Modern Classical Physics

Modern Classical Physics

Optics, Fluids, Plasmas, Elasticity, Relativity, and Statistical Physics

KIP S. THORNE *and* **ROGER D. BLANDFORD**

PRINCETON UNIVERSITY PRESS

Princeton and Oxford

Copyright © 2017 by Princeton University Press

Published by Princeton University Press, 41 William Street, Princeton, New Jersey 08540

In the United Kingdom: Princeton University Press, 6 Oxford Street, Woodstock, Oxfordshire OX20 1TR

press.princeton.edu

All Rights Reserved

Library of Congress Cataloging-in-Publication Data

Names: Thorne, Kip S., author. | Blandford, Roger D., author.
Title: Modern classical physics : optics, fluids, plasmas, elasticity,
   relativity, and statistical physics / Kip S. Thorne and Roger D. Blandford.
Description: Princeton : Princeton University Press, 2017. | Includes
   bibliographical references and index.
Identifiers: LCCN 2014028150 | ISBN 9780691159027 (hardcover : alk. paper) |
   ISBN 0691159025 (hardcover : alk. paper)
Subjects: LCSH: Physics.
Classification: LCC QC21.3 .T46 2015 | DDC 530—dc23
   LC record available at https://lccn.loc.gov/2014028150

British Library Cataloging-in-Publication Data is available

This book has been composed in MinionPro, Whitney, and Ratio Modern using ZzTEX by Windfall Software, Carlisle, Massachusetts

Printed on acid-free paper. ∞

Printed in China

10 9 8 7 6 5 4 3 2 1

To Carolee and Liz

CONTENTS

List of Boxes   xxvii

Preface   xxxi

Acknowledgments   xxxix

PART I FOUNDATIONS   1

1   Newtonian Physics: Geometric Viewpoint   5

1.1   Introduction   5
   1.1.1   The Geometric Viewpoint on the Laws of Physics   5
   1.1.2   Purposes of This Chapter   7
   1.1.3   Overview of This Chapter   7

1.2   Foundational Concepts   8

1.3   Tensor Algebra without a Coordinate System   10

1.4   Particle Kinetics and Lorentz Force in Geometric Language   13

1.5   Component Representation of Tensor Algebra   16
   1.5.1   Slot-Naming Index Notation   17
   1.5.2   Particle Kinetics in Index Notation   19

1.6   Orthogonal Transformations of Bases   20

1.7   Differentiation of Scalars, Vectors, and Tensors; Cross Product and Curl   22

1.8   Volumes, Integration, and Integral Conservation Laws   26
   1.8.1   Gauss's and Stokes' Theorems   27

1.9   The Stress Tensor and Momentum Conservation   29
   1.9.1   Examples: Electromagnetic Field and Perfect Fluid   30
   1.9.2   Conservation of Momentum   31

1.10   Geometrized Units and Relativistic Particles for Newtonian Readers   33
   1.10.1   Geometrized Units   33
   1.10.2   Energy and Momentum of a Moving Particle   34

Bibliographic Note   35

T2   Track Two; see page xxxiv

N   Nonrelativistic (Newtonian) kinetic theory; see page 96

R   Relativistic theory; see page 96

2 Special Relativity: Geometric Viewpoint [T2] 37

- 2.1 Overview 37
- 2.2 Foundational Concepts 38
  - 2.2.1 Inertial Frames, Inertial Coordinates, Events, Vectors, and Spacetime Diagrams 38
  - 2.2.2 The Principle of Relativity and Constancy of Light Speed 42
  - 2.2.3 The Interval and Its Invariance 45
- 2.3 Tensor Algebra without a Coordinate System 48
- 2.4 Particle Kinetics and Lorentz Force without a Reference Frame 49
  - 2.4.1 Relativistic Particle Kinetics: World Lines, 4-Velocity, 4-Momentum and Its Conservation, 4-Force 49
  - 2.4.2 Geometric Derivation of the Lorentz Force Law 52
- 2.5 Component Representation of Tensor Algebra 54
  - 2.5.1 Lorentz Coordinates 54
  - 2.5.2 Index Gymnastics 54
  - 2.5.3 Slot-Naming Notation 56
- 2.6 Particle Kinetics in Index Notation and in a Lorentz Frame 57
- 2.7 Lorentz Transformations 63
- 2.8 Spacetime Diagrams for Boosts 65
- 2.9 Time Travel 67
  - 2.9.1 Measurement of Time; Twins Paradox 67
  - 2.9.2 Wormholes 68
  - 2.9.3 Wormhole as Time Machine 69
- 2.10 Directional Derivatives, Gradients, and the Levi-Civita Tensor 70
- 2.11 Nature of Electric and Magnetic Fields; Maxwell's Equations 71
- 2.12 Volumes, Integration, and Conservation Laws 75
  - 2.12.1 Spacetime Volumes and Integration 75
  - 2.12.2 Conservation of Charge in Spacetime 78
  - 2.12.3 Conservation of Particles, Baryon Number, and Rest Mass 79
- 2.13 Stress-Energy Tensor and Conservation of 4-Momentum 82
  - 2.13.1 Stress-Energy Tensor 82
  - 2.13.2 4-Momentum Conservation 84
  - 2.13.3 Stress-Energy Tensors for Perfect Fluids and Electromagnetic Fields 85
- Bibliographic Note 88

PART II STATISTICAL PHYSICS 91

3 Kinetic Theory 95

- 3.1 Overview 95
- 3.2 Phase Space and Distribution Function 97
  - 3.2.1 Newtonian Number Density in Phase Space, $\mathcal{N}$ 97 [N]
  - 3.2.2 Relativistic Number Density in Phase Space, $\mathcal{N}$ 99 [R][T2]

3.2.3 Distribution Function $f(\mathbf{x}, \mathbf{v}, t)$ for Particles in a Plasma 105
3.2.4 Distribution Function I_ν/ν^3 for Photons 106
3.2.5 Mean Occupation Number η 108
3.3 Thermal-Equilibrium Distribution Functions 111
3.4 Macroscopic Properties of Matter as Integrals over Momentum Space 117
3.4.1 Particle Density n, Flux $\mathbf{S}$, and Stress Tensor $\mathbf{T}$ 117
3.4.2 Relativistic Number-Flux 4-Vector $\vec{S}$ and Stress-Energy Tensor T 118
3.5 Isotropic Distribution Functions and Equations of State 120
3.5.1 Newtonian Density, Pressure, Energy Density, and Equation of State 120
3.5.2 Equations of State for a Nonrelativistic Hydrogen Gas 122
3.5.3 Relativistic Density, Pressure, Energy Density, and Equation of State 125
3.5.4 Equation of State for a Relativistic Degenerate Hydrogen Gas 126
3.5.5 Equation of State for Radiation 128
3.6 Evolution of the Distribution Function: Liouville's Theorem, the Collisionless Boltzmann Equation, and the Boltzmann Transport Equation 132
3.7 Transport Coefficients 139
3.7.1 Diffusive Heat Conduction inside a Star 142
3.7.2 Order-of-Magnitude Analysis 143
3.7.3 Analysis Using the Boltzmann Transport Equation 144
Bibliographic Note 153

4 Statistical Mechanics 155

4.1 Overview 155
4.2 Systems, Ensembles, and Distribution Functions 157
4.2.1 Systems 157
4.2.2 Ensembles 160
4.2.3 Distribution Function 161
4.3 Liouville's Theorem and the Evolution of the Distribution Function 166
4.4 Statistical Equilibrium 168
4.4.1 Canonical Ensemble and Distribution 169
4.4.2 General Equilibrium Ensemble and Distribution; Gibbs Ensemble; Grand Canonical Ensemble 172
4.4.3 Fermi-Dirac and Bose-Einstein Distributions 174
4.4.4 Equipartition Theorem for Quadratic, Classical Degrees of Freedom 177
4.5 The Microcanonical Ensemble 178
4.6 The Ergodic Hypothesis 180
4.7 Entropy and Evolution toward Statistical Equilibrium 181
4.7.1 Entropy and the Second Law of Thermodynamics 181
4.7.2 What Causes the Entropy to Increase? 183
4.8 Entropy per Particle 191
4.9 Bose-Einstein Condensate 193

4.10 Statistical Mechanics in the Presence of Gravity   201
    4.10.1 Galaxies   201
    4.10.2 Black Holes   204
    4.10.3 The Universe   209
    4.10.4 Structure Formation in the Expanding Universe: Violent Relaxation and Phase Mixing   210

4.11 Entropy and Information   211
    4.11.1 Information Gained When Measuring the State of a System in a Microcanonical Ensemble   211
    4.11.2 Information in Communication Theory   212
    4.11.3 Examples of Information Content   214
    4.11.4 Some Properties of Information   216
    4.11.5 Capacity of Communication Channels; Erasing Information from Computer Memories   216

Bibliographic Note   218

5 Statistical Thermodynamics   219

5.1 Overview   219

5.2 Microcanonical Ensemble and the Energy Representation of Thermodynamics   221
    5.2.1 Extensive and Intensive Variables; Fundamental Potential   221
    5.2.2 Energy as a Fundamental Potential   222
    5.2.3 Intensive Variables Identified Using Measuring Devices; First Law of Thermodynamics   223
    5.2.4 Euler's Equation and Form of the Fundamental Potential   226
    5.2.5 Everything Deducible from First Law; Maxwell Relations   227
    5.2.6 Representations of Thermodynamics   228

5.3 Grand Canonical Ensemble and the Grand-Potential Representation of Thermodynamics   229
    5.3.1 The Grand-Potential Representation, and Computation of Thermodynamic Properties as a Grand Canonical Sum   229
    5.3.2 Nonrelativistic van der Waals Gas   232

5.4 Canonical Ensemble and the Physical-Free-Energy Representation of Thermodynamics   239
    5.4.1 Experimental Meaning of Physical Free Energy   241
    5.4.2 Ideal Gas with Internal Degrees of Freedom   242

5.5 Gibbs Ensemble and Representation of Thermodynamics; Phase Transitions and Chemical Reactions   246
    5.5.1 Out-of-Equilibrium Ensembles and Their Fundamental Thermodynamic Potentials and Minimum Principles   248
    5.5.2 Phase Transitions   251
    5.5.3 Chemical Reactions   256

5.6 Fluctuations away from Statistical Equilibrium   260

5.7 Van der Waals Gas: Volume Fluctuations and Gas-to-Liquid Phase Transition 266
5.8 Magnetic Materials 270
    5.8.1 Paramagnetism; The Curie Law 271
    5.8.2 Ferromagnetism: The Ising Model 272
    5.8.3 Renormalization Group Methods for the Ising Model 273
    5.8.4 Monte Carlo Methods for the Ising Model 279
Bibliographic Note 282

6   Random Processes 283

6.1 Overview 283
6.2 Fundamental Concepts 285
    6.2.1 Random Variables and Random Processes 285
    6.2.2 Probability Distributions 286
    6.2.3 Ergodic Hypothesis 288
6.3 Markov Processes and Gaussian Processes 289
    6.3.1 Markov Processes; Random Walk 289
    6.3.2 Gaussian Processes and the Central Limit Theorem; Random Walk 292
    6.3.3 Doob's Theorem for Gaussian-Markov Processes, and Brownian Motion 295
6.4 Correlation Functions and Spectral Densities 297
    6.4.1 Correlation Functions; Proof of Doob's Theorem 297
    6.4.2 Spectral Densities 299
    6.4.3 Physical Meaning of Spectral Density, Light Spectra, and Noise in a Gravitational Wave Detector 301
    6.4.4 The Wiener-Khintchine Theorem; Cosmological Density Fluctuations 303
6.5 2-Dimensional Random Processes 306
    6.5.1 Cross Correlation and Correlation Matrix 306
    6.5.2 Spectral Densities and the Wiener-Khintchine Theorem 307
6.6 Noise and Its Types of Spectra 308
    6.6.1 Shot Noise, Flicker Noise, and Random-Walk Noise; Cesium Atomic Clock 308
    6.6.2 Information Missing from Spectral Density 310
6.7 Filtering Random Processes 311
    6.7.1 Filters, Their Kernels, and the Filtered Spectral Density 311
    6.7.2 Brownian Motion and Random Walks 313
    6.7.3 Extracting a Weak Signal from Noise: Band-Pass Filter, Wiener's Optimal Filter, Signal-to-Noise Ratio, and Allan Variance of Clock Noise 315
    6.7.4 Shot Noise 321
6.8 Fluctuation-Dissipation Theorem 323
    6.8.1 Elementary Version of the Fluctuation-Dissipation Theorem; Langevin Equation, Johnson Noise in a Resistor, and Relaxation Time for Brownian Motion 323
    6.8.2 Generalized Fluctuation-Dissipation Theorem; Thermal Noise in a Laser Beam's Measurement of Mirror Motions; Standard Quantum Limit for Measurement Accuracy and How to Evade It 331

6.9 Fokker-Planck Equation 335
    6.9.1 Fokker-Planck for a 1-Dimensional Markov Process 336
    6.9.2 Optical Molasses: Doppler Cooling of Atoms 340 **T2**
    6.9.3 Fokker-Planck for a Multidimensional Markov Process; Thermal Noise in an Oscillator 343 **T2**
Bibliographic Note 345

PART III OPTICS 347

7 Geometric Optics 351

7.1 Overview 351
7.2 Waves in a Homogeneous Medium 352
    7.2.1 Monochromatic Plane Waves; Dispersion Relation 352
    7.2.2 Wave Packets 354
7.3 Waves in an Inhomogeneous, Time-Varying Medium: The Eikonal Approximation and Geometric Optics 357
    7.3.1 Geometric Optics for a Prototypical Wave Equation 358
    7.3.2 Connection of Geometric Optics to Quantum Theory 362
    7.3.3 Geometric Optics for a General Wave 366
    7.3.4 Examples of Geometric-Optics Wave Propagation 368
    7.3.5 Relation to Wave Packets; Limitations of the Eikonal Approximation and Geometric Optics 369
    7.3.6 Fermat's Principle 371
7.4 Paraxial Optics 375
    7.4.1 Axisymmetric, Paraxial Systems: Lenses, Mirrors, Telescopes, Microscopes, and Optical Cavities 377
    7.4.2 Converging Magnetic Lens for Charged Particle Beam 381
7.5 Catastrophe Optics 384 **T2**
    7.5.1 Image Formation 384 **T2**
    7.5.2 Aberrations of Optical Instruments 395 **T2**
7.6 Gravitational Lenses 396 **T2**
    7.6.1 Gravitational Deflection of Light 396 **T2**
    7.6.2 Optical Configuration 397 **T2**
    7.6.3 Microlensing 398 **T2**
    7.6.4 Lensing by Galaxies 401 **T2**
7.7 Polarization 405
    7.7.1 Polarization Vector and Its Geometric-Optics Propagation Law 405
    7.7.2 Geometric Phase 406 **T2**
Bibliographic Note 409

8   Diffraction   411

8.1   Overview   411
8.2   Helmholtz-Kirchhoff Integral   413
    8.2.1 Diffraction by an Aperture   414
    8.2.2 Spreading of the Wavefront: Fresnel and Fraunhofer Regions   417
8.3   Fraunhofer Diffraction   420
    8.3.1 Diffraction Grating   422
    8.3.2 Airy Pattern of a Circular Aperture: Hubble Space Telescope   425
    8.3.3 Babinet's Principle   428
8.4   Fresnel Diffraction   429
    8.4.1 Rectangular Aperture, Fresnel Integrals, and the Cornu Spiral   430
    8.4.2 Unobscured Plane Wave   432
    8.4.3 Fresnel Diffraction by a Straight Edge: Lunar Occultation of a Radio Source   432
    8.4.4 Circular Apertures: Fresnel Zones and Zone Plates   434
8.5   Paraxial Fourier Optics   436
    8.5.1 Coherent Illumination   437
    8.5.2 Point-Spread Functions   438
    8.5.3 Abbé's Description of Image Formation by a Thin Lens   439
    8.5.4 Image Processing by a Spatial Filter in the Focal Plane of a Lens: High-Pass, Low-Pass, and Notch Filters; Phase-Contrast Microscopy   441
    8.5.5 Gaussian Beams: Optical Cavities and Interferometric Gravitational-Wave Detectors   445
8.6   Diffraction at a Caustic   451
    Bibliographic Note   454

9   Interference and Coherence   455

9.1   Overview   455
9.2   Coherence   456
    9.2.1 Young's Slits   456
    9.2.2 Interference with an Extended Source: Van Cittert-Zernike Theorem   459
    9.2.3 More General Formulation of Spatial Coherence; Lateral Coherence Length   462
    9.2.4 Generalization to 2 Dimensions   463
    9.2.5 Michelson Stellar Interferometer; Astronomical Seeing   464
    9.2.6 Temporal Coherence   472
    9.2.7 Michelson Interferometer and Fourier-Transform Spectroscopy   474
    9.2.8 Degree of Coherence; Relation to Theory of Random Processes   477
9.3   Radio Telescopes   479
    9.3.1 Two-Element Radio Interferometer   479
    9.3.2 Multiple-Element Radio Interferometers   480
    9.3.3 Closure Phase   481
    9.3.4 Angular Resolution   482

9.4 Etalons and Fabry-Perot Interferometers 483
    9.4.1 Multiple-Beam Interferometry; Etalons 483
    9.4.2 Fabry-Perot Interferometer and Modes of a Fabry-Perot Cavity with Spherical Mirrors 490
    9.4.3 Fabry-Perot Applications: Spectrometer, Laser, Mode-Cleaning Cavity, Beam-Shaping Cavity, PDH Laser Stabilization, Optical Frequency Comb 496 [T2]
9.5 Laser Interferometer Gravitational-Wave Detectors 502 [T2]
9.6 Power Correlations and Photon Statistics: Hanbury Brown and Twiss Intensity Interferometer 509
Bibliographic Note 512

10 Nonlinear Optics 513

10.1 Overview 513
10.2 Lasers 515
    10.2.1 Basic Principles of the Laser 515
    10.2.2 Types of Lasers and Their Performances and Applications 519
    10.2.3 Ti:Sapphire Mode-Locked Laser 520
    10.2.4 Free Electron Laser 521
10.3 Holography 521
    10.3.1 Recording a Hologram 522
    10.3.2 Reconstructing the 3-Dimensional Image from a Hologram 525
    10.3.3 Other Types of Holography; Applications 527
10.4 Phase-Conjugate Optics 531
10.5 Maxwell's Equations in a Nonlinear Medium; Nonlinear Dielectric Susceptibilities; Electro-Optic Effects 536
10.6 Three-Wave Mixing in Nonlinear Crystals 540
    10.6.1 Resonance Conditions for Three-Wave Mixing 540
    10.6.2 Three-Wave-Mixing Evolution Equations in a Medium That Is Dispersion-Free and Isotropic at Linear Order 544
    10.6.3 Three-Wave Mixing in a Birefringent Crystal: Phase Matching and Evolution Equations 546 [T2]
10.7 Applications of Three-Wave Mixing: Frequency Doubling, Optical Parametric Amplification, and Squeezed Light 553
    10.7.1 Frequency Doubling 553
    10.7.2 Optical Parametric Amplification 555
    10.7.3 Degenerate Optical Parametric Amplification: Squeezed Light 556
10.8 Four-Wave Mixing in Isotropic Media 558
    10.8.1 Third-Order Susceptibilities and Field Strengths 558
    10.8.2 Phase Conjugation via Four-Wave Mixing in CS_2 Fluid 559
    10.8.3 Optical Kerr Effect and Four-Wave Mixing in an Optical Fiber 562
Bibliographic Note 564

PART IV ELASTICITY 565

11 Elastostatics 567

- 11.1 Overview 567
- 11.2 Displacement and Strain 570
    - 11.2.1 Displacement Vector and Its Gradient 570
    - 11.2.2 Expansion, Rotation, Shear, and Strain 571
- 11.3 Stress, Elastic Moduli, and Elastostatic Equilibrium 577
    - 11.3.1 Stress Tensor 577
    - 11.3.2 Realm of Validity for Hooke's Law 580
    - 11.3.3 Elastic Moduli and Elastostatic Stress Tensor 580
    - 11.3.4 Energy of Deformation 582
    - 11.3.5 Thermoelasticity 584
    - 11.3.6 Molecular Origin of Elastic Stress; Estimate of Moduli 585
    - 11.3.7 Elastostatic Equilibrium: Navier-Cauchy Equation 587
- 11.4 Young's Modulus and Poisson's Ratio for an Isotropic Material: A Simple Elastostatics Problem 589
- 11.5 Reducing the Elastostatic Equations to 1 Dimension for a Bent Beam: Cantilever Bridge, Foucault Pendulum, DNA Molecule, Elastica 592
- 11.6 Buckling and Bifurcation of Equilibria 602
    - 11.6.1 Elementary Theory of Buckling and Bifurcation 602
    - 11.6.2 Collapse of the World Trade Center Buildings 605
    - 11.6.3 Buckling with Lateral Force; Connection to Catastrophe Theory 606
    - 11.6.4 Other Bifurcations: Venus Fly Trap, Whirling Shaft, Triaxial Stars, and Onset of Turbulence 607
- 11.7 Reducing the Elastostatic Equations to 2 Dimensions for a Deformed Thin Plate: Stress Polishing a Telescope Mirror 609
- 11.8 Cylindrical and Spherical Coordinates: Connection Coefficients and Components of the Gradient of the Displacement Vector 614
- 11.9 Solving the 3-Dimensional Navier-Cauchy Equation in Cylindrical Coordinates 619
    - 11.9.1 Simple Methods: Pipe Fracture and Torsion Pendulum 619
    - 11.9.2 Separation of Variables and Green's Functions: Thermoelastic Noise in Mirrors 622
- Bibliographic Note 627

12 Elastodynamics 629

- 12.1 Overview 629
- 12.2 Basic Equations of Elastodynamics; Waves in a Homogeneous Medium 630
    - 12.2.1 Equation of Motion for a Strained Elastic Medium 630
    - 12.2.2 Elastodynamic Waves 636
    - 12.2.3 Longitudinal Sound Waves 637

   12.2.4 Transverse Shear Waves 638
   12.2.5 Energy of Elastodynamic Waves 640
12.3 Waves in Rods, Strings, and Beams 642
   12.3.1 Compression Waves in a Rod 643
   12.3.2 Torsion Waves in a Rod 643
   12.3.3 Waves on Strings 644
   12.3.4 Flexural Waves on a Beam 645
   12.3.5 Bifurcation of Equilibria and Buckling (Once More) 647
12.4 Body Waves and Surface Waves—Seismology and Ultrasound 648
   12.4.1 Body Waves 650
   12.4.2 Edge Waves 654
   12.4.3 Green's Function for a Homogeneous Half-Space 658
   12.4.4 Free Oscillations of Solid Bodies 661
   12.4.5 Seismic Tomography 663
   12.4.6 Ultrasound; Shock Waves in Solids 663
12.5 The Relationship of Classical Waves to Quantum Mechanical Excitations 667   **T2**
   Bibliographic Note 670

PART V FLUID DYNAMICS 671

13 **Foundations of Fluid Dynamics** 675

13.1 Overview 675
13.2 The Macroscopic Nature of a Fluid: Density, Pressure, Flow Velocity; Liquids versus Gases 677
13.3 Hydrostatics 681
   13.3.1 Archimedes' Law 684
   13.3.2 Nonrotating Stars and Planets 686
   13.3.3 Rotating Fluids 689
13.4 Conservation Laws 691
13.5 The Dynamics of an Ideal Fluid 695
   13.5.1 Mass Conservation 696
   13.5.2 Momentum Conservation 696
   13.5.3 Euler Equation 697
   13.5.4 Bernoulli's Theorem 697
   13.5.5 Conservation of Energy 704
13.6 Incompressible Flows 709
13.7 Viscous Flows with Heat Conduction 710
   13.7.1 Decomposition of the Velocity Gradient into Expansion, Vorticity, and Shear 710
   13.7.2 Navier-Stokes Equation 711
   13.7.3 Molecular Origin of Viscosity 713
   13.7.4 Energy Conservation and Entropy Production 714

           13.7.5 Reynolds Number 716
           13.7.6 Pipe Flow 716
　　13.8　Relativistic Dynamics of a Perfect Fluid 719
           13.8.1 Stress-Energy Tensor and Equations of Relativistic Fluid Mechanics 719
           13.8.2 Relativistic Bernoulli Equation and Ultrarelativistic Astrophysical Jets 721
           13.8.3 Nonrelativistic Limit of the Stress-Energy Tensor 723
           Bibliographic Note 726

14　Vorticity 729

14.1　Overview 729

14.2　Vorticity, Circulation, and Their Evolution 731
       14.2.1 Vorticity Evolution 734
       14.2.2 Barotropic, Inviscid, Compressible Flows: Vortex Lines Frozen into Fluid 736
       14.2.3 Tornados 738
       14.2.4 Circulation and Kelvin's Theorem 739
       14.2.5 Diffusion of Vortex Lines 741
       14.2.6 Sources of Vorticity 744

14.3　Low-Reynolds-Number Flow—Stokes Flow and Sedimentation 746
       14.3.1 Motivation: Climate Change 748
       14.3.2 Stokes Flow 749
       14.3.3 Sedimentation Rate 754

14.4　High-Reynolds-Number Flow—Laminar Boundary Layers 757
       14.4.1 Blasius Velocity Profile Near a Flat Plate: Stream Function and Similarity Solution 758
       14.4.2 Blasius Vorticity Profile 763
       14.4.3 Viscous Drag Force on a Flat Plate 763
       14.4.4 Boundary Layer Near a Curved Surface: Separation 764

14.5　Nearly Rigidly Rotating Flows—Earth's Atmosphere and Oceans 766
       14.5.1 Equations of Fluid Dynamics in a Rotating Reference Frame 767
       14.5.2 Geostrophic Flows 770
       14.5.3 Taylor-Proudman Theorem 771
       14.5.4 Ekman Boundary Layers 772

14.6　Instabilities of Shear Flows—Billow Clouds and Turbulence in the Stratosphere 778
       14.6.1 Discontinuous Flow: Kelvin-Helmholtz Instability 778
       14.6.2 Discontinuous Flow with Gravity 782
       14.6.3 Smoothly Stratified Flows: Rayleigh and Richardson Criteria for Instability 784
       Bibliographic Note 786

15　Turbulence 787

15.1　Overview 787

15.2　The Transition to Turbulence—Flow Past a Cylinder 789

15.3 Empirical Description of Turbulence 798
    15.3.1 The Role of Vorticity in Turbulence 799
15.4 Semiquantitative Analysis of Turbulence 800
    15.4.1 Weak-Turbulence Formalism 800
    15.4.2 Turbulent Viscosity 804
    15.4.3 Turbulent Wakes and Jets; Entrainment; the Coanda Effect 805
    15.4.4 Kolmogorov Spectrum for Fully Developed, Homogeneous, Isotropic Turbulence 810
15.5 Turbulent Boundary Layers 817
    15.5.1 Profile of a Turbulent Boundary Layer 818
    15.5.2 Coanda Effect and Separation in a Turbulent Boundary Layer 820
    15.5.3 Instability of a Laminar Boundary Layer 822
    15.5.4 Flight of a Ball 823
15.6 The Route to Turbulence—Onset of Chaos 825
    15.6.1 Rotating Couette Flow 825
    15.6.2 Feigenbaum Sequence, Poincaré Maps, and the Period-Doubling Route to Turbulence in Convection 828
    15.6.3 Other Routes to Turbulent Convection 831
    15.6.4 Extreme Sensitivity to Initial Conditions 832
Bibliographic Note 834

16 Waves 835

16.1 Overview 835
16.2 Gravity Waves on and beneath the Surface of a Fluid 837
    16.2.1 Deep-Water Waves and Their Excitation and Damping 840
    16.2.2 Shallow-Water Waves 840
    16.2.3 Capillary Waves and Surface Tension 844
    16.2.4 Helioseismology 848
16.3 Nonlinear Shallow-Water Waves and Solitons 850
    16.3.1 Korteweg–de Vries (KdV) Equation 850
    16.3.2 Physical Effects in the KdV Equation 853
    16.3.3 Single-Soliton Solution 854
    16.3.4 Two-Soliton Solution 855
    16.3.5 Solitons in Contemporary Physics 856
16.4 Rossby Waves in a Rotating Fluid 858
16.5 Sound Waves 862
    16.5.1 Wave Energy 863
    16.5.2 Sound Generation 865
    16.5.3 Radiation Reaction, Runaway Solutions, and Matched Asymptotic Expansions 869
Bibliographic Note 874

17	**Compressible and Supersonic Flow** 875
17.1	Overview 875
17.2	Equations of Compressible Flow 877
17.3	Stationary, Irrotational, Quasi-1-Dimensional Flow 880
	17.3.1 Basic Equations; Transition from Subsonic to Supersonic Flow 880
	17.3.2 Setting up a Stationary, Transonic Flow 883
	17.3.3 Rocket Engines 887
17.4	1-Dimensional, Time-Dependent Flow 891
	17.4.1 Riemann Invariants 891
	17.4.2 Shock Tube 895
17.5	Shock Fronts 897
	17.5.1 Junction Conditions across a Shock; Rankine-Hugoniot Relations 898
	17.5.2 Junction Conditions for Ideal Gas with Constant γ 904
	17.5.3 Internal Structure of a Shock 906
	17.5.4 Mach Cone 907
17.6	Self-Similar Solutions—Sedov-Taylor Blast Wave 908
	17.6.1 The Sedov-Taylor Solution 909
	17.6.2 Atomic Bomb 912
	17.6.3 Supernovae 914
	Bibliographic Note 916

18	**Convection** 917
18.1	Overview 917
18.2	Diffusive Heat Conduction—Cooling a Nuclear Reactor; Thermal Boundary Layers 918
18.3	Boussinesq Approximation 923
18.4	Rayleigh-Bénard Convection 925
18.5	Convection in Stars 933
18.6	Double Diffusion—Salt Fingers 937
	Bibliographic Note 941

19	**Magnetohydrodynamics** 943
19.1	Overview 943
19.2	Basic Equations of MHD 944
	19.2.1 Maxwell's Equations in the MHD Approximation 946
	19.2.2 Momentum and Energy Conservation 950
	19.2.3 Boundary Conditions 953
	19.2.4 Magnetic Field and Vorticity 957
19.3	Magnetostatic Equilibria 958
	19.3.1 Controlled Thermonuclear Fusion 958
	19.3.2 Z-Pinch 960

		19.3.3 Θ-Pinch 962
		19.3.4 Tokamak 963
19.4	Hydromagnetic Flows 965	
19.5	Stability of Magnetostatic Equilibria 971	
		19.5.1 Linear Perturbation Theory 971
		19.5.2 Z-Pinch: Sausage and Kink Instabilities 975
		19.5.3 The Θ-Pinch and Its Toroidal Analog; Flute Instability; Motivation for Tokamak 978
		19.5.4 Energy Principle and Virial Theorems 980
19.6	Dynamos and Reconnection of Magnetic Field Lines 984	
		19.6.1 Cowling's Theorem 984
		19.6.2 Kinematic Dynamos 985
		19.6.3 Magnetic Reconnection 986
19.7	Magnetosonic Waves and the Scattering of Cosmic Rays 988	
		19.7.1 Cosmic Rays 988
		19.7.2 Magnetosonic Dispersion Relation 989
		19.7.3 Scattering of Cosmic Rays by Alfvén Waves 992
	Bibliographic Note 993	

PART VI PLASMA PHYSICS 995

20 The Particle Kinetics of Plasma 997

20.1	Overview 997
20.2	Examples of Plasmas and Their Density-Temperature Regimes 998
	20.2.1 Ionization Boundary 998
	20.2.2 Degeneracy Boundary 1000
	20.2.3 Relativistic Boundary 1000
	20.2.4 Pair-Production Boundary 1001
	20.2.5 Examples of Natural and Human-Made Plasmas 1001
20.3	Collective Effects in Plasmas—Debye Shielding and Plasma Oscillations 1003
	20.3.1 Debye Shielding 1003
	20.3.2 Collective Behavior 1004
	20.3.3 Plasma Oscillations and Plasma Frequency 1005
20.4	Coulomb Collisions 1006
	20.4.1 Collision Frequency 1006
	20.4.2 The Coulomb Logarithm 1008
	20.4.3 Thermal Equilibration Rates in a Plasma 1010
	20.4.4 Discussion 1012
20.5	Transport Coefficients 1015
	20.5.1 Coulomb Collisions 1015
	20.5.2 Anomalous Resistivity and Anomalous Equilibration 1016

- 20.6 Magnetic Field 1019
  - 20.6.1 Cyclotron Frequency and Larmor Radius 1019
  - 20.6.2 Validity of the Fluid Approximation 1020
  - 20.6.3 Conductivity Tensor 1022
- 20.7 Particle Motion and Adiabatic Invariants 1024
  - 20.7.1 Homogeneous, Time-Independent Magnetic Field and No Electric Field 1025
  - 20.7.2 Homogeneous, Time-Independent Electric and Magnetic Fields 1025
  - 20.7.3 Inhomogeneous, Time-Independent Magnetic Field 1026
  - 20.7.4 A Slowly Time-Varying Magnetic Field 1029
  - 20.7.5 Failure of Adiabatic Invariants; Chaotic Orbits 1030
- Bibliographic Note 1032

21 Waves in Cold Plasmas: Two-Fluid Formalism 1033

- 21.1 Overview 1033
- 21.2 Dielectric Tensor, Wave Equation, and General Dispersion Relation 1035
- 21.3 Two-Fluid Formalism 1037
- 21.4 Wave Modes in an Unmagnetized Plasma 1040
  - 21.4.1 Dielectric Tensor and Dispersion Relation for a Cold, Unmagnetized Plasma 1040
  - 21.4.2 Plasma Electromagnetic Modes 1042
  - 21.4.3 Langmuir Waves and Ion-Acoustic Waves in Warm Plasmas 1044
  - 21.4.4 Cutoffs and Resonances 1049
- 21.5 Wave Modes in a Cold, Magnetized Plasma 1050
  - 21.5.1 Dielectric Tensor and Dispersion Relation 1050
  - 21.5.2 Parallel Propagation 1052
  - 21.5.3 Perpendicular Propagation 1057
  - 21.5.4 Propagation of Radio Waves in the Ionosphere; Magnetoionic Theory 1058
  - 21.5.5 CMA Diagram for Wave Modes in a Cold, Magnetized Plasma 1062
- 21.6 Two-Stream Instability 1065
- Bibliographic Note 1068

22 Kinetic Theory of Warm Plasmas 1069

- 22.1 Overview 1069
- 22.2 Basic Concepts of Kinetic Theory and Its Relationship to Two-Fluid Theory 1070
  - 22.2.1 Distribution Function and Vlasov Equation 1070
  - 22.2.2 Relation of Kinetic Theory to Two-Fluid Theory 1073
  - 22.2.3 Jeans' Theorem 1074
- 22.3 Electrostatic Waves in an Unmagnetized Plasma: Landau Damping 1077
  - 22.3.1 Formal Dispersion Relation 1077
  - 22.3.2 Two-Stream Instability 1079
  - 22.3.3 The Landau Contour 1080
  - 22.3.4 Dispersion Relation for Weakly Damped or Growing Waves 1085

22.3.5 Langmuir Waves and Their Landau Damping   1086
22.3.6 Ion-Acoustic Waves and Conditions for Their Landau Damping to Be Weak   1088
22.4 Stability of Electrostatic Waves in Unmagnetized Plasmas   1090
22.4.1 Nyquist's Method   1091
22.4.2 Penrose's Instability Criterion   1091
22.5 Particle Trapping   1098
22.6 N-Particle Distribution Function   1102
22.6.1 BBGKY Hierarchy   1103
22.6.2 Two-Point Correlation Function   1104
22.6.3 Coulomb Correction to Plasma Pressure   1107
Bibliographic Note   1108

23 Nonlinear Dynamics of Plasmas   1111

23.1 Overview   1111
23.2 Quasilinear Theory in Classical Language   1113
23.2.1 Classical Derivation of the Theory   1113
23.2.2 Summary of Quasilinear Theory   1120
23.2.3 Conservation Laws   1121
23.2.4 Generalization to 3 Dimensions   1122
23.3 Quasilinear Theory in Quantum Mechanical Language   1123
23.3.1 Plasmon Occupation Number η 1123
23.3.2 Evolution of η for Plasmons via Interaction with Electrons   1124
23.3.3 Evolution of f for Electrons via Interaction with Plasmons   1129
23.3.4 Emission of Plasmons by Particles in the Presence of a Magnetic Field   1131
23.3.5 Relationship between Classical and Quantum Mechanical Formalisms   1131
23.3.6 Evolution of η via Three-Wave Mixing   1132
23.4 Quasilinear Evolution of Unstable Distribution Functions—A Bump in the Tail   1136
23.4.1 Instability of Streaming Cosmic Rays   1138
23.5 Parametric Instabilities; Laser Fusion   1140
23.6 Solitons and Collisionless Shock Waves   1142
Bibliographic Note   1149

PART VII GENERAL RELATIVITY   1151

24 From Special to General Relativity   1153

24.1 Overview   1153
24.2 Special Relativity Once Again   1153
24.2.1 Geometric, Frame-Independent Formulation   1154
24.2.2 Inertial Frames and Components of Vectors, Tensors, and Physical Laws   1156
24.2.3 Light Speed, the Interval, and Spacetime Diagrams   1159
24.3 Differential Geometry in General Bases and in Curved Manifolds   1160
24.3.1 Nonorthonormal Bases   1161

24.3.2 Vectors as Directional Derivatives; Tangent Space; Commutators 1165
24.3.3 Differentiation of Vectors and Tensors; Connection Coefficients 1169
24.3.4 Integration 1174
24.4 The Stress-Energy Tensor Revisited 1176
24.5 The Proper Reference Frame of an Accelerated Observer 1180
24.5.1 Relation to Inertial Coordinates; Metric in Proper Reference Frame; Transport Law for Rotating Vectors 1183
24.5.2 Geodesic Equation for a Freely Falling Particle 1184
24.5.3 Uniformly Accelerated Observer 1186
24.5.4 Rindler Coordinates for Minkowski Spacetime 1187
Bibliographic Note 1190

25 Fundamental Concepts of General Relativity 1191

25.1 History and Overview 1191
25.2 Local Lorentz Frames, the Principle of Relativity, and Einstein's Equivalence Principle 1195
25.3 The Spacetime Metric, and Gravity as a Curvature of Spacetime 1196
25.4 Free-Fall Motion and Geodesics of Spacetime 1200
25.5 Relative Acceleration, Tidal Gravity, and Spacetime Curvature 1206
25.5.1 Newtonian Description of Tidal Gravity 1207
25.5.2 Relativistic Description of Tidal Gravity 1208
25.5.3 Comparison of Newtonian and Relativistic Descriptions 1210
25.6 Properties of the Riemann Curvature Tensor 1213
25.7 Delicacies in the Equivalence Principle, and Some Nongravitational Laws of Physics in Curved Spacetime 1217
25.7.1 Curvature Coupling in the Nongravitational Laws 1218 [T2]
25.8 The Einstein Field Equation 1221
25.8.1 Geometrized Units 1224
25.9 Weak Gravitational Fields 1224
25.9.1 Newtonian Limit of General Relativity 1225
25.9.2 Linearized Theory 1227
25.9.3 Gravitational Field outside a Stationary, Linearized Source of Gravity 1231 [T2]
25.9.4 Conservation Laws for Mass, Momentum, and Angular Momentum in Linearized Theory 1237 [T2]
25.9.5 Conservation Laws for a Strong-Gravity Source 1238 [T2]
Bibliographic Note 1239

26 Relativistic Stars and Black Holes 1241

26.1 Overview 1241
26.2 Schwarzschild's Spacetime Geometry 1242
26.2.1 The Schwarzschild Metric, Its Connection Coefficients, and Its Curvature Tensors 1242

- 26.2.2 The Nature of Schwarzschild's Coordinate System, and Symmetries of the Schwarzschild Spacetime 1244
- 26.2.3 Schwarzschild Spacetime at Radii $r \gg M$: The Asymptotically Flat Region 1245
- 26.2.4 Schwarzschild Spacetime at $r \sim M$ 1248

26.3 Static Stars 1250
- 26.3.1 Birkhoff's Theorem 1250
- 26.3.2 Stellar Interior 1252
- 26.3.3 Local Conservation of Energy and Momentum 1255
- 26.3.4 The Einstein Field Equation 1257
- 26.3.5 Stellar Models and Their Properties 1259
- 26.3.6 Embedding Diagrams 1261

26.4 Gravitational Implosion of a Star to Form a Black Hole 1264
- 26.4.1 The Implosion Analyzed in Schwarzschild Coordinates 1264
- 26.4.2 Tidal Forces at the Gravitational Radius 1266
- 26.4.3 Stellar Implosion in Eddington-Finkelstein Coordinates 1267
- 26.4.4 Tidal Forces at $r = 0$—The Central Singularity 1271
- 26.4.5 Schwarzschild Black Hole 1272

26.5 Spinning Black Holes: The Kerr Spacetime 1277 [T2]
- 26.5.1 The Kerr Metric for a Spinning Black Hole 1277 [T2]
- 26.5.2 Dragging of Inertial Frames 1279 [T2]
- 26.5.3 The Light-Cone Structure, and the Horizon 1279 [T2]
- 26.5.4 Evolution of Black Holes—Rotational Energy and Its Extraction 1282 [T2]

26.6 The Many-Fingered Nature of Time 1293 [T2]

Bibliographic Note 1297

27 Gravitational Waves and Experimental Tests of General Relativity 1299

27.1 Overview 1299

27.2 Experimental Tests of General Relativity 1300
- 27.2.1 Equivalence Principle, Gravitational Redshift, and Global Positioning System 1300
- 27.2.2 Perihelion Advance of Mercury 1302
- 27.2.3 Gravitational Deflection of Light, Fermat's Principle, and Gravitational Lenses 1305
- 27.2.4 Shapiro Time Delay 1308
- 27.2.5 Geodetic and Lense-Thirring Precession 1309
- 27.2.6 Gravitational Radiation Reaction 1310

27.3 Gravitational Waves Propagating through Flat Spacetime 1311
- 27.3.1 Weak, Plane Waves in Linearized Theory 1311
- 27.3.2 Measuring a Gravitational Wave by Its Tidal Forces 1315
- 27.3.3 Gravitons and Their Spin and Rest Mass 1319

27.4 Gravitational Waves Propagating through Curved Spacetime 1320
    27.4.1 Gravitational Wave Equation in Curved Spacetime 1321
    27.4.2 Geometric-Optics Propagation of Gravitational Waves 1322
    27.4.3 Energy and Momentum in Gravitational Waves 1324
27.5 The Generation of Gravitational Waves 1327
    27.5.1 Multipole-Moment Expansion 1328
    27.5.2 Quadrupole-Moment Formalism 1330
    27.5.3 Quadrupolar Wave Strength, Energy, Angular Momentum, and Radiation Reaction 1332
    27.5.4 Gravitational Waves from a Binary Star System 1335
    27.5.5 Gravitational Waves from Binaries Made of Black Holes, Neutron Stars, or Both: Numerical Relativity 1341 **T2**
27.6 The Detection of Gravitational Waves 1345
    27.6.1 Frequency Bands and Detection Techniques 1345
    27.6.2 Gravitational-Wave Interferometers: Overview and Elementary Treatment 1347
    27.6.3 Interferometer Analyzed in TT Gauge 1349 **T2**
    27.6.4 Interferometer Analyzed in the Proper Reference Frame of the Beam Splitter 1352 **T2**
    27.6.5 Realistic Interferometers 1355 **T2**
    27.6.6 Pulsar Timing Arrays 1355 **T2**
Bibliographic Note 1358

28 Cosmology 1361

28.1 Overview 1361
28.2 General Relativistic Cosmology 1364
    28.2.1 Isotropy and Homogeneity 1364
    28.2.2 Geometry 1366
    28.2.3 Kinematics 1373
    28.2.4 Dynamics 1376
28.3 The Universe Today 1379
    28.3.1 Baryons 1379
    28.3.2 Dark Matter 1380
    28.3.3 Photons 1381
    28.3.4 Neutrinos 1382
    28.3.5 Cosmological Constant 1382
    28.3.6 Standard Cosmology 1383
28.4 Seven Ages of the Universe 1383
    28.4.1 Particle Age 1384
    28.4.2 Nuclear Age 1387
    28.4.3 Photon Age 1392

- 28.4.4 Plasma Age 1393
- 28.4.5 Atomic Age 1397
- 28.4.6 Gravitational Age 1397
- 28.4.7 Cosmological Age 1400

28.5 Galaxy Formation 1401
- 28.5.1 Linear Perturbations 1401
- 28.5.2 Individual Constituents 1406
- 28.5.3 Solution of the Perturbation Equations 1410
- 28.5.4 Galaxies 1412

28.6 Cosmological Optics 1415
- 28.6.1 Cosmic Microwave Background 1415
- 28.6.2 Weak Gravitational Lensing 1422
- 28.6.3 Sunyaev-Zel'dovich Effect 1428

28.7 Three Mysteries 1431
- 28.7.1 Inflation and the Origin of the Universe 1431
- 28.7.2 Dark Matter and the Growth of Structure 1440
- 28.7.3 The Cosmological Constant and the Fate of the Universe 1444

Bibliographic Note 1447

References 1449

Name Index 1473

Subject Index 1477

BOXES

1.1	Readers' Guide 6	
1.2	Vectors and Tensors in Quantum Theory 18	T2
2.1	Readers' Guide 38	
2.2	Measuring the Speed of Light Without Light 43	
2.3	Propagation Speeds of Other Waves 44	
2.4	Proof of Invariance of the Interval for a Timelike Separation 46	
3.1	Readers' Guide 96	
3.2	Sophisticated Derivation of Relativistic Collisionless Boltzmann Equation 136	R
3.3	Two-Lengthscale Expansions 146	
4.1	Readers' Guide 156	
4.2	Density Operator and Quantum Statistical Mechanics 165	T2
4.3	Entropy Increase Due to Discarding Quantum Correlations 186	
5.1	Readers' Guide 220	
5.2	Two Useful Relations between Partial Derivatives 225	
5.3	Derivation of van der Waals Grand Potential 235	T2
6.1	Readers' Guide 284	
7.1	Readers' Guide 352	
7.2	Bookkeeping Parameter in Two-Lengthscale Expansions 360	
8.1	Readers' Guide 412	
9.1	Readers' Guide 456	
9.2	Astronomical Seeing, Speckle Image Processing, and Adaptive Optics 466	T2
9.3	Modes of a Fabry-Perot Cavity with Spherical Mirrors 491	

10.1	Readers' Guide 514	
10.2	Properties of Some Anisotropic, Nonlinear Crystals 541	
11.1	Readers' Guide 568	
11.2	Irreducible Tensorial Parts of a Second-Rank Tensor in 3-Dimensional Euclidean Space 572	
11.3	Methods of Solving the Navier-Cauchy Equation 590	
11.4	Shear Tensor in Spherical and Cylindrical Coordinates 618	T2
12.1	Readers' Guide 630	
12.2	Wave Equations in Continuum Mechanics 633	
13.1	Readers' Guide 676	
13.2	Thermodynamic Considerations 679	
13.3	Flow Visualization 699	
13.4	Self-Gravity 705	T2
13.5	Terminology Used in Chapter 13 724	
14.1	Readers' Guide 730	
14.2	Movies Relevant to this Chapter 731	
14.3	Swimming at Low and High Reynolds Number: Fish versus Bacteria 747	
14.4	Stream Function for a General, Two-Dimensional, Incompressible Flow 760	T2
14.5	Arbitrariness of Rotation Axis; Ω for Atmospheric and Oceanic Flows 769	
15.1	Readers' Guide 788	
15.2	Movies and Photographs on Turbulence 790	
15.3	Consequences of the Kelvin-Helmholtz Instability 801	
16.1	Readers' Guide 836	
16.2	Movies Relevant to this Chapter 837	
16.3	Nonlinear Shallow-Water Waves with Variable Depth 841	
16.4	Surface Tension 844	
17.1	Readers' Guide 876	
17.2	Movies Relevant to this Chapter 877	
17.3	Velocity Profiles for 1-Dimensional Flow Between Chambers 885	
17.4	Space Shuttle 889	
18.1	Readers' Guide 918	
18.2	Mantle Convection and Continental Drift 932	
19.1	Readers' Guide 944	
20.1	Readers' Guide 998	
21.1	Readers' Guide 1034	

22.1	Readers' Guide 1070	
22.2	Stability of a Feedback-Control System: Analysis by Nyquist's Method 1093	T2
23.1	Readers' Guide 1112	
23.2	Laser Fusion 1141	
24.1	Readers' Guide 1154	
24.2	Stress-Energy Tensor for a Point Particle 1178	T2
24.3	Inertial Guidance Systems 1182	
25.1	Readers' Guide 1192	
25.2	Decomposition of Riemann: Tidal and Frame-Drag Fields 1235	T2
26.1	Readers' Guide 1242	
26.2	Connection Coefficients and Curvature Tensors for Schwarzschild Solution 1243	
26.3	Tendex and Vortex Lines Outside a Black Hole 1295	T2
27.1	Readers' Guide 1300	
27.2	Projecting Out the Gravitational-Wave Field h_{ij}^{TT} 1314	
27.3	Tendex and Vortex Lines for a Gravitational Wave 1318	T2
27.4	Geometrodynamics 1344	T2
28.1	Readers' Guide 1362	

PREFACE

The study of physics (including astronomy) is one of the oldest academic enterprises. Remarkable surges in inquiry occurred in equally remarkable societies—in Greece and Egypt, in Mesopotamia, India and China—and especially in Western Europe from the late sixteenth century onward. Independent, rational inquiry flourished at the expense of ignorance, superstition, and obeisance to authority.

Physics is a constructive and progressive discipline, so these surges left behind layers of understanding derived from careful observation and experiment, organized by fundamental principles and laws that provide the foundation of the discipline today. Meanwhile the detritus of bad data and wrong ideas has washed away. The laws themselves were so general and reliable that they provided foundations for investigation far beyond the traditional frontiers of physics, and for the growth of technology.

The start of the twentieth century marked a watershed in the history of physics, when attention turned to the small and the fast. Although rightly associated with the names of Planck and Einstein, this turning point was only reached through the curiosity and industry of their many forerunners. The resulting quantum mechanics and relativity occupied physicists for much of the succeeding century and today are viewed very differently from each other. Quantum mechanics is perceived as an abrupt departure from the tacit assumptions of the past, while relativity—though no less radical conceptually—is seen as a logical continuation of the physics of Galileo, Newton, and Maxwell. There is no better illustration of this than Einstein's growing special relativity into the general theory and his famous resistance to the quantum mechanics of the 1920s, which others were developing.

This is a book about classical physics—a name intended to capture the pre-quantum scientific ideas, augmented by general relativity. Operationally, it is physics in the limit that Planck's constant $h \to 0$. Classical physics is sometimes used, pejoratively, to suggest that "classical" ideas were discarded and replaced by new principles and laws. Nothing could be further from the truth. The majority of applications of physics today are still essentially classical. This does not imply that physicists or others working in these areas are ignorant or dismissive of quantum physics. It is simply that the issues with which they are confronted are mostly addressed classically. Furthermore, classical physics has not stood still while the quantum world was being explored. In scope and in practice, it has exploded on many fronts and would now be quite unrecognizable to a Helmholtz, a Rayleigh, or a Gibbs. In this book, we have tried to emphasize these contemporary developments and applications at the expense of historical choices, and this is the reason for our seemingly oxymoronic title, *Modern Classical Physics*.

This book is ambitious in scope, but to make it bindable and portable (and so the authors could spend some time with their families), we do not develop classical mechanics, electromagnetic theory, or elementary thermodynamics. We assume the reader has already learned these topics elsewhere, perhaps as part of an undergraduate curriculum. We also assume a normal undergraduate facility with applied mathematics. This allows us to focus on those topics that are less frequently taught in undergraduate and graduate courses.

Another important exclusion is numerical methods and simulation. High-performance computing has transformed modern research and enabled investigations that were formerly hamstrung by the limitations of special functions and artificially imposed symmetries. To do justice to the range of numerical techniques that have been developed—partial differential equation solvers, finite element methods, Monte Carlo approaches, graphics, and so on—would have more than doubled the scope and size of the book. Nonetheless, because numerical evaluations are crucial for physical insight, the book includes many applications and exercises in which user-friendly numerical packages (such as Maple, Mathematica, and Matlab) can be used to produce interesting numerical results without too much effort. We hope that, via this pathway from fundamental principle to computable outcome, our book will bring readers not only physical insight but also enthusiasm for computational physics.

Classical physics as we develop it emphasizes physical phenomena on macroscopic scales: scales where the particulate natures of matter and radiation are secondary to their behavior in bulk; scales where particles' statistical—as opposed to individual—properties are important, and where matter's inherent graininess can be smoothed over.

In this book, we take a journey through spacetime and phase space; through statistical and continuum mechanics (including solids, fluids, and plasmas); and through optics and relativity, both special and general. In our journey, we seek to comprehend the fundamental laws of classical physics in their own terms, and also in relation to quantum physics. And, using carefully chosen examples, we show how the classical laws are applied to important, contemporary, twenty-first-century problems and to everyday phenomena; and we also uncover some deep relationships among the various fundamental laws and connections among the practical techniques that are used in different subfields of physics.

Geometry is a deep theme throughout this book and a very important connector. We shall see how a few geometrical considerations dictate or strongly limit the basic principles of classical physics. Geometry illuminates the character of the classical principles and also helps relate them to the corresponding principles of quantum physics. Geometrical methods can also obviate lengthy analytical calculations. Despite this, long, routine algebraic manipulations are sometimes unavoidable; in such cases, we occasionally save space by invoking modern computational symbol manipulation programs, such as Maple, Mathematica, and Matlab.

This book is the outgrowth of courses that the authors have taught at Caltech and Stanford beginning 37 years ago. Our goal was then and remains now to fill what we saw as a large hole in the traditional physics curriculum, at least in the United States:

- We believe that every masters-level or PhD physicist should be familiar with the basic concepts of all the major branches of classical physics and should have had some experience in applying them to real-world phenomena; this book is designed to facilitate this goal.
- Many physics, astronomy, and engineering graduate students in the United States and around the world use classical physics extensively in their research, and even more of them go on to careers in which classical physics is an essential component; this book is designed to expedite their efforts.
- Many professional physicists and engineers discover, in mid-career, that they need an understanding of areas of classical physics that they had not previously mastered. This book is designed to help them fill in the gaps and see the relationship to already familiar topics.

In pursuit of this goal, we seek, in this book, to *give the reader a clear understanding of the basic concepts and principles of classical physics*. We present these principles in the language of modern physics (not nineteenth-century applied mathematics), and we present them primarily for physicists—though we have tried hard to make the content interesting, useful, and accessible to a much larger community including engineers, mathematicians, chemists, biologists, and so on. As far as possible, we emphasize theory that involves general principles which extend well beyond the particular topics we use to illustrate them.

In this book, we also seek to *teach the reader how to apply the ideas of classical physics*. We do so by presenting contemporary applications from a variety of fields, such as

- fundamental physics, experimental physics, and applied physics;
- astrophysics and cosmology;
- geophysics, oceanography, and meteorology;
- biophysics and chemical physics; and

- engineering, optical science and technology, radio science and technology, and information science and technology.

Why is the range of applications so wide? Because we believe that physicists should have enough understanding of general principles to attack problems that arise in unfamiliar environments. In the modern era, a large fraction of physics students will go on to careers outside the core of fundamental physics. For such students, a broad exposure to non-core applications can be of great value. For those who wind up in the core, such an exposure is of value culturally, and also because ideas from other fields often turn out to have impact back in the core of physics. Our examples illustrate how basic concepts and problem-solving techniques are freely interchanged across disciplines.

We strongly believe that classical physics should *not* be studied in isolation from quantum mechanics and its modern applications. Our reasons are simple:

- Quantum mechanics has primacy over classical physics. Classical physics is an approximation—often excellent, sometimes poor—to quantum mechanics.

- In recent decades, many concepts and mathematical techniques developed for quantum mechanics have been imported into classical physics and there used to enlarge our classical understanding and enhance our computational capability. An example that we shall study is nonlinearly interacting plasma waves, which are best treated as quanta ("plasmons"), despite their being solutions of classical field equations.

- Ideas developed initially for classical problems are frequently adapted for application to avowedly quantum mechanical subjects; examples (not discussed in this book) are found in supersymmetric string theory and in the liquid drop model of the atomic nucleus.

Because of these intimate connections between quantum and classical physics, quantum physics appears frequently in this book.

The amount and variety of material covered in this book may seem overwhelming. If so, keep in mind the key goals of the book: to teach the fundamental concepts, which are not so extensive that they should overwhelm, and to illustrate those concepts. Our goal is not to provide a mastery of the many illustrative applications contained in the book, but rather to convey the spirit of how to apply the basic concepts of classical physics. To help students and readers who feel overwhelmed, we have labeled as "Track Two" sections that can be skipped on a first reading, or skipped entirely—but are sufficiently interesting that many readers may choose to browse or study them. Track-Two sections are labeled by the symbol **T2**. To keep Track One manageable for a one-year course, the Track-One portion of each chapter is rarely longer than 40 pages (including many pages of exercises) and is often somewhat shorter. Track One is designed for a full-year course at the first-year graduate level; that is how we have mostly used it. (Many final-year undergraduates have taken our course successfully, but rarely easily.)

The book is divided into seven parts:

I. **Foundations**—which introduces our book's powerful *geometric* point of view on the laws of physics and brings readers up to speed on some concepts and mathematical tools that we shall need. Many readers will already have mastered most or all of the material in Part I and might find that they can understand most of the rest of the book without adopting our avowedly geometric viewpoint. Nevertheless, we encourage such readers to browse Part I, at least briefly, before moving on, so as to become familiar with this viewpoint. We believe the investment will be repaid. Part I is split into two chapters, Chap. 1 on Newtonian physics and Chap. 2 on special relativity. Since nearly all of Parts II–VI is Newtonian, readers may choose to skip Chap. 2 and the occasional special relativity sections of subsequent chapters, until they are ready to launch into Part VII, General Relativity. Accordingly, Chap. 2 is labeled Track Two, though it becomes Track One when readers embark on Part VII.

II. **Statistical Physics**—including kinetic theory, statistical mechanics, statistical thermodynamics, and the theory of random processes. These subjects underlie some portions of the rest of the book, especially plasma physics and fluid mechanics.

III. **Optics**—by which we mean classical waves of all sorts: light waves, radio waves, sound waves, water waves, waves in plasmas, and gravitational waves. The major concepts we develop for dealing with all these waves include geometric optics, diffraction, interference, and nonlinear wave-wave mixing.

IV. **Elasticity**—elastic deformations, both static and dynamic, of solids. Here we develop the use of tensors to describe continuum mechanics.

V. **Fluid Dynamics**—with flows ranging from the traditional ones of air and water to more modern cosmic and biological environments. We introduce vorticity, viscosity, turbulence, boundary layers, heat transport, sound waves, shock waves, magnetohydrodynamics, and more.

VI. **Plasma Physics**—including plasmas in Earth-bound laboratories and in technological (e.g., controlled-fusion) devices, Earth's ionosphere, and cosmic environments. In addition to magnetohydrodynamics (treated in Part V), we develop two-fluid and kinetic approaches, and techniques of nonlinear plasma physics.

VII. **General Relativity**—the physics of curved spacetime. Here we show how the physical laws that we have discussed in flat spacetime are modified to account for curvature. We also explain how energy and momentum generate this curvature. These ideas are developed for their principal classical applications to neutron stars, black holes, gravitational radiation, and cosmology.

It should be possible to read and teach these parts independently, provided one is prepared to use the cross-references to access some concepts, tools, and results developed in earlier parts.

Five of the seven parts (II, III, V, VI, and VII) conclude with chapters that focus on applications where there is much current research activity and, consequently, there are many opportunities for physicists.

Exercises are a major component of this book. There are five types of exercises:

1. *Practice.* Exercises that provide practice at mathematical manipulations (e.g., of tensors).
2. *Derivation.* Exercises that fill in details of arguments skipped over in the text.
3. *Example.* Exercises that lead the reader step by step through the details of some important extension or application of the material in the text.
4. *Problem.* Exercises with few, if any, hints, in which the task of figuring out how to set up the calculation and get started on it often is as difficult as doing the calculation itself.
5. *Challenge.* Especially difficult exercises whose solution may require reading other books or articles as a foundation for getting started.

We urge readers to try working many of the exercises, especially the examples, which should be regarded as continuations of the text and which contain many of the most illuminating applications. Exercises that we regard as especially important are designated with **.

A few words on units and conventions. In this book we deal with practical matters and frequently need to have a quantitative understanding of the magnitudes of various physical quantities. This requires us to adopt a particular unit system. Physicists use both Gaussian and SI units; units that lie outside both formal systems are also commonly used in many subdisciplines. Both Gaussian and SI units provide a complete and internally consistent set for all of physics, and it is an often-debated issue as to which system is more convenient or aesthetically appealing. We will not enter this debate! One's choice of units should not matter, and a mature physicist should be able to change from one system to another with little thought. However, when learning new concepts, having to figure out "where the 2πs and 4πs go" is a genuine impediment to progress. Our solution to this problem is as follows. For each physics subfield that we study, we consistently use the set of units that seem most natural or that, we judge, constitute the majority usage by researchers in that subfield. We do not pedantically convert cm to m or vice versa at every juncture; we trust that the reader can easily make whatever translation is necessary. However, where the equations are actually different—primarily in electromagnetic theory—we occasionally provide, in brackets or footnotes, the equivalent equations in the other unit system and enough information for the reader to proceed in his or her preferred scheme.

We encourage readers to consult this book's website, http://press.princeton.edu/titles/MCP.html, for information, errata, and various resources relevant to the book.

A large number of people have influenced this book and our viewpoint on the material in it. We list many of them and express our thanks in the Acknowledgments. Many misconceptions and errors have been caught and corrected. However, in a book of this size and scope, others will remain, and for these we take full responsibility. We would be delighted to learn of these from readers and will post corrections and explanations on this book's website when we judge them to be especially important and helpful.

Above all, we are grateful for the support of our wives, Carolee and Liz—and especially for their forbearance in epochs when our enterprise seemed like a mad and vain pursuit of an unreachable goal, a pursuit that we juggled with huge numbers of other obligations, while Liz and Carolee, in the midst of their own careers, gave us the love and encouragement that were crucial in keeping us going.

ACKNOWLEDGMENTS

This book evolved gradually from notes written in 1980–81, through improved notes, then sparse prose, and on into text that ultimately morphed into what you see today. Over these three decades and more, courses based on our evolving notes and text were taught by us and by many of our colleagues at Caltech, Stanford, and elsewhere. From those teachers and their students, and from readers who found our evolving text on the web and dove into it, we have received an extraordinary volume of feedback,[1] and also patient correction of errors and misconceptions as well as help with translating passages that were correct but impenetrable into more lucid and accessible treatments. For all this feedback and to all who gave it, we are extremely grateful. We wish that we had kept better records; the heartfelt thanks that we offer all these colleagues, students, and readers, named and unnamed, are deeply sincere.

Teachers who taught courses based on our evolving notes and text, and gave invaluable feedback, include Professors Richard Blade, Yanbei Chen, Michael Cross, Steven Frautschi, Peter Goldreich, Steve Koonin, Christian Ott, Sterl Phinney, David Politzer, John Preskill, John Schwarz, and David Stevenson at Caltech; Professors Tom Abel, Seb Doniach, Bob Wagoner, and Shoucheng Zhang at Stanford; and Professor Sandor Kovacs at Washington University in St. Louis.

Our teaching assistants, who gave us invaluable feedback on the text, improvements of exercises, and insights into the difficulty of the material for the students, include Jeffrey Atwell, Nate Bode, Yu Cao, Yi-Yuh Chen, Jane Dai, Alexei Dvoretsky, Fernando Echeverria, Jiyu Feng, Eanna Flanagan, Marc Goroff, Dan Grin, Arun Gupta, Alexandr Ikriannikov, Anton Kapustin, Kihong Kim, Hee-Won Lee, Geoffrey Lovelace, Miloje Makivic, Draza Markovic, Keith Matthews, Eric Moranson, Mike Morris, Chung-Yi Mou, Rob Owen, Yi Pan, Jaemo Park, Apoorva Patel, Alexander Putilin, Shuyan Qi, Soo Jong Rey, Fintan Ryan, Bonnie Shoemaker, Paul Simeon,

---

1. Specific applications that were originated by others, to the best of our memory, are acknowledged in the text.

Hidenori Sinoda, Matthew Stevenson, Wai Mo Suen, Marcus Teague, Guodang Wang, Xinkai Wu, Huan Yang, Jimmy Yee, Piljin Yi, Chen Zheng, and perhaps others of whom we have lost track!

Among the students and readers of our notes and text, who have corresponded with us, sending important suggestions and errata, are Bram Achterberg, Mustafa Amin, Richard Anantua, Alborz Bejnood, Edward Blandford, Jonathan Blandford, Dick Bond, Phil Bucksbaum, James Camparo, Conrado Cano, U Lei Chan, Vernon Chaplin, Mina Cho, Ann Marie Cody, Sandro Commandè, Kevin Fiedler, Krzysztof Findeisen, Jeff Graham, Casey Handmer, Ted Jacobson, Matt Kellner, Deepak Kumar, Andrew McClung, Yuki Moon, Evan O'Connor, Jeffrey Oishi, Keith Olive, Zhen Pan, Eric Peterson, Laurence Perreault Levasseur, Vahbod Pourahmad, Andreas Reisenegger, David Reiss, Pavlin Savov, Janet Scheel, Yuki Takahashi, Fun Lim Yee, Yajie Yuan, and Aaron Zimmerman.

For computational advice or assistance, we thank Edward Campbell, Mark Scheel, Chris Mach, and Elizabeth Wood.

Academic support staff who were crucial to our work on this book include Christine Aguilar, JoAnn Boyd, Jennifer Formicelli, and Shirley Hampton.

The editorial and production professionals at Princeton University Press (Peter Dougherty, Karen Fortgang, Ingrid Gnerlich, Eric Henney, and Arthur Werneck) and at Princeton Editorial Associates (Peter Strupp and his freelance associates Paul Anagnostopoulos, Laurel Muller, MaryEllen Oliver, Joe Snowden, and Cyd Westmoreland) have been magnificent, helping us plan and design this book, and transforming our raw prose and primitive figures into a visually appealing volume, with sustained attention to detail, courtesy, and patience as we missed deadline after deadline.

Of course, we the authors take full responsibility for all the errors of judgment, bad choices, and mistakes that remain.

Roger Blandford thanks his many supportive colleagues at Caltech, Stanford University, and the Kavli Institute for Particle Astrophysics and Cosmology. He also acknowledges the Humboldt Foundation, the Miller Institute, the National Science Foundation, and the Simons Foundation for generous support during the completion of this book. And he also thanks the Berkeley Astronomy Department; Caltech; the Institute of Astronomy, Cambridge; and the Max Planck Institute for Astrophysics, Garching, for hospitality.

Kip Thorne is grateful to Caltech—the administration, faculty, students, and staff—for the supportive environment that made possible his work on this book, work that occupied a significant portion of his academic career.

PART I

FOUNDATIONS

In this book, a central theme will be a *Geometric Principle: The laws of physics must all be expressible as geometric (coordinate-independent and reference-frame-independent) relationships between geometric objects (scalars, vectors, tensors, ...) that represent physical entities.*

There are three different conceptual frameworks for the classical laws of physics, and correspondingly, three different geometric arenas for the laws; see Fig. 1. General relativity is the most accurate classical framework; it formulates the laws as geometric relationships among geometric objects in the arena of curved 4-dimensional spacetime. Special relativity is the limit of general relativity in the complete absence of gravity; its arena is flat, 4-dimensional Minkowski spacetime.[1] Newtonian physics is the limit of general relativity when

- gravity is weak but not necessarily absent,
- relative speeds of particles and materials are small compared to the speed of light c, and
- all stresses (pressures) are small compared to the total density of mass-energy.

Its arena is flat, 3-dimensional Euclidean space with time separated off and made universal (by contrast with relativity's reference-frame-dependent time).

In Parts II–VI of this book (covering statistical physics, optics, elasticity, fluid mechanics, and plasma physics), we confine ourselves to the Newtonian formulations of the laws (plus special relativistic formulations in portions of Track Two), and accordingly, our arena will be flat Euclidean space (plus flat Minkowski spacetime in portions of Track Two). In Part VII, we extend many of the laws we have studied into the domain of strong gravity (general relativity)—the arena of curved spacetime.

---

1. This is so-called because Hermann Minkowski (1908) identified the special relativistic invariant interval as defining a metric in spacetime and elucidated the resulting geometry of flat spacetime.

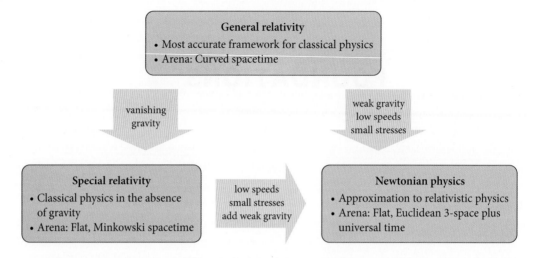

FIGURE 1 The three frameworks and arenas for the classical laws of physics and their relationship to one another.

In Parts II and III (on statistical physics and optics), in addition to confining ourselves to flat space (plus flat spacetime in Track Two), we avoid any sophisticated use of curvilinear coordinates. Correspondingly, when using coordinates in nontrivial ways, we confine ourselves to Cartesian coordinates in Euclidean space (and Lorentz coordinates in Minkowski spacetime).

Part I of this book contains just two chapters. Chapter 1 is an introduction to our geometric viewpoint on Newtonian physics and to all the geometric mathematical tools that we shall need in Parts II and III for Newtonian physics in its arena, 3-dimensional Euclidean space. Chapter 2 introduces our geometric viewpoint on special relativistic physics and extends our geometric tools into special relativity's arena, flat Minkowski spacetime. Readers whose focus is Newtonian physics will have no need for Chap. 2; and if they are already familiar with the material in Chap. 1 but not from our geometric viewpoint, they can successfully study Parts II–VI without reading Chap. 1. However, in doing so, they will miss some deep insights; so we recommend they at least browse Chap. 1 to get some sense of our viewpoint, then return to the chapter occasionally, as needed, when encountering an unfamiliar geometric argument.

In Parts IV, V, and VI, when studying elasticity, fluid dynamics, and plasma physics, we use curvilinear coordinates in nontrivial ways. As a foundation for this, at the beginning of Part IV, we extend our flat-space geometric tools to curvilinear coordinate systems (e.g., cylindrical and spherical coordinates). Finally, at the beginning of Part VII, we extend our geometric tools to the arena of curved spacetime.

Throughout this book, we pay close attention to the relationship between classical physics and quantum physics. Indeed, we often find it powerful to use quantum mechanical language or formalism when discussing and analyzing classical phenomena.

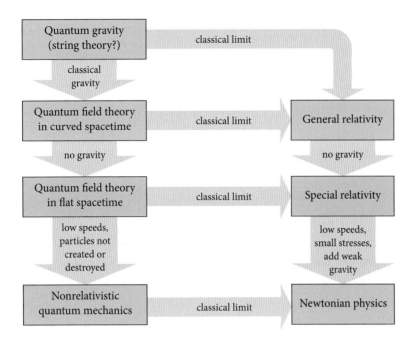

FIGURE 2 The relationship of the three frameworks for classical physics (on right) to four frameworks for quantum physics (on left). Each arrow indicates an approximation. All other frameworks are approximations to the ultimate laws of quantum gravity (whatever they may be—perhaps a variant of string theory).

This quantum power in classical domains arises because quantum physics is primary and classical physics is secondary. Today we see classical physics as arising from quantum physics, though historically the linkage was inverted. The relationship between quantum frameworks and arenas for the laws of physics, and classical frameworks, is sketched in Fig. 2.

CHAPTER ONE

Newtonian Physics: Geometric Viewpoint

> Geometry postulates the solution of these problems from mechanics and teaches the use of the problems thus solved. And geometry can boast that with so few principles obtained from other fields, it can do so much.
>
> ISAAC NEWTON, 1687

1.1 Introduction

1.1.1 The Geometric Viewpoint on the Laws of Physics

In this book, we adopt a different viewpoint on the laws of physics than that in many elementary and intermediate texts. In most textbooks, physical laws are expressed in terms of quantities (locations in space, momenta of particles, etc.) that are measured in some coordinate system. For example, Newtonian vectorial quantities are expressed as triplets of numbers [e.g., $\mathbf{p} = (p_x, p_y, p_z) = (1, 9, -4)$], representing the components of a particle's momentum on the axes of a Cartesian coordinate system; and tensors are expressed as arrays of numbers (e.g.,

$$\mathbf{I} = \begin{bmatrix} I_{xx} & I_{xy} & I_{xz} \\ I_{yx} & I_{yy} & I_{yz} \\ I_{zx} & I_{zy} & I_{zz} \end{bmatrix} \tag{1.1}$$

for the moment of inertia tensor).

By contrast, in this book we express all physical quantities and laws in *geometric forms*, i.e., in forms that are *independent of any coordinate system or basis vectors*. For example, a particle's velocity $\mathbf{v}$ and the electric and magnetic fields $\mathbf{E}$ and $\mathbf{B}$ that it encounters will be vectors described as arrows that live in the 3-dimensional, flat Euclidean space of everyday experience.[1] They require no coordinate system or basis vectors for their existence or description—though often coordinates will be useful. In other words, $\mathbf{v}$ represents the vector itself and is not just shorthand for an ordered list of numbers.

---

1. This interpretation of a vector is close to the ideas of Newton and Faraday. Lagrange, Hamilton, Maxwell, and many others saw vectors in terms of Cartesian components. The vector notation was streamlined by Gibbs, Heaviside, and others, but the underlying coordinate system was still implicit, and $\mathbf{v}$ was usually regarded as shorthand for (v_x, v_y, v_z).

BOX 1.1. READERS' GUIDE

- This chapter is a foundation for almost all of this book.
- Many readers already know the material in this chapter, but from a viewpoint different from our *geometric* one. Such readers will be able to understand almost all of Parts II–VI of this book without learning our viewpoint. Nevertheless, that geometric viewpoint has such power that we encourage them to learn it by browsing this chapter and focusing especially on Secs. 1.1.1, 1.2, 1.3, 1.5, 1.7, and 1.8.
- The stress tensor, introduced and discussed in Sec. 1.9, plays an important role in kinetic theory (Chap. 3) and a crucial role in elasticity (Part IV), fluid dynamics (Part V), and plasma physics (Part VI).
- The integral and differential conservation laws derived and discussed in Secs. 1.8 and 1.9 play major roles throughout this book.
- The Box labeled **T2** is advanced material (Track Two) that can be skipped in a time-limited course or on a first reading of this book.

We insist that the Newtonian laws of physics all obey a *Geometric Principle*: they are all geometric relationships among geometric objects (primarily scalars, vectors, and tensors), expressible without the aid of any coordinates or bases. An example is the Lorentz force law $m d\mathbf{v}/dt = q(\mathbf{E} + \mathbf{v} \times \mathbf{B})$—a (coordinate-free) relationship between the geometric (coordinate-independent) vectors $\mathbf{v}$, $\mathbf{E}$, and $\mathbf{B}$ and the particle's scalar mass m and charge q. As another example, a body's moment of inertia tensor **I** can be viewed as a vector-valued linear function of vectors (a coordinate-independent, basis-independent geometric object). Insert into the tensor **I** the body's angular velocity vector $\mathbf{\Omega}$, and you get out the body's angular momentum vector: $\mathbf{J} = \mathbf{I}(\mathbf{\Omega})$. No coordinates or basis vectors are needed for this law of physics, nor is any description of **I** as a matrix-like entity with components I_{ij} required. Components are secondary; they only exist after one has chosen a set of basis vectors. Components (we claim) are an impediment to a clear and deep understanding of the laws of classical physics. The coordinate-free, component-free description is deeper, and—once one becomes accustomed to it—much more clear and understandable.[2]

---

2. This philosophy is also appropriate for quantum mechanics (see Box 1.2) and, especially, quantum field theory, where it is the invariance of the description under gauge and other symmetry operations that is the powerful principle. However, its implementation there is less direct, simply because the spaces in which these symmetries lie are more abstract and harder to conceptualize.

By adopting this geometric viewpoint, we gain great conceptual power and often also computational power. For example, when we ignore experiment and simply ask what forms the laws of physics can possibly take (what forms are allowed by the requirement that the laws be geometric), we shall find that there is remarkably little freedom. Coordinate independence and basis independence strongly constrain the laws of physics.[3]

This power, together with the elegance of the geometric formulation, suggests that in some deep sense, Nature's physical laws are geometric and have nothing whatsoever to do with coordinates or components or vector bases.

1.1.2 Purposes of This Chapter

The principal purpose of this foundational chapter is to teach the reader this geometric viewpoint.

The mathematical foundation for our geometric viewpoint is *differential geometry* (also called "tensor analysis" by physicists). Differential geometry can be thought of as an extension of the vector analysis with which all readers should be familiar. *A second purpose of this chapter is to develop key parts of differential geometry in a simple form well adapted to Newtonian physics.*

1.1.3 Overview of This Chapter

In this chapter, we lay the geometric foundations for the Newtonian laws of physics in flat Euclidean space. We begin in Sec. 1.2 by introducing some foundational geometric concepts: points, scalars, vectors, inner products of vectors, and the distance between points. Then in Sec. 1.3, we introduce the concept of a tensor as a linear function of vectors, and we develop a number of geometric tools: the tools of coordinate-free tensor algebra. In Sec. 1.4, we illustrate our tensor-algebra tools by using them to describe—without any coordinate system—the kinematics of a charged point particle that moves through Euclidean space, driven by electric and magnetic forces.

In Sec. 1.5, we introduce, for the first time, Cartesian coordinate systems and their basis vectors, and also the components of vectors and tensors on those basis vectors; and we explore how to express geometric relationships in the language of components. In Sec. 1.6, we deduce how the components of vectors and tensors transform when one rotates the chosen Cartesian coordinate axes. (These are the transformation laws that most physics textbooks use to define vectors and tensors.)

In Sec. 1.7, we introduce directional derivatives and gradients of vectors and tensors, thereby moving from tensor algebra to true differential geometry (in Euclidean space). We also introduce the Levi-Civita tensor and use it to define curls and cross

---

3. Examples are the equation of elastodynamics (12.4b) and the Navier-Stokes equation of fluid mechanics (13.69), which are both dictated by momentum conservation plus the form of the stress tensor [Eqs. (11.18), (13.43), and (13.68)]—forms that are dictated by the irreducible tensorial parts (Box 11.2) of the strain and rate of strain.

products, and we learn how to use *index gymnastics* to derive, quickly, formulas for multiple cross products. In Sec. 1.8, we use the Levi-Civita tensor to define vectorial areas, scalar volumes, and integration over surfaces. These concepts then enable us to formulate, in geometric, coordinate-free ways, integral and differential conservation laws. In Sec. 1.9, we discuss, in particular, the law of momentum conservation, formulating it in a geometric way with the aid of a geometric object called the *stress tensor.* As important examples, we use this geometric conservation law to derive and discuss the equations of Newtonian fluid dynamics, and the interaction between a charged medium and an electromagnetic field. We conclude in Sec. 1.10 with some concepts from special relativity that we shall need in our discussions of Newtonian physics.

1.2 Foundational Concepts

In this section, we sketch the foundational concepts of Newtonian physics without using any coordinate system or basis vectors. This is the geometric viewpoint that we advocate.

space and time

The arena for the Newtonian laws of physics is a spacetime composed of the familiar 3-dimensional Euclidean space of everyday experience (which we call *3-space*) and a universal time t. We denote points (locations) in 3-space by capital script letters, such as $\mathcal{P}$ and $\mathcal{Q}$. These points and the 3-space in which they live require no coordinates for their definition.

scalar

A *scalar* is a single number. We are most interested in scalars that directly represent physical quantities (e.g., temperature T). As such, they are real numbers, and when they are functions of location $\mathcal{P}$ in space [e.g., $T(\mathcal{P})$], we call them *scalar fields*. However, sometimes we will work with complex numbers—most importantly in quantum mechanics, but also in various Fourier representations of classical physics.

vector

A *vector* in Euclidean 3-space can be thought of as a straight arrow (or more formally a directed line segment) that reaches from one point, $\mathcal{P}$, to another, $\mathcal{Q}$ (e.g., the arrow $\Delta \mathbf{x}$ in Fig. 1.1a). Equivalently, $\Delta \mathbf{x}$ can be thought of as a direction at $\mathcal{P}$ and a number, the vector's length. Sometimes we shall select one point $\mathcal{O}$ in 3-space as an "origin" and identify all other points, say, $\mathcal{Q}$ and $\mathcal{P}$, by their vectorial separations $\mathbf{x}_\mathcal{Q}$ and $\mathbf{x}_\mathcal{P}$ from that origin.

distance and length

The Euclidean distance $\Delta \sigma$ between two points $\mathcal{P}$ and $\mathcal{Q}$ in 3-space can be measured with a ruler and so, of course, requires no coordinate system for its definition. (If one does have a Cartesian coordinate system, then $\Delta \sigma$ can be computed by the Pythagorean formula, a precursor to the invariant interval of flat spacetime; Sec. 2.2.3.) This distance $\Delta \sigma$ is also the *length* $|\Delta \mathbf{x}|$ of the vector $\Delta \mathbf{x}$ that reaches from $\mathcal{P}$ to $\mathcal{Q}$, and the square of that length is denoted

$$|\Delta \mathbf{x}|^2 \equiv (\Delta \mathbf{x})^2 \equiv (\Delta \sigma)^2. \tag{1.2}$$

Of particular importance is the case when $\mathcal{P}$ and $\mathcal{Q}$ are neighboring points and $\Delta \mathbf{x}$ is a differential (infinitesimal) quantity $d\mathbf{x}$. This *infinitesimal displacement* is a more fundamental physical quantity than the finite $\Delta \mathbf{x}$. To create a finite vector out

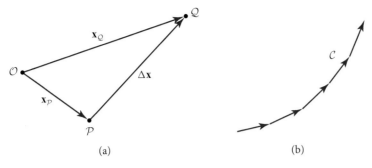

FIGURE 1.1 (a) A Euclidean 3-space diagram depicting two points $\mathcal{P}$ and $\mathcal{Q}$, their respective vectorial separations $\mathbf{x}_\mathcal{P}$ and $\mathbf{x}_\mathcal{Q}$ from the (arbitrarily chosen) origin $\mathcal{O}$, and the vector $\Delta \mathbf{x} = \mathbf{x}_\mathcal{Q} - \mathbf{x}_\mathcal{P}$ connecting them. (b) A curve $\mathcal{P}(\lambda)$ generated by laying out a sequence of infinitesimal vectors, tail-to-tip.

of infinitesimal vectors, one has to add several infinitesimal vectors head to tail, head to tail, and so on, and then take a limit. This involves *translating* a vector from one point to the next. There is no ambiguity about doing this in flat Euclidean space using the geometric notion of parallelism.[4] This simple property of Euclidean space enables us to add (and subtract) vectors at a point. We attach the tail of a second vector to the head of the first vector and then construct the sum as the vector from the tail of the first to the head of the second, or vice versa, as should be quite familiar. The point is that we do not need to add the Cartesian components to sum vectors.

We can also rotate vectors about their tails by pointing them along a different direction in space. Such a rotation can be specified by two angles. The space that is defined by all possible changes of length and direction at a point is called that point's *tangent space*. Again, we generally view the rotation as being that of a physical vector in space, and not, as it is often useful to imagine, the rotation of some coordinate system's basis vectors, with the chosen vector itself kept fixed.

tangent space

We can also construct a path through space by laying down a sequence of infinitesimal $d\mathbf{x}$s, tail to head, one after another. The resulting path is a *curve* to which these $d\mathbf{x}$s are tangent (Fig. 1.1b). The curve can be denoted $\mathcal{P}(\lambda)$, with λ a parameter along the curve and $\mathcal{P}(\lambda)$ the point on the curve whose parameter value is λ, or $\mathbf{x}(\lambda)$ where $\mathbf{x}$ is the vector separation of $\mathcal{P}$ from the arbitrary origin $\mathcal{O}$. The infinitesimal vectors that map the curve out are $d\mathbf{x} = (d\mathcal{P}/d\lambda)\, d\lambda = (d\mathbf{x}/d\lambda)\, d\lambda$, and $d\mathcal{P}/d\lambda = d\mathbf{x}/d\lambda$ is the tangent vector to the curve.

curve

tangent vector

If the curve followed is that of a particle, and the parameter λ is time t, then we have defined the *velocity* $\mathbf{v} \equiv d\mathbf{x}/dt$. In effect we are multiplying the vector $d\mathbf{x}$ by the scalar $1/dt$ and taking the limit. Performing this operation at every point $\mathcal{P}$ in the space occupied by a fluid defines the fluid's *velocity field* $\mathbf{v}(\mathbf{x})$. Multiplying a particle's velocity $\mathbf{v}$ by its scalar mass gives its *momentum* $\mathbf{p} = m\mathbf{v}$. Similarly, the difference $d\mathbf{v}$

---

4. The statement that there is just one choice of line parallel to a given line, through a point not lying on the line, is the famous fifth axiom of Euclid.

of two velocity measurements during a time interval dt, multiplied by $1/dt$, generates the particle's *acceleration* $\mathbf{a} = d\mathbf{v}/dt$. Multiplying by the particle's mass gives the force $\mathbf{F} = m\mathbf{a}$ that produced the acceleration; dividing an electrically produced force by the particle's charge q gives the electric field $\mathbf{E} = \mathbf{F}/q$. And so on.

We can define inner products [see Eq. (1.4a) below] and cross products [Eq. (1.22a)] of pairs of vectors at the same point geometrically; then using those vectors we can define, for example, the rate that work is done by a force and a particle's angular momentum about a point.

These two products can be expressed geometrically as follows. If we allow the two vectors to define a parallelogram, then their cross product is the vector orthogonal to the parallelogram with length equal to the parallelogram's area. If we first rotate one vector through a right angle in a plane containing the other, and then define the parallelogram, its area is the vectors' inner product.

derivatives of scalars and vectors

We can also define spatial derivatives. We associate the difference of a scalar between two points separated by $d\mathbf{x}$ at the same time with a *gradient* and, likewise, go on to define the scalar *divergence* and the vector *curl*. The freedom to translate vectors from one point to the next also underlies the association of a single vector (e.g., momentum) with a group of particles or an extended body. One simply adds all the individual momenta, taking a limit when necessary.

In this fashion (which should be familiar to the reader and will be elucidated, formalized, and generalized below), we can construct all the standard scalars and vectors of Newtonian physics. What is important is that *these physical quantities require no coordinate system for their definition*. They are geometric (coordinate-independent) objects residing in Euclidean 3-space at a particular time.

Geometric Principle

It is a fundamental (though often ignored) principle of physics that *the Newtonian physical laws are all expressible as geometric relationships among these types of geometric objects, and these relationships do not depend on any coordinate system or orientation of axes, nor on any reference frame* (i.e., on any purported velocity of the Euclidean space in which the measurements are made).[5] We call this the *Geometric Principle* for the laws of physics, and we use it throughout this book. It is the Newtonian analog of Einstein's Principle of Relativity (Sec. 2.2.2).

1.3 Tensor Algebra without a Coordinate System

In preparation for developing our geometric view of physical laws, we now introduce, in a coordinate-free way, some fundamental concepts of differential geometry: tensors, the inner product, the metric tensor, the tensor product, and contraction of tensors.

We have already defined a vector $\mathbf{A}$ as a straight arrow from one point, say $\mathcal{P}$, in our space to another, say $\mathcal{Q}$. Because our space is flat, there is a unique and obvious way to

---

5. By changing the velocity of Euclidean space, one adds a constant velocity to all particles, but this leaves the laws (e.g., Newton's $\mathbf{F} = m\mathbf{a}$) unchanged.

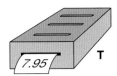

FIGURE 1.2 A rank-3 tensor **T**.

transport such an arrow from one location to another, keeping its length and direction unchanged.[6] Accordingly, we shall regard vectors as unchanged by such transport. This enables us to ignore the issue of where in space a vector actually resides; it is completely determined by its direction and its length.

A *rank-n tensor* **T** is, by definition, a real-valued linear function of n vectors.[7] Pictorially we regard **T** as a box (Fig. 1.2) with n slots in its top, into which are inserted n vectors, and one slot in its end, which prints out a single real number: the value that the tensor **T** has when evaluated as a function of the n inserted vectors. Notationally we denote the tensor by a boldfaced sans-serif character **T**:

tensor

$$\mathbf{T}(\underbrace{\_,\_,\_,\_}_{n \text{ slots in which to put the vectors.}}) \tag{1.3a}$$

This definition of a tensor is very different (and far simpler) than the one found in most standard physics textbooks (e.g., Marion and Thornton, 1995; Jackson, 1999; Griffiths, 1999). There, a tensor is an array of numbers that transform in a particular way under rotations. We shall learn the connection between these definitions in Sec. 1.6 below.

To illustrate this approach, if **T** is a rank-3 tensor (has 3 slots) as in Fig. 1.2, then its value on the vectors **A**, **B**, **C** is denoted **T**(**A**, **B**, **C**). Linearity of this function can be expressed as

$$\mathbf{T}(e\mathbf{E} + f\mathbf{F}, \mathbf{B}, \mathbf{C}) = e\mathbf{T}(\mathbf{E}, \mathbf{B}, \mathbf{C}) + f\mathbf{T}(\mathbf{F}, \mathbf{B}, \mathbf{C}), \tag{1.3b}$$

where e and f are real numbers, and similarly for the second and third slots.

We have already defined the squared length $(\mathbf{A})^2 \equiv \mathbf{A}^2$ of a vector **A** as the squared distance between the points at its tail and its tip. The *inner product* (also called the dot product) $\mathbf{A} \cdot \mathbf{B}$ of two vectors is defined in terms of this squared length by

inner product

$$\boxed{\mathbf{A} \cdot \mathbf{B} \equiv \frac{1}{4}\left[(\mathbf{A} + \mathbf{B})^2 - (\mathbf{A} - \mathbf{B})^2\right].} \tag{1.4a}$$

In Euclidean space, this is the standard inner product, familiar from elementary geometry and discussed above in terms of the area of a parallelogram.

---

6. This is not so in curved spaces, as we shall see in Sec. 24.3.4.
7. This is a different use of the word *rank* than for a matrix, whose rank is its number of linearly independent rows or columns.

1.3 Tensor Algebra without a Coordinate System

One can show that the inner product (1.4a) is a real-valued linear function of each of its vectors. Therefore, we can regard it as a tensor of rank 2. When so regarded, the inner product is denoted **g**(__, __) and is called the *metric tensor*. In other words, the metric tensor **g** is that linear function of two vectors whose value is given by

metric tensor

$$\boxed{\mathbf{g}(\mathbf{A}, \mathbf{B}) \equiv \mathbf{A} \cdot \mathbf{B}.} \tag{1.4b}$$

Notice that, because $\mathbf{A} \cdot \mathbf{B} = \mathbf{B} \cdot \mathbf{A}$, the metric tensor is *symmetric* in its two slots—one gets the same real number independently of the order in which one inserts the two vectors into the slots:

$$\mathbf{g}(\mathbf{A}, \mathbf{B}) = \mathbf{g}(\mathbf{B}, \mathbf{A}). \tag{1.4c}$$

With the aid of the inner product, we can regard any vector **A** as a tensor of rank one: the real number that is produced when an arbitrary vector **C** is inserted into **A**'s single slot is

$$\boxed{\mathbf{A}(\mathbf{C}) \equiv \mathbf{A} \cdot \mathbf{C}.} \tag{1.4d}$$

In Newtonian physics, we rarely meet tensors of rank higher than two. However, second-rank tensors appear frequently—often in roles where one sticks a single vector into the second slot and leaves the first slot empty, thereby producing a single-slotted entity, a vector. An example that we met in Sec. 1.1.1 is a rigid body's moment-of-inertia tensor **I**(__, __), which gives us the body's angular momentum **J**(__) = **I**(__, **Ω**) when its angular velocity **Ω** is inserted into its second slot.[8] Another example is the stress tensor of a solid, a fluid, a plasma, or a field (Sec. 1.9 below).

tensor product

From three vectors **A**, **B**, **C**, we can construct a tensor, their *tensor product* (also called *outer product* in contradistinction to the inner product $\mathbf{A} \cdot \mathbf{B}$), defined as follows:

$$\boxed{\mathbf{A} \otimes \mathbf{B} \otimes \mathbf{C}(\mathbf{E}, \mathbf{F}, \mathbf{G}) \equiv \mathbf{A}(\mathbf{E})\mathbf{B}(\mathbf{F})\mathbf{C}(\mathbf{G}) = (\mathbf{A} \cdot \mathbf{E})(\mathbf{B} \cdot \mathbf{F})(\mathbf{C} \cdot \mathbf{G}).} \tag{1.5a}$$

Here the first expression is the notation for the value of the new tensor, $\mathbf{A} \otimes \mathbf{B} \otimes \mathbf{C}$ evaluated on the three vectors **E**, **F**, **G**; the middle expression is the ordinary product of three real numbers, the value of **A** on **E**, the value of **B** on **F**, and the value of **C** on **G**; and the third expression is that same product with the three numbers rewritten as scalar products. Similar definitions can be given (and should be obvious) for the tensor product of any number of vectors, and of any two or more tensors of any rank; for example, if **T** has rank 2 and **S** has rank 3, then

$$\mathbf{T} \otimes \mathbf{S}(\mathbf{E}, \mathbf{F}, \mathbf{G}, \mathbf{H}, \mathbf{J}) \equiv \mathbf{T}(\mathbf{E}, \mathbf{F})\mathbf{S}(\mathbf{G}, \mathbf{H}, \mathbf{J}). \tag{1.5b}$$

contraction

One last geometric (i.e., frame-independent) concept we shall need is *contraction*. We illustrate this concept first by a simple example, then give the general definition.

---

8. Actually, it doesn't matter which slot, since **I** is symmetric.

Chapter 1. Newtonian Physics: Geometric Viewpoint

From two vectors **A** and **B** we can construct the tensor product $\mathbf{A} \otimes \mathbf{B}$ (a second-rank tensor), and we can also construct the scalar product $\mathbf{A} \cdot \mathbf{B}$ (a real number, i.e., a *scalar*, also known as a *rank-0 tensor*). The process of contraction is the construction of $\mathbf{A} \cdot \mathbf{B}$ from $\mathbf{A} \otimes \mathbf{B}$:

$$\boxed{\text{contraction}(\mathbf{A} \otimes \mathbf{B}) \equiv \mathbf{A} \cdot \mathbf{B}.} \tag{1.6a}$$

One can show fairly easily using component techniques (Sec. 1.5 below) that any second-rank tensor **T** can be expressed as a sum of tensor products of vectors, $\mathbf{T} = \mathbf{A} \otimes \mathbf{B} + \mathbf{C} \otimes \mathbf{D} + \ldots$. Correspondingly, it is natural to define the contraction of **T** to be contraction(**T**) $= \mathbf{A} \cdot \mathbf{B} + \mathbf{C} \cdot \mathbf{D} + \ldots$. Note that this contraction process lowers the rank of the tensor by two, from 2 to 0. Similarly, for a tensor of rank n one can construct a tensor of rank $n - 2$ by contraction, but in this case one must specify which slots are to be contracted. For example, if **T** is a third-rank tensor, expressible as $\mathbf{T} = \mathbf{A} \otimes \mathbf{B} \otimes \mathbf{C} + \mathbf{E} \otimes \mathbf{F} \otimes \mathbf{G} + \ldots$, then the contraction of **T** on its first and third slots is the rank-1 tensor (vector)

$$1\&3\text{contraction}(\mathbf{A} \otimes \mathbf{B} \otimes \mathbf{C} + \mathbf{E} \otimes \mathbf{F} \otimes \mathbf{G} + \ldots) \equiv (\mathbf{A} \cdot \mathbf{C})\mathbf{B} + (\mathbf{E} \cdot \mathbf{G})\mathbf{F} + \ldots. \tag{1.6b}$$

Unfortunately, there is no simple index-free notation for contraction in common use.

All the concepts developed in this section (vector, tensor, metric tensor, inner product, tensor product, and contraction of a tensor) can be carried over, with no change whatsoever, into any vector space[9] that is endowed with a concept of squared length—for example, to the 4-dimensional spacetime of special relativity (next chapter).

1.4 Particle Kinetics and Lorentz Force in Geometric Language

In this section, we illustrate our geometric viewpoint by formulating Newton's laws of motion for particles.

In Newtonian physics, a classical particle moves through Euclidean 3-space as universal time t passes. At time t it is located at some point $\mathbf{x}(t)$ (its *position*). The function $\mathbf{x}(t)$ represents a curve in 3-space, the particle's *trajectory*. The particle's *velocity* $\mathbf{v}(t)$ is the time derivative of its position, its *momentum* $\mathbf{p}(t)$ is the product of its mass m and velocity, its *acceleration* $\mathbf{a}(t)$ is the time derivative of its velocity, and its *kinetic energy* $E(t)$ is half its mass times velocity squared:

trajectory, velocity, momentum, acceleration, and energy

$$\mathbf{v}(t) = \frac{d\mathbf{x}}{dt}, \quad \mathbf{p}(t) = m\mathbf{v}(t), \quad \mathbf{a}(t) = \frac{d\mathbf{v}}{dt} = \frac{d^2\mathbf{x}}{dt^2}, \quad E(t) = \frac{1}{2}m\mathbf{v}^2. \tag{1.7a}$$

---

9. Or, more precisely, any vector space over the real numbers. If the vector space's scalars are complex numbers, as in quantum mechanics, then slight changes are needed.

Since points in 3-space are geometric objects (defined independently of any coordinate system), so also are the trajectory $\mathbf{x}(t)$, the velocity, the momentum, the acceleration, and the energy. (Physically, of course, the velocity has an ambiguity; it depends on one's standard of rest.)

Newton's second law of motion states that the particle's momentum can change only if a force $\mathbf{F}$ acts on it, and that its change is given by

$$d\mathbf{p}/dt = m\mathbf{a} = \mathbf{F}. \tag{1.7b}$$

If the force is produced by an electric field $\mathbf{E}$ and magnetic field $\mathbf{B}$, then this law of motion in SI units takes the familiar Lorentz-force form

$$d\mathbf{p}/dt = q(\mathbf{E} + \mathbf{v} \times \mathbf{B}). \tag{1.7c}$$

(Here we have used the vector cross product, with which the reader should be familiar, and which will be discussed formally in Sec. 1.7.)

The laws of motion (1.7) are geometric relationships among geometric objects. Let us illustrate this using something very familiar, planetary motion. Consider a light planet orbiting a heavy star. If there were no gravitational force, the planet would continue in a straight line with constant velocity $\mathbf{v}$ and speed $v = |\mathbf{v}|$, sweeping out area A at a rate $dA/dt = rv_t/2$, where r is the radius, and v_t is the tangential speed. Elementary geometry equates this to the constant $vb/2$, where b is the impact parameter—the smallest separation from the star. Now add a gravitational force $\mathbf{F}$ and let it cause a small radial impulse. A second application of geometry showed Newton that the product $rv_t/2$ is unchanged to first order in the impulse, and he recovered Kepler's second law ($dA/dt = $ const) without introducing coordinates.[10]

Contrast this approach with one relying on coordinates. For example, one introduces an (r, ϕ) coordinate system, constructs a lagrangian and observes that the coordinate ϕ is ignorable; then the Euler-Lagrange equations immediately imply the conservation of angular momentum, which is equivalent to Kepler's second law. So, which of these two approaches is preferable? The answer is surely "both!" Newton wrote the *Principia* in the language of geometry at least partly for a reason that remains valid today: it brought him a quick understanding of fundamental laws of physics. Lagrange followed his coordinate-based path to the function that bears his name, because he wanted to solve problems in celestial mechanics that would not yield to

---

10. Continuing in this vein, when the force is inverse square, as it is for gravity and electrostatics, we can use Kepler's second law to argue that when the orbit turns through a succession of equal angles $d\theta$, its successive changes in velocity $d\mathbf{v} = \mathbf{a}dt$ (with $\mathbf{a}$ the gravitational acceleration) all have the same magnitude $|d\mathbf{v}|$ and have the same angles $d\theta$ from one to another. So, if we trace the head of the velocity vector in velocity space, it follows a circle. The circle is not centered on zero velocity when the eccentricity is nonzero but there exists a reference frame in which the speed of the planet is constant. This graphical representation is known as a *hodograph*, and similar geometrical approaches are used in fluid mechanics. For Richard Feynman's masterful presentation of these ideas to first-year undergraduates, see Goodstein and Goodstein (1996).

Newton's approach. So it is today. Geometry and analysis are both indispensible. In the domain of classical physics, the geometry is of greater importance in deriving and understanding fundamental laws and has arguably been underappreciated; coordinates hold sway when we apply these laws to solve real problems. Today, both old and new laws of physics are commonly expressed geometrically, using lagrangians, hamiltonians, and actions, for example Hamilton's action principle $\delta \int L dt = 0$ where L is the coordinate-independent lagrangian. Indeed, being able to do this without introducing coordinates is a powerful guide to deriving these laws and a tool for comprehending their implications.

A comment is needed on the famous connection between *symmetry* and *conservation laws*. In our example above, angular momentum conservation followed from axial symmetry which was embodied in the lagrangian's independence of the angle ϕ; but we also deduced it geometrically. This is usually the case in classical physics; typically, we do not need to introduce a specific coordinate system to understand symmetry and to express the associated conservation laws. However, symmetries are sometimes well hidden, for example with a nutating top, and coordinate transformations are then usually the best approach to uncover them.

symmetry and conservation laws

Often in classical physics, real-world factors invalidate or complicate Lagrange's and Hamilton's coordinate-based analytical dynamics, and so one is driven to geometric considerations. As an example, consider a spherical marble rolling on a flat horizontal table. The analytical dynamics approach is to express the height of the marble's center of mass and the angle of its rotation as constraints and align the basis vectors so there is a single horizontal coordinate defined by the initial condition. It is then deduced that linear and angular momenta are conserved. Of course that result is trivial and just as easily gotten without this formalism. However, this model is also used for many idealized problems where the outcome is far from obvious and the approach is brilliantly effective. But consider the real world in which tables are warped and bumpy, marbles are ellipsoidal and scratched, air imposes a resistance, and wood and glass comprise polymers that attract one another. And so on. When one includes these factors, it is to geometry that one quickly turns to understand the real marble's actual dynamics. Even ignoring these effects and just asking what happens when the marble rolls off the edge of a table introduces a *nonholonomic* constraint, and figuring out where it lands and how fast it is spinning are best addressed not by the methods of Lagrange and Hamilton, but instead by considering the geometry of the gravitational and reaction forces. In the following chapters, we shall encounter many examples where we have to deal with messy complications like these.

EXERCISES

Exercise 1.1 *Practice: Energy Change for Charged Particle*
Without introducing any coordinates or basis vectors, show that when a particle with charge q interacts with electric and magnetic fields, its kinetic energy changes at a rate

$$dE/dt = q \mathbf{v} \cdot \mathbf{E}. \qquad (1.8)$$

Exercise 1.2 *Practice: Particle Moving in a Circular Orbit*

Consider a particle moving in a circle with uniform speed $v = |\mathbf{v}|$ and uniform magnitude $a = |\mathbf{a}|$ of acceleration. Without introducing any coordinates or basis vectors, do the following.

(a) At any moment of time, let $\mathbf{n} = \mathbf{v}/v$ be the unit vector pointing along the velocity, and let s denote distance that the particle travels in its orbit. By drawing a picture, show that $d\mathbf{n}/ds$ is a unit vector that points to the center of the particle's circular orbit, divided by the radius of the orbit.

(b) Show that the vector (not unit vector) pointing from the particle's location to the center of its orbit is $(v/a)^2 \mathbf{a}$.

1.5 Component Representation of Tensor Algebra

Cartesian coordinates and orthonormal basis vectors

In the Euclidean 3-space of Newtonian physics, there is a unique set of *orthonormal basis vectors* $\{\mathbf{e}_x, \mathbf{e}_y, \mathbf{e}_z\} \equiv \{\mathbf{e}_1, \mathbf{e}_2, \mathbf{e}_3\}$ associated with any *Cartesian coordinate system* $\{x, y, z\} \equiv \{x^1, x^2, x^3\} \equiv \{x_1, x_2, x_3\}$. (In Cartesian coordinates in Euclidean space, we usually place indices down, but occasionally we place them up. It doesn't matter. By definition, in Cartesian coordinates a quantity is the same whether its index is down or up.) The basis vector $\mathbf{e}_j$ points along the x_j coordinate direction, which is orthogonal to all the other coordinate directions, and it has unit length (Fig. 1.3), so

$$\mathbf{e}_j \cdot \mathbf{e}_k = \delta_{jk}, \tag{1.9a}$$

where δ_{jk} is the Kronecker delta.

Any vector $\mathbf{A}$ in 3-space can be expanded in terms of this basis:

$$\mathbf{A} = A_j \mathbf{e}_j. \tag{1.9b}$$

Einstein summation convention

Here and throughout this book, we adopt the *Einstein summation convention*: repeated indices (in this case j) are to be summed (in this 3-space case over $j = 1, 2, 3$), unless otherwise instructed. By virtue of the orthonormality of the basis, the components A_j of $\mathbf{A}$ can be computed as the scalar product

Cartesian components of a vector

$$A_j = \mathbf{A} \cdot \mathbf{e}_j. \tag{1.9c}$$

[The proof of this is straightforward: $\mathbf{A} \cdot \mathbf{e}_j = (A_k \mathbf{e}_k) \cdot \mathbf{e}_j = A_k(\mathbf{e}_k \cdot \mathbf{e}_j) = A_k \delta_{kj} = A_j$.]

Any tensor, say, the third-rank tensor $\mathbf{T}(\_, \_, \_)$, can be expanded in terms of tensor products of the basis vectors:

$$\mathbf{T} = T_{ijk} \mathbf{e}_i \otimes \mathbf{e}_j \otimes \mathbf{e}_k. \tag{1.9d}$$

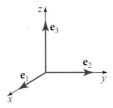

FIGURE 1.3 The orthonormal basis vectors $\mathbf{e}_j$ associated with a Euclidean coordinate system in Euclidean 3-space.

The components T_{ijk} of **T** can be computed from **T** and the basis vectors by the generalization of Eq. (1.9c):

Cartesian components of a tensor

$$T_{ijk} = \mathbf{T}(\mathbf{e}_i, \mathbf{e}_j, \mathbf{e}_k). \tag{1.9e}$$

[This equation can be derived using the orthonormality of the basis in the same way as Eq. (1.9c) was derived.] As an important example, the components of the metric tensor are $g_{jk} = \mathbf{g}(\mathbf{e}_j, \mathbf{e}_k) = \mathbf{e}_j \cdot \mathbf{e}_k = \delta_{jk}$ [where the first equality is the method (1.9e) of computing tensor components, the second is the definition (1.4b) of the metric, and the third is the orthonormality relation (1.9a)]:

$$g_{jk} = \delta_{jk}. \tag{1.9f}$$

The components of a tensor product [e.g., $\mathbf{T}(\_, \_, \_) \otimes \mathbf{S}(\_, \_)$] are easily deduced by inserting the basis vectors into the slots [Eq. (1.9e)]; they are $\mathbf{T}(\mathbf{e}_i, \mathbf{e}_j, \mathbf{e}_k) \otimes \mathbf{S}(\mathbf{e}_l, \mathbf{e}_m) = T_{ijk} S_{lm}$ [cf. Eq. (1.5a)]. In words, the components of a tensor product are equal to the ordinary arithmetic product of the components of the individual tensors.

In component notation, the inner product of two vectors and the value of a tensor when vectors are inserted into its slots are given by

$$\mathbf{A} \cdot \mathbf{B} = A_j B_j, \qquad \mathbf{T}(\mathbf{A}, \mathbf{B}, \mathbf{C}) = T_{ijk} A_i B_j C_k, \tag{1.9g}$$

as one can easily show using previous equations. Finally, the contraction of a tensor [say, the fourth-rank tensor $\mathbf{R}(\_, \_, \_, \_)$] on two of its slots (say, the first and third) has components that are easily computed from the tensor's own components:

$$\text{components of [1\&3 contraction of } \mathbf{R}] = R_{ijik}. \tag{1.9h}$$

Note that R_{ijik} is summed on the i index, so it has only two free indices, j and k, and thus is the component of a second-rank tensor, as it must be if it is to represent the contraction of a fourth-rank tensor.

1.5.1 Slot-Naming Index Notation

We now pause in our development of the component version of tensor algebra to introduce a very important new viewpoint.

> **BOX 1.2. VECTORS AND TENSORS IN QUANTUM THEORY** T2
>
> The laws of quantum theory, like all other laws of Nature, can be expressed as geometric relationships among geometric objects. Most of quantum theory's geometric objects, like those of classical theory, are vectors and tensors: the quantum state $|\psi\rangle$ of a physical system (e.g., a particle in a harmonic-oscillator potential) is a Hilbert-space vector—a generalization of a Euclidean-space vector **A**. There is an inner product, denoted $\langle\phi|\psi\rangle$, between any two states $|\phi\rangle$ and $|\psi\rangle$, analogous to **B** · **A**; but **B** · **A** is a real number, whereas $\langle\phi|\psi\rangle$ is a complex number (and we add and subtract quantum states with complex-number coefficients). The Hermitian operators that represent observables (e.g., the hamiltonian $\hat{H}$ for the particle in the potential) are two-slotted (second-rank), complex-valued functions of vectors; $\langle\phi|\hat{H}|\psi\rangle$ is the complex number that one gets when one inserts ϕ and ψ into the first and second slots of $\hat{H}$. Just as, in Euclidean space, we get a new vector (first-rank tensor) **T**(__, **A**) when we insert the vector **A** into the second slot of **T**, so in quantum theory we get a new vector (physical state) $\hat{H}|\psi\rangle$ (the result of letting $\hat{H}$ "act on" $|\psi\rangle$) when we insert $|\psi\rangle$ into the second slot of $\hat{H}$. In these senses, we can regard **T** as a linear map of Euclidean vectors into Euclidean vectors and $\hat{H}$ as a linear map of states (Hilbert-space vectors) into states.
>
> For the electron in the hydrogen atom, we can introduce a set of orthonormal basis vectors $\{|1\rangle, |2\rangle, |3\rangle, \ldots\}$, that is, the atom's energy eigenstates, with $\langle m|n\rangle = \delta_{mn}$. But by contrast with Newtonian physics, where we only need three basis vectors (because our Euclidean space is 3-dimensional), for the particle in a harmonic-oscillator potential, we need an infinite number of basis vectors (since the Hilbert space of all states is infinite-dimensional). In the particle's quantum-state basis, any observable (e.g., the particle's position $\hat{x}$ or momentum $\hat{p}$) has components computed by inserting the basis vectors into its two slots: $x_{mn} = \langle m|\hat{x}|n\rangle$, and $p_{mn} = \langle m|\hat{p}|n\rangle$. In this basis, the operator $\hat{x}\hat{p}$ (which maps states into states) has components $x_{jk}p_{km}$ (a matrix product), and the noncommutation of position and momentum $[\hat{x}, \hat{p}] = i\hbar$ (an important physical law) is expressible in terms of components as $x_{jk}p_{km} - p_{jk}x_{km} = i\hbar\delta_{jm}$.

Consider the rank-2 tensor **F**(__, __). We can define a new tensor **G**(__, __) to be the same as **F**, but with the slots interchanged: i.e., for any two vectors **A** and **B**, it is true that **G**(**A**, **B**) = **F**(**B**, **A**). We need a simple, compact way to indicate that **F** and **G** are equal except for an interchange of slots. The best way is to give the slots names, say a and b—i.e., to rewrite **F**(__, __) as **F**(__$_a$, __$_b$) or more conveniently as F_{ab}, and then to write the relationship between **G** and **F** as $G_{ab} = F_{ba}$. "NO!" some readers might object. This notation is indistinguishable from our notation for components on a particular basis. "GOOD!" a more astute reader will exclaim. The relation $G_{ab} = F_{ba}$ in a particular basis is a true statement if and only if "**G** = **F** with slots interchanged" is true, so why not use the same notation to symbolize both? In fact, we shall do this. We ask our readers to look at any "index equation," such as $G_{ab} = F_{ba}$, like they would look at an Escher drawing: momentarily think of it as a relationship between components of tensors in a specific basis; then do a quick mind-flip and regard it quite differently, as a relationship between geometric, basis-independent tensors with the indices playing the roles of slot names. This mind-flip approach to tensor algebra will pay substantial dividends.

As an example of the power of this *slot-naming index notation*, consider the contraction of the first and third slots of a third-rank tensor **T**. In any basis the components of 1&3contraction(**T**) are T_{aba}; cf. Eq. (1.9h). Correspondingly, in slot-naming index notation we denote 1&3contraction(**T**) by the simple expression T_{aba}. We can think of the first and third slots as annihilating each other by the contraction, leaving free only the second slot (named b) and therefore producing a rank-1 tensor (a vector).

slot-naming index notation

We should caution that the phrase "slot-naming index notation" is unconventional. You are unlikely to find it in any other textbooks. However, we like it. It says precisely what we want it to say.

1.5.2 Particle Kinetics in Index Notation

1.5.2

As an example of slot-naming index notation, we can rewrite the equations of particle kinetics (1.7) as follows:

$$v_i = \frac{dx_i}{dt}, \quad p_i = mv_i, \quad a_i = \frac{dv_i}{dt} = \frac{d^2x_i}{dt^2},$$

$$E = \frac{1}{2}mv_jv_j, \quad \frac{dp_i}{dt} = q(E_i + \epsilon_{ijk}v_jB_k). \quad (1.10)$$

(In the last equation ϵ_{ijk} is the so-called Levi-Civita tensor, which is used to produce the cross product; we shall learn about it in Sec. 1.7. And note that the scalar energy E must not be confused with the electric field vector E_i.)

Equations (1.10) can be viewed in either of two ways: (i) as the basis-independent geometric laws $\mathbf{v} = d\mathbf{x}/dt$, $\mathbf{p} = m\mathbf{v}$, $\mathbf{a} = d\mathbf{v}/dt = d^2\mathbf{x}/dt^2$, $E = \frac{1}{2}mv^2$, and $d\mathbf{p}/dt = q(\mathbf{E} + \mathbf{v} \times \mathbf{B})$ written in slot-naming index notation; or (ii) as equations for the components of **v**, **p**, **a**, **E**, and **B** in some particular Cartesian coordinate system.

EXERCISES

Exercise 1.3 *Derivation: Component Manipulation Rules*
Derive the component manipulation rules (1.9g) and (1.9h).

Exercise 1.4 *Example and Practice: Numerics of Component Manipulations*
The third-rank tensor **S**(__, __, __) and vectors **A** and **B** have as their only nonzero components $S_{123} = S_{231} = S_{312} = +1$, $A_1 = 3$, $B_1 = 4$, $B_2 = 5$. What are the

components of the vector $\mathbf{C} = \mathbf{S}(\mathbf{A}, \mathbf{B}, \_\_)$, the vector $\mathbf{D} = \mathbf{S}(\mathbf{A}, \_\_, \mathbf{B})$, and the tensor $\mathbf{W} = \mathbf{A} \otimes \mathbf{B}$?

[Partial solution: In component notation, $C_k = S_{ijk} A_i B_j$, where (of course) we sum over the repeated indices i and j. This tells us that $C_1 = S_{231} A_2 B_3$, because S_{231} is the only component of $\mathbf{S}$ whose last index is a 1; this in turn implies that $C_1 = 0$, since $A_2 = 0$. Similarly, $C_2 = S_{312} A_3 B_1 = 0$ (because $A_3 = 0$). Finally, $C_3 = S_{123} A_1 B_2 = +1 \times 3 \times 5 = 15$. Also, in component notation $W_{ij} = A_i B_j$, so $W_{11} = A_1 \times B_1 = 3 \times 4 = 12$, and $W_{12} = A_1 \times B_2 = 3 \times 5 = 15$. Here the $\times$ stands for numerical multiplication, not the vector cross product.]

Exercise 1.5 *Practice: Meaning of Slot-Naming Index Notation*
(a) The following expressions and equations are written in slot-naming index notation. Convert them to geometric, index-free notation: $A_i B_{jk}$, $A_i B_{ji}$, $S_{ijk} = S_{kji}$, $A_i B_i = A_i B_j g_{ij}$.
(b) The following expressions are written in geometric, index-free notation. Convert them to slot-naming index notation: $\mathbf{T}(\_\_, \_\_, \mathbf{A})$, $\mathbf{T}(\_\_, \mathbf{S}(\mathbf{B}, \_\_), \_\_)$.

1.6 Orthogonal Transformations of Bases

Consider two different Cartesian coordinate systems $\{x, y, z\} \equiv \{x_1, x_2, x_3\}$, and $\{\bar{x}, \bar{y}, \bar{z}\} \equiv \{x_{\bar{1}}, x_{\bar{2}}, x_{\bar{3}}\}$. Denote by $\{\mathbf{e}_i\}$ and $\{\mathbf{e}_{\bar{p}}\}$ the corresponding bases. It is possible to expand the basis vectors of one basis in terms of those of the other. We denote the expansion coefficients by the letter R and write

$$\mathbf{e}_i = \mathbf{e}_{\bar{p}} R_{\bar{p}i}, \qquad \mathbf{e}_{\bar{p}} = \mathbf{e}_i R_{i\bar{p}}. \tag{1.11}$$

The quantities $R_{\bar{p}i}$ and $R_{i\bar{p}}$ are not the components of a tensor; rather, they are the elements of transformation matrices

$$[R_{\bar{p}i}] = \begin{bmatrix} R_{\bar{1}1} & R_{\bar{1}2} & R_{\bar{1}3} \\ R_{\bar{2}1} & R_{\bar{2}2} & R_{\bar{2}3} \\ R_{\bar{3}1} & R_{\bar{3}2} & R_{\bar{3}3} \end{bmatrix}, \quad [R_{i\bar{p}}] = \begin{bmatrix} R_{1\bar{1}} & R_{1\bar{2}} & R_{1\bar{3}} \\ R_{2\bar{1}} & R_{2\bar{2}} & R_{2\bar{3}} \\ R_{3\bar{1}} & R_{3\bar{2}} & R_{3\bar{3}} \end{bmatrix}. \tag{1.12a}$$

(Here and throughout this book we use square brackets to denote matrices.) These two matrices must be the inverse of each other, since one takes us from the barred basis to the unbarred, and the other in the reverse direction, from unbarred to barred:

$$R_{\bar{p}i} R_{i\bar{q}} = \delta_{\bar{p}\bar{q}}, \qquad R_{i\bar{p}} R_{\bar{p}j} = \delta_{ij}. \tag{1.12b}$$

The orthonormality requirement for the two bases implies that $\delta_{ij} = \mathbf{e}_i \cdot \mathbf{e}_j = (\mathbf{e}_{\bar{p}} R_{\bar{p}i}) \cdot (\mathbf{e}_{\bar{q}} R_{\bar{q}j}) = R_{\bar{p}i} R_{\bar{q}j} (\mathbf{e}_{\bar{p}} \cdot \mathbf{e}_{\bar{q}}) = R_{\bar{p}i} R_{\bar{q}j} \delta_{\bar{p}\bar{q}} = R_{\bar{p}i} R_{\bar{p}j}$. This says that the transpose of $[R_{\bar{p}i}]$ is its inverse—which we have already denoted by $[R_{i\bar{p}}]$:

$$[R_{i\bar{p}}] \equiv \text{inverse}\left([R_{\bar{p}i}]\right) = \text{transpose}\left([R_{\bar{p}i}]\right). \tag{1.12c}$$

This property implies that the transformation matrix is orthogonal, so the transformation is a reflection or a rotation (see, e.g., Goldstein, Poole, and Safko, 2002). Thus (as should be obvious and familiar), the bases associated with any two Euclidean coordinate systems are related by a reflection or rotation, and the matrices (1.12a) are called *rotation matrices*. Note that Eq. (1.12c) does not say that $[R_{i\bar{p}}]$ is a symmetric matrix. In fact, most rotation matrices are not symmetric [see, e.g., Eq. (1.14)].

orthogonal transformation and rotation

The fact that a vector **A** is a geometric, basis-independent object implies that
$\mathbf{A} = A_i \mathbf{e}_i = A_i(\mathbf{e}_{\bar{p}} R_{\bar{p}i}) = (R_{\bar{p}i} A_i)\mathbf{e}_{\bar{p}} = A_{\bar{p}} \mathbf{e}_{\bar{p}}$:

$$A_{\bar{p}} = R_{\bar{p}i} A_i, \quad \text{and similarly,} \quad A_i = R_{i\bar{p}} A_{\bar{p}}; \tag{1.13a}$$

and correspondingly for the components of a tensor:

$$T_{\bar{p}\bar{q}\bar{r}} = R_{\bar{p}i} R_{\bar{q}j} R_{\bar{r}k} T_{ijk}, \quad T_{ijk} = R_{i\bar{p}} R_{j\bar{q}} R_{k\bar{r}} T_{\bar{p}\bar{q}\bar{r}}. \tag{1.13b}$$

It is instructive to compare the transformation law (1.13a) for the components of a vector with Eqs. (1.11) for the bases. To make these laws look natural, we have placed the transformation matrix on the left in the former and on the right in the latter. In Minkowski spacetime (Chap. 2), the placement of indices, up or down, will automatically tell us the order.

If we choose the origins of our two coordinate systems to coincide, then the vector **x** reaching from the common origin to some point $\mathcal{P}$, whose coordinates are x_j and $x_{\bar{p}}$, has components equal to those coordinates; and as a result, the coordinates themselves obey the same transformation law as any other vector:

$$x_{\bar{p}} = R_{\bar{p}i} x_i, \quad x_i = R_{i\bar{p}} x_{\bar{p}}. \tag{1.13c}$$

The product of two rotation matrices $[R_{i\bar{p}} R_{\bar{p}\bar{s}}]$ is another rotation matrix $[R_{i\bar{s}}]$, which transforms the Cartesian bases $\mathbf{e}_{\bar{s}}$ to $\mathbf{e}_i$. Under this product rule, the rotation matrices form a mathematical *group*: the *rotation group*, whose *group representations* play an important role in quantum theory.

rotation group

EXERCISES

Exercise 1.6 **Example and Practice: Rotation in x-y Plane*
Consider two Cartesian coordinate systems rotated with respect to each other in the x-y plane as shown in Fig. 1.4.

(a) Show that the rotation matrix that takes the barred basis vectors to the unbarred basis vectors is

$$[R_{\bar{p}i}] = \begin{bmatrix} \cos\phi & \sin\phi & 0 \\ -\sin\phi & \cos\phi & 0 \\ 0 & 0 & 1 \end{bmatrix}, \tag{1.14}$$

and show that the inverse of this rotation matrix is, indeed, its transpose, as it must be if this is to represent a rotation.

(b) Verify that the two coordinate systems are related by Eq. (1.13c).

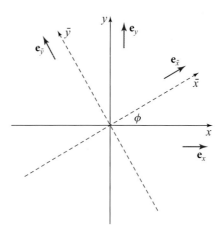

FIGURE 1.4 Two Cartesian coordinate systems $\{x, y, z\}$ and $\{\bar{x}, \bar{y}, \bar{z}\}$ and their basis vectors in Euclidean space, rotated by an angle ϕ relative to each other in the x-y plane. The z- and $\bar{z}$-axes point out of the paper or screen and are not shown.

(c) Let A_j be the components of the electromagnetic vector potential that lies in the x-y plane, so that $A_z = 0$. The two nonzero components A_x and A_y can be regarded as describing the two polarizations of an electromagnetic wave propagating in the z direction. Show that $A_{\bar{x}} + i A_{\bar{y}} = (A_x + i A_y) e^{-i\phi}$. One can show (cf. Sec. 27.3.3) that the factor $e^{-i\phi}$ implies that the quantum particle associated with the wave—the photon—has spin one [i.e., spin angular momentum $\hbar = $ (Planck's constant)$/2\pi$].

(d) Let h_{jk} be the components of a symmetric tensor that is *trace-free* (its contraction h_{jj} vanishes) and is confined to the x-y plane (so $h_{zk} = h_{kz} = 0$ for all k). Then the only nonzero components of this tensor are $h_{xx} = -h_{yy}$ and $h_{xy} = h_{yx}$. As we shall see in Sec. 27.3.1, this tensor can be regarded as describing the two polarizations of a gravitational wave propagating in the z direction. Show that $h_{\bar{x}\bar{x}} + i h_{\bar{x}\bar{y}} = (h_{xx} + i h_{xy}) e^{-2i\phi}$. The factor $e^{-2i\phi}$ implies that the quantum particle associated with the gravitational wave (the graviton) has spin two (spin angular momentum $2\hbar$); cf. Eq. (27.31) and Sec. 27.3.3.

1.7 Differentiation of Scalars, Vectors, and Tensors; Cross Product and Curl

Consider a tensor field $\mathbf{T}(\mathcal{P})$ in Euclidean 3-space and a vector $\mathbf{A}$. We define the *directional derivative* of $\mathbf{T}$ along $\mathbf{A}$ by the obvious limiting procedure

$$\nabla_\mathbf{A} \mathbf{T} \equiv \lim_{\epsilon \to 0} \frac{1}{\epsilon}[\mathbf{T}(\mathbf{x}_\mathcal{P} + \epsilon \mathbf{A}) - \mathbf{T}(\mathbf{x}_\mathcal{P})] \qquad (1.15a)$$

and similarly for the directional derivative of a vector field $\mathbf{B}(\mathcal{P})$ and a scalar field $\psi(\mathcal{P})$. [Here we have denoted points, e.g., $\mathcal{P}$, by the vector $\mathbf{x}_\mathcal{P}$ that reaches from some arbitrary origin to the point, and $\mathbf{T}(\mathbf{x}_\mathcal{P})$ denotes the field's dependence on location in space; $\mathbf{T}$'s slots and dependence on what goes into the slots are suppressed; and the units of ϵ are chosen to ensure that $\epsilon \mathbf{A}$ has the same units as $\mathbf{x}_\mathcal{P}$. There is no other appearance of vectors in this chapter.] In definition (1.15a), the quantity in square brackets is simply the difference between two linear functions of vectors (two tensors), so the quantity on the left-hand side is also a tensor with the same rank as $\mathbf{T}$.

It should not be hard to convince oneself that this directional derivative $\nabla_\mathbf{A} \mathbf{T}$ of any tensor field $\mathbf{T}$ is linear in the vector $\mathbf{A}$ along which one differentiates. Correspondingly, if $\mathbf{T}$ has rank n (n slots), then there is another tensor field, denoted $\nabla \mathbf{T}$, with rank $n+1$, such that

$$\boxed{\nabla_\mathbf{A} \mathbf{T} = \nabla \mathbf{T}(\_, \_, \_, \mathbf{A}).} \tag{1.15b}$$

Here on the right-hand side the first n slots (3 in the case shown) are left empty, and $\mathbf{A}$ is put into the last slot (the "differentiation slot"). The quantity $\nabla \mathbf{T}$ is called the *gradient* of $\mathbf{T}$. In slot-naming index notation, it is conventional to denote this gradient by $T_{abc;d}$, where in general the number of indices preceding the semicolon is the rank of $\mathbf{T}$. Using this notation, the directional derivative of $\mathbf{T}$ along $\mathbf{A}$ reads [cf. Eq. (1.15b)] $T_{abc;j} A_j$.

gradient

It is not hard to show that in any Cartesian coordinate system, the components of the gradient are nothing but the partial derivatives of the components of the original tensor, which we denote by a comma:

$$\boxed{T_{abc;j} = \frac{\partial T_{abc}}{\partial x_j} \equiv T_{abc,j}.} \tag{1.15c}$$

In a non-Cartesian basis (e.g., the spherical and cylindrical bases often used in electromagnetic theory), the components of the gradient typically are not obtained by simple partial differentiation [Eq. (1.15c) fails] because of turning and/or length changes of the basis vectors as we go from one location to another. In Sec. 11.8, we shall learn how to deal with this by using objects called *connection coefficients*. Until then, we confine ourselves to Cartesian bases, so subscript semicolons and subscript commas (partial derivatives) can be used interchangeably.

Because the gradient and the directional derivative are defined by the same standard limiting process as one uses when defining elementary derivatives, they obey the standard (Leibniz) rule for differentiating products:

$$\nabla_\mathbf{A}(\mathbf{S} \otimes \mathbf{T}) = (\nabla_\mathbf{A} \mathbf{S}) \otimes \mathbf{T} + \mathbf{S} \otimes \nabla_\mathbf{A} \mathbf{T},$$
$$\text{or} \quad (S_{ab} T_{cde})_{;j} A_j = (S_{ab;j} A_j) T_{cde} + S_{ab}(T_{cde;j} A_j); \tag{1.16a}$$

and

$$\nabla_\mathbf{A}(f\mathbf{T}) = (\nabla_\mathbf{A} f)\mathbf{T} + f \nabla_\mathbf{A} \mathbf{T}, \quad \text{or} \quad (f T_{abc})_{;j} A_j = (f_{;j} A_j) T_{abc} + f T_{abc;j} A_j. \tag{1.16b}$$

In an orthonormal basis these relations should be obvious: they follow from the Leibniz rule for partial derivatives.

Because the components g_{ab} of the metric tensor are constant in any Cartesian coordinate system, Eq. (1.15c) (which is valid in such coordinates) guarantees that $g_{ab;j} = 0$; i.e., the metric has vanishing gradient:

$$\boldsymbol{\nabla}\mathbf{g} = 0, \quad \text{or} \quad g_{ab;j} = 0. \tag{1.17}$$

From the gradient of any vector or tensor we can construct several other important derivatives by contracting on slots:

1. Since the gradient $\boldsymbol{\nabla}\mathbf{A}$ of a vector field $\mathbf{A}$ has two slots, $\boldsymbol{\nabla}\mathbf{A}(\_,\_)$, we can contract its slots on each other to obtain a scalar field. That scalar field is the *divergence* of $\mathbf{A}$ and is denoted

$$\boldsymbol{\nabla} \cdot \mathbf{A} \equiv (\text{contraction of } \boldsymbol{\nabla}\mathbf{A}) = A_{a;a}. \tag{1.18}$$

divergence

2. Similarly, if $\mathbf{T}$ is a tensor field of rank 3, then $T_{abc;c}$ is its divergence on its third slot, and $T_{abc;b}$ is its divergence on its second slot.

3. By taking the double gradient and then contracting on the two gradient slots we obtain, from any tensor field $\mathbf{T}$, a new tensor field with the same rank,

$$\nabla^2 \mathbf{T} \equiv (\boldsymbol{\nabla} \cdot \boldsymbol{\nabla})\mathbf{T}, \quad \text{or} \quad T_{abc;jj}. \tag{1.19}$$

laplacian

Here and henceforth, all indices following a semicolon (or comma) represent gradients (or partial derivatives): $T_{abc;jj} \equiv T_{abc;j;j}$, $T_{abc,jk} \equiv \partial^2 T_{abc}/\partial x_j \partial x_k$. The operator ∇^2 is called the *laplacian*.

The metric tensor is a fundamental property of the space in which it lives; it embodies the inner product and hence the space's notion of distance. In addition to the metric, there is one (and only one) other fundamental tensor that describes a piece of Euclidean space's geometry: the *Levi-Civita tensor* $\boldsymbol{\epsilon}$, which embodies the space's notion of volume.

Levi-Civita tensor

In a Euclidean space with dimension n, the Levi-Civita tensor $\boldsymbol{\epsilon}$ is a completely antisymmetric tensor with rank n (with n slots). A parallelepiped whose edges are the n vectors $\mathbf{A}, \mathbf{B}, \ldots, \mathbf{F}$ is said to have the *volume*

$$\boxed{\text{volume} = \boldsymbol{\epsilon}(\mathbf{A}, \mathbf{B}, \ldots, \mathbf{F}).} \tag{1.20}$$

volume

(We justify this definition in Sec. 1.8.) Notice that this volume can be positive or negative, and if we exchange the order of the parallelepiped's legs, the volume's sign changes: $\boldsymbol{\epsilon}(\mathbf{B}, \mathbf{A}, \ldots, \mathbf{F}) = -\boldsymbol{\epsilon}(\mathbf{A}, \mathbf{B}, \ldots, \mathbf{F})$ by antisymmetry of $\boldsymbol{\epsilon}$.

It is easy to see (Ex. 1.7) that (i) the volume vanishes unless the legs are all linearly independent, (ii) once the volume has been specified for one parallelepiped (one set of linearly independent legs), it is thereby determined for all parallelepipeds, and therefore, (iii) we require only one number plus antisymmetry to determine $\boldsymbol{\epsilon}$

fully. If the chosen parallelepiped has legs that are orthonormal (all are orthogonal to one another and all have unit length—properties determined by the metric **g**), then it must have unit volume, or more precisely volume ± 1. This is a compatibility relation between **g** and ϵ. It is easy to see (Ex. 1.7) that (iv) ϵ is fully determined by its antisymmetry, compatibility with the metric, and a single sign: the choice of which parallelepipeds have positive volume and which have negative. It is conventional in Euclidean 3-space to give right-handed parallelepipeds positive volume and left-handed ones negative volume: $\epsilon(\mathbf{A}, \mathbf{B}, \mathbf{C})$ is positive if, when we place our right thumb along **C** and the fingers of our right hand along **A**, then bend our fingers, they sweep toward **B** and not $-\mathbf{B}$.

These considerations dictate that in a right-handed orthonormal basis of Euclidean 3-space, the only nonzero components of ϵ are

$$\epsilon_{123} = +1,$$

$$\epsilon_{abc} = \begin{cases} +1 & \text{if } a, b, c \text{ is an even permutation of 1, 2, 3} \\ -1 & \text{if } a, b, c \text{ is an odd permutation of 1, 2, 3} \\ 0 & \text{if } a, b, c \text{ are not all different;} \end{cases} \quad (1.21)$$

and in a left-handed orthonormal basis, the signs of these components are reversed.

The Levi-Civita tensor is used to define the cross product and the curl: **cross product and curl**

$$\mathbf{A} \times \mathbf{B} \equiv \epsilon(\_\_, \mathbf{A}, \mathbf{B}); \quad \text{in slot-naming index notation, } \epsilon_{ijk} A_j B_k; \quad (1.22a)$$

$$\nabla \times \mathbf{A} \equiv (\text{the vector field whose slot-naming index form is } \epsilon_{ijk} A_{k;j}). \quad (1.22b)$$

[Equation (1.22b) is an example of an expression that is complicated if stated in index-free notation; it says that $\nabla \times \mathbf{A}$ is the double contraction of the rank-5 tensor $\epsilon \otimes \nabla \mathbf{A}$ on its second and fifth slots, and on its third and fourth slots.]

Although Eqs. (1.22a) and (1.22b) look like complicated ways to deal with concepts that most readers regard as familiar and elementary, they have great power. The power comes from the following property of the Levi-Civita tensor in Euclidean 3-space [readily derivable from its components (1.21)]:

$$\boxed{\epsilon_{ijm}\epsilon_{klm} = \delta^{ij}_{kl} \equiv \delta^i_k \delta^j_l - \delta^i_l \delta^j_k.} \quad (1.23)$$

Here δ^i_k is the Kronecker delta. Examine the 4-index delta function δ^{ij}_{kl} carefully; it says that either the indices above and below each other must be the same ($i = k$ and $j = l$) with a + sign, or the diagonally related indices must be the same ($i = l$ and $j = k$) with a − sign. [We have put the indices ij of δ^{ij}_{kl} up solely to facilitate remembering this rule. Recall (first paragraph of Sec. 1.5) that in Euclidean space and Cartesian coordinates, it does not matter whether indices are up or down.] With the aid of Eq. (1.23) and the index-notation expressions for the cross product and curl, one can quickly and easily derive a wide variety of useful vector identities; see the very important Ex. 1.8.

EXERCISES

Exercise 1.7 *Derivation: Properties of the Levi-Civita Tensor*
From its complete antisymmetry, derive the four properties of the Levi-Civita tensor, in n-dimensional Euclidean space, that are claimed in the text following Eq. (1.20).

Exercise 1.8 **Example and Practice: Vectorial Identities for the Cross Product and Curl*
Here is an example of how to use index notation to derive a vector identity for the double cross product $\mathbf{A} \times (\mathbf{B} \times \mathbf{C})$: in index notation this quantity is $\epsilon_{ijk} A_j (\epsilon_{klm} B_l C_m)$. By permuting the indices on the second ϵ and then invoking Eq. (1.23), we can write this as $\epsilon_{ijk} \epsilon_{lmk} A_j B_l C_m = \delta^{lm}_{ij} A_j B_l C_m$. By then invoking the meaning of the 4-index delta function [Eq. (1.23)], we bring this into the form $A_j B_i C_j - A_j B_j C_i$, which is the slot-naming index-notation form of $(\mathbf{A} \cdot \mathbf{C})\mathbf{B} - (\mathbf{A} \cdot \mathbf{B})\mathbf{C}$. Thus, it must be that $\mathbf{A} \times (\mathbf{B} \times \mathbf{C}) = (\mathbf{A} \cdot \mathbf{C})\mathbf{B} - (\mathbf{A} \cdot \mathbf{B})\mathbf{C}$. Use similar techniques to evaluate the following quantities.

(a) $\nabla \times (\nabla \times \mathbf{A})$.

(b) $(\mathbf{A} \times \mathbf{B}) \cdot (\mathbf{C} \times \mathbf{D})$.

(c) $(\mathbf{A} \times \mathbf{B}) \times (\mathbf{C} \times \mathbf{D})$.

Exercise 1.9 **Example and Practice: Levi-Civita Tensor in 2-Dimensional Euclidean Space*
In Euclidean 2-space, let $\{\mathbf{e}_1, \mathbf{e}_2\}$ be an orthonormal basis with positive volume.

(a) Show that the components of ϵ in this basis are

$$\epsilon_{12} = +1, \quad \epsilon_{21} = -1, \quad \epsilon_{11} = \epsilon_{22} = 0. \tag{1.24a}$$

(b) Show that

$$\epsilon_{ik} \epsilon_{jk} = \delta_{ij}. \tag{1.24b}$$

1.8 Volumes, Integration, and Integral Conservation Laws

In Cartesian coordinates of 2-dimensional Euclidean space, the basis vectors are orthonormal, so (with a conventional choice of sign) the components of the Levi-Civita tensor are given by Eqs. (1.24a). Correspondingly, the area (i.e., 2-dimensional volume) of a parallelogram whose sides are $\mathbf{A}$ and $\mathbf{B}$ is

$$\text{2-volume} = \epsilon(\mathbf{A}, \mathbf{B}) = \epsilon_{ab} A_a B_b = A_1 B_2 - A_2 B_1 = \det \begin{bmatrix} A_1 & B_1 \\ A_2 & B_2 \end{bmatrix}, \tag{1.25}$$

a relation that should be familiar from elementary geometry. Equally familiar should be the following expression for the 3-dimensional volume of a parallelepiped with legs

A, **B**, and **C** [which follows from the components (1.21) of the Levi-Civita tensor]:

$$\text{3-volume} = \epsilon(\mathbf{A}, \mathbf{B}, \mathbf{C}) = \epsilon_{ijk} A_i B_j C_k = \mathbf{A} \cdot (\mathbf{B} \times \mathbf{C}) = \det \begin{bmatrix} A_1 & B_1 & C_1 \\ A_2 & B_2 & C_2 \\ A_3 & B_3 & C_3 \end{bmatrix}. \quad (1.26)$$

3-volume

Our formal definition (1.20) of volume is justified because it gives rise to these familiar equations.

Equations (1.25) and (1.26) are foundations from which one can derive the usual formulas $dA = dx\, dy$ and $dV = dx\, dy\, dz$ for the area and volume of elementary surface and volume elements with Cartesian side lengths dx, dy, and dz (Ex. 1.10).

In Euclidean 3-space, we define the vectorial surface area of a 2-dimensional parallelogram with legs **A** and **B** to be

$$\boxed{\boldsymbol{\Sigma} = \mathbf{A} \times \mathbf{B} = \epsilon(\_\_, \mathbf{A}, \mathbf{B}).} \quad (1.27)$$

This vectorial surface area has a magnitude equal to the area of the parallelogram and a direction perpendicular to it. Notice that this surface area $\epsilon(\_\_, \mathbf{A}, \mathbf{B})$ can be thought of as an object that is waiting for us to insert a third leg, **C**, so as to compute a 3-volume $\epsilon(\mathbf{C}, \mathbf{A}, \mathbf{B})$—the volume of the parallelepiped with legs **C**, **A**, and **B**.

vectorial surface area

A parallelogram's surface has two faces (two sides), called the *positive face* and the *negative face*. If the vector **C** sticks out of the positive face, then $\boldsymbol{\Sigma}(\mathbf{C}) = \epsilon(\mathbf{C}, \mathbf{A}, \mathbf{B})$ is positive; if **C** sticks out of the negative face, then $\boldsymbol{\Sigma}(\mathbf{C})$ is negative.

1.8.1 Gauss's and Stokes' Theorems

1.8.1

Such vectorial surface areas are the foundation for surface integrals in 3-dimensional space and for the familiar *Gauss's theorem*,

Gauss's and Stokes' theorems

$$\boxed{\int_{\mathcal{V}_3} (\boldsymbol{\nabla} \cdot \mathbf{A}) dV = \int_{\partial \mathcal{V}_3} \mathbf{A} \cdot d\boldsymbol{\Sigma}} \quad (1.28a)$$

(where $\mathcal{V}_3$ is a compact 3-dimensional region, and $\partial \mathcal{V}_3$ is its closed 2-dimensional boundary) and *Stokes' theorem*,

$$\boxed{\int_{\mathcal{V}_2} \boldsymbol{\nabla} \times \mathbf{A} \cdot d\boldsymbol{\Sigma} = \int_{\partial \mathcal{V}_2} \mathbf{A} \cdot d\mathbf{l}} \quad (1.28b)$$

(where $\mathcal{V}_2$ is a compact 2-dimensional region, $\partial \mathcal{V}_2$ is the 1-dimensional closed curve that bounds it, and the last integral is a line integral around that curve); see, e.g., Arfken, Weber, and Harris (2013).

This mathematics is illustrated by the integral and differential conservation laws for electric charge and for particles: The total charge and the total number of particles inside a 3-dimensional region of space $\mathcal{V}_3$ are $\int_{\mathcal{V}_3} \rho_e dV$ and $\int_{\mathcal{V}_3} n dV$, where ρ_e is the charge density and n the number density of particles. The rates that charge and particles flow out of $\mathcal{V}_3$ are the integrals of the current density **j** and the particle flux vector $\mathbf{S}$ over its boundary $\partial \mathcal{V}_3$. Therefore, the *integral laws of charge conservation and particle conservation* are

integral conservation laws

$$\boxed{\frac{d}{dt}\int_{\mathcal{V}_3} \rho_e \, dV + \int_{\partial \mathcal{V}_3} \mathbf{j} \cdot d\boldsymbol{\Sigma} = 0,} \qquad \boxed{\frac{d}{dt}\int_{\mathcal{V}_3} n \, dV + \int_{\partial \mathcal{V}_3} \mathbf{S} \cdot d\boldsymbol{\Sigma} = 0.} \quad (1.29)$$

Pull the time derivative inside each volume integral (where it becomes a partial derivative), and apply Gauss's law to each surface integral; the results are $\int_{\mathcal{V}_3} (\partial \rho_e/\partial t + \nabla \cdot \mathbf{j}) dV = 0$ and similarly for particles. The only way these equations can be true for all choices of $\mathcal{V}_3$ is for the integrands to vanish:

differential conservation laws

$$\boxed{\partial \rho_e/\partial t + \nabla \cdot \mathbf{j} = 0,} \qquad \boxed{\partial n/\partial t + \nabla \cdot \mathbf{S} = 0.} \quad (1.30)$$

These are the *differential conservation laws for charge and for particles*. They have a standard, universal form: the time derivative of the density of a quantity plus the divergence of its flux vanishes.

Note that the integral conservation laws (1.29) and the differential conservation laws (1.30) require no coordinate system or basis for their description, and no coordinate system or basis was used in deriving the differential laws from the integral laws. This is an example of the fundamental principle that *the Newtonian physical laws are all expressible as geometric relationships among geometric objects*.

EXERCISES

Exercise 1.10 *Derivation and Practice: Volume Elements in Cartesian Coordinates*
Use Eqs. (1.25) and (1.26) to derive the usual formulas $dA = dx\,dy$ and $dV = dx\,dy\,dz$ for the 2-dimensional and 3-dimensional integration elements, respectively, in right-handed Cartesian coordinates. [Hint: Use as the edges of the integration volumes $dx\,\mathbf{e}_x$, $dy\,\mathbf{e}_y$, and $dz\,\mathbf{e}_z$.]

Exercise 1.11 *Example and Practice: Integral of a Vector Field over a Sphere*
Integrate the vector field $\mathbf{A} = z\mathbf{e}_z$ over a sphere with radius a, centered at the origin of the Cartesian coordinate system (i.e., compute $\int \mathbf{A} \cdot d\boldsymbol{\Sigma}$). Hints:

(a) Introduce spherical polar coordinates on the sphere, and construct the vectorial integration element $d\boldsymbol{\Sigma}$ from the two legs $a d\theta\, \mathbf{e}_{\hat{\theta}}$ and $a \sin\theta d\phi\, \mathbf{e}_{\hat{\phi}}$. Here $\mathbf{e}_{\hat{\theta}}$ and $\mathbf{e}_{\hat{\phi}}$ are unit-length vectors along the θ and ϕ directions. (Here as in Sec. 1.6 and throughout this book, we use accents on indices to indicate which basis the index is associated with: hats here for the spherical orthonormal basis, bars in Sec. 1.6 for the barred Cartesian basis.) Explain the factors $a d\theta$ and $a \sin\theta d\phi$ in the definitions of the legs. Show that

$$d\boldsymbol{\Sigma} = \boldsymbol{\epsilon}(\underline{\quad}, \mathbf{e}_{\hat{\theta}}, \mathbf{e}_{\hat{\phi}}) a^2 \sin\theta d\theta d\phi. \quad (1.31)$$

(b) Using $z = a\cos\theta$ and $\mathbf{e}_z = \cos\theta \mathbf{e}_{\hat{r}} - \sin\theta \mathbf{e}_{\hat{\theta}}$ on the sphere (where $\mathbf{e}_{\hat{r}}$ is the unit vector pointing in the radial direction), show that

$$\mathbf{A} \cdot d\boldsymbol{\Sigma} = a\cos^2\theta\, \boldsymbol{\epsilon}(\mathbf{e}_{\hat{r}}, \mathbf{e}_{\hat{\theta}}, \mathbf{e}_{\hat{\phi}})\, a^2 \sin\theta d\theta d\phi.$$

(c) Explain why $\epsilon(\mathbf{e}_{\hat{r}}, \mathbf{e}_{\hat{\theta}}, \mathbf{e}_{\hat{\phi}}) = 1$.

(d) Perform the integral $\int \mathbf{A} \cdot d\mathbf{\Sigma}$ over the sphere's surface to obtain your final answer $(4\pi/3)a^3$. This, of course, is the volume of the sphere. Explain pictorially why this had to be the answer.

Exercise 1.12 *Example: Faraday's Law of Induction*
One of Maxwell's equations says that $\nabla \times \mathbf{E} = -\partial \mathbf{B}/\partial t$ (in SI units), where $\mathbf{E}$ and $\mathbf{B}$ are the electric and magnetic fields. This is a geometric relationship between geometric objects; it requires no coordinates or basis for its statement. By integrating this equation over a 2-dimensional surface $\mathcal{V}_2$ with boundary curve $\partial \mathcal{V}_2$ and applying Stokes' theorem, derive Faraday's law of induction—again, a geometric relationship between geometric objects.

1.9 The Stress Tensor and Momentum Conservation

Press your hands together in the *y-z* plane and feel the force that one hand exerts on the other across a tiny area A—say, one square millimeter of your hands' palms (Fig. 1.5). That force, of course, is a vector $\mathbf{F}$. It has a normal component (along the *x* direction). It also has a tangential component: if you try to slide your hands past each other, you feel a component of force along their surface, a "shear" force in the *y* and *z* directions. Not only is the force $\mathbf{F}$ vectorial; so is the 2-surface across which it acts, $\mathbf{\Sigma} = A\,\mathbf{e}_x$. (Here $\mathbf{e}_x$ is the unit vector orthogonal to the tiny area A, and we have chosen the negative side of the surface to be the $-x$ side and the positive side to be $+x$. With this choice, the force $\mathbf{F}$ is that which the negative hand, on the $-x$ side, exerts on the positive hand.)

Now, it should be obvious that the force $\mathbf{F}$ is a linear function of our chosen surface $\mathbf{\Sigma}$. Therefore, there must be a tensor, the *stress tensor*, that reports the force to us when we insert the surface into its second slot:

$$\boxed{\mathbf{F}(\_\_) = \mathbf{T}(\_\_, \mathbf{\Sigma}), \quad \text{or} \quad F_i = T_{ij}\Sigma_j.} \tag{1.32}$$

FIGURE 1.5 Hands, pressed together, exert a force on each other.

Newton's law of action and reaction tells us that the force that the positive hand exerts on the negative hand must be equal and opposite to that which the negative hand exerts on the positive. This shows up trivially in Eq. (1.32): by changing the sign of $\mathbf{\Sigma}$, one reverses which hand is regarded as negative and which positive, and since **T** is linear in $\mathbf{\Sigma}$, one also reverses the sign of the force.

The definition (1.32) of the stress tensor gives rise to the following physical meaning of its components:

meaning of components of stress tensor

$$T_{jk} = \begin{pmatrix} j \text{ component of force per unit area} \\ \text{across a surface perpendicular to } \mathbf{e}_k \end{pmatrix}$$

$$= \begin{pmatrix} j \text{ component of momentum that crosses a unit} \\ \text{area that is perpendicular to } \mathbf{e}_k, \text{ per unit time,} \\ \text{with the crossing being from } -x_k \text{ to } +x_k \end{pmatrix}. \quad (1.33)$$

The stresses inside a table with a heavy weight on it are described by the stress tensor **T**, as are the stresses in a flowing fluid or plasma, in the electromagnetic field, and in any other physical medium. Accordingly, we shall use the stress tensor as an important mathematical tool in our study of force balance in kinetic theory (Chap. 3), elasticity (Part IV), fluid dynamics (Part V), and plasma physics (Part VI).

symmetry of stress tensor

It is not obvious from its definition, but the stress tensor **T** is always symmetric in its two slots. To see this, consider a small cube with side L in any medium (or field) (Fig. 1.6). The medium outside the cube exerts forces, and hence also torques, on the cube's faces. The z-component of the torque is produced by the shear forces on the front and back faces and on the left and right. As shown in the figure, the shear forces on the front and back faces have magnitudes $T_{xy}L^2$ and point in opposite directions, so they exert identical torques on the cube, $N_z = T_{xy}L^2(L/2)$ (where $L/2$ is the distance of each face from the cube's center). Similarly, the shear forces on the left and right faces have magnitudes $T_{yx}L^2$ and point in opposite directions, thereby exerting identical torques on the cube, $N_z = -T_{yx}L^2(L/2)$. Adding the torques from all four faces and equating them to the rate of change of angular momentum, $\frac{1}{6}\rho L^5 d\Omega_z/dt$ (where ρ is the mass density, $\frac{1}{6}\rho L^5$ is the cube's moment of inertia, and Ω_z is the z component of its angular velocity), we obtain $(T_{xy} - T_{yx})L^3 = \frac{1}{6}\rho L^5 d\Omega_z/dt$. Now, let the cube's edge length become arbitrarily small, $L \to 0$. If $T_{xy} - T_{yx}$ does not vanish, then the cube will be set into rotation with an infinitely large angular acceleration, $d\Omega_z/dt \propto 1/L^2 \to \infty$— an obviously unphysical behavior. Therefore, $T_{yx} = T_{xy}$, and similarly for all other components: *the stress tensor is always symmetric under interchange of its two slots.*

1.9.1 Examples: Electromagnetic Field and Perfect Fluid

Two examples will make the concept of the stress tensor more concrete.

- **Electromagnetic field:** See Ex. 1.14.

perfect fluid

- **Perfect fluid:** A *perfect fluid* is a medium that can exert an isotropic pressure P but no shear stresses, so the only nonzero components of its stress tensor

30 Chapter 1. Newtonian Physics: Geometric Viewpoint

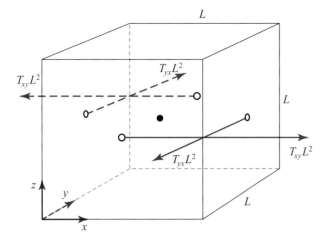

FIGURE 1.6 The shear forces exerted on the left, right, front, and back faces of a vanishingly small cube of side length L. The resulting torque about the z direction will set the cube into rotation with an arbitrarily large angular acceleration unless the stress tensor is symmetric.

in a Cartesian basis are $T_{xx} = T_{yy} = T_{zz} = P$. (Examples of nearly perfect fluids are air and water, but not molasses.) We can summarize this property by $T_{ij} = P\delta_{ij}$ or equivalently, since δ_{ij} are the components of the Euclidean metric, $T_{ij} = Pg_{ij}$. The frame-independent version of this is

$$\mathbf{T} = P\mathbf{g} \quad \text{or, in slot-naming index notation,} \quad T_{ij} = Pg_{ij}. \qquad (1.34)$$

Note that, as always, the formula in slot-naming index notation looks identical to the formula $T_{ij} = Pg_{ij}$ for the components in our chosen Cartesian coordinate system. To check Eq. (1.34), consider a 2-surface $\boldsymbol{\Sigma} = A\mathbf{n}$ with area A oriented perpendicular to some arbitrary unit vector $\mathbf{n}$. The vectorial force that the fluid exerts across $\boldsymbol{\Sigma}$ is, in index notation, $F_j = T_{jk}\Sigma_k = Pg_{jk}An_k = PAn_j$ (i.e., it is a normal force with magnitude equal to the fluid pressure P times the surface area A). This is what it should be.

1.9.2 Conservation of Momentum

The stress tensor plays a central role in the Newtonian law of momentum conservation because (by definition) the force acting across a surface is the same as the rate of flow of momentum, per unit area, across the surface: *the stress tensor is the flux of momentum.*

Consider the 3-dimensional region of space $\mathcal{V}_3$ used above in formulating the integral laws of charge and particle conservation (1.29). The total momentum in $\mathcal{V}_3$ is $\int_{\mathcal{V}_3} \mathbf{G}\,dV$, where $\mathbf{G}$ is the momentum density. This quantity changes as a result of momentum flowing into and out of $\mathcal{V}_3$. The net rate at which momentum flows outward is the integral of the stress tensor over the surface $\partial \mathcal{V}_3$ of $\mathcal{V}_3$. Therefore, by analogy with charge and particle conservation (1.29), *the integral law of momentum conservation* says

integral conservation of momentum

$$\frac{d}{dt}\int_{V_3} \mathbf{G}\, dV + \int_{\partial V_3} \mathbf{T}\cdot d\mathbf{\Sigma} = 0. \quad (1.35)$$

By pulling the time derivative inside the volume integral (where it becomes a partial derivative) and applying the vectorial version of Gauss's law to the surface integral, we obtain $\int_{V_3}(\partial \mathbf{G}/\partial t + \nabla\cdot \mathbf{T})\, dV = 0$. This can be true for all choices of V_3 only if the integrand vanishes:

differential conservation of momentum

$$\frac{\partial \mathbf{G}}{\partial t} + \nabla\cdot \mathbf{T} = 0, \quad \text{or} \quad \frac{\partial G_j}{\partial t} + T_{jk;k} = 0. \quad (1.36)$$

(Because **T** is symmetric, it does not matter which of its slots the divergence acts on.) This is *the differential law of momentum conservation*. It has the standard form for any local conservation law: the time derivative of the density of some quantity (here momentum), plus the divergence of the flux of that quantity (here the momentum flux is the stress tensor), is zero. We shall make extensive use of this Newtonian law of momentum conservation in Part IV (elasticity), Part V (fluid dynamics), and Part VI (plasma physics).

EXERCISES

Exercise 1.13 **Example: *Equations of Motion for a Perfect Fluid*
(a) Consider a perfect fluid with density ρ, pressure P, and velocity $\mathbf{v}$ that vary in time and space. Explain why the fluid's momentum density is $\mathbf{G} = \rho \mathbf{v}$, and explain why its momentum flux (stress tensor) is

$$\mathbf{T} = P\mathbf{g} + \rho \mathbf{v}\otimes \mathbf{v}, \quad \text{or, in slot-naming index notation,} \quad T_{ij} = P g_{ij} + \rho v_i v_j.$$
(1.37a)

(b) Explain why the law of mass conservation for this fluid is

$$\frac{\partial \rho}{\partial t} + \nabla\cdot(\rho \mathbf{v}) = 0. \quad (1.37b)$$

(c) Explain why the derivative operator

$$\frac{d}{dt} \equiv \frac{\partial}{\partial t} + \mathbf{v}\cdot\nabla \quad (1.37c)$$

describes the rate of change as measured by somebody who moves locally with the fluid (i.e., with velocity **v**). This is sometimes called the fluid's *advective time derivative* or *convective time derivative* or *material derivative*.

(d) Show that the fluid's law of mass conservation (1.37b) can be rewritten as

$$\frac{1}{\rho}\frac{d\rho}{dt} = -\nabla \cdot \mathbf{v}, \tag{1.37d}$$

which says that the divergence of the fluid's velocity field is minus the fractional rate of change of its density, as measured in the fluid's local rest frame.

(e) Show that the differential law of momentum conservation (1.36) for the fluid can be written as

$$\frac{d\mathbf{v}}{dt} = -\frac{\nabla P}{\rho}. \tag{1.37e}$$

This is called the fluid's *Euler equation*. Explain why this Euler equation is Newton's second law of motion, $\mathbf{F} = m\mathbf{a}$, written on a per unit mass basis.

In Part V of this book, we use Eqs. (1.37) to study the dynamical behaviors of fluids. For many applications, the Euler equation will need to be augmented by the force per unit mass exerted by the fluid's internal viscosity.

Exercise 1.14 ***Problem: Electromagnetic Stress Tensor*

(a) An electric field $\mathbf{E}$ exerts (in SI units) a pressure $\epsilon_o \mathbf{E}^2/2$ orthogonal to itself and a tension of this same magnitude along itself. Similarly, a magnetic field $\mathbf{B}$ exerts a pressure $\mathbf{B}^2/2\mu_o = \epsilon_o c^2 \mathbf{B}^2/2$ orthogonal to itself and a tension of this same magnitude along itself. Verify that the following stress tensor embodies these stresses:

$$\boxed{\mathbf{T} = \frac{\epsilon_o}{2}\left[(\mathbf{E}^2 + c^2\mathbf{B}^2)\mathbf{g} - 2(\mathbf{E}\otimes\mathbf{E} + c^2\mathbf{B}\otimes\mathbf{B})\right].} \tag{1.38}$$

(b) Consider an electromagnetic field interacting with a material that has a charge density ρ_e and a current density $\mathbf{j}$. Compute the divergence of the electromagnetic stress tensor (1.38) and evaluate the derivatives using Maxwell's equations. Show that the result is the negative of the force density that the electromagnetic field exerts on the material. Use momentum conservation to explain why this has to be so.

1.10 Geometrized Units and Relativistic Particles for Newtonian Readers

Readers who are skipping the relativistic parts of this book will need to know two important pieces of relativity: (i) geometrized units and (ii) the relativistic energy and momentum of a moving particle.

1.10.1 Geometrized Units

The speed of light is independent of one's reference frame (i.e., independent of how fast one moves). This is a fundamental tenet of special relativity, and in the era before 1983, when the meter and the second were defined independently, it was tested and confirmed experimentally with very high precision. By 1983, this constancy had become so universally accepted that it was used to redefine the meter (which is hard to measure precisely) in terms of the second (which is much easier to measure with modern technology).[11] The meter is now related to the second in such a way that the speed of light is precisely $c = 299{,}792{,}458$ m s^{-1} (i.e., 1 meter is the distance traveled by light in 1/299,792,458 seconds). Because of this constancy of the light speed, it is permissible when studying special relativity to set c to unity. Doing so is equivalent to the relationship

$$c = 2.99792458 \times 10^8 \text{ m s}^{-1} = 1 \tag{1.39a}$$

between seconds and centimeters; i.e., equivalent to

$$1\,\text{s} = 2.99792458 \times 10^8 \text{ m}. \tag{1.39b}$$

geometrized units

We refer to units in which $c = 1$ as *geometrized units,* and we adopt them throughout this book when dealing with relativistic physics, since they make equations look much simpler. Occasionally it will be useful to restore the factors of c to an equation, thereby converting it to ordinary (SI or cgs) units. This restoration is achieved easily using dimensional considerations. For example, the equivalence of mass m and relativistic energy $\mathcal{E}$ is written in geometrized units as $\mathcal{E} = m$. In SI units $\mathcal{E}$ has dimensions of joule = kg m^2 s^{-2}, while m has dimensions of kg, so to make $\mathcal{E} = m$ dimensionally correct we must multiply the right side by a power of c that has dimensions m^2 s^{-2} (i.e., by c^2); thereby we obtain $\mathcal{E} = mc^2$.

1.10.2 Energy and Momentum of a Moving Particle

relativistic energy and momentum

A particle with rest mass m, moving with velocity $\mathbf{v} = d\mathbf{x}/dt$ and speed $v = |\mathbf{v}|$, has a relativistic energy $\mathcal{E}$ (including its rest mass), relativistic kinetic energy E (excluding its rest mass), and relativistic momentum $\mathbf{p}$ given by

$$\boxed{\mathcal{E} = \frac{m}{\sqrt{1-v^2}} \equiv \frac{m}{\sqrt{1-v^2/c^2}} \equiv E + m,} \quad \boxed{\mathbf{p} = \mathcal{E}\mathbf{v} = \frac{m\mathbf{v}}{\sqrt{1-v^2}};} \tag{1.40}$$

so $\boxed{\mathcal{E} = \sqrt{m^2 + \mathbf{p}^2}.}$

In the low-velocity (Newtonian) limit, the energy E with rest mass removed (kinetic energy) and the momentum $\mathbf{p}$ take their familiar Newtonian forms:

$$\text{When } v \ll c \equiv 1, \quad E \to \frac{1}{2}mv^2 \quad \text{and } \mathbf{p} \to m\mathbf{v}. \tag{1.41}$$

---

11. The second is defined as the duration of 9,192,631,770 periods of the radiation produced by a certain hyperfine transition in the ground state of a ^{133}Cs atom that is at rest in empty space. Today (2016) all fundamental physical units except mass units (e.g., the kilogram) are defined similarly in terms of fundamental constants of Nature.

A particle with zero rest mass (a photon or a graviton)[12] always moves with the speed of light $v = c = 1$, and like other particles it has momentum $\mathbf{p} = \mathcal{E}\mathbf{v}$, so the magnitude of its momentum is equal to its energy: $|\mathbf{p}| = \mathcal{E}v = \mathcal{E}c = \mathcal{E}$.

When particles interact (e.g., in chemical reactions, nuclear reactions, and elementary-particle collisions) the sum of the particle energies $\mathcal{E}$ is conserved, as is the sum of the particle momenta $\mathbf{p}$.

For further details and explanations, see Chap. 2.

EXERCISES

Exercise 1.15 *Practice: Geometrized Units*
Convert the following equations from the geometrized units in which they are written to SI units.

(a) The "Planck time" t_P expressed in terms of Newton's gravitation constant G and Planck's reduced constant $\hbar$, $t_P = \sqrt{G\hbar}$. What is the numerical value of t_P in seconds? in meters?

(b) The energy $\mathcal{E} = 2m$ obtained from the annihilation of an electron and a positron, each with rest mass m.

(c) The Lorentz force law $md\mathbf{v}/dt = e(\mathbf{E} + \mathbf{v} \times \mathbf{B})$.

(d) The expression $\mathbf{p} = \hbar\omega\mathbf{n}$ for the momentum $\mathbf{p}$ of a photon in terms of its angular frequency ω and direction $\mathbf{n}$ of propagation.

How tall are you, in seconds? How old are you, in meters?

Bibliographic Note

Most of the concepts developed in this chapter are treated, though from rather different viewpoints, in intermediate and advanced textbooks on classical mechanics or electrodynamics, such as Marion and Thornton (1995); Jackson (1999); Griffiths (1999); Goldstein, Poole, and Safko (2002).

Landau and Lifshitz's (1976) advanced text *Mechanics* is famous for its concise and precise formulations; it lays heavy emphasis on symmetry principles and their implications. A similar approach is followed in the next volume in their Course of Theoretical Physics series, *The Classical Theory of Fields* (Landau and Lifshitz, 1975), which is rooted in special relativity and goes on to cover general relativity. We refer to other volumes in this remarkable series in subsequent chapters.

The three-volume *Feynman Lectures on Physics* (Feynman, Leighton, and Sands, 2013) had a big influence on several generations of physicists, and even more so on their teachers. Both of us (Blandford and Thorne) are immensely indebted to Richard Feynman for shaping our own approaches to physics. His insights on the foundations

---

12. We do not know for sure that photons and gravitons are massless, but the laws of physics as currently understood require them to be massless, and there are tight experimental limits on their rest masses.

of classical physics and its relationship to quantum mechanics, and on calculational techniques, are as relevant today as in 1963, when his course was first delivered.

The geometric viewpoint on the laws of physics, which we present and advocate in this chapter, is not common (but it should be because of its great power). For example, the vast majority of mechanics and electrodynamics textbooks, including all those listed above, define a tensor as a matrix-like entity whose components transform under rotations in the manner described by Eq. (1.13b). This is a complicated definition that hides the great simplicity of a tensor as nothing more than a linear function of vectors; it obscures thinking about tensors geometrically, without the aid of any coordinate system or basis.

The geometric viewpoint comes to the physics community from mathematicians, largely by way of relativity theory. By now, most relativity textbooks espouse it. See the Bibliographic Note to Chap. 2. Fortunately, this viewpoint is gradually seeping into the nonrelativistic physics curriculum (e.g., Kleppner and Kolenkow, 2013). We hope this chapter will accelerate that seepage.

CHAPTER TWO

Special Relativity: Geometric Viewpoint T2

> Henceforth space by itself, and time by itself, are doomed to fade away into mere shadows, and only a kind of union of the two will preserve an independent reality.
>
> HERMANN MINKOWSKI, 1908

2.1 Overview

This chapter is a fairly complete introduction to special relativity at an intermediate level. We extend the geometric viewpoint, developed in Chap. 1 for Newtonian physics, to the domain of special relativity; and we extend the tools of differential geometry, developed in Chap. 1 for the arena of Newtonian physics (3-dimensional Euclidean space) to that of special relativity (4-dimensional Minkowski spacetime).

We begin in Sec. 2.2 by defining inertial (Lorentz) reference frames and then introducing fundamental, geometric, reference-frame-independent concepts: events, 4-vectors, and the invariant interval between events. Then in Sec. 2.3, we develop the basic concepts of tensor algebra in Minkowski spacetime (tensors, the metric tensor, the inner product and tensor product, and contraction), patterning our development on the corresponding concepts in Euclidean space. In Sec. 2.4, we illustrate our tensor-algebra tools by using them to describe—without any coordinate system or reference frame—the kinematics (world lines, 4-velocities, 4-momenta) of point particles that move through Minkowski spacetime. The particles are allowed to collide with one another and be accelerated by an electromagnetic field. In Sec. 2.5, we introduce components of vectors and tensors in an inertial reference frame and rewrite our frame-independent equations in slot-naming index notation; then in Sec. 2.6, we use these extended tensorial tools to restudy the motions, collisions, and electromagnetic accelerations of particles. In Sec. 2.7, we discuss Lorentz transformations in Minkowski spacetime, and in Sec. 2.8, we develop spacetime diagrams and use them to study length contraction, time dilation, and simultaneity breakdown. In Sec. 2.9, we illustrate the tools we have developed by asking whether the laws of physics permit a highly advanced civilization to build time machines for traveling backward in time as well as forward. In Sec. 2.10, we introduce directional derivatives, gradients, and the Levi-Civita tensor in Minkowski spacetime, and in Sec. 2.11, we use these tools to discuss Maxwell's equations and the geometric nature of electric and magnetic fields.

BOX 2.1. READERS' GUIDE

- Parts II (Statistical Physics), III (Optics), IV (Elasticity), V (Fluid Dynamics), and VI (Plasma Physics) of this book deal almost entirely with Newtonian physics; only a few sections and exercises are relativistic. Readers who are inclined to skip those relativistic items (which are all labeled Track Two) can skip this chapter and then return to it just before embarking on Part VII (General Relativity). Accordingly, this chapter is Track Two for readers of Parts II–VI and Track One for readers of Part VII—and in this spirit we label it Track Two.

- More specifically, this chapter is a prerequisite for the following: sections on relativistic kinetic theory in Chap. 3, Sec. 13.8 on relativistic fluid dynamics, Ex. 17.9 on relativistic shocks in fluids, many comments in Parts II–VI about relativistic effects and connections between Newtonian physics and relativistic physics, and all of Part VII (General Relativity).

- We recommend that those readers for whom relativity is relevant— and who already have a strong understanding of special relativity— not skip this chapter entirely. Instead, we suggest they browse it, especially Secs. 2.2–2.4, 2.8, and 2.11–2.13, to make sure they understand this book's geometric viewpoint and to ensure their familiarity with such concepts as the stress-energy tensor that they might not have met previously.

In Sec. 2.12, we develop our final set of geometric tools: volume elements and the integration of tensors over spacetime; finally, in Sec. 2.13, we use these tools to define the stress-energy tensor and to formulate very general versions of the conservation of 4-momentum.

2.2 Foundational Concepts

2.2.1 Inertial Frames, Inertial Coordinates, Events, Vectors, and Spacetime Diagrams

Because the nature and geometry of Minkowski spacetime are far less obvious intuitively than those of Euclidean 3-space, we need a crutch in our development of the geometric viewpoint for physics in spacetime. That crutch will be inertial reference frames.

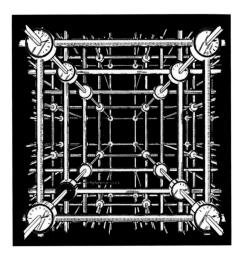

FIGURE 2.1 An inertial reference frame. From Taylor and Wheeler (1966). Used with permission of E. F. Taylor and the estate of J. A. Wheeler.

An inertial reference frame is a 3-dimensional latticework of measuring rods and clocks (Fig. 2.1) with the following properties:

- The latticework is purely conceptual and has arbitrarily small mass, so it does not gravitate.
- The latticework moves freely through spacetime (i.e., no forces act on it) and is attached to gyroscopes, so it is inertially nonrotating.
- The measuring rods form an orthogonal lattice, and the length intervals marked on them are uniform when compared to, for example, the wavelength of light emitted by some standard type of atom or molecule. Therefore, the rods form an orthonormal Cartesian coordinate system with the coordinate x measured along one axis, y along another, and z along the third.
- The clocks are densely packed throughout the latticework so that, ideally, there is a separate clock at every lattice point.
- The clocks tick uniformly when compared to the period of the light emitted by some standard type of atom or molecule (i.e., they are *ideal clocks*).
- The clocks are synchronized by the Einstein synchronization process: if a pulse of light, emitted by one of the clocks, bounces off a mirror attached to another and then returns, the time of bounce t_b, as measured by the clock that does the bouncing, is the average of the times of emission and reception, as measured by the emitting and receiving clock: $t_b = \frac{1}{2}(t_e + t_r)$.[1]

inertial reference frame

ideal clocks and their synchronization

---

1. For a deeper discussion of the nature of ideal clocks and ideal measuring rods see, for example, Misner, Thorne, and Wheeler (1973, pp. 23–29 and 395–399).

T2

(That inertial frames with these properties can exist, when gravity is unimportant, is an empirical fact; it tells us that, in the absence of gravity, spacetime is truly Minkowski.)

event

Our first fundamental, frame-independent relativistic concept is the *event*. An event is a precise location in space at a precise moment of time—a precise location (or *point*) in 4-dimensional spacetime. We sometimes denote events by capital script letters, such as $\mathcal{P}$ and $\mathcal{Q}$—the same notation used for points in Euclidean 3-space.

4-vector

A *4-vector* (also often referred to as a *vector in spacetime* or just a *vector*) is a straight[2] arrow $\Delta\vec{x}$ reaching from one event $\mathcal{P}$ to another, $\mathcal{Q}$. We often deal with 4-vectors and ordinary (3-space) vectors simultaneously, so we shall use different notations for them: boldface Roman font for 3-vectors, $\Delta\mathbf{x}$, and arrowed italic font for 4-vectors, $\Delta\vec{x}$. Sometimes we identify an event $\mathcal{P}$ in spacetime by its vectorial separation $\vec{x}_\mathcal{P}$ from some arbitrarily chosen event in spacetime, the origin $\mathcal{O}$.

An inertial reference frame provides us with a coordinate system for spacetime. The coordinates $(x^0, x^1, x^2, x^3) = (t, x, y, z)$ that it associates with an event $\mathcal{P}$ are $\mathcal{P}$'s location (x, y, z) in the frame's latticework of measuring rods and the time t of $\mathcal{P}$ as measured by the clock that sits in the lattice at the event's location. (Many apparent paradoxes in special relativity result from failing to remember that the time t of an event is always measured by a clock that resides at the event—never by clocks that reside elsewhere in spacetime.)

spacetime diagrams

It is useful to depict events on *spacetime diagrams,* in which the time coordinate $t = x^0$ of some inertial frame is plotted upward; two of the frame's three spatial coordinates, $x = x^1$ and $y = x^2$, are plotted horizontally; and the third coordinate $z = x^3$ is omitted. Figure 2.2 is an example. Two events $\mathcal{P}$ and $\mathcal{Q}$ are shown there, along with their vectorial separations $\vec{x}_\mathcal{P}$ and $\vec{x}_\mathcal{Q}$ from the origin and the vector $\Delta\vec{x} = \vec{x}_\mathcal{Q} - \vec{x}_\mathcal{P}$ that separates them from each other. The coordinates of $\mathcal{P}$ and $\mathcal{Q}$, which are the same as the components of $\vec{x}_\mathcal{P}$ and $\vec{x}_\mathcal{Q}$ in this coordinate system, are $(t_\mathcal{P}, x_\mathcal{P}, y_\mathcal{P}, z_\mathcal{P})$ and $(t_\mathcal{Q}, x_\mathcal{Q}, y_\mathcal{Q}, z_\mathcal{Q})$. Correspondingly, the components of $\Delta\vec{x}$ are

$$\Delta x^0 = \Delta t = t_\mathcal{Q} - t_\mathcal{P}, \qquad \Delta x^1 = \Delta x = x_\mathcal{Q} - x_\mathcal{P},$$
$$\Delta x^2 = \Delta y = y_\mathcal{Q} - y_\mathcal{P}, \qquad \Delta x^3 = \Delta z = z_\mathcal{Q} - z_\mathcal{P}. \qquad (2.1)$$

We denote these components of $\Delta\vec{x}$ more compactly by Δx^α, where the index α and all other lowercased Greek indices range from 0 (for t) to 3 (for z).

When the physics or geometry of a situation being studied suggests some preferred inertial frame (e.g., the frame in which some piece of experimental apparatus is at rest), then we typically use as axes for our spacetime diagrams the coordinates of that preferred frame. By contrast, when our situation provides no preferred inertial frame, or when we wish to emphasize a frame-independent viewpoint, we use as axes

---

2. By "straight" we mean that in any inertial reference frame, the coordinates along $\Delta\vec{x}$ are linear functions of one another.

Chapter 2. Special Relativity: Geometric Viewpoint

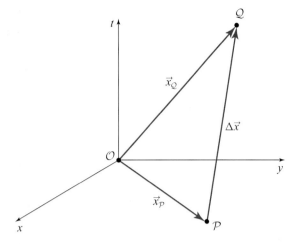

FIGURE 2.2 A spacetime diagram depicting two events $\mathcal{P}$ and $\mathcal{Q}$, their vectorial separations $\vec{x}_\mathcal{P}$ and $\vec{x}_\mathcal{Q}$ from an (arbitrarily chosen) origin $\mathcal{O}$, and the vector $\Delta\vec{x} = \vec{x}_\mathcal{Q} - \vec{x}_\mathcal{P}$ connecting them. The laws of physics cannot involve the arbitrary origin; we introduce it only as a conceptual aid.

the coordinates of a completely arbitrary inertial frame and think of the diagram as depicting spacetime in a coordinate-independent, frame-independent way.

We use the terms *inertial coordinate system* and *Lorentz coordinate system* interchangeably[3] to mean the coordinate system (t, x, y, z) provided by an inertial frame; we also use the term *Lorentz frame* interchangeably with *inertial frame*. A physicist or other intelligent being who resides in a Lorentz frame and makes measurements using its latticework of rods and clocks will be called an *observer*.

Although events are often described by their coordinates in a Lorentz reference frame, and 4-vectors by their components (coordinate differences), it should be obvious that the concepts of an event and a 4-vector need not rely on any coordinate system whatsoever for their definitions. For example, the event $\mathcal{P}$ of the birth of Isaac Newton and the event $\mathcal{Q}$ of the birth of Albert Einstein are readily identified without coordinates. They can be regarded as points in spacetime, and their separation vector is the straight arrow reaching through spacetime from $\mathcal{P}$ to $\mathcal{Q}$. Different observers in different inertial frames will attribute different coordinates to each birth and different components to the births' vectorial separation, but all observers can agree that they are talking about the same events $\mathcal{P}$ and $\mathcal{Q}$ in spacetime and the same separation vector $\Delta\vec{x}$. In this sense, $\mathcal{P}$, $\mathcal{Q}$, and $\Delta\vec{x}$ are *frame-independent, geometric objects* (points and arrows) that reside in spacetime.

inertial coordinates (Lorentz coordinates)

observer

---

3. It was Lorentz (1904) who first wrote down the relationship of one such coordinate system to another: the Lorentz transformation.

2.2 Foundational Concepts

2.2.2 The Principle of Relativity and Constancy of Light Speed

Principle of Relativity

Einstein's Principle of Relativity, stated in modern form, says that *Every (special relativistic) law of physics must be expressible as a geometric, frame-independent relationship among geometric, frame-independent objects* (i.e., such objects as points in spacetime and 4-vectors and tensors, which represent physical quantities, such as events, particle momenta, and the electromagnetic field). This is nothing but our Geometric Principle for physical laws (Chap. 1), lifted from the Euclidean-space arena of Newtonian physics to the Minkowski-spacetime arena of special relativity.

Since the laws are all geometric (i.e., unrelated to any reference frame or coordinate system), they can't distinguish one inertial reference frame from any other. This leads to an alternative form of the Principle of Relativity (one commonly used in elementary textbooks and equivalent to the above): *All the (special relativistic) laws of physics are the same in every inertial reference frame everywhere in spacetime.* This, in fact, is Einstein's own version of his Principle of Relativity; only in the sixty years since his death have we physicists reexpressed it in geometric language.

Because inertial reference frames are related to one another by Lorentz transformations (Sec. 2.7), we can restate Einstein's version of this Principle as *All the (special relativistic) laws of physics are Lorentz invariant.*

A more operational version of this Principle is: Give identical instructions for a specific physics experiment to two different observers in two different inertial reference frames at the same or different locations in Minkowski (i.e., gravity-free) spacetime. The experiment must be self-contained; that is, it must not involve observations of the external universe's properties (the "environment"). For example, an unacceptable experiment would be a measurement of the anisotropy of the universe's cosmic microwave radiation and a computation therefrom of the observer's velocity relative to the radiation's mean rest frame; such an experiment studies the universal environment, not the fundamental laws of physics. An acceptable experiment would be a measurement of the speed of light using the rods and clocks of the observer's own frame, or a measurement of cross sections for elementary particle reactions using particles moving in the reference frame's laboratory. The Principle of Relativity says that in these or any other similarly self-contained experiments, the two observers in their two different inertial frames must obtain identical experimental results—to within the accuracy of their experimental techniques. Since the experimental results are governed by the (nongravitational) laws of physics, this is equivalent to the statement that all physical laws are the same in the two inertial frames.

constancy of light speed

Perhaps the most central of special relativistic laws is the one stating that *The speed of light c in vacuum is frame independent;* that is, it is a constant, independent of the inertial reference frame in which it is measured. In other words, there is no "aether" that supports light's vibrations and in the process influences its speed—a remarkable fact that came as a great experimental surprise to physicists at the end of the nineteenth century.

The constancy of the speed of light, in fact, is built into Maxwell's equations. For these equations to be frame independent, the speed of light, which appears in them, must be frame independent. In this sense, the constancy of the speed of light follows from the Principle of Relativity; it is not an independent postulate. This is illustrated in Box 2.2.

BOX 2.2.   MEASURING THE SPEED OF LIGHT WITHOUT LIGHT

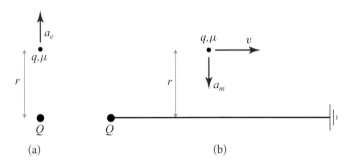

In some inertial reference frame, we perform two thought experiments using two particles, one with a large charge Q; the other, a test particle, with a much smaller charge q and mass μ. In the first experiment, we place the two particles at rest, separated by a distance $|\Delta x| \equiv r$, and measure the electrical repulsive acceleration a_e of q (panel a in the diagram). In Gaussian units (where the speed of light shows up explicitly instead of via $\epsilon_o \mu_o = 1/c^2$), the acceleration is $a_e = qQ/r^2\mu$. In the second experiment, we connect Q to ground by a long wire, and we place q at the distance $|\Delta x| = r$ from the wire and set it moving at speed v parallel to the wire. The charge Q flows down the wire with an e-folding time τ, so the current is $I = dQ/d\tau = (Q/\tau)e^{-t/\tau}$. At early times $0 < t \ll \tau$, this current $I = Q/\tau$ produces a solenoidal magnetic field at q with field strength $B = (2/cr)(Q/\tau)$, and this field exerts a magnetic force on q, giving it an acceleration $a_m = q(v/c)B/\mu = 2vqQ/c^2\tau r\mu$. The ratio of the electric acceleration in the first experiment to the magnetic acceleration in the second experiment is $a_e/a_m = c^2\tau/2rv$. Therefore, we can measure the speed of light c in our chosen inertial frame by performing this pair of experiments; carefully measuring the separation r, speed v, current Q/τ, and accelerations; and then simply computing $c = \sqrt{(2rv/\tau)(a_e/a_m)}$. The Principle of Relativity insists that the result of this pair of experiments should be independent of the inertial frame in which they are performed. Therefore, the speed of light c that appears in Maxwell's equations must be frame independent. In this sense, the constancy of the speed of light follows from the Principle of Relativity as applied to Maxwell's equations.

What makes light so special? What about the propagation speeds of other types of waves? Are they or should they be the same as light's speed? For a digression on this topic, see Box 2.3.

The constancy of the speed of light underlies our ability to use the geometrized units introduced in Sec. 1.10. Any reader who has not studied that section should do so now. We use geometrized units throughout this chapter (and also throughout this book) when working with relativistic physics.

BOX 2.3. PROPAGATION SPEEDS OF OTHER WAVES

Electromagnetic radiation is not the only type of wave in Nature. In this book, we encounter dispersive media, such as optical fibers and plasmas, where electromagnetic signals travel slower than c. We also analyze sound waves and seismic waves, whose governing laws do not involve electromagnetism at all. How do these fit into our special relativistic framework? The answer is simple. Each of these waves involves an underlying medium that is at rest in one particular frame (not necessarily inertial), and the velocity at which the wave's information propagates (the group velocity) is most simply calculated in this frame *from the wave's and medium's fundamental laws.* We can then use the kinematic rules of Lorentz transformations to compute the velocity in another frame. However, if we had chosen to compute the wave speed in the second frame directly, using the same fundamental laws, we would have gotten the same answer, albeit perhaps with greater effort. All waves are in full compliance with the Principle of Relativity. What is special about vacuum electromagnetic waves and, by extension, photons, is that no medium (or "aether," as it used to be called) is needed for them to propagate. Their speed is therefore the same in all frames. (Although some physicists regard the cosmological constant, discussed in Chap. 28, as a modern aether, we must emphaisze that, unlike its nineteenth-century antecedent, its presence does not alter the propagation of photons through Lorentz frames.)

This raises an interesting question. What about other waves that do not require an underlying medium? What about electron de Broglie waves? Here the fundamental wave equation, Schrödinger's or Dirac's, is mathematically different from Maxwell's and contains an important parameter, the electron rest mass. This rest mass allows the fundamental laws of relativistic quantum mechanics to be written in a form that is the same in all inertial reference frames and at the same time allows an electron, considered as either a wave or a particle, to travel at a different speed when measured in a different frame.

(continued)

BOX 2.3. (continued)

Some particles that have been postulated (such as gravitons, the quanta of gravitational waves; Chap. 27) are believed to exist without a rest mass (or an aether!), just like photons. Must these travel at the same speed as photons? The answer, according to the Principle of Relativity, is "yes." Why? Suppose there were two such waves or particles whose governing laws led to different speeds, c and $c' < c$, with each speed claimed to be the same in all reference frames. Such a claim produces insurmountable conundrums. For example, if we move with speed c' in the direction of propagation of the second wave, we will bring it to rest, in conflict with our hypothesis that its speed is frame independent. Therefore, all signals whose governing laws require them to travel with a speed that has no governing parameters (no rest mass and no underlying physical medium) must travel with a unique speed, which we call c. The speed of light is more fundamental to relativity than light itself!

2.2.3 The Interval and Its Invariance

Next we turn to another fundamental concept, the *interval* $(\Delta s)^2$ between the two events $\mathcal{P}$ and $\mathcal{Q}$ whose separation vector is $\Delta \vec{x}$. In a specific but arbitrary inertial reference frame and in geometrized units, $(\Delta s)^2$ is given by

$$(\Delta s)^2 \equiv -(\Delta t)^2 + (\Delta x)^2 + (\Delta y)^2 + (\Delta z)^2 = -(\Delta t)^2 + \sum_{i,j} \delta_{ij} \Delta x^i \Delta x^j;$$

(2.2a)

cf. Eq. (2.1). If $(\Delta s)^2 > 0$, the events $\mathcal{P}$ and $\mathcal{Q}$ are said to have a *spacelike* separation; if $(\Delta s)^2 = 0$, their separation is *null* or *lightlike*; and if $(\Delta s)^2 < 0$, their separation is *timelike*. For timelike separations, $(\Delta s)^2 < 0$ implies that Δs is imaginary; to avoid dealing with imaginary numbers, we describe timelike intervals by

$$(\Delta \tau)^2 \equiv -(\Delta s)^2,$$

(2.2b)

whose square root $\Delta \tau$ is real.

The coordinate separation between $\mathcal{P}$ and $\mathcal{Q}$ depends on one's reference frame: if $\Delta x^{\alpha'}$ and Δx^{α} are the coordinate separations in two different frames, then $\Delta x^{\alpha'} \neq$

Δx^α. Despite this frame dependence, the Principle of Relativity forces the interval $(\Delta s)^2$ to be the same in all frames:

$$(\Delta s)^2 = -(\Delta t)^2 + (\Delta x)^2 + (\Delta y)^2 + (\Delta z)^2$$
$$= -(\Delta t')^2 + (\Delta x')^2 + (\Delta y')^2 + (\Delta z')^2. \tag{2.3}$$

In Box 2.4, we sketch a proof for the case of two events $\mathcal{P}$ and $\mathcal{Q}$ whose separation is timelike.

Because of its frame invariance, the interval $(\Delta s)^2$ can be regarded as a geometric property of the vector $\Delta \vec{x}$ that reaches from $\mathcal{P}$ to $\mathcal{Q}$; we call it the *squared length* $(\Delta \vec{x})^2$ of $\Delta \vec{x}$:

$$(\Delta \vec{x})^2 \equiv (\Delta s)^2. \tag{2.4}$$

BOX 2.4. PROOF OF INVARIANCE OF THE INTERVAL FOR A TIMELIKE SEPARATION

A simple demonstration that the interval is invariant is provided by a thought experiment in which a photon is emitted at event $\mathcal{P}$, reflects off a mirror, and is then detected at event $\mathcal{Q}$. We consider the interval between these events in two reference frames, primed and unprimed, that move with respect to each other. Choose the spatial coordinate systems of the two frames in such a way that (i) their relative motion (with speed β, which will not enter into our analysis) is along the x and x' directions, (ii) event $\mathcal{P}$ lies on the x and x' axes, and (iii) event $\mathcal{Q}$ lies in the x-y and x'-y' planes, as depicted below. Then evaluate the interval between $\mathcal{P}$ and $\mathcal{Q}$ in the unprimed frame by the following construction: Place the mirror parallel to the x-z plane at precisely the height h that permits a photon, emitted from $\mathcal{P}$, to travel along the dashed line to the mirror, then reflect off the mirror and continue along the dashed path, arriving at event $\mathcal{Q}$. If the mirror were placed lower, the photon would arrive at the spatial location of $\mathcal{Q}$ sooner than the time of $\mathcal{Q}$; if placed higher, it would arrive later. Then the distance the photon travels (the length of the two-segment dashed line) is equal to $c \Delta t = \Delta t$, where Δt is the time between events $\mathcal{P}$ and $\mathcal{Q}$ as measured in the unprimed frame. If the mirror had not been present, the photon would have arrived at event $\mathcal{R}$ after time Δt, so $c \Delta t$ is the distance between $\mathcal{P}$ and $\mathcal{R}$. From the diagram, it is easy to see that the height of $\mathcal{R}$ above the x-axis is $2h - \Delta y$, and the Pythagorean theorem then implies that

$$(\Delta s)^2 = -(\Delta t)^2 + (\Delta x)^2 + (\Delta y)^2 = -(2h - \Delta y)^2 + (\Delta y)^2. \tag{1a}$$

The same construction in the primed frame must give the same formula, but with primes:

$$(\Delta s')^2 = -(\Delta t')^2 + (\Delta x')^2 + (\Delta y')^2 = -(2h' - \Delta y')^2 + (\Delta y')^2. \tag{1b}$$

(continued)

BOX 2.4. (continued)

The proof that $(\Delta s')^2 = (\Delta s)^2$ then reduces to showing that the Principle of Relativity requires that distances perpendicular to the direction of relative motion of two frames be the same as measured in the two frames: $h' = h$, $\Delta y' = \Delta y$. We leave it to the reader to develop a careful argument for this (Ex. 2.2).

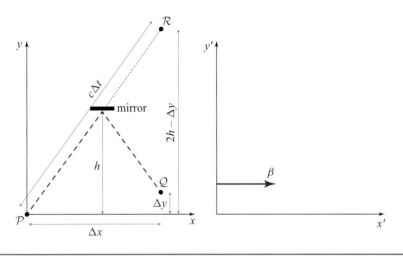

Note that this squared length, despite its name, can be negative (for timelike $\Delta \vec{x}$) or zero (for null $\Delta \vec{x}$) as well as positive (for spacelike $\Delta \vec{x}$).

The invariant interval $(\Delta s)^2$ between two events is as fundamental to Minkowski spacetime as the Euclidean distance between two points is to flat 3-space. Just as the Euclidean distance gives rise to the geometry of 3-space (as embodied, e.g., in Euclid's axioms), so the interval gives rise to the geometry of spacetime, which we shall be exploring. If this spacetime geometry were as intuitively obvious to humans as is Euclidean geometry, we would not need the crutch of inertial reference frames to arrive at it. Nature (presumably) has no need for such a crutch. To Nature (it seems evident), the geometry of Minkowski spacetime, as embodied in the invariant interval, is among the most fundamental aspects of physical law.

EXERCISES

Exercise 2.1 *Practice: Geometrized Units*
Do Ex. 1.15 in Chap. 1.

Exercise 2.2 *Derivation and Example: Invariance of the Interval*
Complete the derivation of the invariance of the interval given in Box 2.4, using the

Principle of Relativity in the form that the laws of physics must be the same in the primed and unprimed frames. Hints (if you need them):

(a) Having carried out the construction in the unprimed frame, depicted at the bottom left of Box 2.4, use the same mirror and photons for the analogous construction in the primed frame. Argue that, independently of the frame in which the mirror is at rest (unprimed or primed), the fact that the reflected photon has (angle of reflection) = (angle of incidence) in its rest frame implies that this is also true for the same photon in the other frame. Thereby conclude that the construction leads to Eq. (1b) in Box 2.4, as well as to Eq. (1a).

(b) Then argue that the perpendicular distance of an event from the common x- and x'-axes must be the same in the two reference frames, so $h' = h$ and $\Delta y' = \Delta y$; whence Eqs. (1b) and (1a) in Box 2.4 imply the invariance of the interval. [Note: For a leisurely version of this argument, see Taylor and Wheeler (1992, Secs. 3.6 and 3.7).]

2.3 Tensor Algebra without a Coordinate System

Having introduced points in spacetime (interpreted physically as events), the invariant interval $(\Delta s)^2$ between two events, 4-vectors (as arrows between two events), and the squared length of a vector (as the invariant interval between the vector's tail and tip), we can now introduce the remaining tools of tensor algebra for Minkowski spacetime in precisely the same way as we did for the Euclidean 3-space of Newtonian physics (Sec. 1.3), with the invariant interval between events playing the same role as the squared length between Euclidean points.

tensor

In particular: a *tensor* **T**(_, _, _) is a real-valued linear function of vectors in Minkowski spacetime. (We use slanted letters **T** for tensors in spacetime and unslanted letters **T** in Euclidean space.) A tensor's *rank* is equal to its number of slots. The *inner product* (also called the dot product) of two 4-vectors is

inner product

$$\vec{A} \cdot \vec{B} \equiv \frac{1}{4}\left[(\vec{A} + \vec{B})^2 - (\vec{A} - \vec{B})^2\right], \quad (2.5)$$

where $(\vec{A} + \vec{B})^2$ is the squared length of this vector (i.e., the invariant interval between its tail and its tip). The *metric tensor* of spacetime is that linear function of 4-vectors whose value is the inner product of the vectors:

metric tensor

$$\mathbf{g}(\vec{A}, \vec{B}) \equiv \vec{A} \cdot \vec{B}. \quad (2.6)$$

Using the inner product, we can regard any vector $\vec{A}$ as a rank-1 tensor: $\vec{A}(\vec{C}) \equiv \vec{A} \cdot \vec{C}$.

tensor product

Similarly, the *tensor product* $\otimes$ is defined precisely as in the Euclidean domain, Eqs. (1.5), as is the *contraction* of two slots of a tensor against each other, Eqs. (1.6), which lowers the tensor's rank by two.

contraction

2.4 Particle Kinetics and Lorentz Force without a Reference Frame

2.4.1 Relativistic Particle Kinetics: World Lines, 4-Velocity, 4-Momentum and Its Conservation, 4-Force

In this section, we illustrate our geometric viewpoint by formulating the special relativistic laws of motion for particles.

An accelerated particle moving through spacetime carries an *ideal clock*. By "ideal" we mean that the clock is unaffected by accelerations: it ticks at a uniform rate when compared to unaccelerated atomic oscillators that are momentarily at rest beside the clock and are well protected from their environments. The builders of inertial guidance systems for airplanes and missiles try to make their clocks as ideal as possible in just this sense. We denote by τ the time ticked by the particle's ideal clock, and we call it the particle's *proper time*.

The particle moves through spacetime along a curve, called its *world line*, which we can denote equally well by $\mathcal{P}(\tau)$ (the particle's spacetime location $\mathcal{P}$ at proper time τ), or by $\vec{x}(\tau)$ (the particle's vector separation from some arbitrarily chosen origin at proper time τ).[4]

We refer to the inertial frame in which the particle is momentarily at rest as its *momentarily comoving inertial frame* or *momentary rest frame*. Now, the particle's clock (which measures τ) is ideal, and so are the inertial frame's clocks (which measure coordinate time t). Therefore, a tiny interval $\Delta\tau$ of the particle's proper time is equal to the lapse of coordinate time in the particle's momentary rest frame $\Delta\tau = \Delta t$. Moreover, since the two events $\vec{x}(\tau)$ and $\vec{x}(\tau + \Delta\tau)$ on the clock's world line occur at the same spatial location in its momentary rest frame ($\Delta x^i = 0$, where $i = 1, 2, 3$) to first order in $\Delta\tau$, the invariant interval between those events is $(\Delta s)^2 = -(\Delta t)^2 + \sum_{i,j} \Delta x^i \Delta x^j \delta_{ij} = -(\Delta t)^2 = -(\Delta \tau)^2$. Thus, *the particle's proper time τ is equal to the square root of the negative of the invariant interval, $\tau = \sqrt{-s^2}$, along its world line.*

Figure 2.3 shows the world line of the accelerated particle in a spacetime diagram where the axes are coordinates of an arbitrary Lorentz frame. This diagram is intended to emphasize the world line as a frame-independent, geometric object. Also shown in the figure is the particle's *4-velocity* $\vec{u}$, which (by analogy with velocity in 3-space) is the time derivative of its position

$$\boxed{\vec{u} \equiv d\mathcal{P}/d\tau = d\vec{x}/d\tau} \quad (2.7)$$

and is the tangent vector to the world line. The derivative is defined by the usual limiting process

$$\frac{d\mathcal{P}}{d\tau} = \frac{d\vec{x}}{d\tau} \equiv \lim_{\Delta\tau \to 0} \frac{\mathcal{P}(\tau + \Delta\tau) - \mathcal{P}(\tau)}{\Delta\tau} = \lim_{\Delta\tau \to 0} \frac{\vec{x}(\tau + \Delta\tau) - \vec{x}(\tau)}{\Delta\tau}. \quad (2.8)$$

---

4. One of the basic ideas in string theory is that an elementary particle is described as a 1-dimensional loop in space rather than a 0-dimensional point. This means that it becomes a cylinder-like surface in spacetime—a world tube.

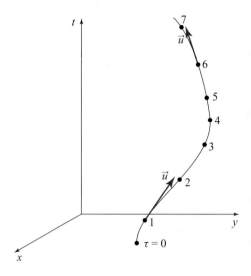

FIGURE 2.3 Spacetime diagram showing the world line $\vec{x}(\tau)$ and 4-velocity $\vec{u}$ of an accelerated particle. Note that the 4-velocity is tangent to the world line.

Here $\mathcal{P}(\tau + \Delta\tau) - \mathcal{P}(\tau)$ and $\vec{x}(\tau + \Delta\tau) - \vec{x}(\tau)$ are just two different ways to denote the same vector that reaches from one point on the world line to another.

The squared length of the particle's 4-velocity is easily seen to be -1:

$$\vec{u}^2 \equiv \mathbf{g}(\vec{u}, \vec{u}) = \frac{d\vec{x}}{d\tau} \cdot \frac{d\vec{x}}{d\tau} = \frac{d\vec{x} \cdot d\vec{x}}{(d\tau)^2} = -1. \tag{2.9}$$

The last equality follows from the fact that $d\vec{x} \cdot d\vec{x}$ is the squared length of $d\vec{x}$, which equals the invariant interval $(\Delta s)^2$ along it, and $(d\tau)^2$ is the negative of that invariant interval.

4-momentum

The particle's *4-momentum* is the product of its 4-velocity and rest mass:

$$\boxed{\vec{p} \equiv m\vec{u} = md\vec{x}/d\tau \equiv d\vec{x}/d\zeta.} \tag{2.10}$$

Here the parameter ζ is a renormalized version of proper time,

$$\zeta \equiv \tau/m. \tag{2.11}$$

affine parameter

This ζ and any other renormalized version of proper time with a position-independent renormalization factor are called *affine parameters* for the particle's world line. Expression (2.10), together with $\vec{u}^2 = -1$, implies that the squared length of the 4-momentum is

$$\boxed{\vec{p}^2 = -m^2.} \tag{2.12}$$

In quantum theory, a particle is described by a relativistic wave function, which, in the geometric optics limit (Chap. 7), has a wave vector $\vec{k}$ that is related to the classical particle's 4-momentum by

$$\vec{k} = \vec{p}/\hbar. \tag{2.13}$$

Chapter 2. Special Relativity: Geometric Viewpoint

The above formalism is valid only for particles with nonzero rest mass, $m \neq 0$. The corresponding formalism for a *particle with zero rest mass* (e.g., a photon or a graviton) can be obtained from the above by taking the limit as $m \to 0$ and $d\tau \to 0$ with the quotient $d\zeta = d\tau/m$ held finite. More specifically, the 4-momentum of a zero-rest-mass particle is well defined (and participates in the conservation law to be discussed below), and it is expressible in terms of the particle's affine parameter ζ by Eq. (2.10):

$$\vec{p} = d\vec{x}/d\zeta. \tag{2.14}$$

By contrast, the particle's 4-velocity $\vec{u} = \vec{p}/m$ is infinite and thus undefined, and proper time $\tau = m\zeta$ ticks vanishingly slowly along its world line and thus is undefined. Because proper time is the square root of the invariant interval along the world line, the interval between two neighboring points on the world line vanishes. Therefore, *the world line of a zero-rest-mass particle is null*. (By contrast, since $d\tau^2 > 0$ and $ds^2 < 0$ along the world line of a particle with finite rest mass, *the world line of a finite-rest-mass particle is timelike*.)

The 4-momenta of particles are important because of the *law of conservation of 4-momentum* (which, as we shall see in Sec. 2.6, is equivalent to the conservation laws for energy and ordinary momentum): If a number of "initial" particles, named $A = 1, 2, 3, \ldots$, enter a restricted region of spacetime $\mathcal{V}$ and there interact strongly to produce a new set of "final" particles, named $\bar{A} = \bar{1}, \bar{2}, \bar{3}, \ldots$ (Fig. 2.4), then the total 4-momentum of the final particles must be the same as the total 4-momentum of the initial ones:

conservation of 4-momentum

$$\boxed{\sum_{\bar{A}} \vec{p}_{\bar{A}} = \sum_A \vec{p}_A.} \tag{2.15}$$

Note that this law of 4-momentum conservation is expressed in frame-independent, geometric language—in accord with Einstein's insistence that all the laws of physics should be so expressible. As we shall see in Part VII, 4-momentum conservation is a consequence of the translation symmetry of flat, 4-dimensional spacetime. In general relativity's curved spacetime, where that translation symmetry is lost, we lose 4-momentum conservation except under special circumstances; see Eq. (25.56) and associated discussion.

If a particle moves freely (no external forces and no collisions with other particles), then its 4-momentum $\vec{p}$ will be conserved along its world line, $d\vec{p}/d\zeta = 0$. Since $\vec{p}$ is tangent to the world line, this conservation means that the direction of the world line in spacetime never changes: the free particle moves along a straight line through spacetime. To change the particle's 4-momentum, one must act on it with a *4-force* $\vec{F}$,

4-force

$$d\vec{p}/d\tau = \vec{F}. \tag{2.16}$$

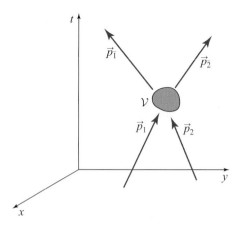

FIGURE 2.4 Spacetime diagram depicting the law of 4-momentum conservation for a situation where two particles, numbered 1 and 2, enter an interaction region V in spacetime, and there interact strongly and produce two new particles, numbered $\bar{1}$ and $\bar{2}$. The sum of the final 4-momenta, $\vec{p}_{\bar{1}} + \vec{p}_{\bar{2}}$, must be equal to the sum of the initial 4-momenta, $\vec{p}_1 + \vec{p}_2$.

If the particle is a fundamental one (e.g., photon, electron, proton), then the 4-force must leave its rest mass unchanged,

$$0 = dm^2/d\tau = -d\vec{p}^2/d\tau = -2\vec{p} \cdot d\vec{p}/d\tau = -2\vec{p} \cdot \vec{F}; \tag{2.17}$$

that is, the 4-force must be orthogonal to the 4-momentum in the 4-dimensional sense that their inner product vanishes.

2.4.2 Geometric Derivation of the Lorentz Force Law

As an illustration of these physical concepts and mathematical tools, we use them to deduce the relativistic version of the Lorentz force law. From the outset, in accord with the Principle of Relativity, we insist that the law we seek be expressible in geometric, frame-independent language, that is, in terms of vectors and tensors.

electromagnetic field tensor

Consider a particle with charge q and rest mass $m \neq 0$ interacting with an electromagnetic field. It experiences an electromagnetic 4-force whose mathematical form we seek. The Newtonian version of the electromagnetic force $\mathbf{F} = q(\mathbf{E} + \mathbf{v} \times \mathbf{B})$ is proportional to q and contains one piece (electric) that is independent of velocity $\mathbf{v}$ and a second piece (magnetic) that is linear in $\mathbf{v}$. It is reasonable to expect that, to produce this Newtonian limit, the relativistic 4-force $\vec{F}$ will be proportional to q and will be linear in the 4-velocity $\vec{u}$. Linearity means there must exist some second-rank tensor $\mathbf{F}(\_, \_)$, the *electromagnetic field tensor*, such that

$$d\vec{p}/d\tau = \vec{F}(\_) = q\mathbf{F}(\_, \vec{u}). \tag{2.18}$$

Because the 4-force $\vec{F}$ must be orthogonal to the particle's 4-momentum and thence also to its 4-velocity, $\vec{F} \cdot \vec{u} \equiv \vec{F}(\vec{u}) = 0$, expression (2.18) must vanish when $\vec{u}$ is inserted into its empty slot. In other words, for all timelike unit-length vectors $\vec{u}$,

$$\mathbf{F}(\vec{u}, \vec{u}) = 0. \tag{2.19}$$

It is an instructive exercise (Ex. 2.3) to show that this is possible only if $\mathbf{F}$ is *antisymmetric*, so the electromagnetic 4-force is

$$\boxed{d\vec{p}/d\tau = q\mathbf{F}(\_\_, \vec{u}), \quad \text{where} \quad \mathbf{F}(\vec{A}, \vec{B}) = -\mathbf{F}(\vec{B}, \vec{A}) \quad \text{for all } \vec{A} \text{ and } \vec{B}.} \tag{2.20}$$

electromagnetic 4-force

Equation (2.20) must be the relativistic form of the Lorentz force law. In Sec. 2.11 we deduce the relationship of the electromagnetic field tensor $\mathbf{F}$ to the more familiar electric and magnetic fields, and the relationship of this relativistic Lorentz force to its Newtonian form (1.7c).

The discussion of particle kinematics and the electromagnetic force in this Sec. 2.4 is elegant but perhaps unfamiliar. In Secs. 2.6 and 2.11, we shall see that it is equivalent to the more elementary (but more complex) formalism based on components of vectors in Euclidean 3-space.

EXERCISES

Exercise 2.3 *Derivation and Example: Antisymmetry of Electromagnetic Field Tensor*
Show that Eq. (2.19) can be true for all timelike, unit-length vectors $\vec{u}$ if and only if $\mathbf{F}$ is antisymmetric. [Hints: (i) Show that the most general second-rank tensor $\mathbf{F}$ can be written as the sum of a symmetric tensor $\mathbf{S}$ and an antisymmetric tensor $\mathbf{A}$, and that the antisymmetric piece contributes nothing to Eq. (2.19), so $\mathbf{S}(\vec{u}, \vec{u})$ must vanish for every timelike $\vec{u}$. (ii) Let $\vec{a}$ be a timelike vector, let $\vec{b}$ be an artibrary vector (timelike, null, or spacelike), and let ϵ be a number small enough that $\vec{A}_\pm \equiv \vec{a} \pm \epsilon \vec{b}$ are both timelike. From the fact that $\mathbf{S}(\vec{A}_+, \vec{A}_+)$, $\mathbf{S}(\vec{A}_-, \vec{A}_-)$, and $\mathbf{S}(\vec{a}, \vec{a})$ all vanish, deduce that $\mathbf{S}(\vec{b}, \vec{b}) = 0$ for the arbitrary vector $\vec{b}$. (iii) From this, deduce that $\mathbf{S}$ vanishes (i.e., it gives zero when any two vectors are inserted into its slots).]

Exercise 2.4 *Problem: Relativistic Gravitational Force Law*
In Newtonian theory, the gravitational potential Φ exerts a force $\mathbf{F} = d\mathbf{p}/dt = -m\nabla\Phi$ on a particle with mass m and momentum $\mathbf{p}$. Before Einstein formulated general relativity, some physicists constructed relativistic theories of gravity in which a Newtonian-like scalar gravitational field Φ exerted a 4-force $\vec{F} = d\vec{p}/d\tau$ on any particle with rest mass m, 4-velocity $\vec{u}$, and 4-momentum $\vec{p} = m\vec{u}$. What must that force law have been for it to (i) obey the Principle of Relativity, (ii) reduce to Newton's law in the nonrelativistic limit, and (iii) preserve the particle's rest mass as time passes?

2.5 Component Representation of Tensor Algebra

2.5.1 Lorentz Coordinates

In Minkowski spacetime, associated with any inertial reference frame (Fig. 2.1 and Sec. 2.2.1), there is a Lorentz coordinate system $\{t, x, y, z\} = \{x^0, x^1, x^2, x^3\}$ generated by the frame's rods and clocks. (Note the use of superscripts.) And associated with these coordinates is a set of *Lorentz basis vectors* $\{\vec{e}_t, \vec{e}_x, \vec{e}_y, \vec{e}_z\} = \{\vec{e}_0, \vec{e}_1, \vec{e}_2, \vec{e}_3\}$. (Note the use of subscripts. The reason for this convention will become clear below.) The basis vector $\vec{e}_\alpha$ points along the x^α coordinate direction, which is orthogonal to all the other coordinate directions, and it has squared length -1 for $\alpha = 0$ (vector pointing in a timelike direction) and $+1$ for $\alpha = 1, 2, 3$ (spacelike):

$$\vec{e}_\alpha \cdot \vec{e}_\beta = \eta_{\alpha\beta}. \tag{2.21}$$

Here $\eta_{\alpha\beta}$ (a spacetime analog of the Kronecker delta) are defined by

$$\eta_{00} \equiv -1, \qquad \eta_{11} \equiv \eta_{22} \equiv \eta_{33} \equiv 1, \qquad \eta_{\alpha\beta} \equiv 0 \text{ if } \alpha \neq \beta. \tag{2.22}$$

Any basis in which $\vec{e}_\alpha \cdot \vec{e}_\beta = \eta_{\alpha\beta}$ is said to be *orthonormal* (by analogy with the Euclidean notion of orthonormality, $\mathbf{e}_j \cdot \mathbf{e}_k = \delta_{jk}$).

Because $\vec{e}_\alpha \cdot \vec{e}_\beta \neq \delta_{\alpha\beta}$, many of the Euclidean-space component-manipulation formulas (1.9b)–(1.9h) do not hold in Minkowski spacetime. There are two approaches to recovering these formulas. One approach, used in many older textbooks (including the first and second editions of Goldstein's *Classical Mechanics* and Jackson's *Classical Electrodynamics*), is to set $x^0 = it$, where $i = \sqrt{-1}$ and correspondingly make the time basis vector be imaginary, so that $\vec{e}_\alpha \cdot \vec{e}_\beta = \delta_{\alpha\beta}$. When this approach is adopted, the resulting formalism does not depend on whether indices are placed up or down; one can place them wherever one's stomach or liver dictates without asking one's brain. However, this $x^0 = it$ approach has severe disadvantages: (i) it hides the true physical geometry of Minkowski spacetime, (ii) it cannot be extended in any reasonable manner to nonorthonormal bases in flat spacetime, and (iii) it cannot be extended in any reasonable manner to the curvilinear coordinates that must be used in general relativity. For these reasons, most modern texts (including the third editions of Goldstein and Jackson) take an alternative approach, one always used in general relativity. This alternative, which we shall adopt, requires introducing two different types of components for vectors (and analogously for tensors): *contravariant components* denoted by superscripts (e.g., $T^{\alpha\beta\gamma}$) and *covariant components* denoted by subscripts (e.g., $T_{\alpha\beta\gamma}$). In Parts I–VI of this book, we introduce these components only for orthonormal bases; in Part VII, we develop a more sophisticated version of them, valid for nonorthonormal bases.

2.5.2 Index Gymnastics

A vector's or tensor's contravariant components are defined as its expansion coefficients in the chosen basis [analogs of Eqs. (1.9b) and (1.9d) in Euclidean 3-space]:

$$\vec{A} \equiv A^\alpha \vec{e}_\alpha, \qquad \mathbf{T} \equiv T^{\alpha\beta\gamma} \vec{e}_\alpha \otimes \vec{e}_\beta \otimes \vec{e}_\gamma. \tag{2.23a}$$

Here and throughout this book, *Greek (spacetime) indices are to be summed when they are repeated with one up and the other down;* this is called the Einstein summation convention.

The covariant components are defined as the numbers produced by evaluating the vector or tensor on its basis vectors [analog of Eq. (1.9e) in Euclidean 3-space]:

covariant components

$$A_\alpha \equiv \vec{A}(\vec{e}_\alpha) = \vec{A} \cdot \vec{e}_\alpha, \qquad T_{\alpha\beta\gamma} \equiv \mathbf{T}(\vec{e}_\alpha, \vec{e}_\beta, \vec{e}_\gamma). \tag{2.23b}$$

(Just as there are contravariant and covariant components A^α and A_α, so also there is a second set of basis vectors $\vec{e}^\alpha$ dual to the set $\vec{e}_\alpha$. However, for economy of notation we delay introducing them until Part VII.)

These definitions have a number of important consequences. We derive them one after another and then summarize them succinctly with equation numbers:

(i) The covariant components of the metric tensor are $g_{\alpha\beta} = \mathbf{g}(\vec{e}_\alpha, \vec{e}_\beta) = \vec{e}_\alpha \cdot \vec{e}_\beta = \eta_{\alpha\beta}$. Here the first equality is the definition (2.23b) of the covariant components, the second equality is the definition (2.6) of the metric tensor, and the third equality is the orthonormality relation (2.21) for the basis vectors.

(ii) The covariant components of any tensor can be computed from the contravariant components by

$$\begin{aligned} T_{\lambda\mu\nu} = \mathbf{T}(\vec{e}_\lambda, \vec{e}_\mu, \vec{e}_\nu) &= T^{\alpha\beta\gamma} \vec{e}_\alpha \otimes \vec{e}_\beta \otimes \vec{e}_\gamma(\vec{e}_\lambda, \vec{e}_\mu, \vec{e}_\nu) \\ &= T^{\alpha\beta\gamma}(\vec{e}_\alpha \cdot \vec{e}_\lambda)(\vec{e}_\beta \cdot \vec{e}_\mu)(\vec{e}_\gamma \cdot \vec{e}_\nu) = T^{\alpha\beta\gamma} g_{\alpha\lambda} g_{\beta\mu} g_{\gamma\nu}. \end{aligned}$$

The first equality is the definition (2.23b) of the covariant components, the second is the expansion (2.23a) of $\mathbf{T}$ on the chosen basis, the third is the definition (1.5a) of the tensor product, and the fourth is one version of our result (i) for the covariant components of the metric.

(iii) This result, $T_{\lambda\mu\nu} = T^{\alpha\beta\gamma} g_{\alpha\lambda} g_{\beta\mu} g_{\gamma\nu}$, together with the numerical values (i) of $g_{\alpha\beta}$, implies that when one lowers a spatial index there is no change in the numerical value of a component, and when one lowers a temporal index, the sign changes: $T_{ijk} = T^{ijk}$, $T_{0jk} = -T^{0jk}$, $T_{0j0} = +T^{0j0}$, $T_{000} = -T^{000}$. We call this the "sign-flip-if-temporal" rule. As a special case, $-1 = g_{00} = g^{00}$, $0 = g_{0j} = -g^{0j}$, $\delta_{jk} = g_{jk} = g^{jk}$—that is, the metric's covariant and contravariant components are numerically identical; they are both equal to the orthonormality values $\eta_{\alpha\beta}$.

raising and lowering indices: sign flip if temporal

components of metric tensor

(iv) It is easy to see that this sign-flip-if-temporal rule for lowering indices implies the same sign-flip-if-temporal rule for raising them, which in turn can be written in terms of metric components as $T^{\alpha\beta\gamma} = T_{\lambda\mu\nu} g^{\lambda\alpha} g^{\mu\beta} g^{\nu\gamma}$.

(v) It is convenient to define *mixed components* of a tensor, components with some indices up and others down, as having numerical values obtained

mixed components

2.5 Component Representation of Tensor Algebra

by raising or lowering some but not all of its indices using the metric, for example, $T^{\alpha}{}_{\mu\nu} = T^{\alpha\beta\gamma}g_{\beta\mu}g_{\gamma\nu} = T_{\lambda\mu\nu}g^{\lambda\alpha}$. Numerically, this continues to follow the sign-flip-if-temporal rule: $T^0{}_{0k} = -T^{00k}$, $T^0{}_{jk} = T^{0jk}$, and it implies, in particular, that the mixed components of the metric are $g^{\alpha}{}_{\beta} = \delta_{\alpha\beta}$ (the Kronecker-delta values; +1 if $\alpha = \beta$ and 0 otherwise).

summary of index gymnastics

These important results can be summarized as follows. *The numerical values of the components of the metric in Minkowski spacetime are expressed in terms of the matrices $[\delta_{\alpha\beta}]$ and $[\eta_{\alpha\beta}]$ as*

$$g_{\alpha\beta} = \eta_{\alpha\beta}, \quad g^{\alpha}{}_{\beta} = \delta_{\alpha\beta}, \quad g_{\alpha}{}^{\beta} = \delta_{\alpha\beta}, \quad g^{\alpha\beta} = \eta_{\alpha\beta}; \tag{2.23c}$$

indices on all vectors and tensors can be raised and lowered using these components of the metric:

$$A_{\alpha} = g_{\alpha\beta}A^{\beta}, \quad A^{\alpha} = g^{\alpha\beta}A_{\beta}, \quad T^{\alpha}{}_{\mu\nu} \equiv g_{\mu\beta}g_{\nu\gamma}T^{\alpha\beta\gamma}, \quad T^{\alpha\beta\gamma} \equiv g^{\beta\mu}g^{\gamma\nu}T^{\alpha}{}_{\mu\nu}, \tag{2.23d}$$

which is equivalent to the sign-flip-if-temporal rule.

This index notation gives rise to formulas for tensor products, inner products, values of tensors on vectors, and tensor contractions that are obvious analogs of those in Euclidean space:

$$[\text{Contravariant components of } \mathbf{T}(\_,\_,\_) \otimes \mathbf{S}(\_,\_)] = T^{\alpha\beta\gamma}S^{\delta\epsilon}, \tag{2.23e}$$

$$\vec{A} \cdot \vec{B} = A^{\alpha}B_{\alpha} = A_{\alpha}B^{\alpha}, \quad \mathbf{T}(\mathbf{A}, \mathbf{B}, \mathbf{C}) = T_{\alpha\beta\gamma}A^{\alpha}B^{\beta}C^{\gamma} = T^{\alpha\beta\gamma}A_{\alpha}B_{\beta}C_{\gamma}, \tag{2.23f}$$

$$\text{Covariant components of } [1\&3\text{contraction of } \mathbf{R}] = R^{\mu}{}_{\alpha\mu\beta},$$

$$\text{Contravariant components of } [1\&3\text{contraction of } \mathbf{R}] = R^{\mu\alpha}{}_{\mu}{}^{\beta}. \tag{2.23g}$$

Notice the very simple pattern in Eqs. (2.23b) and (2.23d), which universally permeates the rules of index gymnastics, a pattern that permits one to reconstruct the rules without any memorization: *Free indices (indices not summed over) must agree in position (up versus down) on the two sides of each equation.* In keeping with this pattern, one can regard the two indices in a pair that is summed as "annihilating each other by contraction," and one speaks of "lining up the indices" on the two sides of an equation to get them to agree. These rules provide helpful checks when performing calculations.

In Part VII, when we use nonorthonormal bases, all these index-notation equations (2.23) will remain valid except for the numerical values [Eq. (2.23c)] of the metric components and the sign-flip-if-temporal rule.

2.5.3 Slot-Naming Notation

In Minkowski spacetime, as in Euclidean space, we can (and often do) use slot-naming index notation to represent frame-independent geometric objects and equations and physical laws. (Readers who have not studied Sec. 1.5.1 on slot-naming index notation should do so now.)

For example, we often write the frame-independent Lorentz force law $d\vec{p}/d\tau = q\mathbf{F}(\_, \vec{u})$ [Eq. (2.20)] as $dp_\mu/d\tau = q F_{\mu\nu} u^\nu$.

Notice that, because the components of the metric in any Lorentz basis are $g_{\alpha\beta} = \eta_{\alpha\beta}$, we can write the invariant interval between two events x^α and $x^\alpha + dx^\alpha$ as

$$ds^2 = g_{\alpha\beta} dx^\alpha dx^\beta = -dt^2 + dx^2 + dy^2 + dz^2. \tag{2.24}$$

This is called the special relativistic *line element*.

line element

EXERCISES

Exercise 2.5 *Derivation: Component Manipulation Rules*
Derive the relativistic component manipulation rules (2.23e)–(2.23g).

Exercise 2.6 *Practice: Numerics of Component Manipulations*
In some inertial reference frame, the vector $\vec{A}$ and second-rank tensor $\mathbf{T}$ have as their only nonzero components $A^0 = 1$, $A^1 = 2$; $T^{00} = 3$, $T^{01} = T^{10} = 2$, $T^{11} = -1$. Evaluate $\mathbf{T}(\vec{A}, \vec{A})$ and the components of $\mathbf{T}(\vec{A}, \_)$ and $\vec{A} \otimes \mathbf{T}$.

Exercise 2.7 *Practice: Meaning of Slot-Naming Index Notation*
(a) Convert the following expressions and equations into geometric, index-free notation: $A^\alpha B_{\gamma\delta}$; $A_\alpha B_\gamma{}^\delta$; $S_\alpha{}^{\beta\gamma} = S^{\gamma\beta}{}_\alpha$; $A^\alpha B_\alpha = A_\alpha B^\beta g^\alpha{}_\beta$.
(b) Convert $\mathbf{T}(\_, \mathbf{S}(\mathbf{R}(\vec{C}, \_), \_), \_)$ into slot-naming index notation.

Exercise 2.8 *Practice: Index Gymnastics*
(a) Simplify the following expression so the metric does not appear in it:

$$A^{\alpha\beta\gamma} g_{\beta\rho} S_{\gamma\lambda} g^{\rho\delta} g^\lambda{}_\alpha.$$

(b) The quantity $g_{\alpha\beta} g^{\alpha\beta}$ is a scalar since it has no free indices. What is its numerical value?

(c) What is wrong with the following expression and equation?

$$A_\alpha{}^{\beta\gamma} S_{\alpha\gamma}; \qquad A_\alpha{}^{\beta\gamma} S_\beta T_\gamma = R_{\alpha\beta\delta} S^\beta.$$

2.6 Particle Kinetics in Index Notation and in a Lorentz Frame

As an illustration of the component representation of tensor algebra, let us return to the relativistic, accelerated particle of Fig. 2.3 and, from the frame-independent equations for the particle's 4-velocity $\vec{u}$ and 4-momentum $\vec{p}$ (Sec. 2.4), derive the component description given in elementary textbooks.

We introduce a specific inertial reference frame and associated Lorentz coordinates x^α and basis vectors $\{\vec{e}_\alpha\}$. In this Lorentz frame, the particle's world line

$\vec{x}(\tau)$ is represented by its coordinate location $x^\alpha(\tau)$ as a function of its proper time τ. The contravariant components of the separation vector $d\vec{x}$ between two neighboring events along the particle's world line are the events' coordinate separations dx^α [Eq. (2.1)]; and correspondingly, the components of the particle's 4-velocity $\vec{u} = d\vec{x}/d\tau$ are

$$u^\alpha = dx^\alpha/d\tau \qquad (2.25a)$$

(the time derivatives of the particle's spacetime coordinates). Note that Eq. (2.25a) implies

$$v^j \equiv \frac{dx^j}{dt} = \frac{dx^j/d\tau}{dt/d\tau} = \frac{u^j}{u^0}. \qquad (2.25b)$$

This relation, together with $-1 = \vec{u}^2 = g_{\alpha\beta}u^\alpha u^\beta = -(u^0)^2 + \delta_{ij}u^i u^j = -(u^0)^2(1 - \delta_{ij}v^i v^j)$, implies that the components of the 4-velocity have the forms familiar from elementary textbooks:

$$u^0 = \gamma, \quad u^j = \gamma v^j, \quad \text{where} \quad \gamma = \frac{1}{(1-\delta_{ij}v^i v^j)^{\frac{1}{2}}}. \qquad (2.25c)$$

ordinary velocity

slice of simultaneity

It is useful to think of v^j as the components of a 3-dimensional vector $\mathbf{v}$, the *ordinary velocity*, that lives in the 3-dimensional Euclidean space $t = \text{const}$ of the chosen Lorentz frame (the green plane in Fig. 2.5). This 3-space is sometimes called the frame's *slice of simultaneity* or *3-space of simultaneity*, because all events lying in it are simultaneous, as measured by the frame's observers. This 3-space is not well defined until a Lorentz frame has been chosen, and correspondingly, $\mathbf{v}$ relies for its existence on a specific choice of frame. However, once the frame has been chosen, $\mathbf{v}$ can be regarded as a coordinate-independent, basis-independent 3-vector lying in the frame's slice of simultaneity. Similarly, the spatial part of the 4-velocity $\vec{u}$ (the part with components u^j in our chosen frame) can be regarded as a 3-vector $\mathbf{u}$ lying in the frame's 3-space; and Eqs. (2.25c) become the component versions of the coordinate-independent, basis-independent 3-space relations

$$\mathbf{u} = \gamma \mathbf{v}, \quad \gamma = \frac{1}{\sqrt{1-\mathbf{v}^2}}, \qquad (2.25d)$$

where $\mathbf{v}^2 = \mathbf{v} \cdot \mathbf{v}$. This γ is called the "Lorentz factor."

The components of the particle's 4-momentum $\vec{p}$ in our chosen Lorentz frame have special names and special physical significances: The time component of the 4-momentum is the particle's (relativistic) *energy* $\mathcal{E}$ as measured in that frame:

relativistic energy

$$\mathcal{E} \equiv p^0 = mu^0 = m\gamma = \frac{m}{\sqrt{1-\mathbf{v}^2}} = \text{(the particle's energy)}$$

$$\simeq m + \frac{1}{2}m\mathbf{v}^2 \quad \text{for } |\mathbf{v}| \ll 1. \qquad (2.26a)$$

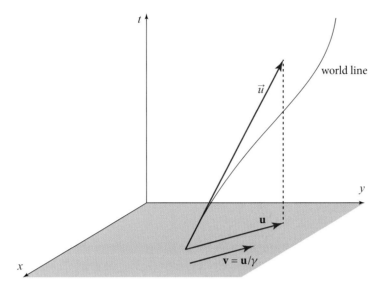

FIGURE 2.5 Spacetime diagram in a specific Lorentz frame, showing the frame's 3-space $t = 0$ (green region), the world line of a particle, the 4-velocity $\vec{u}$ of the particle as it passes through the 3-space, and two 3-dimensional vectors that lie in the 3-space: the spatial part **u** of the particle's 4-velocity and the particle's ordinary velocity **v**.

Note that this energy is the sum of the particle's *rest mass-energy* $m = mc^2$ and its *kinetic energy*

rest mass-energy

$$E \equiv \mathcal{E} - m = m\left(\frac{1}{\sqrt{1-\mathbf{v}^2}} - 1\right)$$

kinetic energy

$$\simeq \frac{1}{2}m\mathbf{v}^2 \quad \text{for } |\mathbf{v}| \ll 1. \tag{2.26b}$$

The spatial components of the 4-momentum, when regarded from the viewpoint of 3-dimensional physics, are the same as the components of the *momentum*, a 3-vector residing in the chosen Lorentz frame's 3-space:

momentum

$$p^j = mu^j = m\gamma v^j = \frac{mv^j}{\sqrt{1-\mathbf{v}^2}} = \mathcal{E}v_j$$

$$= (j \text{ component of particle's momentum}); \tag{2.26c}$$

or, in basis-independent, 3-dimensional vector notation,

$$\mathbf{p} = m\mathbf{u} = m\gamma \mathbf{v} = \frac{m\mathbf{v}}{\sqrt{1-\mathbf{v}^2}} = \mathcal{E}\mathbf{v} = (\text{particle's momentum}). \tag{2.26d}$$

For a zero-rest-mass particle, as for one with finite rest mass, we identify the time component of the 4-momentum, in a chosen Lorentz frame, as the particle's energy, and the spatial part as its momentum. Moreover, if—appealing to quantum theory—we regard a zero-rest-mass particle as a quantum associated with a monochromatic wave, then quantum theory tells us that the wave's angular frequency ω as measured in a chosen Lorentz frame is related to its energy by

$$\mathcal{E} \equiv p^0 = \hbar\omega = \text{(particle's energy)}; \tag{2.27a}$$

and, since the particle has $\vec{p}^2 = -(p^0)^2 + \mathbf{p}^2 = -m^2 = 0$ (in accord with the lightlike nature of its world line), its momentum as measured in the chosen Lorentz frame is

$$\mathbf{p} = \mathcal{E}\mathbf{n} = \hbar\omega\mathbf{n}. \tag{2.27b}$$

Here $\mathbf{n}$ is the unit 3-vector that points in the direction of the particle's travel, as measured in the chosen frame; that is (since the particle moves at the speed of light $v = 1$), $\mathbf{n}$ is the particle's ordinary velocity. Eqs. (2.27a) and (2.27b) are respectively the temporal and spatial components of the geometric, frame-independent relation $\vec{p} = \hbar\vec{k}$ [Eq. (2.13), which is valid for zero-rest-mass particles as well as finite-mass ones].

3+1 split of spacetime into space plus time

The introduction of a specific Lorentz frame into spacetime can be said to produce a 3+1 split of every 4-vector into a 3-dimensional vector plus a scalar (a real number). The 3+1 split of a particle's 4-momentum $\vec{p}$ produces its momentum $\mathbf{p}$ plus its energy $\mathcal{E} = p^0$. Correspondingly, the 3+1 split of the law of 4-momentum conservation (2.15) produces a law of conservation of momentum plus a law of conservation of energy:

$$\sum_{\bar{A}} \mathbf{p}_{\bar{A}} = \sum_{A} \mathbf{p}_A, \qquad \sum_{\bar{A}} \mathcal{E}_{\bar{A}} = \sum_{A} \mathcal{E}_A. \tag{2.28}$$

The unbarred quantities in Eqs. (2.28) are momenta and energies of the particles entering the interaction region, and the barred quantities are those of the particles leaving (see Fig. 2.4).

Because the concept of energy does not even exist until one has chosen a Lorentz frame—and neither does that of momentum—the laws of energy conservation and momentum conservation separately are frame-dependent laws. In this sense, they are far less fundamental than their combination, the frame-independent law of 4-momentum conservation.

By learning to think about the 3+1 split in a geometric, frame-independent way, one can gain conceptual and computational power. As an example, consider a particle with 4-momentum $\vec{p}$, being studied by an observer with 4-velocity $\vec{U}$. In the observer's own Lorentz reference frame, her 4-velocity has components $U^0 = 1$ and $U^j = 0$, and therefore, her 4-velocity is $\vec{U} = U^\alpha \vec{e}_\alpha = \vec{e}_0$; that is, it is identically equal to the time basis vector of her Lorentz frame. Thus the particle energy that she measures is $\mathcal{E} = p^0 = -p_0 = -\vec{p} \cdot \vec{e}_0 = -\vec{p} \cdot \vec{U}$. This equation, derived in the observer's Lorentz frame, is actually a geometric, frame-independent relation: the inner product of two 4-vectors. It says that *when an observer with 4-velocity $\vec{U}$ measures the energy of a* particle with 4-momentum $\vec{p}$, the result she gets (the time part of the 3+1 split of $\vec{p}$ as seen by her) is

$$\boxed{\mathcal{E} = -\vec{p} \cdot \vec{U}.} \quad (2.29)$$

We shall use this equation in later chapters. In Exs. 2.9 and 2.10, the reader can gain experience deriving and interpreting other frame-independent equations for 3+1 splits. Exercise 2.11 exhibits the power of this geometric way of thinking by using it to derive the Doppler shift of a photon.

EXERCISES

Exercise 2.9 **Practice: Frame-Independent Expressions for Energy, Momentum, and Velocity*

An observer with 4-velocity $\vec{U}$ measures the properties of a particle with 4-momentum $\vec{p}$. The energy she measures is $\mathcal{E} = -\vec{p} \cdot \vec{U}$ [Eq. (2.29)].

(a) Show that the particle's rest mass can be expressed in terms of $\vec{p}$ as

$$m^2 = -\vec{p}^2. \quad (2.30a)$$

(b) Show that the momentum the observer measures has the magnitude

$$|\mathbf{p}| = [(\vec{p} \cdot \vec{U})^2 + \vec{p} \cdot \vec{p}]^{\frac{1}{2}}. \quad (2.30b)$$

(c) Show that the ordinary velocity the observer measures has the magnitude

$$|\mathbf{v}| = \frac{|\mathbf{p}|}{\mathcal{E}}, \quad (2.30c)$$

where $|\mathbf{p}|$ and $\mathcal{E}$ are given by the above frame-independent expressions.

(d) Show that the ordinary velocity $\mathbf{v}$, thought of as a 4-vector that happens to lie in the observer's slice of simultaneity, is given by

$$\vec{v} = \frac{\vec{p} + (\vec{p} \cdot \vec{U})\vec{U}}{-\vec{p} \cdot \vec{U}}. \quad (2.30d)$$

Exercise 2.10 **Example: 3-Metric as a Projection Tensor*

Consider, as in Ex. 2.9, an observer with 4-velocity $\vec{U}$ who measures the properties of a particle with 4-momentum $\vec{p}$.

(a) Show that the Euclidean metric of the observer's 3-space, when thought of as a tensor in 4-dimensional spacetime, has the form

$$\boxed{\mathbf{P} \equiv \mathbf{g} + \vec{U} \otimes \vec{U}.} \quad (2.31a)$$

Show, further, that if $\vec{A}$ is an arbitrary vector in spacetime, then $-\vec{A} \cdot \vec{U}$ is the component of $\vec{A}$ along the observer's 4-velocity $\vec{U}$, and

$$\mathbf{P}(\_\,, \vec{A}) = \vec{A} + (\vec{A} \cdot \vec{U})\vec{U} \quad (2.31b)$$

is the projection of $\vec{A}$ into the observer's 3-space (i.e., it is the spatial part of $\vec{A}$ as seen by the observer). For this reason, $\mathbf{P}$ is called a *projection tensor*. In quantum mechanics, the concept of a *projection operator* $\hat{P}$ is introduced as one that satisfies the equation $\hat{P}^2 = \hat{P}$. Show that the projection tensor $\mathbf{P}$ is a projection operator in the same sense:

$$P_{\alpha\mu} P^{\mu}{}_{\beta} = P_{\alpha\beta}. \tag{2.31c}$$

(b) Show that Eq. (2.30d) for the particle's ordinary velocity, thought of as a 4-vector, can be rewritten as

$$\vec{v} = \frac{\mathbf{P}(\_, \vec{p})}{-\vec{p} \cdot \vec{U}}. \tag{2.32}$$

Exercise 2.11 **Example: Doppler Shift Derived without Lorentz Transformations*
(a) An observer at rest in some inertial frame receives a photon that was emitted in direction **n** by an atom moving with ordinary velocity **v** (Fig. 2.6). The photon frequency and energy as measured by the emitting atom are ν_{em} and $\mathcal{E}_{em}$; those measured by the receiving observer are ν_{rec} and $\mathcal{E}_{rec}$. By a calculation carried out solely in the receiver's inertial frame (the frame of Fig. 2.6), and without the aid of any Lorentz transformation, derive the standard formula for the photon's Doppler shift:

$$\frac{\nu_{rec}}{\nu_{em}} = \frac{\sqrt{1-v^2}}{1 - \mathbf{v} \cdot \mathbf{n}}. \tag{2.33}$$

[Hint: Use Eq. (2.29) to evaluate $\mathcal{E}_{em}$ using receiver-frame expressions for the emitting atom's 4-velocity $\vec{U}$ and the photon's 4-momentum $\vec{p}$.]

(b) Suppose that instead of emitting a photon, the emitter produces a particle with finite rest mass m. Using the same method as in part (a), derive an expression for the ratio of received energy to emitted energy, $\mathcal{E}_{rec}/\mathcal{E}_{em}$, expressed in terms of the emitter's ordinary velocity **v** and the particle's ordinary velocity **V** (both as measured in the receiver's frame).

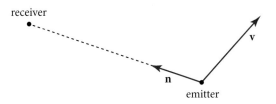

FIGURE 2.6 Geometry for Doppler shift, drawn in a slice of simultaneity of the receiver's inertial frame.

2.7 Lorentz Transformations

Consider two different inertial reference frames in Minkowski spacetime. Denote their Lorentz coordinates by $\{x^\alpha\}$ and $\{x^{\bar{\mu}}\}$ and their bases by $\{\mathbf{e}_\alpha\}$ and $\{\mathbf{e}_{\bar{\mu}}\}$, respectively, and write the transformation from one basis to the other as

$$\vec{e}_\alpha = \vec{e}_{\bar{\mu}} L^{\bar{\mu}}{}_\alpha, \qquad \vec{e}_{\bar{\mu}} = \vec{e}_\alpha L^\alpha{}_{\bar{\mu}}. \tag{2.34}$$

As in Euclidean 3-space (the rotation matrices of Sec 1.6), $L^{\bar{\mu}}{}_\alpha$ and $L^\alpha{}_{\bar{\mu}}$ are elements of two different transformation matrices, and since these matrices operate in opposite directions, they must be the inverse of each other:

transformation matrix

$$L^{\bar{\mu}}{}_\alpha L^\alpha{}_{\bar{\nu}} = \delta^{\bar{\mu}}{}_{\bar{\nu}}, \qquad L^\alpha{}_{\bar{\mu}} L^{\bar{\mu}}{}_\beta = \delta^\alpha{}_\beta. \tag{2.35a}$$

Notice the up/down placement of indices on the elements of the transformation matrices: the first index is always up, and the second is always down. This is just a convenient convention, which helps systematize the index shuffling rules in a way that can easily be remembered. Our rules about summing on the same index when up and down, and matching unsummed indices on the two sides of an equation automatically dictate the matrix to use in each of the transformations (2.34); and similarly for all other equations in this section.

In Euclidean 3-space the orthonormality of the two bases dictates that the transformations must be orthogonal (i.e., must be reflections or rotations). In Minkowski spacetime, orthonormality implies $g_{\alpha\beta} = \vec{e}_\alpha \cdot \vec{e}_\beta = (\vec{e}_{\bar{\mu}} L^{\bar{\mu}}{}_\alpha) \cdot (\vec{e}_{\bar{\nu}} L^{\bar{\nu}}{}_\beta) = L^{\bar{\mu}}{}_\alpha L^{\bar{\nu}}{}_\beta g_{\bar{\mu}\bar{\nu}}$; that is,

$$g_{\bar{\mu}\bar{\nu}} L^{\bar{\mu}}{}_\alpha L^{\bar{\nu}}{}_\beta = g_{\alpha\beta}, \quad \text{and similarly,} \quad g_{\alpha\beta} L^\alpha{}_{\bar{\mu}} L^\beta{}_{\bar{\nu}} = g_{\bar{\mu}\bar{\nu}}. \tag{2.35b}$$

Any matrix whose elements satisfy these equations is a *Lorentz transformation*.

Lorentz transformation

From the fact that vectors and tensors are geometric, frame-independent objects, one can derive the Minkowski-space analogs of the Euclidean transformation laws for components (1.13a) and (1.13b):

$$A^{\bar{\mu}} = L^{\bar{\mu}}{}_\alpha A^\alpha, \qquad T^{\bar{\mu}\bar{\nu}\bar{\rho}} = L^{\bar{\mu}}{}_\alpha L^{\bar{\nu}}{}_\beta L^{\bar{\rho}}{}_\gamma T^{\alpha\beta\gamma}, \tag{2.36a}$$

and similarly in the opposite direction. Notice that here, as elsewhere, these equations can be constructed by lining up indices in accord with our standard rules.

If (as is conventional) we choose the spacetime origins of the two Lorentz coordinate systems to coincide, then the vector $\vec{x}$ extending from the origin to some event $\mathcal{P}$, whose coordinates are x^α and $x^{\bar{\alpha}}$, has components equal to those coordinates. As a result, the transformation law for the coordinates takes the same form as Eq. (2.36a) for the components of a vector:

$$x^\alpha = L^\alpha{}_{\bar{\mu}} x^{\bar{\mu}}, \qquad x^{\bar{\mu}} = L^{\bar{\mu}}{}_\alpha x^\alpha. \tag{2.36b}$$

T2

Lorentz group

The product $L^\alpha{}_{\bar\mu} L^{\bar\mu}{}_{\bar{\bar\rho}}$ of two Lorentz transformation matrices is a Lorentz transformation matrix. Under this product rule, the Lorentz transformations form a mathematical group, the *Lorentz group*, whose representations play an important role in quantum field theory (cf. the rotation group in Sec. 1.6).

An important specific example of a Lorentz transformation is:

$$\left[L^\alpha{}_{\bar\mu}\right] = \begin{bmatrix} \gamma & \beta\gamma & 0 & 0 \\ \beta\gamma & \gamma & 0 & 0 \\ 0 & 0 & 1 & 0 \\ 0 & 0 & 0 & 1 \end{bmatrix}, \quad \left[L^{\bar\mu}{}_\alpha\right] = \begin{bmatrix} \gamma & -\beta\gamma & 0 & 0 \\ -\beta\gamma & \gamma & 0 & 0 \\ 0 & 0 & 1 & 0 \\ 0 & 0 & 0 & 1 \end{bmatrix}, \quad (2.37a)$$

where β and γ are related by

$$|\beta| < 1, \quad \gamma \equiv (1 - \beta^2)^{-\frac{1}{2}}. \quad (2.37b)$$

[Notice that γ is the Lorentz factor associated with β; cf. Eq. (2.25d).]

One can readily verify (Ex. 2.12) that these matrices are the inverses of each other and that they satisfy the Lorentz-transformation relation (2.35b). These transformation matrices produce the following change of coordinates [Eq. (2.36b)]:

Lorentz boost

$$t = \gamma(\bar{t} + \beta\bar{x}), \quad x = \gamma(\bar{x} + \beta\bar{t}), \quad y = \bar{y}, \quad z = \bar{z},$$
$$\bar{t} = \gamma(t - \beta x), \quad \bar{x} = \gamma(x - \beta t), \quad \bar{y} = y, \quad \bar{z} = z. \quad (2.37c)$$

These expressions reveal that any particle at rest in the unbarred frame (a particle with fixed, time-independent x, y, z) is seen in the barred frame to move along the world line $\bar{x} = \text{const} - \beta\bar{t}$, $\bar{y} = \text{const}$, $\bar{z} = \text{const}$. In other words, the unbarred frame is seen by observers at rest in the barred frame to move with uniform velocity $\mathbf{v} = -\beta\mathbf{e}_{\bar{x}}$, and correspondingly the barred frame is seen by observers at rest in the unbarred frame to move with the opposite uniform velocity $\mathbf{v} = +\beta\mathbf{e}_x$. This special Lorentz transformation is called a *pure boost* along the x direction.

EXERCISES

Exercise 2.12 *Derivation: Lorentz Boosts*
Show that the matrices (2.37a), with β and γ satisfying Eq. (2.37b), are the inverses of each other, and that they obey the condition (2.35b) for a Lorentz transformation.

Exercise 2.13 *Example: General Boosts and Rotations*
(a) Show that, if n^j is a 3-dimensional unit vector and β and γ are defined as in Eq. (2.37b), then the following is a Lorentz transformation [i.e., it satisfies Eq. (2.35b)]:

$$L^0{}_{\bar0} = \gamma, \quad L^0{}_{\bar j} = L^j{}_{\bar 0} = \beta\gamma n^j, \quad L^j{}_{\bar k} = L^k{}_{\bar j} = (\gamma - 1)n^j n^k + \delta^{jk}. \quad (2.38)$$

Show, further, that this transformation is a pure boost along the direction $\mathbf{n}$ with speed β, and show that the inverse matrix $L^{\bar\mu}{}_\alpha$ for this boost is the same as $L^\alpha{}_{\bar\mu}$, but with β changed to $-\beta$.

(b) Show that the following is also a Lorentz transformation:

$$[L^{\alpha}{}_{\bar{\mu}}] = \begin{bmatrix} 1 & 0 & 0 & 0 \\ 0 & & & \\ 0 & & [R_{i\bar{j}}] & \\ 0 & & & \end{bmatrix}, \qquad (2.39)$$

where $[R_{i\bar{j}}]$ is a 3-dimensional rotation matrix for Euclidean 3-space. Show, further, that this Lorentz transformation rotates the inertial frame's spatial axes (its latticework of measuring rods) while leaving the frame's velocity unchanged (i.e., the new frame is at rest with respect to the old).

One can show (not surprisingly) that the general Lorentz transformation [i.e., the general solution of Eqs. (2.35b)] can be expressed as a sequence of pure boosts, pure rotations, and pure inversions (in which one or more of the coordinate axes are reflected through the origin, so $x^{\alpha} = -x^{\bar{\alpha}}$).

2.8 Spacetime Diagrams for Boosts

Figure 2.7 illustrates the pure boost (2.37c). Panel a in that figure is a 2-dimensional spacetime diagram, with the y- and z-coordinates suppressed, showing the $\bar{t}$- and $\bar{x}$-axes of the boosted Lorentz frame $\bar{\mathcal{F}}$ in the t, x Lorentz coordinate system of the unboosted frame $\mathcal{F}$. That the barred axes make angles $\tan^{-1}\beta$ with the unbarred axes, as shown, can be inferred from the Lorentz transformation (2.37c). Note that the orthogonality of the $\bar{t}$- and $\bar{x}$-axes to each other ($\vec{e}_{\bar{t}} \cdot \vec{e}_{\bar{x}} = 0$) shows up as the two axes making the same angle $\pi/2 - \beta$ with the null line $x = t$. The invariance of the interval guarantees that, as shown for $a = 1$ or 2 in Fig. 2.7a, the event $\bar{x} = a$ on the $\bar{x}$-axis lies at the intersection of that axis with the dashed hyperbola $x^2 - t^2 = a^2$; and similarly, the event $\bar{t} = a$ on the $\bar{t}$-axis lies at the intersection of that axis with the dashed hyperbola $t^2 - x^2 = a^2$.

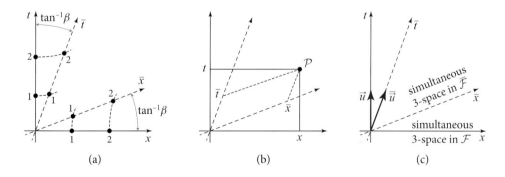

FIGURE 2.7 Spacetime diagrams illustrating the pure boost (2.37c) from one Lorentz reference frame to another.

As shown in Fig. 2.7b, the barred coordinates $\{\bar{t}, \bar{x}\}$ of an event $\mathcal{P}$ can be inferred by projecting from $\mathcal{P}$ onto the $\bar{t}$- and $\bar{x}$-axes, with the projection parallel to the $\bar{x}$- and $\bar{t}$-axes, respectively. Figure 2.7c shows the 4-velocity $\vec{u}$ of an observer at rest in frame $\mathcal{F}$ and that, $\vec{\bar{u}}$, of an observer at rest in frame $\overline{\mathcal{F}}$. The events that observer $\mathcal{F}$ regards as all simultaneous, with time $t = 0$, lie in a 3-space that is orthogonal to $\vec{u}$ and includes the x-axis. This is a slice of simultaneity of reference frame $\mathcal{F}$. Similarly, the events that observer $\overline{\mathcal{F}}$ regards as all simultaneous, with $\bar{t} = 0$, live in the 3-space that is orthogonal to $\vec{\bar{u}}$ and includes the $\bar{x}$-axis (i.e., in a slice of simultaneity of frame $\overline{\mathcal{F}}$).

Exercise 2.14 uses spacetime diagrams similar to those in Fig. 2.7 to deduce a number of important relativistic phenomena, including the contraction of the length of a moving object (length contraction), the breakdown of simultaneity as a universally agreed-on concept, and the dilation of the ticking rate of a moving clock (time dilation). This exercise is extremely important; every reader who is not already familiar with it should study it.

length contraction, time dilation, and breakdown of simultaneity

EXERCISES

Exercise 2.14 **Example: Spacetime Diagrams*
Use spacetime diagrams to prove the following:

(a) Two events that are simultaneous in one inertial frame are not necessarily simultaneous in another. More specifically, if frame $\overline{\mathcal{F}}$ moves with velocity $\vec{v} = \beta \vec{e}_x$ as seen in frame $\mathcal{F}$, where $\beta > 0$, then of two events that are simultaneous in $\overline{\mathcal{F}}$ the one farther "back" (with the more negative value of $\bar{x}$) will occur in $\mathcal{F}$ before the one farther "forward."

(b) Two events that occur at the same spatial location in one inertial frame do not necessarily occur at the same spatial location in another.

(c) If $\mathcal{P}_1$ and $\mathcal{P}_2$ are two events with a timelike separation, then there exists an inertial reference frame in which they occur at the same spatial location, and in that frame the time lapse between them is equal to the square root of the negative of their invariant interval, $\Delta t = \Delta \tau \equiv \sqrt{-(\Delta s)^2}$.

(d) If $\mathcal{P}_1$ and $\mathcal{P}_2$ are two events with a spacelike separation, then there exists an inertial reference frame in which they are simultaneous, and in that frame the spatial distance between them is equal to the square root of their invariant interval, $\sqrt{g_{ij}\Delta x^i \Delta x^j} = \Delta s \equiv \sqrt{(\Delta s)^2}$.

(e) If the inertial frame $\overline{\mathcal{F}}$ moves with speed β relative to the frame $\mathcal{F}$, then a clock at rest in $\overline{\mathcal{F}}$ ticks more slowly as viewed from $\mathcal{F}$ than as viewed from $\overline{\mathcal{F}}$—more slowly by a factor $\gamma^{-1} = (1 - \beta^2)^{\frac{1}{2}}$. This is called *relativistic time dilation*. As one consequence, the lifetimes of unstable particles moving with speed β are increased by the Lorentz factor γ.

(f) If the inertial frame $\overline{\mathcal{F}}$ moves with velocity $\vec{v} = \beta \vec{e}_x$ relative to the frame $\mathcal{F}$, then an object at rest in $\overline{\mathcal{F}}$ as studied in $\mathcal{F}$ appears shortened by a factor $\gamma^{-1} = (1 - \beta^2)^{\frac{1}{2}}$ along the x direction, but its length along the y and z directions is unchanged. This is called *Lorentz contraction*. As one consequence, heavy ions moving at high speeds in a particle accelerator appear to act like pancakes, squashed along their directions of motion.

Exercise 2.15 *Problem: Allowed and Forbidden Electron-Photon Reactions*
Show, using spacetime diagrams and also using frame-independent calculations, that the law of conservation of 4-momentum forbids a photon to be absorbed by an electron, $e + \gamma \to e$, and also forbids an electron and a positron to annihilate and produce a single photon, $e^+ + e^- \to \gamma$ (in the absence of any other particles to take up some of the 4-momentum); but the annihilation to form two photons, $e^+ + e^- \to 2\gamma$, is permitted.

2.9 Time Travel

2.9.1 Measurement of Time; Twins Paradox

Time dilation is one facet of a more general phenomenon: time, as measured by ideal clocks, is a personal thing, different for different observers who move through spacetime on different world lines. This is well illustrated by the infamous "twins paradox," in which one twin, Methuselah, remains forever at rest in an inertial frame and the other, Florence, makes a spacecraft journey at high speed and then returns to rest beside Methuselah.

The twins' world lines are depicted in Fig. 2.8a, a spacetime diagram whose axes are those of Methuselah's inertial frame. The time measured by an ideal clock that Methuselah carries is the coordinate time t of his inertial frame; and its total time lapse, from Florence's departure to her return, is $t_{\text{return}} - t_{\text{departure}} \equiv T_{\text{Methuselah}}$. By contrast, the time measured by an ideal clock that Florence carries is her proper time τ (i.e., the square root of the invariant interval (2.4) along her world line). Thus her total time lapse from departure to return is

$$T_{\text{Florence}} = \int d\tau = \int \sqrt{dt^2 - \delta_{ij} dx^i dx^j} = \int_0^{T_{\text{Methuselah}}} \sqrt{1 - v^2}\, dt. \quad (2.40)$$

Here (t, x^i) are the time and space coordinates of Methuselah's inertial frame, and v is Florence's ordinary speed, $v = \sqrt{\delta_{ij}(dx^i/dt)(dx^j/dt)}$, as measured in Methuselah's frame. Obviously, Eq. (2.40) predicts that T_{Florence} is less than $T_{\text{Methuselah}}$. In fact (Ex. 2.16), even if Florence's acceleration is kept no larger than one Earth gravity throughout her trip, and her trip lasts only $T_{\text{Florence}} = $ (a few tens of years), $T_{\text{Methuselah}}$ can be hundreds or thousands or millions or billions of years.

Does this mean that Methuselah actually "experiences" a far longer time lapse, and actually ages far more than Florence? Yes! The time experienced by humans and the aging of the human body are governed by chemical processes, which in turn are governed by the natural oscillation rates of molecules, rates that are constant to high accuracy when measured in terms of ideal time (or, equivalently, proper time τ).

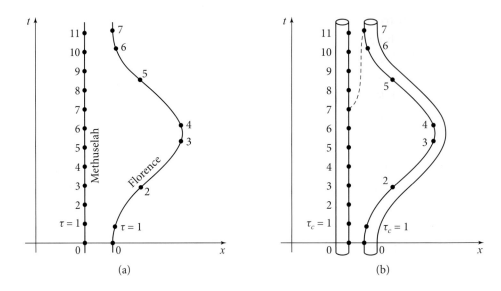

FIGURE 2.8 (a) Spacetime diagram depicting the so-called "twins paradox." Marked along the two world lines are intervals of proper time as measured by the two twins. (b) Spacetime diagram depicting the motions of the two mouths of a wormhole. Marked along the mouths' world tubes are intervals of proper time τ_c as measured by the single clock that sits in the common mouths.

Therefore, a human's experiential time and aging time are the same as the human's proper time—so long as the human is not subjected to such high accelerations as to damage her body.

In effect, then, Florence's spacecraft has functioned as a time machine to carry her far into Methuselah's future, with only a modest lapse of her own proper time (i.e., ideal, experiential, or aging time). This may be a "paradox" in the sense that it is surprising. However, it is in no sense a contradiction. This type of time dilation is routinely measured in high-energy physics storage rings.

2.9.2 Wormholes

Is it also possible, at least in principle, for Florence to construct a time machine that carries her into Methuselah's past—and also her own past? At first sight, the answer would seem to be "yes." Figure 2.8b shows one possible method, using a *wormhole*. [See Frolov and Novikov (1990), Friedman and Higuchi (2006), Everett and Roman (2011) for other approaches.]

Wormholes are hypothetical handles in the topology of space. A simple model of a wormhole can be obtained by taking a flat 3-dimensional space, removing from it the interiors of two identical spheres, and identifying the spheres' surfaces so that if one enters the surface of one of the spheres, one immediately finds oneself exiting through the surface of the other. When this is done, there is a bit of strongly localized spatial curvature at the spheres' common surface, so to analyze such a wormhole properly, one must use general relativity rather than special relativity. In particular, it is the laws of general relativity, combined with the laws of quantum field theory, that tell one how to construct such a wormhole and what kinds of materials are required to hold it open, so things can pass through it. Unfortunately, despite considerable effort, theoretical physicists have not yet deduced definitively whether those laws permit such wormholes to exist and stay open, though indications are pessimistic (Everett and Roman, 2011; Friedman and Higuchi, 2006). However, assuming such wormholes *can* exist, the following special relativistic analysis (Morris et al., 1988) shows how one might be used to construct a machine for backward time travel.

2.9.3 Wormhole as Time Machine

The two identified spherical surfaces are called the wormhole's mouths. Ask Methuselah to keep one mouth with him, forever at rest in his inertial frame, and ask Florence to take the other mouth with her on her high-speed journey. The two mouths' *world tubes* (analogs of world lines for a 3-dimensional object) then have the forms shown in Fig. 2.8b. Suppose that a single ideal clock sits in the wormhole's identified mouths, so that from the external universe one sees it both in Methuselah's wormhole mouth and in Florence's. As seen in Methuselah's mouth, the clock measures his proper time, which is equal to the coordinate time t (see tick marks along the left world tube in Fig. 2.8b). As seen in Florence's mouth, the clock measures her proper time, Eq. (2.40) (see tick marks along the right world tube in Fig. 2.8b). The result should be obvious, if surprising: When Florence returns to rest beside Methuselah, the wormhole has become a time machine. If she travels through the wormhole when the clock reads $\tau_c = 7$, she goes backward in time as seen in Methuselah's (or anyone else's) inertial frame; and then, in fact, traveling along the everywhere timelike world line (dashed in Fig. 2.8b), she is able to meet her younger self before she entered the wormhole.

This scenario is profoundly disturbing to most physicists because of the dangers of science-fiction-type paradoxes (e.g., the older Florence might kill her younger self, thereby preventing herself from making the trip through the wormhole and killing herself). Fortunately perhaps, it seems likely (though far from certain) that vacuum fluctuations of quantum fields will destroy the wormhole at the moment its mouths' motion first makes backward time travel possible. It may be that this mechanism will always prevent the construction of backward-travel time machines, no matter what tools one uses for their construction (Kay et al., 1997; Kim and Thorne, 1991); but see also contrary indications in research reviewed by Everett and Roman (2011) and Friedman and Higuchi (2006). Whether this is so we likely will not know until the laws of quantum gravity have been mastered.

world tube

chronology protection

EXERCISES

Exercise 2.16 *Example: Twins Paradox*
(a) The 4-acceleration of a particle or other object is defined by $\vec{a} \equiv d\vec{u}/d\tau$, where $\vec{u}$ is its 4-velocity and τ is proper time along its world line. Show that, if an observer carries an accelerometer, the magnitude $|\mathbf{a}|$ of the 3-dimensional acceleration $\mathbf{a}$ measured by the accelerometer will always be equal to the magnitude of the observer's 4-acceleration, $|\mathbf{a}| = |\vec{a}| \equiv \sqrt{\vec{a} \cdot \vec{a}}$.

(b) In the twins paradox of Fig. 2.8a, suppose that Florence begins at rest beside Methuselah, then accelerates in Methuselah's x-direction with an acceleration a equal to one Earth gravity, g, for a time $T_{\text{Florence}}/4$ as measured by her, then accelerates in the $-x$-direction at g for a time $T_{\text{Florence}}/2$, thereby reversing her motion; then she accelerates in the $+x$-direction at g for a time $T_{\text{Florence}}/4$, thereby returning to rest beside Methuselah. (This is the type of motion shown in the figure.) Show that the total time lapse as measured by Methuselah is

$$T_{\text{Methuselah}} = \frac{4}{g} \sinh\left(\frac{gT_{\text{Florence}}}{4}\right). \tag{2.41}$$

(c) Show that in the geometrized units used here, Florence's acceleration (equal to acceleration of gravity at the surface of Earth) is $g = 1.033/\text{yr}$. Plot $T_{\text{Methuselah}}$ as a function of T_{Florence}, and from your plot estimate T_{Florence} if $T_{\text{Methuselah}}$ is the age of the Universe, 14 billion years.

Exercise 2.17 *Challenge: Around the World on TWA*
In a long-ago era when an airline named Trans World Airlines (TWA) flew around the world, Josef Hafele and Richard Keating (1972a) carried out a real live twins paradox experiment: They synchronized two atomic clocks and then flew one around the world eastward on TWA, and on a separate trip, around the world westward, while the other clock remained at home at the Naval Research Laboratory near Washington, D.C. When the clocks were compared after each trip, they were found to have aged differently. Making reasonable estimates for the airplane routing and speeds, compute the difference in aging, and compare your result with the experimental data in Hafele and Keating (1972b). [Note: The rotation of Earth is important, as is the general relativistic gravitational redshift associated with the clocks' altitudes; but the gravitational redshift drops out of the difference in aging, if the time spent at high altitude is the same eastward as westward.]

2.10 Directional Derivatives, Gradients, and the Levi-Civita Tensor

Derivatives of vectors and tensors in Minkowski spacetime are defined in precisely the same way as in Euclidean space; see Sec. 1.7. Any reader who has not studied that section should do so now. In particular (in extreme brevity, as the explanations and justifications are the same as in Euclidean space):

The **directional derivative** of a tensor $\mathbf{T}$ along a vector $\vec{A}$ is $\nabla_{\vec{A}} \mathbf{T} \equiv \lim_{\epsilon \to 0}(1/\epsilon)[\mathbf{T}(\vec{x}_{\mathcal{P}} + \epsilon \vec{A}) - \mathbf{T}(\vec{x}_{\mathcal{P}})]$; the **gradient** $\vec{\nabla}\mathbf{T}$ is the tensor that produces the directional derivative when one inserts $\vec{A}$ into its last slot: $\nabla_{\vec{A}} \mathbf{T} = \vec{\nabla}\mathbf{T}(\_, \_, \_, \vec{A})$. In slot-naming index notation (or in components on a basis), the gradient is denoted $T_{\alpha\beta\gamma;\mu}$. In a Lorentz basis (the basis vectors associated with an inertial reference frame), the components of the gradient are simply the partial derivatives of the tensor,

$T_{\alpha\beta\gamma;\mu} = \partial T_{\alpha\beta\gamma}/\partial x^\mu \equiv T_{\alpha\beta\gamma,\mu}$. (The comma means partial derivative in a Lorentz basis, as in a Cartesian basis.)

The gradient and the directional derivative obey all the familiar rules for differentiation of products, for example, $\nabla_{\vec{A}}(\mathbf{S} \otimes \mathbf{T}) = (\nabla_{\vec{A}}\mathbf{S}) \otimes \mathbf{T} + \mathbf{S} \otimes \nabla_{\vec{A}}\mathbf{T}$. The gradient of the metric vanishes, $g_{\alpha\beta;\mu} = 0$. The divergence of a vector is the contraction of its gradient, $\vec{\nabla} \cdot \vec{A} = A_{\alpha;\beta} g^{\alpha\beta} = A^\alpha{}_{;\alpha}$.

Recall that the divergence of the gradient of a tensor in Euclidean space is the Laplacian: $T_{abc;jk} g_{jk} = T_{abc,jk}\delta_{jk} = \partial^2 T_{abc}\partial x^j \partial x^j$. By contrast, in Minkowski spacetime, because $g^{00} = -1$ and $g^{jk} = \delta^{jk}$ in a Lorentz frame, the divergence of the gradient is the wave operator (also called the d'Alembertian):

d'Alembertian

$$T_{\alpha\beta\gamma;\mu\nu}g^{\mu\nu} = T_{\alpha\beta\gamma,\mu\nu}g^{\mu\nu} = -\frac{\partial^2 T_{\alpha\beta\gamma}}{\partial t^2} + \frac{\partial^2 T_{\alpha\beta\gamma}}{\partial x^j \partial x^k}\delta^{jk} = \Box T_{\alpha\beta\gamma}. \quad (2.42)$$

When one sets this to zero, one gets the wave equation.

As in Euclidean space, so also in Minkowski spacetime, there are two tensors that embody the space's geometry: the metric tensor $\mathbf{g}$ and the Levi-Civita tensor $\boldsymbol{\epsilon}$. The Levi-Civita tensor in Minkowski spacetime is the tensor that is completely antisymmetric in all its slots and has value $\boldsymbol{\epsilon}(\vec{A}, \vec{B}, \vec{C}, \vec{D}) = +1$ when evaluated on any *right-handed set of orthonormal 4-vectors*—that is, by definition, any orthonormal set for which $\vec{A}$ is timelike and future directed, and $\{\vec{B}, \vec{C}, \vec{D}\}$ are spatial and right-handed. This means that in any right-handed Lorentz basis, the only nonzero components of $\boldsymbol{\epsilon}$ are

Levi-Civita tensor

$$\epsilon_{\alpha\beta\gamma\delta} = +1 \text{ if } \alpha, \beta, \gamma, \delta \text{ is an even permutation of } 0, 1, 2, 3;$$
$$-1 \text{ if } \alpha, \beta, \gamma, \delta \text{ is an odd permutation of } 0, 1, 2, 3;$$
$$0 \text{ if } \alpha, \beta, \gamma, \delta \text{ are not all different.} \quad (2.43)$$

By the sign-flip-if-temporal rule, $\epsilon_{0123} = +1$ implies that $\epsilon^{0123} = -1$.

2.11 Nature of Electric and Magnetic Fields; Maxwell's Equations

Now that we have introduced the gradient and the Levi-Civita tensor, we can study the relationship of the relativistic version of electrodynamics to the nonrelativistic (Newtonian) version. In doing so, we use Gaussian units (with the speed of light set to 1), as is conventional among relativity theorists, and as does Jackson (1998) in his classic textbook, switching from SI to Gaussian when he moves into the relativistic domain.

Consider a particle with charge q, rest mass m, and 4-velocity $\vec{u}$ interacting with an electromagnetic field $\mathbf{F}(\_\_, \_\_)$. In index notation, the electromagnetic 4-force acting on the particle [Eq. (2.20)] is

$$dp^\alpha/d\tau = q F^{\alpha\beta} u_\beta. \quad (2.44)$$

Let us examine this 4-force in some arbitrary inertial reference frame in which the particle's ordinary-velocity components are $v^j = v_j$ and its 4-velocity components are $u^0 = \gamma, u^j = \gamma v^j$ [Eqs. (2.25c)]. Anticipating the connection with the nonrelativistic viewpoint, we introduce the following notation for the contravariant components of the antisymmetric electromagnetic field tensor:

$$F^{0j} = -F^{j0} = +F_{j0} = -F_{0j} = E_j, \qquad F^{ij} = F_{ij} = \epsilon_{ijk} B_k. \tag{2.45}$$

Inserting these components of **F** and $\vec{u}$ into Eq. (2.44) and using the relationship $dt/d\tau = u^0 = \gamma$ between t and τ derivatives, we obtain for the components of the 4-force $dp_j/d\tau = \gamma dp_j/dt = q(F_{j0} u^0 + F_{jk} u^k) = q u^0 (F_{j0} + F_{jk} v^k) = q\gamma(E_j + \epsilon_{jki} v_k B_i)$ and $dp^0/d\tau = \gamma dp^0/dt = q F^{0j} u_j = q\gamma E_j v_j$. Dividing by γ, converting into 3-space index notation, and denoting the particle's energy by $\mathcal{E} = p^0$, we bring these into the familiar Lorentz-force form

Lorentz force

$$d\mathbf{p}/dt = q(\mathbf{E} + \mathbf{v} \times \mathbf{B}), \qquad d\mathcal{E}/dt = q\mathbf{v} \cdot \mathbf{E}. \tag{2.46}$$

Evidently, **E** is the electric field and **B** the magnetic field as measured in our chosen Lorentz frame.

This may be familiar from standard electrodynamics textbooks (e.g., Jackson, 1999). Not so familiar, but very important, is the following geometric interpretation of **E** and **B**.

The electric and magnetic fields **E** and **B** are spatial vectors as measured in the chosen inertial frame. We can also regard them as 4-vectors that lie in a 3-surface of simultaneity $t = $ const of the chosen frame, i.e., that are orthogonal to the 4-velocity (denote it $\vec{w}$) of the frame's observers (cf. Figs. 2.5 and 2.9). We shall denote this 4-vector version of **E** and **B** by $\vec{E}_{\vec{w}}$ and $\vec{B}_{\vec{w}}$, where the subscript $\vec{w}$ identifies the 4-velocity of the observer who measures these fields. These fields are depicted in Fig. 2.9.

In the rest frame of the observer $\vec{w}$, the components of $\vec{E}_{\vec{w}}$ are $E_{\vec{w}}^0 = 0$, $E_{\vec{w}}^j = E_j = F_{j0}$ [the E_j appearing in Eqs. (2.45)], and similarly for $\vec{B}_{\vec{w}}$; the components of $\vec{w}$ are $w^0 = 1, w^j = 0$. Therefore, in this frame Eqs. (2.45) can be rewritten as

electric and magnetic fields measured by an observer

$$\boxed{E_{\vec{w}}^{\alpha} = F^{\alpha\beta} w_\beta, \qquad B_{\vec{w}}^{\beta} = \frac{1}{2} \epsilon^{\alpha\beta\gamma\delta} F_{\gamma\delta} w_\alpha.} \tag{2.47a}$$

[To verify this, insert the above components of **F** and $\vec{w}$ into these equations and, after some algebra, recover Eqs. (2.45) along with $E_{\vec{w}}^0 = B_{\vec{w}}^0 = 0$.] Equations (2.47a) say that in one special reference frame, that of the observer $\vec{w}$, the components of the 4-vectors on the left and on the right are equal. This implies that in every Lorentz frame the components of these 4-vectors will be equal; that is, Eqs. (2.47a) are true when one regards them as geometric, frame-independent equations written in slot-naming index notation. *These equations enable one to compute the electric and magnetic fields measured by an observer (viewed as 4-vectors in the observer's 3-surface of simultaneity)*

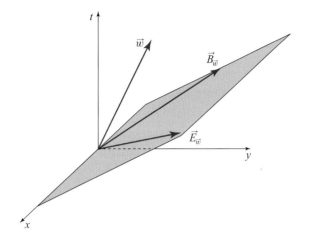

FIGURE 2.9 The electric and magnetic fields measured by an observer with 4-velocity $\vec{w}$, shown as 4-vectors $\vec{E}_{\vec{w}}$ and $\vec{B}_{\vec{w}}$ that lie in the observer's 3-surface of simultaneity (green 3-surface orthogonal to $\vec{w}$).

from the observer's 4-velocity and the electromagnetic field tensor, without the aid of any basis or reference frame.

Equations (2.47a) embody explicitly the following important fact. Although the electromagnetic field tensor **F** is a geometric, frame-independent quantity, the electric and magnetic fields $\vec{E}_{\vec{w}}$ and $\vec{B}_{\vec{w}}$ individually depend for their existence on a specific choice of observer (with 4-velocity $\vec{w}$), that is, a specific choice of inertial reference frame, or in other words, a specific choice of the split of spacetime into a 3-space (the 3-surface of simultaneity orthogonal to the observer's 4-velocity $\vec{w}$) and corresponding time (the Lorentz time of the observer's reference frame). *Only after making such an observer-dependent 3+1 split of spacetime into space plus time do the electric field and the magnetic field come into existence as separate entities.* Different observers with different 4-velocities $\vec{w}$ make this spacetime split in different ways, thereby resolving the frame-independent **F** into different electric and magnetic fields $\vec{E}_{\vec{w}}$ and $\vec{B}_{\vec{w}}$.

By the same procedure as we used to derive Eqs. (2.47a), one can derive the inverse relationship, the following expression for the electromagnetic field tensor in terms of the (4-vector) electric and magnetic fields measured by some observer:

$$F^{\alpha\beta} = w^\alpha E^\beta_{\vec{w}} - E^\alpha_{\vec{w}} w^\beta + \epsilon^{\alpha\beta}{}_{\gamma\delta} w^\gamma B^\delta_{\vec{w}}. \tag{2.47b}$$

Maxwell's equations in geometric, frame-independent form are (in Gaussian units)[5]

$$F^{\alpha\beta}{}_{;\beta} = 4\pi J^\alpha,$$
$$\epsilon^{\alpha\beta\gamma\delta} F_{\gamma\delta;\beta} = 0; \quad \text{i.e.,} \quad F_{\alpha\beta;\gamma} + F_{\beta\gamma;\alpha} + F_{\gamma\alpha;\beta} = 0. \tag{2.48}$$

Maxwell's equations

---

5. In SI units the 4π gets replaced by $\mu_0 = 1/\epsilon_0$, corresponding to the different units for the charge-current 4-vector.

Here $\vec{J}$ is the charge-current 4-vector, which in any inertial frame has components

$$J^0 = \rho_e = \text{(charge density)}, \qquad J^i = j_i = \text{(current density)}. \tag{2.49}$$

Exercise 2.19 describes how to think about this charge density and current density as geometric objects determined by the observer's 4-velocity or 3+1 split of spacetime into space plus time. Exercise 2.20 shows how the frame-independent Maxwell's equations (2.48) reduce to the more familiar ones in terms of **E** and **B**. Exercise 2.21 explores potentials for the electromagnetic field in geometric, frame-independent language and the 3+1 split.

EXERCISES

Exercise 2.18 *Derivation and Practice: Reconstruction of F*
Derive Eq. (2.47b) by the same method as was used to derive Eq. (2.47a). Then show, by a geometric, frame-independent calculation, that Eq. (2.47b) implies Eq. (2.47a).

Exercise 2.19 *Problem: 3+1 Split of Charge-Current 4-Vector*
Just as the electric and magnetic fields measured by some observer can be regarded as 4-vectors $\vec{E}_{\vec{w}}$ and $\vec{B}_{\vec{w}}$ that live in the observer's 3-space of simultaneity, so also the charge density and current density that the observer measures can be regarded as a scalar $\rho_{\vec{w}}$ and 4-vector $\vec{j}_{\vec{w}}$ that live in the 3-space of simultaneity. Derive geometric, frame-independent equations for $\rho_{\vec{w}}$ and $\vec{j}_{\vec{w}}$ in terms of the charge-current 4-vector $\vec{J}$ and the observer's 4-velocity $\vec{w}$, and derive a geometric expression for $\vec{J}$ in terms of $\rho_{\vec{w}}$, $\vec{j}_{\vec{w}}$, and $\vec{w}$.

Exercise 2.20 *Problem: Frame-Dependent Version of Maxwell's Equations*
By performing a 3+1 split on the geometric version of Maxwell's equations (2.48), derive the elementary, frame-dependent version

$$\nabla \cdot \mathbf{E} = 4\pi \rho_e, \qquad \nabla \times \mathbf{B} - \frac{\partial \mathbf{E}}{\partial t} = 4\pi \mathbf{j},$$

$$\nabla \cdot \mathbf{B} = 0, \qquad \nabla \times \mathbf{E} + \frac{\partial \mathbf{B}}{\partial t} = 0. \tag{2.50}$$

Exercise 2.21 *Problem: Potentials for the Electromagnetic Field*
(a) Express the electromagnetic field tensor as an antisymmetrized gradient of a 4-vector potential: in slot-naming index notation

$$F_{\alpha\beta} = A_{\beta;\alpha} - A_{\alpha;\beta}. \tag{2.51a}$$

Show that, whatever may be the 4-vector potential $\vec{A}$, the second of Maxwell's equations (2.48) is automatically satisfied. Show further that the electromagnetic field tensor is unaffected by a gauge change of the form

$$\vec{A}_{\text{new}} = \vec{A}_{\text{old}} + \vec{\nabla}\psi, \tag{2.51b}$$

where ψ is a scalar field (the generator of the gauge change). Show, finally, that it is possible to find a gauge-change generator that enforces *Lorenz gauge*

$$\vec{\nabla} \cdot \vec{A} = 0 \qquad (2.51c)$$

on the new 4-vector potential, and show that in this gauge, the first of Maxwell's equations (2.48) becomes (in Gaussian units)

$$\Box \vec{A} = -4\pi \vec{J}; \quad \text{i.e.,} \quad A^{\alpha;\mu}{}_{\mu} = -4\pi J^{\alpha}. \qquad (2.51d)$$

(b) Introduce an inertial reference frame, and in that frame split **F** into the electric and magnetic fields **E** and **B**, split $\vec{J}$ into the charge and current densities ρ_e and **j**, and split the vector potential into a scalar potential and a 3-vector potential

$$\phi \equiv A_0, \qquad \mathbf{A} = \text{spatial part of } \vec{A}. \qquad (2.51e)$$

Deduce the 3+1 splits of Eqs. (2.51a)–(2.51d), and show that they take the form given in standard textbooks on electromagnetism.

2.12 Volumes, Integration, and Conservation Laws

2.12.1 Spacetime Volumes and Integration

In Minkowski spacetime as in Euclidean 3-space (Sec. 1.8), the Levi-Civita tensor is the tool by which one constructs volumes. The 4-dimensional parallelepiped whose legs are the four vectors $\vec{A}, \vec{B}, \vec{C}, \vec{D}$ has a 4-dimensional volume given by the analog of Eqs. (1.25) and (1.26):

$$4\text{-volume} = \epsilon_{\alpha\beta\gamma\delta} A^{\alpha} B^{\beta} C^{\gamma} D^{\delta} = \epsilon(\vec{A}, \vec{B}, \vec{C}, \vec{D}) = \det \begin{bmatrix} A^0 & B^0 & C^0 & D^0 \\ A^1 & B^1 & C^1 & D^1 \\ A^2 & B^2 & C^2 & D^2 \\ A^3 & B^3 & C^3 & D^3 \end{bmatrix}. \qquad (2.52)$$

Note that this 4-volume is positive if the set of vectors $\{\vec{A}, \vec{B}, \vec{C}, \vec{D}\}$ is right-handed and negative if left-handed [cf. Eq. (2.43)].

Equation (2.52) provides us a way to perform volume integrals over 4-dimensional Minkowski spacetime. To integrate a smooth tensor field **T** over some 4-dimensional region $\mathcal{V}$ of spacetime, we need only divide $\mathcal{V}$ up into tiny parallelepipeds, multiply the 4-volume $d\Sigma$ of each parallelepiped by the value of **T** at its center, add, and take the limit. In any right-handed Lorentz coordinate system, the 4-volume of a tiny parallelepiped whose edges are dx^{α} along the four orthogonal coordinate axes is $d\Sigma = \epsilon(dt\, \vec{e}_0, dx\, \vec{e}_x, dy\, \vec{e}_y, dz\, \vec{e}_z) = \epsilon_{0123}\, dt\, dx\, dy\, dz = dt\, dx\, dy\, dz$ (the analog of $dV = dx\, dy\, dz$). Correspondingly, the integral of **T** over $\mathcal{V}$ can be expressed as

$$\int_{\mathcal{V}} T^{\alpha\beta\gamma} d\Sigma = \int_{\mathcal{V}} T^{\alpha\beta\gamma} dt\, dx\, dy\, dz. \qquad (2.53)$$

vectorial 3-volume

By analogy with the vectorial area (1.27) of a parallelogram in 3-space, any 3-dimensional parallelepiped in spacetime with legs $\vec{A}$, $\vec{B}$, $\vec{C}$ has a vectorial 3-volume $\vec{\Sigma}$ (not to be confused with the scalar 4-volume Σ) defined by

$$\vec{\Sigma}(\_) = \epsilon(\_, \vec{A}, \vec{B}, \vec{C}); \qquad \Sigma_\mu = \epsilon_{\mu\alpha\beta\gamma} A^\alpha B^\beta C^\gamma. \tag{2.54}$$

Here we have written the 3-volume vector both in abstract notation and in slot-naming index notation. This 3-volume vector has one empty slot, ready and waiting for a fourth vector ("leg") to be inserted, so as to compute the 4-volume Σ of a 4-dimensional parallelepiped.

Notice that the 3-volume vector $\vec{\Sigma}$ is orthogonal to each of its three legs (because of the antisymmetry of ϵ), and thus (unless it is null) it can be written as $\vec{\Sigma} = V\vec{n}$, where V is the magnitude of the 3-volume, and $\vec{n}$ is the unit normal to the three legs.

Interchanging any two legs of the parallelepiped reverses the 3-volume's sign. Consequently, the 3-volume is characterized not only by its legs but also by the order of its legs, or equally well, in two other ways: (i) by the direction of the vector $\vec{\Sigma}$ (reverse the order of the legs, and the direction of $\vec{\Sigma}$ will reverse); and (ii) by the *sense* of the 3-volume, defined as follows. Just as a 2-volume (i.e., a segment of a plane) in 3-dimensional space has two sides, so a 3-volume in 4-dimensional spacetime has two sides (Fig. 2.10). Every vector $\vec{D}$ for which $\vec{\Sigma} \cdot \vec{D} > 0$ points out the *positive side* of the 3-volume $\vec{\Sigma}$. Vectors $\vec{D}$ with $\vec{\Sigma} \cdot \vec{D} < 0$ point out its *negative side*. When something moves through, reaches through, or points through the 3-volume from its negative side to its positive side, we say that this thing is moving or reaching or pointing in the "positive sense;" similarly for "negative sense." The examples shown in Fig. 2.10 should make this more clear.

positive and negative sides, and sense of 3-volume

Figure 2.10a shows two of the three legs of the volume vector $\vec{\Sigma} = \epsilon(\_, \Delta x \vec{e}_x, \Delta y \vec{e}_y, \Delta z \vec{e}_z)$, where $\{t, x, y, z\}$ are the coordinates, and $\{\vec{e}_\alpha\}$ is the corresponding right-handed basis of a specific Lorentz frame. It is easy to show that this $\vec{\Sigma}$ can also be written as $\vec{\Sigma} = -\Delta V \vec{e}_0$, where ΔV is the ordinary volume of the parallelepiped as measured by an observer in the chosen Lorentz frame, $\Delta V = \Delta x \Delta y \Delta z$. Thus, the direction of the vector $\vec{\Sigma}$ is toward the past (direction of decreasing Lorentz time t). From this, and the fact that timelike vectors have negative squared length, it is easy to infer that $\vec{\Sigma} \cdot \vec{D} > 0$ if and only if the vector $\vec{D}$ points out of the "future" side of the 3-volume (the side of increasing Lorentz time t); therefore, the positive side of $\vec{\Sigma}$ is the future side. It follows that the vector $\vec{\Sigma}$ points in the negative sense of its own 3-volume.

Figure 2.10b shows two of the three legs of the volume vector $\vec{\Sigma} = \epsilon(\_, \Delta t \vec{e}_t, \Delta y \vec{e}_y, \Delta z \vec{e}_z) = -\Delta t \Delta A \vec{e}_x$ (with $\Delta A = \Delta y \Delta z$). In this case, $\vec{\Sigma}$ points in its own positive sense.

This peculiar behavior is completely general. When the normal to a 3-volume is timelike, its volume vector $\vec{\Sigma}$ points in the negative sense; when the normal is spacelike, $\vec{\Sigma}$ points in the positive sense. And as it turns out, when the normal is null, $\vec{\Sigma}$

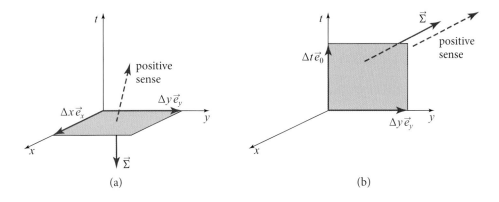

FIGURE 2.10 Spacetime diagrams depicting 3-volumes in 4-dimensional spacetime, with one spatial dimension (that along the z direction) suppressed.

lies in the 3-volume (parallel to its one null leg) and thus points neither in the positive sense nor the negative.[6]

Note the physical interpretations of the 3-volumes of Fig. 2.10. Figure 2.10a shows an instantaneous snapshot of an ordinary, spatial parallelepiped, whereas Fig. 2.10b shows the 3-dimensional region in spacetime swept out during time Δt by the parallelogram with legs $\Delta y \vec{e}_y$, $\Delta z \vec{e}_z$ and with area $\Delta A = \Delta y \Delta z$.

Vectorial 3-volume elements can be used to construct integrals over 3-dimensional volumes (also called 3-dimensional surfaces) in spacetime, for example, $\int_{\mathcal{V}_3} \vec{A} \cdot d\vec{\Sigma}$. More specifically, let (a, b, c) be (possibly curvilinear) coordinates in the 3-surface (3-volume) $\mathcal{V}_3$, and denote by $\vec{x}(a, b, c)$ the spacetime point $\mathcal{P}$ on $\mathcal{V}_3$ whose coordinate values are (a, b, c). Then $(\partial \vec{x}/\partial a)da$, $(\partial \vec{x}/\partial b)db$, $(\partial \vec{x}/\partial c)dc$ are the vectorial legs of the elementary parallelepiped whose corners are at (a, b, c), $(a + da, b, c)$, $(a, b + db, c)$, and so forth; and the spacetime components of these vectorial legs are $(\partial x^\alpha/\partial a)da$, $(\partial x^\alpha/\partial b)db$, $(\partial x^\alpha/\partial c)dc$. The 3-volume of this elementary parallelepiped is $d\vec{\Sigma} = \epsilon(\_\_, (\partial \vec{x}/\partial a)da, (\partial \vec{x}/\partial b)db, (\partial \vec{x}/\partial c)dc)$, which has spacetime components

$$d\Sigma_\mu = \epsilon_{\mu\alpha\beta\gamma} \frac{\partial x^\alpha}{\partial a} \frac{\partial x^\beta}{\partial b} \frac{\partial x^\gamma}{\partial c} da\, db\, dc. \tag{2.55}$$

This is the integration element to be used when evaluating

$$\int_{\mathcal{V}_3} \vec{A} \cdot d\vec{\Sigma} = \int_{\mathcal{V}_3} A^\mu d\Sigma_\mu. \tag{2.56}$$

See Ex. 2.22 for an example.

---

6. This peculiar behavior gets replaced by a simpler description if one uses one-forms rather than vectors to describe 3-volumes; see, for example, Misner, Thorne, and Wheeler (1973, Box 5.2).

2.12 Volumes, Integration, and Conservation Laws

Just as there are Gauss's and Stokes' theorems (1.28a) and (1.28b) for integrals in Euclidean 3-space, so also there are Gauss's and Stokes' theorems in spacetime. Gauss's theorem has the obvious form

$$\boxed{\int_{\mathcal{V}_4} (\vec{\nabla} \cdot \vec{A}) d\Sigma = \int_{\partial \mathcal{V}_4} \vec{A} \cdot d\vec{\Sigma},} \qquad (2.57)$$

where the first integral is over a 4-dimensional region $\mathcal{V}_4$ in spacetime, and the second is over the 3-dimensional boundary $\partial \mathcal{V}_4$ of $\mathcal{V}_4$, with the boundary's positive sense pointing outward, away from $\mathcal{V}_4$ (just as in the 3-dimensional case). We shall not write down the 4-dimensional Stokes' theorem, because it is complicated to formulate with the tools we have developed thus far; easy formulation requires *differential forms* (e.g., Flanders, 1989), which we shall not introduce in this book.

2.12.2 Conservation of Charge in Spacetime

In this section, we use integration over a 3-dimensional region in 4-dimensional spacetime to construct an elegant, frame-independent formulation of the law of conservation of electric charge.

We begin by examining the geometric meaning of the charge-current 4-vector $\vec{J}$. We defined $\vec{J}$ in Eq. (2.49) in terms of its components. The spatial component $J^x = J_x = \vec{J}(\vec{e}_x)$ is equal to the x component of current density j_x: it is the amount Q of charge that flows across a unit surface area lying in the y-z plane in a unit time (i.e., the charge that flows across the unit 3-surface $\vec{\Sigma} = \vec{e}_x$). In other words, $\vec{J}(\vec{\Sigma}) = \vec{J}(\vec{e}_x)$ *is the total charge Q that flows across $\vec{\Sigma} = \vec{e}_x$ in $\vec{\Sigma}$'s positive sense* and similarly for the other spatial directions. The temporal component $J^0 = -J_0 = \vec{J}(-\vec{e}_0)$ is the charge density ρ_e: it is the total charge Q in a unit spatial volume. This charge is carried by particles that are traveling through spacetime from past to future and pass through the unit 3-surface (3-volume) $\vec{\Sigma} = -\vec{e}_0$. Therefore, $\vec{J}(\vec{\Sigma}) = \vec{J}(-\vec{e}_0)$ *is the total charge Q that flows through $\vec{\Sigma} = -\vec{e}_0$ in its positive sense*. This interpretation is the same one we deduced for the spatial components of $\vec{J}$.

This makes it plausible, and indeed one can show, that *for any small 3-surface $\vec{\Sigma}$, $\vec{J}(\vec{\Sigma}) \equiv J^\alpha \Sigma_\alpha$ is the total charge Q that flows across $\vec{\Sigma}$ in its positive sense.*

This property of the charge-current 4-vector is the foundation for our frame-independent formulation of the law of charge conservation. Let $\mathcal{V}$ be a compact 4-dimensional region of spacetime and denote by $\partial \mathcal{V}$ its boundary, a closed 3-surface in 4-dimensional spacetime (Fig. 2.11). The charged media (fluids, solids, particles, etc.) present in spacetime carry electric charge through $\mathcal{V}$, from the past toward the future. The law of charge conservation says that all the charge that enters $\mathcal{V}$ through the past part of its boundary $\partial \mathcal{V}$ must exit through the future part of its boundary. If we choose the positive sense of the boundary's 3-volume element $d\vec{\Sigma}$ to point out of $\mathcal{V}$ (toward the past on the bottom boundary and toward the future on the top), then this *global law of charge conservation* can be expressed as

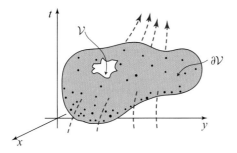

FIGURE 2.11 The 4-dimensional region $\mathcal{V}$ in spacetime and its closed 3-boundary $\partial\mathcal{V}$ (green surface), used in formulating the law of charge conservation. The dashed lines symbolize, heuristically, the flow of charge from the past toward the future.

$$\boxed{\int_{\partial\mathcal{V}} J^\alpha d\Sigma_\alpha = 0.} \quad (2.58)$$

global law of charge conservation

When each tiny charge q enters $\mathcal{V}$ through its past boundary, it contributes negatively to the integral, since it travels through $\partial\mathcal{V}$ in the negative sense (from positive side of $\partial\mathcal{V}$ toward negative side); and when that same charge exits $\mathcal{V}$ through its future boundary, it contributes positively. Therefore, its net contribution is zero, and similarly for all other charges.

In Ex. 2.23 you will show that, when this global law of charge conservation (2.58) is subjected to a 3+1 split of spacetime into space plus time, it becomes the nonrelativistic integral law of charge conservation (1.29).

This global conservation law can be converted into a *local conservation law* with the help of the 4-dimensional Gauss's theorem (2.57), $\int_{\partial\mathcal{V}} J^\alpha d\Sigma_\alpha = \int_{\mathcal{V}} J^\alpha{}_{;\alpha} d\Sigma$. Since the left-hand side vanishes, so must the right-hand side; and for this 4-volume integral to vanish for every choice of $\mathcal{V}$, it is necessary that the integrand vanish everywhere in spacetime:

$$\boxed{J^\alpha{}_{;\alpha} = 0\,; \quad \text{that is,} \quad \vec{\nabla} \cdot \vec{J} = 0.} \quad (2.59)$$

local law of charge conservation

In a specific but arbitrary Lorentz frame (i.e., in a 3+1 split of spacetime into space plus time), Eq. (2.59) becomes the standard differential law of charge conservation (1.30).

2.12.3 Conservation of Particles, Baryon Number, and Rest Mass

Any conserved scalar quantity obeys conservation laws of the same form as those for electric charge. For example, if the number of particles of some species (e.g., electrons, protons, or photons) is conserved, then we can introduce for that species a *number-flux 4-vector* $\vec{S}$ (analog of charge-current 4-vector $\vec{J}$): in any Lorentz frame, S^0 is the number density of particles, also designated n, and S^j is the particle flux. If $\vec{\Sigma}$ is a small

number-flux 4-vector

3-volume (3-surface) in spacetime, then $\vec{S}(\vec{\Sigma}) = S^\alpha \Sigma_\alpha$ is the number of particles that pass through Σ from its negative side to its positive side. The frame-invariant global and local conservation laws for these particles take the same form as those for electric charge:

laws of particle conservation

$$\int_{\partial\mathcal{V}} S^\alpha d\Sigma_\alpha = 0, \quad \text{where } \partial\mathcal{V} \text{ is any closed 3-surface in spacetime,} \quad (2.60a)$$

$$S^\alpha{}_{;\alpha} = 0; \quad \text{that is,} \quad \vec{\nabla}\cdot\vec{S} = 0. \quad (2.60b)$$

When fundamental particles (e.g., protons and antiprotons) are created and destroyed by quantum processes, the total baryon number (number of baryons minus number of antibaryons) is still conserved—or at least this is so to the accuracy of all experiments performed thus far. We shall assume it so in this book. This law of baryon-number conservation takes the forms of Eqs. (2.60), with $\vec{S}$ the number-flux 4-vector for baryons (with antibaryons counted negatively).

It is useful to express this baryon-number conservation law in Newtonian-like language by introducing a universally agreed-on mean rest mass per baryon $\bar{m}_B$. This $\bar{m}_B$ is often taken to be 1/56 the mass of an ^{56}Fe (iron-56) atomic nucleus, since ^{56}Fe is the nucleus with the tightest nuclear binding (i.e., the endpoint of thermonuclear evolution in stars). We multiply the baryon number-flux 4-vector $\vec{S}$ by this mean rest mass per baryon to obtain a rest-mass-flux 4-vector

rest-mass-flux 4-vector

$$\vec{S}_{\text{rm}} = \bar{m}_B \vec{S}, \quad (2.61)$$

which (since $\bar{m}_B$ is, by definition, a constant) satisfies the same conservation laws (2.60) as baryon number.

For such media as fluids and solids, in which the particles travel only short distances between collisions or strong interactions, it is often useful to resolve the particle number-flux 4-vector and the rest-mass-flux 4-vector into a 4-velocity of the medium $\vec{u}$ (i.e., the 4-velocity of the frame in which there is a vanishing net spatial flux of particles), and the particle number density n_o or rest mass density ρ_o as measured in the medium's rest frame:

$$\vec{S} = n_o \vec{u}, \quad \vec{S}_{\text{rm}} = \rho_o \vec{u}. \quad (2.62)$$

See Ex. 2.24.

rest-mass conservation

We make use of the conservation laws $\vec{\nabla}\cdot\vec{S} = 0$ and $\vec{\nabla}\cdot\vec{S}_{\text{rm}} = 0$ for particles and rest mass later in this book (e.g., when studying relativistic fluids); and we shall find the expressions (2.62) for the number-flux 4-vector and rest-mass-flux 4-vector quite useful. See, for example, the discussion of relativistic shock waves in Ex. 17.9.

EXERCISES

Exercise 2.22 *Practice and Example: Evaluation of 3-Surface Integral in Spacetime*
In Minkowski spacetime, the set of all events separated from the origin by a timelike interval a^2 is a 3-surface, the hyperboloid $t^2 - x^2 - y^2 - z^2 = a^2$, where $\{t, x, y, z\}$

are Lorentz coordinates of some inertial reference frame. On this hyperboloid, introduce coordinates $\{\chi, \theta, \phi\}$ such that

$$t = a \cosh \chi, \quad x = a \sinh \chi \sin \theta \cos \phi,$$
$$y = a \sinh \chi \sin \theta \sin \phi, \quad z = a \sinh \chi \cos \theta. \quad (2.63)$$

Note that χ is a radial coordinate and (θ, ϕ) are spherical polar coordinates. Denote by $\mathcal{V}_3$ the portion of the hyperboloid with radius $\chi \leq b$.

(a) Verify that for all values of (χ, θ, ϕ), the points defined by Eqs. (2.63) do lie on the hyperboloid.

(b) On a spacetime diagram, draw a picture of $\mathcal{V}_3$, the $\{\chi, \theta, \phi\}$ coordinates, and the elementary volume element (vector field) $d\vec{\Sigma}$ [Eq. (2.55)].

(c) Set $\vec{A} \equiv \vec{e}_0$ (the temporal basis vector), and express $\int_{\mathcal{V}_3} \vec{A} \cdot d\vec{\Sigma}$ as an integral over $\{\chi, \theta, \phi\}$. Evaluate the integral.

(d) Consider a closed 3-surface consisting of the segment $\mathcal{V}_3$ of the hyperboloid as its top, the hypercylinder $\{x^2 + y^2 + z^2 = a^2 \sinh^2 b, \ 0 < t < a \cosh b\}$ as its sides, and the sphere $\{x^2 + y^2 + z^2 \leq a^2 \sinh^2 b, \ t = 0\}$ as its bottom. Draw a picture of this closed 3-surface on a spacetime diagram. Use Gauss's theorem, applied to this 3-surface, to show that $\int_{\mathcal{V}_3} \vec{A} \cdot d\vec{\Sigma}$ is equal to the 3-volume of its spherical base.

Exercise 2.23 *Derivation and Example: Global Law of Charge Conservation in an Inertial Frame*

Consider the global law of charge conservation $\int_{\partial \mathcal{V}} J^\alpha d\Sigma_\alpha = 0$ for a special choice of the closed 3-surface $\partial \mathcal{V}$: The bottom of $\partial \mathcal{V}$ is the ball $\{t = 0, x^2 + y^2 + z^2 \leq a^2\}$, where $\{t, x, y, z\}$ are the Lorentz coordinates of some inertial frame. The sides are the spherical world tube $\{0 \leq t \leq T, x^2 + y^2 + z^2 = a^2\}$. The top is the ball $\{t = T, x^2 + y^2 + z^2 \leq a^2\}$.

(a) Draw this 3-surface in a spacetime diagram.

(b) Show that for this $\partial \mathcal{V}$, $\int_{\partial \mathcal{V}} J^\alpha d\Sigma_\alpha = 0$ is a time integral of the nonrelativistic integral conservation law (1.29) for charge.

Exercise 2.24 *Example: Rest-Mass-Flux 4-Vector, Lorentz Contraction of Rest-Mass Density, and Rest-Mass Conservation for a Fluid*

Consider a fluid with 4-velocity $\vec{u}$ and rest-mass density ρ_o as measured in the fluid's rest frame.

(a) From the physical meanings of $\vec{u}$, ρ_o, and the rest-mass-flux 4-vector $\vec{S}_{\rm rm}$, deduce Eqs. (2.62).

(b) Examine the components of $\vec{S}_{\rm rm}$ in a reference frame where the fluid moves with ordinary velocity $\mathbf{v}$. Show that $S^0 = \rho_o \gamma$, and $S^j = \rho_o \gamma v^j$, where $\gamma = 1/\sqrt{1 - \mathbf{v}^2}$. Explain the physical interpretation of these formulas in terms of Lorentz contraction.

(c) Show that the law of conservation of rest mass $\vec{\nabla} \cdot \vec{S}_{rm} = 0$ takes the form

$$\frac{d\rho_o}{d\tau} = -\rho_o \vec{\nabla} \cdot \vec{u}, \tag{2.64}$$

where $d/d\tau$ is derivative with respect to proper time moving with the fluid.

(d) Consider a small 3-dimensional volume V of the fluid, whose walls move with the fluid (so if the fluid expands, V increases). Explain why the law of rest-mass conservation must take the form $d(\rho_o V)/d\tau = 0$. Thereby deduce that

$$\vec{\nabla} \cdot \vec{u} = (1/V)(dV/d\tau). \tag{2.65}$$

2.13 Stress-Energy Tensor and Conservation of 4-Momentum

2.13.1 Stress-Energy Tensor

GEOMETRIC DEFINITION

We conclude this chapter by formulating the law of 4-momentum conservation in ways analogous to our laws of conservation of charge, particles, baryon number, and rest mass. This task is not trivial, since 4-momentum is a vector in spacetime, while charge, particle number, baryon number, and rest mass are scalar quantities. Correspondingly, the density-flux of 4-momentum must have one more slot than the density-fluxes of charge, baryon number, and rest mass, $\vec{J}$, $\vec{S}$ and $\vec{S}_{rm}$, respectively; it must be a second-rank tensor. We call it the *stress-energy tensor* and denote it $\mathbf{T}(\_,\_)$. It is a generalization of the Newtonian stress tensor to 4-dimensional spacetime.

stress-energy tensor

Consider a medium or field flowing through 4-dimensional spacetime. As it crosses a tiny 3-surface $\vec{\Sigma}$, it transports a net electric charge $\vec{J}(\vec{\Sigma})$ from the negative side of $\vec{\Sigma}$ to the positive side, and net baryon number $\vec{S}(\vec{\Sigma})$ and net rest mass $\vec{S}_{rm}(\vec{\Sigma})$. Similarly, it transports a net 4-momentum $\mathbf{T}(\_, \vec{\Sigma})$ from the negative side to the positive side:

$$\mathbf{T}(\_, \vec{\Sigma}) \equiv \text{(total 4-momentum } \vec{P} \text{ that flows through } \vec{\Sigma}\text{)}; \quad \text{or } T^{\alpha\beta}\Sigma_\beta = P^\alpha. \tag{2.66}$$

COMPONENTS

From this definition of the stress-energy tensor we can read off the physical meanings of its components on a specific, but arbitrary, Lorentz-coordinate basis: Making use of method (2.23b) for computing the components of a vector or tensor, we see that in a specific, but arbitrary, Lorentz frame (where $\vec{\Sigma} = -\vec{e}_0$ is a volume vector representing a parallelepiped with unit volume $\Delta V = 1$, at rest in that frame, with its positive sense toward the future):

$$-T_{\alpha 0} = \mathbf{T}(\vec{e}_\alpha, -\vec{e}_0) = \vec{P}(\vec{e}_\alpha) = \begin{pmatrix} \alpha \text{ component of 4-momentum that} \\ \text{flows from past to future across a unit} \\ \text{volume } \Delta V = 1 \text{ in the 3-space } t = \text{const} \end{pmatrix}$$

$$= (\alpha \text{ component of density of 4-momentum}). \tag{2.67a}$$

Specializing α to be a time or space component and raising indices, we obtain the specialized versions of (2.67a):

T^{00} = (energy density as measured in the chosen Lorentz frame),

T^{j0} = (density of j component of momentum in that frame). (2.67b)

Similarly, the αx component of the stress-energy tensor (also called the $\alpha 1$ component, since $x = x^1$ and $\vec{e}_x = \vec{e}_1$) has the meaning

$$T_{\alpha 1} \equiv T_{\alpha x} \equiv \mathbf{T}(\vec{e}_\alpha, \vec{e}_x) = \begin{pmatrix} \alpha \text{ component of 4-momentum that crosses} \\ \text{a unit area } \Delta y \Delta z = 1 \text{ lying in a surface of} \\ \text{constant } x, \text{ during unit time } \Delta t, \text{ crossing} \\ \text{from the } -x \text{ side toward the } +x \text{ side} \end{pmatrix}$$

$$= \begin{pmatrix} \alpha \text{ component of flux of 4-momentum} \\ \text{across a surface lying perpendicular to } \vec{e}_x \end{pmatrix}. \quad (2.67c)$$

The specific forms of this for temporal and spatial α are (after raising indices)

$$T^{0x} = \begin{pmatrix} \text{energy flux across a surface perpendicular to } \vec{e}_x, \\ \text{from the } -x \text{ side to the } +x \text{ side} \end{pmatrix}, \quad (2.67d)$$

$$T^{jx} = \begin{pmatrix} \text{flux of } j \text{ component of momentum across a surface} \\ \text{perpendicular to } \vec{e}_x, \text{ from the } -x \text{ side to the } +x \text{ side} \end{pmatrix}$$

$$= \begin{pmatrix} jx \text{ component} \\ \text{of stress} \end{pmatrix}. \quad (2.67e)$$

The αy and αz components have the obvious, analogous interpretations.

These interpretations, restated much more briefly, are:

$$\boxed{\begin{aligned} T^{00} &= \text{(energy density)}, \quad T^{j0} = \text{(momentum density)}, \\ T^{0j} &= \text{(energy flux)}, \quad T^{jk} = \text{(stress)}. \end{aligned}} \quad (2.67f)$$

components of stress-energy tensor

SYMMETRY

Although it might not be obvious at first sight, *the 4-dimensional stress-energy tensor is always symmetric:* in index notation (where indices can be thought of as representing the names of slots, or equally well, components on an arbitrary basis)

$$T^{\alpha\beta} = T^{\beta\alpha}. \quad (2.68)$$

symmetry of stress-energy tensor

This symmetry can be deduced by physical arguments in a specific, but arbitrary, Lorentz frame: Consider, first, the $x0$ and $0x$ components, that is, the x components of momentum density and energy flux. A little thought, symbolized by the following heuristic equation, reveals that they must be equal:

$$T^{x0} = \begin{pmatrix} \text{momentum} \\ \text{density} \end{pmatrix} = \frac{(\Delta \mathcal{E}) dx/dt}{\Delta x \Delta y \Delta z} = \frac{\Delta \mathcal{E}}{\Delta y \Delta z \Delta t} = \begin{pmatrix} \text{energy} \\ \text{flux} \end{pmatrix}, \quad (2.69)$$

2.13 Stress-Energy Tensor and Conservation of 4-Momentum

and similarly for the other space-time and time-space components: $T^{j0} = T^{0j}$. [In the first expression of Eq. (2.69) $\Delta\mathcal{E}$ is the total energy (or equivalently mass) in the volume $\Delta x \Delta y \Delta z$, $(\Delta\mathcal{E})dx/dt$ is the total momentum, and when divided by the volume we get the momentum density. The third equality is just elementary algebra, and the resulting expression is obviously the energy flux.] The space-space components, being equal to the stress tensor, are also symmetric, $T^{jk} = T^{kj}$, by the argument embodied in Fig. 1.6. Since $T^{0j} = T^{j0}$ and $T^{jk} = T^{kj}$, all components in our chosen Lorentz frame are symmetric, $T^{\alpha\beta} = T^{\beta\alpha}$. Therefore, if we insert arbitrary vectors into the slots of **T** and evaluate the resulting number in our chosen Lorentz frame, we find

$$\mathbf{T}(\vec{A}, \vec{B}) = T^{\alpha\beta} A_\alpha B_\beta = T^{\beta\alpha} A_\alpha B_\beta = \mathbf{T}(\vec{B}, \vec{A}); \tag{2.70}$$

that is, **T** is symmetric under interchange of its slots.

Let us return to the physical meanings (2.67f) of the components of the stress-energy tensor. With the aid of **T**'s symmetry, we can restate those meanings in the language of a 3+1 split of spacetime into space plus time: *When one chooses a specific reference frame, that choice splits the stress-energy tensor up into three parts. Its time-time part is the energy density T^{00}, its time-space part $T^{0j} = T^{j0}$ is the energy flux or equivalently the momentum density, and its space-space part T^{jk} is the stress tensor.*

2.13.2 4-Momentum Conservation

Our interpretation of $\vec{J}(\vec{\Sigma}) \equiv J^\alpha \Sigma_\alpha$ as the net charge that flows through a small 3-surface $\vec{\Sigma}$ from its negative side to its positive side gave rise to the global conservation law for charge, $\int_{\partial\mathcal{V}} J^\alpha d\Sigma_\alpha = 0$ [Eq. (2.58) and Fig. 2.11]. Similarly the role of $\mathbf{T}(\_, \vec{\Sigma})$ [$T^{\alpha\beta}\Sigma_\beta$ in slot-naming index notation] as the net 4-momentum that flows through $\vec{\Sigma}$ from its negative side to positive gives rise to the following equation for conservation of 4-momentum:

global law of 4-momentum conservation

$$\boxed{\int_{\partial\mathcal{V}} T^{\alpha\beta} d\Sigma_\beta = 0.} \tag{2.71}$$

(The time component of this equation is energy conservation; the spatial part is momentum conservation.) This equation says that all the 4-momentum that flows into the 4-volume $\mathcal{V}$ of Fig. 2.11 through its 3-surface $\partial\mathcal{V}$ must also leave $\mathcal{V}$ through $\partial\mathcal{V}$; it gets counted negatively when it enters (since it is traveling from the positive side of $\partial\mathcal{V}$ to the negative), and it gets counted positively when it leaves, so its net contribution to the integral (2.71) is zero.

This global law of 4-momentum conservation can be converted into a local law (analogous to $\vec{\nabla} \cdot \vec{J} = 0$ for charge) with the help of the 4-dimensional Gauss's theorem (2.57). Gauss's theorem, generalized in the obvious way from a vectorial integrand to a tensorial one, is:

$$\int_\mathcal{V} T^{\alpha\beta}{}_{;\beta} d\Sigma = \int_{\partial\mathcal{V}} T^{\alpha\beta} d\Sigma_\beta. \tag{2.72}$$

Since the right-hand side vanishes, so must the left-hand side; and for this 4-volume integral to vanish for every choice of $\mathcal{V}$, the integrand must vanish everywhere in spacetime:

$$\boxed{T^{\alpha\beta}{}_{;\beta} = 0; \quad \text{or} \quad \vec{\nabla} \cdot \mathbf{T} = 0.} \tag{2.73a}$$

local law of 4-momentum conservation

In the second, index-free version of this local conservation law, the ambiguity about which slot the divergence is taken on is unimportant, since $\mathbf{T}$ is symmetric in its two slots: $T^{\alpha\beta}{}_{;\beta} = T^{\beta\alpha}{}_{;\beta}$.

In a specific but arbitrary Lorentz frame, the local conservation law (2.73a) for 4-momentum has as its temporal part

$$\frac{\partial T^{00}}{\partial t} + \frac{\partial T^{0k}}{\partial x^k} = 0, \tag{2.73b}$$

that is, the time derivative of the energy density plus the 3-divergence of the energy flux vanishes; and as its spatial part

$$\frac{\partial T^{j0}}{\partial t} + \frac{\partial T^{jk}}{\partial x^k} = 0, \tag{2.73c}$$

that is, the time derivative of the momentum density plus the 3-divergence of the stress (i.e., of momentum flux) vanishes. Thus, as one should expect, the geometric, frame-independent law of 4-momentum conservation includes as special cases both the conservation of energy and the conservation of momentum; and their differential conservation laws have the standard form that one expects both in Newtonian physics and in special relativity: time derivative of density plus divergence of flux vanishes; cf. Eq. (1.36) and associated discussion.

2.13.3 Stress-Energy Tensors for Perfect Fluids and Electromagnetic Fields

As an important example that illustrates the stress-energy tensor, consider a *perfect fluid*—a medium whose stress-energy tensor, evaluated in its *local rest frame* (a Lorentz frame where $T^{j0} = T^{0j} = 0$), has the form

$$T^{00} = \rho, \quad T^{jk} = P\delta^{jk} \tag{2.74a}$$

perfect-fluid stress-energy tensor

[Eq. (1.34) and associated discussion]. Here ρ is a short-hand notation for the energy density T^{00} (density of total mass-energy, including rest mass) as measured in the local rest frame, and the stress tensor T^{jk} in that frame is an isotropic pressure P. From this special form of $T^{\alpha\beta}$ in the local rest frame, one can derive the following geometric, frame-independent expression for the stress-energy tensor in terms of the 4-velocity $\vec{u}$ of the local rest frame (i.e., of the fluid itself), the metric tensor of spacetime $\mathbf{g}$, and the rest-frame energy density ρ and pressure P:

$$\boxed{T^{\alpha\beta} = (\rho + P)u^\alpha u^\beta + P g^{\alpha\beta}; \quad \text{i.e.,} \quad \mathbf{T} = (\rho + P)\vec{u} \otimes \vec{u} + P\mathbf{g}.} \tag{2.74b}$$

See Ex. 2.26.

In Sec. 13.8, we develop and explore the laws of relativistic fluid dynamics that follow from energy-momentum conservation $\vec{\nabla} \cdot \mathbf{T} = 0$ for this stress-energy tensor and from rest-mass conservation $\vec{\nabla} \cdot \vec{S}_{\rm rm} = 0$. By constructing the Newtonian limit of the relativistic laws, we shall deduce the nonrelativistic laws of fluid mechanics, which are the central theme of Part V. Notice, in particular, that the Newtonian limit ($P \ll \rho$, $u^0 \simeq 1$, $u^j \simeq v^j$) of the stress part of the stress-energy tensor (2.74b) is $T^{jk} = \rho v^j v^k + P \delta^{jk}$, which we met in Ex. 1.13.

Another example of a stress-energy tensor is that for the electromagnetic field, which takes the following form in Gaussian units (with $4\pi \to \mu_0 = 1/\epsilon_0$ in SI units):

electromagnetic stress-energy tensor

$$T^{\alpha\beta} = \frac{1}{4\pi}\left(F^{\alpha\mu}F^{\beta}{}_{\mu} - \frac{1}{4}g^{\alpha\beta}F^{\mu\nu}F_{\mu\nu}\right). \tag{2.75}$$

We explore this stress-energy tensor in Ex. 2.28.

EXERCISES

Exercise 2.25 *Example: Global Conservation of Energy in an Inertial Frame*
Consider the 4-dimensional parallelepiped $\mathcal{V}$ whose legs are $\Delta t \vec{e}_t$, $\Delta x \vec{e}_x$, $\Delta y \vec{e}_y$, $\Delta z \vec{e}_z$, where $(t, x, y, z) = (x^0, x^1, x^2, x^3)$ are the coordinates of some inertial frame. The boundary $\partial \mathcal{V}$ of this $\mathcal{V}$ has eight 3-dimensional "faces." Identify these faces, and write the integral $\int_{\partial \mathcal{V}} T^{0\beta} d\Sigma_\beta$ as the sum of contributions from each of them. According to the law of energy conservation, this sum must vanish. Explain the physical interpretation of each of the eight contributions to this energy conservation law. (See Ex. 2.23 for an analogous interpretation of charge conservation.)

Exercise 2.26 **Derivation and Example: Stress-Energy Tensor and Energy-Momentum Conservation for a Perfect Fluid
(a) Derive the frame-independent expression (2.74b) for the perfect fluid stress-energy tensor from its rest-frame components (2.74a).
(b) Explain why the projection of $\vec{\nabla} \cdot \mathbf{T} = 0$ along the fluid 4-velocity, $\vec{u} \cdot (\vec{\nabla} \cdot \mathbf{T}) = 0$, should represent energy conservation as viewed by the fluid itself. Show that this equation reduces to

$$\frac{d\rho}{d\tau} = -(\rho + P)\vec{\nabla} \cdot \vec{u}. \tag{2.76a}$$

With the aid of Eq. (2.65), bring this into the form

$$\frac{d(\rho V)}{d\tau} = -P\frac{dV}{d\tau}, \tag{2.76b}$$

where V is the 3-volume of some small fluid element as measured in the fluid's local rest frame. What are the physical interpretations of the left- and right-hand sides of this equation, and how is it related to the first law of thermodynamics?

(c) Read the discussion in Ex. 2.10 about the tensor $\mathbf{P} = \mathbf{g} + \vec{u} \otimes \vec{u}$ that projects into the 3-space of the fluid's rest frame. Explain why $P_{\mu\alpha}T^{\alpha\beta}{}_{;\beta} = 0$ should represent the law of force balance (momentum conservation) as seen by the fluid. Show that this equation reduces to

$$(\rho + P)\vec{a} = -\mathbf{P} \cdot \vec{\nabla} P, \tag{2.76c}$$

where $\vec{a} = d\vec{u}/d\tau$ is the fluid's 4-acceleration. This equation is a relativistic version of Newton's $\mathbf{F} = m\mathbf{a}$. Explain the physical meanings of the left- and right-hand sides. Infer that $\rho + P$ must be the fluid's inertial mass per unit volume. It is also the enthalpy per unit volume, including the contribution of rest mass; see Ex. 5.5 and Box 13.2.

Exercise 2.27 **Example: Inertial Mass per Unit Volume*
Suppose that some medium has a rest frame (unprimed frame) in which its energy flux and momentum density vanish, $T^{0j} = T^{j0} = 0$. Suppose that the medium moves in the x direction with speed very small compared to light, $v \ll 1$, as seen in a (primed) laboratory frame, and ignore factors of order v^2. The ratio of the medium's momentum density $G_{j'} = T^{j'0'}$ (as measured in the laboratory frame) to its velocity $v_i = v\delta_{ix}$ is called its total *inertial mass per unit volume* and is denoted ρ_{ji}^{inert}:

$$T^{j'0'} = \rho_{ji}^{\text{inert}} v_i. \tag{2.77}$$

In other words, ρ_{ji}^{inert} is the 3-dimensional tensor that gives the momentum density $G_{j'}$ when the medium's small velocity is put into its second slot.

(a) Using a Lorentz transformation from the medium's (unprimed) rest frame to the (primed) laboratory frame, show that

$$\rho_{ji}^{\text{inert}} = T^{00} \delta_{ji} + T_{ji}. \tag{2.78}$$

(b) Give a physical explanation of the contribution $T_{ji} v_i$ to the momentum density.

(c) Show that for a perfect fluid [Eq. (2.74b)] the inertial mass per unit volume is isotropic and has magnitude $\rho + P$, where ρ is the mass-energy density, and P is the pressure measured in the fluid's rest frame:

$$\boxed{\rho_{ji}^{\text{inert}} = (\rho + P)\delta_{ji}.} \tag{2.79}$$

See Ex. 2.26 for this inertial-mass role of $\rho + P$ in the law of force balance.

Exercise 2.28 **Example: Stress-Energy Tensor, and Energy-Momentum Conservation for the Electromagnetic Field*

(a) From Eqs. (2.75) and (2.45) compute the components of the electromagnetic stress-energy tensor in an inertial reference frame (in Gaussian units). Your answer should be the expressions given in electrodynamics textbooks:

$$T^{00} = \frac{\mathbf{E}^2 + \mathbf{B}^2}{8\pi}, \quad \mathbf{G} = T^{0j}\mathbf{e}_j = T^{j0}\mathbf{e}_j = \frac{\mathbf{E} \times \mathbf{B}}{4\pi},$$

$$T^{jk} = \frac{1}{8\pi}\left[(\mathbf{E}^2 + \mathbf{B}^2)\delta_{jk} - 2(E_j E_k + B_j B_k)\right]. \tag{2.80}$$

(In SI units, $4\pi \to \mu_0 = 1/\epsilon_0$.) See also Ex. 1.14 for an alternative derivation of the stress tensor T_{jk}.

(b) Show that the divergence of the stress-energy tensor (2.75) is given by

$$T^{\mu\nu}{}_{;\nu} = \frac{1}{4\pi}(F^{\mu\alpha}{}_{;\nu}F^{\nu}{}_{\alpha} + F^{\mu\alpha}F^{\nu}{}_{\alpha;\nu} - \frac{1}{2}F_{\alpha\beta}{}^{;\mu}F^{\alpha\beta}). \tag{2.81a}$$

(c) Combine this with Maxwell's equations (2.48) to show that

$$\vec{\nabla} \cdot \mathbf{T} = -\mathbf{F}(\_, \vec{J}); \quad \text{i.e., } T^{\alpha\beta}{}_{;\beta} = -F^{\alpha\beta}J_\beta. \tag{2.81b}$$

(d) The matter that carries the electric charge and current can exchange energy and momentum with the electromagnetic field. Explain why Eq. (2.81b) is the rate per unit volume at which that matter feeds 4-momentum into the electromagnetic field, and conversely, $+F^{\alpha\mu}J_\mu$ is the rate per unit volume at which the electromagnetic field feeds 4-momentum into the matter. Show, further, that (as viewed in any reference frame) the time and space components of this quantity are

$$\frac{d\mathcal{E}_{\text{matter}}}{dt\,dV} = F^{0j}J_j = \mathbf{E} \cdot \mathbf{j}, \quad \frac{d\mathbf{p}_{\text{matter}}}{dt\,dV} = \rho_e \mathbf{E} + \mathbf{j} \times \mathbf{B}, \tag{2.81c}$$

where ρ_e is charge density, and $\mathbf{j}$ is current density [Eq. (2.49)]. The first of these equations describes ohmic heating of the matter by the electric field, and the second gives the Lorentz force per unit volume on the matter (cf. Ex. 1.14b).

Bibliographic Note

For an inspiring taste of the history of special relativity, see the original papers by Einstein, Lorentz, and Minkowski, translated into English and archived in Lorentz et al. (1923).

Early relativity textbooks [see the bibliography in Jackson (1999, pp. 566–567)] emphasized the transformation properties of physical quantities, in going from one inertial frame to another, rather than their roles as frame-invariant geometric objects. Minkowski (1908) introduced geometric thinking, but only in recent decades—in large measure due to the influence of John Wheeler—has the geometric viewpoint gained ascendancy.

In our opinion, the best elementary introduction to special relativity is the first edition of Taylor and Wheeler (1966); the more ponderous second edition (Taylor and Wheeler, 1992) is also good. At an intermediate level we strongly recommend the special relativity portions of Hartle (2003).

At a more advanced level, comparable to this chapter, we recommend Goldstein, Poole, and Safko (2002) and the special relativity sections of Misner, Thorne, and Wheeler (1973), Carroll (2004), and Schutz (2009).

These all adopt the geometric viewpoint that we espouse. In this chapter, so far as possible, we have minimized the proliferation of mathematical concepts (avoiding, e.g., differential forms and dual bases). By contrast, the other advanced treatments cited above embrace the richer mathematics.

Much less geometric than these references but still good, in our view, are the special relativity sections of popular electrodynamics texts: Griffiths (1999) at an intermediate level and Jackson (1999) at a more advanced level. We recommend avoiding special relativity treatments that use imaginary time and thereby obfuscate (e.g., earlier editions of Goldstein and of Jackson, and also the more modern and otherwise excellent Zangwill (2013)).

PART II

STATISTICAL PHYSICS

In this second part of the book, we study aspects of classical statistical physics that every physicist should know but that are not usually treated in elementary thermodynamics courses. Our study lays the microphysical (particle-scale) foundations for the continuum physics of Parts III–VII, and it elucidates the intimate connections between relativistic statistical physics and the Newtonian theory, and between quantum statistical physics and the classical theory. Our treatment is both Newtonian and relativistic. Readers who prefer a solely Newtonian treatment can skip the (rather few) relativistic sections. Throughout, we presume that readers are familiar with elementary thermodynamics but not with other aspects of statistical physics.

In Chap. 3, we study *kinetic theory*—the simplest of all formalisms for analyzing systems of huge numbers of particles (e.g., molecules of air, neutrons diffusing through a nuclear reactor, or photons produced in the big-bang origin of the universe). In kinetic theory, the key concept is the distribution function, or number density of particles in phase space, $\mathcal{N}$, that is, the number of particles of some species (e.g., electrons) per unit of physical space and of momentum space. In special relativity, despite first appearances, this $\mathcal{N}$ turns out to be a geometric, reference-frame-independent entity (a scalar field in phase space). This $\mathcal{N}$ and the frame-independent laws it obeys provide us with a means for computing, from microphysics, the macroscopic quantities of continuum physics: mass density, thermal energy density, pressure, equations of state, thermal and electrical conductivities, viscosities, diffusion coefficients, ...

In Chap. 4, we develop the foundations of *statistical mechanics*. Here our statistical study is more sophisticated than in kinetic theory: we deal with ensembles of physical systems. Each ensemble is a (conceptual) collection of a huge number of physical systems that are identical in the sense that they all have the same degrees of freedom, but different in that their degrees of freedom may be in different physical states. For example, the systems in an ensemble might be balloons that are each filled with 10^{23} air molecules so each is describable by 3×10^{23} spatial coordinates (the $\{x, y, z\}$ of all the molecules) and 3×10^{23} momentum coordinates (the $\{p_x, p_y, p_z\}$ of all the molecules). The state of one of the balloons is fully described, then, by 6×10^{23} numbers. We introduce a distribution function $\mathcal{N}$ that is a function of these 6×10^{23} different coordinates (i.e., it is defined in a phase space with 6×10^{23} dimensions). This distribution function tells us how many systems (balloons) in our ensemble lie in a unit volume of that phase space. Using this distribution function, we study such issues as the statistical meaning of entropy, the relationship between entropy and information, the statistical origin of the second law of thermodynamics, the statistical meaning of "thermal equilibrium," and the evolution of ensembles into thermal equilibrium. Our applications include derivations of the Fermi-Dirac distribution for fermions in thermal equilibrium and the Bose-Einstein distribution for bosons, a study of Bose-Einstein condensation in a dilute gas, and explorations of the meaning and role of entropy in gases, black holes, and the universe as a whole.

In Chap. 5, we use the tools of statistical mechanics to study *statistical thermodynamics:* ensembles of systems that are in or near thermal equilibrium (also called statistical equilibrium). Using statistical mechanics, we derive the laws of thermodynamics, and we learn how to use thermodynamic and statistical mechanical tools, hand in hand, to study not only equilibria but also the probabilities for random, spontaneous fluctuations away from equilibrium. Among the applications we study are: (i) chemical and particle reactions, such as ionization equilibrium in a hot gas, and electron-positron pair formation in a still hotter gas and (ii) phase transitions, such as the freezing, melting, vaporization, and condensation of water. We focus special attention on a ferromagnetic phase transition, in which the magnetic moments of atoms spontaneously align with one another as iron is cooled, using it to illustrate two elegant and powerful techniques of statistical physics: the renormalization group and Monte Carlo methods.

In Chap. 6, we develop the theory of random processes (a modern, mathematical component of which is the theory of stochastic differential equations). Here we study the dynamical evolution of processes that are influenced by a huge number of factors over which we have little control and little knowledge, except of their statistical properties. One example is the Brownian motion of a dust particle being buffeted by air molecules; another is the motion of a pendulum used, say, in a gravitational-wave interferometer, where one monitors that motion so accurately that one can see the influences of seismic vibrations and of fluctuating thermal (Nyquist) forces in the pendulum's suspension wire. The position of such a dust particle or pendulum cannot be predicted as a function of time, but one can compute the probability that it will evolve in a given manner. The theory of random processes is a theory of the evolution of the position's probability distribution (and the probability distribution for any other entity driven by random, fluctuating influences). Among the random-process concepts we study are spectral densities, correlation functions, the Fokker-Planck equation (which governs the evolution of probability distributions), and the fluctuation-dissipation theorem (which says that, associated with any kind of friction, there must be fluctuating forces whose statistical properties are determined by the strength of the friction and the temperature of the entities that produce the friction).

The theory of random processes, as treated in Chap. 6, also includes the theory of signals and noise. At first sight this undeniably important topic, which lies at the heart of experimental and observational science, might seem outside the scope of this book. However, we shall discover that it is intimately connected to statistical physics and that similar principles to those used to describe, say, Brownian motion, are appropriate when thinking about, for example, how to detect the electronic signal of a rare particle event against a strong and random background. We study, for example, techniques for extracting weak signals from noisy data by filtering the data, and the limits that noise places on the accuracies of physics experiments and on the reliability of communications channels.

CHAPTER THREE

Kinetic Theory

The gaseous condition is exemplified in the soirée, where the members rush about confusedly, and the only communication is during a collision, which in some instances may be prolonged by button-holing.

JAMES CLERK MAXWELL (1873)

3.1 Overview

In this chapter, we study kinetic theory, the simplest of all branches of statistical physics. Kinetic theory deals with the statistical distribution of a "gas" made from a huge number of "particles" that travel freely, without collisions, for distances (*mean free paths*) long compared to their sizes.

Examples of particles (*italicized*) and phenomena that can be studied via kinetic theory are:

- Whether *neutrons* in a nuclear reactor can survive long enough to maintain a nuclear chain reaction and keep the reactor hot.
- How *galaxies*, formed in the early universe, congregate into clusters as the universe expands.
- How spiral structure develops in the distribution of a galaxy's *stars*.
- How, deep inside a white-dwarf star, relativistic degeneracy influences the equation of state of the star's *electrons and protons*.
- How a supernova explosion affects the evolution of the density and temperature of *interstellar molecules*.
- How anisotropies in the expansion of the universe affect the temperature distribution of the *cosmic microwave photons*—the remnants of the big bang.
- How changes of a metal's temperature affect its thermal and electrical conductivity (with the heat and current carried by *electrons*).

Most of these applications involve particle speeds small compared to that of light and so can be studied with Newtonian theory, but some involve speeds near or at the speed of light and require relativity. Accordingly, we develop both versions of the theory, Newtonian and relativistic, and demonstrate that the Newtonian theory is the low-speed limit of the relativistic theory. As is discussed in the Readers' Guide

> **BOX 3.1. READERS' GUIDE**
>
> - This chapter develops nonrelativistic (Newtonian) kinetic theory and also the relativistic theory. Sections and exercises labeled [N] are Newtonian, and those labeled [R] are relativistic. The [N] material can be read without the [R] material, but the [R] material requires the [N] material as a foundation. The [R] material is all Track Two.
> - This chapter relies on the geometric viewpoint about physics developed in Chap. 1 for Newtonian physics and in Chap. 2 for relativistic physics. It especially relies on
>   - Secs. 1.4 and 1.5.2 on Newtonian particle kinetics,
>   - Secs. 2.4.1 and 2.6 on relativistic particle kinetics,
>   - Sec. 1.8 for the Newtonian conservation laws for particles,
>   - Sec. 2.12.3 for the relativistic number-flux 4-vector and its conservation law,
>   - Sec. 1.9 on the Newtonian stress tensor and its role in conservation laws for momentum,
>   - Sec. 2.13 on the relativistic stress-energy tensor and its role in conservation laws for 4-momentum, and
>   - Sec. 1.10 on aspects of relativity theory that Newtonian readers will need.
> - The Newtonian parts of this chapter are a crucial foundation for the remainder of Part II (Statistical Physics) of this book, for small portions of Part V (Fluid Dynamics; especially equations of state, the origin of viscosity, and the diffusion of heat in fluids), and for half of Part VI (Plasma Physics; Chaps. 22 and 23).

(Box 3.1), the relativistic material is all Track Two and can be skipped by readers who are focusing on the (nonrelativistic) Newtonian theory.

We begin in Sec. 3.2 by introducing the concepts of momentum space, phase space (the union of physical space and momentum space), and the distribution function (number density of particles in phase space). We meet several different versions of the distribution function, all equivalent, but each designed to optimize conceptual thinking or computations in a particular arena (e.g., photons, plasma physics, and the interface with quantum theory). In Sec. 3.3, we study the distribution functions that characterize systems of particles in thermal equilibrium. There are three such equilibrium distributions: one for quantum mechanical particles with half-integral spin (fermions), another for quantum particles with integral spin (bosons), and a third for classical particles. As special applications, we derive the Maxwell velocity distribution for low-speed, classical particles (Ex. 3.4) and its high-speed relativistic analog (Ex. 3.5 and Fig. 3.6 below), and we compute the effects of observers' motions on their measurements of the cosmic microwave radiation created in our universe's big-bang origin (Ex. 3.7). In Sec. 3.4, we learn how to compute macroscopic, physical-space quantities (particle density and flux, energy density, stress tensor, stress-energy tensor, ...) by integrating over the momentum portion of phase space. In Sec. 3.5, we show that, if the distribution function is isotropic in momentum space, in some reference frame, then on macroscopic scales the particles constitute a perfect fluid. We use our momentum-space integrals to evaluate the equations of state of various kinds of perfect fluids: a nonrelativistic hydrogen gas in both the classical, nondegenerate regime and the regime of electron degeneracy (Sec. 3.5.2), a relativistically degenerate gas (Sec. 3.5.4), and a photon gas (Sec. 3.5.5). We use our results to discuss the physical nature of matter as a function of density and temperature (see Fig. 3.7).

In Sec. 3.6, we study the evolution of the distribution function, as described by Liouville's theorem and by the associated collisionless Boltzmann equation when collisions between particles are unimportant, and by the Boltzmann transport equation when collisions are significant. We use a simple variant of these evolution laws to study the heating of Earth by the Sun, and the key role played by the greenhouse effect (Ex. 3.15). Finally, in Sec. 3.7, we learn how to use the Boltzmann transport equation to compute the transport coefficients (diffusion coefficient, electrical conductivity, thermal conductivity, and viscosity) that describe the diffusive transport of particles, charge, energy, and momentum through a gas of particles that collide frequently. We then use the Boltzmann transport equation to study a chain reaction in a nuclear reactor (Ex. 3.21).

Readers who feel overwhelmed by the enormous amount and variety of applications in this chapter (and throughout this book) should remember the authors' goals: We want readers to learn the fundamental concepts of kinetic theory (and other topics in this book), and we want them to meet a variety of applications, so they will understand how the fundamental concepts are used. However, we do not expect readers to become expert in or even remember all these applications. To do so would require much more time and effort than most readers can afford or should expend.

3.2 Phase Space and Distribution Function

3.2.1 Newtonian Number Density in Phase Space, $\mathcal{N}$

In Newtonian, 3-dimensional space (*physical space*), consider a particle with rest mass m that moves along a path $\mathbf{x}(t)$ as universal time t passes (Fig. 3.1a). The particle's time-varying velocity and momentum are $\mathbf{v}(t) = d\mathbf{x}/dt$ and $\mathbf{p}(t) = m\mathbf{v}$. The path $\mathbf{x}(t)$ is a curve in the physical space, and the momentum $\mathbf{p}(t)$ is a time-varying, coordinate-independent vector in the physical space.

physical space

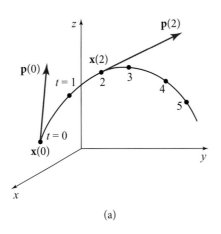
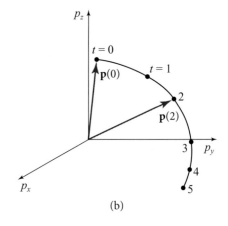

FIGURE 3.1 (a) Euclidean physical space, in which a particle moves along a curve $\mathbf{x}(t)$ that is parameterized by universal time t. In this space, the particle's momentum $\mathbf{p}(t)$ is a vector tangent to the curve. (b) Momentum space, in which the particle's momentum vector $\mathbf{p}$ is placed, unchanged, with its tail at the origin. As time passes, the momentum's tip sweeps out the indicated curve $\mathbf{p}(t)$.

momentum space

It is useful to introduce an auxiliary 3-dimensional space, called *momentum space*, in which we place the tail of $\mathbf{p}(t)$ at the origin. As time passes, the tip of $\mathbf{p}(t)$ sweeps out a curve in momentum space (Fig. 3.1b). This momentum space is "secondary" in the sense that it relies for its existence on the physical space of Fig. 3.1a. Any Cartesian coordinate system of physical space, in which the location $\mathbf{x}(t)$ of the particle has coordinates $\{x, y, z\}$, induces in momentum space a corresponding coordinate system $\{p_x, p_y, p_z\}$. The 3-dimensional physical space and 3-dimensional momentum space together constitute a 6-dimensional *phase space*, with coordinates $\{x, y, z, p_x, p_y, p_z\}$.

phase space

In this chapter, we study a collection of a very large number of identical particles (all with the same rest mass m).[1] As tools for this study, consider a tiny 3-dimensional volume $d\mathcal{V}_x$ centered on some location $\mathbf{x}$ in physical space and a tiny 3-dimensional volume $d\mathcal{V}_p$ centered on location $\mathbf{p}$ in momentum space. Together these make up a tiny 6-dimensional volume

$$d^2\mathcal{V} \equiv d\mathcal{V}_x d\mathcal{V}_p. \tag{3.1}$$

In any Cartesian coordinate system, we can think of $d\mathcal{V}_x$ as being a tiny cube located at (x, y, z) and having edge lengths dx, dy, dz, and similarly for $d\mathcal{V}_p$. Then, as computed in this coordinate system, these tiny volumes are

$$d\mathcal{V}_x = dx\,dy\,dz, \quad d\mathcal{V}_p = dp_x\,dp_y\,dp_z, \quad d^2\mathcal{V} = dx\,dy\,dz\,dp_x\,dp_y\,dp_z. \tag{3.2}$$

---

1. In Ex. 3.2 and Box 3.2, we extend kinetic theory to particles with a range of rest masses.

Denote by dN the number of particles (all with rest mass m) that reside inside $d^2\mathcal{V}$ in phase space (at some moment of time t). Stated more fully: dN is the number of particles that, at time t, are located in the 3-volume $d\mathcal{V}_x$ centered on the location $\mathbf{x}$ in physical space and that also have momentum vectors whose tips at time t lie in the 3-volume $d\mathcal{V}_p$ centered on location $\mathbf{p}$ in momentum space. Denote by

$$\boxed{\mathcal{N}(\mathbf{x}, \mathbf{p}, t) \equiv \frac{dN}{d^2\mathcal{V}} = \frac{dN}{d\mathcal{V}_x \, d\mathcal{V}_p}} \qquad (3.3)$$

distribution function

the *number density of particles at location* $(\mathbf{x}, \mathbf{p})$ *in phase space at time* t. This is also called the *distribution function*.

This distribution function is kinetic theory's principal tool for describing any collection of a large number of identical particles.

In Newtonian theory, the volumes $d\mathcal{V}_x$ and $d\mathcal{V}_p$ occupied by our collection of dN particles are independent of the reference frame that we use to view them. Not so in relativity theory: $d\mathcal{V}_x$ undergoes a Lorentz contraction when one views it from a moving frame, and $d\mathcal{V}_p$ also changes; but (as we shall see in Sec. 3.2.2) their product $d^2\mathcal{V} = d\mathcal{V}_x d\mathcal{V}_p$ is the same in all frames. Therefore, in both Newtonian theory and relativity theory, the distribution function $\mathcal{N} = dN/d^2\mathcal{V}$ is independent of reference frame, and also, of course, independent of any choice of coordinates. It is a coordinate-independent scalar in phase space.

3.2.2 Relativistic Number Density in Phase Space, $\mathcal{N}$ [R] [T2]

SPACETIME

We define the special relativistic distribution function in precisely the same way as the nonrelativistic one, $\mathcal{N}(\mathbf{x}, \mathbf{p}, t) \equiv dN/d^2\mathcal{V} = dN/d\mathcal{V}_x d\mathcal{V}_p$, except that now $\mathbf{p}$ is the relativistic momentum ($\mathbf{p} = m\mathbf{v}/\sqrt{1-\mathbf{v}^2}$ if the particle has nonzero rest mass m). This definition of $\mathcal{N}$ appears, at first sight, to be frame dependent, since the physical 3-volume $d\mathcal{V}_x$ and momentum 3-volume $d\mathcal{V}_p$ do not even exist until we have selected a specific reference frame. In other words, this definition appears to violate our insistence that relativistic physical quantities be described by frame-independent geometric objects that live in 4-dimensional spacetime. In fact, the distribution function defined in this way *is* frame independent, though it does not look so. To elucidate this, we shall develop carefully and somewhat slowly the 4-dimensional spacetime ideas that underlie this relativistic distribution function.

Consider, as shown in Fig. 3.2a, a classical particle with rest mass m moving through spacetime along a world line $\mathcal{P}(\zeta)$, or equivalently $\vec{x}(\zeta)$, where ζ is an affine parameter related to the particle's 4-momentum by

$$\vec{p} = d\vec{x}/d\zeta \qquad (3.4a)$$

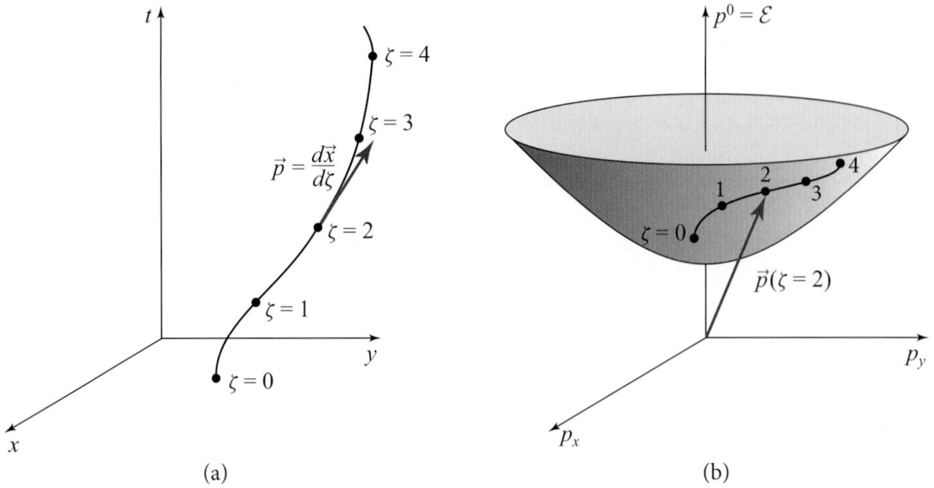

FIGURE 3.2 (a) The world line $\vec{x}(\zeta)$ of a particle in spacetime (with one spatial coordinate, z, suppressed), parameterized by a parameter ζ that is related to the particle's 4-momentum by $\vec{p} = d\vec{x}/d\zeta$. (b) The trajectory of the particle in momentum space. The particle's 4-momentum is confined to the mass hyperboloid, $\vec{p}^2 = -m^2$ (also known as the mass shell).

[as discussed following Eq. (2.10)]. If the particle has nonzero rest mass, then its 4-velocity $\vec{u}$ and proper time τ are related to its 4-momentum and affine parameter by

$$\vec{p} = m\vec{u}, \qquad \zeta = \tau/m \tag{3.4b}$$

[Eqs. (2.10) and (2.11)], and we can parameterize the world line by either τ or ζ. If the particle has zero rest mass, then its world line is null, and τ does not change along it, so we have no choice but to use ζ as the world line's parameter.

MOMENTUM SPACE AND MASS HYPERBOLOID

4-dimensional momentum space

The particle can be thought of not only as living in 4-dimensional spacetime (Fig. 3.2a), but also as living in a 4-dimensional momentum space (Fig. 3.2b). Momentum space, like spacetime, is a geometric, coordinate-independent concept: each point in momentum space corresponds to a specific 4-momentum $\vec{p}$. The tail of the vector $\vec{p}$ sits at the origin of momentum space, and its tip sits at the point representing $\vec{p}$. The momentum-space diagram drawn in Fig. 3.2b has as its coordinate axes the components (p^0, $p^1 = p_1 \equiv p_x$, $p^2 = p_2 \equiv p_y$, $p^3 = p_3 \equiv p_z$) of the 4-momentum as measured in some arbitrary inertial frame. Because the squared length of the 4-momentum is always $-m^2$,

$$\vec{p} \cdot \vec{p} = -(p^0)^2 + (p_x)^2 + (p_y)^2 + (p_z)^2 = -m^2, \tag{3.4c}$$

mass hyperboloid

the particle's 4-momentum (the tip of the 4-vector $\vec{p}$) is confined to a hyperboloid in momentum space. This *mass hyperboloid* requires no coordinates for its existence; it is the frame-independent set of points in momentum space for which $\vec{p} \cdot \vec{p} = -m^2$. If the particle has zero rest mass, then $\vec{p}$ is null, and the mass hyperboloid is a cone with vertex at the origin of momentum space. As in Chap. 2, we often denote the particle's energy p^0 by

$$\mathcal{E} \equiv p^0 \qquad (3.4d)$$

(with the $\mathcal{E}$ in script font to distinguish it from the energy $E = \mathcal{E} - m$ with rest mass removed and its nonrelativistic limit $E = \frac{1}{2}mv^2$), and we embody the particle's spatial momentum in the 3-vector $\mathbf{p} = p_x \mathbf{e}_x + p_y \mathbf{e}_y + p_z \mathbf{e}_z$. Therefore, we rewrite the mass-hyperboloid relation (3.4c) as

$$\mathcal{E}^2 = m^2 + |\mathbf{p}|^2. \qquad (3.4e)$$

If no forces act on the particle, then its momentum is conserved, and its location in momentum space remains fixed. A force (e.g., due to an electromagnetic field) pushes the particle's 4-momentum along some curve in momentum space that lies on the mass hyperboloid. If we parameterize that curve by the same parameter ζ as we use in spacetime, then the particle's trajectory in momentum space can be written abstractly as $\vec{p}(\zeta)$. Such a trajectory is shown in Fig. 3.2b.

Because the mass hyperboloid is 3-dimensional, we can characterize the particle's location on it by just three coordinates rather than four. We typically use as those coordinates the spatial components of the particle's 4-momentum, (p_x, p_y, p_z), or the spatial momentum vector $\mathbf{p}$ as measured in some specific (but usually arbitrary) inertial frame.

PHASE SPACE

Momentum space and spacetime, taken together, constitute the relativistic *phase space*. We can regard phase space as 8-dimensional (four spacetime dimensions plus four momentum space dimensions). Alternatively, if we think of the 4-momentum as confined to the 3-dimensional mass hyperboloid, then we can regard phase space as 7-dimensional. This 7- or 8-dimensional phase space, by contrast with the non-relativistic 6-dimensional phase space, is frame independent. No coordinates or reference frame are actually needed to define spacetime and explore its properties, and none are needed to define and explore 4-momentum space or the mass hyperboloid—though inertial (Lorentz) coordinates are often helpful in practical situations.

phase space

VOLUMES IN PHASE SPACE AND DISTRIBUTION FUNCTION

Now turn attention from an individual particle to a collection of a huge number of identical particles, each with the same rest mass m, and allow m to be finite or zero (it does not matter which). Examine those particles that pass close to a specific event $\mathcal{P}$ (also denoted $\vec{x}$) in spacetime; and *examine them from the viewpoint of a specific observer, who lives in a specific inertial reference frame.* Figure 3.3a is a spacetime diagram drawn in that observer's frame. As seen in that frame, the event $\mathcal{P}$ occurs at time t and spatial location (x, y, z).

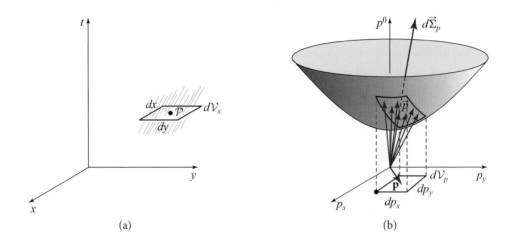

FIGURE 3.3 Definition of the distribution function from the viewpoint of a specific observer in a specific inertial reference frame, whose coordinate axes are used in these drawings. (a) At the event $\mathcal{P}$, the observer selects a 3-volume $d\mathcal{V}_x$ and focuses on the set $\mathcal{S}$ of particles that lie in $d\mathcal{V}_x$. (b) These particles have momenta lying in a region of the mass hyperboloid that is centered on $\vec{p}$ and has 3-momentum volume $d\mathcal{V}_p$. If dN is the number of particles in that set $\mathcal{S}$, then $\mathcal{N}(\mathcal{P}, \vec{p}) \equiv dN/d\mathcal{V}_x d\mathcal{V}_p$.

We ask the observer, at the time t of the chosen event, to define the distribution function $\mathcal{N}$ in identically the same way as in Newtonian theory, except that **p** is the relativistic spatial momentum $\mathbf{p} = m\mathbf{v}/\sqrt{1-v^2}$ instead of the nonrelativistic $\mathbf{p} = m\mathbf{v}$. Specifically, the observer, in her inertial frame, chooses a tiny 3-volume

$$d\mathcal{V}_x = dx\, dy\, dz \tag{3.5a}$$

centered on location $\mathcal{P}$ (little horizontal rectangle shown in Fig. 3.3a) and a tiny 3-volume

$$d\mathcal{V}_p = dp_x\, dp_y\, dp_z \tag{3.5b}$$

centered on **p** in momentum space (little rectangle in the p_x-p_y plane in Fig. 3.3b). Ask the observer to focus on the set $\mathcal{S}$ of particles that lie in $d\mathcal{V}_x$ and have spatial momenta in $d\mathcal{V}_p$ (Fig. 3.3). If there are dN particles in this set $\mathcal{S}$, then the observer will identify

relativistic distribution function

$$\mathcal{N} \equiv \frac{dN}{d\mathcal{V}_x d\mathcal{V}_p} \equiv \frac{dN}{d^2\mathcal{V}} \tag{3.6}$$

as the number density of particles in phase space or *distribution function*.

Notice in Fig. 3.3b that the *4-momenta* of the particles in $\mathcal{S}$ have their tails at the origin of momentum space (as by definition do all 4-momenta) and have their tips in a tiny rectangular box on the mass hyperboloid—a box centered on the 4-momentum $\vec{p}$ whose spatial part is **p** and temporal part is $p^0 = \mathcal{E} = \sqrt{m^2 + \mathbf{p}^2}$.

The momentum volume element $d\mathcal{V}_p$ is the projection of that mass-hyperboloid box onto the horizontal (p_x, p_y, p_z) plane in momentum space. [The mass-hyperboloid box itself can be thought of as a (frame-independent) vectorial 3-volume $d\vec{\Sigma}_p$—the momentum-space version of the vectorial 3-volume introduced in Sec. 2.12.1; see below.]

The number density $\mathcal{N}$ depends on the location $\mathcal{P}$ in spacetime of the 3-volume $d\mathcal{V}_x$ and on the 4-momentum $\vec{p}$ about which the momentum volume on the mass hyperboloid is centered: $\mathcal{N} = \mathcal{N}(\mathcal{P}, \vec{p})$. From the chosen observer's viewpoint, it can be regarded as a function of time t and spatial location $\mathbf{x}$ (the coordinates of $\mathcal{P}$) and of spatial momentum $\mathbf{p}$.

At first sight, one might expect $\mathcal{N}$ to depend also on the inertial reference frame used in its definition (i.e., on the 4-velocity of the observer). If this were the case (i.e., if $\mathcal{N}$ at fixed $\mathcal{P}$ and $\vec{p}$ were different when computed by the above prescription using different inertial frames), then we would feel compelled to seek some other object—one that is frame-independent—to serve as our foundation for kinetic theory. This is because the principle of relativity insists that all fundamental physical laws should be expressible in frame-independent language.

Fortunately, the distribution function (3.6) is frame independent by itself: it is a frame-independent scalar field in phase space, so we need seek no further.

PROOF OF FRAME INDEPENDENCE OF $\mathcal{N} = dN/d^2\mathcal{V}$

To prove the frame independence of $\mathcal{N}$, we shall consider the frame dependence of the spatial 3-volume $d\mathcal{V}_x$, then the frame dependence of the momentum 3-volume $d\mathcal{V}_p$, and finally the frame dependence of their product $d^2\mathcal{V} = d\mathcal{V}_x d\mathcal{V}_p$ and thence of the distribution function $\mathcal{N} = dN/d^2\mathcal{V}$.

The thing that identifies the 3-volume $d\mathcal{V}_x$ and 3-momentum $d\mathcal{V}_p$ is the set of particles $\mathcal{S}$. We select that set once and for all and hold it fixed, and correspondingly, the number of particles dN in the set is fixed. Moreover, we assume that the particles' rest mass m is nonzero and shall deal with the zero-rest-mass case at the end by taking the limit $m \to 0$. Then there is a preferred frame in which to observe the particles $\mathcal{S}$: their own rest frame, which we identify by a prime.

In their rest frame and at a chosen event $\mathcal{P}$, the particles $\mathcal{S}$ occupy the interior of some box with imaginary walls that has some 3-volume $d\mathcal{V}_{x'}$. As seen in some other "laboratory" frame, their box has a Lorentz-contracted volume $d\mathcal{V}_x = \sqrt{1-v^2}\, d\mathcal{V}_{x'}$. Here v is their speed as seen in the laboratory frame. The Lorentz-contraction factor is related to the particles' energy, as measured in the laboratory frame, by $\sqrt{1-v^2} = m/\mathcal{E}$, and therefore $\mathcal{E}\, d\mathcal{V}_x = m\, d\mathcal{V}_{x'}$. The right-hand side is a frame-independent constant m times a well-defined number that everyone can agree on: the particles' rest-frame volume $d\mathcal{V}_{x'}$, i.e.,

$$\boxed{\mathcal{E}\, d\mathcal{V}_x = \text{(a frame-independent quantity)}.} \qquad (3.7a)$$

Thus, the spatial volume $d\mathcal{V}_x$ occupied by the particles is frame dependent, and their energy $\mathcal{E}$ is frame dependent, but the product of the two is independent of reference frame.

Turn now to the frame dependence of the particles' 3-volume $d\mathcal{V}_p$. As one sees from Fig. 3.3b, $d\mathcal{V}_p$ is the projection of the frame-independent mass-hyperboloid region $d\vec{\Sigma}_p$ onto the laboratory's xyz 3-space. Equivalently, it is the time component $d\Sigma_p^0$ of $d\vec{\Sigma}_p$. Now, the 4-vector $d\vec{\Sigma}_p$, like the 4-momentum $\vec{p}$, is orthogonal to the mass hyperboloid at the common point where they intersect it, and therefore $d\vec{\Sigma}_p$ is parallel to $\vec{p}$. This means that, when one goes from one reference frame to another, the time components of these two vectors will grow or shrink in the same manner: $d\vec{\Sigma}_p^0 = d\mathcal{V}_p$ is proportional to $p^0 = \mathcal{E}$, so their ratio must be frame independent:

$$\boxed{\frac{d\mathcal{V}_p}{\mathcal{E}} = \text{(a frame-independent quantity)}.} \tag{3.7b}$$

(If this sophisticated argument seems too slippery to you, then you can develop an alternative, more elementary proof using simpler 2-dimensional spacetime diagrams: Ex. 3.1.)

By taking the product of Eqs. (3.7a) and (3.7b) we see that for our chosen set of particles $\mathcal{S}$,

$$d\mathcal{V}_x d\mathcal{V}_p = d^2\mathcal{V} = \text{(a frame-independent quantity)}; \tag{3.7c}$$

and since the number of particles in the set, dN, is obviously frame-independent, we conclude that

frame independence of relativistic distribution function

$$\boxed{\mathcal{N} = \frac{dN}{d\mathcal{V}_x d\mathcal{V}_p} \equiv \frac{dN}{d^2\mathcal{V}} = \text{(a frame-independent quantity)}.} \tag{3.8}$$

Although we assumed nonzero rest mass ($m \neq 0$) in our derivation, the conclusions that $\mathcal{E}d\mathcal{V}_x$ and $d\mathcal{V}_p/\mathcal{E}$ are frame independent continue to hold if we take the limit as $m \to 0$ and the 4-momenta become null. Correspondingly, Eqs. (3.7a)–(3.8) are valid for particles with zero as well as nonzero rest mass.

EXERCISES

Exercise 3.1 *Derivation and Practice: Frame Dependences of $d\mathcal{V}_x$ and $d\mathcal{V}_p$*
Use the 2-dimensional spacetime diagrams of Fig. 3.4 to show that $\mathcal{E}d\mathcal{V}_x$ and $d\mathcal{V}_p/\mathcal{E}$ are frame independent [Eqs. (3.7a) and (3.7b)].

Exercise 3.2 **Example: Distribution Function for Particles with a Range of Rest Masses*
A galaxy such as our Milky Way contains $\sim 10^{12}$ stars—easily enough to permit a kinetic-theory description of their distribution; each star contains so many atoms ($\sim 10^{56}$) that the masses of the stars can be regarded as continuously distributed, not

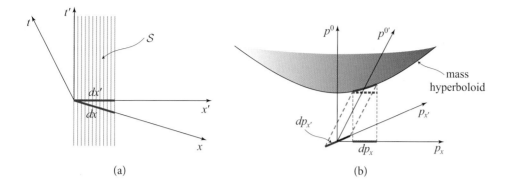

FIGURE 3.4 (a) Spacetime diagram drawn from the viewpoint of the (primed) rest frame of the particles $\mathcal{S}$ for the special case where the laboratory frame moves in the $-x'$ direction with respect to them. (b) Momentum-space diagram drawn from viewpoint of the unprimed observer.

discrete. Almost everywhere in a galaxy, the stars move with speeds small compared to light, but deep in the cores of most galaxies there resides a massive black hole near which the stars move with relativistic speeds. In this exercise we explore the foundations for treating such a system: "particles" with continuously distributed rest masses and relativistic speeds.

(a) For a subset $\mathcal{S}$ of particles like that of Fig. 3.3 and associated discussion, but with a range of rest masses dm centered on some value m, introduce the phase-space volume $d^2\mathcal{V} \equiv d\mathcal{V}_x d\mathcal{V}_p dm$ that the particles $\mathcal{S}$ occupy. Explain why this occupied volume is frame invariant.

(b) Show that this invariant occupied volume can be rewritten as $d^2\mathcal{V} = (d\mathcal{V}_x \, \mathcal{E}/m)(d\mathcal{V}_p d\mathcal{E}) = (d\mathcal{V}_x \, \mathcal{E}/m)(dp^0 dp^x dp^y dp^z)$. Explain the physical meaning of each term in parentheses, and show that each is frame invariant.

If the number of particles in the set $\mathcal{S}$ is dN, then we define the frame-invariant distribution function by

$$\mathcal{N} \equiv \frac{dN}{d^2\mathcal{V}} = \frac{dN}{d\mathcal{V}_x d\mathcal{V}_p dm}. \tag{3.9}$$

This is a function of location $\mathcal{P}$ in 4-dimensional spacetime and location $\vec{p}$ in 4-dimensional momentum space (not confined to the mass hyperboloid), and thus a function of location in 8-dimensional phase space. We explore the evolution of this distribution function in Box 3.2 (near the end of Sec. 3.6).

3.2.3 Distribution Function $f(\mathbf{x}, \mathbf{v}, t)$ for Particles in a Plasma

The normalization that one uses for the distribution function is arbitrary: renormalize $\mathcal{N}$ by multiplying with any constant, and $\mathcal{N}$ will still be a geometric, coordinate-independent, and frame-independent quantity and will still contain the same information as before. In this book, we use several renormalized versions of $\mathcal{N}$, depending on the situation. We now introduce them, beginning with the version used in plasma physics.

In Part VI, when dealing with nonrelativistic plasmas (collections of electrons and ions that have speeds small compared to light), we regard the distribution function as depending on time t, location $\mathbf{x}$ in Euclidean space, and velocity $\mathbf{v}$ (instead of momentum $\mathbf{p} = m\mathbf{v}$), and we denote it by[2]

plasma distribution function

$$f(t, \mathbf{x}, \mathbf{v}) \equiv \frac{dN}{d\mathcal{V}_x \, d\mathcal{V}_v} = \frac{dN}{dx\,dy\,dz \, dv_x \, dv_y \, dv_z} = m^3 \mathcal{N}. \tag{3.10}$$

(This change of viewpoint and notation when transitioning to plasma physics is typical of this textbook. When presenting any subfield of physics, we usually adopt the conventions, notation, and also the system of units that are generally used in that subfield.)

3.2.4 Distribution Function I_ν/ν^3 for Photons

[For readers restricting themselves to the Newtonian portions of this book: Please read Sec. 1.10, which lists a few items of special relativity that you need for the discussion here. As described there, you can deal with photons fairly easily by simply remembering that a photon has zero rest mass, energy $\mathcal{E} = h\nu$, and momentum $\mathbf{p} = (h\nu/c)\mathbf{n}$, where ν is its frequency and $\mathbf{n}$ is a unit vector tangent to the photon's spatial trajectory.]

When dealing with photons or other zero-rest-mass particles, one often expresses $\mathcal{N}$ in terms of the *specific intensity* I_ν. This quantity is defined as follows (see Fig. 3.5). An observer places a CCD (or other measuring device) perpendicular to the photons' propagation direction $\mathbf{n}$—perpendicular as measured in her reference frame. The region of the CCD that the photons hit has surface area dA as measured by her, and because the photons move at the speed of light c, the product of that surface area with c times the time dt that they take to all go through the CCD is equal to the volume they occupy at a specific moment of time:

$$d\mathcal{V}_x = dA \, c \, dt. \tag{3.11a}$$

Focus attention on a set $\mathcal{S}$ of photons in this volume that all have nearly the same frequency ν and propagation direction $\mathbf{n}$ as measured by the observer. Their energies $\mathcal{E}$ and momenta $\mathbf{p}$ are related to ν and $\mathbf{n}$ by

$$\mathcal{E} = h\nu, \quad \mathbf{p} = (h\nu/c)\mathbf{n}, \tag{3.11b}$$

where h is Planck's constant. Their frequencies lie in a range $d\nu$ centered on ν, and they come from a small solid angle $d\Omega$ centered on $-\mathbf{n}$; the volume they occupy in momentum space is related to these quantities by

$$d\mathcal{V}_p = |\mathbf{p}|^2 d\Omega d|\mathbf{p}| = (h\nu/c)^2 d\Omega (h d\nu/c) = (h/c)^3 \nu^2 d\Omega d\nu. \tag{3.11c}$$

---

2. The generalization to relativistic plasmas is straightforward; see, e.g., Ex. 23.12.

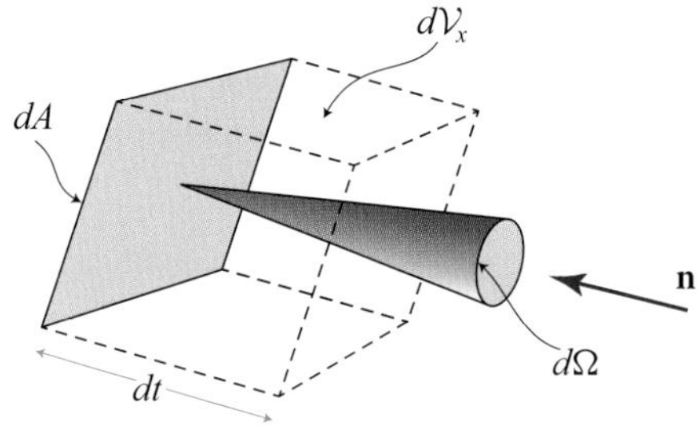

FIGURE 3.5 Geometric construction used in defining the specific intensity I_ν.

The photons' specific intensity, as measured by the observer, is defined to be the total energy

$$d\mathcal{E} = h\nu dN \qquad (3.11d)$$

(where dN is the number of photons) that crosses the CCD per unit area dA, per unit time dt, per unit frequency $d\nu$, and per unit solid angle $d\Omega$ (i.e., per unit everything):

$$I_\nu \equiv \frac{d\mathcal{E}}{dA dt d\nu d\Omega}. \qquad (3.12)$$

specific intensity

(This I_ν is sometimes denoted $I_{\nu\Omega}$.) From Eqs. (3.8), (3.11), and (3.12) we readily deduce the following relationship between this specific intensity and the distribution function:

$$\boxed{\mathcal{N} = \frac{c^2}{h^4}\frac{I_\nu}{\nu^3}.} \qquad (3.13)$$

This relation shows that, with an appropriate renormalization, I_ν/ν^3 is the photons' distribution function.

photon distribution function

Astronomers and opticians regard the specific intensity (or equally well, I_ν/ν^3) as a function of the photon propagation direction **n**, photon frequency ν, location **x** in space, and time t. By contrast, nonrelativistic physicists regard the distribution function $\mathcal{N}$ as a function of the photon momentum **p**, location in space, and time; relativistic physicists regard it as a function of the photon 4-momentum $\vec{p}$ (on the photons' mass hyperboloid, which is the light cone) and of location $\mathcal{P}$ in spacetime. Clearly, the information contained in these three sets of variables, the astronomers' set and the two physicists' sets, is the same.

If two different physicists in two different reference frames at the same event in spacetime examine the same set of photons, they will measure the photons to have different frequencies ν (because of the Doppler shift between their two frames). They will also measure different specific intensities I_ν (because of Doppler shifts of frequencies, Doppler shifts of energies, dilation of times, Lorentz contraction of areas of CCDs, and aberrations of photon propagation directions and thence distortions of solid angles). However, if each physicist computes the ratio of the specific intensity that she measures to the cube of the frequency she measures, that ratio, according to Eq. (3.13), will be the same as computed by the other astronomer: the distribution function I_ν/ν^3 will be frame independent.

3.2.5 Mean Occupation Number η

Although this book is about classical physics, we cannot avoid making frequent contact with quantum theory. The reason is that modern classical physics rests on a quantum mechanical foundation. Classical physics is an approximation to quantum physics, not conversely. Classical physics is derivable from quantum physics, not conversely.

In statistical physics, the classical theory cannot fully shake itself free from its quantum roots; it must rely on them in crucial ways that we shall meet in this chapter and the next. Therefore, rather than try to free it from its roots, we expose these roots and profit from them by introducing a quantum mechanics-based normalization for the distribution function: the mean occupation number η.

As an aid in defining the mean occupation number, we introduce the concept of the *density of states:* Consider a particle of mass m, described quantum mechanically. Suppose that the particle is known to be located in a volume $d\mathcal{V}_x$ (as observed in a specific inertial reference frame) and to have a spatial momentum in the region $d\mathcal{V}_p$ centered on **p**. Suppose, further, that *the particle does not interact with any other particles or fields;* for example, ignore Coulomb interactions. (In portions of Chaps. 4 and 5, we include interactions.) Then how many single-particle quantum mechanical states[3] are available to the free particle? This question is answered most easily by constructing (in some arbitrary inertial frame) a complete set of wave functions for the particle's spatial degrees of freedom, with the wave functions (i) confined to be eigenfunctions of the momentum operator and (ii) confined to satisfy the standard periodic boundary conditions on the walls of a box with volume $d\mathcal{V}_x$. For simplicity, let the box have edge length L along each of the three spatial axes of the Cartesian spatial coordinates, so $d\mathcal{V}_x = L^3$. (This L is arbitrary and will drop out of our analysis shortly.) Then a complete set of wave functions satisfying (i) and (ii) is the set $\{\psi_{j,k,l}\}$ with

$$\psi_{j,k,l}(x, y, z) = \frac{1}{L^{3/2}} e^{i(2\pi/L)(jx+ky+lz)} e^{-i\omega t} \tag{3.14a}$$

[cf., e.g., Cohen-Tannoudji, Diu, and Laloë (1977, pp. 1440–1442), especially the Comment at the end of this page range]. Here the demand that the wave function take

---

3. A quantum mechanical state for a single particle is called an "orbital" in the chemistry literature (where the particle is an election) and in the classic thermal physics textbook by Kittel and Kroemer (1980); we shall use physicists' more conventional but cumbersome phrase "single-particle quantum state," and also, sometimes, "mode."

on the same values at the left and right faces of the box ($x = -L/2$ and $x = +L/2$), at the front and back faces, and at the top and bottom faces (the demand for periodic boundary conditions) dictates that the quantum numbers j, k, and l be integers. The basis states (3.14a) are eigenfunctions of the momentum operator $(\hbar/i)\nabla$ with momentum eigenvalues

$$p_x = \frac{2\pi\hbar}{L}j, \quad p_y = \frac{2\pi\hbar}{L}k, \quad p_z = \frac{2\pi\hbar}{L}l; \tag{3.14b}$$

correspondingly, the wave function's frequency ω has the following values in Newtonian theory **N** and relativity **R**:

$$\boxed{\mathbf{N}} \quad \hbar\omega = E = \frac{\mathbf{p}^2}{2m} = \frac{1}{2m}\left(\frac{2\pi\hbar}{L}\right)^2 (j^2 + k^2 + l^2); \tag{3.14c}$$

$$\boxed{\mathbf{R}} \quad \hbar\omega = \mathcal{E} = \sqrt{m^2 + \mathbf{p}^2} \to m + E \text{ in the Newtonian limit.} \tag{3.14d}$$

Equations (3.14b) tell us that the allowed values of the momentum are confined to lattice sites in 3-momentum space with one site in each cube of side $2\pi\hbar/L$. Correspondingly, the total number of states in the region $d\mathcal{V}_x d\mathcal{V}_p$ of phase space is the number of cubes of side $2\pi\hbar/L$ in the region $d\mathcal{V}_p$ of momentum space:

$$dN_{\text{states}} = \frac{d\mathcal{V}_p}{(2\pi\hbar/L)^3} = \frac{L^3 d\mathcal{V}_p}{(2\pi\hbar)^3} = \frac{d\mathcal{V}_x d\mathcal{V}_p}{h^3}. \tag{3.15}$$

This is true no matter how relativistic or nonrelativistic the particle may be.

Thus far we have considered only the particle's spatial degrees of freedom. Particles can also have an internal degree of freedom called "spin." For a particle with spin s, the number of independent spin states is

$$g_s = \begin{cases} 2s+1 & \text{if } m \neq 0 \text{ (e.g., an electron, proton, or atomic nucleus)} \\ 2 & \text{if } m = 0 \text{ and } s > 0 \text{ [e.g., a photon } (s=1) \text{ or graviton } (s=2)] \\ 1 & \text{if } m = 0 \text{ and } s = 0 \text{ (i.e., a hypothetical massless scalar particle)} \end{cases} \tag{3.16}$$

A notable exception is each species of neutrino or antineutrino, which has nonzero rest mass and spin 1/2, but $g_s = 1$ rather than $g_s = 2s + 1 = 2$.[4] We call this number of internal spin states g_s the particle's *multiplicity*. [It will turn out to play a crucial role in computing the entropy of a system of particles (Chap. 4); thus, it places the imprint of quantum theory on the entropy of even a highly classical system.]

particle's multiplicity

Taking account of both the particle's spatial degrees of freedom and its spin degree of freedom, we conclude that the total number of independent quantum states

---

4. The reason for the exception is the particle's fixed chirality: -1 for neutrinos and $+1$ for antineutrinos; to have $g_s = 2$, a spin-1/2 particle must admit both chiralities.

3.2 Phase Space and Distribution Function

available in the region $d\mathcal{V}_x d\mathcal{V}_p \equiv d^2\mathcal{V}$ of phase space is $dN_{\text{states}} = (g_s/h^3)d^2\mathcal{V}$, and correspondingly the *number density of states in phase space* is

density of states in phase space

$$\mathcal{N}_{\text{states}} \equiv \frac{dN_{\text{states}}}{d^2\mathcal{V}} = \frac{g_s}{h^3}. \tag{3.17}$$

[Relativistic remark: Note that, although we derived this number density of states using a specific inertial frame, it is a frame-independent quantity, with a numerical value depending only on Planck's constant and (through g_s) the particle's rest mass m and spin s.]

The ratio of the number density of particles to the number density of quantum states is obviously the number of particles in each state (the state's *occupation number*) averaged over many neighboring states—but few enough that the averaging region is small by macroscopic standards. In other words, this ratio is the quantum states' *mean occupation number* η:

mean occupation number

$$\eta = \frac{\mathcal{N}}{\mathcal{N}_{\text{states}}} = \frac{h^3}{g_s}\mathcal{N}; \quad \text{i.e.,} \quad \boxed{\mathcal{N} = \mathcal{N}_{\text{states}}\eta = \frac{g_s}{h^3}\eta.} \tag{3.18}$$

The mean occupation number η plays an important role in quantum statistical mechanics, and its quantum roots have a profound impact on classical statistical physics.

fermions and bosons

From quantum theory we learn that the allowed values of the occupation number for a quantum state depend on whether the state is that of a *fermion* (a particle with spin 1/2, 3/2, 5/2, . . .) or that of a *boson* (a particle with spin 0, 1, 2, . . .). For fermions, no two particles can occupy the same quantum state, so the occupation number can only take on the eigenvalues 0 and 1. For bosons, one can shove any number of particles one wishes into the same quantum state, so the occupation number can take on the eigenvalues 0, 1, 2, 3, ... Correspondingly, the mean occupation numbers must lie in the ranges

$$0 \leq \eta \leq 1 \text{ for fermions}, \quad 0 \leq \eta < \infty \text{ for bosons}. \tag{3.19}$$

classical distinguishable particles and classical waves

Quantum theory also teaches us that, when $\eta \ll 1$, the particles, whether fermions or bosons, behave like classical, discrete, distinguishable particles; and when $\eta \gg 1$ (possible only for bosons), the particles behave like a classical wave—if the particles are photons ($s = 1$), like a classical electromagnetic wave; and if they are gravitons ($s = 2$), like a classical gravitational wave. This role of η in revealing the particles' physical behavior will motivate us frequently to use η as our distribution function instead of $\mathcal{N}$.

Of course η, like $\mathcal{N}$, is a function of location in phase space, $\eta(\mathcal{P}, \vec{p})$ in relativity with no inertial frame chosen; or $\eta(t, \mathbf{x}, \mathbf{p})$ in both relativity and Newtonian theory when an inertial frame is in use.

Exercise 3.3 **Practice and Example: Regimes of Particulate and Wave-Like Behavior** N R

(a) Cygnus X-1 is a source of X-rays that has been studied extensively by astronomers. The observations (X-ray, optical, and radio) show that it is a distance $r \sim 6{,}000$ light-years from Earth. It consists of a very hot disk of X-ray-emitting gas that surrounds a black hole with mass $15 M_\odot$, and the hole in turn is in a binary orbit with a heavy companion star. Most of the X-ray photons have energies $\mathcal{E} \sim 2$ keV, their energy flux arriving at Earth is $F \sim 10^{-10}$ W m^{-2}, and the portion of the disk that emits most of them has radius roughly 7 times that of the black hole (i.e., $R \sim 300$ km).[5] Make a rough estimate of the mean occupation number of the X-rays' photon states. Your answer should be in the region $\eta \ll 1$, so the photons behave like classical, distinguishable particles. Will the occupation number change as the photons propagate from the source to Earth?

(b) A highly nonspherical supernova in the Virgo cluster of galaxies (40 million light-years from Earth) emits a burst of gravitational radiation with frequencies spread over the band 0.5–2.0 kHz, as measured at Earth. The burst comes out in a time of about 10 ms, so it lasts only a few cycles, and it carries a total energy of roughly $10^{-3} M_\odot c^2$, where $M_\odot = 2 \times 10^{30}$ kg is the mass of the Sun. The emitting region is about the size of the newly forming neutron-star core (10 km), which is small compared to the wavelength of the waves; so if one were to try to resolve the source spatially by imaging the gravitational waves with a gravitational lens, one would see only a blur of spatial size one wavelength rather than seeing the neutron star. What is the mean occupation number of the burst's graviton states? Your answer should be in the region $\eta \gg 1$, so the gravitons behave like a classical gravitational wave.

3.3 Thermal-Equilibrium Distribution Functions N R

In Chap. 4, we introduce with care and explore in detail the concept of statistical equilibrium—also called "thermal equilibrium." That exploration will lead to a set of distribution functions for particles that are in statistical equilibrium. In this section, we summarize those equilibrium distribution functions, so as to be able to use them for examples and applications of kinetic theory.

If a collection of many identical particles is in thermal equilibrium in the neighborhood of an event $\mathcal{P}$, then, as we shall see in Chap. 4, there is a special inertial reference frame (the *mean rest frame* of the particles near $\mathcal{P}$) in which the distribution function is isotropic, so the mean occupation number η is a function only of the magnitude $|\mathbf{p}|$

---

5. These numbers refer to what astronomers call Cygnus X-1's soft (red) state. It also, sometimes, is seen in a hard (blue) state.

of the particle momentum and does not depend on the momentum's direction. Equivalently, η is a function of the particle's energy. In the relativistic regime, we use two different energies, one denoted $\mathcal{E}$ that includes the contribution of the particle's rest mass and the other denoted E that omits the rest mass and thus represents kinetic energy (cf. Sec. 1.10):

$$E \equiv \mathcal{E} - m = \sqrt{m^2 + \mathbf{p}^2} - m \to \frac{\mathbf{p}^2}{2m} \text{ in the low-velocity, Newtonian limit.}$$

(3.20)

In the nonrelativistic, Newtonian regime we use only $E = \mathbf{p}^2/(2m)$.

Most readers already know that the details of the thermal equilibrium are fixed by two quantities: the mean density of particles and the mean energy per particle, or equivalently (as we shall see) by the *chemical potential* μ and the *temperature* T. By analogy with our treatment of relativistic energy, we use two different chemical potentials: one, $\tilde{\mu}$, that includes rest mass and the other,

$$\mu \equiv \tilde{\mu} - m,$$ (3.21)

that does not. In the Newtonian regime we use only μ.

As we prove by an elegant argument in Chap. 4, in thermal equilibrium the mean occupation number has the following form at all energies, relativistic or nonrelativistic:

Fermi-Dirac and Bose-Einstein distributions

$$\eta = \frac{1}{e^{(E-\mu)/(k_B T)} + 1} \text{ for fermions,}$$ (3.22a)

$$\eta = \frac{1}{e^{(E-\mu)/(k_B T)} - 1} \text{ for bosons.}$$ (3.22b)

Here $k_B = 1.381 \times 10^{-16}$ erg K^{-1} = 1.381×10^{-23} J K^{-1} is Boltzmann's constant. Equation (3.22a) for fermions is the *Fermi-Dirac distribution*; Eq. (3.22b) for bosons is the *Bose-Einstein distribution*. In the relativistic regime, we can also write these distribution functions in terms of the energy $\mathcal{E}$ that includes the rest mass as

$$\eta = \frac{1}{e^{(E-\mu)/(k_B T)} \pm 1} = \frac{1}{e^{(\mathcal{E}-\tilde{\mu})/(k_B T)} \pm 1}.$$ (3.22c)

Notice that the equilibrium mean occupation number (3.22a) for fermions lies in the range 0–1 as required, while that (3.22b) for bosons lies in the range 0 to ∞. In the regime $\mu \ll -k_B T$, the mean occupation number is small compared to unity for all particle energies E (since E is never negative; i.e., $\mathcal{E}$ is never less than m). This is the domain of distinguishable, classical particles, and in it both the Fermi-Dirac and Bose-Einstein distributions become

$$\boxed{\begin{aligned} \eta &\simeq e^{-(E-\mu)/(k_B T)} = e^{-(\mathcal{E}-\tilde{\mu})/(k_B T)} \\ &\text{when } \mu \equiv \tilde{\mu} - m \ll -k_B T \quad \text{(classical particles)}. \end{aligned}} \quad (3.22d)$$

Boltzmann distribution

This limiting distribution is the *Boltzmann distribution*.[6]

By scrutinizing the distribution functions (3.22), one can deduce that the larger the temperature T at fixed μ, the larger will be the typical energies of the particles; the larger the chemical potential μ at fixed T, the larger will be the total density of particles [see Ex. 3.4 and Eqs. (3.39)]. For bosons, μ must always be negative or zero, that is, $\tilde{\mu}$ cannot exceed the particle rest mass m; otherwise, η would be negative at low energies, which is physically impossible. For bosons with μ extremely close to zero, there exist huge numbers of very-low-energy particles, leading quantum mechanically to a *Bose-Einstein condensate*; we study such condensates in Sec. 4.9.

In the special case that the particles of interest can be created and destroyed completely freely, with creation and destruction constrained only by the laws of 4-momentum conservation, the particles quickly achieve a thermal equilibrium in which the relativistic chemical potential vanishes, $\tilde{\mu} = 0$ (as we shall see in Sec. 5.5). For example, inside a box whose walls are perfectly emitting and absorbing and have temperature T, the photons acquire the mean occupation number (3.22b) with zero chemical potential, leading to the standard *blackbody (Planck) form*

$$\eta = \frac{1}{e^{h\nu/(k_B T)} - 1}, \quad \mathcal{N} = \frac{2}{h^3} \frac{1}{e^{h\nu/(k_B T)} - 1}, \quad I_\nu = \frac{(2h/c^2)\nu^3}{e^{h\nu/(k_B T)} - 1}. \quad (3.23)$$

Planck distribution

(Here we have set $\mathcal{E} = h\nu$, where ν is the photon frequency as measured in the box's rest frame, and in the third expression we have inserted the factor c^{-2}, so that I_ν will be in ordinary units.)

By contrast, if one places a fixed number of photons inside a box whose walls cannot emit or absorb them but can scatter them, exchanging energy with them in the process, then the photons will acquire the Bose-Einstein distribution (3.22b) with temperature T equal to that of the walls and with nonzero chemical potential μ fixed by the number of photons present; the more photons there are, the larger will be the chemical potential.

EXERCISES

Exercise 3.4 **Example: Maxwell Velocity Distribution*

Consider a collection of thermalized, classical particles with nonzero rest mass, so they have the Boltzmann distribution (3.22d). Assume that the temperature is low enough ($k_B T \ll mc^2$) that they are nonrelativistic.

(a) Explain why the total number density of particles n in physical space (as measured in the particles' mean rest frame) is given by the integral $n = \int \mathcal{N} d\mathcal{V}_p$. Show

---

6. Lynden-Bell (1967) identifies a fourth type of thermal distribution that occurs in the theory of violent relaxation of star clusters. It corresponds to individually distinguishable, classical particles (in his case stars with a range of masses) that obey the same kind of exclusion principle as fermions.

that $n \propto e^{\mu/k_B T}$, and derive the proportionality constant. [Hint: Use spherical coordinates in momentum space, so $d\mathcal{V}_p = 4\pi p^2 dp$ with $p \equiv |\mathbf{p}|$.] Your answer should be Eq. (3.39a) below.

(b) Explain why the mean energy per particle is given by $\bar{E} = n^{-1} \int (p^2/2m) \mathcal{N} d\mathcal{V}_p$. Show that $\bar{E} = \frac{3}{2} k_B T$.

(c) Show that $P(v) dv \equiv$ (probability that a randomly chosen particle will have speed $v \equiv |\mathbf{v}|$ in the range dv) is given by

$$P(v) = \frac{4}{\sqrt{\pi}} \frac{v^2}{v_o^3} e^{-v^2/v_o^2}, \quad \text{where} \quad v_o = \sqrt{\frac{2k_B T}{m}}. \tag{3.24}$$

This is called the *Maxwell velocity distribution*; it is graphed in Fig. 3.6a. Notice that the peak of the distribution is at speed v_o.

[Side remark: In the normalization of probability distributions such as this one, you will often encounter integrals of the form $\int_0^\infty x^{2n} e^{-x^2} dx$. You can evaluate this quickly via integration by parts, if you have memorized that $\int_0^\infty e^{-x^2} dx = \sqrt{\pi}/2$.]

(d) Consider particles confined to move in a plane or in one dimension (on a line). What is their speed distribution $P(v)$ and at what speed does it peak?

Exercise 3.5 *Problem: Maxwell-Jütner Velocity Distribution for Thermalized, Classical, Relativistic Particles* **R** **T2**

Show that for thermalized, classical relativistic particles the probability distribution for the speed [relativistic version of the Maxwell distribution (3.24)] is

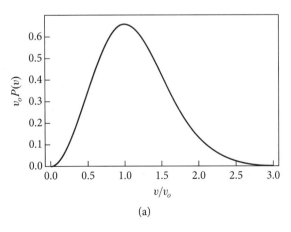

(a)

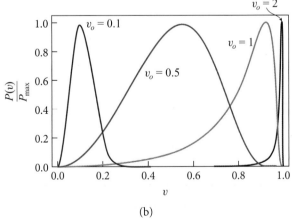

(b)

FIGURE 3.6 (a) Maxwell velocity distribution for thermalized, classical, nonrelativistic particles. (b) Extension of the Maxwell velocity distribution into the relativistic domain. In both plots $v_o = \sqrt{2 k_B T/m}$.

$$P(v) = \frac{2/v_0^2}{K_2(2/v_0^2)} \frac{v^2}{(1-v^2)^{5/2}} \exp\left[-\frac{2/v_o^2}{\sqrt{1-v^2}}\right], \quad \text{where} \quad v_o = \sqrt{\frac{2k_B T}{m}}. \quad (3.25)$$

Where K_2 is the modified Bessel function of the second kind and order 2. This is sometimes called the Maxwell-Jütner distribution, and it is plotted in Fig. 3.6b for a sequence of four temperatures ranging from the nonrelativistic regime $k_B T \ll m$ toward the ultrarelativistic regime $k_B T \gg m$. In the ultrarelativistic regime the particles are (almost) all moving at very close to the speed of light, $v = 1$.

Exercise 3.6 *Example and Challenge: Radiative Processes* R T2

We have described distribution functions for particles and photons and the forms that they have in thermodynamic equilibrium. An extension of these principles can be used to constrain the manner in which particles and photons interact, specifically, to relate the emission and absorption of radiation.

(a) Consider a two-level (two-state) electron system with energy separation $\mathcal{E} = h\nu_0$. Suppose that an electron can transition with a probability per unit time, A, from the upper level (u) to the lower level (l), creating a photon in a specific state. Use the Boltzmann distribution, ignoring degeneracy, Eq. (3.22d), and the expression for the mean photon occupation number, Eq. (3.23), to show that, when the electrons and photons are in thermal equilibrium at the same temperature T, then:

$$An_u + An_u \eta_\gamma = An_l \eta_\gamma, \quad (3.26)$$

where $n_{l,u}$ are the number densities of the electrons in the two states, and η_γ is the photon mean occupation number.

(b) The three terms in Eq. (3.26) are often called the rates per unit volume for *spontaneous emission, stimulated emission,* and *(stimulated) absorption,* respectively. (The second and third terms are sometimes expressed using Einstein B coefficients.) The expressions An_u, $An_u \eta_\gamma$, and $An_l \eta_\gamma$ for the rates of these three types of transition are commonly (and usually correctly) assumed to apply out of thermodynamic equilibrium as well as in equilibrium. Discuss briefly two conditions that need to be satisfied for this to be so: that all three types of transition proceed in a manner that is independent of anything else present, and that the stimulated transitions are time reversible.[7] What additional condition needs to be satisfied if the electron system is more complex than a two-level system? (We return to these issues in the context of cosmology, in Secs. 28.4.4 and 28.6.3.)

(c) Typical transitions have some duration τ, which implies that the photons will be emitted with a distribution of frequencies with width $\Delta \nu \sim \tau^{-1}$ about ν_0. They may also not be emitted isotropically and can carry polarization. Denote the state of a photon by its frequency ν, direction of travel $\widehat{\Omega}$, and polarization $i = 1$ or 2,

---

7. These principles are quite classical in origin but the full justification of the heuristic argument presented here requires quantum electrodynamics.

and denote by $P^i_{\nu\widehat{\Omega}}$ the photons' probability distribution, normalized such that $\Sigma_i \int d\nu\, d\Omega\, P^i_{\nu\widehat{\Omega}} = 1$, with $d\Omega$ an element of solid angle around $\widehat{\Omega}$. Now define a classical emissivity and a classical absorption coefficient by

$$j^i_{\nu\widehat{\Omega}} \equiv n_u h\nu A P^i_{\nu\widehat{\Omega}}, \qquad \kappa^i \equiv n_l A(1 - e^{-h\nu_0/(k_B T_e)}) P^i_{\nu\widehat{\Omega}} c^2/\nu^2, \qquad (3.27)$$

respectively, where T_e is the electron temperature. Interpret these two quantities and prove *Kirchhoff's law*, namely, that $j^i_{\nu\widehat{\Omega}}/\kappa^i$ is the blackbody radiation intensity for one polarization mode. What happens when $h\nu_0 \ll k_B T_e$?

(d) Further generalize the results in part c by assuming that, instead of occupying just two states, the electrons have a continuous distribution of momenta and radiate throughout this distribution. Express the classical emissivity and absorption coefficient as integrals over electron momentum space.

(e) Finally, consider the weak nuclear transformation $\nu + n \to e + p$, where the neutron n and the proton p can be considered as stationary. (This is important in the early universe, as we discuss further in Sec. 28.4.2.) Explain carefully why the rate at which this transformation occurs can be expressed as

$$\frac{dn_p}{dt} = n_n \int \left(\frac{d\mathcal{V}_{p_\nu}}{h^3}\eta_\nu\right)\left(\frac{h^6 W}{m_e^5}\delta(E_\nu + m_n - E_e - m_p)\right)\left(2\frac{d\mathcal{V}_{p_e}}{h^3}(1 - \eta_e)\right) \qquad (3.28)$$

for some W. It turns out that W is a constant (Weinberg, 2008). What is the corresponding rate for the inverse reaction?

Exercise 3.7 **Example: *Observations of Cosmic Microwave Radiation from Earth* R T2

The universe is filled with cosmic microwave radiation left over from the big bang. At each event in spacetime the microwave radiation has a mean rest frame. As seen in that mean rest frame the radiation's distribution function η is almost precisely isotropic and thermal with zero chemical potential:

$$\eta = \frac{1}{e^{h\nu/(k_B T_o)} - 1}, \quad \text{with} \quad T_o = 2.725 \text{ K}. \qquad (3.29)$$

Here ν is the frequency of a photon as measured in the mean rest frame.

(a) Show that the specific intensity of the radiation as measured in its mean rest frame has the *Planck spectrum*, Eq. (3.23). Plot this specific intensity as a function of frequency, and from your plot determine the frequency of the intensity peak.

(b) Show that η can be rewritten in the frame-independent form

$$\eta = \frac{1}{e^{-\vec{p}\cdot\vec{u}_o/(k_B T_o)} - 1}, \qquad (3.30)$$

where $\vec{p}$ is the photon 4-momentum, and $\vec{u}_o$ is the 4-velocity of the mean rest frame. [Hint: See Sec. 2.6 and especially Eq. (2.29).]

(c) In actuality, Earth moves relative to the mean rest frame of the microwave background with a speed v of roughly 400 km s^{-1} toward the Hydra-Centaurus region of the sky. An observer on Earth points his microwave receiver in a direction that makes an angle θ with the direction of that motion, as measured in Earth's frame. Show that the specific intensity of the radiation received is precisely Planckian in form [Eqs. (3.23)], but with a direction-dependent *Doppler-shifted temperature*

$$T = T_o \left(\frac{\sqrt{1-v^2}}{1 - v \cos \theta} \right). \tag{3.31}$$

Note that this Doppler shift of T is precisely the same as the Doppler shift of the frequency of any specific photon [Eq. (2.33)]. Note also that the θ dependence corresponds to an anisotropy of the microwave radiation as seen from Earth. Show that because Earth's velocity is small compared to the speed of light, the anisotropy is very nearly dipolar in form. Measurements by the WMAP satellite give $T_o = 2.725$ K and (averaged over a year) an amplitude of 3.346×10^{-3} K for the dipolar temperature variations (Bennett et al., 2003). What, precisely, is the value of Earth's year-averaged speed v?

3.4 Macroscopic Properties of Matter as Integrals over Momentum Space

3.4.1 Particle Density n, Flux S, and Stress Tensor T

If one knows the Newtonian distribution function $\mathcal{N} = (g_s/h^3)\eta$ as a function of momentum **p** at some location $(\mathbf{x}, t)$ in space and time, one can use it to compute various macroscopic properties of the particles.

From the definition $\mathcal{N} \equiv dN/d\mathcal{V}_x d\mathcal{V}_p$ of the distribution function, it is clear that the number density of particles $n(\mathbf{x}, t)$ in physical space is given by the integral

$$\boxed{n = \frac{dN}{d\mathcal{V}_x} = \int \frac{dN}{d\mathcal{V}_x d\mathcal{V}_p} d\mathcal{V}_p = \int \mathcal{N} d\mathcal{V}_p.} \tag{3.32a}$$

Newtonian particle density

Similarly, the number of particles crossing a unit surface in the y-z plane per unit time (i.e., the x component of the flux of particles) is

$$S_x = \frac{dN}{dy dz dt} = \int \frac{dN}{dx dy dz d\mathcal{V}_p} \frac{dx}{dt} d\mathcal{V}_p = \int \mathcal{N} \frac{p_x}{m} d\mathcal{V}_p,$$

where $dx/dt = p_x/m$ is the x component of the particle velocity. This and the analogous equations for S_y and S_z can be combined into a single geometric, coordinate-independent integral for the vectorial particle flux:

$$\boxed{\mathbf{S} = \int \mathcal{N} \, \mathbf{p} \, \frac{d\mathcal{V}_p}{m}.} \tag{3.32b}$$

Newtonian particle flux

Newtonian momentum density

Notice that, if we multiply this **S** by the particles' mass m, the integral becomes the momentum density:

$$\boxed{\mathbf{G} = m\mathbf{S} = \int \mathcal{N} \mathbf{p} \, d\mathcal{V}_p.} \qquad (3.32c)$$

Finally, since the stress tensor **T** is the flux of momentum [Eq. (1.33)], its j-x component (j component of momentum crossing a unit area in the y-z plane per unit time) must be

$$T_{jx} = \int \frac{dN}{dy\,dz\,dt\,d\mathcal{V}_p} p_j \, d\mathcal{V}_p = \int \frac{dN}{dx\,dy\,dz\,d\mathcal{V}_p} \frac{dx}{dt} p_j \, d\mathcal{V}_p = \int \mathcal{N} p_j \frac{p_x}{m} d\mathcal{V}_p.$$

This and the corresponding equations for T_{jy} and T_{jz} can be collected together into a single geometric, coordinate-independent integral:

Newtonian stress tensor

$$T_{jk} = \int \mathcal{N} p_j p_k \frac{d\mathcal{V}_p}{m}, \quad \text{i.e.,} \quad \boxed{\mathbf{T} = \int \mathcal{N} \mathbf{p} \otimes \mathbf{p} \frac{d\mathcal{V}_p}{m}.} \qquad (3.32d)$$

Notice that the number density n is the zeroth moment of the distribution function in momentum space [Eq. (3.32a)], and aside from factors $1/m$, the particle flux vector is the first moment [Eq. (3.32b)], and the stress tensor is the second moment [Eq. (3.32d)]. All three moments are geometric, coordinate-independent quantities, and they are the simplest such quantities that one can construct by integrating the distribution function over momentum space.

3.4.2 Relativistic Number-Flux 4-Vector $\vec{S}$ and Stress-Energy Tensor **T**

When we switch from Newtonian theory to special relativity's 4-dimensional spacetime viewpoint, we require that all physical quantities be described by geometric, frame-independent objects (scalars, vectors, tensors, . . .) in 4-dimensional spacetime. We can construct such objects as momentum-space integrals over the frame-independent, relativistic distribution function $\mathcal{N}(\mathcal{P}, \vec{p}) = (g_s/h^3)\eta$. The frame-independent quantities that can appear in these integrals are (i) $\mathcal{N}$ itself, (ii) the particle 4-momentum $\vec{p}$, and (iii) the frame-independent integration element $d\mathcal{V}_p/\mathcal{E}$ [Eq. (3.7b)], which takes the form $dp_x dp_y dp_z / \sqrt{m^2 + \mathbf{p}^2}$ in any inertial reference frame. By analogy with the Newtonian regime, the most interesting such integrals are the lowest three moments of the distribution function:

$$R \equiv \int \mathcal{N} \frac{d\mathcal{V}_p}{\mathcal{E}}; \qquad (3.33a)$$

$$\boxed{\vec{S} \equiv \int \mathcal{N} \vec{p} \frac{d\mathcal{V}_p}{\mathcal{E}},} \quad \text{i.e.,} \quad S^\mu \equiv \int \mathcal{N} p^\mu \frac{d\mathcal{V}_p}{\mathcal{E}}; \qquad (3.33b)$$

$$\boxed{\mathbf{T} \equiv \int \mathcal{N} \vec{p} \otimes \vec{p} \frac{d\mathcal{V}_p}{\mathcal{E}},} \quad \text{i.e.,} \quad T^{\mu\nu} \equiv \int \mathcal{N} p^\mu p^\nu \frac{d\mathcal{V}_p}{\mathcal{E}}. \qquad (3.33c)$$

Here and throughout this chapter, relativistic momentum-space integrals are taken over the entire mass hyperboloid unless otherwise specified.

We can learn the physical meanings of each of the momentum-space integrals (3.33) by introducing a specific but arbitrary inertial reference frame and using it to perform a 3+1 split of spacetime into space plus time [cf. the paragraph containing Eq. (2.28)]. When we do this and rewrite $\mathcal{N}$ as $dN/d\mathcal{V}_x d\mathcal{V}_p$, the scalar field R of Eq. (3.33a) takes the form

$$R = \int \frac{dN}{d\mathcal{V}_x d\mathcal{V}_p} \frac{1}{\mathcal{E}} d\mathcal{V}_p \tag{3.34}$$

(where of course $d\mathcal{V}_x = dx\,dy\,dz$ and $d\mathcal{V}_p = dp_x dp_y dp_z$). This is the sum, over all particles in a unit 3-volume, of the inverse energy. Although it is intriguing that this quantity is a frame-independent scalar, it is not a quantity that appears in any important way in the laws of physics.

By contrast, the 4-vector field $\vec{S}$ of Eq. (3.33b) plays a very important role in physics. Its time component in our chosen frame is

$$S^0 = \int \frac{dN}{d\mathcal{V}_x d\mathcal{V}_p} \frac{p^0}{\mathcal{E}} d\mathcal{V}_p = \int \frac{dN}{d\mathcal{V}_x d\mathcal{V}_p} d\mathcal{V}_p \tag{3.35a}$$

(since p^0 and $\mathcal{E}$ are just different notations for the same thing—the relativistic energy $\sqrt{m^2 + \mathbf{p}^2}$ of a particle). Obviously, this S^0 is the number of particles per unit spatial volume as measured in our chosen inertial frame:

$$S^0 = n = \text{(number density of particles)}. \tag{3.35b}$$

The x component of $\vec{S}$ is

$$S^x = \int \frac{dN}{d\mathcal{V}_x d\mathcal{V}_p} \frac{p^x}{\mathcal{E}} d\mathcal{V}_p = \int \frac{dN}{dx\,dy\,dz\, d\mathcal{V}_p} \frac{dx}{dt} d\mathcal{V}_p = \int \frac{dN}{dt\,dy\,dz\, d\mathcal{V}_p} d\mathcal{V}_p, \tag{3.35c}$$

which is the number of particles crossing a unit area in the y-z plane per unit time (i.e., the x component of the particle flux); similarly for other directions j:

$$S^j = (j \text{ component of the particle flux vector } \mathbf{S}). \tag{3.35d}$$

[In Eq. (3.35c), the second equality follows from

$$\frac{p^j}{\mathcal{E}} = \frac{p^j}{p^0} = \frac{dx^j/d\zeta}{dt/d\zeta} = \frac{dx^j}{dt} = (j \text{ component of velocity}), \tag{3.35e}$$

where ζ is the affine parameter such that $\vec{p} = d\vec{x}/d\zeta$.] Since S^0 is the particle number density and S^j is the particle flux, $\vec{S}$ [Eq. (3.33b)] *must be the number-flux 4-vector* introduced and studied in Sec. 2.12.3. Notice that in the Newtonian limit, where $p^0 = \mathcal{E} \to m$, the temporal and spatial parts of the formula $\vec{S} = \int \mathcal{N} \vec{p} \, (d\mathcal{V}_p/\mathcal{E})$ reduce

number-flux 4-vector

to $S^0 = \int \mathcal{N} d\mathcal{V}_p$ and $\mathbf{S} = \int \mathcal{N} \mathbf{p}(d\mathcal{V}_p/m)$, respectively, which are the coordinate-independent expressions (3.32a) and (3.32b) for the Newtonian number density of particles and flux of particles, respectively.

Turn to the quantity **T** defined by the integral (3.33c). When we perform a 3+1 split of it in our chosen inertial frame, we find the following for its various parts:

$$T^{\mu 0} = \int \frac{dN}{d\mathcal{V}_x d\mathcal{V}_p} p^\mu p^0 \frac{d\mathcal{V}_p}{p^0} = \int \frac{dN}{d\mathcal{V}_x d\mathcal{V}_p} p^\mu d\mathcal{V}_p \qquad (3.36a)$$

is the μ component of 4-momentum per unit volume (i.e., T^{00} is the energy density, and T^{j0} is the momentum density). Also,

$$T^{\mu x} = \int \frac{dN}{d\mathcal{V}_x d\mathcal{V}_p} p^\mu p^x \frac{d\mathcal{V}_p}{p^0} = \int \frac{dN}{dx\,dy\,dz\,d\mathcal{V}_p} \frac{dx}{dt} p^\mu d\mathcal{V}_p = \int \frac{dN}{dt\,dy\,dz\,d\mathcal{V}_p} p^\mu d\mathcal{V}_p \qquad (3.36b)$$

is the amount of μ component of 4-momentum that crosses a unit area in the y-z plane per unit time (i.e., it is the x component of flux of μ component of 4-momentum). More specifically, T^{0x} is the x component of energy flux (which is the same as the momentum density T^{x0}), and T^{jx} is the x component of spatial-momentum flux—or, equivalently, the jx component of the stress tensor. These and the analogous expressions and interpretations of $T^{\mu y}$ and $T^{\mu z}$ can be summarized by

$$T^{00} = \text{(energy density)}, \quad T^{j0} = \text{(momentum density)} = T^{0j} = \text{(energy flux)},$$
$$T^{jk} = \text{(stress tensor)}. \qquad (3.36c)$$

stress-energy tensor

Therefore [cf. Eq. (2.67f)], the **T** of Eq. (3.33c) must be the stress-energy tensor introduced and studied in Sec. 2.13. Notice that in the Newtonian limit, where $\mathcal{E} \to m$, the coordinate-independent Eq. (3.33c) for the spatial part of the stress-energy tensor (the stress) becomes $\int \mathcal{N} \mathbf{p} \otimes \mathbf{p}\, d\mathcal{V}_p/m$, which is the same as our coordinate-independent Eq. (3.32d) for the stress tensor.

3.5 Isotropic Distribution Functions and Equations of State

3.5.1 Newtonian Density, Pressure, Energy Density, and Equation of State

Let us return to Newtonian theory.

If the Newtonian distribution function is isotropic in momentum space (i.e., is a function only of the magnitude $p \equiv |\mathbf{p}| = \sqrt{p_x^2 + p_y^2 + p_z^2}$ of the momentum, as is the case, e.g., when the particle distribution is thermalized), then the particle flux **S** vanishes (equal numbers of particles travel in all directions), and the stress tensor is isotropic: $\mathbf{T} = P\mathbf{g}$, or $T_{jk} = P\delta_{jk}$. Thus, it is the stress tensor of a perfect fluid. [Here P is the isotropic pressure, and $\mathbf{g}$ is the metric tensor of Euclidian 3-space, with Cartesian components equal to the Kronecker delta; Eq. (1.9f).] In this isotropic case, the pressure can be computed most easily as 1/3 the trace of the stress tensor (3.32d):

$$P = \frac{1}{3}T_{jj} = \frac{1}{3}\int \mathcal{N}(p_x^2 + p_y^2 + p_z^2)\frac{d\mathcal{V}_p}{m}$$

$$= \frac{1}{3}\int_0^\infty \mathcal{N}p^2 \frac{4\pi p^2 dp}{m} = \frac{4\pi}{3m}\int_0^\infty \mathcal{N}p^4\,dp. \quad (3.37\text{a})$$

Newtonian pressure

Here in the third step we have written the momentum-volume element in spherical polar coordinates as $d\mathcal{V}_p = p^2 \sin\theta\, d\theta\, d\phi\, dp$ and have integrated over angles to get $4\pi p^2 dp$. Similarly, we can reexpress the number density of particles (3.32a) and the corresponding mass density as

$$n = 4\pi \int_0^\infty \mathcal{N}p^2 dp, \quad \rho \equiv mn = 4\pi m \int_0^\infty \mathcal{N}p^2\,dp. \quad (3.37\text{b})$$

Newtonian particle and mass density

Finally, because each particle carries an energy $E = p^2/(2m)$, the energy density in this isotropic case (which we shall denote by U) is 3/2 the pressure:

$$U = \int \frac{p^2}{2m}\mathcal{N}d\mathcal{V}_p = \frac{4\pi}{2m}\int_0^\infty \mathcal{N}p^4 dp = \frac{3}{2}P \quad (3.37\text{c})$$

Newtonian energy density

[cf. Eq. (3.37a)].

If we know the distribution function for an isotropic collection of particles, Eqs. (3.37) give us a straightforward way of computing the collection's number density of particles n, mass density $\rho = nm$, perfect-fluid energy density U, and perfect-fluid pressure P as measured in the particles' mean rest frame. For a thermalized gas, the distribution functions (3.22a), (3.22b), and (3.22d) [with $\mathcal{N} = (g_s/h^3)\eta$] depend on two parameters: the temperature T and chemical potential μ, so this calculation gives n, U, and P in terms of μ and T. One can then invert $n(\mu, T)$ to get $\mu(n, T)$ and insert the result into the expressions for U and P to obtain *equations of state* for thermalized, nonrelativistic particles:

$$U = U(\rho, T), \quad P = P(\rho, T). \quad (3.38)$$

equations of state

For a gas of nonrelativistic, classical particles, the distribution function is Boltzmann [Eq. (3.22d)], $\mathcal{N} = (g_s/h^3)e^{(\mu-E)/(k_B T)}$, with $E = p^2/(2m)$, and this procedure gives, quite easily (Ex. 3.8):

thermalized, classical, nonrelativistic gas

$$\boxed{n = \frac{g_s e^{\mu/(k_B T)}}{\lambda_{T\mathrm{dB}}^3} = \frac{g_s}{h^3}(2\pi m k_B T)^{3/2} e^{\mu/(k_B T)},} \quad (3.39\text{a})$$

$$\boxed{U = \frac{3}{2}nk_B T, \quad P = nk_B T.} \quad (3.39\text{b})$$

Notice that the mean energy per particle is (cf. Ex. 3.4b)

$$\bar{E} = \frac{3}{2}k_B T. \quad (3.39\text{c})$$

In Eq. (3.39a), $\lambda_{T\text{dB}} \equiv h/\sqrt{2\pi m k_B T}$ is the particles' *thermal de Broglie wavelength*: the wavelength of Schrödinger wave-function oscillations for a particle with thermal kinetic energy $E = \pi k_B T$. Note that the classical regime $\eta \ll 1$ (i.e., $\mu/(k_B T) \ll -1$), in which our computation is being performed, corresponds to a mean number of particles in a thermal de Broglie wavelength small compared to 1, $n\lambda_{T\text{dB}}^3 \ll 1$, which should not be surprising.

3.5.2 Equations of State for a Nonrelativistic Hydrogen Gas

As an application, consider ordinary matter. Figure 3.7 shows its physical nature as a function of density and temperature, near and above room temperature (300 K). We study solids (lower right) in Part IV, fluids (lower middle) in Part V, and plasmas (middle) in Part VI.

Our kinetic-theory tools are well suited to any situation where the particles have mean free paths large compared to their sizes. This is generally true in plasmas and sometimes in fluids (e.g., air and other gases, but not water and other liquids), and even sometimes in solids (e.g., electrons in a metal). Here we focus on a nonrelativistic

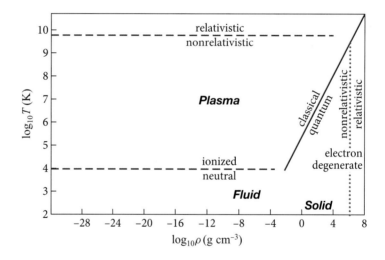

FIGURE 3.7 Physical nature of hydrogen at various densities and temperatures. The plasma regime is discussed in great detail in Part VI, and the equation of state in this regime is Eq. (3.40). The region of relativistic electron degeneracy (to the right of the vertical dotted line) is analyzed in Sec. 3.5.4, and that for the nonrelativistic regime (between slanted solid line and vertical dotted line) in the second half of Sec. 3.5.2. The boundary between the plasma regime and the electron-degenerate regime (slanted solid line) is Eq. (3.41); that between nonrelativistic degeneracy and relativistic degeneracy (vertical dotted line) is Eq. (3.46). The upper relativistic/nonrelativistic boundary is governed by electron-positron pair production (Ex. 5.9 and Fig. 5.7) and is only crudely approximated by the upper dashed line. The ionized-neutral boundary is governed by the Saha equation (Ex. 5.10 and Fig. 20.1) and is crudely approximated by the lower dashed line. For a more accurate and detailed version of this figure, including greater detail on the plasma regime and its boundaries, see Fig. 20.1.

plasma (i.e., the region of Fig. 3.7 that is bounded by the two dashed lines and the slanted solid line). For concreteness and simplicity, we regard the plasma as made solely of hydrogen.[8]

A nonrelativistic hydrogen plasma consists of a mixture of two fluids (gases): free electrons and free protons, in equal numbers. Each fluid has a particle number density $n = \rho/m_p$, where ρ is the total mass density and m_p is the proton mass. (The electrons are so light that they do not contribute significantly to ρ.) Correspondingly, the energy density and pressure include equal contributions from the electrons and protons and are given by [cf. Eqs. (3.39b)]

nondegenerate hydrogen gas

$$U = 3(k_B/m_p)\rho T, \qquad P = 2(k_B/m_p)\rho T. \tag{3.40}$$

In zeroth approximation, the high-temperature boundary of validity for this equation of state is the temperature $T_{\rm rel} = m_e c^2/k_B = 6 \times 10^9$ K, at which the electrons become highly relativistic (top dashed line in Fig. 3.7). In Ex. 5.9, we compute the thermal production of electron-positron pairs in the hot plasma and thereby discover that the upper boundary is actually somewhat lower than this (Figs. 5.7 and 20.1). The bottom dashed line in Fig. 3.7 is the temperature $T_{\rm ion} \sim$ (ionization energy of hydrogen)/(a few k_B) $\sim 10^4$ K, at which electrons and protons begin to recombine and form neutral hydrogen. In Ex. 5.10 on the Saha equation, we analyze the conditions for ionization-recombination equilibrium and thereby refine this boundary (Fig. 20.1). The solid right boundary is the point at which the electrons cease to behave like classical particles, because their mean occupation number η_e ceases to be $\ll 1$. As one can see from the Fermi-Dirac distribution (3.22a), for typical electrons (which have energies $E \sim k_B T$), the regime of classical behavior ($\eta_e \ll 1$; to the left of the solid line) is $\mu_e \ll -k_B T$ and the regime of strong quantum behavior ($\eta_e \simeq 1$; *electron degeneracy;* to the right of the solid line) is $\mu_e \gg +k_B T$. The slanted solid boundary in Fig. 3.7 is thus the location $\mu_e = 0$, which translates via Eq. (3.39a) to

$$\rho = \rho_{\rm deg} \equiv 2m_p/\lambda_{\rm TdB}^3 = (2m_p/h^3)(2\pi m_e k_B T)^{3/2} = 0.00808(T/10^4 \text{ K})^{3/2} \text{ g cm}^{-3}. \tag{3.41}$$

Although the hydrogen gas is *degenerate* to the right of this boundary, we can still compute its equation of state using our kinetic-theory equations (3.37), so long as we use the quantum mechanically correct distribution function for the electrons—the Fermi-Dirac distribution (3.22a).[9] In this electron-degenerate region, $\mu_e \gg k_B T$, the electron mean occupation number $\eta_e = 1/(e^{(E-\mu_e)/(k_B T)} + 1)$ has the form shown

degenerate hydrogen gas

---

8. For both astrophysical and laboratory applications, the non-hydrogen elemental composition often matters, and involves straightforward corrections to purely hydrogen formulae given here.
9. Our kinetic-theory analysis and the Fermi-Dirac distribution ignore Coulomb interactions between the electrons, and between the electrons and the ions. They thereby miss so-called *Coulomb corrections* to the equation of state (Sec. 22.6.3) and other phenomena that are often important in condensed matter physics, but rarely important in astrophysics.

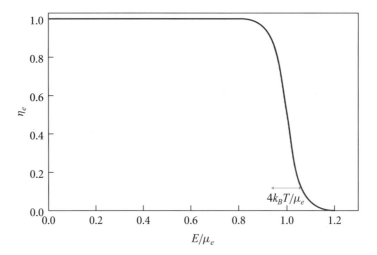

FIGURE 3.8 The Fermi-Dirac distribution function for electrons in the nonrelativistic, degenerate regime $k_B T \ll \mu_e \ll m_e$, with temperature such that $k_B T/\mu_e = 0.03$. Note that η_e drops from near 1 to near 0 over the range $\mu_e - 2k_B T \lesssim E \lesssim \mu_e + 2k_B T$. See Ex. 3.11b.

in Fig. 3.8 and thus can be well approximated by $\eta_e = 1$ for $E = p^2/(2m_e) < \mu_e$ and $\eta_e = 0$ for $E > \mu_e$; or equivalently by

$$\eta_e = 1 \text{ for } p < p_F \equiv \sqrt{2m_e \mu_e}, \qquad \eta_e = 0 \text{ for } p > p_F. \tag{3.42}$$

Fermi momentum

Here p_F is called the *Fermi momentum*. (The word "degenerate" refers to the fact that almost all the quantum states are fully occupied or are empty; i.e., η_e is everywhere nearly 1 or 0.) By inserting this degenerate distribution function [or, more precisely, $\mathcal{N}_e = (2/h^3)\eta_e$] into Eqs. (3.37) and integrating, we obtain $n_e \propto p_F^3$ and $P_e \propto p_F^5$. By then setting $n_e = n_p = \rho/m_p$ and solving for $p_F \propto n_e^{1/3} \propto \rho^{1/3}$ and inserting into the expression for P_e and evaluating the constants, we obtain (Ex. 3.9) the following equation of state for the electron pressure:

$$P_e = \frac{1}{20}\left(\frac{3}{\pi}\right)^{2/3} \frac{m_e c^2}{\lambda_C^3} \left(\frac{\rho}{m_p/\lambda_C^3}\right)^{5/3}. \tag{3.43}$$

Here

$$\lambda_C = h/(m_e c) = 2.426 \times 10^{-10} \text{ cm} \tag{3.44}$$

is the electron Compton wavelength.

The rapid growth $P_e \propto \rho^{5/3}$ of the electron pressure with increasing density is due to the degenerate electrons' being confined by the Pauli Exclusion Principle to regions of ever-shrinking size, causing their zero-point motions and associated pressure to grow. By contrast, the protons, with their far larger rest masses, remain nondegenerate [until their density becomes $(m_p/m_e)^{3/2} \sim 10^5$ times higher than Eq. (3.41)], and so their pressure is negligible compared to that of the electrons: the total pressure is

$$P = P_e = \text{Eq. (3.43)} \tag{3.45}$$

in the regime of nonrelativistic electron degeneracy. This is the equation of state for the interior of a low-mass white-dwarf star and for the outer layers of a high-mass white dwarf—aside from tiny corrections due to Coulomb interactions. In Sec. 13.3.2 we shall see how it can be used to explore the structures of white dwarfs. It is also the equation of state for a neutron star, with m_e replaced by the rest mass of a neutron m_n (since neutron degeneracy pressure dominates over that due to the star's tiny number of electrons and protons) and ρ/m_p replaced by the number density of neutrons— except that for neutron stars there are large corrections due to the strong nuclear force (see, e.g., Shapiro and Teukolsky, 1983).

When the density of hydrogen in this degenerate regime is pushed on upward to

$$\rho_{\text{rel deg}} = \frac{8\pi m_p}{3\lambda_c^3} \simeq 9.8 \times 10^5 \text{ g cm}^{-3} \tag{3.46}$$

(dotted vertical line in Fig. 3.7), the electrons' zero-point motions become relativistically fast (the electron chemical potential μ_e becomes of order $m_e c^2$ and the Fermi momentum p_F of order $m_e c$), so the nonrelativistic, Newtonian analysis fails, and the matter enters a domain of relativistic degeneracy (Sec. 3.5.4). Both domains, nonrelativistic degeneracy ($\mu_e \ll m_e c^2$) and relativistic degeneracy ($\mu_e \gtrsim m_e c^2$), occur for matter inside a massive white-dwarf star—the type of star that the Sun will become when it dies (see Shapiro and Teukolsky, 1983). In Sec. 26.3.5, we shall see how general relativity (spacetime curvature) modifies a star's structure. It also helps force sufficiently massive white dwarfs to collapse (Sec. 6.10 of Shapiro and Teukolsky, 1983).

The (almost) degenerate Fermi-Dirac distribution function shown in Fig. 3.8 has a thermal tail whose energy width is $4k_B T/\mu_e$. As the temperature T is increased, the number of electrons in this tail increases, thereby increasing the electrons' total energy E_{tot}. This increase is responsible for the electrons' *specific heat* (Ex. 3.11)— a quantity of importance for both the electrons in a metal (e.g., a copper wire) and the electrons in a white-dwarf star. The electrons dominate the specific heat when the temperature is sufficiently low; but at higher temperatures it is dominated by the energies of sound waves (see Ex. 3.12, where we use the kinetic theory of phonons to compute the sound waves' specific heat).

3.5.3 Relativistic Density, Pressure, Energy Density, and Equation of State

Now we turn to the relativistic domain of kinetic theory, initially for a single species of particle with rest mass m and then (in the next subsection) for matter composed of electrons and protons.

The relativistic *mean rest frame* of the particles, at some event $\mathcal{P}$ in spacetime, is that frame in which the particle flux $\mathbf{S}$ vanishes. We denote by $\vec{u}_{\text{rf}}$ the 4-velocity of this mean rest frame. As in Newtonian theory (Sec. 3.5.2), we are especially interested in distribution functions $\mathcal{N}$ that are *isotropic* in the mean rest frame: distribution functions that depend on the magnitude $|\mathbf{p}| \equiv p$ of the spatial momentum of a particle but not on its direction—or equivalently, that depend solely on the particles' energy

$$\mathcal{E} = -\vec{u}_{\rm rf} \cdot \vec{p} \quad \text{expressed in frame-independent form [Eq. (2.29)],}$$

$$\mathcal{E} = p^0 = \sqrt{m^2 + p^2} \quad \text{in mean rest frame.} \tag{3.47}$$

Such isotropy is readily produced by particle collisions (Sec. 3.7).

Notice that isotropy in the mean rest frame [i.e., $\mathcal{N} = \mathcal{N}(\mathcal{P}, \mathcal{E})$] does not imply isotropy in any other inertial frame. As seen in some other (primed) frame, $\vec{u}_{\rm rf}$ will have a time component $u_{\rm rf}^{0'} = \gamma$ and a space component $\mathbf{u}'_{\rm rf} = \gamma \mathbf{V}$ [where $\mathbf{V}$ is the mean rest frame's velocity relative to the primed frame, and $\gamma = (1 - \mathbf{V}^2)^{-1/2}$]; and correspondingly, in the primed frame, $\mathcal{N}$ will be a function of

$$\mathcal{E} = -\vec{u}_{\rm rf} \cdot \vec{p} = \gamma[(m^2 + \mathbf{p}'^2)^{\frac{1}{2}} - \mathbf{V} \cdot \mathbf{p}'], \tag{3.48}$$

which is anisotropic: it depends on the direction of the spatial momentum $\mathbf{p}'$ relative to the velocity $\mathbf{V}$ of the particle's mean rest frame. An example is the cosmic microwave radiation as viewed from Earth (Ex. 3.7).

As in Newtonian theory, isotropy greatly simplifies the momentum-space integrals (3.33) that we use to compute macroscopic properties of the particles: (i) The integrands of the expressions $S^j = \int \mathcal{N} p^j (d\mathcal{V}_p/\mathcal{E})$ and $T^{j0} = T^{0j} = \int \mathcal{N} p^j p^0 (d\mathcal{V}_p/\mathcal{E})$ for the particle flux, energy flux, and momentum density are all odd in the momentum-space coordinate p^j and therefore give vanishing integrals: $S^j = T^{j0} = T^{0j} = 0$. (ii) The integral $T^{jk} = \int \mathcal{N} p^j p^k d\mathcal{V}_p/\mathcal{E}$ produces an isotropic stress tensor, $T^{jk} = P g^{jk} = P \delta^{jk}$, whose pressure is most easily computed from its trace, $P = \frac{1}{3} T^{jj}$. Using these results and the relations $|\mathbf{p}| \equiv p$ for the magnitude of the momentum, $d\mathcal{V}_p = 4\pi p^2 dp$ for the momentum-space volume element, and $\mathcal{E} = p^0 = \sqrt{m^2 + p^2}$ for the particle energy, we can easily evaluate Eqs. (3.33) for the particle number density $n = S^0$, the total density of mass-energy T^{00} (which we denote ρ—the same notation as we use for mass density in Newtonian theory), and the pressure P. The results are

relativistic particle density, mass-energy density, and pressure

$$n \equiv S^0 = \int \mathcal{N} d\mathcal{V}_p = 4\pi \int_0^\infty \mathcal{N} p^2 dp, \tag{3.49a}$$

$$\rho \equiv T^{00} = \int \mathcal{N} \mathcal{E} d\mathcal{V}_p = 4\pi \int_0^\infty \mathcal{N} \mathcal{E} p^2 dp, \tag{3.49b}$$

$$P = \frac{1}{3} \int \mathcal{N} p^2 \frac{d\mathcal{V}_p}{\mathcal{E}} = \frac{4\pi}{3} \int_0^\infty \mathcal{N} \frac{p^4 dp}{\sqrt{m^2 + p^2}}. \tag{3.49c}$$

3.5.4 Equation of State for a Relativistic Degenerate Hydrogen Gas

Return to the hydrogen gas whose nonrelativistic equations of state were computed in Sec. 3.5.1. As we deduced there, at densities $\rho \gtrsim 10^5 \text{ g cm}^{-3}$ (near and to the right of the vertical dotted line in Fig. 3.7) the electrons are squeezed into such tiny volumes that their zero-point energies are $\gtrsim m_e c^2$, forcing us to treat them relativistically.

We can do so with the aid of the following approximation for the relativistic Fermi-Dirac mean occupation number $\eta_e = 1/[e^{(\mathcal{E}-\tilde{\mu}_e/(k_B T))} + 1]$:

relativistically degenerate electrons

$$\eta_e \simeq 1 \text{ for } \mathcal{E} < \tilde{\mu}_e \equiv \mathcal{E}_F; \text{ i.e., for } p < p_F = \sqrt{\mathcal{E}_F^2 - m^2}, \tag{3.50}$$

$$\eta_e \simeq 0 \text{ for } \mathcal{E} > \mathcal{E}_F; \text{ i.e., for } p > p_F. \tag{3.51}$$

Here $\mathcal{E}_F$ is called the relativistic *Fermi energy* and p_F the relativistic *Fermi momentum*. By inserting this η_e along with $\mathcal{N}_e = (2/h^3)\eta_e$ into the integrals (3.49) for the electron number density n_e, total density of mass-energy ρ_e, and pressure P_e, and performing the integrals (Ex. 3.10), we obtain results that are expressed most simply in terms of a parameter t (not to be confused with time) defined by

$$\mathcal{E}_F \equiv \tilde{\mu}_e \equiv m_e \cosh(t/4), \qquad p_F \equiv \sqrt{\mathcal{E}_F^2 - m_e^2} \equiv m_e \sinh(t/4). \tag{3.52a}$$

The results are

$$n_e = \frac{8\pi}{3\lambda_C^3}\left(\frac{p_F}{m_e}\right)^3 = \frac{8\pi}{3\lambda_C^3}\sinh^3(t/4), \tag{3.52b}$$

$$\rho_e = \frac{8\pi m_e}{\lambda_C^3}\int_0^{p_F/m_e} x^2\sqrt{1+x^2}\,dx = \frac{\pi m_e}{4\lambda_C^3}[\sinh(t) - t], \tag{3.52c}$$

$$P_e = \frac{8\pi m_e}{\lambda_C^3}\int_0^{p_F/m_e} \frac{x^4}{\sqrt{1+x^2}}\,dx = \frac{\pi m_e}{12\lambda_C^3}[\sinh(t) - 8\sinh(t/2) + 3t]. \tag{3.52d}$$

These parametric relationships for ρ_e and P_e as functions of the electron number density n_e are sometimes called the Anderson-Stoner equation of state, because they were first derived by Wilhelm Anderson and Edmund Stoner in 1930 (see Thorne, 1994, pp. 153–154). They are valid throughout the full range of electron degeneracy, from nonrelativistic up to ultrarelativistic.

In a white-dwarf star, the protons, with their high rest mass, are nondegenerate, the total density of mass-energy is dominated by the proton rest-mass density, and since there is one proton for each electron in the hydrogen gas, that total is

relativistically degenerate hydrogen gas

$$\rho \simeq m_p n_e = \frac{8\pi m_p}{3\lambda_C^3}\sinh^3(t/4). \tag{3.53a}$$

By contrast (as in the nonrelativistic regime), the pressure is dominated by the electrons (because of their huge zero-point motions), not the protons; and so the total pressure is

$$P = P_e = \frac{\pi m_e}{12\lambda_C^3}[\sinh(t) - 8\sinh(t/2) + 3t]. \tag{3.53b}$$

3.5 Isotropic Distribution Functions and Equations of State

In the low-density limit, where $t \ll 1$ so $p_F \ll m_e = m_e c$, we can solve the relativistic equation (3.52b) for t as a function of $n_e = \rho/m_p$ and insert the result into the relativistic expression (3.53b); the result is the nonrelativistic equation of state (3.43).

The dividing line $\rho = \rho_{\text{rel deg}} = 8\pi m_p/(3\lambda_c^3) \simeq 1.0 \times 10^6$ g cm^{-3} [Eq. (3.46)] between nonrelativistic and relativistic degeneracy is the point where the electron Fermi momentum is equal to the electron rest mass [i.e., $\sinh(t/4) = 1$]. The equation of state (3.53a) and (3.53b) implies

$$P_e \propto \rho^{5/3} \quad \text{in the nonrelativistic regime, } \rho \ll \rho_{\text{rel deg}},$$

$$P_e \propto \rho^{4/3} \quad \text{in the relativistic regime, } \rho \gg \rho_{\text{rel deg}}. \tag{3.53c}$$

These asymptotic equations of state turn out to play a crucial role in the structure and stability of white dwarf stars (Secs. 13.3.2 and 26.3.5; Shapiro and Teukolsky, 1983; Thorne, 1994, Chap. 4).

3.5.5 Equation of State for Radiation

thermalized radiation

As was discussed at the end of Sec. 3.3, for a gas of thermalized photons in an environment where photons are readily created and absorbed, the distribution function has the blackbody (Planck) form $\eta = 1/(e^{\mathcal{E}/(k_B T)} - 1)$, which we can rewrite as $1/(e^{p/(k_B T)} - 1)$, since the energy $\mathcal{E}$ of a photon is the same as the magnitude p of its momentum. In this case, the relativistic integrals (3.49) give (see Ex. 3.13)

$$\boxed{n = bT^3, \qquad \rho = aT^4, \qquad P = \frac{1}{3}\rho,} \tag{3.54a}$$

where

$$b = 16\pi \zeta(3) \frac{k_B^3}{h^3 c^3} = 20.28 \text{ cm}^{-3}\text{ K}^{-3}, \tag{3.54b}$$

$$a = \frac{8\pi^5}{15} \frac{k_B^4}{h^3 c^3} = 7.566 \times 10^{-15} \text{ erg cm}^{-3}\text{ K}^{-4} = 7.566 \times 10^{-16} \text{ J m}^{-3}\text{ K}^{-4} \tag{3.54c}$$

are *radiation constants*. Here $\zeta(3) = \sum_{n=1}^{\infty} n^{-3} = 1.2020569\ldots$ is the Riemann zeta function.

As we shall see in Sec. 28.4, when the universe was younger than about 100,000 years, its energy density and pressure were predominantly due to thermalized photons plus neutrinos (which contributed approximately the same as the photons), so its equation of state was given by Eq. (3.54a) with the coefficient changed by a factor of order unity. Einstein's general relativistic field equations (Sec. 25.8 and Chap. 28) relate the energy density ρ of these photons and neutrinos to the age of the universe t as measured in the photons' and neutrinos' mean rest frame:

$$\frac{3}{32\pi G t^2} = \rho \simeq aT^4. \tag{3.55a}$$

Here G is Newton's gravitation constant. Putting in numbers, we find that

$$\rho = \frac{4.5 \times 10^{-10} \text{ g cm}^{-3}}{(\tau/1 \text{ yr})^2}, \qquad T \simeq \frac{2.7 \times 10^6 \text{ K}}{(\tau/1 \text{ yr})^{1/2}}. \tag{3.55b}$$

This implies that, when the universe was 1 minute old, its radiation density and temperature were about 100 g cm^{-3} and 2×10^9 K, respectively. These conditions and the proton density were well suited for burning hydrogen to helium; and, indeed, about 1/4 of all the mass of the universe did get burned to helium at this early epoch. We shall examine this in further detail in Sec. 28.4.2.

EXERCISES

Exercise 3.8 *Derivation and Practice: Equation of State for Nonrelativistic, Classical Gas*
Consider a collection of identical, classical (i.e., with $\eta \ll 1$) particles with a distribution function $\mathcal{N}$ that is thermalized at a temperature T such that $k_B T \ll mc^2$ (nonrelativistic temperature).

(a) Show that the distribution function, expressed in terms of the particles' momenta or velocities in their mean rest frame, is

$$\mathcal{N} = \frac{g_s}{h^3} e^{\mu/(k_B T)} e^{-p^2/(2mk_B T)}, \quad \text{where } p = |\mathbf{p}| = mv, \tag{3.56}$$

with v being the speed of a particle.

(b) Show that the number density of particles in the mean rest frame is given by Eq. (3.39a).

(c) Show that this gas satisfies the equations of state (3.39b).

Note: The following integrals, for nonnegative integral values of q, will be useful:

$$\int_0^\infty x^{2q} e^{-x^2} dx = \frac{(2q-1)!!}{2^{q+1}} \sqrt{\pi}, \tag{3.57}$$

where $n!! \equiv n(n-2)(n-4) \ldots (2 \text{ or } 1)$; and

$$\int_0^\infty x^{2q+1} e^{-x^2} dx = \frac{1}{2} q!. \tag{3.58}$$

Exercise 3.9 *Derivation and Practice: Equation of State for Nonrelativistic, Electron-Degenerate Hydrogen*
Derive Eq. (3.43) for the electron pressure in a nonrelativistic, electron-degenerate hydrogen gas.

Exercise 3.10 *Derivation and Practice: Equation of State for Relativistic, Electron-Degenerate Hydrogen*
Derive the equations of state (3.52) for an electron-degenerate hydrogen gas. (Note: It might be easiest to compute the integrals with the help of symbolic manipulation software, such as Mathematica, Matlab, or Maple.)

Exercise 3.11 *Example: Specific Heat for Nonrelativistic, Degenerate Electrons in White Dwarfs and in Metals*

Consider a nonrelativistically degenerate electron gas at finite but small temperature.

(a) Show that the inequalities $k_B T \ll \mu_e \ll m_e$ are equivalent to the words "nonrelativistically degenerate."

(b) Show that the electron mean occupation number $\eta_e(E)$ has the form depicted in Fig. 3.8: It is near unity out to (nonrelativistic) energy $E \simeq \mu_e - 2k_B T$, and it then drops to nearly zero over a range of energies $\Delta E \sim 4k_B T$.

(c) If the electrons were nonrelativistic but *non*degenerate, their thermal energy density would be $U = \frac{3}{2} n k_B T$, so the total electron energy (excluding rest mass) in a volume V containing $N = nV$ electrons would be $E_{\text{tot}} = \frac{3}{2} N k_B T$, and the electron specific heat, at fixed volume, would be

$$C_V \equiv \left(\frac{\partial E_{\text{tot}}}{\partial T}\right)_V = \frac{3}{2} N k_B \quad \text{(nondegenerate, nonrelativistic)}. \tag{3.59}$$

Using the semiquantitative form of η_e depicted in Fig. 3.8, show that to within a factor of order unity the specific heat of degenerate electrons is smaller than in the nondegenerate case by a factor $\sim k_B T/\mu_e$:

$$C_V \equiv \left(\frac{\partial E_{\text{tot}}}{\partial T}\right)_V \sim \left(\frac{k_B T}{\mu_e}\right) N k_B \quad \text{(degenerate, nonrelativistic)}. \tag{3.60}$$

(d) Compute the multiplicative factor in Eq. (3.60) for C_V. More specifically, show that, to first order in $k_B T/\mu_e$,

$$C_V = \frac{\pi^2}{2} \left(\frac{k_B T}{\mu_e}\right) N k_B. \tag{3.61}$$

(e) As an application, consider hydrogen inside a white dwarf with density $\rho = 10^5$ g cm^{-3} and temperature $T = 10^6$ K. (These are typical values for a white-dwarf interior). What are the numerical values of μ_e/m_e and $k_B T/\mu_e$ for the electrons? What is the numerical value of the dimensionless factor $(\pi^2/2)(k_B T/\mu_e)$ by which degeneracy reduces the electron specific heat?

(f) As a second application, consider the electrons inside a copper wire in a laboratory on Earth at room temperature. Each copper atom donates about one electron to a "gas" of freely traveling (conducting) electrons and keeps the rest of its electrons bound to itself. (We neglect interaction of this electron gas with the ions, thereby missing important condensed matter complexities, such as conduction bands and what distinguishes conducting materials from insulators.)

What are the numerical values of μ_e/m_e and $k_B T/\mu_e$ for the conducting electron gas? Verify that these are in the range corresponding to nonrelativistic degeneracy. What is the value of the factor $(\pi^2/2)(k_B T/\mu_e)$ by which degeneracy reduces the electron specific heat? At room temperature, this electron contribution to the specific heat is far smaller than the contribution from thermal vibrations of the copper atoms (i.e., thermal sound waves, i.e., thermal *phonons*), but at very low temperatures the electron contribution dominates, as we shall see in the next exercise.

Exercise 3.12 *Example: Specific Heat for Phonons in an Isotropic Solid*
In Sec. 12.2 we will study classical sound waves propagating through an isotropic, elastic solid. As we shall see, there are two types of sound waves: *longitudinal* with frequency-independent speed C_L, and *transverse* with a somewhat smaller frequency-independent speed C_T. For each type of wave, $s = L$ or T, the material of the solid undergoes an elastic displacement $\boldsymbol{\xi} = A\mathbf{f}_s \exp(i\mathbf{k}\cdot\mathbf{x} - \omega t)$, where A is the wave amplitude, $\mathbf{f}_s$ is a unit vector (polarization vector) pointing in the direction of the displacement, $\mathbf{k}$ is the wave vector, and ω is the wave frequency. The wave speed is $C_s = \omega/|\mathbf{k}|$ ($= C_L$ or C_T). Associated with these waves are quanta called phonons. As for any wave, each phonon has a momentum related to its wave vector by $\mathbf{p} = \hbar\mathbf{k}$, and an energy related to its frequency by $E = \hbar\omega$. Combining these relations we learn that the relationship between a phonon's energy and the magnitude $p = |\mathbf{p}|$ of its momentum is $E = C_s p$. This is the same relationship as for photons, but with the speed of light replaced by the speed of sound! For longitudinal waves $\mathbf{f}_L$ is in the propagation direction $\mathbf{k}$, so there is just one polarization, $g_L = 1$. For transverse waves $\mathbf{f}_T$ is orthogonal to $\mathbf{k}$, so there are two orthogonal polarizations (e.g., $\mathbf{f}_T = \mathbf{e}_x$ and $\mathbf{f}_T = \mathbf{e}_y$ when $\mathbf{k}$ points in the $\mathbf{e}_z$ direction), $g_T = 2$.

(a) Phonons of both types, longitudinal and transverse, are bosons. Why? [Hint: Each normal mode of an elastic body can be described mathematically as a harmonic oscillator.]

(b) Phonons are fairly easily created, absorbed, scattered, and thermalized. A general argument that we will give for chemical reactions in Sec. 5.5 can be applied to phonon creation and absorption to deduce that, once they reach complete thermal equilibrium with their environment, the phonons will have vanishing chemical potential $\mu = 0$. What, then, will be their distribution functions η and $\mathcal{N}$?

(c) Ignoring the fact that the sound waves' wavelengths $\lambda = 2\pi/|\mathbf{k}|$ cannot be smaller than about twice the spacing between the atoms of the solid, show that the total phonon energy (wave energy) in a volume V of the solid is identical to that for blackbody photons in a volume V, but with the speed of light c replaced by the speed of sound C_s, and with the photon number of spin states, 2, replaced by $g_s = 3$ (2 for transverse waves plus 1 for longitudinal): $E_{\text{tot}} = a_s T^4 V$, with $a_s = g_s(4\pi^5/15)(k_B^4/(h^3 C_s^3))$ [cf. Eqs. (3.54)].

(d) Show that the specific heat of the phonon gas (the sound waves) is $C_V = 4a_s T^3 V$. This scales as T^3, whereas in a metal the specific heat of the degenerate electrons scales as T (previous exercise), so at sufficiently low temperatures the electron specific heat will dominate over that of the phonons.

(e) Show that in the phonon gas, only phonon modes with wavelengths longer than $\sim \lambda_T = C_s h/(k_B T)$ are excited; that is, for $\lambda \ll \lambda_T$ the mean occupation number is $\eta \ll 1$; for $\lambda \sim \lambda_T$, $\eta \sim 1$; and for $\lambda \gg \lambda_T$, $\eta \gg 1$. As T is increased, λ_T gets reduced. Ultimately it becomes of order the interatomic spacing, and our computation fails, because most of the modes that our calculation assumes are thermalized actually don't exist. What is the critical temperature (*Debye temperature*) at which our computation fails and the T^3 law for C_V changes? Show by a roughly one-line argument that above the Debye temperature, C_V is independent of temperature.

Exercise 3.13 *Derivation and Practice: Equation of State for a Photon Gas*
(a) Consider a collection of photons with a distribution function $\mathcal{N}$ that, in the mean rest frame of the photons, is isotropic. Show, using Eqs. (3.49b) and (3.49c), that this photon gas obeys the equation of state $P = \frac{1}{3}\rho$.

(b) Suppose the photons are thermalized with zero chemical potential (i.e., they are isotropic with a blackbody spectrum). Show that $\rho = aT^4$, where a is the radiation constant of Eq. (3.54c). [Note: Do not hesitate to use Mathematica, Matlab, or Maple, or other computer programs to evaluate integrals!]

(c) Show that for this isotropic, blackbody photon gas the number density of photons is $n = bT^3$, where b is given by Eq. (3.54b), and that the mean energy of a photon in the gas is

$$\bar{\mathcal{E}}_\gamma = \frac{\pi^4}{30\zeta(3)} k_B T \simeq 2.701\, k_B T. \tag{3.62}$$

3.6 Evolution of the Distribution Function: Liouville's Theorem, the Collisionless Boltzmann Equation, and the Boltzmann Transport Equation

We now turn to the issue of how the distribution function $\eta(\mathcal{P}, \vec{p})$, or equivalently, $\mathcal{N} = (g_s/h^3)\eta$, evolves from point to point in phase space. We explore the evolution under the simple assumption that between their very brief collisions, the particles all move freely, uninfluenced by any forces. It is straightforward to generalize to a situation where the particles interact with electromagnetic, gravitational, or other fields as they move, and we do so in Box 3.2, and Sec. 4.3. However, in the body of this chapter, we restrict attention to the common situation of free motion between collisions.

Initially we even rule out collisions; only at the end of this section do we restore them by inserting them as an additional term in our collision-free evolution equation for η.

The foundation for the collision-free evolution law will be *Liouville's theorem*. Consider a set $\mathcal{S}$ of particles that are initially all near some location in phase space and initially occupy an infinitesimal (frame-independent) phase-space volume $d^2\mathcal{V} = d\mathcal{V}_x d\mathcal{V}_p$. Pick a particle at the center of the set $\mathcal{S}$ and call it the "fiducial particle." Since all the particles in $\mathcal{S}$ have nearly the same initial position and velocity, they subsequently all move along nearly the same trajectory (world line): they all remain congregated around the fiducial particle. Liouville's theorem says that the phase-space volume occupied by the set of particles $\mathcal{S}$ is conserved along the trajectory of the fiducial particle:

$$\frac{d}{d\ell}(d\mathcal{V}_x d\mathcal{V}_p) = 0. \tag{3.63}$$

Liouville's theorem

Here ℓ is an arbitrary parameter along the trajectory. For example, in Newtonian theory ℓ could be universal time t or distance l traveled, and in relativity it could be proper time τ as measured by the fiducial particle (if its rest mass is nonzero) or the affine parameter ζ that is related to the fiducial particle's 4-momentum by $\vec{p} = d\vec{x}/d\zeta$.

We shall prove Liouville's theorem with the aid of the diagrams in Fig. 3.9. Assume, for simplicity, that the particles have nonzero rest mass. Consider the region in phase space occupied by the particles, as seen in the inertial reference frame (rest frame) of the fiducial particle, and choose for ℓ the time t of that inertial frame (or in Newtonian theory the universal time t). Choose the particles' region $d\mathcal{V}_x d\mathcal{V}_p$ at $t = 0$ to be a rectangular box centered on the fiducial particle (i.e., on the origin $x^j = 0$ of its inertial frame; Fig. 3.9a). Examine the evolution with time t of the 2-dimensional slice $y = p_y = z = p_z = 0$ through the occupied region. The evolution of other slices will be similar. Then, as t passes, the particle at location (x, p_x) moves with velocity $dx/dt = p_x/m$ (where the nonrelativistic approximation to the velocity is used, because all the particles are very nearly at rest in the fiducial particle's inertial frame). Because the particles move freely, each has a conserved p_x, and their motion $dx/dt = p_x/m$ (larger speeds are higher in the diagram) deforms the particles' phase space region into a skewed parallelogram as shown in Fig. 3.9b. Obviously, the area of the occupied region, $\Delta x \Delta p_x$, is conserved.

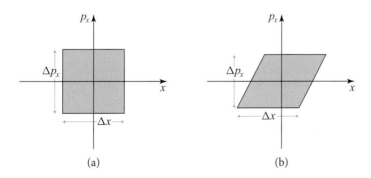

FIGURE 3.9 The phase-space region (x-p_x part) occupied by a set $\mathcal{S}$ of particles with finite rest mass, as seen in the inertial frame of the central, fiducial particle. (a) The initial region. (b) The region after a short time.

3.6 Evolution of the Distribution Function

This same argument shows that the x-p_x area is conserved at *all* values of y, z, p_y, p_z; similarly for the areas in the y-p_y planes and those in the z-p_z planes. As a consequence, the total volume in phase space, $d\mathcal{V}_x d\mathcal{V}_p = \Delta x \Delta p_x \Delta y \Delta p_y \Delta z \Delta p_z$, is conserved.

Although this proof of Liouville's theorem relies on the assumption that the particles have nonzero rest mass, the theorem is also true for particles with zero rest mass—as one can deduce by taking the relativistic limit as the rest mass goes to zero and the particles' 4-momenta become null.

Since, in the absence of collisions or other nongravitational interactions, the number dN of particles in the set $\mathcal{S}$ is conserved, Liouville's theorem immediately implies the conservation of the number density in phase space, $\mathcal{N} = dN/(d\mathcal{V}_x d\mathcal{V}_p)$:

collisionless Boltzmann equation

$$\boxed{\frac{d\mathcal{N}}{d\ell} = 0 \quad \text{along the trajectory of a fiducial particle.}} \tag{3.64}$$

This conservation law is called the *collisionless Boltzmann equation*; in the context of plasma physics (Part VI) it is sometimes called the *Vlasov equation*. Note that it says that *not only is the distribution function $\mathcal{N}$ frame independent; $\mathcal{N}$ also is constant along the phase-space trajectory of any freely moving particle.*

The collisionless Boltzmann equation is actually far more general than is suggested by the above derivation; see Box 3.2, which is best read after finishing this section.

The collisionless Boltzmann equation is most nicely expressed in the frame-independent form Eq. (3.64). For some purposes, however, it is helpful to express the equation in a form that relies on a specific but arbitrary choice of inertial reference frame. Then $\mathcal{N}$ can be regarded as a function of the reference frame's seven phase-space coordinates, $\mathcal{N} = \mathcal{N}(t, x^j, p_k)$, and the collisionless Boltzmann equation (3.64) takes the coordinate-dependent form

$$\frac{d\mathcal{N}}{d\ell} = \frac{dt}{d\ell}\frac{\partial \mathcal{N}}{\partial t} + \frac{dx_j}{d\ell}\frac{\partial \mathcal{N}}{\partial x_j} + \frac{dp_j}{d\ell}\frac{\partial \mathcal{N}}{\partial p_j} = \frac{dt}{d\ell}\left(\frac{\partial \mathcal{N}}{\partial t} + v_j \frac{\partial \mathcal{N}}{\partial x_j}\right) = 0. \tag{3.65}$$

Here we have used the equation of straight-line motion $dp_j/dt = 0$ for the particles and have set dx_j/dt equal to the particle velocity v_j.

Since our derivation of the collisionless Boltzmann equation relies on the assumption that no particles are created or destroyed as time passes, the collisionless Boltzmann equation in turn should guarantee conservation of the number of particles, $\partial n/\partial t + \boldsymbol{\nabla} \cdot \mathbf{S} = 0$ in Newtonian theory (Sec. 1.8), and $\vec{\nabla} \cdot \vec{S} = 0$ relativistically (Sec. 2.12.3). Indeed, this is so; see Ex. 3.14. Similarly, since the collisionless Boltzmann equation is based on the law of momentum (or 4-momentum) conservation for all the individual particles, it is reasonable to expect that the collisionless Boltzmann equation will guarantee the conservation of their total Newtonian momentum [$\partial \mathbf{G}/\partial t + \boldsymbol{\nabla} \cdot \mathbf{T} = 0$, Eq. (1.36)] and their relativistic 4-momentum [$\vec{\nabla} \cdot \boldsymbol{T} = 0$,

Eq. (2.73a)]. And indeed, these conservation laws do follow from the collisionless Boltzmann equation; see Ex. 3.14.

Thus far we have assumed that the particles move freely through phase space with no collisions. If collisions occur, they will produce some nonconservation of $\mathcal{N}$ along the trajectory of a freely moving, noncolliding fiducial particle, and correspondingly, the collisionless Boltzmann equation will be modified to read

$$\boxed{\frac{d\mathcal{N}}{d\ell} = \left(\frac{d\mathcal{N}}{d\ell}\right)_{\text{collisions}}}, \qquad (3.66)$$

Boltzmann transport equation

where the right-hand side represents the effects of collisions. This equation, with collision terms present, is called the *Boltzmann transport equation*. The actual form of the collision terms depends, of course, on the details of the collisions. We meet some specific examples in the next section [Eqs. (3.79), (3.86a), (3.87), and Ex. 3.21] and in our study of plasmas (Chaps. 22 and 23).

When one applies the collisionless Boltzmann equation or Boltzmann transport equation to a given situation, it is helpful to simplify one's thinking in two ways: (i) Adjust the normalization of the distribution function so it is naturally tuned to the situation. For example, when dealing with photons, I_ν/ν^3 is typically best, and if—as is usually the case—the photons do not change their frequencies as they move and only a single reference frame is of any importance, then I_ν alone may do; see Ex. 3.15. (ii) Adjust the differentiation parameter ℓ so it is also naturally tuned to the situation.

EXERCISES

Exercise 3.14 *Derivation and Problem: Collisionless Boltzmann Equation Implies Conservation of Particles and of 4-Momentum*
Consider a collection of freely moving, noncolliding particles that satisfy the collisionless Boltzmann equation $d\mathcal{N}/d\ell = 0$.

(a) Show that this equation guarantees that the Newtonian particle conservation law $\partial n/\partial t + \nabla \cdot \mathbf{S} = 0$ and momentum conservation law $\partial \mathbf{G}/\partial t + \nabla \cdot \mathbf{T} = 0$ are satisfied, where n, $\mathbf{S}$, $\mathbf{G}$, and $\mathbf{T}$ are expressed in terms of the distribution function $\mathcal{N}$ by the Newtonian momentum-space integrals (3.32).

(b) Show that the relativistic Boltzmann equation guarantees the relativistic conservation laws $\vec{\nabla} \cdot \vec{S} = 0$ and $\vec{\nabla} \cdot \mathbf{T} = 0$, where the number-flux 4-vector $\vec{S}$ and the stress-energy tensor $\mathbf{T}$ are expressed in terms of $\mathcal{N}$ by the momentum-space integrals (3.33).

Exercise 3.15 **Example: Solar Heating of Earth: The Greenhouse Effect*
In this example we study the heating of Earth by the Sun. Along the way, we derive some important relations for blackbody radiation.

BOX 3.2. SOPHISTICATED DERIVATION OF RELATIVISTIC COLLISIONLESS BOLTZMANN EQUATION

Denote by $\vec{X} \equiv \{\mathcal{P}, \vec{p}\}$ a point in 8-dimensional phase space. In an inertial frame the coordinates of $\vec{X}$ are $\{x^0, x^1, x^2, x^3, p_0, p_1, p_2, p_3\}$. [We use up (contravariant) indices on x and down (covariant) indices on p, because this is the form required in Hamilton's equations below; i.e., it is p_α and not p^α that is canonically conjugate to x^α.] Regard $\mathcal{N}$ as a function of location $\vec{X}$ in 8-dimensional phase space. Our particles all have the same rest mass, so $\mathcal{N}$ is nonzero only on the mass hyperboloid, which means that as a function of $\vec{X}$, $\mathcal{N}$ entails a delta function. For the following derivation, that delta function is irrelevant; the derivation is valid also for distributions of nonidentical particles, as treated in Ex. 3.2.

A particle in our distribution $\mathcal{N}$ at location $\vec{X}$ moves through phase space along a world line with tangent vector $d\vec{X}/d\zeta$, where ζ is its affine parameter. The product $\mathcal{N} d\vec{X}/d\zeta$ represents the number-flux 8-vector of particles through spacetime, as one can see by an argument analogous to Eq. (3.35c). We presume that, as the particles move through phase space, none are created or destroyed. The law of particle conservation in phase space, by analogy with $\vec{\nabla} \cdot \vec{S} = 0$ in spacetime, takes the form $\vec{\nabla} \cdot (\mathcal{N} d\vec{X}/d\zeta) = 0$. In terms of coordinates in an inertial frame, this conservation law says

$$\frac{\partial}{\partial x^\alpha}\left(\mathcal{N}\frac{dx^\alpha}{d\zeta}\right) + \frac{\partial}{\partial p_\alpha}\left(\mathcal{N}\frac{dp_\alpha}{d\zeta}\right) = 0. \tag{1}$$

The motions of individual particles in phase space are governed by Hamilton's equations

$$\frac{dx^\alpha}{d\zeta} = \frac{\partial \mathcal{H}}{\partial p_\alpha}, \quad \frac{dp_\alpha}{d\zeta} = -\frac{\partial \mathcal{H}}{\partial x^\alpha}. \tag{2}$$

For the freely moving particles of this chapter, a convenient form for the relativistic hamiltonian is [cf. Goldstein, Poole, and Safko (2002, Sec. 8.4) or Misner, Thorne, and Wheeler (1973, p. 489), who call it the super-hamiltonian]

$$\mathcal{H} = \frac{1}{2}(p_\alpha p_\beta g^{\alpha\beta} + m^2). \tag{3}$$

Our derivation of the collisionless Boltzmann equation does not depend on this specific form of the hamiltonian; it is valid for any hamiltonian and thus, for example, for particles interacting with an electromagnetic field or even a relativistic gravitational field (spacetime curvature; Part VII). By inserting Hamilton's equations (2) into the 8-dimensional law of particle conservation (1), we obtain

(continued)

> **BOX 3.2. (continued)**
>
> $$\frac{\partial}{\partial x^\alpha}\left(\mathcal{N}\frac{\partial \mathcal{H}}{\partial p_\alpha}\right) - \frac{\partial}{\partial p_\alpha}\left(\mathcal{N}\frac{\partial \mathcal{H}}{\partial x^\alpha}\right) = 0. \tag{4}$$
>
> Using the rule for differentiating products and noting that the terms involving two derivatives of $\mathcal{H}$ cancel, we bring this into the form
>
> $$0 = \frac{\partial \mathcal{N}}{\partial x^\alpha}\frac{\partial \mathcal{H}}{\partial p_\alpha} - \frac{\partial \mathcal{N}}{\partial p_\alpha}\frac{\partial \mathcal{H}}{\partial x^\alpha} = \frac{\partial \mathcal{N}}{\partial x^\alpha}\frac{dx^\alpha}{d\zeta} - \frac{\partial \mathcal{N}}{\partial p_\alpha}\frac{dp_\alpha}{d\zeta} = \frac{d\mathcal{N}}{d\zeta}, \tag{5}$$
>
> which is the collisionless Boltzmann equation. (To get the second expression we have used Hamilton's equations, and the third follows directly from the formulas of differential calculus.) *Thus, the collisionless Boltzmann equation is a consequence of just two assumptions: conservation of particles and Hamilton's equations for the motion of each particle. This implies that the Boltzmann equation has very great generality.* We extend and explore this generality in the next chapter.

Since we will study photon propagation from the Sun to Earth with Doppler shifts playing a negligible role, there is a preferred inertial reference frame: that of the Sun and Earth with their relative motion neglected. We carry out our analysis in that frame. Since we are dealing with thermalized photons, the natural choice for the distribution function is I_ν/ν^3; and since we use just one unique reference frame, each photon has a fixed frequency ν, so we can forget about the ν^3 and use I_ν.

(a) Assume, as is very nearly true, that each spot on the Sun emits blackbody radiation in all outward directions with a common temperature $T_\odot = 5{,}800$ K. Show, by integrating over the blackbody I_ν, that the total energy flux (i.e., power per unit surface area) F emitted by the Sun is

$$F \equiv \frac{d\mathcal{E}}{dt\,dA} = \sigma T_\odot^4, \quad \text{where} \quad \sigma = \frac{ac}{4} = \frac{2\pi^5}{15}\frac{k_B^4}{h^3 c^2} = 5.67 \times 10^{-5}\frac{\text{erg}}{\text{cm}^2\,\text{s}\,\text{K}^4}. \tag{3.67}$$

(b) Since the distribution function I_ν is conserved along each photon's trajectory, observers on Earth, looking at the Sun, see the same blackbody specific intensity I_ν as they would if they were on the Sun's surface, and similarly for any other star. By integrating over I_ν at Earth [and not by the simpler method of using Eq. (3.67) for the flux leaving the Sun], show that the energy flux arriving at Earth is $F = \sigma T_\odot^4 (R_\odot/r)^2$, where $R_\odot = 696{,}000$ km is the Sun's radius and $r = 1.496 \times 10^8$ km is the mean distance from the Sun to Earth.

(c) Our goal is to compute the temperature $T_\oplus$ of Earth's surface. As a first attempt, assume that all the Sun's flux arriving at Earth is absorbed by Earth's surface, heating it to the temperature $T_\oplus$, and then is reradiated into space as blackbody radiation at temperature $T_\oplus$. Show that this leads to a surface temperature of

$$T_\oplus = T_\odot \left(\frac{R_\odot}{2r}\right)^{1/2} = 280 \text{ K} = 7\,°\text{C}. \tag{3.68}$$

This is a bit cooler than the correct mean surface temperature (287 K = 14 °C).

(d) Actually, Earth has an albedo of $A \simeq 0.30$, which means that 30% of the sunlight that falls onto it gets reflected back into space with an essentially unchanged spectrum, rather than being absorbed. Show that with only a fraction $1 - A = 0.70$ of the solar radiation being absorbed, the above estimate of Earth's temperature becomes

$$T_\oplus = T_\odot \left(\frac{\sqrt{1-A}\, R_\odot}{2r}\right)^{1/2} = 256 \text{ K} = -17\,°\text{C}. \tag{3.69}$$

This is even farther from the correct answer.

(e) The missing piece of physics, which raises the temperature from $-17\,°\text{C}$ to something much nearer the correct 14 °C, is the *greenhouse effect*: The absorbed solar radiation has most of its energy at wavelengths $\sim 0.5\ \mu$m (in the visual band), which pass rather easily through Earth's atmosphere. By contrast, the blackbody radiation that Earth's surface wants to radiate back into space, with its temperature ~ 300 K, is concentrated in the infrared range from $\sim 8\ \mu$m to $\sim 30\ \mu$m. Water molecules, carbon dioxide and methane in Earth's atmosphere absorb about 40% of the energy that Earth tries to reradiate at these energies (Cox, 2000, Sec. 11.22), causing the reradiated energy to be about 60% that of a blackbody at Earth's surface temperature. Show that with this (oversimplified!) greenhouse correction, $T_\oplus$ becomes about 290 K = +17 °C, which is within a few degrees of the true mean temperature. There is overwhelming evidence that the measured increase in the average Earth temperature in recent decades is mostly caused by the measured increase in carbon dioxide and methane, which, in turn, is mostly due to human activity. Although the atmospheric chemistry is not well enough understood to make accurate predictions, mankind is performing a very dangerous experiment.

Exercise 3.16 **Challenge: Olbers' Paradox and Solar Furnace**
Consider a universe (not ours!) in which spacetime is flat and infinite in size and is populated throughout by stars that cluster into galaxies like our own and our neighbors, with interstellar and intergalactic distances similar to those in our neighborhood. Assume that the galaxies are *not* moving apart—there is no universal expansion. Using the collisionless Boltzmann equation for photons, show that Earth's temperature in this universe would be about the same as the surface temperatures of the universe's hotter stars, $\sim 10,000$ K, so we would all be fried. This is Olbers' Paradox. What features of our universe protect us from this fate?

Motivated by this model universe, describe a design for a furnace that relies on sunlight for its heat and achieves a temperature nearly equal to that of the Sun's surface, 5,800 K.

3.7 Transport Coefficients

In this section we turn to a practical application of kinetic theory: the computation of *transport coefficients*. Our primary objective is to illustrate the use of kinetic theory, but the transport coefficients themselves are also of interest: they play important roles in Parts V and VI (Fluid Dynamics and Plasma Physics) of this book.

What are transport coefficients? An example is electrical conductivity κ_e. When an electric field **E** is imposed on a sample of matter, Ohm's law tells us that the matter responds by developing a current density

$$\mathbf{j} = \kappa_e \mathbf{E}. \tag{3.70a}$$

electrical conductivity

The electrical conductivity is high if electrons can move through the material with ease; it is low if electrons have difficulty moving. The impediment to electron motion is scattering off other particles—off ions, other electrons, phonons (sound waves), plasmons (plasma waves), ... Ohm's law is valid when (as almost always) the electrons scatter many times, so they *diffuse* (random-walk their way) through the material. To compute the electrical conductivity, one must analyze, statistically, the effects of the many scatterings on the electrons' motions. The foundation for an accurate analysis is the Boltzmann transport equation (3.66).

Another example of a transport coefficient is thermal conductivity κ, which appears in the law of heat conduction

$$\mathbf{F} = -\kappa \nabla T. \tag{3.70b}$$

thermal conductivity

Here **F** is the diffusive energy flux from regions of high temperature T to low. The impediment to heat flow is scattering of the conducting particles; and, correspondingly, the foundation for accurately computing κ is the Boltzmann transport equation.

Other examples of transport coefficients are (i) the coefficient of shear viscosity η_{shear}, which determines the stress T_{ij} (diffusive flux of momentum) that arises in a shearing fluid [Eq. (13.68)]

coefficient of shear viscosity

$$T_{ij} = -2\eta_{\text{shear}} \sigma_{ij}, \tag{3.70c}$$

where σ_{ij} is the fluid's rate of shear (Ex. 3.19), and (ii) the diffusion coefficient D, which determines the diffusive flux of particles **S** from regions of high particle density n to low (Fick's law):

diffusion coefficient

$$\mathbf{S} = -D\nabla n. \tag{3.70d}$$

There is a *diffusion equation* associated with each of these transport coefficients. For example, the differential law of particle conservation $\partial n/\partial t + \nabla \cdot \mathbf{S} = 0$ [Eq. (1.30)], when applied to material in which the particles scatter many times so $\mathbf{S} = -D\nabla n$, gives the following diffusion equation for the particle number density:

diffusion equation

$$\boxed{\frac{\partial n}{\partial t} = D\nabla^2 n,} \qquad (3.71)$$

where we have assumed that D is spatially constant. In Ex. 3.17, by exploring solutions to this equation, we shall see that the root mean square (rms) distance $\bar{l}$ the particles travel is proportional to the square root of their travel time, $\bar{l} = \sqrt{4Dt}$, a behavior characteristic of diffusive random walks.[10] See Sec. 6.3 for deeper insights into this.

Similarly, the law of energy conservation, when applied to diffusive heat flow $\mathbf{F} = -\kappa\nabla T$, leads to a diffusion equation for the thermal energy density U and thence for temperature [Ex. 3.18 and Eq. (18.4)]. Maxwell's equations in a magnetized fluid, when combined with Ohm's law $\mathbf{j} = \kappa_e \mathbf{E}$, lead to diffusion equation (19.6) for magnetic field lines. And the law of angular momentum conservation, when applied to a shearing fluid with $T_{ij} = -2\eta_{\text{shear}}\sigma_{ij}$, leads to diffusion equation (14.6) for vorticity.

These diffusion equations, and all other physical laws involving transport coefficients, are approximations to the real world—approximations that are valid if and only if (i) many particles are involved in the transport of the quantity of interest (e.g., charge, heat, momentum, particles) and (ii) on average each particle undergoes many scatterings in moving over the length scale of the macroscopic inhomogeneities that drive the transport. This second requirement can be expressed quantitatively in terms of the *mean free path* λ between scatterings (i.e., the mean distance a particle travels between scatterings, as measured in the mean rest frame of the matter) and the *macroscopic inhomogeneity scale* $\mathcal{L}$ for the quantity that drives the transport (e.g., in heat transport that scale is $\mathcal{L} \sim T/|\nabla T|$; i.e., it is the scale on which the temperature changes by an amount of order itself). In terms of these quantities, the second criterion of validity is $\lambda \ll \mathcal{L}$. These two criteria (many particles and $\lambda \ll \mathcal{L}$) together are called *diffusion criteria,* since they guarantee that the quantity being transported (charge, heat, momentum, particles) will diffuse through the matter. If either of the two diffusion criteria fails, then the standard transport law (Ohm's law, the law of heat conduction, the Navier-Stokes equation, or the particle diffusion equation) breaks down and the corresponding transport coefficient becomes irrelevant and meaningless.

mean free path

diffusion criteria

The accuracy with which one can compute a transport coefficient using the Boltzmann transport equation depends on the accuracy of one's description of the scattering. If one uses a high-accuracy collision term $(d\mathcal{N}/d\ell)_{\text{collisions}}$ in the Boltzmann

---

10. Einstein derived this diffusion law $\bar{l} = \sqrt{4Dt}$, and he used it and his formula for the diffusion coefficient D, along with observational data about the diffusion of sugar molecules in water, to demonstrate the physical reality of molecules, determine their sizes, and deduce the numerical value of Avogadro's number; see the historical discussion in Pais (1982, Chap. 5).

equation, one can derive a highly accurate transport coefficient. If one uses a very crude approximation for the collision term, the resulting transport coefficient might be accurate only to within an order of magnitude—in which case, it was probably not worth the effort to use the Boltzmann equation; a simple order-of-magnitude argument would have done just as well. If the interaction between the diffusing particles and the scatterers is electrostatic or gravitational (long-range $1/r^2$ interaction forces), then the particles cannot be idealized as moving freely between collisions, and an accurate computation of transport coefficients requires a more sophisticated analysis: the Fokker-Planck equation developed in Sec. 6.9 and discussed, for plasmas, in Secs. 20.4.3 and 20.5.

In the following three subsections, we compute the coefficient of thermal conductivity κ for hot gas inside a star (where short-range collisions hold sway and the Boltzmann transport equation is highly accurate). We do so first by an order-of-magnitude argument and then by the Boltzmann equation with an accurate collision term. In Exs. 3.19 and 3.20, readers have the opportunity to compute the coefficient of viscosity and the diffusion coefficient for particles using moderately accurate collision terms, and in Ex. 3.21 (neutron diffusion in a nuclear reactor), we will meet diffusion in momentum space, by contrast with diffusion in physical space.

EXERCISES

Exercise 3.17 **Example: Solution of Diffusion Equation in an Infinite Homogeneous Medium**

(a) Show that the following is a solution to the diffusion equation (3.71) for particles in a homogeneous infinite medium:

$$n = \frac{N}{(4\pi Dt)^{3/2}} e^{-r^2/(4Dt)}, \tag{3.72}$$

(where $r \equiv \sqrt{x^2 + y^2 + z^2}$ is radius), and that it satisfies $\int n\, d\mathcal{V}_x = N$, so N is the total number of particles. Note that this is a Gaussian distribution with width $\sigma = \sqrt{4Dt}$. Plot this solution for several values of σ. In the limit as $t \to 0$, the particles are all localized at the origin. As time passes, they random-walk (diffuse) away from the origin, traveling a mean distance $\alpha\sigma = \alpha\sqrt{4Dt}$ after time t, where α is a coefficient of order one. We will meet this square-root-of-time evolution in other random-walk situations elsewhere in this book; see especially Exs. 6.3 and 6.4, and Sec. 6.7.2.

(b) Suppose that the particles have an arbitrary initial distribution $n_o(\mathbf{x})$ at time $t=0$. Show that their distribution at a later time t is given by the following Green's-function integral:

$$n(\mathbf{x}, t) = \int \frac{n_o(\mathbf{x}')}{(4\pi Dt)^{3/2}} e^{-|\mathbf{x}-\mathbf{x}'|^2/(4Dt)} d\mathcal{V}_{x'}. \tag{3.73}$$

(c) What form does the solution take in one dimension? And in two dimensions?

Exercise 3.18 ***Problem: Diffusion Equation for Temperature***
Use the law of energy conservation to show that, when heat diffuses through a homogeneous medium whose pressure is being kept fixed, the evolution of the temperature perturbation $\delta T \equiv T -$ (average temperature) is governed by the diffusion equation

$$\frac{\partial T}{\partial t} = \chi \nabla^2 T, \quad \text{where} \quad \chi = \kappa/(\rho c_P) \tag{3.74}$$

is called the *thermal diffusivity*. Here c_P is the specific heat per unit mass at fixed pressure, and ρc_P is that specific heat per unit volume. For the extension of this to heat flow in a moving fluid, see Eq. (18.4).

3.7.1 Diffusive Heat Conduction inside a Star

The specific transport-coefficient problem we treat here is for heat transport through hot gas deep inside a young, massive star. We confine attention to that portion of the star in which the temperature is $10^7 \text{ K} \lesssim T \lesssim 10^9 \text{ K}$, the mass density is $\rho \lesssim 10 \text{ g cm}^{-3}(T/10^7 \text{ K})^2$, and heat is carried primarily by diffusing photons rather than by diffusing electrons or ions or by convection. (We study convection in Chap. 18.) In this regime the primary impediment to the photons' flow is collisions with electrons. The lower limit on temperature, 10^7 K, guarantees that the gas is almost fully ionized, so there is a plethora of electrons to do the scattering. The upper limit on density, $\rho \sim 10 \text{ g cm}^{-3}(T/10^7 \text{ K})^2$, guarantees that (i) the inelastic scattering, absorption, and emission of photons by electrons accelerating in the Coulomb fields of ions (bremsstrahlung processes) are unimportant as impediments to heat flow compared to scattering off free electrons and (ii) the scattering electrons are nondegenerate (i.e., they have mean occupation numbers η small compared to unity) and thus behave like classical, free, charged particles. The upper limit on temperature, $T \sim 10^9$ K, guarantees that (i) the electrons doing the scattering are moving thermally at much less than the speed of light (the mean thermal energy $\frac{3}{2}k_B T$ of an electron is much less than its rest mass-energy $m_e c^2$) and (ii) the scattering is nearly elastic, with negligible energy exchange between photon and electron, and is describable with good accuracy by the *Thomson scattering cross section*.

In the rest frame of the electron (which to good accuracy will be the same as the mean rest frame of the gas, since the electron's speed relative to the mean rest frame is $\ll c$), the differential cross section $d\sigma$ for a photon to scatter from its initial propagation direction $\mathbf{n}'$ into a unit solid angle $d\Omega$ centered on a new propagation direction $\mathbf{n}$ is

$$\frac{d\sigma(\mathbf{n}' \to \mathbf{n})}{d\Omega} = \frac{3}{16\pi}\sigma_T[1 + (\mathbf{n} \cdot \mathbf{n}')^2]. \tag{3.75a}$$

Here σ_T is the total Thomson cross section [the integral of the differential cross section (3.75a) over solid angle]:

$$\sigma_T = \int \frac{d\sigma(\mathbf{n}' \to \mathbf{n})}{d\Omega} d\Omega = \frac{8\pi}{3}r_o^2 = 0.665 \times 10^{-24} \text{ cm}^2, \tag{3.75b}$$

where $r_o = e^2/(m_e c^2)$ is the classical electron radius. [For a derivation and discussion of the Thomson cross sections (3.75), see, e.g., Jackson (1999, Sec. 14.8).]

3.7.2 Order-of-Magnitude Analysis

Before embarking on any complicated calculation, it is always helpful to do a rough, order-of-magnitude analysis, thereby identifying the key physics and the approximate answer. The first step of a rough analysis of our heat transport problem is to identify the magnitudes of the relevant lengthscales. The inhomogeneity scale $\mathcal{L}$ for the temperature, which drives the heat flow, is the size of the hot stellar core, a moderate fraction of the Sun's radius: $\mathcal{L} \sim 10^5$ km. The mean free path of a photon can be estimated by noting that, since each electron presents a cross section σ_T to the photon and there are n_e electrons per unit volume, the probability of a photon being scattered when it travels a distance l through the gas is of order $n_e \sigma_T l$. Therefore, to build up to unit probability for scattering, the photon must travel a distance

$$\lambda \sim \frac{1}{n_e \sigma_T} \sim \frac{m_p}{\rho \sigma_T} \sim 3 \text{ cm} \left(\frac{1 \text{ g cm}^{-3}}{\rho} \right) \sim 3 \text{ cm}. \tag{3.76}$$

photon mean free path

Here m_p is the proton rest mass, $\rho \sim 1$ g cm^{-3} is the mass density in the core of a young, massive star, and we have used the fact that stellar gas is mostly hydrogen to infer that there is approximately one nucleon per electron in the gas and hence that $n_e \sim \rho/m_p$. Note that $\mathcal{L} \sim 10^5$ km is 3×10^9 times larger than $\lambda \sim 3$ cm, and the number of electrons and photons inside a cube of side $\mathcal{L}$ is enormous, so the diffusion description of heat transport is highly accurate.

In the diffusion description, the heat flux $\mathbf{F}$ as measured in the gas's rest frame is related to the temperature gradient ∇T by the law of diffusive heat conduction $\mathbf{F} = -\kappa \nabla T$. To estimate the thermal conductivity κ, orient the coordinates so the temperature gradient is in the z direction, and consider the rate of heat exchange between a gas layer located near $z = 0$ and a layer one photon mean free path away, at $z = \lambda$ (Fig. 3.10). The heat exchange is carried by photons that are emitted from one layer, propagate nearly unimpeded to the other, and then scatter. Although the individual scatterings are nearly elastic (and we thus are ignoring changes of photon

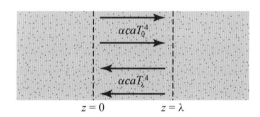

FIGURE 3.10 Heat exchange between two layers of gas separated by a distance of one photon mean free path in the direction of the gas's temperature gradient.

frequency in the Boltzmann equation), tiny changes of photon energy add up over many scatterings to keep the photons nearly in local thermal equilibrium with the gas. Thus, we approximate the photons and gas in the layer at $z = 0$ to have a common temperature T_0 and those in the layer at $z = \lambda$ to have a common temperature $T_\lambda = T_0 + \lambda dT/dz$. Then the photons propagating from the layer at $z = 0$ to that at $z = \lambda$ carry an energy flux

$$F_{0\to\lambda} = \alpha c a (T_0)^4 , \qquad (3.77a)$$

where a is the radiation constant of Eq. (3.54c); $a(T_0)^4$ is the photon energy density at $z = 0$; and α is a dimensionless constant of order $1/4$ that accounts for what fraction of the photons at $z = 0$ are moving rightward rather than leftward, and at what mean angle to the z direction. (Throughout this section, by contrast with early sections of this chapter, we use nongeometrized units, with the speed of light c present explicitly.) Similarly, the flux of energy from the layer at $z = \lambda$ to the layer at $z = 0$ is

$$F_{\lambda\to 0} = -\alpha c a (T_\lambda)^4 ; \qquad (3.77b)$$

and the net rightward flux, the sum of Eqs. (3.77a) and (3.77b), is

$$F = \alpha c a [(T_0)^4 - (T_\lambda)^4] = -4\alpha c a T^3 \lambda \frac{dT}{dz}. \qquad (3.77c)$$

Noting that 4α is approximately 1, inserting expression (3.76) for the photon mean free path, and comparing with the law of diffusive heat flow $\mathbf{F} = -\kappa \nabla T$, we conclude that the thermal conductivity is

thermal conductivity: order of magnitude estimate

$$\kappa \sim a T^3 c \lambda = \frac{a c T^3}{\sigma_T n_e}. \qquad (3.78)$$

3.7.3 Analysis Using the Boltzmann Transport Equation

With these physical insights and rough answer in hand, we turn to a Boltzmann transport analysis of the heat transfer. Our first step is to formulate the Boltzmann transport equation for the photons (including effects of Thomson scattering off the electrons) in the rest frame of the gas. Our second step will be to solve that equation to determine the influence of the heat flow on the distribution function $\mathcal{N}$, and our third step will be to compute the thermal conductivity κ by an integral over that $\mathcal{N}$.

To simplify the analysis we use, as the parameter ℓ in the Boltzmann transport equation $d\mathcal{N}/d\ell = (d\mathcal{N}/d\ell)_{\text{collisions}}$, the distance l that a fiducial photon travels, and we regard the distribution function $\mathcal{N}$ as a function of location $\mathbf{x}$ in space, the photon propagation direction (unit vector) $\mathbf{n}$, and the photon frequency ν: $\mathcal{N}(\mathbf{x}, \mathbf{n}, \nu)$. Because the photon frequency does not change during free propagation or Thomson scattering, it can be treated as a constant when solving the Boltzmann equation.

Along the trajectory of a fiducial photon, $\mathcal{N}(\mathbf{x}, \mathbf{n}, \nu)$ will change as a result of two things: (i) the scattering of photons out of the $\mathbf{n}$ direction and into other directions and (ii) the scattering of photons from other directions $\mathbf{n}'$ into the $\mathbf{n}$ direction. These effects produce the following two collision terms in the Boltzmann transport equation (3.66):

$$\frac{d\mathcal{N}(\mathbf{x}, \mathbf{n}, \nu)}{dl} = -\sigma_T n_e \mathcal{N}(\mathbf{x}, \mathbf{n}, \nu) + \int \frac{d\sigma(\mathbf{n}' \to \mathbf{n})}{d\Omega} n_e \mathcal{N}(\mathbf{x}, \mathbf{n}', \nu) d\Omega'. \quad (3.79)$$

Boltzmann transport equation for diffusing photons

(The second scattering term would be more obvious if we were to use I_ν (which is per unit solid angle) as our distribution function rather than $\mathcal{N}$; but they just differ by a constant.) Because the mean free path $\lambda = 1/(\sigma_T n_e) \sim 3$ cm is so short compared to the length scale $\mathcal{L} \sim 10^5$ km of the temperature gradient, the heat flow will show up as a tiny correction to an otherwise isotropic, perfectly thermal distribution function. Thus, we can write the photon distribution function as the sum of an unperturbed, perfectly isotropic and thermalized piece $\mathcal{N}_0$ and a tiny, anisotropic perturbation $\mathcal{N}_1$:

$$\mathcal{N} = \mathcal{N}_0 + \mathcal{N}_1, \quad \text{where } \mathcal{N}_0 = \frac{2}{h^3} \frac{1}{e^{h\nu/(k_B T)} - 1}. \quad (3.80a)$$

expand $\mathcal{N}$ in powers of $\lambda/\mathcal{L}$

Here the perfectly thermal piece $\mathcal{N}_0(\mathbf{x}, \mathbf{n}, \nu)$ has the standard blackbody form (3.23); it is independent of $\mathbf{n}$, and it depends on $\mathbf{x}$ only through the temperature $T = T(\mathbf{x})$. If the photon mean free path were vanishingly small, there would be no way for photons at different locations $\mathbf{x}$ to discover that the temperature is inhomogeneous; correspondingly, $\mathcal{N}_1$ would be vanishingly small. The finiteness of the mean free path permits $\mathcal{N}_1$ to be finite, and so it is reasonable to expect (and turns out to be true) that the magnitude of $\mathcal{N}_1$ is

$$\mathcal{N}_1 \sim \frac{\lambda}{\mathcal{L}} \mathcal{N}_0. \quad (3.80b)$$

Thus, $\mathcal{N}_0$ is the leading-order term, and $\mathcal{N}_1$ is the first-order correction in an expansion of the distribution function $\mathcal{N}$ in powers of $\lambda/\mathcal{L}$. This is called a *two-lengthscale expansion*; see Box 3.3.

Inserting $\mathcal{N} = \mathcal{N}_0 + \mathcal{N}_1$ into our Boltzmann transport equation (3.79) and using $d/dl = \mathbf{n} \cdot \nabla$ for the derivative with respect to distance along the fiducial photon trajectory, we obtain

$$n_j \frac{\partial \mathcal{N}_0}{\partial x_j} + n_j \frac{\partial \mathcal{N}_1}{\partial x_j} = \left[-\sigma_T n_e \mathcal{N}_0 + \int \frac{d\sigma(\mathbf{n}' \to \mathbf{n})}{d\Omega} n_e c \mathcal{N}_0 d\Omega' \right]$$

$$+ \left[-\sigma_T n_e c \mathcal{N}_1(\mathbf{n}, \nu) + \int \frac{d\sigma(\mathbf{n}' \to \mathbf{n})}{d\Omega} n_e c \mathcal{N}_1(\mathbf{n}', \nu) d\Omega' \right].$$

(3.80c)

perturbative Boltzmann equation for $\mathcal{N}_1$

Because $\mathcal{N}_0$ is isotropic (i.e., is independent of photon direction $\mathbf{n}'$), it can be pulled out of the integral over $\mathbf{n}'$ in the first square bracket on the right-hand side. When this is done, the first and second terms in that square bracket cancel each other. Thus, the unperturbed part of the distribution, $\mathcal{N}_0$, completely drops out of the right-hand side of Eq. (3.80c). On the left-hand side the term involving the perturbation $\mathcal{N}_1$ is tiny compared to that involving the unperturbed distribution $\mathcal{N}_0$, so we shall drop it.

> **BOX 3.3. TWO-LENGTHSCALE EXPANSIONS**
>
> Equation (3.80b) is indicative of the mathematical technique that underlies Boltzmann-transport computations: a perturbative expansion in the dimensionless ratio of two lengthscales, the tiny mean free path λ of the transporter particles and the far larger macroscopic scale $\mathcal{L}$ of the inhomogeneities that drive the transport. Expansions in lengthscale ratios $\lambda/\mathcal{L}$ are called *two-lengthscale expansions* and are widely used in physics and engineering. Most readers will previously have met such an expansion in quantum mechanics: the WKB approximation, where λ is the lengthscale on which the wave function changes, and $\mathcal{L}$ is the scale of changes in the potential $V(x)$ that drives the wave function. Kinetic theory itself is the result of a two-lengthscale expansion: it follows from the more sophisticated statistical-mechanics formalism of Chap. 4, in the limit where the particle sizes are small compared to their mean free paths. In this book we use two-lengthscale expansions frequently—for instance, in the geometric optics approximation to wave propagation (Chap. 7), in the study of boundary layers in fluid mechanics (Secs. 14.4, 14.5.4, and 15.5), in the quasi-linear formalism for plasma physics (Chap. 23), and in the definition of a gravitational wave (Sec. 27.4).

Because the spatial dependence of $\mathcal{N}_0$ is entirely due to the temperature gradient, we can bring the first term and the whole transport equation into the form

$$n_j \frac{\partial T}{\partial x_j} \frac{\partial \mathcal{N}_0}{\partial T} = -\sigma_T n_e \mathcal{N}_1(\mathbf{n}, \nu) + \int \frac{d\sigma(\mathbf{n}' \to \mathbf{n})}{d\Omega} n_e \mathcal{N}_1(\mathbf{n}', \nu) d\Omega'. \quad (3.80d)$$

The left-hand side of this equation is the amount by which the temperature gradient causes $\mathcal{N}_0$ to fail to satisfy the Boltzmann equation, and the right-hand side is the manner in which the perturbation $\mathcal{N}_1$ steps into the breach and enables the Boltzmann equation to be satisfied.

Because the left-hand side is linear in the photon propagation direction n_j (i.e., it has a $\cos\theta$ dependence in coordinates where ∇T is in the z-direction; i.e., it has a dipolar, $l=1$, angular dependence), $\mathcal{N}_1$ must also be linear in n_j (i.e., dipolar), in order to fulfill Eq. (3.80d). Thus, we write $\mathcal{N}_1$ in the dipolar form

$$\mathcal{N}_1 = K_j(\mathbf{x}, \nu) n_j, \quad (3.80e)$$

and we shall solve the transport equation (3.80d) for the function K_j.

multipolar expansion of $\mathcal{N}$

[Side remark: This is a special case of a general situation. When solving the Boltzmann transport equation in diffusion situations, one is performing a power series expansion in $\lambda/\mathcal{L}$; see Box 3.3. The lowest-order term in the expansion, $\mathcal{N}_0$, is isotropic (i.e., it is monopolar in its dependence on the direction of motion of the diffusing particles). The first-order correction, $\mathcal{N}_1$, is down in magnitude by $\lambda/\mathcal{L}$ from $\mathcal{N}_0$ and is dipolar (or sometimes quadrupolar; see Ex. 3.19) in its dependence on the particles' direction of motion. The second-order correction, $\mathcal{N}_2$, is down in magnitude by $(\lambda/\mathcal{L})^2$ from $\mathcal{N}_0$ and its multipolar order is one higher than $\mathcal{N}_1$ (quadrupolar here; octupolar in Ex. 3.19). And so it continues to higher and higher orders.[11]]

When we insert the dipolar expression (3.80e) into the angular integral on the right-hand side of the transport equation (3.80d) and notice that the differential scattering cross section (3.75a) is unchanged under $\mathbf{n}' \to -\mathbf{n}'$, but $K_j n'_j$ changes sign, we find that the integral vanishes. As a result the transport equation (3.80d) takes the simplified form

$$n_j \frac{\partial T}{\partial x_j} \frac{\partial \mathcal{N}_0}{\partial T} = -\sigma_T n_e K_j n_j, \qquad (3.80\text{f})$$

from which we can read off the function K_j and thence $\mathcal{N}_1 = K_j n_j$:

$$\mathcal{N}_1 = -\frac{\partial \mathcal{N}_0/\partial T}{\sigma_T n_e} \frac{\partial T}{\partial x_j} n_j. \qquad (3.80\text{g}) \qquad \text{solution for } \mathcal{N}_1$$

Notice that, as claimed in Eq. (3.80b), the perturbation has a magnitude

$$\frac{\mathcal{N}_1}{\mathcal{N}_0} \sim \frac{1}{\sigma_T n_e} \frac{1}{T} |\nabla T| \sim \frac{\lambda}{\mathcal{L}}. \qquad (3.80\text{h})$$

Having solved the Boltzmann transport equation to obtain $\mathcal{N}_1$, we can now evaluate the energy flux F_i carried by the diffusing photons. Relativity physicists will recognize that flux as the T^{0i} part of the stress-energy tensor and will therefore evaluate it as

$$F_i = T^{0i} = c^2 \int \mathcal{N} p^0 p^i \frac{d\mathcal{V}_p}{p^0} = c^2 \int \mathcal{N} p_i d\mathcal{V}_p \qquad (3.81)$$

[cf. Eq. (3.33c) with the factors of c restored]. Newtonian physicists can deduce this formula by noticing that photons with momentum $\mathbf{p}$ in $d\mathcal{V}_p$ carry energy $E = |\mathbf{p}|c$ and move with velocity $\mathbf{v} = c\mathbf{p}/|\mathbf{p}|$, so their energy flux is $\mathcal{N} E \mathbf{v} \, d\mathcal{V}_p = c^2 \mathcal{N} \mathbf{p} \, d\mathcal{V}_p$; integrating this over momentum space gives Eq. (3.81). Inserting $\mathcal{N} = \mathcal{N}_0 + \mathcal{N}_1$ into this equation and noting that the integral over $\mathcal{N}_0$ vanishes, and inserting Eq. (3.80g) for $\mathcal{N}_1$, we obtain

$$F_i = c^2 \int \mathcal{N}_1 p_i d\mathcal{V}_p = -\frac{c}{\sigma_T n_e} \frac{\partial T}{\partial x_j} \frac{\partial}{\partial T} \int \mathcal{N}_0 c n_j p_i d\mathcal{V}_p. \qquad (3.82\text{a}) \qquad \text{energy flux from } \mathcal{N}_1$$

The relativity physicist will identify this integral as Eq. (3.33c) for the photons' stress tensor T_{ij} (since $n_j = p_j/p_0 = p_j/\mathcal{E}$). The Newtonian physicist, with a little thought, will recognize the integral in Eq. (3.82a) as the j component of the flux of i component

---

11. For full details of this "method-of-moments" analysis in nonrelativistic situations, see, e.g., Grad (1958); and for full relativistic details, see, e.g., Thorne (1981).

of momentum, which is precisely the stress tensor. Since this stress tensor is being computed with the isotropic, thermalized part of $\mathcal{N}$, it is isotropic, $T_{ji} = P\delta_{ji}$, and its pressure has the standard blackbody-radiation form $P = \frac{1}{3}aT^4$ [Eqs. (3.54a)]. Replacing the integral in Eq. (3.82a) by this blackbody stress tensor, we obtain our final answer for the photons' energy flux:

$$F_i = -\frac{c}{\sigma_T n_e} \frac{\partial T}{\partial x_j} \frac{d}{dT}\left(\frac{1}{3}aT^4 \delta_{ji}\right) = -\frac{c}{\sigma_T n_e} \frac{4}{3} aT^3 \frac{\partial T}{\partial x_i}. \tag{3.82b}$$

Thus, *from the Boltzmann transport equation we have simultaneously derived the law of diffusive heat conduction* $\mathbf{q} = -\kappa \nabla T$ *and evaluated the coefficient of thermal conductivity*

thermal conductivity from Boltzmann transport equation

$$\kappa = \frac{4}{3}\frac{acT^3}{\sigma_T n_e}. \tag{3.83}$$

Notice that this heat conductivity is 4/3 times our crude, order-of-magnitude estimate (3.78).

The above calculation, while somewhat complicated in its details, is conceptually fairly simple. The reader is encouraged to go back through the calculation and identify the main conceptual steps [expansion of distribution function in powers of $\lambda/\mathcal{L}$, insertion of zero-order plus first-order parts into the Boltzmann equation, multipolar decomposition of the zero- and first-order parts (with zero-order being monopolar and first-order being dipolar), neglect of terms in the Boltzmann equation that are smaller than the leading ones by factors $\lambda/\mathcal{L}$, solution for the coefficient of the multipolar decomposition of the first-order part, reconstruction of the first-order part from that coefficient, and insertion into a momentum-space integral to get the flux of the quantity being transported]. Precisely these same steps are used to evaluate all other transport coefficients that are governed by classical physics. [For examples of other such calculations, see, e.g., Shkarofsky, Johnston, and Bachynski (1966).]

As an application of the thermal conductivity (3.83), consider a young (main-sequence) 7-solar-mass ($7\,M_\odot$) star as modeled, for example, on pages 480 and 481 of Clayton (1968). Just outside the star's convective core, at radius $r \simeq 0.7 R_\odot \simeq 5 \times 10^5$ km (where $R_\odot$ is the Sun's radius), the density and temperature are $\rho \simeq 5.5$ g cm^{-3} and $T \simeq 1.9 \times 10^7$ K, so the number density of electrons is $n_e \simeq \rho/(1.4 m_p) \simeq 2.3 \times 10^{24}$ cm^{-3} where the 1.4 accounts for the star's chemical composition. For these parameters, Eq. (3.83) gives a thermal conductivity $\kappa \simeq 1.3 \times 10^{18}$ erg s^{-1} cm^{-2} K^{-1}. The lengthscale $\mathcal{L}$ on which the temperature is changing is approximately the same as the radius, so the temperature gradient is $|\nabla T| \sim T/r \sim 4 \times 10^{-4}$ K cm^{-1}. The law of diffusive heat transfer then predicts a heat flux $F = \kappa |\nabla T| \sim 5 \times 10^{14}$ erg s^{-1} cm^{-2}, and thus a total luminosity $L = 4\pi r^2 F \sim 1.5 \times 10^{37}$ erg s$^{-1} \simeq 4000 L_\odot$ (4,000 solar luminosities). (This estimate is a little high. The correct value is about $3,600 L_\odot$.) What a difference the mass makes! The heavier a star, the hotter its core, the faster it burns, and the higher its luminosity will be. Increasing the mass by a factor of 7 drives the luminosity up by 4,000.

Exercise 3.19 **Example: Viscosity of a Monatomic Gas**

Consider a nonrelativistic fluid that, in the neighborhood of the origin, has fluid velocity

$$v_i = \sigma_{ij} x_j, \tag{3.84}$$

with σ_{ij} symmetric and trace-free. As we shall see in Sec. 13.7.1, this represents a purely shearing flow, with no rotation or volume changes of fluid elements; σ_{ij} is called the fluid's *rate of shear*. Just as a gradient of temperature produces a diffusive flow of heat, so the gradient of velocity embodied in σ_{ij} produces a diffusive flow of momentum (i.e., a stress). In this exercise we use kinetic theory to show that, for a monatomic gas with isotropic scattering of atoms off one another, this stress is

$$\boxed{T_{ij} = -2\eta_{\text{shear}} \sigma_{ij},} \tag{3.85a}$$

with the coefficient of shear viscosity

$$\boxed{\eta_{\text{shear}} \simeq \frac{1}{3} \rho \lambda v_{th},} \tag{3.85b}$$

where ρ is the gas density, λ is the atoms' mean free path between collisions, and $v_{th} = \sqrt{3k_B T/m}$ is the atoms' rms speed. Our analysis follows the same route as the analysis of heat conduction in Secs. 3.7.2 and 3.7.3.

(a) Derive Eq. (3.85b) for the shear viscosity, to within a factor of order unity, by an order-of-magnitude analysis like that in Sec. 3.7.2.

(b) Regard the atoms' distribution function $\mathcal{N}$ as being a function of the magnitude p and direction $\mathbf{n}$ of an atom's momentum, and of location $\mathbf{x}$ in space. Show that, if the scattering is isotropic with cross section σ_s and the number density of atoms is n, then the Boltzmann transport equation can be written as

$$\frac{d\mathcal{N}}{dl} = \mathbf{n} \cdot \nabla \mathcal{N} = -\frac{1}{\lambda}\mathcal{N} + \int \frac{1}{4\pi\lambda} \mathcal{N}(p, \mathbf{n}', \mathbf{x}) d\Omega', \tag{3.86a}$$

where $\lambda = 1/n\sigma_s$ is the atomic mean free path (mean distance traveled between scatterings) and l is distance traveled by a fiducial atom.

(c) Explain why, in the limit of vanishingly small mean free path, the distribution function has the following form:

$$\mathcal{N}_0 = \frac{n}{(2\pi m k_B T)^{3/2}} \exp[-(\mathbf{p} - m\boldsymbol{\sigma} \cdot \mathbf{x})^2/(2mk_B T)]. \tag{3.86b}$$

(d) Solve the Boltzmann transport equation (3.86a) to obtain the leading-order correction $\mathcal{N}_1$ to the distribution function at $\mathbf{x} = 0$.

[Answer: $\mathcal{N}_1 = -(\lambda p/(k_B T))\sigma_{ab} n_a n_b \mathcal{N}_0$.]

(e) Compute the stress at $\mathbf{x} = 0$ via a momentum-space integral. Your answer should be Eq. (3.85a) with η_{shear} given by Eq. (3.85b) to within a few tens of percent accuracy. [Hint: Along the way you will need the following angular integral:

$$\int n_a n_b n_i n_j d\Omega = \frac{4\pi}{15}(\delta_{ab}\delta_{ij} + \delta_{ai}\delta_{bj} + \delta_{aj}\delta_{bi}). \quad (3.86c)$$

Derive this by arguing that the integral must have the above delta-function structure, and then computing the multiplicative constant by performing the integral for $a = b = i = j = z$.]

Exercise 3.20 *Example: Diffusion Coefficient in the Collision-Time Approximation*

Consider a collection of identical test particles with rest mass $m \neq 0$ that diffuse through a collection of thermalized scattering centers. (The test particles might be molecules of one species, and the scattering centers might be molecules of a much more numerous species.) The scattering centers have a temperature T such that $k_B T \ll mc^2$, so if the test particles acquire this temperature, they will have thermal speeds small compared to the speed of light as measured in the mean rest frame of the scattering centers. We study the effects of scattering on the test particles using the following collision-time approximation for the collision terms in the Boltzmann equation, which we write in the mean rest frame of the scattering centers:

$$\left(\frac{d\mathcal{N}}{dt}\right)_{\text{collision}} = \frac{1}{\tau}(\mathcal{N}_0 - \mathcal{N}), \quad \text{where} \quad \mathcal{N}_0 \equiv \frac{e^{-p^2/(2mk_B T)}}{(2\pi m k_B T)^{3/2}} n. \quad (3.87)$$

Here the time derivative d/dt is taken moving with a fiducial test particle along its unscattered trajectory, $p = |\mathbf{p}|$ is the magnitude of the test particles' spatial momentum, $n = \int \mathcal{N} d\mathcal{V}_p$ is the number density of test particles, and τ is a constant to be discussed below.

(a) Show that this collision term preserves test particles in the sense that

$$\left(\frac{dn}{dt}\right)_{\text{collision}} \equiv \int \left(\frac{d\mathcal{N}}{dt}\right)_{\text{collision}} dp_x dp_y dp_z = 0. \quad (3.88)$$

(b) Explain why this collision term corresponds to the following physical picture: Each test particle has a probability $1/\tau$ per unit time of scattering; when it scatters, its direction of motion is randomized and its energy is thermalized at the scattering centers' temperature.

(c) Suppose that the temperature T is homogeneous (spatially constant), but the test particles are distributed inhomogeneously, $n = n(\mathbf{x}) \neq \text{const}$. Let $\mathcal{L}$ be the lengthscale on which their number density n varies. What condition must $\mathcal{L}, \tau, T$, and m satisfy for the diffusion approximation to be reasonably accurate? Assume that this condition is satisfied.

(d) Compute, in order of magnitude, the particle flux $\mathbf{S} = -D\nabla n$ produced by the gradient of the number density n, and thereby evaluate the diffusion coefficient D.

(e) Show that the Boltzmann transport equation takes the form (sometimes known as the BKG or Crook model)

$$\frac{\partial \mathcal{N}}{\partial t} + \frac{p_j}{m}\frac{\partial \mathcal{N}}{\partial x_j} = \frac{1}{\tau}(\mathcal{N}_0 - \mathcal{N}). \tag{3.89a}$$

(f) Show that to first order in a small diffusion-approximation parameter, the solution of this equation is $\mathcal{N} = \mathcal{N}_0 + \mathcal{N}_1$, where $\mathcal{N}_0$ is as defined in Eq. (3.87), and

$$\mathcal{N}_1 = -\frac{p_j \tau}{m}\frac{\partial n}{\partial x_j}\frac{e^{-p^2/(2mk_BT)}}{(2\pi m k_B T)^{3/2}}. \tag{3.89b}$$

Note that $\mathcal{N}_0$ is monopolar (independent of the direction of $\mathbf{p}$), while $\mathcal{N}_1$ is dipolar (linear in $\mathbf{p}$).

(g) Show that the perturbation $\mathcal{N}_1$ gives rise to a particle flux given by Eq. (3.70d), with the diffusion coefficient

$$D = \frac{k_B T}{m}\tau. \tag{3.90}$$

How does this compare with your order-of-magnitude estimate in part (d)?

Exercise 3.21 **Example: Neutron Diffusion in a Nuclear Reactor**
A simplified version of a commercial nuclear reactor involves *fissile material* such as enriched uranium[12] and a *moderator* such as graphite, both of which will be assumed in this exercise. Slow (thermalized) neutrons, with kinetic energies ~ 0.1 eV, are captured by the ^{235}U nuclei and cause them to undergo fission, releasing ~ 170 MeV of kinetic energy per fission which appears as heat. Some of this energy is then converted into electric power. The fission releases an average of two or three (assume two) fast neutrons with kinetic energies ~ 1 MeV. (This is an underestimate.) The fast neutrons must be slowed to thermal speeds where they can be captured by ^{235}U atoms and induce further fissions. The slowing is achieved by scattering off the moderator atoms—a scattering in which the crucial effect, energy loss, occurs in momentum space. The momentum-space scattering is elastic and isotropic in the center-of-mass frame, with total cross section (to scatter off one of the moderator's carbon atoms) $\sigma_s \simeq 4.8 \times 10^{-24}$ cm$^2 \equiv 4.8$ barns. Using the fact that in the moderator's rest frame, the incoming neutron has a much higher kinetic energy than the moderator carbon atoms, and using energy and momentum conservation and the isotropy of the scattering, one can show that in the moderator's rest frame, the logarithm of the neutron's

---

12. Natural uranium contains ~ 0.007 of the fissile isotope ^{235}U; the fraction is increased to ~ 0.01–0.05 in enriched uranium.

energy is reduced in each scattering by an average amount ξ that is independent of energy and is given by

$$\xi \equiv -\overline{\Delta \ln E} = 1 + \frac{(A-1)^2}{2A} \ln\left(\frac{A-1}{A+1}\right) \simeq 0.16, \qquad (3.91)$$

a quantity sometimes known as *lethargy*. Here $A \simeq 12$ is the ratio of the mass of the scattering atom to that of the scattered neutron.

There is a dangerous hurdle that the diffusing neutrons must overcome during their slowdown: as the neutrons pass through a critical energy region of about ~ 7 to ~ 6 eV, the ^{238}U atoms can absorb them. The absorption cross section has a huge resonance there, with width ~ 1 eV and resonance integral $\int \sigma_a d \ln E \simeq 240$ barns. For simplicity, we approximate the cross section in this absorption resonance by $\sigma_a \simeq 1600$ barns at 6 eV $< E < 7$ eV, and zero outside this range. To achieve a viable fission chain reaction and keep the reactor hot requires about half of the neutrons (one per original ^{235}U fission) to slow down through this resonant energy without getting absorbed. Those that make it through will thermalize and trigger new ^{235}U fissions (about one per original fission), maintaining the chain reaction.

We idealize the uranium and moderator atoms as homogeneously mixed on lengthscales small compared to the neutron mean free path, $\lambda_s = 1/(\sigma_s n_s) \simeq 2$ cm, where n_s is the number density of moderator (carbon) atoms. Then the neutrons' distribution function $\mathcal{N}$, as they slow down, is isotropic in direction and independent of position; and in our steady-state situation, it is independent of time. It therefore depends only on the magnitude p of the neutron momentum or equivalently, on the neutron kinetic energy $E = p^2/(2m)$: $\mathcal{N} = \mathcal{N}(E)$.

Use the Boltzmann transport equation or other considerations to develop the theory of the slowing of the neutrons in momentum space and of their struggle to pass through the ^{238}U resonance region without getting absorbed. More specifically, do the following.

(a) Use as the distribution function not $\mathcal{N}(E)$ but rather $n_E(E) \equiv dN/d\mathcal{V}_x dE$ = (number of neutrons per unit volume and per unit kinetic energy), and denote by $q(E)$ the number of neutrons per unit volume that slow down through energy E per unit time. Show that outside the resonant absorption region these two quantities are related by

$$q = \sigma_s n_s \xi E\, n_E v, \quad \text{where } v = \sqrt{2mE} \qquad (3.92)$$

is the neutron speed, so q contains the same information as the distribution function n_E. Explain why the steady-state operation of the nuclear reactor requires q to be independent of energy in this nonabsorption region, and infer that $n_E \propto E^{-3/2}$.

(b) Show further that inside the resonant absorption region, 6 eV $< E < 7$ eV, the relationship between q and E is modified:

$$q = (\sigma_s n_s + \sigma_a n_a) \xi E\, n_E v. \qquad (3.93)$$

Here n_s is the number density of scattering (carbon) atoms, and n_a is the number density of absorbing (^{238}U) atoms. [Hint: Require that the rate at which neutrons scatter into a tiny interval of energy $\delta E \ll \xi E$ is equal to the rate at which they leave that tiny interval.] Then show that the absorption causes q to vary with energy according to the following differential equation:

$$\frac{d \ln q}{d \ln E} = \frac{\sigma_a n_a}{(\sigma_s n_s + \sigma_a n_a)\xi}. \tag{3.94}$$

(c) By solving this differential equation in our idealization of constant σ_a over the range 7 to 6 eV, show that the condition to maintain the chain reaction is

$$\frac{n_s}{n_a} \simeq \frac{\sigma_a}{\sigma_s}\left(\frac{\ln(7/6)}{\xi \ln 2} - 1\right) \simeq 0.41 \frac{\sigma_a}{\sigma_s} \simeq 140. \tag{3.95}$$

Thus, to maintain the reaction in the presence of the huge ^{238}U absorption resonance for neutrons, it is necessary that approximately 99% of the reactor volume be taken up by moderator atoms and 1% by uranium atoms.

This is a rather idealized version of what happens inside a nuclear reactor, but it provides insight into some of the important processes and the magnitudes of various relevant quantities. For a graphic example of an additional complexity, see the description of "xenon poisoning" of the chain reaction in the first production-scale nuclear reactor (built during World War II to make plutonium for the first American atomic bombs) in John Archibald Wheeler's autobiography (Wheeler, 2000).

Bibliographic Note

Newtonian kinetic theory is treated in many textbooks on statistical physics. At an elementary level, Kittel and Kroemer (1980, Chap. 14) is rather good. Texts at a more advanced level include Kardar (2007, Chap. 3), Reif (2008, Secs. 7.9–7.13 and Chaps. 12–14), and Reichl (2009, Chap. 11). For a very advanced treatment with extensive applications to electrons and ions in plasmas, and electrons, phonons, and quasi-particles in liquids and solids, see Lifshitz and Pitaevskii (1981).

Relativistic kinetic theory is rarely touched on in statistical-physics textbooks but should be. It is well known to astrophysicists. The treatment in this chapter is easily lifted into general relativity theory (see, e.g., Misner, Thorne, and Wheeler, 1973, Sec. 22.6).

CHAPTER FOUR

Statistical Mechanics

Willard Gibbs did for statistical mechanics and for thermodynamics what Laplace did for celestial mechanics and Maxwell did for electrodynamics, namely, made his field a well-nigh finished theoretical structure.

ROBERT A. MILLIKAN (1938)

4.1 Overview

While kinetic theory (Chap. 3) gives a powerful description of some statistical features of matter, other features are outside its realm and must be treated using the more sophisticated tools of statistical mechanics. Examples include:

- *Correlations:* Kinetic theory's distribution function $\mathcal{N}$ tells us, on average, how many particles will occupy a given phase-space volume, but it says nothing about whether the particles like to clump or prefer to avoid one another. It is therefore inadequate to describe, for example, the distributions of galaxies and stars, which cluster under their mutual gravitational attraction (Sec. 4.10.1), or that of electrons in a plasma, which are mutually repulsive and thus are spatially anticorrelated (Sec. 22.6).

- *Fluctuations:* In experiments to measure a very weak mechanical force (e.g., tests of the equivalence principle and searches for gravitational waves), one typically monitors the motion of a pendulum's test mass, on which the force acts. Molecules of gas hitting the test mass also make it move. Kinetic theory predicts how many molecules will hit in 1 ms, on average, and how strong is the resulting pressure acting in all directions; but kinetic theory's distribution function $\mathcal{N}$ cannot tell us the probability that in 1 ms more molecules will hit one side of the test mass than the other, mimicking the force one is trying to measure. The probability distribution for fluctuations is an essential tool for analyzing the noise in this and any other physical experiment, and it falls in the domain of statistical mechanics, not kinetic theory (Sec. 5.6).

- *Strongly interacting particles:* As should be familiar, the thermal motions of an ionic crystal are best described not in terms of individual atoms (as in the Einstein theory), but instead by decomposing the atoms' motion into

> **BOX 4.1. READERS' GUIDE**
>
> - Relativity enters into portions of this chapter solely via the relativistic energies and momenta of high-speed particles (Sec. 1.10). We presume that all readers are familiar with at least this much relativity and accordingly, we do not provide a Newtonian track through this chapter. We make occasional additional side remarks for the benefit of relativistic readers, but a Newtonian reader's failure to understand them will not compromise mastering all of this chapter's material.
> - This chapter relies in crucial ways on Secs. 3.2, 3.3, 3.5.1, 3.5.2, 3.5.5, and 3.6, and Box 3.3.
> - Chapter 5 is an extension of this chapter. To understand it and portions of Chap. 6, one must master the fundamental concepts of statistical mechanics in this chapter (Secs. 4.2–4.8).
> - Other chapters do not depend strongly on this one.

normal modes (phonons; Debye theory). The thermal excitation of phonons is governed by statistical mechanics [Eq. (4.26)].

- *Microscopic origin of thermodynamic laws:* The laws of classical thermodynamics can be (and often are) derived from a few elementary, macroscopic postulates without any reference to the microscopic, atomic nature of matter. Kinetic theory provides a microscopic foundation for some of thermodynamics' abstract macroscopic ideas (e.g., the first law of thermodynamics) and permits the computation of equations of state. However, a full appreciation of entropy and the second law of thermodynamics (Sec. 4.7), and of behavior at phase transitions (Secs. 4.9, 5.5.2, 5.8.2, and 5.8.4) requires the machinery of statistical mechanics.

In this chapter, we develop the conceptual foundations for classical statistical mechanics and its interface with quantum physics, and we also delve deeply enough into the quantum world to be able to treat a few simple quantum problems. More specifically, in Sec. 4.2, we introduce the concepts of systems, ensembles of systems, and the distribution function for an ensemble. In Sec. 4.3, we use Hamiltonian dynamics to study the evolution of an ensemble's distribution function and derive the statistical mechanical version of Liouville's theorem. In Sec. 4.4, we develop the concept of statistical equilibrium and derive the general form of distribution functions for ensembles of systems that have reached statistical equilibrium (Sec. 4.4.2) and specific forms that depend on what additive macroscopic quantities the systems can exchange with their thermalized environment: energy exchange (canonical distribution, Sec. 4.4.1), energy and volume exchange (Gibbs distribution, Sec. 4.4.2), energy and particle exchange (grand canonical distribution, Sec. 4.4.2), and nothing exchanged (microcanonical distribution, Sec. 4.5).

In Chap. 5, we study these equilibrium distributions in considerable detail, especially their relationship to the laws of thermodynamics. Here in Chap. 4, we use them to explore some fundamental statistical mechanics issues:

1. a derivation of the Bose-Einstein and Fermi-Dirac distributions for the mean occupation number of a single-particle quantum state, which we studied in depth in Chap. 3 (Sec. 4.4.3);

2. a discussion and proof of the equipartition theorem for classical, quadratic degrees of freedom (Sec. 4.4.4);

3. the relationship between the microcanonical ensemble and *ergodicity* (the ergodic evolution of a single, isolated system; Sec. 4.6);

4. the concept of the entropy of an arbitrary ensemble of systems, and the increase of entropy (second law of thermodynamics) as an ensemble of isolated ("closed") systems evolves into its equilibrium, microcanonical form via *phase mixing, coarse graining,* and (quantum mechanically) via *discarding quantum correlations*—it's the physicist's fault!—(Sec. 4.7);

5. the entropy increase when two gases are mixed, and how quantum mechanics resolved the highly classical Gibbs Paradox (Ex. 4.8);

6. the power of entropy per particle as a tool for studying the evolution of physical systems (Sec. 4.8 and Ex. 4.10); and

7. Bose-Einstein condensation of a dilute gas of bosonic atoms (Sec. 4.9).

We conclude with a discussion of statistical mechanics in the presence of gravity (applications to galaxies, black holes, and the universe as a whole; Sec. 4.10) and a brief introduction to the concept of *information* and its connection to entropy (Sec. 4.11).

4.2 Systems, Ensembles, and Distribution Functions

4.2.1 Systems

Systems play the same role in statistical mechanics as is played by particles in kinetic theory. A system is any physical entity. (Obviously, this is an exceedingly general concept!) Examples are a galaxy, the Sun, a sapphire crystal, the fundamental mode of vibration of that crystal, an aluminum atom in that crystal, an electron from that aluminum atom, a quantum state in which that electron could reside, ...

SEMICLOSED SYSTEMS

Statistical mechanics focuses special attention on systems that couple only weakly to the rest of the universe. Stated more precisely, we are interested in systems whose **relevant** internal evolution timescales, τ_{int}, are short compared with the external timescales, τ_{ext}, on which they exchange energy, entropy, particles, and so forth,

semiclosed systems

closed systems

with their surroundings. Such systems are said to be *semiclosed,* and in the idealized limit where one completely ignores their external interactions, they are said to be *closed.* The statistical mechanics formalism for dealing with them relies on the assumption $\tau_{int}/\tau_{ext} \ll 1$; in this sense, it is a variant of a two-lengthscale expansion (Box 3.3).

Some examples will elucidate these concepts. For a galaxy of, say, 10^{11} stars, τ_{int} is the time it takes a star to cross the galaxy, so $\tau_{int} \sim 10^8$ yr. The external timescale is the time since the galaxy's last collision with a neighboring galaxy or the time since it was born by separating from the material that formed neighboring galaxies; both these times are $\tau_{ext} \sim 10^{10}$ yr, so $\tau_{int}/\tau_{ext} \sim 1/100$, and the galaxy is semiclosed. For a small volume of gas inside the Sun (say, 1 m on a side), τ_{int} is the timescale for the constituent electrons, ions, and photons to interact through collisions, $\tau_{int} \lesssim 10^{-10}$ s; this is much smaller than the time for external heat or particles to diffuse from the cube's surface to its center, $\tau_{ext} \gtrsim 10^{-5}$ s, so the cube is semiclosed. An individual atom in a crystal is so strongly coupled to its neighboring atoms by electrostatic forces that $\tau_{int} \sim \tau_{ext}$, which means the atom is not semiclosed. By contrast, for a vibrational mode of the crystal, τ_{int} is the mode's vibration period, and τ_{ext} is the time to exchange energy with other modes and thereby damp the chosen mode's vibrations; quite generally, the damping time is far longer than the period, so the mode is semiclosed. (For a highly polished, cold sapphire crystal weighing several kilograms, τ_{ext} can be $\sim 10^9 \tau_{int}$.) Therefore, it is the crystal's vibrational normal modes and not its atoms that are amenable to the statistical mechanical tools we shall develop.

CLOSED SYSTEMS AND THEIR HAMILTONIANS

When a semiclosed classical system is idealized as closed, so its interactions with the external universe are ignored, then its evolution can be described using Hamiltonian dynamics (see, e.g., Marion and Thornton, 1995; Landau and Lifshitz, 1976; Goldstein, Poole, and Safko, 2002). The system's classical state is described by *generalized coordinates* $\mathbf{q} \equiv \{q_j\}$ and *generalized momenta* $\mathbf{p} \equiv \{p_j\}$, where the index j runs from 1 to $W =$ (the system's number of degrees of freedom). The evolution of $\mathbf{q}, \mathbf{p}$ is governed by *Hamilton's equations*

Hamilton's equations for a closed system

$$\boxed{\frac{dq_j}{dt} = \frac{\partial H}{\partial p_j}, \quad \frac{dp_j}{dt} = -\frac{\partial H}{\partial q_j},} \quad (4.1)$$

where $H(\mathbf{q}, \mathbf{p})$ is the *hamiltonian,* and each equation is really W separate equations. Note that, because the system is idealized as closed, there is no explicit time dependence in the hamiltonian. Of course, not all physical systems (e.g., not those with strong internal dissipation) are describable by Hamiltonian dynamics, though in principle this restriction can usually be circumvented by increasing the number of degrees of freedom to include the cause of the dissipation.[1]

EXAMPLES

Let us return to our examples. For an individual star inside a galaxy, there are three degrees of freedom ($W = 3$), which we might choose to be the motion along three mutually orthogonal Cartesian directions, so $q_1 = x, q_2 = y, q_3 = z$. Because the star's speed is small compared to that of light, its hamiltonian has the standard form for a nonrelativistic particle:

$$H(\mathbf{q}, \mathbf{p}) = \frac{1}{2m}(p_1^2 + p_2^2 + p_3^2) + m\Phi(q_1, q_2, q_3). \tag{4.2}$$

Here m is the stellar mass, and $\Phi(q_1, q_2, q_3)$ is the galaxy's Newtonian gravitational potential (whose sign we take to be negative). Now, consider not just one star, but $K \sim 10^{11}$ of them in a galaxy. There are now $W = 3K$ degrees of freedom, and the hamiltonian is simply the sum of the hamiltonians for each individual star, so long as we ignore interactions between pairs of stars. The great power of the principles that we will develop is that they do not depend on whether $W = 1$ or $W = 3 \times 10^{11}$.

If our system is the fundamental mode of a sapphire crystal, then the number of degrees of freedom is only $W = 1$, and we can take the single generalized coordinate q to be the displacement of one end of the crystal from equilibrium. There will be an effective mass M for the mode (approximately equal to the actual mass of the crystal) such that the mode's generalized momentum is $p = M dq/dt$. The hamiltonian will be the standard one for a harmonic oscillator:

$$H(p, q) = \frac{p^2}{2M} + \frac{1}{2}M\omega^2 q^2, \tag{4.3a}$$

where ω is the mode's angular frequency of oscillation.

If we want to describe a whole crystal weighing several kilograms and having $K \sim 10^{26}$ atoms, then we obtain H by summing over $W = 3K$ oscillator hamiltonians for the crystal's W normal modes and adding an interaction potential H_{int} that accounts for the very weak interactions among modes:

$$H = \sum_{j=1}^{W} \left\{ \frac{p_j^2}{2M_j} + \frac{1}{2}M_j\omega_j^2 q_j^2 \right\} + H_{\text{int}}(q_1, \ldots, q_W, p_1, \ldots, p_W). \tag{4.3b}$$

Here M_j is the effective mass of mode j, and ω_j is the mode's angular frequency. This description of the crystal is preferable to one in which we use, as our generalized coordinates and momenta, the coordinate locations and momentum components of each of the 10^{26} atoms. Why? Because the normal modes are so weakly coupled to one

---

[1] For example, if we add damping to a simple harmonic oscillator, we can either treat the system as (in principle) hamiltonian by allowing for all the internal degrees of freedom associated with the damping, for example friction, or as (in practice) non-hamiltonian with an external heat sink.

another that they are semiclosed subsystems of the crystal, whereas the atoms are so strongly coupled that they are not, individually, semiclosed. As we shall see, there is great power in decomposing a complicated system into semiclosed subsystems.

4.2.2 Ensembles

ensemble of systems

In kinetic theory, we study statistically a collection of a huge number of particles. Similarly, in statistical mechanics, we study statistically a collection or *ensemble* of a huge number of systems. This ensemble is actually only a conceptual device, a foundation for statistical arguments that take the form of thought experiments. As we shall see, there are many different ways that one can imagine forming an ensemble, and this freedom can be used to solve many different types of problems.

In some applications, we require that all the systems in the ensemble be closed and be identical in the sense that they all have the same number of degrees of freedom, W; are governed by hamiltonians with the same functional forms $H(\mathbf{q}, \mathbf{p})$; and have the same volume V and total internal energy E (or $\mathcal{E}$, including rest masses). However, the values of the generalized coordinates and momenta at a specific time t, $\{\mathbf{q}(t), \mathbf{p}(t)\}$, need not be the same (i.e., the systems need not be in the same state at time t). If such a conceptual ensemble of identical closed systems (first studied by Boltzmann) evolves until it reaches statistical equilibrium (Sec. 4.5), it then is called *microcanonical*; see Table 4.1.

microcanonical ensemble

Sometimes we deal with an ensemble of systems that can exchange energy (heat) with their identical surroundings, so the internal energy of each system can fluctuate. If the surroundings (sometimes called *heat baths*) have far greater heat capacity than the individual systems, and if statistical equilibrium has been reached, then we call this sort of ensemble (introduced by Gibbs) *canonical*.

heat baths

canonical ensemble

At the next level of generality, the systems can also expand (i.e., they can exchange volume as well as energy with their identical surroundings). This case was also studied by Gibbs—his text (Gibbs, 1902) still repays reading—and in equilibrium it is known as the *Gibbs ensemble*; its environment is a heat and volume bath. A fourth ensemble in common use is Pauli's *grand canonical ensemble*, in which each system can exchange

Gibbs ensemble

Pauli's grand canonical ensemble

TABLE 4.1: Statistical-equilibrium ensembles used in this chapter

Ensemble	Quantities exchanged with surroundings
Microcanonical	Nothing
Canonical	Energy E
Gibbs	Energy E and volume V
Grand canonical	Energy E and number of particles N_I of various species I

Note: For relativistic systems we usually include the rest masses of all particles in the energy, so E is replaced by $\mathcal{E}$ in the above formulas.

energy and particles (but not volume) with its surroundings; see Table 4.1. We study these equilibrium ensembles and their baths in Secs. 4.4 and 4.5.

4.2.3 Distribution Function

DISTRIBUTION FUNCTION AS A PROBABILITY

In kinetic theory (Chap. 3), we described the statistical properties of a collection of identical particles by a distribution function, and we found it useful to tie that distribution function's normalization to quantum theory: $\eta(t; \mathbf{x}, \mathbf{p}) =$ (mean number of particles that occupy a quantum state at location $\{\mathbf{x}, \mathbf{p}\}$ in 6-dimensional phase space at time t). In statistical mechanics, we use the obvious generalization of this: $\eta =$ (mean number of systems that occupy a quantum state at location $\{\mathbf{q}, \mathbf{p}\}$ in an ensemble's $2W$-dimensional phase space, at time t)—except that we need two modifications:

1. This generalized η is proportional to the number of systems N_{sys} in our ensemble. (If we double N_{sys}, then η will double.) Because our ensemble is only a conceptual device, we don't really care how many systems it contains, so we divide η by N_{sys} to get a renormalized, N_{sys}-independent distribution function, $\rho = \eta/N_{\text{sys}}$, whose physical interpretation is

$$\rho(t; \mathbf{q}, \mathbf{p}) = \begin{pmatrix} \text{probability that a system, drawn randomly} \\ \text{from our ensemble, will be in a quantum state} \\ \text{at location } (\mathbf{q}, \mathbf{p}) \text{ in phase space at time } t \end{pmatrix}. \quad (4.4)$$

probabilistic distribution function

2. If the systems of our ensemble can exchange particles with the external universe (as is the case, for example, in the grand canonical ensemble of Table 4.1), then their number of degrees of freedom, W, can change, so ρ may depend on W as well as on location in the $2W$-dimensional phase space: $\rho(t; W, \mathbf{q}, \mathbf{p})$.

In the sector of the system's phase space with W degrees of freedom, denote the number density of quantum states by

$$\mathcal{N}_{\text{states}}(W, \mathbf{q}, \mathbf{p}) = \frac{dN_{\text{states}}}{d^W q \, d^W p} \equiv \frac{dN_{\text{states}}}{d\Gamma_W}. \quad (4.5)$$

density of states

Here we have used

$$d^W q \equiv dq_1 dq_2 \cdots dq_W, \quad d^W p \equiv dp_1 dp_2 \cdots dp_W, \quad d\Gamma_W \equiv d^W q \, d^W p. \quad (4.6)$$

Then the sum of the occupation probability ρ over all quantum states, which must (by the meaning of probability) be unity, takes the form

$$\sum_n \rho_n = \sum_W \int \rho \mathcal{N}_{\text{states}} d\Gamma_W = 1. \quad (4.7)$$

normalization of distribution function

On the left-hand side n is a formal index that labels the various quantum states $|n\rangle$ available to the ensemble's systems; on the right-hand side the sum is over all possible values of the system's dimensionality W, and the integral is over all of the $2W$-dimensional phase space, with $d\Gamma_W$ a short-hand notation for the phase-space integration element $d^W q \, d^W p$.

GEOMETRICAL VIEWPOINT

Equations (4.4)–(4.7) require some discussion. Just as the events and 4-momenta in relativistic kinetic theory are geometric, frame-independent objects, similarly *location in phase space* in statistical mechanics is a geometric, coordinate-independent concept (though our notation does not emphasize it). The quantities $\{\mathbf{q}, \mathbf{p}\} \equiv \{q_1, q_2, \ldots, q_W, p_1, p_2, \ldots, p_W\}$ are the coordinates of that phase-space location. When one makes a canonical transformation from one set of generalized coordinates and momenta to another (Ex. 4.1), the qs and ps change, but the geometric location in phase space does not. Moreover, just as the individual spatial and momentum volumes $d\mathcal{V}_x$ and $d\mathcal{V}_p$ occupied by a set of relativistic particles in kinetic theory are frame dependent, but their product $d\mathcal{V}_x d\mathcal{V}_p$ is frame-independent [cf. Eqs. (3.7a)–(3.7c)], so also in statistical mechanics the volumes $d^W q$ and $d^W p$ occupied by some chosen set of systems are dependent on the choice of canonical coordinates (they change under a canonical transformation), but the product $d^W q \, d^W p \equiv d\Gamma_W$ (the systems' total volume in phase space) is independent of the choice of canonical coordinates and is unchanged by a canonical transformation. Correspondingly, *the number density of states in phase space* $\mathcal{N}_{\text{states}} = d N_{\text{states}}/d\Gamma_W$ *and the statistical mechanical distribution function* $\rho(t; W, \mathbf{q}, \mathbf{p})$, *like their kinetic-theory counterparts, are geometric, coordinate-independent quantities: they are unchanged by a canonical transformation.* See Ex. 4.1 and references cited there.

ρ and $\mathcal{N}_{\text{states}}$ as geometric objects

DENSITY OF STATES

Classical thermodynamics was one of the crowning achievements of nineteenth-century science. However, thermodynamics was inevitably incomplete and had to remain so until the development of quantum theory. A major difficulty, one that we have already confronted in Chap. 3, was how to count the number of states available to a system. As we saw in Chap. 3, the number density of quantum mechanical states in the 6-dimensional, single-particle phase space of kinetic theory is (ignoring particle spin) $\mathcal{N}_{\text{states}} = 1/h^3$, where h is Planck's constant. Generalizing to the $2W$-dimensional phase space of statistical mechanics, the number density of states turns out to be $1/h^W$ [one factor of $1/h$ for each of the canonical pairs $(q_1, p_1), (q_2, p_2), \cdots, (q_W, p_W)$]. Formally, this follows from the canonical quantization procedure of elementary quantum mechanics.

DISTINGUISHABILITY AND MULTIPLICITY

distinguishability of particles

There was a second problem in nineteenth-century classical thermodynamics, that of *distinguishability*: If we swap two similar atoms in phase space, do we have a new state or not? If we mix two containers of the same gas at the same temperature and pressure, does the entropy increase (Ex. 4.8)? This problem was recognized classically but was not resolved in a completely satisfactory classical manner. When the laws of quantum mechanics were developed, it became clear that all identical particles are indistinguishable, so having particle 1 at location $\mathcal{A}$ in phase space and an identical particle 2 at location $\mathcal{B}$ must be counted as the same state as particle 1 at $\mathcal{B}$ and particle 2 at $\mathcal{A}$. Correspondingly, if we attribute half the quantum state to the classical phase-space location {1 at $\mathcal{A}$, 2 at $\mathcal{B}$} and the other half to {1 at $\mathcal{B}$, 2 at $\mathcal{A}$}, then the classical number density of states per unit volume of phase space must be reduced by a factor of 2—and more generally by some *multiplicity factor* $\mathcal{M}$. In general, therefore, we can write the actual number density of states in phase space as

multiplicity

$$\boxed{\mathcal{N}_{\text{states}} = \frac{dN_{\text{states}}}{d\Gamma_W} = \frac{1}{\mathcal{M}h^W},} \tag{4.8a}$$

density of states

and correspondingly, we can rewrite the normalization condition (4.7) for our probabilistic distribution function as

$$\boxed{\sum_n \rho_n \equiv \sum_W \int \rho \mathcal{N}_{\text{states}} d\Gamma_W = \sum_W \int \rho \frac{d\Gamma_W}{\mathcal{M}h^W} = 1.} \tag{4.8b}$$

This equation can be regarded, in the classical domain, as defining the meaning of the sum over states n. We shall make extensive use of such sums over states.

For N identical and indistinguishable particles with zero spin, it is not hard to see that $\mathcal{M} = N!$. If we include the effects of quantum mechanical spin (and the spin states can be regarded as degenerate), then there are g_s [Eq. (3.16)] more states present in the phase space of each particle than we thought, so an individual state's multiplicity $\mathcal{M}$ (the number of different phase-space locations to be attributed to the state) is reduced to

$$\boxed{\mathcal{M} = \frac{N!}{g_s^N} \quad \text{for a system of N identical particles with spin } s.} \tag{4.8c}$$

This is the quantity that appears in the denominator of the sum over states [Eq. (4.8b)].

Occasionally, for conceptual purposes it is useful to introduce a renormalized distribution function $\mathcal{N}_{\text{sys}}$, analogous to kinetic theory's number density of particles in phase space:

$$\mathcal{N}_{\text{sys}} = N_{\text{sys}} \mathcal{N}_{\text{states}} \rho = \frac{d(\text{number of systems})}{d(\text{volume in } 2W\text{-dimensional phase space})}. \tag{4.9}$$

number density of systems in phase space

However, this version of the distribution function will rarely if ever be useful computationally.

ENSEMBLE AVERAGE

Each system in an ensemble is endowed with a total energy that is equal to its hamiltonian, $E = H(\mathbf{q}, \mathbf{p})$ [or relativistically, $\mathcal{E} = H(\mathbf{q}, \mathbf{p})$]. Because different systems reside at different locations $(\mathbf{q}, \mathbf{p})$ in phase space, they typically will have different energies.

A quantity of much interest is the *ensemble-averaged energy*, which is the average value of E over all systems in the ensemble:

$$\langle E \rangle = \sum_n \rho_n E_n = \sum_W \int \rho \, E \mathcal{N}_{\text{states}} d\Gamma_W = \sum_W \int \rho \, E \frac{d\Gamma_W}{Mh^W}. \quad (4.10a)$$

For any other function $A(\mathbf{q}, \mathbf{p})$ defined on the phase space of a system (e.g., the linear momentum or the angular momentum), one can compute an ensemble average by the obvious analog of Eq. (4.10a):

ensemble average

$$\boxed{\langle A \rangle = \sum_n \rho_n A_n.} \quad (4.10b)$$

Our probabilistic distribution function $\rho_n = \rho(t; W, \mathbf{q}, \mathbf{p})$ has deeper connections to quantum theory than the above discussion reveals. In the quantum domain, even if we start with a system whose wave function ψ is in a *pure* state (ordinary, everyday type of quantum state), the system may evolve into a *mixed state* as a result of (i) interaction with the rest of the universe and (ii) our choice not to keep track of correlations between the universe and the system (Box 4.2 and Sec. 4.7.2). The system's initial, pure state can be described in geometric, basis-independent quantum language by a state vector ("ket") $|\psi\rangle$; but its final, mixed state requires a different kind of quantum description: a *density operator* $\hat{\rho}$. In the classical limit, the quantum mechanical density operator $\hat{\rho}$ becomes our classical probabilistic distribution function $\rho(t, W, \mathbf{q}, \mathbf{p})$; see Box 4.2 for some details.

EXERCISES

Exercise 4.1 *Example: Canonical Transformation*

Canonical transformations are treated in advanced textbooks on mechanics, such as Goldstein, Poole, and Safko (2002, Chap. 9) or, more concisely, Landau and Lifshitz (1976). This exercise gives a brief introduction. For simplicity we assume the hamiltionian is time independent.

Let (q_j, p_k) be one set of generalized coordinates and momenta for a given system. We can transform to another set (Q_j, P_k), which may be more convenient, using a *generating function* that connects the old and new sets. One type of generating function is $F(q_j, P_k)$, which depends on the old coordinates $\{q_j\}$ and new momenta $\{P_k\}$, such that

$$p_j = \frac{\partial F}{\partial q_j}, \quad Q_j = \frac{\partial F}{\partial P_j}. \quad (4.11)$$

(a) As an example, what are the new coordinates and momenta in terms of the old that result from

$$F = \sum_{i=1}^{W} f_i(q_1, q_2, \ldots, q_W) P_i, \quad (4.12)$$

where f_i are arbitrary functions of the old coordinates?

BOX 4.2. DENSITY OPERATOR AND QUANTUM STATISTICAL MECHANICS T2

Here we describe briefly the connection of our probabilistic distribution function ρ to the full quantum statistical theory as laid out, for example, in Sethna (2006, Sec. 7.1.1) and Cohen-Tannoudji, Diu, and Laloë (1977, complement E_{III}), or in much greater detail in Pathria and Beale (2011, Chap. 5) and Feynman (1972).

Consider a single quantum mechanical system that is in a pure state $|\psi\rangle$. One can formulate the theory of such a pure state equally well in terms of $|\psi\rangle$ or the density operator $\hat{\varrho} \equiv |\psi\rangle\langle\psi|$. For example, the expectation value of some observable, described by a Hermitian operator $\hat{A}$, can be expressed equally well as $\langle A \rangle = \langle\psi|\hat{A}|\psi\rangle$ or as $\langle A \rangle = \text{Trace}(\hat{\varrho}\hat{A})$.[2]

If our chosen system interacts with the external universe and we have no knowledge of the correlations that the interaction creates between the system and the universe, then the interaction drives the system into a mixed state, which is describable by a density operator $\hat{\varrho}$ but not by a ket vector $|\psi\rangle$. This $\hat{\varrho}$ can be regarded as a classical-type average of $|\psi\rangle\langle\psi|$ over an ensemble of systems, each of which has interacted with the external universe and then has been driven into a pure state $|\psi\rangle$ by a measurement of the universe. Equivalently, $\hat{\varrho}$ can be constructed from the pure state of universe plus system by "tracing over the universe's degrees of freedom."

If the systems in the ensemble behave nearly classically, then it turns out that in the basis $|\phi_n\rangle$, whose states are labeled by the classical variables $n = \{W, \mathbf{q}, \mathbf{p}\}$, the density matrix $\varrho_{nm} \equiv \langle\phi_n|\hat{\varrho}|\phi_m\rangle$ is very nearly diagonal. The classical probability ρ_n of classical statistical mechanics (and of this book when dealing with classical or quantum systems) is then equal to the diagonal value of this density matrix: $\rho_n = \varrho_{nn}$.

It can be demonstrated that the equation of motion for the density operator $\hat{\varrho}$, when the systems in the quantum mechanical ensemble are all evolving freely (no significant interactions with the external universe), is

$$\frac{\partial \hat{\varrho}}{\partial t} + \frac{1}{i\hbar}[\hat{\varrho}, \hat{H}] = 0. \tag{1}$$

This is the quantum statistical analog of the Liouville equation (4.15), and the quantum mechanical commutator $[\hat{\varrho}, \hat{H}]$ appearing here is the quantum

---

2. In any basis $|\phi_i\rangle$, "Trace" is just the trace of the matrix product $\sum_j \varrho_{ij} A_{jk}$, where $A_{jk} \equiv \langle\phi_j|\hat{A}|\phi_k\rangle$, and $\varrho_{ij} \equiv \langle\phi_i|\hat{\varrho}|\phi_j\rangle$ is called the *density matrix* in that basis. Sometimes $\hat{\rho}$ itself is called the density matrix, even though it is an operator.

(continued)

> BOX 4.2. (continued)
>
> mechanical analog of the Poisson bracket $[\rho, H]_{\mathbf{q},\mathbf{p}}$, which appears in the Liouville equation. If the ensemble's quantum systems are in eigenstates of their hamiltonians, then $\hat{\varrho}$ commutes with $\hat{H}$, so the density matrix is constant in time and there will be no transitions. This is the quantum analog of the classical ρ being constant in time and thus a constant of the motion (Sec. 4.4).

(b) The canonical transformation generated by Eq. (4.11) for arbitrary $F(q_j, P_k)$ leaves unchanged the value, but not the functional form, of the hamiltonian at each point in phase space. In other words, H is a geometric, coordinate-independent function (scalar field) of location in phase space. Show, for the special case of a system with one degree of freedom (one q, one p, one Q, and one P), that if Hamilton's equations (4.1) are satisfied in the old variables (q, p), then they will be satisfied in the new variables (Q, P).

(c) Show, for a system with one degree of freedom, that although $dq \neq dQ$ and $dp \neq dP$, the volume in phase space is unaffected by the canonical transformation: $dp\,dq = dP\,dQ$.

(d) Hence show that for any closed path in phase space, $\oint p\,dq = \oint P\,dQ$. These results are readily generalized to more than one degree of freedom.

4.3 Liouville's Theorem and the Evolution of the Distribution Function

In kinetic theory, the distribution function $\mathcal{N}$ is not only a frame-independent entity, it is also a constant along the trajectory of any freely moving particle, so long as collisions between particles are negligible. Similarly, in statistical mechanics, the probability ρ is not only coordinate-independent (unaffected by canonical transformations); *ρ is also a constant along the phase-space trajectory of any freely evolving system,* so long as the systems in the ensemble are not interacting significantly with the external universe (i.e., so long as they can be idealized as closed). This is the statistical mechanical version of Liouville's theorem, and its proof is a simple exercise in Hamiltonian mechanics, analogous to the "sophisticated" proof of the collisionless Boltzmann equation in Box 3.2.

Since the ensemble's systems are closed, no system changes its dimensionality W during its evolution. This permits us to fix W in the proof. Since no systems are created or destroyed during their evolution, the number density of systems in phase space, $\mathcal{N}_{\text{sys}} = N_{\text{sys}} \mathcal{N}_{\text{states}}\, \rho$ [Eq. (4.9)], must obey the same kind of conservation law as we encountered in Eq. (1.30) for electric charge and particle number in Newtonian physics. For particle number in a fluid, the conservation law is $\partial n/\partial t + \nabla \cdot (n\mathbf{v}) = 0$, where n is the number density of particles in physical space, $\mathbf{v}$ is their common velocity (the fluid's velocity) in physical space, and $n\mathbf{v}$ is their flux. Our ensemble's systems have velocity $dq_j/dt = \partial H/\partial p_j$ in physical space and "velocity" $dp_j/dt = -\partial H/\partial q_j$ in momentum space, so by analogy with particle number in a fluid, the conservation law (valid for ρ as well as for $\mathcal{N}_{\text{sys}}$, since they are proportional to each other) is

$$\frac{\partial \rho}{\partial t} + \frac{\partial}{\partial q_j}\left(\rho \frac{dq_j}{dt}\right) + \frac{\partial}{\partial p_j}\left(\rho \frac{dp_j}{dt}\right) = 0. \tag{4.13}$$

conservation law for systems

Equation (4.13) has an implicit sum, from 1 to W, over the repeated index j (recall the Einstein summation convention, Sec. 1.5). Using Hamilton's equations, we can rewrite this as

$$\begin{aligned}0 &= \frac{\partial \rho}{\partial t} + \frac{\partial}{\partial q_j}\left(\rho \frac{\partial H}{\partial p_j}\right) - \frac{\partial}{\partial p_j}\left(\rho \frac{\partial H}{\partial q_j}\right) \\ &= \frac{\partial \rho}{\partial t} + \frac{\partial \rho}{\partial q_j}\frac{\partial H}{\partial p_j} - \frac{\partial \rho}{\partial p_j}\frac{\partial H}{\partial q_j} = \frac{\partial \rho}{\partial t} + [\rho, H]_{\mathbf{q}, \mathbf{p}},\end{aligned} \tag{4.14}$$

where $[\rho, H]_{\mathbf{q}, \mathbf{p}}$ is the *Poisson bracket* (e.g., Landau and Lifshitz, 1976; Marion and Thornton, 1995; Goldstein, Poole, and Safko, 2002). By using Hamilton's equations once again in the second expression, we discover that this is the time derivative of ρ moving with a fiducial system through the $2W$-dimensional phase space:

$$\boxed{\left(\frac{d\rho}{dt}\right)_{\substack{\text{moving with a}\\\text{fiducial system}}} \equiv \frac{\partial \rho}{\partial t} + \frac{dq_j}{dt}\frac{\partial \rho}{\partial q_j} + \frac{dp_j}{dt}\frac{\partial \rho}{\partial p_j} = \frac{\partial \rho}{\partial t} + [\rho, H]_{\mathbf{q}, \mathbf{p}} = 0.}$$

Liouville equation (collisionless Boltzmann equation)

(4.15)

Therefore, the probability ρ is constant along the system's phase space trajectory, as was to be proved.

We call Eq. (4.15), which embodies this Liouville theorem, the statistical mechanical *Liouville equation* or *collisionless Boltzmann equation*.

As a simple, qualitative example, consider a system consisting of hot gas expanding adiabatically, so its large random kinetic energy is converted into ordered radial motion. If we examine a set $\mathcal{S}$ of such systems very close to one another in phase space, then it is apparent that, as the expansion proceeds, the size of $\mathcal{S}$'s physical-space volume $d^W q$ increases and the size of its momentum-space volume $d^W p$ diminishes, so that the product $d^W q \, d^W p$ remains constant (Fig. 4.1), and correspondingly $\rho \propto \mathcal{N}_{\text{sys}} = dN_{\text{sys}}/d^W q \, d^W p$ is constant.

What happens if the systems being studied interact weakly with their surroundings? We must then include an interaction term on the right-hand side of Eq. (4.15),

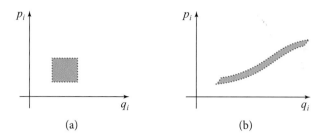

FIGURE 4.1 Liouville's theorem. (a) The region in the q_i-p_i part of phase space (with i fixed) occupied by a set $\mathcal{S}$ of identical, closed systems at time $t = 0$. (b) The region occupied by the same set of systems a short time later, $t > 0$. The hamiltonian-generated evolution of the individual systems has moved them in such a manner as to skew the region they occupy, but the volume $\int dp_i dq_i$ is unchanged.

thereby converting it into the statistical mechanical version of the Boltzmann transport equation:

Boltzmann transport equation

$$\left(\frac{d\rho}{dt}\right)_{\substack{\text{moving with a}\\\text{fiducial system}}} = \left(\frac{d\rho}{dt}\right)_{\text{interactions}}. \tag{4.16}$$

The time derivative on the left is now taken moving through phase space with a fiducial system that does not interact with the external universe.

4.4 Statistical Equilibrium

In this section's discussion of statistical equilibrium, we begin by writing formulas in the relativistic form, where rest masses are included in the energy $\mathcal{E}$; afterward we take the Newtonian limit.

the meaning of temperature

Before we start, we should clarify what we mean by the temperature of a system. The first point to make is that a rigorous definition presumes thermodynamic equilibrium for the system in question, which can be anything from a two-level atom to the early universe. We can then say that two systems that are in equilibrium and whose average properties do not change when they are brought into contact (so they can exchange heat) have the same temperature and that this attribute is transitive. (This is known as the zeroth law of thermodynamics.) This allows us to define the temperature of a reference system as a monotonic function that increases as heat is added to it. This reference system can then be used to assign a temperature to every other body with which it can be brought into equilibrium; cf. Sec. 5.2.3. There are several ways to specify this function. The one that we shall follow [Eqs. (4.19) and following text] is surprisingly general and depends on simple ideas from probability.[3]

---

3. Other approaches to the definition of temperature involve the engineering-inspired Carnot cycle, the mathematical Carathéodory approach, and the fluctuation-dissipation theorem expressed, for example,

4.4.1 Canonical Ensemble and Distribution

STATISTICAL EQUILIBRIUM AND JEANS' THEOREM

Consider an ensemble of identical systems, all of which have the same huge number of degrees of freedom (dimensionality $W \gg 1$). Put all the systems initially in the same state, and then let them exchange heat (but not particles, volume, or anything else) with an external *thermal bath* that has a huge heat capacity and is in thermodynamic equilibrium at some temperature T. (For example, the systems might be impermeable cubes of gas 1 km on a side near the center of the Sun, and the thermal bath might be all the surrounding gas near the Sun's center; or the systems might be identical sapphire crystals inside a huge cryostat, and the thermal bath might be the cryostat's huge store of liquid helium.) After a sufficiently long time, $t \gg \tau_{\text{ext}}$, the ensemble will settle down into equilibrium with the bath (i.e., it will become the canonical ensemble mentioned in Table 4.1 above). In this final, canonical equilibrium state, the probability $\rho(t, \mathbf{q}, \mathbf{p})$ is independent of time t, and it no longer is affected by interactions with the external environment. In other words, the interaction terms in the evolution equation (4.16) have ceased to have any net effect: on average, for each interaction event that feeds energy into a system, there is an interaction event that takes away an equal amount of energy. The distribution function, therefore, satisfies the interaction-free, collisionless Boltzmann equation (4.15) with the time derivative $\partial \rho / \partial t$ removed:

$$[\rho, H]_{\mathbf{q},\mathbf{p}} \equiv \frac{\partial \rho}{\partial q_j} \frac{\partial H}{\partial p_j} - \frac{\partial \rho}{\partial p_j} \frac{\partial H}{\partial q_j} = 0. \qquad (4.17)$$

statistical equilibrium

We use the phrase *statistical equilibrium* to refer to any ensemble whose distribution function has attained such a state and thus satisfies Eq. (4.17).

Equation (4.17) is a well-known equation in Hamiltonian mechanics. It says that ρ is a function solely of constants of the individual systems' hamiltonian-induced motions (e.g., Landau and Lifshitz, 1976; Marion and Thornton, 1995; Goldstein, Poole, and Safko, 2002); in other words, ρ can depend on location $(\mathbf{q}, \mathbf{p})$ in phase space only through those constants of the motion. Sometimes this goes by the name *Jeans' theorem*. Among the constants of motion in typical situations (for typical hamiltonians) are the system's energy $\mathcal{E}$, its linear momentum $\mathbf{P}$, its angular momentum $\mathbf{J}$, its number N_I of conserved particles of various types I (e.g., electrons, protons), and its volume V. Notice that these constants of motion $\mathcal{E}, \mathbf{P}, \mathbf{J}, N_I$, and V are all additive: if we double the size of a system, they each double. We call such additive constants of the hamiltonian-induced motion *extensive variables* (a term borrowed from thermodynamics) and denote them by an enumerated list $K_1, K_2, \ldots$.

Jeans' theorem

extensive variables

Now, the systems we are studying have exchanged energy $\mathcal{E}$ with their environment (the thermal bath) and thereby have acquired some range of $\mathcal{E}$ values; therefore, ρ

---

as Johnson noise (Sec. 6.8.1). These are covered in many thermodynamics texts (e.g., Reif, 2008) and lead to the same temperature scale.

can depend on $\mathcal{E}$. However, the systems have not exchanged anything else with their environment, and they all thus have retained their original (identical) values of the other extensive variables K_A; therefore, ρ must be a delta function in these other variables. We write

$$\rho = \rho(\mathcal{E}) \tag{4.18a}$$

and do not write down the delta functions explicitly.

As an aid in discovering the form of the function $\rho(\mathcal{E})$, let us decompose each system in the ensemble into a huge number of subsystems. For example, each system might be a cube 1 km on a side inside the Sun, and its subsystems might be the 10^9 1-m cubes into which the system can be divided, or the systems might be identical sapphire crystals each containing 10^{26} atoms, and the subsystems might be the crystals' 3×10^{26} normal modes of vibration. We label the subsystems of each system by an integer a in such a way that subsystem a in one system has the same hamiltonian as subsystem a in any other system. (For the sapphire crystals, $a = 1$ could be the fundamental mode of vibration, $a = 2$ the first harmonic, $a = 3$ the second harmonic, etc.) The subsystems with fixed a make up a *subensemble* because of their relationship to the original ensemble.

subensembles

Because the full ensemble is in statistical equilibrium, the subensembles will also be in statistical equilibrium; therefore, their probabilities must be functions of those extensive variables $\mathcal{E}$, K_A that they can exchange with one another:

$$\rho_a = \rho_a(\mathcal{E}_a, K_{1a}, K_{2a}, \ldots). \tag{4.18b}$$

(Although each system can exchange only energy $\mathcal{E}$ with its heat bath, the subsystems may be able to exchange other quantities with one another; for example, if subsystem a is a 1-m cube inside the Sun with permeable walls, then it can exchange energy $\mathcal{E}_a$ and particles of all species I, so $K_{Ia} = N_{Ia}$.)

Since there is such a huge number of subsystems in each system, it is reasonable to expect that in statistical equilibrium there will be no significant correlations at any given time between the actual state of subsystem a and the state of any other subsystem. In other words, the probability $\rho_a(W_a, \mathbf{q}_a, \mathbf{p}_a)$ that subsystem a is in a quantum state with W_a degrees of freedom and with its generalized coordinates and momenta near the values $(\mathbf{q}_a, \mathbf{p}_a)$ is independent of the state of any other subsystem. This lack of correlations, which can be written as

statistical independence

$$\rho(\mathcal{E}) = \prod_a \rho_a, \tag{4.18c}$$

is called *statistical independence*.[4]

---

4. Statistical independence is actually a consequence of a two-lengthscale approximation [Box 3.3]. The size of each subsystem is far smaller than that of the full system, and precise statistical independence arises in the limit as the ratio of these sizes goes to zero.

Statistical independence places a severe constraint on the functional forms of ρ and ρ_a, as the following argument shows. By taking the logarithm of Eq. (4.18c), we obtain

$$\ln \rho(\mathcal{E}) = \sum_a \ln \rho_a(\mathcal{E}_a, K_{1a}, \ldots). \tag{4.18d}$$

We also know, since energy is a linearly additive quantity, that

$$\mathcal{E} = \sum_a \mathcal{E}_a. \tag{4.18e}$$

Now, we have not stipulated the way in which the systems are decomposed into subsystems. For our solar example, the subsystems might have been 2-m or 7-m cubes rather than 1-m cubes. Exploiting this freedom, one can deduce that Eqs. (4.18d) and (4.18e) can be satisfied simultaneously if and only if $\ln \rho$ and $\ln \rho_a$ depend linearly on the energies $\mathcal{E}$ and $\mathcal{E}_a$, with the same proportionality constant $-\beta$:

$$\ln \rho_a = -\beta \mathcal{E}_a + \text{(some function of } K_{1a}, K_{2a}, \ldots), \tag{4.19a}$$

$$\ln \rho = -\beta \mathcal{E} + \text{constant}. \tag{4.19b}$$

The reader presumably will identify β with $1/(k_B T)$, where T is the temperature of the thermal bath, as we have implicitly done in Chap. 3. However, what does this temperature actually mean and how do we define it? One approach is to choose as our subsystem a single identified atom in a classical monatomic perfect gas, with atomic rest mass high enough that the gas is nonrelativistic. The energy $\mathcal{E}$ is then just the atom's kinetic energy (plus rest mass) and the pressure is $nk_B T$. We then repeat this for every other atom. Measuring the pressure—mechanically—in this simple system then defines $k_B T$. The choice of k_B is a mere convention; it is most commonly chosen to set the temperature of the triple point of water as 273.16 K. Much more complex systems can then be assigned the same temperature as the perfect gas with which they would be in equilibrium as outlined above. If the reader protests that gases are not perfect at low or high temperature, then we can substitute free electrons or photons and adopt their distribution functions as we have described in Sec. 3.3 and will derive in Sec. 4.4.3.

definition of temperature

To summarize, an ensemble of identical systems with many degrees of freedom $W \gg 1$, which have reached statistical equilibrium by exchanging energy but nothing else with a huge thermal bath, has the following canonical distribution function:

$$\boxed{\rho_{\text{canonical}} = C \exp(-\mathcal{E}/k_B T),} \quad \boxed{\rho_{\text{canonical}} = C' \exp(-E/k_B T) \text{ nonrelativistically.}} \tag{4.20}$$

canonical distribution

Here $\mathcal{E}(\mathbf{q}, \mathbf{p})$ is the energy of a system at location $\{\mathbf{q}, \mathbf{p}\}$ in phase space, k_B is Boltzmann's constant, T is the temperature of the heat bath, and C is whatever normalization constant is required to guarantee that $\sum_n \rho_n = 1$. The nonrelativistic expression is obtained by removing all the particle rest masses from the total energy $\mathcal{E}$ and then taking the low-temperature, low-thermal-velocities limit.

Actually, we have proved more than Eq. (4.20). Not only must the ensemble of huge systems ($W \gg 1$) have the energy dependence $\rho \propto \exp(-\mathcal{E}/(k_B T))$, so must each subensemble of smaller systems, $\rho_a \propto \exp(-\mathcal{E}_a/(k_B T))$, even if (for example) the subensemble's identical subsystems have only one degree of freedom, $W_a = 1$. Thus, if the subsystems exchanged only heat with their parent systems, then they must have the same canonical distribution (4.20) as the parents. This shows that *the canonical distribution is the equilibrium state independently of the number of degrees of freedom W.*

4.4.2 General Equilibrium Ensemble and Distribution; Gibbs Ensemble; Grand Canonical Ensemble

GENERAL EQUILIBRIUM ENSEMBLE

We can easily generalize the canonical distribution to an ensemble of systems that exchange other additive conserved quantities (extensive variables) $K_1, K_2, \ldots$, in addition to energy $\mathcal{E}$, with a huge, thermalized bath. By an obvious generalization of the argument in Sec. 4.4.1, the resulting statistical equilibrium distribution function must have the form

general equilibrium distribution

$$\rho = C \exp\left(-\beta \mathcal{E} - \sum_A \beta_A K_A\right). \qquad (4.21)$$

extensive variables: energy, volume, particle numbers, momentum, angular momentum

When the extensive variables K_A that are exchanged with the bath (and thus appear explicitly in the distribution function ρ) are energy $\mathcal{E}$, momentum $\mathbf{P}$, angular momentum $\mathbf{J}$, the number N_I of the species I of conserved particles, volume V, or any combination of these quantities, it is conventional to rename the multiplicative factors β and β_A so that ρ takes on the form

$$\rho = C \exp\left[\frac{-\mathcal{E} + \mathbf{U}\cdot\mathbf{P} + \mathbf{\Omega}\cdot\mathbf{J} + \sum_I \tilde{\mu}_I N_I - PV}{k_B T}\right]. \qquad (4.22)$$

bath's intensive variables: temperature, pressure, chemical potentials, velocity, angular velocity

Here T, $\mathbf{U}$, $\mathbf{\Omega}$, $\tilde{\mu}_I$, and P are constants (called *intensive* variables) that are the same for all systems and subsystems (i.e., that characterize the full ensemble and all its subensembles and therefore must have been acquired from the bath); *any extensive variable that is not exchanged with the bath must be omitted from the exponential and be replaced by an implicit delta function.*

As we have seen in Sec. 3.3, T is the temperature that the ensemble and subensembles acquired from the bath (i.e., it is the bath temperature). From the Lorentz transformation law for energy and momentum [$\mathcal{E}' = \gamma(\mathcal{E} - \mathbf{U}\cdot\mathbf{P})$; Eqs. (2.37) and Ex. 2.13] we see that, if we were to transform to a reference frame that moves with velocity $\mathbf{U}$ with respect to our original frame, then the $\exp(\mathbf{U}\cdot\mathbf{P}/(k_B T))$ term in ρ would disappear, and the distribution function would be isotropic in $\mathbf{P}$. Thus $\mathbf{U}$ is the velocity of the bath with respect to our chosen reference frame. By a similar argument, $\mathbf{\Omega}$ is the bath's angular velocity with respect to an inertial frame. By comparison with

Eq. (3.22d), we see that $\tilde{\mu}_I$ is the chemical potential of the conserved species I. Finally, experience with elementary thermodynamics suggests (and it turns out to be true) that P is the bath's pressure.[5] Note that, by contrast with the corresponding extensive variables $\mathcal{E}$, $\mathbf{P}$, $\mathbf{J}$, N_I, and V, the intensive variables T, $\mathbf{U}$, $\mathbf{\Omega}$, $\tilde{\mu}_I$, and P do not double when the size of a system is doubled (i.e., they are not additive); rather, they are properties of the ensemble as a whole and thus are independent of the systems' sizes.

By removing the rest masses m_I of all particles from each system's energy and similarly removing the particle rest mass from each chemical potential,

$$\boxed{E \equiv \mathcal{E} - \sum_I N_I m_I,} \qquad \boxed{\mu_I \equiv \tilde{\mu}_I - m_I} \qquad (4.23)$$

nonrelativistic energy and chemical potentials obtained by removing rest masses

[Eqs. (3.20) and (3.21)], we bring the distribution function into a form that is identical to Eq. (4.22) but with $\mathcal{E} \to E$ and $\tilde{\mu}_I \to \mu_I$:

$$\boxed{\rho = C \exp\left[\frac{-E + \mathbf{U} \cdot \mathbf{P} + \mathbf{\Omega} \cdot \mathbf{J} + \sum_I \mu_I N_I - PV}{k_B T}\right].} \qquad (4.24)$$

general equilibrium distribution

This is the form used in Newtonian theory, but it is also valid relativistically.

SPECIAL EQUILIBRIUM ENSEMBLES

Henceforth (except in Sec. 4.10.2, when discussing black-hole atmospheres), we restrict our baths always to be at rest in our chosen reference frame and to be nonrotating with respect to inertial frames, so that $\mathbf{U} = \mathbf{\Omega} = 0$. The distribution function ρ can then either be a delta function in the system momentum $\mathbf{P}$ and angular momentum $\mathbf{J}$ (if momentum and angular momentum are not exchanged with the bath), or it can involve no explicit dependence on $\mathbf{P}$ and $\mathbf{J}$ (if momentum and angular momentum are exchanged with the bath; cf. Eq. (4.22) with $\mathbf{U} = \mathbf{\Omega} = 0$). In either case, if energy is the only other quantity exchanged with the bath, then the distribution function is the canonical one [Eq. (4.20)]:

$$\boxed{\rho_{\text{canonical}} = C \exp\left[\frac{-\mathcal{E}}{k_B T}\right] = C' \exp\left[\frac{-E}{k_B T}\right],} \qquad (4.25a)$$

where (obviously) the constants C and C' are related by

$$C' = C \exp\left[-\sum_I N_I m_I / k_B T\right].$$

If, in addition to energy, volume can also be exchanged with the bath (e.g., if the systems are floppy bags of gas whose volumes can change and through which heat can

---

5. One can also identify these physical interpretations of T, $\tilde{\mu}_I$, and P by analyzing idealized measuring devices; cf. Sec. 5.2.2.

flow),[6] then the equilibrium is the *Gibbs ensemble,* which has the distribution function

Gibbs distribution

$$\boxed{\rho_{\text{Gibbs}} = C \exp\left[\frac{-(\mathcal{E}+PV)}{k_B T}\right] = C' \exp\left[\frac{-(E+PV)}{k_B T}\right]} \quad (4.25b)$$

enthalpy

(with an implicit delta function in N_I and possibly in **J** and **P**). The combination $\mathcal{E} + PV$ is known as the *enthalpy* H. If the exchanged quantities are energy and particles but not volume (e.g., if the systems are 1-m cubes inside the Sun with totally imaginary walls through which particles and heat can flow), then the equilibrium is the *grand canonical ensemble,* with

grand canonical distribution

$$\boxed{\rho_{\text{grand canonical}} = C \exp\left[\frac{-\mathcal{E} + \sum_I \tilde{\mu}_I N_I}{k_B T}\right] = C \exp\left[\frac{-E + \sum_I \mu_I N_I}{k_B T}\right]}$$

(4.25c)

(with an implicit delta function in V and perhaps in **J** and **P**).

We mention, as a preview of an issue to be addressed in Chap. 5, that an individual system, picked randomly from the ensemble and then viewed as a bath for its own tiny subsystems, will not have the same temperature T, and/or chemical potential $\tilde{\mu}_I$, and/or pressure P as the huge bath with which the ensemble has equilibrated; rather, the individual system's T, $\tilde{\mu}_I$, and/or P can fluctuate a tiny bit around the huge bath's values (around the values that appear in the above probabilities), just as its $\mathcal{E}$, N_I, and/or V fluctuate. We study these fluctuations in Sec. 5.6.

4.4.3 Fermi-Dirac and Bose-Einstein Distributions

The concepts and results developed in this chapter have enormous generality. They are valid (when handled with sufficient care) for quantum systems as well as classical ones, and they are valid for semiclosed or closed systems of any type. The systems need not resemble the examples we have met in the text. They can be radically different, but so long as they are closed or semiclosed, our concepts and results will apply.

SINGLE-PARTICLE QUANTUM STATES (MODES)

As an important example, let each system be a single-particle quantum state of some field. These quantum states can exchange particles (quanta) with one another. As we shall see, in this case the above considerations imply that, in statistical equilibrium at temperature T, the mean number of particles in a state, whose individual particle energies are $\mathcal{E}$, is given by the Fermi-Dirac formula (for fermions) $\eta = 1/(e^{(\mathcal{E}-\tilde{\mu})/(k_B T)} + 1)$ and Bose-Einstein formula (for bosons) $\eta = 1/(e^{(\mathcal{E}-\tilde{\mu})/(k_B T)} - 1)$, which we used in our kinetic-theory studies in the last chapter [Eqs. (3.22a), (3.22b)]. Our derivation of these mean occupation numbers will

---

6. For example, the huge helium-filled balloons made of thin plastic that are used to lift scientific payloads into Earth's upper atmosphere.

illustrate the closeness of classical statistical mechanics and quantum statistical mechanics: the proof is fundamentally quantum mechanical, because the regime $\eta \sim 1$ is quantum mechanical (it violates the classical condition $\eta \ll 1$); nevertheless, the proof makes use of precisely the same concepts and techniques as we have developed for our classical studies.

As a conceptual aid in the derivation, consider an ensemble of complex systems in statistical equilibrium. Each system can be regarded as made up of a large number of fermions (electrons, protons, neutrons, neutrinos, ...) and/or bosons (photons, gravitons, alpha particles, phonons, ...). We analyze each system by identifying a complete set of single-particle quantum states (which we call *modes*) that the particles can occupy[7] (see, e.g., Chandler, 1987, chap. 4). A complete enumeration of modes is the starting point for the *second quantization* formulation of quantum field theory and is also the starting point for our far simpler analysis.

modes

second quantization

Choose one specific mode S [e.g., a nonrelativistic electron plane-wave mode in a box of side L with spin up and momentum $\mathbf{p} = (5, 3, 17)h/L$]. There is one such mode S in each of the systems in our ensemble, and these modes (all identical in their properties) form a subensemble of our original ensemble. Our derivation focuses on this subensemble of identical modes S. Because each of these modes can exchange energy and particles with all the other modes in its system, the subensemble is grand canonically distributed.

The *(many-particle) quantum states* allowed for mode S are states in which S contains a finite number n of particles (quanta). Denote by $\mathcal{E}_S$ the energy of one particle residing in mode S. Then the mode's total energy when it is in the state $|n\rangle$ (when it contains n quanta) is $\mathcal{E}_n = n\mathcal{E}_S$. [For a freely traveling, relativistic electron mode, $\mathcal{E}_S = \sqrt{m^2 + \mathbf{p}^2}$, Eq. (1.40), where $\mathbf{p}$ is the mode's momentum, $p_x = jh/L$ for some integer j and similarly for p_y and p_z; for a phonon mode with angular eigenfrequency of vibration ω, $\mathcal{E}_S = \hbar\omega$.] Since the distribution of the ensemble's modes among the allowed quantum states is grand canonical, the probability ρ_n of being in state $|n\rangle$ is [Eq. (4.25c)]

many-particle quantum states

$$\rho_n = C \, \exp\left(\frac{-\mathcal{E}_n + \tilde{\mu} n}{k_B T}\right) = C \, \exp\left(\frac{n(\tilde{\mu} - \mathcal{E}_S)}{k_B T}\right), \quad (4.26)$$

where $\tilde{\mu}$ and T are the chemical potential and temperature of the bath of other modes, with which the mode S interacts.[8]

---

7. For photons, these modes are the normal modes of the classical electromagnetic field; for phonons in a crystal, they are the normal modes of the crystal's vibrations; for nonrelativistic electrons or protons or alpha particles, they are energy eigenstates of the nonrelativistic Schrödinger equation; for relativistic electrons, they are energy eigenstates of the Dirac equation.
8. Here and throughout Chaps. 4 and 5 we ignore quantum zero point energies, since they are unobservable in this context (though they are observed in the fluctuational forces discussed in Sec. 6.8). Equally well we could include the zero point energies in $\mathcal{E}_n$, $\mathcal{E}_s$, and $\tilde{\mu}$, and they would cancel out in the combinations that appear, e.g., in Eq. (4.26): $\tilde{\mu} - \mathcal{E}_s$, and so forth.

FERMION MODES: FERMI-DIRAC DISTRIBUTION

Suppose that $\mathcal{S}$ is a fermion mode (i.e., its particles have half-integral spin). Then the Pauli exclusion principle dictates that $\mathcal{S}$ cannot contain more than one particle: n can take on only the values 0 and 1. In this case, the normalization constant in the distribution function (4.26) is determined by $\rho_0 + \rho_1 = 1$, which implies that

Fermi-Dirac distribution

$$\rho_0 = \frac{1}{1 + \exp[(\tilde{\mu} - \mathcal{E}_\mathcal{S})/(k_B T)]}, \quad \rho_1 = \frac{\exp[(\tilde{\mu} - \mathcal{E}_\mathcal{S})/(k_B T)]}{1 + \exp[(\tilde{\mu} - \mathcal{E}_\mathcal{S})/(k_B T)]}. \quad (4.27a)$$

This is the explicit form of the grand canonical distribution for a fermion mode. For many purposes (including all those in Chap. 3), this full probability distribution is more than one needs. Quite sufficient instead is the mode's mean occupation number

Fermi-Dirac mean occupation number

$$\boxed{\eta_\mathcal{S} \equiv \langle n \rangle = \sum_{n=0}^{1} n\rho_n = \frac{1}{\exp[(\mathcal{E}_\mathcal{S} - \tilde{\mu})/(k_B T)] + 1} = \frac{1}{\exp[(E_\mathcal{S} - \mu)/(k_B T)] + 1}.}$$

(4.27b)

Here $E_\mathcal{S} = \mathcal{E}_\mathcal{S} - m$ is the energy of a particle in the mode with rest mass removed, and $\mu = \tilde{\mu} - m$ is the chemical potential with rest mass removed—the quantities used in the nonrelativistic (Newtonian) regime.

Equation (4.27b) is the *Fermi-Dirac mean occupation number* asserted in Chap. 3 [Eq. (3.22a)] and studied there for the special case of a gas of freely moving, noninteracting fermions. Because our derivation is completely general, we conclude that this mean occupation number and the underlying grand canonical distribution (4.27a) are valid for any mode of a fermion field—for example, the modes for an electron trapped in an external potential well or a magnetic bottle, and the (single-particle) quantum states of an electron in a hydrogen atom.

BOSON MODES: BOSE-EINSTEIN DISTRIBUTION

Suppose that $\mathcal{S}$ is a boson mode (i.e., its particles have integral spin), so it can contain any nonnegative number of quanta; that is, n can assume the values $0, 1, 2, 3, \ldots$. Then the normalization condition $\sum_{n=0}^{\infty} \rho_n = 1$ fixes the constant in the grand canonical distribution (4.26), resulting in

Bose-Einstein distribution

$$\rho_n = \left[1 - \exp\left(\frac{\tilde{\mu} - \mathcal{E}_\mathcal{S}}{k_B T}\right)\right] \exp\left(\frac{n(\tilde{\mu} - \mathcal{E}_\mathcal{S})}{k_B T}\right). \quad (4.28a)$$

From this grand canonical distribution we can deduce the mean number of bosons in mode $\mathcal{S}$:

Bose-Einstein mean occupation number

$$\boxed{\eta_\mathcal{S} \equiv \langle n \rangle = \sum_{n=1}^{\infty} n\rho_n = \frac{1}{\exp[(\mathcal{E}_\mathcal{S} - \tilde{\mu})/(k_B T)] - 1} = \frac{1}{\exp[(E_\mathcal{S} - \mu)/(k_B T)] - 1},}$$

(4.28b)

in accord with Eq. (3.22b). As for fermions, this *Bose-Einstein mean occupation number* and underlying grand canonical distribution (4.28a) are valid generally, and not solely for the freely moving bosons of Chap. 3.

When the mean occupation number is small, $\eta_S \ll 1$, both the bosonic and the fermionic distribution functions are well approximated by the classical *Boltzmann mean occupation number*

$$\boxed{\eta_S = \exp[-(\mathcal{E}_S - \tilde{\mu})/(k_B T)].} \quad (4.29)$$

Boltzmann mean occupation number

In Sec. 4.9 we explore an important modern application of the Bose-Einstein mean occupation number (4.28b): *Bose-Einstein condensation* of bosonic atoms in a magnetic trap.

4.4.4 Equipartition Theorem for Quadratic, Classical Degrees of Freedom

As a second example of statistical equilibrium distribution functions, we derive the classical equipartition theorem using statistical methods.

To motivate this theorem, consider a diatomic molecule of nitrogen, N_2. To a good approximation, its energy (its hamiltonian) can be written as

$$E = \frac{p_x^2}{2M} + \frac{p_y^2}{2M} + \frac{p_z^2}{2M} + \frac{P_\ell^2}{2M_\ell} + \frac{1}{2} M_\ell \omega_v^2 \ell^2 + \frac{J_x^2}{2I} + \frac{J_y^2}{2I}. \quad (4.30)$$

Here M is the molecule's mass; p_x, p_y, and p_z are the components of its translational momentum; and the first three terms are the molecule's kinetic energy of translation. The next two terms are the molecule's longitudinal vibration energy, with ℓ the change of the molecule's length (change of the separation of its two nuclei) from equilibrium, P_ℓ the generalized momentum conjugate to that length change, ω_v the vibration frequency, and M_ℓ the generalized mass associated with that vibration. The last two terms are the molecule's energy of end-over-end rotation, with J_x and J_y the components of angular momentum associated with this two-dimensional rotator and I its moment of inertia.

Notice that every term in this hamiltonian is quadratic in a generalized coordinate or generalized momentum! Moreover, each of these coordinates and momenta appears only in its single quadratic term and nowhere else, and the density of states is independent of the value of that coordinate or momentum. We refer to such a coordinate or momentum as a *quadratic degree of freedom*.

quadratic degree of freedom

In some cases (e.g., the vibrations and rotations but not the translations), the energy $E_\xi = \alpha \xi^2$ of a quadratic degree of freedom ξ is quantized, with some energy separation ε_0 between the ground state and first excited state (and with energy separations to higher states that are $\lesssim \varepsilon_0$). If (and only if) the thermal energy $k_B T$ is significantly larger than ε_0, then the quadratic degree of freedom ξ will be excited far above its ground state and will behave classically. The equipartition theorem applies only at these high temperatures. For diatomic nitrogen, the rotational degrees of freedom J_x and J_y have $\varepsilon_0 \sim 10^{-4}$ eV and $\varepsilon_0/k_B \sim 1$ K, so temperatures big compared to 1 K are required for J_x and J_y to behave classically. By contrast, the vibrational degrees of freedom ℓ and P_ℓ have $\varepsilon_0 \sim 0.1$ eV and $\varepsilon_0/k_B \sim 1{,}000$ K, so temperatures of a few thousand Kelvins are required for them to behave classically. Above $\sim 10^4$ K, the hamiltonian (4.30) fails: electrons around the nuclei are driven into excited states, and the molecule breaks apart (dissociates into two free atoms of nitrogen).

The equipartition theorem holds for any classical, quadratic degree of freedom [i.e., at temperatures somewhat higher than $T_o = \varepsilon_o/(k_B T)$]. We derive this theorem using the canonical distribution (4.25a). We write the molecule's total energy as $E = \alpha \xi^2 + E'$, where E' does not involve ξ. Then the mean energy associated with ξ is

$$\langle E_\xi \rangle = \frac{\int \alpha \xi^2\, e^{-\beta(\alpha \xi^2 + E')} d\xi\, d(\text{other degrees of freedom})}{\int e^{-\beta(\alpha \xi^2 + E')} d\xi\, d(\text{other degrees of freedom})}. \tag{4.31}$$

Here the exponential is that of the canonical distribution function (4.25a), the denominator is the normalizing factor, and we have set $\beta \equiv 1/(k_B T)$. Because ξ does not appear in the portion E' of the energy, its integral separates out from the others in both numerator and denominator, and the integrals over E' in numerator and denominator cancel. Rewriting $\int \alpha \xi^2 \exp(-\beta \alpha \xi^2)\, d\xi$ as $-d/d\beta[\int \exp(-\beta \alpha \xi^2)\, d\xi]$, Eq. (4.31) becomes

$$\begin{aligned}\langle E_\xi \rangle &= -\frac{d}{d\beta} \ln\left[\int \exp(-\beta \alpha \xi^2)\, d\xi\right] \\ &= -\frac{d}{d\beta} \ln\left[\frac{1}{\sqrt{\beta \alpha}} \int du\, e^{-u^2} du\right] = \frac{1}{2\beta} = \frac{1}{2} k_B T.\end{aligned} \tag{4.32}$$

equipartition theorem

Therefore, *in statistical equilibrium, the mean energy associated with any classical, quadratic degree of freedom is $\frac{1}{2}k_B T$*. This is the equipartition theorem. Note that the factor $\frac{1}{2}$ follows from the quadratic nature of the degrees of freedom.

For our diatomic molecule, at room temperature there are three translational and two rotational classical, quadratic degrees of freedom (p_x, p_y, p_z, J_x, J_y), so the mean total energy of the molecule is $\frac{5}{2}k_B T$. At a temperature of several thousand Kelvins, the two vibrational degrees of freedom, ℓ and P_ℓ, become classical and the molecule's mean total energy is $\frac{7}{2}k_B T$. Above $\sim 10^4$ K the molecule dissociates, and its two parts (the two nitrogen atoms) have only translational quadratic degrees of freedom, so the mean energy per atom is $\frac{3}{2}k_B T$.

The equipartition theorem is valid for any classical, quadratic degree of freedom, whether it is part of a molecule or not. For example, it applies to the generalized coordinate and momentum of any harmonic-oscillator mode of any system: a vibrational mode of Earth or of a crystal, or a mode of an electromagnetic field.

4.5 The Microcanonical Ensemble

Let us now turn from ensembles of systems that interact with an external, thermal bath (as discussed in Sec. 4.4.1) to an ensemble of identical, precisely closed systems (i.e., systems that have no interactions with the external universe). By "identical" we mean that every system in the ensemble has (i) the same set of degrees of freedom, and thus (ii) the same number of degrees of freedom W, (iii) the same hamiltonian, and (iv) the same values for all the additive constants of motion ($\mathcal{E}, K_1, K_2, \ldots$) except perhaps total momentum $\mathbf{P}$ and total angular momentum $\mathbf{J}$.[9]

Suppose that these systems begin with values of $(\mathbf{q}, \mathbf{p})$ that are spread out in some (arbitrary) manner over a hypersurface in phase space that has $H(\mathbf{q}, \mathbf{p})$ equal to the common value of energy $\mathcal{E}$. Of course, we cannot choose systems whose energy is precisely equal to $\mathcal{E}$. For most $\mathcal{E}$ this would be a set of measure zero (see Ex. 4.7). Instead we let the systems occupy a tiny range of energy between $\mathcal{E}$ and $\mathcal{E} + \delta\mathcal{E}$ and then discover (in Ex. 4.7) that our results are highly insensitive to $\delta\mathcal{E}$ as long as it is extremely small compared with $\mathcal{E}$.

It seems reasonable to expect that this ensemble, after evolving for a time much longer than its longest internal dynamical time scale $t \gg \tau_{\text{int}}$, will achieve statistical equilibrium (i.e., will evolve into a state with $\partial\rho/\partial t = 0$). (In the next section we justify this expectation.) The distribution function ρ then satisfies the collisionless Boltzmann equation (4.15) with vanishing time derivative, and therefore is a function only of the hamiltonian's additive constants of the motion $\mathcal{E}$ and K_A. However, we already know that ρ is a delta function in K_A and almost a delta function with a tiny but finite spread in $\mathcal{E}$; and the fact that it cannot depend on any other phase-space quantities then implies *ρ is a constant over the hypersurface in phase space that has the prescribed values of K_A and $\mathcal{E}$, and it is zero everywhere else in phase space.* This equilibrium ensemble is called "microcanonical."

<small>microcanonical distribution</small>

There is a subtle aspect of this microcanonical ensemble that deserves discussion. Suppose we split each system in the ensemble up into a huge number of subsystems that can exchange energy (but for concreteness, nothing else) with one another. We thereby obtain a huge number of subensembles, in the manner of Sec. 4.4.1. The original systems can be regarded as a thermal bath for the subsystems, and correspondingly, the subensembles will have canonical distribution functions, $\rho_a = Ce^{-\mathcal{E}_a/(k_B T)}$. One might also expect the subensembles to be statistically independent, so that $\rho = \prod_a \rho_a$. However, such independence is not possible, since together with additivity of energy $\mathcal{E} = \sum_a \mathcal{E}_a$, it would imply that $\rho = Ce^{-\mathcal{E}/(k_B T)}$ (i.e., that the full ensemble is canonically distributed rather than microcanonical). What is wrong here?

The answer is that there is a tiny correlation between the subensembles: if, at some moment of time, subsystem $a = 1$ happens to have an unusually large energy, then the other subsystems must correspondingly have a little less energy than usual. This very slightly invalidates the statistical-independence relation $\rho = \prod_a \rho_a$, thereby

<small>correlations in the microcanonical ensemble</small>

---

9. Exercise 4.7 below is an example of a microcanonical ensemble where $\mathbf{P}$ and $\mathbf{J}$ are not precisely fixed, though we do not discuss this in the exercise. The gas atoms in that example are contained inside an impermeable box whose walls cannot exchange energy or atoms with the gas, but obviously can and do exchange momentum and angular momentum when atoms collide with the walls. Because the walls are at rest in our chosen reference frame, the distribution function has $\mathbf{U} = \mathbf{\Omega} = 0$ and so is independent of $\mathbf{P}$ and $\mathbf{J}$ [Eq. (4.24)] rather than having precisely defined values of them.

enabling the full ensemble to be microcanonical, even though all its subensembles are canonical. In the language of two-lengthscale expansions, where one expands in the dimensionless ratio (size of subsystems)/(size of full system) [Box 3.3], this correlation is a higher-order correction to statistical independence.

We are now in a position to understand more deeply the nature of the thermalized bath that we have invoked to drive ensembles into statistical equilibrium. That bath can be any huge system that contains the systems we are studying as subsystems; the bath's thermal equilibrium can be either a microcanonical statistical equilibrium or a statistical equilibrium involving exponentials of its extensive variables.

Exercise 4.7 gives a concrete illustration of the microcanonical ensemble, but we delay presenting it until we have developed some additional concepts that it also illustrates.

4.6 The Ergodic Hypothesis

The ensembles we have been studying are almost always just conceptual ones that do not exist in the real universe. We have introduced them and paid so much attention to them not for their own sakes, but because, in the case of statistical-equilibrium ensembles, they can be powerful tools for studying the properties of a single, individual system that really does exist in the universe or in our laboratory.

This power comes about because a sequence of snapshots of the single system, taken at times separated by sufficiently large intervals Δt, has a probability distribution ρ (for the snapshots' instantaneous locations $\{\mathbf{q}, \mathbf{p}\}$ in phase space) that is the same as the distribution function ρ of some conceptual, statistical-equilibrium ensemble. If the single system is closed, so its evolution is driven solely by its own hamiltonian, then the time between snapshots should be $\Delta t \gg \tau_{\text{int}}$, and its snapshots will be (very nearly) microcanonically distributed. If the single system exchanges energy, and only energy, with a thermal bath on a timescale τ_{ext}, then the time between snapshots should be $\Delta t \gg \tau_{\text{ext}}$, and its snapshots will be canonically distributed; similarly for the other types of bath interactions. This property of snapshots is equivalent to the statement that *for the individual system, the long-term time average*[10] *of any function of the system's location in phase space is equal to the statistical-equilibrium ensemble average*:

ergodicity: equality of time average and ensemble average

$$\bar{A} \equiv \lim_{T \to \infty} \frac{1}{T} \int_{-T/2}^{+T/2} A(\mathbf{q}(t), \mathbf{p}(t)) = \langle A \rangle \equiv \sum_n A_n \rho_n. \qquad (4.33)$$

ergodicity

This property comes about because of *ergodicity*: the individual system, as it evolves, visits each accessible quantum state n for a fraction of the time that is equal to the equilibrium ensemble's probability ρ_n. Or, stated more carefully, the system comes

---

10. Physicists often study a system's evolution for too short a time to perform this average, and therefore the system does not reveal itself to be ergodic.

sufficiently close to each state n for a sufficient length of time that, for practical purposes, we can approximate it as spending a fraction ρ_n of its time at n.

At first sight, ergodicity may seem trivially obvious. However, it is not a universal property of all systems. One can easily devise idealized examples of nonergodic behavior (e.g., a perfectly elastic billiard ball bouncing around a square billiard table), and some few-body systems that occur in Nature are nonergodic (e.g., fortunately, planetary systems, such as the Sun's). Moreover, one can devise realistic, nonergodic models of some many-body systems. For examples and a clear discussion see Sethna (2006, Chap. 4). On the other hand, generic closed and semiclosed systems in Nature, whose properties and parameters are not carefully fine tuned, do typically behave ergodically, though to prove so is one of the most difficult problems in statistical mechanics.[11]

We assume throughout this book's discussion of statistical physics that all the systems we study are indeed ergodic; this is called the *ergodic hypothesis*. Correspondingly, sometimes (for ease of notation) we denote the ensemble average with a bar.

ergodic hypothesis

One must be cautious in practical applications of the ergodic hypothesis: it can sometimes require much longer than one might naively expect for a system to wander sufficiently close to accessible states that $\bar{A} = \langle A \rangle$ for observables A of interest.

4.7 Entropy and Evolution toward Statistical Equilibrium

4.7.1 Entropy and the Second Law of Thermodynamics

For any ensemble of systems, whether it is in statistical equilibrium or not, and also whether it is quantum mechanical or not, the ensemble's *entropy* S is defined, in words, by the following awful sentence: S is the mean value (ensemble average) of the natural logarithm of the probability that a random system in the ensemble occupies a given quantum state, summed over states and multiplied by $-k_B$. More specifically, denoting the probability that a system is in state n by ρ_n, the ensemble's entropy S is the following sum over quantum states (or the equivalent integral over phase space):

$$\boxed{S \equiv -k_B \sum_n \rho_n \ln \rho_n.} \qquad (4.34)$$

entropy of an ensemble

If all the systems are in the same quantum state, for example, in the state $n = 17$, then $\rho_n = \delta_{n,17}$ so we know precisely the state of any system pulled at random from the ensemble, and Eq. (4.34) dictates that the entropy vanish. Vanishing entropy thus corresponds to perfect knowledge of the system's quantum state; it corresponds to the quantum state being pure.

Entropy is a measure of our lack of information about the state of any system chosen at random from an ensemble (see Sec. 4.11). In this sense, *the entropy can be*

---

11. The ergodic hypothesis was introduced in 1871 by Boltzmann. For detailed analyses of it from physics viewpoints, see, e.g., ter Haar (1955) and Farquhar (1964). Attempts to understand it rigorously, and the types of systems that do or do not satisfy it, have spawned a branch of mathematical physics called "ergodic theory."

regarded as a property of a random individual system in the ensemble, as well as of the ensemble itself.

By contrast, consider a system in microcanonical statistical equilibrium. In this case, all states are equally likely (ρ is constant), so if there are N_states states available to the system, then $\rho_n = 1/N_\text{states}$, and the entropy (4.34) takes the form[12]

entropy for microcanonical ensemble

$$\boxed{S = k_B \ln N_\text{states}.} \quad (4.35)$$

The entropy, so defined, has some important properties. One is that when the ensemble can be broken up into statistically independent subensembles of subsystems (as is generally the case for big systems in statistical equilibrium), so that $\rho = \prod_a \rho_a$, then the entropy is additive: $S = \sum_a S_a$ (see Ex. 4.3). This permits us to regard the entropy, like the systems' additive constants of motion, as an extensive variable.

second law of thermodynamics

A second very important property is that, as an ensemble of systems evolves, its entropy cannot decrease, and it generally tends to increase. This is the statistical mechanical version of the second law of thermodynamics.

As an example of this second law, consider two different gases (e.g., nitrogen and oxygen) in a container, separated by a thin membrane. One set of gas molecules is constrained to lie on one side of the membrane; the other set lies on the opposite side. The total number of available states N_states is less than if the membrane is ruptured and the two gases are allowed to mix. The mixed state is readily accessible from the partitioned state and not vice versa. When the membrane is removed, the entropy begins to increase in accord with the second law of thermodynamics (cf. Ex. 4.8).

Since any ensemble of identical, closed systems will ultimately, after a time $t \gg \tau_\text{int}$, evolve into microcanonical statistical equilibrium, it must be that the microcanonical distribution function $\rho = $ constant has a larger entropy than any other distribution function that the ensemble could acquire. That this is indeed so can be demonstrated formally as follows.

Consider the class of all distribution functions ρ that: (i) vanish unless the constants of motion have the prescribed values $\mathcal{E}$ (in the tiny range $\delta \mathcal{E}$) and K_A; (ii) can be nonzero anywhere in the region of phase space, which we call $\mathcal{Y}_o$, where the prescribed values $\mathcal{E}, K_A$ are taken; and (iii) are correctly normalized so that

$$\sum_n \rho_n \equiv \int_{\mathcal{Y}_o} \rho \mathcal{N}_\text{states} d\Gamma = 1 \quad (4.36a)$$

[Eq. (4.8b)]. We ask which ρ in this class gives the largest entropy

$$S = -k_B \sum_n \rho_n \ln \rho_n.$$

The requirement that the entropy be extremal (stationary) under variations $\delta \rho$ of ρ that preserve the normalization (4.36a) is embodied in the variational principle (see, e.g., Boas, 2006, Chap. 9):

---

12. This formula, with slightly different notation, can be found on Boltzmann's tomb.

$$\delta S = \delta \int_{\mathcal{Y}_o} (-k_B \rho \ln \rho - \Lambda \rho) \mathcal{N}_{\text{states}} d\Gamma = 0. \tag{4.36b}$$

Here Λ is a Lagrange multiplier that enforces the normalization (4.36a). Performing the variation, we find that

$$\int_{\mathcal{Y}_o} (-k_B \ln \rho - k_B - \Lambda) \delta \rho \mathcal{N}_{\text{states}} d\Gamma = 0, \tag{4.36c}$$

which is satisfied if and only if ρ is a constant, $\rho = e^{-1-\Lambda/k_B}$, independent of location in the allowed region $\mathcal{Y}_o$ of phase space (i.e., if and only if ρ is that of the microcanonical ensemble). This calculation actually only shows that the microcanonical ensemble has stationary entropy. To show it is a maximum, one must perform the second variation (i.e., compute the second-order contribution of $\delta \rho$ to $\delta S = \delta \int (-k_B \rho \ln \rho) \mathcal{N}_{\text{states}} d\Gamma$). That second-order contribution is easily seen to be

$$\delta^2 S = \int_{\mathcal{Y}_o} \left(-k_B \frac{(\delta \rho)^2}{2\rho}\right) \mathcal{N}_{\text{states}} d\Gamma < 0. \tag{4.36d}$$

entropy maximized when ρ is constant (microcanonical distribution)

Thus, the microcanonical distribution does maximize the entropy, as claimed.

4.7.2 What Causes the Entropy to Increase?

There is an apparent paradox at the heart of statistical mechanics, and, at various stages in the development of the subject it has led to confusion and even despair. It still creates controversy (see, e.g., Hawking and Penrose, 2010; Penrose, 1999). Its simplest and most direct expression is to ask: how can the time-reversible, microscopic laws, encoded in a time-independent hamiltonian, lead to the remorseless increase of entropy?

COARSE GRAINING

For insight, first consider a classical, microcanonical ensemble of precisely closed systems (no interaction at all with the external universe). Assume, for simplicity, that at time $t = 0$ all the systems are concentrated in a small but finite region of phase space with volume $\Delta \Gamma$, as shown in Fig. 4.2a, with $\rho = 1/(\mathcal{N}_{\text{states}} \Delta \Gamma)$ in the occupied region and $\rho = 0$ everywhere else. As time passes each system evolves under the action of the systems' common hamiltonian. As depicted in Fig. 4.2b, this evolution distorts the occupied region of phase space; but Liouville's theorem dictates that the occupied region's volume $\Delta \Gamma$ remain unchanged and, correspondingly, that the ensemble's entropy

$$S = -k_B \int (\rho \ln \rho) \mathcal{N}_{\text{states}} d\Gamma = k_B \ln(\mathcal{N}_{\text{states}} \Delta \Gamma) \tag{4.37}$$

remain unchanged.

How can this be so? The ensemble is supposed to evolve into statistical equilibrium, with its distribution function uniformly spread out over that entire portion of

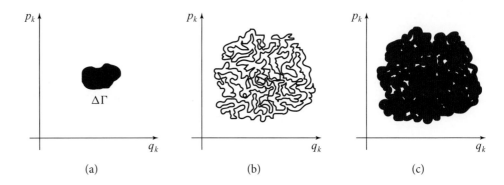

FIGURE 4.2 Evolution of a classical ensemble at $t = 0$ (a) toward statistical equilibrium by means of phase mixing (b) (cf. Fig. 4.1) followed by coarse-graining of one's viewpoint (c).

phase space allowed by the hamiltonian's constants of motion—a portion of phase space far, far larger than $\Delta\Gamma$—and in the process the entropy is supposed to increase.

Figure 4.2b,c resolves the paradox. As time passes, the occupied region becomes more and more distorted. It retains its phase-space volume, but gets strung out into a winding, contorted surface (Fig. 4.2b), which (by virtue of the ergodic hypothesis) ultimately passes arbitrarily close to any given point in the region allowed by the constants of motion. This ergodic wandering is called *phase mixing*. Ultimately, the physicist gets tired of keeping track (or ceases to be able to keep track) of all these contortions of the occupied region and chooses instead to take a *coarse-grained* viewpoint that averages over scales larger than the distance between adjacent portions of the occupied surface, and thereby regards the ensemble as having become spread over the entire allowed region (Fig. 4.2c). More typically, the physicist will perform a coarse-grained smearing out on some given, constant scale at all times. Once the transverse scale of the ensemble's lengthening and narrowing phase-space region drops below the smearing scale, its smeared volume and its entropy start to increase. Thus, *for an ensemble of closed systems it is the physicist's choice (though often a practical necessity) to perform coarse-grain averaging that causes entropy to increase and causes the ensemble to evolve into statistical equilibrium.*

The situation is a bit more subtle for an ensemble of systems interacting with a thermal bath. The evolution toward statistical equilibrium is driven by the interactions. Thus, it might appear at first sight that the physicist is not, this time, to blame for the entropy increase and the achievement of statistical equilibrium. However, a deeper examination reveals the physicist's ultimate culpability. If the physicist were willing to keep track of all those dynamical degrees of freedom of the bath that are influenced by and influence the systems in the ensemble, then the physicist could incorporate those degrees of freedom into the description of the systems and define a phase-space volume that obeys Liouville's theorem and thus does not increase, and an entropy that correspondingly remains constant. However, physicists instead generally choose to ignore the microscopic details of the bath, and that choice forces them to attribute a growing entropy to the ensemble of systems bathed by the bath and regard the ensemble as approaching statistical equilibrium.

DISCARDING CORRELATIONS

When one reexamines these issues in quantum mechanical language, one discovers that the entropy increase is caused by the physicist's discarding the quantum mechanical correlations (the off-diagonal terms in the density matrix of Box 4.2) that get built up through the systems' interaction with the rest of the universe. This discarding of correlations is accomplished through a trace over the external universe's basis states (Box 4.2), and if the state of system plus universe was originally pure, this tracing (discarding of correlations) makes it mixed. From this viewpoint, then, *it is the physicist's choice to discard correlations with the external universe that causes the entropy increase and the evolution toward statistical equilibrium.* Heuristically, we can say that the entropy does not increase until the physicist actually (or figuratively) chooses to let it increase by ignoring the rest of the universe. For a simple example, see Box 4.3 and Ex. 4.9.

discarding quantum correlations (decoherence) causes entropy increase

This viewpoint then raises a most intriguing question. What if we regard the universe as the ultimate microcanonical system? In this case, we might expect that the entropy of the universe will remain identically zero for all time, unless physicists (or other intelligent beings) perform some sort of coarse-graining or discard some sort of correlations. However, such coarse-graining or discarding are made deeply subtle by the fact that the physicists (or other intelligent beings) are themselves part of the system being studied. Further discussion of these questions introduces fascinating, though ill-understood, quantum mechanical and cosmological considerations, which will reappear in a different context in Sec. 28.7.

EXERCISES

Exercise 4.2 *Practice: Estimating Entropy*
Make rough estimates of the entropy of the following systems, assuming they are in statistical equilibrium.

(a) An electron in a hydrogen atom at room temperature.

(b) A glass of wine.

(c) The Pacific ocean.

(d) An ice cube.

(e) The observable universe. [Its entropy is mostly contained in the 3-K microwave background radiation and in black holes (Sec. 4.10). Why?]

Exercise 4.3 *Derivation and Practice: Additivity of Entropy for Statistically Independent Systems*
Consider an ensemble of classical systems with each system made up of a large number of statistically independent subsystems, so $\rho = \prod_a \rho_a$. Show that the entropy of the full ensemble is equal to the sum of the entropies of the subensembles a: $S = \sum_a S_a$.

BOX 4.3. ENTROPY INCREASE DUE TO DISCARDING QUANTUM CORRELATIONS

As an idealized, pedagogical example of entropy increase due to physicists' discarding quantum correlations, consider an electron that interacts with a photon. The electron's initial quantum state is $|\psi_e\rangle = \alpha|\uparrow\rangle + \beta|\downarrow\rangle$, where $|\uparrow\rangle$ is the state with spin up, $|\downarrow\rangle$ is that with spin down, and α and β are complex probability amplitudes with $|\alpha|^2 + |\beta|^2 = 1$. The interaction is so arranged that if the electron spin is up, then the photon is put into a positive helicity state $|+\rangle$, and if down, the photon is put into a negative helicity state $|-\rangle$. Therefore, after the interaction the combined system of electron plus photon is in the state $|\Psi\rangle = \alpha|\uparrow\rangle \otimes |+\rangle + \beta|\downarrow\rangle \otimes |-\rangle$, where $\otimes$ is the tensor product [Eq. (1.5a) generalized to the vector space of quantum states].

The photon flies off into the universe leaving the electron isolated. Suppose that we measure some electron observable $\hat{A}_e$. The expectation values for the measurement before and after the interaction with the photon are

$$\text{Before:} \quad \langle\psi_e|\hat{A}_e|\psi_e\rangle = |\alpha|^2 \langle\uparrow|\hat{A}_e|\uparrow\rangle + |\beta|^2 \langle\downarrow|\hat{A}_e|\downarrow\rangle$$
$$+ \alpha^*\beta \langle\uparrow|\hat{A}_e|\downarrow\rangle + \beta^*\alpha \langle\downarrow|\hat{A}_e|\uparrow\rangle; \quad (1)$$

$$\text{After:} \quad \langle\Psi|\hat{A}_e|\Psi\rangle = |\alpha|^2 \langle\uparrow|\hat{A}_e|\uparrow\rangle \underbrace{\langle+|+\rangle}_{1} + |\beta|^2 \langle\downarrow|\hat{A}_e|\downarrow\rangle \underbrace{\langle-|-\rangle}_{1}$$
$$+ \alpha^*\beta \langle\uparrow|\hat{A}_e|\downarrow\rangle \underbrace{\langle+|-\rangle}_{0} + \beta^*\alpha \langle\downarrow|\hat{A}_e|\uparrow\rangle \underbrace{\langle-|+\rangle}_{0}$$
$$= |\alpha|^2 \langle\uparrow|\hat{A}_e|\uparrow\rangle + |\beta|^2 \langle\downarrow|\hat{A}_e|\downarrow\rangle. \quad (2)$$

Comparing Eqs. (1) and (2), we see that *the correlations with the photon have removed the $\alpha^*\beta$ and $\beta^*\alpha$ quantum interference terms from the expectation value*. The two pieces $\alpha|\uparrow\rangle$ and $\beta|\downarrow\rangle$ of the electron's original quantum state $|\psi_e\rangle$ are said to have *decohered*. Since the outcomes of all measurements can be expressed in terms of expectation values, this quantum decoherence is complete in the sense that no quantum interference between the $\alpha|\uparrow\rangle$ and $\beta|\downarrow\rangle$ pieces of the electron state $|\psi_e\rangle$ will ever be seen again in any measurement on the electron, unless the photon returns, interacts with the electron, and thereby removes its correlations with the electron state.

If physicists are confident the photon will never return and the correlations will never be removed, then they are free to change their mathematical description of the electron state. Instead of describing the postinteraction state as $|\Psi\rangle = \alpha|\uparrow\rangle \otimes |+\rangle + \beta|\downarrow\rangle \otimes |-\rangle$, the physicists can discard the

(continued)

BOX 4.3. (continued)

correlations with the photon and regard the electron as having *classical* probabilities $\rho_\uparrow = |\alpha|^2$ for spin up and $\rho_\downarrow = |\beta|^2$ for spin down (i.e., as being in a *mixed* state). This new, mixed-state viewpoint leads to the same expectation value (2) for all physical measurements as the old, correlated, pure-state viewpoint $|\Psi\rangle$.

The important point for us is that, when discarding the quantum correlations with the photon (with the external universe), the physicist changes the entropy from zero (the value for any pure state including $|\Psi\rangle$) to $S = -k_B(p_\uparrow \ln p_\uparrow + p_\downarrow \ln p_\downarrow) = -k_B(|\alpha|^2 \ln |\alpha|^2 + |\beta|^2 \ln |\beta|^2) > 0$. The physicist's change of viewpoint has increased the entropy.

In Ex. 4.9, this pedagogical example is reexpressed in terms of the density operator discussed in Box 4.2.

Exercise 4.4 ****Example: Entropy of a Thermalized Mode of a Field*
Consider a mode $\mathcal{S}$ of a fermionic or bosonic field, as discussed in Sec. 4.4.3. Suppose that an ensemble of identical such modes is in statistical equilibrium with a heat and particle bath and thus is grand canonically distributed.

(a) Show that if $\mathcal{S}$ is fermionic, then the ensemble's entropy is

$$S_\mathcal{S} = -k_B[\eta \ln \eta + (1-\eta)\ln(1-\eta)]$$
$$\simeq -k_B \eta(\ln \eta - 1) \quad \text{in the classical regime } \eta \ll 1, \qquad (4.38a)$$

where η is the mode's fermionic mean occupation number (4.27b).

(b) Show that if the mode is bosonic, then the entropy is

$$S_\mathcal{S} = k_B[(\eta+1)\ln(\eta+1) - \eta \ln \eta]$$
$$\simeq -k_B \eta(\ln \eta - 1) \quad \text{in the classical regime } \eta \ll 1, \qquad (4.38b)$$

where η is the bosonic mean occupation number (4.28b). Note that in the classical regime, $\eta \simeq e^{-(\mathcal{E}-\tilde{\mu})/(k_B T)} \ll 1$, the entropy is insensitive to whether the mode is bosonic or fermionic.

(c) Explain why the entropy per particle in units of Boltzmann's constant is $\sigma = S_\mathcal{S}/(\eta k_B)$. Plot σ as a function of η for fermions and for bosons. Show analytically that for degenerate fermions ($\eta \simeq 1$) and for the bosons' classical-wave regime ($\eta \gg 1$) the entropy per particle is small compared to unity. See Sec. 4.8 for the importance of the entropy per particle.

Exercise 4.5 *Example: Entropy of Thermalized Radiation Deduced from Entropy per Mode*

Consider fully thermalized electromagnetic radiation at temperature T, for which the mean occupation number has the standard Planck (blackbody) form $\eta = 1/(e^x - 1)$ with $x = h\nu/(k_B T)$.

(a) Show that the entropy per mode of this radiation is
$$S_S = k_B[x/(e^x - 1) - \ln(1 - e^{-x})].$$

(b) Show that the radiation's entropy per unit volume can be written as the following integral over the magnitude of the photon momentum:
$$S/V = (8\pi/h^3) \int_0^\infty S_S \, p^2 dp.$$

(c) By performing the integral (e.g., using Mathematica), show that
$$\frac{S}{V} = \frac{4}{3}\frac{U}{T} = \frac{4}{3}aT^3, \tag{4.39}$$
where $U = aT^4$ is the radiation energy density, and $a = (8\pi^5 k_B^4/15)/(ch)^3$ is the radiation constant [Eqs. (3.54)].

(d) Verify Eq. (4.39) for the entropy density by using the first law of thermodynamics $dE = TdS - PdV$ (which you are presumed to know before reading this book, and which we discuss below and study in the next chapter).

Exercise 4.6 *Problem: Entropy of a Classical, Nonrelativistic, Perfect Gas, Deduced from Entropy per Mode*

Consider a classical, nonrelativistic gas whose particles do not interact and have no excited internal degrees of freedom (a *perfect gas*—not to be confused with perfect fluid). Let the gas be contained in a volume V and be thermalized at temperature T and chemical potential μ. Using the gas's entropy per mode, Ex. 4.4, show that the total entropy in the volume V is
$$S = \left(\frac{5}{2} - \frac{\mu}{k_B T}\right) k_B N, \tag{4.40}$$
where $N = g_s (2\pi m k_B T/h^2)^{3/2} e^{\mu/(k_B T)} V$ is the number of particles in the volume V [Eq. (3.39a), derived in Chap. 3 using kinetic theory], and g_s is each particle's number of spin states.

Exercise 4.7 *Example: Entropy of a Classical, Nonrelativistic, Perfect Gas in a Microcanonical Ensemble*

Consider a microcanonical ensemble of closed cubical cells with volume V. Let each cell contain precisely N particles of a classical, nonrelativistic, perfect gas and contain a nonrelativistic total energy $E \equiv \mathcal{E} - Nmc^2$. For the moment (by contrast with the text's discussion of the microcanonical ensemble), assume that E is precisely fixed instead of being spread over some tiny but finite range.

(a) Explain why the region $\mathcal{Y}_o$ of phase space accessible to each system is

$$|x_A| < L/2, \quad |y_A| < L/2, \quad |z_A| < L/2, \quad \sum_{A=1}^{N} \frac{1}{2m}|\mathbf{p}_A|^2 = E, \quad (4.41\text{a})$$

where A labels the particles, and $L \equiv V^{1/3}$ is the side of the cell.

(b) To compute the entropy of the microcanonical ensemble, we compute the volume $\Delta\Gamma$ in phase space that it occupies, multiply by the number density of states in phase space (which is independent of location in phase space), and then take the logarithm. Explain why

$$\Delta\Gamma \equiv \prod_{A=1}^{N} \int_{\mathcal{Y}_o} dx_A dy_A dz_A dp_A^x dp_A^y dp_A^z \quad (4.41\text{b})$$

vanishes. This illustrates the "set of measure zero" statement in the text (second paragraph of Sec. 4.5), which we used to assert that we must allow the systems' energies to be spread over some tiny but finite range.

(c) Now permit the energies of our ensemble's cells to lie in the tiny but finite range $E_o - \delta E_o < E < E_o$. Show that

$$\Delta\Gamma = V^N [\mathcal{V}_\nu(a) - \mathcal{V}_\nu(a - \delta a)], \quad (4.41\text{c})$$

where $\mathcal{V}_\nu(a)$ is the volume of a sphere of radius a in a Euclidean space with $\nu \gg 1$ dimensions, and where

$$a \equiv \sqrt{2mE_o}, \quad \frac{\delta a}{a} \equiv \frac{1}{2}\frac{\delta E_o}{E_o}, \quad \nu \equiv 3N. \quad (4.41\text{d})$$

It can be shown (and you might want to try to show it) that

$$\mathcal{V}_\nu(a) = \frac{\pi^{\nu/2}}{(\nu/2)!} a^\nu \quad \text{for } \nu \gg 1. \quad (4.41\text{e})$$

(d) Show that, so long as $1 \gg \delta E_o/E_o \gg 1/N$ (where N in practice is an exceedingly huge number),

$$\mathcal{V}_\nu(a) - \mathcal{V}_\nu(a - \delta a) \simeq \mathcal{V}_\nu(a)[1 - e^{-\nu\delta a/a}] \simeq \mathcal{V}_\nu(a), \quad (4.41\text{f})$$

which is independent of δE_o and thus will produce a value for $\Delta\Gamma$ and thence N_{states} and S independent of δE_o, as desired. From this and with the aid of Stirling's approximation (Reif, 2008, Appendix A.6) $n! \simeq (2\pi n)^{1/2}(n/e)^n$ for large n, and taking account of the multiplicity $\mathcal{M} = N!$, show that the entropy of the microcanonically distributed cells is given by

$$S(V, E, N) = Nk_B \ln\left[\frac{V}{N}\left(\frac{E}{N}\right)^{3/2} g_s \left(\frac{4\pi m}{3h^2}\right)^{3/2} e^{5/2}\right]. \quad (4.42)$$

This is known as the *Sackur-Tetrode* equation.

(e) Using Eqs. (3.39), show that this is equivalent to Eq. (4.40) for the entropy, which we derived by a very different method.

Exercise 4.8 **Example: Entropy of Mixing, Indistinguishability of Atoms, and the Gibbs Paradox*

(a) Consider two identical chambers, each with volume V, separated by an impermeable membrane. Into one chamber put energy E and N atoms of helium, and into the other, energy E and N atoms of xenon, with E/N and N/V small enough that the gases are nonrelativistic and nondegenerate. The membrane is ruptured, and the gases mix. Show that this mixing drives the entropy up by an amount $\Delta S = 2 N k_B \ln 2$. [Hint: Use the Sackur-Tetrode equation (4.42).]

(b) Suppose that energy E and N atoms of helium are put into both chambers (no xenon). Show that, when the membrane is ruptured and the gases mix, there is no increase of entropy. Explain why this result is reasonable, and explain its relationship to entropy being an extensive variable.

(c) Suppose that the N helium atoms were distinguishable instead of indistinguishable. Show this would mean that, in the microcanonical ensemble, they have $N!$ more states available to themselves, and their entropy would be larger by $k_B \ln N! \simeq k_B (N \ln N - N)$; and as a result, the Sackur-Tetrode formula (4.42) would be

Distinguishable particles:
$$S(V, E, N) = N k_B \ln \left[V \left(\frac{E}{N} \right)^{3/2} \left(\frac{4\pi m}{3 h^2} \right)^{3/2} e^{3/2} \right]. \quad (4.43)$$

Before the advent of quantum theory, physicists thought that atoms were distinguishable, and up to an additive multiple of N (which they could not compute), they deduced this entropy.

(d) Show that if, as prequantum physicists believed, atoms were distinguishable, then when the membrane between two identical helium-filled chambers is ruptured, there would be an entropy increase identical to that when the membrane between helium and xenon is ruptured: $\Delta S = 2 N k_B \ln 2$ [cf. parts (a) and (b)]. This result, which made prequantum physicists rather uncomfortable, is called the *Gibbs paradox*.

Exercise 4.9 *Problem: Quantum Decoherence and Entropy Increase in Terms of the Density Operator* T2

Reexpress Box 4.3's pedagogical example of quantum decoherence and entropy increase in the language of the quantum mechanical density operator $\hat{\rho}$ (Box 4.2). Use this example to explain the meaning of the various statements made in the next-to-last paragraph of Sec. 4.7.2.

4.8 Entropy per Particle

The entropy per particle in units of Boltzmann's constant,

$$\sigma \equiv S/(Nk_B), \qquad (4.44)$$

entropy per particle

is a very useful concept in both quantitative and order-of-magnitude analyses. (See, e.g., Exs. 4.10, 4.11, and 4.4c, and the discussion of entropy in the expanding universe in Sec. 4.10.3.) One reason is the second law of thermodynamics. Another is that in the real universe σ generally lies somewhere between 0 and 100 and thus is a natural quantity in terms of which to think about and remember the magnitudes of various quantities.

For example, for ionized hydrogen gas in the nonrelativistic, classical domain, the Sackur-Tetrode equation (4.42) for the entropy (derived from the microcanonical ensemble in Ex. 4.7), when specialized either to the gas's protons or to its electrons (both of which, with their spin 1/2, have $g_s = 2$), gives

$$\sigma_p = \frac{5}{2} + \ln\left[\frac{2m_p}{\rho}\left(\frac{2\pi m k_B T}{h^2}\right)^{3/2}\right]. \qquad (4.45)$$

entropy of ionized hydrogen

Here we have set the number density of particles N/V (either protons or electrons) to ρ/m_p, since almost all the mass density ρ is in the protons, and we have set the thermal energy per particle E/N to $\frac{3}{2}k_B T$ [see Eq. (3.39b) derived in Chap. 3 using kinetic theory]. The only way this formula depends on which particle species we are considering, the gas's protons or its electrons, is through the particle mass m; and this factor in Eq. (4.45) tells us that the protons' entropy per proton is a factor $\simeq 10$ higher than the electrons': $\sigma_p - \sigma_e = \frac{3}{2}\ln(m_p/m_e) = 11.27 \simeq 10$. Therefore, for an ionized hydrogen gas most of the entropy is in the protons.

The protons' entropy per proton is plotted as a function of density and temperature in Fig. 4.3. It grows logarithmically as the density ρ decreases, and it ranges from $\sigma \ll 1$ in the regime of extreme proton degeneracy (lower right of Fig. 4.3; see Ex. 4.4) to $\sigma \sim 1$ near the onset of proton degeneracy (the boundary of the classical approximation), to $\sigma \sim 100$ at the lowest density that occurs in the universe, $\rho \sim 10^{-29}$ g cm^{-3}. This range is an example of the fact that the logarithms of almost all dimensionless numbers that occur in Nature lie between approximately -100 and $+100$.

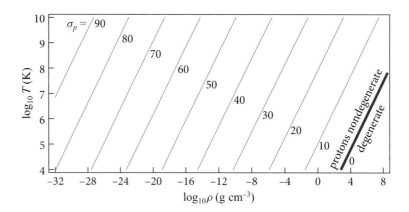

FIGURE 4.3 Proton entropy per proton σ_p for an ionized hydrogen gas. Each line is labeled with its value of σ_p. The electron entropy per electron σ_e is a factor $\simeq 10$ smaller; the electrons are degenerate when $\sigma_e \simeq \sigma_p - 10 \lesssim 1$. The protons are degenerate when $\sigma_p \lesssim 1$.

EXERCISES

Exercise 4.10 ***Problem: Primordial Element Formation*

(a) As we shall see in Sec. 28.4.1, when the early universe was ∼200 s old, its principal constituents were photons, protons, neutrons, electrons, positrons, and (thermodynamically isolated) neutrinos and gravitons. The photon temperature was ∼9×10^8 K, and the baryon density was ∼0.02 kg m^{-3}. The photons, protons, electrons, and positrons were undergoing rapid electromagnetic interactions that kept them in thermodynamic equilibrium. Use the neutron-proton mass difference of 1.3 MeV to argue that, if the neutrons had also been in thermodynamic equilibrium, then their density would have been negligible at this time.

(b) However, the universe expanded too rapidly for weak interactions to keep the neutrons in equilibrium with the other particles, and so the fraction of the baryons at 200 s, in the form of neutrons that did not subsequently decay, was ∼0.14. At this time (Sec. 28.4.2) the neutrons began rapidly combining with protons to form alpha particles through a short chain of reactions, of which the first step—the one that was hardest to make go—was $n + p \rightarrow d + 2.22$ MeV, with the 2.22 MeV going into heat. Only about 10^{-5} of the baryons needed to go into deuterium in order to start and then maintain the full reaction chain. Show, using entropy per particle and the first law of thermodynamics, that at time $t \sim 200$ s after the big bang, this reaction was just barely entropically favorable.

(c) During this epoch, roughly how do you expect the baryon density to have decreased as a function of decreasing photon temperature? (For this you can neglect the heat being released by the nuclear burning and the role of positrons.) Show that before $t \sim 200$ s, the deuterium-formation reaction could not take place, and after $t \sim 200$ s, it rapidly became strongly entropically favorable.

(d) The nuclear reactions shut down when the neutrons had all cycled through deuterium and almost all had been incoporated into α particles, leaving only $\sim 10^{-5}$ of them in deuterium. About what fraction of all the baryons wound up in α particles (Helium 4)? (See Fig. 28.7 for details from a more complete analysis sketched in Sec. 28.4.2.)

Exercise 4.11 *Problem: Reionization of the Universe*
Following the epoch of primordial element formation (Ex. 4.10), the universe continued to expand and cool. Eventually when the temperature of the photons was $\sim$3,000 K, the free electrons and protons combined to form atomic hydrogen; this was the *epoch of recombinaton*. Later, when the photon temperature had fallen to $\sim$30 K, some hot stars and quasars formed and their ultraviolet radiation dissociated the hydrogen; this was the *epoch of reionization*. The details are poorly understood. (For more discussion, see Ex. 28.19.) Making the simplifying assumption that reionization happened rapidly and homogeneously, show that the increase in the entropy per baryon was $\sim 60 k_B$, depending weakly on the temperature of the atomic hydrogen, which you can assume to be $\sim$100 K.

4.9 Bose-Einstein Condensate

In this section, we explore an important modern application of the Bose-Einstein mean occupation number for bosons in statistical equilibrium. Our objectives are (i) to exhibit, in action, the tools developed in this chapter and (ii) to give a nice example of the connections between quantum statistical mechanics (which we use in the first 3/4 of this section) and classical statistical mechanics (which we use in the last 1/4).

For bosons in statistical equilibrium, the mean occupation number $\eta = 1/[e^{(E-\mu)/(k_B T)} - 1]$ diverges as $E \to 0$, if the chemical potential μ vanishes. This divergence is intimately connected to *Bose-Einstein condensation*.

Bose-Einstein condensation

Consider a dilute atomic gas in the form of a large number N of bosonic atoms, spatially confined by a magnetic trap. When the gas is cooled below some critical temperature T_c, μ is negative but gets very close to zero [see Eq. (4.47d)], causing η to become huge near zero energy. This huge η manifests physically as a large number N_0 of atoms collecting into the trap's mode of lowest (vanishing) energy, the Schrödinger equation's ground state [see Eq. (4.49a) and Fig. 4.4a later in this section].

This condensation was predicted by Einstein (1925), but an experimental demonstration was not technologically feasible until 1995, when two research groups independently exhibited it: one at JILA (University of Colorado) led by Eric Cornell and Carl Wieman; the other at MIT led by Wolfgang Ketterle. For these experiments, Cornell, Ketterle, and Wieman were awarded the 2001 Nobel Prize. Bose-Einstein condensates have great promise as tools for precision-measurement technology and nanotechnology.

As a concrete example of Bose-Einstein condensation, we analyze an idealized version of one of the early experiments by the JILA group (Ensher et al., 1996): a gas of 40,000 ^{87}Rb atoms placed in a magnetic trap that we approximate as a spherically symmetric, harmonic oscillator potential:[13]

$$V(r) = \frac{1}{2}m\omega_o^2 r^2 = \frac{1}{2}m\omega_o^2(x^2 + y^2 + z^2). \tag{4.46a}$$

Here x, y, z are Cartesian coordinates, and r is radius. The harmonic-oscillator frequency ω_o and associated temperature $\hbar\omega_o/k_B$, the number N of rubidium atoms trapped in the potential, and the atoms' rest mass m are

$$\omega_o/(2\pi) = 181 \text{ Hz}, \quad \hbar\omega_o/k_B = 8.7 \text{ nK}, \quad N = 40{,}000, \quad m = 1.444 \times 10^{-25} \text{ kg}. \tag{4.46b}$$

Our analysis is adapted, in part, from a review article by Dalfovo et al. (1999).

Each ^{87}Rb atom is made from an even number of fermions [$Z = 37$ electrons, $Z = 37$ protons, and $(A - Z) = 50$ neutrons], and the many-particle wave function Ψ for the system of $N = 40{,}000$ atoms is antisymmetric (changes sign) under interchange of each pair of electrons, each pair of protons, and each pair of neutrons. Therefore, when any pair of atoms is interchanged (entailing interchange of an even number of fermion pairs), there is an even number of sign flips in Ψ. Thus Ψ is symmetric (no sign change) under interchange of atoms (i.e., the atoms behave like bosons and must obey Bose-Einstein statistics).

Repulsive forces between the atoms have a moderate influence on the experiment, but only a tiny influence on the quantities that we compute (see, e.g., Dalfovo et al., 1999). We ignore those forces and treat the atoms as noninteracting.

To make contact with our derivation, in Sec. 4.4.3, of the Bose-Einstein distribution, we must identify the modes (single-atom quantum states) $\mathcal{S}$ available to the atoms. Those modes are the energy eigenstates of the Schrödinger equation for a ^{87}Rb atom in the harmonic-oscillator potential $V(r)$. Solution of the Schrödinger equation (e.g., Cohen-Tannoudji, Diu, and Laloë, 1977, complement B$_{\text{VII}}$) reveals that the energy eigenstates can be labeled by the number of quanta of energy $\{n_x, n_y, n_z\}$ associated with an atom's motion along the x, y, and z directions; the energy of the mode $\{n_x, n_y, n_z\}$ is $E_{n_x, n_y, n_z} = \hbar\omega_o[(n_x + 1/2) + (n_y + 1/2) + (n_z + 1/2)]$. We simplify subsequent formulas by subtracting $\frac{3}{2}\hbar\omega_o$ from all energies and all chemical potentials. This is merely a change in what energy we regard as zero (renormalization), a change under which our statistical formalism is invariant (cf. foot-

---

13. In the actual experiment, the potential was harmonic but prolate spheroidal rather than spherical, i.e., in Cartesian coordinates $V(x, y, z) = \frac{1}{2}m[\omega_\varpi^2(x^2 + y^2) + \omega_z^2 z^2]$, with ω_z somewhat smaller than ω_ϖ. For pedagogical simplicity we treat the potential as spherical, with ω_o set to the geometric mean of the actual frequencies along the three Cartesian axes, $\omega_o = (\omega_\varpi^2 \omega_z)^{1/3}$. This choice of ω_o gives good agreement between our model's predictions and the prolate-spheroidal predictions for the quantities that we compute.

note 8 on page 175). Correspondingly, we attribute to the mode $\{n_x, n_y, n_z\}$ the energy $E_{n_x,n_y,n_z} = \hbar\omega_o(n_x + n_y + n_z)$. Our calculations will be simplified by lumping together all modes that have the same energy, so we switch from $\{n_x, n_y, n_z\}$ to $q \equiv n_x + n_y + n_z = $ (the mode's total number of quanta) as our fundamental quantum number, and we write the mode's energy as

$$E_q = q\hbar\omega_o. \tag{4.47a}$$

It is straightforward to verify that the number of independent modes with q quanta (the number of independent ways to choose $\{n_x, n_y, n_z\}$ such that their sum is q) is $\frac{1}{2}(q+1)(q+2)$. (Of course, one can also derive this same formula in spherical polar coordinates.)

Of special interest is the ground-state mode of the potential, $\{n_x, n_y, n_z\} = \{0, 0, 0\}$. This mode has $q = 0$, and it is unique: $(q+1)(q+2)/2 = 1$. Its energy is $E_0 = 0$, and its Schrödinger wave function is $\psi_o = (\pi\sigma_o^2)^{-3/4} \exp(-r^2/(2\sigma_o^2))$, so for any atom that happens to be in this ground-state mode, the probability distribution for its location is the Gaussian

ground-state mode

$$|\psi_o(r)|^2 = \left(\frac{1}{\pi\sigma_o^2}\right)^{3/2} \exp\left(-\frac{r^2}{\sigma_o^2}\right), \quad \text{where} \quad \sigma_o = \sqrt{\frac{\hbar}{m\omega_o}} = 0.800 \,\mu\text{m}. \tag{4.47b}$$

The entire collection of N atoms in the magnetic trap is a system; it interacts with its environment, which is at temperature T, exchanging energy but nothing else, so a conceptual ensemble consisting of this N-atom system and a huge number of identical systems has the canonical distribution $\rho = C \exp(-E_{\text{tot}}/(k_B T))$, where E_{tot} is the total energy of all the atoms. Each mode (labeled by $\{n_x, n_y, n_z\}$ or by q and two degeneracy parameters that we have not specified) is a subsystem of this system. Because the modes can exchange atoms as well as energy with one another, a conceptual ensemble consisting of any chosen mode and its clones is grand-canonically distributed [Eq. (4.28a)], so its mean occupation number is given by the Bose-Einstein formula (4.28b) with $E_S = q\hbar\omega_o$:

$$\eta_q = \frac{1}{\exp[(q\hbar\omega_o - \mu)/(k_B T)] - 1}. \tag{4.47c}$$

The temperature T is inherited from the environment (heat bath) in which the atoms live. The chemical potential μ is common to all the modes and takes on whatever value is required to guarantee that the total number of atoms in the trap is N—or, equivalently, that the sum of the mean occupation number (4.47c) over all the modes is N.

For the temperatures of interest to us:

1. The number $N_0 \equiv \eta_0$ of atoms in the ground-state mode $q = 0$ will be large, $N_0 \gg 1$, which permits us to expand the exponential in Eq. (4.47c) with $q = 0$ to obtain $N_0 = 1/[e^{-\mu/(k_B T)} - 1] \simeq -k_B T/\mu$, that is,

$$\mu/(k_B T) = -1/N_0. \tag{4.47d}$$

2. The atoms in excited modes will be spread out smoothly over many values of q, so we can approximate the total number of excited-mode atoms by an integral:

$$N - N_0 = \sum_{q=1}^{\infty} \frac{(q+1)(q+2)}{2} \eta_q \simeq \int_0^{\infty} \frac{\frac{1}{2}(q^2 + 3q + 2)dq}{\exp[(q\hbar\omega_o/(k_BT) + 1/N_0)] - 1}. \tag{4.48a}$$

The integral is dominated by large qs, so it is a rather good approximation to keep only the q^2 term in the numerator and to neglect the $1/N_0$ in the exponential:

$$N - N_0 \simeq \int_0^{\infty} \frac{q^2/2}{\exp(q\hbar\omega_o/(k_BT)) - 1} dq = \zeta(3) \left(\frac{k_BT}{\hbar\omega_o}\right)^3. \tag{4.48b}$$

Here $\zeta(3) \simeq 1.202$ is the Riemann zeta function [which also appeared in our study of the equation of state of thermalized radiation, Eq. (3.54b)]. It is useful to rewrite Eq. (4.48b) as

ground-state occupation number

$$N_0 = N \left[1 - \left(\frac{T}{T_c^0}\right)^3\right], \tag{4.49a}$$

where

critical temperature

$$T_c^0 = \frac{\hbar\omega_o}{k_B} \left(\frac{N}{\zeta(3)}\right)^{1/3} = 280 \text{ nK} \gg \hbar\omega_o/k_B = 8.7 \text{ nK} \tag{4.49b}$$

is our leading-order approximation to the critical temperature T_c. Obviously, we cannot have a negative number of atoms in the ground-state mode, so Eq. (4.49a) must fail for $T > T_c^0$. Presumably, N_0 becomes so small there that our approximation (4.47d) fails.

Figure 4.4a, adapted from the review article by Dalfovo et al. (1999), compares our simple prediction (4.49a) for $N_0(T)$ (dashed curve) with the experimental measurements by the JILA group (Ensher et al., 1996). Both theory and experiment show that, when one lowers the temperature T through a critical temperature T_c, the atoms suddenly begin to accumulate in large numbers in the ground-state mode. At $T \simeq 0.8T_c$, **Bose-Einstein condensation** half the atoms have condensed into the ground state (Bose-Einstein condensation); at $T \simeq 0.2T_c$ almost all are in the ground state. The simple formula (4.49a) is remarkably good at $0 < T < T_c$; evidently, T_c^0 [Eq. (4.49b)] is a rather good leading-order approximation to the critical temperature T_c at which the Bose-Einstein condensate begins to form.

Exercise 4.12 and Fig. 4.4b use more accurate approximations to Eq. (4.48a) to explore the onset of condensation as T is gradually lowered through the critical temperature. The onset is actually continuous when viewed on a sufficiently fine temperature scale; but on scales $0.01T_c$ or greater, it appears to be discontinuous.

phase transition

The onset of Bose-Einstein condensation is an example of a *phase transition*: a sudden (nearly) discontinuous change in the properties of a thermalized system. Among

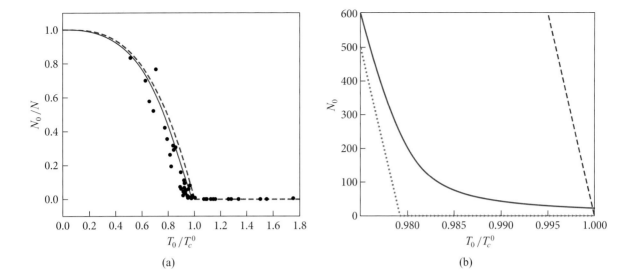

FIGURE 4.4 The number N_0 of atoms in the Bose-Einstein condensate at the center of a magnetic trap as a function of temperature T. (a) Low resolution; (b) High resolution. The dashed curve in each panel is the prediction (4.49a) of the simple theory presented in the text, using the parameters shown in Eq. (4.46b). The dotted curve in panel b is the prediction derived in Ex. 4.12c. The solid curves are our most accurate prediction (4.54) [Ex. 4.12d], including details of the condensation turning on. The large dots are experimental data from Ensher et al. (1996). Panel a is adapted from Dalfovo et al. (1999).

the sudden changes accompanying this phase transition is a (nearly) discontinuous change of the atoms' specific heat (Ex. 4.13). We study some other phase transitions in Chap. 5.

Notice that the critical temperature T_c^0 is larger, by a factor $\sim N^{1/3} = 34$, than the temperature $\hbar\omega_o/k_B = 8.7$ nK associated with the harmonic-oscillator potential. Correspondingly, at the critical temperature there are significant numbers of atoms in modes with q as large as ~ 34—which means that nearly all of the atoms are actually behaving rather classically at $T \simeq T_c^0$, despite our use of quantum mechanical concepts to analyze them!

It is illuminating to compute the spatial distribution of these atoms at the critical temperature using classical techniques. (This distribution could equally well be deduced using the above quantum techniques.) In the near classical, outer region of the trapping potential, the atoms' number density in phase space $\mathcal{N} = (g_s/h^3)\eta$ must have the classical, Boltzmann-distribution form (4.29): $dN/d\mathcal{V}_x d\mathcal{V}_p \propto \exp[-E/(k_B T)] = \exp\{-[V(r) + p^2/(2m)]/(k_B T_c)\}$, where $V(r)$ is the harmonic-oscillator potential (4.46a). Integrating over momentum space, $d\mathcal{V}_p = 4\pi p^2 dp$, we obtain for the number density of atoms $n = dN/d\mathcal{V}_x$

$$n(r) \propto \exp\left(\frac{-V(r)}{k_B T_c}\right) = \exp\left(\frac{-r^2}{a_o^2}\right), \qquad (4.50a)$$

where [using Eqs. (4.49b) for T_c and (4.47b) for ω_o]

$$a_o = \sqrt{\frac{2k_B T_c}{m\omega_o^2}} = \frac{\sqrt{2}}{[\zeta(3)]^{1/6}} N^{1/6} \sigma_o = 1.371 N^{1/6} \sigma_o = 8.02\sigma_o = 6.4 \,\mu\text{m}. \quad (4.50b)$$

Thus, at the critical temperature, the atoms have an approximately Gaussian spatial distribution with radius a_o eight times larger than the trap's 0.80 μm ground-state Gaussian distribution. This size of the distribution gives insight into the origin of the Bose-Einstein condensation: The mean inter-atom spacing at the critical temperature T_c^0 is $a_o/N^{1/3}$. It is easy to verify that this is approximately equal to the typical atom's de Broglie wavelength $\lambda_{T\text{dB}} = h/\sqrt{2\pi m k_B T} = h/(\text{typical momentum})$—which is the size of the region that we can think of each atom as occupying. The atomic separation is smaller than this in the core of the atomic cloud, so the atoms there are beginning to overlap and feel one another's presence, and thereby want to accumulate into the same quantum state (i.e., want to begin condensing). By contrast, the mean separation is larger than λ_{dB} in the outer portion of the cloud, so the atoms there continue to behave classically.

At temperatures below T_c, the N_0 condensed, ground-state-mode atoms have a spatial Gaussian distribution with radius $\sigma_o \sim a_o/8$ (i.e., eight times smaller than the region occupied by the classical, excited-state atoms). Therefore, the condensation is visually manifest by the growth of a sharply peaked core of atoms at the center of the larger, classical, thermalized cloud. In momentum space, the condensed atoms and classical cloud also have Gaussian distributions, with rms momenta $p_{\text{cloud}} \sim 8p_{\text{condensate}}$ (Ex. 4.14) or equivalently, rms speeds $v_{\text{cloud}} \sim 8v_{\text{condensate}}$. In early experiments, the existence of the condensate was observed by suddenly shutting off the trap and letting the condensate and cloud expand ballistically to sizes $v_{\text{condensate}} t$ and $v_{\text{cloud}} t$, and then observing them visually. The condensate was revealed as a sharp Gaussian peak, sticking out of the roughly eight times larger, classical cloud (Fig. 4.5).

EXERCISES

Exercise 4.12 **Example: Onset of Bose-Einstein Condensation

By using more accurate approximations to Eq. (4.48a), explore the onset of the condensation near $T = T_c^0$. More specifically, do the following.

(a) Approximate the numerator in Eq. (4.48a) by $q^2 + 3q$, and keep the $1/N_0$ term in the exponential. Thereby obtain

$$N - N_0 = \left(\frac{k_B T}{\hbar \omega_o}\right)^3 \text{Li}_3(e^{-1/N_0}) + \frac{3}{2}\left(\frac{k_B T}{\hbar \omega_o}\right)^2 \text{Li}_2(e^{-1/N_0}). \quad (4.51)$$

Here

$$\text{Li}_n(u) = \sum_{p=1}^{\infty} \frac{u^p}{p^n} \quad (4.52a)$$

is a special function called the *polylogarithm* (Lewin, 1981), which is known to Mathematica and other symbolic manipulation software and has the properties

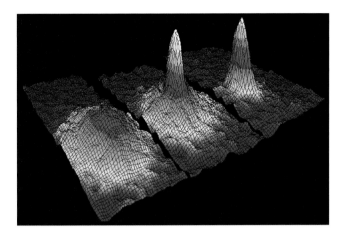

FIGURE 4.5 Velocity distribution of rubidium atoms in a Bose-Einstein condensate experiment by Anderson et al. (1995), as observed by the ballistic expansion method described in the text. In the left frame T is slightly higher than T_c, and there is only the classical cloud. In the center frame T is a bit below T_c, and the condensate sticks up sharply above the cloud. The right frame, at still lower T, shows almost pure condensate. Figure from Cornell (1996).

$$\text{Li}_n(1) = \zeta(n), \qquad \frac{d\text{Li}_n(u)}{du} = \frac{\text{Li}_{n-1}(u)}{u}, \tag{4.52b}$$

where $\zeta(n)$ is the Riemann zeta function.

(b) Show that by setting $e^{-1/N_0} = 1$ and ignoring the second polylogarithm in Eq. (4.51), one obtains the leading-order description of the condensation discussed in the text: Eqs. (4.48b) and (4.49).

(c) By continuing to set $e^{-1/N_0} = 1$ but keeping the second polylogarithm, obtain an improved equation for $N_0(T)$. Your answer should continue to show a discontinuous turn on of the condensation, but at a more accurate, slightly lower critical temperature

$$T_c^1 = T_c^0 \left[1 - \frac{\zeta(2)}{2\zeta(3)^{2/3}} \frac{1}{N^{1/3}} \right] = 0.979 T_c^0. \tag{4.53}$$

This equation illustrates the fact that our approximations are a *large-N expansion* (i.e., an expansion in powers of $1/N$).

(d) By keeping all details of Eq. (4.51) but rewriting it in terms of T_c^0, show that

$$N_0 = N \left[1 - \left(\frac{T}{T_c^0}\right)^3 \frac{\text{Li}_3(e^{-1/N_0})}{\zeta(3)} - \frac{3}{2\zeta(3)^{2/3}} \frac{1}{N^{1/3}} \left(\frac{T}{T_c^0}\right)^2 \text{Li}_2(e^{-1/N_0}) \right]. \tag{4.54}$$

Solve this numerically to obtain $N_0(T/T_c^0)$ for $N = 40{,}000$, and plot your result graphically. It should take the form of the solid curves in Fig. 4.4: a continuous turn on of the condensation over the narrow temperature range $0.98 T_c^0 \lesssim T \lesssim 0.99 T_c^0$ (i.e., a range $\Delta T \sim T_c^0/N^{1/3}$). In the limit of an arbitrarily large number of atoms, the turn on is instantaneous, as described by Eq. (4.49a)—an instantaneous phase transition.

Exercise 4.13 **Problem: Discontinuous Change of Specific Heat*
Analyze the behavior of the atoms' total energy near the onset of condensation, in the limit of arbitrarily large N (i.e., keeping only the leading order in our $1/N^{1/3}$ expansion and approximating the condensation as turning on discontinuously at $T = T_c^0$). More specifically, do the following.

(a) Show that the total energy of the atoms in the magnetic trap is

$$E_{\text{total}} = \frac{3\zeta(4)}{\zeta(3)} N k_B T_c^0 \left(\frac{T}{T_c^0}\right)^4 \quad \text{when } T < T_c^0,$$

$$E_{\text{total}} = \frac{3\text{Li}_4(e^{\mu/(k_B T)})}{\zeta(3)} N k_B T_c^0 \left(\frac{T}{T_c}\right)^4 \quad \text{when } T > T_c^0,$$

(4.55a)

where (at $T > T_c$) $e^{\mu/(k_B T)}$ is a function of N and T determined by $N = \left(k_B T/(\hbar \omega_o)\right)^3 \text{Li}_3(e^{\mu/(k_B T)})$, so $\mu = 0$ at $T = T_c^0$ (see Ex. 4.12). Because $\text{Li}_n(1) = \zeta(n)$, this energy is continuous across the critical temperature T_c.

(b) Show that the specific heat $C = (\partial E_{\text{total}}/\partial T)_N$ is discontinuous across the critical temperature T_c^0:

$$C = \frac{12\zeta(4)}{\zeta(3)} N k_B = 10.80 N k_B \quad \text{as } T \to T_c \text{ from below,}$$

$$C = \left(\frac{12\zeta(4)}{\zeta(3)} - \frac{9\zeta(3)}{\zeta(2)}\right) N k_B = 4.228 N k_B \quad \text{as } T \to T_c \text{ from above.}$$

(4.55b)

Note that for gas contained within the walls of a box, there are two specific heats: $C_V = (\partial E/\partial T)_{N,V}$ when the box volume is held fixed, and $C_P = (\partial E/\partial T)_{N,P}$ when the pressure exerted by the box walls is held fixed. For our trapped atoms, there are no physical walls; the quantity held fixed in place of V or P is the trapping potential $V(r)$.

Exercise 4.14 *Derivation: Momentum Distributions in Condensate Experiments*
Show that in the Bose-Einstein condensate discussed in the text, the momentum distribution for the ground-state-mode atoms is Gaussian with rms momentum $p_{\text{condensate}} = \sqrt{3/2}\hbar/\sigma_o = \sqrt{3\hbar m \omega_o/2}$ and that for the classical cloud it is Gaussian with rms momentum $p_{\text{cloud}} = \sqrt{3 m k_B T_c} \simeq \sqrt{3(N/\zeta(3))^{1/3} \hbar m \omega_o} \simeq 8 p_{\text{condensate}}$.

Chapter 4. Statistical Mechanics

Exercise 4.15 *Problem: Bose-Einstein Condensation in a Cubical Box*

Analyze Bose-Einstein condensation in a cubical box with edge lengths L [i.e., for a potential $V(x, y, z)$ that is zero inside the box and infinite outside it]. In particular, using the analog of the text's simplest approximation, show that the critical temperature at which condensation begins is

$$T_c^0 = \frac{1}{8\pi m k_B} \left[\frac{2\pi\hbar}{L} \left(\frac{N}{\zeta(3/2)} \right)^{1/3} \right]^2, \tag{4.56a}$$

and the number of atoms in the ground-state condensate, when $T < T_c^0$, is

$$N_0 = N \left[1 - \left(\frac{T}{T_c^0} \right)^{3/2} \right]. \tag{4.56b}$$

4.10 Statistical Mechanics in the Presence of Gravity T2

Systems with significant gravity behave quite differently in terms of their statistical mechanics than do systems without gravity. This has led to much controversy as to whether statistical mechanics can really be applied to gravitating systems. Despite that controversy, statistical mechanics has been applied in the presence of gravity in a variety of ways, with great success, resulting in important, fundamental conclusions. In this section, we sketch some of those applications: to galaxies, black holes, the universe as a whole, and the formation of structure in the universe. Our discussion is intended to give just the flavor of these subjects and not full details, so we state some things without derivation. This is necessary in part because many of the phenomena we describe rely for their justification on general relativity (Part VII) and/or quantum field theory in curved spacetime (see, e.g., Parker and Toms, 2009).

4.10.1 Galaxies T2

A galaxy is dominated by a roughly spherical distribution of dark matter (believed to comprise elementary particles with negligible collision cross section) with radius $R_D \sim 3 \times 10^{21}$ m and mass $M_D \sim 10^{42}$ kg. The dark matter and roughly $N \sim 10^{11}$ stars, each with fiducial mass $m \sim 10^{30}$ kg, move in a common gravitational potential well. (As we discuss in Chap. 28, the ratio of regular, or baryonic, matter to dark matter is roughly 1:5 by mass.) The baryons (stars plus gas) are mostly contained within a radius $R \sim 3 \times 10^{20}$ m. The characteristic speed of the dark matter and the stars and gas is $v \sim (GM_D/R_D)^{1/2} \sim (GNm/R)^{1/2} \sim 200$ km s^{-1}. For the moment, focus on the stars, with total mass $M = Nm$, ignoring the dark matter and gas, whose presence does not change our conclusions.

The time it takes stars moving in the dark matter's gravitational potential to cross the baryonic galaxy is $\tau_{\text{int}} \sim 2R/v \sim 10^8$ yr.[14] This time is short compared with the

---

14. 1 yr $\simeq \pi \times 10^7$ s.

age of a galaxy, $\sim 10^{10}$ yr. Galaxies have distant encounters with their neighbors on timescales that can be smaller than their ages but still much longer than τ_{int}; in this sense, they can be thought of as semiclosed systems weakly coupled to their environments. In this subsection, we idealize our chosen galaxy as fully closed (no interaction with its environment). Direct collisions between stars are exceedingly rare, and strong two-star gravitational encounters, which happen when the impact parameter[15] is smaller than $\sim Gm/v^2 \sim R/N$, are also negligibly rare except, sometimes, near the center of a galaxy (which we ignore until the last paragraph of this subsection). We can therefore regard each of the galaxy's stars as moving in a gravitational potential determined by the smoothed-out mass of the dark matter and all the other stars, and can use Hamiltonian dynamics to describe their motions.

Imagine that we have an ensemble of such galaxies, all with the same number of stars N, the same mass M, and the same energy E (in a tiny range δE). We begin our study of that ensemble by making an order-of-magnitude estimate of the probability ρ of finding a chosen galaxy from the ensemble in some chosen quantum state. We compute that probability from the corresponding probabilities for its subsystems, individual stars. The phase-space volume available to each star in the galaxy is $\sim R^3(mv)^3$, the density of single-particle quantum states (modes) in each star's phase space is $1/h^3$, the number of available modes is the product of these, $\sim(Rmv/h)^3$, and the probability of the star occupying the chosen mode, or any other mode, is the reciprocal of this product, $\sim[h/(Rmv)]^3$. The probability of the galaxy occupying a state in its phase space is the product of the probabilities for each of its N stars [Eq. (4.18c)]:

distribution function for stars in a galaxy

$$\rho \sim \left(\frac{h}{Rmv}\right)^{3N} \sim 10^{-2.7 \times 10^{13}}. \tag{4.57}$$

This very small number suggests that it is somewhat silly of us to use quantum mechanics to normalize the distribution function (i.e., silly to use the probabilistic distribution function ρ) when dealing with a system as classical as a whole galaxy. Silly, perhaps; but dangerous, no. The key point is that, so far as classical statistical mechanics is concerned, the only important feature of ρ is that it is proportional to the classical distribution function $\mathcal{N}_{\text{sys}}$; its absolute normalization is usually not important, classically. It was this fact that permitted so much progress to be made in statistical mechanics prior to the advent of quantum mechanics.

Are real galaxies in statistical equilibrium? To gain insight into this question, we estimate the entropy of a galaxy in our ensemble and then ask whether that entropy has any chance of being the maximum value allowed to the galaxy's stars (as it must be if the galaxy is in statistical equilibrium).

Obviously, the stars (by contrast with electrons) are distinguishable, so we can assume multiplicity $\mathcal{M} = 1$ when estimating the galaxy's entropy. Ignoring the (neg-

---

15. The impact parameter is the closest distance between the two stars along their undeflected trajectories.

ligible) correlations among stars, the entropy computed by integrating $\rho \ln \rho$ over the galaxy's full $6N$-dimensional phase space is just N times the entropy associated with a single star, which is $S \sim N k_B \ln(\Delta\Gamma/h^3)$ [Eqs. (4.37) and (4.8a)], where $\Delta\Gamma$ is the phase-space volume over which the star wanders in its ergodic, hamiltonian-induced motion (i.e., the phase space volume available to the star). We express this entropy in terms of the galaxy's total mass M and its total nonrelativistic energy $E \sim -GM^2/(2R)$ as follows. Since the characteristic stellar speed is $v \sim (GM/R)^{1/2}$, the volume of phase space over which the star wanders is $\Delta\Gamma \sim (mv)^3 R^3 \sim (GMm^2R)^{3/2} \sim (-G^2M^3m^2/(2E))^{3/2}$, and the entropy is therefore

$$S_{\text{Galaxy}} \sim (M/m)k_B \ln(\Delta\Gamma/h^3) \sim (3M/(2m))k_B \ln(-G^2M^3m^2/(2Eh^2)). \quad (4.58)$$

Is this the maximum possible entropy available to the galaxy, given the constraints that its mass be M and its nonrelativistic energy be E? No. Its entropy can be made larger by removing a single star from the galaxy to radius $r \gg R$, where the star's energy is negligible. The entropy of the remaining stars will decrease slightly, since the mass M diminishes by m at constant E. However, the entropy associated with the removed star, $\sim (3/2)\ln(GMm^2r/h^2)$, can be made arbitrarily large by making its orbital radius r arbitrarily large. By this thought experiment, we discover that galaxies cannot be in a state of maximum entropy at fixed E and M; they therefore cannot be in a true state of statistical equilibrium.[16] (One might wonder whether there is entropy associated with the galaxy's gravitational field, some of which is due to the stars, and whether that entropy invalidates our analysis. The answer is no. The gravitational field has no randomness, beyond that of the stars themselves, and thus no entropy; its structure is uniquely determined, via Newton's gravitational field equation, by the stars' spatial distribution.)

galaxy never in statistical equilibrium

In a real galaxy or other star cluster, rare near-encounters between stars in the cluster core (ignored in the above discussion) cause individual stars to be ejected from the core into distant orbits or to be ejected from the cluster altogether. These ejections increase the entropy of the cluster plus ejected stars in just the manner of our thought experiment. The core of the galaxy shrinks, a diffuse halo grows, and the total number of stars in the galaxy gradually decreases. This evolution to ever-larger entropy is demanded by the laws of statistical mechanics, but by contrast with systems without gravity, it does not bring the cluster to statistical equilibrium. The long-range influence of gravity prevents a true equilibrium from being reached. Ultimately, the cluster's or galaxy's core may collapse to form a black hole—and, indeed, most large galaxies are observed to have massive black holes in their cores. Despite this somewhat negative conclusion, the techniques of statistical mechanics can be used to understand

to increase entropy, galaxy core shrinks and halo grows

gravity as key to galaxy's behavior

---

16. A true equilibrium can be achieved if the galaxy is enclosed in an idealized spherical box whose walls prevent stars from escaping, or if the galaxy lives in an infinite thermalized bath of stars so that, on average, when one star is ejected into a distant orbit in the bath, another gets injected into the galaxy (see, e.g., Ogorodnikov, 1965; Lynden-Bell, 1967). However, in the real universe galaxies are not surrounded by walls or by thermalized star baths.

galactic dynamics over the comparatively short timescales of interest to astronomers (e.g., Binney and Tremaine, 2003). For a complementary description of a dark matter galaxy in which the stars are ignored, see Ex. 28.7. Further discussion of stellar and dark matter distributions in galaxies is presented in Chap. 28.

4.10.2 Black Holes

Quantum field theory predicts that, near the horizon of a black hole, the vacuum fluctuations of quantized fields behave thermally, as seen by stationary (non-infalling) observers. More specifically, such observers see the horizon surrounded by an atmosphere that is in statistical equilibrium (a thermalized atmosphere) and that rotates with the same angular velocity $\mathbf{\Omega}_H$ as the hole's horizon. This remarkable conclusion, due to Stephen Hawking (1976), William Unruh (1976), and Paul Davies (1977), is discussed pedagogically in books by Thorne, Price, and MacDonald (1986) and Frolov and Zelnikov (2011), and more rigorously in a book by Wald (1994). The atmosphere contains all types of particles that can exist in Nature. Very few of the particles manage to escape from the hole's gravitational pull; most emerge from the horizon, fly up to some maximum height, then fall back down to the horizon. Only if they start out moving almost vertically upward (i.e., with nearly zero angular momentum) do they have any hope of escaping. The few that do escape make up a tiny trickle of *Hawking radiation* (Hawking, 1975) that will ultimately cause the black hole to evaporate, unless it grows more rapidly due to infall of material from the external universe (which it will unless the black hole is far less massive than the Sun).

In discussing the distribution function for the hole's thermalized, rotating atmosphere, one must take account of the fact that the locally measured energy of a particle decreases as it climbs out of the hole's gravitational field (Ex. 26.4). One does so by attributing to the particle the energy that it would ultimately have if it were to escape from the hole's gravitational grip. This is called the particle's "redshifted energy" and is denoted by $\mathcal{E}_\infty$. This $\mathcal{E}_\infty$ is conserved along the particle's world line, as is the projection $\mathbf{j} \cdot \hat{\mathbf{\Omega}}_H$ of the particle's orbital angular momentum $\mathbf{j}$ along the hole's spin axis (unit direction $\hat{\mathbf{\Omega}}_H$).

The hole's horizon behaves like the wall of a blackbody cavity. Into each upgoing mode (single-particle quantum state) a of any and every quantum field that can exist in Nature, it deposits particles that are thermalized with (redshifted) temperature T_H, vanishing chemical potential, and angular velocity $\mathbf{\Omega}_H$. As a result, the mode's distribution function—which is the probability of finding N_a particles in it with net redshifted energy $\mathcal{E}_{a\infty} = N_a \times$ (redshifted energy of one quantum in the mode) and with net axial component of angular momentum $\mathbf{j}_a \cdot \hat{\mathbf{\Omega}}_H = N_a \times$ (angular momentum of one quantum in the mode)—is

$$\rho_a = C \exp\left[\frac{-\mathcal{E}_{a\infty} + \mathbf{\Omega}_H \cdot \mathbf{j}_a}{k_B T_H}\right] \tag{4.59}$$

[see Eq. (4.22) and note that $\mathbf{\Omega}_H = \Omega_H \hat{\mathbf{\Omega}}_H$]. The distribution function for the entire thermalized atmosphere (made of all modes that emerge from the horizon) is, of course, $\rho = \prod_a \rho_a$. (Ingoing modes, which originate at infinity—i.e., far from the black hole—are not thermalized; they contain whatever the universe chooses to send toward the hole.) Because $\mathcal{E}_{a\,\infty}$ is the redshifted energy in mode a, T_H is similarly a redshifted temperature: it is the temperature that the Hawking radiation exhibits when it has escaped from the hole's gravitational grip. Near the horizon, the locally measured atmospheric temperature is gravitationally blue-shifted to much higher values than T_H.

The temperature T_H and angular velocity $\mathbf{\Omega}_H$, like all properties of a black hole, are determined completely by the hole's spin angular momentum $\mathbf{J}_H$ and its mass M_H. To within factors of order unity, they have magnitudes [Ex. 26.16 and Eq. (26.77)]

$$T_H \sim \frac{\hbar}{8\pi k_B G M_H/c^3} \sim \frac{6 \times 10^{-8}\,\text{K}}{M_H/M_\odot}, \quad \mathbf{\Omega}_H \sim \frac{\mathbf{J}_H}{M_H(2GM_H/c^2)^2}. \qquad (4.60)$$

black hole temperature and angular velocity

For a very slowly rotating hole the "$\sim$" becomes an "=" in both equations. Notice how small the hole's temperature is, if its mass is greater than or of order $M_\odot$. For such holes the thermal atmosphere is of no practical interest, though it has deep implications for fundamental physics. Only for tiny black holes (that might conceivably have been formed in the big bang) is T_H high enough to be physically interesting.

Suppose that the black hole evolves much more rapidly by accreting matter than by emitting Hawking radiation. Then the evolution of its entropy can be deduced from the first law of thermodynamics for its atmosphere. By techniques analogous to some developed in the next chapter, one can argue that the atmosphere's equilibrium distribution (4.59) implies the following form for the first law (where we set $c = 1$):

$$dM_H = T_H dS_H + \mathbf{\Omega}_H \cdot d\mathbf{J}_H \qquad (4.61)$$

first law of thermodynamics for black hole

[cf. Eq. (26.92)]. Here dM_H is the change of the hole's mass due to the accretion (with each infalling particle contributing its $\mathcal{E}_\infty$ to dM_H), $d\mathbf{J}_H$ is the change of the hole's spin angular momentum due to the accretion (with each infalling particle contributing its j), and dS_H is the increase of the black hole's entropy.

Because this first law can be deduced using the techniques of statistical mechanics (Chap. 5), it can be argued (e.g., Zurek and Thorne, 1985) that the hole's entropy increase has the standard statistical mechanical origin and interpretation: if N_states is the total number of quantum states that the infalling material could have been in (subject only to the requirement that the total infalling mass-energy be dM_H and total infalling angular momentum be $d\mathbf{J}_H$), then $dS_H = k_B \log N_\text{states}$ [cf. Eq. (4.35)]. In other words, the hole's entropy increases by k_B times the logarithm of the number of quantum mechanically different ways that we could have produced its changes of mass and angular momentum, dM_H and $d\mathbf{J}_H$. Correspondingly, we can regard the hole's total entropy as k_B times the logarithm of the number of ways in which it could have been made. That number of ways is enormous, and correspondingly, the hole's

black hole's entropy

entropy is enormous. This analysis, when carried out in full detail (Zurek and Thorne, 1985), reveals that the entropy is [Eq. (26.93)]

$$S_H = k_B \frac{A_H}{4 L_P^2} \sim 1 \times 10^{77} k_B \left(\frac{M_H}{M_\odot}\right)^2, \tag{4.62}$$

where $A_H \sim 4\pi(2GM_H/c^2)$ is the surface area of the hole's horizon, and $L_P = \sqrt{G\hbar/c^3} = 1.616 \times 10^{-33}$ cm is the Planck length—a result first proposed by Bekenstein (1972) and first proved by Hawking (1975).

What is it about a black hole that leads to this peculiar thermal behavior and enormous entropy? Why is a hole so different from a star or galaxy? The answer lies in the black-hole horizon and the fact that things that fall inward through the horizon cannot get back out. From the perspective of quantum field theory, the horizon produces the thermal behavior. From that of statistical mechanics, the horizon produces the loss of information about how the black hole was made and the corresponding entropy increase. In this sense, the horizon for a black hole plays a role analogous to coarse-graining in conventional classical statistical mechanics.[17]

horizon as key to hole's thermal behavior

The above statistical mechanical description of a black hole's atmosphere and thermal behavior is based on the laws of quantum field theory in curved spacetime—laws in which the atmosphere's fields (electromagnetic, neutrino, etc.) are quantized, but the hole itself is not governed by the laws of quantum mechanics. For detailed but fairly brief analyses along the lines of this section, see Zurek and Thorne (1985); Frolov and Page (1993). For a review of the literature on black-hole thermodynamics and of conundrums swept under the rug in our simple-minded discussion, see Wald (2001).[18]

EXERCISES

Exercise 4.16 *Example: Thermal Equilibria for a Black Hole inside a Box*
This problem was posed and partially solved by Hawking (1976), and fully solved by Page et al. (1977). Place a nonrotating black hole with mass M at the center of a spherical box with volume V that is far larger than the black hole: $V^{1/3} \gg 2GM/c^2$. Put into the box, and outside the black hole, thermalized radiation with energy $\mathcal{E} - Mc^2$ (so the total energy in the box is $\mathcal{E}$). The black hole will emit Hawking

---

17. It seems likely, as of 2017, that the information about what fell into the hole gets retained, in some manner, in the black hole's atmosphere, and we have coarse-grained it away by our faulty understanding of the relevant black-hole quantum mechanics.
18. A much deeper understanding involves string theory—the most promising approach to quantum gravity and to quantization of black holes. Indeed, the thermal properties of black holes, and most especially their entropy, are a powerful testing ground for candidate theories of quantum gravity. A recent, lively debate around the possibility that a classical event horizon might be accompanied by a *firewall* (Almheiri et al., 2013) is a testament to the challenge of these questions, and to the relevance that they have to the relationship between gravity and the other fundamental interactions.

radiation and will accrete thermal radiation very slowly, causing the hole's mass M and the radiation energy $\mathcal{E} - Mc^2$ to evolve. Assume that the radiation always thermalizes during the evolution, so its temperature T is always such that $aT^4V = \mathcal{E} - Mc^2$, where a is the radiation constant. (For simplicity assume that the only form of radiation that the black hole emits and that resides in the box is photons; the analysis is easily extended to the more complicated case where other kinds of particles are present and are emitted.) Discuss the evolution of this system using the second law of thermodynamics. More specifically, do the following.

(a) To simplify your calculations, use so-called *natural units* in which not only is c set equal to unity (geometrized units), but also $G = \hbar = k_B = 1$. If you have never used natural units before, verify that no dimensionless quantities can be constructed from G, c, $\hbar$, and k_B; from this, show that you can always return to cgs or SI units by inserting the appropriate factors of G, c, $\hbar$, and k_B to get the desired units. Show that in natural units the radiation constant is $a = \pi^2/15$.

(b) Show that in natural units the total entropy inside the box is $S = \frac{4}{3}aVT^3 + 4\pi M^2$. The mass M and radiation temperature T will evolve so as to increase this S subject to the constraint that $\mathcal{E} = M + aVT^4$, with fixed $\mathcal{E}$ and V.

(c) Find a rescaling of variables that enables $\mathcal{E}$ to drop out of the problem. [Answer: Set $m = M/\mathcal{E}$, $\Sigma = S/(4\pi\mathcal{E}^2)$, and $\nu = Va/[(3\pi)^4\mathcal{E}^5]$, so m, Σ, and ν are rescaled black-hole mass, entropy in the box, and box volume.] Show that the rescaled entropy is given by

$$\Sigma = m^2 + \nu^{1/4}(1-m)^{3/4}. \qquad (4.63)$$

This entropy is plotted as a function of the black hole mass m in Fig. 4.6 for various ν [i.e., for various $V = (3\pi)^4(\mathcal{E}^5/a)\nu$].

(d) Show that

$$\frac{d\Sigma}{dm} = \frac{3\nu^{1/4}}{4(1-m)^{1/4}}(\tau - 1), \qquad (4.64)$$

where $\tau = T/T_H = 8\pi MT$ is the ratio of the radiation temperature to the black hole's Hawking temperature. Thereby show that (i) when $\tau > 1$, the black hole accretes more energy than it emits, and its mass M grows, thereby increasing the entropy in the box and (ii) when $\tau < 1$, the black hole emits more energy than it accretes, thereby decreasing its mass and again increasing the total entropy.

(e) From the shapes of the curves in Fig. 4.6, it is evident that there are two critical values of the box's volume V (or equivalently, of the rescaled volume ν): V_h and V_g. For $V > V_h$, the only state of maximum entropy is $m = 0$: no black hole. For these large volumes, the black hole will gradually evaporate and disappear. For $V < V_h$, there are two local maxima: one with a black hole whose mass is somewhere between $m = M/\mathcal{E} = 4/5$ and 1; the other with no black hole, $m = 0$.

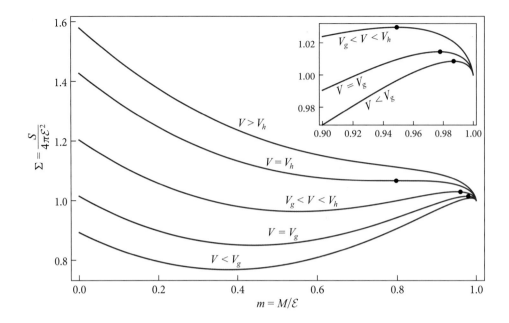

FIGURE 4.6 Total entropy S inside a box of volume V that contains a black hole with mass M and thermalized radiation with energy $\mathcal{E} - Mc^2$ and temperature $T = [(\mathcal{E} - Mc^2)/(aV)]^{1/4}$. For $V < V_h$ there is a statistical equilibrium that contains a black hole, at the local maximum of entropy (large filled circles). At that equilibrium, the radiation temperature T is equal to the black hole's Hawking temperature T_H. See Exercise 4.16e for the definitions of V_g and V_h.

If $V = V_g$, the two states have the same entropy; if $V < V_g$, the state with a black hole has the larger entropy; if $V_g < V < V_h$, the state without a black hole has the larger entropy. Show that

$$V_h = \frac{2^{20}\pi^4}{5^5 a}\mathcal{E}^5 = 4.97 \times 10^4 \left(\frac{\mathcal{E}}{M_P c^2}\right)^2 \left(\frac{G\mathcal{E}}{c^4}\right)^3, \quad V_g = 0.256 V_h, \quad (4.65)$$

where the first expression is in natural units. Here $M_P = \sqrt{\hbar c/G} = 2.177 \times 10^{-5}$ g is the Planck mass, and $G\mathcal{E}/c^4$ is half the radius of a black hole with mass $\mathcal{E}/c^2$. Show that, if $V_g < V < V_h$, and if the box's volume V is made as large as possible—the size of the universe—then the mass of the equilibrium black hole will be roughly 1/100 the mass of the Sun.

When there are two entropy maxima, that is, two states of statistical equilibrium (i.e., for $V < V_h$), an entropy barrier exists between the two equilibria. In principle, quantum fluctuations should be able to carry the system through that barrier, from the equilibrium of lower entropy to that of higher entropy, though the rate will be extremely small for $\mathcal{E} \gg M_P c^2$. Those fluctuations will be governed by the laws of quantum gravity, so studying them is a useful thought experiment in the quest to understand quantum gravity.

4.10.3 The Universe

Observations and theory agree that the universe, when far younger than 1 s old, settled into a very hot, highly thermalized state. All particles except gravitons were in statistical equilibrium at a common, declining temperature, until the dark matter and the neutrinos dropped out of equilibrium and (like the gravitons) became thermodynamically isolated.

During this early relativistic era, the equations of relativistic cosmology imply (as discussed in Sec. 28.4.1) that the temperature T of the universe at age t satisfied $T/T_P \sim (t/t_P)^{-1/2}$. Here $T_P \equiv [\hbar c^5/(Gk_B^2)]^{1/2} \sim 10^{32}$ K is the Planck temperature, and $t_P \equiv (\hbar G/c^5)^{1/2} \sim 10^{-43}$ s is the Planck time. (This approximate T/T_P relationship can be justified on dimensional grounds.) Now the region that was in causal contact at time t (i.e., that was contained within a mutual cosmological horizon) had a volume $\sim (ct)^3$, and thermodynamic considerations imply that the number of relativistic particles that were in causal contact at time t was $N \sim (k_B T t/\hbar)^3 \sim (t/t_P)^{3/2}$. (This remains roughly true today when N has grown to $\sim 10^{91}$, essentially all in microwave background photons.) The associated entropy was then $S \sim Nk_B$ (cf. Sec. 4.8).

number of particles and entropy inside cosmological horizon

Although this seems like an enormous entropy, gravity can do even better. The most efficient way to create entropy, as described in Sec. 4.10.2, is to form massive black holes. Suppose that out of all the relativistic particle mass within the horizon, $M \sim Nk_B T/c^2$, a fraction f has collapsed into black holes of mass M_H. Then, with the aid of Sec. 4.10.2, we estimate that the associated entropy is $S_H \sim f(M_H/M)(t/t_P)^{1/2}S$. If we use the observation that every galaxy has a central black hole with mass in the $\sim 10^6$–10^9 solar mass range, we find that $f \sim 10^{-4}$ and $S_H \sim 10^{11}S$ today!

entropy in black holes

Now it might be claimed that massive black holes are thermodynamically isolated from the rest of the universe because they will take so long to evaporate. That may be so as a practical matter, but more modest gravitational condensations that create stars and starlight can produce large local departures from thermodynamic equilibrium, accompanied by (indeed, driven by) a net increase of entropy and can produce the conditions necessary for life to develop.

Given these considerations, it should not come as a surprise to learn that the behavior of entropy in the expanding universe has provided a foundation for the standard model of cosmology, which we outline in Chap. 28. It has also provided a stimulus for many imaginative proposals addressing fascinating questions that lie beyond the standard model, some of which connect with the nature of life in the universe.[19]

4.10.4 Structure Formation in the Expanding Universe: Violent Relaxation and Phase Mixing

The formation of stars and galaxies ("structure") by gravitational condensation provides a nice illustration of the phase mixing and coarse-graining that underlie the second law of thermodynamics (Sec. 4.7.2).

It is believed that galaxies formed when slight overdensities in the dark matter and gas (presumably irregular in shape) stopped expanding and began to contract under their mutual gravitational attraction. Much of the gas was quickly converted into stars. The dark-matter particles and the stars had very little random motion at this stage relative to their random motions today, $v \sim 200$ km s^{-1}. Correspondingly, although their physical volume $\mathcal{V}_x$ was initially only moderately larger than today, their momentum-space volume $\mathcal{V}_p$ was far smaller than it is today. Translated into the language of an ensemble of N such galaxies, the initial coordinate-space volume $\int d^{3N}x \sim \mathcal{V}_x^N$ occupied by each of the ensemble's galaxies was moderately larger than it is today, while its momentum-space volume $\int d^{3N}p \sim \mathcal{V}_p^N$ was far smaller. The phase-space volume $\mathcal{V}_x^N \mathcal{V}_p^N$ must therefore have increased considerably during the galaxy formation—with the increase due to a big increase in the relative momenta of neighboring stars. For this to occur, it was necessary that the stars changed their relative energies during the contraction, which requires a time-dependent hamiltonian. In other words, the gravitational potential Φ felt by the stars must have varied rapidly, so that the individual stellar energies would vary according to

$$\frac{dE}{dt} = \frac{\partial H}{\partial t} = m\frac{\partial \Phi}{\partial t}. \tag{4.66}$$

violent relaxation of star distribution

The largest changes of energy occurred when the galaxy was contracting dynamically (collapsing), so the potential changed significantly on the timescale it took stars to cross the galaxy, $\tau_{\text{int}} \sim 2R/v$. Numerical simulations show that this energy transfer was highly efficient. This process is known as *violent relaxation*. Although violent relaxation could create the observed stellar distribution functions, it was not by itself a means of diluting the phase-space density, since Liouville's theorem still applied.

phase mixing and coarse-graining of star distribution

The mechanism that changed the phase-space density was phase mixing and coarse-graining (Sec. 4.7.2 above). During the initial collapse, the particles and newly formed stars could be thought of as following highly perturbed radial orbits. The orbits of nearby stars were somewhat similar, though not identical. Therefore small elements of occupied phase space became highly contorted as the particles and stars moved along their phase-space paths.

---

19. For an early and highly influential analysis, see Schrödinger (1944), and for a more recent discussion, see Penrose (2016).

Let us make a simple model of this process by assuming the individual particles and stars initially populate a fraction $f \ll 1$ of the final occupied phase-space volume $\mathcal{V}_{\text{final}}$. After one dynamical timescale $\tau_{int} \sim R/v$, this small volume $f\mathcal{V}_{\text{final}}$ is (presumably) deformed into a convoluted surface that folds back on itself once or twice like dough being kneaded by a baker, while still occupying the same volume $f\mathcal{V}_{\text{final}}$. After n dynamical timescales, there are $\sim 2^n$ such folds (cf. Fig. 4.2b above). After $n \sim -\log_2 f$ dynamical timescales, the spacing between folds becomes comparable with the characteristic thickness of this convoluted surface, and it is no longer practical to distinguish the original distribution function. We expect that coarse-graining has been accomplished for all practical purposes; only a pathological physicist would resist it and insist on trying to continue keeping track of which contorted phase-space regions have the original high density and which do not. For a galaxy we might expect that $f \sim 10^{-3}$ and so this natural coarse-graining can occur in a time approximately equal to $-\log_2 10^{-3} \tau_{int} \sim 10\,\tau_{int} \sim 10^9$ yr, which is 10 times shorter than the present age of galaxies. Therefore it need not be a surprise that the galaxy we know best, our own Milky Way, exhibits little obvious vestigial trace of its initial high-density (low phase-space-volume) distribution function.[20]

4.11 Entropy and Information

4.11.1 Information Gained When Measuring the State of a System in a Microcanonical Ensemble

In Sec. 4.7, we said that entropy is a measure of our lack of information about the state of any system chosen at random from an ensemble. In this section we make this heuristic statement useful by introducing a precise definition of *information*.

Consider a microcanonical ensemble of identical systems. Each system can reside in any one of a finite number, N_{states}, of quantum states, which we label by integers $n = 1, 2, 3, \ldots, N_{\text{states}}$. Because the ensemble is microcanonical, all N_{states} states are equally probable; they have probabilities $\rho_n = 1/N_{\text{states}}$. Therefore the entropy of any system chosen at random from this ensemble is $S = -k_B \sum_n \rho_n \ln \rho_n = k_B \ln N_{\text{states}}$ [Eqs. (4.34) and (4.35)].

Now suppose that we measure the state of our chosen system and find it to be (for example) state number 238 out of the N_{states} equally probable states. How much information have we gained? For this thought experiment, and more generally (see Sec. 4.11.2 below), *the amount of information gained, expressed in bits, is defined to be the minimum number of binary digits required to distinguish the measured state from all the other N_{states} states that the system could have been in*. To evaluate this information gain, we label each state n by the number $n-1$ written in binary code (state $n=1$ is labeled by {000}, state $n=2$ is labeled by {001}, 3 is {010}, 4 is {011}, 5 is {100},

amount of information gained in a measurement for microcanonical ensemble

---

20. However, the dark-matter particles may very well not be fully coarse-grained at the present time, and this influences strategies for detecting it.

6 is {101}, 7 is {110}, 8 is {111}, etc.). If $N_{\text{states}} = 4$, then the number of binary digits needed is 2 (the leading 0 in the enumeration above can be dropped), so in measuring the system's state we gain 2 bits of information. If $N_{\text{states}} = 8$, the number of binary digits needed is 3, so our measurement gives us 3 bits of information. In general, we need $\log_2 N_{\text{states}}$ binary digits to distinguish the states from one another, so *the amount of information gained in measuring the system's state is the base-2 logarithm of the number of states the system could have been in:*

$$I = \log_2 N_{\text{states}} = (1/\ln 2)\ln N_{\text{states}} = 1.4427 \ln N_{\text{states}}. \quad (4.67a)$$

Notice that this information gain is proportional to the entropy $S = k_B \ln N_{\text{states}}$ of the system before the measurement was made:

information gain related to entropy decrease

$$I = S/(k_B \ln 2). \quad (4.67b)$$

The measurement reduces the system's entropy from $S = -k_B \ln N_{\text{states}}$ to zero (and increases the entropy of the rest of the universe by at least this amount), and it gives us $I = S/(k_B \ln 2)$ bits of information about the system. We shall discover below that this entropy/information relationship is true of measurements made on a system drawn from any ensemble, not just a microcanonical ensemble. But first we must develop a more complete understanding of information.

4.11.2 Information in Communication Theory

The definition of "the amount of information I gained in a measurement" was formulated by Claude Shannon (1948) in the context of his laying the foundations of *communication theory.* Communication theory deals (among other things) with the problem of how to encode most efficiently a message as a binary string (a string of 0s and 1s) in order to transmit it across a communication channel that transports binary signals. Shannon defined the information in a message as *the number of bits required, in the most compressed such encoding, to distinguish this message from all other messages that might be transmitted.*

communication theory

general definition of information

symbols and messages

Shannon focused on messages that, before encoding, consist of a sequence of symbols. For an English-language message, each symbol might be a single character (a letter A, B, C, ..., Z or a space; $N = 27$ distinct symbols in all), and a specific message might be the following sequence of length $L = 19$ characters: "I DO NOT UNDERSTAND". Suppose, for simplicity, that in the possible messages, all N distinct symbols appear with equal frequency (this, of course, is not the case for English-language messages), and suppose that the length of some specific message (its number of symbols) is L. Then the number of bits needed to encode this message and distinguish it from all other possible messages of length L is

$$I = \log_2 N^L = L \log_2 N. \quad (4.68a)$$

In other words, the average number of bits per symbol (the average amount of information per symbol) is

$$\bar{I} = \log_2 N. \quad (4.68b)$$

If there are only two possible symbols, we have one bit per symbol in our message. If there are four possible (equally likely) symbols, we have two bits per symbol, and so forth.

It is usually the case that not all symbols occur with the same frequency in the allowed messages. For example, in English messages the letter "A" occurs with a frequency $p_A \simeq 0.07$, while the letter "Z" occurs with the much smaller frequency $p_Z \simeq 0.001$. All English messages, of character length $L \gg N = 27$, constructed by a typical English speaker, will have these frequencies of occurrence for "A" and "Z". Any purported message with frequencies for "A" and "Z" differing substantially from 0.07 and 0.001 will not be real English messages, and thus need not be included in the binary encoding of messages. As a result, it turns out that the most efficient binary encoding of English messages (the most *compressed* encoding) will use an average number of bits per character somewhat less than $\log_2 N = \log_2 27 = 4.755$. In other words, the average information per character in English language messages is somewhat less than $\log_2 27$.

To deduce the average information per character when the characters do not all occur with the same frequency, we begin with a simple example: the number of distinct characters to be used in the message is just $N = 2$ (the characters "B" and "E"), and their frequencies of occurrence in very long allowed messages are $p_B = 3/5$ and $p_E = 2/5$. For example, in the case of messages with length $L = 100$, the message

$$\text{EBBEEBBBBEBBBBBBBBEBEBBBEEEEBBBEB}$$
$$\text{BEEEEEBEEBEEEEEEBBEBBBBBEBBBBBEBBE}$$
$$\text{BBBEBBBEEBBBEBBBBBBBEBBBBEBBEEBEB} \quad (4.69a)$$

contains 63 Bs and 37 Es, and thus (to within statistical variations) has the correct frequencies $p_B \simeq 0.6$, $p_E \simeq 0.4$ to be an allowed message. By contrast, the message

$$\text{BBBBBBBBBBBBBBEBBBBBBBBBBBBBBBBBBBB}$$
$$\text{BBBBBBBBBBBEBBBBBBBBBBBBBBBBBBBBBBB}$$
$$\text{BBBBBBBBBBBBBBEBBBBBBBBBBBBBBBBBB} \quad (4.69b)$$

contains 97 Bs and 3 Es and thus is not an allowed message. To deduce the number of allowed messages and thence the number of bits required to encode them distinguishably, we map this problem of 60% probable Bs and 40% probable Es onto the problem of messages with five equally probable symbols as follows. Let the set of distinct symbols be the letters "a", "b", "c", "y", and "z", all occurring in allowed messages equally frequently: $p_a = p_b = p_c = p_y = p_z = 1/5$. An example of an allowed message is

$$\text{zcczzcaabzccbabcccczaybacyzyzcbbyc}$$
$$\text{ayyyyyayzcyzzzzzcczacbabybbbcczabz}$$
$$\text{bbbybaazybccybaccabazacbzbayycyc} \quad (4.69c)$$

We map each such message from our new message set into one from the previous message set by identifying "a", "b", and "c" as from the beginning of the alphabet (and thus converting them to "B"), and identifying "y" and "z" as from the end of the alphabet (and thus converting them to "E"). Our message (4.69c) from the new message set then maps into the message (4.69a) from the old set. This mapping enables us to deduce the number of bits required to encode the old messages, with their unequal frequencies $p_B = 3/5$ and $p_E = 2/5$, from the number required for the new messages, with their equal frequencies $p_a = p_b = \ldots = p_z = 1/5$.

The number of bits needed to encode the new messages, with length $L \gg N_{\text{new}} = 5$, is $I = L \log_2 5$. The characters "a," "b," and "c" from the beginning of the alphabet occur $\frac{3}{5}L$ times in each new message (in the limit that L is arbitrarily large). When converting the new message to an old one, we no longer need to distinguish between "a", "b", and "c", so we no longer need the $\frac{3}{5}L \log_2 3$ bits that were being used to make those distinctions. Similarly, the number of bits we no longer need when we drop the distinction between our two end-of-alphabet characters "y" and "z" is $\frac{2}{5}L \log_2 2$. As a result, the number of bits still needed to distinguish between old messages (messages with "B" occurring 3/5 of the time and "E" occurring 2/5 of the time) is

$$I = L \log_2 5 - \frac{3}{5}L \log_2 3 - \frac{2}{5} \log_2 2 = L\left[-\frac{3}{5}\log_2 \frac{3}{5} - \frac{2}{5}\log_2 \frac{2}{5}\right]$$
$$= L(-p_B \log_2 p_B - p_E \log_2 p_E). \tag{4.69d}$$

A straightforward generalization of this argument (Ex. 4.17) shows that, *when one constructs messages with very large length $L \gg N$ from a pool of N symbols that occur with frequencies $p_1, p_2, \ldots, p_N$, the minimum number of bits required to distinguish all the allowed messages from one another (i.e., the amount of information in each message) is*

$$I = L \sum_{n=1}^{N} -p_n \log_2 p_n; \tag{4.70}$$

so *the average information per symbol in the message is*

information per symbol in a message

$$\boxed{\bar{I} = \sum_{n=1}^{N} -p_n \log_2 p_n = (1/\ln 2) \sum_{n=1}^{N} -p_n \ln p_n.} \tag{4.71}$$

4.11.3 Examples of Information Content

Notice the similarity of the general information formula (4.70) to the general formula (4.34) for the entropy of an arbitrary ensemble. This similarity has a deep consequence.

INFORMATION FROM MEASURING THE STATE OF ONE SYSTEM IN AN ENSEMBLE
Consider an arbitrary ensemble of systems in statistical mechanics. As usual, label the quantum states available to each system by the integer $n = 1, 2, \ldots, N_{\text{states}}$, and denote by p_n the probability that any chosen system in the ensemble will turn out to be in state n. Now select one system out of the ensemble and measure its quantum state n_1;

select a second system and measure its state, n_2. Continue this process until some large number $L \gg N_{\text{states}}$ of systems have been measured. The sequence of measurement results $\{n_1, n_2, \ldots, n_L\}$ can be regarded as a message. The minimum number of bits needed to distinguish this message from all other possible such messages is given by the general information formula (4.70). This is the total information in the L system measurements. Correspondingly, *the amount of information we get from measuring the state of one system (the average information per measurement) is given by Eq. (4.71).* This acquired information is related to the entropy of the system before measurement [Eq. (4.34)] by the same standard formula (4.67b) as we obtained earlier for the special case of the microcanonical ensemble:

$$\boxed{\bar{I} = S/(k_B \ln 2).} \tag{4.72}$$

information from measurement for a general ensemble

INFORMATION IN AN ENGLISH-LANGUAGE MESSAGE

For another example of information content, we return to English-language messages (Shannon, 1948). Evaluating the information content of a long English message is a very difficult task, since it requires figuring out how to compress the message most compactly. We shall make a series of estimates.

A crude initial estimate of the information per character is that obtained by idealizing all the characters as occurring equally frequently: $\bar{I} = \log_2 27 \simeq 4.76$ bits per character [Eq. (4.68b)]. This is an overestimate, because the 27 characters actually occur with very different frequencies. We could get a better estimate by evaluating $\bar{I} = \sum_{n=1}^{27} -p_n \log_2 p_n$, taking account of the characters' varying frequencies p_n (the result is about $\bar{I} = 4.1$), but we can do even better by converting from characters as our symbols to words. The average number of characters in an English word is about 4.5 letters plus 1 space, or 5.5 characters per word. We can use this number to convert from characters as our symbols to words. The number of words in a typical English speaker's vocabulary is roughly 12,000. If we idealize these 12,000 words as occurring with the same frequencies, then the information per word is $\log_2 12{,}000 \simeq 13.6$, so the information per character is $\bar{I} = (1/5.5) \log_2 12{,}000 \simeq 2.46$. This is much smaller than our previous estimates. A still better estimate is obtained by using Zipf's (1935) approximation $p_n = 0.1/n$ of the frequencies of occurrence of the words in English messages.[21] To ensure that $\sum_{n=1}^{N} p_n = 1$ for Zipf's approximation, we require that the number of words be $N = 12{,}367$. We then obtain, as our improved estimate of the information per word, $\sum_{n=1}^{12,367} (-0.1/n) \log_2(0.1/n) = 9.72$, corresponding to a value of information per character $\bar{I} \simeq 9.72/5.5 = 1.77$. This is substantially smaller than our initial, crudest estimate of 4.76 and is close to more careful estimates $\bar{I} \simeq 1.0$ to 1.5 (Schneier, 1997, Sec. 11.1).

information per character in English message

---

21. The most frequently occurring word is "THE", and its frequency is about 0.1 (1 in 10 words is "THE" in a long message). The next most frequent words are "OF", "AND", and "TO"; their frequencies are about 0.1/2, 0.1/3, and 0.1/4, respectively; and so forth.

4.11.4 Some Properties of Information

Because of the similarity of the general formulas for information and entropy (both proportional to $\sum_n -p_n \ln p_n$), information has very similar properties to entropy. In particular (Ex. 4.18):

1. Information is additive (just as entropy is additive). The information in two successive, independent messages is the sum of the information in each message.

2. If the frequencies of occurrence of the symbols in a message are $p_n = 0$ for all symbols except one, which has $p_n = 1$, then the message contains zero information. This is analogous to the vanishing entropy when all states have zero probability except for one, which has unit probability.

3. For a message L symbols long, whose symbols are drawn from a pool of N distinct symbols, the information content is maximized if the probabilities of the symbols are all equal ($p_n = 1/N$), and the maximal value of the information is $I = L \log_2 N$. This is analogous to the microcanonical ensemble having maximal entropy.

4.11.5 Capacity of Communication Channels; Erasing Information from Computer Memories

NOISELESS COMMUNICATION

A noiseless communication channel has a maximum rate (number of bits per second) at which it can transmit information. This rate is called the *channel capacity* and is denoted C. When one subscribes to a cable internet connection in the United States, one typically pays a monthly fee that depends on the connection's channel capacity; for example, in Pasadena, California, in summer 2011 the fee was \$29.99 per month for a connection with capacity $C = 12$ megabytes/s $= 96$ megabits/s, and \$39.99 for $C = 144$ megabits/s. (This was a 30-fold increase in capacity per dollar since 2003!)

It should be obvious from the way we have defined the information I in a message that the maximum rate at which we can transmit optimally encoded messages, each with information content I, is C/I messages per second.

NOISY COMMUNICATION

When a communication channel is noisy (as all channels actually are), for high-confidence transmission of messages one must put some specially designed redundancy into one's encoding. With cleverness, one can thereby identify and correct errors in a received message, caused by the noise (error-correcting code); see, for example, Shannon, (1948), Raisbeck (1963), and Pierce (2012).[22] The redundancy needed for such error identification and correction reduces the channel's capacity. As an example, consider a *symmetric binary channel*: one that carries messages made of 0s and

---

22. A common form of error-correcting code is based on parity checks.

1s, with equal frequency $p_0 = p_1 = 0.5$, and whose noise randomly converts a 0 into a 1 with some small error probability p_e, and randomly converts a 1 into 0 with that same probability p_e. Then one can show (e.g., Pierce, 2012; Raisbeck, 1963) that the channel capacity is reduced—by the need to find and correct for these errors—by a factor

$$C = C_{\text{noiseless}}[1 - \bar{I}(p_e)], \quad \text{where } \bar{I}(p_e) = -p_e \log_2 p_e - (1 - p_e) \log_2(1 - p_e).$$
(4.73)

reduction of channel capacity due to error correction

Note that the fractional reduction of capacity is by the amount of information per symbol in messages made from symbols with frequencies equal to the probabilities p_e of making an error and $1 - p_e$ of not making an error—a remarkable and nontrivial conclusion! This is one of many important results in communication theory.

MEMORY AND ENTROPY

Information is also a key concept in the theory of computation. As an important example of the relationship of information to entropy, we cite Landauer's (1961, 1991) theorem: In a computer, when one erases L bits of information from memory, one necessarily increases the entropy of the memory and its environment by at least $\Delta S = L k_B \ln 2$ and correspondingly, one increases the thermal energy (heat) of the memory and environment by at least $\Delta Q = T \Delta S = L k_B T \ln 2$ (Ex. 4.21).

Landauer's theorem: entropy increase due to erasure

EXERCISES

Exercise 4.17 *Derivation: Information per Symbol When Symbols Are Not Equally Probable* T2

Derive Eq. (4.70) for the average number of bits per symbol in a long message constructed from N distinct symbols, where the frequency of occurrence of symbol n is p_n. [Hint: Generalize the text's derivation of Eq. (4.69d).]

Exercise 4.18 *Derivation: Properties of Information* T2
Prove the properties of entropy enumerated in Sec. 4.11.4.

Exercise 4.19 *Problem: Information per Symbol for Messages Built from Two Symbols* T2

Consider messages of length $L \gg 2$ constructed from just two symbols ($N = 2$), which occur with frequencies p and $(1 - p)$. Plot the average information per symbol $\bar{I}(p)$ in such messages, as a function of p. Explain why your plot has a maximum $\bar{I} = 1$ when $p = 1/2$, and has $\bar{I} = 0$ when $p = 0$ and when $p = 1$. (Relate these properties to the general properties of information.)

Exercise 4.20 *Problem: Information in a Sequence of Dice Throws* T2
Two dice are thrown randomly, and the sum of the dots showing on the upper faces is computed. This sum (an integer n in the range $2 \leq n \leq 12$) constitutes a symbol,

and the sequence of results of $L \gg 12$ throws is a message. Show that the amount of information per symbol in this message is $\bar{I} \simeq 3.2744$.

Exercise 4.21 *Derivation: Landauer's Theorem* T2
Derive, or at least give a plausibility argument for, Landauer's theorem (stated at the end of Sec. 4.11.5).

Bibliographic Note

Statistical mechanics has inspired a variety of readable and innovative texts. The classic treatment is Tolman (1938). Classic elementary texts are Kittel (2004) and Kittel and Kroemer (1980). Among more modern approaches that deal in much greater depth with the topics covered in this chapter are Lifshitz and Pitaevskii (1980), Chandler (1987), Sethna (2006), Kardar (2007), Reif (2008), Reichl (2009), and Pathria and Beale (2011). The Landau-Lifshitz textbooks (including Lifshitz and Pitaevskii, 1980) are generally excellent after one has already learned the subject at a more elementary level. A highly individual and advanced treatment, emphasizing quantum statistical mechanics, is Feynman (1972). A particularly readable account in which statistical mechanics is used heavily to describe the properties of solids, liquids, and gases is Goodstein (2002). Readable, elementary introductions to information theory are Raisbeck (1963) and Pierce (2012); an advanced text is McEliece (2002).

CHAPTER FIVE

Statistical Thermodynamics

One of the principal objects of theoretical research is to find the point of view from which the subject appears in the greatest simplicty.

J. WILLARD GIBBS (1881)

5.1 Overview

In Chap. 4, we introduced the concept of statistical equilibrium and briefly studied some properties of equilibrated systems. In this chapter, we develop the theory of statistical equilibrium in a more thorough way.

The title of this chapter, "Statistical Thermodynamics," emphasizes two aspects of the theory of statistical equilibrium. The term *thermodynamics* is an ancient one that predates statistical mechanics. It refers to a study of the macroscopic properties of systems that are in or near equilibrium, such as their energy and entropy. Despite paying no attention to the microphysics, classical thermodynamics is a very powerful theory for deriving general relationships among macroscopic properties. Microphysics influences the macroscopic world in a statistical manner, so in the late nineteenth century, Willard Gibbs and others developed statistical mechanics and showed that it provides a powerful conceptual underpinning for classical thermodynamics. The resulting synthesis, *statistical thermodynamics,* adds greater power to thermodynamics by augmenting it with the statistical tools of ensembles and distribution functions.

In our study of statistical thermodynamics, we restrict attention to an ensemble of large systems that are in statistical equilibrium. By "large" is meant a system that can be broken into a large number $N_{\rm ss}$ of subsystems that are all macroscopically identical to the full system except for having $1/N_{\rm ss}$ as many particles, $1/N_{\rm ss}$ as much volume, $1/N_{\rm ss}$ as much energy, $1/N_{\rm ss}$ as much entropy, and so forth. (Note that this definition constrains the energy of interaction between the subsystems to be negligible.) Examples are 1 kg of plasma in the center of the Sun and a 1-kg sapphire crystal.

The equilibrium thermodynamic properties of any type of large system (e.g., an ideal gas)[1] can be derived using any one of the statistical equilibrium ensembles of the last chapter (microcanonical, canonical, grand canonical, or Gibbs). For example,

---

1. An *ideal gas* is one with negligible interactions among its particles.

> **BOX 5.1. READERS' GUIDE**
>
> - Relativity enters into portions of this chapter solely via the relativistic energies and momenta of high-speed particles (Sec. 1.10).
> - This chapter relies in crucial ways on Secs. 3.2 and 3.3 of Chap. 3 and on Secs. 4.2–4.8 of Chap. 4.
> - Portions of Chap. 6 rely on Sec. 5.6 of this chapter. Portions of Part V (Fluid Dynamics) rely on thermodynamic concepts and equations of state treated in this chapter, but most readers will already have met these in a course on elementary thermodynamics.
> - Other chapters do not depend strongly on this one.

each of these ensembles will predict the same equation of state $P = (N/V)k_B T$ for an ideal gas, even though in one ensemble each system's number of particles N is precisely fixed, while in another ensemble N can fluctuate so that strictly speaking, one should write the equation of state as $P = (\overline{N}/V)k_B T$, with $\overline{N}$ the ensemble average of N. (Here and throughout this chapter, for compactness we use bars rather than brackets to denote ensemble averages, i.e., $\overline{N}$ rather than $\langle N \rangle$.) The equations of state are the same to very high accuracy because the fractional fluctuations of N are so extremely small: $\Delta N/N \sim 1/\sqrt{\overline{N}}$ (cf. Ex. 5.11).

Although the thermodynamic properties are independent of the equilibrium ensemble, specific properties are often derived most quickly, and the most insight usually accrues, from the ensemble that most closely matches the physical situation being studied. In Secs. 5.2–5.5, we use the microcanonical, grand canonical, canonical, and Gibbs ensembles to derive many useful results from statistical thermodynamics: fundamental potentials expressed as statistical sums over quantum states, variants of the first law of thermodynamics, equations of state, Maxwell relations, Euler's equation, and others. Table 5.1 summarizes the most important of those statistical-equilibrium results and some generalizations of them. Readers are advised to delay studying this table until they have read further into the chapter.

As we saw in Chap. 4, when systems are out of statistical equilibrium, their evolution toward equilibrium is driven by the law of entropy increase—the second law of thermodynamics. In Sec. 5.5, we formulate the fundamental potential (Gibbs potential) for an out-of-equilibrium ensemble that interacts with a heat and volume bath, and we discover a simple relationship between that fundamental potential and the entropy of system plus bath. From that relationship, we learn that in this case the second law is equivalent to a law of decrease of the Gibbs potential. As applications, we learn how chemical potentials drive chemical reactions and phase transitions. In Sec. 5.6, we discover how the Gibbs potential can be used to study spontaneous fluctuations of a system away from equilibrium, when it is coupled to a heat and particle bath.

TABLE 5.1: Representations and ensembles for systems in statistical equilibrium, in relativistic notation

Representation and ensemble	First law	Quantities exchanged with bath	Distribution function ρ
Energy and microcanonical (Secs. 4.5 and 5.2)	$d\mathcal{E} = TdS + \tilde{\mu}dN - PdV$	None	const $= e^{-S/k_B}$ $\mathcal{E}$ const in $\delta\mathcal{E}$
Enthalpy (Exs. 5.5 and 5.13)	$dH = TdS + \tilde{\mu}dN + VdP$	V and $\mathcal{E}$ $d\mathcal{E} = -PdV$	const $= e^{-S/k_B}$ H const
Physical free energy and canonical (Secs. 4.4.1 and 5.4)	$dF = -SdT + \tilde{\mu}dN - PdV$	$\mathcal{E}$	$e^{(F-\mathcal{E})/(k_B T)}$
Gibbs (Secs. 4.4.2 and 5.5)	$dG = -SdT + \tilde{\mu}dN + VdP$	$\mathcal{E}$ and V	$e^{(G-\mathcal{E}-PV)/(k_B T)}$
Grand canonical (Secs. 4.4.2 and 5.3)	$d\Omega = -SdT - Nd\tilde{\mu} - PdV$	$\mathcal{E}$ and N	$e^{(\Omega-\mathcal{E}+\tilde{\mu}N)/(k_B T)}$

Notes: The nonrelativistic formulas are the same but with the rest masses of particles removed from the chemical potentials ($\tilde{\mu} \to \mu$) and from all fundamental potentials except Ω ($\mathcal{E} \to E$, but no change of notation for H, F, and G). This table will be hard to understand until after reading the sections referenced in column one.

In Sec. 5.7, we employ these tools to explore fluctuations and the gas-to-liquid phase transition for a model of a real gas due to the Dutch physicist Johannes van der Waals. Out-of-equilibrium aspects of statistical mechanics (evolution toward equilibrium and fluctuations away from equilibrium) are summarized in Table 5.2 and discussed in Secs. 5.5.1 and 5.6, not just for heat and volume baths, but for a variety of baths.

Deriving the macroscopic properties of real materials by statistical sums over their quantum states can be formidably difficult. Fortunately, in recent years some powerful approximation techniques have been devised for performing the statistical sums. In Secs. 5.8.3 and 5.8.4, we give the reader the flavor of two of these techniques: the *renormalization group* and *Monte Carlo methods*. We illustrate and compare these techniques by using them to study a phase transition in a simple model for ferromagnetism called the *Ising model*.

5.2 Microcanonical Ensemble and the Energy Representation of Thermodynamics

5.2.1 Extensive and Intensive Variables; Fundamental Potential

Consider a microcanonical ensemble of large, closed systems that have attained statistical equilibrium. We can describe the ensemble macroscopically using a set of thermodynamic variables. These variables can be divided into two classes: *extensive variables* (Sec. 4.4.1), which double if one doubles the system's size, and *intensive variables* (Sec. 4.4.2), whose magnitudes are independent of the system's size. Familiar examples of extensive variables are a system's total energy $\mathcal{E}$, entropy S, volume V, and number of conserved particles of various species N_I. Corresponding examples of

extensive and intensive variables

intensive variables are temperature T, pressure P, and the chemical potentials $\tilde{\mu}_I$ for various species of particles.

For a large, closed system, there is a *complete set of extensive variables that we can specify independently*—usually its volume V, total energy $\mathcal{E}$ or entropy S, and number N_I of particles of each species I. The values of the other extensive variables and all the intensive variables are determined in terms of this complete set by methods that we shall derive.

The particle species I in the complete set must only include those species whose particles are conserved on the timescales of interest. For example, if photons can be emitted and absorbed, then one must not specify N_γ, the number of photons; rather, N_γ will come to an equilibrium value that is governed by the values of the other extensive variables. Also, one must omit from the set $\{I\}$ any conserved particle species whose numbers are automatically determined by the numbers of other, included species. For example, gas inside the Sun is always charge neutral to very high precision, and therefore (neglecting all elements except hydrogen and helium), the number of electrons N_e in a sample of gas is determined by the number of protons N_p and the number of helium nuclei (alpha particles) N_α: $N_e = N_p + 2N_\alpha$. Therefore, if one includes N_p and N_α in the complete set of extensive variables being used, one must omit N_e.

As in Chap. 4, we formulate the theory relativistically correctly but formulate it solely in the mean rest frames of the systems and baths being studied. Correspondingly, in our formulation we generally include the particle rest masses m_I in the total energy $\mathcal{E}$ and in the chemical potentials $\tilde{\mu}_I$. For very nonrelativistic systems, however, we usually replace $\mathcal{E}$ by the nonrelativistic energy $E \equiv \mathcal{E} - \sum_I N_I m_I c^2$ and $\tilde{\mu}_I$ by the nonrelativistic chemical potential $\mu_I \equiv \tilde{\mu}_I - m_I c^2$ (though, as we shall see in Sec. 5.5 when studying chemical reactions, the identification of the appropriate rest mass m_I to subtract is a delicate issue).

5.2.2 Energy as a Fundamental Potential

For simplicity, we temporarily specialize to a microcanonical ensemble of one-species systems, which all have the same values of a complete set of three extensive variables: the energy $\mathcal{E}$,[2] number of particles N, and volume V. Suppose that the microscopic nature of the ensemble's systems is known. Then, at least in principle and often in practice, one can identify from that microscopic nature the quantum states that are available to the system (given its constrained values of $\mathcal{E}$, N, and V), one can count those quantum states, and from their total number N_{states} one can compute the ensemble's total entropy $S = k_B \ln N_{\text{states}}$ [Eq. (4.35)]. The resulting entropy can be regarded as a function of the complete set of extensive variables,

$$S = S(\mathcal{E}, N, V), \quad (5.1)$$

---

2. In practice, as illustrated in Ex. 4.7, one must allow $\mathcal{E}$ to fall in some tiny but finite range $\delta\mathcal{E}$ rather than constraining it precisely, and one must then check to be sure that the results of the analysis are independent of $\delta\mathcal{E}$.

and this equation can then be inverted to give the total energy in terms of the entropy and the other extensive variables:

$$\boxed{\mathcal{E} = \mathcal{E}(S, N, V).} \qquad (5.2)$$

We call the energy $\mathcal{E}$, viewed as a function of S, N, and V, the *fundamental thermodynamic potential for the microcanonical ensemble*. When using this fundamental potential, we regard S, N, and V as our complete set of extensive variables rather than $\mathcal{E}$, N, and V. From the fundamental potential, as we shall see, one can deduce all other thermodynamic properties of the system.

energy as fundamental thermodynamic potential for microcanonical ensemble

5.2.3 Intensive Variables Identified Using Measuring Devices; First Law of Thermodynamics

5.2.3

TEMPERATURE

In Sec. 4.4.1, we used kinetic-theory considerations to identify the thermodynamic temperature T of the canonical ensemble [Eq. (4.20)]. It is instructive to discuss how this temperature arises in the microcanonical ensemble. Our discussion makes use of an idealized *thermometer* consisting of an idealized atom that has only two quantum states, $|0\rangle$ and $|1\rangle$, with energies $\mathcal{E}_0$ and $\mathcal{E}_1 = \mathcal{E}_0 + \Delta\mathcal{E}$. The atom, initially in its ground state, is brought into thermal contact with one of the large systems of our microcanonical ensemble and then monitored over time as it is stochastically excited and deexcited. The ergodic hypothesis (Sec. 4.6) guarantees that the atom traces out a history of excitation and deexcitation that is governed statistically by the canonical ensemble for a collection of such atoms exchanging energy (heat) with our large system (the heat bath). More specifically, if T is the (unknown) temperature of our system, then the fraction of time the atom spends in its excited state, divided by the fraction spent in its ground state, is equal to the canonical distribution's probability ratio:

idealized thermometer

$$\frac{\rho_1}{\rho_0} = \frac{e^{-\mathcal{E}_1/(k_B T)}}{e^{-\mathcal{E}_0/(k_B T)}} = e^{-\Delta\mathcal{E}/(k_B T)} \qquad (5.3a)$$

[cf. Eq. (4.20)].

This ratio can also be computed from the properties of the full system augmented by the two-state atom. This augmented system is microcanonical with a total energy $\mathcal{E} + \mathcal{E}_0$, since the atom was in the ground state when first attached to the full system. Of all the quantum states available to this augmented system, the ones in which the atom is in the ground state constitute a total number $N_0 = e^{S(\mathcal{E}, N, V)/k_B}$; and those with the atom in the excited state constitute a total number $N_1 = e^{S(\mathcal{E} - \Delta\mathcal{E}, N, V)/k_B}$. Here we have used the fact that the number of states available to the augmented system is equal to that of the original, huge system with energy $\mathcal{E}$ or $\mathcal{E} - \Delta\mathcal{E}$ (since the atom, in each of the two cases, is forced to be in a unique state), and we have expressed that number of states of the original system, for each of the two cases, in terms of the original system's entropy function [Eq. (5.1)]. The ratio of the number of states N_1/N_0 is (by the ergodic hypothesis) the ratio of the time that the augmented system spends with the atom excited to the time spent with the atom in its ground state (i.e., it is equal to ρ_1/ρ_0):

$$\frac{\rho_1}{\rho_0} = \frac{N_1}{N_0} = \frac{e^{S(\mathcal{E}-\Delta\mathcal{E},N,V)/k_B}}{e^{S(\mathcal{E},N,V)/k_B}} = \exp\left[-\frac{\Delta\mathcal{E}}{k_B}\left(\frac{\partial S}{\partial \mathcal{E}}\right)_{N,V}\right]. \qquad (5.3b)$$

By equating Eqs. (5.3a) and (5.3b), we obtain an expression for the original system's temperature T in terms of the partial derivative $(\partial \mathcal{E}/\partial S)_{N,V}$ of its fundamental potential $\mathcal{E}(S, N, V)$:

temperature from energy potential $\mathcal{E}(S, N, V)$

$$T = \frac{1}{(\partial S/\partial \mathcal{E})_{N,V}} = \left(\frac{\partial \mathcal{E}}{\partial S}\right)_{N,V}, \qquad (5.3c)$$

where we have used Eq. (3) of Box 5.2.

CHEMICAL POTENTIAL AND PRESSURE

A similar thought experiment—using a highly idealized measuring device that can exchange one particle ($\Delta N = 1$) with the system but cannot exchange any energy with it—gives for the fraction of the time spent with the extra particle in the measuring device ("state 1") and in the system ("state 0"):

$$\frac{\rho_1}{\rho_0} = e^{\tilde{\mu}\Delta N/k_B T}$$

$$= \frac{e^{S(\mathcal{E},N-\Delta N,V)/k_B}}{e^{S(\mathcal{E},N,V)/k_B}} = \exp\left[-\frac{\Delta N}{k_B}\left(\frac{\partial S}{\partial N}\right)_{\mathcal{E},V}\right]. \qquad (5.4a)$$

Here the first expression is computed from the viewpoint of the measuring device's equilibrium ensemble,[3] and the second from the viewpoint of the combined system's microcanonical ensemble. Equating these two expressions, we obtain

chemical potential from energy potential

$$\tilde{\mu} = -T\left(\frac{\partial S}{\partial N}\right)_{\mathcal{E},V} = \left(\frac{\partial \mathcal{E}}{\partial N}\right)_{S,V}. \qquad (5.4b)$$

In the last step we use Eq. (5.3c) and Eq. (4) of Box 5.2. The reader should be able to construct a similar thought experiment involving an idealized pressure transducer (Ex. 5.1), which yields the following expression for the system's pressure:

pressure from energy potential

$$P = -\left(\frac{\partial \mathcal{E}}{\partial V}\right)_{S,N}. \qquad (5.5)$$

FIRST LAW OF THERMODYNAMICS

Having identifed the three intensive variables T, $\tilde{\mu}$, and P as partial derivatives [Eqs. (5.3c), (5.4b), and (5.5)], we now see that the fundamental potential's differential relation

$$d\mathcal{E}(S, N, V) = \left(\frac{\partial \mathcal{E}}{\partial S}\right)_{N,V} dS + \left(\frac{\partial \mathcal{E}}{\partial N}\right)_{S,V} dN + \left(\frac{\partial \mathcal{E}}{\partial V}\right)_{S,N} dV \qquad (5.6)$$

---

3. This ensemble has $\rho = \text{constant } e^{-\tilde{\mu}N/(k_B T)}$, since only particles can be exchanged with the device's heat bath (our system).

> **BOX 5.2. TWO USEFUL RELATIONS BETWEEN PARTIAL DERIVATIVES**
>
> Expand a differential increment in the energy $\mathcal{E}(S, N, V)$ in terms of differentials of its arguments S, N, and V:
>
> $$d\mathcal{E}(S, N, V) = \left(\frac{\partial \mathcal{E}}{\partial S}\right)_{N,V} dS + \left(\frac{\partial \mathcal{E}}{\partial N}\right)_{S,V} dN + \left(\frac{\partial \mathcal{E}}{\partial V}\right)_{S,N} dV. \quad (1)$$
>
> Next expand the entropy $S(\mathcal{E}, N, V)$ similarly, and substitute the resulting expression for dS into the above equation to obtain
>
> $$d\mathcal{E} = \left(\frac{\partial \mathcal{E}}{\partial S}\right)_{N,V} \left(\frac{\partial S}{\partial \mathcal{E}}\right)_{N,V} d\mathcal{E}$$
> $$+ \left[\left(\frac{\partial \mathcal{E}}{\partial S}\right)_{N,V} \left(\frac{\partial S}{\partial N}\right)_{\mathcal{E},V} + \left(\frac{\partial \mathcal{E}}{\partial N}\right)_{S,V}\right] dN$$
> $$+ \left[\left(\frac{\partial \mathcal{E}}{\partial S}\right)_{N,V} \left(\frac{\partial S}{\partial V}\right)_{N,\mathcal{E}} + \left(\frac{\partial \mathcal{E}}{\partial V}\right)_{S,N}\right] dV. \quad (2)$$
>
> Noting that this relation must be satisfied for all values of $d\mathcal{E}$, dN, and dV, we conclude that
>
> $$\left(\frac{\partial \mathcal{E}}{\partial S}\right)_{N,V} = \frac{1}{(\partial S/\partial \mathcal{E})_{N,V}}, \quad (3)$$
>
> $$\left(\frac{\partial \mathcal{E}}{\partial N}\right)_{S,V} = -\left(\frac{\partial \mathcal{E}}{\partial S}\right)_{N,V} \left(\frac{\partial S}{\partial N}\right)_{\mathcal{E},V}, \quad (4)$$
>
> and so forth; similarly for other pairs and triples of partial derivatives.
>
> These equations, and their generalization to other variables, are useful in manipulations of thermodynamic equations.

is nothing more nor less than the ordinary first law of thermodynamics

$$\boxed{d\mathcal{E} = T dS + \tilde{\mu} dN - P dV} \quad (5.7)$$

first law of thermodynamics from energy potential

(cf. Table 5.1 above).

Notice the pairing of intensive and extensive variables in this first law: temperature T is paired with entropy S; chemical potential $\tilde{\mu}$ is paired with number of particles N; and pressure P is paired with volume V. We can think of each intensive variable as a "generalized force" acting on its corresponding extensive variable to change the energy of the system. We can add additional pairs of intensive and extensive variables if appropriate, calling them X_A, Y_A (e.g., an externally imposed magnetic field **B** and a material's magnetization **M**; Sec. 5.8). We can also generalize to a multicomponent system (i.e., one that has several types of conserved particles with numbers N_I and associated chemical potentials $\tilde{\mu}_I$). We can convert to nonrelativistic language by subtracting off the rest-mass contributions (switching from $\mathcal{E}$ to $E \equiv \mathcal{E} - \sum N_I m_I c^2$ and from $\tilde{\mu}_I$ to $\mu_I = \tilde{\mu}_I - m_I c^2$). The result is the nonrelativistic, extended first law:

extended first law

$$dE = TdS + \sum_I \mu_I dN_I - PdV + \sum_A X_A dY_A. \quad (5.8)$$

EXERCISES

Exercise 5.1 *Problem: Pressure-Measuring Device*
For the microcanonical ensemble considered in this section, derive Eq. (5.5) for the pressure using a thought experiment involving a pressure-measuring device.

5.2.4 Euler's Equation and Form of the Fundamental Potential

We can integrate the differential form of the first law to obtain a remarkable—though essentially trivial—relation known as *Euler's equation*. We discuss this for the one-species system whose first law is $d\mathcal{E} = TdS + \tilde{\mu}dN - PdV$. The generalization to other systems should be obvious.

We decompose our system into a large number of subsystems in equilibrium with one another. As they are in equilibrium, they will all have the same values of the intensive variables T, $\tilde{\mu}$, and P; therefore, if we add up all their energies $d\mathcal{E}$ to obtain $\mathcal{E}$, their entropies dS to obtain S, and so forth, we obtain from the first law (5.7)[4]

Euler's equation for energy

$$\mathcal{E} = TS + \tilde{\mu}N - PV. \quad (5.9a)$$

Since the energy $\mathcal{E}$ is itself extensive, Euler's equation (5.9a) must be expressible as

$$\mathcal{E} = Nf(V/N, S/N) \quad (5.9b)$$

for some function f. This is a useful functional form for the fundamental potential $\mathcal{E}(N, V, S)$. For example, for a nonrelativistic, classical, monatomic perfect gas,[5] the

---

4. There are a few (but very few!) systems for which some of the thermodynamic laws, including Euler's equation, take on forms different from those presented in this chapter. A black hole is an example (cf. Sec. 4.10.2). A black hole cannot be divided up into subsystems, so the above derivation of Euler's equation fails. Instead of increasing linearly with the mass $M_H = \mathcal{E}/c^2$ of the hole, the hole's extensive variables $S_H =$ (entropy) and $J_H =$ (spin angular momentum) increase quadratically with M_H. And instead of being independent of the hole's mass, the intensive variables $T_H =$ (temperature) and $\Omega_H =$ (angular velocity) scale as $1/M_H$. See, e.g., Tranah and Landsberg (1980) and see Sec. 4.10.2 for some other aspects of black-hole thermodynamics.
5. Recall (Ex. 4.6) that a perfect gas is one that is ideal (i.e., has negligible interactions among its particles) and whose particles have no excited internal degrees of freedom. The phrase "perfect gas" must not be confused with "perfect fluid" (a fluid whose viscosity is negligible so its stress tensor, in its rest frame, consists solely of an isotropic pressure).

Sackur-Tetrode equation (4.42) can be solved for $E = \mathcal{E} - Nmc^2$ to get the following form of the fundamental potential:

$$E(V, S, N) = N \left(\frac{3h^2}{4\pi m g_s^{2/3}}\right) \left(\frac{V}{N}\right)^{-2/3} \exp\left(\frac{2}{3k_B}\frac{S}{N} - \frac{5}{3}\right). \quad (5.9c)$$

energy potential for nonrelativistic, classical, perfect gas

Here m is the mass of each of the gas's particles, and h is Planck's constant.

5.2.5 Everything Deducible from First Law; Maxwell Relations

There is no need to memorize a lot of thermodynamic relations; *nearly all relations can be deduced almost trivially from the functional form of the first law of thermodynamics,* the main formula shown on the first line of Table 5.1.

For example, in the case of our simple one-species system, the first law $d\mathcal{E} = TdS + \tilde{\mu}dN - PdV$ tells us that the system energy $\mathcal{E}$ should be regarded as a function of the things that appear as differentials on the right-hand side: S, N, and V; that is, the fundamental potential must have the form $\mathcal{E} = \mathcal{E}(S, N, V)$. By thinking about building up our system from smaller systems by adding entropy dS, particles dN, and volume dV at fixed values of the intensive variables, we immediately deduce, from the first law, the Euler equation $\mathcal{E} = TS + \tilde{\mu}N - PV$. By writing out the differential relation (5.6)—which is just elementary calculus—and comparing with the first law, we immediately read off the intensive variables in terms of partial derivatives of the fundamental potential:

$$T = \left(\frac{\partial \mathcal{E}}{\partial S}\right)_{V,N}, \quad \tilde{\mu} = \left(\frac{\partial \mathcal{E}}{\partial N}\right)_{V,S}, \quad P = -\left(\frac{\partial \mathcal{E}}{\partial V}\right)_{S,N}. \quad (5.10a)$$

We can then go on to notice that the resulting $P(V, S, N)$, $T(V, S, N)$, and $\tilde{\mu}(V, S, N)$ are not all independent. The equality of mixed partial derivatives (e.g., $\partial^2 \mathcal{E}/\partial V \partial S = \partial^2 \mathcal{E}/\partial S \partial V$) together with Eqs. (5.10a) implies that they must satisfy the following *Maxwell relations*:

$$\left(\frac{\partial T}{\partial N}\right)_{S,V} = \left(\frac{\partial \tilde{\mu}}{\partial S}\right)_{N,V}, \quad -\left(\frac{\partial P}{\partial S}\right)_{V,N} = \left(\frac{\partial T}{\partial V}\right)_{S,N}, \quad \left(\frac{\partial \tilde{\mu}}{\partial V}\right)_{N,S} = -\left(\frac{\partial P}{\partial N}\right)_{V,S}.$$

Maxwell relations from energy potential

$$(5.10b)$$

Additional relations can be generated using the types of identities proved in Box 5.2—or they can be generated more easily by applying the above procedure to the fundamental potentials associated with other ensembles; see Secs. 5.3–5.5. All equations of state [i.e., all such relations as Eqs. (5.11) between intensive and extensive variables] must satisfy the Maxwell relations. For our simple example of a nonrelativistic, classical, perfect gas, we can substitute the fundamental potential E [Eq. (5.9c)] into Eqs. (5.10a) to obtain

$$T(V, S, N) = \left(\frac{h^2}{2\pi m k_B g_s^{2/3}}\right) \left(\frac{N}{V}\right)^{2/3} \exp\left(\frac{2S}{3k_B N} - \frac{5}{3}\right),$$

$$\mu(V, S, N) = \left(\frac{h^2}{4\pi m g_s^{2/3}}\right) \left(\frac{N}{V}\right)^{2/3} \left(5 - 2\frac{S}{k_B N}\right) \exp\left(\frac{2S}{3k_B N} - \frac{5}{3}\right)$$

$$P(V, S, N) = \left(\frac{h^2}{2\pi m g_s^{2/3}}\right) \left(\frac{N}{V}\right)^{5/3} \exp\left(\frac{2S}{3k_B N} - \frac{5}{3}\right), \quad (5.11)$$

(Ex. 5.2). These clearly do satisfy the Maxwell relations.

EXERCISES

Exercise 5.2 *Derivation: Energy Representation for a Nonrelativistic, Classical, Perfect Gas*

(a) Use the fundamental potential $E(V, S, N)$ for the nonrelativistic, classical, perfect gas [Eq. (5.9c)] to derive Eqs. (5.11) for the gas pressure, temperature, and chemical potential.

(b) Show that these equations of state satisfy Maxwell relations (5.10b).

(c) Combine these equations of state to obtain the ideal-gas equation of state

$$P = \frac{N}{V} k_B T, \quad (5.12)$$

which we derived in Ex. 3.8 using kinetic theory.

5.2.6 Representations of Thermodynamics

The treatment of thermodynamics given in this section is called the *energy representation,* because it is based on the fundamental potential $\mathcal{E}(S, V, N)$ in which the energy is expressed as a function of the complete set of extensive variables $\{S, V, N\}$. As we have seen, this energy representation is intimately related to the microcanonical ensemble. In Sec. 5.3, we meet the grand-potential representation for thermodynamics, which is intimately related to the grand canonical ensemble for systems of volume V in equilibrium with a heat and particle bath that has temperature T and chemical potential $\tilde{\mu}$. Then in Secs. 5.4 and 5.5, we meet the two representations of thermodynamics that are intimately related to the canonical and Gibbs ensembles, and discover their power to handle certain special issues. And in Ex. 5.5, we consider a representation and ensemble based on enthalpy. These five representations and their ensembles are summarized in Table 5.1 above.

5.3 Grand Canonical Ensemble and the Grand-Potential Representation of Thermodynamics

We now turn to the grand canonical ensemble, and its grand-potential representation of thermodynamics, for a semiclosed system that exchanges heat and particles with a thermalized bath. For simplicity, we assume that all particles are identical (just one particle species), but we allow them to be relativistic (speeds comparable to the speed of light) or not and allow them to have nontrivial internal degrees of freedom (e.g., vibrations and rotations; Sec. 5.4.2), and allow them to exert forces on one another via an interaction potential that appears in their hamiltonian (e.g., van der Waals forces; Secs. 5.3.2 and 5.7). We refer to these particles as a gas, though our analysis is more general than gases. The nonrelativistic limit of all our fundamental equations is trivially obtained by removing particle rest masses from the energy ($\mathcal{E}$ gets replaced by $E = \mathcal{E} - mc^2$) and from the chemical potential ($\tilde{\mu}$ gets replaced by $\mu = \tilde{\mu} - mc^2$), but not from the grand potential, as it never has rest masses in it [see, e.g., Eq. (5.18) below].

We begin in Sec. 5.3.1 by deducing the grand potential representation of thermodynamics from the grand canonical ensemble, and by deducing a method for computing the thermodynamic properties of our gas from a grand canonical sum over the quantum states available to the system. In Ex. 5.3, the reader will apply this grand canonical formalism to a relativistic perfect gas. In Sec. 5.3.2, we apply the formalism to a nonrelativistic gas of particles that interact via van der Waals forces, thereby deriving the van der Waals equation of state, which is surprisingly accurate for many nonionized gases.

5.3.1 The Grand-Potential Representation, and Computation of Thermodynamic Properties as a Grand Canonical Sum

Figure 5.1 illustrates the ensemble of systems that we are studying and its bath. Each system is a cell of fixed volume V, with imaginary walls, inside a huge thermal bath of identical particles. Since the cells' walls are imaginary, the cells can and do exchange energy and particles with the bath. The bath is characterized by chemical potential $\tilde{\mu}$ for these particles and by temperature T. Since we allow the particles to be relativistic, we include the rest mass in the chemical potential $\tilde{\mu}$.

We presume that our ensemble of cells has reached statistical equilibrium with the bath, so its probabilistic distribution function has the grand canonical form (4.25c):

$$\boxed{\rho_n = \frac{1}{Z} \exp\left(\frac{-\mathcal{E}_n + \tilde{\mu} N_n}{k_B T}\right) = \exp\left(\frac{\Omega - \mathcal{E}_n + \tilde{\mu} N_n}{k_B T}\right).} \quad (5.13)$$

grand canonical distribution function

Here the index n labels the quantum state $|n\rangle$ of a cell, N_n is the number of particles in that quantum state, $\mathcal{E}_n$ is the total energy of that quantum state (including each particle's rest mass; its energy of translational motion; its internal energy if it has internal vibrations, rotations, or other internal excitations; and its energy of interaction with

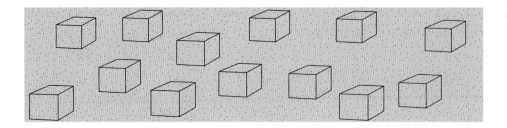

FIGURE 5.1 An ensemble of cells, each with volume V and imaginary walls, inside a heat and particle bath.

other particles), and $1/Z \equiv e^{\Omega/(k_B T)}$ is the normalization constant that guarantees $\sum_n \rho_n = 1$:

grand partition function and grand potential

$$Z \equiv \exp\left(\frac{-\Omega}{k_B T}\right) \equiv \sum_n \exp\left(\frac{-\mathcal{E}_n + \tilde{\mu} N_n}{k_B T}\right). \quad (5.14)$$

This normalization constant, whether embodied in Z or in Ω, is a function of the bath's temperature T and chemical potential $\tilde{\mu}$, and also of the cells' common volume V (which influences the set of available states $|n\rangle$). When regarded as a function of T, $\tilde{\mu}$, and V, the quantity $Z(V, \tilde{\mu}, T)$ is called the gas's *grand partition function*, and $\Omega(T, \tilde{\mu}, V)$ is called its *grand potential*. The following general argument shows that *once one has computed the explicit functional form for the grand potential*

$$\Omega(V, \tilde{\mu}, T) \quad (5.15)$$

[or for the grand partition function $Z(V, \tilde{\mu}, T)$], one can then derive from it all the thermodynamic properties of the thermally equilibrated system. The argument is so general that it applies to every grand canonical ensemble of systems, not just to our chosen gas of identical particles.

As key quantities in the argument, we introduce the mean energy and mean number of particles in the ensemble's systems (i.e., cells of Fig. 5.1):

$$\overline{\mathcal{E}} \equiv \sum_n \rho_n \mathcal{E}_n, \quad \overline{N} \equiv \sum_n \rho_n N_n. \quad (5.16)$$

(We denote these quantities with bars $\overline{\mathcal{E}}$ rather than brackets $\langle \mathcal{E} \rangle$ for ease of notation.) Inserting expression (5.13) for ρ_n into the log term in the definition of entropy $S = -k_B \sum_n \rho_n \ln \rho_n$ and using Eqs. (5.16), we obtain

$$S = -k_B \sum_n \rho_n \ln \rho_n = -k_B \sum_n \rho_n \left(\frac{\Omega - \mathcal{E}_n + \tilde{\mu} N_n}{k_B T}\right) = -\frac{\Omega - \overline{\mathcal{E}} + \tilde{\mu} \overline{N}}{T}; \quad (5.17)$$

or, equivalently,

Legendre transformation between thermodynamic representations

$$\Omega = \overline{\mathcal{E}} - TS - \tilde{\mu} \overline{N}. \quad (5.18)$$

230   Chapter 5. Statistical Thermodynamics

This can be regarded as a *Legendre transformation* that leads from the energy representation of thermodynamics to the grand-potential representation. Legendre transformations are a common tool, for example, in classical mechanics (e.g., Landau and Lifshitz, 1976; Marion and Thornton, 1995; Goldstein, Poole, and Safko, 2002), for switching from one set of independent variables to another. Note that removing rest masses from $\overline{\mathcal{E}} = \bar{E} + Nmc^2$ and from $\tilde{\mu} = \mu + mc^2$ to get a nonrelativistic formula leaves Ω unchanged.

We now ask how the grand potential will change if the temperature T and chemical potential $\tilde{\mu}$ of the bath (and therefore of the ensemble) are slowly altered, and the volumes V of all the ensemble's boxes are slowly altered. Differentiating Eq. (5.18), we obtain $d\Omega = d\overline{\mathcal{E}} - TdS - SdT - \tilde{\mu}d\overline{N} - \overline{N}d\tilde{\mu}$. Expressing $d\mathcal{E}$ in terms of the energy representation's first law of thermodynamics (5.7) (with $\mathcal{E}$ replaced by $\overline{\mathcal{E}}$ and N replaced by $\overline{N}$), we bring this expression into the form

$$\boxed{d\Omega = -PdV - \overline{N}d\tilde{\mu} - SdT.} \tag{5.19}$$

first law in grand potential representation

This is the *grand-potential representation* of the first law of thermodynamics. The quantities P, $\overline{N}$, and S paired with the independent variables V, $\tilde{\mu}$, and T, respectively, can be thought of as generalized forces that push on the independent variables as they change, to produce changes in the grand potential.

generalized forces

From this version of the first law (the key grand canonical equation listed in the last line of Table 5.1), we can easily deduce almost all other equations of the grand-potential representation of thermodynamics. We just follow the same procedure as we used for the energy representation (Sec. 5.2.5).

The grand-potential representation's complete set of independent variables consists of those that appear as differentials on the right-hand side of the first law (5.19): V, $\tilde{\mu}$, and T. From the form (5.19) of the first law we see that Ω is being regarded as a function of these three independent variables: $\Omega = \Omega(V, \tilde{\mu}, T)$. This is the fundamental potential.

The Euler equation of this representation is deduced by building up a system from small pieces that all have the same values of the intensive variables $\tilde{\mu}$, T, and P. The first law (5.19) tells us that this buildup will produce

Euler equation for grand potential

$$\boxed{\Omega = -PV.} \tag{5.20}$$

Thus, if we happen to know P as a function of this representation's independent intensive variables—$P(\tilde{\mu}, T)$ (it must be independent of the extensive variable V)—then we can simply multiply by V to get the functional form of the grand potential: $\Omega(V, \tilde{\mu}, T) = P(\tilde{\mu}, T)V$; see Eqs. (5.24a) and (5.25) as a concrete example.

By comparing the grand-potential version of the first law (5.19) with the elementary calculus equation $d\Omega = (\partial \Omega/\partial V)dV + (\partial \Omega/\partial \tilde{\mu})d\tilde{\mu} + (\partial \Omega/\partial T)dT$, we infer equations for the system's "generalized forces," the pressure P, mean number of particles $\overline{N}$, and entropy S:

$$\overline{N} = -\left(\frac{\partial \Omega}{\partial \tilde{\mu}}\right)_{V,T}, \quad S = -\left(\frac{\partial \Omega}{\partial T}\right)_{V,\tilde{\mu}}, \quad P = -\left(\frac{\partial \Omega}{\partial V}\right)_{\tilde{\mu},T}. \quad (5.21)$$

Maxwell relations from grand potential

By differentiating these relations and equating mixed partial derivatives, we can derive Maxwell relations analogous to those [Eqs. (5.10b)] of the energy representation; for example, $(\partial \overline{N}/\partial T)_{V,\tilde{\mu}} = (\partial S/\partial \tilde{\mu})_{V,T}$. Equations of state are constrained by these Maxwell relations.

If we had begun with a specific functional form of the grand potential as a function of this representation's complete set of independent variables $\Omega(V, T, \tilde{\mu})$ [e.g., Eq. (5.24a) below], then Eqs. (5.21) would give us the functional forms of almost all the other dependent thermodynamic variables. The only one we are missing is the mean energy $\overline{\mathcal{E}}(V, \tilde{\mu}, T)$ in a cell. If we have forgotten Eq. (5.18) (the Legendre transformation) for that quantity, we can easily rederive it from the grand canonical distribution function $\rho = \exp[(\Omega - \mathcal{E} + \tilde{\mu}N)/(k_BT)]$ (the other key equation, besides the first law, on the last line of Table 5.1), via the definition of entropy as $S = -k_B \sum_n \rho_n \ln \rho_n = -k_B \overline{\ln \rho}$, as we did in Eq. (5.17).

This illustrates the power of the sparse information in Table 5.1. From it and little else we can deduce all the thermodynamic equations for each representation of thermodynamics.

It should be easy to convince oneself that *the nonrelativistic versions of all the above equations in this section can be obtained by the simple replacements* $\mathcal{E} \to E$ *(removal of rest masses from total energy) and* $\tilde{\mu} \to \mu$ *(removal of rest mass from chemical potential)*.

5.3.2 Nonrelativistic van der Waals Gas

computing the grand potential from a statistical sum

The statistical sum $Z \equiv e^{-\Omega/(k_BT)} = \sum_n e^{(-\mathcal{E}_n + \tilde{\mu}N_n)/(k_BT)}$ [Eq. (5.14)] is a powerful method for computing the grand potential $\Omega(V, \tilde{\mu}, T)$, a method often used in condensed matter physics. Here we present a nontrivial example: a nonrelativistic, monatomic gas made of atoms or molecules (we call them "particles") that interact with so-called "van der Waals forces." In Ex. 5.3, the reader will explore a simpler example: a relativistic, perfect gas.

We assume that the heat and particle bath that bathes the cells of Fig. 5.1 has (i) sufficiently low temperature that the gas's particles are not ionized (therefore they are also nonrelativistic, $k_BT \ll mc^2$) and (ii) sufficiently low chemical potential that the mean occupation number η of the particles' quantum states is small compared to unity, so they behave classically, $\mu \equiv \tilde{\mu} - mc^2 \ll -k_BT$ [Eq. (3.22d)].

The orbital electron clouds attached to each particle repel one another when the distance r between the particles' centers of mass is smaller than about the diameter r_o of the particles. At larger separations, the particles' electric dipoles (intrinsic or induced) attract one another weakly. The interaction energy (potential energy) $u(r)$

associated with these forces has a form well approximated by the Lennard-Jones potential

$$u(r) = \varepsilon_o \left[\left(\frac{r_o}{r}\right)^{12} - \left(\frac{r_o}{r}\right)^6 \right], \tag{5.22a}$$

where ε_o is a constant energy. When a gradient is taken, the first term gives rise to the small-r repulsive force and the second to the larger-r attractive force. For simplicity of analytic calculations, we use the cruder approximation

$$u(r) = \infty \text{ for } r < r_o, \qquad u(r) = -\varepsilon_o (r_o/r)^6 \text{ for } r > r_o, \tag{5.22b}$$

approximate interaction energy for van der Waals gas

which has an infinitely sharp repulsion at $r = r_o$ (a *hard wall*). For simplicity, we assume that the mean interparticle separation is much larger than r_o (*dilute gas*), so it is highly unlikely that three or more particles are close enough together simultaneously, $r \sim r_o$, to interact (i.e., we confine ourselves to two-particle interactions).[6]

We compute the grand potential $\Omega(V, \mu, T)$ for an ensemble of cells embedded in a bath of these particles (Fig. 5.1), and from $\Omega(V, \mu, T)$ we compute how the particles' interaction energy $u(r)$ alters the gas's equation of state from the form $P = (\bar{N}/V) k_B T$ for an ideal gas [Eqs. (3.39b,c)]. Since this is our objective, any internal and spin degrees of freedom that the particles might have are irrelevant, and we ignore them.

For this ensemble, the nonrelativistic grand partition function $Z = \sum_n \exp[(-E_n + \mu N_n)/(k_B T)]$ is

$$Z = e^{-\Omega/(k_B T)}$$
$$= \sum_{N=0}^{\infty} \frac{e^{\mu N/(k_B T)}}{N!} \int \frac{d^{3N}x \, d^{3N}p}{h^{3N}} \exp\left[-\sum_{i=1}^{N} \frac{\mathbf{p}_i^2}{2mk_B T} - \sum_{i=1}^{N} \sum_{j=i+1}^{N} \frac{u_{ij}}{k_B T} \right].$$
(5.23a)

Here we have used Eq. (4.8b) for the sum over states $\sum_n$ [with $\mathcal{M} = N!$ (the multiplicity factor of Eqs. (4.8)), $W = 3N$, and $d\Gamma_W = d^{3N}x \, d^{3N}p$; cf. Ex. 5.3], and we have written E_n as the sum over the kinetic energies of the N particles in the cell and the interaction energies

$$u_{ij} \equiv u(r_{ij}), \qquad r_{ij} \equiv |\mathbf{x}_i - \mathbf{x}_j| \tag{5.23b}$$

of the $\frac{1}{2} N(N-1)$ pairs of particles.

Evaluation of the integrals and sum in Eq. (5.23a), with the particles' interaction energies given by Eqs. (5.23b) and (5.22b), is a rather complex task, which we relegate to the Track-Two Box 5.3. The result for the grand potential, $\Omega = -k_B T \ln Z$, accurate

---

6. Finding effective ways to tackle many-body problems is a major challenge in many areas of modern theoretical physics.

van der Waals grand potential

to first order in the particles' interactions (in the parameters a and b below), is

$$\Omega = -k_B T V \frac{(2\pi m k_B T)^{3/2}}{h^3} e^{\mu/(k_B T)} \left[1 + \frac{(2\pi m k_B T)^{3/2}}{h^3} e^{\mu/(k_B T)} \left(\frac{a}{k_B T} - b \right) \right].$$

(5.24a)

Here b is four times the volume of each hard-sphere particle, and a is that volume times the interaction energy ε_o of two hard-sphere particles when they are touching:

$$b = \frac{16\pi}{3} \left(\frac{r_o}{2} \right)^3, \qquad a = b\varepsilon_o.$$

(5.24b)

By differentiating this grand potential, we obtain the following expressions for the pressure P and mean number of particles $\overline{N}$ in a volume-V cell:

$$P = -\left(\frac{\partial \Omega}{\partial V} \right)_{\mu,T}$$

$$= k_B T \frac{(2\pi m k_B T)^{3/2}}{h^3} e^{\mu/(k_B T)} \left[1 + \frac{(2\pi m k_B T)^{3/2}}{h^3} e^{\mu/(k_B T)} \left(\frac{a}{k_B T} - b \right) \right],$$

$$\overline{N} = -\left(\frac{\partial \Omega}{\partial \mu} \right)_{V,T}$$

$$= V \frac{(2\pi m k_B T)^{3/2}}{h^3} e^{\mu/(k_B T)} \left[1 + 2 \frac{(2\pi m k_B T)^{3/2}}{h^3} e^{\mu/(k_B T)} \left(\frac{a}{k_B T} - b \right) \right].$$

(5.25)

van der Waals equation of state

Notice that, when the interaction energy is turned off so $a = b = 0$, the second equation gives the standard ideal-gas particle density $\overline{N}/V = (2\pi m k_B T)^{3/2} e^{\mu/(k_B T)}/h^3 = \zeta/\lambda_{T\,dB}^3$ where ζ and $\lambda_{T\,dB}$ are defined in Eq. (2) of Box 5.3. Inserting this into the square bracketed expression in Eqs. (5.25), taking the ratio of expressions (5.25) and multiplying by V and expanding to first order in $a/(k_B T) - b$, we obtain $PV/\overline{N} = k_B T[1 + (\overline{N}/V)(b - a/(k_B T))]$—accurate to first order in $(b - a/(k_B T))$. Bringing the a term to the left-hand side, multiplying both sides by $[1 - (\overline{N}/V)b]$, and linearizing in b, we obtain the standard *van der Waals equation of state*:

$$\boxed{\left(P + \frac{a}{(V/\overline{N})^2} \right) (V/\overline{N} - b) = k_B T.}$$

(5.26)

The quantity $V/\overline{N}$ is the *specific volume* (volume per particle).

A few comments are in order.

1. The factor $(V/\overline{N} - b)$ in the equation of state corresponds to an excluded volume $b = (16\pi/3)(r_o/2)^3$ that is four times larger than the actual volume of each hard-sphere particle (whose radius is $r_o/2$).

2. The term linear in a, $P = -a\overline{N}/V = -b\varepsilon_o \overline{N}/V$, is a pressure reduction due to the attractive force between particles.

BOX 5.3. DERIVATION OF VAN DER WAALS GRAND POTENTIAL T2

In Eq. (5.23a) the momentum integrals and the space integrals separate, and the N momentum integrals are identical, so Z takes the form

$$Z = \sum_{N=0}^{\infty} \frac{e^{\mu N/(k_B T)}}{N! h^{3N}} \left[\int_0^{\infty} 4\pi p^2 dp \exp\left(\frac{-p^2}{2mk_B T}\right) \right]^N J_N \quad (1)$$

$$= \sum_{N=0}^{\infty} \frac{(\zeta/\lambda_{T\,\text{dB}}^3)^N}{N!} J_N,$$

where

$$\zeta \equiv e^{\mu/(k_B T)}, \quad \lambda_{T\,\text{dB}} \equiv \frac{h}{(2\pi m k_B T)^{1/2}},$$

$$J_N = \int d^{3N}x \exp\left[-\sum_{i=1}^{N} \sum_{j=i+1}^{N} \frac{u_{ij}}{k_B T}\right]. \quad (2)$$

Note that λ is the particles' thermal deBroglie wavelength. The Boltzmann factor $e^{-u_{ij}/(k_B T)}$ is unity for large interparticle separations $r_{ij} \gg r_o$, so we write

$$e^{-u_{ij}/(k_B T)} \equiv 1 + f_{ij}, \quad (3)$$

where f_{ij} is zero except when $r_{ij} \lesssim r_o$. Using this definition and rewriting the exponential of a sum as the products of exponentials, we bring J_N into the form

$$J_N = \int d^{3N}x \prod_{i=1}^{N} \prod_{j=i+1}^{N} (1 + f_{ij}). \quad (4)$$

The product contains (i) terms linear in f_{ij} that represent the influence of pairs of particles that are close enough ($r_{ij} \lesssim r_o$) to interact, plus (ii) quadratic terms, such as $f_{14} f_{27}$, that are nonzero only if particles 1 and 4 are near each other and 2 and 7 are near each other (there are so many of these terms that we cannot neglect them!), plus (iii) quadratic terms such as $f_{14} f_{47}$ that are nonzero only if particles 1, 4, and 7 are all within a distance $\sim r_o$ of one another (because our gas is dilute, it turns out these three-particle terms can be neglected), plus (iv) cubic and higher-order terms. At all orders ℓ (linear, quadratic, cubic, quartic, etc.) for our dilute gas, we can ignore terms that require three or more particles to be near one another, so we focus only on terms $f_{ij} f_{mn} \ldots f_{pq}$ where all indices are different. Eq. (4) then becomes

(continued)

BOX 5.3. (continued)

$$J_N = \int d^{3N}x [1 + \underbrace{(f_{12} + f_{13} + \cdots)}_{n_1 \text{ terms}} + \underbrace{(f_{12}f_{34} + f_{13}f_{24} + \cdots)}_{n_2 \text{ terms}}$$
$$+ \underbrace{(f_{12}f_{34}f_{56} + f_{13}f_{24}f_{56} + \cdots)}_{n_3 \text{ terms}} \cdots], \qquad (5)$$

where n_ℓ is the number of terms of order ℓ with all 2ℓ particles different. Denoting

$$V_o \equiv \int f(r) 4\pi r^2 dr, \qquad (6)$$

and performing the integrals, we bring Eq. (5) into the form

$$J_N = \sum_{\ell=0}^{\infty} n_\ell V^{N-\ell} V_o^\ell. \qquad (7)$$

At order ℓ the number of unordered sets of 2ℓ particles that are all different is $N(N-1)\cdots(N-2\ell+1)/\ell!$. The number of ways that these 2ℓ particles can be assembled into unordered pairs is $(2\ell-1)(2\ell-3)(2\ell-5)\cdots 1 \equiv (2\ell-1)!!$. Therefore, the number of terms of order ℓ that appear in Eq. (7) is

$$n_\ell = \frac{N(N-1)\cdots(N-2\ell+1)}{\ell!}(2\ell-1)!!$$
$$= \frac{N(N-1)\cdots(N-2\ell+1)}{2^\ell \ell!}. \qquad (8)$$

Inserting Eqs. (7) and (8) into Eq. (1) for the partition function, we obtain

$$Z = \sum_{N=0}^{\infty} \frac{(\zeta/\lambda^3)^N}{N!} \sum_{\ell=0}^{[N/2]} \frac{N(N-1)\cdots(N-2\ell+1)}{2^\ell \ell!} V^{N-\ell} V_o^\ell, \qquad (9)$$

where $[N/2]$ means the largest integer less than or equal to $N/2$. Performing a little algebra and then reversing the order of the summations, we obtain

$$Z = \sum_{\ell=0}^{\infty} \sum_{N=2\ell}^{\infty} \frac{1}{(N-2\ell)!} \left(\frac{\zeta V}{\lambda^3}\right)^{N-2\ell} \frac{1}{\ell!} \left(\frac{\zeta V}{\lambda^3} \frac{\zeta V_o}{2\lambda^3}\right)^\ell. \qquad (10)$$

By changing the summation index from N to $N' = N - 2\ell$, we decouple the two summations. Each of the sums is equal to an exponential, giving

(continued)

BOX 5.3. (continued)

$$Z = e^{-\Omega/(k_B T)} = \exp\left(\frac{\zeta V}{\lambda^3}\right) \exp\left(\frac{\zeta V}{\lambda^3} \frac{\zeta V_o}{2\lambda^3}\right)$$

$$= \exp\left[\frac{\zeta V}{\lambda^3}\left(1 + \frac{\zeta V_o}{2\lambda^3}\right)\right]. \tag{11}$$

Therefore, the grand potential for our van der Waals gas is

$$\Omega = \frac{-k_B T \zeta V}{\lambda^3}\left(1 + \frac{\zeta V_o}{2\lambda^3}\right). \tag{12}$$

From kinetic theory [Eq. (3.39a)], we know that for an ideal gas, the mean number density is $\overline{N}/V = \zeta/\lambda^3$; this is also a good first approximation for our van der Waals gas, which differs from an ideal gas only by the weakly perturbative interaction energy $u(r)$. Thus $\zeta V_o/(2\lambda^3)$ is equal to $\frac{1}{2} V_o/$(mean volume per particle), which is $\ll 1$ by our dilute-gas assumption. If we had kept three-particle interaction terms, such as $f_{14}f_{47}$, they would have given rise to fractional corrections of order $(\zeta V_o/\lambda^3)^2$, which are much smaller than the leading-order fractional correction $\zeta V_o/(2\lambda^3)$ that we have computed [Eq. (12)]. The higher-order corrections are derived in statistical mechanics textbooks, such as Pathria and Beale (2011, Chap. 10) and Kardar (2007, Chap. 5) using a technique called the "cluster expansion."

For the hard-wall potential (5.22b), f is -1 at $r < r_o$, and assuming that the temperature is high enough that $\varepsilon_o/(k_B T) \ll 1$, then at $r > r_o$, f is very nearly $-u/(k_B T) = (\varepsilon_o/(k_B T))(r_o/r)^6$; therefore we have

$$\frac{V_o}{2} \equiv \frac{1}{2}\int f(r) 4\pi r^2 dr = \frac{a}{k_B T} - b, \quad \text{where } b = \frac{2\pi r_o^3}{3}, \quad a = b\varepsilon_o. \tag{13}$$

Inserting this expression for $V_o/2$ and Eqs. (2) for ζ and λ into Eq. (12), we obtain Eqs. (5.24) for the grand potential of a van der Waals gas.

3. Our derivation is actually only accurate to first order in a and b, so it does not justify the quadratic term $P = ab(\overline{N}/V)^2$ in the equation of state (5.26). However, that quadratic term does correspond to the behavior of real gases: a sharp rise in pressure at high densities due to the short-distance repulsion between particles.

We study this van der Waals equation of state in Sec. 5.7, focusing on the gas-to-liquid phase transition that it predicts and on fluctuations of thermodynamic quantities associated with that phase transition.

In this section we have presented the grand canonical analysis for a van der Waals gas not because such a gas is important (though it is), but rather as a concrete example of how one uses the formalism of statistical mechanics and introduces ingenious approximations to explore the behavior of realistic systems made of interacting particles.

EXERCISES

Exercise 5.3 *Derivation and Example: Grand Canonical Ensemble for a Classical, Relativistic, Perfect Gas*

Consider cells that reside in a heat and particle bath of a classical, relativistic, perfect gas (Fig. 5.1). Each cell has the same volume V and imaginary walls. Assume that the bath's temperature T has an arbitrary magnitude relative to the rest mass-energy mc^2 of the particles (so the thermalized particles might have relativistic velocities), but require $k_B T \ll -\mu$ (so all the particles behave classically). Ignore the particles' spin degrees of freedom, if any. For ease of notation use geometrized units (Sec. 1.10) with $c = 1$.

(a) The number of particles in a chosen cell can be anything from $N = 0$ to $N = \infty$. Restrict attention, for the moment, to a situation in which the cell contains a precise number of particles, N. Explain why the multiplicity is $\mathcal{M} = N!$ even though the density is so low that the particles' wave functions do not overlap, and they are behaving classically (cf. Ex. 4.8).

(b) Still holding fixed the number of particles in the cell, show that the number of degrees of freedom W, the number density of states in phase space $\mathcal{N}_{\text{states}}$, and the energy $\mathcal{E}_N$ in the cell are

$$W = 3N, \quad \mathcal{N}_{\text{states}} = \frac{1}{N! h^{3N}}, \quad \mathcal{E}_N = \sum_{j=1}^{N} (\mathbf{p}_j^2 + m^2)^{\frac{1}{2}}, \quad (5.27a)$$

where $\mathbf{p}_j$ is the momentum of classical particle number j.

(c) Using Eq. (4.8b) to translate from the formal sum over states $\sum_n$ to a sum over $W = 3N$ and an integral over phase space, show that the sum over states (5.14) for the grand partition function becomes

$$Z = e^{-\Omega/(k_B T)} = \sum_{N=0}^{\infty} \frac{V^N}{N! h^{3N}} e^{\tilde{\mu} N/(k_B T)} \left[\int_0^{\infty} \exp\left(-\frac{(p^2 + m^2)^{\frac{1}{2}}}{k_B T}\right) 4\pi p^2 dp \right]^N.$$

(5.27b)

(d) Evaluate the momentum integral in the nonrelativistic limit $k_B T \ll m$, and thereby show that

$$\Omega(T, \mu, V) = -k_B T V \frac{(2\pi m k_B T)^{3/2}}{h^3} e^{\mu/(k_B T)}, \quad (5.28a)$$

where $\mu = \tilde{\mu} - m$ is the nonrelativistic chemical potential. This is the interaction-free limit $V_o = a = b = 0$ of our grand potential (5.24a) for a van der Waals gas.

(e) Show that in the extreme relativistic limit $k_B T \gg m$, Eq. (5.27b) gives

$$\Omega(T, \tilde{\mu}, V) = -\frac{8\pi V (k_B T)^4}{h^3} e^{\tilde{\mu}/(k_B T)}. \tag{5.29}$$

(f) For the extreme relativistic limit use your result (5.29) for the grand potential $\Omega(V, T, \tilde{\mu})$ to derive the mean number of particles $\overline{N}$, the pressure P, the entropy S, and the mean energy $\overline{\mathcal{E}}$ as functions of V, $\tilde{\mu}$, and T. Note that for a photon gas, because of the spin degree of freedom, the correct values of $\overline{N}$, $\overline{\mathcal{E}}$, and S will be twice as large as what you obtain in this calculation. Show that the energy density is $\overline{\mathcal{E}}/V = 3P$ (a relation valid for any ultrarelativistic gas); and that $\overline{\mathcal{E}}/\overline{N} = 3k_B T$ (which is higher than the $2.7011780\ldots k_B T$ for blackbody radiation, as derived in Ex. 3.13, because in the classical regime of $\eta \ll 1$, photons don't cluster in the same states at low frequency; that clustering lowers the mean photon energy for blackbody radiation).

5.4 Canonical Ensemble and the Physical-Free-Energy Representation of Thermodynamics

In this section, we turn to an ensemble of single-species systems that can exchange energy but nothing else with a heat bath at temperature T. The systems thus have variable total energy $\mathcal{E}$, but they all have the same, fixed values of the two remaining extensive variables N and V. We presume that the ensemble has reached statistical equilibrium, so it is canonical with a distribution function (probability of occupying any quantum state of energy $\mathcal{E}$) given by Eq. (4.20):

$$\boxed{\rho_n = \frac{1}{z} e^{-\mathcal{E}_n/(k_B T)} \equiv e^{(F-\mathcal{E}_n)/(k_B T)}.} \tag{5.30}$$

canonical distribution function

Here, as in the grand canonical ensemble [Eq. (5.13)], we have introduced special notations for the normalization constant: $1/z = e^{F/(k_B T)}$, where z (the *partition function*) and F (the *physical free energy* or *Helmholtz free energy*) are functions of the systems' fixed N and V and the bath temperature T. Once the systems' quantum states $|n\rangle$ (with fixed N and V but variable $\mathcal{E}$) have been identified, the functions $z(T, N, V)$ and $F(T, N, V)$ can be computed from the normalization relation $\sum_n \rho_n = 1$:

$$\boxed{e^{-F/(k_B T)} \equiv z(T, N, V) = \sum_n e^{-\mathcal{E}_n/(k_B T)}.} \tag{5.31}$$

partition function and physical free energy

This canonical sum over states, like the grand canonical sum (5.14) that we used for the van der Waals gas, is a powerful tool in statistical mechanics. As an example, in Secs. 5.8.2 and 5.8.3 we use the canonical sum to evaluate the physical free energy F for a model of ferromagnetism and use the resulting F to explore a ferromagnetic phase transition.

Having evaluated $z(T, N, V)$ [or equivalently, $F(T, N, V)$], one can then proceed as follows to determine other thermodynamic properties of the ensemble's systems. The entropy S can be computed from the standard expression $S = -k_B \sum_n \rho_n \ln \rho_n = -k_B \overline{\ln \rho}$, which, with Eq. (5.30) for ρ_n, implies $S = (\overline{\mathcal{E}} - F)/T$. It is helpful to rewrite this as an equation for the physical free energy F:

Legendre transformation

$$\boxed{F = \overline{\mathcal{E}} - TS.} \tag{5.32}$$

This is the Legendre transformation that leads from the energy representation of thermodynamics to the physical-free-energy representation.

Suppose that the canonical ensemble's parameters T, N, and V are changed slightly. By how much will the physical free energy change? Equation (5.32) tells us that $dF = d\overline{\mathcal{E}} - TdS - SdT$. Because macroscopic thermodynamics is independent of the statistical ensemble being studied, we can evaluate $d\overline{\mathcal{E}}$ using the first law of thermodynamics (5.7) with the microcanonical exact energy $\mathcal{E}$ replaced by the canonical mean energy $\overline{\mathcal{E}}$. The result is

first law in physical-free-energy representation

$$\boxed{dF = -SdT + \tilde{\mu}dN - PdV.} \tag{5.33}$$

Equation (5.33) contains the same information as the first law of thermodynamics and can be thought of as the first law rewritten in the physical-free-energy representation. From this form of the first law, we can deduce the other equations of the physical-free-energy representation by the same procedure we used for the energy representation in Sec. 5.2.5 and the grand-potential representation in Sec. 5.3.1.

If we have forgotten our representation's independent variables, we read them off the first law (5.33); they appear as differentials on the right-hand side: T, N, and V. The fundamental potential is the quantity that appears on the left-hand side of the first law: $F(T, N, V)$. By building up a full system from small subsystems that all have the same intensive variables T, $\tilde{\mu}$, and P, we deduce from the first law the Euler relation for this representation:

Euler relation for physical free energy

$$\boxed{F = \tilde{\mu}N - PV.} \tag{5.34}$$

From the first law (5.33) we read off equations of state for this representation's generalized forces [e.g., $-S = (\partial F/\partial T)_{N,V}$]. Maxwell relations can be derived from the equality of mixed partial derivatives.

Thus, as for the energy and grand-potential representations, all the equations of the physical-free-energy representation are easily deducible from the minimal information in Table 5.1.

And as for those representations, *the Newtonian version of this representation's fundamental equations (5.30)–(5.34) is obtained by simply removing rest masses from $\tilde{\mu}$ (which becomes μ), $\mathcal{E}$ (which becomes E), and F (whose notation does not change).*

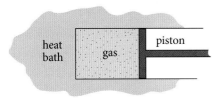

FIGURE 5.2 Origin of the name "physical free energy" for $F(V, T, N)$.

5.4.1 Experimental Meaning of Physical Free Energy

The name "physical free energy" for F can be understood using the idealized experiment shown in Fig. 5.2. Gas is placed in a chamber, one wall of which is a piston, and the chamber comes into thermal equilibrium with a heat bath, with which it can exchange heat but not particles. The volume of the chamber has some initial value V_i; and correspondingly, the gas has some initial physical free energy $F(V_i, T, N)$. The gas is then allowed to push the piston to the right sufficiently slowly for the gas to remain always in thermal equilibrium with the heat bath, at the bath's temperature T. When the chamber has reached its final volume V_f, the total work done on the piston by the gas (i.e., the total energy extracted by the piston from this "engine") is

$$\mathcal{E}_{\text{extracted}} = \int_{V_i}^{V_f} -P dV. \tag{5.35a}$$

Using the first law $dF = -SdT + \tilde{\mu} dN - PdV$ and remembering that T and N are kept constant, Eq. (5.35a) becomes

$$\mathcal{E}_{\text{extracted}} = F(V_f, T, N) - F(V_i, T, N) \equiv \Delta F. \tag{5.35b}$$

physical free energy extracted in isothermal expansion

Thus, *F is the energy that is free to be extracted in an isothermal, physical expansion of the gas*.[7]

If the expansion had been done in a chamber that was perfectly thermally insulated, so no heat could flow in or out of it, then there would have been no entropy change. Correspondingly, with S and N held fixed but V changing during the expansion, the natural way to analyze the expansion would have been in the energy representation; that representation's first law $d\mathcal{E} = -PdV + TdS + \tilde{\mu}dN$ would have told us that the total energy extracted, $\int -PdV$, was the change $\Delta \mathcal{E}$ of the gas's total energy. Such a process, which occurs without any heat flow or entropy increase, is called *adiabatic*. Thus, *the energy $\mathcal{E}$ (or in the nonrelativistic regime, E) measures the amount of energy that can be extracted from an adiabatic engine*, by contrast with F, which measures the energy extracted from an isothermal engine.

---

7. More generally, the phrase "free energy" means the energy that can be extracted in a process that occurs in contact with some sort of environment. The nature of the free energy depends on the nature of the contact. We will meet chemical free energy in Sec. 5.5, and the free energy of a body on which a steady force is acting in Sec. 11.6.1.

5.4.2 Ideal Gas with Internal Degrees of Freedom

As an example of the canonical distribution, we explore the influence of internal molecular degrees of freedom on the properties of a nonrelativistic, ideal gas.[8] This example is complementary to the van der Waals gas that we analyzed in Sec. 5.3.2 using the grand canonical distribution. There we assumed no internal degrees of freedom, but we allowed each pair of particles to interact via an interaction potential $u(r)$ that depended on the particles' separation r. Here, because the gas is ideal, there are no interactions, but we allow for internal degrees of freedom—rotational, vibrational, and electron excitations.

(We have previously studied internal degrees of freedom in Sec. 4.4.4, where we proved the equipartition theorem for those whose generalized coordinates and/or momenta are quadratic in the hamiltonian and are classically excited, such as the vibrations and rotations of a diatomic molecule. Here we allow the internal degrees of freedom to have any form whatsoever and to be arbitrarily excited or nonexcited.)

Our gas is confined to a fixed volume V, it has a fixed number of molecules N, it is in contact with a heat bath with temperature T, and its equilibrium distribution is therefore canonical, $\rho_n = e^{(F-E_n)/(k_B T)}$. The quantum states $|n\rangle$ available to the gas can be characterized by the locations $\{\mathbf{x}_i, \mathbf{p}_i\}$ in phase space of each of the molecules $i = 1, \ldots, N$, and by the state $|K_i\rangle$ of each molecule's internal degrees of freedom. Correspondingly, the partition function and physical free energy are given by [Eq. (5.31)]

$$z = e^{-F/(k_B T)} = \frac{g_s^N}{N!} \int \frac{d^{3N}x \, d^{3N}p}{h^{3N}} \sum_{K_1, K_2, \ldots, K_N} \exp\left[-\sum_{i=1}^{N}\left(\frac{\mathbf{p}_i^2}{2m k_B T} + \frac{E_{K_i}}{k_B T}\right)\right]. \tag{5.36a}$$

It is instructive to compare this with Eq. (5.23a) for the grand partition function of the van der Waals gas. In Eq. (5.36a) there is no interaction energy u_{ij} between molecules, no sum over N, and no $e^{\mu N/(k_B T)}$ (because N is fixed and there is no particle bath). However, we now have sums over the internal states K_i of each of the molecules and a factor g_s in the multiplicity to allow for the molecules' g_s different spin states [cf. Eq. (4.8c)].

Because there are no interactions between molecules, the partition function can be split up into products of independent contributions from each of the molecules. Because there are no interactions between a molecule's internal and translational degrees of freedom, the partition function can be split into a product of translational and internal terms; and because the molecules are all identical, their contributions are all identical, leading to

$$z = e^{-F/(k_B T)} = \frac{1}{N!}\left[g_s \int \frac{d^3x \, d^3p}{h^3} e^{-\mathbf{p}^2/(k_B T)}\right]^N \left[\sum_K e^{-E_K/(k_B T)}\right]^N. \tag{5.36b}$$

---

8. See footnote 5 of this chapter on p. 226 for the meaning of "ideal gas."

The $\int d^3x d^3p \, h^{-3} e^{-\mathbf{p}^2/(k_B T)}$ integral is the same as we encountered in the grand canonical analysis; it gives V/λ^3, where $\lambda = h/(2\pi m k_B T)^{1/2}$ [cf. Eq. (1) in Box 5.3]. The sum over internal states gives a contribution that is some function of temperature:

contribution of internal states to physical free energy of an ideal gas

$$f(T) \equiv \sum_K e^{-E_K/(k_B T)}. \qquad (5.37)$$

Correspondingly [using Stirling's approximation, $N! \simeq (N/e)^N$, to the accuracy needed here], the physical free energy becomes

$$F(N, V, T) = N k_B T \ln\left[\frac{N}{e} \frac{h^3/g_s}{(2\pi m k_B T)^{3/2} V}\right] - N k_B T \ln[f(T)]. \qquad (5.38)$$

Note that because the molecules' translational and internal degrees of freedom are decoupled, their contributions to the free energy are additive. We could have computed them separately and then simply added their free energies.

Because the contribution of the internal degrees of freedom depends only on temperature and not on volume, the ideal gas's pressure

$$P = -(\partial F/\partial V)_{N,T} = (N/V) k_B T \qquad (5.39)$$

is unaffected by the internal degrees of freedom. By contrast, the entropy and the total energy in the box do have internal contributions, which depend on temperature but not on the gas's volume and thence not on its density N/V:

$$S = -(\partial F/\partial T)_{N,V} = S_{\text{translational}} + N k_B (\ln f + d \ln f/d \ln T), \qquad (5.40)$$

where the entropy $S_{\text{translational}}$ can straightforwardly be shown to be equivalent to the Sackur-Tetrode formula (4.42) and

$$\bar{E} = F + TS = N k_B T \left(\frac{3}{2} + \frac{d \ln f}{d \ln T}\right). \qquad (5.41)$$

For degrees of freedom that are classical and quadratic, the internal contribution $N k_B T d \ln f/d \ln T$ gives $\frac{1}{2} k_B T$ for each quadratic term in the hamiltonian, in accord with the equipartition theorem (Sec. 4.4.4).

If there is more than one particle species present (e.g., electrons and protons at high temperatures, so hydrogen is ionized), then the contributions of the species to F, P, S, and E simply add, just as the contributions of internal and translational degrees of freedom added in Eq. (5.38).

EXERCISES

Exercise 5.4 *Example and Derivation: Adiabatic Index for Ideal Gas*
In Part V, when studying fluid dynamics, we shall encounter an *adiabatic index*

$$\Gamma \equiv -\left(\frac{\partial \ln P}{\partial \ln V}\right)_S \qquad (5.42)$$

[Eq. (13.2)] that describes how the pressure P of a fluid changes when it is compressed adiabatically (i.e., compressed at fixed entropy, with no heat being added or removed).

Derive an expression for Γ for an ideal gas that may have internal degrees of freedom (e.g., Earth's atmosphere). More specifically, do the following.

(a) Consider a fluid element (a small sample of the fluid) that contains N molecules. These molecules can be of various species; all species contribute equally to the ideal gas's pressure $P = (N/V)k_B T$ and contribute additively to its energy. Define the fluid element's specific heat at fixed volume to be the amount of heat TdS that must be inserted to raise its temperature by an amount dT while the volume V is held fixed:

$$C_V \equiv T(\partial S/\partial T)_{V,N} = (\partial E/\partial T)_{V,N}. \tag{5.43}$$

Deduce the second equality from the first law of thermodynamics. Show that in an adiabatic expansion the temperature T drops at a rate given by $C_V dT = -PdV$. [Hint: Use the first law of thermodynamics and the fact that for an ideal gas the energy of a fluid element depends only on its temperature and not on its volume (or density); Eq. (5.41).]

(b) Combine the temperature change $dT = (-P/C_V)dV$ for an adiabatic expansion with the equation of state $PV = Nk_B T$ to obtain $\Gamma = (C_V + Nk_B)/C_V$.

(c) To interpret the numerator $C_V + Nk_B$, imagine adding heat to a fluid element while holding its pressure fixed (which requires a simultaneous volume change). Show that in this case the ratio of heat added to temperature change is

$$C_P \equiv T(\partial S/\partial T)_{P,N} = C_V + Nk_B. \tag{5.44}$$

Combining with part (b), conclude that the adiabatic index for an ideal gas is given by

$$\Gamma = \gamma \equiv C_P/C_V, \tag{5.45}$$

a standard result in elementary thermodynamics.

Exercise 5.5 *Example: The Enthalpy Representation of Thermodynamics* **T2**

(a) Enthalpy H is a macroscopic thermodynamic variable defined by

$$\boxed{H \equiv \mathcal{E} + PV.} \tag{5.46}$$

Show that this definition can be regarded as a Legendre transformation that converts from the energy representation of thermodynamics with $\mathcal{E}(V, S, N)$ as the fundamental potential, to an *enthalpy representation* with $H(P, S, N)$ as the fundamental potential. More specifically, show that the first law, reexpressed in terms of H, takes the form

$$\boxed{dH = VdP + TdS + \tilde{\mu}dN;} \tag{5.47}$$

and then explain why this first law dictates that $H(P, S, N)$ be taken as the fundamental potential.

(b) For a nonrelativistic system, it is conventional to remove the particle rest masses from the enthalpy just as one does from the energy, but by contrast with energy, we do not change notation for the enthalpy:

$$H_{\text{nonrelativistic}} \equiv H_{\text{relativistic}} - Nm = E + PV. \tag{5.48}$$

What is the form of the first law (5.47) for the nonrelativistic H?

(c) There is an equilibrium statistical mechanics ensemble associated with the enthalpy representation. Show that each system of this ensemble (fluctuationally) exchanges volume and energy with a surrounding *pressure bath* but does not exchange heat or particles, so the exchanged energy is solely that associated with the exchanged volume, $d\mathcal{E} = -PdV$, and the enthalpy H does not fluctuate. Note that P is the common pressure of the bath and the system.

(d) Show that this ensemble's distribution function is $\rho = e^{-S/k_B} =$ constant for those states in phase space that have a specified number of particles N and a specified enthalpy H. Why do we not need to allow for a small range δH of H, by analogy with the small range $\mathcal{E}$ for the microcanonical ensemble (Sec. 4.5 and Ex. 4.7)?

(e) What equations of state can be read off from the enthalpy first law? What are the Maxwell relations between these equations of state?

(f) What is the Euler equation for H in terms of a sum of products of extensive and intensive variables?

(g) Show that the system's enthalpy is equal to its total (relativistic) inertial mass (multiplied by the speed of light squared); cf. Exs. 2.26 and 2.27.

(h) As another interpretation of the enthalpy, think of the system as enclosed in an impermeable box of volume V. Inject into the box a "sample" of additional material of the same sort as is already there. (It may be helpful to think of the material as a gas.) The sample is to be put into the same thermodynamic state (i.e., macrostate) as that of the box's material (i.e., it is to be given the same values of temperature T, pressure P, and chemical potential $\tilde{\mu}$). Thus, the sample's material is indistinguishable in its thermodynamic properties from the material already in the box, except that its extensive variables (denoted by Δ) are far smaller: $\Delta V/V = \Delta \mathcal{E}/\mathcal{E} = \Delta S/S \ll 1$. Perform the injection by opening up a hole in one of the box's walls, pushing aside the box's material to make a little cavity of volume ΔV equal to that of the sample, inserting the sample into the cavity, and then closing the hole in the wall. The box now has the same volume V as before, but its energy has changed. Show that the energy change (i.e., the energy required to create the sample and perform the injection) is equal to the enthalpy ΔH of the sample. Thus, *enthalpy has the physical interpretation of energy of injection at fixed volume V*. Equivalently, if a sample of material is ejected from the system, the total energy that will come out (including the work done on the sample by the system during the ejection) is the sample's enthalpy ΔH. From this viewpoint, enthalpy is the system's free energy.

5.5 Gibbs Ensemble and Representation of Thermodynamics; Phase Transitions and Chemical Reactions

Next consider systems in which the temperature T and pressure P are both being controlled by an external environment (bath) and thus are treated as independent variables in the fundamental potential. This is the situation in most laboratory experiments and geophysical situations and is common in elementary chemistry but not chemical engineering.

In this case each of the systems has a fixed number of particles N_I for the various independent species I, and it can exchange heat and volume with its surroundings. (We explicitly allow for more than one particle species, because a major application of the Gibbs representation will be to chemical reactions.) There might be a membrane separating each system from its bath—a membrane impermeable to particles but through which heat can pass, and with negligible surface tension so the system and the bath can buffet each other freely, causing fluctuations in the system's volume. For example, this is the case for a so-called "constant-pressure balloon" of the type used to lift scientific payloads into the upper atmosphere. Usually, however, there is no membrane between system and bath. Instead, gravity might hold the system together because it has higher density than the bath (e.g., a liquid in a container); or solid-state forces might hold the system together (e.g., a crystal); or we might just introduce a conceptual, imaginary boundary around the system of interest—one that comoves with some set of particles.

The equilibrium ensemble for this type of system is that of Gibbs, with distribution function

$$\rho_n = e^{G/(k_B T)} e^{-(\mathcal{E}_n + PV_n)/(k_B T)} \qquad (5.49)$$

[Eq. (4.25b), to which we have added the normalization constant $e^{G/(k_B T)}$]. As for the canonical and grand canonical distributions, the quantity G in the normalization constant becomes the fundamental potential for the Gibbs representation of thermodynamics. It is called the *Gibbs potential,* and also, sometimes, the *Gibbs free energy* or *chemical free energy* (see Ex. 5.6). It is a function of the systems' fixed numbers of particles N_I and of the bath's temperature T and pressure P, which appear in the Gibbs distribution function: $G = G(N_I, T, P)$.

The Gibbs potential can be evaluated by a sum over quantum states that follows from $\sum_n \rho_n = 1$:

Gibbs potential as sum over states

$$e^{-G/(k_B T)} = \sum_n e^{-(\mathcal{E}_n + PV_n)/(k_B T)}. \qquad (5.50)$$

See Ex. 5.7 for an example. This sum has proved to be less useful than the canonical and grand canonical sums, so in most statistical mechanics textbooks there is little or no discussion of the Gibbs ensemble. By contrast, the Gibbs representation of thermodynamics is extremely useful, as we shall see, so textbooks pay a lot of attention to it.

We can deduce the equations of the Gibbs representation by the same method as we used for the canonical and grand canonical representations. We begin by writing down a Legendre transformation that takes us from the energy representation to the Gibbs representation. As for the canonical and grand canonical cases, that Legendre transformation can be inferred from the equilibrium ensemble's entropy, $S = -k_B \overline{\ln \rho} = -(G - \overline{\mathcal{E}} - P\overline{V})/T$ [cf. Eq. (5.49) for ρ]. Solving for G, we get

$$G = \overline{\mathcal{E}} + P\overline{V} - TS. \qquad (5.51)$$

Legendre transformation

Once we are in the thermodynamic domain (as opposed to statistical mechanics), we can abandon the distinction between expectation values of quantities and fixed values; that is, we can remove the bars and write this Legendre transformation as $G = \mathcal{E} + PV - TS$.

Differentiating this Legendre transformation and combining with the energy representation's first law (5.8), we obtain the first law in the Gibbs representation:

$$dG = VdP - SdT + \sum_I \tilde{\mu}_I dN_I. \qquad (5.52)$$

first law in Gibbs representation

From this first law we read out the independent variables of the Gibbs representation, namely $\{P, T, N_I\}$ (in case we have forgotten them!), and the values of its generalized forces [equations of state; e.g., $V = (\partial G/\partial P)_{T, N_I}$]. From the equality of mixed partial derivatives, we read off the Maxwell relations. By imagining building up a large system from many tiny subsystems (all with the same, fixed, intensive variables P, T, and $\tilde{\mu}_I$) and applying the first law (5.52) to this buildup, we obtain the Euler relation

$$G = \sum_I \tilde{\mu}_I N_I. \qquad (5.53)$$

Euler relation for Gibbs potential

This Euler relation will be very useful in Sec. 5.5.3, when we discuss chemical reactions.

As with previous representations of thermodynamics, to obtain the Newtonian version of all of this section's equations, we simply remove the particle rest masses from $\tilde{\mu}_I$ (which then becomes μ_I), from $\mathcal{E}$ (which then becomes E), and from G (which does not change notation).

EXERCISES

Exercise 5.6 *Problem: Gibbs Potential Interpreted as Chemical Free Energy*
In Sec. 5.4.1, we explained the experimental meaning of the free energy F for a system

in contact with a heat bath so its temperature is held constant, and in Ex. 5.5h we did the same for contact with a pressure bath. By combining these, give an experimental interpretation of the Gibbs potential G as the free energy for a system in contact with a heat and pressure bath—the "chemical free energy."

Exercise 5.7 *Problem and Practice: Ideal Gas Equation of State from Gibbs Ensemble*
For a nonrelativistic, classical, ideal gas (no interactions between particles), evaluate the statistical sum (5.50) to obtain $G(P, T, N)$, and from it deduce the standard formula for the ideal-gas equation of state $P\bar{V} = Nk_B T$.

5.5.1 Out-of-Equilibrium Ensembles and Their Fundamental Thermodynamic Potentials and Minimum Principles

Despite its lack of usefulness in practical computations of the Gibbs potential G, the Gibbs ensemble plays an important conceptual role in a *minimum principle for G*, which we now derive.

Consider an ensemble of systems, each of which is immersed in an identical heat and volume bath, and assume that the ensemble begins with some arbitrary distribution function ρ_n, one that is not in equilibrium with the baths. As time passes, each system will interact with its bath and will evolve in response to that interaction. Correspondingly, the ensemble's distribution function ρ will evolve. At any moment of time the ensemble's systems will have some mean (ensemble-averaged) energy $\bar{\mathcal{E}} \equiv \sum_n \rho_n \mathcal{E}_n$ and volume $\bar{V} \equiv \sum_n \rho_n V_n$, and the ensemble will have some entropy $S = -k_B \sum_n \rho_n \ln \rho_n$. From these quantities (which are well defined even though the ensemble may be very far from statistical equilibrium), we can compute a Gibbs potential G for the ensemble. This out-of-equilibrium G is defined by the analog of the equilibrium definition (5.51),

out-of-equilibrium Gibbs potential

$$G \equiv \bar{\mathcal{E}} + P_b \bar{V} - T_b S, \qquad (5.54)$$

where P_b and T_b are the pressure and temperature of the identical baths with which the ensemble's systems are interacting.[9] As the evolution proceeds, the total entropy of the baths' ensemble plus the systems' ensemble will continually increase, until equilibrium is reached. Suppose that during a short stretch of evolution the systems' mean energy changes by $\Delta \bar{\mathcal{E}}$, their mean volume changes by $\Delta \bar{V}$, and the entropy of the ensemble

---

9. Notice that, because the number N of particles in the system is fixed (as is the bath temperature T_b), the evolving Gibbs potential is proportional to

$$\frac{G}{Nk_B T_b} = \frac{\bar{E}}{Nk_B T_b} + \frac{P_b \bar{V}}{Nk_B T_b} - \frac{S}{Nk_B}.$$

This quantity is dimensionless and generally of order unity. Note that the last term is the dimensionless entropy per particle [Eq. (4.44) and associated discussion].

changes by ΔS. Then, by conservation of energy and volume, the baths' mean energy and volume must change by

$$\Delta \overline{\mathcal{E}}_b = -\Delta \overline{\mathcal{E}}, \qquad \Delta \bar{V}_b = -\Delta \bar{V}. \tag{5.55a}$$

Because the baths (by contrast with the systems) are in statistical equilibrium, we can apply to them the first law of thermodynamics for equilibrated systems:

$$\Delta \overline{\mathcal{E}}_b = -P_b \Delta \bar{V}_b + T_b \Delta S_b + \sum_I \tilde{\mu}_{Ib} \Delta N_{Ib}. \tag{5.55b}$$

Since the N_{Ib} are not changing (the systems cannot exchange particles with their baths) and since the changes of bath energy and volume are given by Eqs. (5.55a), Eq. (5.55b) tells us that the baths' entropy changes by

$$\Delta S_b = \frac{-\Delta \overline{\mathcal{E}} - P_b \Delta \bar{V}}{T_b}. \tag{5.55c}$$

Correspondingly, the sum of the baths' entropy and the systems' entropy changes by the following amount, which cannot be negative:

$$\Delta S_b + \Delta S = \frac{-\Delta \overline{\mathcal{E}} - P_b \Delta \bar{V} + T_b \Delta S}{T_b} \geq 0. \tag{5.55d}$$

Because the baths' pressure P_b and temperature T_b are not changing (the systems are so tiny compared to the baths that the energy and volume they exchange with the baths cannot have any significant effect on the baths' intensive variables), the numerator of expression (5.55d) is equal to the evolutionary change in the ensemble's out-of-equilibrium Gibbs potential (5.54):

$$\boxed{\Delta S_b + \Delta S = \frac{-\Delta G}{T_b} \geq 0.} \tag{5.56}$$

Thus, the second law of thermodynamics for an ensemble of arbitrary systems in contact with identical heat and volume baths is equivalent to the law that *the systems' out-of-equilibrium Gibbs potential can never increase*. As the evolution proceeds and the entropy of baths plus systems continually increases, the Gibbs potential G will be driven smaller and smaller, until ultimately, when statistical equilibrium with the baths is reached, G will stop at its final, minimum value.

minimum principle for Gibbs potential

The ergodic hypothesis implies that this minimum principle applies not only to an ensemble of systems but also to a single, individual system when that system is averaged over times long compared to its internal timescales τ_{int} (but times that might be short compared to the timescale for interaction with the heat and volume bath). The system's time-averaged energy $\overline{\mathcal{E}}$ and volume $\bar{V}$, and its entropy S (as computed, e.g., by examining the temporal wandering of its state on timescales $\sim \tau_{\text{int}}$), combine with the bath's temperature T_b and pressure P_b to give an out-of-equilibrium

Gibbs potential $G = \overline{\mathcal{E}} + P_b \overline{V} - T_b S$. This G evolves on times long compared to the averaging time used to define it, and that evolution must be one of continually decreasing G. Ultimately, when the system reaches equilibrium with the bath, G achieves its minimum value.

At this point we might ask about the other thermodynamic potentials. Not surprisingly, associated with each of them is an extremum principle analogous to "minimum G":

extremum principles for other thermodynamic potentials

1. For the energy potential $\mathcal{E}(V, S, N)$ (Sec. 5.2), one focuses on closed systems and switches to $S(V, \mathcal{E}, N)$. The extremum principle is then the standard second law of thermodynamics: an ensemble of closed systems of fixed $\mathcal{E}$, V, and N always must evolve toward increasing entropy S; when it ultimately reaches equilibrium, the ensemble will be microcanonical and will have maximum entropy.

2. For the physical free energy (Helmholtz free energy) $F(T_b, V, N)$ (Sec. 5.4), one can derive, in a manner perfectly analogous to the Gibbs derivation, the following minimum principle. *For an ensemble of systems interacting with a heat bath, the out-of-equilibrium physical free energy $F = \overline{\mathcal{E}} - T_b S$ will always decrease, ultimately reaching a minimum when the ensemble reaches its final, equilibrium, canonical distribution.*

3. The grand potential $\Omega(V, T_b, \tilde{\mu}_b)$ (Sec. 5.3) satisfies the analogous minimum principle. *For an ensemble of systems interacting with a heat and particle bath, the out-of-equilibrium grand potential $\Omega = \overline{\mathcal{E}} - \tilde{\mu}_b \overline{N} - T_b S$ will always decrease, ultimately reaching a minimum when the ensemble reaches its final, equilibrium, grand canonical distribution.*

4. For the enthalpy $H(P_b, S, N)$ (Ex. 5.5) the analogous extremum principle is a bit more tricky (see Ex. 5.13). *For an ensemble of systems interacting with a volume bath, as for an ensemble of closed systems, the bath's entropy remains constant, so the systems' entropy S will always increase, ultimately reaching a maximum when the ensemble reaches its final equilibrium distribution.*

Table 5.2 summarizes these extremum principles. The first column lists the quantities that a system exchanges with its bath. The second column shows the out-of-equilibrium fundamental potential for the system, which depends on the bath variables and the system's out-of-equilibrium distribution function ρ (shown explicitly) and also on whatever quantities are fixed for the system (e.g., its volume V and/or number of particles N; not shown explicitly). The third column expresses the total entropy of system plus bath in terms of the bath's out-of-equilibrium fundamental potential. The fourth column expresses the second law of thermodynamics for bath plus system in terms of the fundamental potential. We shall discuss the fifth column in Sec. 5.6, when we study fluctuations away from equilibrium.

TABLE 5.2: Descriptions of out-of-equilibrium ensembles with distribution function ρ

Quantities exchanged with bath	Fundamental potential	Total entropy $S + S_b$	Second law	Fluctuational probability
None	$S(\rho)$ with $\mathcal{E}$ constant	$S +$ const	$dS \geq 0$	$\propto e^{S/k_B}$
Volume and energy with $d\mathcal{E} = -P_b dV$	$S(\rho)$ with $H = \mathcal{E} + P_b V$ constant	$S +$ const (see Ex. 5.13)	$dS \geq 0$	$\propto e^{S/k_B}$
Heat	$F(T_b; \rho) = \overline{\mathcal{E}} - T_b S$	$-F/T_b +$ const	$dF \leq 0$	$\propto e^{-F/(k_B T_b)}$
Heat and volume	$G(T_b, P_b; \rho) = \overline{\mathcal{E}} + P_b \overline{V} - T_b S$	$-G/T_b +$ const	$dG \leq 0$	$\propto e^{-G/(k_B T_b)}$
Heat and particle	$\Omega(T_b, \tilde{\mu}_b, \rho) = \overline{\mathcal{E}} - \tilde{\mu}_b \overline{N} - T_b S$	$-\Omega/T_b +$ const	$d\Omega \leq 0$	$\propto e^{-\Omega/(k_B T_b)}$

Notes: From the distribution function ρ, one computes $S = -k_B \sum_n \rho_n \ln \rho_n$, $\overline{\mathcal{E}} = \sum_n \rho_n \mathcal{E}_n$, $\overline{V} = \sum_n \rho_n V_n$, and $\overline{N}_n = \sum_n \rho_n N_n$. The systems of each ensemble are in contact with the bath shown in column one, and T_b, P_b, and $\tilde{\mu}_b$ are the bath's temperature, pressure, and chemical potential, respectively. For ensembles in statistical equilibrium, see Table 5.1. As in that table, the nonrelativistic formulas are the same as above but with the rest masses of particles removed from the chemical potentials ($\tilde{\mu} \to \mu$) and from all fundamental potentials except Ω ($\mathcal{E} \to E$, but no change of notation for H, F, and G).

5.5.2 Phase Transitions

The minimum principle for the Gibbs potential G is a powerful tool in understanding *phase transitions*. "Phase" here refers to a specific pattern into which the atoms or molecules of a substance organize themselves. The substance H_2O has three familiar phases: water vapor, liquid water, and solid ice. Over one range of pressure P and temperature T, the H_2O molecules prefer to organize themselves into the vapor phase; over another, the liquid phase; and over another, the solid ice phase. It is the Gibbs potential that governs their preferences.

To understand this role of the Gibbs potential, consider a cup of water in a refrigerator (and because the water molecules are highly nonrelativistic, adopt the nonrelativistic viewpoint with the molecules' rest masses removed from their energy E, chemical potential μ_{H_2O}, and Gibbs potential). The refrigerator's air forms a heat and volume bath for the water in the cup (the system). There is no membrane between the air and the water, but none is needed. Gravity, together with the density difference between water and air, serves to keep the water molecules in the cup and the air above the water's surface, for all relevant purposes.

Allow the water to reach thermal and pressure equilibrium with the refrigerator's air, then turn down the refrigerator's temperature slightly and wait for the water to reach equilibrium again, and then repeat the process. Suppose that you are clever enough to compute from first principles the Gibbs potential G for the H_2O at each step of the cooling, using two alternative assumptions: that the H_2O molecules organize themselves into the liquid water phase, and that they organize themselves into the solid ice phase. Your calculations will produce curves for G as a function of the common

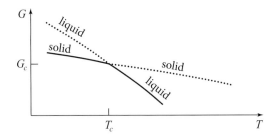

FIGURE 5.3 The Gibbs potential $G(T, P, N)$ for H_2O as a function of temperature T, with fixed P and N, near the freezing point $T = T_c = 273$ K. The solid curves correspond to the actual path traversed by the H_2O if the phase transition is allowed to proceed. The dotted curves correspond to superheated solid ice and supercooled liquid water that are unstable against the phase transition because their Gibbs potentials are higher than those of the other phase. Note that G tends to decrease with increasing temperature. This is caused by the $-TS$ term in $G = E + PV - TS$.

bath and H_2O temperature $T_b = T$ at fixed (atmospheric) pressure, with the shapes shown in Fig. 5.3. At temperatures $T > T_c = 273$ K the liquid phase has the lower Gibbs potential G, and at $T < T_c$ the solid phase has the lower G. Correspondingly, when the cup's temperature sinks slightly below 273 K, the H_2O molecules have a statistical preference for reorganizing themselves into the solid phase. The water freezes, forming ice.

It is a familiar fact that ice floats on water (i.e., ice is less dense than water), even when they are both precisely at the phase-transition temperature of 273 K. Correspondingly, when our sample of water freezes, its volume increases discontinuously by some amount ΔV; that is, when viewed as a function of the Gibbs potential G, the volume V of the statistically preferred phase is discontinuous at the phase-transition point (see Fig. 5.4a). It is also a familiar fact that when water freezes, it releases heat into its surroundings. This is why the freezing requires a moderately long time: the solidifying water can remain at or below its freezing point and continue to solidify only if the surroundings carry away the released heat, and the surroundings typically cannot carry it away quickly. It takes time to conduct heat through the ice and convect it through the water. The heat ΔQ released during the freezing (the *latent heat*) and the volume change ΔV are related to each other in a simple way (see Ex. 5.8, which focuses on the latent heat per unit mass Δq and the density change $\Delta \rho$ instead of on ΔQ and ΔV).

first-order phase transitions

Phase transitions like this one, with finite volume jumps $\Delta V \neq 0$ and finite latent heat $\Delta Q \neq 0$, are called *first-order*. The van der Waals gas (Sec. 5.3.2) provides an analytic model for another first-order phase transition: that from water vapor to liquid water; but we delay studying this model (Sec. 5.7) until we have learned about fluctuations of systems in statistical equilibrium (Sec. 5.6), which the van der Waals gas also illustrates.

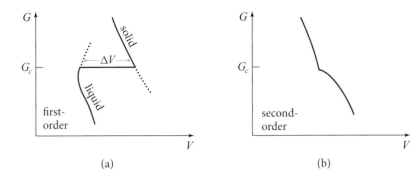

FIGURE 5.4 The changes of volume (plotted rightward) with increasing Gibbs function (plotted upward) at fixed P and N for (a) a first-order phase transition and (b) a second-order phase transition. The critical value of the Gibbs potential at which the transition occurs is G_c.

Less familiar, but also important, are *second-order phase transitions*. In such transitions, the volumes V of the two phases are the same at the transition point, but their rates of change dV/dG are different (and this is so whether one holds P fixed as G decreases, holds T fixed, or holds some combination of P and T fixed); see Fig. 5.4b.

second-order phase transitions

Crystals provide examples of both first-order and second-order phase transitions. A crystal can be characterized as a 3-dimensional repetition of a unit cell, in which ions are distributed in some fixed way. For example, Fig. 5.5a shows the unit cell for a BaTiO$_3$ (barium titanate) crystal at relatively high temperatures. This unit cell has a cubic symmetry. The full crystal can be regarded as made up of such cells stacked side by side and one on another. A first-order phase transition occurs when, with decreasing temperature, the Gibbs potential G of some other ionic arrangement,

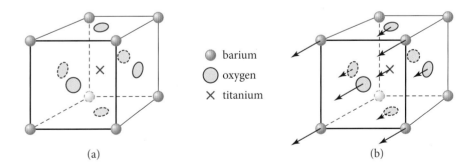

FIGURE 5.5 (a) The unit cell for a BaTiO$_3$ crystal at relatively high temperatures. (b) The displacements of the titanium and oxygen ions relative to the corners of the unit cell that occur in this crystal with falling temperature when it undergoes its second-order phase transition. The magnitudes of the displacements are proportional to the amount $T_c - T$ by which the temperature T drops below the critical temperature T_c, for small $T_c - T$.

with a distinctly different unit cell, drops below the G of the original arrangement. Then the crystal can spontaneously rearrange itself, converting from the old unit cell to the new one with some accompanying release of heat and some discontinuous change in volume.

$BaTiO_3$ does not behave in this way. Rather, as the temperature falls a bit below a critical value, the unit cell begins to elongate parallel to one of its edges (i.e., the cell's atoms get displaced as indicated in Fig. 5.5b). If the temperature is only a tiny bit below critical, they are displaced by only a tiny amount. When the temperature falls further, their displacements increase. If the temperature is raised back up above critical, the ions return to the standard, rigidly fixed positions shown in Fig. 5.5a. The result is a discontinuity, at the critical temperature, in the rate of change of volume dV/dG (Fig. 5.4b), but there is no discontinuous jump of volume and no latent heat.

This $BaTiO_3$ example illustrates a frequent feature of phase transitions: when the transition occurs (i.e., when the atoms start to move), the unit cell's cubic symmetry gets broken. The crystal switches discontinuously to a lower type of symmetry, a tetragonal one in this case. Such spontaneous symmetry breaking is a common occurrence in phase transitions not only in condensed matter physics but also in fundamental particle physics.

Bose-Einstein condensation of a bosonic atomic gas in a magnetic trap (Sec. 4.9) is another example of a phase transition. As we saw in Ex. 4.13, for Bose-Einstein condensation the specific heat of the atoms changes discontinuously (in the limit of an arbitrarily large number of atoms) at the critical temperature; this, or often a mild divergence of the specific heat, is characteristic of second-order phase transitions. Ferromagnetism also exhibits a second-order phase transition, which we explore in Secs. 5.8.3 and 5.8.4 using two powerful computational techniques: the renormalization group and Monte Carlo methods.

EXERCISES

Exercise 5.8 *Example: The Clausius-Clapeyron Equation for Two Phases in Equilibrium with Each Other*

(a) Consider H_2O in contact with a heat and volume bath with temperature T and pressure P. For certain values of T and P the H_2O will be liquid water; for others, ice; for others, water vapor—and for certain values it may be a two- or three-phase mixture of water, ice, and/or vapor. Show, using the Gibbs potential and its Euler equation, that, if two phases a and b are present and in equilibrium with each other, then their chemical potentials must be equal: $\mu_a = \mu_b$. Explain why, for any phase a, μ_a is a unique function of T and P. Explain why the condition $\mu_a = \mu_b$ for two phases to be present implies that the two-phase regions of the T-P plane are lines and the three-phase regions are points (see Fig. 5.6). The three-phase region is called the "triple point." The volume V of the two- or three-phase system will vary, depending on how much of each phase is present, since the density of each phase (at fixed T and P) is different.

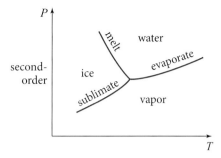

FIGURE 5.6 Phase diagram for H_2O. The temperature of the triple point (where the three phases meet) is 273.16 K and has been used to define the absolute scale of temperature.

(b) Show that the slope of the ice-water interface curve in Fig. 5.6 (the "melting curve") is given by the Clausius-Clapeyron equation

$$\frac{dP_{\text{melt}}}{dT} = \frac{\Delta q_{\text{melt}}}{T}\left(\frac{\rho_{\text{ice}}\,\rho_{\text{water}}}{\rho_{\text{ice}} - \rho_{\text{water}}}\right), \quad (5.57a)$$

where ρ is density (mass per unit volume), and Δq_{melt} is the latent heat per unit mass for melting (or freezing)—the amount of heat required to melt a unit mass of ice, or the amount released when a unit mass of water freezes. Notice that, because ice is less dense than water, the slope of the melting curve is negative. [Hint: Compute dP/dT by differentiating $\mu_a = \mu_b$, and then use the thermodynamic properties of $G_a = \mu_a N_a$ and $G_b = \mu_b N_b$.]

(c) Suppose that a small amount of water is put into a closed container of much larger volume than the water. Initially there is vacuum above the water's surface, but as time passes some of the liquid water evaporates to establish vapor-water equilibrium. The vapor pressure will vary with temperature in accord with the *Clausius-Clapeyron equation*

$$\frac{dP_{\text{vapor}}}{dT} = \frac{\Delta q_{\text{evaporate}}}{T}\left(\frac{\rho_{\text{water}}\,\rho_{\text{vapor}}}{\rho_{\text{water}} - \rho_{\text{vapor}}}\right). \quad (5.57b)$$

Now suppose that a foreign gas (not water vapor) is slowly injected into the container. Assume that this gas does not dissolve in the liquid water. Show that, as the pressure P_{gas} of the foreign gas gradually increases, it does not squeeze water vapor into the water, but rather it induces more water to vaporize:

$$\left(\frac{dP_{\text{vapor}}}{dP_{\text{total}}}\right)_{T\text{ fixed}} = \frac{\rho_{\text{vapor}}}{\rho_{\text{water}}} > 0, \quad (5.57c)$$

where $P_{\text{total}} = P_{\text{vapor}} + P_{\text{gas}}$.

5.5.3 Chemical Reactions

A second important application of the Gibbs potential is to the study of chemical reactions. Here we generalize the term "chemical reactions," to include any change in the constituent particles of the material being studied, including for example the joining of atoms to make molecules, the liberation of electrons from atoms in an ionization process, the joining of two atomic nuclei to make a third kind of nucleus, and the decay of a free neutron to produce an electron and a proton. In other words, the "chemical" of chemical reactions encompasses the reactions studied by nuclear physicists and elementary-particle physicists as well as those studied by chemists. The Gibbs representation is the appropriate one for discussing chemical reactions, because such reactions generally occur in an environment ("bath") of fixed temperature and pressure, with energy and volume being supplied and removed as needed.

As a specific example, in Earth's atmosphere, consider the breakup of two molecules of water vapor to form two hydrogen molecules and one oxygen molecule: $2H_2O \rightarrow 2H_2 + O_2$. The inverse reaction $2H_2 + O_2 \rightarrow 2H_2O$ also occurs in the atmosphere,[10] and it is conventional to write down the two reactions simultaneously in the form

$$2H_2O \leftrightarrow 2H_2 + O_2. \tag{5.58}$$

A chosen (but arbitrary) portion of the atmosphere, with idealized walls to keep all its molecules in, can be regarded as a system. (The walls are unimportant in practice, but are pedagogically useful.) The kinetic motions of this system's molecules reach and maintain statistical equilibrium, at fixed temperature T and pressure P, far more rapidly than chemical reactions can occur. Accordingly, if we view this system on timescales short compared to that τ_{react} for the reactions (5.58) but long compared to the kinetic relaxation time, then we can regard the system as in *partial statistical equilibrium*, with fixed numbers of water molecules N_{H_2O}, hydrogen molecules N_{H_2}, and oxygen molecules N_{O_2}, and with a Gibbs potential whose value is given by the Euler relation (5.53):

$$G = \tilde{\mu}_{H_2O} N_{H_2O} + \tilde{\mu}_{H_2} N_{H_2} + \tilde{\mu}_{O_2} N_{O_2}. \tag{5.59}$$

(Here, even though Earth's atmosphere is highly nonrelativistic, we include rest masses in the chemical potentials and in the Gibbs potential; the reason will become evident at the end of this section.)

When one views the sample over a longer timescale, $\Delta t \sim \tau_{react}$, one discovers that these molecules are not inviolate; they can change into one another via the reactions (5.58), thereby changing the value of the Gibbs potential (5.59). The changes of G are more readily computed from the Gibbs representation of the first law, $dG =$

---

10. In the real world these two reactions are made complicated by the need for free-electron intermediaries, whose availability is influenced by external factors, such as ultraviolet radiation. This, however, does not change the issues of principle discussed here.

$VdP - SdT + \sum_I \tilde{\mu}_I dN_I$, than from the Euler relation (5.59). Taking account of the constancy of P and T and the fact that the reactions entail transforming two water molecules into two hydrogen molecules and one oxygen molecule (or conversely), so that

$$dN_{H_2} = -dN_{H_2O}, \quad dN_{O_2} = -\frac{1}{2}dN_{H_2O}, \tag{5.60a}$$

the first law says

$$dG = (2\tilde{\mu}_{H_2O} - 2\tilde{\mu}_{H_2} - \tilde{\mu}_{O_2})\frac{1}{2}dN_{H_2O}. \tag{5.60b}$$

The reactions (5.58) proceed in both directions, but statistically there is a preference for one direction over the other. The preferred direction, of course, is the one that reduces the Gibbs potential (i.e., increases the entropy of the molecules and their bath). Thus, if $2\tilde{\mu}_{H_2O}$ is larger than $2\tilde{\mu}_{H_2} + \tilde{\mu}_{O_2}$, then water molecules preferentially break up to form hydrogen plus oxygen; but if $2\tilde{\mu}_{H_2O}$ is less than $2\tilde{\mu}_{H_2} + \tilde{\mu}_{O_2}$, then oxygen and hydrogen preferentially combine to form water. As the reactions proceed, the changing N_I values produce changes in the chemical potentials $\tilde{\mu}_I$. [Recall the intimate connection

$$N_I = g_s \frac{(2\pi m_I k_B T)^{3/2}}{h^3} e^{\mu_I/(k_B T)} V \tag{5.61}$$

between $\mu_I = \tilde{\mu}_I - m_I c^2$ and N_I for a gas in the nonrelativistic regime; Eq. (3.39a).] These changes in the N_I and $\tilde{\mu}_I$ values lead ultimately to a macrostate (thermodynamic state) of minimum Gibbs potential G—a state in which the reactions (5.58) can no longer reduce G. In this final state of full statistical equilibrium, the dG of expression (5.60b) must be zero. Correspondingly, the chemical potentials associated with the reactants must balance:

$$2\tilde{\mu}_{H_2O} = 2\tilde{\mu}_{H_2} + \tilde{\mu}_{O_2}. \tag{5.62}$$

The above analysis shows that the "driving force" for the chemical reactions is the combination of chemical potentials in the dG of Eq. (5.60b). Notice that this combination has coefficients in front of the $\tilde{\mu}_I$ terms that are identical to the coefficients in the reactions (5.58) themselves, and the equilibrium relation (5.62) also has the same coefficients as the reactions (5.60b). It is easy to convince oneself that this is true in general. Consider any chemical reaction. Write the reaction in the form

$$\sum_j v_j^L A_j^L \leftrightarrow \sum_j v_j^R A_j^R. \tag{5.63}$$

Here the superscripts L and R denote the "left" and "right" sides of the reaction, the A_js are the names of the species of particle or atomic nucleus or atom or molecule involved in the reaction, and the v_js are the number of such particles (or nuclei or atoms or molecules) involved. Suppose that this reaction is occurring in an environment of fixed temperature and pressure. Then to determine the direction in which the reaction preferentially goes, examine the chemical-potential sums for the two sides of the reaction,

direction of a chemical reaction governed by chemical-potential sums

$$\sum_j v_j^L \tilde{\mu}_j^L, \quad \sum_j v_j^R \tilde{\mu}_j^R. \tag{5.64}$$

The reaction will proceed from the side with the larger chemical-potential sum to the side with the smaller; and ultimately, the reaction will bring the two sides into equality. That final equality is the state of full statistical equilibrium. Exercises 5.9 and 5.10 illustrate this behavior.

rationale for including rest masses in chemical potentials

When dealing with chemical reactions between extremely nonrelativistic molecules and atoms (e.g., water formation and destruction in Earth's atmosphere), one might wish to omit rest masses from the chemical potentials. If one does so, and if one wishes to preserve the criterion that the reaction goes in the direction of decreasing $dG = (2\mu_{H_2O} - 2\mu_{H_2} - \mu_{O_2})\frac{1}{2}dN_{H_2O}$ [Eq. (5.60b) with tildes removed], then one must choose as the "rest masses" to be subtracted values that do not include chemical binding energies; that is, the rest masses must be defined in such a way that $2m_{H_2O} = 2m_{H_2} + m_{O_2}$. This delicacy can be avoided by simply using the relativistic chemical potentials. The derivation of the Saha equation (Ex. 5.10) is an example.

EXERCISES

Exercise 5.9 **Example: Electron-Positron Equilibrium at "Low" Temperatures**
Consider hydrogen gas in statistical equilibrium at a temperature $T \ll m_e c^2/k_B \simeq 6 \times 10^9$ K. Electrons at the high-energy end of the Boltzmann energy distribution can produce electron-positron pairs by scattering off protons:

$$e^- + p \to e^- + p + e^- + e^+. \tag{5.65}$$

(There are many other ways of producing pairs, but in analyzing statistical equilibrium we get all the information we need—a relation among the chemical potentials—by considering just one way.)

(a) In statistical equilibrium, the reaction (5.65) and its inverse must proceed on average at the same rate. What does this imply about the relative magnitudes of the electron and positron chemical potentials $\tilde{\mu}_-$ and $\tilde{\mu}_+$ (with rest masses included)?

(b) Although these reactions require an e^- that is relativistic in energy, almost all the electrons and positrons will have kinetic energies of magnitude $E \equiv \mathcal{E} - m_e c^2 \sim k_B T \ll m_e c^2$, and thus they will have $\mathcal{E} \simeq m_e c^2 + \mathbf{p}^2/(2m_e)$. What are the densities in phase space $\mathcal{N}_\pm = dN_\pm/(d^3x d^3p)$ for the positrons and electrons in terms of $\mathbf{p}$, $\tilde{\mu}_\pm$, and T? Explain why for a hydrogen gas we must have $\tilde{\mu}_- > 0$ and $\tilde{\mu}_+ < 0$.

(c) Assume that the gas is very dilute, so that $\eta \ll 1$ for both electrons and positrons. Then integrate over momenta to obtain the following formula for the number

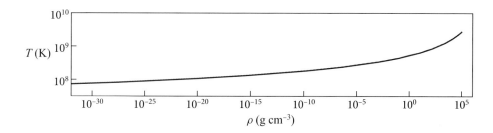

FIGURE 5.7 The temperature T_p at which electron-positron pairs form in a dilute hydrogen plasma, plotted as a function of density ρ. This is the correct upper limit (upper dashed line in Fig. 3.7) on the region where the plasma can be considered fully nonrelativistic. Above this curve, although T may be $\ll m_e c^2 / k_B \simeq 6 \times 10^9$ K, a proliferation of electron-positron pairs radically changes the properties of the plasma.

densities in physical space of electrons and positrons:

$$n_\pm = \frac{2}{h^3}(2\pi m_e k_B T)^{3/2} \exp\left(\frac{\tilde{\mu}_\pm - m_e c^2}{k_B T}\right). \tag{5.66}$$

In cgs units, what does the dilute-gas assumption $\eta \ll 1$ correspond to in terms of $n_\pm$? What region of hydrogen mass density ρ and temperature T is the dilute-gas region?

(d) Let n be the number density of protons. Then by charge neutrality, $n = n_- - n_+$ will also be the number density of "ionization electrons" (i.e., of electrons that have been ionized off of hydrogen). Show that the ratio of positrons (and hence of pairs) to ionization electrons is given by

$$\frac{n_+}{n} = \frac{1}{2y[y + (1+y^2)^{\frac{1}{2}}]}, \tag{5.67a}$$

where

$$y \equiv \frac{1}{4} n \lambda^3 e^{m_e c^2 / (k_B T)}, \quad \text{and} \quad \lambda \equiv \frac{h}{\sqrt{2\pi m_e k_B T}} \tag{5.67b}$$

is the thermal deBroglie wavelength of the electrons. Figure 5.7 shows the temperature T_p at which, according to this formula, $n_+ = n$ (and $y = 0.354$), as a function of mass density $\rho \simeq m_{\text{proton}} n$. This T_p can be thought of as the temperature at which pairs form in a dilute plasma. Somewhat below T_p there are hardly any pairs; somewhat above, the pairs are profuse.

(e) Note that at low densities pairs form at temperatures $T \sim 10^8$ K $\simeq 0.02 m_e c^2 / k_B$. Explain qualitatively in terms of available phase space why the formation temperature is so low.

Exercise 5.10 **Example: Saha Equation for Ionization Equilibrium*
Consider an optically thick hydrogen gas in statistical equilibrium at temperature T. ("Optically thick" means that photons can travel only a small distance compared to

the size of the system before being absorbed, so they are confined by the hydrogen and kept in statistical equilibrium with it.) Among the reactions that are in statistical equilibrium are $H + \gamma \leftrightarrow e + p$ (ionization and recombination of H, with the H in its ground state) and $e + p \leftrightarrow e + p + \gamma$ (emission and absorption of photons by "bremsstrahlung," i.e., by the Coulomb-force-induced acceleration of electrons as they fly past protons). Let $\tilde{\mu}_\gamma, \tilde{\mu}_H, \tilde{\mu}_e$, and $\tilde{\mu}_p$ be the chemical potentials including rest masses; let $m_H, m_e,$ and m_p be the rest masses; denote by $I (\equiv 13.6$ eV$)$ the ionization energy of H, so that $m_H c^2 = m_e c^2 + m_p c^2 - I$; denote $\mu_j \equiv \tilde{\mu}_j - m_j c^2$; and assume that $T \ll m_e c^2 / k_B \simeq 6 \times 10^9$ K and that the density is low enough that the electrons, protons, and H atoms can be regarded as nondegenerate (i.e., as distinguishable, classical particles).

(a) What relationships hold between the chemical potentials $\tilde{\mu}_\gamma, \tilde{\mu}_H, \tilde{\mu}_e$, and $\tilde{\mu}_p$?

(b) What are the number densities $n_H, n_e,$ and n_p expressed in terms of $T, \tilde{\mu}_H, \tilde{\mu}_e,$ and $\tilde{\mu}_p$—taking account of the fact that the electron and proton both have spin $\frac{1}{2}$, and including for H all possible electron and nuclear spin states?

(c) Derive the Saha equation for ionization equilibrium:

$$\frac{n_e n_p}{n_H} = \frac{(2\pi m_e k_B T)^{3/2}}{h^3} e^{-I/(k_B T)}. \qquad (5.68)$$

This equation is widely used in astrophysics and elsewhere.

5.6 Fluctuations away from Statistical Equilibrium

As we saw in Chap. 4, statistical mechanics is built on a distribution function ρ that is equal to the probability of finding a chosen system in a quantum state at some chosen location in the system's phase space. For systems in statistical equilibrium, this probability is given by the microcanonical, canonical, grand canonical, Gibbs, or other distribution, depending on the nature of the system's interactions with its surroundings. Classical thermodynamics makes use of only a tiny portion of the information in this probability distribution: the mean values of a few macroscopic parameters (energy, entropy, volume, pressure, etc.). Also contained in the distribution function, but ignored by classical thermodynamics, is detailed information about fluctuations of a system away from its mean values.

As an example, consider a microcanonical ensemble of boxes, each with volume V and each containing precisely N identical, nonrelativistic, classical gas particles and containing energy (excluding rest mass) between E and $E + \delta E$, where $\delta E \ll E$. (Remember the kludge that was necessary in Ex. 4.7). Consider a quantity y that is not fixed by the set $\{E, V, N\}$. That quantity might be discrete (e.g., the total number

N_R of particles in the right half of the box). Alternatively, it might be continuous (e.g., the total energy E_R in the right half).

In the discrete case, the total number of quantum states that correspond to specific values of y is related to the entropy S by the standard microcanonical relation $N_{\text{states}}(E, V, N; y) = \exp[S(E, V, N; y)/k_B]$; and correspondingly, since all states are equally probable in the microcanonical ensemble, the probability of finding a system of the ensemble to have the specific value y is

$$p(E, V, N; y) = \frac{N_{\text{states}}(E, V, N; y)}{\sum_y N_{\text{states}}(E, V, N; y)} = \text{const} \times \exp\left[\frac{S(E, V, N; y)}{k_B}\right]. \quad (5.69a)$$

probabilities for fluctuations in closed systems

Here the entropy S is to be computed via statistical mechanics (or, when possible, via thermodynamics) not for the original ensemble of boxes in which y was allowed to vary, but for an ensemble in which y is fixed at a chosen value.

The continuous case (e.g., $y = E_R$) can be converted into the discrete case by dividing the range of y into intervals that all have the same infinitesimal width δy. Then the probability of finding y in one of these intervals is $[dp(E, V, N; y \text{ in } \delta y)/dy]\delta y = \text{const} \times \exp[S(E, V, N; y \text{ in } \delta y)]$. Dividing both sides by δy and absorbing δy on the right-hand side into the constant, we obtain

$$\frac{dp(E, V, N; y)}{dy} = \text{const} \times \exp\left[\frac{S(E, V, N; y \text{ in } \delta y)}{k_B}\right]. \quad (5.69b)$$

Obviously, if we are interested in the joint probability for a set of ys, some discrete (e.g., $y_1 = N_R$) and some continuous (e.g., $y_2 = E_R$), that probability will be given by

$$\frac{dp(E, V, N; y_1, y_2, \ldots, y_r)}{dy_q \cdots dy_r} = \text{const} \times \exp\left[\frac{S(E, V, N; y_j)}{k_B}\right], \quad (5.69c)$$

where we keep in mind (but now omit from our notation) the fact that continuous variables are to be given values y_j in some arbitrary but fixed infinitesimal range δy_j.

The probability distribution (5.69c), though exact, is not terribly instructive. To get better insight we expand S in powers of the deviations of the y_j from the values $\bar{y}_j$ that maximize the entropy (these will turn out also to be the means of the distribution). Then for small $|y_j - \bar{y}_j|$, Eq. (5.69c) becomes

$$\frac{dp(E, V, N; y_j)}{dy_q \ldots dy_r} = \text{const} \times \exp\left[\frac{1}{2k_B}\left(\frac{\partial^2 S}{\partial y_j \partial y_k}\right)(y_j - \bar{y}_j)(y_k - \bar{y}_k)\right]. \quad (5.69d)$$

Gaussian approximation to fluctuation probabilities

Here the second partial derivative of the entropy is to be evaluated at the maximum-entropy location, where $y_j = \bar{y}_j$ for all j. Expression (5.69d) is a (multidimensional) Gaussian probability distribution for which the means are obviously $\bar{y}_j$, as predicted. (That this had to be Gaussian follows from the central limit theorem, Sec. 6.3.2.)

The last entry in the first line of Table 5.2 summarizes the above equations: *for a ~sed system, the probability of some fluctuation away from equilibrium is proportional*

5.6 Fluctuations away from Statistical Equilibrium

to e^{S/k_B}, where S is the total entropy for the out-of-equilibrium fluctuational macrostate (e.g., the macrostate with N_R particles in the right half of the box).

For the specific example where $y_1 \equiv N_R =$ (number of perfect-gas particles in right half of box) and $y_2 \equiv E_R =$ (amount of energy in right half of box), we can infer $S(E, V, N; N_R, E_R)$ from the Sackur-Tetrode equation (4.42) as applied to the two halves of the box and then added:[11]

$$S(E, V, N; E_R, N_R) = k_B N_R \ln\left[\left(\frac{4\pi m}{3h^2}\right)^{3/2} e^{5/2} \frac{V}{2} \frac{E_R^{3/2}}{N_R^{5/2}}\right]$$

$$+ k_B(N - N_R) \ln\left[\left(\frac{4\pi m}{3h^2}\right)^{3/2} e^{5/2} \frac{V}{2} \frac{(E - E_R)^{3/2}}{(N - N_R)^{5/2}}\right]. \quad (5.70a)$$

It is straightforward to compute the values $\bar{E}_R$ and $\bar{N}_R$ that maximize this entropy:

$$\bar{E}_R = \frac{E}{2}, \quad \bar{N}_R = \frac{N}{2}. \quad (5.70b)$$

Thus, in agreement with intuition, the mean values of the energy and particle number in the right half-box are equal to half of the box's total energy and particle number. It is also straightforward to compute from expression (5.70a) the second partial derivatives of the entropy with respect to E_R and N_R, evaluate them at $E_R = \bar{E}_R$ and $N_R = \bar{N}_R$, and plug them into the probability distribution (5.69d) to obtain

$$\frac{dp_{N_R}}{dE_R} = \text{const} \times \exp\left(\frac{-(N_R - N/2)^2}{2(N/4)} + \frac{-[(E_R - E/2) - (E/N)(N_R - N/2)]^2}{2(N/6)(E/N)^2}\right).$$

(5.70c)

This Gaussian distribution has the following interpretation. (i) There is a correlation between the energy E_R and the particle number N_R in the right half of the box, as one might have expected: if there is an excess of particles in the right half, then we must expect an excess of energy there as well. (ii) The quantity that is not correlated with N_R is $E_R - (E/N)N_R$, again as one might have expected. (iii) For fixed N_R, dp_{N_R}/dE_R is Gaussian with mean $\bar{E}_R = E/2 + (E/N)(N_R - N/2)$ and with rms fluctuation (standard deviation; square root of variance) $\sigma_{E_R} = (E/N)\sqrt{N/6}$. (iv) After integrating over E_R, we obtain

$$p_{N_R} = \text{const} \times \exp\left[\frac{-(N_R - N/2)^2}{2N/4}\right]. \quad (5.70d)$$

This is Gaussian with mean $\bar{N}_R = N/2$ and rms fluctuation $\sigma_{N_R} = \sqrt{N/4}$. By contrast, if the right half of the box had been in equilibrium with a bath far larger than itself, N_R would have had an rms fluctuation equal to the square root of its mean, $\sigma_{N_R} = \sqrt{N/2}$

---

11. Note that the derivation of Eq. (4.42), as specialized to the right half of the box, requires the same kind of infinitesimal range $\delta y_2 = \delta E_R$ as we used to derive our fluctuational probability equation (5.69d).

(see Ex. 5.11). The fact that the companion of the right half has only the same size as the right half, rather than being far larger, has reduced the rms fluctuation of the number of particles in the right half from $\sqrt{N/2}$ to $\sqrt{N/4}$.

Notice that the probability distributions (5.70c) and (5.70d) are exceedingly sharply peaked about their means. Their standard deviations divided by their means (i.e., the magnitude of their fractional fluctuations) are all of order $1/\sqrt{\overline{N}}$, where $\overline{N}$ is the mean number of particles in a system; and in realistic situations $\overline{N}$ is very large. (For example, $\overline{N}$ is of order 10^{26} for a cubic meter of Earth's atmosphere, and thus the fractional fluctuations of thermodynamic quantities are of order 10^{-13}.) It is this extremely sharp peaking that makes classical thermodynamics insensitive to the choice of type of equilibrium ensemble—that is, sensitive only to means and not to fluctuations about the means.

The generalization of this example to other situations should be fairly obvious; see Table 5.2. When a system is in some out-of-equilibrium macrostate, the total entropy $S + S_b$ of the system and any bath with which it may be in contact is, up to an additive constant, either the system's entropy S or the negative of its out-of-equilibrium potential divided by the bath's temperature ($-F/T_b +$ const, $-G/T_b +$ const, or $-\Omega/T_b +$ const; see column 3 of Table 5.2). Correspondingly, the probability of a fluctuation from statistical equilibrium to this out-of-equilibrium macrostate is proportional to the exponential of this quantity in units of Boltzmann's constant (e^{-S/k_B}, $e^{-F/(k_B T_b)}$, $e^{-G/(k_B T_b)}$, or $e^{-\Omega/(k_B T_b)}$; column 5 of Table 5.2). By expanding the quantity in the exponential around the equilibrium state to second order in the fluctuations, one obtains a Gaussian probability distribution for the fluctuations, like Eq. (5.69d).

probabilities for fluctuations in systems interacting with baths

As examples, in Ex. 5.11 we study fluctuations in the number of particles in a cell that is immersed in a heat and particle bath, so the starting point is the out-of-equilibrium grand potential Ω. And in Ex. 5.12, we study temperature and volume fluctuations for a system in contact with a heat and volume bath; so the starting point is the out-of-equilibrium Gibbs function G.

EXERCISES

Exercise 5.11 *Example: Probability Distribution for the Number of Particles in a Cell*
Consider a cell with volume V, like those of Fig. 5.1, that has imaginary walls and is immersed in a bath of identical, nonrelativistic, classical perfect-gas particles with temperature T_b and chemical potential μ_b. Suppose that we make a large number of measurements of the number of particles in the cell and that from those measurements we compute the probability p_N for that cell to contain N particles.

(a) How widely spaced in time must the measurements be to guarantee that the measured probability distribution is the same as that computed, using the methods of this section, from an ensemble of cells (Fig. 5.1) at a specific moment of time?

(b) Assume that the measurements are widely enough separated for this criterion to be satisfied. Show that p_N is given by

$$p_N \propto \exp\left[\frac{-\Omega(V, T_b, \mu_b; N)}{k_B T_b}\right]$$

$$\equiv \frac{1}{N!}\int \frac{d^{3N}x \, d^{3N}p}{h^{3N}} \exp\left[\frac{-E_n + \mu_b N_n}{k_B T_b}\right] \quad (5.71)$$

$$= \frac{1}{N!}\int \frac{d^{3N}x \, d^{3N}p}{h^{3N}} \exp\left[\frac{-\left(\sum_{i=1}^{N} \mathbf{p}_i^2/(2m)\right) + \mu_b N}{k_B T_b}\right],$$

where $\Omega(V, T_b, \mu_b; N)$ is the grand potential for the ensemble of cells, with each cell constrained to have precisely N particles in it (cf. the last entry in Table 5.2).

(c) By evaluating Eq. (5.71) exactly and then computing the normalization constant, show that the probability p_N for the cell to contain N particles is given by the *Poisson distribution*

$$p_N = e^{-\overline{N}}(\overline{N}^N/N!), \quad (5.72a)$$

where $\overline{N}$ is the mean number of particles in a cell,

$$\overline{N} = (\sqrt{2\pi m k_B T_b}/h)^3 e^{\mu_b/(k_B T_b)} V$$

[Eq. (3.39a)].

(d) Show that for the Poisson distribution (5.72a), the expectation value is $\langle N \rangle = \overline{N}$, and the rms deviation from this is

$$\sigma_N \equiv \langle (N - \overline{N})^2 \rangle^{\frac{1}{2}} = \overline{N}^{\frac{1}{2}}. \quad (5.72b)$$

(e) Show that for $N - \overline{N} \lesssim \sigma_N$, this Poisson distribution is exceedingly well approximated by a Gaussian with mean $\overline{N}$ and variance σ_N.

Exercise 5.12 *Example: Fluctuations of Temperature and Volume in an Ideal Gas*
Consider a gigantic container of gas made of identical particles that might or might not interact. Regard this gas as a bath, with temperature T_b and pressure P_b. Pick out at random a sample of the bath's gas containing precisely N particles, with $N \gg 1$. Measure the volume V of the sample and the temperature T inside the sample. Then pick another sample of N particles, and measure its V and T, and repeat over and over again. Thereby map out a probability distribution $dp/dT \, dV$ for V and T of N-particle samples inside the bath.

(a) Explain in detail why

$$\frac{dp}{dT\,dV} = \text{const} \times \exp\left[-\frac{1}{2k_B T_b}\left(\frac{\partial^2 G}{\partial V^2}(V - \bar{V})^2 + \frac{\partial^2 G}{\partial T^2}(T - T_b)^2 \right.\right.$$
$$\left.\left. + 2\frac{\partial^2 G}{\partial T \partial V}(V - \bar{V})(T - T_b)\right)\right],$$
(5.73a)

where $G(N, T_b, P_b; T, V) = E(T, V, N) + P_b V - T_b S(T, V, N)$ is the out-of-equilibrium Gibbs function for a sample of N particles interacting with this bath (next-to-last line of Table 5.2), $\bar{V}$ is the equilibrium volume of the sample when its temperature and pressure are those of the bath, and the double derivatives in Eq. (5.73a) are evaluated at the equilibrium temperature T_b and pressure P_b.

(b) Show that the derivatives, evaluated at $T = T_b$ and $V = \bar{V}$, are given by

$$\left(\frac{\partial^2 G}{\partial T^2}\right)_{V,N} = \frac{C_V}{T_b}, \quad \left(\frac{\partial^2 G}{\partial V^2}\right)_{T,N} = \frac{1}{\kappa}, \quad \text{and} \quad \left(\frac{\partial^2 G}{\partial T \partial V}\right)_N = 0, \quad (5.73b)$$

where C_V is the gas sample's specific heat at fixed volume and κ is its compressibility at fixed temperature β, multiplied by V/P:

$$C_V \equiv \left(\frac{\partial E}{\partial T}\right)_{V,N} = T\left(\frac{\partial S}{\partial T}\right)_{V,N}, \quad \kappa \equiv \beta V/P = -\left(\frac{\partial V}{\partial P}\right)_{T,N}, \quad (5.73c)$$

both evaluated at temperature T_b and pressure P_b. [Hint: Write $G = G_{\text{eq}} + (P_b - P)V - (T_b - T)S$, where G_{eq} is the equilibrium Gibbs function for the gas samples.] Thereby conclude that

$$\frac{dp}{dT\,dV} = \text{const} \times \exp\left[-\frac{(V - \bar{V})^2}{2k_B T_b \kappa} - \frac{C_V(T - T_b)^2}{2k_B T_b^2}\right]. \quad (5.73d)$$

(c) This probability distribution says that the temperature and volume fluctuations are uncorrelated. Is this physically reasonable? Why?

(d) What are the rms fluctuations of the samples' temperature and volume, σ_T and σ_V? Show that σ_T scales as $1/\sqrt{N}$ and σ_V as $\sqrt{N}$, where N is the number of particles in the samples. Are these physically reasonable? Why?

Exercise 5.13 *Example and Derivation: Evolution and Fluctuations of a System in Contact with a Volume Bath*

Exercise 5.5 explored the enthalpy representation of thermodynamics for an equilibrium ensemble of systems in contact with a volume bath. Here we extend that analysis to an ensemble out of equilibrium. We denote by P_b the bath pressure.

(a) The systems exchange volume but not heat or particles with the bath. Explain why, even though the ensemble may be far from equilibrium, any system's volume change dV must be accompanied by an energy change $d\mathcal{E} = -P_b dV$. This implies that the system's enthalpy $H = \mathcal{E} + P_b V$ is conserved. All systems in the ensemble are assumed to have the same enthalpy H and the same number of particles N.

(b) Using equilibrium considerations for the bath, show that interaction with a system cannot change the bath's entropy.

(c) Show that the ensemble will always evolve toward increasing entropy S, and that when the ensemble finally reaches statistical equilibrium with the bath, its distribution function will be that of the enthalpy ensemble (Table 5.1): $\rho = e^{-S/k_B} = $ const for all regions of phase space that have the specified particle number N and enthalpy H.

(d) Show that fluctuations away from equilibrium are described by the probability distributions (5.69a) and (5.69c), but with the system energy E replaced by the system enthalpy H and the system volume V replaced by the bath pressure P_b (cf. Table 5.2).

5.7 5.7 Van der Waals Gas: Volume Fluctuations and Gas-to-Liquid Phase Transition

The van der Waals gas studied in Sec. 5.3.2 provides a moderately realistic model for real gases such as H_2O and their condensation (phase transition) into liquids, such as water.

The equation of state for a van der Waals gas is

$$\left(P + \frac{a}{v^2}\right)(v - b) = k_B T \tag{5.74}$$

[Eq. (5.26)]. Here a and b are constants, and $v \equiv V/N$ is the specific volume (the inverse of the number density of gas particles). In Fig. 5.8a we depict this equation of state as curves (*isotherms*) of pressure P versus specific volume v at fixed temperature T. Note [as one can easily show from Eq. (5.74)] that there is a critical temperature $T_c = 8a/(27bk_B)$ such that for $T > T_c$ the isotherms are monotonic decreasing; for $T = T_c$ they have an inflection point [at $v = v_c \equiv 3b$ and $P = P_c = a/(27b^2)$]; and for $T < T_c$ they have a maximum and a minimum.

From Eq. (5.73d), derived in Ex. 5.12, we can infer that the probability dp/dv for fluctuations of the specific volume of a portion of this gas containing N particles is

probability for volume fluctuations in van der Waals gas

$$\frac{dp}{dv} = \text{const} \times \exp\left[\frac{N(\partial P/\partial v)_T}{2k_B T}(v - \bar{v})^2\right]. \tag{5.75}$$

This probability is controlled by the slope $(\partial P/\partial v)_T$ of the isotherms. Where the slope is negative, the volume fluctuations are small; where it is positive, the fluid is unstable: its volume fluctuations grow. Therefore, for $T < T_c$, the region of an isotherm between its minimum M and its maximum X (Fig. 5.8b) is unphysical; the fluid cannot exist stably there. Evidently, at $T < T_c$ there are two phases: one with low density ($v > v_X$)

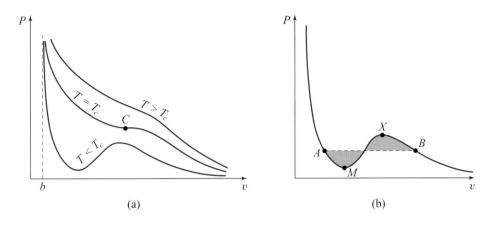

FIGURE 5.8 (a) The van der Waals equation of state $P(N, V, T)$ plotted as pressure P versus specific volume $v \equiv V/N$ at fixed temperature T, for various values of T. (b) The route of a phase transition in the van der Waals gas. The transition is a discontinuous jump from point A to point B.

is gaseous; the other with high density ($v < v_M < v_c = 3b$) is liquid. [Recall, from comment (i) at the end of Sec. 5.3.2, that $b/4$ is the volume of each of the material's particles, so in the high-density phase the particles' separations are not much larger than their diameter; this is characteristic of a fluid.]

Hold the temperature T fixed at $T < T_c$, and gradually increase the density from zero (decrease the specific volume v from infinity). At low densities, the material will be gaseous, and at high densities, it will be liquid. The phase transition from gas to liquid involves a discontinuous jump from some point B in Fig. 5.8b to another point A. The Gibbs potential controls the location of those points.

Since the two phases are in equilibrium with each other at A and B, their Gibbs potential $G = \mu N$ must be the same, which means their chemical potentials must be the same, $\mu_A = \mu_B$—as, of course, must be their temperatures $T_A = T_B$ (they lie on the same isotherm). This in turn implies their pressures must be the same, $P_A = P(\mu_A, T) = P(\mu_B, T) = P_B$. Therefore, the points A and B in Fig 5.8b are connected by a horizontal line (the dashed line in the figure). Let us use the first law of thermodynamics in the Gibbs representation to compute the change in the chemical potential μ as one moves along the isotherm from point A to point B. The first law says $dG = -SdT + VdP + \mu dN$. Focusing on a sample of the material containing N particles, and noting that along the isotherm the sample has $G = \mu N$, $dN = 0$, $dT = 0$, and $V = vN$, we obtain $d\mu = vdP$. Integrating this relation along the isotherm from A to B, we obtain

$$0 = \mu_B - \mu_A = \int_A^B d\mu = \int_A^B vdP. \tag{5.76}$$

This integral is the area of the right green region in Fig. 5.8b minus the area of the left green region. Therefore, these two areas must be equal, which tells us the location of the points A and B that identify the two phases (liquid and gaseous) when the phase transition occurs.

Consider again volume fluctuations. Where an isotherm is flat, $(\partial P/\partial v)_T = 0$, large volume fluctuations occur [Eq. (5.75)]. For $T < T_c$, the isotherm is flat at the minimum M and the maximum X, but these do not occur in Nature—unless the phase transition is somehow delayed as one compresses or expands the material. However, for $T = T_c$, the isotherm is flat at its inflection point $v = v_c$, $P = P_c$ (the material's critical point C in Fig. 5.8a); so the volume fluctuations will be very large there.

At some temperatures T and pressures P, it is possible for two phases, liquid and gas, to exist; at other T and P, only one phase exists. The dividing line in the T-P plane between these two regions is called a *catastrophe*—a term that comes from *catastrophe theory*. We explore this in Ex. 7.16, after first introducing some ideas of catastrophe theory in the context of optics.

EXERCISES

Exercise 5.14 **Example: Out-of-Equilibrium Gibbs Potential for Water; Surface Tension and Nucleation**[12]

Water and its vapor (liquid and gaseous H_2O) can be described moderately well by the van der Waals model, with the parameters $a = 1.52 \times 10^{-48}$ J m^3 and $b = 5.05 \times 10^{-29}$ m^3 determined by fitting to the measured pressure and temperature at the critical point (inflection point C in Fig. 5.8a: $P_c = a/(27b^2) = 22.09$ MPa, $T_c = 8a/(27bk_B) = 647.3$ K). [Note: 1 MPa is 10^6 Pascal; and 1 Pascal is the SI unit of pressure, 1 kg m s^{-2}.]

(a) For an out-of-equilibrium sample of N atoms of H_2O at temperature T and pressure P that has fluctuated to a specific volume v, the van-der-Waals-modeled Gibbs potential is

$$G(N, T, P; v) = N \left[-k_B T + Pv - a/v + k_B T \ln \left(\lambda_{T\,\mathrm{dB}}^3/(v - b) \right) \right],$$

$$\lambda_{T\,\mathrm{dB}} \equiv h/\sqrt{2\pi m k_B T}. \tag{5.77}$$

Verify that this Gibbs potential is minimized when v satisfies the van der Waals equation of state (5.74).

(b) Plot the chemical potential $\mu = G/N$ as a function of v at room temperature, $T = 300$ K, for various pressures in the vicinity of 1 atmosphere = 0.1013 MPa. Adjust the pressure so that the two phases, liquid and gaseous, are in equilibrium (i.e., so the two minima of the curve have the same height). [Answer: The required pressure is about 3.6 atmospheres, and the chemical-potential curve is shown in Fig. 5.9. If the gas is a mixture of air and H_2O rather than pure H_2O, then the required pressure will be lower.]

(c) Compare the actual densities of liquid water and gaseous H_2O with the predictions of Fig. 5.9. They agree moderately well but not very well.

---

12. Exercise adapted from Sethna (2006, Ex. 11.3 and Sec. 11.3).

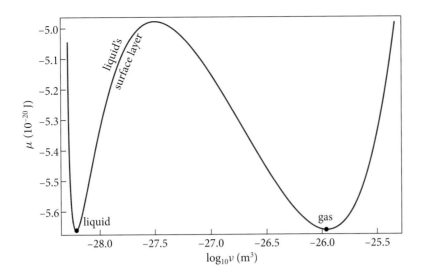

FIGURE 5.9 The out-of-equilibrium chemical potential for a van der Waals gas, with its parameters fitted to the properties of H_2O and with temperature $T = 300$ K and pressure $P = 3.6$ atmospheres at the liquid-gas interface.

(d) At the liquid's surface there is a surface layer a few molecules thick, in which the attractive force between the water molecules, $F = -6\varepsilon_o r_o^6/r^7 = -[27/(2\pi^2)]ab/r^7$ [Eqs. (5.22b) and (5.24b)], produces surface tension. This surface tension is a force per unit length, γ, that the surface molecules on one side of any line lying in the surface exert on the molecules on the other side (Box 16.4). Explain by a simple physical argument why there is an energy γA associated with this surface tension, stored in any area A of the surface. This is a free energy at fixed T, P (a Gibbs free energy) in excess of the free energy that the surface's water molecules would have if they were in the bulk liquid or the bulk gas. This excess free energy shows up in the chemical potential of Fig. 5.9, and the numbers in that figure can be used to estimate the water's surface tension γ. Show that $\gamma \sim \Delta\mu h/v$, where h and v are the thickness and specific volume, respectively, of the surface layer, and $\Delta\mu$ is the difference between the chemical potential in the surface layer and in the bulk water and gas. Estimate γ using numbers from Fig. 5.9 and compare with the measured surface tension, $\gamma \simeq 0.0723$ N/m at $T = 300$ K.

(e) In a cloud or fog, when water vapor is cooled below its equilibrium temperature with liquid water T_e, water drops try to form, that is, *nucleate*. However, there is a potential barrier against nucleation due to the surface tension of an incipient drop. If R is the drop's radius, show that the Gibbs free energy of a droplet (the sum of contributions from its surface layer and its interior) minus the free energy that the droplet's molecules will have if they remain gaseous, is

$$\Delta G = 4\pi R^2 \gamma - \left(\frac{4\pi}{3} R^3\right) \frac{q \Delta T}{v_\ell T_e}. \tag{5.78}$$

Here q is the latent heat per molecule that is released when the vapor liquifies, v_ℓ is the liquid's specific volume, and ΔT is the amount by which the temperature has dropped below the equilibrium point T_e for the two phases. Plot this $\Delta G(R)$, and explain why (i) there is a minimum droplet radius $R_{\min}$ for nucleation to succeed, and (ii) for the droplet to form with this minimum size, thermal fluctuations must put into it some excess energy, B, above statistical equilibrium. Derive these formulas:

$$R_{\min} = \frac{2\gamma T_e v_\ell}{q \Delta T}, \qquad B = \frac{16\pi \gamma^3 T_e^2 v_\ell^2}{3 q^2 \Delta T^2}. \tag{5.79}$$

Explain why the rate at which nucleation occurs must scale as $\exp[-B/(k_B T)]$, which is generally an exceedingly small number. Show that, if the nucleation occurs on a leaf or blade of grass or on the surface of a dust grain, so the drop's interface with the vapor is a hemisphere rather than a sphere, then the energy barrier B is reduced by a factor 8, and the rate of nucleation is enormously increased. That is why nucleation of water droplets almost always occurs on solid surfaces.

5.8 Magnetic Materials T2

The methods we have developed in this chapter can be applied to systems very different from the gases and liquids studied thus far. In this section, we focus on magnetic materials as an example, and we use this example to illustrate two powerful, modern computational techniques: the renormalization group and Monte Carlo methods.

model for magnetic material

We consider, for concreteness, the simplest type of magnetic material: one consisting of a cubic lattice of N identical atoms, each with spin 1/2 and magnetic moment m_o. The material is immersed in a uniform external magnetic field **B**, so each atom (labeled by an index i) has two possible quantum states: one with its spin parallel to **B** (quantum number $s_i = +1$), the other antiparallel to **B** ($s_i = -1$). The energies of these states are $E_{s_i} = -m_o B s_i$. The atoms interact with one another's magnetic fields with a pairwise interaction energy $E_{s_i s_j}$ that we shall make more concrete in Sec. 5.8.2 below. The material's total energy, when the atoms are in the state $|n\rangle = |s_1, s_1, \ldots, s_n\rangle$, is $E_n - M_n B$, where

quantum number s_i for spin orientation

internal energy and magnetization

$$E_n = \sum_{i>j}^{N} \sum_{j=1}^{N} E_{s_i s_j}, \qquad M_n = m_o \sum_{j=1}^{N} s_j \tag{5.80a}$$

are the material's self-interaction energy (internal energy) and *magnetization*.

The atoms interact with a heat bath that has temperature T and with the external magnetic field B, which can be thought of as part of the bath.[13] When they reach sta-

---

13. Arranging the electromagnetic environment to keep B fixed is a choice similar to arranging the thermal environment to keep T fixed. Other choices are possible.

tistical equilibrium with this heat and magnetic bath, the probability for the material (all N atoms) to be in state $|n\rangle = |s_1, s_1, \ldots, s_n\rangle$ is, of course,

$$p_n = e^{G(N,B,T)/(k_B T)} e^{-(E_n - BM_n)/(k_B T)}. \tag{5.80b}$$

Here the first term is the normalization constant, which depends on the number N of atoms in the sample and the bath's B and T, and $G(N, B, T)$ is the fundamental thermodynamic potential for this system, which acts as the normalizing factor for the probability:

$$e^{-G(N,B,T)/(k_B T)} = \sum_n e^{-(E_n - BM_n)/(k_B T)}. \tag{5.80c}$$

Gibbs potential for magnetic material

We have denoted this potential by G, because it is analogous to the Gibbs potential for a gas, but with the gas's volume V_n replaced by minus the magnetization $-M_n$ and the gas bath's pressure P replaced by the material bath's magnetic field strength B. Not surprisingly, the Gibbs thermodynamic formalism for this magnetic material is essentially the same as for a gas, but with $V \to -M$ and $P \to B$.

5.8.1 Paramagnetism; The Curie Law T2

5.8.1

Paramagnetic materials have sufficiently weak self-interaction that we can set $E_n = 0$ and focus solely on the atoms' interaction with the external B field. The magnetic interaction tries to align each atom's spin with B, while thermal fluctuations try to randomize the spins. As a result, the stronger is B (at fixed temperature), the larger will be the mean magnetization $\bar{M}$. From Eq. (5.80b) for the spins' probability distribution we can compute the mean magnetization:

paramagnetic material

$$\bar{M} = \sum_n p_n M_n = e^{G/(k_B T)} \sum_n M_n e^{BM_n/(k_B T)}$$

$$= e^{G/(k_B T)} k_B T \left(\frac{\partial}{\partial B}\right)_{N,T} \sum_n e^{BM_n/(k_B T)}. \tag{5.81a}$$

The last sum is equal to $e^{-G/(k_B T)}$ [Eq. (5.80c) with $E_n = 0$], so Eq. (5.81a) becomes

$$\bar{M} = -\left(\frac{\partial G}{\partial B}\right)_{N,T}. \tag{5.81b}$$

This is obviously our material's analog of $\bar{V} = (\partial G/\partial P)_{N,T}$ for a gas [which follows from the Gibbs representation of the first law, Eq. (5.52)].

To evaluate $\bar{M}$ explicitly in terms of B, we must first compute the Gibbs function from the statistical sum (5.80c) with $E_n = 0$. Because the magnetization M_n in state $|n\rangle = |s_1, s_2, \ldots, s_N\rangle$ is the sum of contributions from individual atoms [Eq. (5.80a)], this sum can be rewritten as the product of identical contributions from each of the N atoms:

$$e^{-G/(k_B T)} = \left(e^{-Bm_o/k_B T} + e^{+Bm_o/k_B T}\right)^N = \left(2\cosh[Bm_o/(k_B T)]\right)^N. \tag{5.81c}$$

(In the second expression, the first term is from state $s_i = -1$ and the second from $s_i = +1$.) Taking the logarithm of both sides, we obtain

$$G(B, T, N) = -N k_B T \ln\left(2 \cosh[B m_o/(k_B T)]\right). \tag{5.82}$$

Differentiating with respect to B [Eq. (5.81b)], we obtain for the mean magnetization

$$\bar{M} = N m_o \tanh\left[B m_o/(k_B T)\right]. \tag{5.83}$$

At high temperatures, $k_B T \gg B m_o$, the magnetization increases linearly with the applied magnetic field (the atoms begin to align with **B**), so the magnetic susceptibility is independent of B:

$$\chi_M \equiv \left(\frac{\partial \bar{M}}{\partial B}\right)_{T,N} \simeq N m_o^2/(k_B T). \tag{5.84}$$

The proportionality $\chi_M \propto 1/T$ for a paramagnetic material at high temperature is called *Curie's law*. At low temperatures ($k_B T \ll B m_o$), the atoms are essentially all aligned with **B**, and the magnetization saturates at $\bar{M} = N m_o$.

5.8.2 Ferromagnetism: The Ising Model

Turn now to a magnetic material for which the spins' interactions are strong, and there is no external B field. In such a material, at high temperatures the spin directions are random, while at low enough temperatures the interactions drive neighboring spins to align with one another, producing a net magnetization. This is called *ferromagnetism*, because it occurs rather strongly in iron. The transition between the two regimes is sharp (i.e., it is a phase transition).

In this section, we introduce a simple model for the spins' interaction: the Ising model.[14] For simplicity, we idealize to two spatial dimensions. The corresponding 3-dimensional model is far more difficult to analyze. Studying the 2-dimensional model carefully brings out many general features of phase transitions, and the intuition it cultivates is very helpful in thinking about experiments and simulations.

In this model, the atoms are confined to a square lattice that lies in the x-y plane, and their spins can point up (along the $+z$ direction) or down. The pairwise interaction energy is nonzero only for nearest neighbor atoms:

$$E_{s_i, s_j} = -J s_i s_j \quad \text{for nearest neighbors;} \tag{5.85}$$

it vanishes for all other pairs. Here J is a positive constant (which depends on the lattice's specific volume $v = V/N$, but that will not be important for us). Note that the interaction energy $-J s_i s_j$ is negative if the spins are aligned ($s_i = s_j$) and positive if they are opposite ($s_i = -s_j$), so like spins attract and opposite spins repel. Although the Ising model does not explicitly include more distant interactions, they are present indirectly: the "knock-on" effect from one spin to the next, as we shall see, introduces

---

14. Named for Ernst Ising, who first investigated it, in 1925.

long-range organization that propagates across the lattice when the temperature is reduced below a critical value T_c, inducing a second-order phase transition. We use the dimensionless parameter

$$K \equiv J/(k_B T) \tag{5.86}$$

to characterize the phase transition. For $K \ll 1$ (i.e., $k_B T \gg J$), the spins will be almost randomly aligned, and the total interaction energy will be close to zero. When $K \gg 1$ (i.e., $k_B T \ll J$), the strong coupling will drive the spins to align over large (2-dimensional) volumes. At some critical intermediate temperature $K_c \sim 1$ [and corresponding temperature $T_c = J/(k_B K_c)$], the phase transition will occur.

critical temperature T_c and critical parameter K_c

We compute this critical K_c, and macroscopic properties of the material near it, using two modern, sophisticated computational techniques: renormalization methods in Sec. 5.8.3 and Monte Carlo methods in Sec. 5.8.4. We examine the accuracy of these methods by comparing our results with an exact solution for the 2-dimensional Ising model, derived in a celebrated paper by Lars Onsager (1944).

5.8.3 Renormalization Group Methods for the Ising Model T2

5.8.3

The key idea behind the renormalization group approach to the Ising model is to try to replace the full lattice by a sparser lattice that has similar thermodynamic properties, and then to iterate, making the lattice more and more sparse; see Fig. 5.10.[15]

idea behind renormalization group

We implement this procedure using the statistical sum (5.80c) for the Gibbs potential, except that here the external magnetic field B vanishes, so the bath is purely thermal and its potential is $F(N, V, T)$—the physical free energy, not G—and the statistical sum (5.80c) reads $e^{-F/(k_B T)} \equiv z = \sum_n e^{-E_n/(k_B T)}$. For our Ising model, with its nearest-neighbor interaction energies (5.85), this becomes

$$e^{-F(N,V,T)/(k_B T)} \equiv z = \sum_{\{s_1=\pm 1, s_2=\pm 1, \ldots\}} e^{K \Sigma^1 s_i s_j}. \tag{5.87a}$$

Here in the exponential Σ^1 means a sum over all pairs of nearest neighbor sites $\{i, j\}$. The dependence on the material's number of atoms N appears in the number of terms in the big sum; the dependence on V/N is via the parameter J, and on T is via the parameter $K = J/(k_B T)$.

The first step in the renormalization group method is to rewrite Eq. (5.87a) so that each of the open-circle spins of Fig. 5.10 (e.g., s_5) appears in only one term in the exponential. Then we explicitly sum each of those spins over ± 1, so they no longer appear in the summations:

$$z = \sum_{\{\ldots, s_4=\pm 1, s_5=\pm 1, s_6=\pm 1, \ldots\}} \cdots e^{K(s_1+s_2+s_3+s_4)s_5} \cdots$$

$$= \sum_{\{\ldots, s_4=\pm 1, s_6=\pm 1, \ldots\}} \cdots \left[e^{K(s_1+s_2+s_3+s_4)} + e^{-K(s_1+s_2+s_3+s_4)} \right] \cdots. \tag{5.87b}$$

---

15. This section is based in part on Maris and Kadanoff (1978) and Chandler (1987, Sec. 5.7).

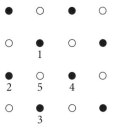

FIGURE 5.10 Partition of a square lattice into two interlaced square lattices (solid circles and open circles). In the renormalization group approach, the open-circle spins are removed from the lattice, and all their interactions are replaced by modified interactions between the remaining solid-circle spins. The new lattice is rotated by $\pi/4$ with respect to the original lattice, and the lattice spacing increases by a factor $\sqrt{2}$.

(This rewriting of z is possible because each open-circle spin interacts only with solid-circle spins.) The partition function z is now a product of terms like those in the square brackets, one for each open-circle lattice site that we have "removed." We would like to rewrite each square-bracketed term in a form involving solely nearest-neighbor interactions of the solid-circle spins, so that we can then iterate our procedure. Such a rewrite, however, is not possible; after some experimentation, one can verify that the rewrite also requires next-nearest-neighbor interactions and four-site interactions:

$$\left[e^{K(s_1+s_2+s_3+s_4)} + e^{-K(s_1+s_2+s_3+s_4)} \right]$$
$$= f(K) e^{[\frac{1}{2}K_1(s_1 s_2 + s_2 s_3 + s_3 s_4 + s_4 s_1) + K_2(s_1 s_3 + s_2 s_4) + K_3 s_1 s_2 s_3 s_4]}. \tag{5.87c}$$

We can determine the four functions $K_1(K)$, $K_2(K)$, $K_3(K)$, $f(K)$ by substituting each of the four possible distinct combinations of $\{s_1, s_2, s_3, s_4\}$ into Eq. (5.87c). Those four combinations, arranged in the pattern of the solid circles of Fig. 5.10, are $\genfrac{}{}{0pt}{}{+}{+}\genfrac{}{}{0pt}{}{+}{+}$, $\genfrac{}{}{0pt}{}{+}{-}\genfrac{}{}{0pt}{}{+}{-}$, $\genfrac{}{}{0pt}{}{+}{-}\genfrac{}{}{0pt}{}{+}{-}$, and $\genfrac{}{}{0pt}{}{+}{+}\genfrac{}{}{0pt}{}{+}{-}$. [Rotating the pattern or changing all signs leaves both sides of Eq. (5.87c) unchanged.] By inserting these combinations into Eq. (5.87c) and performing some algebra, we obtain

$$K_1 = \frac{1}{4} \ln \cosh(4K),$$

$$K_2 = \frac{1}{8} \ln \cosh(4K),$$

$$K_3 = \frac{1}{8} \ln \cosh(4K) - \frac{1}{2} \ln \cosh(2K), \text{ and}$$

$$f(K) = 2[\cosh(2K)]^{1/2} [\cosh(4K)]^{1/8}. \tag{5.87d}$$

By inserting expression (5.87c) and the analogous expressions for the other terms into Eq. (5.87b), we obtain the partition function for our original N-spin lattice of open and closed circles, expressed as a sum over the $(N/2)$-spin lattice of closed circles:

$$z(N, K) = [f(K)]^{N/2} \sum e^{[K_1 \Sigma^1 s_i s_j + K_2 \Sigma^2 s_i s_j + K_3 \Sigma^3 s_i s_j s_k s_l]}. \tag{5.87e}$$

Here the symbol Σ^1 still represents a sum over all nearest neighbors but now in the $N/2$ lattice, Σ^2 is a sum over the four next nearest neighbors, and Σ^3 is a sum over spins located at the vertices of a unit cell. [The reason we defined K_1 with the 1/2 in Eq. (5.87c) was because each nearest neighbor interaction appears in two adjacent squares of the solid-circle lattice, thereby converting the 1/2 to a 1 in Eq. (5.87e).]

So far, what we have done is exact. We now make two drastic approximations that Onsager did not make in his exact treatment, but are designed to simplify the remainder of the calculation and thereby elucidate the renormalization group method. First, in evaluating the partition function (5.87e), we drop completely the quadruple interaction (i.e., we set $K_3 = 0$). This is likely to be decreasingly accurate as we lower the temperature and the spins become more aligned. Second, we assume that near the critical point, in some average sense, the degree of alignment of next nearest neighbors (of which there are as many as nearest neighbors) is "similar" to that of the nearest neighbors, so that we can set $K_2 = 0$ but increase K_1 to

$$K' = K_1 + K_2 = \frac{3}{8}\ln\cosh(4K). \quad (5.88)$$

(If we simply ignored K_2 we would not get a phase transition.) This substitution ensures that the energy of a lattice with $N/2$ *aligned* spins—and therefore N nearest neighbor and N next nearest neighbor bonds, namely, $-(K_1 + K_2)Nk_BT$—is the same as that of a lattice in which we just include the nearest neighbor bonds but strengthen the interaction from K_1 to K'. Clearly this will be unsatisfactory at high temperature, but we only need it to be true near the phase transition's critical temperature.

These approximations bring the partition function (5.87e) into the form

$$z(N, K) = [f(K)]^{N/2} z(N/2, K'), \quad (5.89a)$$

which relates the partition function (5.87a) for our original Ising lattice of N spins and interaction constant K to that of a similar lattice with $N/2$ spins and interaction constant K'.

As the next key step in the renormalization procedure, we note that because the free energy, $F = -k_BT \ln z$, is an extensive variable, $\ln z$ must increase in direct proportion to the number of spins, so that it must have the form

$$-F/(k_BT) \equiv \ln z(N, K) = Ng(K), \quad (5.89b)$$

for some function $g(K)$. By combining Eqs. (5.89a) and (5.89b), we obtain a relation for the function $g(K)$ (the free energy, aside from constants) in terms of the function $f(K)$:

$$g(K') = 2g(K) - \ln f(K), \quad \text{where } f(K) = 2[\cosh(2K)]^{1/2}[\cosh(4K)]^{1/8} \quad (5.90)$$

[cf. Eq. (5.87d)].

Equations (5.88) and (5.90) are the fundamental equations that allow us to calculate thermodynamic properties. They are called the *renormalization group equations,* because their transformations form a mathematical group, and they are a scheme

renormalization group equations

for determining how the effective coupling parameter K changes (gets renormalized) when one views the lattice on larger and larger distance scales. Renormalization group equations like these have been widely applied in elementary-particle theory, condensed-matter theory, and elsewhere. Let us examine these in detail.

The iterative map (5.88) expresses the coupling constant K' for a lattice of enlarged physical size and reduced number of particles $N/2$ in terms of K for the smaller lattice with N particles. [And the associated map (5.90) expresses the free energy when the lattice is viewed on the larger scale in terms of that for a smaller scale.] The map (5.88) has a fixed point that is obtained by setting $K' = K$ [i.e., $K_c = \frac{3}{8} \ln \cosh(4K_c)$], which implies

$$K_c = 0.507. \tag{5.91}$$

This fixed point corresponds to the critical point for the phase transition, with critical temperature T_c such that $K_c = J/(k_B T_c)$.

We can infer that this is the critical point by the following physical argument. Suppose that T is slightly larger than T_c, so K is slightly smaller than K_c. Then, when we make successive iterations, because $dK'/dK > 1$ at $K = K_c$, K decreases with each step, moving farther from K_c; the fixed point is unstable. What this means is that, when $T > T_c$, as we look on larger and larger scales, the effective coupling constant K becomes weaker and weaker, so the lattice becomes more disordered. Conversely, below the critical temperature ($T < T_c$ and $K > K_c$), the lattice becomes more ordered with increasing scale. Only when $K = K_c$ does the lattice appear to be comparably disordered on all scales. It is here that the increase of order with lengthscale changes from greater order at smaller scales (for high temperatures) to greater order at larger scales (for low temperatures).

To demonstrate more explicitly that $K = K_c$ is the location of a phase transition, we compute the lattice's specific heat in the vicinity of K_c. The first step is to compute the lattice's entropy, $S = -(\partial F/\partial T)_{V,N}$. Recalling that $K \propto 1/T$ at fixed V, N [Eq. (5.86)] and using expression (5.89b) for F, we see that

$$S = -\left(\frac{\partial F}{\partial T}\right)_{V,N} = Nk_B \left[g - K\left(\frac{dg}{dK}\right)\right]. \tag{5.92a}$$

The specific heat at constant volume is then, in turn, given by

$$C_V = T\left(\frac{\partial S}{\partial T}\right)_{V,N} = Nk_B K^2 \frac{d^2 g}{dK^2}. \tag{5.92b}$$

Next we note that, because the iteration (5.88) is unstable near K_c, the inverse iteration

$$K = \frac{1}{4}\cosh^{-1}[\exp(8K'/3)] \tag{5.92c}$$

is stable. The corresponding inverse transformation for the function $g(K)$ is obtained from Eq. (5.90), with K in the function f converted to K' using Eq. (5.92c):

$$g(K) = \frac{1}{2}g(K') + \frac{1}{2}\ln\{2\exp(2K'/3)[\cosh(4K'/3)]^{1/4}\}. \tag{5.92d}$$

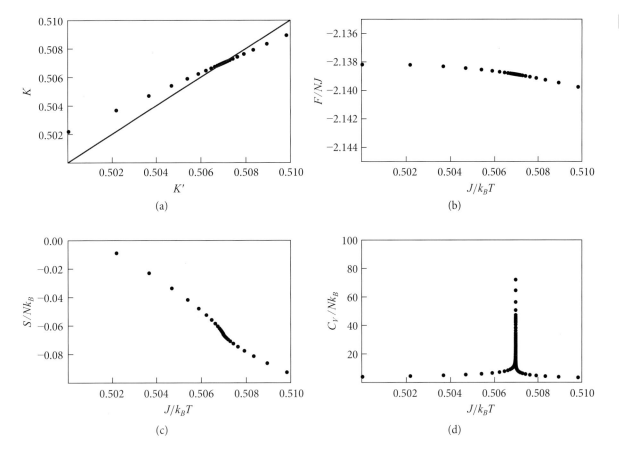

FIGURE 5.11 (a) Iteration map $K(K')$ in the vicinity of the critical point. (b) Free energy per spin. (c) Entropy per spin. (d) Specific heat per spin. Recall that $J/(k_B T) = K$.

Now, we know that at low temperature, $T \ll T_c$ and $K \gg K_c$, all the spins are aligned; correspondingly, in the statistical sum (5.87a) the two terms with all s's identical dominate, giving $z = e^{-F/(k_B T)} = e^{Ng} = 2e^{2NK}$, whence $g(K) \simeq 2K$. Conversely, at high temperature, there is complete disorder, and $K \to 0$. This means that every one of the 2^N terms in the statistical sum (5.87a) is unity, giving $z = e^{Ng} = 2^N$, whence $g(K) \simeq \ln 2$. We can therefore use the iterative map, Eqs. (5.92c) and (5.92d), to approach $K = K_c$ from either side, starting with the high-temperature and low-temperature limits of $g(K)$ and evaluating thermodynamic quantities at each step. More specifically, at each step, we evaluate $g(K)$, dg/dK, and d^2g/dK^2 numerically, and from them we compute F, S, and C_V using Eqs. (5.89b), (5.92a), and (5.92b).

The iterated values of these quantities are plotted as points in Fig. 5.11. Note that the entropy S is continuous at K_c (panel c), but its derivative, the specific heat (panel d), diverges at K_c, as $K \to K_c$ from either side. This is characteristic of a second-order phase transition.

To calculate the explicit form of this divergence, suppose that $g(K)$ is a sum of an analytic (infinitely differentiable) function and a nonanalytic part. Suppose that near the critical point, the nonanalytic part behaves as $g(K) \sim |K - K_c|^{2-\alpha}$ for some "critical exponent" α. This implies that C_V diverges $\propto |K - K_c|^{-\alpha} \propto |T - T_c|^{-\alpha}$. Now, from Eq. (5.92d), we have that

$$|K' - K_c|^{2-\alpha} = 2|K - K_c|^{2-\alpha}, \tag{5.93a}$$

or equivalently,

$$\frac{dK'}{dK} = 2^{1/(2-\alpha)}. \tag{5.93b}$$

Evaluating the derivative at $K = K_c$ from Eq. (5.92c), we obtain

$$\alpha = 2 - \frac{\ln 2}{\ln(dK'/dK)_c} = 0.131, \tag{5.93c}$$

which is consistent with the numerical calculation.

exact analysis of Ising model compared to our approximate analysis

The exact Onsager (1944) analysis of the Ising model gives $K_c = 0.441$ compared to our $K_c = 0.507$, and $C_V \propto -\ln|T - T_c|$ compared to our $C_V \propto |T - T_c|^{-0.131}$. Evidently, our renormalization group approach (formulated by Maris and Kadanoff, 1978) gives a fair approximation to the correct answers but not a good one.

Our approach appears to have a serious problem in that it predicts a negative value for the entropy in the vicinity of the critical point (Fig. 5.11c). This is surely unphysical. (The entropy becomes positive farther away, on either side of the critical point.) This is an artificiality associated with our approach's ansatz [i.e., associated with our setting $K_2 = K_3 = 0$ and $K' = K_1 + K_2$ in Eq. (5.88)]. It does not seem easy to cure this in a simple renormalization-group approach.

Nonetheless, our calculations exhibit the conceptual and physical essentials of the renormalization-group approach to phase transitions.

Why did we bother to go through this cumbersome procedure when Onsager has given us an exact analytical solution to the Ising model? The answer is that it is not possible to generalize the Onsager solution to more complex and realistic problems. In particular, it has not even been possible to find an Onsager-like solution to the 3-dimensional Ising model. However, once the machinery of the renormalization group has been mastered, it can produce approximate answers, with an accuracy that can be estimated, for a variety of problems. In the following section we look at a quite different approach to the same 2-dimensional Ising problem with exactly the same motivation in mind.

EXERCISES

Exercise 5.15 *Example: One-Dimensional Ising Lattice*
(a) Write down the partition function for a 1-dimensional Ising lattice as a sum over terms describing all possible spin organizations.

(b) Show that by separating into even and odd numbered spins, it is possible to factor the partition function and relate $z(N, K)$ exactly to $z(N/2, K')$. Specifically, show that

$$z(N, K) = f(K)^{N/2} z(N/2, K') \tag{5.94}$$

where $K' = \ln[\cosh(2K)]/2$, and $f(K) = 2[\cosh(2K)]^{1/2}$.

(c) Use these relations to demonstrate that the 1-dimensional Ising lattice does not exhibit a second-order phase transition.

5.8.4 Monte Carlo Methods for the Ising Model

In this section, we explore the phase transition of the 2-dimensional Ising model using our second general method: the *Monte Carlo* approach.[16] This method, like the renormalization group, is a powerful tool for a much larger class of problems than just phase transitions.[17]

The Monte Carlo approach is much more straightforward in principle than the renormalization group. We set up a square lattice of atoms and initialize their spins randomly.[18] We imagine that our lattice is in contact with a heat bath with a fixed temperature T (it is one member of a canonical ensemble of systems), and we drive it to approach statistical equilibrium and then wander onward through an equilibrium sequence of states $|n_1\rangle, |n_2\rangle, \ldots$ in a prescribed, ergodic manner. Our goals are to visualize typical equilibrium states (see Fig. 5.12 later in this section) and to compute thermodynamic quantities using $\bar{X} = z^{-1} \sum_n e^{-E_n/(k_B T)} X_n$, where the sum is over the sequence of states $|n_1\rangle, |n_2\rangle, \ldots$. For example, we can compute the specific heat (at constant volume) from

$$C_V = \frac{d\bar{E}}{dT} = \frac{\partial}{\partial T} \left(\frac{\sum_n e^{-E_n/(k_B T)} E_n}{\sum_n e^{-E_n/(k_B T)}} \right) = \frac{\overline{E^2} - \bar{E}^2}{k_B T^2}. \tag{5.95}$$

Monte Carlo method for Ising model

[Note that a singularity in the specific heat at a phase transition will be associated with large fluctuations in the energy, just as it is associated with large fluctuations of temperature; Eq. (5.73d).]

In constructing our sequence of lattice states $|n_1\rangle, |n_2\rangle, \ldots$, we obviously cannot visit all 2^N states even just once, so we must sample them fairly. How can we prescribe the rule for changing the spins when going from one state in our sample to the next, so as to produce a fair sampling? There are many answers to this question; we describe

---

16. The name "Monte Carlo" is a sardonic reference to the casino whose patrons believe they will profit by exploiting random processes.
17. We shall meet it again in Sec. 28.6.1.
18. This and other random steps that follow are performed numerically and require a (pseudo) random number generator. Most programming languages supply this utility, which is mostly used uncritically, occasionally with unintended consequences. Defining and testing "randomness" is an important topic which, unfortunately, we shall not address. See, for example, Press et al. (2007).

and use one of the simplest, due to Metropolis et al. (1953). To understand this, we must appreciate that we don't need to comprehend the detailed dynamics by which a spin in a lattice actually flips. All that is required is that the rule we adopt, for going from one state to the next, should produce a sequence that is in statistical equilibrium.

Let us denote by $p_{nn'}$ the conditional probability that, if the lattice is in state $|n\rangle$, then the next step will take it to state $|n'\rangle$. For statistical equilibrium, it must be that the probability that any randomly observed step takes us out of state $|n\rangle$ is equal to the probability that it takes us into that state:

$$\rho_n \sum_{n'} p_{nn'} = \sum_{n'} \rho_{n'} p_{n'n}. \tag{5.96}$$

(Here ρ_n is the probability that the lattice was in state $|n\rangle$ just before the transition.) We know that in equilibrium, $\rho_{n'} = \rho_n \, e^{(E_n - E_{n'})/(k_B T)}$, so our conditional transition probabilities must satisfy

$$\sum_{n'} p_{nn'} = \sum_{n'} p_{n'n} e^{(E_n - E_{n'})/(k_B T)}. \tag{5.97}$$

Metropolis rule for transition probabilities

The Metropolis rule is simple:

$$\text{if } E_n > E_m, \text{ then } p_{nm} = 1;$$
$$\text{and if } E_n < E_m, \text{ then } p_{nm} = \exp[(E_n - E_m)/(k_B T)] \tag{5.98}$$

up to some normalization constant. It is easy to show that this satisfies the statistical equilibrium condition (5.97) and that it drives an initial out-of-equilibrium system toward equilibrium.

The numerical implementation of the Metropolis rule (5.98) is as follows: Start with the lattice in an initial, random state, and then choose one spin at random to make a trial flip. If the new configuration has a lower energy, we always accept the change. If it has a higher energy, we only accept the change with a probability given by $\exp[-\Delta E/(k_B T)]$, where $\Delta E > 0$ is the energy change.[19] In this way, we produce a sequence of states that will ultimately have the equilibrium distribution function, and we can perform our thermodynamic averages using this sequence in an unweighted fashion. This is a particularly convenient procedure for the Ising problem, because, by changing just one spin at a time, ΔE can only take one of five values ($-4, -2, 0, +2, +4$ in units of J), and it is possible to change from one state to the next very quickly. (It also helps to store the two acceptance probabilities $e^{-2J/(k_B T)}$ and $e^{-4J/(k_B T)}$ for making an energy-gaining transition, so as to avoid evaluating exponentials at every step.)

---

19. There is a small subtlety here. The probability of making a given transition is actually the product of the probability of making the trial flip and of accepting the trial. However, the probability of making a trial flip is the same for all the spins that we might flip ($1/N$), and these trial probabilities cancel, so it is only the ratio of the probabilities of acceptance that matters.

 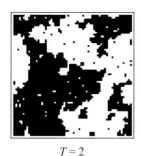 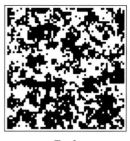

T = 1   T = 2   T = 3

FIGURE 5.12 Typical equilibrium Ising lattices for temperatures $T = 1, 2, 3$ in units of J/k_B. The black regions have spins $s = +1$; the white, $s = -1$.

How big a lattice do we need, and how many states should we consider? The lattice size can be surprisingly small to get qualitatively correct results, if we adopt periodic boundary conditions. That is to say, we imagine a finite tiling of our actual finite lattice, and every time we need to know the spin at a site beyond the tiling's last column (or row), we use the corresponding spin an equal distance beyond the first column (or row). This device minimizes the effects of the boundary on the final answer. Lattices as small as 32×32 can be useful. The length of the computation depends on the required accuracy. (In practice, this is usually implemented the other way around. The time available on a computer of given speed determines the accuracy.) One thing should be clear. It is necessary that we explore a reasonable volume of state space to be able to sample it fairly and compute meaningful estimates of thermodynamic quantities. The final lattice should exhibit no vestigial patterns from the state when the computation was half complete. In practice, it is this consideration that limits the size of the lattice, and it is one drawback of the Metropolis algorithm that the step sizes are necessarily small. There is a large bag of tricks for Monte Carlo simulations that can be used for variance reduction and estimation, but we only concern ourselves here with the general method.

Returning to the Ising model, we show in Fig. 5.12 typical equilibrium states (snapshots) for three temperatures, measured in units of J/k_B. Recall that the critical temperature is $T_c = J/(k_B K_c) = J/(0.441 k_B) = 2.268 J/k_B$. Note the increasingly long-range order as the temperature is reduced below T_c.

Monte Carlo results for Ising model

We have concluded this chapter with an examination of a very simple system that can approach equilibrium according to specified rules and that can exhibit strong fluctuations. In the following chapter, we examine fluctuations more systematically.

EXERCISES

Exercise 5.16 *Practice: Direct Computation of Thermodynamic Integrals*
Estimate how long it would take a personal computer to calculate the partition function for a 32×32 Ising lattice by evaluating every possible state.

5.8 Magnetic Materials

Exercise 5.17 *Example: Monte Carlo Approach to Phase Transition* T2
Write a simple computer program to compute the energy and the specific heat of a 2-dimensional Ising lattice as described in the text. Examine the accuracy of your answers by varying the size of the lattice and the number of states sampled. (You might also try to compute a formal variance estimate.)

Exercise 5.18 *Problem: Ising Lattice with an Applied Magnetic Field* T2
Modify your computer program from Ex. 5.17 to deal with the 2-dimensional Ising model augmented by an externally imposed, uniform magnetic field [Eqs. (5.80)]. Compute the magnetization and the magnetic susceptibility for wisely selected values of $m_o B/J$ and $K = J/(k_B T)$.

Bibliographic Note

Most statistical mechanics textbooks include much detail on statistical thermodynamics. Among those we have found useful at an elementary level are Kittel and Kroemer (1980), and at more advanced levels, Chandler (1987), Sethna (2006), Kardar (2007), Reif (2008), and Pathria and Beale (2011). Chandler (1987) and Sethna (2006) are particularly good for phase transitions. Our treatment of the renormalization group in Sec. 5.8.3 is adapted in part from Chandler, who also covers Monte Carlo methods.

CHAPTER SIX

Random Processes

These motions were such as to satisfy me, after frequently repeated observation, that they arose neither from currents in the fluid, nor from its gradual evaporation, but belonged to the particle itself.

ROBERT BROWN (1828)

6.1 Overview

In this chapter we analyze, among others, the following issues:

- What is the time evolution of the distribution function for an ensemble of systems that begins out of statistical equilibrium and is brought to equilibrium through contact with a heat bath?

- How can one characterize the noise introduced into experiments or observations by noisy devices, such as resistors and amplifiers?

- What is the influence of such noise on one's ability to detect weak signals?

- What filtering strategies will improve one's ability to extract weak signals from strong noise?

- Frictional damping of a dynamical system generally arises from coupling to many other degrees of freedom (a bath) that can sap the system's energy. What is the connection between the fluctuating (noise) forces that the bath exerts on the system and its damping influence?

The mathematical foundation for analyzing such issues is the *theory of random processes* (i.e., of functions that are random and unpredictable but have predictable probabilities for their behavior). A portion of the theory of random processes is the *theory of stochastic differential equations* (equations whose solutions are probability distributions rather than ordinary functions). This chapter is an overview of these topics, sprinkled throughout with applications.

In Sec. 6.2, we introduce the concept of random processes and the various probability distributions that describe them. We introduce restrictions that we shall adhere to—the random processes that we study are stationary and ergodic—and we introduce an example that we return to time and again: a *random-walk* process, of which *Brownian motion* is an example. In Sec. 6.3, we discuss two special classes of random processes: Markov processes and Gaussian processes; we also present two important theorems: the central limit theorem, which explains why random processes so often

BOX 6.1. READERS' GUIDE

- Relativity does not enter into this chapter.
- This chapter does not rely in any major way on previous chapters, but it does make occasional reference to results from Chaps. 4 and 5 about statistical equilibrium and fluctuations in and away from statistical equilibrium.
- No subsequent chapter relies in any major way on this chapter. However:
  - The concepts of spectral density and correlation function, developed in Sec. 6.4, will be used in Ex. 9.8 when treating coherence properties of radiation, in Sec. 11.9.2 when studying thermal noise in solids, in Sec. 15.4 when studying turbulence in fluids, in Sec. 23.2.1 in treating the quasilinear formalism for weak plasma turbulence, and in Sec. 28.6.1 when discussing observations of the anisotropy of the cosmic microwave background radiation.
  - The fluctuation-dissipation theorem, developed in Sec. 6.8, will be used in Sec. 11.9.2 for thermoelastic noise in solids, and in Sec. 12.5 for normal modes of an elastic body.
  - The Fokker-Planck equation, developed in Sec. 6.9, will be referred to or used in Secs. 20.4.3, 20.5.1, and 28.6.3 and Exs. 20.8 and 20.10 when discussing thermal equilibration in a plasma and thermoelectric transport coefficients, and it will be used in Sec. 23.3.3 in developing the quasilinear theory of wave-particle interactions in a plasma.

have Gaussian probability distributions, and Doob's theorem, which says that all the statistical properties of a Markov, Gaussian process are determined by just three parameters. In Sec. 6.4, we introduce two powerful mathematical tools for the analysis of random processes: the correlation function and spectral density, and prove the Wiener-Khintchine theorem, which relates them. As applications of these tools, we use them to prove Doob's theorem and to discuss optical spectra, noise in interferometric gravitational wave detectors, and fluctuations of cosmological mass density and of the distribution of galaxies in the universe. In Secs. 6.6 and 6.7, we introduce another powerful tool, the filtering of a random process, and we use these tools to develop the theory of noise and techniques for extracting weak signals from large noise. As applications we study shot noise (which is important, e.g., in measurements with laser light), frequency fluctuations of atomic clocks, and also the Brownian motion of a dust particle buffeted by air molecules and its connection to random walks. In Sec. 6.8, we develop another powerful tool, the fluctuation-dissipation theorem, which quantifies the relationship between the fluctuations and the dissipation (friction) produced by one and the same heat bath. As examples, we explore Brownian motion (once again), Johnson noise in a resistor and the voltage fluctuations it produces in electric circuits, thermal noise in high-precision optical measurements, and quantum limits on the accuracy of high-precision measurements and how to circumvent them. Finally, in Sec. 6.9 we derive and discuss the Fokker-Planck equation, which governs the evolution of Markov random processes, and we illustrate it with the random motion of an atom that is being cooled by interaction with laser beams (so-called *optical molasses*) and with thermal noise in a harmonic oscillator.

6.2 Fundamental Concepts

In this section we introduce a number of fundamental concepts about random processes.

6.2.1 Random Variables and Random Processes

RANDOM VARIABLE

A (1-dimensional) *random variable* is a (scalar) function $y(t)$, where t is usually time, for which the future evolution is not determined uniquely by any set of initial data—or at least by any set that is knowable to you and me. In other words, *random variable* is just a fancy phrase that means "unpredictable function." Throughout this chapter, we insist for simplicity that our random variables y take on a continuum of *real* values ranging over some interval, often but not always $-\infty$ to $+\infty$. The generalizations to y with complex or discrete (e.g., integer) values, and to independent variables other than time, are straightforward.

Examples of random variables are: (i) the total energy $E(t)$ in a cell of gas that is in contact with a heat bath; (ii) the temperature $T(t)$ at the corner of Main Street and Center Street in Logan, Utah; (iii) the price per share of Google stock $P(t)$; (iv) the mass-flow rate $\dot{M}(t)$ from the Amazon River into the Atlantic Ocean. One can also deal with random variables that are vector or tensor functions of time; in Track-Two portions of this chapter we do so.

RANDOM PROCESS

A (1-dimensional) *random process* (also called "stochastic process") is an ensemble $\mathcal{E}$ of real random variables $y(t)$ that, in a physics context, all represent the same kind of physical entity. For example, each $y(t)$ could be the longitude of a particular oxygen molecule undergoing a random walk in Earth's atmosphere. The individual random variables $y(t)$ in the ensemble $\mathcal{E}$ are often called *realizations* of the random process.

As an example, Fig. 6.1 shows three realizations $y(t)$ of a random process that represents the random walk of a particle in one dimension. For details, see Ex. 6.4, which shows how to generate realizations like these on a computer.

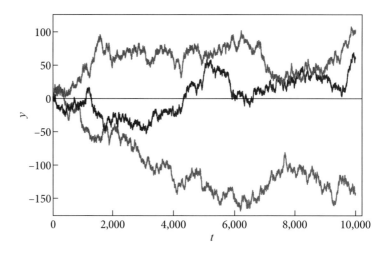

FIGURE 6.1 Three different realizations $y(t)$ of a random process that describes the location y of a particle at time t, when it is undergoing a random walk in 1 dimension (e.g., an atmospheric oxygen molecule's east-west motion).

6.2.2 Probability Distributions

PROBABILITY DISTRIBUTIONS FOR A RANDOM PROCESS

Since the precise time evolution of a random variable $y(t)$ is not predictable, if one wishes to make predictions, one can do so only probabilistically. The foundation for probabilistic predictions is a set of probability functions for the random process (i.e., for the ensemble $\mathcal{E}$ of its realizations).

More specifically, the most general (1-dimensional) random process is fully characterized by the set of probability distributions $p_1, p_2, p_3, \ldots$ defined as

probability distributions for a random process

$$p_n(y_n, t_n; \ldots; y_2, t_2; y_1, t_1) dy_n \ldots dy_2 dy_1. \tag{6.1}$$

Equation (6.1) tells us the probability that a realization $y(t)$, drawn at random from the process (the ensemble $\mathcal{E}$), (i) will take on a value between y_1 and $y_1 + dy_1$ at time t_1, (ii) also will take on a value between y_2 and $y_2 + dy_2$ at a later time $t_2, \ldots$, and (iii) also will take on a value between y_n and $y_n + dy_n$ at a later time t_n. (Note that the subscript n on p_n tells us how many independent values of y appear in p_n, and that earlier times are placed to the right—a practice common for physicists, particularly when dealing with propagators.) If we knew the values of all the process's probability distributions (an infinite number of p_ns!), then we would have full information about its statistical properties. Not surprisingly, it will turn out that, if the process in some sense is in statistical equilibrium, then we can compute all its probability distributions from a very small amount of information. But that comes later; first we must develop more formalism.

ENSEMBLE AVERAGES

From the probability distributions, we can compute ensemble averages (denoted by brackets). For example, the quantities

$$\langle y(t_1) \rangle \equiv \int y_1 p_1(y_1, t_1) dy_1 \quad \text{and} \quad \sigma_y^2(t_1) \equiv \left\langle [y(t_1) - \langle y(t_1) \rangle]^2 \right\rangle \qquad (6.2a)$$

ensemble average and variance

are the ensemble-averaged value of y and the variance of y at time t_1. Similarly,

$$\langle y(t_2) y(t_1) \rangle \equiv \int y_2 y_1 p_2(y_2, t_2; y_1, t_1) dy_2 dy_1 \qquad (6.2b)$$

is the average value of the product $y(t_2) y(t_1)$.

CONDITIONAL PROBABILITIES

Besides the (absolute) probability distributions p_n, we also find useful an infinite series of *conditional* probability distributions $P_2, P_3, \ldots$, defined as

$$P_n(y_n, t_n | y_{n-1}, t_{n-1}; \ldots; y_1, t_1) dy_n. \qquad (6.3)$$

conditional probability distributions

This distribution is the probability that, *if* $y(t)$ took on the values $y_1, y_2, \ldots, y_{n-1}$ at times $t_1, t_2, \ldots, t_{n-1}$, then it will take on a value between y_n and $y_n + dy_n$ at a later time t_n.

It should be obvious from the definitions of the probability distributions that

$$\begin{aligned} &p_n(y_n, t_n; \ldots; y_1, t_1) \\ &= P_n(y_n, t_n | y_{n-1}, t_{n-1}; \ldots; y_1, t_1) p_{n-1}(y_{n-1}, t_{n-1}; \ldots; y_1, t_1). \end{aligned} \qquad (6.4)$$

Using this relation, one can compute all the conditional probability distributions P_n from the absolute distributions $p_1, p_2, \ldots$. Conversely, using this relation recursively, one can build up all the absolute probability distributions p_n from $p_1(y_1, t_1)$ and all the conditional distributions $P_2, P_3, \ldots$.

STATIONARY RANDOM PROCESSES

A random process is said to be *stationary* if and only if its probability distributions p_n depend just on time differences and not on absolute time:

stationary random process

$$p_n(y_n, t_n + \tau; \ldots; y_2, t_2 + \tau; y_1, t_1 + \tau) = p_n(y_n, t_n; \ldots; y_2, t_2; y_1, t_1). \qquad (6.5)$$

If this property holds for the absolute probabilities p_n, then Eq. (6.4) guarantees it also will hold for the conditional probabilities P_n.

Nonstationary random processes arise when one is studying a system whose evolution is influenced by some sort of clock that registers absolute time, not just time differences. For example, the speeds $v(t)$ of all oxygen molecules in downtown St. Anthony, Idaho, make up random processes regulated in part by the atmospheric temperature and therefore by the rotation of Earth and its orbital motion around the Sun. The

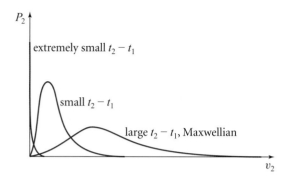

FIGURE 6.2 The probability $P_2(v_2, t_2|0, t_1)$ that a molecule with vanishing speed at time t_1 will have speed v_2 (in a unit interval dv_2) at time t_2. Although the molecular speed is a stationary random process, this probability evolves in time.

influence of these clocks makes $v(t)$ a nonstationary random process. Stationary random processes, by contrast, arise in the absence of any regulating clocks. An example is the speeds $v(t)$ of all oxygen molecules in a room kept at constant temperature.

Stationarity does not mean "no time evolution of probability distributions." For example, suppose one knows that the speed of a specific oxygen molecule vanishes at time t_1, and one is interested in the probability that the molecule will have speed v_2 at time t_2. That probability, $P_2(v_2, t_2|0, t_1)$, is sharply peaked around $v_2 = 0$ for extremely small time differences $t_2 - t_1$ and is Maxwellian for large time differences $t_2 - t_1$ (Fig. 6.2). Despite this evolution, the process is stationary (assuming constant temperature) in the sense that it does not depend on the specific time t_1 at which v happened to vanish, only on the time difference $t_2 - t_1$: $P_2(v_2, t_2|0, t_1) = P_2(v_2, t_2 - t_1|0, 0)$.

Henceforth, throughout this chapter, we restrict attention to random processes that are stationary (at least on the timescales of interest to us); and, accordingly, we use

$$p_1(y) \equiv p_1(y, t_1) \tag{6.6a}$$

for the probability, since it does not depend on the time t_1. We also denote by

$$P_2(y_2, t|y_1) \equiv P_2(y_2, t|y_1, 0) \tag{6.6b}$$

the probability that, if a (realization of a) random process begins with the value y_1, then after the lapse of time t it has the value y_2.

6.2.3 Ergodic Hypothesis

A (stationary) random process (ensemble $\mathcal{E}$ of random variables) is said to satisfy the *ergodic hypothesis* (or, for brevity, it will be called *ergodic*) if and only if it has the following property.

Let $y(t)$ be a random variable in the ensemble $\mathcal{E}$ (i.e., let $y(t)$ be any realization of the process). Construct from $y(t)$ a new ensemble $\mathcal{E}'$ whose members are

$$Y^K(t) \equiv y(t + KT), \tag{6.7}$$

where K runs over all integers, negative and positive, and where T is some very large time interval. Then $\mathcal{E}'$ has the same probability distributions p_n as $\mathcal{E}$; that is, $p_n(Y_n, t_n; \ldots; Y_1, t_1)$ has the same functional form as $p_n(y_n, t_n; \ldots; y_1, t_1)$ for all times such that $|t_i - t_j| < T$.

This is essentially the same ergodic hypothesis as we met in Sec. 4.6.

Henceforth we restrict attention to random processes that satisfy the ergodic hypothesis (i.e., that are ergodic). This, in principle, is a severe restriction. In practice, for a physicist, it is not severe at all. In physics one's objective, when defining random variables that last forever ($-\infty < t < +\infty$) and when introducing ensembles, is usually to acquire computational techniques for dealing with a single, or a small number of, random variables $y(t)$, studied over finite lengths of time. One acquires those techniques by defining conceptual infinite-duration random variables and ensembles in such a way that they satisfy the ergodic hypothesis.

As in Sec. 4.6, because of the ergodic hypothesis, time averages defined using any realization $y(t)$ of a random process are equal to ensemble averages:

$$\bar{F} \equiv \lim_{T \to \infty} \frac{1}{T} \int_{-T/2}^{T/2} F\big(y(t)\big) dt = \langle F(y) \rangle \equiv \int F(y) p_1(y) dy, \tag{6.8}$$

for any function $F = F(y)$. In this sense, each realization of the random process is representative, when viewed over sufficiently long times, of the statistical properties of the process's entire ensemble—and conversely. Correspondingly, we can blur the distinction between the random process and specific realizations of it—and we often do so.

6.3 Markov Processes and Gaussian Processes

6.3.1 Markov Processes; Random Walk

A random process $y(t)$ is said to be *Markov* (also sometimes called "Markovian") if and only if all of its future probabilities are determined by its most recently known value:

Markov random process

$$P_n(y_n, t_n | y_{n-1}, t_{n-1}; \ldots; y_1, t_1) = P_2(y_n, t_n | y_{n-1}, t_{n-1}) \quad \text{for all } t_n \geq \ldots \geq t_2 \geq t_1. \tag{6.9}$$

This relation guarantees that any Markov process (which, of course, we require to be stationary without saying so) is completely characterized by the probabilities

$$p_1(y) \text{ and } P_2(y_2, t | y_1) \equiv \frac{p_2(y_2, t; y_1, 0)}{p_1(y_1)}. \tag{6.10}$$

From $p_1(y)$ and $P_2(y_2, t|y_1)$ one can reconstruct, using the Markov relation (6.9) and the general relation (6.4) between conditional and absolute probabilities, all distribution functions of the process.

Actually, for any random process that satisfies the ergodic hypothesis (which means all random processes considered in this chapter), $p_1(y)$ is determined by the conditional probability $P_2(y_2, t|y_1)$ [Ex. 6.1], so for any Markov (and ergodic) process, all the probability distributions follow from $P_2(y_2, t|y_1)$ alone!

An example of a Markov process is the x component of velocity $v_x(t)$ of a dust particle in an arbitrarily large room,[1] filled with constant-temperature air. Why? Because the molecule's equation of motion is[2] $mdv_x/dt = F'_x(t)$, and the force $F'_x(t)$ is due to random buffeting by other molecules that are uncorrelated (the kick now is unrelated to earlier kicks); thus, there is no way for the value of v_x in the future to be influenced by any earlier values of v_x except the most recent one.

By contrast, the position $x(t)$ of the particle is not Markov, because the probabilities of future values of x depend not just on the initial value of x, but also on the initial velocity v_x—or, equivalently, the probabilities depend on the values of x at two initial, closely spaced times. The pair $\{x(t), v_x(t)\}$ is a 2-dimensional Markov process (see Ex. 6.23).

THE SMOLUCHOWSKI EQUATION

Choose three (arbitrary) times t_1, t_2, and t_3 that are ordered, so $t_1 < t_2 < t_3$. Consider a (realization of an) arbitrary random process that begins with a known value y_1 at t_1, and ask for the probability $P_2(y_3, t_3|y_1)$ (per unit y_3) that it will be at y_3 at time t_3. Since the realization must go through some value y_2 at the intermediate time t_2 (though we don't care what that value is), it must be possible to write the probability to reach y_3 as

$$P_2(y_3, t_3|y_1, t_1) = \int P_3(y_3, t_3|y_2, t_2; y_1, t_1) P_2(y_2, t_2|y_1, t_1) dy_2,$$

where the integration is over all allowed values of y_2. This is not a terribly interesting relation. Much more interesting is its specialization to the case of a Markov process. In that case $P_3(y_3, t_3|y_2, t_2; y_1, t_1)$ can be replaced by $P_2(y_3, t_3|y_2, t_2) = P_2(y_3, t_3 - t_2|y_2, 0) \equiv P_2(y_3, t_3 - t_2|y_2)$, and the result is an integral equation involving only P_2. Because of stationarity, it is adequate to write that equation for the case $t_1 = 0$:

$$\boxed{P_2(y_3, t_3|y_1) = \int P_2(y_3, t_3 - t_2|y_2) P_2(y_2, t_2|y_1) dy_2.} \quad (6.11)$$

Smoluchowski equation

This is the *Smoluchowski equation* (also called *Chapman-Kolmogorov equation*). It is valid for any Markov random process and for times $0 < t_2 < t_3$. We shall discover its power in our derivation of the Fokker-Planck equation in Sec. 6.9.1.

---

1. The room must be arbitrarily large so the effects of the floor, walls, and ceiling can be ignored.
2. By convention, primes are used to identify stochastic forces (i.e., forces that are random processes).

EXERCISES

Exercise 6.1 **Example: Limits of P_2*
Explain why, for any (stationary) random process,

$$\lim_{t \to 0} P_2(y_2, t | y_1) = \delta(y_2 - y_1). \tag{6.12a}$$

Use the ergodic hypothesis to argue that

$$\lim_{t \to \infty} P_2(y_2, t | y_1) = p_1(y_2). \tag{6.12b}$$

Thereby conclude that, for a Markov process, all the probability distributions are determined by the conditional probability $P_2(y_2, t | y_1)$. Give an algorithm for computing them.

Exercise 6.2 *Practice: Markov Processes for an Oscillator*
Consider a harmonic oscillator (e.g., a pendulum), driven by bombardment with air molecules. Explain why the oscillator's position $x(t)$ and velocity $v(t) = dx/dt$ are random processes. Is $x(t)$ Markov? Why? Is $v(t)$ Markov? Why? Is the pair $\{x(t), v(t)\}$ a 2-dimensional Markov process? Why? We study this 2-dimensional random process in Ex. 6.23.

Exercise 6.3 **Example: Diffusion of a Particle; Random Walk*
In Ex. 3.17, we studied the diffusion of particles through an infinite 3-dimensional medium. By solving the diffusion equation, we found that, if the particles' number density at time $t = 0$ was $n_o(\mathbf{x})$, then at time t it has become

$$n(\mathbf{x}, t) = [1/(4\pi Dt)]^{3/2} \int n_o(\mathbf{x}') e^{-(\mathbf{x}-\mathbf{x}')^2/(4Dt)} d^3x',$$

where D is the diffusion coefficient [Eq. (3.73)].

(a) For any one of the diffusing particles, the location $y(t)$ in the y direction (one of three Cartesian directions) is a 1-dimensional random process. From the above $n(\mathbf{x}, t)$, infer that the conditional probability distribution for y is

$$P_2(y_2, t | y_1) = \frac{1}{\sqrt{4\pi Dt}} e^{-(y_2 - y_1)^2/(4Dt)}. \tag{6.13}$$

(b) Verify that the conditional probability (6.13) satisfies the Smoluchowski equation (6.11). [Hint: Consider using symbol-manipulation computer software that quickly can do straightforward calculations like this.]

At first this may seem surprising, since a particle's position y is not Markov. However (as we explore explicitly in Sec. 6.7.2), the diffusion equation from which we derived this P_2 treats as negligibly small the timescale τ_r on which the velocity dy/dt thermalizes. It thereby wipes out all information about what the particle's actual velocity is, making y effectively Markov, and forcing its P_2 to satisfy the Smoluchowski equation. See Ex. 6.10, where we shall also discover that this diffusion is an example of a random walk.

6.3.2 Gaussian Processes and the Central Limit Theorem; Random Walk

GAUSSIAN PROCESSES

Gaussian random process

A random process is said to be Gaussian if and only if all of its (absolute) probability distributions are Gaussian (i.e., have the following form):

$$p_n(y_n, t_n; \ldots; y_2, t_2; y_1, t_1) = A \exp\left[-\sum_{j=1}^{n}\sum_{k=1}^{n} \alpha_{jk}(y_j - \bar{y})(y_k - \bar{y})\right], \quad (6.14a)$$

where (i) A and α_{jk} depend only on the time differences $t_2 - t_1, t_3 - t_1, \ldots, t_n - t_1$; (ii) A is a positive normalization constant; (iii) $[\alpha_{jk}]$ is a *positive-definite*, symmetric matrix (otherwise p_n would not be normalizable); and (iv) $\bar{y}$ is a constant, which one readily can show is equal to the ensemble average of y,

$$\bar{y} \equiv \langle y \rangle = \int y p_1(y)\, dy. \quad (6.14b)$$

Since the conditional probabilities are all computable as ratios of absolute probabilities [Eq. (6.4)], the conditional probabilities of a Gaussian process will be Gaussian.

Gaussian random processes are very common in physics. For example, the total number of particles $N(t)$ in a gas cell that is in statistical equilibrium with a heat bath is a Gaussian random process (Ex. 5.11d); and the primordial fluctuations that gave rise to structure in our universe appear to have been Gaussian (Sec. 28.5.3). In fact, as we saw in Sec. 5.6, macroscopic variables that characterize huge systems in statistical equilibrium always have Gaussian probability distributions. The underlying reason is that, *when a random process is driven by a large number of statistically independent, random influences, its probability distributions become Gaussian*. This general fact is a consequence of the *central limit theorem* of probability. We state and prove a simple variant of this theorem.

CENTRAL LIMIT THEOREM (A SIMPLE VERSION)

central limit theorem

Let y be a random quantity [not necessarily a random variable $y(t)$; there need not be any times involved; however, our applications will be to random variables]. Suppose that y is characterized by an arbitrary probability distribution $p(y)$ (e.g., that of Fig. 6.3a), so the probability of the quantity taking on a value between y and $y + dy$ is $p(y)dy$. Denote by $\bar{y}$ the mean value of y, and by σ_y its standard deviation (also called its rms fluctuation and the square root of its variance):

$$\bar{y} \equiv \langle y \rangle = \int y p(y) dy, \quad (\sigma_y)^2 \equiv \langle (y - \bar{y})^2 \rangle = \langle y^2 \rangle - \bar{y}^2. \quad (6.15a)$$

Randomly draw from this distribution a large number N of values $\{y_1, y_2, \ldots, y_N\}$, and average them to get a number

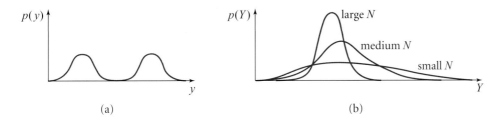

FIGURE 6.3 Example of the central limit theorem. (a) The random variable y with the probability distribution $p(y)$. (b) This variable produces, for various values of N, the variable $Y = (y_1 + \cdots + y_N)/N$ with the probability distributions $p(Y)$. In the limit of very large N, $p(Y)$ is a Gaussian.

$$Y \equiv \frac{1}{N} \sum_{i=1}^{N} y_i. \tag{6.15b}$$

Repeat this process many times, and examine the resulting probability distribution for Y. In the limit of arbitrarily large N, that distribution will be Gaussian with mean and standard deviation

$$\boxed{\bar{Y} = \bar{y} \quad \text{and} \quad \sigma_Y = \frac{\sigma_y}{\sqrt{N}},} \tag{6.15c}$$

that is, it will have the form

$$\boxed{p(Y) = \frac{1}{\sqrt{2\pi \sigma_Y^2}} \exp\left[-\frac{(Y - \bar{Y})^2}{2\sigma_Y^2}\right],} \tag{6.15d}$$

with $\bar{Y}$ and σ_Y given by Eq. (6.15c). See Fig. 6.3b.

Proof of Central Limit Theorem. The key to proving this theorem is the Fourier transform of the probability distribution. (That Fourier transform is called the distribution's *characteristic function*, but in this chapter we do not delve into the details of characteristic functions.) Denote the Fourier transform of $p(y)$ by[3]

$$\tilde{p}_y(f) \equiv \int_{-\infty}^{+\infty} e^{i2\pi f y} p(y) dy = \sum_{n=0}^{\infty} \frac{(i 2\pi f)^n}{n!} \langle y^n \rangle. \tag{6.16a}$$

The second expression follows from a power series expansion of $e^{i2\pi f y}$ in the first. Similarly, since a power series expansion analogous to Eq. (6.16a) must hold for $\tilde{p}_Y(f)$ and since $\langle Y^n \rangle$ can be computed from

$$\langle Y^n \rangle = \langle N^{-n}(y_1 + y_2 + \cdots + y_N)^n \rangle$$
$$= \int N^{-n}(y_1 + \cdots + y_N)^n p(y_1) \ldots p(y_N) dy_1 \ldots dy_N, \tag{6.16b}$$

---

3. See the beginning of Sec. 6.4.2 for the conventions we use for Fourier transforms.

it must be that

$$\tilde{p}_Y(f) = \sum_{n=0}^{\infty} \frac{(i2\pi f)^n}{n!} \langle Y^n \rangle$$

$$= \int \exp[i2\pi f N^{-1}(y_1 + \cdots + y_N)] p(y_1) \ldots p(y_N) dy_1 \ldots dy_n$$

$$= \left[\int e^{i2\pi fy/N} p(y) dy\right]^N = \left[1 + \frac{i2\pi f \bar{y}}{N} - \frac{(2\pi f)^2 \langle y^2 \rangle}{2N^2} + O\left(\frac{1}{N^3}\right)\right]^N$$

$$= \exp\left[i2\pi f \bar{y} - \frac{(2\pi f)^2 (\langle y^2 \rangle - \bar{y}^2)}{2N} + O\left(\frac{1}{N^2}\right)\right]. \tag{6.16c}$$

Here the last equality can be obtained by taking the logarithm of the preceding quantity, expanding in powers of $1/N$, and then exponentiating. By inverting the Fourier transform (6.16c) and using $(\sigma_y)^2 = \langle y^2 \rangle - \bar{y}^2$, we obtain for $p(Y)$ the Gaussian (6.15d). ∎

This proof is a good example of the power of Fourier transforms, a power that we exploit extensively in this chapter. As an important example to which we shall return later, Ex. 6.4 analyzes the simplest version of a random walk.

EXERCISES

Exercise 6.4 **Example: Random Walk with Discrete Steps of Identical Length*
This exercise is designed to make the concept of random processes more familiar and also to illustrate the central limit theorem.

A particle travels in 1 dimension, along the y axis, making a sequence of steps Δy_j (labeled by the integer j), each of which is $\Delta y_j = +1$ with probability 1/2, or $\Delta y_j = -1$ with probability 1/2.

(a) After $N \gg 1$ steps, the particle has reached location $y(N) = y(0) + \sum_{j=1}^{N} \Delta y_j$. What does the central limit theorem predict for the probability distribution of $y(N)$? What are its mean and its standard deviation?

(b) Viewed on lengthscales $\gg 1$, $y(N)$ looks like a continuous random process, so we shall rename $N \equiv t$. Using the (pseudo)random number generator from your favorite computer software language, compute a few concrete realizations of $y(t)$ for $0 < t < 10^4$ and plot them.[4] Figure 6.1 above shows one realization of this random process.

(c) Explain why this random process is Markov.

---

[4]. If you use Mathematica, the command `RandomInteger[]` generates a pseudorandom number that is 0 with probability 1/2 or 1 with probability 1/2. Therefore, the following simple script will carry out the desired computation: `y = Table[0, {10000}]; For[t = 1, t < 10000, t++, y[[t + 1]] = y[[t]] + 2 RandomInteger[] - 1]; ListPlot[y, Joined -> True]`. This was used to generate Fig. 6.1.

(d) Use the central limit theorem to infer that the conditional probability P_2 for this random process is

$$P_2(y_2, t|y_1) = \frac{1}{\sqrt{2\pi t}} \exp\left[-\frac{(y_2 - y_1)^2}{2t}\right]. \quad (6.17)$$

(e) Notice that this is the same probability distribution as we encountered in the diffusion exercise (Ex. 6.3) but with $D = 1/2$. Why did this have to be the case?

(f) Using an extension of the computer program you wrote in part (b), evaluate $y(t = 10^4)$ for 1,000 realizations of this random process, each with $y(0) = 0$, then bin the results in bins of width $\delta y = 10$, and plot the number of realizations $y(10^4)$ that wind up in each bin. Repeat for 10,000 realizations. Compare your plots with the probability distribution (6.17).

6.3.3 Doob's Theorem for Gaussian-Markov Processes, and Brownian Motion

A large fraction of the random processes that one meets in physics are Gaussian, and many are Markov. Therefore, the following remarkable theorem is very important. *Any 1-dimensional random process $y(t)$ that is both Gaussian and Markov has the following form for its conditional probability distribution P_2:*

$$\boxed{P_2(y_2, t|y_1) = \frac{1}{[2\pi \sigma_{y_t}^2]^{\frac{1}{2}}} \exp\left[-\frac{(y_2 - \bar{y}_t)^2}{2\sigma_{y_t}^2}\right],} \quad (6.18a)$$

where the mean $\bar{y}_t$ and variance $\sigma_{y_t}^2$ at time t are given by

$$\boxed{\bar{y}_t = \bar{y} + e^{-t/\tau_r}(y_1 - \bar{y}), \quad \sigma_{y_t}^2 = (1 - e^{-2t/\tau_r})\sigma_y^2.} \quad (6.18b)$$

Here $\bar{y}$ and σ_y^2 are respectively the process's equilibrium mean and variance (the values at $t \to \infty$), and τ_r is its *relaxation time*. This result is *Doob's theorem*.[5] We shall prove it in Ex. 6.5, after we have developed some necessary tools.

Note the great power of Doob's theorem: Because $y(t)$ is Markov, all of its probability distributions are computable from this P_2 (Ex. 6.1), which in turn is determined by $\bar{y}, \sigma_y$, and τ_r. Correspondingly, all statistical properties of a Gaussian-Markov process are determined by just three parameters: its (equilibrium) mean $\bar{y}$ and variance σ_y^2,

---

5. It is so named because it was first formulated and proved by J. L. Doob (1942).

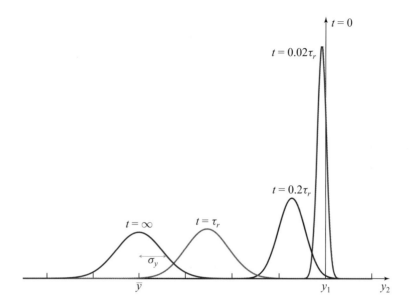

FIGURE 6.4 Evolution of the conditional probability $P_2(y_2, t|y_1)$ for a Gaussian-Markov random process [Eq. (6.18a)], as predicted by Doob's theorem. The correlation function and spectral density for this process are shown later in the chapter in Fig. 6.8.

and its relaxation time τ_r. As an example, the first absolute probability distribution is

$$p_1(y) = \lim_{t \to \infty} P_2(y, t|y_1) = \frac{1}{\sqrt{2\pi \sigma_y^2}} \exp\left[-\frac{(y - \bar{y})^2}{2\sigma_y^2}\right]. \quad (6.18c)$$

The time evolution of P_2 [Eqs. (6.18a,b)] is plotted in Fig. 6.4. At $t = 0$ it is a delta function at y_1, in accord with Eq. (6.12a). As t increases, its peak (its mean) moves toward $\bar{y}$, and it spreads out. Ultimately, at $t = \infty$, its peak asymptotes to $\bar{y}$, and its standard deviation (half-width) asymptotes to σ_y, so $P_2 \to p_1$—in accord with Eqs. (6.12b) and (6.18c).

Brownian motion

An example that we explore in Sec. 6.7.2 is a dust particle being buffeted by air molecules in a large, constant-temperature room (Brownian motion). As we discussed near the beginning of Sec. 6.3.1, any Cartesian component v of the dust particle's velocity is a Markov process. It is also Gaussian (because its evolution is influenced solely by the independent forces of collisions with a huge number of independent air molecules), so $P_2(v, t|v_1)$ is given by Doob's theorem. In equilibrium, positive and negative values of the Cartesian velocity component v are equally probable, so $\bar{v} = 0$, which means that $\frac{1}{2}m\sigma_v^2 = \frac{1}{2}m\overline{v^2}$, which is the equilibrium mean kinetic energy— a quantity we know to be $\frac{1}{2}k_B T$ from the equipartition theorem (Sec. 4.4.4); thus, $\bar{v} = 0$, and $\sigma_v = \sqrt{k_B T/m}$. The relaxation time τ_r is the time required for the particle to change its velocity substantially, due to collisions with air molecules; we compute it in Sec. 6.8.1 using the fluctuation-dissipation theorem; see Eq. (6.78).

6.4 Correlation Functions and Spectral Densities

6.4.1 Correlation Functions; Proof of Doob's Theorem

Let $y(t)$ be a (realization of a) random process with time average $\bar{y}$. Then the correlation function of $y(t)$ is defined by

$$C_y(\tau) \equiv \overline{[y(t) - \bar{y}][y(t+\tau) - \bar{y}]} \equiv \lim_{T \to \infty} \frac{1}{T} \int_{-T/2}^{+T/2} [y(t) - \bar{y}][y(t+\tau) - \bar{y}] dt. \quad \text{correlation function}$$

(6.19)

This quantity, as its name suggests, is a measure of the extent to which the values of y at times t and $t + \tau$ tend to be correlated. The quantity τ is sometimes called the *delay time*, and by convention it is taken to be positive. [One can easily see that, if one also defines $C_y(\tau)$ for negative delay times τ by Eq. (6.19), then $C_y(-\tau) = C_y(\tau)$. Thus nothing is lost by restricting attention to positive delay times.]

As an example, for a Gaussian-Markov process with P_2 given by Doob's formula (6.18a) (Fig. 6.4), we can compute $C(\tau)$ by replacing the time average in Eq. (6.19) with an ensemble average: $C_y(\tau) = \int y_2 y_1 p_2(y_2, \tau; y_1) dy_1 dy_2$. If we use $p_2(y_2, \tau; y_1) = P_2(y_2, \tau; y_1) p_1(y_1)$ [Eq. (6.10)], insert P_2 and p_1 from Eqs. (6.18), and perform the integrals, we obtain

$$C_y(\tau) = \sigma_y^2 e^{-\tau/\tau_r}. \quad (6.20)$$

This correlation function has two properties that are quite general:

1. The following is true for all (ergodic and stationary) random processes: properties of correlation function

$$\boxed{C_y(0) = \sigma_y^2,} \quad (6.21a)$$

   as one can see by replacing time averages with ensemble averages in definition (6.19); in particular, $C_y(0) \equiv \overline{(y - \bar{y})^2} = \langle (y - \bar{y})^2 \rangle$, which by definition is the variance σ_y^2 of y.

2. In addition, we have that

$$\boxed{C_y(\tau) \text{ asymptotes to zero for } \tau \gg \tau_r,} \quad (6.21b)$$

   where τ_r is the process's *relaxation time* or *correlation time* (see Fig. 6.5). This is true for all ergodic, stationary random processes, since our definition of ergodicity in Sec. 6.2.3 relies on each realization $y(t)$ losing its memory of earlier values after some sufficiently long time T. Otherwise, it would not be possible to construct the ensemble $\mathcal{E}'$ of random variables $Y^K(t)$ [Eq. (6.7)] and have them behave like independent random variables.

relaxation time

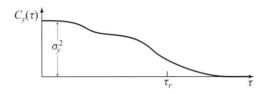

FIGURE 6.5 Properties (6.21) of correlation functions.

As an example of how one can use correlation functions, in Ex. 6.5 we use them to prove Doob's theorem.

EXERCISES

Exercise 6.5 *Derivation: Proof of Doob's Theorem*
Prove Doob's theorem. More specifically, for any Gaussian-Markov random process, show that $P_2(y_2, t|y_1)$ is given by Eqs. (6.18a,b).

[Hint: For ease of notation, set $y_{\text{new}} = (y_{\text{old}} - \bar{y}_{\text{old}})/\sigma_{y_{\text{old}}}$, so $\bar{y}_{\text{new}} = 0$ and $\sigma_{y_{\text{new}}} = 1$. If the theorem is true for y_{new}, then by the rescalings inherent in the definition of $P_2(y_2, t|y_1)$, it will also be true for y_{old}.]

(a) Show that the Gaussian process y_{new} has probability distributions

$$p_1(y) = \frac{1}{\sqrt{2\pi}} e^{-y^2/2}, \tag{6.22a}$$

$$p_2(y_2, t_2; y_1, t_1) = \frac{1}{\sqrt{(2\pi)^2(1 - C_{21}^2)}} \exp\left[-\frac{y_1^2 + y_2^2 - 2C_{21}y_1 y_2}{2(1 - C_{21}^2)}\right]; \tag{6.22b}$$

and show that the constant C_{21} that appears here is the correlation function $C_{21} = C_y(t_2 - t_1)$.

(b) From the relationship between absolute and conditional probabilities [Eq. (6.4)], show that

$$P_2(y_2, t_2|y_1, t_1) = \frac{1}{\sqrt{2\pi(1 - C_{21}^2)}} \exp\left[-\frac{(y_2 - C_{21}y_1)^2}{2(1 - C_{21}^2)}\right]. \tag{6.22c}$$

(c) Show that for any three times $t_3 > t_2 > t_1$,

$$C_{31} = C_{32} C_{21}; \quad \text{i.e.,} \quad C_y(t_3 - t_1) = C_y(t_3 - t_2) C_y(t_2 - t_1). \tag{6.22d}$$

To show this, you could (i) use the relationship between absolute and conditional probabilities and the Markov nature of the random process to infer that

$$p_3(y_3, t_3; y_2, t_2; y_1, t_1) = P_3(y_3, t_3|y_2, t_2; y_1, t_1) p_2(y_2, t_2; y_1, t_1)$$
$$= P_2(y_3, t_3|y_2, t_2) p_2(y_2, t_2; y_1, t_1);$$

then (ii) compute the last expression explicitly, getting

$$\frac{1}{\sqrt{2\pi(1-C_{32}^2)}} \exp\left[-\frac{(y_3 - C_{32}y_2)^2}{2(1-C_{32}^2)}\right]$$

$$\times \frac{1}{\sqrt{(2\pi)^2(1-C_{21}^2)}} \exp\left[-\frac{(y_1^2 + y_2^2 - 2C_{21}y_1y_2)}{2(1-C_{21}^2)}\right];$$

(iii) then using this expression, evaluate

$$C_y(t_3 - t_1) \equiv C_{31} \equiv \langle y(t_3)y(t_1)\rangle = \int p_3(y_3, t_3; y_2, t_2; y_1, t_1) y_3 y_1 dy_3 dy_2 dy_1.$$

(6.22e)

The result should be $C_{31} = C_{32}C_{21}$.

(d) Argue that the unique solution to this equation, with the "initial condition" that $C_y(0) = \sigma_y^2 = 1$, is $C_y(\tau) = e^{-\tau/\tau_r}$, where τ_r is a constant (which we identify as the relaxation time). Correspondingly, $C_{21} = e^{-(t_2-t_1)/\tau_r}$.

(e) By inserting this expression into Eq. (6.22c), complete the proof for $y_{\text{new}}(t)$, and thence conclude that Doob's theorem is also true for our original, unrescaled $y_{\text{old}}(t)$.

6.4.2 Spectral Densities

There are several different normalization conventions for Fourier transforms. In this chapter, we adopt a normalization that is commonly (though not always) used in the theory of random processes and that differs from the one common in quantum theory. Specifically, instead of using the angular frequency ω, we use the ordinary frequency $f \equiv \omega/(2\pi)$. We define the Fourier transform of a function $y(t)$ and its inverse by

$$\boxed{\tilde{y}(f) \equiv \int_{-\infty}^{+\infty} y(t)e^{i2\pi ft}dt, \qquad y(t) \equiv \int_{-\infty}^{+\infty} \tilde{y}(f)e^{-i2\pi ft}df.}$$

(6.23)   **Fourier transform**

Notice that with this set of conventions, there are no factors of $1/(2\pi)$ or $1/\sqrt{2\pi}$ multiplying the integrals. Those factors have been absorbed into the df of Eq. (6.23), since $df = d\omega/(2\pi)$.

The integrals in Eq. (6.23) are not well defined as written because a random process $y(t)$ is generally presumed to go on forever so its Fourier transform $\tilde{y}(f)$ is divergent. One gets around this problem by crude trickery. From $y(t)$ construct, by truncation, the function

$$y_T(t) \equiv \begin{cases} y(t) & \text{if } -T/2 < t < +T/2, \\ 0 & \text{otherwise.} \end{cases}$$

(6.24a)

Then the Fourier transform $\tilde{y}_T(f)$ is finite, and by Parseval's theorem (e.g., Arfken, Weber, and Harris, 2013) it satisfies

$$\int_{-T/2}^{+T/2} [y(t)]^2 dt = \int_{-\infty}^{+\infty} [y_T(t)]^2 dt = \int_{-\infty}^{+\infty} |\tilde{y}_T(f)|^2 df = 2\int_0^{\infty} |\tilde{y}_T(f)|^2 df. \tag{6.24b}$$

In the last equality we have used the fact that because $y_T(t)$ is real, $\tilde{y}_T^*(f) = \tilde{y}_T(-f)$, where * denotes complex conjugation. Consequently, the integral from $-\infty$ to 0 of $|\tilde{y}_T(f)|^2$ is the same as the integral from 0 to $+\infty$. Now, the quantities on the two sides of (6.24b) diverge in the limit as $T \to \infty$, and it is obvious from the left-hand side that they diverge linearly as T. Correspondingly, the limit

$$\lim_{T \to \infty} \frac{1}{T} \int_{-T/2}^{+T/2} [y(t)]^2 dt = \lim_{T \to \infty} \frac{2}{T} \int_0^{\infty} |\tilde{y}_T(f)|^2 df \tag{6.24c}$$

is convergent.

These considerations motivate the following definition of the *spectral density* (also sometimes called the *power spectrum*) $S_y(f)$ of the random process $y(t)$:

spectral density

$$\boxed{S_y(f) \equiv \lim_{T \to \infty} \frac{2}{T} \left| \int_{-T/2}^{+T/2} [y(t) - \bar{y}] e^{i2\pi f t} dt \right|^2.} \tag{6.25}$$

Notice that the quantity inside the absolute value sign is just $\tilde{y}_T(f)$, but with the mean of y removed before computation of the Fourier transform. (The mean is removed to avoid an uninteresting delta function in $S_y(f)$ at zero frequency.) Correspondingly, by virtue of our motivating result (6.24c), the spectral density satisfies $\int_0^{\infty} S_y(f) df = \lim_{T\to\infty} \frac{1}{T} \int_{-T/2}^{+T/2} [y(t)-\bar{y}]^2 dt = \overline{(y-\bar{y})^2} = \sigma_y^2$, or

integral of spectral density

$$\boxed{\int_0^{\infty} S_y(f) df = C_y(0) = \sigma_y^2.} \tag{6.26}$$

Thus the integral of the spectral density of y over all positive frequencies is equal to the variance of y.

By convention, our spectral density is defined only for nonnegative frequencies f. This is because, were we to define it also for negative frequencies, the fact that $y(t)$ is real would imply that $S_y(f) = S_y(-f)$, so the negative frequencies contain no new information. Our insistence that f be positive goes hand in hand with the factor 2 in the $2/T$ of definition (6.25): that factor 2 folds the negative-frequency part onto the positive-frequency part. This choice of convention is called the *single-sided spectral density*. Sometimes one encounters a *double-sided spectral density*,

$$S_y^{\text{double-sided}}(f) = \frac{1}{2} S_y(|f|), \tag{6.27}$$

in which f is regarded as both positive and negative, and frequency integrals generally run from $-\infty$ to $+\infty$ instead of 0 to ∞ (see, e.g., Ex. 6.7).

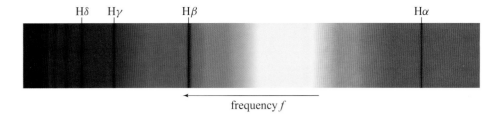

FIGURE 6.6 A spectrum obtained by sending light through a diffraction grating. The intensity of the image is proportional to $d\mathcal{E}/dtdf$, which, in turn, is proportional to the spectral density $S_E(f)$ of the electric field $E(t)$ of the light that entered the diffraction grating.

Notice that the spectral density has units of y^2 per unit frequency; or, more colloquially (since frequency f is usually measured in Hertz, i.e., cycles per second), its units are y^2/Hz.

6.4.3 Physical Meaning of Spectral Density, Light Spectra, and Noise in a Gravitational Wave Detector

We can infer the physical meaning of the spectral density from previous experience with light spectra. Specifically, consider the scalar electric field[6] $E(t)$ of a plane-polarized light wave entering a telescope from a distant star, galaxy, or nebula. (We must multiply this $E(t)$ by the polarization vector to get the vectorial electric field.) This $E(t)$ is a superposition of emission from an enormous number of atoms, molecules, and high-energy particles in the source, so it is a Gaussian random process. It is not hard to convince oneself that $E(t)$'s spectral density $S_E(f)$ is proportional to the light power per unit frequency $d\mathcal{E}/dtdf$ (the light's power spectrum) entering the telescope. When we send the light through a diffraction grating, we get this power spectrum spread out as a function of frequency f in the form of spectral lines superposed on a continuum, as in Fig. 6.6. The amount of light power in this spectrum, in some narrow bandwidth Δf centered on some frequency f, is $(d\mathcal{E}/dtdf)\Delta f \propto S_E(f)\Delta f$ (assuming S_E is nearly constant over that band).

Another way to understand this role of the spectral density $S_E(f)$ is by examining the equation for the variance of the oscillating electric field E as an integral over frequency, $\sigma_E^2 = \int_0^\infty S_E(f)df$. If we filter the light so only that portion at frequency f, in a very narrow bandwidth Δf, gets through the filter, then the variance of the filtered, oscillating electric field will obviously be the portion of the integral coming from this frequency band. The rms value of the filtered electric field will be the square root of this—and similarly for any other random process $y(t)$:

$$\left(\begin{array}{c}\text{rms value of } y\text{'s oscillations} \\ \text{at frequency } f \text{ in a very narrow bandwidth } \Delta f\end{array}\right) \simeq \sqrt{S_y(f)\Delta f}. \quad (6.28)$$

rms oscillation

---

6. In this section, and only here, E represents the electric field rather than (nonrelativistic) energy.

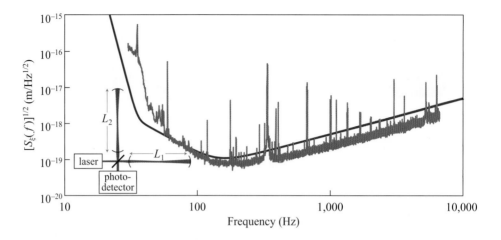

FIGURE 6.7 The square root of the spectral density of the time-varying arm-length difference $\xi(t) = L_1 - L_2$ (see inset), in the Laser Interferometer Gravitational-wave Observatory (LIGO) interferometer at Hanford, Washington, as measured on February 22, 2010. See Sec. 9.5 and Fig. 9.13. The dark blue curve is the noise that was specified as this instrument's goal. The narrow spectral lines (sharp spikes in the spectrum produced by internal resonances in the instrument) contain negligible power, and so can be ignored for our purposes. At high frequencies, $f \gtrsim 150$ Hz, the noise is due to randomness in arrival times of photons used to measure the mirror motions (photon shot noise, Sec. 6.7.4). At intermediate frequencies, 40 Hz $\lesssim f \lesssim 150$ Hz, it is primarily thermal noise (end of Sec. 6.8.2). At low frequencies, $f \lesssim 40$ Hz, it is primarily mechanical vibrations that sneak through a vibration isolation system ("seismic" noise).

(In Sec. 6.7.1, we develop a mathematical formalism to describe this type of filtering).

As a practical example, consider the output of an interferometric gravitational wave detector (to be further discussed in Secs. 9.5 and 27.6). The gravitational waves from some distant source (e.g., two colliding black holes) push two mirrors (hanging by wires) back and forth with respect to each other. Laser interferometry is used to monitor the difference $\xi(t) = L_1 - L_2$ between the two arm lengths. (Here L_1 is the separation between the mirrors in one arm of the interferometer, and L_2 is that in the other arm; see inset in Fig. 6.7.) The measured $\xi(t)$ is influenced by noise in the instrument as well as by gravitational waves. Figure 6.7 shows the square root of the spectral density of the noise-induced fluctuations in $\xi(t)$. Note that this $\sqrt{S_\xi(f)}$ has units of meters/$\sqrt{\text{Hertz}}$ (since ξ has units of meters).

The minimum of the noise power spectrum is at $f \simeq 150$ Hz. If one is searching amidst this noise for a broadband gravitational-wave signal, then one might filter the interferometer output so one's data analysis sees only a frequency band of order the frequency of interest: $\Delta f \simeq f$. Then the rms noise in this band will be $\sqrt{S_\xi(f) \times f} \simeq 10^{-19}$ m/$\sqrt{\text{Hz}} \times \sqrt{150 \text{ Hz}} \simeq 10^{-18}$ m, which is $\sim 1/1{,}000$ the diameter of a proton. If a gravitational wave with frequency ~ 150 Hz changes the mirrors' separations by much more than this miniscule amount, it should be detectable!

6.4.4 The Wiener-Khintchine Theorem; Cosmological Density Fluctuations

The Wiener-Khintchine theorem says that, for any random process $y(t)$, the correlation function $C_y(\tau)$ and the spectral density $S_y(f)$ are the cosine transforms of each other and thus contain precisely the same information:

Wiener-Khintchine theorem

$$\boxed{C_y(\tau) = \int_0^\infty S_y(f)\cos(2\pi f\tau)df, \qquad S_y(f) = 4\int_0^\infty C_y(\tau)\cos(2\pi f\tau)d\tau.}$$

(6.29)

The factor 4 results from our folding negative frequencies into positive in our definition of the spectral density.

Proof of Wiener-Khintchine Theorem. This theorem is readily proved as a consequence of Parseval's theorem: Assume, from the outset, that the mean has been subtracted from $y(t)$, so $\bar{y} = 0$. (This is not really a restriction on the proof, since C_y and S_y are insensitive to the mean of y.) Denote by $y_T(t)$ the truncated y of Eq. (6.24a) and by $\tilde{y}_T(f)$ its Fourier transform. Then the generalization of Parseval's theorem[7]

$$\int_{-\infty}^{+\infty}(gh^* + hg^*)dt = \int_{-\infty}^{+\infty}(\tilde{g}\tilde{h}^* + \tilde{h}\tilde{g}^*)df \qquad (6.30a)$$

[with $g = y_T(t)$ and $h = y_T(t+\tau)$ both real and with $\tilde{g} = \tilde{y}_T(f), \tilde{h} = \tilde{y}_T(f)e^{-i2\pi f\tau}$], states

$$\int_{-\infty}^{+\infty} y_T(t)y_T(t+\tau)dt = \int_{-\infty}^{+\infty} \tilde{y}_T^*(f)\tilde{y}_T(f)e^{-i2\pi f\tau}df. \qquad (6.30b)$$

By dividing by T, taking the limit as $T \to \infty$, and using Eqs. (6.19) and (6.25), we obtain the first equality of Eqs. (6.29). The second follows from the first by Fourier inversion. ∎

The Wiener-Khintchine theorem implies (Ex. 6.6) the following formula for the ensemble averaged self-product of the Fourier transform of the random process $y(t)$:

spectral density as mean square of random process's Fourier transform

$$\boxed{2\langle\tilde{y}(f)\tilde{y}^*(f')\rangle = S_y(f)\delta(f-f').} \qquad (6.31)$$

This equation quantifies the strength of the infinite value of $|\tilde{y}(f)|^2$, which motivated our definition (6.25) of the spectral density.

As an application of the Wiener-Khintchine theorem, we can deduce the spectral density $S_y(f)$ for any Gaussian-Markov process by performing the cosine transform of its correlation function $C_y(\tau) = \sigma_y^2 e^{-\tau/\tau_r}$ [Eq. (6.20)]. The result is

$$\boxed{S_y(f) = \frac{(4/\tau_r)\sigma_y^2}{(2\pi f)^2 + (1/\tau_r)^2};} \qquad (6.32)$$

see Fig. 6.8.

---

7. This follows by subtracting Parseval's theorem for g and for h from Parseval's theorem for $g+h$.

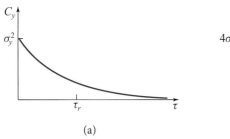
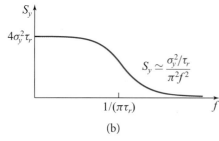

FIGURE 6.8 (a) The correlation function (6.20) and (b) the spectral density (6.32) for a Gaussian-Markov process. The conditional probability $P_2(y_2, \tau | y_1)$ for this process is shown in Fig. 6.4.

As a second application, in Ex. 6.7 we explore fluctuations in the density of galaxies in the universe, caused by gravity pulling them into clusters.

EXERCISES

Exercise 6.6 *Derivation: Spectral Density as the Mean Square of Random Process's Fourier Transform*

Derive Eq. (6.31).

[Hint: Write $\langle \tilde{y}(f)\tilde{y}^*(f') \rangle = \int_{-\infty}^{+\infty} \int_{-\infty}^{+\infty} \langle y(t)y(t') \rangle e^{i2\pi ft} e^{-i2\pi f't'} dt dt'$. Then set $t' = t + \tau$, and express the expectation value as $C_y(\tau)$, and use an expression for the Dirac delta function in terms of Fourier transforms.]

Exercise 6.7 ***Example: Cosmological Density Fluctuations*[8]

Random processes can be stochastic functions of some variable or variables other than time. For example, it is conventional to describe fractional fluctuations in the large-scale distribution of mass in the universe, or the distribution of galaxies, using the quantity

$$\delta(\mathbf{x}) \equiv \frac{\rho(\mathbf{x}) - \langle \rho \rangle}{\langle \rho \rangle}, \quad \text{or} \quad \delta(\mathbf{x}) \equiv \frac{n(\mathbf{x}) - \langle n \rangle}{\langle n \rangle} \tag{6.33}$$

(not to be confused with the Dirac delta function). Here $\rho(\mathbf{x})$ is mass density and $n(\mathbf{x})$ is the number density of galaxies, which we assume, for didactic purposes, to have equal mass and to be distributed in the same fashion as the dark matter (Sec. 28.3.2). This $\delta(\mathbf{x})$ is a function of 3-dimensional position rather than 1-dimensional time, and $\langle \cdot \rangle$ is to be interpreted conceptually as an ensemble average and practically as a volume average (ergodic hypothesis!).

(a) Define the Fourier transform of δ over some large averaging volume V as

$$\tilde{\delta}_V(\mathbf{k}) = \int_V e^{i\mathbf{k} \cdot \mathbf{x}} \delta(\mathbf{x}) d^3x, \tag{6.34a}$$

---

8. Discussed further in Sec. 28.5.4.

and define its spectral density (Sec. 28.5.4) by

$$P_\delta(\mathbf{k}) \equiv \lim_{V \to \infty} \frac{1}{V} |\tilde{\delta}_V(\mathbf{k})|^2. \tag{6.34b}$$

(Note that we here use cosmologists' "double-sided" normalization for P_δ, which is different from our normalization for a random process in time; we do not fold negative values of the Cartesian components k_j of $\mathbf{k}$ onto positive values.) Show that the two-point correlation function for cosmological density fluctuations, defined by

$$\xi_\delta(\mathbf{r}) \equiv \langle \delta(\mathbf{x})\delta(\mathbf{x}+\mathbf{r}) \rangle, \tag{6.34c}$$

is related to $P_\delta(\mathbf{k})$ by the following version of the Wiener-Khintchine theorem:

$$\xi_\delta(\mathbf{r}) = \int P_\delta(\mathbf{k}) e^{-i\mathbf{k}\cdot\mathbf{r}} \frac{d^3k}{(2\pi)^3} = \int_0^\infty P_\delta(k) \, \text{sinc}(kr) \frac{k^2 dk}{2\pi^2}, \tag{6.35a}$$

$$P_\delta(\mathbf{k}) = \int \xi_\delta(\mathbf{r}) e^{i\mathbf{k}\cdot\mathbf{r}} d^3x = \int_0^\infty \xi_\delta(r) \, \text{sinc}(kr) 4\pi r^2 dr, \tag{6.35b}$$

where $\text{sinc } x \equiv \sin x / x$. In deriving these expressions, use the fact that the universe is isotropic to infer that ξ_δ can depend only on the distance r between points and not on direction, and P_δ can depend only on the magnitude k of the wave number and not on its direction.

(b) Figure 6.9 shows observational data for the galaxy correlation function $\xi_\delta(r)$. These data are rather well approximated at $r < 20$ Mpc by

$$\xi_\delta(r) = (r_o/r)^\gamma, \quad r_o \simeq 7 \text{ Mpc}, \quad \gamma \simeq 1.8. \tag{6.36}$$

(Here 1 Mpc means 1×10^6 parsecs or about 3×10^6 light-years.) Explain why this implies that galaxies are strongly correlated (they cluster together strongly) on lengthscales $r \lesssim r_o \simeq 7$ Mpc. (Recall that the distance between our Milky Way galaxy and the nearest other large galaxy, Andromeda, is about 0.8 Mpc.) Use the Wiener-Khintchine theorem to compute the spectral density $P_\delta(k)$ and then the rms fractional density fluctuations at wave number k in bandwidth $\Delta k = k$. From your answer, infer that the density fluctuations are very large on lengthscales $\lambda = 1/k < r_o$.

(c) As a more precise measure of these density fluctuations, show that the variance of the total number $N(R)$ of galaxies inside a sphere of radius R is

$$\sigma_N^2 = \langle n \rangle^2 \int_0^\infty \frac{dk}{2\pi^2} k^2 P_\delta(k) \, W^2(kR), \tag{6.37a}$$

where

$$W(x) = \frac{3(\text{sinc } x - \cos x)}{x^2}. \tag{6.37b}$$

Evaluate this for the spectral density $P_\delta(r)$ that you computed in part (b). [Hint: Although it is straightforward to calculate this directly, it is faster to regard the

6.4 Correlation Functions and Spectral Densities

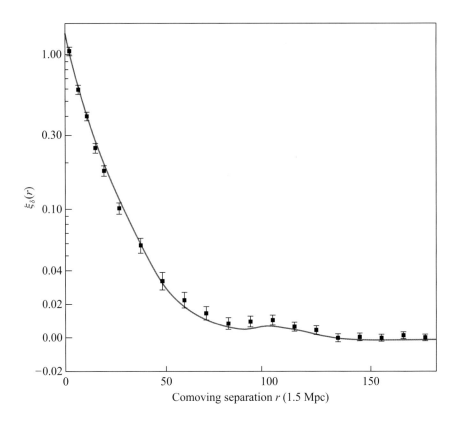

FIGURE 6.9 The galaxy correlation function $\xi_\delta(r)$ [defined in Eq. (6.34c)], as measured in the Sloan Digital Sky Survey. Notice that the vertical scale is linear for $\xi_\delta \lesssim 0.04$ and logarithmic for larger ξ_δ. Adapted from Eisenstein et al. (2005).

distribution of galaxies inside the sphere as an infinite distribution multiplied by a step function in radius and to invoke the convolution theorem.]

6.5 2-Dimensional Random Processes T2

Sometimes two (or more) random processes are closely related, and one wants to study their connections. An example is the position $x(t)$ and momentum $p(t)$ of a harmonic oscillator (Ex. 6.23), and we will encounter other examples in Secs. 28.5 and 28.6. Such pairs can be regarded as a 2-dimensional random process. In this Track-Two section, we generalize the concepts of correlation function and spectral density to such related processes.

6.5.1 Cross Correlation and Correlation Matrix T2

If $x(t)$ and $y(t)$ are two random processes, then by analogy with the correlation function $C_y(\tau)$ we define their *cross correlation* as

$$C_{xy}(\tau) \equiv \overline{x(t)y(t+\tau)}. \tag{6.38a}$$

cross correlation

When $x = y$, the cross correlation function becomes the autocorrelation function, $C_{yy}(\tau)$ interchangeable here with $C_y(\tau)$. The matrix

$$\begin{bmatrix} C_{xx}(\tau) & C_{xy}(\tau) \\ C_{yx}(\tau) & C_{yy}(\tau) \end{bmatrix} \equiv \begin{bmatrix} C_x(\tau) & C_{xy}(\tau) \\ C_{yx}(\tau) & C_y(\tau) \end{bmatrix} \quad (6.38b)$$

correlation matrix

can be regarded as a correlation matrix for the 2-dimensional random process $\{x(t), y(t)\}$. Notice that the elements of this matrix satisfy

$$C_{ab}(-\tau) = C_{ba}(\tau). \quad (6.39)$$

6.5.2 Spectral Densities and the Wiener-Khintchine Theorem

If $x(t)$ and $y(t)$ are two random processes, then by analogy with the spectral density $S_y(f)$ we define their *cross spectral density* as

$$S_{xy}(f) = \lim_{T \to \infty} \frac{2}{T} \int_{-T/2}^{+T/2} [x(t) - \bar{x}] e^{-2\pi i f t} dt \int_{-T/2}^{+T/2} [y(t') - \bar{y}] e^{+2\pi i f t'} dt'. \quad (6.40a)$$

cross spectral density

Notice that the cross spectral density of a random process with itself is equal to its spectral density, $S_{yy}(f) = S_y(f)$, and is real, but if $x(t)$ and $y(t)$ are different random processes, then $S_{xy}(f)$ is generally complex, with

$$S_{xy}^*(f) = S_{xy}(-f) = S_{yx}(f). \quad (6.40b)$$

This relation allows us to confine attention to positive f without any loss of information. The Hermitian matrix

$$\begin{bmatrix} S_{xx}(f) & S_{xy}(f) \\ S_{yx}(f) & S_{yy}(f) \end{bmatrix} = \begin{bmatrix} S_x(f) & S_{xy}(f) \\ S_{yx}(f) & S_y(f) \end{bmatrix} \quad (6.40c)$$

spectral density matrix

can be regarded as a spectral density matrix that describes how the power in the 2-dimensional random process $\{x(t), y(t)\}$ is distributed over frequency.

A generalization of the 1-dimensional Wiener-Khintchine Theorem (6.29) states that *for any two random processes $x(t)$ and $y(t)$, the cross correlation function $C_{xy}(\tau)$ and the cross spectral density $S_{xy}(f)$ are Fourier transforms of each other and thus contain precisely the same information*:

Wiener-Khintchine theorem for cross spectral density

$$C_{xy}(\tau) = \frac{1}{2} \int_{-\infty}^{+\infty} S_{xy}(f) e^{-i2\pi f \tau} df = \frac{1}{2} \int_0^{\infty} \left[S_{xy}(f) e^{-i2\pi f \tau} + S_{yx}(f) e^{+i2\pi f \tau} \right] df,$$

$$S_{xy}(f) = 2 \int_{-\infty}^{\infty} C_{xy}(\tau) e^{i2\pi f \tau} d\tau = 2 \int_0^{\infty} \left[C_{xy}(f) e^{+i2\pi f \tau} + C_{yx}(f) e^{-i2\pi f \tau} \right] df.$$

$$(6.41)$$

The factors 1/2 and 2 in these formulas result from folding negative frequencies into positive in our definitions of the spectral density. Equations (6.41) can be proved by the same Parseval-theorem-based argument as we used for the 1-dimensional Wiener-Khintchine theorem (Sec. 6.4.4).

> **T2**
> cross spectral density as mean product of random processes' Fourier transforms

The Wiener-Khintchine theorem implies the following formula for the ensemble averaged product of the Fourier transform of the random processes $x(t)$ and $y(t)$:

$$2\langle \tilde{x}(f)\tilde{y}^*(f')\rangle = S_{xy}(f)\delta(f - f'). \tag{6.42}$$

This can be proved by the same argument as we used in Ex. 6.6 to prove its single-process analog, $2\langle \tilde{y}(f)\tilde{y}^*(f')\rangle = S_y(f)\delta(f - f')$ [Eq. (6.31)].

EXERCISES

Exercise 6.8 *Practice: Spectral Density of the Sum of Two Random Processes* **T2**
Let u and v be two random processes. Show that

$$S_{u+v}(f) = S_u(f) + S_v(f) + S_{uv}(f) + S_{vu}(f) = S_u(f) + S_v(f) + 2\Re S_{uv}(f), \tag{6.43}$$

where $\Re$ denotes the real part of the argument.

6.6 Noise and Its Types of Spectra

Experimental physicists and engineers encounter random processes in the form of noise that is superposed on signals they are trying to measure. Examples include:

1. In radio communication, static on the radio is noise.

2. When modulated laser light is used for optical communication, random fluctuations in the arrival times of photons always contaminate the signal; the effects of such fluctuations are called "shot noise" and will be studied in Sec. 6.6.1.

3. Even the best of atomic clocks fail to tick with absolutely constant angular frequencies ω. Their frequencies fluctuate ever so slightly relative to an ideal clock, and those fluctuations can be regarded as noise.

Sometimes the signal that one studies amidst noise is actually itself some very special noise. (One person's noise is another person's signal.) An example is the light passing through an optical telescope and diffraction grating, discussed in Sec. 6.4.3. There the electric field $E(t)$ of the light from a star is a random process whose spectral density the astronomer measures as a function of frequency, studying with great interest features in the spectral lines and continuum. When the source is dim, the astronomer must try to separate its spectral density from those of noise in the photodetector and noise of other sources in the sky.

6.6.1 Shot Noise, Flicker Noise, and Random-Walk Noise; Cesium Atomic Clock

Physicists, astronomers, and engineers give names to certain shapes of noise spectra:

> white, flicker, and random-walk spectra

$$\boxed{S_y(f) \text{ independent of } f\text{—white noise spectrum,}} \tag{6.44a}$$

$$\boxed{S_y(f) \propto 1/f\text{—flicker noise spectrum,}} \tag{6.44b}$$

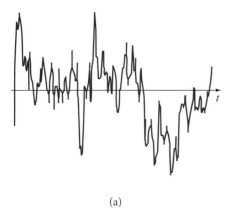 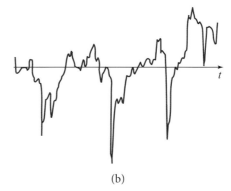

(a) (b)

FIGURE 6.10 Examples of two random processes that have flicker noise spectra, $S_y(f) \propto 1/f$. Adapted from Press (1978).

$$S_y(f) \propto 1/f^2 \text{—random-walk spectrum.} \qquad (6.44c)$$

White noise, S_y independent of f, is called "white" because it has equal amounts of power per unit frequency S_y at all frequencies, just as white light has roughly equal powers at all light frequencies. Put differently, if $y(t)$ has a white-noise spectrum, then its rms fluctuations in fixed bandwidth Δf are independent of frequency f (i.e., $\sqrt{S_y(f)\Delta f}$ is independent of f). white noise

Flicker noise, $S_y \propto 1/f$, gets its name from the fact that, when one looks at the time evolution $y(t)$ of a random process with a flicker-noise spectrum, one sees fluctuations ("flickering") on all timescales, and the rms amplitude of flickering is independent of the timescale one chooses. Stated more precisely, choose any timescale Δt and then choose a frequency $f \sim 3/\Delta t$, so one can fit roughly three periods of oscillation into the chosen timescale. Then the rms amplitude of the fluctuations observed will be $\sqrt{S_y(f)f/3}$, which is a constant independent of f when the spectrum is that of flicker noise, $S_y \propto 1/f$. In other words, flicker noise has the same amount of power in each octave of frequency. Figure 6.10 is an illustration: both graphs shown there depict random processes with flicker-noise spectra. (The differences between the two graphs will be explained in Sec. 6.6.2.) No matter what time interval one chooses, these processes look roughly periodic with one, two, or three oscillations in that time interval; and the amplitudes of those oscillations are independent of the chosen time interval. Flicker noise occurs widely in the real world, at low frequencies, for instance, in many electronic devices, in some atomic clocks, in geophysics (the flow rates of rivers, ocean currents, etc.), in astrophysics (the light curves of quasars, sunspot numbers, etc.); even in classical music. For an interesting discussion, see Press (1978). flicker noise

Random-walk noise, $S_y \propto 1/f^2$, arises when a random process $y(t)$ undergoes a random walk. In Sec. 6.7.2, we explore an example: the time evolving position $x(t)$ of a dust particle buffeted by air molecules—the phenomenon of Brownian motion. random-walk noise

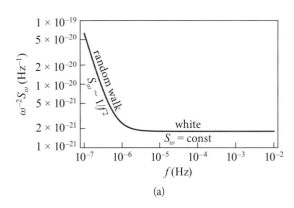
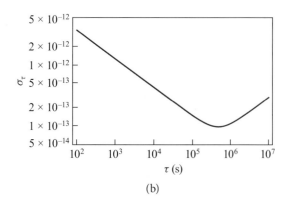

FIGURE 6.11 (a) Spectral density of the fluctuations in angular frequency ω of a typical cesium atomic clock. (b) Square root of the Allan variance for the same clock; see Ex. 6.13. Adapted from Galleani (2012). The best cesium clocks in 2016 (e.g., the U.S. primary time and frequency standard) have amplitude noise, $\sqrt{S_\omega}$ and σ_τ, 1000 times lower than this.

Notice that for a Gaussian-Markov process, the spectrum [Eq. (6.32) and Fig. 6.8b] is white at frequencies $f \ll 1/(2\pi \tau_r)$, where τ_r is the relaxation time, and it is random-walk at frequencies $f \gg 1/(2\pi \tau_r)$. This is typical: random processes encountered in the real world tend to have one type of spectrum over one large frequency interval and then switch to another type over another large interval. The angular frequency ω of ticking of a cesium atomic clock is another example.[9] It fluctuates slightly with time, $\omega = \omega(t)$, with the fluctuation spectral density shown in Fig. 6.11. At low frequencies, $f \lesssim 10^{-6}$ Hz (over long timescales $\Delta t \gtrsim 2$ weeks), ω exhibits random-walk noise. At higher frequencies, $f \gtrsim 10^{-6}$ Hz (timescales $\Delta t \lesssim 2$ weeks), it exhibits white noise—which is just the opposite of a Gaussian-Markov process (see, e.g., Galleani, 2012).

6.6.2 Information Missing from Spectral Density

In experimental studies of noise, attention focuses heavily on the spectral density $S_y(f)$ and on quantities that can be computed from it. In the special case of a Gaussian-Markov process, the spectrum $S_y(f)$ and the mean $\bar{y}$ together contain full information about all statistical properties of the random process. However, most random processes that one encounters are not Markov (though most are Gaussian). (When the spectrum deviates from the special form shown in Fig. 6.8, one can be sure the process is not Gaussian-Markov.) Correspondingly, for most processes the spectrum contains only a tiny part of the statistical information required to characterize

---

9. The U.S. national primary time and frequency standard is currently (2016) a cesium atomic clock; but it might be replaced, in a few years, by an atomic clock that oscillates at optical frequencies rather than the cesium clock's microwave frequencies. In their current (2016) experimental form, such clocks have achieved frequency stabilities and accuracies as good as a few $\times 10^{-18}$, nearly 100 times better than the current U.S. standard. Optical-frequency combs (Sec. 9.4.3) can be used to lock microwave-frequency oscillators to such optical-frequency clocks.

the process. The two random processes shown in Fig. 6.10 are good examples. They were constructed on a computer as superpositions of pulses $F(t - t_o)$ with random arrival times t_o and with identical forms

$$F(t) = 0 \text{ for } t < 0, \qquad F(t) = K/\sqrt{t} \text{ for } t > 0 \tag{6.45}$$

(cf. Sec. 6.7.4). The two $y(t)$s look very different, because the first (Fig. 6.10a) involves frequent small pulses, while the second (Fig. 6.10b) involves less frequent, larger pulses. These differences are obvious to the eye in the time evolutions $y(t)$. However, they do not show up at all in the spectra $S_y(f)$, which are identical: both are flicker spectra (Ex. 6.15). Moreover, the differences do not show up in $p_1(y_1)$ or in $p_2(y_2, t_2; y_1, t_1)$, because the two processes are both superpositions of many independent pulses and thus are Gaussian, and for Gaussian processes p_1 and p_2 are determined fully by the mean and the correlation function, or equivalently by the mean and spectral density, which are the same for the two processes. Thus, the differences between the two processes show up only in the probabilities p_n of third order and higher, $n \geq 3$ [as defined in Eq. (6.1)].

6.7 Filtering Random Processes

6.7.1 Filters, Their Kernels, and the Filtered Spectral Density

In experimental physics and engineering, one often takes a signal $y(t)$ or a random process $y(t)$ and filters it to produce a new function $w(t)$ that is a *linear functional* of $y(t)$:

$$w(t) = \int_{-\infty}^{+\infty} K(t - t') y(t') dt'. \tag{6.46}$$

The quantity $y(t)$ is called the filter's *input*; $K(t - t')$ is the filter's *kernel*, and $w(t)$ is its *output*. We presume throughout this chapter that the kernel depends only on the time difference $t - t'$ and not on absolute time. When this is so, the filter is said to be *stationary*; and when it is violated so $K = K(t, t')$ depends on absolute time, the filter is said to be nonstationary. Our restriction to stationary filters goes hand-in-hand with our restriction to stationary random processes, since if $y(t)$ is stationary and the filter is stationary (as we require), then the filtered process $w(t) = \int_{-\infty}^{+\infty} K(t - t') y(t') dt'$ is also stationary.

a filter and its input, output, and kernel

Some examples of kernels and their filtered outputs are

$$\begin{aligned} K(\tau) &= \delta(\tau): & w(t) &= y(t), \\ K(\tau) &= \delta'(\tau): & w(t) &= dy/dt, \\ K(\tau) &= 0 \text{ for } \tau < 0 \text{ and } 1 \text{ for } \tau > 0: & w(t) &= \int_{-\infty}^{t} y(t') dt'. \end{aligned} \tag{6.47}$$

Here $\delta'(\tau)$ denotes the derivative of the Dirac δ-function.

As with any function, a knowledge of the kernel $K(\tau)$ is equivalent to a knowledge of its Fourier transform:

Fourier transform of filter's kernel

$$\boxed{\tilde{K}(f) \equiv \int_{-\infty}^{+\infty} K(\tau) e^{i2\pi f\tau} d\tau.} \tag{6.48}$$

This Fourier transform plays a central role in the theory of filtering (also called the theory of *linear signal processing*): the convolution theorem of Fourier transform theory states that, if $y(t)$ is a function whose Fourier transform $\tilde{y}(f)$ exists (converges), then the Fourier transform of the filter's output $w(t)$ [Eq. (6.46)] is given by

$$\tilde{w}(f) = \tilde{K}(f)\tilde{y}(f). \tag{6.49}$$

Similarly, by virtue of the definition (6.25) of spectral density in terms of Fourier transforms, if $y(t)$ is a random process with spectral density $S_y(f)$, then the filter's output $w(t)$ will be a random process with spectral density

spectral density of filter's output

$$\boxed{S_w(f) = |\tilde{K}(f)|^2 S_y(f).} \tag{6.50}$$

[Note that, although $\tilde{K}(f)$, like all Fourier transforms, is defined for both positive and negative frequencies, when its modulus is used in Eq. (6.50) to compute the effect of the filter on a spectral density, only positive frequencies are relevant; spectral densities are strictly positive-frequency quantities.]

easy way to compute kernel's squared Fourier transform

The quantity $|\tilde{K}(f)|^2$ that appears in the very important relation (6.50) is most easily computed not by evaluating directly the Fourier transform (6.48) and then squaring, but rather by sending the function $e^{i2\pi ft}$ through the filter (i.e., by computing the output w that results from the special input $y = e^{i2\pi ft}$), and then squaring the output: $|\tilde{K}(f)|^2 = |w|^2$. To see that this works, notice that the result of sending $y = e^{i2\pi ft}$ through the filter is

$$w = \int_{-\infty}^{+\infty} K(t-t') e^{i2\pi ft'} dt' = \tilde{K}^*(f) e^{i2\pi ft}, \tag{6.51}$$

which differs from $\tilde{K}(f)$ by complex conjugation and a change of phase, and which thus has absolute value squared of $|w|^2 = |\tilde{K}(f)|^2$.

For example, if $w(t) = d^n y/dt^n$, then when we set $y = e^{i2\pi ft}$, we get $w = d^n(e^{i2\pi ft})/dt^n = (i2\pi f)^n e^{i2\pi ft}$; and, accordingly, $|\tilde{K}(f)|^2 = |w|^2 = (2\pi f)^{2n}$, whence, for any random process $y(t)$, the quantity $w(t) = d^n y/dt^n$ will have $S_w(f) = (2\pi f)^{2n} S_y(f)$.

how differentiation and integration change spectral density

This example also shows that by differentiating a random process once, one changes its spectral density by a multiplicative factor $(2\pi f)^2$; for example, one can thereby convert random-walk noise into white noise. Similarly, by integrating a random process once in time (the inverse of differentiating), one multiplies its spectral density by $(2\pi f)^{-2}$. If instead one wants to multiply by f^{-1}, one can achieve that using the filter whose kernel is

$$K(\tau) = 0 \text{ for } \tau < 0, \qquad K(\tau) = \sqrt{\frac{2}{\tau}} \text{ for } \tau > 0; \tag{6.52a}$$

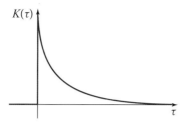

FIGURE 6.12 The kernel (6.52a), whose filter multiplies the spectral density by a factor $1/f$, thereby converts white noise into flicker noise and flicker noise into random-walk noise.

see Fig. 6.12. Specifically, it is easy to show, by sending a sinusoid through this filter, that

$$w(t) \equiv \int_{-\infty}^{t} \sqrt{\frac{2}{t-t'}} y(t') dt' \tag{6.52b}$$

has spectral density

$$S_w(f) = \frac{1}{f} S_y(f). \tag{6.52c}$$

Thus by filtering in this way, one can convert white noise into flicker noise and flicker noise into random-walk noise.

Exercise 6.9 *Derivations and Practice: Examples of Filters*

(a) Show that the kernels $K(\tau)$ in Eq. (6.47) produce the indicated outputs $w(t)$. Deduce the ratio $S_w(f)/S_y(f) = |\tilde{K}(f)|^2$ in two ways: (i) by Fourier transforming each $K(\tau)$; (ii) by setting $y = e^{i2\pi ft}$, deducing the corresponding filtered output w directly from the expression for w in terms of y, and then squaring to get $|\tilde{K}(f)|^2$.

(b) Derive Eqs. (6.52b) and (6.52c) for the kernel (6.52a).

6.7.2 Brownian Motion and Random Walks

As an example of the uses of filtering, consider the motion of an organelle derived from pollen (henceforth "dust particle") being buffeted by thermalized air molecules—the phenomenon of Brownian motion, named for the Scottish botanist Robert Brown (1828), one of the first to observe it in careful experiments. As we discussed in Sec. 6.3.1 and in greater detail at the end of Sec. 6.3.3, any Cartesian component $v(t)$ of the particle's velocity is a Gaussian-Markov process, whose statistical properties are all determined by its equilibrium mean $\bar{v} = 0$ and standard deviation $\sigma_v = \sqrt{k_B T/m}$, and its relaxation time τ_r (which we compute in Sec. 6.8.1). Here m is the particle's mass, and T is the temperature of the air molecules that buffet it. The conditional probability distribution P_2 for v is given by Doob's theorem:

$$P_2(v_2, t|v_1) = \frac{e^{-(v_2-\bar{v}_t)^2/(2\sigma_{v_t}^2)}}{[2\pi\sigma_{v_t}^2]^{\frac{1}{2}}}, \quad \bar{v}_t = v_1 e^{-t/\tau_r},$$

$$\sigma_{v_t}^2 = (1 - e^{-2t/\tau_r})\sigma_v^2, \quad \sigma_v = \sqrt{\frac{k_B T}{m}} \tag{6.53a}$$

[Eqs. (6.18)], and its corresponding correlation function and spectral density have the standard forms (6.20) and (6.32) for a Gaussian-Markov process:

correlation function and spectral densities for Brownian motion

$$C_v(\tau) = \sigma_v^2 e^{-\tau/\tau_r}, \quad S_v(f) = \frac{4\sigma_v^2/\tau_r}{(2\pi f)^2 + (1/\tau_r)^2}. \tag{6.53b}$$

The Cartesian coordinate (position) of the dust particle, $x(t) = \int v\, dt$, is of special interest. Its spectral density can be deduced by applying the time-integral filter $|\tilde{K}(f)|^2 = 1/(2\pi f)^2$ to $S_v(f)$. The result, using Eq. (6.53b), is

$$S_x(f) = \frac{4\tau_r \sigma_v^2}{(2\pi f)^2[1 + (2\pi f \tau_r)^2]}. \tag{6.53c}$$

Notice that at frequencies $f \ll 1/\tau_r$ (corresponding to times long compared to the relaxation time), our result (6.53c) reduces to the random-walk spectrum $S_x = 4\sigma_v^2\tau_r/(2\pi f)^2$. From this spectrum, we can compute the rms distance $\sigma_{\Delta x}$ in the x direction that the dust particle travels in a time interval $\Delta \tau \gg \tau_r$. That $\sigma_{\Delta x}$ is the standard deviation of the random process $\Delta x(t) \equiv x(t + \Delta \tau) - x(t)$. The filter that takes $x(t)$ into $\Delta x(t)$ has

$$|\tilde{K}(f)|^2 = |e^{i2\pi f(t+\Delta\tau)} - e^{i2\pi ft}|^2 = 4\sin^2(\pi f \Delta\tau). \tag{6.54a}$$

Correspondingly, $\Delta x(t)$ has spectral density

$$S_{\Delta x}(f) = |\tilde{K}(f)|^2 S_x(f) = 4\sigma_v^2 \tau_r (\Delta\tau)^2 \operatorname{sinc}^2(\pi f \Delta\tau) \tag{6.54b}$$

(where, again, $\operatorname{sinc} u \equiv \sin u/u$), so the variance of Δx (i.e., the square of the rms distance traveled) is

$$(\sigma_{\Delta x})^2 = \int_0^\infty S_{\Delta x}(f)\, df = 2(\sigma_v \tau_r)^2 \frac{\Delta\tau}{\tau_r}. \tag{6.54c}$$

This equation has a simple physical interpretation. The damping time τ_r is the time required for collisions to change substantially the dust particle's momentum, so we can think of it as the duration of a single step in the particle's random walk. The particle's mean speed is roughly $\sqrt{2}\sigma_v$, so the distance traveled during each step (the particle's mean free path) is roughly $\sqrt{2}\sigma_v \tau_r$. (The $\sqrt{2}$ comes from our analysis; this physical argument could not have predicted it.) Therefore, during a time interval $\Delta\tau$ long compared to a single step τ_r, the rms distance traveled in the x direction by the random-walking dust particle is about one mean-free path $\sqrt{2}\sigma_v \tau_r$, multiplied by the square root of the mean number of steps taken, $\sqrt{\Delta\tau/\tau_r}$:

$$\sigma_{\Delta x} = \sqrt{2}\sigma_v \tau_r \sqrt{\Delta\tau/\tau_r}. \qquad (6.55)$$

rms distance traveled in Brownian motion

This "square root of the number of steps taken" behavior is a universal rule of thumb for random walks; one meets it time and again in science, engineering, and mathematics. We have met it previously in our studies of diffusion (Exs. 3.17 and 6.3) and of the elementary "unit step" random walk problem that we studied using the central limit theorem in Ex. 6.4. We could have guessed Eq. (6.55) from this rule of thumb, up to an unknown multiplicative factor of order unity. Our analysis has told us that factor: $\sqrt{2}$.

Exercise 6.10 *Example: Position, Viewed on Timescales* $\Delta\tau \gg \tau_r$, *as a Markov Process*

EXERCISES

(a) Explain why, physically, when the Brownian motion of a particle (which starts at $x = 0$ at time $t = 0$) is observed only on timescales $\Delta\tau \gg \tau_r$ corresponding to frequencies $f \ll 1/\tau_r$, its position $x(t)$ must be a Gaussian-Markov process with $\bar{x} = 0$. What are the spectral density of $x(t)$ and its relaxation time in this case?

(b) Use Doob's theorem to compute the conditional probability $P_2(x_2, \tau | x_1)$. Your answer should agree with the result deduced in Ex. 6.3 from the diffusion equation, and in Ex. 6.4 from the central limit theorem for a random walk.

6.7.3 Extracting a Weak Signal from Noise: Band-Pass Filter, Wiener's Optimal Filter, Signal-to-Noise Ratio, and Allan Variance of Clock Noise

6.7.3

In experimental physics and engineering, one often meets a random process $Y(t)$ that consists of a sinusoidal signal on which is superposed noise $y(t)$:

$$Y(t) = \sqrt{2}Y_s \cos(2\pi f_o t + \delta_o) + y(t). \qquad (6.56a)$$

[The factor $\sqrt{2}$ is included in Eq. (6.56a) because the time average of the square of the cosine is 1/2; correspondingly, with the factor $\sqrt{2}$ present, Y_s is the rms signal amplitude.] We assume that the frequency f_o and phase δ_o of the signal are known, and we want to determine the signal's rms amplitude Y_s. The noise $y(t)$ is an impediment to the determination of Y_s. To reduce that impediment, we can send $Y(t)$ through a *band-pass filter* centered on the signal frequency f_o (i.e., a filter with a shape like that of Fig. 6.13).

band-pass filter

For such a filter, with central frequency f_o and with bandwidth $\Delta f \ll f_o$, the bandwidth is defined by

bandwidth

$$\Delta f \equiv \frac{\int_0^\infty |\tilde{K}(f)|^2 df}{|\tilde{K}(f_o)|^2}. \qquad (6.56b)$$

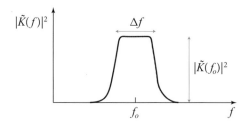

FIGURE 6.13 A band-pass filter centered on frequency f_o with bandwidth Δf.

The output $W(t)$ of such a filter, when the input is $Y(t)$, has the form

$$W(t) = |\tilde{K}(f_o)|\sqrt{2}Y_s \cos(2\pi f_o t + \delta_1) + w(t), \tag{6.56c}$$

where the first term is the filtered signal and the second is the filtered noise. The output signal's phase δ_1 may be different from the input signal's phase δ_o, but that difference can be evaluated in advance for one's filter and can be taken into account in the measurement of Y_s; thus it is of no interest to us. Assuming (as we shall) that the input noise $y(t)$ has spectral density S_y that varies negligibly over the small bandwidth of the filter, the filtered noise $w(t)$ will have spectral density

spectral density of band-pass filter's output

$$S_w(f) = |\tilde{K}(f)|^2 S_y(f_o). \tag{6.56d}$$

This means that $w(t)$ consists of a random superposition of sinusoids all with nearly—but not quite—the same frequency f_o; their frequency spread is Δf. Now, when one superposes two sinusoids with frequencies that differ by $\Delta f \ll f_o$, the two beat against each other, producing a modulation with period $1/\Delta f$. Correspondingly, with its random superposition of many such sinusoids, the noise $w(t)$ will have the form

$$w(t) = w_o(t) \cos[2\pi f_o t + \phi(t)], \tag{6.56e}$$

with amplitude $w_o(t)$ and phase $\phi(t)$ that fluctuate randomly on timescales

$$\boxed{\Delta t \sim 1/\Delta f,} \tag{6.56f}$$

but that are nearly constant on timescales $\Delta t \ll 1/\Delta f$.

band-pass filter's output in time domain

The filter's net output $W(t)$ thus consists of a precisely sinusoidal signal at frequency f_o, with known phase δ_1 and with an amplitude that we wish to determine, plus noise $w(t)$ that is also sinusoidal at frequency f_o but with amplitude and phase that wander randomly on timescales $\Delta t \sim 1/\Delta f$. The rms output signal is

$$S \equiv |\tilde{K}(f_o)|Y_s \tag{6.56g}$$

[Eq. (6.56c)], while the rms output noise is

$$N \equiv \sigma_w = \left[\int_0^\infty S_w(f)df\right]^{\frac{1}{2}}$$

$$= \sqrt{S_y(f_o)} \left[\int_0^\infty |\tilde{K}(f)|^2 df\right]^{\frac{1}{2}} = |\tilde{K}(f_o)|\sqrt{S_y(f_o)\Delta f}, \quad (6.56h)$$

where the first integral follows from Eq. (6.26), the second from Eq. (6.56d), and the third from the definition (6.56b) of the bandwidth Δf. The ratio of the rms signal (6.56g) to the rms noise (6.56h) after filtering is

$$\boxed{\frac{S}{N} = \frac{Y_s}{\sqrt{S_y(f_o)\Delta f}}.} \quad (6.57)$$

signal-to-noise ratio for band-pass filter

Thus, the rms output $S + N$ of the filter is the signal amplitude to within an rms fractional error N/S given by the reciprocal of (6.57). Notice that the narrower the filter's bandwidth, the more accurate will be the measurement of the signal. In practice, of course, one does not know the signal frequency with complete precision in advance; correspondingly, one does not want to make one's filter so narrow that the signal might be lost from it.

A simple example of a band-pass filter is the following *finite Fourier transform filter*:

$$w(t) = \int_{t-\Delta t}^{t} \cos[2\pi f_o(t-t')] y(t')dt', \quad \text{where } \Delta t \gg 1/f_o. \quad (6.58a)$$

In Ex. 6.11 it is shown that this is indeed a band-pass filter, and that the integration time Δt used in the Fourier transform is related to the filter's bandwidth by

$$\Delta f = 1/\Delta t. \quad (6.58b)$$

Often the signal one seeks amidst noise is not sinusoidal but has some other known form $s(t)$. In this case, the optimal way to search for it is with a so-called *Wiener filter* (an alternative to the band-pass filter); see the very important Ex. 6.12.

EXERCISES

Exercise 6.11 *Derivation and Example: Bandwidths of a Finite-Fourier-Transform Filter and an Averaging Filter*

(a) If y is a random process with spectral density $S_y(f)$, and $w(t)$ is the output of the finite-Fourier-transform filter (6.58a), what is $S_w(f)$?

(b) Sketch the filter function $|\tilde{K}(f)|^2$ for this finite-Fourier-transform filter, and show that its bandwidth is given by Eq. (6.58b).

(c) An "averaging filter" is one that averages its input over some fixed time interval Δt:

$$w(t) \equiv \frac{1}{\Delta t} \int_{t-\Delta t}^{t} y(t')dt'. \quad (6.59a)$$

What is $|\tilde{K}(f)|^2$ for this filter? Draw a sketch of this $|\tilde{K}(f)|^2$.

(d) Suppose that $y(t)$ has a spectral density that is very nearly constant at all frequencies, $f \lesssim 1/\Delta t$, and that this y is put through the averaging filter (6.59a). Show that the rms fluctuations in the averaged output $w(t)$ are

$$\sigma_w = \sqrt{S_y(0)\Delta f}, \tag{6.59b}$$

where Δf, interpretable as the bandwidth of the averaging filter, is

$$\Delta f = \frac{1}{2\Delta t}. \tag{6.59c}$$

(Recall that in our formalism we insist that f be nonnegative.) Why is there a factor $1/2$ here and not one in the equation for an averaging filter [Eq. (6.58b)]? Because here, with f restricted to positive frequencies and the filter centered on zero frequency, we see only the right half of the filter: $f \geq f_o = 0$ in Fig. 6.13.

Exercise 6.12 **Example: Wiener's Optimal Filter*
Suppose that you have a noisy receiver of weak signals (e.g., a communications receiver). You are expecting a signal $s(t)$ with finite duration and known form to come in, beginning at a predetermined time $t = 0$, but you are not sure whether it is present. If it is present, then your receiver's output will be

$$Y(t) = s(t) + y(t), \tag{6.60a}$$

where $y(t)$ is the receiver's noise, a random process with spectral density $S_y(f)$ and with zero mean, $\bar{y} = 0$. If it is absent, then $Y(t) = y(t)$. A powerful way to find out whether the signal is present is by passing $Y(t)$ through a filter with a carefully chosen kernel $K(t)$. More specifically, compute the quantity

$$W \equiv \int_{-\infty}^{+\infty} K(t)Y(t)dt. \tag{6.60b}$$

If $K(t)$ is chosen optimally, then W will be maximally sensitive to the signal $s(t)$ in the presence of the noise $y(t)$. Correspondingly, if W is large, you infer that the signal was present; if it is small, you infer that the signal was either absent or so weak as not to be detectable. This exercise derives the form of the *optimal filter, $K(t)$* (i.e., the filter that will most effectively discern whether the signal is present). As tools in the derivation, we use the quantities S and N defined by

$$S \equiv \int_{-\infty}^{+\infty} K(t)s(t)dt, \quad N \equiv \int_{-\infty}^{+\infty} K(t)y(t)dt. \tag{6.60c}$$

Note that S is the filtered signal, N is the filtered noise, and $W = S + N$. Since $K(t)$ and $s(t)$ are precisely defined functions, S is a number. But since $y(t)$ is a random process, the value of N is not predictable, and instead is given by some probability distribution $p_1(N)$. We shall also need the Fourier transform $\tilde{K}(f)$ of the kernel $K(t)$.

(a) In the measurement being done one is not filtering a function of time to get a new function of time; rather, one is just computing a number, $W = S + N$.

Nevertheless, as an aid in deriving the optimal filter it is helpful to consider the time-dependent output of the filter that results when noise $y(t)$ is fed continuously into it:

$$N(t) \equiv \int_{-\infty}^{+\infty} K(t-t')y(t')dt'. \tag{6.61a}$$

Show that this random process has a mean squared value

$$\overline{N^2} = \int_0^\infty |\tilde{K}(f)|^2 S_y(f)df. \tag{6.61b}$$

Explain why this quantity is equal to the average of the number N^2 computed using (6.60c) in an ensemble of many experiments:

$$\overline{N^2} = \langle N^2 \rangle \equiv \int p_1(N)N^2 dN = \int_0^\infty |\tilde{K}(f)|^2 S_y(f)df. \tag{6.61c}$$

(b) Show that of all choices of the kernel $K(t)$, the one that will give the largest value of

$$\frac{S}{\langle N^2 \rangle^{\frac{1}{2}}} \tag{6.61d}$$

is Norbert Wiener's (1949) *optimal filter*, whose kernel has the Fourier transform

$$\boxed{\tilde{K}(f) = \text{const} \times \frac{\tilde{s}(f)}{S_y(f)},} \tag{6.62a}$$

Wiener's optimal filter

where $\tilde{s}(f)$ is the Fourier transform of the signal $s(t)$, and $S_y(f)$ is the spectral density of the noise. Note that when the noise is white, so $S_y(f)$ is independent of f, this optimal kernel is just $K(t) = \text{const} \times s(t)$ (i.e., one should simply multiply the known signal form into the receiver's output and integrate). By contrast, when the noise is not white, the optimal kernel (6.62a) is a distortion of const $\times s(t)$ in which frequency components at which the noise is large are suppressed, while frequency components at which the noise is small are enhanced.

(c) Show that when the optimal kernel (6.62a) is used, the square of the signal-to-noise ratio is

$$\boxed{\frac{S^2}{\langle N^2 \rangle} = 4 \int_0^\infty \frac{|\tilde{s}(f)|^2}{S_y(f)}df.} \tag{6.62b}$$

signal-to-noise ratio for Wiener's optimal filter

(d) As an example, suppose the signal consists of n cycles of some complicated waveform with frequencies spread out over the range $f_o/2$ to $2f_o$ with amplitude $\sim A$ for its entire duration. Also suppose that S_y is approximately constant (near white noise) over this frequency band. Show that $S/\langle N^2 \rangle^{1/2} \sim 2nA/\sqrt{f_o S_y(f_o)}$, so the amplitude signal to noise increases linearly with the number of cycles in the signal.

(e) Suppose (as an idealization of searches for gravitational waves in noisy LIGO data) that (i) we do not know the signal $s(t)$ in advance, but we do know that it is from a set of N distinct signals, all of which have frequency content concentrated around some f_o; (ii) we do not know when the signal will arrive, but we search for it for a long time τ_s (say, a year); and (iii) the noise superposed on the signal is Gaussian. Show that, in order to have 99% confidence that any signal found is real, it must have amplitude signal-to-noise ratio of $S/\langle N^2\rangle^{1/2} \gtrsim [2\ln(H/\sqrt{2\ln H})]^{1/2}$, where $H = 100 N f_o \tau_s$. For $N \sim 10^4$, $f_o \sim 100$ Hz, $\tau_s \sim 1$ yr, this says $S/\langle N^2\rangle^{1/2} \gtrsim 8.2$. This is so small because the Gaussian probability distribution falls off so rapidly. If the noise is non-Gaussian, then the minimum detectable signal will be larger than this, possibly much larger.

Exercise 6.13 **Example: Allan Variance of Clocks*
Highly stable clocks (e.g., cesium clocks, hydrogen maser clocks, or quartz crystal oscillators) have angular frequencies ω of ticking that tend to wander so much over very long timescales that their variances diverge. For example, a cesium clock has random-walk noise on very long timescales (low frequencies)

$$S_\omega(f) \propto 1/f^2 \quad \text{at low } f; \tag{6.63a}$$

and correspondingly,

$$\sigma_\omega^2 = \int_0^\infty S_\omega(f) df = \infty \tag{6.63b}$$

(cf. Fig. 6.11 and associated discussion). For this reason, clock makers have introduced a special technique for quantifying the frequency fluctuations of their clocks. Using the phase

$$\phi(t) = \int_0^t \omega(t') dt', \tag{6.64a}$$

they define the quantity

$$\Phi_\tau(t) = \frac{[\phi(t+2\tau) - \phi(t+\tau)] - [\phi(t+\tau) - \phi(t)]}{\sqrt{2}\bar\omega\tau}, \tag{6.64b}$$

where $\bar\omega$ is the mean frequency. Aside from the $\sqrt{2}$, this $\Phi_\tau(t)$ is the fractional difference of clock readings for two successive intervals of duration τ. (In practice the measurement of t is made by a clock more accurate than the one being studied; or, if a more accurate clock is not available, by a clock or ensemble of clocks of the same type as is being studied.)

(a) Show that the spectral density of $\Phi_\tau(t)$ is related to that of $\omega(t)$ by

$$S_{\Phi_\tau}(f) = \frac{2}{\bar\omega^2}\left[\frac{\cos 2\pi f\tau - 1}{2\pi f\tau}\right]^2 S_\omega(f)$$

$$\propto f^2 S_\omega(f) \text{ at } f \ll 1/(2\pi\tau) \tag{6.65}$$

$$\propto f^{-2} S_\omega(f) \text{ at } f \gg 1/(2\pi\tau).$$

Note that $S_{\Phi_\tau}(f)$ is much better behaved (more strongly convergent when integrated) than $S_\omega(f)$ is, both at low frequencies and at high.

(b) The *Allan variance* of the clock is defined as

$$\sigma_\tau^2 \equiv [\text{variance of } \Phi_\tau(t)] = \int_0^\infty S_{\Phi_\tau}(f)df. \qquad (6.66)$$

Show that

$$\sigma_\tau = \left[\alpha \frac{S_\omega(1/(2\tau))}{\bar{\omega}^2} \frac{1}{2\tau}\right]^{\frac{1}{2}}, \qquad (6.67)$$

where α is a constant of order unity that depends on the spectral shape of $S_\omega(f)$ near $f = 1/(2\tau)$. Explain why, aside from the factor α, the right-hand side of Eq. (6.67) is the rms fractional fluctuation of ω at frequency $1/(2\tau)$ in bandwidth $1/(2\tau)$.

(c) Show that, if ω has a white-noise spectrum, then the clock stability is better for long averaging times than for short; if ω has a flicker-noise spectrum, then the clock stability is independent of averaging time; and if ω has a random-walk spectrum, then the clock stability is better for short averaging times than for long. (See Fig. 6.11.)

6.7.4 Shot Noise

A specific kind of noise that one frequently meets and frequently wants to filter is *shot noise*. A random process $y(t)$ is said to consist of shot noise if it is a random superposition of a large number of finite-duration pulses. In this chapter, we restrict attention to a simple version of shot noise in which the pulses all have the same shape, $F(\tau)$ (e.g., Fig. 6.14a), but their arrival times t_i are random

$$y(t) = \sum_i F(t - t_i), \qquad (6.68a)$$

as may be the case for individual photons arriving at a photodetector. We denote by $\mathcal{R}$ the mean rate of pulse arrivals (the mean number per second). From the definition (6.25) of spectral density, it is straightforward to see that the spectral density of y is

$$\boxed{S_y(f) = 2\mathcal{R}|\tilde{F}(f)|^2,} \qquad (6.68b)$$

where $\tilde{F}(f)$ is the Fourier transform of $F(\tau)$ (Fig. 6.14). See Ex. 6.14 for proof. If the pulses are broadband bursts without much substructure in them (as in Fig. 6.14a), then the duration τ_p of the pulse is related to the frequency $f_{\max}$ at which the spectral density starts to cut off by $f_{\max} \sim 1/\tau_p$. Since the correlation function is the cosine transform of the spectral density, the correlation's relaxation time is $\tau_r \sim 1/f_{\max} \sim \tau_p$ (Ex. 6.14).

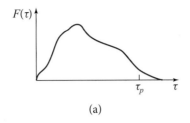
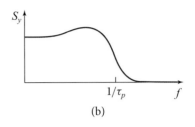

FIGURE 6.14 (a) A broadband pulse that produces shot noise by arriving at random times. (b) The spectral density of the shot noise produced by that pulse.

In the common (but not universal) case that many pulses are on at once on average ($\mathcal{R}\tau_p \gg 1$), then at any moment of time $y(t)$ is the sum of many random processes. Correspondingly, the central limit theorem guarantees that y is a Gaussian random process. Over time intervals smaller than $\tau_p \sim \tau_r$ the process will not generally be Markov, because a knowledge of both $y(t_1)$ and $y(t_2)$ gives some rough indication of how many pulses happen to be on and how many new ones have turned on during the time interval between t_1 and t_2 and thus are still in their early stages at time t_3; this knowledge helps one predict $y(t_3)$ with greater confidence than if one knew only $y(t_2)$. In other words, $P_3(y_3, t_3 | y_2, t_2; y_1, t_1)$ is not equal to $P_2(y_3, t_3 | y_2, t_2)$, which implies non-Markov behavior.

By contrast, if many pulses are on at once, and if one takes a coarse-grained view of time, never examining time intervals as short as τ_p or shorter, then a knowledge of $y(t_1)$ is of no help in predicting $y(t_2)$. All correlations between different times are lost, so the process is Markov, and (because it is a random superposition of many independent influences) it is also Gaussian—an example of the central limit theorem at work. Thus it must have the standard Gaussian-Markov spectral density (6.32) with vanishing correlation time τ_r (i.e., it must be white). Indeed, it is: For $f \ll 1/\tau_p$, the limit of Eq. (6.68b) for S_y and the corresponding correlation function are

$$S_y(f) = 2\mathcal{R}|\tilde{F}(0)|^2, \quad C_y(\tau) = \mathcal{R}|\tilde{F}(0)|^2 \delta(\tau). \tag{6.68c}$$

This formula remains true if the pulses have different shapes, so long as their Fourier transforms at zero frequency, $\tilde{F}_j(0) = \int_{-\infty}^{\infty} F_j dt$, are all the same; see Ex. 6.14b.

As an important example, consider a (nearly) monochromatic beam of light with angular frequency $\omega_o \sim 10^{15}\,\text{s}^{-1}$ and with power (energy per unit time) $W(t)$ that is being measured by a photodetector. The arriving light consists of discrete photons, each with its own pulse shape $W_j(t - t_j)$,[10] which lasts for a time τ_p long compared to the light's period ($\sim 3 \times 10^{-15}$ s) but short compared to the inverse frequency f^{-1} at which we measure the photon shot noise. The Fourier transform of W_j at zero

---

10. For a single photon, $W_j(t)$ is the probability per unit time for the photon's arrival, times the photon's energy $\hbar\omega_o$.

frequency is just $\tilde{W}_j(0) = \int_0^\infty W_j dt = \hbar\omega$ (the total energy carried by the photon), which is the same for all pulses; the rate of arrival of photons is $\mathcal{R} = \overline{W}/\hbar\omega_o$. Therefore the spectral density of the intensity measured by the photodetector is

$$S_W(f) = 2\overline{W}\,\hbar\omega. \tag{6.69}$$

In the LIGO interferometer, whose noise power spectrum is shown in Fig. 6.7, this photon shot noise dominates in the frequency band $f \gtrsim 200$ Hz. (Although S_W for the laser light has white noise, when passed through the interferometer as a filter, it produces $S_x \propto f^2$.)

EXERCISES

Exercise 6.14 *Derivation: Shot Noise*
(a) Show that for shot noise, $y(t) = \sum_i F(t - t_i)$, the spectral density $S_y(f)$ is given by Eq. (6.68b). Show that the relaxation time appearing in the correlation function is approximately the duration τ_p of $F(t)$.
(b) Suppose the shapes of $F_j(t - t_j)$ are all different instead of being identical but all last for times $\lesssim \tau_p$, and all have the same Fourier transform at zero frequency, $\tilde{F}_j(0) = \int_{-\infty}^\infty F_j dt = \tilde{F}(0)$. Show that the shot noise at frequencies $f \ll 1/\tau_p$ is still given by Eq. (6.68c).

Exercise 6.15 *Example: Shot Noise with Flicker Spectrum*
(a) Show that for shot noise with identical pulses that have the infinitely sharply peaked shape of Eq. (6.45), the power spectrum has the flicker form $S_y \propto 1/f$ for all f.
(b) Construct realizations of shot noise with flicker spectrum [Eq. (6.68a) with pulse shape (6.45)] that range from a few large pulses in the time interval observed to many small pulses, and describe the visual differences; cf. Fig. 6.10 and the discussion in Sec. 6.6.2.

6.8 Fluctuation-Dissipation Theorem

6.8.1 Elementary Version of the Fluctuation-Dissipation Theorem; Langevin Equation, Johnson Noise in a Resistor, and Relaxation Time for Brownian Motion

Friction is generally caused by interaction with the huge number of degrees of freedom of some sort of bath (e.g., the molecules of air against which a moving ball or dust particle pushes). Those degrees of freedom also produce fluctuating forces. In this section, we study the relationship between the friction and the fluctuating forces when the bath is thermalized at some temperature T (so it is a heat bath).

For simplicity, we restrict ourselves to a specific generalized coordinate q of the system that will interact with a bath (e.g., the x coordinate of the ball or dust particle).

We require just one special property for q: its time derivative $\dot{q} = dq/dt$ must appear in the system's lagrangian as a kinetic energy,

$$E_{\text{kinetic}} = \frac{1}{2}m\dot{q}^2, \tag{6.70}$$

and in no other way. Here m is a (generalized) mass associated with q. Then the equation of motion for q will have the simple form of Newton's second law, $m\ddot{q} = F$, where F includes contributions $\mathcal{F}$ from the system itself (e.g., a restoring force in the case of a normal mode), plus a force F_{bath} due to the heat bath (i.e., due to all the degrees of freedom in the bath). This F_{bath} is a random process whose time average is a frictional (damping) force proportional to $\dot{q}$:

$$\bar{F}_{\text{bath}} = -R\dot{q}, \quad F_{\text{bath}} \equiv \bar{F}_{\text{bath}} + F'. \tag{6.71}$$

Here R is the coefficient of friction. The fluctuating part F' of F_{bath} is responsible for driving q toward statistical equilibrium.

examples:

Three specific examples, to which we shall return below, are as follows.

dust particle

1. The system might be a dust particle with q its x coordinate and m its mass. The heat bath might be air molecules at temperature T, which buffet the dust particle, producing Brownian motion.

electric circuit

2. The system might be an L-C-R circuit (i.e., an electric circuit containing an inductance L, a capacitance C, and a resistance R) with q the total electric charge on the top plate of the capacitor. The bath in this case would be the many mechanical degrees of freedom in the resistor. For such a circuit, the "equation of motion" is

$$L\ddot{q} + C^{-1}q = F_{\text{bath}}(t) = -R\dot{q} + F', \tag{6.72}$$

so the effective mass is the inductance L; the coefficient of friction is the resistance (both denoted R); $-R\dot{q} + F'$ is the total voltage across the resistor; and F' is the fluctuating voltage produced by the resistor's internal degrees of freedom (the bath) and so might better be denoted V'.

normal mode of crystal

3. The system might be the fundamental mode of a 10-kg sapphire crystal with q its generalized coordinate (cf. Sec. 4.2.1). The heat bath might be all the other normal modes of vibration of the crystal, with which the fundamental mode interacts weakly.

LANGEVIN EQUATION

In general, the equation of motion for the generalized coordinate $q(t)$ under the joint action of (i) the bath's damping force $-R\dot{q}$, (ii) the bath's fluctuating forces F', and (iii) the system's internal force $\mathcal{F}$ will take the form [cf. Eq. (6.71)]

$$m\ddot{q} + R\dot{q} = \mathcal{F} + F'(t). \tag{6.73}$$

The internal force $\mathcal{F}$ is derived from the system's hamiltonian or lagrangian in the absence of the heat bath. For the L-C-R circuit of Eq. (6.72) that force is $\mathcal{F} = -C^{-1}q$; for the dust particle, if the particle were endowed with a charge Q and were in an external electric field with potential $\Phi(t, x, y, z)$, it would be $\mathcal{F} = -Q\partial\Phi/\partial x$; for the normal mode of a crystal, it is $\mathcal{F} = -m\omega^2 q$, where ω is the mode's eigenfrequency.

Because the equation of motion (6.73) involves a driving force $F'(t)$ that is a random process, one cannot solve it to obtain $q(t)$. Instead, one must solve it in a statistical way to obtain the evolution of q's probability distributions $p_n(q_1, t_1; \ldots; q_n, t_n)$. This and other evolution equations involving random-process driving terms are called by modern mathematicians *stochastic differential equations,* and there is an extensive body of mathematical formalism for solving them. In statistical physics the specific stochastic differential equation (6.73) is known as the *Langevin equation*.

<small>stochastic differential equation</small>

<small>Langevin equation</small>

ELEMENTARY FLUCTUATION-DISSIPATION THEOREM

Because the damping force $-R\dot{q}$ and the fluctuating force F' both arise from interaction with the same heat bath, there is an intimate connection between them. For example, the stronger the coupling to the bath, the stronger will be the coefficient of friction R and the stronger will be F'. The precise relationship between the dissipation embodied in R and the fluctuations embodied in F' is given by the following *fluctuation-dissipation theorem:* At frequencies

$$f \ll 1/\tau_r, \tag{6.74a}$$

<small>elementary fluctuation-dissipation theorem</small>

where τ_r is the (very short) relaxation time for the fluctuating force F', the fluctuating force has the spectral density

$$\boxed{S_{F'}(f) = 4R\left(\frac{1}{2}hf + \frac{hf}{e^{hf/(k_B T)} - 1}\right) \quad \text{in general,}} \tag{6.74b}$$

$$\boxed{S_{F'}(f) = 4Rk_B T \quad \text{in the classical domain, } k_B T \gg hf.} \tag{6.74c}$$

Here T is the temperature of the bath, and h is Planck's constant.

Notice that in the classical domain, $k_B T \gg hf$, the spectral density has a white-noise spectrum. In fact, since we are restricting attention to frequencies at which F' has no self-correlations ($f^{-1} \gg \tau_r$), F' is Markov; and since it is produced by interaction with the huge number of degrees of freedom of the bath, F' is also Gaussian. Thus, in the classical domain F' is a Gaussian-Markov, white-noise process.

<small>classical fluctuations accompanying $-R\dot{q}$ damping are Gaussian, Markov, white-noise processes</small>

At frequencies $f \gg k_B T/h$ (quantum domain), in Eq. (6.74b) the term $S_{F'} = 4R\frac{1}{2}hf$ is associated with vacuum fluctuations of the degrees of freedom that make up the heat bath (one-half quantum of fluctuations per mode as for any quantum mechanical simple harmonic oscillator). In addition, the second term, $S_{F'}(f) = 4Rhf e^{-hf/(k_B T)}$, associated with thermal excitations of the bath's degrees of freedom, is exponentially suppressed because at these high frequencies, the bath's modes have exponentially small probabilities of containing any quanta at all. Since in this quantum domain $S_{F'}(f)$ does not have the standard Gaussian-Markov frequency dependence (6.32), in the quantum domain F' is not a Gaussian-Markov process.

Proof of the Fluctuation-Dissipation Theorem.

proof of elementary fluctuation-dissipation theorem

In principle, we can alter the system's internal restoring force $\mathcal{F}$ without altering its interactions with the heat bath [i.e., without altering R or $S_{F'}(f)$]. For simplicity, we set $\mathcal{F}$ to zero so q becomes the coordinate of a free mass. The basic idea of our proof is to choose a frequency f_o at which to evaluate the spectral density of F', and then, in an idealized thought experiment, very weakly couple a harmonic oscillator with eigenfrequency f_o to q. Through that coupling, the oscillator is indirectly damped by the resistance R of q and is indirectly driven by R's associated fluctuating force F', which arises from a bath with temperature T. After a long time, the oscillator will reach thermal equilibrium with that bath and will then have the standard thermalized mean kinetic energy ($\bar{E} = k_B T$ in the classical regime). We shall compute that mean energy in terms of $S_{F'}(f_o)$ and thereby deduce $S_{F'}(f_o)$.

The Langevin equation (6.73) and equation of motion for the coupled free mass and harmonic oscillator are

$$m\ddot{q} + R\dot{q} = -\kappa Q + F'(t), \quad M\ddot{Q} + M\omega_o^2 Q = -\kappa q. \quad (6.75a)$$

Here M, Q, and $\omega_o = 2\pi f_o$ are the oscillator's mass, coordinate, and angular eigenfrequency, and κ is the arbitrarily small coupling constant. (The form of the coupling terms $-\kappa Q$ and $-\kappa q$ in the two equations can be deduced from the coupling's interaction hamiltonian $H_I = \kappa q Q$.) Equations (6.75a) can be regarded as a filter to produce from the fluctuating-force input $F'(t)$ a resulting motion of the oscillator, $Q(t) = \int_{-\infty}^{+\infty} K(t-t')F'(t')dt'$. The squared Fourier transform $|\tilde{K}(f)|^2$ of this filter's kernel $K(t-t')$ is readily computed by the standard method [Eq. (6.51) and associated discussion] of inserting a sinusoid $e^{-i\omega t}$ (with $\omega = 2\pi f$) into the filter [i.e., into the differential equations (6.75a)] in place of F', then solving for the sinusoidal output Q, and then setting $|\tilde{K}|^2 = |Q|^2$. The resulting $|\tilde{K}|^2$ is the ratio of the spectral densities of input and output. We carefully manipulate the resulting $|\tilde{K}|^2$ so as to bring it into the following standard resonant form:

$$S_q(f) = |\tilde{K}(f)|^2 S_{F'}(f) = \frac{|B|^2}{(\omega - \omega_o')^2 + (2M\omega_o^2 R|B|^2)^2]} S_{F'}(f). \quad (6.75b)$$

Here $B = \kappa/[2M\omega_o(m\omega_o^2 + iR\omega_o)]$ is arbitrarily small because κ is arbitrarily small; and $\omega_o'^2 = \omega_o^2 + 4mM\omega_o^4|B|^2$ is the oscillator's squared angular eigenfrequency after coupling to q, and is arbitrarily close to ω_o^2 because $|B|^2$ is arbitrarily small. In these equations we have replaced ω by ω_o everywhere except in the resonance term $(\omega - \omega_o')^2$ because $|\tilde{K}|^2$ is negligibly small everywhere except near resonance, $\omega \cong \omega_o$.

The mean energy of the oscillator, averaged over an arbitrarily long timescale, can be computed in either of two ways.

1. Because the oscillator is a mode of some boson field and is in statistical equilibrium with a heat bath, its mean occupation number must have the standard Bose-Einstein value $\eta = 1/[e^{\hbar\omega_o/(k_BT)} - 1]$, and since each quantum carries an energy $\hbar\omega_o$, the mean energy is

$$\bar{E} = \frac{\hbar\omega_o}{e^{\hbar\omega_o/(k_BT)} - 1} + \frac{1}{2}\hbar\omega_o. \tag{6.75c}$$

Here we have included the half-quantum of energy associated with the mode's vacuum fluctuations.

2. Because on average the energy is half potential and half kinetic, and the mean potential energy is $\frac{1}{2}m\omega_o^2\overline{Q^2}$, and because the ergodic hypothesis tells us that time averages are the same as ensemble averages, it must be that

$$\bar{E} = 2\frac{1}{2}M\omega_o^2\langle Q^2\rangle = M\omega_o^2 \int_0^\infty S_Q(f)\,df. \tag{6.75d}$$

By inserting the spectral density (6.75b) and performing the frequency integral with the help of the narrowness of the resonance, we obtain

$$\bar{E} = \frac{S_{F'}(f_o)}{4R}. \tag{6.75e}$$

Equating this to our statistical-equilibrium expression (6.75c) for the mean energy, we see that at the frequency $f_o = \omega_o/(2\pi)$ the spectral density $S_{F'}(f_o)$ has the form (6.74b) claimed in the fluctuation-dissipation theorem. Moreover, since f_o can be chosen to be any frequency in the range (6.74a), the spectral density $S_{F'}(f)$ has the claimed form anywhere in this range. ∎

Let us discuss two examples of the elementary fluctuation-dissipation theorem (6.74):

JOHNSON NOISE IN A RESISTOR

For the L-C-R circuit of Eq. (6.72), $R\dot{q}$ is the dissipative voltage across the resistor, and $F'(t)$ is the fluctuating voltage [more normally denoted $V'(t)$] across the resistor. The fluctuating voltage is called *Johnson noise*, and the fluctuation-dissipation relationship $S_V(f) = 4Rk_BT$ (classical regime) is called *Nyquist's theorem*, because John Johnson (1928) discovered the voltage fluctuations $V'(t)$ experimentally, and Harry Nyquist (1928) derived the fluctuation-dissipation relationship for a resistor to explain them. The fluctuation-dissipation theorem as formulated here is a generalization of Nyquist's original theorem to any system with kinetic energy $\frac{1}{2}m\dot{q}^2$ associated with a generalized coordinate q and with frictional dissipation produced by a heat bath.

Nyquist's theorem for Johnson noise

BROWNIAN MOTION

In Secs. 6.3.3 and 6.7.2, we have studied the Brownian motion of a dust particle being buffeted by air molecules, but until now we omitted any attempt to deduce the motion's relaxation time τ_r. We now apply the fluctuation-dissipation theorem to deduce τ_r, using a model in which the particle is idealized as a sphere with mass m and radius a that, of course, is far larger than the air molecules.

The equation of motion for the dust particle, when we ignore the molecules' fluctuating forces, is $m dv/dt = -Rv$. Here the resistance (friction) R due to interaction with the molecules has a form that depends on whether the molecules' mean free path λ is small or large compared to the particle. From the kinetic-theory formula $\lambda = 1/(n\sigma_{\mathrm{mol}})$, where n is the number density of molecules and σ_{mol} is their cross section to scatter off each other (roughly their cross sectional area), we can deduce that for air $\lambda \sim 0.1\,\mu$m. This is tiny compared to a dust particle's radius $a \sim 10$ to $1{,}000\,\mu$m. This means that, when interacting with the dust particle, the air molecules will behave like a fluid. As we shall learn in Chap. 15, the friction for a fluid depends on whether a quantity called the Reynolds number, $\mathrm{Re} = va/\nu$, is small or large compared to unity; here $\nu \sim 10^{-5}\,\mathrm{m^2\,s^{-1}}$ is the kinematic viscosity of air. Inserting numbers, we see that $\mathrm{Re} \sim (v/0.1\,\mathrm{m\,s^{-1}})(a/100\,\mu\mathrm{m})$. The speeds v of dust particles being buffeted by air are far smaller than $0.1\,\mathrm{m\,s^{-1}}$ as anyone who has watched them in a sunbeam knows, or as you can estimate from Eq. (6.53a). Therefore, the Reynolds number is small. From an analysis carried out in Sec. 14.3.2, we learn that in this low-Re fluid regime, the resistance (friction) on our spherical particle with radius a is [Eq. (14.34)]

$$R = 6\pi \rho \nu a, \qquad (6.76)$$

where $\rho \sim 1\,\mathrm{kg\,m^{-3}}$ is the density of air. (Notice that this resistance is proportional to the sphere's radius a or circumference; if λ were $\gg a$, then R would be proportional to the sphere's cross sectional area, i.e., to a^2.)

When we turn on the molecules' fluctuating force F', the particle's equation of motion becomes $m dv/dt + Rv = F'$. Feeding $e^{i2\pi ft}$ through this equation in place of F', we get the output $v = 1/(R + i 2\pi f m)$, whose modulus squared then is the ratio of S_v to $S_{F'}$. In this obviously classical regime, the fluctuation-dissipation theorem states that $S_{F'} = 4 R k_B T$. Therefore, we have

$$S_v = \frac{S_{F'}}{R^2 + (2\pi f m)^2} = \frac{4 R k_B T}{R^2 + (2\pi f m)^2} = \frac{4 R k_B T/m^2}{(2\pi f)^2 + (R/m)^2}. \qquad (6.77)$$

relaxation time for Brownian motion

By comparing with the S_v that we derived from Doob's theorem, Eq. (6.53b), we can read off the particle's rms velocity (in one dimension, x or y or z), $\sigma_v = \sqrt{k_B T/m}$—which agrees with Eq. (6.53a) as it must—and we can also read off the particle's relaxation time (not to be confused with the bath's relaxation time),

$$\tau_r = m/R = m/(6\pi \rho \nu a). \qquad (6.78)$$

If we had tried to derive this relaxation time by analyzing the buffeting of the particle directly, we would have had great difficulty. The fluctuation-dissipation theorem,

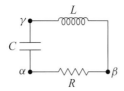

FIGURE 6.15 An L-C-R circuit. See Ex. 6.16.

Doob's theorem, and the fluid-mechanics analysis of friction on a sphere have made the task straightforward.

EXERCISES

Exercise 6.16 *Practice: Noise in an L-C-R Circuit*

Consider an L-C-R circuit as shown in Fig. 6.15. This circuit is governed by the differential equation (6.72), where F' is the fluctuating voltage produced by the resistor's microscopic degrees of freedom (so we shall rename it V'), and $F \equiv V$ vanishes, since there is no driving voltage in the circuit. Assume that the resistor has temperature $T \gg \hbar\omega_o/k$, where $\omega_o = (LC)^{-1/2}$ is the circuit's resonant angular frequency, $\omega_o = 2\pi f_o$, and also assume that the circuit has a large quality factor (weak damping) so $R \ll 1/(\omega_o C) \simeq \omega_o L$.

(a) Initially consider the resistor R decoupled from the rest of the circuit, so current cannot flow across it. What is the spectral density $V_{\alpha\beta}$ of the voltage across this resistor?

(b) Now place the resistor into the circuit as shown in Fig. 6.15. The fluctuating voltage V' will produce a fluctuating current $I = \dot{q}$ in the circuit (where q is the charge on the capacitor). What is the spectral density of I? And what, now, is the spectral density $V_{\alpha\beta}$ across the resistor?

(c) What is the spectral density of the voltage $V_{\alpha\gamma}$ between points α and γ? and of $V_{\beta\gamma}$?

(d) The voltage $V_{\alpha\beta}$ is averaged from time $t = t_0$ to $t = t_0 + \tau$ (with $\tau \gg 1/f_o$), giving some average value U_0. The average is measured once again from t_1 to $t_1 + \tau$, giving U_1. A long sequence of such measurements gives an ensemble of numbers $\{U_0, U_1, \ldots, U_n\}$. What are the mean $\bar{U}$ and rms deviation $\Delta U \equiv \langle (U - \bar{U})^2 \rangle^{\frac{1}{2}}$ of this ensemble?

Exercise 6.17 ***Example: Detectability of a Sinusoidal Force that Acts on an Oscillator with Thermal Noise*

To measure a very weak sinusoidal force, let the force act on a simple harmonic oscillator with eigenfrequency at or near the force's frequency, and measure the oscillator's response. Examples range in physical scale from nanomechanical oscillators ($\sim 1\,\mu$m in size) with eigenfrequency ~ 1 GHz that might play a role in future quantum information technology (e.g., Chan et al., 2011), to the fundamental mode of a ~ 10-kg sapphire crystal, to a ~ 40-kg LIGO mirror on which light pressure produces a restoring force, so its center of mass oscillates mechanically at frequency ~ 100 Hz (e.g., Abbott et al., 2009a). The oscillator need not be mechanical; for example, it could be an L-C-R circuit, or a mode of an optical (Fabry-Perot) cavity.

The displacement $x(t)$ of any such oscillator is governed by the driven-harmonic-oscillator equation

$$m(\ddot{x} + \frac{2}{\tau_*}\dot{x} + \omega^2 x) = F(t) + F'(t). \tag{6.79}$$

Here m, ω, and τ_* are respectively the effective mass, angular frequency, and amplitude damping time associated with the oscillator; $F(t)$ is an external driving force; and $F'(t)$ is the fluctuating force associated with the dissipation that gives rise to τ_*. Assume that $\omega \tau_* \gg 1$ (weak damping).

(a) Weak coupling to other modes is responsible for the damping. If the other modes are thermalized at temperature T, what is the spectral density $S_{F'}(f)$ of the fluctuating force F'? What is the spectral density $S_x(f)$ of x?

(b) A very weak sinusoidal force drives the fundamental mode precisely on resonance:

$$F = \sqrt{2} F_s \cos \omega t. \tag{6.80}$$

Here F_s is the rms signal. What is the $x(t)$ produced by this signal force?

(c) A sensor with negligible noise monitors this $x(t)$ and feeds it through a narrow-band filter with central frequency $f = \omega/(2\pi)$ and bandwidth $\Delta f = 1/\hat{\tau}$ (where $\hat{\tau}$ is the averaging time used by the filter). Assume that $\hat{\tau} \gg \tau_*$. What is the rms thermal noise σ_x after filtering? Show that the strength F_s of the signal force that produces a signal $x(t) = \sqrt{2} x_s \cos(\omega t + \delta)$ with rms amplitude x_s equal to σ_x and phase δ is

$$F_s = \sqrt{\frac{8 m k_B T}{\hat{\tau} \tau_*}}. \tag{6.81}$$

This is the minimum detectable force at the "one-σ level."

(d) Suppose that the force acts at a frequency ω_o that differs from the oscillator's eigenfrequency ω by an amount $|\omega - \omega_o| \lesssim 1/\tau_*$. What, then, is the minimum detectable force strength F_s? What might be the advantages and disadvantages of operating off resonance in this way, versus on resonance?

6.8.2 Generalized Fluctuation-Dissipation Theorem; Thermal Noise in a Laser Beam's Measurement of Mirror Motions; Standard Quantum Limit for Measurement Accuracy and How to Evade It T2

Not all generalized coordinates q have kinetic energy $\frac{1}{2}m\dot{q}^2$. An important example (Levin, 1998) arises when one measures the location of the front of a mirror by bouncing a laser beam perpendicularly off of it—a common and powerful tool in modern technology. If the mirror moves along the beam's optic axis by Δz, the distance of the bouncing light's travel changes by $2\Delta z$, and the light acquires a phase shift $(2\pi/\lambda)2\Delta z$ (with λ the light's wavelength) that can be read out via interferometry (Chap. 9). Because the front of the mirror can deform, Δz is actually the change in a spatial average of the mirror front's location $z(r, \phi; t)$, an average weighted by the number of photons that hit a given region. In other words, the (time varying) mirror position monitored by the light is

$$q(t) = \int z(r, \phi; t) \frac{e^{-(r/r_o)^2}}{\pi r_o^2} r d\phi \, dr. \tag{6.82}$$

Here (r, ϕ) are cylindrical coordinates centered on the laser beam's optic axis, and $e^{-(r/r_o)^2}$ is the Gaussian distribution of the beam's energy flux, so

$$\left[e^{-(r/r_o)^2}/(\pi r_o^2)\right] r d\phi dr$$

is the probability that a photon of laser light will hit the mirror at (r, ϕ) in the range $(dr, d\phi)$.

Because the mirror front's deformations $z(r, \phi; t)$ can be expanded in normal modes, this q is a linear superposition of the generalized coordinates $q_j(t)$ of the mirror's normal modes of oscillation and its center-of-mass displacement $q_0(t)$: $q(t) = q_0(t) + \sum_j Q_j(r, \phi) q_j(t)$, where $Q_j(r, \phi)$ is mode j's displacement eigenfunction evaluated at the mirror's face. Each of the generalized coordinates q_0 and q_j has a kinetic energy proportional to $\dot{q}_j^2$, but this q does not. Therefore, the elementary version of the fluctuation-dissipation theorem, treated in the previous section, is not valid for this q.

Fortunately, there is a remarkably powerful generalized fluctuation-dissipation theorem due to Callen and Welton (1951) that works for this q and all other generalized coordinates that are coupled to a heat bath. To formulate this theorem, we must first introduce the *complex impedance* $Z(\omega)$ for a generalized coordinate.

Let a sinusoidal external force $F = F_o e^{-i\omega t}$ act on the generalized coordinate q [so q's canonically conjugate momentum p is being driven as $(dp/dt)_{\text{drive}} = F_o e^{-i\omega t}$]. Then the velocity of the resulting sinuosoidal motion will be

$$\boxed{\dot{q} \equiv \frac{dq}{dt} = -i\omega q = \frac{1}{Z(\omega)} F_o e^{-i\omega t},} \tag{6.83a}$$

T2

complex impedance for a generalized coordinate

where the real part of each expression is to be taken. This equation can be regarded as the definition of q's complex impedance $Z(\omega)$ (ratio of force to velocity); it is determined by the system's details. If the system were completely conservative, then the impedance would be perfectly imaginary, $Z = iI$, where I is real. For example, for a freely moving dust particle in vacuum, driven by a sinusoidal force, the momentum is $p = m\dot{q}$ (where m is the particle's mass), the equation of motion is $F_o e^{-i\omega t} = dp/dt = m(d/dt)\dot{q} = m(-i\omega)\dot{q}$, and so the impedance is $Z = -im\omega$, which is purely imaginary.

The bath prevents the system from being conservative—energy can be fed back and forth between the generalized coordinate q and the bath's many degrees of freedom. This energy coupling influences the generalized coordinate q in two important ways. First, it changes the impedance $Z(\omega)$ from purely imaginary to complex,

real and imaginary parts of complex impedance; resistance $R(\omega)$

$$Z(\omega) = iI(\omega) + R(\omega), \tag{6.83b}$$

where R is the *resistance* experienced by q. Correspondingly, when the sinusoidal force $F = F_o e^{-i\omega t}$ is applied, the resulting motions of q feed energy into the bath, dissipating power at a rate $W_{\text{diss}} = \langle \Re(F)\Re(\dot{q})\rangle = \langle \Re(F_o e^{-i\omega t})\Re(F_o e^{-i\omega t}/Z)\rangle = \langle F_o \cos\omega t\, \Re(1/Z)\, F_o \cos\omega t\rangle$; that is,

power dissipation rate

$$W_{\text{diss}} = \frac{1}{2}\frac{R}{|Z|^2}F_o^2. \tag{6.84}$$

Second, the thermal motions of the bath exert a randomly fluctuating force $F'(t)$ on q, driving its generalized momentum as $(dp/dt)_{\text{drive}} = F'$.

As an example, consider the L-C-R circuit of Eq. (6.72). We can identify the generalized momentum by shutting off the bath (the resistor and its fluctuating voltage); writing down the lagrangian for the resulting L-C circuit, $\mathcal{L} = \frac{1}{2}L\dot{q}^2 - \frac{1}{2}q^2/C$; and computing $p = \partial\mathcal{L}/\partial\dot{q} = L\dot{q}$. [Equally well, we can identify p from one of Hamilton's equations for the hamiltonian $H = p^2/(2L) + q^2/(2C)$.] We evaluate the impedance $Z(\omega)$ from the equation of motion for this lagrangian with the bath's resistance restored (but not its fluctuating voltage) and with a sinusoidal voltage $V = V_o e^{-i\omega t}$ imposed:

$$L\frac{d\dot{q}}{dt} - \frac{q}{C} + R\dot{q} = \left(-i\omega L + \frac{1}{-i\omega C} + R\right)\dot{q} = V_o e^{-i\omega t}. \tag{6.85a}$$

Evidently, $V = V_o e^{-i\omega t}$ is the external force F that drives the generalized momentum $p = L\dot{q}$, and the complex impedance (ratio of force to velocity) is

$$Z(\omega) = \frac{V}{\dot{q}} = -i\omega L + \frac{1}{-i\omega C} + R. \tag{6.85b}$$

This is identical to the impedance as defined in the standard theory of electrical circuits (which is what motivates our $Z = F/\dot{q}$ definition of impedance), and as expected, the real part of this impedance is the circuit's resistance R.

Returning to our general q, the fluctuating force F' (equal to fluctuating voltage V' in the case of the circuit) and the resistance R to an external force both arise from interaction with the same heat bath. Therefore, it should not be surprising that they are connected by the generalized fluctuation-dissipation theorem:

$$S_{F'}(f) = 4R(f)\left(\frac{1}{2}hf + \frac{hf}{e^{hf/(k_BT)} - 1}\right) \quad \text{in general,} \tag{6.86a}$$

generalized fluctuation-dissipation theorem

$$S_{F'}(f) = 4R(f)k_BT \quad \text{in the classical domain, } k_BT \gg hf, \tag{6.86b}$$

which is valid at all frequencies

$$f \ll 1/\tau_r, \tag{6.87}$$

where τ_r is the (very short) relaxation time for the bath's fluctuating forces F'. Here T is the temperature of the bath, h is Planck's constant, and we have written the resistance as $R(f)$ to emphasize that it can depend on frequency $f = \omega/(2\pi)$.

A derivation of this generalized fluctuation-dissipation theorem is sketched in Ex. 6.18.

One is usually less interested in the spectral density of the bath's force F' than in that of the generalized coordinate q. The definition (6.83a) of impedance implies $-i\omega\tilde{q} = \tilde{F}'/Z(\omega)$ for Fourier transforms, whence $S_q = S_{F'}/[(2\pi f)^2|Z|^2]$. When combined with Eqs. (6.86) and (6.84), this implies

spectral density for generalized coordinate

$$S_q(f) = \frac{8W_{\text{diss}}}{(2\pi f)^2 F_o^2}\left(\frac{1}{2}hf + \frac{hf}{e^{hf/(k_BT)} - 1}\right) \quad \text{in general,} \tag{6.88a}$$

$$S_q(f) = \frac{8W_{\text{diss}}k_BT}{(2\pi f)^2 F_o^2} \quad \text{in the classical domain, } k_BT \gg hf. \tag{6.88b}$$

Therefore, to evaluate $S_q(f)$, one does not need to know the complex impedance $Z(\omega)$. Rather, one only needs the power dissipation W_{diss} that results when a sinusoidal force F_o is applied to the generalized momentum p that is conjugate to the coordinate q of interest.

The light beam bouncing off a mirror (beginning of this section) is a good example. To couple the sinusoidal force $F(t) = F_o e^{-i\omega t}$ to the mirror's generalized coordinate q, we add an interaction term $H_I = -F(t)q$ to the mirror's hamiltonian H_{mirror}. Hamilton's equation for the evolution of the momentum conjugate to q then becomes $dp/dt = -\partial[H_{\text{mirror}} - F(t)q]/\partial q = \partial H_{\text{mirror}}/\partial t + F(t)$. Thus, $F(t)$ drives p as desired. The form of the interaction term is, by Eq. (6.82) for q,

$$H_I = -F(t)q = -\int z(r,\phi)\frac{F(t)e^{-(r/r_o)^2}}{\pi r_o^2}r d\phi\, dr. \tag{6.89}$$

6.8 Fluctuation-Dissipation Theorem

This is the mathematical description of a time varying *pressure*

$$P = F_o e^{-i\omega t} e^{-(r/r_o)^2}/(\pi r_o^2)$$

applied to the mirror face, which has coordinate location $z(r, \phi)$. Therefore, to compute the spectral density of the mirror's light-beam-averaged displacement q, at frequency $f = \omega/(2\pi)$, we can

Levin's method for computing spectral density of mirror's light-averaged displacement

1. apply to the mirror's front face a pressure with spatial shape the same as that of the light beam's energy flux (a Gaussian in our example) and with total force $F_o e^{-i\omega t}$;

2. evaluate the power dissipation W_{diss} produced by this sinusoidally oscillating pressure; and then

3. insert the ratio W_{diss}/F_o^2 into Eq. (6.88a) or Eq. (6.88b). This is called Levin's (1998) method.

In practice, in this thought experiment the power can be dissipated at many locations: in the mirror coating (which makes the mirror reflective), in the substrate on which the coating is placed (usually glass, i.e., fused silica), in the attachment of the mirror to whatever supports it (usually a wire or glass fiber), and in the supporting structure (the wire or fiber and the solid object to which it is attached). The dissipations W_{diss} at each of these locations add together, and therefore the fluctuating noises from the various dissipation locations are additive. Correspondingly, one speaks of "coating thermal noise," "substrate thermal noise," and so forth; physicists making delicate optical measurements deduce each through a careful computation of its corresponding dissipation W_{diss}.

In the LIGO interferometer, whose noise power spectrum is shown in Fig. 6.7, these thermal noises dominate in the intermediate frequency band 40 Hz $\lesssim f \lesssim$ 150 Hz.

EXERCISES

Exercise 6.18 *Derivation: Generalized Fluctuation-Dissipation Theorem*
By a method analogous to that used for the elementary fluctuation-dissipation theorem (Sec. 6.8.1), derive the generalized fluctuation-dissipation theorem [Eqs. (6.86)].

Hints: Consider a thought experiment in which the system's generalized coordinate q is very weakly coupled to an external oscillator that has a mass M and an angular eigenfrequency ω_o, near which we wish to derive the fluctuation-dissipation formulas (6.86). Denote by Q and P the external oscillator's generalized coordinate and momentum and by κ the arbitrarily weak coupling constant between the oscillator and q, so the hamiltonian of system plus oscillator plus fluctuating force F' acting on q is

$$H = H_{\text{system}}(q, p, \ldots) + \frac{P^2}{2M} + \frac{1}{2}M\omega_o^2 Q^2 + \kappa Q q - F'(t)\, q. \tag{6.90}$$

Here the "..." refers to the other degrees of freedom of the system, some of which might be strongly coupled to q and p (as is the case, e.g., for the laser-measured mirror discussed in the text).

(a) By combining Hamilton's equations for q and its conjugate momentum p [which give Eq. (6.83a) with the appropriate driving force] with those for the external oscillator (Q, P), derive an equation that shows quantitatively how the force F', acting through q, influences the oscillator's coordinate Q:

$$\left[M(-\omega^2 + {\omega'_o}^2) - \frac{i\kappa^2 R}{\omega |Z|^2}\right]\tilde{Q} = \frac{\kappa}{i\omega Z}\tilde{F'}. \tag{6.91}$$

Here the tildes denote Fourier transforms; $\omega = 2\pi f$ is the angular frequency at which the Fourier transforms are evaluated; and ${\omega'_o}^2 = \omega_o^2 - \kappa^2 I/(\omega|Z|^2)$, with $Z = R + iI$, is the impedance of q at angular frequency ω.

(b) Show that

$$S_Q = \frac{(\kappa/\omega|Z|)^2 S_{F'}}{M^2(-\omega^2 + {\omega'_o}^2)^2 + \kappa^4 R^2/(\omega|Z|^2)^2}. \tag{6.92}$$

(c) Make the resonance in this equation arbitrarily sharp by choosing the coupling constant κ arbitrarily small. Then show that the mean energy in the oscillator is

$$\bar{E} = M\omega_o^2 \int_0^\infty S_Q(f)df = \frac{S_{F'}(f = \omega_o/(2\pi))}{4R}. \tag{6.93}$$

(d) By equating this to expression (6.75c) for the mean energy of any oscillator coupled to a heat bath, deduce the desired generalized fluctuation-dissipation equations (6.86).

6.9 Fokker-Planck Equation

In statistical physics, we often want to know the collective influence of many degrees of freedom (a bath) on a single (possibly vectorial) degree of freedom q. The bath might or might not be thermalized. The forces it exerts on q might have short range (as in molecular collisions buffeting an air molecule or dust particle) or long range (as in Coulomb forces from many charged particles in a plasma pushing stochastically on an electron that interests us, or gravitational forces from many stars pulling on a single star of interest). There might also be long-range, macroscopic forces that produce anisotropies and/or inhomogeneities (e.g., applied electric or magnetic fields). We might want to compute how the bath's many degrees of freedom influence, for example, the diffusion of a particle as embodied in its degree of freedom q. Or we might want to compute the statistical properties of q for a representative electron in a plasma and from them deduce the plasma's transport coefficients (diffusivity, heat conductivity, and thermal conductivity). Or we might want to know how the gravitational pulls of many stars in the vicinity of a black hole drive the collective evolution of the stars' distribution function.

The Fokker-Planck equation is a powerful tool in such situations. To apply it, we must identify a (possibly vectorial) degree of freedom q to analyze that is Markov. For the types of problems described here, this is typically the velocity (or a component of velocity) of a representative particle or star. The Fokker-Planck equation is then a differential equation for the evolution of the conditional probability distribution P_2 [Eq. (6.6b)], or other distribution function, for that Markov degree of freedom. In Sec. 6.9.1, we present the simplest, 1-dimensional example. Then in Sec. 6.9.3, we generalize to several dimensions.

crucial assumption of a Markov Process

6.9.1 Fokker-Planck for a 1-Dimensional Markov Process

For a 1-dimensional Markov process $y(t)$ (e.g., the x component of the velocity of a particle) being driven by a bath (not necessarily thermalized!) with many degrees of freedom, the *Fokker-Planck equation*[11] states

1-dimensional Fokker-Planck equation

$$\boxed{\frac{\partial}{\partial t} P_2 = -\frac{\partial}{\partial y}[A(y)P_2] + \frac{1}{2}\frac{\partial^2}{\partial y^2}[B(y)P_2].} \qquad (6.94)$$

Here $P_2 = P_2(y, t|y_o)$ is to be regarded as a function of the variables y and t with y_o fixed; that is, Eq. (6.94) is to be solved subject to the initial condition

$$\boxed{P_2(y, 0|y_o) = \delta(y - y_o).} \qquad (6.95)$$

As we shall see later, this Fokker-Planck equation is a generalized diffusion equation for the probability P_2: as time passes, the probability diffuses away from its initial location, $y = y_o$, spreading gradually out over a wide range of values of y.

In the Fokker-Planck equation (6.94) the function $A(y)$ produces a motion of the mean away from its initial location, while the function $B(y)$ produces a diffusion of the probability. If one can deduce the evolution of P_2 for very short times by some other method [e.g., in the case of a dust particle being buffeted by air molecules, by solving statistically the Langevin equation $mdv/dt + Rv = F'(t)$], then from that short-time evolution one can compute the functions $A(y)$ and $B(y)$:

$$A(y) = \lim_{\Delta t \to 0} \frac{1}{\Delta t} \int_{-\infty}^{+\infty} (y' - y) P_2(y', \Delta t|y) dy', \qquad (6.96a)$$

$$B(y) = \lim_{\Delta t \to 0} \frac{1}{\Delta t} \int_{-\infty}^{+\infty} (y' - y)^2 P_2(y', \Delta t|y) dy'. \qquad (6.96b)$$

---

11. A very important generalization of this equation is to replace the probability P_2 by a particle distribution function and the Markov process $y(t)$ by the 3-dimensional velocity or momentum of the particles. The foundations for this generalization are laid in the Track-Two Sec. 6.9.3, and an application is in Ex. 20.8.

(These equations can be deduced by reexpressing the limit as an integral of the time derivative $\partial P_2/\partial t$ and then inserting the Fokker-Planck equation and integrating by parts; Ex. 6.19.) Note that the integral (6.96a) for $A(y)$ is the mean change $\overline{\Delta y}$ in the value of y that occurs in time Δt, if at the beginning of Δt (at $t = 0$) the value of the process is precisely y; moreover (since the integral of yP_2 is just equal to y, which is a constant), $A(y)$ is also the rate of change of the mean, $d\bar{y}/dt$. Correspondingly we can write Eq. (6.96a) in the more suggestive form

$$\boxed{A(y) = \lim_{\Delta t \to 0}\left(\frac{\overline{\Delta y}}{\Delta t}\right) = \left(\frac{d\bar{y}}{dt}\right)_{t=0}.} \qquad (6.97a)$$

Similarly, the integral (6.96b) for $B(y)$ is the mean-squared change in y, $\overline{(\Delta y)^2}$, if at the beginning of Δt the value of the process is precisely y; and (as one can fairly easily show; Ex. 6.19) it is also the rate of change of the variance $\sigma_y^2 = \int (y' - \bar{y})^2 P_2 dy'$. Correspondingly, Eq. (6.96b) can be written as

$$\boxed{B(y) = \lim_{\Delta t \to 0}\left(\frac{\overline{(\Delta y)^2}}{\Delta t}\right) = \left(\frac{d\sigma_y^2}{dt}\right)_{t=0}.} \qquad (6.97b)$$

It may seem surprising that $\overline{\Delta y}$ and $\overline{(\Delta y)^2}$ can both increase linearly in time for small times [cf. the Δt in the denominators of both Eq. (6.97a) and Eq. (6.97b)], thereby both giving rise to finite functions $A(y)$ and $B(y)$. In fact, this is so: the linear evolution of $\overline{\Delta y}$ at small t corresponds to the motion of the mean (i.e., of the peak of the probability distribution), while the linear evolution of $\overline{(\Delta y)^2}$ corresponds to the diffusive broadening of the probability distribution.

DERIVATION OF THE FOKKER-PLANCK EQUATION (6.94)

Because y is Markov, it satisfies the Smoluchowski equation (6.11), which we rewrite here with a slight change of notation:

$$P_2(y, t+\tau|y_o) = \int_{-\infty}^{+\infty} P_2(y - \xi, t|y_o) P_2(y - \xi + \xi, \tau|y - \xi) d\xi. \qquad (6.98a)$$

Take τ to be small so only small ξ will contribute to the integral, and expand in a Taylor series in τ on the left-hand side of (6.98a) and in the ξ of $y - \xi$ on the right-hand side:

$$P_2(y, t|y_o) + \sum_{n=1}^{\infty} \frac{1}{n!}\left[\frac{\partial^n}{\partial t^n} P_2(y, t|y_o)\right]\tau^n$$

$$= \int_{-\infty}^{+\infty} P_2(y, t|y_o) P_2(y + \xi, \tau|y) d\xi$$

$$+ \sum_{n=1}^{\infty} \frac{1}{n!} \int_{-\infty}^{+\infty} (-\xi)^n \frac{\partial^n}{\partial y^n} [P_2(y, t|y_o) P_2(y + \xi, \tau|y)] d\xi. \qquad (6.98b)$$

In the first integral on the right-hand side the first term is independent of ξ and can be pulled out from under the integral, and the second term then integrates to one; thereby the first integral on the right reduces to $P_2(y, t|y_o)$, which cancels the first term on the left. The result is then

$$\sum_{n=1}^{\infty} \frac{1}{n!} \left[\frac{\partial^n}{\partial t^n} P_2(y, t|y_o) \right] \tau^n$$

$$= \sum_{n=1}^{\infty} \frac{(-1)^n}{n!} \frac{\partial^n}{\partial y^n} \left[P_2(y, t|y_o) \int_{-\infty}^{+\infty} \xi^n P_2(y+\xi, \tau|y) \, d\xi \right]. \quad (6.98c)$$

Divide by τ, take the limit $\tau \to 0$, and set $\xi \equiv y' - y$ to obtain

$$\frac{\partial}{\partial t} P_2(y, t|y_o) = \sum_{n=1}^{\infty} \frac{(-1)^n}{n!} \frac{\partial^n}{\partial y^n} [M_n(y) P_2(y, t|y_o)], \quad (6.99a)$$

where

$$M_n(y) \equiv \lim_{\Delta t \to 0} \frac{1}{\Delta t} \int_{-\infty}^{+\infty} (y' - y)^n P_2(y', \Delta t|y) \, dy' \quad (6.99b)$$

is the nth moment of the probability distribution P_2 after time Δt. This is a form of the Fokker-Planck equation that has slightly wider validity than Eq. (6.94). Almost always, however, the only nonvanishing functions $M_n(y)$ are $M_1 \equiv A$, which describes the linear motion of the mean, and $M_2 \equiv B$, which describes the linear growth of the variance. Other moments of P_2 grow as higher powers of Δt than the first power, and correspondingly, their M_ns vanish. Thus, almost always[12] (and always, so far as we are concerned), Eq. (6.99a) reduces to the simpler version (6.94) of the Fokker-Planck equation.

TIME-INDEPENDENT FOKKER-PLANCK EQUATION

If, as we assume in this chapter, y is ergodic, then $p_1(y)$ can be deduced as the limit of $P_2(y, t|y_o)$ for arbitrarily large times t. Then (and in general) p_1 can be deduced from the time-independent Fokker-Planck equation:

time-independent Fokker-Planck equation

$$\boxed{-\frac{\partial}{\partial y}[A(y) p_1(y)] + \frac{1}{2} \frac{\partial^2}{\partial y^2}[B(y) p_1(y)] = 0.} \quad (6.100)$$

GAUSSIAN-MARKOV PROCESS

For a Gaussian-Markov process, the mathematical form of $P_2(y_2, t|y_1)$ is known from Doob's theorem: Eqs. (6.18). In the notation of those equations, the Fokker-Planck functions A and B are

$$A(y_1) = (d\bar{y}_t/dt)_{t=0} = -(y_1 - \bar{y})/\tau_r, \quad \text{and} \quad B(y_1) = (d\sigma_{y_t}^2/dt)_{t=0} = 2\sigma_y^2/\tau_r.$$

---

12. In practice, when there are important effects not captured by A and B (e.g., in the mathematical theory of finance; Hull, 2014), they are usually handled by adding other terms to Eq. (6.94), including sometimes integrals.

Translating back to the notation of this section, we have

$$A(y) = -(y - \bar{y})/\tau_r, \qquad B(y) = 2\sigma_y^2/\tau_r. \qquad (6.101)$$

Thus, if we can compute $A(y)$ and $B(y)$ explicitly for a Gaussian-Markov process, then from them we can read off the process's relaxation time τ_r, long-time mean $\bar{y}$, and long-time variance σ_y^2. As examples, in Ex. 6.22 we revisit Brownian motion of a dust particle in air and in the next section, we analyze laser cooling of atoms. A rather different example is the evolution of a photon distribution function under Compton scattering (Sec. 28.6.3).

EXERCISES

Exercise 6.19 *Derivation: Equations for A and B*
Derive Eqs. (6.96) for A and B from the Fokker-Planck equation (6.94), and then from Eqs. (6.96) derive Eqs. (6.97).

Exercise 6.20 *Problem: Fokker-Planck Equation as Conservation Law for Probability*
Show that the Fokker-Planck equation can be interpreted as a conservation law for probability. What is the probability flux in this conservation law? What is the interpretation of each of its two terms?

Exercise 6.21 *Example: Fokker-Planck Coefficients When There Is Direct Detailed Balance*
Consider an electron that can transition between two levels by emitting or absorbing a photon; and recall (as discussed in Ex. 3.6) that we have argued that the stimulated transitions should be microscopically reversible. This is an example of a general principle introduced by Boltzmann called *detailed balance*. In the context of classical physics, it is usually considered in the context of the time reversibility of the underlying physical equations. For example, if two molecules collide elastically and exchange energy, the time-reversed process happens with the same probability per unit time when the colliding particles are in the time-reversed initial states. However, this does not necessarily imply that the probability of a single molecule changing its velocity from $\mathbf{v}$ to $\mathbf{v}'$ is the same as that of the reverse change. (A high-energy molecule is more likely to lose energy when one averages over all collisions.)

An important simplification happens when the probability of a change in y is equal to the probability of the opposite change. An example might be a light particle colliding with a heavy particle for which the recoil can be ignored. Under this more restrictive condition, we can write that $P_2(y', \Delta t | y) = P_2(y, \Delta t | y')$. Show that the Fokker-Planck equation then simplifies (under the usual assumptions) to a standard diffusion equation:

$$\frac{\partial}{\partial t} P_2 = \frac{1}{2} \frac{\partial}{\partial y} B(y) \frac{\partial}{\partial y} P_2. \qquad (6.102)$$

Of course, there can be other contributions to the total Fokker-Planck coefficients that do not satisfy this condition, but this simplification can be very instructive. An example described in Sec. 23.3.3 is the quasilinear interaction between waves and particles in a plasma.

Exercise 6.22 *Example: Solution of Fokker-Planck Equation for Brownian Motion of a Dust Particle*

(a) Write down the explicit form of the Langevin equation for the x component of velocity $v(t)$ of a dust particle interacting with thermalized air molecules.

(b) Suppose that the dust particle has velocity v at time t. By integrating the Langevin equation, show that its velocity at time $t + \Delta t$ is $v + \Delta v$, where

$$m\Delta v + Rv\Delta t + O[(\Delta t)^2] = \int_t^{t+\Delta t} F'(t')dt', \quad (6.103a)$$

with R the frictional resistance and m the particle's mass. Take an ensemble average of this and use $\overline{F'} = 0$ to conclude that the function $A(v)$ appearing in the Fokker-Planck equation (6.94) has the form

$$A(v) \equiv \lim_{\Delta t \to 0} \frac{\overline{\Delta v}}{\Delta t} = -\frac{Rv}{m}. \quad (6.103b)$$

Compare this expression with the first of Eqs. (6.101) to conclude that the mean and relaxation time are $\bar{v} = 0$ and $\tau_r = m/R$, respectively, in agreement with the second of Eqs. (6.53a) in the limit $\tau \to \infty$ and with Eq. (6.78).

(c) From Eq. (6.103a) show that

$$(\Delta v)^2 = \left[-\frac{v}{\tau_r}\Delta t + O[(\Delta t)^2] + \frac{1}{m}\int_t^{t+\Delta t} F'(t')dt' \right]^2. \quad (6.103c)$$

Take an ensemble average of this expression, and use $\overline{F'(t_1)F'(t_2)} = C_{F'}(t_2 - t_1)$—together with the Wiener-Khintchine theorem—to evaluate the terms involving F' in terms of $S_{F'}$, which in turn is known from the fluctuation-dissipation theorem. Thereby obtain

$$B(v) = \lim_{\Delta t \to 0} \frac{\overline{(\Delta v)^2}}{\Delta t} = \frac{2Rk_B T}{m^2}. \quad (6.103d)$$

Combine with Eq. (6.101) and $\tau_r = m/R$ [from part (b)], to conclude that $\sigma_v^2 = k_B T/m$, in accord with the last of Eqs. (6.53a).

6.9.2 Optical Molasses: Doppler Cooling of Atoms T2

The 1997 Nobel Prize was awarded to Steven Chu, Claude Cohen-Tannoudji, and William D. Phillips for the "development of methods to cool and trap atoms with laser light" (Chu et al., 1998). In this section, we use the Fokker-Planck equation to analyze

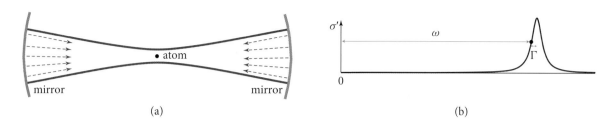

FIGURE 6.16 Doppler cooling of an atom in a Fabry-Perot cavity. (a) The cavity formed by two mirrors with laser light bouncing back and forth between them, and the sodium atom at the center. (b) The cross section $\sigma' = d\sigma/d\omega$ for the atom in its ground state to absorb a photon of laser light. The laser angular frequency ω is tuned to the off-resonance inflection point (steepest slope) of σ', indicated by the dot.

one of the most important methods they developed: *Doppler cooling*, also called *laser cooling* or *optical molasses*.

A neutral sodium atom is placed near the center (waist) of a Fabry-Perot optical cavity (Sec. 9.4.2), so it is bathed by laser light traveling in both the $+z$ and $-z$ directions (Fig. 6.16a). The atom absorbs and reemits photons and their momenta, resulting in a stochastic evolution of its z component of velocity, v. Using the Fokker-Planck equation to analyze this evolution, we shall discover that, if the light frequency and power are tuned appropriately, there is a strong resisting force ("optical molasses") on the atom as well as a randomizing force. The net effect, after the atom relaxes into equilibrium with the photon field, is a very low effective temperature ($\sim 100~\mu$K) for the atom's motion in the z direction.

The atom has a large cross section $\sigma' \equiv d\sigma/d\omega$ to absorb a photon with angular frequency $\omega \simeq 3.20 \times 10^{15}$ s^{-1} (yellow light), thereby getting excited into a state with energy $\hbar\omega \simeq 2.11$ eV. The absorption cross section $\sigma'(\omega)$ has a narrow resonance (Lorentzian line shape $\propto [1 + (\omega - \omega_o)^2/\Gamma^2]^{-1}$; Fig. 6.16b) with half-width $\Gamma \simeq$ 10 MHz; and, correspondingly, the excited atom has a half-life $1/\Gamma$ to reemit a photon and return to its ground state. The laser power is adjusted to make the excitation rate $\mathcal{R}$ equal to Γ,

$$\mathcal{R} = \Gamma \simeq 10^7~\text{s}^{-1}, \tag{6.104a}$$

thereby maximizing the rate of excitations. (At a higher power, the excitation rate will saturate at $1/\Gamma$, because the atom spends most of its time excited and waiting to reemit.)

The laser frequency is tuned to the resonance's inflection point (point of greatest slope $d\sigma'/d\omega$ on one side of the line), so that, when an atom is moving rightward with velocity v, the Doppler shift $\delta\omega/\omega = v/c$ produces a maximal fractional increase in the cross section and rate for absorbing leftward-moving photons and decrease in those for rightward-moving photons:

$$\frac{\delta\mathcal{R}}{\mathcal{R}} = \frac{\delta\sigma'}{\sigma'} = \frac{1}{\sigma'}\frac{d\sigma'}{d\omega}\left(\omega\frac{v}{c}\right) \sim \frac{\omega}{\Gamma}\frac{v}{c}. \tag{6.104b}$$

6.9 Fokker-Planck Equation

(Here and henceforth "$\sim$" means accurate to within a factor of order unity.) This results in a net resisting force $F \sim \delta R\, \hbar k$ on the atom, due to the imbalance in absorption rates for leftward and rightward photons; here $k = \omega/c = 10.7~\mu\text{m}^{-1}$ is the photons' wave number, and $\hbar k$ is the momentum absorbed from each photon. This slow-down force produces a rate of change of the atom's mean velocity

$$A = \frac{d\bar{v}}{dt} \sim -\frac{\delta R\, \hbar k}{m} \sim -\frac{\hbar k^2}{m} v. \tag{6.104c}$$

Here we have used Eqs. (6.104b) and (6.104a), and $\omega/c = k$; and we have set the slow-down rate equal to the coefficient A in the Fokker-Planck equation for v [Eq. (6.97a)].

There are two sources of randomness in the atom's velocity, both of the same magnitude: statistical randomness ($\sqrt{N}$) in the number of photons absorbed from the two directions, and randomness in the direction of reemission of photons and thence in the recoil direction. During a short time interval Δt, the mean number of absorptions and reemissions is $\sim \mathcal{R}\Delta t$, so the rms fluctuation in the momentum transfer to the atom (along the z direction) is $\sim \hbar k \sqrt{\mathcal{R}\Delta t}$, whence the change of the variance of the velocity is $\overline{(\Delta v)^2} \sim (\hbar k)^2 \mathcal{R}\Delta t/m^2$. (Here $m \simeq 3.82 \times 10^{-26}$ kg is the sodium atom's mass.) Correspondingly, the B coefficient (6.103d) in the Fokker-Planck equation for v is

$$B = \frac{\overline{(\Delta v)^2}}{\Delta t} \sim \left(\frac{\hbar k}{m}\right)^2 \mathcal{R} = \left(\frac{\hbar k}{m}\right)^2 \Gamma. \tag{6.104d}$$

From the A and B coefficients [Eqs. (6.104c) and (6.104d)] we infer, with the aid of Eqs. (6.101), the relaxation time, long-term mean, and long-term variance of the atom's velocity along the z direction, and also an effective temperature associated with the variance:[13]

$$\tau_r \sim \frac{m}{(\hbar k^2)} = 3~\mu\text{s}, \quad \bar{v} = 0, \quad \sigma_v^2 \sim \frac{\hbar \Gamma}{m} = (0.17~\text{ms}^{-1})^2,$$

$$T_{\text{eff}} = \frac{m \sigma_v^2}{k_B} \sim \frac{\hbar \Gamma}{k_B} \sim 8~\mu\text{K}. \tag{6.105}$$

It is remarkable how effective this optical molasses can be!

If one wants to cool all components of velocity, one can either impose counter-propagating laser beams along all three Cartesian axes, or put the atom into a potential well (inside the Fabry Perot cavity) that deflects its direction of motion on a timescale much less than τ_r.

This optical molasses technique is widely used today in atomic physics, for example, when cooling ensembles of atoms to produce Bose-Einstein condensates

---

13. The atom's long-term, ergodically wandering velocity distribution is Gaussian rather than Maxwellian, so it is not truly thermalized. However, it has the same velocity variance as a thermal distribution with temperature $\sim \hbar \Gamma / k_B$, so we call this its "effective temperature."

(Sec. 4.9), and for cooling atoms to be used as the ticking mechanisms of atomic clocks (Fig. 6.11, footnote 9 in Sec. 6.6.1, Ex. 6.13, and associated discussions).

6.9.3 Fokker-Planck for a Multidimensional Markov Process; Thermal Noise in an Oscillator [T2]

Few 1-dimensional random processes are Markov, so only a few can be treated using the 1-dimensional Fokker-Planck equation. However, it is frequently the case that, if one augments additional variables into the random process, it becomes Markov. An important example is a harmonic oscillator driven by a Gaussian random force (Ex. 6.23). Neither the oscillator's position $x(t)$ nor its velocity $v(t)$ is Markov, but the pair $\{x, v\}$ is a 2-dimensional Markov process.

For such a process, and more generally for any n-dimensional Gaussian-Markov process $\{y_1(t), y_2(t), \ldots, y_n(t)\} \equiv \{\mathbf{y}(t)\}$, the conditional probability distribution $P_2(\mathbf{y}, t | \mathbf{y}_o)$ satisfies the following Fokker-Planck equation [the obvious generalization of Eq. (6.94)]:

$$\frac{\partial}{\partial t} P_2 = -\frac{\partial}{\partial y_j}[A_j(y) P_2] + \frac{1}{2} \frac{\partial^2}{\partial y_j \partial y_k}[B_{jk}(y) P_2]. \qquad (6.106a)$$

multidimensional Fokker-Planck equation

Here the functions A_j and B_{jk}, by analogy with Eqs. (6.96) and (6.97), are

$$A_j(\mathbf{y}) = \lim_{\Delta t \to 0} \frac{1}{\Delta t} \int (y'_j - y_j) P_2(\mathbf{y}', \Delta t | \mathbf{y}) d^n y' = \lim_{\Delta t \to 0} \left(\frac{\overline{\Delta y_j}}{\Delta t} \right), \qquad (6.106b)$$

$$B_{jk}(\mathbf{y}) = \lim_{\Delta t \to 0} \frac{1}{\Delta t} \int (y'_j - y_j)(y'_k - y_k) P_2(\mathbf{y}', \Delta t | \mathbf{y}) d^n y' = \lim_{\Delta t \to 0} \left(\frac{\overline{\Delta y_j \Delta y_k}}{\Delta t} \right).$$
$$(6.106c)$$

In Ex. 6.23 we use this Fokker-Planck equation to explore how a harmonic oscillator settles into equilibrium with a dissipative heat bath. In Ex. 20.8 we apply it to Coulomb collisions in an ionized plasma.

The multidimensional Fokker-Planck equation can be used to solve the Boltzmann transport equation (3.66) for the kinetic-theory distribution function $\mathcal{N}(\mathbf{p}, t)$ or (in the conventions of plasma physics) for the velocity distribution $f(\mathbf{v}, t)$ (Chap. 20). The reasons are (i) $\mathcal{N}(\mathbf{p}, t)$ and $f(\mathbf{v}, t)$ are the same kind of probability distribution as P_2—probabilities for a Markov momentum or velocity—with the exception that $\mathcal{N}(\mathbf{p}, t)$ and $f(\mathbf{v}, t)$ usually have different initial conditions at time $t = 0$ than P_2's delta function [in fact, P_2 can be regarded as a Green's function for $\mathcal{N}(\mathbf{p}, t)$ and $f(\mathbf{v}, t)$] and (ii) the initial conditions played no role in our derivation of the Fokker-Planck equation. In Sec. 20.4.3, we discuss the use of the Fokker-Planck equation to deduce how long-range Coulomb interactions drive the equilibration of the distribution functions $f(\mathbf{v}, t)$ for the velocities of electrons and ions in a plasma. In Sec. 23.3.3, we use the Fokker-Planck equation to study the interaction of electrons and ions with plasma waves (plasmons).

EXERCISES

Exercise 6.23 **Example: Solution of Fokker-Planck Equation for Thermal Noise in an Oscillator** T2

Consider a classical simple harmonic oscillator (e.g., the nanomechanical oscillator, LIGO mass on an optical spring, L-C-R circuit, or optical resonator briefly discussed in Ex. 6.17). Let the oscillator be coupled weakly to a dissipative heat bath with temperature T. The Langevin equation for the oscillator's generalized coordinate x is Eq. (6.79). The oscillator's coordinate $x(t)$ and momentum $p(t) \equiv m\dot{x}$ together form a 2-dimensional Gaussian-Markov process and thus obey the 2-dimensional Fokker-Planck equation (6.106a). As an aid to solving this Fokker-Planck equation, change variables from $\{x, p\}$ to the real and imaginary parts X_1 and X_2 of the oscillator's complex amplitude:

$$x = \Re[(X_1 + iX_2)e^{-i\omega t}] = X_1(t) \cos \omega t + X_2(t) \sin \omega t. \qquad (6.107)$$

Then $\{X_1, X_2\}$ is a Gaussian-Markov process that evolves on a timescale $\sim \tau_r$.

(a) Show that X_1 and X_2 obey the Langevin equation

$$-2\omega(\dot{X}_1 + X_1/\tau_r) \sin \omega t + 2\omega(\dot{X}_2 + X_2/\tau_r) \cos \omega t = F'/m. \qquad (6.108a)$$

(b) To compute the functions $A_j(\mathbf{X})$ and $B_{jk}(\mathbf{X})$ that appear in the Fokker-Planck equation (6.106a), choose the timescale Δt to be short compared to the oscillator's damping time τ_r but long compared to its period $2\pi/\omega$. By multiplying the Langevin equation successively by $\sin \omega t$ and $\cos \omega t$ and integrating from $t = 0$ to $t = \Delta t$, derive equations for the changes ΔX_1 and ΔX_2 produced during Δt by the fluctuating force $F'(t)$ and its associated dissipation. [Neglect fractional corrections of order $1/(\omega \Delta t)$ and of order $\Delta t/\tau_r$]. Your equations should be analogous to Eq. (6.103a).

(c) By the same technique as was used in Ex. 6.22, obtain from the equations derived in part (b) the following forms of the Fokker-Planck functions:

$$A_j = \frac{-X_j}{\tau_r}, \qquad B_{jk} = \frac{2k_B T}{m\omega^2 \tau_r} \delta_{jk}. \qquad (6.108b)$$

(d) Show that the Fokker-Planck equation, obtained by inserting functions (6.108b) into Eq. (6.106a), has the following Gaussian solution:

$$P_2(X_1, X_2, t | X_1^{(o)}, X_2^{(o)}) = \frac{1}{2\pi \sigma^2} \exp\left[-\frac{(X_1 - \bar{X}_1)^2 + (X_2 - \bar{X}_2)^2}{2\sigma^2}\right], \qquad (6.109a)$$

where the means and variance of the distribution are

$$\bar{X}_j = X_j^{(o)} e^{-t/\tau_r}, \qquad \sigma^2 = \frac{k_B T}{m\omega^2}\left(1 - e^{-2t/\tau_r}\right) \simeq \begin{cases} \frac{k_B T}{m\omega^2} \frac{2t}{\tau_r} & \text{for } t \ll \tau_r \\ \frac{k_B T}{m\omega^2} & \text{for } t \gg \tau_r. \end{cases} \qquad (6.109b)$$

(e) Discuss the physical meaning of the conditional probability (6.109a). Discuss its implications for the physics experiment described in Ex. 6.17, when the signal force acts for a time short compared to τ_r rather than long.

Bibliographic Note

Random processes are treated in many standard textbooks on statistical physics, typically under the rubric of fluctuations or nonequilibrium statistical mechanics (and sometimes not even using the phrase "random process"). We like Kittel (2004), Sethna (2006), Reif (2008), and Pathria and Beale (2011). A treatise on signal processing that we recommend, despite its age, is Wainstein and Zubakov (1962). There are a number of textbooks on random processes (also called "stochastic processes" in book titles), usually aimed at mathematicians, engineers, or finance folks (who use the theory of random processes to try to make lots of money, and often succeed). But we do not like any of those books as well as the relevant sections in the above statistical mechanics texts. Nevertheless, you might want to peruse Lax et al. (2006), Van Kampen (2007), and Paul and Baschnagel (2010).

PART III

OPTICS

Prior to the twentieth century's quantum mechanics and opening of the electromagnetic spectrum observationally, the study of optics was concerned solely with visible light.

Reflection and refraction of light were first described by the Greeks and Arabs and further studied by such medieval scholastics as Roger Bacon (thirteenth century), who explained the rainbow and used refraction in the design of crude magnifying lenses and spectacles. However, it was not until the seventeenth century that there arose a strong commercial interest in manipulating light, particularly via the telescope and compound microscope, improved and famously used by Galileo and Newton.

The discovery of Snell's law in 1621 and observations of diffractive phenomena by Grimaldi in 1665 stimulated serious speculation about the physical nature of light. The wave and corpuscular theories were propounded by Huygens in 1678 and Newton in 1704, respectively. The corpuscular theory initially held sway, for 100 years. However, observational studies of interference by Young in 1803 and the derivation of a wave equation for electromagnetic disturbances by Maxwell in 1865 then seemed to settle the matter in favor of the undulatory theory, only for the debate to be resurrected in 1887 with the discovery of the photoelectric effect by Hertz. After quantum mechanics was developed in the 1920s, the dispute was abandoned, the wave and particle descriptions of light became "complementary," and Hamilton's optics-inspired formulation of classical mechanics was modified to produce the Schrödinger equation.

Many physics students are all too familiar with this potted history and may consequently regard optics as an ancient precursor to modern physics, one that has been completely subsumed by quantum mechanics. Not so! Optics has developed dramatically and independently from quantum mechanics in recent decades and is now a major branch of classical physics. And it is no longer concerned primarily with light. The principles of optics are routinely applied to all types of wave propagation: for example, all parts of the electromagnetic spectrum, quantum mechanical waves (e.g., of electrons and neutrinos), waves in elastic solids (Part IV of this book), fluids (Part V), plasmas (Part VI), and the geometry of spacetime (Part VII). There is a commonality, for instance, to seismology, oceanography, and radio physics that allows ideas to be freely interchanged among these different disciplines. Even the study of visible light has seen major developments: the invention of the laser has led to the modern theory of coherence and has begotten the new field of nonlinear optics.

An even greater revolution has occurred in optical technology. From the credit card and white-light hologram to the laser scanner at a supermarket checkout, from laser printers to CDs, DVDs, and BDs, from radio telescopes capable of nanoradian angular resolution to Fabry-Perot systems that detect displacements smaller than the size of an elementary particle, we are surrounded by sophisticated optical devices in our everyday and scientific lives. Many of these devices turn out to be clever and direct applications of the fundamental optical principles that we discuss in this part of the book.

Our treatment of optics in this part differs from that found in traditional texts, in that we assume familiarity with basic classical mechanics and quantum mechanics and, consequently, fluency in the language of Fourier transforms. This inversion of the historical development reflects contemporary priorities and allows us to emphasize those aspects of the subject that involve fresh concepts and modern applications.

In Chap. 7, we discuss optical (wave-propagation) phenomena in the geometric optics approximation. This approximation is accurate when the wavelength and the wave period are short compared with the lengthscales and timescales on which the wave amplitude and the waves' environment vary. We show how a wave equation can be solved approximately, with optical rays becoming the classical trajectories of quantum particles (photons, phonons, plasmons, and gravitons) and the wave field propagating along these trajectories. We also show how, in general, these trajectories develop singularities or caustics where the geometric optics approximation breaks down, and we must revert to the wave description.

In Chap. 8, we develop the theory of diffraction that arises when the geometric optics approximation fails, and the waves' energy spreads in a non-particle-like way. We analyze diffraction in two limiting regimes, called "Fresnel" and "Fraunhofer" (after the physicists who discovered them), in which the wavefronts are approximately planar or spherical, respectively. As we are working with a linear theory of wave propagation, we make heavy use of Fourier methods and show how elementary applications of Fourier transforms can be used to design powerful optics instruments.

Most elementary diffractive phenomena involve the superposition of an infinite number of waves. However, in many optical applications, only a small number of waves from a common source are combined. This is known as interference and is the subject of Chap. 9. In this chapter, we also introduce the notion of coherence, which is a quantitative measure of the distributions of the combining waves and their capacity to interfere constructively.

The final chapter on optics, Chap. 10, is concerned with nonlinear phenomena that arise when waves, propagating through a medium, become sufficiently strong to couple to one another. These nonlinear phenomena can occur for all types of waves (we meet them for fluid waves in Sec. 16.3 and plasma waves in Chap. 23). For light (the focus of Chap. 10), nonlinearities have become especially important in recent years; the nonlinear effects that arise when laser light is shone through certain crystals are having a strong impact on technology and on fundamental scientific research. We explore several examples.

CHAPTER SEVEN

Geometric Optics

Solar rays parallel to OB and passing through this solid are refracted at the hyperbolic surface, and the refracted rays converge at A.

IBN SAHL (984)

7.1 Overview

Geometric optics, the study of "rays," is the oldest approach to optics. It is an accurate description of wave propagation when the wavelengths and periods of the waves are far smaller than the lengthscales and timescales on which the wave amplitude and the medium supporting the waves vary.

After reviewing wave propagation in a homogeneous medium (Sec. 7.2), we begin our study of geometric optics in Sec. 7.3. There we derive the geometric-optics propagation equations with the aid of the eikonal approximation, and we elucidate the connection to Hamilton-Jacobi theory (which we assume the reader has already encountered). This connection is made more explicit by demonstrating that a classical, geometric-optics wave can be interpreted as a flux of quanta. In Sec. 7.4, we specialize the geometric-optics formalism to any situation where a bundle of nearly parallel rays is being guided and manipulated by some sort of apparatus. This is the paraxial approximation, and we illustrate it with a magnetically focused beam of charged particles and show how matrix methods can be used to describe the particle (i.e., ray) trajectories. In Sec. 7.5, we explore how imperfect optics can produce multiple images of a distant source, and that as one moves from one location to another, the images appear and disappear in pairs. Locations where this happens are called "caustics" and are governed by catastrophe theory, a topic we explore briefly. In Sec. 7.6, we describe gravitational lenses, remarkable astronomical phenomena that illustrate the formation of multiple images and caustics. Finally, in Sec. 7.7, we turn from scalar waves to the vector waves of electromagnetic radiation. We deduce the geometric-optics propagation law for the waves' polarization vector and explore the classical version of a phenomenon called geometric phase.

> **BOX 7.1. READERS' GUIDE**
>
> - This chapter does not depend substantially on any previous chapter, but it does assume familiarity with classical mechanics, quantum mechanics, and classical electromagnetism.
> - Secs. 7.1–7.4 are foundations for the remaining optics chapters, 8, 9, and 10.
> - The discussion of caustics in Sec. 7.5 is a foundation for Sec. 8.6 on diffraction at a caustic.
> - Secs. 7.2 and 7.3 (monochromatic plane waves and wave packets in a homogeneous, time-independent medium; the dispersion relation; and the geometric-optics equations) are used extensively in subsequent parts of this book, including
>   - Chap. 12 for elastodynamic waves,
>   - Chap. 16 for waves in fluids,
>   - Sec. 19.7 and Chaps. 21–23 for waves in plasmas, and
>   - Chap. 27 for gravitational waves.
>   - Sec. 28.6.2 for weak gravitational lensing.

7.2 Waves in a Homogeneous Medium

7.2.1 Monochromatic Plane Waves; Dispersion Relation

Consider a monochromatic plane wave propagating through a homogeneous medium. Independently of the physical nature of the wave, it can be described mathematically by

$$\psi = A e^{i(\mathbf{k}\cdot\mathbf{x} - \omega t)} \equiv A e^{i\varphi}, \tag{7.1}$$

plane wave: complex amplitude, phase, angular frequency, wave vector, wavelength, and propagation direction

where ψ is any oscillatory physical quantity associated with the wave, for example, the y component of the magnetic field associated with an electromagnetic wave. If, as is usually the case, the physical quantity is real (not complex), then we must take the real part of Eq. (7.1). In Eq. (7.1), A is the wave's *complex amplitude*; $\varphi = \mathbf{k}\cdot\mathbf{x} - \omega t$ is the wave's *phase*; t and $\mathbf{x}$ are time and location in space; $\omega = 2\pi f$ is the wave's *angular frequency*; and $\mathbf{k}$ is its *wave vector* (with $k \equiv |\mathbf{k}|$ its *wave number*, $\lambda = 2\pi/k$ its *wavelength*, $\lambdabar = \lambda/(2\pi)$ its *reduced wavelength*, and $\hat{\mathbf{k}} \equiv \mathbf{k}/k$ its *propagation direction*). Surfaces of constant phase φ are orthogonal to the propagation direction $\hat{\mathbf{k}}$ and move in the $\hat{\mathbf{k}}$ direction with the *phase velocity*

phase velocity

$$\mathbf{V}_{\mathrm{ph}} \equiv \left(\frac{\partial \mathbf{x}}{\partial t}\right)_{\varphi} = \frac{\omega}{k}\hat{\mathbf{k}} \tag{7.2}$$

(cf. Fig. 7.1). The frequency ω is determined by the wave vector $\mathbf{k}$ in a manner that depends on the wave's physical nature; the functional relationship

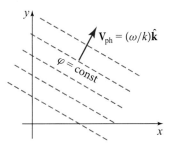

FIGURE 7.1 A monochromatic plane wave in a homogeneous medium.

$$\omega = \Omega(\mathbf{k}) \quad (7.3)$$

dispersion relation

is called the wave's *dispersion relation,* because (as we shall see in Ex. 7.2) it governs the dispersion (spreading) of a wave packet that is constructed by superposing plane waves.

Some examples of plane waves that we study in this book are:

examples:

1. Electromagnetic waves propagating through an isotropic dielectric medium with index of refraction n [Eq. 10.20)], for which ψ could be any Cartesian component of the electric or magnetic field or vector potential and the dispersion relation is

electromagnetic waves

$$\omega = \Omega(\mathbf{k}) = Ck \equiv C|\mathbf{k}|, \quad (7.4)$$

with $C = c/\mathrm{n}$ the phase speed and c the speed of light in vacuum.

2. Sound waves propagating through a solid (Sec. 12.2.3) or fluid (liquid or vapor; Secs. 7.3.1 and 16.5), for which ψ could be the pressure or density perturbation produced by the sound wave (or it could be a potential whose gradient is the velocity perturbation), and the dispersion relation is the same as for electromagnetic waves, Eq. (7.4), but with C now the sound speed.

sound waves

3. Waves on the surface of a deep body of water (depth $\gg \lambda$; Sec. 16.2.1), for which ψ could be the height of the water above equilibrium, and the dispersion relation is [Eq. (16.9)]:

water waves

$$\omega = \Omega(\mathbf{k}) = \sqrt{gk} = \sqrt{g|\mathbf{k}|}, \quad (7.5)$$

with g the acceleration of gravity.

4. Flexural waves on a stiff beam or rod (Sec. 12.3.4), for which ψ could be the transverse displacement of the beam from equilibrium, and the dispersion relation is

flexural waves

$$\omega = \Omega(\mathbf{k}) = \sqrt{\frac{D}{\Lambda}k^2} = \sqrt{\frac{D}{\Lambda}\mathbf{k}\cdot\mathbf{k}}, \quad (7.6)$$

7.2 Waves in a Homogeneous Medium

with Λ the rod's mass per unit length and D its "flexural rigidity" [Eq. (12.33)].

Alfvén waves

5. Alfvén waves in a magnetized, nonrelativistic plasma (bending waves of the plasma-laden magnetic field lines; Sec. 19.7.2), for which ψ could be the transverse displacement of the field and plasma, and the dispersion relation is [Eq. (19.75)]

$$\omega = \Omega(\mathbf{k}) = \mathbf{a} \cdot \mathbf{k}, \tag{7.7}$$

with $\mathbf{a} = \mathbf{B}/\sqrt{\mu_o \rho}$, $[= \mathbf{B}/\sqrt{4\pi\rho}]^1$ the Alfvén speed, $\mathbf{B}$ the (homogeneous) magnetic field, μ_o the magnetic permittivity of the vacuum, and ρ the plasma mass density.

gravitational waves

6. Gravitational waves propagating across the universe, for which ψ can be a component of the waves' metric perturbation which describes the waves' stretching and squeezing of space; these waves propagate nondispersively at the speed of light, so their dispersion relation is Eq. (7.4) with C replaced by the vacuum light speed c.

In general, one can derive the dispersion relation $\omega = \Omega(\mathbf{k})$ by inserting the plane-wave ansatz (7.1) into the dynamical equations that govern one's physical system [e.g., Maxwell's equations, the equations of elastodynamics (Chap. 12), or the equations for a magnetized plasma (Part VI)]. We shall do so time and again in this book.

7.2.2 Wave Packets

7.2.2

Waves in the real world are not precisely monochromatic and planar. Instead, they occupy wave packets that are somewhat localized in space and time. Such wave packets can be constructed as superpositions of plane waves:

wave packet

$$\boxed{\psi(\mathbf{x}, t) = \int A(\mathbf{k}) e^{i\alpha(\mathbf{k})} e^{i(\mathbf{k}\cdot\mathbf{x} - \omega t)} \frac{d^3k}{(2\pi)^3}, \tag{7.8a}}$$
$$\text{where } A(\mathbf{k}) \text{ is concentrated around some } \mathbf{k} = \mathbf{k}_o.$$

Here A and α (both real) are the modulus and phase of the complex amplitude $Ae^{i\alpha}$, and the integration element is $d^3k \equiv d\mathcal{V}_k \equiv dk_x dk_y dk_z$ in terms of components of $\mathbf{k}$ along Cartesian axes x, y, and z. In the integral (7.8a), the contributions from adjacent $\mathbf{k}$s will tend to cancel each other except in that region of space and time where the oscillatory phase factor changes little with changing $\mathbf{k}$ (when $\mathbf{k}$ is near $\mathbf{k}_o$). This is the spacetime region in which the wave packet is concentrated, and its center is where $\nabla_\mathbf{k}$(phase factor) $= 0$:

$$\left(\frac{\partial \alpha}{\partial k_j} + \frac{\partial}{\partial k_j}(\mathbf{k} \cdot \mathbf{x} - \omega t)\right)_{\mathbf{k}=\mathbf{k}_o} = 0. \tag{7.8b}$$

---

1. Gaussian unit equivalents will be given with square brackets.

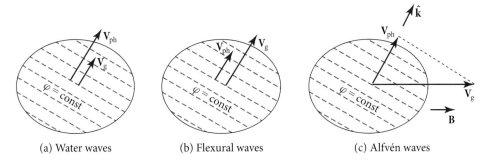

(a) Water waves   (b) Flexural waves   (c) Alfvén waves

FIGURE 7.2 (a) A wave packet of waves on a deep body of water. The packet is localized in the spatial region bounded by the ellipse. The packet's (ellipse's) center moves with the group velocity $\mathbf{V}_g$. The ellipse expands slowly due to wave-packet dispersion (spreading; Ex. 7.2). The surfaces of constant phase (the wave's oscillations) move twice as fast as the ellipse and in the same direction, $\mathbf{V}_{ph} = 2\mathbf{V}_g$ [Eq. (7.11)]. This means that the wave's oscillations arise at the back of the packet and move forward through the packet, disappearing at the front. The wavelength of these oscillations is $\lambda = 2\pi/k_o$, where $k_o = |\mathbf{k}_o|$ is the wave number about which the wave packet is concentrated [Eq. (7.8a) and associated discussion]. (b) A flexural wave packet on a beam, for which $\mathbf{V}_{ph} = \frac{1}{2}\mathbf{V}_g$ [Eq. (7.12)], so the wave's oscillations arise at the packet's front and, traveling more slowly than the packet, disappear at its back. (c) An Alfvén wave packet. Its center moves with a group velocity $\mathbf{V}_g$ that points along the direction of the background magnetic field [Eq. (7.13)], and its surfaces of constant phase (the wave's oscillations) move with a phase velocity $\mathbf{V}_{ph}$ that can be in any direction $\hat{\mathbf{k}}$. The phase speed is the projection of the group velocity onto the phase propagation direction, $|\mathbf{V}_{ph}| = \mathbf{V}_g \cdot \hat{\mathbf{k}}$ [Eq. (7.13)], which implies that the wave's oscillations remain fixed inside the packet as the packet moves; their pattern inside the ellipse does not change. (An even more striking example is provided by the Rossby wave, discussed in Sec. 16.4, in which the group velocity is equal and oppositely directed to the phase velocity.)

Evaluating the derivative with the aid of the wave's dispersion relation $\omega = \Omega(\mathbf{k})$, we obtain for the location of the wave packet's center

$$x_j - \left(\frac{\partial \Omega}{\partial k_j}\right)_{\mathbf{k}=\mathbf{k}_o} t = -\left(\frac{\partial \alpha}{\partial k_j}\right)_{\mathbf{k}=\mathbf{k}_o} = \text{const.} \qquad (7.8c)$$

This tells us that the *wave packet* moves with the *group velocity*

$$\boxed{\mathbf{V}_g = \nabla_{\mathbf{k}} \Omega, \quad \text{i.e.,} \quad V_{g\,j} = \left(\frac{\partial \Omega}{\partial k_j}\right)_{\mathbf{k}=\mathbf{k}_o}.} \qquad (7.9) \qquad \textbf{group velocity}$$

When, as for electromagnetic waves in a dielectric medium or sound waves in a solid or fluid, the dispersion relation has the simple form of Eq. (7.4), $\omega = \Omega(\mathbf{k}) = Ck$ with $k \equiv |\mathbf{k}|$, then the group and phase velocities are the same,

$$\mathbf{V}_g = \mathbf{V}_{ph} = C\hat{\mathbf{k}}, \qquad (7.10)$$

and the waves are said to be *dispersionless*. If the dispersion relation has any other form, then the group and phase velocities are different, and the wave is said to exhibit *dispersion;* cf. Ex. 7.2. Examples are (see Fig. 7.2 and the list in Sec. 7.2.1, from which our numbering is taken):

3. Waves on a deep body of water [dispersion relation (7.5); Fig. 7.2a], for which

$$\mathbf{V}_g = \frac{1}{2}\mathbf{V}_{\text{ph}} = \frac{1}{2}\sqrt{\frac{g}{k}}\,\hat{\mathbf{k}}; \tag{7.11}$$

4. Flexural waves on a stiff beam or rod [dispersion relation (7.6); Fig. 7.2b], for which

$$\mathbf{V}_g = 2\mathbf{V}_{\text{ph}} = 2\sqrt{\frac{D}{\Lambda}}\,k\hat{\mathbf{k}}; \tag{7.12}$$

5. Alfvén waves in a magnetized and nonrelativistic plasma [dispersion relation (7.7); Fig. 7.2c], for which

$$\mathbf{V}_g = \mathbf{a}, \qquad \mathbf{V}_{\text{ph}} = (\mathbf{a} \cdot \hat{\mathbf{k}})\hat{\mathbf{k}}. \tag{7.13}$$

Notice that, depending on the dispersion relation, the group speed $|\mathbf{V}_g|$ can be less than or greater than the phase speed, and if the homogeneous medium is anisotropic (e.g., for a magnetized plasma), the group velocity can point in a different direction than the phase velocity.

Physically, it should be obvious that the energy contained in a wave packet must remain always with the packet and cannot move into the region outside the packet where the wave amplitude vanishes. Correspondingly, the wave packet's energy must propagate with the group velocity $\mathbf{V}_g$ and not with the phase velocity $\mathbf{V}_{\text{ph}}$. When one examines the wave packet from a quantum mechanical viewpoint, its quanta must move with the group velocity $\mathbf{V}_g$. Since we have required that the wave packet have its wave vectors concentrated around $\mathbf{k}_o$, the energy and momentum of each of the packet's quanta are given by the standard quantum mechanical relations:

$$\boxed{\mathcal{E} = \hbar\Omega(\mathbf{k}_o), \quad \text{and} \quad \mathbf{p} = \hbar\mathbf{k}_o.} \tag{7.14}$$

EXERCISES

Exercise 7.1 *Practice: Group and Phase Velocities*
Derive the group and phase velocities (7.10)–(7.13) from the dispersion relations (7.4)–(7.7).

Exercise 7.2 **Example: Gaussian Wave Packet and Its Dispersion*
Consider a 1-dimensional wave packet, $\psi(x,t) = \int A(k)e^{i\alpha(k)}e^{i(kx-\omega t)}dk/(2\pi)$, with dispersion relation $\omega = \Omega(k)$. For concreteness, let $A(k)$ be a narrow Gaussian peaked around k_o: $A \propto \exp[-\kappa^2/(2(\Delta k)^2)]$, where $\kappa = k - k_o$.

(a) Expand α as $\alpha(k) = \alpha_o - x_o\kappa$ with x_o a constant, and assume for simplicity that higher order terms are negligible. Similarly, expand $\omega \equiv \Omega(k)$ to quadratic order, and explain why the coefficients are related to the group velocity V_g at $k = k_o$ by $\Omega = \omega_o + V_g \kappa + (dV_g/dk)\kappa^2/2$.

(b) Show that the wave packet is given by

$$\psi \propto \exp[i(\alpha_o + k_o x - \omega_o t)] \int_{-\infty}^{+\infty} \exp[i\kappa(x - x_o - V_g t)] \quad (7.15a)$$

$$\times \exp\left[-\frac{\kappa^2}{2}\left(\frac{1}{(\Delta k)^2} + i\frac{dV_g}{dk}t\right)\right] d\kappa.$$

The term in front of the integral describes the phase evolution of the waves inside the packet; cf. Fig. 7.2.

(c) Evaluate the integral analytically (with the help of a computer, if you wish). From your answer, show that the modulus of ψ satisfies

$$\boxed{|\psi| \propto \frac{1}{L^{1/2}} \exp\left[-\frac{(x - x_o - V_g t)^2}{2L^2}\right], \quad \text{where } L = \frac{1}{\Delta k}\sqrt{1 + \left(\frac{dV_g}{dk}(\Delta k)^2 t\right)^2}}$$

(7.15b)

is the packet's half-width.

(d) Discuss the relationship of this result at time $t = 0$ to the uncertainty principle for the localization of the packet's quanta.

(e) Equation (7.15b) shows that the wave packet spreads (i.e., disperses) due to its containing a range of group velocities [Eq. (7.11)]. How long does it take for the packet to enlarge by a factor 2? For what range of initial half-widths can a water wave on the ocean spread by less than a factor 2 while traveling from Hawaii to California?

7.3 Waves in an Inhomogeneous, Time-Varying Medium: The Eikonal Approximation and Geometric Optics

Suppose that the medium in which the waves propagate is spatially inhomogeneous and varies with time. If the lengthscale $\mathcal{L}$ and timescale $\mathcal{T}$ for substantial variations are long compared to the waves' reduced wavelength and period,

$$\mathcal{L} \gg \lambdabar = 1/k, \quad \mathcal{T} \gg 1/\omega, \quad (7.16)$$

then the waves can be regarded locally as planar and monochromatic. The medium's inhomogeneities and time variations may produce variations in the wave vector $\mathbf{k}$ and frequency ω, but those variations should be substantial only on scales $\gtrsim \mathcal{L} \gg 1/k$ and $\gtrsim \mathcal{T} \gg 1/\omega$. This intuitively obvious fact can be proved rigorously using a two-lengthscale expansion (i.e., an expansion of the wave equation in powers of $\lambdabar/\mathcal{L} = 1/k\mathcal{L}$ and $1/\omega\mathcal{T}$). Such an expansion, in this context of wave propagation, is called

two-lengthscale expansion

eikonal approximation

WKB approximation

the *geometric-optics approximation* or the *eikonal approximation* (after the Greek word ϵικων, meaning image). When the waves are those of elementary quantum mechanics, it is called the *WKB approximation*.[2] The eikonal approximation converts the laws of wave propagation into a remarkably simple form, in which the waves' amplitude is transported along trajectories in spacetime called *rays*. In the language of quantum mechanics, these rays are the world lines of the wave's quanta (photons for light, phonons for sound, plasmons for Alfvén waves, and gravitons for gravitational waves), and the law by which the wave amplitude is transported along the rays is one that conserves quanta. These ray-based propagation laws are called the laws of *geometric optics*.

In this section we develop and study the eikonal approximation and its resulting laws of geometric optics. We begin in Sec. 7.3.1 with a full development of the eikonal approximation and its geometric-optics consequences for a prototypical dispersion-free wave equation that represents, for example, sound waves in a weakly inhomogeneous fluid. In Sec. 7.3.3, we extend our analysis to cover all other types of waves. In Sec. 7.3.4 and a number of exercises we explore examples of geometric-optics waves, and in Sec. 7.3.5 we discuss conditions under which the eikonal approximation breaks down and some non-geometric-optics phenomena that result from the breakdown. Finally, in Sec. 7.3.6 we return to nondispersive light and sound waves, deduce Fermat's principle, and explore some of its consequences.

7.3.1

7.3.1 Geometric Optics for a Prototypical Wave Equation

Our prototypical wave equation is

prototypical wave equation in a slowly varying medium

$$\frac{\partial}{\partial t}\left(W\frac{\partial \psi}{\partial t}\right) - \nabla \cdot (WC^2 \nabla \psi) = 0. \tag{7.17}$$

Here $\psi(\mathbf{x}, t)$ is the quantity that oscillates (the *wave field*), $C(\mathbf{x}, t)$ will turn out to be the wave's slowly varying *propagation speed*, and $W(\mathbf{x}, t)$ is a slowly varying *weighting function* that depends on the properties of the medium through which the wave propagates. As we shall see, W has no influence on the wave's dispersion relation or on its geometric-optics rays, but it does influence the law of transport for the waves' amplitude.

The wave equation (7.17) describes sound waves propagating through a static, isentropic, inhomogeneous fluid (Ex. 16.13), in which case ψ is the wave's pressure perturbation δP, $C(\mathbf{x}) = \sqrt{(\partial P/\partial \rho)_s}$ is the adabiatic sound speed, and the weighting function is $W(\mathbf{x}) = 1/(\rho C^2)$, with ρ the fluid's unperturbed density. This wave equation also describes waves on the surface of a lake or pond or the ocean, in the limit that the slowly varying depth of the undisturbed water $h_o(\mathbf{x})$ is small compared

---

2. Sometimes called "JWKB," adding Jeffreys to the attribution, though Carlini, Liouville, and Green used it a century earlier.

to the wavelength (shallow-water waves; e.g., tsunamis); see Ex. 16.3. In this case ψ is the perturbation of the water's depth, $W = 1$, and $C = \sqrt{gh_o}$ with g the acceleration of gravity. In both cases—sound waves in a fluid and shallow-water waves—if we turn on a slow time dependence in the unperturbed fluid, then additional terms enter the wave equation (7.17). For pedagogical simplicity we leave those terms out, but in the analysis below we do allow W and C to be slowly varying in time, as well as in space: $W = W(\mathbf{x}, t)$ and $C = C(\mathbf{x}, t)$.

Associated with the wave equation (7.17) are an energy density $U(\mathbf{x}, t)$ and energy flux $\mathbf{F}(\mathbf{x}, t)$ given by

$$U = W\left[\frac{1}{2}\left(\frac{\partial \psi}{\partial t}\right)^2 + \frac{1}{2}C^2(\nabla \psi)^2\right], \quad \mathbf{F} = -WC^2 \frac{\partial \psi}{\partial t}\nabla \psi; \quad (7.18)$$

energy density and flux

see Ex. 7.4. It is straightforward to verify that, if C and W are independent of time t, then the scalar wave equation (7.17) guarantees that the U and $\mathbf{F}$ of Eq. (7.18) satisfy the law of energy conservation:

$$\frac{\partial U}{\partial t} + \nabla \cdot \mathbf{F} = 0; \quad (7.19)$$

cf. Ex. 7.4.[3]

We now specialize to a weakly inhomogeneous and slowly time-varying fluid and to nearly plane waves, and we seek a solution of the wave equation (7.17) that locally has approximately the plane-wave form $\psi \simeq Ae^{i\mathbf{k}\cdot\mathbf{x}-\omega t}$. Motivated by this plane-wave form, (i) we express the waves in the eikonal approximation as the product of a real amplitude $A(\mathbf{x}, t)$ that varies slowly on the length- and timescales $\mathcal{L}$ and $\mathcal{T}$, and the exponential of a complex phase $\varphi(\mathbf{x}, t)$ that varies rapidly on the timescale $1/\omega$ and lengthscale λ:

eikonal approximated wave: amplitude, phase, wave vector, and angular frequency

$$\psi(\mathbf{x}, t) = A(\mathbf{x}, t)e^{i\varphi(\mathbf{x},t)}; \quad (7.20)$$

and (ii) we define the wave vector (field) and angular frequency (field) by

$$\mathbf{k}(\mathbf{x}, t) \equiv \nabla \varphi, \quad \omega(\mathbf{x}, t) \equiv -\partial \varphi / \partial t. \quad (7.21)$$

In addition to our two-lengthscale requirement, $\mathcal{L} \gg 1/k$ and $\mathcal{T} \gg 1/\omega$, we also require that A, $\mathbf{k}$, and ω vary slowly (i.e., vary on lengthscales $\mathcal{R}$ and timescales $\mathcal{T}'$ long compared to $\lambda = 1/k$ and $1/\omega$).[4] This requirement guarantees that the waves are locally planar, $\varphi \simeq \mathbf{k} \cdot \mathbf{x} - \omega t + \text{constant}$.

---

3. Alternatively, one can observe that a stationary medium will not perform work.
4. Note that these variations can arise both (i) from the influence of the medium's inhomogeneity (which puts limits $\mathcal{R} \lesssim \mathcal{L}$ and $\mathcal{T}' \lesssim \mathcal{T}$ on the wave's variations) and (ii) from the chosen form of the wave. For example, the wave might be traveling outward from a source and so have nearly spherical phase fronts with radii of curvature $r \simeq$ (distance from source); then $\mathcal{R} = \min(r, \mathcal{L})$.

> **BOX 7.2. BOOKKEEPING PARAMETER IN TWO-LENGTHSCALE EXPANSIONS**
>
> When developing a two-lengthscale expansion, it is sometimes helpful to introduce a bookkeeping parameter σ and rewrite the ansatz (7.20) in a fleshed-out form:
>
> $$\psi = (A + \sigma B + \ldots)e^{i\varphi/\sigma}. \qquad (1)$$
>
> The numerical value of σ is unity, so it can be dropped when the analysis is finished. We use σ to tell us how various terms scale when $\lambdabar$ is reduced at fixed $\mathcal{L}$ and $\mathcal{R}$. The amplitude A has no attached σ and so scales as $\lambdabar^0$, B is multiplied by σ and so scales proportional to $\lambdabar$, and φ is multiplied by σ^{-1} and so scales as $\lambdabar^{-1}$. When one uses these factors of σ in the evaluation of the wave equation, the first term on the right-hand side of Eq. (7.22) gets multiplied by σ^{-2}, the second term by σ^{-1}, and the omitted terms by σ^0. These factors of σ help us to quickly group together all terms that scale in a similar manner and to identify which of the groupings is leading order, and which subleading, in the two-lengthscale expansion. In Eq. (7.22) the omitted σ^0 terms are the first ones in which B appears; they produce a propagation law for B, which can be regarded as a post-geometric-optics correction.
>
> Occasionally the wave equation itself will contain terms that scale with $\lambdabar$ differently from one another (e.g., Ex. 7.9). One should always look out for this possibility.

We now insert the eikonal-approximated wave field (7.20) into the wave equation (7.17), perform the differentiations with the aid of Eqs. (7.21), and collect terms in a manner dictated by a two-lengthscale expansion (see Box 7.2):

$$0 = \frac{\partial}{\partial t}\left(W\frac{\partial \psi}{\partial t}\right) - \nabla \cdot (WC^2 \nabla \psi) \qquad (7.22)$$

$$= \left(-\omega^2 + C^2 k^2\right) W\psi$$

$$+ i\left[-2\left(\omega \frac{\partial A}{\partial t} + C^2 k_j A_{,j}\right)W - \frac{\partial(W\omega)}{\partial t}A - (WC^2 k_j)_{,j} A\right] e^{i\varphi} + \cdots.$$

The first term on the right-hand side, $(-\omega^2 + C^2 k^2)W\psi$, scales as $\lambdabar^{-2}$ when we make the reduced wavelength $\lambdabar$ shorter and shorter while holding the macroscopic lengthscales $\mathcal{L}$ and $\mathcal{R}$ fixed; the second term (in square brackets) scales as $\lambdabar^{-1}$; and the omitted terms scale as $\lambdabar^0$. This is what we mean by "collecting terms in a manner dictated by a two-lengthscale expansion." Because of their different scaling, the first, second, and omitted terms must vanish separately; they cannot possibly cancel one another.

The vanishing of the first term in the eikonal-approximated wave equation (7.22) implies that the waves' frequency field $\omega(\mathbf{x}, t) \equiv -\partial\varphi/\partial t$ and wave-vector field $\mathbf{k} \equiv \nabla\varphi$ satisfy the dispersionless dispersion relation,

dispersion relation

$$\omega = \Omega(\mathbf{k}, \mathbf{x}, t) \equiv C(\mathbf{x}, t)k, \tag{7.23}$$

where (as throughout this chapter) $k \equiv |\mathbf{k}|$. Notice that, as promised, this dispersion relation is independent of the weighting function W in the wave equation. Notice further that this dispersion relation is identical to that for a precisely plane wave in a homogeneous medium, Eq. (7.4), except that the propagation speed C is now a slowly varying function of space and time. This will always be so.

One can always deduce the geometric-optics dispersion relation by (i) considering a precisely plane, monochromatic wave in a precisely homogeneous, time-independent medium and deducing $\omega = \Omega(\mathbf{k})$ in a functional form that involves the medium's properties (e.g., density) and then (ii) allowing the properties to be slowly varying functions of $\mathbf{x}$ and t. The resulting dispersion relation [e.g., Eq. (7.23)] then acquires its $\mathbf{x}$ and t dependence from the properties of the medium.

The vanishing of the second term in the eikonal-approximated wave equation (7.22) dictates that the wave's real amplitude A is transported with the group velocity $\mathbf{V}_g = C\hat{\mathbf{k}}$ in the following manner:

$$\frac{dA}{dt} \equiv \left(\frac{\partial}{\partial t} + \mathbf{V}_g \cdot \nabla\right) A = -\frac{1}{2W\omega}\left[\frac{\partial(W\omega)}{\partial t} + \nabla \cdot (WC^2\mathbf{k})\right] A. \tag{7.24}$$

propagation law for amplitude

This propagation law, by contrast with the dispersion relation, does depend on the weighting function W. We return to this propagation law shortly and shall understand more deeply its dependence on W, but first we must investigate in detail the directions along which A is transported.

The time derivative $d/dt = \partial/\partial t + \mathbf{V}_g \cdot \nabla$ appearing in the propagation law (7.24) is similar to the derivative with respect to proper time along a world line in special relativity, $d/d\tau = u^0 \partial/\partial t + \mathbf{u} \cdot \nabla$ (with u^α the world line's 4-velocity). This analogy tells us that the waves' amplitude A is being propagated along some sort of world lines (trajectories). Those world lines (the waves' rays), in fact, are governed by Hamilton's equations of particle mechanics with the dispersion relation $\Omega(\mathbf{x}, t, \mathbf{k})$ playing the role of the hamiltonian and $\mathbf{k}$ playing the role of momentum:

rays

$$\boxed{\frac{dx_j}{dt} = \left(\frac{\partial\Omega}{\partial k_j}\right)_{\mathbf{x},t} \equiv V_{gj},} \quad \boxed{\frac{dk_j}{dt} = -\left(\frac{\partial\Omega}{\partial x_j}\right)_{\mathbf{k},t},} \quad \boxed{\frac{d\omega}{dt} = \left(\frac{\partial\Omega}{\partial t}\right)_{\mathbf{x},\mathbf{k}}.} \tag{7.25}$$

Hamilton's equations for rays

The first of these Hamilton equations is just our definition of the group velocity, with which [according to Eq. (7.24)] the amplitude is transported. The second tells us how

7.3 Waves in an Inhomogeneous, Time-Varying Medium

the wave vector $\mathbf{k}$ changes along a ray, and together with our knowledge of $C(\mathbf{x}, t)$, it tells us how the group velocity $\mathbf{V}_g = C\hat{\mathbf{k}}$ for our dispersionless waves changes along a ray, and thence defines the ray itself. The third tells us how the waves' frequency changes along a ray.

To deduce the second and third of these Hamilton equations, we begin by inserting the definitions $\omega = -\partial\varphi/\partial t$ and $\mathbf{k} = \nabla\varphi$ [Eqs. (7.21)] into the dispersion relation $\omega = \Omega(\mathbf{x}, t; \mathbf{k})$ for an arbitrary wave, thereby obtaining

$$\boxed{\frac{\partial\varphi}{\partial t} + \Omega(\mathbf{x}, t; \nabla\varphi) = 0.} \tag{7.26a}$$

eikonal equation and Hamilton-Jacobi equation

This equation is known in optics as the *eikonal equation*. It is formally the same as the Hamilton-Jacobi equation of classical mechanics (see, e.g., Goldstein, Poole, and Safko, 2002), if we identify Ω with the hamiltonian and φ with Hamilton's principal function (cf. Ex. 7.9). This suggests that, to derive the second and third of Eqs. (7.25), we can follow the same procedure as is used to derive Hamilton's equations of motion. We take the gradient of Eq. (7.26a) to obtain

$$\frac{\partial^2\varphi}{\partial t \partial x_j} + \frac{\partial\Omega}{\partial k_l}\frac{\partial^2\varphi}{\partial x_l \partial x_j} + \frac{\partial\Omega}{\partial x_j} = 0, \tag{7.26b}$$

where the partial derivatives of Ω are with respect to its arguments $(\mathbf{x}, t; \mathbf{k})$; we then use $\partial\varphi/\partial x_j = k_j$ and $\partial\Omega/\partial k_l = V_{g\,l}$ to write Eq. (7.26b) as $dk_j/dt = -\partial\Omega/\partial x_j$. This is the second of Hamilton's equations (7.25), and it tells us how the wave vector changes along a ray. The third Hamilton equation, $d\omega/dt = \partial\Omega/\partial t$ [Eq. (7.25)], is obtained by taking the time derivative of the eikonal equation (7.26a).

Not only is the waves' amplitude A propagated along the rays, so also is their phase:

propagation equation for phase

$$\frac{d\varphi}{dt} = \frac{\partial\varphi}{\partial t} + \mathbf{V}_g \cdot \nabla\varphi = -\omega + \mathbf{V}_g \cdot \mathbf{k}. \tag{7.27}$$

Since our dispersionless waves have $\omega = Ck$ and $\mathbf{V}_g = C\hat{\mathbf{k}}$, this vanishes. Therefore, for the special case of dispersionless waves (e.g., sound waves in a fluid and electromagnetic waves in an isotropic dielectric medium), the phase is constant along each ray:

$$\boxed{d\varphi/dt = 0.} \tag{7.28}$$

7.3.2 Connection of Geometric Optics to Quantum Theory

Although the waves $\psi = Ae^{i\varphi}$ are classical and our analysis is classical, their propagation laws in the eikonal approximation can be described most nicely in quantum mechanical language.[5] Quantum mechanics insists that, associated with any wave in

---

5. This is intimately related to the fact that quantum mechanics underlies classical mechanics; the classical world is an approximation to the quantum world, often a very good approximation.

the geometric-optics regime, there are real quanta: the wave's quantum mechanical particles. If the wave is electromagnetic, the quanta are photons; if it is gravitational, they are gravitons; if it is sound, they are phonons; if it is a plasma wave (e.g., Alfvén), they are plasmons. When we multiply the wave's $\mathbf{k}$ and ω by $\hbar$, we obtain the particles' momentum and energy:

$$\boxed{\mathbf{p} = \hbar \mathbf{k}, \qquad \mathcal{E} = \hbar \omega.} \tag{7.29}$$

momentum and energy of quanta

Although the originators of the nineteenth-century theory of classical waves were unaware of these quanta, once quantum mechanics had been formulated, the quanta became a powerful conceptual tool for thinking about classical waves.

In particular, we can regard the rays as the world lines of the quanta, and by multiplying the dispersion relation by $\hbar$, we can obtain the hamiltonian for the quanta's world lines:

$$\boxed{H(\mathbf{x}, t; \mathbf{p}) = \hbar \Omega(\mathbf{x}, t; \mathbf{k} = \mathbf{p}/\hbar).} \tag{7.30}$$

hamiltonian for quanta

Hamilton's equations (7.25) for the rays then immediately become Hamilton's equations for the quanta: $dx_j/dt = \partial H/\partial p_j$, $dp_j/dt = -\partial H/\partial x_j$, and $d\mathcal{E}/dt = \partial H/\partial t$.

Return now to the propagation law (7.24) for the waves' amplitude, and examine its consequences for the waves' energy. By inserting the ansatz $\psi = \Re(Ae^{i\varphi}) = A\cos(\varphi)$ into Eqs. (7.18) for the energy density U and energy flux $\mathbf{F}$ and averaging over a wavelength and wave period (so $\overline{\cos^2 \varphi} = \overline{\sin^2 \varphi} = 1/2$), we find that

$$U = \frac{1}{2} W C^2 k^2 A^2 = \frac{1}{2} W \omega^2 A^2, \qquad \mathbf{F} = U(C\hat{\mathbf{k}}) = U\mathbf{V}_g. \tag{7.31}$$

Inserting these into the expression $\partial U/\partial t + \nabla \cdot \mathbf{F}$ for the rate at which energy (per unit volume) fails to be conserved and using the propagation law (7.24) for A, we obtain

$$\frac{\partial U}{\partial t} + \nabla \cdot \mathbf{F} = U \frac{\partial \ln C}{\partial t}. \tag{7.32}$$

Thus, as the propagation speed C slowly changes at a fixed location in space due to a slow change in the medium's properties, the medium slowly pumps energy into the waves or removes it from them at a rate per unit volume of $U \partial \ln C/\partial t$.

This slow energy change can be understood more deeply using quantum concepts. The number density and number flux of quanta are

$$\boxed{n = \frac{U}{\hbar\omega}, \qquad \mathbf{S} = \frac{\mathbf{F}}{\hbar\omega} = n\mathbf{V}_g.} \tag{7.33}$$

number density and flux for quanta

By combining these equations with the energy (non)conservation equation (7.32), we obtain

$$\frac{\partial n}{\partial t} + \nabla \cdot \mathbf{S} = n \left[\frac{\partial \ln C}{\partial t} - \frac{d \ln \omega}{dt} \right]. \tag{7.34}$$

7.3 Waves in an Inhomogeneous, Time-Varying Medium

The third Hamilton equation (7.25) tells us that

$$d\omega/dt = (\partial\Omega/\partial t)_{x,k} = [\partial(Ck)/\partial t]_{x,k} = k\partial C/\partial t,$$

whence $d \ln \omega/dt = \partial \ln C/\partial t$, which, when inserted into Eq. (7.34), implies that the quanta are conserved:

conservation of quanta

$$\boxed{\frac{\partial n}{\partial t} + \nabla \cdot \mathbf{S} = 0.} \quad (7.35a)$$

Since $\mathbf{S} = n\mathbf{V}_g$ and $d/dt = \partial/\partial t + \mathbf{V}_g \cdot \nabla$, we can rewrite this conservation law as a propagation law for the number density of quanta:

$$\boxed{\frac{dn}{dt} + n\nabla \cdot \mathbf{V}_g = 0.} \quad (7.35b)$$

The propagation law for the waves' amplitude, Eq. (7.24), can now be understood much more deeply: *The amplitude propagation law is nothing but the law of conservation of quanta in a slowly varying medium, rewritten in terms of the amplitude. This is true quite generally, for any kind of wave (Sec. 7.3.3); and the quickest route to the amplitude propagation law is often to express the wave's energy density U in terms of the amplitude and then invoke conservation of quanta,* Eq. (7.35b).

In Ex. 7.3 we show that the conservation law (7.35b) is equivalent to

$$\boxed{\frac{d(nC\mathcal{A})}{dt} = 0, \quad \text{i.e., } nC\mathcal{A} \text{ is a constant along each ray.}} \quad (7.35c)$$

Here $\mathcal{A}$ is the cross sectional area of a bundle of rays surrounding the ray along which the wave is propagating. Equivalently, by virtue of Eqs. (7.33) and (7.31) for the number density of quanta in terms of the wave amplitude A, we have

$$\frac{d}{dt}A\sqrt{CW\omega\mathcal{A}} = 0, \quad \text{i.e., } A\sqrt{CW\omega\mathcal{A}} \text{ is a constant along each ray.} \quad (7.35d)$$

In Eqs. (7.33) and (7.35), we have boxed those equations that are completely general (because they embody conservation of quanta) and have not boxed those that are specialized to our prototypical wave equation.

EXERCISES

Exercise 7.3 ** *Derivation and Example: Amplitude Propagation for Dispersionless Waves Expressed as Constancy of Something along a Ray*

(a) In connection with Eq. (7.35b), explain why $\nabla \cdot \mathbf{V}_g = d \ln \mathcal{V}/dt$, where $\mathcal{V}$ is the tiny volume occupied by a collection of the wave's quanta.

(b) Choose for the collection of quanta those that occupy a cross sectional area $\mathcal{A}$ orthogonal to a chosen ray, and a longitudinal length Δs along the ray, so $\mathcal{V} = \mathcal{A}\Delta s$. Show that $d \ln \Delta s/dt = d \ln C/dt$ and correspondingly, $d \ln \mathcal{V}/dt = d \ln(C\mathcal{A})/dt$.

(c) Given part (b), show that the conservation law (7.35b) is equivalent to the constancy of $nC\mathcal{A}$ along a ray, Eq. (7.35c).

(d) From the results of part (c), derive the constancy of $A\sqrt{CW\omega\mathcal{A}}$ along a ray (where A is the wave's amplitude), Eq. (7.35d).

Exercise 7.4 ***Example: Energy Density and Flux, and Adiabatic Invariant, for a Dispersionless Wave*

(a) Show that the prototypical scalar wave equation (7.17) follows from the variational principle

$$\delta \int \mathcal{L} \, dt \, d^3x = 0, \qquad (7.36a)$$

where $\mathcal{L}$ is the lagrangian density

$$\mathcal{L} = W\left[\frac{1}{2}\left(\frac{\partial \psi}{\partial t}\right)^2 - \frac{1}{2}C^2 (\nabla \psi)^2\right] \qquad (7.36b)$$

(not to be confused with the lengthscale $\mathcal{L}$ of inhomogeneities in the medium).

(b) For any scalar-field lagrangian density $\mathcal{L}(\psi, \partial\psi/\partial t, \nabla\psi, \mathbf{x}, t)$, the energy density and energy flux can be expressed in terms of the lagrangian, in Cartesian coordinates, as

$$U(\mathbf{x}, t) = \frac{\partial \psi}{\partial t}\frac{\partial \mathcal{L}}{\partial \psi/\partial t} - \mathcal{L}, \qquad F_j = \frac{\partial \psi}{\partial t}\frac{\partial \mathcal{L}}{\partial \psi/\partial x_j} \qquad (7.36c)$$

(Goldstein, Poole, and Safko, 2002, Sec. 13.3). Show, from the Euler-Lagrange equations for $\mathcal{L}$, that these expressions satisfy energy conservation, $\partial U/\partial t + \nabla \cdot \mathbf{F} = 0$, if $\mathcal{L}$ has no explicit time dependence [e.g., for the lagrangian (7.36b) if $C = C(\mathbf{x})$ and $W = W(\mathbf{x})$ do not depend on time t].

(c) Show that expression (7.36c) for the field's energy density U and its energy flux F_j agree with Eqs. (7.18).

(d) Now, regard the wave amplitude ψ as a generalized (field) coordinate. Use the lagrangian $L = \int \mathcal{L} d^3x$ to define a field momentum Π conjugate to this ψ, and then compute a *wave action*,

$$J \equiv \int_0^{2\pi/\omega} \int \Pi(\partial\psi/\partial t) d^3x \, dt, \qquad (7.36d)$$

which is the continuum analog of Eq. (7.43) in Sec. 7.3.6. The temporal integral is over one wave period. Show that this J is proportional to the wave energy divided by the frequency and thence to the number of quanta in the wave.

It is shown in standard texts on classical mechanics that, for approximately periodic oscillations, the particle action (7.43), with the integral limited to one period of oscillation of q, is an *adiabatic invariant*. By the extension of that proof to continuum physics, the wave action (7.36d) is also an adiabatic invariant. This means that the wave action and hence the number of quanta in the waves are conserved when the medium [in our case the index of refraction n(**x**)] changes very slowly in time—a result asserted in the text, and one that also follows from quantum mechanics. We study the particle version (7.43) of this adiabatic invariant in detail when we analyze charged-particle motion in a slowly varying magnetic field in Sec. 20.7.4.

Exercise 7.5 *Problem: Propagation of Sound Waves in a Wind*
Consider sound waves propagating in an atmosphere with a horizontal wind. Assume that the sound speed C, as measured in the air's local rest frame, is constant. Let the wind velocity $\mathbf{u} = u_x \mathbf{e}_x$ increase linearly with height z above the ground: $u_x = Sz$, where S is the constant shearing rate. Consider only rays in the x-z plane.

(a) Give an expression for the dispersion relation $\omega = \Omega(\mathbf{x}, t; \mathbf{k})$. [Hint: In the local rest frame of the air, Ω should have its standard sound-wave form.]

(b) Show that k_x is constant along a ray path, and then demonstrate that sound waves will not propagate when

$$\left| \frac{\omega}{k_x} - u_x(z) \right| < C. \tag{7.37}$$

(c) Consider sound rays generated on the ground that make an angle θ to the horizontal initially. Derive the equations describing the rays, and use them to sketch the rays, distinguishing values of θ both less than and greater than $\pi/2$. (You might like to perform this exercise numerically.)

7.3.3 Geometric Optics for a General Wave

With the simple case of nondispersive sound waves (Secs. 7.3.1 and 7.3.2) as our model, we now study an arbitrary kind of wave in a weakly inhomogeneous and slowly time varying medium (e.g., any of the examples in Sec. 7.2.1: light waves in a dielectric medium, deep water waves, flexural waves on a stiff beam, or Alfvén waves). Whatever the wave may be, we seek a solution to its wave equation using the eikonal approximation $\psi = A e^{i\varphi}$ with slowly varying amplitude A and rapidly varying phase φ. Depending on the nature of the wave, ψ and A might be a scalar (e.g., sound waves), a vector (e.g., light waves), or a tensor (e.g., gravitational waves).

When we insert the ansatz $\psi = A e^{i\varphi}$ into the wave equation and collect terms in the manner dictated by our two-lengthscale expansion [as in Eq. (7.22) and Box 7.2], the leading-order term will arise from letting every temporal or spatial derivative act on the $e^{i\varphi}$. This is precisely where the derivatives would operate in the case of a plane wave in a homogeneous medium, and here, as there, the result of each differentiation is $\partial e^{i\varphi}/\partial t = -i\omega e^{i\varphi}$ or $\partial e^{i\varphi}/\partial x_j = i k_j e^{i\varphi}$. Correspondingly, the leading-order terms in the wave equation here will be identical to those in the homogeneous plane wave case: they will be the dispersion relation multiplied by something times the wave,

$$[-\omega^2 + \Omega^2(\mathbf{x}, t; \mathbf{k})] \times (\text{something}) A e^{i\varphi} = 0, \tag{7.38a}$$

with the spatial and temporal dependence of Ω^2 entering through the medium's properties. This guarantees that (as we claimed in Sec. 7.3.1) the dispersion relation can be obtained by analyzing a plane, monochromatic wave in a homogeneous, time-independent medium and then letting the medium's properties, in the dispersion relation, vary slowly with $\mathbf{x}$ and t.

dispersion relation for general wave

Each next-order ("subleading") term in the wave equation will entail just one of the wave operator's derivatives acting on a slowly varying quantity (A, a medium property, ω, or $\mathbf{k}$) and all the other derivatives acting on $e^{i\varphi}$. The subleading terms that interest us, for the moment, are those in which the one derivative acts on A, thereby propagating it. Therefore, the subleading terms can be deduced from the leading-order terms (7.38a) by replacing just one $i\omega A e^{i\varphi} = -A(e^{i\varphi})_{,t}$ by $-A_{,t} e^{i\varphi}$, and replacing just one $ik_j A e^{i\varphi} = A(e^{i\varphi})_{,j}$ by $A_{,j} e^{i\varphi}$ (where the subscript commas denote partial derivatives in Cartesian coordinates). A little thought then reveals that the equation for the vanishing of the subleading terms must take the form [deducible from the leading terms (7.38a)]:

$$-2i\omega \frac{\partial A}{\partial t} - 2i\Omega(\mathbf{k}, \mathbf{x}, t) \frac{\partial \Omega(\mathbf{k}, \mathbf{x}, t)}{\partial k_j} \frac{\partial A}{\partial x_j} = \text{terms proportional to } A. \tag{7.38b}$$

Using the dispersion relation $\omega = \Omega(\mathbf{x}, t; \mathbf{k})$ and the group velocity (first Hamilton equation) $\partial \Omega / \partial k_j = V_{g\,j}$, we bring this into the "propagate A along a ray" form:

$$\frac{dA}{dt} \equiv \frac{\partial A}{\partial t} + \mathbf{V}_g \cdot \nabla A = \text{terms proportional to } A. \tag{7.38c}$$

Let us return to the leading-order terms (7.38a) in the wave equation [i.e., to the dispersion relation $\omega = \Omega(\mathbf{x}, t; k)$]. For our general wave, as for the prototypical dispersionless wave of the previous two sections, the argument embodied in Eqs. (7.26) shows that the rays are determined by Hamilton's equations (7.25),

$$\boxed{\frac{dx_j}{dt} = \left(\frac{\partial \Omega}{\partial k_j}\right)_{\mathbf{x},t} \equiv V_{g\,j}, \quad \frac{dk_j}{dt} = -\left(\frac{\partial \Omega}{\partial x_j}\right)_{\mathbf{k},t}, \quad \frac{d\omega}{dt} = \left(\frac{\partial \Omega}{\partial t}\right)_{\mathbf{x},\mathbf{k}},} \tag{7.39}$$

Hamilton's equations for general wave

but using the general wave's dispersion relation $\Omega(\mathbf{k}, \mathbf{x}, t)$ rather than $\Omega = C(\mathbf{x}, t)k$. These Hamilton equations include propagation laws for $\omega = -\partial \varphi / \partial t$ and $k_j = \partial \varphi / \partial x_j$, from which we can deduce the propagation law (7.27) for φ along the rays:

$$\boxed{\frac{d\varphi}{dt} = -\omega + \mathbf{V}_g \cdot \mathbf{k}.} \tag{7.40}$$

propagation law for phase of general wave

For waves with dispersion, by contrast with sound in a fluid and other waves that have $\Omega = Ck$, φ will not be constant along a ray.

For our general wave, as for dispersionless waves, the Hamilton equations for the rays can be reinterpreted as Hamilton's equations for the world lines of the waves' quanta [Eq. (7.30) and associated discussion]. And for our general wave, as for dispersionless waves, the medium's slow variations are incapable of creating or destroying wave quanta.[6] Correspondingly, if one knows the relationship between the waves' energy density U and their amplitude A, and thence the relationship between the waves' quantum number density $n = U/\hbar\omega$ and A, then *from the quantum conservation law* [boxed Eqs. (7.35)]

conservation of quanta and propagation of amplitude for general wave

$$\boxed{\frac{\partial n}{\partial t} + \nabla \cdot (n\mathbf{V}_g) = 0, \quad \frac{dn}{dt} + n\nabla \cdot \mathbf{V}_g = 0, \quad \text{or} \quad \frac{d(nC\mathcal{A})}{dt} = 0,} \quad (7.41)$$

one can deduce the propagation law for A—and the result must be the same propagation law as one obtains from the subleading terms in the eikonal approximation.

7.3.4 Examples of Geometric-Optics Wave Propagation

SPHERICAL SOUND WAVES

As a simple example of these geometric-optics propagation laws, consider a sound wave propagating radially outward through a homogeneous fluid from a spherical source (e.g., a radially oscillating ball; cf. Sec. 16.5.3). The dispersion relation is Eq. (7.4): $\Omega = Ck$. It is straightforward (Ex. 7.6) to integrate Hamilton's equations and learn that the rays have the simple form $\{r = Ct + \text{constant}, \theta = \text{constant}, \phi = \text{constant}, \mathbf{k} = (\omega/C)\mathbf{e}_r\}$ in spherical polar coordinates, with $\mathbf{e}_r$ the unit radial vector. Because the wave is dispersionless, its phase φ must be conserved along a ray [Eq. (7.28)], so φ must be a function of $Ct - r, \theta,$ and ϕ. For the waves to propagate radially, it is essential that $\mathbf{k} = \nabla\varphi$ point very nearly radially, which implies that φ must be a rapidly varying function of $Ct - r$ and a slowly varying one of θ and ϕ. The law of conservation of quanta in this case reduces to the propagation law $d(rA)/dt = 0$ (Ex. 7.6), so rA is also a constant along the ray; we call it $\mathcal{B}$. Putting this all together, we conclude that the sound waves' pressure perturbation $\psi = \delta P$ has the form

$$\psi = \frac{\mathcal{B}(Ct - r, \theta, \phi)}{r} e^{i\varphi(Ct-r,\theta,\phi)}, \quad (7.42)$$

where the phase φ is rapidly varying in $Ct - r$ and slowly varying in the angles, and the amplitude $\mathcal{B}$ is slowly varying in $Ct - r$ and the angles.

FLEXURAL WAVES

As another example of the geometric-optics propagation laws, consider flexural waves on a spacecraft's tapering antenna. The dispersion relation is $\Omega = k^2\sqrt{D/\Lambda}$ [Eq. (7.6)] with $D/\Lambda \propto h^2$, where h is the antenna's thickness in its direction of bend (or the

---

6. This is a general feature of quantum theory; creation and destruction of quanta require imposed oscillations at the high frequency and short wavelength of the waves themselves, or at some submultiple of them (in the case of nonlinear creation and annihilation processes; Chap. 10).

antenna's diameter, if it has a circular cross section); cf. Eq. (12.33). Since Ω is independent of t, as the waves propagate from the spacecraft to the antenna's tip, their frequency ω is conserved [third of Eqs. (7.39)], which implies by the dispersion relation that $k = (D/\Lambda)^{-1/4}\omega^{1/2} \propto h^{-1/2}$; hence the wavelength decreases as $h^{1/2}$. The group velocity is $V_g = 2(D/\Lambda)^{1/4}\omega^{1/2} \propto h^{1/2}$. Since the energy per quantum $\hbar\omega$ is constant, particle conservation implies that the waves' energy must be conserved, which in this 1-dimensional problem means that the energy flowing through a segment of the antenna per unit time must be constant along the antenna. On physical grounds this constant energy flow rate must be proportional to $A^2 V_g h^2$, which means that the amplitude A must increase $\propto h^{-5/4}$ as the flexural waves approach the antenna's end. A qualitatively similar phenomenon is seen in the cracking of a bullwhip (where the speed of the end can become supersonic).

LIGHT THROUGH A LENS AND ALFVÉN WAVES

Figure 7.3 sketches two other examples: light propagating through a lens and Alfvén waves propagating in the magnetosphere of a planet. In Sec. 7.3.6 and the exercises we explore a variety of other applications, but first we describe how the geometric-optics propagation laws can fail (Sec. 7.3.5).

Exercise 7.6 *Derivation and Practice: Quasi-Spherical Solution to Vacuum Scalar Wave Equation*

Derive the quasi-spherical solution (7.42) of the vacuum scalar wave equation $-\partial^2\psi/\partial t^2 + \nabla^2\psi = 0$ from the geometric-optics laws by the procedure sketched in the text.

EXERCISES

7.3.5 Relation to Wave Packets; Limitations of the Eikonal Approximation and Geometric Optics

The form $\psi = Ae^{i\varphi}$ of the waves in the eikonal approximation is remarkably general. At some initial moment of time, A and φ can have any form whatsoever, so long as the two-lengthscale constraints are satisfied [A, $\omega \equiv -\partial\varphi/\partial t$, $\mathbf{k} \equiv \nabla\varphi$, and dispersion relation $\Omega(\mathbf{k}; \mathbf{x}, t)$ all vary on lengthscales long compared to $\lambdabar = 1/k$ and on timescales long compared to $1/\omega$]. For example, ψ could be as nearly planar as is allowed by the inhomogeneities of the dispersion relation. At the other extreme, ψ could be a moderately narrow wave packet, confined initially to a small region of space (though not too small; its size must be large compared to its mean reduced wavelength). In either case, the evolution will be governed by the above propagation laws.

Of course, the eikonal approximation is an approximation. Its propagation laws make errors, though when the two-lengthscale constraints are well satisfied, the errors will be small for sufficiently short propagation times. Wave packets provide an important example. Dispersion (different group velocities for different wave vectors) causes wave packets to spread (disperse) as they propagate; see Ex. 7.2. This spreading

phenomena missed by geometric optics

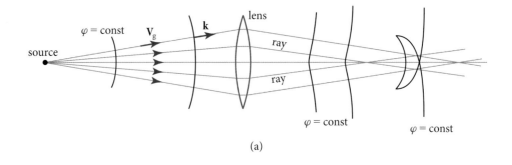

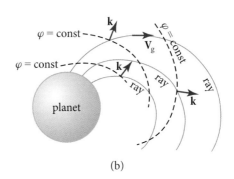

FIGURE 7.3 (a) The rays and the surfaces of constant phase φ at a fixed time for light passing through a converging lens [dispersion relation $\Omega = ck/\mathfrak{n}(\mathbf{x})$, where $\mathfrak{n}$ is the index of refraction]. In this case the rays (which always point along $\mathbf{V}_g$) are parallel to the wave vector $\mathbf{k} = \nabla\varphi$ and thus are also parallel to the phase velocity $\mathbf{V}_{ph}$, and the waves propagate along the rays with a speed $V_g = V_{ph} = c/\mathfrak{n}$ that is independent of wavelength. The strange self-intersecting shape of the last phase front is due to caustics; see Sec. 7.5. (b) The rays and surfaces of constant phase for Alfvén waves in the magnetosphere of a planet [dispersion relation $\Omega = \mathbf{a}(\mathbf{x}) \cdot \mathbf{k}$]. In this case, because $\mathbf{V}_g = \mathbf{a} \equiv \mathbf{B}/\sqrt{\mu_0\rho}$, the rays are parallel to the magnetic field lines and are not parallel to the wave vector, and the waves propagate along the field lines with speeds V_g that are independent of wavelength; cf. Fig. 7.2c. As a consequence, if some electric discharge excites Alfvén waves on the planetary surface, then they will be observable by a spacecraft when it passes magnetic field lines on which the discharge occurred. As the waves propagate, because $\mathbf{B}$ and ρ are time independent and hence $\partial\Omega/\partial t = 0$, the frequency ω and energy $\hbar\omega$ of each quantum is conserved, and conservation of quanta implies conservation of wave energy. Because the Alfvén speed generally diminishes with increasing distance from the planet, conservation of wave energy typically requires the waves' energy density and amplitude to increase as they climb upward.

is not included in the geometric-optics propagation laws; it is a fundamentally wave-based phenomenon and is lost when one goes to the particle-motion regime. In the limit that the wave packet becomes very large compared to its wavelength or that the packet propagates for only a short time, the spreading is small (Ex. 7.2). This is the geometric-optics regime, and geometric optics ignores the spreading.

Many other wave phenomena are missed by geometric optics. Examples are diffraction (e.g., at a geometric-optics caustic; Secs. 7.5 and 8.6), nonlinear wave-wave coupling (Chaps. 10 and 23, and Sec. 16.3), and parametric amplification of waves by rapid time variations of the medium (Sec. 10.7.3)—which shows up in quantum mechanics as particle production (i.e., a breakdown of the law of conservation of quanta). In Sec. 28.7.1, we will encounter such particle production in inflationary models of the early universe.

7.3.6 Fermat's Principle

Hamilton's equations of optics allow us to solve for the paths of rays in media that vary both spatially and temporally. When the medium is time independent, the rays $\mathbf{x}(t)$ can be computed from a variational principle due to Fermat. This is the optical analog of the classical dynamics principle of least action,[7] which states that, when a particle moves from one point to another through a time-independent potential (so its energy, the hamiltonian, is conserved), then the path $\mathbf{q}(t)$ that it follows is one that extremizes the action

$$J = \int \mathbf{p} \cdot d\mathbf{q} \qquad (7.43)$$

principle of least action

(where $\mathbf{q}$ and $\mathbf{p}$ are the particle's generalized coordinates and momentum), subject to the constraint that the paths have a fixed starting point, a fixed endpoint, and constant energy. The proof (e.g., Goldstein, Poole, and Safko, 2002, Sec. 8.6) carries over directly to optics when we replace the hamiltonian by Ω, $\mathbf{q}$ by $\mathbf{x}$, and $\mathbf{p}$ by $\mathbf{k}$. The resulting Fermat principle, stated with some care, has the following form.

Consider waves whose hamiltonian $\Omega(\mathbf{k}, \mathbf{x})$ is independent of time. Choose an initial location $\mathbf{x}_{\text{initial}}$ and a final location $\mathbf{x}_{\text{final}}$ in space, and consider the rays $\mathbf{x}(t)$ that connect these two points. The rays (usually only one) are those paths that satisfy the variational principle

Fermat's principle

$$\boxed{\delta \int \mathbf{k} \cdot d\mathbf{x} = 0.} \qquad (7.44)$$

In this variational principle, $\mathbf{k}$ must be expressed in terms of the trial path $\mathbf{x}(t)$ using Hamilton's equation $dx^j/dt = -\partial\Omega/\partial k_j$; the rate that the trial path is traversed (i.e., the magnitude of the group velocity) must be adjusted to keep Ω constant along the trial path (which means that the total time taken to go from $\mathbf{x}_{\text{initial}}$ to $\mathbf{x}_{\text{final}}$ can differ from one trial path to another). And of course, the trial paths must all begin at $\mathbf{x}_{\text{initial}}$ and end at $\mathbf{x}_{\text{final}}$.

PATH INTEGRALS

Notice that, once a ray has been identified by this action principle, it has $\mathbf{k} = \nabla\varphi$, and therefore the extremal value of the action $\int \mathbf{k} \cdot d\mathbf{x}$ along the ray is equal to the waves'

---

[7]. This is commonly attributed to Maupertuis, though others, including Leibniz and Euler, understood it earlier or better. This "action" and the rules for its variation are different from those in play in Hamilton's principle.

phase difference $\Delta\varphi$ between $\mathbf{x}_{\text{initial}}$ and $\mathbf{x}_{\text{final}}$. Correspondingly, for any trial path, we can think of the action as a phase difference along that path,

$$\Delta\varphi = \int \mathbf{k} \cdot d\mathbf{x}, \tag{7.45a}$$

and we can think of Fermat's principle as saying that the particle travels along a path of extremal phase difference $\Delta\varphi$. This can be reexpressed in a form closely related to *Feynman's path-integral formulation of quantum mechanics* (Feynman, 1966). We can regard all the trial paths as being followed with equal probability. For each path, we are to construct a probability amplitude $e^{i\Delta\varphi}$, and we must then add together these amplitudes,

$$\sum_{\text{all paths}} e^{i\Delta\varphi}, \tag{7.45b}$$

to get the net complex amplitude for quanta associated with the waves to travel from $\mathbf{x}_{\text{initial}}$ to $\mathbf{x}_{\text{final}}$. The contributions from almost all neighboring paths will interfere destructively. The only exceptions are those paths whose neighbors have the same values of $\Delta\varphi$, to first order in the path difference. These are the paths that extremize the action (7.44): they are the wave's rays, the actual paths of the quanta.

SPECIALIZATION TO $\Omega = C(\mathbf{x})k$

Fermat's principle takes on an especially simple form when not only is the hamiltonian $\Omega(\mathbf{k}, \mathbf{x})$ time independent, but it also has the simple dispersion-free form $\Omega = C(\mathbf{x})k$—a form valid for the propagation of light through a time-independent dielectric, and sound waves through a time-independent, inhomogeneous fluid, and electromagnetic or gravitational waves through a time-independent, Newtonian gravitational field (Sec. 7.6). In this $\Omega = C(\mathbf{x})k$ case, the hamiltonian dictates that for each trial path, $\mathbf{k}$ is parallel to $d\mathbf{x}$, and therefore $\mathbf{k} \cdot d\mathbf{x} = k\,ds$, where s is distance along the path. Using the dispersion relation $k = \Omega/C$ and noting that Hamilton's equation $dx^j/dt = \partial\Omega/\partial k_j$ implies $ds/dt = C$ for the rate of traversal of the trial path, we see that $\mathbf{k} \cdot d\mathbf{x} = k\,ds = \Omega\,dt$. Since the trial paths are constrained to have Ω constant, Fermat's principle (7.44) becomes a principle of extremal time: The rays between $\mathbf{x}_{\text{initial}}$ and $\mathbf{x}_{\text{final}}$ are those paths along which

principle of extreme time for dispersionless wave

$$\boxed{\int dt = \int \frac{ds}{C(\mathbf{x})} = \int \frac{\mathrm{n}(\mathbf{x})}{c} ds} \tag{7.46}$$

is extremal. In the last expression we have adopted the convention used for light in a dielectric medium, that $C(\mathbf{x}) = c/\mathrm{n}(\mathbf{x})$, where c is the speed of light in vacuum, and n is the medium's index of refraction. Since c is constant, the rays are paths of extremal optical path length $\int \mathrm{n}(\mathbf{x}) ds$.

index of refraction

We can use Fermat's principle to demonstrate that, if the medium contains no opaque objects, then there will always be at least one ray connecting any two points. This is because there is a lower bound on the optical path between any two points, given by $n_{min}L$, where n_{min} is the lowest value of the refractive index anywhere in the medium, and L is the distance between the two points. This means that for some path the optical path length must be a minimum, and that path is then a ray connecting the two points.

From the principle of extremal time, we can derive the Euler-Lagrange differential equation for the ray. For ease of derivation, we write the action principle in the form

$$\delta \int n(\mathbf{x}) \sqrt{\frac{d\mathbf{x}}{ds} \cdot \frac{d\mathbf{x}}{ds}} \, ds, \tag{7.47}$$

where the quantity in the square root is identically one. Performing a variation in the usual manner then gives

$$\boxed{\frac{d}{ds}\left(n \frac{d\mathbf{x}}{ds}\right) = \nabla n, \quad \text{i.e.,} \quad \frac{d}{ds}\left(\frac{1}{C}\frac{d\mathbf{x}}{ds}\right) = \nabla \left(\frac{1}{C}\right).} \tag{7.48}$$

ray equation for dispersionless wave

This is equivalent to Hamilton's equations for the ray, as one can readily verify using the hamiltonian $\Omega = kc/n$ (Ex. 7.7).

Equation (7.48) is a second-order differential equation requiring two boundary conditions to define a solution. We can either choose these to be the location of the start of the ray and its starting direction, or the start and end of the ray. A simple case arises when the medium is stratified [i.e., when $n = n(z)$, where (x, y, z) are Cartesian coordinates]. Projecting Eq. (7.48) perpendicular to $\mathbf{e}_z$, we discover that ndy/ds and ndx/ds are constant, which implies

$$\boxed{n \sin \theta = \text{constant},} \tag{7.49}$$

Snell's law

where θ is the angle between the ray and $\mathbf{e}_z$. This is a variant of Snell's law of refraction. Snell's law is just a mathematical statement that the rays are normal to surfaces (wavefronts) on which the eikonal (phase) φ is constant (cf. Fig. 7.4).[8] Snell's law is valid not only when $n(\mathbf{x})$ varies slowly but also when it jumps discontinuously, despite the assumptions underlying geometric optics failing at a discontinuity.

EXERCISES

Exercise 7.7 *Derivation: Hamilton's Equations for Dispersionless Waves; Fermat's Principle*
Show that Hamilton's equations for the standard dispersionless dispersion relation (7.4) imply the same ray equation (7.48) as we derived using Fermat's principle.

---

8. Another important application of this general principle is to the design of optical instruments, where it is known as the *Abbé condition*. See, e.g., Born and Wolf (1999).

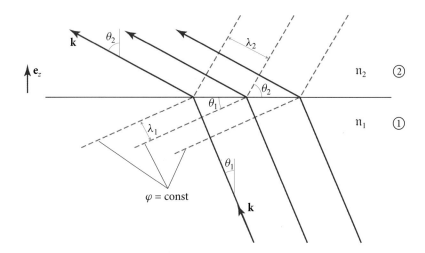

FIGURE 7.4 Illustration of Snell's law of refraction at the interface between two media, for which the refractive indices are n_1 and n_2 (assumed less than n_1). As the wavefronts must be continuous across the interface, simple geometry tells us that $\lambda_1/\sin\theta_1 = \lambda_2/\sin\theta_2$. This and the fact that the wavelengths are inversely proportional to the refractive index, $\lambda_j \propto 1/n_j$, imply that $n_1 \sin\theta_1 = n_2 \sin\theta_2$, in agreement with Eq. (7.49).

Exercise 7.8 *Example: Self-Focusing Optical Fibers*

Optical fibers in which the refractive index varies with radius are commonly used to transport optical signals. When the diameter of the fiber is many wavelengths, we can use geometric optics. Let the refractive index be

$$n = n_0(1 - \alpha^2 r^2)^{1/2}, \tag{7.50a}$$

where n_0 and α are constants, and r is radial distance from the fiber's axis.

(a) Consider a ray that leaves the axis of the fiber along a direction that makes an angle β to the axis. Solve the ray-transport equation (7.48) to show that the radius of the ray is given by

$$r = \frac{\sin\beta}{\alpha} \left| \sin\left(\frac{\alpha z}{\cos\beta}\right) \right|, \tag{7.50b}$$

where z measures distance along the fiber.

(b) Next consider the propagation time T for a light pulse propagating along the ray with $\beta \ll 1$, down a long length L of fiber. Show that

$$T = \frac{n_0 L}{C}[1 + O(\beta^4)], \tag{7.50c}$$

and comment on the implications of this result for the use of fiber optics for communication.

Chapter 7. Geometric Optics

Exercise 7.9 **Example: Geometric Optics for the Schrödinger Equation*

Consider the nonrelativistic Schrödinger equation for a particle moving in a time-dependent, 3-dimensional potential well:

$$-\frac{\hbar}{i}\frac{\partial \psi}{\partial t} = \left[\frac{1}{2m}\left(\frac{\hbar}{i}\nabla\right)^2 + V(\mathbf{x}, t)\right]\psi. \quad (7.51)$$

(a) Seek a geometric-optics solution to this equation with the form $\psi = Ae^{iS/\hbar}$, where A and V are assumed to vary on a lengthscale $\mathcal{L}$ and timescale $\mathcal{T}$ long compared to those, $1/k$ and $1/\omega$, on which S varies. Show that the leading-order terms in the two-lengthscale expansion of the Schrödinger equation give the Hamilton-Jacobi equation

$$\frac{\partial S}{\partial t} + \frac{1}{2m}(\nabla S)^2 + V = 0. \quad (7.52a)$$

Our notation $\varphi \equiv S/\hbar$ for the phase φ of the wave function ψ is motivated by the fact that the geometric-optics limit of quantum mechanics is classical mechanics, and the function $S = \hbar\varphi$ becomes, in that limit, "Hamilton's principal function," which obeys the Hamilton-Jacobi equation (see, e.g., Goldstein, Poole, and Safko, 2002, Chap. 10). [Hint: Use a formal parameter σ to keep track of orders (Box 7.2), and argue that terms proportional to $\hbar^n$ are of order σ^n. This means there must be factors of σ in the Schrödinger equation (7.51) itself.]

(b) From Eq. (7.52a) derive the equation of motion for the rays (which of course is identical to the equation of motion for a wave packet and therefore is also the equation of motion for a classical particle):

$$\frac{d\mathbf{x}}{dt} = \frac{\mathbf{p}}{m}, \quad \frac{d\mathbf{p}}{dt} = -\nabla V, \quad (7.52b)$$

where $\mathbf{p} = \nabla S$.

(c) Derive the propagation equation for the wave amplitude A and show that it implies

$$\frac{d|A|^2}{dt} + |A|^2 \frac{\nabla \cdot \mathbf{p}}{m} = 0. \quad (7.52c)$$

Interpret this equation quantum mechanically.

7.4 Paraxial Optics

It is quite common in optics to be concerned with a bundle of rays that are almost parallel (i.e., for which the angle the rays make with some reference ray can be treated as small). This approximation is called *paraxial optics,* and it permits one to linearize the geometric-optics equations and use matrix methods to trace their

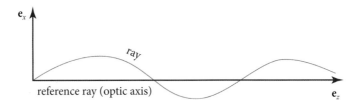

FIGURE 7.5 A reference ray (the z-axis) and an adjacent ray identified by its transverse distances $x(z)$ and $y(z)$, from the reference ray.

rays. The resulting matrix formalism underlies the first-order theory of simple optical instruments (e.g., the telescope and the microscope).

We develop the paraxial optics formalism for waves whose dispersion relation has the simple, time-independent, nondispersive form $\Omega = kc/\mathfrak{n}(\mathbf{x})$. This applies to light in a dielectric medium—the usual application. As we shall see, it also applies to charged particles in a storage ring or electron microscope (Sec. 7.4.2) and to light being lensed by a weak gravitational field (Sec. 7.6).

We restrict ourselves to a situation where there exists a ray that is a straight line, except when it reflects off a mirror or other surface. We choose this as a *reference ray* (also called the *optic axis*) for our formalism, and we orient the z-axis of a Cartesian coordinate system along it (Fig. 7.5). Let the 2-dimensional vector $\mathbf{x}(z)$ be the transverse displacement of some other ray from this reference ray, and denote by $(x, y) = (x_1, x_2)$ the Cartesian components of $\mathbf{x}$.

Under paraxial conditions, $|\mathbf{x}|$ is small compared to the z lengthscales of the propagation, so we can Taylor expand the refractive index $\mathfrak{n}(\mathbf{x}, z)$ in (x_1, x_2):

$$\mathfrak{n}(\mathbf{x}, z) = \mathfrak{n}(0, z) + x_i \mathfrak{n}_{,i}(0, z) + \frac{1}{2} x_i x_j \mathfrak{n}_{,ij}(0, z) + \ldots. \tag{7.53a}$$

Here the subscript commas denote partial derivatives with respect to the transverse coordinates, $\mathfrak{n}_{,i} \equiv \partial \mathfrak{n}/\partial x_i$. The linearized form of the ray-propagation equation (7.48) is then given by

$$\frac{d}{dz}\left(\mathfrak{n}(0, z)\frac{dx_i}{dz}\right) = \mathfrak{n}_{,i}(0, z) + x_j \mathfrak{n}_{,ij}(0, z). \tag{7.53b}$$

In order for the reference ray $x_i = 0$ to satisfy this equation, $\mathfrak{n}_{,i}(0, z)$ must vanish, so Eq. (7.53b) becomes a linear, homogeneous, second-order equation for the path of a nearby ray, $\mathbf{x}(z)$:

paraxial ray equation

$$\boxed{\left(\frac{d}{dz}\right)\left(\frac{\mathfrak{n}\, dx_i}{dz}\right) = x_j \mathfrak{n}_{,ij}.} \tag{7.54}$$

Here $\mathfrak{n}$ and $\mathfrak{n}_{,ij}$ are evaluated on the reference ray. It is helpful to regard z as "time" and think of Eq. (7.54) as an equation for the 2-dimensional motion of a particle (the ray) in a quadratic potential well. We can solve Eq. (7.54) given starting values $\mathbf{x}(z')$ and $\dot{\mathbf{x}}(z')$, where the dot denotes differentiation with respect to z, and z' is the starting location. The solution at some later point z is linearly related to the starting values. We can capitalize on this linearity by treating $\{\mathbf{x}(z), \dot{\mathbf{x}}(z)\}$ as a 4-dimensional vector $V_i(z)$, with

$$V_1 = x, \quad V_2 = \dot{x}, \quad V_3 = y, \quad \text{and} \quad V_4 = \dot{y}, \tag{7.55a}$$

and embodying the linear transformation [linear solution of Eq. (7.54)] from location z' to location z in a *transfer matrix* $J_{ab}(z, z')$:

$$V_a(z) = J_{ab}(z, z') \cdot V_b(z'), \tag{7.55b}$$

paraxial transfer matrix

where there is an implied sum over the repeated index b. The transfer matrix contains full information about the change of position and direction of all rays that propagate from z' to z. As is always the case for linear systems, the transfer matrix for propagation over a large interval, from z' to z, can be written as the product of the matrices for two subintervals, from z' to z'' and from z'' to z:

$$J_{ac}(z, z') = J_{ab}(z, z'') J_{bc}(z'', z'). \tag{7.55c}$$

7.4.1 Axisymmetric, Paraxial Systems: Lenses, Mirrors, Telescopes, Microscopes, and Optical Cavities

7.4.1

If the index of refraction is everywhere axisymmetric, so $\mathfrak{n} = \mathfrak{n}(\sqrt{x^2 + y^2}, z)$, then there is no coupling between the motions of rays along the x and y directions, and the equations of motion along x are identical to those along y. In other words, $J_{11} = J_{33}, J_{12} = J_{34}, J_{21} = J_{43}$, and $J_{22} = J_{44}$ are the only nonzero components of the transfer matrix. This reduces the dimensionality of the propagation problem from 4 dimensions to 2: V_a can be regarded as either $\{x(z), \dot{x}(z)\}$ or $\{y(z), \dot{y}(z)\}$, and in both cases the 2×2 transfer matrix J_{ab} is the same.

Let us illustrate the paraxial formalism by deriving the transfer matrices of a few simple, axisymmetric optical elements. In our derivations it is helpful conceptually to focus on rays that move in the x-z plane (i.e., that have $y = \dot{y} = 0$). We write the 2-dimensional V_i as a column vector:

axisymmetric transfer matrices

$$V_a = \begin{pmatrix} x \\ \dot{x} \end{pmatrix}. \tag{7.56a}$$

The simplest case is a straight section of length d extending from z' to $z = z' + d$. The components of V will change according to

$$x = x' + \dot{x}'d,$$
$$\dot{x} = \dot{x}',$$

7.4 Paraxial Optics

so

for straight section

$$\boxed{J_{ab} = \begin{pmatrix} 1 & d \\ 0 & 1 \end{pmatrix} \text{ for a straight section of length } d,} \quad (7.56b)$$

where $x' = x(z')$, and so forth. Next, consider a thin lens with focal length f. The usual convention in optics is to give f a positive sign when the lens is converging and a negative sign when diverging. A thin lens gives a deflection to the ray that is linearly proportional to its displacement from the optic axis, but does not change its transverse location. Correspondingly, the transfer matrix in crossing the lens (ignoring its thickness) is

for thin lens

$$\boxed{J_{ab} = \begin{pmatrix} 1 & 0 \\ -f^{-1} & 1 \end{pmatrix} \text{ for a thin lens with focal length } f.} \quad (7.56c)$$

Similarly, a spherical mirror with radius of curvature R (again adopting a positive sign for a converging mirror and a negative sign for a diverging mirror) has a transfer matrix

for spherical mirror

$$\boxed{J_{ab} = \begin{pmatrix} 1 & 0 \\ -2R^{-1} & 1 \end{pmatrix} \text{ for a spherical mirror with radius of curvature } R.}$$

$$(7.56d)$$

(Recall our convention that z always increases along a ray, even when the ray reflects off a mirror.)

As a simple illustration, we consider rays that leave a point source located a distance u in front of a converging lens of focal length f, and we solve for the ray positions a distance v behind the lens (Fig. 7.6). The total transfer matrix is the transfer matrix (7.56b) for a straight section, multiplied by the product of the lens transfer matrix (7.56c) and a second straight-section transfer matrix:

$$J_{ab} = \begin{pmatrix} 1 & v \\ 0 & 1 \end{pmatrix} \begin{pmatrix} 1 & 0 \\ -f^{-1} & 1 \end{pmatrix} \begin{pmatrix} 1 & u \\ 0 & 1 \end{pmatrix} = \begin{pmatrix} 1 - vf^{-1} & u + v - uvf^{-1} \\ -f^{-1} & 1 - uf^{-1} \end{pmatrix}. \quad (7.57)$$

When the 1-2 element (upper right entry) of this transfer matrix vanishes, the position of the ray after traversing the optical system is independent of the starting direction. In other words, rays from the point source form a point image. When this happens, the planes containing the source and the image are said to be conjugate. The condition for this to occur is

conjugate planes

thin-lens equations

$$\boxed{\frac{1}{u} + \frac{1}{v} = \frac{1}{f}.} \quad (7.58)$$

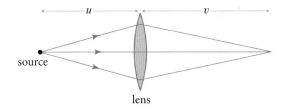

FIGURE 7.6 Simple converging lens used to illustrate the use of transfer matrices. The total transfer matrix is formed by taking the product of the straight-section transfer matrix with the lens matrix and another straight-section matrix.

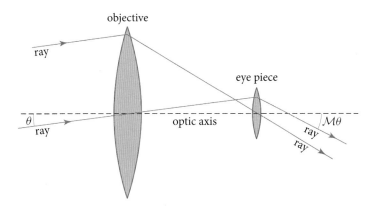

FIGURE 7.7 Simple refracting telescope. By convention $\theta > 0$ and $\mathcal{M}\theta < 0$, so the image is inverted.

This is the standard thin-lens equation. The linear magnification of the image is given by $\mathcal{M} = J_{11} = 1 - v/f$, that is,

$$\mathcal{M} = -\frac{v}{u}, \qquad (7.59)$$

where the negative sign means that the image is inverted. Note that, if a ray is reversed in direction, it remains a ray, but with the source and image planes interchanged; u and v are exchanged, Eq. (7.58) is unaffected, and the magnification (7.59) is inverted: $\mathcal{M} \to 1/\mathcal{M}$.

Exercise 7.10 *Problem: Matrix Optics for a Simple Refracting Telescope*

Consider a simple refracting telescope (Fig. 7.7) that comprises two converging lenses, the *objective* and the *eyepiece*. This telescope takes parallel rays of light from distant stars, which make an angle $\theta \ll 1$ with the optic axis, and converts them into parallel rays making a much larger angle $\mathcal{M}\theta$. Here $\mathcal{M}$ is the magnification with $\mathcal{M}$ negative,

EXERCISES

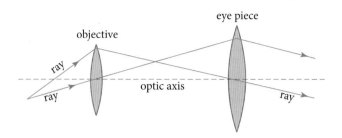

FIGURE 7.8 Simple microscope.

$|\mathcal{M}| \gg 1$, and $|\mathcal{M}\theta| \ll 1$. (The parallel output rays are then focused by the lens of a human's eye, to a point on the eye's retina.)

(a) Use matrix methods to investigate how the output rays depend on the separation of the two lenses, and hence find the condition that the output rays are parallel when the input rays are parallel.

(b) How does the magnification $\mathcal{M}$ depend on the ratio of the focal lengths of the two lenses?

(c) If, instead of looking through the telescope with one's eye, one wants to record the stars' image on a photographic plate or CCD, how should the optics be changed?

Exercise 7.11 *Problem: Matrix Optics for a Simple Microscope*
A microscope takes light rays from a point on a microscopic object, very near the optic axis, and transforms them into parallel light rays that will be focused by a human eye's lens onto the eye's retina (Fig. 7.8). Use matrix methods to explore the operation of such a microscope. A single lens (magnifying glass) could do the same job (rays from a point converted to parallel rays). Why does a microscope need two lenses? What focal lengths and lens separations are appropriate for the eye to resolve a bacterium 100 μm in size?

Exercise 7.12 *Example: Optical Cavity—Rays Bouncing between Two Mirrors*
Consider two spherical mirrors, each with radius of curvature R, separated by distance d so as to form an optical cavity (Fig. 7.9). A laser beam bounces back and forth between the two mirrors. The center of the beam travels along a geometric-optics ray. (We study such beams, including their diffractive behavior, in Sec. 8.5.5.)

(a) Show, using matrix methods, that the central ray hits one of the mirrors (either one) at successive locations $\mathbf{x}_1, \mathbf{x}_2, \mathbf{x}_3, \ldots$ (where $\mathbf{x} \equiv (x, y)$ is a 2-dimensional vector in the plane perpendicular to the optic axis), which satisfy the difference equation

$$\mathbf{x}_{k+2} - 2b\mathbf{x}_{k+1} + \mathbf{x}_k = 0, \qquad (7.60a)$$

FIGURE 7.9 An optical cavity formed by two mirrors, and a light beam bouncing back and forth inside it.

where

$$b = 1 - \frac{4d}{R} + \frac{2d^2}{R^2}. \quad (7.60b)$$

Explain why this is a difference-equation analog of the simple-harmonic-oscillator equation.

(b) Show that this difference equation has the general solution

$$\mathbf{x}_k = \mathbf{A} \cos(k \cos^{-1} b) + \mathbf{B} \sin(k \cos^{-1} b). \quad (7.60c)$$

Obviously, $\mathbf{A}$ is the transverse position $\mathbf{x}_0$ of the ray at its 0th bounce. The ray's 0th position $\mathbf{x}_0$ and its 0th direction of motion $\dot{\mathbf{x}}_0$ together determine $\mathbf{B}$.

(c) Show that if $0 \leq d \leq 2R$, the mirror system is stable. In other words, all rays oscillate about the optic axis. Similarly, show that if $d > 2R$, the mirror system is unstable and the rays diverge from the optic axis.

(d) For an appropriate choice of initial conditions $\mathbf{x}_0$ and $\dot{\mathbf{x}}_0$, the laser beam's successive spots on the mirror lie on a circle centered on the optic axis. When operated in this manner, the cavity is called a *Harriet delay line*. How must d/R be chosen so that the spots have an angular step size θ? (There are two possible choices.)

7.4.2 Converging Magnetic Lens for Charged Particle Beam

Since geometric optics is the same as particle dynamics, matrix equations can be used to describe paraxial motions of electrons or ions in a storage ring. (Note, however, that the hamiltonian for such particles is dispersive, since it does not depend linearly on the particle momentum, and so for our simple matrix formalism to be valid, we must confine attention to a monoenergetic beam of particles.)

The simplest practical lens for charged particles is a quadrupolar magnet. Quadrupolar magnetic fields are used to guide particles around storage rings. If we orient our axes appropriately, the magnet's magnetic field can be expressed in the form

$$\mathbf{B} = \frac{B_0}{r_0}(y\mathbf{e}_x + x\mathbf{e}_y) \quad \text{independent of } z \text{ within the lens} \quad (7.61)$$

quadrupolar magnetic field

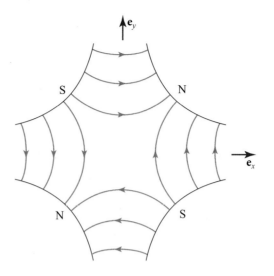

FIGURE 7.10 Quadrupolar magnetic lens. The magnetic field lines lie in a plane perpendicular to the optic axis. Positively charged particles moving along $\mathbf{e}_z$ converge when $y = 0$ and diverge when $x = 0$.

(Fig. 7.10). Particles traversing this magnetic field will be subjected to a Lorentz force that curves their trajectories. In the paraxial approximation, a particle's coordinates satisfy the two differential equations

$$\ddot{x} = -\frac{x}{\lambda^2}, \qquad \ddot{y} = \frac{y}{\lambda^2}, \tag{7.62a}$$

where the dots (as above) mean $d/dz = v^{-1}d/dt$, and

$$\lambda = \left(\frac{pr_0}{qB_0}\right)^{1/2} \tag{7.62b}$$

[cf. Eq. (7.61)], with q the particle's charge (assumed positive) and p its momentum. The motions in the x and y directions are decoupled. It is convenient in this case to work with two 2-dimensional vectors, $\{V_{x1}, V_{x2}\} \equiv \{x, \dot{x}\}$ and $\{V_{y1}, V_{y2}\} = \{y, \dot{y}\}$. From the elementary solutions to the equations of motion (7.62a), we infer that the transfer matrices from the magnet's entrance to its exit are $J_{x\,ab}$, $J_{y\,ab}$, where

transfer matrices for quadrupolar magnetic lens

$$J_{x\,ab} = \begin{pmatrix} \cos\phi & \lambda \sin\phi \\ -\lambda^{-1}\sin\phi & \cos\phi \end{pmatrix}, \tag{7.63a}$$

$$J_{y\,ab} = \begin{pmatrix} \cosh\phi & \lambda \sinh\phi \\ \lambda^{-1}\sinh\phi & \cosh\phi \end{pmatrix}, \tag{7.63b}$$

and

$$\phi = L/\lambda, \tag{7.63c}$$

with L the distance from entrance to exit (i.e., the lens thickness).

The matrices $J_{x\,ab}$ and $J_{y\,ab}$ can be decomposed as follows:

$$J_{x\,ab} = \begin{pmatrix} 1 & \lambda\tan\phi/2 \\ 0 & 1 \end{pmatrix} \begin{pmatrix} 1 & 0 \\ -\lambda^{-1}\sin\phi & 1 \end{pmatrix} \begin{pmatrix} 1 & \lambda\tan\phi/2 \\ 0 & 1 \end{pmatrix} \qquad (7.63d)$$

$$J_{y\,ab} = \begin{pmatrix} 1 & \lambda\tanh\phi/2 \\ 0 & 1 \end{pmatrix} \begin{pmatrix} 1 & 0 \\ \lambda^{-1}\sinh\phi & 1 \end{pmatrix} \begin{pmatrix} 1 & \lambda\tanh\phi/2 \\ 0 & 1 \end{pmatrix} \qquad (7.63e)$$

Comparing with Eqs. (7.56b) and (7.56c), we see that the action of a single magnet is equivalent to the action of a straight section, followed by a thin lens, followed by another straight section. Unfortunately, if the lens is focusing in the x direction, it must be defocusing in the y direction and vice versa. However, we can construct a lens that is focusing along both directions by combining two magnets that have opposite polarity but the same focusing strength $\phi = L/\lambda$.

combining two magnets to make a converging lens

Consider first the particles' motion in the x direction. Let

$$f_+ = \lambda/\sin\phi \quad \text{and} \quad f_- = -\lambda/\sinh\phi \qquad (7.64)$$

be the equivalent focal lengths of the first converging lens and the second diverging lens. If we separate the magnets by a distance s, this must be added to the two effective lengths of the two magnets to give an equivalent separation of $d = \lambda\tan(\phi/2) + s + \lambda\tanh(\phi/2)$ for the two equivalent thin lenses. The combined transfer matrix for the two thin lenses separated by this distance d is then

$$\begin{pmatrix} 1 & 0 \\ -f_-^{-1} & 1 \end{pmatrix} \begin{pmatrix} 1 & d \\ 0 & 1 \end{pmatrix} \begin{pmatrix} 1 & 0 \\ -f_+^{-1} & 1 \end{pmatrix} = \begin{pmatrix} 1 - df_+^{-1} & d \\ -f_*^{-1} & 1 - df_-^{-1} \end{pmatrix}, \qquad (7.65a)$$

transfer matrix for converging magnetic lens

where

$$\frac{1}{f_*} = \frac{1}{f_-} + \frac{1}{f_+} - \frac{d}{f_- f_+} = \frac{\sin\phi}{\lambda} - \frac{\sinh\phi}{\lambda} + \frac{d\sin\phi\sinh\phi}{\lambda^2}. \qquad (7.65b)$$

If we assume that $\phi \ll 1$ and $s \ll L$, then we can expand as a Taylor series in ϕ to obtain

$$f_* \simeq \frac{3\lambda}{2\phi^3} = \frac{3\lambda^4}{2L^3}. \qquad (7.66)$$

The effective focal length f_* of the combined magnets is positive, and so the lens has a net focusing effect. From the symmetry of Eq. (7.65b) under interchange of f_+ and f_-, it should be clear that f_* is independent of the order in which the magnets are encountered. Therefore, if we were to repeat the calculation for the motion in the y direction, we would get the same focusing effect. (The diagonal elements of the transfer matrix are interchanged, but as they are both close to unity, this difference is rather small.)

The combination of two quadrupole lenses of opposite polarity can therefore imitate the action of a converging lens. Combinations of magnets like this are used to collimate particle beams in storage rings, particle accelerators, and electron microscopes.

7.4 Paraxial Optics

7.5 Catastrophe Optics

7.5.1 Image Formation

CAUSTICS

Many simple optical instruments are carefully made to form point images from point sources. However, naturally occurring optical systems, and indeed precision optical instruments when examined in fine detail, bring light to a focus not at a point, but instead on a 2-dimensional surface—an envelope formed by the rays—called a *caustic*. Caustics are often seen in everyday life. For example, when bright sunlight is reflected by the inside of an empty coffee mug some of the rays are reflected *specularly* (angle of incidence equals angle of reflection) and some of the rays are reflected *diffusely* (in all directions due to surface irregularity and multiple reflections and refractions beneath the surface). The specular reflection by the walls—a cylindrical mirror— forms a caustic surface. The intersection of this surface with the bottom forms caustic lines that can be seen in diffuse reflection.[9] These caustic lines are observed to meet in a point. When the optical surfaces are quite irregular (e.g., the water surface in a swimming pool[10] or the type of glass used in bathrooms), then a *caustic network* forms. Caustic lines and points are seen, just the same as with the mug (Fig. 7.11).

What may be surprising is that caustics like these, formed under quite general conditions, can be classified into a rather small number of types, called *catastrophes*, possessing generic properties and scaling laws (Thom, 1994). The scaling laws are reminiscent of the renormalization group discussed in Sec. 5.8.3. Although we focus on catastrophes in the context of optics (e.g., Berry and Upstill, 1980), where they are caustics, the phenomenon is quite general and crops up in other subfields of physics, especially dynamics (e.g., Arnol'd, 1992) and thermodynamics (e.g., Ex. 7.16). It has also been invoked, often quite controversially, in fields outside physics (e.g., Poston and Stewart, 2012). Catastrophes can be found whenever we have a physical system whose states are determined by extremizing a function, such as energy. Our treatment will be quite heuristic, but the subject does have a formal mathematical foundation that connects it to bifurcations and *Morse theory* (see Sec. 11.6; see also, e.g., Petters et al., 2001).

STATE VARIABLES AND CONTROL PARAMETERS

Let us start with a specific, simple example. Suppose that there is a distant *source* S and a *detector* D separated by free space. If we consider all the paths from S to

---

9. The curve that is formed is called a "nephroid."
10. The optics is quite complicated. Some rays from the Sun are reflected specularly by the surface of the water, creating multiple images of the Sun. As the Sun is half a degree in diameter, these produce thickened caustic lines. Some rays are refracted by the water, forming caustic surfaces that are intersected by the bottom of the pool to form a caustic pattern. Some light from this pattern is reflected diffusely before being refracted a second time on the surface of the water and ultimately detected by a retina or a CCD. Other rays are reflected multiple times.

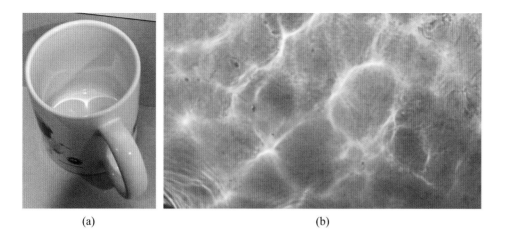

FIGURE 7.11 Photographs of caustics. (a) Simple caustic pattern formed by a coffee mug. (b) Caustic network formed in a swimming pool. The generic structure of these patterns comprises *fold* lines meeting at *cusp* points.

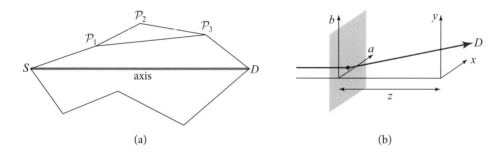

FIGURE 7.12 (a) Alternative paths make up a sequence of straight segments from a source S to a detector D. The path S-$\mathcal{P}_1$-$\mathcal{P}_2$-$\mathcal{P}_3$-D can be simplified to a shorter path S-$\mathcal{P}_1$-$\mathcal{P}_3$-D, and this process can be continued until we have the minimum number of segments needed to exhibit the catastrophe. The (true) ray, with the smallest phase difference, is the axis S-D. (b) A single path from a distant source intersecting a screen at $\{a, b\}$ and ending at detector D with coordinates $\{x, y, z\}$.

D there is a single extremum—a minimum—in the phase difference, $\Delta\varphi = \omega t = \int \mathbf{k} \cdot d\mathbf{x}$. By Fermat's principle, this is the (true) ray—the *axis*—connecting the two points. There are an infinite number of alternative paths that could be defined by an infinite set of parameters, but wherever else the rays go, the phase difference is larger. Take one of these alternative paths connecting S and D and break it down into a sequence of connected segments (Fig. 7.12a). We can imagine replacing two successive segments with a single segment connecting their endpoints. This will reduce the phase difference. The operation can be repeated until we are left with the minimum number of segments, specified by the minimum number of necessary variables that we need to exhibit the catastrophe. The order in which we do this does not matter, and the final variables characterizing the path can be chosen for convenience. These variables are known as *state variables*.

state variables

Next, introduce a screen perpendicular to the S–D axis and close to D (Fig. 7.12b). Consider a path from S, nearly parallel to the axis and intersecting the screen at a point with Cartesian coordinates $\{a, b\}$ measured from the axis. There let it be deflected toward D. In this section and the next, introduce the *delay* $t \equiv \Delta\varphi/\omega$, subtracting off the constant travel time along the axis in the absence of the screen, to measure the phase. The additional geometric delay associated with this ray is given approximately by

$$t_{\text{geo}} = \frac{a^2 + b^2}{2zc}, \qquad (7.67)$$

control parameters

where $z \gg \{a, b\}$ measures the distance from the screen to D, parallel to the axis, and c is the speed of light. The coordinates $\{a, b\}$ act as state variables, and the true ray is determined by differentiating with respect to them. Next, move D off the axis and give it Cartesian coordinates $\{x, y, z\}$ with the x-y plane parallel to the a-b plane, and the transverse coordinates measured from the original axis. As these coordinates specify one of the endpoints, they do not enter into the variation that determines the true ray, but they do change t_{geo} to $[(a-x)^2 + (b-y)^2]/(2zc)$. These $\{x, y, z\}$ parameters are examples of *control parameters*. In general, the number of control parameters that we use is also the minimum needed to exhibit the catastrophe, and the choice is usually determined by algebraic convenience.

FOLD CATASTROPHE

Now replace the screen with a thin lens of refractive index $\mathfrak{n}$ and thickness $w(a, b)$. This introduces an additional contribution to the delay, $t_{\text{lens}} = (\mathfrak{n} - 1)w/c$. The true ray will be fixed by the variation of the sum of the geometric and lens delays with respect to a and b plus any additional state variables that are needed. Suppose that the lens is cylindrical so rays are bent only in the x direction and one state variable, a, suffices. Let us use an analytically tractable example, $t_{\text{lens}} = s^2(1 - 2a^2/s^2 + a^4/s^4)/(4fc)$, for $|a| < s$, where $f \gg s$ is the focal length (Fig. 7.13). Place the detector D on the axis with $z < f$. The delay along a path is:

$$t \equiv t_{\text{geo}} + t_{\text{lens}} = \frac{a^2}{2fzc}\left(f - z + \frac{a^2 z}{2s^2}\right), \qquad (7.68)$$

dropping a constant. This leaves the single minimum and the true ray at $a = 0$.

Now displace D perpendicular to the axis a distance x with z fixed. t_{geo} becomes $(a - x)^2/(2zc)$, and the true ray will be parameterized by the single real value of a that minimizes t and therefore solves

$$x = \frac{a}{f}\left(f - z + z\frac{a^2}{s^2}\right). \qquad (7.69)$$

The first two terms, $f - z$, represent a perfect thin lens [cf. J_{11} in Eq. (7.57)]; the third represents an imperfection.

A human eye at D [coordinates (x, y, z)] focuses the ray through D and adjacent rays onto its retina, producing there a point image of the point source at S. The

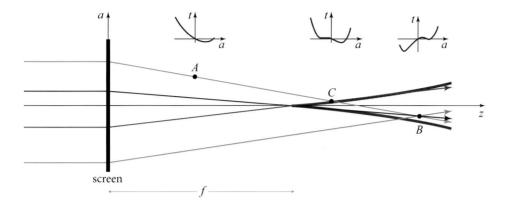

FIGURE 7.13 Light from a distant source is normally incident on a thin, phase-changing lens. The phase change depends solely on the distance a from the axis. The delay t along paths encountering a detector D located at (x, z) can be calculated using an equation such as Eq. (7.68). The true rays are located where t is extremized. The envelope created by these rays comprises two fold caustic curves (red) that meet at a cusp point. When D lies outside the caustic, for example at A, there is only one true ray (cyan) where t has its single minimum. When D lies inside the caustic, for example at point B, there are three true rays associated with two minima (orange, cyan) and one maximum (purple) of t. When D lies on the caustic, for example at C, there is one minimum (cyan) and a point of inflection (green). The magnification is formally infinite on the caustic. The cusp at the end of the caustic is the focus, a distance f from the screen.

power $P = d\mathcal{E}/dt$ in that image is the energy flux at D times the area inside the eye's iris, so the image power is proportional to the energy flux. Since the energy flux is proportional to the square of the field amplitude A^2 and $A \propto 1/\sqrt{\mathcal{A}}$, where $\mathcal{A}$ is the area of a bundle of rays [Eq. (7.35d)], the image power is $P \propto 1/\mathcal{A}$.

Consider a rectangular ray bundle that passes through $\{a, b\}$ on the screen with edges da and db, and area $\mathcal{A}_{screen} = da\, db$. The bundle's area when it arrives at D is $\mathcal{A}_D = dx\, dy$. Because the lens is cylindrical, $dy = db$, so the ratio of power seen by an eye at D to that seen by an eye at the screen (i.e., the lens's *magnification*) is $\mathcal{M} = dP_D/dP_{screen} = \mathcal{A}_{screen}/\mathcal{A}_D = da/dx$. Using Eq. (7.69), we find

$$\mathcal{H} \equiv \mathcal{M}^{-1} = \frac{dx}{da} = zc\left(\frac{d^2 t}{da^2}\right) = \left(\frac{f-z}{f} + 3\frac{a^2 z}{s^2 f}\right). \quad (7.70)$$

If the curvature $\mathcal{H}$ of the delay in the vicinity of a true ray is decreased, the magnification is increased. When $z < f$, the curvature is positive. When $z > f$ the curvature for $a = x = 0$ is negative, and the magnification is $\mathcal{M} = -f/(z - f)$, corresponding to an inverted image. However, there are now two additional images with $a = \pm z^{-1/2}(z - f)^{1/2} s$ at minima with associated magnifications $\mathcal{M} = \frac{1}{2} f(z - f)^{-1}$.

Next move the detector D farther away from the axis. A maximum and a minimum in t will become a point of inflection and, equivalently, two of the images will merge when $a = a_f = \pm 3^{-1/2} z^{-1/2}(z - f)^{1/2} s$ or

$$x = x_f \equiv \mp 2 f^{-1} z^{-1/2}(z - f)^{3/2} s. \quad (7.71)$$

This is the location of the caustic; in 3-dimensional space it is the surface shown in the first panel of Fig. 7.15 below.

fold catastrophe

The magnification of the two images will diverge at the caustic and, expanding $\mathcal{H}$ to linear order about zero, we find that $\mathcal{M} = \pm 2^{-1} 3^{-1/4} f^{1/2} s^{1/2} z^{-1/4} (z-f)^{-1/4} |x - x_f|^{-1/2}$ for each image. However, when $|x| > x_f$, the two images vanish. This abrupt change in the optics—two point images becoming infinitely magnified and then vanishing as the detector is moved—is an example of a *fold catastrophe* occurring at a caustic. (Note that the algebraic sum of the two magnifications is zero in the limit.)

It should be pointed out that the divergence of the magnification does not happen in practice for two reasons. The first is that a point source is only an idealization, and if we allow the source to have finite size, different parts will produce caustics at slightly different locations. The second is that geometric optics, on which our analysis is based, pretends that the wavelength of light is vanishingly small. In actuality, the wavelength is always nonzero, and near a caustic its finiteness leads to diffraction effects, which also limit the magnification to a finite value (Sec. 8.6).

Although we have examined one specific and stylized example, the algebraic details can be worked out for any configuration governed by geometric optics. However, they are less important than the scaling laws—for example, $\mathcal{M} \propto |x - x_f|^{-1/2}$—that become increasingly accurate as the catastrophe (caustic) is approached. For this reason, catastrophes are commonly given a *standard form* chosen to exhibit these features and only valid very close to the catastrophe. We discuss this here just in the context of geometrical optics, but the basic scalings are useful in other physics applications (see, e.g., Ex. 7.16).

standard form for catastrophes

First we measure the state variables $a, b, \ldots$ in units of some appropriate scale. Next we do likewise for the control parameters, $x, y, \ldots$. We call the new state variables and control parameters $\tilde{a}, \tilde{b}, \ldots$ and $\tilde{x}, \tilde{y} \ldots$, respectively. We then Taylor expand the delay about the catastrophe. In the case of the fold, we want to be able to find up to two extrema. This requires a cubic equation in $\tilde{a}$. The constant is clearly irrelevant, and an overall multiplying factor will not change the scalings. We are also free to change the origin of $\tilde{a}$, allowing us to drop either the linear or the quadratic term (we choose the latter, so that the coefficient is linearly related to x). If we adjust the scaled delay in the vicinity of a fold catastrophe, Eq. (7.68) can be written in the standard form:

for fold catastrophe

$$\tilde{t}_{\text{fold}} = \frac{1}{3}\tilde{a}^3 - \tilde{x}\tilde{a}, \tag{7.72}$$

where the coefficients are chosen for algebraic convenience. The maximum number of rays involved in the catastrophe is two, and the number of control parameters required is one, which we can think of as being used to adjust the difference in $\tilde{t}$ between two stationary points. The scaled magnifications are now given by $\widetilde{\mathcal{M}} \equiv (d\tilde{x}/d\tilde{a})^{-1} = \pm\frac{1}{2}\tilde{x}^{-1/2}$, and the combined, scaled magnification is $\widetilde{\mathcal{M}} = \tilde{x}^{-1/2}$ for $\tilde{x} > 0$.

CUSP CATASTROPHE

So far, we have only allowed D to move perpendicular to the axis along x. Now move it along the axis toward the screen. We find that x_f decreases with decreasing z until it vanishes at $z = f$. At this point, the central maximum in t merges simultaneously with both minima, leaving a single image. This is an example of a *cusp catastrophe*. Working in 1-dimensional state-variable space with two control parameters and applying the same arguments as we just used with the fold, the standard form for the cusp can be written as

$$\tilde{t}_{\text{cusp}} = \frac{1}{4}\tilde{a}^4 - \frac{1}{2}\tilde{z}\tilde{a}^2 - \tilde{x}\tilde{a}. \qquad (7.73)$$

for cusp catastrophe

The parameter $\tilde{x}$ is still associated with a transverse displacement of D, and we can quickly persuade ourselves that $\tilde{z} \propto z - f$ by inspecting the quadratic term in Eq. (7.68).

The cusp then describes a transition between one and three images, one of which must be inverted with respect to the other two. The location of the image for a given $\tilde{a}$ and $\tilde{z}$ is

$$\tilde{x} = \tilde{a}^3 - \tilde{z}\tilde{a}. \qquad (7.74)$$

Conversely, for a given $\tilde{x}$ and $\tilde{z}$, there are one or three real solutions for $\tilde{a}(\tilde{x}, \tilde{z})$ and one or three images. The equation satisfied by the fold lines where the transition occurs is

$$\tilde{x} = \pm\frac{2}{3^{3/2}}\tilde{z}^{3/2}. \qquad (7.75)$$

These are the two branches of a semi-cubical parabola (the caustic surface in 3 dimensions depicted in the second panel of Fig. 7.15 below), and they meet at the cusp catastrophe where $\tilde{x} = \tilde{z} = 0$.

The scaled magnification at the cusp is

$$\widetilde{\mathcal{M}}(\tilde{x}, \tilde{z}) = \left(\frac{\partial \tilde{x}}{\partial \tilde{a}}\right)_{\tilde{z}}^{-1} = [3\tilde{a}(\tilde{x}, \tilde{z})^2 - z]^{-1}. \qquad (7.76)$$

SWALLOWTAIL CATASTROPHE

Now let the rays propagate in 3 dimensions, so that there are three control variables, x, y, and z, where the y-axis is perpendicular to the x- and z-axes. The fold, which was a point in 1 dimension and a line in 2, becomes a surface in 3 dimensions, and the point cusp in 2 dimensions becomes a line in 3 (see Fig. 7.15 below). If there is still only one state variable, then $\tilde{t}$ should be a quintic with up to four extrema. (In general, a catastrophe involving as many as N images requires $N - 1$ control parameters for its full description. These parameters can be thought of as independently changing the relative values of $\tilde{t}$ at the extrema.) The resulting, four-image catastrophe is called a *swallowtail*. (In practice, this catastrophe only arises when there are two state variables, and additional images are always present. However, these are not involved in the catastrophe.) Again following our procedure, we can write the standard form of the swallowtail catastrophe as

$$\tilde{t}_{\text{swallowtail}} = \frac{1}{5}\tilde{a}^5 - \frac{1}{3}\tilde{z}\tilde{a}^3 - \frac{1}{2}\tilde{y}\tilde{a}^2 - \tilde{x}\tilde{a}. \tag{7.77}$$

There are two cusp lines in the half-space $\tilde{z} > 0$, and these meet at the catastrophe where $\tilde{x} = \tilde{y} = \tilde{z} = 0$ (see Fig. 7.15 below). The relationship between $\tilde{x}$, $\tilde{y}$, and $\tilde{z}$ and x, y, and z in this or any other example is not simple, and so the variation of the magnification in the vicinity of the swallowtail catastrophe depends on the details.

HYPERBOLIC UMBILIC CATASTROPHE

Next increase the number of essential state variables to two. We can choose these to be $\tilde{a}$ and $\tilde{b}$. To see what is possible, sketch contours of Δt in the $\tilde{a}$-$\tilde{b}$ plane for fixed values of the control variables. The true rays will be associated with maxima, minima, or saddle points, and each distinct catastrophe corresponds to a different way to nest the contours (Fig. 7.14). The properties of the fold, cusp, and swallowtail are essentially unchanged by the extra dimension. We say that they are *structurally stable*. However, a little geometric experimentation uncovers a genuinely 2-dimensional nesting. The *hyperbolic umbilic* catastrophe has two saddles, one maximum and one minimum. Further algebraic experiment produces a standard form:

$$\tilde{t} = \frac{1}{3}(\tilde{a}^3 + \tilde{b}^3) - \tilde{z}\tilde{a}\tilde{b} - \tilde{x}\tilde{a} - \tilde{y}\tilde{b}. \tag{7.78}$$

This catastrophe can be exhibited by a simple generalization of our example. We replace the cylindrical lens described by Eq. (7.68) with a nearly circular lens where the focal length f_a for rays in the a-z plane differs from the focal length f_b for rays in the b-z plane.

$$t = \frac{a^2}{2f_a z c}\left(f_a - z + \frac{a^2 z}{2s_a^2}\right) + \frac{b^2}{2f_b z c}\left(f_b - z + \frac{b^2 z}{2s_b^2}\right). \tag{7.79}$$

This is an example of *astigmatism*. A pair of fold surfaces is associated with each of these foci. These surfaces can cross, and when this happens a cusp line associated with one fold surface can transfer onto the other fold surface. The point where this happens is the hyperbolic umbilic catastrophe.

This example also allows us to illustrate a simple feature of magnification. When the source and the detector are both treated as 2-dimensional, then we generalize the curvature to the *Hessian* matrix

$$\widetilde{\mathcal{H}} = \widetilde{\mathcal{M}}^{-1} = \begin{pmatrix} \frac{\partial \tilde{x}}{\partial \tilde{a}} & \frac{\partial \tilde{y}}{\partial \tilde{a}} \\ \frac{\partial \tilde{x}}{\partial \tilde{b}} & \frac{\partial \tilde{y}}{\partial \tilde{b}} \end{pmatrix}. \tag{7.80}$$

The magnification matrix $\widetilde{\mathcal{M}}$, which describes the mapping from the source plane to the image plane, is simply the inverse of $\widetilde{\mathcal{H}}$. The four matrix elements also describe the deformation of the image. As we describe in more detail when discussing elastostatics (Sec. 11.2.2), the antisymmetric part of $\widetilde{\mathcal{M}}$ describes the rotation of the image, and

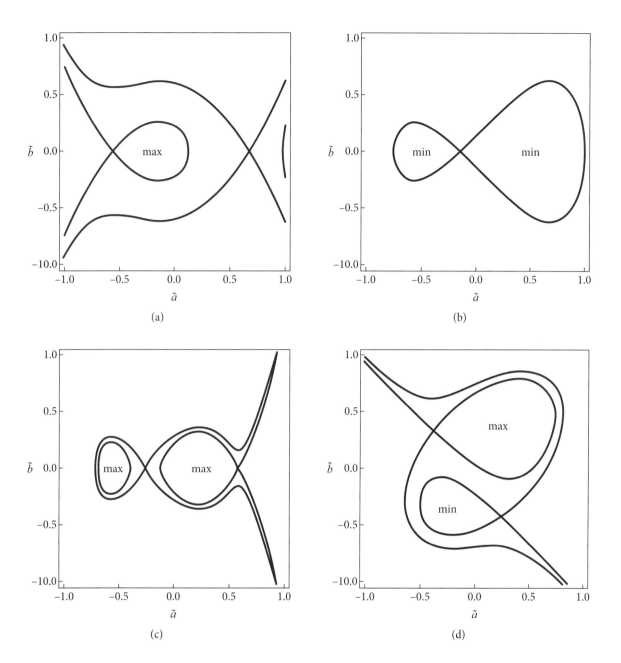

FIGURE 7.14 Distinct nestings of contours of $\tilde{t}$ in state-variable space. (a) A 1-dimensional arrangement of two saddle points and a maximum in the vicinity of a cusp. The locations of the extrema and their curvatures change as the control parameters change. (b) A cusp formed by two minima and a saddle. Although the nestings look different in 2 dimensions, this is essentially the same catastrophe when considered in 1 dimension, which is all that is necessary to determine its salient properties. These are the only contour nestings possible with two state variables and three extrema (or rays). (c) When we increase the number of extrema to four, two more nestings are possible. The swallowtail catastrophe is essentially a cusp with an additional extremum added to the end, requiring three control parameters to express. It, too, is essentially 1-dimensional. (d) The hyperbolic umbilic catastrophe is essentially 2-dimensional and is associated with a maximum, a minimum, and two saddles. A distinct nesting of contours with three saddle points and one extremum occurs in the elliptic umbilic catastrophe (Ex. 7.13).

7.5 Catastrophe Optics

the symmetric part its magnification and stretching or *shear*. Both eigenvalues of $\widetilde{\mathcal{M}}$ are positive at a minimum, and the image is a distorted version of the source. At a saddle, one eigenvalue is positive, the other negative, and the image is inverted; at a maximum, they are both negative, and the image is doubly inverted so that it appears to have been rotated through a half-turn.

ELLIPTIC UMBILIC CATASTROPHE

There is a second standard form that can describe the nesting of contours just discussed—a distinct catastrophe called the *elliptic umbilic catastrophe* (Ex. 7.13b):

standard form for elliptic umbilic catastrophe

$$\tilde{t} = \frac{1}{3}\tilde{a}^3 - \tilde{a}\tilde{b}^2 - \tilde{z}(\tilde{a}^2 + \tilde{b}^2) - \tilde{x}\tilde{a} - \tilde{y}\tilde{b}. \qquad (7.81)$$

The caustic surfaces in three dimensions $(\tilde{x}, \tilde{y}, \tilde{z})$ for the five elementary catastrophes discussed here are shown in Fig. 7.15. Additional types of catastrophe are found with more control parameters, for example, time (e.g., Poston and Stewart, 2012). This is relevant, for example, to the twinkling of starlight in the geometric-optics limit.

EXERCISES

Exercise 7.13 *Derivation and Problem: Cusps and Elliptic Umbilics*

(a) Work through the derivation of Eq. (7.73) for the scaled time delay in the vicinity of the cusp caustic for our simple example [Eq. (7.68)], with the aid of a suitable change of variables (Goodman, Romani, Blandford, and Narayan, 1987, Appendix B).

(b) Sketch the nesting of the contours for the elliptic umbilic catastrophe as shown for the other four catastrophes in Fig. 7.14. Verify that Eq. (7.81) describes this catastrophe.

Exercise 7.14 *Problem: Cusp Scaling Relations*

Consider a cusp catastrophe created by a screen as in the example and described by a standard cusp potential, Eq. (7.73). Suppose that a detector lies between the folds, so that there are three images of a single point source with state variables $\tilde{a}_i$.

(a) Explain how, in principle, it is possible to determine $\tilde{a}$ for a single image by measurements at D.

(b) Make a 3-dimensional plot of the location of the image(s) in $\tilde{a}$-$\tilde{x}$-$\tilde{y}$ space and explain why the names "fold" and "cusp" were chosen.

(c) Prove as many as you can of the following scaling relations, valid in the limit as the cusp catastrophe is approached:

$$\sum_{i=1}^{3}\tilde{a}_i = 0, \quad \sum_{i=1}^{3}\frac{1}{\tilde{a}_i} = -\frac{\tilde{z}}{\tilde{x}}, \quad \sum_{i=1}^{3}\widetilde{\mathcal{M}}_i = 0, \quad \sum_{i=1}^{3}\tilde{a}_i\widetilde{\mathcal{M}}_i = 0,$$

$$\sum_{i=1}^{3}\tilde{a}_i^2\widetilde{\mathcal{M}}_i = 1, \quad \sum_{i=1}^{3}\tilde{a}_i^3\widetilde{\mathcal{M}}_i = 0 \quad \text{and} \quad \sum_{i=1}^{3}\tilde{a}_i^4\widetilde{\mathcal{M}}_i = \tilde{z}. \qquad (7.82)$$

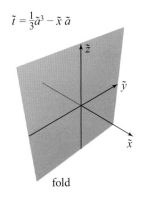
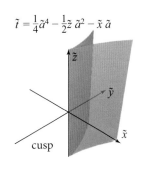
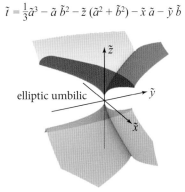
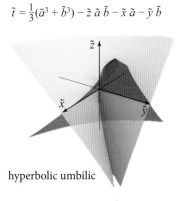
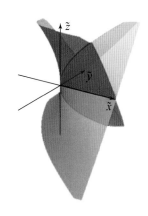

FIGURE 7.15 The five elementary catastrophes (caustic structures) that are possible for a set of light rays specified by one or two state varables $\{\tilde{a}, \tilde{b}\}$ in 3-dimensional space with coordinates (control parameters) $\{\tilde{x}, \tilde{y}, \tilde{z}\}$. The surfaces represent the loci of points of infinite magnification assuming a point source and geometric optics. The actual caustic surfaces will be deformed versions of these basic shapes. The hyperbolic umbilic surfaces are shown from two different viewpoints.

7.5 Catastrophe Optics

[Hint: You must retain the sign of the magnification.] Of course, not all of these are useful. However, relations like these exist for all catastrophes and are increasingly accurate as the separation of the images becomes much smaller than the scale of variation of $\tilde{t}$.

Exercise 7.15 *Problem: Wavefronts* T2

As we have emphasized, representing light using wavefronts is complementary to treating it in terms of rays. Sketch the evolution of the wavefronts after they propagate through a phase-changing screen and eventually form caustics. Do this for a 2-dimensional cusp, and then consider the formation of a hyperbolic umbilic catastrophe by an astigmatic lens.

Exercise 7.16 ***Example: Van der Waals Catastrophe* T2

The van der Waals equation of state $(P + a/v^2)(v - b) = k_B T$ for H_2O relates the pressure P and specific volume (volume per molecule) v to the temperature T; see Sec. 5.7. Figure 5.8 makes it clear that, at some temperatures T and pressures P, there are three allowed volumes $v(T, P)$, one describing liquid water, one water vapor, and the third an unstable phase that cannot exist in Nature. At other values of T and P, there is only one allowed v. The transition between three allowed v values and one occurs along some curve in the T-P plane—a catastrophe curve.

(a) This curve must correspond to one of the elementary catastrophes explored in the previous exercise. Based on the number of solutions for $v(T, P)$, which catastrophe must it be?

(b) Change variables in the van der Waals equation of state to $p = P/P_c - 1$, $\tau = T/T_c - 1$, and $\rho = v_c/v - 1$, where $T_c = 8a/(27bk_B)$, $P_c = a/(27b^2)$, and $v_c = 3b$ are the temperature, pressure, and specific volume at the critical point C of Fig. 5.8. Show that this change of variables brings the van der Waals equation of state into the form

$$\rho^3 - z\rho - x = 0, \tag{7.83}$$

where $z = -(p/3 + 8\tau/3)$ and $x = 2p/3 - 8\tau/3$.

(c) This equation $\rho^3 - z\rho - x$ is the equilibrium surface associated with the catastrophe-theory potential $t(\rho; x, z) = \frac{1}{4}\rho^4 - \frac{1}{2}z\rho^2 - x\rho$ [Eq. (7.73)]. Correspondingly, the catastrophe [the boundary between three solutions $v(T, P)$ and one] has the universal cusp form $x = \pm 2(z/3)^{2/3}$ [Eq. (7.75)]. Plot this curve in the temperature-pressure plane.

Note that we were guaranteed by catastrophe theory that the catastrophe curve would have this form near its cusp point. However, it is a surprise and quite unusual that, for the van der Waals case, the cusp shape $x = \pm 2(z/3)^{2/3}$ is not confined to the vicinity of the cusp point but remains accurate far from that point.

7.5.2 Aberrations of Optical Instruments

Much computational effort is expended in the design of expensive optical instruments prior to prototyping and fabrication. This is conventionally discussed in terms of *aberrations,* which provide a perturbative description of rays that complements the singularity-based approach of catastrophe theory. While it is possible to design instruments that take all the rays from a point source S and focus them geometrically onto a point detector D,[11] this is not what is demanded of them in practice. Typically, they have to map an extended image onto an extended surface, for example, a CCD detector. Sometimes the source is large, and the instrument must achieve a large *field of view;* sometimes it is small, and image fidelity close to the axis matters. Sometimes light levels are low, and transmission losses must be minimized. Sometimes the bandwidth of the light is large, and the variation of the imaging with frequency must be minimized. Sometimes diffractive effects are important. The residual imperfections of an instrument are known as *aberrations.*

As we have shown, any (geometric-optics) instrument will map, one to many, source points onto detector points. This mapping is usually expanded in terms of a set of basis functions, and several choices are in use, for example, those due to Seidel and Zernike (e.g., Born and Wolf, 1999, Secs. 5.3, 9.2). If we set aside effects caused by the variation of the refractive index with wavelength, known as *chromatic aberration,* there are five common types of geometrical aberration. *Spherical aberration* is the failure to bring a point on the optic axis to a single focus. Instead, an axisymmetric cusp/fold caustic is created. We have already exhibited *astigmatism* in our discussion of the hyperbolic umbilic catastrophe with a non-axisymmetric lens and an axial source (Sec. 7.5.1). It is not hard to make axisymmetric lenses and mirrors, so this does not happen much in practice. However, as soon as we consider off-axis surfaces, we break the symmetry, and astigmatism is unavoidable. *Curvature* arises when the surface on which the rays from point sources are best brought to a focus lies on a curved surface, not on a plane. It is sometimes advantageous to accept this aberration and to curve the detector surface.[12] To understand *coma,* consider a small pencil of rays from an off-axis source that passes through the center of an instrument and is brought to a focus. Now consider a cone of rays about this pencil that passes through the periphery of the lens. When there is coma, these rays will on average be displaced

---

11. A simple example is to make the interior of a prolate ellipsoidal detector perfectly reflecting and to place S and D at the two foci, as crudely implemented in whispering galleries.
12. For example, in a traditional Schmidt telescope.

radially. Coma can be ameliorated by reducing the aperture. Finally, there is *distortion*, in which the sides of a square in the source plane are pushed in (*pin cushion*) or out (*barrel*) in the image plane.

7.6 Gravitational Lenses

7.6.1 Gravitational Deflection of Light

Albert Einstein's general relativity theory predicts that light rays should be deflected by the gravitational field of the Sun (Ex. 27.3; Sec. 27.2.3). Newton's law of gravity combined with his corpuscular theory of light also predicts this deflection, but through an angle half as great as relativity predicts. A famous measurement, during a 1919 solar eclipse, confirmed the relativistic prediction, thereby making Einstein world famous.

The deflection of light by gravitational fields allows a cosmologically distant galaxy to behave like a crude lens and, in particular, to produce multiple images of a more distant quasar. Many examples of this phenomenon have been observed. The optics of these gravitational lenses provides an excellent illustration of the use of Fermat's principle (e.g., Blandford and Narayan, 1992; Schneider, Ehlers, and Falco, 1992). We explore these issues in this section.

The action of a gravitational lens can only be understood properly using general relativity. However, when the gravitational field is weak, there exists an equivalent Newtonian model, due to Eddington (1919), that is adequate for our purposes. In this model, curved spacetime behaves as if it were spatially flat and endowed with a refractive index given by

$$\mathfrak{n} = 1 - \frac{2\Phi}{c^2}, \tag{7.84}$$

where Φ is the Newtonian gravitational potential, normalized to vanish far from the source of the gravitational field and chosen to have a negative sign (so, e.g., the field at a distance r from a point mass M is $\Phi = -GM/r$). Time is treated in the Newtonian manner in this model. In Sec. 27.2.3, we use a general relativistic version of Fermat's principle to show that for static gravitational fields this index-of-refraction model gives the same predictions as general relativity, up to fractional corrections of order $|\Phi|/c^2$, which are $\lesssim 10^{-5}$ for the lensing examples in this chapter.

A second Newtonian model gives the same predictions as this index-of-refraction model to within its errors, $\sim |\Phi|/c^2$. We deduce it by rewriting the ray equation (7.48) in terms of Newtonian time t using $ds/dt = C = c/\mathfrak{n}$. The resulting equation is $(\mathfrak{n}^3/c^2)d^2\mathbf{x}/dt^2 = \nabla \mathfrak{n} - 2(\mathfrak{n}/c)^2(d\mathfrak{n}/dt)d\mathbf{x}/dt$. The second term changes the length of the velocity vector $d\mathbf{x}/dt$ by a fractional amount of order $|\Phi|/c^2 \lesssim 10^{-5}$ (so as to keep the length of $d\mathbf{x}/ds$ unity). This is of no significance for our Newtonian model, so we drop this term. The factor $\mathfrak{n}^3 \simeq 1 - 6\Phi/c^2$ produces a fractional correction to $d^2\mathbf{x}/dt^2$ that is of the same magnitude as the fractional errors in our index of refraction model, so we replace this factor by one. The resulting equation of motion for the ray is

$$\frac{d^2\mathbf{x}}{dt^2} = c^2 \nabla \mathfrak{n} = -2\nabla\Phi. \tag{7.85}$$

Equation (7.85) says that the photons that travel along rays feel a Newtonian gravitational potential that is twice as large as the potential felt by low-speed particles; the photons, moving at speed c (aside from fractional changes of order $|\Phi|/c^2$), respond to that doubled Newtonian field in the same way as any Newtonian particle would. The extra deflection is attributable to the geometry of the spatial part of the metric being non-Euclidean (Sec. 27.2.3).

7.6.2 Optical Configuration

To understand how gravitational lenses work, we adopt some key features from our discussion of optical catastrophes formed by an intervening screen. However, there are some essential differences.

- The source is not assumed to be distant from the screen (which we now call a lens, L).
- Instead of tracing rays emanating from a point source S, we consider a *congruence* of rays emanating from the observer O (i.e., us) and propagating backward in time past the lens to the sources. This is because there are many stars and galaxies whose images will be distorted by the lens. The caustics envelop the sources.
- The universe is expanding, which makes the optics formally time-dependent. However, as we discuss in Sec. 28.6.2, we can work in comoving coordinates and still use Fermat's principle. For the moment, we introduce three distances: d_{OL} for distance from the observer to the lens, d_{OS} for the distance from the observer to the source, and d_{LS} for the distance from the lens to the source. We evaluate these quantities cosmologically in Sec. 28.6.2.
- Instead of treating a and b as the state variables that describe rays (Sec. 7.5.1), we use a 2-dimensional (small) angular vector $\boldsymbol{\theta}$ measuring the image position on the sky. We also replace the control parameters x and y with the 2-dimensional angle $\boldsymbol{\beta}$, which measures the location that the image of the source would have in the absence of the lens. We can also treat the distance d_{OS} as a third control parameter replacing z.

The Hessian matrix, replacing Eq. (7.80), is now the Jacobian of the vectorial angles that a small, finite source would subtend in the absence and in the presence of the lens:

$$\mathcal{H} = \mathcal{M}^{-1} = \frac{\partial \boldsymbol{\beta}}{\partial \boldsymbol{\theta}}. \tag{7.86}$$

As the specific intensity $I_\nu = dE/dA\,dt\,d\nu\,d\Omega$ is conserved along a ray (see Sec. 3.6), the determinant of $\mathcal{H}$ is just the ratio of the flux of energy per unit frequency without the lens to the flux with the lens, and correspondingly the determinant of $\mathcal{M}$ (the scalar magnification) is the ratio of flux with the lens to that without the lens.

7.6.3 Microlensing

Our first example of a gravitational lens is a point mass—specifically, a star. This phenomenon is known as *microlensing,* because the angles of deflection are typically microarcseconds.[13] The source is also usually another star, which we also treat as a point.

We first compute the deflection of a Newtonian particle with speed v passing by a mass M with impact parameter b. By computing the perpendicular impulse, it is straightforward to show that the deflection angle is $2GM/v^2$. Replacing v by c and doubling the answer gives the small deflection angle for light:

microlensing deflection angle

$$\boldsymbol{\alpha} = \frac{4GM}{bc^2} = 1.75 \left(\frac{M}{M_\odot}\right)\left(\frac{b}{R_\odot}\right)^{-1} \hat{\mathbf{b}} \text{ arcsec}, \tag{7.87}$$

where $\hat{\mathbf{b}}$ is a unit vector along the impact parameter, which allows us to treat the deflection as a 2-dimensional vector like $\boldsymbol{\theta}$ and $\boldsymbol{\beta}$. $M_\odot$ and $R_\odot$ are the solar mass and radius, respectively.

microlensing lens equation

The imaging geometry can be expressed as a simple vector equation called the *lens equation* (Fig. 7.16):

$$\boldsymbol{\theta} = \boldsymbol{\beta} + \frac{d_{LS}}{d_{OS}}\boldsymbol{\alpha}. \tag{7.88}$$

A point mass exhibits circular symmetry, so we can treat this equation as a scalar equation and rewrite it in the form

$$\theta = \beta + \frac{\theta_E^2}{\theta}, \tag{7.89}$$

where

$$\theta_E = \left(\frac{4GM}{d_{\text{eff}}c^2}\right)^{1/2} = 903 \left(\frac{M}{M_\odot}\right)^{1/2}\left(\frac{d_{\text{eff}}}{10\text{ kpc}}\right)^{-1/2} \mu \text{ arcsec} \tag{7.90}$$

is the *Einstein radius,* and

$$d_{\text{eff}} = \frac{d_{OL}d_{OS}}{d_{LS}} \tag{7.91}$$

is the *effective distance.* (Here 10 kpc means 10 kiloparsecs, about 30,000 light years.)

---

13. Interestingly, Newton speculated that light rays could be deflected by gravity, and the underlying theory of microlensing was worked out correctly by Einstein in 1912, before he realized that the deflection was twice the Newtonian value.

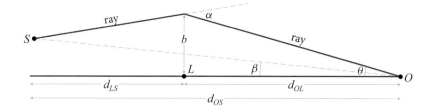

FIGURE 7.16 Geometry for microlensing of a stellar source S by a stellar lens L observed at O.

The solutions to this quadratic equation are

$$\theta_\pm = \frac{\beta}{2} \pm \sqrt{\theta_E^2 + \left(\frac{\beta}{2}\right)^2}. \qquad (7.92)$$ image locations

The magnification of the two images can be computed directly by evaluating the reciprocal of the determinant of $\mathcal{H}$ from Eq. (7.86). However, it is quicker to exploit the circular symmetry and note that the element of source solid angle in polar coordinates is $\beta d\beta d\phi$, while the element of image solid angle is $\theta d\theta d\phi$, so that

$$\mathcal{M} = \frac{\theta}{\beta}\frac{d\theta}{d\beta} = \frac{1}{1-(\theta_E/\theta)^4}. \qquad (7.93)$$ image magnifications

The eigenvalues of the magnification matrix are $[1+(\theta_E/\theta)^2]^{-1}$ and $[1-(\theta_E/\theta)^2]^{-1}$. The former describes the radial magnification, the latter the tangential magnification. As $\beta \to 0$, $\theta \to \theta_E$ for both images. If we consider a source of finite angular size, when β approaches this size then the tangential stretching is pronounced and two nearly circular arcs are formed on opposite sides of the lens, one with $\theta > \theta_E$; the other, inverted, with $\theta < \theta_E$. When β is reduced even more, the arcs join up to form an *Einstein ring* (Fig. 7.17).   Einstein ring

Astronomers routinely observe microlensing events when stellar sources pass behind stellar lenses. They are unable to distinguish the two images and so measure a combined magnification $\mathcal{M} = |\mathcal{M}_+| + |\mathcal{M}_-|$. If we substitute β for θ, then we have

$$\mathcal{M} = \frac{(\theta_E^2 + \frac{1}{2}\beta^2)}{(\theta_E^2 + \frac{1}{4}\beta^2)^{1/2}\beta}. \qquad (7.94)$$

Note that in the limit of large magnifications, $\mathcal{M} \sim \theta_E/\beta$, and the probability that the magnification exceeds $\mathcal{M}$ is proportional to the cross sectional area $\pi\beta^2 \propto \mathcal{M}^{-2}$ (cf. Fig. 7.16).

If the speed of the source relative to the continuation of the O-L line is v, and the closest approach to this line is h, which happens at time $t=0$, then $\beta = (h^2 + v^2 t^2)^{1/2}/d_{OS}$, and then there is a one-parameter family of magnification curves (shown in Fig. 7.18). The characteristic variation of the magnification can be used to distinguish this phenomenon from intrinsic stellar variation. This behavior is very

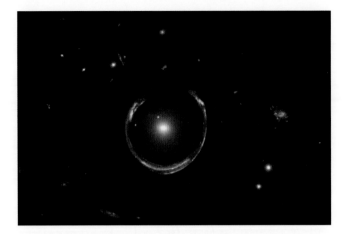

FIGURE 7.17 The source LRG 3-757, as imaged by the Hubble Space Telescope. The blue Einstein ring is the image of two background galaxies formed by the gravitational field associated with the intervening central (yellow) lens galaxy. The accurate alignment of the lens and source galaxies is quite unusual. (ESA/Hubble and NASA.)

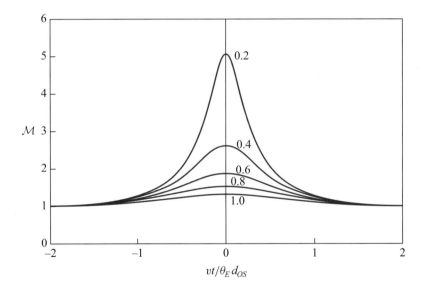

FIGURE 7.18 Variation of the combined magnifications of a stellar source as it passes behind a lens star. The time t is measured in units of $d_{OS}\theta_E/v$, and the parameter that labels the curves is $h/(d_{OS}\theta_E)$.

different from that at a generic caustic (Sec. 7.5.1) and is not structurally stable: if the axisymmetry is broken, then the behavior will change significantly. Nevertheless, for finite-sized sources of light and stars or nearly circular galaxies, stable Einstein rings are commonly formed.

Exercise 7.17 *Example: Microlensing Time Delay* T2

An alternative derivation of the lens equation for a point-mass lens, Eq. (7.88), evaluates the time delay along a path from the source to the observer and finds the true ray by extremizing it with respect to variations of θ [cf. Eq. (7.68)].

(a) Show that the geometric time delay is given by

$$t_{\text{geo}} = \frac{1}{2c} d_{\text{eff}} (\boldsymbol{\theta} - \boldsymbol{\beta})^2. \tag{7.95}$$

(b) Next show that the lens time delay can be expressed as

$$t_{\text{lens}} = -(4GM/c^3) \ln b + \text{const},$$

where b is the impact parameter. (It will be helpful to evaluate the difference in delays between two rays with differing impact parameters.) This is known as the *Shapiro delay* and is discussed further in Sec. 27.2.4.

(c) Show that the lens delay can also be written as

$$t_{\text{lens}} = -\frac{2}{c^3} \int dz\, \Phi = -\frac{2}{c^3} \Phi_2, \tag{7.96}$$

where Φ_2 is the surface gravitational potential obtained by integrating the 3-dimensional potential Φ along the path. The surface potential is only determined up to an unimportant, divergent constant, which is acceptable because we are only interested in dt_{lens}/db which is finite.

(d) By minimizing $t_{\text{geo}} + t_{\text{lens}}$, derive the lens equation (7.88).

Exercise 7.18 *Derivation: Microlensing Variation* T2
Derive Eq. (7.94).

Exercise 7.19 *Problem: Magnification by a Massive Black Hole* T2
Suppose that a large black hole forms two images of a background source separated by an angle θ. Let the fluxes of the two images be F_+ and $F_- < F_+$. Show that the flux from the source would be $F_+ - F_-$ if there were no lens and that the black hole should be located an angular distance $[1 + (F_-/F_+)^{-1/2}]^{-1}\theta$ along the line from the brighter image to the fainter one. (Only consider small angle deflections.)

7.6.4 Lensing by Galaxies T2

Most observed gravitational lenses are galaxies. Observing these systems brings out new features of the optics and proves useful for learning about galaxies and the universe. Galaxies comprise dark matter and stars, and the dispersion $\langle v_\parallel^2 \rangle$ in the stars' velocities along the line of sight can be measured using spectroscopy. The virial theorem (Goldstein, Poole, and Safko, 2002) tells us that the kinetic energy of the matter in the galaxy is half the magnitude of its gravitational potential energy Φ. We can therefore make an order of magnitude estimate of the ray deflection angle caused by a galaxy by using Eq. (7.87):

$$\alpha \sim \frac{4GM}{bc^2} \sim \frac{4|\Phi|}{c^2} \sim 4 \times 2 \times \frac{3}{2} \times \frac{\langle v_\parallel^2 \rangle}{c^2} \sim \frac{12 \langle v_\parallel^2 \rangle}{c^2}. \tag{7.97}$$

This evaluates to $\alpha \sim 2$ arcsec for a typical galaxy velocity dispersion of ~ 300 km s^{-1}. The images can typically be resolved using radio and optical telescopes, but their separations are much less than the full angular sizes of distant galaxies and so the imaging is sensitive to the lens galaxy's structure. As the lens is typically far more complicated than a point mass, it is now convenient to measure the angle $\boldsymbol{\theta}$ relative to the reference ray that connects us to the source.

We can describe the optics of a galaxy gravitational lens by adapting the formalism that we developed for a point-mass lens in Ex. 7.17. We assume that there is a point source at $\boldsymbol{\beta}$ and consider paths designated by $\boldsymbol{\theta}$. The geometrical time delay is unchanged. In Ex. 7.17, we showed that the lens time delay for a point mass was proportional to the surface gravitational potential Φ_2. A distributed lens is handled by adding the potentials associated with all the point masses out of which it can be considered as being composed. In other words, we simply use the surface potential for the distributed mass in the galaxy,

$$t = t_{\text{geo}} + t_{\text{lens}} = \frac{d_{\text{eff}}}{2c}(\boldsymbol{\theta} - \boldsymbol{\beta})^2 - \frac{2}{c^3}\Phi_2 = \frac{d_{\text{eff}} \tilde{t}}{c}, \tag{7.98}$$

scaled time delay for lensing by galaxies

where the *scaled time delay t* is defined by $\tilde{t}(\boldsymbol{\theta}; \boldsymbol{\beta}) = \frac{1}{2}(\boldsymbol{\theta} - \boldsymbol{\beta})^2 - \Psi(\boldsymbol{\theta})$, and $\Psi = 2\Phi_2/(c^2 d_{\text{eff}})$. The quantity Ψ satisfies the 2-dimensional Poisson equation:

$$\nabla_{2,\theta}^2 \Psi = \frac{8\pi G \Sigma}{d_{\text{eff}} c^2}, \tag{7.99}$$

where Σ is the density of matter per unit solid angle, and the 2-dimensional laplacian describes differentiation with respect to the components of $\boldsymbol{\theta}$.[14]

As written, Eq. (7.98) describes all paths for a given source position, only a small number of which correspond to true rays. However, if, instead, we set $\boldsymbol{\beta} = 0$ and choose any convenient origin for $\boldsymbol{\theta}$ so that

---

14. The minimum surface density (expressed as mass per area) of a cosmologically distant lens needed to produce multiple images of a background source turns out to be ~ 1 g cm^{-2}. It is remarkable that such a seemingly small surface density operating on these scales can make such a large difference to our view of the universe. It is tempting to call this a "rule of thumb," because it is roughly the column density associated with one's thumb!

$$\tilde{t}(\boldsymbol{\theta}) = \frac{1}{2}\theta^2 - \Psi(\boldsymbol{\theta}), \qquad (7.100)$$

then three useful features of $\tilde{t}$ emerge:

uses of scaled time delay

- if there is an image at $\boldsymbol{\theta}$, then the source position is simply $\boldsymbol{\beta} = \nabla_\theta \tilde{t}$ [cf. Eq. (7.88)];
- the magnification tensor associated with this image can be calculated by taking the inverse of the Hessian matrix $\mathcal{H} = \nabla_\theta \nabla_\theta \tilde{t}$ [cf. Eq. (7.86)]; and
- the measured differences in the times of variation observed in multiple images of the same source are just the differences in $d_{\text{eff}}\tilde{t}/c$ evaluated at the image positions.[15]

Computing $\tilde{t}$ for a model of a putative lens galaxy allows one to assess whether background sources are being multiply imaged and, if so, to learn about the lens as well as the source.

EXERCISES

Exercise 7.20 *Problem: Catastrophe Optics of an Elliptical Gravitational Lens* T2
Consider an elliptical gravitational lens where the potential Ψ is modeled by

$$\Psi(\boldsymbol{\theta}) = (1 + A\theta_x^2 + 2B\theta_x\theta_y + C\theta_y^2)^q; \quad 0 < q < 1/2. \qquad (7.101)$$

Determine the generic form of the caustic surfaces, the types of catastrophe encountered, and the change in the number of images formed when a point source crosses these surfaces. Note that it is in the spirit of catastrophe theory *not* to compute exact expressions but to determine scaling laws and to understand the qualitative features of the images.

Exercise 7.21 *Challenge: Microlensing in a Galaxy* T2
Our discussion of microlensing assumed a single star and a circularly symmetric potential about it. This is usually a good approximation for stars in our galaxy. However, when the star is in another galaxy and the source is a background quasar (Figs. 7.19, 7.20), it is necessary to include the gravitational effects of the galaxy's other stars and its dark matter. Recast the microlensing analysis (Sec. 7.6.3) in the potential formulation of Eq. (7.98) and add *external magnification* and *external shear* contributions to

---

15. These differences can be used to measure the size and age of the universe. To order of magnitude, the relative delays are the times it takes light to cross the universe (or equivalently, the age of the universe, roughly 10 Gyr) times the square of the scattering angle (roughly 2 arcsec or $\sim 10^{-5}$ radians), which is roughly 1 year. This is very convenient for astronomers (see Fig. 7.20).

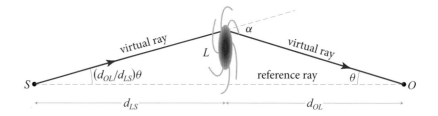

FIGURE 7.19 Geometry for gravitational lensing of a quasar source S by a galaxy lens L observed at O.

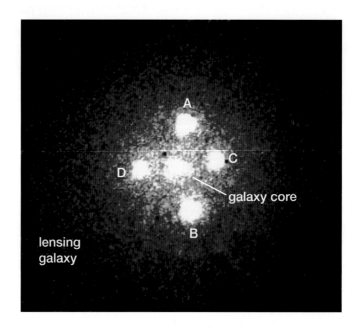

FIGURE 7.20 Gravitational lens in which a distant quasar, Q2237+0305, is quadruply imaged by an intervening galaxy. The four quasar images are denoted A, B, C, and D. The galaxy is much larger than the separation of the images (1–1.5 arcsec) but its bright central core is labeled. There is also a fifth, faint image coincident with this core. There are many examples of gravitational lenses like this, where the source region is very compact and variable so that the delay in the variation seen in the individual images can be used to measure the distance of the source and the size and age of the universe. In addition, microlensing-like variations in the images are induced by individual stars in the lens galaxy moving in front of the quasar. Analyzing these changes can be used to measure the proportion of stars and dark matter in galaxies. Adapted from image by Hubble Space Telescope. (NASA, ESA, STScI.)

Ψ that are proportional to $\theta_x^2 + \theta_y^2$ and $\theta_x^2 - \theta_y^2$, respectively. The latter will break the circular symmetry, and structurally stable caustics will be formed. Explore the behavior of these caustics as you vary the strength and sign of the magnification and shear contributions. Plot a few flux variations that might be observed.

7.7 Polarization

In our geometric-optics analyses thus far, we have either dealt with a scalar wave (e.g., a sound wave) or simply supposed that individual components of vector or tensor waves can be treated as scalars. For most purposes, this is indeed the case, and we continue to use this simplification in the following chapters. However, there are some important wave properties that are unique to vector (or tensor) waves. Most of these come under the heading of *polarization* effects. In Secs. 27.3, 27.4, and 27.5, we study polarization effects for (tensorial) gravitational waves. Here and in Secs. 10.5 and 28.6.1, we examine them for electromagnetic waves.

An electromagnetic wave's two polarizations are powerful tools for technology, engineering, and experimental physics. However, we forgo any discussion of this in the present chapter. Instead we focus solely on the geometric-optics propagation law for polarization (Sec. 7.7.1) and an intriguing aspect of it—the geometric phase (Sec. 7.7.2).

7.7.1 Polarization Vector and Its Geometric-Optics Propagation Law

A plane electromagnetic wave in a vacuum has its electric and magnetic fields **E** and **B** perpendicular to its propagation direction $\hat{\mathbf{k}}$ and perpendicular to each other. In a medium, **E** and **B** may or may not remain perpendicular to $\hat{\mathbf{k}}$, depending on the medium's properties. For example, an Alfvén wave has its vibrating magnetic field perpendicular to the background magnetic field, which can make an arbitrary angle with respect to $\hat{\mathbf{k}}$. By contrast, in the simplest case of an isotropic dielectric medium, where the dispersion relation has our standard dispersion-free form $\Omega = (c/\mathfrak{n})k$, the group and phase velocities are parallel to $\hat{\mathbf{k}}$, and **E** and **B** turn out to be perpendicular to $\hat{\mathbf{k}}$ and to each other—as in a vacuum. In this section, we confine attention to this simple situation and to linearly polarized waves, for which **E** oscillates linearly along a unit polarization vector $\hat{\mathbf{f}}$ that is perpendicular to $\hat{\mathbf{k}}$:

polarization vector

$$\mathbf{E} = Ae^{i\varphi}\hat{\mathbf{f}}, \qquad \hat{\mathbf{f}} \cdot \hat{\mathbf{k}} \equiv \hat{\mathbf{f}} \cdot \nabla\varphi = 0. \tag{7.102}$$

In the eikonal approximation, $Ae^{i\varphi} \equiv \psi$ satisfies the geometric-optics propagation laws of Sec. 7.3, and the polarization vector $\hat{\mathbf{f}}$, like the amplitude A, will propagate along the rays. The propagation law for $\hat{\mathbf{f}}$ can be derived by applying the eikonal approximation to Maxwell's equations, but it is easier to infer that law by simple physical reasoning:

1. Since $\hat{\mathbf{f}}$ is orthogonal to $\hat{\mathbf{k}}$ for a plane wave, it must also be orthogonal to $\hat{\mathbf{k}}$ in the eikonal approximation (which, after all, treats the wave as planar on lengthscales long compared to the wavelength).

2. If the ray is straight, then the medium, being isotropic, is unable to distinguish a slow right-handed rotation of $\hat{\mathbf{f}}$ from a slow left-handed rotation, so there will be no rotation at all: $\hat{\mathbf{f}}$ will continue always to point in the same direction (i.e., $\hat{\mathbf{f}}$ will be kept parallel to itself during transport along the ray).

3. If the ray bends, so $d\hat{\mathbf{k}}/ds \neq 0$ (where s is distance along the ray), then $\hat{\mathbf{f}}$ will have to change as well, so as always to remain perpendicular to $\hat{\mathbf{k}}$. The direction of $\hat{\mathbf{f}}$'s change must be $\hat{\mathbf{k}}$, since the medium, being isotropic, cannot provide any other preferred direction for the change. The magnitude of the change is determined by the requirement that $\hat{\mathbf{f}} \cdot \hat{\mathbf{k}}$ remain zero all along the ray and that $\hat{\mathbf{k}} \cdot \hat{\mathbf{k}} = 1$. This immediately implies that the propagation law for $\hat{\mathbf{f}}$ is

propagation law for polarization vector

$$\boxed{\frac{d\hat{\mathbf{f}}}{ds} = -\hat{\mathbf{k}}\left(\hat{\mathbf{f}} \cdot \frac{d\hat{\mathbf{k}}}{ds}\right).} \qquad (7.103)$$

This equation states that the vector $\hat{\mathbf{f}}$ is parallel-transported along the ray (cf. Fig. 7.5 in Sec. 24.3.3). Here "parallel transport" means: (i) Carry $\hat{\mathbf{f}}$ a short distance along the ray, keeping it parallel to itself in 3-dimensional space. Because of the bending of the ray and its tangent vector $\hat{\mathbf{k}}$, this will cause $\hat{\mathbf{f}}$ to no longer be perpendicular to $\hat{\mathbf{k}}$. (ii) Project $\hat{\mathbf{f}}$ perpendicular to $\hat{\mathbf{k}}$ by adding onto it the appropriate multiple of $\hat{\mathbf{k}}$. (The techniques of differential geometry for curved lines and surfaces, which we develop in Chaps. 24 and 25 in preparation for studying general relativity, give powerful mathematical tools for analyzing this parallel transport.)

7.7.2 Geometric Phase [T2]

We use the polarization propagation law (7.103) to illustrate a quite general phenomenon known as the *geometric phase*, or sometimes as the *Berry phase*, after Michael Berry who elucidated it. For further details and some history of this concept, see Berry (1990).

As a simple context for the geometric phase, consider a linearly polarized, monochromatic light beam that propagates in an optical fiber. Focus on the evolution of the polarization vector along the fiber's optic axis. We can imagine bending the fiber into any desired shape, thereby controlling the shape of the ray. The ray's shape in turn will control the propagation of the polarization via Eq. (7.103).

If the fiber and ray are straight, then the propagation law (7.103) keeps $\hat{\mathbf{f}}$ constant. If the fiber and ray are circular, then Eq. (7.103) causes $\hat{\mathbf{f}}$ to rotate in such a way as to always point along the generator of a cone, as shown in Fig. 7.21a. This polarization behavior, and that for any other ray shape, can be deduced with the aid of a unit sphere on which we plot the ray direction $\hat{\mathbf{k}}$ (Fig. 7.21b). For example, the ray directions at ray locations C and H of panel a are as shown in panel b of the figure. Notice that the trajectory of $\hat{\mathbf{k}}$ around the unit sphere is a great circle.

On the unit sphere we also plot the polarization vector $\hat{\mathbf{f}}$—one vector at each point corresponding to a ray direction. Because $\hat{\mathbf{f}} \cdot \hat{\mathbf{k}} = 0$, the polarization vectors are always tangent to the unit sphere. Notice that each $\hat{\mathbf{f}}$ on the unit sphere is identical in length and direction to the corresponding one in the physical space of Fig. 7.21a.

The parallel-transport law (7.103) keeps constant the angle α between $\hat{\mathbf{f}}$ and the trajectory of $\hat{\mathbf{k}}$ (i.e., the great circle in panel b of the figure). Translated back to

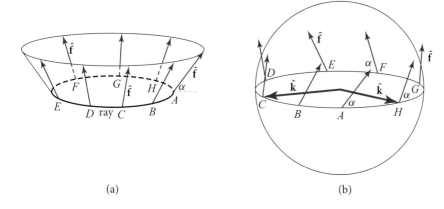

FIGURE 7.21 (a) The ray along the optic axis of a circular loop of optical fiber, and the polarization vector $\hat{\mathbf{f}}$ that is transported along the ray by the geometric-optics transport law $d\hat{\mathbf{f}}/ds = -\hat{\mathbf{k}}(\hat{\mathbf{f}} \cdot d\hat{\mathbf{k}}/ds)$. (b) The polarization vector $\hat{\mathbf{f}}$ drawn on the unit sphere. The vector from the center of the sphere to each of the points $A, B, \ldots, H$ is the ray's propagation direction $\hat{\mathbf{k}}$, and the polarization vector (which is orthogonal to $\hat{\mathbf{k}}$ and thus tangent to the sphere) is identical to that in the physical space of the ray (panel a).

panel a, this constancy of α implies that the polarization vector points always along the generators of the cone, whose opening angle is $\pi/2 - \alpha$, as shown.

Next let the fiber and its central axis (the ray) be helical as shown in Fig. 7.22a. In this case, the propagation direction $\hat{\mathbf{k}}$ rotates, always maintaining the same angle θ to the vertical direction, and correspondingly its trajectory on the unit sphere of Fig. 7.22b is a circle of constant polar angle θ. Therefore (as one can see, e.g., with the aid of a large globe of Earth and a pencil transported around a circle of latitude $90° - \theta$), the parallel-transport law dictates that the angle α between $\hat{\mathbf{f}}$ and the circle *not* remain constant, but instead rotate at the rate

$$d\alpha/d\phi = \cos\theta. \tag{7.104}$$

Here ϕ is the angle (longitude on the globe) around the circle. This is the same propagation law as for the direction of swing of a Foucault pendulum as Earth turns (cf. Box 14.5), and for the same reason: the gyroscopic action of the Foucault pendulum is described by parallel transport of its plane along Earth's spherical surface.

In the case where θ is arbitrarily small (a nearly straight ray), Eq. (7.104) says $d\alpha/d\phi = 1$. This is easily understood: although $\hat{\mathbf{f}}$ remains arbitrarily close to constant, the trajectory of $\hat{\mathbf{k}}$ turns rapidly around a tiny circle about the pole of the unit sphere, so α changes rapidly—by a total amount $\Delta\alpha = 2\pi$ after one trip around the pole, $\Delta\phi = 2\pi$; whence $d\alpha/d\phi = \Delta\alpha/\Delta\phi = 1$. For any other helical pitch angle θ, Eq. (7.104) says that during one round trip, α will change by an amount $2\pi \cos\theta$ that lags behind its change for a tiny circle (nearly straight ray) by the lag angle $\alpha_{\text{lag}} = 2\pi(1 - \cos\theta)$, which is also the solid angle $\Delta\Omega$ enclosed by the path of $\hat{\mathbf{k}}$ on the unit sphere:

$$\alpha_{\text{lag}} = \Delta\Omega. \tag{7.105}$$

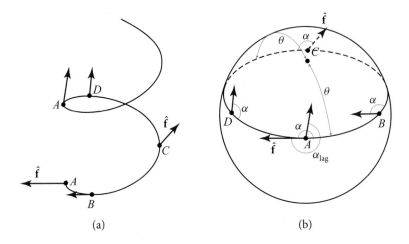

FIGURE 7.22 (a) The ray along the optic axis of a helical loop of optical fiber, and the polarization vector $\hat{\mathbf{f}}$ that is transported along this ray by the geometric-optics transport law $d\hat{\mathbf{f}}/ds = -\hat{\mathbf{k}}(\hat{\mathbf{f}} \cdot d\hat{\mathbf{k}}/ds)$. The ray's propagation direction $\hat{\mathbf{k}}$ makes an angle $\theta = 73°$ to the vertical direction. (b) The trajectory of $\hat{\mathbf{k}}$ on the unit sphere (a circle with polar angle $\theta = 73°$), and the polarization vector $\hat{\mathbf{f}}$ that is parallel transported along that trajectory. The polarization vectors in panel a are deduced from the parallel-transport law demonstrated in panel b. The lag angle $\alpha_{\text{lag}} = 2\pi(1 - \cos\theta) = 1.42\pi$ is equal to the solid angle contained inside the trajectory of $\hat{\mathbf{k}}$ (the $\theta = 73°$ circle).

(For the circular ray of Fig. 7.21, the enclosed solid angle is $\Delta\Omega = 2\pi$ steradians, so the lag angle is 2π radians, which means that $\hat{\mathbf{f}}$ returns to its original value after one trip around the optical fiber, in accord with the drawings in the figure.)

lag angle for polarization vector in an optical fiber

Remarkably, Eq. (7.105) is true for light propagation along an optical fiber of any shape: if the light travels from one point on the fiber to another at which the tangent vector $\hat{\mathbf{k}}$ has returned to its original value, then the lag angle is given by the enclosed solid angle on the unit sphere, Eq. (7.105).

By itself, the relationship $\alpha_{\text{lag}} = \Delta\Omega$ is merely a cute phenomenon. However, it turns out to be just one example of a very general property of both classical and quantum mechanical systems when they are forced to make slow, *adiabatic* changes described by circuits in the space of parameters that characterize them. In the more general case, one focuses on a phase lag rather than a direction-angle lag. We can easily translate our example into such a phase lag.

geometric phase change

The apparent rotation of $\hat{\mathbf{f}}$ by the lag angle $\alpha_{\text{lag}} = \Delta\Omega$ can be regarded as an advance of the phase of one circularly polarized component of the wave by $\Delta\Omega$ and a phase retardation of the other circular polarization by the same amount. Thus the phase of a circularly polarized wave will change, after one circuit around the fiber's helix, by an amount equal to the usual phase advance $\Delta\varphi = \int \mathbf{k} \cdot d\mathbf{x}$ (where $d\mathbf{x}$ is displacement along the fiber) plus an extra *geometric* phase change $\pm\Delta\Omega$, where the sign is given by the sense of circular polarization. This type of geometric phase change is found quite generally, when classical vector or tensor waves propagate through

408   Chapter 7. Geometric Optics

backgrounds that change slowly, either temporally or spatially. The phases of the wave functions of quantum mechanical particles with spin behave similarly.

Exercise 7.22 *Derivation: Parallel Transport* T2
Use the parallel-transport law (7.103) to derive the relation (7.104).

Exercise 7.23 *Problem: Martian Rover* T2
A Martian Rover is equipped with a single gyroscope that is free to pivot about the direction perpendicular to the plane containing its wheels. To climb a steep hill on Mars without straining its motor, it must circle the summit in a decreasing spiral trajectory. Explain why there will be an error in its measurement of North after it has reached the summit. Could it be programmed to navigate correctly? Will a stochastic error build up as it traverses a rocky terrain?

Bibliographic Note

Modern textbooks on optics deal with the geometric-optics approximation only for electromagnetic waves propagating through a dispersion-free medium. Accordingly, they typically begin with Fermat's principle and then treat in considerable detail the paraxial approximation, applications to optical instruments, and sometimes the human eye. There is rarely any mention of the eikonal approximation or of multiple images and caustics. Examples of texts of this sort that we like are Bennett (2008), Ghatak (2010), and Hecht (2017). For a far more thorough treatise on geometric optics of scalar and electromagnetic waves in isotropic and anisotropic dielectric media, see Kravtsov (2005). A good engineering-oriented text with many contemporary applications is Iizuka (1987).

We do not know of textbooks that treat the eikonal approximation to the degree of generality used in this chapter, though some should, since it has applications to all types of waves (many of which are explored later in this book). For the eikonal approximation specialized to Maxwell's equations, see Kravtsov (2005) and the classic treatise on optics by Born and Wolf (1999), which in this new edition has modern updates by a number of other authors. For the eikonal approximation specialized to the Schrödinger equation and its connection to Hamilton-Jacobi theory, see most any quantum mechanics textbook (e.g., Griffiths, 2004).

Multiple-image formation and caustics are omitted from most standard optics textbooks, except for a nice but out-of-date treatment in Born and Wolf (1999). Much better are the beautiful review by Berry and Upstill (1980) and the much more thorough treatments in Kravtsov (2005) and Nye (1999). For an elementary mathematical treatment of catastrophe theory, we like Saunders (1980). For a pedagogical treatise on gravitational lenses, see Schneider, Ehlers, and Falco (1992). Finally, for some history and details of the geometric phase, see Berry (1990).

CHAPTER EIGHT

Diffraction

I have seen—without any illusion—three broad stripes in the spectrum of Sirius, which seem to have no similarity to those of sunlight.

JOSEPH VON FRAUNHOFER (1814-1815)

8.1 Overview

The previous chapter was devoted to the classical mechanics of wave propagation. We showed how a classical wave equation can be solved in the short-wavelength (eikonal) approximation to yield Hamilton's dynamical equations for its rays. When the medium is time independent (as we require in this chapter), we showed that the frequency of a wave packet is constant, and we imported a result from classical mechanics—the principle of stationary action—to show that the true geometric-optics rays coincide with paths along which the action (the phase) is stationary [Eqs. (7.44) and (7.45a) and associated discussion]. Our physical interpretation of this result was that the waves do indeed travel along every path, from some source to a point of observation, where they are added together, but they only give a significant net contribution when they can add coherently in phase—along the true rays [Eq. (7.45b)]. Essentially, this is Huygens' model of wave propagation, or, in modern language, a *path integral*.

Huygens' principle asserts that every point on a wavefront acts as a source of secondary waves that combine so their envelope constitutes the advancing wavefront. This principle must be supplemented by two ancillary conditions: that the secondary waves are only formed in the forward direction, not backward, and that a $\pi/2$ phase shift be introduced into the secondary wave. The reason for the former condition is obvious, that for the latter, less so. We discuss both together with the formal justification of Huygens' construction in Sec. 8.2.

We begin our exploration of the "wave mechanics" of optics in this chapter and continue it in Chaps. 9 and 10. Wave mechanics differs increasingly from geometric optics as the reduced wavelength λ increases relative to the lengthscales $\mathcal{R}$ of the phase fronts and $\mathcal{L}$ of the medium's inhomogeneities. The number of paths that can combine constructively increases, and the rays that connect two points become blurred. In quantum mechanics, we recognize this phenomenon as the uncertainty principle, and it is just as applicable to photons as to electrons.

> **BOX 8.1. READERS' GUIDE**
>
> - This chapter depends substantially on Secs. 7.1–7.4 (geometric optics).
> - In addition, Sec. 8.6 of this chapter (on diffraction at a caustic) depends on Sec. 7.5.
> - Chapters 9 and 10 depend substantially on Secs. 8.1–8.5 of this chapter.
> - Nothing else in this book relies on this chapter.

Solving the wave equation exactly is very hard, except in simple circumstances. Geometric optics is one approximate method of solving it—a method that works well in the short-wavelength limit. In this chapter and the next two, we develop approximate techniques that work when the wavelength becomes longer and geometric optics fails.

In this book, we make a somewhat artificial distinction between phenomena that arise when an effectively infinite number of propagation paths are involved (which we call *diffraction* and describe in this chapter) and those when a few paths, or, more correctly, a few tight bundles of rays are combined (which we term *interference,* and whose discussion we defer to the next chapter).

In Sec. 8.2, we present the Fresnel-Helmholtz-Kirchhoff theory that underlies most elementary discussions of diffraction. We then distinguish between Fraunhofer diffraction (the limiting case when spreading of the wavefront mandated by the uncertainty principle is important) and Fresnel diffraction (where wavefront spreading is a modest effect, and geometric optics is beginning to work, at least roughly). In Sec. 8.3, we illustrate Fraunhofer diffraction by computing the angular resolution of the Hubble Space Telescope, and in Sec. 8.4, we analyze Fresnel diffraction and illustrate it using zone plates and lunar occultation of radio waves.

Many contemporary optical devices can be regarded as linear systems that take an input wave signal and transform it into a linearly related output. Their operation, particularly as image-processing devices, can be considerably enhanced by processing the signal in the spatial Fourier domain, a procedure known as spatial filtering. In Sec. 8.5 we introduce a tool for analyzing such devices: paraxial Fourier optics—a close analog of the paraxial geometric optics of Sec. 7.4. Using paraxial Fourier optics, we develop the theory of image processing by spatial filters and use it to discuss various types of filters and the phase-contrast microscope. We also use Fourier optics to develop the theory of Gaussian beams—the kind of light beam produced by lasers when (as is usual) their optically resonating cavities have spherical mirrors. Finally, in Sec. 8.6 we analyze diffraction near a caustic of a wave's phase field, a location where geometric optics predicts a divergent magnification of the wave (Sec. 7.5). As we shall see, diffraction keeps the magnification finite and produces an oscillating energy-flux pattern (interference fringes).

8.2 Helmholtz-Kirchhoff Integral

In this section, we derive a formalism for describing diffraction. We restrict attention to the simplest (and, fortunately, the most widely useful) case: a monochromatic scalar wave,

$$\Psi = \psi(\mathbf{x})e^{-i\omega t}, \tag{8.1a}$$

with field variable ψ that satisfies the Helmholtz equation

$$\nabla^2 \psi + k^2 \psi = 0, \quad \text{with } k = \omega/c, \tag{8.1b}$$

Helmholtz equation

except at boundaries. Generally, Ψ will represent a real-valued physical quantity, but for mathematical convenience we give it a complex representation and take the real part of Ψ when making contact with physical measurements. This is in contrast to a quantum mechanical wave function satisfying the Schrödinger equation, which is an intrinsically complex function. We assume that the wave in Eqs. (8.1) is monochromatic (constant ω) and nondispersive, and that the medium is isotropic and homogeneous (phase speed equal to group speed, and both with a constant value C, so k is also constant). Each of these assumptions can be relaxed, but with some technical penalty.

The scalar formalism that we develop based on Eq. (8.1b) is fully valid for weak sound waves in a homogeneous fluid or solid (e.g., air; Secs. 12.2 and 16.5). It is also quite accurate, but not precisely so, for the most widely used application of diffraction theory: electromagnetic waves in a vacuum or in a medium with a homogeneous dielectric constant. In this case ψ can be regarded as one of the Cartesian components of the electric field vector, such as E_x (or equally well, a Cartesian component of the vector potential or the magnetic field vector). In a vacuum or in a homogeneous dielectric medium, Maxwell's equations imply that this $\psi = E_x$ satisfies the scalar wave equation exactly and hence, for fixed frequency, the Helmholtz equation (8.1b). However, when the wave hits a boundary of the medium (e.g., the edge of an aperture, or the surface of a mirror or lens), its interaction with the boundary can couple the various components of $\mathbf{E}$, thereby invalidating the simple scalar theory we develop in this chapter. Fortunately, this polarizational coupling is usually weak in the paraxial (small angle) limit, and also under a variety of other circumstances, thereby making our simple scalar formalism quite accurate.[1]

applications of scalar diffraction formalism

---

1. For a formulation of diffraction that takes account of these polarization effects, see, e.g., Born and Wolf (1999, Chap. 11).

The Helmholtz equation (8.1b) is an elliptic, linear, partial differential equation, and we can thus express the value $\psi_\mathcal{P}$ of ψ at any point $\mathcal{P}$ inside some closed surface $\mathcal{S}$ as an integral over $\mathcal{S}$ of some linear combination of ψ and its normal derivative; see Fig. 8.1. To derive such an expression, we proceed as follows. First, we introduce in the interior of $\mathcal{S}$ a second solution of the Helmholtz equation:

spherical wave

$$\psi_0 = \frac{e^{ikr}}{r}. \tag{8.2}$$

This is a spherical wave originating from the point $\mathcal{P}$, and r is the distance from $\mathcal{P}$ to the point where ψ_0 is evaluated. Next we apply Gauss's theorem, Eq. (1.28a), to the vector field $\psi \nabla \psi_0 - \psi_0 \nabla \psi$ and invoke Eq. (8.1b), thereby obtaining

$$\int_{\mathcal{S} \cup \mathcal{S}_o} (\psi \nabla \psi_0 - \psi_0 \nabla \psi) \cdot d\mathbf{\Sigma} = -\int_\mathcal{V} (\psi \nabla^2 \psi_0 - \psi_0 \nabla^2 \psi) dV$$
$$= 0. \tag{8.3}$$

Here we have introduced a small sphere $\mathcal{S}_o$ of radius r_o surrounding $\mathcal{P}$ (Fig. 8.1); $\mathcal{V}$ is the volume between the two surfaces $\mathcal{S}_o$ and $\mathcal{S}$ (so $\mathcal{S} \cup \mathcal{S}_o$ is the boundary $\partial \mathcal{V}$ of $\mathcal{V}$). For future convenience we have made an unconventional choice of direction for the integration element $d\mathbf{\Sigma}$: it points into $\mathcal{V}$ instead of outward, thereby producing the minus sign in the second expression in Eq. (8.3). As we let the radius r_o decrease to zero, we find that $\psi \nabla \psi_0 - \psi_0 \nabla \psi \to -\psi_\mathcal{P}/r_o^2 \, \mathbf{e}_r + O(1/r_o)$, and so the integral over $\mathcal{S}_o$ becomes $-4\pi \psi_\mathcal{P}$ (where $\psi_\mathcal{P}$ is the value of ψ at $\mathcal{P}$). Rearranging, we obtain

Helmholtz-Kirchhoff integral

$$\boxed{\psi_\mathcal{P} = \frac{1}{4\pi} \int_\mathcal{S} \left(\psi \nabla \frac{e^{ikr}}{r} - \frac{e^{ikr}}{r} \nabla \psi \right) \cdot d\mathbf{\Sigma}.} \tag{8.4}$$

Equation (8.4), known as the *Helmholtz-Kirchhoff integral*, is the promised expression for the field ψ at some point $\mathcal{P}$ in terms of a linear combination of its value and normal derivative on a surrounding surface. The specific combination of ψ and $d\mathbf{\Sigma} \cdot \nabla \psi$ that appears in this formula is perfectly immune to contributions from any wave that might originate at $\mathcal{P}$ and pass outward through $\mathcal{S}$ (any outgoing wave). The integral thus is influenced only by waves that enter $\mathcal{V}$ through $\mathcal{S}$, propagate through $\mathcal{V}$, and then leave through $\mathcal{S}$. (There cannot be sources inside $\mathcal{S}$, except conceivably at $\mathcal{P}$, because we assumed ψ satisfies the source-free Helmholtz equation throughout $\mathcal{V}$.) If $\mathcal{P}$ is many wavelengths away from the boundary $\mathcal{S}$, then to high accuracy, the integral is influenced by the waves ψ only when they are entering through $\mathcal{S}$ (when they are incoming) and not when they are leaving (outgoing). This fact is important for applications, as we shall see.

8.2.1 Diffraction by an Aperture

Now let us suppose that some aperture $\mathcal{Q}$, with size much larger than a wavelength but much smaller than the distance to $\mathcal{P}$, is illuminated by a distant wave source (left side

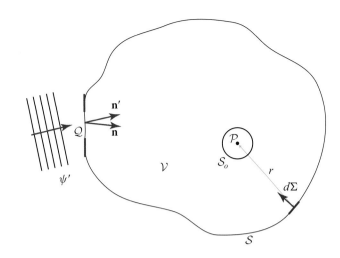

FIGURE 8.1 Geometry for the Helmholtz-Kirchhoff integral (8.4), which expresses the field $\psi_\mathcal{P}$ at the point $\mathcal{P}$ in terms of an integral of the field and its normal derivative over the surrounding surface $\mathcal{S}$. The small sphere $\mathcal{S}_o$ is centered on the observation point $\mathcal{P}$, and $\mathcal{V}$ is the volume bounded by $\mathcal{S}$ and $\mathcal{S}_o$. (The aperture $\mathcal{Q}$, the vectors $\mathbf{n}$ and $\mathbf{n}'$ at the aperture, the incoming wave ψ', and the point $\mathcal{P}'$ are irrelevant to the formulation of the Helmholtz-Kirchhoff integral, but appear in applications later in this chapter—initially in Sec. 8.2.1.)

of Fig. 8.1).[2] Let the surface $\mathcal{S}$ pass through the aperture $\mathcal{Q}$, and denote by ψ' the wave incident on $\mathcal{Q}$. We assume that the diffracting aperture has a local and linear effect on ψ'. More specifically, we suppose that the wave transmitted through the aperture is given by

$$\psi_\mathcal{Q} = \mathsf{t}\, \psi', \tag{8.5}$$

where t is a complex transmission function that varies over the aperture. In practice, t is usually zero (completely opaque region) or unity (completely transparent region). However, t can also represent a variable phase factor when, for example, the aperture consists of a medium (lens) of variable thickness and of different refractive index from that of the homogeneous medium outside the aperture—as is the case in microscopes, telescopes, eyes, and other optical devices.

complex transmission function t

What this formalism does not allow is that $\psi_\mathcal{Q}$ at any point on the aperture be influenced by the wave's interaction with other parts of the aperture. For this reason, not only the aperture, but also any structure that it contains must be many wavelengths across. To give a specific example of what might go wrong, suppose that electromagnetic radiation is normally incident on a wire grid. A surface current will

limitations of this scalar diffraction formalism

---

2. If the aperture were comparable to a wavelength in size, or if part of it were only a few wavelengths from $\mathcal{P}$, then polarizational coupling effects at the aperture would be large (Born and Wolf, 1999). Our assumption avoids this complication.

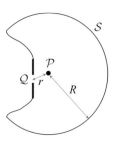

FIGURE 8.2 Geometry for deriving Eq. (8.6) for diffraction by an aperture.

be induced in each wire by the wave's electric field, and that current will produce a secondary wave that cancels the primary wave immediately behind the wire, thereby "eclipsing" the wave. If the wire separation is not large compared to a wavelength, then the secondary wave from the current flowing in one wire will drive currents in adjacent wires, thereby modifying their secondary waves and disturbing the eclipse in a complex, polarization-dependent manner. Such modifications are negligible if the wires are many wavelengths apart.

Let us now use the Helmholtz-Kirchhoff integral (8.4) to compute the field at $\mathcal{P}$ due to the wave $\psi_\mathcal{Q} = \mathfrak{t}\,\psi'$ transmitted through the aperture. Let the (imaginary) surface $\mathcal{S}$ in Fig. 8.1 comprise the aperture $\mathcal{Q}$, a sphere of radius $R \gg r$ centered on $\mathcal{P}$, and an extension of the aperture to meet the sphere as sketched in Fig. 8.2. Assume that the only incoming waves are those passing through the aperture. Then, for the reason discussed in the paragraph following Eq. (8.4), when the incoming waves subsequently pass on outward through the spherical part of $\mathcal{S}$, they contribute negligibly to the integral (8.4). Also, with the extension from aperture to sphere swinging back as drawn (Fig. 8.2), the contribution from the extension will also be negligible. Therefore, the only contribution is from the aperture itself.[3]

Because $kr \gg 1$, we can write $\nabla(e^{ikr}/r) \simeq -ik\mathbf{n}e^{ikr}/r$ on the aperture. Here $\mathbf{n}$ is a unit vector pointing toward $\mathcal{P}$ (Fig. 8.1). Similarly, we write $\nabla\psi \simeq ik\mathfrak{t}\,\mathbf{n}'\psi'$, where $\mathbf{n}'$ is a unit vector along the direction of propagation of the incident wave (and where our assumption that anything in the aperture varies on scales long compared to $\lambdabar = 1/k$ permits us to ignore the gradient of $\mathfrak{t}$). Inserting these gradients into the Helmholtz-Kirchhoff integral, we obtain

---

3. Actually, the incoming waves will diffract around the edge of the aperture onto the back side of the screen that bounds the aperture (i.e., the side facing $\mathcal{P}$), and this diffracted wave will contribute to the Helmholtz-Kirchhoff integral in a polarization-dependent way (see Born and Wolf, 1999, Chap. 11). However, because the diffracted wave decays along the screen with an e-folding length of order a wavelength, its contribution will be negligible if the aperture is many wavelengths across and $\mathcal{P}$ is many wavelengths away from the edge of the aperture, as we have assumed.

$$\boxed{\psi_\mathcal{P} = -\frac{ik}{2\pi} \int_\mathcal{Q} d\mathbf{\Sigma} \cdot \left(\frac{\mathbf{n}+\mathbf{n}'}{2}\right) \frac{e^{ikr}}{r} t\, \psi'.} \quad (8.6)$$

wave field behind an aperture

Equation (8.6) can be used to compute the wave from a small aperture $\mathcal{Q}$ at any point $\mathcal{P}$ in the far field. It has the form of an integral transform of the incident field variable ψ', where the integral is over the area of the aperture. The kernel of the transform is the product of several factors. The factor $1/r$ guarantees that the wave's energy flux falls off as the inverse square of the distance to the aperture, as we might have expected. The phase factor $-ie^{ikr}$ advances the phase of the wave by an amount equal to the optical path length between the element $d\mathbf{\Sigma}$ of the aperture and $\mathcal{P}$, minus $\pi/2$ (the phase of $-i$). The amplitude and phase of the wave ψ' can also be changed by the transmission function t. Finally there is the geometric factor $d\hat{\mathbf{\Sigma}} \cdot (\mathbf{n}+\mathbf{n}')/2$ (with $d\hat{\mathbf{\Sigma}}$ the unit vector normal to the aperture). This is known as the *obliquity factor*, and it ensures that the waves from the aperture propagate only forward with respect to the original wave and not backward (i.e., not in the direction $\mathbf{n} = -\mathbf{n}'$). More specifically, this factor prevents the backward-propagating secondary wavelets in a Huygens construction from reinforcing one another to produce a back-scattered wave. When dealing with paraxial Fourier optics (Sec. 8.5), we can usually set the obliquity factor to unity.

obliquity factor

It is instructive to specialize to a point source seen through a small diffracting aperture. If we suppose that the source has unit strength and is located at $\mathcal{P}'$, a distance r' before $\mathcal{Q}$ (Fig. 8.1), then $\psi' = -e^{ikr'}/(4\pi r')$ (our definition of unit strength), and $\psi_\mathcal{P}$ can be written in the symmetric form

$$\boxed{\psi_\mathcal{P} = \int_\mathcal{Q} \left(\frac{e^{ikr}}{4\pi r}\right) it\, (\mathbf{k}+\mathbf{k}') \cdot d\mathbf{\Sigma} \left(\frac{e^{ikr'}}{4\pi r'}\right).} \quad (8.7)$$

propagator through an aperture

We can think of this expression as the Green's function response at $\mathcal{P}$ to a delta function source at $\mathcal{P}'$. Alternatively, we can regard it as a *propagator* from $\mathcal{P}'$ to $\mathcal{P}$ by way of the aperture (co-opting a concept that was first utilized in quantum mechanics but is also useful in classical optics).

8.2.2 Spreading of the Wavefront: Fresnel and Fraunhofer Regions

8.2.2

Equation (8.6) [or Eq. (8.7)] gives a general prescription for computing the diffraction pattern from an illuminated aperture. It is commonly used in two complementary limits, called "Fraunhofer" and "Fresnel."

Suppose that the aperture has linear size a (as in Fig. 8.3) and is roughly centered on the geometric ray from the source point $\mathcal{P}'$ to the field point $\mathcal{P}$, so $\mathbf{n} \cdot d\hat{\mathbf{\Sigma}} = \mathbf{n}' \cdot d\hat{\mathbf{\Sigma}} \simeq 1$. Consider the variations of the phase φ of the contributions to $\psi_\mathcal{P}$ that come from various places in the aperture. Using elementary trigonometry, we can estimate that locations on the aperture's opposite edges produce phases at $\mathcal{P}$ that differ by $\Delta \varphi =$

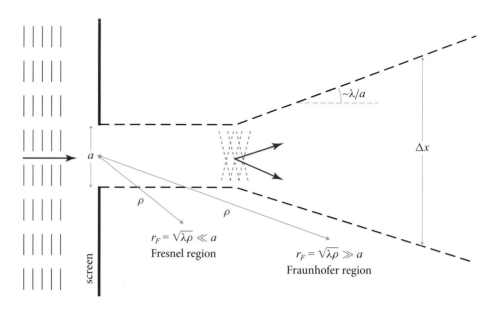

FIGURE 8.3 Fraunhofer and Fresnel diffraction. The dashed line is an approximation to the edge of the aperture's shadow.

$k(\rho_2 - \rho_1) \sim ka^2/2\rho$, where ρ_1 and ρ_2 are the distances of $\mathcal{P}$ from the two edges of the aperture, and ρ is the distance from the center of the aperture. There are two limiting regions for ρ, depending on whether $\mathcal{P}$'s so-called *Fresnel length*,

Fresnel length

$$\boxed{r_F \equiv \left(\frac{2\pi\rho}{k}\right)^{1/2} = (\lambda\rho)^{1/2}} \qquad (8.8)$$

regions of Fraunhofer and Fresnel diffraction

(a surrogate for the distance ρ), is large or small compared to the aperture. Notice that $(a/r_F)^2 = ka^2/(2\pi\rho) \sim \Delta\varphi/\pi$. Therefore, when $r_F \gg a$ (field point far from the aperture), the phase variation $\Delta\varphi$ across the aperture is $\ll \pi$ and can be ignored, so the contributions at $\mathcal{P}$ from different parts of the aperture are essentially in phase with one another. This is the *Fraunhofer* region. When $r_F \ll a$ (near the aperture), the phase variation is $\Delta\varphi \gg \pi$ and therefore is of utmost importance in determining the observed energy-flux pattern $F \propto |\psi_\mathcal{P}|^2$. This is the *Fresnel* region; see Fig. 8.3.

We can use an argument familiar, perhaps, from quantum mechanics to deduce the qualitative form of the flux patterns in these two regions. For simplicity, let the incoming wave be planar [r' huge in Eq. (8.7)]; let it propagate perpendicular to the aperture (as shown in Fig. 8.3); and let the aperture be empty, so $t = 1$ inside the aperture. Then geometric optics (photons treated like classical particles) would predict that the aperture's edge will cast a sharp shadow; the wave leaves the plane of the aperture as a beam with a sharp edge. However, wave optics insists that the transverse localization of the wave into a region of size $\Delta x \sim a$ must produce a spread in its transverse wave vector, $\Delta k_x \sim 1/a$ (a momentum uncertainty $\Delta p_x = \hbar\Delta k_x \sim \hbar/a$ in the language of the Heisenberg uncertainty principle). This uncertain transverse wave vector produces, after propagating a distance ρ, a corresponding uncertainty $(\Delta k_x/k)\rho \sim r_F^2/a$ in the beam's transverse size. This uncertainty superposes incoherently on the aperture-induced size a of the beam to produce a net transverse beam size of

$$\Delta x \sim \sqrt{a^2 + (r_F^2/a)^2}$$
$$\sim \begin{cases} a & \text{if } r_F \ll a \text{ (i.e., } \rho \ll a^2/\lambda; \text{ Fresnel region)}, \\ \left(\frac{\lambda}{a}\right)\rho & \text{if } r_F \gg a \text{ (i.e., } \rho \gg a^2/\lambda; \text{ Fraunhofer region)}. \end{cases} \quad (8.9)$$

Therefore, in the nearby, Fresnel region, the aperture creates a beam whose edges have the same shape and size as the aperture itself, and are reasonably sharp (but with some oscillatory blurring, associated with wavepacket spreading, that we analyze below); see Fig. 8.4. Thus, in the Fresnel region the field behaves approximately as one would predict using geometric optics. By contrast, in the more distant, Fraunhofer region, wavefront spreading causes the transverse size of the entire beam to grow linearly with distance; as illustrated in Fig. 8.4, the flux pattern differs markedly from the aperture's shape. We analyze the distant (Fraunhofer) region in Sec. 8.3 and the near (Fresnel) region in Sec. 8.4.

Fresnel and Fraunhofer diffraction regions and patterns

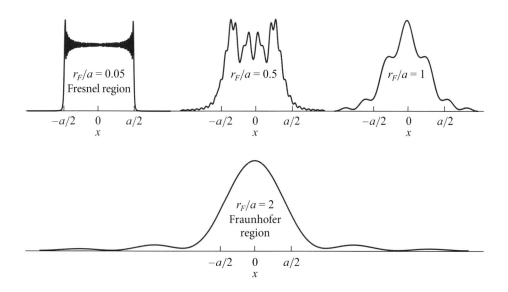

FIGURE 8.4 The 1-dimensional energy-flux diffraction pattern $|\psi|^2$ produced on a screen a distance z from a slit with $\mathfrak{t}(x) = 1$ for $|x| < a/2$ and $\mathfrak{t}(x) = 0$ for $|x| > a/2$. Four patterns are shown, each for a different value of $r_F/a = \sqrt{\lambda z}/a$. For $r_F/a = 0.05$ (very near the slit; extreme Fresnel region), the flux distribution resembles the slit itself: sharp edges at $x = \pm a/2$, but with damped oscillations (interference fringes) near the edges. For $r_F/a = 1$ (beginning of Fraunhofer region) there is a bright central peak and low-brightness, oscillatory side bands. As r_F/a increases from 0.05 to 2, the pattern transitions (quite rapidly between 0.5 and 2) from the Fresnel to the Fraunhofer pattern. These flux distributions are derived in Ex. 8.8.

8.2 Helmholtz-Kirchhoff Integral

8.3 Fraunhofer Diffraction

Consider the Fraunhofer region of strong wavefront spreading, $r_F \gg a$, and for simplicity specialize to the case of an incident plane wave with wave vector **k** orthogonal to the aperture plane; see Fig. 8.5. Regard the line along **k** through the center of the aperture $\mathcal{Q}$ as the optic axis and identify points in the aperture by their transverse 2-dimensional vectorial separation **x** from that axis. Identify $\mathcal{P}$ by its distance ρ from the aperture center and its 2-dimensional transverse separation $\rho\boldsymbol{\theta}$ from the optic axis, and restrict attention to small-angle diffraction $|\boldsymbol{\theta}| \ll 1$. Then the geometric path length between $\mathcal{P}$ and a point **x** on $\mathcal{Q}$ [the length denoted r in Eq. (8.6)] can be expanded as

$$\text{Path length} = r = (\rho^2 - 2\rho\mathbf{x}\cdot\boldsymbol{\theta} + x^2)^{1/2} \simeq \rho - \mathbf{x}\cdot\boldsymbol{\theta} + \frac{x^2}{2\rho} + \ldots; \quad (8.10)$$

cf. Fig. 8.5. The first term in this expression, ρ, just contributes an **x**-independent phase $e^{ik\rho}$ to the $\psi_\mathcal{P}$ of Eq. (8.6). The third term, $x^2/(2\rho)$, contributes a phase variation that is $\ll 1$ in the Fraunhofer region (but that will be important in the Fresnel region; Sec. 8.4). Therefore, in the Fraunhofer region we can retain just the second term, $-\mathbf{x}\cdot\boldsymbol{\theta}$, and write Eq. (8.6) in the form

Fraunhofer diffraction of a plane wave

$$\boxed{\psi_\mathcal{P}(\boldsymbol{\theta}) \propto \int e^{-ik\mathbf{x}\cdot\boldsymbol{\theta}} \mathfrak{t}(\mathbf{x})\, d\Sigma \equiv \tilde{\mathfrak{t}}(\boldsymbol{\theta}),} \quad (8.11a)$$

where $d\Sigma$ is the surface-area element in the aperture plane, and we have dropped a constant phase factor and constant multiplicative factors. Thus, $\psi_\mathcal{P}(\boldsymbol{\theta})$ in the Fraunhofer region is given by the 2-dimensional Fourier transform, denoted $\tilde{\mathfrak{t}}(\boldsymbol{\theta})$, of the transmission function $\mathfrak{t}(\mathbf{x})$, with **x** made dimensionless in the transform by multiplying by $k = 2\pi/\lambda$. [Note that with this optics convention for the Fourier transform, the inverse transform is

optics convention for Fourier transform

$$\mathfrak{t}(\mathbf{x}) = \left(\frac{k}{2\pi}\right)^2 \int e^{ik\mathbf{x}\cdot\boldsymbol{\theta}} \tilde{\mathfrak{t}}(\boldsymbol{\theta})\, d\Omega, \quad (8.11b)$$

where $d\Omega = d\theta_x d\theta_y$ is the solid angle.]

The flux distribution of the diffracted wave [Eq. (8.11a)] is

Fraunhofer diffracted flux

$$\boxed{F(\boldsymbol{\theta}) = \overline{(\mathfrak{R}[\psi_\mathcal{P}(\boldsymbol{\theta})e^{-i\omega t}])^2} = \tfrac{1}{2}|\psi_\mathcal{P}(\boldsymbol{\theta})|^2 \propto |\tilde{\mathfrak{t}}(\boldsymbol{\theta})|^2,} \quad (8.12)$$

where $\mathfrak{R}$ means take the real part, and the bar means average over time.

As an example, the bottom curve in Fig. 8.4 (the curve $r_F = 2a$) shows the flux distribution $F(\theta)$ from a slit

$$\mathfrak{t}(x) = H_1(x) \equiv \begin{cases} 1 & |x| < a/2 \\ 0 & |x| > a/2, \end{cases} \quad (8.13a)$$

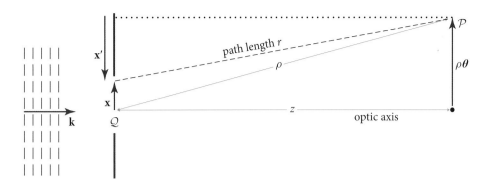

FIGURE 8.5 Geometry for computing the path length between a point $\mathcal{Q}$ in the aperture and the point of observation $\mathcal{P}$. The transverse vector $\mathbf{x}$ is used to identify $\mathcal{Q}$ in our Fraunhofer analysis (Sec. 8.3), and $\mathbf{x}'$ is used in the Fresnel analysis (Sec. 8.4).

for which

$$\psi_\mathcal{P}(\theta) \propto \tilde{H}_1 \propto \int_{-a/2}^{a/2} e^{ikx\theta} dx \propto \mathrm{sinc}\left(\tfrac{1}{2}ka\theta\right), \tag{8.13b}$$

$$F(\theta) \propto \mathrm{sinc}^2\left(\tfrac{1}{2}ka\theta\right). \tag{8.13c}$$

Fraunhofer diffraction from a slit with width a

Here $\mathrm{sinc}(\xi) \equiv \sin(\xi)/\xi$. The bottom flux curve in Fig. 8.4 is almost—but not quite—described by Eq. (8.13c); the differences (e.g., the not-quite-zero value of the minimum between the central peak and the first side lobe) are due to the field point not being fully in the Fraunhofer region, $r_F/a = 2$ rather than $r_F/a \gg 1$.

It is usually uninteresting to normalize a Fraunhofer diffraction pattern. On those rare occasions when the absolute value of the observed flux is needed, rather than just the pattern's angular shape, it typically can be derived most easily from conservation of the total wave energy. This is why we have ignored the proportionality factors in the diffraction patterns.

All techniques for handling Fourier transforms (which should be familiar from quantum mechanics and elsewhere, though with different normalizations) can be applied to derive Fraunhofer diffraction patterns. In particular, the *convolution theorem* turns out to be very useful. It says that the Fourier transform of the convolution

$$f_2 \otimes f_1 \equiv \int_{-\infty}^{+\infty} f_2(\mathbf{x} - \mathbf{x}') f_1(\mathbf{x}') d\Sigma' \tag{8.14}$$

convolution

of two functions f_1 and f_2 is equal to the product $\tilde{f}_2(\boldsymbol{\theta}) \tilde{f}_1(\boldsymbol{\theta})$ of their Fourier transforms, and conversely, the Fourier transform of a product is equal to the convolution. [Here and throughout this chapter, we use the optics version of a Fourier transform, in which 2-dimensional transverse position $\mathbf{x}$ is made dimensionless via the wave number k; Eq. (8.11a).]

8.3 Fraunhofer Diffraction

As an example of the convolution theorem's power, we now compute the diffraction pattern produced by a diffraction grating.

8.3.1 Diffraction Grating

A diffraction grating[4] can be modeled as a finite series of alternating transparent and opaque, long, parallel strips. Let there be N transparent and opaque strips each of width $a \gg \lambda$ (Fig. 8.6a), and idealize them as infinitely long, so their diffraction pattern is 1-dimensional. We outline how to use the convolution theorem to derive the grating's Fraunhofer diffraction pattern. The details are left as an exercise for the reader (Ex. 8.2).

using convolution to build transmission function

Our idealized N-slit grating can be regarded as an infinite series of delta functions with separation $2a$, convolved with the transmission function H_1 [Eq. (8.13a)] for a single slit of width a,

$$\int_{-\infty}^{\infty} \left[\sum_{n=-\infty}^{+\infty} \delta(\xi - 2an) \right] H_1(x - \xi) d\xi, \tag{8.15a}$$

and then multiplied by an aperture function with width $2Na$:

$$H_{2N}(x) \equiv \begin{cases} 1 & |x| < Na \\ 0 & |x| > Na. \end{cases} \tag{8.15b}$$

More explicitly, we have

$$\mathfrak{t}(x) = \left(\int_{-\infty}^{\infty} \left[\sum_{n=-\infty}^{+\infty} \delta(\xi - 2an) \right] H_1(x - \xi) d\xi \right) H_{2N}(x), \tag{8.16}$$

which is shown graphically in Fig. 8.6b.

using the convolution theorem to compute Fraunhofer diffraction

Let us use the convolution theorem to evaluate the Fourier transform $\tilde{\mathfrak{t}}(\theta)$ of expression (8.16), thereby obtaining the diffraction pattern $\psi_\mathcal{P}(\theta) \propto \tilde{\mathfrak{t}}(\theta)$ for our transmission grating. The Fourier transform of the infinite series of delta functions with spacing $2a$ is itself an infinite series of delta functions with reciprocal spacing $2\pi/(2ka) = \lambda/(2a)$ (see the hint in Ex. 8.2). This must be multiplied by the Fourier transform $\tilde{H}_1(\theta) \propto \text{sinc}(\frac{1}{2}ka\theta)$ of the single narrow slit and then convolved with the Fourier transform $\tilde{H}_{2N}(\theta) \propto \text{sinc}(Nka\theta)$ of the aperture (wide slit). The result is shown schematically in Fig. 8.6c. (Each of the transforms is real, so the 1-dimensional functions shown in the figure fully embody them.)

The resulting diffracted energy flux, $F \propto |\mathfrak{t}(\theta)|^2$ (as computed in Ex. 8.2), is shown in Fig. 8.6d. The grating has channeled the incident radiation into a few equally spaced beams with directions $\theta = \pi p/(ka) = p\lambda/(2a)$, where p is an integer known as the *order* of the beam. Each of these beams has a shape given by $|\tilde{H}_{2N}(\theta)|^2$: a sharp central peak with half-width (distance from center of peak to first null of the flux) $\lambda/(2Na)$, followed by a set of *side lobes* whose intensities are $\propto N^{-1}$.

---

4. Diffraction gratings were first fabricated by David Rittenhouse in 1785 by winding hair around a pair of screws.

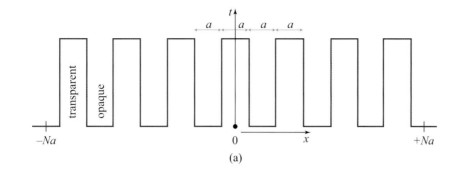

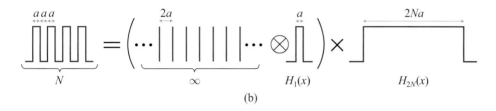

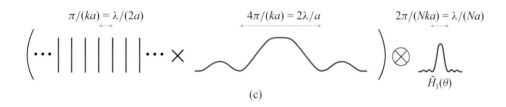

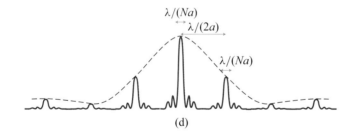

FIGURE 8.6 (a) Diffraction grating $t(x)$ formed by N alternating transparent and opaque strips each of width a. (b) Decomposition of this finite grating into an infinite series of equally spaced delta functions that are convolved (the symbol $\otimes$) with the shape of an individual transparent strip (i.e., a slit) and then multiplied (the symbol $\times$) by a large aperture function covering N such slits; Eq. (8.16). (c) The resulting Fraunhofer diffraction pattern $\tilde{t}(\theta)$ is shown schematically as the Fourier transform of a series of delta functions multiplied by the Fourier transform of a single slit and then convolved with the Fourier transform of the aperture. (d) The energy flux $F \propto |\tilde{t}(\theta)|^2$ of this diffraction pattern.

8.3 Fraunhofer Diffraction

The deflection angles $\theta = p\lambda/(2a)$ of these beams are proportional to λ; this underpins the use of diffraction gratings for spectroscopy (different wavelengths deflected into beams at different angles). It is of interest to ask what the wavelength resolution of such an idealized grating might be. If one focuses attention on the pth-order beams at two wavelengths λ and $\lambda + \delta\lambda$ [which are located at $\theta = p\lambda/(2a)$ and $p(\lambda + \delta\lambda)/(2a)$], then one can distinguish the beams from each other when their separation $\delta\theta = p\delta\lambda/(2a)$ is at least as large as the angular distance $\lambda/(2Na)$ between the maximum of each beam's diffraction pattern and its first minimum, that is, when

diffraction grating's chromatic resolving power

$$\frac{\lambda}{\delta\lambda} \lesssim \mathcal{R} \equiv Np. \tag{8.17}$$

(Recall that N is the total number of slits in the grating, and p is the order of the diffracted beam.) This $\mathcal{R}$ is called the grating's *chromatic resolving power*.

Real gratings are not this simple. First, they usually work not by modulating the amplitude of the incident radiation in this simple manner, but instead by modulating the phase. Second, the manner in which the phase is modulated is such as to channel most of the incident power into a particular order, a technique known as *blazing*. Third, gratings are often used in reflection rather than transmission. Despite these complications, the principles of a real grating's operation are essentially the same as our idealized grating. Manufactured gratings typically have $N \gtrsim 10{,}000$, giving a wavelength resolution for visual light that can be as small as $\lambda/10^5 \sim 10$ pm (i.e., 10^{-11} m).

EXERCISES

Exercise 8.1 *Practice: Convolutions and Fourier Transforms*
(a) Calculate the 1-dimensional Fourier transforms [Eq. (8.11a) reduced to 1 dimension] of the functions $f_1(x) \equiv e^{-x^2/2\sigma^2}$, and $f_2 \equiv 0$ for $x < 0$, $f_2 \equiv e^{-x/h}$ for $x \geq 0$.
(b) Take the inverse transforms of your answers to part (a) and recover the original functions.
(c) Convolve the exponential function f_2 with the Gaussian function f_1, and then compute the Fourier transform of their convolution. Verify that the result is the same as the product of the Fourier transforms of f_1 and f_2.

Exercise 8.2 *Derivation: Diffraction Grating*
Use the convolution theorem to carry out the calculation of the Fraunhofer diffraction pattern from the grating shown in Fig. 8.6. [Hint: To show that the Fourier transform of the infinite sequence of equally spaced delta functions is a similar sequence of delta functions, perform the Fourier transform to get $\sum_{n=-\infty}^{+\infty} e^{i2kan\theta}$ (aside from a multiplicative factor); then use the formulas for a Fourier *series* expansion, and its inverse, for any function that is periodic with period $\pi/(ka)$ to show that $\sum_{n=-\infty}^{+\infty} e^{i2kan\theta}$ is a sequence of delta functions.]

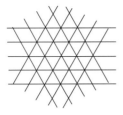

FIGURE 8.7 Diffraction grating formed from three groups of parallel lines.

Exercise 8.3 *Problem: Triangular Diffraction Grating*
Sketch the Fraunhofer diffraction pattern you would expect to see from a diffraction grating made from three groups of parallel lines aligned at angles of 120° to one another (Fig. 8.7).

8.3.2 Airy Pattern of a Circular Aperture: Hubble Space Telescope

The Hubble Space Telescope was launched in April 1990 to observe planets, stars, and galaxies above Earth's atmosphere. One reason for going into space is to avoid the irregular refractive-index variations in Earth's atmosphere, known, generically, as *seeing*, which degrade the quality of the images. (Another reason is to observe the ultraviolet part of the spectrum, which is absorbed in Earth's atmosphere.) Seeing typically limits the angular resolution of Earth-bound telescopes to $\sim 0.5''$ at visual wavelengths (see Box 9.2). We wish to compute how much the angular resolution improves by going into space. As we shall see, the computation is essentially an exercise in Fraunhofer diffraction theory.

The essence of the computation is to idealize the telescope as a circular aperture with diameter equal to that of the primary mirror. Light from this mirror is actually reflected onto a secondary mirror and then follows a complex optical path before being focused on a variety of detectors. However, this path is irrelevant to the angular resolution. The purposes of the optics are merely (i) to bring the Fraunhofer-region light to a focus close to the mirror [Eq. (8.31) and subsequent discussion], in order to produce an instrument that is compact enough to be launched, and (ii) to match the sizes of stars' images to the pixel size on the detector. In doing so, however, the optics leaves the angular resolution unchanged; the resolution is the same as if we were to observe the light, which passes through the primary mirror's circular aperture, far beyond the mirror, in the Fraunhofer region.

If the telescope aperture were very small, for example, a pin hole, then the light from a point source (a very distant star) would create a broad diffraction pattern, and the telescope's angular resolution would be correspondingly poor. As we increase the diameter of the aperture, we still see a diffraction pattern, but its angular width diminishes.

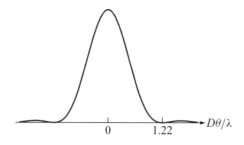

FIGURE 8.8 Airy diffraction pattern produced by a circular aperture.

Using these considerations, we can compute how well the telescope can distinguish neighboring stars. We do not expect it to resolve them fully if they are closer together on the sky than the angular width of the diffraction patttern. Of course, optical imperfections and pointing errors in a real telescope may degrade the image quality even further, but this is the best that can be done, limited only by the uncertainty principle.

The calculation of the Fraunhofer amplitude far from the aperture, via the Fourier transform (8.11a), is straightforward (Ex. 8.4):

Airy diffraction pattern from circular aperture

$$\psi(\theta) \propto \int_{\text{Disk with diameter } D} e^{-i k \mathbf{x} \cdot \boldsymbol{\theta}} d\Sigma$$

$$\propto \text{jinc}\left(\frac{kD\theta}{2}\right), \tag{8.18}$$

where D is the diameter of the aperture (i.e., of the telescope's primary mirror), $\theta \equiv |\boldsymbol{\theta}|$ is the angle from the optic axis, and $\text{jinc}(x) \equiv J_1(x)/x$, with J_1 the Bessel function of the first kind and of order one. The flux from the star observed at angle θ is therefore $\propto \text{jinc}^2(kD\theta/2)$. This energy-flux pattern, known as the *Airy pattern*, is shown in Fig. 8.8. The image appears as a central "Airy disk" surrounded by a circle where the flux vanishes, and then further surrounded by a series of concentric rings whose flux diminishes with radius. Only 16% of the total light falls outside the central Airy disk. The angular radius θ_A of the Airy disk (i.e., the radius of the dark circle surrounding it) is determined by the first zero of $J_1(kD\theta/2) = J_1(\theta \pi D/\lambda)$:

$$\boxed{\theta_A = 1.22\lambda/D.} \tag{8.19}$$

A conventional, though essentially arbitrary, criterion for angular resolution is to say that two point sources can be distinguished if they are separated in angle by more than θ_A. For the Hubble Space Telescope, $D = 2.4$ m and $\theta_A \sim 0.05''$ at visual wavelengths, which is more than 10 times better than is achievable on the ground with conventional (nonadaptive) optics.

Initially, there was a serious problem with Hubble's telescope optics. The hyperboloidal primary mirror was ground to the wrong shape, so rays parallel to the optic axis did not pass through a common focus after reflection off a convex hyperboloidal secondary mirror. This defect, known as *spherical aberration* (Sec. 7.5.2), created blurred images. However, it was possible to correct this error in subsequent instruments in the optical train, and the Hubble Space Telescope became the most successful optical telescope of all time, transforming our view of the universe.

spherical aberration in the Hubble Space Telescope

The Hubble Space Telescope should be succeeded in 2018 by the James Webb Space Telescope. Webb will have a diameter $D \simeq 6.5$ m (2.7 times larger than Hubble) but its wavelengths of observation are somewhat longer (0.6 μm to 28.5 μm), so its angular resolution will be, on average, roughly the same as Hubble's.

EXERCISES

Exercise 8.4 *Derivation: Airy Pattern*
Derive and plot the Airy diffraction pattern [Eq. (8.18)] and show that 84% of the light is contained within the Airy disk.

Exercise 8.5 *Problem: Pointillist Painting*
The neoimpressionist painter Georges Seurat was a member of the pointillist school. His paintings consisted of an enormous number of closely spaced dots of pure pigment (of size ranging from ∼0.4 mm in his smaller paintings to ∼4 mm in his largest paintings, such as *Gray Weather, Grande Jatte,* Fig. 8.9). The illusion of color mixing was produced only in the eye of the observer. How far from the painting should one stand to obtain the desired blending of color?

(a) (b)

FIGURE 8.9 (a) Georges Seurat's painting *Gray Weather, Grande Jatte.* When viewed from a sufficient distance, adjacent dots of paint with different colors blend together in the eye to form another color. (b) Enlargement of the boat near the center of the painting. In this enlargement, one sees clearly the individual dots of paint. *Gray Weather, Grande Jatte* by Georges Seurat (ca. 1886–88); The Metropolitan Museum of Art (www.metmuseum.org); The Walter H. and Leonore Annenberg Collection, Gift of Walter H. and Leonore Annenberg, 2002, Bequest of Walter H. Annenberg, 2002.

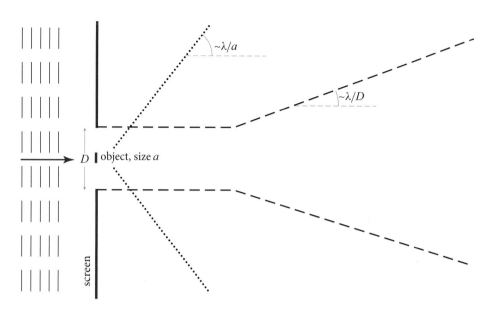

FIGURE 8.10 Geometry for Babinet's principle. The beam produced by the large aperture D is confined between the long-dashed lines. Outside this beam, the energy-flux pattern $F(\boldsymbol{\theta}) \propto |\mathfrak{t}(\boldsymbol{\theta})|^2$ produced by a small object (size a) and its complement are the same, Eqs. (8.20). This diffracted flux pattern is confined between the dotted lines.

8.3.3 8.3.3 Babinet's Principle

Suppose that monochromatic light falls normally onto a large circular aperture with diameter D. At distances $z \lesssim D^2/\lambda$ (i.e., $r_F \lesssim D$), the transmitted light will be collimated into a beam with diameter D, and at larger distances, the beam will become conical with opening angle λ/D (Fig. 8.3) and flux distribution given by the Airy diffraction pattern of Fig. 8.8.

Now place into this aperture a significantly smaller object (size $a \ll D$; Fig. 8.10) with transmissivity $\mathfrak{t}_1(\mathbf{x})$—for example, an opaque star-shaped object. This object will produce a Fraunhofer diffraction pattern with opening angle $\lambda/a \gg \lambda/D$ that extends well beyond the large aperture's beam. Outside that beam, the diffraction pattern will be insensitive to the shape and size of the large aperture, because only the small object can diffract light to these large angles; so the diffracted flux will be $F_1(\boldsymbol{\theta}) \propto |\tilde{\mathfrak{t}}_1(\theta)|^2$.

Next, suppose that we replace the small object by one with a *complementary transmissivity* $\mathfrak{t}_2$, complementary in the sense that

$$\boxed{\mathfrak{t}_1(\mathbf{x}) + \mathfrak{t}_2(\mathbf{x}) = 1.} \tag{8.20a}$$

Babinet's principle

For example, we replace a small, opaque star-shaped object by an opaque screen that fills the original, large aperture except for a star-shaped hole. This complementary object will produce a diffraction pattern $F_2(\boldsymbol{\theta}) \propto |\tilde{\mathfrak{t}}_2(\theta)|^2$. Outside the large aperture's beam, this pattern again is insensitive to the size and shape of the large aperture, that is, it is insensitive to the 1 in $\mathfrak{t}_2 = 1 - \mathfrak{t}_1$ (which sends light solely inside the large aperture's beam); so at these large angles, $\tilde{t}_2(\boldsymbol{\theta}) = -\tilde{t}_1(\boldsymbol{\theta})$, which implies that the energy-flux diffraction pattern of the original object and the new, complementary object will be the same outside the large aperture's beam:

$$F_2(\boldsymbol{\theta}) \propto |\tilde{t}_2(\boldsymbol{\theta})|^2 = |\tilde{t}_1(\boldsymbol{\theta})|^2 \propto F_1(\boldsymbol{\theta}). \tag{8.20b}$$

This is called *Babinet's principle*.

EXERCISES

Exercise 8.6 *Problem: Light Scattering by a Large, Opaque Particle*
Consider the scattering of light by an opaque particle with size $a \gg 1/k$. Neglect any scattering via excitation of electrons in the particle. Then the scattering is solely due to diffraction of light around the particle. With the aid of Babinet's principle, do the following.

(a) Explain why the scattered light is confined to a cone with opening angle $\Delta\theta \sim \pi/(ka) \ll 1$.

(b) Show that the total power in the scattered light, at very large distances from the particle, is $P_S = FA$, where F is the incident energy flux, and A is the cross sectional area of the particle perpendicular to the incident wave.

(c) Explain why the total "extinction" (absorption plus scattering) cross section is equal to $2A$ independent of the shape of the opaque particle.

Exercise 8.7 *Problem: Thickness of a Human Hair*
Conceive and carry out an experiment using light diffraction to measure the thickness of a hair from your head, accurate to within a factor of ~ 2. [Hint: Make sure the source of light that you use is small enough—e.g., a very narrow laser beam—that its finite size has negligible influence on your result.]

8.4 Fresnel Diffraction

We next turn to the Fresnel region of observation points $\mathcal{P}$ with $r_F = \sqrt{\lambda \rho}$ much smaller than the aperture. In this region, the field at $\mathcal{P}$ arriving from different parts of the aperture has significantly different phase $\Delta\varphi \gg 1$. We again specialize to incoming wave vectors that are approximately orthogonal to the aperture plane and to small diffraction angles, so we can ignore the obliquity factor $d\hat{\boldsymbol{\Sigma}} \cdot (\mathbf{n} + \mathbf{n}')/2$ in Eq. (8.6). By contrast with the Fraunhofer case, we now identify $\mathcal{P}$ by its distance z from the aperture plane instead of its distance ρ from the aperture center, and we use as our integration variable in the aperture $\mathbf{x}' \equiv \mathbf{x} - \rho\boldsymbol{\theta}$ (Fig. 8.5), thereby writing the dependence of the phase at $\mathcal{P}$ on $\mathbf{x}$ in the form

$$\Delta\varphi \equiv k \times [(\text{path length from } \mathbf{x} \text{ to } \mathcal{P}) - z] = \frac{k\mathbf{x}'^2}{2z} + O\left(\frac{k x'^4}{z^3}\right). \tag{8.21}$$

In the Fraunhofer region (Sec. 8.3) only the linear term $-k\mathbf{x}\cdot\boldsymbol{\theta}$ in $k\mathbf{x}'^2/(2z) \simeq k(\mathbf{x}-r\boldsymbol{\theta})^2/r$ was significant. In the Fresnel region the term quadratic in $\mathbf{x}$ is also significant (and we have changed variables to $\mathbf{x}'$ so as to simplify it), but the $O(x'^4)$ term is negligible. Therefore, in the Fresnel region the diffraction pattern (8.6) is

$$\psi_\mathcal{P} = \frac{-ike^{ikz}}{2\pi z}\int e^{i\Delta\varphi}\mathfrak{t}\psi'd\Sigma' = \frac{-ike^{ikz}}{2\pi z}\int e^{ik\mathbf{x}'^2/(2z)}\mathfrak{t}(\mathbf{x}')\psi'(\mathbf{x}')d\Sigma', \quad (8.22)$$

where in the denominator we have replaced r by z to excellent approximation.

Let us consider the Fresnel diffraction pattern formed by an empty ($\mathfrak{t}=1$) simple aperture of arbitrary shape, illuminated by a normally incident plane wave. It is convenient to introduce transverse Cartesian coordinates $\mathbf{x}' = (x', y')$ and define

$$\sigma = \left(\frac{k}{\pi z}\right)^{1/2} x', \quad \tau = \left(\frac{k}{\pi z}\right)^{1/2} y'. \quad (8.23a)$$

[Notice that $(k/[\pi z])^{1/2}$ is $\sqrt{2}/r_F$; cf. Eq. (8.8).] We can thereby rewrite Eq. (8.22) in the form

Fresnel diffraction pattern from arbitrary empty aperture $\mathcal{Q}$

$$\psi_\mathcal{P} = -\frac{i}{2}\int\int_\mathcal{Q} e^{i\pi\sigma^2/2}e^{i\pi\tau^2/2}\psi_\mathcal{Q} e^{ikz}d\sigma d\tau. \quad (8.23b)$$

Here we have changed notation for the field ψ' impinging on the aperture $\mathcal{Q}$ to $\psi_\mathcal{Q}$.

Equations (8.23) depend on the transverse location of the observation point $\mathcal{P}$ through the origin used to define $\mathbf{x}' = (x', y')$; see Fig. 8.5. We use these rather general expressions in Sec. 8.5, when discussing paraxial Fourier optics, as well as in Secs. 8.4.1–8.4.4 on Fresnel diffraction.

8.4.1 Rectangular Aperture, Fresnel Integrals, and the Cornu Spiral

In this and the following three sections, we explore the details of the Fresnel diffraction pattern for an incoming plane wave that falls perpendicularly on the aperture, so $\psi_\mathcal{Q}$ is constant over the aperture.

Fresnel diffraction by rectangular aperture

For simplicity, we initially confine attention to a rectangular aperture with edges along the x' and y' directions. Then the two integrals have limits that are independent of each other, and the integrals can be expressed in the form $\mathcal{E}(\sigma_{\max}) - \mathcal{E}(\sigma_{\min})$ and $\mathcal{E}(\tau_{\max}) - \mathcal{E}(\tau_{\min})$, so

$$\psi_\mathcal{P} = \frac{-i}{2}[\mathcal{E}(\sigma_{\max}) - \mathcal{E}(\sigma_{\min})][\mathcal{E}(\tau_{\max}) - \mathcal{E}(\tau_{\min})]\psi_\mathcal{Q} e^{ikz} \equiv \frac{-i}{2}\Delta\mathcal{E}_\sigma \Delta\mathcal{E}_\tau \psi_\mathcal{Q} e^{ikz}, \quad (8.24a)$$

where the arguments are the limits of integration (the two edges of the aperture), and where

$$\mathcal{E}(\xi) \equiv \int_0^\xi e^{i\pi\sigma^2/2}d\sigma \equiv C(\xi) + iS(\xi). \quad (8.24b)$$

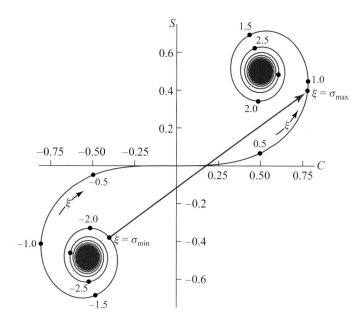

FIGURE 8.11 Cornu spiral in the complex plane; the real part of $\mathcal{E}(\xi) = C(\xi) + iS(\xi)$ is plotted horizontally, and the imaginary part vertically; the point $\xi = 0$ is at the origin, positive ξ is in the upper right quadrant, and negative ξ is in the lower left quadrant. The diffracted energy flux is proportional to the squared length of the arrow reaching from $\xi = \sigma_{\min}$ to $\xi = \sigma_{\max}$.

Here

$$C(\xi) \equiv \int_0^\xi d\sigma \, \cos(\pi \sigma^2/2), \qquad S(\xi) \equiv \int_0^\xi d\sigma \, \sin(\pi \sigma^2/2) \qquad (8.24c)$$

Fresnel integrals

are known as *Fresnel integrals* and are standard functions tabulated in many books and can be found in Mathematica, Matlab, and Maple. Notice that the energy-flux distribution is

$$F \propto |\psi_\mathcal{P}|^2 \propto |\Delta\mathcal{E}_\sigma|^2 |\Delta\mathcal{E}_\tau|^2. \qquad (8.24d)$$

It is convenient to exhibit the Fresnel integrals graphically using a *Cornu spiral* (Fig. 8.11). This is a graph of the parametric equation $[C(\xi), S(\xi)]$, or equivalently, a graph of $\mathcal{E}(\xi) = C(\xi) + iS(\xi)$ in the complex plane. The two terms $\Delta\mathcal{E}_\sigma$ and $\Delta\mathcal{E}_\tau$ in Eq. (8.24a) can be represented in amplitude and phase by arrows in the C-S plane reaching from $\xi = \sigma_{\min}$ on the Cornu spiral to $\xi = \sigma_{\max}$ (Fig. 8.11), and from $\xi = \tau_{\min}$ to $\xi = \tau_{\max}$. Correspondingly, the flux F [Eq. (8.24d)] is proportional to the product of the squared lengths of these two vectors.

Cornu spiral

As the observation point $\mathcal{P}$ moves around in the observation plane (Fig. 8.5), $x'_{\min}$ and $x'_{\max}$ change, and hence $\sigma_{\min} = [k/(\pi z)]^{1/2} x'_{\min}$ and $\sigma_{\max} = [k/(\pi z)]^{1/2} x'_{\max}$ change [i.e., the tail and tip of the arrow in Fig. 8.11 move along the Cornu spiral, thereby changing the diffracted flux, which is $\propto$ (length of arrow)^2].

8.4 Fresnel Diffraction

8.4.2 Unobscured Plane Wave

The simplest illustration of Fresnel integrals is the totally unobscured plane wave. In this case, the limits of both integrations extend from $-\infty$ to $+\infty$, which, as we see in Fig. 8.11, is an arrow of length $2^{1/2}$ and phase $\pi/4$. Therefore, ψ_P is equal to $(2^{1/2}e^{i\pi/4})^2(-i/2)\psi_Q e^{ikz} = \psi_Q e^{ikz}$, as we could have deduced simply by solving the Helmholtz equation (8.1b) for a plane wave.

This unobscured-wavefront calculation elucidates three issues that we have already met. First, it illustrates our interpretation of Fermat's principle in geometric optics. In the limit of short wavelengths, the paths that contribute to the wave field are just those along which the phase is stationary to small variations. Our present calculation shows that, because of the tightening of the Cornu spiral as one moves toward a large argument, the paths that contribute significantly to ψ_P are those that are separated from the geometric-optics path by less than a few Fresnel lengths at Q. (For a laboratory experiment with light and $z \sim 2$ m, the Fresnel length is $\sqrt{\lambda z} \sim 1$ mm.)

for Fresnel diffraction dominant paths are near geometric optics path

A second, and related, point is that, when computing the Fresnel diffraction pattern from a more complicated aperture, we need only perform the integral (8.6) in the immediate vicinity of the geometric-optics ray. We can ignore the contribution from the extension of the aperture Q to meet the "sphere at infinity" (the surface S in Fig. 8.2), even when the wave is unobstructed there. The rapid phase variation makes the contribution from that extension and from S sum to zero.

Third, when integrating over the whole area of the wavefront at Q, we have summed contributions with increasingly large phase differences that add in such a way that the total has a net extra phase of $\pi/2$ relative to the geometric-optics ray. This phase factor cancels exactly the prefactor $-i$ in the Helmholtz-Kirchhoff formula (8.6). (This phase factor is unimportant in the limit of geometric optics.)

8.4.3 Fresnel Diffraction by a Straight Edge: Lunar Occultation of a Radio Source

The next-simplest case of Fresnel diffraction is the pattern formed by a straight edge. As a specific example, consider a cosmologically distant source of radio waves that is occulted by the Moon. If we treat the lunar limb as a straight edge, then the radio source will create a changing diffraction pattern as it passes behind the Moon, and the diffraction pattern can be measured by a radio telescope on Earth. We orient our coordinates so the Moon's edge is along the y' direction (τ direction). Then in Eq. (8.24a) $\Delta \mathcal{E}_\tau \equiv \mathcal{E}(\tau_{\max}) - \mathcal{E}(\tau_{\min}) = \sqrt{2i}$ is constant, and $\Delta \mathcal{E}_\sigma \equiv \mathcal{E}(\sigma_{\max}) - \mathcal{E}(\sigma_{\min})$ is described by the Cornu spiral of Fig. 8.12b.

Long before the occultation, we can approximate $\sigma_{\min} = -\infty$ and $\sigma_{\max} = +\infty$ (Fig. 8.12a), so $\Delta \mathcal{E}_\sigma$ is given by the arrow from $(-1/2, -1/2)$ to $(1/2, 1/2)$ in Fig. 8.12b (i.e., $\Delta \mathcal{E}_\sigma = \sqrt{2i}$). The observed wave amplitude, Eq. (8.24a), is therefore $\psi_Q e^{ikz}$. When the Moon approaches occultation of the radio source, the upper bound on the Fresnel integral begins to diminish from $\sigma_{\max} = +\infty$, and the complex vector on the Cornu spiral begins to oscillate in length (e.g., from A to B in Fig. 8.12b,c) and in phase. The observed flux also oscillates, more and more strongly as geometric

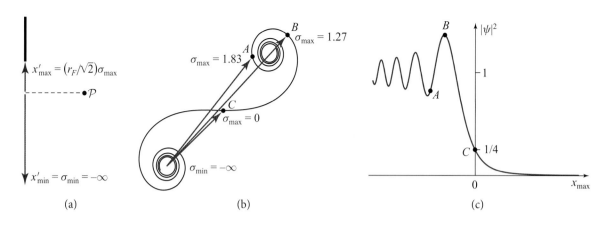

FIGURE 8.12 Diffraction from a straight edge. (a) The straight edge, onto which a plane wave impinges orthogonally, the observation point $\mathcal{P}$, and the vectors that reach to the lower and upper limits of integration, $x'_{min} = -\infty$ and x'_{max} for computation of the diffraction integral (8.23b). (b) The Cornu spiral, showing the arrows that represent the contribution $\Delta \mathcal{E}_\sigma$ to the diffracted field $\psi_\mathcal{P}$. (c) The energy-flux diffraction pattern formed by the straight edge. The flux $|\psi|^2 \propto |\Delta \mathcal{E}_\sigma|^2$ is proportional to the squared length of the arrow on the Cornu spiral in panel b.

occultation is approached. At the point of geometric occultation (point C in Fig. 8.12b,c), the complex vector extends from $(-1/2, -1/2)$ to $(0, 0)$, and so the observed wave amplitude is one-half the unocculted value, and the flux is reduced to one-fourth. As the occultation proceeds, the length of the complex vector and the observed flux decrease monotonically to zero, while the phase continues to oscillate.

Historically, diffraction of a radio source's waves by the Moon led to the discovery of quasars—the hyperactive nuclei of distant galaxies. In the early 1960s, a team of Australian and British radio observers led by Cyril Hazard knew that the Moon would occult the powerful radio source 3C273, so they set up their telescope to observe the development of the diffraction pattern as the occultation proceeded. From the pattern's observed times of ingress (passage into the Moon's shadow) and egress (emergence from the Moon's shadow), Hazard determined the coordinates of 3C273 on the sky and did so with remarkable accuracy, thanks to the oscillatory features in the diffraction pattern. These coordinates enabled Maarten Schmidt at the 200-inch telescope on Palomar Mountain to identify 3C273 optically and discover (from its optical redshift) that it was surprisingly distant and consequently had an unprecedented luminosity. It was the first example of a quasar—a previously unknown astrophysical object.

discovery of quasars

In Hazard's occultation measurements, the observing wavelength was $\lambda \sim 0.2$ m. Since the Moon is roughly $z \sim 400{,}000$ km distant, the Fresnel length was about $r_F = \sqrt{\lambda z} \sim 10$ km. The Moon's orbital speed is $v \sim 200$ m s^{-1}, so the diffraction pattern took a time $\sim 5 r_F / v \sim 4$ min to pass through the telescope.

The straight-edge diffraction pattern of Fig. 8.12c occurs universally along the edge of the shadow of any object, so long as the source of light is sufficiently small

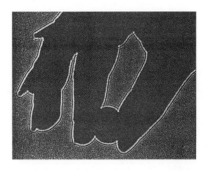

FIGURE 8.13 Fresnel diffraction pattern in the shadow of *Mary's Hand Holding a Dime*—a photograph by Eugene Hecht, from the first figure in Hecht (2017, Chap. 10).

and the shadow's edge bends on lengthscales long compared to the Fresnel length $r_F = \sqrt{\lambda z}$. Examples are the diffraction patterns on the two edges of a slit's shadow in the upper-left curve in Fig. 8.4 (cf. Ex. 8.8) and the diffraction pattern along the edge of a shadow cast by a person's hand in Fig. 8.13.

EXERCISES

Exercise 8.8 *Example: Diffraction Pattern from a Slit*
Derive a formula for the energy-flux diffraction pattern $F(x)$ of a slit with width a, as a function of distance x from the center of the slit, in terms of Fresnel integrals. Plot your formula for various distances z from the slit's plane (i.e., for various values of $r_F/a = \sqrt{\lambda z/a^2}$), and compare with Fig. 8.4.

8.4.4 Circular Apertures: Fresnel Zones and Zone Plates

We have seen how the Fresnel diffraction pattern for a plane wave can be thought of as formed by waves that derive from a patch in the diffracting object's plane a few Fresnel lengths in size. This notion can be made quantitatively useful by reanalyzing an unobstructed wavefront in circular polar coordinates. More specifically, consider a plane wave incident on an aperture $\mathcal{Q}$ that is infinitely large (no obstruction), and define $\varpi \equiv |\mathbf{x}'|/r_F = \sqrt{\frac{1}{2}(\sigma^2 + \tau^2)}$. Then the phase factor in Eq. (8.23b) is $\Delta\varphi = \pi\varpi^2$, so the observed wave coming from the region inside a circle of radius $|\mathbf{x}'| = r_F \varpi$ will be given by

$$\psi_{\mathcal{P}} = -i \int_0^{\varpi} 2\pi \varpi' d\varpi' e^{i\pi\varpi'^2} \psi_{\mathcal{Q}} e^{ikz}$$

$$= (1 - e^{i\pi\varpi^2}) \psi_{\mathcal{Q}} e^{ikz}. \qquad (8.25)$$

Now, this integral does not appear to converge as $\varpi \to \infty$. We can see what is happening if we sketch an amplitude-and-phase diagram (Fig. 8.14).

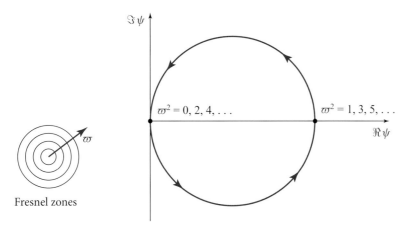

FIGURE 8.14 Amplitude-and-phase diagram for an unobstructed plane wavefront, decomposed into Fresnel zones; Eq. (8.25).

As we integrate outward from $\varpi = 0$, the complex vector has the initial phase retardation of $\pi/2$ but then moves on a semi-circle, so that by the time we have integrated out to a radius of $|\mathbf{x}'| = r_F$ ($\varpi = 1$), the contribution to the observed wave is $\psi_\mathcal{P} = 2\psi_\mathcal{Q}$ in phase with the incident wave. When the integration has been extended to $\sqrt{2}\, r_F$, ($\varpi = \sqrt{2}$), the circle has been completed, and $\psi_\mathcal{P} = 0$! The integral continues on around the same circle over and over again, as the upper-bound radius is further increased; see Fig. 8.14.

Of course, the field must actually have a well-defined value as $\varpi \to \infty$, despite this apparent failure of the integral to converge. To understand how the field becomes well defined, imagine splitting the aperture $\mathcal{Q}$ up into concentric annular rings, known as *Fresnel half-period zones,* with outer radii $\sqrt{n}\, r_F$, where $n = 1, 2, 3 \ldots$. The integral fails to converge because the contribution from each odd-numbered ring cancels that from an adjacent even-numbered ring. However, the thickness of these rings decreases as $1/\sqrt{n}$, and eventually we must allow for the fact that the incoming wave is not exactly planar; or, equivalently and more usefully, we must allow for the fact that the wave's distant source has some finite angular size. The finite size causes different pieces of the source to have their Fresnel rings centered at slightly different points in the aperture plane, which causes our computation of $\psi_\mathcal{P}$ to begin averaging over rings. This averaging forces the tip of the complex vector ($\Re\psi, \Im\psi$), where $\Re$ and $\Im$ correspond to the real and imaginary parts, to asymptote to the center of the circle in Fig. 8.14. Correspondingly, due to the averaging, the observed energy flux asymptotes to $|\psi_\mathcal{Q}|^2$ [Eq. (8.25) with the exponential $e^{i\pi\varpi^2}$ going to zero].

half-period zones

Although this may not seem to be a particularly wise way to decompose a plane wavefront, it does allow a striking experimental verification of our theory of diffraction. Suppose that we fabricate an aperture (called a *zone plate*) in which, for a chosen wavelength and observation point $\mathcal{P}$ on the optic axis, alternate half-period

zone plate

8.4 Fresnel Diffraction

zones are obscured. Then the wave observed at $\mathcal{P}$ will be the linear sum of several diameters of the circle in Fig. 8.14 and therefore will be far larger than ψ_Q. This strong amplification is confined to our chosen spot on the optic axis; most everywhere else the field's energy flux is reduced, thereby conserving energy. Thus the zone plate behaves like a lens. Because the common area of the half-period zones is $A = n\pi r_F^2 - (n-1)\pi r_F^2 = \pi r_F^2 = \pi \lambda z$, if we construct a zone plate with fixed area A for the zones, its focal length will be $f = z = A/(\pi\lambda)$. For $A =$ (a few square millimeters)—the typical choice—and $\lambda \sim 500$ nm (optical wavelengths), we have $f \sim$ a few meters.

secondary foci

Zone plates are only good lenses when the radiation is monochromatic, since the focal length is wavelength dependent, $f \propto \lambda^{-1}$. They have the further interesting property that they possess secondary foci, where the fields from 3 contiguous zones, or 5 or 7 or so forth, add up coherently (Ex. 8.9).

EXERCISES

Exercise 8.9 *Problem: Zone Plate*
(a) Use an amplitude-and-phase diagram to explain why a zone plate has secondary foci at distances of $f/3, f/5, f/7 \ldots$.
(b) An opaque, perfectly circular disk of diameter D is placed perpendicular to an incoming plane wave. Show that, at distances r such that $r_F \ll D$, the disk casts a rather sharp shadow, but at the precise center of the shadow there must be a bright spot.[5] How bright?

Exercise 8.10 *Problem: Spy Satellites*
Telescopes can also look down through the same atmospheric irregularities as those mentioned in Sec. 8.3.2 (see also Box 9.2). In what important respects will the optics differ from those for ground-based telescopes looking upward?

8.5

8.5 Paraxial Fourier Optics

We have developed a linear theory of wave optics that has allowed us to calculate diffraction patterns in the Fraunhofer and Fresnel limiting regions. That these calculations agree with laboratory measurements provides some vindication of the theory and the assumptions implicit in it. We now turn to practical applications of these ideas, specifically, to the acquisition and processing of images by instruments operating throughout the electromagnetic spectrum. As we shall see, these instruments rely

---

5. Siméon Poisson predicted the existence of this spot as a consequence of Fresnel's wave theory of light, in order to demonstrate that Fresnel's theory was wrong. However, François Arago (who was briefly, in 1848, premier of France) quickly demonstrated experimentally that the bright spot existed. It is now called the Poisson spot (despite Poisson's skepticism) or the Arago spot.

on an extension of paraxial geometric optics (Sec. 7.4) to situations where diffraction effects are important. Because of the central role played by Fourier transforms in diffraction [e.g., Eq. (8.11a)], the theory underlying these instruments is called *paraxial Fourier optics,* or just *Fourier optics.*

Although the conceptual framework and mathematical machinery for image processing by Fourier optics were developed in the nineteenth century, Fourier optics was not widely exploited until the second half of the twentieth century. Its maturation was driven in part by a growing recognition of similarities between optics and communication theory—for example, the realization that a microscope is simply an *image-processing system.* The development of electronic computation has also triggered enormous strides; computers are now seen as extensions of optical devices, and vice versa. It is a matter of convenience, accuracy, economics, and practicality to decide which parts of the image processing are carried out with mirrors, lenses, and the like, and which parts are performed numerically.

image processing with mirrors, lenses, etc. plus computing

One conceptually simple example of optical image processing is an improvement in one's ability to identify a faint star in the Fraunhofer diffraction rings ("fringes") of a much brighter star. As we shall see [Eq. (8.31) and subsequent discussion], the bright image of a source in a telescope's or microscope's focal plane has the same Airy diffraction pattern as we met in Eq. (8.18) and Fig. 8.8. If the shape of that image could be changed from the ring-endowed Airy pattern to a Gaussian, then it would be far easier to identify a nearby feature or faint star. One way to achieve this would be to attenuate the incident radiation at the telescope aperture in such a way that, immediately after passing through the aperture, it has a Gaussian profile instead of a sharp-edged profile. Its Fourier transform (the diffraction pattern in the focal plane) would then also be Gaussian. Such a Gaussian-shaped attenuation is difficult to achieve in practice, but it turns out—as we shall see—that there are easier options.

Gaussian aperture

Before exploring these options, we must lay some foundations, beginning with the concept of coherent illumination in Sec. 8.5.1, and then moving on to point-spread functions in Sec. 8.5.2.

8.5.1 Coherent Illumination

8.5.1

If the radiation arriving at the input of an optical system derives from a single source (e.g., a point source that has been collimated into a parallel beam by a converging lens), then the radiation is best described by its complex amplitude ψ (as we are doing in this chapter). An example might be a biological specimen on a microscope slide, illuminated by an external point source, for which the phases of the waves leaving different parts of the slide are strongly correlated with one another. This is called *coherent illumination.* If, by contrast, the source is self-luminous and of nonnegligible size, with the atoms or molecules in its different parts radiating independently—for example, a cluster of stars—then the phases of the radiation from different parts are uncorrelated, and it may be the radiation's energy flux, not its complex amplitude, that obeys well-defined (nonprobabilistic) evolution laws. This is called *incoherent illumination*. In this chapter we develop Fourier optics for a coherently illuminating source (the kind of illumination tacitly assumed in previous sections of the chapter). A parallel theory with a similar vocabulary can be developed for incoherent illumination, and some of the foundations for it are laid in Chap. 9. In Chap. 9, we also develop a more precise formulation of the concept of *coherence*.

8.5.2 Point-Spread Functions

In our treatment of paraxial geometric optics (Sec. 7.4), we showed how it is possible to regard a group of optical elements as a sequence of linear devices and relate the output rays to the input by linear operators (i.e., matrices). This chapter's theory of diffraction is also linear, and so a similar approach can be followed. As in Sec. 7.4, we restrict attention to small angles relative to some optic axis ("paraxial Fourier optics"). We describe the wave field at some distance z_j along the optic axis by the function $\psi_j(\mathbf{x})$, where $\mathbf{x}$ is a 2-dimensional vector perpendicular to the optic axis (as in Fig. 8.5). If we consider a single linear optical device, then we can relate the output field ψ_2 at z_2 to the input ψ_1 at z_1 using a Green's function denoted $P_{21}(\mathbf{x}_2, \mathbf{x}_1)$:[6]

Green's function, point-spread function, propagator

$$\psi_2(\mathbf{x}_2) = \int P_{21}(\mathbf{x}_2, \mathbf{x}_1) d\Sigma_1 \psi_1. \tag{8.26}$$

If ψ_1 were a delta function, then the output would be simply given by the function P_{21}, up to normalization. For this reason, P_{21} is usually known as the *point-spread function*. Alternatively, we can think of it as a *propagator*. If we now combine two optical devices sequentially, so the output ψ_2 of the first device is the input to the second, then the point-spread functions combine in the natural manner of any linear propagator to give a total point-spread function

combining point-spread functions

$$P_{31}(\mathbf{x}_3, \mathbf{x}_1) = \int P_{32}(\mathbf{x}_3, \mathbf{x}_2) d\Sigma_2 P_{21}(\mathbf{x}_2, \mathbf{x}_1). \tag{8.27}$$

Just as the simplest matrix for paraxial, geometric-optics propagation is that for free propagation through some distance d, so also the simplest point-spread function is that for free propagation. From Eq. (8.22) we see that it is given by

free-propagation point-spread function

$$P_{21} = \frac{-ik}{2\pi d} e^{ikd} \exp\left(\frac{ik(\mathbf{x}_2 - \mathbf{x}_1)^2}{2d}\right) \tag{8.28}$$
$$\text{for free propagation through a distance } d = z_2 - z_1.$$

---

6. The approach followed here has an interesting history. It originated in the treatment of Markov processes by Norbert Wiener and others (see Sec 6.3). It was then cleverly adopted for use in quantum mechanics and quantum field theory by Paul Dirac, Richard Feynman, and others, where it constitutes the path integral formulation. This, in turn, suggested the simpler adaptation to problems in classical optics laid out here.

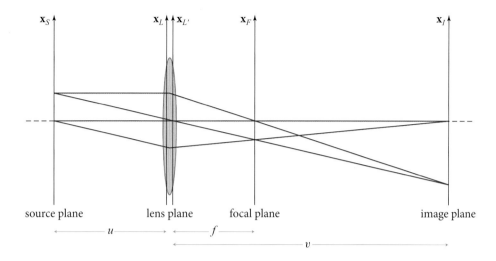

FIGURE 8.15 Wave theory of a single converging lens. The focal plane is a distance f (lens focal length) from the lens plane; and the image plane is a distance $v = fu/(u - f)$ from the lens plane.

Note that this P_{21} depends only on $\mathbf{x}_2 - \mathbf{x}_1$ as it should, and not on $\mathbf{x}_2$ or $\mathbf{x}_1$ individually, because there is joint translational invariance in the transverse $\mathbf{x}_1$, $\mathbf{x}_2$ planes.

A thin lens adds or subtracts an extra phase $\Delta\varphi$ to the wave, and $\Delta\varphi$ depends quadratically on distance $|\mathbf{x}|$ from the optic axis, so the angle of deflection, which is proportional to the gradient of the extra phase, will depend linearly on $\mathbf{x}$. Correspondingly, the point-spread function for a thin lens with focal length f is

$$\boxed{P_{21} = \exp\left(\frac{-ik|\mathbf{x}_1|^2}{2f}\right)\delta(\mathbf{x}_2 - \mathbf{x}_1).} \quad (8.29)$$

thin-lens point-spread function

For a converging lens, f is positive; for a diverging one, it is negative.

8.5.3 Abbé's Description of Image Formation by a Thin Lens

We can use the two point-spread functions (8.28) and (8.29) to give a wave description of the production of images by a single converging lens, in parallel to the geometric-optics description of Fig. 7.6. This description was formulated by Ernst Abbé in 1873. We develop Abbé's description in two stages. First, we propagate the wave from the source plane S a distance u in front of the lens, through the lens L, and to its focal plane F a distance f behind the lens (Fig. 8.15). Then we propagate the wave a further distance $v - f$ from the focal plane to the image plane. We know from geometric optics that $v = fu/(u - f)$ [Eq. (7.58)]. We restrict ourselves to $u > f$, so v is positive and the lens forms a real, inverted image.

Using Eqs. (8.27)–(8.29), we obtain the following expression for the propagator from the source plane S through the lens plane L to the focal plane F:

$$P_{FS} = \int P_{FL'} d\Sigma_{L'} P_{L'L} d\Sigma_L P_{LS}$$

$$= \int \frac{-ik}{2\pi f} e^{ikf} \exp\left(\frac{ik(\mathbf{x}_F - \mathbf{x}'_L)^2}{2f}\right) d\Sigma_{L'} \delta(\mathbf{x}_{L'} - \mathbf{x}_L) \exp\left(\frac{-ik|\mathbf{x}_L|^2}{2f}\right)$$

$$\times d\Sigma_L \frac{-ik}{2\pi u} e^{iku} \exp\left(\frac{ik(\mathbf{x}_L - \mathbf{x}_S)^2}{2u}\right)$$

$$= \frac{-ik}{2\pi f} e^{ik(f+u)} \exp\left(\frac{-ikx_F^2}{2(v-f)}\right) \exp\left(\frac{-ik\mathbf{x}_F \cdot \mathbf{x}_S}{f}\right). \tag{8.30}$$

Here we have extended all integrations to $\pm\infty$ and have used the values of the Fresnel integrals at infinity, $\mathcal{E}(\pm\infty) = \pm(1+i)/2$, to get the expression on the last line. The wave in the focal plane is given by $\psi_F(\mathbf{x}_F) = \int P_{FS} d\Sigma_S \psi_S(\mathbf{x}_S)$, which integrates to

wave field in focal plane of a thin lens

$$\boxed{\psi_F(\mathbf{x}_F) = -\frac{ik}{2\pi f} e^{ik(f+u)} \exp\left(\frac{-ikx_F^2}{2(v-f)}\right) \tilde{\psi}_S(\mathbf{x}_F/f).} \tag{8.31}$$

Here

$$\tilde{\psi}_S(\boldsymbol{\theta}) = \int d\Sigma_S \psi_S(\mathbf{x}_S) e^{-ik\boldsymbol{\theta} \cdot \mathbf{x}_S}. \tag{8.32}$$

Fraunhofer diffraction in focal plane

Equation (8.31) states that the field in the focal plane is, apart from a phase factor, proportional to the Fourier transform of the field in the source plane [recall our optics convention Eq. (8.11a) for the Fourier transform]. In other words, *the focal-plane field is the Fraunhofer diffraction pattern of the input wave.* That this has to be the case can be understood from Fig. 8.15. The focal plane F is where the converging lens brings parallel rays from the source plane to a focus. By doing so, the lens in effect brings in from "infinity" the Fraunhofer diffraction pattern of the source [Eq. (8.11a)][7] and places it into the focal plane.

It now remains to propagate the final distance from the focal plane to the image plane. We do so with the free-propagation point-spread function (8.28): $\psi_I = \int P_{IF} d\Sigma_F \psi_F$, which integrates to

wave field in image plane of a thin lens

$$\boxed{\psi_I(\mathbf{x}_I) = -\left(\frac{u}{v}\right) e^{ik(u+v)} \exp\left(\frac{ikx_I^2}{2(v-f)}\right) \psi_S(\mathbf{x}_S = -\mathbf{x}_I u/v).} \tag{8.33}$$

---

7. In Eq. (8.11a), the input wave at the system's entrance aperture is $\psi_S = \psi'\, t(\mathbf{x}) \propto t(\mathbf{x})$ [Eq. (8.6) and Fig. 8.2], the Fraunhofer diffraction pattern is $\psi_\mathcal{P} \propto \tilde{t}(\boldsymbol{\theta})$, and the lens produces the focal-plane field $\tilde{\psi}_F \propto \tilde{t}(\mathbf{x}_F/f)$.

Chapter 8. Diffraction

Thus (again ignoring a phase factor) the wave in the image plane is just a magnified and inverted version of the wave in the source plane, as we might have expected from geometric optics. In words, the lens acts by taking the Fourier transform of the source and then takes the Fourier transform again to recover the source structure.

8.5.4 Image Processing by a Spatial Filter in the Focal Plane of a Lens: High-Pass, Low-Pass, and Notch Filters; Phase-Contrast Microscopy

The focal plane of a lens is a convenient place to process an image by altering its Fourier transform. This process, known as *spatial filtering*, is a powerful technique. We shall gain insight into its power via several examples.

spatial filtering

In each of these examples, we assume for simplicity the one-lens system of Fig. 8.15, for which we worked out Abbé's description in the previous section. If the source wave has planar phase fronts parallel to the source plane so ψ_S is real, then the output wave in the image plane has spherical phase fronts, embodied in the phase factor $\exp[ikx_I^2/(2(v-f))]$. If, instead, one wants the output wave to have the same planar phase fronts as the input wave, one can achieve that by using a two-lens system with the lenses separated by the sum of the lenses' focal lengths and altering the Fourier transform occurring in the common focal plane between them (Ex. 8.11).

LOW-PASS FILTER: CLEANING A LASER BEAM

In low-pass filtering, a small circular aperture or "stop" is introduced into the focal plane, thereby allowing only the low-order spatial Fourier components (long-wavelength components) to be transmitted to the image plane. This produces a considerable smoothing of the wave. An application is to the output beam from a laser (Chap. 10), which ought to be smooth but in practice has high spatial frequency structure (high transverse wave numbers, short wavelengths) on account of noise and imperfections in the optics. A low-pass filter can be used to clean the beam. In the language of Fourier transforms, if we multiply the transform of the source, in the focal plane, by a small-diameter circular aperture function, we will thereby convolve the image with a broad Airy-disk smoothing function.

aperture stop to remove high spatial-frequency noise

HIGH-PASS FILTER: ACCENTUATING AN IMAGE'S FEATURES

Conversely, we can exclude the low spatial frequencies with a high-pass filter (e.g., by placing an opaque circular disk in the focal plane, centered on the optic axis). This accentuates boundaries and discontinuities in the source's image and can be used to highlight features where the gradient of the brightness is large.

high-pass filter to accentuate edges

NOTCH FILTER: REMOVING PIXELATION FROM AN IMAGE

Another type of filter is used when the image is pixelated and thus has unwanted structure with wavelength equal to pixel size. A narrow range of frequencies centered around this spatial frequency is removed by putting an appropriate filter in the focal plane; this is sometimes called a "notch filter."

notch filter to remove pixelations

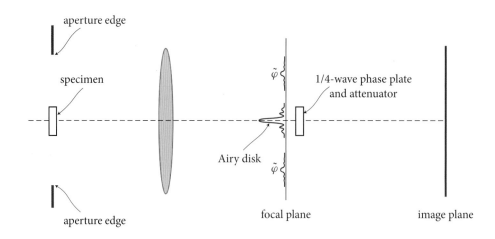

FIGURE 8.16 Schematic phase-contrast microscope.

PHASE-CONTRAST MICROSCOPY

Phase-contrast microscopy (Fig. 8.16) is a useful technique for studying small objects, such as transparent biological specimens, that modify the phase of coherent illuminating light but not its amplitude. Suppose that the differential phase change in the object is small, $|\varphi| \ll 1$, as often is the case for biological specimens. We can then write the field just after it passes through the specimen as

$$\psi_S(\mathbf{x}) = H(\mathbf{x})e^{i\varphi(\mathbf{x})} \simeq H(\mathbf{x}) + i\varphi(\mathbf{x})H(\mathbf{x}). \tag{8.34}$$

Here H is the microscope's aperture function, unity for $|\mathbf{x}| < D/2$ and zero for $|\mathbf{x}| > D/2$, with D the aperture diameter. The energy flux is not modulated, and therefore the effect of the specimen on the wave is hard to observe unless one is clever.

Equation (8.34) and the linearity of the Fourier transform imply that the wave in the focal plane is the sum of (i) the Fourier transform of the aperture function [Eq. (8.18) and Ex. 8.4] and (ii) the transform of the phase function convolved with that of the aperture [in which the fine-scale variations of the phase function dominate and push $\tilde\varphi$ to large radii in the focal plane (Fig. 8.16), where the aperture has little influence]:

$$\psi_F \sim \mathrm{jinc}\left(\frac{kD|\mathbf{x}_F|}{2f}\right) + i\tilde\varphi\left(\frac{\mathbf{x}_F}{f}\right), \quad \text{where} \quad \mathrm{jinc}(x) = J_1(x)/x. \tag{8.35}$$

converting phase changes to amplitude changes

If a high-pass filter is used to remove the Airy disk completely, then the remaining wave field in the image plane will be essentially φ magnified by v/u. However, the energy flux $F \propto (\varphi v/u)^2$ will be quadratic in the phase, and so the contrast in the image will still be small. A better technique is to phase shift the Airy disk in the focal plane by $\pi/2$ so that the two terms in Eq. (8.35) are in phase. The flux variations, $F \sim (1 \pm \varphi)^2 \simeq 1 \pm 2\varphi$, will now be linear in the phase φ. An even better procedure is

442   Chapter 8. Diffraction

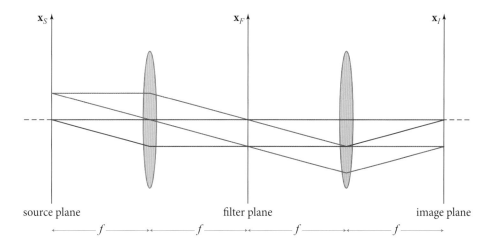

FIGURE 8.17 Two-lens system for spatial filtering.

to attenuate the Airy disk until its amplitude is comparable with the rms value of φ and also phase shift it by $\pi/2$ (as indicated by the "1/4-wave phase plate and attenuator" in Fig. 8.16). This will maximize the contrast in the final image. Analogous techniques are used in communications to interconvert amplitude-modulated and phase-modulated signals.

Exercise 8.11 **Example: Two-Lens Spatial Filter*

EXERCISES

Figure 8.17 depicts a two-lens system for spatial filtering (also sometimes called a "$4f$ system," since it involves five special planes separated by four intervals with lengths equal to the common focal length f of the two lenses). Develop a description, patterned after Abbé's, of image formation with this system. Most importantly, do the following.

(a) Show that the field at the filter plane is

$$\psi_F(\mathbf{x}_F) = -\frac{ik}{2\pi f} e^{2ikf} \tilde{\psi}_S(\mathbf{x}_F/f). \tag{8.36a}$$

This is like Eq. (8.31) for the one-lens system but with the spatially dependent phase factor $\exp[-ikx_F^2/(2(v-f))]$ gone, so aside from a multiplicative constant, the filter-plane field is precisely the Fourier transform of the source-plane field.

(b) In the filter plane we place a filter whose transmissivity we denote $\tilde{K}(-\mathbf{x}_F/f)$, so it is (proportional to) the filter-plane field that would be obtained from some source-plane field $K(-\mathbf{x}_S)$. Using the optics conventions (8.11) for the Fourier transform and its inverse, show that the image-plane field, with the filter present, is

$$\psi_I(\mathbf{x}_I) = -e^{4ikf} \Psi_S(\mathbf{x}_S = -\mathbf{x}_I). \tag{8.36b}$$

8.5 Paraxial Fourier Optics

Here Ψ_S is the convolution of the source field and the filter function

$$\Psi_S(\mathbf{x}_S) = \int K(\mathbf{x}_S - \mathbf{x}')\psi_S(\mathbf{x}')d\Sigma'. \tag{8.36c}$$

In the absence of the filter, Ψ_S is equal to ψ_S, so Eq. (8.36b) is like the image-plane field (8.33) for the single-lens system, but with the spatially dependent phase factor $\exp[-ikx_F^2/(2(v-f))]$ gone. Thus ψ_I is precisely the same as the inverted ψ_S, aside from an overall phase.

Exercise 8.12 *Problem: Convolution via Fourier Optics*
(a) Suppose that you have two thin sheets with transmission functions $t = g(x, y)$ and $t = h(x, y)$, and you wish to compute via Fourier optics the convolution

$$g \otimes h(x_o, y_o) \equiv \int\int g(x, y)h(x_o - x, y_o - y)\, dx\, dy. \tag{8.37}$$

Devise a method for doing so using Fourier optics. [Hint: Use several lenses and a projection screen with a pinhole through which passes light whose energy flux is proportional to the convolution; place the two sheets at strategically chosen locations along the optic axis, and displace one of the two sheets transversely with respect to the other.]

(b) Suppose you wish to convolve a large number of different pairs of 1-dimensional functions $\{g_1 h_1\}, \{g_2 h_2\}, \ldots$ simultaneously; that is, you want to compute

$$g_j \otimes h_j(x_o) \equiv \int g_j(x)h_j(x_o - x)dx \tag{8.38}$$

for $j = 1, 2, \ldots$. Devise a way to do this via Fourier optics using appropriately constructed transmissive sheets and cylindrical lenses.

Exercise 8.13 **Example: Transmission Electron Microscope*
In a transmission electron microscope, electrons, behaving as waves, are prepared in near-plane-wave quantum wave-packet states with transverse sizes large compared to the object ("sample") being imaged. The sample, placed in the source plane of Fig. 8.15, is sufficiently thin—with a transmission function t(**x**)—that the electron waves can pass through it, being diffracted in the process. The diffracted waves travel through the lens shown in Fig. 8.15 (the *objective lens*), then onward to and through a second lens, called the *projector lens*, which focuses them onto a fluorescent screen that shines with an energy flux proportional to the arriving electron flux. At least two lenses are needed, as in the simplest of optical microscopes and telescopes (Figs. 7.7 and 7.8), and for the same reason: to make images far larger than could be achieved with a single lens.

(a) The electrons all have the same kinetic energy $E \sim 200$ keV, to within an energy spread $\Delta E \sim 1$ eV. What is the wavelength λ of their nearly plane-wave quantum wave functions, and what is their fractional wavelength spread $\Delta\lambda/\lambda$? Your answer for λ (∼a few picometers) is so small that electron microscopy can be used to study atoms and molecules. Contrast this with light's million-fold longer wavelength ∼1 μm, which constrains it to imaging objects a million times larger than atoms.

(b) Explain why the paraxial Fourier-optics formalism that we developed in Sec. 8.5 can be used without change to analyze this electron microscope. This is true even though the photons of an ordinary microscope are in states with mean occupation numbers η huge compared to unity while the electrons have $\eta \ll 1$, and the photons have zero rest mass while the electrons have finite rest mass m with roughly the same magnitude as their kinetic energies, $E \sim mc^2$.

(c) Suppose that each magnetic lens in the electron microscope is made of two transverse magnetic quadrupoles, as described in Sec. 7.4.2. Show that, although these quadrupoles are far from axisymmetric, their combined influence on each electron's wave function is given by the axisymmetric thin-lens point-spread function (8.29) with focal length (7.66), to within the accuracy of the analysis of Sec. 7.4.2. [In practice, higher-order corrections make the combined lens sufficiently nonaxisymmetric that electron microscopes do not use this type of magnetic lens. Instead they use a truly axisymmetric lens in which the magnetic field lines lie in planes of constant azimuthal angle ϕ, and the field lines first bend the electron trajectories into helices (give them motion in the ϕ direction), then bend them radially, and then undo the helical motion.]

(d) By appropriate placement of the projector lens, the microscope can produce, in the fluorescing plane, either a vastly enlarged image of the source-plane sample, $|t(\mathbf{x})|^2$, or a large image of the modulus of that object's Fourier transform (the object's diffraction pattern), $|\tilde{t}(\mathbf{k})|^2$. Explain how each of these is achieved.

8.5.5 Gaussian Beams: Optical Cavities and Interferometric Gravitational-Wave Detectors

The mathematical techniques of Fourier optics enable us to analyze the structure and propagation of light beams that have Gaussian profiles. Such Gaussian beams are the natural output of ideal lasers; they are the real output of spatially filtered lasers; and they are widely used for optical communication, interferometry, and other practical applications. Moreover, they are the closest one can come in the real world of wave optics to the idealization of a geometric-optics pencil beam.

Consider a beam that is precisely plane-fronted, with a Gaussian profile, at location $z = 0$ on the optic axis:

$$\psi_0 = \exp\left(\frac{-\varpi^2}{\sigma_0^2}\right); \tag{8.39}$$

here $\varpi = |\mathbf{x}|$ is radial distance from the optic axis. The form of this same wave at a distance z farther down the optic axis can be computed by propagating this ψ_0 using the point-spread function (8.28) (with the distance d replaced by z). The result is

freely propagating Gaussian beam

$$\boxed{\psi_z = \frac{\sigma_0}{\sigma_z} \exp\left(\frac{-\varpi^2}{\sigma_z^2}\right) \exp\left[i\left(\frac{k\varpi^2}{2R_z} + kz - \arctan\frac{z}{z_0}\right)\right],} \qquad (8.40a)$$

where

$$\boxed{z_0 = \frac{k\sigma_0^2}{2} = \frac{\pi\sigma_0^2}{\lambda}, \quad \sigma_z = \sigma_0(1 + z^2/z_0^2)^{1/2}, \quad R_z = z(1 + z_0^2/z^2),}$$

(8.40b)

and a subscript z indicates that this quantity is a function of z. These equations for the freely propagating Gaussian beam are valid for negative z as well as positive.

From these equations we learn the following properties of the beam:

- The beam's cross sectional energy-flux distribution

$$F \propto |\psi_z|^2 \propto \exp(-\varpi^2/\sigma_z^2)$$

remains a Gaussian as the wave propagates, with a beam radius

beam radius

$$\sigma_z = \sigma_0\sqrt{1 + z^2/z_0^2}$$

that is a minimum at $z = 0$ (the beam's *waist*) and grows away from the waist, both forward and backward, in just the manner to be expected from our uncertainty-principle discussion of wavefront spreading [Eq. (8.9)]. At distances $|z| \ll z_0$ from the waist location (corresponding to a Fresnel length $r_F = \sqrt{\lambda|z|} \ll \sqrt{\pi}\sigma_0$), the beam radius is nearly constant; this is the Fresnel region. At distances $z \gg z_0$ ($r_F \gg \sqrt{\pi}\sigma_0$), the beam radius increases linearly [i.e., the beam spreads with an opening angle $\theta = \sigma_0/z_0 = \lambda/(\pi\sigma_0)$]; this is the Fraunhofer region.

- The beam's wavefronts (surfaces of constant phase) have phase

$$\varphi = k\varpi^2/(2R_z) + kz - \arctan(z/z_0) = \text{constant}.$$

Gouy phase

The arctan term (called the wave's *Gouy phase*) varies far far more slowly with changing z than does the kz term, so the wavefronts are almost precisely $z = -\varpi^2/2R_z + \text{const}$, which is a segment of a sphere of radius R_z. Thus, the wavefronts are spherical, with radii of curvature $R_z = z(1 + z_0^2/z^2)$, which is infinite (flat phase fronts) at the waist $z = 0$, decreases to a minimum of $2z_0$ at $z = z_0$ (boundary between Fresnel and Fraunhofer regions and beginning of substantial wavefront spreading), and then increases as z (gradual flattening of spreading wavefronts) when one moves deep into the Fraunhofer region.

- The Gaussian beam's form [Eqs. (8.40)] at some arbitrary location is fully characterized by three parameters: the wavelength $\lambda = 2\pi/k$, the distance z to the waist, and the beam radius σ_0 at the waist [from which one can compute the local beam radius σ_z and the local wavefront radius of curvature R_z via Eqs. (8.40b)].

One can easily compute the effects of a thin lens on a Gaussian beam by folding the ψ_z at the lens's location into the lens point-spread function (8.29). The result is a phase change that preserves the general Gaussian form of the wave but alters the distance z to the waist and the radius σ_0 at the waist. Thus, by judicious placement of lenses or mirrors and with judicious choices of the lenses' and mirrors' focal lengths, one can tailor the parameters of a Gaussian beam to fit whatever optical device one is working with. For example, if one wants to send a Gaussian beam into a self-focusing optical fiber (Exs. 7.8 and 8.14), one should place its waist at the entrance to the fiber and adjust its waist size there to coincide with that of the fiber's Gaussian mode of propagation (the mode analyzed in Ex. 8.14). The beam will then enter the fiber smoothly and will propagate steadily along the fiber, with the effects of the transversely varying index of refraction continually compensating for the effects of diffraction so as to keep the phase fronts flat and the waist size constant.

tailoring the beam parameters

Gaussian beams are used (among many other places) in interferometric gravitational-wave detectors, such as LIGO (the Laser Interferometer Gravitational-Wave Observatory). We shall learn how these *GW interferometers* work in Sec. 9.5. For the present, all we need to know is that a GW interferometer entails an optical cavity formed by mirrors facing each other, as in Fig. 7.9. A Gaussian beam travels back and forth between the two mirrors, its light superposing on itself coherently after each round trip—the light *resonates* in the cavity formed by the two mirrors. Each mirror hangs from an overhead support, and when a gravitational wave passes, it pushes the hanging mirrors back and forth with respect to each other, causing the cavity to lengthen and shorten by a tiny fraction of a light wavelength. This puts a minuscule phase shift on the resonating light, which is measured by allowing some of the light to leak out of the cavity and interfere with light from another, similar cavity (see Sec. 9.5).

use of Gaussian beams in LIGO

For the light to resonate in the cavity, the mirrors' surfaces must coincide with the Gaussian beam's wavefronts. Suppose that the mirrors are identical, with radii of curvature R, and are separated by a distance $L = 4$ km, as in LIGO. Then the beam must be symmetric around the center of the cavity, so its waist must be halfway between the mirrors. What is the smallest that the beam radius can be at the mirrors' locations $z = \pm L/2 = \pm 2$ km? From $\sigma_z = \sigma_0(1 + z^2/z_0^2)^{1/2}$ together with $z_0 = \pi \sigma_0^2/\lambda$, we see that $\sigma_{L/2}$ is minimized when $z_0 = L/2 = 2$ km. If the wavelength is $\lambda = 1.064$ μm (Nd:YAG—neodymium-doped yttrium aluminum garnet—laser light, as in LIGO), then the beam radii at the waist and at the mirrors are $\sigma_0 = \sqrt{\lambda z_0/\pi} = \sqrt{\lambda L/2\pi} = 2.6$ cm, and $\sigma_{L/2} = \sqrt{2}\sigma_0 = 3.7$ cm, respectively, and the mirrors' radii of curvature are $R_{L/2} = L = 4$ km. This was approximately the regime of parameters used for

LIGO's initial GW interferometers, which carried out a 2-year-long search for gravitational waves from autumn 2005 to autumn 2007 and then, after some sensitivity improvements, a second long search in 2009 and 2010.

advanced LIGO

A new generation of GW interferometers, called "Advanced LIGO" (LIGO Scientific Collaboration, 2015), began operating in summer 2015 and shortly thereafter discovered gravitational waves. In these GW interferometers, the spot sizes on the mirrors are enlarged, so as to reduce thermal noise by averaging over a much larger spatial sampling of thermal fluctuations of the mirror surfaces (cf. Secs. 6.8.2 and 11.9.2). How were the spot sizes enlarged? From Eqs. (8.40b) we see that, in the limit $z_0 = \pi \sigma_0^2/\lambda \to 0$, the mirrors' radii of curvature approach the cavity half-length, $R_{L/2} \to L/2$, and the beam radii on the mirrors diverge as $\sigma_{L/2} \to L\lambda/(2\pi\sigma_0) \to \infty$. This is the same instability as we discovered in the geometric-optics limit (Ex. 7.12). Advanced LIGO takes advantage of this instability by moving toward the near-unstable regime, causing the beams on the mirrors to enlarge. In Advanced LIGO's semifinal design, the mirrors' radii of curvature were set at $R_{L/2} = 2.076$ km, just 4% above the unstable point $R = L/2 = 2$ km; and Eqs. (8.40b) then tell us that $\sigma_0 = 1.15$ cm, $z_0 = 0.389$ km $\ll L/2 = 2$ km, and σ_z was pushed up by nearly a factor of two, to $\sigma_z = 6.01$ cm. The mirrors are deep into the Fraunhofer, wavefront-spreading region. In the final design the optical cavity was made slightly asymmetric with mirror radii of curvature a bit below and a bit above 2.076 km (Table 1 of LIGO Scientific Collaboration, 2015).

EXERCISES

Exercise 8.14 *Problem: Guided Gaussian Beams*
Consider a self-focusing optical fiber discussed in Ex. 7.8, in which the refractive index is

$$\mathfrak{n}(\mathbf{x}) = \mathfrak{n}_o (1 - \alpha^2 r^2)^{1/2}, \tag{8.41}$$

where $r = |\mathbf{x}|$.

(a) Write down the Helmholtz equation in cylindrical polar coordinates and seek an axisymmetric mode for which $\psi = R(r)Z(z)$, where R and Z are functions to be determined, and z measures distance along the fiber. In particular, show that there exist modes with a Gaussian radial profile that propagate along the fiber without spreading.

(b) Compute the group and phase velocities along the fiber for this mode.

Exercise 8.15 *Example: Noise Due to Scattered Light in LIGO*
In LIGO and other GW interferometers, one potential source of noise is scattered light. When the Gaussian beam in one of LIGO's cavities reflects off a mirror, a small portion of the light gets scattered toward the walls of the cavity's vacuum tube. Some of this scattered light can reflect or scatter off the tube wall and then propagate toward the distant mirror, where it scatters back into the Gaussian beam; see Fig. 8.18a (without

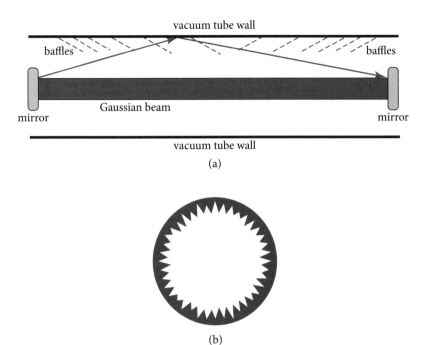

FIGURE 8.18 (a) Scattered light in LIGO's beam tube. (b) Cross section of a baffle used to reduce the noise due to scattered light.

the baffles, which are shown dashed). This is troublesome because the tube wall vibrates from sound-wave excitations and seismic excitations, and those vibrations put a phase shift on the scattered light. Although the fraction of all the light that scatters in this way is tiny, the phase shift is huge compared to that produced in the Gaussian beam by gravitational waves. When the tiny amount of scattered light with its huge oscillating phase shift recombines into the Gaussian beam, it produces a net Gaussian-beam phase shift that can be large enough to mask a gravitational wave. This exercise explores some aspects of this scattered-light noise and its control.

(a) The scattering of Gaussian-beam light off the mirror is caused by bumps in the mirror surface (imperfections). Denote by $h(\mathbf{x})$ the height of the mirror surface relative to the desired shape (a segment of a sphere with radius of curvature that matches the Gaussian beam's wavefronts). Show that, if the Gaussian-beam field emerging from a perfect mirror is $\psi^G(\mathbf{x})$ [Eq. (8.40a)] at the mirror plane, then the beam emerging from the actual mirror is $\psi'(\mathbf{x}) = \psi^G(\mathbf{x}) \exp[-i2kh(\mathbf{x})]$. The magnitude of the mirror irregularities is very small compared to a wavelength, so $|2kh| \ll 1$, and the wave field emerging from the mirror is $\psi'(\mathbf{x}) = \psi^G(\mathbf{x})[1 - i2kh(\mathbf{x})]$. Explain why the factor 1 does not contribute at all to the scattered light (where does its light go?), so the scattered light field emerging from the mirror is

$$\psi^S(\mathbf{x}) = -i\psi^G(\mathbf{x})2kh(\mathbf{x}). \tag{8.42}$$

(b) Assume that, when arriving at the vacuum-tube wall, the scattered light is in the Fraunhofer region. (You will justify this later in this example.) Then at the tube wall, the scattered light field is given by the Fraunhofer formula

$$\psi^S(\boldsymbol{\theta}) \propto \int \psi^G(\mathbf{x}) k h(\mathbf{x}) e^{i k \mathbf{x} \cdot \boldsymbol{\theta}} d\Sigma. \tag{8.43}$$

Show that the light that hits the tube wall at an angle $\theta = |\boldsymbol{\theta}|$ to the optic axis arises from irregularities in the mirror that have spatial wavelengths $\lambda_{\text{mirror}} \sim \lambda/\theta$. The radius of the beam tube is $\mathcal{R} = 60$ cm in LIGO, and the length of the tube (distance between cavity mirrors) is $L = 4$ km. What is the spatial wavelength of the mirror irregularities that scatter light to the tube wall at distances $z \sim L/2$ (which can then reflect or scatter off the wall toward the distant mirror and there scatter back into the Gaussian beam)? Show that for these irregularities, the tube wall is indeed in the Fraunhofer region. [Hint: The irregularities have a coherence length of only a few wavelengths λ_{mirror}.]

(c) In the initial LIGO interferometers, the mirrors' scattered light consisted of two components: one peaked strongly toward small angles so it hit the distant tube wall (e.g., at $z \sim L/2$), and the other roughly isotropically distributed. What is the size of the irregularities that produced the isotropic component?

(d) To reduce substantially the amount of scattered light reaching the distant mirror via reflection or scattering from the tube wall, a set of baffles was installed in the tube, in such a way as to hide the wall from scattered light (dashed lines in Fig. 8.18). The baffles have an angle of 35° to the tube wall, so when light hits a baffle, it reflects at a steep angle, $\sim 70°$ toward the opposite tube wall, and after a few bounces gets absorbed. However, a small portion of the scattered light can now *diffract* off the top of each baffle and then propagate to the distant mirror and scatter back into the main beam. Especially troublesome is the case of a mirror in the center of the beam tube's cross section, because light that scatters off such a mirror travels nearly the same total distance from the mirror to the top of some baffle and then to the distant mirror, independent of the azimuthal angle ϕ on the baffle at which it diffracts. There is then a danger of *coherent superposition* of all scattered light that diffracts off all angular locations around any given baffle—and coherent superposition means a much enlarged net noise (a variant of the Poisson or Arago spot discussed in the footnote to Ex. 8.9). To protect against any such coherence, the baffles in the LIGO beam tubes are serrated (i.e., they have sawtooth edges), and the heights of the teeth are drawn from a random (Gaussian) probability distribution (Fig. 8.18b). The typical tooth heights are large enough to extend through about six Fresnel zones. How wide is each Fresnel zone at the baffle location, and correspondingly, how high must be the typical baffle tooth? By approximately how much do the random serrations reduce the light-scattering noise, relative to what it would be with no serrations and with coherent scattering? [Hint: See part e. There are two ways that the noise is reduced: (i) The breaking of coherence of scattered light by the randomness of serrated tooth height, which causes the phase of the scattered light diffracting off different teeth to be randomly different. (ii) The reduction in energy flux of the scattered light due to the teeth reaching through six Fresnel zones, on average.]

(e) To aid you in answering part d, show that the propagator (point-spread function) for light that begins at the center of one mirror, travels to the edge of a baffle, and then propagates to the center of the distant mirror is

$$P \propto \int_0^{2\pi} \exp\left(\frac{ikR^2(\phi)}{2\ell_{\text{red}}}\right) d\phi, \quad \text{where} \quad \frac{1}{\ell_{\text{red}}} = \frac{1}{\ell} + \frac{1}{L-\ell}. \quad (8.44)$$

Here $R(\phi)$ is the radial distance of the baffle edge from the beam tube axis at azimuthal angle ϕ around the beam tube, and at distance ℓ down the tube from the scattering mirror. Note that ℓ_{red} is the "reduced baffle distance" analogous to the reduced mass in a binary system. One can show that the time-varying part of the scattered-light amplitude (i.e., the part whose time dependence is produced by baffle vibrations) is proportional to this propagator. Explain why this is plausible. Then explain how the baffle serrations, embodied in the ϕ dependence of $R(\phi)$, produce the reduction of scattered-light amplitude in the manner described in part c.

8.6 Diffraction at a Caustic

In Sec. 7.5, we described how caustics can be formed in general in the geometric-optics limit (e.g., on the bottom of a swimming pool when the water's surface is randomly rippled, or behind a gravitational lens). We chose as an example an imperfect lens illuminated by a point source (Fig. 7.13) and showed how a pair of images would merge as the transverse distance x of the observer from the caustic decreases to zero. That merging was controlled by maxima and minima of the scaled time delay $\tilde{t}$ or equivalently, the phase φ of the light that originates at transverse location a just before the lens and arrives at transverse location x in the observation plane. We argued that for a fold caustic, this phase, when expressed as a Taylor series in a, has the standard form $\varphi(\tilde{a}, \tilde{x}) = \tilde{a}^3/3 - \tilde{x}\tilde{a}$, where $\tilde{a}$ and $\tilde{x}$ are rescaled a and x [Eq. (7.72)]. Using this φ, we showed that the magnification $\mathcal{M}$ of the images diverged $\propto \tilde{x}^{-1/2}$ as the caustic was approached ($\tilde{x} \to 0$), and then $\mathcal{M}$ crashed to zero just past the caustic (the two images disappeared). This singular behavior raised the question of what happens when we take into account the finite wavelength of the wave while still assuming the source has negligible size.

geometric optics divergence at a fold caustic

We are now in a position to answer this question. We simply use the Helmholtz-Kirchhoff formula (8.6) to write the expression for the amplitude measured at position $\tilde{x}$ in the form (Ex. 8.17)

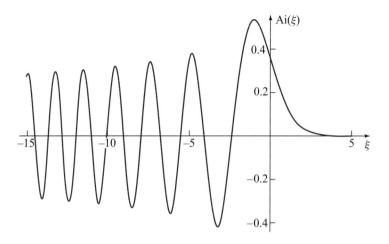

FIGURE 8.19 The Airy function Ai(ξ) describing diffraction at a fold caustic. The argument is $\xi = -\tilde{x}$, where $\tilde{x}$ is rescaled distance from the caustic.

$$\psi(\tilde{x}) \propto \int d\tilde{a}\, e^{i\varphi(\tilde{a},\tilde{x})} = \int d\tilde{a}(\cos\varphi + i\sin\varphi), \quad \text{where} \quad \varphi = \frac{\tilde{a}^3}{3} - \tilde{x}\tilde{a}. \tag{8.45}$$

The phase φ varies rapidly with location $\tilde{a}$ in the lens at large $|\tilde{a}|$, so we can treat the limits of integration as $\pm\infty$. Because $\varphi(\tilde{a}, \tilde{x})$ is odd in $\tilde{a}$, the sine term integrates to zero, and the cosine integral turns out to be the Airy function

diffraction at a caustic: Airy function

$$\psi \propto \int_{-\infty}^{\infty} d\tilde{a}\, \cos(\tilde{a}^3/3 - \tilde{x}\tilde{a}) = 2\pi\, \text{Ai}(-\tilde{x}). \tag{8.46}$$

The Airy function Ai(ξ) is displayed in Fig. 8.19.

The asymptotic behavior of Ai(ξ) is

$$\text{Ai}(\xi) \sim \pi^{-1/2}|\xi|^{-1/4}\sin(2|\xi|^{3/2}/3 + \pi/4), \quad \text{for } \xi \to -\infty$$

$$\sim \frac{e^{-2\xi^{3/2}/3}}{2\pi^{1/2}\xi^{1/4}}, \quad \text{for } \xi \to \infty. \tag{8.47}$$

diffractive field structure at a fold caustic

We see that the amplitude ψ remains finite as the caustic is approached (as $\tilde{x} = -\xi \to 0$) instead of diverging as in the geometric-optics limit, and it decreases smoothly toward zero when the caustic is passed, instead of crashing instantaneously to zero. For $\tilde{x} > 0$ ($\xi = -\tilde{x} < 0$; left part of Fig. 8.19), where an observer sees two geometric-optics images, the envelope of ψ diminishes $\propto \tilde{x}^{-1/4}$, so the energy flux $|\psi|^2$ decreases $\propto \tilde{x}^{-1/2}$ just as in the geometric-optics limit. What is actually seen is a series of bands of alternating dark and light with their spacing calculated by using $\Delta(2\tilde{x}^{3/2}/3) = \pi$ or $\Delta\tilde{x} \propto \tilde{x}^{-1/2}$. At sufficient distance from the caustic, it is not possible to resolve these bands, and a uniform illumination of average flux is observed, so we recover the geometric-optics limit.

These near-caustic scalings and others in Ex. 8.16, like the geometric-optics scalings (Sec. 7.5), are a universal property of this type of caustic (the simplest caustic of all, the "fold").

There is a helpful analogy, familiar from quantum mechanics. Consider a particle in a harmonic potential well in a highly excited state. Its wave function is given in the usual way using Hermite polynomials of large order. Close to the classical turning point, these functions change from being oscillatory to having exponential decay, just like the Airy function (and if we were to expand about the turning point, we would recover Airy functions). Of course, what is happening is that the probability density of finding the particle close to its turning point diverges classically, because it is moving vanishingly slowly at the turning point; the oscillations stem from interference between waves associated with the particle moving in opposite directions at the turning point.

analogy with quantum wave function in harmonic oscillator potential

For light near a caustic, if we consider the transverse component of the photon motion, then we have essentially the same problem. The field's oscillations stem from interference of the waves associated with the motions of the photons in two geometric-optics beams coming from slightly different directions and thus having slightly different transverse photon speeds.

This is our first illustration of the formation of large-contrast interference fringes when only a few beams are combined. We shall meet other examples of such interference in the next chapter.

EXERCISES

Exercise 8.16 *Problem: Wavelength Scaling at a Fold Caustic*
For the fold caustic discussed in the text, assume that the phase change introduced by the imperfect lens is nondispersive, so that the $\varphi(\tilde{a}, \tilde{x})$ in Eq. (8.45) satisfies $\varphi \propto \lambda^{-1}$. Show that the peak magnification of the interference fringes at the caustic scales with wavelength, $\propto \lambda^{-4/3}$. Also show that the spacing Δx of the fringes near a fixed observing position x is $\propto \lambda^{2/3}$. Discuss qualitatively how the fringes will be affected if the source has a finite size.

Exercise 8.17 *Problem: Diffraction at Generic Caustics*
In Sec. 7.5, we explored five elementary (generic) caustics that can occur in geometric optics. Each is described by its phase $\varphi(\tilde{a}, \tilde{b}; \tilde{x}, \tilde{y}, \tilde{z})$ for light arriving at an observation point with Cartesian coordinates $\{\tilde{x}, \tilde{y}, \tilde{z}\}$ along paths labeled by $(\tilde{a}, \tilde{b})$.

(a) Suppose the (monochromatic) wave field $\psi(\tilde{x}, \tilde{y}, \tilde{z})$ that exhibits one of these caustics is produced by plane-wave light that impinges orthogonally on a phase-shifting surface on which are laid out Cartesian coordinates $(\tilde{a}, \tilde{b})$. Using the Helmholtz-Kirchhoff diffraction integral (8.6), show that the field near a caustic is given by

$$\psi(\tilde{x}, \tilde{y}, \tilde{z}) \propto \int\!\!\int d\tilde{a}\, d\tilde{b}\, e^{i\varphi(\tilde{a}, \tilde{b}; \tilde{x}, \tilde{y}, \tilde{z})}. \tag{8.48}$$

(b) In the text we evaluated this near-caustic diffraction integral for a fold caustic, obtaining the Airy function. For the higher-order elementary caustics, the integral cannot be evaluated analytically in terms of standard functions. To get insight into the influence of finite wavelength, evaluate the integral numerically for the case of a cusp caustic, $\varphi = \frac{1}{4}\tilde{a}^4 - \frac{1}{2}\tilde{z}\tilde{a}^2 - \tilde{x}\tilde{a}$, and plot the real and imaginary parts of ψ ($\Re\psi$ and $\Im\psi$). Before doing so, though, guess what these parts will look like. As foundations for this guess, (i) pay attention to the shape $\tilde{x} = \pm 2(\tilde{z}/3)^{3/2}$ of the caustic in the geometric-optics approximation, (ii) notice that away from the cusp point, each branch of the caustic is a fold, whose ψ is the Airy function (Fig. 8.19), and (iii) note that the oscillating ψ associated with each branch interferes with that associated with the other branch. The numerical computation may take awhile, so make a wise decision from the outset as to the range of $\tilde{z}$ and $\tilde{x}$ to include in your computations and plots. If your computer is really slow, you may want to prove that the integral is symmetric in $\tilde{x}$ and so restrict yourself to positive $\tilde{x}$, and argue that the qualitative behaviors of $\Re\psi$ and $\Im\psi$ must be the same, and so restrict yourself to $\Re\psi$.

Bibliographic Note

Hecht (2017) gives a pedagogically excellent treatment of diffraction at roughly the same level as this chapter, but in much more detail, with many illustrations and intuitive explanations. Other nice treatments at about our level will be found in standard optics textbooks, including Jenkins and White (1976), Brooker (2003), Sharma (2006), Bennett (2008), Ghatak (2010), and, from an earlier era, Longhurst (1973) and Welford (1988). The definitive treatment of diffraction at an advanced and thorough level is that of Born and Wolf (1999). For an excellent and thorough treatment of paraxial Fourier optics and spatial filtering, see Goodman (2005). The standard textbooks say little or nothing about diffraction at caustics, though they should; for this, we recommend the brief treatment by Berry and Upstill (1980) and the thorough treatment by Nye (1999).

CHAPTER NINE

Interference and Coherence

When two Undulations, from different Origins, coincide either perfectly or very nearly in Direction, their joint effect is a Combination of the Motions belonging to each.

THOMAS YOUNG (1802)

9.1 Overview

In the last chapter, we considered superpositions of waves that pass through a (typically large) aperture. The foundation for our analysis was the Helmholtz-Kirchhoff expression (8.4) for the field at a chosen point $\mathcal{P}$ as a sum of contributions from all points on a closed surface surrounding $\mathcal{P}$. The spatially varying field pattern resulting from this superposition of many different contributions is known as diffraction.

In this chapter, we continue our study of superposition, but for the special case where only two (or at most, several) discrete beams are being superposed. For this special case one uses the term *interference* rather than diffraction. Interference is important in a wide variety of practical instruments designed to measure or use the spatial and temporal structures of electromagnetic radiation. However, interference is not just of practical importance. Attempting to understand it forces us to devise ways of describing the radiation field that are independent of the field's origin and of the means by which it is probed. Such descriptions lead us naturally to the fundamental concept of coherence (Sec. 9.2).

The light from a distant, monochromatic point source is effectively a plane wave; we call it "perfectly coherent" radiation. In fact, there are two different types of coherence present: *lateral or spatial coherence* (coherence in the angular structure of the radiation field), and *temporal or longitudinal coherence* (coherence in the field's temporal structure, which clearly must imply something also about its frequency structure). We shall see in Sec. 9.2 that for both types of coherence there is a measurable quantity, called the *degree of coherence*, that is the Fourier transform of either the radiation's angular intensity distribution $I(\boldsymbol{\alpha})$ (energy flux per unit angle or solid angle, as a function of direction $\boldsymbol{\alpha}$) or its spectral energy flux $F_\omega(\omega)$ (energy flux per unit angular frequency ω, as a function of angular frequency).

Interspersed with our development of the theory of coherence are two historical devices with modern applications: (i) the stellar interferometer (Sec. 9.2.5), by which Michelson measured the diameters of Jupiter's moons and several bright stars using

> **BOX 9.1. READERS' GUIDE**
>
> - This chapter depends substantially on
>   - Secs. 8.2, 8.3, and 8.5.5 and
>   - correlation functions, spectral densities, and the Wiener-Khintchine theorem for random processes (Sec. 6.4).
> - The concept of coherence length or coherence time, as developed in this chapter, is used in Chaps. 10, 15, 16, and 23.
> - Interferometry as developed in this chapter, especially in Sec. 9.5, is a foundation for the discussion of gravitational-wave detection in Sec. 27.6.
> - Nothing else in this book relies substantially on this chapter.

spatial coherence; and (ii) the Michelson interferometer and its practical implementation in a Fourier-transform spectrometer (Sec. 9.2.7), which use temporal coherence to measure electromagnetic spectra (e.g., the spectral energy flux of the cosmic microwave background radiation). After developing our full formalism for coherence, we go on in Sec. 9.3 to apply it to the operation of radio telescope arrays, which function by measuring the spatial coherence of the radiation field.

In Sec. 9.4, we turn to multiple-beam interferometry, in which incident radiation is split many times into several different paths and then recombined. A simple example is an etalon, made from two parallel, reflecting surfaces. A Fabry-Perot cavity interferometer, in which light is trapped between two highly reflecting mirrors (e.g., in a laser), is essentially a large-scale etalon. In Secs. 9.4.3 and 9.5, we discuss a number of applications of Fabry-Perot interferometers, including lasers, their stabilization, manipulations of laser light, the optical frequency comb, and laser interferometer gravitational-wave detectors.

Finally, in Sec. 9.6, we turn to the intensity interferometer. This has not proved especially powerful in application but does illustrate some quite subtle issues of physics and, in particular, highlights the relationship between the classical and quantum theories of light.

9.2 Coherence

9.2.1 Young's Slits

Young's slits

The most elementary example of interference is provided by Young's slits. Suppose two long, narrow, parallel slits are illuminated coherently by monochromatic light from a distant source that lies on the perpendicular bisector of the line joining the slits (the

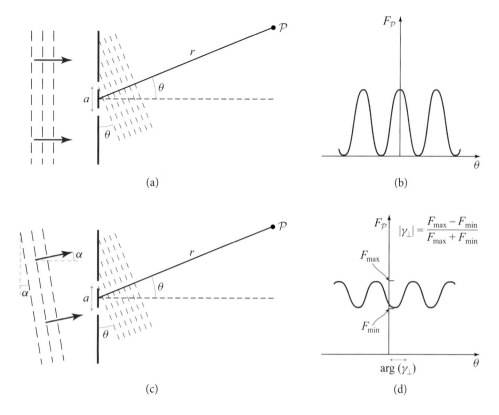

FIGURE 9.1 (a) Young's slits. (b) Interference fringes observed in a transverse plane [Eq. (9.1b)]. (c) The propagation direction of the incoming waves is rotated to make an angle α to the optic axis; as a result, the angular positions of the interference fringes in panel b are shifted by $\Delta\theta = \alpha$ [Eq. (9.3); not shown]. (d) Interference fringes observed from an extended source [Eq. (9.8)].

optic axis), so an incident wavefront reaches the slits simultaneously (Fig. 9.1a). This situation can be regarded as having only one lateral dimension because of translation invariance in the other. The waves from the slits (effectively, two 1-dimensional beams) fall onto a screen in the distant, Fraunhofer region, and there they interfere. The Fraunhofer interference pattern observed at a point $\mathcal{P}$, whose position is specified using the polar coordinates $\{r, \theta\}$ shown in Fig. 9.1, is proportional to the spatial Fourier transform of the transmission function [Eq. (8.11a)]. If the slits are very narrow, we can regard the transmission function as two delta functions, separated by the slit spacing a, and its Fourier transform will be

$$\psi(\theta) \propto e^{-ika\theta/2} + e^{ika\theta/2} \propto \cos\left(\frac{ka\theta}{2}\right), \qquad (9.1a)$$

where $k = 2\pi/\lambda$ is the light's wave number, and a is the slit's separation. (That we can sum the wave fields from the two slits in this manner is a direct consequence of the linearity of the underlying wave equation.) The *energy flux* (energy per unit time crossing a unit area) at $\mathcal{P}$ (at angle θ to the optic axis) will be

$$\boxed{F(\theta) \propto |\psi|^2 \propto \cos^2(ka\theta/2);} \qquad (9.1b)$$

interference fringes from Young's slits

cf. Fig. 9.1b. The alternating regions of dark and bright illumination in this flux distribution are known as *interference fringes*. Notice that the flux falls to zero between the bright fringes. This will be very nearly so even if (as is always the case in practice) the field is slightly nonmonochromatic, that is, even if the field hitting the slits has the form $e^{i[\omega_o t + \delta\varphi(t)]}$, where $\omega_o = c/k$ is the light's average angular frequency, and $\delta\varphi(t)$ is a phase [not to be confused with the light's full phase $\varphi = \omega_o t + \delta\varphi(t)$], which varies randomly on a timescale extremely long compared to $1/\omega_o$.[1] Notice also that there are many fringes, symmetrically disposed with respect to the optic axis. [If we were to take account of the finite width $w \ll a$ of the two slits, then we would find, by contrast with Eq. (9.1b), that the actual number of fringes is finite, in fact of order a/w; cf. Fig. 8.6 and associated discussion.] This type of interferometry is sometimes known as *interference by division of the wavefront*.

interference by division of the wavefront

Young's slits in quantum mechanics

Of course, this Young's slits experiment is familiar from quantum mechanics, where it is often used as a striking example of the nonparticulate behavior of electrons (e.g., Feynman, Leighton, and Sands, 2013, Vol. III, Chap. 1). Just as for electrons, so also for photons it is possible to produce interference fringes even if only one photon is in the apparatus at any time, as was demonstrated in a famous experiment performed by G. I. Taylor in 1909. However, our concerns in this chapter are with the classical limit, where many photons are present simultaneously and their fields can be described by Maxwell's equations. In the next subsection we depart from the usual quantum mechanical treatment by asking what happens to the fringes when the source of radiation is spatially extended.

EXERCISES

Exercise 9.1 *Problem: Single-Mirror Interference*
X-rays with wavelength 8.33 Å (0.833 nm) coming from a point source can be reflected at shallow angles of incidence from a plane mirror. The direct ray from a point source

---

1. More precisely, if $\delta\varphi(t)$ wanders by $\sim \pi$ on a timescale $\tau_c \gg 2\pi/\omega_o$ (the waves' *coherence time*), then the waves are contained in a bandwidth $\Delta\omega_o \sim 2\pi/\tau_c \ll \omega_o$ centered on ω_o, k is in a band $\Delta k \sim k\Delta\omega/\omega_o$, and the resulting superposition of precisely monochromatic waves has fringe minima with fluxes $F_{\min}$ that are smaller than the maxima by $F_{\min}/F_{\max} \sim (\pi \Delta\omega/\omega_o)^2 \ll 1$. [One can see this in order of magnitude by superposing the flux (9.1b) with wave number k and the same flux with wave number $k + \Delta k$.] Throughout this section, until Eq. (9.15), we presume that the waves have such a small bandwidth (such a long coherence time) that this $F_{\min}/F_{\max}$ is completely negligible; for example, $1 - F_{\min}/F_{\max}$ is far closer to unity than any fringe visibility V [Eq. (9.8)] that is of interest to us. This can be achieved in practice by either controlling the waves' source or by band-pass filtering the measured signals just before detecting them.

to a detector 3 m away interferes with the reflected ray to produce fringes with spacing 25 μm. Calculate the distance of the X-ray source from the mirror plane.

9.2.2 Interference with an Extended Source: Van Cittert-Zernike Theorem

We approach the topic of extended sources in steps. Our first step was taken in the last subsection, where we dealt with an idealized, single, incident plane wave, such as might be produced by an ideal, distant laser. We have called this type of radiation "perfectly coherent," which we have implicitly taken to mean that the field oscillates with a fixed angular frequency ω_o and a randomly but very slowly varying phase $\delta\varphi(t)$ (see footnote 1), and thus, for all practical purposes, there is a time-independent phase difference between any two points in the region under consideration.

perfect coherence

As our second step, we keep the incoming waves perfectly coherent and perfectly planar, but change their incoming direction in Fig. 9.1 so it makes a small angle α to the optic axis (and correspondingly, its wavefronts make an angle α to the plane of the slits) as shown in Fig. 9.1c. Then the distribution of energy flux in the Fraunhofer diffraction pattern on the screen will be modified to

$$F(\theta) \propto |e^{-ika(\theta-\alpha)/2} + e^{+ika(\theta-\alpha)/2}|^2 \propto \cos^2\left(\frac{ka(\theta-\alpha)}{2}\right)$$

interference fringes for perfectly coherent waves from angle α

$$\propto \{1 + \cos[ka(\theta-\alpha)]\}. \tag{9.2}$$

Notice that, as the direction α of the incoming waves is varied, the locations of the bright and dark fringes change by $\Delta\theta = \alpha$, but the fringes remain fully sharp (their minima remain essentially zero; cf. footnote 1). Thus, the positions of the fringes carry information about the direction to the source.

Now, in our third and final step, we deal with an extended source (i.e., one whose radiation comes from a finite range of angles α), with (for simplicity) $|\alpha| \ll 1$. We assume that the source is monochromatic (and in practice we can make it very nearly monochromatic by band-pass filtering the waves just before detection). However, in keeping with how all realistic monochromatic sources (including band-pass filtered sources) behave, we give it a randomly fluctuating phase $\delta\varphi(t)$ [and amplitude $A(t)$], and require that the timescale on which the phase and amplitude wander (the waves' coherence time) be long compared to the waves' period $2\pi/\omega_o$; cf. footnote 1.

We shall also assume that the sources of the light propagating in different directions are independent and uncorrelated. Typically, they are separate electrons, ions, atoms, or molecules. To make this precise, we write the field in the form[2]

$$\Psi(x,z,t) = e^{i(kz-\omega_o t)} \int \psi(\alpha,t) e^{ik\alpha x} d\alpha, \tag{9.3}$$

wave field from extended source

---

2. As in Chap. 8, we denote the full field by Ψ and reserve ψ to denote the portion of the field from which a monochromatic part $e^{-i\omega_o t}$ or $e^{i(kz-\omega_o t)}$ has been factored out.

where $\psi(\alpha, t) = Ae^{-i\delta\varphi}$ is the slowly wandering complex amplitude of the waves from direction α. When we consider the total flux arriving at a given point (x, z) from two different directions α_1 and α_2 and average it over times long compared to the waves' coherence time, then we lose all interference between the two contributions:

incoherent superposition of radiation

$$\overline{|\psi(\alpha_1, t) + \psi(\alpha_2, t)|^2} = \overline{|\psi(\alpha_1, t)|^2} + \overline{|\psi(\alpha_2, t)|^2}. \tag{9.4}$$

Such radiation is said to be incoherent in the incoming angle α, and we say that the contributions from different directions *superpose incoherently*. This is just a fancy way of saying that their intensities (averaged over time) add linearly.

The angularly incoherent light from our extended source is sent through two Young's slits and produces fringes on a screen in the distant Fraunhofer region. We assume that the coherence time for the light from each source point is very long compared to the difference in light travel time to the screen via the two different slits. Then the light from each source point in the extended source forms the sharp interference fringes described by Eq. (9.2). However, because contributions from different source directions add incoherently, the flux distribution on the screen is a linear sum of the fluxes from source points:

$$F(\theta) \propto \int d\alpha\, I(\alpha)\{1 + \cos[ka(\theta - \alpha)]\}. \tag{9.5}$$

Here $I(\alpha)d\alpha \propto \overline{|\psi(\alpha, t)|^2}d\alpha$ is the flux incident on the plane of the slits from the infinitesimal range $d\alpha$ of directions, so $I(\alpha)$ is the radiation's *intensity* (its energy per unit time falling on a unit area and coming from a unit angle). The remainder of the integrand, $1 + \cos[ka(\theta - \alpha)]$, is the Fraunhofer diffraction pattern [Eq. (9.2)] for coherent radiation from direction α.

We presume that the range of angles present in the waves, $\Delta\alpha$, is large compared to their fractional bandwidth $\Delta\alpha \gg \Delta\omega/\omega_o$; so, whereas the finite but tiny bandwidth produced negligible smearing out of the interference fringes (see footnote 1 in this chapter), the finite but small range of directions may produce significant smearing [i.e., the minima of $F(\theta)$ might not be very sharp]. We quantify the fringes' non-sharpness and their locations by writing the slit-produced flux distribution (9.5) in the form

interference fringes from extended source

$$\boxed{F(\theta) = F_S[1 + \Re\{\gamma_\perp(ka)e^{-ika\theta}\}],} \tag{9.6a}$$

where

$$\boxed{F_S \equiv \int d\alpha\, I(\alpha)} \tag{9.6b}$$

(subscript S for "source") is the total flux arriving at the slits from the source, and

degree of lateral coherence

$$\boxed{\gamma_\perp(ka) \equiv \frac{\int d\alpha\, I(\alpha)e^{ika\alpha}}{F_S}} \tag{9.7a}$$

is defined as the radiation's *degree of spatial (or lateral) coherence*.[3] The phase of $\gamma_\perp$ determines the angular locations of the fringes; its modulus determines their depth (the amount of their smearing due to the source's finite angular size).

The nonzero value of $\gamma_\perp(ka)$ reflects the existence of some degree of relative coherence between the waves arriving at the two slits, whose separation is a. The radiation can have this finite spatial coherence, despite its complete lack of angular coherence, because each angle contributes coherently to the field at the two slits. The lack of coherence for different angles reduces the net spatial coherence (smears the fringes), but it does not drive the coherence all the way to zero (does not completely destroy the fringes).

Equation (9.7a) states that the degree of lateral coherence of the radiation from an extended, angularly incoherent source is the Fourier transform of the source's angular intensity pattern. Correspondingly, if one knows the degree of lateral coherence as a function of the (dimensionless) distance ka, from it one can reconstruct the source's angular intensity pattern by Fourier inversion:

$$\boxed{I(\alpha) = F_S \int \frac{d(ka)}{2\pi} \gamma_\perp(ka) e^{-ika\alpha}.} \qquad (9.7b)$$

The two Fourier relations (9.7a) and (9.7b) make up the *van Cittert-Zernike theorem*. In Ex. 9.8, we shall see that this theorem is a complex-variable version of Chap. 6's *Wiener-Khintchine theorem* for random processes.

van Cittert-Zernike theorem for lateral coherence

Because of its Fourier-transform relationship to the source's angular intensity pattern $I(\alpha)$, the degree of spatial coherence $\gamma_\perp(ka)$ is of great practical importance. For a given choice of ka (a given distance between the slits), $\gamma_\perp$ is a complex number that one can read off the interference fringes of Eq. (9.6a) and Fig. 9.1d as follows. Its modulus is

$$\boxed{|\gamma_\perp| \equiv V = \frac{F_{\max} - F_{\min}}{F_{\max} + F_{\min}},} \qquad (9.8)$$

where $F_{\max}$ and $F_{\min}$ are the maximum and minimum values of the flux F on the screen; and its phase $\arg(\gamma_\perp)$ is ka times the displacement $\Delta\theta$ of the centers of the bright fringes from the optic axis. The modulus (9.8) is called the *fringe visibility*, or simply the *visibility*, because it measures the fractional contrast in the fringes [Eq. (9.8)], and this name is the reason for the symbol V. Analogously, the complex quantity $\gamma_\perp$ (or a close relative) is sometimes known as the *complex fringe visibility*. Notice that V can lie anywhere in the range from zero (no contrast; fringes completely undetectable) to unity (monochromatic plane wave; contrast as large as possible).

fringe visibility, or visibility

complex fringe visibility

---

3. In Sec. 9.2.6, we introduce the degree of temporal coherence (also known as longitudinal coherence) to describe the correlation along the direction of propagation. The correlation in a general, lateral and longitudinal, direction is called the degree of coherence (Sec. 9.2.8).

phase of $\gamma_\perp$

When the phase $\arg(\gamma_\perp)$ of the complex visibility (degree of coherence) is zero, there is a bright fringe precisely on the optic axis. This will be the case for a source that is symmetric about the optic axis, for example. If the symmetry point of such a source is gradually moved off the optic axis by an angle $\delta\alpha$, the fringe pattern will shift correspondingly by $\delta\theta = \delta\alpha$, which will show up as a corresponding shift in the argument of the fringe visibility, $\arg(\gamma_\perp) = ka\delta\alpha$.

The above analysis shows that Young's slits, even when used virtually, are nicely suited to measuring both the modulus and the phase of the complex fringe visibility (the degree of spatial coherence) of the radiation from an extended source.

9.2.3 More General Formulation of Spatial Coherence; Lateral Coherence Length

It is not necessary to project the light onto a screen to determine the contrast and angular positions of the fringes. For example, if we had measured the field at the locations of the two slits, we could have combined the signals electronically and cross correlated them numerically to determine what the fringe pattern would be with slits. All we are doing with the Young's slits is sampling the wave field at two different points, which we now label 1 and 2. Observing the fringes corresponds to adding a phase φ ($= ka\theta$) to the field at one of the points and then adding the fields and measuring the flux $\propto |\psi_1 + \psi_2 e^{i\varphi}|^2$ averaged over many periods. Now, since the source is far away, the rms value of the wave field will be the same at the two slits: $\overline{|\psi_1|^2} = \overline{|\psi_2|^2} \equiv \overline{|\psi|^2}$. We can therefore express this time-averaged flux in the symmetric-looking form

$$F(\varphi) \propto \overline{(\psi_1 + \psi_2 e^{i\varphi})(\psi_1^* + \psi_2^* e^{-i\varphi})}$$

$$\propto 1 + \Re\left(\frac{\overline{\psi_1 \psi_2^*}}{\overline{|\psi|^2}} e^{-i\varphi}\right). \tag{9.9}$$

degree of spatial coherence

Here a bar denotes an average over times long compared to the coherence times for ψ_1 and ψ_2. Comparing with Eq. (9.6a) and using $\varphi = ka\theta$, we identify

$$\boxed{\gamma_{\perp 12} = \frac{\overline{\psi_1 \psi_2^*}}{\overline{|\psi|^2}}} \tag{9.10}$$

as the *degree of spatial coherence* in the radiation field between the two points 1 and 2. Equation (9.10) is the general definition of degree of spatial coherence. Equation (9.6a) is the special case for points separated by a lateral distance a.

If the radiation field is strongly correlated between the two points, we describe it as having strong spatial or lateral coherence. Correspondingly, we shall define a field's **spatial or lateral coherence length** *lateral coherence length* $l_\perp$ as the linear size of a region over which the field is strongly correlated (has $V = |\gamma_\perp| \sim 1$). If the angle subtended by the source is $\sim \delta\alpha$, then by virtue of the van Cittert-Zernike theorem [Eqs. (9.7)] and the usual reciprocal relation for Fourier transforms, the radiation field's lateral coherence length will be

$$\boxed{l_\perp \sim \frac{2\pi}{k\,\delta\alpha} = \frac{\lambda}{\delta\alpha}.} \qquad (9.11)$$

This relation has a simple physical interpretation. Consider two beams of radiation coming from opposite sides of the brightest portion of the source. These beams are separated by the incoming angle $\delta\alpha$. As one moves laterally in the plane of the Young's slits, one sees a varying relative phase delay between these two beams. The coherence length $l_\perp$ is the distance over which the variations in that relative phase delay are of order 2π: $k\,\delta\alpha\,l_\perp \sim 2\pi$.

9.2.4 Generalization to 2 Dimensions

We have so far just considered a 1-dimensional intensity distribution $I(\alpha)$ observed through the familiar Young's slits. However, most sources will be 2-dimensional, so to investigate the full radiation pattern, we should allow the waves to come from 2-dimensional angular directions $\boldsymbol{\alpha}$:

$$\Psi = e^{i(kz-\omega_0 t)} \int \psi(\boldsymbol{\alpha},t)e^{ik\boldsymbol{\alpha}\cdot\mathbf{x}}d^2\alpha \equiv e^{i(kz-\omega_0 t)}\psi(\mathbf{x},t) \qquad (9.12a)$$

[where $\psi(\boldsymbol{\alpha},t)$ is slowly varying in time], and we should use several pairs of slits aligned along different directions. Stated more generally, we should sample the wave field (9.12a) at a variety of points separated by a variety of 2-dimensional vectors $\mathbf{a}$ transverse to the direction of wave propagation. The complex visibility (degree of spatial coherence) will then be a function of $k\mathbf{a}$,

$$\boxed{\gamma_\perp(k\mathbf{a}) = \frac{\overline{\psi(\mathbf{x},t)\psi^*(\mathbf{x}+\mathbf{a},t)}}{\overline{|\psi|^2}},} \qquad (9.12b)$$

complex visibility, or degree of spatial coherence

and the van Cittert-Zernike theorem (9.7) will take the 2-dimensional form

$$\boxed{\gamma_\perp(k\mathbf{a}) = \frac{\int d\Omega_\alpha I(\boldsymbol{\alpha})e^{ik\mathbf{a}\cdot\boldsymbol{\alpha}}}{F_S},} \qquad (9.13a)$$

2-dimensional van Cittert-Zernike theorem

$$\boxed{I(\boldsymbol{\alpha}) = F_S \int \frac{d^2(k a)}{(2\pi)^2}\gamma_\perp(k\mathbf{a})e^{-ik\mathbf{a}\cdot\boldsymbol{\alpha}}.} \qquad (9.13b)$$

Here $I(\boldsymbol{\alpha}) \propto \overline{|\psi(\boldsymbol{\alpha},t)|^2}$ is the source's *intensity* (energy per unit time crossing a unit area from a unit solid angle $d\Omega_\alpha$); $F_S = \int d\Omega_\alpha I(\boldsymbol{\alpha})$ is the source's total energy flux; and $d^2(ka) = k^2 d\Sigma_a$ is a (dimensionless) surface area element in the lateral plane.

EXERCISES

Exercise 9.2 *Problem: Lateral Coherence of Solar Radiation*
How closely separated must a pair of Young's slits be to see strong fringes from the Sun (angular diameter $\sim 0.5°$) at visual wavelengths? Suppose that this condition is

just satisfied, and the slits are 10 μm in width. Roughly how many fringes would you expect to see?

Exercise 9.3 *Problem: Degree of Coherence for a Source
with Gaussian Intensity Distribution*
A circularly symmetric light source has an intensity distribution $I(\alpha) = I_0 \exp[-\alpha^2/(2\alpha_0^2)]$, where α is the angular radius measured from the optic axis. Compute the degree of spatial coherence. What is the lateral coherence length? What happens to the degree of spatial coherence and the interference fringe pattern if the source is displaced from the optic axis?

9.2.5 Michelson Stellar Interferometer; Astronomical Seeing

The classic implementation of Young's slits for measuring spatial coherence is Michelson's stellar interferometer, which Albert Michelson and Francis Pease (1921) used for measuring the angular diameters of Betelgeuse and several other bright stars in 1920.[4] The starlight was sampled at two small mirrors separated by a variable distance $a \leq 6$ m and was then reflected into the 100-inch (2.5-m) telescope on Mount Wilson, California, to form interference fringes (Fig. 9.2). (As we have emphasized, the way in which the fringes are formed is unimportant; all that matters is the two locations where the light is sampled, i.e., the first two mirrors in Fig. 9.2.) As Michelson and Pease increased the separation a between the mirrors, the fringe visibility V decreased. Michelson and Pease modeled Betelgeuse (rather badly, in fact) as a circular disk of uniform brightness, $I(\alpha) = 1$ for $|\alpha| < \alpha_r$ and 0 for $|\alpha| > \alpha_r$, so its visibility was given, according to Eq. (9.13a), as

$$V = \gamma_\perp = 2\,\mathrm{jinc}(ka\alpha_r) \qquad (9.14)$$

where α_r is the star's true angular radius, and $\mathrm{jinc}(\xi) = J_1(\xi)/\xi$. They identified the separation $a \simeq 3$ m, where the fringes disappeared, with the first zero of the function $\mathrm{jinc}(ka\alpha_r)$. From this and the mean wavelength $\lambda = 575$ nm of the starlight, they inferred that the angular radius of Betelgeuse is $\alpha_r \sim 0.02$ arcsec, which at Betelgeuse's then-estimated distance of 60 pc (180 light-years) corresponds to a physical radius ~ 300 times larger than that of the Sun. The modern parallax-measured distance is 200 pc, so Betelgeuse's physical radius is actually $\sim 1{,}000$ times larger than the Sun.

This technique only works for big, bright stars and is very difficult to use, because turbulence in Earth's atmosphere causes the fringes to keep moving around; see Box 9.2 and Ex. 9.4 for details.

---

4. Similar principles are relevant to imaging by Earth-observing satellites where one looks down, not up. In this case, there is the important distinction that most of the diffraction happens relatively close to the source, not the detector.

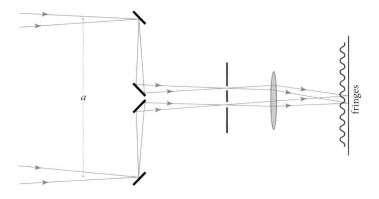

FIGURE 9.2 Schematic illustration of a Michelson stellar interferometer.

EXERCISES

Exercise 9.4 *Example and Derivation: Time-Averaged Visibility and Image for a Distant Star Seen through Earth's Turbulent Atmosphere* **T2**
Fill in the details of the analysis of time-averaged seeing in Box 9.2. More specifically, do the following. If you have difficulty, Roddier (1981) may be helpful.

(a) Give an order-of-magnitude derivation of Eq. (4a) in Box 9.2 for the mean-square phase fluctuation of light induced by propagation through a thin, turbulent layer of atmosphere. [Hint: Consider turbulent cells of size a, each of which produces some $\delta\varphi$, and argue that the contributions add up as a random walk.]

(b) Deduce the factor 2.91 in Eq. (4a) in Box 9.2 by evaluating

$$D_{\delta\varphi} = k^2 \left\langle \left\{ \int_z^{z+\delta h} [\delta\mathfrak{n}(\mathbf{x}+\mathbf{a}, z, t) - \delta\mathfrak{n}(\mathbf{x}, z, t)] \, dz \right\}^2 \right\rangle.$$

(c) Derive Eq. (4b) in Box 9.2 for the time-averaged complex visibility after propagating through the thin layer. [Hint: Because $\zeta \equiv \delta\varphi(\mathbf{x}, t) - \delta\varphi(\mathbf{x}+\mathbf{a}, t)$ is the result of contributions from a huge number of independent turbulent cells, the central limit theorem (Sec. 6.3.2) suggests it is a Gaussian random variable though, in fact intermittency (Sec. 15.3) can make it rather non-Gaussian. Idealizing it as Gaussian, evaluate $\gamma_\perp = \langle e^{i\zeta} \rangle = \langle \int_{-\infty}^{\infty} p(\zeta) e^{i\zeta} \, d\zeta \rangle$ with $p(\zeta)$ the Gaussian distribution.]

(d) Use the point-spread function (8.28) for free propagation of the light field ψ to show that, under free propagation, the complex visibility $\gamma_\perp(\mathbf{a}, z, t) = \langle \psi(\mathbf{x}+\mathbf{a}, z, t) \psi^*(\mathbf{x}, z, t) \rangle$ (with averaging over $\mathbf{x}$ and t) is constant (i.e., independent of height z).

(e) By combining parts c and d, deduce Eqs. (5) in Box 9.2 for the mean-square phase fluctuations and spacetime-averaged visibility on the ground.

(f) Perform a numerical Fourier transform of $\bar{\gamma}_\perp(\mathbf{a})$ [Eq. (5b) in Box 9.2] to get the time-averaged intensity distribution $I(\alpha)$. Construct a log-log plot of it, and compare with panel b of the first figure in Box 9.2. What is r_o for the observational data shown in that figure?

BOX 9.2. ASTRONOMICAL SEEING, SPECKLE IMAGE PROCESSING, AND ADAPTIVE OPTICS T2

When light from a star passes through turbulent layers of Earth's atmosphere, the turbulently varying index of refraction $\mathfrak{n}(\mathbf{x}, t)$ diffracts the light in a random, time varying way. One result is "twinkling" (fluctuations in the flux observed by eye on the ground, with fluctuational frequencies $f_o \sim 100$ Hz). Another is astronomical seeing: the production of many images of the star (i.e., *speckles*) as seen through a large optical telescope (panel a of the box figure), with the image pattern fluctuating at $\sim$100 Hz.

Here and in Ex. 9.4 we quantify astronomical seeing using the theory of 2-dimensional lateral coherence. We do this not because seeing is important (though it is), but rather because our analysis provides an excellent illustration of three fundamental concepts working together: (i) turbulence in fluids and its Kolmogorov spectrum (Chap. 15), (ii) random processes (Chap. 6), and (iii) coherence of light (this chapter).

We begin by deriving, for a star with arbitrarily small angular diameter, the time-averaged complex visibility $\gamma_\perp$ observed on the ground and the visibility's Fourier transform [the observed intensity distribution averaged

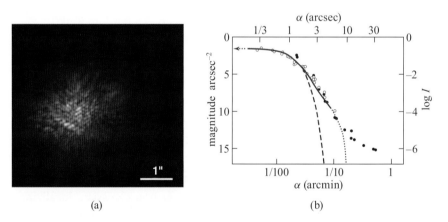

(a) (b)

(a) Picture of a bright star with a dimmer companion star, as seen through the Russian 6-m telescope in the Caucasus Mountains, in an exposure shorter than 10 ms. Atmospheric turbulence creates a large number of images of each star (speckles) spread over a region with angular diameter of order 2 arcsec. (b) The theory discussed in the text and in Ex. 9.4 predicts the solid curve for the time-averaged intensity distribution for a single bright star. Notice the logarithmic axes. The dotted curve is an estimate of the influence of the small-scale cutoff of the turbulence, and the dashed curve is a Gaussian. The circles are observational data. Panel (a), Gerd Weigelt; panel (b), adapted from Roddier (1981).

(continued)

BOX 9.2. (continued)

over the speckles, $\bar{I}(\boldsymbol{\alpha})$]. Then we briefly discuss the temporally fluctuating speckle pattern and techniques for extracting information from it.

TIME-AVERAGED VISIBILITY AND ANGULAR DISTRIBUTION OF INTENSITY

When analyzing light propagation through a turbulent atmosphere, it is convenient to describe the turbulent fluctuations of the index of refraction by their spatial correlation function $C_{\mathfrak{n}}(\boldsymbol{\xi}) \equiv \langle \delta\mathfrak{n}(\mathbf{X}, t)\delta\mathfrak{n}(\mathbf{X}+\boldsymbol{\xi}, t)\rangle$ (discussed in Sec. 6.4.1); or, better yet, by $\mathfrak{n}$'s *mean-square fluctuation on the lengthscale* ξ,

$$D_{\mathfrak{n}}(\boldsymbol{\xi}) \equiv \langle [\delta\mathfrak{n}(\mathbf{X}+\boldsymbol{\xi}, t) - \delta\mathfrak{n}(\mathbf{X}, t)]^2\rangle = 2[\sigma_{\mathfrak{n}}^2 - C_{\mathfrak{n}}(\boldsymbol{\xi})], \quad (1)$$

which is called $\mathfrak{n}$'s *structure function*. Here $\delta\mathfrak{n}$ is the perturbation of the index of refraction, $\mathbf{X}$ is location in 3-dimensional space, t is time, $\langle\cdot\rangle$ denotes a spacetime average, and $\sigma_{\mathfrak{n}}^2 \equiv \langle \delta\mathfrak{n}^2\rangle = C_{\mathfrak{n}}(0)$ is the variance of the fluctuations.

In Sec. 15.4.4, we show that, for strong and isotropic turbulence, $D_{\mathfrak{n}}$ has the functional form $D_{\mathfrak{n}} \propto \xi^{2/3}$ (where $\xi \equiv |\boldsymbol{\xi}|$), with a multiplicative coefficient $d_{\mathfrak{n}}^2$ that characterizes the strength of the perturbations:

$$D_{\mathfrak{n}}(\boldsymbol{\xi}) = d_{\mathfrak{n}}^2 \xi^{2/3} \quad (2)$$

[Eq. (15.29)]. The 2/3 power is called the *Kolmogorov spectrum* for the turbulence.

When light from a very distant star (a point source), directly overhead for simplicity, reaches Earth's atmosphere, its phase fronts lie in horizontal planes, so the frequency ω component of the electric field is $\psi = e^{ikz}$, where z increases downward. (Here we have factored out the field's overall amplitude.) When propagating through a thin layer of turbulent atmosphere of thickness δh, the light acquires the phase fluctuation

$$\delta\varphi(\mathbf{x}, t) = k \int_z^{z+\delta h} \delta\mathfrak{n}(\mathbf{x}, z, t)dz. \quad (3)$$

Here $\mathbf{x}$ is the transverse (i.e., horizontal) location, and Eq. (3) follows from $d\varphi = kdz$, with $k = (\mathfrak{n}/c)\omega$ and $\mathfrak{n} \simeq 1$.

In Ex. 9.4, we derive some spacetime-averaged consequences of the phase fluctuations (3):

1. When the light emerges from the thin, turbulent layer, it has acquired a mean-square phase fluctuation on transverse lengthscale a given by

$$D_{\delta\varphi}(\mathbf{a}) \equiv \langle [\delta\varphi(\mathbf{x}+\mathbf{a}, t) - \delta\varphi(\mathbf{x}, t)]^2\rangle = 2.91 d_{\mathfrak{n}}^2 \,\delta h\, k^2 a^{5/3} \quad (4a)$$

(continued)

BOX 9.2. (continued)

[Eq. (2)], and a spacetime-averaged complex visibility given by

$$\bar{\gamma}_\perp(\mathbf{a}) = \langle \psi(\mathbf{x},t)\psi^*(\mathbf{x}+\mathbf{a},t)\rangle = \langle \exp\{i\,[\delta\varphi(\mathbf{x},t) - \delta\varphi(\mathbf{x}+\mathbf{a},t)]\}\rangle$$

$$= \exp\left[-\frac{1}{2}D_{\delta\varphi}(a)\right] = \exp\left[-1.455\,d_\mathfrak{n}^2\,\delta h k^2 a^{5/3}\right]. \quad (4b)$$

2. Free propagation (including free-propagator diffraction effects, which are important for long-distance propagation) preserves the spacetime-averaged complex visibility: $d\bar{\gamma}_\perp/dz = 0$.

3. Therefore, not surprisingly, when the turbulence is spread out vertically in some arbitrary manner, the net mean-square phase shift and time-averaged complex visibility observed on the ground are $D_\varphi(\mathbf{a}) = 2.91\left[\int d_\mathfrak{n}^2(z)dz\right]k^2 a^{5/3}$, and $\bar{\gamma}_\perp(\mathbf{a}) = \exp\left[-\frac{1}{2}D_\varphi(a)\right]$.

It is conventional to introduce a transverse lengthscale

$$r_o \equiv \left[0.423 k^2 \int d_\mathfrak{n}^2(z)dz\right]^{-3/5}$$

called the *Fried parameter*, in terms of which D_φ and $\bar{\gamma}_\perp$ are

$$D_\varphi(\mathbf{a}) = 6.88(a/r_o)^{5/3}, \quad (5a)$$

$$\bar{\gamma}_\perp(\mathbf{a}) = \exp\left[-\frac{1}{2}D_\varphi(a)\right] = \exp\left[-3.44(a/r_o)^{5/3}\right]. \quad (5b)$$

This remarkably simple result provides opportunities to test the Kolmogorov power law. For light from a distant star, one can use a large telescope to measure $\bar{\gamma}_\perp(a)$ and then plot $\ln\bar{\gamma}_\perp$ as a function of a. Equation (5b) predicts a slope 5/3 for this plot, and observations confirm that prediction.

Notice that the Fried parameter r_o is the lengthscale on which the rms phase fluctuation $\varphi_{\rm rms} = \sqrt{D_\varphi(a=r_o)}$ is $\sqrt{6.88} = 2.62$ radians: r_o is the transverse lengthscale beyond which the turbulence-induced phase fluctuations are large compared to unity. These large random phase fluctuations drive $\bar{\gamma}_\perp$ rapidly toward zero with increasing distance a [Eq. (5b)], that is, they cause the light field to become spatially decorrelated with itself for distances $a \gtrsim r_o$. Therefore, r_o is (approximately) the time-averaged light field's spatial correlation length on the ground. Moreover, since $\bar{\gamma}_\perp$ is preserved under free propagation from the turbulent region to the ground, r_o must be the transverse correlation length of the light as it exits the turbulent region that

(continued)

BOX 9.2. (continued)

produces the seeing. A correlated region with transverse size r_o is called an *isoplanatic patch*.

The observed time-averaged intensity $\bar{I}(\alpha)$ from the point-source star is the Fourier transform of the complex visibility (5b); see Eq. (9.13b). This transform cannot be performed analytically, but a numerical computation gives the solid curve in panel b of the figure above, which agrees remarkably well with observations out to $\sim 10^{-4}$ of the central intensity, where the Kolmogorov power law is expected to break down. Notice that the intensity distribution has a large-radius tail with far larger intensity than a Gaussian distribution (the dashed curve). This large-radius light is produced by large-angle diffraction, which is caused by very small-spatial-scale fluctuations (eddies) in the index of refraction.

Astronomers attribute to this time-averaged $I(\alpha)$ a full width at half maximum (FWHM) angular diameter $\alpha_{\text{Kol}}^{\text{FWHM}} = 0.98\lambda/r_o$ (Ex. 9.4). For blue light in very good seeing conditions, r_o is about 20 cm and $\alpha_{\text{Kol}}^{\text{FWHM}}$ is about 0.5 arcsec. Much more common is $r_o \sim 10$ cm and $\alpha_{\text{Kol}}^{\text{FWHM}} \sim 1$ arcsec.

SPECKLE PATTERN AND ITS INFORMATION

The speckle pattern seen on short timescales, $\lesssim 1/f_o \sim 0.01$ s, can be understood in terms of the turbulence's isoplanatic patches (see the drawing below). When the light field exits the turbulent region, at a height $h \lesssim 1$ km, the isoplanatic patches on its wavefronts, with transverse size r_o, are planar to within roughly a reduced wavelength $\bar{\lambda} = 1/k = \lambda/(2\pi)$ (since the rms phase variation across a patch is just 2.62 radians). Each patch carries an image of the star, or whatever other object the astronomer is observing. The patch's light rays make an angle $\theta \lesssim \alpha_{\text{Kol}}^{\text{FWHM}} = 0.98\lambda/r_o$ to the vertical. The patch's Fresnel length from the ground is $r_F = \sqrt{\bar{\lambda}h} \lesssim 2$ cm (since $\lambda \sim 0.5\ \mu$m and $h \lesssim 1$ km). This is significantly smaller than the patch size $r_o \sim 10$ to 20 cm; so there is little diffraction in the trip to ground. When these patches reach

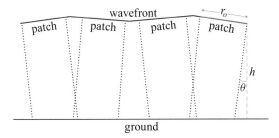

(continued)

BOX 9.2. (continued)

a large telescope (one with diameter $D \gg r_o$), each is focused to produce an image of the object centered on the angular position θ of its rays (dotted lines in the drawing). These images are the speckles seen in panel a of the first box figure.

The speckle pattern varies rapidly, because winds at high altitude sweep the isoplanatic patches through the star's light rays. For a wind speed $u \sim 20$ m s^{-1}, the frequency of the pattern's fluctuations is $f_o \sim u/r_o \sim 100$ Hz, in agreement with observations.

To study the speckles and extract information about the object's above-atmosphere intensity distribution $I_o(\boldsymbol{\alpha})$, one must observe them on timescales $\lesssim 1/f_o \sim 10$ ms. The first observations of this sort were the measurements of a few stellar diameters by Michelson and Pease, using the Michelson stellar interferometer (Sec. 9.2.7). The fringes they saw were produced by the speckles, and because the phase of each speckle's fringes was random, the many speckles contributed incoherently to produce a far smaller net fringe visibility V than in the absence of atmospheric turbulence. Moreover, because the speckle pattern varied at $f_o \sim 100$ Hz, the net visibility and its phase also varied at $f_o \sim 100$ Hz. Fortunately, the human eye can discern things that vary this fast, so Michelson and Pease were able to see the fringes.

In the modern era of CCDs, fast electronics, and powerful computers, a variety of more sophisticated techniques have been devised and implemented for observing these speckles and extracting their information. Two common techniques are speckle-image processing and adaptive optics.

In *speckle image processing*, which is really only usable for bright sources, one makes optical measurements of the speckle pattern (sometimes with multi-pinhole masks) on timescales $\lesssim 0.01$ s for which the speckles are unchanging. One then uses optical or computational techniques to construct fourth-order or sixth-order correlations of the light field, for example, $\int \gamma_\perp(\mathbf{a} - \mathbf{a}') \gamma_\perp^*(\mathbf{a}') d^2 a'$ (which is fourth-order in the field), from which a good approximation to the source's above-atmosphere intensity distribution $I_o(\boldsymbol{\alpha})$ can be computed.

In *adaptive optics*, one focuses not on the speckles themselves, but on the turbulence-induced distortions of the wavefronts arriving at the large telescope's mirror. The simplest implementation uses *natural guide stars* to act as point sources. Unfortunately the incidence of sufficiently bright stars only allows a small fraction of the sky to be examined using this technique.

(continued)

> **BOX 9.2. (continued)**
>
> Therefore, *laser guide stars* are created by shining collimated laser light close to the direction of the astronomical sources under study. The laser light generates the guide star via either Rayleigh scattering by molecules at modest altitude (∼20 km) or resonant scattering from a layer of sodium atoms at high altitude (∼90 km).
>
> The guide star must be within an angular distance $\lesssim r_o/h \sim 3$ arcsec of the astronomical object one is observing (where $h \sim 10$ km is the height of the highest turbulent layers that contribute significantly to the seeing). This $\lesssim 3$ arcsec separation guarantees that light rays arriving at the same spot on the telescope mirror from the object and from the artificial star will have traversed the same isoplanatic patches of turbulent atmosphere and thus have experienced the same phase delay and acquired the same wavefront distortions. One measures the wavefront distortions of the artificial star's light and dynamically reshapes the telescope mirror so as to compensate for them. This removes the distortions not only from the artificial star's wavefronts but also from the astronomical object's wavefronts. Thereby one converts the speckle pattern into the object's true intensity distribution $I_o(\boldsymbol{\alpha})$.
>
> Two recent successes of adaptive optics are to observe the stars orbiting the massive black hole in our galactic center and thereby measure its mass (Ghez et al., 2008; Genzel, Eisenhauer, and Gillessen, 2010); and to image exoplanets directly by masking out the light—typically millions of times brighter than the planet—from the stars that they orbit (Macintosh et al., 2014).
>
> The techniques described here also find application to the propagation of radio waves through the turbulent interplanetary and interstellar media. The refractive index is due to the presence of free electrons in a plasma (see Sec. 21.4.1). Interestingly, this turbulence is commonly characterized by a Kolmogorov spectrum, though the presence of a magnetic field makes it anisotropic.
>
> These techniques are also starting to find application in ophthalmology and industry.

(g) Reexpress the turbulence-broadened image's FWHM as $\alpha_{\text{Kol}}^{\text{FWHM}} = 0.98\lambda/r_o$. Show that the Airy intensity distribution for light for a circular aperture of diameter D [Eq. (8.18)] has FWHM $\alpha_{\text{Airy}}^{\text{FWHM}} = 1.03\lambda/D$.

(h) The fact that the coefficients in these two expressions for α^{FWHM} are both close to unity implies that when the diameter $D \lesssim r_o$, the seeing is determined by the telescope. Conversely, when $D \gtrsim r_o$ (which is true for essentially all ground-based research optical telescopes), it is the atmosphere that determines the image quality. Large telescopes act only as "light buckets," unless some additional correctives are applied, such as speckle image processing or adaptive optics (Box 9.2). A common measure of the performance of a telescope with or without this correction is the *Strehl ratio*, S, which is the ratio of the peak intensity in the actual image of a point source on the telescope's optic axis to the peak intensity in the Airy disk for the telescope's aperture. Show that without correction, $S = 1.00(r_o/D)^2$. Modern adaptive optics systems on large telescopes can achieve $S \sim 0.5$.

9.2.6 Temporal Coherence

In addition to the degree of spatial (or lateral) coherence, which measures the correlation of the field transverse to the direction of wave propagation, we can also measure the *degree of temporal coherence*, also called the *degree of longitudinal coherence*. This describes the correlation at a given time at two points separated by a distance s along the direction of propagation. Equivalently, it measures the field sampled at a fixed position at two times differing by $\tau = s/c$. When (as in our discussion of spatial coherence) the waves are nearly monochromatic so the field arriving at the fixed position has the form $\Psi = \psi(t)e^{-i\omega_o t}$, then the degree of temporal coherence is complex and has a form completely analogous to the transverse case [Eq. (9.12b)]:

degree of temporal or longitudinal coherence for nearly monochromatic radiation

$$\gamma_\parallel(\tau) = \frac{\overline{\psi(t)\psi^*(t+\tau)}}{\overline{|\psi|^2}} \quad \text{for nearly monochromatic radiation.} \quad (9.15)$$

Here the average is over sufficiently long times t for the averaged value to settle down to an unchanging value.

When studying temporal coherence, one often wishes to deal with waves that contain a wide range of frequencies—such as the nearly Planckian (blackbody) cosmic microwave radiation emerging from the very early universe (Ex. 9.6). In this case, one should not factor any $e^{-i\omega_o t}$ out of the field Ψ, and one gains nothing by regarding $\Psi(t)$ as complex, so the *temporal coherence*

degree of temporal coherence for broadband radiation

$$\gamma_\parallel(\tau) = \frac{\overline{\Psi(t)\Psi(t+\tau)}}{\overline{\Psi^2}} \quad \text{for real } \Psi \text{ and broadband radiation} \quad (9.16)$$

is also real. We use this real $\gamma_\parallel$ throughout this subsection and the next. It obviously is the correlation function of Ψ [Eq. (6.19)] renormalized so $\gamma_\parallel(0) = 1$.

As τ is increased, $\gamma_\parallel$ typically remains near unity until some critical value τ_c is reached, and then it begins to fall off toward zero. The critical value τ_c, the longest time over which the field is strongly coherent, is the coherence time, of which we have already spoken: If the wave is roughly monochromatic, so $\Psi(t) \propto \cos[\omega_o t + \delta\varphi(t)]$, with ω_o fixed and the phase $\delta\varphi$ randomly varying in time, then it should be clear that

Chapter 9. Interference and Coherence

the mean time for $\delta\varphi$ to change by an amount of order unity is the coherence time τ_c at which $\gamma_\parallel$ begins to fall significantly.

The uncertainty principle dictates that a field with coherence time τ_c, when Fourier analyzed in time, must contain significant power over a bandwidth $\Delta f = \Delta\omega/(2\pi) \sim 1/\tau_c$. Correspondingly, if we define the field's *longitudinal coherence length* by

$$l_\parallel \equiv c\tau_c, \quad (9.17)$$

longitudinal coherence length

then $l_\parallel$ for broadband radiation will be only a few times the peak wavelength, but for a narrow spectral line of width $\Delta\lambda$, it will be $\lambda^2/\Delta\lambda$.

These relations between the coherence time or longitudinal coherence length and the field's spectral energy flux are order-of-magnitude consequences not only of the uncertainty relation, but also of the temporal analog of the van Cittert-Zernike theorem. That analog is just the Wiener-Khintchine theorem in disguise, and it can be derived by the same methods as we used in the transverse spatial domain. In that theorem the degree of lateral coherence $\gamma_\perp$ is replaced by the degree of temporal coherence $\gamma_\parallel$, and the angular intensity distribution $I(\alpha)$ (distribution of energy over angle) is replaced by the field's spectral energy flux $F_\omega(\omega)$ (the energy crossing a unit area per unit time and per unit angular frequency ω)—which is also called its *spectrum*.[5] The theorem takes the explicit form

spectrum or spectral energy flux $F_\omega(\omega)$

$$\gamma_\parallel(\tau) = \frac{\int_{-\infty}^{\infty} d\omega F_\omega(\omega) e^{i\omega\tau}}{F_S} = \frac{2\int_0^\infty d\omega F_\omega(\omega) \cos\omega\tau}{F_S} \quad (9.18a)$$
for real $\Psi(t)$, valid for broadband radiation

temporal analog of van Cittert-Zernike theorem

and

$$F_\omega(\omega) = F_S \int_{-\infty}^{\infty} \frac{d\tau}{2\pi} \gamma_\parallel(\tau) e^{-i\omega\tau} = 2F_S \int_0^\infty \frac{d\tau}{2\pi} \gamma_\parallel(\tau) \cos\omega\tau. \quad (9.18b)$$

[Here the normalization of our Fourier transform and the sign of its exponential are those conventionally used in optics, and differ from those used in the theory of random processes (Chap. 6). Also, because we have chosen Ψ to be real, $F_\omega(-\omega) = F_\omega(+\omega)$ and $\gamma_\parallel(-\tau) = \gamma_\parallel(+\tau)$.] One can measure $\gamma_\parallel$ by combining the radiation from two points displaced longitudinally to produce interference fringes just as we did when measuring spatial coherence. This type of interference is sometimes called *interference by division of the amplitude*, in contrast with "interference by division of the wavefront" for a Young's-slit-type measurement of lateral spatial coherence (next-to-last paragraph of Sec. 9.2.1).

interference by division of the amplitude

---

5. Note that the spectral energy flux (spectrum) is simply related to the spectral density of the field: if the field Ψ is so normalized that the energy density is $U = \beta\,\overline{\Psi_{,t}\Psi_{,t}}$ with β some constant, then $F_\omega(\omega) = \beta c\omega^2/(2\pi)S_\Psi(f)$, with $f = \omega/(2\pi)$.

EXERCISES

Exercise 9.5 *Problem: Longitudinal Coherence of Radio Waves*
An FM radio station has a carrier frequency of 91.3 MHz and transmits heavy metal rock music in frequency-modulated side bands of the carrier. Estimate the coherence length of the radiation.

9.2.7 Michelson Interferometer and Fourier-Transform Spectroscopy

The classic instrument for measuring the degree of longitudinal coherence is the Michelson interferometer of Fig. 9.3 (not to be confused with the Michelson stellar interferometer). In the simplest version, incident light (e.g., in the form of a Gaussian beam; Sec. 8.5.5) is split by a beam splitter into two beams, which are reflected off different plane mirrors and then recombined. The relative positions of the mirrors are adjustable so that the two light paths can have slightly different lengths. (An early version of this instrument was used in the famous Michelson-Morley experiment.) There are two ways to view the fringes. One way is to tilt one of the reflecting mirrors slightly so there is a range of path lengths in one of the arms. Light and dark interference bands (fringes) can then be seen across the circular cross section of the recombined beam. The second method is conceptually more direct but requires aligning the mirrors sufficiently accurately so the phase fronts of the two beams are parallel after recombination and the recombined beam has no banded structure. The end mirror in one arm of the interferometer is then slowly moved backward or forward, and as it moves, the recombined light slowly changes from dark to light to dark and so on.

It is interesting to interpret this second method in terms of the Doppler shift. One beam of light undergoes a Doppler shift on reflection off the moving mirror. There is then a beat wave produced when it is recombined with the unshifted radiation of the other beam.

Whichever method is used (tilted mirror or longitudinal motion of mirror), the visibility $\gamma_\parallel$ of the interference fringes measures the beam's degree of longitudinal coherence, which is related to the spectral energy flux (spectrum) F_ω by Eqs. (9.18).

Let us give an example. Suppose we observe a spectral line with rest angular frequency ω_0 that is broadened by random thermal motions of the emitting atoms. Then the line profile is

$$F_\omega \propto \exp\left(-\frac{(\omega_0 - \omega)^2}{2(\Delta\omega)^2}\right). \tag{9.19a}$$

The width of the line is given by the formula for the Doppler shift,

$$\Delta\omega \sim \omega_0 [k_B T/(mc^2)]^{1/2},$$

where T is the temperature of the emitting atoms, and m is their mass. (We ignore other sources of line broadening, e.g., natural broadening and pressure broadening,

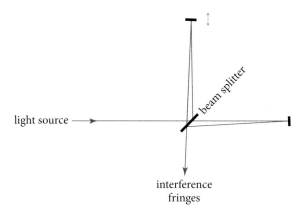

FIGURE 9.3 Michelson interferometer.

which actually dominate under normal conditions.) For example, for hydrogen at $T = 10^3$ K, the Doppler-broadened line width is $\Delta\omega \sim 10^{-5}\omega_0$.

By Fourier transforming this line profile, using the well-known result that the Fourier transform of a Gaussian is another Gaussian and invoking the fundamental relations (9.18) between the spectrum and temporal coherence, we obtain

$$\gamma_\parallel(\tau) = \exp\left(-\frac{\tau^2(\Delta\omega)^2}{2}\right)\cos\omega_0\tau. \tag{9.19b}$$

If we had used the nearly monochromatic formalism with the field written as $\Psi = \psi(t)e^{-i\omega_0 t}$, then we would have obtained

$$\gamma_\parallel(\tau) = \exp\left(-\frac{\tau^2(\Delta\omega)^2}{2}\right)e^{i\omega_0\tau}, \tag{9.19c}$$

the real part of which is our broadband formalism's $\gamma_\parallel$. In either case, $\gamma_\parallel$ oscillates with angular frequency ω_0, and the amplitude of this oscillation is the *fringe visibility* V:

fringe visibility

$$V = \exp\left(-\frac{\tau^2(\Delta\omega)^2}{2}\right). \tag{9.19d}$$

The variation $V(\tau)$ of this visibility with lag time τ is sometimes called an *interferogram*. For time lags $\tau \ll (\Delta\omega)^{-1}$, the line appears to be monochromatic, and fringes with unit visibility should be seen. However, for lags $\tau \gtrsim (\Delta\omega)^{-1}$, the fringe visibility will decrease exponentially with increasing τ^2. In our Doppler-broadened hydrogen-line example with $\Delta\omega \sim 10^{-5}\omega_0$, the rest angular frequency is $\omega_0 \sim 3 \times 10^{15}$ rad s^{-1}, so the longitudinal coherence length is $l_\parallel = c\tau_c \sim 10$ mm. No fringes will be seen when the radiation is combined from points separated by much more than this distance.

interferogram

This procedure is an example of *Fourier transform spectroscopy,* in which, by measuring the degree of temporal coherence $\gamma_\parallel(\tau)$ and then Fourier transforming

Fourier transform spectroscopy

it [Eq. (9.18)], one infers the shape of the radiation's spectrum, or in this case, the width of a specific spectral line.

When (as in Ex. 9.6) the waves are very broad band, the degree of longitudinal coherence $\gamma_\parallel(\tau)$ will not have the form of a sinusoidal oscillation (regular fringes) with slowly varying amplitude (visibility). Nevertheless, the broadband van Cittert-Zernike theorem [Eqs. (9.18)] still guarantees that the spectrum (spectral energy flux) will be the Fourier transform of the coherence $\gamma_\parallel(\tau)$, which can be measured by a Michelson interferometer.

EXERCISES

Exercise 9.6 *Problem: COBE Measurement of the Cosmic Microwave Background Radiation*

An example of a Michelson interferometer is the Far Infrared Absolute Spectrophotometer (FIRAS) carried by the Cosmic Background Explorer satellite (COBE). COBE studied the spectrum and anisotropies of the cosmic microwave background radiation (CMB) that emerged from the very early, hot phase of our universe's expansion (Sec. 28.3.3). One of the goals of the COBE mission was to see whether the CMB spectrum really has the shape of 2.7 K blackbody (Planckian) radiation, or if it is highly distorted, as some measurements made on rocket flights had suggested. COBE's spectrophotometer used Fourier transform spectroscopy to meet this goal: it compared accurately the degree of longitudinal coherence $\gamma_\parallel$ of the CMB radiation with that of a calibrated source on board the spacecraft, which was known to be a blackbody at about 2.7 K. The comparison was made by alternately feeding radiation from the microwave background and radiation from the calibrated source into the same Michelson interferometer and comparing their fringe spacings. The result (Mather et al., 1994) was that the background radiation has a spectrum that is Planckian with temperature 2.726 ± 0.010 K over the wavelength range 0.5–5.0 mm, in agreement with simple cosmological theory that we shall explore in the last chapter of this book.

(a) Suppose that the CMB had had a Wien spectrum

$$F_\omega \propto \omega^3 \exp[-\hbar\omega/(k_B T)].$$

Show that the visibility of the fringes would have been

$$V = |\gamma_\parallel| \propto \frac{|s^4 - 6s_0^2 s^2 + s_0^4|}{(s^2 + s_0^2)^4} \quad (9.20)$$

where $s = c\tau$ is longitudinal distance, and calculate a numerical value for s_0.

(b) Compute the interferogram $V(\tau)$ for a Planck function either analytically (perhaps with the help of a computer) or numerically using a fast Fourier transform. Compare graphically the interferogram for the Wien and Planck spectra.

9.2.8 Degree of Coherence; Relation to Theory of Random Processes

Having separately discussed spatial and temporal coherence, we now can easily perform a final generalization and define the full degree of coherence of the radiation field between two points separated both laterally by a vector **a** and longitudinally by a distance s (or equivalently, by a time $\tau = s/c$). If we restrict ourselves to nearly monochromatic waves and use the complex formalism so the waves are written as $\Psi = e^{i(kz-\omega_0 t)}\psi(\mathbf{x}, t)$ [Eq. (9.12a)], then we have

$$\gamma_{12}(k\mathbf{a}, \tau) \equiv \frac{\overline{\psi(\mathbf{x}_1, t)\psi^*(\mathbf{x}_1 + \mathbf{a}, t + \tau)}}{\left[\overline{|\psi(\mathbf{x}_1, t)|^2}\,\overline{|\psi(\mathbf{x}_1 + \mathbf{a}, t)|^2}\right]^{1/2}} = \frac{\overline{\psi(\mathbf{x}_1, t)\psi^*(\mathbf{x}_1 + \mathbf{a}, t + \tau)}}{\overline{|\psi|^2}}. \quad (9.21)$$

full (3-dimensional) degree of coherence

In the denominator of the second expression we have used the fact that, because the source is far away, $\overline{|\psi|^2}$ is independent of the spatial location at which it is evaluated, in the region of interest. Consistent with the definition (9.21), we can define a *volume of coherence* $\mathcal{V}_c$ as the product of the longitudinal coherence length $l_\parallel = c\tau_c$ and the square of the transverse coherence length $l_\perp^2$: $\mathcal{V}_c = l_\perp^2 c\tau_c$.

volume of coherence

The 3-dimensional version of the van Cittert-Zernike theorem relates the complex degree of coherence (9.21) to the radiation's *specific intensity*, $I_\omega(\boldsymbol{\alpha}, \omega)$, also called its spectral intensity (i.e., the energy crossing a unit area per unit time per unit solid angle and per unit angular frequency, or energy "per unit everything"). (Since the frequency ν and the angular frequency ω are related by $\omega = 2\pi\nu$, the specific intensity I_ω of this chapter and that I_ν of Chap. 3 are related by $I_\nu = 2\pi I_\omega$.) The 3-dimensional van Cittert-Zernike theorem states that

specific intensity

$$\boxed{\gamma_{12}(k\mathbf{a}, \tau) = \frac{\int d\Omega_\alpha d\omega I_\omega(\boldsymbol{\alpha}, \omega) e^{i(k\mathbf{a}\cdot\boldsymbol{\alpha} + \omega\tau)}}{F_S},} \quad (9.22a)$$

and

3-dimensional van Cittert-Zernike theorem

$$\boxed{I_\omega(\boldsymbol{\alpha}, \omega) = F_S \int \frac{d\tau d^2 ka}{(2\pi)^3} \gamma_{12}(k\mathbf{a}, \tau) e^{-i(k\mathbf{a}\cdot\boldsymbol{\alpha} + \omega\tau)}.} \quad (9.22b)$$

There obviously must be an intimate relationship between the theory of random processes, as developed in Chap. 6, and the theory of a wave's coherence, as we have developed it in Sec. 9.2. That relationship is explained in Ex. 9.8.

EXERCISES

Exercise 9.7 *Problem: Decomposition of Degree of Coherence*
We have defined the degree of coherence $\gamma_{12}(\mathbf{a}, \tau)$ for two points in the radiation field separated laterally by a distance **a** and longitudinally by a time τ. Under what conditions will this be given by the product of the spatial and temporal degrees of coherence?

$$\gamma_{12}(\mathbf{a}, \tau) = \gamma_\perp(\mathbf{a})\gamma_\parallel(\tau). \quad (9.23)$$

Exercise 9.8 ****Example: Complex Random Processes and the van Cittert-Zernike Theorem**

In Chap. 6 we developed the theory of real-valued random processes that vary randomly with time t (i.e., that are defined on a 1-dimensional space in which t is a coordinate). Here we generalize a few elements of that theory to a complex-valued random process $\Phi(\mathbf{x})$ defined on a (Euclidean) space with n dimensions. We assume the process to be stationary and to have vanishing mean (cf. Chap. 6 for definitions). For $\Phi(\mathbf{x})$ we define a complex-valued correlation function by

$$C_\Phi(\boldsymbol{\xi}) \equiv \overline{\Phi(\mathbf{x})\Phi^*(\mathbf{x}+\boldsymbol{\xi})} \tag{9.24a}$$

(where $*$ denotes complex conjugation) and a real-valued spectral density by

$$S_\Phi(\mathbf{k}) = \lim_{L\to\infty} \frac{1}{L^n} |\tilde{\Phi}_L(\mathbf{k})|^2. \tag{9.24b}$$

Here Φ_L is Φ confined to a box of side L (i.e., set to zero outside that box), and the tilde denotes a Fourier transform defined using the conventions of Chap. 6:

$$\tilde{\Phi}_L(\mathbf{k}) = \int \Phi_L(\mathbf{x}) e^{-i\mathbf{k}\cdot\mathbf{x}} d^n x, \quad \Phi_L(\mathbf{x}) = \int \tilde{\Phi}_L(\mathbf{k}) e^{+i\mathbf{k}\cdot\mathbf{x}} \frac{d^n k}{(2\pi)^n}. \tag{9.25}$$

Because Φ is complex rather than real, $C_\Phi(\boldsymbol{\xi})$ is complex; and as we shall see below, its complexity implies that [although $S_\Phi(\mathbf{k})$ is real], $S_\Phi(-\mathbf{k}) \neq S_\Phi(\mathbf{k})$. This fact prevents us from folding negative $\mathbf{k}$ into positive $\mathbf{k}$ and thereby making $S_\Phi(\mathbf{k})$ into a "single-sided" spectral density as we did for real random processes in Chap. 6. In this complex case we must distinguish $-\mathbf{k}$ from $+\mathbf{k}$ and similarly $-\boldsymbol{\xi}$ from $+\boldsymbol{\xi}$.

(a) The complex Wiener-Khintchine theorem [analog of Eq. (6.29)] states that

$$S_\Phi(\mathbf{k}) = \int C_\Phi(\boldsymbol{\xi}) e^{+i\mathbf{k}\cdot\boldsymbol{\xi}} d^n\xi, \tag{9.26a}$$

$$C_\Phi(\boldsymbol{\xi}) = \int S_\Phi(\mathbf{k}) e^{-i\mathbf{k}\cdot\boldsymbol{\xi}} \frac{d^n k}{(2\pi)^n}. \tag{9.26b}$$

Derive these relations. [Hint: Use Parseval's theorem in the form $\int A(\mathbf{x}) B^*(\mathbf{x}) d^n x = \int \tilde{A}(\mathbf{k}) \tilde{B}^*(\mathbf{k}) d^n k/(2\pi)^n$ with $A(\mathbf{x}) = \Phi_L(\mathbf{x})$ and $B(\mathbf{x}) = \Phi_L(\mathbf{x}+\boldsymbol{\xi})$, and then take the limit as $L \to \infty$.] Because $S_\Phi(\mathbf{k})$ is real, this Wiener-Khintchine theorem implies that $C_\Phi(-\boldsymbol{\xi}) = C_\Phi^*(\boldsymbol{\xi})$. Show that this is so directly from the definition (9.24a) of $C_\Phi(\boldsymbol{\xi})$. Because $C_\Phi(\boldsymbol{\xi})$ is complex, the Wiener-Khintchine theorem implies that $S_\Phi(\mathbf{k}) \neq S_\Phi(-\mathbf{k})$.

(b) Let $\psi(\mathbf{x},t)$ be the complex-valued wave field defined in Eq. (9.12a), and restrict $\mathbf{x}$ to range only over the two transverse dimensions so ψ is defined on a 3-dimensional space. Define $\Phi(\mathbf{x},t) \equiv \psi(\mathbf{x},t)/\left[\overline{|\psi(\mathbf{x},t)|^2}\right]^{1/2}$. Show that

$$C_\Phi(\mathbf{a},\tau) = \gamma_{12}(k\mathbf{a},\tau), \quad S_\Phi(-\alpha k, -\omega) = \text{const} \times \frac{I_\omega(\boldsymbol{\alpha},\omega)}{F_S}, \tag{9.27}$$

and that the complex Wiener-Khintchine theorem (9.26) is the van Cittert-Zernike theorem (9.22). (Note: The minus signs in S_Φ result from the difference in Fourier transform conventions between the theory of random processes [Eq. (9.25) and Chap. 6] and the theory of optical coherence [this chapter].) Evaluate the constant in Eq. (9.27).

9.3 Radio Telescopes

The interferometry technique pioneered by Michelson for measuring the angular sizes of stars at visual wavelengths has been applied to great effect in radio astronomy.

A modern radio telescope is a large surface that reflects radio waves onto a "feed," where the waves' fluctuating electric field creates a tiny electric voltage that subsequently can be amplified and measured electronically. Modern radio telescopes have diameters D that range from ~ 10 m to the ~ 300 m of the Arecibo telescope in Puerto Rico. A typical observing wavelength might be $\lambda \sim 6$ cm. This implies an angular resolution $\theta_A \sim \lambda/D \sim 2 \text{ arcmin}(\lambda/6 \text{ cm})(D/100 \text{ m})^{-1}$ [Eq. (8.18) and subsequent discussion]. However, many of the most interesting cosmic sources are much smaller than this. To achieve much better angular resolution, the technique of radio interferometry was pioneered in the 1950s and has been steadily developing since then.[6]

radio telescope interferometry

9.3.1 Two-Element Radio Interferometer

If we have two radio telescopes, then we can think of them as two Young's slits, and we can link them using a combination of waveguides and electric cables, as shown in Fig. 9.4. When they are both pointed at a source, they both measure the electric field in radio waves from that source. We combine their signals by narrowband filtering their voltages to make them nearly monochromatic and then either adding the filtered voltages and measuring the power, or multiplying the two voltages directly. In either case a measurement of the degree of coherence [Eq. (9.10)] can be achieved. [If the source is not vertically above the two telescopes, one obtains some nonlateral component of the full degree of coherence $\gamma_{12}(\mathbf{a}, \tau)$. However, by introducing a time delay into one of the signals, as in Fig. 9.4, one can measure the degree of lateral coherence $\gamma_\perp(\mathbf{a})$, which is what the astronomer usually needs.]

The objective is usually to produce an image of the radio waves' source. This is achieved by Fourier inverting the lateral degree of coherence $\gamma_\perp(\mathbf{a})$ [Eq. (9.13b)], which therefore must be measured for a variety of values of the relative separation vector $\mathbf{a}$ of the telescopes perpendicular to the source's direction. As Earth rotates, the separation vector will trace out half an ellipse in the 2-dimensional $\mathbf{a}$ plane every 12 hours. [The source intensity is a real quantity, so we can use Eq. (9.13a) to deduce

---

6. This type of interferometry is also developing fast at optical wavelengths. This was not possible until optical technology became good enough to monitor the phase of light as well as its amplitude, at separate locations, and then produce interference.

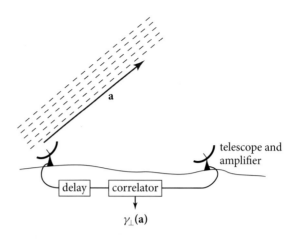

FIGURE 9.4 Two-element radio interferometer.

that $\gamma_\perp(-\mathbf{a}) = \gamma_\perp^*(\mathbf{a})$, which gives the other half of the ellipse.] By changing the spacing between the two telescopes daily and collecting data for a number of days, the degree of coherence can be well sampled. This technique[7] is known as *Earth-rotation aperture synthesis*, because the telescopes are being made to have the angular resolution of a giant telescope as big as their maximum separation, with the aid of Earth's rotation. They do not, of course, have the sensitivity of this giant telescope.

9.3.2 Multiple-Element Radio Interferometers

In practice, a modern radio interferometer has many more than two telescopes. For example, the Karl G. Jansky Very Large Array (JVLA) in New Mexico (USA) has 27 individual telescopes arranged in a Y pattern and operating simultaneously, with a maximum baseline of 36 km and a minimum observing wavelength of 7 mm. The degree of coherence can thus be measured simultaneously over $27 \times 26/2 = 351$ different relative separations. The results of these measurements can then be interpolated to give values of $\gamma_\perp(\mathbf{a})$ on a regular grid of points (usually $2^N \times 2^N$ for some integer N). This is then suitable for applying the fast Fourier transform algorithm to infer the source structure $I(\boldsymbol{\alpha})$.

The Atacama Large Millimeter Array (ALMA) being constructed in Chile is already (2016) operational. It comprises 66 telescopes observing with wavelengths between 0.3 and 9.6 mm and baselines as long as 16 km. Future ambitions at longer radio wavelengths are centered on the proposed Square Kilometer Array (SKA) to be built in South Africa and Australia comprising thousands of dishes with a combined collecting area of a square kilometer.

---

7. For which Martin Ryle was awarded the Nobel Prize.

9.3.3 Closure Phase

Among the many technical complications of interferometry is one that brings out an interesting point about Fourier methods. It is usually much easier to measure the modulus than the phase of the complex degree of coherence. This is partly because it is hard to introduce the necessary delays in the electronics accurately enough to know where the zero of the fringe pattern should be located and partly because unknown, fluctuating phase delays are introduced into the phase of the field as the wave propagates through the upper atmosphere and ionosphere. [This is a radio variant of the problem of "seeing" for optical telescopes (cf. Box 9.2), and it also plagues the Michelson stellar interferometer.] It might therefore be thought that we would have to make do with just the modulus of the degree of coherence (i.e., the fringe visibility) to perform the Fourier inversion for the source structure. This is not so.

Consider a three-element interferometer measuring fields ψ_1, ψ_2, and ψ_3, and suppose that at each telescope there are unknown phase errors, $\delta\varphi_1$, $\delta\varphi_2$, and $\delta\varphi_3$ (Fig. 9.5). For baseline $\mathbf{a}_{12}$, we measure the degree of coherence $\gamma_{\perp 12} \propto \overline{\psi_1 \psi_2^*}$, a complex number with phase $\Phi_{12} = \varphi_{12} + \delta\varphi_1 - \delta\varphi_2$, where φ_{12} is the phase of $\gamma_{\perp 12}$ in the absence of phase errors. If we also measure the degrees of coherence for the other two pairs of telescopes in the triangle and derive their phases Φ_{23} and Φ_{31}, we can then calculate the quantity

$$C_{123} = \Phi_{12} + \Phi_{23} + \Phi_{31}$$
$$= \varphi_{12} + \varphi_{23} + \varphi_{31}, \quad (9.28)$$

from which the phase errors cancel out.

The quantity C_{123}, known as the *closure phase*, can be measured with high accuracy. In the JVLA, there are $27 \times 26 \times 25/6 = 2{,}925$ such closure phases, and they can all be measured with considerable redundancy. Although absolute phase information cannot be recovered, 93% of the telescopes' relative phases can be inferred in this manner and used to construct an image far superior to what could be achieved without any phase information.

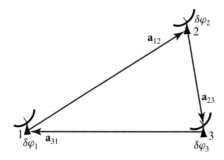

FIGURE 9.5 Closure-phase measurement using a triangle of telescopes.

9.3.4 Angular Resolution

When the telescope spacings are well sampled and the source is bright enough to carry out these image-processing techniques, an interferometer can have an angular resolving power approaching that of an equivalent filled aperture as large as the maximum telescope spacing. For the JVLA the best angular resolution is 50 milliarcsec and at ALMA it will be as fine as 4 milliarcsec, thousands of times better than single dishes.

Even greater angular resolution is achieved with a technique known as very long baseline interferometry (VLBI). Here the telescopes can be located on different continents and instead of linking them directly, the oscillating field amplitudes $\psi(t)$ are stored electronically. Then they are combined digitally long after the observation to compute the complex degree of coherence and thence the source structure $I(\boldsymbol{\alpha})$. In this way angular resolutions more than 300 times better than those achievable by the JVLA have been obtained. Structure smaller than a milliarcsec, corresponding to a few light-years at cosmological distances, can be measured in this manner. A most exciting prospect is to use submillimeter VLBI to resolve marginally the event horizons of massive black holes, specifically, the four-million-solar-mass hole at the center of our galaxy and the six-billion-solar-mass hole at the center of the nearby galaxy M87. Existing observations by a collaboration called the "event horizon telescope" have already (2016) measured interference fringes on angular scales ~5 times larger than these black holes' gravitational radii, i.e. ~$10\, GM/c^2$.

EXERCISES

Exercise 9.9 *Example: Radio Interferometry from Space*

The longest radio-telescope separation available in 2016 is that between telescopes on Earth's surface and a 10-m diameter radio telescope in the Russian RadioAstron satellite, which was launched into a highly elliptical orbit around Earth in summer 2011, with perigee ~10,000 km (1.6 Earth radii) and apogee ~350,000 km (55 Earth radii).

(a) Radio astronomers conventionally describe the specific intensity $I_\omega(\boldsymbol{\alpha}, \omega)$ of a source in terms of its brightness temperature. This is the temperature $T_b(\omega)$ that a blackbody would have to emit, in the Rayleigh-Jeans (low-frequency) end of its spectrum, to produce the same specific intensity as the source. Show that for a single (linear or circular) polarization, if the solid angle subtended by a source is $\Delta\Omega$ and the spectral energy flux measured from the source is $F_\omega \equiv \int I_\omega d\Omega = I_\omega \Delta\Omega$, then the brightness temperature is

$$T_b = \frac{(2\pi)^3 c^2 I_\omega}{k_B \omega^2} = \frac{(2\pi)^3 c^2 F_\omega}{k_B \omega^2 \Delta\Omega}, \qquad (9.29)$$

where k_B is Boltzmann's constant.

(b) The brightest quasars emit radio spectral fluxes of about $F_\omega = 10^{-25}$ W m^{-2} Hz^{-1}, independent of frequency. The smaller such a quasar is, the larger will be its brightness temperature. Thus, one can characterize the smallest sources that a radio-telescope system can resolve by the highest brightness temperatures it can measure. Show that the maximum brightness temperature measurable by the Earth-to-orbit RadioAstron interferometer is independent of the frequency at which the observation is made, and estimate its numerical value.

9.4 Etalons and Fabry-Perot Interferometers

We have shown how a Michelson interferometer (Fig. 9.3) can be used as a Fourier-transform spectrometer: one measures the complex fringe visibility as a function of the two arms' optical path difference and then takes the visibility's Fourier transform to obtain the spectrum of the radiation. The inverse process is also powerful. One can drive a Michelson interferometer with radiation with a known, steady spectrum (usually close to monochromatic), and look for time variations of the positions of its fringes caused by changes in the relative optical path lengths of the interferometer's two arms. This was the philosophy of the famous Michelson-Morley experiment to search for ether drift, and it is also the underlying principle of a laser interferometer ("interferometric") gravitational-wave detector.

To reach the sensitivity required for gravitational-wave detection, one must modify the Michelson interferometer by making the light travel back and forth in each arm many times, thereby amplifying the phase shift caused by changes in the arm lengths. This is achieved by converting each arm into a Fabry-Perot interferometer. In this section, we study Fabry-Perot interferometers and some of their other applications, and in Sec. 9.5, we explore their use in gravitational-wave detection.

9.4.1 Multiple-Beam Interferometry; Etalons

Fabry-Perot interferometry is based on trapping monochromatic light between two highly reflecting surfaces. To understand such trapping, let us consider the concrete situation where the reflecting surfaces are flat and parallel to each other, and the transparent medium between the surfaces has one index of refraction n, while the medium outside the surfaces has another index n' (Fig. 9.6). Such a device is sometimes called an *etalon*. One example is a glass slab in air ($n \simeq 1.5$, $n' \simeq 1$); another is a vacuum maintained between two glass mirrors ($n = 1$, $n' \simeq 1.5$). For concreteness, we discuss the slab case, though all our formulas are equally valid for a vacuum between mirrors or for any other etalon.

etalon

Suppose that a monochromatic plane wave (i.e., with parallel rays) with angular frequency ω is incident on one of the slab's reflecting surfaces, where it is partially reflected and partially transmitted with refraction. The transmitted wave will propagate through to the second surface, where it will be partially reflected and partially transmitted. The reflected portion will return to the first surface, where it too will be split, and so on (Fig. 9.6a). The resulting total fields in and outside the slab can be computed by summing the series of sequential reflections and transmissions (Ex. 9.12). Alternatively, they can be computed as follows.

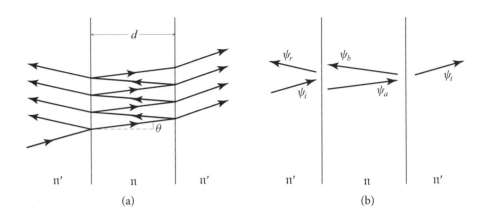

FIGURE 9.6 Multiple-beam interferometry using a type of Fabry-Perot etalon.

Assume, for pedagogical simplicity, that there is translational invariance along the slab (i.e., the slab and incoming wave are perfectly planar). Then the series, if summed, would lead to the five waves shown in Fig. 9.6b: an incident wave (ψ_i), a reflected wave (ψ_r), a transmitted wave (ψ_t), and two internal waves (ψ_a and ψ_b).

amplitude reflection and transmission coefficients

We introduce amplitude reflection and transmission coefficients, denoted $\mathfrak{r}$ and $\mathfrak{t}$, for waves incident on the slab surface from outside. Likewise, we introduce coefficients $\mathfrak{r}'$, $\mathfrak{t}'$ for waves incident on the slab from inside. These coefficients are functions of the angles of incidence and the light's polarization. They can be computed using electromagnetic theory (e.g., Hecht, 2017, Sec. 4.6.2), but this will not concern us here.

Armed with these definitions, we can express the reflected and transmitted waves at the first surface (location A in Fig. 9.7) in the form

$$\psi_r = \mathfrak{r}\psi_i + \mathfrak{t}'\psi_b,$$
$$\psi_a = \mathfrak{t}\psi_i + \mathfrak{r}'\psi_b, \tag{9.30a}$$

where ψ_i, ψ_a, ψ_b, and ψ_r are the values of ψ at A for waves impinging on or leaving the surface along the paths i, a, b, and r, respectively, depicted in Fig. 9.7. Simple geometry shows that the waves at the second surface are as depicted in Fig. 9.7. Correspondingly, the relationships between the ingoing and outgoing waves there are

$$\psi_b e^{-iks_1} = \mathfrak{r}'\psi_a e^{ik(s_1-s_2)},$$
$$\psi_t = \mathfrak{t}'\psi_a e^{iks_1}, \tag{9.30b}$$

where $k = n\omega/c$ is the wave number in the slab, and (as is shown in the figure) s_1 and s_2 are defined as

$$s_1 = d \sec\theta, \qquad s_2 = 2d \tan\theta \sin\theta, \tag{9.30c}$$

with d the thickness of the slab, and θ the angle that the wavefronts inside the slab make to the slab's faces.

In solving Eqs. (9.30) for the net transmitted and reflected waves ψ_t and ψ_r in terms of the incident wave ψ_i, we need *reciprocity relations* between the reflection

Chapter 9. Interference and Coherence

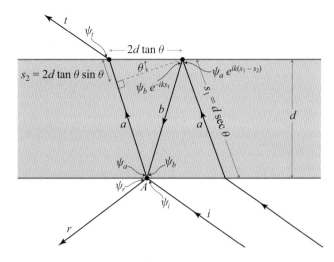

FIGURE 9.7 Construction for calculating the phase differences across the slab for the two internal waves in an etalon.

and transmission coefficients $\mathfrak{r}$ and $\mathfrak{t}$ for waves that hit the reflecting surfaces from one side, and those between $\mathfrak{r}'$ and $\mathfrak{t}'$ for waves from the other side. These reciprocity relations are analyzed quite generally in Ex. 9.10. To derive the reciprocity relations in our case of sharp boundaries between homogeneous media, consider the limit in which the slab thickness $d \to 0$. This is allowed because the wave equation is linear, and the solution for one surface can be superposed on that for the other surface. In this limit $s_1 = s_2 = 0$ and the slab must become transparent, so

$$\psi_r = 0, \qquad \psi_t = \psi_i. \tag{9.31}$$

Equations (9.30a), (9.30b), and (9.31) are then six homogeneous equations in the five wave amplitudes ψ_i, ψ_r, ψ_t, ψ_a, and ψ_b, from which we can extract the two desired reciprocity relations:

$$\boxed{\mathfrak{r}' = -\mathfrak{r}, \qquad \mathfrak{t}\mathfrak{t}' - \mathfrak{r}\mathfrak{r}' = 1.} \tag{9.32}$$

reciprocity relations

Since there is no mechanism to produce a phase shift as the waves propagate across a perfectly sharp boundary, it is reasonable to expect $\mathfrak{r}$, $\mathfrak{r}'$, $\mathfrak{t}$, and $\mathfrak{t}'$ all to be real, as indeed they are (Ex. 9.10). [If the interface has a finite thickness, it is possible to adjust the spatial origins on the two sides of the interface so as to make $\mathfrak{r}$, $\mathfrak{r}'$, $\mathfrak{t}$, and $\mathfrak{t}'$ all be real, leading to the reciprocity relations (9.32), but a price will be paid; see Ex. 9.10.]

Now return to the case of finite slab thickness. By solving Eqs. (9.30) for the reflected and transmitted fields and invoking the reciprocity relations (9.32), we obtain

$$\frac{\psi_r}{\psi_i} \equiv \mathfrak{r}_e = \frac{\mathfrak{r}(1 - e^{i\varphi})}{1 - \mathfrak{r}^2 e^{i\varphi}}, \qquad \frac{\psi_t}{\psi_i} \equiv \mathfrak{t}_e = \frac{(1 - \mathfrak{r}^2)e^{i\varphi/(2\cos^2\theta)}}{1 - \mathfrak{r}^2 e^{i\varphi}}. \tag{9.33a}$$

9.4 Etalons and Fabry-Perot Interferometers

Here $\mathfrak{r}_e$ and $\mathfrak{t}_e$ are the etalon's reflection and transmission coefficients, and $\varphi = k(2s_1 - s_2)$, which reduces to

$$\varphi = 2\mathfrak{n}\omega d \cos\theta/c, \tag{9.33b}$$

is the light's round-trip phase shift (along path a and then b) inside the etalon, relative to the phase of the incoming light that it meets at location A. If φ is a multiple of 2π, the round-trip light will superpose coherently on the new, incoming light.

We are particularly interested in the *reflectivity and transmissivity* for the energy flux—the coefficients that tell us what fraction of the total flux (and therefore also the total power) incident on the etalon is reflected by it and what fraction emerges from its other side:

etalon's reflectivity and transmissivity

$$R = |\mathfrak{r}_e|^2 = \frac{|\psi_r|^2}{|\psi_i|^2} = \frac{2\mathfrak{r}^2(1-\cos\varphi)}{1 - 2\mathfrak{r}^2\cos\varphi + \mathfrak{r}^4}, \quad T = |\mathfrak{t}_e|^2 = \frac{|\psi_t|^2}{|\psi_i|^2} = \frac{(1-\mathfrak{r}^2)^2}{1 - 2\mathfrak{r}^2\cos\varphi + \mathfrak{r}^4}. \tag{9.33c}$$

From these expressions, we see that

energy conservation

$$\boxed{R + T = 1,} \tag{9.33d}$$

which says that the energy flux reflected from the slab plus that transmitted is equal to that impinging on the slab (energy conservation). It is actually the reciprocity relations (9.32) for the amplitude reflection and transmission coefficients that enforce this energy conservation. If they had contained a provision for absorption or scattering of light in the interfaces, $R + T$ would have been less than one.

We discuss the etalon's reflectivity and transmissivity, Eq. (9.33c), at length in Sec. 9.4.2. But first, in a set of example exercises, we clarify some important issues related to the above analysis.

EXERCISES

Exercise 9.10 *Example: Reciprocity Relations for a Locally Planar Optical Device*
Modern mirrors, etalons, beam splitters, and other optical devices are generally made of glass or fused silica (quartz) and have dielectric coatings on their surfaces. The coatings consist of alternating layers of materials with different dielectric constants, so the index of refraction $\mathfrak{n}$ varies periodically. If, for example, the period of $\mathfrak{n}$'s variations is half a wavelength of the radiation, then waves reflected from successive dielectric layers build up coherently, producing a large net reflection coefficient; the result is a highly reflecting mirror.

In this exercise, we use a method due to Stokes to derive the reciprocity relations for devices with dielectric coatings, and in fact for much more general devices. Specifically, our derivation will be valid for locally plane-fronted, monochromatic waves

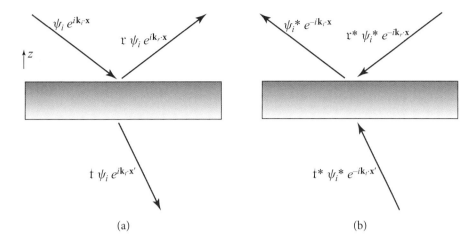

FIGURE 9.8 Construction for deriving reciprocity relations for amplitude transmission and reflection coefficients.

impinging on an arbitrary, locally planar, lossless optical device.[8] The device could be a mirror, a surface with an antireflection coating (Ex. 9.13 below), an etalon, or any sequence of such objects with locally parallel surfaces.

Let a plane, monochromatic wave $\psi_i e^{i\mathbf{k}_i \cdot \mathbf{x}} e^{-i\omega t}$ impinge on the optical device from above, and orient the device so its normal is in the z direction and it is translation invariant in the x and y directions; see Fig. 9.8a. Then the reflected and transmitted waves are as shown in the figure. Because the medium below the device can have a different index of refraction from that above, the waves' propagation direction below may be different from that above, as shown. For reasons explained in part (e), we denote position below the device by $\mathbf{x}'$ and position above the device by $\mathbf{x}$. Some arbitrary choice has been made for the locations of the vertical origins $z = 0$ and $z' = 0$ on the two sides of the device.

(a) Consider a thought experiment in which the waves of Fig. 9.8a are time-reversed, so they impinge on the device from the original reflection and transmission directions and emerge toward the original input direction, as shown in Fig. 9.8b. If the device had been lossy, the time-reversed waves would not satisfy the field's wave equation; the absence of losses guarantees they do. Show that mathematically, the time reversal can be achieved by complex conjugating the spatial part of the waves while leaving the temporal part $e^{-i\omega t}$ unchanged. (Such phase conjugation can be achieved in practice using techniques of nonlinear optics, as we

---

8. By "locally" plane-fronted and planar, we mean that transverse variations are on scales sufficiently long compared to the wavelength of light that we can use the plane-wave analysis sketched here; for example, the spherical mirrors and Gaussian beams of an interferometric gravitational-wave detector (see Fig. 9.13 in Sec. 9.5) easily satisfy this requirement. By lossless we mean that there is no absorption or scattering of the light.

shall see in the next chapter.) Show, correspondingly, that the spatial part of the time-reversed waves is described by the formulas shown in Fig. 9.8b.

(b) Use the reflection and transmission coefficients to compute the waves produced by the inputs of Fig. 9.8b. From the requirement that the wave emerging from the device's upward side must have the form shown in the figure, conclude that

$$\boxed{1 = \mathfrak{r}\mathfrak{r}^* + \mathfrak{t}'\mathfrak{t}^*.} \tag{9.34a}$$

Similarly, from the requirement that no wave emerge from the device's downward side, conclude that

$$\boxed{0 = \mathfrak{t}\mathfrak{r}^* + \mathfrak{t}^*\mathfrak{r}'.} \tag{9.34b}$$

Eqs. (9.34) are the most general form of the reciprocity relations for lossless, planar devices.

(c) For a sharp interface between two homogeneous media, combine these general reciprocity relations with the ones derived in the text [Eqs. (9.32)] to show that $\mathfrak{t}$, $\mathfrak{t}'$, $\mathfrak{r}$, and $\mathfrak{r}'$ are all real (as was asserted in the text).

(d) For the etalon of Figs. 9.6 and 9.7, $\mathfrak{r}_e$ and $\mathfrak{t}_e$ are given by Eqs. (9.33a). What do the reciprocity relations tell us about the coefficients for light propagating in the opposite direction, $\mathfrak{r}'_e$ and $\mathfrak{t}'_e$?

(e) Show that for a general optical device, the reflection and transmission coefficients can all be made real by appropriate, independent *adjustments of the origins of the vertical coordinates z* (for points above the device) and *z'* (for points below the device). More specifically, show that by setting $z_{\text{new}} = z_{\text{old}} + \delta z$ and $z'_{\text{new}} = z'_{\text{old}} + \delta z'$ and choosing δz and $\delta z'$ appropriately, one can make $\mathfrak{t}$ and $\mathfrak{r}$ real. Show further that the reciprocity relations (9.34a) and (9.34b) then imply that $\mathfrak{t}'$ and $\mathfrak{r}'$ are also real. Finally, show that this adjustment of origins brings the real reciprocity relations into the same form (9.32) as for a sharp interface between two homogeneous media.

As attractive as it may be to have these coefficients real, one must keep in mind some disadvantages: (i) the displaced origins for z and z' in general will depend on frequency, and correspondingly, (ii) frequency-dependent information (most importantly, frequency-dependent phase shifts of the light) is lost by making the coefficients real. If the phase shifts depend only weakly on frequency over the band of interest (as is typically the case for the dielectric coating of a mirror face), then these disadvantages are unimportant and it is conventional to choose the coefficients real. If the phase shifts depend strongly on frequency over the band of interest [e.g., for the etalon of Eqs. (9.33a), when its two faces are highly reflecting and its round-trip phase φ is near a multiple of 2π], the disadvantages are severe. One then should leave the origins frequency independent, and correspondingly leave the device's $\mathfrak{r}$, $\mathfrak{r}'$, $\mathfrak{t}$, and $\mathfrak{t}'$ complex [as we have for the etalon in Eqs. (9.33a)].

Exercise 9.11 **Example: Transmission and Reflection Coefficients for an Interface between Dielectric Media*

Consider monochromatic electromagnetic waves that propagate from a medium with index of refraction n_1 into a medium with index of refraction n_2. Let z be a Cartesian coordinate perpendicular to the planar interface between the media.

(a) From the Helmholtz equation $[-\omega^2 + (c^2/n^2)\nabla^2]\psi = 0$, show that both ψ and $\psi_{,z}$ must be continuous across the interface.

(b) Using these continuity requirements, show that for light propagating orthogonal to the interface (z direction), the reflection and transmission coefficients, in going from medium 1 to medium 2, are

$$\boxed{\mathfrak{r} = \frac{n_1 - n_2}{n_1 + n_2}, \quad \mathfrak{t} = \frac{2n_1}{n_1 + n_2}.} \tag{9.35}$$

Notice that these $\mathfrak{r}$ and $\mathfrak{t}$ are both real.

(c) Use the reciprocity relations (9.34) to deduce the reflection and transmission coefficients $\mathfrak{r}'$ and $\mathfrak{t}'$ for a wave propagating in the opposite direction, from medium 2 to medium 1.

Exercise 9.12 **Example: Etalon's Light Fields Computed by Summing the Contributions from a Sequence of Round Trips*

Study the step-by-step buildup of the field inside an etalon and the etalon's transmitted field, when the input field is suddenly turned on. More specifically, carry out the following steps.

(a) When the wave first turns on, the transmitted field inside the etalon, at point A of Fig. 9.7, is $\psi_a = \mathfrak{t}\psi_i$, which is very small if the reflectivity is high so that $|\mathfrak{t}| \ll 1$. Show (with the aid of Fig. 9.7) that, after one round-trip-travel time in the etalon, the transmitted field at A is $\psi_a = \mathfrak{t}\psi_i + (\mathfrak{r}')^2 e^{i\varphi}\mathfrak{t}\psi_i$. Show that for high reflectivity and on resonance, the tiny transmitted field has doubled in amplitude and its energy flux has quadrupled.

(b) Compute the transmitted field ψ_a at A after more and more round trips, and watch it build up. Sum the series to obtain the steady-state field ψ_a. Explain the final, steady-state amplitude: why is it not infinite, and why, physically, does it have the value you have derived?

(c) Show that, at any time during this buildup, the field transmitted out the far side of the etalon is $\psi_t = \mathfrak{t}'\psi_a e^{iks_1}$ [Eq. (9.30b)]. What is the final, steady-state transmitted field? Your answer should be Eqs. (9.33a).

Exercise 9.13 **Example: Anti-Reflection Coating*

A common technique used to reduce the reflection at the surface of a lens is to coat it with a quarter wavelength of material with refractive index equal to the geometric mean of the refractive indices of air and glass.

(a) Show that this does indeed lead to perfect transmission of normally incident light.

(b) Roughly how thick must the layer be to avoid reflection of blue light? Estimate the energy-flux reflection coefficient for red light in this case.

[Hint: The amplitude reflection coefficients at an interface are given by Eqs. (9.35).]

Exercise 9.14 *Problem: Oil Slick*
When a thin layer of oil lies on top of water, one sometimes sees beautiful, multicolored, irregular bands of light reflecting off the oil layer. Explain qualitatively what causes this.

9.4.2 Fabry-Perot Interferometer and Modes of a Fabry-Perot Cavity with Spherical Mirrors

When an etalon's two faces are highly reflecting (reflection coefficient $\mathfrak{r}$ near unity), we can think of them as mirrors, between which the light resonates. The etalon is then a special case of a *Fabry-Perot interferometer*. The general case is any device in which light resonates between two high-reflectivity mirrors. The mirrors need not be planar and need not have the same reflectivities, and the resonating light need not be plane fronted.

A common example is the optical cavity of Fig. 7.9, formed by two mirrors that are segments of spheres, which we studied using geometric optics in Ex. 7.12. Because the phase fronts of a Gaussian beam (Sec. 8.5.5) are also spherical, such a beam can resonate in the optical cavity if (i) the beam's waist location and waist radius are adjusted so its phase-front radii of curvature, at the mirrors, are the same as the mirrors' radii of curvature, and (ii) the light's frequency is adjusted so a half-integral number of wavelengths fit perfectly inside the cavity. Box 9.3 gives details for the case where the two mirrors have identical radii of curvature. In that box we also learn that the Gaussian beams are not the only eigenmodes that can resonate inside such a cavity. Other, "higher-order" modes can also resonate. They have more complex transverse distributions of the light. There are two families of such modes: one with rectangular transverse light distributions, and the other with wedge-shaped, spoke-like light distributions.

For any Fabry-Perot interferometer with identical mirrors, driven by light with a transverse cross section that matches one of the interferometer's modes, one can study the interferometer's response to the driving light by the same kind of analysis as we used for an etalon in the previous section. And the result will be the same: the interferometer's reflected and transmitted light, at a given transverse location $\{x, y\}$, will be given by

$$\boxed{\frac{\psi_r}{\psi_i} \equiv \mathfrak{r}_{\mathrm{FP}} = \frac{\mathfrak{r}(1 - e^{i\varphi})}{1 - \mathfrak{r}^2 e^{i\varphi}}, \quad \frac{\psi_t}{\psi_i} \equiv \mathfrak{t}_{\mathrm{FP}} = \frac{(1 - \mathfrak{r}^2) e^{i\varphi/2}}{1 - \mathfrak{r}^2 e^{i\varphi}}} \qquad (9.36)$$

BOX 9.3. MODES OF A FABRY-PEROT CAVITY WITH SPHERICAL MIRRORS

Consider a Fabry-Perot cavity whose spherical mirrors have the same radius of curvature R and are separated by a distance L. Introduce (i) Cartesian coordinates with $z = 0$ at the cavity's center, and (ii) the same functions we used for Gaussian beams [Eqs. (8.40b)]:

$$z_0 = \frac{k\sigma_0^2}{2} = \frac{\pi \sigma_0^2}{\lambda}, \quad \sigma_z = \sigma_0(1 + z^2/z_0^2)^{1/2}, \quad R_z = z(1 + z_0^2/z^2), \quad (1)$$

with k the wave number and σ_0 a measure of the transverse size of the beam at the cavity's center. Then it is straightforward to verify that the following functions (i) satisfy the Helmholtz equation, (ii) are orthonormal when integrated over their transverse Cartesian coordinates x and y, and (iii) have phase fronts (surfaces of constant phase) that are spheres with radius of curvature R_z:

$$\psi_{mn}(x, y, z) = \frac{e^{-(x^2+y^2)/\sigma_z^2}}{\sqrt{2^{m+n-1}\pi \, m!n!}\, \sigma_z} H_m\left(\frac{\sqrt{2}\, x}{\sigma_z}\right) H_n\left(\frac{\sqrt{2}\, y}{\sigma_z}\right)$$

$$\times \exp\left\{i\left[\frac{k(x^2+y^2)}{2R_z} + kz - (n+m+1)\tan^{-1}\frac{z}{z_0}\right]\right\}. \quad (2)$$

Here $H_n(\xi) = (-1)^n e^{\xi^2} d^n e^{-\xi^2}/d\xi^n$ is the Hermite polynomial of index n, and m and n range over nonnegative integers. By adjusting σ_0, we can make the phase-front radius of curvature R_z match that, R, of the mirrors at the mirror locations, $z = \pm L/2$. Then the u_{mn} are a transversely orthonormal set of modes for the light field inside the cavity. Their flux distribution $|u_{mn}|^2$ on each mirror consists of an $m+1$ by $n+1$ matrix of discrete spots; see panel b in the box figure. The mode with $m = n = 0$ (panel a) is the Gaussian beam explored in the previous chapter: Eqs. (8.40). A given mode, specified by $\{m, n, k\}$, cannot resonate inside the cavity unless its wave number matches the cavity length, in the sense that the total phase shift in traveling from the cavity center $z = 0$ to the cavity end ($z = \pm L/2$) is an integral multiple of $\pi/2$; that is, $kL/2 - (n+m+1)\tan^{-1}[L/(2z_0)] = N\pi/2$ for some integer N.

There is a second family of modes that can resonate in the cavity, one whose eigenfunctions separate in circular polar coordinates:

$$\psi_{pm}(\varpi, \phi, z) = \frac{2p!\, e^{-\varpi^2/\sigma_z^2}}{\sqrt{1+\delta_{m0}}\, \pi(p+m)!\, \sigma_z} \left(\frac{\sqrt{2}\varpi}{\sigma_z}\right)^m L_p^m\left(\frac{2\varpi^2}{\sigma_z^2}\right) \begin{pmatrix} \cos m\phi \text{ or} \\ \sin m\phi \end{pmatrix}$$

$$\times \exp\left\{i\left[\frac{k\varpi^2}{2R_z} + kz - (2p+m+1)\tan^{-1}\frac{z}{z_0}\right]\right\}, \quad (3)$$

(continued)

BOX 9.3. (continued)

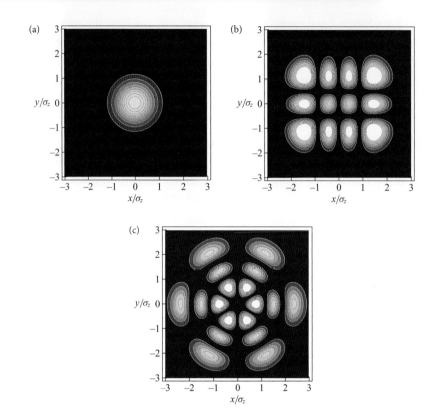

Energy flux distributions for (a) the Gaussian mode, (b) the Hermite mode of order 3,2, (c) the associated Laguerre mode of order 2,3. The contours are at 90%, 80%, ..., 10% of maximum.

where $L_p^m(\xi) = (1/p!)\, e^\xi\, \xi^{-m}\, d^p/d\xi^p(e^{-\xi}\, \xi^{p+m})$ is the associated Laguerre polynomial, and the indices p and m range over nonnegative integers. These modes make spots on the mirrors shaped like azimuthal wedges cut radially by circles; see panel c in the box figure. Again, they can resonate only if the phase change from the center of the cavity to an end mirror at $z = \pm L/2$ is an integral multiple of $\pi/2$: $kL/2 - (2p + m + 1)\tan^{-1}[L/(2z_0)] = N\pi/2$.

As one goes to larger mode numbers m, n (Hermite modes) or p, m (associated Laguerre modes), the region with substantial light power gets larger (see the box figure). As a result, more light gets lost off the edges of the cavity's mirrors. So unless the mirrors are made very large, high-order modes have large losses and do not resonate well.

For further details on these modes, see, for example, Yariv and Yeh (2007, Secs. 2.5–2.8 and 4.3).

[Eqs. (9.33a) with $\theta = 0$ so the light rays are orthogonal to the mirrors]. Here $\mathfrak{r}$ is the mirrors' reflection coefficient, and the round-trip phase is now

$$\boxed{\varphi = 2\pi\omega/\omega_f + \varphi_G, \quad \text{where } \omega_f = 2\pi/\tau_{\text{rt}}.} \quad (9.37)$$

Here τ_{rt} is the time required for a high-frequency photon to travel round-trip in the interferometer along the optic axis, from one mirror to the other; ω_f (called the *free spectral range*) is, as we shall see, the angular-frequency separation between the interferometer's resonances; and φ_G is an additive contribution (the Gouy phase), caused by the curvature of the phase fronts [e.g., the $\tan^{-1}(z/z_o)$ term in Eq. (8.40a) for a Gaussian beam and in Eqs. (2) and (3) of Box 9.3 for higher-order modes]. Because φ_G is of order one while $2\pi\omega/\omega_f$ is huge compared to one, and because φ_G changes very slowly with changing light frequency, it is unimportant in principle (and we henceforth shall ignore it). However, it is important in practice: it causes modes with different transverse light distributions (e.g., the Gaussian and higher-order modes in Box 9.3), which have different Gouy phases, to resonate at different frequencies.

free spectral range

Gouy phase

The Fabry-Perot interferometer's power transmissivity T and reflectivity R are given by Eqs. (9.33c), which we can rewrite in the following simplified form:

$$\boxed{T = 1 - R = \frac{1}{1 + (2\mathcal{F}/\pi)^2 \sin^2 \tfrac{1}{2}\varphi}.} \quad (9.38)$$

Here $\mathcal{F}$, called the interferometer's *finesse*, is defined by

$$\boxed{\mathcal{F} \equiv \pi\mathfrak{r}/(1 - \mathfrak{r}^2).} \quad (9.39)$$

finesse

[This finesse should not be confused with the *coefficient of finesse* $F = (2\mathcal{F}/\pi)^2$, which is sometimes used in optics, but which we eschew to avoid confusion.]

In Fig. 9.9 we plot, as functions of the round-trip phase $\varphi = 2\pi\omega/\omega_f$ (ignoring φ_G), the interferometer's power reflectivity and transmissivity T and R, and the phase changes $\arg(\mathfrak{t}_{\text{FP}})$ and $\arg(\mathfrak{r}_{\text{FP}})$ of the light that is transmitted and reflected by the interferometer.

Notice in Fig. 9.9a that, when the finesse $\mathcal{F}$ is large compared to unity, the interferometer exhibits sharp resonances at frequencies separated by the free spectral range ω_f. On resonance, the interferometer is perfectly transmitting ($T = 1$); away from resonance, it is nearly perfectly reflecting ($R \simeq 1$). The *full width at half maximum* (FWHM) of each sharp transmission resonance is given by

$$\delta\varphi_{1/2} = \frac{2\pi}{\mathcal{F}}, \quad \delta\omega_{1/2} = \frac{\omega_f}{\mathcal{F}}. \quad (9.40a)$$

resonance FWHM

In other words, if the frequency ω of the light is swept slowly through resonance, the transmission will be within 50% of its peak value (unity) over a bandwidth $\delta\omega_{1/2} = \omega_f/\mathcal{F}$. Notice also, in Fig. 9.9b, that for large finesse, the phase of the reflected and

9.4 Etalons and Fabry-Perot Interferometers

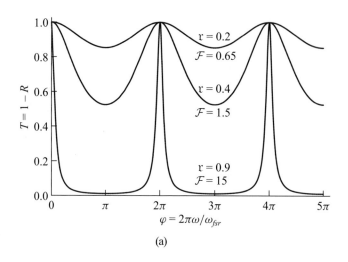

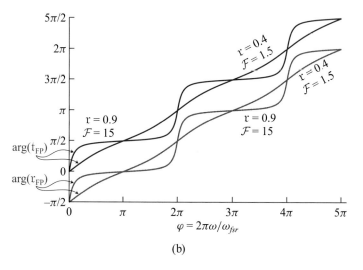

FIGURE 9.9 (a) Power transmissivity and reflectivity [Eq. (9.38)] for a Fabry-Perot interferometer with identical mirrors that have reflection coefficients $\mathfrak{r}$, as a function of the round-trip phase shift φ inside the interferometer. (b) The phase of the light transmitted (red) or reflected (blue) from the interferometer, relative to the input phase [Eqs. (9.36)]. The interferometer's finesse $\mathcal{F}$ is related to the mirrors' reflectivity by Eq. (9.39).

transmitted light near resonance changes very rapidly with a change in frequency of the driving light. Precisely on resonance, that rate of change is

resonance rate of change of phase

$$\left(\frac{d\,\arg(\mathfrak{t}_{FP})}{d\omega}\right)_{\text{on resonance}} = \left(\frac{d\,\arg(\mathfrak{r}_{FP})}{d\omega}\right)_{\text{on resonance}} = \frac{2\mathcal{F}}{\omega_f} = \frac{2}{\delta\omega_{1/2}}. \quad (9.40b)$$

The large transmissivity at resonance for large finesse can be understood by considering what happens when one first turns on the incident wave. Since the reflectivity of the first (input) mirror is near unity, the incoming wave has a large amplitude for reflection, and correspondingly, only a tiny amplitude for transmission into the optical cavity. The tiny bit that gets transmitted travels through the first mirror, gets strongly reflected from the second mirror, and returns to the first precisely in phase with the incoming wave (because φ is an integer multiple of 2π). Correspondingly, it superposes coherently on the tiny field being transmitted by the incoming wave, and so the net wave inside the cavity is doubled. After one more round trip inside the slab, this wave returns to the first face again in phase with the tiny field being transmitted by the incoming wave; again they superpose coherently, and the internal wave now has a three-times-larger amplitude than it began with. This process continues until a very strong field has built up inside the cavity (Ex. 9.12). As it builds up, that field begins to leak out of the cavity's first mirror with just such a phase as to destructively interfere with the wave being reflected there. The net reflected wave is thereby driven close to zero. The field leaking out of the second mirror has no other wave to interfere with. It remains strong, so the interferometer settles down into a steady state with strong net transmission. Heuristically, one can say that, because the wave inside the cavity is continually constructively superposing on itself, the cavity sucks almost all the incoming wave into itself, and then ejects it out the other side. Quantum mechanically, this sucking is due to the photons' Bose-Einstein statistics: the photons "want" to be in the same quantum state. We shall study this phenomenon in the context of plasmons that obey Bose-Einstein statistics in Sec. 23.3.2.

superposition of internal waves on resonance

Bose-Einstein behavior on resonance

This discussion makes it clear that, when the properties of the input light are changed, a high-finesse Fabry-Perot interferometer will change its response rather slowly—on a timescale approximately equal to the inverse of the resonance FWHM (i.e., the finesse times the round-trip travel time):

$$\tau_{\text{response}} \sim \frac{2\pi}{\delta\omega_{1/2}} = \mathcal{F}\,\tau_{\text{rt}}. \quad (9.40c)$$

These properties of a high-finesse Fabry-Perot interferometer are similar to those of a high-Q mechanical or electrical oscillator. The similarity arises because, in both cases, energy is being stored in a resonant, sinusoidal manner inside the device (the oscillator or the interferometer). For the interferometer, the light's round-trip travel time τ_{rt} is analogous to the oscillator's period, the interferometer's free spectral range ω_f is analogous to the oscillator's resonant angular frequency, and the interferometer's finesse $\mathcal{F}$ is analogous to the oscillator's quality factor Q. However, there are some major differences between an ordinary oscillator and a Fabry-Perot interferometer. Perhaps the most important is that the interferometer has several large families of resonant modes (families characterized by the number of longitudinal nodes between the mirrors and by the 2-dimensional transverse distributions of the light), whereas an oscillator has just one mode. This gives an interferometer much greater versatility than a simple oscillator possesses.

similarity between Fabry-Perot interferometer and an oscillator

9.4.3 Fabry-Perot Applications: Spectrometer, Laser, Mode-Cleaning Cavity, Beam-Shaping Cavity, PDH Laser Stabilization, Optical Frequency Comb

Just as mechanical and electrical oscillators have a wide variety of important applications in science and technology, so also do Fabry-Perot interferometers. In this section, we sketch a few of them.

SPECTROMETER

In the case of a Fabry-Perot etalon (highly reflecting parallel mirrors; plane-parallel light beam), the resonant transmission enables the etalon to be used as a spectrometer. The round-trip phase change $\varphi = 2\mathfrak{n}\omega d \cos\theta/c$ inside the etalon varies linearly with the wave's angular frequency ω, but only waves with round-trip phase φ near an integer multiple of 2π will be transmitted efficiently. The etalon can be tuned to a particular frequency by varying either the slab width d or the angle of incidence of the radiation (and thence the angle θ inside the etalon). Either way, impressively good chromatic resolving power can be achieved. We say that waves with two nearby frequencies can just be resolved by an etalon when the half-power point of the transmission coefficient of one wave coincides with the half-power point of the transmission coefficient of the other. Using Eq. (9.38) we find that the phases for the two frequencies must differ by $\delta\varphi \simeq 2\pi/\mathcal{F}$. Correspondingly, since $\varphi = 2\mathfrak{n}\omega d \cos\theta/c$, the *chromatic resolving power* is

chromatic resolving power of a spectrometer

$$\mathcal{R} = \frac{\omega}{\delta\omega} = \frac{4\pi \mathfrak{n} d \cos\theta}{\lambda_{\text{vac}} \delta\varphi} = \frac{2 \mathfrak{n} d \cos\theta \mathcal{F}}{\lambda_{\text{vac}}}. \tag{9.41}$$

Here $\lambda_{\text{vac}} = 2\pi c/\omega$ is the wavelength in vacuum (i.e., outside the etalon).

LASER

Fabry-Perot interferometers are exploited in the construction of many types of lasers. For example, in a gas-phase laser, the atoms are excited to emit a spectral line. This radiation is spontaneously emitted isotropically over a wide range of frequencies. Placing the gas between the mirrors of a Fabry-Perot interferometer allows one or more highly collimated and narrowband modes to be trapped and, while trapped, to be amplified by stimulated emission (i.e., to lase). See Sec. 10.2.1.

MODE CLEANER FOR A MESSY LASER BEAM

The output beam from a laser often has a rather messy cross sectional profile, for example because it contains multiple modes of excitation of the Fabry-Perot interferometer in which the lasing material resides (see the discussion of possible modes in Box 9.3). For many applications, one needs a much cleaner laser beam (e.g., one with a Gaussian profile). To clean the beam, one can send it into a high-finesse Fabry-Perot cavity with identical spherical mirrors, whose mirror curvatures and cavity length are adjusted so that, among the modes present in the beam, only the desired Gaussian mode will resonate and thereby be transmitted (see the sharp transmission peaks in Fig. 9.9a). The beam's unwanted modes will not resonate in the cavity, and therefore will be reflected backward off its input mirror, leaving the transmitted beam clean.

cleaning a light beam

Chapter 9. Interference and Coherence

BEAM-SHAPING CAVITY

In some applications one wants a light beam whose cross sectional distribution of flux $F(x, y)$ is different from any of the modes that resonate in a spherical-mirror cavity—for example, one might want a circular, flat-topped flux distribution $F(\varpi)$ with steeply dropping edges, like the shape of a circular mesa in a North American desert. One can achieve the desired light distribution (or something approximating it) as follows. Build a Fabry-Perot cavity with identical mirrors that are shaped in such a way that there is a cavity mode with the desired distribution. Then drive the cavity with a Gaussian beam. That portion of the beam with the desired flux distribution will resonate in the interferometer and leak out of the other mirror as the desired beam; the rest of the input beam will be rejected by the cavity.

reshaping a light beam

LASER STABILIZATION

There are two main ways to stabilize the frequency of a laser. One is to lock it onto the frequency of some fundamental atomic or molecular transition. The other is to lock it onto a resonant frequency of a mechanically stable Fabry-Perot cavity—a technique called *Pound-Drever-Hall* (PDH) *locking*.

PDH locking

In PDH locking to a cavity with identical mirrors, one passes the laser's output light (with frequency ω) through a device that modulates its frequency,[9] so ω becomes $\omega + \delta\omega$ with $\delta\omega = \sigma \cos(\Omega t)$ and $\sigma \ll \delta\omega_{1/2}$, the cavity resonance's FWHM. One then sends the modulated light into the cavity and monitors the reflected light power. Assume, for simplicity, that the modulation is slow compared to the cavity response time, $\Omega \ll 1/\tau_{\text{response}}$. Then the cavity's response at any moment will be that for steady light, that is, the reflected power will be $P_i R(\omega + \delta\omega)$, where R is the reflectivity at frequency $\omega + \delta\omega$. Using Eq. (9.38) for the reflectivity, specialized to frequencies very near resonance so the denominator is close to one, and using Eqs. (9.37) and (9.40a), we bring the reflected power into the form

$$P_r = P_i R(\omega + \delta\omega) = P_i \times \left[R(\omega) + \frac{dR}{d\omega} \delta\omega(t) \right]$$

$$= P_i \left[R(\omega) + \frac{8\sigma(\omega - \omega_o)}{(\delta\omega_{1/2})^2} \cos\Omega t \right], \qquad (9.42)$$

where ω_o is the cavity's resonant frequency (at which φ is an integral multiple of 2π).

The modulated part of the reflected power has an amplitude directly proportional to the laser's frequency error, $\omega - \omega_o$. In the PDH technique, one monitors this modulated power with a photodetector, followed by a band-pass filter on the photodetector's output electric current to get rid of the unmodulated part of the signal [arising from $P_i R(\omega)$]. The amplitude of the resulting, modulated output current is proportional

---

9. Actually, one sends it through a phase modulator called a *Pockels cell*, consisting of a crystal whose index of refraction is modulated by applying an oscillating voltage to it. The resulting phase modulation, $\delta\phi \propto \sin\Omega t$, is equivalent to a frequency modulation, $\delta\omega = d\delta\phi/dt \propto \cos\Omega t$.

to $\omega - \omega_o$ and is used to control a *feedback circuit* that drives the laser back toward the desired, cavity-resonant frequency ω_o. (See, e.g., Black, 2001, for details.)

In Ex. 9.15 it is shown that, if one needs a faster feedback and therefore requires a modulation frequency $\Omega \gtrsim 1/\tau_{\text{response}}$, this PDH locking technique still works.

This technique was invented by Ronald Drever for use in interferometric gravitational-wave detectors, relying on earlier ideas of Robert Pound, and it was first demonstrated experimentally by Drever and John Hall. It is now used widely in many areas of science and technology.

OPTICAL FREQUENCY COMB

John Hall and Theodor Hänsch were awarded the 2005 Nobel Prize for development of the optical frequency comb. This powerful tool is based on an optical cavity of length L, filled with a lasing medium that creates and maintains a sharply pulsed internal light field with the following form (Fig. 9.10):

$$\psi = \psi_o(z - V_g t) \exp[i k_p (z - V_p t)]. \tag{9.43}$$

Here

1. we use a z coordinate that increases monotonically along the optic axis as one moves rightward from mirror 1 to mirror 2, then leftward from 2 to 1, then rightward from 1 to 2, and so forth;
2. $\exp[i k_p(z - V_p t)]$ is the pth longitudinal monochromatic mode of the cavity, with wave number $k_p \equiv p\pi/L$, phase velocity V_p, and angular frequency $k_p V_p$ lying in the optical range $\sim 10^{15}$ Hz;
3. $\psi_o(z - V_g t)$ is the envelope of a wave packet so narrow that only about one wavelength of the mode p can fit inside it;
4. the envelope travels with group velocity V_g and does not spread.

For the wave packet not to spread, V_g must have the same (constant) value over all frequencies contained in the packet, which means that the dispersion relation must have $0 = (\partial V_g/\partial k) = \partial^2 \omega/\partial k^2$; ω must be linear in k:

$$\omega = V_g(k + \kappa) \tag{9.44}$$

for some constant κ, which is typically considerably smaller than the wave numbers k contained in the packet.

It was a huge technical challenge to build a lasing cavity that creates and sustains a very narrow wave packet of this sort. Two of the keys to this achievement were (i) using a nonlinear lasing medium that amplifies light more strongly at high energy fluxes $|\psi|^2$ than at low ones and thereby tends to produce intense, short pulses rather than long, monochromatic waves and (ii) using some trickery to ensure that the lasing medium and anything else in the cavity jointly give rise to the linear dispersion relation (9.44) over a sufficiently wide frequency band. For some of the details, see Sec. 10.2.3. Because the sharp wave packet (9.43) has fixed relationships between the phases of the

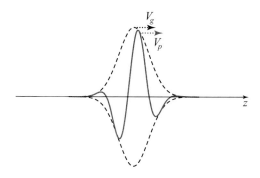

FIGURE 9.10 The sharply pulsed electric field [Eq. (9.43)] inside a Fabry-Perot cavity. The envelope is shown dotted; the red curve is the full field ψ.

various monochromatic modes that make it up, the lasing optical cavity that creates it is called a *mode-locked laser*.

As the internal field's pulse hits mirror 2 time and again, it transmits through the mirror a sequence of outgoing pulses separated by the wave packet's round-trip travel time in the cavity, $\tau_{\rm rt} = 2L/V_g$. Assuming, for pedagogical clarity, a Gaussian shape for each pulse, the oscillating internal field (9.43) produces the outgoing field $\psi \propto \sum_n \exp[-\sigma^2(t - z/c - n\tau_{\rm rt})^2/2] \exp[-ik_p V_p(t - z/c)]$. Here $1/\sigma$ is the pulse length in time, and we have assumed vacuum-light-speed propagation outside the cavity. It is helpful to rewrite the frequency $k_p V_p$ of the oscillatory piece of this field as the sum of its nearest multiple of the cavity's free spectral range, $\omega_f = 2\pi/\tau_{\rm rt}$, plus a frequency shift ω_s: $k_p V_p = q\omega_f + \omega_s$. The integer $q =$ (largest integer contained in $k_p V_p/\omega_f$) will typically be quite close to p, and ω_s is guaranteed to lie in the interval $0 \leq \omega_s < \omega_f$. The emerging electric field is then

$$\psi \propto \sum_n \exp[-\sigma^2(t - z/c - n\tau_{\rm rt})^2/2] \exp[-i(q\omega_f + \omega_s)(t - z/c)] \quad (9.45{\rm a})$$

(Fig. 9.11a).

This entire emerging field is periodic in $t - z/c$ with period $\tau_{\rm rt} = 2\pi/\omega_f$, except for the frequency-shift term $\exp[-i\omega_s(t - z/c)]$. The periodic piece can be expanded as a sum over discrete frequencies that are multiples of $\omega_f = 2\pi/\tau_{\rm rt}$. Since the Fourier transform of a Gaussian is a Gaussian, this sum, augmented by the frequency shift term, turns out to be

$$\psi \propto \sum_{m=-\infty}^{+\infty} \exp\left[\frac{-(m-q)^2 \omega_f^2}{2\sigma^2}\right] \exp[-i(m\omega_f + \omega_s)(t - z/c)]. \quad (9.45{\rm b})$$

The set of discrete frequencies (spectral lines) appearing in this outgoing field is the *frequency comb* displayed in Fig. 9.11b.

9.4 Etalons and Fabry-Perot Interferometers

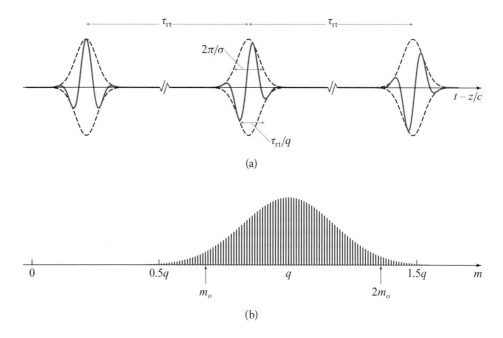

FIGURE 9.11 Optical frequency comb. (a) The pulsed electric field (9.45a) emerging from the cavity. (b) The field's comb spectrum. Each line, labeled by m, has angular frequency $m\omega_f$, and is shown with height proportional to its power, $\propto \exp\left[-(m-q)^2\omega_f^2/\sigma^2\right]$ [cf. Eq. (9.45b)].

Some concrete numbers make clear how very remarkable this pulsed electric field (9.45a) and its frequency comb (9.45b) are:

1. The Fabry-Perot cavity typically has a length L somewhere between ~ 3 cm and ~ 3 m, so (with the group velocity V_g of order the vacuum light speed) the round-trip travel time and free spectral range are $\tau_{\rm rt} = 2L/V_g \sim 0.3$ to 30 ns; and $\omega_f/2\pi \sim 1/\tau_{\rm rt} \sim 30$ MHz to 3 GHz, which are radio and microwave frequencies, respectively. Since the shift frequency is $\omega_s < \omega_f$, it is also in the radio or microwave band.

2. The comb's central frequency is in the optical, $q\omega_f/2\pi \sim 3 \times 10^{14}$ Hz, so the harmonic number of the central frequency is $q \sim 10^5$ to 10^7, roughly a million.

3. The pulse width $\sim 2/\sigma$ contains roughly one period ($2\pi/q\omega_f$) of the central frequency, so $\sigma \sim q\omega_f/3$, which means that most of the comb's power is contained in the range $m \sim 2q/3$ to $m \sim 4q/3$ (i.e., there are roughly a million strong teeth in the comb).

It is possible to lock the comb's free spectral range ω_f to a very good cesium atomic clock, whose oscillation frequency ~ 9 GHz is stable to $\delta\omega/\omega \sim 10^{-13}$ (Fig. 6.11), so ω_f has that same phenomenal stability. One can then measure the shift frequency ω_s

and calibrate the comb (identify the precise mode number m of each frequency in the comb) as follows.

calibrating the optical frequency comb

1. Arbitrarily choose a tooth at the low-frequency end of the comb, $m_o \simeq 2q/3$ (Fig. 9.11b); it has frequency $\omega_o = m_o \omega_f + \omega_s$.
2. Separate the light in that tooth from light in the other teeth, and send a beam of that tooth's light through a frequency doubler (to be discussed in Sec. 10.7.1), thereby getting a beam with frequency $2\omega_o = 2(m_o \omega_f + \omega_s)$.
3. By beating this beam against the light in teeth at $m \sim 4q/3$, identify the tooth that most closely matches this beam's frequency. It will have frequency $2m_o \omega_f + \omega_s$, and the frequency difference (beat frequency) will be ω_s.

This reveals ω_s to very high accuracy, and one can count the number of teeth ($m_o - 1$) between this tooth $2m_o$ and its undoubled parent m_o, thereby learning the precise numerical value of m_o. From this, by tooth counting, one learns the precise mode numbers m of all the optical-band teeth in the comb, and also their frequencies $m\omega_f + \omega_s$.

With the comb now calibrated, it can be used to measure the frequency of any other beam of light in the optical band in terms of the ticking frequency of the cesium clock, to which the entire comb has been locked. The optical frequency accuracies thereby achieved are orders of magnitude better than were possible before this optical frequency comb was developed. And in the near future, as optical-frequency atomic clocks become much more accurate and stable than the microwave-frequency cesium clock (see the footnote in Sec. 6.6.1), this comb will be used to calibrate microwave and radio frequencies in terms of the ticking rates of optical frequency clocks.

some applications

For further details about optical frequency combs, see the review articles by Cundiff (2002) and Cundiff and Ye (2003).

EXERCISES

Exercise 9.15 *Problem: PDH Laser Stabilization*
Show that the PDH method for locking a laser's frequency to an optical cavity works for modulations faster than the cavity's response time, $\Omega \gtrsim 1/\tau_{\text{response}}$, and even works for $\Omega \gg 1/\tau_{\text{response}}$. More specifically, show that the reflected power still contains the information needed for feedback to the laser. [For a quite general analysis and some experimental details, see Black (2001).]

Exercise 9.16 *Derivation: Optical Frequency Comb*
Fill in the details of the derivation of all the equations in the section describing the optical frequency comb.

Exercise 9.17 ***Problem: Sagnac Interferometer*
A Sagnac interferometer is a rudimentary version of a laser gyroscope for measuring rotation with respect to an inertial frame. The optical configuration is shown in

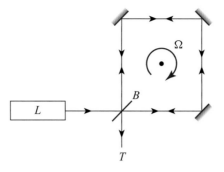

FIGURE 9.12 Sagnac interferometer used as a type of laser gyro.

Fig. 9.12. Light from a laser L is split by a beam splitter B and travels both clockwise and counterclockwise around the optical circuit, reflecting off three plane mirrors. The light is then recombined at B, and interference fringes are viewed through the telescope T. The whole assembly rotates with angular velocity Ω. Calculate the difference in the time it takes light to traverse the circuit in the two directions, and show that the consequent fringe shift (total number of fringes that enter the telescope during one round trip of the light in the interferometer) can be expressed as $\Delta N = 4A\Omega/(c\lambda)$, where λ is the wavelength, and A is the area bounded by the beams. Show further that, for a square Sagnac interferometer with side length L, the rate at which fringes enter the telescope is $\Omega L/\lambda$.

9.5 Laser Interferometer Gravitational-Wave Detectors

As we discuss in Chap. 27, gravitational waves are predicted to exist by general relativity theory, and their emission by the binary neutron-star system PSR B1913+16 has been monitored since 1974, via their back-action on the binary's orbital motion. As orbital energy is lost to gravitational waves, the binary gradually spirals inward, so its orbital angular velocity gradually increases. The measured rate of increase agrees with general relativity's predictions to within the experimental accuracy of a fraction of a percent (for which Russell Hulse and Joseph Taylor received the 1993 Nobel Prize in physics). Unfortunately, the gravitational analog of Hertz's famous laboratory emission and detection of electromagnetic waves has not yet been performed, and cannot be in the authors' lifetime because of the waves' extreme weakness. For waves strong enough to be detectable on Earth, one must turn to violent astrophysical events, such as the collision and coalescence of two neutron stars or black holes.

When the gravitational waves from such an event reach Earth and pass through a laboratory, general relativity predicts that they will produce tiny relative accelerations of free test masses. The resulting oscillatory variation of the distance between two such masses can be measured optically using a Michelson interferometer, in which (to increase the signal strength) each of the two arms is operated as a Fabry-Perot

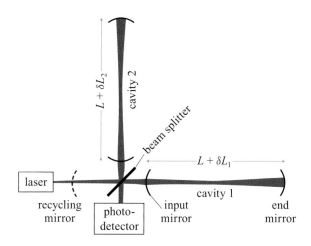

FIGURE 9.13 Schematic design of an initial gravitational-wave interferometer operated in LIGO (at Livingston, Louisiana, and Hanford, Washington, USA) during 2005–2010.

cavity. Two such instruments, called *interferometric gravitational wave detectors* [for which Rainer Weiss (1972) was the primary inventor, with important contributions from Ronald Drever and many others] made the first discovery of gravitational waves arriving at Earth on September 14, 2015 (Abbott et al., 2016a). The waves came from the last few orbits of inspiral and the collision of two black holes with masses $29M_\odot$ and $36M_\odot$, where $M_\odot$ is the Sun's mass.

The observatory that discovered these waves is called LIGO, the Laser Interferometer Gravitational Wave Observatory, and its interferometric detectors were second generation instruments, called advanced LIGO. These advanced LIGO detectors are more complex than is appropriate to analyze in this chapter, so we shall forcus instead on the first generation detectors called initial LIGO (Abbott et al., 2009b), and then near the end of this section we shall briefly describe how advanced LIGO differs from initial LIGO.

In each of the initial LIGO gravitational-wave detectors, the two cavities are aligned along perpendicular directions, as shown in Fig. 9.13. A Gaussian beam of light (Sec. 8.5.5) from a laser passes through a beam splitter, creating two beams with correlated phases. The beams excite the two cavities near resonance. Each cavity has an end mirror with extremely high reflectivity,[10] $1 - \mathfrak{r}_e^2 < 10^{-4}$, and a corner mirror ("input mirror") with a lower reflectivity, $1 - \mathfrak{r}_i^2 \sim 0.03$. Because of this lower reflectivity, by contrast with the etalons discussed in previous sections, the resonant light leaks out through the input mirror instead of through the end mirror. This reflectivity of the input mirror is so adjusted that the typical photon is stored in

interferometric gravitational wave detector

advanced LIGO

initial LIGO

mirror reflectivity

---

10. Because LIGO operates with monochromatic light, it is convenient to adjust the phases of the mirrors' reflection and transmission coefficients so $\mathfrak{r}$ and $\mathfrak{t}$ are both real; cf. Ex. 9.10e. We do so here.

the cavity for roughly half the period of the expected gravitational waves (a few milliseconds), which means that the input mirror's reflectivity $\mathfrak{r}_i^2$, the arm length L, and the gravitational-wave angular frequency ω_{gw} are related by

$$\frac{4L}{c(1-\mathfrak{r}_i^2)} \sim \frac{1}{\omega_{\text{gw}}}. \tag{9.46}$$

The light emerging from the cavity, like that transmitted by an etalon, has a phase that is highly sensitive to the separation between the mirrors: a tiny change δL in their separation produces a change in the outcoming phase

$$\delta\varphi_o \simeq \frac{8\omega\delta L}{c}\frac{1}{(1-\mathfrak{r}_i^2)} \sim \frac{2\omega}{\omega_{\text{gw}}}\frac{\delta L}{L} \tag{9.47}$$

in the limit $1-\mathfrak{r}_i \ll 1$; see Ex. 9.18. The outcoming light beams from the two cavities return to the beam splitter and are recombined there. The relative distances from the beam splitter to the cavities are adjusted so that, in the absence of any perturbations of the cavity lengths, almost all the interfered light goes back toward the laser, and only a tiny (but nonzero) amount goes toward the photodetector of Fig. 9.13, which monitors the output. Perturbations δL_1 and δL_2 in the cavity lengths then produce a change

phase shift induced by mirror motions

$$\delta\varphi_{o1} - \delta\varphi_{o2} \sim \frac{2\omega}{\omega_{\text{gw}}}\frac{(\delta L_1 - \delta L_2)}{L} \tag{9.48}$$

in the relative phases at the beam splitter, and this in turn produces a change of the light power entering the photodetector. By using two cavities in this way and keeping their light-storage times (and hence response times) the same, one makes the light power entering the photodetector be insensitive to fluctuations in the laser frequency. This is crucial for obtaining the high sensitivities that gravitational-wave detection requires.

The mirrors at the ends of each cavity are suspended as pendula, and when a gravitational wave with dimensionless amplitude h (discussed in Chap. 27) passes, it moves the mirrors back and forth, producing changes

mirror motions produced by gravitational wave

$$\delta L_1 - \delta L_2 = hL \tag{9.49}$$

in the arm-length difference. The resulting change in the relative phases of the two beams returning to the beam splitter,

$$\delta\varphi_{o1} - \delta\varphi_{o2} \sim \frac{2\omega}{\omega_{\text{gw}}}h, \tag{9.50}$$

is monitored via the changes in power that it produces for the light going into the photodetector. If one builds the entire detector optimally and uses the best possible photodetector, these phase changes can be measured with a photon shot-noise-limited precision of $\sim 1/\sqrt{N}$. Here $N \sim [W_\ell/(\hbar\omega)](\pi/\omega_{\text{gw}})$ is the number of photons put into

detectable phase shift

the detector by the laser (with power W_ℓ) during half a gravitational-wave period.[11] By combining this with Eq. (9.50), we see that the weakest wave that can be detected (at signal-to-noise ratio 1) is

$$h \sim \left(\frac{\hbar \omega_{gw}^3}{4\pi \omega W_\ell}\right)^{1/2}. \qquad (9.51)$$

estimate of weakest detectable gravitational wave

For a laser power $W_\ell \sim 5$ W, and $\omega_{gw} \sim 10^3$ s^{-1}, $\omega \sim 2 \times 10^{15}$ s^{-1}, this gravitational-wave sensitivity (noise level) is $h \sim 1 \times 10^{-21}$.

When operated in this manner, about 97% of the light returns toward the laser from the beam splitter and the other 3% goes out the end mirror, goes into the photodetector, gets absorbed, or is scattered due to imperfections in the optics. In LIGO's initial detectors, the 97% returning toward the laser was recycled back into the interferometer, in phase with new laser light, by placing a mirror between the laser and the beam splitter. This "power recycling mirror" (shown dashed in Fig. 9.13) made the entire optical system into a big optical resonator with two subresonators (the arms' Fabry-Perot cavities), and the practical result was a 50-fold increase in the input light power, from 5 W to 250 W—and an optical power in each arm of about $\frac{1}{2} \times 250$ W $\times 4/(1 - \mathfrak{r}_i^2) \sim 15$ kW; see Abbott et al. (2009b, Fig. 3). When operated in this manner, the interferometer achieved a noise level $h \sim 1 \times 10^{-21}/\sqrt{50} \sim 1 \times 10^{-22}$. For a more accurate analysis of the sensitivity, see Exs. 9.18 and 9.19.

power recycling

This estimate of sensitivity is actually the rms noise in a bandwidth equal to frequency at the minimum of LIGO's noise curve. Figure 6.7 shows the noise curve as the square root of the spectral density of the measured arm-length difference $\xi \equiv L_1 - L_2$, $\sqrt{S_\xi(f)}$. Since the waves produce a change of ξ given by $\delta\xi = hL$ [Eq. (9.49)], the corresponding noise-induced fluctuations in the measured h have $S_h = S_\xi/L^2$, and the rms noise fluctuations in a bandwidth equal to frequency f are $h_{rms} = \sqrt{S_h f} = (1/L)\sqrt{S_\xi f}$. Inserting $\sqrt{S_\xi} \simeq 10^{-19}$ m Hz$^{-1/2}$ and $f \simeq 100$ Hz from Fig. 6.7, and $L = 4$ km for the LIGO arm length, we obtain $h_{rms} \simeq 2.5 \times 10^{-22}$, a factor 2.5 larger than our 1×10^{-22} estimate—in part because thermal noise is roughly as large as photon shot noise at 100 Hz.

measured noise in initial LIGO

There are enormous obstacles to achieving such high sensitivity. Here we name just a few. Imperfections in the optics will absorb some of the high light power, heating the mirrors and beam splitter and causing them to deform. Even without such heating, the mirrors and beam splitter must be exceedingly smooth and near perfectly shaped to minimize the scattering of light from them (which causes noise; Ex. 8.15). Thermal noise in the mirrors and their suspensions (described by the fluctuation-dissipation

experimental challenges

---

11. This measurement accuracy is related to the Poisson distribution of the photons entering the interferometer's two arms: if N is the mean number of photons during a half gravitational-wave period, then the variance is $\sqrt{N}$, and the fractional fluctuation is $1/\sqrt{N}$. The interferometer's shot noise is actually caused by a beating of quantum electrodynamical vacuum fluctuations against the laser's light; for details see Caves (1980).

theorem) will cause the mirrors to move in manners that simulate the effects of a gravitational wave (Secs. 6.8.2 and 11.9.2), as will seismic- and acoustic-induced vibrations of the mirror suspensions. LIGO's arms must be long (4 km) to minimize the effects of these noise sources. While photon shot noise dominates above the noise curve's minimum, $f \gtrsim 100$ Hz, these and other noises dominate at lower frequencies.

The initial LIGO detectors operated from 2005 to 2010, carrying out gravitational-wave searches, much of the time in collaboration with international partners (the French-Italian VIRGO and British/German GEO600 detectors). From 2010 to 2015, the second generation advanced LIGO detectors were installed in the LIGO vacuum system, with amplitude design sensitivity 10-fold higher than the initial detectors. These advanced detectors discovered gravitational waves in September 2015, just before their first search officially began, and when their sensitivity at the noise-curve minimum was about 3 times worse than their design but 3 times better than initial LIGO.

One of several major changes, in going from initial LIGO to advanced LIGO, was the insertion of a "signal recycling mirror" between the beam splitter and the photodetector (Abbott et al., 2016b). The output signal light, in the cavity bounded by this new mirror and the two input mirrors, "sucks" signal light out of the interferometer arms due to Bose-Einstein statistics. This permits laser light to be stored in the arm cavities far longer than half a gravitational wave period (and thereby build up to higher intensity), while the signal light gets extracted in roughly a half period. This added complication is why we chose not to analyze the advanced detectors.

EXERCISES

Exercise 9.18 *Derivation and Problem: Phase Shift in LIGO Arm Cavity*
In this exercise and the next, simplify the analysis by treating each Gaussian light beam as though it were a plane wave. The answers for the phase shifts will be the same as for a true Gaussian beam, because on the optic axis, the Gaussian beam's phase [Eq. (8.40a) with $\varpi = 0$] is the same as that of a plane wave, except for the Gouy phase $\tan^{-1}(z/z_0)$, which is very slowly changing and thus is irrelevant.

(a) For the interferometric gravitational-wave detector depicted in Fig. 9.13 (with the arms' input mirrors having amplitude reflectivities $\mathfrak{r}_i$ close to unity and the end mirrors idealized as perfectly reflecting), analyze the light propagation in cavity 1 by the same techniques as used for an etalon in Sec. 9.4.1. Show that, if ψ_{i1} is the light field impinging on the input mirror, then the total reflected light field ψ_{r1} is

$$\psi_{r1} = e^{i\varphi_1} \frac{1 - \mathfrak{r}_i e^{-i\varphi_1}}{1 - \mathfrak{r}_i e^{i\varphi_1}} \psi_{i1}, \quad \text{where } \varphi_1 = 2kL_1. \tag{9.52a}$$

(b) From this, infer that the reflected flux $|\psi_{r1}|^2$ is identical to the cavity's input flux $|\psi_{i1}|^2$, as it must be, since no light can emerge through the perfectly reflecting end mirror.

(c) The arm cavity is operated on resonance, so φ_1 is an integer multiple of 2π. From Eq. (9.52a) infer that (up to fractional errors of order $1 - \mathfrak{r}_i$) a change δL_1 in the length of cavity 1 produces a change

$$\delta\varphi_{r1} = \frac{8k\delta L_1}{1 - \mathfrak{r}_i^2}. \tag{9.52b}$$

With slightly different notation, this is Eq. (9.47), which we used in the text's order-of-magnitude analysis of LIGO's sensitivity. In this exercise and the next, we carry out a more precise analysis.

Exercise 9.19 *Example: Photon Shot Noise in LIGO* T2

This exercise continues the preceding one. We continue to treat the light beams as plane waves.

(a) Denote by ψ_ℓ the light field from the laser that impinges on the beam splitter and gets split in two, with half going into each arm (Fig. 9.13). Using the equations from Ex. 9.18 and Sec. 9.5, infer that the light field returning to the beam splitter from arm 1 is $\psi_{s1} = \frac{1}{\sqrt{2}}\psi_\ell e^{i\varphi_1}(1 + i\delta\varphi_{r1})$, where φ_1 is some net accumulated phase that depends on the separation between the beam splitter and the input mirror of arm 1.

(b) Using the same formula for the field ψ_{s2} from arm 2, and assuming that the phase changes between beam splitter and input mirror are almost the same in the two arms, so $\varphi_o \equiv \varphi_1 - \varphi_2$ is small compared to unity (mod 2π), show that the light field that emerges from the beam splitter, traveling toward the photodetector, is

$$\psi_{pd} = \frac{1}{\sqrt{2}}(\psi_{s1} - \psi_{s2}) = \frac{i}{2}(\varphi_o + \delta\varphi_{r1} - \delta\varphi_{r2})\psi_\ell \tag{9.53a}$$

to first order in the small phases. Show that the condition $|\varphi_o| \ll 1$ corresponds to the experimenters' having adjusted the positions of the input mirrors in such a way that almost all the light returns toward the laser, and only a small fraction goes toward the photodetector.

(c) For simplicity, let the gravitational wave travel through the interferometer from directly overhead and have an optimally oriented polarization. Then, as we shall see in Chap. 27 [Eq. (27.81)], the dimensionless gravitational-wave field $h(t)$ produces the arm-length changes $\delta L_1 = -\delta L_2 = \frac{1}{2}h(t)L$, where L is the unperturbed arm length. Show, then, that the field traveling toward the photodetector is

$$\psi_{pd} = \frac{i}{2}(\varphi_o + \delta\varphi_{gw})\psi_\ell, \quad \text{where} \quad \delta\varphi_{gw} = \frac{8kL}{1-\mathfrak{r}_i^2}h(t) = \frac{16\pi L/\lambda}{1-\mathfrak{r}_i^2}h(t). \tag{9.53b}$$

The experimenter adjusts φ_o so it is large compared to the tiny $\delta\varphi_{gw}$.

(d) Actually, this equation has been derived assuming, when analyzing the arm cavities [Eq. (9.52a)], that the arm lengths are static. Explain why it should still be nearly valid when the gravitational waves are moving the mirrors, so long as the gravitational-wave half-period $1/(2f) = \pi/\omega_{\rm gw}$ is somewhat longer than the mean time that a photon is stored inside an arm cavity, or so long as $f \ll f_o$, where

$$f_o \equiv \frac{1-\mathfrak{r}_i^2}{4\pi} \frac{c}{2L}. \tag{9.54}$$

Assume that this is so. For the initial LIGO detectors, $1 - \mathfrak{r}_i^2 \sim 0.03$ and $L = 4$ km, so $f_o \sim 90$ Hz.

(e) Show that, if W_ℓ is the laser power impinging on the beam splitter (proportional to $|\psi_\ell|^2$), then the steady-state light power going toward the photodetector is $W_{\rm pd} = (\varphi_o/2)^2 W_\ell$, and the time variation in that light power due to the gravitational wave (the gravitational-wave signal) is

$$W_{\rm gw}(t) = \sqrt{W_\ell W_{\rm pd}} \, \frac{16\pi L/\lambda}{1-\mathfrak{r}_i^2} h(t). \tag{9.55a}$$

The photodetector monitors these changes $W_{\rm gw}(t)$ in the light power $W_{\rm pd}$ and from them infers the gravitational-wave field $h(t)$. This is called a "DC" or "homodyne" readout system; it works by beating the gravitational-wave signal field ($\propto \delta\varphi_{\rm gw}$) against the steady light field ("local oscillator," $\propto \varphi_o$) to produce the signal light power $W_{\rm gw}(t) \propto h(t)$.

(f) Shot noise in the interferometer's output light power $W_{\rm pd}$ gives rise to noise in the measured gravitational-wave field $h(t)$. From Eq. (9.55a) show that the spectral density of the noise in the measured $h(t)$ is

$$S_h(f) = \left(\frac{(1-\mathfrak{r}_i^2)\lambda}{16\pi L}\right)^2 \frac{S_{W_{\rm pd}}}{W_\ell W_{\rm pd}}. \tag{9.55b}$$

In Sec. 6.7.4, we derived the formula $S_{W_{\rm pd}} = 2\mathcal{R}(\hbar\omega)^2 = 2W_{\rm pd}\hbar\omega$ [Eq. (6.69)] for the (frequency-independent) spectral density of a steady, monochromatic light beam's power fluctuations due to shot noise; here $\mathcal{R} = W_{\rm pd}/(\hbar\omega)$ is the average rate of arrival of photons. Combining with Eq. (9.55b), deduce your final formula for the spectral density of the noise in the inferred gravitational-wave signal:

$$S_h(f) = \left(\frac{(1-\mathfrak{r}_i^2)\lambda}{16\pi L}\right)^2 \frac{2}{W_\ell/(\hbar\omega)}. \tag{9.56a}$$

From this deduce the rms noise in a bandwidth equal to frequency

$$h_{\rm rms} = \sqrt{f S_h} = \left(\frac{(1-\mathfrak{r}_i^2)\lambda}{16\pi L \sqrt{N}}\right), \quad \text{where} \quad N = \frac{W_\ell}{\hbar\omega} \frac{1}{2f} \tag{9.56b}$$

is the number of photons that impinge on the beam splitter, from the laser, in half a gravitational-wave period.

(g) In Ex. 9.20 we shall derive (as a challenge) the modification to the spectral density that arises at frequencies $f \gtrsim f_o$. The signal strength that gets through the interferometer is reduced because the arm length is increasing, then decreasing, then increasing again (and so forth) while the typical photon is in an arm cavity. The result of the analysis is an increase of $S_h(f)$ by $1 + (f/f_o)^2$, so we have

$$S_h(f) = \left(\frac{(1-\mathfrak{r}_i^2)\lambda}{16\pi L}\right)^2 \frac{2}{W_\ell/\hbar\omega}\left(1 + \frac{f^2}{f_o^2}\right). \tag{9.57}$$

Compare this with the measured noise, at frequencies above $f_o \sim 90$ Hz in the initial LIGO detectors (Fig. 6.7 with $\xi = hL$), using the initial LIGO parameters: $\lambda = 1.06\ \mu$m, $\omega = 2\pi c/\lambda \simeq 2 \times 10^{15}$ s^{-1}, $L = 4$ km, $W_\ell = 250$ W, $1 - \mathfrak{r}_i^2 = 0.03$. It should agree fairly well with the measured noise at frequencies $f \gtrsim f_o$, where most of the noise is due to photon shot noise. Also compare the noise (9.57) in a bandwidth equal to frequency $\sqrt{f S_h}$ and evaluated at frequency $f = f_o$ with the estimate (9.51) worked out in the text.

Exercise 9.20 *Challenge: LIGO Shot Noise at $f \gtrsim f_o$* T2
Derive the factor $1 + (f/f_o)^2$ by which the spectral density of the shot noise is increased at frequencies $f \gtrsim f_o$. [Hint: Redo the analysis of the arm cavity fields, part (a) of Ex. 9.19, using an arm length that varies sinusoidally at frequency f due to a sinusoidal gravitational wave, and then use the techniques of Ex. 9.19 to deduce $S_h(f)$.]

9.6 Power Correlations and Photon Statistics: Hanbury Brown and Twiss Intensity Interferometer

A type of interferometer that is rather different from those studied earlier in this chapter was proposed and constructed by Robert Hanbury Brown and Richard Q. Twiss in 1956. In this interferometer, light powers, rather than amplitudes, are combined to measure the degree of coherence of the radiation field. This is often called an *intensity interferometer*, because the optics community often uses the word "intensity" to mean energy flux (power per unit area).

In their original experiment, Hanbury Brown and Twiss divided light from an incandescent mercury lamp and sent it along two paths of variable length before detecting photons in each beam separately using a photodetector (see Fig. 9.14). The electrical output from each photodetector measures the rate of arrival of its beam's photons, or equivalently, its beam's power $W(t)$, which we can write as $K(\Re\Psi)^2$,

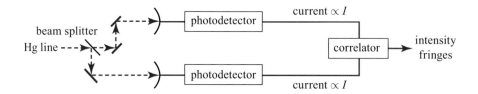

FIGURE 9.14 Hanbury Brown and Twiss intensity interferometer.

fluctuations of power W in each beam

where K is a constant. This W exhibits fluctuations δW about its mean value $\overline{W}$, and it was found that the fluctuations in the two beams were correlated. How can this be?

The light that was detected originated from many random and independent emitters and therefore obeys Gaussian statistics, according to the central limit theorem (Sec. 6.3.2). This turns out to mean that the fourth-order correlations of the wave field Ψ with itself can be expressed in terms of the second-order correlations (i.e., in terms of the degree of coherence $\gamma_\|$).

More specifically, continuing to treat the wave field Ψ as a scalar, we (i) write each beam's power as the sum over a set of Fourier components Ψ_j with precise frequencies ω_j and slowly wandering, complex amplitudes $W(t) = (\sum_j \Re \Psi_j)^2$; (ii) form the product $W(t)W(t+\tau)$; (iii) keep only terms that will have nonzero averages by virtue of containing products of the form $\propto e^{+i\omega_j t} e^{-i\omega_j t} e^{+i\omega_k t} e^{-i\omega_k t}$ (where j and k are generally not the same); and (iv) average over time. Thereby we obtain

$$\overline{W(t)W(t+\tau)} = K^2 \overline{\Psi(t)\Psi^*(t) \times \Psi(t+\tau)\Psi^*(t+\tau)}$$
$$+ K^2 \overline{\Psi(t)\Psi^*(t+\tau) \times \Psi^*(t)\Psi(t+\tau)}$$
$$= \overline{W}^2 [1 + |\gamma_\|(\tau)|^2] \qquad (9.58)$$

[cf. Eq. (9.16) with Ψ allowed to be complex]. If we now measure the relative fluctuations, we find that

longitudinal correlation of beam power

$$\frac{\overline{\delta W(t)\delta W(t+\tau)}}{\overline{W(t)}^2} = \frac{\overline{W(t)W(t+\tau)} - \overline{W(t)}^2}{\overline{W(t)}^2} = |\gamma_\|(\tau)|^2. \qquad (9.59)$$

[Note: This analysis is only correct if the radiation comes from many uncorrelated sources—the many independently emitting mercury atoms in Fig. 9.14—and therefore has Gaussian statistics.]

Equation (9.59) tells us that the power as well as the amplitude of coherent radiation must exhibit a positive longitudinal correlation, and the degree of coherence for the fluxes is equal to the squared modulus of the degree of coherence for the amplitudes. Although this result was rather controversial at the time the experiments were first performed, it is easy to interpret qualitatively if we think in terms of photons rather than classical waves. Photons are bosons and are therefore positively correlated even in thermal equilibrium; cf. Chaps. 3 and 4. When they arrive at the beam splitter of Fig. 9.14, they clump more than would be expected for a random distribution of classical particles, a phenomenon called photon bunching.[12] In fact, treating the problem from the point of view of photon statistics gives an answer equivalent to Eq. (9.59).

explanation of power correlation by Bose-Einstein statistics: photon bunching

Some practical considerations should be mentioned. The first is that Eq. (9.59), derived for a scalar wave, is really only valid for electromagnetic waves if they are completely polarized. If the incident waves are unpolarized, then the intensity fluctuations are reduced by a factor of two. The second point is that, in the Hanbury Brown and Twiss experiments, the photon counts were actually averaged over longer times than the correlation time of the incident radiation. This reduced the magnitude of the measured effect further.

Nevertheless, after successfully measuring temporal power correlations, Hanbury Brown and Twiss constructed a *stellar intensity interferometer,* with which they were able to measure the angular diameters of bright stars. This method had the advantage that it did not depend on the phase of the incident radiation, so the results were insensitive to atmospheric fluctuations (seeing), one of the drawbacks of the Michelson stellar interferometer (Sec. 9.2.5). Indeed, it is not even necessary to use accurately ground mirrors to measure the effect. The method has the disadvantage that it can only measure the modulus of the degree of coherence; the phase is lost. It was the first example of using fourth-order correlations of the light field to extract image information from light that has passed through Earth's turbulent atmosphere (Box 9.2).

stellar intensity interferometer

EXERCISES

Exercise 9.21 *Derivation: Power Correlations*
By expressing the field as either a Fourier sum or a Fourier integral complete the argument that leads to Eq. (9.58).

Exercise 9.22 *Problem: Electron Intensity Interferometry*
Is it possible to construct an intensity interferometer (i.e., a number-flux interferometer) to measure the coherence properties of a beam of electrons? What qualitative differences do you expect there to be from a photon-intensity interferometer? What do you expect Eq. (9.59) to become?

---

12. Pions emerging from high-energy, heavy ion collisions exhibit similar bunching because pions, like photons, are bosons.

Bibliographic Note

For pedagogical introductions to interference and coherence at an elementary level in greater detail than this chapter, see Klein and Furtak (1986) and Hecht (2017). For more advanced treatments, we like Pedrotti, Pedrotti, and Pedrotti (2007), Saleh and Teich (2007), Ghatak (2010), and especially Brooker (2003). For a particularly deep and thorough discussion of coherence, see Goodman (1985). For modern applications of interferometry (including the optical frequency comb), see Yariv and Yeh (2007), and at a more elementary level, Francon and Willmans (1966) and Hariharan (2007).

CHAPTER TEN

Nonlinear Optics

The development of the maser and laser . . . followed no script except to hew to the nature of humans groping to understand, to explore, and to create . . .

CHARLES H. TOWNES (2002)

10.1 Overview

Communication technology is undergoing a revolution, and computer technology may do so soon—a revolution in which the key devices used (e.g., switches and communication lines) are changing from radio and microwave frequencies to optical frequencies. This revolution has been made possible by the invention and development of lasers (especially semiconductor diode lasers) and other technology developments, such as nonlinear media whose polarization P_i is a nonlinear function of the applied electric field, $P_i = \epsilon_0(\chi_{ij}E^j + 2d_{ijk}E^jE^k + 4\chi_{ijkl}E^jE^kE^l + \cdots)$. In this chapter we study lasers, nonlinear media, and various nonlinear optics applications that are based on them.

Most courses in elementary physics idealize the world as linear. From the simple harmonic oscillator to Maxwell's equations to the Schrödinger equation, most elementary physical laws one studies are linear, and most of the applications studied make use of this linearity. In the real world, however, nonlinearities abound, creating such phenomena as avalanches, breaking ocean waves, holograms, optical switches, and neural networks; and in the past three decades nonlinearities and their applications have become major themes in physics research, both basic and applied. This chapter, with its exploration of nonlinear effects in optics, serves as a first introduction to some fundamental nonlinear phenomena and their present and future applications. In later chapters, we revisit some of these phenomena and meet others, in the context of fluids (Chaps. 16 and 17), plasmas (Chap. 23), and spacetime curvature (Chaps. 25–28).

Since highly coherent and monochromatic laser light is one of the key foundations on which modern nonlinear optics has been built, we begin in Sec. 10.2 with a review of the basic physics principles that underlie the laser: the pumping of an active medium to produce a molecular population inversion and the stimulated emission of radiation from the inverted population. Then we briefly describe the wide variety of lasers now available, how a few of them are pumped, and the characteristics of their light. As

> **BOX 10.1. READERS' GUIDE**
>
> - This chapter depends substantially on Secs. 7.2, 7.3, and 7.7.1 (geometric optics).
> - Sec. 10.6, on wave-wave mixing, is an important foundation for Chap. 23 on the nonlinear dynamics of plasmas, and (to a lesser extent) for the discussions of solitary waves (solitons) in Secs. 16.3 and 23.6. Nothing else in this book relies substantially on this chapter.

an important example (crucial for the optical frequency combs of Sec. 9.4.3), we give details about mode-locked lasers.

In Sec. 10.3, we meet our first example of an application of nonlinear optics: holography. In the simplest variant of holography, a 3-dimensional, monochromatic image of an object is produced by a two-step process: recording a hologram of the image, and then passing coherent light through the hologram to reconstruct the image. We analyze this recording and reconstruction and then describe a few of the many variants of holography now available and some of their practical applications.

Holography differs from more modern nonlinear optics applications in not being a real-time process. Real-time processes have been made possible by nonlinear media and other new technologies. In Sec. 10.4, we study an example of a real-time, nonlinear-optics process: phase conjugation of light by a phase-conjugating mirror (though we delay a detailed discussion of how such mirrors work until Sec. 10.8.2). In Sec. 10.4, we also see how phase conjugation can be used to counteract the distortion of images and signals by media through which they travel.

In Sec. 10.5, we introduce nonlinear media and formulate Maxwell's equations for waves propagating through such media. As an example, we briefly discuss electro-optic effects, where a slowly changing electric field modulates the optical properties of a nonlinear crystal, thereby modulating light waves that propagate through it. Then in Sec. 10.6, we develop a detailed description of how such a nonlinear crystal couples two optical waves to produce a new, third wave—so-called "three-wave mixing." Three-wave mixing has many important applications in modern technology. In Sec. 10.7, we describe and analyze several: frequency doubling (e.g., in a green laser pointer), optical parametric amplification of signals, and driving light into a squeezed state (e.g., the squeezed vacuum of quantum electrodynamics).

In an isotropic medium, three-wave mixing is suppressed, but a new, fourth wave can be produced by three incoming waves. In Sec. 10.8, we describe and analyze this four-wave mixing and how it is used in phase-conjugate mirrors and produces problems in the optical fibers widely used to transmit internet, television, and telephone signals.

These topics just scratch the surface of the exciting field of nonlinear optics, but they will give the reader an overview and some major insights into this field, and into nonlinear phenomena in the physical world.

10.2 Lasers

10.2.1 Basic Principles of the Laser

In quantum mechanics one identifies three different types of interaction of light with material systems (atoms, molecules, atomic nuclei, electrons, etc.): (i) *spontaneous emission*, in which a material system in an excited state spontaneously drops into a state of lesser excitation and emits a photon in the process; (ii) *absorption*, in which an incoming photon is absorbed by a material system, exciting it; and (iii) *stimulated emission*, in which a material system, initially in some excited state, is "tickled" by passing photons, and this tickling stimulates it to emit a photon of the same sort (in the same state) as the photons that tickled it.

spontaneous emission, absorption, and stimulated emission

As peculiar as stimulated emission may seem at first sight, in fact it is easily understood and analyzed classically. It is nothing but "negative absorption": In classical physics, when a light beam with electric field $E = \Re[A e^{i(kz-\omega t+\varphi)}]$ travels through an absorbing medium, its real amplitude A decays exponentially with the distance propagated, $A \propto e^{-\mu z/2}$ (corresponding to an energy-flux decay $F \propto e^{-\mu z}$), while its frequency ω, wave number k, and phase φ remain nearly constant. For normal materials, the absorption rate $\mu = F^{-1} dF/dz$ is positive, and the energy lost goes ultimately into heat. However, one can imagine a material with an internally stored energy that amplifies a passing light beam. Such a material would have a negative absorption rate, $\mu < 0$, and correspondingly, the amplitude of the passing light would grow with the distance traveled, $A \propto e^{+|\mu|z/2}$, while its frequency, wave number, and phase would remain nearly constant. Such materials do exist; they are called "active media," and their amplification of passing waves is stimulated emission.

active medium

This elementary, classical description of stimulated emission is equivalent to the quantum mechanical description in the domain where the stimulated emission is strong: the domain of large photon-occupation numbers $\eta \gg 1$ (which, as we learned in Sec. 3.2.5, is the domain of classical waves).

The classical description of stimulated emission takes for granted the existence of an active medium. To understand the nature of such a medium, we must turn to quantum mechanics.

As a first step toward such understanding, consider a beam of monochromatic light with frequency ω that impinges on a collection of molecules (or atoms or charged particles) that are all in the same quantum mechanical state $|1\rangle$. Suppose the molecules have a second state $|2\rangle$ with energy $E_2 = E_1 + \hbar\omega$. Then the light will resonantly excite the molecules from their initial state $|1\rangle$ to the higher state $|2\rangle$, and in the process photons will be absorbed (Fig. 10.1a). The strength of the interaction is proportional to the beam's energy flux F. Stated more precisely, the rate of absorption of photons

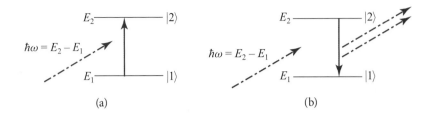

FIGURE 10.1 (a) *Photon absorption.* A photon with energy $\hbar\omega = E_2 - E_1$ excites a molecule from its ground state, with energy E_1 to an excited state with energy E_2 (as depicted by an energy-level diagram). (b) *Stimulated emission.* The molecule is initially in its excited state, and the incoming photon stimulates it to deexcite into its ground state, emitting a photon identical to the incoming one.

is proportional to the number flux of photons in the beam, $dn/dA dt = F/(\hbar\omega)$, and thence is proportional to F, in accord with the classical description of absorption.

Next suppose that when the light beam first arrives, the atoms are all in the higher state $|2\rangle$ rather than the lower state $|1\rangle$. There will still be a resonant interaction, but this time the interaction will deexcite the atoms, with an accompanying emission of photons (Fig. 10.1b). As in the absorption case, the strength of the interaction is proportional to the flux of the incoming beam (i.e., the rate of emission of new photons is proportional to the number flux of photons that the beam already has), and thence it is also proportional to the beam's energy flux F. A quantum mechanical analysis shows that the photons from this stimulated emission come out in the same quantum state as is occupied by the photons of the incoming beam (Bose-Einstein statistics: photons, being bosons, tend to congregate in the same state). Correspondingly, when viewed classically, the beam's flux will be amplified at a rate proportional to its initial flux, with no change of its frequency, wave number, or phase.

In Nature, molecules usually have their energy levels populated in accord with the laws of statistical (thermodynamic) equilibrium. Such thermalized populations, as we saw at the end of Sec. 4.4.1, entail a ratio $N_2/N_1 = \exp[-(E_2 - E_1)/(k_B T)] < 1$ for the number N_2 of molecules in state $|2\rangle$ to the number N_1 in state $|1\rangle$. Here T is the molecular temperature, and for simplicity it is assumed that the states are nondegenerate. Since there are more molecules in the lower state $|1\rangle$ than the higher one $|2\rangle$, an incoming light beam will experience more absorption than stimulated emission.

population inversion

By contrast, occasionally in Nature and often in the laboratory a collection of molecules develops a "population inversion" in which $N_2 > N_1$. The two states can then be thought of as having a negative temperature with respect to each other. Light propagating through population-inverted molecules will experience more stimulated emission than absorption (i.e., it will be amplified). The result is "light amplification by stimulated emission," or "laser" action.

laser

This basic principle underlying the laser has been known since the early years of quantum mechanics, but only in the 1950s did physicists succeed in designing, constructing, and operating real lasers. The first proposals for practical devices were

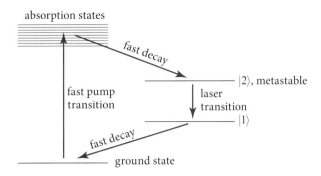

FIGURE 10.2 The mechanism for creating the population inversion that underlies laser action. The horizontal lines and band represent energy levels of a molecule, and the arrows represent transitions in which the molecules are excited by pumping or decay by emission of photons.

made, independently, in the United States by Weber (1953) and Gordon, Zeiger, and Townes (1954), and in the Soviet Union by Basov and Prokhorov (1954, 1955). The first successful construction and operation of a laser was by Gordon, Zeiger, and Townes (1954, 1955), and soon thereafter by Basov and Prokhorov—though these first lasers actually used radiation not at optical frequencies but rather at microwave frequencies and were based on a population inversion of ammonia molecules (see Feynman, Leighton, and Sands, 2013, Chap. 9), and thus were called *masers*. The first optical frequency laser, one based on a population inversion of chromium ions in a ruby crystal, was constructed and operated by Maiman (1960).

The key to laser action is the population inversion. Population inversions are incompatible with thermodynamic equilibrium; thus, to achieve them, one must manipulate the molecules in a nonequilibrium way. This is usually done by some concrete variant of the process shown in the energy-level diagram of Fig. 10.2. Some sort of pump mechanism (to be discussed in the next section) rapidly excites molecules from the ground state into some group of absorption states. The molecules then decay rapidly from the absorption states into the state $|2\rangle$, which is metastable (i.e., has a long lifetime against spontaneous decay), so the molecules linger there. The laser transition is from state $|2\rangle$ to state $|1\rangle$. Once a molecule has decayed into state $|1\rangle$, it quickly decays on down to the ground state and then may be quickly pumped back up into the absorption states. This is called "four-level pumping." It is instead called "three-level pumping" if state $|1\rangle$ is the ground state.

pump mechanism

If the pump acts suddenly and briefly, this process produces a temporary population inversion of states $|2\rangle$ and $|1\rangle$, with which an incoming, weak burst of "seed" light can interact to produce a burst of amplification. The result is a pulsed laser. If the pump acts continuously, the result may be a permanently maintained population inversion with which continuous seed light can interact to produce continuous-wave laser light.

pulsed lasers and continuous-wave lasers

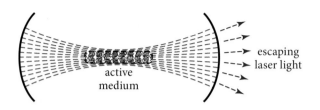

FIGURE 10.3 The use of a Fabry-Perot cavity to enhance the interaction of the light in a laser with its active medium.

As the laser beam travels through the active medium (the population-inverted molecules), its flux F builds up with distance z as $dF/dz = F/\ell_o$, so $F(z) = F_o e^{z/\ell_o}$. Here F_o is the initial flux, and $\ell_o \equiv 1/|\mu|$ (the e-folding length) depends on the strength of the population inversion and the strength of the coupling between the light and the active medium. Typically, ℓ_o is so long that strong lasing action cannot be achieved by a single pass through the active medium. In this case, the lasing action is enhanced by placing the active medium inside a Fabry-Perot cavity (Fig. 10.3 and Sec. 9.4.3). The length L of the cavity is adjusted to maximize the output power, which occurs when the lasing transition frequency $\omega = (E_2 - E_1)/\hbar$ is an eigenfrequency of the cavity. The lasing action then excites a standing wave mode of the cavity, from which the light leaks out through one or both cavity mirrors. If $\mathcal{F}$ is the cavity's finesse [approximately the average number of times a photon bounces back and forth inside the cavity before escaping through a mirror; cf. Eq. (9.39)], then the cavity increases the distance that typical photons travel through the active medium by a factor $\sim \mathcal{F}$, thereby increasing the energy flux of the light output by a factor $\sim e^{\mathcal{F}L/\ell_o}$.

Typically, many modes of the Fabry-Perot cavity are excited, so the laser's output is multimodal and contains a mixture of polarizations. When a single mode and polarization are desired, the polarization is made pure by oblique optical elements at the ends of the laser that transmit only the desired polarization, and all the modes except one are removed from the output light by a variety of techniques (e.g., filtering with a second Fabry-Perot cavity; Sec. 9.4.3).

coherent-state laser light

For an ideal laser (one, e.g., with a perfectly steady pump maintaining a perfectly steady population inversion that in turn maintains perfectly steady lasing), the light comes out in the most perfectly classical state that quantum mechanics allows. This state, called a *quantum mechanical coherent state*, has a perfectly sinusoidally oscillating electric field on which is superimposed the smallest amount of noise (the smallest wandering of phase and amplitude) allowed by quantum mechanics: the noise of quantum electrodynamical vacuum fluctuations. The value of the oscillations' well-defined phase is determined by the phase of the seed field from which the coherent state was built up by lasing. Real lasers have additional noise due to a variety of practical factors, but nevertheless, their outputs are usually highly coherent, with long coherence times.

10.2.2 Types of Lasers and Their Performances and Applications

Lasers can have continuous, nearly monochromatic output, or they can be pulsed. Their active media can be liquids, gases (ionized or neutral), or solids (semiconductors, glasses, or crystals; usually carefully doped with impurities). Lasers can be pumped by radiation (e.g., from a flash tube), by atomic collisions that drive the lasing atoms into their excited states, by nonequilibrium chemical reactions, or by electric fields associated with electric currents (e.g., in semiconductor diode lasers that can be powered by ordinary batteries and are easily modulated for optical communication).

laser pump mechanisms

Lasers can be made to pulse by turning the pump on and off, by mode-locked operation (Sec. 10.2.3), or by Q-switching (turning off the lasing, e.g., by inserting into the Fabry-Perot cavity an electro-optic material that absorbs light until the pump has produced a huge population inversion, and then suddenly applying an electric field to the absorber, which makes it transparent and restores the lasing).

pulsed lasers

Laser pulses can be as short as a few femtoseconds (thus enabling experimental investigations of fast chemical reactions) and they can carry as much as 20,000 J with duration of a few picoseconds and pulse power $\sim 10^{15}$ W (at the U.S. National Ignition Facility in Livermore, California, for controlled fusion).

The most powerful continuous laser in the United States is the Mid-Infrared Advanced Chemical Laser (MIRACL), developed by the Navy to shoot down missiles and satellites, with ~ 1 MW power in a 14×14 cm^2 beam lasting ~ 70 s. Continuous CO_2 lasers with powers ~ 3 kW are used industrially for cutting and welding metal.

continuous lasers

The beam from a Q-switched CO_2 laser with ~ 1 GW power can be concentrated into a region with transverse dimensions as small as one wavelength (~ 1 μm), yielding a local energy flux of 10^{21} W m^{-2}, an rms magnetic field strength of ~ 3 kT, an electric field of ~ 1 TV m^{-1}, and an electrical potential difference across a wavelength of ~ 1 MeV. It then should not be surprising that high-power lasers can create electron-positron-pair plasmas!

For many applications, large power is irrelevant or undesirable, but high frequency stability (a long coherence time) is often crucial. By locking the laser frequency to an optical frequency atomic transition (e.g., in the Al$^+$ atomic clock; see footnote in Sec. 6.6.1), one can achieve a frequency stability $\Delta f/f \sim 10^{-17}$, so $\Delta f \sim 3$ mHz, for hours or longer, corresponding to coherence times of ~ 100 s and coherence lengths of $\sim 3 \times 10^7$ km. By locking the frequency to a mode of a physically highly stable Fabry-Perot cavity (e.g., PDH locking; Sec. 9.4.3), stabilities have been achieved as high as $\Delta f/f \sim 10^{-16}$ for times of ~ 1 hr in a physically solid cavity (the superconducting cavity stabilized oscillator), and $\Delta f/f \sim 10^{-22}$ for a few ms in LIGO's 4-km-long cavity with freely hanging mirrors and sophisticated seismic isolation (Sec. 9.5).

locking a laser's frequency

When first invented, lasers were called "a solution looking for a problem." Now they permeate everyday life and high technology. Examples are supermarket bar-code readers, laser pointers, DVD players, eye surgery, laser printers, laser cutting and welding, laser gyroscopes (which are standard on commercial aircraft), laser-based surveying, Raman spectroscopy, laser fusion, optical communication, optically based computers, holography, maser amplifiers, and atomic clocks.

10.2.3 Ti:Sapphire Mode-Locked Laser

As a concrete example of a modern, specialized laser, consider the titanium-sapphire (Ti:sapphire) mode-locked laser (Cundiff, 2002) that is used to generate the optical frequency comb described in Sec. 9.4.3. Recall that this laser's light must be concentrated in a very short pulse that travels unchanged back and forth between its Fabry-Perot mirrors. The pulse is made from phase-locked (Gaussian) modes of the optical cavity that extend over a huge frequency band, $\Delta\omega \sim \omega$. Among other things, this mode-locked laser illustrates the use of an optical nonlinearity called the "Kerr effect," whose underlying physics we describe later in this chapter (Sec. 10.8.3).

As we discussed in Sec. 9.4.3, this mode-locked laser must (i) more strongly amplify modes with high energy flux than with low (this pushes the light into the short, high-flux pulse), and (ii) its (Gaussian) modes must have a group velocity V_g that is independent of frequency over the frequency band $\Delta\omega \sim \omega$ (this enables the pulse to stay short rather than disperse).

Figure 10.4 illustrates the Ti:sapphire laser that achieves this. The active medium is a sapphire crystal doped with titanium ions. This medium exhibits the optical Kerr effect, which means that its index of refraction n is a sum of two terms, one independent of the light's energy flux; the other proportional to the flux [Eq. (10.69)]. The flux-dependent term slows the light's speed near the beam's center and thereby focuses the beam, making its cross section smaller. A circular aperture attenuates large light beams but not small. As a result, the lasing is stronger the smaller the beam (which means the higher its flux). This drives the lasing light into the desired short, high-flux pulse.

The Ti:sapphire crystal has a group velocity that increases with frequency. The two prisms and mirror (Fig. 10.4) compensate this. The first prism bends low-frequency light more than high, so the high-frequency light traverses more glass in the second prism and is slowed. By adjusting the mirror tilt, one adjusts the amount of slowing to keep the round-trip-averaged phase velocity the same at high frequencies as at low.

mode-locked laser produces frequency comb

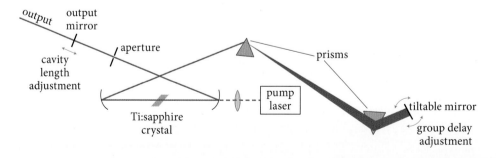

FIGURE 10.4 The Ti:sapphire mode-locked laser. Adapted from Cundiff (2002).

The laser then generates multimode, high-intensity, pulsed light (with pulse durations that can be as short as attoseconds, when augmented by a technique called chirped pulse amplification), resulting in the optical frequency comb of Sec. 9.4.3.

10.2.4 Free Electron Laser

The free electron laser is quite different from the lasers we have been discussing. A collimated beam of essentially monoenergetic electrons (the population inversion) passes through an undulator comprising an alternating transverse magnetic field, which causes the electrons to radiate X-rays along the direction of the beam. From a classical perspective, this X-ray-frequency electromagnetic wave induces correlated electron motions, which emit coherently, enhancing the X-ray wave. In other words, the undulating electron beam lases.

As an example, the LINAC Coherent Light Source (LCLS) at SLAC National Accelerator Laboratory can produce femtosecond pulses of keV X-rays with pulse frequency 120 Hz and peak intensity $\sim 10^9$ times higher than that achievable from a conventional synchrotron. An upgrade, LCLS-II, is planned to start operating in 2020 with a pulse frequency ~ 100 kHz and an increase in average brightness of $\sim 10^3$.

Exercise 10.1 *Challenge: Nuclear Powered X-Ray Laser*
A device much ballyhooed in the United States during the presidency of Ronald Reagan, but thankfully never fully deployed, was a futuristic, superpowerful X-ray laser pumped by a nuclear explosion. As part of Reagan's Strategic Defense Initiative ("Star Wars"), this laser was supposed to shoot down Soviet missiles.

How would you design a nuclear powered X-ray laser? The energy for the pump comes from a nuclear explosion that you set off in space above Earth. You want to use that energy to create a population inversion in an active medium that will lase at X-ray wavelengths, and you want to focus the resulting X-ray beam onto an intercontinental ballistic missile that is rising out of Earth's atmosphere. What would you use for the active medium? How would you guarantee that a population inversion is created in the active medium? How would you focus the resulting X-ray beam? (Note: This is a highly nontrivial exercise, intended more as a stimulus for thought than as a test of one's understanding of things taught in this book.)

10.3 Holography

Holography is an old[1] and well-explored example of nonlinear optics—an example in which the nonlinear interaction of light with itself is produced not in real time, but rather by means of a recording followed by a later readout (e.g., Cathey, 1974; Iizuka, 1987).

---

1. Holography was invented by Dennis Gabor (1948) for use in electron microscopy. The first practical optical holograms were made in 1962, soon after lasers became available.

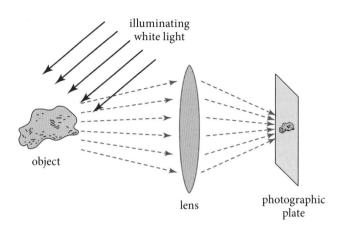

FIGURE 10.5 Ordinary photography.

By contrast with ordinary photography (Fig. 10.5), which produces a colored, 2-dimensional image of 3-dimensional objects, holography (Figs. 10.6 and 10.8) normally produces a monochromatic 3-dimensional image of 3- dimensional objects. Roughly speaking, the two processes contain the same amount of information, two items at each location in the image. For ordinary photography, they are the energy flux and color; for holography, the energy flux and phase of monochromatic light.

It is the phase, lost from an ordinary photograph but preserved in holography, that carries the information about the third dimension. Our brain deduces the distance to a point on an object from the difference in the directions of propagation of the point's light as it arrives at our two eyes. Those propagation directions are encoded in the light as variations of the phase with transverse location [see, e.g., the point-spread function for a thin lens, Eq. (8.29)].

ordinary photography

In an ordinary photograph (Fig. 10.5), white light scatters off an object, with different colors scattering at different strengths. The resulting colored light is focused through a lens to form a colored image on a photographic plate or a CCD. The plate or CCD records the color and energy flux at each point or pixel in the focal plane, thereby producing the ordinary photograph.

hologram

In holography, one records a hologram with flux and phase information (see Fig. 10.6), and one then uses the hologram to reconstruct the 3-dimensional, monochromatic, holographic image (see Fig. 10.8).

10.3.1 Recording a Hologram

Consider, first, the recording of the hologram. Monochromatic, linearly polarized plane-wave light with electric field

$$E = \Re[\psi(x, y, z)e^{-i\omega t}], \tag{10.1}$$

angular frequency ω, and wave number $k = \omega/c$ illuminates the object and also a mirror, as shown in Fig. 10.6. The light must be spatially coherent over the entire

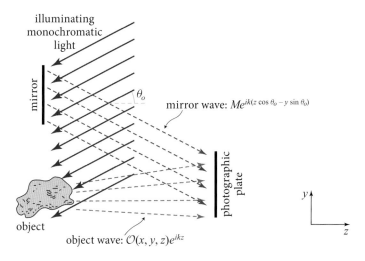

FIGURE 10.6 Recording a hologram.

region of mirror plus object. The propagation vector **k** of the illuminating light lies in the y-z plane at some angle θ_o to the z-axis, and the mirror lies in the x-y plane. The mirror reflects the illuminating light, producing a so-called "reference beam," which we call the *mirror wave*:

reference beam or mirror wave

$$\psi_{\text{mirror}} = Me^{ik(z\cos\theta_o - y\sin\theta_o)}, \tag{10.2}$$

where M is a real constant. The object (shown in Fig. 10.6) scatters the illuminating light, producing a wave that propagates in the z direction toward the recording medium (a photographic plate, for concreteness). We call this the *object wave* and denote it

object wave

$$\psi_{\text{object}} = \mathcal{O}(x, y, z)e^{ikz}. \tag{10.3}$$

It is the slowly varying complex amplitude $\mathcal{O}(x, y, z)$ of this object wave that carries the 3-dimensional, but monochromatic, information about the object's appearance. Thus it is this $\mathcal{O}(x, y, z)$ that will be reconstructed in the second step of holography.

In the first step (Fig. 10.6), the object wave propagates along the z direction to the photographic plate (or, more commonly today, a photoresist) at $z = 0$, where it interferes with the mirror wave to produce the transverse pattern of energy flux

$$\begin{aligned} F(x, y) &\propto |\mathcal{O} + Me^{-iky\sin\theta_o}|^2 \\ &= M^2 + |\mathcal{O}(x, y, z=0)|^2 + \mathcal{O}(x, y, z=0)Me^{iky\sin\theta_o} \\ &\quad + \mathcal{O}^*(x, y, z=0)Me^{-iky\sin\theta_o}. \end{aligned} \tag{10.4}$$

(Here and throughout this chapter, * denotes complex conjugation.) The plate is blackened at each point in proportion to this flux. The plate is then developed, and a positive or negative print (it doesn't matter which because of Babinet's principle,

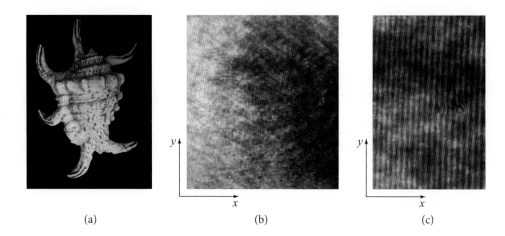

FIGURE 10.7 (a) Ordinary photograph of an object. (b) Hologram of the same object. (c) Magnification of the hologram. Photos courtesy Jason Sapan, Holographic Studios, New York City.

Sec. 8.3.3) is made on a transparent sheet of plastic or glass. This print, the *hologram*, has a transmissivity as a function of x and y that is proportional to the flux distribution (10.4):

hologram's transmissivity

$$\mathfrak{t}(x, y) \propto M^2 + |\mathcal{O}(x, y, z=0)|^2 + \mathcal{O}(x, y, z=0)M e^{iky \sin\theta_o}$$
$$+ \mathcal{O}^*(x, y, z=0)M e^{-iky \sin\theta_o}. \tag{10.5}$$

hologram's nonlinearity

In this transmissivity we meet our first example of nonlinearity: $\mathfrak{t}(x, y)$ contains a nonlinear superposition of the mirror wave and the object wave. Stated more precisely, the superposition is not a linear sum of wave fields, but instead is a sum of products of one wave field with the complex conjugate of another wave field. A further nonlinearity will arise in the reconstruction of the holographic image [Eq. (10.7)].

Figure 10.7 shows an example. Figure 10.7a is an ordinary photograph of an object, Fig. 10.7b is a hologram of the same object, and Fig. 10.7c is a blow-up of a portion of that hologram. The object is not at all recognizable in the hologram, because the object wave $\mathcal{O}$ was not focused to form an image at the plane of the photographic plate. Rather, light from each region of the object was scattered to and recorded by all regions of the photographic plate. Nevertheless, the plate contains the full details of the scattered light $\mathcal{O}(x, y, z=0)$, including its phase. That information is recorded in the piece $M(\mathcal{O}e^{iky \sin\theta_o} + \mathcal{O}^*e^{-iky \sin\theta_o}) = 2M \, \Re(\mathcal{O}e^{iky \sin\theta_o})$ of the hologram's transmissivity. This piece oscillates sinusoidally in the y direction with wavelength $2\pi/(k \sin\theta_o)$; and the amplitude and phase of its oscillations are modulated by the object wave $\mathcal{O}(x, y, z=0)$. Those modulated oscillations show up clearly when one magnifies the hologram (Fig. 10.7c); they turn the hologram into a sort of diffraction grating, with the object wave $\mathcal{O}(x, y, z=0)$ encoded as variations of the darkness and spacings of the grating lines.

hologram's information encoding

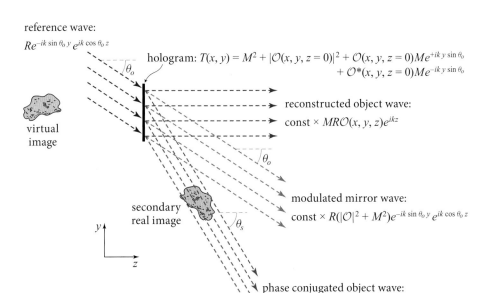

FIGURE 10.8 Reconstructing the holographic image from the hologram. Note that $\sin\theta_s = 2\sin\theta_o$.

What about the other pieces of the transmissivity (10.5), which superpose linearly on the diffraction grating? One piece, $t \propto M^2$, is spatially uniform and thus has no effect except to make the lightest parts of the hologram slightly gray rather than leaving it absolutely transparent (since this hologram is a negative rather than a positive). The other piece, $t \propto |\mathcal{O}|^2$, is the flux of the object's unfocused, scattered light. It produces a graying and whitening of the hologram (Fig. 10.7b) that varies on lengthscales long compared to the grating's wavelength $2\pi/(k\sin\theta_o)$ and thus blots out the diffraction grating a bit here and there, but it does not change the amplitude or phase of the grating's modulation.

10.3.2 Reconstructing the 3-Dimensional Image from a Hologram

To reconstruct the object's 3-dimensional wave, $\mathcal{O}(x, y, z)e^{ikz}$, one sends through the hologram monochromatic, plane-wave light identical to the mirror light used in making the hologram (Fig. 10.8). If, for pedagogical simplicity, we place the hologram at the same location $z = 0$ as was previously occupied by the photographic plate, then the incoming light has the same form (10.2) as the original mirror wave, but with an amplitude that we denote as R, corresponding to the phrase *reference beam* that is used to describe this incoming light:

$$\psi_{\text{reference}} = Re^{ik(z\cos\theta_o - y\sin\theta_o)}. \tag{10.6}$$

In passing through the hologram at $z = 0$, this reference beam is partially absorbed and partially transmitted. The result, immediately on exiting from the hologram, is a reconstructed light-wave field whose value and normal derivative are given by [cf. Eq. (10.5)]

reconstructed wave

$$\psi_{\text{reconstructed}}\Big|_{z=0} \equiv \mathcal{R}(x, y, z=0) = \mathfrak{t}(x, y) R e^{-iky \sin\theta_o}$$

$$= \left[M^2 + |\mathcal{O}(x, y, z=0)|^2\right] R e^{-iky \sin\theta_o}$$

$$+ MR\mathcal{O}(x, y, z=0)$$

$$+ MR\mathcal{O}^*(x, y, z=0) e^{-i2ky \sin\theta_o};$$

$$\psi_{\text{reconstructed},z}\Big|_{z=0} \equiv \mathcal{Z}(x, y, z=0) = ik \cos\theta_o \mathcal{R}(x, y, z=0). \tag{10.7}$$

This field and normal derivative act as initial data for the subsequent evolution of the reconstructed wave. Note that the field and derivative, and thus also the reconstructed wave, are triply nonlinear: each term in Eq. (10.7) is a product of (i) the original mirror wave M used to construct the hologram or the original object wave $\mathcal{O}$, (ii) $\mathcal{O}^*$ or $M^* = M$, and (iii) the reference wave R that is being used in the holographic reconstruction.

The evolution of the reconstructed wave beyond the hologram (at $z > 0$) can be computed by combining the initial data [Eq. (10.7)] for $\psi_{\text{reconstructed}}$ and $\psi_{\text{reconstructed},z}$ at $z=0$ with the Helmholtz-Kirchhoff formula (8.4); see Exs. 10.2 and 10.6. From the four terms in Eq. (10.7) [which arise from the four terms in the hologram's transmissivity $\mathfrak{t}(x, y)$, Eq. (10.5)], the reconstruction produces four wave fields; see Fig. 10.8. The direction of propagation of each of these waves can easily be inferred from the vertical spacing of its phase fronts along the outgoing face of the hologram, or equivalently, from the relation $\partial \psi_{\text{reconstructed}}/\partial y = ik_y \psi_{\text{reconstructed}} = -ik \sin\theta \psi$, where θ is the angle of propagation relative to the horizontal z direction. Since, immediately in front of the hologram, $\psi_{\text{reconstructed}} = \mathcal{R}$, the propagation angle is

$$\sin\theta = \frac{\partial \mathcal{R}/\partial y}{-ik\mathcal{R}}. \tag{10.8}$$

Comparing with Eqs. (10.5) and (10.7), we see that the first two, slowly spatially varying, terms in the transmissivity, $\mathfrak{t} \propto M^2$ and $T \propto |\mathcal{O}|^2$, both produce waves that propagate in the same direction as the reference wave, $\theta = \theta_o$. This combined wave has an uninteresting, smoothly and slowly varying energy-flux pattern.

object wave and its 3-dimensional image

The two diffraction-grating terms in the hologram's transmissivity produce two interesting waves. One, arising from $\mathfrak{t} \propto \mathcal{O}(x, y, z=0) M e^{iky \sin\theta_o}$ [and produced by the $MR\mathcal{O}$ term of the initial conditions (10.7)], is *precisely the same object wave* $\psi_{\text{object}} = \mathcal{O}(x, y, z) e^{ikz}$ (aside from overall amplitude) *as one would have seen while making the hologram if one had replaced the photographic plate by a window and looked through it*. This object wave, carrying [encoded in $\mathcal{O}(x, y, z)$] the holographic image with full 3-dimensionality, propagates in the z direction, $\theta = 0$.

526   Chapter 10. Nonlinear Optics

The transmissivity's second diffraction-grating term,

$$t \propto \mathcal{O}^*(x, y, z=0) M e^{-iky \sin \theta_o},$$

acting via the $MR\mathcal{O}^*$ term of the initial conditions (10.7), gives rise to a secondary wave, which [according to Eq. (10.8)] propagates at an angle θ_s to the z-axis, where

$$\sin \theta_s = 2 \sin \theta_o. \tag{10.9}$$

secondary (phase conjugate) wave

(If $\theta_o > 30°$, then $2 \sin \theta_s > 1$, which means θ_s cannot be a real angle, and there will be no secondary wave.) *This secondary wave, if it exists, carries an image that is encoded in the complex conjugate $\mathcal{O}^*(x, y, z=0)$ of the transverse (i.e., x, y) part of the original object wave.* Since complex conjugation of an oscillatory wave just reverses the sign of the wave's phase, this wave in some sense is a "phase conjugate" of the original object wave.

When one recalls that the electric and magnetic fields that make up an electromagnetic wave are actually real rather than complex, and that we are using complex wave fields to describe electromagnetic waves only for mathematical convenience, one then realizes that this phase conjugation of the object wave is actually a highly nonlinear process. There is no way, by linear manipulations of the real electric and magnetic fields, to produce the phase-conjugated wave from the original object wave.

In Sec. 10.4 we develop in detail the theory of phase-conjugated waves, and in Ex. 10.6, we relate our holographically constructed secondary wave to that theory. As we shall see, our secondary wave is not quite the same as the phase-conjugated object wave, but it is the same aside from some distortion along the y direction and a change in propagation direction. More specifically, *if one looks into the object wave with one's eyes* (i.e., if one focuses it onto one's retinas), *one sees the original object in all its 3-dimensional glory, though single colored, sitting behind the hologram at the object's original position* (shown in Fig. 10.8). Because the image one sees is behind the hologram, it is called a *virtual image*. *If, instead, one looks into the secondary wave with one's eyes, one sees the original 3-dimensional object, sitting in front of the hologram but turned inside out and distorted* (also shown in the figure). For example, if the object is a human face, the secondary image looks like the interior of a mask made from that human face, with distortion along the y direction. Because this secondary image appears to be in front of the hologram, it is called a *real image*—even though one can pass one's hands through it and feel nothing but thin air.

object wave's image is virtual

secondary wave's image is real

10.3.3 Other Types of Holography; Applications

There are many variants on the basic holographic technique depicted in Figs. 10.6–10.8. These include the following.

PHASE HOLOGRAPHY

Instead of darkening the high-flux regions of the hologram as in photography, one produces a phase-shifting screen, whose phase shift (due to thickening of the hologram's material) is proportional to the incoming flux. Such a phase hologram transmits more of the reference-wave light than a standard, darkened hologram, thus making a brighter reconstructed image.

VOLUME HOLOGRAPHY

The hologram is a number of wavelengths deep rather than being just 2-dimensional. For example, it could be made from a thick photographic emulsion, in which the absorption length for light is longer than the thickness. Such a hologram has a 3-dimensional grating structure (grating "surfaces" rather than grating "lines"), with two consequences. When one reconstructs the holographic image from it in the manner of Fig. 10.8, (i) the third dimension of the grating suppresses the secondary wave while enhancing the (desired) object wave, so more power goes into it; and (ii) the reference wave's incoming angle θ_o must be controlled much more precisely, as modest errors suppress the reconstructed object wave. This second consequence enables one to record multiple images in a volume hologram, each using its own angle θ_o for the illuminating light and reference wave.

3-dimensional gratings

REFLECTION HOLOGRAPHY

One reads out the hologram by reflecting light off of it rather than transmitting light through it, and the hologram's diffraction grating produces a 3-dimensional holographic image by the same process as in transmission; see Ex. 10.3.

WHITE-LIGHT HOLOGRAPHY

The hologram is recorded with monochromatic light as usual, but it is optimized for reading out with white light. Even for the simple 2-dimensional hologram of Fig. 10.8, if one sends in white light at the angle θ_o, one will get a 3-dimensional object wave: the hologram's grating will diffract various wavelengths in various directions. In the direction of the original object wave (the horizontal direction in Fig. 10.8), one gets a 3-dimensional reconstructed image of the same color as was used when constructing the hologram. When one moves away from that direction (vertically in Fig. 10.8), one sees the color of the 3-dimensional image continuously change (Ex. 10.2b). White-light reflection holograms are used on credit cards and money as impediments to counterfeiting; they have even been used on postage stamps.

colored holograms

COMPUTATIONAL HOLOGRAMS

Just as one can draw 2-dimensional pictures numerically, pixel-by-pixel, so one can also create and modify holograms numerically, then read them out optically.

FULL-COLOR HOLOGRAPHY

A full-color holographic image of an object can be constructed by superposing three monochromatic holographic images with the three primary colors—red, green, and

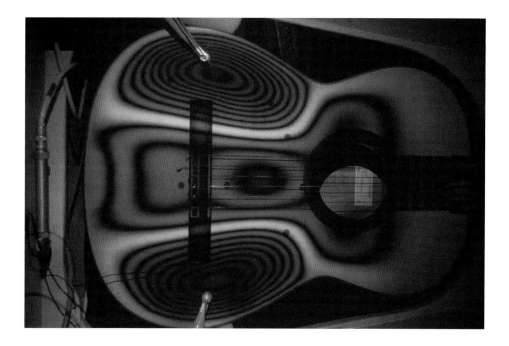

FIGURE 10.9 In-and-out vibrations of a guitar body visualized via holographic interferometry with green light. The dark and bright curves are a contour map, in units of the green light's wavelength, of the amplitude of vibration. Image courtesy Bernard Richardson, Cardiff University.

blue. One way to achieve this is to construct a single volume hologram using illuminating light from red, green, and blue laser beams, each arriving from a different 2-dimensional direction θ_o. Each beam produces a diffraction grating in the hologram with a different orientation and with spatial wave number corresponding to the beam's color. The 3-dimensional image can then be reconstructed using three white-light reference waves, one from each of the original three directions θ_o. The hologram will pick out of each beam the appropriate primary color, and produce the desired three overlapping images, which the eye will interpret as having approximately the true colors of the original object.

combining red, green, and blue holograms

HOLOGRAPHIC INTERFEROMETRY

One can observe changes in the shape of a surface at about the micrometer level by constructing two holograms, one of the original surface and the other of the changed surface, and then interfering the reconstructed light from the two holograms. Among other things, this holographic interferometry is used to observe small strains and vibrations of solid bodies—for example, sonic vibrations of a guitar in Fig. 10.9.

HOLOGRAPHIC LENSES

Instead of designing a hologram to reconstruct a 3-dimensional image, one can design it to manipulate light beams in most any way one wishes. As a simple example (Ex. 10.4c) of such a holographic lens, one can construct a holographic lens that splits one beam into two and focuses each of the two beams on a different spot. Holographic lenses are widely used in everyday technology, for example, to read bar codes in supermarket checkouts and to read the information off CDs, DVDs, and BDs (Ex. 10.5).

FUTURE APPLICATIONS

Major applications of holography that are under development include (i) dynamically changing volume holograms for 3-dimensional movies (which, of course, will require no eye glasses), and (ii) volume holograms for storage of large amounts of data—up to terabytes cm^{-3}.

EXERCISES

Exercise 10.2 *Derivation and Problem: Holographically Reconstructed Wave*
(a) Use the Helmholtz-Kirchhoff integral (8.4) or (8.6) to compute all four pieces of the holographically reconstructed wave field. Show that the piece generated by

$$t \propto \mathcal{O}(x, y, z = 0) M e^{iky \sin\theta_o}$$

is the same (aside from overall amplitude) as the field $\psi_{\text{object}} = \mathcal{O}(x, y, z)e^{-i\omega t}$ that would have resulted if when making the hologram (Fig. 10.6), the mirror wave had been absent and the photographic plate replaced by a window. Show that the other pieces have the forms and propagation directions indicated heuristically in Fig. 10.8. We shall examine the secondary wave, generated by $t \propto M\mathcal{O}^* e^{-iky \sin\theta_o}$, in Ex. 10.6.

(b) Suppose that plane-parallel white light is used in the holographic reconstruction of Fig. 10.8. Derive an expression for the direction in which one sees the object's 3-dimensional image have a given color (or equivalently, wave number). Assume that the original hologram was made with green light and $\theta_o = 45°$. What are the angles at which one sees the image as violet, green, yellow, and red?

Exercise 10.3 *Problem: Recording a Reflection Hologram*
How would you record a hologram if you want to read it out via reflection? Draw diagrams illustrating this, similar to Figs. 10.6 and 10.8. [Hint: The mirror wave and object wave can impinge on the photographic plate from either side; it's your choice.]

Exercise 10.4 *Example: Holographic Lens to Split and Focus a Light Beam*
A holographic lens, like any other hologram, can be described by its transmissivity $t(x, y)$.

(a) What $t(x, y)$ will take a reference wave, impinging from the θ_o direction (as in Fig. 10.8) and produce from it a primary object wave that converges on the spot

$(x, y, z) = (0, 0, d)$? [Hint: Consider, at the hologram's plane, a superposition of the incoming mirror wave and the point-spread function (8.28), which represents a beam that diverges from a point source. Then phase conjugate the point-spread function, so it converges to a point instead of diverging.]

(b) Draw a contour plot of the transmissivity $t(x, y)$ of the lens in part (a). Notice the resemblance to the Fresnel zone plate of Sec. 8.4.4. Explain the connection of the two, paying attention to how the holographic lens changes when one alters the chosen angle θ_o of the reference wave.

(c) What $t(x, y)$ will take a reference wave, impinging from the θ_o direction, and produce from it a primary wave that splits in two, with equal light powers converging on the spots $(x, y, z) = (-a, 0, d)$ and $(x, y, z) = (+a, 0, d)$?

Exercise 10.5 **Problem: CDs, DVDs, and BDs*
Information on CDs, DVDs, and BDs (compact, digital video, and blu-ray disks) is recorded and read out using holographic lenses, but it is not stored holographically. Rather, it is stored in a linear binary code consisting of pits and no-pits (for 0 and 1) along a narrow spiraling track. In each successive generation of storage device, the laser light has been pushed to a shorter wavelength ($\lambda = 780$ nm for CDs, 650 nm for DVDs, and 405 nm for BDs), and in each generation, the efficiency of the information storage has been improved. In CDs, the information is stored in a single holographic layer on the surface of the disk; in DVDs and BDs, it is usually stored in a single layer but can also be stored in as many as four layers, one above the other, though with a modest price in access time.

(a) Explain why one can expect to record in a disk's recording layer, at the very most, (close to) four bits of information per square wavelength of the recording light.

(b) The actual storage capacities are up to 900 MB for CDs, 4.7 GB for DVDs, and 25 GB for BDs. How efficient are each of these technologies relative to the maximum given in part (a)?

(c) Estimate the number of volumes of the *Encyclopedia Britannica* that can be recorded on a CD, on a DVD, and on a BD.

10.4 Phase-Conjugate Optics

Nonlinear optical techniques make it possible to phase conjugate an optical wave in real time, by contrast with holography, where the phase conjugation requires recording a hologram and then reconstructing the wave later. In this section, we explore the properties of phase-conjugated waves of any sort (light, sound, plasma waves, etc.), and in the next section, we discuss technology by which real-time phase conjugation is achieved for light.

The basic ideas and foundations for phase conjugation of waves were laid in Moscow by Boris Yakovovich Zel'dovich[2] and his colleagues (1972) and at Caltech by Amnon Yariv (1978).

phase conjugation

Phase conjugation is the process of taking a monochromatic wave

$$\Psi_O = \Re[\psi(x,y,z)e^{-i\omega t}] = \frac{1}{2}(\psi e^{-i\omega t} + \psi^* e^{+i\omega t}), \qquad (10.10a)$$

and from it constructing the wave

$$\Psi_{PC} = \Re[\psi^*(x,y,z)e^{-i\omega t}] = \frac{1}{2}(\psi^* e^{-i\omega t} + \psi e^{+i\omega t}). \qquad (10.10b)$$

phase-conjugating mirror

Notice that the phase-conjugated wave Ψ_{PC} is obtainable from the original wave Ψ_O by time reversal, $t \to -t$. This has a number of important consequences. One is that Ψ_{PC} propagates in the opposite direction to Ψ_O. Others are explained most clearly with the help of a *phase-conjugating mirror*.

Consider a wave Ψ_O with spatial modulation (i.e., a wave that carries a picture or a signal of some sort). Let the wave propagate in the z direction (rightward in Fig. 10.10), so that

incoming wave

$$\psi = \mathcal{A}(x,y,z)e^{i(kz-\omega t)}, \qquad (10.11)$$

where $\mathcal{A} = A e^{i\varphi}$ is a complex amplitude whose modulus A and phase φ change slowly in x, y, and z (slowly compared to the wave's wavelength $\lambda = 2\pi/k$). Suppose that this wave propagates through a time-independent medium with slowly varying physical properties [e.g., a dielectric medium with slowly varying index of refraction $\mathfrak{n}(x,y,z)$]. These slow variations will distort the wave's complex amplitude as it propagates. The wave equation for the real, classical field $\Psi = \Re[\psi e^{-i\omega t}]$ will have the form $\mathcal{L}\Psi - \partial^2\Psi/\partial t^2 = 0$, where $\mathcal{L}$ is a real spatial differential operator that depends on the medium's slowly varying physical properties. This wave equation implies that the complex field ψ satisfies

$$\mathcal{L}\psi + \omega^2\psi = 0 \qquad (10.12)$$

[which is the Helmholtz equation (8.1b) if $\mathcal{L}$ is the vacuum wave operator]. Equation (10.12) is the evolution equation for the wave's complex amplitude.

phase conjugating mirror and its phase conjugated wave Ψ_{PC}

Let the distorted, rightward propagating wave Ψ_O reflect off a mirror located at $z = 0$. If the mirror is a phase-conjugating one, then very near it (at z near zero) the reflected wave will have the form

$$\Psi_{PC} = \Re[\mathcal{A}^*(x,y,z=0)e^{i(-kz-\omega t)}], \qquad (10.13)$$

---

2. Zel'dovich is the famous son of a famous Russian/Jewish physicist, Yakov Borisovich Zel'dovich, who with Andrei Dmitrievich Sakharov fathered the Soviet hydrogen bomb and then went on to become a dominant figure internationally in astrophysics and cosmology.

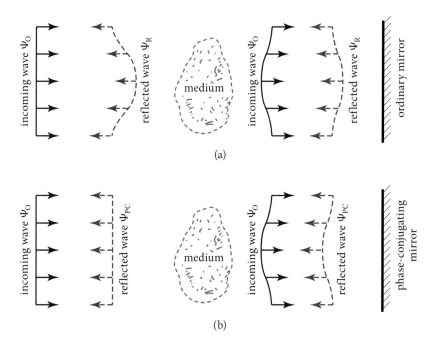

FIGURE 10.10 A rightward propagating wave and the reflected wave produced by (a) an ordinary mirror and (b) a phase-conjugating mirror. In both cases the waves propagate through a medium with spatially variable properties, which distorts their phase fronts. In case (a) the distortion is reinforced by the second passage through the variable medium; in case (b) the distortion is removed by the second passage.

while if it is an ordinary mirror, then the reflected wave will be

$$\Psi_R = \Re[\pm\mathcal{A}(x, y, z = 0)e^{i(-kz-\omega t)}]. \tag{10.14}$$

ordinary mirror and its reflected wave Ψ_R

(Here the sign depends on the physics of the wave. For example, if Ψ is the transverse electric field of an electromagnetic wave and the mirror is a perfect conductor, the sign will be minus to guarantee that the total electric field, original plus reflected, vanishes at the mirror's surface.)

These two waves, the phase-conjugated one Ψ_{PC} and the ordinary reflected one Ψ_R, have very different surfaces of constant phase (*phase fronts*). The phase of the incoming wave Ψ_O [Eqs. (10.10a) and (10.11)] as it nears the mirror ($z = 0$) is $\varphi + kz$, so (taking account of the fact that φ is slowly varying) the surfaces of constant phase are $z = -\varphi(x, y, z = 0)/k$. Similarly, the phase of the wave Ψ_R [Eq. (10.14)] reflected from the ordinary mirror is $\varphi - kz$, so its surfaces of constant phase near the mirror are $z = +\varphi(x, y, z = 0)/k$, which are reversed from those of the incoming wave, as shown in the upper right of Fig. 10.10. Finally, the phase of the wave Ψ_{PC} [Eq. (10.13)] reflected from the phase-conjugating mirror is $-\varphi - kz$, so its surfaces of constant phase near the mirror are $z = -\varphi(x, y, z = 0)/k$, which are the same as those of the

phase fronts of Ψ_{PC} and Ψ_R

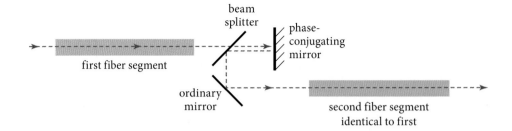

FIGURE 10.11 The use of a phase-conjugating mirror in an optical transmission line to prevent the fiber from distorting an optical image. The distortions put onto the image as it propagates through the first segment of fiber are removed during propagation through the second segment.

incoming wave (lower right of Fig. 10.10), even though the two waves are propagating in opposite directions.

The phase fronts of the original incoming wave and the phase-conjugated wave are the same not only near the phase-conjugating mirror; they are the same everywhere. More specifically, as the phase-conjugated wave Ψ_{PC} propagates away from the mirror [near which it is described by Eq. (10.13)], the propagation equation (10.12) forces it to evolve in such a way as to remain always the phase conjugate of the incoming wave:

$$\Psi_{PC} = \Re[\mathcal{A}^*(x, y, z)e^{-ikz}e^{-i\omega t}]. \tag{10.15}$$

This should be obvious: because the differential operator $\mathcal{L}$ in the propagation equation (10.12) for $\psi(x, y, z) = \mathcal{A}e^{ikz}$ is real, $\psi^*(x, y, z) = \mathcal{A}^*e^{-ikz}$ will satisfy this propagation equation when $\psi(x, y, z)$ does.

distortion removal in Ψ_{PC}

That the reflected wave Ψ_{PC} always remains the phase conjugate of the incoming wave Ψ_O means that *the distortions put onto the incoming wave, as it propagates rightward through the inhomogeneous medium, get removed from the phase-conjugated wave as it propagates back leftward* (Fig. 10.10).

This removal of distortions has a number of important applications. One is for image transmission in optical fibers. Normally when an optical fiber is used to transmit an optical image, the transverse spatial variations $\mathfrak{n}(x, y)$ of the fiber's index of refraction (which are required to hold the light in the fiber; Ex. 7.8) distort the image somewhat. The distortions can be eliminated by using a sequence of identical segments of optical fibers separated by phase-conjugating mirrors (Fig. 10.11). A few other applications include (i) real-time holography; (ii) removal of phase distortions in Fabry-Perot cavities by making one of the mirrors a phase-conjugating one, with a resulting improvement in the shape of the beam that emerges from the cavity; (iii) devices that can memorize an optical image and compare it to other images; (iv) the production of squeezed light (Ex. 10.16); and (v) improved focusing of laser light for laser fusion.

As we shall see in the next section, phase-conjugating mirrors rely crucially on the sinusoidal time evolution of the wave field; they integrate that sinusoidal evolution coherently over some timescale $\hat{\tau}$ (typically microseconds to nanoseconds) to produce the phase-conjugated wave. Correspondingly, if an incoming wave varies on timescales τ long compared to this $\hat{\tau}$ (e.g., if it carries a temporal modulation with bandwidth $\Delta\omega \sim 1/\tau$ small compared to $1/\hat{\tau}$), then the wave's temporal modulations will *not* be time reversed by the phase-conjugating mirror. For example, if the wave impinging on a phase-conjugating mirror has a frequency that is ω_a initially, and then gradually, over a time τ, increases to $\omega_b = \omega_a + 2\pi/\tau$, then the phase-conjugated wave will not emerge from the mirror with frequency ω_b first and ω_a later. Rather, it will emerge with ω_a first and ω_b later (same order as for the original wave). When the incoming wave's temporal variations are fast compared to the mirror's integration time, $\tau \ll \hat{\tau}$, the mirror encounters a variety of frequencies during its integration time and ceases to function properly. Thus, *even though phase conjugation is equivalent to time reversal in a formal sense, a phase-conjugating mirror cannot time reverse a temporal signal. It only time reverses monochromatic waves (which might carry a spatial signal).*

phase conjugation and time reversal

EXERCISES

Exercise 10.6 *Derivation and Example: Secondary Wave in Holography*
Consider the secondary wave generated by $t \propto \mathcal{M}\mathcal{O}^* e^{-iky\sin\theta_o}$ in the holographic reconstruction process of Fig. 10.8, Eq. (10.7), and Ex. 10.2.

(a) Assume, for simplicity, that the mirror and reference waves propagate nearly perpendicular to the hologram, so $\theta_o \ll 90°$ and $\theta_s \simeq 2\theta_o \ll 90°$; but assume that θ_s is still large enough that fairly far from the hologram the object wave and secondary waves separate cleanly from each other. Then, taking account of the fact that the object wave field has the form $\mathcal{O}(x, y, z)e^{ikz}$, show that the secondary wave is the phase-conjugated object wave defined in this section, except that it is propagating in the $+z$ direction rather than $-z$ (i.e., it has been reflected through the $z = 0$ plane). Then use this and the discussion of phase conjugation in the text to show that the secondary wave carries an image that resides in front of the hologram and is turned inside out, as discussed near the end of Sec. 10.3. Show, further, that if θ_o is not $\ll 90°$ (but is $< 30°$, so θ_s is a real angle, and the secondary image actually exists), then the secondary image is changed by a distortion along the y direction. What is the nature of the distortion, a squashing or a stretch?

(b) Suppose that a hologram has been made with $\theta_o < 30°$. Show that it is possible to perform image reconstruction with a modified reference wave (different from Fig. 10.8) in such a manner that the secondary, phase-conjugated wave emerges precisely perpendicular to the hologram and undistorted.

10.5 Maxwell's Equations in a Nonlinear Medium; Nonlinear Dielectric Susceptibilities; Electro-Optic Effects

In nonlinear optics, one is often concerned with media that are electrically polarized with *polarization* (electric dipole moment per unit volume) **P** but have no free charges or currents and are unmagnetized. In such a medium the charge and current densities associated with the polarization are

$$\rho_P = -\nabla \cdot \mathbf{P}, \qquad \mathbf{j}_P = \frac{\partial \mathbf{P}}{\partial t}, \qquad (10.16a)$$

and Maxwell's equations in SI units take the form

$$\nabla \cdot \mathbf{E} = \frac{\rho_P}{\epsilon_0}, \quad \nabla \cdot \mathbf{B} = 0, \quad \nabla \times \mathbf{E} = -\frac{\partial \mathbf{B}}{\partial t}, \quad \nabla \times \mathbf{B} = \mu_0 \left(\mathbf{j}_P + \epsilon_0 \frac{\partial \mathbf{E}}{\partial t} \right), \quad (10.16b)$$

which should be familiar. When rewritten in terms of the electric displacement vector

$$\mathbf{D} \equiv \epsilon_0 \mathbf{E} + \mathbf{P}, \qquad (10.17)$$

these Maxwell equations take the alternative form

$$\nabla \cdot \mathbf{D} = 0, \qquad \nabla \cdot \mathbf{B} = 0, \qquad \nabla \times \mathbf{E} = -\frac{\partial \mathbf{B}}{\partial t}, \qquad \nabla \times \mathbf{B} = \mu_0 \frac{\partial \mathbf{D}}{\partial t}, \qquad (10.18)$$

which should also be familiar. By taking the curl of the third equation (10.16b), using the relation $\nabla \times \nabla \times \mathbf{E} = -\nabla^2 \mathbf{E} + \nabla(\nabla \cdot \mathbf{E})$, and combining with the time derivative of the fourth equation (10.16b) and with $\epsilon_0 \mu_0 = 1/c^2$ and $\mathbf{j}_P = \partial \mathbf{P}/\partial t$, we obtain the following wave equation for the electric field, sourced by the medium's polarization:

wave equation for light in a polarizable medium

$$\boxed{\nabla^2 \mathbf{E} - \nabla(\nabla \cdot \mathbf{E}) = \frac{1}{c^2} \frac{\partial^2 (\mathbf{E} + \mathbf{P}/\epsilon_0)}{\partial t^2}.} \qquad (10.19)$$

If the electric field is sufficiently weak and the medium is homogeneous and isotropic (the case treated in most textbooks on electromagnetic theory), the polarization **P** is proportional to the electric field: $\mathbf{P} = \epsilon_0 \chi_0 \mathbf{E}$, where χ_0 is the medium's electrical susceptibility. In this case the medium does not introduce any nonlinearities into Maxwell's equations, the right-hand side of Eq. (10.19) becomes $[(1 + \chi_0)/c^2]\partial^2 \mathbf{E}/\partial t^2$, the divergence of Eq. (10.19) implies that the divergence of **E** vanishes, and therefore Eq. (10.19) becomes the standard dispersionless wave equation:

$$\nabla^2 \mathbf{E} - \frac{\mathfrak{n}^2}{c^2} \frac{\partial^2 \mathbf{E}}{\partial t^2} = 0, \quad \text{with} \quad \mathfrak{n}^2 = 1 + \chi_0. \qquad (10.20)$$

In many dielectric media, however, a strong electric field can produce a polarization that is nonlinear in the field. In such nonlinear media, the general expression for the (real) polarization in terms of the (real) electric field is

polarization in a nonlinear medium

$$\boxed{P_i = \epsilon_0 (\chi_{ij} E_j + 2 d_{ijk} E_j E_k + 4 \chi_{ijkl} E_j E_k E_l + \ldots),} \qquad (10.21)$$

where we sum over repeated indices. Here χ_{ij}, the linear susceptibility, is proportional to the 3-dimensional metric, $\chi_{ij} = \chi_0 g_{ij} = \chi_0 \delta_{ij}$, if the medium is isotropic (i.e., if all directions in it are equivalent), but otherwise it is tensorial; and the d_{ijk} and χ_{ijkl} are nonlinear susceptibilities, referred to as *second-order* and *third-order*, respectively, because of the two and three **E** terms that multiply them in Eq. (10.21). The normalizations used for these second- and third-order susceptibilities differ from one researcher to another: sometimes the factor ϵ_0 is omitted in Eq. (10.21); occasionally the factors of 2 and 4 are omitted. A compressed 2-index notation is sometimes used for the components of d_{ijk}; see Box 10.2 in Sec. 10.6.

second-order and third-order nonlinear susceptibilities

With **P** given by Eq. (10.21), the wave equation (10.19) becomes

$$\nabla^2 \mathbf{E} - \nabla(\nabla \cdot \mathbf{E}) - \frac{1}{c^2}\boldsymbol{\epsilon} \cdot \frac{\partial^2 \mathbf{E}}{\partial t^2} = \frac{1}{c^2 \epsilon_0} \frac{\partial^2 \mathbf{P}^{\text{NL}}}{\partial t^2}, \quad \text{where} \quad \epsilon_{ij} = \delta_{ij} + \chi_{ij}$$

dielectric tensor ϵ_{ij}

(10.22a)

is the "dielectric tensor," and $\mathbf{P}^{\text{NL}}$ is the nonlinear part of the polarization:

$$P_i^{\text{NL}} = \epsilon_0 (2 d_{ijk} E_j E_k + 4 \chi_{ijkl} E_j E_k E_l + \ldots).$$

(10.22b)

When $\mathbf{P}^{\text{NL}}$ is strong enough to be important and a monochromatic wave at frequency ω enters the medium, the nonlinearities lead to harmonic generation—the production of secondary waves with frequencies 2ω, 3ω, ...; see Secs. 10.7.1 and 10.8.1. As a result, an electric field in the medium cannot oscillate at just one frequency, and each of the electric fields in expression (10.22b) for the nonlinear polarization must be a sum of pieces with different frequencies. Because *the susceptibilities can depend on frequency*, this means that, when using expression (10.21), one sometimes must break up P_i and each E_i into its frequency components and use different values of the susceptibility to couple the different frequencies together. For example, one of the terms in Eq. (10.22b) will become

harmonic generation by nonlinearities

$$P_i^{(4)} = 4\epsilon_0 \chi_{ijkl} E_j^{(1)} E_k^{(2)} E_l^{(3)},$$

(10.23)

where $E_j^{(n)}$ oscillates at frequency ω_n, $P_i^{(4)}$ oscillates at frequency ω_4, and χ_{ijkl} depends on the four frequencies $\omega_1, \ldots, \omega_4$. Although this is complicated in the general case, in most practical applications, resonant coupling (or equivalently, energy and momentum conservation for photons) guarantees that only a single set of frequencies is important, and the resulting analysis simplifies substantially (see, e.g., Sec. 10.6.1).

Because all the tensor indices on the susceptibilities except the first index get contracted into the electric field in expression (10.21), we are free (and it is conventional) to define the susceptibilities as symmetric under interchange of any pair of indices that does not include the first. When [as has been tacitly assumed in Eq. (10.21)] there is no hysteresis in the medium's response to the electric field, the energy density of interaction between the polarization and the electric field is

polarizational energy density

$$U = \epsilon_0 \left(\frac{\chi_{ij} E_i E_j}{2} + \frac{2 d_{ijk} E_i E_j E_k}{3} + \frac{4 \chi_{ijkl} E_i E_j E_k E_l}{4} + \cdots \right), \quad (10.24a)$$

and the polarization is related to this energy of interaction, in Cartesian coordinates, by

$$P_i = \frac{\partial U}{\partial E_i}, \quad (10.24b)$$

which agrees with Eq. (10.21) *providing the susceptibilities are symmetric under interchange of all pairs of indices, including the first.* We shall assume such symmetry.[3] If the crystal is isotropic (as will be the case if it has cubic symmetry and reflection symmetry), then each of its tensorial susceptibilities is constructable from the metric $g_{ij} = \delta_{ij}$ and a single scalar susceptibility (see Ex. 10.7):[4]

susceptibilities for isotropic crystal

$$\chi_{ij} = \chi_0 g_{ij}, \quad d_{ijk} = 0,$$
$$\chi_{ijkl} = \tfrac{1}{3} \chi_4 (g_{ij} g_{kl} + g_{ik} g_{jl} + g_{il} g_{jk}), \quad \chi_{ijklm} = 0, \ldots. \quad (10.25)$$

A simple model of a crystal that explains how nonlinear susceptibilities can arise is the following. Imagine each ion in the crystal as having a valence electron that can oscillate in response to a sinusoidal electric field. The electron can be regarded as residing in a potential well, which, for low-amplitude oscillations, is very nearly harmonic (potential energy quadratic in displacement; restoring force proportional to displacement; "spring constant" independent of displacement). However, if the electron's displacement from equilibrium becomes a significant fraction of the interionic distance, it will begin to feel the electrostatic attraction of the neighboring ions, and its spring constant will weaken. So the potential the electron sees is really not that of a harmonic oscillator, but rather that of an *anharmonic oscillator*, $V(x) = \alpha x^2 - \beta x^3 + \cdots$, where x is the electron's displacement from equilibrium. The nonlinearities in this potential cause the electron's amplitude of oscillation, when driven by a sinusoidal electric field,

---

3. For further details see, e.g., Sharma (2006, Sec. 14.3) or Yariv (1989, Secs. 16.2 and 16.3). In a lossy medium, symmetry on the first index is lost; see Yariv and Yeh (2007, Sec. 8.1).
4. There is a caveat to these symmetry arguments. When the nonlinear susceptibilities depend significantly on the frequencies of the three or four waves, then these simple symmetries can be broken. For example, the third-order susceptibility χ_{ijkl} for an isotropic medium depends on which of the three input waves is paired with the output wave in Eq. (10.25); so when one orders the input waves with wave 1 on the j index, 2 on the k index, and 3 on the l index (and output 4 on the i index), the three terms in χ_{ijkl} [Eq. (10.25)] have different scalar coefficients. We ignore this subtlety in the remainder of this chapter. (For details, see, e.g., Sharma, 2006, Sec. 14.3.)

to be nonlinear in the field strength, and that nonlinear displacement causes the crystal's polarization to be nonlinear (e.g., Yariv, 1989, Sec. 16.3). For most crystals, the spatial arrangement of the ions causes the electron's potential energy V to be different for displacements in different directions, which causes the nonlinear susceptibilities to be anisotropic.

Because the total energy required to liberate the electron from its lattice site is roughly 1 eV and the separation between lattice sites is $\sim 10^{-10}$ m, the characteristic electric field for strong instantaneous nonlinearities is $\sim 1\,\text{V}(10^{-10}\,\text{m})^{-1} = 10^{10}\,\text{V m}^{-1} = 1\,\text{V}\,(100\,\text{pm})^{-1}$. Correspondingly, since d_{ijk} has dimensions 1/(electric field) and χ_{ijkl} has dimensions 1/(electric field)^2, rough upper limits on their Cartesian components are

$$\boxed{d_{ijk} \lesssim 100\,\text{pm V}^{-1}, \qquad \chi_4 \sim \chi_{ijkl} \lesssim \left(100\,\text{pm V}^{-1}\right)^2.} \qquad (10.26)$$

magnitudes of nonlinear susceptibilities

For comparison, because stronger fields will pull electrons out of solids, the strongest continuous-wave electric fields that occur in practical applications are $E \sim 10^6\,\text{V m}^{-1}$, corresponding to maximum intensities $F \sim 1\,\text{kW mm}^{-2} = 1\,\text{GW m}^{-2}$. These numbers dictate that, unless the second-order d_{ijk} are suppressed by isotropy, they will produce much larger effects than the third-order χ_{ijkl}, which in turn will dominate over all higher orders.

In the next few sections, we explore how the nonlinear susceptibilities produce nonlinear couplings of optical waves. There is, however, another application that we must mention in passing. When a slowly changing, non-wave electric field E_k is applied to a nonlinear medium, it can be thought of as producing a change in the linear dielectric tensor for waves $\Delta \chi_{ij} = 2(d_{ijk} + d_{ikj})E_k +$ quadratic terms [cf. Eq. (10.22b)]. This is an example (Boyd, 2008) of an *electro-optic effect*: the modification of optical properties of a medium by an applied electric field. Electro-optic effects are important in modern optical technology. For example, Pockels cells (used to modulate Gaussian light beams), optical switches (used in Q-switched lasers), and liquid-crystal displays (used for computer screens and television screens) are based on electro-optic effects. For some details of several important electro-optic effects and their applications, see, for example, Yariv and Yeh (2007, Chap. 9).

electro-optic effects

EXERCISES

Exercise 10.7 *Derivation and Example: Nonlinear Susceptibilities for an Isotropic Medium*

Explain why the nonlinear susceptibilities for an isotropic medium have the forms given in Eq. (10.25). [Hint: Use the facts that the χs must be symmetric in all their indices, and that, because the medium is isotropic, they must be constructable from the only isotropic tensors available to us, the (symmetric) metric tensor g_{ij} and the (antisymmetric) Levi-Civita tensor ϵ_{ijk}.] What are the corresponding forms, in an isotropic medium, of χ_{ijklmn} and $\chi_{ijklmnp}$? [Note: We will encounter an argument similar to this, in Ex. 28.1, for the form of the Riemann tensor in an isotropic universe.]

10.6 Three-Wave Mixing in Nonlinear Crystals

10.6.1 Resonance Conditions for Three-Wave Mixing

When a beam of light is sent through a nonlinear crystal, the nonlinear susceptibilities produce wave-wave mixing. The mixing due to the second-order susceptibility d_{ijk} is called *three-wave mixing*, because three electric fields appear in the polarization-induced interaction energy, Eq. (10.24a). The mixing produced by the third-order χ_{ijkl} is similarly called *four-wave mixing*. Three-wave mixing dominates in an anisotropic medium, but it is suppressed when the medium is isotropic, leaving four-wave mixing as the leading-order nonlinearity.

For use in our analyses of three-wave mixing, in Box 10.2 we list the second-order susceptibilities and some other properties of several specific nonlinear crystals.

Let us examine three-wave mixing in a general anisotropic crystal. Because the nonlinear susceptibilities are so small [i.e., because the input wave will generally be far weaker than 10^{10} V m^{-1} = 1 V(100 pm)$^{-1}$], the nonlinearities can be regarded as small perturbations. Suppose that two waves, labeled $n = 1$ and $n = 2$, are injected into the anisotropic crystal, and let their wave vectors be $\mathbf{k}_n$ when one ignores the (perturbative) nonlinear susceptibilities but keeps the large linear χ_{ij}. Because χ_{ij} is an anisotropic function of frequency, the dispersion relation $\Omega(\mathbf{k})$ for these waves (ignoring the nonlinearities) will typically be anisotropic. The frequencies of the two input waves satisfy the dispersion relation, $\omega_n = \Omega(\mathbf{k}_n)$, and the waves' forms are

$$E_j^{(n)} = \Re\left(\mathcal{A}_j^{(n)} e^{i(\mathbf{k}_n \cdot \mathbf{x} - \omega_n t)}\right) = \frac{1}{2}\left(\mathcal{A}_j^{(n)} e^{i(\mathbf{k}_n \cdot \mathbf{x} - \omega_n t)} + \mathcal{A}_j^{(n)*} e^{i(-\mathbf{k}_n \cdot \mathbf{x} + \omega_n t)}\right), \quad (10.27)$$

where we have denoted their vectorial complex amplitudes by $\mathcal{A}_j^{(n)}$. We adopt the convention that wave 1 is the wave with the larger frequency, so $\omega_1 - \omega_2 \geq 0$.

These two input waves couple, via the second-order nonlinear susceptibility d_{ijk}, to produce the following contribution to the medium's nonlinear polarization vector:

$$P_i^{(3)} = 2\epsilon_0 d_{ijk} 2 E_j^{(1)} E_k^{(2)}$$
$$= \epsilon_0 d_{ijk}\left(\mathcal{A}_j^{(1)} \mathcal{A}_k^{(2)} e^{i(\mathbf{k}_1+\mathbf{k}_2)\cdot\mathbf{x}} e^{i(\omega_1+\omega_2)t} + \mathcal{A}_j^{(1)} \mathcal{A}_k^{(2)*} e^{i(\mathbf{k}_1-\mathbf{k}_2)\cdot\mathbf{x}} e^{i(\omega_1-\omega_2)t} + \text{cc}\right),$$

(10.28)

BOX 10.2. PROPERTIES OF SOME ANISOTROPIC, NONLINEAR CRYSTALS

NOTATION FOR SECOND-ORDER SUSCEPTIBILITIES

In tabulations of the second-order nonlinear susceptibilities d_{ijk}, a compressed two-index notation d_{ab} is often used, with the indices running over

a: $1 = x$, $2 = y$, $3 = z$,

b: $1 = xx$, $2 = yy$, $3 = zz$, $4 = yz = zy$, $5 = xz = zx$, $6 = xy = yx$.

$$\tag{1}$$

CRYSTALS WITH LARGE SECOND-ORDER SUSCEPTIBILITIES

The following crystals have especially large second-order susceptibilities:

$$\text{Te: tellurium} \quad d_{11} = d_{xxx} = 650 \text{ pm V}^{-1}$$
$$\text{CdGeAs}_2 \quad d_{36} = d_{zyx} = 450 \text{ pm V}^{-1}$$
$$\text{Se: selenium} \quad d_{11} = d_{xxx} = 160 \text{ pm V}^{-1}. \tag{2}$$

However, they are not widely used in nonlinear optics, because some of their other properties are unfavorable. By contrast, glasses containing tellurium or selenium have moderately large nonlinearities and are useful.

KH$_2$PO$_4$

Potassium dihydrogen phosphate (KDP) is among the most widely used nonlinear crystals in 2016, not because of its nonlinear susceptibilities (which are quite modest) but because (i) it can sustain large electric fields without suffering damage, (ii) it is highly birefringent (different light speeds in different directions and for different polarizations, which as we shall see in Sec. 10.6.3 is useful for phase matching), and (iii) it has large electro-optic coefficients (end of Sec. 10.5). At linear order, it is axisymmetric around the z-axis, and its indices of refraction and susceptibilities have the following dependence on wavelength λ (measured in microns), which we use in Sec. 10.6.3 and Fig. 10.12a:

$$(n_o)^2 = 1 + \chi_{xx} = 1 + \chi_{yy} = 2.259276 + \frac{0.01008956}{\lambda^2 - 0.012942625} + \frac{13.005522\lambda^2}{\lambda^2 - 400},$$

$$(n_e)^2 = 1 + \chi_{zz} = 2.132668 + \frac{0.008637494}{\lambda^2 - 0.012281043} + \frac{3.2279924\lambda^2}{\lambda^2 - 400}. \tag{3}$$

The second-order nonlinearities break the axisymmetry of KDP, giving rise to

$$d_{36} = d_{zyx} = 0.44 \text{ pm V}^{-1}. \tag{4}$$

Although this is three orders of magnitude smaller than the largest nonlinearities available, its smallness is compensated for by its ability to sustain large electric fields.

(continued)

> **BOX 10.2. (continued)**
>
> KTiOPO$_4$
>
> Potassium titanyl phosphate (KTP) is quite widely used in 2016 (e.g., in green laser pointers; Ex. 10.13). At linear order it is nonaxisymmetric, but with only modest birefringence: its indices of refraction along its three principal axes, at the indicated wavelengths, are
>
> $$1{,}064 \text{ nm: } n_x = \sqrt{1+\chi_{xx}} = 1.740, \quad n_y = \sqrt{1+\chi_{yy}} = 1.747,$$
> $$n_z = \sqrt{1+\chi_{zz}} = 1.830;$$
> $$532 \text{ nm: } n_x = \sqrt{1+\chi_{xx}} = 1.779, \quad n_y = \sqrt{1+\chi_{yy}} = 1.790,$$
> $$n_z = \sqrt{1+\chi_{zz}} = 1.887. \tag{5}$$
>
> Its third-order nonlinearities are moderately large. In units of pm V^{-1}, they are
>
> $$d_{31} = d_{zxx} = 6.5, \quad d_{32} = d_{zyy} = 5.0, \quad d_{33} = d_{zzz} = 13.7,$$
> $$d_{24} = d_{xyz} = d_{xzy} = 7.6, \quad d_{15} = d_{xxz} = d_{xzx} = 6.1. \tag{6}$$
>
> Notice that symmetry on the first index is modestly broken: $d_{zxx} = 6.5 \neq d_{xxz} = 7.6$. This symmetry breaking is caused by the crystal's dissipating a small portion of the light power that drives it.
>
> **EVOLUTION OF MATERIALS**
>
> Over the past three decades materials scientists have found and developed nonlinear crystals with ever-improving properties. By the time you read this book, the most widely used crystals are likely to have changed.

where "cc" means complex conjugate.[5] This sinusoidally oscillating polarization produces source terms in Maxwell's equations (10.16b) and the wave equation (10.19): an oscillating, polarization-induced charge density $\rho_P = -\nabla \cdot \mathbf{P}^{(3)}$ and current density $\mathbf{j}_P = \partial \mathbf{P}^{(3)}/\partial t$. This polarization charge and current, like $\mathbf{P}^{(3)}$ itself [Eq. (10.28)], consist of two traveling waves, one with frequency and wave vector

resonance conditions and dispersion relation for new, third wave

$$\boxed{\omega_3 = \omega_1 + \omega_2, \quad \mathbf{k}_3 = \mathbf{k}_1 + \mathbf{k}_2;} \tag{10.29a}$$

the other with frequency and wave vector

$$\boxed{\omega_3 = \omega_1 - \omega_2, \quad \mathbf{k}_3 = \mathbf{k}_1 - \mathbf{k}_2.} \tag{10.29b}$$

---

5. The reason for the factor 2 in the definition $P_i = 2\epsilon_0 d_{ijk} E_j E_k$ is to guarantee a factor unity in Eq. (10.28) and in the resulting coupling constant κ of Eq. (10.38).

If either of these (ω_3, $\mathbf{k}_3$) satisfies the medium's dispersion relation $\omega = \Omega(\mathbf{k})$, then the polarization will generate an electromagnetic wave $E_j^{(3)}$ that propagates along in resonance with its polarization-vector source in the wave equation

$$\nabla^2 \mathbf{E}^{(3)} - \boldsymbol{\nabla}(\boldsymbol{\nabla} \cdot \mathbf{E}^{(3)}) + \frac{\omega_3^2}{c^2}\boldsymbol{\epsilon} \cdot \mathbf{E}^{(3)} = \frac{1}{c^2 \epsilon_0} \frac{\partial^2 \mathbf{P}^{(3)}}{\partial t^2} \qquad (10.30)$$

[the frequency-ω_3 part of Eq. (10.22a)]. Therefore, this new electromagnetic wave, with frequency ω_3 and wave vector $\mathbf{k}_3$, will grow as it propagates.

For most choices of the input waves—most choices of $\{\mathbf{k}_1, \omega_1 = \Omega(\mathbf{k}_1), \mathbf{k}_2, \omega_2 = \Omega(\mathbf{k}_2)\}$—neither of the polarizations $\mathbf{P}^{(3)}$ will have a frequency $\omega_3 = \omega_1 \pm \omega_2$ and wave vector $\mathbf{k}_3 = \mathbf{k}_1 \pm \mathbf{k}_2$ that satisfy the medium's dispersion relation, and thus neither will be able to create a third electromagnetic wave resonantly; the wave-wave mixing is ineffective. However, for certain special choices of the input waves, resonant coupling will be achieved, and a strong third wave will be produced. See Sec. 10.6.3 for details.

In nonlinear optics, enforcing the resonance conditions (10.29), with all three waves satisfying their dispersion relations, is called *phase matching*, because it guarantees that the new wave propagates along in phase with the polarization produced by the two old waves.

phase matching

The resonance conditions (10.29) have simple quantum mechanical interpretations—a fact that is not at all accidental: quantum mechanics underlies the classical theory that we are developing. Each classical wave is carried by photons that have discrete energies $\mathcal{E}_n = \hbar\omega_n$ and discrete momenta $\mathbf{p}_n = \hbar\mathbf{k}_n$. The input waves are able to produce resonantly waves with $\omega_3 = \omega_1 \pm \omega_2$ and $\mathbf{k}_3 = \mathbf{k}_1 \pm \mathbf{k}_2$, if those waves satisfy the dispersion relation. Restated in quantum mechanical terms, the condition of resonance with the "+" sign rather than the "−" is

quantum description of resonance conditions

$$\mathcal{E}_3 = \mathcal{E}_1 + \mathcal{E}_2, \qquad \mathbf{p}_3 = \mathbf{p}_1 + \mathbf{p}_2. \qquad (10.31a)$$

one photon created from two

This has the quantum mechanical meaning that one photon of energy $\mathcal{E}_1$ and momentum $\mathbf{p}_1$, and another of energy $\mathcal{E}_2$ and momentum $\mathbf{p}_2$ combine together, via the medium's nonlinearities, and are annihilated (in the language of quantum field theory). By their annihilation they create a new photon with energy $\mathcal{E}_3 = \mathcal{E}_1 + \mathcal{E}_2$ and momentum $\mathbf{p}_3 = \mathbf{p}_1 + \mathbf{p}_2$. Thus the classical condition of resonance is the quantum mechanical condition of energy-momentum conservation for the sets of photons involved in a process of quantum annihilation and creation. For this process to proceed, not only must energy-momentum conservation be satisfied, but also all three photons must have energies and momenta that obey the photons' semiclassical hamiltonian relation $\mathcal{E} = H(\mathbf{p})$ (i.e., the dispersion relation $\omega = \Omega(\mathbf{k})$ with $H = \hbar\Omega$, $\mathcal{E} = \hbar\omega$, and $\mathbf{p} = \hbar\mathbf{k}$).

Similarly, the classical conditions of resonance with the "−" sign rather than the "+" can be written (after bringing photon 2 to the left-hand side) as

two photons created from one

$$\mathcal{E}_3 + \mathcal{E}_2 = \mathcal{E}_1, \qquad \mathbf{p}_3 + \mathbf{p}_2 = \mathbf{p}_1. \tag{10.31b}$$

This has the quantum mechanical meaning that one photon of energy $\mathcal{E}_1$ and momentum $\mathbf{p}_1$ is annihilated, via the medium's nonlinearities, and from its energy and momentum two photons are created, with energies $\mathcal{E}_2$, $\mathcal{E}_3$ and momenta $\mathbf{p}_2$, $\mathbf{p}_3$ that satisfy energy-momentum conservation.

Resonance conditions play a major role in other areas of physics, whenever one deals with nonlinear wave-wave coupling or wave-particle coupling. In this book we meet them again in both classical language and quantum language when studying excitations of plasmas (Chap. 23).

10.6.2 Three-Wave-Mixing Evolution Equations in a Medium That Is Dispersion-Free and Isotropic at Linear Order

Consider the simple, idealized case where the linear part of the susceptibility χ_{jk} is isotropic and frequency independent, $\chi_{jk} = \chi_0 g_{jk}$; correspondingly, Maxwell's equations imply $\nabla \cdot \mathbf{E} = 0$. The Track-One part of this chapter will be confined to this idealized case. In Sec. 10.6.3 (Track Two), we treat the more realistic case, which has dispersion and anisotropy at linear order.

In our idealized case the dispersion relation, ignoring the nonlinearities, takes the simple, nondispersive form [which follows from Eq. (10.20)]:

$$\omega = \frac{c}{\mathfrak{n}} k, \quad \text{where} \quad k = |\mathbf{k}|, \ \mathfrak{n} = \sqrt{1 + \chi_0}. \tag{10.32}$$

Consider three-wave mixing for waves 1, 2, and 3 that all propagate in the same z direction with wave numbers that satisfy the resonance condition $k_3 = k_1 + k_2$. The dispersion-free dispersion relation (10.32) guarantees that the frequencies will also resonate: $\omega_3 = \omega_1 + \omega_2$. If we write the new wave as $E_i^{(3)} = \Re(\mathcal{A}_i^{(3)} e^{i(k_3 z - \omega_3 t)}) = \frac{1}{2}\mathcal{A}_i^{(3)} e^{i(k_3 z - \omega_3 t)} + \text{cc}$, then its evolution equation (10.30), when combined with Eqs. (10.27) and (10.28), takes the form

$$\nabla^2 \left(\mathcal{A}_i^{(3)} e^{i(k_3 z - \omega_3 t)} \right) + \frac{\mathfrak{n}^2 \omega_3^2}{c^2} \mathcal{A}_i^{(3)} e^{i(k_3 z - \omega_3 t)} = -2 \frac{\omega_3^2}{c^2} d_{ijk} \mathcal{A}_j^{(1)} \mathcal{A}_k^{(2)} e^{i(k_3 z - \omega_3 t)}. \tag{10.33}$$

Using the dispersion relation (10.32) and the fact that the lengthscale on which wave 3 changes is long compared to its wavelength (which is always the case, because the fields are always much weaker than 10^{10} V m^{-1}), the left-hand side becomes $2i k_3 d\mathcal{A}_i^{(3)}/dz$, and Eq. (10.33) then becomes (with the aid of the dispersion relation) $d\mathcal{A}_i^{(3)}/dz = i(k_3/\mathfrak{n}^2) d_{ijk} \mathcal{A}_j^{(1)} \mathcal{A}_k^{(2)}$. This and similar computations for evolution of the other two waves (Ex. 10.8) give the following equations for the rates of change of the three waves' complex amplitudes:

$$\frac{d\mathcal{A}_i^{(3)}}{dz} = i\frac{k_3}{\mathfrak{n}^2} d_{ijk} \mathcal{A}_j^{(1)} \mathcal{A}_k^{(2)} \quad \text{at } \omega_3 = \omega_1 + \omega_2, \quad k_3 = k_1 + k_2; \quad (10.34\text{a})$$

$$\frac{d\mathcal{A}_i^{(1)}}{dz} = i\frac{k_1}{\mathfrak{n}^2} d_{ijk} \mathcal{A}_j^{(3)} \mathcal{A}_k^{(2)*} \quad \text{at } \omega_1 = \omega_3 - \omega_2, \quad k_1 = k_3 - k_2; \quad (10.34\text{b})$$

$$\frac{d\mathcal{A}_i^{(2)}}{dz} = i\frac{k_2}{\mathfrak{n}^2} d_{ijk} \mathcal{A}_j^{(3)} \mathcal{A}_k^{(1)*} \quad \text{at } \omega_2 = \omega_3 - \omega_1, \quad k_2 = k_3 - k_1. \quad (10.34\text{c})$$

Therefore, each wave's amplitude changes with distance z traveled at a rate proportional to the product of the field strengths of the other two waves.

It is instructive to rewrite the evolution equations (10.34) in terms of *renormalized scalar amplitudes* $\mathfrak{A}_n$ and *unit-normed polarization vectors* $f_j^{(n)}$ for the three waves $n = 1, 2, 3$:

$$\mathcal{A}_j^{(n)} = \sqrt{\frac{2k_n}{\epsilon_0 \mathfrak{n}^2}} \, \mathfrak{A}_n \, f_j^{(n)} = \sqrt{\frac{2\omega_n}{\epsilon_0 c \, \mathfrak{n}}} \, \mathfrak{A}_n \, f_j^{(n)}. \quad (10.35)$$

renormalized wave amplitudes

This renormalization is motivated by the fact that $|\mathfrak{A}_n|^2$ is proportional to the flux of quanta $dN_n/dA dt$ associated with wave n. Specifically, the energy density in wave n is (neglecting nonlinearities) $U = \epsilon_o(1 + \chi_o)\overline{\mathbf{E}^2} = \frac{1}{2}\epsilon_o \mathfrak{n}^2 |\mathcal{A}^{(n)}|^2$ (where the bar means time average); the energy flux is this U times the wave speed $c/\mathfrak{n}$:

$$\boxed{F_n = \frac{1}{2}\epsilon_o \mathfrak{n} c |\mathcal{A}^{(n)}|^2 = \omega_n |\mathfrak{A}_n|^2;} \quad (10.36)$$

and the flux of quanta is this F_n divided by the energy $\mathcal{E}_n = \hbar \omega_n$ of each quantum: $dN_n/dA dt = |\mathfrak{A}_n|^2/\hbar$, where dA is a unit area orthogonal to $\mathbf{k}_n$.

The three-wave-mixing evolution equations (10.34), rewritten in terms of the renormalized amplitudes, take the simple form

$$\frac{d\mathfrak{A}_3}{dz} = i\kappa \, \mathfrak{A}_1 \mathfrak{A}_2, \quad \frac{d\mathfrak{A}_1}{dz} = i\kappa \, \mathfrak{A}_3 \mathfrak{A}_2^*, \quad \frac{d\mathfrak{A}_2}{dz} = i\kappa \, \mathfrak{A}_3 \mathfrak{A}_1^*;$$

$$\kappa = \sqrt{\frac{2\omega_1 \omega_2 \omega_3}{\epsilon_0 c^3 \mathfrak{n}^3}} \, d_{ijk} \, f_i^{(1)} f_j^{(2)} f_k^{(3)}. \quad (10.37)$$

three-wave mixing evolution equations in an isotropic, dispersion-free medium

It is straightforward to verify that these evolution equations guarantee energy conservation $d(F_1 + F_2 + F_3)/dz = 0$, with F_n given by Eq. (10.36). Therefore, at least one wave will grow and at least one wave will decay due to three-wave mixing.

When waves 1 and 2 are the same wave, the three-wave mixing leads to frequency doubling: $\omega_3 = 2\omega_1$. In this case, the nonlinear polarization that produces the third wave is $P_i = \epsilon_0 d_{ijk} E_j^{(1)} E_k^{(1)}$, by contrast with that when waves 1 and 2 are different, $P_i = 2\epsilon_0 d_{ijk} E_j^{(1)} E_k^{(2)}$ [Eq. (10.28)]. [In the latter case the factor 2 arises because we are dealing with cross terms in $(E_j^{(1)} + E_j^{(2)})(E_k^{(1)} + E_k^{(2)})$.] Losing the factor 2 and

frequency doubling

making wave 2 the same as wave 1 leads to an obvious modification of the evolution equations (10.37):

evolution equations for frequency doubling

$$\frac{d\mathfrak{A}_3}{dz} = \frac{i\kappa}{2}(\mathfrak{A}_1)^2, \quad \frac{d\mathfrak{A}_1}{dz} = i\kappa \mathfrak{A}_3 \mathfrak{A}_1^*; \quad \kappa = \sqrt{\frac{2\omega_1^2 \omega_3}{\epsilon_0 c^3 \mathfrak{n}^3}} \, d_{ijk} \, f_i^{(1)} f_j^{(1)} f_k^{(3)}. \quad (10.38)$$

Once again, it is easy to verify energy conservation, $d(F_1 + F_3)/dz = 0$. We discuss frequency doubling in Sec. 10.7.1.

EXERCISES

Exercise 10.8 *Derivation: Evolution Equations in Idealized Three-Wave Mixing*
Derive Eqs. (10.34b) and (10.34c) for the amplitudes of waves 1 and 2 produced by three-wave mixing.

10.6.3 Three-Wave Mixing in a Birefringent Crystal: Phase Matching and Evolution Equations T2

ORDINARY WAVES, EXTRAORDINARY WAVES, AND DISPERSION RELATIONS

In reality, all nonlinear media have frequency-dependent dispersion relations and most are anisotropic at linear order and therefore birefringent (different wave speeds in different directions). An example is the crystal KDP (Box 10.2), which is symmetric around the z-axis and has indices of refraction[6]

birefringent crystal

$$\mathfrak{n}_o = \sqrt{1 + \chi_{xx}} = \sqrt{1 + \chi_{yy}}, \quad \mathfrak{n}_e = \sqrt{1 + \chi_{zz}} \quad (10.39)$$

that depend on the light's wave number $k = 2\pi/\lambda$ in the manner shown in Fig. 10.12a and in Eq. (3) of Box 10.2. The subscript o stands for *ordinary*; e, for *extraordinary*; see the next paragraph.

Maxwell's equations imply that, in this crystal, for plane, monochromatic waves propagating in the x-z plane at an angle θ to the symmetry axis [$\mathbf{k} = k(\sin\theta \mathbf{e}_x + \cos\theta \mathbf{e}_z)$], there are two dispersion relations corresponding to the two polarizations of the electric field:

1. If $\mathbf{E}$ is orthogonal to the symmetry axis, then (as is shown in Ex. 10.9), it must also be orthogonal to the propagation direction (i.e., must point in the $\mathbf{e}_y$ direction), and the dispersion relation is

$$\frac{\omega/k}{c} = \text{(phase speed in units of speed of light)} = \frac{1}{\mathfrak{n}_o}, \quad (10.40a)$$

---

[6]. For each wave, the index of refraction is the ratio of light's vacuum speed c to the wave's phase speed $V_{\rm ph} = \omega/k$.

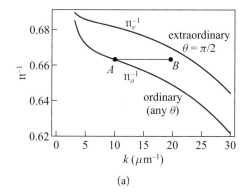
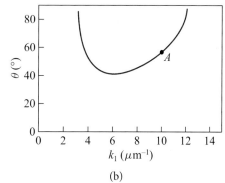

FIGURE 10.12 (a) The inverse of the index of refraction $\mathfrak{n}^{-1}$ (equal to the phase speed in units of the speed of light) for electromagnetic waves propagating at an angle θ to the symmetry axis of a KDP crystal, as a function of wave number k in reciprocal microns. See Eq. (10.40a) for the lower curve and Eq. (10.40b) with $\theta = \pi/2$ for the upper curve. For extraordinary waves propagating at an arbitrary angle θ to the crystal's symmetry axis, $\mathfrak{n}^{-1}$ is a mean [Eq. (10.40b)] of the two plotted curves. The plotted curves are fit by the analytical formulas (3) of Box 10.2. (b) The angle θ to the symmetry axis at which ordinary waves with wave number k_1 (e.g., point A) must propagate for three-wave mixing to be able to produce frequency-doubled or phase-conjugated extraordinary waves (e.g., point B).

independent of the angle θ. These waves are called *ordinary,* and their phase speed (10.40a) is the lower curve in Fig. 10.12a; at $k = 10\ \mu\mathrm{m}^{-1}$ (point A), the phase speed is $0.663c$, while at $k = 20\ \mu\mathrm{m}^{-1}$, it is $0.649c$.

ordinary waves

2. If $\mathbf{E}$ is not orthogonal to the symmetry axis, then (Ex. 10.9) it must lie in the plane formed by $\mathbf{k}$ and the symmetry axis (the x-z plane, with $E_x/E_z = -(\mathfrak{n}_e/\mathfrak{n}_o)^2 \cot\theta$ (which means that $\mathbf{E}$ is not orthogonal to the propagation direction unless the crystal is isotropic, $\mathfrak{n}_e = \mathfrak{n}_o$, which it is not); and the dispersion relation is

$$\frac{\omega/k}{c} = \frac{1}{\mathfrak{n}} = \sqrt{\frac{\cos^2\theta}{\mathfrak{n}_o^2} + \frac{\sin^2\theta}{\mathfrak{n}_e^2}}. \tag{10.40b}$$

In this case the waves are called *extraordinary.*[7] As the propagation direction varies from parallel to the symmetry axis ($\cos\theta = 1$) to perpendicular ($\sin\theta = 1$), this extraordinary phase speed varies from $c/\mathfrak{n}_o$ (the lower curve in Fig. 10.12; $0.663c$ at $k = 10\ \mu\mathrm{m}^{-1}$), to $c/\mathfrak{n}_e$ (the upper curve; $0.681c$ at $k = 10\ \mu\mathrm{m}^{-1}$).

extraordinary waves

7. When studying perturbations of a cold, magnetized plasma (Sec. 21.5.3) we will meet two wave modes that have these same names: ordinary and extraordinary. However, because of the physical differences between an axially symmetric magnetized plasma and an axially symmetric birefringent crystal, the physics of those plasma modes (e.g., the direction of their oscillating electric field) is rather different from that of the crystal modes studied here.

10.6 Three-Wave Mixing in Nonlinear Crystals

PHASE MATCHING FOR FREQUENCY DOUBLING IN A KDP CRYSTAL

This birefringence enables one to achieve phase matching (satisfy the resonance conditions) in three-wave mixing. As an example, consider the resonance conditions for a *frequency-doubling device* (discussed in further detail in Sec. 10.7.1): one in which the two input waves are identical, so $\mathbf{k}_1 = \mathbf{k}_2$ and $\mathbf{k}_3 = 2\mathbf{k}_1$ point in the same direction. Let this common propagation direction be at an angle θ to the symmetry axis. Then the resonance conditions reduce to the demands that the output wave number be twice the input wave number, $k_3 = 2k_1$, and the output phase speed be the same as the input phase speed, $\omega_3/k_3 = \omega_1/k_1$. Now, for waves of the same type (both ordinary or both extraordinary), the phase speed is a monotonic decreasing function of wave number [Fig. 10.12a and Eqs. (10.40)], so there is no choice of propagation angle θ that enables these resonance conditions to be satisfied. The only way to satisfy them is by using ordinary input waves and extraordinary output waves, and then only for a special, frequency-dependent propagation direction. This technique is called *type I phase matching*; "type I" because there are other techniques for phase matching (i.e., for arranging that the resonance conditions be satisfied; see, e.g., Table 8.4 of Yariv and Yeh, 2007).

As an example, if the input waves are ordinary, with $k_1 = 10~\mu\text{m}^{-1}$ (approximately the value for light from a ruby laser; point A in Fig. 10.12a), then the output waves must be extraordinary and must have the same phase speed as the input waves (same height in Fig. 10.12a) and have $k_3 = 2k_1 = 20~\mu\text{m}^{-1}$ (i.e., point B). This phase speed is between $c/\mathfrak{n}_e(2k_1)$ and $c/\mathfrak{n}_o(2k_1)$, and thus can be achieved for a special choice of propagation angle: $\theta = 56.7°$ (point A in Fig. 10.12b). In general, Eqs. (10.40) imply that the unique propagation direction θ at which the resonance conditions can be satisfied is the following function of the input wave number k_1:

$$\sin^2\theta = \frac{1/\mathfrak{n}_o^2(k_1) - 1/\mathfrak{n}_o^2(2k_1)}{1/\mathfrak{n}_e^2(2k_1) - 1/\mathfrak{n}_o^2(2k_1)}. \tag{10.41}$$

This resonance angle is plotted as a function of wave number for KDP in Fig. 10.12b.

This special case of identical input waves illustrates a general phenomenon: *at fixed input frequencies, the resonance conditions can be satisfied only for special, discrete input and output directions.*

For our frequency-doubling example, the extraordinary dispersion relation (10.40b) for the output wave can be rewritten as

$$\omega = \frac{ck}{\mathfrak{n}} = \Omega_e(\mathbf{k}) = c\sqrt{\frac{k_z^2}{\mathfrak{n}_o(k)^2} + \frac{k_x^2}{\mathfrak{n}_e(k)^2}}, \quad \text{where} \quad k = \sqrt{k_x^2 + k_z^2}. \tag{10.42}$$

Correspondingly, the group velocity[8] $V_g^j = \partial\Omega/\partial k_j$ for the output wave has components

$$V_g^x = V_{\text{ph}} \sin\theta \left(\frac{n^2}{n_e^2} - \frac{n^2 \cos^2\theta}{n_o^2} \frac{d \ln n_o}{d \ln k} - \frac{n^2 \sin^2\theta}{n_e^2} \frac{d \ln n_e}{d \ln k} \right),$$

$$V_g^z = V_{\text{ph}} \cos\theta \left(\frac{n^2}{n_o^2} - \frac{n^2 \cos^2\theta}{n_o^2} \frac{d \ln n_o}{d \ln k} - \frac{n^2 \sin^2\theta}{n_e^2} \frac{d \ln n_e}{d \ln k} \right), \qquad (10.43)$$

where $V_{\text{ph}} = \omega/k = c/n$ is the phase velocity. For an ordinary input wave with $k_1 = 10\ \mu\text{m}^{-1}$ (point A in Fig. 10.12) and an extraordinary output wave with $k_3 = 20\ \mu\text{m}^{-1}$ (point B), these formulas give for the direction of the output group velocity (direction along which the output waves grow) $\theta_g = \arctan(V_g^x/V_g^z) = 58.4°$, compared to the direction of the common input-output phase velocity $\theta = 56.7°$. They give for the magnitude of the group velocity $V_g = 0.628c$, compared to the common phase velocity $v_{\text{ph}} = 0.663c$. Thus, the differences between the group velocity and the phase velocity are small, but they do differ.

EVOLUTION EQUATIONS

Once one has found wave vectors and frequencies that satisfy the resonance conditions, the evolution equations for the two (or three) coupled waves have the same form as in the idealized dispersion-free, isotropic case [Eqs. (10.38) or (10.37)], but with the following minor modifications.

Let planar input waves impinge on a homogeneous, nonlinear crystal at some plane $z = 0$ and therefore (by symmetry) have energy fluxes inside the crystal that evolve as functions of z only: $\mathbf{F}_n = \mathbf{F}_n(z)$ for waves $n = 1$ and 3 in the case of frequency doubling (or 1, 2, and 3 in the case of three different waves). Then energy conservation dictates that

$$\frac{d}{dz} \sum_n F_{n\,z} = 0, \qquad (10.44)$$

where $F_{n\,z}(z)$ is the z component of the energy flux for wave n. It is convenient to define a complex amplitude $\mathcal{A}_n$ for wave n that is related to the wave's complex electric field amplitude by an analog of Eq. (10.35):

$$\mathcal{A}_j^{(n)} = \zeta_n \sqrt{\frac{2\omega_n}{\epsilon_0 c\, \mathfrak{n}_n}}\, \mathcal{A}_n f_j^{(n)}. \qquad (10.45)$$

renormalized amplitude in birefringent crystal

Here $f_j^{(n)}$ is the wave's polarization vector [Eq. (10.35)], $\mathfrak{n}_n$ is its index of refraction (defined by $\omega_n/k_n = c/\mathfrak{n}_n$), and ζ_n is some positive real constant that depends on the

---

8. Since a wave's energy travels with the group velocity, it must be that $\mathbf{V}_g = \mathbf{E} \times \mathbf{H}/U$, where U is the wave's energy density, $\mathbf{E} \times \mathbf{H}$ is its Poynting vector (energy flux), and $\mathbf{H} = \mathbf{B}/\mu_0$ (in our dielectric medium). It can be shown explicitly that, indeed, this is the case.

relative directions of $\mathbf{k}_n$, $\mathbf{f}^{(n)}$, and $\mathbf{e}_z$ and has a value ensuring that

$$F_{n\,z} = \omega_n |\mathcal{A}_n|^2 \qquad (10.46)$$

[same as Eq. (10.36) but with F_n replaced by $F_{n\,z}$]. Since the energy flux is $\hbar\omega_n$ times the photon-number flux, this equation tells us that $|\mathcal{A}_n|^2/\hbar$ is the photon-number flux (just like the idealized case).

Because the evolution equations involve the same photon creation and annihilation processes as in the idealized case, they must have the same mathematical form as in that case [Eqs. (10.38) or (10.37)], except for the magnitude of the coupling constant. (For a proof, see Ex. 10.10.) Specifically, for frequency doubling of a wave 1 to produce wave 3 ($\omega_3 = 2\omega_1$), the resonant evolution equations and coupling constant are

three-wave mixing evolution equations in a birefringent crystal

$$\frac{d\mathcal{A}_3}{dz} = \frac{i\kappa}{2}(\mathcal{A}_1)^2, \quad \frac{d\mathcal{A}_1}{dz} = i\kappa\,\mathcal{A}_3\mathcal{A}_1^*; \quad \kappa = \beta\sqrt{\frac{2\omega_1^2\omega_3}{\epsilon_0 c^3 n_1^2 n_3}}\,d_{ijk}\,f_i^{(1)}f_j^{(1)}f_k^{(3)}$$

$$(10.47)$$

[cf. Eqs. (10.38) for the idealized case]. For resonant mixing of three different waves ($\omega_3 = \omega_1 + \omega_2$), they are

$$\frac{d\mathcal{A}_3}{dz} = i\kappa\,\mathcal{A}_1\mathcal{A}_2, \quad \frac{d\mathcal{A}_1}{dz} = i\kappa\,\mathcal{A}_3\mathcal{A}_2^*, \quad \frac{d\mathcal{A}_2}{dz} = i\kappa\,\mathcal{A}_3\mathcal{A}_1^*;$$

$$\kappa = \beta'\sqrt{\frac{2\omega_1\omega_2\omega_3}{\epsilon_0 c^3 n_1 n_2 n_3}}\,d_{ijk}\,f_i^{(1)}f_j^{(2)}f_k^{(3)}$$

$$(10.48)$$

[cf. Eqs. (10.37) for the idealized case]. Here β and β' are constants of order unity that depend on the relative directions of $\mathbf{e}_z$ and the wave vectors $\mathbf{k}_n$ and polarization vectors $\mathbf{f}^{(n)}$; see Ex. 10.10.

It is useful to keep in mind the following magnitudes of the quantities that appear in these three-wave-mixing equations (Ex. 10.11):

$$F_n \lesssim 1\,\text{GW m}^{-2}, \quad |\mathcal{A}_n| \lesssim 10^{-3}\,\text{J}^{1/2}\,\text{m}^{-1}, \quad \kappa \lesssim 10^5\,\text{J}^{-1/2}, \quad |\kappa\,\mathcal{A}_n| \lesssim 1\,\text{cm}^{-1}.$$

$$(10.49)$$

We use the evolution equations (10.47) and (10.48) in Sec. 10.7 to explore several applications of three-wave mixing.

One can reformulate the equations of three-wave mixing in fully quantum mechanical language, with a focus on the mean occupation numbers of the wave modes. This is commonly done in plasma physics; in Sec. 23.3.6 we discuss the example of coupled electrostatic waves in a plasma.

Exercise 10.9 **Example: Dispersion Relation for an Anisotropic Medium** T2

Consider a wave propagating through a dielectric medium that is anisotropic, but not necessarily—for the moment—axisymmetric. Let the wave be sufficiently weak that nonlinear effects are unimportant. Then the nonlinear wave equation (10.22a) takes the linear form

$$-\nabla^2 \mathbf{E} + \nabla(\nabla \cdot \mathbf{E}) = -\frac{1}{c^2}\boldsymbol{\epsilon} \cdot \frac{\partial^2 \mathbf{E}}{\partial t^2}. \qquad (10.50)$$

(a) Specialize to a monochromatic plane wave with angular frequency ω and wave vector $\mathbf{k}$. Show that the wave equation (10.50) reduces to

$$L_{ij}E_j = 0, \quad \text{where } L_{ij} = k_i k_j - k^2 \delta_{ij} + \frac{\omega^2}{c^2}\epsilon_{ij}. \qquad (10.51a)$$

This equation says that E_j is an eigenvector of L_{ij} with vanishing eigenvalue, which is possible if and only if

$$\det ||L_{ij}|| = 0. \qquad (10.51b)$$

This vanishing determinant is the waves' dispersion relation. We use it in Chap. 21 to study waves in plasmas.

(b) Next specialize to an axisymmetric medium, and orient the symmetry axis along the z direction, so the only nonvanishing components of the dielectric tensor ϵ_{ij} are $\epsilon_{11} = \epsilon_{22}$ and ϵ_{33}. Let the wave propagate in a direction $\hat{\mathbf{k}}$ that makes an angle θ to the symmetry axis. Show that in this case L_{ij} has the form

$$||L_{ij}|| = k^2 \begin{Vmatrix} (n_o/n)^2 - \cos^2\theta & 0 & \sin\theta\cos\theta \\ 0 & (n_o/n)^2 - 1 & 0 \\ \sin\theta\cos\theta & 0 & (n_e/n)^2 - \sin^2\theta \end{Vmatrix}, \qquad (10.52a)$$

and the dispersion relation (10.51b) reduces to

$$\left(\frac{1}{n^2} - \frac{1}{n_o^2}\right)\left(\frac{1}{n^2} - \frac{\cos^2\theta}{n_o^2} - \frac{\sin^2\theta}{n_e^2}\right) = 0, \qquad (10.52b)$$

where $1/n = \omega/kc$, $n_o = \sqrt{\epsilon_{11}} = \sqrt{\epsilon_{22}}$, and $n_e = \sqrt{\epsilon_{33}}$, in accord with Eq. (10.39).

(c) Show that this dispersion relation has the two solutions (ordinary and extraordinary) discussed in the text, Eqs. (10.40a) and (10.40b), and show that the electric fields associated with these two solutions point in the directions described in the text.

Exercise 10.10 **Derivation and Example: Evolution Equations for Realistic Wave-Wave Mixing** T2

Derive the evolution equations (10.48) for three-wave mixing. [The derivation of those (10.47) for frequency doubling is similar.] You could proceed as follows.

(a) Insert expressions (10.27) and (10.28) into the general wave equation (10.30) and extract the portions with frequency $\omega_3 = \omega_1 + \omega_2$, thereby obtaining the generalization of Eq. (10.33):

$$\nabla^2 \left(A_i^{(3)} e^{i(k_3 z - \omega_3 t)} \right) - \frac{\partial^2}{\partial x^i \partial x^j} \left(A_j^{(3)} e^{i(k_3 z - \omega_3 t)} \right) + \frac{\omega_3^2}{c^2} \epsilon_{ij} A_j^{(3)} e^{i(k_3 z - \omega_3 t)}$$

$$= -2 \frac{\omega_3^2}{c^2} d_{ijk} A_j^{(1)} A_k^{(2)} e^{i k_3 z - \omega_3 t}. \tag{10.53}$$

(b) Explain why $e^{i(\mathbf{k}_3 \cdot \mathbf{x} - \omega_3 t)} \mathbf{f}^{(3)}$ satisfies the homogeneous wave equation (10.50). Then, splitting each wave into its scalar field and polarization vector, $\mathcal{A}_i^{(n)} \equiv \mathcal{A}^{(n)} f_i^{(n)}$, and letting each $\mathcal{A}^{(n)}$ be a function of z (because of the boundary condition that the three-wave mixing begins at the crystal face $z = 0$), show that Eq. (10.53) reduces to

$$\alpha_3 d\mathcal{A}^{(3)}/dz = i(k_3/\mathfrak{n}_3^2) d_{ijk} f_j^{(1)} f_k^{(2)} \mathcal{A}^{(1)} \mathcal{A}^{(2)},$$

where α_3 is a constant of order unity that depends on the relative orientations of the unit vectors $\mathbf{e}_z$, $\mathbf{f}^{(3)}$, and $\hat{\mathbf{k}}_3$. Note that, aside from α_3, this is the same evolution equation as for our idealized isotropic, dispersion-free medium, Eq. (10.34a). Show that, similarly, $\mathcal{A}^{(1)}(z)$ and $\mathcal{A}^{(2)}(z)$ satisfy the same equations (10.34b) and (10.34c) as in the idealized case, aside from multiplicative constants α_1 and α_2.

(c) Adopting the renormalizations $\mathcal{A}^{(n)} = \zeta_n \sqrt{2\omega_n/(\epsilon_0 c \, \mathfrak{n}_n)} \, \mathfrak{A}_n$ [Eq. (10.45)] with ζ_n so chosen that the photon-number flux for wave n is proportional to $|\mathfrak{A}_n|^2$, show that your evolution equations for $\mathcal{A}_n$ become Eqs. (10.48), except that the factor β' and thence the value of κ might be different for each equation.

(d) Since the evolution entails one photon with frequency ω_1 and one with frequency ω_2 annihilating to produce a photon with frequency ω_3, it must be that $d|\mathfrak{A}_1|^3/dz = d|\mathfrak{A}_2|^3/dz = -d|\mathfrak{A}_3|^2/dz$. (These are called *Manley-Rowe relations*.) By imposing this on your evolution equations in part (c), deduce that all three coupling constants κ must be the same, and thence also that all three β' must be the same; therefore the evolution equations take precisely the claimed form, Eqs. (10.48).

Exercise 10.11 **Derivation: Magnitudes of Three-Wave-Mixing Quantities**
Derive Eqs. (10.49). [Hint: The maximum energy flux in a wave arises from the limit $E \lesssim 10^6$ V m^{-1} on the wave's electric field to ensure that it not pull electrons out of the surface of the nonlinear medium. The maximum coupling constant κ arises from the largest values $|d_{ijk}| \lesssim 10$ pm V^{-1} for materials typically used in three-wave mixing (Box 10.2).]

10.7 Applications of Three-Wave Mixing: Frequency Doubling, Optical Parametric Amplification, and Squeezed Light

10.7.1 Frequency Doubling

Frequency doubling (also called *second harmonic generation*) is one of the most important applications of wave-wave mixing. As we have already discussed briefly in Secs. 10.6.2 (Track One) and 10.6.3 (Track Two), it can be achieved by passing a single wave (which plays the role of both wave $n=1$ and wave $n=2$) through a nonlinear crystal, with the propagation direction chosen to satisfy the resonance conditions. As we have also seen in the previous section (Track Two), the crystal's birefringence and dispersion have little influence on the growth of the output wave, $n=3$ with $\omega_3 = 2\omega_1$; it grows with distance inside the crystal at a rate given by Eqs. (10.47), which is the same as in the Track-One case of a medium that is isotropic at linear order, Eqs. (10.38), aside from the factor β of order unity in the coupling constant κ. By doing a sufficiently good job of phase matching (satisfying the resonance conditions) and choosing the thickness of the crystal appropriately, one can achieve close to 100% conversion of the input-wave energy into frequency-doubled energy. More specifically, if wave 1 enters the crystal at $z=0$ with $\mathfrak{A}_1(0) = \mathfrak{A}_{1o}$, which we choose (without loss of generality) to be real, and if there is no incoming wave 3 so $\mathfrak{A}_3(0) = 0$, then the solution to the evolution equations (10.47) or (10.38) is

second harmonic generation (frequency doubling)

efficiency of frequency doubling

$$\mathfrak{A}_3 = \frac{i}{\sqrt{2}} \mathfrak{A}_{1o} \tanh\left(\frac{\kappa}{\sqrt{2}} \mathfrak{A}_{1o} z\right), \quad \mathfrak{A}_1 = \mathfrak{A}_{1o} \operatorname{sech}\left(\frac{\kappa}{\sqrt{2}} \mathfrak{A}_{1o} z\right). \quad (10.54)$$

wave amplitude evolution

It is easy to see that this solution has the following properties. (i) It satisfies energy conservation, $2|\mathfrak{A}_3|^2 + |\mathfrak{A}_1|^2 = |\mathfrak{A}_{1o}|^2$. (ii) At a depth $z = 1.246/(\kappa \mathfrak{A}_{1o})$ in the crystal, half the initial energy has been frequency doubled. (iii) As z increases beyond this half-doubling depth, the light asymptotes to fully frequency doubled.

One might expect the frequency doubling to proceed onward to $4\omega_1$, and so forth. However, it typically does not, because these higher-frequency waves typically fail to satisfy the crystal's dispersion relation.

As an example, the neodymium:YAG (Nd^{3+}:YAG) laser, which is based on an yttrium-aluminum-garnet crystal with trivalent neodymium impurities, is among the most attractive of all lasers for a combination of high frequency stability, moderately high power, and high efficiency. However, this laser operates in the infrared, at a wavelength of 1.064 microns. For many purposes, one wants optical light. This can be achieved by frequency doubling the laser's output, thereby obtaining 0.532-micron (green) light. This is how green laser pointers, used in lecturing, work (though in 2016 they are typically driven not by Nd:YAG but rather a relative; see Ex. 10.13).

Nd: YAG laser

Frequency doubling also plays a key role in laser fusion, where intense, pulsed laser beams, focused on a pellet of fusion fuel, compress and heat the pellet to high densities and temperatures. Because the beam's energy flux is inversely proportional to the area of its focused cross section—and because the larger the wavelength, the more seriously diffraction impedes making the cross section small—in order to achieve efficient compression of the pellet, it is important to give the beam a very short wavelength. This is achieved by multiple frequency doublings, which can and do occur in the experimental setup of laser fusion.

EXERCISES

Exercise 10.12 *Derivation: Saturation in Frequency Doubling*
Derive the solution (10.54) to the evolution equations (10.47) for frequency doubling, and verify that it has the claimed properties.

Exercise 10.13 **Example: Frequency Doubling in a Green Laser Pointer*
Green laser pointers, popular in 2016, have the structure shown in Fig. 10.13. A battery-driven infrared diode laser puts out 808-nm light that pumps a Nd:YVO$_4$ laser crystal (neodymium-doped yttrium vanadate; a relative of Nd:YAG). The 1,064-nm light beam from this Nd:YVO$_4$ laser is frequency doubled by a KTP crystal, resulting in 532-nm green light. An infrared filter removes all the 880-nm and 1,064-nm light from the output, leaving only the green.

(a) To make the frequency doubling as efficient as possible, the light is focused to as small a beam radius ϖ_o as diffraction allows as it travels through the KTP crystal. Assuming that the crystal length is $L \simeq 3$ mm, show that $\varpi_o \simeq \sqrt{\lambda L/(4\pi n_1)} \simeq$

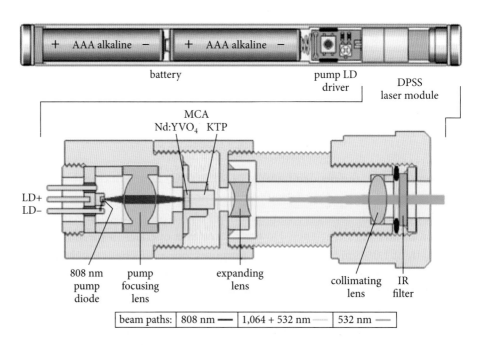

FIGURE 10.13 Structure of a green laser pointer, circa 2012. Adapted with minor changes from a drawing copyright by Samuel M. Goldwasser (Sam's Laser FAQ at http://www.repairfaq.org/lasersam.htm), and printed here with his permission.

12 μm (about 12 times larger than the 1,064-nm wavelength). [Hint: Use the properties of Gaussian beams; Sec. 8.5.5 adjusted for propagation in a medium with index of refraction $\mathfrak{n}_1$.]

(b) The 1,064-nm beam has an input power $W_{1o} \simeq 100$ mW as it enters the KTP crystal. Show that its energy flux and its electric field strength are $F \simeq 230$ MW m^{-2} and $\mathcal{A}^{(1)} \simeq 400$ kV m^{-1}.

(c) Assuming that phase matching has been carried out successfully (i.e., photon energy and momentum conservation have been enforced), explain why it is reasonable to expect the quantity $\beta d_{ijk} f_i^{(1)} f_j^{(1)} f_k^{(3)}$ in the coupling constant κ to be roughly 4 pm/V [cf. Eq. (6) of Box 10.2]. Then show that the green output beam at the end of the KTP crystal has $|\mathfrak{A}_3|^2 \sim 0.7 \times 10^{-4} \mathfrak{A}_{1o}^2$, corresponding to an output power $W_3 \sim 1.5 \times 10^{-4} W_{1o} \simeq 0.015$ mW. This is far below the output power, 5 mW, of typical green laser pointers. How do you think the output power is boosted by a factor $\sim 5/0.015 \simeq 300$?

(d) The answer is (i) to put reflective coatings on the two ends of the KTP crystal so it becomes a Fabry-Perot resonator for the 1.064 μm input field; and also (ii) make the input face (but not the output face) reflective for the 0.532 μm green light. Show that, if the 1.064 μm resonator has a finesse $\mathcal{F} \simeq 30$, then the green-light output power will be increased to $\simeq 5$ mW.

(e) Explain why this strategy makes the output power sensitive to the temperature of the KTP crystal. To minimize this sensitivity, the crystal is oriented so that its input and output faces are orthogonal to its (approximate) symmetry axis—the z-axis—for which the thermal expansion coefficient is very small (0.6×10^{-6}/C, by contrast with $\simeq 10 \times 10^{-6}$/C along other axes. Show that, in this case, a temperature increase or reduction of 6 C from the pointer's optimal 22 C (room temperature) will reduce the output power from 5 mW to much less than 1 mW. Astronomers complain that green laser pointers stop working outdoors on cool evenings.

10.7.2 Optical Parametric Amplification

In optical parametric amplification, the energy of a *pump wave* is used to amplify an initially weak *signal wave* and also amplify an uninteresting *idler wave*. The waves satisfy the resonance conditions with $\omega_p = \omega_s + \omega_i$. The pump wave and signal wave are fed into an anisotropic nonlinear crystal, propagating in (nearly) the same direction, with nonzero renormalized amplitudes $\mathfrak{A}_{po}$ and $\mathfrak{A}_{so}$ at $z = 0$. The idler wave has $\mathfrak{A}_{io} = 0$ at the entry plane. Because the pump wave is so strong, it is negligibly influenced by the three-wave mixing (i.e., $\mathfrak{A}_p$ remains constant inside the crystal).

The evolution equations for the (renormalized) signal and idler amplitudes are

$$\frac{d\mathfrak{A}_s}{dz} = i\kappa \mathfrak{A}_p \mathfrak{A}_i^*, \qquad \frac{d\mathfrak{A}_i}{dz} = i\kappa \mathfrak{A}_p \mathfrak{A}_s^* \qquad (10.55)$$

[Eqs. (10.48) or (10.37)]. For the initial conditions of weak signal wave and no idler wave, the solution to these equations is

wave amplitude evolution

$$\mathfrak{A}_s = \mathfrak{A}_{so} \cosh(|\gamma|z), \qquad \mathfrak{A}_i = \frac{\gamma}{|\gamma|} \mathfrak{A}_{so}^* \sinh(|\gamma|z); \qquad \gamma \equiv i\kappa \mathfrak{A}_p. \quad (10.56)$$

Thus the signal field grows exponentially, after an initial pause, with an e-folding length $1/|\gamma|$, which for strong three-wave nonlinearities is of order 1 cm [Ex. (10.14)].

EXERCISES

Exercise 10.14 *Derivation: e-Folding Length for an Optical Parametric Amplifier*
Estimate the magnitude of the e-folding length for an optical parametric amplifier that is based on a strong three-wave nonlinearity.

10.7.3 Degenerate Optical Parametric Amplification: Squeezed Light

degenerate optical parametric amplification

Consider optical parametric amplification with the signal and idler frequencies identical, so the idler field is the same as the signal field, and the pump frequency is twice the signal frequency: $\omega_p = 2\omega_s$. This condition is called *degenerate*. Adjust the phase of the pump field so that $\gamma = i\kappa \mathfrak{A}_p$ is real and positive. Then the equation of evolution for the signal field is the same as appears in frequency doubling [Eqs. (10.47) or (10.38)]:

$$d\mathfrak{A}_s/dz = \gamma \mathfrak{A}_s^*. \quad (10.57)$$

The resulting evolution is most clearly understood by decomposing $\mathfrak{A}_s$ into its real and imaginary parts (as we did in Ex. 6.23 when studying thermal noise in an oscillator): $\mathfrak{A}_s = X_1 + iX_2$. Then the time-evolving electric field is

$$E \propto \Re(\mathfrak{A}_s e^{i(k_s z - \omega_s t)}) = X_1 \cos(k_s z - \omega_s t) + X_2 \sin(k_s z - \omega_s t). \quad (10.58)$$

Therefore, X_1 is the amplitude of the field's cosine quadrature, and X_2 is the amplitude of its sine quadrature. Equation (10.57) then says that $dX_1/dz = \gamma X_1$, $dX_2/dz = -\gamma X_2$, so we have

$$X_1 = X_{1o} e^{\gamma z}, \qquad X_2 = X_{2o} e^{-\gamma z}. \quad (10.59)$$

squeezing

Therefore, the wave's cosine quadrature gets amplified as the wave propagates, and its sine quadrature is attenuated. This is called *squeezing*, because X_2 is reduced (squeezed) while X_1 is increased. It is a phenomenon known to children who swing; see Ex. 10.15.

Squeezing is especially interesting when it is applied to noise. Typically, a wave has equal amounts of noise in its two quadratures (i.e., the standard deviations ΔX_1 and ΔX_2 of the two quadratures are equal, as was the case in Ex. 6.23). When such a wave is squeezed, its two standard deviations get altered in such a way that their product is unchanged:

$$\boxed{\Delta X_1 = \Delta X_{1o}\, e^{\gamma z}; \qquad \Delta X_2 = \Delta X_{2o}\, e^{-\gamma z}, \qquad \Delta X_1 \Delta X_2 = \text{const}} \quad (10.60)$$

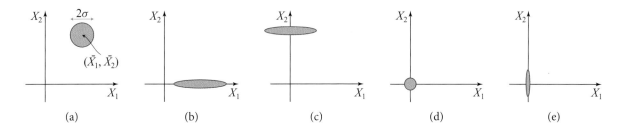

FIGURE 10.14 Error boxes in the complex amplitude plane for several different electromagnetic waves: (a) Classical light. (b) Phase-squeezed light. (c) Amplitude-squeezed light. (d) The quantum electrodynamical vacuum. (e) The squeezed vacuum.

(see Fig. 10.14). When, as here, the standard deviations of two quadratures differ, the light is said to be in a *squeezed state*.

In quantum theory, X_1 and X_2 are complementary observables; they are described by Hermitian operators that do not commute. The uncertainty principle associated with their noncommutation implies that their product $\Delta X_1 \Delta X_2$ has some minimum possible value. This minimum is achieved by the wave's vacuum state, which has $\Delta X_1 = \Delta X_2$ with values corresponding to a half quantum of energy (vacuum fluctuations) in the field mode that we are studying. When this "quantum electrodynamic vacuum" is fed into a degenerate optical parametric amplifier, the vacuum noise gets squeezed in the same manner [Eq. (10.59)] as any other noise.

Squeezed states of light, including this *squeezed vacuum*, have great value for fundamental physics experiments and technology. Most importantly, they can be used to reduce the photon shot noise of an interferometer below the standard quantum limit of $\Delta N = \sqrt{N}$ (Poisson statistics), thereby improving the signal-to-noise ratio in certain communications devices and in laser interferometer gravitational-wave detectors such as LIGO (Caves, 1981; McClelland et al., 2011; Oelker et al., 2016). We explore some properties of squeezed light in Ex. 10.16.

squeezed state of light

squeezed vacuum

EXERCISES

Exercise 10.15 **Example: Squeezing by Children Who Swing*
A child, standing in a swing, bends her knees and then straightens them twice per swing period, making the distance ℓ from the swing's support to her center of mass oscillate as $\ell = \ell_0 + \ell_1 \sin 2\omega_0 t$. Here $\omega_0 = \sqrt{g\ell_0}$ is the swing's mean angular frequency.

(a) Show that the swing's angular displacement from vertical, θ, obeys the equation of motion

$$\frac{d^2\theta}{dt^2} + \omega_0^2 \theta = -\omega_1^2 \sin(2\omega_0 t)\theta, \quad (10.61)$$

where $\omega_1 = \sqrt{g\ell_1}$, and θ is assumed to be small, $\theta \ll 1$.

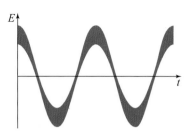

FIGURE 10.15 The error band for the electric field $E(t)$, as measured at a fixed location in space, when phase-squeezed light passes by.

(b) Write $\theta = X_1 \cos \omega_0 t + X_2 \sin \omega_0 t$. Assuming that $\ell_1 \ll \ell_0$ so $\omega_1 \ll \omega_0$, show that the child's knee bending (her "pumping" the swing) squeezes X_1 and amplifies X_2 (parametric amplification):

$$X_1(t) = X_1(0)e^{-[\omega_1^2/(2\omega_0)]t}, \qquad X_2(t) = X_2(0)e^{+[\omega_1^2/(2\omega_0)]t} \qquad (10.62)$$

(c) Explain how this squeezing is related to the child's conscious manipulation of the swing (i.e., to her strategy for increasing the swing's amplitude when she starts up, and her strategy for reducing the amplitude when she wants to quit swinging).

Exercise 10.16 **Example: Squeezed States of Light*

Consider a plane, monochromatic electromagnetic wave with angular frequency ω, whose electric field is expressed in terms of its complex amplitude $X_1 + iX$ by Eq. (10.58). Because the field (inevitably) is noisy, its quadrature amplitudes X_1 and X_2 are random processes with means $\bar{X}_1, \bar{X}_2$ and variances $\Delta X_1, \Delta X_2$.

(a) Normal, classical light has equal amounts of noise in its two quadratures. Explain why it can be represented by Fig. 10.14a.

(b) Explain why Fig. 10.14b represents *phase-squeezed light,* and show that its electric field as a function of time has the form shown in Fig. 10.15.

(c) Explain why Fig. 10.14c represents *amplitude-squeezed light*, and construct a diagram of its electric field as a function of time analogous to Fig. 10.15.

(d) Figure 10.14d represents the vacuum state of light's frequency-ω plane-wave mode. Give a formula for the diameter of the mode's circular error box. Construct a diagram of the electric field as a function of time analogous to Fig. 10.15.

(e) Figure 10.14e represents the squeezed vacuum. Construct a diagram of its electric field as a function of time analogous to Fig. 10.15.

10.8 Four-Wave Mixing in Isotropic Media

10.8.1 Third-Order Susceptibilities and Field Strengths

The nonlinear polarization for four-wave mixing, $P_i^{(4)} = 4\epsilon_0 \chi_{ijkl} E_j E_k E_l$ [Eq. (10.21)], is typically smaller than that, $P_i^{(3)} = 2\epsilon_0 d_{ijk} E_j E_k$, for three-wave mix-

TABLE 10.1: Materials used in four-wave mixing

Material	Wavelength (μm)	n	χ_{1111} (pm^2 V^{-2})	n_2 (10^{-20} m^2 W^{-1})
Fused silica	0.694	1.455	56	3
SF$_6$ glass	1.06	1.77	590	21
CS$_2$ liquid	1.06	1.594	6,400	290
2-methyl-4-nitroaniline (MNA) organic crystal[a]		1.8	1.7×10^5	5,800
PTS polydiacetylene polymeric crystal[a]		1.88	5.5×10^5	1.8×10^4

a. Also has large d_{ijk}.

Notes: At the indicated light wavelength, n is the index of refraction, χ_{1111} is the third-order nonlinear susceptibility, and n_2 is the Kerr coefficient of Eq. (10.69). Adapted from Yariv and Yeh (2007, Table 8.8), whose χ_{1111} is $1/\epsilon_0$ times ours.

ing by $\sim E|\chi/d| \sim (10^6 \text{ V m}^{-1})(100 \text{ pm V}^{-1}) \sim 10^{-4}$. (Here we have used the largest electric field that solids typically can support.) Therefore (as we have already discussed), only when d_{ijk} is greatly suppressed by isotropy of the nonlinear material does χ_{ijkl} and four-wave mixing become the dominant nonlinearity. And in that case, we expect the propagation lengthscale for strong, cumulative four-wave mixing to be $\sim 10^4$ larger than that (~ 1 cm) for the strongest three-wave mixing (i.e., $\ell_{4w} \gtrsim 100$ m).

order of magnitude estimate for strength of four-wave mixing

In reality, as we shall see in Sec. 10.8.2, this estimate is overly pessimistic. In special materials, ℓ_{4w} can be less than a meter (though still much bigger than the three-wave mixing's centimeter). Two factors enable this. (i) If the nonlinear material is a fluid (e.g., CS$_2$) confined by solid walls, then it can support somewhat larger electric field strengths than a nonlinear crystal's maximum, 10^6 V m^{-1}. (ii) If the nonlinear material is made of molecules significantly larger than 10^{-10} m (e.g., organic molecules), then the molecular electric dipoles induced by an electric field can be significantly larger than our estimates (10.26); correspondingly, $|\chi_{ijkl}|$ can significantly exceed $(100 \text{ pm V}^{-1})^2$; see Table 10.1.

In Secs. 10.8.2 and 10.8.3, we give (i) an example with strong four-wave mixing: phase conjugation by a half-meter-long cell containing CS$_2$ liquid and then (ii) an example with weak but important four-wave mixing: light propagation in a multikilometer-long fused-silica optical fiber.

10.8.2 Phase Conjugation via Four-Wave Mixing in CS$_2$ Fluid

phase conjugation by four-wave mixing

As an example of four-wave mixing, we discuss phase conjugation in a rectangular cell that contains carbon disulfide (CS$_2$) liquid (Fig. 10.16a). The fluid is pumped by two

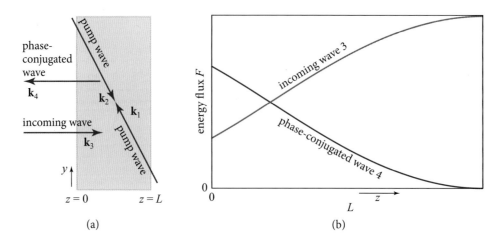

FIGURE 10.16 (a) A phase-conjugating mirror based on four-wave mixing. (b) The evolution of the incoming wave's flux and the phase-conjugated wave's flux inside the mirror (the nonlinear medium).

strong waves, 1 and 2, propagating in opposite directions with the same frequency as the incoming wave 3 that is to be phase conjugated. The pump waves are planar without modulation, but wave 3 has a spatial modulation (slow compared to the wave number) that carries, for example, a picture; $\mathcal{A}_3 = \mathcal{A}_3(x, y; z)$. As we shall see, nonlinear interaction of the two pump waves 1 and 2 and the incoming wave 3 produces outgoing wave 4, which is the phase conjugate of wave 3. All four waves propagate in planes of constant x and have their electric fields in the x direction, so the relevant component of the third-order nonlinearity is $\chi_{xxxx} = \chi_{1111}$.

resonance conditions for four-wave mixing

The resonance conditions (photon energy and momentum conservation) for this four-wave mixing process are $\omega_4 = \omega_1 + \omega_2 - \omega_3$ and $\mathbf{k}_4 = \mathbf{k}_1 + \mathbf{k}_2 - \mathbf{k}_3$. Since the three input waves all have the same frequency, $\omega_1 = \omega_2 = \omega_3 = \omega$, the output wave 4 will also have $\omega_4 = \omega$, so this is *fully degenerate four-wave mixing*. The pump waves propagate in opposite directions, so they satisfy $\mathbf{k}_1 = -\mathbf{k}_2$, whence the output wave has $\mathbf{k}_4 = -\mathbf{k}_3$. That is, it propagates in the opposite direction to the input wave 3 and has the same frequency, as it must, if it is to be (as we claim) the phase conjugate of 3.

The nonlinear polarization that generates wave 4 is

$$P_x^{(4)} = 4\epsilon_0 \chi_{1111}(E_x^{(1)} E_x^{(2)} E_x^{(3)} + E_x^{(2)} E_x^{(3)} E_x^{(1)} + \ldots).$$

There are six terms in the sum (six ways to order the three waves), so

$$P_x^{(4)} = 24\epsilon_0 \chi_{1111} E_x^{(1)} E_x^{(2)} E_x^{(3)}.$$

Inserting

$$E_x^{(n)} = \frac{1}{2}(\mathcal{A}^{(n)} e^{i(\mathbf{k}_n \cdot \mathbf{x} - \omega_n t)} + \mathcal{A}^{(n)*} e^{i(-\mathbf{k}_n + \omega_n t)})$$

into this $P_x^{(4)}$ and plucking out the relevant term for our phase-conjugation process (with the signal wave 3 phase conjugated and the pump waves not), we obtain

$$P_x^{(4)} = 3\epsilon_0 \chi_{1111} \mathcal{A}^{(1)} \mathcal{A}^{(2)} \mathcal{A}^{(3)*} e^{i(\mathbf{k}_4 \cdot \mathbf{x} - \omega_4 t)}. \tag{10.63}$$

Inserting this expression into the wave equation for wave 4 in an isotropic medium, $(\nabla^2 + \mathrm{n}^2 \omega_4^2/c^2)\,(\mathcal{A}^{(4)} e^{i(\mathbf{k}_4 \cdot \mathbf{x} - \omega_4 t)}) = -(\omega_4^2/c^2) P_x^{(4)}$ [analog of the isotropic-medium wave equation (10.33) for three-wave mixing] and making use of the fact that the lengthscale on which wave 4 changes is long compared to its wavelength, we obtain the following evolution equation for wave 4, which we augment with that for wave 3, obtained in the same way:

$$\frac{d\mathcal{A}^{(4)}}{dz} = -\frac{3ik}{\mathrm{n}^2} \chi_{1111} \mathcal{A}^{(1)} \mathcal{A}^{(2)} \mathcal{A}^{(3)*}, \quad \frac{d\mathcal{A}^{(3)}}{dz} = -\frac{3ik}{\mathrm{n}^2} \chi_{1111} \mathcal{A}^{(1)} \mathcal{A}^{(2)} \mathcal{A}^{(4)*}. \tag{10.64}$$

Here we have dropped subscripts from k, since all waves have the same scalar wave number, and we have used the dispersion relation $\omega/k = c/\mathrm{n}$ common to all four waves (since they all have the same frequency). We have not written down the evolution equations for the pump waves, because in practice they are very much stronger than the incoming and phase-conjugated waves, so they change hardly at all during the evolution.

It is convenient to change normalizations of the wave fields, as in three-wave mixing, from $\mathcal{A}^{(n)}$ (the electric field) to $\mathfrak{A}_n$ (the square root of energy flux divided by frequency, $\mathfrak{A}_n = \sqrt{F_n/\omega_n}$):

$$\mathcal{A}^{(n)} = \sqrt{\frac{2k}{\epsilon_0 \mathrm{n}^2}} \mathfrak{A}_n = \sqrt{\frac{2\omega_n}{\epsilon_0 c \mathrm{n}}} \mathfrak{A}_n \tag{10.65}$$

renormalized wave amplitudes

[Eq. (10.35)]. Inserting these into the evolution equations (10.64) and absorbing the constant pump-wave amplitudes into a coupling constant, we obtain

$$\boxed{\frac{d\mathfrak{A}_4}{dz} = -i\kappa \mathfrak{A}_3^*, \quad \frac{d\mathfrak{A}_3}{dz} = i\kappa \mathfrak{A}_4^*; \quad \kappa = \frac{6\omega^2}{c^2 \mathrm{n}^2 \epsilon_0} \chi_{1111} \mathfrak{A}_1 \mathfrak{A}_2.} \tag{10.66}$$

evolution equations for fully degenerate four-wave mixing in an isotropic medium

Equations (10.66) are our final, simple equations for the evolution of the input and phase-conjugate waves in our isotropic, nonlinear medium (CS_2 liquid). Inserting the index of refraction $\mathrm{n} = 1.594$ and nonlinear susceptibility $\chi_{1111} = 6{,}400$ (pm/V)^2 for CS_2 (Table 10.1), the angular frequency corresponding to 1.064 μm wavelength light from, say, a Nd:YAG laser, and letting both pump waves $n = 1$ and 2 have energy fluxes $F_n = \omega |\mathfrak{A}_n|^2 = 5 \times 10^{10}$ W m^{-2} (corresponding to electric field amplitudes 6.1×10^6 V m^{-1}, six times larger than good nonlinear crystals can support), we obtain for the magnitude of the coupling constant $|\kappa| = 1/0.59$ m^{-1}. Thus, the CS_2 cell of Fig. 10.16a need only be a half meter thick to produce strong phase conjugation.

For an input wave $\mathfrak{A}_{3_0}(x, y)$ at the cell's front face $z = 0$ and no input wave 4 at $z = L$, the solution to the evolution equations (10.66) is easily found to be, in the interior of the cell:

$$\mathfrak{A}_4(x, y, z) = \frac{-i\kappa}{|\kappa|} \left(\frac{\sin[|\kappa|(z-L)]}{\cos[|\kappa|L]} \right) \mathfrak{A}_{3_0}^*(x, y),$$

$$\mathfrak{A}_3(x, y, z) = \left(\frac{\cos[|\kappa|(z-L)]}{\cos[|\kappa|L]} \right) \mathfrak{A}_{3_0}(x, y). \tag{10.67}$$

The corresponding energy fluxes, $F_n = \omega|\mathfrak{A}_n|^2$ are plotted in Fig. 10.16b, for a crystal thickness $L = 1/|\kappa| = 0.59$ m. Notice that the pump waves amplify the rightward propagating input wave, so it grows from the crystal front to the crystal back. At the same time, the interaction of the input wave with the pump waves generates the leftward propagating phase-conjugated wave, which begins with zero strength at the back of the crystal and grows (when $L \sim 1/|\kappa|$) to be stronger than the input wave at the crystal front.

EXERCISES

Exercise 10.17 **Problem: Photon Creation and Annihilation in a Phase-Conjugating Mirror*
Describe the creation and annihilation of photons that underlies a phase-conjugating mirror's four-wave mixing. Specifically, how many photons of each wave are created or annihilated? [Hint: See the discussion of photon creation and annihilation for three-wave mixing at the end of Sec. 10.6.1.]

Exercise 10.18 **Problem: Spontaneous Oscillation in Four-Wave Mixing*
Suppose the thickness of the nonlinear medium of the text's four-wave mixing analysis is $L = \pi/(2\kappa)$, so the denominators in Eqs. (10.67) are zero. Explain the physical nature of the resulting evolution of waves 3 and 4.

Exercise 10.19 *Problem: Squeezed Light Produced by Phase Conjugation*
Suppose a light beam is split in two by a beam splitter. One beam is reflected off an ordinary mirror and the other off a phase-conjugating mirror. The beams are then recombined at the beam splitter. Suppose that the powers returning to the beam splitter are nearly the same; they differ by a fractional amount $\Delta P/P = \epsilon \ll 1$. Show that the recombined light is in a strongly squeezed state, and discuss how one can guarantee it is phase squeezed, or (if one prefers) amplitude squeezed.

10.8.3 Optical Kerr Effect and Four-Wave Mixing in an Optical Fiber

Suppose that an isotropic, nonlinear medium is driven by a single input plane wave polarized in the x direction, $E_x = \Re[Ae^{-(kz-\omega t)}]$. This input wave produces the following polarization that propagates along with itself in resonance (Ex. 10.20):

$$P_x = \epsilon_0 \chi_0 E_x + 6\epsilon_0 \chi_{1111} \overline{E^2} E_x. \tag{10.68}$$

Here the second term is due to four-wave mixing, and $\overline{E^2}$ is the time average of the square of the electric field, which can be expressed in terms of the energy flux as $\overline{E^2} = F/(\epsilon_0 \mathfrak{n} c)$. The four-wave-mixing term can be regarded as a nonlinear correction to χ_0: $\Delta \chi_0 = 6\chi_{1111}\overline{E^2} = [6\chi_{1111}/(\mathfrak{n} c \epsilon_0)]F$. Since the index of refraction is $\mathfrak{n} = \sqrt{1 + \chi_0}$, this corresponds to a fractional change of index of refraction given by

$$\Delta \mathfrak{n} = n_2 F, \quad \text{where } n_2 = \frac{3\chi_{1111}}{\mathfrak{n}^2 c \epsilon_0}. \tag{10.69}$$

This nonlinear change of $\mathfrak{n}$ is called the *optical Kerr effect*, and the coefficient n_2 is the *Kerr coefficient* and has dimensions of 1/(energy flux), or m² W⁻¹. Values for n_2 for several materials are listed in Table 10.1.

optical Kerr effect; Kerr coefficient

We have already briefly discussed an important application of the optical Kerr effect: the self-focusing of a laser beam, which plays a key role in mode locked lasers (Sec. 10.2.3) and also in laser fusion.

The optical Kerr effect is also important in the optical fibers used in modern communication (e.g., to carry telephone, television, and internet signals to your home). Such fibers are generally designed to support just one spatial mode of propagation: the fundamental Gaussian mode of Sec. 8.5.5 or some analog of it. Their light-carrying cores are typically made from fused silica doped with particular impurities, so their Kerr coefficients are $n_2 \simeq 3 \times 10^{-20}$ m² W⁻¹ (Table 10.1). Although the fibers are not spatially homogeneous and the wave is not planar, one can show (and it should not be surprising) that the fibers nonetheless exhibit the optical Kerr effect, with $\Delta \mathfrak{n} = n_2 F_{\text{eff}}$. Here F_{eff}, the *effective energy flux*, is the light beam's power P divided by an effective cross sectional area, $\pi \sigma_0^2$, with σ_0 the Gaussian beam's radius, defined in Eq. (8.39): $F_{\text{eff}} = P/\pi \sigma_0^2$; see Yariv and Yeh (2007, Sec. 14.1).

As a realistic indication of the importance of the optical Kerr effect in communication fibers, consider a signal beam with mean power $P = 10$ mW and a beam radius $\sigma_0 = 5\ \mu$m and thence an effective energy flux $F_{\text{eff}} = 127$ MW m⁻². If the wavelength is $2\pi/k = 0.694\ \mu$m, then Table 10.1 gives $\mathfrak{n} = 1.455$ and $n_2 = 3 \times 10^{-20}$ m² W⁻¹. When this beam travels a distance $L = 50$ km along the fiber, its light experiences a phase shift

$$\Delta \phi = \frac{\Delta n}{n} kL = \frac{n_2}{\mathfrak{n}} FkL \simeq 1.2 \text{ rad}. \tag{10.70}$$

A phase shift of this size or larger can cause significant problems for optical communication. For example:

1. Variations of the flux, when pulsed signals are being transmitted, cause time-varying phase shifts that modify the signals' phase evolution (*self-phase modulation*). One consequence of this is the broadening of each pulse; another is a nonlinearly induced chirping of each pulse (slightly lower frequency at beginning and higher at end).

2. Fibers generally carry many channels of communication with slightly different carrier frequencies, and the time-varying flux of one channel can modify the phase evolution of another (*cross-phase modulation*).

Various techniques have been developed to deal with these issues (see, e.g., Yariv and Yeh, 2007, Chap. 14).

In long optical fibers, pulse broadening due to the nonlinear optical Kerr effect can be counterbalanced by a narrowing of a pulse due to linear dispersion (dependence of group velocity on frequency). The result is an *optical soliton:* a pulse of light with a special shape that travels down the fiber without any broadening or narrowing (see, e.g., Yariv and Yeh, 2007, Sec. 14.5). In Sec. 16.3, we study in full mathematical detail this soliton phenomenon for nonlinear waves on the surface of water; in Sec. 23.6, we study it for nonlinear waves in plasmas.

EXERCISES

Exercise 10.20 *Derivation: Optical Kerr Effect*

(a) Derive Eq. (10.68) for the polarization induced in an isotropic medium by a linearly polarized electromagnetic wave.

(b) Fill in the remaining details of the derivation of Eq. (10.69) for the optical Kerr effect.

Bibliographic Note

For a lucid and detailed discussion of lasers and their applications, see Saleh and Teich (2007), and at a more advanced level, Yariv and Yeh (2007). For less detailed but clear discussions, see standard optics textbooks, such as Jenkins and White (1976), Ghatak (2010), and Hecht (2017).

For a lucid and detailed discussion of holography and its applications, see Goodman (2005). Most optics textbooks contain less detailed but clear discussions. We like Jenkins and White (1976), Brooker (2003), Sharma (2006), Ghatak (2010), and Hecht (2017).

Wave-wave mixing in nonlinear media is discussed in great detail and with many applications by Yariv and Yeh (2007). Some readers might find an earlier book by Yariv (1989) pedagogically easier; it was written when the subject was less rich, but the foundations were already in place, and it has more of a quantum mechanical focus. The fundamental concepts of wave-wave mixing and its underlying physical processes are treated especially nicely by Boyd (2008). A more elementary treatment with focus on applications is given by Saleh and Teich (2007). Among treatments in standard optics texts, we like Sharma (2006).

PART IV

ELASTICITY

Although ancient civilizations built magnificent pyramids, palaces, and cathedrals and presumably developed insights into how to avoid their collapse, mathematical models for this (the theory of elasticity) were not developed until the seventeenth century and later.

The seventeenth-century focus was on a beam (e.g., a vertical building support) under compression or tension. Galileo initiated this study in 1632, followed most notably by Robert Hooke[1] in 1660. Bent beams became the focus with the work of Edme Mariotte in 1680. Bending and compression came together with Leonhard Euler's 1744 theory of the buckling of a compressed beam and his derivation of the complex shapes of very thin wires, whose ends are pushed toward each other (*elastica*). These ideas were extended to 2-dimensional thin plates by Marie-Sophie Germain and Joseph-Louis Lagrange in 1811–1816 in a study that was brought to full fruition by Augustus Edward Hugh Love in 1888. The full theory of 3-dimensional, stressed, elastic objects was developed by Claude-Louis Navier and by Augustin-Louis Cauchy in 1821–1822. A number of great mathematicians and natural philosophers then developed techniques for solving the Navier-Cauchy equations, particularly for phenomena relevant to railroads and construction. In 1956, with the advent of modern digital computers, M. J. Turner, R. W. Clough, H. C. Martin, and L. J. Topp pioneered finite-element methods for numerically modeling stressed bodies. Finite-element numerical simulations are now a standard tool for designing mechanical structures and devices, and, more generally, for solving difficult elasticity problems.

These historical highlights cannot begin to do justice to the history of elasticity research. For much more detail see, for example, the (out-of-date) long introduction in Love (1927); and for far more detail than most anyone wants, see the (even more out-of-date) two-volume work by Todhunter and Pearson (1886).

---

1. One of whose many occupations was architect.

Despite its centuries-old foundations, elasticity remains of great importance today, and its modern applications include some truly interesting phenomena. Among those applications, most of which we shall touch on in this book, are

1. the design of bridges, skyscrapers, automobiles, and other structures and mechanical devices and the study of their structural failure;
2. the development and applications of new materials, such as carbon nanotubes, which are so light and strong that one could aspire to use them to build a tether connecting a geostationary satellite to Earth's surface;
3. the design of high-precision physics experiments with torsion pendula and microcantilevers, including brane-worlds-motivated searches for gravitational evidence of macroscopic higher dimensions of space;
4. the creation of nano-scale cantilever probes in atomic-force microscopes;
5. the study of biophysical systems, such as DNA molecules, cell walls, and the Venus fly trap plant; and
6. the study of plate tectonics, quakes, seismic waves, and seismic tomography in Earth and other planets. Much of the modern, geophysical description of mountain building involves buckling and fracture within the lithosphere through viscoelastic processes that combine the principles of elasticity with those of fluid dynamics discussed in Part V.

Indeed, elastic solids are so central to everyday life and to modern science that a basic understanding of their behavior should be part of the repertoire of every physicist. That is the goal of Part IV of the book.

We devote just two chapters to elasticity. The first (Chap. 11) will focus on *elastostatics:* the properties of materials and solid objects that are in static equilibrium, with all forces and torques balancing out. The second (Chap. 12) will focus on *elastodynamics*: the dynamical behavior of materials and solid objects that are perturbed away from equilibrium.

CHAPTER ELEVEN

Elastostatics

Ut tensio, sic vis
ROBERT HOOKE (1678)

11.1 Overview

From the viewpoint of continuum mechanics, a *solid* is a substance that recovers its original shape after the application and removal of any small stress. Note the requirement that this be true for *any* small stress. Many fluids (e.g., water) satisfy our definition as long as the applied stress is isotropic, but they will deform permanently under a shear stress. Other materials (e.g., Earth's crust) are only solid for limited times but undergo plastic flow when a small stress is applied for a long time.

We focus our attention in this chapter on solids whose deformation (quantified by a *tensorial strain*) is linearly proportional to the applied, small, *tensorial stress*. This linear, 3-dimensional stress-strain relationship, which we develop and explore in this chapter, generalizes Hooke's famous 1-dimensional law, which states that if an elastic wire or rod is stretched by an applied force F (Fig. 11.1a), its fractional change of length (its strain) is proportional to the force, $\Delta \ell / \ell \propto F$.

Hooke's law

Hooke's law turns out to be one component of a 3-dimensional stress-strain relation, but to understand it deeply in that language, we must first define and understand the *strain tensor* and the *stress tensor*. Our approach to these tensors follows the geometric, frame-independent philosophy introduced in Chap. 1. Some readers may wish to review that philosophy and mathematics by rereading or browsing Chap. 1.

We begin our development of elasticity theory in Sec. 11.2 by introducing, in a frame-independent way, the vectorial displacement field $\boldsymbol{\xi}(\mathbf{x})$ inside a stressed body (Fig. 11.1b), and its gradient $\boldsymbol{\nabla \xi}$, whose symmetric part is the strain tensor **S**. We then express the strain tensor as the sum of an expansion Θ that represents volume changes and a shear $\boldsymbol{\Sigma}$ that represents shape changes.

In Sec. 11.3.1, we introduce the stress tensor, and in Sec. 11.3.2, we discuss the realms in which there is a linear relationship between stress and strain, and ways in which linearity can fail. In Sec. 11.3.3, assuming linearity, we discuss how the material resists volume change by developing an opposing isotropic stress, with a stress/strain ratio that is equal to the *bulk modulus K*. We discuss how the material also

> **BOX 11.1. READERS' GUIDE**
>
> - This chapter relies heavily on the geometric view of Newtonian physics (including vector and tensor analysis) laid out in Chap. 1.
> - Chapter 12 (Elastodynamics) is an extension of this chapter; to understand it, this chapter must be mastered.
> - The idea of the irreducible tensorial parts of a tensor, and its most important example, decomposition of the gradient of a displacement vector into expansion, rotation, and shear (Sec. 11.2.2 and Box 11.2) will be encountered again in Part V (Fluid Dynamics), Part VI (Plasma Physics), and Part VII (General Relativity).
> - Differentiation of vectors and tensors with the help of connection coefficients (Sec. 11.8; Track Two) will be used occasionally in Part V (Fluid Dynamics) and Part VI (Plasma Physics), and will be generalized to nonorthonormal bases in Part VII (General Relativity), where it will become Track One and will be used extensively.
> - No other portions of this chapter are important for subsequent parts of this book.

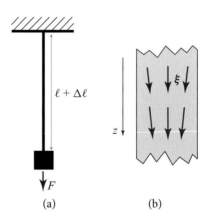

FIGURE 11.1 (a) Hooke's 1-dimensional law for a rod stretched by a force F: $\Delta\ell/\ell \propto F$. (b) The 3-dimensional displacement vector $\boldsymbol{\xi}(\mathbf{x})$ inside the stretched rod.

resists a shear-type strain by developing an opposing shear stress with a stress/strain ratio equal to twice the shear modulus 2μ. In Sec. 11.3.4, we evaluate the energy density stored in elastostatic strains; in Sec. 11.3.5, we explore the influence of thermal expansion on the stress-strain relationship; and in Sec. 11.3.6, we discuss the atomic-force origin of the elastostatic stresses and use atomic considerations to estimate the magnitudes of the bulk and shear moduli. Then in Sec. 11.3.7, we compute the elastic force density inside a linear material as the divergence of the sum of its elastic stresses, and we formulate the law of elastostatic stress balance (the Navier-Cauchy equation) as the vanishing sum of the material's internal elastic force density and any other force densities that may act (usually a gravitational force density due to the weight of the elastic material). We discuss the analogy between this elastostatic stress-balance equation and Maxwell's electrostatic and magnetostatic equations. We describe how mathematical techniques common in electrostatics can also be applied to solve the Navier-Cauchy equation, subject to boundary conditions that describe external forces.

In Sec. 11.4, as a simple example, we use our 3-dimensional formulas to deduce Hooke's law for the 1-dimensional longitudinal stress and strain in a stretched wire.

When the elastic body that one studies is very thin in two dimensions compared to the third (e.g., a wire or rod), we can reduce the 3-dimensional elastostatic equations to a set of coupled 1-dimensional equations by taking moments of the elastostatic equations. We illustrate this technique in Sec. 11.5, where we treat the bending of beams and other examples.

Elasticity theory, as developed in this chapter, is an example of a common (some would complain far too common) approach to physics problems, namely, to linearize them. Linearization may be acceptable when the distortions are small. However, when deformed by sufficiently strong forces, elastic media may become unstable to small displacements, which can then grow to large amplitude, causing rupture. We study an example of this in Sec. 11.6: the buckling of a beam when subjected to a sufficiently large longitudinal stress. Buckling is associated with *bifurcation of equilibria,* a phenomenon that is common to many physical systems, not just elastostatic ones. We illustrate bifurcation in Sec. 11.6 using our beam under a compressive load, and we explore its connection to catastrophe theory.

In Sec. 11.7, we discuss dimensional reduction by the method of moments for bodies that are thin in only 1 dimension, not two, such as plates and thin mirrors. In such bodies, the 3-dimensional elastostatic equations are reduced to 2 dimensions. We illustrate our 2-dimensional formalism by the stress polishing of telescope mirrors.

Because elasticity theory entails computing gradients of vectors and tensors, and practical calculations are often best performed in cylindrical or spherical coordinate systems, we present a mathematical digression in Track-Two Sec. 11.8—an introduction to how one can perform practical calculations of gradients of vectors and tensors in the orthonormal bases associated with curvilinear coordinate systems, using the concept of a connection coefficient.

As illustrative examples of both connection coefficients and elastostatic force balance, in Track-Two Sec. 11.9 and various exercises, we give practical examples of solutions of the elastostatic force-balance equation in cylindrical coordinates using two common techniques of elastostatics and electrostatics: separation of variables (text of Sec. 11.9.2) and Green's functions (Ex. 11.27).

11.2 Displacement and Strain

We begin our study of elastostatics by introducing the elastic displacement vector, its gradient, and the irreducible tensorial parts of its gradient. We then identify the strain as the symmetric part of the displacement's gradient.

11.2.1 Displacement Vector and Its Gradient

Elasticity provides a major application of the tensorial techniques we developed in Chap. 1. Label the position of a *point* (a tiny bit of solid) in an unstressed body, relative to some convenient origin in the body, by its position vector $\mathbf{x}$. Let a force be applied, so the body deforms and the point moves from $\mathbf{x}$ to $\mathbf{x} + \boldsymbol{\xi}(\mathbf{x})$; we call $\boldsymbol{\xi}$ the point's *displacement vector* (Fig. 11.1b). If $\boldsymbol{\xi}$ were constant (i.e., if its components in a Cartesian coordinate system were independent of location in the body), then the body would simply be translated and would undergo no deformation. To produce a deformation, we must make the displacement $\boldsymbol{\xi}$ change from one location to another. The most simple, coordinate-independent way to quantify those changes is by the gradient of $\boldsymbol{\xi}$, $\nabla \boldsymbol{\xi}$. This gradient is a second-rank tensor field, which we denote by $\mathbf{W}$:

displacement vector

$$\boxed{\mathbf{W} \equiv \nabla \boldsymbol{\xi}.} \tag{11.1a}$$

This tensor is a geometric object, defined independently of any coordinate system in the manner described in Sec. 1.7. In slot-naming index notation (Sec. 1.5), it is denoted

$$W_{ij} = \xi_{i;j}, \tag{11.1b}$$

where the index j after the semicolon is the name of the gradient slot.

In a Cartesian coordinate system the components of the gradient are always just partial derivatives [Eq. (1.15c)], and therefore the Cartesian components of $\mathbf{W}$ are

$$W_{ij} = \frac{\partial \xi_i}{\partial x_j} = \xi_{i,j}. \tag{11.1c}$$

(Recall that indices following a comma represent partial derivatives.) In Sec. 11.8, we learn how to compute the components of the gradient in cylindrical and spherical coordinates.

In any small neighborhood of any point $\mathbf{x}_o$ in a deformed body, we can reconstruct the displacement vector $\boldsymbol{\xi}$ from its gradient $\mathbf{W}$ up to an additive constant. Specifically, in Cartesian coordinates, by virtue of a Taylor-series expansion, $\boldsymbol{\xi}$ is given by

$$\begin{aligned}\xi_i(\mathbf{x}) &= \xi_i(\mathbf{x}_o) + (x_j - x_{oj})(\partial \xi_i / \partial x_j) + \ldots \\ &= \xi_i(\mathbf{x}_o) + (x_j - x_{oj}) W_{ij} + \ldots.\end{aligned} \tag{11.2}$$

If we place the origin of Cartesian coordinates at $\mathbf{x}_o$ and let the origin move with the point there as the body deforms [so $\boldsymbol{\xi}(\mathbf{x}_o) = 0$], then Eq. (11.2) becomes

$$\xi_i = W_{ij} x_j \quad \text{when } |\mathbf{x}| \text{ is sufficiently small.} \tag{11.3}$$

We have derived this as a relationship between components of $\boldsymbol{\xi}$, $\mathbf{x}$, and $\mathbf{W}$ in a Cartesian coordinate system. However, the indices can also be thought of as the names of slots (Sec. 1.5) and correspondingly, Eq. (11.3) can be regarded as a geometric, coordinate-independent relationship among the vectors and tensor $\boldsymbol{\xi}$, $\mathbf{x}$, and $\mathbf{W}$.

In Ex. 11.2, we use Eq. (11.3) to gain insight into the displacements associated with various parts of the gradient $\mathbf{W}$.

11.2.2 Expansion, Rotation, Shear, and Strain

In Box 11.2, we introduce the concept of the irreducible tensorial parts of a tensor, and we state that in physics, when one encounters an unfamiliar tensor, it is often useful to identify the tensor's irreducible parts. The gradient of the displacement vector, $\mathbf{W} = \nabla \boldsymbol{\xi}$, is an important example. It is a second-rank tensor. Therefore, as discussed in Box 11.2, its irreducible tensorial parts are its trace $\Theta \equiv \text{Tr}(\mathbf{W}) = W_{ii} = \nabla \cdot \boldsymbol{\xi}$, which is called the deformed body's *expansion* (for reasons we shall explore below); its symmetric, trace-free part $\boldsymbol{\Sigma}$, which is called the body's *shear*; and its antisymmetric part $\mathbf{R}$, which is called the body's *rotation*:

irreducible tensorial parts of a tensor

expansion, shear, and rotation

$$\Theta = W_{ii} = \nabla \cdot \boldsymbol{\xi}, \tag{11.4a}$$

$$\Sigma_{ij} = \frac{1}{2}(W_{ij} + W_{ji}) - \frac{1}{3}\Theta g_{ij} = \frac{1}{2}(\xi_{i;j} + \xi_{j;i}) - \frac{1}{3}\xi_{k;k}\, g_{ij}, \tag{11.4b}$$

$$R_{ij} = \frac{1}{2}(W_{ij} - W_{ji}) = \frac{1}{2}(\xi_{i;j} - \xi_{j;i}). \tag{11.4c}$$

Here g_{ij} is the metric, which has components $g_{ij} = \delta_{ij}$ (Kronecker delta) in Cartesian coordinates, and repeated indices [the ii in Eq. (11.4a)] are to be summed [Eq. (1.9b) and subsequent discussion].

We can reconstruct $\mathbf{W} = \nabla \boldsymbol{\xi}$ from these irreducible tensorial parts in the following manner [Eq. (4) of Box 11.2, rewritten in abstract notation]:

$$\nabla \boldsymbol{\xi} = \mathbf{W} = \frac{1}{3}\Theta \mathbf{g} + \boldsymbol{\Sigma} + \mathbf{R}. \tag{11.5}$$

Let us explore the physical effects of the three separate parts of $\mathbf{W}$ in turn. To understand expansion, consider a small 3-dimensional piece $\mathcal{V}$ of a deformed body (a *volume element*). When the deformation $\mathbf{x} \to \mathbf{x} + \boldsymbol{\xi}$ occurs, a much smaller element of area[1] $d\boldsymbol{\Sigma}$ on the surface $\partial \mathcal{V}$ of $\mathcal{V}$ gets displaced through the vectorial distance $\boldsymbol{\xi}$ and in the process sweeps out a volume $\boldsymbol{\xi} \cdot d\boldsymbol{\Sigma}$. Therefore, the change in the volume element's volume, produced by $\boldsymbol{\xi}$, is

$$\delta V = \int_{\partial \mathcal{V}} d\boldsymbol{\Sigma} \cdot \boldsymbol{\xi} = \int_{\mathcal{V}} dV \nabla \cdot \boldsymbol{\xi} = \nabla \cdot \boldsymbol{\xi} \int_{\mathcal{V}} dV = (\nabla \cdot \boldsymbol{\xi})\, V. \tag{11.6}$$

---

1. Note that we use $\boldsymbol{\Sigma}$ for a vectorial area and $\boldsymbol{\Sigma}$ for the shear tensor. There should be no confusion.

BOX 11.2. IRREDUCIBLE TENSORIAL PARTS OF A SECOND-RANK TENSOR IN 3-DIMENSIONAL EUCLIDEAN SPACE

In quantum mechanics, an important role is played by the *rotation group*: the set of all rotation matrices, viewed as a mathematical entity called a group (e.g., Mathews and Walker, 1970, Chap. 16). Each tensor in 3-dimensional Euclidean space, when rotated, is said to generate a specific *representation* of the rotation group. Tensors that are "big" (in a sense to be discussed later in this box) can be broken down into a sum of several tensors that are "as small as possible." These smallest tensors are said to generate *irreducible representations* of the rotation group. All this mumbo-jumbo is really simple, when one thinks about tensors as geometric, frame-independent objects.

As an example, consider an arbitrary second-rank tensor W_{ij} in 3-dimensional, Euclidean space. In the text W_{ij} is the gradient of the displacement vector. From this tensor we can construct the following "smaller" tensors by linear operations that involve only W_{ij} and the metric g_{ij}. (As these smaller tensors are enumerated, the reader should think of the notation used as the basis-independent, frame-independent, slot-naming index notation of Sec. 1.5.1.) The smaller tensors are the contraction (i.e., trace) of W_{ij},

$$\Theta \equiv W_{ij} g_{ij} = W_{ii}; \tag{1}$$

the antisymmetric part of W_{ij},

$$R_{ij} \equiv \frac{1}{2}(W_{ij} - W_{ji}); \tag{2}$$

and the symmetric, trace-free part of W_{ij},

$$\Sigma_{ij} \equiv \frac{1}{2}(W_{ij} + W_{ji}) - \frac{1}{3} g_{ij} W_{kk}. \tag{3}$$

It is straightforward to verify that the original tensor W_{ij} can be reconstructed from these three smaller tensors plus the metric g_{ij} as follows:

$$W_{ij} = \frac{1}{3} \Theta g_{ij} + R_{ij} + \Sigma_{ij}. \tag{4}$$

One way to see the sense in which Θ, R_{ij}, and Σ_{ij} are smaller than W_{ij} is by counting the number of independent real numbers required to specify their components in an arbitrary basis. (Think of the index notation as components on a chosen basis.) The original tensor W_{ij} has three × three = nine components ($W_{11}, W_{12}, W_{13}, W_{21}, \ldots W_{33}$), all of which are independent. By contrast, the scalar Θ has just one. The antisymmetric

(continued)

BOX 11.2. (continued)

tensor R_{ij} has just three independent components, R_{12}, R_{23}, and R_{31}. Finally, the nine components of Σ_{ij} are not independent; symmetry requires that $\Sigma_{ij} \equiv \Sigma_{ji}$, which reduces the number of independent components from nine to six; being trace-free, $\Sigma_{ii} = 0$, reduces it further from six to five. Therefore, (five independent components in Σ_{ij}) + (three independent components in R_{ij}) + (one independent component in Θ) = 9 = (number of independent components in W_{ij}).

The number of independent components (one for Θ, three for R_{ij}, and five for Σ_{ij}) is a geometric, basis-independent concept: It is the same, regardless of the basis used to count the components; and for each of the smaller tensors that make up W_{ij}, it is easily deduced without introducing a basis at all (think here in slot-naming index notation): The scalar Θ is clearly specified by just one real number. The antisymmetric tensor R_{ij} contains precisely the same amount of information as the vector

$$\phi_i \equiv -\frac{1}{2}\epsilon_{ijk}R_{jk}, \tag{5}$$

as can be seen from the fact that Eq. (5) can be inverted to give

$$R_{ij} = -\epsilon_{ijk}\phi_k; \tag{6}$$

and the vector ϕ_i can be characterized by its direction in space (two numbers) plus its length (a third). The symmetric, trace-free tensor Σ_{ij} can be characterized geometrically by the ellipsoid $(g_{ij} + \varepsilon\Sigma_{ij})\zeta_i\zeta_j = 1$, where ε is an arbitrary number $\ll 1$, and ζ_i is a vector whose tail sits at the center of the ellipsoid and whose head moves around on the ellipsoid's surface. Because Σ_{ij} is trace-free, this ellipsoid has unit volume. Therefore, it is specified fully by the direction of its longest principal axis (two numbers) plus the direction of a second principal axis (a third number) plus the ratio of the length of the second axis to the first (a fourth number) plus the ratio of the length of the third axis to the first (a fifth number).

Each of the tensors Θ, R_{ij} (or equivalently, ϕ_i), and Σ_{ij} is irreducible in the sense that one cannot construct any smaller tensors from it, by any linear operation that involves only it, the metric, and the Levi-Civita tensor. Irreducible tensors in 3-dimensional Euclidean space always have an odd number of components. It is conventional to denote this number by $2l + 1$, where the integer l is called the "order of the irreducible representation of the

(continued)

> **BOX 11.2. (continued)**
>
> rotation group" that the tensor generates. For Θ, R_{ij} (or equivalently, ϕ_i), and Σ_{jk}, l is 0, 1, and 2, respectively. These three tensors can be mapped into the spherical harmonics of order $l = 0, 1, 2$; and their $2l + 1$ components correspond to the $2l + 1$ values of the quantum number $m = -l, -l+1 \ldots, l-1, l$. (For details see, e.g., Thorne, 1980, Sec. II.C.)
>
> In physics, when one encounters a new tensor, it is often useful to identify the tensor's irreducible parts. They almost always play important, independent roles in the physical situation one is studying. We meet one example in this chapter, another when we study fluid dynamics (Chap. 13), and a third in general relativity (Box 25.2).

Here we have invoked Gauss's theorem in the second equality, and in the third we have used the smallness of $\mathcal{V}$ to infer that $\nabla \cdot \boldsymbol{\xi}$ is essentially constant throughout $\mathcal{V}$ and so can be pulled out of the integral. Therefore, the fractional change in volume is equal to the trace of the stress tensor (i.e., the expansion):

expansion as fractional volume change

$$\boxed{\frac{\delta V}{V} = \nabla \cdot \boldsymbol{\xi} = \Theta.} \tag{11.7}$$

See Fig. 11.2 for a simple example.

shearing displacements

The shear tensor $\boldsymbol{\Sigma}$ produces the shearing displacements illustrated in Figs. 11.2 and 11.3. As the tensor has zero trace, there is no volume change when a body undergoes a pure shear deformation. The shear tensor has five independent components (Box 11.2). However, by rotating our Cartesian coordinates appropriately, we can transform away all the off-diagonal elements, leaving only the three diagonal elements Σ_{xx}, Σ_{yy}, and Σ_{zz}, which must sum to zero. This is known as a *principal-axis transformation*. Each element produces a stretch ($\Sigma_{..} > 0$) or squeeze ($\Sigma_{..} < 0$) along its

shear's stretch and squeeze along principal axes

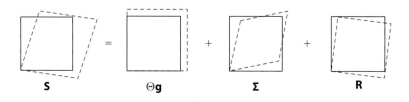

FIGURE 11.2 A simple example of the decomposition of a 2-dimensional distortion **S** of a square body into an expansion Θ, a shear $\boldsymbol{\Sigma}$, and a rotation **R**.

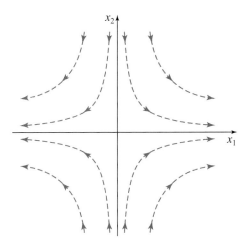

FIGURE 11.3 Shear in 2 dimensions. The displacement of points in a solid undergoing pure shear is the vector field $\boldsymbol{\xi}(\mathbf{x})$ given by Eq. (11.3) with W_{ji} replaced by Σ_{ji}: $\xi_j = \Sigma_{ji} x_i = \Sigma_{j1} x_1 + \Sigma_{j2} x_2$. The integral curves of this vector field are plotted in this figure. The figure is drawn using principal axes, which are Cartesian, so $\Sigma_{12} = \Sigma_{21} = 0$ and $\Sigma_{11} = -\Sigma_{22}$, which means that $\xi_1 = \Sigma_{11} x_1$ and $\xi_2 = -\Sigma_{11} x_2$; or, equivalently, $\xi_x = \Sigma_{xx} x$ and $\xi_y = -\Sigma_{xx} y$. The integral curves of this simple vector field are the hyperbolas shown. Note that the displacement increases linearly with distance from the origin. The shear shown in Fig. 11.2 is the same as this, but with the axes rotated counterclockwise by 45°.

axis,[2] and their vanishing sum (the vanishing trace of $\boldsymbol{\Sigma}$) means that there is no net volume change. The components of the shear tensor in any Cartesian coordinate system can be written down immediately from Eq. (11.4b) by substituting the Kronecker delta δ_{ij} for the components of the metric tensor g_{ij} and treating all derivatives as partial derivatives:

$$\Sigma_{xx} = \frac{2}{3} \frac{\partial \xi_x}{\partial x} - \frac{1}{3} \left(\frac{\partial \xi_y}{\partial y} + \frac{\partial \xi_z}{\partial z} \right), \quad \Sigma_{xy} = \frac{1}{2} \left(\frac{\partial \xi_x}{\partial y} + \frac{\partial \xi_y}{\partial x} \right), \quad (11.8)$$

and similarly for the other components. The analogous equations in spherical and cylindrical coordinates are given in Sec. 11.8.

The third term **R** in Eq. (11.5) describes a pure rotation, which does not deform the solid. To verify this, write $\boldsymbol{\xi} = \boldsymbol{\phi} \times \mathbf{x}$, where $\boldsymbol{\phi}$ is a small rotation of magnitude ϕ about an axis parallel to the direction of $\boldsymbol{\phi}$. Using Cartesian coordinates in 3-dimensional Euclidean space, we can demonstrate by direct calculation that the symmetric part of $\mathbf{W} = \nabla \boldsymbol{\xi}$ vanishes (i.e., $\Theta = \boldsymbol{\Sigma} = 0$) and that

rotation vector

$$R_{ij} = -\epsilon_{ijk} \phi_k, \quad \phi_i = -\frac{1}{2} \epsilon_{ijk} R_{jk}. \quad (11.9a)$$

2. More explicitly, $\Sigma_{xx} > 0$ produces a stretch along the x-axis, $\Sigma_{yy} < 0$ produces a squeeze along the y-axis, etc.

Therefore, the elements of the tensor **R** in a Cartesian coordinate system just involve the vectorial rotation angle $\boldsymbol{\phi}$. Note that expression (11.9a) for $\boldsymbol{\phi}$ and expression (11.4c) for R_{ij} imply that $\boldsymbol{\phi}$ is half the curl of the displacement vector:

$$\boldsymbol{\phi} = \frac{1}{2}\nabla \times \boldsymbol{\xi}. \tag{11.9b}$$

A simple example of rotation is shown in the last picture in Fig. 11.2.

Elastic materials resist expansion Θ and shear $\boldsymbol{\Sigma}$, but they don't mind at all having their orientation in space changed (i.e., they do not resist rotations **R**). Correspondingly, in elasticity theory a central focus is on expansion and shear. For this reason the symmetric part of the gradient of $\boldsymbol{\xi}$,

strain tensor

$$S_{ij} \equiv \frac{1}{2}(\xi_{i;j} + \xi_{j;i}) = \Sigma_{ij} + \frac{1}{3}\Theta g_{ij}, \tag{11.10}$$

which includes the expansion and shear but omits the rotation, is given a special name—the *strain*—and is paid great attention.

Let us consider some examples of strains that arise in physical systems.

1. Understanding how materials deform under various *loads* (externally applied forces) is central to mechanical, civil, and structural engineering. As we learn in Sec. 11.3.2, all Hookean materials (materials with strain proportional to stress when the stress is small) crack or break when the load is so great that any component of their strain exceeds ~ 0.1, and almost all crack or break at strains ~ 0.001. For this reason, in our treatment of elasticity theory (this chapter and the next), we focus on strains that are small compared to unity.

2. Continental drift can be measured on the surface of Earth using very long baseline interferometry, a technique in which two or more radio telescopes are used to detect interferometric fringes using radio waves from an astronomical point source. (A similar technique uses the Global Positioning System to achieve comparable accuracy.) By observing the fringes, it is possible to detect changes in the spacing between the telescopes as small as a fraction of a wavelength (~ 1 cm). As the telescopes are typically 1,000 km apart, this means that dimensionless strains $\sim 10^{-8}$ can be measured. The continents drift apart on a timescale $\lesssim 10^8$ yr, so it takes roughly a year for these changes to grow large enough to be measured. Such techniques are also useful for monitoring earthquake faults.

3. The smallest time-varying strains that have been measured so far involve laser interferometer gravitational-wave detectors, such as LIGO. In each arm of a LIGO interferometer, two mirrors hang freely, separated by 4 km. In 2015 their separations were monitored (at frequencies of ~ 100 Hz) to $\sim 4 \times 10^{-19}$ m, four ten-thousandths the radius of a nucleon. The associated strain is 1×10^{-22}. Although this strain is not associated with an elastic solid, it does indicate the high accuracy of optical measurement techniques.

Exercise 11.1 *Derivation and Practice: Reconstruction of a Tensor from Its Irreducible Tensorial Parts*

Using Eqs. (1), (2), and (3) of Box 11.2, show that $\frac{1}{3}\Theta g_{ij} + \Sigma_{ij} + R_{ij}$ is equal to W_{ij}.

Exercise 11.2 *Example: Displacement Vectors Associated with Expansion, Rotation, and Shear*

(a) Consider a $\mathbf{W} = \nabla \boldsymbol{\xi}$ that is pure expansion: $W_{ij} = \frac{1}{3}\Theta g_{ij}$. Using Eq. (11.3) show that, in the vicinity of a chosen point, the displacement vector is $\xi_i = \frac{1}{3}\Theta x_i$. Draw this displacement vector field.

(b) Similarly, draw $\boldsymbol{\xi}(\mathbf{x})$ for a $\mathbf{W}$ that is pure rotation. [Hint: Express $\boldsymbol{\xi}$ in terms of the vectorial angle $\boldsymbol{\phi}$ with the aid of Eq. (11.9b).]

(c) Draw $\boldsymbol{\xi}(\mathbf{x})$ for a $\mathbf{W}$ that is pure shear. To simplify the drawing, assume that the shear is confined to the x-y plane, and make your drawing for a shear whose only nonzero components are $\Sigma_{xx} = -\Sigma_{yy}$. Compare your drawing with Fig. 11.3.

11.3 Stress, Elastic Moduli, and Elastostatic Equilibrium

11.3.1 Stress Tensor

The forces acting in an elastic solid are measured by a second-rank tensor, the *stress tensor* introduced in Sec. 1.9. Let us recall the definition of this stress tensor.

Consider two small, contiguous regions in a solid. If we take a small element of area $d\boldsymbol{\Sigma}$ in the contact surface with its positive sense[3] (same as the direction of $d\boldsymbol{\Sigma}$ viewed as a vector) pointing from the first region toward the second, then the first region exerts a force $d\mathbf{F}$ (not necessarily normal to the surface) on the second through this area. The force the second region exerts on the first (through the area $-d\boldsymbol{\Sigma}$) will, by Newton's third law, be equal and opposite to that force. The force and the area of contact are both vectors, and there is a linear relationship between them. (If we double the area, we double the force.) The two vectors therefore will be related by a second-rank tensor, the stress tensor $\mathbf{T}$:

$$d\mathbf{F} = \mathbf{T} \cdot d\boldsymbol{\Sigma} = \mathbf{T}(\ldots, d\boldsymbol{\Sigma}); \quad dF_i = T_{ij} d\Sigma_j. \quad (11.11)$$

Thus the tensor $\mathbf{T}$ is the net (vectorial) force per unit (vectorial) area that a body exerts on its surroundings. Be aware that many books on elasticity (e.g., Landau and Lifshitz, 1986) define the stress tensor with the opposite sign to that in Eq. (11.11). Also be careful not to confuse the shear tensor Σ_{jk} with the vectorial infinitesimal surface area $d\Sigma_j$.

---

[3]. For a discussion of area elements including their positive sense, see Sec. 1.8.

We often need to compute the total elastic force acting on some finite volume $\mathcal{V}$. To aid in this, we make an important assumption, discussed in Sec. 11.3.6: the stress is determined by local conditions and can be computed from the local arrangement of atoms. If this assumption is valid, then (as we shall see in Sec. 11.3.6), we can compute the total force acting on the volume element by integrating the stress over its surface $\partial \mathcal{V}$:

$$\mathbf{F} = -\int_{\partial \mathcal{V}} \mathbf{T} \cdot d\mathbf{\Sigma} = -\int_{\mathcal{V}} \mathbf{\nabla} \cdot \mathbf{T} \, dV, \tag{11.12}$$

where we have invoked Gauss's theorem, and the minus sign is included because by convention, for a closed surface $\partial \mathcal{V}$, $d\mathbf{\Sigma}$ points out of $\mathcal{V}$ instead of into it.

Equation (11.12) must be true for arbitrary volumes, so we can identify the *elastic force density* $\mathbf{f}$ acting on an elastic solid as

elastic force density

$$\boxed{\mathbf{f} = -\mathbf{\nabla} \cdot \mathbf{T}.} \tag{11.13}$$

In elastostatic equilibrium, this force density must balance all other volume forces acting on the material, most commonly the gravitational force density, so

force balance equation

$$\boxed{\mathbf{f} + \rho \mathbf{g} = 0,} \tag{11.14}$$

where $\mathbf{g}$ is the gravitational acceleration. (Again, there should be no confusion between the vector $\mathbf{g}$ and the metric tensor $\mathbf{g}$.) There are other possible external forces, some of which we shall encounter later in the context of fluids (e.g., an electromagnetic force density). These can be added to Eq. (11.14).

Just as for the strain, the stress tensor $\mathbf{T}$ can be decomposed into its irreducible tensorial parts, a pure trace (the *pressure P*) plus a symmetric trace-free part (the *shear stress*):

pressure and shear stress

$$\mathbf{T} = P\mathbf{g} + \mathbf{T}^{\text{shear}}; \quad P = \frac{1}{3}\,\text{Tr}(\mathbf{T}) = \frac{1}{3}T_{ii}. \tag{11.15}$$

There is no antisymmetric part, because the stress tensor is symmetric, as we saw in Sec. 1.9. Fluids at rest exert isotropic stresses: $\mathbf{T} = P\mathbf{g}$. They cannot exert shear stress when at rest, though when moving and shearing, they can exert a viscous shear stress, as we discuss extensively in Part V (initially in Sec. 13.7.2).

In SI units, stress is measured in units of Pascals, denoted Pa:

Pascal

$$1\,\text{Pa} = 1\,\text{N m}^{-2} = 1\frac{\text{kg m s}^{-2}}{\text{m}^2}, \tag{11.16}$$

or sometimes in GPa = 10^9 Pa. In cgs units, stress is measured in dyne cm^{-2}. Note that 1 Pa = 10 dyne cm^{-2}.

Now let us consider some examples of stresses.

1. Atmospheric pressure is equal to the weight of the air in a column of unit area extending above the surface of Earth, and thus is roughly $P \sim \rho g H \sim$

10^5 Pa, where $\rho \simeq 1$ kg m^{-3} is the density of air, $g \simeq 10$ m s^{-2} is the acceleration of gravity at Earth's surface, and $H \simeq 10$ km is the atmospheric scale height $[H \equiv (d \ln P/dz)^{-1}$, with z the vertical distance]. Thus 1 atmosphere is $\sim 10^5$ Pa (or, more precisely, 1.01325×10^5 Pa). The stress tensor is isotropic.

2. Suppose we hammer a nail into a block of wood. The hammer might weigh $m \sim 0.3$ kg and be brought to rest from a speed of $v \sim 10$ m s^{-1} in a distance of, say, $d \sim 3$ mm. Then the average force exerted on the wood by the nail, as it is driven, is $F \sim mv^2/d \sim 10^4$ N. If this is applied over an effective area $A \sim 1$ mm^2, then the magnitude of the typical stress in the wood is $\sim F/A \sim 10^{10}$ Pa $\sim 10^5$ atmosphere. There is a large shear component to the stress tensor, which is responsible for separating the fibers in the wood as the nail is hammered.

3. Neutron stars are as massive as the Sun, $M \sim 2 \times 10^{30}$ kg, but have far smaller radii, $R \sim 10$ km. Their surface gravities are therefore $g \sim GM/R^2 \sim 10^{12}$ m s^{-2}, 10 billion times that encountered on Earth. They have solid crusts of density $\rho \sim 10^{16}$ kg m^{-3} that are about 1 km thick. In the crusts, the main contribution to the pressure is from the degeneracy of relativistic electrons (see Sec. 3.5.3). The magnitude of the stress at the base of a neutron-star crust is $P \sim \rho g H \sim 10^{31}$ Pa! The crusts are solid, because the free electrons are neutralized by a lattice of ions. However, a crust's shear modulus is only a few percent of its bulk modulus.

4. As we discuss in Sec. 28.7.1, a popular cosmological theory called *inflation* postulates that the universe underwent a period of rapid, exponential expansion during its earliest epochs. This expansion was driven by the stress associated with a *false vacuum*. The action of this stress on the universe can be described quite adequately using a classical stress tensor. If the interaction energy is $E \sim 10^{15}$ GeV, the supposed scale of grand unification, and the associated lengthscale is the Compton wavelength associated with that energy, $l \sim \hbar c/E$, then the magnitude of the stress is $\sim E/l^3 \sim 10^{97}(E/10^{15}$ GeV$)^4$ Pa.

5. Elementary particles interact through forces. Although it makes no sense to describe this interaction using classical elasticity, it is reasonable to make order-of-magnitude estimates of the associated stress. One promising model of these interactions involves *strings* with mass per unit length $\mu = g_s^2 c^2/(8\pi G) \sim 1$ Megaton/fermi (where Megaton is not the TNT equivalent!), and cross section of order the Planck length squared, $L_P^2 = \hbar G/c^3 \sim 10^{-70}$ m^2, and tension (negative pressure) $T_{zz} \sim \mu c^2/L_P^2 \sim 10^{110}$ Pa. Here $\hbar$, G, and c are Planck's reduced constant, Newton's gravitation constant, and the speed of light, and $g_s^2 \sim 0.025$ is the string coupling constant.

6. The highest possible stress is presumably associated with spacetime singularities, for example at the birth of the universe or inside a black hole. Here the characteristic energy is the Planck energy $E_P = (\hbar c^5/G)^{1/2} \sim 10^{19}$ GeV, the lengthscale is the Planck length $L_P = (\hbar G/c^3)^{1/2} \sim 10^{-35}$ m, and the associated ultimate stress is $\sim E_P/L_P^3 \sim 10^{114}$ Pa.

11.3.2 Realm of Validity for Hooke's Law

In elasticity theory, motivated by Hooke's Law (Fig. 11.1), we assume a linear relationship between a material's stress and strain tensors. Before doing so, however, we discuss the realm in which this linearity is true and some ways in which it can fail.

For this purpose, consider again the stretching of a rod by an applied force (Fig. 11.1a, shown again in Fig. 11.4a). For a sufficiently small stress $T_{zz} = F/A$ (with A the cross sectional area of the rod), the strain $S_{zz} = \Delta \ell/\ell$ follows Hooke's law (straight red line in Fig. 11.4b). However, at some point, called the *proportionality limit* (first big dot in Fig. 11.4b), the strain begins to depart from Hooke's law. Despite this deviation, if the stress is removed, the rod returns to its original length. At a somewhat larger stress, called the *elastic limit*, that ceases to be true; the rod is permanently stretched. At a still larger stress, called the *yield limit* or *yield point*, little or no increase in stress causes a large increase in strain, usually because the material begins to flow plasticly. At an even larger stress, the *rupture point*, the rod cracks or breaks. For a ductile substance like polycrystalline copper, the proportionality limit and elastic limit both occur at about the same rather low strain $\Delta \ell/\ell \sim 10^{-4}$, but yield and rupture do not occur until $\Delta \ell/\ell \sim 10^{-3}$. For a more resilient material like cemented tungsten carbide, strains can be proportional and elastic up to $\sim 3 \times 10^{-3}$. Rubber is non-Hookean (stress is not proportional to strain) at essentially all strains; its proportionality limit is exceedingly small, but it returns to its original shape from essentially all nonrupturing deformations, which can be as large as $\Delta \ell/\ell \sim 8$ (the yield and rupture points).[4] Especially significant is that in almost all solids except rubber, the proportionality, elastic, and yield limits are all small compared to unity.

11.3.3 Elastic Moduli and Elastostatic Stress Tensor

In realms where Hooke's law is valid, there is a corresponding linear relationship between the material's stress tensor and its strain tensor. The most general linear equation relating two second-rank tensors involves a fourth-rank tensor known as the *elastic modulus tensor* **Y**. In slot-naming index notation,

$$T_{ij} = -Y_{ijkl} S_{kl}. \tag{11.17}$$

---

4. Rubber is made of long, polymeric molecules, and its elasticity arises from uncoiling of the molecules when a force is applied, which is a different mechanism than is found in crystalline materials (Xing, Goldbart, and Radzihovsky, 2007).

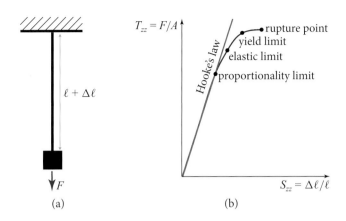

FIGURE 11.4 The stress-strain relation for a rod, showing special points at which the behavior of the rod's material changes.

Now, a general fourth-rank tensor in 3 dimensions has $3^4 = 81$ independent components. Elasticity can get complicated! However, the situation need not be so dire. There are several symmetries that we can exploit. Let us look first at the general case. As the stress and strain tensors are both symmetric, **Y** is symmetric in its first pair of slots, and we are free to choose it symmetric in its second pair: $Y_{ijkl} = Y_{jikl} = Y_{ijlk}$. There are therefore 6 independent components Y_{ijkl} for variable i, j and fixed k, l, and vice versa. In addition, as we will show, **Y** is symmetric under an interchange of its first and second pairs of slots: $Y_{ijkl} = Y_{klij}$. There are therefore $(6 \times 7)/2 = 21$ independent components in **Y**. This is an improvement over 81. Many substances, notably crystals, exhibit additional symmetries, which can reduce the number of independent components considerably.

The simplest, and in fact most common, case arises when the medium is *isotropic*. In other words, there are no preferred directions in the material. This occurs when the solid is polycrystalline or amorphous and completely disordered on a scale large compared with the atomic spacing, but small compared with the solid's inhomogeneity scale.

If a medium is isotropic, then its elastic properties must be describable by scalars that relate the irreducible parts P and $\mathbf{T}^{\text{shear}}$ of the stress tensor **T** to those, Θ and $\boldsymbol{\Sigma}$, of the strain tensor **S**. The only mathematically possible, linear, coordinate-independent relationship between $\{P, \mathbf{T}^{\text{shear}}\}$ and $\{\Theta, \boldsymbol{\Sigma}\}$ involving solely scalars is $P = -K\Theta$, $\mathbf{T}^{\text{shear}} = -2\mu\boldsymbol{\Sigma}$, corresponding to a total stress tensor

$$\boxed{\mathbf{T} = -K\Theta \mathbf{g} - 2\mu\boldsymbol{\Sigma}.} \tag{11.18}$$

bulk modulus, shear modulus, and stress tensor for isotropic elastic medium

Here K is called the *bulk modulus* and μ the *shear modulus*, and the factor 2 is included for purely historical reasons. The first minus sign (with $K > 0$) ensures that the isotropic part of the stress, $-K\Theta\mathbf{g}$, resists volume changes; the second minus sign (with $\mu > 0$) ensures that the symmetric, trace-free part, $-2\mu\boldsymbol{\Sigma}$, resists shape changes (resists shearing).

11.3 Stress, Elastic Moduli, and Elastostatic Equilibrium

Hooke's law (Figs. 11.1 and 11.4) can be expressed in this same stress-proportional-to-strain form. The stress, when the rod is stretched, is the force F that does the stretching divided by the rod's cross sectional area A, the strain is the rod's fractional change of length $\Delta\ell/\ell$, and so Hooke's law takes the form

$$F/A = -E\Delta\ell/\ell, \qquad (11.19)$$

with E an elastic coefficient called *Young's modulus*. In Sec. 11.4, we show that E is a combination of the bulk and shear moduli: $E = 9\mu K/(3K + \mu)$.

In many treatments and applications of elasticity, the shear tensor $\mathbf{\Sigma}$ is paid little attention. The focus instead is on the strain S_{ij} and its trace $S_{kk} = \Theta$, and the elastic stress tensor (11.18) is written as $\mathbf{T} = -\lambda\Theta\mathbf{g} - 2\mu\mathbf{S}$, where $\lambda \equiv K - \frac{2}{3}\mu$. In these treatments μ and λ are called the *first and second Lamé coefficients* and are used in place of μ and K. We shall not adopt this viewpoint.

Lamé coefficients

11.3.4 Energy of Deformation

Take a wire of length ℓ and cross sectional area A, and stretch it (e.g., via the "Hooke's-law experiment" of Figs. 11.1 and 11.4) by an amount ζ' that grows gradually from 0 to $\Delta\ell$. When the stretch is ζ', the force that does the stretching is [by Eq. (11.19)] $F' = EA(\zeta'/\ell) = (EV/\ell^2)\zeta'$; here $V = A\ell$ is the wire's volume, and E is its Young's modulus. As the wire is gradually lengthened, the stretching force F' does work

$$W = \int_0^{\Delta\ell} F'd\zeta' = \int_0^{\Delta\ell} (EV/\ell^2)\zeta'd\zeta'$$
$$= \frac{1}{2}EV(\Delta\ell/\ell)^2.$$

This tells us that the stored elastic energy per unit volume is

$$U = \frac{1}{2}E(\Delta\ell/\ell)^2. \qquad (11.20)$$

To generalize this formula to a strained, isotropic, 3-dimensional medium, consider an arbitrary but small region $\mathcal{V}$ inside a body that has already been strained by a displacement vector field ξ_i and is thus already experiencing an elastic stress $T_{ij} = -K\Theta\delta_{ij} - 2\mu\Sigma_{ij}$ [Eq. (11.18)]. Imagine building up this displacement gradually from zero at the same rate everywhere in and around $\mathcal{V}$, so at some moment during the buildup the displacement field is $\xi'_i = \xi_i\epsilon$ (with the parameter ϵ gradually growing from 0 to 1). At that moment, the stress tensor (by virtue of the linearity of the stress-strain relation) is $T'_{ij} = T_{ij}\epsilon$. On the boundary $\partial\mathcal{V}$ of the region $\mathcal{V}$, this stress exerts a force $\Delta F'_i = -T'_{ij}\Delta\Sigma_j$ across any surface element $\Delta\Sigma_j$, from the exterior of $\partial\mathcal{V}$ to its interior. As the displacement grows, this surface force does the following amount of work on $\mathcal{V}$:

$$\Delta W_{\text{surf}} = \int \Delta F'_i d\xi'_i = \int (-T'_{ij}\Delta\Sigma_j)d\xi'_i = -\int_0^1 T_{ij}\epsilon\Delta\Sigma_j\xi'_i d\epsilon = -\frac{1}{2}T_{ij}\Delta\Sigma_j\xi_i.$$

$$(11.21)$$

The total amount of work done can be computed by adding up the contributions from all the surface elements of $\partial\mathcal{V}$:

$$W_{\text{surf}} = -\frac{1}{2}\int_{\partial\mathcal{V}} T_{ij}\xi_i d\Sigma_j = -\frac{1}{2}\int_{\mathcal{V}} (T_{ij}\xi_i)_{;j} dV = -\frac{1}{2}(T_{ij}\xi_i)_{;j} V. \quad (11.22)$$

In the second step we have used Gauss's theorem, and in the third step we have used the smallness of the region $\mathcal{V}$ to infer that the integrand is very nearly constant and the integral is the integrand times the total volume V of $\mathcal{V}$.

Does W_{surf} equal the elastic energy stored in $\mathcal{V}$? The answer is "no," because we must also take account of the work done in the interior of $\mathcal{V}$ by gravity or any other nonelastic force that may be acting. Although it is not easy in practice to turn gravity off and then on, we must do so in the following thought experiment. In the volume's final deformed state, the divergence of its elastic stress tensor is equal to the gravitational force density, $\nabla \cdot \mathbf{T} = \rho \mathbf{g}$ [Eqs. (11.13) and (11.14)]; and in the initial, undeformed and unstressed state, $\nabla \cdot \mathbf{T}$ must be zero, whence so must $\mathbf{g}$. Therefore, we must imagine growing the gravitational force proportional to ϵ just like we grow the displacement, strain, and stress. During this growth, with $\mathbf{g}' = \epsilon \mathbf{g}$, the gravitational force $\rho \mathbf{g}'V$ does the following amount of work on our tiny region $\mathcal{V}$:

$$W_{\text{grav}} = \int \rho V \mathbf{g}' \cdot d\boldsymbol{\xi}' = \int_0^1 \rho V \mathbf{g}\epsilon \cdot \boldsymbol{\xi} d\epsilon = \frac{1}{2}\rho V \mathbf{g} \cdot \boldsymbol{\xi} = \frac{1}{2}(\nabla \cdot \mathbf{T}) \cdot \boldsymbol{\xi} V = \frac{1}{2}T_{ij;j}\xi_i V. \quad (11.23)$$

The total work done to deform $\mathcal{V}$ is the sum of the work done by the elastic force (11.22) on its surface and the gravitational force (11.23) in its interior, $W_{\text{surf}} + W_{\text{grav}} = -\frac{1}{2}(\xi_i T_{ij})_{;j} V + \frac{1}{2} T_{ij;j}\xi_i V = -\frac{1}{2} T_{ij}\xi_{i;j} V$. This work gets stored in $\mathcal{V}$ as elastic energy, so the energy density is $U = -\frac{1}{2} T_{ij}\xi_{i;j}$. Inserting (for an isotropic material) $T_{ij} = -K\Theta g_{ij} - 2\mu \Sigma_{ij}$ and $\xi_{i;j} = \frac{1}{3}\Theta g_{ij} + \Sigma_{ij} + R_{ij}$ in this equation for U and performing some simple algebra that relies on the symmetry properties of the expansion, shear, and rotation (Ex. 11.3), we obtain

$$\boxed{U = \frac{1}{2}K\Theta^2 + \mu\Sigma_{ij}\Sigma_{ij}.} \quad (11.24)$$

elastic energy density at fixed temperature

Note that this elastic energy density is always positive if the elastic moduli are positive, as they must be for matter to be stable against small perturbations, and note that it is independent of the rotation R_{ij}, as it should be on physical grounds.

For the more general, anisotropic case, expression (11.24) becomes [by virtue of the stress-strain relation $T_{ij} = -Y_{ijkl}\xi_{k;l}$, Eq. (11.17)]

$$U = \frac{1}{2}\xi_{i;j} Y_{ijkl} \xi_{k;l}. \quad (11.25)$$

The volume integral of the elastic energy density given by Eq. (11.24) or (11.25) can be used as an action from which to compute the stress, by varying the displacement (Ex. 11.4). Since only the part of $\mathbf{Y}$ that is symmetric under interchange of the first and second pairs of slots contributes to U, only that part can affect the action-principle-derived stress. Therefore, it must be that $Y_{ijkl} = Y_{klij}$. This is the symmetry we asserted earlier.

EXERCISES

Exercise 11.3 *Derivation and Practice: Elastic Energy*
Beginning with $U = -\frac{1}{2}T_{ij}\xi_{i;j}$ [text following Eq. (11.23)], derive Eq. (11.24) for the elastic energy density inside a body.

Exercise 11.4 *Derivation and Practice: Action Principle for Elastic Stress*
For an anisotropic, elastic medium with elastic energy density $U = \frac{1}{2}\xi_{i;j}Y_{ijkl}\xi_{k;l}$, integrate this energy density over a 3-dimensional region $\mathcal{V}$ (not necessarily small) to get the total elastic energy E. Now consider a small variation $\delta\xi_i$ in the displacement field. Evaluate the resulting change δE in the elastic energy without using the relation $T_{ij} = -Y_{ijkl}\xi_{k;l}$. Convert to a surface integral over $\partial\mathcal{V}$, and thence infer the stress-strain relation $T_{ij} = -Y_{ijkl}\xi_{k;l}$.

11.3.5 Thermoelasticity

In our discussion of deformation energy, we tacitly assumed that the temperature of the elastic medium was held fixed during the deformation (i.e., we ignored the possibility of any thermal expansion). Correspondingly, the energy density U that we computed is actually the physical free energy per unit volume $\mathcal{F}$, at some chosen temperature T_0 of a heat bath. If we increase the bath's and material's temperature from T_0 to $T = T_0 + \delta T$, then the material wants to expand by $\Theta = \delta V/V = 3\alpha\delta T$ (i.e., it will have vanishing expansional elastic energy if Θ has this value). Here α is its **coefficient of linear thermal expansion**. (The factor 3 is because there are three directions into which it can expand: x, y, and z.) Correspondingly, the physical-free-energy density at temperature $T = T_0 + \delta T$ is

$$\mathcal{F} = \mathcal{F}_0(T) + \frac{1}{2}K(\Theta - 3\alpha\delta T)^2 + \mu\Sigma_{ij}\Sigma_{ij}. \tag{11.26}$$

The stress tensor in this heated and strained state can be computed from $T_{ij} = -\partial\mathcal{F}/\partial S_{ij}$ [a formula most easily inferred from Eq. (11.25) with U reinterpreted as $\mathcal{F}$ and $\xi_{i;j}$ replaced by its symmetrization, S_{ij}]. Reexpressing Eq. (11.26) in terms of S_{ij} and computing the derivative, we obtain (not surprisingly!)

$$T_{ij} = -\frac{\partial\mathcal{F}}{\partial S_{ij}} = -K(\Theta - 3\alpha\delta T)\delta_{ij} - 2\mu\Sigma_{ij}. \tag{11.27}$$

What happens if we allow our material to expand *adiabatically* rather than at fixed temperature? Adiabatic expansion means expansion at fixed entropy S. Consider a small sample of material that contains mass M and has volume $V = M/\rho$. Its entropy is $S = -[\partial(\mathcal{F}V)/\partial T]_V$ [cf. Eq. (5.33)], which, using Eq. (11.26), becomes

$$S = S_0(T) + 3\alpha K \Theta V. \quad (11.28)$$

Here we have neglected the term $-9\alpha^2 K \delta T$, which can be shown to be negligible compared to the temperature dependence of the elasticity-independent term $S_0(T)$. If our sample expands adiabatically by an amount $\Delta V = V \Delta \Theta$, then its temperature must go down by the amount $\Delta T < 0$ that keeps S fixed (i.e., that makes $\Delta S_0 = -3\alpha K V \Delta \Theta$). Noting that $T \Delta S_0$ is the change of the sample's thermal energy, which is $\rho c_V \Delta T$ (c_V is the specific heat per unit mass), we see that the temperature change is

$$\frac{\Delta T}{T} = \frac{-3\alpha K \Delta \Theta}{\rho c_V} \quad \text{for adiabatic expansion.} \quad (11.29)$$

temperature change in adiabatic expansion

This temperature change, accompanying an adiabatic expansion, alters slightly the elastic stress [Eq. (11.27)] and thence the bulk modulus K (i.e., it gives rise to an adiabatic bulk modulus that differs slightly from the isothermal bulk modulus K introduced in previous sections). However, the differences are so small that they are generally ignored. For further details, see Landau and Lifshitz (1986, Sec. 6).

11.3.6 Molecular Origin of Elastic Stress; Estimate of Moduli

It is important to understand the microscopic origin of the elastic stress. Consider an ionic solid in which singly ionized ions (e.g., positively charged sodium and negatively charged chlorine) attract their nearest (opposite-species) neighbors through their mutual Coulomb attraction and repel their *next* nearest (same-species) neighbors, and so on. Overall, there is a net electrostatic attraction on each ion, which is balanced by the short-range repulsion of its bound electrons against its neighbors' bound electrons. Now consider a thin slice of material of thickness intermediate between the inter-atomic spacing and the solid's inhomogeneity scale (Fig. 11.5).

Although the electrostatic force between individual pairs of ions is long range, the material is electrically neutral on the scale of several ions; as a result, when averaged

FIGURE 11.5 A thin slice of an ionic solid (between the dark lines) that interacts electromagnetically with ions outside it. The electrostatic force on the slice is dominated by interactions between ions lying in the two thin shaded areas, a few atomic layers thick, one on each side of the slice. The force is effectively a surface force rather than a volume force. In elastostatic equilibrium, the forces on the two sides are equal and opposite, if the slice is sufficiently thin.

TABLE 11.1: Density ρ; bulk, shear, and Young's moduli K, μ, and E, respectively; Poisson's ratio ν; and yield strain S_Y under tension, for various materials

Substance	ρ (kg m^{-3})	K (GPa)	μ (GPa)	E (GPa)	ν	S_Y	c_L (km s^{-1})	c_T (km s^{-1})
Carbon nanotube	1,300			~1,000		0.05		
Steel	7,800	170	81	210	0.29	0.003	5.9	3.2
Copper	8,960	130	45	120	0.34	0.0006	4.6	2.2
Rock	3,000	70	40	100	0.25	0.001	6.0	3.5
Glass	2,500	47	28	70	0.25	0.0005	5.8	3.3
Rubber	1,200	10	0.0007	0.002	0.50	~8	1.0	0.03
DNA molecule				0.3		~0.1		

Notes: The final two columns are the longitudinal and transverse sound speeds c_L, c_T, defined in Chap. 12. The DNA molecule is discussed in Ex. 11.12.

over many ions, the net electric force is short range (Fig. 11.5). We can therefore treat the net force acting on the thin slice as a surface force, governed by local conditions in the material. This is essential if we are to be able to write down a localized linear stress-strain relation $T_{ij} = -Y_{ijkl}S_{kl}$ or $T_{ij} = -K\Theta\delta_{ij} - 2\mu\Sigma_{ij}$. This need not have been the case; there are other circumstances where the net electrostatic force is long range, not short. One example occurs in certain types of crystal (e.g., tourmaline), which develop internal, long-range *piezoelectric* fields when strained.

Our treatment so far has implicitly assumed that matter is continuous on all scales and that derivatives are mathematically well defined. Of course, this is not the case. In fact, we not only need to acknowledge the existence of atoms, we must also use them to compute the elastic moduli.

magnitudes of elastic moduli

We can estimate the elastic moduli in ionic or metallic materials by observing that, if a crystal lattice were to be given a dimensionless strain of order unity, then the elastic stress would be of order the electrostatic force between adjacent ions divided by the area associated with each ion. If the lattice spacing is $a \sim 2$ Å $= 0.2$ nm and the ions are singly charged, then K and $\mu \sim e^2/4\pi\epsilon_0 a^4 \sim 100$ GPa. This is about a million atmospheres. Covalently bonded compounds are less tightly bound and have somewhat smaller elastic moduli; exotic carbon nanotubes have larger moduli. See Table 11.1.

On the basis of this argument, it might be thought that crystals can be subjected to strains of order unity before they attain their elastic limits. However, as discussed in Sec. 11.3.2, most materials are only elastic for strains $\lesssim 10^{-3}$. The reason for this difference is that crystals are generally imperfect and are laced with *dislocations*. Relatively small stresses suffice for the dislocations to move through the solid and for the crystal thereby to undergo permanent deformation (Fig. 11.6).

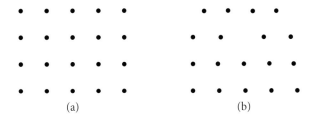

FIGURE 11.6 The ions in one layer of a crystal. In subsequent layers, going into each picture, the ion distribution is the same. (a) This perfect crystal, in which the atoms are organized in a perfectly repeating lattice, can develop very large shear strains without yielding. (b) Real materials contain dislocations that greatly reduce their rigidity. The simplest type of dislocation, shown here, is the *edge dislocation* (with the central vertical atomic layer having a terminating edge that extends into the picture). The dislocation will move transversely, and the crystal thereby will undergo inelastic deformation when the strain is typically greater than $\sim 10^{-3}$, which is $\sim 1\%$ of the yield shear strain for a perfect crystal.

EXERCISES

Exercise 11.5 *Problem: Order-of-Magnitude Estimates*

(a) What is the maximum size of a nonspherical asteroid? [Hint: If the asteroid is too large, its gravity will deform it into a spherical shape.]

(b) What length of steel wire can hang vertically without breaking? What length of carbon nanotube? What are the prospects for creating a tether that hangs to Earth's surface from a geostationary satellite?

(c) Can a helium balloon lift the tank used to transport its helium gas? (Purcell, 1983).

Exercise 11.6 *Problem: Jumping Heights*

Explain why all animals, from fleas to humans to elephants, can jump to roughly the same height. The field of science that deals with topics like this is called *allometry* (Ex. 11.18).

11.3.7 Elastostatic Equilibrium: Navier-Cauchy Equation

It is commonly the case that the elastic moduli K and μ are constant (i.e., independent of location in the medium), even though the medium is stressed in an inhomogeneous way. (This is because the strains are small and thus perturb the material properties by only small amounts.) If so, then from the elastic stress tensor $\mathbf{T} = -K\Theta\mathbf{g} - 2\mu\boldsymbol{\Sigma}$ and expressions (11.4a) and (11.4b) for the expansion and shear in terms of the displacement vector, we can deduce the following expression for the elastic force density $\mathbf{f}$ [Eq. (11.13)] inside the body:

elastic force density

$$\mathbf{f} = -\nabla \cdot \mathbf{T} = K\nabla\Theta + 2\mu\nabla\cdot\boldsymbol{\Sigma} = \left(K + \frac{1}{3}\mu\right)\nabla(\nabla\cdot\boldsymbol{\xi}) + \mu\nabla^2\boldsymbol{\xi}; \quad (11.30)$$

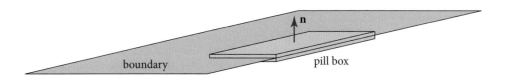

FIGURE 11.7 Pill box used to derive boundary conditions in electrostatics and elastostatics.

see Ex. 11.7. Here $\nabla \cdot \mathbf{\Sigma}$ in index notation is $\Sigma_{ij;j} = \Sigma_{ji;j}$. Extra terms must be added if we are dealing with anisotropic materials. However, in this book Eq. (11.30) will be sufficient for our needs.

If no other countervailing forces act in the interior of the material (e.g., if there is no gravitational force), and if, as in this chapter, the material is in a static, equilibrium state rather than vibrating dynamically, then this force density will have to vanish throughout the material's interior. This vanishing of $\mathbf{f} \equiv -\nabla \cdot \mathbf{T}$ is just a fancy version of Newton's law for static situations, $\mathbf{F} = m\mathbf{a} = 0$. If the material has density ρ and is pulled on by a gravitational acceleration $\mathbf{g}$, then the sum of the elastostatic force per unit volume and gravitational force per unit volume must vanish, $\mathbf{f} + \rho \mathbf{g} = 0$:

Navier-Cauchy equation for elastostatic equilibrium

$$\boxed{\mathbf{f} + \rho \mathbf{g} = \left(K + \frac{1}{3}\mu\right) \nabla(\nabla \cdot \boldsymbol{\xi}) + \mu \nabla^2 \boldsymbol{\xi} + \rho \mathbf{g} = 0.} \quad (11.31)$$

This is often called the *Navier-Cauchy equation*.[5]

When external forces are applied to the surface of an elastic body (e.g., when one pushes on the face of a cylinder) and gravity acts on the interior, the distribution of the strain $\boldsymbol{\xi}(\mathbf{x})$ inside the body can be computed by solving the Navier-Cauchy equation (11.31) subject to boundary conditions provided by the applied forces.

In electrostatics, one can derive boundary conditions by integrating Maxwell's equations over the interior of a thin box (a "pill box") with parallel faces that snuggle up to the boundary (Fig. 11.7). For example, by integrating $\nabla \cdot \mathbf{E} = \rho_e/\epsilon_o$ over the interior of the pill box and then applying Gauss's law to convert the left-hand side to a surface integral, we obtain the junction condition that the discontinuity in the normal component of the electric field is equal $1/\epsilon_o$ times the surface charge density. Similarly, in elastostatics one can derive boundary conditions by integrating the elastostatic equation $\nabla \cdot \mathbf{T} = 0$ over the pill box of Fig. 11.7 and then applying Gauss's law:

$$0 = \int_\mathcal{V} \nabla \cdot \mathbf{T}\, dV = \int_{\partial \mathcal{V}} \mathbf{T} \cdot d\mathbf{\Sigma} = \int_{\partial \mathcal{V}} \mathbf{T} \cdot \mathbf{n}\, dA = [(\mathbf{T} \cdot \mathbf{n})_{\text{upper face}} - (\mathbf{T} \cdot \mathbf{n})_{\text{lower face}}]\, A. \quad (11.32)$$

Here in the next-to-last expression we have used $d\mathbf{\Sigma} = \mathbf{n}\, dA$, where dA is the scalar area element, and $\mathbf{n}$ is the unit normal to the pill-box face. In the last term we have

---

5. It was first written down by Claude-Louis Navier (in 1821) and in a more general form by Augustin-Louis Cauchy (in 1822).

assumed the pill box has a small face, so **T** · **n** can be treated as constant and be pulled outside the integral. The result is the boundary condition that

$$\mathbf{T} \cdot \mathbf{n} \quad \text{must be continuous across any boundary;} \tag{11.33}$$

boundary conditions for Navier-Cauchy equation

in index notation, $T_{ij}n_j$ is continuous.

Physically, this is nothing but the law of force balance across the boundary: the force per unit area acting from the lower side to the upper side must be equal and opposite to that acting from upper to lower. As an example, if the upper face is bounded by vacuum, then the solid's stress tensor must satisfy $T_{ij}n_j = 0$ at the surface. If a normal pressure P is applied by some external agent at the upper face, then the solid must respond with a normal force equal to P: $n_i T_{ij} n_j = P$. If a vectorial force per unit area $\mathcal{F}_i$ is applied at the upper face by some external agent, then it must be balanced: $T_{ij}n_j = -\mathcal{F}_i$.

Solving the Navier-Cauchy equation (11.32) for the displacement field $\boldsymbol{\xi}(\mathbf{x})$, subject to specified boundary conditions, is a problem in elastostatics analogous to solving Maxwell's equations for an electric field subject to boundary conditions in electrostatics, or for a magnetic field subject to boundary conditions in magnetostatics. The types of solution techniques used in electrostatics and magnetostatics can also be used here. See Box 11.3.

EXERCISES

Exercise 11.7 *Derivation and Practice: Elastic Force Density*
From Eq. (11.18) derive expression (11.30) for the elastostatic force density inside an elastic body.

Exercise 11.8 ***Practice: Biharmonic Equation*
A homogeneous, isotropic, elastic solid is in equilibrium under (uniform) gravity and applied surface stresses. Use Eq. (11.30) to show that the displacement inside it, $\boldsymbol{\xi}(\mathbf{x})$, is biharmonic, i.e., it satisfies the differential equation

$$\boxed{\nabla^2 \nabla^2 \boldsymbol{\xi} = 0.} \tag{11.34a}$$

Show also that the expansion Θ satisfies the Laplace equation

$$\boxed{\nabla^2 \Theta = 0.} \tag{11.34b}$$

11.4 Young's Modulus and Poisson's Ratio for an Isotropic Material: A Simple Elastostatics Problem

As a simple example of an elastostatics problem, we explore the connection between our 3-dimensional theory of stress and strain and the 1-dimensional Hooke's law (Fig. 11.1).

> **BOX 11.3. METHODS OF SOLVING THE NAVIER-CAUCHY EQUATION**
>
> Many techniques have been devised to solve the Navier-Cauchy equation (11.31), or other equations equivalent to it, subject to appropriate boundary conditions. Among them are:
>
> - Separation of variables. See Sec. 11.9.2.
> - Green's functions. See Ex. 11.27 and Johnson (1985).
> - Variational principles. See Marsden and Hughes (1986, Chap. 5) and Slaughter (2002, Chap. 10).
> - Saint-Venant's principle. One changes the boundary conditions to something simpler, for which the Navier-Cauchy equation can be solved analytically, and then one uses linearity of the Navier-Cauchy equation to compute an approximate, additive correction that accounts for the difference in boundary conditions.[1]
> - Dimensional reduction. This method reduces the theory to 2 dimensions in the case of thin plates (Sec. 11.7), and to 1 dimension for rods and for translation-invariant plates (Sec. 11.5).
> - Complex variable methods. These are particularly useful in solving the 2-dimensional equations (Boresi and Chong, 1999, Appendix 5B).
> - Numerical simulations on computers. These are usually carried out by the method of finite elements, in which one approximates stressed objects by a finite set of elementary, interconnected physical elements, such as rods; thin, triangular plates; and tetrahedra (Ugural and Fenster, 2012, Chap. 7).
> - Replace Navier-Cauchy by equivalent equations. For example, and widely used in the engineering literature, write force balance $T_{ij;j} = 0$ in terms of the strain tensor S_{ij}, supplement this with an equation that guarantees S_{ij} can be written as the symmetrized gradient of a vector field (the displacement vector), and develop techniques to solve these coupled equations plus boundary conditions for S_{ij} [Ugural and Fenster (2012, Sec. 2.4); also large parts of Boresi and Chong (1999) and Slaughter (2002)].
> - Mathematica or other computer software. These software packages can be used to perform complicated analytical analyses. One can then explore their predictions numerically (Constantinescu and Korsunsky, 2007).
>
> 1. In 1855 Barré de Saint-Venant had the insight to realize that, under suitable conditions, the correction will be significant only locally (near the altered boundary) and not globally. (See Boresi and Chong, 1999, pp. 288ff; Ugural and Fenster, 2012, Sec. 2.16, and references therein.)

Consider a thin rod of square cross section hanging along the $\mathbf{e}_z$ direction of a Cartesian coordinate system (Fig. 11.1). Subject the rod to a stretching force applied normally and uniformly at its ends. (It could just as easily be a rod under compression.) Its sides are free to expand or contract transversely, since no force acts on them: $dF_i = T_{ij}d\Sigma_j = 0$. As the rod is slender, vanishing of dF_i at its x and y sides implies to high accuracy that the stress components T_{ix} and T_{iy} will vanish throughout the interior; otherwise there would be a very large force density $T_{ij;j}$ inside the rod. Using $T_{ij} = -K\Theta g_{ij} - 2\mu\Sigma_{ij}$, we then obtain

$$T_{xx} = -K\Theta - 2\mu\Sigma_{xx} = 0, \tag{11.35a}$$

$$T_{yy} = -K\Theta - 2\mu\Sigma_{yy} = 0, \tag{11.35b}$$

$$T_{yz} = -2\mu\Sigma_{yz} = 0, \tag{11.35c}$$

$$T_{xz} = -2\mu\Sigma_{xz} = 0, \tag{11.35d}$$

$$T_{xy} = -2\mu\Sigma_{xy} = 0, \tag{11.35e}$$

$$T_{zz} = -K\Theta - 2\mu\Sigma_{zz}. \tag{11.35f}$$

From the first two of these equations and $\Sigma_{xx} + \Sigma_{yy} + \Sigma_{zz} = 0$, we obtain a relationship between the expansion and the nonzero components of the shear,

$$K\Theta = \mu\Sigma_{zz} = -2\mu\Sigma_{xx} = -2\mu\Sigma_{yy}; \tag{11.36}$$

and from this and Eq. (11.35f), we obtain $T_{zz} = -3K\Theta$. The decomposition of S_{ij} into its irreducible tensorial parts tells us that $S_{zz} = \xi_{z;z} = \Sigma_{zz} + \frac{1}{3}\Theta$, which becomes, on using Eq. (11.36), $\xi_{z;z} = [(3K + \mu)/(3\mu)]\Theta$. Combining with $T_{zz} = -3K\Theta$, we obtain Hooke's law and an expression for Young's modulus E in terms of the bulk and shear moduli:

$$\frac{-T_{zz}}{\xi_{z;z}} = \frac{9\mu K}{3K + \mu} = E. \tag{11.37}$$

Hooke's law and Young's modulus

It is conventional to introduce *Poisson's ratio* ν, which is minus the ratio of the lateral strain to the longitudinal strain during a deformation of this type, where the transverse motion is unconstrained. It can be expressed as a ratio of elastic moduli as follows:

$$\nu \equiv -\frac{\xi_{x;x}}{\xi_{z;z}} = -\frac{\xi_{y;y}}{\xi_{z;z}} = -\frac{\Sigma_{xx} + \frac{1}{3}\Theta}{\Sigma_{zz} + \frac{1}{3}\Theta} = \frac{3K - 2\mu}{2(3K + \mu)}, \tag{11.38}$$

Poisson's ratio

where we have used Eq. (11.36). We tabulate these relations and their inverses for future use:

$$\boxed{E = \frac{9\mu K}{3K + \mu}, \quad \nu = \frac{3K - 2\mu}{2(3K + \mu)}; \quad K = \frac{E}{3(1 - 2\nu)}, \quad \mu = \frac{E}{2(1 + \nu)}.}$$

(11.39)

We have already remarked that mechanical stability of a solid requires that $K, \mu > 0$. Using Eq. (11.39), we observe that this imposes a restriction on Poisson's ratio, namely that $-1 < \nu < 1/2$. For metals, Poisson's ratio is typically about 1/3, and the shear modulus is roughly half the bulk modulus. For a substance that is easily sheared but not easily compressed, like rubber (or neutron star crusts; Sec. 11.3.6), the bulk modulus is relatively high and $\nu \simeq 1/2$ (cf. Table 11.1). For some exotic materials, Poisson's ratio can be negative (cf. Yeganeh-Haeri, Weidner, and Parise, 1992).

Although we derived them for a square strut under extension, our expressions for Young's modulus and Poisson's ratio are quite general. To see this, observe that the derivation would be unaffected if we combined many parallel, square fibers together. All that is necessary is that the transverse motion be free, so that the only applied force is uniform and normal to a pair of parallel faces.

11.5 Reducing the Elastostatic Equations to 1 Dimension for a Bent Beam: Cantilever Bridge, Foucault Pendulum, DNA Molecule, Elastica

When dealing with bodies that are much thinner in 2 dimensions than the third (e.g., rods, wires, and beams), one can use the *method of moments* to reduce the 3-dimensional elastostatic equations to ordinary differential equations in 1 dimension (a process called *dimensional reduction*). We have already met an almost trivial example of this in our discussion of Hooke's law and Young's modulus (Sec. 11.4 and Fig. 11.1). In this section, we discuss a more complicated example, the bending of a beam through a small displacement angle. In Ex. 11.13, we shall analyze a more complicated example: the bending of a long, elastic wire into a complicated shape called an *elastica*.

Our beam-bending example is motivated by a common method of bridge construction, which uses cantilevers. (A famous historical example is the old bridge over the Firth of Forth in Scotland that was completed in 1890 with a main span of half a kilometer.)

The principle is to attach two independent beams to the two shores as cantilevers, and allow them to meet in the middle. (In practice the beams are usually supported at the shores on piers and strengthened along their lengths with trusses.) Similar cantilevers, with lengths of order a micron or less, are used in atomic force microscopes and other nanotechnology applications, including quantum-information experiments.

Let us make a simple model of a cantilever (Fig. 11.8). Consider a beam clamped rigidly at one end, with length ℓ, horizontal width w, and vertical thickness h. Introduce local Cartesian coordinates with $\mathbf{e}_x$ pointing along the beam and $\mathbf{e}_z$ pointing vertically upward. Imagine the beam extending horizontally in the absence of gravity. Now let it sag under its own weight, so that each element is displaced through a small distance $\boldsymbol{\xi}(\mathbf{x})$. The upper part of the beam is stretched, while the lower part is compressed, so there must be a *neutral surface* where the horizontal strain $\xi_{x,x}$ vanishes.

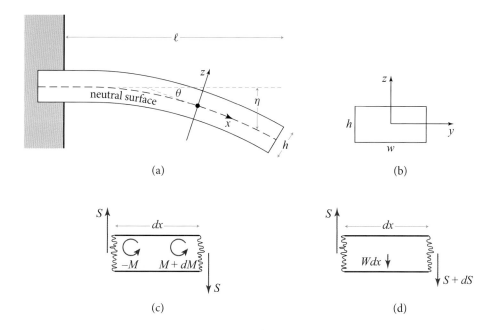

FIGURE 11.8 Bending of a cantilever. (a) A beam is held rigidly at one end and extends horizontally with the other end free. We introduce an orthonormal coordinate system (x, y, z) with $\mathbf{e}_x$ extending along the beam. We only consider small departures from equilibrium. The bottom of the beam will be compressed, the upper portion extended. There is therefore a neutral surface $z = 0$ on which the strain $\xi_{x,x}$ vanishes. (b) The beam has a rectangular cross section with horizontal width w and vertical thickness h; its length is ℓ. (c) The bending torque M must be balanced by the torque exerted by the vertical shear force S. (d) The shear force S must vary along the beam so as to support the beam's weight per unit length, W.

This neutral surface must itself be curved downward. Let its downward displacement from the horizontal plane that it occupied before sagging be $\eta(x)$ (> 0), let a plane tangent to the neutral surface make an angle $\theta(x)$ (also > 0) with the horizontal, and adjust the x and z coordinates so x runs along the slightly curved neutral plane and z is orthogonal to it (Fig. 11.8). The longitudinal strain is then given to first order in small quantities by

$$\xi_{x,x} = \frac{z}{\mathcal{R}} = z\frac{d\theta}{dx} \simeq z\frac{d^2\eta}{dx^2}, \qquad (11.40a)$$ longitudinal strain

where $\mathcal{R} = dx/d\theta > 0$ is the radius of curvature of the beam's bend, and we have chosen $z = 0$ at the neutral surface. The 1-dimensional displacement $\eta(x)$ will be the focus for dimensional reduction of the elastostatic equations.

As in our discussion of Hooke's law for a stretched rod (Sec. 11.4), we can regard the beam as composed of a bundle of long, parallel fibers, stretched or squeezed along their length and free to contract transversely. The longitudinal stress is therefore

$$T_{xx} = -E\xi_{x,x} = -Ez\frac{d^2\eta}{dx^2}. \qquad (11.40b)$$

We can now compute the horizontal force density, which must vanish in elastostatic equilibrium:[6]

$$f_x = -T_{xx,x} - T_{xz,z} = Ez\frac{d^3\eta}{dx^3} - T_{xz,z} = 0. \qquad (11.40c)$$

method of moments

This is a partial differential equation. We convert it into a 1-dimensional ordinary differential equation by the *method of moments*: We multiply it by z and integrate over z (i.e., we compute its "first moment"). Integrating the second term, $\int zT_{xz,z}dz$, by parts and using the boundary condition $T_{xz} = 0$ on the upper and lower surfaces of the beam, we obtain

$$\frac{Eh^3}{12}\frac{d^3\eta}{dx^3} = -\int_{-h/2}^{h/2} T_{xz}\,dz. \qquad (11.40d)$$

Using $T_{xz} = T_{zx}$, notice that the integral, when multiplied by the beam's width w in the y direction, is the vertical shear force $S(x)$ in the beam:

shear force, S

$$\boxed{S = \int T_{zx}dydz = w\int_{-h/2}^{h/2} T_{zx}dz = -D\frac{d^3\eta}{dx^3}.} \qquad (11.41a)$$

Here

flexural rigidity or bending modulus of an elastic beam, D

$$\boxed{D \equiv E\int z^2 dydz \equiv EA\,r_g^2 = Ewh^3/12} \qquad (11.41b)$$

is called the beam's *flexural rigidity*, or its *bending modulus*. Notice that, quite generally, D is the beam's Young's modulus E times the second moment of the beam's cross sectional area A along the direction of bend. Engineers call that second moment $A\,r_g^2$ and call r_g the radius of gyration. For our rectangular beam, this D is $Ewh^3/12$.

As an aside, we can gain some insight into Eq. (11.41a) by examining the torques that act on a segment of the beam with length dx. As shown in Fig. 11.8c, the shear forces on the two ends of the segment exert a clockwise torque $2S(dx/2) = Sdx$. This is balanced by a counterclockwise torque due to the stretching of the upper half of the segment and compression of the lower half (i.e., due to the bending of the beam). This *bending torque* is

bending torque, M

$$\boxed{M \equiv \int T_{xx}zdydz = -D\frac{d^2\eta}{dx^2}} \qquad (11.41c)$$

---

6. Because the coordinates are slightly curvilinear rather than precisely Cartesian, our Cartesian-based analysis makes small errors. Track-Two readers who have studied Sec. 11.8 can evaluate those errors using connection-coefficient terms that were omitted from this equation: $-\Gamma_{xjk}T_{jk} - \Gamma_{jkj}T_{xk}$. Each Γ has magnitude $1/\mathcal{R}$, so these terms are of order $T_{jk}/\mathcal{R}$, whereas the terms kept in Eq. (11.40c) are of order T_{xx}/ℓ and T_{xz}/h. Since the thickness h and length ℓ of the beam are small compared to the beam's radius of curvature $\mathcal{R}$, the connection-coefficient terms are negligible.

on the right end of the segment and minus this on the left, so torque balance says $(dM/dx)dx = Sdx$:

$$\boxed{S = dM/dx;} \tag{11.42}$$

see Fig. 11.8c. This is precisely Eq. (11.41a).

Equation (11.41a) [or equivalently, Eq. (11.42)] embodies half of the elastostatic equations. It is the x component of force balance $f_x = 0$, converted to an ordinary differential equation by evaluating its lowest nonvanishing moment: its first moment, $\int z f_x dy dz = 0$ [Eq. (11.40d)]. The other half is the z component of stress balance, which we can write as

$$T_{zx,x} + T_{zz,z} + \rho g = 0 \tag{11.43}$$

(vertical elastic force balanced by gravitational pull on the beam). We can convert this to a 1-dimensional ordinary differential equation by taking its lowest nonvanishing moment, its zeroth moment (i.e., by integrating over y and z). The result is

$$\boxed{\frac{dS}{dx} = -W,} \tag{11.44}$$

where $W = g\rho wh$ is the beam's weight per unit length (Fig. 11.8d).

weight per unit length, W

Combining our two dimensionally reduced components of force balance, Eqs. (11.41a) and (11.44), we obtain a fourth-order differential equation for our 1-dimensional displacement $\eta(x)$:

$$\boxed{\frac{d^4\eta}{dx^4} = \frac{W}{D}.} \tag{11.45}$$

elastostatic force balance equation for bent beam

(Fourth-order differential equations are characteristic of elasticity.)

Equation (11.45) can be solved subject to four appropriate boundary conditions. However, before we solve it, notice that *for a beam of a fixed length ℓ, the deflection η is inversely proportional to the flexural rigidity*. Let us give a simple example of this scaling. Floors in U.S. homes are conventionally supported by wooden joists of 2" by 6" lumber with the 6" side vertical. Suppose an inept carpenter installed the joists with the 6" side horizontal. The flexural rigidity of the joist would be reduced by a factor 9, and the center of the floor would be expected to sag 9 times as much as if the joists had been properly installed—a potentially catastrophic error.

Also, before solving Eq. (11.45), let us examine the approximations that we have made. First, we have assumed that the sag is small compared with the length of the beam, when making the small-angle approximation in Eq. (11.40a); we have also assumed the beam's radius of curvature is large compared to its length, when treating our slightly curved coordinates as Cartesian.[7] These assumptions will usually be valid, but are not so for the elastica studied in Ex. 11.13. Second, by using the method of moments rather than solving for the complete local stress tensor field, we have ignored the effects of some components of the stress tensor. In particular, when evaluating the bending torque [Eq. (11.41c)] we have ignored the effect of the T_{zx} component of the stress tensor. This is $O(h/\ell)T_{xx}$, and so our equations can only be accurate for fairly slender beams. Third, the extension above the neutral surface and the compression below the neutral surface lead to changes in the cross sectional shape of the beam. The fractional error here is of order the longitudinal shear, which is small for real materials.

The solution to Eq. (11.45) is a fourth-order polynomial with four unknown constants, to be set by boundary conditions. In this problem, the beam is held horizontal at the fixed end, so that $\eta(0) = \eta'(0) = 0$, where the prime denotes d/dx. At the free end, T_{zx} and T_{xx} must vanish, so the shear force S must vanish, whence $\eta'''(\ell) = 0$ [Eq. (11.41a)]; the bending torque M [Eq. (11.41c)] must also vanish, whence [by Eq. (11.42)] $\int S dx \propto \eta''(\ell) = 0$. By imposing these four boundary conditions $\eta(0) = \eta'(0) = \eta''(\ell) = \eta'''(\ell) = 0$ on the solution of Eq. (11.45), we obtain for the beam shape

displacement of a clamped cantilever

$$\eta(x) = \frac{W}{D}\left(\frac{1}{4}\ell^2 x^2 - \frac{1}{6}\ell x^3 + \frac{1}{24}x^4\right). \tag{11.46a}$$

Therefore, the end of the beam sags by

$$\eta(\ell) = \frac{W\ell^4}{8D}. \tag{11.46b}$$

Problems in which the beam rests on supports rather than being clamped can be solved in a similar manner. The boundary conditions will be altered, but the differential equation (11.45) will be unchanged.

Now suppose that we have a cantilever bridge of constant vertical thickness h and total span $2\ell \sim 100$ m made of material with density $\rho \sim 8 \times 10^3$ kg m^{-3} (e.g., steel) and Young's modulus $E \sim 100$ GPa. Suppose further that we want the center of the bridge to sag by no more than $\eta \sim 1$ m. According to Eq. (11.46b), the thickness of the beam must satisfy

$$h \gtrsim \left(\frac{3\rho g\ell^4}{2E\eta}\right)^{1/2} \sim 2.8 \text{ m.} \tag{11.47}$$

This estimate makes no allowance for all the extra strengthening and support present in real structures (e.g., via trusses and cables), and so it is an overestimate.

---

7. In more technical language, when neglecting the connection-coefficient terms discussed in the previous footnote.

Exercise 11.9 *Derivation: Sag in a Cantilever*
(a) Verify Eqs. (11.46) for the sag in a horizontal beam clamped at one end and allowed to hang freely at the other end.
(b) Now consider a similar beam with constant cross section and loaded with weights, so that the total weight per unit length is $W(x)$. What is the sag of the free end, expressed as an integral over $W(x)$, weighted by an appropriate Green's function?

Exercise 11.10 *Example: Microcantilever*
A microcantilever, fabricated from a single crystal of silicon, is being used to test the inverse square law of gravity on micron scales (Weld et al., 2008). It is clamped horizontally at one end, and its horizontal length is $\ell = 300$ μm, its horizontal width is $w = 12$ μm, and its vertical height is $h = 1$ μm. (The density and Young's modulus for silicon are $\rho = 2{,}000$ kg m^{-3} and $E = 100$ GPa, respectively.) The cantilever is loaded at its free end with a $m = 10$ μg gold mass.
(a) Show that the static deflection of the end of the cantilever is $\eta(\ell) = mg\ell^3/(3D) = 9$ μm, where $g = 10$ m s^{-2} is the acceleration due to gravity. Explain why it is permissible to ignore the weight of the cantilever.
(b) Next suppose the mass is displaced slightly vertically and then released. Show that the natural frequency of oscillation of the cantilever is $f = 1/(2\pi)\sqrt{g/\eta(\ell)} \simeq 170$ Hz.
(c) A second, similar gold mass is placed 100 μm away from the first. Estimate roughly the Newtonian gravitational attraction between these two masses, and compare with the attraction of Earth. Suggest a method that exploits the natural oscillation of the cantilever to measure the tiny gravitational attraction between the two gold masses.

The motivation for developing this technique was to seek departures from Newton's inverse-square law of gravitation on ∼micron scales, which had been predicted if our universe is a membrane ("brane") in a higher-dimensional space ("bulk") with at least one macroscopic extra dimension. No such departures have been found as of 2016.

Exercise 11.11 *Example: Foucault Pendulum*
In any high-precision Foucault pendulum, it is important that the pendular restoring force be isotropic, since anisotropy will make the swinging period different in different planes and thereby cause precession of the plane of swing.
(a) Consider a pendulum of mass m and length ℓ suspended (as shown in Fig. 11.9) by a rectangular wire with thickness h in the plane of the bend (X-Z plane) and thickness w orthogonal to that plane (Y direction). Explain why the force that the wire exerts on the mass is approximately $-\mathbf{F} = -(mg\cos\theta_o + m\ell\dot{\theta}_o^2)\mathbf{e}_x$, where g is the acceleration of gravity, θ_o is defined in the figure, and $\dot{\theta}_o$ is the time derivative of θ_o due to the swinging of the pendulum. In the second term we have

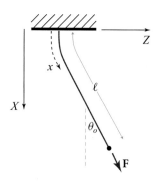

FIGURE 11.9 Foucault pendulum.

assumed that the wire is long compared to its region of bend. Express the second term in terms of the amplitude of swing $\theta_o^{\max}$, and show that for small amplitudes $\theta_o^{\max} \ll 1$, $\mathbf{F} \simeq -mg\mathbf{e}_x$. Use this approximation in the subsequent parts.

(b) Assuming that all along the wire, its angle $\theta(x)$ to the vertical is small, $\theta \ll 1$, show that

$$\theta(x) = \theta_o\left(1 - e^{-x/\lambda}\right), \tag{11.48a}$$

where λ (not to be confused with the second Lamé coefficient) is

$$\lambda = \frac{h}{(12\epsilon)^{1/2}}, \tag{11.48b}$$

$\epsilon = \xi_{x,x}$ is the longitudinal strain in the wire, and h is the wire's thickness in the plane of its bend. [Hint: The solution to Ex. 11.9 might be helpful.] Note that the bending of the wire is concentrated near the support, so this is where dissipation will be most important and where most of the suspension's thermal noise will arise (cf. Sec. 6.8 for discussion of thermal noise).

(c) Hence show that the shape of the wire is given in terms of Cartesian coordinates by

$$Z = [X - \lambda(1 - e^{-X/\lambda})]\theta_o \tag{11.48c}$$

to leading order in λ, and that the pendulum period is

$$P = 2\pi \left(\frac{\ell - \lambda}{g}\right)^{1/2}. \tag{11.48d}$$

(d) Finally, show that the pendulum periods when swinging along $\mathbf{e}_x$ and $\mathbf{e}_y$ differ by

$$\frac{\delta P}{P} = \left(\frac{h-w}{\ell}\right)\left(\frac{1}{48\epsilon}\right)^{1/2}. \tag{11.48e}$$

From Eq. (11.48e) one can determine how accurately the two thicknesses h and w must be equal to achieve a desired degree of isotropy in the period. A similar analysis can be carried out for the more realistic case of a slightly elliptical wire.

Exercise 11.12 *Example: DNA Molecule—Bending, Stretching, Young's Modulus, and Yield Point*

A DNA molecule consists of two long strands wound around each other as a helix, forming a cylinder with radius $a \simeq 1$ nm. In this exercise, we explore three ways of measuring the molecule's Young's modulus E. For background and further details, see Marko and Cocco (2003) and Nelson (2008, Chap. 9).

(a) Show that if a segment of DNA with length ℓ is bent into a segment of a circle with radius R, its elastic energy is $E_{el} = D\ell/(2R^2)$, where $D = (\pi/4)a^4 E$ is the molecule's flexural rigidity.

(b) Biophysicists define the DNA's *persistence length* ℓ_p as that length which, when bent through an angle of 90°, has elastic energy $E_{el} = k_B T$, where k_B is Boltzmann's constant and T is the temperature of the molecule's environment. Show that $\ell_p \simeq D/(k_B T)$. Explain why, in a thermalized environment, segments much shorter than ℓ_p will be more or less straight, and segments with length $\sim \ell_p$ will be randomly bent through angles of order 90°.

(c) Explain why a DNA molecule with total length L will usually be found in a random coil with diameter $d \simeq \ell_p \sqrt{L/\ell_p} = \sqrt{L\ell_p}$. Observations at room temperature with $L \simeq 17\ \mu$m reveal that $d \simeq 1\ \mu$m. From this show that the persistence length is $\ell_p \simeq 50$ nm at room temperature, and thence evaluate the molecule's flexural rigidity and from it, show that the molecule's Young's modulus is $E \simeq 0.3$ GPa; cf. Table 11.1.

(d) When the ends of a DNA molecule are attached to glass beads and the beads are pulled apart with a gradually increasing force F, the molecule begins to uncoil, just like rubber. To understand this semiquantitatively, think of the molecule as like a chain made of N links, each with length ℓ_p, whose interfaces can bend freely. If the force acts along the z direction, explain why the probability that any chosen link will make an angle θ to the z axis is $dP/d\cos\theta \propto \exp[+F\ell_p \cos\theta/(k_B T)]$. [Hint: This is analogous to the probability $dP/dV \propto \exp[-PV/(k_B T)]$ for the volume V of a system in contact with a bath that has pressure P and temperature T [Eq. (5.49)]; see also the discussion preceding Eq. (11.56).] Infer that when the force is F, the molecule's length along the force's direction is $\bar{L} \simeq L(\coth\alpha - 1/\alpha)$, where $\alpha = F\ell_p/(k_B T)$ and $L = N\ell_p$ is the length of the uncoiled molecule. Infer, further, that for $\alpha \ll 1$ (i.e., $F \ll k_B T/\ell_p \sim 0.1$ pN), our model predicts $\bar{L} \simeq \alpha L/3$, i.e. a linear force-length relation $F = (3k_B T/\ell_p)\bar{L}/L$, with a strongly temperature dependent spring constant, $3k_B T/\ell_p \propto T^2$. The measured value of this spring constant, at room temperature, is about 0.13 pN (Fig. 9.5 of Nelson, 2008). From this infer a value 0.5 GPa for the molecule's Young's modulus. This agrees surprisingly well with the 0.3 GPa deduced in part (c), given the crudeness of the jointed chain model.

(e) Show that when $F \gg k_B T/\ell_p \sim 0.1$ pN, our crude model predicts (correctly) that the molecule is stretched to its full length $L = N\ell_p$. At this point, its true elasticity should take over and allow genuine stretching. That true elasticity turns out to dominate only for forces $\gtrsim 10$ pN. [For details of what happens between 0.1 and 10 pN, see, e.g., Nelson (2008), Secs. 9.1–9.4.] For a force between ~ 10 and ~ 80 pN, the molecule is measured to obey Hooke's law, with a Young's modulus $E \simeq 0.3$ GPa that agrees with the value inferred in part (c) from its random-coil diameter. When the applied force reaches $\simeq 65$ pN, the molecule's double helix suddenly stretches greatly with small increases of force, changing its structure, so this is the molecule's yield point. Show that the strain at this yield point is $\Delta \ell / \ell \sim 0.1$; cf. Table 11.1.

Exercise 11.13 ***Example: Elastica*

Consider a slender wire of rectangular cross section with horizontal thickness h and vertical thickness w that is resting on a horizontal surface, so gravity is unimportant. Let the wire be bent in the horizontal plane as a result of equal and opposite forces F that act at its ends; Fig. 11.10. The various shapes the wire can assume are called *elastica;* they were first computed by Euler in 1744 and are discussed in Love (1927), pp. 401–404). The differential equation that governs the wire's shape is similar to that for the cantilever [Eq. (11.45)], with the simplification that the wire's weight does not enter the problem and the complication that the wire is long enough to deform through large angles.

It is convenient (as in the cantilever problem, Fig. 11.8) to introduce curvilinear coordinates with coordinate x measuring distance along the neutral surface, z measuring distance orthogonal to x in the plane of the bend (horizontal plane), and y measured perpendicular to the bending plane (vertically). The unit vectors along the x, y, and z directions are $\mathbf{e}_x$, $\mathbf{e}_y$, $\mathbf{e}_z$ (Fig. 11.10). Let $\theta(x)$ be the angle between $\mathbf{e}_x$ and the applied force $\mathbf{F}$; $\theta(x)$ is determined, of course, by force and torque balance.

(a) Show that force balance along the x and z directions implies

$$F \cos \theta = \int T_{xx} dy dz, \qquad F \sin \theta = \int T_{zx} dy dz \equiv S. \qquad (11.49a)$$

(b) Show that torque balance for a short segment of wire implies

$$S = \frac{dM}{dx}, \qquad (11.49b)$$

where $M(x) \equiv \int z T_{xx} dy dz$ is the bending torque.

(c) Show that the stress-strain relation in the wire implies

$$M = -D \frac{d\theta}{dx}, \qquad (11.49c)$$

where $D = Ewh^3/12$ is the flexural rigidity [Eq. (11.41b)].

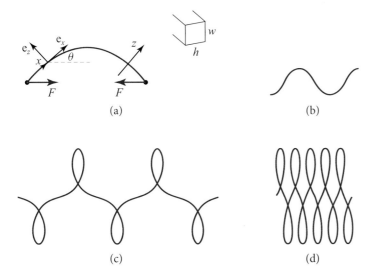

FIGURE 11.10 Elastica. (a) A bent wire is in elastostatic equilibrium under the action of equal and opposite forces applied at its two ends. x measures distance along the neutral surface; z measures distance orthogonal to the wire in the plane of the bend. (b)–(d) Examples of the resulting shapes.

(d) From the relations in parts (a)–(c), derive the following differential equation for the shape of the wire:

$$\frac{d^2\theta}{dx^2} = -\frac{F \sin \theta}{D}. \tag{11.49d}$$

This is the same equation as describes the motion of a simple pendulum!

(e) For Track-Two readers who have studied Sec. 11.8: Go back through your analysis and identify any place that connection coefficients would enter into a more careful computation, and explain why the connection-coefficient terms are negligible.

(f) Find one nontrivial solution of the elastica equation (11.49d) either analytically using elliptic integrals or numerically. (The general solution can be expressed in terms of elliptic integrals.)

(g) Solve analytically or numerically for the shape adopted by the wire corresponding to your solution in part (f), in terms of precisely Cartesian coordinates (X, Z) in the bending (horizontal) plane. Hint: Express the curvature of the wire, $1/\mathcal{R} = d\theta/dx$, as

$$\frac{d\theta}{dx} = -\frac{d^2 X}{dZ^2}\left[1 + \left(\frac{dX}{dZ}\right)^2\right]^{-3/2}. \tag{11.49e}$$

(h) Obtain a uniform piece of wire and adjust the force **F** to compare your answer with experiment.

11.6 Buckling and Bifurcation of Equilibria

So far, we have considered stable elastostatic equilibria and have implicitly assumed that the only reason for failure of a material is exceeding the yield limit. However, anyone who has built a house of cards knows that mechanical equilibria can be unstable, with startling consequences. In this section, we explore a specific, important example of a mechanical instability: *buckling*—the theory of which was developed long ago, in 1744 by Leonard Euler.

A tragic example of buckling was the collapse of the World Trade Center's Twin Towers on September 11, 2001. We discuss it near the end of this section, after first developing the theory in the context of a much simpler and cleaner example.

11.6.1 Elementary Theory of Buckling and Bifurcation

Take a new playing card and squeeze it between your finger and thumb (Fig. 11.11). When you squeeze gently, the card remains flat. But when you gradually increase the compressive force F past a critical value F_{crit}, the card suddenly buckles (i.e., bends), and the curvature of the bend then increases rather rapidly with increasing applied force.

To understand quantitatively the sudden onset of buckling, we derive an eigenequation for the transverse displacement η as a function of distance x from one end of the card. (Although the card is effectively 2-dimensional, it has translation symmetry along its transverse dimension, so we can use the 1-dimensional equations of Sec. 11.5.) We suppose that the ends are free to pivot but not move transversely, so

$$\eta(0) = \eta(\ell) = 0. \tag{11.50}$$

For small displacements, the bending torque of our dimensionally reduced 1-dimensional theory is [Eq. (11.41c)]

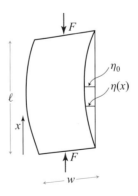

FIGURE 11.11 A playing card of length ℓ, width w, and thickness h is subjected to a compressive force F applied at both ends. The ends of the card are free to pivot.

$$M(x) = -D\frac{d^2\eta}{dx^2}, \quad (11.51)$$

where $D = wh^3E/12$ is the flexural rigidity [Eq. (11.41b)]. As the card is very light (negligible gravity), the total torque around location x, acting on a section of the card from x to one end, is the bending torque $M(x)$ acting at x plus the torque $-F\eta(x)$ associated with the applied force. This sum must vanish:

$$D\frac{d^2\eta}{dx^2} + F\eta = 0. \quad (11.52)$$

The eigensolutions of Eq. (11.52) satisfying boundary conditions (11.50) are

$$\eta = \eta_0 \sin kx, \quad (11.53a)$$

with eigenvalues

$$k = \left(\frac{F}{D}\right)^{1/2} = \frac{n\pi}{\ell} \quad \text{for nonnegative integers } n. \quad (11.53b)$$

Therefore, there is a critical force (first derived by Leonhard Euler in 1744), given by

$$F_{\text{crit}} = \frac{\pi^2 D}{\ell^2} = \frac{\pi^2 wh^3 E}{12\ell^2}. \quad (11.54) \qquad \text{critical force for buckling}$$

When $F < F_{\text{crit}}$, there is no solution except $\eta = 0$ (an unbent card). When $F = F_{\text{crit}}$, the unbent card is still a solution, and there suddenly is the additional, arched solution (11.53) with $n = 1$, depicted in Fig. 11.11.

The linear approximation we have used cannot tell us the height η_0 of the arch as a function of F for $F \geq F_{\text{crit}}$; it reports, incorrectly, that for $F = F_{\text{crit}}$ all arch heights are allowed, and that for $F > F_{\text{crit}}$ there is no solution with $n = 1$. However, when nonlinearities are taken into account (Ex. 11.14), the $n = 1$ solution continues to exist for $F > F_{\text{crit}}$, and the arch height η_0 is related to F by

$$F = F_{\text{crit}} \left\{ 1 + \frac{1}{2}\left(\frac{\pi\eta_0}{2\ell}\right)^2 + O\left[\left(\frac{\pi\eta_0}{2\ell}\right)^4\right] \right\}. \quad (11.55)$$

The sudden appearance of the arched equilibrium state as F is increased through F_{crit} is called a *bifurcation of equilibria*. This bifurcation also shows up in the elastodynamics of the playing card, as we deduce in Sec. 12.3.5. When $F < F_{\text{crit}}$, small perturbations of the card's unbent shape oscillate stably. When $F = F_{\text{crit}}$, the unbent card is neutrally stable, and its zero-frequency motion leads the card from its unbent equilibrium state to its $n = 1$, arched equilibrium. When $F > F_{\text{crit}}$, the straight card is an unstable equilibrium: its $n = 1$ perturbations grow in time, driving the card toward the $n = 1$ arched equilibrium state.

A nice way of looking at this bifurcation is in terms of free energy. Consider candidate equilibrium states labeled by the height η_0 of their arch. For each value of η_0, give the card (for concreteness) the $n = 1$ sine-wave shape $\eta = \eta_0 \sin(\pi x/\ell)$. Compute the total elastic energy $E(\eta_0)$ associated with the card's bending, and subtract off the

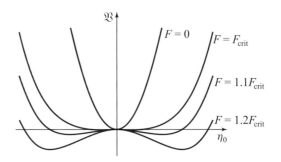

FIGURE 11.12 Representation of bifurcation by a potential energy function $\mathfrak{V}(\eta_0)$. When the applied force is small ($F \leq F_{\text{crit}}$), there is only one stable equilibrium. As the applied force F is increased, the bottom of the potential well flattens, and eventually (for $F > F_{\text{crit}}$) the number of equilibria increases from one to three, of which only two are stable.

work $F\delta X$ done on the card by the applied force F when the card arches from $\eta_0 = 0$ to height η_0. [Here $\delta X(\eta_0)$ is the arch-induced decrease in straight-line separation between the card's ends.] The resulting quantity, $\mathfrak{V}(\eta_0) = E - F\delta X$, is the card's *free energy*—analogous to the physical free energy $F = E - TS$ for a system in contact with a heat bath (Secs. 5.4.1 and 11.3.5), the enthalpic free energy when in contact with a pressure bath (Ex. 5.5h), and the Gibbs (chemical) free energy $G = E - TS + PV$ when in contact with a heat and pressure bath (Sec. 5.5). It is the relevant energy for analyzing the card's equilibrium and dynamics when the force F is continually being applied at the two ends. In Ex. 11.15 we deduce that this free energy is

free energy of a bent card or beam

$$\mathfrak{V} = \left(\frac{\pi\eta_0}{2\ell}\right)^2 \ell \left[(F_{\text{crit}} - F) + \frac{1}{4}F_{\text{crit}}\left(\frac{\pi\eta_0}{2\ell}\right)^2\right] + O\left[F_{\text{crit}}\ell\left(\frac{\pi\eta_0}{2\ell}\right)^6\right], \quad (11.56)$$

which we depict in Fig. 11.12.

At small values of the compressive force $F < F_{\text{crit}}$, the free energy has only one minimum $\eta_0 = 0$ corresponding to a single stable equilibrium, the straight card. However, as the force is increased through F_{crit}, the potential minimum flattens out and then becomes a maximum flanked by two new minima (e.g., the curve $F = 1.2F_{\text{crit}}$). The maximum for $F > F_{\text{crit}}$ is the unstable, zero-displacement (straight-card) equilibrium, and the two minima are the stable, finite-amplitude equilibria with positive and negative η_0 given by Eq. (11.55).

This procedure of representing a continuous system with an infinite number of degrees of freedom by just one or a few coordinates and finding the equilibrium by minimizing a free energy is quite common and powerful.

Thus far, we have discussed only two of the card's equilibrium shapes (11.53): the straight shape $n = 0$ and the single-arch shape $n = 1$. If the card were con-

higher order equilibria

strained, by gentle, lateral stabilizing forces, to remain straight beyond $F = F_{\text{crit}}$, then at $F = n^2 F_{\text{crit}}$ for each $n = 2, 3, 4, \ldots$, the nth-order perturbative mode, with

$\eta = \eta_0 \sin(n\pi x/\ell)$, would become unstable, and a new, stable equilibrium with this shape would bifurcate from the straight equilibrium. You can easily explore this for $n = 2$ using a new playing card.

These higher-order modes are rarely of practical importance. In the case of a beam with no lateral constraints, as F increases above F_{crit}, it will buckle into its single-arched shape. For beam dimensions commonly used in construction, a fairly modest further increase of F will bend it enough that its yield point and then rupture point are reached. To experience this yourself, take a thin meter stick, compress its ends between your two hands, and see what happens.

11.6.2 Collapse of the World Trade Center Buildings

Now we return to the example with which we began this section. On September 11, 2001, al-Qaeda operatives hijacked two Boeing 767 passenger airplanes and crashed them into the 110-story Twin Towers of the World Trade Center in New York City, triggering the towers' collapse a few hours later, with horrendous loss of life.

The weight of a tall building such as the towers is supported by vertical steel beams, called "columns." The longer the column is, the lower the weight it can support without buckling, since $F_{\text{crit}} = \pi^2 D/\ell^2 = \pi^2 E A (r_g/\ell)^2$, with A the beam's cross sectional area, r_g its radius of gyration along its bending direction, and ℓ its length [Eqs. (11.54) and (11.41b)].[8] The column lengths are typically chosen such that the critical stress for buckling, $F_{\text{crit}}/A = E(\pi r_g/\ell)^2$, is roughly the same as the yield stress, $F_{\text{yield}} \simeq 0.003 E$ (cf. Table 11.1), which means that the columns' *slenderness ratio* is $\ell/r_g \sim 50$. The columns are physically far longer than $50 r_g$, but they are anchored to each other laterally every $\sim 50 r_g$ by beams and girders in the floors, so their effective length for buckling is $\ell \sim 50 r_g$. The columns' radii of gyration r_g are generally made large, without using more steel than needed to support the overhead weight, by making the columns hollow, or giving them H-shaped cross sections. In the Twin Towers, the thinnest beams had $r_g \sim 13$ cm, and they were anchored in every floor, with floor separations $\ell \simeq 3.8$ m, so their slenderness ratio was actually $\ell/r_g \simeq 30$.

According to a detailed investigation (NIST, 2005, especially Secs. 6.14.2 and 6.14.3), the crashing airplanes ignited fires in and near floors 93–99 of the North Tower and 78–83 of the South Tower, where the airplanes hit. The fires were most intense in the floors and around uninsulated central steel columns. The heated central columns lost their rigidity and began to sag, and trusses then transferred some of the weight above to the outer columns. In parallel, the heated floor structures began

description of failure modes

---

8. As noted in the discussion, after Eq. (11.41b), $A r_g^2$ is really the second moment of the column's cross sectional area, along its direction of bend. If the column is supported at its ends against movement in both transverse directions, then the relevant second moment is the transverse tensor $\int x^i x^j dx dy$, and the direction of buckling (if it occurs) will be the eigendirection of this tensor that has the smallest eigenvalue (the column's narrowest direction).

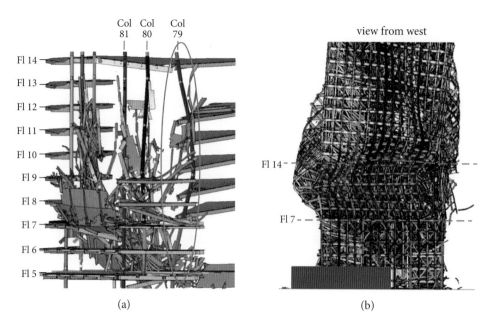

FIGURE 11.13 (a) The buckling of column 79 in building WTC7 at the World Trade Center, based on a finite-element simulation informed by all available observational data. (b) The subsequent buckling of the building's core. From NIST (2008).

to sag, pulling inward on the buildings' exterior steel columns, which bowed inward and then buckled, initiating the buildings' collapse. [This is a somewhat oversimplified description of a complex situation; for full complexities, see the report, NIST (2005).]

This column buckling was somewhat different from the buckling of a playing card because of the inward pull of the sagging floors. Much more like our playing-card buckle was the fate of an adjacent, 47-story building called WTC7. When the towers collapsed, they injected burning debris onto and into WTC7. About 7 hours later, fire-induced thermal expansion triggered a cascade of failures in floors 13–16, which left column 79 with little stabilizing lateral support, so its effective length ℓ was increased far beyond $50 r_g$. It then quickly buckled (Fig. 11.13a) in much the same manner as our playing card, followed by column 80, then 81, and subsequently columns 77, 78, and 76 (NIST, 2008, especially Sec. 2.4). Within seconds, the building's entire core was buckling (Fig. 11.13b).

11.6.3 Buckling with Lateral Force; Connection to Catastrophe Theory

Returning to the taller Twin Towers, we can crudely augment the inward pull of the sagging floors into our free-energy description of buckling, by adding a term $-F_{\text{lat}} \eta_0$, which represents the energy inserted into a bent column by a lateral force F_{lat} when its center has been displaced laterally through the distance η_0. Then the free energy

(11.56), made dimensionless and with its terms rearranged, takes the form

$$\varphi \equiv \frac{\mathfrak{V}}{F_{\text{crit}}\ell} = \frac{1}{4}\left(\frac{\pi\eta_0}{2\ell}\right)^4 - \frac{1}{2}\left(\frac{2(F - F_{\text{crit}})}{F_{\text{crit}}}\right)\left(\frac{\pi\eta_0}{2\ell}\right)^2 - \left(\frac{2F_{\text{lat}}}{\pi F_{\text{crit}}}\right)\left(\frac{\pi\eta_0}{2\ell}\right). \tag{11.57}$$

Notice that this equation has the canonical form $\varphi = \frac{1}{4}a^4 - \frac{1}{2}za^2 - xa$ for the potential that governs a cusp catastrophe, whose state variable is $a = \pi\eta_0/(2\ell)$ and control variables are $z = 2(F - F_{\text{crit}})/F_{\text{crit}}$ and $x = (2/\pi)F_{\text{lat}}/F_{\text{crit}}$; see Eq. (7.72).[9] From the elementary mathematics of this catastrophe, as worked out in Sec. 7.5.1, we learn that although the lateral force F_{lat} will make the column bend, it will not induce a bifurcation of equilibria until the control-space cusp $x = \pm 2(z/3)^{3/2}$ is reached:

interpretation in terms of catastrophe theory

$$\frac{F_{\text{lat}}}{F_{\text{crit}}} = \pm\pi\left(\frac{2(F - F_{\text{crit}})}{3F_{\text{crit}}}\right)^{3/2}. \tag{11.58}$$

Notice that the lateral force F_{lat} actually delays the bifurcation to a higher vertical force, $F > F_{\text{crit}}$. However, this is not significant for the physical buckling, since the column in this case is bent from the outset, and as F_{lat} increases, it stops carrying its share of the building's weight and moves smoothly toward its yield point and rupture; Ex. 11.16.

11.6.4 Other Bifurcations: Venus Fly Trap, Whirling Shaft, Triaxial Stars, and Onset of Turbulence

This bifurcation of equilibria, associated with the buckling of a column, is just one of many bifurcations that occur in physical systems. Another is a buckling type bifurcation that occurs in the 2-dimensional leaves of the Venus fly trap plant; the plant uses the associated instability to snap together a pair of leaves in a small fraction of a second, thereby capturing insects for it to devour; see Fortere et al. (2005). Yet another is the onset of a lateral bend in a shaft (rod) that spins around its longitudinal axis (see Love, 1927, Sec. 286). This is called *whirling*; it is an issue in drive shafts for automobiles and propellers, and a variant of it occurs in spinning DNA molecules during replication—see Wolgemuth, Powers, and Goldstein (2000). One more example is the development of triaxiality in self-gravitating fluid masses (i.e., stars) when their rotational kinetic energy reaches a critical value, about 1/4 of their gravitational energy; see Chandrasekhar (1962). Bifurcations also play a major role in the onset of turbulence in fluids and in the route to chaos in other dynamical systems; we study turbulence and chaos in Sec. 15.6.

whirling shaft

---

9. The lateral force F_{lat} makes the bifurcation *structurally stable*, in the language of catastrophe theory (discussed near the end of Sec. 7.5) and thereby makes it describable by one of the generic catastrophes. Without F_{lat}, the bifurcation is not structurally stable.

For further details on the mathematics of bifurcations with emphasis on elastostatics and elastodynamics, see, for example, Marsden and Hughes (1986, Chap. 7). For details on buckling from an engineering viewpoint, see Ugural and Fenster (2012, Chap. 11).

EXERCISES

Exercise 11.14 *Derivation and Example: Bend as a Function of Applied Force*
Derive Eq. (11.55) relating the angle $\theta_o = (d\eta/dx)_{x=0} = k\eta_o = \pi\eta_o/\ell$ to the applied force F when the card has an $n = 1$, arched shape. [Hint: Consider the card as comprising many thin parallel wires and use the elastica differential equation $d^2\theta/dx^2 = -(F/D)\sin\theta$ [Eq. (11.49d)] for the angle between the card and the applied force at distance x from the card's end. The $\sin\theta$ becomes θ in the linear approximation used in the text; the nonlinearities embodied in the sine give rise to the desired relation. The following steps along the way toward a solution are mathematically the same as used when computing the period of a pendulum as a function of its amplitude of swing.]

(a) Derive the first integral of the elastica equation

$$(d\theta/dx)^2 = 2(F/D)(\cos\theta - \cos\theta_o), \tag{11.59}$$

where θ_o is an integration constant. Show that the boundary condition of no bending torque (no inflection of the card's shape) at the card ends implies $\theta = \theta_o$ at $x = 0$ and $x = \ell$; whence $\theta = 0$ at the card's center, $x = \ell/2$.

(b) Integrate the differential equation (11.59) to obtain

$$\frac{\ell}{2} = \sqrt{\frac{D}{2F}} \int_0^{\theta_o} \frac{d\theta}{\sqrt{\cos\theta - \cos\theta_o}}. \tag{11.60}$$

(c) Perform the change of variable $\sin(\theta/2) = \sin(\theta_o/2)\sin\phi$ and thereby bring Eq. (11.60) into the form

$$\ell = 2\sqrt{\frac{D}{F}} \int_0^{\pi/2} \frac{d\phi}{\sqrt{1 - \sin^2(\theta_o/2)\sin^2\phi}} = 2\sqrt{\frac{D}{F}} K[\sin^2(\theta_o/2)]. \tag{11.61}$$

Here $K(y)$ is the complete elliptic integral of the first type, with the parameterization used by Mathematica (which differs from that of many books).

(d) Expand Eq. (11.61) in powers of $\theta_o/2$ to obtain

$$F = F_{\text{crit}} \frac{4}{\pi^2} K^2[\sin^2(\theta_o/2)] = F_{\text{crit}} \left[1 + \frac{1}{2}\left(\frac{\theta_o/2}{2}\right)^2 + \cdots\right], \tag{11.62}$$

from which deduce our desired result, Eq. (11.55).

Exercise 11.15 *Problem: Free Energy of a Bent, Compressed Beam*
Derive Eq. (11.56) for the free energy $\mathfrak{V}$ of a beam that is compressed with a force F and has a critical compression $F_{\text{crit}} = \pi^2 D/\ell^2$, where D is its flexural rigidity. [Hint: It

must be that $\partial V/\partial \eta_0 = 0$ gives Eq. (11.55) for the beam's equilibrium bend amplitude η_0 as a function of $F - F_{\text{crit}}$. Use this to reduce the number of terms in $\mathfrak{V}(\eta_0)$ in Eq. (11.56) that you need to derive.]

Exercise 11.16 *Problem: Bent Beam with Lateral Force*
Explore numerically the free energy (11.57) of a bent beam with a compressive force F and lateral force F_{lat}. Examine how the extrema (equilibrium states) evolve as F and F_{lat} change, and deduce the physical consequences.

Exercise 11.17 **Problem: Applications of Buckling—Mountains and Pipes*
Buckling plays a role in many natural and human-caused phenomena. Explore the following examples.

(a) *Mountain building.* When two continental plates are in (very slow) collision, the compressional force near their interface drives their crustal rock to buckle upward, producing mountains. Estimate how high such mountains can be on Earth and on Mars, and compare your estimates with their actual heights. Read about such mountain building in books or on the web.

(b) *Thermal expansion of pipes.* When a segment of pipe is heated (e.g., by the rising sun in the morning), it will expand. If its ends are held fixed, this can easily produce a longitudinal stress large enough to buckle the pipe. How would you deal with this in an oil pipeline on Earth's surface? In a long vacuum tube? Compare your answers with standard engineering solutions, which you can find in books or on the web.

Exercise 11.18 *Example: Allometry*
Allometry is the study of biological scaling laws that relate various features of an animal to its size or mass. One example concerns the ratio of the width to the length of leg bones. Explain why the width to the length of a thigh bone in a quadruped might scale as the square root of the stress that it has to support. Compare elephants with cats in this regard. (The density of bone is roughly 1.5 times that of water, and its Young's modulus is ~ 20 GPa.)

11.7 Reducing the Elastostatic Equations to 2 Dimensions for a Deformed Thin Plate: Stress Polishing a Telescope Mirror

The world's largest optical telescopes (as of 2016) are the Gran Telescopio Canarias in the Canary Islands and the two Keck telescopes on Mauna Kea in Hawaii, which are all about 10 m in diameter. It is very difficult to support traditional, monolithic mirrors so that the mirror surfaces maintain their shape (their "figure") as the telescope slews, because they are so heavy, so for Keck a new method of fabrication was sought. The solution devised by Jerry Nelson and his colleagues was to construct the telescope out of 36 separate hexagons, each 0.9 m on a side. However, this posed a second problem: how to grind each hexagon's reflecting surface to the required hyperboloidal shape. For this, a novel technique called *stressed mirror polishing* was developed. This technique relies on the fact that it is relatively easy to grind a surface to a spherical shape, but technically highly challenging to create a nonaxisymmetric shape. So during the grinding, stresses are applied around the boundary of the mirror to deform it, and a spherical surface is produced. The stresses are then removed, and the mirror springs into the desired nonspherical shape. Computing the necessary stresses is a problem in classical elasticity theory and, in fact, is a good example of a large number of applications where the elastic body can be approximated as a thin plate and its shape analyzed using elasticity equations that are reduced from 3 dimensions to 2 by the method of moments.

For stress polishing of mirrors, the applied stresses are so large that we can ignore gravitational forces (at least in our simplified treatment). We suppose that the hexagonal mirror has a uniform thickness h and idealize it as a circle of radius R, and we introduce Cartesian coordinates with (x, y) in the horizontal plane (the plane of the mirror before deformation and polishing begin), and z vertical. The mirror is deformed as a result of a net vertical force per unit area (pressure) $P(x, y)$. This pressure is applied at the lower surface when upward (positive) and the upper surface when downward (negative). In addition, there are shear forces and bending torques applied around the rim of the mirror.

As in our analysis of a cantilever in Sec. 11.5, we assume the existence of a neutral surface in the deformed mirror, where the horizontal strain vanishes, $T_{ab} = 0$. (Here and below we use letters from the early part of the Latin alphabet for horizontal components $x = x^1$ and $y = x^2$.) We denote the vertical displacement of the neutral surface by $\eta(x, y)$. By applying the method of moments to the 3-dimensional stress-balance equation $T_{jk,k} = 0$ in a manner similar to our cantilever analysis, we obtain the following 2-dimensional equation for the mirror's shape (Ex. 11.19):

$$\boxed{\nabla^2(\nabla^2 \eta) = P(x, y)/D.} \quad (11.63a)$$

Here ∇^2 is the horizontal Laplacian: $\nabla^2 \eta \equiv \eta_{,aa} = \eta_{,xx} + \eta_{,yy}$. Equation (11.63a) is the 2-dimensional analog of the equation $d^4\eta/dx^4 = W(x)/D$ for the shape of a cantilever on which a downward force per unit length $W(x)$ acts [Eq. (11.45)]. The 2-dimensional flexural rigidity that appears in Eq. (11.63a) is

$$\boxed{D = \frac{Eh^3}{12(1 - \nu^2)},} \quad (11.63b)$$

where E is the mirror's Young's modulus, h is its thickness, and ν is its Poisson's ratio. The operator $\nabla^2 \nabla^2$ acting on η in the shape equation (11.63a) is called the *biharmonic operator;* it also appears in 3-dimensional form in the biharmonic equation (11.34a)

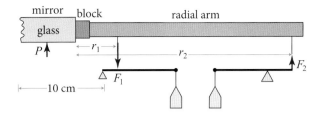

FIGURE 11.14 Schematic showing the mirror rim, a radial arm attached to it via a block, and a lever assembly used to apply shear forces and bending torques to the rim during stress polishing. (F_1 need not equal F_2, as there is a pressure P applied to the back surface of the mirror and forces applied at 23 other points around its rim.) The shear force on the mirror rim is $S = F_2 - F_1$, and the bending torque is $M = r_2 F_2 - r_1 F_1$.

for the displacement inside a homogeneous, isotropic body to which surface stresses are applied.

The shape equation (11.63a) must be solved subject to boundary conditions around the mirror's rim: the applied shear forces and bending torques.

The individual Keck mirror segments were constructed out of a ceramic material with Young's modulus $E = 89$ GPa and Poisson's ratio $\nu = 0.24$ (similar to glass; cf. Table 11.1). A mechanical jig was constructed to apply the shear forces and bending torques at 24 uniformly spaced points around the rim of the mirror (Fig. 11.14). The maximum stress was applied for the six outermost mirrors and was 2.4×10^6 N m^{-2}, 12% of the material's breaking tensile strength (2×10^7 N m^{-2}).

This stress polishing worked beautifully, and the Keck telescopes have become highly successful tools for astronomical research.

EXERCISES

Exercise 11.19 **Derivation and Example: Dimensionally Reduced Shape Equation for a Stressed Plate*

Use the method of moments (Sec. 11.5) to derive the 2-dimensional shape equation (11.63a) for the stress-induced deformation of a thin plate, and expression (11.63b) for the 2-dimensional flexural rigidity. Here is a step-by-step guide, in case you want or need it.

(a) Show on geometrical grounds that the in-plane strain is related to the vertical displacement by [cf. Eq. (11.40a)]

$$\xi_{a,b} = -z\eta_{,ab}. \tag{11.64a}$$

(b) Derive an expression for the horizontal components of the stress, T_{ab}, in terms of double derivatives of the displacement function $\eta(x, y)$ [analog of $T_{xx} = -Ezd^2\eta/dx^2$, Eq. (11.40b), for a stressed rod]. This can be done (i) by arguing on physical grounds that the vertical component of stress, T_{zz}, is much smaller than the horizontal components and therefore can be approximated as zero [an

11.7 Stress Polishing a Telescope Mirror

approximation to be checked in part (f)]; (ii) by expressing $T_{zz} = 0$ in terms of the strain and thence displacement and using Eqs. (11.39) to obtain

$$\Theta = -\left(\frac{1-2\nu}{1-\nu}\right) z \nabla^2 \eta, \tag{11.64b}$$

where ∇^2 is the horizontal Laplacian; and (iii) by then writing T_{ab} in terms of Θ and $\xi_{a,b}$ and combining with Eqs. (11.64a) and (11.64b) to get the desired equation:

$$T_{ab} = Ez \left[\frac{\nu}{(1-\nu^2)} \nabla^2 \eta \, \delta_{ab} + \frac{\eta_{,ab}}{(1+\nu)} \right]. \tag{11.64c}$$

(c) With the aid of Eq. (11.64c), write the horizontal force density in the form

$$f_a = -T_{ab,b} - T_{az,z} = -\frac{Ez}{1-\nu^2} \nabla^2 \eta_{,a} - T_{az,z} = 0. \tag{11.64d}$$

Then, as in the cantilever analysis [Eq. (11.40d)], reduce the dimensionality of this force equation by the method of moments. The zeroth moment (integral over z) vanishes. Why? Therefore, the lowest nonvanishing moment is the first (multiply f_a by z and integrate). Show that this gives

$$S_a \equiv \int T_{za} dz = D \nabla^2 \eta_{,a}, \tag{11.64e}$$

where D is the 2-dimensional flexural rigidity (11.63b). The quantity S_a is the vertical shear force per unit length acting perpendicular to a line in the mirror whose normal is in the direction a; it is the 2-dimensional analog of a stressed rod's shear force S [Eq. (11.41a)].

(d) For physical insight into Eq. (11.64e), define the bending torque per unit length (bending torque density) as

$$M_{ab} \equiv \int z T_{ab} dz, \tag{11.64f}$$

and show with the aid of Eq. (11.64c) that (11.64e) is the *law of torque balance* $S_a = M_{ab,b}$—the 2-dimensional analog of a stressed rod's $S = dM/dx$ [Eq. (11.42)].

(e) Compute the total vertical shear force acting on a small area of the plate as the line integral of S_a around its boundary, and by applying Gauss's theorem, deduce that the vertical shear force per unit area is $S_{a,a}$. Argue that this must be balanced by the net pressure P applied to the face of the plate, and thereby deduce the *law of vertical force balance*:

$$S_{a,a} = P. \tag{11.64g}$$

By combining this equation with the law of torque balance (11.64e), obtain the plate's bending equation $\nabla^2(\nabla^2 \eta) = P/D$ [Eq. (11.63a)—the final result we were seeking].

(f) Use this bending equation to verify the approximation made in part (b) that T_{zz} is small compared to the horizontal stresses. Specifically, show that $T_{zz} \simeq P$ is $O(h/R)^2 T_{ab}$, where h is the plate thickness, and R is the plate radius.

Exercise 11.20 *Example: Paraboloidal Mirror*

Show how to construct a paraboloidal mirror of radius R and focal length f by stress polishing.

(a) Adopt a strategy of polishing the stressed mirror into a segment of a sphere with radius of curvature equal to that of the desired paraboloid at its center, $r = 0$. By comparing the shape of the desired paraboloid to that of the sphere, show that the required vertical displacement of the stressed mirror during polishing is

$$\eta(r) = \frac{r^4}{64 f^3}, \tag{11.64h}$$

where r is the radial coordinate, and we only retain terms of leading order.

(b) Hence use Eq. (11.63a) to show that a uniform force per unit area

$$P = \frac{D}{f^3}, \tag{11.64i}$$

where D is the flexural rigidity, must be applied to the bottom of the mirror. (Ignore the weight of the mirror.)

(c) Based on the results of part (b), show that if there are N equally spaced levers attached at the rim, the vertical force applied at each of them must be

$$F_z = \frac{\pi D R^2}{N f^3}, \tag{11.64j}$$

and the applied bending torque must be

$$M = \frac{\pi D R^3}{2 N f^3}. \tag{11.64k}$$

(d) Show that the radial displacement inside the mirror is

$$\xi_r = -\frac{r^3 z}{16 f^3}, \tag{11.64l}$$

where z is the vertical distance from the neutral surface, halfway through the mirror.

(e) Hence evaluate the expansion Θ and the components of the shear tensor Σ, and show that the maximum stress in the mirror is

$$T_{\max} = \frac{(3 - 2\nu) R^2 h E}{32(1 - 2\nu)(1 + \nu) f^3}, \tag{11.64m}$$

where h is the mirror thickness. Comment on the limitations of this technique for making a thick, "fast" (i.e., $2R/f$ large) mirror.

11.8 Cylindrical and Spherical Coordinates: Connection Coefficients and Components of the Gradient of the Displacement Vector

Thus far in our discussion of elasticity, we have restricted ourselves to Cartesian coordinates. However, many problems in elasticity are most efficiently solved using cylindrical or spherical coordinates, so in this section, we develop some mathematical tools for those coordinate systems. In doing so, we follow the vectorial conventions of standard texts on electrodynamics and quantum mechanics (e.g., Jackson, 1999; Cohen-Tannoudji, Diu, and Laloë, 1977). We introduce an *orthonormal* set of basis vectors associated with each of our curvilinear coordinate systems; the coordinate lines are orthogonal to one another, and the basis vectors have unit lengths and point along the coordinate lines. In our study of continuum mechanics (Part IV, Elasticity; Part V, Fluid Dynamics; and Part VI, Plasma Physics), we follow this practice. When studying General Relativity (Part VII), we introduce and use basis vectors that are *not* orthonormal.

Our notation for cylindrical coordinates is (ϖ, ϕ, z); ϖ (pronounced "pomega") is distance from the z-axis, and ϕ is the angle around the z-axis:

$$\varpi = \sqrt{x^2 + y^2}, \qquad \phi = \arctan(y/x). \tag{11.65a}$$

The unit basis vectors that point along the coordinate axes are denoted $\mathbf{e}_\varpi$, $\mathbf{e}_\phi$, and $\mathbf{e}_z$, and are related to the Cartesian basis vectors by

$$\mathbf{e}_\varpi = (x/\varpi)\mathbf{e}_x + (y/\varpi)\mathbf{e}_y, \qquad \mathbf{e}_\phi = -(y/\varpi)\mathbf{e}_x + (x/\varpi)\mathbf{e}_y,$$
$$\mathbf{e}_z = \text{Cartesian } \mathbf{e}_z. \tag{11.65b}$$

Our notation for spherical coordinates is (r, θ, ϕ), with (as should be very familiar)

$$r = \sqrt{x^2 + y^2 + z^2}, \qquad \theta = \arccos(z/r), \qquad \phi = \arctan(y/x). \tag{11.66a}$$

The unit basis vectors associated with these coordinates are

$$\mathbf{e}_r = \frac{x}{r}\mathbf{e}_x + \frac{y}{r}\mathbf{e}_y + \frac{z}{r}\mathbf{e}_z, \qquad \mathbf{e}_\theta = \frac{z}{r}\mathbf{e}_\varpi - \frac{\varpi}{r}\mathbf{e}_z, \qquad \mathbf{e}_\phi = -\frac{y}{\varpi}\mathbf{e}_x + \frac{x}{\varpi}\mathbf{e}_y. \tag{11.66b}$$

Because our bases are orthonormal, the components of the metric of 3-dimensional space retain the Kronecker-delta values

$$\boxed{g_{jk} \equiv \mathbf{e}_j \cdot \mathbf{e}_k = \delta_{jk},} \tag{11.67}$$

which permits us to keep all vector and tensor indices down, by contrast with spacetime, where we must distinguish between up and down; cf. Sec. 2.5.[10]

---

[10] Occasionally—e.g., in the useful equation $\epsilon_{ijm}\epsilon_{klm} = \delta^{ij}_{kl} \equiv \delta^i_k \delta^j_l - \delta^i_l \delta^j_k$ [Eq. (1.23)]—it is convenient to put some indices up. In our orthonormal basis, any component with an index up is equal to that same component with an index down: e.g., $\delta^i_k \equiv \delta_{ik}$.

In Jackson (1999), Cohen-Tannoudji, Diu, and Laloë (1977), and other standard texts, formulas are written down for the gradient and Laplacian of a scalar field, and the divergence and curl of a vector field, in cylindrical and spherical coordinates; one uses these formulas over and over again. In elasticity theory, we deal largely with second-rank tensors and will need formulas for their various derivatives in cylindrical and spherical coordinates. In this book we introduce a mathematical tool, *connection coefficients* Γ_{ijk}, by which those formulas can be derived when needed.

connection coefficients

The connection coefficients quantify the turning of the orthonormal basis vectors as one moves from point to point in Euclidean 3-space: they tell us how the basis vectors at one point in space are *connected to* (related to) those at another point. More specifically, we define Γ_{ijk} by the two equivalent relations

$$\boxed{\nabla_k \mathbf{e}_j = \Gamma_{ijk} \mathbf{e}_i; \qquad \Gamma_{ijk} = \mathbf{e}_i \cdot (\nabla_k \mathbf{e}_j).} \tag{11.68}$$

Here $\nabla_k \equiv \nabla_{\mathbf{e}_k}$ is the directional derivative along the orthonormal basis vector $\mathbf{e}_k$; cf. Eq. (1.15a). Notice that (as is true quite generally; cf. Sec. 1.7) the differentiation index comes *last* on Γ; and notice that the middle index of Γ names the basis vector that is differentiated. Because our basis is orthonormal, it must be that $\nabla_k(\mathbf{e}_i \cdot \mathbf{e}_j) = 0$. Expanding this expression out using the standard rule for differentiating products, we obtain $\mathbf{e}_j \cdot (\nabla_k \mathbf{e}_i) + \mathbf{e}_i \cdot (\nabla_k \mathbf{e}_j) = 0$. Then invoking the definition (11.68) of the connection coefficients, we see that Γ_{ijk} is antisymmetric on its first two indices:

$$\boxed{\Gamma_{ijk} = -\Gamma_{jik}.} \tag{11.69}$$

In Part VII, when we use nonorthonormal bases, this antisymmetry will break down.

It is straightforward to compute the connection coefficients for cylindrical and spherical coordinates from (i) the definition (11.68); (ii) expressions (11.65b) and (11.66b) for the cylindrical and spherical basis vectors in terms of the Cartesian basis vectors; and (iii) the fact that *in Cartesian coordinates the connection coefficients vanish* ($\mathbf{e}_x$, $\mathbf{e}_y$, and $\mathbf{e}_z$ do not rotate as one moves through Euclidean 3-space). One can also deduce the cylindrical and spherical connection coefficients by drawing pictures of the basis vectors and observing how they change from point to point. As an example, for cylindrical coordinates we see from Fig. 11.15 that $\nabla_\phi \mathbf{e}_\varpi = \mathbf{e}_\phi/\varpi$. A similar pictorial calculation (which the reader is encouraged to do) reveals that $\nabla_\phi \mathbf{e}_\phi = -\mathbf{e}_\varpi/\varpi$. All other derivatives vanish. By comparing with Eq. (11.68), we see that *the only nonzero connection coefficients in cylindrical coordinates are*

connection coefficients for orthonormal bases of cylindrical and spherical coordinates

$$\boxed{\Gamma_{\varpi\phi\phi} = -\frac{1}{\varpi}, \quad \Gamma_{\phi\varpi\phi} = \frac{1}{\varpi},} \tag{11.70}$$

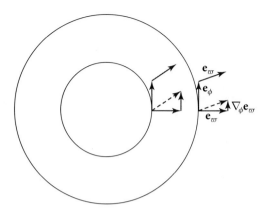

FIGURE 11.15 Pictorial evaluation of $\Gamma_{\phi\varpi\phi}$. In the rightmost assemblage of vectors we compute $\nabla_\phi \mathbf{e}_\varpi$ as follows. We draw the vector to be differentiated, $\mathbf{e}_\varpi$, at the tail of $\mathbf{e}_\phi$ (the vector along which we differentiate) and also at its head. We then subtract $\mathbf{e}_\varpi$ at the head from that at the tail; this difference is $\nabla_\phi \mathbf{e}_\varpi$. It obviously points in the $\mathbf{e}_\phi$ direction. When we perform the same calculation at a radius ϖ that is smaller by a factor 2 (left assemblage of vectors), we obtain a result, $\nabla_\phi \mathbf{e}_\varpi$, that is twice as large. Therefore, the length of this vector must scale as $1/\varpi$. By looking quantitatively at the length at some chosen radius ϖ, one can see that the multiplicative coefficient is unity: $\nabla_\phi \mathbf{e}_\varpi = \mathbf{e}_\phi/\varpi$. Comparing with Eq. (11.68), we deduce that $\Gamma_{\phi\varpi\phi} = 1/\varpi$.

which have the required antisymmetry [Eq. (11.69)]. Likewise, for spherical coordinates (Ex. 11.22), we have

$$\boxed{\Gamma_{\theta r\theta} = \Gamma_{\phi r\phi} = -\Gamma_{r\theta\theta} = -\Gamma_{r\phi\phi} = \frac{1}{r}, \quad \Gamma_{\phi\theta\phi} = -\Gamma_{\theta\phi\phi} = \frac{\cot\theta}{r}.} \quad (11.71)$$

These connection coefficients are the keys to differentiating vectors and tensors. Consider the gradient of the displacement, $\mathbf{W} = \nabla\boldsymbol{\xi}$. Applying the product rule for differentiation, we obtain

$$\nabla_k(\xi_j \mathbf{e}_j) = (\nabla_k \xi_j)\mathbf{e}_j + \xi_j(\nabla_k \mathbf{e}_j) = \xi_{j,k}\mathbf{e}_j + \xi_j \Gamma_{ljk}\mathbf{e}_l. \quad (11.72)$$

directional derivative along basis vector

Here the comma denotes the directional derivative, along a basis vector, of the components treated as scalar fields. For example, in cylindrical coordinates we have

$$\xi_{i,\varpi} = \frac{\partial \xi_i}{\partial \varpi}, \quad \xi_{i,\phi} = \frac{1}{\varpi}\frac{\partial \xi_i}{\partial \phi}, \quad \xi_{i,z} = \frac{\partial \xi_i}{\partial z}; \quad (11.73)$$

and in spherical coordinates we have

$$\xi_{i,r} = \frac{\partial \xi_i}{\partial r}, \quad \xi_{i,\theta} = \frac{1}{r}\frac{\partial \xi_i}{\partial \theta}, \quad \xi_{i,\phi} = \frac{1}{r\sin\theta}\frac{\partial \xi_i}{\partial \phi}. \quad (11.74)$$

Taking the ith component of Eq. (11.72) we obtain

$$\boxed{W_{ik} = \xi_{i;k} = \xi_{i,k} + \Gamma_{ijk}\xi_j.} \tag{11.75}$$

Here $\xi_{i;k}$ are the nine components of the gradient of the vector field $\boldsymbol{\xi}(\mathbf{x})$.

We can use Eq. (11.75) to evaluate the expansion $\Theta = \text{Tr } \mathbf{W} = \nabla \cdot \boldsymbol{\xi}$. Using Eqs. (11.70) and (11.71), we obtain

$$\Theta = \nabla \cdot \boldsymbol{\xi} = \frac{\partial \xi_\varpi}{\partial \varpi} + \frac{1}{\varpi}\frac{\partial \xi_\phi}{\partial \phi} + \frac{\partial \xi_z}{\partial z} + \frac{\xi_\varpi}{\varpi}$$

$$= \frac{1}{\varpi}\frac{\partial}{\partial \varpi}(\varpi \xi_\varpi) + \frac{1}{\varpi}\frac{\partial \xi_\phi}{\partial \phi} + \frac{\partial \xi_z}{\partial z} \tag{11.76}$$

in cylindrical coordinates and

$$\Theta = \nabla \cdot \boldsymbol{\xi} = \frac{\partial \xi_r}{\partial r} + \frac{1}{r}\frac{\partial \xi_\theta}{\partial \theta} + \frac{1}{r \sin\theta}\frac{\partial \xi_\phi}{\partial \phi} + \frac{2\xi_r}{r} + \frac{\cot\theta\, \xi_\theta}{r}$$

$$= \frac{1}{r^2}\frac{\partial}{\partial r}(r^2 \xi_r) + \frac{1}{r \sin\theta}\frac{\partial}{\partial \theta}(\sin\theta\, \xi_\theta) + \frac{1}{r \sin\theta}\frac{\partial \xi_\phi}{\partial \phi} \tag{11.77}$$

in spherical coordinates, in agreement with formulas in standard textbooks, such as the flyleaf of Jackson (1999).

The components of the rotation are most easily deduced using $R_{ij} = -\epsilon_{ijk}\phi_k$ with $\boldsymbol{\phi} = \frac{1}{2}\nabla \times \boldsymbol{\xi}$, and the standard expressions for the curl in cylindrical and spherical coordinates (Jackson, 1999). Since the rotation does not enter into elasticity theory in a significant way, we refrain from writing down the results. The components of the shear are computed in Box 11.4.

By a computation analogous to Eq. (11.72), we can construct an expression for the gradient of a tensor of any rank. For a second-rank tensor $\mathbf{T} = T_{ij}\mathbf{e}_i \otimes \mathbf{e}_j$ we obtain (Ex. 11.21)

$$\boxed{T_{ij;k} = T_{ij,k} + \Gamma_{ilk}T_{lj} + \Gamma_{jlk}T_{il}.} \tag{11.78}$$

components of gradient of a tensor

Equation (11.78) for the components of the gradient can be understood as follows. In cylindrical or spherical coordinates, the components T_{ij} can change from point to point as a result of two things: a change of the tensor $\mathbf{T}$ or the turning of the basis vectors. The two connection coefficient terms in Eq. (11.78) remove the effects of the basis turning, leaving in $T_{ij;k}$ only the influence of the change of $\mathbf{T}$ itself. There are two correction terms corresponding to the two slots (indices) of $\mathbf{T}$; the effects of basis turning on each slot get corrected one after another. If $\mathbf{T}$ had had n slots, then there would have been n correction terms, each with the form of the two in Eq. (11.78).

These expressions for derivatives of tensors are not required for dealing with the vector fields of introductory electromagnetic theory or quantum theory, but they are essential for manipulating the tensor fields encountered in elasticity. As we shall see in Sec. 24.3, with one further generalization, we can go on to differentiate tensors in

BOX 11.4. SHEAR TENSOR IN SPHERICAL AND CYLINDRICAL COORDINATES T2

Using our rules (11.75) for forming the gradient of a vector, we can derive a general expression for the shear tensor:

$$\Sigma_{ij} = \frac{1}{2}(\xi_{i;j} + \xi_{j;i}) - \frac{1}{3}\delta_{ij}\xi_{k;k}$$

$$= \frac{1}{2}(\xi_{i,j} + \xi_{j,i} + \Gamma_{ilj}\xi_l + \Gamma_{jli}\xi_l) - \frac{1}{3}\delta_{ij}(\xi_{k,k} + \Gamma_{klk}\xi_l). \quad (1)$$

Evaluating this in cylindrical coordinates using the connection coefficients (11.70), we obtain

$$\Sigma_{\varpi\varpi} = \frac{2}{3}\frac{\partial \xi_\varpi}{\partial \varpi} - \frac{1}{3}\frac{\xi_\varpi}{\varpi} - \frac{1}{3\varpi}\frac{\partial \xi_\phi}{\partial \phi} - \frac{1}{3}\frac{\partial \xi_z}{\partial z},$$

$$\Sigma_{\phi\phi} = \frac{2}{3\varpi}\frac{\partial \xi_\phi}{\partial \phi} + \frac{2}{3}\frac{\xi_\varpi}{\varpi} - \frac{1}{3}\frac{\partial \xi_\varpi}{\partial \varpi} - \frac{1}{3}\frac{\partial \xi_z}{\partial z},$$

$$\Sigma_{zz} = \frac{2}{3}\frac{\partial \xi_z}{\partial z} - \frac{1}{3}\frac{\partial \xi_\varpi}{\partial \varpi} - \frac{1}{3}\frac{\xi_\varpi}{\varpi} - \frac{1}{3\varpi}\frac{\partial \xi_\phi}{\partial \phi},$$

$$\Sigma_{\phi z} = \Sigma_{z\phi} = \frac{1}{2\varpi}\frac{\partial \xi_z}{\partial \phi} + \frac{1}{2}\frac{\partial \xi_\phi}{\partial z},$$

$$\Sigma_{z\varpi} = \Sigma_{\varpi z} = \frac{1}{2}\frac{\partial \xi_\varpi}{\partial z} + \frac{1}{2}\frac{\partial \xi_z}{\partial \varpi},$$

$$\Sigma_{\varpi\phi} = \Sigma_{\phi\varpi} = \frac{1}{2}\frac{\partial \xi_\phi}{\partial \varpi} - \frac{\xi_\phi}{2\varpi} + \frac{1}{2\varpi}\frac{\partial \xi_\varpi}{\partial \phi}. \quad (2)$$

Likewise, in spherical coordinates using the connection coefficients (11.71), we obtain

$$\Sigma_{rr} = \frac{2}{3}\frac{\partial \xi_r}{\partial r} - \frac{2}{3r}\xi_r - \frac{\cot\theta}{3r}\xi_\theta - \frac{1}{3r}\frac{\partial \xi_\theta}{\partial \theta} - \frac{1}{3r\sin\theta}\frac{\partial \xi_\phi}{\partial \phi},$$

$$\Sigma_{\theta\theta} = \frac{2}{3r}\frac{\partial \xi_\theta}{\partial \theta} + \frac{\xi_r}{3r} - \frac{1}{3}\frac{\partial \xi_r}{\partial r} - \frac{\cot\theta\,\xi_\theta}{3r} - \frac{1}{3r\sin\theta}\frac{\partial \xi_\phi}{\partial \phi},$$

$$\Sigma_{\phi\phi} = \frac{2}{3r\sin\theta}\frac{\partial \xi_\phi}{\partial \phi} + \frac{2\cot\theta\,\xi_\theta}{3r} + \frac{\xi_r}{3r} - \frac{1}{3}\frac{\partial \xi_r}{\partial r} - \frac{1}{3r}\frac{\partial \xi_\theta}{\partial \theta},$$

$$\Sigma_{\theta\phi} = \Sigma_{\phi\theta} = \frac{1}{2r}\frac{\partial \xi_\phi}{\partial \theta} - \frac{\cot\theta\,\xi_\phi}{2r} + \frac{1}{2r\sin\theta}\frac{\partial \xi_\theta}{\partial \phi},$$

$$\Sigma_{\phi r} = \Sigma_{r\phi} = \frac{1}{2r\sin\theta}\frac{\partial \xi_r}{\partial \phi} + \frac{1}{2}\frac{\partial \xi_\phi}{\partial r} - \frac{\xi_\phi}{2r},$$

$$\Sigma_{r\theta} = \Sigma_{\theta r} = \frac{1}{2}\frac{\partial \xi_\theta}{\partial r} - \frac{\xi_\theta}{2r} + \frac{1}{2r}\frac{\partial \xi_r}{\partial \theta}. \quad (3)$$

any basis (orthonormal or nonorthonormal) in a curved spacetime, as is needed to perform calculations in general relativity.

Although the algebra of evaluating the components of derivatives such as in Eq. (11.78) in explicit form (e.g., in terms of $\{r, \theta, \phi\}$) can be long and tedious when done by hand, in the modern era of symbolic manipulation using computers (e.g., Mathematica, Matlab, or Maple), the algebra can be done quickly and accurately to obtain expressions such as Eqs. (3) of Box 11.4.

EXERCISES

Exercise 11.21 *Derivation and Practice: Gradient of a Second-Rank Tensor* T2
By a computation analogous to Eq. (11.72), derive Eq. (11.78) for the components of the gradient of a second-rank tensor in any orthonormal basis.

Exercise 11.22 *Derivation and Practice: Connection in Spherical Coordinates* T2
(a) By drawing pictures analogous to Fig. 11.15, show that

$$\nabla_\phi \mathbf{e}_r = \frac{1}{r}\mathbf{e}_\phi, \quad \nabla_\theta \mathbf{e}_r = \frac{1}{r}\mathbf{e}_\theta, \quad \nabla_\phi \mathbf{e}_\theta = \frac{\cot\theta}{r}\mathbf{e}_\phi. \tag{11.79}$$

(b) From these relations and antisymmetry on the first two indices [Eq. (11.69)], deduce the connection coefficients (11.71).

Exercise 11.23 *Derivation and Practice: Expansion in Cylindrical and Spherical Coordinates* T2
Derive Eqs. (11.76) and (11.77) for the divergence of the vector field $\boldsymbol{\xi}$ in cylindrical and spherical coordinates using the connection coefficients (11.70) and (11.71).

11.9 Solving the 3-Dimensional Navier-Cauchy Equation in Cylindrical Coordinates T2

11.9.1 Simple Methods: Pipe Fracture and Torsion Pendulum T2

As an example of an elastostatic problem with cylindrical symmetry, consider a cylindrical pipe that carries a high-pressure fluid (e.g., water, oil, natural gas); Fig. 11.16. How thick must the pipe's wall be to ensure that it will not burst due to the fluid's pressure? We sketch the solution, leaving the details to Ex. 11.24.

We suppose, for simplicity, that the pipe's length is held fixed by its support system: it does not lengthen or shorten when the fluid pressure is changed. Then by symmetry, the displacement field in the pipe wall is purely radial and depends only on radius: its only nonzero component is $\xi_\varpi(\varpi)$. The radial dependence is governed by radial force balance,

$$f_\varpi = K\Theta_{;\varpi} + 2\mu\Sigma_{\varpi j;j} = 0 \tag{11.80}$$

[Eq. (11.30)].

FIGURE 11.16 A pipe whose wall has inner and outer radii ϖ_1 and ϖ_2.

Because ξ_ϖ is independent of ϕ and z, the expansion [Eq. (11.76)] is given by

$$\Theta = \frac{d\xi_\varpi}{d\varpi} + \frac{\xi_\varpi}{\varpi}. \tag{11.81}$$

The second term in the radial force balance equation (11.80) is proportional to $\Sigma_{\varpi j;j}$ which—using Eq. (11.78) and noting that the only nonzero connection coefficients are $\Gamma_{\varpi\phi\phi} = -\Gamma_{\phi\varpi\phi} = -1/\varpi$ [Eq. (11.70)] and that symmetry requires the shear tensor to be diagonal—becomes

$$\Sigma_{\varpi j;j} = \Sigma_{\varpi\varpi,\varpi} + \Gamma_{\varpi\phi\phi}\Sigma_{\phi\phi} + \Gamma_{\phi\varpi\phi}\Sigma_{\varpi\varpi}. \tag{11.82}$$

Inserting the components of the shear tensor from Eqs. (2) of Box 11.4 and the values of the connection coefficients and comparing the result with expression (11.81) for the expansion, we obtain the remarkable result that $\Sigma_{\varpi j;j} = \frac{2}{3}\partial\Theta/\partial\varpi$. Inserting this into the radial force balance equation (11.80), we obtain

$$f_\varpi = \left(K + \frac{4\mu}{3}\right)\frac{d\Theta}{d\varpi} = 0. \tag{11.83}$$

Thus, inside the pipe wall, the expansion is independent of radius ϖ, and correspondingly, the radial displacement must have the form [cf. Eq. (11.81)]

$$\xi_\varpi = A\varpi + \frac{B}{\varpi} \tag{11.84}$$

for some constants A and B, whence $\Theta = 2A$ and $\Sigma_{\varpi\varpi} = \frac{1}{3}A - B/\varpi^2$. The values of A and B are fixed by the boundary conditions at the inner and outer faces of the pipe wall: $T_{\varpi\varpi} = P$ at $\varpi = \varpi_1$ (inner wall) and $T_{\varpi\varpi} = 0$ at $\varpi = \varpi_2$ (outer wall). Here P is the pressure of the fluid that the pipe carries, and we have neglected the atmosphere's pressure on the outer face by comparison. Evaluating $T_{\varpi\varpi} = -K\Theta - 2\mu\Sigma_{\varpi\varpi}$ and then imposing these boundary conditions, we obtain

$$A = \frac{P}{2(K + \mu/3)}\frac{\varpi_1^2}{(\varpi_2^2 - \varpi_1^2)}, \qquad B = \frac{P}{2\mu}\frac{\varpi_1^2\varpi_2^2}{(\varpi_2^2 - \varpi_1^2)}. \tag{11.85}$$

The only nonvanishing components of the strain then work out to be

$$S_{\varpi\varpi} = \frac{\partial \xi_\varpi}{\partial \varpi} = A - \frac{B}{\varpi^2}, \quad S_{\phi\phi} = \frac{\xi_\varpi}{\varpi} = A + \frac{B}{\varpi^2}. \tag{11.86}$$

This strain is maximal at the inner wall of the pipe; expressing it in terms of the ratio $\zeta \equiv \varpi_2/\varpi_1$ of the outer to the inner pipe radius and using the values of $K = 180$ GPa and $\mu = 81$ GPa for steel, we bring this maximum strain into the form

$$S_{\varpi\varpi} \simeq -\frac{P}{\mu} \frac{5\zeta^2 - 2}{10(\zeta^2 - 1)}, \quad S_{\phi\phi} \simeq \frac{P}{\mu} \frac{5\zeta^2 + 2}{10(\zeta^2 - 1)}. \tag{11.87}$$

Note that $|S_{\phi\phi}| > |S_{\varpi\varpi}|$.

The pipe will fracture at a strain $\sim 10^{-3}$; for safety it is best to keep the actual strain smaller than this by an order of magnitude, $|S_{ij}| \lesssim 10^{-4}$. A typical pressure for an oil pipeline is $P \simeq 10$ atmospheres ($\simeq 10^6$ Pa), compared to the shear modulus of steel $\mu = 81$ GPa, so $P/\mu \simeq 1.2 \times 10^{-5}$. Inserting this number into Eq. (11.87) with $|S_{\phi\phi}| \lesssim 10^{-4}$, we deduce that the ratio of the pipe's outer radius to its inner radius must be $\zeta = \varpi_2/\varpi_1 \gtrsim 1.04$. If the pipe has a diameter of a half meter, then its wall thickness should be about 1 cm or more. This is typical of oil pipelines.

criterion for safety against fracture

Exercise 11.25 presents a second fairly simple example of elastostatics in cylindrical coordinates: a computation of the period of a torsion pendulum.

EXERCISES

Exercise 11.24 *Derivation and Practice: Fracture of a Pipe* T2
Fill in the details of the text's analysis of the deformation of a pipe carrying a high-pressure fluid, and the wall thickness required to protect the pipe against fracture. (See Fig. 11.16.)

Exercise 11.25 *Practice: Torsion Pendulum* T2
A torsion pendulum is a very useful tool for testing the equivalence principle (Sec. 25.2), for seeking evidence for hypothetical fifth (not to mention sixth!) forces, and for searching for deviations from gravity's inverse-square law on submillimeter scales, which could arise from gravity being influenced by macroscopic higher spatial dimensions. (See, e.g., Kapner et al., 2008; Wagner et al., 2012.) It would be advantageous to design a torsion pendulum with a 1-day period (Fig. 11.17). Here we estimate whether this is possible. The pendulum consists of a thin cylindrical wire of length ℓ and radius a. At the bottom of the wire are suspended three masses at the corners of an equilateral triangle at a distance b from the wire.

(a) Show that the longitudinal strain is

$$\xi_{z;z} = \frac{3mg}{\pi a^2 E}. \tag{11.88a}$$

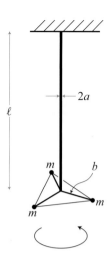

FIGURE 11.17 Torsion pendulum.

(b) What component of shear is responsible for the restoring force in the wire, which causes the torsion pendulum to oscillate?

(c) Show that the pendulum undergoes torsional oscillations with period

$$P = 2\pi \left(\frac{\ell}{g}\right)^{1/2} \left(\frac{2b^2 E \xi_{z;z}}{a^2 \mu}\right)^{1/2}. \tag{11.88b}$$

(d) Do you think you could design a pendulum that attains the goal of a 1-day period?

11.9.2 Separation of Variables and Green's Functions: Thermoelastic Noise in Mirrors

In complicated situations that have moderate amounts of symmetry, the elastostatic equations can be solved by the same kinds of sophisticated mathematical techniques as one uses in electrostatics: separation of variables, Green's functions, complex potentials, or integral transform methods (see, e.g., Gladwell, 1980). We provide an example in this section, focusing on separation of variables and Green's functions.

MOTIVATION

Our example is motivated by an important issue in high-precision measurements with light, including, among others, gravitational-wave detectors and quantum-optics experiments in which photons and atoms are put into entangled nonclassical states by coupling them to one another inside Fabry-Perot cavities.

In these situations, noise due to thermal motions of the mirror is a serious issue. It can hide a gravitational wave, and it can cause decoherence of the atom/photon quantum states. In Sec. 6.8.2, we formulated a generalized fluctuation-dissipation theorem by which this mirror thermal noise can be computed (Levin, 1998).

Specifically, in a thought experiment one applies to the mirror face a force F_o that oscillates at some frequency f at which one wants to evaluate the thermal noise. This force has the same transverse pressure distribution as the light beam—say, for concreteness, a Gaussian distribution:

$$T_{zz}^{\text{applied}} = \frac{e^{-\varpi^2/\varpi_o^2}}{\pi \varpi_o^2} F_o \cos(2\pi f t). \tag{11.89}$$

This applied pressure induces a strain distribution **S** inside the mirror, and that oscillating strain interacts with imperfections to dissipate energy at some rate $W_{\text{diss}}(f)$. The fluctuation-dissipation theorem states that in the real experiment, where the light beam bounces off the mirror, the reflected light will encode a noisy transverse-averaged position q for the mirror face, and the noise spectral density for q will be

$$S_q(f) = \frac{8 W_{\text{diss}}(f) k_B T}{(2\pi f)^2 F_o^2} \tag{11.90}$$

[Eq. (6.88b)].

Even if one could make the mirror perfect (no dislocations or impurities), so there is no dissipation due to imperfections, there will remain one other source of dissipation in this thought experiment: the applied pressure (11.89) will produce a spatially inhomogeneous expansion $\Theta(\mathbf{x}, t)$ inside the mirror, which in turn will produce the thermoelastic temperature change $\Delta T/T = -[3\alpha K/(\rho c_V)]\Theta$ [Eq. (11.29)]. The gradient of this temperature will induce heat flow, with a thermal energy flux $\mathbf{F}_{\text{th}} = -\kappa \nabla \Delta T$, where κ is the thermal conductivity. When an amount Q of this thermal energy flows from a region with temperature T to a region of lower temperature $T - dT$, it produces an entropy increase $dS = Q/(T - dT) - Q/T = Q dT/T^2$; and correspondingly, there is a rate of entropy increase per unit volume given by $d^2 S/dV dt = -\mathbf{F}_{\text{th}} \cdot \nabla \Delta T/T^2 = \kappa (\nabla \Delta T)^2/T^2$. This entropy increase has an accompanying energy dissipation rate $W_{\text{diss}} = \int T(d^2 S/dt dV) dV = \int (\kappa/T)(\nabla \Delta T)^2 dV$. Expressing ΔT in terms of the expansion that drives it via $\Delta T/T = -[3\alpha K/(\rho c_V)]\Theta$ and inserting that into Eq. (11.90) and using the third of Eqs. (11.39), we obtain the thermal noise spectral density that the experimenters must contend with:

$$S_q(f) = \frac{8\kappa E^2 \alpha^2 k_B T^2}{(1 - 2\nu)^2 c_V^2 \rho^2 F_o^2 (2\pi f)^2} \left\langle \int (\nabla \Theta)^2 \varpi \, d\phi \, d\varpi \, dz \right\rangle. \tag{11.91}$$

Here $\langle \cdot \rangle$ means average over time as Θ oscillates due to the oscillation of the driving force $F_o \cos(2\pi f t)$. Because the dissipation producing this noise is due to heat flowing down a thermoelastic temperature gradient, it is called *thermoelastic noise*.

This is the motivation for an elasticity problem that we shall solve to illustrate separation of variables: to evaluate this thermoelastic noise, we must compute the expansion $\Theta(\mathbf{x}, t)$ inside a mirror, produced by the oscillating pressure (11.89) on the mirror face; and we must then perform the integral (11.91).

T2

SOLUTION FOR Θ VIA SEPARATION OF VARIABLES

The frequencies f at which we wish to evaluate the thermal noise are low compared to the inverse sound travel time across the mirror, so when computing Θ we can regard the force as oscillating very slowly (i.e., we can use our elastostatic equations rather than dynamical equations of the next chapter). Also, the size of the light spot on the mirror is usually small compared to the mirror's transverse size and thickness, so we can idealize the mirror as being infinitely large and thick—a homogeneous half-space of isotropic, elastic material.

Because the applied stress is axially symmetric, the induced strain and expansion will also be axially symmetric and are thus computed most easily using cylindrical coordinates. Our challenge, then, is to solve the Navier-Cauchy equation $\mathbf{f} = (K + \frac{1}{3}\mu)\nabla(\nabla \cdot \boldsymbol{\xi}) + \mu \nabla^2 \boldsymbol{\xi} = 0$ for the cylindrical components $\xi_\varpi(z, \varpi)$ and $\xi_z(z, \varpi)$ of the displacement, and then evaluate the expansion $\Theta = \nabla \cdot \boldsymbol{\xi}$. (The component ξ_ϕ vanishes by symmetry.)

Equations of elasticity in cylindrical coordinates, and their homogeneous solution

It is straightforward, using the techniques of Sec. 11.8, to compute the cylindrical components of $\mathbf{f}$. Reexpressing the bulk K and shear μ moduli in terms of Young's modulus E and Poisson's ratio ν [Eqs. (11.39)] and setting the internal forces to zero, we obtain

elastostatic force balance in cylindrical coordinates

$$f_\varpi = \frac{E}{2(1+\nu)(1-2\nu)} \left[2(1-\nu)\left(\frac{\partial^2 \xi_\varpi}{\partial \varpi^2} + \frac{1}{\varpi}\frac{\partial \xi_\varpi}{\partial \varpi} - \frac{\xi_\varpi}{\varpi^2}\right) \right.$$
$$\left. + (1-2\nu)\frac{\partial^2 \xi_\varpi}{\partial z^2} + \frac{\partial^2 \xi_z}{\partial z \partial \varpi} \right] = 0, \quad (11.92a)$$

$$f_z = \frac{E}{2(1+\nu)(1-2\nu)} \left[(1-2\nu)\left(\frac{\partial^2 \xi_z}{\partial \varpi^2} + \frac{1}{\varpi}\frac{\partial \xi_z}{\partial \varpi}\right) \right.$$
$$\left. + 2(1-\nu)\frac{\partial^2 \xi_z}{\partial z^2} + \frac{\partial^2 \xi_\varpi}{\partial z \partial \varpi} + \frac{1}{\varpi}\frac{\partial \xi_\varpi}{\partial z} \right] = 0. \quad (11.92b)$$

These are two coupled, linear, second-order differential equations for the two unknown components of the displacement vector. As with the analogous equations of electrostatics and magnetostatics, these can be solved by separation of variables, that is, by setting $\xi_\varpi = R_\varpi(\varpi) Z_\varpi(z)$ and $\xi_z = R_z(\varpi) Z_z(z)$, and inserting into Eq. (11.92a). We seek the general solution that dies out at large ϖ and z. The general solution of this sort, to the complicated-looking Eqs. (11.92), turns out to be (really!!)

separation-of-variables solution of force-balance equation $f_j = 0$

$$\xi_\varpi = \int_0^\infty [\alpha(k) - (2 - 2\nu - kz)\beta(k)] e^{-kz} J_1(k\varpi) \, dk,$$

$$\xi_z = \int_0^\infty [\alpha(k) + (1 - 2\nu + kz)\beta(k)] e^{-kz} J_0(k\varpi) \, dk. \quad (11.93)$$

624 Chapter 11. Elastostatics

Here J_0 and J_1 are Bessel functions of order 0 and 1, and $\alpha(k)$ and $\beta(k)$ are arbitrary functions.

Boundary conditions

The functions $\alpha(k)$ and $\beta(k)$ are determined by boundary conditions on the face of the test mass. The force per unit area exerted across the face by the strained test-mass material, T_{zj} at $z=0$ with $j = \{\varpi, \phi, z\}$, must be balanced by the applied force per unit area, T_{zj}^{applied} [Eq. (11.89)]. The (shear) forces in the ϕ direction, $T_{z\phi}$ and $T_{z\phi}^{\text{applied}}$, vanish because of cylindrical symmetry and thus provide no useful boundary condition. The (shear) force in the ϖ direction, which must vanish at $z=0$ since $T_{z\varpi}^{\text{applied}} = 0$, is given by [cf. Eq. (2) in Box 11.4]

$$T_{z\varpi}(z=0) = -2\mu \Sigma_{z\varpi} = -\mu \left(\frac{\partial \xi_z}{\partial \varpi} + \frac{\partial \xi_\varpi}{\partial z} \right)$$

$$= -\mu \int_0^\infty [\beta(k) - \alpha(k)] J_1(kz) k \, dk = 0, \quad (11.94)$$

shear-force boundary condition at $z=0$

which implies that $\beta(k) = \alpha(k)$. The (normal) force in the z direction, which must balance the applied pressure (11.89), is $T_{zz} = -K\Theta - 2\mu\Sigma_{zz}$; using Eq. (2) in Box 11.4 and Eqs. (11.39), (11.76), (11.93) and (11.89), this reduces to

$$T_{zz}(z=0) = 2\mu \int_0^\infty \alpha(k) J_0(k\varpi) k \, dk = T_{zz}^{\text{applied}} = \frac{e^{-\varpi^2/\varpi_o^2}}{\pi \varpi_o^2} F_o \cos(2\pi f t),$$

longitudinal-force boundary condition at $z=0$

$$(11.95)$$

which can be inverted[11] to give

$$\alpha(k) = \beta(k) = \frac{1}{4\pi\mu} e^{-k^2 \varpi_o^2/4} F_o \cos(2\pi f t). \quad (11.96)$$

solution for expansion coefficients

Inserting this equation into the Eqs. (11.93) for the displacement and then evaluating the expansion $\Theta = \nabla \cdot \boldsymbol{\xi} = \xi_{z,z} + \varpi^{-1}(\varpi \xi_\varpi)_{,\varpi}$, we obtain

$$\Theta = 2(1 - 2\nu) \int_0^\infty \alpha(k) e^{-kz} J_0(k\varpi) k \, dk. \quad (11.97)$$

As in electrostatics and magnetostatics, so also in elasticity theory, one can solve an elastostatics problem using Green's functions instead of separation of variables. We explore this option for our applied Gaussian force in Ex. 11.27. For greater detail on

---

11. The inversion and the subsequent evaluation of the integral of $(\nabla \Theta)^2$ are aided by the following expressions for the Dirac delta function:

$$\delta(k - k') = k \int_0^\infty J_0(k\varpi) J_0(k'\varpi) \varpi \, d\varpi = k \int_0^\infty J_1(k\varpi) J_1(k'\varpi) \varpi \, d\varpi.$$

11.9 Solving the 3-Dimensional Navier-Cauchy Equation

Green's functions in elastostatics and their applications from an engineer's viewpoint, see Johnson (1985). For other commonly used solution techniques, see Box 11.3.

THERMOELASTIC NOISE SPECTRAL DENSITY

Let us return to the mirror-noise problem that motivated our calculation. It is straightforward to compute the gradient of the expansion (11.97), and square and integrate it to get the spectral density $S_q(f)$ [Eq. (11.91)]. The result is (Braginsky, Gorodetsky, and Vyatchanin, 1999; Liu and Thorne, 2000)

$$S_q(f) = \frac{8(1+\nu)^2 \kappa \alpha^2 k_B T^2}{\sqrt{2\pi} c_V^2 \rho^2 \varpi_o^3 (2\pi f)^2}. \tag{11.98}$$

Early plans for advanced LIGO gravitational wave detectors (Sec. 9.5; LIGO Scientific Collaboration, 2015) called for mirrors made of high-reflectivity dielectric coatings on sapphire crystal substrates. Sapphire was chosen because it can be grown in giant crystals with very low impurities and dislocations, resulting in low thermal noise. However, the thermoelastic noise (11.98) in sapphire turns out to be uncomfortably high. Using sapphire's $\nu = 0.29, \kappa = 40$ W m^{-1} K$^{-1}, \alpha = 5.0 \times 10^{-6}$ K^{-1}, $c_V = 790$ J kg^{-1} K^{-1}, $\rho = 4{,}000$ kg m^{-3}, and a light-beam radius $\varpi_o = 4$ cm and room temperature $T = 300$ K, Eq. (11.98) gives the following for the noise in a bandwidth equal to frequency:

$$\sqrt{f S_q(f)} = 5 \times 10^{-20} \text{ m} \sqrt{\frac{100 \text{ Hz}}{f}}. \tag{11.99}$$

Because this was uncomfortably high at low frequencies, $f \sim 10$ Hz, and because of the birefringence of sapphire, which could cause technical problems, a decision was made to switch to fused silica for the advanced LIGO mirrors.

EXERCISES

Exercise 11.26 *Derivation and Practice: Evaluation of Elastostatic Force in Cylindrical Coordinates*

Derive Eqs. (11.92) for the cylindrical components of the internal elastostatic force per unit volume $\mathbf{f} = (K + \frac{1}{3}\mu)\nabla(\nabla \cdot \boldsymbol{\xi}) + \mu \nabla^2 \boldsymbol{\xi}$ in a cylindrically symmetric situation.

Exercise 11.27 ***Example: Green's Function for Normal Force on a Half-Infinite Body*

Suppose that a stress $T_{zj}^{\text{applied}}(\mathbf{x}_o)$ is applied on the face $z=0$ of a half-infinite elastic body (one that fills the region $z > 0$). Then by virtue of the linearity of the elastostatics equation $\mathbf{f} = (K + \frac{1}{3}\mu)\nabla(\nabla \cdot \boldsymbol{\xi}) + \mu \nabla^2 \boldsymbol{\xi} = 0$ and the linearity of its boundary conditions, $T_{zj}^{\text{internal}} = T_{zj}^{\text{applied}}$, there must be a Green's function $G_{jk}(\mathbf{x} - \mathbf{x}_o)$ such that the body's internal displacement $\boldsymbol{\xi}(\mathbf{x})$ is given by

$$\boxed{\xi_j(\mathbf{x}) = \int G_{jk}(\mathbf{x} - \mathbf{x}_0) T_{kz}^{\text{applied}}(\mathbf{x}_o) d^2 x_o.} \tag{11.100}$$

Here the integral is over all points $\mathbf{x}_o$ on the face of the body ($z = 0$), and $\mathbf{x}$ can be anywhere inside the body, $z \geq 0$.

(a) Show that if a force F_j is applied on the body's surface at a single point (the origin of coordinates), then the displacement inside the body is

$$\xi_j(\mathbf{x}) = G_{jk}(\mathbf{x}) F_k. \tag{11.101}$$

Thus, the Green's function can be thought of as the body's response to a point force on its surface.

(b) As a special case, consider a point force F_z directed perpendicularly into the body. The resulting displacement turns out to have cylindrical components[12]

$$\xi_z = G_{zz}(\varpi, z) F_z = \frac{(1+\nu)}{2\pi E} \left[\frac{2(1-\nu)}{r} + \frac{z^2}{r^3} \right] F_z,$$

$$\xi_\varpi = G_{\varpi z}(\varpi, z) F_z = -\frac{(1+\nu)}{2\pi E} \left[\frac{(1-2\nu)\varpi}{r(r+z)} - \frac{\varpi z}{r^3} \right] F_z, \tag{11.102}$$

where $r = \sqrt{\varpi^2 + z^2}$. It is straightforward to show that this displacement does satisfy the elastostatics equations (11.92). Show that it also satisfies the required boundary condition $T_{z\varpi}(z=0) = -2\mu \Sigma_{z\varpi} = 0$.

(c) Show that for this displacement [Eq. (11.102)], $T_{zz} = -K\Theta - 2\mu\Sigma_{zz}$ vanishes everywhere on the body's surface $z = 0$ except at the origin $\varpi = 0$ and is infinite there. Show that the integral of this normal stress over the surface is F_z, and therefore, $T_{zz}(z=0) = F_z \delta_2(\mathbf{x})$, where δ_2 is the 2-dimensional Dirac delta function on the surface. This is the second required boundary condition.

(d) Plot the integral curves of the displacement vector $\boldsymbol{\xi}$ (i.e., the curves to which $\boldsymbol{\xi}$ is parallel) for a reasonable choice of Poisson's ratio ν. Explain physically why the curves have the form you find.

(e) One can use the Green's function (11.102) to compute the displacement $\boldsymbol{\xi}$ induced by the Gaussian-shaped pressure (11.89) applied to the body's face, and to then evaluate the induced expansion and thence the thermoelastic noise; see Braginsky, Gorodetsky, and Vyatchanin (1999) and Liu and Thorne (2000). The results agree with Eqs. (11.97) and (11.98) deduced using separation of variables.

Bibliographic Note

Elasticity theory was developed in the eighteenth, nineteenth, and early twentieth centuries. The classic advanced textbook from that era is Love (1927). An outstanding, somewhat more modern advanced text is Landau and Lifshitz (1986)—originally

---

12. For the other components of the Green's function, written in Cartesian coordinates (since a non-normal applied force breaks the cylindrical symmetry), see Landau and Lifshitz [1986, Eqs. (8.18)].

written in the 1950s and revised in a third edition shortly before Lifshitz's death. This is among the most readable textbooks that Landau and Lifshitz wrote and is still widely used by physicists in the early twenty-first century.

Some significant new insights, both mathematical and physical, have been developed in recent decades, for example, catastrophe theory and its applications to bifurcations and stability, practical insights from numerical simulations, and practical applications based on new materials (e.g., carbon nanotubes). For a modern treatment that deals with these and much else from an engineering viewpoint, we strongly recommend Ugural and Fenster (2012). For a fairly brief and elementary modern treatment, we recommend Lautrup (2005, Part III). Other good texts that focus particularly on solving the equations for elastostatic equilibrium include Southwell (1941), Timoshenko and Goodier (1970), Gladwell (1980), Johnson (1985), Boresi and Chong (1999), and Slaughter (2002); see also the discussion and references in Box 11.3.

CHAPTER TWELVE

Elastodynamics

. . . logarithmic plots are a device of the devil.

CHARLES RICHTER (1980)

12.1 Overview

In the previous chapter we considered elastostatic equilibria, in which the forces acting on elements of an elastic solid were balanced, so the solid remained at rest. When this equilibrium is disturbed, the solid will undergo accelerations. This is the subject of this chapter—*elastodynamics*.

In Sec. 12.2, we derive the equations of motion for elastic media, paying particular attention to the underlying conservation laws and focusing especially on elastodynamic waves. We show that there are two distinct wave modes that propagate in a uniform, isotropic solid—longitudinal waves and shear waves—and both are nondispersive (their phase speeds are independent of frequency).

A major use of elastodynamics is in structural engineering, where one encounters vibrations (usually standing waves) on the beams that support buildings and bridges. In Sec. 12.3, we discuss the types of waves that propagate on bars, rods, and beams and find that the boundary conditions at the free transverse surfaces make the waves dispersive. We also return briefly to the problem of bifurcation of equilibria (treated in Sec. 11.6) and show how, as the parameters controlling an equilibrium are changed so it passes through a bifurcation point, the equilibrium becomes unstable; the instability drives the system toward the new, stable equilibrium state.

A second application of elastodynamics is to seismology (Sec. 12.4). Earth is mostly a solid body through which waves can propagate. The waves can be excited naturally by earthquakes or artificially using human-made explosions. Understanding how waves propagate through Earth is important for locating the sources of earthquakes, for diagnosing the nature of an explosion (was it an illicit nuclear bomb test?), and for analyzing the structure of Earth. We briefly describe some of the wave modes that propagate through Earth and some of the inferences about Earth's structure that have been drawn from studying their propagation. In the process, we gain some experience in applying the tools of geometric optics to new types of waves, and we learn

> **BOX 12.1. READERS' GUIDE**
>
> - This chapter is a companion to Chap. 11 (Elastostatics) and relies heavily on it.
> - This chapter also relies rather heavily on geometric-optics concepts and formalism, as developed in Secs. 7.2 and 7.3, especially phase velocity, group velocity, dispersion relation, rays, and the propagation of waves and information and energy along them, the role of the dispersion relation as a hamiltonian for the rays, and ray tracing.
> - The discussion of continuum-mechanics wave equations in Box 12.2 underlies this book's treatment of waves in fluids (Chap. 16), especially in plasmas (Part VI) and in general relativity (Chap. 27).
> - The experience that the reader gains in this chapter with waves in solids will be useful when we encounter much more complicated waves in plasmas in Part VI.
> - No other portions of this chapter are important for subsequent parts of this book.

how rich can be the Green's function for elastodynamic waves, even when the medium is as simple as a homogeneous half-space. We briefly discuss how the methods by which geophysicists probe Earth's structure using seismic waves also find application in ultrasonic technology: imaging solid structures and the human body using high-frequency sound waves.

Finally, in Sec. 12.5, we return to physics to consider the quantum theory of elastodynamic waves. We compare the classical theory with the quantum theory, specializing to quantized vibrations in an elastic solid: phonons.

12.2 Basic Equations of Elastodynamics; Waves in a Homogeneous Medium

In this section, we derive a vectorial equation that governs the dynamical displacement $\boldsymbol{\xi}(\mathbf{x}, t)$ of a dynamically disturbed elastic medium. We then consider monochromatic plane wave solutions and explore wave propagation through inhomogeneous media in the geometric optics limit. This general approach to wave propagation in continuum mechanics will be taken up again in Chap. 16 for waves in fluids, Part VI for waves in plasmas, and Chap. 27 for general relativistic gravitational waves. We conclude this section with a discussion of the energy density and energy flux of these waves.

12.2.1 Equation of Motion for a Strained Elastic Medium

In Chap. 11, we learned that when an elastic medium undergoes a displacement $\boldsymbol{\xi}(\mathbf{x})$, it builds up a strain $S_{ij} = \frac{1}{2}(\xi_{i;j} + \xi_{j;i})$, which in turn produces an internal stress

$\mathbf{T} = -K\Theta\mathbf{g} - 2\mu\mathbf{\Sigma}$. Here $\Theta \equiv \nabla \cdot \boldsymbol{\xi}$ is the expansion, and $\mathbf{\Sigma} \equiv$ (the trace-free part of $\mathbf{S}$) is the shear; see Eqs. (11.4) and (11.18). The stress $\mathbf{T}$ produces an elastic force per unit volume

$$\mathbf{f} = -\nabla \cdot \mathbf{T} = \left(K + \frac{1}{3}\mu\right)\nabla(\nabla \cdot \boldsymbol{\xi}) + \mu\nabla^2\boldsymbol{\xi} \qquad (12.1)$$

[Eq. (11.30)], where K and μ are the bulk and shear moduli.

In Chap. 11, we restricted ourselves to elastic media that are in elastostatic equilibrium, so they are static. This equilibrium requires that the net force per unit volume acting on the medium vanish. If the only force is elastic, then $\mathbf{f}$ must vanish. If the pull of gravity is also significant, then $\mathbf{f} + \rho\mathbf{g}$ vanishes, where ρ is the medium's mass density and $\mathbf{g}$ the acceleration of gravity.

In this chapter, we focus on dynamical situations, in which an unbalanced force per unit volume causes the medium to move—with the motion taking the form of an elastodynamic wave. For simplicity, we assume unless otherwise stated that the only significant force is elastic (i.e., the gravitational force is negligible by comparison). In Ex. 12.1, we show that this is the case for elastodynamic waves in most media on Earth when the wave frequency $\omega/(2\pi)$ is higher than about 0.001 Hz (which is usually the case in practice). Stated more precisely, in a homogeneous medium we can ignore the gravitational force when the elastodynamic wave's angular frequency ω is much larger than g/C, where g is the acceleration of gravity, and C is the wave's propagation speed.

when gravity can be ignored

Consider, then, a dynamical, strained medium with elastic force per unit volume (12.1) and no other significant force (negligible gravity), and with velocity

$$\boxed{\mathbf{v} = \frac{\partial \boldsymbol{\xi}}{\partial t}.} \qquad (12.2\text{a})$$

The law of momentum conservation states that the force per unit volume $\mathbf{f}$, if nonzero, must produce a rate of change of momentum per unit volume $\rho\mathbf{v}$ according to the equation[1]

$$\frac{\partial(\rho\mathbf{v})}{\partial t} = \mathbf{f} = -\nabla \cdot \mathbf{T} = \left(K + \frac{1}{3}\mu\right)\nabla(\nabla \cdot \boldsymbol{\xi}) + \mu\nabla^2\boldsymbol{\xi}. \qquad (12.2\text{b})$$

elastodynamic momentum conservation

Notice that, when rewritten in the form

$$\frac{\partial(\rho\mathbf{v})}{\partial t} + \nabla \cdot \mathbf{T} = 0, \qquad (12.2\text{b}')$$

---

1. In Sec. 13.5, we learn that the motion of the medium produces a stress $\rho\mathbf{v} \otimes \mathbf{v}$ that must be included in this equation if the velocities are large. However, this subtle dynamical stress is always negligible in elastodynamic waves, because the displacements and hence velocities $\mathbf{v}$ are tiny and $\rho\mathbf{v} \otimes \mathbf{v}$ is second order in the displacement. For this reason, we delay studying this subtle nonlinear effect until Chap. 13. A similar remark applies to Eq. (12.2c). In general the conservation laws of continuum mechanics are nonlinear, as we discuss in Box 12.2.

this is the version of the law of momentum conservation discussed in Chap. 1 [Eq. (1.36)]. It has the standard form for a conservation law (time derivative of density of something plus divergence of flux of that something vanishes; see end of Sec. 1.8); $\rho \mathbf{v}$ is the density of momentum, and the stress tensor **T** is by definition the flux of momentum. Equations (12.2a) and (12.2b), together with the law of mass conservation [the obvious analog of Eqs. (1.30) for conservation of charge and particle number],

mass conservation

$$\boxed{\frac{\partial \rho}{\partial t} + \nabla \cdot (\rho \mathbf{v}) = 0,} \qquad (12.2c)$$

are a complete set of equations for the evolution of the displacement $\boldsymbol{\xi}(\mathbf{x}, t)$, the velocity $\mathbf{v}(\mathbf{x}, t)$, and the density $\rho(\mathbf{x}, t)$.

To derive a linear wave equation, we must find some small parameter in which to expand. The obvious choice in elastodynamics is the magnitude of the components of the strain $S_{ij} = \frac{1}{2}(\xi_{i;j} + \xi_{j;i})$, which are less than about 10^{-3} so as to remain below the proportionality limit (i.e., to remain in the realm where stress is proportional to strain; Sec. 11.3.2). Equally well, we can regard the displacement $\boldsymbol{\xi}$ itself as our small parameter (or more precisely, $\xi/\lambdabar$, the magnitude of $\boldsymbol{\xi}$ divided by the reduced wavelength of its perturbations).

If the medium's equilibrium state were homogeneous, the linearization would be trivial. However, we wish to be able to treat perturbations of inhomogeneous equilibria, such as seismic waves in Earth, or perturbations of slowly changing equilibria, such as vibrations of a pipe or mirror that is gradually changing temperature. In most practical situations the lengthscale $\mathcal{L}$ and timescale $\mathcal{T}$ on which the medium's equilibrium properties (ρ, K, μ) vary are extremely large compared to the lengthscale and timescale of the dynamical perturbations [their reduced wavelength $\lambdabar =$ wavelength$/(2\pi)$ and $1/\omega =$ period$/(2\pi)$]. This permits us to perform a two-lengthscale expansion (like the one that underlies geometric optics; Sec. 7.3) alongside our small-strain expansion.

density perturbation

When analyzing a dynamical perturbation of an equilibrium state, we use $\boldsymbol{\xi}(\mathbf{x}, t)$ to denote the dynamical displacement (i.e., we omit from it the equilibrium's static displacement, and similarly we omit from **S** the equilibrium strain). We write the density as $\rho + \delta\rho$, where $\rho(\mathbf{x})$ is the equilibrium density distribution, and $\delta\rho(\mathbf{x}, t)$ is the dynamical density perturbation, which is first order in the dynamical displacement $\boldsymbol{\xi}$. Inserting these into the equation of mass conservation (12.2c), we obtain $\partial\delta\rho/\partial t + \nabla \cdot [(\rho + \delta\rho)\mathbf{v}] = 0$. Because $\mathbf{v} = \partial\boldsymbol{\xi}/\partial t$ is first order, the term $(\delta\rho)\mathbf{v}$ is second order and can be dropped, resulting in the linearized equation $\partial\delta\rho/\partial t + \nabla \cdot (\rho\mathbf{v}) = 0$. Because ρ varies on a much longer lengthscale than does $\mathbf{v}$ ($\mathcal{L}$ versus $\lambdabar$), we can pull ρ out of the derivative. Setting $\mathbf{v} = \partial\boldsymbol{\xi}/\partial t$ and interchanging the time derivative and divergence, we then obtain $\partial\delta\rho/\partial t + \rho\partial(\nabla \cdot \boldsymbol{\xi})/\partial t = 0$. Noting that ρ varies on a

BOX 12.2. WAVE EQUATIONS IN CONTINUUM MECHANICS

Here we make an investment for future chapters by considering wave equations in some generality. Most wave equations arise as approximations to the full set of equations that govern a dynamical physical system. It is usually possible to arrange those full equations as a set of first-order partial differential equations that describe the dynamical evolution of a set of n physical quantities, V_A, with $A = 1, 2, \ldots, n$:

$$\frac{\partial V_A}{\partial t} + F_A(V_B) = 0. \tag{1}$$

[For elastodynamics there are $n = 7$ quantities V_A: $\{\rho, \rho v_x, \rho v_y, \rho v_z, \xi_x, \xi_y, \xi_z\}$ (in Cartesian coordinates); and the seven equations (1) are mass conservation, momentum conservation, and $\partial \xi_j / \partial t = v_j$; Eqs. (12.2).]

Most dynamical systems are intrinsically nonlinear (Maxwell's equations in a vacuum being a conspicuous exception), and it is usually quite hard to find nonlinear solutions. However, it is generally possible to make a perturbation expansion in some small physical quantity about a time-independent equilibrium and retain only those terms that are linear in this quantity. We then have a set of n linear partial differential equations that are much easier to solve than the nonlinear ones—and that usually turn out to have the character of wave equations (i.e., to be *hyperbolic*; see Arfken, Weber, and Harris, 2013). Of course, the solutions will only be a good approximation for small-amplitude waves. [In elastodynamics, we justify linearization by requiring that the strains be below the proportionality limit and linearize in the strain or displacement of the dynamical perturbation. The resulting linear wave equation is $\rho \partial^2 \boldsymbol{\xi}/\partial t^2 = (K + \frac{1}{3}\mu)\boldsymbol{\nabla}(\boldsymbol{\nabla} \cdot \boldsymbol{\xi}) + \mu \nabla^2 \boldsymbol{\xi}$; Eq. (12.4b).]

BOUNDARY CONDITIONS

In some problems (e.g., determining the normal modes of vibration of a building during an earthquake, or analyzing the sound from a violin or the vibrations of a finite-length rod), the boundary conditions are intricate and have to be incorporated as well as possible to have any hope of modeling the problem. The situation is rather similar to that familiar from elementary quantum mechanics. The waves are often localized in some region of space, like bound states, in such a way that the eigenfrequencies are discrete (e.g., standing-wave modes of a plucked string). In other problems the volume in which the wave propagates is essentially infinite (e.g., waves on the surface of the ocean, or seismic waves propagating through Earth), as happens with

(continued)

BOX 12.2. (continued)

unbound quantum states. Then the only boundary condition is essentially that the wave amplitude remain finite at large distances. In this case, the wave spectrum is usually continuous.

GEOMETRIC-OPTICS LIMIT AND DISPERSION RELATIONS

The solutions to the wave equation will reflect the properties of the medium through which the wave is propagating, as well as its boundaries. If the medium and boundaries have a finite number of discontinuities but are otherwise smoothly varying, there is a simple limiting case: waves of short enough wavelength and high enough frequency can be analyzed in the geometric-optics approximation (Chap. 7).

The key to geometric optics is the dispersion relation, which (as we learned in Sec. 7.3) acts as a hamiltonian for the propagation. Recall from Chap. 7 that although the medium may actually be inhomogeneous and might even be changing with time, when deriving the dispersion relation, we can approximate it as precisely homogeneous and time independent and can resolve the waves into plane-wave modes [i.e., modes in which the perturbations vary $\propto \exp i(\mathbf{k} \cdot \mathbf{x} - \omega t)$]. Here $\mathbf{k}$ is the wave vector and ω is the angular frequency. This allows us to remove all the temporal and spatial derivatives and converts our set of partial differential equations into a set of homogeneous, linear algebraic equations. When we do this, we say that our normal modes are *local*. If, instead, we were to go to the trouble of solving the partial differential wave equation with its attendant boundary conditions, the modes would be referred to as *global*.

The linear algebraic equations for a local problem can be written in the form $M_{AB} V_B = 0$, where V_A is the vector of n dependent variables, and the elements M_{AB} of the $n \times n$ matrix $||M_{AB}||$ depend on $\mathbf{k}$ and ω as well as on parameters p_α that describe the local conditions of the medium. [See, e.g., Eq. (12.6).] This set of equations can be solved in the usual manner by requiring that the determinant of $||M_{AB}||$ vanish. Carrying through this procedure yields a polynomial, usually of nth order, for $\omega(\mathbf{k}, p_\alpha)$. This polynomial is the dispersion relation. It can be solved (analytically in simple cases and numerically in general) to yield a number of complex values for ω (the eigenfrequencies), with $\mathbf{k}$ regarded as real. (Some problems involve complex $\mathbf{k}$ and real ω, but for concreteness, we shall take $\mathbf{k}$ to be real.) Armed with the complex eigenfrequencies, we can solve for the associated eigenvectors V_A. The eigenfrequencies and eigenvectors fully characterize the

(continued)

> **BOX 12.2. (continued)**
>
> solution of the local problem and can be used to solve for the waves' temporal evolution from some given initial conditions in the usual manner. (As we shall see several times, especially when we discuss Landau damping in Sec. 22.3, there are some subtleties that can arise.)
>
> What does a complex value of the angular frequency ω mean? We have posited that all small quantities vary $\propto \exp[i(\mathbf{k} \cdot \mathbf{x} - \omega t)]$. If ω has a positive imaginary part, then the small perturbation quantities will grow exponentially with time. Conversely, if it has a negative imaginary part, they will decay. Now, polynomial equations with real coefficients have complex conjugate solutions. Therefore, if there is a decaying mode there must also be a growing mode. Growing modes correspond to instability, a topic that we shall encounter often.

timescale $\mathcal{T}$ long compared to the $1/\omega$ of $\boldsymbol{\xi}$ and $\delta\rho$, we can integrate this to obtain the linear relation

$$\boxed{\frac{\delta\rho}{\rho} = -\boldsymbol{\nabla} \cdot \boldsymbol{\xi}.} \quad (12.3)$$

This linearized equation for the fractional perturbation of density could equally well have been derived by considering a small volume V of the medium that contains mass $M = \rho V$ and noting that the dynamical perturbations lead to a volume change $\delta V/V = \Theta = \boldsymbol{\nabla} \cdot \boldsymbol{\xi}$ [Eq. (11.7)]. Then conservation of mass requires $0 = \delta M = \delta(\rho V) = V \delta\rho + \rho \delta V = V \delta\rho + \rho V \boldsymbol{\nabla} \cdot \boldsymbol{\xi}$, which implies $\delta\rho/\rho = -\boldsymbol{\nabla} \cdot \boldsymbol{\xi}$. This is the same as Eq. (12.3).

The equation of momentum conservation (12.2b) can be handled similarly. By setting $\rho \to \rho(\mathbf{x}) + \delta\rho(\mathbf{x}, t)$, then linearizing (i.e., dropping the $\delta\rho\, \mathbf{v}$ term) and pulling the slowly varying $\rho(\mathbf{x})$ out from the time derivative, we convert $\partial(\rho\mathbf{v})/\partial t$ into $\rho \partial \mathbf{v}/\partial t = \rho \partial^2 \boldsymbol{\xi}/\partial t^2$. Inserting this into Eq. (12.2b), we obtain the linear wave equation

$$\rho \frac{\partial^2 \boldsymbol{\xi}}{\partial t^2} = -\boldsymbol{\nabla} \cdot \mathbf{T}_{\text{el}}, \quad (12.4a)$$

where $\mathbf{T}_{\text{el}}$ is the elastic contribution to the stress tensor. Expanding it, we obtain

$$\boxed{\rho \frac{\partial^2 \boldsymbol{\xi}}{\partial t^2} = \left(K + \frac{1}{3}\mu\right) \boldsymbol{\nabla}(\boldsymbol{\nabla} \cdot \boldsymbol{\xi}) + \mu \nabla^2 \boldsymbol{\xi}.} \quad (12.4b)$$

elastodynamic wave equation

In this equation, terms involving a derivative of K or μ have been omitted, because the two-lengthscale assumption $\mathcal{L} \gg \lambda$ makes them negligible compared to the terms we have kept.

Equation (12.4b) is the first of many wave equations we shall encounter in elastodynamics, fluid mechanics, plasma physics, and general relativity.

12.2.2 Elastodynamic Waves

Continuing to follow our general procedure for deriving and analyzing wave equations as outlined in Box 12.2, we next derive dispersion relations for two types of waves (longitudinal and transverse) that are jointly incorporated into the general elastodynamic wave equation (12.4b).

Recall from Sec. 7.3.1 that, although a dispersion relation can be used as a hamiltonian for computing wave propagation through an inhomogeneous medium, one can derive the dispersion relation most easily by specializing to monochromatic plane waves propagating through a medium that is precisely homogeneous. Therefore, we seek a plane-wave solution, that is, a solution of the form

$$\boldsymbol{\xi}(\mathbf{x}, t) \propto e^{i(\mathbf{k}\cdot\mathbf{x}-\omega t)}, \tag{12.5}$$

to the wave equation (12.4b) with ρ, K, and μ regarded as homogeneous (constant). (To deal with more complicated perturbations of a homogeneous medium, we can think of this wave as being an individual Fourier component and linearly superpose many such waves as a Fourier integral.) Since our wave is planar and monochromatic, we can remove the derivatives in Eq. (12.4b) by making the substitutions $\nabla \to i\mathbf{k}$ and $\partial/\partial t \to -i\omega$ (the first of which implies $\nabla^2 \to -k^2$, $\nabla \cdot \to i\mathbf{k}\cdot$, $\nabla \times \to i\mathbf{k}\times$). We thereby reduce the partial differential equation (12.4b) to a vectorial algebraic equation:

elastodynamic eigenequation for plane waves

$$\rho\omega^2 \boldsymbol{\xi} = \left(K + \frac{1}{3}\mu\right) \mathbf{k}(\mathbf{k} \cdot \boldsymbol{\xi}) + \mu k^2 \boldsymbol{\xi}. \tag{12.6}$$

[This reduction is only possible because the medium is uniform, or in the geometric-optics limit of near uniformity; otherwise, we must solve the second-order partial differential equation (12.4b).]

How do we solve this equation? The sure way is to write it as a 3×3 matrix equation $M_{ij}\xi_j = 0$ for the vector $\boldsymbol{\xi}$ and set the determinant of M_{ij} to zero (Box 12.2 and Ex. 12.3). This is not hard for small or sparse matrices. However, some wave equations are more complicated, and it often pays to think about the waves in a geometric, coordinate-independent way before resorting to brute force.

The quantity that oscillates in the elastodynamic waves of Eq. (12.6) is the vector field $\boldsymbol{\xi}$. The nature of its oscillations is influenced by the scalar constants ρ, μ, K, and ω and by just one quantity that has directionality: the constant vector $\mathbf{k}$. It seems reasonable to expect the description (12.6) of the oscillations to simplify, then, if we resolve the oscillations into a "longitudinal" component (or "mode") along $\mathbf{k}$ and a "transverse" component (or "mode") perpendicular to $\mathbf{k}$, as shown in Fig. 12.1:

decomposition into longitudinal and transverse modes

$$\boxed{\boldsymbol{\xi} = \boldsymbol{\xi}_L + \boldsymbol{\xi}_T, \quad \boldsymbol{\xi}_L = \xi_L \hat{\mathbf{k}}, \quad \boldsymbol{\xi}_T \cdot \hat{\mathbf{k}} = 0.} \tag{12.7a}$$

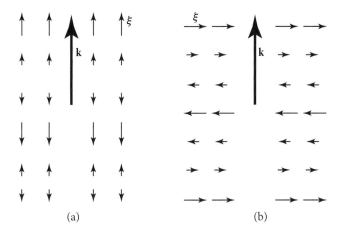

FIGURE 12.1 Displacements in an isotropic, elastic solid, perturbed by (a) a longitudinal mode and (b) a transverse mode.

Here $\hat{\mathbf{k}} \equiv \mathbf{k}/k$ is the unit vector along the propagation direction. It is easy to see that the longitudinal mode $\boldsymbol{\xi}_L$ has nonzero expansion $\Theta \equiv \nabla \cdot \boldsymbol{\xi}_L \neq 0$ but vanishing rotation $\boldsymbol{\phi} = \frac{1}{2} \nabla \times \boldsymbol{\xi}_L = 0$. It can therefore be written as the gradient of a scalar potential:

$$\boxed{\boldsymbol{\xi}_L = \nabla \psi.} \tag{12.7b}$$

By contrast, the transverse mode has zero expansion but nonzero rotation and can thus be written as the curl of a vector potential:

$$\boxed{\boldsymbol{\xi}_T = \nabla \times \mathbf{A};} \tag{12.7c}$$

cf. Ex. 12.2.

12.2.3 Longitudinal Sound Waves

For the longitudinal mode the algebraic wave equation (12.6) reduces to the following simple relation [as one can easily see by inserting $\boldsymbol{\xi} \equiv \boldsymbol{\xi}_L = \xi_L \hat{\mathbf{k}}$ into Eq. (12.6), or, alternatively, by taking the divergence of Eq. (12.6), which is equivalent to taking the scalar product with $\mathbf{k}$]:

$$\omega^2 = \frac{K + \frac{4}{3}\mu}{\rho} k^2; \quad \text{or} \quad \boxed{\omega = \Omega(\mathbf{k}) = \left(\frac{K + \frac{4}{3}\mu}{\rho}\right)^{1/2} k.} \tag{12.8}$$

longitudinal dispersion relation

Here $k = |\mathbf{k}|$ is the wave number (the magnitude of $\mathbf{k}$). This relation between ω and $\mathbf{k}$ is the longitudinal mode's dispersion relation.

From the geometric-optics analysis in Sec. 7.3 we infer that if K, μ, and ρ vary spatially on an inhomogeneity lengthscale $\mathcal{L}$ large compared to $1/k = \lambdabar$, and vary temporally on a timescale $\mathcal{T}$ large compared to $1/\omega$, then the dispersion relation (12.8)—with Ω now depending on $\mathbf{x}$ and t through K, μ, and ρ—serves as a hamiltonian for the wave propagation. In Sec. 12.4 and Fig. 12.5 below we use this hamiltonian to deduce details of the propagation of seismic waves through Earth's inhomogeneous interior.

As discussed in great detail in Sec. 7.2, associated with any wave mode is its phase velocity $\mathbf{V}_{\text{ph}} = (\omega/k)\hat{\mathbf{k}}$ and its phase speed $V_{\text{ph}} = \omega/k$. The dispersion relation (12.8) implies that for longitudinal elastodynamic modes, the phase speed is

longitudinal wave speed

$$\boxed{C_L = \frac{\omega}{k} = \left(\frac{K + \frac{4}{3}\mu}{\rho}\right)^{1/2}.} \tag{12.9a}$$

As Eq. (12.9a) does not depend on the wave number $k \equiv |\mathbf{k}|$, the mode is nondispersive. As it does not depend on the direction $\hat{\mathbf{k}}$ of propagation through the medium, the phase speed is also isotropic, naturally enough, and the group velocity $V_{gj} = \partial\Omega/\partial k_j$ is equal to the phase velocity:

$$\boxed{\mathbf{V}_g = \mathbf{V}_{\text{ph}} = C_L\hat{\mathbf{k}}.} \tag{12.9b}$$

Elastodynamic longitudinal modes are similar to sound waves in a fluid. However, in a fluid, as we can infer from Eq. (16.48), the sound waves travel with phase speed $V_{\text{ph}} = (K/\rho)^{1/2}$ [the limit of Eq. (12.9a) when the shear modulus vanishes].[2] This fluid sound speed is lower than the C_L of a solid with the same bulk modulus, because the longitudinal displacement necessarily entails shear (note that in Fig. 12.1a the motions are not an isotropic expansion), and in a solid there is a restoring shear stress (proportional to μ) that is absent in a fluid.

Because the longitudinal phase velocity is independent of frequency, we can write down general planar longitudinal-wave solutions to the elastodynamic wave equation (12.4b) in the following form:

general longitudinal plane wave

$$\boldsymbol{\xi} = \xi_L\hat{\mathbf{k}} = F(\hat{\mathbf{k}} \cdot \mathbf{x} - C_Lt)\hat{\mathbf{k}}, \tag{12.10}$$

where $F(x)$ is an arbitrary function. This describes a wave propagating in the (arbitrary) direction $\hat{\mathbf{k}}$ with an arbitrary profile determined by the function F.

12.2.4 Transverse Shear Waves

To derive the dispersion relation for a transverse wave we can simply make use of the transversality condition $\mathbf{k} \cdot \boldsymbol{\xi}_T = 0$ in Eq. (12.6); or, equally well, we can take the curl of Eq. (12.6) (multiply it by $i\mathbf{k}\times$), thereby projecting out the transverse piece, since

---

2. Equation (16.48) below gives the fluid sound speed as $C = \sqrt{(\partial P/\partial\rho)_s}$ (i.e., the square root of the derivative of the fluid pressure with respect to density at fixed entropy). In the language of elasticity theory, the fractional change of density is related to the expansion Θ by $\delta\rho/\rho = -\Theta$ [Eq. (12.3)], and the accompanying change of pressure is $\delta P = -K\Theta$ [sentence preceding Eq. (11.18)], so $\delta P = K(\delta\rho/\rho)$. Therefore, the fluid mechanical sound speed is $C = \sqrt{\delta P/\delta\rho} = \sqrt{K/\rho}$ [see passage following Eq. (12.43)].

the longitudinal part of $\boldsymbol{\xi}$ has vanishing curl. The result is

$$\omega^2 = \frac{\mu}{\rho} k^2, \quad \text{or} \quad \boxed{\omega = \Omega(\mathbf{k}) \equiv \left(\frac{\mu}{\rho}\right)^{1/2} k.} \qquad (12.11)$$

transverse dispersion relation

This dispersion relation $\omega = \Omega(\mathbf{k})$ serves as a geometric-optics hamiltonian for wave propagation when μ and ρ vary slowly with $\mathbf{x}$ and/or t. It also implies that the transverse waves propagate with a phase speed C_T and phase and group velocities given by

$$\boxed{C_T = \left(\frac{\mu}{\rho}\right)^{1/2};} \qquad (12.12a)$$

transverse wave speed

$$\boxed{\mathbf{V}_{\text{ph}} = \mathbf{V}_{\text{g}} = C_T \hat{\mathbf{k}}.} \qquad (12.12b)$$

As $K > 0$, the shear wave speed C_T is always less than the speed C_L of longitudinal waves [Eq. (12.9a)].

These transverse modes are known as *shear waves*, because they are driven by the shear stress; cf. Fig. 12.1b. There is no expansion and therefore no change in volume associated with shear waves. They do not exist in fluids, but they are close analogs of the transverse vibrations of a string.

Longitudinal waves can be thought of as scalar waves, since they are fully describable by a single component ξ_L of the displacement $\boldsymbol{\xi}$: that along $\hat{\mathbf{k}}$. Shear waves, by contrast, are inherently vectorial. Their displacement $\boldsymbol{\xi}_T$ can point in any direction orthogonal to $\mathbf{k}$. Since the directions orthogonal to $\mathbf{k}$ form a 2-dimensional space, once $\mathbf{k}$ has been chosen, there are two independent states of polarization for the shear wave. These two polarization states, together with the single one for the scalar, longitudinal wave, make up the three independent degrees of freedom in the displacement $\boldsymbol{\xi}$.

polarization

In Ex. 12.3 we deduce these properties of $\boldsymbol{\xi}$ using matrix techniques.

EXERCISES

Exercise 12.1 ***Problem: Influence of Gravity on Wave Speed*
Modify the wave equation (12.4b) to include the effect of gravity. Assume that the medium is homogeneous and the gravitational field is constant. By comparing the orders of magnitude of the terms in the wave equation, verify that the gravitational terms can be ignored for high-enough frequency elastodynamic modes: $\omega \gg g/C_{L,T}$. For wave speeds ~ 3 km s^{-1}, this gives $\omega/(2\pi) \gg 0.0005$ Hz. Seismic waves are mostly in this regime.

Exercise 12.2 *Example: Scalar and Vector Potentials for Elastic Waves in a Homogeneous Solid*
Just as in electromagnetic theory, it is sometimes useful to write the displacement $\boldsymbol{\xi}$ in terms of scalar and vector potentials:

$$\boldsymbol{\xi} = \nabla \psi + \nabla \times \mathbf{A}. \qquad (12.13)$$

(The vector potential **A** is, as usual, only defined up to a gauge transformation, $\mathbf{A} \to \mathbf{A} + \nabla\varphi$, where φ is an arbitrary scalar field.) By inserting Eq. (12.13) into the general elastodynamic wave equation (12.4b), show that the scalar and vector potentials satisfy the following wave equations in a homogeneous solid:

$$\frac{\partial^2 \psi}{\partial t^2} = C_L^2 \nabla^2 \psi, \quad \frac{\partial^2 \mathbf{A}}{\partial t^2} = C_T^2 \nabla^2 \mathbf{A}. \tag{12.14}$$

Thus, the scalar potential ψ generates longitudinal waves, while the vector potential **A** generates transverse waves.

Exercise 12.3 *Example: Solving the Algebraic Wave Equation by Matrix Techniques*
By using the matrix techniques discussed in the next-to-the-last paragraph of Box 12.2, deduce that the general solution to the algebraic wave equation (12.6) is the sum of a longitudinal mode with the properties deduced in Sec. 12.2.3 and two transverse modes with the properties deduced in Sec. 12.2.4. [Note: This matrix technique is necessary and powerful when the algebraic dispersion relation is complicated, such as for plasma waves; Secs. 21.4.1 and 21.5.1. Elastodynamic waves are simple enough that we did not need this matrix technique in the text.] Specifically, do the following.

(a) Rewrite the algebraic wave equation in the matrix form $M_{ij}\xi_j = 0$, obtaining thereby an explicit form for the matrix $||M_{ij}||$ in terms of ρ, K, μ, ω and the components of **k**.

(b) This matrix equation from part (a) has a solution if and only if the determinant of the matrix $||M_{ij}||$ vanishes. (Why?) Show that $\det||M_{ij}|| = 0$ is a cubic equation for ω^2 in terms of k^2, and that one root of this cubic equation is $\omega = C_L k$, while the other two roots are $\omega = C_T k$ with C_L and C_T given by Eqs. (12.9a) and (12.12a).

(c) Orient Cartesian axes so that **k** points in the z direction. Then show that, when $\omega = C_L k$, the solution to $M_{ij}\xi_j = 0$ is a longitudinal wave (i.e., a wave with $\boldsymbol{\xi}$ pointing in the z direction, the same direction as **k**).

(d) Show that, when $\omega = C_T k$, there are two linearly independent solutions to $M_{ij}\xi_j = 0$, one with $\boldsymbol{\xi}$ pointing in the x direction (transverse to **k**) and the other in the y direction (also transverse to **k**).

12.2.5 Energy of Elastodynamic Waves

Elastodynamic waves transport energy, just like waves on a string. The waves' kinetic energy density is obviously $\frac{1}{2}\rho\mathbf{v}^2 = \frac{1}{2}\rho\dot{\boldsymbol{\xi}}^2$, where the dot means $\partial/\partial t$. The elastic energy density is given by Eq. (11.24), so the total energy density is

$$U = \frac{1}{2}\rho\dot{\boldsymbol{\xi}}^2 + \frac{1}{2}K\Theta^2 + \mu\Sigma_{ij}\Sigma_{ij}. \tag{12.15a}$$

In Ex. 12.4 we show that (as one might expect) the elastodynamic wave equation (12.4b) can be derived from an action whose lagrangian density is the kinetic energy density minus the elastic energy density. We also show that associated with the waves is an energy flux **F** (not to be confused with a force for which we use the same notation) given by

$$F_i = -K\Theta\dot{\xi}_i - 2\mu\Sigma_{ij}\dot{\xi}_j = T_{ij}\dot{\xi}_j. \tag{12.15b}$$

energy density and flux for general elastodynamic wave in isotropic medium

As the waves propagate, energy sloshes back and forth between the kinetic and the elastic parts, with the time-averaged kinetic energy being equal to the time-averaged elastic energy (equipartition of energy). For the planar, monochromatic, longitudinal mode, the time-averaged energy density and flux are

$$\boxed{U_L = \rho\langle\dot{\xi}_L^2\rangle, \qquad \mathbf{F}_L = U_L C_L \hat{\mathbf{k}},} \tag{12.16}$$

energy density and flux for plane, monochromatic elastodynamic wave

where $\langle\cdot\rangle$ denotes an average over one period or wavelength of the wave. Similarly, for the planar, monochromatic, transverse mode, the time-averaged density and flux of energy are

$$\boxed{U_T = \rho\langle\dot{\xi}_T^2\rangle, \quad \mathbf{F}_T = U_T C_T \hat{\mathbf{k}}} \tag{12.17}$$

(Ex. 12.4). Thus, elastodynamic waves transport energy at the same speed $C_{L,T}$ as the waves propagate, and in the same direction $\hat{\mathbf{k}}$. This is the same behavior as electromagnetic waves in vacuum, whose Poynting flux and energy density are related by $\mathbf{F}_{\text{EM}} = U_{\text{EM}} c \hat{\mathbf{k}}$ with c the speed of light, and the same as all forms of dispersion-free scalar waves [e.g., sound waves in a medium; cf. Eq. (7.31)]. Actually, this is the dispersion-free limit of the more general result that the energy of any wave, in the geometric-optics limit, is transported with the wave's group velocity $\mathbf{V}_g$; see Sec. 7.2.2.

EXERCISES

Exercise 12.4 *Example: Lagrangian and Energy for Elastodynamic Waves*
Derive the energy-density, energy-flux, and lagrangian properties of elastodynamic waves given in Sec. 12.2.5. Specifically, do the following.

(a) For ease of calculation (and for greater generality), consider an elastodynamic wave in a possibly anisotropic medium, for which

$$T_{ij} = -Y_{ijkl}\xi_{k;l}, \tag{12.18}$$

with Y_{ijkl} the tensorial modulus of elasticity, which is symmetric under interchange (i) of the first two indices ij, (ii) of the last two indices kl, and (iii) of the first pair ij with the last pair kl [Eq. (11.17) and associated discussion]. Show that for an isotropic medium

$$Y_{ijkl} = \left(K - \frac{2}{3}\mu\right)g_{ij}g_{kl} + \mu(g_{ik}g_{jl} + g_{il}g_{jk}). \tag{12.19}$$

(Recall that in the orthonormal bases to which we have confined ourselves, the components of the metric are $g_{ij} = \delta_{ij}$, i.e., the Kronecker delta.)

(b) For these waves the elastic energy density is $\frac{1}{2}Y_{ijkl}\xi_{i;j}\xi_{k;l}$ [Eq. (11.25)]. Show that the kinetic energy density minus the elastic energy density,

$$\mathcal{L} = \frac{1}{2}\rho\, \dot{\xi}_i \dot{\xi}_i - \frac{1}{2}Y_{ijkl}\xi_{i;j}\xi_{k;l}, \tag{12.20}$$

is a lagrangian density for the waves; that is, show that the vanishing of its variational derivative $\delta\mathcal{L}/\delta\xi_j \equiv \partial\mathcal{L}/\partial\xi_j - (\partial/\partial t)(\partial\mathcal{L}/\partial\dot{\xi}_j) = 0$ is equivalent to the elastodynamic equations $\rho\ddot{\boldsymbol{\xi}} = -\nabla \cdot \mathbf{T}$.

(c) The waves' energy density and flux can be constructed by the vector-wave analog of the canonical procedure of Eq. (7.36c):

$$U = \frac{\partial\mathcal{L}}{\partial\dot{\xi}_i}\dot{\xi}_i - \mathcal{L} = \frac{1}{2}\rho\,\dot{\xi}_i\dot{\xi}_i + \frac{1}{2}Y_{ijkl}\xi_{i;j}\xi_{k;l},$$

$$F_j = \frac{\partial\mathcal{L}}{\partial\xi_{i;j}}\dot{\xi}_i = -Y_{ijkl}\dot{\xi}_i\xi_{k;l}. \tag{12.21}$$

Verify that these density and flux values satisfy the energy conservation law, $\partial U/\partial t + \nabla \cdot \mathbf{F} = 0$. Using Eq. (12.19), verify that for an isotropic medium, expressions (12.21) for the energy density and flux become the expressions (12.15) given in the text.

(d) Show that in general (for an arbitrary mixture of wave modes), the time average of the total kinetic energy in some huge volume is equal to that of the total elastic energy. Show further that for an individual longitudinal or transverse, planar, monochromatic mode, the time-averaged kinetic energy density and time-averaged elastic energy density are both independent of spatial location. Combining these results, infer that for a single mode, the time-averaged kinetic and elastic energy densities are equal, and therefore the time-averaged total energy density is equal to twice the time-averaged kinetic energy density. Show that this total time-averaged energy density is given by the first of Eqs. (12.16) and (12.17).

(e) Show that the time average of the energy flux (12.15b) for the longitudinal and transverse modes is given by the second of Eqs. (12.16) and (12.17), so the energy propagates with the same speed and direction as the waves themselves.

12.3 Waves in Rods, Strings, and Beams

Before exploring applications (Sec. 12.4) of the longitudinal and transverse waves just described, we discuss how the wave equations and wave speeds are modified when the medium through which they propagate is bounded. Despite this situation being formally "global" in the sense of Box 12.2, elementary considerations enable us to derive the relevant dispersion relations without much effort.

12.3.1 Compression Waves in a Rod

First consider a longitudinal wave propagating along a light (negligible gravity), thin, unstressed rod. We shall call this a *compression wave*. Introduce a Cartesian coordinate system with the x-axis parallel to the rod. When there is a small displacement ξ_x independent of y and z, the restoring stress is given by $T_{xx} = -E\partial\xi_x/\partial x$, where E is Young's modulus [cf. Eq. (11.37)]. Hence the restoring force density $\mathbf{f} = -\nabla \cdot \mathbf{T}$ is $f_x = E\partial^2\xi_x/\partial x^2$. The wave equation then becomes

$$\frac{\partial^2 \xi_x}{\partial t^2} = \left(\frac{E}{\rho}\right)\frac{\partial^2 \xi_x}{\partial x^2}, \tag{12.22}$$

and so the sound speed for compression waves in a long straight rod is

$$\boxed{C_C = \left(\frac{E}{\rho}\right)^{\frac{1}{2}}.} \tag{12.23}$$

speed of compression wave in a rod

Referring to Table 11.1 we see that a typical value of Young's modulus in a solid is $\sim$100 GPa. If we adopt a typical density of $\sim 3 \times 10^3$ kg m^{-3}, then we estimate the compressional sound speed to be $\sim$5 km s^{-1}. This is roughly 15 times the sound speed in air.

As this compressional wave propagates, in regions where the rod is compressed longitudinally, it bulges laterally by an amount given by Poisson's ratio; where it is expanded longitudinally, the rod shrinks laterally. By contrast, for a longitudinal wave in a homogeneous medium, transverse forces do not cause lateral bulging and shrinking. This accounts for the different propagation speeds, $C_C \neq C_L$; see Ex. 12.6.

12.3.2 Torsion Waves in a Rod

Next consider a rod with circular cross section of radius a, subjected to a twisting force (Fig. 12.2). Let us introduce an angular displacement $\Delta\phi \equiv \varphi(x)$ that depends on distance x down the rod. The only nonzero component of the displacement vector is $\xi_\phi = \varpi\varphi(x)$, where $\varpi = \sqrt{y^2 + z^2}$ is the cylindrical radius. We can calculate the total torque, exerted by the portion of the rod to the left of x on the portion to the right, by integrating over a circular cross section. For small twists, there is no expansion, and the only components of the shear tensor are

$$\Sigma_{\phi x} = \Sigma_{x\phi} = \frac{1}{2}\xi_{\phi,x} = \frac{\varpi}{2}\frac{\partial \varphi}{\partial x}. \tag{12.24}$$

The torque contributed by an annular ring of radius ϖ and thickness $d\varpi$ is $\varpi \cdot T_{\phi x} \cdot 2\pi\varpi d\varpi$, and we substitute $T_{\phi x} = -2\mu\Sigma_{\phi x}$ to obtain the total torque of the rod to the left of x on that to the right:

$$N = -\int_0^a 2\pi\mu\varpi^3 d\varpi \frac{\partial\varphi}{\partial x} = -\frac{\mu}{\rho}I\frac{\partial\varphi}{\partial x}, \quad \text{where} \quad I = \frac{\pi}{2}\rho a^4 \tag{12.25}$$

is the moment of inertia per unit length.

12.3 Waves in Rods, Strings, and Beams

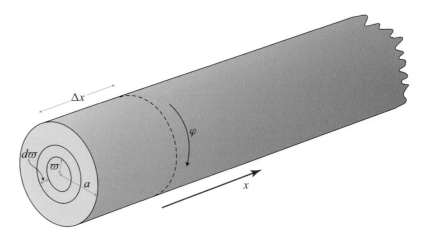

FIGURE 12.2 When a rod with circular cross section is twisted, there will be a restoring torque.

Equating the net torque on a segment with length Δx to the rate of change of its angular momentum, we obtain

$$-\frac{\partial N}{\partial x}\Delta x = I \frac{\partial^2 \varphi}{\partial t^2}\Delta x, \tag{12.26}$$

which, using Eq. (12.25), becomes the wave equation for torsional waves:

$$\left(\frac{\mu}{\rho}\right)\frac{\partial^2 \varphi}{\partial x^2} = \frac{\partial^2 \varphi}{\partial t^2}. \tag{12.27}$$

The speed of torsional waves is thus

speed of torsion wave in a rod

$$\boxed{c_T = \left(\frac{\mu}{\rho}\right)^{\frac{1}{2}}.} \tag{12.28}$$

Note that this is the same speed as that of transverse shear waves in a uniform medium. This might have been anticipated, as there is no change in volume in a torsional oscillation and so only the shear stress acts to produce a restoring force.

12.3.3 Waves on Strings

This example is surely all too familiar. When a string under a tensile force T (*not* force per unit area) is plucked, there will be a restoring force proportional to the curvature of the string. If $\xi_z \equiv \eta$ is the transverse displacement (in the same notation as we used for rods in Secs. 11.5 and 11.6), then the wave equation will be

$$T\frac{\partial^2 \eta}{\partial x^2} = \Lambda \frac{\partial^2 \eta}{\partial t^2}, \tag{12.29}$$

where Λ is the mass per unit length. The wave speed is thus

$$\boxed{C_S = \left(\frac{T}{\Lambda}\right)^{1/2}.} \qquad (12.30)$$

speed of wave in a string under tension

12.3.4 Flexural Waves on a Beam

Now consider the small-amplitude, transverse displacement of a horizontal rod or beam that can be flexed. In Sec. 11.5 we showed that such a flexural displacement produces a net elastic restoring force per unit length given by $D\partial^4\eta/\partial x^4$, and we considered a situation where that force was balanced by the beam's weight per unit length, $W = \Lambda g$ [Eq. (11.45)]. Here

$$D = \frac{1}{12}Ewh^3 \qquad (12.31)$$

is the flexural rigidity [Eq. (11.41b)], h is the beam's thickness in the direction of bend, w is its width, $\eta = \xi_z$ is the transverse displacement of the neutral surface from the horizontal, Λ is the mass per unit length, and g is Earth's acceleration of gravity. The solution of the resulting force-balance equation, $-D\partial^4\eta/\partial x^4 = W = \Lambda g$, is the quartic (11.46a), which describes the equilibrium beam shape.

When gravity is absent and the beam is allowed to bend dynamically, the acceleration of gravity g gets replaced by the beam's dynamical acceleration, $\partial^2\eta/\partial t^2$. The result is a wave equation for flexural waves on the beam:

$$-D\frac{\partial^4 \eta}{\partial x^4} = \Lambda \frac{\partial^2 \eta}{\partial t^2}. \qquad (12.32)$$

(This derivation of the wave equation is an elementary illustration of the principle of equivalence—the equivalence of gravitational and inertial forces, or gravitational and inertial accelerations—which underlies Einstein's general relativity theory; see Sec. 25.2.)

The wave equations we have encountered so far in this chapter have all described nondispersive waves, for which the wave speed is independent of the frequency. Flexural waves, by contrast, are dispersive. We can see this by assuming that $\eta \propto \exp[i(kx - \omega t)]$ and thereby deducing from Eq. (12.32) the dispersion relation:

dispersion relation for flexural waves on a beam

$$\boxed{\omega = \sqrt{D/\Lambda}\, k^2 = \sqrt{Eh^2/(12\rho)}\, k^2.} \qquad (12.33)$$

Here we have used Eq. (12.31) for the flexural rigidity D and $\Lambda = \rho w h$ for the mass per unit length.

Before considering the implications of this dispersion relation, we complicate the equilibrium a little. Let us suppose that, in addition to the net shearing force per

dispersion relation for flexural waves on a stretched beam

unit length $-D\partial^4\eta/\partial x^4$, the beam is also held under a tensile force T. We can then combine the two wave equations (12.29) and (12.32) to obtain

$$-D\frac{\partial^4\eta}{\partial x^4} + T\frac{\partial^2\eta}{\partial x^2} = \Lambda\frac{\partial^2\eta}{\partial t^2}, \tag{12.34}$$

for which the dispersion relation is

$$\omega^2 = C_S^2 k^2 \left(1 + \frac{k^2}{k_c^2}\right), \tag{12.35}$$

where $C_S = \sqrt{T/\Lambda}$ is the wave speed when the flexural rigidity D is negligible so the beam is string-like, and

$$k_c = \sqrt{T/D} \tag{12.36}$$

is a critical wave number. If the average strain induced by the tension is $\epsilon = T/(Ewh)$, then $k_c = (12\epsilon)^{1/2}h^{-1}$, where h is the thickness of the beam, w is its width, and we have used Eq. (12.31). [Notice that k_c is equal to 1/(the lengthscale on which a pendulum's support wire—analog of our beam—bends), as discussed in Ex. 11.11.] For short wavelengths $k \gg k_c$, the shearing force dominates, and the beam behaves like a tension-free beam; for long wavelengths $k \ll k_c$, it behaves like a string.

phase and group velocities for flexural waves on a stretched beam

A consequence of dispersion is that waves with different wave numbers k propagate with different speeds, and correspondingly, the group velocity $V_g = d\omega/dk$ with which wave packets propagate differs from the phase velocity $V_{\rm ph} = \omega/k$ with which a wave's crests and troughs move (see Sec. 7.2.2). For the dispersion relation (12.35), the phase and group velocities are

$$V_{\rm ph} \equiv \omega/k = C_S(1 + k^2/k_c^2)^{1/2},$$
$$V_g \equiv d\omega/dk = C_S(1 + 2k^2/k_c^2)(1 + k^2/k_c^2)^{-1/2}. \tag{12.37}$$

As we discussed in detail in Sec. 7.2.2 and Ex. 7.2, for dispersive waves such as this one, the fact that different Fourier components in the wave packet propagate with different speeds causes the packet to gradually spread; we explore this quantitatively for longitudinal waves on a beam in Ex. 12.5.

EXERCISES

Exercise 12.5 *Derivation: Dispersion of Flexural Waves*
Verify Eqs. (12.35) and (12.37). Sketch the dispersion-induced evolution of a Gaussian wave packet as it propagates along a stretched beam.

Exercise 12.6 *Problem: Speeds of Elastic Waves*
Show that the sound speeds for the following types of elastic waves in an isotropic material are in the ratios $1 : (1-\nu^2)^{-1/2} : \left(\frac{1-\nu}{(1+\nu)(1-2\nu)}\right)^{1/2} : [2(1+\nu)]^{-1/2} : [2(1+\nu)]^{-1/2}$. The elastic waves are (i) longitudinal waves along a rod, (ii) longitudinal waves along a sheet, (iii) longitudinal waves along a rod embedded in an

incompressible fluid, (iv) shear waves in an extended solid, and (v) torsional waves along a rod. [Note: Here and elsewhere in this book, if you encounter grungy algebra (e.g., frequent conversions from $\{K, \mu\}$ to $\{E, \nu\}$), do not hesitate to use Mathematica, Matlab, Maple, or other symbolic manipulation software to do the algebra!]

Exercise 12.7 *Problem: Xylophones*
Consider a beam of length ℓ, whose weight is negligible in the elasticity equations, supported freely at both ends (so the slope of the beam is unconstrained at the ends). Show that the frequencies of standing flexural waves satisfy

$$\omega = \left(\frac{n\pi}{\ell}\right)^2 \left(\frac{D}{\rho A}\right)^{1/2},$$

where A is the cross sectional area, and n is an integer. Now repeat the exercise when the ends are clamped. Based on your result, explain why xylophones don't have clamped ends.

12.3.5 Bifurcation of Equilibria and Buckling (Once More)

We conclude this discussion by returning to the problem of buckling, which we introduced in Sec. 11.6. The example we discussed there was a playing card compressed until it wants to buckle. We can analyze small dynamical perturbations of the card, $\eta(x, t)$, by treating the tension T of the previous section as negative, $T = -F$, where F is the compressional force applied to the card's two ends in Fig. 11.11. Then the equation of motion (12.34) becomes

$$-D\frac{\partial^4 \eta}{\partial x^4} - F\frac{\partial^2 \eta}{\partial x^2} = \Lambda\frac{\partial^2 \eta}{\partial t^2}. \tag{12.38}$$

We seek solutions for which the ends of the playing card are held fixed (as shown in Fig. 11.11): $\eta = 0$ at $x = 0$ and at $x = \ell$. Solving Eq. (12.38) by separation of variables, we see that

$$\eta = \eta_n \sin\left(\frac{n\pi}{\ell}x\right) e^{-i\omega_n t}. \tag{12.39}$$

Here $n = 1, 2, 3, \ldots$ labels the card's modes of oscillation, $n - 1$ is the number of nodes in the card's sinusoidal shape for mode n, η_n is the amplitude of deformation for mode n, and the mode's eigenfrequency ω_n (of course) satisfies the same dispersion relation (12.35) as for waves on a long, stretched beam, with $T \to -F$ and $k \to n\pi/\ell$:

eigenfrequencies for normal modes of a compressed beam

$$\omega_n^2 = \frac{1}{\Lambda}\left(\frac{n\pi}{\ell}\right)^2\left[\left(\frac{n\pi}{\ell}\right)^2 D - F\right] = \frac{1}{\Lambda}\left(\frac{n\pi}{\ell}\right)^2 \left(n^2 F_{\text{crit}} - F\right), \tag{12.40}$$

where $F_{\text{crit}} = \pi^2 D/\ell^2$ is the critical force that we introduced in Chap. 11 [Eq. (11.54)].

Consider the lowest normal mode, $n = 1$, for which the playing card is bent in the single-arch manner of Fig. 11.11 as it oscillates. When the compressional force F is small, ω_1^2 is positive, so ω_1 is real and the normal mode oscillates sinusoidally and stably. But for $F > F_{\text{crit}} = \pi^2 D/\ell^2$, ω_1^2 is negative, so ω_1 is imaginary and there are two normal-mode solutions, one decaying exponentially with time, $\eta \propto \exp(-|\omega_1|t)$, and the other increasing exponentially with time, $\eta \propto \exp(+|\omega_1|t)$, signifying an instability against buckling.

onset of buckling for a compressed beam

Notice that the onset of instability occurs at the same compressional force, $F = F_{\text{crit}}$, as the bifurcation of equilibria [Eq. (11.54)], where a new (bent) equilibrium state for the playing card comes into existence. Notice, moreover, that the card's $n = 1$ normal mode has zero frequency, $\omega_1 = 0$, at this onset of instability and bifurcation of equilibria; the card can bend by an amount that grows linearly in time, $\eta = A \sin(\pi x/\ell) \, t$, with no restoring force or exponential growth. This zero-frequency motion leads the card from its original, straight equilibrium shape, to its new, bent equilibrium shape.

This is an example of a very general phenomenon, which we shall meet again in fluid mechanics (Sec. 15.6.1). For mechanical systems without dissipation (no energy losses to friction, viscosity, radiation, or anything else), as one gradually changes some "control parameter" (in this case the compressional force F), there can occur bifurcation of equilibria. At each bifurcation point, a normal mode of the original equilibrium becomes unstable, and at its onset of instability the mode has zero frequency and represents a motion from the original equilibrium (which is becoming unstable) to the new, stable equilibrium.

zero-frequency mode at bifurcation of equilibria

In our simple playing-card example, we see this phenomenon repeated again and again as the control parameter F is increased. One after another, at $F = n^2 F_{\text{crit}}$, the modes $n = 1, n = 2, n = 3, \ldots$ become unstable. At each onset of instability, ω_n vanishes, and the zero-frequency mode (with $n - 1$ nodes in its eigenfunction) leads from the original, straight-card equilibrium to the new, stable, $(n - 1)$-noded, bent equilibrium.

12.4 Body Waves and Surface Waves—Seismology and Ultrasound

In Sec. 12.2, we derived the dispersion relations $\omega = C_L k$ and $\omega = C_T k$ for longitudinal and transverse elastodynamic waves in uniform media. We now consider how the waves are modified in an inhomogeneous, finite body: Earth. Earth is well approximated as a sphere of radius $R \sim 6{,}000$ km and mean density $\bar{\rho} \sim 6{,}000$ kg m^{-3}. The outer crust, extending down to 5–10 km below the ocean floor and comprising rocks of high tensile strength, rests on a more malleable mantle, the two regions being separated by the famous Mohorovičić (or Moho for short) discontinuity. Underlying the mantle is an outer core mainly composed of liquid iron, which itself surrounds a denser, solid inner core of mostly iron; see Table 12.1 and Fig. 12.5 below.

The pressure in Earth's interior is much larger than atmospheric, and the rocks are therefore quite compressed. Their atomic structure cannot be regarded as a small

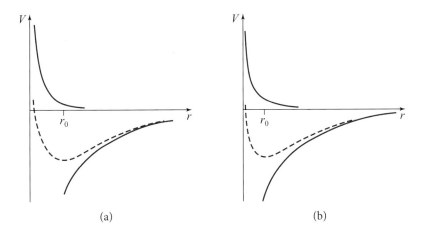

FIGURE 12.3 Potential energy curves (dashed) for nearest neighbors in a crystal lattice. (a) At atmospheric (effectively zero) pressure, the equilibrium spacing is set by the minimum in the potential energy, which is a combination of hard electrostatic repulsion by the nearest neighbors (upper solid curve) and a softer overall attraction associated with all the nearby ions (lower solid curve). (b) At much higher pressure, the softer, attractive component is moved inward, and the equilibrium spacing is greatly reduced. The bulk modulus is proportional to the curvature of the potential energy curve at its minimum and so is considerably increased.

perturbation from their structure in a vacuum. Nevertheless, we can still use linear elasticity theory to discuss small perturbations about their equilibrium. This is because the crystal lattice has had plenty of time to establish a new equilibrium with a much smaller lattice spacing than at atmospheric pressure (Fig. 12.3). The density of lattice defects and dislocations will probably not differ appreciably from those at atmospheric pressure, so the proportionality limit and yield stress should be about the same as for rocks near Earth's surface. The linear stress-strain relation will still apply below the proportionality limit, although the elastic moduli are much greater than those measured at atmospheric pressure.

We can estimate the magnitude of the pressure P in Earth's interior by idealizing the planet as an isotropic medium with negligible shear stress, so its stress tensor is like that of a fluid, $\mathbf{T} = P\mathbf{g}$ (where $\mathbf{g}$ is the metric tensor). Then the equation of static equilibrium takes the form

$$\frac{dP}{dr} = -g\rho, \qquad (12.41)$$

where ρ is density, and $g(r)$ is the acceleration of gravity at radius r. This equation can be approximated by

$$P \sim \bar{\rho} g R \sim 300 \text{ GPa} \sim 3 \times 10^6 \text{ atmospheres}, \qquad (12.42)$$

pressure at Earth's center

where g is now the acceleration of gravity at Earth's surface $r = R$, and $\bar{\rho}$ is Earth's mean density. This agrees well numerically with the accurate value of 360 GPa

12.4 Body Waves and Surface Waves—Seismology and Ultrasound

bulk modulus related to equation of state

at Earth's center. The bulk modulus produces the isotropic pressure $P = -K\Theta$ [Eq. (11.18)]; and since $\Theta = -\delta\rho/\rho$ [cf. Eq. (12.3)], the bulk modulus can be expressed as

$$K = \frac{dP}{d\ln\rho}. \qquad (12.43)$$

[Strictly speaking, we should distinguish between adiabatic and isothermal variations in Eq. (12.43), but the distinction is small for solids; see Sec. 11.3.5. It is significant for gases.] Typically, the bulk modulus inside Earth is 4 to 5 times the pressure, and the shear modulus in the crust and mantle is about half the bulk modulus.

12.4.1 Body Waves

Virtually all our direct information about the internal structure of Earth comes from measurements of the propagation times of elastic waves that are generated by earthquakes and propagate through Earth's interior (*body waves*). There are two fundamental kinds of body waves: the longitudinal and transverse modes of Sec. 12.2. These are known in seismology as P-modes (P for pressure) and S-modes (S for shear), respectively. The two polarizations of the transverse shear waves are designated SH and SV, where H and V stand for "horizontal" and "vertical" displacements (i.e., displacements orthogonal to **k** that are fully horizontal, or that are obtained by projecting the vertical direction $\mathbf{e}_z$ orthogonal to $\hat{\mathbf{k}}$).

P-modes and S-modes

We shall first be concerned with what seismologists call high-frequency (of order 1 Hz) modes, which leads to three related simplifications. As typical wave speeds lie in the range 3–14 km s^{-1}, the wavelengths lie in the range ~1 to 10 km, which is generally small compared with the distance over which gravity causes the pressure to change significantly—the *pressure scale height*. It turns out that we then can ignore the effects of gravity on the propagation of small perturbations. In addition, we can regard the medium locally as effectively homogeneous and so use the local dispersion relations $\omega = C_{L,T} k$. Finally, as the wavelengths are short, we can trace rays through Earth using geometrical optics (Sec. 7.3).

Now, Earth is quite inhomogeneous globally, and the sound speeds therefore vary significantly with radius; see Table 12.1. To a fair approximation, Earth is horizontally stratified below the outer crust (whose thickness is irregular). Two types of variation can be distinguished, the abrupt and the gradual. There are several abrupt changes in the crust and mantle (including the Moho discontinuity at the interface between crust and mantle), and also at the transitions between mantle and outer core, and between outer core and inner core. At these *surfaces of discontinuity*, the density and elastic constants apparently change over distances short compared with a wavelength. Seismic waves incident on these discontinuities behave like light incident on the surface of a glass plate; they can be reflected and refracted. In addition, as there are now two different waves with different phase speeds, it is possible to generate SV waves from pure P waves and vice versa at a discontinuity (Fig. 12.4). However, this

surfaces of discontinuity

TABLE 12.1: Typical outer radii (R), densities (ρ), bulk moduli (K), shear moduli (μ), P-wave speeds (C_P), and S-wave speeds (C_S) in different zones of Earth

Zone	R (10^3 km)	ρ (10^3 kg m^{-3})	K (GPa)	μ (GPa)	C_P (km s^{-1})	C_S (km s^{-1})
Inner core	1.2	13	1,400	160	11	2
Outer core	3.5	10–12	600–1,300	—	8–10	—
Mantle	6.35	3–5	100–600	70–250	8–14	5–7
Crust	6.37	3	50	30	6–7	3–4
Ocean	6.37	1	2	—	1.5	—

Notes: Note the absence of shear waves (denoted by a —) in the fluid regions. Adapted from Stacey (1977).

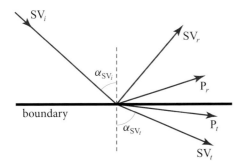

FIGURE 12.4 An incident shear wave polarized in the vertical direction (SV$_i$), incident from above on a surface of discontinuity, produces both a longitudinal (P) wave and an SV wave in reflection and in transmission. If the wave speeds increase across the boundary (the case shown), then the transmitted waves, SV$_t$, P$_t$, will be refracted away from the vertical. A shear mode, SV$_r$, will be reflected at the same angle as the incident wave. However, the reflected P mode, P$_r$, will be reflected at a greater angle to the vertical, as it has greater speed than the incident wave.

wave-wave mixing is confined to SV and P; the SH waves do not mix with SV or P; see Ex. 12.9.

The junction conditions that control this wave mixing and all other details of the waves' behavior at a surface of discontinuity are: (i) the displacement $\boldsymbol{\xi}$ must be continuous across the discontinuity (otherwise infinite strain and infinite stress would develop there); and (ii) the net force acting on an element of surface must be zero (otherwise the surface, having no mass, would have infinite acceleration), so the force per unit area acting from the front face of the discontinuity to the back must be balanced by that acting from the back to the front. If we take the unit normal to the horizontal discontinuity to be $\mathbf{e}_z$, then these boundary conditions become

$$\boxed{[\xi_j] = [T_{jz}] = 0,} \qquad (12.44)$$

wave boundary conditions at interfaces

12.4 Body Waves and Surface Waves—Seismology and Ultrasound

where the notation $[X]$ signifies the difference in X across the boundary, and the j is a vector index. (For an alternative, more formal derivation of $[T_{jz}] = 0$, see Ex. 12.8.)

One consequence of these boundary conditions is Snell's law for the directions of propagation of the waves. Since these continuity conditions must be satisfied all along the surface of discontinuity and at all times, the phase $\phi = \mathbf{k} \cdot \mathbf{x} - \omega t$ of the wave must be continuous across the surface at all locations $\mathbf{x}$ on it and all times, which means that the phase ϕ must be the same on the surface for all transmitted waves and all reflected waves as for the incident waves. This is possible only if the frequency ω, the horizontal wave number $k_H = k \sin\alpha$, and the horizontal phase speed $C_H = \omega/k_H = \omega/(k \sin\alpha)$ are the same for all the waves. (Here $k_H = k \sin\alpha$ is the magnitude of the horizontal component of a wave's propagation vector, and α is the angle between its propagation direction and the vertical; cf. Fig. 12.4.) Thus we arrive at Snell's law: for every reflected or transmitted wave J, the horizontal phase speed must be the same as for the incident wave:

Snell's law for waves at interfaces

$$\boxed{\frac{C_J}{\sin\alpha_J} = C_H \text{ is the same for all } J.} \tag{12.45}$$

It is straightforward though tedious to compute the reflection and transmission coefficients (e.g., the strength of a transmitted P-wave produced by an incident SV wave) for the general case using the boundary conditions (12.44) and (12.45) (see, e.g., Eringen and Suhubi, 1975, Sec. 7.7). For the very simplest of examples, see Ex. 12.10.

In the regions between the discontinuities, the pressures (and consequently, the elastic moduli) increase steadily, over many wavelengths, with depth. The elastic moduli generally increase more rapidly than the density does, so the wave speeds generally also increase with depth (i.e., $dC/dr < 0$). This radial variation in C causes the rays along which the waves propagate to bend. The details of this bending are governed by Hamilton's equations, with the hamiltonian $\Omega(\mathbf{x}, \mathbf{k})$ determined by the simple nondispersive dispersion relation $\Omega = C(\mathbf{x})k$ (Sec. 7.3.1). Hamilton's equations in this case reduce to the simple ray equation (7.48), which (since the index of refraction is $\propto 1/C$) can be rewritten as

geometric optics ray equation for elastodynamic waves in inhomogeneous elastic medium

$$\frac{d}{ds}\left(\frac{1}{C}\frac{d\mathbf{x}}{ds}\right) = \boldsymbol{\nabla}\left(\frac{1}{C}\right). \tag{12.46}$$

Here s is distance along the ray, so $d\mathbf{x}/ds = \mathbf{n}$ is the unit vector tangent to the ray. This ray equation can be reexpressed in the following form:

$$d\mathbf{n}/ds = -(\boldsymbol{\nabla}\ln C)_\perp, \tag{12.47}$$

where the subscript $\perp$ means "projected perpendicular to the ray;" and this in turn means that the ray bends *away* from the direction in which C increases (i.e., it bends upward inside Earth, since C increases downward) with the radius of curvature of the bend given by

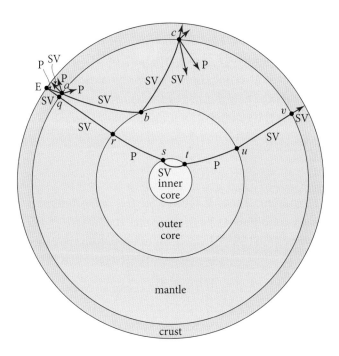

FIGURE 12.5 Seismic wave propagation in a schematic Earth model. A P wave made by an earthquake, E, propagates to the crust-mantle boundary at a where it generates two transmitted waves (SV and P) and two reflected waves (SV and P). The transmitted SV wave propagates along a ray that bends upward a bit (geometric-optics bending) and hits the mantle–outer-core boundary at b. There can be no transmitted SV wave at b, because the outer core is fluid; there can be no transmitted or reflected P wave, because the angle of incidence of the SV wave is too great. So the SV wave is perfectly reflected. It then travels along an upward curving ray, to the crust-mantle interface at c, where it generates four waves, two of which hit Earth's surface. The earthquake E also generates an SV wave traveling almost radially inward, through the crust-mantle interface at q, to the mantle–outer-core interface at r. Because the outer core is liquid, it cannot support an SV wave, so only a P wave is transmitted into the outer core at r. That P wave propagates to the interface with the inner core at s, where it regenerates an SV wave along with the transmitted and reflected P waves (not shown). The SV wave refracts back upward in the inner core and generates a P wave at the interface with the outer core t. That P wave propagates through the liquid outer core to u, where it generates an SV wave along with its transmitted and reflected P waves (not shown); that SV wave travels nearly radially outward, through v and to Earth's surface.

$$\mathcal{R} = \frac{1}{|(\nabla \ln C)_\perp|} = \frac{1}{|(d \ln C/dr) \sin \alpha|}. \tag{12.48}$$

Here α is the angle between the ray and the radial direction.

Figure 12.5 shows schematically the propagation of seismic waves through Earth. At each discontinuity in Earth's material, Snell's law governs the directions of the reflected and transmitted waves. As an example, note from Eq. (12.45) that an SV mode incident on a boundary cannot generate any P mode when its angle of incidence exceeds $\sin^{-1}(C_{Ti}/C_{Lt})$. (Here we use the standard notation C_T for the phase speed of an S wave and C_L for that of a P wave.) This is what happens at point b in Fig. 12.5.

seismic wave propagation

12.4 Body Waves and Surface Waves—Seismology and Ultrasound

EXERCISES

Exercise 12.8 *Derivation: Junction Condition at a Discontinuity*
Derive the junction condition $[T_{jz}] = 0$ at a horizontal discontinuity between two media by the same method as one uses in electrodynamics to show that the normal component of the magnetic field must be continuous: Integrate the equation of motion $\rho d\mathbf{v}/dt = -\nabla \cdot \mathbf{T}$ over the volume of an infinitesimally thin pill box centered on the boundary (see Fig. 11.7), and convert the volume integral to a surface integral via Gauss's theorem.

Exercise 12.9 *Derivation: Wave Mixing at a Surface of Discontinuity*
Using the boundary conditions (12.44), show that at a surface of discontinuity inside Earth, SV and P waves mix, but SH waves do not mix with the other waves.

Exercise 12.10 *Example: Reflection and Transmission of Normal, Longitudinal Waves at a Boundary*
Consider a longitudinal elastic wave incident normally on the boundary between two media, labeled 1 and 2. By matching the displacement and the normal component of stress at the boundary, show that the ratio of the transmitted wave amplitude to the incident amplitude is given by

$$t = \frac{2Z_1}{Z_1 + Z_2},$$

where $Z_{1,2} = [\rho_{1,2}(K_{1,2} + 4\mu_{1,2}/3)]^{1/2}$ is known as the *acoustic impedance*. (The impedance is independent of frequency and is just a characteristic of the material.) Likewise, evaluate the amplitude reflection coefficient, and verify that wave energy flux is conserved.

12.4.2 Edge Waves

One phenomenon that is important in seismology (and also occurs in plasmas; Ex. 21.17) but is absent for many other types of wave motion is *edge waves*, i.e., waves that propagate along a discontinuity in the elastic medium. An important example is *surface waves*, which propagate along the surface of a medium (e.g., Earth's surface) and that decay exponentially with depth. Waves with such exponential decay are sometimes called *evanescent*.

The simplest type of surface wave is the *Rayleigh wave*, which propagates along the surface of an elastic medium. We analyze Rayleigh waves for the idealization of a *plane semi-infinite solid*—also sometimes called a *homogeneous half-space*. When it is applied to Earth, this discussion must be modified to allow for both the density stratification and (if the wavelength is sufficiently long) the surface curvature. However, the qualitative character of the mode is unchanged.

Rayleigh waves are an intertwined mixture of P and SV waves. When analyzing them, it is useful to resolve their displacement vector $\boldsymbol{\xi}$ into a sum of a (longitudinal) P-wave component $\boldsymbol{\xi}^L$ and a (transverse) SV-wave component $\boldsymbol{\xi}^T$.

Consider a semi-infinite elastic medium, and introduce a local Cartesian coordinate system with $\mathbf{e}_z$ normal to the surface, $\mathbf{e}_x$ lying in the surface, and the propagation vector $\mathbf{k}$ in the $\mathbf{e}_z$-$\mathbf{e}_x$ plane. The propagation vector has a real component along the horizontal ($\mathbf{e}_x$) direction, corresponding to true propagation, and an imaginary component along the $\mathbf{e}_z$ direction, corresponding to an exponential decay of the amplitude as one goes down into the medium. For the longitudinal (P-wave) and transverse (SV-wave) parts of the wave to remain in phase with each other as they propagate along the boundary, they must have the same values of the frequency ω and horizontal wave number k_x. However, there is no reason why their vertical e-folding lengths should be the same (i.e., why their imaginary k_z values should be the same). We therefore denote their imaginary k_zs by $-iq_L$ for the longitudinal (P-wave) component and $-iq_T$ for the transverse (S-wave) component, and we denote their common k_x by k.

First focus attention on the longitudinal part of the wave. Its displacement can be written as the gradient of a scalar [Eq. (12.7b)], $\boldsymbol{\xi}_L = \nabla(\psi_0 e^{q_L z + i(kx - \omega t)})$:

$$\xi_x^L = ik\psi_0 e^{q_L z + i(kx-\omega t)}, \quad \xi_z^L = q_L \psi_0 e^{q_L z + i(kx-\omega t)}, \quad z \leq 0. \qquad (12.49)$$

Substituting into the general dispersion relation $\omega^2 = C_L^2 \mathbf{k}^2$ for longitudinal waves, we obtain

$$q_L = \left(k^2 - \frac{\omega^2}{C_L^2}\right)^{1/2}. \qquad (12.50)$$

Because the transverse part of the wave is divergence free, the wave's expansion comes entirely from the longitudinal part, $\Theta = \nabla \cdot \boldsymbol{\xi}_L$, and is given by

$$\Theta = (q_L^2 - k^2)\psi_0 e^{q_L z + i(kx-\omega t)}. \qquad (12.51)$$

The transverse (SV) part of the wave can be written as the curl of a vector potential [Eq. (12.7c)], which we can take to point in the y direction, $\boldsymbol{\xi}_T = \nabla \times (A_0 \mathbf{e}_y e^{q_T z + i(kx-\omega t)})$:

$$\xi_x^T = -q_T A_0 e^{q_T z + i(kx-\omega t)}, \quad \xi_z^T = ik A_0 e^{q_T z + i(kx-\omega t)}, \quad z \leq 0. \qquad (12.52)$$

The dispersion relation $\omega^2 = C_T^2 \mathbf{k}^2$ for transverse waves tells us that

$$q_T = \left(k^2 - \frac{\omega^2}{C_T^2}\right)^{1/2}. \qquad (12.53)$$

We next impose boundary conditions at the surface. Since the surface is free, there will be no force acting on it:

$$\mathbf{T} \cdot \mathbf{e}_z|_{z=0} = 0, \qquad (12.54)$$

which is a special case of the general boundary condition (12.44). (Note that we can evaluate the stress at the unperturbed surface rather than at the displaced surface as we are only working to linear order.) The tangential stress is $T_{xz} = -2\mu(\xi_{z,x} + \xi_{x,z}) = 0$, so its vanishing is equivalent to

$$\xi_{z,x} + \xi_{x,z} = 0 \quad \text{at } z = 0. \tag{12.55}$$

Inserting $\xi = \xi_L + \xi_T$ and Eqs. (12.49) and (12.52) for the components and solving for the ratio of the transverse amplitude to the longitudinal amplitude, we obtain

$$\frac{A_0}{\psi_0} = \frac{2ikq_L}{k^2 + q_T^2}. \tag{12.56}$$

The normal stress is $T_{zz} = -K\Theta - 2\mu(\xi_{z,z} - \frac{1}{3}\Theta) = 0$, so from the values $C_L^2 = (K + \frac{4}{3}\mu)/\rho$ and $C_T^2 = \mu/\rho$ of the longitudinal and transverse sound speeds, which imply $K/\mu = (C_L/C_T)^2 - \frac{4}{3}$, we deduce that vanishing T_{zz} is equivalent to

$$(1 - 2\kappa)\Theta + 2\kappa\xi_{z,z} = 0 \quad \text{at } z = 0, \quad \text{where} \quad \kappa \equiv \frac{C_T^2}{C_L^2} = \frac{1 - 2\nu}{2(1 - \nu)}. \tag{12.57}$$

Here we have used Eqs. (11.39) to express the speed ratio in terms of Poisson's ratio ν. Inserting Θ from Eq. (12.51) and $\xi_z = \xi_z^L + \xi_z^T$ with components from Eqs. (12.49) and (12.52), using the relation $q_L^2 - k^2 + 2\kappa k^2 = \kappa(q_T^2 + k^2)$ [which follows from Eqs. (12.50), (12.53), and (12.57)], and solving for the ratio of amplitudes, we obtain

$$\frac{A_0}{\psi_0} = \frac{k^2 + q_T^2}{-2ikq_T}. \tag{12.58}$$

The dispersion relation for Rayleigh waves is obtained by equating the two expressions (12.56) and (12.58) for the amplitude ratio:

$$4k^2 q_T q_L = (k^2 + q_T^2)^2. \tag{12.59}$$

We can express this dispersion relation more explicitly in terms of the ratio of the Rayleigh-wave speed $C_R = \omega/k$ to the transverse-wave speed C_T:

$$\zeta = \left(\frac{\omega}{C_T k}\right)^2 = \left(\frac{C_R}{C_T}\right)^2. \tag{12.60}$$

By inserting $q_T = k\sqrt{1 - \zeta}$ [from Eqs. (12.53) and (12.60)] and $q_L = k\sqrt{1 - \kappa\zeta}$ [from Eqs. (12.50), (12.57), and (12.60)], and expressing κ in terms of Poisson's ratio via Eq. (12.57), we bring Eq. (12.59) into the explicit form

dispersion relation for Rayleigh wave

$$\zeta^3 - 8\zeta^2 + 8\left(\frac{2 - \nu}{1 - \nu}\right)\zeta - \frac{8}{(1 - \nu)} = 0. \tag{12.61}$$

phase speed of Rayleigh wave

This dispersion relation is a third-order polynomial in $\zeta \propto \omega^2$ with just one positive real root $\zeta(\nu)$, which we plot in Fig. 12.6. From that plot, we see that for a Poisson's ratio characteristic of rocks, $0.2 \lesssim \nu \lesssim 0.3$, *the phase speed of a Rayleigh wave is roughly 0.9 times the speed of a pure shear wave*; cf. Fig. 12.6.

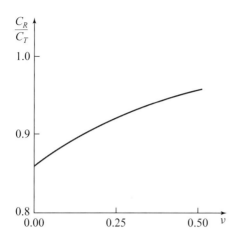

FIGURE 12.6 Solution $C_R/C_T = \sqrt{\zeta}$ of the dispersion relation (12.61) as a function of Poisson's ratio ν.

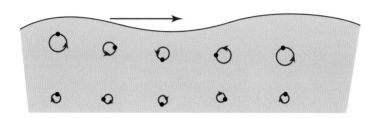

FIGURE 12.7 Rayleigh waves in a semi-infinite elastic medium.

The displacement vector for Rayleigh waves is the sum of Eqs. (12.49) and (12.52) with the amplitudes related by (12.56):

$$\xi_x = ik\psi_o \left[e^{q_L z} - \frac{2q_L q_T}{k^2 + q_T^2} e^{q_T z} \right] e^{i(kx-\omega t)},$$

$$\xi_z = q_L \psi_o \left[e^{q_L z} - \frac{2k^2}{k^2 + q_T^2} e^{q_T z} \right] e^{i(kx-\omega t)}. \quad (12.62)$$

material displacement in Rayleigh wave

Equation (12.62) represents a backward rotating, elliptical motion for each fluid element near the surface (as depicted in Fig. 12.7), reversing to a forward rotation at depths where the sign of ξ_x has flipped.

Rayleigh waves propagate around the surface of Earth rather than penetrate its interior. However, our treatment is inadequate, because their wavelengths—typically 1–10 km if generated by an earthquake—are not necessarily small compared with the scale heights in the outer crust over which C_S and C_T vary. Our wave equation has to be modified to include these vertical gradients.

Love waves

This vertical stratification has an important additional consequence. Ignoring these gradients, if we attempt to find an orthogonal surface mode just involving SH waves, we find that we cannot simultaneously satisfy the surface boundary conditions on displacement and stress with a single evanescent wave. We need two modes to do this. However, when we allow for stratification, the strong refraction allows an SH surface wave to propagate. This is known as a *Love wave*. The reason for its practical importance is that seismic waves are also created by underground nuclear explosions, and it is important to be able to distinguish explosion-generated waves from earthquake waves. An earthquake is usually caused by the transverse slippage of two blocks of crust across a fault line. It is therefore an efficient generator of shear modes, including Love waves. By contrast, explosions involve radial motions away from the point of explosion and are inefficient emitters of Love waves. This allows these two sources of seismic disturbance to be distinguished.

EXERCISES

Exercise 12.11 *Example: Earthquakes*

The magnitude M of an earthquake, on modern variants of the *Richter scale*, is a quantitative measure of the strength of the seismic waves it creates. The earthquake's seismic-wave energy release can be estimated using a rough semi-empirical formula due to Båth (1966):

earthquake magnitude
$$E = 10^{5.24+1.44M} \text{J}. \tag{12.63}$$

The largest earthquakes have magnitude ~ 9.5.

One type of earthquake is caused by slippage along a fault deep in the crust. Suppose that most of the seismic power in an earthquake with $M \sim 8.5$ is emitted at frequencies ~ 1 Hz and that the quake lasts for a time $T \sim 100$ s. If C is an average wave speed, then it is believed that the stress is relieved over an area of fault of length $\sim CT$ and a depth of order one wavelength (Fig. 12.8). By comparing the stored elastic energy with the measured energy release, make an estimate of the minimum strain prior to the earthquake. Is your answer reasonable? Hence estimate the typical displacement during the earthquake in the vicinity of the fault. Make an order-of-magnitude estimate of the acceleration measurable by a seismometer in the next state and in the next continent. (Ignore the effects of density stratification, which are actually quite significant.)

12.4.3 Green's Function for a Homogeneous Half-Space

To gain insight into the combination of waves generated by a localized source, such as an explosion or earthquake, it is useful to examine the Green's function for excitations in a homogeneous half-space. Physicists define the Green's function $G_{jk}(\mathbf{x}, t; \mathbf{x}', t')$ to be the displacement response $\xi_j(\mathbf{x}, t)$ to a unit delta-function force in the $\mathbf{e}_k$ direction at location $\mathbf{x}'$ and time t': $\mathbf{F} = \delta(\mathbf{x} - \mathbf{x}')\delta(t - t')\mathbf{e}_k$. Geophysicists sometimes find

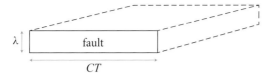

FIGURE 12.8 Earthquake: The region of a fault that slips (solid rectangle), and the volume over which the strain is relieved, on one side of the fault (dashed region).

it useful to work, instead, with the "Heaviside Green's function," $G_{jk}^H(\mathbf{x}, t; \mathbf{x}', t')$, which is the displacement response $\xi_j(\mathbf{x}, t)$ to a unit step-function force (one that turns on to unit strength and remains forever constant afterward) at $\mathbf{x}'$ and t': $\mathbf{F} = \delta(\mathbf{x} - \mathbf{x}')H(t - t')\mathbf{e}_k$. Because $\delta(t - t')$ is the time derivative of the Heaviside step function $H(t - t')$, *the Heaviside Green's function is the time integral of the physicists' Green's function*. The Heaviside Green's function has the advantage that one can easily see the size of the step functions it contains, by contrast with the size of the delta functions contained in the physicists' Green's function.

Heaviside Green's function for elastodynamic waves

It is a rather complicated task to compute the Heaviside Green's function, and geophysicists have devoted much effort to doing so. We shall not give the details of such computations, but merely show the function graphically in Fig. 12.9 for an instructive situation: the displacement produced by a step-function force in a homogeneous half-space with the observer at the surface and the force at two different locations: a point nearly beneath the observer (Fig. 12.9a), and a point close to the surface and some distance away in the x direction (Fig. 12.9b).

Several features of this Heaviside Green's function deserve note:

Properties of Heaviside Green's function

- Because of their relative propagation speeds, the P waves arrive at the observer first, then (about twice as long after the excitation) the S waves, and shortly thereafter the Rayleigh waves. From the time interval ΔT between the first P waves and first S waves, one can estimate the distance to the source: $\ell \simeq (C_P - C_S)\Delta T \sim 3(\Delta T/\text{s})$ km.
- For the source nearly beneath the observer (Fig. 12.9a), there is no sign of any Rayleigh wave, whereas for the source close to the surface, the Rayleigh wave is the strongest feature in the x and z (longitudinal and vertical) displacements but is absent from the y (transverse) displacement. From this, one can infer the waves' propagation direction.
- The y (transverse) component of force produces a transverse displacement that is strongly concentrated in the S wave.
- The x and z (longitudinal and vertical) components of force produce x and z displacements that include P waves, S waves, and (for the source near the surface) Rayleigh waves.

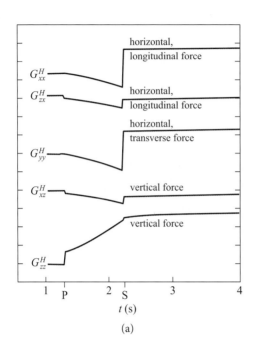
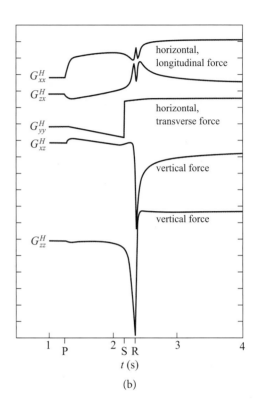

FIGURE 12.9 The Heaviside Green's function (displacement response to a step-function force) in a homogeneous half-space. The observer is at the surface. The force is applied at a point in the x-z plane, with a direction stated in words and also given by the second index of G^H; the displacement direction is given by the first index of G^H. The longitudinal and transverse speeds are $C_L = 8.00$ km s^{-1} and $C_S = 4.62$ km s^{-1}, and the density is 3.30×10^3 kg m^{-3}. For a force of 1 Newton, a division on the vertical scale is 10^{-16} m. The moments of arrival of the P wave, S wave, and Rayleigh (R) wave from the moment the force is turned on are indicated on the horizontal axis. (a) The source is nearly directly beneath the observer, so the waves propagate nearly vertically upward. More specifically, the source is at 10 km depth and is 2 km distant along the horizontal x direction. (b) The source is close to the surface, and the waves propagate nearly horizontally (in the x direction). More specifically, the source is at 2 km depth and is 10 km distant along the horizontal x direction. Adapted from Johnson (1974, Figs. 2, 4).

- The gradually changing displacements that occur between the arrival of the turn-on P wave and the turn-on S wave are due to P waves that hit the surface some distance from the observer, and from there diffract to the observer as a mixture of P and S waves. Similarly for gradual changes of displacement after the turn-on S wave.

The complexity of seismic waves arises in part from the richness of features in this homogeneous-half-space Heaviside Green's function, in part from the influences of Earth's inhomogeneities, and in part from the complexity of an earthquake's or explosion's forces.

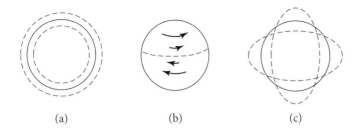

FIGURE 12.10 Displacements associated with three types of global modes for an elastic sphere, such as Earth. The polar axis points upward. (a) A radial mode shown on a cross section through the sphere's center; matter originally on the solid circle moves in and out between the two dashed circles. (b) An $l = 1$, $m = 0$ torsional mode; the arrows are proportional to the displacement vector on the sphere's surface at one moment of time. (c) An $l = 2$, $m = 0$ spheroidal mode shown on a cross section through the sphere's center; matter originally on the solid circle moves between the two ellipses.

12.4.4 Free Oscillations of Solid Bodies

When computing the dispersion relations for body (P- and S-wave) modes and surface (Rayleigh-wave) modes, we have assumed that the wavelength is small compared with Earth's radius; therefore the modes have a continuous frequency spectrum. However, it is also possible to excite global modes in which the whole Earth "rings" with a discrete spectrum. If we approximate Earth as spherically symmetric and ignore its rotation (whose period is long compared to a normal-mode period), then we can isolate three types of global modes: *radial*, *torsional*, and *spheroidal*.

types of normal modes of an elastic sphere

To compute the details of these modes, we can solve by separation of variables the equations of elastodynamics for the displacement vector in spherical polar coordinates. This is much like solving the Schrödinger equation for a central potential. See Ex. 12.12 for a relatively simple example. In general (as in that example), each of the three types of modes has a displacement vector $\boldsymbol{\xi}$ characterized by its own type of spherical harmonic.

The *radial modes* have spherically symmetric, purely radial displacements $\xi_r(r)$ and so have spherical harmonic order $l = 0$; see Fig. 12.10a and Ex. 12.12a.

radial modes

The *torsional modes* have vanishing radial displacements, and their nonradial displacements are proportional to the vector spherical harmonic $\hat{\mathbf{L}} Y_l^m(\theta, \phi)$, where θ, ϕ are spherical coordinates, Y_l^m is the scalar spherical harmonic, and $\hat{\mathbf{L}} = \mathbf{x} \times \boldsymbol{\nabla}$ is the angular momentum operator (aside from an omitted factor $\hbar/i$), which plays a major role in quantum theory. Spherical symmetry guarantees that these modes' eigenfrequencies are independent of m (because by reorienting the polar axis, the various m are mixed among one another), so $m = 0$ is representative. In this case, $\boldsymbol{\nabla} Y_l^0 \propto \boldsymbol{\nabla} P_l(\cos\theta)$ is in the θ direction, so $\hat{\mathbf{L}} Y_l^0$ is in the ϕ direction. Hence the only nonzero component of the displacement vector is

torsional modes

$$\xi_\phi \propto \partial P_l(\cos\theta)/\partial\theta = \sin\theta\, P_l'(\cos\theta). \tag{12.64}$$

(Here P_l is the Legendre polynomial and P_l' is the derivative of P_l with respect to its argument.) Therefore, in these modes alternate zones of different latitude oscillate in opposite directions (clockwise or counterclockwise at some chosen moment of time) in such a way as to conserve total angular momentum. See Fig. 12.10b and Ex. 12.12b. In the high-frequency limit, the torsional modes become SH waves (since their displacements are horizontal).

spheroidal modes

The *spheroidal modes* have radial displacements proportional to $Y_l^m(\theta, \phi)\mathbf{e}_r$, and they have nonradial components proportional to $\boldsymbol{\nabla} Y_l^m$. These two displacements can combine into a single mode, because they have the same parity (and opposite parity from the torsional modes) as well as the same spherical-harmonic orders l and m. The eigenfrequencies again are independent of m and thus can be studied by specializing to $m = 0$, in which case the displacements become

$$\xi_r \propto P_l(\cos\theta), \qquad \xi_\theta \propto \sin\theta\, P_l'(\cos\theta). \tag{12.65}$$

These displacements deform the sphere in a spheroidal manner for the special case $l = 2$ (Fig. 12.10c and Ex. 12.12c), whence their name "spheroidal." The radial modes are the special case $l = 0$ of these spheroidal modes. It is sometimes mistakenly asserted that there are no $l = 1$ spheroidal modes because of conservation of momentum. In fact, $l = 1$ modes do exist: for example, the central regions of the sphere can move up, while the outer regions move down. For Earth, the lowest-frequency $l = 2$ spheroidal mode has a period of 53 minutes and can ring for about 1,000 periods (i.e., its quality factor is $Q \sim 1{,}000$). This is typical for solid planets. In the high-frequency limit, the spheroidal modes become a mixture of P and SV waves.

solving elastodynamic equations for modes

When one writes the displacement vector $\boldsymbol{\xi}$ for a general vibration of Earth as a sum over these various types of global modes and inserts that sum into the wave equation (12.4b) (augmented, for greater realism, by gravitational forces), spherical symmetry of unperturbed Earth guarantees that the various modes will separate from one another. For each mode the wave equation will give a radial wave equation analogous to that for a hydrogen atom in quantum mechanics. The boundary condition $\mathbf{T}\cdot\mathbf{e}_r = 0$

mode spectra

at Earth's surface constrains the solutions of the radial wave equation for each mode to be a discrete set, which one can label by the number n of radial nodes that the mode possesses (just as for the hydrogen atom). The frequencies of the modes increase with both n and l. See Ex. 12.12 for details in a relatively simple case.

For small values of the quantum numbers l and n, the modes are quite sensitive to the model assumed for Earth's structure. For example, they are sensitive to whether one correctly includes the gravitational restoring force in the wave equation. However, for large l and n, the spheroidal and toroidal modes become standing combinations of P, SV, SH, Rayleigh, and Love waves, and therefore they are rather insensitive to the effects of gravity.

12.4.5 Seismic Tomography

Observations of all of these types of seismic waves clearly code much information about Earth's structure. Inverting the measurements to infer this structure has become a highly sophisticated and numerically intensive branch of geophysics—and also of oil exploration! The travel times of the P and S body waves can be measured between various pairs of points over Earth's surface and essentially allow C_L and C_T (and hence K/ρ and μ/ρ) to be determined as functions of radius inside Earth. Travel times are $\lesssim 1$ hour. Using this type of analysis, seismologists can infer the presence of hot and cold regions in the mantle and then infer how the rocks are circulating under the crust.

It is also possible to combine the observed travel times with Earth's equation of elastostatic equilibrium

$$\frac{dP}{dr} = -g(r)\rho, \quad g(r) = \frac{4\pi G}{r^2}\int_0^r r'^2 \rho(r')dr', \quad (12.66)$$

elastostatic equilibrium equation for gravitating, elastic sphere (e.g., Earth)

where $g(r)$ is the gravitational acceleration, to determine the distributions of density, pressure, and elastic constants. Measurements of Rayleigh and Love waves can be used to probe the surface layers. The results of this procedure are then input to obtain free oscillation frequencies (global mode frequencies), which compare well with the observations. The damping rates for the free oscillations furnish information on the interior viscosity.

12.4.6 Ultrasound; Shock Waves in Solids

Sound waves at frequencies above 20,000 Hz (where human hearing ends) are widely used in modern technology, especially for imaging and tomography. This is much like exploring Earth with seismic waves.

Just as seismic waves can travel from Earth's solid mantle through its liquid outer core and into its solid inner core, so these *ultrasonic waves* can travel through both solids and liquids, with reflection, refraction, and transmission similar to those in Earth.

Applications of ultrasound include, among others, medical imaging (e.g., of structures in the eye and of a human fetus during pregnancy), inspection of materials (e.g., of cracks, voids, and welding joints), acoustic microscopy at frequencies up to $\sim$3 GHz (e.g., by scanning acoustic microscopes), ultrasonic cleaning (e.g., of jewelry), and ultrasonic welding (with sonic vibrations creating heat at the interface of the materials to be joined).

When an ultrasonic wave (or any other sound wave) in an elastic medium reaches a sufficiently high amplitude, its high-amplitude regions propagate faster than its low-amplitude regions. This causes it to develop a *shock front*, in which the compression increases almost discontinuously. This nonlinear mechanism for shock formation in a solid is similar to that in a fluid, where we shall study the mechanism in detail (Sec. 17.4.1). The theory of the elastodynamic shock itself, especially *jump conditions across the shock*, is also similar to that in a fluid (Sec. 17.5.1). For details of the theory in solids, including effects of plastic flow at sufficiently high compressions, see Davison (2010). Ultrasonic shock waves are used to break up kidney stones in the human body. In laboratories, shocks from the impact of a rapidly moving projectile on a material specimen, or from intense pulsed laser-induced ablation (e.g., at the U.S. National Ignition Facility, Sec. 10.2.2), are used to compress the specimen to pressures similar to those in Earth's core or higher and thereby explore the specimen's high-density properties. Strong shock waves from meteorite impacts alter the properties of the rock through which they travel and can be used to infer aspects of the equation of state for planetary interiors.

EXERCISES

Exercise 12.12 **Example: Normal Modes of a Homogeneous, Elastic Sphere[3]*
Show that, for frequency-ω oscillations of a homogeneous elastic sphere with negligible gravity, the displacement vector can everywhere be written as $\boldsymbol{\xi} = \boldsymbol{\nabla}\psi + \boldsymbol{\nabla} \times \mathbf{A}$, where ψ is a scalar potential that satisfies the longitudinal-wave Helmholtz equation, and $\mathbf{A}$ is a divergence-free vector potential that satisfies the transverse-wave Helmholtz equation:

$$\boldsymbol{\xi} = \boldsymbol{\nabla}\psi + \boldsymbol{\nabla} \times \mathbf{A}, \quad (\nabla^2 + k_L^2)\psi = 0,$$

$$(\nabla^2 + k_T^2)\mathbf{A} = 0, \quad \boldsymbol{\nabla} \cdot \mathbf{A} = 0; \quad k_L = \frac{\omega}{C_L}, \quad k_T = \frac{\omega}{C_T}. \quad (12.67)$$

[Hint: See Ex. 12.2, and make use of gauge freedom in the vector potential.]

(a) *Radial Modes.* The radial modes are purely longitudinal; their scalar potential [general solution of $(\nabla^2 + k_L^2)\psi = 0$ that is regular at the origin] is the spherical Bessel function of order zero: $\psi = j_0(k_L r) = \sin(k_L r)/(k_L r)$. The corresponding displacement vector $\boldsymbol{\xi} = \boldsymbol{\nabla}\psi$ has as its only nonzero component

$$\xi_r = j_0'(k_L r), \quad (12.68)$$

where the prime on j_0 and on any other function in this exercise means the derivative with respect to its argument. (Here we have dropped a multiplicative factor k_L.) Explain why the only boundary condition that need be imposed is $T_{rr} = 0$ at the sphere's surface. By computing T_{rr} for the displacement vector of

---

3. For a detailed and extensive treatment of this problem and many references, see Eringen and Suhubi (1975, Secs. 8.13, 8.14). Our approach to the mathematics is patterned after that of Ashby and Dreitlein (1975, Sec. III), and our numerical evaluations are for the same cases as Love (1927, Secs. 195, 196) and as the classic paper on this topic, Lamb (1882). It is interesting to compare the mathematics of our analysis with the nineteenth-century mathematics of Lamb and Love, which uses radial functions $\psi_l(r) \sim r^{-l} j_l(r)$ and *solid harmonics* that are sums over m of $r^l Y_l^m(\theta, \phi)$ and satisfy Laplace's equation. Here j_l is the spherical Bessel function of order l, and Y_l^m is the spherical harmonic. Solid harmonics can be written as $\mathcal{F}_{ij\ldots q} x_i x_j \ldots x_q$, with $\mathcal{F}$ a symmetric, trace-free tensor of order l. A variant of them is widely used today in multipolar expansions of gravitational radiation (see, e.g., Thorne, 1980, Sec. II).

Eq. (12.68) and setting it to zero, deduce the following eigenequation for the wave numbers k_L and thence frequencies $\omega = C_L k_L$ of the radial modes:

$$\frac{\tan x_L}{x_L} = \frac{1}{1 - (x_T/2)^2}, \qquad (12.69)$$

where

$$x_L \equiv k_L R = \frac{\omega R}{C_L} = \omega R \sqrt{\frac{\rho}{K + 4\mu/3}}, \quad x_T \equiv k_T R = \frac{\omega R}{C_T} = \omega R \sqrt{\frac{\rho}{\mu}}, \qquad (12.70)$$

with R the radius of the sphere. For $(x_T/x_L)^2 = 3$, which corresponds to a Poisson's ratio $\nu = 1/4$ (about that of glass and typical rock; see Table 11.1), numerical solution of this eigenequation gives the spectrum for modes $\{l = 0; n = 0, 1, 2, 3, \ldots\}$:

$$\frac{x_L}{\pi} = \frac{\omega}{\pi C_L/R} = \{0.8160, 1.9285, 2.9539, 3.9658, \ldots\}. \qquad (12.71)$$

Note that these eigenvalues get closer and closer to integers as one ascends the spectrum; this also will be true for any other physically reasonable value of x_T/x_L. Explain why.

(b) **T2** *Torsional Modes.* The general solution to the scalar Helmholtz equation $(\nabla^2 + k^2)\psi$ that is regular at the origin is a sum over eigenfunctions of the form $j_l(kr)Y_l^m(\theta, \phi)$. Show that the angular momentum operator $\hat{\mathbf{L}} = \mathbf{x} \times \nabla$ commutes with the laplacian ∇^2, and thence infer that $\hat{\mathbf{L}}[j_l Y_l^m]$ satisfies the vector Helmholtz equation. Verify further that it is divergence free, which means it must be expressible as the curl of a vector potential that is also divergence free and satisfies the vector Helmholtz equation. This means that $\boldsymbol{\xi} = \hat{\mathbf{L}}[j_l(k_T r)Y_l^m(\theta, \phi)]$ is a displacement vector that satisfies the elastodynamic equation for transverse waves. Since $\hat{\mathbf{L}}$ differentiates only transversely (in the θ and ϕ directions), we can rewrite this expression as $\boldsymbol{\xi} = j_l(k_T r)\hat{\mathbf{L}} Y_l^m(\theta, \phi)$. To simplify computing the eigenfrequencies (which are independent of m), specialize to $m = 0$, and show that the only nonzero component of the displacement is

$$\xi_\phi \propto j_l(k_T r) \sin\theta\, P_l'(\theta). \qquad (12.72)$$

Equation (12.72), for our homogeneous sphere, is the torsional mode discussed in the text [Eq. (12.64)]. Show that the only boundary condition that must be imposed on this displacement function is $T_{\phi r} = 0$ at the sphere's surface, $r = R$. Compute this component of the stress tensor (with the aid of Box 11.4 for the shear), and by setting it to zero, derive the following eigenequation for the torsional-mode frequencies:

$$x_T\, j_l'(x_T) - j_l(x_T) = 0. \qquad (12.73)$$

For $l = 1$ (the case illustrated in Fig. 12.10b), this eigenequation reduces to $(3 - x_T^2) \tan x_T = 3x_T$, and the lowest few eigenfrequencies are

$$\frac{x_T}{\pi} = \frac{\omega}{\pi C_T/R} = \{1.8346, 2.8950, 3.9225, 4.9385, \ldots\}. \qquad (12.74)$$

As for the radial modes, these eigenvalues get closer and closer to integers as one ascends the spectrum. Explain why.

(c) **T2** *Ellipsoidal Modes.* The displacement vector for the ellipsoidal modes is the sum of a longitudinal piece $\nabla\psi$ [with $\psi = j_l(k_L r)Y_l^m(\theta, \phi)$ satisfying the longitudinal wave Helmholtz equation] and a transverse piece $\nabla \times \mathbf{A}$ [with $\mathbf{A} = j_l(k_T r)\hat{\mathbf{L}}\, Y_l^m(\theta, \phi)$, which as we saw in part (b) satisfies the transverse wave Helmholtz equation and is divergence free]. Specializing to $m = 0$ to derive the eigenequation (which is independent of m), show that the components of the displacement vector are

$$\xi_r = \left[\frac{\alpha}{k_L} j_l'(k_L r) + \frac{\beta}{k_T} l(l+1) \frac{j_l(k_T r)}{k_T r}\right] P_l(\cos\theta),$$

$$\xi_\theta = \left[-\frac{\alpha}{k_L} \frac{j_l(k_L r)}{k_L r} + \frac{\beta}{k_T}\left(j'(k_T r) + \frac{j(k_T r)}{k_T r}\right)\right] \sin\theta\, P_l'(\cos\theta), \qquad (12.75)$$

where α/k_L and β/k_T are constants that determine the weightings of the longitudinal and transverse pieces, and we have included the k_L and k_T to simplify the stress tensor derived below. (To get the β term in ξ_r, you will need to use a differential equation satisfied by P_l.) Show that the boundary conditions we must impose on these eigenfunctions are $T_{rr} = 0$ and $T_{r\theta} = 0$ at the sphere's surface, $r = R$. By evaluating these [using the shear components in Box 11.4, the differential equation satisfied by $j_l(x)$, and $(K + \frac{4}{3}\mu)/\mu = (x_T/x_L)^2$], obtain the following expressions for these components of the stress tensor at the surface:

$$T_{rr} = -\mu P_l(\cos\theta)\left[\alpha\left\{2j_l''(x_L) - [(x_T/x_L)^2 - 2]j_l(x_L)\right\} + \beta\, 2l(l+1)f_1(x_T)\right] = 0,$$

$$T_{r\theta} = \mu \sin\theta\, P_l'(\cos\theta)\left[\alpha\, 2f_1(x_L) + \beta\left\{j_l''(x_T) + [l(l+1) - 2]f_0(x_T)\right\}\right] = 0,$$

where $f_0(x) \equiv j_l(x)/x^2$ and $f_1(x) \equiv (j_l(x)/x)'$. $\qquad (12.76)$

These simultaneous equations for the ratio α/β have a solution if and only if the determinant of their coefficients vanishes:

$$\left\{2j_l''(x_L) - [(x_T/x_L)^2 - 2]j_l(x_L)\right\}\left\{j_l''(x_T) + (l+2)(l-1)f_0(x_T)\right\}$$

$$- 4l(l+1)f_1(x_L)f_1(x_T) = 0. \qquad (12.77)$$

This is the eigenequation for the ellipsoidal modes. For $l = 2$ and $(x_T/x_L)^2 = 3$, it predicts for the lowest four ellipsoidal eigenfrequencies

$$\frac{x_T}{\pi} = \frac{\omega}{\pi C_T/R} = \{0.8403, 1.5487, 2.6513, 3.1131, \ldots\}. \qquad (12.78)$$

Notice that these are significantly smaller than the torsional frequencies. Show that, in the limit of an incompressible sphere ($K \to \infty$) and for $l = 2$, the eigenequation becomes $(4 - x_T^2)[j_2''(x_T) + 4f_0(x_T)] - 24 f_1(x_T) = 0$, which predicts for the lowest four eigenfrequencies

$$\frac{x_T}{\pi} = \frac{\omega}{\pi C_T / R} = \{0.8485, 1.7421, 2.8257, 3.8709, \ldots\}. \tag{12.79}$$

These are modestly larger than the compressible case, Eq. (12.78).

12.5 The Relationship of Classical Waves to Quantum Mechanical Excitations

In the previous chapter, we identified the effects of atomic structure on the continuum approximation for elastic solids. Specifically, we showed that atomic structure accounts for the magnitude of the elastic moduli and explains why most solids yield under comparatively small strain. A quite different connection of the continuum theory to atomic structure is provided by the normal modes of vibration of a finite solid body (e.g., the sphere treated in Sec. 12.4.4 and Ex. 12.12).

For any such body, one can solve the vector wave equation (12.4b) [subject to the vanishing-surface-force boundary condition $\mathbf{T} \cdot \mathbf{n} = 0$, Eq. (11.33)] to find the body's normal modes, as we did in Ex. 12.12 for the sphere. In this section, we label the normal modes by a single index N (encompassing $\{l, m, n\}$ in the case of a sphere) and denote the eigenfrequency of mode N by ω_N and its (typically complex) eigenfunction by $\boldsymbol{\xi}_N$. Then any general, small-amplitude disturbance in the body can be decomposed into a linear superposition of these normal modes:

normal modes of a general elastic body

$$\boxed{\boldsymbol{\xi}(\mathbf{x}, t) = \Re \sum_N a_N(t) \boldsymbol{\xi}_N(\mathbf{x}), \qquad a_N = A_N \exp(-i\omega_N t).} \tag{12.80}$$

displacement expanded in normal modes

Here $\Re$ means to take the real part, a_N is the *complex generalized coordinate* of mode N, and A_N is its complex amplitude. It is convenient to normalize the eigenfunctions so that

$$\int \rho |\boldsymbol{\xi}_N|^2 dV = M, \tag{12.81}$$

where M is the mass of the body; A_N then measures the mean physical displacement in mode N.

Classical electromagnetic waves in a vacuum are described by linear Maxwell equations; so after they have been excited, they will essentially propagate forever. This is not so for elastic waves, where the linear wave equation is only an approximation. Nonlinearities—and most especially impurities and defects in the structure of the

T2

body's material—will cause the different modes to interact weakly and also damp, so that their complex amplitudes A_N change slowly with time according to a damped simple harmonic oscillator differential equation of the form

equation of motion for a normal mode

$$\ddot{a}_N + (2/\tau_N)\dot{a}_N + \omega_N^2 a_N = F'_N/M. \qquad (12.82)$$

Here the second term on the left-hand side is a damping term (due to frictional heating and weak coupling with other modes) that will cause the mode to decay as long as $\tau_N > 0$; F'_N is a fluctuating or *stochastic* force on mode N (also caused by weak coupling to the other modes). Equation (12.82) is the Langevin equation that we studied in Sec. 6.8.1. The spectral density of the fluctuating force F'_N is proportional to $1/\tau_N$ and is determined by the fluctuation-dissipation theorem, Eqs. (6.74) or (6.86). If the modes are thermalized at temperature T, then the fluctuating forces maintain an average energy of kT in each mode.

What happens quantum mechanically? The ions and electrons in an elastic solid interact so strongly that it is difficult to analyze them directly. A quantum mechanical treatment is much easier if one makes a canonical transformation from the coordinates and momenta of the individual ions or atoms to new, generalized coordinates $\hat{x}_N$ and momenta $\hat{p}_N$ that represent weakly interacting normal modes. These coordinates and momenta are Hermitian operators, and they are related to the quantum mechanical complex generalized coordinate $\hat{a}_N$ by

quantization of a normal mode

$$\hat{x}_N = \frac{1}{2}(\hat{a}_N + \hat{a}_N^\dagger), \qquad (12.83a)$$

$$\hat{p}_N = \frac{M\omega_N}{2i}(\hat{a}_N - \hat{a}_N^\dagger), \qquad (12.83b)$$

where the dagger denotes the Hermitian adjoint. We can transform back to obtain an expression for the displacement of the ith ion:

$$\hat{\mathbf{x}}_i = \frac{1}{2}\Sigma_N[\hat{a}_N \boldsymbol{\xi}_N(\mathbf{x}_i) + \hat{a}_N^\dagger \boldsymbol{\xi}_N^*(\mathbf{x}_i)] \qquad (12.84)$$

[a quantum version of Eq. (12.80)].

The hamiltonian can be written in terms of these coordinates as

$$\hat{H} = \Sigma_N \left(\frac{\hat{p}_N^2}{2M} + \frac{1}{2}M\omega_N^2 \hat{x}_N^2 \right) + \hat{H}_{\text{int}}, \qquad (12.85)$$

where the first term is a sum of simple harmonic oscillator hamiltonians for individual modes; and $\hat{H}_{\text{int}}$ is the perturbative interaction hamiltonian, which takes the place of the combined damping and stochastic forcing terms in the classical Langevin

Chapter 12. Elastodynamics

equation (12.82). When the various modes are thermalized, the mean energy in mode N takes on the standard Bose-Einstein form:

$$\bar{E}_N = \hbar\omega_N \left[\frac{1}{2} + \frac{1}{\exp[\hbar\omega_N/(k_B T)] - 1} \right] \quad (12.86)$$

thermalized normal modes

[Eq. (4.28b) with vanishing chemical potential and augmented by a "zero-point energy" of $\frac{1}{2}\hbar\omega$], which reduces to $k_B T$ in the classical limit $\hbar \to 0$.

As the unperturbed hamiltonian for each mode is identical to that for a particle in a harmonic oscillator potential well, it is sensible to think of each wave mode as analogous to such a particle-in-well. Just as the particle-in-well can reside in any one of a series of discrete energy levels lying above the zero-point energy of $\hbar\omega/2$ and separated by $\hbar\omega$, so each wave mode with frequency ω_N must have an energy $(n + 1/2)\hbar\omega_N$, where n is an integer. The operator that causes the energy of the mode to decrease by $\hbar\omega_N$ is the *annihilation operator* for mode n:

$$\boxed{\hat{\alpha}_N = \left(\frac{M\omega_N}{2\hbar} \right)^{1/2} \hat{a}_N,} \quad (12.87)$$

creation and annihilation operators for normal-mode quanta (phonons)

the operator that causes an increase in the energy by $\hbar\omega_N$ is its Hermitian conjugate, the *creation operator* $\hat{\alpha}_N^\dagger$, and their commutator is $[\hat{\alpha}_N, \hat{\alpha}_N^\dagger] = 1$, corresponding to $[\hat{x}_N, \hat{p}_N] = i\hbar$; see for example Chap. 5 of Cohen-Tannoudji, Diu, and Laloë (1977). It is useful to think of each increase or decrease of a mode's energy as the creation or annihilation of an individual quantum or "particle" of energy, so that when the energy in mode N is $(n + 1/2)\hbar\omega_N$, there are n quanta (particles) present. These particles are called *phonons*. Because phonons can coexist in the same state (the same mode), they are bosons. They have individual energies and momenta which must be conserved in their interactions with one another and with other types of particles (e.g., electrons). This conservation law shows up classically as resonance conditions in mode-mode mixing (cf. the discussion in nonlinear optics, Sec. 10.6.1).

phonons are bosons

The important question is, given an elastic solid at finite temperature, do we think of its thermal agitation as a superposition of classical modes, or do we regard it as a gas of quanta? The answer depends on what we want to do. From a purely fundamental viewpoint, the quantum mechanical description takes precedence. However, for many problems where the number of phonons per mode (the mode's mean occupation number) $\eta_N \sim k_B T/(\hbar\omega_N)$ is large compared to one, the classical description is amply adequate and much easier to handle. We do not need a quantum treatment when computing the normal modes of a vibrating building excited by an earthquake or when trying to understand how to improve the sound quality of a violin. Here the difficulty often is in accommodating the boundary conditions so as to determine the normal modes. All this was expected. What comes as more of a surprise is that often, for purely classical problems (where $\hbar$ is quantitatively irrelevant), the fastest way to

relation of classical and quantum theories for normal modes

analyze a practical problem formally is to follow the quantum route and then take the limit $\hbar \to 0$. We shall see this graphically demonstrated when we discuss nonlinear plasma physics in Chap. 23.

Bibliographic Note

The classic textbook treatments of elastodynamics from a physicist's viewpoint are Landau and Lifshitz (1986) and—in nineteenth-century language—Love (1927). For a lovely and very readable introduction to the basic concepts, with a focus on elastodynamic waves, see Kolsky (1963). Our favorite advanced textbook and treatise is Eringen and Suhubi (1975).

By contrast with elastostatics, where there are a number of good, twenty-first-century engineering-oriented textbooks at the elementary and intermediate levels, there are none that we know of for elastodynamics.

However, we do know two good, advanced engineering-oriented books on methods to solve the elastodynamic equations in nontrivial situations: Poruchikov, Khokhryakov, and Groshev (1993) and Kausel (2006). And for a compendium of practical engineering lore about vibrations of engineering structures, from building foundations to bell towers to suspension bridges, see Bachman (1994).

For seismic waves and their geophysical applications, we recommend the textbooks by Stein and Wysession (2003) and by Shearer (2009).

PART V

FLUID DYNAMICS

Having studied elasticity theory, we now turn to a second branch of continuum mechanics: *fluid dynamics*. Three of the four states of matter (gases, liquids, and plasmas) can be regarded as fluids, so it is not surprising that interesting fluid phenomena surround us in our everyday lives. Fluid dynamics is an experimental discipline; much of our current understanding has come in response to laboratory investigations. Fluid dynamics finds experimental application in engineering, physics, biophysics, chemistry, and many other fields. The observational sciences of oceanography, meteorology, astrophysics, and geophysics, in which experiments are less frequently performed, also rely heavily on fluid dynamics. Many of these fields have enhanced our appreciation of fluid dynamics by presenting flows under conditions that are inaccessible to laboratory study.

Despite this rich diversity, the fundamental principles are common to all these applications. The key assumption that underlies the equations governing the motion of a fluid is that the length- and timescales associated with the flow are long compared with the corresponding microscopic scales, so the continuum approximation can be invoked.

The fundamental equations of fluid dynamics are, in some respects, simpler than the corresponding laws of elastodynamics. However, as with particle dynamics, simplicity of equations does not imply the solutions are simple. Indeed, they are not! One reason is that fluid displacements are usually not small (by contrast with elastodynamics, where the elastic limit keeps them small), so most fluid phenomena are immediately nonlinear.

Relatively few problems in fluid dynamics admit complete, closed-form, analytic solutions, so progress in describing fluid flows has usually come from introducing clever physical models and using judicious mathematical approximations. Semi-empirical scaling laws are also common, especially for engineering applications. In more recent years, numerical fluid dynamics has come of age, and in many areas of fluid dynamics, computer simulations are complementing and even supplanting laboratory experiments and measurements. For example, most design work for airplanes and automobiles is now computational.

In fluid dynamics, considerable insight accrues from visualizing the flow. This is true of fluid experiments, where much technical skill is devoted to marking the fluid so it can be imaged; it is also true of numerical simulations, where frequently more time is devoted to computer graphics than to solving the underlying partial differential equations. Indeed, obtaining an analytic solution to the equations of fluid dynamics is not the same as understanding the flow; as a tool for understanding, at the very least it is usually a good idea to sketch the flow pattern.

We present the fundamental concepts of fluid dynamics in Chap. 13, focusing particularly on the underlying physical principles and the conservation laws for mass, momentum, and energy. We explain why, when flow velocities are very subsonic, a fluid's density changes very little (i.e., it is effectively incompressible), and we specialize the fundamental principles and equations to incompressible flows.

Vorticity plays major roles in fluid dynamics. In Chap. 14, we focus on those roles for incompressible flows, both in the fundamental equations of fluid dynamics and in applications. Our applications include, among others, tornados and whirlpools, boundary layers abutting solid bodies, the influence of boundary layers on bulk flows, and how wind drives ocean waves and is ultimately responsible for deep-ocean currents.

Viscosity has a remarkably strong influence on fluid flows, even when the viscosity is weak. When strong, it keeps a flow laminar (smooth); when weak, it controls details of the turbulence that pervades the bulk flow (the flow away from boundary layers). In Chap. 15, we describe turbulence, a phenomenon so difficult to handle theoretically that semiquantitative ideas and techniques pervade its theoretical description, even in the incompressible approximation (to which we adhere). The onset of turbulence is especially intriguing. We illuminate it by exploring a closely related phenomenon: chaotic behavior in mathematical maps.

In Chap. 16, we focus on waves in fluids, beginning with waves on the surface of water, where we shall see, for shallow water, how nonlinear effects and dispersion together give rise to "solitary waves" (solitons) that hold themselves together as they propagate. In this chapter, we abandon the incompressible approximation, which has permeated Part V thus far, to study sound waves. Radiation reaction in sound generation is much simpler than in, for example, electrodynamics, so we use sound waves to elucidate the physical origin of radiation reaction and the nonsensical nature of pre-acceleration.

In Chap. 17, we turn to transonic and supersonic flows, in which density changes are of major importance. Here we meet some beautiful and powerful mathematical tools: characteristics and their associated Riemann invariants. We focus especially on flow through rocket nozzles and other constrictions, and on shock fronts, with applications to explosions (bombs and supernovae).

Convection is another phenomenon in which density changes are crucial—though here the density changes are induced by thermal expansion rather than by physical compression. We study convection in Chap. 18, paying attention to the (usually small but sometimes large) influence of diffusive heat conduction and the diffusion of chemical constituents (e.g., salt).

When a fluid is electrically conducting and has an embedded magnetic field, the exchange of momentum between the field and the fluid can produce remarkable phenomena (e.g., dynamos that amplify a seed magnetic field, a multitude of instabilities, and Alfvén waves and other magnetosonic waves). This is the realm of magnetohydrodynamics, which we explore in Chap. 19. The most common application of magnetohydrodynamics is to a highly ionized plasma, the topic of Part VI of this book, so Chap. 19 serves as a transition from fluid dynamics (Part V) to plasma physics (Part VI).

CHAPTER THIRTEEN

Foundations of Fluid Dynamics

εὕρηκα

ARCHIMEDES (CA. 250 BC)

13.1 Overview

In this chapter, we develop the fundamental concepts and equations of fluid dynamics, first in the flat-space venue of Newtonian physics (Track One) and then in the Minkowski spacetime venue of special relativity (Track Two). Our relativistic treatment is rather brief. This chapter contains a large amount of terminology that may be unfamiliar to readers. A glossary of terminology is given in Box 13.5, near the end of the chapter.

We begin in Sec. 13.2 with a discussion of the physical nature of a fluid: the possibility of describing it by a piecewise continuous density, velocity, and pressure and the relationship between density changes and pressure changes. Then in Sec. 13.3, we discuss hydrostatics (density and pressure distributions of a static fluid in a static gravitational field); this parallels our discussion of elastostatics in Chap. 11. After explaining the physical basis of Archimedes' law, we discuss stars, planets, Earth's atmosphere, and other applications.

Our foundation for moving from hydrostatics to hydrodynamics will be conservation laws for mass, momentum, and energy. To facilitate that transition, in Sec. 13.4, we examine in some depth the physical and mathematical origins of these conservation laws in Newtonian physics.

The stress tensor associated with most fluids can be decomposed into an isotropic pressure and a viscous term linear in the rate of shear (i.e., in the velocity gradient). Under many conditions the viscous stress can be neglected over most of the flow, and diffusive heat conductivity is negligible. The fluid is then called *ideal* or *perfect*.[1] We study the laws governing ideal fluids in Sec. 13.5. After deriving the relevant

ideal fluid (perfect fluid)

---

[1] An ideal fluid (also called a *perfect fluid*) should not be confused with an ideal or perfect gas—one whose pressure is due solely to kinetic motions of particles and thus is given by $P = nk_B T$, with n the particle number density, k_B Boltzmann's constant, and T temperature, and that may (ideal gas) or may not (perfect gas) have excited internal molecular degrees of freedom; see Box 13.2.

> **BOX 13.1. READERS' GUIDE**
>
> - This chapter relies heavily on the geometric view of Newtonian physics (including vector and tensor analysis) laid out in Chap. 1.
> - This chapter also relies on some elementary thermodynamic concepts treated in Chap. 5 and on the concepts of expansion, shear, and rotation (the irreducible tensorial parts of the gradient of displacement) introduced in Chap. 11.
> - Our brief introduction to relativistic fluid dynamics (Track Two of this chapter) relies heavily on the geometric viewpoint on special relativity developed in Chap. 2.
> - Chapters 14–19 (fluid dynamics including magnetohydrodynamics) are extensions of this chapter; to understand them, the Track-One parts of this chapter must be mastered.
> - Portions of Part VI, Plasma Physics (especially Chap. 21 on the "two-fluid formalism"), rely heavily on Track-One parts of this chapter.
> - Portions of Part VII, General Relativity, entail relativistic fluids in curved spacetime, for which the Track-Two Sec. 13.8 serves as a foundation.

conservation laws and equation of motion (the Euler equation), we derive and discuss Bernoulli's theorem for an ideal fluid and explain how it can simplify the description of many flows. In flows for which the fluid velocities are much smaller than the speed of sound and gravity is too weak to compress the fluid much, the fractional changes in fluid density are small. It can then be a good approximation to treat the fluid as *incompressible*, which leads to considerable simplification, also addressed in Sec. 13.5. As we shall see in Sec. 13.6, incompressibility can be a good approximation not just for liquids (which tend to have large bulk moduli) but also, more surprisingly, for gases. It is so widely applicable that we restrict ourselves to the incompressible approximation throughout Chaps. 14 and 15.

In Sec. 13.7, we augment our basic equations with terms describing viscous stresses and also heat conduction. This allows us to derive the famous Navier-Stokes equation and illustrate its use by analyzing the flow of a fluid through a pipe, and then use this to make a crude model of blood flowing through an artery. Much of our study of fluids in future chapters will focus on this Navier-Stokes equation.

In our study of fluids we often deal with the influence of a uniform gravitational field (e.g., Earth's, on lengthscales small compared to Earth's radius). Occasionally, however, we consider inhomogeneous gravitational fields produced by the fluid whose motion we study. For such situations it is useful to introduce gravitational contributions to the stress tensor and energy density and flux. We present and discuss these in Box 13.4, where they will not impede the flow of the main stream of ideas.

We conclude this chapter in Sec. 13.8 with a brief, Track-Two overview of relativistic fluid mechanics for a perfect (ideal) fluid. As an important application, we explore the structure of a relativistic astrophysical jet: the conversion of internal thermal energy into the energy of organized bulk flow as the jet travels outward from the nucleus of a galaxy into intergalactic space and widens. We also explore how the fundamental equations of Newtonian fluid mechanics arise as low-velocity limits of the relativistic equations.

13.2 The Macroscopic Nature of a Fluid: Density, Pressure, Flow Velocity; Liquids versus Gases

The macroscopic nature of a fluid follows from two simple observations. The first is that in most flows the macroscopic continuum approximation is valid. Because the molecular mean free paths in a fluid are small compared to macroscopic lengthscales, we can define a mean local velocity $\mathbf{v}(\mathbf{x}, t)$ of the fluid's molecules, which varies smoothly both spatially and temporally; we call this the fluid's velocity. For the same reason, other quantities that characterize the fluid [e.g., the density $\rho(\mathbf{x}, t)$] also vary smoothly on macroscopic scales. This need not be the case everywhere in the flow. An important exception is a shock front, which we study in Chap. 17; there the flow varies rapidly, over a length of order the molecules' mean free path for collisions. In this case, the continuum approximation is only piecewise valid, and we must perform a matching at the shock front. One might think that a second exception is a turbulent flow, where it seems plausible that the average molecular velocity will vary rapidly on whatever lengthscale we choose to study, all the way down to intermolecular distances, so averaging becomes problematic. As we shall see in Chap. 15, this is not the case; turbulent flows generally have a lengthscale far larger than intermolecular distances, below which the flow varies smoothly.

The second observation is that fluids do not oppose a steady shear strain. This is easy to understand on microscopic grounds, as there is no lattice to deform, and the molecular velocity distribution remains locally isotropic in the presence of a static shear. By kinetic theory considerations (Chap. 3), we therefore expect that a fluid's stress tensor $\mathbf{T}$ will be isotropic in the local rest frame of the fluid (i.e., in a frame where $\mathbf{v} = 0$). Because of viscosity, this is not quite true when the shear varies with time. However, we neglect viscosity as well as diffusive heat flow until Sec. 13.7 (i.e., we restrict ourselves to ideal fluids). This assumption allows us to write $\mathbf{T} = P\mathbf{g}$ in the local rest frame, where P is the fluid's pressure, and $\mathbf{g}$ is the metric (with Kronecker delta components, $g_{ij} = \delta_{ij}$, in Cartesian coordinates).

The laws of fluid mechanics as we develop them are equally valid for liquids, gases, and (under many circumstances) plasmas. In a liquid, as in a solid, the molecules are packed side by side (but can slide over one another easily). In a gas or plasma, the molecules are separated by distances large compared to their sizes. This difference leads to different behaviors under compression.

liquid

gas or plasma

For a liquid (e.g., water in a lake), the molecules resist strongly even a small compression; as a result, it is useful to characterize the pressure increase by a bulk modulus K, as in an elastic solid (Chap. 11):

pressure changes in a liquid

$$\boxed{\delta P = -K\Theta = K\frac{\delta\rho}{\rho} \quad \text{for a liquid.}} \qquad (13.1)$$

(Here we have used the fact that the expansion Θ is the fractional increase in volume, or equivalently by mass conservation, it is the fractional decrease in density.) The bulk modulus for water is about 2.2 GPa, so as one goes downward in a lake far enough to double the pressure from one atmosphere (10^5 Pa to 2×10^5 Pa), the fractional change in density is only $\delta\rho/\rho = (2 \times 10^5/2.2 \times 10^9) \simeq 1$ part in 10,000.

By contrast, gases and plasmas are much less resistant to compression. Due to the large distance between molecules, a doubling of the pressure requires, in order of magnitude, a doubling of the density:

pressure changes in a gas or plasma

$$\boxed{\frac{\delta P}{P} = \Gamma\frac{\delta\rho}{\rho} \quad \text{for a gas,}} \qquad (13.2)$$

where Γ is a proportionality factor of order unity. The numerical value of Γ depends on the physical situation. If the gas is ideal [so $P = \rho k_B T/(\mu m_p)$ in the notation of Box 13.2, Eq. (4)] and the temperature T is being held fixed by thermal contact with some heat source as the density changes (*isothermal process*), then $\delta P \propto \delta\rho$, and $\Gamma = 1$. Alternatively, and much more commonly, a fluid element's entropy may remain constant, because no significant heat can flow in or out of it during the density change. In this case Γ is called the *adiabatic index,* and (continuing to assume ideality), it can be shown using the laws of thermodynamics that

adiabatic index

$$\boxed{\Gamma = \gamma \equiv c_P/c_V \quad \text{for adiabatic processes in an ideal gas.}} \qquad (13.3)$$

specific heats per unit mass

Here c_P and c_V are the specific heats at constant pressure and volume; see Ex. 5.4 in Chap. 5.[2]

---

2. In fluid dynamics, our specific heats, and other extensive variables, such as energy, entropy, and enthalpy, are defined per unit mass and are denoted by lowercased letters. So $c_P = T(\partial s/\partial T)_P$ is the amount of heat that must be added to a unit mass of the fluid to increase its temperature by one unit, and similarly for $c_V = T(\partial s/\partial T)_\rho$. By contrast, in statistical thermodynamics (Chap. 5) our extensive variables are defined for some chosen sample of material and are denoted by capital letters, e.g., $C_P = T(\partial S/\partial T)_P$.

BOX 13.2. THERMODYNAMIC CONSIDERATIONS

One feature of fluid dynamics (especially gas dynamics) that distinguishes it from elastodynamics is that the thermodynamic properties of the fluid are often important, so we must treat energy conservation explicitly. In this box we review, from Chap. 5 or any book on thermodynamics (e.g., Kittel and Kroemer, 1980), the thermodynamic concepts needed in our study of fluids. We have no need for partition functions, ensembles, and other statistical aspects of thermodynamics. Instead, we only need elementary thermodynamics.

We begin with the nonrelativistic first law of thermodynamics (5.8) for a sample of fluid with energy E, entropy S, volume V, number N_I of molecules of species I, temperature T, pressure P, and chemical potential μ_I for species I:

$$dE = TdS - PdV + \sum_I \mu_I dN_I. \tag{1}$$

Almost everywhere in our treatment of fluid dynamics (and throughout this chapter), we assume that the term $\sum_I \mu_I dN_I$ vanishes. Physically this holds because (i) all relevant nuclear reactions are frozen (occur on timescales τ_{react} far longer than the dynamical timescales τ_{dyn} of interest to us), so $dN_I = 0$, and (ii) each chemical reaction is either frozen $dN_I = 0$ or goes so rapidly ($\tau_{\text{react}} \ll \tau_{\text{dyn}}$) that it and its inverse are in local thermodynamic equilibrium: $\sum_I \mu_I dN_I = 0$ for those species involved in the reactions. In the rare intermediate situation, where some relevant reaction has $\tau_{\text{react}} \sim \tau_{\text{dyn}}$, we would have to carefully keep track of the relative abundances of the chemical or nuclear species and their chemical potentials.

Consider a small fluid element with mass Δm, internal energy per unit mass u, entropy per unit mass s, and volume per unit mass $1/\rho$. Then inserting $E = u\Delta m$, $S = s\Delta m$, and $V = \Delta m/\rho$ into the first law $dE = TdS - PdV$, we obtain the form of the first law that we use in almost all of our fluid-dynamics studies:

$$\boxed{du = Tds - Pd\left(\frac{1}{\rho}\right).} \tag{2}$$

The internal energy (per unit mass) u comprises the random translational energy of the molecules that make up the fluid, together with the energy associated with their internal degrees of freedom (rotation, vibration, etc.) and with their intermolecular forces. The term Tds represents some amount of heat (per unit mass) that may be injected into a fluid element (e.g., by

(continued)

BOX 13.2. (continued)

viscous heating; Sec. 13.7) or may be removed (e.g., by radiative cooling). The term $-P d(1/\rho)$ represents work done on the fluid.

In fluid mechanics it is useful to introduce the enthalpy $H = E + PV$ of a fluid element (cf. Ex. 5.5) and the corresponding enthalpy per unit mass $h = u + P/\rho$. Inserting $u = h - P/\rho$ into the left-hand side of the first law (2), we obtain the first law in the enthalpy representation [Eq. (5.47)]:

$$dh = T ds + \frac{dP}{\rho}. \tag{3}$$

Because we assume that all reactions are frozen or are in local thermodynamic equilibrium, the relative abundances of the various nuclear and chemical species are fully determined by a fluid element's density ρ and temperature T (or by any two other variables in the set ρ, T, s, and P). Correspondingly, the thermodynamic state of a fluid element is completely determined by any two of these variables. To calculate all features of that state from two variables, we must know the relevant equations of state, such as $P(\rho, T)$ and $s(\rho, T)$; $P(\rho, s)$ and $T(\rho, s)$; or the fluid's fundamental thermodynamic potential (Table 5.1), from which follow the equations of state.

We often deal with ideal gases (i.e., gases in which intermolecular forces and the volume occupied by the molecules are treated as totally negligible). For any ideal gas, the pressure arises solely from the kinetic motions of the molecules, and so the equation of state $P(\rho, T)$ is

$$P = \frac{\rho k_B T}{\mu m_p}. \tag{4}$$

Here μ is the *mean molecular weight*, and m_p is the proton mass [cf. Eqs. (3.39b), with the number density of particles n reexpressed as $\rho/\mu m_p$]. The mean molecular weight μ is the mean mass per gas molecule in units of the proton mass (e.g., $\mu = 1$ for hydrogen, $\mu = 32$ for O_2, and $\mu = 28.8$ for air). This μ should not be confused with the chemical potential of species I, μ_I (which will rarely if ever be used in our fluid dynamics analyses). [The concept of an ideal gas must not be confused with an ideal fluid; see footnote 1.]

An idealization that is often accurate in fluid dynamics is that the fluid is *adiabatic:* no heating or cooling from dissipative processes, such as viscosity,

(continued)

> **BOX 13.2. (continued)**
>
> thermal conductivity, or the emission and absorption of radiation. When this is a good approximation, the entropy per unit mass s of a fluid element is constant.
>
> In an adiabatic flow, there is only one thermodynamic degree of freedom, so we can write $P = P(\rho, s) = P(\rho)$. Of course, this function will be different for fluid elements that have different s. In the case of an ideal gas, a standard thermodynamic argument (Ex. 5.4) shows that the pressure in an adiabatically expanding or contracting fluid element varies with density as $\delta P/P = \gamma \delta \rho/\rho$, where $\gamma = c_P/c_V$ is the adiabatic index [Eqs. (13.2) and (13.3)]. If, as is often the case, the adiabatic index remains constant over several doublings of the pressure and density, then we can integrate this expression to obtain the equation of state
>
> $$\boxed{P = K(s)\rho^\gamma,} \tag{5}$$
>
> where $K(s)$ is some function of the entropy. Equation (5) is sometimes called the *polytropic* equation of state, and a *polytropic index n* (not to be confused with number density of particles!) is defined by $\gamma = 1 + 1/n$. See the discussion of stars and planets in Sec. 13.3.2 and Ex. 13.4. A special case of adiabatic flow is *isentropic* flow. In this case, the entropy is constant everywhere, not just inside individual fluid elements.
>
> When the pressure can be regarded as a function of the density alone (the same function everywhere), the fluid is called *barotropic*.

From Eqs. (13.1) and (13.2), we see that $K = \Gamma P$; so why use K for liquids and Γ for gases and plasmas? Because in a liquid K remains nearly constant when P changes by large fractional amounts $\delta P/P \gtrsim 1$, while in a gas or plasma it is Γ that remains nearly constant.

For other thermodynamic aspects of fluid dynamics that will be relevant as we proceed, see Box 13.2.

13.3 Hydrostatics

Just as we began our discussion of elasticity with a treatment of elastostatics, so we introduce fluid mechanics by discussing hydrostatic equilibrium.

The *equation of hydrostatic equilibrium* for a fluid at rest in a gravitational field $\mathbf{g}$ is the same as the equation of elastostatic equilibrium with a vanishing shear stress, so $\mathbf{T} = P\mathbf{g}$:

$$\boxed{\nabla \cdot \mathbf{T} = \nabla P = \rho \mathbf{g}} \tag{13.4}$$

equation of hydrostatic equilibrium

[Eq. (11.14) with $\mathbf{f} = -\nabla \cdot \mathbf{T}$]. Here $\mathbf{g}$ is the acceleration of gravity (which need not be constant; e.g., it varies from location to location inside the Sun). It is often useful to express $\mathbf{g}$ as the gradient of the Newtonian gravitational potential Φ:

Newtonian gravitational potential

$$\mathbf{g} = -\nabla \Phi. \tag{13.5}$$

Note our sign convention: Φ is negative near a gravitating body and zero far from all bodies, and it is determined by Newton's field equation for gravity:

Newtonian field equation for gravity

$$\nabla^2 \Phi = -\nabla \cdot \mathbf{g} = 4\pi G \rho. \tag{13.6}$$

We can draw some immediate and important inferences from Eq. (13.4). Take the curl of Eq. (13.4) and use Eq. (13.5) to obtain

$$\nabla \Phi \times \nabla \rho = 0. \tag{13.7}$$

theorems about hydrostatic structure

Equation (13.7) tells us that in hydrostatic equilibrium, the contours of constant density (isochores) coincide with the equipotential surfaces: $\rho = \rho(\Phi)$. Equation (13.4) itself, with (13.5), then tells us that, as we move from point to point in the fluid, the changes in P and Φ are related by $dP/d\Phi = -\rho(\Phi)$. This, in turn, implies that the difference in pressure between two equipotential surfaces Φ_1 and Φ_2 is given by

$$\Delta P = -\int_{\Phi_1}^{\Phi_2} \rho(\Phi) d\Phi. \tag{13.8}$$

Moreover, as $\nabla P \propto \nabla \Phi$, the surfaces of constant pressure (the *isobars*) coincide with the gravitational equipotentials. This is all true whether $\mathbf{g}$ varies inside the fluid or is constant.

pressure is weight of overlying fluid

The gravitational acceleration $\mathbf{g}$ is actually constant to high accuracy in most non-astrophysical applications of fluid dynamics, for example, on the surface of Earth. In this case the pressure at a point in a fluid is, from Eq. (13.8), equal to the total weight of fluid per unit area above the point,

$$P(z) = g \int_z^\infty \rho dz, \tag{13.9}$$

where the integral is performed by integrating upward in the gravitational field (cf. Fig. 13.1). For example, the deepest point in the world's oceans is the bottom of the Mariana Trench in the Pacific, 11 km below sea level. Adopting a density $\rho \simeq 10^3 \text{ kg m}^{-3}$ for water and $g \simeq 10 \text{ m s}^{-2}$, we obtain a pressure of $\simeq 10^8$ Pa or $\simeq 10^3$ atmospheres. This is comparable with the yield stress of the strongest materials. It should therefore come as no surprise that the record for the deepest dive ever recorded by a submersible—a depth of 10.91 km (just $\sim$100 m shy of the lowest point in the trench) achieved by the *Trieste* in 1960—remained unbroken for more than half a century. Only in 2012 was that last 100 m conquered and the trench's bottom reached, by the filmmaker James Cameron in the *Deep Sea Challenger*. Since the bulk modulus

Chapter 13. Foundations of Fluid Dynamics

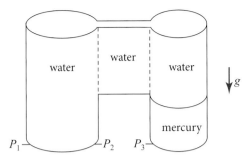

FIGURE 13.1 Elementary demonstration of the principles of hydrostatic equilibrium. Water and mercury, two immiscible fluids of different density, are introduced into a container with two connected chambers as shown. In each chamber, isobars (surfaces of constant pressure) coincide with surfaces of constant $\Phi = -gz$, and so are horizontal. The pressure at each point on the flat bottom of a container is equal to the weight per unit area of the overlying fluid [Eq. (13.9)]. The pressures P_1 and P_2 at the bottom of the left chamber are equal, but because of the density difference between mercury and water, they differ from the pressure P_3 at the bottom of the right chamber.

of sea water is $K = 2.3$ GPa, at the bottom of the trench the water is compressed by $\delta\rho/\rho = P/K \simeq 0.05$.

EXERCISES

Exercise 13.1 *Example: Earth's Atmosphere*

As mountaineers know, it gets cooler as you climb. However, the rate at which the temperature falls with altitude depends on the thermal properties of air. Consider two limiting cases.

(a) In the lower stratosphere (Fig. 13.2), the air is isothermal. Use the equation of hydrostatic equilibrium (13.4) to show that the pressure decreases exponentially with height z:

$$P \propto \exp(-z/H), \tag{13.10a}$$

where the scale height H is given by

$$H = \frac{k_B T}{\mu m_p g}, \tag{13.10b}$$

with μ the mean molecular weight of air and m_p the proton mass. Evaluate this numerically for the lower stratosphere, and compare with the stratosphere's thickness. By how much does P drop between the bottom and top of the isothermal region?

(b) Suppose that the air is isentropic, so that $P \propto \rho^\gamma$ [Eq. (5) of Box 13.2], where γ is the specific heat ratio. (For diatomic gases like nitrogen and oxygen, $\gamma \sim 1.4$; see Ex. 17.1.) Show that the temperature gradient satisfies

$$\frac{dT}{dz} = -\left(\frac{\gamma - 1}{\gamma}\right)\frac{g\mu m_p}{k}. \tag{13.10c}$$

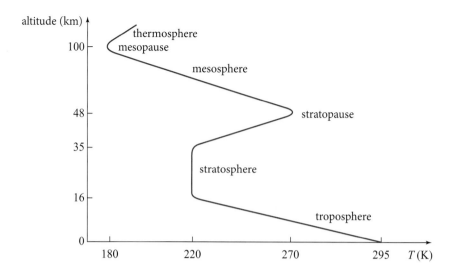

FIGURE 13.2 Actual temperature variation of Earth's mean atmosphere at temperate latitudes.

Note that the temperature gradient vanishes when $\gamma \to 1$. Evaluate the temperature gradient, also known at low altitudes as the *lapse rate*. The average lapse rate is measured to be ~ 6 K km^{-1} (Fig. 13.2). Show that this is intermediate between the two limiting cases of isentropic and isothermal lapse rates.

13.3.1 Archimedes' Law

Archimedes' law

The law of Archimedes states that, *when a solid body is totally or partially immersed in a fluid in a uniform gravitational field* $\mathbf{g} = -g\mathbf{e}_z$, *the total buoyant upward force of the fluid on the body is equal to the weight of the displaced fluid.*

Proof of Archimedes' Law. A formal proof can be made as follows (Fig. 13.3). The fluid, pressing inward on the body across a small element of the body's surface $d\mathbf{\Sigma}$, exerts a force $d\mathbf{F}^{\text{buoy}} = \mathbf{T}(\_, -d\mathbf{\Sigma})$ [Eq. (1.32)], where $\mathbf{T}$ is the fluid's stress tensor, and the minus sign is because by convention, $d\mathbf{\Sigma}$ points out of the body rather than into it. Converting to index notation and integrating over the body's surface $\partial \mathcal{V}$, we obtain for the net buoyant force

$$F_i^{\text{buoy}} = -\int_{\partial \mathcal{V}} T_{ij} d\Sigma_j. \tag{13.11a}$$

Now, imagine removing the body and replacing it by fluid that has the same pressure $P(z)$ and density $\rho(z)$, at each height z as the surrounding fluid; this is the fluid that was originally displaced by the body. Since the fluid stress on $\partial \mathcal{V}$ has not changed, the buoyant force will be unchanged. Use Gauss's law to convert the surface integral (13.11a) into a volume integral over the interior fluid (the originally displaced fluid):

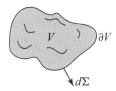

FIGURE 13.3 Derivation of Archimedes' law.

$$F_i^{\text{buoy}} = -\int_\mathcal{V} T_{ij;j} dV. \qquad (13.11b)$$

The displaced fluid obviously is in hydrostatic equilibrium with the surrounding fluid, and its equation of hydrostatic equilibrium $T_{ij;j} = \rho g_i$ [Eq. (13.4)], when inserted into Eq. (13.11b), implies that

$$\mathbf{F}^{\text{buoy}} = -\mathbf{g} \int_\mathcal{V} \rho dV = -M\mathbf{g}, \qquad (13.12)$$

where M is the mass of the displaced fluid. Thus, the upward buoyant force on the original body is equal in magnitude to the weight Mg of the displaced fluid. Clearly, if the body has a higher density than the fluid, then the downward gravitational force on it (its weight) will exceed the weight of the displaced fluid and thus exceed the buoyant force it feels, and the body will fall. If the body's density is less than that of the fluid, the buoyant force will exceed its weight and it will be pushed upward. ∎

A key piece of physics underlying Archimedes' law is the fact that the intermolecular forces acting in a fluid, like those in a solid (cf. Sec. 11.3.6), are short range. If, instead, the forces were long range, Archimedes' law could fail. For example, consider a fluid that is electrically conducting, with currents flowing through it that produce a magnetic field and resulting long-range magnetic forces (the magnetohydrodynamic situation studied in Chap. 19). If we then substitute an insulating solid for some region $\mathcal{V}$ of the conducting fluid, the force that acts on the solid will be different from the force that acted on the displaced fluid.

short-range versus long-range forces

EXERCISES

Exercise 13.2 *Practice: Weight in Vacuum*
How much more would you weigh in a vacuum?

Exercise 13.3 *Problem: Stability of Boats*
Use Archimedes' law to explain qualitatively the conditions under which a boat floating in still water will be stable to small rolling motions from side to side. [Hint: You might want to define and introduce a *center of buoyancy* and a *center of gravity* inside

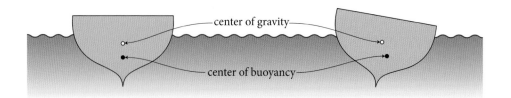

FIGURE 13.4 The center of gravity and center of buoyancy of a boat when it is upright (left) and tilted (right).

the boat, and pay attention to the change in the center of buoyancy when the boat tilts. See Fig. 13.4.]

13.3.2 Nonrotating Stars and Planets

Stars and massive planets—if we ignore their rotation—are self-gravitating fluid spheres. We can model the structure of such a nonrotating, spherical, self-gravitating fluid body by combining the equation of hydrostatic equilibrium (13.4) in spherical polar coordinates,

$$\frac{dP}{dr} = -\rho \frac{d\Phi}{dr}, \tag{13.13}$$

with Poisson's equation,

$$\nabla^2 \Phi = \frac{1}{r^2} \frac{d}{dr}\left(r^2 \frac{d\Phi}{dr}\right) = 4\pi G \rho, \tag{13.14}$$

to obtain

$$\frac{1}{r^2} \frac{d}{dr}\left(\frac{r^2}{\rho} \frac{dP}{dr}\right) = -4\pi G \rho. \tag{13.15}$$

Equation (13.15) can be integrated once radially with the aid of the boundary condition $dP/dr = 0$ at $r = 0$ (pressure cannot have a cusp-like singularity) to obtain

equations of structure for nonrotating star or planet

$$\boxed{\frac{dP}{dr} = -\rho \frac{Gm}{r^2},} \tag{13.16a}$$

where

$$\boxed{m = m(r) \equiv \int_0^r 4\pi \rho r^2 dr} \tag{13.16b}$$

is the total mass inside radius r. Equation (13.16a) is an alternative form of the equation of hydrostatic equilibrium (13.13) at radius r inside the body: Gm/r^2 is the gravitational acceleration g at r, $\rho(Gm/r^2) = \rho g$ is the downward gravitational force per unit volume on the fluid, and dP/dr is the upward buoyant force per unit volume.

Equations (13.13)–(13.16b) are a good approximation for solid planets (e.g., Earth) as well as for stars and fluid planets (e.g., Jupiter) because at the enormous stresses encountered in the interior of a solid planet, the strains are so large that plastic flow occurs. In other words, the shear stresses are much smaller than the isotropic part of the stress tensor.

Let us make an order-of-magnitude estimate of the interior pressure in a star or planet of mass M and radius R. We use the equation of hydrostatic equilibrium (13.4) or (13.16a), approximating m by M, the density ρ by M/R^3, and the gravitational acceleration by GM/R^2; the result is

$$P \sim \frac{GM^2}{R^4}. \tag{13.17}$$

To improve this estimate, we must solve Eq. (13.15). For that, we need a prescription relating the pressure to the density (i.e., an equation of state). A common idealization is the polytropic relation:

$$P \propto \rho^{1+1/n}, \tag{13.18}$$

polytropic equation of state

where n is the polytropic index (cf. last part of Box 13.2). [This finesses the issue of the generation and flow of heat in stellar interiors, which determines the temperature $T(r)$ and thence the pressure $P(\rho, T)$.] Low-mass white-dwarf stars are well approximated as $n = 1.5$ polytropes [Eq. (3.53c)], and red-giant stars are somewhat similar in structure to $n = 3$ polytropes. The giant planets, Jupiter and Saturn, are mainly composed of an H-He fluid that can be approximated by an $n = 1$ polytrope, and the density of a small planet like Mercury is roughly constant ($n = 0$).

To solve Eqs. (13.16), we also need boundary conditions. We can choose some density ρ_c and corresponding pressure $P_c = P(\rho_c)$ at the star's center $r = 0$, then integrate Eqs. (13.16) outward until the pressure P drops to zero, which will be the star's (or planet's) surface. The values of r and m there will be the star's radius R and mass M. For mathematical details of polytropic stellar models constructed in this manner, see Ex. 13.4. This exercise is particularly important as an example of the power of converting to dimensionless variables, a procedure we use frequently in this part of the book.

boundary conditions for stellar/planetary structure

We can easily solve the equations of hydrostatic equilibrium (13.16) for a planet with constant density ($n = 0$) to obtain $m = (4\pi/3)\rho r^3$ and

$$P = P_0 \left(1 - \frac{r^2}{R^2}\right), \tag{13.19}$$

where the central pressure is

$$P_0 = \left(\frac{3}{8\pi}\right) \frac{GM^2}{R^4}, \tag{13.20}$$

pressure in constant-density planet

consistent with our order-of-magnitude estimate (13.17).

13.3 Hydrostatics

EXERCISES

Exercise 13.4 **Example: Polytropes—The Power of Dimensionless Variables*

When dealing with differential equations describing a physical system, it is often helpful to convert to dimensionless variables. *Polytropes* (nonrotating, spherical fluid bodies with the polytropic equation of state $P = K\rho^{1+1/n}$) are a nice example.

(a) Combine the two equations of stellar structure (13.16) to obtain a single second-order differential equation for P and ρ as functions of r.

(b) In the equation from part (a) set $P = K\rho^{1+1/n}$ to obtain a nonlinear, second-order differential equation for $\rho(r)$.

(c) It is helpful to change dependent variables from $\rho(r)$ to some other variable, call it $\theta(r)$, so chosen that the quantity being differentiated is linear in θ and the only θ nonlinearity is in the driving term. Show that choosing $\rho \propto \theta^n$ achieves this.

(d) It is helpful to choose the proportionality constant in $\rho \propto \theta^n$ in such a way that θ is dimensionless and takes the value 1 at the polytrope's center and 0 at its surface. This is achieved by setting

$$\rho = \rho_c \theta^n, \tag{13.21a}$$

where ρ_c is the polytrope's (possibly unknown) central density.

(e) Similarly, it is helpful to make the independent variable r dimensionless by setting $r = a\xi$, where a is a constant with dimensions of length. The value of a should be chosen wisely, so as to simplify the differential equation as much as possible. Show that the choice

$$r = a\xi, \quad \text{where } a = \left[\frac{(n+1)K\rho_c^{(1/n-1)}}{4\pi G}\right]^{1/2}, \tag{13.21b}$$

brings the differential equation into the form

$$\frac{1}{\xi^2}\frac{d}{d\xi}\xi^2\frac{d\theta}{d\xi} = -\theta^n. \tag{13.22}$$

Equation (13.22) is called the *Lane-Emden equation of stellar structure*, after Jonathan Homer Lane and Jacob Robert Emden, who introduced and explored it near the end of the nineteenth century. There is an extensive literature on solutions of the Lane-Emden equation (see, e.g., Chandrasekhar, 1939, Chap. 4; Shapiro and Teukolsky, 1983, Sec. 3.3).

(f) Explain why the Lane-Emden equation (13.22) must be solved subject to the following boundary conditions (where $\theta' \equiv d\theta/d\xi$):

$$\theta = 1 \quad \text{and} \quad \theta' = 0 \quad \text{at } \xi = 0. \tag{13.23}$$

(g) One can integrate the Lane-Emden equation, numerically or analytically, outward from $\xi = 0$ until some radius ξ_1 at which θ (and thus also ρ and P) goes to zero. That is the polytrope's surface. Its physical radius is then $R = a\xi_1$, and its mass is $M = \int_0^R 4\pi\rho r^2 dr$, which is readily shown to be $M = 4\pi a^3 \rho_c \xi_1^2 |\theta'(\xi_1)|$.

Then, using the value of a given in Eq. (13.21b), we have:

$$R = \left[\frac{(n+1)K}{4\pi G}\right]^{1/2} \rho_c^{(1-n)/2n} \xi_1,$$

$$M = 4\pi \left[\frac{(n+1)K}{4\pi G}\right]^{3/2} \rho_c^{(3-n)/2n} \xi_1^2 |\theta'(\xi_1)|, \quad (13.24a)$$

whence

$$M = 4\pi R^{(3-n)/(1-n)} \left[\frac{(n+1)K}{4\pi G}\right]^{n/(n-1)} \xi_1^{(n+1)/(n-1)} |\theta'(\xi_1)|. \quad (13.24b)$$

(h) When one converts a problem into dimensionless variables that satisfy some differential or algebraic equation(s) and then expresses physical quantities in terms of the dimensionless variables, the resulting expressions describe how the physical quantities scale with one another. As an example, Jupiter and Saturn are both made up of an H-He fluid that is well approximated by a polytrope of index $n = 1$, $P = K\rho^2$, with the same constant K. Use the information that $M_J = 2 \times 10^{27}$ kg, $R_J = 7 \times 10^4$ km, and $M_S = 6 \times 10^{26}$ kg to estimate the radius of Saturn. For $n = 1$, the Lane-Emden equation has a simple analytical solution: $\theta = \sin\xi/\xi$. Compute the central densities of Jupiter and Saturn.

13.3.3 Rotating Fluids

The equation of hydrostatic equilibrium (13.4) and the applications of it discussed above are valid only when the fluid is static in a nonrotating reference frame. However, they are readily extended to bodies that rotate rigidly with some uniform angular velocity $\boldsymbol{\Omega}$ relative to an inertial frame. In a frame that corotates with the body, the fluid will have vanishing velocity $\mathbf{v}$ (i.e., will be static), and the equation of hydrostatic equilibrium (13.4) will be changed only by the addition of the centrifugal force per unit volume:

$$\boxed{\nabla P = \rho(\mathbf{g} + \mathbf{g}_{\text{cen}}) = -\rho\nabla(\Phi + \Phi_{\text{cen}}).} \quad (13.25)$$

hydrostatic equilibrium in corotating reference frame

Here

$$\boxed{\mathbf{g}_{\text{cen}} = -\boldsymbol{\Omega} \times (\boldsymbol{\Omega} \times \mathbf{r}) = -\nabla\Phi_{\text{cen}}} \quad (13.26)$$

is the centrifugal acceleration, $\rho\mathbf{g}_{\text{cen}}$ is the centrifugal force per unit volume, and

$$\boxed{\Phi_{\text{cen}} = -\frac{1}{2}(\boldsymbol{\Omega} \times \mathbf{r})^2} \quad (13.27)$$

is a *centrifugal potential* whose gradient is equal to the centrifugal acceleration when $\boldsymbol{\Omega}$ is constant. The centrifugal potential can be regarded as an augmentation of the gravitational potential Φ. Indeed, *in the presence of uniform rotation, all hydrostatic theorems* [e.g., Eqs. (13.7) and (13.8)] *remain valid in the corotating reference frame with Φ replaced by $\Phi + \Phi_{\text{cen}}$.*

hydrostatic theorems for rotating body

We can illustrate this by considering the shape of a spinning fluid planet. Let us suppose that almost all the planet's mass is concentrated in its core, so the gravitational potential $\Phi = -GM/r$ is unaffected by the rotation. The surface of the planet must be an equipotential of $\Phi + \Phi_{\text{cen}}$ [coinciding with the zero-pressure isobar; cf. the sentence following Eq. (13.8), with $\Phi \to \Phi + \Phi_{\text{cen}}$]. The contribution of the centrifugal potential at the equator is $-\Omega^2 R_e^2/2$, and at the pole it is zero. The difference in the gravitational potential Φ between the equator and the pole is $\simeq g(R_e - R_p)$ where R_e and R_p are the equatorial and polar radii, respectively, and g is the gravitational acceleration at the planet's surface. Therefore, adopting this centralized-mass model and requiring that $\Phi + \Phi_{\text{cen}}$ be the same at the equator as at the pole, we estimate the difference between the polar and equatorial radii to be

centrifugal flattening

$$R_e - R_p \simeq \frac{\Omega^2 R^2}{2g}. \tag{13.28a}$$

Earth, although not a fluid, is unable to withstand large shear stresses, because its shear strain cannot exceed the yield strain of rock, ~ 0.001; see Sec. 11.3.2 and Table 11.1. Since the heights of the tallest mountains are also governed by the yield strain, Earth's surface will not deviate from its equipotential by more than the maximum height of a mountain, $\simeq 9$ km.

If, for Earth, we substitute $g \simeq 10$ m s^{-2}, $R \simeq 6 \times 10^6$ m, and $\Omega \simeq 7 \times 10^{-5}$ rad s^{-1} into Eq. (13.28a), we obtain $R_e - R_p \simeq 10$ km, about half the correct value of 21 km. The reason for this discrepancy lies in our assumption that all the planet's mass resides at its center. In fact, the mass is distributed fairly uniformly in radius and, in particular, some mass is found in the equatorial bulge. This deforms the gravitational equipotential surfaces from spheres to ellipsoids, which accentuates the flattening. If, following Newton (in his *Principia Mathematica*, published in 1687), we assume that Earth has uniform density, then the flattening estimate is 2.5 times larger than our centralized-mass estimate (Ex. 13.5) (i.e., $R_e - R_p \simeq 25$ km), in fairly good agreement with Earth's actual shape.

EXERCISES

Exercise 13.5 *Example: Shape of a Constant-Density, Spinning Planet*
(a) Show that the spatially variable part of the gravitational potential for a uniform-density, nonrotating planet can be written as $\Phi = 2\pi G\rho r^2/3$, where ρ is the density.
(b) Hence argue that the gravitational potential for a slowly spinning planet can be written in the form

$$\Phi = \frac{2\pi G\rho r^2}{3} + Ar^2 P_2(\mu),$$

where A is a constant, and P_2 is the Legendre polynomial with argument $\mu = \sin(\text{latitude})$. What happens to the P_1 term?

(c) Give an equivalent expansion for the potential outside the planet.

(d) Now transform into a frame spinning with the planet, and add the centrifugal potential to give a total potential.

(e) By equating the potential and its gradient at the planet's surface, show that the difference between the polar and the equatorial radii is given by

$$R_e - R_p \simeq \frac{5\Omega^2 R^2}{4g}, \qquad (13.28b)$$

where g is the gravitational acceleration at the surface. Note that this is 5/2 times the answer for a planet whose mass is concentrated at its center [Eq. (13.28a)].

Exercise 13.6 *Problem: Shapes of Stars in a Tidally Locked Binary System*
Consider two stars with the same mass M orbiting each other in a circular orbit with diameter (separation between the stars' centers) a. Kepler's laws tell us that the stars' orbital angular velocity is $\Omega = \sqrt{2GM/a^3}$. Assume that each star's mass is concentrated near its center, so that everywhere except near a star's center the gravitational potential, in an inertial frame, is $\Phi = -GM/r_1 - GM/r_2$ with r_1 and r_2 the distances of the observation point from the center of star 1 and star 2. Suppose that the two stars are "tidally locked": tidal gravitational forces have driven them each to rotate with rotational angular velocity equal to the orbital angular velocity Ω. (The Moon is tidally locked to Earth, which is why it always keeps the same face toward Earth.) Then in a reference frame that rotates with angular velocity Ω, each star's gas will be at rest, $\mathbf{v} = 0$.

(a) Write down the total potential $\Phi + \Phi_{cen}$ for this binary system in the rotating frame.

(b) Using Mathematica, Maple, Matlab, or some other computer software, plot the equipotentials $\Phi + \Phi_{cen} = $ const for this binary in its orbital plane, and use these equipotentials to describe the shapes that these stars will take if they expand to larger and larger radii (with a and M held fixed). You should obtain a sequence in which the stars, when compact, are well separated and nearly round. As they grow, tidal gravity elongates them ultimately into tear-drop shapes, followed by merger into a single, highly distorted star. With further expansion, the merged star starts flinging mass off into the surrounding space (a process not included in this hydrostatic analysis).

13.4 Conservation Laws

As a foundation for the transition from hydrostatics to hydrodynamics [i.e., to situations with nonzero fluid velocity $\mathbf{v}(\mathbf{x}, t)$], we give a general discussion of Newtonian conservation laws, focusing especially on the conservation of mass and of linear momentum.

We begin with the differential law of mass conservation,

mass conservation: differential form

$$\frac{\partial \rho}{\partial t} + \nabla \cdot (\rho \mathbf{v}) = 0, \tag{13.29}$$

which we met and used in our study of elastic media [Eq. (12.2c)]. Eq. (13.29) is the obvious analog of the laws of conservation of charge, $\partial \rho_e/\partial t + \nabla \cdot \mathbf{j} = 0$, and of particles, $\partial n/\partial t + \nabla \cdot \mathbf{S} = 0$, which we met in Chap. 1 [Eqs. (1.30)]. In each case the law has the form $(\partial/\partial t)$(density of something) $+\nabla \cdot$ (flux of that something) $= 0$. In fact, this form is universal for a differential conservation law.

Each Newtonian differential conservation law has a corresponding integral conservation law (Sec. 1.8), which we obtain by integrating the differential law over some arbitrary 3-dimensional volume $\mathcal{V}$, for example, the volume used in Fig. 13.3 to discuss Archimedes' law: $(d/dt) \int_\mathcal{V} \rho dV = \int_\mathcal{V} (\partial \rho/\partial t) dV = -\int_\mathcal{V} \nabla \cdot (\rho \mathbf{v}) dV$. Applying Gauss's law to the last integral, we obtain

mass conservation: integral form

$$\frac{d}{dt} \int_\mathcal{V} \rho dV = -\int_{\partial \mathcal{V}} \rho \mathbf{v} \cdot d\mathbf{\Sigma}, \tag{13.30}$$

where $\partial \mathcal{V}$ is the closed surface bounding $\mathcal{V}$. The left-hand side is the rate of change of mass inside the region $\mathcal{V}$. The right-hand side is the rate at which mass flows into $\mathcal{V}$ through $\partial \mathcal{V}$ (since $\rho \mathbf{v}$ is the mass flux, and the inward pointing surface element is $-d\mathbf{\Sigma}$). This is the same argument, connecting differential to integral conservation laws, as we gave when deriving Eqs. (1.29) and (1.30) for electric charge and for particles, but going in the opposite direction. And this argument depends in no way on whether the flowing material is a fluid or not. The mass conservation laws (13.29) and (13.30) are valid for any kind of material.

Writing the differential conservation law in the form (13.29), where we monitor the changing density at a given location in space rather than moving with the material, is called the *Eulerian* approach. There is an alternative *Lagrangian* approach to mass conservation, in which we focus on changes of density as measured by somebody who moves, locally, with the material (i.e., with velocity $\mathbf{v}$). We obtain this approach by differentiating the product $\rho \mathbf{v}$ in Eq. (13.29) to obtain

mass conservation: alternative differential form

$$\frac{d\rho}{dt} = -\rho \nabla \cdot \mathbf{v}, \tag{13.31}$$

where

convective (advective) time derivative

$$\frac{d}{dt} \equiv \frac{\partial}{\partial t} + \mathbf{v} \cdot \nabla. \tag{13.32}$$

The operator d/dt is known as the *convective time derivative* (or *advective time derivative*) and crops up often in continuum mechanics. Its physical interpretation is very simple. Consider first the partial derivative $(\partial/\partial t)_{\mathbf{x}}$. This is the rate of change of some quantity [the density ρ in Eq. (13.31)] at a fixed point $\mathbf{x}$ in space in some reference frame. In other words, if there is motion, $\partial/\partial t$ compares this quantity at the same point $\mathcal{P}$ in space for two different points in the material: one that is at $\mathcal{P}$ at time $t + dt$; the other that was at $\mathcal{P}$ at the earlier time t. By contrast, the convective time derivative d/dt follows the motion, taking the difference in the value of the quantity at successive times at the same point in the moving matter. It is the time derivative for the Lagrangian approach.

For a fluid, the Lagrangian approach can also be expressed in terms of fluid elements. Consider a small fluid element with a bounding surface attached to the fluid, and denote its volume by V. The mass inside the fluid element is $M = \rho V$. As the fluid flows, this mass must be conserved, so $dM/dt = (d\rho/dt)V + \rho(dV/dt) = 0$, which we can rewrite as

$$\frac{d\rho}{dt} = -\rho \frac{dV/dt}{V}. \tag{13.33}$$

Lagrangian formulation of mass conservation

Comparing with Eq. (13.31), we see that

$$\boxed{\nabla \cdot \mathbf{v} = \frac{dV/dt}{V}.} \tag{13.34}$$

Thus, the divergence of $\mathbf{v}$ is the fractional rate of increase of a fluid element's volume. Notice that this is just the time derivative of our elastostatic equation $\Delta V/V = \nabla \cdot \boldsymbol{\xi} = \Theta$ [Eq. (11.7)] (since $\mathbf{v} = d\boldsymbol{\xi}/dt$). Correspondingly we denote

$$\boxed{\nabla \cdot \mathbf{v} \equiv \theta = d\Theta/dt,} \tag{13.35}$$

rate of expansion of fluid

and call it the fluid's *rate of expansion*.

Equation (13.29), $\partial \rho/\partial t + \nabla \cdot (\rho \mathbf{v}) = 0$, is our model for Newtonian conservation laws. It says that there is a quantity (in this case mass) with a certain density (ρ), and a certain flux ($\rho \mathbf{v}$), and this quantity is neither created nor destroyed. The temporal derivative of the density (at a fixed point in space) added to the divergence of the flux must vanish. Of course, not all physical quantities have to be conserved. If there were sources or sinks of mass, then these would be added to the right-hand side of Eq. (13.29).

Now turn to momentum conservation. The (Newtonian) law of momentum conservation must take the standard conservation-law form $(\partial/\partial t)$(momentum density) $+\nabla \cdot$ (momentum flux) $= 0$.

If we just consider the *mechanical momentum* associated with the motion of mass, its density is the vector field $\rho \mathbf{v}$. There can also be other forms of momentum density (e.g., electromagnetic), but these do not enter into Newtonian fluid mechanics. For fluids, as for an elastic medium (Chap. 12), the momentum density is simply $\rho \mathbf{v}$.

13.4 Conservation Laws

The momentum flux is more interesting and rich. Quite generally it is, by definition, the stress tensor **T**, and the differential conservation law states

momentum conservation: Eulerian formulation

$$\frac{\partial(\rho \mathbf{v})}{\partial t} + \nabla \cdot \mathbf{T} = 0 \qquad (13.36)$$

[Eq. (1.36)]. For an elastic medium, $\mathbf{T} = -K\Theta\mathbf{g} - 2\mu\boldsymbol{\Sigma}$ [Eq. (11.18)], and the conservation law (13.36) gives rise to the elastodynamic phenomena that we explored in Chap. 12. For a fluid we build up **T** piece by piece.

We begin with the rate $d\mathbf{p}/dt$ that mechanical momentum flows through a small element of surface area $d\boldsymbol{\Sigma}$, from its back side to its front. The rate that mass flows through is $\rho \mathbf{v} \cdot d\boldsymbol{\Sigma}$, and we multiply that mass by its velocity **v** to get the momentum flow rate: $d\mathbf{p}/dt = (\rho \mathbf{v})(\mathbf{v} \cdot d\boldsymbol{\Sigma})$. This rate of flow of momentum is the same thing as a force $\mathbf{F} = d\mathbf{p}/dt$ acting across $d\boldsymbol{\Sigma}$; so it can be computed by inserting $d\boldsymbol{\Sigma}$ into the second slot of a "mechanical" stress tensor $\mathbf{T}_m$: $d\mathbf{p}/dt = \mathbf{T}_m(\_\_, d\boldsymbol{\Sigma})$ [cf. the definition (1.32) of the stress tensor]. By writing these two expressions for the momentum flow in index notation, $dp_i/dt = (\rho v_i)v_j d\Sigma_j = T_{m\,ij}d\Sigma_j$, we read off the mechanical stress tensor: $T_{m\,ij} = \rho v_i v_j$:

mechanical stress tensor

$$\mathbf{T}_m = \rho \mathbf{v} \otimes \mathbf{v}. \qquad (13.37)$$

This tensor is symmetric (as any stress tensor must be; Sec. 1.9), and it obviously is the flux of mechanical momentum, since it has the form (momentum density) $\otimes$ (velocity).

force f per unit volume acting on fluid

Let us denote by **f** the net force per unit volume that acts on the fluid. Then instead of writing momentum conservation in the usual Eulerian differential form (13.36), we can write it as

$$\frac{\partial(\rho \mathbf{v})}{\partial t} + \nabla \cdot \mathbf{T}_m = \mathbf{f} \qquad (13.38)$$

(conservation law with a source on the right-hand side). Inserting $\mathbf{T}_m = \rho \mathbf{v} \otimes \mathbf{v}$ into this equation, converting to index notation, using the rule for differentiating products, and combining with the law of mass conservation, we obtain the *Lagrangian law*

momentum conservation: Lagrangian formulation

$$\rho \frac{d\mathbf{v}}{dt} = \mathbf{f}. \qquad (13.39)$$

Here $d/dt = \partial/\partial t + \mathbf{v} \cdot \nabla$ is the convective time derivative (i.e., the time derivative moving with the fluid); so this equation is just Newton's $\mathbf{F} = m\mathbf{a}$, per unit volume. For the equivalent equations (13.38) and (13.39) of momentum conservation to also be equivalent to the Eulerian formulation (13.36), there must be a stress tensor $\mathbf{T}_f$ such that

stress tensor for the force f

$$\mathbf{f} = -\nabla \cdot \mathbf{T}_f; \quad \text{and} \quad \mathbf{T} = \mathbf{T}_m + \mathbf{T}_f. \qquad (13.40)$$

Then Eq. (13.38) becomes the Eulerian conservation law (13.36).

Evidently, a knowledge of the stress tensor $\mathbf{T}_f$ for some material is equivalent to a knowledge of the force density $\mathbf{f}$ that acts on the material. It often turns out to be much easier to figure out the form of the stress tensor, for a given situation, than the form of the force. Correspondingly, as we add new pieces of physics to our fluid analysis (e.g., isotropic pressure, viscosity, gravity, magnetic forces), an efficient way to proceed at each stage is to insert the relevant physics into the stress tensor $\mathbf{T}_f$ and then evaluate the resulting contribution $\mathbf{f} = -\boldsymbol{\nabla} \cdot \mathbf{T}_f$ to the force density and thence to the Lagrangian law of force balance (13.39). At each step, we get out in the form $\mathbf{f} = -\boldsymbol{\nabla} \cdot \mathbf{T}_f$ the physics that we put into $\mathbf{T}_f$.

stress tensor for force density f

force density as divergence of stress tensor

There may seem something tautological about the procedure (13.40) by which we went from the Lagrangian $\mathbf{F} = m\mathbf{a}$ equation (13.39) to the Eulerian conservation law (13.36), (13.40). The $\mathbf{F} = m\mathbf{a}$ equation makes it look like mechanical momentum is not conserved in the presence of the force density $\mathbf{f}$. But we make it be conserved by introducing the momentum flux $\mathbf{T}_f$. It is almost as if we regard conservation of momentum as a principle to be preserved at all costs, and so every time there appears to be a momentum deficit, we simply define it as a bit of the momentum flux. However, this is not the whole story. What is important is that the force density $\mathbf{f}$ can always be expressed as the divergence of a stress tensor; that fact is central to the nature of force and of momentum conservation. An erroneous formulation of the force would not necessarily have this property, and so no differential conservation law could be formulated. Therefore, the fact that we can create elastostatic, thermodynamic, viscous, electromagnetic, gravitational, etc. contributions to some grand stress tensor (that go to zero outside the regions occupied by the relevant matter or fields)—as we shall do in the coming chapters—is significant. It affirms that our physical model is complete at the level of approximation to which we are working.

We can proceed in the same way with energy conservation as we have with momentum. There is an energy density $U(\mathbf{x}, t)$ for a fluid and an energy flux $\mathbf{F}(\mathbf{x}, t)$, and they obey a conservation law with the standard form

energy density U, energy flux F, and energy conservation

$$\boxed{\frac{\partial U}{\partial t} + \boldsymbol{\nabla} \cdot \mathbf{F} = 0.} \qquad (13.41)$$

At each stage in our buildup of fluid dynamics (adding, one by one, the influences of compressional energy, viscosity, gravity, and magnetism), we can identify the relevant contributions to U and $\mathbf{F}$ and then grind out the resulting conservation law (13.41). At each stage we get out the physics that we put into U and $\mathbf{F}$.

13.5 The Dynamics of an Ideal Fluid

We now use the general conservation laws of the previous section to derive the fundamental equations of fluid dynamics. We do so in several stages. In this section and Sec. 13.6, we confine our attention to ideal fluids—flows for which it is safe to ignore dissipative processes (viscosity and thermal conductivity) and thus for which the entropy of a fluid element remains constant with time. In Sec. 13.7, we introduce the effects of viscosity and diffusive heat flow.

13.5.1 Mass Conservation

As we have seen, mass conservation takes the (Eulerian) form $\partial\rho/\partial t + \nabla \cdot (\rho \mathbf{v}) = 0$ [Eq. (13.29)], or equivalently the (Lagrangian) form $d\rho/dt = -\rho \nabla \cdot \mathbf{v}$ [Eq. (13.31)], where $d/dt = \partial/\partial t + \mathbf{v} \cdot \nabla$ is the convective time derivative [i.e., the time derivative moving with the fluid; Eq. (13.32)].

As we shall see in Sec. 13.6, when flow speeds are small compared to the speed of sound and the effects of gravity are sufficiently modest, the density of a fluid element remains nearly constant: $|(1/\rho)d\rho/dt| = |\nabla \cdot \mathbf{v}| \ll 1/\tau$, where τ is the fluid flow's characteristic timescale. It is then a good approximation to rewrite the law of mass conservation as $\nabla \cdot \mathbf{v} = 0$, which is the *incompressible approximation*.

incompressible approximation

13.5.2 Momentum Conservation

For an ideal fluid, the only forces that can act are those of gravity and of the fluid's isotropic pressure P. We have already met and discussed the contribution of P to the stress tensor, $\mathbf{T} = P\mathbf{g}$, when dealing with elastic media (Chap. 11) and in hydrostatics (Sec. 13.3). The gravitational force density $\rho\mathbf{g}$ is so familiar that it is easier to write it down than the corresponding gravitational contribution to the stress. Correspondingly, we can most easily write momentum conservation in the form

momentum conservation for an ideal fluid

$$\frac{\partial(\rho\mathbf{v})}{\partial t} + \nabla \cdot \mathbf{T} = \rho\mathbf{g}; \quad \text{or} \quad \frac{\partial(\rho\mathbf{v})}{\partial t} + \nabla \cdot (\rho\mathbf{v} \otimes \mathbf{v} + P\mathbf{g}) = \rho\mathbf{g}, \quad (13.42)$$

where the stress tensor is given by

stress tensor for an ideal fluid

$$\boxed{\mathbf{T} = \rho\mathbf{v} \otimes \mathbf{v} + P\mathbf{g} \quad \text{for an ideal fluid}} \quad (13.43)$$

[cf. Eqs. (13.37), (13.38), and (13.4)]. The first term, $\rho\mathbf{v} \otimes \mathbf{v}$, is the mechanical momentum flux (also called the *kinetic* stress), and the second, $P\mathbf{g}$, is that associated with the fluid's pressure.

In most of our Newtonian applications, the gravitational field $\mathbf{g}$ will be externally imposed (i.e., it will be produced by some object, e.g., Earth that is different from the fluid we are studying). However, the law of momentum conservation remains the same [Eq. (13.42)], independently of what produces gravity—the fluid, or an external body, or both. And independently of its source, one can write the stress tensor $\mathbf{T}_g$ for the gravitational field $\mathbf{g}$ in a form presented and discussed in the Track-Two Box 13.4 later in the chapter—a form that has the required property $-\nabla \cdot \mathbf{T}_g = \rho\mathbf{g} =$ (the gravitational force density).

13.5.3 Euler Equation

The Euler equation is the equation of motion that one gets out of the momentum conservation law (13.42) for an ideal fluid by performing the differentiations and invoking mass conservation (13.29):[3]

$$\boxed{\frac{d\mathbf{v}}{dt} \equiv \frac{\partial \mathbf{v}}{\partial t} + (\mathbf{v} \cdot \nabla)\mathbf{v} = -\frac{\nabla P}{\rho} + \mathbf{g} \quad \text{for an ideal fluid.}} \quad (13.44)$$

Euler equation (momentum conservation) for an ideal fluid

The Euler equation has a simple physical interpretation: $d\mathbf{v}/dt$ is the convective derivative of the velocity (i.e., the derivative moving with the fluid), which means it is the acceleration felt by the fluid. This acceleration has two causes: gravity, $\mathbf{g}$, and the pressure gradient, ∇P. In a hydrostatic situation, $\mathbf{v} = 0$, the Euler equation reduces to the equation of hydrostatic equilibrium: $\nabla P = \rho \mathbf{g}$ [Eq. (13.4)].

In Cartesian coordinates, the Euler equation (13.44) and mass conservation $d\rho/dt + \rho \nabla \cdot \mathbf{v} = 0$ [Eq. (13.31)] comprise four equations in five unknowns, ρ, P, v_x, v_y, and v_z. The remaining fifth equation gives P as a function of ρ. For an ideal fluid, this equation comes from the fact that the entropy of each fluid element is conserved (because there is no mechanism for dissipation):

$$\frac{ds}{dt} = 0, \quad (13.45)$$

entropy conservation

together with an equation of state for the pressure in terms of the density and the entropy: $P = P(\rho, s)$. In practice, the equation of state is often well approximated by incompressibility, $\rho = $ const, or by a polytropic relation, $P = K(s)\rho^{1+1/n}$ [Eq. (13.18)].

13.5.4 Bernoulli's Theorem

Bernoulli's theorem is well known. Less well appreciated are the conditions under which it is true. To deduce these, we must first introduce a kinematic quantity known as the *vorticity*:

$$\boxed{\boldsymbol{\omega} \equiv \nabla \times \mathbf{v}.} \quad (13.46)$$

vorticity

The physical interpretation of vorticity is simple. Consider a small fluid element. As it moves and deforms over a tiny period of time δt, each bit of fluid inside it undergoes a tiny displacement $\boldsymbol{\xi} = \mathbf{v}\delta t$. The gradient of that displacement field can be decomposed into an expansion, rotation, and shear (as we discussed in the context of an elastic medium in Sec. 11.2.2). The vectorial angle of the rotation is $\boldsymbol{\phi} = \frac{1}{2}\nabla \times \boldsymbol{\xi}$ [Eq. (11.9b)]. The time derivative of that vectorial angle, $d\boldsymbol{\phi}/dt = \frac{1}{2}d\boldsymbol{\xi}/dt = \frac{1}{2}\nabla \times \mathbf{v}$, is obviously the fluid element's rotational angular velocity; hence *the vorticity $\boldsymbol{\omega} = \nabla \times \mathbf{v}$ is twice*

angular velocity of a fluid element

---

[3]. This equation was first derived in 1757 by the Swiss mathematician and physicist Leonhard Euler—the same Euler who formulated the theory of buckling of a compressed beam (Sec. 11.6.1).

the angular velocity of rotation of a fluid element. Vorticity plays a major role in fluid mechanics, as we shall see in Chap. 14.

To derive Bernoulli's theorem (with the aid of vorticity), we begin with the Euler equation for an ideal fluid, $d\mathbf{v}/dt = -(1/\rho)\nabla P + \mathbf{g}$. We express $\mathbf{g}$ as $-\nabla\Phi$ and convert the convective derivative of velocity (i.e., the acceleration) into its two parts $d\mathbf{v}/dt = \partial\mathbf{v}/\partial t + (\mathbf{v}\cdot\nabla)\mathbf{v}$. Then we rewrite $(\mathbf{v}\cdot\nabla)\mathbf{v}$ using the vector identity

$$\mathbf{v}\times\boldsymbol{\omega}\equiv\mathbf{v}\times(\nabla\times\mathbf{v})=\frac{1}{2}\nabla v^2-(\mathbf{v}\cdot\nabla)\mathbf{v}. \tag{13.47}$$

The result is

Bernoulli's theorem: most general version

$$\boxed{\frac{\partial\mathbf{v}}{\partial t}+\nabla\left(\frac{1}{2}v^2+\Phi\right)+\frac{\nabla P}{\rho}-\mathbf{v}\times\boldsymbol{\omega}=0.} \tag{13.48}$$

This is just the Euler equation written in a new form, but it is also *the most general version of Bernoulli's theorem*—valid for any ideal fluid. Two special cases are of interest.

BERNOULLI'S THEOREM FOR STEADY FLOW OF AN IDEAL FLUID

Since the fluid is ideal, dissipation (due to viscosity and heat flow) can be ignored, so the entropy is constant following the flow: $ds/dt = (\mathbf{v}\cdot\nabla)s = 0$. When, in addition, the flow is steady, meaning $\partial(\text{everything})/\partial t = 0$, the thermodynamic identity $dh = Tds + dP/\rho$ [Eq. (3) of Box 13.2] combined with $ds/dt = 0$ implies

$$(\mathbf{v}\cdot\nabla)P=\rho(\mathbf{v}\cdot\nabla)h. \tag{13.49}$$

Dotting the velocity $\mathbf{v}$ into the most general Bernoulli theorem (13.48) and invoking Eq. (13.49) and $\partial\mathbf{v}/\partial t = 0$, we obtain

$$\frac{dB}{dt}=(\mathbf{v}\cdot\nabla)B=0, \tag{13.50}$$

where

Bernoulli function

$$\boxed{B\equiv\frac{1}{2}v^2+h+\Phi.} \tag{13.51}$$

Bernoulli's theorem for steady flow of an ideal fluid

Equation (13.50) states that *in a steady flow of an ideal fluid, the Bernoulli function B, like the entropy, is conserved moving with a fluid element.* This is the most elementary form of the Bernoulli theorem.

Let us define *streamlines*, analogous to lines of force of a magnetic field, by the differential equations

streamlines

$$\frac{dx}{v_x}=\frac{dy}{v_y}=\frac{dz}{v_z}. \tag{13.52}$$

In the language of Sec. 1.5, these are just the integral curves of the (steady) velocity field; they are also the spatial world lines of fluid elements. Equation (13.50) states: *In a steady flow of an ideal fluid, the Bernoulli function B is constant along streamlines.*

BOX 13.3. FLOW VISUALIZATION

Various methods are used to visualize fluid flows. One way is via *streamlines*, which are the integral curves of the velocity field **v** at a given time [Eq. (13.52)]. Streamlines are the analog of magnetic field lines. They coincide with the *paths* of individual fluid elements if the flow is steady, but not if the flow is time dependent. In general, the paths are the solutions of the equation $d\mathbf{x}/dt = \mathbf{v}(\mathbf{x}, t)$. These paths are the analog of particle trajectories in mechanics.

Another type of flow line is a *streak*. Monitoring streaks is a common way of visualizing a flow experimentally. Streaks are usually produced by introducing some colored or fluorescent tracer into the flow continuously at some fixed release point $\mathbf{x}_r$, and observing the locus of the tracer at some fixed time, say, t_0. Each point on the streak can be parameterized by the common release point $\mathbf{x}_r$, the common time of observation t_0, and the time t_r at which its marker was released, $\mathbf{x}(\mathbf{x}_r, t_r; t_0)$; so the streak is the parameterized curve $\mathbf{x}(t_r) = \mathbf{x}(\mathbf{x}_r, t_r; t_0)$.

Examples of streamlines, paths, and streaks are sketched below.

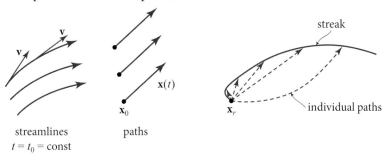

Streamlines are a powerful way to visualize fluid flows. There are other ways, sketched in Box 13.3.

The Bernoulli function $B = \frac{1}{2}v^2 + h + \Phi = \frac{1}{2}v^2 + u + P/\rho + \Phi$ has a simple physical meaning. It is the fluid's total energy density (kinetic plus internal plus potential) per unit mass, plus the work $P(1/\rho)$ that must be done to inject a unit mass of fluid (with volume $1/\rho$) into surrounding fluid that has pressure P. This goes hand in hand with the enthalpy $h = u + P/\rho$ being the *injection energy* (per unit mass) in the absence of kinetic and potential energy; see the last part of Ex. 5.5. This meaning of B leads to the following physical interpretation of the constancy of B for a stationary, ideal flow.

In a steady flow of an ideal fluid, consider a *stream tube* made of a bundle of streamlines (Fig. 13.5). A fluid element with unit mass occupies region $\mathcal{A}$ of the stream

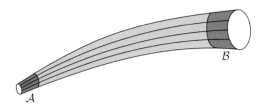

FIGURE 13.5 A stream tube used to explain the Bernoulli theorem for stationary flow of an ideal fluid.

tube at some early time and has moved into region $\mathcal{B}$ at some later time. When it vacates region $\mathcal{A}$, the fluid element carries an energy B [including the energy $P(1/\rho)$ it acquires by being squeezed out of $\mathcal{A}$ by the pressure of the surrounding fluid]. When it moves into region $\mathcal{B}$, it similarly carries the total injection energy B. Because the flow is steady, the energy it extracts from $\mathcal{A}$ must be precisely what it needs to occupy $\mathcal{B}$ (i.e., B must be constant along the stream tube, and hence also along each streamline in the tube).

The most immediate consequence of Bernoulli's theorem for steady flow is that, if gravity is having no significant effect, then the enthalpy falls when the speed increases, and conversely. This is just the conversion of internal (injection) energy into bulk kinetic energy, and conversely. For our ideal fluid, entropy must be conserved moving with a fluid element, so the first law of thermodynamics says $dh = T ds + dP/\rho = dP/\rho$. Therefore, as the speed increases and h decreases, P will also decrease.

This behavior is the foundation for the *Pitot tube*, a simple device used to measure the air speed of an aircraft (Fig. 13.6). The Pitot tube extends out from the side of the aircraft, all the way through a boundary layer of slow-moving air and into the bulk flow. There it bends into the flow. The Pitot tube is actually two tubes: (i) an outer tube with several orifices along its sides, past which the air flows with its incoming speed V and pressure P, so the pressure inside that tube is also P; and (ii) an inner tube with a small orifice at its end, where the flowing air is brought essentially to rest (to *stagnation*). At this stagnation point and inside the orifice's inner tube, the pressure, by Bernoulli's theorem with Φ and ρ both essentially constant, is the *stagnation pressure*: $P_{\text{stag}} = P + \frac{1}{2}\rho V^2$. The pressure difference $\Delta P = P_{\text{stag}} - P$ between the two tubes is measured by an instrument called a *manometer*, from which the air speed is computed as $V = (2\Delta P/\rho)^{1/2}$. If $V \sim 100$ m s^{-1} and $\rho \sim 1$ kg m^{-3}, then $\Delta P \sim 5{,}000$ N m$^{-3} \sim 0.05$ atmospheres.

In this book, we shall meet many other applications of the Bernoulli theorem for steady, ideal flows.

BERNOULLI'S THEOREM FOR IRROTATIONAL FLOW OF AN IDEAL, ISENTROPIC FLUID

An even more specialized type of flow is one that is *isentropic* (so s is the same everywhere) and *irrotational* (meaning its vorticity vanishes everywhere), as well as ideal.

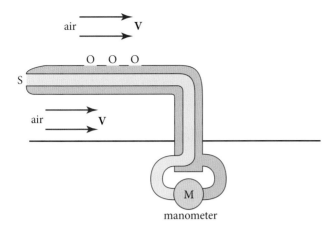

FIGURE 13.6 Schematic illustration of a Pitot tube used to measure air speed. The manometer M measures the pressure difference between the stagnation point S and the orifices O.

(In Sec. 14.2, we shall learn that if an incompressible flow initially is irrotational and it encounters no walls and experiences no significant viscous stresses, then it remains irrotational.) As $\boldsymbol{\omega} = \nabla \times \mathbf{v}$ vanishes, we can follow the electrostatic precedent and introduce a *velocity potential* $\psi(\mathbf{x}, t)$, so that at any time,

$$\boxed{\mathbf{v} = \nabla \psi \quad \text{for an irrotational flow.}} \tag{13.53}$$

velocity potential for irrotational (vorticity-free) flow

The first law of thermodynamics [Eq. (3) of Box 13.2] implies that $\nabla h = T\nabla s + (1/\rho)\nabla P$. Therefore, in an isentropic flow, $\nabla P = \rho \nabla h$. Imposing these conditions on Eq. (13.48), we obtain, for a (possibly unsteady) isentropic, irrotational flow:

$$\nabla \left(\frac{\partial \psi}{\partial t} + B \right) = 0. \tag{13.54}$$

Bernoulli's theorem for isentropic, irrotational flow of an ideal fluid

Thus *in an isentropic, irrotational flow of an ideal fluid, the quantity $\partial \psi/\partial t + B$ is constant everywhere.* (If $\partial \psi/\partial t + B$ is a function of time, we can absorb that function into ψ without affecting $\mathbf{v}$, leaving it constant in time as well as in space.) Of course, if the flow is steady, so $\partial(\text{everything})/\partial t = 0$, then B itself is constant.

EXERCISES

Exercise 13.7 *Problem: A Hole in My Bucket*
There's a hole in my bucket. How long will it take to empty? (Try an experiment, and if the time does not agree with the estimate, explain why not.)

Exercise 13.8 *Problem: Rotating Planets, Stars, and Disks*
Consider a stationary, axisymmetric planet, star, or disk differentially rotating under the action of a gravitational field. In other words, the motion is purely in the azimuthal direction.

(a) Suppose that the fluid has a *barotropic* equation of state $P = P(\rho)$. Write down the equations of hydrostatic equilibrium, including the centrifugal force, in cylindrical polar coordinates. Hence show that the angular velocity must be constant on surfaces of constant cylindrical radius. This is called *von Zeipel's theorem*. (As an application, Jupiter is differentially rotating and therefore might be expected to have similar rotation periods at the same latitudes in the north and the south. This is only roughly true, suggesting that the equation of state is not completely barotropic.)

(b) Now suppose that the structure is such that the surfaces of constant entropy per unit mass and angular momentum per unit mass coincide. (This state of affairs can arise if slow convection is present.) Show that the Bernoulli function (13.51) is also constant on these surfaces. [Hint: Evaluate ∇B.]

Exercise 13.9 ***Problem: Crocco's Theorem*

(a) Consider steady flow of an ideal fluid. The Bernoulli function (13.51) is conserved along streamlines. Show that the variation of B across streamlines is given by Crocco's theorem:

$$\nabla B = T\nabla s + \mathbf{v} \times \boldsymbol{\omega}. \tag{13.55}$$

(b) As an example, consider the air in a tornado. In the tornado's core, the velocity vanishes; it also vanishes beyond the tornado's outer edge. Use Crocco's theorem to show that the pressure in the core is substantially different from that at the outer edge. Is it lower, or is it higher? How does this explain the ability of a tornado to make the walls of a house explode? For more detail, see Ex. 14.5.

Exercise 13.10 *Problem: Cavitation (Suggested by P. Goldreich)*
A hydrofoil moves with speed V at a depth $D = 3$ m below the surface of a lake; see Fig. 13.7. Estimate how fast V must be to make the water next to the hydrofoil boil. [This boiling, which is called *cavitation*, results from the pressure P trying to go negative (see, e.g., Batchelor, 2000, Sec. 6.12; Potter, Wiggert, and Ramadan, 2012, Sec. 8.3.4).] [Note: For a more accurate value of the speed V that triggers cavitation,

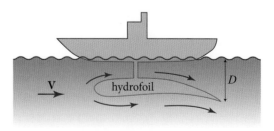

FIGURE 13.7 Water flowing past a hydrofoil as seen in the hydrofoil's rest frame.

one would have to compute the velocity field $\mathbf{v}(\mathbf{x})$ around the hydrofoil—for example, using the method of Ex. 14.17—and identify the maximum value of $v = |\mathbf{v}|$ near the hydrofoil's surface.]

Exercise 13.11 *Example: Collapse of a Bubble*
Suppose that a spherical bubble has just been created in the water above the hydrofoil in the previous exercise. Here we analyze its collapse—the decrease of the bubble's radius $R(t)$ from its value R_o at creation, using the incompressible approximation (which is rather good in this situation). This analysis is an exercise in solving the Euler equation.

(a) Introduce spherical polar coordinates with origin at the center of the bubble, so the collapse entails only radial fluid motion, $\mathbf{v} = v(r, t)\mathbf{e}_r$. Show that the incompressibility approximation $\nabla \cdot \mathbf{v} = 0$ implies that the radial velocity can be written in the form $v = w(t)/r^2$. Then use the radial component of the Euler equation (13.44) to show that

$$\frac{1}{r^2}\frac{dw}{dt} + v\frac{\partial v}{\partial r} + \frac{1}{\rho}\frac{\partial P}{\partial r} = 0.$$

At fixed time t, integrate this outward from the bubble surface at radius $R(t)$ to a large enough radius that the bubble's influence is no longer felt. Thereby obtain

$$\frac{-1}{R}\frac{dw}{dt} + \frac{1}{2}\dot{R}^2(R) = \frac{P_0}{\rho},$$

where P_0 is the ambient pressure and $-\dot{R}(R)$ is the speed of collapse of the bubble's surface when its radius is R. Assuming vanishing collapse speed when the bubble is created, $\dot{R}(R_o) = 0$, show that

$$\dot{R}(R) = -\left(\frac{2P_0}{3\rho}\right)^{1/2}\left[\left(\frac{R_0}{R}\right)^3 - 1\right]^{1/2},$$

which can be integrated to get $R(t)$.

(b) Suppose that bubbles formed near the pressure minimum on the surface of the hydrofoil are swept back onto a part of the surface where the pressure is much larger. By what factor R_o/R must the bubbles collapse if they are to create stresses that inflict damage on the hydrofoil?

Pistol shrimp can create collapsing bubbles and use the shock waves to stun their prey. A modification of this solution is important in interpreting the fascinating phenomenon of *sonoluminescence* (Brenner, Hilgenfeldt, and Lohse, 2002), which arises when fluids are subjected to high-frequency acoustic waves that create oscillating bubbles. The temperatures inside these bubbles can get so large that the air becomes ionized and radiates.

13.5.5 Conservation of Energy

As well as imposing conservation of mass and momentum, we must also address energy conservation in its general form (by contrast with the specialized version of energy conservation inherent in Bernoulli's theorem for a stationary, ideal flow).

In general, energy conservation is needed for determining the temperature T of a fluid, which in turn is needed to compute the pressure $P(\rho, T)$. So far in our treatment of fluid dynamics, we have finessed this issue by either postulating some relationship between the pressure P and the density ρ (e.g., the polytropic relation $P = K\rho^\gamma$) or by focusing on the flow of ideal fluids, where the absence of dissipation guarantees the entropy is constant moving with the flow, so that $P = P(\rho, s)$ with constant s. In more general situations, one cannot avoid confronting energy conservation. Moreover, even for ideal fluids, understanding how energy is conserved is often useful for gaining physical insight—as we have seen in our discussion of Bernoulli's theorem.

The most fundamental formulation of the law of energy conservation is Eq. (13.41): $\partial U/\partial t + \nabla \cdot \mathbf{F} = 0$. To explore its consequences for an ideal fluid, we must insert the appropriate ideal-fluid forms of the energy density U and energy flux $\mathbf{F}$.

When (for simplicity) the fluid is in an externally produced gravitational field Φ, its energy density is obviously

energy density for ideal fluid with external gravity

$$\boxed{U = \rho \left(\frac{1}{2}v^2 + u + \Phi\right)} \quad \text{for ideal fluid with external gravity.} \quad (13.56)$$

Here the three terms are kinetic, internal, and gravitational energy. When the fluid participates in producing gravity and one includes the energy of the gravitational field itself, the energy density is a bit more subtle; see the Track-Two Box 13.4.

In an external gravitational field, one might expect the energy flux to be $\mathbf{F} = U\mathbf{v}$, but this is not quite correct. Consider a bit of surface area dA orthogonal to the direction in which the fluid is moving (i.e., orthogonal to $\mathbf{v}$). The fluid element that crosses dA during time dt moves through a distance $dl = vdt$, and as it moves, the fluid behind this element exerts a force PdA on it. That force, acting through the distance dl, feeds an energy $dE = (PdA)dl = PvdAdt$ across dA; the corresponding energy flux across dA has magnitude $dE/dAdt = Pv$ and points in the $\mathbf{v}$ direction, so it contributes $P\mathbf{v}$ to the energy flux $\mathbf{F}$. This contribution is missing from our initial guess $\mathbf{F} = U\mathbf{v}$. We explore its importance at the end of this subsection. When it is added to our guess, we obtain for the total energy flux

energy flux for ideal fluid with external gravity

$$\boxed{\mathbf{F} = \rho \mathbf{v}\left(\frac{1}{2}v^2 + h + \Phi\right)} \quad \text{for ideal fluid with external gravity.} \quad (13.57)$$

Here $h = u + P/\rho$ is the enthalpy per unit mass (cf. Box 13.2). Inserting Eqs. (13.56) and (13.57) into the law of energy conservation (13.41), and requiring that the external

BOX 13.4. SELF-GRAVITY T2

In the text, we mostly treat the gravitational field as externally imposed and independent of the fluid. This approximation is usually a good one. However, it is inadequate for planets and stars, whose self-gravity is crucial. It is easiest to discuss the modifications due to the fluid's self-gravitational effects by amending the conservation laws.

As long as we work in the domain of Newtonian physics, the mass conservation equation (13.29) is unaffected by self-gravity. However, we included the gravitational force per unit volume $\rho \mathbf{g}$ as a source of momentum in the momentum conservation law (13.42). It would fit much more neatly in our formalism if we could express it as the divergence of a gravitational stress tensor $\mathbf{T}_g$. To see that this is indeed possible, use Poisson's equation $\nabla \cdot \mathbf{g} = -4\pi G \rho$ (which embodies self-gravity) to write

$$\nabla \cdot \mathbf{T}_g = -\rho \mathbf{g} = \frac{(\nabla \cdot \mathbf{g})\mathbf{g}}{4\pi G} = \frac{\nabla \cdot [\mathbf{g} \otimes \mathbf{g} - \frac{1}{2}g^2 \mathbf{g}]}{4\pi G},$$

so

$$\boxed{\mathbf{T}_g = \frac{\mathbf{g} \otimes \mathbf{g} - \frac{1}{2}g^2 \mathbf{g}}{4\pi G}.} \qquad (1)$$

Readers familiar with classical electromagnetic theory will notice an obvious and understandable similarity to the Maxwell stress tensor [Eqs. (1.38) and (2.80)], whose divergence equals the Lorentz force density.

What of the gravitational momentum density? We expect that it can be related to the gravitational energy density using a Lorentz transformation. That is to say, it is $O(v/c^2)$ times the gravitational energy density, where v is some characteristic speed. However, in the Newtonian approximation, the speed of light c is regarded as infinite, and so we should expect the gravitational momentum density to be identically zero in Newtonian theory—and indeed it is. We therefore can write the full equation of motion (13.42), including gravity, as a conservation law:

$$\frac{\partial(\rho \mathbf{v})}{\partial t} + \nabla \cdot \mathbf{T}_{\text{total}} = 0, \qquad (2)$$

where $\mathbf{T}_{\text{total}}$ includes $\mathbf{T}_g$.

Now consider energy conservation. We have seen in the text that in a constant, external gravitational field, the fluid's total energy density U and flux $\mathbf{F}$ are given by Eqs. (13.56) and (13.57), respectively. In a general

(continued)

BOX 13.4. (continued)

situation, we must add to these some field energy density and flux. On dimensional grounds, these must be $U_{\text{field}} \propto \mathbf{g}^2/G$ and $\mathbf{F}_{\text{field}} \propto \Phi_{,t}\mathbf{g}/G$ (where $\mathbf{g} = -\nabla\Phi$). The proportionality constants can be deduced by demanding that for an ideal fluid in the presence of gravity, the law of energy conservation when combined with mass conservation, momentum conservation, and the first law of thermodynamics, lead to $ds/dt = 0$ (no dissipation in, so no dissipation out); see Eq. (13.59) and associated discussion. The result (Ex. 13.13) is

$$U = \rho\left(\frac{1}{2}v^2 + u + \Phi\right) + \frac{g^2}{8\pi G}, \tag{3}$$

$$\mathbf{F} = \rho\mathbf{v}\left(\frac{1}{2}v^2 + h + \Phi\right) + \frac{1}{4\pi G}\frac{\partial\Phi}{\partial t}\mathbf{g}. \tag{4}$$

Actually, there is an ambiguity in how the gravitational energy is localized. This ambiguity arises physically because one can transform away the gravitational acceleration $\mathbf{g}$, at any point in space, by transforming to a reference frame that falls freely there. Correspondingly, it turns out, one can transform away the gravitational energy density at any desired point in space. This possibility is embodied mathematically in the possibility of adding to the energy flux $\mathbf{F}$ the time derivative of $\alpha\Phi\nabla\Phi/(4\pi G)$ and adding to the energy density U minus the divergence of this quantity (where α is an arbitrary constant), while preserving energy conservation $\partial U/\partial t + \nabla \cdot \mathbf{F} = 0$. Thus the following choice of energy density and flux is just as good as Eqs. (3) and (4); both satisfy energy conservation:

$$U = \rho\left(\frac{1}{2}v^2 + u + \Phi\right) + \frac{g^2}{8\pi G} - \alpha\nabla\cdot\left(\frac{\Phi\nabla\Phi}{4\pi G}\right)$$

$$= \rho\left[\frac{1}{2}v^2 + u + (1-\alpha)\Phi\right] + (1-2\alpha)\frac{g^2}{8\pi G}, \tag{5}$$

$$\mathbf{F} = \rho\mathbf{v}\left(\frac{1}{2}v^2 + h + \Phi\right) + \frac{1}{4\pi G}\frac{\partial\Phi}{\partial t}\mathbf{g} + \alpha\frac{\partial}{\partial t}\left(\frac{\Phi\nabla\Phi}{4\pi G}\right)$$

$$= \rho\mathbf{v}\left(\frac{1}{2}v^2 + h + \Phi\right) + (1-\alpha)\frac{1}{4\pi G}\frac{\partial\Phi}{\partial t}\mathbf{g} - \frac{\alpha}{4\pi G}\Phi\frac{\partial\mathbf{g}}{\partial t}. \tag{6}$$

[Here we have used the gravitational field equation $\nabla^2\Phi = 4\pi G\rho$ and $\mathbf{g} = -\nabla\Phi$.] Note that the choice $\alpha = 1/2$ puts all the energy density into the

(continued)

BOX 13.4. (continued)

$\rho\Phi$ term, while the choice $\alpha = 1$ puts all the energy density into the field term $\mathbf{g}^2$. In Ex. 13.14 it is shown that the total gravitational energy of an isolated system is independent of the arbitrary parameter α, as it must be on physical grounds.

A full understanding of the nature and limitations of the concept of gravitational energy requires the general theory of relativity (Part VII). The relativistic analog of the arbitrariness of Newtonian energy localization is an arbitrariness in the gravitational "stress-energy pseudotensor" (see, e.g., Misner, Thorne, and Wheeler, 1973, Sec. 20.3).

---

gravity be static (time independent), so the work it does on the fluid is conservative, we obtain the following ideal-fluid equation of energy balance:

$$\frac{\partial}{\partial t}\left[\rho\left(\frac{1}{2}v^2 + u + \Phi\right)\right] + \nabla \cdot \left[\rho\mathbf{v}\left(\frac{1}{2}v^2 + h + \Phi\right)\right] = 0$$

energy conservation for ideal fluid with external gravity

for an ideal fluid and static external gravity. (13.58)

When the gravitational field is dynamical or is being generated by the fluid itself (or both), we must use a more complete gravitational energy density and stress; see Box 13.4.

By combining the law of energy conservation (13.58) with the corresponding laws of momentum (13.29) and mass conservation (13.42), and using the first law of thermodynamics $dh = Tds + (1/\rho)dP$, we obtain the remarkable result that the entropy per unit mass is conserved moving with the fluid:

$$\boxed{\frac{ds}{dt} = 0 \quad \text{for an ideal fluid.}} \tag{13.59}$$

entropy conservation for ideal fluid

The same conclusion can be obtained when the gravitational field is dynamical and not external (cf. Box 13.4 and Ex. 13.13), so no statement about gravity is included with this equation. This entropy conservation should not be surprising. If we put no dissipative processes into the energy density, energy flux, or stress tensor, then we get no dissipation out. Moreover, the calculation that leads to Eq. (13.59) ensures that, *so long as we take full account of mass and momentum conservation, then the full and sole content of the law of energy conservation for an ideal fluid is $ds/dt = 0$.*

Table 13.1 summarizes our formulas for the density and flux of mass, momentum, and energy in an ideal fluid with externally produced gravity.

TABLE 13.1: Densities and fluxes of mass, momentum, and energy for an ideal fluid in an externally produced gravitational field

Quantity	Density	Flux
Mass	ρ	$\rho \mathbf{v}$
Momentum	$\rho \mathbf{v}$	$\mathbf{T} = P\mathbf{g} + \rho \mathbf{v} \otimes \mathbf{v}$
Energy	$U = (\tfrac{1}{2}v^2 + u + \Phi)\rho$	$\mathbf{F} = (\tfrac{1}{2}v^2 + h + \Phi)\rho \mathbf{v}$

EXERCISES

Exercise 13.12 *Example: Joule-Kelvin Cooling*

A good illustration of the importance of the Pv term in the energy flux is provided by the *Joule-Kelvin method* commonly used to cool gases (Fig. 13.8). Gas is driven from a high-pressure chamber 1 through a nozzle or porous plug into a low-pressure chamber 2, where it expands and cools.

(a) Using the energy flux (13.57), including the Pv term contained in h, show that a mass ΔM ejected through the nozzle carries a total energy ΔE_1 that is equal to the enthalpy ΔH_1 that this mass had while in chamber 1.

(b) This ejected gas expands and crashes into the gas of chamber 2, temporarily going out of statistical (thermodynamic) equilibrium. Explain why, after it has settled down into statistical equilibrium as part of the chamber-2 gas, the total energy it has deposited into chamber 2 is its equilibrium enthalpy ΔH_2. Thereby conclude that the enthalpy per unit mass is the same in the two chambers, $h_1 = h_2$.

(c) From $h_1 = h_2$, show that the temperature drop between the two chambers is

$$\Delta T = \int_{P_1}^{P_2} \mu_{\text{JK}} dP, \qquad (13.60)$$

where $\mu_{\text{JK}} \equiv (\partial T/\partial P)_h$ is the Joule-Kelvin coefficient (also called Joule-Thomson). A straightforward thermodynamic calculation yields the identity

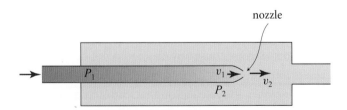

FIGURE 13.8 Schematic illustration of Joule-Kelvin cooling of a gas. Gas flows steadily through a nozzle from a chamber at high pressure P_1 to one at low pressure P_2. The flow proceeds at constant enthalpy. Work done against attractive intermolecular forces leads to cooling. The efficiency of cooling can be enhanced by exchanging heat between the two chambers. Gases can also be liquified in this manner.

$$\mu_{\text{JK}} \equiv \left(\frac{\partial T}{\partial P}\right)_h = -\frac{1}{\rho^2 c_p}\left(\frac{\partial(\rho T)}{\partial T}\right)_P. \qquad (13.61)$$

(d) Show that the Joule-Kelvin coefficient of an ideal gas vanishes. Therefore, the cooling must arise because of the attractive forces (van der Waals forces; Sec. 5.3.2) between the molecules, which are absent in an ideal gas. When a real gas expands, work is done against these forces, and the gas therefore cools.

Exercise 13.13 *Derivation: No Dissipation "in" Means No Dissipation "out" and Verification of the Claimed Gravitational Energy Density and Flux* T2
Consider an ideal fluid interacting with a (possibly dynamical) gravitational field that the fluid itself generates via $\nabla^2\Phi = 4\pi G\rho$. For this fluid, take the law of energy conservation, $\partial U/\partial t + \nabla \cdot \mathbf{F} = 0$, and from it subtract the scalar product of $\mathbf{v}$ with the law of momentum conservation, $\mathbf{v} \cdot [\partial(\rho\mathbf{v})/\partial t + \nabla \cdot \mathbf{T}]$; then simplify using the law of mass conservation and the first law of thermodynamics, to obtain $\rho ds/dt = 0$. In your computation, use for U and $\mathbf{F}$ the expressions given in Eqs. (3) and (4) of Box 13.4. This calculation tells us two things. (i) The law of energy conservation for an ideal fluid reduces simply to conservation of entropy moving with the fluid; we have put no dissipative physics into the fluxes of momentum and energy, so we get no dissipation out. (ii) The gravitational energy density and flux contained in Eqs. (3) and (4) of Box 13.4 must be correct, since they guarantee that gravity does not alter this "no dissipation in, no dissipation out" result.

Exercise 13.14 *Example: Gravitational Energy* T2
Integrate the energy density U of Eq. (5) of Box 13.4 over the interior and surroundings of an isolated gravitating system to obtain the system's total energy. Show that the gravitational contribution to this total energy (i) is independent of the arbitrariness (parameter α) in the energy's localization, and (ii) can be written in the following forms:

$$\boxed{E_g = \int dV \frac{1}{2}\rho\Phi = -\frac{1}{8\pi G}\int dV\, g^2 = \frac{-G}{2}\int\int dV\, dV' \frac{\rho(\mathbf{x})\rho(\mathbf{x}')}{|\mathbf{x}-\mathbf{x}'|}.} \qquad (13.62)$$

Interpret each of these expressions physically.

13.6 Incompressible Flows

A common assumption made when discussing the fluid dynamics of highly subsonic flows is that the density is constant—that the fluid is incompressible. This is a natural approximation to make when dealing with a liquid like water, which has a very large bulk modulus. It is a bit of a surprise that it is also useful for flows of gases, which are far more compressible under static conditions.

To see its validity, suppose that we have a flow in which the characteristic length L over which the fluid variables P, ρ, v, and so forth vary is related to the characteristic timescale T over which they vary by $L \lesssim vT$. In this case, we can compare the magnitude of the various terms in the Euler equation (13.44) to obtain an estimate of the magnitude of the pressure variation:

$$\underbrace{\frac{\partial \mathbf{v}}{\partial t}}_{v/T} + \underbrace{(\mathbf{v} \cdot \nabla)\mathbf{v}}_{v^2/L} = -\underbrace{\frac{\nabla P}{\rho}}_{\delta P/\rho L} - \underbrace{\nabla \Phi}_{\delta \Phi/L}. \tag{13.63}$$

Multiplying through by L and using $L/T \lesssim v$, we obtain $\delta P/\rho \sim v^2 + |\delta \Phi|$. The variation in pressure will be related to the variation in density by $\delta P \sim C^2 \delta \rho$, where $C = \sqrt{(\partial P/\partial \rho)_s}$ is the sound speed (Sec. 16.5), and we drop constants of order unity when making these estimates. Inserting this into our expression for δP, we obtain the estimate for the fractional density fluctuation:

$$\boxed{\frac{\delta \rho}{\rho} \sim \frac{v^2}{C^2} + \frac{\delta \Phi}{C^2}.} \tag{13.64}$$

incompressible approximation

Therefore, *if the fluid speeds are highly subsonic* ($v \ll C$) *and the gravitational potential does not vary greatly along flow lines* ($|\delta \Phi| \ll C^2$), *then we can ignore the density variations moving with the fluid when solving for the velocity field.* More specifically, since $\rho^{-1} d\rho/dt = -\nabla \cdot \mathbf{v} = -\theta$ [Eq. (13.35)], we can make the incompressible approximation

$$\nabla \cdot \mathbf{v} \simeq 0 \tag{13.65}$$

(which means that the velocity field is *solenoidal*, like a magnetic field; i.e., is expressible as the curl of some potential). This argument breaks down when we are dealing with sound waves for which $L \sim CT$.

For air at atmospheric temperature, the speed of sound is $C \sim 300$ m/s, which is very fast compared to most flow speeds one encounters, so most flows are incompressible.

It should be emphasized, though, that the incompressible approximation for the velocity field, $\nabla \cdot \mathbf{v} \simeq 0$, does not imply that the density variation can be neglected in all other contexts. A particularly good example is provided by convection flows, which are driven by buoyancy, as we shall discuss in Chap. 18.

Incompressibility is a weaker condition than requiring the density to be constant everywhere; for example, the density varies substantially from Earth's center to its surface, but if the material inside Earth were moving more or less on surfaces of constant radius, the flow would be incompressible.

We restrict ourselves to incompressible flows throughout the next two chapters and then abandon incompressibility in subsequent chapters on fluid dynamics.

13.7 Viscous Flows with Heat Conduction

13.7.1 Decomposition of the Velocity Gradient into Expansion, Vorticity, and Shear

It is an observational fact that many fluids, when they flow, develop a shear stress (also called a *viscous stress*). Honey pouring off a spoon is a nice example. Most fluids, however, appear to flow quite freely; for example, a cup of tea appears to offer little resistance to stirring other than the inertia of the water. In such cases, it might be thought that viscous effects only produce a negligible correction to the flow's details. However, this is not so.

One of the main reasons is that most flows touch solid bodies, at whose surfaces the velocity must vanish. This leads to the formation of boundary layers, whose thickness and behavior are controlled by viscous forces. The boundary layers in turn can exert a controlling influence on the bulk flow (where the viscosity is negligible); for example, they can trigger the development of turbulence in the bulk flow—witness the stirred tea. For details see Chaps. 14 and 15.

We must therefore augment our equations of fluid dynamics to include viscous stresses. Our formal development proceeds in parallel to that used in elasticity, with the velocity field $\mathbf{v} = d\boldsymbol{\xi}/dt$ replacing the displacement field $\boldsymbol{\xi}$. We decompose the velocity gradient tensor $\nabla \mathbf{v}$ into its irreducible tensorial parts: a *rate of expansion* θ; a symmetric, trace-free *rate of shear* tensor $\boldsymbol{\sigma}$; and an antisymmetric *rate of rotation* tensor $\mathbf{r}$:

$$\nabla \mathbf{v} = \frac{1}{3}\theta \mathbf{g} + \boldsymbol{\sigma} + \mathbf{r}. \tag{13.66}$$

Note that we use lowercased symbols to distinguish the fluid case from its elastic counterpart: $\theta = d\Theta/dt$, $\boldsymbol{\sigma} = d\boldsymbol{\Sigma}/dt$, $\mathbf{r} = d\mathbf{R}/dt$. Proceeding directly in parallel to the treatment in Sec. 11.2.2 and Box 11.2, we can invert Eq. (13.66) to obtain

$$\theta = \nabla \cdot \mathbf{v}, \tag{13.67a}$$

rates of expansion, shear, and rotation

$$\sigma_{ij} = \frac{1}{2}(v_{i;j} + v_{j;i}) - \frac{1}{3}\theta g_{ij}, \tag{13.67b}$$

$$r_{ij} = \frac{1}{2}(v_{i;j} - v_{j;i}) = -\frac{1}{2}\epsilon_{ijk}\omega^k, \tag{13.67c}$$

where $\boldsymbol{\omega} = 2d\boldsymbol{\phi}/dt$ is the vorticity, which we introduced and discussed in Sec. 13.5.4.

EXERCISES

Exercise 13.15 **Example: Kinematic Interpretation of Vorticity*
Consider a velocity field with nonvanishing curl. Define a locally orthonormal basis at a point in the velocity field, so that one basis vector, $\mathbf{e}_x$, is parallel to the vorticity. Now imagine the remaining two basis vectors as being frozen into the fluid. Show that they will both rotate about the axis defined by $\mathbf{e}_x$ and that the vorticity will be the sum of their angular velocities (i.e., twice the average of their angular velocities).

13.7.2 Navier-Stokes Equation

Although, as we have emphasized, a fluid at rest does not exert a shear stress, and this distinguishes it from an elastic solid, a fluid in motion can resist shear in the

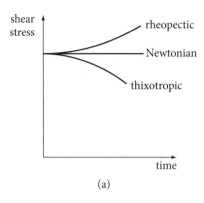
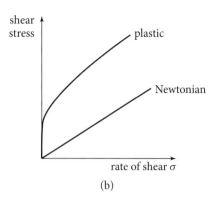

FIGURE 13.9 Some examples of non-Newtonian behavior in fluids. (a) In a Newtonian fluid the shear stress is proportional to the rate of shear σ and does not vary with time when σ is constant as here. However, some substances, such as paint, flow more freely with time and are said to be *thixotropic*. Microscopically, the long, thin paint molecules gradually become aligned with the flow, which reduces their resistance to shear. The opposite behavior is exhibited by *rheopectic* substances such as ink and some lubricants. (b) An alternative type of non-Newtonian behavior is exhibited by various plastics, where a threshold stress is needed before flow will commence.

velocity field. It has been found experimentally that in most fluids the magnitude of this shear stress is linearly related to the velocity gradient. This law, due to Hooke's contemporary, Isaac Newton, is the analog of the linear relation between stress and strain that we used in our discussion of elasticity. Fluids that obey this law are known as *Newtonian*. (Some examples of non-Newtonian fluid behavior are shown in Fig. 13.9.) We analyze only Newtonian fluids in this book.

Fluids are usually isotropic. (Important exceptions include *smectic* liquid crystals.) In this book we restrict ourselves to isotropic fluids. By analogy with the theory of elasticity, we describe the linear relation between stress and rate of strain using two constants called the coefficients of *bulk* and *shear* viscosity, denoted ζ and η, respectively. We write the viscous contribution to the stress tensor as

coefficients of bulk and shear viscosity

viscous stress tensor

$$\mathbf{T}_{\text{vis}} = -\zeta \theta \mathbf{g} - 2\eta \boldsymbol{\sigma}, \tag{13.68}$$

by analogy to Eq. (11.18), $\mathbf{T}_{\text{elas}} = -K\Theta\mathbf{g} - 2\mu\boldsymbol{\Sigma}$, for an elastic solid. Here as there, shear-free rotation about a point does not produce a resistive stress.

If we include this viscous contribution in the stress tensor, then the law of momentum conservation $\partial(\rho\mathbf{v})/\partial t + \boldsymbol{\nabla} \cdot \mathbf{T} = \rho\mathbf{g}$ gives the following generalization of Euler's equation (13.44):

Navier-Stokes equation: general form

$$\rho \frac{d\mathbf{v}}{dt} = -\boldsymbol{\nabla} P + \rho\mathbf{g} + \boldsymbol{\nabla}(\zeta\theta) + 2\boldsymbol{\nabla} \cdot (\eta\boldsymbol{\sigma}). \tag{13.69}$$

This is the *Navier-Stokes equation*, and the last two terms are the viscous force density.

For incompressible flows (e.g., when the flow is highly subsonic; Sec. 13.6), θ can be approximated as zero, so the bulk viscosity can be ignored. The viscosity coefficient

TABLE 13.2: Approximate kinematic viscosity for common fluids

Quantity	Kinematic viscosity v (m² s⁻¹)
Water	10^{-6}
Air	10^{-5}
Glycerine	10^{-3}
Blood	3×10^{-6}

η generally varies in space far more slowly than the shear $\boldsymbol{\sigma}$, and so can be taken outside the divergence. In this case, Eq. (13.69) simplifies to

$$\boxed{\frac{d\mathbf{v}}{dt} = -\frac{\nabla P}{\rho} + \mathbf{g} + v\nabla^2 \mathbf{v},} \qquad (13.70)$$

Navier-Stokes equation for incompressible flow

where

$$\boxed{v \equiv \frac{\eta}{\rho}} \qquad (13.71)$$

kinematic shear viscosity coefficient

is known as the *kinematic viscosity*, by contrast to η, which is often called the *dynamic viscosity*. Equation (13.70) is the commonly quoted form of the Navier-Stokes equation; it is the form that we shall almost always use. Approximate values of the kinematic viscosity for common fluids are given in Table 13.2.

13.7.3 Molecular Origin of Viscosity

13.7.3

We can distinguish gases from liquids microscopically. In a gas, a molecule of mass m travels a distance of order its *mean free path* λ before it collides. If there is a shear in the fluid (Fig. 13.10), then the molecule, traveling in the y direction, on average

estimate of shear viscosity

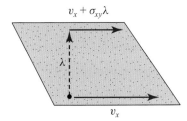

FIGURE 13.10 Molecular origin of viscosity in a gas. A molecule travels a distance λ in the y direction between collisions. Its mean x velocity at its point of origin is that of the fluid there, v_x, which differs from the mean x velocity at its next collision by $-\sigma_{xy}\lambda$. As a result, it transports a momentum $-m\sigma_{xy}\lambda$ to the location of its next collision.

13.7 Viscous Flows with Heat Conduction

will transfer an x momentum of about $-m\lambda\sigma_{xy}$ between collision points. If there are n molecules per unit volume traveling with mean thermal speeds v_{th}, then the transferred momentum crossing a unit area in unit time is $T_{xy} \sim -nmv_{\text{th}}\lambda\sigma_{xy}$, from which, by comparison with Eq. (13.68), we can extract an estimate of the coefficient of shear viscosity:

$$\boxed{\eta \simeq \frac{1}{3}\rho v_{\text{th}}\lambda.} \tag{13.72}$$

Here the numerical coefficient of 1/3 (which arises from averaging over molecular directions and speeds) has been inserted to agree with a proper kinetic-theory calculation; see Ex. 3.19 in Chap. 3. Note from Eq. (13.72) that in a gas, where the mean thermal kinetic energy $\frac{3}{2}k_B T$ is $\sim m v_{\text{th}}^2$, the coefficient of viscosity will increase with temperature as $\nu \propto T^{1/2}$.

In a liquid, where the molecules are less mobile, it is the close intermolecular attraction that produces the shear stress. The ability of molecules to slide past one another increases rapidly with their thermal activation, causing typical liquid viscosity coefficients to fall dramatically with rising temperature.

EXERCISES

Exercise 13.16 *Problem: Mean Free Path*
Estimate the collision mean free path of the air molecules around you. Hence verify the estimate for the kinematic viscosity of air given in Table 13.2.

13.7.4 Energy Conservation and Entropy Production

The viscous stress tensor represents an additional momentum flux that can do work on the fluid at a rate $\mathbf{T}_{\text{vis}} \cdot \mathbf{v}$ per unit area. Therefore a contribution

$$\boxed{\mathbf{F}_{\text{vis}} = \mathbf{T}_{\text{vis}} \cdot \mathbf{v}} \tag{13.73}$$

is made to the energy flux, just like the term $P\mathbf{v}$ appearing (as part of the $\rho\mathbf{v}h$) in Eq. (13.57). Diffusive heat flow (thermal conductivity) can also contribute to the energy flux; its contribution is [Eq. (3.70b)]

viscous energy flux

energy flux for heat conduction

$$\boxed{\mathbf{F}_{\text{cond}} = -\kappa\nabla T,} \tag{13.74}$$

coefficient of thermal conductivity κ

where κ is the coefficient of thermal conductivity. The molecules or particles that produce the viscosity and the heat flow also carry energy, but their energy density already is included in u, the total internal energy per unit mass, and their energy flux in $\rho\mathbf{v}h$. The total energy flux, including these contributions, is shown in Table 13.3, along with the energy density and the density and flux of momentum.

We see most clearly the influence of the dissipative viscous forces and heat conduction on energy conservation by inserting the energy density and flux from Table 13.3 into the law of energy conservation $\partial U/\partial t + \nabla \cdot \mathbf{F} = 0$, subtracting $\mathbf{v} \cdot [\partial(\rho\mathbf{v})/\partial t +$

TABLE 13.3: Densities and fluxes of mass, momentum, and energy for a dissipative fluid in an externally produced gravitational field

Quantity	Density	Flux
Mass	ρ	$\rho \mathbf{v}$
Momentum	$\rho \mathbf{v}$	$\mathbf{T} = \rho \mathbf{v} \otimes \mathbf{v} + P\mathbf{g} - \zeta\theta\mathbf{g} - 2\eta\boldsymbol{\sigma}$
Energy	$U = (\frac{1}{2}v^2 + u + \Phi)\rho$	$\mathbf{F} = (\frac{1}{2}v^2 + h + \Phi)\rho\mathbf{v} - \zeta\theta\mathbf{v} - 2\eta\boldsymbol{\sigma}\cdot\mathbf{v} - \kappa\nabla T$

Note: For self-gravitating systems, see Box 13.4.

$\nabla \cdot \mathbf{T} = \rho \mathbf{g}$] ($\mathbf{v}$ dotted into momentum conservation), and simplifying using mass conservation and the first law of thermodynamics. The result (Ex. 13.17) is the following equation for the evolution of entropy:

$$\boxed{T\left[\rho\left(\frac{ds}{dt}\right) + \nabla\cdot\left(\frac{\mathbf{F}_{\text{cond}}}{T}\right)\right] = \zeta\theta^2 + 2\eta\boldsymbol{\sigma}:\boldsymbol{\sigma} + \frac{\kappa}{T}(\nabla T)^2,} \quad (13.75)$$

Lagrangian equation for entropy evolution, for viscous, heat-conducting fluid

where $\boldsymbol{\sigma}:\boldsymbol{\sigma}$ is the double scalar product $\sigma_{ij}\sigma_{ij}$. The term in square brackets on the left-hand side represents an increase of entropy per unit volume moving with the fluid due to dissipation (the total increase minus that due to heat flowing conductively into a unit volume); multiplied by T, this is the dissipative increase in entropy density. This increase of random, thermal energy is being produced, on the right-hand side, by viscous heating (first two terms), and by the flow of heat $\mathbf{F}_{\text{cond}} = -\kappa\nabla T$ down a temperature gradient $-\nabla T$ (third term).

The dissipation equation (13.75) is the full content of the law of energy conservation for a dissipative fluid, when one takes account of mass conservation, momentum conservation, and the first law of thermodynamics.

We can combine this Lagrangian rate of viscous dissipation with the equation of mass conservation (13.29) to obtain an Eulerian differential equation for the entropy increase:

$$\boxed{\frac{\partial(\rho s)}{\partial t} + \nabla\cdot(\rho s\mathbf{v} - \kappa\nabla \ln T) = \frac{1}{T}\left(\zeta\theta^2 + 2\eta\boldsymbol{\sigma}:\boldsymbol{\sigma} + \frac{\kappa}{T}(\nabla T)^2\right).} \quad (13.76)$$

Eulerian equation for entropy evolution

The left-hand side of this equation describes the rate of change of entropy density plus the divergence of entropy flux. The right-hand side is therefore the rate of production of entropy per unit volume. Invoking the second law of thermodynamics, this quantity must be positive definite. Therefore the two coefficients of viscosity, like the bulk and shear moduli, must be positive, as must the coefficient of thermal conductivity κ (heat must flow from hotter regions to cooler ones).

In most laboratory and geophysical flows, thermal conductivity is unimportant, so we largely ignore it until our discussion of convection in Chap. 18.

13.7 Viscous Flows with Heat Conduction

EXERCISES

Exercise 13.17 *Derivation: Entropy Increase*

(a) Derive the Lagrangian equation (13.75) for the rate of increase of entropy in a dissipative fluid by carrying out the steps in the sentence preceding that equation. [Hint: If you have already done the analogous problem (Ex. 13.13) for an ideal fluid, then you need only compute the new terms that arise from the dissipative momentum flux $\mathbf{T}_{\mathrm{vis}} = -\zeta \theta \mathbf{g} - 2\eta \boldsymbol{\sigma}$ and dissipative energy fluxes $\mathbf{F}_{\mathrm{vis}} = \mathbf{T}_{\mathrm{vis}} \cdot \mathbf{v}$ and $\mathbf{F}_{\mathrm{cond}} = -\kappa \nabla T$. The sum of these new contributions, when you subtract $\mathbf{v} \cdot$ (momentum conservation) from energy conservation, is $\nabla \cdot \mathbf{F}_{\mathrm{cond}} + \nabla \cdot (\mathbf{T}_{\mathrm{vis}} \cdot \mathbf{v}) - \mathbf{v} \cdot (\nabla \cdot \mathbf{T}_{\mathrm{vis}})$; this must be added to the left-hand side of the result $\rho T ds/dt = 0$, Eq. (13.59), for an ideal fluid. When doing the algebra, it may be useful to decompose the gradient of the velocity into its irreducible tensorial parts, Eq. (13.66).]

(b) From the Lagrangian equation of entropy increase (13.75) derive the corresponding Eulerian equation (13.76).

13.7.5 Reynolds Number

The kinematic viscosity ν has dimensions length^2/time. This suggests that we quantify the importance of viscosity in a fluid flow by comparing ν with the product of the flow's characteristic velocity V and its characteristic lengthscale L. The dimensionless combination

Reynolds number

$$\mathrm{Re} = \frac{LV}{\nu} \tag{13.77}$$

is known as the *Reynolds number* and is the first of many dimensionless numbers we shall encounter in our study of fluid mechanics. Flows with Reynolds number much less than unity are dominated by viscosity. Large Reynolds number flows can also be strongly influenced by viscosity (as we shall see in later chapters), especially when the viscosity acts near boundaries—even though the viscous stresses are negligible over most of the flow's volume.

13.7.6 Pipe Flow

Let us now consider a simple example of viscous stresses at work, namely, the steady-state flow of blood down an artery.[4] We model the artery as a cylindrical pipe of radius a, through which the blood is forced by a time-independent pressure gradient. This is an example of what is called *pipe flow*.

Because gravity is unimportant and the flow is time independent, the Navier-Stokes equation (13.70) reduces to

---

4. We approximate the blood as a Newtonian fluid although, in reality, its shear viscosity η decreases at high rates of shear σ.

$$(\mathbf{v} \cdot \nabla)\mathbf{v} = -\frac{\nabla P}{\rho} + \nu \nabla^2 \mathbf{v}. \tag{13.78}$$

We assume that the flow is *laminar* (smooth, as is usually the case for blood in arteries), so **v** points solely along the z direction and is only a function of cylindrical radius ϖ. (This restriction is very important. As we discuss in Chap. 15, in other types of pipe flow, e.g., in crude oil pipelines, it often fails because the flow becomes turbulent, which has a major impact on the flow. In arteries, turbulence occasionally occurs and can lead to blood clots and a stroke!) Writing Eq. (13.78) in cylindrical coordinates, and denoting by $v(\varpi)$ the z component of velocity (the only nonvanishing component), we deduce that the nonlinear $\mathbf{v} \cdot \nabla \mathbf{v}$ term vanishes, and the pressure P is a function of z only and not of ϖ:

laminar flow contrasted with turbulent flow

$$\frac{1}{\varpi}\frac{d}{d\varpi}\left(\varpi \frac{dv}{d\varpi}\right) = \frac{1}{\eta}\frac{dP}{dz}. \tag{13.79}$$

Here dP/dz (which is negative) is the pressure gradient along the pipe, and $\eta = \nu\rho$ is the dynamic viscosity. This differential equation must be solved subject to the boundary conditions that the velocity gradient vanish at the center of the pipe and the velocity vanish at its walls. The solution is

velocity profile for laminar pipe flow

$$v(\varpi) = -\frac{dP}{dz}\frac{a^2 - \varpi^2}{4\eta}. \tag{13.80}$$

Using this velocity field, we can evaluate the pipe's total *flow rate*—volume per unit time—for (incompressible) blood volume:

Poiseuille's law for pipe flow

$$\mathcal{F} = \int_0^a v 2\pi \varpi \, d\varpi = -\frac{dP}{dz}\frac{\pi a^4}{8\eta}. \tag{13.81}$$

This relation is known as *Poiseuille's law* and because of the parabolic shape of the velocity profile (13.80), this pipe flow is sometimes called *parabolic Poiseuille flow*.

Now let us apply this result to a human body. The healthy adult heart, beating at about 60 beats per minute, pumps $\mathcal{F} \sim 5$ L min^{-1} (liters per minute) of blood into a circulatory system of many branching arteries that reach into all parts of the body and then return. This circulatory system can be thought of as like an electric circuit, and Poiseuille's law (13.81) is like the current-voltage relation for a small segment of wire in the circuit. The flow rate $\mathcal{F}$ in an arterial segment plays the role of the electric current I in the wire segment, the pressure gradient dP/dz is the voltage drop per unit length dV/dz, and $dR/dz \equiv 8\eta/\pi a^4$ is the resistance per unit length. Thus $-dP/dz = \mathcal{F}\, dR/dz$ [Eq. (13.81)] is equivalent to the voltage-current relation $-dV/dz = I\, dR/dz$. Moreover, just as the total current is conserved at a circuit branch point (sum of currents in equals sum of currents out), so also the total blood flow rate is conserved at an arterial branch point. These identical conservation laws and identical pressure-and-voltage-drop equations imply that the analysis of pressure changes and flow distributions in the body's many-branched circulatory system is the same as that of voltage changes and current distributions in an equivalent many-branched electrical circuit.

blood flow in human body

13.7 Viscous Flows with Heat Conduction

Because of the heart's periodic pumping, blood flow is *pulsatile* (pulsed; periodic) in the great vessels leaving the heart (the aorta and its branches); see Ex. 13.19. However, as the vessels divide into smaller and smaller arterial branches, the pulsatility becomes lost, so the flow is steady in the smallest vessels, the *arterioles*. Since a vessel's resistance per unit length scales with its radius as $dR/dz \propto 1/a^4$, and hence its pressure drop per unit length $-dP/dz = \mathcal{F}dR/dz$ also scales as $1/a^4$ [Eq. (13.81)], it should not be surprising that in a healthy human the circulatory system's pressure drop occurs primarily in the tiny arterioles, which have radii $a \sim 5$–$50 \, \mu$m.

The walls of these arterioles have circumferentially oriented smooth muscle structures, which are capable of changing the vessel radius a by as much as a factor ~ 2 or 3 in response to various stimuli (exercise, cold, stress, etc.). Note that a factor 3 radius increase means a factor $3^4 \sim 100$ decrease in pressure gradient at fixed flow rate! Accordingly, drugs designed to lower blood pressure do so by triggering radius changes in the arterioles. And anything you can do to keep your arteries from hardening, narrowing, or becoming blocked will help keep your blood pressure down. Eat salads!

EXERCISES

Exercise 13.18 *Problem: Steady Flow between Two Plates*
A viscous fluid flows steadily (no time dependence) in the z direction, with the flow confined between two plates that are parallel to the x-z plane and are separated by a distance $2a$. Show that the flow's velocity field is

$$v_z = -\frac{dP}{dz}\frac{a^2}{2\eta}\left[1 - \left(\frac{y}{a}\right)^2\right], \tag{13.82a}$$

and the mass flow rate (the discharge) per unit width of the plates is

$$\frac{dm}{dtdx} = -\frac{dP}{dz}\frac{2\rho a^3}{3\eta}. \tag{13.82b}$$

Here dP/dz (which is negative) is the pressure gradient along the direction of flow. (In Sec. 19.4 we return to this problem, augmented by a magnetic field and electric current, and discover great added richness.)

Exercise 13.19 *Example: Pulsatile Blood Flow*
Consider the pulsatile flow of blood through one of the body's larger arteries. The pressure gradient $dP/dz = P'(t)$ consists of a steady term plus a term that is periodic, with the period of the heart's beat.

(a) Assuming laminar flow with **v** pointing in the z direction and being a function of radius and time, $\mathbf{v} = v(\varpi, t)\mathbf{e}_z$, show that the Navier-Stokes equation reduces to $\partial v/\partial t = -P'/\rho + \nu\nabla^2 v$.

(b) Explain why $v(\varpi, t)$ is the sum of a steady term produced by the steady (time-independent) part of P', plus terms at angular frequencies $\omega_0, 2\omega_0, \ldots$, produced by parts of P' that have these frequencies. Here $\omega_0 \equiv 2\pi/$(heart's beat period).

(c) Focus on the component with angular frequency $\omega = n\omega_0$ for some integer n. For what range of ω do you expect the ϖ dependence of v to be approximately Poiseuille [Eq. (13.80)], and what ϖ dependence do you expect in the opposite extreme, and why?

(d) By solving the Navier-Stokes equation for the frequency-ω component, which is driven by the pressure-gradient term $dP/dz = \Re(P'_\omega e^{-i\omega t})$, and by imposing appropriate boundary conditions at $\varpi = 0$ and $\varpi = a$, show that

$$v = \Re\left[\frac{P'_\omega e^{-i\omega t}}{i\omega\rho}\left(1 - \frac{J_0(\sqrt{i}\,W\varpi/a)}{J_0(\sqrt{i}\,W)}\right)\right]. \tag{13.83}$$

Here $\Re$ means take the real part, a is the artery's radius, J_0 is the Bessel function, i is $\sqrt{-1}$, and $W \equiv \sqrt{\omega a^2/\nu}$ is called the (dimensionless) *Womersley number.*

(e) Plot the pieces of this $v(\varpi)$ that are in phase and out of phase with the driving pressure gradient. Compare with the prediction you made in part (b). Explain the phasing physically. Notice that in the extreme non-Poiseuille regime, there is a boundary layer attached to the artery's wall, with sharply changing flow velocity. What is its thickness in terms of a and the Womersley number? We study boundary layers like this one in Sec. 14.4 and especially Ex. 14.18.

13.8 Relativistic Dynamics of a Perfect Fluid T2

When a fluid's speed $v = |\mathbf{v}|$ becomes comparable to the speed of light c, or $P/(\rho c^2)$ or u/c^2 become of order unity, Newtonian fluid mechanics breaks down and must be replaced by a relativistic treatment. In this section, we briefly sketch the resulting laws of relativistic fluid dynamics for an ideal (perfect) fluid. For the extension to a fluid with dissipation (viscosity and heat conductivity), see, e.g., Misner, Thorne, and Wheeler (1973, Ex. 22.7).

Our treatment takes off from the brief description of an ideal, relativistic fluid in Secs. 2.12.3 and 2.13.3. As done there, we shall use geometrized units in which the speed of light is set to unity: $c = 1$ (see Sec. 1.10).

13.8.1 Stress-Energy Tensor and Equations of Relativistic Fluid Mechanics T2

For relativistic fluids we use ρ to denote the total density of mass-energy (including rest mass and internal energy) in the fluid's local rest frame; it is sometimes written as

$$\rho = \rho_o(1+u), \quad \text{where} \quad \rho_o = \bar{m}_B n \tag{13.84}$$

rest-mass density and total density of mass-energy for relativistic fluid

is the density of rest mass, $\bar{m}_B$ is some standard mean rest mass per baryon, n is the number density of baryons, ρ_o is the density of rest mass, and u is the specific internal energy (Sec. 2.12.3).

T2

stress-energy tensor for relativistic, ideal fluid

The stress-energy tensor $T^{\alpha\beta}$ for a relativistic, ideal fluid takes the form [Eq. (2.74b)]

$$T^{\alpha\beta} = (\rho + P)u^\alpha u^\beta + P g^{\alpha\beta}, \tag{13.85}$$

where P is the fluid's pressure (as measured in its local rest frame), u^α is its 4-velocity, and $g^{\alpha\beta}$ is the spacetime metric. In the fluid's local rest frame, where $u^0 = 1$ and $u^j = 0$, the components of this stress-energy tensor are, of course, $T^{00} = \rho$, $T^{j0} = T^{0j} = 0$, and $T^{jk} = P g^{jk} = P \delta^{jk}$.

The dynamics of our relativistic, ideal fluid are governed by five equations. The first equation is the law of *rest-mass conservation*, $(\rho_o u^\alpha)_{;\alpha} = 0$, which can be rewritten in the form [Eqs. (2.64) and (2.65) of Ex. 2.24]

rest-mass conservation

$$\frac{d\rho_o}{d\tau} = -\rho_o \vec{\nabla} \cdot \vec{u}, \quad \text{or} \quad \frac{d(\rho_o V)}{d\tau} = 0, \tag{13.86a}$$

where $d/d\tau = \vec{u} \cdot \vec{\nabla}$ is the derivative with respect to proper time moving with the fluid, $\vec{\nabla} \cdot \vec{u} = (1/V)(dV/d\tau)$ is the divergence of the fluid's 4-velocity, and V is the volume of a fluid element. The second equation is *energy conservation*, in the form of the vanishing divergence of the stress-energy tensor projected onto the fluid 4-velocity, $u_\alpha T^{\alpha\beta}{}_{;\beta} = 0$, which, when combined with the law of rest-mass conservation, reduces to [Eqs. (2.76)]

energy conservation

$$\frac{d\rho}{d\tau} = -(\rho + P)\vec{\nabla} \cdot \vec{u}, \quad \text{or} \quad \frac{d(\rho V)}{d\tau} = -P \frac{dV}{d\tau}. \tag{13.86b}$$

The third equation follows from the first law of thermodynamics moving with the fluid, $d(\rho V)/d\tau = -P dV/d\tau + T d(\rho_o V s)/d\tau$, combined with rest-mass conservation (13.86a) and energy conservation (13.86b), to yield *conservation of the entropy per unit rest mass s* (adiabaticity of the flow):

entropy conservation for relativistic, ideal fluid

$$\frac{ds}{d\tau} = 0. \tag{13.86c}$$

As in Newtonian theory, the ultimate source of this adiabaticity is our restriction to an ideal fluid (i.e., one without any dissipation). The fourth equation is *momentum conservation*, which we obtain by projecting the vanishing divergence of the stress-energy tensor orthogonally to the fluid's 4-velocity, $P_{\alpha\mu} T^{\mu\nu}{}_{;\nu} = 0$, resulting in [Eq. (2.76c)]

momentum conservation (relativistic Euler equation)

$$(\rho + P)\frac{du^\alpha}{d\tau} = -P^{\alpha\mu} P_{;\mu}, \quad \text{where } P^{\alpha\mu} = g^{\alpha\mu} + u^\alpha u^\mu. \tag{13.86d}$$

This is the *relativistic Euler equation*, and $P^{\alpha\mu}$ (not to be confused with the fluid pressure P or its gradient $P_{;\mu}$) is the tensor that projects orthogonally to $\vec{u}$. Note that the inertial mass per unit volume is $\rho + P$ (Ex. 2.27), and that the pressure gradient produces a force that is orthogonal to $\vec{u}$. The fifth equation is an *equation of state*, for example, in the form

equation of state

$$P = P(\rho_o, s). \tag{13.86e}$$

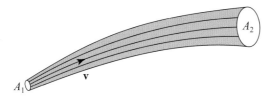

FIGURE 13.11 A tube generated by streamlines for a stationary flow. The tube ends are orthogonal to the streamlines and have areas A_1 and A_2.

Equations (13.86) are four independent scalar equations and one vector equation for the four scalars ρ_o, ρ, P, and s, and one vector $\vec{u}$.

As an example of an equation of state, one that we use below, consider a fluid so hot that its pressure and energy density are dominated by thermal motions of relativistic particles (photons, electron-positron pairs, etc.), so $P = \rho/3$ [Eqs. (3.54a)]. Then from the first law of thermodynamics for a fluid element, $d(\rho V) = -P dV$, and the law of rest-mass conservation, $d(\rho_o V) = 0$, one can deduce the relativistic polytropic equation of state

$$P = \frac{1}{3}\rho = K(s)\rho_o^{4/3}. \tag{13.87}$$

13.8.2 Relativistic Bernoulli Equation and Ultrarelativistic Astrophysical Jets

When the relativistic flow is steady (independent of time t in some chosen inertial frame), the law of energy conservation in that frame, $T^{0\mu}{}_{;\mu} = 0$, and the law of mass conservation, $(\rho_o u^\mu)_{;\mu} = 0$, together imply that the relativistic Bernoulli function B is conserved along flow lines; specifically:

$$\frac{dB}{d\tau} = \gamma v_j \frac{\partial B}{\partial x^j} = 0, \quad \text{where } B = \frac{(\rho + P)\gamma}{\rho_o}. \tag{13.88}$$

Bernoulli theorem for steady flow of relativistic, ideal fluid

Here $\gamma = u^0 = 1/\sqrt{1-\mathbf{v}^2}$. A direct proof is left as an exercise (Ex. 13.20). The following more indirect geometric proof provides useful insight.

Consider a narrow tube in space, whose walls are generated by steady flow lines (streamlines), which are tangent to the steady velocity field $\mathbf{v}$ (Fig. 13.11). Denote the tube's interior by $\mathcal{V}$ and its boundary by $\partial\mathcal{V}$. Because the flow is steady, the law of mass conservation, $(\rho_o u^\alpha)_{;\alpha} = 0$, reduces to the vanishing spatial divergence: $(\rho_o u^j)_{;j} = (\rho_o \gamma v^j)_{;j} = 0$. Integrate this equation over the tube's interior, and apply Gauss's law to obtain $\int_{\partial\mathcal{V}} \rho_o \gamma v^j d\Sigma_j = 0$. Because the walls of the tube are parallel to v^j, they make no contribution. The contribution from the ends is $(\rho_o \gamma v)_2 A_2 - (\rho_o \gamma v)_1 A_1 = 0$. In other words, *the product $\rho_o \gamma v A$—the discharge—is constant along the stream tube.*

Similarly, for our steady flow the law of energy conservation, $T^{0\alpha}{}_{;\alpha} = 0$, reduces to the vanishing spatial divergence, $T^{0j}{}_{;j} = [(\rho + P)\gamma^2 v^j]_{;j} = 0$, which, when integrated over the tube's interior and converted to a surface integral using Gauss's theorem, implies that *the product $(\rho + P)\gamma^2 v A$—the power—is constant along the stream tube.*

The ratio of these two constants, $(\rho + P)\gamma/\rho_o = B$, *must also be constant along the stream tube;* equivalently (since it is independent of the area of the narrow tube), *B must be constant along a streamline,* which is Bernoulli's theorem.

Important venues for relativistic fluid mechanics are the early universe (Secs. 28.4 and 28.5), and also the narrow relativistic jets that emerge from some galactic nuclei and gamma ray bursts. The flow velocities in some of these jets are measured to be so close to the speed of light that $\gamma \sim 1{,}000$. For the moment, ignore dissipation and electromagnetic contributions to the stress-energy tensor (which are quite important in practice), and assume that the gas pressure P and mass-energy density ρ are dominated by relativistic particles, so the equation of state is $P = \rho/3 = K\rho_o^{4/3}$ [Eq. (13.87)]. Then we can use the above, italicized conservation laws for the discharge, power, and Bernoulli function to learn how ρ, P, and γ evolve along such a jet.

From the relativistic Bernoulli theorem (the ratio B of the two constants, discharge and power) and the equation of state, we deduce that $\gamma \propto B\rho_o/(\rho + P) \propto \rho_o/\rho_o^{4/3} \propto \rho_o^{-1/3} \propto \rho^{-1/4}$. This describes a conversion of the jet's internal mass-energy ($\sim \rho$) into bulk flow energy ($\sim \gamma$): as the internal energy (energy of random thermal motions of the jet's particles) goes down, the bulk flow energy (energy of organized flow) goes up.

This energy exchange is actually driven by changes in the jet's cross sectional area, which (in some jets) is controlled by competition between the inward pressure of surrounding, slowly moving gas and outward pressure from the jet itself. Since $\rho_o \gamma v A \simeq \rho_o \gamma A$ is constant along the jet, we have $A \propto 1/(\rho_o \gamma) \propto \gamma^2 \propto \rho^{-1/2}$. Therefore, as the jet's cross sectional area A increases, internal energy ($\rho \propto 1/A^2$) goes down, and the energy of bulk flow ($\gamma \propto A^{1/2}$) goes up. Another way to think about this is in terms of the relativistic distribution function of the constituent particles. As the volume of configuration space $d\mathcal{V}_x$ that they occupy increases, their volume of momentum space $d\mathcal{V}_p$ decreases (Secs. 3.2.2 and 3.6).

We explore the origin of relativistic jets in Sec. 26.5.

EXERCISES

Exercise 13.20 *Derivation: Relativistic Bernoulli Theorem*
By manipulating the differential forms of the law of rest-mass conservation and the law of energy conservation, derive the constancy of $B = (\rho + P)\gamma/\rho_o$ along steady flow lines, Eq. (13.88).

Exercise 13.21 *Example: Relativistic Momentum Conservation*
Give an expression for the change in the *thrust*—the momentum crossing a surface perpendicular to the tube per unit time—along a slender stream tube when the discharge and power are conserved. Explain why the momentum has to change.

13.8.3 Nonrelativistic Limit of the Stress-Energy Tensor T2

It is instructive to evaluate the nonrelativistic limit of the perfect-fluid stress-energy tensor, $T^{\alpha\beta} = (\rho + P)u^\alpha u^\beta + P g^{\alpha\beta}$, and verify that it has the form we deduced in our study of nonrelativistic fluid mechanics (see Table 13.1) with vanishing gravitational potential $\Phi = 0$.

In the nonrelativistic limit, the fluid is nearly at rest in the chosen Lorentz reference frame. It moves with ordinary velocity $\mathbf{v} = d\mathbf{x}/dt$ that is small compared to the speed of light, so the temporal part of its 4-velocity $u^0 = 1/\sqrt{1-v^2}$ and the spatial part $\mathbf{u} = u^0 \mathbf{v}$ can be approximated as

$$u^0 \simeq 1 + \frac{1}{2}v^2, \quad \mathbf{u} \simeq \left(1 + \frac{1}{2}v^2\right)\mathbf{v}. \quad (13.89a)$$

nonrelativistic limit of 4-velocity

We write $\rho = \rho_o(1 + u)$ [Eq. (13.84)], where u is the specific internal energy (not to be confused with the fluid 4-velocity $\vec{u}$ or its spatial part $\mathbf{u}$). In our chosen Lorentz frame the volume of each fluid element is Lorentz contracted by the factor $\sqrt{1-v^2}$, and therefore the rest mass density is increased from ρ_o to $\rho_o/\sqrt{1-v^2} = \rho_o u^0$. Correspondingly the rest-mass flux is increased from $\rho_o \mathbf{v}$ to $\rho_o u^0 \mathbf{v} = \rho_o \mathbf{u}$ [Eq. (2.62)], and the law of rest-mass conservation becomes $\partial(\rho_o u^0)/\partial t + \partial(\rho_o u^j)/\partial x^j = 0$. When taking the Newtonian limit, we should identify the Newtonian mass ρ_N with the low-velocity limit of this Lorentz-contracted rest-mass density:

$$\rho_N = \rho_o u^0 \simeq \rho_o\left(1 + \frac{1}{2}v^2\right). \quad (13.89b)$$

nonrelativistic limit of Lorentz-contracted mass density

In the nonrelativistic limit the specific internal energy u, the kinetic energy per unit mass $\frac{1}{2}v^2$, and the ratio of pressure to rest-mass density P/ρ_o are of the same order of smallness:

$$u \sim \frac{1}{2}v^2 \sim \frac{P}{\rho_o} \ll 1, \quad (13.90)$$

and the momentum density T^{j0} is accurate to first order in $v \equiv |\mathbf{v}|$, the momentum flux (stress) T^{jk} and the energy density T^{00} are both accurate to second order in v, and the energy flux T^{0j} is accurate to third order in v. To these accuracies, the perfect-fluid stress-energy tensor (13.85), when combined with Eqs. (13.84) and (13.89), takes the following form:

nonrelativistic limit of stress-energy tensor for ideal (perfect) fluid

$$T^{j0} = \rho_N v^j, \quad T^{jk} = P g^{jk} + \rho_N v^j v^k,$$

$$T^{00} = \rho_N + \frac{1}{2}\rho_N v^2 + \rho_N u, \quad T^{0j} = \rho_N v^j + \left(\frac{1}{2}v^2 + u + \frac{P}{\rho_N}\right)\rho_N v^j; \quad (13.91)$$

see Ex. 13.22. These are precisely the same as the nonrelativistic momentum density, momentum flux, energy density, and energy flux in Table 13.1, aside from (i) the notational change $\rho \to \rho_N$ from there to here, (ii) including the rest mass-energy, $\rho_N = \rho_N c^2$, in T_{00} here but not there, and (iii) including the rest-mass-energy flux $\rho_N v^j$ in T^{0j} here but not there.

EXERCISES

Exercise 13.22 *Derivation: Nonrelativistic Limit of Perfect-Fluid Stress-Energy Tensor* T2

(a) Show that in the nonrelativistic limit, the components of the perfect-fluid stress-energy tensor (13.85) take on the forms (13.91), and verify that these agree with the densities and fluxes of energy and momentum that are used in nonrelativistic fluid mechanics (Table 13.1).

(b) Show that the contribution of the pressure P to the relativistic density of inertial mass causes the term $(P/\rho_N)\rho_N \mathbf{v} = P\mathbf{v}$ to appear in the nonrelativistic energy flux.

BOX 13.5. TERMINOLOGY USED IN CHAPTER 13

This chapter introduces a large amount of terminology. We list much of it here.

adiabatic A process in which each fluid element conserves its entropy.

adiabatic index The parameter Γ that relates pressure and density changes, $\delta P/P = \Gamma \delta\rho/\rho$, in an adiabatic process. For an ideal gas, it is the ratio of specific heats: $\Gamma = \gamma \equiv C_P/C_V$.

advective time derivative The time derivative moving with the fluid: $d/dt = \partial/\partial t + \mathbf{v} \cdot \nabla$. Also called the convective time derivative.

barotropic A process or equation in which pressure can be regarded as a function solely of density: $P = P(\rho)$.

Bernoulli function, also sometimes called Bernoulli constant: $B = \rho(\tfrac{1}{2}v^2 + h + \Phi)$.

bulk viscosity, coefficient of The proportionality constant ζ relating rate of expansion to viscous stress: $\mathbf{T}_{\text{vis}} = -\zeta \theta \mathbf{g}$.

convective time derivative See advective time derivative.

dissipation A process that increases the entropy. Viscosity and diffusive heat flow (heat conduction) are forms of dissipation.

dynamic viscosity The coefficient of shear viscosity, η.

(continued)

BOX 13.5. (continued)

equation of state In this chapter, where chemical and nuclear reactions do not occur: relations of the form $u(\rho, s)$, $P(\rho, s)$ or $u(\rho, T)$, $P(\rho, T)$.

Euler equation Newton's second law for an ideal fluid: $\rho d\mathbf{v}/dt = -\nabla P + \rho \mathbf{g}$.

Eulerian changes Changes in a quantity measured at fixed locations in space; cf. Lagrangian changes.

expansion, rate of Fractional rate of increase of a fluid element's volume: $\theta = \nabla \cdot \mathbf{v}$.

gas A fluid in which the separations between molecules are large compared to the molecular sizes, and no long-range forces act among molecules except gravity; cf. liquid.

ideal flow A flow in which there is no dissipation.

ideal fluid A fluid in which there are no dissipative processes (also called "perfect fluid").

ideal gas A gas in which the sizes of the molecules and (nongravitational) forces among them are neglected, so the pressure is due solely to the molecules' kinetic motions: $P = nk_B T = [\rho/(\mu m_p)]k_B T$.

incompressible A process or fluid in which the fractional changes of density are small, $\delta\rho/\rho \ll 1$, so the velocity can be approximated as divergence free: $\nabla \cdot \mathbf{v} = 0$.

inviscid Having negligible viscosity.

irrotational A flow or fluid with vanishing vorticity.

isentropic A process or fluid in which the entropy per unit rest mass s is the same everywhere.

isobar A surface of constant pressure.

isothermal A process or fluid in which the temperature is the same everywhere.

kinematic viscosity The ratio of the coefficient of shear viscosity to the density: $\nu \equiv \eta/\rho$.

Lagrangian changes Changes measured moving with the fluid; cf. Eulerian changes.

laminar flow A nonturbulent flow.

(continued)

> **BOX 13.5. (continued)**
>
> **liquid** A fluid in which the molecules are packed side by side (e.g., water); contrast this with a gas.
>
> **mean molecular weight** The average mass of a molecule in a fluid, divided by the mass of a proton: μ.
>
> **Navier-Stokes equation** Newton's second law for a viscous, incompressible fluid: $d\mathbf{v}/dt = -(1/\rho)\boldsymbol{\nabla} P + \nu\nabla^2\mathbf{v} + \mathbf{g}$.
>
> **Newtonian fluid** A (i) nonrelativistic fluid, or (ii) a fluid in which the shear-stress tensor is proportional to the rate of shear $\boldsymbol{\sigma}$ and is time-independent when $\boldsymbol{\sigma}$ is constant.
>
> **perfect fluid** See ideal fluid.
>
> **perfect gas** An ideal gas (with $P = [\rho/(\mu m_p)]k_B T$) that has negligible excitation of internal molecular degrees of freedom.
>
> **polytropic** A barotropic pressure-density relation of the form $P \propto \rho^{1+1/n}$ for some constant n called the *polytropic index*. The proportionality constant is usually a function of entropy.
>
> **Reynolds number** The ratio $\text{Re} = LV/\nu$, where L is the characteristic lengthscale of a flow, V is the characteristic velocity, and ν is the kinematic viscosity. In order of magnitude it is the ratio of inertial acceleration $(\mathbf{v} \cdot \boldsymbol{\nabla})\mathbf{v}$ to viscous acceleration $\nu\nabla^2\mathbf{v}$ in the Navier-Stokes equation.
>
> **rotation, rate of** Antisymmetric part of the gradient of velocity; vorticity converted into an antisymmetric tensor using the Levi-Civita tensor.
>
> **shear, rate of** Symmetric, trace-free part of the gradient of velocity: $\boldsymbol{\sigma}$.
>
> **shear viscosity, coefficient of** The proportionality constant η relating rate of shear to viscous stress: $\mathbf{T}_{\text{vis}} = -\eta\boldsymbol{\sigma}$.
>
> **steady flow** Flow that is independent of time in some chosen reference frame.
>
> **turbulent flow** Flow characterized by chaotic fluid motions.
>
> **vorticity** The curl of the velocity field: $\boldsymbol{\omega} = \boldsymbol{\nabla} \times \mathbf{v}$.

Bibliographic Note

There are many good texts on fluid mechanics. Among those with a physicist's perspective, we particularly like Acheson (1990) and Lautrup (2005) at an elementary level, and Lighthill (1986) and Batchelor (2000) at a more advanced level. Landau and Lifshitz (1959), as always, is terse but good for physicists who already have some knowledge of the subject. Tritton (1987) takes an especially physical approach to the subject with lots of useful diagrams and photographs of fluid flows. Faber (1995) gives a more deductive yet still physical approach, including a broader range of flows. A general graduate text covering many topics that we discuss (including blood flow) is Kundu, Cohen, and Dowling (2012). For relativistic fluid mechanics we recommend Rezzolla and Zanotti (2013).

Given the importance of fluids to modern engineering and technology, it should not be surprising that there are many more texts with an engineering perspective than with a physics one. Those we particularly like include Potter, Wiggert, and Ramadan (2012), which has large numbers of useful examples, illustrations, and exercises; also recommended are Munson, Young, and Okiishi (2006) and White (2008).

Physical intuition is very important in fluid mechanics and is best developed with the aid of visualizations—both movies and photographs. In recent years many visualizations have been made available on the web. Movies that we have found especially useful are those produced by the National Committee for Fluid Mechanics Films (Shapiro 1961a) and those produced by Hunter Rouse (1963a–f).

The numerical solution of the equations of fluid dynamics on computers (computational fluid dynamics, or CFD) is a mature field of science in its own right. CFD simulations are widely used in engineering, geophysics, astrophysics, and the movie industry. We do not treat CFD in this book. For an elementary introduction, see Lautrup (2005, Chap. 21) and Kundu, Cohen, and Dowling (2012, Chap. 11). For more thorough pedagogical treatments see, for example, Fletcher (1991) and Toro (2010); in the relativistic domain, see Rezzolla and Zanotti (2013).

CHAPTER FOURTEEN

Vorticity

The flow of wet water

RICHARD FEYNMAN (1964, VOLUME 2, CHAP. 41)

14.1 Overview

In the last chapter, we introduced an important quantity called *vorticity*, which is the principal subject of the present chapter. Although the most mathematically simple flows are potential, with velocity $\mathbf{v} = \boldsymbol{\nabla}\psi$ for some ψ so the vorticity $\boldsymbol{\omega} = \boldsymbol{\nabla} \times \mathbf{v}$ vanishes, most naturally occurring flows are *vortical*, with $\boldsymbol{\omega} \neq 0$. By studying vorticity, we shall develop an intuitive understanding of how flows evolve. We shall also see that computing the vorticity can be a powerful step along the path to determining a flow's full velocity field.

We all think we can recognize a vortex. The most hackneyed example is water disappearing down a drainhole in a bathtub or shower. The angular velocity around the drain increases inward, because the angular momentum per unit mass is conserved when the water moves radially slowly, in addition to rotating. Remarkably, angular momentum conservation means that the product of the circular velocity v_ϕ and the radius ϖ is independent of radius, which in turn implies that $\boldsymbol{\nabla} \times \mathbf{v} = 0$. So this is a vortex without vorticity! (Except, as we shall see, a delta-function spike of vorticity right at the drainhole's center; see Sec. 14.2 and Ex. 14.24.) Vorticity is a precise physical quantity defined by $\boldsymbol{\omega} = \boldsymbol{\nabla} \times \mathbf{v}$, not just any vaguely circulatory motion.

In Sec. 14.2, we introduce two tools for analyzing and using vorticity: vortex lines and circulation. Vorticity is a vector field and therefore has integral curves obtained by solving $d\mathbf{x}/d\lambda = \boldsymbol{\omega}$ for some parameter λ. These integral curves are the *vortex lines*; they are analogous to magnetic field lines. The flux of vorticity $\int_S \boldsymbol{\omega} \cdot d\boldsymbol{\Sigma}$ across a surface S is equal to the integral of the velocity field, $\Gamma \equiv \int_{\partial S} \mathbf{v} \cdot d\mathbf{x}$, around the surface's boundary ∂S (by Stokes' theorem). We call this Γ the *circulation* around ∂S; it is analogous to magnetic-field flux. In fact, the analogy with magnetic fields turns out to be extremely useful. Vorticity, like a magnetic field, automatically has vanishing divergence, which means that the vortex lines are continuous, just like magnetic field lines. Vorticity, again like a magnetic field, is an axial vector and thus can be written

> **BOX 14.1. READERS' GUIDE**
>
> - This chapter relies heavily on Chap. 13.
> - Chapters 15–19 (fluid mechanics and magnetohydrodynamics) are extensions of this chapter; to understand them, this chapter must be mastered.
> - Portions of Part VI, Plasma Physics (especially Chap. 21 on the two-fluid formalism), rely on this chapter.

as the curl of a polar vector potential, the velocity $\mathbf{v}$.[1] Vorticity has the interesting property that it evolves in a perfect fluid (ideal fluid) in such a manner that the flow carries the vortex lines along with it; we say that the vortex lines are "frozen into the fluid." When viscous stresses make the fluid imperfect, then the vortex lines diffuse through the moving fluid with a diffusion coefficient that is equal to the kinematic viscosity ν.

In Sec. 14.3, we study a classical problem that illustrates both the action and the propagation of vorticity: the *creeping* flow of a low-Reynolds-number fluid around a sphere. (Low-Reynolds-number flow arises when the magnitude of the viscous-acceleration term in the equation of motion is much larger than the magnitude of the inertial acceleration.) The solution to this problem finds contemporary application in the sedimentation rates of soot particles in the atmosphere.

In Sec. 14.4, we turn to high-Reynolds-number flows, in which the viscous stress is quantitatively weak over most of the fluid. Here, the action of vorticity can be concentrated in relatively thin *boundary layers* in which the vorticity, created at a wall, diffuses away into the main body of the flow. Boundary layers arise because in real fluids, intermolecular attraction requires that the component of the fluid velocity parallel to the boundary (not just the normal component) vanish. The vanishing of both components of velocity distinguishes real fluid flow at high Reynolds number (i.e., low viscosity) from the solutions obtained assuming no vorticity. Nevertheless, it is often (but sometimes not) a good approximation to seek a solution to the equations of fluid dynamics in which vortex-free fluid slips freely past the solid and then match it to a boundary-layer solution near the solid.

Stirred water in a tea cup and Earth's oceans and atmosphere rotate nearly rigidly, so they are most nicely analyzed in a co-rotating reference frame. In Sec. 14.5, we use

---

1. Pursuing the electromagnetic analogy further, we can ask the question, "Given a specified vorticity field $\boldsymbol{\omega}(\mathbf{x}, t)$, can I solve uniquely for the velocity $\mathbf{v}(\mathbf{x}, t)$?" The answer, of course, is "No." There is gauge freedom, so many solutions exist. Interestingly, if we specify that the flow be incompressible $\boldsymbol{\nabla} \cdot \mathbf{v} = 0$ (i.e., be the analog of the Coulomb gauge), then $\mathbf{v}(\mathbf{x}, t)$ is unique up to an additive constant.

> **BOX 14.2. MOVIES RELEVANT TO THIS CHAPTER**
>
> In the 1960s, Asher Shapiro and the National Committee for Fluid Mechanics Films and Hunter Rouse at the University of Iowa produced movies that are pedagogically powerful and still fully relevant a half-century later. Those most germane to this chapter are:
>
> - Shapiro (1961b), relevant to the entire chapter;
> - Taylor (1964), relevant to Sec. 14.3;
> - Abernathy (1968), the portion dealing with (nonturbulent) laminar boundary layers, relevant to Sec. 14.4;
> - Rouse (1963e), relevant to Sec. 14.4;
> - Fultz (1969), relevant to Sec. 14.5; and
> - Taylor (1968), relevant to Sec. 14.5.4.
>
> Also relevant are many segments of the movies produced at the University of Iowa, such as Rouse (1963a–f).

such an analysis to discover novel phenomena produced by Coriolis forces—including winds around pressure depressions; Taylor columns of fluid that hang together like a rigid body; Ekman boundary layers with spiral-shaped velocity fields; gyres (humps of water, e.g., the Sargasso Sea), around which ocean currents (e.g., the Gulf Stream) circulate; and tea leaves that accumulate at the bottom center of a tea cup.

When a flow has a large amount of shear, Nature often finds ways to tap the relative kinetic energy of neighboring stream tubes. In Sec. 14.6 we explore the resulting instabilities, focusing primarily on horizontally stratified fluids with relative horizontal velocities, which have vorticity concentrated in regions where the velocity changes sharply. The instabilities we encounter show up, in Nature, as (among other things) billow clouds and clear-air turbulence in the stratosphere. These phenomena provide motivation for the principal topic of the next chapter: turbulence.

Physical insight into the phenomena of this chapter is greatly aided by movies of fluid flows. The reader is urged to view relevant movies in parallel with reading this chapter; see Box 14.2.

14.2 Vorticity, Circulation, and Their Evolution

In Sec. 13.5.4, we defined the vorticity as the curl of the velocity field, $\boldsymbol{\omega} = \boldsymbol{\nabla} \times \mathbf{v}$, analogous to defining the magnetic field as the curl of a vector potential. To get insight into vorticity, consider the three simple 2-dimensional flows shown in Fig. 14.1.

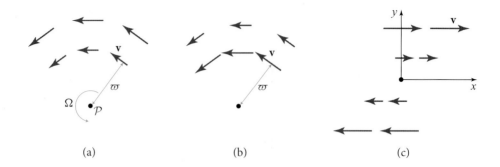

FIGURE 14.1 Vorticity in three 2-dimensional flows. The vorticity vector points in the z direction (orthogonal to the plane of the flow) and so can be thought of as a scalar ($\omega = \omega_z$). (a) Constant angular velocity Ω. If we measure radius ϖ from the center $\mathcal{P}$, the circular velocity satisfies $v = \Omega\varpi$. This flow has vorticity $\omega = 2\Omega$ everywhere. (b) Constant angular momentum per unit mass j, with $v = j/\varpi$. This flow has zero vorticity except at its center, $\omega = 2\pi j \delta(\mathbf{x})$. (c) Shearing flow in a laminar boundary layer, $v_x = -\omega y$ with $\omega < 0$. The vorticity is $\omega = -v_x/y$, and the rate of shear is $\sigma_{xy} = \sigma_{yx} = -\frac{1}{2}\omega$.

Figure 14.1a shows *uniform (rigid) rotation*, with constant angular velocity $\mathbf{\Omega} = \Omega \mathbf{e}_z$. The velocity field is $\mathbf{v} = \mathbf{\Omega} \times \mathbf{x}$, where $\mathbf{x}$ is measured from the rotation axis. Taking its curl, we discover that $\boldsymbol{\omega} = 2\mathbf{\Omega}$ everywhere.

Figure 14.1b shows a flow in which the angular momentum per unit mass $\mathbf{j} = j\mathbf{e}_z$ is constant, because it was approximately conserved as the fluid gradually drifted inward to create this flow. In this case the rotation is *differential* (radially changing angular velocity), with $\mathbf{v} = \mathbf{j} \times \mathbf{x}/\varpi^2$ (where $\varpi = |\mathbf{x}|$ and $\mathbf{j} = $ const). This is the kind of flow that occurs around a bathtub vortex and around a tornado—but outside the vortex's or tornado's core. The vorticity is $\boldsymbol{\omega} = 2\pi j \delta(\mathbf{x})$ (where $\delta(\mathbf{x})$ is the 2-dimensional Dirac delta function in the plane of the flow), so it vanishes everywhere except at the center, $\mathbf{x} = 0$ (or, more precisely, except in the vortex's or tornado's core). Anywhere in the flow, two neighboring fluid elements, separated tangentially, rotate about each other with an angular velocity $+\mathbf{j}/\varpi^2$, but when the two elements are separated radially, their relative angular velocity is $-\mathbf{j}/\varpi^2$; see Ex. 14.1. The average of these two angular velocities vanishes, which seems reasonable, since the vorticity vanishes.

vortex without vorticity except at its center

The vanishing vorticity in this case is an illustration of a simple geometrical description of vorticity in any 2-dimensional flow (Ex. 13.15): If we orient the $\mathbf{e}_z$ axis of a Cartesian coordinate system along the vorticity, then

$$\boldsymbol{\omega} = \left(\frac{\partial v_y}{\partial x} - \frac{\partial v_x}{\partial y}\right)\mathbf{e}_z. \tag{14.1}$$

vorticity measured by a vane with orthogonal fins

This expression implies that the vorticity at a point is the sum of the angular velocities of a pair of mutually perpendicular, infinitesimal lines passing through that point (one along the x direction, the other along the y direction) and moving with the fluid; for example, these lines could be thin straws suspended in the fluid. If we float a little vane with orthogonal fins in the flow, with the vane parallel to $\boldsymbol{\omega}$, then the vane will rotate with an angular velocity that is the average of the flow's angular velocities at its fins, which is half the vorticity. Equivalently, the vorticity is twice the rotation rate of the vane. In the case of constant-angular-momentum flow in Fig. 14.1b, the average of the two angular velocities is zero, the vane doesn't rotate, and the vorticity vanishes.

Figure 14.1c shows the flow in a plane-parallel shear layer. In this case, a line in the flow along the x direction does not rotate, while a line along the y direction rotates with angular velocity ω. The sum of these two angular velocities, $0 + \omega = \omega$, is the vorticity. Evidently, curved streamlines are not a necessary condition for vorticity.

EXERCISES

Exercise 14.1 *Practice: Constant-Angular-Momentum Flow—Relative Motion of Fluid Elements*

Verify that for the constant-angular-momentum flow of Fig. 14.1b, with $\mathbf{v} = \mathbf{j} \times \mathbf{x}/\varpi^2$, two neighboring fluid elements move around each other with angular velocity $+j/\varpi^2$ when separated tangentially and $-j/\varpi^2$ when separated radially. [Hint: If the fluid elements' separation vector is $\boldsymbol{\xi}$, then their relative velocity is $\nabla_{\boldsymbol{\xi}} \mathbf{v} = \boldsymbol{\xi} \cdot \nabla \mathbf{v}$. Why?]

Exercise 14.2 *Practice: Vorticity and Incompressibility*

Sketch the streamlines for the following stationary 2-dimensional flows, determine whether the flow is compressible, and evaluate its vorticity. The coordinates are Cartesian in parts (a) and (b), and are circular polar with orthonormal bases $\{\mathbf{e}_\varpi, \mathbf{e}_\phi\}$ in (c) and (d).

(a) $v_x = 2xy$, $v_y = x^2$,
(b) $v_x = x^2$, $v_y = -2xy$,
(c) $v_\varpi = 0$, $v_\phi = \varpi$,
(d) $v_\varpi = 0$, $v_\phi = \varpi^{-1}$.

Exercise 14.3 **Example: Rotating Superfluids*

At low temperatures certain fluids undergo a phase transition to a superfluid state. A good example is ^{4}He, for which the transition temperature is 2.2 K. As a superfluid has no viscosity, it cannot develop vorticity. How then can it rotate? The answer (e.g., Feynman 1972, Chap. 11) is that not all the fluid is in a superfluid state; some of it is normal and can have vorticity. When the fluid rotates, all the vorticity is concentrated in microscopic vortex cores of normal fluid that are parallel to the rotation axis and have quantized circulations $\Gamma = h/m$, where m is the mass of the atoms and h is Planck's constant. The fluid external to these vortex cores is irrotational (has vanishing vorticity). These normal fluid vortices may be pinned at the walls of the container.

(a) Explain, using a diagram, how the vorticity of the macroscopic velocity field, averaged over many vortex cores, is twice the mean angular velocity of the fluid.

(b) Make an order-of-magnitude estimate of the spacing between these vortex cores in a beaker of superfluid helium on a turntable rotating at 10 rpm.

(c) Repeat this estimate for a neutron star, which mostly comprises superfluid neutron pairs at the density of nuclear matter and spins with a period of order a millisecond. (The mass of the star is roughly 3×10^{30} kg.)

14.2.1 Vorticity Evolution

By analogy with magnetic field lines, we define a flow's *vortex lines* to be parallel to the vorticity vector $\boldsymbol{\omega}$ and to have a line density proportional to $\omega = |\boldsymbol{\omega}|$. These vortex lines are always continuous throughout the fluid, because the vorticity field, like the magnetic field, is a curl and therefore is necessarily solenoidal ($\boldsymbol{\nabla} \cdot \boldsymbol{\omega} = 0$). However, vortex lines can begin and end on solid surfaces, as the equations of fluid dynamics no longer apply there. Figure 14.2 shows an example: vortex lines that emerge from the wingtip of a flying airplane.

Vorticity and its vortex lines depend on the velocity field at a particular instant and evolve with time as the velocity field evolves. We can determine how by manipulating the Navier-Stokes equation.

In this chapter and the next one, we restrict ourselves to flows that are incompressible in the sense that $\boldsymbol{\nabla} \cdot \mathbf{v} = 0$. As we saw in Sec. 13.6, this is the case when the flow is substantially subsonic and gravitational potential differences are not too extreme. We also require (as is almost always the case) that the shear viscosity vary spatially far

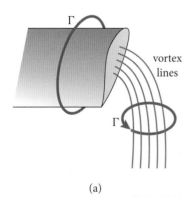  

(a) (b) (c)

FIGURE 14.2 (a) Sketch of the wing of a flying airplane and the vortex lines that emerge from the wing tip and sweep backward behind the plane. The lines are concentrated in a region with small cross section, a vortex of whirling air. The closed red curves encircle the wing and the vortex; the integral of the velocity field around these curves, $\Gamma = \int \mathbf{v} \cdot d\mathbf{x}$, is the circulation contained in the wing and its boundary layers, and in the vortex; see Sec. 14.2.4 and especially Ex. 14.8. (b) Photograph of the two vortices emerging from the wingtips of an Airbus, made visible by light scattering off water droplets in the vortex cores (Ex. 14.6). Photo © Daniel Umaña. (c) Sketch of vortex lines (dashed) in the wingtip vortices of a flying bird and the flow lines (solid) of air around them. Sketch from Vogel, 1994, Fig. 12.7c, reprinted by permission.

more slowly than the shear itself. These restrictions allow us to write the Navier-Stokes equation in its simplest form:

$$\frac{d\mathbf{v}}{dt} \equiv \frac{\partial \mathbf{v}}{\partial t} + (\mathbf{v} \cdot \nabla)\mathbf{v} = -\frac{\nabla P}{\rho} - \nabla \Phi + \nu \nabla^2 \mathbf{v} \qquad (14.2)$$

[Eq. (13.70) with $\mathbf{g} = -\nabla \Phi$].

To derive the desired evolution equation for vorticity, we take the curl of Eq. (14.2) and use the vector identity $(\mathbf{v} \cdot \nabla)\mathbf{v} = \nabla(v^2)/2 - \mathbf{v} \times \boldsymbol{\omega}$ (easily derivable using the Levi-Civita tensor and index notation) to obtain

$$\frac{\partial \boldsymbol{\omega}}{\partial t} = \nabla \times (\mathbf{v} \times \boldsymbol{\omega}) - \frac{\nabla P \times \nabla \rho}{\rho^2} + \nu \nabla^2 \boldsymbol{\omega}. \qquad (14.3)$$

Although the flow is assumed incompressible, $\nabla \cdot \mathbf{v} = 0$, the density can vary spatially due to a varying chemical composition (e.g., some regions might be oil and others water) or varying temperature and associated thermal expansion. Therefore, we must not omit the $\nabla P \times \nabla \rho$ term.

It is convenient to rewrite the vorticity evolution equation (14.3) with the aid of the relation (again derivable using the Levi-Civita tensor)

$$\nabla \times (\mathbf{v} \times \boldsymbol{\omega}) = (\boldsymbol{\omega} \cdot \nabla)\mathbf{v} + \mathbf{v}(\nabla \cdot \boldsymbol{\omega}) - \boldsymbol{\omega}(\nabla \cdot \mathbf{v}) - (\mathbf{v} \cdot \nabla)\boldsymbol{\omega}. \qquad (14.4)$$

Inserting this into Eq. (14.3), using $\nabla \cdot \boldsymbol{\omega} = 0$ and $\nabla \cdot \mathbf{v} = 0$, and introducing a new type of time derivative[2]

$$\frac{D\boldsymbol{\omega}}{Dt} \equiv \frac{\partial \boldsymbol{\omega}}{\partial t} + (\mathbf{v} \cdot \nabla)\boldsymbol{\omega} - (\boldsymbol{\omega} \cdot \nabla)\mathbf{v} = \frac{d\boldsymbol{\omega}}{dt} - (\boldsymbol{\omega} \cdot \nabla)\mathbf{v}, \qquad (14.5)$$

we bring Eq. (14.3) into the following form:

$$\frac{D\boldsymbol{\omega}}{Dt} = -\frac{\nabla P \times \nabla \rho}{\rho^2} + \nu \nabla^2 \boldsymbol{\omega}. \qquad (14.6)$$

vorticity evolution equation for incompressible flow

This is our favorite form for the vorticity evolution equation for an incompressible flow, $\nabla \cdot \mathbf{v} = 0$. If there are additional accelerations acting on the fluid, then their curls must be added to the right-hand side. The most important examples are the Coriolis acceleration $-2\boldsymbol{\Omega} \times \mathbf{v}$ in a reference frame that rotates rigidly with angular velocity $\boldsymbol{\Omega}$ (Sec. 14.5.1), and the Lorentz-force acceleration $\mathbf{j} \times \mathbf{B}/\rho$ when the fluid has an internal electric current density $\mathbf{j}$ and an immersed magnetic field $\mathbf{B}$ (Sec. 19.2.1); then Eq. (14.6) becomes

$$\frac{D\boldsymbol{\omega}}{Dt} = -\frac{\nabla P \times \nabla \rho}{\rho^2} + \nu \nabla^2 \boldsymbol{\omega} - 2\nabla \times (\boldsymbol{\Omega} \times \mathbf{v}) + \nabla \times (\mathbf{j} \times \mathbf{B}/\rho). \qquad (14.7)$$

influence of Coriolis and Lorentz forces on vorticity

---

2. The combination of spatial derivatives appearing here is called the *Lie derivative* and is denoted $\mathcal{L}_\mathbf{v} \boldsymbol{\omega} \equiv (\mathbf{v} \cdot \nabla)\boldsymbol{\omega} - (\boldsymbol{\omega} \cdot \nabla)\mathbf{v}$; it is also the *commutator* of $\mathbf{v}$ and $\boldsymbol{\omega}$ and is denoted $[\mathbf{v}, \boldsymbol{\omega}]$. It is often encountered in differential geometry.

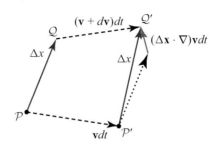

FIGURE 14.3 Equation of motion for an infinitesimal vector $\Delta\mathbf{x}$ connecting two fluid elements. As the fluid elements at $\mathcal{P}$ and $\mathcal{Q}$ move to $\mathcal{P}'$ and $\mathcal{Q}'$ in a time interval dt, the vector changes by $(\Delta\mathbf{x}\cdot\nabla)\mathbf{v}dt$.

In the remainder of Sec. 14.2, we explore the predictions of our favorite form [Eq. (14.6)] of the vorticity evolution equation.

The operator D/Dt [defined by Eq. (14.5) when acting on a vector and by $D/Dt = d/dt$ when acting on a scalar] is called the *fluid derivative*. (Warning: The notation D/Dt is used in some older texts for the convective derivative d/dt.) The geometrical meaning of the fluid derivative can be understood from Fig. 14.3. Denote by $\Delta\mathbf{x}(t)$ the vector connecting two points $\mathcal{P}$ and $\mathcal{Q}$ that are moving with the fluid. Then the figure shows that the convective derivative $d\Delta\mathbf{x}/dt$ is the relative velocity of these two points, namely $(\Delta\mathbf{x}\cdot\nabla)\mathbf{v}$. Therefore, by the second equality in Eq. (14.5), the fluid derivative of $\Delta\mathbf{x}$ vanishes:

$$\frac{D\Delta\mathbf{x}}{Dt} = 0. \tag{14.8}$$

meaning of the fluid derivative D/Dt

Correspondingly, *the fluid derivative of any vector is its rate of change relative to a vector, such as $\Delta\mathbf{x}$, whose tail and head move with the fluid.*

14.2.2 Barotropic, Inviscid, Compressible Flows: Vortex Lines Frozen into Fluid

To understand the vorticity evolution law (14.6) physically, we explore various special cases in this and the next few subsections.

Here we specialize to a barotropic $[P = P(\rho)]$, inviscid ($\nu = 0$) fluid flow. (This kind of flow often occurs in Earth's atmosphere and oceans, well away from solid boundaries.) Then the right-hand side of Eq. (14.6) vanishes, leaving $D\boldsymbol{\omega}/Dt = 0$.

For generality, we temporarily (this subsection only) abandon our restriction to incompressible flow, $\nabla\cdot\mathbf{v} = 0$, but keep the flow barotropic and inviscid. Then it is straightforward to deduce, from the curl of the Euler equation (13.44), that

$$\frac{D\boldsymbol{\omega}}{Dt} = -\boldsymbol{\omega}\nabla\cdot\mathbf{v} \tag{14.9}$$

(Ex. 14.4). This equation shows that the vorticity has a fluid derivative parallel to itself: the fluid slides along its vortex lines, or equivalently, the vortex lines are frozen into

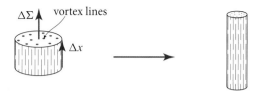

FIGURE 14.4 Simple demonstration of the kinematics of vorticity propagation in a compressible, barotropic, inviscid flow. A short, thick cylindrical fluid element with generators parallel to the local vorticity is deformed, by the flow, into a long, slender cylinder. By virtue of Eq. (14.10), we can think of the vortex lines as being convected with the fluid, with no creation of new lines or destruction of old ones, so that the number of vortex lines passing through the cylinder (through its end surface $\Delta\mathbf{\Sigma}$) remains constant.

(i.e., are carried by) the moving fluid. The wingtip vortex lines of Fig. 14.2 are an example. They are carried backward by the air that flowed over the wingtips, and they endow that air with vorticity that emerges from a wingtip.

We can actually make the fluid derivative vanish by substituting $\boldsymbol{\nabla}\cdot\mathbf{v} = -\rho^{-1} d\rho/dt$ (the equation of mass conservation) into Eq. (14.9); the result is

$$\boxed{\frac{D}{Dt}\left(\frac{\boldsymbol{\omega}}{\rho}\right) = 0 \quad \text{for barotropic, inviscid flow.}} \tag{14.10}$$

Therefore, the quantity $\boldsymbol{\omega}/\rho$ evolves according to the same equation as the separation $\Delta\mathbf{x}$ of two points in the fluid. To see what this implies, consider a small cylindrical fluid element whose symmetry axis is parallel to $\boldsymbol{\omega}$ (Fig. 14.4). Denote its vectorial length by $\Delta\mathbf{x}$, its vectorial cross sectional area by $\Delta\mathbf{\Sigma}$, and its conserved mass by $\Delta M = \rho\Delta\mathbf{x}\cdot\Delta\mathbf{\Sigma}$. Then, since $\boldsymbol{\omega}/\rho$ points along $\Delta\mathbf{x}$ and both are frozen into the fluid, it must be that $\boldsymbol{\omega}/\rho = \text{const} \times \Delta\mathbf{x}$. Therefore, the fluid element's conserved mass is $\Delta M = \rho\Delta\mathbf{x}\cdot\Delta\mathbf{\Sigma} = \text{const}\times\boldsymbol{\omega}\cdot\Delta\mathbf{\Sigma}$, so $\boldsymbol{\omega}\cdot\Delta\mathbf{\Sigma}$ is conserved as the cylindrical fluid element moves and deforms. We thereby conclude that the fluid's vortex lines, with number per unit area proportional to $|\boldsymbol{\omega}|$, are convected by our barotropic, inviscid fluid, without being created or destroyed.

for barotropic, inviscid flow: vortex lines are convected (frozen into the fluid)

Now return to an incompressible flow, $\boldsymbol{\nabla}\cdot\mathbf{v} = 0$ (which includes, of course, Earth's oceans and atmosphere), so the vorticity evolution equation becomes $D\boldsymbol{\omega}/Dt = 0$. Suppose that the flow is 2 dimensional (as it commonly is to moderate accuracy when averaged over transverse scales large compared to the thickness of the atmosphere and oceans), so $\mathbf{v}$ is in the x and y directions and is independent of z. Then $\boldsymbol{\omega} = \omega\mathbf{e}_z$, and we can regard the vorticity as the scalar ω. Then Eq. (14.5) with $(\boldsymbol{\omega}\cdot\boldsymbol{\nabla})\mathbf{v} = 0$ implies that the vorticity obeys the simple propagation law

$$\frac{d\omega}{dt} = 0. \tag{14.11}$$

Thus, *in a 2-dimensional, incompressible, barotropic, inviscid flow, the scalar vorticity is conserved when convected,* just like entropy per unit mass in an adiabatic fluid.

EXERCISES

Exercise 14.4 *Derivation: Vorticity Evolution in a Compressible, Barotropic, Inviscid Flow*

By taking the curl of the Euler equation (13.44), derive the vorticity evolution equation (14.9) for a compressible, barotropic, inviscid flow.

14.2.3 Tornados

A particularly graphic illustration of the behavior of vorticity is provided by a tornado. Tornados in North America are most commonly formed at a front where cold, dry air from the north meets warm, moist air from the south, and huge, cumulonimbus thunderclouds form. A strong updraft of the warm, moist air creates rotational motion about a horizontal axis, and updraft of the central part of the rotation axis itself makes it somewhat vertical. A low-pressure vortical core is created at the center of this spinning fluid (recall Crocco's theorem, Ex. 13.9), and the spinning region lengthens under the action of up- and downdrafts. Now, consider this process in the context of vorticity propagation. As the flow (to first approximation) is incompressible, a lengthening of the spinning region's vortex lines corresponds to a reduction in the cross section and a strengthening of the vorticity. This in turn corresponds to an increase in the tornado's circulatory speeds. (Speeds in excess of 450 km/hr have been reported.) If and when the tornado touches down to the ground and its very-low-pressure core passes over the walls and roof of a building, the far larger, normal atmospheric pressure inside the building can cause the building to explode. Further details are explored in Exs. 13.9 and 14.5.

EXERCISES

Exercise 14.5 *Problem: Tornado*

(a) Figure 14.5 shows photographs of two particularly destructive tornados and one waterspout (a tornado sucking water from the ocean). For the tornados the wind speeds near the ground are particularly high: about 450 km/hr. Estimate the wind speeds at the top, where the tornados merge with the clouds. For the water spout, the wind speed near the water is about 150 km/hr. Estimate the wind speed at the top.

(b) Estimate the air pressure in atmospheres in the cores of these tornados and water spout. (Hint: Use Crocco's theorem, Ex. 13.9.)

FIGURE 14.5 (a,b) Two destructive tornados and (c) a waterspout. (a) Eric Nguyen / Science Source; (b) Justin James Hobson, licensed under Creative Commons-ShareAlike 3.0 Unported (CC BY-SA 3.0); (c) NOAA; http://www.spc.noaa.gov/faq/tornado/wtrspout.htm.

Exercise 14.6 *Problem: Visualizing a Wingtip Vortex*
Explain why the pressure and temperature of the core of a wingtip vortex are significantly lower than the pressure and temperature of the ambient air. Under what circumstances will this lead to condensation of tiny water droplets in the vortex core, off which light can scatter, as in Fig. 14.2b?

14.2.4 Circulation and Kelvin's Theorem

Intimately related to vorticity is a quantity called *circulation* Γ; it is defined as the line integral of the velocity around a closed curve ∂S lying in the fluid:

$$\Gamma \equiv \int_{\partial S} \mathbf{v} \cdot d\mathbf{x}; \qquad (14.12a)$$

it can be regarded as a property of the closed curve ∂S. We can invoke Stokes' theorem to convert this circulation into a surface integral of the vorticity passing through a surface S bounded by ∂S:

$$\Gamma = \int_{S} \boldsymbol{\omega} \cdot d\boldsymbol{\Sigma}. \qquad (14.12b)$$

[Note, though, that Eq. (14.12b) is only valid if the area bounded by the contour is simply connected. If the area enclosed contains a solid body, this equation may fail.] Equation (14.12b) states that the circulation Γ around ∂S is the flux of vorticity through S, or equivalently, it is proportional to the number of vortex lines passing through S. Circulation is thus the fluid counterpart of magnetic flux.

Kelvin's theorem tells us the rate of change of the circulation associated with a particular contour $\partial \mathcal{S}$ that is attached to the moving fluid. Let us evaluate this rate directly using the convective derivative of Γ. We do this by differentiating the two vector quantities inside the integral (14.12a):

$$\frac{d\Gamma}{dt} = \int_{\partial \mathcal{S}} \frac{d\mathbf{v}}{dt} \cdot d\mathbf{x} + \int_{\partial \mathcal{S}} \mathbf{v} \cdot d\left(\frac{d\mathbf{x}}{dt}\right)$$

$$= -\int_{\partial \mathcal{S}} \frac{\nabla P}{\rho} \cdot d\mathbf{x} - \int_{\partial \mathcal{S}} \nabla \Phi \cdot d\mathbf{x} + \nu \int_{\partial \mathcal{S}} (\nabla^2 \mathbf{v}) \cdot d\mathbf{x} + \int_{\partial \mathcal{S}} d\frac{1}{2}v^2, \quad (14.13)$$

where we have used the Navier-Stokes equation (14.2) with $\nu =$ constant. The second and fourth terms on the right-hand side of Eq. (14.13) vanish around a closed curve, and the first can be rewritten in different notation to give

Kelvin's theorem for evolution of circulation

$$\boxed{\frac{d\Gamma}{dt} = -\int_{\partial \mathcal{S}} \frac{dP}{\rho} + \nu \int_{\partial \mathcal{S}} (\nabla^2 \mathbf{v}) \cdot d\mathbf{x}.} \quad (14.14)$$

This is Kelvin's theorem for the evolution of circulation. It is an integral version of our evolution equation (14.6) for vorticity. In a rotating reference frame it must be augmented by the integral of the Coriolis acceleration $-2\mathbf{\Omega} \times \mathbf{v}$ around the closed curve $\partial \mathcal{S}$, and if the fluid is electrically conducting with current density $\mathbf{j}$ and possesses a magnetic field $\mathbf{B}$, it must be augmented by the integral of the Lorentz force per unit mass $(\mathbf{j} \times \mathbf{B})/\rho$ around $\partial \mathcal{S}$:

$$\frac{d\Gamma}{dt} = -\int_{\partial \mathcal{S}} \frac{dP}{\rho} + \nu \int_{\partial \mathcal{S}} (\nabla^2 \mathbf{v}) \cdot d\mathbf{x} - 2\int_{\partial \mathcal{S}} \mathbf{\Omega} \times \mathbf{v} \cdot d\mathbf{x} + \int_{\partial \mathcal{S}} \frac{\mathbf{j} \times \mathbf{B}}{\rho} \cdot d\mathbf{x}. \quad (14.15)$$

This is the integral form of Eq. (14.7).

If the fluid is barotropic, $P = P(\rho)$, and the effects of viscosity are negligible (and the coordinates are inertial and there is no magnetic field and no electric current), then the right-hand sides of Eqs. (14.14) and (14.15) vanish, and Kelvin's theorem takes the simple form

conservation of circulation in a barotropic inviscid flow

$$\boxed{\frac{d\Gamma}{dt} = 0 \quad \text{for barotropic, inviscid flow.}} \quad (14.16)$$

Eq. (14.16) is the global version of our result that the circulation $\boldsymbol{\omega} \cdot \Delta \mathbf{\Sigma}$ of an infinitesimal fluid element is conserved.

The qualitative content of Kelvin's theorem is that vorticity in a fluid is long lived. A fluid's vorticity and circulation (or lack thereof) will persist, unchanged, unless and until viscosity or a $(\nabla P \times \nabla \rho)/\rho^2$ term (or a Coriolis or Lorentz force term) comes into play in the vorticity evolution equation (14.3). We now explore these sources and modifications of vorticity and circulation.

14.2.5 Diffusion of Vortex Lines

First consider the action of viscous stresses on an existing vorticity distribution. For an incompressible, barotropic fluid with nonnegligible viscosity, the vorticity evolution law (14.6) becomes

$$\boxed{\frac{D\boldsymbol{\omega}}{Dt} = \nu \nabla^2 \boldsymbol{\omega} \quad \text{for an incompressible, barotropic fluid.}} \tag{14.17}$$

diffusion equation for vorticity

This is a *convective vectorial diffusion equation*: the viscous term $\nu\nabla^2\boldsymbol{\omega}$ causes the vortex lines to diffuse through the moving fluid, and the kinematic viscosity ν is the diffusion coefficient for the vorticity. When viscosity is negligible, the vortex lines are frozen into the flow. When viscosity is significant and no boundaries impede the vorticity's diffusion and the vorticity initially is concentrated in a thin vortex, then as time t passes the vorticity diffuses outward into a cross sectional area $\sim \nu t$ in the moving fluid (Ex. 14.7; i.e., the vortex expands). Thus the kinematic viscosity not only has the dimensions of a diffusion coefficient, it also actually controls the diffusion of vortex lines relative to the moving fluid.

diffusion of vortex lines

As a simple example of the spreading of vortex lines, consider an infinite plate moving parallel to itself relative to a fluid at rest. Transform to the rest frame of the plate (Fig. 14.6a) so the fluid moves past it. Suppose that at time $t = 0$ the velocity has

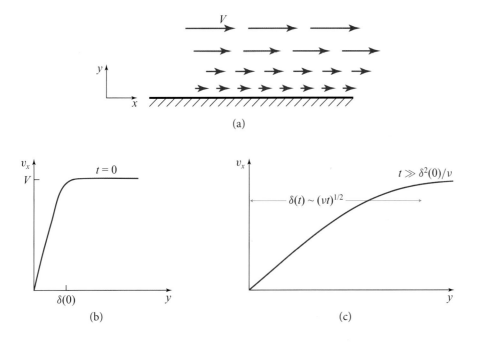

FIGURE 14.6 A simple shear layer that is translation invariant along the flow direction x. Vorticity diffuses away from the static plate at $y = 0$, under the action of viscous torques, in much the same way that heat diffuses away from a heated surface.

only a component v_x parallel to the plate, which depends solely on the distance y from the plate, so the flow is translation invariant in the x direction; then it will continue always to be translation invariant. Suppose further that, at $t = 0$, v_x is constant, $v_x = V$, except in a thin boundary layer near the plate, where it drops rapidly to 0 at $y = 0$ (as it must, because of the plate's no-slip boundary condition; Fig. 14.6b). As the flow is a function only of y (and t), and $\mathbf{v}$ and $\boldsymbol{\omega}$ point in directions orthogonal to $\mathbf{e}_y$, in this flow the fluid derivative D/Dt [Eq. (14.5)] reduces to $\partial/\partial t$, and the convective diffusion equation (14.17) becomes an ordinary scalar diffusion equation for the only nonzero component of the vorticity: $\partial \omega_z / \partial t = \nu \nabla^2 \omega_z$. Let the initial thickness of the boundary layer at time $t = 0$ be $\delta(0)$. Then our experience with the diffusion equation (e.g., Exs. 3.17 and 6.3) tells us that the viscosity will diffuse through the fluid under the action of viscous stress, and as a result, the boundary-layer thickness will increase with time as

$$\delta(t) \sim (\nu t)^{\frac{1}{2}} \quad \text{for} \quad t \gtrsim \delta(0)^2/\nu; \tag{14.18}$$

see Fig. 14.6c. We compute the evolving structure $v_x(y, t)$ of the expanding boundary layer in Sec. 14.4.

EXERCISES

Exercise 14.7 **Example: Diffusive Expansion of a Vortex*
At time $t = 0$, a 2-dimensional barotropic flow has a velocity field, in circular polar coordinates, $\mathbf{v} = (j/\varpi)\mathbf{e}_\phi$ (Fig. 14.1b); correspondingly, its vorticity is $\boldsymbol{\omega} = 2\pi j \delta(x) \delta(y) \mathbf{e}_z$: it is a delta-function vortex. In this exercise you will solve for the full details of the subsequent evolution of the flow.

(a) Solve the vorticity evolution equation (14.6) to determine the vorticity as a function of time. From your solution, show that the area in which the vorticity is concentrated (the cross sectional area of the vortex) at time t is $A \sim \nu t$, and show that the vorticity is becoming smoothed out—it is evolving toward a state of uniform vorticity, where the viscosity will no longer have any influence.

(b) From your computed vorticity in part (a), plus circular symmetry, compute the velocity field as a function of time.

(c) From the Navier-Stokes equation (or equally well, from Crocco's theorem), compute the evolution of the pressure distribution $P(\varpi, t)$.

Remark: This exercise illustrates a frequent phenomenon in fluid mechanics: The pressure adjusts itself to whatever it needs to be to accommodate the flow. One can often solve for the pressure distribution in the end, after having worked out other details. This happens here because, when one takes the curl of the Navier-Stokes equation for a barotropic fluid (which we did to get the evolution equation for vorticity), the pressure drops out—it decouples from the evolution equation for vorticity and hence velocity.

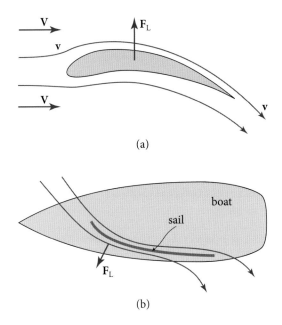

FIGURE 14.7 (a) Air flow around an airfoil viewed in cross section. The solid green lines are flow lines with velocity **v**; the incoming velocity is **V**. (b) Air flow around a sail. The shaded body represents the hull of a boat seen from above.

Exercise 14.8 **Example: The Lift on an Airplane Wing, Wingtip Vortices, Sailing Upwind, and Fish Locomotion*

When an appropriately curved airfoil (e.g., an airplane wing) is introduced into a steady flow of air, the air has to flow faster along the upper surface than along the lower surface, which can create a lifting force (Fig. 14.7a). In this situation, compressibility and gravity are usually unimportant for the flow.

(a) Show that the pressure difference across the airfoil is given approximately by $\Delta P = \frac{1}{2}\rho\Delta(v^2) = \rho v \Delta v$. Hence show that the lift exerted by the air on an airfoil of length L is given approximately by

$$\boxed{F_L = L\int \Delta P\, dx = \rho V L \Gamma,} \tag{14.19}$$

where Γ is the circulation around the airfoil, and V is the air's incoming speed in the airfoil's rest frame. This is known as *Kutta-Joukowski's theorem*. Interpret this result in terms of the conservation of linear momentum, and sketch the overall flow pattern.

(b) Explain why the circulation around an airplane wing (left orange curve in Fig. 14.2a) is the same as that around the wingtip's vortex (right orange curve in

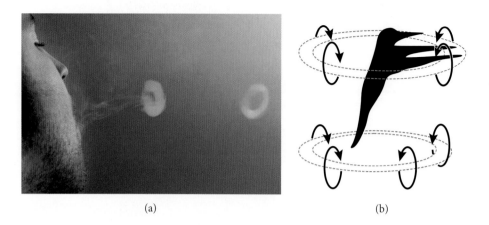

FIGURE 14.8 (a) Smoke rings blown by a man travel away from his mouth. Photo ©Andrew Vargas. (b) A ring-shaped wingtip vortex is produced by each half beat of the wings of a hovering hummingbird; the vortices travel downward. Sketch from Vogel, 1994, Fig. 12.7a, reprinted by permission.

Fig. 14.2a), and correspondingly explain why wingtip vortices are essential features of an airplane's flight. Without them, an airplane could not take off.

(c) How might birds' wingtip vortices (Fig. 14.2c) be related to the V-shaped configuration of birds in a flying flock? (For discussion, see Vogel, 1994, p. 288.)

(d) Explain how the same kind of lift as occurs on an airplane wing propels a sailboat forward when sailing upwind, as in Fig. 14.7b.

(e) Snakes, eels, and most fish undulate their bodies and/or fins as they swim. Draw pictures that explain how the same principle that propels a sailboat pushes these animals forward as well.

Exercise 14.9 *Problem: Vortex Rings*

Smoke rings (ring-shaped vortices) blown by a person (Fig. 14.8a) propagate away from him. Similarly, a hovering hummingbird produces ring-shaped vortices that propagate downward (Fig. 14.8b). Sketch the velocity field of such a vortex and explain how it propels itself through the ambient air. For the hovering hummingbird, discuss the role of the vortices in momentum conservation.

14.2.6 Sources of Vorticity

Having discussed how vorticity is conserved in simple inviscid flows and how it diffuses away under the action of viscosity, we now consider its sources. The most important source is a solid surface. When fluid suddenly encounters a solid surface, such as the leading edge of an airplane wing or a spoon stirring coffee, intermolecular forces act to decelerate the fluid rapidly in a thin boundary layer along the surface. This deceleration introduces circulation and consequently vorticity into the flow, where

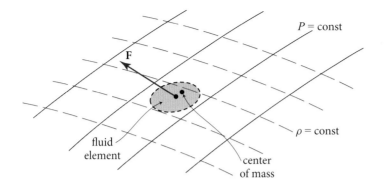

FIGURE 14.9 Mechanical explanation for the creation of vorticity in a nonbarotropic fluid. The net pressure gradient force **F**, acting at the geometric center of a small fluid element, is normal to the isobars (solid lines) and does not pass through the center of mass of the element; thereby a torque is produced.

none existed before; that vorticity then diffuses into the bulk flow, thickening the boundary layer (Sec. 14.4).

If the fluid is nonbarotropic (usually due to spatially variable chemical composition or spatially variable temperature), then pressure gradients can also create vorticity, as described by the first term on the right-hand side of the vorticity evolution law (14.6): $(-\nabla P \times \nabla \rho)/\rho^2$. Physically, when the surfaces of constant pressure (*isobars*) do not coincide with the surfaces of constant density (*isochors*), then the net pressure force on a small fluid element does not pass through its center of mass. The pressure therefore exerts a torque on the fluid element, introducing some rotational motion and vorticity (Fig. 14.9). Nonbarotropic pressure gradients can therefore create vorticity within the body of the fluid. Note that because the vortex lines must be continuous, any fresh ones that are created in the fluid must be created as loops that expand from a point or a line.

<small>vorticity generated by nonbarotropic pressure gradients</small>

There are three other common sources of vorticity in fluid dynamics:

1. Coriolis forces, when one's reference frame is rotating rigidly;
2. Lorentz forces, when the fluid is magnetized and electrically conducting [last two terms in Eqs. (14.7) and (14.15)]; and
3. curving shock fronts (when the fluid speed is supersonic).

<small>vorticity generated by Coriolis forces, Lorentz forces, and curving shock fronts</small>

We discuss these sources in Sec. 14.5 and Chaps. 19 and 17, respectively.

EXERCISES

Exercise 14.10 *Problem: Vortices Generated by a Spatula*
Fill a bathtub with water and sprinkle baby powder liberally over the water's surface to aid in viewing the motion of the surface water. Then take a spatula, insert it gently into the water, move it slowly and briefly perpendicular to its flat face, then extract it

gently from the water. Twin vortices will have been generated. Observe the vortices' motions. Explain (i) the generation of the vortices, and (ii) the sense in which the velocity field of each vortex convects the other vortex through the ambient water. Use your bathtub or a swimming pool to perform other experiments on the generation and propagation of vortices.

Exercise 14.11 *Example: Vorticity Generated by Heating*
Rooms are sometimes heated by radiators (hot surfaces) that have no associated blowers or fans. Suppose that, in a room whose air is perfectly still, a radiator is turned on to high temperature. The air will begin to circulate (convect), and that air motion contains vorticity. Explain how the vorticity is generated in terms of the $-\int dP/\rho$ term of Kelvin's theorem (14.14) and the $(-\nabla P \times \nabla \rho)/\rho^2$ term of the vorticity evolution equation (14.6).

14.3 Low-Reynolds-Number Flow—Stokes Flow and Sedimentation

In the previous chapter, we defined the Reynolds number Re to be the product of the characteristic speed V and lengthscale a of a flow divided by its kinematic viscosity $\nu = \eta/\rho$: $\text{Re} \equiv Va/\nu$. The significance of the Reynolds number follows from the fact that, in the Navier-Stokes equation (14.2), the ratio of the magnitude of the inertial acceleration $|(\mathbf{v} \cdot \nabla)\mathbf{v}|$ to the viscous acceleration $|\nu \nabla^2 \mathbf{v}|$ is approximately equal to Re. Therefore, when $\text{Re} \ll 1$, the inertial acceleration can often be ignored, and the velocity field is determined by balancing the pressure gradient against the viscous stress. The velocity then scales linearly with the magnitude of the pressure gradient and vanishes when the pressure gradient vanishes.

This has the amusing consequence that a low-Reynolds-number flow driven by a solid object moving through a fluid at rest is effectively reversible. An example, depicted in the movie by Taylor (1964), is a rotating sphere. If it is rotated slowly in a viscous fluid for N revolutions in one direction, then rotated in reverse for N revolutions, the fluid elements will return almost to their original positions. This phenomenon is easily understood by examining the Navier-Stokes equation (14.2) with the inertial acceleration $d\mathbf{v}/dt$ neglected and gravity omitted: $\nabla P = \eta \nabla^2 \mathbf{v}$; pressure gradient balances viscous force density. No time derivatives appear here, and the equation is linear. When the direction of the sphere's rotation is reversed, the pressure gradients reverse and the velocity field reverses, bringing the fluid back to its original state.

From the magnitudes of viscosities of real fluids (Table 13.2), it follows that the low-Reynolds-number limit is appropriate either for small-scale flows (e.g., the motion of microorganisms in water; see Box 14.3) or for very viscous large scale fluids (e.g., Earth's mantle; Ex. 14.13).

BOX 14.3. SWIMMING AT LOW AND HIGH REYNOLDS NUMBER: FISH VERSUS BACTERIA

Swimming provides insight into the differences between flows with low and high Reynolds numbers. In water, the flow around a swimming fish has Re $\sim 10^{+6}$, while that around a bacterium has Re $\sim 10^{-5}$. In the simplest variant of fish locomotion, the fish wags its tail fin back and forth nearly rigidly, pushing water backward and itself forward with each stroke [schematic drawing (a) below]. And a simple variant of bacterial propulsion is that of the *E. coli* bacterium: it has a rigid corkscrew-shaped tail (made from several flagella), which rotates, pushing water backward via friction (viscosity) and itself forward [schematic drawing (b), adapted from Nelson (2008)].

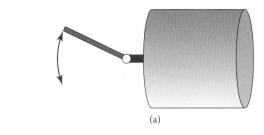

(a)

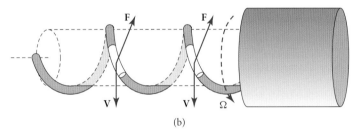

(b)

A fish's tail-wagging propulsion would fail at low Reynolds number, because the flow would be reversible: after each back-forth wagging cycle, the fluid would return to its original state—a consequence of the linear, no-time-derivatives balance of pressure and viscous forces: $\nabla P = \eta \nabla^2 \mathbf{v}$ (see second paragraph of Sec. 14.3). However, for the fish with high Reynolds number, aside from a very thin boundary layer near the fin's surface, viscosity is negligible, and so the flow is governed by Euler's equation: $d\mathbf{v}/dt = (\partial/\partial t + \mathbf{v} \cdot \nabla)\mathbf{v} = -\nabla P/\rho$ (and of course mass conservation). The time derivatives and nonlinearity make the fluid's motion nonreversible in this case. With each back-forth cycle of wag, the tail feeds substantial net backward momentum into the water, and the fish acquires net forward momentum (which will be counteracted by friction in boundary layers along the fish's body if the fish is moving fast enough).

(continued)

> **BOX 14.3. (continued)**
>
> *E. coli*'s corkscrew propulsion would fail at high Reynolds number, because its tail's flagella are so thin ($d \sim 20$ nm) that they could not push any noticeable amount of water inertially. However, at *E. coli*'s low Reynolds number, the amount of water entrained by the tail's viscous friction is almost independent of the tail's thickness. To understand the resulting frictional propulsion, consider a segment of the tail shown white in drawing (b). It moves laterally with velocity **V**. If the segment's length ℓ is huge compared to its thickness d, then the water produces a drag force **F** on it of magnitude $F \sim 2\pi \eta V \ell / \ln(\nu/Vd)$ (Ex. 14.12) that points *not* opposite to **V** but rather somewhat forward of that direction, as shown in the drawing. The reason is that this segment of a thin rod has a drag that is larger (by about a factor two) when pulled perpendicular to its long axis than when pulled along its long axis. Hence the drag is a tensorial function of **V**, $F_i = H_{ij} V_j$, and is not parallel to **V**. In drawing (b) the transverse component of the drag cancels out when one integrates it along the winding tail, but the forward component adds coherently along the tail, giving the bacterium a net forward force.
>
> For further discussion of fish as well as bacteria, see, for example, Vogel (1994) and Nelson (2008).

14.3.1 Motivation: Climate Change

An important example of a small-scale flow arises when we ask whether cooling of Earth due to volcanic explosions can mitigate global warming.

The context is concern about anthropogenic (human-made) climate change. Earth's atmosphere is a subtle and fragile protector of the environment that allows life to flourish. Especially worrisome is the increase in atmospheric carbon dioxide—by nearly 25% over the past 50 years to a mass of 3×10^{15} kg. As an important greenhouse gas, carbon dioxide traps solar radiation. Increases in its concentration are contributing to the observed increase in mean surface temperature, the rise of sea levels, and the release of oceanic carbon dioxide with potential runaway consequences.

These effects are partially mitigated by volcanos like Krakatoa, which exploded in 1883,[3] releasing roughly 250 megatons or $\sim 10^{18}$ J of energy and nearly 10^{14} kg of small particles (aerosols, ash, soot, etc.), of which $\sim 10^{12}$ kg was raised into the stratosphere,

---

[3]. A more recent example was the Pinatubo volcano in 1991, which released roughly a tenth the mass of Krakatoa. Studying the consequences of this explosion provided important calibration for models of greater catastrophes.

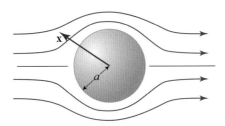

FIGURE 14.10 Flow lines for Stokes flow around a sphere.

where much of it remained for several years. These micron-sized particles absorb light with roughly their geometrical cross section. The area of Earth's surface is roughly 5×10^{14} m^2, and the density of the particles is roughly 2000 kg m^{-3}, so that 10^{12} kg of aerosols is sufficient to blot out the Sun. More specifically, the micron-sized particles absorb solar optical and ultraviolet radiation while remaining reasonably transparent to infrared radiation escaping from Earth's surface. The result is a noticeable global cooling of Earth for as long as the particles remain suspended in the atmosphere.[4]

A key issue in assessing how our environment is likely to change over the next century is how long particles of size 1–10 μm will remain in the atmosphere after volcanic explosions (i.e., their rate of sedimentation). This problem is one of low-Reynolds-number flow.

We model the sedimentation by computing the speed at which a spherical soot particle falls through quiescent air when the Reynolds number is small. The speed is governed by a balance between the downward force of gravity and the speed-dependent upward drag force of the air. We compute this sedimentation speed by first evaluating the force of the air on the moving particle ignoring gravity, and then, at the end of the calculation, inserting the influence of gravity.

14.3.2 Stokes Flow

We model the soot particle as a sphere with radius a. The low-Reynolds-number flow of a viscous fluid past such a sphere is known as *Stokes flow*. We calculate the flow's velocity field, and then from it, the force of the fluid on the sphere. (This calculation also finds application in the famous Millikan oil-drop experiment and is a prototype for many more complex calculations of low-Reynolds-number flow.)

SOLUTION FOR VELOCITY FIELD AND PRESSURE

It is easiest to tackle this problem in the frame of the sphere, where the flow is stationary (time-independent; Fig. 14.10). We seek a solution to the Navier-Stokes equation in which the flow velocity $\mathbf{v}(\mathbf{x})$ tends to a constant value $\mathbf{V}$ (the velocity of the sphere

---

4. Similar effects could follow the explosion of nuclear weapons in a major nuclear war according to Turco et al. (1986), a phenomenon they called *nuclear winter*.

through the fluid) at large distances from the sphere's center. We presume the asymptotic flow velocity **V** to be highly subsonic, so the flow is effectively incompressible: $\nabla \cdot \mathbf{v} = 0$.

insignificance of inertial acceleration

We define the Reynolds number for this flow by $\mathrm{Re} = \rho V a / \eta = V a / \nu$. As this is assumed to be small, in the Navier-Stokes equation (14.2) we can ignore the inertial term, which is $O(V \Delta v / a)$, in comparison with the viscous term, which is $O(\nu \Delta v / a^2)$; here $\Delta v \sim V$ is the total velocity variation. The time-independent Navier-Stokes equation (14.2) can thus be well approximated by $\nabla P = \rho \mathbf{g} + \eta \nabla^2 \mathbf{v}$. The uniform gravitational force density $\rho \mathbf{g}$ is obviously balanced by a uniform pressure gradient (hydrostatic equilibrium). Removing these uniform terms from both sides of the equation, we get

$$\nabla P = \eta \nabla^2 \mathbf{v}, \tag{14.20}$$

where ∇P is now just the nonuniform part of the pressure gradient required to balance the viscous force density. The full details of the flow are governed by this force-balance equation, the flow's incompressibility

$$\nabla \cdot \mathbf{v} = 0, \tag{14.21}$$

and the boundary conditions $\mathbf{v} = 0$ at $r = a$ and $\mathbf{v} \to \mathbf{V}$ at $r \to \infty$.

From force balance (14.20) we infer that in order of magnitude the difference between the fluid's pressure on the front of the sphere and that on the back is $\Delta P \sim \eta V / a$. We also expect a viscous drag stress along the sphere's sides of magnitude $T_{r\theta} \sim \eta V / a$, where V/a is the magnitude of the shear. These two stresses, acting on the sphere's surface area $\sim a^2$, will produce a net drag force $F \sim \eta V a$. Our goal is to verify this order-of-magnitude estimate, compute the force more accurately, then balance this force against gravity (adjusted for the much smaller Archimedes buoyancy force produced by the uniform part of the pressure gradient), and thereby infer the speed of fall V of a soot particle.

For a highly accurate analysis of the flow, we could write the full solution as a perturbation expansion in powers of the Reynolds number Re. We compute only the leading term in this expansion; the next term, which corrects for inertial effects, will be smaller than our solution by a factor $O(\mathrm{Re})$.

some general solution ideas

Our solution to this classic problem is based on some general ideas that ought to be familiar from other areas of physics. First, we observe that the quantities in which we are interested are the pressure P, the velocity $\mathbf{v}$, and the vorticity $\boldsymbol{\omega}$—a scalar, a polar vector, and an axial vector, respectively—and to first order in the Reynolds number they should be linear in **V**. The only scalar we can form that is linear in **V** is $\mathbf{V} \cdot \mathbf{x}$, so we expect the variable part of the pressure to be proportional to this combination. For the polar-vector velocity we have two choices, a part $\propto \mathbf{V}$ and a part $\propto (\mathbf{V} \cdot \mathbf{x})\mathbf{x}$; both terms are present. Finally, for the axial-vector vorticity, our only option is a term $\propto \mathbf{V} \times \mathbf{x}$. We use these combinations below.

Now take the divergence of Eq. (14.20), and conclude that the pressure must satisfy Laplace's equation: $\nabla^2 P = 0$. The solution should be axisymmetric about $\mathbf{V}$, and we know that axisymmetric solutions to Laplace's equation that decay as $r \to \infty$ can be expanded as a sum over Legendre polynomials, $\sum_{\ell=0}^{\infty} P_\ell(\mu)/r^{\ell+1}$, where μ is the cosine of the angle θ between $\mathbf{V}$ and $\mathbf{x}$, and r is $|\mathbf{x}|$. Since the variable part of P is proportional to $\mathbf{V} \cdot \mathbf{x}$, the dipolar ($\ell = 1$) term [for which $P_1(\mu) = \mu = \mathbf{V} \cdot \mathbf{x}/(Vr)$] is all we need; the higher-order polynomials will be higher-order in $\mathbf{V}$ and thus must arise at higher orders of the Reynolds-number expansion. We therefore write

$$P = P_\infty + \frac{k\eta(\mathbf{V} \cdot \mathbf{x})a}{r^3}. \tag{14.22}$$

Here k is a numerical constant that must be determined, we have introduced a factor η to make k dimensionless, and P_∞ is the pressure far from the sphere.

Next consider the vorticity. Since it is proportional to $\mathbf{V} \times \mathbf{x}$ and cannot depend in any other way on $\mathbf{V}$, it must be expressible as

$$\boldsymbol{\omega} = \frac{\mathbf{V} \times \mathbf{x}}{a^2} f(r/a). \tag{14.23}$$

The factor a appears in the denominator to make the unknown function f dimensionless. We determine this unknown function by rewriting Eq. (14.20) in the form

$$\nabla P = -\eta \nabla \times \boldsymbol{\omega} \tag{14.24}$$

[which we can do because $\nabla \times \boldsymbol{\omega} = -\nabla^2 \mathbf{v} + \nabla(\nabla \cdot \mathbf{v})$, and $\nabla \cdot \mathbf{v} = 0$], and then inserting Eqs. (14.22) and (14.23) into (14.24) to obtain $f(\xi) = k\xi^{-3}$; hence we have

$$\boldsymbol{\omega} = \frac{k(\mathbf{V} \times \mathbf{x})a}{r^3}. \tag{14.25}$$

Equation (14.25) for the vorticity looks familiar. It has the form of the Biot-Savart law for the magnetic field from a current element. We can therefore write down immediately a formula for its associated "vector potential," which in this case is the velocity:

electromagnetic analogy

$$\mathbf{v}(\mathbf{x}) = \frac{ka\mathbf{V}}{r} + \nabla \psi. \tag{14.26}$$

The addition of the $\nabla \psi$ term corresponds to the familiar gauge freedom in defining the vector potential. However, in the case of fluid dynamics, where the velocity is a directly observable quantity, the choice of the scalar ψ is fixed by the boundary conditions instead of being free. As ψ is a scalar linear in $\mathbf{V}$, it must be expressible in terms of a second dimensionless function $g(\xi)$ as

$$\psi = g(r/a)\mathbf{V} \cdot \mathbf{x}. \tag{14.27}$$

14.3 Low-Reynolds-Number Flow—Stokes Flow and Sedimentation

Next we recall that the flow is incompressible: $\nabla \cdot \mathbf{v} = 0$. Substituting Eq. (14.27) into Eq. (14.26) and setting the divergence expressed in spherical polar coordinates to zero, we obtain an ordinary differential equation for g:

$$\frac{d^2 g}{d\xi^2} + \frac{4}{\xi}\frac{dg}{d\xi} - \frac{k}{\xi^3} = 0. \tag{14.28}$$

This has the solution

$$g(\xi) = A - \frac{k}{2\xi} + \frac{B}{\xi^3}, \tag{14.29}$$

where A and B are integration constants. As $\mathbf{v} \to \mathbf{V}$ far from the sphere, the constant $A = 1$. The constants B and k can be found by imposing the boundary condition $\mathbf{v} = 0$ for $r = a$. We thereby obtain $B = -1/4$ and $k = -3/2$. After substituting these values into Eq. (14.26), we obtain for the velocity field:

$$\mathbf{v} = \left[1 - \frac{3}{4}\left(\frac{a}{r}\right) - \frac{1}{4}\left(\frac{a}{r}\right)^3\right]\mathbf{V} - \frac{3}{4}\left(\frac{a}{r}\right)^3\left[1 - \left(\frac{a}{r}\right)^2\right]\frac{(\mathbf{V} \cdot \mathbf{x})\mathbf{x}}{a^2}. \tag{14.30}$$

velocity, pressure, and vorticity in Stokes flow

The associated pressure and vorticity, from Eqs. (14.22) and (14.25), are given by

$$P = P_\infty - \frac{3\eta a(\mathbf{V} \cdot \mathbf{x})}{2r^3},$$

$$\boldsymbol{\omega} = \frac{3a(\mathbf{x} \times \mathbf{V})}{2r^3}. \tag{14.31}$$

The pressure is seen to be largest on the upstream hemisphere, as expected. However, the vorticity, which points in the direction of $\mathbf{e}_\phi$, is seen to be symmetric between the front and the back of the sphere. This is because our low-Reynolds-number approximation neglects the advection of vorticity by the velocity field and only retains the diffusive term. Vorticity is generated on the front surface of the sphere and diffuses into the surrounding flow; then, after the flow passes the sphere's equator, the vorticity diffuses back inward and is absorbed onto the sphere's back face.

An analysis that includes higher orders in the Reynolds number would show that not all of the vorticity is reabsorbed; a small portion is left in the fluid downstream from the sphere.

We have been able to obtain a simple solution for low-Reynolds-number flow past a sphere. Although closed-form solutions like this are not common, the methods used to derive it are of widespread applicability. Let us recall them. First, we approximated the equation of motion by omitting the subdominant inertial term and invoked a symmetry argument. We used our knowledge of elementary electrostatics to write the pressure in the form of Eq. (14.22). We then invoked a second symmetry argument to solve for the vorticity and drew on another analogy with electromagnetic theory to derive a differential equation for the velocity field, which was solved subject to the no-slip boundary condition on the surface of the sphere.

Having obtained a solution for the velocity field and pressure, it is instructive to reexamine our approximations. The first point to notice is that the velocity perturbation, given by Eq. (14.30), dies off slowly—it is inversely proportional to distance r from the sphere. Thus for our solution to be valid, the region through which the sphere is moving must be much larger than the sphere; otherwise, the boundary conditions at $r \to \infty$ have to be modified. This is not a concern for a soot particle in the atmosphere. A second, related point is that, if we compare the sizes of the inertial term (which we neglected) and the pressure gradient (which we kept) in the full Navier-Stokes equation, we find that

$$|(\mathbf{v}\cdot\nabla)\mathbf{v}| \sim \frac{V^2 a}{r^2}, \quad \left|\frac{\nabla P}{\rho}\right| \sim \frac{\eta a V}{\rho r^3}. \qquad (14.32)$$

At $r = a$ their ratio is $Va\rho/\eta = Va/\nu$, which is the (small) Reynolds number. However, at a distance $r \sim \eta/\rho V = a/\text{Re}$ from the sphere's center, the inertial term becomes comparable to the pressure term. Correspondingly, to improve on our zero-order solution, we must perform a second expansion at large r including inertial effects and then match it asymptotically to our near-zone expansion (see, e.g., Panton, 2005, Sec. 21.9). This technique of *matched asymptotic expansions* (Panton, 2005, Chap. 15) is a very powerful and general way of finding approximate solutions valid over a wide range of lengthscales, where the dominant physics changes from one scale to the next. We present an explicit example of such a matched asymptotic expansion in Sec. 16.5.3.

matched asymptotic expansions

DRAG FORCE

Let us return to the problem that motivated this calculation: computing the drag force on the sphere. It can be computed by integrating the stress tensor $\mathbf{T} = P\mathbf{g} - 2\eta\boldsymbol{\sigma}$ over the sphere's surface. If we introduce a local orthonormal basis $\{\mathbf{e}_r, \mathbf{e}_\theta, \mathbf{e}_\phi\}$ with polar axis ($\theta = 0$) along the flow direction $\mathbf{V}$, then we readily see that the only nonzero viscous contribution to the surface stress tensor is $T_{r\theta} = T_{\theta r} = \eta \partial v_\theta/\partial r$. The net resistive force along the direction of the velocity (drag force) is then given by

$$F = \int_{r=a} \frac{d\boldsymbol{\Sigma}\cdot\mathbf{T}\cdot\mathbf{V}}{V}$$

$$= \int_0^{2\pi} 2\pi a^2 \sin\theta\, d\theta \left[-P_\infty \cos\theta + \frac{3\eta V \cos^2\theta}{2a} + \frac{3\eta V \sin^2\theta}{2a}\right], \qquad (14.33)$$

where the first two terms are from the fluid's pressure on the sphere and the third is from its viscous stress. The integrals are easy and give $F = 6\pi \eta a V$ for the force, in the direction of the flow. In the rest frame of the fluid, the sphere moves with velocity $\mathbf{V}_{\text{sphere}} = -\mathbf{V}$, so the drag force that the sphere experiences is

$$\boxed{\mathbf{F} = -6\pi \eta a \mathbf{V}_{\text{sphere}}.} \qquad (14.34)$$

Stokes' law for drag force in low-Re flow

Eq. (14.34) is Stokes' law for the drag force in low-Reynolds-number flow. Two-thirds of the force comes from the viscous stress and one third from the pressure. When the influence of inertial forces at $r \gtrsim a/\mathrm{Re}$ is taken into account via matched asymptotic expansions, one obtains a correction to the drag force:

correction to drag force

$$\mathbf{F} = -6\pi \eta a \mathbf{V}_{\text{sphere}}\left(1 + \frac{3aV}{8\nu}\right) = -6\pi \eta a \mathbf{V}_{\text{sphere}}\left(1 + \frac{3\,\mathrm{Re}_d}{16}\right), \quad (14.35)$$

where (as is common) the Reynolds number Re_d is based on the sphere's diameter $d = 2a$ rather than its radius.

EXERCISES

Exercise 14.12 *Problem: Stokes Flow around a Cylinder: Stokes' Paradox*
Consider low-Reynolds-number flow past an infinite cylinder whose axis coincides with the z-axis. Try to repeat the analysis we used for a sphere to obtain an order-of-magnitude estimate for the drag force per unit length. [Hint: You might find it useful to write $\mathbf{v} = \nabla \times (\zeta \mathbf{e}_z)$, which guarantees $\nabla \cdot \mathbf{v} = 0$ (cf. Box 14.4); then show that the scalar *stream function* $\zeta(\varpi, \phi)$ satisfies the biharmonic equation $\nabla^2 \nabla^2 \zeta = 0$.] You will encounter difficulty in finding a solution for $\mathbf{v}$ that satisfies the necessary boundary conditions at the cylinder's surface $\varpi = a$ and at large radii $\varpi \gg a$. This difficulty is called *Stokes' paradox*, and the resolution to it by including inertial forces at large radii was given by Carl Wilhelm Oseen (see, e.g., Panton, 2005, Sec. 21.10). The result for the drag force per unit length is $\mathbf{F} = -2\pi \eta \mathbf{V}(\alpha^{-1} - 0.87\alpha^{-3} + \ldots)$, where $\alpha = \ln(3.703/\mathrm{Re}_d)$, and $\mathrm{Re}_d = 2aV/\nu$ is the Reynolds number computed from the cylinder's diameter $d = 2a$. The logarithmic dependence on the Reynolds number and thence on the cylinder's diameter is a warning of the subtle mixture of near-cylinder viscous flow and far-distance inertial flow that influences the drag.

14.3.3 Sedimentation Rate

Now we return to the problem that motivated our study of Stokes flow: the rate of sedimentation of soot particles (the rate at which they sink to the ground) after a gigantic volcanic eruption. To analyze this, we must restore gravity to our analysis. We can do so by restoring to the Navier-Stokes equation the uniform pressure gradient and balancing gravitational term that we removed just before Eq. (14.20). Gravity and the buoyancy (Archimedes) force from the uniform pressure gradient exert a net downward force $(4\pi a^3/3)(\rho_s - \rho)g$ on the soot particle, which must balance the upward resistive force (14.34). Here $\rho_s \sim 2{,}000\,\mathrm{kg\,m^{-3}}$ is the density of soot and $\rho \sim 1\,\mathrm{kg\,m^{-3}}$ is the far smaller (and here negligible) density of air. Equating these forces, we obtain

sedimentation speed

$$V = \frac{2\rho_s a^2 g}{9\eta}. \quad (14.36)$$

The kinematic viscosity of air at sea level is, according to Table 13.2, $\nu \sim 10^{-5}\,\mathrm{m^2\,s^{-1}}$, and the density is $\rho_a \sim 1\,\mathrm{kg\,m^{-3}}$, so the coefficient of viscosity is $\eta = \rho_a \nu \sim 10^{-5}\,\mathrm{kg\,m^{-1}\,s^{-1}}$. This viscosity is proportional to the square root of temperature and is independent of the density [cf. Eq. (13.72)]; however, the temperature does not vary by more than about 25% up to the stratosphere (Fig. 13.2), so for an approximate calculation, we can use its value at sea level. Substituting the above values into Eq. (14.36), we obtain an equilibrium sedimentation speed

$$V \sim 0.5(a/1\,\mu\text{m})^2 \text{ mm s}^{-1}. \tag{14.37}$$

For self-consistency we should also estimate the Reynolds number:

$$\text{Re} \sim \frac{2\rho_a V a}{\eta} \sim 10^{-4}\left(\frac{a}{1\,\mu\text{m}}\right)^3. \tag{14.38}$$

Our analysis is therefore only likely to be adequate for particles of radius $a \lesssim 10\,\mu\text{m}$.

There is also a lower bound to the size of the particle for validity of this analysis: the mean free path of the nitrogen and oxygen molecules must be smaller than the particle. The mean free path is $\sim 0.3\,\mu\text{m}$, so the resistive force is reduced when $a \lesssim 1\,\mu\text{m}$.[5]

limitations on validity of analysis

The sedimentation speed (14.37) is much smaller than wind speeds in the upper atmosphere: $v_{\text{wind}} \sim 30$ m s^{-1}. However, as the stratosphere is reasonably stratified, the net vertical motion due to the winds is quite small,[6] and so we can estimate the settling time by dividing the stratosphere's height ~ 30 km by the speed [Eq. (14.37)] to obtain

$$t_{\text{settle}} \sim 6 \times 10^7 \left(\frac{a}{1\,\mu\text{m}}\right)^{-2} \text{s} \sim 2\left(\frac{a}{1\,\mu\text{m}}\right)^{-2} \text{ years}. \tag{14.39}$$

This calculation is a simple model for more serious and complex analyses of sedimentation after volcanic eruptions, and the resulting mitigation of global warming. Of course, huge volcanic eruptions are rare, so no matter the result of reliable future analyses, we cannot count on volcanos to save humanity from runaway global warming. And the consequences of "geo-engineering" fixes in which particles are deliberately introduced into the atmosphere are correspondingly uncertain.

EXERCISES

Exercise 14.13 *Problem: Viscosity of Earth's Mantle*
Episodic glaciation subjects Earth's crust to loading and unloading by ice. The last major ice age was 10,000 years ago, and the subsequent unloading produces a nontidal contribution to the acceleration of Earth's rotation rate of order $|\Omega|/|\dot{\Omega}| \simeq 6 \times 10^{11}$ yr, detectable from observing the positions of distant stars. Corresponding changes in Earth's oblateness produce a decrease in the rate of nodal line regression of the geodetic satellites LAGEOS.

(a) Estimate the speed with which the polar regions (treated as spherical caps of radius $\sim 1{,}000$ km) are rebounding now. Do you think the speed was much greater in the past?

---

5. A dimensionless number—the ratio of the mean free path to the lengthscale of the flow (in this case the radius of the particle), called the *Knudsen number* or Kn—has been defined to describe this situation. Corrections to Stokes' law for the drag force are needed when Kn $\gtrsim 1$.
6. Brownian motion also affects the sedimentation rate for very small particles.

(b) Geological evidence suggests that a particular glaciated region of radius about 1,000 km sank in ~3,000 yr during the last ice age. By treating this as a low-Reynolds-number viscous flow, make an estimate of the coefficient of viscosity for the mantle.

Exercise 14.14 *Example: Undulatory Locomotion in Microorganisms*
Many microorganisms propel themselves, at low Reynolds number, using undulatory motion. Examples include the helical motion of E. coli's corkscrew tail (Box 14.3), and undulatory waves in a forest of cilia attached to an organism's surface, or near a bare surface itself. As a 2-dimensional model for locomotion via surface waves, we idealize the organism's undisturbed surface as the plane $y = 0$, and we assume the surface undulates in and out with displacement $\delta y = (-u/kC) \sin[k(x - Ct)]$ and hence velocity $V_y = u \cos[k(x - Ct)]$. Here u is the amplitude, k the wave number, and C the surface speed in the organism's rest frame. Derive the velocity field $\mathbf{v}$ for the fluid at $y > 0$ produced by this wall motion, and from it deduce the velocity of the organism through the fluid. You could proceed as follows.

(a) The velocity field must satisfy the incompressible, low-Reynolds-number equations $\nabla P = \eta \nabla^2 \mathbf{v}$ and $\nabla \cdot \mathbf{v} = 0$ [Eqs. (14.20) and (14.21)]. Explain why $\mathbf{v}$ (which must point in the x and y directions) can be expressed as the curl of a vector potential that points in the z direction: $\mathbf{v} = \nabla \times (\zeta \mathbf{e}_z)$; and show that $\zeta(t, x, y)$ (the *stream function*; see Box 14.4 in Sec. 14.4.1) satisfies the biharmonic equation: $\nabla^2 \nabla^2 \zeta = 0$.

(b) Show that the following ζ satisfies the biharmonic equation and satisfies the required boundary conditions at the organism's surface, $v_x[x, \delta y(x, t), t] = 0$ and $v_y[x, \delta y(x, t), t] = V_y(x, t)$, and the appropriate boundary conditions at $y \to \infty$:

$$\zeta = -\frac{u}{k}(1 + ky) \exp(-ky) \sin[k(x - Ct)]$$

$$+ \frac{u^2}{2C} y \{1 - \exp(-2ky) \cos[2k(x - Ct)]\} + O(u^3). \quad (14.40)$$

Explain how $\nabla P = \eta \nabla^2 \mathbf{v}$ is then easily satisfied.

(c) Show that the streamlines (tangent to $\mathbf{v}$) are surfaces of constant ζ. Plot these streamlines at $t = 0$ and discuss how they change as t passes. Explain why they are physically reasonable.

(d) Show that at large y the fluid moves with velocity $\mathbf{v} = [u^2/(2C)]\mathbf{e}_x$, and that therefore, in the asymptotic rest frame of the fluid, the organism moves with velocity $-[u^2/(2C)]\mathbf{e}_x$, opposite to $\mathbf{k}$. Is this physically reasonable? Why does the organism's inertia (and hence its mass) not influence this velocity? That the organism's velocity is second order in the velocity of its surface waves illustrates the difficulty of locomotion at low Reynolds number.

(e) Now consider what happens if the undulation is longitudinal with motion lying in the plane of the undulating surface. Sketch the flow pattern, and show that the organism will now move in the direction of **k**.

14.4 High-Reynolds-Number Flow—Laminar Boundary Layers

As we have described, flow near a solid surface creates vorticity, and consequently, the velocity field near the surface cannot be derived from a scalar potential, $\mathbf{v} = \boldsymbol{\nabla}\psi$. However, if the Reynolds number is high, then the vorticity may be localized in a thin boundary layer adjacent to the surface, as in Fig. 14.6. Then the flow may be very nearly of potential form $\mathbf{v} = \boldsymbol{\nabla}\psi$ outside that boundary layer. In this section, we use the equations of hydrodynamics to model the flow in the simplest example of such a boundary layer: that formed when a long, thin plate is placed in a steady, uniform flow $\mathbf{v} = V\mathbf{e}_x$ with its surface parallel to the flow (Fig. 14.11).

boundary layers in high-Re flow

If the plate is not too long, then the flow will be *laminar,* that is, steady and 2-dimensional—a function only of the distances x along the plate's length and y perpendicular to the plate (both being measured from an origin at the plate's front). We assume the flow to be highly subsonic, so it can be regarded as incompressible. As the viscous stress decelerates the fluid close to the plate, it must therefore be deflected away from the plate to avoid accumulating, thereby producing a small y component of velocity along with the larger x component. As the velocity is uniform well away from the plate, the pressure is constant outside the boundary layer. We use this condition to motivate the approximation that P is also constant in the boundary layer. After solving for the flow, we will check the self-consistency of this ansatz (guess). With

laminar flow

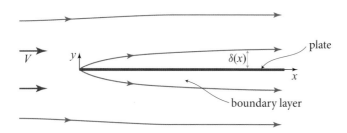

FIGURE 14.11 Laminar boundary layer formed by a long, thin plate in a flow with asymptotic speed V. The length ℓ of the plate must give a Reynolds number $\mathrm{Re}_\ell \equiv V\ell/\nu$ in the range $10 \lesssim \mathrm{Re}_\ell \lesssim 10^6$. If Re_ℓ is much less than 10, the plate will be in or near the regime of low-Reynolds-number flow (Sec. 14.3), and the boundary layer will be so thick everywhere that our analysis will fail. If Re_ℓ is much larger than 10^6, then at sufficiently great distances x down the plate ($\mathrm{Re}_x = Vx/\nu \gtrsim 10^6$), a portion of the boundary layer will become turbulently unstable and its simple laminar structure will be destroyed (see Chap. 15).

P = constant and the flow stationary, only the inertial and viscous terms remain in the Navier-Stokes equation (14.2):

$$(\mathbf{v} \cdot \nabla)\mathbf{v} \simeq \nu \nabla^2 \mathbf{v}. \tag{14.41}$$

This equation must be solved in conjunction with $\nabla \cdot \mathbf{v} = 0$ and the boundary conditions $\mathbf{v} \to V\mathbf{e}_x$ as $y \to \infty$ and $\mathbf{v} \to 0$ as $y \to 0$.

The fluid first encounters the no-slip boundary condition at the front of the plate, $x = y = 0$. The flow there abruptly decelerates to vanishing velocity, creating a steep velocity gradient that contains a sharp spike of vorticity. This is the birth of the boundary layer.

vorticity created at beginning of plate; circulation conserved thereafter

Farther downstream, the total flux of vorticity inside the rectangle C of Fig. 14.12, $\int \boldsymbol{\omega} \cdot d\boldsymbol{\Sigma}$, is equal to the circulation $\Gamma_C = \int_C \mathbf{v} \cdot d\mathbf{x}$ around C. The flow velocity is zero on the bottom leg of C, and it is (very nearly) orthogonal to C on the vertical legs, so the only nonzero contribution is from the top leg, which gives $\Gamma_C = V\Delta x$. Therefore, the circulation per unit length (flux of vorticity per unit length $\Gamma_C/\Delta x$) is V everywhere along the plate. This means that there is no new vorticity acquired, and none is lost after the initial spike at the front of the plate.

As the fluid flows down the plate, from $x = 0$ to larger x, the spike of vorticity, created at the plate's leading edge, gradually diffuses outward from the wall into the flow, thickening the boundary layer.

Let us compute the order of magnitude of the boundary layer's thickness $\delta(x)$ as a function of distance x down the plate. Incompressibility, $\nabla \cdot \mathbf{v} = 0$, implies that $v_y \sim v_x \delta/x$. Using this to estimate the relative magnitudes of the various terms in the x component of the force-balance equation (14.41), we see that the dominant inertial term (left-hand side) is $\sim V^2/x$ and the dominant viscous term (right-hand side) is $\sim \nu V/\delta^2$. We therefore obtain the estimate $\delta \sim \sqrt{\nu x/V}$. This motivates us to define the function

boundary layer thickens due to viscous diffusion of vorticity

$$\boxed{\delta(x) \equiv \sqrt{\frac{\nu x}{V}}} \tag{14.42}$$

for use in our quantitative analysis. Our analysis will reveal that the actual thickness of the boundary layer is several times larger than this $\delta(x)$.

Equation (14.42) shows that the boundary layer has a parabolic shape: $y \sim \delta(x) = \sqrt{\nu x/V}$. To keep the analysis manageable, we confine ourselves to the region, not too close to the front of the plate, where the layer is thin, $\delta \ll x$, and the velocity is nearly parallel to the plate, $v_y \sim (\delta/x)v_x \ll v_x$.

14.4.1 Blasius Velocity Profile Near a Flat Plate: Stream Function and Similarity Solution

To proceed further, we use a technique of widespread applicability in fluid mechanics: we make a *similarity ansatz*, whose validity we elucidate near the end of the calculation.

similarity ansatz

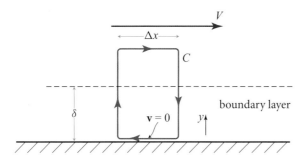

FIGURE 14.12 A rectangle C used in showing that a boundary layer's circulation per unit length $\Gamma_C/\Delta x$ is equal to the flow speed V just above the boundary layer.

We suppose that, once the boundary layer has become thin ($\delta \ll x$), the cross sectional shape of the flow is independent of distance x down the plate (it is "similar" at all x, also called "self-similar"). Stated more precisely, we assume that $v_x(x, y)$ (which has magnitude $\sim V$) and $(x/\delta)v_y$ (which also has magnitude $\sim V$) are functions only of the single transverse, dimensionless variable:

$$\xi = \frac{y}{\delta(x)} = y\sqrt{\frac{V}{\nu x}}. \tag{14.43}$$

Then our task is to compute $\mathbf{v}(\xi)$ subject to the boundary conditions $\mathbf{v} = 0$ at $\xi = 0$, and $\mathbf{v} = V\mathbf{e}_x$ at $\xi \gg 1$. We do so with the aid of a second, very useful calculational device. Recall that any vector field [$\mathbf{v}(\mathbf{x})$ in our case] can be expressed as the sum of the gradient of a scalar potential and the curl of a vector potential: $\mathbf{v} = \nabla\psi + \nabla \times \mathbf{A}$. If our flow were irrotational ($\boldsymbol{\omega} = 0$), we would need only $\nabla\psi$, but it is not; the vorticity in the boundary layer is large. On the other hand, to high accuracy the flow is incompressible, $\theta = \nabla \cdot \mathbf{v} = 0$, which means we need only the vector potential: $\mathbf{v} = \nabla \times \mathbf{A}$. And because the flow is 2-dimensional (depends only on x and y and has $\mathbf{v}$ pointing only in the x and y directions), the vector potential need only have a z component: $\mathbf{A} = A_z \mathbf{e}_z$. We denote its nonvanishing component by $A_z \equiv \zeta(x, y)$ and give it the name *stream function*, since it governs how the laminar flow streams. In terms of the stream function, the relation $\mathbf{v} = \nabla \times A$ takes the simple form

stream function

$$v_x = \frac{\partial \zeta}{\partial y}, \quad v_y = -\frac{\partial \zeta}{\partial x}. \tag{14.44}$$

Equation (14.44) automatically satisfies $\nabla \cdot \mathbf{v} = 0$. Notice that $\mathbf{v} \cdot \nabla\zeta = v_x \partial\zeta/\partial x + v_y \partial\zeta/\partial y = -v_x v_y + v_y v_x = 0$. Thus *the stream function is constant along streamlines.* (As an aside that often will be useful, e.g., in Exs. 14.12 and 14.20, we generalize this stream function in Box 14.4.)

BOX 14.4. STREAM FUNCTION FOR A GENERAL, TWO-DIMENSIONAL, INCOMPRESSIBLE FLOW T2

Consider any orthogonal coordinate system in flat 3-dimensional space, for which the metric coefficients are independent of one of the coordinates, say, x_3:

$$ds^2 = g_{11}(x_1, x_2)\, dx_1^2 + g_{22}(x_1, x_2)\, dx_2^2 + g_{33}(x_1, x_2)\, dx_3^2. \tag{1}$$

The most common examples are Cartesian coordinates $\{x, y, z\}$ with $g_{11} = g_{22} = g_{33} = 1$; cylindrical coordinates $\{\varpi, z, \phi\}$ with $g_{11} = g_{22} = 1$ and $g_{33} = \varpi^2$; and spherical coordinates $\{r, \theta, \phi\}$ with $g_{11} = 1$, $g_{22} = r^2$, and $g_{33} = r^2 \sin^2 \theta$. Suppose the velocity field is also independent of x_3, so it is effectively 2-dimensional (translation invariant for Cartesian coordinates; axisymmetric for cylindrical or spherical coordinates).

Because the flow is incompressible, $\nabla \cdot \mathbf{v} = 0$, we can write the velocity as the curl of a vector potential: $\mathbf{v} = \nabla \times \mathbf{A}(t, x_1, x_2)$. By imposing the Lorenz gauge on the vector potential (i.e., making it divergence free, as is commonly done in electromagnetism), we can ensure that its only nonvanishing component is $A_3 = \mathbf{A} \cdot \mathbf{e}_3$, where $\mathbf{e}_3$ is the unit vector pointing in the x_3 direction. Now, a special role is played by the vector that generates local translations along the x_3 direction (i.e., that generates the flow's symmetry). If we write a location $\mathcal{P}$ in space as a function of the coordinates $\mathcal{P}(x_1, x_2, x_3)$, then this generator is $\partial \mathcal{P}/\partial x_3 = \sqrt{g_{33}}\, \mathbf{e}_3$. We define the flow's stream function by

$$\boxed{\zeta(t, x_1, x_2) \equiv \mathbf{A} \cdot \partial \mathcal{P}/\partial x_3,} \tag{2}$$

which implies that the only nonzero component of the vector potential is $A_3 = \zeta/\sqrt{g_{33}}$.

Then it is straightforward to show (Ex. 14.19) the following. (i) The orthonormal components of the velocity field $\mathbf{v} = \nabla \times \mathbf{A}$ are

$$\boxed{v_1 = \frac{1}{\sqrt{g_{22}g_{33}}} \frac{\partial \zeta}{\partial x_2}, \quad v_2 = \frac{-1}{\sqrt{g_{11}g_{33}}} \frac{\partial \zeta}{\partial x_1}.} \tag{3}$$

These expressions enable one to reduce an analysis of the flow to solving for three scalar functions of (t, x_1, x_2): the stream function ζ, the pressure P, and the density ρ. (ii) The stream function is a constant along flow lines: $\mathbf{v} \cdot \nabla \zeta = 0$. (iii) The stream function is proportional to the flow rate (the amount of fluid volume crossing a surface per unit time). More specifically,

(continued)

BOX 14.4. (continued)

consider a segment of some curve reaching from point $\mathcal{A}$ to point $\mathcal{B}$ in a surface of constant x_3, and expand this curve into a segment of a 2-dimensional surface by translating it through some Δx_3 along the symmetry generator $\partial \mathcal{P}/\partial x_3$. Then the flow rate across this surface is

$$\mathcal{F} = \int_{\mathcal{A}}^{\mathcal{B}} \mathbf{v} \cdot \left(\Delta x_3 \frac{\partial \mathcal{P}}{\partial x_3} \times d\mathbf{x} \right)$$

$$= \int_{\mathcal{A}}^{\mathcal{B}} \mathbf{v} \cdot (\sqrt{g_{33}} \Delta x_3 \mathbf{e}_3 \times d\mathbf{x}) = [\zeta(\mathcal{B}) - \zeta(\mathcal{A})] \Delta x_3. \quad (4)$$

Since the stream function varies on the lengthscale δ, to produce a velocity field with magnitude $\sim V$, it must have magnitude $\sim V\delta$. This motivates us to guess that it has the functional form

$$\zeta = V\delta(x) f(\xi), \quad (14.45)$$

where $f(\xi)$ is some dimensionless function of order unity. This guess will be good if, when inserted into Eq. (14.44), it produces a self-similar flow [i.e., one with v_x and $(x/\delta)v_y$ depending only on ξ]. Indeed, inserting Eq. (14.45) into Eq. (14.44), we obtain

mathematical form of self-similar boundary layer

$$v_x = Vf', \quad v_y = \frac{\delta(x)}{2x} V(\xi f' - f), \quad (14.46\text{a})$$

where the prime denotes $d/d\xi$. These equations have the desired self-similar form.

By inserting these self-similar v_x and v_y into the x component of the force-balance equation $\mathbf{v} \cdot \nabla \mathbf{v} = \nu \nabla^2 \mathbf{v}$ [Eq. (14.41)] and neglecting $\partial^2 v_x/\partial x^2$ compared to $\partial^2 v_x/\partial y^2$, we obtain a nonlinear third-order differential equation for $f(\xi)$:

$$\frac{d^3 f}{d\xi^3} + \frac{f}{2} \frac{d^2 f}{d\xi^2} = 0. \quad (14.46\text{b})$$

That this equation involves x and y only in the combination $\xi = y\sqrt{V/\nu x}$ confirms that our self-similar ansatz was a good one. Equation (14.46b) must be solved subject to the boundary condition that the velocity vanish at the surface and approach V as $y \to \infty$ ($\xi \to \infty$) [cf. Eqs. (14.46a)]:

$$f(0) = f'(0) = 0, \quad f'(\infty) = 1. \quad (14.46\text{c})$$

Not surprisingly, Eq. (14.46b) does not admit an analytic solution. However, it is simple to compute a numerical solution with the boundary conditions (14.46c).

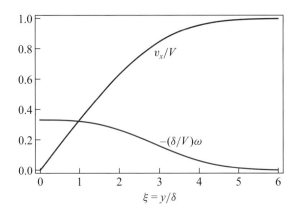

FIGURE 14.13 Laminar boundary layer near a flat plate: the Blasius velocity profile $v_x/V = f'(\xi)$ (blue curve) and vorticity profile $(\delta/V)\omega = -f'''(\xi)$ (red curve) as functions of scaled perpendicular distance $\xi = y/\delta$. Note that the flow speed is 90% of V at a distance of 3δ from the surface, so δ is a good measure of the thickness of the boundary layer.

Blasius profile for self-similar boundary layer

The result for $v_x/V = f'(\xi)$ is shown in Fig. 14.13. This solution, the *Blasius profile*, qualitatively has the form we expected: the velocity v_x rises from 0 to V in a smooth manner as one moves outward from the plate, achieving a sizable fraction of V at a distance several times larger than $\delta(x)$.

This Blasius profile is not only our first example (aside from Ex. 14.12) of the use of a stream function ($v_x = \partial\zeta/\partial y$, $v_y = -\partial\zeta/\partial x$); it is also our first example of another common procedure in fluid dynamics: *taking account of a natural scaling in the problem to make a self-similar ansatz and thereby transform the partial differential fluid equations into ordinary differential equations*. Solutions of this type are known as *similarity solutions* and also as *self-similar solutions*.

self-similar solutions

The motivation for using similarity solutions is obvious. The nonlinear partial differential equations of fluid dynamics are much harder to solve, even numerically, than are ordinary differential equations. Elementary similarity solutions are especially appropriate for problems where there is no intrinsic characteristic lengthscale or timescale associated with the relevant physical quantities except those explicitly involving the spatial and temporal coordinates. High-Reynolds-number flow past a large plate has a useful similarity solution, whereas flow with $\mathrm{Re}_\ell \sim 1$, where the size of the plate is clearly a significant scale in the problem, does not. We shall encounter more examples of similarity solutions in the following chapters.

Now that we have a solution for the flow, we must examine a key approximation that underlies it: constancy of the pressure P. To do this, we begin with the y component of the force-balance equation (14.41) (a component that we never used explicitly in our analysis). The inertial and viscous terms are both $O(V^2\delta/x^2)$, so if we reinstate a term $-\boldsymbol{\nabla} P/\rho \sim -\Delta P/\rho\delta$, it can be no larger than $\sim V^2\delta/x^2$. From this we estimate that the pressure difference across the boundary layer is $\Delta P \lesssim \rho V^2 \delta^2/x^2$. Using this estimate in the x component of force balance (14.41) (the component on which our analysis was based), we verify that the pressure gradient term is smaller than those we kept by a factor $\lesssim \delta^2/x^2 \ll 1$. For this reason, *when the boundary layer is thin, we can indeed neglect pressure gradients across it when computing its structure from longitudinal force balance.*

14.4.2 Blasius Vorticity Profile

It is illuminating to consider the structure of the Blasius boundary layer in terms of its vorticity. Since the flow is 2-dimensional with velocity $\mathbf{v} = \nabla \times (\zeta \mathbf{e}_z)$, its vorticity is $\boldsymbol{\omega} = \nabla \times \nabla \times (\zeta \mathbf{e}_z) = -\nabla^2(\zeta \mathbf{e}_z)$, which has as its only nonzero component

$$\omega \equiv \omega_z = -\nabla^2 \zeta = -\frac{V}{\delta} f''(\xi), \tag{14.47}$$

aside from fractional corrections of order δ^2/x^2. This vorticity is exhibited in Fig. 14.13.

From Eq. (14.46b), we observe that the gradient of vorticity vanishes at the plate. This means that the vorticity is not diffusing out from the plate's surface. Neither is it being convected away from the plate's surface, as the perpendicular velocity vanishes there. This confirms what we already learned from Fig. 14.12: the flux of vorticity per unit length is conserved along the plate, once it has been created as a spike at the plate's leading edge.

If we transform to a frame moving with an intermediate speed $\sim V/2$, and measure time t since passing the leading edge, the vorticity will diffuse a distance $\sim (\nu t)^{1/2} \sim (\nu x/V)^{1/2} = \delta(x)$ away from the surface after time t; see the discussion of vorticity diffusion in Sec. 14.2.5. This behavior exhibits the connection between that diffusion discussion and the similarity solution for the boundary layer in this section.

vorticity diffusion in boundary layer

14.4.3 Viscous Drag Force on a Flat Plate

It is of interest to compute the total drag force exerted on the plate. Let ℓ be the plate's length and $w \gg \ell$ be its width. Noting that the plate has two sides, the drag force produced by the viscous stress acting on the plate's surface is

$$F = 2\int T_{xy}^{\text{vis}} dx\,dz = 2\int (-2\eta \sigma_{xy}) dx\,dz = 2w \int_0^\ell \rho\nu \left(\frac{\partial v_x}{\partial y}\right)_{y=0} dx. \tag{14.48}$$

Inserting $\partial v_x/\partial y = (V/\delta)f''(0) = V\sqrt{V/(\nu x)}\, f''(0)$ from Eqs. (14.46a) and performing the integral, we obtain

$$\boxed{F = \frac{1}{2}\rho V^2 \times (2\ell w) \times C_D,} \tag{14.49}$$

where

$$\boxed{C_D = 4f''(0)\operatorname{Re}_\ell^{-1/2}.} \tag{14.50}$$

drag coefficient

Here we have introduced an often-used notation for expressing the drag force of a fluid on a solid body. We have written it as half the incoming fluid's kinetic stress ρV^2 times the surface area of the body $2\ell w$ on which the drag force acts, times a dimensionless drag coefficient C_D. We have expressed the drag coefficient in terms of the Reynolds number

$$\boxed{\mathrm{Re}_\ell = \frac{V\ell}{\nu}} \tag{14.51}$$

formed from the body's relevant dimension, ℓ, and the speed and viscosity of the incoming fluid.

From Fig. 14.13, we estimate that $f''(0) \simeq 0.3$ (an accurate numerical value is 0.332), and so $C_D \simeq 1.328 R_\ell^{-1/2}$. Note that the drag coefficient decreases as the viscosity decreases and the Reynolds number increases. However, as we discuss in Sec. 15.5, this model breaks down for very large Reynolds numbers $\mathrm{Re}_\ell \gtrsim 10^6$, because a portion of the boundary layer becomes turbulent (cf. caption of Fig. 14.11).

14.4.4 Boundary Layer Near a Curved Surface: Separation

Next consider flow past a nonplanar surface (e.g., the aircraft wings of Fig. 14.2a,b). In this case, in general a longitudinal pressure gradient exists along the boundary layer, which cannot be ignored, in contrast to the transverse pressure gradient across the boundary layer. If the pressure decreases along the flow, the flow will accelerate, and so more vorticity will be created at the surface and will diffuse away from the surface.

adverse pressure gradient

However, if there is an "adverse" pressure gradient causing the flow to decelerate, then negative vorticity must be created at the wall. For a sufficiently adverse gradient, the negative vorticity gets so strong that it cannot diffuse fast enough into and through the boundary layer to maintain a simple boundary-layer-type flow. Instead, the boundary

boundary-layer separation

layer *separates* from the surface, as shown in Fig. 14.14, and a backward flow is generated beyond the separation point by the negative vorticity. This phenomenon can occur on an aircraft when the wings' angle of attack (i.e., the inclination of the wings to the horizontal) is too great. An adverse pressure gradient develops on the

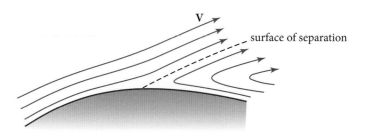

FIGURE 14.14 Separation of a boundary layer in the presence of an adverse pressure gradient.

upper wing surfaces, the flow separates, and the plane stalls. The designers of wings make great efforts to prevent this, as we discuss briefly in Sec. 15.5.2.

EXERCISES

Exercise 14.15 *Problem: Reynolds Numbers*
Estimate the Reynolds numbers for the following flows. Make sketches of the flow fields, pointing out any salient features.

(a) A hang glider in flight.
(b) Plankton in the ocean.
(c) A physicist waving her hands.

Exercise 14.16 ***Problem: Fluid Dynamical Scaling*
An auto manufacturer wishes to reduce the drag force on a new model by changing its design. She does this by building a one-sixth scale model and putting it into a wind tunnel. How fast must the air travel in the wind tunnel to simulate the flow at 40 mph on the road?

[Remark: This is our first example of "scaling" relations in fluid dynamics, a powerful concept that we develop and explore in later chapters.]

Exercise 14.17 *Example: Potential Flow around a Cylinder (D'Alembert's Paradox)*
Consider stationary incompressible flow around a cylinder of radius a with sufficiently large Reynolds number that viscosity may be ignored except in a thin boundary layer, which is assumed to extend all the way around the cylinder. The velocity is assumed to have the uniform value **V** at large distances from the cylinder.

(a) Show that the velocity field outside the boundary layer can be derived from a scalar *velocity potential* (introduced in Sec. 13.5.4), $\mathbf{v} = \boldsymbol{\nabla}\psi$, that satisfies Laplace's equation: $\nabla^2\psi = 0$.

(b) Write down suitable boundary conditions for ψ.

(c) Write the velocity potential in the form
$$\psi = \mathbf{V}\cdot\mathbf{x} + f(\mathbf{x}),$$
and solve for f. Sketch the streamlines and equipotentials.

(d) Use Bernoulli's theorem to compute the pressure distribution over the surface and the net drag force given by this solution. Does your drag force seem reasonable? It did not seem reasonable to d'Alembert in 1752, and it came to be called *d'Alembert's paradox*.

(e) Finally, consider the effect of the pressure distribution on the boundary layer. How do you think this will make the real flow different from the potential solution? How will the drag change?

Exercise 14.18 *Example: Stationary Laminar Flow down a Long Pipe*

Fluid flows down a long cylindrical pipe of length b much larger than radius a, from a reservoir maintained at pressure P_0 (which connects to the pipe at $x = 0$) to a free end at large x, where the pressure is negligible. In this problem, we try to understand the velocity field $v_x(\varpi, x)$ as a function of radius ϖ and distance x down the pipe, for a given discharge (i.e., mass flow per unit time) $\dot{M}$. Assume that the Reynolds number is small enough for the flow to be treated as laminar all the way down the pipe.

(a) Close to the entrance of the pipe (small x), the boundary layer will be thin, and the velocity will be nearly independent of radius. What is the fluid velocity outside the boundary layer in terms of its density and $\dot{M}$?

(b) How far must the fluid travel along the pipe before the vorticity diffuses into the center of the flow and the boundary layer becomes as thick as the radius? An order-of-magnitude calculation is adequate, and you may assume that the pipe is much longer than your estimate.

(c) At a sufficiently great distance down the pipe, the profile will cease evolving with x and settle down into the Poiseuille form derived in Sec. 13.7.6, with the discharge $\dot{M}$ given by the Poiseuille formula. Sketch how the velocity profile changes along the pipe, from the entrance to this final Poiseuille region.

(d) Outline a procedure for computing the discharge in a long pipe of arbitrary cross section.

Exercise 14.19 *Derivation: Stream Function in General* T2

Derive results (i), (ii), and (iii) in the last paragraph of Box 14.4. [Hint: The derivation is simplest if one works in a coordinate basis (Sec. 24.3) rather than in the orthonormal bases that we use throughout Parts I–VI of this book.]

Exercise 14.20 *Problem: Stream Function for Stokes Flow around a Sphere* T2

Consider low-Reynolds-number flow around a sphere. Derive the velocity field (14.30) using the stream function of Box 14.4. This method is more straightforward but less intuitive than that used in Sec. 14.3.2.

14.5 Nearly Rigidly Rotating Flows—Earth's Atmosphere and Oceans

In Nature one often encounters fluids that rotate nearly rigidly (i.e., fluids with a nearly uniform distribution of vorticity). Earth's oceans and atmosphere are important examples, where the rotation is forced by the underlying rotation of Earth. Such rotating fluids are best analyzed in a rotating reference frame, in which the unperturbed fluid is at rest and the perturbations are influenced by Coriolis forces, resulting in surprising phenomena. We explore some of these phenomena in this section.

14.5.1 Equations of Fluid Dynamics in a Rotating Reference Frame

As a foundation for this exploration, we transform the Navier-Stokes equation from the inertial frame in which it was derived to a uniformly rotating frame: the mean rest frame of the flows we study.

We begin by observing that the Navier-Stokes equation has the same form as Newton's second law for particle motion:

$$\frac{d\mathbf{v}}{dt} = \mathbf{f}, \tag{14.52}$$

where the force per unit mass is $\mathbf{f} = -\nabla P/\rho - \nabla \Phi + \nu \nabla^2 \mathbf{v}$. We transform to a frame rotating with uniform angular velocity $\mathbf{\Omega}$ by adding "fictitious" Coriolis and centrifugal accelerations, given by $-2\mathbf{\Omega} \times \mathbf{v}$ and $-\mathbf{\Omega} \times (\mathbf{\Omega} \times \mathbf{x})$, respectively, and expressing the force $\mathbf{f}$ in rotating coordinates. The fluid velocity transforms as

$$\mathbf{v} \to \mathbf{v} + \mathbf{\Omega} \times \mathbf{x}. \tag{14.53}$$

Coriolis and centrifugal accelerations in rigidly rotating reference frame

It is straightforward to verify that this transformation leaves the expression for the viscous acceleration, $\nu \nabla^2 \mathbf{v}$, unchanged. Therefore the expression for the force $\mathbf{f}$ is unchanged, and the Navier-Stokes equation in rotating coordinates becomes

$$\frac{d\mathbf{v}}{dt} = -\frac{\nabla P}{\rho} - \nabla \Phi + \nu \nabla^2 \mathbf{v} - 2\mathbf{\Omega} \times \mathbf{v} - \mathbf{\Omega} \times (\mathbf{\Omega} \times \mathbf{x}). \tag{14.54}$$

Navier-Stokes equation in rotating reference frame

The centrifugal acceleration $-\mathbf{\Omega} \times (\mathbf{\Omega} \times \mathbf{x})$ can be expressed as the gradient of a centrifugal potential, $\nabla[\frac{1}{2}(\mathbf{\Omega} \times \mathbf{x})^2] = \nabla[\frac{1}{2}(\Omega \varpi)^2]$, where ϖ is distance from the rotation axis. (The location of the rotation axis is actually arbitrary, aside from the requirement that it be parallel to $\mathbf{\Omega}$; see Box 14.5.) For simplicity we confine ourselves to an incompressible fluid, so that ρ is constant. This allows us to define an *effective pressure*

$$\boxed{P' = P + \rho \left[\Phi - \frac{1}{2} (\mathbf{\Omega} \times \mathbf{x})^2 \right]} \tag{14.55}$$

effective pressure for incompressible fluid in rotating reference frame

that includes the combined effects of the real pressure, gravity, and the centrifugal force. In terms of P' the Navier-Stokes equation in the rotating frame becomes

$$\boxed{\frac{d\mathbf{v}}{dt} = -\frac{\nabla P'}{\rho} + \nu \nabla^2 \mathbf{v} - 2\mathbf{\Omega} \times \mathbf{v}.} \tag{14.56a}$$

Navier-Stokes equation for incompressible fluid in rotating reference frame

The quantity P' will be constant if the fluid is at rest in the rotating frame, $\mathbf{v} = 0$, in contrast to the true pressure P, which does have a gradient. Equation (14.56a) is the most useful form for the Navier-Stokes equation in a rotating frame. In keeping with our assumptions that ρ is constant and the flow speeds are very low in comparison with the speed of sound, we augment Eq. (14.56a) by the incompressibility condition

$\nabla \cdot \mathbf{v} = 0$, which is left unchanged by the transformation (14.53) to a rotating reference frame:

$$\nabla \cdot \mathbf{v} = 0. \tag{14.56b}$$

It should be evident from Eq. (14.56a) that two dimensionless numbers characterize rotating fluids. The first is the *Rossby number*,

Rossby number

$$\mathrm{Ro} = \frac{V}{\Omega L}, \tag{14.57}$$

where V is a characteristic velocity of the flow relative to the rotating frame, and L is a characteristic length. Ro measures the relative strength of the inertial acceleration and the Coriolis acceleration:

$$\mathrm{Ro} \sim \frac{|(\mathbf{v} \cdot \nabla)\mathbf{v}|}{|2\mathbf{\Omega} \times \mathbf{v}|} \sim \frac{\text{inertial force}}{\text{Coriolis force}}. \tag{14.58}$$

The second dimensionless number is the *Ekman number*,

Ekman number

$$\mathrm{Ek} = \frac{\nu}{\Omega L^2}, \tag{14.59}$$

which analogously measures the relative strengths of the viscous and Coriolis accelerations:

$$\mathrm{Ek} \sim \frac{|\nu \nabla^2 \mathbf{v}|}{|2\mathbf{\Omega} \times \mathbf{v}|} \sim \frac{\text{viscous force}}{\text{Coriolis force}}. \tag{14.60}$$

Notice that Ro/Ek = Re is the Reynolds number.

storms, ocean currents, and tea cups

The three traditional examples of rotating flows are large-scale storms and other weather patterns on rotating Earth, deep currents in Earth's oceans, and water in a stirred tea cup.

For a typical storm, the wind speed might be $V \sim 25$ mph (~ 10 m s^{-1}), and a characteristic lengthscale might be $L \sim 1{,}000$ km. The effective angular velocity at a temperate latitude is (see Box 14.5) $\Omega_\star = \Omega_\oplus \sin 45° \sim 10^{-4}$ rad s^{-1}, where $\Omega_\oplus$ is Earth's rotational angular velocity. As the air's kinematic viscosity is $\nu \sim 10^{-5}$ m^2 s^{-1}, we find that Ro ~ 0.1 and Ek $\sim 10^{-13}$. This tells us immediately that Coriolis forces are important but not totally dominant, compared to inertial forces, in controlling the weather, and that viscous forces are unimportant except in thin boundary layers.

For deep ocean currents, such as the Gulf Stream, V ranges from ~ 0.01 to ~ 1 m s^{-1}, so we use $V \sim 0.1$ m s^{-1}, lengthscales are $L \sim 1{,}000$ km, and for water $\nu \sim 10^{-6}$ m^2 s^{-1}, so Ro $\sim 10^{-3}$ and Ek $\sim 10^{-14}$. Thus, Coriolis accelerations are far more important than inertial forces, and viscous forces are important only in thin boundary layers.

BOX 14.5. ARBITRARINESS OF ROTATION AXIS; $\mathbf{\Omega}$ FOR ATMOSPHERIC AND OCEANIC FLOWS

ARBITRARINESS OF ROTATION AXIS

Imagine yourself on the rotation axis $\mathbf{x} = 0$ of a rigidly rotating flow. All fluid elements circulate around you with angular velocity $\mathbf{\Omega}$. Now move perpendicular to the rotation axis to a new location $\mathbf{x} = \mathbf{a}$, and ride with the flow there. All other fluid elements will still rotate around you with angular velocity $\mathbf{\Omega}$! The only way you can tell you have moved (if all fluid elements look identical) is that you will now experience a centrifugal force $\mathbf{\Omega} \times (\mathbf{\Omega} \times \mathbf{a})$.

This shows up mathematically in the rotating-frame Navier-Stokes equation (14.54). When we set $\mathbf{x} = \mathbf{x}_{\text{new}} - \mathbf{a}$, the only term that changes is the centrifugal force; it becomes $-\mathbf{\Omega} \times (\mathbf{\Omega} \times \mathbf{x}_{\text{new}}) + \mathbf{\Omega} \times (\mathbf{\Omega} \times \mathbf{a})$. If we absorb the new, constant, centrifugal term $\mathbf{\Omega} \times (\mathbf{\Omega} \times \mathbf{a})$ into the gravitational acceleration $\mathbf{g} = -\nabla \Phi$, then the Navier-Stokes equation is completely unchanged. In this sense, the choice of rotation axis is arbitrary.

ANGULAR VELOCITY $\mathbf{\Omega}$ FOR LARGE-SCALE FLOWS IN EARTH'S ATMOSPHERE AND OCEANS

For large-scale flows in Earth's atmosphere and oceans (e.g., storms), the rotation of the unperturbed fluid is that due to the rotation of Earth. One might think that this means we should take as the angular velocity $\mathbf{\Omega}$ in the Coriolis term of the Navier-Stokes equation (14.56a) Earth's angular velocity $\mathbf{\Omega}_\oplus$. Not so. The atmosphere and ocean are so thin vertically that vertical motions cannot achieve small Rossby numbers: Coriolis forces are unimportant for vertical motions. Correspondingly, the only component of Earth's angular velocity $\mathbf{\Omega}_\oplus$ that is important for Coriolis forces is that which couples horizontal flows to horizontal flows: the vertical component $\Omega_* = \Omega_\oplus \sin(\text{latitude})$. (A similar situation occurs for a Foucault pendulum.) Thus, in the Coriolis term of the Navier-Stokes equation, we must set $\mathbf{\Omega} = \Omega_* \mathbf{e}_z = \Omega_\oplus \sin(\text{latitude}) \mathbf{e}_z$, where $\mathbf{e}_z$ is the vertical unit vector. By contrast, in the centrifugal potential $\frac{1}{2}(\mathbf{\Omega} \times \mathbf{x})^2$, $\mathbf{\Omega}$ remains the full angular velocity of Earth, $\mathbf{\Omega}_\oplus$—unless (as is commonly done) we absorb a portion of it into the gravitational potential as when we change rotation axes, in which case we can use $\mathbf{\Omega} = \Omega_* \mathbf{e}_z$ in the centrifugal potential.

For water stirred in a tea cup (with parameters typical of many flows in the laboratory), $L \sim 10$ cm, $\Omega \sim V/L \sim 10$ rad s^{-1}, and $\nu \sim 10^{-6}$ m^2 s^{-1} giving Ro ~ 1 and Ek $\sim 10^{-5}$. Coriolis and inertial forces are comparable in this case, and viscous forces again are confined to boundary layers, but the layers are much thicker relative to the bulk flow than in the atmospheric and oceanic cases.

Notice that for all these flows—atmospheric, oceanic, and tea cup—the (effective) rotation axis is vertical: $\mathbf{\Omega}$ is vertically directed (cf. Box 14.5). This will be the case for all nearly rigidly rotating flows considered in this chapter.

14.5.2 Geostrophic Flows

Stationary flows $\partial \mathbf{v}/\partial t = 0$ in which both the Rossby and Ekman numbers are small (i.e., with Coriolis forces big compared to inertial and viscous forces) are called *geostrophic*, even in the laboratory. Geostrophic flow is confined to the bulk of the fluid, well away from all boundary layers, since viscosity will become important in those layers. For such geostrophic flows, the Navier-Stokes equation (14.56a) reduces to

$$2\mathbf{\Omega} \times \mathbf{v} = -\frac{\nabla P'}{\rho}. \qquad (14.61)$$

This equation states that the velocity $\mathbf{v}$ (measured in the rotating frame) is orthogonal to the body force $\nabla P'$, which drives it. Correspondingly, the streamlines are perpendicular to the gradient of the generalized pressure (i.e., they lie on surfaces of constant P').

An example of geostrophic flow is the motion of atmospheric winds around a low pressure region or *depression*. [Since $P' = P + \rho(\Phi - \frac{1}{2}\Omega^2 \varpi^2)$, when the actual pressure P goes up or down at some fixed location, P' goes up or down by the same amount, so a depression of P' is a depression of P.] The geostrophic equation (14.61) tells us that such winds must be counterclockwise in the northern hemisphere as seen from a satellite, and clockwise in the southern hemisphere. For a flow with speed $v \sim 10$ m s^{-1} around a $\sim$1,000-km depression, the drop in effective pressure at the depression's center is $\Delta P' = \Delta P \sim 1$ kPa ~ 10 mbar ~ 0.01 atmosphere ~ 0.3 inches of mercury $\sim$4 inches of water. Around a high-pressure region, winds circulate in the opposite direction.

It is here that we can see the power of introducing the effective pressure P'. In the case of atmospheric and oceanic flows, the true pressure P changes significantly vertically, and the pressure scale height is generally much shorter than the horizontal lengthscale. However, the effective pressure will be almost constant vertically, any small variation being responsible for minor updrafts and downdrafts, which we generally ignore when describing the wind or current flow pattern. It is the horizontal pressure gradients that are responsible for driving the flow. When pressures are quoted, they must therefore be referred to some reference equipotential surface: $\Phi - \frac{1}{2}(\mathbf{\Omega} \times \mathbf{x})^2 = $ const. The convenient one to use is the equipotential associated

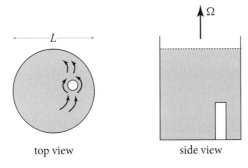

FIGURE 14.15 Taylor column. A solid cylinder (white) is placed in a large container of water, which is then spun up on a turntable to a high enough angular velocity Ω that the Ekman number is small: $\mathrm{Ek} = \nu/(\Omega L^2) \ll 1$. A slow, steady flow relative to the cylinder is then induced. [The flow's velocity $\mathbf{v}$ in the rotating frame must be small enough to keep the Rossby number $\mathrm{Ro} = v/(\Omega L) \ll 1$.] The water in the bottom half of the container flows around the cylinder. The water in the top half does the same as if there were an invisible cylinder present. This is an illustration of the Taylor-Proudman theorem, which states that there can be no vertical gradients in the velocity field. The effect can also be demonstrated with vertical velocity: If the cylinder is slowly made to rise, then the fluid immediately above it will also be pushed upward rather than flow past the cylinder—except at the water's surface, where the geostrophic flow breaks down. The fluid above the cylinder, which behaves as though it were rigidly attached to the cylinder, is called a *Taylor column*.

with the surface of the ocean, usually called "mean sea level." This is the pressure that appears on a meteorological map.

14.5.3 Taylor-Proudman Theorem

There is a simple theorem due to Taylor and Proudman that simplifies the description of 3-dimensional, geostrophic flows. Take the curl of Eq. (14.61) and use $\nabla \cdot \mathbf{v} = 0$; the result is

$$\boxed{(\boldsymbol{\Omega} \cdot \nabla)\mathbf{v} = 0.} \tag{14.62}$$

Taylor-Proudman theorem for geostrophic flow

Thus, there can be no vertical gradient (gradient along $\boldsymbol{\Omega}$) of the velocity under geostrophic conditions. This result provides a good illustration of the stiffness of vortex lines: the vortex lines associated with the rigid rotation $\boldsymbol{\Omega}$ are frozen in the fluid under geostrophic conditions (where other contributions to the vorticity are small), and they refuse to be bent. The simplest demonstration of this is the Taylor column of Fig. 14.15.

It is easy to see that any vertically constant, divergence-free velocity field $\mathbf{v}(x, y)$ can be a solution to the geostrophic equation (14.61). The generalized pressure P' can be adjusted to make it a solution (see the discussion of pressure adjusting itself to whatever the flow requires, in Ex. 14.7). However, one must keep in mind that to guarantee it is also a true (approximate) solution of the full Navier-Stokes equation (14.56a), its Rossby and Ekman numbers must be $\ll 1$.

14.5.4 Ekman Boundary Layers

As we have seen, Ekman numbers are typically small in the bulk of a rotating fluid. However, as was true in the absence of rotation, the no-slip condition at a solid surface generates a boundary layer that can significantly influence the global velocity field, albeit indirectly.

When the Rossby number is less than one, the structure of a laminar boundary layer is dictated by a balance between viscous and Coriolis forces rather than between viscous and inertial forces. Balancing the relevant terms in Eq. (14.56a), we obtain an estimate of the boundary-layer thickness:

Ekman boundary layer and its thickness

$$\boxed{\text{thickness} \sim \delta_E \equiv \sqrt{\frac{\nu}{\Omega}}.} \tag{14.63}$$

In other words, the thickness of the boundary layer is that which makes the layer's Ekman number unity: $\text{Ek}(\delta_E) = \nu/(\Omega \delta_E^2) = 1$.

Consider such an "Ekman boundary layer" at the bottom or top of a layer of geostrophically flowing fluid. For the same reasons as for ordinary laminar boundary layers (Sec. 14.4), the generalized pressure P' will be nearly independent of height z through the Ekman layer, that is, it will have the value dictated by the flow just outside the layer: $\nabla P' = -2\rho \mathbf{\Omega} \times \mathbf{V} = \text{const}$. Here $\mathbf{V}$ is the velocity just outside the layer (the velocity of the bulk flow), which we assume to be constant on scales $\sim \delta_E$. Since $\mathbf{\Omega}$ is vertical, $\nabla P'$, like $\mathbf{V}$, will be horizontal (i.e., they both will lie in the x-y plane). To simplify the analysis, we introduce the fluid velocity relative to the bulk flow,

$$\mathbf{w} \equiv \mathbf{v} - \mathbf{V}, \tag{14.64}$$

which goes to zero outside the boundary layer. When rewritten in terms of $\mathbf{w}$, the Navier-Stokes equation (14.56a) [with $\nabla P'/\rho = -2\mathbf{\Omega} \times \mathbf{V}$ and with $d\mathbf{v}/dt = \partial \mathbf{v}/\partial t + (\mathbf{v} \cdot \nabla)\mathbf{v} = 0$, because the flow in the thin boundary layer is steady and $\mathbf{v}$ is horizontal and varies only vertically] takes the simple form $d^2\mathbf{w}/dz^2 = (2/\nu)\mathbf{\Omega} \times \mathbf{w}$. Choosing Cartesian coordinates with an upward vertical z direction, assuming $\mathbf{\Omega} = +\Omega \mathbf{e}_z$ (as is the case for the oceans and atmosphere in the northern hemisphere), and introducing the complex quantity

complex velocity field

$$w = w_x + i w_y \tag{14.65}$$

to describe the horizontal velocity field, we can rewrite $d^2\mathbf{w}/dz^2 = (2/\nu)\mathbf{\Omega} \times \mathbf{w}$ as

$$\frac{d^2 w}{dz^2} = \frac{2i}{\delta_E^2} w = \left(\frac{1+i}{\delta_E}\right)^2 w. \tag{14.66}$$

Equation (14.66) must be solved subject to $w \to 0$ far from the water's boundary and some appropriate condition at the boundary.

For a first illustration of an Ekman layer, consider the effects of a wind blowing in the $\mathbf{e}_x$ direction above a still ocean, $\mathbf{V} = 0$, and set $z = 0$ at the ocean's surface. The

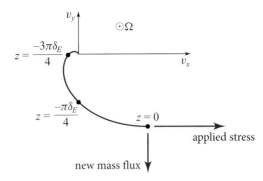

FIGURE 14.16 Ekman spiral (water velocity as a function of depth) at the ocean surface, where the wind exerts a stress.

wind will exert, through a turbulent boundary layer of air, a stress $-T_{xz} > 0$ on the ocean's surface. This stress must be balanced by an equal, viscous stress $\nu\rho\, dw_x/dz$ at the top of the water's boundary layer, $z = 0$. Thus there must be a velocity gradient, $dw_x/dz = -T_{xz}/\nu\rho = |T_{xz}|/\nu\rho$, in the water at $z = 0$. (This replaces the no-slip boundary condition that we have when the boundary is a solid surface.) Imposing this boundary condition along with $w \to 0$ as $z \to -\infty$ (down into the ocean), we find from Eqs. (14.66) and (14.65):

$$v_x = w_x = \left(\frac{|T_{xz}|\delta_E}{\sqrt{2}\,\nu\rho}\right) e^{z/\delta_E} \cos(z/\delta_E - \pi/4),$$

$$v_y = w_y = \left(\frac{|T_{xz}|\delta_E}{\sqrt{2}\,\nu\rho}\right) e^{z/\delta_E} \sin(z/\delta_E - \pi/4), \qquad (14.67)$$

wind-driven Ekman boundary layer at top of a geostrophic flow

for $z \leq 0$ (Fig. 14.16). As a function of depth, this velocity field has the form of a spiral—the so-called *Ekman spiral*. When $\boldsymbol{\Omega}$ points toward us (as in Fig. 14.16), the spiral is clockwise and tightens as we move away from the boundary ($z = 0$ in the figure) into the bulk flow.

structure of the boundary layer: Ekman spiral

By integrating the mass flux $\rho\mathbf{w}$ over z, we find for the total mass flowing per unit time per unit length of the ocean's surface

$$\mathbf{F} = \rho \int_{-\infty}^{0} \mathbf{w}\, dz = -\frac{\delta_E^2}{2\nu}|T_{xz}|\mathbf{e}_y; \qquad (14.68)$$

see Fig. 14.16. Thus the wind, blowing in the $\mathbf{e}_x$ direction, causes a net mass flow in the direction of $\mathbf{e}_x \times \boldsymbol{\Omega}/\Omega = -\mathbf{e}_y$. This response (called "Ekman pumping") may seem less paradoxical if one recalls how a gyroscope responds to applied forces.

Ekman pumping

This mechanism is responsible for the creation of *gyres* in the oceans (Ex. 14.21 and Fig. 14.18 below).

gyres

As a second illustration of an Ekman boundary layer, we consider a geostrophic flow with nonzero velocity $\mathbf{V} = V\mathbf{e}_x$ in the bulk of the fluid, and examine this flow's

14.5 Nearly Rigidly Rotating Flows—Earth's Atmosphere and Oceans

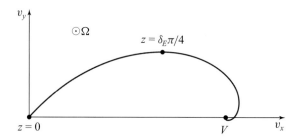

FIGURE 14.17 Ekman spiral (water velocity as a function of height) in the bottom boundary layer, when the bulk flow above it moves geostrophically with speed $V\mathbf{e}_x$.

interaction with a static, solid surface at its bottom. We set $z = 0$ at the bottom, with z increasing upward, in the $\boldsymbol{\Omega}$ direction. The structure of the boundary layer on the bottom is governed by the same differential equation (14.66) as at the wind-blown surface, but with altered boundary conditions. The solution is

Ekman boundary layer at bottom of a geostrophic flow

$$v_x - V = w_x = -V\exp(-z/\delta_E)\cos(z/\delta_E),$$
$$v_y = w_y = +V\exp(-z/\delta_E)\sin(z/\delta_E). \tag{14.69}$$

This solution is shown in Fig. 14.17.

Recall that we have assumed $\boldsymbol{\Omega}$ points in the upward $+z$ direction, which is appropriate for the ocean and atmosphere in the northern hemisphere. If, instead, $\boldsymbol{\Omega}$ points downward (as in the southern hemisphere), then the handedness of the Ekman spiral is reversed.

Ekman boundary layers drive circulation in rotating fluids

Ekman boundary layers are important, because they can circulate rotating fluids faster than viscous diffusion can. Suppose we have a nonrotating container (e.g., a tea cup) of radius L containing a fluid that rotates with angular velocity Ω (e.g., due to stirring; cf. Ex. 14.22). As you will see in your analysis of Ex. 14.22, the Ekman layer at the container's bottom experiences a pressure difference between the wall and the container's center given by $\Delta P \sim \rho L^2 \Omega^2$. This drives a fluid circulation in the Ekman layer, from the wall toward the center, with radial speed $V \sim \Omega L$. The circulating fluid must upwell at the bottom's center from the Ekman layer into the bulk fluid. This produces a poloidal mixing of the fluid on a timescale given by

$$t_E \sim \frac{L^3}{L\delta_E V} \sim \frac{L\delta_E}{\nu}. \tag{14.70}$$

This timescale is shorter than that for simple diffusion of vorticity, $t_\nu \sim L^2/\nu$, by a factor $t_E/t_\nu \sim \sqrt{\mathrm{Ek}}$, which (as we have seen) can be very small. This circulation and mixing are key to the piling up of tea leaves at the bottom center of a stirred tea cup, and to the mixing of the tea or milk into the cup's hot water (Ex. 14.22).

The circulation driven by an Ekman layer is an example of a *secondary flow*—a weakly perturbative bulk flow that is produced by interaction with a boundary. For other examples of secondary flows, see Taylor (1968).

secondary flows

EXERCISES

Exercise 14.21 **Example: Winds and Ocean Currents in the North Atlantic*
The north Atlantic Ocean exhibits the pattern of winds and ocean currents shown in Fig. 14.18. Westerly winds blow from west to east at 40° latitude. Trade winds blow from east to west at 20° latitude. In between, around 30° latitude, is the Sargasso Sea: a 1.5-m-high gyre (raised hump of water). The gyre is created by ocean surface currents, extending down to a depth of only about 30 m, that flow northward from the trade-wind region and southward from the westerly wind region (upper inset in Fig. 14.18). A deep ocean current, extending from the surface down to near the bottom, circulates around the Sargasso Sea gyre in a clockwise manner. This current goes under different names in different regions of the ocean: Gulf Stream, West Wind Drift, Canaries Current, and North Equatorial Current. Explain both qualitatively and semiquantitatively (in terms of order of magnitude) how the winds are ultimately responsible for all these features of the ocean. More specifically, do the following.

(a) Explain the surface currents in terms of an Ekman layer at the top of the ocean, with thickness δ_E about 30 m. From this measured δ_E compute the kinematic viscosity ν in the boundary layer. Your result, $\nu \sim 0.03 \text{ m}^2 \text{ s}^{-1}$, is far larger than

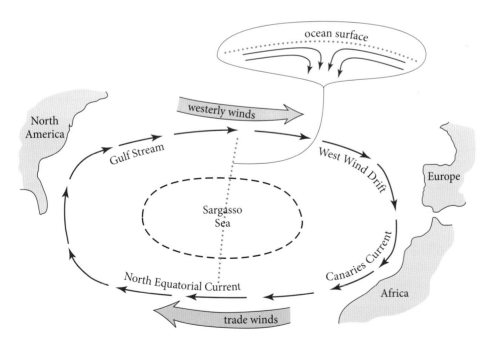

FIGURE 14.18 Winds and ocean currents in the north Atlantic. The upper inset shows the surface currents, along the dotted north-south line, that produce the Sargasso Sea gyre.

14.5 Nearly Rigidly Rotating Flows—Earth's Atmosphere and Oceans the molecular viscosity of water, $\sim 10^{-6}$ m^2 s^{-1} (Table 13.2). The reason is that the boundary layer is turbulent, and its eddies produce this large viscosity (see Sec. 15.4.2).

(b) Explain why the height of the gyre that the surface currents produce in the Sargasso Sea is about 1.5 m.

(c) Explain the deep ocean current (Gulf Stream, etc.) in terms of a geostrophic flow, and estimate the speed of this current. This current, like circulation in a tea cup, is an example of a "secondary flow."

(d) If there were no continents on Earth, but only an ocean of uniform depth, what would be the flow pattern of this deep current—its directions of motion at various locations around Earth, and its speeds? The continents (North America, Europe, and Africa) must be responsible for the deviation of the actual current (Gulf Stream, etc.) from this continent-free flow pattern. How do you think the continents give rise to the altered flow pattern?

Exercise 14.22 **Example: Circulation in a Tea Cup*

Place tea leaves and water in a tea cup, glass, or other larger container. Stir the water until it is rotating uniformly, and then stand back and watch the motion of the water and leaves. Notice that the tea leaves tend to pile up at the cup's center. An Ekman boundary layer on the bottom of the cup is responsible for this phenomenon. In this exercise you explore the origin and consequences of this Ekman layer.

(a) Evaluate the pressure distribution $P(\varpi, z)$ in the bulk flow (outside all boundary layers), assuming that it rotates rigidly. (Here z is height and ϖ is distance from the water's rotation axis.) Perform your evaluation in the water's rotating reference frame. From this $P(\varpi, z)$ deduce the shape of the top surface of the water. Compare your deduced shape with the actual shape in your experiment.

(b) Estimate the thickness of the Ekman layer at the bottom of your container. It is very thin. Show, using the Ekman spiral diagram (Fig. 14.17), that the water in this Ekman layer flows inward toward the container's center, causing the tea leaves to pile up at the center. Estimate the radial speed of this Ekman-layer flow and the mass flux that it carries.

(c) To get a simple physical understanding of this inward flow, examine the radial gradient $\partial P/\partial \varpi$ of the pressure P in the bulk flow just above the Ekman layer. Explain why $\partial P/\partial \varpi$ in the Ekman layer will be the same as in the rigidly rotating flow above it. Then apply force balance in an inertial frame to deduce that the water in the Ekman layer will be accelerated inward toward the center.

(d) Using geostrophic-flow arguments, deduce the fate of the boundary-layer water after it reaches the center of the container's bottom. Where does it go? What is the large-scale circulation pattern that results from the "driving force" of the Ekman layer's mass flux? What is the Rossby number for this large-scale circulation pattern? How and where does water from the bulk, nearly rigidly rotating flow, enter the bottom boundary layer so as to be swept inward toward the center?

(e) Explain how this large-scale circulation pattern can mix much of the water through the boundary layer in the time t_E of Eq. (14.70). What is the value of this t_E for the water in your container? Explain why this, then, must also be the time for the angular velocity of the bulk flow to slow substantially. Compare your computed value of t_E with the observed slow-down time for the water in your container.

Exercise 14.23 **Problem: Water down a Drain in Northern and Southern Hemispheres*

One often hears the claim that water in a bathtub or basin swirls down a drain clockwise in the northern hemisphere and counterclockwise in the southern hemisphere. In fact, on YouTube you are likely to find video demonstrations of this (e.g., by searching on "water down drain at equator"). Show that for Earth-rotation centrifugal forces to produce this effect, it is necessary that the water in the basin initially be moving with a speed smaller than

$$v_{max} \sim a\Omega_* \sim a\Omega_\oplus \frac{\ell}{R_\oplus} \sim \frac{30 \text{ cm}}{\text{yr}} \left(\frac{a}{1 \text{ m}}\right)\left(\frac{\ell}{1 \text{ km}}\right). \tag{14.71}$$

Here $\Omega_\oplus$ is Earth's rotational angular velocity, Ω_* is its vertical component (Box 14.5), $R_\oplus$ is Earth's radius, a is the radius of the basin, and ℓ is the distance of the basin from the equator. Even for a basin in Europe or North America, this maximum speed is $\sim$3 mm min^{-1} for a 1-m-diameter basin—exceedingly difficult to achieve. Therefore, the residual initial motion of the water in any such basin will control the direction in which the water swirls down the drain. There is effectively no difference between northern and southern hemispheres.

Exercise 14.24 **Example: Water down Drain: Experiment*

(a) In a shower or bathtub with the drain somewhere near the center, not the wall, set water rotating so a whirlpool forms over the drain. Perform an experiment to see where the water going down the drain comes from: the surface of the water, its bulk, or its bottom. For example, you could sprinkle baby powder on top of the water, near the whirlpool, and measure how fast the powder is pulled inward and down the drain; put something neutrally buoyant in the bulk and watch its motion; and put sand on the bottom of the shower near the whirlpool and measure how fast the sand is pulled inward and down the drain.

(b) Explain the result of your experiment in part (a). How is it related to the tea cup problem, Ex. 14.22?

(c) Compute the shape of the surface of the water near and in the whirlpool.

14.6 Instabilities of Shear Flows—Billow Clouds and Turbulence in the Stratosphere

Kelvin-Helmholtz instability

Here we explore the stability of a variety of shear flows. We begin with the simplest case of two incompressible fluids, one above the other, that move with different uniform speeds, when gravity is negligible. Such a flow has a delta-function spike of vorticity at the interface between the fluids, and, as we shall see, the interface is always unstable against growth of so-called *internal waves*. This is the *Kelvin-Helmholtz instability*. We then explore the ability of gravity to suppress this instability. If the densities of the two fluids are nearly the same, there is no suppression, which is why the Kelvin-Helmholtz instability is seen in a variety of places in Earth's atmosphere and oceans. If the densities are substantially different, then gravity easily suppresses the instability, unless the two flow speeds are very different. Finally, we allow the density and horizontal velocity to change continuously in the vertical direction (e.g., in Earth's stratosphere), and for such a flow we deduce the *Richardson criterion for instability*, which is often satisfied in the stratosphere and leads to turbulence. Along the way we briefly visit several other instabilities that occur in stratified fluids.

14.6.1 Discontinuous Flow: Kelvin-Helmholtz Instability

A particularly interesting and simple type of vorticity distribution is one where the vorticity is confined to a thin, plane interface between two immiscible fluids. In other words, one fluid is in uniform motion relative to the other. This type of flow arises quite frequently; for example, when the wind blows over the ocean or when smoke from a smokestack discharges into the atmosphere (a flow that is locally but not globally planar).

We analyze the stability of such flows, initially without gravity (this subsection), then with gravity present (next subsection). Our analysis provides another illustration of the behavior of vorticity and an introduction to techniques that are commonly used to analyze fluid instabilities.

We restrict attention to the simplest version of this flow: an equilibrium with a fluid of density ρ_+ moving horizontally with speed V above a second fluid, which is at rest, with density ρ_-. Let x be a Cartesian coordinate measured along the planar interface in the direction of the flow, and let y be measured perpendicular to it. The equilibrium contains a sheet of vorticity lying in the plane $y = 0$, across which the velocity changes discontinuously. Now, this discontinuity ought to be treated as a boundary layer, with a thickness determined by the viscosity. However, in this problem we analyze disturbances with lengthscales much greater than the thickness of the boundary layer, and so we can ignore it. As a corollary, we can also ignore viscous stresses in the body of the flow. In addition, we specialize to very subsonic speeds, for which the flow can be treated as incompressible; we also ignore the effects of surface tension as well as gravity.

A full description of this flow requires solving the full equations of fluid dynamics, which are quite nonlinear and, as it turns out, for this problem can only be solved numerically. However, we can make progress analytically on an important subproblem. This is the issue of whether this equilibrium flow is stable to small perturbations, and if unstable, the nature of the growing modes. To answer this question, we linearize the fluid equations in the amplitude of the perturbations.

We consider a small vertical perturbation $\delta y = \xi(x, t)$ in the location of the interface (see Fig. 14.19a below). We denote the associated perturbations to the pressure and velocity by δP and $\delta \mathbf{v}$. That is, we write

$$P(\mathbf{x}, t) = P_0 + \delta P(\mathbf{x}, t), \quad \mathbf{v} = V H(y)\mathbf{e}_x + \delta \mathbf{v}(\mathbf{x}, t), \tag{14.72}$$

where P_0 is the constant pressure in the equilibrium flow about which we are perturbing, V is the constant speed of the flow above the interface, and $H(y)$ is the Heaviside step function (1 for $y > 0$ and 0 for $y < 0$). We substitute these $P(\mathbf{x}, t)$ and $\mathbf{v}(\mathbf{x}, t)$ into the governing equations: the incompressibility relation,

$$\nabla \cdot \mathbf{v} = 0, \tag{14.73}$$

and the viscosity-free Navier-Stokes equation (i.e., the Euler equation),

$$\frac{d\mathbf{v}}{dt} = \frac{-\nabla P}{\rho}. \tag{14.74}$$

We then subtract off the equations satisfied by the equilibrium quantities to obtain, for the perturbed variables,

$$\nabla \cdot \delta\mathbf{v} = 0, \tag{14.75}$$

$$\frac{d\delta\mathbf{v}}{dt} = -\frac{\nabla \delta P}{\rho}. \tag{14.76}$$

Combining these two equations, we find, as for Stokes flow (Sec. 14.3.2), that the pressure satisfies Laplace's equation:

$$\nabla^2 \delta P = 0. \tag{14.77}$$

We now follow the procedure used in Sec. 12.4.2 when treating Rayleigh waves on the surface of an elastic medium: we seek an internal wave mode, in which the perturbed quantities vary $\propto \exp[i(kx - \omega t)] f(y)$, with $f(y)$ dying out away from the interface. From Laplace's equation (14.77), we infer an exponential falloff with $|y|$:

$$\delta P = \delta P_0 e^{-k|y| + i(kx - \omega t)}, \tag{14.78}$$

where δP_0 is a constant.

Our next step is to substitute this δP into the perturbed Euler equation (14.76) to obtain

$$\delta v_y = \frac{i k \delta P}{(\omega - kV)\rho_+} \text{ for } y > 0, \quad \delta v_y = \frac{-i k \delta P}{\omega \rho_-} \text{ for } y < 0. \tag{14.79}$$

14.6 Instabilities of Shear Flows—Billow Clouds and Turbulence in the Stratosphere

We must impose two boundary conditions at the interface between the fluids: continuity of the vertical displacement ξ of the interface (the tangential displacement need not be continuous, since we are examining scales large compared to the boundary layer), and continuity of the pressure P across the interface. [See Eq. (12.44) and associated discussion for the analogous boundary conditions at a discontinuity in an elastic medium.] The vertical interface displacement ξ is related to the velocity perturbation by $d\xi/dt = \delta v_y(y=0)$, which implies that

$$\xi = \frac{i\delta v_y}{(\omega - kV)} \quad \text{at } y = 0_+ \quad \text{(immediately above the interface),}$$

$$\xi = \frac{i\delta v_y}{\omega} \quad \text{at } y = 0_- \quad \text{(immediately below the interface).} \tag{14.80}$$

Then, by virtue of Eqs. (14.78), (14.80), and (14.79), the continuity of pressure and vertical displacement at $y = 0$ imply that

$$\rho_+(\omega - kV)^2 + \rho_-\omega^2 = 0, \tag{14.81}$$

where ρ_+ and ρ_- are the densities of the fluid above and below the interface, respectively. Solving for frequency ω as a function of horizontal wave number k, we obtain the following dispersion relation for internal wave modes localized at the interface, which are also called *linear Kelvin-Helmholtz modes*:

dispersion relation for linear Kelvin-Helmholtz modes at an interface

$$\boxed{\omega = kV\left(\frac{\rho_+ \pm i(\rho_+\rho_-)^{1/2}}{\rho_+ + \rho_-}\right).} \tag{14.82}$$

This dispersion relation can be used to describe both a sinusoidal perturbation whose amplitude grows in time, and a time-independent perturbation that grows spatially:

TEMPORAL GROWTH

Suppose that we create some small, localized disturbance at time $t = 0$. We can Fourier analyze the disturbance in space, and—as we have linearized the problem— can consider the temporal evolution of each Fourier component separately. What we ought to do is solve the initial value problem carefully, taking account of the initial conditions. However, when there are growing modes, we can usually infer the long-term behavior by ignoring the transients and just considering the growing solutions. In our case, we infer from Eqs. (14.78)–(14.80) and the dispersion relation (14.82) that a mode with spatial frequency k must grow as

temporal growth of mode with wavelength $2\pi/k$

$$\delta P, \xi \propto \exp\left[\left(\frac{kV(\rho_+\rho_-)^{1/2}}{(\rho_+ + \rho_-)}\right)t + ik\left(x - V\frac{\rho_+}{\rho_+ + \rho_-}t\right)\right]. \tag{14.83}$$

Thus, this mode grows exponentially with time (Fig. 14.19a). Note that the mode is nondispersive, and if the two densities are equal, it e-folds in $1/(2\pi)$ times a period. This means that the fastest modes to grow are those with the shortest periods and hence the shortest wavelengths. (However, the wavelength must not approach the

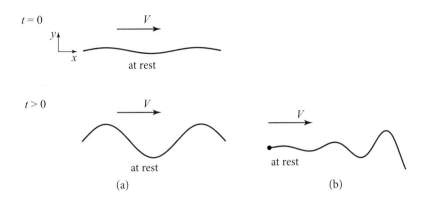

FIGURE 14.19 Kelvin-Helmholtz instability. (a) Temporally growing mode. (b) Spatially growing mode.

thickness of the boundary layer and thereby compromise our assumption that the effects of viscosity are negligible.)

We can understand this growing mode somewhat better by transforming to the center-of-momentum frame, which moves with speed $\rho_+ V/(\rho_+ + \rho_-)$ relative to the frame in which the lower fluid is at rest. In this (primed) frame, the velocity of the upper fluid is $V' = \rho_- V/(\rho_+ + \rho_-)$, and so the perturbations evolve as

$$\delta P, \xi \propto \exp[kV'(\rho_+/\rho_-)^{1/2}t]\cos(kx'). \tag{14.84}$$

In this frame the wave is purely growing, whereas in our original frame it oscillated with time as it grew.

SPATIAL GROWTH

An alternative type of mode is one in which a small perturbation is excited temporally at some point where the shear layer begins (Fig. 14.19b). In this case we regard the frequency ω as real and look for the mode with negative imaginary k corresponding to spatial growth. Using Eq. (14.82), we obtain

$$k = \frac{\omega}{V}\left[1 - i\left(\frac{\rho_-}{\rho_+}\right)^{1/2}\right]. \tag{14.85}$$

spatial growth of mode with angular frequency ω

The mode therefore grows exponentially with distance from the point of excitation.

PHYSICAL INTERPRETATION OF THE INSTABILITY

We have performed a normal mode analysis of a particular flow and discovered that there are unstable internal wave modes. However much we calculate the form of the growing modes, though, we cannot be said to understand the instability until we can explain it physically. In the case of this Kelvin-Helmholtz instability, this task is simple.

physical origin of Kelvin-Helmholtz instability

The flow pattern is shown schematically in Fig. 14.20. The upper fluid will have to move faster when passing over a crest in the water wave, because the cross sectional

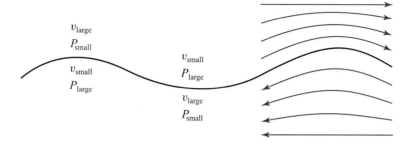

FIGURE 14.20 Physical explanation for the Kelvin-Helmholtz instability.

area of a flow tube diminishes and the flux of fluid must be conserved. By Bernoulli's theorem, the upper pressure will be lower than ambient at this point, and so the crest will rise even higher. Conversely, in the trough of the wave the upper fluid will travel slower, and its pressure will increase. The pressure differential will push the trough downward, making it grow.

Equivalently, we can regard the boundary layer as a plane containing parallel vortex lines that interact with one another, much like magnetic field lines exert pressure on one another. When the vortex lines all lie strictly in a plane, they are in equilibrium, because the repulsive force exerted by one on its neighbor is balanced by an opposite force exerted by the opposite neighbor. However, when this equilibrium is disturbed, the forces become unbalanced and the vortex sheet effectively buckles.

More generally, when there is a large amount of relative kinetic energy available in a high-Reynolds-number flow, there exists the possibility of unstable modes that can tap this energy and convert it into a spectrum of growing modes. These modes can then interact nonlinearly to create fluid turbulence, which ultimately is dissipated as heat. However, the availability of free kinetic energy does not necessarily imply that the flow is unstable; sometimes it is stable. Instability must be demonstrated—often a very difficult task.

14.6.2 Discontinuous Flow with Gravity

What happens to this Kelvin-Helmholtz instability if we turn on gravity? To learn the answer, we insert a downward gravitational acceleration **g** into the above analysis. The result (Ex. 14.25) is the following modification of the dispersion relation (14.82):

$$\frac{\omega}{k} = \frac{\rho_+ V}{\rho_+ + \rho_-} \pm \left[\frac{g}{k}\left(\frac{\rho_- - \rho_+}{\rho_+ + \rho_-}\right) - \frac{\rho_+ \rho_-}{(\rho_+ + \rho_-)^2} V^2 \right]^{1/2}. \quad (14.86)$$

gravity suppresses Kelvin-Helmholtz instability

If the lower fluid is sufficiently more dense than the upper fluid (ρ_- sufficiently bigger than ρ_+), then gravity g will change the sign of the quantity inside the square root, making ω/k real, which means the Kelvin-Helmholtz instability is suppressed. In other words, *for the flow to be Kelvin-Helmholtz unstable in the presence of gravity, the two fluids must have nearly the same density:*

FIGURE 14.21 Billow clouds above San Francisco. The cloud structure is generated by the Kelvin-Helmholtz instability. These types of billow clouds may have inspired the swirls in Vincent van Gogh's famous painting "Starry Night." Courtesy of Science Source.

$$\frac{\rho_- - \rho_+}{\rho_+ + \rho_-} < \frac{\rho_+\rho_-}{(\rho_+ + \rho_-)^2}\frac{kV^2}{g}. \tag{14.87}$$

This is the case for many interfaces in Nature, which is why the Kelvin-Helmholtz instability is often seen. An example is the interface between a water-vapor-laden layer of air under a fast-moving, drier layer. The result is the so-called "billow clouds" shown in Fig. 14.21. Other examples are flow interfaces in the ocean and the edges of dark smoke pouring out of a smoke stack.

As another application of the dispersion relation (14.86), consider the excitation of ocean waves by a laminar-flowing wind. In this case, the "+" fluid is air, and the "−" fluid is water, so the densities are very different: $\rho_+/\rho_- \simeq 0.001$. The instability criterion (14.87) tells us the minimum wind velocity V required to make the waves grow:

$$V_{\min} \simeq \left(\frac{g}{k}\frac{\rho_-}{\rho_+}\right)^{1/2} \simeq 450 \text{ km h}^{-1}\sqrt{\lambda/10 \text{ m}}, \tag{14.88}$$

where $\lambda = 2\pi/k$ is the waves' wavelength.

Obviously, this answer is physically wrong. Water waves are easily driven by winds that are far slower than this. Evidently, some other mechanism of interaction between wind and water must drive the waves much more strongly. Observations of wind over water reveal the answer: The winds near the sea's surface are typically quite turbulent (Chap. 15), not laminar. The randomly fluctuating pressures in a turbulent wind are far more effective than the smoothly varying pressure of a laminar wind in driving ocean waves. For two complementary models of this, see Phillips (1957) and Miles (1993).

As another, very simple application of the dispersion relation (14.86), set the speed V of the upper fluid to zero. In this case, the interface is unstable if and only if the upper

Rayleigh-Taylor instability

fluid has higher density than the lower fluid: $\rho_+ > \rho_-$. This is called the *Rayleigh-Taylor instability* for incompressible fluids.

EXERCISES

Exercise 14.25 *Problem: Discontinuous Flow with Gravity*
Insert gravity into the analysis of the Kelvin-Helmholtz instability (with the uniform gravitational acceleration **g** pointing perpendicularly to the fluid interface, from the upper "+" fluid to the lower "−" fluid). Thereby derive the dispersion relation (14.86).

14.6.3 Smoothly Stratified Flows: Rayleigh and Richardson Criteria for Instability

Sometimes one can diagnose a fluid instability using simple physical arguments rather than detailed mathematical analysis. We conclude this chapter with two examples.

First, we consider *rotating Couette flow*: the azimuthal flow of an incompressible, effectively inviscid fluid, rotating axially between two coaxial cylinders (Fig. 14.22).

We explore the stability of this flow to purely axisymmetric perturbations by using a thought experiment in which we interchange two fluid rings. As there are no azimuthal forces (no forces in the ϕ direction), the interchange will occur at constant angular momentum per unit mass. Suppose that the ring that moves outward in radius ϖ has lower specific angular momentum j than the surroundings into which it has moved; then it will have less centrifugal force per unit mass j^2/ϖ^3 than its surroundings and will thus experience a restoring force that drives it back to its original position. Conversely, if the surroundings' angular momentum per unit mass decreases outward, then the displaced ring will continue to expand. We conclude on this basis that *Couette and similar flows are unstable when the angular momentum per unit mass decreases outward*. This is known as the *Rayleigh criterion*. We return to Couette flow in Sec. 15.6.1.

Compact stellar objects (black holes, neutron stars, and white dwarfs) are sometimes surrounded by orbiting *accretion disks* of gas (Exs. 26.17 and 26.18). The gas in these disks orbits with its angular velocity approximately dictated by Kepler's laws. Therefore, the specific angular momentum of the gas increases radially outward, approximately proportional to the square root of the orbital radius. Consequently, accretion disks are stable to this type of instability.

For our second example of a simple physical diagnosis of instability, consider the situation analyzed in Ex. 14.25 (Kelvin-Helmholtz instability with gravity present) but with the density and velocity changing continuously instead of discontinuously, as one moves upward. More specifically, focus on Earth's stratosphere, which extends from ~10 km height to ~40 km. The density in the stratosphere decreases upward faster than if the stratosphere were isentropic. This means that, when a fluid element moves upward adiabatically, its pressure-induced density reduction is smaller than the density decrease in its surroundings. Since it is now more dense than its surroundings, the downward pull of gravity on the fluid element will exceed the upward buoyant

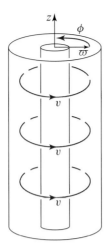

FIGURE 14.22 Rotating Couette flow.

force, so the fluid element will be pulled back down. Therefore, the stratosphere is stably stratified. (We shall return to this phenomenon, in the context of stars, in Sec. 18.5.)

However, it may be possible for the stratosphere to tap the relative kinetic energy in its horizontal winds, so as to mix the air vertically. Consider the pedagogical example of two thin stream tubes of horizontally flowing air in the stratosphere, separated vertically by a distance $\delta\ell$ large compared to their thicknesses. The speed of the upper stream will exceed that of the lower stream by $\delta V = V'\delta\ell$, where $V' = dV/dz$ is the velocity gradient. In the center of velocity frame, the streams each have speed $\delta V/2$, and they differ in density by $\delta\rho = |\rho'\delta\ell|$. To interchange these streams requires doing work per unit mass[7] $\delta W = g[\delta\rho/(2\rho)]\delta\ell$ against gravity (where the factor 2 comes from the fact that there are two unit masses involved in the interchange, one going up and the other coming down). This work can be supplied by the streams' kinetic energy, if the available kinetic energy per unit mass, $\delta E_k = (\delta V/2)^2/2$, exceeds the required work. *A necessary condition for instability* is then that

$$\delta E_k = (\delta V)^2/8 > \delta W = g|\rho'/(2\rho)|\delta\ell^2, \tag{14.89}$$

or

$$\boxed{\mathrm{Ri} = \frac{|\rho'|g}{\rho V'^2} < \frac{1}{4}}, \tag{14.90}$$

Richardson criterion for instability of a stratified shear flow

---

7. Here, for simplicity we idealize the streams' densities as not changing when they move vertically—an idealization that makes gravity be more effective at resisting the instability. In the stratosphere, where the temperature is constant (Fig. 13.2) so the density drops vertically much faster than if it were isentropic, this idealization is pretty good.

where $V' = dV/dz$ is the velocity gradient. This is known as the *Richardson criterion for instability*, and Ri is the *Richardson number*. Under a wide variety of circumstances this criterion turns out to be sufficient for instability.

The density scale height in the stratosphere is $\rho/|\rho'| \sim 10$ km. Therefore the maximum velocity gradient allowed by the Richardson criterion is

$$V' \lesssim 60 \frac{\text{m s}^{-1}}{\text{km}}. \tag{14.91}$$

Larger velocity gradients are rapidly disrupted by instabilities. This instability is responsible for much of the clear air turbulence encountered by airplanes, and it is to a discussion of turbulence that we turn in the next chapter.

Bibliographic Note

Vorticity is so fundamental to fluid mechanics that it and its applications are treated in detail in all fluid-dynamics textbooks. Among those with a physicist's perspective, we particularly like Acheson (1990) and Lautrup (2005) at an elementary level, and Landau and Lifshitz (1959), Lighthill (1986), and Batchelor (2000) at a more advanced level. Tritton (1987) is especially good for physical insight. Panton (2005) is almost encyclopedic and nicely bridges the viewpoints of physicists and engineers. For an engineering emphasis, we like Munson, Young, and Okiishi (2006) and Potter, Wiggert, and Ramadan (2012); see also the more advanced text on viscous flow, White (2006). For the viewpoint of an applied mathematician, see Majda and Bertozzi (2002).

To build up physical intuition, we recommend Tritton (1987) and the movies listed in Box 14.2. For a textbook treatment of rotating flows and weak perturbations of them, we recommend Greenspan (1973). For geophysical applications at an elementary level, we like Cushman-Roisin and Beckers (2011), and for a more mathematical treatment Pedlosky (1987). A more encyclopedic treatment at an elementary level (including some discussion of simulations) can be found in Gill (1982) and Vallis (2006).

CHAPTER FIFTEEN

15

Turbulence

> Big whirls have little whirls, which feed on their velocity.
> Little whirls have lesser whirls, and so on to viscosity.
> LEWIS RICHARDSON (1922)

15.1 Overview

In Sec. 13.7.6, we derived the Poiseuille formula for the flow of a viscous fluid down a pipe by assuming that the flow is laminar (i.e., that it has a velocity parallel to the pipe wall). We showed how balancing the stress across a cylindrical surface led to a parabolic velocity profile and a rate of flow proportional to the fourth power of the pipe diameter d. We also defined the Reynolds number; for pipe flow it is $\text{Re}_d \equiv \bar{v}d/\nu$, where $\bar{v}$ is the mean speed in the pipe, and ν is the kinematic viscosity. Now, it turns out experimentally that the pipe flow only remains laminar up to a critical Reynolds number that has a value in the range $\sim 10^3$–10^5, depending on the smoothness of the pipe's entrance and roughness of its walls. If the pressure gradient is increased further (and thence the mean speed $\bar{v}$ and Reynolds number Re_d are increased), then the velocity field in the pipe becomes irregular both temporally and spatially, a condition known as *turbulence*.

Turbulence is common in high-Reynolds-number flows. Much of our experience with fluids involves air or water, for which the kinematic viscosities are $\sim 10^{-5}$ and 10^{-6} m^2 s^{-1}, respectively. For a typical everyday flow with a characteristic speed of $v \sim 10$ m s^{-1} and a characteristic length of $d \sim 1$ m, the Reynolds number is huge: $\text{Re} = vd/\nu \sim 10^6 - 10^7$. It is therefore not surprising that we see turbulent flows all around us. Smoke in a smokestack, a cumulus cloud, and the wake of a ship are examples.

In Sec. 15.2 we illustrate the phenomenology of the transition to turbulence as the Reynolds number increases using a particularly simple example: the flow of a fluid past a circular cylinder oriented perpendicular to the flow's incoming velocity. We shall see how the flow pattern is dictated by the Reynolds number, and how the velocity changes from steady creeping flow at low Re to fully developed turbulence at high Re.

> **BOX 15.1. READERS' GUIDE**
>
> - This chapter relies heavily on Chaps. 13 and 14.
> - The remaining chapters on fluid mechanics and magnetohydrodynamics (Chaps. 16–19) do not rely significantly on this chapter, nor do any of the remaining chapters in this book.

What is turbulence? Fluid dynamicists can recognize it, but they have a hard time defining it precisely[1] and an even harder time describing it quantitatively.[2] So typically for a definition they rely on empirical, qualitative descriptions of its physical properties (Sec. 15.3). Closely related to this description is the crucial role of vorticity in driving turbulent energy from large scales to small (Sec. 15.3.1).

At first glance, a quantitative description of turbulence appears straightforward. Decompose the velocity field into Fourier components just as is done for the electromagnetic field when analyzing electromagnetic radiation. Then recognize that the equations of fluid dynamics are nonlinear, so there will be coupling between different modes (akin to wave-wave coupling between optical modes in a nonlinear crystal, discussed in Chap. 10). Analyze that coupling perturbatively. The resulting *weak-turbulence formalism* is sketched in Secs. 15.4.1 and 15.4.2 and in Ex. 15.5.[3]

However, most turbulent flows come under the heading of *fully developed* or *strong turbulence* and cannot be well described by weak-turbulence models. Part of the problem is that the $(\mathbf{v} \cdot \nabla)\mathbf{v}$ term in the Navier-Stokes equation is a strong nonlinearity, not a weak coupling between linear modes. As a consequence, eddies of size ℓ persist for typically no more than one turnover timescale $\sim \ell/v$ before they are broken up, and so they do not behave like weakly coupled normal modes.

In the absence of a decent quantitative theory of strong turbulence, fluid dynamicists sometimes simply push the weak-turbulence formalism into the strong-turbulence regime and use it there to gain qualitative or semiquantitative insights (e.g., Fig. 15.7 below and associated discussion in the text). A simple alternative (which we explore in Sec. 15.4.3 in the context of wakes and jets and in Sec. 15.5 for turbulent boundary layers) is intuitive, qualitative, and semiquantitative approaches to the *physical* description of turbulence. We emphasize the adjective "physical," because our goal is to start to comprehend the underlying physical character of turbulence, going beyond empirical descriptions of its consequences on the one hand and uninstruc-

---

1. The analogy to Justice Potter Stewart's definition of pornography should be resisted.
2. Werner Heisenberg's dissertation was "On the Stability and Turbulence of Fluid Flow." He was disappointed with his progress and was glad to change to a more tractable problem.
3. Another weak-turbulence formalism that is developed along similar lines is the *quasilinear* theory of nonlinear plasma interactions, which we discuss in Chap. 23.

tive mathematical expansions on the other. Much modern physics has this character. An important feature that we meet in Sec. 15.4.3, when we discuss wakes and jets, is *entrainment*. This leads to irregular boundaries of turbulent flows caused by giant eddies and to dramatic time dependence, including *intermittency*.

One triumph of this approach (Sec. 15.4.4) is the Kolmogorov analysis of the shape of the time-averaged turbulent energy spectrum (the turbulent energy per unit wave number as a function of wave number) in a stationary turbulent flow. This spectrum has been verified experimentally under many different conditions. The arguments used to justify it are characteristic of many semiempirical derivations of scaling relations that find confident practical application in the world of engineering.

In the context of turbulent boundary layers, our physical approach will reveal semiquantitatively the structures of such boundary layers (Sec. 15.5.1), and it will explain why turbulent boundary layers generally exert more shear stress on a surface than do laminar boundary layers, but nevertheless usually produce less total drag on airplane wings, baseballs, etc. (Sec. 15.5.2).

Whether or not a flow becomes turbulent can have a major influence on how fast chemical reactions occur in liquids and gases. In Sec. 15.5.3, we briefly discuss how turbulence can arise through instability of a laminar boundary layer.

One can gain additional insight into turbulence by a technique that is often useful when struggling to understand complex physical phenomena: Replace the system being studied by a highly idealized model system that is much simpler than the original one, both conceptually and mathematically, but that retains at least one central feature of the original system. Then analyze the model system completely, with the hope that the quantitative insight so gained will be useful in understanding the original problem. Since the 1970s, new insights into turbulence have come from studying idealized dynamical systems that have very few degrees of freedom but have the same kinds of nonlinearities as produce turbulence in fluids (e.g., Ott, 1982; Ott, 1993). We examine several such low-dimensional dynamical systems and the insights they give in Sec. 15.6.

The most useful of those insights deal with the onset of weak turbulence and the observation that it seems to have much in common with the onset of chaos (irregular and unpredictable dynamical behavior) in a wide variety of other dynamical systems. A great discovery of modern classical physics/mathematics has been that there exist organizational principles that govern the behavior of these seemingly quite different chaotic physical systems.

In parallel with studying this chapter, to build up physical intuition the reader is urged to watch movies and study photographs that deal with turbulence; see Box 15.2.

15.2 The Transition to Turbulence—Flow Past a Cylinder

We illustrate qualitatively how a flow (and especially its transition to turbulence) depends on its Reynolds number by considering a specific problem: the flow of a

> **BOX 15.2. MOVIES AND PHOTOGRAPHS ON TURBULENCE**
>
> The most relevant movies for this chapter are:
>
> - Stewart (1968)—demonstrates key features of turbulent flows.
> - Rouse (1963d)—exhibits the transition from laminar to turbulent flow.
>
> Also useful are photographs of turbulence (e.g., in Van Dyke, 1982).

uniformly moving fluid past a cylinder oriented transversely to the flow's incoming velocity (Fig. 15.1). We assume that the flow velocity is small compared with the speed of sound, so the effects of compressibility can be ignored. Let the cylinder diameter be d, and choose this as the characteristic length in the problem. Similarly, let the velocity far upstream be V and choose this as the characteristic velocity, so the Reynolds number is[4]

Reynolds number

$$\mathrm{Re}_d = \frac{Vd}{\nu}. \tag{15.1}$$

equations governing stationary, unperturbed flows

Initially, we assume that the flow is stationary (no turbulence) as well as incompressible, and the effects of gravity are negligible. Then the equations governing the flow are incompressibility,

$$\nabla \cdot \mathbf{v} = 0, \tag{15.2a}$$

and the time-independent Navier-Stokes equation (13.69) with $\partial \mathbf{v}/\partial t = 0$:

$$(\mathbf{v} \cdot \nabla)\mathbf{v} = -\frac{\nabla P}{\rho} + \nu \nabla^2 \mathbf{v}. \tag{15.2b}$$

These four equations (one for incompressibility, three for the components of Navier-Stokes) can be solved for the pressure and the three components of velocity subject to the velocity vanishing on the surface of the cylinder and becoming uniform far upstream.

From the parameters of the flow (the cylinder's diameter d; the fluid's incoming velocity V, its density ρ, and its kinematic viscosity ν) we can construct only one dimensionless number, the Reynolds number $\mathrm{Re}_d = Vd/\nu$. (If the flow speed were high enough that incompressibility fails, then the sound speed c_s would also be a relevant parameter, and a second dimensionless number could be formed: the Mach number, $M = V/c_s$; Chap. 17.) With Re_d the only dimensionless number, we are

---

4. The subscript d is just a reminder that in this instance we have chosen the diameter as the characteristic lengthscale; for other applications, the length of the cylinder might be more relevant.

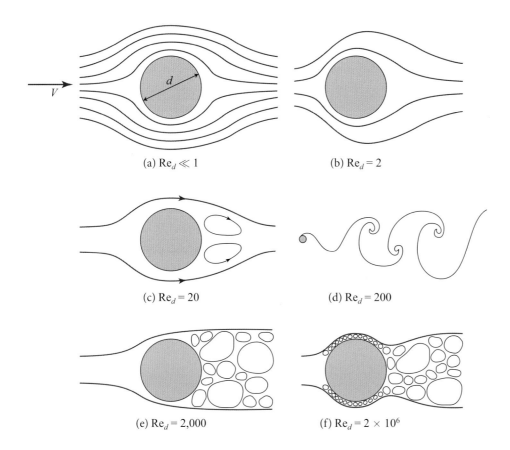

FIGURE 15.1 Schematic depiction of flow past a cylinder for steadily increasing values of the Reynolds number $\text{Re}_d = Vd/\nu$ as labeled. There are many photographs, drawings, and simulations of this flow on the web, perhaps best found by doing a search on "Kármán vortex street."

guaranteed on dimensional grounds that the solution to the flow equations can be expressed as

$$\mathbf{v}/V = \mathbf{U}(\mathbf{x}/d, \text{Re}_d). \tag{15.3}$$

dimensional analysis gives functional form of flow

Here $\mathbf{U}$ is a dimensionless function of the dimensionless $\mathbf{x}/d$, and it can take wildly different forms, depending on the value of the Reynolds number Re_d (cf. Fig. 15.1, which we discuss below).

The functional form of $\mathbf{v}$ [Eq. (15.3)] has important implications. If we compute the flow for specific values of the upstream velocity V, the cylinder's diameter d, and the kinematic viscosity ν and then double V and d and quadruple ν so that Re_d is unchanged, then the new solution will be *similar* to the original one. It can be produced from the original by rescaling the flow velocity to the new upstream velocity and the distance to the new cylinder diameter. [For this reason, Eq. (15.3) is sometimes called a *scaling relation*.] By contrast, if we had only doubled the kinematic

scaling relation

15.2 The Transition to Turbulence—Flow Past a Cylinder

viscosity, the Reynolds number would have also doubled, and we could be dealing with a qualitatively different flow.

When discussing flow past the cylinder, a useful concept is the *stagnation pressure* in the upstream flow. This is the pressure the fluid would have, according to the Bernoulli principle ($v^2/2 + \int dP/\rho = \text{const}$), if it were brought to rest at the leading edge of the cylinder without significant action of viscosity. Ignoring the effects of compressibility (so ρ is constant), this stagnation pressure is

stagnation pressure

$$\boxed{P_{\text{stag}} = P_0 + \frac{1}{2}\rho V^2}, \tag{15.4}$$

where P_0 is the upstream pressure. Suppose that this stagnation pressure were to act over the whole front face of the cylinder, while the pressure P_0 acted on the downstream face. The net force on the cylinder per unit length, F_D, would then be $\frac{1}{2}\rho V^2 d$. This is a first rough estimate for the drag force. It is conventional to define a *drag coefficient* as the ratio of the actual drag force per unit length to this rough estimate:

drag coefficient

$$C_D \equiv \frac{F_D}{\frac{1}{2}\rho V^2 d}. \tag{15.5}$$

drag coefficient is function of Reynolds number

This drag coefficient, being a dimensionless feature of the flow (15.3), can depend only on the dimensionless Reynolds number Re_d: $C_D = C_D(\text{Re}_d)$; see Fig. 15.2. Similarly for flow past a 3-dimensional body with cross sectional area A perpendicular to the flow and with any other shape, the drag coefficient

$$\boxed{C_D \equiv \frac{F_D}{\frac{1}{2}\rho V^2 A}} \tag{15.6}$$

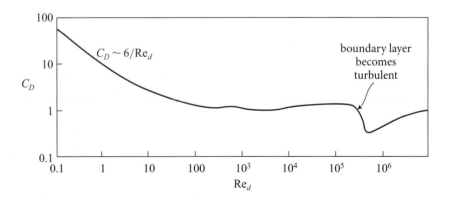

FIGURE 15.2 Drag coefficient C_D for flow past a cylinder as a function of Reynolds number $\text{Re}_d = Vd/\nu$. This graph, adapted from Tritton (1987, Fig. 3.15), is based on experimental measurements.

Chapter 15. Turbulence

will be a function only of Re. However, the specific functional form of $C_D(\text{Re})$ will depend on the body's shape and orientation.

Now, turn to the details of the flow around a cylinder as described in Figs. 15.1 and 15.2. At low Reynolds number, $\text{Re}_d \ll 1$, there is creeping flow (Fig. 15.1a) just like that analyzed in detail for a spherical obstacle in Sec. 14.3.2. As you might have surmised by tackling Ex. 14.12, the details of low-Reynolds-number flow past a long object, such as a cylinder, are subtly different from those of flow past a short one, such as a sphere. This is because, for a cylinder, inertial forces become comparable with viscous and pressure forces at distances $\sim d/\text{Re}_d$, where the flow is still significantly perturbed from uniform motion, while for short objects inertial forces become significant only at much larger radii, where the flow is little perturbed by the object's presence. Despite this, the flow streamlines around a cylinder at $\text{Re}_d \ll 1$ (Fig. 15.1a) are similar to those for a sphere (Fig. 14.10) and are approximately symmetric between upstream and downstream. The fluid is decelerated by viscous stresses as it moves past the cylinder along these streamlines, and the pressure is higher on the cylinder's front face than on its back. Both effects contribute to the net drag force acting on the cylinder. The momentum removed from the flow is added to the cylinder. At cylindrical radius $\varpi \ll d/\text{Re}_d$ the viscous stress dominates over the fluid's inertial stress, and the fluid momentum therefore is being transferred largely to the cylinder at a rate per unit area $\sim \rho V^2$. In contrast, for $\varpi \gtrsim d/\text{Re}_d$ the viscous and inertial stresses are comparable and balance each other, and the flow's momentum is not being transferred substantially to the cylinder. This implies that the effective cross sectional width over which the cylinder extracts the fluid's momentum is $\sim d/\text{Re}_d$, and correspondingly, the net drag force per unit length is $F \sim \rho V^2 d/\text{Re}_d$, which implies [cf. Eq. (15.5)] a drag coefficient $\sim 1/\text{Re}_d$ at low Reynolds numbers ($\text{Re}_d \ll 1$). A more careful analysis gives $C_D \sim 6/\text{Re}_d$, as shown experimentally in Fig. 15.2.

at $\text{Re}_d \ll 1$: creeping flow and $C_D \sim 6/\text{Re}_d$

As the Reynolds number is increased to ~ 1 (Fig. 15.1b), the effective cross section gets reduced to roughly the cylinder's diameter d; correspondingly, the drag coefficient decreases to $C_D \sim 1$. At this Reynolds number, $\text{Re}_d \sim 1$, the velocity field begins to appear asymmetric from front to back.

With a further increase in Re_d, a laminar boundary layer of thickness $\delta \sim d/\sqrt{\text{Re}_d}$ starts to form. The viscous force per unit length due to this boundary layer is $F \sim \rho V^2 d/\sqrt{\text{Re}_d}$ [cf. Eqs. (14.49)–(14.51) divided by the transverse length w, and with $\ell \sim d$ and $v_o = V$]. It might therefore be thought that the drag would continue to decrease as $C_D \sim 1/\sqrt{\text{Re}_d}$, when Re_d increases substantially above unity, making the boundary layer thin and the external flow start to resemble potential flow. However, this does *not* happen. Instead, at $\text{Re}_d \sim 5$, the flow begins to separate from the back side of the cylinder and is there replaced by two retrograde eddies (Fig. 15.1c). As described in Sec. 14.4.4, this separation occurs because an adverse pressure gradient $(\mathbf{v} \cdot \nabla) P > 0$ develops outside the boundary layer, near the cylinder's downstream face, and causes the separated boundary layer to be replaced by these two counter-rotating eddies. The pressure in these eddies, and thus also on the cylinder's back face,

at $\text{Re}_d \sim 1$: $C_D \sim 1$ and laminar boundary layer starts to form

at $\text{Re}_d \sim 5$: separation begins

is of order the flow's incoming pressure P_0 and is significantly less than the stagnation pressure $P_{\text{stag}} = P_0 + \frac{1}{2}\rho V^2$ at the cylinder's front face, so the drag coefficient stabilizes at $C_D \sim 1$.

As the Reynolds number increases above $\text{Re}_d \sim 5$, the size of the two eddies increases until, at $\text{Re}_d \sim 100$, the eddies are shed dynamically, and the flow becomes nonstationary. The eddies tend to be shed alternately in time, first one and then the other, producing a beautiful pattern of alternating vortices downstream known as a *Kármán vortex street* (Fig. 15.1d).

at $\text{Re}_d \sim 100$: eddies shed dynamically; form Kármán vortex street

When $\text{Re}_d \sim 1{,}000$, the downstream vortices are no longer visible, and the wake behind the cylinder contains a velocity field irregular on all macroscopic scales (Fig. 15.1e). This downstream flow has become turbulent. Finally, at $\text{Re}_d \sim 3 \times 10^5$, the boundary layer itself, which has been laminar up to this point, becomes turbulent (Fig. 15.1f), reducing noticeably the drag coefficient (Fig. 15.2). We explore the cause of this reduction in Sec. 15.5. [The physically relevant Reynolds number for onset of turbulence in the boundary layer is that computed not from the cylinder diameter d, $\text{Re}_d = V d / \nu$, but rather from the boundary layer thickness $\delta \sim d / \text{Re}_d^{1/2}$:

at $\text{Re}_d \sim 1{,}000$: turbulent wake

at $\text{Re}_d \sim 3 \times 10^5$: turbulent boundary layer

$$\text{Re}_\delta = \frac{V \delta}{\nu} \sim \frac{V d \text{Re}_d^{-1/2}}{\nu} = \sqrt{\text{Re}_d}. \tag{15.7}$$

The onset of boundary-layer turbulence is at $\text{Re}_\delta \sim \sqrt{3 \times 10^5} \sim 500$, about the same as the $\text{Re}_d \sim 1{,}000$ for the onset of turbulence in the wake.]

An important feature of this changing flow pattern is that at $\text{Re}_d \ll 1{,}000$ (Figs. 15.1a–d), before any turbulence sets in, the flow (whether steady or dynamical) is translation symmetric—it is independent of distance z down the cylinder (i.e., it is 2-dimensional). This is true even of the Kármán vortex street. By contrast, the turbulent velocity field at $\text{Re}_d \gtrsim 1{,}000$ is fully 3-dimensional. At these high Reynolds numbers, small, nontranslation-symmetric perturbations of the translation-symmetric flow grow into vigorous, 3-dimensional turbulence. This is a manifestation of the inability of 2-dimensional flows to exhibit all the chaotic motion associated with 3-dimensional turbulence (Sec. 15.4.4).

turbulence is 3-dimensional

The most important feature of this family of flows—one that is characteristic of most such families—is that there is a critical Reynolds number for the onset of turbulence. That critical number can range from ~ 30 to $\sim 10^5$, depending on the geometry of the flow and on precisely what length and speed are used to define the Reynolds number.

critical Reynolds number for onset of turbulence

EXERCISES

Exercise 15.1 **Example: *The 2-Dimensional Laminar Wake behind a Cylinder*
In Sec. 15.4.3, we explore the structure of the wake behind the cylinder when the Reynolds number is high enough that the flow is turbulent. For comparison, here we

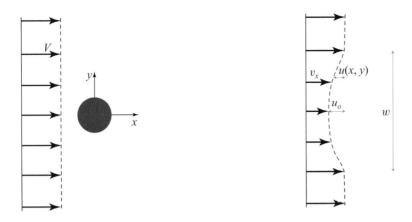

FIGURE 15.3 The 2-dimensional laminar wake behind an infinitely long cylinder. (This figure also describes a turbulent wake, at high Reynolds numbers, if v_x is replaced by the time-averaged $\bar{v}_x$; Sec. 15.4.3.)

compute the wake's structure at lower Reynolds numbers, when the wake is laminar. This computation is instructive: using order-of-magnitude estimates first, followed by detailed calculations, it illustrates the power of momentum conservation. It is our first encounter with the velocity field in a wake.

(a) Begin with an order-of-magnitude estimate. Characterize the wake by its width $w(x)$ at distance x downstream from the cylinder and by the reduction in the flow velocity (the "velocity deficit"), $u_o(x) \equiv V - v_x(x)$, at the center of the wake; see Fig. 15.3. From the diffusion of vorticity show that $w \simeq 2\sqrt{\nu x/V}$.

(b) Explain why momentum conservation requires that the force per unit length on the cylinder, $F_D = C_D \tfrac{1}{2}\rho V^2 d$ [Eq. (15.5)], equals the transverse integral $\int T_{xx} dy$ of the fluid's kinetic stress ($T_{xx} = \rho v_x v_x$) before the fluid reaches the sphere, minus that integral at distance x after the sphere. Use this requirement to show that the fractional velocity deficit at the center of the wake is $u_o/V \simeq \tfrac{1}{4} C_D d/w \simeq C_D \sqrt{d\,\mathrm{Re}_d/(64x)}$.

(c) For a more accurate description of the flow, solve the Navier-Stokes equation to obtain the profile of the velocity deficit, $u(x, y) \equiv V - v_x(x, y)$. [Hint: Ignoring the pressure gradient, which is negligible (Why?), the Navier-Stokes equation should reduce to the 1-dimensional diffusion equation, which we have met several times previously in this book.] Your answer should be

$$u = u_o e^{-(2y/w)^2}, \quad w = 4\left(\frac{\nu x}{V}\right)^{1/2}, \quad u_o = V C_D \left(\frac{d\,\mathrm{Re}_d}{16\pi x}\right)^{1/2}, \quad (15.8)$$

where w and u_o are more accurate values of the wake's width and its central velocity deficit.

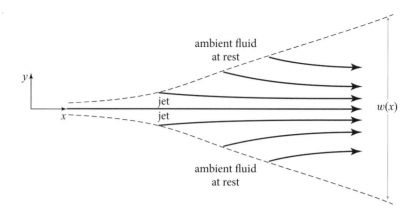

FIGURE 15.4 2-dimensional laminar jet. As the jet widens, it entrains ambient fluid.

Exercise 15.2 *Problem: The 3-Dimensional Laminar Wake behind a Sphere*
Repeat Ex. 15.1 for the 3-dimensional laminar wake behind a sphere.

Exercise 15.3 *Example: Structure of a 2-Dimensional Laminar Jet; Entrainment*
Consider a narrow, 2-dimensional, incompressible (i.e., subsonic) jet emerging from a 2-dimensional nozzle into ambient fluid at rest with the same composition and pressure. (By 2-dimensional we mean that the nozzle and jet are translation symmetric in the third dimension.) Let the Reynolds number be low enough for the flow to be laminar; we study the turbulent regime in Ex. 15.7. We want to understand how rapidly this laminar jet spreads.

(a) Show that the pressure forces far downstream from the nozzle are likely to be much smaller than the viscous forces and can therefore be ignored.

(b) Let the jet's thrust per unit length (i.e., the momentum per unit time per unit length flowing through the nozzle) be $\mathcal{F}$. Introduce Cartesian coordinates x, y, with x parallel to and y perpendicular to the jet (cf. Fig. 15.4). As in Ex. 15.1, use vorticity diffusion (or the Navier-Stokes equation) and momentum conservation to estimate the speed v_x of the jet and its width w as functions of distance x downstream.

(c) Use the scalings from part (b) to modify the self-similarity analysis of the Navier-Stokes equation that we used for the laminar boundary layer in Sec. 14.4, and thereby obtain the following approximate solution for the jet's velocity profile:

$$v_x = \left(\frac{3\mathcal{F}^2}{32\rho^2 \nu x} \right)^{1/3} \operatorname{sech}^2 \left[\left(\frac{\mathcal{F}}{48\rho \nu^2 x^2} \right)^{1/3} y \right]. \quad (15.9)$$

(d) Equation (15.9) shows that the jet width w increases downstream as $x^{2/3}$. As the jet widens, it scoops up (entrains) ambient fluid, as depicted in Fig. 15.4. This entrainment actually involves pulling fluid inward in a manner described by the y component of velocity, v_y. Solve the incompressibility equation $\nabla \cdot \mathbf{v} = 0$ to obtain the following expression for v_y:

$$v_y = -\frac{1}{3x}\left(\frac{3\mathcal{F}^2}{32\rho^2 \nu x}\right)^{1/3} \tag{15.10}$$

$$\times \left\{\left(\frac{48\rho\nu^2 x^2}{\mathcal{F}}\right)^{1/3} \tanh\left[\left(\frac{\mathcal{F}}{48\rho\nu^2 x^2}\right)^{1/3} y\right] - 2y \ \text{sech}^2\left[\left(\frac{\mathcal{F}}{48\rho\nu^2 x^2}\right)^{1/3} y\right]\right\}$$

$$\simeq -\left(\frac{1}{6}\frac{\mathcal{F}\nu}{\rho x^2}\right)^{1/3} \text{sign}(y) \quad \text{for} \quad |y| \gg \frac{1}{2}w(x) = \left(\frac{48\rho\nu^2 x^2}{\mathcal{F}}\right)^{1/3}.$$

Thus ambient fluid is pulled inward from both sides to satisfy the jet's entrainment appetite.

Exercise 15.4 *Example: Marine Animals*

One does not have to be a biologist to appreciate the strong evolutionary advantage that natural selection confers on animals that can reduce their drag coefficients. It should be no surprise that the shapes and skins of many animals are highly streamlined. This is particularly true for aquatic animals. Of course, an animal minimizes drag while developing efficient propulsion and lift, which change the flow pattern. Comparisons of the properties of flows past different stationary (e.g., dead or towed) animals are, therefore, of limited value. The species must also organize its internal organs in 3 dimensions, which is an important constraint. The optimization varies from species to species and is suited to the environment. There are some impressive performers in the animal world.

(a) First, idealize our animal as a thin rectangle with thickness t, length ℓ, and width w such that $t \ll \ell \ll w$.[5] Let the animal be aligned with the flow parallel to the ℓ direction. Assuming an area of $A = 2\ell w$, the drag coefficient is $C_D = 1.33 \text{Re}_\ell^{-0.5}$ [Eq. (14.50)]. This assumes that the flow is laminar. The corresponding result for a turbulent flow is $C_D \simeq 0.072 \text{Re}_\ell^{-0.2}$. Show that the drag can be considerably reduced if the transition to turbulence takes place at high Re_ℓ. Estimate the effective Reynolds number for an approximately flat fish like a flounder of size ~ 0.3 m that can move with a speed ~ 0.3 m s^{-1}, and then compute the drag force. Express your answer as a stopping length.

(b) One impressive performer is the mackerel (a highly streamlined fish), for which the reported drag coefficient is 0.0043 at $\text{Re} \sim 10^5$. Compare this with a thin plate and a sphere. (The drag coefficient for a sphere is $C_D \sim 0.5$, assuming a reference area equal to the total area of the sphere. The drag decreases abruptly when there is a transition to turbulence, just as we found with the cylinder.)

---

5. The aquatic counterpart to the famous Spherical Cow!

(c) Another fine swimmer is the California sea lion, which has a drag coefficient of $C_D = 0.0041$ at Re $\sim 2 \times 10^6$. How does this compare with a plate and a sphere?

Further details can be found in Vogel (1994) and references therein.

15.3 Empirical Description of Turbulence

Empirical studies of turbulent flows have revealed some universal properties that are best comprehended through movies (Box 15.2). Here we simply list the most important of them and comment on them briefly. We revisit most of them in more detail in the remainder of the chapter. Throughout, we restrict ourselves to turbulence with velocities that are very subsonic, and thus the fluid is incompressible.

characteristics of turbulence

Turbulence is characterized by:

- *Disorder, irreproducible in detail but with rich, nonrandom structure.* This disorder is intrinsic to the flow. It appears to arise from a variety of instabilities. No forcing by external agents is required to produce it. If we try to resolve the flow into modes, however, we find that the phases of the modes are not fully random, either spatially or temporally: there are strong correlations.[6] Correspondingly, if we look at a snapshot of a turbulent flow, we frequently observe large, well-defined coherent structures like eddies and jets, which suggests that the flow is more organized than a purely random superposition of modes, just as the light reflected from the surface of a painting differs from that emitted by a blackbody. If we monitor the time variation of some fluid variable, such as one component of the velocity at a given point in the flow, we observe *intermittency*—the irregular starting and ceasing of strong turbulence. Again, this effect is so pronounced that more than a random-mode superposition is at work, reminiscent of the distinction between noise and music (at least some music). [A major consequence that we shall have to face is that strong turbulence is *not* well treated by perturbation theory. As an alternative, semiquantitative techniques of analysis must be devised.]

- *A wide range of interacting scales.* When the fluid velocity and pressure are Fourier analyzed, one finds them varying strongly over many decades of wave number and frequency. We can think of these variations as due to *eddies* with a huge range of sizes. These eddies interact strongly. Typically, large eddies appear to feed their energy to smaller eddies, which in turn feed energy to still smaller eddies, and so forth. Occasionally, amazingly, the flow of energy appears to reverse: small-scale turbulent structures give rise

---

6. As we discuss in Chap. 28, many cosmologists suspected that the primordial fluctuations, out of which galaxies and stars eventually grew, would share this characteristic with turbulence. Insofar as we can measure, this is not the case, and the fluctuations appear to be quite random and uncorrelated.

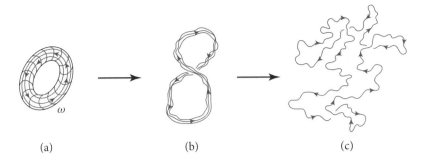

FIGURE 15.5 Schematic illustration of the propagation of turbulence by the stretching of vortex lines. The tube of vortex lines in (a) is stretched and thereby forced into a reduced cross section by the turbulent evolution from (a) to (b) to (c). The reduced cross section means an enhanced vorticity on smaller scales.

to large-scale structures,[7] resulting in intermittency. A region of the flow that appears to have calmed down may suddenly and unexpectedly become excited again.

- *Vorticity, irregularly distributed in three dimensions.* This vorticity varies in magnitude and direction over the same wide range of scales as for the fluid velocity. It appears to play a central role in coupling large scales to small; see Sec. 15.3.1.
- *Large dissipation.* Typically, turbulent energy is fed from large scales to small in just one turnover time of a large eddy. This is extremely fast. The energy cascades down to smaller and smaller lengthscales in shorter and shorter timescales until it reaches eddies so small that their shear, coupled to molecular viscosity, converts the turbulent energy into heat.
- *Efficient mixing and transport.* Most everything that can be transported is efficiently mixed and moved: momentum, heat, salt, chemicals, and so forth.

15.3.1 The Role of Vorticity in Turbulence

Turbulent flows contain tangled vorticity. As we discussed in Sec. 14.2.2, when viscosity is unimportant, vortex lines are frozen into the fluid and can be stretched by the action of neighboring vortex lines. As a bundle of vortex lines is stretched and twisted (Fig. 15.5), the incompressibility of the fluid causes the bundle's cross section to decrease and correspondingly causes the magnitude of its vorticity to increase and the lengthscale on which the vorticity changes to decrease (cf. Sec. 14.2). The continuous lengthening and twisting of fluid elements therefore creates vorticity on progressively smaller lengthscales.

evolution of vorticity in turbulence: toward progressively smaller scales

---

7. The creation of larger structures from smaller ones—sometimes called an *inverse cascade*—is a strong feature of 2-dimensional fluid turbulence.

Note that, when the flow is 2-dimensional (i.e., has translation symmetry), the vortex lines point in the translational direction, so they do not get stretched, and there is thus no inexorable driving of the turbulent energy to smaller and smaller lengthscales. This is one reason why true turbulence does not occur in 2 dimensions, only in 3.

However, something akin to turbulence but with much less richness and small-scale structure, *does* occur in 2 dimensions (e.g., in 2-dimensional simulations of the Kelvin-Helmholtz instability; Box 15.3). But in the real world, once the Kelvin-Helmholtz instability is fully developed, 3-dimensional instabilities grow strong, vortex-line stretching increases, and the flow develops full 3-dimensional turbulence. The same happens in other ostensibly 2-dimensional flows (e.g., the "2-dimensional" wake behind a long circular cylinder and "2-dimensional" jets and boundary layers; Secs. 15.4.3 and 15.5).

15.4 Semiquantitative Analysis of Turbulence

In this section, we develop a semiquantitative mathematical analysis of turbulence and explore a few applications. This analysis is fairly good for weak turbulence. However, for the much more common strong turbulence, it is at best semiquantitative—but nonetheless widely used for lack of anything simple that is much better.

weak turbulence versus strong turbulence

15.4.1 Weak-Turbulence Formalism

The meaning of weak turbulence can be explained in terms of interacting eddies (a concept we exploit in Sec. 15.4.4 when studying the flow of turbulent energy from large scales to small). One can regard turbulence as weak if the timescale τ_* for a large eddy to feed most of its energy to smaller eddies is long compared to the large eddy's turnover time τ (i.e., its rotation period). The weak-turbulence formalism (model) that we sketch here can be thought of as an expansion in τ/τ_*.

weak turbulence

Unfortunately, for most turbulence seen in fluids, the large eddies' energy-loss time is of order its turnover time, $\tau/\tau_* \sim 1$, which means the eddy loses its identity in roughly one turnover time and the turbulence is strong. In this case, the weak-turbulence formalism that we sketch here is only semiquantitatively accurate.

Our formalism for weak turbulence (with gravity negligible and the flow highly subsonic, so it can be regarded as incompressible) is based on the standard incompressibility equation and the time-dependent Navier-Stokes equation, which we write as

$$\nabla \cdot \mathbf{v} = 0, \quad (15.11a)$$

$$\rho \frac{\partial \mathbf{v}}{\partial t} + \nabla \cdot (\rho \mathbf{v} \otimes \mathbf{v}) = -\nabla P + \rho \nu \nabla^2 \mathbf{v}, \quad (15.11b)$$

assuming for simplicity that ρ is constant. [Equation (15.11b) is equivalent to (15.2b) with $\partial \mathbf{v}/\partial t$ added to account for time dependence and with the inertial force term rewritten using $\nabla \cdot (\rho \mathbf{v} \otimes \mathbf{v}) = \rho(\mathbf{v} \cdot \nabla)\mathbf{v}$, or equivalently in index notation, $(\rho v_i v_j)_{;i} = \rho_{,i} v_i v_j + \rho(v_{i;i} v_j + v_i v_{j;i}) = \rho v_i v_{j;i}$.] Equations (15.11) are four scalar

> **BOX 15.3. CONSEQUENCES OF THE KELVIN-HELMHOLTZ INSTABILITY**
>
> The Kelvin-Helmholtz instability arises when two fluids with nearly the same density are in uniform motion past each other (Sec. 14.6.1). Their interface (a vortex sheet) develops corrugations that grow (Figs. 14.19 and 14.20). That growth bends the corrugations more and more sharply. Along the corrugated interface the fluids on each side are still sliding past each other, so the instability arises again on this smaller scale—and again and again, somewhat like the cascade of turbulent energy from large scales to small.
>
> In the real world, 3-dimensional instabilities also arise, and the flow becomes fully turbulent. However, much insight is gained into the difference between 2- and 3-dimensional flow by artificially constraining the Kelvin-Helmholtz flow to remain 2-dimensional. This is easily done in numerical simulations. Movies of such simulations abound on the web (e.g., on the Wikipedia web page for the Kelvin-Helmholtz instability; the following picture is a still from that movie).
>
>
>
> By Bdubb12 (Own work) [Public domain], via Wikimedia Commons.
> https://commons.wikimedia.org/wiki/File%3AKHI.gif.
>
> Although the structures in this simulation are complex, they are much less rich than those that appear in fully 3-dimensional turbulence—perhaps largely due to the absence of stretching and twisting of vortex lines when the flow is confined to 2 dimensions. (Not surprisingly, the structures in this simulation resemble some in Jupiter's atmosphere, which also arise from a Kelvin-Helmholtz instability.)

equations for four unknowns, $P(\mathbf{x}, t)$ and the three components of $\mathbf{v}(\mathbf{x}, t)$; ρ and ν can be regarded as constants.

To obtain the weak-turbulence versions of these equations, we split the velocity field $\mathbf{v}(\mathbf{x}, t)$ and pressure $P(\mathbf{x}, t)$ into steady parts $\bar{\mathbf{v}}$, $\bar{P}$, plus fluctuating parts, $\delta \mathbf{v}$, δP:

foundations for weak-turbulence theory

$$\boxed{\mathbf{v} = \bar{\mathbf{v}} + \delta \mathbf{v}, \qquad P = \bar{P} + \delta P.} \qquad (15.12)$$

We can think of (or, in fact, define) $\bar{\mathbf{v}}$ and $\bar{P}$ as the time averages of $\mathbf{v}$ and P, and define $\delta \mathbf{v}$ and δP as the difference between the exact quantities and the time-averaged quantities.

The time-averaged variables $\bar{\mathbf{v}}$ and $\bar{P}$ are governed by the time-averaged incompressibility and Navier-Stokes equations (15.11). Because the incompressibility equation is linear, its time average,

$$\boxed{\nabla \cdot \bar{\mathbf{v}} = 0,} \tag{15.13a}$$

entails no coupling of the steady variables to the fluctuating ones. By contrast, the nonlinear inertial term $\nabla \cdot (\rho \mathbf{v} \otimes \mathbf{v})$ in the Navier-Stokes equation gives rise to such a coupling in the (time-independent) time-averaged equation:

$$\boxed{\rho(\bar{\mathbf{v}} \cdot \nabla)\bar{\mathbf{v}} = -\nabla \bar{P} + \rho \nu \nabla^2 \bar{\mathbf{v}} - \nabla \cdot \mathbf{T}_R.} \tag{15.13b}$$

time-averaged Navier-Stokes equation: couples turbulence (via Reynolds stress tensor) to time-averaged flow

Here

$$\boxed{\mathbf{T}_R \equiv \rho \overline{\delta \mathbf{v} \otimes \delta \mathbf{v}}} \tag{15.13c}$$

is known as the *Reynolds stress tensor*. It serves as a driving term in the time-averaged Navier-Stokes equation (15.13b)—a term by which the fluctuating part of the flow acts back on and so influences the time-averaged flow.

This Reynolds stress $\mathbf{T}_R$ can be regarded as an additional part of the total stress tensor, analogous to the gas pressure computed in kinetic theory,[8] $P = \frac{1}{3}\rho \overline{v^2}$, where v is the molecular speed. $\mathbf{T}_R$ will be dominated by the largest eddies present, and it can be anisotropic, especially when the largest-scale turbulent velocity fluctuations are distorted by interaction with an averaged shear flow [i.e., when $\bar{\sigma}_{ij} = \frac{1}{2}(\bar{v}_{i;j} + \bar{v}_{j;i})$ is large].

Reynolds stress for stationary, homogeneous turbulence

If the turbulence is both stationary and homogeneous (a case we specialize to below when studying the Kolmogorov spectrum), then the Reynolds stress tensor can be written in the form $\mathbf{T}_R = P_R \mathbf{g}$, where P_R is the Reynolds pressure, which is independent of position, and $\mathbf{g}$ is the metric, so $g_{ij} = \delta_{ij}$. In this case, the turbulence exerts no force density on the mean flow, so $\nabla \cdot \mathbf{T}_R = \nabla P_R$ will vanish in the time-averaged Navier-Stokes equation (15.13b). By contrast, near the edge of a turbulent region (e.g., near the edge of a turbulent wake, jet, or boundary layer), the turbulence is inhomogeneous, and thereby (as we shall see in Sec. 15.4.2) exerts an important influence on the time-independent, averaged flow.

correlation functions in turbulence

Notice that the Reynolds stress tensor is the tensorial *correlation function* (also called "autocorrelation function") of the velocity fluctuation field at zero time delay (multiplied by density ρ; cf. Secs. 6.4.1 and 6.5.1). Notice also that it involves the

---

8. Deducible from Eq. (3.37c) or from Eqs. (3.39b) and (3.39c) with mean energy per particle $\bar{E} = \frac{1}{2}m\overline{v^2}$.

temporal cross-correlation function of components of the velocity fluctuation [e.g., $\overline{\delta v_x(\mathbf{x}, t)\delta v_y(\mathbf{x}, t)}$; Sec. 6.5.1]. It is possible to extend this weak-turbulence formalism so it probes the statistical properties of turbulence more deeply, with the aid of correlation functions with finite time delays and correlation functions of velocity components (or other relevant physical quantities) at two different points in space simultaneously; see, for example, Sec. 28.5.3. (It is relatively straightforward experimentally to measure these correlation functions.) As we discuss in greater detail in Sec. 15.4.4 (and also saw for 1- and 2-dimensional random processes in Secs. 6.4.4 and 6.5.2, and for multidimensional, complex random processes in Ex. 9.8), the Fourier transforms of these correlation functions give the spatial and temporal spectral densities of the fluctuating quantities.

Just as the structure of the time-averaged flow is governed by the time-averaged incompressibility and Navier-Stokes equations (15.13) (with the fluctuating variables acting on the time-averaged flow through the Reynolds stress), so also the fluctuating part of the flow is governed by the fluctuating (difference between instantaneous and time-averaged) incompressibility and Navier-Stokes equations. For details, see Ex. 15.5. This exercise is important; it exhibits the weak-turbulence formalism in action and underpins the application to spatial energy flow in a 2-dimensional, turbulent wake in Fig. 15.7 below.

EXERCISES

Exercise 15.5 **Example: Reynolds Stress; Fluctuating Part of Navier-Stokes Equation in Weak Turbulence*

(a) Derive the time-averaged Navier-Stokes equation (15.13b) from the time-dependent form [Eq. (15.11b)], and thereby infer the definition (15.13c) for the Reynolds stress. Equation (15.13b) shows how the Reynolds stress affects the evolution of the mean velocity. However, it does not tell us how the Reynolds stress evolves.

(b) Explain why an equation for the evolution of the Reynolds stress must involve averages of triple products of the velocity fluctuation. Similarly, the time evolution of the averaged triple products will involve averaged quartic products, and so on (cf. the BBGKY hierarchy of equations in plasma physics, Sec. 22.6.1). How do you think you might "close" this sequence of equations (i.e., terminate it at some low order) and get a fully determined system of equations? [Answer: The simplest way is to use the concept of turbulent viscosity discussed in Sec. 15.4.2.]

(c) Show that the fluctuating part of the Navier-Stokes equation (the difference between the exact Navier-Stokes equation and its time average) takes the following form:

$$\boxed{\begin{aligned}\frac{\partial \delta \mathbf{v}}{\partial t} + (\bar{\mathbf{v}} \cdot \nabla)\delta \mathbf{v} + (\delta \mathbf{v} \cdot \nabla)\bar{\mathbf{v}} + [(\delta \mathbf{v} \cdot \nabla)\delta \mathbf{v} - \overline{(\delta \mathbf{v} \cdot \nabla)\delta \mathbf{v}}] = \\ -\frac{1}{\rho}\nabla \delta P + \nu \nabla^2(\delta \mathbf{v})\end{aligned}} \quad (15.14a)$$

This equation and the fluctuating part of the incompressibility equation

$$\nabla \cdot \delta \mathbf{v} = 0 \tag{15.14b}$$

govern the evolution of the fluctuating variables $\delta\mathbf{v}$ and δP. [The challenge, of course, is to devise ways to solve these equations despite the nonlinearities and the coupling to the mean flow $\bar{\mathbf{v}}$ that show up strongly in Eq. (15.14a).]

(d) By dotting $\delta\mathbf{v}$ into Eq. (15.14a) and then taking its time average, derive the following law for the spatial evolution of the turbulent energy density $\frac{1}{2}\rho\overline{\delta v^2}$:

spatial evolution equation for turbulent energy density

$$\bar{\mathbf{v}} \cdot \nabla(\tfrac{1}{2}\rho\overline{\delta v^2}) + \nabla \cdot \overline{\left(\tfrac{1}{2}\rho\delta v^2 \delta\mathbf{v} + \delta P\,\delta\mathbf{v}\right)} = -T_R^{ij}\bar{v}_{i;j} + \nu\rho\overline{\delta\mathbf{v}\cdot(\nabla^2\delta\mathbf{v})}. \tag{15.15}$$

Here $T_R^{ij} = \rho\overline{\delta v_i \delta v_j}$ is the Reynolds stress [Eq. (15.13c)]. Interpret each term in this equation. [The four interpretations will be discussed below, for a 2-dimensional turbulent wake, in Sec. 15.4.3.]

(e) Now derive a similar law for the spatial evolution of the energy density of ordered motion, $\frac{1}{2}\rho\bar{\mathbf{v}}^2$. Show that the energy lost by the ordered motion is compensated for by the energy gained by the turbulent motion.

15.4.2 Turbulent Viscosity

Additional tools that are often introduced in the theory of weak turbulence come from taking the analogy with the kinetic theory of gases one stage further and defining *turbulent transport coefficients*, most importantly a turbulent viscosity that governs the turbulent transport of momentum. These turbulent transport coefficients are derived by simple analogy with the kinetic-theory transport coefficients (Sec. 3.7).

Momentum, heat, and so forth are transported most efficiently by the largest turbulent eddies in the flow; therefore, when estimating the transport coefficients, we replace the particle mean free path by the size ℓ of the largest eddies and the mean particle speed by the magnitude v_ℓ of the fluctuations of velocity in the largest eddies. The result, for momentum transport, is a model turbulent viscosity:

turbulent viscosity determined by largest eddies

$$\nu_t \simeq \tfrac{1}{3} v_\ell \ell \tag{15.16}$$

[cf. Eq. (13.72) for molecular viscosity, with $\nu = \eta/\rho$]. The Reynolds stress is then approximated as a turbulent shear stress of the standard form:

$$\mathbf{T}_R \simeq -2\rho\nu_t \bar{\boldsymbol{\sigma}}. \tag{15.17}$$

Here $\bar{\boldsymbol{\sigma}}$ is the rate of shear tensor (13.67b) evaluated using the mean velocity field $\bar{\mathbf{v}}$. Note that the turbulent kinematic viscosity defined in this manner, ν_t, is a property of the turbulent flow and not an intrinsic property of the fluid; it differs from molecular viscosity in this important respect.

We have previously encountered turbulent viscosity in our study of the physical origin of the Sargasso Sea gyre in the north Atlantic Ocean and the gyre's role in generating the Gulf Stream (Ex. 14.21). The gyre is produced by water flowing in a wind-driven Ekman boundary layer at the ocean's surface. From the measured thickness of that boundary layer, $\delta_E \sim 30$ m, we deduced that the boundary layer's viscosity is $\nu \sim 0.03$ m^2 s^{-1}, which is 30,000 times larger than water's molecular viscosity; it is the turbulent viscosity of Eq. (15.16).

By considerations similar to those above for turbulent viscosity, one can define and estimate a turbulent thermal conductivity for the spatial transport of time-averaged heat (cf. Sec. 3.7.2) and a turbulent diffusion coefficient for the spatial transport of one component of a time-averaged fluid through another, for example, an odor crossing a room (cf. Ex. 3.20).

turbulent thermal conductivity and turbulent diffusion coefficient

The turbulent viscosity ν_t and the other turbulent transport coefficients can be far larger than their kinetic-theory counterparts. Besides the Sargasso Sea gyre, another example is air in a room subjected to typical uneven heating and cooling. The air may circulate with an average largest eddy velocity of $v_\ell \sim 1$ cm s^{-1} and an associated eddy size of $\ell \sim 3$ m. (The values can be estimated by observing the motion of illuminated dust particles.) The turbulent viscosity ν_t and the turbulent diffusion coefficient D_t (Ex. 3.20) associated with these motions are $\nu_t \sim D_t \sim 10^{-2}$ m^2 s^{-1}, some three orders of magnitude larger than the molecular values.

15.4.3 Turbulent Wakes and Jets; Entrainment; the Coanda Effect

As instructive applications of turbulent viscosity and related issues, we now explore in an order-of-magnitude way the structures of turbulent wakes and jets. The more complicated extension to a turbulent boundary layer will be explored in Sec. 15.5. In the text of this section we focus on the 2-dimensional wake behind a cylinder. In Exs. 15.6, 15.7, and 15.8, we study the 3-dimensional wake behind a sphere and 2- and 3-dimensional jets.

TWO-DIMENSIONAL TURBULENT WAKE: ORDER-OF-MAGNITUDE COMPUTATION OF WIDTH AND VELOCITY DEFICIT; ENTRAINMENT

For the turbulent wake behind a cylinder at high Reynolds number (see Fig. 15.3), we begin by deducing the turbulent viscosity $\nu_t \sim \frac{1}{3} v_\ell \ell$. It is reasonable to expect (and observations confirm) that the largest eddies in a turbulent wake, at distance x past the cylinder, extend transversely across nearly the entire width $w(x)$ of the wake, so their size is $\ell \sim w(x)$. What is the largest eddies' circulation speed v_ℓ? Because these eddies' energies are fed into smaller eddies in (roughly) one eddy turnover time, these eddies must be continually regenerated by interaction between the wake and the uniform flow at its transverse boundaries. This means that the wake's circulation speed v_ℓ cannot be sensitive to the velocity difference V between the incoming flow and the cylinder far upstream; the wake has long since lost memory of that difference. The only characteristic speed that influences the wake is the difference between its own mean downstream speed $\bar{v}_x$ and the speed V of the uniform flow at its boundaries. It seems physically reasonable that the interaction between these two flows will drive the eddy to circulate with that difference speed, the deficit u_o depicted in Fig. 15.3, which observations show to be true. Thus we have $v_\ell \sim u_o$. This means that the turbulent viscosity is $v_t \sim \tfrac{1}{3} v_\ell \ell \sim \tfrac{1}{3} u_o(x) w(x)$.

wake's eddy circulation speed and turbulent viscosity

Knowing the viscosity, we can compute our two unknowns, the wake's velocity deficit $u_o(x)$ and its width $w(x)$, from vorticity diffusion and momentum conservation, as we did for a laminar wake in Ex. 15.1a,b.

First, consider momentum conservation. Here, as in Ex. 15.1b, the drag force per unit length on the cylinder, $F_D = C_D \tfrac{1}{2} \rho V^2 d$, must be equal to the difference between the momentum per unit length ($\int \rho V^2 dy$) in the water impinging on the cylinder and that ($\int \rho v_x^2 dy$) at any chosen distance x behind the cylinder. That difference is easily seen, from Fig. 15.3, to be $\sim \rho V u_o w$. Equating this to F_D, we obtain the product $u_o w$ and thence $v_t \sim \tfrac{1}{3} u_o w \sim \tfrac{1}{6} C_D V d$.

Thus, remarkably, the turbulent viscosity is independent of distance x downstream from the cylinder. This means that the wake's vorticity (which is contained primarily in its large eddies) will diffuse transversely in just the manner we have encountered several times before [Sec. 14.2.5, Eq. (14.42), and Ex. 15.1a], causing its width to grow as $w \sim \sqrt{v_t x / V}$. Inserting $v_t \sim \tfrac{1}{6} C_D V d \sim \tfrac{1}{3} u_o w$, we obtain

average structure of turbulent wake

$$w \sim \left(\frac{C_D d}{6} x\right)^{1/2}, \quad u_o \sim V \left(\frac{3 C_D d}{2x}\right)^{1/2} \qquad (15.18)$$

for the width and velocity deficit in the turbulent wake.

entrainment and its role in a turbulent wake

In this analysis, our appeal to vorticity diffusion obscures the physical mechanism by which the turbulent wake widens downstream. That mechanism is entrainment—the capture of fluid from outside the wake into the wake (a phenomenon we met for a laminar jet in Ex. 15.3d). The inertia of the largest eddies enables them to sweep outward into the surrounding flow with a transverse speed that is a sizable fraction of their turnover speed, say, $\tfrac{1}{6} v_\ell$ (the 1/6 ensures agreement with the diffusion argument). This means that, moving with the flow at speed V, the eddy widens by entrainment at a rate $dw/dt = V dw/dx \sim \tfrac{1}{6} v_\ell \sim \tfrac{1}{6} u_o$. Inserting $u_o \sim \tfrac{1}{2} V C_D d/w$ from momentum conservation and solving the resulting differential equation, we obtain the same wake width $w \sim \sqrt{C_D d \, x/6}$ as we got from our diffusion argument.

DISTRIBUTION OF VORTICITY IN THE WAKE; IRREGULARITY OF THE WAKE'S EDGE

The wake's fluid can be cleanly distinguished from the exterior, ambient fluid by vorticity: the ambient flow velocity $\mathbf{v}(\mathbf{x}, t)$ is vorticity free; the wake has nonzero vorticity. If we look at the actual flow velocity and refrain from averaging, the only way a fluid element in the wake can acquire vorticity is by molecular diffusion. Molecular diffusion is so slow that the boundary between a region with vorticity and one without

FIGURE 15.6 Contours of constant magnitude of vorticity $|\boldsymbol{\omega}|$ in the 2-dimensional turbulent wake between and behind a pair of cylinders. The outer edge of the blue region is the edge of the wake. From a numerical simulation by the research group of Sanjiva K. Lele at Stanford University, http://flowgallery.stanford.edu/research.html.

(the boundary between the wake fluid and the ambient fluid) is very sharp. Sharp, yes, but straight, no! The wake's eddies, little as well as big, drive the boundary into a very convoluted shape (Fig. 15.6). Correspondingly, the order-of-magnitude wake width $w \sim \sqrt{C_D d\, x/6}$ that we have derived is only the width of a thoroughly averaged flow and is not at all the instantaneous, local width of the wake.

Intermittency is one consequence of the wake's highly irregular edge. A fixed location behind the cylinder that is not too close to the wake's center will sometimes be outside the wake and sometimes inside it. This will show up in one's measurement of any flow variable [e.g., $v_y(t)$, $\omega_x(t)$, or pressure P] at the fixed location. When outside the wake, the measured quantities will be fairly constant; when inside, they will change rapidly and stochastically. The quiet epochs combined with interspersed stochastic epochs are a form of intermittency.

intermittency

AVERAGED ENERGY FLOW IN THE WAKE

The weak-turbulence formalism of Sec. 15.4.1 and Ex. 15.5 can be used to explore the generation and flow of turbulent energy in the 2-dimensional wake behind a cylinder. This formalism, even when extended, is not good enough to make definitive predictions, but it *can* be used to deduce the energy flow from measurements of the mean (time-averaged) flow velocity $\bar{\mathbf{v}}$, the turbulent velocity $\delta \mathbf{v}$, and the turbulent pressure δP. A classic example of this was carried out long ago by Townsend (1949) and is summarized in Fig. 15.7.

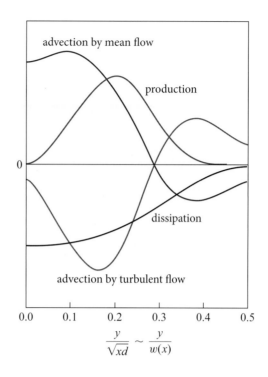

FIGURE 15.7 The four terms in the rate of change of the time-averaged turbulent energy density [Eq. (15.15)] for the 2-dimensional turbulent wake behind a cylinder. The horizontal axis measures the distance y across the wake in units of the wake's mean width $w(x)$. The vertical axis shows the numerical value of each term. For a discussion of the four terms, see the four bulleted points in the text. Energy conservation and stationarity of the averaged flow guarantee that the sum of the four terms vanishes at each y [Eq. (15.15)]. Adapted from Tritton (1987, Fig. 21.8), which is based on measurements and analysis by Townsend (1949) at $\mathrm{Re}_d = 1{,}360$ and $x/d > 500$.

averaged energy flow in a wake: production, advection, and dissipation

The time-averaged turbulent energy density changes due to four processes that are graphed in that figure as a function of distance y across the wake:

- *Production.* Energy in the organized bulk flow (mean flow) $\bar{\mathbf{v}}$ is converted into turbulent energy by the interaction of the mean flow's shear with the turbulence's Reynolds stress, at a rate per unit volume of $T_R^{ij} \bar{v}_{i;j}$. This production vanishes at the wake's center ($y = 0$), because the shear of the mean flow vanishes there; it also vanishes at the edge of the wake, because both the mean-flow shear and the Reynolds stress go to zero there.

- *Advection by mean flow.* Once produced, the turbulent energy is advected across the wake by the mean flow. This causes an increase in turbulent energy density in the center of the wake and a decrease in the wake's outer parts at the rate $\nabla \cdot (\tfrac{1}{2}\rho \overline{\delta v^2} \bar{\mathbf{v}}) = (\bar{\mathbf{v}} \cdot \nabla)(\tfrac{1}{2}\rho \overline{\delta v^2})$.

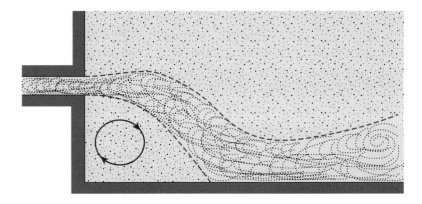

FIGURE 15.8 The Coanda effect. A turbulent jet emerging from an orifice in the left wall is attracted to the solid bottom wall.

- *Advection by turbulent flow.* The turbulent energy also is advected by the turbulent part of the flow, causing a decrease of turbulent energy density in the central regions of the wake and an increase in the outer regions at the rate $\nabla \cdot (\tfrac{1}{2}\rho \overline{\delta v^2 \delta \mathbf{v}} + \overline{\delta P \delta \mathbf{v}})$.
- *Dissipation.* The turbulent energy is converted to heat by molecular viscosity at a rate per unit volume given by $-\nu \rho \overline{\delta \mathbf{v} \cdot (\nabla^2 \delta \mathbf{v})}$. This dissipation is largest at the wake's center and falls off gradually toward the wake's averaged edge.

Energy conservation plus stationarity of the averaged flow guarantees that the sum of these four terms vanishes at all locations in the wake (all y of Fig. 15.7). This is the physical content of Eq. (15.15) and is confirmed in Fig. 15.7 by the experimental data.

ENTRAINMENT AND COANDA EFFECT

Notice how much wider the (averaged) turbulent wake is than the corresponding laminar wake of Ex. 15.1. The ratio of their widths [Eqs. (15.18) and (15.8)] is $w_t/w_l \sim \sqrt{C_D V d/\nu}$. For $C_D \sim 1$ in the turbulent wake, $V \sim 1$ m s^{-1}, $d \sim 1$ m, and water's kinematic viscosity $\nu \sim 10^{-6}$ m^2 s^{-1}, the ratio is $w_t/w_l \sim 10^3$, independent of distance x downstream. In this sense, entrainment in the turbulent wake is a thousand times stronger than entrainment in the laminar wake. Turbulent wakes and jets have voracious appetites for ambient fluid!

strong entrainment in turbulent jets and wakes

Entrainment is central to the *Coanda effect,* depicted in Fig. 15.8. Consider a turbulent flow (e.g., the jet of Fig. 15.8) that is widening by entrainment of surrounding fluid. The jet normally widens downstream by pulling surrounding fluid into itself, and the inflow toward the jet extends far beyond the jet's boundaries (see, e.g., Ex. 15.3d). However, when a solid wall is nearby, so there is no source for inflowing ambient fluid, the jet's entrainment causes a drop in the pressure of the ambient fluid near the wall. The resulting pressure gradient pushes the jet toward the wall as depicted in Fig. 15.8.

Similarly, if a turbulent flow is already close to a wall and the wall begins to curve away from the flow, the flow develops a pressure gradient that tends to keep the turbulent region attached to the wall. In other words, turbulent flows are attracted to solid surfaces and tend to stick to them. This is the Coanda effect.

Coanda effect

The Coanda effect also occurs for laminar flows, but because entrainment is typically orders of magnitude weaker in laminar flows than in turbulent ones, the effect is also orders of magnitude weaker.

The Coanda effect is important in aeronautics; for example, it is exploited to prevent the separation of the boundary layer from the upper surface of a wing, thereby improving the wing's lift and reducing its drag, as we discuss in Sec. 15.5.2.

EXERCISES

Exercise 15.6 *Problem: Turbulent Wake behind a Sphere*
Compute the width $w(x)$ and velocity deficit $u_o(x)$ for the 3-dimensional turbulent wake behind a sphere.

Exercise 15.7 *Problem: Turbulent Jets in 2 and 3 Dimensions*
Consider a 2-dimensional turbulent jet emerging into an ambient fluid at rest, and contrast it to the laminar jet analyzed in Ex. 15.3.

(a) Find how the mean jet velocity and the jet width scale with distance downstream from the nozzle.

(b) Repeat the exercise for a 3-dimensional jet.

Exercise 15.8 *Problem: Entrainment and Coanda Effect in a 3-Dimensional Jet*
(a) Evaluate the scaling of the rate of mass flow (discharge) $\dot{M}(x)$ along the 3-dimensional turbulent jet of the previous exercise. Show that $\dot{M}$ increases with distance from the nozzle, so that mass must be entrained in the flow and become turbulent.

(b) Compare the entrainment rate for a turbulent jet with that for a laminar jet (Ex. 15.3). Do you expect the Coanda effect to be stronger for a turbulent or a laminar jet?

15.4.4 Kolmogorov Spectrum for Fully Developed, Homogeneous, Isotropic Turbulence

When a fluid exhibits turbulence over a large volume that is well removed from any solid bodies, there will be no preferred directions and no substantial gradients in the statistically averaged properties of the turbulent velocity field. This suggests that the statistical properties of the turbulence will be stationary and isotropic. We derive a semiquantitative description of some of these statistical properties, proceeding in two steps: first we analyze the turbulence's velocity field and then the turbulent distribution of quantities that are transported by the fluid.

TURBULENT VELOCITY FIELD

Our analysis will be based on the following simple physical model. We idealize the turbulent velocity field as made of a set of large eddies, each of which contains a set of smaller eddies, and so on. We suppose that each eddy splits into eddies roughly half its size after a few turnover times. This process can be described mathematically as nonlinear or triple velocity correlation terms [terms like the second one in Eq. (15.15)] producing, in the law of energy conservation, an energy transfer (a "cascade" of energy) from larger eddies to smaller ones. Now, for large enough eddies, we can ignore the effects of molecular viscosity in the flow. However, for sufficiently small eddies, viscous dissipation will convert the eddy bulk kinetic energy into heat. This simple model enables us to derive a remarkably successful formula (the *Kolmogorov spectrum*) for the distribution of turbulent energy over eddy size.

physical model underlying Kolmogorov spectrum for turbulence: cascading eddies

We must first introduce and define the turbulent energy per unit wave number and per unit mass, $u_k(k)$. For this purpose, we focus on a volume $\mathcal{V}$ much larger than the largest eddies. At some moment of time t, we compute the spatial Fourier transform of the fluctuating part of the velocity field $\delta \mathbf{v}(\mathbf{x})$, confined to this volume [with $\delta \mathbf{v}(\mathbf{x})$ set to zero outside $\mathcal{V}$], and also write down the inverse Fourier transform:

$$\boxed{\delta\tilde{\mathbf{v}}(\mathbf{k}) = \int_\mathcal{V} d^3x \, \delta\mathbf{v}(\mathbf{x}) e^{-i\mathbf{k}\cdot\mathbf{x}}, \qquad \delta\mathbf{v} = \int \frac{d^3k}{(2\pi)^3} \delta\tilde{\mathbf{v}} e^{i\mathbf{k}\cdot\mathbf{x}} \quad \text{inside } \mathcal{V}.} \qquad (15.19)$$

The total energy per unit mass u in the turbulence, averaged over the box $\mathcal{V}$, is then

$$u = \int \frac{d^3x}{\mathcal{V}} \frac{1}{2} \overline{|\delta\mathbf{v}|^2} = \int \frac{d^3k}{(2\pi)^3} \frac{\overline{|\delta\tilde{\mathbf{v}}|^2}}{2\mathcal{V}} \equiv \int_0^\infty dk \, u_k(k), \qquad (15.20)$$

where we have used Parseval's theorem in the second equality, we have used $d^3k = 4\pi k^2 dk$, and we have defined

$$u_k(k) \equiv \frac{\overline{|\delta\tilde{\mathbf{v}}|^2} k^2}{4\pi^2 \mathcal{V}}. \qquad (15.21)$$

spectral energy per unit mass in turbulence

Here the bars denote a time average, k is the magnitude of the wave vector: $k \equiv |\mathbf{k}|$ (i.e., it is the wave number or equivalently 2π divided by the wavelength), and $u_k(k)$ is called the *spectral energy per unit mass* of the turbulent velocity field $\delta\mathbf{v}$. In the third equality in Eq. (15.20), we have assumed that the turbulence is isotropic, so the integrand depends only on wave number k and not on the direction of $\mathbf{k}$. Correspondingly, we have defined $u_k(k)$ as the energy per unit wave number rather than an energy per unit volume of $\mathbf{k}$-space. This means that $u_k(k)dk$ is the average kinetic energy per unit mass associated with modes that have k lying in the interval dk; we treat k as positive.

In Chap. 6, we introduced the concepts of a random process and its spectral density. The Cartesian components of the fluctuating velocity δv_x, δv_y, and δv_z obviously are random processes that depend on vectorial location in space $\mathbf{x}$ rather than on time as in Chap. 6. It is straightforward to show that their *double-sided* spectral densities are related to $u_k(k)$ by

spectral densities of velocity field

$$S_{v_x}(\mathbf{k}) = S_{v_y}(\mathbf{k}) = S_{v_z}(\mathbf{k}) = \frac{(2\pi)^2}{3k^2} \times u_k(k). \tag{15.22}$$

If we fold negative k_x into positive, and similarly for k_y and k_z, so as to get the kind of single-sided spectral density (Sec. 6.4.2) that we used in Chap. 6, then these spectral densities should be multiplied by $2^3 = 8$.

We now use our physical model of turbulence to derive an expression for $u_k(k)$. Denote by $k_{\min} = 2\pi/\ell$ the wave number of the largest eddies, and by $k_{\max}$ that of the smallest ones (those in which viscosity dissipates the cascading, turbulent energy). Our derivation will be valid (and the result valid) only when $k_{\max}/k_{\min} \gg 1$: only when there is a long sequence of eddies from the largest to half the largest to a quarter the largest and so forth down to the smallest.

eddy turnover speed and time

As a tool in computing $u_k(k)$, we introduce the root-mean-square turbulent turnover speed of the eddies with wave number k: $v(k) \equiv v$. Ignoring factors of order unity, we treat the size of these eddies as k^{-1}. Then their turnover time is $\tau(k) \sim k^{-1}/v(k) = 1/[kv(k)]$. Our model presumes that in this same time τ (to within a factor of order unity), each eddy of size k^{-1} splits into eddies of half this size (i.e., the turbulent energy cascades from k to $2k$). In other words, our model presumes the turbulence is strong. Since the energy cascade is presumed stationary (i.e., no energy is accumulating at any wave number), the energy per unit mass that cascades in a unit time from k to $2k$ must be independent of k. Denote by q that k-independent, cascading energy per unit mass per unit time. Since the energy per unit mass in the eddies of size k^{-1} is v^2 (aside from a factor 2, which we neglect), and the cascade time is $\tau \sim 1/(kv)$, then $q \sim v^2/\tau \sim v^3 k$. This tells us that the rms turbulent velocity is

$$v(k) \sim (q/k)^{1/3}. \tag{15.23}$$

Our model lumps together all eddies with wave numbers in a range $\Delta k \sim k$ around k and treats them all as having wave number k. The total energy per unit mass in these eddies is $u_k(k)\Delta k \sim k u_k(k)$ when expressed in terms of the sophisticated quantity $u_k(k)$; it is $\sim v(k)^2$ when expressed in terms of our simple model. Thus our model predicts that $u_k(k) \sim v(k)^2/k$, which by Eqs. (15.23) and (15.22) implies

Kolmogorov spectrum for stationary, isotropic, incompressible turbulence

$$u_k(k) \sim q^{2/3} k^{-5/3}, \quad S_{v_j}(\mathbf{k}) \sim q^{2/3} k^{-11/3} \quad \text{for } k_{\min} \ll k \ll k_{\max}; \tag{15.24}$$

see Fig. 15.9. This is the *Kolmogorov spectrum* for the spectral energy density of stationary, isotropic, incompressible turbulence. It is valid only in the range $k_{\min} \ll k \ll k_{\max}$, because only in this range are the turbulent eddies continuously receiving energy from larger lengthscales and passing it on to smaller scales. At the ends of the range, the spectrum has to be modified in the manner illustrated qualitatively in Fig. 15.9.

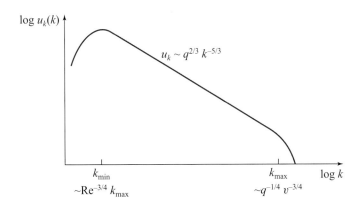

FIGURE 15.9 The Kolmogorov spectral energy per unit mass for stationary, homogeneous turbulence.

At the smallest lengthscales present, $k_{\max}^{-1}$, the molecular viscous forces become competitive with inertial forces in the Navier-Stokes equation; these viscous forces convert the cascading energy into heat. Since the ratio of inertial forces to viscous forces is the Reynolds number, the smallest eddies have a Reynolds number of order unity: $\mathrm{Re}_{k_{\max}} = v(k_{\max}) k_{\max}^{-1}/\nu \sim 1$. Inserting Eq. (15.23) for $v(k)$, we obtain

$$k_{\max} \sim q^{1/4} \nu^{-3/4}. \tag{15.25}$$

The largest eddies have sizes $\ell \sim k_{\min}^{-1}$ and turnover speeds $v_\ell = v(k_{\min}) \sim (q/k_{\min})^{1/3}$. By combining these relations with Eq. (15.25), we see that the ratio of the largest wave numbers present in the turbulence to the smallest is

$$\boxed{\frac{k_{\max}}{k_{\min}} \sim \left(\frac{v_\ell \ell}{\nu}\right)^{3/4} = \mathrm{Re}_\ell^{3/4}.} \tag{15.26}$$

range of wave numbers in turbulence

Here Re_ℓ is the Reynolds number for the flow's largest eddies.

Let us now take stock of our results. If we know the scale ℓ of the largest eddies and their rms turnover speeds v_ℓ (and, of course, the viscosity of the fluid), then we can compute their Reynolds number Re_ℓ; from that, Eq. (15.26), and $k_{\min} \sim \ell^{-1}$, we can compute the flow's maximum and minimum wave numbers; and from $q \sim v_\ell^3/\ell$ and Eq. (15.24), we can compute the spectral energy per unit mass in the turbulence.

We can also compute the total time required for energy to cascade from the largest eddies to the smallest. Since $\tau(k) \sim 1/(kv) \sim 1/(q^{1/3} k^{2/3})$, each successive set of eddies feeds its energy downward in a time $2^{-2/3}$ shorter than the preceding set did. As a result, it takes roughly the same amount of time for energy to pass from the second-largest eddies (size $\ell/2$) to the very smallest (size $k_{\max}^{-1}$) as it takes for the second-largest to extract the energy from the very largest. The total cascade occurs in a time of several ℓ/v_ℓ (during which time, of course, the mean flow has fed new energy into the largest eddies and they are sending it downward).

These results are accurate only to within factors of order unity—with one major exception. The $-5/3$ power law in the Kolmogorov spectrum is very accurate. That this ought to be so one can verify in two equivalent ways: (i) Repeat the above derivation inserting arbitrary factors of order unity at every step. These factors will influence the final multiplicative factor in the Kolmogorov spectrum but will not influence the $-5/3$ power. (ii) Use dimensional analysis. Specifically, notice that the only dimensional entities that can influence the spectrum in the region $k_{\min} \ll k \ll k_{\max}$ are the energy cascade rate q and the wave number k. Then notice that the only quantity with the dimensions of $u_k(k)$ (energy per unit mass per unit wave number) that can be constructed from q and k is $q^{2/3} k^{-5/3}$. Thus, aside from a multiplicative factor of order unity, this must be the form of $u_k(k)$.

phenomena missed by Kolmogorov spectrum

intermittency

Let us now review and critique the assumptions that went into our derivation of the Kolmogorov spectrum. First, we assumed that the turbulence is stationary and homogeneous. Real turbulence is neither of these, since it exhibits intermittency (Sec. 15.3), and smaller eddies tend to occupy less volume overall than do larger eddies and so cannot be uniformly distributed in space. Second, we assumed that the energy source is large-lengthscale motion and that the energy transport is local in k-space from the large lengthscales to steadily smaller ones. In the language of a Fourier decomposition into normal modes, we assumed that nonlinear coupling between modes with wave number k causes modes with wave number of order $2k$ to grow but does not significantly enhance modes with wave number $100k$ or $0.01k$. Again this behavior is not completely in accord with observations, which reveal the development of *coherent structures*—large-scale regions with correlated vorticity in the flow. These structures are evidence for a reversed flow of energy in k-space from small scales to large ones, and they play a major role in another feature of real turbulence, entrainment—the spreading of an organized motion (e.g., a jet) into the surrounding fluid (Sec. 15.4.3).

coherent structures, flow of energy from small scales to large, and entrainment

Despite these issues with its derivation, the Kolmogorov law is surprisingly accurate and useful. It has been verified in many laboratory flows, and it describes many naturally occurring instances of turbulence. For example, the twinkling of starlight is caused by refractive index fluctuations in Earth's atmosphere, whose power spectrum we can determine optically. The underlying turbulence spectrum turns out to be of Kolmogorov form; see Box 9.2.

KOLMOGOROV SPECTRUM FOR QUANTITIES TRANSPORTED BY THE FLUID

In applications of the Kolmogorov spectrum, including twinkling starlight, one often must deal with such quantities as the index of refraction $\mathfrak{n}$ that, in the turbulent cascade, are transported passively with the fluid, so that $d\mathfrak{n}/dt = 0$.[9] As the cascade

---

9. In Earth's atmospheric turbulence, $\mathfrak{n}$ is primarily a function of the concentration of water molecules and the air density, which is controlled by temperature via thermal expansion and contraction. Both water concentration and temperature are carried along by the fluid in the turbulent cascade, whence so is $\mathfrak{n}$.

strings out and distorts fluid elements, it will also stretch and distort any inhomogeneities of n, driving them toward smaller lengthscales (larger k).

What is the k-dependence of the resulting spectral density $S_n(k)$? The quickest route to an answer is dimensional analysis. This S_n can be influenced only by the quantities that characterize the velocity field (its wave number k and its turbulent-energy cascade rate q) plus a third quantity: the rate $q_n = d\sigma_n^2/dt$ at which the variance $\sigma_n^2 = \int S_n \, d^3k/(2\pi)^3$ of n would die out if the forces driving the turbulence were turned off. (This q_n is the analog, for n, of q for velocity.) The only combination of k, q, and q_n that has the same dimensions as S_n is (Ex. 15.9)

$$S_n \sim q_n q^{-1/3} k^{-11/3}. \tag{15.27}$$

Kolmogorov spectrum for quantities transported with the fluid in its turbulent cascade

This is the same $-11/3$ power law as for S_{v_j} [Eq. (15.24)], and it will hold true also for any other quantity Θ that is transported with the fluid ($d\Theta/dt = 0$) in the turbulent cascade.

As for the velocity's inhomogeneities, so also for the inhomogeneities of n, there is some small lengthscale, $1/k_{n\,\text{max}}$, at which diffusion wipes them out, terminating the cascade. Because n is a function of temperature and water concentration, the relevant diffusion is a combination of thermal diffusion and water-molecule diffusion, which will proceed at a (perhaps modestly) different rate from the viscous diffusion for velocity inhomogeneities. Therefore, the maximum k in the cascade may be different for the index of refraction (and for any other quantity transported by the fluid) than for the velocity.

In applications, one often needs a spatial description of the turbulent spectrum rather than a wave-number description. This is normally given by the Fourier transform of the spectral density, which is the correlation function $C_n(\xi) = \langle \delta n(x) \delta n(x + \xi) \rangle$ (where δn is the perturbation away from the mean and $\langle \cdot \rangle$ is the ensemble average or average over space); see Sec. 6.4. For turbulence, the correlation function has the unfortunate property that $C_n(0) = \langle \delta n^2 \rangle \equiv \sigma_n^2$ is the variance, which is dominated by the largest lengthscales, where the Kolmogorov power-law spectrum breaks down. To avoid this problem it is convenient, when working with turbulence, to use in place of the correlation function the so-called *structure function*:

$$D_n(\xi) = \langle [\delta n(\mathbf{x} + \boldsymbol{\xi}) - \delta n(\mathbf{x})]^2 \rangle = 2[\sigma_n^2 - C_n(\xi)]. \tag{15.28}$$

structure function for spatial structure of turbulence

By Fourier transforming the spectral density (15.27), or more quickly by dimensional analysis, one can infer the Kolmogorov spectrum for this structure function:

$$D_n(\xi) \sim (q_n/q^{1/3}) \xi^{2/3} \quad \text{for} \quad k_{\text{max}}^{-1} \lesssim \xi \lesssim k_{\text{min}}^{-1}. \tag{15.29}$$

In Box 9.2, we use this version of the spectrum to analyze the twinkling of starlight due to atmospheric turbulence.

15.4 Semiquantitative Analysis of Turbulence

EXERCISES

Exercise 15.9 *Derivation and Practice: Kolmogorov Spectrum for Quantities Transported by the Fluid*

(a) Fill in the details of the dimensional-analysis derivation of the Kolmogorov spectrum (15.27) for a quantity such as $\mathfrak{n}$ that is transported by the fluid and thus satisfies $d\mathfrak{n}/dt = 0$. In particular, convince yourself (or refute!) that the only quantities $S_\mathfrak{n}$ can depend on are k, q, and $q_\mathfrak{n}$, identify their dimensions and the dimension of $S_\mathfrak{n}$, and use those dimensions to deduce Eq. (15.27).

(b) Derive the spatial version (15.29) of the Kolmogorov spectrum for a transported quantity by two methods: (i) Fourier transforming the wave-number version (15.27), and (ii) dimensional analysis.

Exercise 15.10 *Example: Excitation of Earth's Normal Modes by Atmospheric Turbulence*[10]

Earth has normal modes of oscillation, many of which are in the milli-Hertz frequency range. Large earthquakes occasionally excite these modes strongly, but the quakes are usually widely spaced in time compared to the ringdown time of a particular mode (typically a few days). There is evidence of a background level of continuous excitation of these modes, with an rms ground acceleration per mode of $\sim 10^{-10}$ cm s^{-2} at seismically "quiet" times. The excitation mechanism is suspected to be stochastic forcing by the pressure fluctuations associated with atmospheric turbulence. This exercise deals with some aspects of this hypothesis.

(a) Estimate the rms pressure fluctuations $P(f)$ at frequency f, in a bandwidth equal to frequency $\Delta f = f$, produced on Earth's surface by atmospheric turbulence, assuming a Kolmogorov spectrum for the turbulent velocities and energy. Make your estimate using two methods: (i) using dimensional analysis (what quantity can you construct from the energy cascade rate q, atmospheric density ρ, and frequency f that has dimensions of pressure?) and (ii) using the kinds of arguments about eddy sizes and speeds developed in Sec. 15.4.4.

(b) Your answer using method (i) in part (a) should scale with frequency as $P(f) \propto 1/f$. In actuality, the measured pressure spectra have a scaling law more nearly like $P(f) \propto 1/f^{2/3}$, not $P(f) \propto 1/f$ (e.g., Tanimoto and Um, 1999, Fig. 2a). Explain this discrepancy [i.e., what is wrong with the argument in method (i) and how can you correct it to give $P(f) \propto 1/f^{2/3}$?].

(c) The low-frequency cutoff for the $P(f) \propto 1/f^{2/3}$ pressure spectrum is about 0.5 mHz, and at 1 mHz, $P(f)$ has the value $P(f = 1\,\text{mHz}) \sim 0.3$ Pa, which is

---

[10] Problem devised by David Stevenson, based in part on Tanimoto and Um (1999), who, however, used the pressure spectrum deduced in method (i) of part (a) rather than the more nearly correct spectrum of part (b). The difference in spectra does not much affect their conclusions.

about 3 × 10^{-6} of atmospheric pressure. Assuming that 0.5 mHz corresponds to the largest eddies, which have a lengthscale of a few kilometers (a little less than the scale height of the atmosphere), derive an estimate for the eddies' turbulent viscosity ν_t in the lower atmosphere. By how many orders of magnitude does this exceed the molecular viscosity? What fraction of the Sun's energy input to Earth ($\sim 10^6$ erg cm^{-2} s^{-1}) goes into maintaining this turbulence (assumed to be distributed over the lowermost 10 km of the atmosphere)?

(d) At $f = 1$ mHz, what is the characteristic spatial scale (wavelength) of the relevant normal modes of Earth? [Hint: The relevant modes have few or no nodes in the radial direction. All you need to answer this question is a typical wave speed for seismic shear waves, which you can take to be 5 km s^{-1}.] What is the characteristic spatial scale (eddy size) of the atmospheric pressure fluctuations at this same frequency, assuming isotropic turbulence? Suggest a plausible estimate for the rms amplitude of the pressure fluctuation averaged over a surface area equal to one square wavelength of Earth's normal modes. (You must keep in mind the random spatially and temporally fluctuating character of the turbulence.)

(e) **Challenge:** Using your answer from part (d) and a characteristic shear and bulk modulus for Earth's deformations of $K \sim \mu \sim 10^{12}$ dyne cm^{-2}, comment on how the observed rms normal-mode acceleration (10^{-10} cm s^{-2}) compares with that expected from stochastic forcing due to atmospheric turbulence. You may need to review Chaps. 11 and 12, and think about the relationship between surface force and surface deformation. [Note: Several issues emerge when doing this assessment accurately that have not been dealt with in this exercise (e.g., number of modes in a given frequency range), so don't expect to be able to get an answer more accurate than an order of magnitude.]

15.5 Turbulent Boundary Layers

Great interest surrounds the projection of spheres of cork, rubber, leather, and string by various parts of the human anatomy, with and without the mechanical advantage of levers of willow, ceramic, and the finest Kentucky ash. As is well known, changing the surface texture, orientation, and spin of a sphere in various sports can influence its trajectory markedly. Much study has been made of ways to do this both legally and illegally. Some procedures used by professional athletes are pure superstition, but many others find physical explanations that are good examples of the behavior of boundary layers. Many sports involve the motion of balls at Reynolds numbers where the boundary layers can transition between laminar and turbulent, which allows opportunities for controlling the flow. With the goal of studying this transition, let us now consider the structure of a turbulent boundary layer—first along a straight wall and later along a ball's surface.

15.5.1 Profile of a Turbulent Boundary Layer

In Sec. 14.4.1, we derived the Blasius profile for a laminar boundary layer and showed that its thickness a distance x downstream from the start of the boundary layer was roughly $3\delta = 3(\nu x/V)^{1/2}$, where V is the free-stream speed (cf. Fig. 14.13). As we have described, when the Reynolds number is large enough—$\mathrm{Re}_d = Vd/\nu \sim 3 \times 10^5$ or $\mathrm{Re}_\delta \sim \sqrt{\mathrm{Re}_d} \sim 500$ in the case of flow past a cylinder (Figs. 15.1 and 15.2)—the boundary layer becomes turbulent.

A turbulent boundary layer consists of a thin laminar sublayer of thickness δ_{ls} close to the wall and a much thicker turbulent zone of thickness δ_t (Fig. 15.10).

turbulent boundary layer's laminar sublayer and turbulent zone

In the following paragraphs we use the turbulence concepts developed in the previous sections to compute the order of magnitude of the structures of the laminar sublayer and the turbulent zone, and the manner in which those structures evolve along the boundary. We let the wall have any shape, so long as its radius of curvature is large compared to the boundary layer's thickness, so it looks locally planar. We also use locally Cartesian coordinates with distance y measured perpendicular to the boundary and distance x along it, in the direction of the near-wall flow.

One key to the structure of the boundary layer is that, in the x component of the time-averaged Navier-Stokes equation, the stress-divergence term $T_{xy,y}$ has the *potential* to be so large (because of the boundary layer's small thickness) that no other term can compensate for it. This is true in the turbulent zone, where T_{xy} is the huge Reynolds stress; it is also true in the laminar sublayer, where T_{xy} is the huge viscous stress produced by a large shear that results from the thinness of the layer. (One can check at the end of the following analysis that, for the computed boundary-layer structure, other terms in the x component of the Navier-Stokes equation are indeed so small that they could not compensate a significantly nonzero $T_{xy,y}$.) This potential dominance by $T_{xy,y}$ implies that the flow must adjust itself to make $T_{xy,y}$ nearly zero, so T_{xy} must be (very nearly) independent of distance y from the boundary.

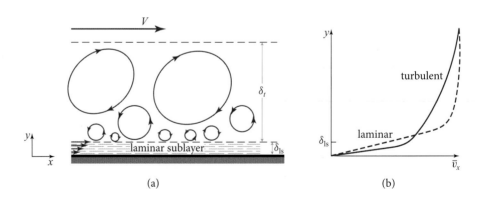

FIGURE 15.10 (a) Physical structure of a turbulent boundary layer. (b) Mean flow speed $\bar{v}_x$ as a function of distance from the wall for the turbulent boundary layer [solid curve, Eqs. (15.30)] and for a laminar boundary layer [dashed curve; the Blasius profile, Eqs. (14.46)].

In the turbulent zone, T_{xy} is the Reynolds stress, ρv_ℓ^2, where v_ℓ is the turbulent velocity of the largest eddies at a distance y from the wall; therefore, constancy of T_{xy} implies constancy of v_ℓ. The largest eddies at y have a size ℓ of order the distance y from the wall; correspondingly, the turbulent viscosity is $v_t \sim v_\ell y/3$. Equating the expression ρv_ℓ^2 for the Reynolds stress to the alternative expression $2\rho v_t \frac{1}{2} \bar{v}_{,y}$ (where $\bar{v}$ is the mean-flow speed at y, and $\frac{1}{2}\bar{v}_{,y}$ is the shear), and using $v_t \sim v_\ell y/3$ for the turbulent viscosity, we discover that in the turbulent zone the mean flow speed varies logarithmically with distance from the wall: $\bar{v} \sim v_\ell \ln y + \text{constant}$. Since the turbulence is created at the inner edge of the turbulent zone, $y \sim \delta_{ls}$ (Fig. 15.10) by interaction of the mean flow with the laminar sublayer, the largest turbulent eddies there must have their turnover speeds v_ℓ equal to the mean-flow speed there: $\bar{v} \sim v_\ell$ at $y \sim \delta_{ls}$. This tells us the normalization of the logarithmically varying mean-flow speed:

$$\boxed{\bar{v} \sim v_\ell [1 + \ln(y/\delta_{ls})] \quad \text{at } y \gtrsim \delta_{ls}.} \tag{15.30a}$$

profile of mean flow speed in turbulent zone

Turn next to the structure of the laminar sublayer. There the constant shear stress is viscous, $T_{xy} = \rho v \bar{v}_{,y}$. Stress balance at the interface between the laminar sublayer and the turbulent zone requires that this viscous stress be equal to the turbulent zone's ρv_ℓ^2. This equality implies a linear profile for the mean-flow speed in the laminar sublayer, $\bar{v} = (v_\ell^2/v)y$. The thickness of the sublayer is then fixed by continuity of $\bar{v}$ at its outer edge: $(v_\ell^2/v)\delta_{ls} = v_\ell$. Combining these last two relations, we obtain the following profile and laminar-sublayer thickness:

$$\boxed{\bar{v} \sim v_\ell \left(\frac{y}{\delta_{ls}}\right) \quad \text{at } y \lesssim \delta_{ls} \sim v/v_\ell.} \tag{15.30b}$$

mean flow velocity in laminar sublayer

Having deduced the internal structure of the boundary layer, we turn to the issue of what determines the y-independent turbulent velocity v_ℓ of the largest eddies. This v_ℓ is fixed by matching the turbulent zone to the free-streaming region outside it. The free-stream velocity V (which may vary slowly with x due to curvature of the wall) must be equal to the mean flow velocity $\bar{v}$ [Eq. (15.30a)] at the outer edge of the turbulent zone. The logarithmic term dominates, so $V = v_\ell \ln(\delta_t/\delta_{ls})$. Introducing an overall Reynolds number for the boundary layer,

$$\boxed{\text{Re}_\delta \equiv V\delta_t/v,} \tag{15.31}$$

and noting that turbulence requires a huge value ($\gtrsim 1{,}000$) of this Re_δ, we can reexpress V as $V \sim v_\ell \ln \text{Re}_\delta$. This should actually be regarded as an equation for the turbulent velocity of the largest-scale eddies in terms of the free-stream velocity:

$$\boxed{v_\ell \sim \frac{V}{\ln \text{Re}_\delta}.} \tag{15.32}$$

turbulent velocity of largest-scale eddies in turbulent zone

15.5 Turbulent Boundary Layers

If the free-stream velocity $V(x)$ and the thickness $\delta_t + \delta_{ls} \simeq \delta_t$ of the entire boundary layer are given, then Eq. (15.31) determines the boundary layer's Reynolds number, Eq. (15.32) then determines the turbulent velocity, and Eqs. (15.30) determine the layer's internal structure.

Finally, we turn to the issue of how the boundary layer thickness δ_t evolves with distance x down the wall (and, correspondingly, how the rest of the boundary layer's structure, which is fixed by δ_t, evolves). The key to the evolution of δ_t is entrainment, which we met in our discussion of turbulent wakes and jets (Sec. 15.4.3). At the turbulent zone's outer edge, the largest turbulent eddies move with speed $\sim v_\ell$ into the free-streaming fluid, entraining that fluid into themselves. Correspondingly, the thickness grows at a rate

thickness of turbulent boundary layer

$$\boxed{\frac{d\delta_t}{dx} \sim \frac{v_\ell}{V} \sim \frac{1}{\ln \mathrm{Re}_\delta}.} \tag{15.33}$$

Since $\ln \mathrm{Re}_\delta$ depends only extremely weakly on δ_t, the turbulent boundary layer expands essentially linearly with distance x, by contrast with a laminar boundary layer's $\delta \propto x^{1/2}$.

15.5.2 Coanda Effect and Separation in a Turbulent Boundary Layer

One can easily verify that not only does the turbulent boundary layer expand more rapidly than the corresponding laminar boundary layer would (if it were stable) but also that the turbulent layer is thicker at all locations down the wall. Physically, this distinction can be traced, in part, to the fact that the turbulent boundary layer involves a 3-dimensional velocity field, whereas the corresponding laminar layer would involve only a 2-dimensional field. The enhanced thickness and expansion contribute to the Coanda effect for a turbulent boundary layer—the layer's ability to stick to the wall under adverse conditions (Sec. 15.4.3).

However, there is a price to be paid for this benefit. Since the velocity gradient is increased near the surface, the actual surface shear stress exerted by the turbulent layer, through its laminar sublayer, is significantly larger than in the corresponding laminar boundary layer. As a result, if the entire boundary layer were to remain laminar, the portion that would adhere to the surface would produce less viscous drag than the near-wall laminar sublayer of the turbulent boundary layer. Correspondingly, in a long, straight pipe, the drag on the pipe wall goes up when the boundary layer becomes turbulent.

influence of turbulence on separation of boundary layer and thence on drag

However, for flow around a cylinder or other confined body, the drag goes down! (Cf. Fig. 15.2.) The reason is that in the separated, laminar boundary layer the dominant source of drag is not viscosity but rather a pressure differential between the front face of the cylinder (where the layer adheres) and the back face (where the reverse eddies circulate). The pressure is much lower in the back-face eddies than in the front-face boundary layer, and that pressure differential gives rise to a significant drag, which is reduced when the boundary layer goes turbulent and adheres to the back

Chapter 15. Turbulence

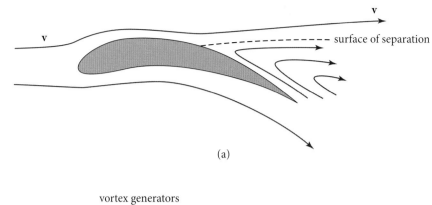

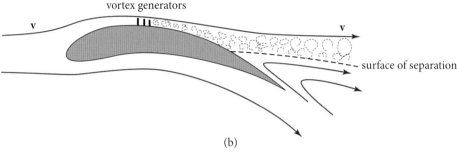

FIGURE 15.11 (a) A laminar boundary layer separating from an airplane wing due to an adverse pressure gradient in the wing's back side. (b) Vortex generators attached to the wing's top face generate turbulence. The turbulent boundary layer sticks to the wing more effectively than does the laminar boundary layer (the Coanda effect). Separation from the wing is delayed, and the wing's lift is increased and drag is decreased.

face. Therefore, if one's goal is to reduce the overall drag and the laminar flow is prone to separation, a nonseparating (or delayed-separation) turbulent layer is preferred to the laminar layer. Similarly (and for essentially the same reason), for an airplane wing, if one's goal is to maintain a large lift, then a nonseparating (or delayed-separation) turbulent layer is superior to a separating, laminar one.[11]

For this reason, steps are often taken in engineering flows to ensure that boundary layers become and remain turbulent. A crude but effective example is provided by the vortex generators that are installed on the upper surfaces of some airplane wings (Fig. 15.11). These structures are small obstacles on the wing that penetrate through a laminar boundary layer into the free flow. By changing the pressure distribution, they force air into the boundary layer and initiate 3-dimensional vortical motion in the boundary layer, forcing it to become partially turbulent. This turbulence improves the

vortex generators on airplane wings

---

11. Another example of separation occurs in "lee waves," which can form when wind blows over a mountain range. These consist of standing-wave eddies in the separated boundary layer, somewhat analogous to the Kármán vortex street of Fig. 15.1d. Lee waves are sometimes used by glider pilots to regain altitude.

15.5.3 Instability of a Laminar Boundary Layer

Much work has been done on the linear stability of laminar boundary layers. The principles of such stability analyses should now be familiar, although the technical details are formidable. In the simplest case, an equilibrium flow is identified (e.g., the Blasius profile), and the equations governing the time evolution of small perturbations are written down. The individual Fourier components are assumed to vary in space and time as $\exp i(\mathbf{k} \cdot \mathbf{x} - \omega t)$, and we seek modes that have zero velocity perturbation on the solid surface past which the fluid flows and that decay to zero in the free stream. We ask whether there are unstable modes (i.e., modes with real $\mathbf{k}$ for which the imaginary part of ω is positive, so they grow exponentially in time). Such exponential growth drives the perturbation to become nonlinear, which then typically triggers turbulence.

The results can generally be expressed in the form of a diagram like that in Fig. 15.12. As shown in that figure, there is generally a critical Reynolds number $\mathrm{Re}_{\mathrm{crit}} \sim 500$ at which one mode becomes linearly unstable. At higher values of the Reynolds number, modes with a range of k-vectors are linearly unstable. One interesting result of these calculations is that in the absence of viscous forces (i.e., in the limit $\mathrm{Re}_\delta \to \infty$), the boundary layer is unstable if and only if there is a point of inflection in the velocity profile (a point where $d^2 v_x / dy^2$ changes sign; cf. Fig. 15.12 and Ex. 15.11).

dynamically stable and unstable modes in laminar boundary layer

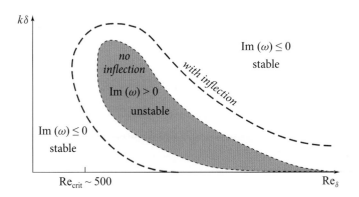

FIGURE 15.12 Values of wave number k for stable and unstable wave modes in a laminar boundary layer with thickness δ, as a function of the boundary layer's Reynolds number $\mathrm{Re}_\delta = V\delta/\nu$. If the unperturbed velocity distribution $v_x(y)$ has no inflection point (i.e., if $d^2 v_x / dy^2 < 0$ everywhere, as is the case for the Blasius profile; Fig. 14.13), then the unstable modes are confined to the shaded region. If there is an inflection point (so $d^2 v_x / dy^2$ is positive near the wall but becomes negative farther from the wall, as is the case near a surface of separation; Fig. 14.14), then the unstable region is larger (dashed boundary) and does not asymptote to $k = 0$ as $\mathrm{Re}_\delta \to \infty$.

Although, in the absence of an inflection, an inviscid flow $v_x(y)$ is stable, for some such profiles even the slightest viscosity can trigger instability. Physically, this is because viscosity can tap the relative kinetic energies of adjacent flow lines. Viscous-triggered instabilities of this sort are sometimes called *secular instabilities* by contrast with the *dynamical instabilities* that arise in the absence of viscosity. Secular instabilities are quite common in fluid mechanics.

secular instabilities contrasted with dynamical instabilities

---

EXERCISES

Exercise 15.11 *Problem: Tollmien-Schlichting Waves*
Consider an inviscid ($\nu = 0$), incompressible flow near a plane wall where a laminar boundary layer is established. Introduce coordinates x parallel to the wall and y perpendicular to it. Let the components of the equilibrium velocity be $v_x(y)$.

(a) Show that a weak propagating-wave perturbation in the velocity, $\delta v_y \propto \exp ik(x - Ct)$, with k real and frequency Ck possibly complex, satisfies the differential equation

$$\frac{\partial^2 \delta v_y}{\partial y^2} = \left[\frac{1}{(v_x - C)} \frac{d^2 v_x}{dy^2} + k^2 \right] \delta v_y. \tag{15.34}$$

These are called *Tollmien-Schlichting waves*.

(b) Hence argue that a sufficient condition for unstable wave modes [Im(C) > 0] is that the velocity field possess a point of inflection (i.e., a point where $d^2 v_x/dy^2$ changes sign; cf. Fig. 15.12). The boundary layer can also be unstable in the absence of a point of inflection, but viscosity must then trigger the instability.

15.5.4 Flight of a Ball

Having developed some insights into boundary layers and their stability, we now apply those insights to the balls used in various sports.

physics of sports balls

The simplest application is to the dimples on a golf ball (Fig. 15.13a). The dimples provide finite-amplitude disturbances in the flow that can trigger growing wave modes and then turbulence in the boundary layer. The adherence of the boundary layer to the ball is improved, and separation occurs farther behind the ball, leading to a lower drag coefficient and a greater range of flight; see Figs. 15.2 and 15.13a.

golf ball: turbulent boundary layer reduces drag

A variant on this mechanism is found in the game of cricket, which is played with a ball whose surface is polished leather with a single equatorial seam of rough stitching. When the ball is "bowled" in a nonspinning way with the seam inclined to the direction of motion, a laminar boundary layer exists on the smooth side and a turbulent one on the side with the rough seam (Fig. 15.13b). These two boundary layers separate at different points behind the flow, leading to a net deflection of the air. The ball therefore swerves toward the side with the leading seam. (The effect is strongest when the ball is new and still shiny and on days when the humidity is high, so the thread in the seam swells and is more efficient at making turbulence.)

cricket ball: two boundary layers, turbulent and laminar, produce deflection

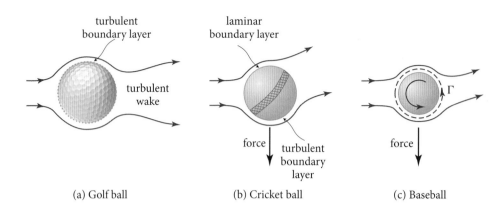

FIGURE 15.13 Boundary layers around (a) golf ball, (b) cricket ball, and (c) baseball as they move leftward relative to the air (i.e., as the air flows rightward as seen in their rest frames).

baseball and table tennis: circulation produces deflection

This mechanism is different from that used to throw a slider or curveball in baseball, in which the pitcher causes the ball to spin about an axis roughly perpendicular to the direction of motion. In the slider the axis is vertical; for a curveball it is inclined at about 45° to the vertical. The spin of the ball creates circulation (in a nonrotating, inertial frame) like that around an airfoil. The pressure forces associated with this circulation produce a net sideways force in the direction of the baseball's rotational velocity on its leading hemisphere (i.e., as seen by the hitter; Fig. 15.13c). The physical origin of this effect is actually quite complex and is only properly described with reference to experimental data. The major effect is that separation is delayed on the side of the ball where the rotational velocity is in the same direction as the airflow and happens sooner on the opposite side (Fig. 15.13c), leading to a pressure differential. The reader may be curious as to how this circulation can be established in view of Kelvin's theorem, Eq. (14.16), which states that if we use a closed curve ∂S, Eq. (14.12a), that is so far from the ball and its wake that viscous forces cannot cause the vorticity to diffuse to it, then the circulation must be zero. What actually happens is similar to an airplane wing (Fig. 14.2b). When the flow is initiated, starting vortices are shed by the baseball and are then convected downstream, leaving behind the net circulation Γ that passes through the ball (Fig. 15.14). This effect is very much larger in 2 dimensions with a rotating cylinder than in 3 dimensions, because the magnitude of the shed vorticity is much larger. It goes by the name of *Magnus effect* in 2 dimensions and *Robins effect* in 3, and it underlies Kutta-Joukowski's theorem for the lift on an airplane wing (Ex. 14.8).

Magnus and Robins effects

In table tennis, a drive is often hit with *topspin*, so that the ball rotates about a horizontal axis perpendicular to the direction of motion. In this case, the net force is downward, and the ball falls faster toward the table, the effect being largest after it has somewhat decelerated. This allows a ball to be hit hard over the net and bounce before passing the end of the table, increasing the margin for errors in the direction of the hit.

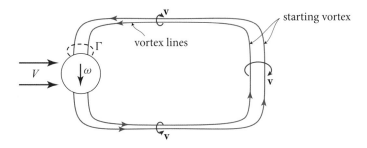

FIGURE 15.14 Vortex lines passing through a spinning ball. The starting vortex is created and shed when the ball is thrown and is carried downstream by the flow as seen in the ball's frame of reference. The vortex lines connecting this starting vortex to the ball lengthen as the flow continues.

Those wishing to improve their curveballs or cure a bad slice are referred to the monographs by Adair (1990), Armenti (1992), and Lighthill (1986).

Exercise 15.12 *Problem: Effect of Drag*

A well-hit golf ball travels about 300 yards. A fast bowler or fastball pitcher throws a cricket ball or baseball at more than 90 mph (miles per hour). A table-tennis player can hit a forehand return at about 30 mph. The masses and diameters of each of these four types of balls are $m_g \sim 46$ g, $d_g \sim 43$ mm; $m_c \sim 160$ g, $d_c \sim 70$ mm; $m_b \sim 140$ g, $d_b \sim 75$ mm; and $m_{tt} \sim 2.5$ g, $d_{tt} \sim 38$ mm.

(a) For golf, cricket (or baseball), and table tennis, estimate the Reynolds number of the flow and infer the drag coefficient C_D. (The variation of C_D with Re_d can be assumed to be similar to that in flow past a cylinder; Fig. 15.2.)

(b) Hence estimate the importance of aerodynamic drag in determining the range of a ball in each of these three cases.

15.6 The Route to Turbulence—Onset of Chaos

15.6.1 Rotating Couette Flow

Let us examine qualitatively how a viscous flow becomes turbulent. A good example is rotating Couette flow between two long, concentric, differentially rotating cylinders, as introduced in Sec. 14.6.3 and depicted in Fig. 15.15a. This flow was studied in a classic set of experiments by Gollub, Swinney, and collaborators (Fenstermacher, Swinney, and Gollub, 1979; Gorman and Swinney, 1982, and references therein). In these experiments, the inner cylinder rotates and the outer one does not, and the fluid is a liquid whose viscosity is gradually decreased by heating it, so the Reynolds number gradually increases. At low Reynolds numbers, the equilibrium flow is stable,

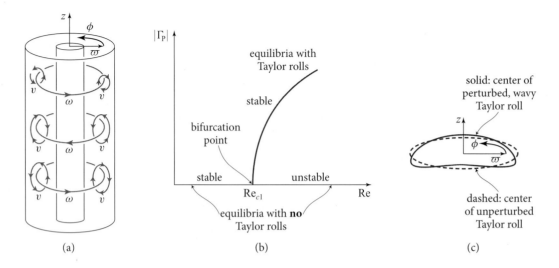

FIGURE 15.15 Bifurcation of equilibria in rotating Couette flow. (a) Equilibrium flow with Taylor rolls. (b) Bifurcation diagram in which the amplitude of the poloidal circulation $|\Gamma_P|$ in a Taylor roll is plotted against the Reynolds number. At low Reynolds numbers (Re < Re_{c1}), the only equilibrium flow configuration is smooth, azimuthal flow. At larger Reynolds numbers (Re_{c1} < Re < Re_{c2}), there are two equilibria, one with Taylor rolls and stable; the other with smooth, azimuthal flow, which is unstable. (c) Shape of a Taylor roll at Re_{c1} < Re < Re_{c2} (dashed ellipse) and at higher values, Re_{c2} < Re < Re_{c3} (solid, wavy curve).

stationary, and azimuthal (strictly in the ϕ direction; Fig. 15.15a). At very high Reynolds numbers, the flow is unstable, according to the Rayleigh criterion (angular momentum per unit mass decreases outward; Sec. 14.6.3). Therefore, as the Reynolds number is gradually increased, at some critical value Re_{c1}, the flow becomes unstable to the growth of small perturbations. These perturbations drive a transition to a new stationary equilibrium whose form is what one might expect from the Rayleigh-criterion thought experiment (Sec. 14.6.3): it involves poloidal circulation (quasi-circular motions in the r and z directions, called *Taylor rolls*; see Fig. 15.15a).

Taylor rolls

bifurcation to a lower symmetry state

Thus an equilibrium with a high degree of symmetry has become unstable, and a new, lower-symmetry, stable equilibrium has been established; see Fig. 15.15b. Translational invariance along the cylinder axis has been lost from the flow, even though the boundary conditions remain translationally symmetric. This transition is another example of the bifurcation of equilibria that we discussed when treating the buckling of beams and playing cards (Secs. 11.6 and 12.3.5).

As the Reynolds number is increased further, this process repeats. At a second critical Reynolds number Re_{c2}, a second bifurcation of equilibria occurs in which the azimuthally smooth Taylor rolls become unstable and are replaced by new, azimuthally wavy Taylor rolls; see Fig. 15.15c. Again, an equilibrium with higher symmetry (rotation invariance) has been replaced at a bifurcation point by one of lower symmetry (no rotation invariance). A fundamental frequency f_1 shows up in the fluid's velocity $\mathbf{v}(\mathbf{x}, t)$ as the wavy Taylor rolls circulate around the central cylinder.

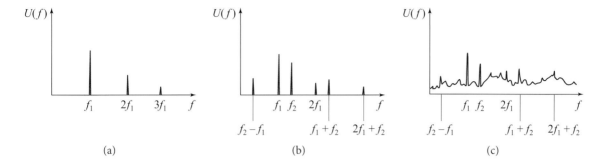

FIGURE 15.16 The energy spectrum of velocity fluctuations in rotating Couette flow (schematic). (a) Spectrum for a moderate Reynolds number, $Re_{c2} < Re < Re_{c3}$, at which the stable equilibrium flow is that with the wavy Taylor rolls of Fig. 15.15c. (b) Spectrum for a higher Reynolds number, $Re_{c3} < Re < Re_{c4}$, at which the stable flow has wavy Taylor rolls with two incommensurate fundamental frequencies present. (c) Spectrum for a still higher Reynolds number, $Re > Re_{c4}$, at which turbulence has set in.

Since the waves are nonlinearly large, harmonics of this fundamental are also seen when the velocity field is Fourier decomposed (cf. Fig. 15.16a). When the Reynolds number is increased still further to some third critical value Re_{c3}, there is yet another bifurcation. The Taylor rolls now develop a second set of waves, superimposed on the first, with a corresponding new fundamental frequency f_2 that is incommensurate with f_1. In the energy spectrum one now sees various harmonics of f_1 and of f_2, as well as sums and differences of these two fundamentals (Fig. 15.16b).

<small>wavy Taylor rolls: two frequencies</small>

It is exceedingly difficult to construct an experimental apparatus that is clean enough—and sufficiently free from the effects of finite lengths of the cylinders—to reveal what happens next as the Reynolds number increases. However, despite the absence of clean experiments, it seemed obvious before the 1970s what would happen. The sequence of bifurcations would continue, with ever-decreasing intervals of Reynolds number ΔRe between them, eventually producing such a complex maze of frequencies, harmonics, sums, and differences, as to be interpreted as turbulence. Indeed, one finds the onset of turbulence described in just this manner in the classic fluid mechanics textbook of Landau and Lifshitz (1959), based on earlier research by Landau (1944).

The 1970s and 1980s brought a major breakthrough in our understanding of the onset of turbulence. This breakthrough came from studies of model dynamical systems with only a few degrees of freedom, in which nonlinear effects play similar roles to the nonlinearities of the Navier-Stokes equation. These studies revealed only a handful of routes to irregular or unpredictable behavior known as chaos, and none were of the Landau type. However, one of these routes starts out in the same manner as does rotating Couette flow: As a control parameter (the Reynolds number in the case of Couette flow) is gradually increased, first oscillations with one fundamental frequency f_1 and its harmonics turn on; then a second frequency f_2 (incommensurate with the first) and its harmonics turn on, along with sums and differences of f_1

<small>route to turbulence in rotating Couette flow: one frequency, two frequencies, then turbulence (chaos)</small>

15.6 The Route to Turbulence—Onset of Chaos

and f_2; and then, suddenly, chaos sets in. Moreover, the chaos is clearly not being produced by a complicated superposition of new frequencies; it is fundamentally different from that.

Remarkably, the experiments of Gollub, Swinney, and colleagues gave convincing evidence that the onset of turbulence in rotating Couette flow takes precisely this route (Fig. 15.16c).

15.6.2 Feigenbaum Sequence, Poincaré Maps, and the Period-Doubling Route to Turbulence in Convection

The simplest of systems in which one can study the several possible routes to chaos are 1-dimensional mathematical maps. A lovely example is the *Feigenbaum sequence*, explored by Mitchell Feigenbaum (1978).

The Feigenbaum sequence is a sequence $\{x_1, x_2, x_3, \ldots\}$ of values of a real variable x, given by the rule (sometimes called the *logistic equation*)[12]

logistic equation and its Feigenbaum sequence

$$x_{n+1} = 4ax_n(1 - x_n). \tag{15.35}$$

Here a is a fixed control parameter. It is easy to compute Feigenbaum sequences $\{x_1, x_2, x_3, \ldots\}$ for different values of a on a personal computer (Ex. 15.13). What is found is that there are critical parameters $a_1, a_2, \ldots$ at which the character of the sequence changes sharply. For $a < a_1$, the sequence asymptotes to a stable fixed point. For $a_1 < a < a_2$, the sequence asymptotes to stable, periodic oscillations between two fixed points. If we increase the parameter further, so that $a_2 < a < a_3$, the sequence becomes a periodic oscillation between four fixed points. The period of the oscillation has doubled. This *period doubling* (not to be confused with frequency doubling) happens again: When $a_3 < a < a_4$, x asymptotes to regular motion between eight fixed points. Period doubling increases with shorter and shorter intervals of a until at some value a_∞, the period becomes infinite, and the sequence does not repeat. Chaos has set in.

period doubling route to chaos for Feigenbaum sequence

This period doubling is a second route to chaos, very different in character from the one-frequency, two-frequencies, chaos route that appears to occur in rotating Couette flow. Remarkably, fluid dynamical turbulence can set in by this second route as well as by the first. It does so in certain very clean experiments on convection. We return to this phenomenon at the end of this section and then again in Sec. 18.4.

How can so starkly simple and discrete a model as a 1-dimensional map bear any relationship to the continuous solutions of the fluid dynamical differential equations? The answer is quite remarkable.

---

12. This equation first appeared in discussions of population biology (Verhulst, 1838). If we consider x_n as being proportional to the number of animals in a species (traditionally rabbits), the number in the next season should be proportional to the number of animals already present and to the availability of resources, which will decrease as x_n approaches some maximum value, in this case unity. Hence the terms x_n and $1 - x_n$ in Eq. (15.35).

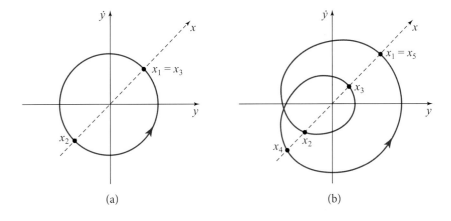

FIGURE 15.17 (a) Representation of a single periodic oscillation as motion in phase space. (b) Motion in phase space after period doubling. The behavior of the system may also be described by using the coordinate x of the Poincaré map.

Consider a steady flow in which one parameter a (e.g., the Reynolds number) can be adjusted. Now, as we change a and approach turbulence, the flow may develop a periodic oscillation with a single frequency f_1. We could measure this by inserting some probe at a fixed point in the flow to measure a fluid variable y (e.g., one component of the velocity). We can detect the periodicity either by inspecting the readout $y(t)$ or its Fourier transform $\tilde{y}$. However, there is another way, which may be familiar from classical mechanics. This is to regard $\{y, \dot{y}\}$ as the two coordinates of a 2-dimensional phase space. (Of course, instead one could measure many variables and their time derivatives, resulting in an arbitrarily large phase space, but let us keep matters as simple as possible.) For a single periodic oscillation, the system will follow a closed path in this phase space (Fig. 15.17a). As we increase a further, a period doubling may occur, and the trajectory in phase space may look like Fig. 15.17b. As we are primarily interested in the development of the oscillations, we need only keep one number for every fundamental period $P_1 = 1/f_1$. Let us do this by taking a section through phase space and introducing a coordinate x on this section, as shown in Fig. 15.17. The nth time the trajectory crosses this section, its crossing point is x_n, and the mapping from x_n to x_{n+1} can be taken as a representative characterization of the flow. When only the frequency f_1 is present, the map reads $x_{n+2} = x_n$ (Fig. 15.17a). When f_1 and $f_2 = \frac{1}{2}f_1$ are present, the map reads $x_{n+4} = x_n$ (Fig. 15.17b). (These specific maps are overly simple compared to what one may encounter in a real flow, but they illustrate the idea.)

phase-space representation of dynamics

To reiterate, instead of describing the flow by the full solution $\mathbf{v}(\mathbf{x}, t)$ to the Navier-Stokes equation and the flow's boundary conditions, we can construct the simple map $x_n \to x_{n+1}$ to characterize the flow. This procedure is known as a *Poincaré map*. The mountains have labored and brought forth a mouse! However, this mouse turns out to be all that we need. For some convection experiments, the same period-

reducing the dynamics of a fluid's route to turbulence to a 1-dimensional map: the Poincaré map doubling behavior and approach to chaos are present in these maps as in the 2-dimensional phase-space diagram and in the full solution to the fluid dynamical equations. Furthermore, when observed in the Poincaré maps, the transition looks qualitatively the same as in the Feigenbaum sequence. It is remarkable that for a system with so many degrees of freedom, chaotic behavior can be observed by suppressing almost all of them.

If, in the period-doubling route to chaos, we compute the limiting ratio of successive critical parameters,

Feigenbaum number for period doubling route to chaos, and its universality

$$\mathcal{F} = \lim_{j \to \infty} \frac{a_j - a_{j-1}}{a_{j+1} - a_j}, \tag{15.36}$$

we find that it has the value 4.66920160910.... This *Feigenbaum number* seems to be a universal constant characteristic of most period-doubling routes to chaos, independent of the particular map that was used. For example, if we had used

$$x_{n+1} = a \sin \pi x_n \tag{15.37}$$

we would have gotten the same constant.

period doubling route to turbulence in Libchaber convection experiments

The most famous illustration of the period-doubling route to chaos is a set of experiments by Libchaber and colleagues on convection in liquid helium, water, and mercury, culminating with the mercury experiment by Libchaber, Laroche, and Fauve (1982). In each experiment (depicted in Fig. 18.1), the temperature at a point was monitored with time as the temperature difference ΔT between the fluid's bottom and top surfaces was slowly increased. In each experiment, at some critical temperature difference ΔT_1 the temperature began to oscillate with a single period; then at some ΔT_2 that oscillation was joined by another at twice the period; at some ΔT_3 another period doubling occurred; at ΔT_4 another; and at ΔT_5 yet another. The frequency doubling could not be followed beyond this because the signal was too weak, but shortly thereafter the convection became turbulent. In each experiment it was possible to estimate the Feigenbaum ratio (15.36), with a_j being the jth critical ΔT. For the highest-accuracy (mercury) experiment, the experimental result agreed with the Feigenbaum number 4.669... to within the experimental accuracy (about 6%); the helium and water experiments also agreed with Feigenbaum to within their experimental accuracies. This result was remarkable. *There truly is a deep universality in the route to chaos, a universality that extends even to fluid-dynamical convection!* This work won Libchaber and Feigenbaum together the prestigious 1986 Wolf Prize.

universality in the route to chaos

EXERCISES

Exercise 15.13 *Problem: Feigenbaum Number for the Logistic Equation*
Use a computer to calculate the first five critical parameters a_j for the sequence of numbers generated by the logistic equation (15.35). Hence verify that the ratio of successive differences tends toward the Feigenbaum number $\mathcal{F}$ quoted in Eq. (15.36). (Hint: To find suitable starting values x_1 and starting parameter a, you might find it

helpful to construct a graph.) For insights into the universality of the Feigenbaum number, based in part on the renormalization group (Sec. 5.8.3), see Sethna (2006, Ex. 12.9).

15.6.3 Other Routes to Turbulent Convection

Some other routes to turbulence have been seen in convection experiments, in addition to the Feigenbaum/Libchaber period-doubling route. Particularly impressive were convection experiments by Gollub and Benson (1980), which showed—depending on the experimental parameters and the nature of the initial convective flow—four different routes:

four different routes to turbulence in convection experiments

1. The period-doubling route.

2. A variant of the one-frequency, two-frequencies, chaos route. In this variant, a bit after (at higher ΔT) the second frequency appears, incommensurate with the first, the two frequencies become commensurate and their modes become phase locked (one entrains the other); and then a bit later (higher ΔT) the phase lock is broken and simultaneously turbulence sets in.

3. A one-frequency, two-frequencies, three-frequencies, chaos route.

4. An intermittency route, in which, as ΔT is increased, the fluid oscillates between a state with two or three incommensurate modes and a state with turbulence.

These four routes to chaos are all seen in simple mathematical maps or low-dimensional dynamical systems; for example, the intermittency route is seen in the Lorenz equations of Ex. 15.16.

Note that the convective turbulence that is triggered by each of these routes is *weak*; the control parameter ΔT must be increased further to drive the fluid into a fully developed, strong-turbulence state.

weakness and confinement of the turbulence that follows these routes

Also important is the fact that these experimental successes, which compare the onset of turbulence with the behaviors of simple mathematical maps or low-dimensional dynamical systems, all entail fluids that are confined by boundaries and have a modest aspect ratio (ratio of the largest to the smallest dimension), 20:1 for rotating Couette flow and 5:1 for convection. For confined fluids with much larger aspect ratios, and for unconfined fluids, there has been no such success. It may be that the successes are related to the small number of relevant modes in a system with modest aspect ratio.

These successes occurred in the 1970s and 1980s. Much subsequent research focuses on the mechanism by which the final transition to turbulence occurs [e.g., the role of the Navier-Stokes nonlinear term $(\mathbf{v} \cdot \boldsymbol{\nabla})\mathbf{v}$]. This research involves a mixture of analytical work and numerical simulations, plus some experiment (see, e.g., Grossman, 2000).

15.6.4 Extreme Sensitivity to Initial Conditions

The evolution of a dynamical system becomes essentially incalculable after the onset of chaos. This is because, as can be shown mathematically, the state of the system (as measured by the value of a map, or in a fluid by the values of a set of fluid variables) at some future time becomes highly sensitive to the assumed initial state. Paths (in the map or in phase space) that start arbitrarily close together diverge from each other exponentially rapidly with time.

It is important to distinguish this unpredictability of classical chaos from unpredictability in the evolution of a quantum mechanical system. A classical system evolves under precisely deterministic differential equations. Given a full characterization of the system at any time t, the system is fully specified at a later time $t + \Delta t$, for any Δt. However, what characterizes chaos is that the evolution of two identical systems in neighboring initial states will eventually evolve so that they follow totally different histories. The time it takes for this to happen is called the *Lyapunov time*. The practical significance of this essentially mathematical feature is that if, as will always be the case, we can only specify the initial state up to a given accuracy (due to practical considerations, not issues of principle), then the true initial state could be any one of those lying in some region, so we have no way of predicting what the state will be after a few Lyapunov times have elapsed.

Lyapunov time: characterizes sensitivity of a classical system to initial conditions

quantum indeterminacy

Quantum mechanical indeterminacy is different. If we can prepare a system in a given state described by a wave function, then the wave function's evolution will be governed fully deterministically by the time-dependent Schrödinger equation. However, if we choose to make a measurement of an observable, many quite distinct outcomes are immediately possible, and (for a high-precision measurement) the system will be left in an eigenstate corresponding to the actual measured outcome. The quantum mechanical description of classical chaos is the subject of *quantum chaos* (e.g., Gutzwiller, 1990).

The realization that many classical systems have an intrinsic unpredictability despite being deterministic from instant to instant has been widely publicized in popularizations of research on chaos. However, the concept is not particularly new. It was well understood, for example, by Poincaré around 1900, and watching the weather report on the nightly news bears witness to its dissemination into popular culture! What *is* new and intriguing is the manner in which a system transitions from a deterministic to an unpredictable evolution.

examples of chaotic behavior

Chaotic behavior is well documented in a variety of physical dynamical systems: electrical circuits, nonlinear pendula, dripping faucets, planetary dynamics, and so on. The extent to which the principles that have been devised to describe chaos in these systems can also be applied to general fluid turbulence remains a matter for debate. There is no question that similarities exist, and (as we have seen) quantitative success has been achieved by applying chaos results to particular forms of turbulent convection. However, most forms of turbulence are not so easily described, and

EXERCISES

Exercise 15.14 *Example: Lyapunov Exponent*
Consider the logistic equation (15.35) for the special case $a = 1$, which is large enough to ensure that chaos has set in.

(a) Make the substitution $x_n = \sin^2 \pi \theta_n$, and show that the logistic equation can be expressed in the form $\theta_{n+1} = 2\theta_n$ (mod 1); that is, θ_{n+1} equals the fractional part of $2\theta_n$.

(b) Write θ_n as a "binimal" (binary decimal). For example, $11/16 = 1/2 + 0/4 + 1/8 + 1/16$ has the binary decimal form 0.1011. Explain what happens to this number in each successive iteration.

(c) Now suppose that an error is made in the ith digit of the starting binimal. When will it cause a major error in the predicted value of x_n?

(d) If the error after n iterations is written ϵ_n, show that the Lyapunov exponent p defined by

$$p = \lim_{n \to \infty} \frac{1}{n} \ln \left| \frac{\epsilon_n}{\epsilon_0} \right| \qquad (15.38)$$

is ln 2 (so $\epsilon_n \simeq 2^n \epsilon_0$ for large enough n). Lyapunov exponents play an important role in the theory of dynamical systems.

Exercise 15.15 *Example: Strange Attractors*
Another interesting 1-dimensional map is provided by the recursion relation

$$x_{n+1} = a \left(1 - 2 \left| x_n - \frac{1}{2} \right| \right). \qquad (15.39)$$

(a) Consider the asymptotic behavior of the variable x_n for different values of the parameter a, with both x_n and a being confined to the interval [0, 1]. In particular, find that for $0 < a < a_{\text{crit}}$ (for some a_{crit}), the sequence x_n converges to a stable fixed point, but for $a_{\text{crit}} < a < 1$, the sequence wanders chaotically through some interval $[x_{\min}, x_{\max}]$.

(b) Using a computer, calculate the value of a_{crit} and the interval $[x_{\min}, x_{\max}]$ for $a = 0.8$.

(c) The interval $[x_{\min}, x_{\max}]$ is an example of a *strange attractor*. It has the property that if we consider sequences with arbitrarily close starting values, their values of x_n in this range will eventually diverge. Show that the attractor is strange by computing the sequences with $a = 0.8$ and starting values $x_1 = 0.5, 0.51, 0.501$, and 0.5001. Determine the number of iterations n_ϵ required to produce significant divergence as a function of $\epsilon = x_1 - 0.5$. It is claimed that $n_\epsilon \sim -\ln_2(\epsilon)$. Can you verify this? Note that the onset of chaos at $a = a_{\text{crit}}$ is quite sudden in this case, unlike the behavior exhibited by the Feigenbaum sequence. See Ruelle (1989) for more on strange attractors.

Exercise 15.16 *Problem: Lorenz Equations*

One of the first discoveries of chaos in a mathematical model was by Lorenz (1963), who made a simple model of atmospheric convection. In this model, the temperature and velocity field are characterized by three variables, x, y, and z, which satisfy the coupled, nonlinear differential equations

$$dx/dt = 10(y - x),$$
$$dy/dt = -xz + 28x - y,$$
$$dz/dt = xy - 8z/3. \quad (15.40)$$

(The precise definitions of x, y, and z need not concern us here.) Integrate these equations numerically to show that x, y, and z follow nonrepeating orbits in the 3-dimensional phase space that they span, and quickly asymptote to a 2-dimensional strange attractor. (It may be helpful to plot out the trajectories of pairs of the dependent variables.)

These Lorenz equations are often studied with the numbers 10, 28, 8/3 replaced by parameters σ, ρ, and β. As these parameters are varied, the behavior of the system changes.

Bibliographic Note

For physical insight into turbulence, we strongly recommend the movies cited in Box 15.2 and the photographs in Van Dyke (1982) and Tritton (1987).

Many fluid mechanics textbooks have good treatments of turbulence. We particularly like White (2008), and also recommend Lautrup (2005) and Panton (2005). Some treatises on turbulence go into the subject much more deeply (though the deeper treatments often have somewhat limited applicability); among these, we particularly like Tennekes and Lumley (1972) and also recommend the more up-to-date Davidson (2005) and Pope (2000). Standard fluid mechanics textbooks for engineers focus particularly on turbulence in pipe flow and in boundary layers (see, e.g., Munson, Young, and Okiishi, 2006; White, 2006; Potter, Wiggert, and Ramadan, 2012).

For the influence of boundary layers and turbulence on the flight of balls of various sorts, see Lighthill (1986), Adair (1990), and Armenti (1992).

For the onset of turbulence in unstable laminar flows, we particularly like Sagdeev, Usikov, and Zaslovsky (1988, Chap. 11) and Acheson (1990, Chap. 9). For the route to chaos in low-dimensional dynamical systems with explicit connections to the onset of turbulence, see Ruelle (1989); for chaos theory with little or no discussion of turbulence, Baker and Gollub (1990), Alligood, Sauer, and Yorke (1996), and Strogatz (2008).

CHAPTER SIXTEEN

Waves

*An ocean traveller has even more vividly the impression
that the ocean is made of waves than that it is made of water.*

ARTHUR EDDINGTON (1927)

16.1 Overview

In the preceding chapters, we have derived the basic equations of fluid dynamics and developed a variety of techniques to describe stationary flows. We have also demonstrated how, even if there exists a rigorous, stationary solution of these equations for a time-steady flow, instabilities may develop, in which the amplitude of an oscillatory disturbance grows with time. These unstable modes of an unstable flow can usually be thought of as waves that interact strongly with the flow and extract energy from it. Of course, wave modes can also be stable and can be studied as independent, individual modes.

Fluid dynamical waves come in a wide variety of forms. They can be driven by a combination of gravitational, pressure, rotational, and surface-tension stresses and also by mechanical disturbances, such as water rushing past a boat or air passing through a larynx. In this chapter, we describe a few examples of wave modes in fluids, chosen to illustrate general wave properties.

The most familiar types of wave are probably *gravity waves* on the surface of a large body of water (Sec. 16.2), such as ocean waves and waves on lakes and rivers. We consider them in the linear approximation and find that they are dispersive in general, though they become nondispersive in the long-wavelength (shallow-water) limit (i.e., when they are influenced by the water's bottom). We also examine the effects of surface tension on gravity waves, which converts them into *capillary waves*, and in this connection we develop a mathematical description of surface tension (see Box 16.4). Boundary conditions can give rise to a discrete spectrum of normal modes, which we illustrate by *helioseismology*: the study of coherent-wave modes excited in the Sun by convective overturning motions.

By contrast with the elastodynamic waves of Chap. 12, waves in fluids often develop amplitudes large enough that nonlinear effects become important (Sec. 16.3). The nonlinearities can cause the front of a wave to steepen and then break—a phenomenon we have all seen at the sea shore. It turns out that, at least under some

> **BOX 16.1. READERS' GUIDE**
>
> - This chapter relies heavily on Chaps. 13 and 14, and less heavily on geometric-optics concepts introduced in Secs. 7.2 and 7.3.
> - Chap. 17 relies to some extent on Secs. 16.2, 16.3, and 16.5 of this chapter.
> - Sec. 16.3 on solitons on water is a foundation for Sec. 23.6 on solitons in plasmas.
> - The remaining chapters of this book do not rely significantly on this chapter.

restrictive conditions, nonlinear waves have very surprising properties. There exist *soliton* or *solitary-wave* modes, in which the front-steepening due to nonlinearity is stably held in check by dispersion, so particular wave profiles are quite robust and propagate for long intervals of time without breaking or dispersing. We demonstrate this behavior by studying flow in a shallow channel. We also explore the remarkable behaviors of such solitons when they pass through each other.

In a nearly rigidly rotating fluid, there are remarkable waves in which the restoring force is the Coriolis effect; they have the unusual property that their group and phase velocities are oppositely directed. These so-called *Rossby waves,* studied in Sec. 16.4, are important in both the oceans and the atmosphere.

The simplest fluid waves of all are small-amplitude *sound waves*—a paradigm for scalar waves. They are nondispersive, just like electromagnetic waves, and are therefore sometimes useful for human communication. We shall study sound waves in Sec. 16.5 and use them to explore (i) the radiation reaction force that acts back on a wave-emitting object (a fundamental physics issue) and (ii) matched asymptotic expansions (a mathematical physics technique). We also describe how sound waves can be produced by fluid flows. This process will be illustrated with the problem of sound generation by high-speed turbulent flows—a problem that provides a good starting point for the topic of the following chapter, compressible flows.

Other examples of fluid waves are treated elsewhere in Part V: Kelvin-Helmholtz waves at the interface between two fluids that move relative to each other (Sec. 14.6.1); Tollmien-Schlichting waves in a laminar boundary layer (Ex. 15.11); lee waves in a separated boundary layer (footnote in Sec. 15.5.2); wavy Taylor rolls in rotating Couette flow (Sec. 15.6.1); shock waves (Sec. 17.5); Sedov-Taylor blast waves (Sec. 17.6); hydraulic jumps (Ex. 17.10); internal waves, which propagate when fluid is stratified in a gravitational field (Ex. 18.8); and various types of magnetohydrodynamic waves (Chap. 19).

> **BOX 16.2. MOVIES RELEVANT TO THIS CHAPTER**
>
> We strongly recommend that the reader view the following movies dealing with waves:
>
> - Bryson (1964)—waves in fluids.
> - Fultz (1969)—relevant to Rossby waves, Sec. 16.4.
> - Rouse (1963c)—relevant to gravity waves on the surface of water, Sec. 16.2.

As in Chaps. 14 and 15, readers are urged to watch movies in parallel with reading this chapter; see Box 16.2.

16.2 Gravity Waves on and beneath the Surface of a Fluid

Gravity waves[1] are waves on and beneath the surface of a fluid, for which the restoring force is the downward pull of gravity. Familiar examples are ocean waves and the waves produced on the surface of a pond when a pebble is thrown in. Less familiar examples are "g modes" of vibration inside the Sun, discussed at the end of this section.

Consider a small-amplitude wave propagating along the surface of a flat-bottomed lake with depth h_o, as shown in Fig. 16.1. As the water's displacement is small, we can describe the wave as a linear perturbation about equilibrium. The equilibrium water is at rest (i.e., it has velocity $\mathbf{v} = 0$). The water's perturbed motion is essentially inviscid and incompressible, so $\boldsymbol{\nabla} \cdot \mathbf{v} = 0$. A simple application of the equation of vorticity transport, Eq. (14.6), assures us that, since the water is static and thus irrotational before and after the wave passes, it must also be irrotational in the wave. Therefore, we can describe the wave inside the water by a velocity potential ψ whose gradient is the velocity field:

$$\mathbf{v} = \boldsymbol{\nabla}\psi. \quad (16.1)$$

Incompressibility, $\boldsymbol{\nabla} \cdot \mathbf{v} = 0$, applied to this equation, implies that ψ satisfies Laplace's equation:

$$\nabla^2 \psi = 0. \quad (16.2)$$

We introduce horizontal coordinates x, y and a vertical coordinate z measured upward from the lake's equilibrium surface (Fig. 16.1). For simplicity we confine

---

1. Not to be confused with gravitational waves, which are waves in the relativistic gravitational field (spacetime curvature) that propagate at the speed of light, and which we meet in Chap. 27.

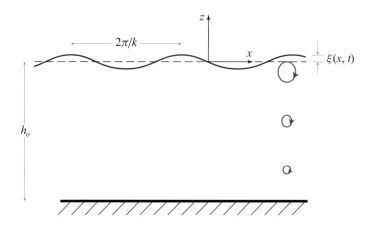

FIGURE 16.1 Gravity waves propagating horizontally across a lake with constant depth h_o.

attention to a sinusoidal wave propagating in the x direction with angular frequency ω and wave number k. Then ψ and all other perturbed quantities have the form $f(z) \exp[i(kx - \omega t)]$ for some function $f(z)$. More general disturbances can be expressed as a superposition of many of these elementary wave modes propagating in various horizontal directions (and in the limit, as a Fourier integral over modes). All the properties of such superpositions follow straightforwardly from those of our elementary plane-wave mode (see Secs. 7.2.2 and 7.3), so we continue to focus on it.

We can use Laplace's equation (16.2) to solve for the vertical variation $f(z)$ of the velocity potential. As the horizontal variation at a particular time is $\propto \exp(ikx)$, direct substitution into Eq. (16.2) gives two possible vertical variations: $\psi \propto \exp(\pm kz)$. The precise linear combination of these two forms is dictated by the boundary conditions. The boundary condition we need is that the vertical component of velocity $v_z = \partial \psi / \partial z$ vanish at the bottom of the lake ($z = -h_o$). The only combination that can vanish is a sinh function. Its integral, the velocity potential, therefore involves a cosh function:

boundary condition at bottom of lake

velocity potential for gravity waves

$$\psi = \psi_0 \cosh[k(z + h_o)] \exp[i(kx - \omega t)]. \tag{16.3}$$

An alert reader might note at this point that, for this ψ, the horizontal component of velocity $v_x = \psi_{,x} = ik\psi$ does not vanish at the lake bottom, in violation of the no-slip boundary condition. In fact, as we discussed in Sec. 14.4, a thin, viscous boundary layer along the bottom of the lake will join our potential-flow solution (16.3) to nonslip fluid at the bottom. We ignore the boundary layer under the (justifiable) assumption that for our oscillating waves, it is too thin to affect much of the flow.

Returning to the potential flow, we must also impose a boundary condition at the surface. This can be obtained from Bernoulli's law. The version of Bernoulli's law that we need is that for an irrotational, isentropic, time-varying flow:

$$\mathbf{v}^2/2 + h + \Phi + \partial \psi / \partial t = \text{const everywhere in the flow} \tag{16.4}$$

[Eqs. (13.51), (13.54)]. We apply this law at the surface of the perturbed water. Let us examine each term:

Boundary condition at surface of water

1. The term $\mathbf{v}^2/2$ is quadratic in a perturbed quantity and therefore can be dropped.
2. The enthalpy $h = u + P/\rho$ (cf. Box 13.2) is a constant, since u and ρ are constants throughout the fluid and P is constant on the surface (equal to the atmospheric pressure).[2]
3. The gravitational potential at the fluid surface is $\Phi = g\xi$, where $\xi(x, t)$ is the surface's vertical displacement from equilibrium, and we ignore an additive constant.
4. The constant on the right-hand side, which could depend on time $[C(t)]$, can be absorbed in the velocity potential term $\partial \psi/\partial t$ without changing the physical observable $\mathbf{v} = \boldsymbol{\nabla} \psi$.

Bernoulli's law applied at the surface therefore simplifies to give

$$g\xi + \frac{\partial \psi}{\partial t} = 0. \tag{16.5}$$

The vertical component of the surface velocity in the linear approximation is just $v_z(z = 0, t) = \partial \xi/\partial t$. Expressing v_z in terms of the velocity potential, we then obtain

$$\frac{\partial \xi}{\partial t} = v_z = \frac{\partial \psi}{\partial z}. \tag{16.6}$$

Combining this expression with the time derivative of Eq. (16.5), we obtain an equation for the vertical gradient of ψ in terms of its time derivative:

$$g \frac{\partial \psi}{\partial z} = -\frac{\partial^2 \psi}{\partial t^2}. \tag{16.7}$$

Finally, substituting Eq. (16.3) into Eq. (16.7) and setting $z = 0$ [because we derived Eq. (16.7) only at the water's surface], we obtain the dispersion relation[3] for linearized gravity waves:

$$\boxed{\omega^2 = gk \, \tanh(kh_o).} \tag{16.8}$$

dispersion relation for gravity waves in water of arbitrary depth

How do the individual elements of fluid move in a gravity wave? We can answer this question (Ex. 16.1) by computing the vertical and horizontal components of the velocity $v_x = \psi_{,x}$, and $v_z = \psi_{,z}$, with ψ given by Eq. (16.3). We find that the fluid elements undergo forward-rotating elliptical motion, as depicted in Fig. 16.1, similar

---

2. Actually, a slight variation of the surface pressure is caused by the varying weight of the air above the surface, but as the density of air is typically $\sim 10^{-3}$ that of water, this correction is very small.
3. For a discussion of the dispersion relation, phase velocity, and group velocity for waves, see Sec. 7.2.

to that for Rayleigh waves on the surface of a solid (Sec. 12.4.2). However, in gravity waves, the sense of rotation is the same (forward) at all depths, in contrast to reversals with depth found in Rayleigh waves [cf. the discussion following Eq. (12.62)].

We now consider two limiting cases: deep water and shallow water.

16.2.1 Deep-Water Waves and Their Excitation and Damping

When the water is deep compared to the wavelength of the waves, $kh_o \gg 1$, the dispersion relation (16.8) becomes

dispersion relation for gravity waves in deep water

$$\boxed{\omega = \sqrt{gk}.} \tag{16.9}$$

Thus deep-water waves are dispersive (see Sec. 7.2); their group velocity $V_g \equiv d\omega/dk = \frac{1}{2}\sqrt{g/k}$ is half their phase velocity, $V_{\rm ph} \equiv \omega/k = \sqrt{g/k}$. [Note that we could have deduced the deep-water dispersion relation (16.9), up to a dimensionless multiplicative constant, by dimensional arguments: The only frequency that can be constructed from the relevant variables g, k, and ρ is $\sqrt{gk}$.]

dispersion relation from dimensional argument

The quintessential example of deep-water waves is waves on an ocean or a lake before they near the shore, so the water's depth is much greater than their wavelength. Such waves, of course, are excited by wind. We have discussed this excitation in Sec. 14.6.2. There we found that the Kelvin-Helmholtz instability (where air flowing in a laminar fashion over water would raise waves) is strongly suppressed by gravity, so this mechanism does not work. Instead, the excitation is by a turbulent wind's randomly fluctuating pressures pounding on the water's surface. Once the waves have been generated, they can propagate great distances before viscous dissipation damps them; see Ex. 16.2.

16.2.2 Shallow-Water Waves

For shallow-water waves ($kh_o \ll 1$), the dispersion relation (16.8) becomes

dispersion relation for gravity waves in shallow water

$$\boxed{\omega = \sqrt{gh_o}\, k.} \tag{16.10}$$

Thus these waves are nondispersive; their phase and group velocities are equal: $V_{\rm ph} = V_g = \sqrt{gh_o}$.

Later, when studying solitons, we shall need two special properties of shallow-water waves. First, when the depth of the water is small compared with the wavelength, but not very small, the waves will be slightly dispersive. We can obtain a correction to Eq. (16.10) by expanding the tanh function of Eq. (16.8) as $\tanh x = x - x^3/3 + \dots$. The dispersion relation then becomes

$$\omega = \sqrt{gh_o}\left(1 - \frac{1}{6}k^2 h_o^2\right) k. \tag{16.11}$$

Second, by computing $\mathbf{v} = \boldsymbol{\nabla}\psi$ from Eq. (16.3), we find that in the shallow-water limit the water's horizontal motions are much larger than its vertical motions and are essentially independent of depth. The reason, physically, is that the fluid acceleration is produced almost entirely by a horizontal pressure gradient (caused by spatially variable water depth) that is independent of height; see Ex. 16.1.

Often shallow-water waves have heights ξ that are comparable to the water's undisturbed depth h_o, and h_o changes substantially from one region of the flow to another. A familiar example is an ocean wave nearing a beach. In such cases, the wave equation is modified by nonlinear and height-dependent effects. In Box 16.3 we derive the equations that govern such waves, and in Ex. 16.3 and Sec. 16.3 we explore properties of these waves.

BOX 16.3. NONLINEAR SHALLOW-WATER WAVES WITH VARIABLE DEPTH

Consider a nonlinear shallow-water wave propagating on a body of water with variable depth. Let $h_o(x, y)$ be the depth of the undisturbed water at location (x, y), and let $\xi(x, y, t)$ be the height of the wave, so the depth of the water in the presence of the wave is $h = h_o + \xi$. As in the linear-wave case, the transverse fluid velocity (v_x, v_y) inside the water is nearly independent of height z, so the wave is characterized by three functions $\xi(x, y, t)$, $v_x(x, y, t)$, and $v_y(x, y, t)$. These functions are governed by the law of mass conservation and the inviscid Navier-Stokes equation (i.e., the Euler equation).

The mass per unit area is $\rho h = \rho(h_o + \xi)$, and the corresponding mass flux (mass crossing a unit length per unit time) is $\rho h \mathbf{v} = \rho(h_o + \xi)\mathbf{v}$, where $\mathbf{v}$ is the 2-dimensional, horizontal vectorial velocity $\mathbf{v} = v_x \mathbf{e}_x + v_y \mathbf{v}_y$. Mass conservation, then, requires that $\partial[\rho(h_o + \xi)]/\partial t + {}^{(2)}\mathbf{\nabla} \cdot [\rho(h_o + \xi)\mathbf{v}] = 0$, where ${}^{(2)}\mathbf{\nabla}$ is the 2-dimensional gradient operator that acts solely in the horizontal (x-y) plane. Since ρ is constant, and h_o is time independent, this expression becomes

$$\partial \xi/\partial t + {}^{(2)}\mathbf{\nabla} \cdot [(h_o + \xi)\mathbf{v}] = 0. \tag{1a}$$

The Euler equation for $\mathbf{v}$ at an arbitrary height z in the water is $\partial \mathbf{v}/\partial t + (\mathbf{v} \cdot {}^{(2)}\mathbf{\nabla})\mathbf{v} = -(1/\rho){}^{(2)}\mathbf{\nabla} P$, and hydrostatic equilibrium gives the pressure as the weight per unit area of the overlying water: $P = g(\xi - z)\rho$ (where height z is measured from the water's undisturbed surface). Combining these equations, we obtain

$$\partial \mathbf{v}/\partial t + (\mathbf{v} \cdot {}^{(2)}\mathbf{\nabla})\mathbf{v} + g\, {}^{(2)}\mathbf{\nabla}\xi = 0. \tag{1b}$$

Equations (1) are used, for example, in theoretical analyses of tsunamis (Ex. 16.3).

EXERCISES

Exercise 16.1 *Example: Fluid Motions in Gravity Waves*

(a) Show that in a gravity wave in water of arbitrary depth (deep, shallow, or in between), each fluid element undergoes forward-rolling elliptical motion as shown in Fig. 16.1. (Assume that the amplitude of the water's displacement is small compared to a wavelength.)

(b) Calculate the longitudinal diameter of the motion's ellipse, and the ratio of vertical to longitudinal diameters, as functions of depth.

(c) Show that for a deep-water wave, $kh_o \gg 1$, the ellipses are all circles with diameters that die out exponentially with depth.

(d) We normally think of a circular motion of fluid as entailing vorticity, but a gravity wave in water has vanishing vorticity. How can this vanishing vorticity be compatible with the circular motion of fluid elements?

(e) Show that for a shallow-water wave, $kh_o \ll 1$, the motion is (nearly) horizontal and is independent of height z.

(f) Compute the fluid's pressure perturbation $\delta P(x, z, t)$ inside the fluid for arbitrary depth. Show that, for a shallow-water wave, the pressure is determined by the need to balance the weight of the overlying fluid, but for greater depth, vertical fluid accelerations alter this condition of weight balance.

Exercise 16.2 *Problem: Viscous Damping of Deep-Water Waves*

(a) Show that viscosity damps a monochromatic deep-water wave with an amplitude e-folding time $\tau_* = (2\nu k^2)^{-1}$, where k is the wave number, and ν is the kinematic viscosity. [Hint: Compute the energy E in the wave and the rate of loss of energy to viscous heating $\dot{E}$, and argue that $\tau_* = -2E/\dot{E}$. Recall the discussions of energy in Sec. 13.5.5 and viscous heating in Sec. 13.7.4.]

(b) As an example, consider the ocean waves that one sees at an ocean beach when the surf is "up" (large-amplitude waves). These are usually generated by turbulent winds in storms in the distant ocean, 1,000 km or farther from shore. The shortest wavelengths present should be those for which the damping length $C\tau_*$ is about 1,000 km; shorter wavelengths than that will have been dissipated before reaching shore. Using the turbulent viscosity $\nu \sim 0.03$ m² s^{-1} that we deduced from the observed thicknesses of wind-driven Ekman layers in the ocean (Ex. 14.21), compute this shortest wavelength for the large-amplitude waves, and compare with the wavelengths you have observed at an ocean beach. You should get pretty good agreement, thanks to a weak sensitivity of the wavelength to the rather uncertain turbulent viscosity.

(c) Make similar comparisons of theory and observation for (i) the choppy, short-wavelength ocean waves that one sees when (and only when) a local wind is blowing, and (ii) waves on a small lake.

Exercise 16.3 *Example: Shallow-Water Waves with Variable Depth; Tsunamis*[4]
Consider small-amplitude (linear) shallow-water waves in which the height of the bottom boundary varies, so the unperturbed water's depth is variable: $h_o = h_o(x, y)$.

(a) Using the theory of nonlinear shallow-water waves with variable depth (Box 16.3), show that the wave equation for the perturbation $\zeta(x, y, t)$ of the water's height takes the form

$$\frac{\partial^2 \zeta}{\partial t^2} - {}^{(2)}\nabla \cdot (g h_o {}^{(2)}\nabla \zeta) = 0. \tag{16.12}$$

Here ${}^{(2)}\nabla$ is the 2-dimensional gradient operator that acts in the horizontal (i.e., x-y) plane. Note that $g h_o$ is the square of the wave's propagation speed C^2 (phase speed and group speed), so this equation takes the same form as Eq. (7.17) from the geometric-optics approximation in Sec. 7.3.1, with $W = 1$.

(b) Describe what happens to the direction of propagation of a wave as the depth h_o of the water varies (either as a set of discrete jumps in h_o or as a slowly varying h_o). As a specific example, how must the propagation direction change as waves approach a beach (but when they are sufficiently far out from the beach that nonlinearities have not yet caused them to begin to break)? Compare with your own observations at a beach.

(c) Tsunamis are gravity waves with enormous wavelengths ($\sim$100 km or so) that propagate on the deep ocean. Since the ocean depth is typically $h_o \sim 4$ km, tsunamis are governed by the shallow-water wave equation (16.12). What would you have to do to the ocean floor to create a lens that would focus a tsunami, generated by an earthquake near Japan, so that it destroys Los Angeles? (For simulations of tsunami propagation, see, e.g., http://bullard.esc.cam.ac.uk/~taylor/Tsunami.html.)

(d) The height of a tsunami, when it is in the ocean with depth $h_o \sim 4$ km, is only $\sim$1 m or less. Use the geometric-optics approximation (Sec. 7.3) to show that the tsunami's wavelength decreases as $\lambda \propto \sqrt{h_o}$ and its amplitude increases as $\max(\zeta) \propto 1/h_o^{1/4}$ as the tsunami nears land and the water's depth h_o decreases.

How high [$\max(\zeta)$] does the tsunami get when nonlinearities become strongly important? (Assume a height of 1 m in the deep ocean.) How does this compare with the heights of historically disastrous tsunamis when they hit land? From your answer you should conclude that the nonlinearities must play a major role in raising the height. Equations (16.11) in Box 16.3 are used by geophysicists to analyze this nonlinear growth of the tsunami height. If the wave breaks, then these equations fail, and ideas developed (in rudimentary form) in Ex. 17.10 must be used.

---

[4]. Exercise courtesy of David Stevenson.

16.2.3 Capillary Waves and Surface Tension

When the wavelength is short (so k is large), we must include the effects of *surface tension* on the surface boundary condition. Surface tension can be treated as an isotropic force per unit length, γ, that lies in the surface and is unaffected by changes in the shape or size of the surface; see Box 16.4. In the case of a gravity wave traveling in the x direction, this tension produces on the fluid's surface a net downward force per unit area $-\gamma d^2\xi/dx^2 = \gamma k^2 \xi$, where k is the horizontal wave number. [This downward force is like that on a violin string; cf. Eq. (12.29) and associated discussion.] This additional force must be included as an augmentation of $\rho g \xi$. Correspondingly, the effect of surface tension on a mode with wave number k is simply to change the true acceleration of gravity to an effective acceleration of gravity:

$$g \to g + \frac{\gamma k^2}{\rho}. \tag{16.13}$$

The remainder of the derivation of the dispersion relation for deep-water gravity waves carries over unchanged, and the dispersion relation becomes

---

BOX 16.4. SURFACE TENSION

In a water molecule, the two hydrogen atoms stick out from the larger oxygen atom somewhat like Mickey Mouse's ears, with an H-O-H angle of 105°. This asymmetry of the molecule gives rise to a large electric dipole moment. In the interior of a body of water the dipole moments are oriented rather randomly, but near the water's surface they tend to be parallel to the surface and bond with one another to create surface tension—a macroscopically isotropic, 2-dimensional tension force (force per unit length) γ that is confined to the water's surface.

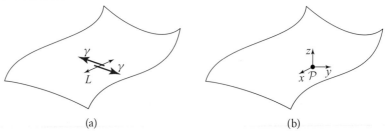

(a)   (b)

More specifically, consider a line L of unit length in the water's surface [drawing (a)]. The surface water on one side of L exerts a tension (pulling) force on the surface water on the other side. The magnitude of this force is γ, and it is orthogonal to the line L regardless of L's orientation. This behavior is analogous to an isotropic pressure P in 3 dimensions, which acts orthogonally across any unit area.

(continued)

BOX 16.4. (continued)

Choose a point $\mathcal{P}$ in the water's surface, and introduce local Cartesian coordinates there with x and y lying in the surface and z orthogonal to it [drawing (b)]. In this coordinate system, the 2-dimensional stress tensor associated with surface tension has components $^{(2)}T_{xx} = {}^{(2)}T_{yy} = -\gamma$, analogous to the 3-dimensional stress tensor for an isotropic pressure: $T_{xx} = T_{yy} = T_{zz} = P$. We can also use a 3-dimensional stress tensor to describe the surface tension: $T_{xx} = T_{yy} = -\gamma \delta(z)$; all other $T_{jk} = 0$. If we integrate this 3-dimensional stress tensor through the water's surface, we obtain the 2-dimensional stress tensor: $\int T_{jk} dz = {}^{(2)}T_{jk}$ (i.e., $\int T_{xx} dz = \int T_{yy} dz = -\gamma$). The 2-dimensional metric of the surface is $^{(2)}\mathbf{g} = \mathbf{g} - \mathbf{e}_z \otimes \mathbf{e}_z$; in terms of this 2-dimensional metric, the surface tension's 3-dimensional stress tensor is $\mathbf{T} = -\gamma \delta(z) {}^{(2)}\mathbf{g}$.

Water is not the only fluid that exhibits surface tension; all fluids do so, at the interfaces between themselves and other substances. For a thin film (e.g., a soap bubble), there are two interfaces (the top and bottom faces of the film), so if we ignore the film's thickness, its stress tensor is twice as large as for a single surface, $\mathbf{T} = -2\gamma \delta(z) {}^{(2)}\mathbf{g}$.

The hotter the fluid, the more randomly its surface molecules will be oriented (and hence the smaller the fluid's surface tension γ will be). For water, γ varies from 75.6 dyne/cm at $T = 0\,°C$, to 72.0 dyne/cm at $T = 25\,°C$, to 58.9 dyne/cm at $T = 100\,°C$.

In Exs. 16.4–16.6, we explore some applications of surface tension. In Sec. 16.2.3 and Exs. 16.7 and 16.8, we consider the influence of surface tension on water waves. In Ex. 5.14, we study the statistical thermodynamics of surface tension and its role in the nucleation of water droplets in clouds and fog.

$$\omega^2 = gk + \frac{\gamma k^3}{\rho} \quad (16.14)$$

dispersion relation for gravity waves in deep water with surface tension

[cf. Eqs. (16.9) and (16.13)]. When the second term dominates, the waves are sometimes called *capillary waves*. In Exs. 16.7 and 16.8 we explore some aspects of capillary waves. In Exs. 16.4–16.6 we explore some other aspects of surface tension.

capillary waves

EXERCISES

Exercise 16.4 *Problem: Maximum Size of a Water Droplet*
What is the maximum size of water droplets that can form by water very slowly dripping out of a syringe? Out of a water faucet (whose opening is far larger than that of a syringe)?

Exercise 16.5 *Problem: Force Balance for an Interface between Two Fluids*
Consider a point $\mathcal{P}$ in the curved interface between two fluids. Introduce Cartesian coordinates at $\mathcal{P}$ with x and y parallel to the interface and z orthogonal [as in diagram (b) in Box 16.4], and orient the x- and y-axes along the directions of the interface's principal curvatures, so the local equation for the interface is

$$z = \frac{x^2}{2R_1} + \frac{y^2}{2R_2}. \tag{16.15}$$

Here R_1 and R_2 are the surface's principal radii of curvature at $\mathcal{P}$; note that each of them can be positive or negative, depending on whether the surface bends up or down along their directions. Show that, in equilibrium, stress balance, $\nabla \cdot \mathbf{T} = 0$, for the surface implies that the pressure difference across the surface is

$$\boxed{\Delta P = \gamma \left(\frac{1}{R_1} + \frac{1}{R_2} \right),} \tag{16.16}$$

where γ is the surface tension.

Exercise 16.6 *Challenge: Minimum Area of a Soap Film*
For a soap film that is attached to a bent wire (e.g., to the circular wire that a child uses to blow a bubble), the air pressure on the film's two sides is the same. Therefore, Eq. (16.16) (with γ replaced by 2γ, since the film has two faces) tells us that at every point in the film, its two principal radii of curvature must be equal and opposite: $R_1 = -R_2$. It is an interesting exercise in differential geometry to show that this requirement means that the soap film's surface area is an extremum with respect to variations of the film's shape, holding its boundary on the wire fixed. If you know enough differential geometry, prove this extremal-area property of soap films, and then show that for the film's shape to be stable, its extremal area must actually be a minimum.

Exercise 16.7 *Problem: Capillary Waves*
Consider deep-water gravity waves of short enough wavelength that surface tension must be included, so the dispersion relation is Eq. (16.14). Show that there is a minimum value of the group velocity, and find its value together with the wavelength of the associated wave. Evaluate these for water ($\gamma \sim 0.07$ N m^{-1} = 70 dyne/cm). Try performing a crude experiment to verify this phenomenon.

Exercise 16.8 *Example: Boat Waves*
A toy boat moves with uniform velocity $\mathbf{u}$ across a deep pond (Fig. 16.2). Consider the wave pattern (time-independent in the boat's frame) produced on the water's surface at distances large compared to the boat's size. Both gravity waves and surface-tension (*capillary*) waves are excited. Show that capillary waves are found both ahead of and behind the boat, whereas gravity waves occur solely inside a trailing wedge. More specifically, do the following.

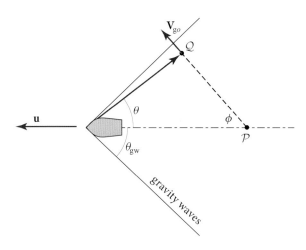

FIGURE 16.2 Capillary and gravity waves excited by a small boat (Ex. 16.8).

(a) In the rest frame of the water, the waves' dispersion relation is Eq. (16.14). Change notation so that ω is the waves' angular velocity as seen in the boat's frame, and ω_o in the water's frame, so the dispersion relation becomes $\omega_o^2 = gk + (\gamma/\rho)k^3$. Use the Doppler shift (i.e., the transformation between frames) to derive the boat-frame dispersion relation $\omega(k)$.

(b) The boat radiates a spectrum of waves in all directions. However, only those with vanishing frequency in the boat's frame, $\omega = 0$, contribute to the time-independent (stationary) pattern. As seen in the water's frame and analyzed in the geometric-optics approximation of Chap. 7, these waves are generated by the boat (at points along its horizontal dash-dot trajectory in Fig. 16.2) and travel outward with the group velocity $\mathbf{V}_{go}$. Regard Fig. 16.2 as a snapshot of the boat and water at a particular moment of time. Consider a wave that was generated at an earlier time, when the boat was at location $\mathcal{P}$, and that traveled outward from there with speed V_{go} at an angle ϕ to the boat's direction of motion. (You may restrict yourself to $0 \leq \phi \leq \pi/2$.) Identify the point $\mathcal{Q}$ that this wave has reached, at the time of the snapshot, by the angle θ shown in the figure. Show that θ is given by

$$\tan \theta = \frac{V_{go}(k) \sin \phi}{u - V_{go}(k) \cos \phi}, \qquad (16.17a)$$

where k is determined by the dispersion relation $\omega_0(k)$ together with the vanishing ω condition:

$$\omega_0(k, \phi) = uk \cos \phi. \qquad (16.17b)$$

(c) Specialize to capillary waves $[k \gg \sqrt{g\rho/\gamma}]$. Show that

$$\tan \theta = \frac{3 \tan \phi}{2 \tan^2 \phi - 1}. \qquad (16.18)$$

Demonstrate that the capillary-wave pattern is present for all values of θ (including in front of the boat, $\pi/2 < \theta < \pi$, and behind it, $0 \leq \theta \leq \pi/2$).

(d) Next, specialize to gravity waves, and show that

$$\tan\theta = \frac{\tan\phi}{2\tan^2\phi + 1}. \tag{16.19}$$

Demonstrate that the gravity-wave pattern is confined to a trailing wedge with angles $\theta < \theta_{\rm gw} = \sin^{-1}(1/3) = 19.47°$ (cf. Fig. 16.2). You might try to reproduce these results experimentally.

16.2.4 Helioseismology

The Sun provides an excellent example of the excitation of small-amplitude waves in a fluid body. In the 1960s, Robert Leighton and colleagues at Caltech discovered that the surface of the Sun oscillates vertically with a period of roughly 5 min and a speed of ~ 1 km s^{-1}. This motion was thought to be an incoherent surface phenomenon until it was shown that the observed variation was, in fact, the superposition of thousands of highly coherent wave modes excited in the Sun's interior—normal modes of the Sun. Present-day techniques allow surface velocity amplitudes as small as 2 mm s^{-1} to be measured, and phase coherence for intervals as long as a year has been observed. Studying the frequency spectrum and its variation provides a unique probe of the Sun's interior structure, just as the measurement of conventional seismic waves (Sec. 12.4) probes Earth's interior.

properties of the Sun that influence its helioseismic modes

The description of the Sun's normal modes requires some modification of our treatment of gravity waves. We eschew the details and just outline the principles—which are rather similar to those for normal modes of a homogeneous elastic sphere (Sec. 12.4.4 and Ex. 12.12). First, the Sun is (very nearly) spherical. We therefore work in spherical polar coordinates rather than Cartesian coordinates. Second, the Sun is made of hot gas, and it is no longer a good approximation to assume that the fluid is incompressible. We must therefore replace the equation $\nabla \cdot \mathbf{v} = 0$ with the full equation of continuity (mass conservation) together with the equation of energy conservation, which governs the relationship between the perturbations of density and pressure. Third, the Sun is not uniform. The pressure and density in the unperturbed gas vary with radius in a known manner and must be included. Fourth, the Sun has a finite surface area. Instead of assuming a continuous spectrum of waves, we must now anticipate that the boundary conditions will lead to a discrete spectrum of normal modes. Allowing for these complications, it is possible to derive a differential equation for the perturbations to replace Eq. (16.7). It turns out that a convenient dependent variable (replacing the velocity potential ψ) is the pressure perturbation. The boundary conditions are that the displacement vanish at the center of the Sun and the pressure perturbation vanish at the surface.

At this point the problem is reminiscent of the famous solution for the eigenfunctions of the Schrödinger equation for a hydrogen atom in terms of associated Laguerre

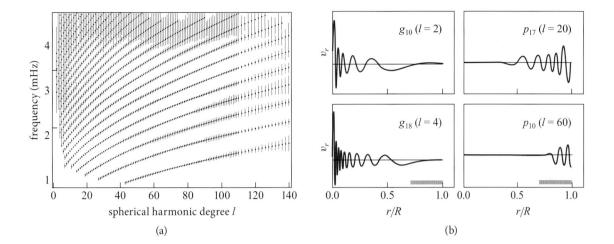

FIGURE 16.3 (a) Measured frequency spectrum for solar p modes with different values of the quantum numbers n and l. The error bars are magnified by a factor of 1,000. The lowest ridge is $n = 0$, the next is $n = 1, \ldots$. (b) Sample eigenfunctions for g and p modes labeled by n (subscripts) and l (in parentheses). The ordinate is the radial velocity, and the abscissa is fractional radial distance from the Sun's center to its surface. The solar convection zone is the shaded region at the bottom. Adapted from Libbrecht and Woodard (1991).

polynomials. The wave frequencies of the Sun's normal modes are given by the eigenvalues of the differential equation. The corresponding eigenfunctions can be classified using three quantum numbers, n, l, m, where n counts the number of radial nodes in the eigenfunction, and the angular variation of the pressure perturbation is proportional to the spherical harmonic $Y_l^m(\theta, \phi)$. If the Sun were precisely spherical, the modes with the same n and l but different m would be degenerate, just as is the case for an atom when there is no preferred direction in space. However, the Sun rotates with a latitude-dependent period in the range $\sim$25–30 days, which breaks the degeneracy, just as an applied magnetic field in an atom breaks the degeneracy of the atom's states (the Zeeman effect). From the observed splitting of the solar-mode spectrum, it is possible to learn about the distribution of rotational angular momentum inside the Sun.

When this problem is solved in detail, it turns out that there are two general classes of modes. One class is similar to gravity waves, in the sense that the forces that drive the gas's motions are produced primarily by gravity (either directly, or indirectly via the weight of overlying material producing pressure that pushes on the gas). These are called *g modes*. In the second class (known as *p modes*),[5] the pressure forces arise mainly from the compression of the fluid just like in sound waves (which we study in Sec. 16.5). It turns out that the g modes have large amplitudes in the middle of the Sun, whereas the p modes are dominant in the outer layers (Fig. 16.3b). The reasons for this are relatively easy to understand and introduce ideas to which we shall return.

g modes of the Sun or other gravitating fluid spheres

p modes of the Sun or other gravitating fluid spheres

---

5. There are also formally distinguishable f-modes, which for our purposes are just a subset of the p-modes.

properties of p modes

The Sun is a hot body, much hotter at its center ($T \sim 1.5 \times 10^7$ K) than on its surface ($T \sim 6{,}000$ K). The sound speed C is therefore much greater in its interior, and so p modes of a given frequency ω can carry their energy flux $\sim \rho \xi^2 \omega^2 C$ (Sec. 16.5) with much smaller amplitudes ξ than near the surface. Therefore the p-mode amplitudes are much smaller in the center of the Sun than near its surface.

properties of g modes

The g modes are controlled by different physics and thus behave differently. The outer $\sim$30% (by radius) of the Sun is *convective* (Chap. 18), because the diffusion of heat is inadequate to carry the huge amount of nuclear power being generated in the solar core. The convection produces an equilibrium variation of pressure and density with radius that are just such as to keep the Sun almost neutrally stable, so that regions that are slightly hotter (cooler) than their surroundings will rise (sink) in the solar gravitational field. Therefore there cannot be much of a mechanical restoring force that would cause these regions to oscillate about their average positions, and so the g modes (which are influenced almost solely by gravity) have little restoring force and thus are evanescent in the convection zone; hence their amplitudes decay quickly with increasing radius there.

We should therefore expect only p modes to be seen in the surface motions, which is indeed the case. Furthermore, we should not expect the properties of these modes to be very sensitive to the physical conditions in the core. A more detailed analysis bears this out.

16.3 Nonlinear Shallow-Water Waves and Solitons

In recent decades, *solitons* or solitary waves have been studied intensively in many different areas of physics. However, fluid dynamicists became familiar with them in the nineteenth century. In an oft-quoted passage, John Scott-Russell (1844) described how he was riding along a narrow canal and watched a boat stop abruptly. This deceleration launched a single smooth pulse of water which he followed on horseback for 1 or 2 miles, observing it "rolling on a rate of some eight or nine miles an hour, preserving its original figure some thirty feet long and a foot to a foot and a half in height." This was a soliton—a 1-dimensional, nonlinear wave with fixed profile traveling with constant speed. Solitons can be observed fairly readily when gravity waves are produced in shallow, narrow channels. We use the particular example of a shallow, nonlinear gravity wave to illustrate solitons in general.

16.3.1 Korteweg–de Vries (KdV) Equation

The key to a soliton's behavior is a robust balance between the effects of dispersion and those of nonlinearity. When one grafts these two effects onto the wave equation for shallow-water waves, then to leading order in the strengths of the dispersion and nonlinearity one gets the *Korteweg–de Vries* (KdV) equation for solitons. Since a completely rigorous derivation of the KdV equation is quite lengthy, we content ourselves with a somewhat heuristic derivation that is based on this grafting process and is designed to emphasize the equation's physical content.

We choose as the dependent variable in our wave equation the height ξ of the water's surface above its quiescent position, and we confine ourselves to a plane wave that propagates in the horizontal x direction, so $\xi = \xi(x, t)$.

In the limit of very weak waves, $\xi(x, t)$ is governed by the shallow-water dispersion relation, $\omega = \sqrt{gh_o}\, k$, where h_o is the depth of the quiescent water. This dispersion relation implies that $\xi(x, t)$ must satisfy the following elementary wave equation [cf. Eq. (16.12)]:

$$0 = \frac{\partial^2 \xi}{\partial t^2} - gh_o \frac{\partial^2 \xi}{\partial x^2} = \left(\frac{\partial}{\partial t} - \sqrt{gh_o}\frac{\partial}{\partial x}\right)\left(\frac{\partial}{\partial t} + \sqrt{gh_o}\frac{\partial}{\partial x}\right)\xi. \tag{16.20}$$

In the second expression, we have factored the wave operator into two pieces, one that governs waves propagating rightward, and the other for those moving leftward. To simplify our derivation and the final wave equation, we confine ourselves to rightward-propagating waves; correspondingly, we can simply remove the left-propagation operator, obtaining

$$\frac{\partial \xi}{\partial t} + \sqrt{gh_o}\frac{\partial \xi}{\partial x} = 0. \tag{16.21}$$

(Leftward-propagating waves are described by this same equation with a change of sign on one of the terms.)

We now graft the effects of dispersion onto this rightward-wave equation. The dispersion relation, including the effects of dispersion to leading order, is $\omega = \sqrt{gh_o}\, k(1 - \frac{1}{6}k^2 h_o^2)$ [Eq. (16.11)]. Now, this dispersion relation ought to be derivable by assuming a variation $\xi \propto \exp[i(kx - \omega t)]$ and substituting into a generalization of Eq. (16.21) with corrections that take account of the finite depth of the channel. We take a short cut and reverse this process to obtain the generalization of Eq. (16.21) from the dispersion relation. The result is

$$\frac{\partial \xi}{\partial t} + \sqrt{gh_o}\frac{\partial \xi}{\partial x} = -\frac{1}{6}\sqrt{gh_o}\, h_o^2 \frac{\partial^3 \xi}{\partial x^3}, \tag{16.22}$$

as a simple calculation confirms. This is the linearized KdV equation. It incorporates weak dispersion associated with the finite depth of the channel but is still a linear equation, only useful for small-amplitude waves.

Now let us set aside the dispersive correction and tackle the nonlinearity using the equations derived in Box 16.3. Denoting the depth of the disturbed water by $h = h_o + \xi$, the nonlinear law of mass conservation [Eq. (1a) of Box 16.3] becomes

$$\frac{\partial h}{\partial t} + \frac{\partial (hv)}{\partial x} = 0, \tag{16.23a}$$

and the Euler equation [Eq. (1b) of Box 16.3] becomes

$$\frac{\partial v}{\partial t} + v\frac{\partial v}{\partial x} + g\frac{\partial h}{\partial x} = 0. \tag{16.23b}$$

Here we have specialized the equations in Box 16.3 to a 1-dimensional wave in the channel and to a constant depth h_o of the channel's undisturbed water. Equations (16.23a) and (16.23b) can be combined to obtain

$$\frac{\partial (v - 2\sqrt{gh})}{\partial t} + \left(v - \sqrt{gh}\right) \frac{\partial (v - 2\sqrt{gh})}{\partial x} = 0. \tag{16.23c}$$

This equation shows that the quantity $v - 2\sqrt{gh}$ is constant along characteristics that propagate with speed $v - \sqrt{gh}$. (This constant quantity is a special case of a *Riemann invariant*, a concept that we study in Sec. 17.4.1.) When (as we require below) the nonlinearities are modest, so h does not differ greatly from h_o and the water speed v is small, these characteristics propagate leftward, which implies that for rightward-propagating waves they begin at early times in undisturbed fluid, where $v = 0$ and $h = h_o$. Therefore, the constant value of $v - 2\sqrt{gh}$ is $-2\sqrt{gh_o}$, and correspondingly in regions of disturbed fluid we have

$$v = 2\left(\sqrt{gh} - \sqrt{gh_o}\right). \tag{16.24}$$

Substituting this into Eq. (16.23a), we obtain

$$\frac{\partial h}{\partial t} + \left(3\sqrt{gh} - 2\sqrt{gh_o}\right) \frac{\partial h}{\partial x} = 0. \tag{16.25}$$

We next substitute $\xi = h - h_o$ and expand to second order in ξ to obtain the final form of our wave equation with nonlinearities but no dispersion:

$$\frac{\partial \xi}{\partial t} + \sqrt{gh_o}\frac{\partial \xi}{\partial x} = -\frac{3\xi}{2}\sqrt{\frac{g}{h_o}}\frac{\partial \xi}{\partial x}, \tag{16.26}$$

where the term on the right-hand side is the nonlinear correction.

We now have separate dispersive corrections (16.22) and nonlinear corrections (16.26) to the rightward-wave equation (16.21). Combining the two corrections into a single equation, we obtain

$$\frac{\partial \xi}{\partial t} + \sqrt{gh_o}\left[\left(1 + \frac{3\xi}{2h_o}\right)\frac{\partial \xi}{\partial x} + \frac{h_o^2}{6}\frac{\partial^3 \xi}{\partial x^3}\right] = 0. \tag{16.27}$$

Finally, we substitute

rightward moving spatial coordinate

$$\boxed{\chi \equiv x - \sqrt{gh_o}\, t} \tag{16.28}$$

to transform to a frame moving rightward with the speed of small-amplitude gravity waves. The result is the full *Korteweg–de Vries* or KdV equation:

Korteweg–de Vries (KdV) equation

$$\boxed{\frac{\partial \xi}{\partial t} + \frac{3}{2}\sqrt{\frac{g}{h_o}}\left(\xi \frac{\partial \xi}{\partial \chi} + \frac{1}{9}h_o^3 \frac{\partial^3 \xi}{\partial \chi^3}\right) = 0.} \tag{16.29}$$

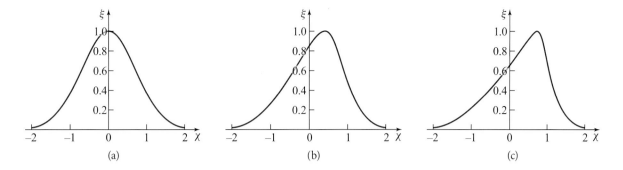

FIGURE 16.4 Steepening of a Gaussian wave profile by the nonlinear term in the KdV equation. The increase of wave speed with amplitude causes the leading part of the profile to steepen with time (going left to right) and the trailing part to flatten. In the full KdV equation (16.29), this effect can be balanced by the effect of dispersion, which causes the high-frequency Fourier components in the wave to travel slightly slower than the low-frequency ones. This allows stable solitons to form.

16.3.2 Physical Effects in the KdV Equation

Before exploring solutions to the KdV equation (16.29), let us consider the physical effects of its nonlinear and dispersive terms. The second (nonlinear) term $\frac{3}{2}\sqrt{g/h_o}\,\xi\,\partial\xi/\partial\chi$ derives from the nonlinearity in the $(\mathbf{v}\cdot\nabla)\mathbf{v}$ term of the Euler equation. The effect of this nonlinearity is to steepen the leading edge of a wave profile and flatten the trailing edge (Fig. 16.4). Another way to understand the effect of this term is to regard it as a nonlinear coupling of linear waves. Since it is nonlinear in the wave amplitude, it can couple waves with different wave numbers k. For example, if we have a purely sinusoidal wave $\propto \exp(ikx)$, then this nonlinearity leads to the growth of a first harmonic $\propto \exp(2ikx)$. Similarly, when two linear waves with spatial frequencies k and k' are superposed, this term describes the production of new waves at the sum and difference of spatial frequencies. We have already met such wave-wave coupling in our study of nonlinear optics (Chap. 10), and in the route to turbulence for rotating Couette flow (Fig. 15.16). We meet it again in nonlinear plasma physics (Chap. 23).

The third term in Eq. (16.29), $\frac{1}{6}\sqrt{g/h_o}\,h_o^3\,\partial^3\xi/\partial\chi^3$, is linear and is responsible for a weak dispersion of the wave. The higher-frequency Fourier components travel with slower phase velocities than do the lower-frequency components. This has two effects. One is an overall spreading of a wave in a manner qualitatively familiar from elementary quantum mechanics (cf. Ex. 7.2). For example, in a Gaussian wave packet with width Δx, the range of wave numbers k contributing significantly to the profile is $\Delta k \sim 1/\Delta x$. The spread in the group velocity is then $\Delta v_g \sim \Delta k\, \partial^2\omega/\partial k^2 \sim (gh_o)^{1/2} h_o^2 k\, \Delta k$ [cf. Eq. (16.11)]. The wave packet will then double in size in a time

$$t_{\text{spread}} \sim \frac{\Delta x}{\Delta v_g} \sim \left(\frac{\Delta x}{h_o}\right)^2 \frac{1}{k\sqrt{gh_o}}. \tag{16.30}$$

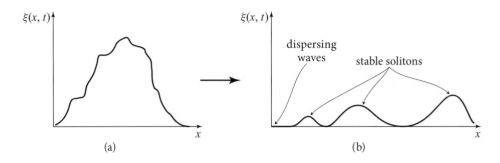

FIGURE 16.5 Production of stable solitons out of an irregular initial wave profile.

The second effect is that since the high-frequency components travel somewhat slower than the low-frequency ones, the profile tends to become asymmetric, with the leading edge less steep than the trailing edge.

Given the opposite effects of these two corrections (nonlinearity makes the wave's leading edge steeper; dispersion reduces its steepness), it should not be too surprising in hindsight that it is possible to find solutions to the KdV equation in which non-linearity balances dispersion, so there is no change of shape as the wave propagates and no spreading. What is quite surprising, though, is that these solutions, called *solitons*, are very robust and arise naturally out of random initial data. That is to say, if we solve an initial value problem numerically starting with several peaks of random shape and size, then although much of the wave will spread and disappear due to dispersion, we will typically be left with several smooth soliton solutions, as in Fig. 16.5.

solitons in which nonlinearity balances dispersion

16.3.3 Single-Soliton Solution

We can discard some unnecessary algebraic luggage in the KdV equation (16.29) by transforming both independent variables using the substitutions

$$\zeta = \frac{\xi}{h_o}, \quad \eta = \frac{3\chi}{h_o} = \frac{3(x - \sqrt{gh_o}\, t)}{h_o}, \quad \tau = \frac{9}{2}\sqrt{\frac{g}{h_o}}\, t. \tag{16.31}$$

The KdV equation then becomes

simplified form of KdV equation

$$\frac{\partial \zeta}{\partial \tau} + \zeta \frac{\partial \zeta}{\partial \eta} + \frac{\partial^3 \zeta}{\partial \eta^3} = 0. \tag{16.32}$$

There are well-understood mathematical techniques (see, e.g., Whitham, 1974) for solving equations like the KdV equation. However, here we just quote solutions and explore their properties. The simplest solution to the dimensionless KdV equation (16.32) is

single-soliton solution of KdV equation

$$\zeta = \zeta_0 \operatorname{sech}^2\left[\left(\frac{\zeta_0}{12}\right)^{1/2}\left(\eta - \frac{1}{3}\zeta_0 \tau\right)\right]. \tag{16.33}$$

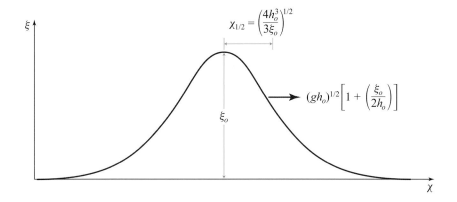

FIGURE 16.6 Profile of the single-soliton solution [Eqs. (16.33) and (16.31)] of the KdV equation. The half-width $\chi_{1/2}$ is inversely proportional to the square root of the peak height ξ_o.

This solution, depicted in Fig. 16.6, describes a one-parameter family of stable solitons. For each such soliton (each ζ_0), the soliton maintains its shape while propagating at speed $d\eta/d\tau = \zeta_0/3$ relative to a weak wave. By transforming to the rest frame of the unperturbed water using Eqs. (16.28) and (16.31), we find for the soliton's speed there:

$$\frac{dx}{dt} = \sqrt{gh_o}\left(1 + \frac{\xi_o}{2h_o}\right). \qquad (16.34) \qquad \text{speed of soliton}$$

The first term is the propagation speed of a weak (linear) wave. The second term is the nonlinear correction, proportional to the wave amplitude $\xi_o = h_o\zeta_o$. A "half-width" of the wave may be defined by setting the argument of the hyperbolic secant to unity. It is $\eta_{1/2} = (12/\zeta_o)^{1/2}$, corresponding to

$$x_{1/2} = \chi_{1/2} = \left(\frac{4h_o^3}{3\xi_o}\right)^{1/2}. \qquad (16.35) \qquad \text{half-width of soliton}$$

The larger the wave amplitude, the narrower its width will be, and the faster it will propagate (cf. Fig. 16.6).

Let us return to Scott-Russell's soliton (start of Sec. 16.3). Converting to SI units, the observed speed was about 4 m s^{-1}, giving an estimate of the depth of the canal of $h_o \sim 1.6$ m. Using the observed half-width $x_{1/2} \sim 5$ m, we obtain a peak height $\xi_o \sim 0.22$ m, somewhat smaller than quoted but within the errors allowing for the uncertainty in the definition of the width.

16.3.4 Two-Soliton Solution

One of the most fascinating properties of solitons is the way that two or more waves interact. The expectation, derived from physics experience with weakly coupled normal modes, might be that, if we have two well-separated solitons propagating in the

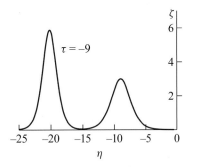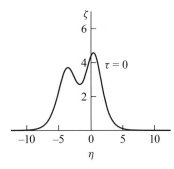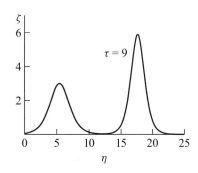

FIGURE 16.7 Two-soliton solution to the dimensionless KdV equation (16.32). This solution describes two waves well separated at $\tau \to -\infty$ that coalesce and then separate, producing the original two waves in reverse order at $\tau \to +\infty$. The notation is that of Eq. (16.36); the values of the parameters in that equation are $\eta_1 = \eta_2 = 0$ (so the solitons will merge at position $\eta = 0$), $\alpha_1 = 1$, and $\alpha_2 = 1.4$.

same direction with the larger wave chasing the smaller one, then the larger will eventually catch up with the smaller, and nonlinear interactions between the two waves will essentially destroy both, leaving behind a single, irregular pulse that will spread and decay after the interaction. However, this is not what happens. Instead, the two waves pass through each other unscathed and unchanged, except that they emerge from the interaction a bit sooner than they would have had they moved with their original speeds during the interaction. See Fig. 16.7. We shall not pause to explain why the two waves survive unscathed, except to remark that there are topological invariants in the solution that must be preserved. However, we can exhibit one such two-soliton solution analytically:

two-soliton solution of KdV equation

$$\zeta = \frac{\partial^2}{\partial \eta^2}[12 \ln F(\eta, \tau)],$$

$$\text{where } F = 1 + f_1 + f_2 + \left(\frac{\alpha_2 - \alpha_1}{\alpha_2 + \alpha_1}\right)^2 f_1 f_2,$$

$$\text{and } f_i = \exp[-\alpha_i(\eta - \eta_i) + \alpha_i^3 \tau], \quad i = 1, 2; \tag{16.36}$$

here α_i and η_i are constants. This solution is depicted in Fig. 16.7.

16.3.5 Solitons in Contemporary Physics

Solitons were rediscovered in the 1960s when they were found in numerical simulations of plasma waves. Their topological properties were soon understood, and general methods to generate solutions were derived. Solitons have been isolated in such different subjects as the propagation of magnetic flux in a Josephson junction, elastic waves in anharmonic crystals, quantum field theory (as *instantons*), and classical general relativity (as solitary, nonlinear gravitational waves). Most classical solitons

some equations with soliton solutions

are solutions to one of a relatively small number of nonlinear partial differential equations, including the KdV equation, the *nonlinear Schrödinger equation* (which governs solitons in optical fibers; Sec. 10.8.3), *Burgers equation,* and the *sine-Gordon* equation. Unfortunately, it has proved difficult to generalize these equations and their soliton solutions to 2 and 3 spatial dimensions.

Just like research into chaos (Sec. 15.6), studies of solitons have taught physicists that nonlinearity need not lead to maximal disorder in physical systems, but instead can create surprisingly stable, ordered structures.

EXERCISES

Exercise 16.9 *Example: Breaking of a Dam*
Consider the flow of water along a horizontal channel of constant width after a dam breaks. Sometime after the initial transients have died away[6] the flow may be described by the nonlinear, unidirectional, shallow-water wave equations (16.23a) and (16.23b):

$$\frac{\partial h}{\partial t} + \frac{\partial (hv)}{\partial x} = 0, \quad \frac{\partial v}{\partial t} + v\frac{\partial v}{\partial x} + g\frac{\partial h}{\partial x} = 0. \tag{16.37}$$

Here h is the height of the flow, v is the horizontal speed of the flow, and x is distance along the channel measured from the location of the dam. Solve for the flow, assuming that initially (at $t=0$) $h = h_o$ for $x < 0$ and $h = 0$ for $x > 0$ (no water). Your solution should have the form shown in Fig. 16.8. What is the speed of the front of the water? [Hint: From the parameters of the problem we can construct only one velocity, $\sqrt{gh_o}$, and no length except h_o. It therefore is a reasonable guess that the solution has the self-similar form $h = h_o \tilde{h}(\xi)$, $v = \sqrt{gh_o}\, \tilde{v}(\xi)$, where $\tilde{h}$ and $\tilde{v}$ are dimensionless functions of the similarity variable

$$\xi = \frac{x/t}{\sqrt{gh_o}}. \tag{16.38}$$

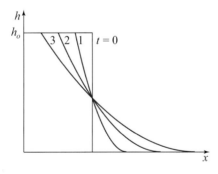

FIGURE 16.8 The water's height $h(x, t)$ after a dam breaks.

---

6. In the idealized case that the dam is removed instantaneously, there are no transients, and Eqs. (16.37) describe the flow from the outset.

Using this ansatz, convert the partial differential equations (16.37) into a pair of ordinary differential equations that can be solved so as to satisfy the initial conditions.]

Exercise 16.10 *Derivation: Single-Soliton Solution*
Verify that expression (16.33) does indeed satisfy the dimensionless KdV equation (16.32).

Exercise 16.11 *Derivation: Two-Soliton Solution*
(a) Verify, using symbolic-manipulation computer software (e.g., Maple, Matlab, or Mathematica) that the two-soliton expression (16.36) satisfies the dimensionless KdV equation. (Warning: Considerable algebraic travail is required to verify this by hand directly.)

(b) Verify analytically that the two-soliton solution (16.36) has the properties claimed in the text. First consider the solution at early times in the spatial region where $f_1 \sim 1$, $f_2 \ll 1$. Show that the solution is approximately that of the single soliton described by Eq. (16.33). Demonstrate that the amplitude is $\zeta_{01} = 3\alpha_1^2$, and find the location of its peak. Repeat the exercise for the second wave and for late times.

(c) Use a computer to follow numerically the evolution of this two-soliton solution as time η passes (thereby filling in timesteps between those shown in Fig. 16.7).

16.4 Rossby Waves in a Rotating Fluid

Coriolis force as restoring force for Rossby waves

In a nearly rigidly rotating fluid with the rotational angular velocity $\mathbf{\Omega}$ parallel or antiparallel to the acceleration of gravity $\mathbf{g} = -g\mathbf{e}_z$, the Coriolis effect observed in the co-rotating reference frame (Sec. 14.5) provides the restoring force for an unusual type of wave motion called *Rossby waves*. These waves are seen in Earth's oceans and atmosphere [with $\mathbf{\Omega} = $ (Earth's rotational angular velocity) $\sin(\text{latitude})\mathbf{e}_z$; see Box 14.5].

For a simple example, we consider the sea above a sloping seabed; Fig. 16.9. We assume the unperturbed fluid has vanishing velocity $\mathbf{v} = 0$ in Earth's rotating frame, and we study weak waves in the sea with oscillating velocity $\mathbf{v}$. (Since the fluid is at rest in the equilibrium state about which we are perturbing, we write the perturbed velocity as $\mathbf{v}$ rather than $\delta\mathbf{v}$.) We assume that the wavelengths are long enough that viscosity and surface tension are negligible. In this case we also restrict attention to small-amplitude waves, so that nonlinear terms can be dropped from the dynamical equations. The perturbed Navier-Stokes equation (14.56a) then becomes (after linearization)

$$\frac{\partial \mathbf{v}}{\partial t} + 2\mathbf{\Omega} \times \mathbf{v} = \frac{-\nabla \delta P'}{\rho}. \tag{16.39}$$

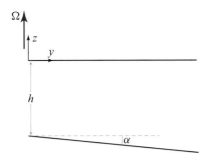

FIGURE 16.9 Geometry of the sea for Rossby waves.

Here, as in Sec. 14.5, $\delta P'$ is the perturbation in the effective pressure [which includes gravitational and centrifugal effects: $P' = P + \rho\Phi - \frac{1}{2}\rho(\mathbf{\Omega} \times \mathbf{x})^2$]. Taking the curl of Eq. (16.39), we obtain for the time derivative of the waves' vorticity:

$$\frac{\partial \boldsymbol{\omega}}{\partial t} = 2(\mathbf{\Omega} \cdot \nabla)\mathbf{v}. \tag{16.40}$$

We seek a wave mode with angular frequency ω (not to be confused with vorticity $\boldsymbol{\omega}$) and wave number k, in which the horizontal fluid velocity oscillates in the x direction and (in accord with the Taylor-Proudman theorem; Sec. 14.5.3) is independent of z, so

$$v_x \text{ and } v_y \propto \exp[i(kx - \omega t)], \quad \frac{\partial v_x}{\partial z} = \frac{\partial v_y}{\partial z} = 0. \tag{16.41}$$

The only allowed vertical variation is in the vertical velocity v_z; differentiating $\nabla \cdot \mathbf{v} = 0$ with respect to z, we obtain

$$\frac{\partial^2 v_z}{\partial z^2} = 0. \tag{16.42}$$

The vertical velocity therefore varies linearly between the surface and the sea floor (Fig. 16.9). One boundary condition is that the vertical velocity must vanish at the sea's surface. The other is that at the sea floor $z = -h$, we must have $v_z(-h) = -\alpha v_y$, where α is the tangent of the angle of inclination of the sea floor. The solution to Eq. (16.42) satisfying these boundary conditions is

$$v_z = \frac{\alpha z}{h} v_y. \tag{16.43}$$

Taking the vertical component of Eq. (16.40) and evaluating $\omega_z = v_{y,x} - v_{x,y} = ikv_y$, we obtain

$$\omega k v_y = 2\Omega \frac{\partial v_z}{\partial z} = \frac{2\Omega \alpha v_y}{h}. \tag{16.44}$$

The dispersion relation therefore has the quite unusual form $\omega \propto 1/k$:

dispersion relation for Rossby waves

$$\boxed{\omega k = \frac{2\Omega \alpha}{h}.} \tag{16.45}$$

properties of Rossby waves

Rossby waves have interesting properties. They can only propagate in one direction—parallel to the intersection of the sea floor with the horizontal (our $\mathbf{e}_x$ direction). Their phase velocity $\mathbf{V}_{\text{ph}}$ and group velocity $\mathbf{V}_g$ are equal in magnitude but opposite in direction:

$$\mathbf{V}_{\text{ph}} = -\mathbf{V}_g = \frac{2\Omega\alpha}{k^2 h}\mathbf{e}_x. \tag{16.46}$$

Using $\nabla \cdot \mathbf{v} = 0$, we discover that the two components of horizontal velocity are in quadrature: $v_x = i\alpha v_y/(kh)$. This means that the fluid circulates in the opposite sense to the angular velocity Ω.

Rossby waves play an important role in the circulation of Earth's oceans (see, e.g., Chelton and Schlax, 1996). A variant of these Rossby waves in air can be seen as undulations in the atmosphere's jet stream, produced when the stream goes over a sloping terrain, such as that of the Rocky Mountains. Another variant found in neutron stars, called "r modes," generates gravitational waves (ripples of spacetime curvature) that are a promising source for ground-based gravitational-wave detectors, such as LIGO.

EXERCISES

Exercise 16.12 *Example: Rossby Waves in a Cylindrical Tank with a Sloping Bottom*
In the film Fultz (1969), about 20 min 40 s into the film, an experiment is described in which Rossby waves are excited in a rotating cylindrical tank with inner and outer vertical walls and a sloping bottom. Figure 16.10a is a photograph of the tank from the side, showing its bottom, which slopes upward toward the center, and a hump on the bottom that generates the Rossby waves. The tank is filled with water, then set into rotation with an angular velocity Ω. The water is given time to settle down into rigid rotation with the cylinder. Then the cylinder's angular velocity is reduced by a small amount, so the water is rotating at angular velocity $\Delta\Omega \ll \Omega$ relative to the cylinder. As the water passes over the hump on the tank bottom, the hump generates Rossby waves. Those waves are made visible by injecting dye at a fixed radius through a syringe attached to the tank. Figure 16.10b is a photograph of the dye trace as seen looking down on the tank from above. If there were no Rossby waves present, the trace would be circular. The Rossby waves make it pentagonal. In this exercise you will work out the details of the Rossby waves, explore their physics, and explain the shape of the trace.

Because the slope of the bottom is cylindrical rather than planar, this is somewhat different from the situation discussed in the text (see Fig. 16.9). However, we can deduce the details of the waves in this cylindrical case from those for the planar case by using geometric-optics considerations (Sec. 7.3), making modest errors because the wavelength of the waves is not all that small compared to the circumference of the tank.

(a) Using geometric optics, show that the rays along which the waves propagate are circles centered on the tank's symmetry axis.

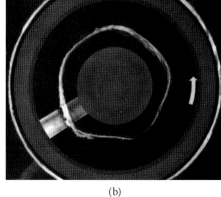

FIGURE 16.10 Rossby waves in a rotating cylinder with sloping bottom. (a) Side view. (b) Top view, showing dye trace. Images from *NCFMF Book of Film Notes,* 1972; The MIT Press with Education Development Center, Inc. © 2014 Education Development Center, Inc. Reprinted with permission with all other rights reserved.

(b) Focus on the ray that is halfway between the inner and outer walls of the tank. Let its radius be a, the depth of the water there be h, and the slope angle of the tank floor be α. Introduce quasi-Cartesian coordinates $x = a\phi$ and $y = -\varpi$, where $\{\varpi, \phi, z\}$ are cylindrical coordinates. By translating the Cartesian-coordinate waves of the text into quasi-Cartesian coordinates and noting from Fig. 16.10b that five wavelengths must fit into the circumference around the cylinder, show that the velocity field has the form $v_\varpi, v_\phi, v_z \propto e^{i(5\phi+\omega t)}$, and deduce the ratios of the three components of velocity to one another. The solution has nonzero radial velocity at the walls—a warning that edge effects will modify the waves somewhat. This analysis ignores those edge effects.

(c) Because the waves are generated by the ridge on the bottom of the tank, the wave pattern must remain at rest relative to that ridge, which means it must rotate relative to the fluid's frame with the angular velocity $d\phi/dt = -\Delta\Omega$. From the waves' dispersion relation, deduce $\Delta\Omega/\Omega$ (the fractional slowdown of the tank that had to be imposed to generate the observed pentagonal wave).

(d) Compute the displacement field $\delta\mathbf{x}(\varpi, \phi, z, t)$ of a fluid element whose undisplaced location (in the rigidly rotating cylindrical coordinates) is (ϖ, ϕ, z). Explain the pentagonal shape of the movie's dye lines in terms of this displacement field.

(e) Compute the wave's vertical vorticity field ω_z (relative to the rigidly rotating flow), and show that as a fluid element moves and the vertical vortex line through it shortens or lengthens due to the changing water depth, ω_z changes proportionally to the vortex line's length.

16.5 Sound Waves

So far our discussion of fluid dynamics has mostly been concerned with flows sufficiently slow that the density can be treated as constant. We now introduce the effects of compressibility in the context of sound waves (in a nonrotating reference frame). Sound waves are prototypical scalar waves and therefore are simpler in many respects than vector electromagnetic waves and tensor gravitational waves.

Consider a small-amplitude sound wave propagating through a homogeneous, time-independent fluid. The wave's oscillations are generally quick compared to the time for heat to diffuse across a wavelength, so the pressure and density perturbations are adiabatically related:

$$\delta P = C^2 \delta \rho, \tag{16.47}$$

where

adiabatic sound speed

$$\boxed{C \equiv \left[\left(\frac{\partial P}{\partial \rho}\right)_s\right]^{1/2},} \tag{16.48}$$

which will turn out to be the wave's propagation speed—the speed of sound. The perturbation of the fluid velocity (which we denote $\mathbf{v}$, since the unperturbed fluid is static) is related to the pressure perturbation by the linearized Euler equation:

$$\frac{\partial \mathbf{v}}{\partial t} = -\frac{\nabla \delta P}{\rho}. \tag{16.49a}$$

A second relation between $\mathbf{v}$ and δP can be obtained by combining the linearized law of mass conservation, $\partial \rho / \partial t = -\rho \nabla \cdot \mathbf{v}$, with the adiabatic pressure-density relation (16.47):

$$\nabla \cdot \mathbf{v} = -\frac{1}{\rho C^2} \frac{\partial \delta P}{\partial t}. \tag{16.49b}$$

By equating the divergence of Eq. (16.49a) to the time derivative of Eq. (16.49b), we obtain a simple, dispersion-free wave equation for the pressure perturbation:

wave equation for sound waves

$$\left(\frac{\partial^2}{\partial t^2} - C^2 \nabla^2\right) \delta P = 0. \tag{16.50}$$

Thus, as claimed, C is the wave's propagation speed.

For a perfect gas, this adiabatic sound speed is $C = (\gamma P/\rho)^{1/2}$, where γ is the ratio of specific heats (see Ex. 5.4). The sound speed in air at 20 °C is 343 m s^{-1}. In water under atmospheric conditions, it is about 1.5 km s^{-1} (not much different from sound speeds in solids).

Because the vorticity of the unperturbed fluid vanishes and the wave contains no vorticity-producing forces, the wave's vorticity vanishes: $\nabla \times \mathbf{v} = 0$. This permits us to express the wave's velocity perturbation as the gradient of a velocity potential: $\mathbf{v} = \nabla \psi$.

Inserting this expression into the perturbed Euler equation (16.49a), we express the pressure perturbation in terms of ψ:

$$\boxed{\delta P = -\rho \frac{\partial \psi}{\partial t}, \quad \text{where} \quad \mathbf{v} = \nabla \psi.} \tag{16.51}$$

velocity potential for sound waves

The first of these relations guarantees that ψ satisfies the same wave equation as δP:

$$\boxed{\left(\frac{\partial^2}{\partial t^2} - C^2 \nabla^2\right) \psi = 0.} \tag{16.52}$$

It is sometimes useful to describe the wave by its oscillating pressure δP and sometimes by its oscillating potential ψ.

The general solution of the wave equation (16.52) for plane sound waves propagating in the $\pm x$ directions is

$$\psi = f_1(x - Ct) + f_2(x + Ct), \tag{16.53}$$

where f_1 and f_2 are arbitrary functions.

EXERCISES

Exercise 16.13 *Problem: Sound Wave in an Inhomogeneous Fluid*
Consider a sound wave propagating through a static, inhomogeneous fluid with no gravity. Explain why the unperturbed fluid has velocity $\mathbf{v} = 0$ and pressure $P_o = $ constant, but can have variable density and sound speed, $\rho_o(\mathbf{x})$ and $C(\mathbf{x}, t)$. By repeating the analysis in Eqs. (16.47)–(16.50), show that the wave equation is $\partial^2 \delta P/\partial t^2 = C^2 \rho_o \nabla \cdot (\rho_o^{-1} \nabla \delta P)$, which can be rewritten as

$$W \frac{\partial^2 \delta P}{\partial t^2} - \nabla \cdot (W C^2 \nabla \delta P) = 0, \tag{16.54}$$

where $W = (C^2 \rho_o)^{-1}$. [Hint: It may be helpful to employ the concept of Lagrangian versus Eulerian perturbations, as described by Eq. (19.44).] Equation (16.54) is an example of the prototypical wave equation (7.17) that we used in Sec. 7.3.1 to illustrate the geometric-optics formalism. The functional form of W and the placement of W and C^2 (inside versus outside the derivatives) have no influence on the wave's dispersion relation or its rays or phase in the geometric-optics limit, but they do influence the propagation of the wave's amplitude. See Sec. 7.3.1.

16.5.1 Wave Energy

In Sec. 7.3.1 and Ex. 7.4, we used formal mathematical techniques to derive the energy density U and energy flux $\mathbf{F}$ [Eqs. (7.18)] associated with waves satisfying the prototypical wave equation (16.54). In this section, we rederive U and $\mathbf{F}$ for sound waves using a physical, fluid dynamical analysis. We get precisely the same expressions up to a constant multiplicative factor ρ^2. Because of the formal nature of the arguments leading to Eqs. (7.18), we only had a right to expect the same answer up to some multiplicative constant.

The fluid's energy density is $U = (\frac{1}{2}v^2 + u)\rho$ (Table 13.1 with $\Phi = 0$). The first term is the fluid's kinetic energy density; the second is its internal energy density. The internal energy density can be evaluated by a Taylor expansion in the wave's density perturbation:

$$u\rho = [u\rho] + \left[\left(\frac{\partial(u\rho)}{\partial \rho}\right)_s\right]\delta\rho + \frac{1}{2}\left[\left(\frac{\partial^2(u\rho)}{\partial \rho^2}\right)_s\right]\delta\rho^2, \qquad (16.55)$$

where the three coefficients in square brackets are evaluated at the equilibrium density. The first term in Eq. (16.55) is the energy of the background fluid, so we drop it. The second term averages to zero over a wave period, so we also drop it. The third term can be simplified using the first law of thermodynamics in the form $du = Tds - Pd(1/\rho)$ (which implies $[\partial(u\rho)/\partial\rho]_s = u + P/\rho$). We then apply the definition $h = u + P/\rho$ of enthalpy density, followed by the first law in the form $dh = Tds + dP/\rho$, and then followed by expression (16.48) for the speed of sound. The result is

$$\left(\frac{\partial^2(u\rho)}{\partial \rho^2}\right)_s = \left(\frac{\partial h}{\partial \rho}\right)_s = \frac{C^2}{\rho}. \qquad (16.56)$$

Inserting this relation into the third term of Eq. (16.55) and averaging over a wave period and wavelength, we obtain for the wave energy per unit volume $U = \frac{1}{2}\rho\overline{v^2} + [C^2/(2\rho)]\overline{\delta\rho^2}$. Using $\mathbf{v} = \nabla\psi$ [the second of Eqs. (16.51)] and $\delta\rho = (\rho/C^2)\partial\psi/\partial t$ [from $\delta\rho = (\partial\rho/\partial P)_s \delta P = \delta P/C^2$ and the first of Eqs. (16.51)], we bring the equation for U into the form

energy density for sound waves

$$\boxed{U = \frac{1}{2}\rho\left[\overline{(\nabla\psi)^2} + \frac{1}{C^2}\overline{\left(\frac{\partial\psi}{\partial t}\right)^2}\right] = \rho\overline{(\nabla\psi)^2}.} \qquad (16.57)$$

The second equality can be deduced by multiplying the wave equation (16.52) by ψ and averaging. Thus, energy is equipartitioned between the kinetic and internal energy terms.

The energy flux is $\mathbf{F} = (\frac{1}{2}v^2 + h)\rho\mathbf{v}$ (Table 13.1 with $\Phi = 0$). The kinetic energy flux (first term) is third order in the velocity perturbation and therefore vanishes on average. For a sound wave, the internal energy flux (second term) can be brought into a more useful form by expanding the enthalpy per unit mass:

$$h = [h] + \left[\left(\frac{\partial h}{\partial P}\right)_s\right]\delta P = [h] + \frac{\delta P}{\rho}. \qquad (16.58)$$

Here we have used the first law of thermodynamics $dh = Tds + (1/\rho)dP$ and adiabaticity of the perturbation ($s = $ const); the terms in square brackets are unperturbed quantities. Inserting Eq. (16.58) into $\mathbf{F} = h\rho\mathbf{v}$, expressing δP and $\mathbf{v}$ in terms of the velocity potential [Eqs. (16.51)], and averaging over a wave period and wavelength, we obtain for the energy flux $\mathbf{F} = \overline{\rho h \mathbf{v}} = \overline{\delta P \mathbf{v}}$, which becomes

$$\mathbf{F} = -\rho \overline{\left(\frac{\partial \psi}{\partial t}\right) \nabla \psi}. \tag{16.59}$$

energy flux for soundwaves

Aside from a multiplicative constant factor ρ^2, this equation and Eq. (16.57) agree with Eqs. (7.18) [with ψ there being this chapter's velocity potential ψ, and with $W = (C^2 \rho)^{-1}$; Eq. (16.54)], which we derived by formal techniques in Sec. 7.3.1 and Ex. 7.4.

For a locally plane wave with $\psi = \psi_o \cos(\mathbf{k} \cdot \mathbf{x} - \omega t + \varphi)$ (where φ is an arbitrary phase), the energy density (16.57) is $U = \frac{1}{2}\rho \psi_o^2 k^2$, and the energy flux (16.59) is $\mathbf{F} = \frac{1}{2}\rho \psi_o^2 \omega \mathbf{k}$. Since for this dispersion-free wave, the phase and group velocities are both $\mathbf{V} = (\omega/k)\hat{\mathbf{k}} = C\hat{\mathbf{k}}$ (where $\hat{\mathbf{k}} = \mathbf{k}/k$ is the unit vector pointing in the wave-propagation direction), the energy density and flux are related by

$$\boxed{\mathbf{F} = U\mathbf{V} = UC\hat{\mathbf{k}}.} \tag{16.60}$$

The energy flux is therefore the product of the energy density and the wave velocity, as it must be [Eq. (7.31), where we see that, if the waves were to have dispersion, it would be the group velocity that appears in this expression].

The energy flux carried by sound is conventionally measured in dB (decibels). The flux in decibels, F_{dB}, is related to the flux F in W m^{-2} by

$$\boxed{F_{\text{dB}} = 120 + 10 \log_{10}(F).} \tag{16.61}$$

Sound that is barely audible is about 1 dB. Normal conversation is about 50–60 dB. Jet aircraft, rock concerts, and volcanic eruptions can cause exposure to more than 120 dB, with consequent damage to the ear.

16.5.2 Sound Generation

So far in this book we have been concerned with describing how different types of waves propagate. It is also important to understand how they are generated. We now outline some aspects of the theory of sound generation.

The reader should be familiar with the theory of electromagnetic wave emission (e.g., Jackson, 1999, Chap. 9). For electromagnetic waves one considers a localized region containing moving charges and varying currents. The source can be described as a sum over electric and magnetic multipoles, and each multipole produces a characteristic angular variation of the distant radiation field. The radiation-field amplitude decays inversely with distance from the source, and so the Poynting flux varies with the inverse square of the distance. Integrating over a large sphere gives the total power radiated by the source, broken down into the power radiated by each multipolar component. The ratio of the power in successive multipole pairs [e.g., (magnetic dipole power)/(electric dipole power) $\sim$ (electric quadrupole power)/(electric dipole power)] is typically $\sim (b/\lambdabar)^2$, where b is the size of the source, and $\lambdabar = 1/k$ is the waves' reduced wavelength. When $\lambdabar$ is large compared to b (a situation referred to as *slow motion*, since the source's charges then generally move at speeds $\sim(b/\lambdabar)c$ small compared to the speed of light c), the most powerful radiating multipole is the electric dipole $\mathbf{d}(t)$, unless it happens to be suppressed. The dipole's average emitted power is given by the Larmor formula:

$$\mathcal{P} = \frac{\overline{\ddot{\mathbf{d}}^2}}{6\pi\epsilon_0 c^3}, \tag{16.62}$$

where $\ddot{\mathbf{d}}$ is the second time derivative of $\mathbf{d}$, the bar denotes a time average, ϵ_0 is the permittivity of free space, and c is the speed of light.

This same procedure can be followed when describing sound generation. However, as we are dealing with a scalar wave, sound can have a monopolar source. As a pedagogical example, let us set a small, spherical, elastic ball, surrounded by fluid, into radial oscillation (not necessarily sinusoidal) with oscillation frequencies of order ω, so the emitted waves have reduced wavelengths of order $\lambdabar = C/\omega$. Let the surface of the ball have radius $a + \xi(t)$, and impose the slow-motion and small-amplitude conditions that

$$\lambdabar \gg a \gg |\xi|. \tag{16.63}$$

As the waves will be spherical, the relevant outgoing-wave solution of the wave equation (16.52) is

$$\psi = \frac{f(t - r/C)}{r}, \tag{16.64}$$

where f is a function to be determined. Since the fluid's velocity at the ball's surface must match that of the ball, we have (to first order in $\mathbf{v}$ and ψ):

$$\dot{\xi}\mathbf{e}_r = \mathbf{v}(a, t) = \boldsymbol{\nabla}\psi \simeq -\frac{f(t - a/C)}{a^2}\mathbf{e}_r \simeq -\frac{f(t)}{a^2}\mathbf{e}_r, \tag{16.65}$$

where in the third equality we have used the slow-motion condition $\lambdabar \gg a$. Solving for $f(t)$ and inserting into Eq. (16.64), we see that

$$\psi(r, t) = -\frac{a^2\dot{\xi}(t - r/C)}{r}. \tag{16.66}$$

It is customary to express the radial velocity perturbation v in terms of an oscillating fluid *monopole moment*

$$\boxed{q = 4\pi\rho a^2 \dot{\xi}.} \tag{16.67}$$

monopole moment for spherical sound waves from an oscillating ball

Physically, this is the total radial discharge of air mass (i.e., mass per unit time) crossing an imaginary fixed spherical surface of radius slightly larger than that of the oscillating ball. In terms of q, we have $\dot{\xi}(t) = q(t)/[4\pi\rho a^2]$. Using this expression and Eq. (16.66), we compute for the power radiated as sound waves [Eq. (16.59) integrated over a sphere centered on the ball]:

$$\boxed{\mathcal{P} = \frac{\overline{\dot{q}^2}}{4\pi\rho C}.} \qquad (16.68)$$

power emitted into spherical sound waves

Note that the power is inversely proportional to the signal speed, which is characteristic of monopolar emission and is in contrast to the inverse-cube variation for dipolar emission [Eq. (16.62)].

The emission of monopolar waves requires that the volume of the emitting solid body oscillate. When the solid simply oscillates without changing its volume (e.g., the reed in a musical instrument), dipolar emission usually dominates. We can think of this as two monopoles of size a in antiphase separated by some displacement $b \sim a$. The velocity potential in the far field is then the sum of two monopolar contributions, which almost cancel. Making a Taylor expansion, we obtain

$$\frac{\psi_{\text{dipole}}}{\psi_{\text{monopole}}} \sim \frac{b}{\lambdabar} \sim \frac{\omega b}{C}, \qquad (16.69)$$

typical multipolar sound-wave strengths

where ω and $\lambdabar$ are the characteristic magnitudes of the angular frequency and reduced wavelength of the waves (which we have not assumed to be precisely sinusoidal).

This reduction of ψ by the slow-motion factor $b/\lambdabar$ implies that the dipolar power emission is weaker than the monopolar power by a factor $\sim(b/\lambdabar)^2$ for similar frequencies and amplitudes of motion—the same factor as for electromagnetic waves (see the start of this subsection). However, to emit dipole radiation, momentum must be given to and removed from the fluid. In other words, the fluid must be forced by a solid body. In the absence of such a solid body, the lowest multipole that can be radiated effectively is quadrupolar radiation, which is weaker by yet one more factor of $(b/\lambdabar)^2$.

lowest order sound wave emitted in absence of solid-body forcing: quadrupolar

These considerations are important for understanding how noise is produced by the intense turbulence created by jet engines, especially close to airports. We expect that the sound emitted by the free turbulence in the wake just behind the engine will be quadrupolar and will be dominated by emission from the largest (and hence fastest) turbulent eddies. (See the discussion of turbulent eddies in Sec. 15.4.4.) Denote by ℓ and v_ℓ the size and turnover speed of these largest eddies. Then the characteristic size of the sound's source is $a \sim b \sim \ell$, the mass discharge is $q \sim \rho\ell^2 v_\ell$, the characteristic frequency is $\omega \sim v_\ell/\ell$, the reduced wavelength of the sound waves is $\lambdabar = C/\omega \sim \ell C/v_\ell$, and the slow-motion parameter is $b/\lambdabar \sim \omega b/C \sim v_\ell/C$. The quadrupolar power radiated per unit volume [Eq. (16.68) divided by the volume ℓ^3 of an eddy and reduced by $\sim(b/\lambdabar)^4$] is therefore

sound-wave power from free turbulence

$$\frac{d\mathcal{P}}{d^3x} \sim \rho \frac{v_\ell^3}{\ell} \left(\frac{v_\ell}{C}\right)^5, \qquad (16.70)$$

and this power is concentrated around frequency $\omega \sim v_\ell/\ell$. For air of fixed sound speed and lengthscale ℓ of the largest eddies, and for which the largest eddy speed v_ℓ is proportional to some characteristic speed V (e.g., the average speed of the air leaving

16.5 Sound Waves

Lighthill's law for sound generation by turbulence

the engine), Eq. (16.70) says the sound generation increases proportional to the eighth power of the Mach number $M = V/C$. This is known as Lighthill's law (Lighthill 1952, 1954). The implications for the design of jet engines should be obvious.

EXERCISES

Exercise 16.14 *Problem: Attenuation of Sound Waves*
Viscosity and thermal conduction will attenuate sound waves. For the moment just consider a monatomic gas where the bulk viscosity can be neglected.

(a) Consider the entropy equation (13.75), and evaluate the influence of the heat flux on the relationship between the pressure and the density perturbations. [Hint: Assume that all quantities vary as $e^{i(kx-\omega t)}$, where k is real and ω complex.]

(b) Consider the momentum equation (13.69), and include the viscous term as a perturbation.

(c) Combine these two relations in parts (a) and (b), together with the equation of mass conservation, to solve for the imaginary part of ω in the linear regime.

(d) Substitute kinetic-theory expressions for the coefficient of shear viscosity and the coefficient of thermal conductivity (Secs. 13.7.3 and 3.7) to obtain a simple expression for the attenuation length involving the wave's wavelength and the atoms' collisional mean free path.

(e) How do you think the wave attenuation will be affected if the fluid is air or is turbulent, or both?

See Faber (1995) for more details.

Exercise 16.15 *Example: Plucked Violin String*
Consider the G string (196 Hz) of a violin. It is ∼30 cm from bridge to nut (the fixed endpoints), and the tension in the string is ∼40 N.

(a) Infer the mass per unit length in the string and estimate its diameter. Hence estimate the strain in the string before being plucked. Estimate the strain's increase if its midpoint is displaced through 3 mm.

(b) Now suppose that the string is released. Estimate the speed with which it moves as it oscillates back and forth.

(c) Estimate the dipolar sound power emitted and the distance out to which the note can be heard (when its intensity is a few decibels). Do your answers seem reasonable? What factors, omitted from this calculation, might change your answers?

Exercise 16.16 *Example: Trumpet*
Idealize the trumpet as a bent pipe of length 1.2 m from the mouthpiece (a node of the air's displacement) to the bell (an antinode). The lowest note is a first overtone and should correspond to B flat (233 Hz). Does it?

Exercise 16.17 *Problem: Aerodynamic Sound Generation*

Consider the emission of quadrupolar sound waves by a Kolmogorov spectrum of free turbulence (Sec. 15.4.4). Show that the power radiated per unit frequency interval has a spectrum

$$\mathcal{P}_\omega \propto \omega^{-7/2}.$$

Also show that the total power radiated is roughly a fraction M^5 of the power dissipated in the turbulence, where M is the Mach number.

16.5.3 Radiation Reaction, Runaway Solutions, and Matched Asymptotic Expansions [T2]

Let us return to our idealized example of sound waves produced by a radially oscillating, spherical ball. We use this example to illustrate several deep issues in theoretical physics: the *radiation-reaction force* that acts back on a source due to its emission of radiation, a spurious *runaway solution* to the source's equation of motion caused by the radiation-reaction force, and *matched asymptotic expansions*, a mathematical technique for solving field equations when there are two different regions of space in which the equations have rather different behaviors.[7] These issues also arise, in a rather more complicated way, in analyses of the electromagnetic radiation reaction force on an accelerated electron (the "Abraham-Lorentz force"), and the radiation-reaction force caused by emission of gravitational waves; see the derivation, by Burke (1971), of gravitational results quoted in Sec. 27.5.3.

For our oscillating ball, the two different regions of space that we match to each other are the *near zone*, $r \ll \lambdabar$, and the *wave zone*, $r \gtrsim \lambdabar$.

We consider, first, the near zone, and we redo, from a new point of view, the analysis of the matching of the near-zone fluid velocity to the ball's surface velocity and the computation of the pressure perturbation. Because the region near the ball is small compared to $\lambdabar$ and the fluid speeds are small compared to C, the flow is very nearly incompressible, $\nabla \cdot \mathbf{v} = \nabla^2 \psi = 0$; see the discussion of conditions for incompressibility in Sec. 13.6. (The near-zone equation $\nabla^2 \psi = 0$ is analogous to $\nabla^2 \Phi = 0$ for the Newtonian gravitational potential in the weak-gravity near zone of a gravitational-wave source; Sec. 27.5.)

The general monopolar (spherical) solution to $\nabla^2 \psi = 0$ is

$$\psi = \frac{A(t)}{r} + B(t). \quad (16.71)$$

Matching the fluid's radial velocity $v = \partial \psi/\partial r = -A/r^2$ at $r = a$ to the ball's radial velocity $\dot\xi$, we obtain

$$A(t) = -a^2 \dot\xi(t). \quad (16.72)$$

---

7. Our treatment is based on Burke (1970).

From the point of view of near-zone physics, no mechanism exists for generating a nonzero spatially constant term $B(t)$ in ψ [Eq. (16.71)], so if one were unaware of the emitted sound waves and their action back on the source, one would be inclined to set $B(t)$ to zero. [This line of reasoning is analogous to that of a Newtonian physicist, who would be inclined to write the quadrupolar contribution to an axisymmetric source's external gravitational field in the form $\Phi = P_2(\cos\theta)[A(t)r^{-3} + B(t)r^2]$ and then, being unaware of gravitational waves and their action back on the source, would set $B(t)$ to zero.] Taking this near-zone viewpoint, with $B = 0$, we infer that the fluid's pressure perturbation acting on the ball's surface is

$$\delta P = -\rho \frac{\partial \psi(a,t)}{\partial t} = -\rho \frac{\dot{A}}{a} = \rho a \ddot{\xi} \tag{16.73}$$

[Eqs. (16.51) and (16.72)].

The motion $\xi(t)$ of the ball's surface is controlled by the elastic restoring forces in its interior and the fluid pressure perturbation δP on its surface. In the absence of δP the surface would oscillate sinusoidally with some angular frequency ω_o, so $\ddot{\xi} + \omega_o^2 \xi = 0$. The pressure will modify this expression to

$$m(\ddot{\xi} + \omega_o^2 \xi) = -4\pi a^2 \delta P, \tag{16.74}$$

where m is an effective mass, roughly equal to the ball's true mass, and the right-hand side is the integral of the radial component of the pressure perturbation force over the sphere's surface. Inserting the near-zone viewpoint's pressure perturbation (16.73), we obtain

$$(m + 4\pi a^3 \rho)\ddot{\xi} + m\omega_o^2 \xi = 0. \tag{16.75}$$

Evidently, the fluid increases the ball's effective inertial mass (it *loads* additional mass on the ball) and thereby reduces its frequency of oscillation to

$$\omega = \frac{\omega_o}{\sqrt{1+\kappa}}, \quad \text{where } \kappa = \frac{4\pi a^3 \rho}{m} \tag{16.76}$$

is a measure of the coupling strength between the ball and the fluid. In terms of this loaded frequency, the equation of motion becomes

$$\ddot{\xi} + \omega^2 \xi = 0. \tag{16.77}$$

This near-zone viewpoint is not quite correct, just as the standard Newtonian viewpoint is not quite correct for the near-zone gravity of a gravitational-wave source (Sec. 27.5.3). To improve on this viewpoint, we temporarily move out into the wave zone and identify the general, outgoing-wave solution to the sound wave equation:

$$\psi = \frac{f(t - \epsilon r/C)}{r} \tag{16.78}$$

[Eq. (16.64)]. Here f is a function to be determined by matching to the near zone, and ϵ is a parameter that has been inserted to trace the influence of the outgoing-wave boundary condition. For outgoing waves (the real, physical, situation), $\epsilon = +1$; if the waves were ingoing, we would have $\epsilon = -1$.

This wave-zone solution remains valid into the near zone. In the near zone we can perform a slow-motion expansion to bring it into the same form as the near-zone velocity potential (16.71):

slow-motion expansion

$$\psi = \frac{f(t)}{r} - \epsilon \frac{\dot{f}(t)}{C} + \cdots. \quad (16.79)$$

The second term is sensitive to whether the waves are outgoing or incoming and thus must ultimately be responsible for the radiation-reaction force that acts back on the oscillating ball; for this reason we call it the *radiation-reaction potential*.

radiation-reaction potential

Equating the first term of this ψ to the first term of Eq. (16.71) and using the value in Eq. (16.72) of $A(t)$, which was obtained by matching the fluid velocity to the ball velocity, we obtain

$$f(t) = A(t) = -a^2 \dot{\xi}(t). \quad (16.80)$$

This equation tells us that the wave field $f(t - r/C)/r$ generated by the ball's surface displacement $\xi(t)$ is given by $\psi = -a^2 \dot{\xi}(t - r/C)/r$ [Eq. (16.66)]—the result we derived more quickly in the previous section. We can regard Eq. (16.80) as matching the near-zone solution outward onto the wave-zone solution to determine the wave field as a function of the source's motion.

Equating the second term of Eq. (16.79) to the second term of the near-zone velocity potential (16.71), we obtain

$$B(t) = -\epsilon \frac{\dot{f}(t)}{C} = \epsilon \frac{a^2}{C} \ddot{\xi}(t). \quad (16.81)$$

This is the term in the near-zone velocity potential $\psi = A/r + B$ that is responsible for radiation reaction. *We can regard this radiation-reaction potential $\psi^{RR} = B(t)$ as having been generated by matching the wave zone's outgoing ($\epsilon = +1$) or ingoing ($\epsilon = -1$) wave field back into the near zone.* [A similar matching analysis by Burke (1971) led him to the gravitational radiation-reaction potential (27.64).]

how radiation-reaction potential is generated

This pair of matchings, outward then inward (Fig. 16.11), is a special, almost trivial example of the technique of matched asymptotic expansions—a technique developed by applied mathematicians to deal with much more complicated matching problems than this one (see, e.g., Cole, 1974).

matched asymptotic expansions

The radiation-reaction potential $\psi^{RR} = B(t) = \epsilon(a^2/C)\ddot{\xi}(t)$ gives rise to a radiation-reaction contribution to the pressure on the ball's surface: $\delta P^{RR} = -\rho \dot{\psi}^{RR} = -\epsilon(\rho a^2/C)\dddot{\xi}$. Inserting this into the equation of motion (16.74) along with the loading pressure (16.73) and performing the same algebra as before, we get the following radiation-reaction-modified form of Eq. (16.77):

radiation-reaction pressure on ball's surface

$$\boxed{\ddot{\xi} + \omega^2 \xi = \epsilon \tau \dddot{\xi},} \quad \text{where} \quad \tau = \frac{\kappa}{1+\kappa} \frac{a}{C} \quad (16.82)$$

ball's equation of motion with radiation reaction

16.5 Sound Waves 871

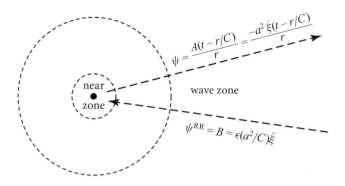

FIGURE 16.11 Matched asymptotic expansions for an oscillating ball emitting sound waves. The near-zone expansion feeds the radiation field $\psi = \frac{1}{r}A(t-r/C) = -\frac{1}{r}a^2\dot{\xi}(t-r/C)$ into the wave zone. The wave-zone expansion then feeds the radiation-reaction field $\psi^{RR} = B = \epsilon(a^2/C)\dddot{\xi}$ back into the near zone, where it produces the radiation-reaction pressure $\delta P^{RR} = -\rho\dot{\psi}^{RR}$ on the ball's surface.

is less than the fluid's sound travel time to cross the ball's radius, a/C. The term $\epsilon\tau\dddot{\xi}$ in the equation of motion is the ball's *radiation-reaction acceleration,* as we see from the fact that it would change sign if we switched from outgoing waves ($\epsilon = +1$) to incoming waves ($\epsilon = -1$).

radiation-reaction induced damping

In the absence of radiation reaction, the ball's surface oscillates sinusoidally in time: $\xi = e^{\pm i\omega t}$. The radiation-reaction term produces a weak damping of these oscillations:

$$\xi \propto e^{\pm i\omega t}e^{-\sigma t}, \quad \sigma = \frac{1}{2}\epsilon(\omega\tau)\omega, \qquad (16.83)$$

where σ is the radiation-reaction-induced damping rate (with $\epsilon = +1$). Note that in order of magnitude the ratio of the damping rate to the oscillation frequency is $\sigma/\omega \sim \omega\tau \lesssim \omega a/C = a/\lambda$, which is small compared to unity by virtue of the slow-motion assumption. If the waves were incoming rather than outgoing, $\epsilon = -1$, the fluid's oscillations would grow. In either case, outgoing waves or ingoing waves, the radiation-reaction force removes energy from the ball or adds it at the same rate as the sound waves carry energy off or bring it in. The total energy, wave plus ball, is conserved.

Expression (16.83) is two linearly independent solutions to the equation of motion (16.82), one with the plus sign and the other with the minus sign. Since this equation of motion has been made third order by the radiation-reaction term, there must be a third independent solution. It is easy to see that, up to a tiny fractional correction, that third solution is

$$\xi \propto e^{\epsilon t/\tau}. \qquad (16.84)$$

runaway solution to ball's equation of motion

For outgoing waves, $\epsilon = +1$, this solution grows exponentially in time on an extremely rapid timescale, $\tau \lesssim a/C$; it is called a *runaway solution.*

Such runaway solutions are ubiquitous in equations of motion with radiation reaction. For example, a computation of the electromagnetic radiation reaction on a small, classical, electrically charged, spherical particle gives the Abraham-Lorentz equation of motion:

$$m(\ddot{\mathbf{x}} - \tau \dddot{\mathbf{x}}) = \mathbf{F}_{\text{ext}} \qquad (16.85)$$

runaway solutions for an electrically charged particle

(Rohrlich, 1965; Jackson, 1999, Sec. 16.2). Here $\mathbf{x}(t)$ is the particle's world line, $\mathbf{F}_{\text{ext}}$ is the external force that causes the particle to accelerate, and the particle's inertial mass m includes an electrostatic contribution analogous to $4\pi a^3 \rho$ in our fluid problem. The timescale τ, like that in our fluid problem, is very short, and when the external force is absent, there is a runaway solution $\mathbf{x} \propto e^{t/\tau}$.

Much human heat and confusion were generated, in the early and mid-twentieth century, over these runaway solutions (see, e.g., Rohrlich, 1965). For our simple model problem, there need be little heat or confusion. One can easily verify that the runaway solution (16.84) violates the slow-motion assumption $a/\lambda \ll 1$ that underlies our derivation of the radiation-reaction acceleration. It therefore is a spurious solution.

Our model problem is sufficiently simple that one can dig deeper into it and learn that the runaway solution arises from the slow-motion approximation failing to reproduce a genuine, rapidly damped solution and getting the sign of the damping wrong (Ex. 16.19 and Burke, 1970).

origin of runaway solution

EXERCISES

Exercise 16.18 *Problem: Energy Conservation for Radially Oscillating Ball Plus Sound Waves*
For the radially oscillating ball as analyzed in Sec. 16.5.3, verify that the radiation-reaction acceleration removes energy from the ball, plus the fluid loaded onto it, at the same rate as the sound waves carry energy away. See Ex. 27.12 for the analogous gravitational-wave result.

Exercise 16.19 *Problem: Radiation Reaction without the Slow-Motion Approximation*
Redo the computation of radiation reaction for a radially oscillating ball immersed in a fluid without imposing the slow-motion assumption and approximation. Thereby obtain the following coupled equations for the radial displacement $\xi(t)$ of the ball's surface and the function $\Phi(t) \equiv a^{-2} f(t - \epsilon a/C)$, where $\psi = r^{-1} f(t - \epsilon r/C)$ is the sound-wave field:

$$\ddot{\xi} + \omega_o^2 \xi = \kappa \dot{\Phi}, \quad \dot{\xi} = -\Phi - \epsilon(a/C)\dot{\Phi}. \qquad (16.86)$$

Show that in the slow-motion regime, this equation of motion has two weakly damped solutions of the same form as we derived using the slow-motion approximation [Eq. (16.83)], and one rapidly damped solution: $\xi \propto \exp(-\epsilon \kappa t/\tau)$. Burke (1970) shows that the runaway solution (16.84) obtained using the slow-motion approximation is caused by that approximation's futile attempt to reproduce this genuine, rapidly damped solution.

Exercise 16.20 *Problem: Sound Waves from a Ball Undergoing Quadrupolar Oscillations*

Repeat the analysis of sound-wave emission, radiation reaction, and energy conservation—as given in Sec. 16.5.3 and Ex. 16.18—for axisymmetric, quadrupolar oscillations of an elastic ball: $r_{\text{ball}} = a + \xi(t) P_2(\cos\theta)$.

Comment: Since the lowest multipolar order for gravitational waves is quadrupolar, this exercise is closer to the analogous problem of gravitational wave emission (Secs. 27.5.2 and 27.5.3) than is the monopolar analysis in the text.

[Hint: If ω is the frequency of the ball's oscillations, then the sound waves have the form

$$\psi = K \Re\left[e^{-i\omega t} \left(\frac{n_2(\omega r/C) - i\epsilon j_2(\omega r/C)}{r} \right) \right] P_2(\cos\theta), \qquad (16.87)$$

where K is a constant; $\Re(X)$ is the real part of X; ϵ is $+1$ for outgoing waves and -1 for ingoing waves; j_2 and n_2 are the spherical Bessel and spherical Neuman functions of order 2, and P_2 is the Legendre polynomial of order 2. In the distant wave zone, $x \equiv \omega r/C \gg 1$, we have

$$n_2(x) - i\epsilon j_2(x) = \frac{e^{i\epsilon x}}{x}; \qquad (16.88)$$

in the near zone $x = \omega r/C \ll 1$, we have

$$n_2(x) = -\frac{3}{x^3}\left(1 \, \& \, x^2 \, \& \, x^4 \, \& \, \dots \right), \quad j_2(x) = \frac{x^2}{15}\left(1 \, \& \, x^2 \, \& \, x^4 \, \& \, \dots \right). \qquad (16.89)$$

Here "$\& \, x^n$" means "+ (some constant) x^n".]

Bibliographic Note

For physical insight into waves in fluids, we recommend the movies discussed in Box 16.2. Among fluid-dynamics textbooks, those that we most like for their treatment of waves are Acheson (1990), Lautrup (2005), and Kundu, Cohen, and Dowling (2012). For greater depth and detail, we recommend two books solely devoted to waves: Lighthill (2001) and Whitham (1974).

For Rossby waves (which are omitted from most fluid-dynamics texts), we recommend the very physical descriptions and analyses in Tritton (1987). For solitons, we like Whitham (1974, Chap. 17); also Drazin and Johnson (1989), which focuses on the mathematics of the Korteweg–de Vries equation; Ablowitz (2011), which treats the mathematics of solitons plus applications to fluids and nonlinear optics; and Dauxois and Peyrard (2010), which treats the mathematics and applications to plasmas and condensed-matter physics.

The mathematics of matched asymptotic expansions is nicely developed by Cole (1974) and Lagerstrom (1988). Radiation reaction in wave emission is not treated pedagogically in any textbook that we know of except ours; for pedagogical original literature, we like Burke (1970).

CHAPTER SEVENTEEN

Compressible and Supersonic Flow

> Rocket science is tough, and rockets have a way of failing.
> PHYSICIST, ASTRONAUT, AND EDUCATOR SALLY RIDE (2012)

17.1 Overview

So far we have mainly been concerned with flows that are slow enough that they may be treated as incompressible. We now consider flows in which the velocity approaches or even exceeds the speed of sound and in which changes of density along streamlines cannot be ignored. Such flows are common in aeronautics and astrophysics. For example, the motion of a rocket through the atmosphere is faster than the speed of sound in air. In other words, it is *supersonic*. Therefore, if we transform into the frame of the rocket, the flow of air past the rocket is also supersonic.

When the flow speed exceeds the speed of sound in some reference frame, it is not possible for a pressure pulse to travel upstream in that frame and change the direction of the flow. However, if there is a solid body in the way (e.g., a rocket or aircraft), the flow direction must change (Fig. 17.1). In a supersonic flow, this change happens nearly discontinuously, through the formation of *shock fronts,* at which the flow suddenly decelerates from supersonic to subsonic. Shock fronts are an inevitable feature of supersonic flows.

In another example of supersonic flow, a rocket itself is propelled by the thrust created by its nozzle's escaping hot gases. These hot gases move through the rocket nozzle at supersonic speeds, expanding and cooling as they accelerate. In this manner, the random thermal motion of the gas molecules is converted into an organized bulk motion that carries negative momentum away from the rocket and pushes it forward.

The solar wind furnishes yet another example of a supersonic flow. This high-speed flow of ionized gas is accelerated in the solar corona and removes a fraction of $\sim 10^{-14}$ of the Sun's mass every year. Its own pressure accelerates it to supersonic speeds of ~ 400 km s^{-1}. When the outflowing solar wind encounters a planet, it is rapidly decelerated to subsonic speed by passing through a strong discontinuity, known as a *bow shock,* that surrounds the planet (Fig. 17.2). The bulk kinetic energy in the solar wind, built up during acceleration, is rapidly and irreversibly transformed into heat as it passes through this shock front.

BOX 17.1. READERS' GUIDE

- This chapter relies heavily on Chap. 13 and on Secs. 16.2, 16.3, and 16.5.
- No subsequent chapters rely substantially on this one.

FIGURE 17.1 Complex pattern of shock fronts formed around a model aircraft in a wind tunnel with air moving 10% faster than the speed of sound (i.e., with Mach number $M = 1.1$). Image from NASA/Ames Imaging Library System.

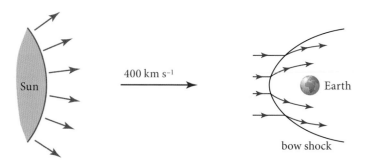

FIGURE 17.2 The supersonic solar wind forms a type of shock front known as a bow shock when it passes by a planet. Earth is ~ 200 solar radii from the Sun.

> **BOX 17.2. MOVIES RELEVANT TO THIS CHAPTER**
>
> We strongly recommend that the reader view the following movies dealing with compressible and supersonic flows:
>
> - Rouse (1963f)—covers most of this chapter except Riemann invariants.
> - Coles (1965)—focuses on shock fronts and quasi-1-dimensional flow through throats.
> - Bryson (1964)—includes segments on shock fronts and hydraulic jumps.

In this chapter, we study some properties of supersonic flows. After restating the basic equations of compressible fluid dynamics (Sec. 17.2), we analyze three important, simple cases: stationary, quasi-1-dimensional flow (Sec. 17.3); time-dependent, 1-dimensional flow (Sec. 17.4); and normal adiabatic shock fronts (Sec. 17.5). In these sections, we apply the results of our analyses to some contemporary examples, including the Space Shuttle (Box 17.4); rocket engines; shock tubes; and the Mach cone, N-wave, and sonic booms produced by supersonic projectiles and aircraft. In Sec. 17.6, we develop similarity-solution techniques for supersonic flows and apply them to supernovae, underwater depth charges, and nuclear-bomb explosions in Earth's atmosphere.

As in our previous fluid-dynamics chapters, we strongly encourage readers to view relevant movies in parallel with reading this chapter. See Box 17.2.

17.2 Equations of Compressible Flow

In Chap. 13, we derived the equations of fluid dynamics, allowing for compressibility. We expressed them as laws of mass conservation [Eq. (13.29)], momentum conservation [$\partial(\rho \mathbf{v})/\partial t + \nabla \cdot \mathbf{T} = 0$, with $\mathbf{T}$ as given in Table 13.3], energy conservation [$\partial U/\partial t + \nabla \cdot \mathbf{F} = 0$, with U and $\mathbf{F}$ as given in Table 13.3], and also an evolution law for entropy [Eq. (13.76)]. When, as in this chapter, heat conduction is negligible ($\kappa \to 0$) and the gravitational field is a time-independent, external one (not generated by the flowing fluid), these equations become

equations of compressible flow

$$\boxed{\frac{\partial \rho}{\partial t} + \nabla \cdot (\rho \mathbf{v}) = 0,} \qquad (17.1a)$$

mass conservation

$$\boxed{\frac{\partial(\rho \mathbf{v})}{\partial t} + \nabla \cdot (P\mathbf{g} + \rho \mathbf{v} \otimes \mathbf{v} - 2\eta \boldsymbol{\sigma} - \zeta \theta \mathbf{g}) = \rho \mathbf{g},} \qquad (17.1b)$$

momentum conservation

energy conservation
$$\frac{\partial}{\partial t}\left[\left(\tfrac{1}{2}v^2 + u + \Phi\right)\rho\right] + \nabla \cdot \left[\left(\tfrac{1}{2}v^2 + h + \Phi\right)\rho\mathbf{v} - 2\eta\boldsymbol{\sigma}\cdot\mathbf{v} - \zeta\theta\mathbf{v}\right] = 0, \quad (17.1c)$$

entropy evolution
$$\frac{\partial(\rho s)}{\partial t} + \nabla \cdot (\rho s \mathbf{v}) = \frac{1}{T}\left(2\eta\boldsymbol{\sigma}:\boldsymbol{\sigma} + \zeta\theta^2\right). \quad (17.1d)$$

Here $\boldsymbol{\sigma}:\boldsymbol{\sigma}$ is index-free notation for $\sigma_{ij}\sigma_{ij}$.

Some comments are in order. Equation (17.1a) is the complete mass-conservation equation (continuity equation), assuming that matter is neither added to nor removed from the flow, for example, no electron-positron pair creation. Equation (17.1b) expresses the conservation of momentum allowing for one external force, gravity. Other external forces (e.g., electromagnetic) can be added. Equation (17.1c), expressing energy conservation, includes a viscous contribution to the energy flux. If there are sources or sinks of fluid energy, then these must be included on the right-hand side of this equation. Possible sources of energy include chemical or nuclear reactions; possible energy sinks include cooling by emission of radiation. Equation (17.1d) expresses the evolution of entropy and will also need modification if there are additional contributions to the energy equation. The right-hand side of this equation is the rate of increase of entropy due to viscous heating. This equation is not independent of the preceding equations and the laws of thermodynamics, but it is often convenient to use. In particular, one often uses it (together with the first law of thermodynamics) in place of energy conservation (17.1c).

These equations must be supplemented with an equation of state in the form

equation of state

$P(\rho, T)$ or $P(\rho, s)$. For simplicity, we often focus on an ideal gas (one with $P \propto \rho k_B T$) that undergoes adiabatic evolution with constant specific-heat ratio (adiabatic index γ; Ex. 5.4), so the equation of state has the simple polytropic form (Box 13.2)

adiabatic index

polytropic equation of state
$$P = K(s)\rho^\gamma. \quad (17.2a)$$

Here $K(s)$ is a function of the entropy per unit mass s and is thus constant during adiabatic evolution, but it will change across shocks, because the entropy increases in a shock (Sec. 17.5). The value of γ depends on the number of thermalized internal degrees of freedom of the gas's constituent particles (Ex. 17.1). For a gas of free particles (e.g., fully ionized hydrogen), the value is $\gamma = 5/3$; for Earth's atmosphere at temperatures between about 10 K and 400 K, it is $\gamma = 7/5 = 1.4$ (Ex. 17.1).

polytropic relations

For a polytropic gas with $P = K(s)\rho^\gamma$, we can integrate the first law of thermodynamics (Box 13.2) to obtain a formula for the internal energy per unit mass:

$$u = \frac{P}{(\gamma - 1)\rho}, \quad (17.2b)$$

where we have assumed that the internal energy vanishes as the temperature $T \to 0$ and thence $P \to 0$. It will prove convenient to express the density ρ, the internal

Chapter 17. Compressible and Supersonic Flow

energy per unit mass u, and the enthalpy per unit mass h in terms of the sound speed:

$$C = \sqrt{\left(\frac{\partial P}{\partial \rho}\right)_S} = \sqrt{\frac{\gamma P}{\rho}} \qquad (17.2c)$$

[Eq. (16.48)]. A little algebra gives

$$\rho = \left(\frac{C^2}{\gamma K}\right)^{1/(\gamma-1)}, \quad u = \frac{C^2}{\gamma(\gamma-1)}, \quad h = u + \frac{P}{\rho} = \frac{C^2}{\gamma-1} = \frac{\gamma P}{(\gamma-1)\rho}. \qquad (17.2d)$$

EXERCISES

Exercise 17.1 **Example: Values of γ*

Consider an ideal gas consisting of several different particle species (e.g., diatomic oxygen molecules and nitrogen molecules in the case of Earth's atmosphere). Consider a sample of this gas with volume V, containing N_A particles of various species A, all in thermodynamic equilibrium at a temperature T sufficiently low that we can ignore the effects of special relativity. Let species A have ν_A internal degrees of freedom with the hamiltonian quadratic in their generalized coordinates (e.g., rotation and vibration), and assume that those degrees of freedom are sufficiently thermally excited to have reached energy equipartition. Then the equipartition theorem (Sec. 4.4.4) dictates that each such particle has $\frac{3}{2}k_B T$ of translational energy plus $\frac{1}{2}\nu_A k_B T$ of internal energy, and because the gas is ideal, each particle contributes $k_B T/V$ to the pressure. Correspondingly, the sample's total energy E and pressure P are

$$E = \sum_A \left(\frac{3}{2} + \frac{\nu_A}{2}\right) N_A k_B T, \quad P = \frac{1}{V} \sum_A N_A k_B T. \qquad (17.3a)$$

(a) Use the laws of thermodynamics to show that the specific heats at fixed volume and pressure are

$$C_V \equiv \left(T\frac{\partial S}{\partial T}\right)_{V,N_A} = \frac{E}{T} = \sum_A \left(\frac{3}{2} + \frac{\nu_A}{2}\right) N_A k_B,$$

$$C_P = \left(T\frac{\partial S}{\partial T}\right)_{P,N_A} = C_V + \frac{PV}{T}, \qquad (17.3b)$$

so the ratio of specific heats is

$$\gamma = \frac{C_P}{C_V} = 1 + \frac{\sum_A N_A}{\sum_A N_A \left(\frac{3}{2} + \frac{\nu_A}{2}\right)} \qquad (17.3c)$$

(cf. Ex. 5.4).

(b) If there are no thermalized internal degrees of freedom, $\nu_A = 0$ (e.g., for a fully ionized, nonrelativistic gas), then $\gamma = 5/3$. For Earth's atmosphere, at temperatures between about 10 K and 400 K, the rotational degrees of freedom of the O_2 and N_2 molecules are thermally excited, but the temperature is too low to excite their vibrational degrees of freedom. Explain why this means that $\nu_{O_2} = \nu_{N_2} = 2$,

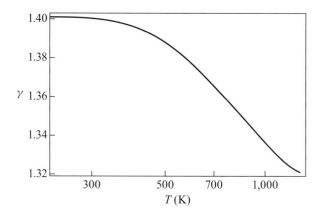

FIGURE 17.3 The ratio of specific heats γ for air as a function of temperature.

which implies $\gamma = 7/5 = 1.4$. [Hint: There are just two orthogonal axes around which the diatomic molecule can rotate.]

(c) Between about 1,300 K and roughly 10,000 K the vibrational degrees of freedom are thermalized, but the molecules have not dissociated substantially into individual atoms, nor have they become substantially ionized. Explain why this means that $\nu_{O_2} = \nu_{N_2} = 4$ in this temperature range, which implies $\gamma = 9/7 \simeq 1.29$. [Hint: An oscillator has kinetic energy and potential energy.]

(d) At roughly 10,000 K the two oxygen atoms in O_2 dissociate, the two nitrogen atoms in N_2 dissociate, and electrons begin to ionize. Explain why this drives γ up toward $5/3 \simeq 1.67$.

The actual value of γ as a function of temperature for the range 200 K to 1,300 K is shown in Fig. 17.3. Evidently, as stated, $\gamma = 1.4$ is a good approximation only up to about 400 K, and the transition toward $\gamma = 1.29$ occurs gradually between about 400 K and 1,400 K as the vibrational degrees of freedom gradually become thermalized and begin to obey the equipartition theorem (Sec. 4.4.4).

17.3 Stationary, Irrotational, Quasi-1-Dimensional Flow

17.3.1 Basic Equations; Transition from Subsonic to Supersonic Flow

In their full generality, the fluid dynamic equations (17.1) are quite unwieldy. To demonstrate some of the novel features of supersonic flow, we proceed as in earlier chapters: we specialize to a simple type of flow in which the physical effects of interest are strong, and extraneous effects are negligible.

In particular, in this section, we seek insight into smooth transitions between subsonic and supersonic flow by restricting ourselves to a stationary ($\partial/\partial t = 0$),

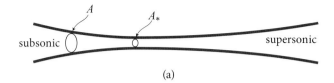

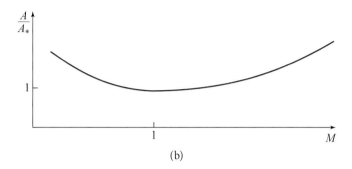

FIGURE 17.4 Stationary, transonic flow in a converging and then diverging streamtube. (a) The streamtube. (b) The flow's Mach number $M = v/C$ (horizontal axis) as a function of the streamtube's area A (vertical axis). The flow is subsonic to the left of the streamtube's throat $A = A_*$, sonic at the throat, and supersonic to the right.

irrotational ($\nabla \times \mathbf{v} = 0$) flow in which gravity and viscosity are negligible ($\Phi = \mathbf{g} = \eta = \zeta = 0$), as are various effects not included in our general equations: chemical reactions, thermal conductivity, and radiative losses. (We explore some effects of gravity in Ex. 17.5.) The vanishing viscosity implies [from the entropy evolution equation (17.1d)] that the entropy per baryon s is constant along each flow line. We assume that s is the same on all flow lines, so the flow is fully isentropic (s is constant everywhere), and the pressure $P = P(\rho, s)$ can thus be regarded as a function only of the density: $P = P(\rho)$. When we need a specific form for $P(\rho)$, we will use the polytropic form $P = K(s)\rho^\gamma$ for an ideal gas with constant specific-heat ratio γ [Eqs. (17.2); Ex. 17.1], but much of our analysis is done for a general isentropic $P(\rho)$. We make one further approximation: the flow is almost 1-dimensional. In other words, the velocity vectors all make small angles with one another in the region of interest.

These drastic simplifications are actually appropriate for many cases of practical interest. Granted these simplifications, we can consider a narrow bundle of streamlines—which we call a *streamtube*—and introduce as a tool for analysis its cross sectional area A normal to the flow (Fig. 17.4a).

As the flow is stationary, the equation of mass conservation (17.1a) states that the rate $\dot{m}$ at which mass passes through the streamtube's cross section must be independent of position along the tube:

$$\rho v A = \dot{m} = \text{const}; \qquad (17.4a)$$

here v is the speed of the fluid in the streamtube. Rewriting this in differential form, we obtain

$$\frac{dA}{A} + \frac{d\rho}{\rho} + \frac{dv}{v} = 0. \tag{17.4b}$$

Because the flow is stationary and inviscid, the law of energy conservation (17.1c) reduces to Bernoulli's theorem [Eqs. (13.51), (13.50)]:

$$h + \frac{1}{2}v^2 = h_1 = \text{const} \tag{17.4c}$$

along each streamline and thus along our narrow streamtube. Here h_1 is the specific enthalpy at a location where the flow velocity v vanishes (e.g., in chamber 1 of Fig. 17.5a below). Since the flow is adiabatic, we can use the first law of thermodynamics (Box 13.2) $dh = dP/\rho + Tds = dP/\rho = C^2 d\rho/\rho$ [where C is the speed of sound; Eq. (17.2c)] to write Eq. (17.4c) in the differential form

$$\frac{d\rho}{\rho} + \frac{vdv}{C^2} = 0. \tag{17.4d}$$

Finally and most importantly, we combine Eqs. (17.4b) and (17.4d) to obtain

$$\boxed{\frac{dv}{v} = \frac{dA/A}{M^2 - 1}, \quad \frac{d\rho}{\rho} = \frac{dA/A}{M^{-2} - 1},} \tag{17.5}$$

where

Mach number

$$\boxed{M \equiv v/C} \tag{17.6}$$

is the *Mach number*. This Mach number is an important dimensionless number that is used to characterize compressible flows. When the Mach number is less than 1, the flow is *subsonic*; when $M > 1$, it is *supersonic*. By contrast with the Reynolds, Rossby, and Ekman numbers, which are usually defined using a single set of (characteristic) values of the flow parameters (V, ν, Ω, and L) and thus have a single value for any given flow, the Mach number by convention is defined at each point in the flow and thus is a flow variable, $M(\mathbf{x})$, similar to $v(\mathbf{x})$ and $\rho(\mathbf{x})$.

subsonic and supersonic flow

properties of subsonic and supersonic flow

Equations (17.5) make remarkable predictions, which we illustrate in Fig. 17.4 for a particular flow called "transonic":

1. The only locations along a streamtube at which M can be unity ($v = C$) are those where A is an extremum—for example, for the streamtube in Fig. 17.4, the minimum $A = A_*$ (the tube's throat).

2. At points along a streamtube where the flow is subsonic, $M < 1$ (left side of the streamtube in Fig. 17.4), v increases when A decreases, in accord with everyday experience.

3. At points where the flow is supersonic, $M > 1$ (right side of Fig. 17.4), v increases when A increases—just the opposite of everyday experience.

These conclusions are useful when analyzing stationary, high-speed flows.

17.3.2 Setting up a Stationary, Transonic Flow

At this point the reader may wonder whether it is easy to set up a transonic flow in which the speed of the fluid changes continuously from subsonic to supersonic, as in Fig. 17.4. The answer is quite illuminating. We can illustrate the answer using two chambers maintained at different pressures, P_1 and P_2, and connected through a narrow channel, along which the cross sectional area passes smoothly through a minimum $A = A_*$, the channel's throat (Fig. 17.5a). When $P_2 = P_1$, no flow occurs between the two chambers. When we decrease P_2 slightly below P_1, a slow subsonic flow moves through the channel (curves 1 in Fig. 17.5b,c). As we decrease P_2 further, there comes a point ($P = P_2^{\rm crit}$) at which the flow becomes transonic at the channel's throat $A = A_*$ (curves 2). For all pressures $P_2 < P_2^{\rm crit}$, the flow is also transonic at the throat and has a universal form to the left of and near the throat, independent of the value of P_2 (curves 2)—including a universal value $\dot{m}_{\rm crit}$ for the rate of mass flow through the throat! This universal flow is supersonic to the right of the throat (curves 2b), but it must be brought to rest in chamber 2, since there is a hard wall at the chamber's end. How is it brought to rest? Through a shock front, where it is driven subsonic almost discontinuously (curves 3 and 4; see Sec. 17.5).

inducing transonic flow by increasing pressure difference

How, physically, is it possible for the transonic flow to have a universal form to the left of the shock? The key is that, in any supersonic region of the flow, disturbances are unable to propagate upstream, so the upstream fluid has no way of knowing what the pressure P_2 is in chamber 2. Although the flow to the left of the shock is universal, the location of the shock and the nature of the subsonic, post-shock flow are affected by P_2, since information can propagate upstream through that subsonic flow, from chamber 2 to the shock.

The reader might now begin to suspect that the throat, in the transonic case, is a special location. It is, and that location is known as a *critical point* of the stationary flow. From a mathematical point of view, critical points are singular points of the equations (17.4) and (17.5) of stationary flow. This singularity shows up in the solutions to the equations, as depicted in Fig. 17.5c. The universal solution that passes transonically through the critical point (solution 2) joins onto two different solutions to the right of the throat: solution 2a, which is supersonic, and solution 2b, which is subsonic. Which solution occurs in practice depends on conditions downstream. Other solutions that are arbitrarily near this universal solution (dashed curves in Fig. 17.5c) are either double valued and consequently unphysical, or are everywhere subsonic or everywhere supersonic (in the absence of shocks); see Box 17.3 and Ex. 17.2.

critical point of a stationary flow

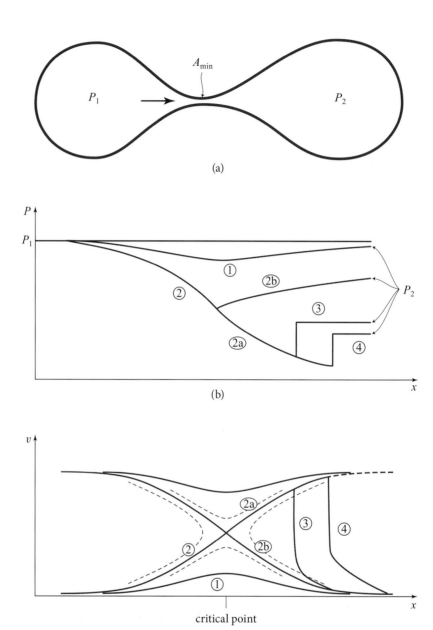

FIGURE 17.5 Stationary flow through a channel between two chambers maintained at different pressures P_1 and P_2. (a) Setup of the chambers. When the pressure difference $P_1 - P_2$ is large enough [see curves in panel (b)], the flow is subsonic to the left of the channel's throat and supersonic to the right [see curves in panel (c)]. As it nears or enters the second chamber, the supersonic flow must encounter a strong shock, where it decelerates abruptly to subsonic speed [panels (b) and (c)]. The forms of the various velocity profiles $v(x)$ in panel (c) are explained in Box 17.3.

BOX 17.3. VELOCITY PROFILES FOR 1-DIMENSIONAL FLOW BETWEEN CHAMBERS

Consider the adiabatic, stationary flow of an isentropic, polytropic fluid $P = K\rho^\gamma$ between the two chambers shown in Fig. 17.5. Describe the channel between chambers by its cross sectional area $A(x)$ as a function of distance x, and describe the flow by the fluid's velocity $v(x)$ and its sound speed $C(x)$. There are two coupled algebraic equations for $v(x)$ and $C(x)$: mass conservation $\rho v A = \dot{m}$ [Eq. (17.4a)] and the Bernoulli theorem $h + \frac{1}{2}v^2 = h_1$ [Eq. (17.4c)], which, for our polytropic fluid, become [see Eqs. (17.2d)]:

$$C^{2/(\gamma-1)} v = (\gamma K)^{1/(\gamma-1)} \dot{m}/A, \qquad \frac{C^2}{\gamma-1} + \frac{v^2}{2} = \frac{C_1^2}{\gamma-1}. \qquad (1)$$

These equations are graphed in diagrams below for three different mass flow rates $\dot{m}$. Mass conservation [the first of Eqs. (1)] is a set of generalized hyperbolas, one for each value of the channel's area $A_* < A_a < A_b < A_c$. The Bernoulli theorem [the second of Eqs. (1)] is a single ellipse. On a chosen diagram (for a chosen $\dot{m}$), the dot at the intersection of the ellipse with a hyperbola tells us the flow velocity v and speed of sound C at each of the two points in the channel where the area A has the hyperbola's value.

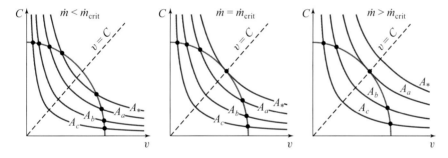

There is a critical mass-flow rate $\dot{m}_{\rm crit}$ (central diagram above), such that the hyperbola $A = A_*$ (the channel's throat) is tangent to the ellipse at the point $v = C$, so the flow is Mach 1 ($v = C$). For this $\dot{m}$, the sequence of dots along the ellipse, moving from lower right to upper left, represents the transonic flow, which begins as subsonic and becomes supersonic (upward swooping solid curve in the drawing below); and the same sequence of dots, moving in the opposite direction from upper left to lower right, represents a flow that begins as supersonic and smoothly transitions to subsonic (downward swooping solid curve, below). When $\dot{m} < \dot{m}_{\rm crit}$ (left diagram above; top and bottom quadrants below), the sequence of dots beginning at lower

(continued)

BOX 17.3. (continued)

right reaches the throat $A = A_*$ at a subsonic velocity, so the solution climbs up along the ellipse to that point and then descends back down, mapping out a fully subsonic solution $v(x)$ below—and similarly for the dots on the upper branch of the ellipse, which map out a fully supersonic solution below. When $\dot{m} > \dot{M}_{\rm crit}$ (right diagram above), the dots map out curves in the left and right quadrants below that never reach the sonic point and are double valued for $v(x)$—and are thus unphysical. Therefore, the mass flow rate $\dot{m}$ can never exceed $\dot{m}_{\rm crit}$.

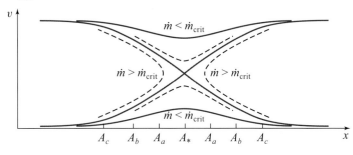

The existence of critical points is a price we must pay, mathematically, for not allowing our equations to be time dependent. If we were to solve the time-dependent equations (which would then be partial differential equations), we would find that they change from elliptic to hyperbolic as the flow passes through a critical point.

properties of critical points

From a physical point of view, critical points are the places where a sound wave propagating upstream remains at rest in the flow. They are therefore the one type of place from which time-dependent transients, associated with setting up the flow in the first place, cannot decay away (if the equations are dissipation-free, i.e., inviscid). Thus, even the time-dependent equations can display peculiar behaviors at a critical point. However, when dissipation, for example due to viscosity, is introduced, these peculiarities get smeared out.

dissipation smears out critical points

EXERCISES

Exercise 17.2 *Problem: Explicit Solution for Flow between Chambers When $\gamma = 3$*
For $\gamma = 3$ and for a channel with $A = A_*(1 + x^2)$, solve the flow equations (1) of Box 17.3 analytically and explicitly for $v(x)$, and verify that the solutions have the qualitative forms depicted in the last figure of Box 17.3.

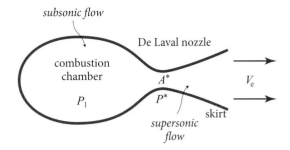

FIGURE 17.6 Schematic illustration of a rocket engine. Note the skirt, which increases the thrust produced by the escaping exhaust gases.

17.3.3 Rocket Engines

We have shown that, to push a quasi-1-dimensional flow from subsonic to supersonic, one must send it through a throat. This result is exploited in the design of rocket engines and jet engines.

In a rocket engine, hot gas is produced by controlled burning of fuel in a large chamber, and the gas then escapes through a *converging-diverging* (also known as a *De Laval*[1]) nozzle, as shown in Fig. 17.6. The nozzle is designed with a skirt so the flow becomes supersonic smoothly when it passes through the nozzle's throat.

To analyze this flow in some detail, let us approximate it as precisely steady, isentropic, and quasi-1-dimensional, and the gas as ideal and inviscid with a constant ratio of specific heats γ. In this case, the enthalpy is $h = C^2/(\gamma - 1)$ [Eqs. (17.2d)], so Bernoulli's theorem (17.4c) reduces to

$$\frac{C^2}{(\gamma - 1)} + \frac{1}{2}v^2 = \frac{C_1^2}{(\gamma - 1)}. \tag{17.7}$$

Here C is the sound speed in the flow and C_1 is the *stagnation* sound speed (i.e., the sound speed evaluated in the rocket chamber where $v = 0$). Dividing this Bernoulli theorem by C^2 and manipulating, we learn how the sound speed varies with Mach number $M = v/C$:

$$C = C_1 \left[1 + \frac{\gamma - 1}{2} M^2 \right]^{-1/2}. \tag{17.8}$$

From mass conservation [Eq. (17.4a)], we know that the cross sectional area A varies as $A \propto \rho^{-1} v^{-1} \propto \rho^{-1} M^{-1} C^{-1} \propto M^{-1} C^{(\gamma+1)/(1-\gamma)}$, where we have used $\rho \propto C^{2/(\gamma-1)}$ [Eqs. (17.2d)]. Combining with Eq. (17.8), and noting that $M = 1$ where $A = A_*$ (i.e.,

---

1. First used in a steam turbine in 1882 by the Swedish engineer Gustaf de Laval. De Laval is most famous for developing a device to separate cream from milk, centrifugally!

how Mach number varies with area

the flow is sonic at the throat), we find that

$$\frac{A}{A_*} = \frac{1}{M}\left[\frac{2}{\gamma+1} + \left(\frac{\gamma-1}{\gamma+1}\right)M^2\right]^{\frac{(\gamma+1)}{2(\gamma-1)}}. \quad (17.9)$$

The pressure P_* at the throat can be deduced from $P \propto \rho^\gamma \propto C^{2\gamma/(\gamma-1)}$ [Eqs. (17.2a) and (17.2d)] together with Eq. (17.8) with $M = 0$ and $P = P_1 =$ (stagnation pressure) in the chamber and $M = 1$ at the throat:

$$P_* = P_1 \left(\frac{2}{\gamma+1}\right)^{\frac{\gamma}{\gamma-1}}. \quad (17.10)$$

We use these formulas in Box 17.4 and Ex. 17.3 to evaluate, numerically, some features of the Space Shuttle and its rocket engines.

Bernoulli's theorem is a statement that the fluid's energy is conserved along a streamtube. (For conceptual simplicity we regard the entire interior of the nozzle as a single streamtube.) By contrast with energy, the fluid's momentum is not conserved, since it pushes against the nozzle wall as it flows. As the subsonic flow accelerates down the nozzle's converging region, the area of its streamtube diminishes, and the momentum flowing per second in the streamtube, $(P + \rho v^2)A$, decreases; momentum is being transferred to the nozzle wall. If the rocket did not have a skirt but instead opened up completely to the outside world at its throat, the rocket thrust would be

momentum flow inside nozzle

$$T_* = (\rho_* v_*^2 + P_*)A_* = (\gamma + 1)P_* A_*. \quad (17.11)$$

This is much less than if momentum had been conserved along the subsonic, accelerating streamtubes.

role of a rocket engine's skirt

Much of the "lost" momentum is regained, and the thrust is made significantly larger than T_*, by the force of the skirt on the stream tube in the diverging part of the nozzle (Fig. 17.6). The nozzle's skirt keeps the flow quasi-1-dimensional well beyond the throat, driving it to be more and more strongly supersonic. In this accelerating, supersonic flow the tube's rate of momentum flow $(\rho v^2 + P)A$ increases downstream, with a compensating increase of the rocket's forward thrust. This skirt-induced force accounts for a significant fraction of the thrust of a well-designed rocket engine.

matching exit pressure to external pressure

Rockets work most efficiently when the exit pressure of the gas, as it leaves the base of the skirt, matches the external pressure in the surrounding air. When the pressure in the exhaust is larger than the external pressure, the flow is termed *underexpanded* and a pulse of low pressure, known as a *rarefaction*, will be driven into the escaping gases, causing them to expand and increasing their speed. However, the exhaust will now be pushing on the surrounding air, rather than on the rocket. More thrust could have been exerted on the rocket if the flow had not been underexpanded. By contrast, when the exhaust has a smaller pressure than the surrounding air (i.e., is *overexpanded*), shock fronts will form near the exit of the nozzle, affecting the fluid flow and sometimes causing separation of the flow from the nozzle's walls. It is important that the nozzle's skirt be shaped so that the exit flow is neither seriously over- nor underexpanded.

BOX 17.4. SPACE SHUTTLE

NASA's (now retired) Space Shuttle provides many nice examples of the behavior of supersonic flows. At launch, the shuttle and fuel had a mass $\sim 2 \times 10^6$ kg. The maximum thrust, $T \sim 3 \times 10^7$ N, occurred at lift-off and gave the rocket an initial acceleration relative to the ground of $\sim 0.5g$. This increased to $\sim 3g$ as the fuel was burned and the total mass diminished. Most of the thrust was produced by two solid-fuel boosters that burned fuel at a combined rate of $\dot{m} \sim 10{,}000$ kg s^{-1} over a 2-min period. Their combined thrust was $T \sim 2 \times 10^7$ N averaged over the 2 minutes, from which we can estimate the speed of the escaping gases as they left the nozzles' skirts. Assuming this speed was quite supersonic (so $P_e \ll \rho_e v_e^2$), we estimate that $v_e \sim T/\dot{m} \sim 2$ km s^{-1}. The combined exit areas of the two skirts was $A_e \sim 20$ m^2, roughly four times the combined throat area, A_*. Using Eq. (17.9) with $\gamma \sim 1.29$, we deduce that the exit Mach number was $M_e \sim 3$.

From $T \sim \rho_e v_e^2 A_e$ and $P_e = C_e^2 \rho_e / \gamma$, we deduce the exit pressure, $P_e \sim T/(\gamma M_e^2 A_e) \sim 8 \times 10^4$ N m^{-2}, about atmospheric. The stagnation pressure in the combustion region was [combine Eqs. (17.2a), (17.2d), and (17.8)]

$$P_1 \sim P_e \left[1 + \frac{(\gamma - 1) M_e^2}{2} \right]^{\frac{\gamma}{\gamma - 1}} \sim 35 \text{ atmospheres.} \tag{1}$$

Of course, the actual operation was far more complex than this. For example, to optimize the final altitude, one must allow for the decreasing mass and atmospheric pressure as well as the 2-dimensional gas flow through the nozzle.

The Space Shuttle can also be used to illustrate the properties of shock waves (Sec. 17.5). When the shuttle reentered the atmosphere, it was highly supersonic, and therefore was preceded by a strong shock front that heated the onrushing air and consequently heated the shuttle. The shuttle continued moving supersonically down to an altitude of 15 km, and until this time it created a shock-front pattern that could be heard on the ground as a sonic boom. The maximum heating rate occurred at 70 km. Here, the shuttle moved at $V \sim 7$ km s^{-1}, and the sound speed is about 280 m s^{-1}, giving a Mach number of 25. For the specific-heat ratio $\gamma \sim 1.5$ and mean molecular weight $\mu \sim 10$ appropriate to dissociated air, the strong-shock Rankine-Hugoniot relations (17.37) (see Sec. 17.5.2), together with $P = [\rho/(\mu m_p)] k_B T$ and $C^2 = \gamma P/\rho$, predict a post-shock temperature of

$$T \sim \frac{2(\gamma - 1) \mu m_p V^2}{(\gamma + 1)^2 k_B} \sim 9{,}000 \text{ K.} \tag{2}$$

Exposure to gas at this high temperature heated the shuttle's nose to $\sim 1{,}800$ K.

(continued)

BOX 17.4. (continued)

There is a second, well-known consequence of this high temperature: it is sufficient to ionize the air partially as well as dissociate it. As a result, during reentry the shuttle was surrounded by a sheath of plasma, which, as we shall discover in Chap. 19, prevented radio communication. The blackout lasted for about 12 minutes.

EXERCISES

Exercise 17.3 *Problem: Space Shuttle's Solid-Fuel Boosters*
Use the rough figures in Box 17.4 to estimate the energy released per unit mass in burning the fuel. Does your answer seem reasonable?

Exercise 17.4 *Problem: Relativistic 1-Dimensional Flow* T2
Use the development of relativistic gas dynamics in Sec. 13.8.2 to show that the cross sectional area of a relativistic 1-dimensional flow tube is also minimized when the flow is transonic. Assume that the equation of state is $P = \frac{1}{3}\rho c^2$. For details see Blandford and Rees (1974).

Exercise 17.5 **Example: Adiabatic, Spherical Accretion of Gas onto a Black Hole or Neutron Star*
Consider a black hole or neutron star with mass $\mathcal{M}$ at rest in interstellar gas that has constant ratio of specific heats γ. In this exercise you will derive some features of the adiabatic, spherical accretion of the gas onto the hole or star, a problem first solved by Bondi (1952b). This exercise shows how gravity can play a role analogous to a De Laval nozzle: it can trigger a transition of the flow from subsonic to supersonic. Although, near the black hole or neutron star, spacetime is significantly curved and the flow becomes relativistic, we shall confine ourselves to a Newtonian treatment.

(a) Let ρ_∞ and C_∞ be the density and sound speed in the gas far from the hole (at radius $r = \infty$). Use dimensional analysis to estimate the rate of accretion of mass $\dot{\mathcal{M}}$ onto the star or hole in terms of the parameters of the system: $\mathcal{M}, \gamma, \rho_\infty, C_\infty$, and Newton's gravitation constant G. [Hint: Dimensional considerations alone cannot give the answer. Why? Augment your dimensional considerations by a knowledge of how the answer should scale with one of the parameters (e.g., the density ρ_∞).]

(b) Give a simple physical argument, devoid of dimensional considerations, that produces the same answer for $\dot{\mathcal{M}}$, to within a multiplicative factor of order unity, as you deduced in part (a).

(c) Because the neutron star and black hole are both very compact with intense gravity near their surfaces, the inflowing gas is guaranteed to accelerate to supersonic speeds as it falls in. Explain why the speed will remain supersonic in the case of the hole, but must transition through a shock to subsonic flow near the surface of the neutron star. If the star has the same mass $\mathcal{M}$ as the hole, will the details of its accretion flow $[\rho(r), C(r), v(r)]$ be the same as or different from those for the hole, outside the star's shock? Will the mass accretion rates $\dot{M}$ be the same or different? Justify your answers physically.

(d) By combining the Euler equation for $v(r)$ with the equation of mass conservation, $\dot{M} = 4\pi r^2 \rho v$, and with the sound-speed equation $C^2 = (\partial P/\partial \rho)_s$, show that

$$(v^2 - C^2)\frac{1}{\rho}\frac{d\rho}{dr} = \frac{GM}{r^2} - \frac{2v^2}{r}, \qquad (17.12)$$

and the flow speed v_s, sound speed C_s, and radius r_s at the *sonic point* (the transition from subsonic to supersonic; the flow's critical point) are related by

$$v_s^2 = C_s^2 = \frac{GM}{2r_s}. \qquad (17.13)$$

(e) By combining with the Bernoulli equation (with the effects of gravity included), deduce that the sound speed at the sonic point is related to that at infinity by

$$C_s^2 = \frac{2C_\infty^2}{5 - 3\gamma} \qquad (17.14)$$

and that the radius of the sonic point is

$$r_s = \frac{(5-3\gamma)}{4}\frac{GM}{C_\infty^2}. \qquad (17.15)$$

Thence also deduce a precise value for the mass accretion rate $\dot{M}$ in terms of the parameters of the problem. Compare with your estimate of $\dot{M}$ in parts (a) and (b). [Comment: For $\gamma = 5/3$, which is the value for hot, ionized gas, this analysis places the sonic point at an arbitrarily small radius. In this limiting case, (i) general relativistic effects strengthen the gravitational field (Sec. 26.2), thereby moving the sonic point well outside the star or hole, and (ii) the answer for $\dot{M}$ has a finite value close to the general relativistic prediction.]

(f) Much of the interstellar medium is hot and ionized, with a density of about 1 proton per cubic centimeter and temperature of about 10^4 K. In such a medium, what is the mass accretion rate onto a 10-solar-mass hole, and approximately how long does it take for the hole's mass to double?

17.4 1-Dimensional, Time-Dependent Flow

17.4.1 Riemann Invariants

Let us turn now to time-dependent flows. Again we confine our attention to the simplest situation that illustrates the physics—in this case, truly 1-dimensional motion of an isentropic fluid in the absence of viscosity, thermal conductivity, and gravity, so

1-dimensional flow of an isentropic fluid without viscosity, thermal conductivity, or gravity the flow is adiabatic as well as isentropic (entropy constant in time as well as space). The motion of the gas in such a flow is described by the equation of continuity (17.1a) and the Euler equation (17.1b) specialized to 1 dimension:

$$\frac{d\rho}{dt} = -\rho \frac{\partial v}{\partial x}, \quad \frac{dv}{dt} = -\frac{1}{\rho}\frac{\partial P}{\partial x}, \tag{17.16}$$

where

$$\frac{d}{dt} = \frac{\partial}{\partial t} + v\frac{\partial}{\partial x} \tag{17.17}$$

is the convective (advective) time derivative—the time derivative moving with the fluid.

Given an isentropic equation of state $P = P(\rho)$ that relates the pressure to the density, these two nonlinear equations can be combined into a single second-order differential equation in the velocity. However, it is more illuminating to work with the first-order set. As the gas is isentropic, the density ρ and sound speed $C = (dP/d\rho)^{1/2}$ can both be regarded as functions of a single thermodynamic variable, which we choose to be the pressure.

Taking linear combinations of Eqs. (17.16), we obtain two partial differential equations:

$$\frac{\partial v}{\partial t} \pm \frac{1}{\rho C}\frac{\partial P}{\partial t} + (v \pm C)\left(\frac{\partial v}{\partial x} \pm \frac{1}{\rho C}\frac{\partial P}{\partial x}\right) = 0, \tag{17.18}$$

which together are equivalent to Eqs. (17.16). We can rewrite these equations in terms of *Riemann invariants*

Riemann invariants

$$\boxed{J_\pm \equiv v \pm \int \frac{dP}{\rho C}} \tag{17.19}$$

and *characteristic speeds*

characteristic speeds

$$\boxed{V_\pm \equiv v \pm C} \tag{17.20}$$

in the following way:

flow equations

$$\boxed{\left(\frac{\partial}{\partial t} + V_\pm \frac{\partial}{\partial x}\right) J_\pm = 0.} \tag{17.21}$$

Equation (17.21) tells us that the convective derivative of each Riemann invariant $J_\pm$ vanishes for an observer who moves, not with the fluid speed, but, instead, with the speed $V_\pm$. We say that each Riemann invariant is conserved along its characteristic (denoted by $\mathcal{C}_\pm$), which is a path through spacetime satisfying

Riemann invariant conserved along its characteristic

$$\boxed{\mathcal{C}_\pm: \quad \frac{dx}{dt} = v \pm C.} \tag{17.22}$$

Note that in these equations, both v and C are functions of x and t.

Chapter 17. Compressible and Supersonic Flow

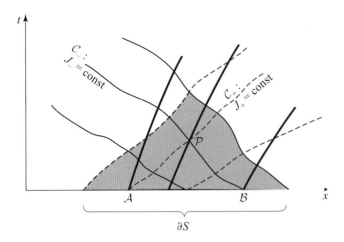

FIGURE 17.7 Spacetime diagram showing the characteristics (thin solid and dashed lines) for a 1-dimensional adiabatic flow of an isentropic gas. The paths of the fluid elements are shown as thick solid lines. Initial data are presumed to be specified over some interval ∂S of x at time $t = 0$. The Riemann invariant J_+ is constant along each characteristic $\mathcal{C}_+$ (thin dashed line) and thus at point $\mathcal{P}$ it has the same value, unchanged, as at point $\mathcal{A}$ in the initial data. Similarly J_- is invariant along each characteristic $\mathcal{C}_-$ (thin solid line) and thus at $\mathcal{P}$ it has the same value as at $\mathcal{B}$. The shaded area of spacetime is the domain of dependence S of ∂S.

The characteristics have a natural interpretation. They describe the motion of small disturbances traveling backward and forward relative to the fluid at the local sound speed. As seen in the fluid's local rest frame $v = 0$, two neighboring events in the flow, separated by a small time interval Δt and a space interval $\Delta x = +C\Delta t$—so that they lie on the same $\mathcal{C}_+$ characteristic—will have small velocity and pressure differences satisfying $\Delta v = -\Delta P/(\rho C)$ [as one can deduce from Eqs. (17.16) with $v = 0$, $d/dt = \partial/\partial t$, and $C^2 = dP/d\rho$]. Now, for a linear sound wave propagating along the positive x direction, Δv and ΔP will separately vanish. However, in a nonlinear wave, only the combination $\Delta J_+ = \Delta v + \Delta P/(\rho C)$ will vanish along $\mathcal{C}_+$. Integrating over a finite interval of time, we recover the constancy of J_+ along the characteristic $\mathcal{C}_+$ [Eq. (17.21)].

The Riemann invariants provide a general method for deriving the details of the flow from initial conditions. Suppose that the fluid velocity and the thermodynamic variables are specified over an interval of x, designated ∂S, at an initial time $t = 0$ (Fig. 17.7). This means that $J_\pm$ are also specified over this interval. We can then determine $J_\pm$ at any point $\mathcal{P}$ in the *domain of dependence* S of ∂S (i.e., at any point linked to ∂S by two characteristics $\mathcal{C}_\pm$) by simply propagating each of $J_\pm$ unchanged along its characteristic. From these values of $J_\pm$ at $\mathcal{P}$, we can solve algebraically for all the other flow variables (v, P, ρ, etc.) at $\mathcal{P}$. To learn the evolution outside the domain of dependence S, we must specify the initial conditions outside ∂S.

physical interpretation of characteristics

using characteristics and Riemann invariants to solve for details of flow in domain of dependence

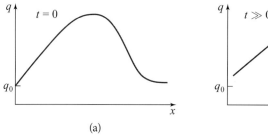 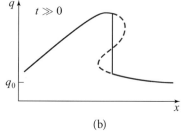

FIGURE 17.8 Evolution of a nonlinear sound wave. (a) The fluid at the crest of the wave moves faster than the fluid in the trough. (b) Mathematically, the flow eventually becomes triple-valued (dashed curve). Physically, a shock wave develops (vertical solid line).

In practice, we do not actually know the characteristics $\mathcal{C}_\pm$ until we have solved for the flow variables, so we must solve for the characteristics as part of the solution process. This means, in practice, that the solution involves algebraic manipulations of (i) the equation of state and the relations $J_\pm = v \pm \int dP/(\rho C)$, which give $J_\pm$ in terms of v and C; and (ii) the conservation laws that $J_\pm$ are constant along $\mathcal{C}_\pm$ (i.e., along curves $dx/dt = v \pm C$). These algebraic manipulations have the goal of deducing $C(x, t)$ and $v(x, t)$ from the initial conditions on ∂S. We exhibit a specific example in Sec. 17.4.2.

how a nonlinear sound wave evolves to form a shock wave

We can use Riemann invariants to understand qualitatively how a nonlinear sound wave evolves with time. If the wave propagates in the positive x direction into previously undisturbed fluid (fluid with $v = 0$), then the J_- invariant, propagating backward along $\mathcal{C}_-$, is constant everywhere, so $v = \int dP/(\rho C) + \text{const}$. Let us use $q \equiv \int dP/(\rho C)$ as our wave variable. For an ideal gas with a constant ratio of specific heats γ, we have $q = 2C/(\gamma - 1)$, so our oscillating wave variable is essentially the oscillating sound speed. Constancy of J_- then says that $v = q - q_0$, where q_0 is the stagnation value of q (i.e., the value of q in the undisturbed fluid in front of the wave).

Now, $J_+ = v + q$ is conserved on each rightward characteristic $\mathcal{C}_+$, and so both v and q are separately conserved on each $\mathcal{C}_+$. If we sketch a profile of the wave pulse as in Fig. 17.8 and measure its amplitude using the quantity q, then the relation $v = q - q_0$ requires that the fluid at the crest of the wave moves faster than the fluid in a trough. This causes the leading edge of the wave to steepen, a process we have already encountered in our discussion of shallow-water solitons (Fig. 16.4). Now, sound waves, by contrast with shallow-water waves (where dispersion counteracts the steepening), are nondispersive so the steepening will continue until $|dv/dx| \to \infty$ (Fig. 17.8). When the velocity gradient becomes sufficiently large, viscosity and dissipation become strong, producing an increase of entropy and a breakdown of our isentropic flow. This breakdown and entropy increase will occur in an extremely thin region—a shock wave, which we study in Sec. 17.5.

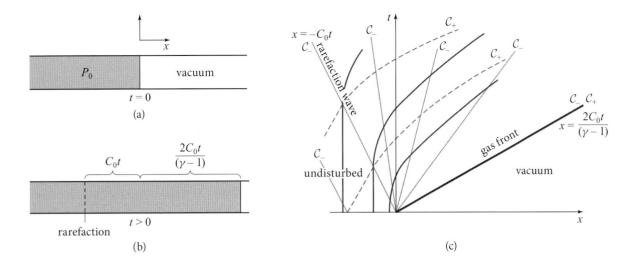

FIGURE 17.9 Shock tube. (a) At $t \leq 0$ gas is held at rest at high pressure P_0 in the left half of the tube. (b) At $t > 0$ the high-pressure gas moves rightward down the tube at high speed, and a rarefaction wave propagates leftward at the sound speed. (c) Space-time diagram showing the flow's characteristics (C_+: thin dashed lines; C_-: thin solid lines) and fluid paths (thick solid lines). To the left of the rarefaction wave, $x < -C_0 t$, the fluid is undisturbed. To the right of the gas front, $x > [2/(\gamma - 1)]C_0 t$, resides undisturbed (near) vacuum.

17.4.2 Shock Tube

We have shown how 1-dimensional isentropic flows can be completely analyzed by propagating the Riemann invariants along characteristics. Let us illustrate this in more detail by analyzing a shock tube, a laboratory device for creating supersonic flows and studying the behavior of shock waves. In a shock tube, high-pressure gas is retained at rest in the left half of a long tube by a thin membrane (Fig. 17.9a). At time $t = 0$, the membrane is ruptured by a laser beam, and the gas rushes into the tube's right half, which has usually been evacuated. Diagnostic photographs and velocity and pressure measurements are synchronized with the onset of the flow.

Let us idealize the operation of a shock tube by assuming, once more, that the gas is ideal with constant γ, so that $P \propto \rho^\gamma$. For times $t \leq 0$, we suppose that the gas has uniform density ρ_0 and pressure P_0 (and consequently, uniform sound speed C_0) at $x \leq 0$, and that $\rho = P = 0$ at $x \geq 0$. At time $t = 0$, the barrier is ruptured, and the gas flows toward positive x. The first Riemann invariant J_+ is conserved on C_+, which originates in the static gas, so it has the value

$$J_+ = v + \frac{2C}{\gamma - 1} = \frac{2C_0}{\gamma - 1}. \tag{17.23}$$

Note that in this case, the invariant is the same on all rightward characteristics (i.e., throughout the flow), so that

$$v = \frac{2(C_0 - C)}{\gamma - 1} \quad \text{everywhere.} \tag{17.24}$$

The second invariant is

$$J_- = v - \frac{2C}{\gamma - 1}. \tag{17.25}$$

Its constant values are not so easy to identify, because those characteristics $\mathcal{C}_-$ that travel through the perturbed flow all emerge from the origin, where v and C are indeterminate (cf. Fig. 17.9c). However, by combining Eq. (17.24) with Eq. (17.25), we deduce that v and C are separately constant on each characteristic $\mathcal{C}_-$. This enables us, trivially, to solve the differential equation $dx/dt = v - C$ for the leftward characteristics $\mathcal{C}_-$, obtaining

$$\mathcal{C}_-: \quad x = (v - C)t. \tag{17.26}$$

Here we have set the constant of integration equal to zero to obtain all the characteristics that propagate through the perturbed fluid. (For those in the unperturbed fluid, $v = 0$ and $C = C_0$, so $x = x_0 - C_0 t$, with $x_0 < 0$ the characteristic's initial location.)

flow in a shock tube

Now Eq. (17.26) is true on each characteristic in the perturbed fluid. Therefore, it is true throughout the perturbed fluid. We can then combine Eqs. (17.24) and (17.26) to solve for $v(x, t)$ and $C(x, t)$ throughout the perturbed fluid. That solution, together with the obvious solution (same as initial data) to the left and right of the perturbed fluid, is:

$$v = 0, \quad C = C_0 \quad \text{at } x < -C_0 t,$$

$$v = \frac{2}{\gamma + 1}\left(C_0 + \frac{x}{t}\right), \quad C = \frac{2C_0}{\gamma + 1} - \left(\frac{\gamma - 1}{\gamma + 1}\right)\frac{x}{t} \quad \text{at } -C_0 t < x < \frac{2C_0}{\gamma - 1}t,$$

$$\text{vacuum prevails at } x > \frac{2C_0}{\gamma - 1}t. \tag{17.27}$$

rarefaction wave as flow initiator

In this solution, notice that the gas at $x < 0$ remains at rest until a *rarefaction wave* from the origin reaches it. Thereafter, it is accelerated rightward by the local pressure gradient, and as it accelerates it expands and cools, so its sound speed C goes down; asymptotically it reaches zero temperature, as exhibited by $C = 0$ and an asymptotic speed $v = 2C_0/(\gamma - 1)$ [cf. Eq. (17.23)]; see Fig. 17.9b,c. In the expansion, the internal random velocity of the gas molecules is transformed into an ordered velocity, just as in a rocket's exhaust. However, the total energy per unit mass in the stationary gas is $u = C_0^2/[\gamma(\gamma - 1)]$ [Eq. (17.2d)], which is less than the asymptotic kinetic energy per unit mass of $2C_0^2/(\gamma - 1)^2$. The additional energy has come from the gas in the rarefaction wave, which is pushing the asymptotic gas rightward.

In the more realistic case, where there initially is some low-density gas in the evacuated half of the tube, the expanding driver gas creates a strong shock as it plows into the low-density gas; hence the name "shock tube." In the next section we explore the structure of this and other shock fronts.

Exercise 17.6 *Problem: Fluid Paths in Free Expansion*

We have computed the velocity field for a freely expanding gas in 1 dimension, Eqs. (17.27). Use this result to show that the path of an individual fluid element, which begins at $x = x_0 < 0$, is

$$x = \frac{2C_0 t}{\gamma - 1} + \left(\frac{\gamma + 1}{\gamma - 1}\right) x_0 \left(\frac{-C_0 t}{x_0}\right)^{\frac{2}{\gamma+1}} \quad \text{at} \quad 0 < -\frac{x_0}{C_0} < t.$$

Exercise 17.7 *Problem: Riemann Invariants for Shallow-Water Flow; Breaking of a Dam*

Consider the 1-dimensional flow of shallow water in a straight, narrow channel, neglecting dispersion and boundary layers. The equations governing the flow, as derived and discussed in Box 16.3 and Eqs. (16.23), are

$$\frac{\partial h}{\partial t} + \frac{\partial (hv)}{\partial x} = 0, \quad \frac{\partial v}{\partial t} + v\frac{\partial v}{\partial x} + g\frac{\partial h}{\partial x} = 0. \tag{17.28}$$

Here $h(x, t)$ is the height of the water, and $v(x, t)$ is its depth-independent velocity.

(a) Find two Riemann invariants $J_\pm$ for these equations, and find two conservation laws for these $J_\pm$ that are equivalent to the shallow-water equations (17.28).

(b) Use these Riemann invariants to demonstrate that shallow-water waves steepen in the manner depicted in Fig. 16.4, a manner analogous to the peaking of the nonlinear sound wave in Fig. 17.8.

(c) Use these Riemann invariants to solve for the flow of water $h(x, t)$ and $v(x, t)$ after a dam breaks (the problem posed in Ex. 16.9, and there solved via similarity methods). The initial conditions (at $t = 0$) are $v = 0$ everywhere, and $h = h_o$ at $x < 0$, $h = 0$ (no water) at $x > 0$.

17.5 Shock Fronts

We have just demonstrated that in an ideal gas with constant adiabatic index γ, large perturbations to fluid dynamical variables inevitably evolve to form a divergently large velocity gradient—a "shock front," or a "shock wave," or simply a "shock." When the velocity gradient becomes large, we can no longer ignore the viscous stress, because the viscous terms in the Navier-Stokes equation involve second derivatives of **v** in space, whereas the inertial term involves only first derivatives. As in turbulence and in boundary layers, so also in a shock front, the viscous stresses convert the fluid's ordered, bulk kinetic energy into microscopic kinetic energy (i.e., thermal energy). The ordered fluid velocity **v** thereby is rapidly—almost discontinuously—reduced from supersonic to subsonic, and the fluid is heated.

terminology for shocks

The cooler, supersonic region of incoming fluid is said to be *ahead of* or *upstream from* the shock, and it hits the shock's *front side;* the hotter, subsonic region of outgoing fluid is said to be *behind* or *downstream from* the shock, and it emerges from the shock's *back side;* see Fig. 17.10 below.

17.5.1 Junction Conditions across a Shock; Rankine-Hugoniot Relations

role of viscosity in a shock

Viscosity is crucial to the internal structure of the shock, but it is just as negligible in the downstream flow behind the shock as in the upstream flow ahead of the shock, since there velocity gradients are modest again. Remarkably, if (as is usually the case) the shock front is very thin compared to the length scales in the upstream and downstream flows, and the time for the fluid to pass through the shock is short compared to the upstream and downstream timescales, then we can deduce the net influence of the shock on the flow without any reference to the viscous processes that operate in the shock and without reference to the shock's detailed internal structure. We do so by treating the shock as a discontinuity across which certain junction conditions must be satisfied. This is similar to electromagnetic theory, where the junction conditions for the electric and magnetic fields across a material interface are independent of the detailed structure of the interface.

The keys to the shock's junction conditions are the conservation laws for mass, momentum, and energy: The fluxes of mass, momentum, and energy must usually be the same in the downstream flow, emerging from the shock, as in the upstream flow, entering it. To understand this, we first note that, because the time to pass through the shock is so short, mass, momentum, and energy cannot accumulate in the shock, so the flow can be regarded as stationary. In a stationary flow, the mass flux is always constant, as there is no way to create new mass or destroy old mass. Its continuity across the shock can be written as

mass conservation across shock

$$[\rho \mathbf{v} \cdot \mathbf{n}] = 0, \tag{17.29a}$$

where $\mathbf{n}$ is the unit normal to the shock front, and the square bracket means the difference in the values on the downstream and upstream sides of the shock. Similarly, the total momentum flux $\mathbf{T} \cdot \mathbf{n}$ must be conserved in the absence of external forces. Now $\mathbf{T}$ has both a mechanical component, $P\mathbf{g} + \rho \mathbf{v} \otimes \mathbf{v}$, and a viscous component, $-\zeta \theta \mathbf{g} - 2\eta \boldsymbol{\sigma}$. However, the viscous component is negligible in the upstream and downstream flows, which are being matched to each other, so the mechanical component by itself must be conserved across the shock front:

momentum conservation

$$[(P\mathbf{g} + \rho \mathbf{v} \otimes \mathbf{v}) \cdot \mathbf{n}] = 0. \tag{17.29b}$$

Similar remarks apply to the energy flux, though here we must be slightly more restrictive. There are three ways that a change in the energy flux could occur. First, energy may be added to the flow by chemical or nuclear reactions that occur in the shock front. Second, the gas may be heated to such a high temperature that it will lose energy in the shock front through the emission of radiation. Third, energy may be conducted far upstream by suprathermal particles so as to preheat the incoming gas. Any of these processes will thicken the shock front and may make it so thick that it can no longer sensibly be approximated as a discontinuity. If any of these processes occurs, we must check to see whether it is strong enough to significantly influence energy conservation across the shock. What such a check often reveals is that preheating is negligible, and the lengthscales over which the chemical and nuclear reactions and radiation emission operate are much greater than the length over which viscosity acts. In this case we can conserve energy flux across the viscous shock and then follow the evolutionary effects of reactions and radiation (if significant) in the downstream flow.

A shock with negligible preheating—and with negligible radiation emission and chemical and nuclear reactions inside the shock—will have the same energy flux in the departing, downstream flow as in the entering, upstream flow, so it will satisfy

$$\boxed{\left[\left(\frac{1}{2}v^2 + h\right)\rho \mathbf{v}\cdot\mathbf{n}\right] = 0.} \quad (17.29c) \quad \text{energy conservation}$$

Shocks that satisfy the conservation laws of mass, momentum, and energy, Eqs. (17.29), are said to be *adiabatic*.

By contrast to mass, momentum, and energy, the entropy will not be conserved across a shock front, since viscosity and other dissipative processes increase the entropy as the fluid flows through the shock. So far, the only type of dissipation that we have discussed is viscosity, and this is sufficient by itself to produce a shock front and keep it thin. However, heat conduction (Sec. 18.2) and electrical resistivity, which is important in magnetic shocks (Chap. 19), can also contribute to the dissipation and can influence the detailed structure of the shock front.

For an adiabatic shock, the three requirements of mass, momentum, and energy conservation [Eqs. (17.29)], known collectively as the *Rankine-Hugoniot relations*, enable us to relate the downstream flow and its thermodynamic variables to their upstream counterparts.[2]

Let us work in a reference frame where the incoming flow is normal to the shock front and the shock is at rest, so the flow is stationary in the shock's vicinity. Then the conservation of tangential momentum—the tangential component of Eq. (17.29b)—tells us that the outgoing flow is also normal to the shock in our chosen reference frame. We say that the shock is *normal*, not *oblique*.

We use the subscripts 1, 2 to denote quantities measured ahead of and behind the shock, respectively (i.e., 1 denotes the incoming flow, and 2 denotes the outgoing flow;

---

2. The existence of shocks was actually understood quite early on, more or less in this way, by Stokes. However, he was persuaded by his former student Rayleigh and others that such discontinuities were impossible, because they would violate energy conservation. With a deference that professors traditionally show their students, Stokes believed Rayleigh. They were both making an error in their analysis of energy conservation, due to an inadequate understanding of thermodynamics in that era.

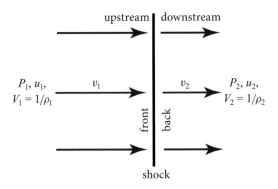

FIGURE 17.10 Terminology and notation for a shock front and the flow into and out of it.

Fig. 17.10). The Rankine-Hugoniot relations (17.29) then take the forms:

$$\rho_2 v_2 = \rho_1 v_1 = j, \tag{17.30a}$$

$$P_2 + \rho_2 v_2^2 = P_1 + \rho_1 v_1^2, \tag{17.30b}$$

$$h_2 + \frac{1}{2}v_2^2 = h_1 + \frac{1}{2}v_1^2, \tag{17.30c}$$

where j is the mass flux, which is determined by the upstream flow.

These equations can be brought into a more useful form by replacing the density ρ with the specific volume $V \equiv 1/\rho$, replacing the specific enthalpy h by its value in terms of P and V, $h = u + P/\rho = u + PV$, and performing some algebra. The result is

Rankine-Hugoniot jump conditions across a shock

$$\boxed{u_2 - u_1 = \frac{1}{2}(P_1 + P_2)(V_1 - V_2),} \tag{17.31a}$$

$$\boxed{j^2 = \frac{P_2 - P_1}{V_1 - V_2},} \tag{17.31b}$$

$$\boxed{v_1 - v_2 = [(P_2 - P_1)(V_1 - V_2)]^{1/2}.} \tag{17.31c}$$

This is the most widely used form of the Rankine-Hugoniot relations. It must be augmented by an equation of state in the form

equation of state

$$\boxed{u = u(P, V).} \tag{17.32}$$

Some of the physical content of these Rankine-Hugoniot relations is depicted in Fig. 17.11. The thermodynamic state of the upstream (incoming) fluid is the point **shock adiabat** (V_1, P_1) in this volume-pressure diagram. The thick solid curve, called the *shock adiabat*, is the set of all possible downstream (outgoing) fluid states. This shock adiabat can be computed by combining Eq. (17.31a) with the equation of state (17.32). Those

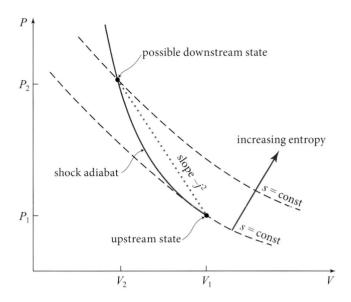

FIGURE 17.11 Shock adiabat. The pressure and specific volume $V = 1/\rho$ in the upstream flow are P_1 and V_1, and in the downstream flow they are P_2 and V_2. The dashed curves are ordinary adiabats (curves of constant entropy per unit mass s). The thick curve is the shock adiabat, the curve of allowed downstream states (V_2, P_2) for a given upstream state (V_1, P_1). The actual location of the downstream state on this adiabat is determined by the mass flux j flowing through the shock: the slope of the dotted line connecting the upstream and downstream states is $-j^2$.

equations will actually give a curve that extends away from (V_1, P_1) in both directions, up-leftward and down-rightward. Only the up-leftward portion is compatible with an increase of entropy across the shock; the down-rightward portion requires an entropy decrease, which is forbidden by the second law of thermodynamics and therefore is not drawn in Fig. 17.11. The actual location (V_2, P_2) of the downstream state along the shock adiabat is determined by Eq. (17.31b) in a simple way: the slope of the dotted line connecting the upstream and downstream states is $-j^2$, where j is the mass flux passing through the shock. When one thereby has learned (V_2, P_2), one can compute the downstream speed v_2 from Eq. (17.31c).

It can be shown that the pressure and density always increase across a shock (as is the case in Fig. 17.11), and the fluid always decelerates:

$$P_2 > P_1, \quad V_2 < V_1, \quad v_2 < v_1; \qquad (17.33)$$

pressure and density increase, and velocity decrease across shock

see Ex. 17.8. It also can be demonstrated in general—and will be verified in a particular case below, that the Rankine-Hugoniot relations require the flow to be supersonic with respect to the shock front upstream $v_1 > C_1$ and subsonic downstream, $v_2 < C_2$. Physically, this requirement is sensible (as we have seen): When the fluid approaches the shock supersonically, it is not possible to communicate a pressure pulse upstream from the shock (via a Riemann invariant moving at the speed of sound relative to the flow) and thereby cause the flow to decelerate; therefore, to slow the flow, a shock must develop.[3] By contrast, the shock front can and does respond to changes in the downstream conditions, since it is in causal contact with the downstream flow; sound waves and a Riemann invariant can propagate upstream, through the downstream flow, to the shock.

summary of shocks

To summarize, shocks are machines that decelerate a normally incident upstream flow to a subsonic speed, so it can be in causal contact with conditions downstream. In the process, bulk momentum flux ρv^2 is converted into pressure, bulk kinetic energy is converted into internal energy, and entropy is manufactured by the dissipative processes at work in the shock front. For given upstream conditions, the downstream conditions are fixed by the conservation laws of mass, momentum, and energy and are independent of the detailed dissipation mechanisms.

EXERCISES

Exercise 17.8 *Derivation and Challenge: Signs of Change across a Shock*

(a) Almost all equations of state satisfy the condition $(\partial^2 V/\partial P^2)_s > 0$. Show that, when this condition is satisfied, the Rankine-Hugoniot relations and the law of entropy increase imply that the pressure and density must increase across a shock and the fluid must decelerate: $P_2 > P_1$, $V_2 < V_1$, and $v_2 < v_1$.

(b) Show that in a fluid that violates $(\partial^2 V/\partial P^2)_s > 0$, the pressure and density must still increase and the fluid decelerate across a shock, as otherwise the shock would be unstable.

For a solution to this exercise, see Landau and Lifshitz (1959, Sec. 84).

Exercise 17.9 *Problem: Relativistic Shock* **T2**

In astrophysics (e.g., in supernova explosions and in jets emerging from the vicinities of black holes), one sometimes encounters shock fronts for which the flow speeds relative to the shock approach the speed of light, and the internal energy density is comparable to the fluid's rest-mass density.

(a) Show that the relativistic Rankine-Hugoniot equations for such a shock take the following form:

$$\boxed{\eta_2^2 - \eta_1^2 = (P_2 - P_1)(\eta_1 V_1 + \eta_2 V_2),} \qquad (17.34\text{a})$$

$$\boxed{j^2 = \frac{P_2 - P_1}{\eta_1 V_1 - \eta_2 V_2},} \qquad (17.34\text{b})$$

$$\boxed{v_2 \gamma_2 = j V_2, \qquad v_1 \gamma_1 = j V_1.} \qquad (17.34\text{c})$$

---

3. Of course, if there is some faster means of communication, for example, photons or, in an astrophysical context, cosmic rays or neutrinos, then there may be a causal contact between the shock and the inflowing gas, which can either prevent shock formation or lead to a more complex shock structure.

Here,

(i) We use units in which the speed of light is 1 (as in Chap. 2).

(ii) The volume per unit rest mass is $V \equiv 1/\rho_o$, and ρ_o is the rest-mass density (equal to some standard rest mass per baryon times the number density of baryons; cf. Sec. 2.12.3).

(iii) We denote the total density of mass-energy including rest mass by ρ_R (it was denoted ρ in Chap. 2) and the internal energy per unit rest mass by u, so $\rho_R = \rho_o(1+u)$. In terms of these the quantity $\eta \equiv (\rho_R + P)/\rho_o = 1 + u + P/\rho_o = 1 + h$ is the relativistic enthalpy per unit rest mass (i.e., the enthalpy per unit rest mass, including the rest-mass contribution to the energy) as measured in the fluid rest frame.

(iv) The pressure as measured in the fluid rest frame is P.

(v) The flow velocity in the shock's rest frame is v, and $\gamma \equiv 1/\sqrt{1-v^2}$ (*not* the adiabatic index!), so $v\gamma$ is the spatial part of the flow 4-velocity.

(vi) The rest-mass flux is j (rest mass per unit area per unit time) entering and leaving the shock.

(b) Use a pressure-volume diagram to discuss these relativistic Rankine-Hugoniot equations in a manner analogous to Fig. 17.11.

(c) Show that in the nonrelativistic limit, the relativistic Rankine-Hugoniot equations (17.34) reduce to the nonrelativistic ones (17.31).

(d) It can be shown (Thorne, 1973) that relativistically, just as for nonrelativistic shocks, in general $P_2 > P_1$, $V_2 < V_1$, and $v_2 < v_1$. Consider, as an example, a relativistic shock propagating through a fluid in which the mass density due to radiation greatly exceeds that due to matter (a *radiation-dominated fluid*), so $P = \rho_R/3$ (Sec. 3.5.5). Show that $v_1 v_2 = 1/3$, which implies $v_1 > 1/\sqrt{3}$ and $v_2 < 1/\sqrt{3}$. Show further that $P_2/P_1 = (9v_1^2 - 1)/[3(1 - v_1^2)]$.

Exercise 17.10 ****Problem: Hydraulic Jumps and Breaking Ocean Waves**

Run water at a high flow rate from a kitchen faucet onto a dinner plate (Fig. 17.12). What you see is called a hydraulic jump. It is the kitchen analog of a breaking ocean wave, and the shallow-water-wave analog of a shock front in a compressible gas. In this exercise you will develop the theory of hydraulic jumps (and breaking ocean waves) using the same tools as those used for shock fronts.

(a) Recall that for shallow-water waves, the water motion, below the water's surface, is nearly horizontal, with speed independent of depth z (Ex. 16.1). The same is true of the water in front of and behind a hydraulic jump. Apply the conservation of mass and momentum to a hydraulic jump, in the jump's rest frame, to obtain equations for the height of the water h_2 and water speed v_2 behind the jump (emerging from it) in terms of those in front of the jump, h_1 and v_1. These are the analog of the Rankine-Hugoniot relations for a shock front. [Hint: For

FIGURE 17.12 Hydraulic jump on a dinner plate under a kitchen tap.

momentum conservation you will need to use the pressure P as a function of height in front of and behind the jump.]

(b) You did not use energy conservation across the jump in your derivation, but it was needed in the analysis of a shock front. Why?

(c) Show that the upstream speed v_1 is greater than the speed $\sqrt{gh_1}$ of small-amplitude, upstream gravity waves [shallow-water waves; Eq. (16.10) and associated discussion]; thus the upstream flow is supersonic. Similarly, show that the downstream flow speed v_2 is slower than the speed $\sqrt{gh_2}$ of small-amplitude, downstream gravity waves (i.e., the downstream flow is subsonic).

(d) We normally view a breaking ocean wave in the rest frame of the quiescent upstream water. Use your hydraulic-jump equations to show that the speed of the breaking wave as seen in this frame is related to the depths h_1 and h_2 in front of and behind the breaking wave by

$$v_{\text{break}} = \left[\frac{g(h_1 + h_2)h_2}{2h_1} \right]^{1/2};$$

see Fig. 17.13.

17.5.2 Junction Conditions for Ideal Gas with Constant γ

shock Mach number

To make the shock junction conditions more explicit, let us again specialize to an ideal gas with constant specific-heat ratio γ (a polytropic gas), so the equation of state is

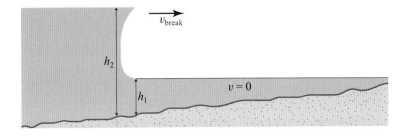

FIGURE 17.13 An ocean wave breaking on a gradually sloping beach. The depth of water ahead of the wave is h_1, and the depth behind the wave is h_2.

$u = PV/(\gamma - 1)$, and the sound speed is $C = \sqrt{\gamma P/\rho} = \sqrt{\gamma PV}$ [Eqs. (17.2)]. We measure the strength of the shock using the *shock Mach number M*, which is defined to be the Mach number in the upstream flow relative to the shock:

$$M \equiv M_1 = v_1/C_1 = \sqrt{v_1^2/(\gamma P_1 V_1)}. \tag{17.35}$$

With the aid of this equation of state and Mach number, we can bring the Rankine-Hugoniot relations (17.31) into the form:

$$\boxed{\frac{\rho_1}{\rho_2} = \frac{V_2}{V_1} = \frac{v_2}{v_1} = \frac{\gamma - 1}{\gamma + 1} + \frac{2}{(\gamma + 1)M^2},} \tag{17.36a}$$

Rankine-Hugoniot relations for polytropic equation of state

$$\boxed{\frac{P_2}{P_1} = \frac{2\gamma M^2}{\gamma + 1} - \frac{\gamma - 1}{\gamma + 1},} \tag{17.36b}$$

$$\boxed{M_2^2 = \frac{2 + (\gamma - 1)M^2}{2\gamma M^2 - (\gamma - 1)}.} \tag{17.36c}$$

Here $M_2 \equiv v_2/c_2$ is the downstream Mach number.

The results for this equation of state illustrate a number of general features of shocks. The density and pressure increase across the shock, the flow speed decreases, and the downstream flow is subsonic—all discussed previously—and one important new feature: a shock weakens as its Mach number M decreases. In the limit that $M \to 1$, the jumps in pressure, density, and speed vanish, and the shock disappears.

In the *strong-shock limit*, $M \gg 1$, the jumps are

$$\boxed{\frac{\rho_1}{\rho_2} = \frac{V_2}{V_1} = \frac{v_2}{v_1} \simeq \frac{\gamma - 1}{\gamma + 1},} \tag{17.37a}$$

Rankine-Hugoniot relations for strong polytropic shock

$$\boxed{\frac{P_2}{P_1} \simeq \frac{2\gamma M^2}{\gamma + 1}.} \tag{17.37b}$$

Thus the density jump is always of order unity, but the pressure jump grows ever larger as M increases. Air has $\gamma \simeq 1.4$ (Ex. 17.1), so the density compression ratio for a strong shock in air is $\rho_2/\rho_1 = 6$, and the pressure ratio is $P_2/P_1 = 1.2M^2$. The Space Shuttle's reentry provides a nice example of these strong-shock Rankine-Hugoniot relations; see the bottom half of Box 17.4.

EXERCISES

Exercise 17.11 *Problem: Shock Tube*
Consider a shock tube as discussed in Sec. 17.4.2 and Fig. 17.9. High density "driver" gas with sound speed C_0 and specific heat ratio γ is held in place by a membrane that separates it from target gas with very low density, sound speed C_1, and the same specific-heat ratio γ. When the membrane is ruptured, a strong shock propagates into the target gas. Show that the Mach number of this shock is given approximately by

$$M = \left(\frac{\gamma+1}{\gamma-1}\right)\left(\frac{C_0}{C_1}\right). \tag{17.38}$$

[Hint: Think carefully about which side of the shock is supersonic and which side subsonic.]

17.5.3 Internal Structure of a Shock

Although they are often regarded as discontinuities, shocks, like boundary layers, do have structure. The simplest case is that of a gas in which the shear-viscosity coefficient is molecular in origin and is given by $\eta = \rho\nu \sim \rho\lambda v_{\text{th}}/3$, where λ is the molecular mean free path, and $v_{\text{th}} \sim C$ is the mean thermal speed of the molecules (Ex. 3.19). In this case for 1-dimensional flow $v = v_x(x)$, the viscous stress $T_{xx} = -\zeta\theta - 2\eta\sigma_{xx}$ is $-(\zeta + 4\eta/3)dv/dx$, where ζ is the coefficient of bulk viscosity, which can be of the same order as the coefficient of shear viscosity. In the shock, this viscous stress must roughly balance the total kinetic momentum flux $\sim\rho v^2$. If we estimate the velocity gradient dv/dx by v_1/δ_S, where δ_S is a measure of the shock thickness, and we estimate the sound speed in the shock front by $C \sim v_1$, then we deduce that the shock thickness δ_S is roughly equal to λ, the collision mean free path in the gas. For air at standard temperature and pressure, the mean free path is $\lambda \sim (\sqrt{2}n\pi\sigma^2)^{-1} \sim 70$ nm, where n is the molecular density, and σ is the molecular diameter. This is very small! Microscopically, it makes sense that $\delta_S \sim \lambda$, as an individual molecule only needs a few collisions to randomize its ordered motion perpendicular to the shock front. However, this estimate raises a problem, as it brings into question our use of the continuum approximation (cf. Sec. 13.2). It turns out that, when a more careful calculation of the shock structure is carried out incorporating heat conduction, the shock thickness is several mean free paths, fluid dynamics is acceptable for an approximate theory, and the results are in rough accord with measurements of the velocity profiles of shocks with modest Mach numbers. Despite this, for an accurate description of the shock structure, a kinetic treatment is usually necessary.

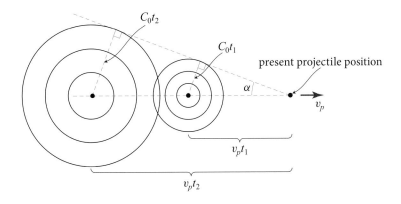

FIGURE 17.14 Construction for the Mach cone formed by a supersonic projectile. The cone angle is $\alpha = \sin^{-1}(1/M)$, where $M = v_p/C_0$ is the Mach number of the projectile.

So far we have assumed that the shocked fluid is made of uncharged molecules. A more complicated type of shock can arise in an ionized gas (i.e., a plasma; Part VI). Shocks in the solar wind are examples. In this case the collision mean free paths are enormous; in fact, they are comparable to the transverse size of the shock, and therefore one might expect the shocks to be so thick that the Rankine-Hugoniot relations fail. However, spacecraft measurements reveal solar-wind shocks that are relatively thin—far thinner than the collisional mean free paths of the plasma's electrons and ions. In this case, it turns out that collisionless, collective electromagnetic and charged-particle interactions in the plasma are responsible for the viscosity and dissipation. (The particles create plasma waves, which in turn deflect the particles.) These processes are so efficient that thin shock fronts can occur without individual particles having to hit one another. Since the shocks are thin, they must satisfy the Rankine-Hugoniot relations. We discuss these collisionless shocks further in Sec. 23.6.

dissipation mechanisms for shocks in collisionless plasmas

17.5.4 Mach Cone

The shock waves formed by a supersonically moving body are quite complex close to the body and depend on its detailed shape, Reynolds' number, and so forth (see, e.g., Fig. 17.1). However, far from the body, the leading shock has the form of the *Mach cone* shown in Fig. 17.14. We can understand this cone by the construction shown in the figure. The shock is the boundary between the fluid that is in sound-based causal contact with the projectile and the fluid that is not. This boundary is mapped out by (conceptual) sound waves that propagate into the fluid from the projectile at the ambient sound speed C_0. When the projectile is at the indicated position, the envelope of the circles is the shock front and has the shape of the Mach cone, with opening angle (the *Mach angle*)

$$\alpha = \sin^{-1}(1/M). \tag{17.39}$$

Mach cone

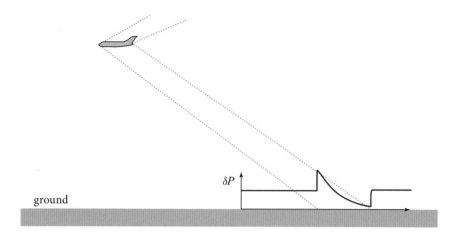

FIGURE 17.15 Double shock created by a supersonic body and its associated N-wave pressure distribution on the ground.

Usually, there are two such shock cones, one attached to the projectile's bow shock and the other formed out of the complex shock structure in its tail region. The pressure must jump twice, once across each of these shocks, and therefore forms an *N wave*, which propagates cylindrically away from the projectile, as shown in Fig. 17.15. Behind the first shock, the density and pressure drop off gradually by more than the first shock's compression. As a result, the fluid flowing into the second shock has a lower pressure, density, and sound speed than that flowing into the first (Fig. 17.15). This causes the Mach number of the second shock to be higher than that of the first, and its Mach angle thus to be smaller. As a result, the separation between the shocks increases as they travel as $\varpi^{1/2}$, it turns out. The pressure jumps across the shocks decrease as $\varpi^{-3/4}$. Here ϖ is the perpendicular distance of the point of observation from the projectile's trajectory. Often a double boom can be heard on the ground. For a detailed analysis, see Whitham (1974, Sec. 9.3).

EXERCISES

Exercise 17.12 *Problem: Sonic Boom from the Space Shuttle*
Use the quoted scaling of N-wave amplitude with cylindrical radius ϖ to make an order-of-magnitude estimate of the flux of acoustic energy produced by the Space Shuttle flying at Mach 2 at an altitude of 20 km. Give your answer in decibels [Eq. (16.61)].

17.6 Self-Similar Solutions—Sedov-Taylor Blast Wave

Strong explosions can generate shock waves. Examples include atmospheric nuclear explosions, supernova explosions, and depth charges. The debris from a strong explosion is at much higher pressure than the surrounding gas and therefore drives a strong spherical shock into its surroundings. Initially, this shock wave travels at roughly the radial speed of the expanding debris. However, the mass of fluid swept up by the shock eventually exceeds that of the explosion debris. The shock then decelerates and the energy of the explosion is transferred to the swept-up fluid. It is of obvious importance to be able to calculate how fast and how far the shock front will travel.

17.6.1 The Sedov-Taylor Solution

We first make an order-of-magnitude estimate. Let the total energy of the explosion be E and the density of the surrounding fluid (assumed uniform) be ρ_0. Then after time t, when the shock radius is $R(t)$, the mass of swept-up fluid will be $m \sim \rho_0 R^3$. The fluid velocity behind the shock will be roughly the radial velocity of the shock front, $v \sim \dot{R} \sim R/t$, and so the kinetic energy of the swept-up gas will be $\sim mv^2 \sim \rho_0 R^5/t^2$. There will also be internal energy in the post-shock flow, with energy density roughly equal to the post-shock pressure: $\rho u \sim P \sim \rho_0 \dot{R}^2$ [cf. the strong-shock jump condition (17.37b) with $P_1 \sim \rho_0 C_0^2$, so $P_1 M^2 \sim \rho_0 v^2 \sim \rho_0 \dot{R}^2$]. The total internal energy behind the expanding shock will then be $\sim \rho \dot{R}^2 R^3 \sim \rho_0 R^5/t^2$, equal in order of magnitude to the kinetic energy. Equating this to the total energy E of the explosion, we obtain the estimate

$$E = \kappa \rho_0 R^5 t^{-2}, \tag{17.40}$$

where κ is a numerical constant of order unity. This expression implies that at time t the shock front has reached the radius

$$R = \left(\frac{E}{\kappa \rho_0}\right)^{1/5} t^{2/5}. \tag{17.41}$$

radius of a strong, spherical shock as a function of time

This scaling should hold roughly from the time that the mass of the swept-up gas is of order that of the exploding debris, to the time that the shock weakens to a Mach number of order unity so we can no longer use the strong-shock value $\sim \rho_0 \dot{R}^2$ for the post-shock pressure.

Note that we could have obtained Eq. (17.41) by a purely dimensional argument: E and ρ_0 are the only significant controlling parameters in the problem, and $E^{1/5} \rho_0^{-1/5} t^{2/5}$ is the only quantity with dimensions of length that can be constructed from E, ρ_0, and t. However, it is usually possible and always desirable to justify any such dimensional argument on the basis of the governing equations.

If, as we shall assume, the motion remains radial and the gas is ideal with constant specific-heat ratio γ, then we can solve for the details of the flow behind the shock front by integrating the radial flow equations:

$$\frac{\partial \rho}{\partial t} + \frac{1}{r^2} \frac{\partial}{\partial r}(r^2 \rho v) = 0, \tag{17.42a}$$

$$\frac{\partial v}{\partial t} + v \frac{\partial v}{\partial r} + \frac{1}{\rho} \frac{\partial P}{\partial r} = 0, \tag{17.42b}$$

$$\frac{\partial}{\partial t}\left(\frac{P}{\rho^\gamma}\right) + v \frac{\partial}{\partial r}\left(\frac{P}{\rho^\gamma}\right) = 0. \tag{17.42c}$$

The first two equations are the familiar continuity equation and Euler equation written for a spherical flow. The third equation is energy conservation expressed as the adiabatic-expansion relation, $P/\rho^\gamma = $ const moving with a fluid element. Although P/ρ^γ is time independent for each fluid element, its value will change from element to element. Gas that has passed through the shock more recently will be given a smaller entropy than gas that was swept up when the shock was stronger, and thus it will have a smaller value of P/ρ^γ.

Given suitable initial conditions, the partial differential equations (17.42) can be integrated numerically. However, there is a practical problem: it is not easy to determine the initial conditions in an explosion! Fortunately, at late times, when the initial debris mass is far less than the swept-up mass, the fluid evolution is independent of the details of the initial expansion and in fact can be understood analytically as a *similarity solution*. By this term we mean that the shape of the radial profiles of pressure, density, and velocity are independent of time.

examples of similarity solutions

We have already met three examples of similarity solutions: the Blasius structure of a laminar boundary layer (Sec. 14.4.1), the structure of a turbulent jet (Ex. 15.3), and the flow of water following the sudden rupture of a dam (Ex. 16.9). The one we explored in greatest detail was the Blasius boundary layer (Sec. 14.4.1). There we argued on the basis of mass and momentum conservation (or, equally well, by dimensional analysis) that the thickness of the boundary layer as a function of distance x downstream would be $\sim \delta = (\nu x/V)^{1/2}$, where V is the speed of the flow above the boundary layer. This motivated us to introduce the dimensionless variable $\xi = y/\delta$ and argue that the boundary layer's speed $v_x(x, y)$ would be equal to the free-stream velocity V times some universal function $f'(\xi)$. This ansatz converted the fluid's partial differential equations into an ordinary differential equation for $f(\xi)$, which we solved numerically.

Our explosion problem is somewhat similar. The characteristic scaling length in the explosion is the radius $R(t) = [E/(\kappa \rho_0)]^{1/5} t^{2/5}$ of the shock [Eq. (17.41)], with κ an as-yet-unknown constant, so the fluid and thermodynamic variables should be expressible as some characteristic values multiplying universal functions of

$$\xi \equiv r/R(t). \tag{17.43}$$

similarity solution behind strong, spherical shock

Our thermodynamic variables are P, ρ, and u, and natural choices for their characteristic values are the values P_2, ρ_2, and v_2 immediately behind the shock. If we assume the shock is strong, then we can use the strong-shock jump conditions (17.37) to determine those values, and then write

$$P = \frac{2}{\gamma + 1} \rho_0 \dot{R}^2 \tilde{P}(\xi), \tag{17.44a}$$

$$\rho = \frac{\gamma + 1}{\gamma - 1} \rho_0 \, \tilde{\rho}(\xi), \tag{17.44b}$$

$$v = \frac{2}{\gamma + 1} \dot{R} \, \tilde{v}(\xi), \tag{17.44c}$$

Chapter 17. Compressible and Supersonic Flow

with $\tilde{P}(1) = \tilde{\rho}(1) = \tilde{v}(1) = 1$, since $\xi = 1$ is the shock's location. Note that the velocity v is scaled to the post-shock velocity v_2 measured in the inertial frame in which the upstream fluid is at rest, rather than in the noninertial frame in which the decelerating shock is at rest. The self-similarity ansatz (17.44) and resulting similarity solution for the flow are called the *Sedov-Taylor blast-wave solution,* since L. I. Sedov and G. I. Taylor independently developed it (Sedov, 1946, 1993; Taylor, 1950).

The partial differential equations (17.42) can now be transformed into ordinary differential equations by inserting the ansatz (17.44) together with expression (17.41) for $R(t)$, changing the independent variables from r and t to R and ξ and using

$$\left(\frac{\partial}{\partial t}\right)_r = -\left(\frac{\xi \dot{R}}{R}\right)\left(\frac{\partial}{\partial \xi}\right)_R + \dot{R}\left(\frac{\partial}{\partial R}\right)_\xi = -\left(\frac{2\xi}{5t}\right)\left(\frac{\partial}{\partial \xi}\right)_R + \frac{2R}{5t}\left(\frac{\partial}{\partial R}\right)_\xi, \tag{17.45}$$

$$\left(\frac{\partial}{\partial r}\right)_t = \left(\frac{1}{R}\right)\left(\frac{\partial}{\partial \xi}\right)_R. \tag{17.46}$$

Mass conservation, the Euler equation, and the equation of adiabatic expansion become, in that order:

$$0 = 2\tilde{\rho}\tilde{v}' - (\gamma + 1)\xi \tilde{\rho}' + \tilde{v}\left(2\tilde{\rho}' + \frac{4}{\xi}\tilde{\rho}\right), \tag{17.47a}$$

$$0 = \tilde{\rho}\tilde{v}[3(\gamma + 1) - 4\tilde{v}'] + 2(\gamma + 1)\xi \tilde{\rho}\tilde{v}' - 2(\gamma - 1)\tilde{P}', \tag{17.47b}$$

$$3 = \left(\frac{2\tilde{v}}{\gamma + 1} - \xi\right)\left(\frac{\tilde{P}'}{\tilde{P}} - \gamma\frac{\tilde{\rho}'}{\tilde{\rho}}\right). \tag{17.47c}$$

differential equations for similarity solution

These self-similarity equations can be solved numerically, subject to the boundary conditions that $\tilde{v}$, $\tilde{\rho}$, and $\tilde{P}$ are all zero at $\xi = 0$ and 1 at $\xi = 1$. Remarkably, Sedov and Taylor independently found an analytic solution (also given in Landau and Lifshitz, 1959, Sec. 99). The solutions for an explosion in air ($\gamma = 1.4$) are exhibited in Fig. 17.16.

Armed with the solution for $\tilde{v}(\xi)$, $\tilde{\rho}(\xi)$, and $\tilde{P}(\xi)$ (numerical or analytic), we can evaluate the flow's energy E, which is equal to the explosion's total energy during the time interval when this similarity solution is accurate. The energy E is given by the integral

$$E = \int_0^R 4\pi r^2 dr \rho \left(\frac{1}{2}v^2 + u\right)$$

$$= \frac{4\pi \rho_0 R^3 \dot{R}^2 (\gamma + 1)}{(\gamma - 1)} \int_0^1 d\xi \xi^2 \tilde{\rho}\left(\frac{2\tilde{v}^2}{(\gamma + 1)^2} + \frac{2\tilde{P}}{(\gamma + 1)^2 \tilde{\rho}}\right). \tag{17.48}$$

explosion's total energy

Here we have used Eqs. (17.44) and substituted $u = P/[\rho(\gamma - 1)]$ for the internal energy [Eq. (17.2b)]. The energy E appears not only on the left-hand side of this equation but also on the right, in the terms $\rho_o R^3 \dot{R}^2 = (4/25)E/\kappa$. Thus, E cancels

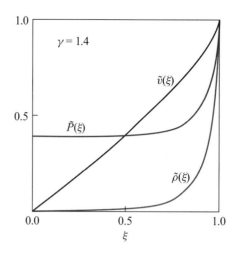

FIGURE 17.16 Scaled pressure, density, and velocity as a function of scaled radius behind a Sedov-Taylor blast wave in air with $\gamma = 1.4$.

out, and Eq. (17.48) becomes an equation for the unknown constant κ. Evaluating that equation numerically, we find that κ varies from $\kappa = 0.85$ for $\gamma = 1.4$ (air) to $\kappa = 0.49$ for $\gamma = 1.67$ (monatomic gas or fully ionized plasma).

It is enlightening to see how the fluid behaves in this blast-wave solution. The fluid that passes through the shock is compressed, so that it mostly occupies a fairly thin spherical shell immediately behind the shock [see the spike in $\tilde{\rho}(\xi)$ in Fig. 17.16]. The fluid in this shell moves somewhat more slowly than the shock [$v = 2\dot{R}/(\gamma + 1)$; Eq. (17.44c) and Fig. 17.16]; it flows from the shock front through the high-density shell (fairly slowly relative to the shell, which remains attached to the shock), and on into the lower density post-shock region. Since the post-shock flow is subsonic, the pressure in the blast wave is fairly uniform [see the curve $\tilde{P}(\xi)$ in Fig. 17.16]; in fact the central pressure is typically about half the maximum pressure immediately behind the shock. This pressure pushes on the spherical shell, thereby accelerating the freshly swept-up fluid.

17.6.2 Atomic Bomb

The first atomic bomb was exploded in New Mexico in 1945, and photographs released in 1947 (Fig. 17.17) showed the radius of the blast wave as a function of time. The pictures were well fit by $R \sim 37(t/1\,\text{ms})^{0.4}$ m up to about $t = 100$ ms, when the shock Mach number fell to about unity (Fig. 17.17). Combining this information with the similarity solution (Sec. 17.6.1), that they had earlier derived independently, the Russian physicist L. I. Sedov and the British physicist G. I. Taylor were both able to infer the total energy released, which was an official American secret at the time. Their analyses of the data were published later (Taylor, 1950; Sedov, 1957, 1993).

Adopting the specific-heat ratio $\gamma = 1.4$ of air, the corresponding value $\kappa = 0.85$, and the measured $R \sim 37(t/1\,\text{ms})^{0.4}$ m, we obtain from Eq. (17.48) the estimate

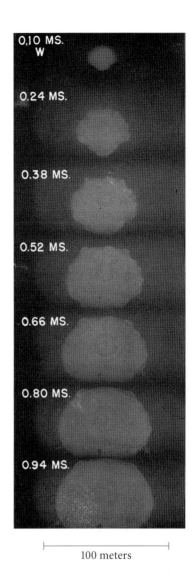
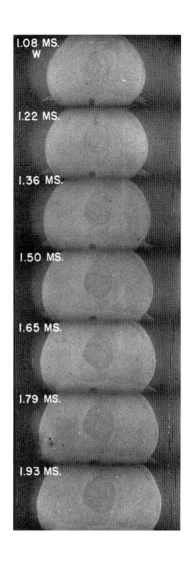

100 meters

FIGURE 17.17 Photographs of the fireball (very hot post-shock gas) from the first atomic bomb explosion, at Almagordo, New Mexico, on July 16, 1945. Courtesy of the Los Alamos National Laboratory Archives.

$E \sim 7.2 \times 10^{13}$ J, which is about the same energy release as 17 kilotons of TNT. This estimate is close to the Los Alamos scientists' estimate of 18–20 kilotons. (Hydrogen bombs have been manufactured that are more than a thousand times more energetic than this—as much as 57 megatons—but such awesome weapons have not been deemed militarily useful, so today's arsenals contain bombs that are typically *only* ~ 1 megaton!)

We can use the Sedov-Taylor solution to infer some further features of the explosion. The post-shock gas is at density $\sim (\gamma+1)/(\gamma-1) \sim 5$ times the ambient density $\rho_0 \sim 1$ kg m^{-3}. Similarly, using the ideal gas law $P = [\rho/(m_p \mu)] k_B T$ with

17.6 Self-Similar Solutions—Sedov-Taylor Blast Wave

a mean molecular weight $\mu \sim 10$ and the strong-shock jump conditions (17.37), the post-shock temperature can be computed:

$$T_2 = \frac{m_p \mu}{\rho_2 k_B} P_2 \sim 4 \times 10^4 \left(\frac{t}{1 \text{ ms}}\right)^{-1.2} \text{ K}. \tag{17.49}$$

At early times, this temperature is high enough to ionize the gas.

17.6.3 Supernovae

The evolution of most massive stars ends in a supernova explosion (like the one observed in 1987 in the Large Magellanic Cloud), in which a neutron star of mass $m \sim 3 \times 10^{30}$ kg is usually formed. This neutron star has a gravitational binding energy of about $0.1mc^2 \sim 3 \times 10^{46}$ J. Most of this binding energy is released in the form of neutrinos in the collapse that forms the neutron star, but an energy $E \sim 10^{44}$ J drives off the outer envelope of the pre-supernova star, a mass $M_0 \sim 10^{31}$ kg. This stellar material escapes with an rms speed $V_0 \sim (2E/M_0)^{1/2} \sim 5{,}000$ km s^{-1}. The expanding debris drives a blast wave into the surrounding interstellar medium, which has density $\rho_0 \sim 10^{-21}$ kg m^{-3}. The expansion of the blast wave can be modeled using the Sedov-Taylor solution after the mass of the swept-up interstellar gas has become large enough to dominate the blast debris, so the star-dominated initial conditions are no longer important—after a time $\sim (3M_0/4\pi\rho_0)^{1/3}/V_0 \sim 1{,}000$ yr. The blast wave then decelerates as for a Sedov-Taylor similarity solution until the shock speed nears the sound speed in the surrounding gas; this takes about 100,000 yr. *Supernova remnants* of this sort are efficient emitters of radio waves and X-rays, and several hundred have been observed in our Milky Way galaxy.

In some of the younger examples, like Tycho's remnant (Fig. 17.18), it is possible to determine the expansion speed, and the effects of deceleration can be measured. The observations from remnants like this are close to the prediction of the Sedov-Taylor solution; namely, the radius variation satisfies $d \ln R / d \ln t = 0.4$. (For Tycho's remnant, a value of 0.54 ± 0.05 is found, and using the estimated density of the interstellar medium plus the kinematics, an explosion energy of $E \sim 5 \times 10^{43}$ J is inferred.)

EXERCISES

Exercise 17.13 *Problem: Underwater Explosions*

A simple analytical solution to the Sedov-Taylor similarity equations can be found for the particular case $\gamma = 7$. This is a fair approximation to the behavior of water under explosive conditions, as it will be almost incompressible.

(a) Make the ansatz (whose self-consistency we will check later) that the velocity in the post-shock flow varies linearly with radius from the origin to the shock: $\tilde{v}(\xi) = \xi$. Use Eq. (17.45) to transform the equation of continuity into an ordinary differential equation and hence solve for the density function $\tilde{\rho}(\xi)$.

FIGURE 17.18 This remnant of the supernova explosion observed by the astronomer Tycho Brahe in 1572 is roughly 7,000 light-years from us. (Recent observations of a *light echo* have demonstrated that the supernova was of the same type used to discover the acceleration of the universe, which is discussed in Ex. 28.6.) This image was made by the Chandra X-ray Observatory and shows the emission from the hot (in red) ($\sim 1 - 10 \times 10^7$ K) gas heated by the outer and inner shock fronts as well as the even more energetic (in blue) X-rays that delineate the position of the outer shock front. Shock fronts like this are believed to be the sites for acceleration of galactic cosmic rays (cf. Sec. 19.7, Ex. 23.8). NASA/CXC/Rutgers/J. Warren and J. Hughes, et al.

(b) Next use the equation of motion to discover that $\tilde{P}(\xi) = \xi^3$.

(c) Verify that your solutions for the functions $\tilde{P}$, $\tilde{\rho}$, and $\tilde{v}$ satisfy the remaining entropy equation, thereby vindicating the original ansatz.

(d) Substitute your results from parts (a) and (b) into Eq. (17.48) to show that

$$E = \frac{2\pi R^5 \rho_0}{225 t^2}.$$

(e) An explosive charge weighing 100 kg with an energy release of 10^8 J kg^{-1} is detonated underwater. For what range of shock radii do you expect the Sedov-Taylor similarity solution to be valid?

Exercise 17.14 *Problem: Stellar Winds*

Many stars possess powerful stellar winds that drive strong spherical shock waves into the surrounding interstellar medium. If the strength of the wind remains constant, the kinetic and internal energy of the swept-up interstellar medium will increase linearly with time.

(a) Modify the text's analysis of a point explosion to show that in this case the speed of the shock wave at time t is $\frac{3}{5}R(t)/t$, where R is the associated shock radius. What is the speed of the post-shock gas?

(b) Now suppose that the star explodes as a supernova, and the blast wave expands into the relatively slowly moving stellar wind. Suppose that before the explosion the rate at which mass left the star and the speed of the wind were constant for a long time. How do you expect the density of gas in the wind to vary with radius? Modify the Sedov-Taylor analysis again to show that the expected speed of the shock wave at time t is now $\frac{2}{3}R(t)/t$.

Exercise 17.15 *Problem: Similarity Solution for a Shock Tube*
Use a similarity analysis to derive the solution (17.27) for the shock-tube flow depicted in Fig. 17.9.

Bibliographic Note

For physical insight into compressible flows and shock waves, we recommend the movies cited in Box 17.2. For textbook treatments, we recommend the relevant sections of Landau and Lifshitz (1959); Thompson (1984); Liepmann and Roshko (2002); Anderson (2003); and, at a more elementary and sometimes cursory level, Lautrup (2005). Whitham (1974) is superb on all aspects, from an applied mathematician's viewpoint.

Engineering-oriented textbooks on fluid mechanics generally contain detailed treatments of quasi-1-dimensional flows, shocks, hydraulic jumps, and their real-world applications. We like Munson, Young, and Okiishi (2006), White (2006), and Potter, Wiggert, and Ramadan (2012).

The two-volume treatise Zel'dovich and Raizer (2002) is a compendium of insights into shock waves and high-temperature hydrodynamics by an author (Yakov Borisovich Zel'dovich) who had a huge influence on the design of atomic and thermonuclear weapons in the Soviet Union and later on astrophysics and cosmology. Stanyukovich (1960) is a classic treatise on nonstationary flows, shocks and detonation waves. Sedov (1993)—the tenth edition of a book whose first was in 1943—is a classic and insightful treatise on similarity methods in physics.

CHAPTER EIGHTEEN

Convection

Approximation methods derived from physical intuition are frequently more reliable than rigorous mathematical methods, because in the case of the latter it is easier for errors to creep into the fundamental assumptions.

WERNER HEISENBERG (1969)

18.1 Overview

In Chaps. 13 and 14, we demonstrated that viscosity can exert a major influence on subsonic fluid flows. When the viscosity ν is large and the Reynolds number (Re $= LV/\nu$) is low, viscous stresses transport momentum directly, and the fluid's behavior can be characterized by the diffusion of the vorticity ($\boldsymbol{\omega} = \boldsymbol{\nabla} \times \mathbf{v}$) through the fluid [cf. Eq. (14.3)]. As the Reynolds number increases, the advection of the vorticity becomes more important. In the limit of large Reynolds number, we think of the vortex lines as being frozen into the flow. However, as we learned in Chap. 15, this insight is only qualitatively helpful, because high-Reynolds-number flows are invariably turbulent. Large, irregular, turbulent eddies transport shear stress very efficiently. This is particularly in evidence in turbulent boundary layers.

When viewed microscopically, heat conduction is a similar transport process to viscosity, and it is responsible for analogous physical effects. If a viscous fluid has high viscosity, then vorticity diffuses through it rapidly; similarly, if a fluid has high thermal conductivity, then heat diffuses through it rapidly. In the other extreme, when viscosity is low (i.e., when the Reynolds number is high), instabilities produce turbulence, which transports vorticity far more rapidly than diffusion could possibly do. Analogously, in heated fluids with low conductivity, the local accumulation of heat drives the fluid into convective motion, and the heat is transported much more efficiently by this motion than by thermal diffusion. As the convective heat transport increases, the fluid motion becomes more vigorous, and if the viscosity is sufficiently low, the thermally driven flow can also become turbulent. These effects are very much in evidence near solid boundaries, where thermal boundary layers can be formed, analogous to viscous boundary layers.

In addition to thermal effects that resemble the effects of viscosity, there are also unique thermal effects—particularly the novel and subtle combined effects of gravity and heat. Heat, unlike vorticity, causes a fluid to expand and thus, in the presence

> **BOX 18.1. READERS' GUIDE**
>
> - This chapter relies heavily on Chap. 13.
> - No subsequent chapters rely substantially on this one.

of gravity, to become buoyant; this buoyancy can drive thermal circulation or *free convection* in an otherwise stationary fluid. (Free convection should be distinguished from *forced convection*, in which heat is carried passively in a flow driven by externally imposed pressure gradients, e.g., when you blow on hot food to cool it, or stir soup on a hot stove.)

The transport of heat is a fundamental characteristic of many flows. It dictates the form of global weather patterns and ocean currents. It is also of great technological importance and is studied in detail, for example, in the cooling of nuclear reactors and the design of automobile engines. From a more fundamental perspective, as we have already discussed, the analysis and experimental studies of convection have led to major insights into the route to chaos (Sec. 15.6).

In this chapter, we describe some flows where thermal effects are predominant. We begin in Sec. 18.2 by writing down and then simplifying the equations of fluid mechanics with heat conduction. Then in Sec. 18.3, we discuss the *Boussinesq approximation*, which is appropriate for modest-scale flows where buoyancy is important. This allows us in Sec. 18.4 to derive the conditions under which convection is initiated. Unfortunately, this Boussinesq approximation sometimes breaks down. In particular, as we discuss in Sec. 18.5, it is inappropriate for convection in stars and planets, where circulation takes place over several gravitational scale heights. For such cases we have to use alternative, more heuristic arguments to derive the relevant criterion for convective instability, known as the *Schwarzschild criterion*, and to quantify the associated heat flux. We shall apply this theory to the solar convection zone.

Finally, in Sec. 18.6 we return to simple buoyancy-driven convection in a stratified fluid to consider *double diffusion*, a quite general type of instability that can arise when the diffusion of two physical quantities (in our case heat and the concentration of salt) render a fluid unstable even though the fluid would be stably stratified if there were only concentration gradients of one of these quantities.

18.2 Diffusive Heat Conduction—Cooling a Nuclear Reactor; Thermal Boundary Layers

So long as the mean free path of heat-carrying particles is small compared to the fluid's inhomogeneity lengthscales (as is almost always the case), and the fractional temperature change in one mean free path is small (as is also almost always true), the energy flux of heat flow takes the thermal-diffusion form

$$\mathbf{F}_{\text{cond}} = -\kappa \nabla T; \tag{18.1}$$

see Secs. 3.7 and 13.7.4. Here κ is the thermal conductivity.

For a viscous, heat-conducting fluid flowing in an external gravitational field, the most general governing equations are the fundamental thermodynamic potential $u(\rho, s)$; the first law of thermodynamics, Eq. (2) or (3) of Box 13.2; the law of mass conservation [Eq. (13.29) or (13.31)]; the Navier-Stokes equation (13.69); and the law of dissipative entropy production (13.75):

evolution equations for a viscous, heat-conducting fluid in an external gravitational field

$$u = u(\rho, s), \tag{18.2a}$$

$$\frac{du}{dt} = T\frac{ds}{dt} - P\frac{d(1/\rho)}{dt}, \tag{18.2b}$$

$$\frac{d\rho}{dt} = -\rho \nabla \cdot \mathbf{v}, \tag{18.2c}$$

$$\rho \frac{d\mathbf{v}}{dt} = -\nabla P + \rho \mathbf{g} + \nabla(\zeta \theta) + 2\nabla \cdot (\eta \boldsymbol{\sigma}), \tag{18.2d}$$

$$T\left[\rho\left(\frac{ds}{dt}\right) + \nabla \cdot \left(\frac{-\kappa \nabla T}{T}\right)\right] = \zeta \theta^2 + 2\eta \boldsymbol{\sigma} : \boldsymbol{\sigma} + \frac{\kappa}{T}(\nabla T)^2. \tag{18.2e}$$

These are four scalar equations and one vector equation for four scalar and one vector variables: the density ρ, internal energy per unit mass u, entropy per unit mass s, pressure P, and velocity $\mathbf{v}$. The thermal conductivity κ and coefficients of shear viscosity ζ and bulk viscosity $\eta = \rho \nu$ are presumed to be functions of ρ and s (or equally well, ρ and T).

This set of equations is far too complicated to solve, except via massive numerical simulations, unless some strong simplifications are imposed. We therefore introduce approximations. Our first approximation (already implicit in the above equations) is that the thermal conductivity κ is constant, as are the coefficients of viscosity. For most real applications this approximation is close to true, and no significant physical effects are missed by assuming it. Our second approximation, which does limit somewhat the type of problem we can address, is that the fluid motions are very slow—slow enough that, not only can the flow be regarded as incompressible ($\theta = \nabla \cdot \mathbf{v} = 0$), but also the squares of the shear $\boldsymbol{\sigma}$ and expansion θ (which are quadratic in the fluid speed) are negligibly small, and we thus can ignore viscous dissipation. These approximations bring the last three of the fluid evolution equations (18.2) into the simplified form

approximations

$$\nabla \cdot \mathbf{v} \simeq 0, \quad d\rho/dt \simeq 0, \tag{18.3a}$$

$$\frac{d\mathbf{v}}{dt} = -\frac{\nabla P}{\rho} + \mathbf{g} + \nu \nabla^2 \mathbf{v}, \tag{18.3b}$$

$$\rho T \frac{ds}{dt} = \kappa \nabla^2 T. \tag{18.3c}$$

[Our reasons for using "$\simeq$" in Eqs. (18.3a) become clear in Sec. 18.3, in connection with buoyancy.] Note that Eq. (18.3b) is the standard form of the Navier-Stokes equation for incompressible flows, which we have used extensively in the past several chapters. Equation (18.3c) is an elementary law of energy conservation: the temperature times the rate of increase of entropy density moving with the fluid is equal to minus the divergence of the conductive energy flux, $\mathbf{F}_{\text{heat}} = -\kappa \nabla T$.

We can convert the entropy evolution equation (18.3c) into an evolution equation for temperature by expressing the changes ds/dt of entropy per unit mass in terms of changes dT/dt of temperature. The usual way to do this is to note that $T\,ds$ (the amount of heat deposited in a unit mass of fluid) is given by $c\,dT$, where c is the fluid's specific heat per unit mass. However, the specific heat depends on what one holds fixed during the energy deposition: the fluid element's volume or its pressure. As we have assumed that the fluid motions are slow, the fractional pressure fluctuations will be correspondingly small. (This assumption does not preclude significant temperature perturbations, provided they are compensated by density fluctuations of opposite sign.) Therefore, the relevant specific heat for a slowly moving fluid is the one at constant pressure, c_P, and we must write $T\,ds = c_P dT$.[1] Equation (18.3c) then becomes a linear partial differential equation for the temperature:

temperature diffusion equation

$$\boxed{\frac{dT}{dt} \equiv \frac{\partial T}{\partial t} + \mathbf{v} \cdot \nabla T = \chi \nabla^2 T,} \tag{18.4}$$

where

thermal diffusivity

$$\boxed{\chi = \kappa/(\rho\, c_P)} \tag{18.5}$$

is known as the *thermal diffusivity*, and we have again taken the easiest route in treating c_P and ρ as constant. When the fluid moves so slowly that the advective term $\mathbf{v} \cdot \nabla T$ is negligible, then Eq. (18.4) shows that the heat simply diffuses through the fluid, with the thermal diffusivity χ being the diffusion coefficient for temperature.

Prandtl number: vorticity diffusion over heat diffusion

The diffusive transport of heat by thermal conduction is similar to the diffusive transport of vorticity by viscous stress [Eq. (14.3)], and the thermal diffusivity χ is the direct analog of the kinematic viscosity ν. This observation motivates us to introduce a new dimensionless number known as the *Prandtl number*, which measures the relative importance of viscosity and heat conduction (in the sense of their relative abilities to diffuse vorticity and heat):

$$\boxed{\Pr = \frac{\nu}{\chi}.} \tag{18.6}$$

---

1. See, e.g., Turner (1973) for a more formal justification of the use of the specific heat at constant pressure rather than at constant volume.

TABLE 18.1: Order-of-magnitude estimates for kinematic viscosity ν, thermal diffusivity χ, and Prandtl number $\mathrm{Pr} = \nu/\chi$ for earth, fire, air, and water

Fluid	ν (m² s⁻¹)	χ (m² s⁻¹)	Pr
Earth's mantle	10^{17}	10^{-6}	10^{23}
Solar interior	10^{-2}	10^{2}	10^{-4}
Atmosphere	10^{-5}	10^{-5}	1
Ocean	10^{-6}	10^{-7}	10

For gases, both ν and χ are given to order of magnitude by the product of the mean molecular speed and the mean free path, and so Prandtl numbers are typically of order unity. (For air, $\mathrm{Pr} \sim 0.7$.) By contrast, in liquid metals the free electrons carry heat very efficiently compared with the transport of momentum (and vorticity) by diffusing ions, and so their Prandtl numbers are small. This is why liquid sodium is used as a coolant in nuclear power reactors. At the other end of the spectrum, water is a relatively poor thermal conductor with $\mathrm{Pr} \sim 6$, and Prandtl numbers for oils, which are quite viscous and are poor conductors, measure in the thousands. Other Prandtl numbers are given in Table 18.1.

typical Prandtl numbers

One might think that, when the Prandtl number is small (so κ is large compared to ν), one should necessarily include heat flow in the fluid equations and pay attention to thermally induced buoyancy (Sec. 18.3). Not so. In some low-Prandtl-number flows the heat conduction is so effective that the fluid becomes essentially isothermal, and buoyancy effects are minimized. Conversely, in some large-Prandtl-number flows the large viscous stress reduces the velocity gradient so that slow, thermally driven circulation takes place, and thermal effects are very important. In general the kinematic viscosity is of direct importance in controlling the transport of momentum, and hence in establishing the velocity field, whereas heat conduction affects the velocity field only indirectly (Sec. 18.3). We must therefore examine each flow on its individual merits.

Another dimensionless number is commonly introduced when discussing thermal effects: the *Péclet number*. It is defined, by analogy with the Reynolds number, as

$$\boxed{\mathrm{Pe} = \frac{LV}{\chi},} \qquad (18.7)$$

Péclet number: heat advection over heat conduction

where L is a characteristic length scale of the flow, and V is a characteristic speed. The Péclet number measures the relative importance of advection and heat conduction.

18.2 Diffusive Heat Conduction—Cooling a Nuclear Reactor; Thermal Boundary Layers

EXERCISES

Exercise 18.1 *Problem: Fukushima-Daiichi Power Plant*

After the earthquake that triggered the catastrophic failure of the Fukushima-Daiichi nuclear power plant on March 11, 2011, reactor operation was immediately stopped. However, the subsequent tsunami disabled the cooling system needed to remove the decay heat, which was still being generated. This system failure led to a meltdown of three reactors and escape of radioactive material.

The *boiling water reactors* that were in use each generated $\sim$500 MW of heat, under normal operation, and the decay-heat production amounted to $\sim$30 MW. Suppose that this decay heat was being carried off equally by a large number of cylindrical pipes of length $L \sim 10$ m and inner radius $R \sim 10$ mm, taking it from the reactor core, where the water temperature was $T_0 \sim 550$ K and the pressure was $P_0 \sim 10^7$ N m^{-2}, to a heat exchanger. Suppose, initially, that the flow was laminar, so that the fluid velocity had the parabolic Poiseuille profile:

$$v = 2\bar{v}\left(1 - \frac{\varpi^2}{R^2}\right) \quad (18.8)$$

[Eq. (13.80) and associated discussion]. Here, ϖ is the cylindrical radial coordinate measured from the axis of the pipe, and $\bar{v}$ is the mean speed along the pipe. As the goal was to carry the heat away efficiently during normal operation, the pipe was thermally well insulated, so its inner wall was at nearly the same temperature as the core of the fluid (at $\varpi = 0$). The total temperature drop ΔT down the length L was then $\Delta T \ll T_0$, and the longitudinal temperature gradient was constant, so the temperature distribution in the pipe had the form:

$$T = T_0 - \Delta T \frac{z}{L} + f(\varpi). \quad (18.9)$$

(a) Use Eq. (18.4) to show that

$$f = \frac{\bar{v} R^2 \Delta T}{2 \chi L}\left[\frac{3}{4} - \frac{\varpi^2}{R^2} + \frac{1}{4}\frac{\varpi^4}{R^4}\right]. \quad (18.10)$$

(b) Derive an expression for the conductive heat flux through the walls of the pipe and show that the ratio of the heat escaping through the walls to that advected by the fluid was $\Delta T/T$. (Ignore the influence of the temperature gradient on the velocity field, and treat the thermal diffusivity and specific heat as constant throughout the flow.)

(c) The flow would only remain laminar so long as the Reynolds number was Re $\lesssim$ Re$_{\text{crit}} \sim 2{,}000$, the critical value for transition to turbulence. Show that the maximum power that could be carried by a laminar flow was:

$$\dot{Q}_{\max} \sim \frac{\pi}{2}\rho R \nu \text{Re}_{\text{crit}} c_p T_0 \sim \frac{\pi}{4}\left(\frac{\rho P_0 R^5 \text{Re}_{\text{crit}}}{L}\right)^{1/2} c_p T_0. \quad (18.11)$$

Estimate the mean velocity $\bar{v}$, the temperature drop ΔT, and the number of cooling pipes needed in normal operation, when the flow was about to transition to turbulence. Assume that $\chi = 10^{-7}$ m^2 s^{-1}, $c_p = 6$ kJ kg^{-1} K^{-1}, and $\rho v = \eta = 1 \times 10^{-4}$ Pa s^{-1}.

(d) Describe qualitatively what would happen if the flow became turbulent (as it did under normal operation).

Exercise 18.2 *Problem: Thermal Boundary Layers*
In Sec. 14.4, we introduced the notion of a laminar boundary layer by analyzing flow past a thin plate. Now suppose that this same plate is maintained at a different temperature from the free flow. A thermal boundary layer will form, in addition to the viscous boundary layer, which we presume to be laminar. These two boundary layers both extend outward from the wall but (usually) have different thicknesses.

(a) Explain why their relative thicknesses depend on the Prandtl number.

(b) Using Eq. (18.4), show that in order of magnitude the thickness δ_T of the thermal boundary layer is given by

$$v(\delta_T)\delta_T^2 = \ell\chi,$$

where $v(\delta_T)$ is the fluid velocity parallel to the plate at the outer edge of the thermal boundary layer, and ℓ is the distance downstream from the leading edge.

(c) Let V be the free-stream fluid velocity and ΔT be the temperature difference between the plate and the body of the flow. Estimate δ_T in the limits of large and small Prandtl numbers.

(d) What will be the boundary layer's temperature profile when the Prandtl number is exactly unity? [Hint: Seek a self-similar solution to the relevant equations. For the solution, see Lautrup (2005, Sec. 31.1).]

18.3 Boussinesq Approximation

When heat fluxes are sufficiently small, we can use Eq. (18.4) to solve for the temperature distribution in a given velocity field, ignoring the feedback of thermal effects on the velocity. However, if we imagine increasing the flow's temperature differences, so the heat fluxes also increase, at some point thermal feedback effects begin to influence the velocity significantly. Typically, the first feedback effect to occur is *buoyancy*, the tendency of the hotter (and hence lower-density) fluid to rise in a gravitational field and the colder (and hence denser) fluid to descend.[2] In this section, we describe the

---

[2] This effect is put to good use in a domestic "gravity-fed" warm-air circulation system. The furnace generally resides in the basement, not the attic!

Boussinesq approximation for describing heat-driven buoyancy effects

effects of buoyancy as simply as possible. The minimal approach, which is adequate surprisingly often, is called the *Boussinesq approximation*. Leading to Eqs. (18.12), (18.18), and (18.19), it can be used to describe many heat-driven laboratory flows and atmospheric flows, and some geophysical flows.

The types of flows for which the Boussinesq approximation is appropriate are those in which the fractional density changes are small ($|\Delta\rho| \ll \rho$). By contrast, the velocity can undergo large changes, though it remains constrained by the incompressibility relation (18.3a):

$$\boxed{\nabla \cdot \mathbf{v} = 0 \quad \text{Boussinesq (1)}.} \tag{18.12}$$

One might think that this constraint implies constant density moving with a fluid element, since mass conservation requires $d\rho/dt = -\rho \nabla \cdot v$. However, thermal expansion causes small density changes, with tiny corresponding violations of Eq. (18.12); this explains the "$\simeq$" that we used in Eqs. (18.3a). The key point is that, for these types of flows, the density is controlled to high accuracy by thermal expansion, and the velocity field is divergence free to high accuracy.

When discussing thermal expansion, it is convenient to introduce a *reference density* ρ_0 and *reference temperature* T_0, equal to some mean of the density and temperature in the region of fluid that one is studying. We shall denote by

$$\boxed{\tau \equiv T - T_0} \tag{18.13}$$

the perturbation of the temperature away from its reference value. The thermally perturbed density can then be written as

thermal expansion

$$\boxed{\rho = \rho_0(1 - \alpha\tau),} \tag{18.14}$$

where α is the thermal expansion coefficient for volume[3] [evaluated at constant pressure for the same reason as the specific heat was chosen as at constant pressure in the paragraph following Eq. (18.3c)]:

$$\boxed{\alpha = -\left(\frac{\partial \ln \rho}{\partial T}\right)_P.} \tag{18.15}$$

Equation (18.14) enables us to eliminate density perturbations as an explicit variable and replace them by temperature perturbations.

Now turn to the Navier-Stokes equation (18.3b) in a uniform external gravitational field. We expand the pressure-gradient term as

$$-\frac{\nabla P}{\rho} \simeq -\frac{\nabla P}{\rho_0}(1 + \alpha\tau) = \frac{\nabla P}{\rho_0} - \alpha\tau\mathbf{g}, \tag{18.16}$$

---

3. Note that α is three times larger than the thermal expansion coefficient for the linear dimensions of the fluid.

where we have used hydrostatic equilibrium for the unperturbed flow. As in our analysis of rotating flows [Eq. (14.55)], we introduce an *effective pressure* designed to compensate for the first-order effects of the uniform gravitational field:

$$P' = P + \rho_0 \Phi = P - \rho_0 \mathbf{g} \cdot \mathbf{x}. \tag{18.17}$$

(Notice that P' measures the amount the pressure differs from the value it would have in supporting a hydrostatic atmosphere of the fluid at the reference density.) The Navier-Stokes equation (18.3b) then becomes

$$\frac{d\mathbf{v}}{dt} = -\frac{\nabla P'}{\rho_0} - \alpha \tau \mathbf{g} + \nu \nabla^2 \mathbf{v} \quad \text{Boussinesq (2)}, \tag{18.18}$$

dropping the small term $O(\alpha P')$. In words, a fluid element accelerates in response to a buoyancy force [the sum of the first and second terms on the right-hand side of Eq. (18.18)] and a viscous force.

To solve this equation, we must be able to solve for the temperature perturbation τ. This evolves according to the standard equation of heat diffusion [Eq. (18.4)]:

$$\frac{d\tau}{dt} = \chi \nabla^2 \tau \quad \text{Boussinesq (3)}. \tag{18.19}$$

Equations (18.12), (18.18), and (18.19) are the equations of fluid flow in the Boussinesq approximation; they control the coupled evolution of the velocity $\mathbf{v}$ and the temperature perturbation τ. We now use them to discuss free convection in a laboratory apparatus.

Boussinesq equations

18.4 Rayleigh-Bénard Convection

In a relatively simple laboratory experiment to demonstrate convection, a fluid is confined between two rigid plates a distance d apart, each maintained at a fixed temperature, with the upper plate cooler than the lower by ΔT. When ΔT is small, viscous stresses, together with the no-slip boundary conditions at the plates, inhibit circulation; so, despite the upward buoyancy force on the hotter, less-dense fluid near the bottom plate, the fluid remains stably at rest with heat being conducted diffusively upward. If the plates' temperature difference ΔT is gradually increased, the buoyancy becomes gradually stronger. At some critical ΔT it will overcome the restraining viscous forces, and the fluid starts to circulate (convect) between the two plates. Our goal is to determine the critical temperature difference ΔT_{crit} for the onset of this *Rayleigh-Bénard* convection.

We now make some physical arguments to simplify the calculation of ΔT_{crit}. From our experience with earlier instability calculations, especially those involving elastic bifurcations (Secs. 11.6.1 and 12.3.5), we anticipate that for $\Delta T < \Delta T_{\text{crit}}$ the response of the equilibrium to small perturbations will be oscillatory (i.e., will have positive squared eigenfrequency ω^2), while for $\Delta T > \Delta T_{\text{crit}}$, perturbations will grow

Rayleigh-Bénard convection

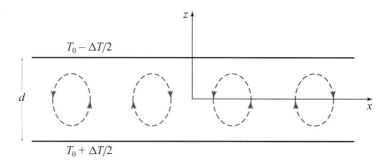

FIGURE 18.1 Rayleigh-Bénard convection. A fluid is confined between two horizontal surfaces separated by a vertical distance d. When the temperature difference between the two plates ΔT is increased sufficiently, the fluid starts to convect heat vertically. The reference effective pressure P'_0 and reference temperature T_0 are the values of P' and T measured at the midplane $z = 0$.

exponentially (i.e., will have negative ω^2). Correspondingly, at $\Delta T = \Delta T_{\text{crit}}$, ω^2 for some mode will be zero. This zero-frequency mode marks the bifurcation of equilibria from one with no fluid motions to one with slow, convective motions. We search for ΔT_{crit} by searching for a solution to the Boussinesq equations (18.12), (18.18), and (18.19) that represents this zero-frequency mode. In those equations, we choose for the reference temperature T_0, density ρ_0, and effective pressure P_0 the values at the midplane between the plates, $z = 0$ (cf. Fig. 18.1).

The unperturbed equilibrium, when $\Delta T = \Delta T_{\text{crit}}$, is a solution of the Boussinesq equations (18.12), (18.18), and (18.19) with vanishing velocity, a time-independent vertical temperature gradient $dT/dz = -\Delta T/d$, and a compensating, time-independent, vertical pressure gradient:

$$\mathbf{v} = 0, \quad \tau = T - T_0 = -\frac{\Delta T}{d} z, \quad P' = P'_0 + g\rho_0 \alpha \frac{\Delta T}{d} \frac{z^2}{2}. \tag{18.20}$$

When the zero-frequency mode is present, the velocity $\mathbf{v}$ is nonzero, and the temperature and effective pressure have additional perturbations $\delta \tau$ and $\delta P'$:

$$\mathbf{v} \neq 0, \quad \tau = T - T_0 = -\frac{\Delta T}{d} z + \delta \tau, \quad P' = P'_0 + g\rho_0 \alpha \frac{\Delta T}{d} \frac{z^2}{2} + \delta P'. \tag{18.21}$$

The perturbations $\mathbf{v}$, $\delta \tau$, and $\delta P'$ are governed by the Boussinesq equations and the boundary conditions at the plates ($z = \pm d/2$): $\mathbf{v} = 0$ (no-slip) and $\delta \tau = 0$. We manipulate these equations and boundary conditions in such a way as to get a partial differential equation for the scalar temperature perturbation $\delta \tau$ by itself, decoupled from the velocity and the pressure perturbation.

First consider the result of inserting expressions (18.21) into the Boussinesq-approximated Navier-Stokes equation (18.18). Because the perturbation mode has zero frequency, $\partial \mathbf{v}/\partial t$ vanishes; and because $\mathbf{v}$ is extremely small, we can neglect the quadratic advective term $\mathbf{v} \cdot \nabla \mathbf{v}$, thereby bringing Eq. (18.18) into the form:

$$\frac{\nabla \delta P'}{\rho_0} = \nu \nabla^2 \mathbf{v} - \mathbf{g}\alpha\delta\tau. \tag{18.22}$$

We want to eliminate $\delta P'$ from this equation. The other Boussinesq equations are of no help for this, since $\delta P'$ is absent from them. One might be tempted to eliminate δP using the equation of state $P = P(\rho, T)$; but in the present analysis our Boussinesq approximation insists that the only significant changes of density are those due to thermal expansion (i.e., it neglects the influence of pressure on density), so the equation of state cannot help us. Lacking any other way to eliminate $\delta P'$, we employ a common trick: we take the curl of Eq. (18.22). As the curl of a gradient vanishes, $\delta P'$ drops out. We then take the curl one more time and use $\nabla \cdot \mathbf{v} = 0$ to obtain

$$\nu \nabla^2 (\nabla^2 \mathbf{v}) = \alpha \mathbf{g} \nabla^2 \delta\tau - \alpha (\mathbf{g} \cdot \nabla)\nabla \delta\tau. \tag{18.23}$$

Next turn to the Boussinesq version of the equation of heat transport [Eq. (18.19)]. Inserting into it Eqs. (18.21) for τ and $\mathbf{v}$, setting $\partial \delta\tau/\partial t$ to zero (because our perturbation has zero frequency), linearizing in the perturbation, and using $\mathbf{g} = -g\mathbf{e}_z$, we obtain

$$\frac{v_z \Delta T}{d} = -\chi \nabla^2 \delta\tau. \tag{18.24}$$

This is an equation for the vertical velocity v_z in terms of the temperature perturbation $\delta\tau$. By inserting this v_z into the z component of Eq. (18.23), we achieve our goal of a scalar equation for $\delta\tau$ alone:

$$\boxed{\nu\chi \nabla^2 \nabla^2 \nabla^2 \delta\tau = \frac{\alpha g \Delta T}{d}\left(\frac{\partial^2 \delta\tau}{\partial x^2} + \frac{\partial^2 \delta\tau}{\partial y^2}\right).} \tag{18.25}$$

linearized perturbation equation for Rayleigh-Bénard convection

This is a sixth-order differential equation, even more formidable than the fourth-order equations that arise in the elasticity calculations of Chaps. 11 and 12. We now see how prudent it was to make simplifying assumptions at the outset!

The differential equation (18.25) is, however, linear, so we can seek solutions using separation of variables. As the equilibrium is unbounded horizontally, we look for a single horizontal Fourier component with some wave number k; that is, we seek a solution of the form

$$\delta\tau \propto \exp(ikx) f(z), \tag{18.26}$$

where $f(z)$ is some unknown function. Such a $\delta\tau$ will be accompanied by motions $\mathbf{v}$ in the x and z directions (i.e., $v_y = 0$) that also have the form $v_j \propto \exp(ikx) f_j(z)$ for some other functions $f_j(z)$.

18.4 Rayleigh-Bénard Convection

The ansatz (18.26) converts the partial differential equation (18.25) into the single ordinary differential equation

$$\left(\frac{d^2}{dz^2} - k^2\right)^3 f + \frac{\text{Ra}\, k^2 f}{d^4} = 0, \tag{18.27}$$

where we have introduced yet another dimensionless number

Rayleigh number: buoyancy force over viscous force

$$\boxed{\text{Ra} = \frac{\alpha g \Delta T d^3}{\nu \chi}} \tag{18.28}$$

called the *Rayleigh number*. By virtue of relation (18.24) between v_z and $\delta\tau$, the Rayleigh number is a measure of the ratio of the strength of the buoyancy term $-\alpha\delta\tau\mathbf{g}$ to the viscous term $\nu\nabla^2\mathbf{v}$ in the Boussinesq version [Eq. (18.18)] of the Navier-Stokes equation:

$$\boxed{\text{Ra} \sim \frac{\text{buoyancy force}}{\text{viscous force}}.} \tag{18.29}$$

The general solution of Eq. (18.27) is an arbitrary, linear combination of three sine functions and three cosine functions:

$$f = \sum_{n=1}^{3} A_n \cos(\mu_n k z) + B_n \sin(\mu_n k z), \tag{18.30}$$

where the dimensionless numbers μ_n are given by

$$\mu_n = \left[\left(\frac{\text{Ra}}{k^4 d^4}\right)^{1/3} e^{2\pi n i/3} - 1\right]^{1/2}; \quad n = 1, 2, 3, \tag{18.31}$$

which involves the three cube roots of unity, $e^{2\pi n i/3}$. The values of five of the coefficients A_n, B_n are fixed in terms of the sixth (an overall arbitrary amplitude) by five boundary conditions at the bounding plates. A sixth boundary condition then determines the critical temperature difference ΔT_{crit} (or equivalently, the critical Rayleigh number Ra_{crit}) at which convection sets in.

boundary conditions for Rayleigh-Bénard convection

The six boundary conditions are paired as follows:

1. The requirement that the fluid temperature be the same as the plate temperature at each plate, so $\delta\tau = 0$ at $z = \pm d/2$.

2. The no-slip boundary condition $v_z = 0$ at each plate, which by virtue of Eq. (18.24) and $\delta\tau = 0$ at the plates, translates to $\delta\tau_{,zz} = 0$ at $z = \pm d/2$ (where the indices after the comma are partial derivatives).

3. The no-slip boundary condition $v_x = 0$, which by virtue of incompressibility $\nabla \cdot \mathbf{v} = 0$ implies $v_{z,z} = 0$ at the plates, which in turn by Eq. (18.24) implies $\delta\tau_{,zzz} + \delta\tau_{,xxz} = 0$ at $z = \pm d/2$.

It is straightforward but computationally complex to impose these six boundary conditions and from them deduce the critical Rayleigh number for onset of convection (Pellew and Southwell, 1940). Rather than present the nasty details, we switch to a toy problem in which the boundary conditions are adjusted to give a simpler solution but one with the same qualitative features as for the real problem. Specifically, we replace the no-slip condition (3) ($v_x = 0$ at the plates) by a condition of no shear:

toy problem for Rayleigh-Bénard convection

3′. $v_{x,z} = 0$ at the plates. By virtue of incompressibility $\nabla \cdot \mathbf{v} = 0$, the x derivative of this condition translates to $v_{z,zz} = 0$, which by Eq. (18.24) becomes $\delta\tau_{,zzxx} + \delta\tau_{,zzzz} = 0$.

To recapitulate, we seek a solution of the form (18.30) and (18.31) that satisfies the boundary conditions (1), (2), and (3′).

The terms in Eq. (18.30) with $n = 1, 2$ always have complex arguments and thus always have z dependences that are products of hyperbolic and trigonometric functions with real arguments. For $n = 3$ and a large enough Rayleigh number, μ_3 is positive, and the solutions are pure sines and cosines. Let us just consider the $n = 3$ terms alone, in this regime, and impose boundary condition (1): $\delta\tau = 0$ at the plates. The cosine term by itself,

$$\delta\tau = \text{const} \times \cos(\mu_3 k z)\, e^{ikx}, \tag{18.32}$$

satisfies this boundary condition, if we set

$$\frac{\mu_3 k d}{2} \equiv \left[\left(\frac{\text{Ra}}{k^4 d^4}\right)^{1/3} - 1\right]^{1/2} \frac{kd}{2} = \left(m + \frac{1}{2}\right)\pi, \tag{18.33}$$

where m is an integer. It is straightforward to show, remarkably, that Eqs. (18.32) and (18.33) also satisfy boundary conditions (2) and (3′), so they solve the toy version of our problem.

As ΔT is gradually increased from zero, the Rayleigh number gradually grows, passing through the sequence of values given by Eq. (18.33) with $m = 0, 1, 2, \ldots$ (for any chosen k). At each of these values there is a zero-frequency, circulatory mode of fluid motion with horizontal wave number k, which is passing from stability to instability. The first of these, $m = 0$, represents the onset of circulation for the chosen k, and the Rayleigh number at this onset [Eq. (18.33) with $m = 0$] is

$$\text{Ra} = \frac{(k^2 d^2 + \pi^2)^3}{k^2 d^2}. \tag{18.34}$$

This Ra(k) relation is plotted as a thick curve in Fig. 18.2.

Notice in Fig. 18.2 that there is a critical Rayleigh number Ra_{crit} below which all modes are stable, independent of their wave numbers, and above which modes in some range $k_{\min} < k < k_{\max}$ are unstable. From Eq. (18.34) we deduce that for our toy problem, $\text{Ra}_{\text{crit}} = 27\pi^4/4 \simeq 658$.

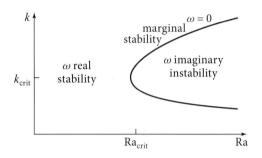

FIGURE 18.2 Horizontal wave number k of the first mode to go unstable, as a function of Rayleigh number. Along the solid curve the mode has zero frequency; to the left of the curve it is stable, to the right it is unstable. Ra_{crit} is the minimum Rayleigh number for convective instability.

When one imposes the correct boundary conditions (1), (2), and (3) [instead of our toy choice (1), (2), and (3′)] and works through the nasty details of the computation, one obtains a relation for $Ra(k)$ that looks qualitatively the same as Fig. 18.2. One deduces that convection should set in at $Ra_{crit} = 1{,}708$, which agrees reasonably well with experiment. One can carry out the same computation with the fluid's upper surface free to move (e.g., due to placing air rather than a solid plate at $z = d/2$). Such a computation predicts that convection begins at $Ra_{crit} \simeq 1{,}100$, though in practice surface tension is usually important, and its effect must be included.

<small>critical Rayleigh number for onset of Rayleigh-Bénard convection</small>

One feature of these critical Rayleigh numbers is very striking. Because the Rayleigh number is an estimate of the ratio of buoyancy forces to viscous forces [Eq. (18.29)], an order-of-magnitude analysis suggests that convection should set in at $Ra \sim 1$—which is wrong by three orders of magnitude! This provides a vivid reminder that order-of-magnitude estimates can be quite inaccurate. In this case, the main reason for the discrepancy is that the convective onset is governed by a sixth-order differential equation (18.25) and thus is highly sensitive to the lengthscale d used in the order-of-magnitude analysis. If we choose d/π rather than d as the length scale, then an order-of-magnitude estimate could give $Ra \sim \pi^6 \sim 1{,}000$, a much more satisfactory value.

<small>how order-of-magnitude analyses can fail</small>

Once convection has set in, the unstable modes grow until viscosity and non-linearities stabilize them, at which point they carry far more heat upward between the plates than does conduction. The convection's velocity pattern depends in practice on the manner in which the heat is applied, the temperature dependence of the viscosity, and the fluid's boundaries. For a limited range of Rayleigh numbers near Ra_{crit}, it is possible to excite a hexagonal pattern of *convection cells* as is largely but not entirely the case in Fig. 18.3; however other patterns can also be excited.

<small>patterns of convection cells in Rayleigh-Bénard convection</small>

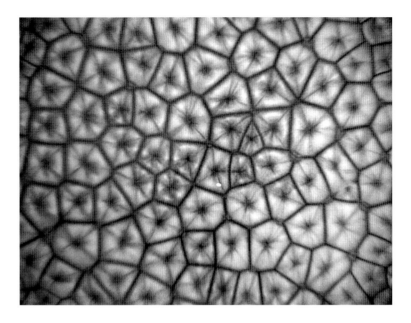

FIGURE 18.3 Convection cells in Rayleigh-Bénard convection. The fluid, which is visualized using aluminum powder, rises at the center of each cell and falls around its edges. From Maroto, Perez-Munuzuri, and Romero-Cano (2007).

Drazin and Reid (2004) suggest a kitchen experiment for observing convection cells. Place a 2-mm layer of corn or canola oil on the bottom of a skillet, and sprinkle cocoa or Ovaltine or other powder over it. Heat the skillet bottom gently and uniformly. The motion of the powder particles will reveal the convection cells, with upwelling at the cell centers and surface powder collecting and falling at the edges.

kitchen experiment

In Rayleigh-Bénard convection experiments, as the Rayleigh number is increased beyond the onset of convection, one or another sequences of equilibrium bifurcations leads to weak turbulence (see Secs. 15.6.2 and 15.6.3). When the Rayleigh number becomes very large, the convection becomes strongly turbulent.

bifurcation of Rayleigh-Bénard equilibria leading to convective turbulence

Free convection, like that in these laboratory experiments, also occurs in meteorological and geophysical flows. For example, for air in a room, the relevant parameter values are $\alpha = 1/T \sim 0.003 \text{ K}^{-1}$ (Charles' Law), and $\nu \sim \chi \sim 10^{-5} \text{ m}^2 \text{ s}^{-1}$, so the Rayleigh number is Ra $\sim 3 \times 10^8 (\Delta T / 1 \text{ K})(d/1 \text{ m})^3$. Convection in a room thus occurs extremely readily, even for small temperature differences. In fact, so many modes of convective motion can be excited that heat-driven air flow is invariably turbulent. It is therefore common in everyday situations to describe heat transport using a phenomenological turbulent thermal conductivity (Sec. 15.4.2; White, 2008, Sec. 6.10.1).

A second example, convection in Earth's mantle, is described in Box 18.2.

18.4 Rayleigh-Bénard Convection

BOX 18.2. MANTLE CONVECTION AND CONTINENTAL DRIFT

As is now well known, the continents drift over the surface of the globe on a timescale of roughly 100 million years. Despite the clear geographical evidence that the continents fit together, some geophysicists were, for a long while, skeptical that this occurred, because they were unable to identify the forces responsible for overcoming the visco-elastic resilience of the crust. It is now known that these motions are in fact slow convective circulation of the mantle driven by internally generated heat from the radioactive decay of unstable isotopes, principally uranium, thorium, and potassium.

When the heat is generated in the convective layer (which has radial thickness d), rather than passively transported from below, we must modify our definition of the Rayleigh number. Let the heat generated per unit mass per unit time be Q. In the analog of our laboratory analysis, where the fluid is assumed marginally unstable to convective motions, this Q will generate a heat flux $\sim \rho Q d$, which must be carried diffusively. Equating this flux to $\kappa \Delta T / d$, we can solve for the temperature difference ΔT between the lower and upper edges of the convective mantle: $\Delta T \sim \rho Q d^2 / \kappa$. Inserting this ΔT into Eq. (18.28), we obtain a modified expression for the Rayleigh number

$$\text{Ra}' = \frac{\alpha \rho g Q d^5}{\kappa \chi \nu}. \tag{1}$$

Let us now estimate the value of Ra' for Earth's mantle. The mantle's kinematic viscosity can be measured by post-glacial rebound studies (cf. Ex. 14.13) to be $\nu \sim 10^{17}$ m^2 s^{-1}. We can use the rate of attenuation of diurnal and annual temperature variation with depth in surface rock to estimate a thermal diffusivity $\chi \sim 10^{-6}$ m^2 s^{-1}. Direct experiment furnishes an expansion coefficient, $\alpha \sim 3 \times 10^{-5}$ K^{-1} and thermal conductivity $\kappa \sim 4$ W m^{-1} K^{-1}. The thickness of the upper mantle is $d \sim 700$ km, and the rock density is $\rho \sim 4{,}000$ kg m^{-3}. The rate of heat generation can be estimated both by chemical analysis and direct measurement at Earth's surface and turns out to be $Q \sim 10^{-11}$ W kg^{-1}. Combining these quantities, we obtain an estimated Rayleigh number $\text{Ra}' \sim 10^7$, well in excess of the critical value for convection under free slip conditions, which evaluates to $\text{Ra}'_{\text{crit}} = 868$ (Turcotte and Schubert, 1982). For this reason, it is now believed that continental drift is driven primarily by mantle convection.

Exercise 18.3 *Problem: Critical Rayleigh Number*

Use the Rayleigh criterion to estimate the temperature difference that would have to be maintained for 2 mm of corn/canola oil, or water, or mercury in a skillet to start convecting. Look up the relevant physical properties and comment on your answers. Do not perform this experiment with mercury.

Exercise 18.4 *Problem: Width of a Thermal Plume*

Consider a knife on its back, so its sharp edge points in the upward, z direction. The edge (idealized as extending infinitely far in the y direction) is hot, and by heating adjacent fluid, it creates a rising thermal plume. Introduce a temperature deficit $\Delta T(z)$ that measures the typical difference in temperature between the plume and the surrounding, ambient fluid at height z above the knife edge, and let $\delta_p(z)$ be the width of the plume at height z.

(a) Show that energy conservation implies the constancy of $\delta_p \Delta T \bar{v}_z$, where $\bar{v}_z(z)$ is the plume's mean vertical speed at height z.

(b) Make an estimate of the buoyancy acceleration, and use it to estimate $\bar{v}_z$.

(c) Use Eq. (18.19) to relate the width of the plume to the speed. Hence, show that the width of the plume scales as $\delta_p \propto z^{2/5}$ and the temperature deficit as $\Delta T \propto z^{-3/5}$.

(d) Repeat this exercise for a 3-dimensional plume above a hot spot.

18.5 Convection in Stars

The Sun and other stars generate heat in their interiors by nuclear reactions. In most stars the internal energy is predominantly in the form of hot hydrogen and helium ions and their electrons, while the thermal conductivity is due primarily to diffusing photons (Sec. 3.7.1), which have much longer mean free paths than the ions and electrons. When the photon mean free path becomes small due to high opacity (as happens in the outer 30% of the Sun; Fig. 18.4), the thermal conductivity goes down, so to transport the heat from nuclear burning, the star develops an increasingly steep

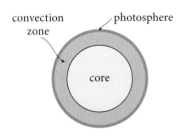

FIGURE 18.4 A convection zone occupies the outer 30% of a solar-type star.

temperature gradient. The star may then become convectively unstable and transport its energy far more efficiently by circulating its hot gas than it could have done by photon diffusion. Describing this convection is a key step in understanding the interiors of the Sun and other stars.

A heuristic argument provides the basis for a surprisingly simple description of this convection. As a foundation for our argument, let us identify the relevant physics:

1. The pressure in stars varies through many orders of magnitude (a factor $\sim 10^{12}$ for the Sun). Therefore, we cannot use the Boussinesq approximation; instead, as a fluid element rises or descends, we must allow for its density to change in response to large changes of the surrounding pressure.

2. The convection involves circulatory motions on such large scales that the attendant shears are small, and viscosity is thus unimportant.

3. Because the convection is driven by the ineffectiveness of conduction, we can idealize each fluid element as retaining its heat as it moves, so the flow is adiabatic.

4. The convection is usually well below sonic, as subsonic motions are easily sufficient to transport the nuclear-generated heat, except very close to the solar surface.

heuristic analysis of the onset of convection in stars

Our heuristic argument, then, focuses on convecting fluid blobs that move through the star's interior very subsonically, adiabatically, and without viscosity. As the motion is subsonic, each blob remains in pressure equilibrium with its surroundings. Now, suppose we make a virtual interchange between two blobs at different heights (Fig. 18.5). The blob that rises (blob B in the figure) experiences a decreased pressure and thus expands, so its density diminishes. If its density after rising is lower than that of its surroundings, then it is buoyant and continues to rise. Conversely, if the raised blob is denser than its surroundings, then it will sink back to its original location. Therefore, a criterion for convective instability is that the raised blob has lower density than its surroundings. Since the blob and its surroundings have the same pressure, and since the larger is the entropy s per unit mass of gas, the lower is its density (there being more phase space available to its particles), the fluid is convectively unstable if the raised blob has a higher entropy than its surroundings. Now, the blob's motion was adiabatic, so its entropy per unit mass s is the same after it rises as before. Therefore, the fluid is convectively unstable if the entropy per unit mass s at the location where the blob began (lower in the star) is greater than that at the location to which it rose (higher in the star); that is, *the star is convectively unstable if its entropy per unit mass decreases outward: $ds/dr < 0$*. For small blobs, this instability will be counteracted by both viscosity and heat conduction. But for large blobs, viscosity and conduction are ineffective, and the convection proceeds.

When building stellar models, astrophysicists find it convenient to determine whether a region of a model is convectively unstable by computing what its struc-

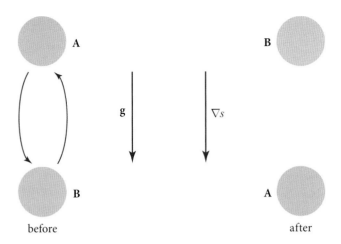

FIGURE 18.5 Convectively unstable interchange of two blobs in a star whose entropy per unit mass increases downward. Blob B rises to the former position of blob A and expands adiabatically to match the surrounding pressure. The entropy per unit mass of the blob is higher than that of the surrounding gas, and so the blob has a lower density. It will therefore be buoyant and continue to rise. Similarly, blob A will continue to sink.

ture would be without convection (i.e., with all its heat carried radiatively). That computation gives some temperature gradient dT/dr. If this computed dT/dr is *superadiabatic*, that is, if

$$-\frac{d \ln T}{d \ln r} > \left(\frac{\partial \ln T}{\partial \ln P}\right)_s \left(-\frac{d \ln P}{d \ln r}\right) \equiv -\left(\frac{d \ln T}{d \ln r}\right)_s, \qquad (18.35)$$

then the entropy s decreases outward, and the star is convectively unstable. This is known as the *Schwarzschild criterion for convection*, since it was formulated by the same Karl Schwarzschild who discovered the Schwarzschild solution to Einstein's equations (which describes a nonrotating black hole; Chap. 26).

Schwarzschild criterion for convection in stars

In practice, if the star is convective, then the convection is usually so efficient at transporting heat that the actual temperature gradient is only slightly superadiabatic, that is, the entropy s is nearly independent of radius—it decreases outward only very slightly. (Of course, the entropy can increase significantly outward in a convectively stable zone, where radiative diffusion is adequate to transport heat.)

actual temperature gradient in convective region of a star

We can demonstrate the efficiency of convection by estimating the convective heat flux when the temperature gradient is slightly superadiabatic, that is, when $\Delta|\nabla T| \equiv |(dT/dr)| - |(dT/dr)_s|$ is slightly positive. As a tool in our estimate, we introduce the concept of the *mixing length*, denoted by l—the typical distance a blob travels before breaking up. As the blob is in pressure equilibrium, we can estimate its fractional density difference from its surroundings by $\Delta\rho/\rho \sim \Delta T/T \sim \Delta|\nabla T|l/T$. Invoking Archimedes' law, we estimate the blob's acceleration to be $\sim g\Delta\rho/\rho \sim g\Delta|\nabla T|l/T$ (where g is the local acceleration of gravity); hence the average speed with which a

mixing length

18.5 Convection in Stars

blob rises or sinks is $\bar{v} \sim (g\Delta|\nabla T|/T)^{1/2} l$. The convective heat flux is then given by

estimate of convective heat flux

$$F_{\text{conv}} \sim c_P \rho \bar{v} l \Delta |\nabla T|$$
$$\sim c_P \rho (g/T)^{1/2} (\Delta |\nabla T|)^{3/2} l^2. \qquad (18.36)$$

estimate of amount by which temperature gradient exceeds adiabatic in convective region

We can bring this expression into a more useful form, accurate to within factors of order unity, by (i) setting the mixing length equal to the pressure scale height $l \sim H = |dr/d\ln P|$ (as is usually the case in the outer parts of a star); (ii) setting $c_P \sim h/T$, where h is the enthalpy per unit mass [cf. the first law of thermodynamics, Eq. (3) of Box 13.2]; (iii) setting $g = -(P/\rho) d\ln P/dr \sim C^2 |d\ln P/dr|$ [cf. the equation of hydrostatic equilibrium (13.13) and Eq. (16.48) for the speed of sound C]; and (iv) setting $|\nabla T| \equiv |dT/dr| \sim T|d\ln P/dr|$. The resulting expression for F_{conv} can then be inverted to give

$$\frac{|\Delta \nabla T|}{|\nabla T|} \sim \left(\frac{F_{\text{conv}}}{h\rho C}\right)^{2/3} \sim \left(\frac{F_{\text{conv}}}{\frac{5}{2} P \sqrt{k_B T/m_p}}\right)^{2/3}. \qquad (18.37)$$

Here the last expression is obtained from the fact that the gas is fully ionized, so its enthalpy is $h = \frac{5}{2} P/\rho$, and its speed of sound is about the thermal speed of its protons (the most numerous massive particle), $C \sim \sqrt{k_B T/m_p}$ (with k_B Boltzmann's constant and m_p the proton rest mass).

solar convection zone

It is informative to apply this estimate to the convection zone of the Sun (the outer $\sim$30% of its radius; Fig. 18.4). The luminosity of the Sun is $\sim 4 \times 10^{26}$ W, and its radius is 7×10^5 km, so its convective energy flux is $F_{\text{conv}} \sim 10^8$ W m^{-2}. First consider the convection zone's base. The pressure there is $P \sim 1$ TPa, and the temperature is $T \sim 10^6$ K, so Eq. (18.37) predicts $|\Delta \nabla T|/|\nabla T| \sim 3 \times 10^{-7}$, so the temperature gradient at the base of the convection zone need only be superadiabatic by a few parts in 10 million to carry the solar energy flux.

By contrast, at the top of the convection zone (which is nearly at the solar surface), the gas pressure is only $\sim$10 kPa, and the sound speed is $\sim$10 km s^{-1}, so $h\rho c \sim 10^8$ W m^{-2}, and $|\Delta \nabla T|/|\nabla T| \sim 1$; that is, the temperature gradient must depart significantly from the adiabatic gradient to carry the heat. Moreover, the convective elements, in their struggle to carry the heat, move with a significant fraction of the sound speed, so it is no longer true that they are in pressure equilibrium with their surroundings. A more sophisticated theory of convection is therefore necessary near the solar surface.

Convection is important in some other types of stars. It is the primary means of heat transport in the cores of stars with high mass and high luminosity, and throughout very young stars before they start to burn their hydrogen in nuclear reactions.

Exercise 18.5 *Problem: Radiative Transport*

The density and temperature in the deep interior of the Sun are roughly 0.1 kg m^{-3} and 1.5×10^7 K.

(a) Estimate the central gas pressure and radiation pressure and their ratio.

(b) The mean free path of the radiation is determined almost equally by Thomson scattering, bound-free absorption, and free-free absorption. Estimate numerically the photon mean free path and hence estimate the photon escape time and the luminosity. How well do your estimates compare with the known values for the Sun?

Exercise 18.6 *Problem: Bubbles*

Consider a small bubble of air rising slowly in a large expanse of water. If the bubble is large enough for surface tension to be ignored, then it will form an irregular cap of radius r. Show that the speed with which the bubble rises is roughly $(gr)^{1/2}$. (A more refined estimate gives a numerical coefficient of 2/3.)

18.6 Double Diffusion—Salt Fingers [T2]

As we have described it so far, convection is driven by the presence of an unbalanced buoyancy force in an equilibrium distribution of fluid. However, it can also arise as a higher-order effect, even if the fluid initially is stably stratified (i.e., if the density gradient is in the same direction as gravity). An example is *salt fingering*, a rapid mixing that can occur when warm, salty water lies at rest above colder fresh water. The higher temperature of the upper fluid outbalances the weight of its salt, making it more buoyant than the fresh water below. However, in a small, localized, downward perturbation of the warm, salty water, heat diffuses laterally into the colder surrounding water faster than salt diffuses, increasing the perturbation's density, so it will continue to sink.

salt fingering and the mechanism behind it

It is possible to describe this instability using a local perturbation analysis. The setup is somewhat similar to the one we used in Sec. 18.4 to analyze Rayleigh-Bénard convection. We consider a stratified fluid in an equilibrium state, in which there is a vertical gradient of the temperature, and as before, we measure its departure from a reference temperature T_0 at a midplane ($z = 0$) by $\tau \equiv T - T_0$. We presume that in the equilibrium state τ varies linearly with z, so $\nabla \tau = (d\tau/dz)\mathbf{e}_z$ is constant. Similarly, we characterize the salt concentration by $\mathcal{C} \equiv$ (concentration) $-$ (equilibrium concentration at the midplane); and we assume that in the equilibrium state, $\mathcal{C}$, like τ, varies linearly with height, so $\nabla \mathcal{C} = (d\mathcal{C}/dz)\mathbf{e}_z$ is constant. The density ρ is equal to the equilibrium density at the midplane plus corrections due to thermal expansion and salt concentration:

$$\rho = \rho_0 - \alpha \rho_0 \tau + \beta \rho_0 \mathcal{C} \qquad (18.38)$$

[cf. Eq. (18.14)]. Here β is a constant for concentration analogous to the thermal expansion coefficient α for temperature. In this problem, by contrast with Rayleigh-Bénard convection, it is easier to work directly with the pressure than with the modified pressure. In equilibrium, hydrostatic equilibrium dictates that its gradient be $\nabla P = -\rho \mathbf{g}$.

Now, let us perturb about this equilibrium state and write down the linearized equations for the evolution of the perturbations. We denote the perturbation of temperature (relative to the reference temperature) by $\delta\tau$, of salt concentration by $\delta\mathcal{C}$, of density by $\delta\rho$, of pressure by δP, and of velocity by simply $\mathbf{v}$ (since the unperturbed state has $\mathbf{v} = 0$). We do not ask about the onset of instability, but rather (because we expect our situation to be generically unstable) we seek a dispersion relation $\omega(\mathbf{k})$ for the perturbations. Correspondingly, in all our perturbation equations we replace $\partial/\partial t$ with $-i\omega$ and ∇ with $i\mathbf{k}$, except for the equilibrium $\nabla\mathcal{C}$ and $\nabla\tau$, which are constants.

perturbation equations

The first of our perturbation equations is the linearized Navier-Stokes equation (18.3b):

$$-i\omega\rho_0\mathbf{v} = -i\mathbf{k}\delta P + \mathbf{g}\delta\rho - \nu k^2 \rho_0 \mathbf{v}, \tag{18.39a}$$

where we have kept the viscous term, because we expect the Prandtl number to be of order unity (for water, $\Pr \sim 6$). Low velocity implies incompressibity $\nabla \cdot \mathbf{v} = 0$, which becomes

$$\mathbf{k} \cdot \mathbf{v} = 0. \tag{18.39b}$$

The density perturbation follows from the perturbed form of Eq. (18.38):

$$\delta\rho = -\alpha\rho_0\delta\tau + \beta\rho_0\delta\mathcal{C}. \tag{18.39c}$$

The temperature perturbation is governed by Eq. (18.19), which linearizes to

$$-i\omega\delta\tau + (\mathbf{v} \cdot \nabla)\tau = -\chi k^2 \delta\tau. \tag{18.39d}$$

Assuming that the timescale for the salt to diffuse is much longer than that for the temperature to diffuse, we can ignore salt diffusion altogether, so that $d\delta\mathcal{C}/dt = 0$:

$$-i\omega\delta\mathcal{C} + (\mathbf{v} \cdot \nabla)\mathcal{C} = 0. \tag{18.39e}$$

Equations (18.39) are five equations for the five unknowns δP, $\delta\rho$, $\delta\mathcal{C}$, $\delta\tau$, and $\mathbf{v}$, one of which is a three-component vector! Unless we are careful, we will end up with a seventh-order algebraic equation. Fortunately, there is a way to keep the algebra manageable. First, we eliminate the pressure perturbation by taking the curl of Eq. (18.39a) [or equivalently, by crossing $\mathbf{k}$ into Eq. (18.39a)]:

$$(-i\omega + \nu k^2)\rho_0 \mathbf{k} \times \mathbf{v} = \mathbf{k} \times \mathbf{g}\delta\rho. \tag{18.40a}$$

Taking the curl of this equation again allows us to incorporate incompressibility (18.39b). Then dotting into $\mathbf{g}$, we obtain

$$(i\omega - \nu k^2)\rho_0 k^2 \mathbf{g} \cdot \mathbf{v} = [(\mathbf{k} \cdot \mathbf{g})^2 - k^2 g^2]\delta\rho. \tag{18.40b}$$

Since **g** points vertically, Eq. (18.40b) is one equation for the density perturbation in terms of the vertical velocity perturbation v_z. We can obtain a second equation of this sort by inserting Eq. (18.39d) for $\delta\tau$ and Eq. (18.39e) for $\delta\mathcal{C}$ into Eq. (18.39c); the result is

$$\delta\rho = -\left(\frac{\alpha\rho_0}{i\omega - \chi k^2}\right)(\mathbf{v}\cdot\nabla)\tau + \frac{\beta\rho_0}{i\omega}(\mathbf{v}\cdot\nabla)\mathcal{C}. \tag{18.40c}$$

Since the unperturbed gradients of temperature and salt concentration are both vertical, Eq. (18.40c), like Eq. (18.40b), involves only v_z and not v_x or v_y. Solving both Eqs. (18.40b) and (18.40c) for the ratio $\delta\rho/v_z$ and equating these two expressions, we obtain the following dispersion relation for our perturbations:

$$\omega(\omega + i\nu k^2)(\omega + i\chi k^2) + \left[1 - \frac{(\mathbf{k}\cdot\mathbf{g})^2}{k^2 g^2}\right][\omega\alpha(\mathbf{g}\cdot\nabla)\tau - (\omega + i\chi k^2)\beta(\mathbf{g}\cdot\nabla)\mathcal{C}] = 0. \tag{18.41}$$

dispersion relation for salt-fingering perturbations

When **k** is real, as we shall assume, we can write this dispersion relation as a cubic equation for $p = -i\omega$ with real coefficients. The roots for p are either all real or one real and two complex conjugates, and growing modes have the real part of p positive. When the constant term in the cubic is negative, that is, when

$$(\mathbf{g}\cdot\nabla)\mathcal{C} < 0, \tag{18.42}$$

we are guaranteed that there is at least one positive, real root p and this root corresponds to an unstable, growing mode. Therefore, *a sufficient condition for instability is that the concentration of salt increase with height!*

sufficient condition for salt-fingering instability

By inspecting the dispersion relation, we conclude that the growth rate will be maximal when $\mathbf{k}\cdot\mathbf{g} = 0$ (i.e., when the wave vector is horizontal). What is the direction of the velocity **v** for these fastest-growing modes? The incompressibility equation (18.39b) shows that **v** is orthogonal to the horizontal **k**; Eq. (18.40a) states that $\mathbf{k}\times\mathbf{v}$ points in the same direction as $\mathbf{k}\times\mathbf{g}$, which is horizontal, since **g** is vertical. These two conditions imply that **v** points vertically. Therefore, these fastest modes represent *fingers* of salty water descending past rising fingers of fresh water (Fig. 18.6). For large k (narrow fingers), the dispersion relation (18.41) predicts a growth rate given approximately by

fastest salt-fingering modes

$$p = -i\omega \sim \frac{\beta(-\mathbf{g}\cdot\nabla)\mathcal{C}}{\nu k^2}. \tag{18.43}$$

Thus the growth of narrow fingers is driven by the concentration gradient and retarded by viscosity. For larger fingers, the temperature gradient participates in the retardation, since the heat must diffuse to break the buoyant stability.

Now let us turn to the nonlinear development of this instability. Although we have just considered a single Fourier mode, the fingers that grow are roughly cylindrical rather than sheet-like. They lengthen at a rate that is slow enough for the heat to diffuse horizontally, though not so slow that the salt can diffuse. Let the diffusion coefficient for the salt be χ_C by analogy with χ for temperature. If the length of the fingers is L

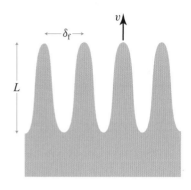

FIGURE 18.6 Salt fingers in a fluid in which warm, salty water lies on top of cold, fresh water.

and their width is δ_f, then to facilitate heat diffusion and prevent salt diffusion, the vertical speed v must satisfy

$$\frac{\chi_C L}{\delta_f^2} \ll v \ll \frac{\chi L}{\delta_f^2}. \tag{18.44}$$

Balancing the viscous acceleration $\nu v/\delta_f^2$ by the buoyancy acceleration $g\beta\delta C$, we obtain

$$v \sim \frac{g\beta\delta C \delta_f^2}{\nu}. \tag{18.45}$$

We can therefore rewrite Eq. (18.44) as

width of nonlinear salt-fingering modes

$$\left(\frac{\chi_C \nu L}{g\beta\delta C}\right)^{1/4} \ll \delta_f \ll \left(\frac{\chi \nu L}{g\beta\delta C}\right)^{1/4}. \tag{18.46}$$

Typically, $\chi_C \sim 0.01\chi$, so Eq. (18.46) implies that the widths of the fingers lie in a narrow range, as is verified by laboratory experiments.

Salt fingering can occur naturally, for example, in an estuary where cold river water flows beneath sea water warmed by the Sun. However, the development of salt fingers is quite slow, and in practice it only leads to mixing when the equilibrium velocity field is very small.

This instability is one example of a quite general type of instability known as *double diffusion*, which can arise when two physical quantities can diffuse through a fluid at different rates. Other examples include the diffusion of two different solutes and the diffusion of vorticity and heat in a rotating flow.

EXERCISES

Exercise 18.7 *Problem: Laboratory Experiment with Salt Fingers*
Make an order-of-magnitude estimate of the size of the fingers and the time it takes for them to grow in a small transparent jar. You might like to try an experiment.

Exercise 18.8 *Problem: Internal Waves* **T2**

Consider a stably stratified fluid at rest with a small (negative) vertical density gradient $d\rho/dz$.

(a) By modifying the analysis in this section, ignoring the effects of viscosity, heat conduction, and concentration gradients, show that small-amplitude linear waves, which propagate in a direction making an angle θ to the vertical, have an angular frequency given by $\omega = N|\sin\theta|$, where $N \equiv [(\mathbf{g}\cdot\nabla)\ln\rho]^{1/2}$ is known as the *Brunt-Väisälä frequency*. These waves are called *internal waves*. They can also be found at abrupt discontinuities as well as in the presence of a slow variation in the background medium. They are analogous to the Love and Rayleigh waves we have already met in our discussion of seismology (Sec. 12.4.2). Another type of internal wave is the Kelvin-Helmholtz wave (Sec. 14.6.1).

(b) Show that the group velocity of these waves is orthogonal to the phase velocity, and interpret this result physically.

Bibliographic Note

For pedagogical treatments of almost all the topics in this chapter plus much more related material, we particularly like Tritton (1987), whose phenomenological approach is lucid and appealing; and also Turner (1973), which is a thorough treatise on the influence of buoyancy (thermally induced and otherwise) on fluid motions.

Lautrup (2005) treats very nicely all this chapter's topics except convection in stars, salt fingers, and double diffusion. Landau and Lifshitz (1959, Chaps. 5 and 6) give a fairly succinct treatment of diffusive heat flow in fluids, the onset of convection in several different physical situations, and the concepts underlying double diffusion. Chandrasekhar (1961, Chaps. 2–6) gives a thorough and rich treatment of the influence of a wide variety of phenomena on the onset of convection, and on the types of fluid motions that can occur near the onset of convection. For a few pages on strongly turbulent convective heat transfer, see White (2006, Sec. 6-10).

Engineering-oriented textbooks typically say little about convection. For an engineer's viewpoint and engineering issues in convection, we recommend more specialized texts, such as Bejan (2013). For an applied mathematician's viewpoint, we suggest the treatise Pop and Ingham (2001).

CHAPTER NINETEEN

19

Magnetohydrodynamics

> ... it is only the plasma itself which does not 'understand' how beautiful the theories are
> and absolutely refuses to obey them.
>
> HANNES ALFVÉN (1970)

19.1 Overview

In preceding chapters we have described the consequences of incorporating viscosity and thermal conductivity into the description of a fluid. We now turn to our final embellishment of fluid mechanics, in which the fluid is electrically conducting and moves in a magnetic field. The study of flows of this type is known as *magnetohydrodynamics,* or MHD for short. In our discussion, we eschew full generality and with one exception just use the basic Euler equation (no viscosity, no heat diffusion, etc.) augmented by magnetic terms. This approach suffices to highlight peculiarly magnetic effects and is adequate for many applications.

The simplest example of an electrically conducting fluid is a liquid metal, for example, mercury or liquid sodium. However, the major application of MHD is in plasma physics—discussed in Part VI. (A plasma is a hot, ionized gas containing free electrons and ions.) It is by no means obvious that plasmas can be regarded as fluids, since the mean free paths for Coulomb-force collisions between a plasma's electrons and ions are macroscopically long. However, as we shall learn in Sec. 20.5, collective interactions between large numbers of plasma particles can isotropize the particles' velocity distributions in some local mean reference frame, thereby making it sensible to describe the plasma macroscopically by a mean density, velocity, and pressure. These mean quantities can then be shown to obey the same conservation laws of mass, momentum, and energy as we derived for fluids in Chap. 13. As a result, a fluid description of a plasma is often reasonably accurate. We defer to Part VI further discussion of this point, asking the reader to take it on trust for the moment. In MHD, we also implicitly assume that the average velocity of the ions is nearly the same as the average velocity of the electrons. This is usually a good approximation; if it were not so, then the plasma would carry an unreasonably large current density.

Two serious technological applications of MHD may become very important in the future. In the first, strong magnetic fields are used to confine rings or columns of hot plasma that (it is hoped) will be held in place long enough for thermonuclear fusion to occur and for net power to be generated. In the second, which is directed toward a

> **BOX 19.1. READERS' GUIDE**
>
> - This chapter relies heavily on Chap. 13 and somewhat on the treatment of vorticity transport in Sec. 14.2.
> - Part VI, Plasma Physics (Chaps. 20–23), relies heavily on this chapter.

similar goal, liquid metals or plasmas are driven through a magnetic field to generate electricity. The study of magnetohydrodynamics is also motivated by its widespread application to the description of space (in the solar system) and astrophysical plasmas (beyond the solar system). We illustrate the principles of MHD using examples drawn from all these areas.

After deriving the basic equations of MHD (Sec. 19.2), we elucidate magnetostatic (also called "hydromagnetic") equilibria by describing a *tokamak* (Sec. 19.3). This is currently the most popular scheme for the magnetic confinement of hot plasma. In our second application (Sec. 19.4) we describe the flow of conducting liquid metals or plasma along magnetized ducts and outline its potential as a practical means of electrical power generation and spacecraft propulsion. We then return to the question of magnetostatic confinement of hot plasma and focus on the stability of equilibria (Sec. 19.5). This issue of stability has occupied a central place in our development of fluid mechanics, and it will not come as a surprise to learn that it has dominated research on thermonuclear fusion in plasmas. When a magnetic field plays a role in the equilibrium (e.g., for magnetic confinement of a plasma), the field also makes possible new modes of oscillation, and some of these MHD modes can be unstable to exponential growth. Many magnetic-confinement geometries exhibit such instabilities. We demonstrate this qualitatively by considering the physical action of the magnetic field, and also formally by using variational methods.

In Sec. 19.6, we turn to a geophysical problem, the origin of Earth's magnetic field. It is generally believed that complex fluid motions in Earth's liquid core are responsible for regenerating the field through dynamo action. We use a simple model to illustrate this process.

When magnetic forces are added to fluid mechanics, a new class of waves, called magnetosonic waves, can propagate. We conclude our discussion of MHD in Sec. 19.7 by deriving the properties of these wave modes in a homogeneous plasma and discussing how they control the propagation of cosmic rays in the interplanetary and interstellar media.

As in previous chapters, we encourage our readers to view films; on magnetohydrodynamics, for example, Shercliff (1965).

19.2 Basic Equations of MHD

The equations of MHD describe the motion of a conducting fluid in a magnetic field. This fluid is usually either a liquid metal or a plasma. In both cases, the conductivity, strictly speaking, should be regarded as a tensor (Sec. 20.6.3) if the electrons' cyclotron frequency (Sec. 20.6.1) exceeds their collision frequency (the inverse of the mean time between collisions; Sec. 20.4.1). (If there are several collisions per cyclotron orbit, then the influence of the magnetic field on the transport coefficients will be minimal.) However, to keep the mathematics simple, we treat the conductivity as a constant scalar, κ_e. In fact, it turns out that for many of our applications, it is adequate to take the conductivity as infinite, and it does not matter whether that infinity is a scalar or a tensor!

Two key physical effects occur in MHD, and understanding them well is key to developing physical intuition. The first effect arises when a good conductor moves into a magnetic field (Fig. 19.1a). Electric current is induced in the conductor, which, by Lenz's law, creates its own magnetic field. This induced magnetic field tends to cancel the original, externally supported field, thereby in effect excluding the magnetic field lines from the conductor. Conversely, when the magnetic field penetrates the conductor and the conductor is moved out of the field, the induced field reinforces the applied field. The net result is that the lines of force appear to be dragged along with the conductor—they "go with the flow." Naturally, if the conductor is a fluid with complex motions, the ensuing magnetic field distribution can become quite complex, and the current builds up until its growth is balanced by Ohmic dissipation.

The second key effect is dynamical. When currents are induced by a motion of a conducting fluid through a magnetic field, a Lorentz (or $\mathbf{j} \times \mathbf{B}$) force acts on the fluid and modifies its motion (Fig. 19.1b). In MHD, the motion modifies the field, and the field, in turn, reacts back and modifies the motion. This behavior makes the theory highly nonlinear.

Before deriving the governing equations of MHD, we should consider the choice of primary variables. In electromagnetic theory, we specify the spatial and temporal variation of either the electromagnetic field or its source, the electric charge density and current density. One choice is computable (at least in principle) from the other

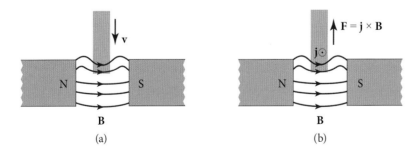

FIGURE 19.1 The two key physical effects that occur in MHD. (a) A moving conductor modifies the magnetic field by dragging the field lines with it. When the conductivity is infinite, the field lines are frozen in the moving conductor. (b) When electric current, flowing in the conductor, crosses magnetic field lines, a Lorentz force is generated that accelerates the fluid.

two key physical effects in MHD

using Maxwell's equations, augmented by suitable boundary conditions. So it is with MHD, and the choice depends on convenience. It turns out that for the majority of applications, it is most instructive to deal with the magnetic field as primary, and to use Maxwell's equations

in MHD the magnetic field is the primary variable

Maxwell's equations

$$\nabla \cdot \mathbf{E} = \frac{\rho_e}{\epsilon_0}, \tag{19.1a}$$

$$\nabla \cdot \mathbf{B} = 0, \tag{19.1b}$$

$$\nabla \times \mathbf{E} = -\frac{\partial \mathbf{B}}{\partial t}, \tag{19.1c}$$

$$\nabla \times \mathbf{B} = \mu_0 \mathbf{j} + \mu_0 \epsilon_0 \frac{\partial \mathbf{E}}{\partial t} \tag{19.1d}$$

to express the electric field $\mathbf{E}$, the current density $\mathbf{j}$, and the charge density ρ_e in terms of the magnetic field (next subsection).

19.2.1 Maxwell's Equations in the MHD Approximation

As normally formulated, Ohm's law is valid only in the rest frame of the conductor. In particular, for a conducting fluid, Ohm's law relates the current density $\mathbf{j}'$ measured in the fluid's local rest frame to the electric field $\mathbf{E}'$ measured there:

$$\mathbf{j}' = \kappa_e \mathbf{E}', \tag{19.2}$$

where κ_e is the scalar electric conductivity. Because the fluid is generally accelerated, $d\mathbf{v}/dt \neq 0$, its local rest frame is generally not inertial. Since it would produce a terrible headache to have to transform time and again from some inertial frame to the continually changing local rest frame when applying Ohm's law, it is preferable to reformulate Ohm's law in terms of the fields $\mathbf{E}$, $\mathbf{B}$, and $\mathbf{j}$ measured in an inertial frame. To facilitate this (and for completeness), we explore the frame dependence of all our electromagnetic quantities $\mathbf{E}$, $\mathbf{B}$, $\mathbf{j}$, and ρ_e.

Throughout our development of magnetohydrodynamics, we assume that the fluid moves with a nonrelativistic speed $v \ll c$ relative to our chosen reference frame. We can then express the rest-frame electric field in terms of the inertial-frame electric and magnetic fields as

$$\mathbf{E}' = \mathbf{E} + \mathbf{v} \times \mathbf{B}; \quad E' = |\mathbf{E}'| \ll E, \quad \text{so } \mathbf{E} \simeq -\mathbf{v} \times \mathbf{B}. \tag{19.3a}$$

In the first equation we have set the Lorentz factor $\gamma \equiv 1/\sqrt{1 - v^2/c^2}$ to unity, consistent with our nonrelativistic approximation. The second equation follows from the high conductivity of the fluid, which guarantees that current will quickly flow in whatever manner it must to annihilate any electric field $\mathbf{E}'$ that might be formed in the fluid's local rest frame. By contrast with the extreme frame dependence (19.3a) of the electric field, the magnetic field is essentially the same in the fluid's local rest frame as in the laboratory. More specifically, the analog of Eq. (19.3a) is $\mathbf{B}' = \mathbf{B} - (\mathbf{v}/c^2) \times \mathbf{E}$; and since $E \sim vB$, the second term is of magnitude $(v/c)^2 B$, which is negligible, giving

$$\mathbf{B}' \simeq \mathbf{B}. \tag{19.3b}$$

Because $\mathbf{E}$ is highly frame dependent, so is its divergence, the electric charge density ρ_e. In the laboratory frame, where $E \sim vB$, Gauss's and Ampère's laws [Eqs. (19.1a,d)] imply that $\rho_e \sim \epsilon_0 vB/L \sim (v/c^2) j$, where L is the lengthscale on which $\mathbf{E}$ and $\mathbf{B}$ vary; and the relation $E' \ll E$ with Gauss's law implies $|\rho'_e| \ll |\rho_e|$:

$$\rho_e \sim j\, v/c^2, \qquad |\rho'_e| \ll |\rho_e|. \tag{19.3c}$$

By transforming the current density between frames and approximating $\gamma \simeq 1$, we obtain $\mathbf{j}' = \mathbf{j} + \rho_e \mathbf{v} = \mathbf{j} + O(v/c)^2 j$; so in the nonrelativistic limit (first order in v/c) we can ignore the charge density and write

$$\mathbf{j}' = \mathbf{j}. \tag{19.3d}$$

To recapitulate, in nonrelativistic magnetohydrodynamic flows, the magnetic field and current density are frame independent up to fractional corrections of order $(v/c)^2$, while the electric field and charge density are highly frame dependent and are generally small in the sense that $E/c \sim (v/c)B \ll B$ and $\rho_e \sim (v/c^2) j \ll j/c$ [in Gaussian cgs units we have $E \sim (v/c)B \ll B$ and $\rho_e c \sim (v/c) j \ll j$].

in MHD, magnetic field and current density are approximately frame independent; electric field and charge density are small and frame dependent

Combining Eqs. (19.2), (19.3a), and (19.3d), we obtain the nonrelativistic form of Ohm's law in terms of quantities measured in our chosen inertial, laboratory frame:

$$\mathbf{j} = \kappa_e (\mathbf{E} + \mathbf{v} \times \mathbf{B}). \tag{19.4}$$

Ohm's law

We are now ready to derive explicit equations for the (inertial-frame) electric field and current density in terms of the (inertial-frame) magnetic field. In our derivation, we denote by L the lengthscale on which the magnetic field changes.

We begin with Ampère's law written as $\nabla \times \mathbf{B} - \mu_0 \mathbf{j} = \mu_0 \epsilon_0 \partial \mathbf{E}/\partial t = (1/c^2) \partial \mathbf{E}/\partial t$, and we notice that the time derivative of $\mathbf{E}$ is of order $Ev/L \sim Bv^2/L$ (since $E \sim vB$). Therefore, the right-hand side is $O[Bv^2/(c^2 L)]$ and thus can be neglected compared to the $O(B/L)$ term on the left, yielding:

$$\boxed{\mathbf{j} = \frac{1}{\mu_0} \nabla \times \mathbf{B}.} \tag{19.5a}$$

current density in terms of magnetic field

We next insert this expression for $\mathbf{j}$ into the inertial-frame Ohm's law (19.4), thereby obtaining

$$\boxed{\mathbf{E} = -\mathbf{v} \times \mathbf{B} + \frac{1}{\kappa_e \mu_0} \nabla \times \mathbf{B}.} \tag{19.5b}$$

electric field in terms of magnetic field

If we happen to be interested in the charge density (which is rare in MHD), we can compute it by taking the divergence of this electric field:

charge density in terms of magnetic field

$$\rho_e = -\epsilon_0 \nabla \cdot (\mathbf{v} \times \mathbf{B}). \tag{19.5c}$$

Equations (19.5) express all the secondary electromagnetic variables in terms of our primary one, $\mathbf{B}$. This has been possible because of the high electric conductivity κ_e and our choice to confine ourselves to nonrelativistic (low-velocity) situations; it would not be possible otherwise.

We next derive an evolution law for the magnetic field by taking the curl of Eq. (19.5b), using Maxwell's equation $\nabla \times \mathbf{E} = -\partial \mathbf{B}/\partial t$ and the vector identity $\nabla \times (\nabla \times \mathbf{B}) = \nabla(\nabla \cdot \mathbf{B}) - \nabla^2 \mathbf{B}$, and using $\nabla \cdot \mathbf{B} = 0$. The result is

evolution law for magnetic field

$$\frac{\partial \mathbf{B}}{\partial t} = \nabla \times (\mathbf{v} \times \mathbf{B}) + \left(\frac{1}{\mu_0 \kappa_e}\right) \nabla^2 \mathbf{B}, \tag{19.6}$$

which, using Eqs. (14.4) and (14.5) with $\boldsymbol{\omega}$ replaced by $\mathbf{B}$, can also be written as

$$\frac{D\mathbf{B}}{Dt} = -\mathbf{B} \nabla \cdot \mathbf{v} + \left(\frac{1}{\mu_0 \kappa_e}\right) \nabla^2 \mathbf{B}, \tag{19.7}$$

where D/Dt is the fluid derivative defined in Eq. (14.5). When the flow is incompressible (as it often will be), the $\nabla \cdot \mathbf{v}$ term vanishes.

Equation (19.6)—or equivalently, Eq. (19.7)—is called the *induction equation* and describes the temporal evolution of the magnetic field. It is the same in form as the propagation law for vorticity $\boldsymbol{\omega}$ in a flow with $\nabla P \times \nabla \rho = 0$ [Eq. (14.3), or (14.6) with $\boldsymbol{\omega} \nabla \cdot \mathbf{v}$ added in the compressible case]. The $\nabla \times (\mathbf{v} \times \mathbf{B})$ term in Eq. (19.6) dominates when the conductivity is large and can be regarded as describing the freezing of magnetic field lines in the fluid in the same way as the $\nabla \times (\mathbf{v} \times \boldsymbol{\omega})$ term describes the freezing of vortex lines in a fluid with small viscosity ν (Fig. 19.2). By analogy with Eq. (14.10), when flux-freezing dominates, the fluid derivative of $\mathbf{B}/\rho$ can be written as

for large conductivity: freezing of magnetic field into the fluid

$$\frac{D}{Dt}\left(\frac{\mathbf{B}}{\rho}\right) \equiv \frac{d}{dt}\left(\frac{\mathbf{B}}{\rho}\right) - \left(\frac{\mathbf{B}}{\rho} \cdot \nabla\right) \mathbf{v} = 0, \tag{19.8}$$

where ρ is mass density (not to be confused with charge density ρ_e). Equation (19.8) states that $\mathbf{B}/\rho$ evolves in the same manner as the separation $\Delta \mathbf{x}$ between two points in the fluid (cf. Fig. 14.4 and associated discussion).

The term $[1/(\mu_0 \kappa_e)]\nabla^2 \mathbf{B}$ in the B-field evolution equation (19.6) or (19.7) is analogous to the vorticity diffusion term $\nu \nabla^2 \boldsymbol{\omega}$ in the vorticity evolution equation (14.3) or (14.6). Therefore, when κ_e is not too large, magnetic field lines will diffuse through the fluid. The effective diffusion coefficient (analogous to ν) is

magnetic diffusion coefficient

$$D_M = 1/(\mu_0 \kappa_e). \tag{19.9a}$$

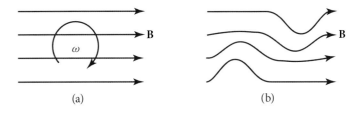

FIGURE 19.2 Pictorial representation of the evolution of the magnetic field in a fluid endowed with infinite electrical conductivity. (a) A uniform magnetic field at time $t = 0$ in a vortex. (b) At a later time, when the fluid has rotated through $\sim 30°$, the circulation has stretched and distorted the magnetic field.

Earth's magnetic field provides an example of field diffusion. That field is believed to be supported by electric currents flowing in Earth's iron core. Now, we can estimate the electric conductivity of iron under these conditions and from it deduce a value for the diffusivity, $D_M \sim 1 \, \text{m}^2 \, \text{s}^{-1}$. The size of Earth's core is $L \sim 10^4$ km, so if there were no fluid motions, then we would expect the magnetic field to diffuse out of the core and escape from Earth in a time

$$\tau_M \sim \frac{L^2}{D_M} \quad (19.9\text{b}) \quad \text{magnetic decay time}$$

~ 3 million years, which is much shorter than the age of Earth, ~ 5 billion years. The reason for this discrepancy, as we discuss in Sec. 19.6, is that there are internal circulatory motions in the liquid core that are capable of regenerating the magnetic field through dynamo action.

Although Eq. (19.6) describes a genuine diffusion of the magnetic field, to compute with confidence the resulting magnetic decay time, one must solve the complete boundary value problem. To give a simple illustration, suppose that a poor conductor (e.g., a weakly ionized column of plasma) is surrounded by an excellent conductor (e.g., the metal walls of the container in which the plasma is contained), and that magnetic field lines supported by wall currents thread the plasma. The magnetic field will only diminish after the wall currents undergo Ohmic dissipation, which can take much longer than the diffusion time for the plasma column alone.

It is customary to introduce a dimensionless number called the *magnetic Reynolds number*, R_M, directly analogous to the fluid Reynolds number Re, to describe the relative importance of flux freezing and diffusion. The fluid Reynolds number can be regarded as the ratio of the magnitude of the vorticity-freezing term, $\nabla \times (\mathbf{v} \times \boldsymbol{\omega}) \sim (V/L)\omega$, in the vorticity evolution equation, $\partial \boldsymbol{\omega}/\partial t = \nabla \times (\mathbf{v} \times \boldsymbol{\omega}) + \nu \nabla^2 \boldsymbol{\omega}$, to the magnitude of the diffusion term, $\nu \nabla^2 \boldsymbol{\omega} \sim (\nu/L^2)\omega$: $\text{Re} = (V/L)(\nu/L^2)^{-1} = VL/\nu$. Here V is a characteristic speed, and L a characteristic lengthscale of the flow. Similarly, the magnetic Reynolds number is the ratio of the magnitude of the

TABLE 19.1: Characteristic magnetic diffusivities D_M, decay times τ_M, and magnetic Reynolds numbers R_M for some common MHD flows with characteristic length scales L and velocities V

Substance	L (m)	V (m s^{-1})	D_M (m^2 s^{-1})	τ_M (s)	R_M
Mercury	0.1	0.1	1	0.01	0.01
Liquid sodium	0.1	0.1	0.1	0.1	0.1
Laboratory plasma	1	100	10	0.1	10
Earth's core	10^7	0.1	1	10^{14}	10^6
Interstellar gas	10^{17}	10^3	10^3	10^{31}	10^{17}

magnetic-flux-freezing term, $\nabla \times (\mathbf{v} \times \mathbf{B}) \sim (V/L)B$, to the magnitude of the magnetic-flux-diffusion term, $D_M \nabla^2 \mathbf{B} = [1/(\mu_o \kappa_e)] \nabla^2 \mathbf{B} \sim B/(\mu_o \kappa_e L^2)$, in the induction equation (19.6):

magnetic Reynolds number and magnetic field freezing

$$R_M = \frac{V/L}{D_M/L^2} = \frac{VL}{D_M} = \mu_o \kappa_e V L. \quad (19.9c)$$

When $R_M \gg 1$, the field lines are effectively frozen in the fluid; when $R_M \ll 1$, Ohmic dissipation is dominant, and the field lines easily diffuse through the fluid.

Magnetic Reynolds numbers and diffusion times for some typical MHD flows are given in Table 19.1. For most laboratory conditions, R_M is modest, which means that electric resistivity $1/\kappa_e$ is significant, and the magnetic diffusivity D_M is rarely

perfect MHD: infinite conductivity and magnetic field freezing

negligible. By contrast, in space physics and astrophysics, R_M is usually very large, $R_M \gg 1$, so the resistivity can be ignored almost always and everywhere. This limiting case, when the electric conductivity is treated as infinite, is often called *perfect MHD*.

The phrase "almost always and everywhere" needs clarification. Just as for large-Reynolds-number fluid flows, so also here, boundary layers and discontinuities can be formed, in which the gradients of physical quantities are automatically large enough

magnetic reconnection and its influence

to make $R_M \sim 1$ locally. An important example discussed in Sec. 19.6.3 is *magnetic reconnection*. This occurs when regions magnetized along different directions are juxtaposed, for example, when the solar wind encounters Earth's magnetosphere. In such discontinuities and boundary layers, the current density is high, and magnetic diffusion and Ohmic dissipation are important. As in ordinary fluid mechanics, these dissipative layers and discontinuities can control the character of the overall flow despite occupying a negligible fraction of the total volume.

19.2.2 Momentum and Energy Conservation

The fluid dynamical aspects of MHD are handled by adding an electromagnetic force term to the Euler or Navier-Stokes equation. The magnetic force density $\mathbf{j} \times \mathbf{B}$ is the sum of the Lorentz forces acting on all the fluid's charged particles in a unit volume.

There is also an electric force density $\rho_e \mathbf{E}$, but this is smaller than $\mathbf{j} \times \mathbf{B}$ by a factor $O(v^2/c^2)$ by virtue of Eqs. (19.5), so we ignore it. When $\mathbf{j} \times \mathbf{B}$ is added to the Euler equation (13.44) (or equivalently, to the Navier-Stokes equation with the viscosity neglected as unimportant in the situations we shall study), it takes the following form:

$$\rho \frac{d\mathbf{v}}{dt} = \rho \mathbf{g} - \nabla P + \mathbf{j} \times \mathbf{B} = \rho \mathbf{g} - \nabla P + \frac{(\nabla \times \mathbf{B}) \times \mathbf{B}}{\mu_0}. \quad (19.10)$$

MHD equation of motion for fluid

Here we have used expression (19.5a) for the current density in terms of the magnetic field. This is our basic MHD force equation. In Sec. 20.6.2 we will generalize it to situations where, due to electron cyclotron motion, the pressure P is anisotropic.

Like all other force densities in this equation, the magnetic one $\mathbf{j} \times \mathbf{B}$ can be expressed as minus the divergence of a stress tensor, the magnetic portion of the Maxwell stress tensor:

$$\mathbf{T}_M = \frac{B^2 \mathbf{g}}{2\mu_0} - \frac{\mathbf{B} \otimes \mathbf{B}}{\mu_0}; \quad (19.11)$$

magnetic stress tensor

see Ex. 19.1. By virtue of $\mathbf{j} \times \mathbf{B} = -\nabla \cdot \mathbf{T}_M$ and other relations explored in Sec. 13.5 and Box 13.4, we can convert the force-balance equation (19.10) into the conservation law for momentum [generalization of Eq. (13.42)]:

momentum conservation

$$\frac{\partial (\rho \mathbf{v})}{\partial t} + \nabla \cdot (P\mathbf{g} + \rho \mathbf{v} \otimes \mathbf{v} + \mathbf{T}_g + \mathbf{T}_M) = 0. \quad (19.12)$$

Here $\mathbf{T}_g$ is the gravitational stress tensor [Eq. (1) of Box 13.4], which resembles the magnetic one:

$$\mathbf{T}_g = -\frac{g^2 \mathbf{g}}{8\pi G} + \frac{\mathbf{g} \otimes \mathbf{g}}{4\pi G}; \quad (19.13)$$

it is generally unimportant in laboratory plasmas but can be quite important in and near stars and black holes.

The two terms in the magnetic Maxwell stress tensor [Eq. (19.11)] can be identified as the "push" of an isotropic magnetic pressure of $B^2/(2\mu_0)$ that acts just like the gas pressure P, and the "pull" of a tension B^2/μ_0 that acts parallel to the magnetic field. The combination of the tension and the isotropic pressure give a net tension $B^2/(2\mu_0)$ along the field and a net pressure $B^2/(2\mu_0)$ perpendicular to the field lines (Ex. 1.14).

The magnetic force density

magnetic force density

$$\mathbf{f}_m = -\nabla \cdot \mathbf{T}_M = \mathbf{j} \times \mathbf{B} = \frac{(\nabla \times \mathbf{B}) \times \mathbf{B}}{\mu_0} \quad (19.14)$$

can be rewritten, using standard vector identities, as

$$\mathbf{f}_m = -\nabla \left(\frac{B^2}{2\mu_0} \right) + \frac{(\mathbf{B} \cdot \nabla) \mathbf{B}}{\mu_0} = -\left[\nabla \left(\frac{B^2}{2\mu_0} \right) \right]_\perp + \left[\frac{(\mathbf{B} \cdot \nabla) \mathbf{B}}{\mu_0} \right]_\perp. \quad (19.15)$$

19.2 Basic Equations of MHD

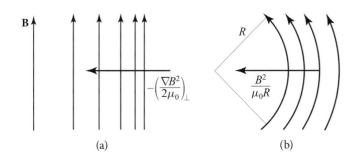

FIGURE 19.3 Contributions to the electromagnetic force density acting on a conducting fluid in a nonuniform magnetic field. A magnetic-pressure force density $-[\nabla B^2/(2\mu_0)]_\perp$ acts perpendicularly to the field. And a magnetic-curvature force density $[(\mathbf{B}\cdot\nabla)\mathbf{B}/\mu_0]_\perp$, which is also perpendicular to the magnetic field and lies in the plane of the field's bend, points toward its center of curvature. The magnitude of this curvature force density is $B^2/(\mu_0 R)$, where R is the radius of curvature.

Here "$\perp$" means keep only the components perpendicular to the magnetic field; the fact that $\mathbf{f}_m = \mathbf{j}\times\mathbf{B}$ guarantees that the net force parallel to $\mathbf{B}$ must vanish, so we can throw away the component along $\mathbf{B}$ in each term. This transversality of $\mathbf{f}_m$ means that the magnetic force neither inhibits nor promotes motion of the fluid along the magnetic field. Instead, fluid elements are free to slide along the field like beads that slide without friction along a magnetic "wire."

The "$\perp$" expressions in Eq. (19.15) indicate that the magnetic force density has two parts: first, the negative of the 2-dimensional gradient of the magnetic pressure $B^2/(2\mu_0)$ orthogonal to $\mathbf{B}$ (Fig. 19.3a), and second, an orthogonal *curvature force* $(\mathbf{B}\cdot\nabla)\mathbf{B}/\mu_0$, which has magnitude $B^2/(\mu_0 R)$, where R is the radius of curvature of a field line. This curvature force acts toward the field line's center of curvature (Fig. 19.3b) and is the magnetic-field-line analog of the force that acts on a curved wire or curved string under tension.

Just as the magnetic force density dominates and the electric force is negligible [$O(v^2/c^2)$] in our nonrelativistic situation, so also the electromagnetic contribution to the energy density is predominantly due to the magnetic term $U_M = B^2/(2\mu_0)$ with negligible electric contribution. The electromagnetic energy flux is just the Poynting flux $\mathbf{F}_M = \mathbf{E}\times\mathbf{B}/\mu_0$, with $\mathbf{E}$ given by Eq. (19.5b). Inserting these expressions into the law of energy conservation (13.58) (and continuing to neglect viscosity), we obtain

energy conservation

$$\frac{\partial}{\partial t}\left[\left(\frac{1}{2}v^2 + u + \Phi\right)\rho + \frac{B^2}{2\mu_0}\right] + \nabla\cdot\left[\left(\frac{1}{2}v^2 + h + \Phi\right)\rho\mathbf{v} + \frac{\mathbf{E}\times\mathbf{B}}{\mu_0}\right] = 0.$$

(19.16)

When the fluid's self-gravity is important, we must augment this equation with the gravitational energy density and flux, as discussed in Box 13.4.

As in Sec. 13.7.4, we can combine this energy conservation law with mass conservation and the first law of thermodynamics to obtain an equation for the evolution of entropy: Eqs. (13.75) and (13.76) are modified to read

$$\frac{\partial(\rho s)}{\partial t} + \nabla \cdot (\rho s \mathbf{v}) = \rho \frac{ds}{dt} = \frac{j^2}{\kappa_e T}. \quad (19.17)$$

entropy evolution; Ohmic dissipation

Thus, just as viscosity increases entropy through viscous dissipation, and thermal conductivity increases entropy through diffusive heat flow [Eqs. (13.75) and (13.76)], so also electrical resistivity (formally, κ_e^{-1}) increases entropy through Ohmic dissipation. From Eq. (19.17) we see that our fourth transport coefficient κ_e, like our previous three (the two coefficients of viscosity $\eta \equiv \rho \nu$ and ζ and the thermal conductivity κ), is constrained to be positive by the second law of thermodynamics.

Exercise 19.1 *Derivation: Basic Equations of MHD*

EXERCISES

(a) Verify that $-\nabla \cdot \mathbf{T}_M = \mathbf{j} \times \mathbf{B}$, where $\mathbf{T}_M$ is the magnetic stress tensor (19.11).

(b) Take the scalar product of the fluid velocity $\mathbf{v}$ with the equation of motion (19.10) and combine with mass conservation to obtain the energy conservation equation (19.16).

(c) Combine energy conservation (19.16) with the first law of thermodynamics and mass conservation to obtain Eq. (19.17) for the evolution of the entropy.

19.2.3 Boundary Conditions

The equations of MHD must be supplemented by boundary conditions at two different types of interfaces. The first is a *contact discontinuity* (i.e., the interface between two distinct media that do not mix; e.g., the surface of a liquid metal or a rigid wall of a plasma containment device). The second is a shock front that is being crossed by the fluid. Here the boundary is between shocked and unshocked fluid.

types of interfaces: contact discontinuity and shock front

We can derive the boundary conditions by transforming into a primed frame in which the interface is instantaneously at rest (not to be confused with the fluid's local rest frame) and then transforming back into our original unprimed inertial frame. In the primed frame, we resolve the velocity and magnetic and electric vectors into components normal and tangential to the surface. If $\mathbf{n}$ is a unit vector normal to the surface, then the normal and tangential components of velocity in either frame are

$$v_n = \mathbf{n} \cdot \mathbf{v}, \quad \mathbf{v}_t = \mathbf{v} - (\mathbf{n} \cdot \mathbf{v})\mathbf{n}, \quad (19.18)$$

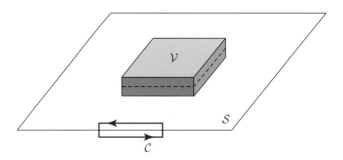

FIGURE 19.4 Elementary pill box $\mathcal{V}$ and elementary circuit $\mathcal{C}$ used in deriving the MHD junction conditions at a surface $\mathcal{S}$.

and similarly for the **E** and **B**. At a contact discontinuity, we have

boundary conditions at interfaces: Eqs. (19.19)

$$v'_n = v_n - v_{sn} = 0 \tag{19.19a}$$

on both sides of the interface surface; here v_{sn} is the normal velocity of the surface. At a shock front, mass flux across the surface is conserved [cf. Eq. (17.29a)]:

junction condition for mass flux

$$[\rho v'_n] = [\rho(v_n - v_{sn})] = 0. \tag{19.19b}$$

Here, as in Sec. 17.5, we use the notation $[X]$ to signify the difference in some quantity X across the interface, that is, the *junction condition* for X.

When we consider the magnetic field, it does not matter which frame we use, since **B** is unchanged to the Galilean order at which we are working. Let us construct a thin "pill box" $\mathcal{V}$ (Fig. 19.4) and integrate the equation $\nabla \cdot \mathbf{B} = 0$ over its volume, invoke the divergence theorem, and let the box thickness diminish to zero; thereby we see that

electromagnetic junction conditions

$$[B_n] = 0. \tag{19.19c}$$

By contrast, the tangential component of the magnetic field can be discontinuous across an interface because of surface currents: by integrating $\nabla \times \mathbf{B} = \mu_0 \mathbf{j}$ across the shock front, we can deduce that

$$[\mathbf{B}_t] = -\mu_0 \mathbf{n} \times \mathbf{J}, \tag{19.19d}$$

where **J** is the surface current density.

We deduce the junction condition on the electric field by integrating Maxwell's equation $\nabla \times \mathbf{E} = -\partial \mathbf{B}/\partial t$ over the area bounded by the circuit $\mathcal{C}$ in Fig. 19.4 and using Stokes' theorem, letting the two short legs of the circuit vanish. We thereby obtain

$$[\mathbf{E}'_t] = [\mathbf{E}_t] + [(\mathbf{v}_s \times \mathbf{B})_t] = 0, \tag{19.19e}$$

where $\mathbf{v}_s$ is the velocity of a frame that moves with the surface. Note that only the normal component of the velocity contributes to this expression, so we can replace $\mathbf{v}_s$ by $v_{sn}\mathbf{n}$. The normal component of the electric field, like the tangential component of the magnetic field, can be discontinuous, as there may be surface charge at the interface.

There are also dynamical junction conditions that can be deduced by integrating the laws of momentum conservation (19.12) and energy conservation (19.16) over the pill box and using Gauss's theorem to convert the volume integral of a divergence to a surface integral. The results, naturally, are the requirements that the normal fluxes of momentum $\mathbf{T}\cdot\mathbf{n}$ and energy $\mathbf{F}\cdot\mathbf{n}$ be continuous across the surface. Here $\mathbf{T}$ is the total stress [i.e., the quantity inside the divergence in Eq. (19.12)], and $\mathbf{F}$ is the total energy flux [i.e., the quantity inside the divergence in Eq. (19.16)]; see Eqs. (17.29)–(17.31) and associated discussion. The normal and tangential components of $[\mathbf{T}\cdot\mathbf{n}] = 0$ read

$$\boxed{\left[P + \rho(v_n - v_{sn})^2 + \frac{B_t^2}{2\mu_0}\right] = 0,} \qquad (19.19\mathrm{f})$$

dynamical junction conditions

$$\boxed{\left[\rho(v_n - v_{sn})(\mathbf{v}_t - \mathbf{v}_{st}) - \frac{B_n \mathbf{B}_t}{\mu_0}\right] = 0,} \qquad (19.19\mathrm{g})$$

where we have omitted the gravitational stress, since it will always be continuous in situations studied in this chapter (no surface layers of mass). Similarly, continuity of the energy flux $[\mathbf{F}\cdot\mathbf{n}] = 0$ reads

$$\boxed{\left[\left(\frac{1}{2}v^2 + h\right)\rho(v_n - v_{sn}) + \frac{\mathbf{n}\cdot[(\mathbf{E} + \mathbf{v}_s\times\mathbf{B})\times\mathbf{B}]}{\mu_0}\right] = 0.} \qquad (19.19\mathrm{h})$$

When the interface has plasma on one side and a vacuum magnetic field on the other, as in devices for magnetic confinement of plasmas (Sec. 19.3), the vacuum electromagnetic field, like that in the plasma, has small time derivatives: $\partial/\partial t \sim v\partial/\partial x^j$. As a result, the vacuum displacement current $\epsilon_0 \partial\mathbf{E}/\partial t$ is very small, and the vacuum Maxwell equations reduce to the same form as those in the MHD plasma but with ρ_e and $\mathbf{j}$ zero. As a result, the boundary conditions at the vacuum-plasma interface (Ex. 19.2) are those discussed above [Eqs. (19.19)], but with ρ and P vanishing on the vacuum side.

EXERCISES

Exercise 19.2 *Example and Derivation: Perfect MHD Boundary Conditions at a Fluid-Vacuum Interface*

When analyzing the stability of configurations for magnetic confinement of a plasma (Sec. 19.5), one needs boundary conditions at the plasma-vacuum interface for the special case of perfect MHD (electrical conductivity idealized as arbitrarily large).

Denote by a tilde ($\tilde{\mathbf{B}}$ and $\tilde{\mathbf{E}}$) the magnetic and electric fields in the vacuum, and reserve non-tilde symbols for quantities on the plasma side of the interface.

(a) Show that the normal-force boundary condition (19.19f) reduces to an equation for the vacuum region's tangential magnetic field:

$$\frac{\tilde{B}_t^2}{2\mu_0} = P + \frac{B_t^2}{2\mu_0}. \tag{19.20a}$$

(b) By combining Eqs. (19.19c) and (19.19g) and noting that $v_n - v_{sn}$ must vanish (why?), and assuming that the magnetic confinement entails surface currents on the interface, show that the normal component of the magnetic field must vanish on both sides of the interface:

$$\tilde{B}_n = B_n = 0. \tag{19.20b}$$

(c) When analyzing energy flow across the interface, it is necessary to know the tangential electric field. On the plasma side $\tilde{\mathbf{E}}_t$ is a secondary quantity fixed by projecting tangentially the relation $\mathbf{E} + \mathbf{v} \times \mathbf{B} = 0$. On the vacuum side it is fixed by the boundary condition (19.19e). By combining these two relations, show that

$$\tilde{\mathbf{E}}_t + \mathbf{v}_{sn} \times \tilde{\mathbf{B}}_t = 0. \tag{19.20c}$$

Exercise 19.3 *Problem: Diffusion of Magnetic Field*

Consider an infinitely long cylinder of plasma with constant electric conductivity, surrounded by vacuum. Assume that the cylinder initially is magnetized uniformly parallel to its length, and assume that the field decays quickly enough that the plasma's inertia keeps it from moving much during the decay (so $\mathbf{v} \simeq 0$).

(a) Show that the reduction of magnetic energy as the field decays is compensated by the Ohmic heating of the plasma plus energy lost to outgoing electromagnetic waves (which will be negligible if the decay is slow).

(b) Compute the approximate magnetic profile after the field has decayed to a small fraction of its original value. Your answer should be expressible in terms of a Bessel function.

Exercise 19.4 *Example: Shock with Transverse Magnetic Field*

Consider a normal shock wave ($\mathbf{v}$ perpendicular to the shock front), in which the magnetic field is parallel to the shock front, analyzed in the shock front's rest frame.

(a) Show that the junction conditions across the shock are the vanishing of all the following quantities:

$$[\rho v] = [P + \rho v^2 + B^2/(2\mu_0)] = [h + v^2/2 + B^2/(\mu_0 \rho)] = [vB] = 0. \tag{19.21}$$

(b) Specialize to a fluid with equation of state $P \propto \rho^\gamma$. Show that these junction conditions predict no compression, $[\rho] = 0$, if the upstream velocity is $v_1 = C_f \equiv \sqrt{(B_1^2/\mu_0 + \gamma P_1)/\rho_1}$. For v_1 greater than this C_f, the fluid gets compressed.

(c) Explain why the result in part (b) means that the speed of sound perpendicular to the magnetic field must be C_f. As we shall see in Sec. 19.7.2, this indeed is the case: C_f is the speed [Eq. (19.77)] of a *fast magnetosonic wave*, the only kind of sound wave that can propagate perpendicular to **B**.

Exercise 19.5 *Problem: Earth's Bow Shock*

The solar wind is a supersonic, hydromagnetic flow of plasma originating in the solar corona. At the radius of Earth's orbit, the wind's density is $\rho \sim 6 \times 10^{-21}$ kg m^{-3}, its velocity is $v \sim 400$ km s^{-1}, its temperature is $T \sim 10^5$ K, and its magnetic field strength is $B \sim 1$ nT.

(a) By balancing the wind's momentum flux with the magnetic pressure exerted by Earth's dipole magnetic field, estimate the radius above Earth at which the solar wind passes through a bow shock (Fig. 17.2).

(b) Consider a strong perpendicular shock at which the magnetic field is parallel to the shock front. Show that the magnetic field strength will increase by the same ratio as the density, when crossing the shock front. Do you expect the compression to increase or decrease as the strength of the field is increased, keeping all of the other flow variables constant?

19.2.4 Magnetic Field and Vorticity

We have already remarked on how the magnetic field and the vorticity are both axial vectors that can be written as the curl of a polar vector and that they satisfy similar transport equations. It is not surprising that they are physically intimately related. To explore this relationship in full detail would take us beyond the scope of this book. However, we can illustrate their interaction by showing how they can create each other. In brief: vorticity can twist a magnetic field, amplifying it; and an already twisted field, trying to untwist itself, can create vorticity.

vorticity–magnetic-field interactions

First, consider a simple vortex through which passes a uniform magnetic field (Fig. 19.2a). If the magnetic Reynolds number is large enough, then the magnetic field is carried with the flow and is wound up like spaghetti on the end of a fork (Fig. 19.2b, continued for a longer time). This process increases the magnetic energy in the vortex, though not the mean flux of the magnetic field. This amplification continues until either the field gradient is large enough that the field decays through Ohmic dissipation, or the field strength is large enough to react back on the flow and stop it from spinning.

Second, consider an irrotational flow containing a twisted magnetic field (Fig. 19.5a). Provided that the magnetic Reynolds number is sufficiently large, the magnetic stress, attempting to untwist the field, will act on the flow and induce vorticity (Fig. 19.5b). We can describe this formally by taking the curl of the equation

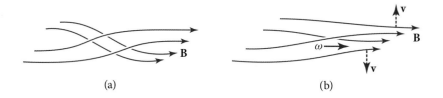

FIGURE 19.5 (a) A twisted magnetic field is frozen in an irrotational flow. (b) The field tries to untwist and in the process creates vorticity.

of motion (19.10). Assuming, for simplicity, that the density ρ is constant and the electric conductivity is infinite, we obtain

$$\frac{\partial \boldsymbol{\omega}}{\partial t} - \boldsymbol{\nabla} \times (\mathbf{v} \times \boldsymbol{\omega}) = \frac{\boldsymbol{\nabla} \times [(\boldsymbol{\nabla} \times \mathbf{B}) \times \mathbf{B}]}{\mu_0 \rho}. \qquad (19.22)$$

The term on the right-hand side of this equation changes the number of vortex lines threading the fluid, just like the $-\boldsymbol{\nabla} P \times \boldsymbol{\nabla}\rho/\rho^2$ term on the right-hand side of Eq. (14.3). However, because the divergence of the vorticity is zero, any fresh vortex lines that are made must be created as continuous curves that grow out of points or lines where the vorticity vanishes.

19.3 Magnetostatic Equilibria

19.3.1 Controlled Thermonuclear Fusion

GLOBAL POWER DEMAND

We start this section with an oversimplified discussion of the underlying problem. Earth's population has quadrupled over the past century to its present (2016) value of 7.4 billion and is still rising. We consume ~ 16 TW of power more or less equally for manufacture, transportation, and domestic use. The power is derived from oil (~ 5 TW), coal (~ 4 TW), gas (~ 4 TW), nuclear (fission) reactors (~ 1 TW), hydroelectric turbines (~ 1 TW), and alternative sources, such as solar, wind, wave, and biomass (~ 1 TW). The average power consumption is ~ 2 kW per person, with Canada and the United States in the lead, consuming ~ 10 kW per person. Despite conservation efforts, the demand for energy still appears to be rising.

motivation for controlled fusion program

Meanwhile, the burning of coal, oil, and gas produces carbon dioxide at a rate of ~ 1 Gg s^{-1}, about 10% of the total transfer rate in Earth's biomass–atmosphere–ocean *carbon cycle*. This disturbance of the carbon-cycle equilibrium has led to an increase in the atmospheric concentration of carbon dioxide by about a third over the past century and it is currently growing at an average rate of about a half percent per year. There is strong evidence to link this increase in carbon dioxide and other greenhouse gases to climate change, as exemplified by an increase in the globally averaged mean temperature of ~ 1 K over the past century. Given the long time constants associated with the three components of the carbon cycle, future projections of climate change are alarming. These considerations strongly motivate the rapid deployment of low-carbon sources of power—renewables and nuclear—and conservation.

THERMONUCLEAR FUSION

For more than 60 years, plasma physicists have striven to address this problem by releasing nuclear energy in a controlled, peaceful manner through confining plasma at a temperature in excess of 100 million degrees using strong magnetic fields. In the most widely studied scheme, deuterium and tritium combine according to the reaction

$$d + t \rightarrow \alpha + n + 22.4 \text{ MeV}. \tag{19.23}$$

d, t fusion reaction

The energy release is equivalent to $\sim$400 TJ kg^{-1}. The fast neutrons can be absorbed in a surrounding blanket of lithium, and the heat can then be used to drive a generator.

PLASMA CONFINEMENT

At first this task seemed quite simple. However, it eventually became clear that it is very difficult to confine hot plasma with a magnetic field, because most confinement geometries are unstable. In this book we restrict our attention to a few simple confinement devices, emphasizing the one that is the basis of most modern efforts, the *tokamak*.[1] In this section, we treat equilibrium configurations; in Sec. 19.5, we consider their stability.

In our discussions of both equilibrium and stability, we treat the plasma as a magnetized fluid in the MHD approximation. At first sight, treating the plasma as a fluid might seem rather unrealistic, because we are dealing with a dilute gas of ions and electrons that undergo infrequent Coulomb collisions. However, as we discuss in Sec. 20.5.2 and justify in Chaps. 22 and 23, collective effects produce a sufficiently high effective collision frequency to make the plasma behave like a fluid, so MHD is usually a good approximation for describing these equilibria and their rather slow temporal evolution.

Let us examine some numbers that characterize the regime in which a successful controlled-fusion device must operate.

PLASMA PRESSURE

The ratio of plasma pressure to magnetic pressure

$$\boxed{\beta \equiv \frac{P}{B^2/(2\mu_0)}} \tag{19.24}$$

pressure ratio β for controlled fusion

plays a key role. For the magnetic field to have any chance of confining the plasma, its pressure must exceed that of the plasma (i.e., β must be less than one). The most successful designs achieve $\beta \sim 0.2$. The largest field strengths that can be safely

---

1. Originally proposed in the Soviet Union by Andrei Sakharov and Igor Tamm in 1950. The word is a Russian abbreviation for "toroidal magnetic field."

sustained in the laboratory are $B \sim 10$ T $= 100$ kG, so $\beta \lesssim 0.2$ limits the gas pressure to $P \lesssim 10^7$ Pa ~ 100 atmospheres.

LAWSON CRITERION

Plasma fusion can only be economically feasible if more power is released by nuclear reactions than is lost to radiative cooling. Both heating and cooling are proportional to the square of the number density of hydrogen ions, n^2. However, while the radiative cooling rate increases comparatively slowly with temperature, the nuclear reaction rate increases very rapidly. (This is because, as the mean energy of the ions increases, the number of ions in the Maxwellian tail of the distribution function that are energetic enough to penetrate the Coulomb barrier increases exponentially.) Thus for the rate of heat production to greatly exceed the rate of cooling, the temperature need only be modestly higher than that required for the rates to be equal—which is a minimum temperature essentially fixed by atomic and nuclear physics. In the case of a d-t plasma, this is $T_{\min} \sim 10^8$ K. The maximum hydrogen density that can be confined is therefore $n_{\max} = P/(2k_B T_{\min}) \sim 3 \times 10^{21}$ m^{-3}. (The factor 2 comes from the electrons, which produce the same pressure as the ions.)

Now, if a volume V of plasma is confined at a given number density n and temperature $T_{\min}$ for a time τ, then the amount of nuclear energy generated will be proportional to $n^2 V \tau$, while the energy to heat the plasma up to $T_{\min}$ is $\propto nV$. Therefore, there is a minimum value of the product $n\tau$ that must be attained before net energy is produced. This condition is known as the *Lawson criterion*. Numerically, the plasma must be confined for

$$\tau \sim (n/10^{20} \text{ m}^{-3})^{-1} \text{ s}, \tag{19.25}$$

typically ~ 30 ms. The sound speed at these temperatures is $\sim 1 \times 10^6$ m s^{-1}, and so an unconfined plasma would hit the few-meter-sized walls of the vessel in which it is held in a few μs. Therefore, the magnetic confinement must be effective for typically 10^4–10^5 dynamical timescales (sound-crossing times). It is necessary that the plasma be confined and confined well if we want to build a viable fusion reactor.

19.3.2 Z-Pinch

Before discussing plasma confinement by tokamaks, we describe a simpler confinement geometry known as the *Z-pinch* and often called the Bennett pinch (Fig. 19.6a). In a Z-pinch, electric current is induced to flow along a cylinder of plasma. This current creates a toroidal magnetic field whose tension prevents the plasma from expanding radially, much like hoops on a barrel prevent it from exploding. Let us assume that the cylinder has a radius R and is surrounded by vacuum.

Now, in static equilibrium we must balance the plasma pressure gradient by a Lorentz force:

$$\nabla P = \mathbf{j} \times \mathbf{B}. \tag{19.26}$$

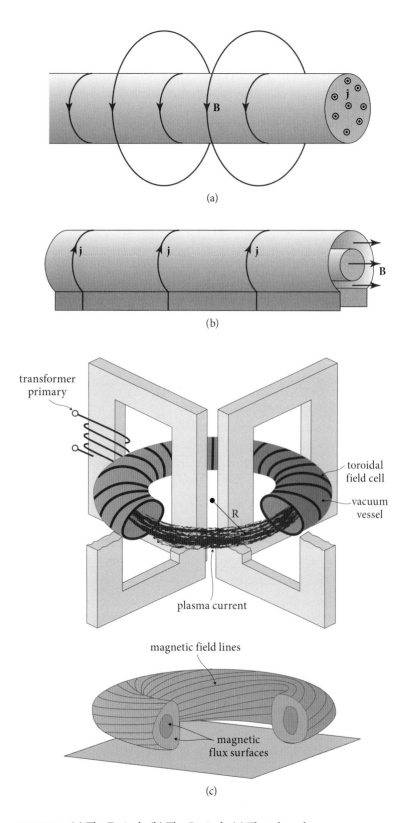

FIGURE 19.6 (a) The Z-pinch. (b) The Θ-pinch. (c) The tokamak.

(Gravitational forces can safely be ignored.) Equation (19.26) implies immediately that $\mathbf{B} \cdot \nabla P = \mathbf{j} \cdot \nabla P = 0$, so both the magnetic field and the current density lie on constant pressure (or *isobaric*) surfaces. An equivalent version of the force-balance equation (19.26), obtained using Eq. (19.15) and Fig. 19.3, is

$$\frac{d}{d\varpi}\left(P + \frac{B^2}{2\mu_0}\right) = -\frac{B^2}{\mu_0 \varpi}, \tag{19.27}$$

where ϖ is the radial cylindrical coordinate. Equation (19.27) exhibits the balance between the gradient of plasma and magnetic pressure on the left, and the magnetic tension (the "hoop force") on the right. Treating it as a differential equation for B^2 and integrating it, assuming that P falls to zero at the surface of the column, we obtain for the surface magnetic field

$$B^2(R) = \frac{4\mu_0}{R^2} \int_0^R P \varpi \, d\varpi. \tag{19.28}$$

We can reexpress the surface toroidal field in terms of the total current flowing along the plasma as $B(R) = \mu_0 I / (2\pi R)$ (Ampère's law); and assuming that the plasma is primarily hydrogen (so its ion density n and electron density are equal), we can write the pressure as $P = 2nk_B T$. Inserting these expressions into Eq. (19.28), integrating, and solving for the current, we obtain

$$I = \left(\frac{16\pi N k_B T}{\mu_0}\right)^{1/2}, \tag{19.29}$$

where N is the number of ions per unit length. For a column of plasma with diameter $2R \sim 1$ m, hydrogen density $n \sim 10^{20}$ m^{-3}, and temperature $T \sim 10^8$ K, Eq. (19.29) indicates that currents ~ 1 MA are required for confinement.

The most promising Z-pinch experiments to date have been carried out at Sandia National Laboratories in Albuquerque, New Mexico. The experimenters have impulsively compressed a column of gas into a surprisingly stable cylinder of diameter ~ 1 mm for a time ~ 100 ns, using a current of ~ 30 MA from giant capaciter banks. A transient field of ~ 1 kT was created.

19.3.3 Θ-Pinch

Θ-pinch configuration for plasma confinement

There is a complementary equilibrium for a cylindrical plasma, in which the magnetic field lies parallel to the axis and the current density encircles the cylinder (Fig. 19.6b). This configuration is called the Θ-*pinch*. It is usually established by making a cylindrical metal tube with a small gap, so that current can flow around it as shown in the figure. The tube is filled with cold plasma, and then the current is turned on quickly, producing a quickly growing longitudinal field in the tube (as in a solenoid). Since the plasma is highly conducting, the field lines cannot penetrate the plasma column but instead exert a pressure on its surface, causing it to shrink radially and rapidly. The plasma heats up due to both the radial work done on it and Ohmic heating. Equilibrium is established when the plasma's pressure P balances the magnetic pressure $B^2/(2\mu_0)$ at the plasma's surface.

Despite early promise, interest in Θ-pinches has waned, and mirror machines (Sec. 19.5.3) have become more popular.

19.3.4 Tokamak

One of the problems with the Θ- and Z-pinches (and we shall find other problems below!) is that they have ends through which plasma can escape. This is readily addressed by replacing the cylinder with a torus. The most stable geometry, called the *tokamak*, combines features of both Z- and Θ-pinches; see Fig. 19.6c. If we introduce spherical coordinates (r, θ, ϕ), then magnetic field lines and currents that lie in an r-θ plane (orthogonal to $\mathbf{e}_\phi$) are called *poloidal*, whereas their ϕ components are called *toroidal*. In a tokamak, the toroidal magnetic field is created by external poloidal current windings. However, the poloidal field is mostly created as a consequence of toroidal current induced to flow in the plasma torus. The resulting net field lines wrap around the plasma torus in a helical manner, defining a magnetic surface on which the pressure is constant. The number of poloidal transits around the torus during one toroidal transit is denoted $\iota/(2\pi)$; ι is called the *rotational transform* and is a property of the magnetic surface on which the field line resides. If $\iota/(2\pi)$ is a rational number, then the field line will close after a finite number of circuits. However, in general, $\iota/(2\pi)$ will not be rational, so a single field line will cover the whole magnetic surface ergodically. This allows the plasma to spread over the whole surface rapidly. The rotational transform is a measure of the toroidal current flowing inside the magnetic surface and of course increases as we move out from the innermost magnetic surface, while the pressure decreases.

The best performers to date (2016) include the MIT Alcator C-Mod ($n \sim 2 \times 10^{20}$ m^{-3}, $T \sim 35$ MK, $B \sim 6$ T, $\tau \sim 2$ s), and the larger Joint European Torus or JET (Keilhacker and the JET team, 1998). JET generated 16 MW of nuclear power with $\tau \sim 1$ s by burning d-t fuel, but its input power was ~ 25 MW and so, despite being a major step forward, JET fell short of achieving "break even."

The largest device, currently under construction, is the International Thermonuclear Experimental Reactor (ITER) (whose acronym means "journey" in Latin). ITER is a tokamak-based experimental fusion reactor being constructed in France by a large international consortium (see http://www.iter.org/). The outer diameter of the device is ~ 20 m, and the maximum magnetic field produced by its superconducting magnets will be ~ 14 T. Its goal is to use d-t fuel to convert an input power of ~ 50 MW into an output power of ~ 500 MW, sustained for $\sim 3{,}000$ s.[2] However, many engineering, managerial, financial, and political challenges remain to be addressed before mass production of economically viable, durable, and safe fusion reactors can begin.

---

2. Note that it would require of order 10,000 facilities with ITER's projected peak power operating continuously to supply, say, one-third of the current global power demand.

EXERCISES

Exercise 19.6 *Problem: Strength of Magnetic Field in a Magnetic Confinement Device*
The currents that are sources for strong magnetic fields have to be held in place by solid conductors. Estimate the limiting field that can be sustained using normal construction materials.

Exercise 19.7 *Problem: Force-Free Equilibria*
In an equilibrium state of a very low-β plasma, the plasma's pressure force density $-\nabla P$ is ignorably small, and so the Lorentz force density $\mathbf{j} \times \mathbf{B}$ must vanish [Eq. (19.10)]. Such a plasma is said to be "force-free." As a result, the current density is parallel to the magnetic field, so $\nabla \times \mathbf{B} = \alpha \mathbf{B}$. Show that α must be constant along a field line, and that if the field lines eventually travel everywhere, then α must be constant everywhere.

Exercise 19.8 *Example and Challenge: Spheromak*
Another magnetic confinement device which brings out some important principles is the *spheromak*. Spheromaks can be made in the laboratory (Bellan, 2000) and have also been proposed as the basis of a fusion reactor.[3] It is simplest to consider a spheromak in the limit when the plasma pressure is ignorable (low β) and the magnetic field distribution is force-free (Ex. 19.7). We just describe the simplest example of this regime.

(a) As in the previous exercise, assume that α is constant everywhere and, without loss of generality, set it equal to unity. Show that the magnetic field—and also the current density and vector potential, adopting the Coulomb gauge—satisfy the vector Helmholtz equation: $\nabla^2 \mathbf{B} + \alpha^2 \mathbf{B} = 0$.

(b) Introduce a scalar χ such that $\mathbf{B} = \alpha \mathbf{r} \times \nabla \chi + \nabla \times (\mathbf{r} \times \nabla \chi)$, with $\mathbf{r}$ the radial vector pointing out of the spheromak's center, and show that χ satisfies the scalar Helmholtz equation: $\nabla^2 \chi + \alpha^2 \chi = 0$.

(c) The Helmholtz equation in part (b) separates in spherical coordinates (r, θ, ϕ). Show that it has a nonsingular solution $\chi = j_l(\alpha r) Y_{lm}(\theta, \phi)$, where $j_l(\alpha r)$ is a spherical Bessel function, and $Y_{lm}(\theta, \phi)$ is a spherical harmonic. Evaluate this for the simplest example, the spheromak, with $l = 2$ and $m = 0$.

(d) Calculate expressions for the associated magnetic field in part (c) and either sketch or plot it.

(e) Show that the magnetic field's radial component vanishes on the surface of a sphere of radius equal to the first zero of j_l. Hence explain why a spheromak may be confined in a conducting sphere or, alternatively, by a current-free field that is uniform at large distance.

---

3. Spheromak configurations of magnetic fields have even been associated with ball lightning!

Exercise 19.9 *Problem: Magnetic Helicity*

(a) A physical quantity that turns out to be useful in describing the evolution of magnetic fields in confinement devices is *magnetic helicity*. This is defined by $H = \int dV \mathbf{A} \cdot \mathbf{B}$, where $\mathbf{A}$ is the vector potential, and the integral should be performed over all the space visited by the field lines to remove a dependence on the electromagnetic gauge. Compute the partial derivative of $\mathbf{A}$ and $\mathbf{B}$ with respect to time to show that H is conserved if $\mathbf{E} \cdot \mathbf{B} = 0$.

(b) The helicity H primarily measures the topological linkage of the magnetic field lines. Therefore, it should not be a surprise that H turns out to be relatively well-preserved, even when the plasma is losing energy through resistivity and radiation. To discuss magnetic helicity properly would take us too far into the domain of classical electromagnetic theory, but some indication of its value follows from computing it for the case of two rings of magnetic field containing fluxes Φ_1 and Φ_2. Start with the rings quite separate, and show that $H = 0$. Then allow the rings to be linked while not sharing any magnetic field lines. Show that now $H = 2\Phi_1\Phi_2$ and that $dH/dt = -2\int dV \mathbf{E} \cdot \mathbf{B}$.[4]

19.4 Hydromagnetic Flows

We now introduce fluid motions into our applications of magnetohydrodynamics. Specifically, we explore a simple class of *stationary hydromagnetic flows*: the flow of an electrically conducting fluid along a duct of constant cross section perpendicular to a uniform magnetic field B_0 (see Fig. 19.7). This is sometimes known as *Hartmann flow*. The duct has two insulating walls (top and bottom, as shown in the figure), separated by a distance $2a$ that is much smaller than the separation of short side walls, which are electrically conducting.

Hartmann flow and its applications

To relate Hartmann flow to magnetic-free Poiseuille flow (viscous, laminar flow between plates; Ex. 13.18), we reinstate the viscous force in the equation of motion. For simplicity we assume that the time-independent flow ($\partial \mathbf{v}/\partial t = 0$) has traveled sufficiently far down the duct (x direction) to have reached a z-independent form, so $\mathbf{v} \cdot \nabla \mathbf{v} = 0$ and $\mathbf{v} = \mathbf{v}(y, z)$. We also assume that gravitational forces are unimportant. Then the flow's equation of motion takes the form

$$\nabla P = \mathbf{j} \times \mathbf{B} + \eta \nabla^2 \mathbf{v}, \tag{19.30}$$

where $\eta = \rho \nu$ is the coefficient of dynamical viscosity. The magnetic (Lorentz) force $\mathbf{j} \times \mathbf{B}$ alters the balance between the Poiseuille flow's viscous force $\eta \nabla^2 \mathbf{v}$ and the

---

4. For further discussion, see Bellan (2000).

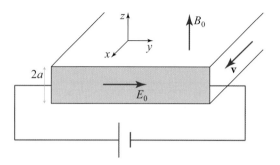

FIGURE 19.7 Hartmann flow with average speed v along a duct of thickness $2a$, perpendicular to an applied magnetic field of strength B_0. The short side walls are conducting and the two long horizontal walls are electrically insulating.

four versions of Hartmann flow

pressure gradient ∇P. The details of that altered balance and the resulting magnetic-influenced flow depend on how the walls are connected electrically. Let us consider four possibilities chosen to bring out the essential physics.

ELECTROMAGNETIC BRAKE

We short circuit the electrodes, so a current $\mathbf{j}$ can flow (Fig. 19.8a). The magnetic field lines are partially dragged by the fluid, bending them (as embodied in $\nabla \times \mathbf{B} = \mu_0 \mathbf{j}$), so they can exert a decelerating tension force $\mathbf{j} \times \mathbf{B} = (\nabla \times \mathbf{B}) \times \mathbf{B}/\mu_0 = \mathbf{B} \cdot \nabla \mathbf{B}/\mu_0$ on the flow (Fig. 19.3b). This configuration is an electromagnetic brake. The pressure gradient, which is trying to accelerate the fluid, is balanced by the magnetic tension and viscosity. The work being done (per unit volume) by the pressure gradient, $\mathbf{v} \cdot (-\nabla P)$, is converted into heat through viscous and Ohmic dissipation.

MHD POWER GENERATOR

The MHD power generator is similar to the electromagnetic brake except that an external load is added to the circuit (Fig. 19.8b). Useful power can be extracted from the flow. Variants of this configuration were developed in the 1970s–1990s in many countries, but they are currently not seen to be economically competitive with other power-generation methods.

FLOW METER

When the electrodes in a flow meter are on an open circuit, the induced electric field produces a measurable potential difference across the duct (Fig. 19.8c). This voltage will increase monotonically with the rate of flow of fluid through the duct and therefore can provide a measurement of the flow.

ELECTROMAGNETIC PUMP

Finally, we can attach a battery to the electrodes and allow a current to flow (Figs. 19.7 and 19.8d). This produces a Lorentz force which either accelerates or decelerates the

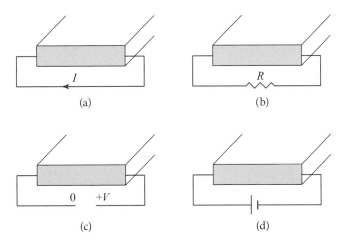

FIGURE 19.8 Four variations on Hartmann flow: (a) Electromagnetic brake. (b) MHD power generator. (c) Flow meter. (d) Electromagnetic pump.

flow, depending on the direction of the magnetic field. This method can be used to pump liquid sodium coolant around a nuclear reactor. It has also been proposed as a means of spacecraft propulsion in interplanetary space.

We consider in more detail two limiting cases of the electromagnetic pump. When there is a constant pressure gradient $Q = -dP/dx$ but no magnetic field, a flow with modest Reynolds number will be approximately laminar with velocity profile (Ex. 13.18):

$$v_x(z) = \frac{Qa^2}{2\eta}\left[1 - \left(\frac{z}{a}\right)^2\right], \tag{19.31}$$

where a is the half-width of the channel. This flow is the 1-dimensional version of the Poiseuille flow in a pipe, such as a blood artery, which we studied in Sec. 13.7.6 [cf. Eq. (13.82a)]. Now suppose that uniform electric and magnetic fields E_0 and B_0 are applied along the $\mathbf{e}_y$ and $\mathbf{e}_z$ directions, respectively (Fig. 19.7). The resulting magnetic force $\mathbf{j} \times \mathbf{B}$ can either reinforce or oppose the fluid's motion. When the applied magnetic field is small, $B_0 \ll E_0/v_x$, the effect of the magnetic force will be similar to that of the pressure gradient, and Eq. (19.31) must be modified by replacing $Q \equiv -dP/dx$ by $-dP/dx + j_y B_z = -dP/dy + \kappa_e E_0 B_0$. [Here $j_y = \kappa_e(E_y - v_x B_z) \simeq \kappa_e E_0$.]

If the strength of the magnetic field is increased sufficiently, then the magnetic force will dominate the viscous force, except in thin boundary layers near the walls. Outside the boundary layers, in the bulk of the flow, the velocity will adjust so that the electric field vanishes in the rest frame of the fluid (i.e., $v_x = E_0/B_0$). In the boundary layers v_x drops sharply from E_0/B_0 to zero at the walls, and correspondingly, a strong viscous force $\eta\nabla^2\mathbf{v}$ develops. Since the pressure gradient ∇P must be essentially the same in the boundary layer as in the adjacent bulk flow and thus cannot balance this

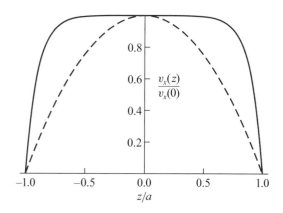

FIGURE 19.9 Velocity profiles [Eq. (19.35)] for flow in an electromagnetic pump of width $2a$ with small and large Hartmann number scaled to the velocity at the center of the channel. The dashed curve shows the almost parabolic profile for Ha $= 0.1$ [Eq. (19.31)]. The solid curve shows the almost flat-topped profile for Ha $= 10$.

large viscous force, it must be balanced instead by the magnetic force: $\mathbf{j} \times \mathbf{B} + \eta \nabla^2 \mathbf{v} = 0$ [Eq. (19.30)], with $\mathbf{j} = \kappa_e(\mathbf{E} + \mathbf{v} \times \mathbf{B}) \sim \kappa_e v_x B_0 \mathbf{e}_y$. We thereby see that the thickness of the boundary layer is given by

$$\delta_H \sim \left(\frac{\eta}{\kappa_e B_0^2}\right)^{1/2}. \tag{19.32}$$

This suggests a new dimensionless number to characterize the flow,

Hartmann number
$$\boxed{\mathrm{Ha} = \frac{a}{\delta_H} = B_0 a \left(\frac{\kappa_e}{\eta}\right)^{1/2},} \tag{19.33}$$

called the *Hartmann number*. The square of the Hartmann number, Ha^2, is essentially the ratio of the magnetic force, $|\mathbf{j} \times \mathbf{B}| \sim \kappa_e v_x B_0^2$, to the viscous force, $\sim \eta v_x/a^2$, assuming a lengthscale a rather than δ_H for variations of the velocity.

The detailed velocity profile $v_x(z)$ away from the vertical side walls is computed in Ex. 19.10 and is shown for low and high Hartmann numbers in Fig. 19.9. Notice that at low Hartmann numbers, the plotted profile is nearly parabolic as expected, and at high Hartmann numbers it consists of boundary layers at $z \sim -a$ and $z \sim a$, and a uniform flow in between.

In Exs. 19.11 and 19.12 we explore two important astrophysical examples of hydromagnetic flow: the magnetosphere of a rotating, magnetized star or other body, and the solar wind.

EXERCISES

Exercise 19.10 *Example: Hartmann Flow*
Compute the velocity profile of a conducting fluid in a duct of thickness $2a$ perpendicular to externally generated, uniform electric and magnetic fields ($E_0 \mathbf{e}_y$

and $B_0 \mathbf{e}_z$) as shown in Fig. 19.7. Away from the vertical sides of the duct, the velocity v_x is just a function of y, and the pressure can be written in the form $P = -Qx + p(z)$, where Q is the longitudinal pressure gradient.

(a) Show that the velocity field satisfies the differential equation

$$\frac{d^2 v_x}{dz^2} - \frac{\kappa_e B_0^2}{\eta} v_x = -\frac{(Q + \kappa_e B_0 E_0)}{\eta}. \tag{19.34}$$

(b) Impose suitable boundary conditions at the bottom and top walls of the channel, and solve this differential equation to obtain the following velocity field:

$$v_x = \frac{Q + \kappa_e B_0 E_0}{\kappa_e B_0^2} \left[1 - \frac{\cosh(\mathrm{Ha}\, z/a)}{\cosh(\mathrm{Ha})} \right], \tag{19.35}$$

where Ha is the Hartmann number (see Fig. 19.9).

Exercise 19.11 **Example: Rotating Magnetospheres*

Many self-gravitating cosmic bodies are both spinning and magnetized. Examples are Earth, the Sun, black holes surrounded by highly conducting accretion disks (which hold a magnetic field on the hole), neutron stars (pulsars), and magnetic white dwarfs. As a consequence of the magnetic field's spin-induced motion, large electric fields are produced outside the rotating body. The divergence of these electric fields must be balanced by free electric charge, which implies that the region around the body cannot be a vacuum. It is usually filled with plasma and is called a *magnetosphere*. MHD provides a convenient formalism for describing the structure of this magnetosphere. Magnetospheres are found around most planets and stars. Magnetospheres surrounding neutron stars and black holes are believed to be responsible for the emissions from pulsars and quasars.

As a model of a rotating magnetosphere, consider a magnetized and infinitely conducting star, spinning with angular frequency Ω_*. Suppose that the magnetic field is stationary and axisymmetric with respect to the spin axis and that the magnetosphere, like the star, is perfectly conducting.

(a) Show that the azimuthal component E_ϕ of the magnetospheric electric field must vanish if the magnetic field is to be stationary. Hence show that there exists a function $\mathbf{\Omega}(\mathbf{r})$ that must be parallel to $\mathbf{\Omega}_*$ and must satisfy

$$\mathbf{E} = -(\mathbf{\Omega} \times \mathbf{r}) \times \mathbf{B}. \tag{19.36}$$

Show that if the motion of the magnetosphere's conducting fluid is simply a rotation, then its angular velocity must be $\mathbf{\Omega}$.

(b) Use the induction equation (magnetic-field transport law) to show that

$$(\mathbf{B} \cdot \nabla)\mathbf{\Omega} = 0. \tag{19.37}$$

(c) Use the boundary condition at the surface of the star to show that the magnetosphere corotates with the star (i.e., $\Omega = \Omega_*$). This is known as *Ferraro's law of isorotation*.

Exercise 19.12 *Example: Solar Wind*

The solar wind is a magnetized outflow of plasma that emerges from the solar corona. We make a simple model of it by generalizing the results from the previous exercise that emphasize its hydromagnetic features while ignoring gravity and thermal pressure (which are also important in practice). We consider stationary, axisymmetric motion in the equatorial plane, and idealize the magnetic field as having the form $B_r(r)$, $B_\phi(r)$. (If this were true at all latitudes, the Sun would have to contain magnetic monopoles!)

(a) Use the results from the previous exercise plus the perfect MHD relation, $\mathbf{E} = -\mathbf{v} \times \mathbf{B}$, to argue that the velocity field can be written in the form

$$\mathbf{v} = \frac{\kappa \mathbf{B}}{\rho} + (\mathbf{\Omega} \times \mathbf{r}), \tag{19.38}$$

where κ and $\mathbf{\Omega}$ are constant along a field line. Interpret this relation kinematically.

(b) Resolve the velocity and the magnetic field into radial and azimuthal components, v_r, v_ϕ, and B_r, B_ϕ, and show that $\rho v_r r^2$ and $B_r r^2$ are constant.

(c) Use the induction equation to show that

$$\frac{v_r}{v_\phi - \Omega r} = \frac{B_r}{B_\phi}. \tag{19.39}$$

(d) Use the equation of motion to show that the specific angular momentum, including both the mechanical and the magnetic contributions,

$$\Lambda = r v_\phi - \frac{r B_r B_\phi}{\mu_0 \rho v_r}, \tag{19.40}$$

is constant.

(e) Combine Eqs. (19.39) and (19.40) to argue that

$$v_\phi = \frac{\Omega r [M_A^2 \Lambda/(\Omega r^2) - 1]}{M_A^2 - 1}, \tag{19.41}$$

where

$$M_A = \frac{v_r (\mu_0 \rho)^{1/2}}{B_r} \tag{19.42}$$

is the Alfvén Mach number [cf. Eq. (19.73)]. Show that the solar wind must pass through a critical point (Sec. 17.3.2) where its radial speed equals the Alfvén speed.

(f) Suppose that the critical point in part (e) is located at 20 solar radii and that $v_r = 100$ km s^{-1}, $v_\phi = 20$ km s^{-1}, and $B_r = 400$ nT there. Calculate values for Λ, B_ϕ, ρ and use these to estimate the fraction of the solar mass and the solar angular momentum that could have been carried off by the solar wind over its ~5 Gyr lifetime. Comment on your answer.

(g) Suppose that there is no poloidal current density so that $B_\phi \propto r^{-1}$, and deduce values for the velocity and the magnetic field in the neighborhood of Earth where $r \sim 200$ solar radii. Sketch the magnetic field and the path of the solar wind as it flows from r_c to Earth.

The solar mass, radius and rotation period are $\sim 2 \times 10^{30}$ kg, $\sim 7 \times 10^8$ m, and ~ 25 d.

19.5 Stability of Magnetostatic Equilibria

Having used the MHD equation of motion to analyze some simple flows, we return to the problem of magnetic confinement and demonstrate a procedure to analyze the stability of the confinement's magnetostatic equilibria. We first perform a straightforward linear perturbation analysis about equilibrium, obtaining an eigenequation for the perturbation's oscillation frequencies ω. For sufficiently simple equilibria, this eigenequation can be solved analytically, but most equilibria are too complex for this approach, so the eigenequation must be solved numerically or by other approximation techniques. This is rather similar to the task one faces in attempting to solve the Schrödinger equation for multi-electron atoms. It will not be a surprise to learn that variational methods are especially practical and useful, and we develop a suitable formalism for them.

We develop the perturbation theory, eigenequation, and variational formalism in some detail not only because of their importance for the stability of magnetostatic equilibria, but also because essentially the same techniques (with different equations) are used in studying the stability of other equilibria. One example is the oscillations and stability of stars, in which the magnetic field is unimportant, while self-gravity is crucial [see, e.g., Shapiro and Teukolsky (1983, Chap. 6), and Sec. 16.2.4 of this book, on helioseismology]. Another example is the oscillations and stability of elastostatic equilibria, in which **B** is absent but shear stresses are important (Secs. 12.3 and 12.4).

19.5.1 Linear Perturbation Theory

Consider a perfectly conducting, isentropic fluid at rest in equilibrium, with pressure gradients that balance magnetic forces—for example, the Z-pinch, Θ-pinch, and tokamak configurations of Fig. 19.6. For simplicity, we ignore gravity. (This is usually justified in laboratory situations.) The equation of equilibrium then reduces to

$$\nabla P = \mathbf{j} \times \mathbf{B} \qquad (19.43)$$

[Eq. (19.10)].

equation of magnetostatic equilibrium without gravity

We now perturb slightly about our chosen equilibrium and ignore the (usually negligible) effects of viscosity and magnetic-field diffusion, so $\eta = \rho \nu \simeq 0$ and $\kappa_e \simeq \infty$. It is useful and conventional to describe the perturbations in terms of two different types of quantities: (i) The change in a quantity (e.g., the fluid density) moving with the fluid, which is called a *Lagrangian* perturbation and is denoted by the symbol Δ (e.g., the lagrangian density perturbation $\Delta \rho$). (ii) The change at a fixed location in

Lagrangian and Eulerian perturbations and how they are related

space, which is called an *Eulerian* perturbation and is denoted by the symbol δ (e.g., the Eulerian density perturbation $\delta\rho$). The fundamental variable used in the theory is the fluid's *Lagrangian displacement* $\Delta \mathbf{x} \equiv \boldsymbol{\xi}(\mathbf{x}, t)$ (i.e., the change in a fluid element's location). A fluid element whose location is $\mathbf{x}$ in the unperturbed equilibrium is moved to location $\mathbf{x} + \boldsymbol{\xi}(\mathbf{x}, t)$ by the perturbations. From their definitions, one can see that the Lagrangian and Eulerian perturbations are related by

$$\boxed{\Delta = \delta + \boldsymbol{\xi} \cdot \boldsymbol{\nabla} \quad (\text{e.g.,}\ \Delta\rho = \delta\rho + \boldsymbol{\xi} \cdot \boldsymbol{\nabla}\rho).} \tag{19.44}$$

Now consider the transport law for the magnetic field in the limit of infinite electrical conductivity, $\partial \mathbf{B}/\partial t = \boldsymbol{\nabla} \times (\mathbf{v} \times \mathbf{B})$ [Eq. (19.6)]. To linear order, the fluid velocity is $\mathbf{v} = \partial \boldsymbol{\xi}/\partial t$. Inserting this into the transport law and setting the full magnetic field at fixed $\mathbf{x}$ and t equal to the equilibrium field plus its Eulerian perturbation $\mathbf{B} \to \mathbf{B} + \delta\mathbf{B}$, we obtain $\partial \delta\mathbf{B}/\partial t = \boldsymbol{\nabla} \times [(\partial \boldsymbol{\xi}/\partial t) \times (\mathbf{B} + \delta\mathbf{B})]$. Linearizing in the perturbation, and integrating in time, we obtain for the Eulerian perturbation of the magnetic field:

$$\delta\mathbf{B} = \boldsymbol{\nabla} \times (\boldsymbol{\xi} \times \mathbf{B}). \tag{19.45a}$$

Since the current and the field are related, in general, by the linear equation $\mathbf{j} = \boldsymbol{\nabla} \times \mathbf{B}/\mu_0$, their Eulerian perturbations are related in the same way:

$$\delta\mathbf{j} = \boldsymbol{\nabla} \times \delta\mathbf{B}/\mu_0. \tag{19.45b}$$

In the equation of mass conservation, $\partial\rho/\partial t + \boldsymbol{\nabla} \cdot (\rho\mathbf{v}) = 0$, we replace the density by its equilibrium value plus its Eulerian perturbation, $\rho \to \rho + \delta\rho$, and replace $\mathbf{v}$ by $\partial\boldsymbol{\xi}/\partial t$. We then linearize in the perturbation to obtain

$$\delta\rho + \rho\boldsymbol{\nabla} \cdot \boldsymbol{\xi} + \boldsymbol{\xi} \cdot \boldsymbol{\nabla}\rho = 0. \tag{19.45c}$$

The lagrangian density perturbation, obtained from this via Eq. (19.44), is

$$\Delta\rho = -\rho\boldsymbol{\nabla} \cdot \boldsymbol{\xi}. \tag{19.45d}$$

We assume that, as it moves, the fluid gets compressed or expanded adiabatically (no Ohmic or viscous heating, or radiative cooling). Then the Lagrangian change of pressure ΔP in each fluid element (moving with the fluid) is related to the Lagrangian change of density by

$$\Delta P = \left(\frac{\partial P}{\partial \rho}\right)_S \Delta\rho = \frac{\gamma P}{\rho}\Delta\rho = -\gamma P \boldsymbol{\nabla} \cdot \boldsymbol{\xi}, \tag{19.45e}$$

where γ is the fluid's adiabatic index (ratio of specific heats), which might or might not be independent of position in the equilibrium configuration. Correspondingly, the Eulerian perturbation of the pressure (perturbation at fixed location) is

$$\delta P = \Delta P - (\boldsymbol{\xi} \cdot \boldsymbol{\nabla})P = -\gamma P(\boldsymbol{\nabla} \cdot \boldsymbol{\xi}) - (\boldsymbol{\xi} \cdot \boldsymbol{\nabla})P. \tag{19.45f}$$

This is the pressure perturbation that appears in the fluid's equation of motion.

By replacing $\mathbf{v} \to \partial\boldsymbol{\xi}/\partial t$, $P \to P + \delta P$, $\mathbf{B} \to \mathbf{B} + \delta\mathbf{B}$, and $\mathbf{j} \to \mathbf{j} + \delta\mathbf{j}$ in the fluid's equation of motion (19.10) and neglecting gravity, and by then linearizing in the perturbation, we obtain

$$\boxed{\rho \frac{\partial^2 \boldsymbol{\xi}}{\partial t^2} = \mathbf{j} \times \delta\mathbf{B} + \delta\mathbf{j} \times \mathbf{B} - \nabla\delta P \equiv \hat{\mathbf{F}}[\boldsymbol{\xi}].} \quad (19.46)$$

dynamical equation for adiabatic perturbations of a magnetostatic equilibrium

Here $\hat{\mathbf{F}}$ is a real, linear differential operator, whose form one can deduce by substituting expressions (19.45a), (19.45b), and (19.45f) for $\delta\mathbf{B}$, $\delta\mathbf{j}$, and δP, and by substituting $\nabla \times \mathbf{B}/\mu_0$ for $\mathbf{j}$. Performing these substitutions and carefully rearranging the terms, we eventually convert the operator $\hat{\mathbf{F}}$ into the following form, expressed in slot-naming index notation:

$$\boxed{\begin{aligned}\hat{F}_i[\boldsymbol{\xi}] = &\left\{\left[(\gamma-1)P + \frac{B^2}{2\mu_0}\right]\xi_{k;k} + \frac{B_j B_k}{\mu_0}\xi_{j;k}\right\}_{;i} \\ &+ \left[\left(P + \frac{B^2}{2\mu_0}\right)\xi_{j;i} + \frac{B_j B_k}{\mu_0}\xi_{i;k} + \frac{B_i B_j}{\mu_0}\xi_{k;k}\right]_{;j}\end{aligned}} \quad (19.47)$$

Honestly! Here the semicolons denote gradients (partial derivatives in Cartesian coordinates; connection coefficients (Sec. 11.8) are required in curvilinear coordinates).

We write the operator $\hat{F}_i$ in the explicit form (19.47) because of its power for demonstrating that $\hat{F}_i$ is self-adjoint (Hermitian, with real variables rather than complex): by introducing the Kronecker-delta components of the metric, $g_{ij} = \delta_{ij}$, we can easily rewrite Eq. (19.47) in the form

force operator $\hat{F}_i[\boldsymbol{\xi}]$ is self-adjoint

$$\hat{F}_i[\boldsymbol{\xi}] = (T_{ijkl}\xi_{k;l})_{;j}, \quad (19.48)$$

where T_{ijkl} are the components of a fourth-rank tensor that is symmetric under interchange of its first and second pairs of indices: $T_{ijkl} = T_{klij}$.

Now, our magnetic-confinement equilibrium configuration (e.g., Fig. 19.6) typically consists of a plasma-filled interior region $\mathcal{V}$ surrounded by a vacuum magnetic field (which in turn may be surrounded by a wall). Our MHD equations with force operator $\hat{\mathbf{F}}$ are valid only in the plasma region $\mathcal{V}$, and not in vacuum, where Maxwell's equations with small displacement current prevail. We use Eq. (19.48) to prove that $\hat{\mathbf{F}}$ is a self-adjoint operator when integrated over $\mathcal{V}$, with the appropriate boundary conditions at the vacuum interface.

Specifically, we contract a vector field $\boldsymbol{\zeta}$ into $\hat{\mathbf{F}}[\boldsymbol{\xi}]$, integrate over $\mathcal{V}$, and perform two integrations by parts to obtain

$$\int_\mathcal{V} \boldsymbol{\zeta} \cdot \mathbf{F}[\boldsymbol{\xi}] dV = \int_\mathcal{V} \zeta_i (T_{ijkl}\xi_{k;l})_{;j} dV = -\int_\mathcal{V} T_{ijkl}\zeta_{i;j}\xi_{k;l} dV = \int_\mathcal{V} \xi_i(T_{ijkl}\zeta_{k;l})_{;j} dV$$

$$= \int_\mathcal{V} \boldsymbol{\xi} \cdot \mathbf{F}[\boldsymbol{\zeta}] dV. \quad (19.49)$$

Here we have discarded the integrals of two divergences, which by Gauss's theorem can be expressed as surface integrals at the fluid-vacuum interface $\partial \mathcal{V}$. Those unwanted surface integrals vanish if $\boldsymbol{\xi}$ and $\boldsymbol{\zeta}$ satisfy

$$T_{ijkl}\xi_{k;l}\zeta_i n_j = 0, \quad \text{and} \quad T_{ijkl}\zeta_{k;l}\xi_i n_j = 0, \tag{19.50}$$

with n_j the normal to the interface ∂V.

boundary conditions for perturbations

Now, $\boldsymbol{\xi}$ and $\boldsymbol{\zeta}$ are physical displacements of the MHD fluid, and as such, they must satisfy the appropriate boundary conditions at the boundary $\partial \mathcal{V}$ of the plasma region $\mathcal{V}$. In the simplest, idealized case, the conducting fluid would extend far beyond the region where the disturbances have appreciable amplitude, so $\boldsymbol{\xi} = \boldsymbol{\zeta} = 0$ at the distant boundary, and Eqs. (19.50) are satisfied. More reasonably, the fluid might butt up against rigid walls at $\partial \mathcal{V}$, where the normal components of $\boldsymbol{\xi}$ and $\boldsymbol{\zeta}$ vanish, guaranteeing again that Eqs. (19.50) are satisfied. This configuration is fine for liquid mercury or sodium, but not for a hot plasma, which would quickly destroy the walls. For confinement by a surrounding vacuum magnetic field, no current flows outside $\mathcal{V}$, and the displacement current is negligible, so $\nabla \times \delta \mathbf{B} = 0$ there. By combining this with the rest of Maxwell's equations and paying careful attention to the motion of the interface and boundary conditions [Eqs. (19.20)] there, one can show, once again, that Eqs. (19.50) are satisfied (see, e.g., Goedbloed and Poedts, 2004, Sec. 6.6.2). Therefore, in all these cases Eq. (19.49) is also true, which demonstrates the self-adjointness (Hermiticity) of $\hat{\mathbf{F}}$.[5] We use this property below.

Returning to our perturbed MHD system, we seek its normal modes by assuming a harmonic time dependence, $\boldsymbol{\xi} \propto e^{-i\omega t}$. The first-order equation of motion (19.46) then becomes

Sturm-Liouville type eigenequation for perturbations

$$\boxed{\hat{\mathbf{F}}[\boldsymbol{\xi}] + \rho \omega^2 \boldsymbol{\xi} = 0.} \tag{19.51}$$

This is an eigenequation for the fluid's Lagrangian displacement $\boldsymbol{\xi}$, with eigenvalue ω^2. It must be augmented by the boundary conditions (19.20) at the edge $\partial \mathcal{V}$ of the fluid.

By virtue of the elegant, self-adjoint mathematical form (19.48) of the differential operator $\hat{\mathbf{F}}$, our eigenequation (19.51) is of a very special and powerful type, called *Sturm-Liouville*; see any good text on mathematical physics (e.g., Mathews and Walker, 1970; Arfken, Weber, and Harris, 2013; Hassani, 2013). From the general (rather simple) theory of Sturm-Liouville equations, we can infer that all the eigenvalues ω^2 are real, so the normal modes are purely oscillatory ($\omega^2 > 0$, $\boldsymbol{\xi} \propto e^{\pm i|\omega|t}$) or are purely exponentially growing or decaying ($\omega^2 < 0$, $\boldsymbol{\xi} \propto e^{\pm |\omega|t}$). Exponentially growing modes represent instabilities. Sturm-Liouville theory also implies that all

properties of eigenfrequencies and eigenfunctions

---

5. Self-adjointness can also be deduced from energy conservation without getting entangled in detailed boundary conditions (see, e.g., Bellan, 2006, Sec. 10.4.2).

eigenfunctions [labeled by indices "(n)"] with different eigenfrequencies are orthogonal to one another, in the sense that $\int_V \rho \boldsymbol{\xi}^{(n)} \cdot \boldsymbol{\xi}^{(m)} dV = 0$.

EXERCISES

Exercise 19.13 *Derivation: Properties of Eigenmodes*
Derive the properties of the eigenvalues and eigenfunctions for perturbations of an MHD equilibrium that are asserted in the last paragraph of Sec. 19.5.1, namely, the following.

(a) For each normal mode the eigenfrequency ω_n is either real or imaginary.

(b) Eigenfunctions $\boldsymbol{\xi}^{(m)}$ and $\boldsymbol{\xi}^{(n)}$ that have different eigenvalues $\omega_m \neq \omega_n$ are orthogonal to each other: $\int \rho\, \boldsymbol{\xi}^{(m)} \cdot \boldsymbol{\xi}^{(m)} dV = 0$.

19.5.2 Z-Pinch: Sausage and Kink Instabilities

We illustrate MHD stability theory using a simple, analytically tractable example: a variant of the Z-pinch configuration of Fig. 19.6a. We consider a long, cylindrical column of a conducting, incompressible liquid (e.g., mercury) with column radius R and fluid density ρ. The column carries a current I longitudinally along its surface (rather than in its interior as in Fig. 19.6a), so $\mathbf{j} = [I/(2\pi R)]\delta(\varpi - R)\mathbf{e}_z$, and the liquid is confined by the resulting external toroidal magnetic field $B_\phi \equiv B$. The interior of the plasma is field free and at constant pressure P_0. From $\nabla \times \mathbf{B} = \mu_0 \mathbf{j}$, we deduce that the exterior magnetic field is

$$B_\phi \equiv B = \frac{\mu_0 I}{2\pi \varpi} \quad \text{at } \varpi \geq R. \tag{19.52}$$

Here (ϖ, ϕ, z) are the usual cylindrical coordinates. This is a variant of the Z-pinch, because the z-directed current on the column's surface creates the external toroidal field B, which pinches the column until its internal pressure is balanced by the field's pressure:

$$P_0 = \left(\frac{B^2}{2\mu_0}\right)_{\varpi=R}. \tag{19.53}$$

It is quicker and more illuminating to analyze the stability of this Z-pinch equilibrium directly instead of by evaluating $\hat{\mathbf{F}}$, and the outcome is the same. (For a treatment based on $\hat{\mathbf{F}}$, see Ex. 19.16.) Treating only the most elementary case, we consider small, axisymmetric perturbations with an assumed variation $\boldsymbol{\xi} \propto e^{i(kz-\omega t)}\mathbf{f}(\varpi)$ for some function $\mathbf{f}$. As the magnetic field interior to the column vanishes, the equation of motion $\rho d\mathbf{v}/dt = -\nabla(P + \delta P)$ becomes

$$-\omega^2 \rho \xi_\varpi = -\delta P', \quad -\omega^2 \rho \xi_z = -ik\delta P, \tag{19.54a}$$

where the prime denotes differentiation with respect to radius ϖ. Combining these two equations, we obtain

$$\xi_z' = ik\xi_\varpi. \tag{19.54b}$$

Because the fluid is incompressible, it satisfies $\nabla \cdot \boldsymbol{\xi} = 0$:

$$\varpi^{-1}(\varpi \xi_\varpi)' + ik\xi_z = 0, \tag{19.54c}$$

which, with Eq. (19.54b), leads to

$$\xi_z'' + \frac{\xi_z'}{\varpi} - k^2 \xi_z = 0. \tag{19.54d}$$

The solution of this equation that is regular at $\varpi = 0$ is

$$\xi_z = A I_0(k\varpi) \quad \text{at } \varpi \leq R, \tag{19.54e}$$

where A is a constant, and $I_n(x)$ is the modified Bessel function: $I_n(x) = i^{-n} J_n(ix)$. From Eq. (19.54b) and $dI_0(x)/dx = I_1(x)$, we obtain

$$\xi_\varpi = -i A I_1(k\varpi). \tag{19.54f}$$

Next we consider the region exterior to the fluid column. As this is vacuum, it must be current free; and as we are dealing with a purely axisymmetric perturbation, the ϖ component of Maxwell's equation $\nabla \times \delta \mathbf{B} = 0$ (with negligible displacement current) reads

$$\frac{\partial \delta B_\phi}{\partial z} = ik \delta B_\phi = 0. \tag{19.54g}$$

The ϕ component of the magnetic perturbation therefore vanishes outside the column.

The interior and exterior solutions must be connected by the law of force balance, that is, by the boundary condition (19.19f) [or equivalently, Eq. (19.20a) with the tildes removed] at the plasma's surface. Allowing for the displacement of the surface and retaining only linear terms, this becomes

$$P_0 + \Delta P = P_0 + (\boldsymbol{\xi} \cdot \nabla) P_0 + \delta P = \frac{(B + \Delta B_\phi)^2}{2\mu_0} = \frac{B^2}{2\mu_0} + \frac{B}{\mu_0}(\boldsymbol{\xi} \cdot \nabla) B + \frac{B \delta B_\phi}{\mu_0}, \tag{19.54h}$$

where all quantities are evaluated at $\varpi = R$. The equilibrium force-balance condition gives us $P_0 = B^2/(2\mu_0)$ [Eq. (19.53)] and $\nabla P_0 = 0$. In addition, we have shown that $\delta B_\phi = 0$. Therefore, Eq. (19.54h) becomes simply

$$\delta P = \frac{B B'}{\mu_0} \xi_\varpi. \tag{19.54i}$$

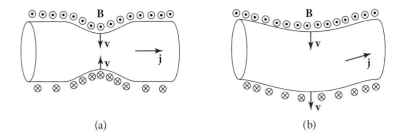

FIGURE 19.10 Physical interpretation of (a) sausage and (b) kink instabilities.

Substituting δP from Eqs. (19.54a) and (19.54e), B from Eq. (19.52), and ξ_ϖ from Eq. (19.54f), we obtain the dispersion relation

$$\omega^2 = \frac{-\mu_0 I^2}{4\pi^2 R^4 \rho} \frac{kR I_1(kR)}{I_0(kR)}$$

$$\simeq \frac{-\mu_0 I^2}{8\pi^2 R^2 \rho} k^2; \quad \text{for } k \ll R^{-1}$$

$$\simeq \frac{-\mu_0 I^2}{4\pi^2 R^3 \rho} k; \quad \text{for } k \gg R^{-1}, \tag{19.55}$$

where we have used $I_0(x) \sim 1$, $I_1(x) \sim x/2$ as $x \to 0$ and $I_1(x)/I_0(x) \to 1$ as $x \to \infty$.

Because I_0 and I_1 are positive for all $kR > 0$, for every wave number k this dispersion relation shows that ω^2 is negative. Therefore, ω is imaginary, the perturbation grows exponentially with time, and so *the Z-pinch configuration is dynamically unstable.* If we define a characteristic Alfvén speed by $a = B(R)/(\mu_0 \rho)^{1/2}$ [see Eq. (19.73)], then we see that the growth time for modes with wavelengths comparable to the column diameter is roughly an Alfvén crossing time $2R/a$. This is fast!

This instability is sometimes called a *sausage instability,* because its eigenfunction $\xi_\varpi \propto e^{ikz}$ consists of oscillatory pinches of the column's radius that resemble the pinches between sausages in a link. This sausage instability has a simple physical interpretation (Fig. 19.10a), one that illustrates the power of the concepts of flux freezing and magnetic tension for developing intuition. If we imagine an inward radial motion of the fluid, then the toroidal loops of magnetic field will be carried inward, too, and will therefore shrink. As the external field is unperturbed, $\delta B_\phi = 0$, we have $B_\phi \propto 1/\varpi$, whence the surface field at the inward perturbation increases, leading to a larger "hoop" stress or, equivalently, a larger $\mathbf{j} \times \mathbf{B}$ Lorentz force, which accelerates the inward perturbation (see Fig. 19.10a).

sausage instability of Z-pinch

So far, we have only considered axisymmetric perturbations. We can generalize our analysis by allowing the perturbations to vary as $\boldsymbol{\xi} \propto \exp(im\phi)$. (Our sausage instability corresponds to $m = 0$.) Modes with $m \geq 1$, like $m = 0$, are also generically unstable. The $m = 1$ modes are known as *kink* modes. In this case the column bends,

kink instability of Z-pinch

19.5 Stability of Magnetostatic Equilibria

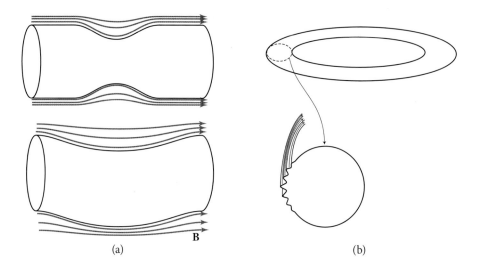

FIGURE 19.11 (a) Stabilizing magnetic fields for Θ-pinch configuration. (b) Flute instability for Θ-pinch configuration made into a torus.

and the field strength is intensified along the inner face of the bend and reduced on the outer face, thereby amplifying the instability (Fig. 19.10b). The incorporation of compressibility, as is appropriate for plasma instead of mercury, introduces only algebraic complexity; the conclusions are unchanged. The column is still highly unstable. It remains so if we distribute the longitudinal current throughout the column's interior, thereby adding a magnetic field to the interior, as in Fig. 19.6a.

MHD instabilities such as these have bedeviled attempts to confine plasma long enough to bring about nuclear fusion. Indeed, considerations of MHD stability were one of the primary motivations for the tokamak, the most consistently successful of experimental fusion devices.

19.5.3 The Θ-Pinch and Its Toroidal Analog; Flute Instability; Motivation for Tokamak

full stability of Θ-pinch

By contrast with the extreme instability of the Z-pinch configuration, the Θ-pinch configuration (Sec. 19.3.3 and Fig. 19.6b) is fully stable against MHD perturbations! (See Ex. 19.14.) This is easily understood physically (Fig. 19.11a). When the plasma cylinder is pinched or bent, at outward displaced regions of the cylinder, the external longitudinal magnetic field lines are pushed closer together, thereby strengthening the magnetic field and its pressure and thence creating an inward restoring force. Similarly, at inward displaced regions, the field lines are pulled apart, weakening their pressure and creating an outward restoring force.

toroidal Θ-pinch: flute instability

Unfortunately, the Θ-pinch configuration cannot confine plasma without closing its ends. The ends can be partially closed by a pair of magnetic mirrors, but there remain losses out the ends that cause problems. The ends can be fully closed by bending the column into a closed torus, but, sadly, the resulting toroidal Θ-pinch configuration exhibits a new MHD "flute" instability (Fig. 19.11b).

This flute instability arises on and near the outermost edge of the torus. That edge is curved around the torus's symmetry axis in just the same way as the face of the Z-pinch configuration is curved around its symmetry axis—with the magnetic field in both cases forming closed loops around the axis. As a result, this outer edge is subject to a sausage-type instability, similar to that of the Z-pinch. The resulting corrugations are translated along the torus, so they resemble the flutes on a Greek column that supports an architectural arch or roof (hence the name "flute instability"). This fluting can also be understood as a flux-tube interchange instability (Ex. 19.15).

The flute instability can be counteracted by endowing the torus with an internal magnetic field that twists (shears) as one goes radially inward (the tokamak configuration of Fig. 19.6c). Adjacent magnetic surfaces (isobars), with their different field directions, counteract each other's MHD instabilities. The component of the magnetic field along the plasma torus provides a pressure that stabilizes against sausage instabilities and a tension that stabilizes against kink-type instabilities; the component around the torus's guiding circle acts to stabilize its flute modes. In addition, the formation of image currents in the conducting walls of a tokamak vessel can also have a stabilizing influence.

how tokamak configuration counteracts instabilities

EXERCISES

Exercise 19.14 *Problem and Challenge: Stability of Θ-Pinch*
Derive the dispersion relation $\omega^2(k)$ for axisymmetric perturbations of the Θ-pinch configuration when the magnetic field is confined to the cylinder's exterior, and conclude from it that the Θ-pinch is stable against axisymmetric perturbations. [Hint: The analysis of the interior of the cylinder is the same as for the Z-pinch analyzed in the text.] Repeat your analysis for a general, variable-separated perturbation of the form $\boldsymbol{\xi} \propto e^{i(m\theta + kz - \omega t)}$, and thereby conclude that the Θ-pinch is fully MHD stable.

Exercise 19.15 *Example: Flute Instability Understood by Flux-Tube Interchange*
Carry out an analysis of the flute instability patterned after that for rotating Couette flow (first long paragraph of Sec. 14.6.3) and that for convection in stars (Fig. 18.5): Imagine exchanging two plasma-filled magnetic flux tubes that reside near the outermost edge of the torus. Argue that the one displaced outward experiences an unbalanced outward force, and the one displaced inward experiences an unbalanced inward force. [Hint: (i) To simplify the analysis, let the equilibrium magnetic field rise from zero continuously in the outer layers of the torus rather than arising discontinuously at its surface, and consider flux tubes in that outer, continuous region. (ii) Argue that the unbalanced force per unit length on a displaced flux tube is $[-\boldsymbol{\nabla}(P + B^2/(2\mu_0)) - (B_{\text{tube}}^2/\varpi)\mathbf{e}_{\varpi}]A$. Here A is the tube's cross sectional area, ϖ is distance from the torus's symmetry axis, $\mathbf{e}_{\varpi}$ is the unit vector pointing away from the symmetry axis, B_{tube} is the field strength inside the displaced tube, and $\boldsymbol{\nabla}(P + B^2/(2\mu_0))$ is the net surrounding pressure gradient at the tube's

displaced location.] Argue further that the innermost edge of the torus is stable against flux-tube interchange, so the flute instability is confined to the torus's outer face.

19.5.4 Energy Principle and Virial Theorems T2

For the perturbation eigenequation (19.46) and its boundary conditions, analytical or even numerical solutions are only readily obtained in the most simple of geometries and for the simplest fluids. However, as the eigenequation is self-adjoint, it is possible to write down a variational principle and use it to derive approximate stability criteria. This variational principle has been a powerful tool for analyzing the stability of tokamak and other magnetostatic configurations.

To derive the variational principle, we begin by multiplying the fluid velocity $\dot{\boldsymbol{\xi}} = \partial \boldsymbol{\xi}/\partial t$ into the eigenequation (equation of motion) $\rho \partial^2 \boldsymbol{\xi}/\partial t^2 = \hat{\mathbf{F}}[\boldsymbol{\xi}]$. We then integrate over the plasma-filled region $\mathcal{V}$, and use the self-adjointness of $\hat{\mathbf{F}}$ to write $\int_\mathcal{V} dV \dot{\boldsymbol{\xi}} \cdot \hat{\mathbf{F}}[\boldsymbol{\xi}] = \frac{1}{2}\int_\mathcal{V} dV (\dot{\boldsymbol{\xi}} \cdot \hat{\mathbf{F}}[\boldsymbol{\xi}] + \boldsymbol{\xi} \cdot \hat{\mathbf{F}}[\dot{\boldsymbol{\xi}}])$. We thereby obtain

conserved energy for adiabatic perturbations of magnetostatic configurations

$$\frac{dE}{dt} = 0, \quad \text{where } E = T + W, \tag{19.56a}$$

$$T = \int_\mathcal{V} dV \frac{1}{2}\rho \dot{\boldsymbol{\xi}}^2, \quad \text{and} \quad W = W[\boldsymbol{\xi}] \equiv -\frac{1}{2}\int_\mathcal{V} dV \boldsymbol{\xi} \cdot \hat{\mathbf{F}}[\boldsymbol{\xi}]. \tag{19.56b}$$

The integrals T and W are the perturbation's kinetic and potential energy, and $E = T + W$ is the conserved total energy.

Any solution of the equation of motion $\partial^2 \boldsymbol{\xi}/\partial t^2 = \hat{\mathbf{F}}[\boldsymbol{\xi}]$ can be expanded in terms of a complete set of normal modes $\boldsymbol{\xi}^{(n)}(\mathbf{x})$ with eigenfrequencies ω_n: $\boldsymbol{\xi} = \sum_n A_n \boldsymbol{\xi}^{(n)} e^{-i\omega_n t}$. Because $\hat{\mathbf{F}}$ is a real, self-adjoint operator, these normal modes can all be chosen to be real and orthogonal, even when some of their frequencies are degenerate. As the perturbation evolves, its energy sloshes back and forth between kinetic T and potential W, so time averages of T and W are equal: $\overline{T} = \overline{W}$. This implies, for each normal mode,

$$\boxed{\omega_n^2 = \frac{W[\boldsymbol{\xi}^{(n)}]}{\int_\mathcal{V} dV \frac{1}{2}\rho \boldsymbol{\xi}^{(n)2}}.} \tag{19.57}$$

energy principle (Rayleigh principle) for stability

As the denominator is positive definite, we conclude that *a magnetostatic equilibrium is stable against small perturbations if and only if the potential energy $W[\boldsymbol{\xi}]$ is a positive-definite functional of the perturbation $\boldsymbol{\xi}$*. This is sometimes called the *Rayleigh principle* for a general Sturm-Liouville problem. In the MHD context, it is known as the *energy principle*.

action principle for eigenfrequencies

It is straightforward to verify, by virtue of the self-adjointness of $\hat{\mathbf{F}}[\boldsymbol{\xi}]$, that expression (19.57) serves as an action principle for the eigenfrequencies: If one inserts into Eq. (19.57) a trial function $\boldsymbol{\xi}_{\text{trial}}$ in place of $\boldsymbol{\xi}^{(n)}$, then the resulting value of the equation will be stationary under small variations of $\boldsymbol{\xi}_{\text{trial}}$ if and only if $\boldsymbol{\xi}_{\text{trial}}$ is equal to some eigenfunction $\boldsymbol{\xi}^{(n)}$; and the stationary value of Eq. (19.57) is that eigenfunction's squared eigenfrequency ω_n^2. This action principle is most useful for estimating the lowest few squared frequencies ω_n^2. Because first-order differences between $\boldsymbol{\xi}_{\text{trial}}$ and $\boldsymbol{\xi}^{(n)}$ produce second-order errors in ω_n^2, relatively crude trial eigenfunctions can furnish surprisingly accurate eigenvalues.

Whatever may be our chosen trial function $\boldsymbol{\xi}_{\text{trial}}$, the computed value of the action (19.57) will always be larger than ω_0^2, the squared eigenfrequency of the most unstable mode. Therefore, if we compute a negative value of Eq. (19.57) using some trial eigenfunction, we know that the equilibrium must be even more unstable.

The MHD energy principle and action principle are special cases of the general conservation law and action principle for Sturm-Liouville differential equations (see, e.g., Mathews and Walker, 1970; Arfken, Weber, and Harris, 2013; Hassani, 2013). For further insights into the energy and action principles, see the original MHD paper by Bernstein et al. (1958), in which these ideas were developed; also see Mikhailovskii (1998), Goedbloed and Poedts (2004, Chap. 6), and Bellan (2006, Chap. 10).

EXERCISES

Exercise 19.16 *Example: Reformulation of the Energy Principle; Application to Z-Pinch* T2

The form of the potential energy functional derived in the text [Eq. (19.47)] is optimal for demonstrating that the operator $\hat{\mathbf{F}}$ is self-adjoint. However, there are several simpler, equivalent forms that are more convenient for practical use.

(a) Use Eq. (19.46) to show that

$$\boldsymbol{\xi} \cdot \hat{\mathbf{F}}[\boldsymbol{\xi}] = \mathbf{j} \times \mathbf{b} \cdot \boldsymbol{\xi} - \mathbf{b}^2/\mu_0 - \gamma P (\nabla \cdot \boldsymbol{\xi})^2 - (\nabla \cdot \boldsymbol{\xi})(\boldsymbol{\xi} \cdot \nabla) P$$
$$+ \nabla \cdot [(\boldsymbol{\xi} \times \mathbf{B}) \times \mathbf{b}/\mu_0 + \gamma P \boldsymbol{\xi}(\nabla \cdot \boldsymbol{\xi}) + \boldsymbol{\xi}(\boldsymbol{\xi} \cdot \nabla) P], \quad (19.58)$$

where $\mathbf{b} \equiv \delta \mathbf{B}$ is the Eulerian perturbation of the magnetic field.

(b) Insert Eq. (19.58) into expression (19.56b) for the potential energy $W[\boldsymbol{\xi}]$ and convert the volume integral of the divergence into a surface integral. Then impose the boundary condition of a vanishing normal component of the magnetic field at $\partial \mathcal{V}$ [Eq. (19.20b)] to show that

$$W[\boldsymbol{\xi}] = \frac{1}{2} \int_{\mathcal{V}} dV \left[-\mathbf{j} \times \mathbf{b} \cdot \boldsymbol{\xi} + \mathbf{b}^2/\mu_0 + \gamma P (\nabla \cdot \boldsymbol{\xi})^2 + (\nabla \cdot \boldsymbol{\xi})(\boldsymbol{\xi} \cdot \nabla) P \right]$$
$$- \frac{1}{2} \int_{\partial \mathcal{V}} d\boldsymbol{\Sigma} \cdot \boldsymbol{\xi} \left[\gamma P (\nabla \cdot \boldsymbol{\xi}) + \boldsymbol{\xi} \cdot \nabla P - \mathbf{B} \cdot \mathbf{b}/\mu_0 \right]. \quad (19.59)$$

(c) Consider axisymmetric perturbations of the cylindrical Z-pinch of an incompressible fluid, as discussed in Sec. 19.5.2, and argue that the surface integral vanishes.

(d) Adopt a simple trial eigenfunction, and obtain a variational estimate of the growth rate of the sausage instability's fastest growing mode.

Exercise 19.17 *Problem: Potential Energy in Its Most Physically Interpretable Form*
(a) Show that the potential energy (19.59) can be transformed into the following form:

$$W[\boldsymbol{\xi}] = \frac{1}{2}\int_{\mathcal{V}} dV \left[-\mathbf{j} \times \mathbf{b} \cdot \boldsymbol{\xi} + \frac{\mathbf{b}^2}{\mu_0} + \delta P \frac{\Delta\rho}{\rho}\right]$$

$$+ \frac{1}{2}\int_{\partial\mathcal{V}} d\boldsymbol{\Sigma} \cdot \nabla\left(\frac{\tilde{B}^2}{2\mu_0} - P - \frac{B^2}{2\mu_0}\right)\xi_n^2 + \frac{1}{2}\int_{\text{vacuum}} dV \frac{\tilde{\mathbf{b}}^2}{\mu_0}. \quad (19.60)$$

Here symbols without tildes represent quantities in the plasma region, and those with tildes are in the vacuum region; ξ_n is the component of the fluid displacement orthogonal to the vacuum-plasma interface.

(b) Explain the physical interpretation of each term in the expression for the potential energy in part (a). Notice that, although our original expression for the potential energy, $W = -\frac{1}{2}\int_{\mathcal{V}} \boldsymbol{\xi} \cdot \mathbf{F}[\boldsymbol{\xi}]dV$, entails an integral only over the plasma region, it actually includes the vacuum region's magnetic energy.

Exercise 19.18 *Example: Virial Theorems*
Additional mathematical tools that are useful in analyzing MHD equilibria and their stability—and are also useful in astrophysics—are the *virial theorems*. In this exercise and the next, you will deduce time-dependent and time-averaged virial theorems for any system for which the law of momentum conservation can be written in the form

$$\frac{\partial(\rho v_j)}{\partial t} + T_{jk;k} = 0. \quad (19.61)$$

Here ρ is mass density, v_j is the material's velocity, ρv_j is momentum density, and T_{jk} is the stress tensor. We have met this formulation of momentum conservation in elastodynamics [Eq. (12.2b)], in fluid mechanics with self-gravity [Eq. (2) of Box 13.4], and in magnetohydrodynamics with self-gravity [Eq. (19.12)].

The virial theorems involve integrals over any region $\mathcal{V}$ for which there is no mass flux or momentum flux across its boundary: $\rho v_j n_j = T_{jk}n_k = 0$ everywhere on $\partial\mathcal{V}$, where n_j is the normal to the boundary.[6] For simplicity we use Cartesian coordinates, so there are no connection coefficients to worry about, and momentum conservation becomes $\partial(\rho v_j) + \partial T_{jk}/\partial x_k = 0$.

(a) Show that mass conservation, $\partial\rho/\partial t + \nabla \cdot (\rho \mathbf{v}) = 0$, implies that for any field f (scalar, vector, or tensor), we have

$$\frac{d}{dt}\int_{\mathcal{V}} \rho f dV = \int_{\mathcal{V}} \rho \frac{df}{dt} dV. \quad (19.62)$$

---

6. If self-gravity is included, then the boundary will have to be at spatial infinity (i.e., no boundary), as $T_{jk}^{\text{grav}}n_k$ cannot vanish everywhere on a finite enclosing wall.

(b) Use Eq. (19.62) to show that

$$\frac{d^2 I_{jk}}{dt^2} = 2 \int_{\mathcal{V}} T_{jk} dV, \qquad (19.63a)$$

where I_{jk} is the second moment of the mass distribution:

$$I_{jk} = \int_{\mathcal{V}} \rho x_j x_k dV \qquad (19.63b)$$

with x_j the distance from a chosen origin. This is the *time-dependent tensor virial theorem*. (Note that the system's mass quadrupole moment is the trace-free part of I_{jk}, $\mathcal{I}_{jk} = I_{jk} - \frac{1}{3} I g_{jk}$, and the system's moment of inertia tensor is $\mathfrak{I}_{jk} = I_{jk} - I g_{jk}$, where $I = I_{jj}$ is the trace of I_{jk}.)

(c) If the time integral of $d^2 I_{jk}/dt^2$ vanishes, then the time-averaged stress tensor satisfies

$$\int_{\mathcal{V}} \bar{T}_{jk} dV = 0. \qquad (19.64)$$

This is the *time-averaged tensor virial theorem*. Under what circumstances is this theorem true?

Exercise 19.19 *Example: Scalar Virial Theorems* T2

(a) By taking the trace of the time-dependent tensorial virial theorem and specializing to an MHD plasma with (or without) self-gravity, show that

$$\frac{1}{2}\frac{d^2 I}{dt^2} = 2 E_{\text{kin}} + E_{\text{mag}} + E_{\text{grav}} + 3 E_P, \qquad (19.65a)$$

where I is the trace of I_{jk}, E_{kin} is the system's kinetic energy, E_{mag} is its magnetic energy, E_P is the volume integral of its pressure, and E_{grav} is its gravitational self-energy:

$$I = \int_{\mathcal{V}} \rho \mathbf{x}^2 dV, \quad E_{\text{kin}} = \int_{\mathcal{V}} \frac{1}{2} \rho v^2 dV, \quad E_{\text{mag}} = \int \frac{\mathbf{B}^2}{2\mu_0}, \quad E_P = \int_{\mathcal{V}} P dV,$$

$$E_{\text{grav}} = \frac{1}{2}\int_{\mathcal{V}} \rho \Phi = -\int_{\mathcal{V}} \frac{1}{8\pi G}(\nabla \Phi)^2 = -\frac{1}{2}\int_{\mathcal{V}}\int_{\mathcal{V}} G \frac{\rho(\mathbf{x})\rho(\mathbf{x}')}{|\mathbf{x}-\mathbf{x}'|} dV' dV, \quad (19.65b)$$

with Φ the gravitational potential energy [cf. Eq. (13.62)].

(b) When the time integral of $d^2 I / dt^2$ vanishes, then the time average of the right-hand side of Eq. (19.65a) vanishes:

$$2 \bar{E}_{\text{kin}} + \bar{E}_{\text{mag}} + \bar{E}_{\text{grav}} + 3 \bar{E}_P = 0. \qquad (19.66)$$

This is the *time-averaged scalar virial theorem*. Give examples of circumstances in which it holds.

(c) Equation (19.66) is a continuum analog of the better-known scalar virial theorem, $2\bar{E}_{\rm kin} + \bar{E}_{\rm grav} = 0$, for a system consisting of particles that interact via their self-gravity—for example, the solar system (see, e.g., Goldstein, Poole, and Safko, 2002, Sec. 3.4).

(d) As a simple but important application of the time-averaged scalar virial theorem, show—neglecting self-gravity—that it is impossible for internal currents in a plasma to produce a magnetic field that confines the plasma to a finite volume: external currents (e.g., in solenoids) are necessary.

(e) For applications to the oscillation and stability of self-gravitating systems, see Chandrasekhar (1961, Sec. 118).

19.6 Dynamos and Reconnection of Magnetic Field Lines

As we have already remarked, the timescale for Earth's magnetic field to decay is estimated to be roughly 1 million years. Since Earth is far older than that, some process inside Earth must be regenerating the magnetic field. This process is known as a *dynamo*. Generally speaking, in a dynamo the motion of the fluid stretches the magnetic field lines, thereby increasing their magnetic energy density, which compensates for the decrease in magnetic energy associated with Ohmic decay. The details of how this happens inside Earth are not well understood. However, some general principles of dynamo action have been formulated, and their application to the Sun is somewhat better understood (Exs. 19.20 and 19.21).

19.6.1 Cowling's Theorem

It is simple to demonstrate the impossibility of dynamo action in any time-independent, axisymmetric flow. Suppose that there were such a dynamo configuration, and the time-independent, poloidal (meridional) field had—for concreteness—the form sketched in Fig. 19.12. Then there must be at least one neutral point marked $\mathcal{P}$ (actually a circle about the symmetry axis), where the poloidal field vanishes. However, the curl of the magnetic field does not vanish at $\mathcal{P}$, so a toroidal current j_ϕ must exist there. Now, in the presence of finite resistivity, there must also be a toroidal electric field at $\mathcal{P}$, since

$$j_\phi = \kappa_e[E_\phi + (\mathbf{v}_P \times \mathbf{B}_P)_\phi] = \kappa_e E_\phi. \tag{19.67}$$

Here $\mathbf{v}_P$ and $\mathbf{B}_P$ are the poloidal components of $\mathbf{v}$ and $\mathbf{B}$. The nonzero E_ϕ in turn implies, via $\nabla \times \mathbf{E} = -\partial \mathbf{B}/\partial t$, that the amount of poloidal magnetic flux threading the circle at $\mathcal{P}$ must change with time, violating our original supposition that the magnetic field distribution is stationary.

We therefore conclude that any time-independent, self-perpetuating dynamo must be more complicated than a purely axisymmetric magnetic field. This result is known as *Cowling's theorem*.

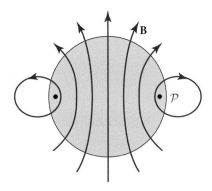

FIGURE 19.12 Impossibility of an axisymmetric dynamo.

19.6.2 Kinematic Dynamos

The simplest types of dynamo to consider are those in which we specify a particular velocity field and allow the magnetic field to evolve according to the transport law (19.6). Under certain circumstances, this can produce dynamo action. Note that we do not consider in our discussion the dynamical effect of the magnetic field on the velocity field.

The simplest type of motion is one in which a *dynamo cycle* occurs. In this cycle, there is one mechanism for creating a toroidal magnetic field from a poloidal field and a separate mechanism for regenerating the poloidal field. The first mechanism is usually differential rotation. The second is plausibly magnetic buoyancy, in which a toroidal magnetized loop is lighter than its surroundings and therefore rises in the gravitational field. As the loop rises, Coriolis forces twist the flow, causing a poloidal magnetic field to appear, which completes the dynamo cycle.

Small-scale, turbulent velocity fields may also be responsible for dynamo action. In this case, it can be shown on general symmetry grounds that the velocity field must contain *hydrodynamic helicity*—a nonzero mean value of $\mathbf{v} \cdot \boldsymbol{\omega}$ [which is an obvious analog of magnetic helicity (Ex. 19.9), the volume integral or mean value of $\mathbf{A} \cdot \mathbf{B} = \mathbf{A} \cdot (\nabla \times \mathbf{A})$].

If the magnetic field strength grows, then its dynamical effect will eventually react back on the flow and modify the velocity field. A full description of a dynamo must include this back reaction. Dynamos are a prime target for numerical simulations of MHD, and in recent years, significant progress has been made using these simulations to understand the terrestrial dynamo and other specialized problems.

EXERCISES

Exercise 19.20 *Problem: Differential Rotation in the Solar Dynamo*
This problem shows how differential rotation leads to the production of a toroidal magnetic field from a poloidal field.

(a) Verify that for a fluid undergoing differential rotation around a symmetry axis with angular velocity $\Omega(r, \theta)$, the ϕ component of the induction equation reads

$$\frac{\partial B_\phi}{\partial t} = \sin\theta \left(B_\theta \frac{\partial \Omega}{\partial \theta} + B_r r \frac{\partial \Omega}{\partial r} \right), \tag{19.68}$$

where θ is the co-latitude. (The resistive term can be ignored.)

(b) It is observed that the angular velocity on the solar surface is largest at the equator and decreases monotonically toward the poles. Analysis of solar oscillations (Sec. 16.2.4) has shown that this variation $\Omega(\theta)$ continues inward through the convection zone (cf. Sec. 18.5). Suppose that the field of the Sun is roughly poloidal. Sketch the appearance of the toroidal field generated by the poloidal field.

Exercise 19.21 *Problem: Buoyancy in the Solar Dynamo* T2

Consider a slender flux tube with width much less than its length which, in turn, is much less than the external pressure scale height H. Also assume that the magnetic field is directed along the tube so there is negligible current along the tube.

(a) Show that the requirement of magnetostatic equilibrium implies that inside the flux tube

$$\nabla \left(P + \frac{B^2}{2\mu_0} \right) \simeq 0. \tag{19.69}$$

(b) Consider a segment of this flux tube that is horizontal and has length ℓ. Holding both ends fixed, bend it vertically upward so that the radius of curvature of its center line is $R \gg \ell$. Assume that the fluid is isentropic with adiabatic index γ. By balancing magnetic tension against buoyancy, show that magnetostatic equilibrium is possible if $R \simeq 2\gamma H$. Do you think this equilibrium could be stable?

(c) In the solar convection zone (cf. Sec. 18.5), small entropy differences are important in driving the convective circulation. Following Ex. 19.20(b), suppose that a length of toroidal field is carried upward by a convecting "blob." Consider the action of the Coriolis force due to the Sun's rotation (cf. Sec. 14.2.1) on a single blob, and argue that it will rotate. What will this do to the magnetic field? Sketch the generation of a large-scale poloidal field from a toroidal field through the combined effects of many blobs. What do you expect to observe when a flux tube breaks through the solar surface (known as the photosphere)?

The combined effect of differential rotation, magnetic stress, and buoyancy, as outlined in Exs. 19.20 and 19.21, is thought to play an important role in sustaining the solar dynamo cycle. See Goedbloed and Poedts (2004, Secs. 8.2, 8.3) for further insights.

19.6.3 Magnetic Reconnection T2

So far, our discussion of the evolution of the magnetic field has centered on the induction equation (19.6) (the magnetic transport law). We have characterized the

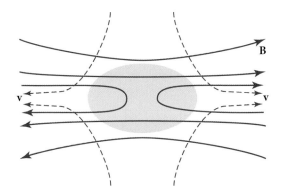

FIGURE 19.13 Magnetic reconnection. In the shaded reconnection region, Ohmic diffusion is important and allows magnetic field lines to "exchange partners," changing the overall field topology. Magnetic field components perpendicular to the plane of the illustration do not develop large gradients and so do not inhibit the reconnection process.

magnetized fluid by a magnetic Reynolds number using some characteristic length L associated with the flow, and have found that, when $R_M \gg 1$, Ohmic dissipation and field-line diffusion in the transport law are unimportant. This is reminiscent of the procedure we followed when discussing vorticity. However, for vorticity we discovered a very important exception to an uncritical neglect of viscosity, dissipation, and vortex-line diffusion at large Reynolds numbers, namely, boundary layers (Sec. 14.4). In particular, we found that large-Reynolds-number flows near solid surfaces develop large velocity gradients on account of the no-slip boundary condition, and that the local Reynolds number can thereby decrease to near unity, allowing viscous stress to change the character of the flow completely. Something similar, called *magnetic reconnection*, can happen in hydromagnetic flows with large R_M, even without the presence of solid surfaces.

Consider two oppositely magnetized regions of conducting fluid moving toward each other (the upper and lower regions in Fig. 19.13). Mutual magnetic attraction of the two regions occurs, as magnetic energy would be reduced if the two sets of field lines were superposed. However, strict flux freezing prevents superposition. Something has to give! What happens is a compromise. The attraction causes large magnetic gradients to develop, accompanied by a buildup of large current densities, until Ohmic diffusion ultimately allows the magnetic field lines to slip sideways through the fluid and to *reconnect* with the field in the other region (the sharply curved field lines in Fig. 19.13).

how magnetic reconnection occurs at high R_M

This reconnection mechanism can be clearly observed at work in tokamaks and in Earth's magnetopause, where the solar wind's magnetic field meets Earth's magnetosphere. However, the details of the reconnection mechanism are quite complex, involving plasma instabilities, anisotropic electrical conductivity, and shock fronts.

Large, inductive electric fields can also develop when the magnetic geometry undergoes rapid change. This can happen in the reversing magnetic field in Earth's *magnetotail*,[7] leading to the acceleration of charged particles that impact Earth during a *magnetic substorm*. Like dynamo action, reconnection has a major role in determining how magnetic fields actually behave in both laboratory and space plasmas.

For further detail on the physics and observations of reconnection, see, for example, Birn and Priest (2007) and Forbes and Priest (2007).

19.7 Magnetosonic Waves and the Scattering of Cosmic Rays

In Sec. 19.5, we discussed global perturbations of a nonuniform magnetostatic plasma and described how they may be unstable. We now consider a different, particularly simple example of dynamical perturbations: planar, monochromatic, propagating waves in a uniform, magnetized, conducting medium. These are called *magnetosonic waves*. They can be thought of as sound waves that are driven not just by gas pressure but also by magnetic pressure and tension.

Although magnetosonic waves have been studied experimentally under laboratory conditions, there the magnetic Reynolds numbers are generally quite small, so the waves damp quickly by Ohmic dissipation. No such problem arises in space plasmas, where magnetosonic waves are routinely studied by the many spacecraft that monitor the solar wind and its interaction with planetary magnetospheres. It appears that these modes perform an important function in space plasmas: they control the transport of cosmic rays. Let us describe some properties of cosmic rays before giving a formal derivation of the magnetosonic-wave dispersion relation.

19.7.1 Cosmic Rays

Cosmic rays are the high-energy particles, primarily protons, that bombard Earth's magnetosphere from outer space. They range in energy from ~ 1 MeV to $\sim 3 \times 10^{11}$ GeV $= 0.3$ ZeV ~ 50 J. (The highest cosmic-ray energy ever measured was ~ 50 J. Thus, naturally occurring particle accelerators are far more impressive than their terrestrial counterparts, which can only reach ~ 10 TeV $= 10^4$ GeV!) Most subrelativistic particles originate in the solar system. Their relativistic counterparts, up to energies ~ 100 TeV, are believed to come mostly from interstellar space, where they are accelerated by expanding shock waves created by supernova explosions (cf. Sec. 17.6.3). The origin of the highest energy particles, greater than ~ 100 TeV, is an intriguing mystery.

The distribution of cosmic-ray arrival directions at Earth is inferred to be quite isotropic (to better than 1 part in 10^4 at an energy of 10 GeV). This is somewhat surprising, because their sources, both in and beyond the solar system, are believed to be distributed anisotropically, so the isotropy needs to be explained. Part of the reason for the isotropy is that the interplanetary and interstellar media are magnetized, and

---

7. The magnetotail is the region containing trailing field lines on the night side of Earth.

the particles gyrate around the magnetic field with the gyro (relativistic cyclotron) frequency $\omega_G = eBc^2/\mathcal{E}$, where $\mathcal{E}$ is the (relativistic) particle energy including rest mass, and B is the magnetic field strength. The gyro (Larmor) radii of the non-relativistic particle orbits are typically small compared to the size of the solar system, and those of the relativistic particles are typically small compared to characteristic lengthscales in the interstellar medium. Therefore, this gyrational motion can effectively erase any azimuthal asymmetry around the field direction. However, this does not stop the particles from streaming away from their sources along the magnetic field, thereby producing anisotropy at Earth. So something else must be impeding this along-line flow, by scattering the particles and causing them to effectively diffuse along and across the field through interplanetary and interstellar space.

evidence for scattering

As we verify in Sec. 20.4, Coulomb collisions are quite ineffective (even if they were effective, they would cause huge cosmic-ray energy losses, in violation of observations). We therefore seek some means of changing a cosmic ray's momentum without altering its energy significantly. This is reminiscent of the scattering of electrons in metals, where it is phonons (elastic waves in the crystal lattice) that are responsible for much of the scattering. It turns out that in the interstellar medium, magnetosonic waves can play a role analogous to phonons and can scatter the cosmic rays. As an aid to understanding this phenomenon, we now derive the waves' dispersion relation.

19.7.2 Magnetosonic Dispersion Relation

19.7.2

Our procedure for deriving the dispersion relation (last paragraph of Sec. 7.2.1) should be familiar by now. We consider a uniform, isentropic, magnetized fluid at rest; we perform a linear perturbation and seek monochromatic plane-wave solutions varying as $e^{i(\mathbf{k} \cdot \mathbf{x} - \omega t)}$. We ignore gravity and dissipative processes (specifically, viscosity, thermal conductivity, and electrical resistivity), as well as gradients in the equilibrium, which can all be important in some circumstances.

It is convenient to use the velocity perturbation $\delta \mathbf{v}$ as the independent variable. The perturbed and linearized equation of motion (19.10) then takes the form

$$-i\rho\omega\delta\mathbf{v} = -iC^2\mathbf{k}\delta\rho + \frac{(i\mathbf{k} \times \delta\mathbf{B}) \times \mathbf{B}}{\mu_0}. \tag{19.70}$$

Here C is the sound speed [$C^2 = (\partial P/\partial \rho)_s = \gamma P/\rho$, not to be confused with the speed of light c], and $\delta P = C^2 \delta \rho$ is the Eulerian pressure perturbation for our homogeneous equilibrium.[8] (Note that $\nabla P = \nabla \rho = 0$, so Eulerian and Lagrangian perturbations are the same.) The perturbed equation of mass conservation, $\partial \rho/\partial t + \nabla \cdot (\rho \mathbf{v}) = 0$, becomes

$$\omega \delta \rho = \rho \mathbf{k} \cdot \delta \mathbf{v}, \tag{19.71}$$

---

8. Note that we are assuming that (i) the equilibrium pressure tensor is isotropic and (ii) the perturbations to the pressure are also isotropic. This is unlikely to be the case for a collisionless plasma, so our treatment there must be modified. See Sec. 20.6.2 and Ex. 21.1.

and the MHD law of magnetic-field transport with dissipation ignored, $\partial \mathbf{B}/\partial t = \nabla \times (\mathbf{v} \times \mathbf{B})$, becomes

$$\omega \delta \mathbf{B} = -\mathbf{k} \times (\delta \mathbf{v} \times \mathbf{B}). \tag{19.72}$$

Alfvén velocity

We introduce the Alfvén velocity

$$\boxed{\mathbf{a} \equiv \frac{\mathbf{B}}{(\mu_0 \rho)^{1/2}},} \tag{19.73}$$

and insert $\delta\rho$ [Eq. (19.71)] and $\delta\mathbf{B}$ [Eq. (19.72)] into Eq. (19.70) to obtain

eigenequation for magnetosonic waves

$$\{\mathbf{k} \times [\mathbf{k} \times (\delta \mathbf{v} \times \mathbf{a})]\} \times \mathbf{a} + C^2 (\mathbf{k} \cdot \delta \mathbf{v}) \mathbf{k} = \omega^2 \delta \mathbf{v}. \tag{19.74}$$

Equation (19.74) is an eigenequation for the wave's squared frequency ω^2 and eigendirection $\delta\mathbf{v}$. The straightforward way to solve it is to rewrite it in the standard matrix form $M_{ij}\delta v_j = \omega^2 \delta v_i$ and then use standard matrix (determinant) methods. It is quicker, however, to seek the three eigendirections $\delta\mathbf{v}$ and eigenfrequencies ω one by one, by projection along preferred directions.

Alfvén mode and its dispersion relation

We first seek a solution to Eq. (19.74) for which $\delta\mathbf{v}$ is orthogonal to the plane formed by the unperturbed magnetic field and the wave vector, $\delta\mathbf{v} = \mathbf{a} \times \mathbf{k}$ up to a multiplicative constant. Inserting this $\delta\mathbf{v}$ into Eq. (19.74), we obtain the dispersion relation

$$\boxed{\omega = \pm \mathbf{a} \cdot \mathbf{k}; \qquad \frac{\omega}{k} = \pm a \cos\theta,} \tag{19.75}$$

where θ is the angle between $\mathbf{k}$ and the unperturbed field, and $a \equiv |\mathbf{a}| = B/(\mu_0\rho)^{1/2}$ is the Alfvén speed. This type of wave is known as the *intermediate magnetosonic mode* and also as the *Alfvén mode*. Its phase speed $\omega/k = a\cos\theta$ is plotted as the larger figure-8 curve in Fig. 19.14. The velocity and magnetic perturbations $\delta\mathbf{v}$ and $\delta\mathbf{B}$ are both along the direction $\mathbf{a} \times \mathbf{k}$, so the Alfvén wave is fully transverse. There is no compression ($\delta\rho = 0$), which accounts for the absence of the sound speed C in the dispersion relation. This Alfvén mode has a simple physical interpretation in the limiting case when $\mathbf{k}$ is parallel to $\mathbf{B}$. We can think of the magnetic field lines as strings with tension B^2/μ_0 and inertia ρ, which are plucked transversely. Their transverse oscillations then propagate with speed $\sqrt{\text{tension/inertia}} = B/\sqrt{\mu_0\rho} = a$. For details and delicacies, see Ex. 21.8.

Alfvén-mode properties

The dispersion relations for the other two modes can be deduced by projecting the eigenequation (19.74) successively along $\mathbf{k}$ and along $\mathbf{a}$ to obtain two scalar equations:

$$k^2(\mathbf{k} \cdot \mathbf{a})(\mathbf{a} \cdot \delta \mathbf{v}) = [(a^2 + C^2)k^2 - \omega^2](\mathbf{k} \cdot \delta \mathbf{v}),$$
$$C^2(\mathbf{k} \cdot \mathbf{a})(\mathbf{k} \cdot \delta \mathbf{v}) = \omega^2(\mathbf{a} \cdot \delta \mathbf{v}). \tag{19.76}$$

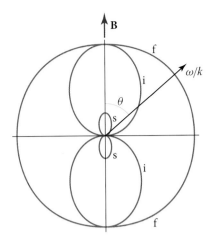

FIGURE 19.14 Phase-velocity surfaces for the three types of magnetosonic modes: fast (f), intermediate (i), and slow (s). The three curves are polar plots of the wave phase velocity ω/k in units of the Alfvén speed $a = B/\sqrt{\mu_0 \rho}$. In the particular example shown, the sound speed C is half the Alfvén speed.

Combining these equations, we obtain the dispersion relation

$$\left(\frac{\omega}{k}\right)^2 = \frac{1}{2}(a^2 + C^2)\left[1 \pm \left(1 - \frac{4C^2 a^2 \cos^2\theta}{(a^2 + C^2)^2}\right)^{1/2}\right]. \quad (19.77)$$

dispersion relation for fast (+) and slow (−) magnetosonic modes

(By inserting this dispersion relation, with the upper or lower sign, back into Eqs. (19.76), we can deduce the mode's eigendirection $\delta \mathbf{v}$.) This dispersion relation tells us that ω^2 is positive, so no unstable modes exist, which seems reasonable, as there is no source of free energy. (The same is true, of course, for the Alfvén mode.)

These waves are compressive, with the gas being moved by a combination of gas pressure and magnetic pressure and tension. The modes can be seen to be non-dispersive, which is also to be expected, as we have introduced neither a characteristic timescale nor a characteristic length into the problem.

The mode with the plus signs in Eq. (19.77) is called the *fast magnetosonic mode*; its phase speed is depicted by the outer, quasi-circular curve in Fig. 19.14. A good approximation to its phase speed when $a \gg C$ or $a \ll C$ is $\omega/k \simeq \pm(a^2 + C^2)^{1/2}$. When propagating perpendicularly to $\mathbf{B}$, the fast mode can be regarded as simply a longitudinal sound wave in which the gas pressure is augmented by the magnetic pressure $B^2/(2\mu_0)$ (adopting a specific-heat ratio γ for the magnetic field of 2, as $B \propto \rho$ and so $P_{\text{mag}} \propto \rho^2$ under perpendicular compression).

fast magnetosonic-mode properties

The mode with the minus signs in Eq. (19.77) is called the *slow magnetosonic mode*. Its phase speed (depicted by the inner figure-8 curve in Fig. 19.14) can be approximated by $\omega/k = \pm aC \cos\theta/(a^2 + C^2)^{1/2}$ when $a \gg C$ or $a \ll C$. Note that

slow magnetosonic-mode properties

slow mode—like the intermediate mode but unlike the fast mode—is incapable of propagating perpendicularly to the unperturbed magnetic field; see Fig. 19.14. In the limit of vanishing Alfvén speed or vanishing sound speed, the slow mode ceases to exist for all directions of propagation.

In Chap. 21, we will discover that MHD is a good approximation to the behavior of plasmas only at frequencies below the "ion cyclotron frequency," which is a rather low frequency. For this reason, magnetosonic modes are usually regarded as low-frequency modes.

19.7.3 Scattering of Cosmic Rays by Alfvén Waves

mechanism for Alfvén waves to scatter cosmic-ray particles

Now let us return to the issue of cosmic-ray propagation, which motivated our investigation of magnetosonic modes. Let us consider 100-GeV particles in the interstellar medium. The electron (and ion, mostly proton) density and magnetic field strength are typically $n \sim -3 \times 10^4 \, \text{m}^{-3}$ and $B \sim 300 \, \text{pT}$. The Alfvén speed is then $a \sim 30 \, \text{km s}^{-1}$, much slower than the speeds of the cosmic rays. When analyzing cosmic-ray propagation, a magnetosonic wave can therefore be treated as essentially a magnetostatic perturbation. A relativistic cosmic ray of energy $\mathcal{E}$ has a gyro (relativistic Larmor) radius of $r_G = \mathcal{E}/eBc$, in this case $\sim 3 \times 10^{12}$ m. Cosmic rays will be unaffected by waves with wavelength either much greater than or much less than r_G. However, magnetosonic waves (especially Alfvén waves, which are responsible for most of the scattering), with wavelength matched to the gyro radius, will be able to change the particle's *pitch angle* α (the angle its momentum makes with the mean magnetic field direction). See Sec. 23.4.1 for some details. If the Alfvén waves in this wavelength range have rms dimensionless amplitude $\delta B/B \ll 1$, then the particle's pitch angle will change by an amount $\delta\alpha \sim \delta B/B$ for every wavelength. Now, if the wave spectrum is broadband, individual waves can be treated as uncorrelated, so the particle pitch angle changes stochastically. In other words, the particle diffuses in pitch angle. The effective diffusion coefficient is

$$D_\alpha \sim \left(\frac{\delta B}{B}\right)^2 \omega_G, \tag{19.78}$$

where $\omega_G = c/r_G$ is the gyro frequency (relativistic analog of cyclotron frequency ω_c). The particle is therefore scattered by roughly a radian in pitch angle every time it traverses a distance $\ell \sim (B/\delta B)^2 r_G$. This is effectively the particle's collisional mean free path. Associated with this mean free path is a spatial diffusion coefficient

$$D_x \sim \frac{\ell c}{3}. \tag{19.79}$$

estimated and measured cosmic-ray anisotropy

It is thought that $\delta B/B \sim 10^{-1}$ in the relevant wavelength range in the interstellar medium. An estimate of the collision mean free path is then $\ell(100 \, \text{GeV}) \sim 10^{14}$ m. Now, the thickness of our galaxy's interstellar disk of gas is roughly $L \sim 3 \times 10^{18}$ m $\sim 10^4 \ell$. Therefore, an estimate of the cosmic-ray anisotropy is

$\sim \ell/L \sim 3 \times 10^{-5}$, roughly compatible with the measurements. Although this discussion is an oversimplification, it does demonstrate that the cosmic rays in the interplanetary, interstellar, and intergalactic media can be scattered efficiently by magnetosonic waves. This allows the particle transport to be impeded without much loss of energy, so that the theory of cosmic-ray acceleration (Ex. 23.12) and scattering (this section) together can account for the particle fluxes observed as a function of energy and direction at Earth.

A good question to ask at this point is "Where do the Alfvén waves come from?" The answer turns out to be that they are maintained as part of a turbulence spectrum and created by the cosmic rays themselves through the growth of plasma instabilities. To proceed further and give a more quantitative description of this interaction, we must go beyond a purely fluid description and explore the motions of individual particles. This is where we shall turn in the next few chapters, culminating with a return to the interaction of cosmic rays with Alfvén waves in Sec. 23.4.1.

Bibliographic Note

For intuitive insight into magnetohydrodynamics, we recommend Shercliff (1965).

For textbook introductions to magnetohydrodynamics, we recommend the relevant chapters of Landau, Pitaevskii, and Lifshitz (1979), Schmidt (1979), Boyd and Sanderson (2003), and Bellan (2006). For far greater detail, we recommend a textbook that deals solely with magnetohydrodynamics: Goedbloed and Poedts (2004) and its advanced supplement Goedbloed, Keppens, and Poedts (2010). For a very readable treatment from the viewpoint of an engineer, with applications to engineering and metallurgy, see Davidson (2001).

For the theory of MHD instabilities and applications to magnetic confinement, see the above references, and also Bateman (1978) and the collection of early papers edited by Jeffrey and Taniuti (1966). For applications to astrophysics and space physics, see Parker (1979), Parks (2004), and Kulsrud (2005).

PART VI

PLASMA PHYSICS

A *plasma* is a gas that is significantly ionized (through heating or photoionization) and thus is composed of electrons and ions, and that has a low enough density to behave classically (i.e., to obey Maxwell-Boltzmann statistics rather than Fermi-Dirac or Bose-Einstein). Plasma physics originated in the nineteenth century, with the study of gas discharges (Crookes, 1879). In 1902, Heaviside and Kennelly realized that plasma is also the key to understanding the propagation of radio waves across the Atlantic. The subject received a further boost in the early 1950s, with the start of the controlled (and the uncontrolled) thermonuclear fusion program. The various confinement devices described in the preceding chapter are intended to hold plasma at temperatures as high as $\sim 10^8$ K, high enough for fusion to begin; the difficulty of this task has turned out to be an issue of plasma physics as much as of MHD. The next new venue for plasma research was extraterrestrial. Although it was already understood that Earth was immersed in a tenuous outflow of ionized hydrogen known as the solar wind, the dawn of the space age in 1957 also initiated experimental *space plasma physics*. More recently, the interstellar and intergalactic media beyond the solar system as well as exotic astronomical objects like quasars and pulsars have allowed us to observe plasmas under quite extreme conditions, irreproducible in any laboratory experiment.

The dynamical behavior of a plasma is more complex than the dynamics of the gases and fluids we have met so far. This dynamical complexity has two main origins:

1. The dominant form of interparticle interaction in a plasma, Coulomb scattering, is so weak that the mean free paths of the electrons and ions are often larger than the plasma's macroscopic lengthscales. This allows the particles' momentum distribution functions to deviate seriously from their equilibrium Maxwellian forms and, in particular, to be highly anisotropic.

2. The electromagnetic fields in a plasma are of long range. This allows charged particles to couple to one another electromagnetically and act in concert as modes of excitation (plasma waves) that behave like single dynamical entities

(plasmons). Much of plasma physics consists of the study of the properties and interactions of these modes.

The dynamical behavior of a plasma depends markedly on frequency. At the lowest frequencies, the ions and electrons are locked together by electrostatic forces and behave like an electrically conducting fluid; this is the regime of MHD (Chap. 19). At somewhat higher frequencies, the electrons and the ions can move relative to each other, behaving like two separate, interpenetrating fluids; we study this two-fluid regime in Chap. 21. At still higher frequencies, complex dynamics are supported by momentum space anisotropies and can be analyzed using a variant of the kinetic-theory collisionless Boltzmann equation that we introduced in Chap. 3. We study such dynamics in Chap. 22. In the two-fluid and collisionless-Boltzmann analyses of Chaps. 21 and 22, we focus on phenomena that can be treated as linear perturbations of an equilibrium state. However, the complexities and long mean free paths of plasmas also produce rich nonlinear phenomena; we study some of these in Chap. 23. But first, as a foundation for the dynamical studies in Chaps. 21, 22, and 23, we develop in Chap. 20 detailed insights into the microscopic structure of a plasma.

CHAPTER TWENTY

The Particle Kinetics of Plasma

> The study of individual particles frequently gives insight
> into the behavior of an ionized gas.
> LYMAN SPITZER (1962)

20.1 Overview

The preceding chapter, Chap. 19, can be regarded as a transition from fluid mechanics to plasma physics. In the context of a magnetized plasma, it describes equilibrium and low-frequency dynamical phenomena using fluid-mechanics techniques. In this chapter, we prepare for more sophisticated descriptions of a plasma by introducing a number of elementary foundational concepts peculiar to plasmas and by exploring a plasma's structure on the scale of individual particles using elementary techniques from kinetic theory.

Specifically, in Sec. 20.2, we identify the region of densities and temperatures in which matter, in statistical equilibrium, takes the form of a plasma, and we meet specific examples of plasmas that occur in Nature and in the laboratory. Then in Sec. 20.3, we study two phenomena that are important for plasmas: the collective manner in which large numbers of electrons and ions shield out the electric field of a charge in a plasma (Debye shielding), and the oscillations of a plasma's electrons relative to its ions (plasma oscillations).

In Sec. 20.4, we study the Coulomb scattering by which a plasma's electrons and ions deflect an individual charged particle from straight-line motion and exchange energy with it. We then examine the statistical properties of large numbers of such Coulomb scatterings—most importantly, the rates (inverse timescales) for the velocities of a plasma's electrons and ions to isotropize, and the rates for them to thermalize. Our calculations reveal that Coulomb scattering is so weak that, in most plasmas encountered in Nature, it is unlikely to produce isotropized or thermalized velocity distributions. In Sec. 20.5, we give a brief preview of the fact that in real plasmas the scattering of electrons and ions off collective plasma excitations (*plasmons*) often isotropizes and thermalizes their velocities far faster than would Coulomb scattering. Thus many real plasmas are far more isotropic and thermalized than our Coulomb-scattering analyses suggest. (We explore this "anomalous" behavior in Chaps. 22 and 23.) In Sec. 20.5, we also use the statistical properties

> **BOX 20.1. READERS' GUIDE**
>
> - This chapter relies significantly on portions of nonrelativistic kinetic theory as developed in Chap. 3.
> - It also relies a bit (but not greatly) on portions of MHD as developed in Chap. 19.
> - Chapters 21–23 of Part VI rely heavily on this chapter.

of Coulomb scatterings to derive a plasma's transport coefficients—specifically, its electrical and thermal conductivities—for situations where Coulomb scattering dominates over particle-plasmon scattering.

Most plasmas are significantly magnetized. This introduces important new features into their dynamics, which we describe in Sec. 20.6: cyclotron motion (the spiraling of particles around magnetic field lines), a resulting anisotropy of the plasma's pressure (different pressures along and orthogonal to the field lines), and the split of a plasma's adiabatic index into four different adiabatic indices for four different types of compression. Finally, in Sec. 20.7, we examine the motion of an individual charged particle in a slightly inhomogeneous and slowly time-varying magnetic field, and we describe *adiabatic invariants*, which control that motion in easily understood ways.

20.2 Examples of Plasmas and Their Density-Temperature Regimes

The density-temperature regime in which matter behaves as a nonrelativistic plasma is shown in Fig. 20.1. In this figure, and in most of Part VI, we confine our attention to pure hydrogen plasma comprising protons and electrons. Many plasmas contain large fractions of other ions, which can have larger charges and do have greater masses than protons. This generalization introduces few new issues of principle so, for simplicity, we eschew it and focus on a hydrogen plasma.

The boundaries of the plasma regime in Fig. 20.1 are dictated by the following considerations.

20.2.1 Ionization Boundary

We are mostly concerned with fully ionized plasmas, even though partially ionized plasmas, such as the ionosphere, are often encountered in physics, astronomy, and engineering, and more complex plasmas (such as dusty plasmas) can also be important. The plasma regime's ionization boundary is the bottom curve in Fig. 20.1, at a temperature of a few thousand Kelvins. This boundary is dictated by chemical equilibrium for the reaction

$$H \leftrightarrow p + e, \tag{20.1}$$

ionization boundary

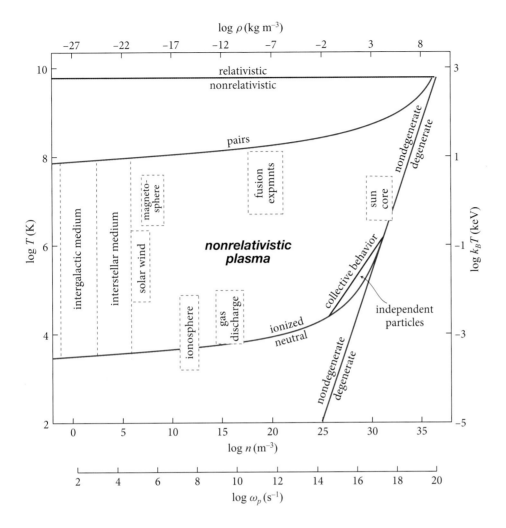

FIGURE 20.1 The density-temperature regime in which matter, made largely of hydrogen, behaves as a nonrelativistic plasma. The densities and temperatures of specific examples of plasmas are indicated by dashed lines. The number density n of electrons (and also of protons) is shown horizontally at the bottom, and the corresponding mass density ρ is shown at the top. The temperature T is shown at the left in Kelvins, and at the right $k_B T$ is shown in keV, thousands of electron volts. The bottom-most scale is the plasma frequency, discussed in Sec. 20.3.3. This figure is a more detailed version of Fig. 3.7.

as described by the Saha equation (Ex. 5.10):

$$\frac{n_e n_p}{n_H} = \frac{(2\pi m_e k_B T)^{3/2}}{h^3} e^{-I_P/(k_B T)}. \tag{20.2}$$

Here n_e, n_p, and n_H are the number densities of electrons, protons, and neutral hydrogen atoms (at the relevant temperatures hydrogen molecules have dissociated into individual atoms), respectively; T is temperature; m_e is the electron rest mass; h is Planck's constant; k_B is Boltzmann's constant; and $I_P = 13.6$ eV is the ionization energy of hydrogen (i.e., the binding energy of its ground state). Notice that the prefactor in Eq. (20.2) is λ_{TdB}^{-3} with $\lambda_{TdB} \equiv h/\sqrt{2\pi m_e k_B T}$, the electrons' thermal de Broglie wavelength. The bottom boundary in Fig. 20.1 is that of 50% ionization [i.e., $n_e = n_p = n_H = \rho/(2m_H)$, with m_H the mass of a hydrogen atom]; but because of the exponential factor in Eq. (20.2), the line of 90% ionization is virtually indistinguishable from that of 50% ionization on the scale of the figure. Using the rough equivalence 1 eV $\cong 10^4$ K, we might have expected that the ionization boundary would correspond to a temperature $T \sim I_P/k_B \sim 10^5$ K. However, this is true only near the degeneracy boundary (see below). When the plasma is strongly nondegenerate, ionization occurs at a significantly lower temperature due to the vastly greater number of states available to an electron when free than when bound in a hydrogen atom. Equivalently, at low densities, once a hydrogen atom has been broken up into an electron plus a proton, the electron (or proton) must travel a large distance before encountering another proton (or electron) with which to recombine, making a new hydrogen atom; as a result, equilibrium occurs at a lowered temperature, where the ionization rate is thereby lowered to match the smaller recombination rate.

20.2.2 Degeneracy Boundary

degeneracy boundary

The electrons, with their small rest masses, become degenerate more easily than the protons or hydrogen atoms. The slanting line on the right side of Fig. 20.1 is the plasma's boundary of electron degeneracy. This boundary is determined by the demand that the mean occupation number of the electrons' single-particle quantum states not be $\ll 1$. In other words, the volume of phase space per electron—the product of the volumes of real space $\sim n_e^{-1}$ and of momentum space $\sim (m_e k_B T)^{3/2}$ occupied by each electron—should be comparable with the elementary quantum mechanical phase-space volume given by the uncertainty principle, h^3. Inserting the appropriate factors of order unity [cf. Eq. (3.41)], this relation becomes the boundary equation:

$$n_e \simeq 2\frac{(2\pi m_e k_B T)^{3/2}}{h^3}. \tag{20.3}$$

When the electrons become degenerate (rightward of the degeneracy line in Fig. 20.1), as they do in a metal or a white dwarf, the electron de Broglie wavelength becomes large compared with the mean interparticle spacing, and quantum mechanical considerations are of paramount importance.

20.2.3 Relativistic Boundary

relativistic boundary

Another important limit arises when the electron thermal speeds become relativistic. This occurs when

$$T \sim m_e c^2/k_B \sim 6 \times 10^9 \text{ K} \tag{20.4}$$

(top horizontal line in Fig. 20.1). Although we shall not consider them much further, the properties of relativistic plasmas (above this line) are mostly analogous to those of nonrelativistic plasmas (below it).

20.2.4 Pair-Production Boundary

Finally, for plasmas in statistical equilibrium, electron-positron pairs are created in profusion at high enough temperatures. In Ex. 5.9 we showed that, for $k_B T \ll m_e c^2$ but T high enough that pairs begin to form, the density of positrons divided by that of protons is

$$\frac{n_+}{n_p} = \frac{1}{2y[y + (1 + y^2)^{1/2}]}, \quad \text{where } y \equiv \frac{1}{4} n_e \left(\frac{h}{\sqrt{2\pi m_e k_B T}}\right)^3 e^{m_e c^2/(k_B T)}. \quad (20.5)$$

Setting this expression to unity gives the pair-production boundary. This boundary curve, labeled "pairs" in Fig. 20.1, is similar in shape to the ionization boundary but shifted in temperature by $\sim 2 \times 10^4 \sim \alpha_F^{-2}$, where α_F is the fine structure constant. This is because we are now effectively "ionizing the vacuum" rather than a hydrogen atom, and the "ionization potential of the vacuum" is $\sim 2 m_e c^2 = 4 I_P / \alpha_F^2$.

pair-production boundary

We shall encounter a plasma above the pair-production boundary, and thus with a profusion of electron-positron pairs, in our discussion of the early universe in Secs. 28.4.1 and 28.4.2.

20.2.5 Examples of Natural and Human-Made Plasmas

Figure 20.1 and Table 20.1 show the temperature-density regions for the following plasmas:

- *Laboratory gas discharge.* The plasmas created in the laboratory by electric currents flowing through hot gas (e.g., in vacuum tubes, spark gaps, welding arcs, and neon and fluorescent lights).
- *Controlled thermonuclear fusion experiments.* The plasmas in which experiments for controlled thermonuclear fusion are carried out (e.g., in tokamaks).
- *Ionosphere.* The part of Earth's upper atmosphere (at heights of $\sim$50–300 km) that is partially photoionized by solar ultraviolet radiation.
- *Magnetosphere.* The plasma of high-speed electrons and ions that are locked on Earth's dipolar magnetic field and slide around on its field lines at several Earth radii.
- *Sun's core.* The plasma at the center of the Sun, where fusion of hydrogen to form helium generates the Sun's heat.
- *Solar wind.* The wind of plasma that blows off the Sun and outward through the region between the planets.
- *Interstellar medium.* The plasma, in our galaxy, that fills the region between the stars. This plasma exhibits a fairly wide range of density and temperature as a result of such processes as heating by photons from stars, heating and compression by shock waves from supernovae, and cooling by thermal emission of radiation.

TABLE 20.1: Representative electron number densities n, temperatures T, and magnetic field strengths B, together with derived plasma parameters in a variety of environments

Plasma	n (m^{-3})	T (K)	B (T)	λ_D (m)	N_D	ω_p (s^{-1})	ν_{ee} (s^{-1})	ω_c (s^{-1})	r_L (m)
Gas discharge	10^{16}	10^4	—	10^{-4}	10^4	10^{10}	10^5	—	—
Fusion experiments	10^{20}	10^8	10	10^{-4}	10^8	10^{12}	10^4	10^{12}	10^{-5}
Ionosphere	10^{12}	10^3	10^{-5}	10^{-3}	10^5	10^8	10^3	10^6	10^{-1}
Magnetosphere	10^7	10^7	10^{-8}	10^2	10^{10}	10^5	10^{-8}	10^3	10^4
Sun's core	10^{32}	10^7	—	10^{-11}	1	10^{18}	10^{16}	—	—
Solar wind	10^6	10^5	10^{-9}	10	10^{11}	10^5	10^{-6}	10^2	10^4
Interstellar medium	10^5	10^4	10^{-10}	10	10^{10}	10^4	10^{-5}	10	10^4
Intergalactic medium	10^{-1}	10^6	—	10^5	10^{15}	10^2	10^{-12}	—	—

Notes: The derived parameters, discussed later in this chapter, are: λ_D, Debye length; N_D, Debye number; ω_p, plasma frequency; ν_{ee}, equilibration rate for electron energy; ω_c, electron cyclotron frequency; and r_L, electron Larmor radius. Values are given to order of magnitude, as all of these environments are quite inhomogeneous. — indicates this quantity is irrelevant.

- *Intergalactic medium.* The plasma that fills the space outside galaxies and clusters of galaxies (the locale of almost all the universe's plasma). We shall meet the properties and evolution of this intergalactic plasma in our study of cosmology, Sec. 28.3.1.

Characteristic plasma properties in these various environments are collected in Table 20.1. In the next three chapters we study applications from all these environments.

EXERCISES

Exercise 20.1 *Derivation: Boundary of Degeneracy*
Show that the condition $n_e \ll (2\pi m_e k_B T)^{3/2}/h^3$ [cf. Eq. (20.3)] that electrons be nondegenerate is equivalent to the following statements.

(a) The mean separation between electrons, $l \equiv n_e^{-1/3}$, is large compared to the electrons' thermal de Broglie wavelength, $\lambda_{T\,\mathrm{dB}} = h/\sqrt{2\pi m_e k_B T}$.

(b) The uncertainty in the location of an electron drawn at random from the thermal distribution is small compared to the average inter-electron spacing.

(c) The quantum mechanical zero-point energy associated with squeezing each electron into a region of size $l = n_e^{-1/3}$ is small compared to the electron's mean thermal energy $k_B T$.

20.3 Collective Effects in Plasmas—Debye Shielding and Plasma Oscillations

In this section, we introduce two key ideas that are associated with most of the collective effects in plasma dynamics: Debye shielding (also called Debye screening) and plasma oscillations.

20.3.1 Debye Shielding

Any charged particle inside a plasma attracts other particles with opposite charge and repels those with the same charge, thereby creating a net cloud of opposite charges around itself. This cloud shields the particle's own charge from external view (i.e., it causes the particle's Coulomb field to fall off exponentially at large radii, rather than falling off as $1/r^2$).[1]

mechanism of Debye shielding

This Debye shielding (or screening) of a particle's charge can be demonstrated and quantified as follows. Consider a single fixed test charge Q surrounded by a plasma of protons and electrons. Denote the average densities of protons and electrons—considered as smooth functions of radius r from the test charge—by $n_p(r)$ and $n_e(r)$, and let the mean densities of electrons and protons (averaged over a large volume) be $\bar{n}$. (The mean electron and proton densities must be equal because of overall charge neutrality, which is enforced by the electrostatic restoring force that we explore in Sec. 20.3.3.) Then the electrostatic potential $\Phi(r)$ outside the particle satisfies Poisson's equation, which we write in SI units:[2]

$$\nabla^2 \Phi = -\frac{(n_p - n_e)e}{\epsilon_0} - \frac{Q}{\epsilon_0}\delta(\mathbf{r}). \qquad (20.6)$$

Poisson's equation

(We denote the positive charge of a proton by $+e$ and the negative charge of an electron by $-e$.)

A proton at radius r from the particle has an electrostatic potential energy $e\Phi(r)$. Correspondingly, the number density of protons at radius r is altered from $\bar{n}$ by the Boltzmann factor $\exp[-e\Phi/(k_B T)]$; and, similarly, the density of electrons is altered by $\exp[+e\Phi/(k_B T)]$:

$$n_p = \bar{n}\exp[-e\Phi/(k_B T)] \simeq \bar{n}[1 - e\Phi/(k_B T)],$$
$$n_e = \bar{n}\exp[+e\Phi/(k_B T)] \simeq \bar{n}[1 + e\Phi/(k_B T)], \qquad (20.7)$$

where we have made a Taylor expansion of the Boltzmann factor valid for $|e\Phi| \ll k_B T$. By inserting the linearized versions of Eqs. (20.7) into Eq. (20.6), we obtain

$$\nabla^2 \Phi = \frac{2e^2 \bar{n}}{\epsilon_0 k_B T}\Phi - \frac{Q}{\epsilon_0}\delta(\mathbf{r}). \qquad (20.8)$$

---

1. Analogous effects are encountered in condensed-matter physics and quantum electrodynamics.
2. For those who prefer Gaussian units, the translation is most easily effected by the transformations $4\pi\epsilon_0 \to 1$ and $\mu_0/(4\pi) \to 1$, and inserting factors of c by inspection using dimensional analysis. It is also useful to recall that $1\,\text{T} \equiv 10^4$ Gauss and that the charge on an electron is $-1.602 \times 10^{-19}\,\text{C} \equiv -4.803 \times 10^{-10}$ esu.

The spherically symmetric solution to this equation,

$$\Phi = \frac{Q}{4\pi \epsilon_0 r} e^{-\sqrt{2}\, r/\lambda_D}, \tag{20.9}$$

has the form of a Coulomb field with an exponential cutoff. The characteristic length-scale of the exponential cutoff,

Debye length

$$\lambda_D \equiv \left(\frac{\epsilon_0 k_B T}{\bar{n} e^2}\right)^{1/2} = 69 \left(\frac{T/1\,\mathrm{K}}{\bar{n}/1\,\mathrm{m}^{-3}}\right)^{1/2}\,\mathrm{m}, \tag{20.10}$$

is called the *Debye length*. It is a rough measure of the size of the Debye shielding cloud that the charged particle carries with itself.

The charged particle could be some foreign charged object (not a plasma electron or proton), or equally well, it could be one of the plasma's own electrons or protons. Thus, we can think of each electron in the plasma as carrying with itself a positively charged Debye shielding cloud of size λ_D, and each proton as carrying a negatively charged cloud. Each electron and proton not only carries its own cloud; it also plays a role as one of the contributors to the clouds around other electrons and protons.

20.3.2 Collective Behavior

A charged particle's Debye cloud is almost always made of a huge number of electrons and very nearly the same number of protons. It is only a tiny, time-averaged excess of electrons over protons (or protons over electrons) that produces the cloud's net charge and the resulting exponential decay of the electrostatic potential. Ignoring this tiny excess, the mean number of electrons in the cloud and the mean number of protons are roughly

Debye number

$$N_D \equiv \bar{n}\, \frac{4\pi}{3} \lambda_D^3 = 1.4 \times 10^6 \frac{(T/1\,\mathrm{K})^{3/2}}{(\bar{n}/1\,\mathrm{m}^{-3})^{1/2}}. \tag{20.11}$$

This *Debye number* is large compared to unity throughout the density-temperature regime of plasmas, except for the tiny lower right-hand corner of Fig. 20.1. The boundary of that region (labeled "collective behavior"/"independent particles") is given by $N_D = 1$. The upper left-hand side of that boundary has $N_D \gg 1$ and is called the regime of collective behavior, because a huge number of particles are collectively responsible for the Debye cloud; this leads to a variety of collective dynamical phenomena in the plasma. The lower right-hand side has $N_D < 1$ and is called the regime of independent particles, because in it collective phenomena are of small importance. (In this regime Φ on the right-hand side of Eq. (20.8) is replaced by $(kT/e)\sinh[e\Phi/(k_B T)]$.) In this book we restrict ourselves to the extensive regime of collective behavior and ignore the tiny regime of independent particles.

Characteristic values for the Debye length and Debye number in a variety of environments are listed in Table 20.1.

20.3.3 Plasma Oscillations and Plasma Frequency

Of all the dynamical phenomena that can occur in a plasma, perhaps the most important is a relative oscillation of the plasma's electrons and protons. A simple, idealized version of this *plasma oscillation* is depicted in Fig. 20.2. Suppose for the moment that the protons are all fixed, and displace the electrons rightward (in the x direction) with respect to the protons by an amount ξ, thereby producing a net negative charge per unit area $-e\bar{n}\xi$ at the right end of the plasma, a net positive charge per unit area $+e\bar{n}\xi$ at the left end, and a corresponding electric field $E = e\bar{n}\xi/\epsilon_0$ in the x direction throughout the plasma. The electric field pulls on the plasma's electrons and protons, giving the electrons an acceleration $d^2\xi/dt^2 = -eE/m_e$ and the protons an acceleration smaller by $m_e/m_p = 1/1,836$, which we neglect. The result is an equation of motion for the electrons' collective displacement:

$$\frac{d^2\xi}{dt^2} = -\frac{e}{m_e}E = -\frac{e^2\bar{n}}{\epsilon_0 m_e}\xi. \tag{20.12}$$

Since Eq. (20.12) is a harmonic-oscillator equation, the electrons oscillate sinusoidally, $\xi = \xi_o \cos(\omega_p t)$, at the *plasma (angular) frequency*

$$\boxed{\omega_p \equiv \left(\frac{\bar{n}e^2}{\epsilon_0 m_e}\right)^{1/2} = 56.3\left(\frac{\bar{n}}{1\,\mathrm{m}^{-3}}\right)^{1/2}\,\mathrm{s}^{-1}.} \tag{20.13}$$

plasma frequency, for plasma oscillations

Notice that this frequency of plasma oscillations depends only on the plasma density $\bar{n}$ and not on its temperature. Note that, if we define the electron thermal speed to be $v_e \equiv (k_B T_e/m_e)^{1/2}$, then $\omega_p \equiv v_e/\lambda_D$. In other words, a thermal electron travels roughly a Debye length in a plasma oscillation period. We can think of the Debye length as the electrostatic correlation length, and the plasma period as the electrostatic correlation time.

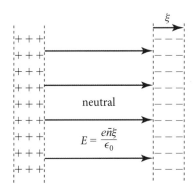

FIGURE 20.2 Idealized depiction of the displacement of electrons relative to protons, which occurs during plasma oscillations.

Characteristic values for the plasma frequency in a variety of environments are listed in Table 20.1 and depicted on the bottom-most scale of Fig. 20.1.

20.4 Coulomb Collisions

In this section and the next, we study transport coefficients (electrical and thermal conductivities) and the establishment of local thermodynamic equilibrium in a plasma under the hypothesis that Coulomb collisions are the dominant source of scattering for both electrons and protons. In fact, as we shall see later, Coulomb scattering is usually a less-effective scattering mechanism than *collisionless* processes mediated by fluctuating electromagnetic fields (plasmons).

20.4.1 Collision Frequency

Consider first, as we did in our discussion of Debye shielding, a single test particle—let it be an electron—interacting with background field particles—let these be protons for the moment. The test electron moves with speed v_e. The field protons will move much more slowly if the electron and protons are near thermodynamic equilibrium (as the proton masses are much greater than that of the electron), so the protons can be treated, for the moment, as at rest. When the electron flies by a single proton, we can characterize the encounter using an *impact parameter b*, which is what the distance of closest approach would have been if the electron were not deflected; see Fig. 20.3. The electron will be scattered by the Coulomb field of the proton, a process sometimes called *Rutherford scattering*. If the deflection angle is small, $\theta_D \ll 1$, we can approximate its value by computing the perpendicular impulse exerted by the proton's Coulomb field, integrating along the electron's unperturbed straight-line trajectory:

$$m_e v_e \theta_D = \int_{-\infty}^{+\infty} \frac{e^2 b}{4\pi \epsilon_o (b^2 + v_e^2 t^2)^{3/2}} dt = \frac{e^2}{2\pi \epsilon_o v_e b}, \quad (20.14a)$$

which implies that

$$\theta_D = b_o / b \text{ for } b \gg b_o, \quad (20.14b)$$

where

$$b_o \equiv \frac{e^2}{2\pi \epsilon_0 m_e v_e^2}. \quad (20.14c)$$

When $b \lesssim b_o$, this approximation breaks down, and the deflection angle is of order 1 radian.[3]

Below we shall need to know how much energy the electron loses, for the large-impact-parameter case. That energy loss, $-\Delta E$, is equal to the energy gained by the

---

[3]. A more careful calculation gives $2 \tan(\theta_D/2) = b_o/b$; see, e.g., Bellan (2006, Assignment 1 in Sec. 1.13).

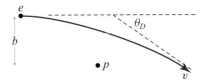

FIGURE 20.3 The geometry of a Coulomb collision.

proton. Since the proton is initially at rest, and since momentum conservation implies it gains a momentum $\Delta p = m_e v_e \theta_D$, ΔE must be

$$\Delta E = -\frac{(\Delta p)^2}{2m_p} = -\frac{m_e}{m_p}\left(\frac{b_o}{b}\right)^2 E, \quad \text{for } b \gg b_o. \tag{20.15}$$

Here $E = \frac{1}{2}m_e v_e^2$ is the electron's initial energy.

We turn from an individual Coulomb collision to the net, statistically averaged effect of many collisions. The first thing we compute is the mean time t_D required for the orbit of the test electron to be deflected by an angle of order 1 radian from its initial direction, and the inverse of t_D, which we call the "deflection rate" or "deflection frequency" and denote $\nu_D = 1/t_D$. If the dominant source of this deflection were a single large-angle scattering event, then the relevant cross section would be $\sigma = \pi b_o^2$ (since all impact parameters $\lesssim b_o$ produce large-angle scatterings), and the mean deflection time and frequency would be

$$\nu_D \equiv \frac{1}{t_D} = n\sigma v_e = n\pi b_o^2 v_e. \tag{20.16}$$

time and frequency for a single large-angle deflection of an electron by a proton

Here n is the proton number density, which is the same as the electron number density in our hydrogen plasma.

The cumulative, random-walk effects of many small-angle scatterings off field protons actually produce a net deflection of order 1 radian in a time shorter than this. As the directions of the individual scatterings are random, the mean deflection angle after many scatterings vanishes. However, the mean-square deflection angle, $\langle \Theta^2 \rangle = \sum_{\text{all encounters}} \theta_D^2$, will not vanish. That mean-square deflection angle, during a time t, accumulates to

$$\langle \Theta^2 \rangle = \int_{b_{\min}}^{b_{\max}} \left(\frac{b_o}{b}\right)^2 n v_e t\, 2\pi b\, db = n 2\pi b_o^2 v_e t\, \ln\left(\frac{b_{\max}}{b_{\min}}\right). \tag{20.17}$$

mean-square deflection from small-angle scatterings

Here the factor $(b_o/b)^2$ in the integrand is the squared deflection angle θ_D^2 for impact parameter b, and the remaining factor $nv_e t\, 2\pi b\, db$ is the number of encounters that occur with impact parameters between b and $b + db$ during time t. The integral diverges logarithmically at both its lower limit $b_{\min}$ and its upper limit $b_{\max}$. In Sec. 20.4.2 we discuss the physical origins of and values of the cutoffs $b_{\min}$ and $b_{\max}$. The value of t that makes the mean-square deflection angle $\langle \Theta^2 \rangle$ equal to unity is, to within factors of order unity, the deflection time t_D (and inverse deflection frequency ν_D^{-1}):

time and frequency for large net deflection of an electron by protons

$$\nu_D^{ep} = \frac{1}{t_D^{ep}} = n 2\pi b_o^2 v_e \ln \Lambda = \frac{n e^4 \ln \Lambda}{2\pi \epsilon_0^2 m_e^2 v_e^3}, \quad \text{where } \Lambda = b_{\max}/b_{\min}. \quad (20.18a)$$

Here the superscript ep indicates that the test particle is an electron and the field particles are protons. Notice that this deflection frequency is larger, by a factor $2 \ln \Lambda$, than the frequency (20.16) for a single large-angle scattering.

We must also consider the repulsive collisions of our test electron with field electrons. Although we are no longer justified in treating the field electrons as being at rest, the impact parameter for a large angle deflection is still $\sim b_0$, so Eq. (20.18a) is also appropriate to this case, in order of magnitude:

time and frequency for large net deflection of an electron by electrons

$$\nu_D^{ee} = \frac{1}{t_D^{ee}} \sim \nu_D^{ep} = n 2\pi b_0^2 v_e \ln \Lambda = \frac{n e^4 \ln \Lambda}{2\pi \epsilon_0^2 m_e^2 v_e^3}. \quad (20.18b)$$

Finally, and in the same spirit, we can compute the collision frequency for the protons. Because electrons are so much lighter than protons, proton-proton collisions will be more effective in deflecting protons than are proton-electron collisions. Therefore, the proton collision frequency is given by Eq. (20.18b) with the electron subscripts replaced by proton subscripts:

time and frequency for large net deflection of a proton by protons

$$\nu_D^{pp} = \frac{1}{t_D^{pp}} \sim \frac{n e^4 \ln \Lambda}{2\pi \epsilon_0^2 m_p^2 v_p^3}. \quad (20.18c)$$

20.4.2 The Coulomb Logarithm

The maximum impact parameter $b_{\max}$, which appears in $\Lambda \equiv b_{\max}/b_{\min}$, is the Debye length λ_D, since for impact parameters $b \gg \lambda_D$ the Debye shielding screens out a field particle's Coulomb field, while for $b \ll \lambda_D$ Debye shielding is unimportant.

The minimum impact parameter $b_{\min}$ has different values, depending on whether quantum mechanical wave-packet spreading is important for the test particle during the collision. Because of wave-packet spreading, the nearest the test particle can come to a field particle is the test particle's de Broglie wavelength: $b_{\min} = \hbar/mv$. However, if the de Broglie wavelength is smaller than b_0, then the effective value of $b_{\min}$ will be simply b_0. Therefore, in $\ln \Lambda = \ln(b_{\max}/b_{\min})$ (the *Coulomb logarithm*), we have

Coulomb logarithm

$$b_{\min} = \max[b_o = e^2/(2\pi \epsilon_0 m_e v_e^2), \hbar/(m_e v_e)], \quad \text{and} \quad b_{\max} = \lambda_D \quad \text{for test electrons;}$$
$$b_{\min} = \max[b_o = e^2/(2\pi \epsilon_0 m_p v_p^2), \hbar/(m_p v_p)], \quad \text{and} \quad b_{\max} = \lambda_D \quad \text{for test protons.}$$
$$(20.19)$$

Over most of the accessible range of density and temperature for a plasma, we have $3 \lesssim \ln \Lambda \lesssim 30$. Therefore, if we set

$$\boxed{\ln \Lambda \simeq 10,} \qquad (20.20)$$

our estimate is good to a factor ~ 3. For numerical values of $\ln \Lambda$ in a thermalized plasma, see Spitzer (1962, Table 5.1).

Logarithmic terms, closely related to the Coulomb logarithm, arise in a variety of other situations where one is dealing with a field whose potential varies as $1/r$ and force as $1/r^2$—and for the same reason: one encounters integrals of the form $\int (1/r) dr$ that diverge logarithmically. Specific examples are (i) the *Gaunt factors*, which arise in the theory of bremsstrahlung radiation (electromagnetic waves from Coulomb scattering of charged particles), (ii) *Coulomb wave functions* (solutions to the Schrödinger equation for the quantum mechanical theory of Coulomb scattering), and (iii) the Shapiro time delay for electromagnetic waves that travel past a general relativistic gravitating body, such as the Sun (Sec. 27.2.4).

EXERCISES

Exercise 20.2 *Problem: The Coulomb Logarithm*

(a) Express the Coulomb logarithm in terms of the Debye number, N_D, in the classical regime, where $b_{\min} \sim b_0$.

(b) What range of electron temperatures corresponds to the classical regime, and what range to the quantum regime?

(c) Use the representative parameters from Table 20.1 to evaluate Coulomb logarithms for the Sun's core, a tokamak, and the interstellar medium, and verify that they lie in the range $3 \lesssim \ln \Lambda \lesssim 30$.

Exercise 20.3 *Example: Parameters for Various Plasmas*

Estimate the Debye length λ_D, the Debye number N_D, the plasma frequency $f_p \equiv \omega_p/2\pi$, and the electron deflection timescale $t_D^{ee} \sim t_D^{ep}$, for the following plasmas.

(a) An atomic bomb explosion in Earth's atmosphere 1 ms after the explosion. (Use the Sedov-Taylor similarity solution for conditions behind the bomb's shock wave; Sec. 17.6.)

(b) The ionized gas that enveloped the Space Shuttle (Box 17.4) as it reentered Earth's atmosphere.

(c) The expanding universe during its early evolution, just before it became cool enough for electrons and protons to combine to form neutral hydrogen (i.e., just before ionization "turned off"). (As we shall discuss in Chap. 28, the universe today is filled with blackbody radiation, produced in the big bang, that has a temperature $T = 2.7$ K, and the universe today has a mean density of hydrogen $\rho \sim 4 \times 10^{-28}$ kg m^{-3}. Extrapolate backward in time to infer the density and temperature at the epoch just before ionization turned off.)

20.4.3 Thermal Equilibration Rates in a Plasma

Suppose that a hydrogen plasma is heated in some violent way (e.g., by a shock wave). Such heating will typically give the plasma's electrons and protons a non-Maxwellian velocity distribution. Coulomb collisions then, as time passes (and in the absence of more violent disruptions), will force the particles to exchange energy in random ways, gradually driving them into thermal equilibrium. As we shall see, thermal equilibration is achieved at different rates for the electrons and the protons. Correspondingly, the following three timescales are all different:

$$t_{ee}^{\text{eq}} \equiv \begin{pmatrix} \text{time required for electrons to equilibrate with one another,} \\ \text{achieving a near-Maxwellian velocity distribution} \end{pmatrix},$$

$$t_{pp}^{\text{eq}} \equiv (\text{time for protons to equilibrate with one another}),$$

$$t_{ep}^{\text{eq}} \equiv (\text{time for electrons to equilibrate with protons}).$$

(20.21)

In this section we compute these three equilibration times.

ELECTRON-ELECTRON EQUILIBRATION

To evaluate t_{ee}^{eq}, we assume that the electrons begin with typical individual energies of order $k_B T_e$, where T_e is the temperature to which they are going to equilibrate, but their initial velocity distribution is rather non-Maxwellian. Then we can choose a typical electron as the "test particle." We have argued that Coulomb interactions with electrons and protons are comparably effective in deflecting test electrons. However, they are not comparably effective in transferring energy. When the electron collides with a stationary proton, the energy transfer is

$$\frac{\Delta E}{E} \simeq -\frac{m_e}{m_p}\theta_D{}^2 \quad (20.22)$$

[Eq. (20.15)]. This is smaller than the typical energy transfer in an electron-electron collision by the ratio m_e/m_p. Therefore, the collisions between electrons are responsible for establishing an electron Maxwellian distribution function.

The alert reader may spot a problem at this point. According to Eq. (20.22), electrons always lose energy to protons and never gain it. This would cause the electron temperature to continue to fall below the proton temperature, in clear violation of the second law of thermodynamics. Actually what happens in practice is that, if we allow for nonzero proton velocities, then the electrons can gain energy from some electron-proton collisions. This is also the case for the electron-electron collisions of immediate concern.

The most accurate formalism for dealing with this situation is the Fokker-Planck formalism, discussed in Sec. 6.9 and made explicit for Coulomb scattering in Ex. 20.8. Fokker-Planck is appropriate because, as we have shown, many weak scatterings dominate the few strong scatterings. If we use the Fokker-Planck approach to compute an energy equilibration time for a nearly Maxwellian distribution of electrons with temperature T_e, then it turns out that a simple estimate, based on combining the deflection time (20.16) with the typical energy transfer (20.22) (with $m_p \to m_e$), and on assuming a typical electron velocity $v_e = (3k_B T_e/m_e)^{1/2}$, gives an answer good to a factor of 2. It is actually convenient to express this electron energy equilibration timescale using its reciprocal, the *electron-electron equilibration rate*, v_{ee}. This facilitates comparison with the other frequencies characterizing the plasma. The true Fokker-Planck result for electrons near equilibrium is then

$$\boxed{\begin{aligned} v_{ee} = \frac{1}{t_{ee}} &= \frac{n\sigma_T c \ln \Lambda}{2\pi^{1/2}} \left(\frac{k_B T_e}{m_e c^2}\right)^{-3/2} \\ &= 2.6 \times 10^{-5} \left(\frac{n}{1\,\mathrm{m}^{-3}}\right) \left(\frac{T_e}{1\,\mathrm{K}}\right)^{-3/2} \left(\frac{\ln \Lambda}{10}\right) \mathrm{s}^{-1} \end{aligned}} \quad (20.23a)$$

electron-electron equilibration rate

where we have used the Thomson cross section

$$\sigma_T = \frac{8\pi}{3}\left(\frac{e^2}{4\pi\epsilon_0 m_e c^2}\right)^2 = 6.65 \times 10^{-29}\,\mathrm{m}^2. \quad (20.23b)$$

PROTON-PROTON EQUILIBRATION

As for proton deflections [Eq. (20.18c)], so also for proton energy equilibration: the light electrons are far less effective at influencing the protons than are other protons. Therefore, the protons achieve a thermal distribution by equilibrating with one another, and their *proton-proton equilibration rate* can be written down immediately from Eq. (20.23a) by replacing the electron masses and temperatures with the proton values:

$$\boxed{\begin{aligned} v_{pp} = \frac{1}{t_{pp}} &= \frac{n\sigma_T c \ln \Lambda}{2\pi^{1/2}} \left(\frac{m_e}{m_p}\right)^{1/2} \left(\frac{k_B T_p}{m_e c^2}\right)^{-3/2} \\ &= 6.0 \times 10^{-7} \left(\frac{n}{1\,\mathrm{m}^{-3}}\right) \left(\frac{T_p}{1\,\mathrm{K}}\right)^{-3/2} \left(\frac{\ln \Lambda}{10}\right) \mathrm{s}^{-1} \end{aligned}} \quad (20.23c)$$

proton-proton equilibration rate

ELECTRON-PROTON EQUILIBRATION

Finally, if the electrons and protons have different temperatures, we should compute the timescale for the two species to equilibrate with each other. This again is easy to estimate using the energy-transfer equation (20.22): $t_{ep} \simeq (m_p/m_e) t_{ee}$. The more accurate Fokker-Planck result for the *electron-proton equilibration rate* is again close to this value and is given by

$$\boxed{\begin{aligned} v_{ep} = \frac{1}{t_{ep}} &= \frac{2n\sigma_T c \ln \Lambda}{\pi^{1/2}} \left(\frac{m_e}{m_p}\right) \left(\frac{k_B T_e}{m_e c^2}\right)^{-3/2} \\ &= 5.6 \times 10^{-8} \left(\frac{n}{1\,\mathrm{m}^{-3}}\right) \left(\frac{T_e}{1\,\mathrm{K}}\right)^{-3/2} \left(\frac{\ln \Lambda}{10}\right) \mathrm{s}^{-1} \end{aligned}} \quad (20.23d)$$

electron-proton equilibration rate

Thus, at the same density and temperature, protons require $\sim (m_p/m_e)^{1/2} = 43$ times longer to reach thermal equilibrium among themselves than do the electrons, and proton-electron equilibration takes a time $\sim (m_p/m_e) = 1{,}836$ longer than electron-electron equilibration.

20.4.4 Discussion

Table 20.1 lists the electron-electron equilibration rates for a variety of plasma environments. Generally, they are very small compared with the plasma frequencies. For example, if we take parameters appropriate to fusion experiments (e.g., a tokamak), we find that $\nu_{ee} \sim 10^{-8} \omega_p$ and $\nu_{ep} \sim 10^{-11} \omega_p$. In fact, the equilibration time is comparable, to order of magnitude, with the total plasma confinement time ~ 0.1 s (cf. Sec. 19.3). The disparity between ν_{ee} and ω_p is even greater in the interstellar medium. For this reason most plasmas are well described as collisionless, and we must anticipate that the particle distribution functions will depart significantly from Maxwellian.

EXERCISES

Exercise 20.4 *Derivation: Electron-Electron Equilibration Rate*
Using the non-Fokker-Planck arguments outlined in the text, compute an estimate of the electron-electron equilibration rate, and show that it agrees with the Fokker-Planck result, Eq. (20.23a), to within a factor of 2.

Exercise 20.5 *Problem: Dependence of Thermal Equilibration on Charge and Mass*
Compute the ion equilibration rate for a pure He^3 plasma with electron density 10^{20} m^{-3} and temperature 10^8 K.

Exercise 20.6 *Example: Stopping of α-Particles*
A 100-MeV α-particle is incident on a plastic object. Estimate the distance that it will travel before coming to rest. This is known as the particle's *range*. [Hints: (i) Debye shielding does not occur in a plastic. (Why?) (ii) The α-particle loses far more energy to Coulomb scattering off electrons than off atomic nuclei. (Why?) Estimate the electron density as $n_e = 2 \times 10^{29}$ m^{-3}. (iii) Ignore relativistic corrections and refinements, such as the so-called *density effect*. (iv) Consider the appropriate values of $b_{\max}$ and $b_{\min}$.]

Exercise 20.7 *Problem: Equilibration Time for a Globular Star Cluster*
Collections of stars have many similarities to a plasma of electrons and ions. These similarities arise from the fact that in both cases the interaction between the individual particles (stars, or ions and electrons) is a radial, $1/r^2$ force. The principal difference is that the force between stars is always attractive, so there is no analog of Debye shielding. One consequence of this difference is that a plasma can be spatially homogeneous and static, when one averages over lengthscales large compared to the interparticle separation; but a collection of stars cannot be—the stars congregate into clusters that are held together by the stars' mutual gravity.

A globular star cluster is an example. A typical globular cluster is a nearly spherical swarm of stars with the following parameters: cluster radius $\equiv R = 10$ light-years;

total number of stars in the cluster $\equiv N = 10^6$; and mass of a typical star $\equiv m = 0.4$ solar masses $= 8 \times 10^{32}$ g. Each star moves on an orbit of the average, "smeared out" gravitational field of the entire cluster. Since that smeared-out gravitational field is independent of time, each star conserves its total energy (kinetic plus gravitational) as it moves. Actually, the total energy is only approximately conserved. Just as in a plasma, gravitational "Coulomb collisions" of the star with other stars produce changes in the star's energy.

(a) What is the mean time t_E for a typical star in a globular cluster to change its energy substantially? Express your answer, accurate to within a factor of ~ 3, in terms of N, R, and m. Evaluate it numerically and compare it with the age of the universe.

(b) The cluster evolves substantially on the timescale t_E. What type of evolution would you expect to occur? What type of stellar energy distribution would you expect to result from this evolution?[4]

Exercise 20.8 **Example and Challenge: Fokker-Planck Formalism for Coulomb Collisions** T2

Consider two families of particles that interact via Coulomb collisions (e.g., electrons and protons, or electrons and electrons). Regard one family as test particles, with masses m and charges q and a kinetic-theory distribution function $f(\mathbf{v}) = dN/d\mathcal{V}_x d\mathcal{V}_v$ that is homogeneous in space and thus only a function of the test particles' velocities $\mathbf{v}$. (For discussion of this distribution function, see Sec. 3.2.3, and in greater detail, Sec. 22.2.1.) Regard the other family as field particles, with masses m' and charges q' and a spatially homogeneous distribution function $f_F(\mathbf{v}')$.

The Fokker-Planck equation (Sec. 6.9.3) can be used to give a rather accurate description of how the field particles' distribution function $f(\mathbf{v}, t)$ evolves due to Coulomb collisions with the test particles. In this exercise, you will work out explicit versions of this Fokker-Planck equation. These explicit versions are concrete foundations for deducing Eqs. (20.23) for the electron and proton equilibration rates, Eqs. (20.26) for the thermal and electrical conductivities of a plasma, and other evolutionary effects of Coulomb collisions.

(a) Explain why, for a spatially homogeneous plasma with the only electromagnetic fields present being the Coulomb fields of the plasma particles, the collisionless Boltzmann transport equation (3.65) (with $\mathcal{N}$ replaced by f and p_j by v_j) becomes simply $\partial f/\partial t = 0$. Then go on to explain why the collision terms in the collisional Boltzmann transport equation (3.66) are of Fokker-Planck form, Eq. (6.106a):

$$\frac{\partial f}{\partial t} = -\frac{\partial}{\partial v_j}\left[A_j(\mathbf{v}) f(\mathbf{v}, t)\right] + \frac{1}{2}\frac{\partial^2}{\partial v_j \partial v_k}\left[B_{jk}(\mathbf{v}) f(\mathbf{v}, t)\right], \quad (20.24a)$$

---

4. For a detailed discussion, see, e.g., Binney and Tremaine (2003).

where A_j and B_{jk} are, respectively, the slowdown rate and velocity diffusion rate of the test particles due to Coulomb scatterings off the field particles:

$$A_j = \lim_{\Delta t \to 0} \left(\frac{\overline{\Delta v_j}}{\Delta t} \right), \quad B_{jk} = \lim_{\Delta t \to 0} \left(\frac{\overline{\Delta v_j \Delta v_k}}{\Delta t} \right)$$

[Eqs. (6.106b) and (6.106c)].

(b) Analyze, in the center-of-mass reference frame, the small-angle Coulomb scattering of a test particle off a field particle (with some chosen impact parameter b and relative velocity $\mathbf{v}_{\rm rel} = \mathbf{v} - \mathbf{v}' =$ difference between lab-frame velocities), and then transform to the laboratory frame where the field particles on average are at rest. Then add up the scattering influences of all field particles on the chosen test particle (i.e., integrate over b and over $\mathbf{v}'$) to show that the test particle's rate of change of velocity is

$$A_j = \Gamma \frac{\partial}{\partial v_j} h(\mathbf{v}). \tag{20.24b}$$

Here the constant Γ and the *first Rosenbluth potential h* are

$$\Gamma = \frac{q^2 q'^2 \ln \Lambda}{4\pi \epsilon_0^2 m^2}, \quad h(\mathbf{v}) = \frac{m}{\mu} \int \frac{f_F(\mathbf{v}')}{|\mathbf{v} - \mathbf{v}'|} d\mathcal{V}_{v'}. \tag{20.24c}$$

In addition, $\mu = mm'/(m+m')$ is the reduced mass, which appears in the center-of-mass-frame analysis of the Coulomb collision; $\Lambda = b_{\rm max}/b_{\rm min}$ is the ratio of the maximum to the minimum possible impact parameters as in Sec. 20.4.2; and $d\mathcal{V}_{v'} = dv'_x dv'_y dv'_z$ is the volume element in velocity space. Although $b_{\rm min}$ depends on the particles' velocities, without significant loss of accuracy we can evaluate it for some relevant mean velocity and pull it out from under the integral, as we have done. This is because $\Lambda = b_{\rm max}/b_{\rm min}$ appears only in the logarithm.

(c) By the same method as in part (b), show that the velocity diffusion rate is

$$B_{jk}(\mathbf{v}) = \Gamma \frac{\partial^2}{\partial v_j \partial v_k} g(\mathbf{v}), \tag{20.24d}$$

where the *second Rosenbluth potential* is

$$g(\mathbf{v}) = \int |\mathbf{v} - \mathbf{v}'| f_F(\mathbf{v}') d\mathcal{V}_{v'}. \tag{20.24e}$$

(d) Show that, with A_j and B_{jk} expressed in terms of the Rosenbluth potentials [Eqs. (20.24b) and (20.24d)], the Fokker-Planck equation (20.24a) can be brought into the following alternative form:

$$\frac{\partial f}{\partial t} = \frac{\Gamma m}{2} \frac{\partial}{\partial v_j} \int \frac{\partial^2 |\mathbf{v} - \mathbf{v}'|}{\partial v_j \partial v_k} \left[\frac{f_F(\mathbf{v}')}{m} \frac{\partial f(\mathbf{v})}{\partial v_k} - \frac{f(\mathbf{v})}{m'} \frac{\partial f_F(\mathbf{v}')}{\partial v'_k} \right] d\mathcal{V}_{v'}. \tag{20.25}$$

[Hint: Manipulate the partial derivatives in a wise way and perform integrations by parts.]

(e) Using Eq. (20.25), show that, if the two species—test and field—are thermalized at the same temperature, then the test-particle distribution function will remain unchanged: $\partial f / \partial t = 0$.

For details of the solution to this exercise see, for example, the original paper by Rosenbluth, MacDonald, and Judd (1957), or the pedagogical treatments in standard plasma physics textbooks (e.g., Boyd and Sanderson, 2003, Sec. 8.4; Bellan, 2006, Sec. 13.2).

The explicit form (20.24) of the Fokker-Planck equation has been solved approximately for idealized situations. It has also been integrated numerically for more realistic situations, so as to evolve the distribution functions of interacting particle species and, for example, watch them thermalize. (See, e.g., Bellan, 2006, Sec. 13.2; Boyd and Sanderson, 2003, Secs. 8.5, 8.6; and Shkarofsky, Johnston, and Bachynski, 1966, Secs. 7.5–7.11.) An extension to a plasma with an applied electric field and/or temperature gradient (using techniques developed in Sec. 3.7.3) has been used to deduce the electrical and thermal conductivities discussed in the next section (see, e.g., Shkarofsky, Johnston, and Bachynski, 1966, Chap. 8).

20.5 Transport Coefficients

Because electrons have far lower masses than ions, they have far higher typical speeds at fixed temperature and are much more easily accelerated (i.e., they are much more mobile). As a result, it is the motion of the electrons, not the ions, that is responsible for the transport of heat and charge through a plasma. In the spirit of the discussion above, we can compute transport properties, such as the electric conductivity and thermal conductivity, on the presumption that it is Coulomb collisions that determine the electron mean free paths and that magnetic fields are unimportant. (Later we will see that collisionless effects—scattering off plasmons—usually provide a more serious impediment to charge and heat flow than Coulomb collisions and thus dominate the conductivities.)

20.5.1 Coulomb Collisions

First consider an electron exposed to a constant, accelerating electric field E. The electron's typical field-induced velocity is along the direction of the electric field and has magnitude $-eE/(m_e \nu_D)$, where ν_D is the deflection frequency (rate) evaluated in Eqs. (20.18a) and (20.18b). We call this a *drift velocity*, because it is superposed on the electrons' collision-induced isotropic velocity distribution. The associated current density is $j \sim ne^2 E/(m_e \nu_D)$, and the electrical conductivity therefore is $\kappa_e \sim ne^2/(m_e \nu_D)$. (Note that electron-electron collisions conserve momentum and thus do not impede the flow of current, so electron-proton collisions, which happen about as frequently, produce all the electrical resistance and are thus responsible for this κ_e.)

The thermal conductivity can likewise be estimated by noting that a typical electron travels a mean distance $\ell \sim v_e/\nu_D$ between large net deflections, from an initial location where the average temperature is different from the final location's temperature by an amount $\Delta T \sim \ell |\nabla T|$. The heat flux transported by the electrons is therefore $\sim n v_e k_B \Delta T$, which should be equated to $-\kappa \nabla T$. We thereby obtain the electron contribution to the thermal conductivity as $\kappa \sim n k_B^2 T / (m_e \nu_D)$.

Computations based on the Fokker-Planck approach (Spitzer and Harm, 1953; see also Ex. 20.8) produce equations for the electrical and thermal conductivities that agree with the above estimates within factors of order ten:

electrical and thermal conductivities when Coulomb collisions are responsible for the resistivity

$$\boxed{\kappa_e = 4.9 \left(\frac{e^2}{\sigma_T c \ln \Lambda m_e} \right) \left(\frac{k_B T_e}{m_e c^2} \right)^{3/2} = 1.5 \times 10^{-3} \left(\frac{T_e}{1\,\text{K}} \right)^{3/2} \left(\frac{\ln \Lambda}{10} \right)^{-1} \Omega^{-1}\,\text{m}^{-1},}$$

(20.26a)

$$\boxed{\kappa = 19.1 \left(\frac{k_B c}{\sigma_T \ln \Lambda} \right) \left(\frac{k_B T_e}{m_e c^2} \right)^{5/2} = 4.4 \times 10^{-11} \left(\frac{T_e}{1\,\text{K}} \right)^{5/2} \left(\frac{\ln \Lambda}{10} \right)^{-1} \text{W}\,\text{m}^{-1}\,\text{K}^{-1}.}$$

(20.26b)

Here σ_T is the Thomson cross section, Eq. (20.23b). Note that neither transport coefficient depends explicitly on the density; increasing the number of charge or heat carriers is compensated by the reduction in their mean free paths.

20.5.2 Anomalous Resistivity and Anomalous Equilibration

We have demonstrated that the theoretical Coulomb interaction between charged particles gives very long mean free paths. Correspondingly, the electrical and thermal conductivities (20.26) are very large in practical, geophysical, and astrophysical applications. Is this the way that real plasmas behave? The answer is almost always "no."

mechanism for anomalous resistivity and equilibration

As we shall show in the next three chapters, a plasma can support a variety of modes of collective excitation (plasmons), in which large numbers of electrons and/or ions move in collective, correlated fashions that are mediated by electromagnetic fields that they create. When the modes of a plasma are sufficiently excited (which is common), the electromagnetic fields carried by the excitations can be much more effective than Coulomb scattering at deflecting the orbits of individual electrons and ions and at feeding energy into or removing it from the electrons and ions. Correspondingly, the electrical and thermal conductivities will be reduced. The reduced transport coefficients are termed *anomalous*, as is the scattering by plasmons that controls them. Providing quantitative calculations of the anomalous scattering and these anomalous transport coefficients is one of the principal tasks of nonlinear plasma physics—as we start to discuss in Chap. 22.

Exercise 20.9 *Example: Bremsstrahlung*

Bremsstrahlung (or *free-free* radiation) arises when electrons are accelerated by ions and the changing electric dipole moment that they create leads to the emission of photons. This process can only be fully described using quantum mechanics, though the low-frequency spectrum can be calculated accurately using classical electromagnetism. Here we make a "back of the envelope" estimate of the absorption coefficient and convert it into an emission coefficient.

(a) Consider an electron with average speed v oscillating in an electromagnetic wave of electric field amplitude E and (slow) angular frequency ω. Show that its energy of oscillation is $E_{\mathrm{osc}} \sim e^2 E^2 m^{-1} \omega^{-2}$.

(b) Let the electron encounter an ion and undergo a rapid deflection. Explain why the electron will, on average, gain energy of $\sim E_{\mathrm{osc}}$ as a result of the encounter. Hence estimate the average rate of electron heating. (You may ignore the cumulative effects of distant encounters.)

(c) Recognizing that classical absorption at low frequency is the difference between quantum absorption and stimulated emission (e.g., Sec. 10.2.1), convert the result from part (b) into a spontaneous emission rate, assuming that the electron velocity is typical for a gas in thermal equilibrium at temperature T.

(d) Hence show that the free-free cooling rate (power radiated per unit volume) of an electron-proton plasma with electron density n_e and temperature T is $\sim n_e^2 \alpha_F \sigma_T c^2 (k_B T m_e)^{1/2}$, where $\alpha_F = e^2/(4\pi\epsilon_0 \hbar c) = 7.3 \times 10^{-3}$ is the fine structure constant, and $\sigma_T = 6.65 \times 10^{-29}$ m^2 is the Thomson cross section. In a more detailed analysis, the long-range ($1/r^2$) character of the electric force gives rise to a multiplicative logarithmic factor, called the Gaunt factor, analogous to the Coulomb logarithm of Sec. 20.4.2.

This approach is useful when the plasma is so dense that the emission rate is significantly different from the rate in a vacuum (Boyd and Sanderson, 2003, Chap. 9).

Exercise 20.10 *Challenge and Example: Thermoelectric Transport Coefficients*

(a) Consider a plasma in which the magnetic field is so weak that it presents little impediment to the flow of heat and electric current. Suppose that the plasma has a gradient ∇T_e of its electron temperature and also has an electric field **E**. It is a familiar fact that the temperature gradient will cause heat to flow and the electric field will create an electric current. Not so familiar, but somewhat obvious if one stops to think about it, is that the temperature gradient also creates an electric current and the electric field also causes heat flow. Explain in physical terms why this is so.

(b) So long as the mean free path of an electron between substantial deflections by electrons and protons, $\ell_{D,e} = (3k_B T_e/m_e)^{1/2} t_{D,e}$, is short compared to the lengthscale for substantial temperature change, $T_e/|\nabla T_e|$, and short compared to the lengthscale for the electrons to be accelerated to near the speed of light by the electric field, $m_e c^2/(eE)$, the fluxes of heat $\mathbf{q}$ and of electric charge $\mathbf{j}$ will be governed by nonrelativistic electron diffusion and will be linear in ∇T and $\mathbf{E}$:

$$\mathbf{q} = -\kappa \nabla T - \beta \mathbf{E}, \quad \mathbf{j} = \kappa_e \mathbf{E} + \alpha \nabla T. \tag{20.27}$$

The coefficients κ (heat conductivity), κ_e (electrical conductivity), β, and α are called *thermoelectric transport coefficients*. Use kinetic theory (Chap. 3), in a situation where $\nabla T = 0$, to derive the conductivity equations $\mathbf{j} = \kappa_e \mathbf{E}$ and $\mathbf{q} = -\beta \mathbf{E}$, and the following approximate formulas for the transport coefficients:

$$\kappa_e \sim \frac{ne^2 t_{D,e}}{m_e}, \quad \beta \sim \frac{k_B T}{e} \kappa_e. \tag{20.28a}$$

Show that, aside from a coefficient of order unity, this κ_e, when expressed in terms of the plasma's temperature and density, reduces to the Fokker-Planck result Eq. (20.26a).

(c) Use kinetic theory, in a situation where $\mathbf{E} = 0$ and the plasma is near thermal equilibrium at temperature T, to derive the conductivity equations $\mathbf{q} = -\kappa \nabla T$ and $\mathbf{j} = \alpha \nabla T$, and the approximate formulas

$$\kappa \sim k_B n \frac{k_B T}{m_e} t_{D,e}, \quad \alpha \sim \frac{e}{k_B T} \kappa. \tag{20.28b}$$

Show that, aside from a coefficient of order unity, this κ reduces to the Fokker-Planck result Eq. (20.26b).

(d) It can be shown (Spitzer and Harm, 1953) that for a hydrogen plasma

$$\frac{\alpha \beta}{\kappa_e \kappa} = 0.581. \tag{20.28c}$$

By studying the entropy-governed probability distributions for fluctuations away from statistical equilibrium, one can derive another relation among the thermoelectric transport coefficients, the *Onsager relation* (Kittel, 2004, Secs. 33, 34; Reif, 2008, Sec. 15.8):

$$\beta = \alpha T + \frac{5}{2} \frac{k_B T_e}{e} \kappa_e; \tag{20.28d}$$

Eqs. (20.28c) and (20.28d) determine α and β in terms of κ_e and κ. Show that your approximate values of the transport coefficients, Eqs. (20.28a) and (20.28b), are in rough accord with Eqs. (20.28c) and (20.28d).

(e) If a temperature gradient persists for sufficiently long, it will give rise to sufficient charge separation in the plasma to build up an electric field (called a "secondary field") that prevents further charge flow. Show that this cessation of charge flow suppresses the heat flow. The total heat flux is then $\mathbf{q} = -\kappa_{T\,\text{effective}} \nabla T$, where

$$\kappa_{T\,\text{effective}} = \left(1 - \frac{\alpha \beta}{\kappa_e \kappa}\right) \kappa = 0.419 \kappa. \tag{20.29}$$

20.6 Magnetic Field

20.6.1 Cyclotron Frequency and Larmor Radius

Many of the plasmas that we will encounter are endowed with a strong magnetic field. This causes the charged particles to travel along helical orbits about the field direction rather than move rectilinearly between collisions.

If we denote the magnetic field by **B**, then the equation of motion for a nonrelativistic electron becomes

$$m_e \frac{d\mathbf{v}}{dt} = -e\, \mathbf{v} \times \mathbf{B}, \tag{20.30}$$

which gives rise to a constant speed $v_\parallel$ parallel to the magnetic field and a circular motion perpendicular to the field with angular velocity sometimes denoted ω_{ce} and sometimes ω_c:

$$\boxed{\omega_{ce} = \omega_c = \frac{eB}{m_e} = 1.76 \times 10^{11} \left(\frac{B}{1\,\mathrm{T}}\right)\,\mathrm{s}^{-1}.} \tag{20.31}$$

(electron) cyclotron frequency

This angular velocity is called the *electron cyclotron frequency,* or simply the *cyclotron frequency;* and it is sometimes called the *gyro frequency.* Notice that this cyclotron frequency depends only on the magnetic field strength B and not on the plasma's density n or the electron velocity (i.e., the plasma temperature T). Nor does it depend on the angle between **v** and **B** (the *pitch angle, α*).

The radius of the electron's gyrating (spiraling) orbit, projected perpendicular to the direction of the magnetic field, is called the *Larmor radius* and is given by

$$\boxed{r_L = \frac{v_\perp}{\omega_{ce}} = \frac{v \sin\alpha}{\omega_{ce}} = 5.7 \times 10^{-9} \left(\frac{v_\perp}{1\,\mathrm{km\,s}^{-1}}\right)\left(\frac{B}{1\,\mathrm{T}}\right)^{-1}\,\mathrm{m},} \tag{20.32}$$

Larmor radius

where $v_\perp$ is the electron's velocity projected perpendicular to the field. Protons (and other ions) in a plasma also undergo cyclotron motion. Because the proton mass is larger by $m_p/m_e = 1{,}836$ than the electron mass, its angular velocity

$$\boxed{\omega_{cp} = \frac{eB}{m_p} = 0.96 \times 10^8\,\mathrm{s}^{-1}\left(\frac{B}{1\,\mathrm{T}}\right)} \tag{20.33}$$

proton cyclotron frequency

is 1,836 times lower. The quantity ω_{cp} is called the *proton cyclotron frequency* or *ion cyclotron frequency.* The sense of gyration is, of course, opposite to that of the electrons. If the protons have similar temperatures to the electrons, their speeds are typically $\sim\sqrt{m_p/m_e} = 43$ times smaller than those of the electrons, and their typical Larmor radii are ~ 43 times larger than those of the electrons.

We demonstrated in Sec. 20.3.3 that all the electrons in a plasma can oscillate in phase at the plasma frequency. The electrons' cyclotron motions can also be coherent. Such coherent motions are called *cyclotron resonances* or *cyclotron oscillations,* and we shall study them in Chap. 21. Ion cyclotron resonances can also occur. Characteristic electron cyclotron frequencies and Larmor radii are tabulated in Table 20.1. As shown

cyclotron resonances (oscillations)

there, the cyclotron frequency, like the plasma frequency, is typically far larger than the rates for Coulomb-mediated energy equilibration.

20.6.2 Validity of the Fluid Approximation

MAGNETOHYDRODYNAMICS

In Chap. 19, we developed the magnetohydrodynamic (MHD) description of a magnetized plasma. We described the plasma by its density and temperature (or equivalently, its pressure). Under what circumstances, for a plasma, is this description accurate? The answer to this question turns out to be quite complex, and a full discussion would go well beyond the scope of this book. Some aspects, however, are easy to describe. A fluid description ought to be acceptable when (i) the timescales τ that characterize the macroscopic flow are long compared with the time required to establish Maxwellian equilibrium (i.e., $\tau \gg \nu_{ep}^{-1}$), and (ii) the excitation level of collective wave modes is so small that the modes do not interfere seriously with the influence of Coulomb collisions. Unfortunately, these conditions rarely apply. (One type of plasma where they might be a quite good approximation is that in the interior of the Sun.)

conditions for validity of MHD approximation

Magnetohydrodynamics can still provide a moderately accurate description of a plasma, even if the electrons and ions are not fully equilibrated, when the electrical conductivity can be treated as very large and the thermal conductivity as very small. Then we can treat the magnetic Reynolds number [Eq. (19.9c)] as effectively infinite and the plasma as an adiabatic perfect fluid (as we assumed in much of Chap. 19). It is not so essential that the actual particle distribution functions be Maxwellian, merely that they have second moments that can be associated with a (roughly defined) common temperature.

Quite often in plasma physics almost all of the dissipation is localized—for example, to the vicinity of a shock front or a site of magnetic-field-line reconnection—and the remainder of the flow can be treated using MHD. The MHD description then provides a boundary condition for a fuller plasma physical analysis of the dissipative region. This approach simplifies the analysis of such situations.

GENERALIZATIONS OF MHD

The great advantage of fluid descriptions, and the reason physicists abandon them with such reluctance, is that they are much simpler than other descriptions of a plasma. One only has to cope with the fluid pressure, density, and velocity and does not have to deal with an elaborate statistical description of the positions and velocities of all the particles. Generalizations of the simple fluid approximation have therefore been devised that can extend the domain of validity of simple MHD ideas.

two-fluid approximation

One extension, which we develop in the following chapter, is to treat the protons and the electrons as two separate fluids and derive dynamical equations that describe their (coupled) evolution. Another extension, which we describe now, is to acknowledge that, in most plasmas:

1. the cyclotron period is very short compared with the Coulomb collision time (and with the anomalous scattering time), and
2. the timescale on which energy is transferred back and forth among the electrons, the protons, and the electromagnetic field is intermediate between ω_c^{-1} and ν_{ee}^{-1}.

Intuitively, these assumptions allow the electron and proton velocity distributions to become axisymmetric with respect to the magnetic field direction, though not fully isotropic. In other words, we can characterize the plasma using a density and two separate components of pressure, one associated with motion along the direction of the magnetic field, and the other with gyration around the field lines.

anisotropic fluid in presence of magnetic field

For simplicity, let us just consider the electrons and their stress tensor, which we can write as

$$T_e^{jk} = \int \mathcal{N}_e \, p^j p^k \frac{d\mathcal{V}_p}{m} \tag{20.34}$$

[Eq. (3.32d)], where $\mathcal{N}_e$ is the electron number density in phase space and $d\mathcal{V}_p = dp_x dp_y dp_z$ is the volume element in phase space. If we orient Cartesian axes so that the direction of $\mathbf{e}_z$ is parallel to the local magnetic field, then

$$\boxed{||T_e^{jk}|| = \begin{bmatrix} P_{e\perp} & 0 & 0 \\ 0 & P_{e\perp} & 0 \\ 0 & 0 & P_{e\|} \end{bmatrix},} \tag{20.35}$$

anisotropic pressure

where $P_{e\perp}$ is the electron pressure perpendicular to $\mathbf{B}$, and $P_{e\|}$ is the electron pressure parallel to $\mathbf{B}$. Now suppose that there is a compression or expansion on a timescale that is long compared with the cyclotron period but short compared with the Coulomb collision and anomalous scattering timescales. Then we should not expect $P_{e\perp}$ to be equal to $P_{e\|}$, and we anticipate that they will evolve with density according to different laws.

The adiabatic indices governing $P_\perp$ and $P_\|$ in such a situation are easily derived from kinetic-theory arguments (Ex. 20.11). For compression perpendicular to $\mathbf{B}$ and no change of length along $\mathbf{B}$, we have

$$\boxed{\gamma_\perp \equiv \left(\frac{\partial \ln P_\perp}{\partial \ln \rho} \right)_s = 2, \quad \gamma_\| \equiv \left(\frac{\partial \ln P_\|}{\partial \ln \rho} \right)_s = 1;} \tag{20.36a}$$

collisionless, anistotropic adiabatic indices

for compression parallel to $\mathbf{B}$ and no change of transverse area, we have

$$\boxed{\gamma_\perp \equiv \left(\frac{\partial \ln P_\perp}{\partial \ln \rho} \right)_s = 1, \quad \gamma_\| \equiv \left(\frac{\partial \ln P_\|}{\partial \ln \rho} \right)_s = 3.} \tag{20.36b}$$

By contrast, if the expansion is sufficiently slow that Coulomb collisions are effective (though not so slow that heat conduction can operate), then we expect the velocity distribution to maintain isotropy and both components of pressure to evolve according to the law appropriate to a monatomic gas:

collisional, isotropic adiabatic index

$$\gamma = \left(\frac{\partial \ln P_\perp}{\partial \ln \rho}\right)_s = \left(\frac{\partial \ln P_\parallel}{\partial \ln \rho}\right)_s = \frac{5}{3}. \quad (20.37)$$

20.6.3 Conductivity Tensor

As is evident from the foregoing remarks, if we are in a regime where Coulomb scattering really does determine the particle mean free path, then an extremely small magnetic field strength suffices to ensure that individual particles complete gyrational orbits before they collide. Specifically, for electrons, the deflection time t_D, given by Eq. (20.18a,b), exceeds ω_c^{-1} if

$$B \gtrsim 10^{-16} \left(\frac{n}{1\,\text{m}^{-3}}\right) \left(\frac{T_e}{1\,\text{K}}\right)^{-3/2} \text{T}. \quad (20.38)$$

This is almost always the case. It is also almost always true for the ions.

When inequality (20.38) is satisfied and also anomalous scattering is negligible on the timescale ω_c^{-1}, the transport coefficients must be generalized to form tensors. Let us compute the electrical conductivity tensor for a plasma in which a steady electric field **E** is applied. Once again orienting our coordinate system so that the magnetic field is parallel to $\mathbf{e}_z$, we can write down an equation of motion for the electrons by balancing the electromagnetic acceleration with the average rate of loss of momentum due to collisions or anomalous scattering:

electron equation of motion

$$-e(\mathbf{E} + \mathbf{v} \times \mathbf{B}) - m_e \nu_D \mathbf{v} = 0. \quad (20.39)$$

Solving for the velocity, we obtain

$$\begin{pmatrix} v_x \\ v_y \\ v_z \end{pmatrix} = -\frac{e}{m_e \nu_D (1+\omega_c^2/\nu_D^2)} \begin{pmatrix} 1 & \omega_c/\nu_D & 0 \\ -\omega_c/\nu_D & 1 & 0 \\ 0 & 0 & 1+\omega_c^2/\nu_D^2 \end{pmatrix} \begin{pmatrix} E_x \\ E_y \\ E_z \end{pmatrix}. \quad (20.40)$$

As the current density is $\mathbf{j}_e = -ne\mathbf{v} = \kappa_e \mathbf{E}$, the electrical conductivity tensor is given by

anisotropic electric conductivity

$$\kappa_e = \frac{ne^2}{m_e \nu_D (1+\omega_c^2/\nu_D^2)} \begin{pmatrix} 1 & \omega_c/\nu_D & 0 \\ -\omega_c/\nu_D & 1 & 0 \\ 0 & 0 & 1+\omega_c^2/\nu_D^2 \end{pmatrix}. \quad (20.41)$$

It is apparent from the form of this conductivity tensor that, when $\omega_c \gg \nu_D$ (as is almost always the case), the conductivity perpendicular to the magnetic field is greatly inhibited, whereas that along the magnetic field is unaffected. Similar remarks apply to the flow of heat. It is therefore often assumed that only transport parallel to the field is effective. However, as is made clear in the next section, if the plasma is inhomogeneous, then cross-field transport can be quite rapid in practice.

EXERCISES

Exercise 20.11 *Example and Derivation: Adiabatic Indices for Compression of a Magnetized Plasma*

Consider a plasma in which, in the local mean rest frame of the electrons, the electron stress tensor has the form (20.35) with $\mathbf{e}_z$ the direction of the magnetic field. The following analysis for the electrons can be carried out independently for the ions, resulting in the same formulas.

(a) Show that

$$P_{e\parallel} = nm_e \langle v_\parallel^2 \rangle, \quad P_{e\perp} = \frac{1}{2} nm_e \langle |\mathbf{v}_\perp|^2 \rangle, \quad (20.42)$$

where $\langle v_\parallel^2 \rangle$ is the mean-square electron velocity parallel to $\mathbf{B}$, and $\langle |\mathbf{v}_\perp|^2 \rangle$ is the mean-square velocity orthogonal to $\mathbf{B}$. (The velocity distributions are not assumed to be Maxwellian.)

(b) Consider a fluid element with length l along the magnetic field and cross sectional area A orthogonal to the field. Let $\bar{\mathbf{v}}$ be the mean velocity of the electrons ($\bar{\mathbf{v}} = 0$ in the mean electron rest frame), and let θ and σ_{jk} be the expansion and shear of the mean electron motion as computed from $\bar{\mathbf{v}}$ (Sec. 13.7.1). Show that

$$\frac{dl/dt}{l} = \frac{1}{3}\theta + \sigma^{jk} b_j b_k, \quad \frac{dA/dt}{A} = \frac{2}{3}\theta - \sigma^{jk} b_j b_k, \quad (20.43)$$

where $\mathbf{b} = \mathbf{B}/|\mathbf{B}| = \mathbf{e}_z$ is a unit vector in the direction of the magnetic field.

(c) Assume that the timescales for compression and shearing are short compared with those for Coulomb scattering and anomalous scattering: $\tau \ll t_{D,e}$. Show, using the laws of energy and particle conservation, that

$$\frac{1}{\langle v_\parallel^2 \rangle} \frac{d\langle v_\parallel^2 \rangle}{d} = -\frac{2}{l} \frac{dl}{dt},$$

$$\frac{1}{\langle v_\perp^2 \rangle} \frac{d\langle v_\perp^2 \rangle}{dt} = -\frac{1}{A} \frac{dA}{dt}, \quad (20.44)$$

$$\frac{1}{n} \frac{dn}{dt} = -\frac{1}{l} \frac{dl}{dt} - \frac{1}{A} \frac{dA}{dt}.$$

(d) Show that

$$\frac{1}{P_{e\parallel}} \frac{dP_{e\parallel}}{dt} = -3\frac{dl/dt}{l} - \frac{dA/dt}{A} = -\frac{5}{3}\theta - 2\sigma^{jk} b_j b_k,$$

$$\frac{1}{P_{e\perp}} \frac{dP_{e\perp}}{dt} = -\frac{dl/dt}{l} - 2\frac{dA/dt}{A} = -\frac{5}{3}\theta + \sigma^{jk} b_j b_k. \quad (20.45)$$

(e) Show that, when the fluid is expanding or compressing entirely perpendicular to **B**, with no expansion or compression along **B**, the pressures change in accord with the adiabatic indices of Eq. (20.36a). Show, similarly, that when the fluid expands or compresses along **B**, with no expansion or compression in the perpendicular direction, the pressures change in accord with the adiabatic indices of Eq. (20.36b).

(f) Hence derive the so-called *double adiabatic* or *CGL* (Chew, Goldberger, and Low, 1956) equations of state:

$$P_\parallel \propto n^3/B^2, \quad P_\perp \propto nB, \qquad (20.46)$$

valid for changes on timescales long compared with the cyclotron period but short compared with all Coulomb collision and anomalous scattering times.

Exercise 20.12 *Problem: Relativistic Electron Motion*

Use the relativistic equation of motion to show that the relativistic electron cyclotron frequency is $\omega_c = eB/(\gamma m_e)$, where $\gamma = 1/\sqrt{1-(v/c)^2}$ is the electron Lorentz factor. What is the relativistic electron Larmor radius? What is the relativistic proton cyclotron frequency and Larmor radius?

Exercise 20.13 *Example: Ultra-High-Energy Cosmic Rays*

The most energetic (ultra-high-energy) cosmic rays are probably created with energies up to ~ 1 ZeV $= 10^{21}$ eV in sources roughly 100 million light-years away.

(a) Show that they start with the "energy of a well-hit baseball and the momentum of a snail." By the time they arrive at Earth, their energies are likely to have been reduced by a factor of ~ 3.

(b) Assuming that the intergalactic magnetic field is $\sim 10^{-13}$ T $\sim 10^{-9}$ G, compute the cosmic rays' Larmor radii, assuming that they are protons and iron nuclei.

(c) Do you expect their arrival directions to point back to their sources?

(d) Suppose that an ultra-high-energy cosmic ray collides with a stationary nitrogen nucleus in our atmosphere. How much energy becomes available in the center-of-mass frame? Compare this with the energy at the Large Hadron Collider, where protons, each with energy ~ 7 TeV, collide head on.

20.7 Particle Motion and Adiabatic Invariants

In the next three chapters we shall meet a variety of plasma phenomena that can be understood in terms of the orbital motions of individual electrons and ions. These phenomena typically entail motions in an electromagnetic field that is nearly but not quite spatially homogeneous on the scale of the Larmor radius r_L and nearly but not quite constant during a cyclotron period $2\pi/\omega_c$. In this section, in preparation for the next three chapters, we review charged-particle motion in such nearly homogeneous, nearly time-independent fields.

Since the motions of electrons are usually of greater interest than those of ions, we presume throughout this section that the charged particle is an electron. We denote its charge by $-e$ and its mass by m_e.

20.7.1 Homogeneous, Time-Independent Magnetic Field and No Electric Field

From the nonrelativistic version of the Lorentz force equation, $d\mathbf{v}/dt = -(e/m_e)\mathbf{v} \times \mathbf{B}$, one readily deduces that an electron in a homogeneous, time-independent magnetic field $\mathbf{B}$ moves with uniform velocity $\mathbf{v}_\parallel$ parallel to the field, and it moves perpendicular to the field in a circular orbit with the cyclotron frequency $\omega_c = eB/m_e$ and Larmor radius $r_L = m_e v_\perp/(eB)$. Here $v_\perp$ is the electron's time-independent transverse speed (speed perpendicular to $\mathbf{B}$).

20.7.2 Homogeneous, Time-Independent Electric and Magnetic Fields

Suppose that the homogeneous magnetic field $\mathbf{B}$ is augmented by a homogeneous electric field $\mathbf{E}$, and assume initially that $|\mathbf{E} \times \mathbf{B}| < B^2 c$. Then examine the electric and magnetic fields in a new reference frame, one that moves with the velocity

$$\boxed{\mathbf{v}_D = \frac{\mathbf{E} \times \mathbf{B}}{B^2},} \quad (20.47)$$

relative to the original frame. Note that the moving frame's velocity $\mathbf{v}_D$ is perpendicular to both the magnetic field and the electric field. From the Lorentz transformation law for the electric field, $\mathbf{E}' = \gamma(\mathbf{E} + \mathbf{v}_D \times \mathbf{B})$, we infer that in the moving frame the electric field and the magnetic field are parallel to each other. As a result, in the moving frame the electron's motion perpendicular to the magnetic field is purely circular; and, correspondingly, in the original frame its perpendicular motion consists of a *drift* with velocity $\mathbf{v}_D$, and superposed on that drift, a circular motion (Fig. 20.4). In other words, the electron moves in a circle whose center (the electron's *guiding center*) drifts with velocity $\mathbf{v}_D$. Notice that the drift velocity (20.47) is independent of the electron's charge and mass, and thus is the same for ions as for electrons. This drift is called the "$\mathbf{E} \times \mathbf{B}$ drift." guiding-center approximation; $\mathbf{E} \times \mathbf{B}$ drift

When the component of the electric field orthogonal to $\mathbf{B}$ is so large that the drift velocity computed from Eq. (20.47) exceeds the speed of light, the electron's guiding center, of course, cannot move with that velocity. Instead, the electric field drives the electron up to higher and higher velocities as time passes, but in a sinusoidally modulated manner. Ultimately, the electron velocity becomes arbitrarily close to the speed of light. breakdown of guiding-center approximation when B is large

When a uniform, time-independent gravitational field $\mathbf{g}$ accompanies a uniform, time-independent magnetic field $\mathbf{B}$, its effect on an electron will be the same as that guiding-center approximation in gravitational field

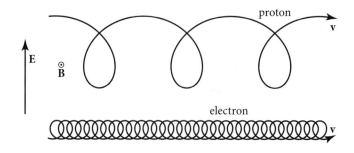

FIGURE 20.4 The proton motion (upper trajectory) and electron motion (lower trajectory) orthogonal to the magnetic field, when there are constant electric and magnetic fields with $|\mathbf{E} \times \mathbf{B}| < B^2 c$. Each electron and proton moves in a circle with a superposed drift velocity $\mathbf{v}_D$ given by Eq. (20.47).

of an electric field $\mathbf{E}_{\text{equivalent}} = -(m_e/e)\mathbf{g}$: The electron's guiding center will acquire a drift velocity

$$\mathbf{v}_D = -\frac{m_e}{e}\frac{\mathbf{g} \times \mathbf{B}}{B^2}, \tag{20.48}$$

and similarly for a proton. This *gravitational drift* velocity is typically very small.

20.7.3 Inhomogeneous, Time-Independent Magnetic Field

When the electric field vanishes, but the magnetic field is spatially inhomogeneous and time-independent, and the inhomogeneity scale is large compared to the Larmor radius r_L of the electron's orbit, the electron motion again is nicely described in terms of a guiding center.

First consider the effects of a curvature of the field lines (Fig. 20.5a). Suppose that the speed of the electron along the field lines is $v_\parallel$. We can think of this as a longitudinal guiding-center motion. As the field lines bend in, say, the direction of the unit vector $\mathbf{n}$ with radius of curvature R, this longitudinal guiding-center motion experiences the acceleration $\mathbf{a} = v_\parallel^2 \mathbf{n}/R$. That acceleration is equivalent to the effect of an electric field $\mathbf{E}_{\text{effective}} = (-m_e/e)v_\parallel^2 \mathbf{n}/R$, and it therefore produces a transverse drift of the guiding center with $\mathbf{v}_D = (\mathbf{E}_{\text{effective}} \times \mathbf{B})/B^2$. Since the curvature R of the field line and the direction $\mathbf{n}$ of its bend are given by $B^{-2}(\mathbf{B} \cdot \nabla)\mathbf{B} = \mathbf{n}/R$, this *curvature drift* velocity is

curvature drift

$$\boxed{\mathbf{v}_D = -\frac{m_e v_\parallel^2}{e}\mathbf{B} \times \frac{(\mathbf{B} \cdot \nabla)\mathbf{B}}{B^4}.} \tag{20.49}$$

Notice that the magnitude of this drift is

$$v_D = \frac{r_L}{R}\frac{v_\parallel}{v_\perp}v_\parallel. \tag{20.50}$$

This particle drift and others discussed below also show up as fluid drifts in a magnetized plasma; see Ex. 21.1.

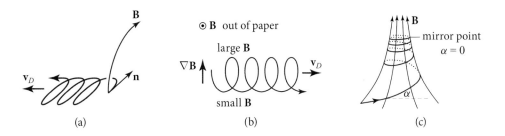

FIGURE 20.5 An electron's motion in a time-independent, inhomogeneous magnetic field. (a) The drift induced by the curvature of the field lines. (b) The drift induced by a transverse gradient of the magnitude of the magnetic field. (c) The change in electron pitch angle induced by a longitudinal gradient of the magnitude of the magnetic field.

A second kind of inhomogeneity is a transverse spatial gradient ∇B of the magnitude of $\mathbf{B}$. As is shown in Fig. 20.5b, such a gradient causes the electron's circular motion to be tighter (smaller radius of curvature of the circle) in the region of larger B than in the region of smaller B; this difference in radii of curvature clearly induces a drift. It is straightforward to show that the resulting *gradient drift* velocity is

gradient drift

$$\mathbf{v}_D = \frac{-m_e v_\perp^2}{2e} \frac{\mathbf{B} \times \nabla B}{B^3}. \tag{20.51}$$

The third and final kind of inhomogeneity is a longitudinal gradient of the magnitude of $\mathbf{B}$ (Fig. 20.5c). Such a gradient results from the magnetic field lines converging toward (or diverging away from) one another. The effect of this convergence (or divergence) is most easily inferred in a frame that moves longitudinally with the electron. In such a frame the magnetic field changes with time, $\partial \mathbf{B}'/\partial t \neq 0$, and correspondingly, the resultant electric field satisfies $\nabla \times \mathbf{E}' = -\partial \mathbf{B}'/\partial t$. The kinetic energy of the electron as measured in this longitudinally moving frame is the same as the transverse energy $\frac{1}{2} m_e v_\perp^2$ in the original frame. This kinetic energy is forced to change by the electric field $\mathbf{E}'$. The change in energy during one circuit around the magnetic field is

effects of longitudinal gradient of B

$$\Delta \left(\frac{1}{2} m_e v_\perp^2 \right) = -e \oint \mathbf{E}' \cdot d\mathbf{l} = e \int \frac{\partial \mathbf{B}'}{\partial t} \cdot d\mathbf{A} = e \left(\frac{\omega_c}{2\pi} \Delta B \right) \pi r_L^2 = \frac{m_e v_\perp^2}{2} \frac{\Delta B}{B}. \tag{20.52}$$

Here the second expression in the equation involves a line integral once around the electron's circular orbit. The third expression involves a surface integral over the interior of the orbit and has $\partial \mathbf{B}'/\partial t$ parallel to $d\mathbf{A}$. In the fourth the time derivative of the magnetic field has been expressed as $(\omega_c/(2\pi))\Delta B$, where ΔB is the change in magnetic field strength along the electron's guiding center during one circular orbit.

Equation (20.52) can be rewritten as a conservation law along the world line of the electron's guiding center:

$$\frac{m_e v_\perp^2}{2B} = \text{const.} \tag{20.53}$$

20.7 Particle Motion and Adiabatic Invariants

conservation of enclosed flux (i.e., of magnetic moment)

Notice that the conserved quantity $m_e v_\perp^2/(2B)$ is proportional to the total magnetic flux threading the electron's circular orbit, $\pi r_L^2 B$. Thus the electron moves along the field lines in such a manner as to keep constant the magnetic flux enclosed in its orbit; see Fig. 20.5c. A second interpretation of Eq. (20.53) is in terms of the magnetic moment created by the electron's circulatory motion. That moment is $\boldsymbol{\mu} = -m_e v_\perp^2/(2B^2)\mathbf{B}$, and its magnitude is the conserved quantity

$$\boxed{\mu = \frac{m_e v_\perp^2}{2B} = \text{const.}} \tag{20.54}$$

An important consequence of the conservation law (20.54) is a gradual change in the electron's pitch angle,

pitch angle of electron orbit

$$\alpha \equiv \tan^{-1}(v_\parallel/v_\perp), \tag{20.55}$$

as it spirals along the converging field lines. Because there is no electric field in the original frame, the electron's total kinetic energy is conserved in that frame:

$$E_{\text{kin}} = \frac{1}{2} m_e (v_\parallel^2 + v_\perp^2) = \text{const.} \tag{20.56}$$

This expression, together with the constancy of $\mu = m_e v_\perp^2/(2B)$ and the definition of the electron pitch angle [Eq. (20.55)], implies that the pitch angle varies with magnetic field strength as

variation of pitch angle along guiding center

$$\boxed{\tan^2 \alpha = \frac{E_{\text{kin}}}{\mu B} - 1.} \tag{20.57}$$

Notice that as the field lines converge, B increases in magnitude, and α decreases. Ultimately, when B reaches a critical value $B_{\text{crit}} = E_{\text{kin}}/\mu$, the pitch angle α goes to zero. The electron then "reflects" off the strong-field region and starts moving back toward weak fields, with increasing pitch angle. The location at which the electron reflects is called the electron's *mirror point*.

mirror point

magnetic bottle

Figure 20.6 shows two examples of this mirroring. The first example is a "magnetic bottle" (Ex. 20.14). Electrons whose pitch angles at the center of the bottle are sufficiently small have mirror points in the bottle and thus cannot leak out. The second example is the van Allen belts of Earth. Electrons (and also ions) travel up and down the magnetic field lines of the van Allen belts, reflecting at mirror points.

It is not hard to show that the gradient of $\mathbf{B}$ can be split up into the three pieces we have studied: a curvature with no change of $B = |\mathbf{B}|$ (Fig. 20.5a), a change of B orthogonal to the magnetic field (Fig. 20.5b), and a change of B along the magnetic field (Fig. 20.5c). When (as we have assumed) the lengthscales of these changes are far greater than the electron's Larmor radius, their effects on the electron's motion superpose linearly.

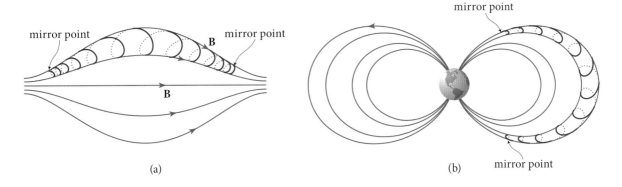

FIGURE 20.6 Two examples of the mirroring of particles in an inhomogeneous magnetic field. (a) A magnetic bottle. (b) Earth's van Allen belts.

20.7.4 A Slowly Time-Varying Magnetic Field

When the magnetic field changes on timescales long compared to the cyclotron period $2\pi/\omega_c$, its changes induce alterations of the electron's orbit that can be deduced with the aid of *adiabatic invariants*—quantities that are nearly constant when the field changes adiabatically, in other words, slowly (see, e.g., Lifshitz and Pitaevskii, 1981; Northrop, 1963). The conserved magnetic moment $\mu = m_e v_\perp^2/(2B)$ associated with an electron's transverse, circular motion is an example of an adiabatic invariant. We proved its invariance in Eqs. (20.52) and (20.53) (in Sec. 20.7.3, where we were working in a reference frame in which the magnetic field changed slowly, and associated with that change was a weak electric field). This adiabatic invariant can be shown to be, aside from a constant multiplicative factor $2\pi m_e/e$, the action associated with the electron's circular motion: $J_\phi = \oint p_\phi d\phi$. Here ϕ is the angle around the circular orbit, and $p_\phi = (m_e v_\perp - e A_\phi) r_L$ is the ϕ component of the electron's canonical momentum. The action J_ϕ is a well-known adiabatic invariant.

adiabatic invariants

magnetic moment as circular-motion action

When a slightly inhomogeneous magnetic field varies slowly in time, not only is $\mu = m_e v_\perp^2/(2B)$ adiabatically invariant (conserved), so also are two other actions. One is the action associated with motion from one mirror point of the magnetic field to another and back:

$$J_\| = \oint \mathbf{p}_\| \cdot d\mathbf{l}. \qquad (20.58)$$

longitudinal, mirror-point action

Here $\mathbf{p}_\| = m_e \mathbf{v}_\| - e\mathbf{A}_\| = m_e \mathbf{v}_\|$ is the generalized (canonical) momentum along the field line, and $d\mathbf{l}$ is distance along the field line. Thus the adiabatic invariant is the spatial average $\langle v_\| \rangle$ of the longitudinal speed of the electron, multiplied by m_e and twice the distance Δl between mirror points: $J_\| = 2m_e \langle v_\| \rangle \Delta l$.

The other (third) adiabatic invariant is the action associated with the drift of the guiding center: an electron mirroring back and forth along the field lines drifts sideways, and by its drift it traces out a 2-dimensional surface to which the magnetic

guiding-center-drift action

field is tangent (e.g., the surface of the center of the magnetic bottle in Fig. 20.6a, rotated around the horizontal axis). The action of the electron's drift around this magnetic surface turns out to be proportional to the total magnetic flux enclosed by the surface. Thus, if the field geometry changes slowly, the magnetic flux enclosed by the magnetic surface on which the electron's guiding center moves is adiabatically conserved.

<small>accuracy of adiabatic invariants</small>

How nearly constant are the adiabatic invariants? The general theory of adiabatic invariants shows that, so long as the temporal changes of the magnetic field structure are smooth enough to be described by analytic functions of time, the fractional failures of the adiabatic invariants to be conserved are of order $e^{-\tau/P}$. Here τ is the timescale on which the field changes, and P is the period of the motion associated with the adiabatic invariant. (This period P is $2\pi/\omega_c$ for the invariant μ; it is the mirroring period for the longitudinal action, and it is the drift period for the magnetic flux enclosed in the electron's magnetic surface.) Because the exponential $e^{-\tau/P}$ dies out so quickly with increasing timescale τ, the adiabatic invariants are conserved to very high accuracy when $\tau \gg P$.

20.7.5 Failure of Adiabatic Invariants; Chaotic Orbits

When the magnetic field changes as fast as or more rapidly than a cyclotron orbit (in space or in time), then the adiabatic invariants fail, and the charged-particle orbits may even be chaotic in some cases.

<small>failure of adiabatic invariants and chaotic motion near a magnetic neutral line</small>

An example is charged-particle motion near a neutral line (also called an X-line) of a magnetic field. Near an X-line, the field has the hyperbolic shape $\mathbf{B} = B_0(y\,\mathbf{e}_x + \gamma\,x\,\mathbf{e}_y)$ with constants B_0 and γ; see Fig. 19.13 for an example that occurs in magnetic-field-line reconnection. The X-line is the z-axis, $(x, y) = (0, 0)$, on which the field vanishes. Near there the Larmor radius becomes arbitrarily large, far larger than the scale on which the field changes. Correspondingly, no adiabatic invariants exist near the X-line, and it turns out that some charged-particle orbits near there are chaotic in the sense of extreme sensitivity to initial conditions (Sec. 15.6.4). An electric field in the $\mathbf{e}_z$ direction (orthogonal to $\mathbf{B}$ and parallel to the X-line) enhances the chaos (for details, see, e.g., Martin, 1986).

EXERCISES

Exercise 20.14 *Example: Mirror Machine*
One method for confining hot plasma is to arrange electric coils so as to make a mirror machine in which the magnetic field has the geometry sketched in Fig. 20.6a. Suppose that the magnetic field in the center is 1 T and the field strength at the two necks is 10 T, and that plasma is introduced with an isotropic velocity distribution near the center of the bottle.

(a) What fraction of the plasma particles will escape?

(b) Sketch the pitch-angle distribution function for the particles that remain.

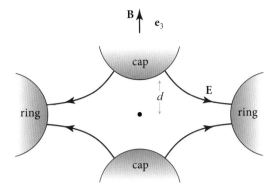

FIGURE 20.7 Penning trap for localizing individual charged particles. The magnetic field is uniform and parallel to the vertical axis of symmetry $\mathbf{e}_z$. The electric field is maintained between a pair of hyperboloidal caps and a hyperboloidal ring.

(c) Suppose that Coulomb collisions cause particles to diffuse in pitch angle α with a diffusion coefficient

$$D_{\alpha\alpha} \equiv \left\langle \frac{\Delta\alpha^2}{\Delta t} \right\rangle = t_D^{-1}. \tag{20.59}$$

Estimate how long it will take most of the plasma to escape the mirror machine.

(d) What do you suspect will happen in practice?

Exercise 20.15 *Challenge: Penning Trap*
A clever technique for studying the behavior of individual electrons or ions is to entrap them using a combination of electric and magnetic fields. One of the simplest and most useful devices is the *Penning trap* (see, e.g., Brown and Gabrielse, 1986). Basically this comprises a uniform magnetic field B combined with a hyperboloidal electrostatic field that is maintained between hyperboloidal electrodes as shown in Fig. 20.7. The electrostatic potential has the form $\Phi(\mathbf{x}) = \Phi_0(z^2 - x^2/2 - y^2/2)/(2d^2)$, where Φ_0 is the potential difference maintained across the electrodes, and d is the minimum axial distance from the origin to the hyperboloidal cap (as well as being $1/\sqrt{2}$ times the minimum radius of the ring electrode).

(a) Show that the potential satisfies Laplace's equation, as it must.
(b) Now consider an individual charged particle in the trap. Show that three separate oscillations can be excited:
   (i) cyclotron orbits in the magnetic field with angular frequency ω_c,
   (ii) "magnetron" orbits produced by $\mathbf{E} \times \mathbf{B}$ drift around the axis of symmetry with angular frequency ω_m (which you should compute), and

(iii) axial oscillations parallel to the magnetic field with angular frequency ω_z (which you should also compute).

Assume that $\omega_m \ll \omega_z \ll \omega_c$, and show that $\omega_z^2 \simeq 2\omega_m \omega_c$.

(c) Typically, the potential difference across the electrodes is ~ 10 V, the magnetic field strength is $B \sim 6$ T, and the radius of the ring and the height of the caps above the center of the traps are ~ 3 mm. Estimate the three independent angular frequencies for electrons and ions, verifying the ordering $\omega_m \ll \omega_z \ll \omega_c$. Also estimate the maximum velocities associated with each of these oscillations if the particle is to be retained in the trap.

(d) Solve the classical equation of motion exactly, and demonstrate that the magnetron motion is formally unstable.

Penning traps have been used to perform measurements of the electron-proton mass ratio and the magnetic moment of the electron with unprecedented precision.

Bibliographic Note

For a thorough treatment of the particle kinetics of plasmas, see Shkarofsky, Johnston, and Bachynski (1966). For less detailed treatments, we recommend the relevant portions of Spitzer (1962), Bittencourt (2004), and Bellan (2006).

For applications of the Fokker-Planck equation to Coulomb scattering in plasmas, we like Shkarofsky, Johnston, and Bachynski (1966, Chaps. 7 and 8), Boyd and Sanderson (2003, Chap. 8), Kulsrud (2005, Chap. 8), and Bellan (2006, Chap. 13).

For charged-particle motion in inhomogeneous and time-varying magnetic fields, we recommend Northrop (1963); Jackson (1999, Chap. 12), which is formulated using special relativity; Bittencourt (2004, Chaps. 2–4); and Bellan (2006, Chap. 3).

CHAPTER TWENTY-ONE

Waves in Cold Plasmas: Two-Fluid Formalism

> Waves from moving sources: Adagio. Andante. Allegro moderato.
> OLIVER HEAVISIDE (1912)

21.1 Overview

The growth of plasma physics came about, in the early twentieth century, through studies of oscillations in electric discharges and the contemporaneous development of means to broadcast radio waves over great distances by reflecting them off Earth's ionosphere. It is therefore not surprising that most early plasma-physics research was devoted to describing the various modes of wave propagation. Even in the simplest, linear approximation for a plasma sufficiently cold that thermal effects are unimportant, we will see that the variety of possible modes is immense.

In the previous chapter, we introduced several length- and timescales, most importantly the Larmor (gyro) radius, the Debye length, the plasma period, the cyclotron (gyro) period, the collision time (inverse collision frequency), and the equilibration times (inverse collision rates). To these we must now add the wavelength and period of the wave under study. The wave's characteristics are controlled by the relative sizes of these parameters, and in view of the large number of parameters, there is a bewildering number of possibilities. If we further recognize that plasmas are collisionless, so there is no guarantee that the particle distribution functions can be characterized by a single temperature, then the possibilities multiply further.

Fortunately, the techniques needed to describe the propagation of linear wave perturbations in a particular equilibrium configuration of a plasma are straightforward and can be amply illustrated by studying a few simple cases. In this chapter, we follow this course by restricting our attention to one class of modes, those where we can either ignore completely the thermal motions of the ions and electrons that compose the plasma (i.e., treat these species as *cold*) or include them using just a velocity dispersion or temperature. We can then apply our knowledge of fluid dynamics by treating the ions and electrons separately as fluids acted on by electromagnetic forces. This is called the *two-fluid formalism* for plasmas. In the next chapter, we explore when and how waves are sensitive to the actual distribution of particle speeds by developing

> **BOX 21.1. READERS' GUIDE**
>
> - This chapter relies significantly on:
>   - Chap. 20 on the particle kinetics of plasmas;
>   - the basic concepts of fluid dynamics, Secs. 13.4 and 13.5;
>   - magnetosonic waves, Sec. 19.7; and
>   - the basic concepts of geometric optics, Secs. 7.2 and 7.3.
> - The remaining Chaps. 22 and 23 of Part VI rely heavily on this chapter.

the more sophisticated *kinetic-theory formalism* and using it to study waves in *warm plasmas*.

We begin our two-fluid study of plasma waves in Sec. 21.2 by deriving a very general wave equation, which governs weak waves in a homogeneous plasma that may or may not have a magnetic field and also governs electromagnetic waves in any other dielectric medium. That wave equation and the associated dispersion relation for the wave modes depend on a dielectric tensor, which must be derived from an examination of the motions of the electrons and protons (or other charge carriers) inside the wave. Those motions are described, in our two-fluid (electron-fluid and proton-fluid) model by equations that we write down in Sec. 21.3.

In Sec. 21.4, we use those two-fluid equations to derive the dielectric tensor, and then we combine with our general wave equation from Sec. 21.2 to obtain the dispersion relation for wave modes in a uniform, unmagnetized plasma. The modes fall into two classes: (i) *Transverse or electromagnetic waves,* with the electric field **E** perpendicular to the wave's propagation direction. These are modified versions of electromagnetic waves in vacuum. As we shall see, they can propagate only at frequencies above the plasma frequency; at lower frequencies they become evanescent. (ii) *Longitudinal waves*, with **E** parallel to the propagation direction, which come in two species: *Langmuir waves* and *ion-acoustic waves*. Longitudinal waves are a melded combination of sound waves in a fluid and electrostatic plasma oscillations; their restoring force is a mixture of thermal pressure and electrostatic forces.

In Sec. 21.5, we explore how a uniform magnetic field changes the character of these waves. The **B** field makes the plasma anisotropic but axially symmetric. As a result, the dielectric tensor, dispersion relation, and wave modes have much in common with those in an anisotropic but axially symmetric dielectric crystal, which we studied in the context of nonlinear optics in Chap. 10. A plasma, however, has a much richer set of characteristic frequencies than does a crystal (electron plasma frequency, electron cyclotron frequency, ion cyclotron frequency, etc.). As a result, even in the regime of weak linear waves and a cold plasma (no thermal pressure), the plasma has a far greater richness of modes than does a crystal.

In Sec. 21.5.1, we derive the general dispersion relation that encompasses all these cold-magnetized-plasma modes, and in Secs. 21.5.2 and 21.5.3, we explore the special cases of modes that propagate parallel to and perpendicular to the magnetic field. Then in Sec. 21.5.4, we examine a practical example: the propagation of radio waves in Earth's ionosphere, where it is a good approximation to ignore the ion motion and work with a one-fluid (i.e., electron-fluid) theory. Having gained insight into simple cases (parallel modes, perpendicular modes, and one-fluid modes), we return in Sec. 21.5.5 to the full class of linear modes in a cold, magnetized, two-fluid plasma and briefly describe some tools by which one can make sense of them all.

Finally, in Sec. 21.6, we turn to the question of plasma stability. In Sec. 14.6.1 and Chap. 15, we saw that fluid flows that have sufficient shear are unstable; perturbations can feed off the relative kinetic energy of adjacent regions of the fluid and use that energy to power an exponential growth. In plasmas, with their long mean free paths, there can similarly exist kinetic energies of relative, ordered motion in velocity space; and perturbations, feeding off those energies, can grow exponentially. To study this phenomenon in full requires the kinetic-theory description of a plasma, which we develop in Chap. 22; but in Sec. 21.6 we gain insight into a prominent example of such a velocity-space instability by analyzing two cold plasma streams moving through each other. We illustrate the resulting *two-stream instability* by a short discussion of particle beams that are created in disturbances on the surface of the Sun and propagate out through the solar wind.

21.2 Dielectric Tensor, Wave Equation, and General Dispersion Relation

We begin our study of waves in plasmas by deriving a general wave equation that applies equally well to electromagnetic waves in unmagnetized plasmas; in magnetized plasmas; and in any other kind of dielectric medium, such as an anisotropic crystal. This wave equation is the same one we used in our study of nonlinear optics in Chap. 10 [Eqs. (10.50) and (10.51a)], and the derivation is essentially a linearized variant of the one we gave in Chap. 10 [Eqs. (10.16a)–(10.22b)].

When a wave propagates through a plasma (or other dielectric), it entails a relative motion of electrons and protons (or other charge carriers). Assuming the wave has small enough amplitude to be linear, those charge motions can be embodied in an oscillating polarization (electric dipole moment per unit volume) $\mathbf{P}(\mathbf{x}, t)$, which is related to the plasma's (or dielectric's) varying charge density ρ_e and current density $\mathbf{j}$ in the usual way:

polarization vector

$$\rho_e = -\nabla \cdot \mathbf{P}, \qquad \mathbf{j} = \frac{\partial \mathbf{P}}{\partial t}. \tag{21.1}$$

(These relations enforce charge conservation, $\partial \rho_e/\partial t + \nabla \cdot \mathbf{j} = 0$.) When these ρ_e and $\mathbf{j}$ are inserted into the standard Maxwell equations for $\mathbf{E}$ and $\mathbf{B}$, one obtains

Maxwell equations

$$\boxed{\nabla \cdot \mathbf{E} = -\frac{\nabla \cdot \mathbf{P}}{\epsilon_0}, \quad \nabla \cdot \mathbf{B} = 0, \quad \nabla \times \mathbf{E} = -\frac{\partial \mathbf{B}}{\partial t}, \quad \nabla \times \mathbf{B} = \mu_0 \frac{\partial \mathbf{P}}{\partial t} + \frac{1}{c^2} \frac{\partial \mathbf{E}}{\partial t}.}$$

(21.2)

If the plasma is endowed with a uniform magnetic field $\mathbf{B}_o$, that field can be left out of these equations, as its divergence and curl are guaranteed to vanish. Thus we can regard $\mathbf{E}$, $\mathbf{B}$, and $\mathbf{P}$ in these Maxwell equations as the perturbed quantities associated with the waves.

From a detailed analysis of the response of the charge carriers to the oscillating $\mathbf{E}$ and $\mathbf{B}$ fields, one can deduce a linear (frequency-dependent and wave-vector-dependent) relationship between the waves' electric field $\mathbf{E}$ and the polarization $\mathbf{P}$:

$$\boxed{P_j = \epsilon_0 \chi_{jk} E_k.}$$

(21.3)

tensorial electric susceptibility

Here ϵ_0 is the vacuum permittivity, and χ_{jk} is a dimensionless, tensorial electric susceptibility [cf. Eq. (10.21)]. A different, but equivalent, viewpoint on the relationship between $\mathbf{P}$ and $\mathbf{E}$ can be deduced by taking the time derivative of Eq. (21.3); setting $\partial \mathbf{P}/\partial t = \mathbf{j}$; assuming a sinusoidal time variation $e^{-i\omega t}$ so that $\partial \mathbf{E}/\partial t = -i\omega \mathbf{E}$; and then reinterpreting the result as Ohm's law with a tensorial electric conductivity κ_{ejk}:

tensorial electric conductivity

$$\boxed{j_j = \kappa_{ejk} E_k, \qquad \kappa_{ejk} = -i\omega \epsilon_0 \chi_{jk}.}$$

(21.4)

Evidently, for sinusoidal waves the electric susceptibility χ_{jk} and the electric conductivity κ_{ejk} embody the same information about the wave-particle interactions.

dielectric tensor

That information is also embodied in a third object: the dimensionless dielectric tensor ϵ_{jk}, which relates the electric displacement $\mathbf{D}$ to the electric field $\mathbf{E}$:

$$\boxed{D_j \equiv \epsilon_0 E_j + P_j = \epsilon_0 \epsilon_{jk} E_k, \qquad \epsilon_{jk} = \delta_{jk} + \chi_{jk} = \delta_{jk} + \frac{i}{\epsilon_0 \omega} \kappa_{ejk}.}$$

(21.5)

In the next section, we derive the value of the dielectric tensor ϵ_{jk} for waves in an unmagnetized plasma, and in Sec. 21.4.1, we derive it for a magnetized plasma.

Using the definition $\mathbf{D} = \epsilon_0 \mathbf{E} + \mathbf{P}$, we can eliminate $\mathbf{P}$ from Eqs. (21.2), thereby obtaining the familiar form of Maxwell's equations for dielectric media with no nonpolarization-based charges or currents:

$$\nabla \cdot \mathbf{D} = 0, \quad \nabla \cdot \mathbf{B} = 0, \quad \nabla \times \mathbf{E} = -\frac{\partial \mathbf{B}}{\partial t}, \quad \nabla \times \mathbf{B} = \mu_0 \frac{\partial \mathbf{D}}{\partial t}.$$

(21.6)

general wave equation in a dielectric

By taking the curl of the third of these equations and combining with the fourth and with $D_j = \epsilon_0 \epsilon_{jk} E_k$, we obtain the wave equation that governs the perturbations:

$$\boxed{\nabla^2 \mathbf{E} - \nabla(\nabla \cdot \mathbf{E}) - \boldsymbol{\epsilon} \cdot \frac{1}{c^2} \frac{\partial^2 \mathbf{E}}{\partial t^2} = 0,} \qquad (21.7)$$

where $\boldsymbol{\epsilon}$ is our index-free notation for ϵ_{jk}. [This is the same as the linearized approximation Eq. (10.50) to our nonlinear-optics wave equation (10.22a).] Specializing to a plane-wave mode with wave vector $\mathbf{k}$ and angular frequency ω, so $\mathbf{E} \propto e^{i\mathbf{k}\mathbf{x}}e^{-i\omega t}$, we convert this wave equation into a homogeneous, algebraic equation for the Cartesian components of the electric vector E_j (cf. Box 12.2):

$$\boxed{L_{ij} E_j = 0,} \qquad (21.8)$$

algebratized wave equation

where

$$\boxed{L_{ij} = k_i k_j - k^2 \delta_{ij} + \frac{\omega^2}{c^2} \epsilon_{ij}.} \qquad (21.9)$$

algebratized wave operator

We call Eq. (21.8) the *algebratized wave equation*, and L_{ij} the *algebratized wave operator*.

The algebratized wave equation (21.8) can have a solution only if the determinant of the 3-dimensional matrix L_{ij} vanishes:

$$\boxed{\det||L_{ij}|| \equiv \det \left\| k_i k_j - k^2 \delta_{ij} + \frac{\omega^2}{c^2} \epsilon_{ij} \right\|.} \qquad (21.10)$$

general dispersion relation for electromagnetic waves in a dielectric medium

This is a polynomial equation for the angular frequency as a function of the wave vector (with ω and $\mathbf{k}$ appearing not only explicitly in L_{ij} but also implicitly in the functional form of ϵ_{jk}). Each solution $\omega(\mathbf{k})$ of this equation is the dispersion relation for a particular wave mode. Therefore, we can regard Eq. (21.10) as the general dispersion relation for plasma waves—and for linear electromagnetic waves in any other kind of dielectric medium.

To obtain an explicit form of the dispersion relation (21.10), we must give a prescription for calculating the dielectric tensor $\epsilon_{ij}(\omega, \mathbf{k})$, or equivalently [cf. Eq. (21.5)] the conductivity tensor $\kappa_{e\,ij}$ or the susceptibility tensor χ_{ij}. The simplest prescription involves treating the electrons and ions as independent fluids; so we digress, briefly, from our discussion of waves to present the two-fluid formalism for plasmas.

21.3 Two-Fluid Formalism

21.3

A plasma necessarily contains rapidly moving electrons and ions, and their individual responses to an applied electromagnetic field depend on their velocities. In the simplest model of these responses, we average over all the particles in a species (electrons or protons in this case) and treat them collectively as a fluid. Now, the fact that all the electrons are treated as one fluid does not mean that they have to collide with one

two-fluid formalism

another. In fact, as we have already emphasized in Chap. 20, electron-electron collisions are quite rare, and we can usually ignore them. Nevertheless, we can still define a mean fluid velocity for both the electrons and the protons by averaging over their total velocity distribution functions just as we would for a gas:

mean fluid velocity

$$\mathbf{u}_s = \langle \mathbf{v} \rangle_s; \quad s = p, e, \tag{21.11}$$

where the subscripts p and e refer to protons and electrons. Similarly, for each fluid we define a pressure tensor using the fluid's dispersion of particle velocities:

pressure tensor

$$\mathbf{P}_s = n_s m_s \langle (\mathbf{v} - \mathbf{u}_s) \otimes (\mathbf{v} - \mathbf{u}_s) \rangle \tag{21.12}$$

[cf. Eqs. (20.34) and (20.35)].

The density n_s and mean velocity $\mathbf{u}_s$ of each species s must satisfy the equation of continuity (particle conservation):

particle conservation

$$\frac{\partial n_s}{\partial t} + \nabla \cdot (n_s \mathbf{u}_s) = 0. \tag{21.13a}$$

They must also satisfy an equation of motion: the law of momentum conservation (i.e., the Euler equation with the Lorentz force added to the right-hand side):

momentum conservation

$$n_s m_s \left(\frac{\partial \mathbf{u}_s}{\partial t} + (\mathbf{u}_s \cdot \nabla) \mathbf{u}_s \right) = -\nabla \cdot \mathbf{P}_s + n_s q_s (\mathbf{E} + \mathbf{u}_s \times \mathbf{B}). \tag{21.13b}$$

Here we have neglected the tiny influence of collisions between the two species. In these equations and below, $q_s = \pm e$ is the particles' charge (positive for protons and negative for electrons). Note that, as collisions are ineffectual, we cannot assume that the pressure tensors are isotropic.

Although the particle- and momentum-conservation equations (21.13) for each species (electron or proton) are formally decoupled from the equations for the other species, there is actually a strong physical coupling induced by the electromagnetic field. The two species together produce $\mathbf{E}$ and $\mathbf{B}$ through their joint charge density and current density:

charge and current density

$$\rho_e = \sum_s q_s n_s, \quad \mathbf{j} = \sum_s q_s n_s \mathbf{u}_s, \tag{21.14}$$

and those $\mathbf{E}$ and $\mathbf{B}$ fields strongly influence the electron and proton fluids' dynamics via their equations of motion (21.13b).

EXERCISES

Exercise 21.1 *Problem: Fluid Drifts in a Magnetized Plasma*
We developed a one-fluid (MHD) description of plasma in Chap. 19, and in Chap. 20 we showed how to describe the orbits of individual charged particles in a magnetic field that varies slowly compared with the particles' orbital periods and radii. In this

chapter, we describe the plasma as two or more cold fluids. We relate these three approaches in this exercise.

(a) Generalize Eq. (21.13b) to a single fluid as:

$$\rho \mathbf{a} = \rho \mathbf{g} - \nabla \cdot \mathbf{P} + \rho_e \mathbf{E} + \mathbf{j} \times \mathbf{B}, \qquad (21.15)$$

where $\mathbf{a}$ is the fluid acceleration, and $\mathbf{g}$ is the acceleration of gravity, and write the pressure tensor as $\mathbf{P} = P_\perp \mathbf{g} + (P_\parallel - P_\perp) \mathbf{B} \otimes \mathbf{B}/B^2$, where we suppress the subscript s.[1] Show that the component of current density perpendicular to the local magnetic field is

$$\mathbf{j}_\perp = \frac{\mathbf{B} \times \nabla \cdot \mathbf{P}}{B^2} + \rho_e \frac{\mathbf{E} \times \mathbf{B}}{B^2} + \rho \frac{(\mathbf{g} - \mathbf{a}) \times \mathbf{B}}{B^2}. \qquad (21.16a)$$

(b) Use vector identities to rewrite Eq. (21.16a) in the form:

$$\mathbf{j}_\perp = P_\parallel \frac{\mathbf{B} \times (\mathbf{B} \cdot \nabla) \mathbf{B}}{B^4} + P_\perp \frac{\mathbf{B} \times \nabla B}{B^3} - (\nabla \times \mathbf{M})_\perp + \rho_e \frac{\mathbf{E} \times \mathbf{B}}{B^2} + \rho \frac{(\mathbf{g} - \mathbf{a}) \times \mathbf{B}}{B^2}, \qquad (21.16b)$$

where

$$\mathbf{M} = P_\perp \frac{\mathbf{B}}{B^3} \qquad (21.16c)$$

is the *magnetization*.

(c) Identify the first term of Eq. (21.16b) with the curvature drift (20.49) and the second term with the gradient drift (20.51).

(d) Using a diagram, explain how the magnetization (20.54)—the magnetic moment per unit volume—can contribute to the current density. In particular, consider what might happen at the walls of a cavity containing plasma.[2] Argue that there should also be a local magnetization current parallel to the magnetic field, even when there is no net drift of the particles.

(e) Associate the final two terms of Eq. (21.16b) with the "$\mathbf{E} \times \mathbf{B}$" drift (20.47) and the gravitational drift (20.48). Explain the presence of the acceleration in the gravitational drift.

(f) Discuss how to combine these contributions to rederive the standard formulation of MHD, and specify some circumstances under which MHD might be a poor approximation.

---

1. Note that when writing the pressure tensor this way, we are implicitly assuming that it has only two components, along and perpendicular to the local magnetic field. Physically, this is reasonable if the particles that make up the fluid are *magnetized*, i.e., the particles' orbital periods and radii are short compared to the scales on which the magnetic field changes.
2. This is an illustration of a general theorem in statistical mechanics due to Niels Bohr in 1911, which essentially shows that magnetization cannot arise in classical physics. This theorem may have had a role in the early development of quantum mechanics.

Note that this single-fluid formalism does not describe the component of the current parallel to the magnetic field, $\mathbf{j}_\parallel$. For this, we need to introduce an effective collision frequency for the electrons and ions either with waves (Chap. 23) or with other particles (Chap. 20). If we assume that the electrical conductivity is perfect, then $\mathbf{E} \cdot \mathbf{B} = 0$, and $\mathbf{j}_\parallel$ is essentially fixed by the boundary conditions.

21.4 Wave Modes in an Unmagnetized Plasma

We now specialize to waves in a homogeneous, unmagnetized electron-proton plasma.

unperturbed plasma

First consider an unperturbed plasma in the absence of a wave, and work in a frame in which the proton fluid velocity $\mathbf{u}_p$ vanishes. By assumption, the equilibrium is spatially uniform. If there were an electric field, then charges would quickly flow to neutralize it; so there can be no electric field, and hence (since $\mathbf{\nabla} \cdot \mathbf{E} = \rho_e/\epsilon_0$) no net charge density. Therefore, the electron density must equal the proton density. Furthermore, there can be no net current, as this would lead to a magnetic field; so since the proton current $e\, n_p \mathbf{u}_p$ vanishes, the electron current $= -e\, n_e \mathbf{u}_e$ must also vanish. Hence the electron fluid velocity $\mathbf{u}_e$ must vanish in our chosen frame. Thus in an equilibrated homogeneous, unmagnetized plasma, $\mathbf{u}_e$, $\mathbf{u}_p$, $\mathbf{E}$, and $\mathbf{B}$ all vanish in the protons' mean rest frame.

Now apply an electromagnetic perturbation. This will induce a small, oscillating fluid velocity $\mathbf{u}_s$ in both the proton and electron fluids. It should not worry us that the fluid velocity is small compared with the random speeds of the constituent particles; the same is true in any subsonic gas dynamical flow, but the fluid description remains good there and also here.

21.4.1 Dielectric Tensor and Dispersion Relation for a Cold, Unmagnetized Plasma

Continuing to keep the plasma unmagnetized, let us further simplify matters (until Sec. 21.4.3) by restricting ourselves to a cold plasma, so the tensorial pressures vanish: $\mathbf{P}_s = 0$. As we are only interested in linear wave modes, we rewrite the two-fluid equations (21.13), just retaining terms that are first order in perturbed quantities [i.e., dropping the $(\mathbf{u}_s \cdot \mathbf{\nabla})\mathbf{u}_s$ and $\mathbf{u}_s \times \mathbf{B}$ terms]. Then, focusing on a plane-wave mode, $\propto \exp[i(\mathbf{k} \cdot \mathbf{x} - \omega t)]$, we bring the equation of motion (21.13b) into the form

linearized perturbative equation of motion

$$\boxed{-i\omega n_s m_s \mathbf{u}_s = q_s n_s \mathbf{E}} \qquad (21.17)$$

for each species, $s = p, e$. From this, we can immediately deduce the linearized current density:

$$\mathbf{j} = \sum_s n_s q_s \mathbf{u}_s = \sum_s \frac{i n_s q_s^2}{m_s \omega} \mathbf{E}, \qquad (21.18)$$

from which we infer that the conductivity tensor κ_e has Cartesian components

$$\kappa_{eij} = \sum_s \frac{in_s q_s^2}{m_s \omega} \delta_{ij}, \tag{21.19}$$

where δ_{ij} is the Kronecker delta. Note that the conductivity is purely imaginary, which means that the current oscillates out of phase with the applied electric field, which in turn implies that there is no time-averaged ohmic energy dissipation: $\langle \mathbf{j} \cdot \mathbf{E} \rangle = 0$. Inserting the conductivity tensor (21.19) into the general equation (21.5) for the dielectric tensor, we obtain

$$\epsilon_{ij} = \delta_{ij} + \frac{i}{\epsilon_0 \omega} \kappa_{eij} = \left(1 - \frac{\omega_p^2}{\omega^2}\right) \delta_{ij}. \tag{21.20}$$

Here and throughout this chapter, the plasma frequency ω_p is slightly different from that used in Chap. 20: it includes a tiny (1/1,836) correction due to the motion of the protons, which we neglected in our analysis of plasma oscillations in Sec. 20.3.3:

$$\boxed{\omega_p^2 = \sum_s \frac{n_s q_s^2}{m_s \epsilon_0} = \frac{ne^2}{m_e \epsilon_0}\left(1 + \frac{m_e}{m_p}\right) = \frac{ne^2}{\mu \epsilon_0},} \tag{21.21}$$

plasma frequency

where μ is the reduced mass $\mu = m_e m_p / (m_e + m_p)$. Note that because there is no physical source of a preferred direction in the plasma, the dielectric tensor (21.20) is isotropic.

Now, without loss of generality, let the waves propagate in the z direction, so $\mathbf{k} = k\mathbf{e}_z$. Then the algebratized wave operator (21.9), with $\boldsymbol{\epsilon}$ given by Eq. (21.20), takes the following form:

$$L_{ij} = \frac{\omega^2}{c^2} \begin{pmatrix} 1 - \frac{c^2 k^2}{\omega^2} - \frac{\omega_p^2}{\omega^2} & 0 & 0 \\ 0 & 1 - \frac{c^2 k^2}{\omega^2} - \frac{\omega_p^2}{\omega^2} & 0 \\ 0 & 0 & 1 - \frac{\omega_p^2}{\omega^2} \end{pmatrix}. \tag{21.22}$$

algebratized wave operator and general dispersion relation for waves in a cold, unmagnetized plasma

The corresponding dispersion relation $\det\|L_{jk}\| = 0$ [Eq. (21.10)] becomes

$$\left(1 - \frac{c^2 k^2}{\omega^2} - \frac{\omega_p^2}{\omega^2}\right)^2 \left(1 - \frac{\omega_p^2}{\omega^2}\right) = 0. \tag{21.23}$$

This is a polynomial equation of order 6 for ω as a function of k, so formally there are six solutions corresponding to three pairs of modes propagating in opposite directions.

Two of the pairs of modes are degenerate with frequency

$$\boxed{\omega = \sqrt{\omega_p^2 + c^2 k^2}.} \tag{21.24}$$

21.4 Wave Modes in an Unmagnetized Plasma

plasma electromagnetic modes

These are called *plasma electromagnetic modes,* and we study them in the next subsection. The remaining pair of modes exist at a single frequency:

$$\boxed{\omega = \omega_p.} \tag{21.25}$$

electrostatic plasma oscillations

These must be the electrostatic plasma oscillations that we studied in Sec. 20.3.3 (though now with an arbitrary wave number k, while in Sec. 20.3.3 the wave number was assumed to be zero). In Sec. 21.4.3, we show that this is so and explore how these plasma oscillations are modified by finite-temperature effects.

21.4.2 Plasma Electromagnetic Modes

properties of plasma electromagnetic modes

To learn the physical nature of the modes with dispersion relation $\omega = \sqrt{\omega_p^2 + c^2 k^2}$ [Eq. (21.24)], we must examine the details of their electric-field oscillations, magnetic-field oscillations, and electron and proton motions. A key to this is the algebratized wave equation $L_{ij} E_j = 0$, with L_{ij} specialized to the dispersion relation (21.24): $\|L_{ij}\| = \text{diag}[0, 0, (\omega^2 - \omega_p^2)/c^2]$. In this case, the general solution to $L_{ij} E_j = 0$ is an electric field that lies in the x-y plane (transverse plane) and that therefore is orthogonal to the waves' propagation vector $\mathbf{k} = k \mathbf{e}_z$. The third of the Maxwell equations (21.2) implies that the magnetic field is

transverse electric and magnetic fields

$$\mathbf{B} = (\mathbf{k}/\omega) \times \mathbf{E}, \tag{21.26}$$

which also lies in the transverse plane and is orthogonal to $\mathbf{E}$. Evidently, these modes are close analogs of electromagnetic waves in vacuum; correspondingly, they are known as the plasma's *electromagnetic modes*. The electron and proton motions in these modes, as given by Eq. (21.17), are oscillatory displacements in the direction of $\mathbf{E}$ but are out of phase with $\mathbf{E}$. The amplitudes of the fluid motions, at fixed electric-field amplitude, vary as $1/\omega$; when ω decreases, the fluid amplitudes grow.

The dispersion relation for these modes, Eq. (21.24), implies that they can only propagate (i.e., have real angular frequency when the wave vector is real) if ω exceeds the plasma frequency. As ω is decreased toward ω_p, k approaches zero, so these modes become electrostatic plasma oscillations with arbitrarily long wavelength orthogonal to the oscillation direction (i.e., they become a spatially homogeneous variant of the plasma oscillations studied in Sec. 20.3.3). At $\omega < \omega_p$ these modes become evanescent.

In their regime of propagation, $\omega > \omega_p$, these cold-plasma electromagnetic waves have a phase velocity given by

phase velocity

$$\mathbf{V}_{\text{ph}} = \frac{\omega}{k} \hat{\mathbf{k}} = c \left(1 - \frac{\omega_p^2}{\omega^2}\right)^{-1/2} \hat{\mathbf{k}}, \tag{21.27a}$$

where $\hat{\mathbf{k}} \equiv \mathbf{k}/k$ is a unit vector in the propagation direction. Although this phase velocity exceeds the speed of light, causality is not violated, because information (and energy) propagate at the group velocity, not the phase velocity. The group velocity is readily shown to be

$$\mathbf{V}_g = \nabla_{\mathbf{k}}\omega = \frac{c^2\mathbf{k}}{\omega} = c\left(1 - \frac{\omega_p^2}{\omega^2}\right)^{1/2}\hat{\mathbf{k}}, \qquad (21.27b)$$

group velocity

which is less than c as it must be.

These cold-plasma electromagnetic modes transport energy and momentum just like wave modes in a fluid. There are three contributions to the waves' mean (time-averaged) energy density: the electric, the magnetic, and the kinetic-energy densities. (If we had retained the pressure, then an additional contribution would come from the internal energy.) To compute these mean energy densities, we must form the time average of products of physical quantities. Now, we have used the complex representation to denote each of our oscillating quantities (e.g., E_x), so we must be careful to remember that $A = ae^{i(\mathbf{k}\cdot\mathbf{x}-\omega t)}$ is an abbreviation for the real part of this quantity—which is the physical A. It is easy to show (Ex. 21.3) that the time-averaged value of the physical A times the physical B (which we shall denote by $\langle AB \rangle$) is given in terms of their complex amplitudes by

$$\langle AB \rangle = \frac{AB^* + A^*B}{4}, \qquad (21.28)$$

time-averaged product in terms of complex amplitudes

where $*$ denotes a complex conjugate.

Using Eqs. (21.26) and (21.27a), we can write the magnetic energy density in the form $\langle B^2 \rangle/(2\mu_0) = (1 - \omega_p^2/\omega^2)\epsilon_0\langle E^2\rangle/2$. Using Eq. (21.17), the particle kinetic energy is $\sum_s n_s m_s \langle u_s^2 \rangle/2 = (\omega_{pe}^2/\omega^2)\epsilon_0\langle E^2\rangle/2$. Summing these contributions and using Eq. (21.28), we obtain

$$U = \frac{\epsilon_0 EE^*}{4} + \frac{BB^*}{4\mu_0} + \sum_s \frac{n_s m_s u_s u_s^*}{4}$$

$$= \frac{\epsilon_0 EE^*}{2}. \qquad (21.29a)$$

energy density

The mean energy flux in the wave is carried (to quadratic order) by the electromagnetic field and is given by the Poynting flux. (The kinetic energy flux vanishes to this order.) A straightforward calculation gives

$$\mathbf{F}_{\rm EM} = \langle \mathbf{E} \times \mathbf{B} \rangle = \frac{\mathbf{E} \times \mathbf{B}^* + \mathbf{E}^* \times \mathbf{B}}{4} = \frac{EE^*\mathbf{k}}{2\mu_0\omega}$$

$$= U\mathbf{V}_g, \qquad (21.29b)$$

energy flux

where we have used $\mu_0 = c^{-2}\epsilon_0^{-1}$. We therefore find that the energy flux is the product of the energy density and the group velocity, as is true quite generally (cf. Sec. 6.3). (If it were not true, then a localized wave packet, which propagates at the group velocity, would move along a different trajectory from its energy, and we would wind up with energy in regions with vanishing amplitude!)

EXERCISES

Exercise 21.2 *Derivation: Phase and Group Velocities for Electromagnetic Modes*
Derive Eqs. (21.27) for the phase and group velocities of electromagnetic modes in a plasma.

Exercise 21.3 *Derivation: Time-Averaging Formula*
Verify Eq. (21.28).

Exercise 21.4 *Problem: Collisional Damping in an Electromagnetic Wave Mode*
Consider a transverse electromagnetic wave mode propagating in an unmagnetized, partially ionized gas in which the electron-neutral collision frequency is ν_e. Include the effects of collisions in the electron equation of motion (21.17), by introducing a term $-n_e m_e \nu_e \mathbf{u}_e$ on the right-hand side. Ignore ion motion and electron-ion and electron-electron collisions.

Derive the dispersion relation when $\omega \gg \nu_e$, and show by explicit calculation that the rate of loss of energy per unit volume ($-\nabla \cdot \mathbf{F}_{EM}$, where $\mathbf{F}_{EM}$ is the Poynting flux) is balanced by the Ohmic heating of the plasma. [Hint: It may be easiest to regard ω as real and $\mathbf{k}$ as complex.]

21.4.3 Langmuir Waves and Ion-Acoustic Waves in Warm Plasmas

longitudinal electrostatic plasma oscillations

For our case of a cold, unmagnetized plasma, the third pair of modes embodied in the dispersion relation (21.23) only exists at a single frequency, the plasma frequency: $\omega = \omega_p$. These modes' wave equation $L_{ij} E_j = 0$ with $||L_{ij}|| = \text{diag}(-k^2, -k^2, 0)$ [Eq. (21.22) with $\omega^2 = \omega_p^2$] implies that $\mathbf{E}$ points in the z direction (i.e., along $\mathbf{k}$, which is to say the longitudinal direction). Maxwell's equations then imply $\mathbf{B} = 0$, and the equation of motion (21.17) implies that the fluid displacements are also in the direction of $\mathbf{E}$—the longitudinal direction. Clearly, these modes, like electromagnetic modes in the limit $k = 0$ and $\omega = \omega_p$, are electrostatic plasma oscillations. However, in this case, where the spatial variations of $\mathbf{E}$ and $\mathbf{u}_s$ are along the direction of oscillation instead of perpendicular to it, k is not constrained to vanish; instead, all wave numbers are allowed. This means that the plasma can undergo plane-parallel oscillations at $\omega = \omega_p$ with displacements in some Cartesian z direction and with any arbitrary z-dependent amplitude that one might wish. But these oscillations cannot transport energy: because ω is independent of $\mathbf{k}$, their group velocity, $\mathbf{V}_g = \nabla_{\mathbf{k}} \omega$, vanishes. Their phase velocity, $\mathbf{V}_{ph} = (\omega_p/k)\hat{\mathbf{k}}$, by contrast, is finite.

finite temperature converts longitudinal plasma oscillations into Langmuir waves

So far, we have confined ourselves to wave modes in cold plasmas. When thermal motions are turned on, the resulting thermal pressure gradients convert longitudinal plasma oscillations, at finite wave number k, into propagating, energy-transporting, longitudinal modes called *Langmuir waves*.[3] As we have already intimated, because

---

3. The chemist Irving Langmuir observed these waves and introduced the name "plasma" for ionized gas in 1927.

the plasma is collisionless, we must turn to kinetic theory (Chap. 22) to understand the thermal effects fully. However, using the present chapter's two-fluid formalism and with the guidance of physical arguments, we can deduce the leading-order effects of finite temperature.

In our physical arguments, we assume that the electrons are thermalized with one another at a temperature T_e, the protons are thermalized at temperature T_p, and T_e and T_p may differ (because the timescale for electrons and protons to exchange energy is so much longer than the timescales for electrons to exchange energy among themselves and for protons to exchange energy among themselves; see Sec. 20.4.3).

Physically, the key to the Langmuir waves' propagation is the warm electrons' thermal pressure. (The proton pressure is unimportant here, because the protons oscillate electrostatically with an amplitude that is tiny compared to the electron amplitude; nevertheless, as we shall see below, the proton pressure is important in other ways.)

In an adiabatic sound wave *in a fluid* (where the particle mean free paths are small compared to the wavelength), we relate the pressure perturbation to the density perturbation by assuming that the entropy is held constant in each fluid element. In other words, we write $\nabla P = C^2 m \nabla n$, where $C = [\gamma P/(nm)]^{1/2}$ is the adiabatic sound speed, n is the particle density, m is the particle mass, and γ is the adiabatic index [which is equal to the specific heat ratio $\gamma = c_P/c_V$ (Ex. 5.4)], whose value is 5/3 for a monatomic gas.

However, the electron gas *in the plasma we are considering* is collisionless on the short timescale of a perturbation period, and we are only interested in the tensorial pressure gradient parallel to **k** (which we take to point in the z direction), $\delta P_{e\,zz,z}$. We can therefore ignore all electron motion perpendicular to the wave vector, as it is not coupled to the parallel motion. The electron motion is then effectively 1-dimensional, since there is only one (translational) degree of freedom. The relevant specific heat at constant volume is therefore just $k_B/2$ per electron, while that at constant pressure is $3k_B/2$, giving $\gamma = 3$.[4] The effective sound speed for the electron gas is then $C = (3k_B T_e/m_e)^{1/2}$, and correspondingly, the perturbations of longitudinal electron pressure and electron density are related by

properties of Langmuir waves:

$$\frac{\delta P_{e\,zz}}{m_e \delta n_e} = C^2 = \frac{3k_B T_e}{m_e}. \quad (21.30a)$$

perturbations of pressure, electron density, electron fluid velocity, and electric field

This is one of the equations governing Langmuir waves. The others are the linearized equation of continuity (21.13a), which relates the electrons' density perturbation to the longitudinal component of their fluid velocity perturbation:

$$\delta n_e = n_e \frac{k}{\omega} u_{e\,z}, \quad (21.30b)$$

---

4. We derived this longitudinal adiabatic index $\gamma = 3$ by a different method in Ex. 20.11e [Eq. (20.36b)] in the context of a plasma with a magnetic field along the longitudinal (or z) direction. It is valid also in our unmagnetized case, because the magnetic field has no influence on longitudinal electron motions.

the linearized equation of motion (21.13b), which relates $u_{e\,z}$ and $\delta P_{e\,zz}$ to the longitudinal component of the oscillating electric field:

$$-i\omega n_e m_e u_{e\,z} = -ik\delta P_{e\,zz} - n_e e E_z, \quad (21.30c)$$

and the linearized form of Poisson's equation $\nabla \cdot \mathbf{E} = \rho_e/\epsilon_0$, which relates E_z to δn_e:

$$ik E_z = -\frac{\delta n_e e}{\epsilon_0}. \quad (21.30d)$$

Equations (21.30) are four equations for three ratios of the perturbed quantities. By combining these equations, we obtain a condition that must be satisfied for them to have a solution:

Bohm-Gross dispersion relation

$$\boxed{\omega^2 = \omega_{pe}^2 + \frac{3k_B T_e}{m_e}k^2 = \omega_{pe}^2(1 + 3k^2\lambda_D^2).} \quad (21.31)$$

Here $\lambda_D = \sqrt{\epsilon_0 k_B T_e/(n_e e^2)}$ is the Debye length [Eq. (20.10)]. Equation (21.31) is the *Bohm-Gross* dispersion relation for Langmuir waves.

From this dispersion relation, we deduce the phase speed of a Langmuir wave:

phase speed

$$V_{\text{ph}} = \frac{\omega}{k} = \left(\frac{k_B T_e}{m_e}\right)^{1/2}\left(3 + \frac{1}{k^2\lambda_D^2}\right)^{1/2}. \quad (21.32)$$

Evidently, when the reduced wavelength $\lambda/(2\pi) = 1/k$ is less than or of order the Debye length ($k\lambda_D \gtrsim 1$), the phase speed becomes comparable with the electron thermal speed. It is then possible for individual electrons to transfer energy between adjacent compressions and rarefactions in the wave. As we shall see in Sec. 22.3, when we recover Eq. (21.31) from a kinetic treatment, the resulting energy transfers damp the wave; this is called *Landau damping*. Therefore, the Bohm-Gross dispersion relation is only valid for reduced wavelengths much longer than the Debye length (i.e., $k\lambda_D \ll 1$; Fig. 21.1).

Landau damping

In our analysis of Langmuir waves we have ignored the proton motion. This is justified as long as the proton thermal speeds are small compared to the electron thermal speeds (i.e., $T_p \ll m_p T_e/m_e$), which will almost always be the case. Proton motion is, however, not ignorable in a second type of plasma wave that owes its existence to finite temperature: *ion-acoustic waves* (also called ion-sound waves). These are waves that propagate with frequencies far below the electron plasma frequency—frequencies so low that the electrons remain locked electrostatically to the protons, keeping the plasma charge neutral and preventing electromagnetic fields from participating in the oscillations. As for Langmuir waves, we can derive the ion-acoustic dispersion relation using fluid theory combined with physical arguments.

properties of ion-acoustic waves:

low frequencies; electrons locked to protons

Using kinetic theory in the next chapter, we shall see that ion-acoustic waves can propagate only when the proton temperature is small compared with the electron temperature: $T_p \ll T_e$; otherwise they are damped. (Such a temperature disparity

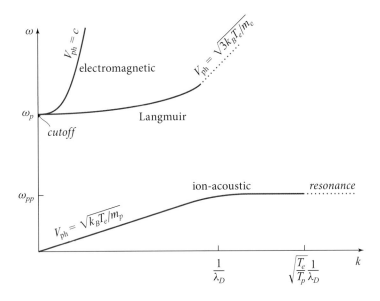

FIGURE 21.1 Dispersion relations for electromagnetic waves, Langmuir waves, and ion-acoustic waves in an unmagnetized plasma, whose electrons are thermalized with one another at temperature T_e, and whose protons are thermalized at a temperature T_p that might not be equal to T_e. In the dotted regions the waves are strongly damped, according to kinetic-theory analyses (Chap. 22). Ion-acoustic waves are wiped out by that damping at all k unless $T_p \ll T_e$ (as is assumed on the horizontal axis), in which case they survive on the non-dotted part of their curve.

can be produced, e.g., when a plasma passes through a shock wave, and it can be maintained for a long time because Coulomb collisions are so ineffective at restoring $T_p \sim T_e$; cf. Sec. 20.4.3.) Because $T_p \ll T_e$, the proton pressure can be ignored, and the waves' restoring force is provided by electron pressure. Now, in an ion-acoustic wave—by contrast with a Langmuir wave—the individual thermal electrons can travel over many wavelengths during a single wave period, so the electrons remain isothermal as their mean motion oscillates in lock-step with the protons' mean motion. Correspondingly, the electrons' effective (1-dimensional) specific-heat ratio is $\gamma_{\rm eff} = 1$.

Although the electrons provide the ion-acoustic waves' restoring force, the inertia of the electrostatically locked electrons and protons is almost entirely that of the heavy protons. Correspondingly, the waves' phase velocity is

require low proton temperature; restoring force is electron pressure and inertia is proton mass

$$\mathbf{V}_{ia} = \left(\frac{\gamma_{\rm eff} P_e}{n_p m_p}\right)^{1/2} \hat{k} = \left(\frac{k_B T_e}{m_p}\right)^{1/2} \hat{k} \qquad (21.33)$$

phase velocity

(cf. Ex. 21.5), and the dispersion relation is $\omega = V_{\rm ph} k = (k_B T_e/m_p)^{1/2} k$.

From this phase velocity and our physical description of these ion-acoustic waves, it should be evident that they are the magnetosonic waves of MHD theory (Sec. 19.7.2), in the limit that the plasma's magnetic field is turned off.

21.4 Wave Modes in an Unmagnetized Plasma

In Ex. 21.5, we show that the character of these waves is modified when their wavelength becomes of order the Debye length (i.e., when $k\lambda_D \sim 1$). The dispersion relation then becomes

dispersion relation

$$\omega = \left(\frac{k_B T_e/m_p}{1+\lambda_D^2 k^2}\right)^{1/2} k, \qquad (21.34)$$

which means that for $k\lambda_D \gg 1$, the waves' frequency approaches the *proton plasma frequency* $\omega_{pp} \equiv \sqrt{ne^2/(\epsilon_0 m_p)} \simeq \sqrt{m_e/m_p}\,\omega_p$. A kinetic-theory treatment reveals that these waves are strongly damped when $k\lambda_D \gtrsim \sqrt{T_e/T_p}$. These features of the ion-acoustic dispersion relation are shown in Fig. 21.1.

proton plasma frequency

The regime in which ion-acoustic (magnetosonic) waves can propagate, $\omega < \omega_{pp}$, is quite generally the regime in which the electrons and ions are locked to each other, making magnetohydrodynamics a good appoximation to the plasma's dynamics.

MHD regime

EXERCISES

Exercise 21.5 *Derivation: Ion-Acoustic Waves*

Ion-acoustic waves can propagate in an unmagnetized plasma when the electron temperature T_e greatly exceeds the ion temperature T_p. In this limit, the electron density n_e can be approximated by $n_e = n_0 \exp[e\Phi/(k_B T_e)]$, where n_0 is the mean electron density, and Φ is the electrostatic potential.

(a) Show that for ion-acoustic waves that propagate in the z direction, the nonlinear equations of continuity, the motion for the ion (proton) fluid, and Poisson's equation for the potential take the form

$$\frac{\partial n}{\partial t} + \frac{\partial (nu)}{\partial z} = 0,$$

$$\frac{\partial u}{\partial t} + u\frac{\partial u}{\partial z} = -\frac{e}{m_p}\frac{\partial \Phi}{\partial z},$$

$$\frac{\partial^2 \Phi}{\partial z^2} = -\frac{e}{\epsilon_0}\left(n - n_0 e^{e\Phi/(k_B T_e)}\right). \qquad (21.35)$$

Here n is the proton density, and u is the proton fluid velocity (which points in the z direction).

(b) Linearize Eqs. (21.35), and show that the dispersion relation for small-amplitude ion-acoustic modes is

$$\omega = \omega_{pp}\left(1 + \frac{1}{\lambda_D^2 k^2}\right)^{-1/2} = \left(\frac{k_B T_e/m_p}{1+\lambda_D^2 k^2}\right)^{1/2} k, \qquad (21.36)$$

where λ_D is the Debye length.

Exercise 21.6 *Challenge: Ion-Acoustic Solitons*

In this exercise we explore nonlinear effects in ion-acoustic waves (Ex. 21.5), and show that they give rise to solitons that obey the same KdV equation as governs solitonic water waves (Sec. 16.3). This version of the solitons is only mildly nonlinear. In Sec. 23.6 we will generalize to strong nonlinearity.

(a) Introduce a bookkeeping expansion parameter ε whose numerical value is unity,[5] and expand the ion density, ion velocity, and potential in the forms

$$n = n_0(1 + \varepsilon n_1 + \varepsilon^2 n_2 + \ldots),$$
$$u = (k_B T_e/m_p)^{1/2}(\varepsilon u_1 + \varepsilon^2 u_2 + \ldots),$$
$$\Phi = (k_B T_e/e)(\varepsilon \Phi_1 + \varepsilon^2 \Phi_2 + \ldots). \quad (21.37)$$

Here n_1, u_1, and Φ_1 are small compared to unity, and the factors of ε tell us that, as the wave amplitude is decreased, these quantities scale proportionally to one another, while n_2, u_2, and Φ_2 scale proportionally to the squares of n_1, u_1, and Φ_1, respectively. Change independent variables from (t, z) to (τ, η), where

$$\eta = \sqrt{2}\varepsilon^{1/2}\lambda_D^{-1}[z - (k_B T_e/m_p)^{1/2} t],$$
$$\tau = \sqrt{2}\varepsilon^{3/2}\omega_{pp} t. \quad (21.38)$$

Explain, now or at the end, the chosen powers $\varepsilon^{1/2}$ and $\varepsilon^{3/2}$. By substituting Eqs. (21.37) and (21.38) into the nonlinear equations (21.35), equating terms of the same order in ε, and then setting $\varepsilon = 1$ (bookkeeping parameter!), show that n_1, u_1, and Φ_1 *each* satisfy the KdV equation (16.32):

$$\frac{\partial \zeta}{\partial \tau} + \zeta \frac{\partial \zeta}{\partial \eta} + \frac{\partial^3 \zeta}{\partial \eta^3} = 0. \quad (21.39)$$

(b) In Sec. 16.3 we discussed the exact, single-soliton solution (16.33) to this KdV equation. Show that for an ion-acoustic soliton, this solution propagates with the physical speed $(1 + n_{1o})(k_B T_e/m_p)^{1/2}$ (where n_{1o} is the value of n_1 at the peak of the soliton), which is greater the larger is the wave's amplitude.

21.4.4 Cutoffs and Resonances

Electromagnetic waves, Langmuir waves, and ion-acoustic waves in an unmagnetized plasma provide examples of *cutoffs* and *resonances*.

A *cutoff* is a frequency at which a wave mode ceases to propagate because its wave number k there becomes zero. Langmuir and electromagnetic waves at $\omega \to \omega_p$ are examples; see their dispersion relations in Fig. 21.1. For concreteness, consider a monochromatic radio-frequency electromagnetic wave propagating upward into Earth's ionosphere at some nonzero angle to the vertical (left side of Fig. 21.2), and neglect the effects of Earth's magnetic field. As the wave moves deeper (higher) into the ionosphere, it encounters a rising electron density n and correspondingly, a rising plasma frequency ω_p. The wave's wavelength will typically be small compared to the

cutoff, where k goes to zero

---

5. See Box 7.2 for a discussion of such bookkeeping parameters in a different context.

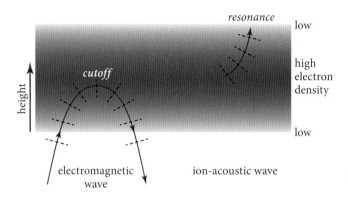

FIGURE 21.2 Cutoff and resonance illustrated by wave propagation in Earth's ionosphere. The thick, arrowed curves are rays, and the thin, dashed lines are phase fronts. The electron density is proportional to the darkness of the shading.

inhomogeneity scale for ω_p, so the wave propagation can be analyzed using geometric optics (Sec. 7.3). Across a phase front, the portion of the wave that is higher in the ionosphere will have a smaller k and thus a larger wavelength and phase speed; it thus has a greater distance between phase fronts (dashed lines). Therefore, the rays, which are orthogonal to the phase fronts, will bend away from the vertical (left side of Fig. 21.2); that is, the wave will be reflected away from the cutoff, at which $\omega_p \to \omega$ and $k \to 0$. Clearly, this behavior is quite general. Wave modes are generally reflected from regions in which slowly changing plasma conditions give rise to cutoffs.

waves are reflected by cutoff regions

A *resonance* is a frequency at which a wave mode ceases to propagate because its wave number k there becomes infinite (i.e., its wavelength goes to zero). Ion-acoustic waves provide an example; see their dispersion relation in Fig. 21.1. For concreteness, consider an ion-acoustic wave deep in the ionosphere, where ω_{pp} is larger than the wave's frequency ω (right side of Fig. 21.2). As the wave propagates toward the upper edge of the ionosphere, at some nonzero angle to the vertical, the portion of a phase front that is higher sees a smaller electron density and thus a smaller ω_{pp}; thence it has a larger k and shorter wavelength and thus a shorter distance between phase fronts (dashed lines). This causes the rays to bend toward the vertical (right side of Fig. 21.2). The wave is "attracted" to the region of the resonance, $\omega \to \omega_{pp}$, $k \to \infty$, where it is "Landau damped" (Chap. 22) and dies. This behavior is quite general. Wave modes are generally attracted to regions in which slowly changing plasma conditions give rise to resonances, and on reaching a resonance, they die.

resonance, where k becomes infinite

waves are attracted to resonance regions, and damped

We study wave propagation in the ionosphere in greater detail in Sec. 21.5.4.

21.5 Wave Modes in a Cold, Magnetized Plasma

21.5.1 Dielectric Tensor and Dispersion Relation

We now complicate matters somewhat by introducing a uniform magnetic field $\mathbf{B}_0$ into the unperturbed plasma. To avoid additional complications, we make the plasma cold (i.e., we omit thermal effects). The linearized equation of motion (21.13b) for each species s then becomes

$$-i\omega \mathbf{u}_s = \frac{q_s \mathbf{E}}{m_s} + \frac{q_s}{m_s} \mathbf{u}_s \times \mathbf{B}_0. \quad (21.40)$$

It is convenient to multiply this equation of motion by $n_s q_s / \epsilon_0$ and introduce for each species a scalar plasma frequency and scalar and vectorial cyclotron frequencies:

$$\boxed{\omega_{ps} = \left(\frac{n_s q_s^2}{\epsilon_0 m_s}\right)^{1/2}, \quad \omega_{cs} = \frac{q_s B_0}{m_s}, \quad \boldsymbol{\omega}_{cs} = \omega_{cs} \hat{\mathbf{B}}_0 = \frac{q_s \mathbf{B}_0}{m_s}} \quad (21.41)$$

electron and proton plasma frequencies, and cyclotron frequencies

[so $\omega_{pp} = \sqrt{(m_e/m_p)}\,\omega_{pe} \simeq \omega_{pe}/43$, $\omega_p = \sqrt{\omega_{pe}^2 + \omega_{pp}^2}$, $\omega_{ce} < 0$, $\omega_{cp} > 0$, and $\omega_{cp} = (m_e/m_p)|\omega_{ce}| \simeq |\omega_{ce}|/1{,}836$]. Thereby we bring the equation of motion (21.40) into the form

$$-i\omega\left(\frac{n_s q_s}{\epsilon_0}\mathbf{u}_s\right) + \boldsymbol{\omega}_{cs} \times \left(\frac{n_s q_s}{\epsilon_0}\mathbf{u}_s\right) = \omega_{ps}^2 \mathbf{E}. \quad (21.42)$$

By combining this equation with $\boldsymbol{\omega}_{cs} \times$ (this equation) and $\boldsymbol{\omega}_{cs} \cdot$ (this equation), we can solve for the fluid velocity of species s as a linear function of the electric field $\mathbf{E}$:

properties of wave modes in a cold, magnetized plasma:

$$\boxed{\frac{n_s q_s}{\epsilon_0}\mathbf{u}_s = -i\left(\frac{\omega \omega_{ps}^2}{\omega_{cs}^2 - \omega^2}\right)\mathbf{E} - \frac{\omega_{ps}^2}{(\omega_{cs}^2 - \omega^2)}\boldsymbol{\omega}_{cs} \times \mathbf{E} + \boldsymbol{\omega}_{cs}\frac{i\omega_{ps}^2}{(\omega_{cs}^2 - \omega^2)\omega}\boldsymbol{\omega}_{cs}\cdot \mathbf{E}.}$$

fluid velocities

$$(21.43)$$

(This relation is useful for deducing the physical properties of wave modes.) From this fluid velocity we can read off the current $\mathbf{j} = \sum_s n_s q_s \mathbf{u}_s$ as a linear function of $\mathbf{E}$; by comparing with Ohm's law, $\mathbf{j} = \boldsymbol{\kappa}_e \cdot \mathbf{E}$, we then obtain the tensorial conductivity $\boldsymbol{\kappa}_e$, which we insert into Eq. (21.20) to get the following expression for the dielectric tensor (in which $\mathbf{B}_0$ and hence $\boldsymbol{\omega}_{cs}$ are taken to be along the z-axis):

$$\boldsymbol{\epsilon} = \begin{pmatrix} \epsilon_1 & -i\epsilon_2 & 0 \\ i\epsilon_2 & \epsilon_1 & 0 \\ 0 & 0 & \epsilon_3 \end{pmatrix}, \quad (21.44)$$

dielectric tensor

where

$$\boxed{\epsilon_1 = 1 - \sum_s \frac{\omega_{ps}^2}{\omega^2 - \omega_{cs}^2}, \quad \epsilon_2 = \sum_s \frac{\omega_{ps}^2 \omega_{cs}}{\omega(\omega^2 - \omega_{cs}^2)}, \quad \epsilon_3 = 1 - \sum_s \frac{\omega_{ps}^2}{\omega^2}.}$$

$$(21.45)$$

Let the wave propagate in the x-z plane at an angle θ to the z-axis (i.e., to the magnetic field). Then the algebratized wave operator (21.9) takes the form

$$||L_{ij}|| = \frac{\omega^2}{c^2} \begin{pmatrix} \epsilon_1 - \mathfrak{n}^2\cos^2\theta & -i\epsilon_2 & \mathfrak{n}^2\sin\theta\cos\theta \\ i\epsilon_2 & \epsilon_1 - \mathfrak{n}^2 & 0 \\ \mathfrak{n}^2\sin\theta\cos\theta & 0 & \epsilon_3 - \mathfrak{n}^2\sin^2\theta \end{pmatrix}, \quad (21.46)$$

where

index of refraction

$$\boxed{\mathfrak{n} = \frac{ck}{\omega}} \quad (21.47)$$

is the wave's index of refraction (i.e., the wave's phase velocity is $V_{\rm ph} = \omega/k = c/\mathfrak{n}$). (Note: $\mathfrak{n}$ must not be confused with the number density of particles n.) The algebratized wave operator (21.46) will be needed when we explore the physical nature of modes, in particular, the directions of their electric fields, which satisfy $L_{ij}E_j = 0$.

From the wave operator (21.46), we deduce the waves' dispersion relation $\det ||L_{ij}|| = 0$. Some algebra brings this into the form

dispersion relation

$$\boxed{\tan^2\theta = \frac{-\epsilon_3(\mathfrak{n}^2 - \epsilon_R)(\mathfrak{n}^2 - \epsilon_L)}{\epsilon_1(\mathfrak{n}^2 - \epsilon_3)\left(\mathfrak{n}^2 - \epsilon_R\epsilon_L/\epsilon_1\right)},} \quad (21.48)$$

where

$$\boxed{\epsilon_L = \epsilon_1 - \epsilon_2 = 1 - \sum_s \frac{\omega_{ps}^2}{\omega(\omega - \omega_{cs})}, \quad \epsilon_R = \epsilon_1 + \epsilon_2 = 1 - \sum_s \frac{\omega_{ps}^2}{\omega(\omega + \omega_{cs})}.}$$

(21.49)

21.5.2 Parallel Propagation

waves propagating parallel to the magnetic field

As a first step in making sense out of the general dispersion relation (21.48) for waves in a cold, magnetized plasma, let us consider wave propagation along the magnetic field, so $\theta = 0$. The dispersion relation (21.48) then factorizes to give three solutions:

dispersion relations for three modes: left, right, and plasma oscillations; Fig. 21.3

$$\mathfrak{n}^2 \equiv \frac{c^2 k^2}{\omega^2} = \epsilon_L, \quad \mathfrak{n}^2 \equiv \frac{c^2 k^2}{\omega^2} = \epsilon_R, \quad \epsilon_3 = 0. \quad (21.50)$$

LEFT AND RIGHT MODES; PLASMA OSCILLATIONS

Consider the first solution in Eq. (21.50): $\mathfrak{n}^2 = \epsilon_L$. The algebratized wave equation $L_{ij}E_j = 0$ [with L_{ij} given by Eq. (21.46)] in this case with the wave propagating along the $\mathbf{B}$ ($\mathbf{e}_z$) direction, requires that the electric field direction be $\mathbf{E} \propto (\mathbf{e}_x - i\mathbf{e}_y)e^{-i\omega t}$, which we define to be a left-hand circular polarized mode. The second solution in (21.50), $\mathfrak{n}^2 = \epsilon_R$, is the corresponding right-hand circular polarized mode. From Eqs. (21.49) we see that these two modes propagate with different phase velocities (but only slightly different, if ω is far from the electron cyclotron frequency and far

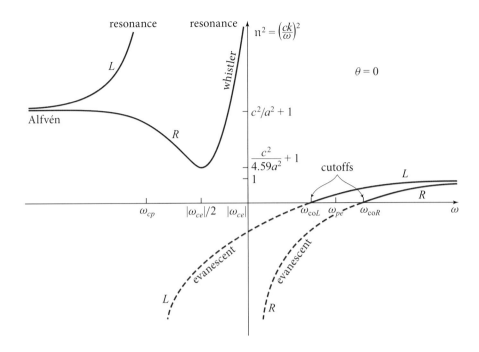

FIGURE 21.3 Square of wave refractive index for circularly polarized waves propagating along the static magnetic field in a proton-electron plasma with $\omega_{pe} > \omega_{ce}$. The horizontal, angular-frequency scale is logarithmic. L means left-hand circularly polarized ("left mode"), and R, right-hand circularly polarized ("right mode").

from the proton cyclotron frequency). The third solution in Eq. (21.50), $\epsilon_3 = 0$, is just the electrostatic plasma oscillation in which the electrons and protons oscillate parallel to and are unaffected by the static magnetic field.

As an aid to exploring the frequency dependence of the left and right modes, Fig. 21.3 shows the refractive index $\mathfrak{n} = ck/\omega$ squared as a function of ω.

HIGH-FREQUENCIES: FARADAY ROTATION

In the high-frequency limit, the refractive index for both modes is slightly less than unity and approaches the index for an unmagnetized plasma, $\mathfrak{n} = ck/\omega \simeq 1 - \frac{1}{2}\omega_p^2/\omega^2$ [cf. Eq. (21.27a)], but with a small difference between the modes given, to leading order, by

$$\mathfrak{n}_L - \mathfrak{n}_R \simeq -\frac{\omega_{pe}^2 \omega_{ce}}{\omega^3}. \tag{21.51}$$

This difference is responsible for an important effect known as *Faraday rotation*.

Suppose that a linearly polarized wave is incident on a magnetized plasma and propagates parallel to the magnetic field. We can deduce the behavior of the polarization by expanding the mode as a linear superposition of the two circularly polarized eigenmodes, left and right. These two modes propagate with slightly different phase

Faraday rotation at high frequency: difference between left and right modes

velocities, so after propagating some distance through the plasma, they acquire a relative phase shift $\Delta\phi$. When one then reconstitutes the linear polarized mode from the circular eigenmodes, this phase shift is manifest in a rotation of the plane of polarization through an angle $\Delta\phi/2$ (for small $\Delta\phi$). This—together with the difference in refractive indices [Eq. (21.51)], which determines $\Delta\phi$—implies a Faraday rotation rate for the plane of polarization given by (Ex. 21.7)

rate of rotation

$$\frac{d\chi}{dz} = -\frac{\omega_{pe}^2 \omega_{ce}}{2\omega^2 c}. \tag{21.52}$$

INTERMEDIATE FREQUENCIES: CUTOFFS

As the wave frequency is reduced, the refractive index decreases to zero, first for the right circular wave, then for the left one (Fig. 21.3). Vanishing of n at a finite frequency corresponds to vanishing of k and infinite wavelength, that is, it signals a cutoff (Fig. 21.2 and associated discussion). When the frequency is lowered further, the squared refractive index becomes negative, and the wave mode becomes evanescent. Correspondingly, when a circularly polarized electromagnetic wave with constant real frequency propagates into an inhomogeneous plasma parallel to its density gradient and parallel to a magnetic field, then beyond the spatial location of the wave's cutoff, its wave number k becomes purely imaginary, and the wave dies out with distance (gradually at first, then more rapidly).

The cutoff frequencies are different for the two modes and are given by

cutoff frequencies

$$\omega_{\text{co}R,L} = \frac{1}{2}\left[\left\{(\omega_{ce} + \omega_{cp})^2 + 4(\omega_{pe}^2 + \omega_{pp}^2)\right\}^{1/2} \pm (|\omega_{ce}| - \omega_{cp})\right]$$

$$\simeq \omega_{pe} \pm \frac{1}{2}|\omega_{ce}| \quad \text{if } \omega_{pe} \gg |\omega_{ce}|, \quad \text{as is often the case.} \tag{21.53}$$

LOW FREQUENCIES: RESONANCES; WHISTLER MODES

As we lower the frequency further (Fig. 21.3), first the right mode and then the left regain the ability to propagate. When the wave frequency lies between the proton and electron gyro frequencies, $\omega_{cp} < \omega < |\omega_{ce}|$, only the right mode propagates.

whistler mode at intermediate frequencies

This mode is sometimes called a *whistler*. As the mode's frequency increases toward the electron gyro frequency $|\omega_{ce}|$ (where it first recovered the ability to propagate), its refractive index and wave vector become infinite—a signal that $\omega = |\omega_{ce}|$ is a resonance for the whistler (Fig. 21.2 and associated discussion). The physical origin

resonance with electron gyrations

of this resonance is that the wave frequency becomes resonant with the gyrational frequency of the electrons that are orbiting the magnetic field in the same sense as the wave's electric vector rotates. To quantify the strong wave absorption that occurs at this resonance, one must carry out a kinetic-theory analysis that takes account of the electrons' thermal motions (Chap. 22).

large dispersion near resonance

Another feature of the whistler is that it is highly dispersive close to resonance; its dispersion relation there is given approximately by

$$\omega \simeq \frac{|\omega_{ce}|}{1 + \omega_{pe}^2/(c^2 k^2)}. \tag{21.54}$$

The group velocity, obtained by differentiating Eq. (21.54), is given approximately by

$$\mathbf{V}_g = \nabla_\mathbf{k}\, \omega \simeq \frac{2\omega_{ce} c}{\omega_{pe}} \left(1 - \frac{\omega}{|\omega_{ce}|}\right)^{3/2} \hat{\mathbf{B}}_0. \quad (21.55)$$

This velocity varies extremely rapidly close to resonance, so waves of slightly different frequency propagate at very different speeds.

This is the physical origin of the phenomenon by which whistlers were discovered. They were encountered in World War One by radio operators who heard, in their earphones, strange tones with rapidly changing pitch. These turned out to be whistler modes excited by lightning in the southern hemisphere; they propagated along Earth's magnetic field through the magnetosphere to the northern hemisphere. Only modes below the lowest electron gyro frequency on the waves' path (their geometric-optics ray) could propagate, and these were highly dispersed, with the lower frequencies arriving first.

There is also a resonance associated with the left circularly polarized wave, which propagates below the proton cyclotron frequency; see Fig. 21.3.

VERY LOW FREQUENCIES: ALFVÉN MODES

Finally, let us examine the very low-frequency limit of these waves (Fig. 21.3). We find that both dispersion relations $\mathfrak{n}^2 = \epsilon_L$ and $\mathfrak{n}^2 = \epsilon_R$ asymptote, at arbitrarily low frequencies, to

$$\omega = ak \left(1 + \frac{a^2}{c^2}\right)^{-1/2}. \quad (21.56)$$

Here $a = B_0[\mu_0 n_e(m_p + m_e)]^{-1/2}$ is the Alfvén speed that arose in our discussion of magnetohydrodynamics [Eq. (19.75)]. In fact at very low frequencies, both modes, left and right, have become the Alfvén waves that we studied using MHD in Sec. 19.7.2. However, our two-fluid formalism reports a phase speed $\omega/k = a/\sqrt{1+a^2/c^2}$ for these Alfvén waves that is slightly lower than the speed $\omega/k = a$ predicted by our MHD formalism. The $1/\sqrt{1+a^2/c^2}$ correction could not be deduced using non-relativistic MHD, because that formalism neglects the displacement current. (Relativistic MHD includes the displacement current and predicts precisely this correction factor.)

Alfvén modes at very low frequencies

relativistic correction to phase speed

We can understand the physical origin of this correction by examining the particles' motions in a very-low-frequency Alfvén wave; see Fig. 21.4. Because the wave frequency is far below both the electron and the proton cyclotron frequencies, both types of particle orbit the field $\mathbf{B}_0$ many times in a wave period. When the wave's slowly changing electric field is applied, the guiding centers of both types of orbits acquire the same slowly changing drift velocity $\mathbf{v} = \mathbf{E} \times \mathbf{B}_0/B_0^2$, so the two fluid velocities also drift at this rate, and the currents associated with the proton and electron drifts cancel. However, when we consider corrections to the guiding-center response that are of higher order in ω/ω_{cp} and ω/ω_{ce}, we find that the ions drift slightly faster than the electrons, which produces a net current that modifies the magnetic field and gives rise to the $1/\sqrt{1+a^2/c^2}$ correction to the Alfvén wave's phase speed.

interpretation as drift motion

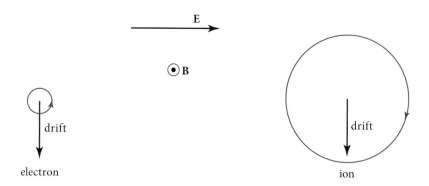

FIGURE 21.4 Gyration of electrons and ions in a very-low-frequency Alfvén wave. Although the electrons and ions gyrate with opposite senses about the magnetic field, their $\mathbf{E} \times \mathbf{B}$ drifts are similar. It is only in the next highest order of approximation that a net ion current is produced parallel to the applied electric field. Note that, if the electrons and ions have the same temperature, then the ratio of radii of the orbits is $\sim (m_e/m_i)^{1/2}$, or ~ 0.02 for protons.

interpretation as magnetic contribution to inertia

A second way to understand this correction is by the contribution of the magnetic field to the plasma's inertial mass per unit volume (Ex. 21.8).

EXERCISES

Exercise 21.7 *Derivation: Faraday Rotation*
Derive Eq. (21.52) for Faraday rotation.

Exercise 21.8 *Example: Alfvén Waves as Plasma-Laden, Plucked Magnetic Field Lines*
A narrow bundle of magnetic field lines with cross sectional area A, together with the plasma attached to them, can be thought of as like a stretched string. When such a string is plucked, waves travel down it with phase speed $\sqrt{T/\Lambda}$, where T is the string's tension, and Λ is its mass per unit length (Sec. 12.3.3). The plasma analog is Alfvén waves propagating parallel to the plasma-laden magnetic field.

(a) Analyzed nonrelativistically, the tension for our bundle of field lines is $T = [B^2/(2\mu_0)]A$ and the mass per unit length is $\Lambda = \rho A$, so we expect a phase velocity $\sqrt{T/\Lambda} = \sqrt{B^2/(2\mu_0 \rho)}$, which is $1/\sqrt{2}$ of the correct result. Where is the error? [Hint: In addition to the restoring force on bent field lines, due to tension along the field, there is also the curvature force $(\mathbf{B} \cdot \nabla)\mathbf{B}/\mu_0$; Eq. (19.15).]

(b) In special relativity, the plasma-laden magnetic field has a tensorial inertial mass per unit volume that is discussed in Ex. 2.27. Explain why, when the field lines (which point in the z direction) are plucked so they vibrate in the x direction, the inertial mass per unit length that resists this motion is $\Lambda = (T^{00} + T^{xx})A = [\rho + B^2/(\mu_0 c^2)]A$. (In the first expression for Λ, T^{00} is the mass-energy density of plasma and magnetic field, T^{xx} is the magnetic pressure along the x direction, and the speed of light is set to unity as in Chap. 2; in the second expression, the speed of light has been restored to the equation using dimensional arguments.) Show that the magnetic contribution to this inertial mass gives the relativistic correction $1/\sqrt{1+a^2/c^2}$ to the Alfvén waves' phase speed, Eq. (21.56).

21.5.3 Perpendicular Propagation

21.5.3 waves propagating perpendicular to the magnetic field

Turn next to waves that propagate perpendicular to the static magnetic field: $\mathbf{k} = k\mathbf{e}_x$; $\mathbf{B}_0 = B_0 \mathbf{e}_z$; $\theta = \pi/2$. In this case our general dispersion relation (21.48) again has three solutions corresponding to three modes:

$$\mathfrak{n}^2 \equiv \frac{c^2 k^2}{\omega^2} = \epsilon_3, \qquad \mathfrak{n}^2 \equiv \frac{c^2 k^2}{\omega^2} = \frac{\epsilon_R \epsilon_L}{\epsilon_1}, \qquad \epsilon_1 = 0. \qquad (21.57)$$

The first solution

$$\mathfrak{n}^2 = \epsilon_3 = 1 - \frac{\omega_p^2}{\omega^2} \qquad (21.58)$$

ordinary mode—E and v parallel to $\mathbf{B}_0$

has the same index of refraction as for an electromagnetic wave in an unmagnetized plasma [cf. Eq. (21.24)], so this is called the *ordinary mode*. In this mode the electric vector and velocity perturbation are parallel to the static magnetic field, so the field has no influence on the wave. The wave is identical to an electromagnetic wave in an unmagnetized plasma.

The second solution in Eq. (21.57),

$$\mathfrak{n}^2 = \epsilon_R \epsilon_L / \epsilon_1 = \frac{\epsilon_1^2 - \epsilon_2^2}{\epsilon_1}, \qquad (21.59)$$

extraordinary mode—E orthogonal to $\mathbf{B}_0$

is known as the *extraordinary mode* and has an electric field that is orthogonal to $\mathbf{B}_0$ but not to $\mathbf{k}$. (Note that the names "ordinary" and "extraordinary" are used differently here than for waves in a nonlinear crystal in Sec. 10.6.)

The refractive indices for the ordinary and extraordinary modes are plotted as functions of frequency in Fig. 21.5. The ordinary-mode curve is dull; it is just like that in an unmagnetized plasma. The extraordinary-mode curve is more interesting. It has two cutoffs, with frequencies (ignoring the ion motions)

cutoffs and resonances

$$\omega_{co1,2} \simeq \left(\omega_{pe}^2 + \frac{1}{4}\omega_{ce}^2\right)^{1/2} \pm \frac{1}{2}\omega_{ce}, \qquad (21.60)$$

and two resonances with strong absorption, at frequencies known as the *upper hybrid* (UH) and *lower hybrid* (LH) frequencies. These frequencies are given approximately by

$$\omega_{\text{UH}} \simeq (\omega_{pe}^2 + \omega_{ce}^2)^{1/2},$$

$$\omega_{\text{LH}} \simeq \left[\frac{(\omega_{pe}^2 + |\omega_{ce}|\omega_{cp})|\omega_{ce}|\omega_{cp}}{\omega_{pe}^2 + \omega_{ce}^2}\right]^{1/2}. \qquad (21.61)$$

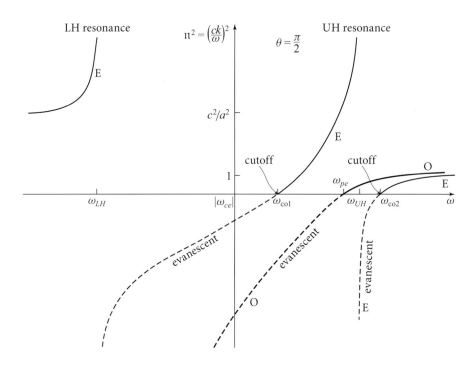

FIGURE 21.5 Square of wave refractive index n as a function of frequency ω, for wave propagation perpendicular to the magnetic field in an electron ion plasma with $\omega_{pe} > \omega_{ce}$. The ordinary mode is designated by O, the extraordinary mode by E.

very low frequencies: fast magnetosonic mode and Alfvén mode

In the limit of very low frequency, the extraordinary, perpendicularly propagating mode has the same dispersion relation $\omega = ak/\sqrt{1 + a^2/c^2}$ as the parallel propagating modes [Eq. (21.56)]. In this regime the mode has become the fast magnetosonic wave, propagating perpendicularly to the static magnetic field (Sec. 19.7.2), while the parallel waves have become the Alfvén modes.

The third solution in Eq. (21.57), $\epsilon_1 = 0$, has vanishing eigenvector **E** (i.e., $L_{ij} E_j = 0$ for $\epsilon_1 = 0$ implies $E_j = 0$), so it does not represent a physical oscillation of any sort.

21.5.4 Propagation of Radio Waves in the Ionosphere; Magnetoionic Theory

magnetoionic theory for radio waves in ionosphere

The discovery that radio waves could be reflected off the ionosphere, and thereby could be transmitted over long distances,[6] revolutionized communications and stimulated intensive research on radio-wave propagation in a magnetized plasma. In this section, we discuss radio-wave propagation in the ionosphere for waves whose prop-

---

6. This effect was predicted independently by Heaviside and Kennelly in 1902. Appleton demonstrated its existence in 1924, for which he received the Nobel prize. It was claimed by some physicists that the presence of a wave speed > c violated special relativity until it was appreciated that it is the group velocity that is relevant.

agation vectors make arbitrary angles θ to the magnetic field. The approximate formalism we develop is sometimes called *magneto-ionic theory*.

The ionosphere is a dense layer of partially ionized gas between 50 and 300 km above the surface of Earth. The ionization is due to incident solar ultraviolet radiation. Although the ionization fraction increases with height, the actual density of free electrons passes through a maximum whose height rises and falls with the position of the Sun.

properties of ionosphere

The electron gyro frequency varies from ~ 0.5 to ~ 1 MHz in the ionosphere, and the plasma frequency increases from effectively zero to a maximum that can be as high as 100 MHz; so typically, but not everywhere, $\omega_{pe} \gg |\omega_{ce}|$. We are interested in wave propagation at frequencies above the electron plasma frequency, which in turn is well in excess of the ion plasma frequency and the ion gyro frequency. It is therefore a good approximation to ignore ion motions altogether. In addition, at the altitudes of greatest interest for radio-wave propagation, the temperature is low, $T_e \sim 200\text{–}600$ K, so the cold plasma approximation is well justified. A complication that one must sometimes face in the ionosphere is the influence of collisions (see Ex. 21.4), but in this section we ignore it.

It is conventional in magneto-ionic theory to introduce two dimensionless parameters:

$$X = \frac{\omega_{pe}^2}{\omega^2}, \quad Y = \frac{|\omega_{ce}|}{\omega}, \tag{21.62}$$

in terms of which (ignoring ion motions) the components (21.45) of the dielectric tensor (21.44) are

$$\epsilon_1 = 1 + \frac{X}{Y^2 - 1}, \quad \epsilon_2 = \frac{XY}{Y^2 - 1}, \quad \epsilon_3 = 1 - X. \tag{21.63}$$

In this case it is convenient to rewrite the dispersion relation $\det ||L_{ij}|| = 0$ in a form different from Eq. (21.48)—a form derivable, for example, by computing explicitly the determinant of the matrix (21.46), setting

$$x = \frac{X - 1 + \mathfrak{n}^2}{1 - \mathfrak{n}^2}, \tag{21.64}$$

solving the resulting quadratic in x, and then solving for $\mathfrak{n}^2$. The result is the *Appleton-Hartree* dispersion relation:

$$\boxed{\mathfrak{n}^2 = 1 - \frac{X}{1 - \frac{Y^2 \sin^2 \theta}{2(1-X)} \pm \left[\frac{Y^4 \sin^4 \theta}{4(1-X)^2} + Y^2 \cos^2 \theta\right]^{1/2}}.} \tag{21.65}$$

Appleton-Hartree dispersion relation

There are two commonly used approximations to this dispersion relation. The first is the *quasi-longitudinal* approximation, which is used when $\mathbf{k}$ is approximately parallel to the static magnetic field (i.e., when θ is small). In this case, just retaining

quasi-longitudinal approximation

21.5 Wave Modes in a Cold, Magnetized Plasma

the dominant terms in the dispersion relation, we obtain

$$\mathfrak{n}^2 \simeq 1 - \frac{X}{1 \pm Y \cos \theta}. \tag{21.66}$$

This is just the dispersion relation (21.50) for the left and right modes in strictly parallel propagation, with the substitution $B_0 \to B_0 \cos \theta$. By comparing the magnitude of the terms that we dropped from the full dispersion relation in deriving Eq. (21.66) with those that we retained, one can show that the quasi-longitudinal approximation is valid when

$$Y \sin^2 \theta \ll 2(1 - X) \cos \theta. \tag{21.67}$$

quasi-transverse approximation

The second approximation is the *quasi-transverse* approximation; it is appropriate when inequality (21.67) is reversed. In this case the two modes are generalizations of the precisely perpendicular ordinary and extraordinary modes, and their approximate dispersion relations are

$$\mathfrak{n}_O^2 \simeq 1 - X,$$
$$\mathfrak{n}_X^2 \simeq 1 - \frac{X(1 - X)}{1 - X - Y^2 \sin^2 \theta}. \tag{21.68}$$

The ordinary-mode dispersion relation (subscript O) is unchanged from the strictly perpendicular one, Eq. (21.58); the extraordinary dispersion relation (subscript X) is obtained from the strictly perpendicular one [Eq. (21.59)] by the substitution $B_0 \to B_0 \sin \theta$.

The quasi-longitudinal and quasi-transverse approximations simplify the problem of tracing rays through the ionosphere.

ionospheric reflection of AM and SW radio waves

Commercial radio stations operate in the AM (amplitude-modulated) band (0.5–1.6 MHz), the SW (short-wave) band (2.3–18 MHz), and the FM (frequency-modulated) band (88–108 MHz). Waves in the first two bands are reflected by the ionosphere and can therefore be transmitted over large surface areas (Ex. 21.12). FM waves, with their higher frequencies, are not reflected and must therefore be received as "ground waves" (waves that propagate directly, near the ground). However, they have the advantage of a larger bandwidth and consequently a higher fidelity audio output. As the altitude of the reflecting layer rises at night, short-wave communication over long distances becomes easier.

EXERCISES

Exercise 21.9 *Derivation: Appleton-Hartree Dispersion Relation*
Derive Eq. (21.65).

Exercise 21.10 *Example: Dispersion and Faraday Rotation of Pulsar Pulses*
A radio pulsar emits regular pulses at 1-s intervals, which propagate to Earth through the ionized interstellar plasma with electron density $n_e \simeq 3 \times 10^4$ m^{-3}. The pulses ob-

served at $f = 100$ MHz are believed to be emitted at the same time as those observed at much higher frequency, but they arrive with a delay of 100 ms.

(a) Explain briefly why pulses travel at the group velocity instead of the phase velocity, and show that the expected time delay of the $f = 100$-MHz pulses relative to the high-frequency pulses is given by

$$\Delta t = \frac{e^2}{8\pi^2 m_e \epsilon_0 f^2 c} \int n_e dx, \qquad (21.69)$$

where the integral is along the waves' propagation path. Hence compute the distance to the pulsar.

(b) Now suppose that the pulses are linearly polarized and that their propagation is accurately described by the quasi-longitudinal approximation. Show that the plane of polarization will be Faraday rotated through an angle

$$\Delta \chi = \frac{e \Delta t}{m_e} \langle B_\| \rangle, \qquad (21.70)$$

where $\langle B_\| \rangle = \int n_e \mathbf{B} \cdot d\mathbf{x} / \int n_e dx$. The plane of polarization of the pulses emitted at 100 MHz is believed to be the same as the emission plane for higher frequencies, but when the pulses arrive at Earth, the 100-MHz polarization plane is observed to be rotated through 3 radians relative to that at high frequencies. Calculate the mean parallel component of the interstellar magnetic field.

Exercise 21.11 *Challenge: Faraday Rotation Paradox*
Consider a wave mode propagating through a plasma—for example, the ionosphere—in which the direction of the background magnetic field is slowly changing. We have just demonstrated that so long as $\mathbf{B}$ is not almost perpendicular to $\mathbf{k}$, we can use the quasi-longitudinal approximation, the difference in phase velocity between the two eigenmodes is $\propto \mathbf{B} \cdot \mathbf{k}$, and the integral for the magnitude of the rotation of the plane of polarization is $\propto \int n_e \mathbf{B} \cdot d\mathbf{x}$.

Now, suppose that the parallel component of the magnetic field changes sign. It has been implicitly assumed that the faster eigenmode, which is circularly polarized in the quasi-longitudinal approximation, becomes the slower eigenmode (and vice versa) when the field is reversed, and the Faraday rotation is undone. However, if we track the modes using the full dispersion relation, we find that the faster quasi-longitudinal eigenmode remains the faster eigenmode in the quasi-perpendicular regime, and it becomes the faster eigenmode with opposite sense of circular polarization in the field-reversed quasi-longitudinal regime. Now, let there be a second field reversal where an analogous transition occurs. Following this logic, the net rotation should be $\propto \int n_e |\mathbf{B} \cdot d\mathbf{x}|$. What is going on?

Exercise 21.12 *Example: Reflection of Short Waves by the Ionosphere*
The free electron density in the night-time ionosphere increases exponentially from 10^9 m^{-3} to 10^{11} m^{-3} as the altitude increases from 100 to 200 km, and the density

diminishes above this height. Use Snell's law [Eq. (7.49)] to calculate the maximum range of 10-MHz waves transmitted from Earth's surface, assuming a single ionospheric reflection.

21.5.5 CMA Diagram for Wave Modes in a Cold, Magnetized Plasma

Magnetized plasmas are anisotropic, just like many nonlinear crystals (Chap. 10). This implies that the phase speed of a propagating wave mode depends on the angle between the direction of propagation and the magnetic field. Two convenient ways are used to exhibit this anisotropy diagrammatically. The first method, due originally to Fresnel, is to construct *phase-velocity surfaces* (also called *wave-normal surfaces*), which are polar plots of the wave phase velocity $V_{\rm ph} = \omega/k$, at fixed frequency ω, as a function of the angle θ that the wave vector **k** makes with the magnetic field; see Fig. 21.6a.

The second type of surface, used originally by Hamilton, is the *refractive-index surface*. This is a polar plot of the refractive index $\mathfrak{n} = ck/\omega$ for a given frequency, again as a function of the wave vector's angle θ to **B**; see Fig. 21.6b. This plot has the important property that the group velocity is perpendicular to the surface

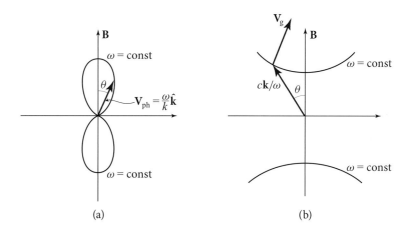

(a)   (b)

FIGURE 21.6 (a) *Wave-normal surface* (i.e., phase-velocity surface) for a whistler mode propagating at an angle θ with respect to the magnetic field direction. This diagram plots the phase velocity $\mathbf{V}_{\rm ph} = (\omega/k)\hat{\mathbf{k}}$ as a vector from the origin, with the direction of the magnetic field chosen as upward. When we fix the frequency ω of the wave, the tip of the phase velocity vector sweeps out the figure-8 curve as its angle θ to the magnetic field changes. This curve should be thought of as rotated around the vertical (magnetic-field) direction to form the figure-8 wave-normal surface. Note that there are some directions where no mode can propagate. (b) *Refractive-index surface* for the same whistler mode. Here we plot ck/ω as a vector from the origin, and as its direction changes with fixed ω, this vector sweeps out the two hyperboloid-like surfaces. Since the length of the vector is $ck/\omega = \mathfrak{n}$, this figure can be thought of as a polar plot of the refractive index $\mathfrak{n}$ as a function of wave-propagation direction θ for fixed ω; hence the name "refractive-index surface." The group velocity $\mathbf{V}_g$ is orthogonal to the refractive-index surface (Ex. 21.13). Note that for this whistler mode, the energy flow (along $\mathbf{V}_g$) is focused toward the direction of the magnetic field.

(Ex. 21.13). As discussed above, the energy flow is along the direction of the group velocity and, in a magnetized plasma, this velocity can make a large angle with the wave vector.

A particularly useful application of these ideas is to a graphical representation of the various types of wave modes that can propagate in a cold, magnetized plasma (Fig. 21.7). This representation is known as the *Clemmow-Mullaly-Allis* or *CMA* diagram. The character of waves of a given frequency ω depends on the ratio of this frequency to the plasma frequency and the cyclotron frequency. This suggests defining two dimensionless numbers, $\omega_p^2/\omega^2 \equiv (\omega_{pe}^2 + \omega_{pp}^2)/\omega^2$ and $|\omega_{ce}|\omega_{cp}/\omega^2$, which are plotted on the horizontal and vertical axes of the CMA diagram. [Recall that $\omega_{pp} = \omega_{pe}\sqrt{m_e/m_p}$ and $\omega_{cp} = \omega_{ce}(m_e/m_p)$.] The CMA space defined by these two dimensionless parameters can be subdivided into sixteen regions, in each of which the propagating modes have a distinctive character. The mode properties are indicated by sketching the topological form of the *wave-normal surfaces* associated with each region.

CMA diagram, Fig. 21.7

horizontal and vertical axes

sixteen regions; topological form of wave-normal surfaces

The form of the wave-normal surface in each region can be deduced from the general dispersion relation (21.48). To deduce it, one must solve the dispersion relation for $1/\mathfrak{n} = \omega/kc = V_{ph}/c$ as a function of θ and ω, and then generate the polar plot of $V_{ph}(\theta)$.

On the CMA diagram's wave-normal curves, the characters of the parallel and perpendicular modes are indicated by labels: R and L for right and left parallel modes ($\theta = 0$), and O and X for ordinary and extraordinary perpendicular modes ($\theta = \pi/2$). As one moves across a boundary from one region to another, there is often a change of which parallel mode gets deformed continuously, with increasing θ, into which perpendicular mode. In some regions a wave-normal surface has a figure-8 shape, indicating that the waves can propagate only over a limited range of angles, $\theta < \theta_{max}$. In some regions there are two wave-normal surfaces, indicating that—at least in some directions θ—two modes can propagate; in other regions there is just one wave-normal surface, so only one mode can propagate; and in the bottom-right two regions there are no wave-normal surfaces, since no waves can propagate at these high densities and low magnetic-field strengths.

EXERCISES

Exercise 21.13 *Derivation: Refractive-Index Surface*
Verify that the group velocity of a wave mode is perpendicular to the refractive-index surface (Fig. 21.6b).

Exercise 21.14 *Problem: Exploration of Modes in the CMA Diagram*
For each of the following modes studied earlier in this chapter, identify in the CMA diagram the phase speed, as a function of frequency ω, and verify that the turning on and cutting off of the modes, and the relative speeds of the modes, are in accord with the CMA diagram's wave-normal curves.

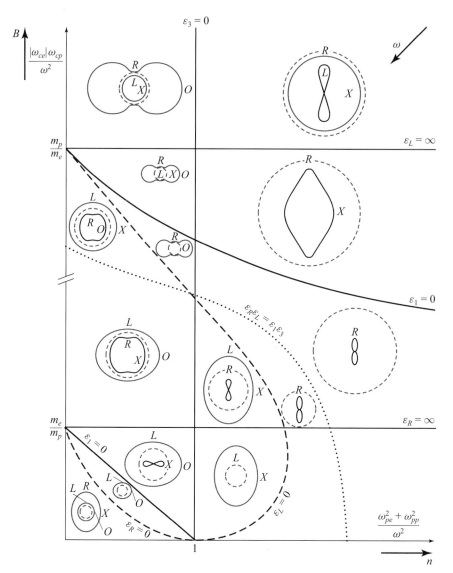

FIGURE 21.7 CMA diagram for wave modes with frequency ω propagating in a plasma with plasma frequencies ω_{pe}, ω_{pp} and gyro frequencies ω_{ce}, ω_{cp}. Plotted upward is the dimensionless quantity $|\omega_{ce}|\omega_{cp}/\omega^2$, which is proportional to B^2, so magnetic field strength also increases upward. Plotted rightward is the dimensionless quantity $(\omega_{pe}^2 + \omega_{pp}^2)/\omega^2$, which is proportional to n, so the plasma number density also increases rightward. Since both the ordinate and the abscissa scale as $1/\omega^2$, ω increases in the left-down direction. This plane is split into sixteen regions by a set of curves on which various dielectric components have special values. In each of the sixteen regions are shown two, one, or no wave-normal surfaces (phase-velocity surfaces) at fixed ω (cf. Fig. 21.6a). These surfaces depict the types of wave modes that can propagate for that region's values of frequency ω, magnetic field strength B, and electron number density n. In each wave-normal diagram the dashed circle indicates the speed of light; a point outside that circle has phase velocity $V_{\rm ph}$ greater than c; inside the circle, $V_{\rm ph} < c$. The topologies of the wave-normal surfaces and speeds relative to c are constant throughout each of the sixteen regions; they change as one moves between regions. Adapted from Boyd and Sanderson (2003, Fig. 6.12), which in turn is adapted from Allis, Buchsbaum, and Bers (1963).

(a) Electromagnetic modes in an unmagnetized plasma.

(b) Left and right modes for parallel propagation in a magnetized plasma.

(c) Ordinary and extraordinary modes for perpendicular propagation in a magnetized plasma.

21.6 Two-Stream Instability

When considered on large enough scales, plasmas behave like fluids and are subject to a wide variety of fluid dynamical instabilities. However, as we are discovering, plasmas have internal degrees of freedom associated with their velocity distributions, which offers additional opportunities for unstable wave modes to grow and for free energy to be released. A full description of velocity-space instabilities is necessarily kinetic and must await the following chapter. However, it is instructive to consider a particularly simple example, the *two-stream instability*, using cold-plasma theory, as this brings out several features of the more general theory in a particularly simple manner.

We will apply our results in a slightly unusual way, to the propagation of fast electron beams through the slowly outflowing solar wind. These electron beams are created by coronal disturbances generated on the surface of the Sun (specifically, those associated with "Type III" radio bursts). The observation of these fast electron beams was initially a puzzle, because plasma physicists knew that they should be unstable to the exponential growth of electrostatic waves. What we do in this section is demonstrate the problem. What we will not do is explain what is thought to be its resolution, since that involves nonlinear plasma-physics considerations beyond the scope of this book (see, e.g., Melrose, 1980, or Sturrock, 1994).

Consider a simple, cold (i.e., with negligible thermal motions) electron-proton plasma at rest. Ignore the protons for the moment. We can write the dispersion relation for electron plasma oscillations in the form

$$\frac{\omega_{pe}^2}{\omega^2} = 1. \tag{21.71}$$

Now allow the ions also to oscillate about their mean positions. The dispersion relation is slightly modified to

$$\frac{\omega_p^2}{\omega^2} = \frac{\omega_{pe}^2}{\omega^2} + \frac{\omega_{pp}^2}{\omega^2} = 1 \tag{21.72}$$

[cf. Eq. (21.23)]. If we were to add other components (e.g., helium ions), that would simply add extra terms to Eq. (21.72).

Next, return to Eq. (21.71) and look at it in a reference frame in which the electrons are moving with speed u. The observed wave frequency is then Doppler shifted, and so the dispersion relation becomes

$$\frac{\omega_{pe}^2}{(\omega - ku)^2} = 1, \tag{21.73}$$

dispersion relation for electron plasma oscillations in a frame where electrons move with speed u

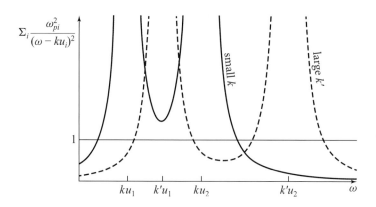

FIGURE 21.8 Left-hand side of the dispersion relation (21.74) for two cold plasma streams and two different choices of wave vector, k (small) and k' (large). For small enough k, there are only two real roots for ω.

where ω is now the angular frequency measured in this new frame. It should be evident from this how to generalize Eq. (21.72) to the case of several cold streams moving with different speeds u_j. We simply add the terms associated with each component using angular frequencies that have been appropriately Doppler shifted:

dispersion relation for electron plasma oscillations in a plasma with counterstreaming beams

$$\frac{\omega_{p1}^2}{(\omega - ku_1)^2} + \frac{\omega_{p2}^2}{(\omega - ku_2)^2} + \ldots = 1. \tag{21.74}$$

(This procedure will be justified via kinetic theory in the next chapter.)

The left-hand side of the dispersion relation (21.74) is plotted in Fig. 21.8 for the case of two cold plasma streams. Equation (21.74) in this case is a quartic in ω, and so it should have four roots. However, for small enough k only two of these roots will be real (solid curves in Fig. 21.8). The remaining two roots must be a complex-conjugate pair, and the root with the positive imaginary part corresponds to a growing mode. We have therefore shown that for small enough k the two-stream plasma will be unstable. Small electrostatic disturbances will grow exponentially to large amplitude and ultimately react back on the plasma. As we add more cold streams to the plasma, we add more modes, some of which will be unstable. This simple example demonstrates how easy it is for a plasma to tap the free energy residing in anisotropic particle-distribution functions.

two-stream instability: tapping free energy of anisotropic particle-distribution functions

two-stream instability for electron beams in solar wind

Let us return to our solar-wind application and work in the rest frame of the wind ($u_1 = 0$), where the plasma frequency is $\omega_{p1} = \omega_p$. If the beam density is a small fraction α of the solar-wind density, so $\omega_{p2}^2 = \alpha \omega_p^2$, and the beam velocity (as seen in the wind's rest frame) is $u_2 = V$, then by differentiating Eq. (21.74), we find that the local minimum of the left-hand side occurs at $\omega = kV/(1 + \alpha^{1/3})$. The value of

Chapter 21. Waves in Cold Plasmas: Two-Fluid Formalism

the left-hand side at that minimum is $\omega_p^2(1+\alpha^{1/3})/\omega^2$. This minimum exceeds unity (thereby making two roots of the dispersion relation complex) for

$$k < \frac{\omega_p}{V}(1+\alpha^{1/3})^{3/2}. \tag{21.75}$$

This is therefore the condition for the existence of a growing mode. The maximum value for the growth rate can be found simply by varying k. It occurs at $k = \omega_p/V$ and is

$$\omega_i = \frac{3^{1/2}\alpha^{1/3}\omega_p}{2^{4/3}}. \tag{21.76}$$

For the solar wind near Earth, we have $\omega_p \sim 10^5$ rad s^{-1}, $\alpha \sim 10^{-3}$, and $V \sim 10^4$ km s^{-1}. We therefore find that the instability should grow, thereby damping the fast electron beam, in a length of 30 km, which is much less than the distance from the Sun (1.5×10^8 km)! This describes the problem that we will not resolve here.

EXERCISES

Exercise 21.15 *Derivation: Two-Stream Instability*
Verify Eq. (21.76).

Exercise 21.16 *Problem: Relativistic Two-Stream Instability*
In a very strong magnetic field, we can consider electrons as constrained to move in 1 dimension along the direction of the magnetic field. Consider a beam of relativistic protons propagating with density n_b and speed $u_b \sim c$ through a cold electron-proton plasma along **B**. Generalize the dispersion relation (21.74) for modes with $\mathbf{k} \parallel \mathbf{B}$.

Exercise 21.17 *Example: Drift Waves*
Another type of wave mode that can be found from a fluid description of a plasma (but requires a kinetic treatment to understand completely) is a *drift wave*. Just as the two-stream instability provides a mechanism for plasmas to erase nonuniformity in velocity space, so drift waves can rapidly remove spatial irregularities.

The limiting case that we consider here is a modification of an ion-acoustic mode in a strongly magnetized plasma with a density gradient. Suppose that the magnetic field is uniform and points in the $\mathbf{e}_z$ direction. Let there be a gradient in the equilibrium density of both the electrons and the protons: $n_0 = n_0(x)$. In the spirit of our description of ion-acoustic modes in an unmagnetized, homogeneous plasma [cf. Eq. (21.33)], treat the proton fluid as cold, but allow the electrons to be warm and isothermal with temperature T_e. We seek modes of frequency ω propagating perpendicular to the density gradient [i.e., with $\mathbf{k} = (0, k_y, k_z)$].

(a) Consider the equilibrium of the warm electron fluid, and show that there must be a fluid drift velocity along the direction $\mathbf{e}_y$ of magnitude

$$V_{de} = -\frac{V_{ia}^2}{\omega_{ci}}\frac{1}{n_0}\frac{dn_0}{dx}, \tag{21.77}$$

where $V_{ia} = (k_B T_e/m_p)^{1/2}$ is the ion-acoustic speed. Explain in physical terms the origin of this drift and why we can ignore the equilibrium drift motion for the ions (protons).

(b) We limit our attention to low-frequency electrostatic modes that have phase velocities below the Alfvén speed. Under these circumstances, perturbations to the magnetic field can be ignored, and the electric field can be written as $\mathbf{E} = -\nabla \Phi$. Write down the three components of the linearized proton equation of motion in terms of the perturbation to the proton density n, the proton fluid velocity $\mathbf{u}$, and the electrostatic potential Φ.

(c) Write down the linearized equation of proton continuity, including the gradient in n_0, and combine with the equation of motion to obtain an equation for the fractional proton density perturbation at low frequencies:

$$\frac{\delta n}{n_0} = \left(\frac{(\omega_{cp}^2 k_z^2 - \omega^2 k^2) V_{ia}^2 + \omega_{cp}^2 \omega k_y V_{de}}{\omega^2 (\omega_{cp}^2 - \omega^2)} \right) \left(\frac{e\Phi}{k_B T_e} \right). \quad (21.78)$$

(d) Argue that the fractional electron-density perturbation follows a linearized Boltzmann distribution, so that

$$\frac{\delta n_e}{n_0} = \frac{e\Phi}{k_B T_e}. \quad (21.79)$$

(e) Use both the proton- and the electron-density perturbations in Poisson's equation to obtain the electrostatic drift wave dispersion relation in the low-frequency ($\omega \ll \omega_{cp}$), long-wavelength ($k\lambda_D \ll 1$) limit:

$$\omega = \frac{k_y V_{de}}{2} \pm \frac{1}{2} \left(k_y^2 V_{de}^2 + 4 k_z^2 V_{ia}^2 \right)^{1/2}. \quad (21.80)$$

Describe the physical character of the mode in the additional limit $k_z \to 0$. A proper justification of this procedure requires a kinetic treatment, which also shows that, under some circumstances, drift waves can be unstable and grow exponentially.

Bibliographic Note

The definitive monograph on waves in plasmas is Stix (1992); also very good, and with a controlled-fusion orientation, is Swanson (2003).

For an elementary and lucid textbook treatment, which makes excellent contact with laboratory experiments, see Chap. 4 of Chen (1974) or Chen (2016). For more sophisticated and detailed textbook treatments, we especially like the relevant chapters of Clemmow and Dougherty (1969), Krall and Trivelpiece (1973), Lifshitz and Pitaevskii (1981), Sturrock (1994), Boyd and Sanderson (2003), Bittencourt (2004), and Bellan (2006). For treatments that focus on astrophysical plasmas, see Melrose (1980) and Parks (2004).

CHAPTER TWENTY-TWO

Kinetic Theory of Warm Plasmas

> Any complete theory of the kinetics of a plasma
> cannot ignore the plasma oscillations.
> IRVING LANGMUIR (1928)

22.1 Overview

At the end of Chap. 21, we showed how to generalize cold-plasma two-fluid theory to accommodate several distinct plasma beams, and thereby we discovered the two-stream instability. If the beams are not individually monoenergetic (i.e., cold), as we assumed there, but instead have broad velocity dispersions that overlap in velocity space (i.e., if the beams are *warm*), then the two-fluid approach of Chap. 21 cannot be used, and a more powerful, kinetic-theory description of the plasma is required.

Chapter 21's approach entailed specifying the positions and velocities of specific groups of particles (the "fluids"); this is an example of a *Lagrangian* description. It turns out that the most robust and powerful method for developing the kinetic theory of warm plasmas is an *Eulerian* one in which we specify how many particles are to be found in a fixed volume of one-particle phase space.

In this chapter, using this Eulerian approach, we develop the kinetic theory of plasmas. We begin in Sec. 22.2 by introducing kinetic theory's one-particle distribution function $f(\mathbf{v}, \mathbf{x}, t)$ and recovering its evolution equation (the collisionless Boltzmann equation, also called the Vlasov equation), which we have met previously in Chap. 3. We then use this *Vlasov equation* to derive the two-fluid formalism used in Chap. 21 and to deduce some physical approximations that underlie the two-fluid description of plasmas.

In Sec. 22.3, we explore the application of the Vlasov equation to Langmuir waves—the 1-dimensional electrostatic modes in an unmagnetized plasma that we studied in Chap. 21 using the two-fluid formalism. Using kinetic theory, we rederive Sec. 21.4.3's Bohm-Gross dispersion relation for Langmuir waves, and as a bonus we uncover a physical damping mechanism, called *Landau damping*, that did not and cannot emerge from the two-fluid analysis. This subtle process leads to the transfer of energy from a wave to those particles that can "surf" or "phase-ride" the wave (i.e., those whose velocity projected parallel to the wave vector is slightly less than the wave's phase speed). We show that Landau damping works because there are usually fewer

> **BOX 22.1. READERS' GUIDE**
>
> - This chapter relies significantly on:
>   - Portions of Chap. 3 on kinetic theory: Secs. 3.2.1 and 3.2.3 on the distribution function, and Sec. 3.6 on Liouville's theorem and the collisionless Boltzmann equation.
>   - Section 20.3 on Debye shielding, collective behavior of plasmas, and plasma oscillations.
>   - Portions of Chap. 21: Sec. 21.2 on the wave equation and dispersion relation for dielectrics, Sec. 21.3 on the two-fluid formalism, Sec. 21.4 on Langmuir and ion-acoustic waves, and Sec. 21.6 on the two-stream instability.
> - Chapter 23 on nonlinear dynamics of plasmas relies heavily on this chapter.

particles traveling faster than the wave and losing energy to it than those traveling slower and extracting energy from it. However, in a collisionless plasma, the particle distributions need not be Maxwellian. In particular, it is possible for a plasma to possess an "inverted" particle distribution with more fast than slow ones; then there is a net injection of particle energy into the waves, which creates an instability. In Sec. 22.4, we use kinetic theory to derive a necessary and sufficient criterion for this instability.

In Sec. 22.5, we examine in greater detail the physics of Landau damping and show that it is an intrinsically nonlinear phenomenon; and we give a semi-quantitative discussion of *nonlinear Landau damping*, preparatory to a more detailed treatment of some other nonlinear plasma effects in the following chapter.

Although the kinetic-theory, Vlasov description of a plasma that is developed and used in this chapter is a great improvement on the two-fluid description of Chap. 21, it is still an approximation, and some situations require more accurate descriptions. We conclude this chapter in Sec. 22.6 by introducing greater accuracy via *N-particle distribution functions*, and as applications we use them (i) to explore the approximations underlying the Vlasov description, and (ii) to explore two-particle correlations that are induced in a plasma by Coulomb interactions and the influence of those correlations on a plasma's equation of state.

22.2 Basic Concepts of Kinetic Theory and Its Relationship to Two-Fluid Theory

22.2.1 Distribution Function and Vlasov Equation

In Chap. 3, we introduced the number density of particles in phase space, the distribution function $\mathcal{N}(\mathbf{p}, \mathbf{x}, t)$. We showed that this quantity is Lorentz invariant, and that it satisfies the collisionless Boltzmann equation (3.64) and (3.65). We interpreted this equation as $\mathcal{N}$ being constant along the phase-space trajectory of any freely moving particle.

To comply with the conventions of the plasma-physics community, we use the name *Vlasov equation* in place of *collisionless Boltzmann equation*.[1] We change notation in a manner described in Sec. 3.2.3: we use a particle's velocity $\mathbf{v}$ rather than its momentum $\mathbf{p}$ as an independent variable, and we define the distribution function f to be the number density of particles in physical and velocity space:

Vlasov equation

$$f(\mathbf{v}, \mathbf{x}, t) = \frac{dN}{d\mathcal{V}_x d\mathcal{V}_v} = \frac{dN}{dx\,dy\,dz\,dv_x\,dv_y\,dv_z}. \quad (22.1)$$

distribution function in plasma physics conventions

Note that the integral of f over velocity space is the number density $n(\mathbf{x}, t)$ of particles in physical space:

$$\boxed{\int f(\mathbf{v}, \mathbf{x}, t)\, d\mathcal{V}_v = n(\mathbf{x}, t),} \quad (22.2)$$

where $d\mathcal{V}_v \equiv dv_x dv_y dv_z$ is the 3-dimensional volume element of velocity space. (For simplicity, we also restrict ourselves to nonrelativistic speeds; the generalization to relativistic plasma theory is straightforward.)

This one-particle distribution function $f(\mathbf{v}, \mathbf{x}, t)$ and its resulting kinetic theory give a good description of a plasma in the regime of large Debye number, $N_D \gg 1$—which includes almost all plasmas that occur in the universe (cf. Sec. 20.3.2 and Fig. 20.1). The reason is that, when $N_D \gg 1$, we can define $f(\mathbf{v}, \mathbf{x}, t)$ by averaging over a physical-space volume that is large compared to the average interparticle spacing and that thus contains many particles, but is still small compared to the Debye length. By such an average—the starting point of kinetic theory—the electric fields of individual particles are made unimportant, as are the Coulomb-interaction-induced correlations between pairs of particles. We explore this issue in detail in Sec. 22.6.2, using a two-particle distribution function.

In Chap. 3, we showed that, in the absence of collisions (a good assumption for plasmas!), the distribution function evolves in accord with the Vlasov equation (3.64) and (3.65). We now rederive that Vlasov equation beginning with the law of conservation of particles for each species $s = e$ (electrons) and p (protons):

$$\frac{\partial f_s}{\partial t} + \boldsymbol{\nabla} \cdot (f_s \mathbf{v}) + \boldsymbol{\nabla}_v \cdot (f_s \mathbf{a}) \equiv \frac{\partial f_s}{\partial t} + \frac{\partial (f_s v_j)}{\partial x_j} + \frac{\partial (f_s a_j)}{\partial v_j} = 0. \quad (22.3)$$

particle conservation for species s (electron or proton)

---

1. This equation was introduced and explored in 1913 by James Jeans in the context of stellar dynamics, and then rediscovered and explored by Anatoly Alexandrovich Vlasov in 1938 in the context of plasma physics. Plasma physicists have honored Vlasov by naming the equation after him. For details of this history, see Hénon (1982).

Here

$$\mathbf{a} = \frac{d\mathbf{v}}{dt} = \frac{q_s}{m_s}(\mathbf{E} + \mathbf{v} \times \mathbf{B}) \tag{22.4}$$

is the electromagnetic acceleration of a particle of species s, which has mass m_s and charge q_s, and $\mathbf{E}$ and $\mathbf{B}$ are the electric and magnetic fields averaged over the same volume as is used in constructing f. Equation (22.3) has the standard form for a conservation law: the time derivative of a density (in this case density of particles in phase space, not just physical space), plus the divergence of a flux (in this case the spatial divergence of the particle flux, $f\mathbf{v} = f d\mathbf{x}/dt$, in the physical part of phase space, plus the velocity divergence of the particle flux, $f\mathbf{a} = f d\mathbf{v}/dt$, in velocity space) is equal to zero.

Now $\mathbf{x}$ and $\mathbf{v}$ are independent variables, so that $\partial x_i/\partial v_j = 0$ and $\partial v_i/\partial x_j = 0$. In addition, $\mathbf{E}$ and $\mathbf{B}$ are functions of $\mathbf{x}$ and t but not of $\mathbf{v}$, and the term $\mathbf{v} \times \mathbf{B}$ is perpendicular to $\mathbf{v}$. Therefore, we have

$$\nabla_v \cdot (\mathbf{E} + \mathbf{v} \times \mathbf{B}) = 0. \tag{22.5}$$

These facts permit us to pull $\mathbf{v}$ and $\mathbf{a}$ out of the derivatives in Eq. (22.3), thereby obtaining

Vlasov equation for species s

$$\frac{\partial f_s}{\partial t} + (\mathbf{v} \cdot \nabla) f_s + (\mathbf{a} \cdot \nabla_v) f_s \equiv \frac{\partial f_s}{\partial t} + \frac{dx_j}{dt}\frac{\partial f_s}{\partial x_j} + \frac{dv_j}{dt}\frac{\partial f_s}{\partial v_j} = 0. \tag{22.6}$$

We recognize this as the statement that f_s is a constant along the trajectory of a particle in phase space:

$$\frac{df_s}{dt} = 0, \tag{22.7}$$

which is the *Vlasov equation* for species s.

Equation (22.7) tells us that, when the space density near a given particle increases, the velocity-space density must decrease, and vice versa. Of course, if we find that other forces or collisions are important in some situation, we can represent them by extra terms added to the right-hand side of the Vlasov equation (22.7) in the manner of the Boltzmann transport equation (3.66) (cf. Sec. 3.6).

So far, we have treated the electromagnetic field as being somehow externally imposed. However, it is actually produced by the net charge and current densities associated with the two particle species. These are expressed in terms of the distribution functions by

charge density and current density as integrals over distribution functions

$$\rho_e = \sum_s q_s \int f_s \, d\mathcal{V}_v, \quad \mathbf{j} = \sum_s q_s \int f_s \mathbf{v} \, d\mathcal{V}_v. \tag{22.8}$$

Equations (22.8), together with Maxwell's equations and the Vlasov equation (22.6), with $\mathbf{a} = d\mathbf{v}/dt$ given by the Lorentz force law (22.4), form a complete set of equations for the structure and dynamics of a plasma. They constitute the kinetic theory of plasmas.

22.2.2 Relation of Kinetic Theory to Two-Fluid Theory

Before developing techniques to solve the Vlasov equation, we first relate it to the two-fluid approach used in the previous chapter. We begin by constructing the moments of the distribution function f_s, defined by

$$n_s = \int f_s \, d\mathcal{V}_v,$$

$$\mathbf{u}_s = \frac{1}{n_s} \int f_s \mathbf{v} \, d\mathcal{V}_v,$$

$$\mathbf{P}_s = m_s \int f_s (\mathbf{v} - \mathbf{u}_s) \otimes (\mathbf{v} - \mathbf{u}_s) \, \mathcal{V}_v. \tag{22.9}$$

macroscopic, two-fluid quantities as integrals over distribution functions

These are respectively the density, the mean fluid velocity, and the pressure tensor for species s. (Of course, $\mathbf{P}_s$ is just the 3-dimensional stress tensor $\mathbf{T}_s$ [Eq. (3.32d)] evaluated in the rest frame of the fluid.)

By integrating the Vlasov equation (22.6) over velocity space and using

$$\int (\mathbf{v} \cdot \nabla) f_s \, d\mathcal{V}_v = \int \nabla \cdot (f_s \mathbf{v}) \, d\mathcal{V}_v = \nabla \cdot \int f_s \mathbf{v} \, d\mathcal{V}_v,$$

$$\int (\mathbf{a} \cdot \nabla_v) f_s \, d\mathcal{V}_v = -\int (\nabla_v \cdot \mathbf{a}) f_s \, d\mathcal{V}_v = 0, \tag{22.10}$$

together with Eq. (22.9), we obtain the continuity equation,

$$\frac{\partial n_s}{\partial t} + \nabla \cdot (n_s \mathbf{u}_s) = 0, \tag{22.11}$$

continuity equation

for each particle species s. [It should not be surprising that the Vlasov equation implies the continuity equation, since the Vlasov equation is equivalent to the conservation of particles in phase space [Eq. (22.3)], while the continuity equation is just the conservation of particles in physical space.]

The continuity equation is the first of the two fundamental equations of two-fluid theory. The second is the equation of motion (i.e., the evolution equation for the fluid velocity $\mathbf{u}_s$). To derive this, we multiply the Vlasov equation (22.6) by the particle velocity $\mathbf{v}$ and then integrate over velocity space (i.e., we compute the Vlasov equation's first moment). The details are a useful exercise for the reader (Ex. 22.1); the result is

$$n_s m_s \left(\frac{\partial \mathbf{u}_s}{\partial t} + (\mathbf{u}_s \cdot \nabla) \mathbf{u}_s \right) = -\nabla \cdot \mathbf{P}_s + n_s q_s (\mathbf{E} + \mathbf{u}_s \times \mathbf{B}), \tag{22.12}$$

equation of motion

which is identical with Eq. (21.13b).

constructing two-fluid equations from kinetic theory by method of moments

A difficulty now presents itself in the two-fluid approximation to kinetic theory. We can use Eqs. (22.8) to compute the charge and current densities from n_s and $\mathbf{u}_s$, which evolve via the fluid equations (22.11) and (22.12). However, we do not yet know how to compute the pressure tensor $\mathbf{P}_s$ in the two-fluid approximation. We could derive a fluid equation for its evolution by taking the second moment of the Vlasov equation (i.e., multiplying it by $\mathbf{v} \otimes \mathbf{v}$ and integrating over velocity space), but that evolution equation would involve an unknown third moment of f_s on the right-hand side, $\mathbf{M}_3 = \int f_s \mathbf{v} \otimes \mathbf{v} \otimes \mathbf{v} \, d\mathcal{V}_v$, which is related to the heat-flux tensor. To determine the evolution of this $\mathbf{M}_3$, we would have to construct the third moment of the Vlasov equation, which would involve the fourth moment of f_s as a driving term, and so on. Clearly, this procedure will never terminate unless we introduce some additional relationship between the moments. Such a relationship, called a *closure relation*, permits us to build a self-contained theory involving only a finite number of moments.

the need for a closure relation

For the two-fluid theory of Chap. 21, the closure relation that we implicitly used was the same idealization that one makes when regarding a fluid as perfect, namely, that the heat-flux tensor vanishes. This idealization is less well justified in a collisionless plasma, with its long mean free paths, than in a normal gas or liquid (which has short mean free paths).

An example of an alternative closure relation is one that is appropriate if radiative processes thermostat the plasma to a particular temperature, so $T_s = $ const; then we can set $\mathbf{P}_s = n_s k_B T_s \mathbf{g} \propto n_s$, where $\mathbf{g}$ is the metric tensor. Clearly, a fluid theory of plasmas can be no more accurate than its closure relation.

22.2.3 Jeans' Theorem

Let us now turn to the difficult task of finding solutions to the Vlasov equation. There is an elementary (and, after the fact, obvious) method to write down a class of solutions that is often useful. This is based on Jeans' theorem (named after the astronomer who first drew attention to it in the context of stellar dynamics; Jeans, 1929).

Suppose that we know the particle acceleration $\mathbf{a}$ as a function of $\mathbf{v}$, $\mathbf{x}$, and t. (We assume this for pedagogical purposes; it is not necessary for our final conclusion.) Then for any particle with phase-space coordinates $(\mathbf{x}_0, \mathbf{v}_0)$ specified at time t_0, we can (at least in principle) compute the particle's future motion: $\mathbf{x} = \mathbf{x}(\mathbf{x}_0, \mathbf{v}_0, t)$, $\mathbf{v} = \mathbf{v}(\mathbf{x}_0, \mathbf{v}_0, t)$. These particle trajectories are the *characteristics* of the Vlasov equation, analogous to the characteristics of the equations for 1-dimensional supersonic fluid flow that we studied in Sec. 17.4 (see Fig. 17.7). For many choices of the acceleration $\mathbf{a}(\mathbf{v}, \mathbf{x}, t)$, there are *constants of the motion*, also known as *integrals of the motion*, that are preserved along the particle trajectories. Simple examples, familiar from elementary mechanics, include the energy (for a time-independent plasma) and the angular momentum (for a spherically symmetric plasma). These integrals can be expressed in terms of the initial coordinates $(\mathbf{x}_0, \mathbf{v}_0)$. If we know n constants of

constants of motion for a particle

the motion, then only $6 - n$ additional variables need be chosen from $(\mathbf{x}_0, \mathbf{v}_0)$ to completely specify the motion of the particle.

The Vlasov equation tells us that f_s is constant along a trajectory in $\mathbf{x}$-$\mathbf{v}$ space. Therefore, in general f_s must be expressible as a function of $(\mathbf{x}_0, \mathbf{v}_0)$. Equivalently, it can be rewritten as a function of the n constants of the motion and the remaining $6 - n$ initial phase-space coordinates. However, there is no requirement that it actually depend on all of these variables. In particular, any function of the integrals of motion alone that is independent of the remaining initial coordinates will satisfy the Vlasov equation (22.6). This is *Jeans' theorem*. In words, *functions of constants of the motion take constant values along actual dynamical trajectories in phase space and therefore satisfy the Vlasov equation.*

Jeans' theorem: solution of Vlasov equation based on constants of particle motion

Of course, a situation may be so complex that no integrals of the particles' equation of motion can be found, in which case Jeans' theorem won't help us find solutions of the Vlasov equation. Alternatively, there may be integrals but the initial conditions may be sufficiently complex that extra variables are required to determine f_s. However, it turns out that in a wide variety of applications, particularly those with symmetries (e.g., time independence $\partial f_s/\partial t = 0$), simple functions of simple constants of the motion suffice to describe a plasma's distribution functions.

We have already met and used a variant of Jeans' theorem in our analysis of statistical equilibrium in Sec. 4.4. There the statistical mechanics distribution function ρ was found to depend only on the integrals of the motion.

We have also, unknowingly, used Jeans' theorem in our discussion of Debye shielding in a plasma (Sec. 20.3.1). To understand this, let us suppose that we have a single isolated positive charge at rest in a stationary plasma ($\partial f_s/\partial t = 0$), and we want to know the electron distribution function in its vicinity. Let us further suppose that the electron distribution at large distances from the charge is known to be Maxwellian with temperature T: $f_e(\mathbf{v}, \mathbf{x}, t) \propto \exp[-\frac{1}{2}m_e v^2/(k_B T)]$. Now, the electrons have an energy integral, $E = \frac{1}{2}m_e v^2 - e\Phi$, where Φ is the electrostatic potential. As Φ becomes constant at large distances from the charge, we can therefore write $f_e \propto \exp[-E/(k_B T)]$ at large distances. However, the particles near the charge must have traveled there from a large distance and so must have this same distribution function. Therefore, close to the charge, we have

$$f_e \propto e^{-E/k_B T} = e^{-[(m_e v^2/2 - e\Phi)/(k_B T)]}, \tag{22.13}$$

and the electron density is obtained by integration over velocity:

$$n_e = \int f_e \, d\mathcal{V}_v \propto e^{[e\Phi/(k_B T)]}. \tag{22.14}$$

This is just the Boltzmann distribution that we asserted to be appropriate in Sec. 20.3.1.

EXERCISES

Exercise 22.1 *Derivation: Two-Fluid Equation of Motion*
Derive the two-fluid equation of motion (22.12) by multiplying the Vlasov equation (22.6) by **v** and integrating over velocity space.

Exercise 22.2 *Example: Positivity of Distribution Function*
The particle distribution function $f(\mathbf{v}, \mathbf{x}, t)$ ought not to become negative if it is to remain physical. Show that this is guaranteed if it initially is everywhere nonnegative and it evolves by the collisionless Vlasov equation.

Exercise 22.3 *Problem and Challenge: Jeans' Theorem in Stellar Dynamics*
Jeans' theorem is of great use in studying the motion of stars in a galaxy. The stars are also almost collisionless and can be described by a distribution function $f(\mathbf{v}, \mathbf{x}, t)$. However, there is only one sign of "charge" and no possibility of screening. In this problem, we make a model of a spherical galaxy composed of identical-mass stars.[2]

(a) The simplest type of distribution function is a function of one integral of a star's motion, the energy per unit mass: $E = \frac{1}{2}v^2 + \Phi$, where Φ is the gravitational potential. A simple example is $f(E) \propto |-E|^{7/2}$ for negative energy and zero for positive energy. Verify that the associated mass density satisfies: $\rho \propto [1 + (r/s)^2]^{-5/2}$, where r is the radius, and s is a scale that you should identify.

(b) This density profile does not agree with observations, so we need to find an algorithm for computing the distribution function that gives a specified density profile. Show that for a general $f(E)$ satisfying appropriate boundary conditions,

$$\frac{d\rho}{d\Phi} = -8^{1/2}\pi\, m \int_\Phi^0 dE \frac{f(E)}{(E-\Phi)^{1/2}}, \tag{22.15a}$$

where m is the mass of a star.

(c) Equation (22.15a) is an Abel integral equation. Confirm that it can be inverted to give the desired algorithm:

$$f(E) = \frac{1}{8^{1/2}\pi^2\, m}\frac{d}{dE}\int_E^0 d\Phi \frac{d\rho/d\Phi}{(\Phi-E)^{1/2}}. \tag{22.15b}$$

(d) Compute the distribution function that is paired with the *Jaffe* profile, which looks more like a real galaxy:

$$\rho \propto \frac{1}{r^2(r+s)^2}. \tag{22.16}$$

---

2. We now know that the outer parts of galaxies are dominated by *dark matter* (Sec. 28.3.2), which may comprise weakly interacting elementary particles that should satisfy the Vlasov equation just as stars do.

(e) We can construct *two-integral* spherical models using $f(E, L)$, where L is a star's total angular momentum per unit mass. What extra feature can we hope to capture using this broader class of distribution functions?

22.3 Electrostatic Waves in an Unmagnetized Plasma: Landau Damping

As our principal application of the kinetic theory of plasmas, we explore its predictions for the dispersion relations; stability; and damping of longitudinal, electrostatic waves in an unmagnetized plasma—Langmuir waves and ion-acoustic waves. When studying these waves in Sec. 21.4.3 using two-fluid theory, we alluded time and again to properties of the waves that could not be derived by fluid techniques. Our goal now is to elucidate those properties using kinetic theory. As we shall see, their origin lies in the plasma's velocity-space dynamics.

22.3.1 Formal Dispersion Relation

Consider an electrostatic wave propagating in the z direction. Such a wave is 1-dimensional in that the electric field points in the z direction, $\mathbf{E} = E\mathbf{e}_z$, and varies as $e^{i(kz-\omega t)}$, so it depends only on z and not on x or y; the distribution function similarly varies as $e^{i(kz-\omega t)}$ and is independent of x, y; and the Vlasov, Maxwell, and Lorentz force equations produce no coupling of particle velocities v_x and v_y to the z direction. These properties suggest the introduction of 1-dimensional distribution functions, obtained by integration over v_x and v_y:

$$F_s(v, z, t) \equiv \int f_s(v_x, v_y, v = v_z, z, t) dv_x dv_y. \quad (22.17)$$

1-dimensional distribution function for plasma interacting with a longitudinal electrostatic wave (Langmuir or ion-acoustic)

Here and throughout we suppress the subscript z on v_z.

Restricting ourselves to weak waves, so nonlinearities can be neglected, we linearize the 1-dimensional distribution functions:

$$F_s(v, z, t) \simeq F_{s0}(v) + F_{s1}(v, z, t). \quad (22.18)$$

Here $F_{s0}(v)$ is the distribution function of the unperturbed particles ($s = e$ for electrons and $s = p$ for protons) in the absence of the wave, and F_{s1} is the perturbation induced by and linearly proportional to the electric field E. The evolution of F_{s1} is governed by the linear approximation to the Vlasov equation (22.6):

$$\frac{\partial F_{s1}}{\partial t} + v \frac{\partial F_{s1}}{\partial z} + \frac{q_s E}{m_s} \frac{dF_{s0}}{dv} = 0. \quad (22.19)$$

Here E is a first-order quantity, so in its term we keep only the zero-order dF_{s0}/dv.

We seek a monochromatic plane-wave solution to this Vlasov equation, so $\partial/\partial t \to -i\omega$ and $\partial/\partial z \to ik$ in Eq. (22.19). Solving the resulting equation for F_{s1}, we obtain

solution of linearized Vlasov equation

$$F_{s1} = \frac{-iq_s}{m_s(\omega - kv)} \frac{dF_{s0}}{dv} E. \tag{22.20}$$

This equation implies that the charge density associated with the wave is related to the electric field by

$$\rho_e = \sum_s q_s \int_{-\infty}^{+\infty} F_{s1} dv = \left(\sum_s \frac{-iq_s^2}{m_s} \int_{-\infty}^{+\infty} \frac{F'_{s0} dv}{\omega - kv} \right) E, \tag{22.21}$$

where the prime denotes a derivative with respect to v: $F'_{s0} = dF_{s0}/dv$.

A quick route from here to the waves' dispersion relation is to insert this charge density into Poisson's equation, $\nabla \cdot \mathbf{E} = ikE = \rho_e/\epsilon_0$, and note that both sides are proportional to E, so a solution is possible only if

$$1 + \sum_s \frac{q_s^2}{m_s \epsilon_0 k} \int_{-\infty}^{+\infty} \frac{F'_{s0} dv}{\omega - kv} = 0. \tag{22.22}$$

An alternative route, which makes contact with the general analysis of waves in a dielectric medium (Sec. 21.2), is developed in Ex. 22.4. This route reveals that the dispersion relation is given by the vanishing of the zz component of the dielectric tensor, which we denoted ϵ_3 in Chap. 21 [Eq. (21.45)], and it shows that ϵ_3 is given by expression (22.22):

$$\epsilon_3(\omega, k) = 1 + \sum_s \frac{q_s^2}{m_s \epsilon_0 k} \int_{-\infty}^{+\infty} \frac{F'_{s0} dv}{\omega - kv} = 0. \tag{22.23}$$

Since $\epsilon_3 = \epsilon_{zz}$ is the only component of the dielectric tensor that we meet in this chapter, we simplify notation henceforth by omitting the subscript 3 (i.e., by denoting $\epsilon_{zz} = \epsilon$).

The form of the dispersion relation (22.23) suggests that we combine the unperturbed electron and proton distribution functions $F_{e0}(v)$ and $F_{p0}(v)$ to produce a single, unified distribution function:

unified distribution function

$$F(v) \equiv F_{e0}(v) + \frac{m_e}{m_p} F_{p0}(v), \tag{22.24}$$

in terms of which the dispersion relation takes the form

general dispersion relation for electrostatic waves—derived by Fourier techniques

$$\epsilon(\omega, k) = 1 + \frac{e^2}{m_e \epsilon_0 k} \int_{-\infty}^{+\infty} \frac{F' dv}{\omega - kv} = 0. \tag{22.25}$$

Note that each proton is weighted less heavily than each electron by a factor $m_e/m_p = 1/1{,}836$ in the unified distribution function (22.24) and the dispersion relation (22.25). This is due to the protons' greater inertia and corresponding weaker response to an applied electric field; it causes the protons to be of no importance in Langmuir waves (Sec. 22.3.5). However, in ion-acoustic waves (Sec. 22.3.6), the protons can play an important role, because large numbers of them may move with thermal speeds that are close to the waves' phase velocity, and thereby they can interact resonantly with the waves.

EXERCISES

Exercise 22.4 *Example: Dielectric Tensor and Dispersion Relation for Longitudinal, Electrostatic Waves*

Derive expression (22.23) for the zz component of the dielectric tensor in a plasma excited by a weak electrostatic wave, and show that the wave's dispersion relation is $\epsilon_3 = 0$. [Hint: Notice that the z component of the plasma's electric polarization P_z is related to the charge density by $\nabla \cdot \mathbf{P} = ikP_z = -\rho_e$ [Eq. (21.1)]. Combine this with Eq. (22.21) to get a linear relationship between P_z and $E_z = E$. Argue that the only nonzero component of the plasma's electric susceptibility is χ_{zz}, and deduce its value by comparing the above result with Eq. (21.3). Then construct the dielectric tensor ϵ_{ij} from Eq. (21.5) and the algebratized wave operator L_{ij} from Eq. (21.9), and deduce that the dispersion relation $\det||L_{ij}|| = 0$ takes the form $\epsilon_{zz} \equiv \epsilon_3 = 0$, where ϵ_3 is given by Eq. (22.23).]

22.3.2 Two-Stream Instability

As a first application of the general dispersion relation (22.25), we use it to rederive the dispersion relation (21.74) associated with the cold-plasma two-stream instability of Sec. 21.6.

We begin by performing an integration by parts on the general dispersion relation (22.25), obtaining:

$$\frac{e^2}{m_e \epsilon_0} \int_{-\infty}^{+\infty} \frac{F\, dv}{(\omega - kv)^2} = 1. \qquad (22.26)$$

multi-stream approach to two-stream instability

We then presume, as in Sec. 21.6, that the fluid consists of two or more streams of cold particles (protons or electrons) moving in the z direction with different fluid speeds $u_1, u_2, \ldots$, so $F(v) = n_1 \delta(v - u_1) + n_2 \delta(v - u_2) + \ldots$. Here n_j is the number density of particles in stream j if the particles are electrons, and m_e/m_p times the number density if they are protons. Inserting this $F(v)$ into Eq. (22.26) and noting that $n_j e^2/(m_e \epsilon_0)$ is the squared plasma frequency ω_{pj}^2 of stream j, we obtain the dispersion relation

$$\frac{\omega_{p1}^2}{(\omega - ku_1)^2} + \frac{\omega_{p2}^2}{(\omega - ku_2)^2} + \cdots = 1, \qquad (22.27)$$

which is identical to the dispersion relation (21.74) used in our analysis of the two-stream instability.

It should be evident that the general dispersion relation (22.26) [or equally well, Eq. (22.25)] provides us with a tool for exploring how the two-stream instability is influenced by a warming of the plasma (i.e., by a spread of particle velocities around the mean, fluid velocity of each stream). We explore this in Sec. 22.4.

22.3.3 The Landau Contour

The general dispersion relation (22.25) has a troubling feature: for real ω and k its integrand becomes singular at $v = \omega/k =$ (the waves' phase velocity) unless dF/dv vanishes there, which is generically unlikely. This tells us that if k is real (as we shall assume), then ω cannot be real, except perhaps for a nongeneric mode whose phase velocity happens to coincide with a velocity for which $dF/dv = 0$.

ambiguity in general dispersion relation

With ω/k complex, we must face the possibility of some subtlety in how the integral over v in the dispersion relation (22.25) is performed—the possibility that we may have to make v complex in the integral and follow some special route in the complex velocity plane from $v = -\infty$ to $v = +\infty$. Indeed, there is such a subtlety, as Landau (1946) has shown. Our simple derivation of the dispersion relation in Sec. 22.3.1 cannot reveal this subtlety—and, indeed, is suspicious, since in going from Eq. (22.19) to Eq. (22.20), our derivation entailed dividing by $\omega - kv$, which vanishes when $v = \omega/k$, and dividing by zero is always a suspicious practice. Faced with this conundrum, Landau developed a more sophisticated derivation of the dispersion relation. It is based on posing generic initial data for electrostatic waves, then evolving those data forward in time and identifying the plasma's electrostatic modes by their late-time sinusoidal behaviors, and finally reading off the dispersion relation for the modes from the equations for the late-time evolution. In the remainder of this section, we present a variant of Landau's analysis.

derivation of general dispersion relation by Landau's method

This analysis is very important, including the portion (Ex. 22.5) assigned for the reader to work out. The reader is encouraged to read through this section slowly, with care, so as to understand clearly what is going on.

For simplicity, from the outset we restrict ourselves to plane waves propagating in the z direction with some fixed, real wave number k, so the linearized 1-dimensional distribution function and the electric field have the forms

$$F_s(v, z, t) = F_{s0}(v) + F_{s1}(v, t)e^{ikz}, \quad E(z, t) = E(t)e^{ikz}. \qquad (22.28)$$

Chapter 22. Kinetic Theory of Warm Plasmas

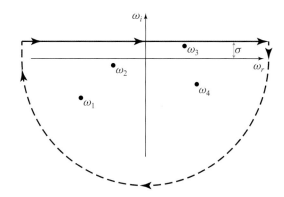

FIGURE 22.1 Contour of integration for evaluating $E(t)$ [Eq. (22.29)] as a sum over residues of the integrand's poles ω_n—the complex frequencies of the plasma's modes.

At $t = 0$ we pose initial data $F_{s1}(v, 0)$ for the electron and proton velocity distributions; these data determine the initial electric field $E(0)$ via Poisson's equation. We presume that these initial distributions [and also the unperturbed plasma's velocity distribution $F_{s0}(v)$] are analytic functions of velocity v, but aside from this constraint, the $F_{s1}(v, 0)$ are generic. (A Maxwellian distribution is analytic, and most any physically reasonable initial distribution can be well approximated by an analytic function.)

We then evolve these initial data forward in time. The ideal tool for such evolution is the Laplace transform, and *not* the Fourier transform. The power of the Laplace transform is much appreciated by engineers and is underappreciated by many physicists. Those readers who are not intimately familiar with evolution via Laplace transforms should work carefully through Ex. 22.5. That exercise uses Laplace transforms, followed by conversion of the final answer into Fourier language, to derive the following formula for the time-evolving electric field in terms of the initial velocity distributions $F_{s1}(v, 0)$:

key to the derivation— evolve arbitrary initial data using Laplace transform

$$E(t) = \int_{i\sigma-\infty}^{i\sigma+\infty} \frac{e^{-i\omega t}}{\epsilon(\omega, k)} \left[\sum_s \frac{q_s}{2\pi\epsilon_0 k} \int_{-\infty}^{+\infty} \frac{F_{s1}(v, 0)}{\omega - kv} dv \right] d\omega. \qquad (22.29)$$

Here the integral in frequency space is along the solid horizontal line at the top of Fig. 22.1, with the imaginary part of ω held fixed at $\omega_i = \sigma$ and the real part ω_r varying from $-\infty$ to $+\infty$. The Laplace techniques used to derive this formula are carefully designed to avoid any divergences and any division by zero. This careful design leads to the requirement that the height σ of the integration line above the real frequency axis be greater than the e-folding rate $\Im(\omega)$ of the plasma's most rapidly growing mode (or, if none grow, still larger than zero and thus larger than $\Im(\omega)$ for the most slowly decaying mode):

$$\sigma > p_o \equiv \max_n \Im(\omega_n), \quad \text{and} \quad \sigma > 0. \qquad (22.30)$$

22.3 Electrostatic Waves in an Unmagnetized Plasma: Landau Damping

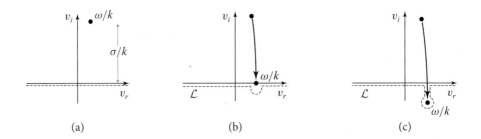

FIGURE 22.2 Derivation of the Landau contour $\mathcal{L}$. (a) The dielectric function $\epsilon(\omega, k)$ is originally defined, in Eq. (22.31), solely for the pole (the point labeled ω/k) lying in the upper half plane; that is, for $\omega_i/k = \sigma/k > 0$. (b,c) Since $\epsilon(\omega, k)$ must be an analytic function of ω at fixed k and thus must vary continuously as ω is continuously changed, the dashed contour of integration in Eq. (22.31) must be kept always below the pole $v = \omega/k$.

Here $n = 1, 2, \ldots$ labels the modes, ω_n is the complex frequency of mode n, and $\Im$ means take the imaginary part of. We shall see that the ω_n are the zeros of the dielectric function $\epsilon(\omega, k)$ that appears in Eq. (22.29), that is, they are poles of the integrand of the frequency integral.

Equation (22.29) also entails a velocity integral. In the Laplace-based analysis (Ex. 22.5) that leads to this formula, there is never any question about the nature of the velocity v: it is always real, so the integral is over real v running from $-\infty$ to $+\infty$. However, because all the frequencies ω appearing in Eq. (22.29) have imaginary parts $\omega_i = \sigma > 0$, there is no possibility in the velocity integral of any divergence of the integrand.

In Eq. (22.29) for the evolving field, $\epsilon(\omega, k)$ is the same dielectric function (22.25) as we deduced in our previous analysis (Sec. 22.3.1):

$$\epsilon(\omega, k) = 1 + \frac{e^2}{m_e \epsilon_0 k} \int_{-\infty}^{+\infty} \frac{F' \, dv}{\omega - kv}, \quad \text{where } F(v) = F_{e0}(v) + \frac{m_e}{m_p} F_{p0}. \quad (22.31)$$

However, by contrast with Sec. 22.3.1, our derivation here dictates unequivocally how to handle the v integration—the same way as in Eq. (22.29): v is strictly real, and the only frequencies appearing in the evolution equations have $\omega_i = \sigma > 0$, so the v integral, running along the real velocity axis, passes under the integrand's pole at $v = \omega/k$, as shown in Fig. 22.2a.

To read off the modal frequencies from the evolving field $E(t)$ at times $t > 0$, we use techniques from complex-variable theory. It can be shown that, because (by hypothesis) $F_{s1}(v, 0)$ and $F_{s0}(v)$ are analytic functions of v, the integrand of the ω integral in Eq. (22.29) is meromorphic: when the integrand is analytically continued throughout the complex frequency plane, its only singularities are poles. This permits us to evaluate the frequency integral, at times $t > 0$, by closing the integration contour in the lower-half frequency plane, as shown by the dashed curve in Fig. 22.1. Because of the exponential factor $e^{-i\omega t}$, the contribution from the dashed part of the contour vanishes, which means that the integral around the contour is equal to $E(t)$ (the contribution from the solid horizontal part). Complex-variable theory tells us that this integral is given by a sum over the residues R_n of the integrand at the poles (labeled $n = 1, 2, \ldots$):

$$E(t) = 2\pi i \sum_n R_n = \sum_n A_n e^{-i\omega_n t}. \tag{22.32}$$

Here ω_n is the frequency at pole n, and A_n is $2\pi i R_n$ with its time dependence $e^{-i\omega_n t}$ factored out. It is evident, then, that each pole of the analytically continued integrand of Eq. (22.29) corresponds to a mode of the plasma, and the pole's complex frequency is the mode's frequency.

Now, for special choices of the initial data $F_{s1}(v, 0)$, there may be poles in the square-bracketed term in Eq. (22.29), but for generic initial data there will be none, and the only poles will be the zeros of $\epsilon(\omega, k)$. Therefore, generically, the modes' frequencies are the zeros of $\epsilon(\omega, k)$—when that function (which was originally defined only for ω along the line $\omega_i = \sigma$) has been analytically extended throughout the complex frequency plane.

So how do we compute the analytically extended dielectric function $\epsilon(\omega, k)$? Imagine holding k fixed and real, and exploring the (complex) value of ϵ, thought of as a function of ω/k, by moving around the complex ω/k plane (same as the complex velocity plane). In particular, imagine computing ϵ from Eq. (22.31) at one point after another along the arrowed path shown in Fig. 22.2b,c. This path begins at an initial location ω/k, where $\omega_i/k = \sigma/k > 0$, and travels down to some other location below the real axis. At the starting point, the discussion in the paragraph following Eq. (22.30) tells us how to handle the velocity integral: just integrate v along the real axis. As ω/k is moved continuously (with k held fixed), $\epsilon(\omega, k)$ must vary continuously, because it is analytic. When ω/k crosses the real velocity axis, if the integration contour in Eq. (22.31) were to remain on the velocity axis, then the contour would jump over the integral's moving pole $v = \omega/k$, and the function $\epsilon(\omega, k)$ would jump discontinuously at the moment of crossing, which is not possible. To avoid such a discontinuous jump, it is necessary that the contour of integration be kept below the pole, $v = \omega/k$, as that pole moves into the lower half of the velocity plane (Fig. 22.2b,c).

The rule that the integration contour must always pass beneath the pole $v = \omega/k$ as shown in Fig. 22.2 is called the *Landau prescription;* the contour is called the *Landau contour* and is denoted $\mathcal{L}$. Our final formula for the dielectric function (and for its vanishing at the modal frequencies—the dispersion relation) is

Landau contour and the unambiguous general dispersion relation for electrostatic waves in an unmagnetized plasma

$$\boxed{\epsilon(\omega, k) = 1 + \frac{e^2}{m_e \epsilon_0 k} \int_{\mathcal{L}} \frac{F' dv}{\omega - kv} = 0,} \quad \text{where} \quad \boxed{F(v) = F_e(v) + \frac{m_e}{m_p} F_p(v).}$$

(22.33)

For future use we have omitted the subscript 0 from the unperturbed distribution functions F_s, as there should be no confusion in future contexts. We refer to Eq. (22.33) as the general dispersion relation for electrostatic waves in an unmagnetized plasma.

EXERCISES

Exercise 22.5 **Example: Electric Field for Electrostatic Wave Deduced Using Laplace Transforms*

Use Laplace-transform techniques to derive Eqs. (22.29)–(22.31) for the time-evolving electric field of electrostatic waves with fixed wave number k and initial velocity perturbations $F_{s1}(v, 0)$. A sketch of the solution follows.

(a) When the initial data are evolved forward in time, they produce $F_{s1}(v, t)$ and $E(t)$. Construct the Laplace transforms (see, e.g., Arfken, Weber, and Harris, 2013, Chap. 20, or Mathews and Walker, 1970, Sec. 4-3) of these evolving quantities:

$$\tilde{F}_{s1}(v, p) = \int_0^\infty dt\, e^{-pt} F_{s1}(v, t), \qquad \tilde{E}(p) = \int_0^\infty dt\, e^{-pt} E(t). \quad (22.34)$$

To ensure that the time integral is convergent, insist that $\Re(p)$ be greater than $p_0 \equiv \max_n \Im(\omega_n) \equiv$ (the e-folding rate of the most strongly growing mode—or, if none grow, then the most weakly damped mode). Also, to simplify the subsequent analysis, insist that $\Re(p) > 0$. Below, in particular, we need the Laplace transforms for $\Re(p) = $ (some fixed value σ that satisfies $\sigma > p_o$ and $\sigma > 0$).

(b) By constructing the Laplace transform of the 1-dimensional Vlasov equation (22.19) and integrating by parts the term involving $\partial F_{s1}/\partial t$, obtain an equation for a linear combination of $\tilde{F}_{s1}(v, p)$ and $\tilde{E}(p)$ in terms of the initial data $F_{s1}(v, t=0)$. By then combining with the Laplace transform of Poisson's equation, show that

$$\tilde{E}(p) = \frac{1}{\epsilon(ip, k)} \sum_s \frac{q_s}{k\epsilon_0} \int_{-\infty}^\infty \frac{F_{s1}(v, 0)}{ip - kv} dv. \quad (22.35)$$

Here $\epsilon(ip, k)$ is the dielectric function (22.25) evaluated for frequency $\omega = ip$, with the integral running along the real v-axis, and [as we noted in part (a)] with $\Re(p)$ greater than p_0, the largest ω_i of any mode, and greater than 0. This situation for the dielectric function is the one depicted in Fig. 22.2a.

(c) Laplace-transform theory tells us that the time-evolving electric field (with wave number k) can be expressed in terms of its Laplace transform (22.35) by

$$E(t) = \int_{\sigma-i\infty}^{\sigma+i\infty} \tilde{E}(p)\, e^{pt}\, \frac{dp}{2\pi i}, \quad (22.36)$$

where σ [as introduced in part (a)] is any real number larger than p_0 and larger than 0. Combine this equation with expression (22.35) for $\tilde{E}(p)$, and set $p = -i\omega$. Thereby arrive at the desired result, Eq. (22.29).

22.3.4 Dispersion Relation for Weakly Damped or Growing Waves

In most practical situations, electrostatic waves are weakly damped or weakly unstable: $|\omega_i| \ll \omega_r$ (where ω_r and ω_i are the real and imaginary parts of the wave frequency ω), so the amplitude changes little in one wave period. In this case, the dielectric function (22.33) can be evaluated at $\omega = \omega_r + i\omega_i$ using the first term in a Taylor series expansion away from the real axis:

$$\epsilon(\omega_r + i\omega_i, k) \simeq \epsilon(\omega_r, k) + \omega_i \left(\frac{\partial \epsilon_r}{\partial \omega_i} + i \frac{\partial \epsilon_i}{\partial \omega_i} \right)_{\omega_i=0}$$

$$= \epsilon(\omega_r, k) + \omega_i \left(-\frac{\partial \epsilon_i}{\partial \omega_r} + i \frac{\partial \epsilon_r}{\partial \omega_r} \right)_{\omega_i=0}$$

$$\simeq \epsilon(\omega_r, k) + i\omega_i \left(\frac{\partial \epsilon_r}{\partial \omega_r} \right)_{\omega_i=0}. \quad (22.37)$$

Here ϵ_r and ϵ_i are the real and imaginary parts of ϵ. In going from the first line to the second we have assumed that $\epsilon(\omega, k)$ is an analytic function of ω near the real axis and hence have used the Cauchy-Riemann equations for the derivatives (see, e.g., Arfken, Weber, and Harris, 2013). In going from the second line to the third we have used the fact that $\epsilon_i \to 0$ when the velocity distribution is one that produces $\omega_i \to 0$ [cf. Eq. (22.39)], so the middle term on the second line is second order in ω_i and can be neglected.

Equation (22.37) expresses the dielectric function slightly away from the real axis in terms of its value and derivative on and along the real axis. The on-axis value can be computed from Eq. (22.33) by breaking the Landau contour depicted in Fig. 22.2b into three pieces—two lines from $\pm\infty$ to a small distance δ from the pole, plus a semicircle of radius δ under and around the pole—and by then taking the limit $\delta \to 0$. The first two terms (the two straight lines) together produce the Cauchy principal value of the integral (denoted $\int_P$ below), and the third produces $2\pi i$ times half the residue of the pole at $v = \omega_r/k$, so Eq. (22.33) becomes:

$$\epsilon(\omega_r, k) = 1 - \frac{e^2}{m_e \epsilon_0 k^2} \left[\int_P \frac{F' \, dv}{v - \omega_r/k} dv + i\pi F'(v = \omega_r/k) \right]. \quad (22.38)$$

Inserting this equation and its derivative with respect to ω_r into Eq. (22.37), and setting the result to zero, we obtain

$$\epsilon(\omega_r + i\omega_i, k) \simeq 1 - \frac{e^2}{m_e \epsilon_0 k^2} \left[\int_P \frac{F' \, dv}{v - \omega_r/k} + i\pi F'(\omega_r/k) \right.$$

$$\left. + i\omega_i \frac{\partial}{\partial \omega_r} \int_P \frac{F' \, dv}{v - \omega_r/k} \right] = 0. \quad (22.39)$$

Notice that the vanishing of ϵ_r determines the real part ω_r of the frequency:

$$\boxed{1 - \frac{e^2}{m_e \epsilon_0 k^2} \int_P \frac{F'}{v - \omega_r/k} dv = 0,} \quad (22.40a)$$

and the vanishing of ϵ_i determines the imaginary part:

$$\boxed{\omega_i = \frac{\pi F'(\omega_r/k)}{-\frac{\partial}{\partial \omega_r} \int_P \frac{F'}{v - \omega_r/k} dv}.} \qquad (22.40\text{b})$$

general dispersion relation for weakly damped elastostatic waves in an unmagnetized plasma

Equations (22.40) are the dispersion relation in the limit $|\omega_i| \ll \omega_r$. We refer to this as the small-$|\omega_i|$ dispersion relation for electrostatic waves in an unmagnetized plasma.

Notice that the sign of ω_i is influenced by the sign of $F' = dF/dv$ at $v = \omega_r/k = V_{\rm ph} =$ (the waves' phase velocity). As we shall see, this has a simple physical origin and important physical consequences. Usually, but not always, the denominator of Eq. (22.40b) is positive, so the sign of ω_i is the same as the sign of $F'(\omega_r/k)$.

22.3.5 Langmuir Waves and Their Landau Damping

We now apply the small-$|\omega_i|$ dispersion relation (22.40) to Langmuir waves in a thermalized plasma. Langmuir waves typically move so fast that the slow ions cannot interact with them, so their dispersion relation is influenced significantly only by the electrons. Therefore, we ignore the ions and include only the electrons in $F(v)$. We obtain $F(v)$ by integrating out v_y and v_z in the 3-dimensional Boltzmann distribution [Eq. (3.22d) with $E = \tfrac{1}{2}m_e(v_x^2 + v_y^2 + v_z^2)$]; the result, correctly normalized so that $\int F(v) dv = n$, is

$$F \simeq F_e = n \left(\frac{m_e}{2\pi k_B T}\right)^{1/2} e^{-[m_e v^2/(2k_B T)]}, \qquad (22.41)$$

where T is the electron temperature.

Now, as we saw in Eq. (22.40b), ω_i is proportional to $F'(v = \omega_r/k)$ with a proportionality constant that is usually positive. Physically, this proportionality arises from the manner in which electrons surf on the waves. Those electrons moving slightly faster than the waves' phase velocity $V_{\rm ph} = \omega_r/k$ (usually) lose energy to the waves on average, while those moving slightly slower (usually) extract energy from the waves on average. Therefore,

energy conservation in Landau damping

1. if there are more slightly slower particles than slightly faster [$F'(v = \omega_r/k) < 0$], then the particles on average gain energy from the waves and the waves are damped [$\omega_i < 0$];

2. if there are more slightly faster particles than slightly slower [$F'(v = \omega_r/k) > 0$], then the particles on average lose energy to the waves and the waves are amplified [$\omega_i > 0$]; and

3. the bigger the disparity between the number of slightly faster electrons and the number of slightly slower electrons [i.e., the bigger $|F'(\omega_r/k)|$], the larger will be the damping or growth of wave energy (i.e., the larger will be $|\omega_i|$).

Quantitatively, it turns out that if the waves' phase velocity ω_r/k is anywhere near the steepest point on the side of the electron velocity distribution (i.e., if ω_r/k is of order the electron thermal velocity $\sqrt{k_B T/m_e}$), then the waves will be strongly damped: $\omega_i \sim -\omega_r$. Since our dispersion relation (22.41) is valid only when the waves are weakly damped, we must restrict ourselves to waves with $\omega_r/k \gg \sqrt{k_B T/m_e}$ (a physically allowed regime) or to $\omega_r/k \ll \sqrt{k_B T/m_e}$ (a regime that does not occur in Langmuir waves; cf. Fig. 21.1).

Requiring, then, that $\omega_r/k \gg \sqrt{k_B T/m_e}$ and noting that the integral in Eq. (22.39) gets its dominant contribution from velocities $v \lesssim \sqrt{k_B T/m_e}$, we can expand $1/(v - \omega_r/k)$ in the integrand as a power series in vk/ω_r, obtaining

$$\int_P \frac{F' \, dv}{v - \omega_r/k} = -\int_{-\infty}^{\infty} dv F' \left[\frac{k}{\omega_r} + \frac{k^2 v}{\omega_r^2} + \frac{k^3 v^2}{\omega_r^3} + \frac{k^4 v^3}{\omega_r^4} + \cdots \right]$$

$$= \frac{nk^2}{\omega_r^2} + \frac{3n\langle v^2 \rangle k^4}{\omega_r^4} + \cdots$$

$$= \frac{nk^2}{\omega_r^2}\left(1 + \frac{3 k_B T k^2}{m_e \omega_r^2} + \cdots \right)$$

$$\simeq \frac{nk^2}{\omega_r^2}\left(1 + 3k^2 \lambda_D^2 \frac{\omega_p^2}{\omega_r^2}\right). \quad (22.42)$$

Substituting Eqs. (22.41) and (22.42) into Eqs. (22.40a) and (22.40b), and noting that $\omega_r/k \gg \sqrt{k_B T/m_e} \equiv \omega_p \lambda_D$ implies $k\lambda_D \ll 1$ and $\omega_r \simeq \omega_p$, we obtain

$$\boxed{\omega_r = \omega_p (1 + 3k^2 \lambda_D^2)^{1/2},} \quad (22.43a)$$

$$\boxed{\omega_i = -\left(\frac{\pi}{8}\right)^{1/2} \frac{\omega_p}{k^3 \lambda_D^3} \exp\left(-\frac{1}{2k^2 \lambda_D^2} - \frac{3}{2}\right).} \quad (22.43b)$$

dispersion relation for Langmuir waves in an unmagnetized, thermalized plasma

The real part of this dispersion relation, $\omega_r = \omega_p \sqrt{1 + 3k^2 \lambda_D^2}$, reproduces the Bohm-Gross result (21.31) that we derived using the two-fluid theory in Sec. 21.4.3 and plotted in Fig. 21.1. The imaginary part reveals the damping of these Langmuir waves by surfing electrons—so-called *Landau damping*. The two-fluid theory could not predict this Landau damping, because the damping is a result of internal dynamics in the electrons' velocity space, of which that theory is oblivious.

Landau damping of Langmuir waves

Notice that, as the waves' wavelength is decreased (i.e., as k increases), the waves' phase velocity decreases toward the electron thermal velocity and the damping becomes stronger, as is expected from our discussion of the number of electrons that can surf on the waves. In the limit $k \to 1/\lambda_D$ (where our dispersion relation has broken down and so is only an order-of-magnitude guide), the dispersion relation predicts that $\omega_r/k \sim \sqrt{k_B T/m_e}$ and $\omega_i/\omega_r \sim 1/10$.

In the opposite regime of large wavelength $k\lambda_D \ll 1$ (where our dispersion relation should be quite accurate), the Landau damping is very weak—so weak that ω_i decreases to zero with increasing k faster than any power of k.

22.3.6 Ion-Acoustic Waves and Conditions for Their Landau Damping to Be Weak

As we saw in Sec. 21.4.3, ion-acoustic waves are the analog of ordinary sound waves in a fluid: they occur at low frequencies where the mean (fluid) electron velocity is very nearly locked to the mean (fluid) proton velocity, so the electric polarization is small; the restoring force is due to thermal pressure and not to the electrostatic field; and the inertia is provided by the heavy protons. It was asserted in Sec. 21.4.3 that to avoid these waves being strongly damped, the electron temperature must be much higher than the proton temperature: $T_e \gg T_p$. We can now understand this condition in terms of particle surfing and Landau damping.

requirement that $T_e \gg T_p$ for ion-acoustic waves

Suppose that the electrons and protons have Maxwellian velocity distributions but possibly with different temperatures. Because of their far greater inertia, the protons will have a far smaller mean thermal speed than the electrons, $\sqrt{k_B T_p/m_p} \ll \sqrt{k_B T_e/m_e}$, so the net 1-dimensional distribution function $F(v) = F_e(v) + (m_e/m_p)F_p(v)$ [Eq. (22.24)] that appears in the kinetic-theory dispersion relation has the form shown in Fig. 22.3. Now, if $T_e \sim T_p$, then the contributions of the electron pressure and proton pressure to the waves' restoring force will be comparable, and the waves' phase velocity will therefore be $\omega_r/k \sim \sqrt{k_B(T_e+T_p)/m_p} \sim \sqrt{k_B T_p/m_p} = v_{\text{th},p}$, which is the thermal proton velocity and also is the speed at which the proton contribution to $F(v)$ has its steepest slope (see the left tick mark on the horizontal axis in Fig. 22.3); so $|F'(v = \omega_r/k)|$ is large. This means large numbers of protons can surf on the waves and a disparity develops between the number moving slightly slower than the waves (which extract energy from the waves) and the number moving slightly faster (which give energy to the waves). The result will be strong Landau damping by the protons.

ion-acoustic waves' strong Landau damping if electrons and ions are thermalized at same temperature

This strong Landau damping is avoided if $T_e \gg T_p$. Then the waves' phase velocity will be $\omega_r/k \sim \sqrt{k_B T_e/m_p}$, which is large compared to the proton thermal velocity $v_{\text{th},p} = \sqrt{k_B T_p/m_p}$ and so is way out on the tail of the proton velocity distribution, where there are few protons that can surf and damp the waves; see the right tick mark on the horizontal axis in Fig. 22.3. Thus Landau damping by protons has been shut down by raising the electron temperature.

What about Landau damping by electrons? The phase velocity $\omega_r/k \sim \sqrt{k_B T_e/m_p}$ is small compared to the electron thermal velocity $v_{\text{th},e} = \sqrt{k_B T_e/m_e}$, so the waves reside near the peak of the electron velocity distribution, where $F_e(v)$ is large, so many electrons can surf with the waves. But $F'_e(v)$ is small, so there are nearly equal numbers of faster and slower electrons, and the surfing produces little net Landau damping. Thus $T_e/T_p \gg 1$ leads to successful propagation of ion acoustic waves.

A detailed computation, based on our small-ω_i kinetic-theory dispersion relation (22.40) makes this physical argument quantitative. The details are carried out in

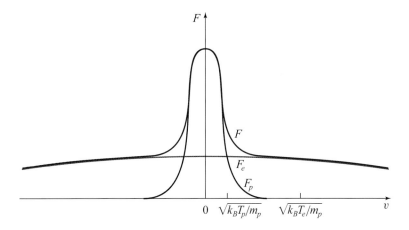

FIGURE 22.3 Electron and ion contributions to the net distribution function $F(v)$ in a thermalized plasma. When $T_e \sim T_p$, the phase speed of ion acoustic waves is near the left tick mark on the horizontal axis—a speed at which surfing protons have maximum ability to Landau-damp the waves, and the waves are strongly damped. When $T_e \gg T_p$, the phase speed is near the right tick mark—far out on the tail of the proton velocity distribution—so few protons can surf and damp the waves. The phase speed is near the peak of the electron distribution, so the number of electrons moving slightly slower than the waves is nearly the same as the number moving slightly faster, and there is little net damping by the electrons. In this case the waves can propagate.

Ex. 22.6 under the assumptions that $T_e \gg T_p$ and $\sqrt{k_B T_p/m_p} \ll \omega_r/k \ll \sqrt{k_B T_e/m_e}$ (corresponding to the above discussion). The result is

$$\boxed{\frac{\omega_r}{k} = \sqrt{\frac{k_B T_e/m_p}{1+k^2\lambda_D^2}},} \quad (22.44a)$$

$$\boxed{\frac{\omega_i}{\omega_r} = -\frac{\sqrt{\pi/8}}{(1+k^2\lambda_D^2)^{3/2}}\left[\sqrt{\frac{m_e}{m_p}} + \left(\frac{T_e}{T_p}\right)^{3/2} \exp\left(\frac{-T_e/T_p}{2(1+k^2\lambda_D^2)} - \frac{3}{2}\right)\right].} \quad (22.44b)$$

dispersion relation for ion-acoustic waves in an unmagnetized plasma with electrons and ions thermalized at different temperatures, $T_e \gg T_p$

The real part of this dispersion relation was plotted in Fig. 21.1. As is shown there and in the above formulas, for $k\lambda_D \ll 1$ the waves' phase speed is $\sqrt{k_B T_e/m_p}$, and the waves are only weakly damped: they can propagate for roughly $\sqrt{m_p/m_e} \sim 43$ periods before damping has a strong effect. This damping is independent of the proton temperature, so it must be due to surfing electrons. When the wavelength is decreased (k is increased) into the regime $k\lambda_D \gtrsim 1$, the waves' frequency asymptotes toward $\omega_r = \omega_{pp}$, the proton plasma frequency. Then the phase velocity decreases, so more protons can surf the waves, and the Landau damping increases. Equations (22.44) show us that the damping becomes very strong when $k\lambda_D \sim \sqrt{T_e/T_p}$, and that this is also the point at which ω_r/k has decreased to the proton thermal velocity $\sqrt{k_B T_p/m_p}$—which is in accord with our physical arguments about proton surfing.

When T_e/T_p is decreased from $\gg 1$ toward unity, the ion damping becomes strong regardless of how small may be k [cf. the second term of Eq. (22.44b)]. This is also in accord with our physical reasoning.

Ion-acoustic waves are readily excited at Earth's bow shock, where Earth's magnetosphere impacts the solar wind. It is observed that these waves are not able to propagate very far away from the shock, by contrast with Alfvén waves, which are much less rapidly damped.

EXERCISES

Exercise 22.6 *Derivation: Ion-Acoustic Dispersion Relation*
Consider a plasma in which the electrons have a Maxwellian velocity distribution with temperature T_e, the protons are Maxwellian with temperature T_p, and $T_p \ll T_e$. Consider an ion acoustic mode in this plasma for which $\sqrt{k_B T_p/m_p} \ll \omega_r/k \ll \sqrt{k_B T_e/m_e}$ (i.e., the wave's phase velocity is between the left and right tick marks in Fig. 22.3). As was argued in the text, for such a mode it is reasonable to expect weak damping: $|\omega_i| \ll \omega_r$. Making approximations based on these "$\ll$" inequalities, show that the small-$|\omega_i|$ dispersion relation (22.40) reduces to Eqs. (22.44).

Exercise 22.7 *Problem: Dispersion Relations for a Non-Maxwellian Distribution Function*
Consider a plasma with cold protons and hot electrons with a 1-dimensional distribution function proportional to $1/(v_0^2 + v^2)$, so the full 1-dimensional distribution function is

$$F(v) = \frac{nv_0}{\pi(v_0^2 + v^2)} + n\frac{m_e}{m_p}\delta(v). \tag{22.45}$$

(a) Show that the dispersion relation for Langmuir waves, with phase speeds large compared to v_0, is

$$\omega \simeq \omega_{pe} - ikv_0. \tag{22.46}$$

(b) Compute the dispersion relation for ion-acoustic waves, assuming that their phase speeds are much less than v_0 but are large compared to the cold protons' thermal velocities (so the contribution from proton surfing can be ignored). Your result should be

$$\omega \simeq \frac{kv_0(m_e/m_p)^{1/2}}{[1 + (kv_0/\omega_{pe})^2]^{1/2}} - \frac{ikv_0(m_e/m_p)}{[1 + (kv_0/\omega_{pe})^2]^2}. \tag{22.47}$$

22.4 Stability of Electrostatic Waves in Unmagnetized Plasmas

Our small-ω_i dispersion relation (22.40) implies that the sign of F' at resonance dictates the sign of the imaginary part of ω. This raises the interesting possibility that distribution functions that increase with velocity over some range of positive v might be unstable to the exponential growth of electrostatic waves. In fact, the criterion for instability turns out to be a little more complex than this [as one might suspect from the fact that the sign of the denominator of Eq. (22.40b) is not obvious], and deriving it is an elegant exercise in complex-variable theory.

We carry out our derivation in two steps. We first introduce a general method, due to Harry Nyquist, for diagnosing instabilities of dynamical systems. Then we apply Nyquist's method explicitly to electrostatic waves in an unmagnetized plasma and thereby deduce an instability criterion due to Oliver Penrose.

22.4.1 Nyquist's Method

Consider any dynamical system whose modes of oscillation have complex eigenfreqencies ω that are zeros of some function $\mathcal{D}(\omega)$. [In our case the dynamical system is electrostatic waves in an unmagnetized plasma with some chosen wave number k, and because the waves' dispersion relation is $\epsilon(\omega, k) = 0$, Eq. (22.33), the function $\mathcal{D}$ can be chosen as $\mathcal{D}(\omega) = \epsilon(\omega, k)$.] Unstable modes are zeros of $\mathcal{D}(\omega)$ that lie in the upper half of the complex-ω plane.

Assume that $\mathcal{D}(\omega)$ is analytic in the upper half of the ω plane. Then a well-known theorem in complex-variable theory[3] says that the number N_z of zeros of $\mathcal{D}(\omega)$ in the upper half-plane, minus the number N_p of poles, is equal to the number of times that $\mathcal{D}(\omega)$ encircles the origin counterclockwise, in the complex-$\mathcal{D}$ plane, as ω travels counterclockwise along the closed path $\mathcal{C}$ that encloses the upper half of the frequency plane; see Fig. 22.4.

Nyquist diagram and method for deducing the number of unstable modes of a dynamical system

If one knows the number of poles of $\mathcal{D}(\omega)$ in the upper half of the frequency plane, then one can infer, from the Nyquist diagram, the number of unstable modes of the dynamical system.

In Sec. 22.4.2, we use this Nyquist method to derive the Penrose criterion for instability of electrostatic modes of an unmagnetized plasma. As a second example, in Box 22.2, we show how it can be used to diagnose the stability of a feedback control system.

22.4.2 Penrose's Instability Criterion

The straightforward way to apply Nyquist's method to electrostatic waves would be to set $\mathcal{D}(\omega) = \epsilon(\omega, k)$. However, to reach our desired instability criterion more quickly, we set $\mathcal{D} = k^2 \epsilon$; then the zeros of $\mathcal{D}$ are still the electrostatic waves' modes. From Eq. (22.33) for ϵ, we see that

$$\mathcal{D}(\omega) = k^2 - Z(\omega/k), \qquad (22.48a)$$

---

3. This is variously called "the principle of the argument" or "Cauchy's theorem," and it follows from the theorem of residues (e.g., Copson, 1935, Sec. 6.2; Arfken, Weber, and Harris, 2013, Chap. 11).

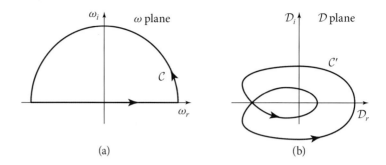

FIGURE 22.4 Nyquist diagram for stability of a dynamical system. (a) The curve C in the complex-ω plane extends along the real frequency axis from $\omega_r = -\infty$ to $\omega_r = +\infty$, then closes up along the semicircle at $|\omega| = \infty$, so it encloses the upper half of the frequency plane. (b) As ω travels along C, $\mathcal{D}(\omega)$ travels along the closed curve C' in the complex-$\mathcal{D}$ plane, which counterclockwise encircles the origin twice. Thus the number of zeros of the analytic function $\mathcal{D}(\omega)$ in the upper half of the frequency plane minus the number of poles is $N_z - N_p = 2$.

where

$$Z(\zeta) \equiv \frac{e^2}{m_e \epsilon_0} \int_{\mathcal{L}} \frac{F'}{v - \zeta} dv, \tag{22.48b}$$

with $\zeta = \omega/k$ the waves' phase velocity.

These equations have several important consequences.

1. For ζ in the upper half-plane—the region that concerns us—we can choose the Landau contour $\mathcal{L}$ to travel along the real v-axis from $-\infty$ to $+\infty$, and the resulting $\mathcal{D}(\omega)$ is analytic in the upper half of the frequency plane, as required for Nyquist's method.

2. For all ζ in the upper half-plane, and for all distribution functions $F(v)$ that are nonnegative and normalizable and thus physically acceptable, the velocity integral is finite, so there are no poles of $\mathcal{D}(\omega)$ in the upper half-plane. Thus there is an unstable mode, for fixed k, if and only if $\mathcal{D}(\omega)$ encircles the origin at least once, as ω travels around the curve C of Fig. 22.4. Note that the encircling is guaranteed to be counterclockwise, since there are no poles.

3. The wave frequency ω traveling along the curve C in the complex-frequency plane is equivalent to ζ traveling along the same curve in the complex-phase-velocity plane; and $\mathcal{D}$ encircling the origin of the complex-$\mathcal{D}$ plane is equivalent to $Z(\zeta)$ encircling the point $Z = k^2$ on the positive real axis of the complex-Z plane.

4. For every point on the semicircular segment of the curve C at $|\zeta| \to \infty$ (Fig. 22.4), $Z(\zeta)$ vanishes, so the curve C can be regarded as going just along the real axis from $-\infty$ to $+\infty$, during which $Z(\zeta)$ emerges from the origin, travels around some curve, and returns to the origin.

BOX 22.2. STABILITY OF A FEEDBACK-CONTROL SYSTEM: ANALYSIS BY NYQUIST'S METHOD T2

A control system can be described quite generally by the following block diagram.

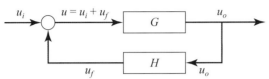

An input signal $u_i(t)$ and the feedback signal $u_f(t)$ are added, then fed through a filter G to produce an output signal $u_o(t) = \int_{-\infty}^{+\infty} G(t - t')u(t')dt'$; or, in the Fourier domain, $\tilde{u}_o(\omega) = \tilde{G}(\omega)\tilde{u}(\omega)$. [See Sec. 6.7 for filtering of signals. Here, for consistency with this plasma-physics chapter, we adopt the opposite sign convention for Fourier transforms from that in Sec. 6.7.] Then the output signal is fed through a filter H to produce the feedback signal $u_f(t)$.

As an example, consider the following simple model for an automobile's cruise control. The automobile's speed v is to be locked to some chosen value V by measuring v and applying a suitable feedback acceleration. To simplify the analysis, we focus on the difference $u \equiv v - V$, which is to be locked to zero. The input to the control system is the speed $u_i(t)$ that the automobile would have in the absence of feedback, plus the speed change $u_f(t)$ due to the feedback acceleration. Their sum, $u = u_i + u_f$, is measured by averaging over a short time interval, with the average exponentially weighted toward the present (in our simple model), so the output of the measurement is $u_o(t) = (1/\tau) \int_{-\infty}^{t} e^{(t'-t)/\tau} u(t')dt'$. By comparing with $u_o = \int_{-\infty}^{+\infty} G(t - t')u(t')dt'$ to infer the measurement filter's kernel $G(t)$, then Fourier transforming, we find that $\tilde{G}(\omega) \equiv \int_{-\infty}^{+\infty} G(t)e^{i\omega t}dt = 1/(1 - i\omega\tau)$. If $u_o(t)$ is positive, we apply to the automobile a negative feedback acceleration $a_f(t)$ proportional to it; if u_o is negative, we apply a positive feedback acceleration; so in either case, $a_f = -K u_o$ for some positive constant K. The feedback speed u_f is the time integral of this acceleration: $u_f(t) = -K \int_{-\infty}^{t} u_o(t')dt'$. Setting this to $\int_{-\infty}^{\infty} H(t - t')u_o(t')dt'$, reading off the kernel H, and computing its Fourier transform, we find $\tilde{H}(\omega) = -K/(i\omega)$.

From the block diagram, we see, fully generally, that in the Fourier domain $\tilde{u}_o = \tilde{G}(\tilde{u}_i + \tilde{u}_f) = \tilde{G}(\tilde{u}_i + \tilde{H}\tilde{u}_o)$; so the output in terms of the input is

$$\tilde{u}_o = \frac{\tilde{G}}{1 + \tilde{G}\tilde{H}}\tilde{u}_i. \qquad (1)$$

(continued)

BOX 22.2. (continued)

Evidently, the feedback system will undergo self-excited oscillations, with no input, at any complex frequency ω that is a zero of $\mathcal{D}(\omega) \equiv 1 + \tilde{G}(\omega)\tilde{H}(\omega)$. If that ω is in the lower half of the complex frequency plane, the oscillations will die out and so are not a problem; but if it is in the upper half-plane, they will grow exponentially with time. Thus the zeros of $\mathcal{D}(\omega)$ in the upper half of the ω plane represent unstable modes of self-excitation and must be avoided in the design of any feedback-control system.

For our cruise-control example, $\mathcal{D}$ is $1 + \tilde{G}\tilde{H} = 1 - K[i\omega(1-i\omega\tau)]^{-1}$, which can be brought into a more convenient form by introducing the dimensionless frequency $z = \omega\tau$ and dimensionless feedback strength $\kappa = K\tau$: $\mathcal{D} = 1 - \kappa[iz(1-iz)]^{-1}$. The Nyquist diagram for this $\mathcal{D}$ has the following form.

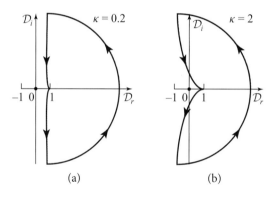

As $z = \omega\tau$ travels around the upper half of the frequency plane (curve $\mathcal{C}$ in Fig. 22.4a), $\mathcal{D}$ travels along the left curve (for feedback strength $\kappa = 0.2$), or the right curve (for $\kappa = 2$) in the above diagram. These curves do not encircle the origin at all—nor does the curve for any other $\kappa > 0$, so the number of zeros minus the number of poles in the upper half-plane is $N_z - N_p = 0$. Moreover, $\mathcal{D} = 1 - \kappa[iz(1-iz)]^{-1}$ has no poles in the upper half-plane, so $N_p = 0$ and $N_z = 0$: our cruise-control feedback system is stable. For further details, see Ex. 22.10.

When designing control systems, it is important to have a significant margin of protection against instability. As an example, consider a control system for which $\tilde{G}\tilde{H} = -\kappa(1+iz)[iz(1-iz)]^{-1}$ (Ex. 22.11). The Nyquist diagrams take the common form shown here.

(continued)

BOX 22.2. (continued)

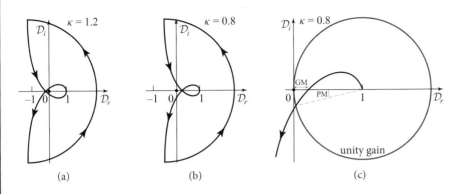

There are no poles in the upper half-plane; and for $\kappa > 1$ (drawing a) the origin is encircled twice, while for $\kappa < 1$ it is not encircled at all (drawing b). Therefore, the control system is unstable for $\kappa > 1$ and stable for $\kappa < 1$. One often wants to push κ as high as possible to achieve one's stabilization goals but must maintain a margin of safety against instability. That margin is quantified by either or both of two quantities: (i) The *phase margin* (labeled PM in diagram c): the amount by which the phase of $\tilde{G}\tilde{H} = \mathcal{D} - 1$ exceeds $180°$ at the unity gain point, $|\tilde{G}\tilde{H}| = 1$ (red curve). (ii) The *gain margin* GM: the amount by which the gain $|\tilde{G}\tilde{H}|$ is less than 1 when the phase of $\tilde{G}\tilde{H}$ reaches $180°$. As κ is increased toward the onset of instability, $\kappa = 1$, both PM and GM approach zero.

For the theory of control systems, see, for example, Franklin, Powell, and Emami-Naeini (2005); Dorf and Bishop (2012).

In view of these facts, Nyquist's method tells us the following: *there will be an unstable mode, for one or more values of k, if and only if, as ζ travels from $-\infty$ to $+\infty$, $Z(\zeta)$ encloses one or more points on the positive real Z-axis. In addition, the wave numbers for any resulting unstable modes are $k = \pm\sqrt{Z}$, for all Z on the positive real axis that are enclosed.*

In Fig. 22.5, we show three examples of $Z(\zeta)$ curves. For Fig. 22.5a, no points on the positive real axis are enclosed, so all electrostatic modes are stable for all wave numbers k. For Fig. 22.5b,c, a segment of the positive real axis is enclosed, so there are unstable modes; those unstable modes have $k = \pm\sqrt{Z}$ for all Z on the enclosed line segment.

Notice that in Fig. 22.5a,b, the rightmost crossing of the real axis is at positive Z, and the curve $\mathcal{C}'$ moves upward as it crosses. A little thought reveals that this must

Nyquist's method for diagnosing unstable electrostatic modes in an unmagnetized plasma

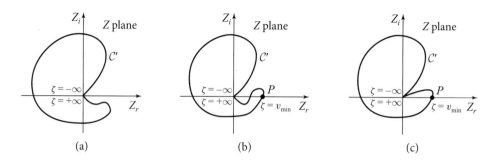

FIGURE 22.5 Nyquist diagrams for electrostatic waves. As a mode's real phase velocity ζ increases from $-\infty$ to $+\infty$, $Z(\zeta)$ travels, in the complex-Z plane, around the closed curve C', which always begins and ends at the origin. (a) The curve C' encloses no points on the positive real axis, so there are no unstable electrostatic modes. (b,c) The curve does enclose a set of points on the positive real axis, so there are unstable modes.

always be the case: $Z(\zeta)$ will encircle, counterclockwise, points on the positive-Z axis if and only if it somewhere crosses the positive-Z axis traveling upward.

alternative Nyquist criterion for instability

Therefore, *an unstable electrostatic mode exists in an unmagnetized plasma if and only if, as ζ travels along its real axis from $-\infty$ to $+\infty$, $Z(\zeta)$ crosses some point P on its positive real axis, traveling upward*. This version of the Nyquist criterion enables us to focus on the small-ω_i domain (while still treating the general case)—for which $\epsilon(\zeta, k)$ is given by Eq. (22.39). From that expression and real ζ (our case), we infer that $Z(\zeta) = k^2[1 - \epsilon]$ is given by

$$Z(\zeta) = \frac{e^2}{m_e \epsilon_0} \left[\int_P \frac{F' \, dv}{v - \zeta} + i\pi F'(\zeta) \right], \quad (22.49a)$$

where P means the Cauchy principal value of the integral. This means that $Z(\zeta)$ crosses its real axis at any ζ where $F'(\zeta) = 0$, and it crosses moving upward if and only if $F''(\zeta) > 0$ at that crossing point. These two conditions together say that, at the crossing point P, $\zeta = v_{\min}$, a particle speed at which $F(v)$ has a minimum. Moreover, Eq. (22.49a) says that $Z(\zeta)$ crosses its positive real axes (rather than negative) if and only if $\int_P [F'/(v - v_{\min})] \, dv > 0$. We can evaluate this integral using an integration by parts:

$$\int_P \frac{F'}{v - v_{\min}} dv = \int_P \frac{d[F(v) - F(v_{\min})]/dv}{v - v_{\min}} dv$$

$$= \int_P \frac{[F(v) - F(v_{\min})]}{(v - v_{\min})^2} dv + \lim_{\delta \to 0} \left[\frac{F(v_{\min} - \delta) - F(v_{\min})}{-\delta} \right. \quad (22.49b)$$

$$\left. - \frac{F(v_{\min} + \delta) - F(v_{\min})}{\delta} \right].$$

The $\lim_{\delta \to 0}$ terms inside the square bracket vanish since $F'(v_{\min}) = 0$, and in the first $\int_P$ term we do not need the Cauchy principal value, because F is a minimum at $v_{\min}$. Therefore, our requirement is that

$$\boxed{\int_{-\infty}^{+\infty} \frac{[F(v) - F(v_{\min})]}{(v - v_{\min})^2} dv > 0.} \qquad (22.50)$$

Thus, *a necessary and sufficient condition for an unstable mode is that there exist some velocity $v_{\min}$ at which the distribution function $F(v)$ has a minimum, and that in addition the minimum be deep enough that the integral (22.50) is positive.* This is called the *Penrose criterion* for instability (Penrose, 1960).

Penrose's criterion for instability of an electrostatic wave in an unmagnetized plasma

For a more in-depth, pedagogical derivation and discussion of the Penrose criterion, see, for example, Krall and Trivelpiece (1973, Sec. 9.6).

EXERCISES

Exercise 22.8 *Example: Penrose Criterion*
Consider an unmagnetized electron plasma with a 1-dimensional distribution function:

$$F(v) \propto [(v - v_0)^2 + u^2]^{-1} + [(v + v_0)^2 + u^2]^{-1}, \qquad (22.51)$$

where v_0 and u are constants. Show that this distribution function possesses a minimum if $v_0 > 3^{-1/2}u$, but the minimum is not deep enough to cause instability unless $v_0 > u$.

Exercise 22.9 *Problem: Range of Unstable Wave Numbers*
Consider a plasma with a distribution function $F(v)$ that has precisely two peaks, at $v = v_1$ and $v = v_2$ [with $F(v_2) \geq F(v_1)$], and a minimum between them at $v = v_{\min}$, and assume that the minimum is deep enough to satisfy the Penrose criterion for instability [Eq. (22.50)]. Show that there will be at least one unstable mode for every wave number k in the range $k_{\min} < k < k_{\max}$, where

$$k_{\min}^2 = \frac{e^2}{\epsilon_0 m_e} \int_{-\infty}^{+\infty} \frac{F(v) - F(v_1)}{(v - v_1)^2} dv, \quad k_{\max}^2 = \frac{e^2}{\epsilon_0 m_e} \int_{-\infty}^{+\infty} \frac{F(v) - F(v_{\min})}{(v - v_{\min})^2} dv. \qquad (22.52)$$

Show, further, that the marginally unstable mode at $k = k_{\max}$ has phase velocity $\omega/k = v_{\min}$, and the marginally unstable mode at $k = k_{\min}$ has $\omega/k = v_1$. [Hint: Use a Nyquist diagram like those in Fig. 22.5.]

Exercise 22.10 *Example and Derivation: Cruise-Control System* **T2**
(a) Show that the cruise-control feedback system described at the beginning of Box 22.2 has $\tilde{G}(z) = 1/(1 - iz)$ and $\tilde{H} = -\kappa/(iz)$, with $z = \omega\tau$ and $\kappa = K\tau$, as claimed.
(b) Show that the Nyquist diagram has the forms shown in the second figure in Box 22.2, and that this control system is stable for all feedback strengths $\kappa > 0$.
(c) Solve explicitly for the zeros of $\mathcal{D} = 1 + \tilde{G}(z)\tilde{H}(z)$, and verify that none are in the upper half of the frequency plane.

(d) To understand the stability from a different viewpoint, imagine that the automobile's speed v is oscillating with an amplitude δv and a real frequency ω around the desired speed V, $v = V + \delta v \sin(\omega t)$, and that the feedback is turned off. Show that the output of the control system is $u_o = [\delta v/\sqrt{1+\omega^2\tau^2}] \sin(\omega t - \Delta\varphi)$, with a phase delay $\Delta\varphi = \arctan(\omega\tau)$ relative to the oscillations of v. Now turn on the feedback, but at a low strength, so it only weakly changes the speed's oscillations in one period. Show that, because $\Delta\varphi < \pi/2$, $d(\delta v^2)/dt$ is negative, so the feedback damps the oscillations. Show that an instability would arise if the phase delay were in the range $\pi/2 < |\Delta\varphi| < 3\pi/2$. For high-frequency oscillations, $\omega\tau \gg 1$, $\Delta\varphi$ approaches $\pi/2$, so the cruise-control system is only marginally stable.

Exercise 22.11 *Derivation: Phase Margin and Gain Margin for a Feedback-Control System* **T2**

Consider the control system discussed in the last two long paragraphs of Box 22.2. It has $\tilde{G}\tilde{H} = -\kappa(1+iz)[iz(1-iz)]^{-1}$, with $z = \omega\tau$ a dimensionless frequency and τ some time constant.

(a) Show that there are no poles of $\mathcal{D} = 1 + \tilde{G}\tilde{H}$ in the upper half of the complex frequency plane (z plane).

(b) Construct the Nyquist diagram for various feedback strengths κ. Show that for $\kappa > 1$ the curve encircles the origin twice (diagram a in the third figure in Box 22.2), so the control system is unstable, while for $\kappa < 1$, it does not encircle the origin (diagram b in the same figure), so the control system is stable.

(c) Show that the phase margin and gain margin, defined in diagram c in the third figure in Box 22.2, approach zero as κ increases toward the instability point, $\kappa = 1$.

(d) Compute explicitly the zeros of $\mathcal{D} = 1 + \tilde{G}\tilde{H}$, and plot their trajectories in the complex frequency plane as κ increases from zero through one to ∞. Verify that two zeros enter the upper half of the frequency plane as κ increases through one, and they remain in the upper half-plane for all $\kappa > 1$, as is guaranteed by the Nyquist diagrams.

22.5 Particle Trapping

We now return to the description of Landau damping. Our treatment so far has been essentially linear in the wave amplitude (or equivalently, in the perturbation to the distribution function). What happens when the wave amplitude is not infinitesimally small?

Consider a single Langmuir wave mode as observed in a frame moving with the mode's phase velocity and, for the moment, ignore its growth or damping. Then in this frame the electrostatic field oscillates spatially, $E = E_0 \sin kz$, but has no time

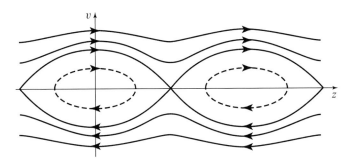

FIGURE 22.6 Phase-space orbits for trapped (dashed lines) and untrapped (solid lines) electrons.

dependence. Figure 22.6 shows some phase-space orbits of electrons in this oscillatory potential. The solid curves are orbits of particles that move fast enough to avoid being trapped in the potential wells at $kz = 0, 2\pi, 4\pi, \ldots$. The dashed curves are orbits of trapped particles. As seen in another frame, these trapped particles are surfing on the wave, with their velocities performing low-amplitude oscillations around the wave's phase velocity ω/k.

trapped and untrapped electrons in a Langmuir wave

The equation of motion for an electron trapped in the minimum $z = 0$ has the form

$$\ddot{z} = \frac{-eE_0 \sin kz}{m_e}$$
$$\simeq -\omega_b^2 z, \qquad (22.53)$$

where we have assumed small-amplitude oscillations and approximated $\sin kz \simeq kz$, and where

$$\omega_b = \left(\frac{eE_0 k}{m_e}\right)^{1/2} \qquad (22.54)$$

is known as the *bounce frequency*. Since the potential well is actually anharmonic, the trapped particles will mix in phase quite rapidly.

The growth or damping of the wave is characterized by a growth or damping of E_0, and correspondingly by a net acceleration or deceleration of untrapped particles, when averaged over a wave cycle. It is this net feeding of energy into and out of the untrapped particles that causes the wave's Landau damping or growth.

Landau damping associated with accelerating or decelerating untrapped particles

Now suppose that the amplitude E_0 of this particular wave mode is changing on a time scale τ due to interactions with the electrons, or possibly (as we shall see in Chap. 23) due to interactions with other waves propagating through the plasma. The potential well then changes on this same timescale, and we expect that τ is also a measure of the maximum length of time a particle can be trapped in the potential well. Evidently, nonlinear wave-trapping effects should only be important when the bounce period $\sim \omega_b^{-1}$ is short compared with τ, that is, when $E_0 \gg m_e/(ek\tau^2)$.

22.5 Particle Trapping

Electron trapping can cause particles to be bunched together at certain preferred phases of a growing wave. This can have important consequences for the radiative properties of the plasma. Suppose, for example, that the plasma is magnetized. Then the electrons gyrate around the magnetic field and emit cyclotron radiation. If their gyrational phases are random, then the total power that they radiate will be the sum of their individual particle powers. However, if N electrons are localized at the same gyrational phase due to being trapped in a potential well of a wave, then they will radiate like one giant electron with a charge $-Ne$. As the radiated power is proportional to the square of the charge carried by the radiating particle, the total power radiated by the bunched electrons will be N times the power radiated by the same number of unbunched electrons. Very large amplification factors are thereby possible both in the laboratory and in Nature, for example, in the Jovian magnetosphere.

radiation from bunched, trapped electrons: example of a nonlinear plasma effect

This brief discussion suggests that there may be much more richness in plasma waves than is embodied in our dispersion relations with their simple linear growth and decay, even when the wave amplitude is small enough that the particle motion is only slightly perturbed by its interaction with the wave. This motivates us to discuss more systematically nonlinear plasma physics, which is the topic of our next chapter.

EXERCISES

Exercise 22.12 *Challenge: BGK Waves*
Consider a steady, 1-dimensional, large-amplitude electrostatic wave in an unmagnetized, proton-electron plasma. Write down the Vlasov equation for each particle species in a frame moving with the wave [i.e., a frame in which the electrostatic potential is a time-independent function of z, $\Phi = \Phi(z)$, not necessarily precisely sinusoidal].

(a) Use Jeans' theorem to argue that proton and electron distribution functions that are just functions of the energy,

$$F_s = F_s(W_s), \qquad W_s = \frac{m_s v_s^2}{2} + q_s \Phi(z), \qquad (22.55a)$$

satisfy the Vlasov equation; here $s = p, e$, and as usual $q_p = e$, $q_e = -e$. Then show that Poisson's equation for the potential Φ can be rewritten in the form

$$\frac{1}{2}\left(\frac{d\Phi}{dz}\right)^2 + V(\Phi) = \text{const}, \qquad (22.55b)$$

where the potential V is $-2/\epsilon_0$ times the kinetic-energy density of the particles:

$$V = \frac{-2}{\epsilon_0} \sum_s \int \frac{1}{2} m_s v_s^2 F_s \, dv_s, \qquad (22.55c)$$

which depends on Φ. (Yes, it is weird to construct a potential V from kinetic energy, and a "kinetic energy" term from the electrical potential Φ, but it also is very clever.)

(b) It is possible to find many electrostatic potential profiles $\Phi(z)$ and distribution functions $F_s[W_s(v, \Phi)]$ that satisfy Eqs. (22.55). These are called BGK waves after

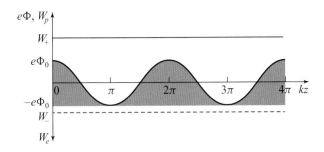

FIGURE 22.7 BGK waves as analyzed in Ex. 22.12. Two quantitites are plotted upward: $e\Phi$, where $\Phi(z)$ is the 1-dimensional electrostatic potential, and W_p, the proton total energy. Values $e\Phi_0$ and $-e\Phi_0$ of $e\Phi$, and W_+ of W_p are marked on the vertical axis. The electron total energy W_e is plotted downward, and its value W_- is marked on the axis. The shaded region shows the range of bound electron energies as a function of location z; the solid line W_+ and dashed line W_- are the energies of the unbound protons and electrons.

Bernstein, Greene, and Kruskal (1957), who first analyzed them. Explain how, in principle, one can solve for (nonunique) BGK distribution functions F_s in a large-amplitude wave of given electrostatic potential profile $\Phi(z)$.

(c) Carry out this procedure, assuming that the potential profile is of the form $\Phi(z) = \Phi_0 \cos kz$, with $\Phi_0 > 0$. Assume also that the protons are monoenergetic, with $W_p = W_+ > e\Phi_0$, and move along the positive z direction. In addition, assume that there are both monoenergetic (with $W_e = W_- > -e\Phi_0$), untrapped electrons (also moving along the positive z direction), and trapped electrons with distribution $F_e(W_e)$, $-e\Phi_0 \leq W_e < e\Phi_0$; see Fig. 22.7. Show that the density of trapped electrons must vanish at the wave troughs [at $z = (2n+1)\pi/k; n = 0, 1, 2, 3\ldots$]. Let the proton density at the troughs be n_{p0}, and assume that there is no net current. Show that the total electron density can then be written as

$$n_e(z) = \left[\frac{m_e(W_+ + e\Phi_0)}{m_p(W_- + e\Phi)}\right]^{1/2} n_{p0} + \int_{-e\Phi}^{e\Phi_0} \frac{dW_e F_e(W_e)}{[2m_e(W_e + e\Phi)]^{1/2}}. \quad (22.56)$$

(d) Use Poisson's equation to show that

$$\int_{-e\Phi}^{e\Phi_0} \frac{dW_e F_e(W_e)}{[2m_e(W_e + e\Phi)]^{1/2}}$$
$$= -\frac{\epsilon_0 k^2 \Phi}{e} + n_{p0} \left[\left(\frac{W_+ + e\Phi_0}{W_+ - e\Phi}\right)^{1/2} - \left(\frac{m_e(W_+ + e\Phi_0)}{m_p(W_- + e\Phi)}\right)^{1/2}\right]. \quad (22.57)$$

(e) Solve the integral equation in part (d) for $F_e(W_e)$. [Hint: It is of Abel type.]
(f) Exhibit some solutions of Eq. (22.57) graphically.

22.6 N-Particle Distribution Function

particle correlations are missed by the Vlasov formalism

Before turning to nonlinear phenomena in plasmas (the next chapter), let us digress briefly and explore ways to study correlations among particles, of which our Vlasov formalism is oblivious.

The Vlasov formalism treats each particle as independent, migrating through phase space in response to the local mean electromagnetic field and somehow unaffected by individual electrostatic interactions with neighboring particles. As we discussed in Chap. 20, this approximation is likely to be good in a collisionless plasma because of the influence of Debye screening—except in the tiny "independent-particle" region of Fig. 20.1. However, we would like to have some means of quantifying this and of deriving an improved formalism that takes into account the correlations among individual particles.

One environment where this formalism may be relevant is the interior of the Sun. Here, although the gas is fully ionized, the Debye number is not particularly large (i.e., one is near the independent-particle region of Fig. 20.1). As a result, Coulomb corrections to the perfect-gas equation of state may be responsible for measurable changes in the Sun's internal structure, as deduced, for example, using helioseismological analysis (cf. Sec. 16.2.4). In this application our task is simpler than in the general case, because the gas will locally be in thermodynamic equilibrium at some temperature T. It turns out that the general case, where the plasma departs significantly from thermal equilibrium, is extremely hard to treat.

k-particle distribution functions

The one-particle distribution function that we use in the Vlasov formalism is the first member of a hierarchy of k-particle distribution functions: $f^{(k)}(\mathbf{x}_1, \mathbf{x}_2, \ldots, \mathbf{x}_k, \mathbf{v}_1, \mathbf{v}_2, \ldots, \mathbf{v}_k, t)$. For example, $f^{(2)}(\mathbf{x}_1, \mathbf{x}_2, \mathbf{v}_1, \mathbf{v}_2, t)d\mathbf{x}_1 d\mathbf{v}_1 d\mathbf{x}_2 d\mathbf{v}_2 \equiv$ [the probability of finding a particle (any particle) in a volume $d\mathbf{x}_1 d\mathbf{v}_1 \equiv dx_1 dy_1 dz_1 dv_{x_1} dv_{y_1} dv_{z_1}$ of phase space, and another particle (any particle) in volume $d\mathbf{x}_2 d\mathbf{v}_2$ of phase space].[4] This definition and its obvious generalization dictate the normalization

$$\int f^{(k)} d\mathbf{x}_1 d\mathbf{v}_1 \cdots d\mathbf{x}_k d\mathbf{v}_k = N^k, \tag{22.58}$$

where $N \gg k \geq 1$ is the total number of particles.

It is useful to relate the distribution functions $f^{(k)}$ to the concepts of statistical mechanics, which we developed in Chap. 4. Suppose we have an ensemble of N-electron plasmas, and let the probability that a member of this ensemble is in a volume $d^N\mathbf{x}\, d^N\mathbf{v}$ of the $6N$-dimensional phase space of all its particles be $f_{\text{all}} d^N\mathbf{x}\, d^N\mathbf{v}$ (N is

N-particle distribution function f_{all} in N-particle plasma

---

4. In this section—and only in this section—we adopt the volume notation commonly used in this multi-particle subject: we use $d\mathbf{x}_j \equiv dx_j dy_j dz_j$ and $d\mathbf{v}_j \equiv dv_{x_j} dv_{y_j} dv_{z_j}$ to denote the volume elements for particle j. Warning: Despite the boldface notation, $d\mathbf{x}_j$ and $d\mathbf{v}_j$ are not vectors! Also, in this section we do *not* study waves, so k represents the number of particles in a distribution function, rather than a wave number.

a very large number!). Of course, f_{all} satisfies the Liouville equation

$$\frac{\partial f_{\text{all}}}{\partial t} + \sum_{i=1}^{N}\left[(\mathbf{v}_i \cdot \boldsymbol{\nabla}_i)f_{\text{all}} + (\mathbf{a}_i \cdot \boldsymbol{\nabla}_{\mathbf{v}i})f_{\text{all}}\right] = 0, \qquad (22.59)$$

Liouville equation for f_{all}

where $\mathbf{a}_i$ is the electromagnetic acceleration of the ith particle, and $\boldsymbol{\nabla}_i$ and $\boldsymbol{\nabla}_{\mathbf{v}i}$ are gradients with respect to the position and velocity of particle i, respectively. We can construct the k-particle "reduced" distribution function from the statistical-mechanics distribution function f_{all} by integrating over all but k of the particles:

$$f^{(k)}(\mathbf{x}_1, \mathbf{x}_2, \ldots, \mathbf{x}_k, \mathbf{v}_1, \mathbf{v}_2, \ldots, \mathbf{v}_k, t)$$
$$= N^k \int d\mathbf{x}_{k+1}\ldots d\mathbf{x}_N d\mathbf{v}_{k+1}\ldots d\mathbf{v}_N f_{\text{all}}(\mathbf{x}_1,\ldots,\mathbf{x}_N,\mathbf{v}_1,\ldots,\mathbf{v}_N). \qquad (22.60)$$

k-particle distribution function constructed from f_{all}

(Note that k is typically a very small number, by contrast with N; below we shall only be concerned with $k = 1, 2, 3$.) The reason for the prefactor N^k in Eq. (22.60) is that, whereas f_{all} refers to the probability of finding particle 1 in $d\mathbf{x}_1 d\mathbf{v}_1$, particle 2 in $d\mathbf{x}_2 d\mathbf{v}_2$, and so forth, the reduced distribution function $f^{(k)}$ describes the probability of finding *any* of the N identical (though distinguishable) particles in $d\mathbf{x}_1 d\mathbf{v}_1$ and so on. (As long as we are dealing with nondegenerate plasmas we can count the electrons classically.) As $N \gg k$, the number of possible ways we can choose k particles for k locations in phase space is approximately N^k.

For simplicity, suppose that the protons are immobile and form a uniform, neutralizing background of charge, so we need only consider the electron distribution and its correlations. Let us further suppose that the forces associated with the mean electromagnetic fields produced by external charges and currents can be ignored. We can then restrict our attention to direct electron-electron electrostatic interactions. The acceleration of an electron is then

$$\mathbf{a}_i = \frac{e}{m_e}\sum_j \boldsymbol{\nabla}_i \Phi_{ij}, \qquad (22.61)$$

where $\Phi_{ij}(x_{ij}) = -e/(4\pi\epsilon_0 x_{ij})$ is the electrostatic potential of electron i at the location of electron j, and $x_{ij} \equiv |\mathbf{x}_i - \mathbf{x}_j|$.

22.6.1 BBGKY Hierarchy

We can now derive the so-called BBGKY hierarchy of kinetic equations (Bogolyubov, 1962; Born and Green, 1949; Kirkwood, 1946; Yvon, 1935) which relate the k-particle distribution function to integrals over the $(k + 1)$-particle distribution function. The first equation in this hierarchy is given by integrating the Liouville equation (22.59) over $d\mathbf{x}_2 \ldots d\mathbf{x}_N d\mathbf{v}_2 \ldots d\mathbf{v}_N$. If we assume that the distribution function decreases to zero at large distances, then integrals of the type $\int d\mathbf{x}_i \mathbf{V}_i f_{\text{all}}$ vanish, and the one-particle distribution function evolves according to

$$\frac{\partial f^{(1)}}{\partial t} + (\mathbf{v}_1 \cdot \mathbf{\nabla}_1) f^{(1)} = \frac{-eN}{m_e} \int d\mathbf{x}_2 \ldots d\mathbf{x}_N d\mathbf{v}_2 \ldots d\mathbf{v}_N \Sigma_j \mathbf{\nabla}_1 \Phi_{1j} \cdot \mathbf{\nabla}_{\mathbf{v}_1} f_{\text{all}}$$

$$= \frac{-eN^2}{m_e} \int d\mathbf{x}_2 \ldots d\mathbf{x}_N d\mathbf{v}_2 \ldots d\mathbf{v}_N \mathbf{\nabla}_1 \Phi_{12} \cdot \mathbf{\nabla}_{\mathbf{v}_1} f_{\text{all}}$$

$$= \frac{-e}{m_e} \int d\mathbf{x}_2 d\mathbf{v}_2 \left(\mathbf{\nabla}_{v1} f^{(2)} \cdot \mathbf{\nabla}_1 \right) \Phi_{12}, \tag{22.62}$$

where, in the second line, we have replaced the probability of having any particle at a location in phase space by N times the probability of having one specific particle there. The left-hand side of Eq. (22.62) describes the evolution of independent particles, and the right-hand side takes account of their pairwise mutual correlation.

The evolution equation for $f^{(2)}$ can similarly be derived by integrating the Liouville equation (22.59) over $d\mathbf{x}_3 \ldots d\mathbf{x}_N d\mathbf{v}_3 \ldots d\mathbf{v}_N$:

$$\frac{\partial f^{(2)}}{\partial t} + (\mathbf{v}_1 \cdot \mathbf{\nabla}_1) f^{(2)} + (\mathbf{v}_2 \cdot \mathbf{\nabla}_2) f^{(2)} + \frac{e}{m_e} \left[\left(\mathbf{\nabla}_{v1} f^{(2)} \cdot \mathbf{\nabla}_1 \right) \Phi_{12} + \left(\mathbf{\nabla}_{v2} f^{(2)} \cdot \mathbf{\nabla}_2 \right) \Phi_{12} \right]$$

$$= \frac{-e}{m_e} \int d\mathbf{x}_3 d\mathbf{v}_3 \left[\left(\mathbf{\nabla}_{v1} f^{(3)} \cdot \mathbf{\nabla}_1 \right) \Phi_{13} + \left(\mathbf{\nabla}_{v2} f^{(3)} \cdot \mathbf{\nabla}_2 \right) \Phi_{23} \right]. \tag{22.63}$$

BBGKY hierarchy of evolution equations for k-particle distribution functions, derived by integrating the Liouville equation

Generalizing and allowing for the presence of a mean electromagnetic field (in addition to the inter-electron electrostatic field) causing an acceleration $\mathbf{a}^{\text{ext}} = -(e/m_e)(\mathbf{E} + \mathbf{v} \times \mathbf{B})$, we obtain the BBGKY hierarchy of kinetic equations:

$$\frac{\partial f^{(k)}}{\partial t} + \sum_{i=1}^{k} \left[(\mathbf{v}_i \cdot \mathbf{\nabla}_i) f^{(k)} + (\mathbf{a}_i^{\text{ext}} \cdot \mathbf{\nabla}_{\mathbf{v}i}) f^{(k)} + \frac{e}{m_e} (\mathbf{\nabla}_{\mathbf{v}_i} f^{(k)} \cdot \mathbf{\nabla}_i) \sum_{j \neq i}^{k} \Phi_{ij} \right]$$

$$= \frac{-e}{m_e} \int d\mathbf{x}_{k+1} d\mathbf{v}_{k+1} \sum_{i=1}^{k} \left(\mathbf{\nabla}_{\mathbf{v}_i} f^{(k+1)} \cdot \mathbf{\nabla}_i \right) \Phi_{ik+1}. \tag{22.64}$$

This kth equation in the hierarchy shows explicitly how we require knowledge of the $(k+1)$-particle distribution function to determine the evolution of the k-particle distribution function.

22.6.2 Two-Point Correlation Function

It is convenient to define the *two-point correlation function*, $\xi_{12}(\mathbf{x}_1, \mathbf{v}_1, \mathbf{x}_2, \mathbf{v}_2, t)$ for particles 1 and 2, by

two-point correlation function

$$\boxed{f^{(2)}(\mathbf{x}_1, \mathbf{v}_1, \mathbf{x}_2, \mathbf{v}_2, t) = f_1 f_2 (1 + \xi_{12}),} \tag{22.65}$$

where we introduce the notation $f_1 = f^{(1)}(\mathbf{x}_1, \mathbf{v}_1, t)$, and $f_2 = f^{(1)}(\mathbf{x}_2, \mathbf{v}_2, t)$. For the analysis that follows to be accurate, we require $|\xi_{12}| \ll 1$. We restrict attention to a plasma in thermal equilibrium at temperature T. In this case, f_1 and f_2 are

Maxwellian distribution functions, independent of $\mathbf{x}$ and t. Now, let us make an ansatz, namely, that ξ_{12} is just a function of the electrostatic interaction energy between the two electrons, and therefore it does not involve the electron velocities. (It is, actually, possible to justify this directly for an equilibrium distribution of particles interacting electrostatically, but we make do with showing that our final answer for ξ_{12} is just a function of $x_{12} = |\mathbf{x}_1 - \mathbf{x}_2|$, in accord with our ansatz.) As Debye screening should be effective at large distances, we anticipate that $\xi_{12} \to 0$ as $x_{12} \to \infty$.

Now turn to Eq. (22.62), and introduce the simplest imaginable closure relation: $\xi_{12} = 0$. In other words, completely ignore all correlations. We can then replace $f^{(2)}$ by $f_1 f_2$ and perform the integral over $\mathbf{x}_2$ and $\mathbf{v}_2$ to recover the collisionless Vlasov equation (22.6). Therefore, we see explicitly that particle-particle correlations are indeed ignored in the simple Vlasov approach.

For the three-particle distribution function, we expect that, when electron 1 is distant from both electrons 2 and 3, then $f^{(3)} \sim f_1 f_2 f_3 (1 + \xi_{23})$, and so forth. Summing over all three pairs, we write:

$$f^{(3)} = f_1 f_2 f_3 (1 + \xi_{23} + \xi_{31} + \xi_{12} + \chi_{123}), \tag{22.66}$$

three-particle distribution function and three-point correlation function

where χ_{123} is the *three-point correlation function* that ought to be significant when all three particles are close together. The function χ_{123} is, of course, determined by the next equation in the BBGKY hierarchy.

We next make the closure relation $\chi_{123} = 0$, that is to say, we ignore the influence of third bodies on pair interactions. This is plausible because close, three-body encounters are even less frequent than close two-body encounters. We can now derive an equation for ξ_{12} by seeking a steady-state solution to Eq. (22.63), that is, a solution with $\partial f^{(2)}/\partial t = 0$. We substitute Eqs. (22.65) and (22.66) into Eq. (22.63) (with $\chi_{123} = 0$) to obtain

closure relation for computing two-point correlation function

$$f_1 f_2 \left[(\mathbf{v}_1 \cdot \nabla_1) \xi_{12} + (\mathbf{v}_2 \cdot \nabla_2) \xi_{12} - \frac{e(1+\xi_{12})}{k_B T} \left\{ (\mathbf{v}_1 \cdot \nabla_1) \Phi_{12} + (\mathbf{v}_2 \cdot \nabla_2) \Phi_{12} \right\} \right]$$
$$= \frac{e f_1 f_2}{k_B T} \int d\mathbf{x}_3 d\mathbf{v}_3 f_3 \left(1 + \xi_{23} + \xi_{31} + \xi_{12} \right) \left[(\mathbf{v}_1 \cdot \nabla_1) \Phi_{13} + (\mathbf{v}_2 \cdot \nabla_2) \Phi_{23} \right],$$
$$\tag{22.67}$$

where we have used the relation

$$\nabla_{\mathbf{v}1} f_1 = -\frac{m_e \mathbf{v}_1 f_1}{k_B T}, \tag{22.68}$$

valid for an unperturbed Maxwellian distribution function. We can rewrite this equation using the relations

$$\nabla_1 \Phi_{12} = -\nabla_2 \Phi_{12}, \quad \nabla_1 \xi_{12} = -\nabla_2 \xi_{12}, \quad \xi_{12} \ll 1, \quad \int d\mathbf{v}_3 f_3 = n, \tag{22.69}$$

to obtain

$$(\mathbf{v}_1 - \mathbf{v}_2) \cdot \nabla_1 \left(\xi_{12} - \frac{e\Phi_{12}}{k_B T} \right)$$
$$= \frac{ne}{k_B T} \int d\mathbf{x}_3 (1 + \xi_{23} + \xi_{31} + \xi_{12})[(\mathbf{v}_1 \cdot \nabla_1)\Phi_{13} + (\mathbf{v}_2 \cdot \nabla_2)\Phi_{23}]. \quad (22.70)$$

Now, symmetry considerations tell us that

$$\int d\mathbf{x}_3 (1 + \xi_{31})\nabla_1 \Phi_{13} = 0, \quad \int d\mathbf{x}_3 (1 + \xi_{23})\nabla_2 \Phi_{23} = 0, \quad (22.71)$$

and, in addition,

$$\int d\mathbf{x}_3 \xi_{12} \nabla_1 \Phi_{13} = -\xi_{12} \int d\mathbf{x}_3 \nabla_3 \Phi_{13} = 0,$$

$$\int d\mathbf{x}_3 \xi_{12} \nabla_2 \Phi_{23} = -\xi_{12} \int d\mathbf{x}_3 \nabla_3 \Phi_{23} = 0. \quad (22.72)$$

Therefore, we end up with

$$(\mathbf{v}_1 - \mathbf{v}_2) \cdot \nabla_1 \left(\xi_{12} - \frac{e\Phi_{12}}{k_B T} \right) = \frac{ne}{k_B T} \int d\mathbf{x}_3 [\xi_{23}(\mathbf{v}_1 \cdot \nabla_1)\Phi_{13} + \xi_{31}(\mathbf{v}_2 \cdot \nabla_2)\Phi_{23}].$$
$$(22.73)$$

As this equation must be true for arbitrary velocities, we can set $\mathbf{v}_2 = 0$ and obtain

$$\nabla_1 (k_B T \xi_{12} - e\Phi_{12}) = ne \int d\mathbf{x}_3 \xi_{23} \nabla_1 \Phi_{13}. \quad (22.74)$$

We take the divergence of Eq. (22.74) and use Poisson's equation, $\nabla_1^2 \Phi_{12} = e\delta(\mathbf{x}_{12})/\epsilon_0$, to obtain

$$\nabla_1^2 \xi_{12} - \frac{\xi_{12}}{\lambda_D^2} = \frac{e^2}{\epsilon_0 k_B T} \delta(\mathbf{x}_{12}), \quad (22.75)$$

where $\lambda_D = [k_B T \epsilon_0/(ne^2)]^{1/2}$ is the Debye length [Eq. (20.10)]. The solution of Eq. (22.75) is

$$\boxed{\xi_{12} = \frac{-e^2}{4\pi \epsilon_0 k_B T} \frac{e^{-x_{12}/\lambda_D}}{x_{12}}.} \quad (22.76)$$

two-point correlation function for electrons in an unmagnetized, thermalized plasma

Note that the sign is negative, because the electrons repel one another. Note also that, to order of magnitude, $\xi_{12}(x_{12} = \lambda_D) \sim -N_D^{-1}$, which is $\ll 1$ in magnitude if the Debye number N_D is much greater than unity. At the mean interparticle spacing, we have $\xi_{12}(x_{12} = n^{-1/3}) \sim -N_D^{-2/3}$. Only for distances $x_{12} \lesssim e^2/(\epsilon_0 k_B T)$ will the correlation effects become large and our expansion procedure and truncation ($\chi_{123} = 0$) become invalid. This analysis justifies the use of the Vlasov equation when $N_D \gg 1$; see the discussion at the end of Sec. 22.6.3.

Exercise 22.13 *Problem: Correlations in a Tokamak Plasma* T2

For a tokamak plasma, compute, to order of magnitude, the two-point correlation function for two electrons separated by

(a) a Debye length, and

(b) the mean interparticle spacing.

22.6.3 Coulomb Correction to Plasma Pressure T2

Let us now turn to the problem of computing the Coulomb correction to the pressure of a thermalized ionized gas. It is easiest to begin by computing the Coulomb correction to the internal energy density. Once again ignoring the protons, this is simply given by

$$U_c = \frac{-e}{2} \int d\mathbf{x}_1 n_1 n_2 \xi_{12} \Phi_{12}, \tag{22.77}$$

where the factor 1/2 compensates for double counting the interactions. Substituting Eq. (22.76) and performing the integral, we obtain

$$U_c = \frac{-ne^2}{8\pi \epsilon_0 \lambda_D} = -\frac{(e^2 n/\epsilon_0)^{3/2}}{8\pi (k_B T)^{1/2}}, \tag{22.78}$$

where n is the number density of electrons. The pressure can be obtained from this energy density using elementary thermodynamics. From the definition (5.32) of the physical free energy converted to a per-unit-volume basis and the first law of thermodynamics [Eq. (5.33)], the volume density of Coulomb free energy, $\mathcal{F}_c$, is given by integrating

$$U_c = -T^2 \left(\frac{\partial (\mathcal{F}_c / T)}{\partial T} \right)_n. \tag{22.79}$$

From this expression, we obtain $\mathcal{F}_c = \frac{2}{3} U_c$. The Coulomb contribution to the pressure is then given by $P = -\partial F/\partial V$ [Eq. (5.33)], rewritten as

$$P_c = n^2 \left(\frac{\partial (\mathcal{F}_c / n)}{\partial n} \right)_T = \frac{1}{3} U_c. \tag{22.80}$$

Therefore, including the Coulomb interaction decreases the pressure at a given density and temperature.

So far we have only allowed the electrons to move and have kept the protons fixed, and we found a Coulomb pressure and energy density independent of the mass of the electron. Therefore, if we had only allowed the protons to move, we would have gotten the same answer. If (as is the case in reality) we allow both protons and electrons to move, we must include the proton-electron attractions as well; but because U_c and P_c are proportional to e^2, these contribute with identical magnitude and sign to the

T2

electron-electron repulsions. Therefore the electrons and protons play completely equivalent roles, and the full electron-proton U_c and P_c are simply obtained by replacing n by the total density of particles, $2n$. Inserting $\lambda_D = \sqrt{\epsilon_0 k_B T/(ne^2)}$, the end result for the Coulomb correction to the pressure is

Coulomb correction to pressure in a thermalized plasma

$$P_c = \frac{-n^{3/2} e^3}{2^{3/2} 3\pi \epsilon_0^{3/2} (k_B T)^{1/2}}, \tag{22.81}$$

where n is still the number density of electrons. Numerically, the gas pressure for a perfect electron-proton gas is

$$P = 1.6 \times 10^{13} (\rho/1{,}000 \text{ kg m}^{-3})(T/10^6 \text{ K}) \text{ N m}^{-2}, \tag{22.82}$$

and the Coulomb correction to this pressure is

$$P_c = -7.3 \times 10^{11} (\rho/1{,}000 \text{ kg m}^{-3})^{3/2} (T/10^6 \text{ K})^{-1/2} \text{ N m}^{-2}. \tag{22.83}$$

In the interior of the Sun this is about 1% of the total pressure. In denser, cooler stars, it is significantly larger.

By contrast, for most of the plasmas that one encounters, our analysis implies that the order of magnitude of the two-point correlation function ξ_{12} is $\sim N_D^{-1}$ across a Debye sphere and only $\sim N_D^{-2/3}$ at the distance of the mean interparticle spacing (see end of Sec. 22.6.2). Only those particles that are undergoing large deflections, through angles ~ 1 radian, are close enough together for $\xi_{12} = O(1)$. This is the ultimate justification for treating plasmas as collisionless and for using mean electromagnetic fields in the Vlasov description.

EXERCISES

Exercise 22.14 *Derivation: Thermodynamic Identities* **T2**
Verify Eqs. (22.79) and (22.80).

Exercise 22.15 *Problem: Thermodynamics of a Coulomb Plasma* **T2**
Compute the entropy of a proton-electron plasma in thermal equilibrium at temperature T including the Coulomb correction.

Bibliographic Note

The kinetic theory of warm plasmas and its application to electrostatic waves and their stability are treated in nearly all texts on plasma physics. For maximum detail and good pedagogy, we particularly like Krall and Trivelpiece (1973, Chaps. 7, 8) (but beware of typographical errors). We also recommend, in Bellan (2006): the early parts of Chap. 2, and all of Chap. 5, and for the extension to magnetized plasmas, Chap. 8.

Also useful are Schmidt (1979, Chaps. 3, 7), Lifshitz and Pitaevskii (1981, Chap. 3), Stix (1992, Chaps. 8, 10), Boyd and Sanderson (2003, Chap. 8), and Swanson (2003, Chap. 4).

For brief discussions of the BBGKY hierarchy of N-particle distribution functions and their predicted correlations in a plasma, see Boyd and Sanderson (2003, Chap. 12) and Swanson (2003, Sec. 4.1.3). For detailed discussions, see the original literature cited in Sec. 22.6.1.

The theory of galaxy dynamics is very clearly developed in Binney and Tremaine (2003). The reader will find many parallels between warm plasmas and self-gravitating stellar distributions.

CHAPTER TWENTY-THREE

Nonlinear Dynamics of Plasmas

Plasma seems to have the kinds of properties one would like for life. It's somewhat like liquid water—unpredictable and thus able to behave in an enormously complex fashion. It could probably carry as much information as DNA does. It has at least the potential for organizing itself in interesting ways.

FREEMAN DYSON (1986)

23.1 Overview

In Sec. 21.6 we met our first example of a velocity-space instability: the two-stream instability, which illustrated the general principle that in a collisionless plasma, departures from Maxwellian equilibrium in velocity space can be unstable and lead to exponential growth of small-amplitude waves. This is similar to the Kelvin-Helmholtz instability in a fluid (Sec. 14.6.1), where a spatially varying fluid velocity produces an instability. In Chap. 22, we derived the dispersion relation for electrostatic waves in a warm, unmagnetized plasma. We discovered Landau damping of the waves when the phase-space density of the resonant particles diminishes with increasing speed, and we showed that in the opposite case of an increasing phase-space density, the waves can grow at the expense of the energies of near-resonant particles (a generalization of the two-stream instability). In this chapter, we explore the back-reaction of the waves on the near-resonant particles. This back-reaction is a (weakly) nonlinear process, so we have to extend our analysis of wave-particle interactions to include the leading nonlinearity.

This extension is called *quasilinear theory* or *the theory of weak plasma turbulence*, and it allows us to follow the time development of the waves and the near-resonant particles simultaneously. We develop this formalism in Sec. 23.2 and verify that it enforces the laws of particle conservation, energy conservation, and momentum conservation. Our initial development of the formalism is entirely in classical language and meshes nicely with the kinetic theory of electrostatic waves as presented in Chap. 22. In Sec. 23.3, we reformulate the theory in terms of the emission, absorption, and scattering of wave quanta, which are called *plasmons*. Although waves in plasmas almost always entail large quantum occupation numbers and thus are highly classical, this quantum formulation of the classical theory has great computational and heuristic power. And, as one might expect, despite the presence of Planck's reduced constant $\hbar$ in the formalism, $\hbar$ nowhere appears in the final answers to problems.

> **BOX 23.1. READERS' GUIDE**
>
> - This chapter relies significantly on:
>   - portions of Chap. 3 on kinetic theory: Secs. 3.2.1 and 3.2.3 on the distribution function, Sec. 3.2.5 on the mean occupation number, and Sec. 3.6 on Liouville's theorem and the collisionless Boltzmann equation;
>   - portions of Secs. 7.2 and 7.3 on geometric optics;
>   - Sec. 20.3 on Debye shielding, collective behavior of plasmas and plasma oscillations; and
>   - Secs. 22.1–22.5 on kinetic theory of warm plasmas.
> - This chapter also relies to some extent but not greatly on:
>   - the concept of spectral density as developed in Sec. 6.4, and
>   - Sec. 6.9 on the Fokker-Planck equation.
> - No subsequent material in this book relies significantly on this chapter.

Our initial derivation and development of the formalism is restricted to the interaction of electrons with electrostatic waves, but in Sec. 23.3 we also describe how the formalism can be generalized to describe a multitude of wave modes and particle species interacting with one another.

We also describe circumstances in which this formalism can fail because the resonant particles couple strongly, *not* to a broad-band distribution of incoherent waves (as the formalism presumes) but instead to one or a few individual, coherent modes. (In Sec. 23.6, we explore an example.)

In Sec. 23.4, we give two illustrative applications of quasilinear theory. For a warm electron beam that propagates through a stable plasma, generating Langmuir plasmons, we show how the plasmons act back on the beam's particle distribution function so as to shut down the plasmon production. For our second application, we describe how galactic cosmic rays (relativistic charged particles) generate Alfvén waves, which in turn scatter the cosmic rays, isotropizing their distribution functions and confining them to our galaxy's interior much longer than one might expect.

In the remainder of this chapter, we describe a few of the many other nonlinear phenomena that can be important in plasmas. In Sec. 23.5, we consider parametric instabilities in plasmas (analogs of the optical parametric amplification that we met in nonlinear crystals in Sec. 10.7). These instabilities are important for the absorption of laser light in experimental studies of the compression of small deuterium-tritium pellets—a possible forerunner of a commercial nuclear fusion reactor. In Sec. 23.6, we return to ion-acoustic solitons (which we studied, for small amplitudes, in Ex. 21.6) and explore how they behave when their amplitudes are so large that nonlinearities are strong. We discover that dissipation can convert such a soliton into a collisionless shock, similar to the bow shock that forms where Earth's magnetic field meets the solar wind.

23.2 Quasilinear Theory in Classical Language

23.2.1 Classical Derivation of the Theory

In Chap. 22, we discovered that hot, thermalized electrons or ions can Landau damp a wave mode. We also showed that some nonthermal particle velocity distributions lead to exponential growth of the waves. Either way there is energy transfer between the waves and the particles. We now turn to the back-reaction of the waves on the near-resonant particles that damp or amplify them. For simplicity, we derive and explore the back-reaction equations ("quasilinear theory") in the special case of electrons interacting with electrostatic Langmuir waves. Then we assert the (rather obvious) generalization to protons or other ions and (less obviously) to interactions with other types of wave modes.

TWO-LENGTHSCALE EXPANSION

We begin with the electrons' 1-dimensional distribution function $F_e(v, z, t)$ [Eq. (22.17)]. As in Sec. 22.3.1, we split F_e into two parts, but we must do so more carefully here than we did there. The foundation for our split is a two-lengthscale expansion of the same sort as we used in developing geometric optics (Sec. 7.3). We introduce two disparate lengthscales, the short one being the typical reduced wavelength of a Langmuir wave $\lambdabar \sim 1/k$, and the long one being a scale $L \gg \lambdabar$ over which we perform spatial averages. Later, when applying our formalism to an inhomogeneous plasma, L must be somewhat shorter than the spatial inhomogeneity scale but still $\gg \lambdabar$. In our present situation of a homogeneous background plasma, there can still be large-scale spatial inhomogeneities caused by the growth or damping of the wave modes via their interaction with the electrons. We must choose L somewhat smaller than the growth or damping length but still large compared to $\lambdabar$.

Our split of F_e is into the spatial average of F_e over the length L (denoted F_0) plus a rapidly varying part that averages to zero (denoted F_1):

$$\boxed{F_0 \equiv \langle F_e \rangle, \quad F_1 \equiv F_e - F_0, \quad F_e = F_0 + F_1.} \quad (23.1)$$

[For simplicity, we omit the subscript e from F_0 and F_1. Therefore—by contrast with Chap. 22, where $F_0 = F_{e0} + (m_e/m_p)F_{p0}$—we here include only F_{e0} in F_0.]

The time evolution of F_e, and hence of F_0 and F_1, is governed by the 1-dimensional Vlasov equation, in which we assume a uniform neutralizing ion (proton) background, no background magnetic field, and interaction with 1-dimensional (planar) electrostatic waves. We cannot use the linearized Vlasov equation (22.19), which formed the foundation for all of Chap. 22, because the processes we wish to study are nonlinear. Rather, we must use the fully nonlinear Vlasov equation [Eq. (22.6) integrated over the irrelevant components v_x and v_y of velocity, as in Eq. (22.17)]:

$$\frac{\partial F_e}{\partial t} + v\frac{\partial F_e}{\partial z} - \frac{eE}{m_e}\frac{\partial F_e}{\partial v} = 0. \tag{23.2}$$

Here E is the rapidly varying electric field (pointing in the z direction) associated with the waves.

Inserting $F_e = F_0 + F_1$ into this Vlasov equation, and arranging the terms in an order that will make sense below, we obtain

$$\frac{\partial F_0}{\partial t} + v\frac{\partial F_0}{\partial z} - \frac{e}{m_e}\frac{\partial F_1}{\partial v}E + \frac{\partial F_1}{\partial t} + v\frac{\partial F_1}{\partial z} - \frac{e}{m_e}\frac{\partial F_0}{\partial v}E = 0. \tag{23.3}$$

We then split this equation into two parts: its average over the large lengthscale L and its remaining time-varying part.

The averaged part gets contributions only from the first three terms (since the last three are linear in F_1 and E, which have vanishing averages):

evolution of slowly varying part F_0 of electron distribution function

$$\boxed{\frac{\partial F_0}{\partial t} + v\frac{\partial F_0}{\partial z} - \frac{e}{m_e}\left\langle\frac{\partial F_1}{\partial v}E\right\rangle = 0.} \tag{23.4}$$

This is an evolution equation for the averaged distribution F_0; the third, nonlinear term drives the evolution. This driving term is the only nonlinearity that we keep in the quasilinear Vlasov equation.

RAPID EVOLUTION OF F_1

The rapidly varying part of the Vlasov equation (23.3) is just the last three terms of Eq. (23.3):

evolution of rapidly varying part F_1 of electron distribution function

$$\boxed{\frac{\partial F_1}{\partial t} + v\frac{\partial F_1}{\partial z} - \frac{e}{m_e}\frac{\partial F_0}{\partial v}E = 0,} \tag{23.5}$$

plus a nonlinear term

$$-\frac{e}{m_e}\left(\frac{\partial F_1}{\partial v}E - \left\langle\frac{\partial F_1}{\partial v}E\right\rangle\right), \tag{23.6}$$

three-wave mixing

which we discard as being far smaller than the linear ones. If we were to keep this term, we would find that it can produce a "three-wave mixing" (analogous to that for light in a nonlinear crystal, Sec. 10.6), in which two electrostatic waves with different wave numbers k_1 and k_2 interact weakly to try to generate a third electrostatic wave with wave number $k_3 = k_1 \pm k_2$. We discuss such three-wave mixing in Sec. 23.3.6; for the moment we ignore it and correspondingly discard the nonlinearity (23.6).

Equation (23.5) is the same linear evolution equation for F_1 as we developed and studied in Chap. 22. Here, as there, we bring its physics to the fore by decomposing into monochromatic modes; but here we are dealing with many modes and sums over the effects of many modes, so we must do the decomposition a little more carefully than in Chap. 22. The foundation for the decomposition is a spatial Fourier transform inside our averaging "box" of length L:

$$\tilde{F}_1(v, k, t) = \int_0^L e^{-ikz} F_1(v, z, t) dz, \quad \tilde{E}(k, t) = \int_0^L e^{-ikz} E(z, t) dz. \quad (23.7)$$

Fourier analysis inside an averaging box

We take F_1 and E to represent the physical quantities and thus to be real; this implies that $\tilde{F}_1(-k) = \tilde{F}_1^*(k)$ and similarly for $\tilde{E}$, so the inverse Fourier transforms are

$$F_1(v, z, t) = \int_{-\infty}^{\infty} e^{ikz} \tilde{F}_1(v, k, t) \frac{dk}{2\pi},$$
$$E(z, t) = \int_{-\infty}^{\infty} e^{ikz} \tilde{E}(v, k, t) \frac{dk}{2\pi}, \quad \text{for } 0 < z < L. \quad (23.8)$$

(This choice of how to do the mathematics corresponds to idealizing F_1 and E as vanishing outside the box; alternatively, we could treat them as though they were periodic with period L and replace Eq. (23.8) by a sum over discrete values of k—multiples of $2\pi/L$.)

From our study of linearized waves in Chap. 21, we know that a mode with real wave number k will oscillate in time with some complex frequency $\omega(k)$:

conventions for linear modes

$$\tilde{F}_1 \propto e^{-i\omega(k)t}, \quad \tilde{E} \propto e^{-i\omega(k)t}. \quad (23.9)$$

For simplicity, we assume that the mode propagates in the $+z$ direction; when studying modes traveling in the opposite direction, we just turn our z-axis around. (In Sec. 23.2.4, we generalize to 3-dimensional situations and include all directions of propagation simultaneously.) For simplicity, we also assume that for each wave number k there is at most one mode type present (i.e., only a Langmuir wave or only an ion-acoustic wave). With these simplifications, $\omega(k)$ is a unique function with its real part positive ($\omega_r > 0$) when $k > 0$. Notice that the reality of $E(z, t)$ implies [from the second of Eqs. (23.7)] that $\tilde{E}(-k, t) = \tilde{E}^*(k, t)$ for all t; in other words [cf. Eqs. (23.9)], $\tilde{E}(-k, 0)e^{-i\omega(-k)t} = \tilde{E}^*(k, 0)e^{+i\omega^*(k)t}$ for all t, which in turn implies

$$\omega(-k) = -\omega^*(k); \quad \text{or} \quad \omega_r(-k) = -\omega_r(k), \quad \omega_i(-k) = \omega_i(k). \quad (23.10)$$

These equations should be obvious physically: they say that, for our chosen conventions, both the negative k and positive k contributions to Eq. (23.8) propagate in the $+z$ direction, and both grow or are damped in time at the same rate. In general, $\omega(k)$ is determined by the Landau-contour dispersion relation (22.33). However, throughout Secs. 23.2–23.4, we specialize to weakly damped or growing Langmuir waves with phase velocities ω_r/k large compared to the rms electron speed:

$$v_{\text{ph}} = \frac{\omega_r}{k} \gg v_{\text{rms}} = \sqrt{\frac{1}{n} \int v^2 F_0(v) dv}. \quad (23.11)$$

23.2 Quasilinear Theory in Classical Language

For such waves, from Eqs. (22.40a), (22.40b), and the first two lines of Eq. (22.42), we deduce the following explicit forms for the real and imaginary parts of ω:

dispersion relation for slowly evolving Langmuir waves

$$\boxed{\omega_r^2 = \omega_p^2 \left(1 + 3\frac{v_{\rm rms}^2}{(\omega_p/k)^2}\right), \quad \text{for } k > 0,} \tag{23.12a}$$

$$\boxed{\omega_i = \frac{\pi e^2}{2\epsilon_0 m_e} \frac{\omega_r}{k^2} F_0'(\omega_r/k), \quad \text{for } k > 0.} \tag{23.12b}$$

The linearized Vlasov equation (23.5) implies that the modes' amplitudes $\tilde{F}_1(v, k, t)$ and $\tilde{E}(k, t)$ are related by

$$\tilde{F}_1 = \frac{ie}{m_e} \frac{\partial F_0/\partial v}{(\omega - kv)} \tilde{E}. \tag{23.13}$$

This is just Eq. (22.20) with d/dv replaced by $\partial/\partial v$, because F_0 now varies slowly in space and time as well as varying with v.

SPECTRAL ENERGY DENSITY $\mathcal{E}_k$

Now leave the rapidly varying quantities F_1 and E and their Vlasov equation (23.5), dispersion relation (23.12a), and damping rate (23.12b), and turn to the spatially averaged distribution function F_0 and its spatially averaged Vlasov equation (23.4). We shall bring this Vlasov equation's nonlinear term into a more useful form. The key quantity in this nonlinear term is the average of the product of the rapidly varying quantities F_1 and E. Parseval's theorem permits us to rewrite this as

$$\langle EF_1 \rangle = \frac{1}{L}\int_0^L EF_1 dz = \frac{1}{L}\int_{-\infty}^{\infty} EF_1 dz = \int_{-\infty}^{\infty} \frac{dk}{2\pi} \frac{\tilde{E}^* \tilde{F}_1}{L}, \tag{23.14}$$

where in the second step we have used our mathematical idealization that F_1 and E vanish outside our averaging box, and the third equality is Parseval's theorem. Inserting Eq. (23.13), we bring Eq. (23.14) into the form

$$\langle EF_1 \rangle = \frac{e}{m_e} \int_{-\infty}^{\infty} \frac{dk}{2\pi} \frac{\tilde{E}^* \tilde{E}}{L} \frac{i}{\omega - kv} \frac{\partial F_0}{\partial v}. \tag{23.15}$$

The quantity $\tilde{E}^*\tilde{E}/L$ is a function of wave number k, time t, and also the location and size L of the averaging box. For Eq. (23.15) to be physically and computationally useful, it is essential that this quantity not fluctuate wildly as k, t, L, and the box location are varied. In most circumstances, if the box is chosen to be far larger than $\lambda = 1/k$, then $\tilde{E}^*\tilde{E}/L$ indeed will not fluctuate wildly. When one develops the quasilinear theory with greater care and rigor than we can do in so brief a treatment, one discovers that this nonfluctuation is a consequence of the *random-phase approximation* or RPA for short—an approximation that the phase of $\tilde{E}$ varies randomly with k, t, L, and

random-phase approximation (RPA)

the box location on suitably short lengthscales.[1] Like ergodicity (Secs. 4.6 and 6.2.3), although the RPA is often valid, sometimes it can fail. Sometimes there is an organized bunching of the particles in phase space that induces nonrandom phases on the plasma waves. Quasilinear theory requires that RPA be valid, and for the moment we assume it is, but in Sec. 23.6 we meet an example for which it fails: strong ion-acoustic solitons.

The RPA implies that, as we increase the length L of our averaging box, $\tilde{E}^*\tilde{E}/L$ will approach a well-defined limit. This limit is half the spectral density $S_E(k)$ of the random process $E(z,t)$ at fixed time t [cf. Eq. (6.25)]. Correspondingly, it is natural to express quasilinear theory in the language of spectral densities. We shall do so, but with a normalization of the spectral density that is tied to the physical energy density and differs slightly from that used in Chap. 6: in place of $S_E(k)$, we use the *Langmuir-wave spectral energy density* $\mathcal{E}_k$. We follow plasma physicists' conventions by defining this quantity to include the oscillatory kinetic energy in the electrons, as well as the electrical energy to which it is, on average, equal. As in the theory of random processes (Chap. 6), we add the energy at $-k$ to that at $+k$, so that all the energy is regarded as residing at positive wave number, and $\int_0^\infty dk\,\mathcal{E}_k$ is the total wave energy per unit volume in the plasma, averaged over length L.

Invoking the RPA, we can use Parseval's theorem to compute the electrical energy density:

$$\frac{\epsilon_0\langle E^2\rangle}{2} = \int_{-\infty}^{\infty}\frac{dk}{2\pi}\epsilon_0\left\langle\frac{\tilde{E}\tilde{E}^*}{2L}\right\rangle = \int_0^{\infty}\frac{dk}{2\pi}\epsilon_0\left\langle\frac{\tilde{E}\tilde{E}^*}{L}\right\rangle, \qquad (23.16)$$

where we have used $\tilde{E}(k)\tilde{E}^*(k) = \tilde{E}(-k)\tilde{E}^*(-k)$. We double this quantity to account for the wave energy in the oscillating electrons and then read off the spectral energy density as the integrand:

$$\boxed{\mathcal{E}_k = \frac{\epsilon_0\langle\tilde{E}\tilde{E}^*\rangle}{\pi L}.} \qquad (23.17)$$

spectral energy density of Langmuir waves

This wave energy density can be regarded as a function either of wave number k or wave phase velocity $v_{\rm ph} = \omega_r/k$. It is useful to plot $\mathcal{E}_k(v_{\rm ph})$ on the same graph as the averaged electron velocity distribution $F_0(v)$. Figure 23.1 is such a plot. It shows the physical situation we are considering: approximately thermalized electrons with (possibly) a weak beam (bump) of additional electrons at velocities $v \gg v_{\rm rms}$, and a distribution of Langmuir waves with phase velocities $v_{\rm ph} \gg v_{\rm rms}$.

EVOLUTION OF WAVE SPECTRAL ENERGY DENSITY $\mathcal{E}_K$

There is an implicit time dependence associated with the growth or decay of the waves, so $\mathcal{E}_k \propto e^{2\omega_i t}$. Now, the waves' energy density $\mathcal{E}_k$ travels through phase space (physical

---

1. For detailed discussions, see Davidson (1972) and Pines and Schrieffer (1962).

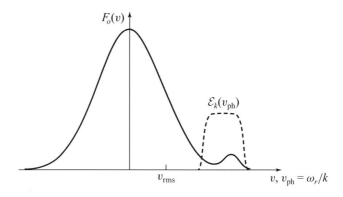

FIGURE 23.1 The spatially averaged electron velocity distribution $F_0(v)$ (solid curve) and wave spectral energy density $\mathcal{E}_k(v_{\rm ph})$ (dashed curve) for the situation treated in Secs. 23.2–23.4.

space and wave-vector space) on the same trajectories as a wave packet (i.e., along the geometric-optics rays discussed in Sec. 7.3), which for our waves propagating in the z direction are given by

derivatives moving with Langmuir-wave rays (wave packets)

$$\boxed{\left(\frac{dz}{dt}\right)_{\rm wp} = \frac{\partial \omega_r}{\partial k_j}, \qquad \left(\frac{dk}{dt}\right)_{\rm wp} = -\frac{\partial \omega_r}{\partial z}} \qquad (23.18)$$

[Eqs. (7.25)]. Correspondingly, the waves' growth or decay actually occurs along these rays (with the averaging boxes also having to be carried along the rays). Thus the equation of motion for the waves' energy density is

evolution of Langmuir waves' spectral energy density along a ray

$$\boxed{\frac{d\mathcal{E}_k}{dt} \equiv \frac{\partial \mathcal{E}_k}{\partial t} + \left(\frac{dz}{dt}\right)_{\rm wp}\frac{\partial \mathcal{E}_k}{\partial z} + \left(\frac{dk}{dt}\right)_{\rm wp}\frac{\partial \mathcal{E}_k}{\partial k} = 2\omega_i \mathcal{E}_k.} \qquad (23.19)$$

Here we have used the fact that the electrostatic waves are presumed to propagate in the z direction, so the only nonzero component of $\mathbf{k}$ is $k_z \equiv k$, and the only nonzero component of the group velocity is $V_{gz} = (dz/dt)_{\rm wp} = (\partial \omega_r/\partial k)_z$. For weakly damped, high-phase-speed Langmuir waves, $\omega_r(\mathbf{x}, k)$ is given by Eq. (23.12a), with the $\mathbf{x}$ dependence arising from the slowly spatially varying electron density $n(\mathbf{x})$, which induces a slow spatial variation in the plasma frequency: $\omega_p = \sqrt{ne^2/(\epsilon_0 m_e)}$.

SLOW EVOLUTION OF F_0

evolution of slowly varying part F_0 of electron distribution function

The context in which the quantity $\tilde{E}(k)^*\tilde{E}(k) = (\pi L/\epsilon_0)\mathcal{E}_k$ arose was our evaluation of the nonlinear term in the Vlasov equation (23.4) for the electrons' averaged distribution function F_0. By inserting Eqs. (23.15) and (23.17) into Eq. (23.4), we bring that nonlinear Vlasov equation into the form

$$\boxed{\frac{\partial F_0}{\partial t} + v\frac{\partial F_0}{\partial z} = \frac{\partial}{\partial v}\left(D\frac{\partial F_0}{\partial v}\right),} \qquad (23.20)$$

where

$$D(v) = \frac{e^2}{2\epsilon_0 m_e^2} \int_{-\infty}^{\infty} dk\, \mathcal{E}_k \frac{i}{\omega - kv}$$

$$= \frac{e^2}{\epsilon_0 m_e^2} \int_0^{\infty} dk\, \mathcal{E}_k \frac{\omega_i}{(\omega_r - kv)^2 + \omega_i^2}. \quad (23.21)$$

diffusion coefficient for electrons

In the second equality we have used Eq. (23.10).

Equation (23.20) says that $F_0(v, z, t)$ is transported in physical space with the electron speed v and diffuses in velocity space with the *velocity diffusion coefficient* $D(v)$. Notice that $D(v)$ is manifestly real, and a major contribution to it comes from waves whose phase speeds ω_r/k nearly match the particle speed v (i.e., from resonant waves).

The two-lengthscale approximation that underlies quasilinear theory requires that the waves grow or damp on a lengthscale long compared to a wavelength, and equivalently, that $|\omega_i|$ be much smaller than ω_r. This allows us, for each v, to split the integral in Eq. (23.21) into a piece due to modes that can resonate with electrons of that speed because $\omega_r/k \simeq v$ plus a piece that cannot so resonate. We consider these two pieces in turn.

The resonant piece of the diffusion coefficient can be written, in the limit $|\omega_i| \ll \omega_r$, by approximating the resonance in Eq. (23.21) as a delta function:

$$\boxed{D^{\text{res}} \simeq \frac{e^2 \pi}{\epsilon_0 m_e^2} \int_0^{\infty} dk\, \mathcal{E}_k \delta(\omega_r - kv).} \quad (23.22\text{a})$$

diffusion coefficient for electrons that are resonant with the waves

In the diffusion equation (23.20) this influences $F_0(v)$ only at those velocities v of resonating waves with substantial wave energy (i.e., on the tail of the electron velocity distribution, under the dashed $\mathcal{E}_k$ curve of Fig. 23.1). We refer to electrons in this region as the *resonant electrons*. In Sec. 23.4, we explore the dynamical influence of this resonant diffusion coefficient on the velocity distribution $F_0(v)$ of the resonant electrons.

Nearly all the electrons reside at velocities $|v| \lesssim v_{\text{rms}}$, where there are no waves (because waves there get damped very quickly). For these *nonresonant electrons*, the denominator in Eq. (23.21) for the diffusion coefficient is approximately equal to $\omega_r^2 \simeq \omega_p^2 = e^2 n/(\epsilon_0 m_e)$, and correspondingly, the diffusion coefficient has the form

$$\boxed{D^{\text{non-res}} \simeq \frac{1}{nm_e} \int_0^{\infty} \omega_i \mathcal{E}_k dk,} \quad (23.22\text{b})$$

diffusion coefficient for nonresonant electrons

which is independent of the electron velocity v.

The nonresonant electrons at $v \lesssim v_{\text{rms}}$ are the ones that participate in the wave motions and account for the waves' oscillating charge density (via F_1). The time-averaged kinetic energy of these nonresonant electrons (embodied in F_0) thus must include a conserved part independent of the waves plus a part associated with the waves. The wave part must be equal to the waves' electrical energy and so to half the waves' total energy, $\frac{1}{2}\int_0^\infty dk \mathcal{E}_k$, and it thus must change at a rate $2\omega_i$ times that energy. Correspondingly, we expect the nonresonant piece of $D(v)$ to produce a rate change of the time-averaged electron energy (embodied in F_0) given by

nonresonant energy conservation

$$\frac{\partial U_e}{\partial t} + \frac{\partial \mathcal{F}_{ez}}{\partial z} = \frac{1}{2}\int_0^\infty dk\, 2\omega_i \mathcal{E}_k, \qquad (23.23)$$

where $\mathcal{F}_{ez}$ is the electron energy flux. Indeed, this is the case; see Ex. 23.1. Because we have already accounted for this electron contribution to the wave energy in our definition of $\mathcal{E}_k$, we ignore it henceforth in the evolution of $F_0(v)$, and correspondingly, for weakly damped or growing waves we focus solely on the resonant part of the diffusion coefficient, Eq. (23.22a).

EXERCISES

Exercise 23.1 *Problem: Nonresonant Particle Energy in Wave*
Show that the nonresonant part of the diffusion coefficient in velocity space, Eq. (23.22b), produces a rate of change of electron kinetic energy given by Eq. (23.23).

23.2.2 Summary of Quasilinear Theory

All the fundamental equations of quasilinear theory are now in hand. They are:

summary: fundamental equations of quasilinear theory for electrons interacting with Langmuir waves

1. The general dispersion relation (22.40) for the (real part of the) waves' frequency $\omega_r(k)$ and their growth rate $\omega_i(k)$ [which, for the high-speed Langmuir waves on which we are focusing, reduces to Eqs. (23.12)]; this dispersion relation depends on the electrons' slowly evolving, time-averaged velocity distribution $F_0(v, z, t)$.

2. The equation of motion (23.19) for the waves' slowly evolving spectral energy density $\mathcal{E}_k(k, z, t)$, in which appear $\omega_r(k)$ and $\omega_i(k)$.

3. Equations (23.21) or (23.22) for the diffusion coefficient $D(v)$ in terms of $\mathcal{E}_k$.

4. The diffusive evolution equation (23.20) for the slow evolution of $F_0(v, z, t)$.

The fundamental functions in this theory are $\mathcal{E}_k(k, z, t)$ for the waves and $F_0(v, z, t)$ for the electrons.

Quasilinear theory sweeps under the rug and ignores the details of the oscillating electric field $E(z, t)$ and the oscillating part of the distribution function $F_1(v, z, t)$. Those quantities were needed in deriving the quasilinear equations, but they are needed no longer—except, sometimes, as an aid to physical understanding.

23.2.3 Conservation Laws

It is instructive to verify that the quasilinear equations enforce the conservation of particles, momentum, and energy.

We begin with particles (electrons). The number density of electrons is $n = \int F_0 dv$, and the z component of particle flux is $S_z = \int n v dv$ (where F_1 contributes nothing because it is oscillatory, and here and below all velocity integrals go from $-\infty$ to $+\infty$). Therefore, by integrating the diffusive evolution equation (23.20) for F_0 over velocity, we obtain

$$\frac{\partial n}{\partial t} + \frac{\partial S_z}{\partial z} = \int \left(\frac{\partial}{\partial t} F_0 + v \frac{\partial}{\partial z} F_0 \right) dv = \int \frac{\partial}{\partial v} \left(D \frac{\partial F_0}{\partial v} \right) dv = 0, \quad (23.24)$$

particle conservation

which is the law of particle conservation for our 1-dimensional situation where nothing depends on x or y.

The z component of electron momentum density is $G_z^e \equiv \int m_e v F_0 dv$, and the zz component of electron momentum flux (stress) is $T_{zz}^e = \int m_e v^2 F_0 dv$; so evaluating the first moment of the evolution equation (23.20) for F_0, we obtain

$$\frac{\partial G_z^e}{\partial t} + \frac{\partial T_{zz}^e}{\partial z} = m_e \int v \left(\frac{\partial F_0}{\partial t} + v \frac{\partial F_0}{\partial z} \right) dv$$

$$= m_e \int v \frac{\partial}{\partial v} \left(D \frac{\partial F_0}{\partial v} \right) dv = -m_e \int D \frac{\partial F_0}{\partial v} dv, \quad (23.25)$$

where we have used integration by parts in the last step. The waves influence the momentum of the resonant electrons through the delta-function part of the diffusion coefficient D [Eq. (23.22a)], and the momentum of the nonresonant electrons through the nonresonant part of the diffusion coefficient (23.22b). Because we have included the evolving part of the nonresonant electrons' momentum and energy as part of the waves' momentum and energy, we must restrict attention in Eq. (23.25) to the resonant electrons (see last sentence in Sec. 23.2.1). We therefore insert the delta-function part of D [Eq. (23.22a)] into Eq. (23.25), thereby obtaining

$$\frac{\partial G_z^e}{\partial t} + \frac{\partial T_{zz}^e}{\partial z} = -\frac{\pi e^2}{\epsilon_0 m_e} \int dv \int dk \mathcal{E}_k \delta(\omega_r - kv) \frac{\partial F_0}{\partial v} = -\frac{\pi e^2}{\epsilon_0 m_e} \int \mathcal{E}_k F_0'(\omega_r/k) \frac{dk}{k}. \quad (23.26)$$

Here we have changed the order of integration, integrated out v, and set $F_0'(v) \equiv \partial F_0/\partial v$. Assuming, for definiteness, high-speed Langmuir waves, we can rewrite the last expression in terms of ω_i with the aid of Eq. (23.12b):

$$\frac{\partial G_z^e}{\partial t} + \frac{\partial T_{zz}^e}{\partial z} = -2 \int dk \omega_i \frac{\mathcal{E}_k k}{\omega_r} = -\frac{\partial}{\partial t} \int dk \frac{\mathcal{E}_k k}{\omega_r} - \frac{\partial}{\partial z} \int dk \frac{\mathcal{E}_k k}{\omega_r} \frac{\partial \omega_r}{\partial k}$$

conservation of total momentum in quasilinear theory

$$= -\frac{\partial G_z^w}{\partial t} - \frac{\partial T_{zz}^w}{\partial z}. \quad (23.27)$$

The second equality follows from Eq. (23.19) together with the fact that for high-speed Langmuir waves, $\omega_r \simeq \omega_p$ so $\partial \omega_r / \partial z \simeq \frac{1}{2}(\omega_p / n_e)\partial n_e / \partial z$ is independent of k. After the second equality in Eq. (23.19), $\int dk\, \mathcal{E}_k k/\omega_r = G_z^w$ is the waves' density of the z component of momentum [as one can see from the fact that each plasmon (wave quantum) carries a momentum $p_z = \hbar k$ and an energy $\hbar \omega_r$; cf. Secs. 7.3.2 and 23.3]. Similarly, since the waves' momentum and energy travel with the group velocity $d\omega_r / dk$, $\int dk\, \mathcal{E}_k (k/\omega_r)(\partial \omega_r / \partial k) = T_{zz}^w$ is the waves' flux of momentum. Obviously, Eq. (23.27) represents the conservation of total momentum, that of the resonant electrons plus that of the waves (which includes the evolving part of the nonresonant electron momentum).

conservation of total energy in quasilinear theory

Energy conservation can be handled in a similar fashion; see Ex. 23.2.

EXERCISES

Exercise 23.2 *Problem: Energy Conservation*
Show that the quasilinear evolution equations guarantee conservation of total energy—that of the resonant electrons plus that of the waves. Pattern your analysis after that for momentum, Eqs. (23.25)–(23.27).

23.2.4 Generalization to 3 Dimensions

quasilinear equations in 3 dimensions

So far, we have restricted our attention to Langmuir waves propagating in one direction, $+\mathbf{e}_z$. The generalization to 3 dimensions is straightforward. The waves' wave number k is replaced by the wave vector $\mathbf{k} = k\hat{\mathbf{k}}$, where $\hat{\mathbf{k}}$ is a unit vector in the direction of the phase velocity. The waves' spectral energy density becomes $\mathcal{E}_\mathbf{k}$, which depends on $\mathbf{k}$, varies slowly in space $\mathbf{x}$ and time t, and is related to the waves' total energy density by

waves' spectral energy density

$$U_w = \int \mathcal{E}_\mathbf{k} d\mathcal{V}_k, \quad (23.28)$$

where $d\mathcal{V}_k \equiv dk_x dk_y dk_z$ is the volume integral in wave-vector space.

Because the plasma is isotropic, the dispersion relation $\omega(\mathbf{k}) = \omega(|\mathbf{k}|)$ has the same form as in the 1-dimensional case, and the group and phase velocities point in the same direction $\hat{\mathbf{k}}$: $\mathbf{V}_{\text{ph}} = (\omega/k)\hat{\mathbf{k}}$, $\mathbf{V}_g = (d\omega_r/dk)\hat{\mathbf{k}}$. The evolution equation (23.19) for the waves' spectral energy density moving along a ray (a wave-packet trajectory) becomes

evolution of waves' spectral energy density

$$\boxed{\frac{d\mathcal{E}_\mathbf{k}}{dt} \equiv \frac{\partial \mathcal{E}_\mathbf{k}}{\partial t} + \left(\frac{dx_j}{dt}\right)_{\text{wp}} \frac{\partial \mathcal{E}_j}{\partial x_j} + \left(\frac{dk_j}{dt}\right)_{\text{wp}} \frac{\partial \mathcal{E}_\mathbf{k}}{\partial k_j} = 2\omega_i \mathcal{E}_\mathbf{k},} \quad (23.29a)$$

and the rays themselves are given by

rays

$$\boxed{\left(\frac{dx_j}{dt}\right)_{\text{wp}} = \frac{\partial \omega_r}{\partial k_j} = V_{g\,j}, \quad \left(\frac{dk_j}{dt}\right)_{\text{wp}} = -\frac{\partial \omega_r}{\partial x_j}.} \quad (23.29b)$$

We shall confine ourselves to the resonant part of the diffusion coefficient, as the nonresonant part is only involved in coupling a wave's electrostatic part to its oscillating, nonresonant-electron part, and we already understand that issue [Eqs. (23.22b), (23.23), and associated discussion] and shall not return to it.

The 3-dimensional velocity diffusion coefficient acts only along the direction of the waves; that is, its $\hat{\mathbf{k}} \otimes \hat{\mathbf{k}}$ component has the same resonant form [Eq. (23.22a)] as in the 1-dimensional case, and components orthogonal to $\hat{\mathbf{k}}$ vanish, so

$$\boxed{\mathbf{D} = \frac{\pi e^2}{\epsilon_0 m_e^2} \int \mathcal{E}_{\mathbf{k}}\, \hat{\mathbf{k}} \otimes \hat{\mathbf{k}}\, \delta(\omega_r - \mathbf{k} \cdot \mathbf{v}) d\mathcal{V}_{\mathbf{k}}.} \quad (23.29c)$$

tensorial diffusion coefficient for resonant electrons

Because the waves will generally propagate in a variety of directions, the net $\mathbf{D}$ is not unidirectional. This diffusion coefficient enters into the obvious generalization of the Fokker-Planck-like evolution equation (23.20) for the averaged distribution function $f_0(\mathbf{v}, \mathbf{x}, t)$—in which we suppress the subscript 0 for ease of notation:

$$\boxed{\frac{\partial f}{\partial t} + \mathbf{v} \cdot \nabla f = \nabla_v \cdot (\mathbf{D} \cdot \nabla_v f), \quad \text{or} \quad \frac{\partial f}{\partial t} + v_j \frac{\partial f}{\partial x_j} = \frac{\partial}{\partial v_i} \left(D_{ij} \frac{\partial f}{\partial v_j} \right).}$$

evolution of electrons' averaged distribution function

$$(23.29d)$$

Here ∇_v (in index notation, $\partial/\partial v_j$) is the gradient in velocity space, not to be confused with the spatial derivative along a vector $\mathbf{v}$, $\nabla_{\mathbf{v}} \equiv \mathbf{v} \cdot \nabla = v_j \partial/\partial x_j$.

23.3 Quasilinear Theory in Quantum Mechanical Language

The attentive reader will have noticed a familiar structure to our quasilinear theory. It is reminiscent of the geometric-optics formalism that we introduced in Chap. 7. Here, as there, we can reinterpret the formalism in terms of quanta carried by the waves. At the most fundamental level, we could (second) quantize the field of plasma waves into quanta (usually called "plasmons"), and describe their creation and annihilation using quantum mechanical transition probabilities. However, there is no need to go through the rigors of the quantization procedure, since the basic concepts of creation, annihilation, and transition probabilities should already be familiar to most readers in the context of photons coupled to atomic systems. Those concepts can be carried over essentially unchanged to plasmons, and by doing so, we recover our quasilinear theory rewritten in quantum language.

plasmons: quantized plasma wave's particles

23.3.1 Plasmon Occupation Number η

A major role in the quantum theory is played by the *occupation number* for electrostatic wave modes, which are just the single-particle states of quantum theory. Our electrostatic waves have spin zero, since there is no polarization freedom (the direction of $\mathbf{E}$ is unique: it must point along $\mathbf{k}$). In other words, there is only one polarization state for each $\mathbf{k}$, so the number of modes (i.e., of quantum states) in a volume

$dV_x dV_k = dxdydzdk_xdk_ydk_z$ of phase space is $dN_{\text{states}} = dV_x dV_k/(2\pi)^3$, and correspondingly, the number density of states in phase space is

$$dN_{\text{states}}/dV_x dV_k = 1/(2\pi)^3 \tag{23.30}$$

(cf. Sec. 3.2.5 with $\mathbf{p} = \hbar\mathbf{k}$). The density of energy in phase space is $\mathcal{E}_\mathbf{k}$, and the energy of an individual plasmon is $\hbar\omega_r$, so the number density of plasmons in phase space is

$$dN_{\text{plasmons}}/dV_x dV_k = \mathcal{E}_\mathbf{k}/\hbar\omega_r. \tag{23.31}$$

Therefore, the states' occupation number is given by

Langmuir plasmons' mean occupation number

$$\boxed{\eta(\mathbf{k}, \mathbf{x}, t) = \frac{dN_{\text{plasmons}}/dV_x dV_k}{dN_{\text{states}}/dV_x dV_k} = \frac{dN_{\text{plasmons}}}{dV_x\, dV_k/(2\pi)^3} = \frac{(2\pi)^3 \mathcal{E}_\mathbf{k}}{\hbar\omega_r}.} \tag{23.32}$$

This is actually the mean occupation number; the occupation numbers of individual states will fluctuate statistically around this mean. In this chapter (as in most of our treatment of statistical physics in Part II of this book), we do not deal with the individual occupation numbers, since quasilinear theory is oblivious to them and deals only with the mean. Thus, without any danger of ambiguity, we simplify our terminology by suppressing the word "mean."

Equation (23.32) says that $\eta(\mathbf{k}, \mathbf{x}, t)$ and $\mathcal{E}_\mathbf{k}$ are the same quantity, aside from normalization. In the classical formulation of quasilinear theory we use $\mathcal{E}_\mathbf{k}$; in the equivalent quantum formulation we use η. We can think of η equally well as a function of the state's wave number $\mathbf{k}$ or of the momentum $\mathbf{p} = \hbar\mathbf{k}$ of the individual plasmons that reside in the state.

The third expression in Eq. (23.32) allows us to think of η as the number density in $\mathbf{x}$-$\mathbf{k}$ phase space, with the relevant phase-space volume renormalized from dV_k to $dV_k/(2\pi)^3 = (dk_x/2\pi)(dk_y/2\pi)(dk_z/2\pi)$. The factors of 2π appearing here are the same ones as appear in the relationship between a spatial Fourier transform and its inverse [e.g., Eq. (23.8)]. We shall see the quantity $dV_k/(2\pi)^3$ appearing over and over again in the quantum mechanical theory, and it can generally be traced to that Fourier transform relationship.

23.3.2 Evolution of η for Plasmons via Interaction with Electrons

CLASSICAL FORMULA FOR EVOLUTION OF η

The motion of individual plasmons is governed by Hamilton's equations with the hamiltonian determined by their dispersion relation:

$$H(\mathbf{p}, \mathbf{x}, t) = \hbar\omega_r(\mathbf{k} = \mathbf{p}/\hbar, \mathbf{x}, t); \tag{23.33}$$

see Sec. 7.3.2. The plasmon trajectories in phase space are, of course, identical to the wave-packet trajectories (the rays) of the classical wave theory, since

$$\frac{dx_j}{dt} = \frac{\partial H}{\partial p_j} = \frac{\partial \omega_r}{\partial k_j}, \quad \frac{dp_j}{dt} = \hbar\frac{\partial k_j}{\partial t} = -\frac{\partial H}{\partial x_j} = -\hbar\frac{\partial \omega_r}{\partial x_j}. \tag{23.34}$$

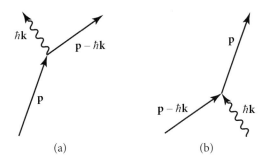

FIGURE 23.2 Feynman diagrams showing (a) creation and (b) annihilation of a plasmon (with momentum $\hbar\mathbf{k}$) by an electron (with momentum $\mathbf{p}$).

Interactions between the waves and resonant electrons cause $\mathcal{E}_\mathbf{k}$ to evolve as $d\mathcal{E}_\mathbf{k}/dt = 2\omega_i \mathcal{E}_\mathbf{k}$ [Eq. (23.29a)]. Therefore, the plasmon occupation number will also vary as

$$\frac{d\eta}{dt} \equiv \frac{\partial \eta}{\partial t} + \frac{dx_j}{dt}\frac{\partial \eta}{\partial x_j} + \frac{dp_j}{dt}\frac{\partial \eta}{\partial p_j} = 2\omega_i \eta, \qquad (23.35)$$

evolution of plasmon (mean) occupation number

where dx_j/dt and dp_j/dt are given by Hamilton's equations (23.34).

QUANTUM FORMULA FOR EVOLUTION OF η

The fundamental process that we are dealing with in Eq. (23.35) is the creation (or annihilation) of a plasmon by an electron (Fig. 23.2). The kinematics of this process is simple: energy and momentum must be conserved in the interaction. In plasmon creation (Fig. 23.2a), the plasma gains one quantum of energy $\hbar\omega_r$ so the electron must lose this same energy: $\hbar\omega_r = -\Delta(m_e v^2/2) \simeq -\Delta\mathbf{p}\cdot\mathbf{v}$, where $\Delta\mathbf{p}$ is the electron's change of momentum, and $\mathbf{v}$ is its velocity; by "$\simeq$" we assume that the electron's fractional change of energy is small (an assumption inherent in quasilinear theory). Since the plasmon momentum change, $\hbar\mathbf{k}$, is minus the electron momentum change, we conclude that

$$\hbar\omega_r = -\Delta(m_e v^2/2) \simeq -\Delta\mathbf{p}\cdot\mathbf{v} = \hbar\mathbf{k}\cdot\mathbf{v}. \qquad (23.36)$$

conservation of energy and momentum in the creation of plasmons: electron resonance and surfing

This is just the resonance condition contained in the delta function $\delta(\omega_r - \mathbf{k}\cdot\mathbf{v})$ of Eq. (23.29c). Thus, energy and momentum conservation in the fundamental plasmon creation process imply that the electron producing the plasmon must resonate with the plasmon's mode (i.e., the component of the electron's velocity along $\mathbf{k}$ must be the same as the mode's phase speed). In other words, the electron must "surf" with the wave mode in which it is creating the plasmon, always remaining in the same trough or crest of the mode.

A fundamental quantity in the quantum description is the probability per unit time for an electron with velocity $\mathbf{v}$ to *spontaneously* emit a Langmuir plasmon into a volume $\Delta\mathcal{V}_k$ in $\mathbf{k}$ space (i.e., the number of $\mathbf{k}$ plasmons emitted by a single velocity-$\mathbf{v}$

electron per unit time into the volume $\Delta \mathcal{V}_k$ centered on some wave vector $\mathbf{k}$). This probability is expressed in the following form:

spontaneous emission of Langmuir plasmons: fundamental probability W

$$\boxed{\left(\frac{dN_{\text{plasmons}}}{dt}\right)_{\text{from one electron}} \equiv W(\mathbf{v}, \mathbf{k}) \frac{\Delta \mathcal{V}_k}{(2\pi)^3}.} \qquad (23.37)$$

In a volume $\Delta \mathcal{V}_x$ of physical space and $\Delta \mathcal{V}_v$ of electron-velocity space, there are $dN_e = f(\mathbf{v}) \Delta \mathcal{V}_x \Delta \mathcal{V}_v$ electrons. Equation (23.37) tells us that these electrons increase the number of plasmons in $\Delta \mathcal{V}_x$, and with $\mathbf{k}$ in $\Delta \mathcal{V}_k$, by

$$\left(\frac{dN_{\text{plasmons}}}{dt}\right)_{\text{from all electrons in } \Delta \mathcal{V}_x \Delta \mathcal{V}_v} \equiv W(\mathbf{v}, \mathbf{k}) \frac{\Delta \mathcal{V}_k}{(2\pi)^3} f(\mathbf{v}) \Delta \mathcal{V}_x \Delta \mathcal{V}_v. \qquad (23.38)$$

Dividing by $\Delta \mathcal{V}_x$ and by $\Delta \mathcal{V}_k/(2\pi)^3$, and using the second expression for η in Eq. (23.32), we obtain for the rate of change of the plasmon occupation number produced by electrons with velocity $\mathbf{v}$ in $\Delta \mathcal{V}_v$:

$$\left(\frac{d\eta(\mathbf{k})}{dt}\right)_{\text{from all electrons in } \Delta \mathcal{V}_v} = W(\mathbf{v}, \mathbf{k}) f(\mathbf{v}) \Delta \mathcal{V}_v. \qquad (23.39)$$

Integrating over all of velocity space, we obtain a final expression for the influence of spontaneous plasmon emission, by electrons of all velocities, on the plasmon occupation number:

spontaneous plasmon emission

$$\left(\frac{d\eta(\mathbf{k})}{dt}\right)_s = \int W(\mathbf{v}, \mathbf{k}) f(\mathbf{v}) d\mathcal{V}_v. \qquad (23.40)$$

(Here and below the subscript s means "spontaneous.") Our introduction of the factor $(2\pi)^3$ in the definition (23.37) of $W(\mathbf{v}, \mathbf{k})$ was designed to avoid a factor $(2\pi)^3$ in this equation for the evolution of the occupation number.

Below [Eq. (23.45)] we deduce the fundamental emission rate W for high-speed Langmuir plasmons by comparing with our classical formulation of quasilinear theory.

Because the plasmons have spin zero, they obey Bose-Einstein statistics, which means that the rate for *induced* emission of plasmons is larger than that for spontaneous emission by the occupation number η of the state that receives the plasmons. Furthermore, the principle of detailed balance ("unitarity" in quantum mechanical language) tells us that W is also the relevant transition probability for the inverse process of absorption of a plasmon in a transition between the same two electron momentum states (Fig. 23.2b). This permits us to write down a *master equation* for the evolution of the plasmon occupation number in a homogeneous plasma:

master equation for evolution of plasmon occupation number

$$\boxed{\frac{d\eta}{dt} = \int W(\mathbf{v}, \mathbf{k}) \left\{ f(\mathbf{v})[1 + \eta(\mathbf{k})] - f(\mathbf{v} - \hbar \mathbf{k}/m_e) \eta(\mathbf{k}) \right\} d\mathcal{V}_v.} \qquad (23.41)$$

The $f(\mathbf{v})$ term in the integrand is the contribution from spontaneous emission, the $f(\mathbf{v})\eta(\mathbf{k})$ term is that from induced emission, and the $f(\mathbf{v} - \hbar\mathbf{k}/m_e)\eta(\mathbf{k})$ term is from absorption.

COMPARISON OF CLASSICAL AND QUANTUM FORMULAS FOR $d\eta/dt$; THE FUNDAMENTAL QUANTUM EMISSION RATE W

The master equation (23.41) is actually the evolution law (23.35) for η in disguise, with the e-folding rate ω_i written in a fundamental quantum mechanical form. To make contact with Eq. (23.35), we first notice that in our classical development of quasilinear theory, for which $\eta \gg 1$, we neglected spontaneous emission, so we drop it from Eq. (23.41). In the absorption term, the momentum of the plasmon is so much smaller than the electron momentum that we can make a Taylor expansion:

$$f(\mathbf{v} - \hbar\mathbf{k}/m_e) \simeq f(\mathbf{v}) - (\hbar/m_e)(\mathbf{k} \cdot \nabla_v)f. \qquad (23.42)$$

Inserting this into Eq. (23.41) and removing the spontaneous-emission term, we obtain

$$\boxed{\frac{d\eta}{dt} \simeq \eta \int W \frac{\hbar}{m_e}(\mathbf{k} \cdot \nabla_v) f \, d\mathcal{V}_v.} \qquad (23.43)$$

For comparison, Eq. (23.35), with ω_i given by the classical high-speed Langmuir relation (23.12b) and converted to 3-dimensional notation, becomes

$$\frac{d\eta}{dt} = \eta \int \frac{\pi e^2 \omega_r}{\epsilon_0 k^2 m_e} \delta(\omega_r - \mathbf{k} \cdot \mathbf{v}) \mathbf{k} \cdot \nabla_v f \, d\mathcal{V}_v. \qquad (23.44)$$

Comparing Eqs. (23.43) and (23.44), we infer that the fundamental quantum emission rate for plasmons must be

$$\boxed{W = \frac{\pi e^2 \omega_r}{\epsilon_0 k^2 \hbar} \delta(\omega_r - \mathbf{k} \cdot \mathbf{v}).} \qquad (23.45)$$

fundamental plasmon emission rate deduced from classical limit

Note that this emission rate is inversely proportional to $\hbar$ and is therefore a very large number under the classical conditions of our quasilinear theory.

This computation has shown that the classical absorption rate $-\omega_i$ is the difference between the quantum mechanical absorption rate and induced emission rate. Under normal conditions, when $\mathbf{k} \cdot \nabla_v f < 0$ ($\partial F_0/\partial v < 0$ in 1-dimensional language), the absorption dominates over emission, so the absorption rate $-\omega_i$ is positive, describing Landau damping. However (as we saw in Chap. 22), when this inequality is reversed, there can be wave growth [subject of course to there being a suitable mode into which the plasmons can be emitted, as guaranteed when the Penrose criterion (22.50) is fulfilled].

SPONTANEOUS EMISSION OF LANGMUIR PLASMONS AS CERENKOV RADIATION

Although spontaneous emission was absent from our classical development of quasilinear theory, it nevertheless can be a classical process and therefore must be added

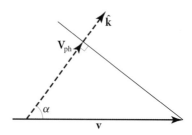

FIGURE 23.3 Geometry for Cerenkov emission. An electron moving with velocity **v** and speed $v = |\mathbf{v}|$ emits waves with phase speed $V_{\text{ph}} < v$ along a direction $\hat{\mathbf{k}}$ that makes an angle $\alpha = \cos^{-1}(V_{\text{ph}}/v)$ with the direction of the electron's motion. The slanted solid line is a phase front.

plasmon emission as Cerenkov radiation

to the quasilinear formalism. Classically or quantum mechanically, the spontaneous emission is a form of *Cerenkov radiation*, since (as for Cerenkov light emitted by electrons moving through a dielectric medium), the plasmons are produced when an electron moves through the plasma faster than the waves' phase velocity. More specifically, only when $v > V_{\text{ph}} = \omega_r/k$ can there be an angle α of **k** relative to **v** along which the resonance condition is satisfied: $\mathbf{v} \cdot \hat{\mathbf{k}} = v \cos \alpha = \omega_r/k$. The plasmons are emitted at this angle α to the electron's direction of motion (Fig. 23.3).

The spontaneous Cerenkov emission rate (23.40) takes the following form when we use the Langmuir expression (23.45) for W:

$$\left(\frac{d\eta}{dt}\right)_s = \frac{\pi e^2}{\epsilon_0 \hbar} \frac{\omega_r}{k^2} \int f(\mathbf{v}) \delta(\omega_r - \mathbf{k} \cdot \mathbf{v}) d\mathcal{V}_v. \tag{23.46}$$

Translated into classical language via Eq. (23.32), this Cerenkov emission rate is

$$\boxed{\left(\frac{d\mathcal{E}_\mathbf{k}}{dt}\right)_s = \frac{e^2}{8\pi^2 \epsilon_0} \frac{\omega_r^2}{k^2} \int f(\mathbf{v}) \delta(\omega_r - \mathbf{k} \cdot \mathbf{v}) d\mathcal{V}_v.} \tag{23.47}$$

Note that Planck's constant is absent from the classical expression, but present in the quantum one.

In the above analysis, we computed the fundamental emission rate W by comparing the quantum induced emission rate minus absorption rate with the classical growth rate for plasma energy. An alternative route to Eq. (23.45) for W would have been to use classical plasma considerations to compute the classical Cerenkov emission rate (23.47), then convert to quantum language using $\eta = (2\pi)^3 \mathcal{E}_\mathbf{k}/\hbar \omega_r$, thereby obtaining Eq. (23.46), and then compare with the fundamental formula (23.40).

conditions for spontaneous Cerenkov emission to be ignorable

By comparing Eqs. (23.44) and (23.46) and assuming a thermal (Maxwellian) distribution for the electron velocities, we see that the spontaneous Cerenkov emission is ignorable in comparison with Landau damping when the electron temperature T_e is smaller than $\eta \hbar \omega_r/k_B$. Sometimes it is convenient to define a classical brightness temperature $T_B(\mathbf{k})$ for the plasma waves given implicitly by

$$\eta(\mathbf{k}) \equiv [e^{\hbar\omega_r/[k_B T_B(\mathbf{k})]} - 1]^{-1} \quad (23.48)$$

$$\simeq \frac{k_B T_B(\mathbf{k})}{\hbar\omega_r} \quad \text{when } \eta(\mathbf{k}) \gg 1, \text{ as in the classical regime.}$$

In this language, spontaneous emission of plasmons with wave vector $\mathbf{k}$ is generally ignorable when the wave brightness temperature exceeds the electron kinetic temperature—as one might expect on thermodynamic grounds. In a plasma in strict thermal equilibrium, Cerenkov emission is balanced by Landau damping, so as to maintain a thermal distribution of Langmuir waves with a temperature equal to that of the electrons: $T_B(\mathbf{k}) = T_e$ for all $\mathbf{k}$.

the balance of Cerenkov emission by Landau damping in thermal equilibrium

EXERCISES

Exercise 23.3 *Problem: Cerenkov Power in Electrostatic Waves*
Show that the Langmuir wave power radiated by an electron moving with speed v in a plasma with plasma frequency ω_p is given by

$$P \simeq \frac{e^2 \omega_p^2}{4\pi \epsilon_0 v} \ln\left(\frac{k_{\max} v}{\omega_p}\right), \quad (23.49)$$

where $k_{\max}$ is the largest wave number at which the waves can propagate. (For larger k the waves are strongly Landau damped.)

23.3.3 Evolution of f for Electrons via Interaction with Plasmons

Now turn from the evolution of the plasmon distribution $\eta(\mathbf{k}, \mathbf{x}, t)$ to that of the particle distribution $f(\mathbf{v}, \mathbf{x}, t)$. Classically, f evolves via the velocity-space diffusion equation (23.29d). We shall write down a fundamental quantum mechanical evolution equation (the "kinetic equation") that appears at first sight to differ remarkably from Eq. (23.29d), but then we will recover (23.29d) in the classical limit.

To derive the electron kinetic equation, we must consider three electron velocity states, $\mathbf{v}$ and $\mathbf{v} \pm \hbar\mathbf{k}/m_e$ (Fig. 23.4). Momentum conservation says that an electron can move between these states by emission or absorption of a plasmon with wave vector $\mathbf{k}$. The fundamental probability for these transitions is the same one, W, as for plasmon

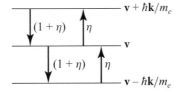

FIGURE 23.4 Three-level system for understanding the electron kinetic equation.

evolution of averaged electron distribution function due to plasmon emission and absorption—fundamental form of equation

emission, since these transitions are merely plasmon emissions and absorptions as seen from the electron's viewpoint. Therefore, the electron kinetic equation must take the form

$$\frac{df(\mathbf{v})}{dt} = \int \frac{d\mathcal{V}_k}{(2\pi)^3} \{(1+\eta)[W(\mathbf{v}+\hbar\mathbf{k}/m_e, \mathbf{k})f(\mathbf{v}+\hbar\mathbf{k}/m_e) - W(\mathbf{v},\mathbf{k})f(\mathbf{v})]$$
$$-\eta[W(\mathbf{v}+\hbar\mathbf{k}/m_e, \mathbf{k})f(\mathbf{v}) - W(\mathbf{v},\mathbf{k})f(\mathbf{v}-\hbar\mathbf{k}/m_e)]\}. \quad (23.50)$$

The four terms can be understood by inspection of Fig. 23.4. The two downward transitions in that diagram entail plasmon emission and thus are weighted by $(1+\eta)$, where the 1 is the spontaneous contribution and the η is the induced emission. In the first of these $(1+\eta)$ terms in Eq. (23.50), the $\mathbf{v}$ electron state gets augmented, so the sign is positive; in the second it gets depleted, so the sign is negative. The two upward transitions entail plasmon absorption and thus are weighted by η; the sign in Eq. (23.50) is plus when the final electron state has velocity $\mathbf{v}$, and minus when the initial state is $\mathbf{v}$.

In the domain of classical quasilinear theory, the momentum of each emitted or absorbed plasmon must be small compared to that of the electron, so we can expand the terms in Eq. (23.50) in powers of $\hbar\mathbf{k}/m_e$. Carrying out that expansion to second order and retaining those terms that are independent of $\hbar$ and therefore classical, we obtain (Ex. 23.4) the *quasilinear electron kinetic equation*:

evolution of averaged electron distribution in quasilinear domain—Fokker-Planck equation

$$\boxed{\frac{df}{dt} = \nabla_v \cdot [\mathbf{R}(\mathbf{v})f + \mathbf{D}(\mathbf{v}) \cdot \nabla_v f],} \quad (23.51a)$$

where ∇_v is the gradient in velocity space (and not $\mathbf{v} \cdot \nabla$), and where

$$\mathbf{R}(\mathbf{v}) = \int \frac{d\mathcal{V}_k}{(2\pi)^3} \frac{W(\mathbf{v},\mathbf{k})\hbar\mathbf{k}}{m_e}, \quad (23.51b)$$

$$\mathbf{D}(\mathbf{v}) = \int \frac{d\mathcal{V}_k}{(2\pi)^3} \frac{\eta(\mathbf{k})W(\mathbf{v},\mathbf{k})\hbar\mathbf{k}\otimes\hbar\mathbf{k}}{m_e^2}. \quad (23.51c)$$

Note that $\mathbf{R}$ is a resistive coefficient associated with spontaneous emission, and $\mathbf{D}$ is a diffusive coefficient associated with plasmon absorption and induced emission in the limit that electron recoil due to spontaneous emission can be ignored. Following Eq. (6.106c), we can rewrite Eq. (23.51c) for $\mathbf{D}$ as

$$\mathbf{D} = \left\langle \frac{\Delta\mathbf{v}\otimes\Delta\mathbf{v}}{\Delta t} \right\rangle \quad (23.51d)$$

with $\Delta\mathbf{v} = -\hbar\mathbf{k}/m_e$.

Our electron kinetic equation (23.51a) has the standard Fokker-Planck form (6.94) derived in Sec. 6.9.1 (generalized to three dimensions), except that the resistive coefficient there is $A = -R + \partial D/\partial v$ and the diffusion coefficient there is $B = 2D$. Note that the diffusive term here has the form $(\partial/\partial v)(D\,\partial f/\partial v)$ discussed in Ex. 6.21 [Eq. (6.102)], as it should, because our formalism does not admit electron recoil (so the probability of a change in **v** due to plasmon absorption is the same as for the opposite change due to plasmon emission).

In our classical, quasilinear analysis, we ignored spontaneous emission and thus had no resistive term **R** in the evolution equation (23.29d) for f. We can recover that evolution equation and its associated **D** by dropping the resistive term from the quantum kinetic equation (23.51a) and inserting expression (23.45) for W into Eq. (23.51c) for the quantum **D**. The results agree with the classical equations (23.29c) and (23.29d).

EXERCISES

Exercise 23.4 *Derivation: Electron Fokker-Planck Equation*
Fill in the missing details in the derivation of the electron Fokker-Planck equation (23.51a). It may be helpful to use index notation.

23.3.4 Emission of Plasmons by Particles in the Presence of a Magnetic Field

Let us return briefly to Cerenkov emission by an electron or ion. In the presence of a background magnetic field **B**, the resonance condition for Cerenkov emission must be modified. Only the momentum parallel to the magnetic field need be conserved, not the total vectorial momentum. The unbalanced components of momentum perpendicular to **B** are compensated by a reaction force from **B** itself and thence by the steady currents that produce **B**; correspondingly, the unbalanced perpendicular momentum ultimately does work on those currents. In this situation, one can show that the Cerenkov resonance condition $\omega_r - \mathbf{k} \cdot \mathbf{v} = 0$ is modified to

$$\omega_r - \mathbf{k}_\parallel \cdot \mathbf{v}_\parallel = 0, \tag{23.52}$$

where $\parallel$ means the component parallel to **B**. If we allow for the electron gyrational motion as well, then some number N of gyrational (cyclotron) quanta can be fed into each emitted plasmon, so Eq. (23.52) gets modified to read

$$\omega_r - \mathbf{k}_\parallel \cdot \mathbf{v}_\parallel = N\omega_{ce}, \tag{23.53}$$

Cerenkov resonance condition in presence of a magnetic field

where N is an integer. For nonrelativistic electrons, the strongest resonance is for $N = 1$. We use this modified Cerenkov resonance condition in Sec. 23.4.1.

23.3.5 Relationship between Classical and Quantum Mechanical Formalisms

We have demonstrated how the structure of the classical quasilinear equations is mandated by quantum mechanics. In developing the quantum equations, we had to rely on one classical calculation, Eq. (23.44), which gave us the emission rate W. However, even this was not strictly necessary, since with significant additional effort we could have calculated the relevant quantum mechanical matrix elements and then computed W directly from Fermi's golden rule.[2] This has to be the case, because

---

2. See, for example, Cohen-Tannoudji, Diu, and Laloë (1977), and, in the plasma physics context, Melrose (2008, 2012).

quantum mechanics is the fundamental physical theory and thus must be applicable to plasma excitations just as it is applicable to atoms. Of course, if we are only interested in classical processes, as is usually the case in plasma physics, then we end up taking the limit $\hbar \to 0$ in all observable quantities, and the classical rate is all we need.

using quantum calculations in the classical domain

This raises an important point of principle. Should we perform our calculations classically or quantum mechanically? The best answer is to be pragmatic. Many calculations in nonlinear plasma theory are so long and arduous that we need all the help we can get to complete them. We therefore combine both classical and quantum considerations (confident that both must be correct throughout their overlapping domain of applicability), in whatever proportion minimizes our computational effort.

23.3.6 Evolution of η via Three-Wave Mixing

We have discussed plasmon emission and absorption both classically and quantum mechanically. Our classical and quantum formalisms can be generalized straightforwardly to encompass other nonlinear processes.

Among the most important other processes are three-wave interactions (in which two waves coalesce to form a third wave or one wave splits up into two) and scattering processes (in which waves are scattered off particles without creating or destroying plasmons). In this section we focus on three-wave mixing. We present the main ideas in the text but leave most of the details to Exs. 23.5 and 23.6.

three-wave mixing in general

In three-wave mixing, where waves A and B combine to create wave C (Fig. 23.5a), the equation for the growth of the amplitude of wave C will contain nonlinear driving terms that combine the harmonic oscillations of waves A and B, that is, driving terms proportional to $\exp[i(\mathbf{k}_A \cdot \mathbf{x} - \omega_A t)] \exp[i(\mathbf{k}_B \cdot \mathbf{x} - \omega_B t)]$. For wave C to build up coherently over many oscillation periods, it is necessary that the spacetime dependence of these driving terms be the same as that of wave C, $\exp[i(\mathbf{k}_C \cdot \mathbf{x} - \omega_C t)]$; in other words, it is necessary that

$$\mathbf{k}_C = \mathbf{k}_A + \mathbf{k}_B, \qquad \omega_C = \omega_A + \omega_B. \tag{23.54}$$

Quantum mechanically, these expressions can be recognized as momentum and energy conservation for the waves' plasmons. We have met three-wave mixing previously, for electromagnetic waves in a nonlinear dielectric crystal (Sec. 10.6). There, as here, the conservation laws (23.54) were necessary for the mixing to proceed; see Eqs. (10.29) and associated discussion.

When the three waves are all electrostatic, the three-wave mixing arises from the nonlinear term (23.6) in the rapidly varying part [wave part, Eq. (23.5)] of the Vlasov equation, which we discarded in our quasilinear analysis. Generalized to 3 dimensions with this term treated as a driver, that Vlasov equation takes the form

three-wave mixing for electrostatic waves: driving term from quasilinear analysis

$$\frac{\partial f_1}{\partial t} + \mathbf{v} \cdot \nabla f_1 - \frac{e}{m_e} \mathbf{E} \cdot \nabla_v f_0 = \frac{e}{m_e} \left(\mathbf{E} \cdot \nabla_v f_1 - \langle \mathbf{E} \cdot \nabla_v f_1 \rangle \right). \tag{23.55}$$

In the driving term (right-hand side of this equation), $\mathbf{E}$ could be the electric field of wave A and f_1 could be the perturbed velocity distribution of wave B or vice versa,

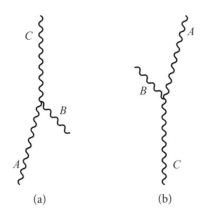

FIGURE 23.5 (a) A three-wave process in which two plasmons A and B interact nonlinearly to create a third plasmon C. Conserving energy and linear momentum, we obtain $\omega_C = \omega_A + \omega_B$ and $\mathbf{k}_C = \mathbf{k}_A + \mathbf{k}_B$. For example, A and C might be transverse electromagnetic plasmons satisfying the dispersion relation (21.24), and B might be a longitudinal plasmon (Langmuir or ion acoustic); or A and C might be Langmuir plasmons, and B might be an ion-acoustic plasmon—the case treated in the text and in Exs. 23.5 and 23.6. (b) The time-reversed three-wave process in which plasmon C generates plasmon B by an analog of Cerenkov emission, and while doing so recoils into plasmon state A.

and the $\mathbf{E}$ and f_1 terms on the left side could be those of wave C. If the wave vectors and frequencies are related by Eq. (23.54), then via this equation waves A and B will coherently generate wave C.

The dispersion relations for Langmuir and ion-acoustic waves permit the conservation law (23.54) to be satisfied if A is Langmuir (so $\omega_A \sim \omega_{pe}$), B is ion acoustic (so $\omega_B \lesssim \omega_{pp} \ll \omega_A$), and C is Langmuir. By working out the detailed consequences of the driving term (23.55) in the quasilinear formalism and comparing with the quantum equations for three-wave mixing (Ex. 23.5), one can deduce the fundamental rate for the process $A + B \to C$ [Fig. 23.5a; Eqs. (23.56)]. Detailed balance (unitarity) guarantees that the time-reversed process $C \to A + B$ (Fig. 23.5b) will have identically the same fundamental rate. This time-reversed process has a physical interpretation analogous to the emission of a Cerenkov plasmon by a high-speed, resonant electron: C is a "high-energy" Langmuir plasmon ($\omega_C \sim \omega_{pe}$) that can be thought of as Cerenkov-emitting a "low-energy" ion-acoustic plasmon ($\omega_B \lesssim \omega_{pp} \ll \omega_C$) and in the process recoiling slightly into Langmuir state A.

electrostatic three-wave mixing—two Langmuir waves (A and C) and one ion-acoustic wave (B)

The fundamental rate that one obtains for this wave-wave Cerenkov process $A + B \to C$ and its time reversal $C \to A + B$, when the plasma's electrons are thermalized at temperature T_e, is (Ex. 23.5)[3]

$$W_{AB\leftrightarrow C} = R_{AB\leftrightarrow C}(\mathbf{k}_A, \mathbf{k}_B, \mathbf{k}_C)\delta(\mathbf{k}_A + \mathbf{k}_B - \mathbf{k}_C)\delta(\omega_A + \omega_B - \omega_C), \quad (23.56a)$$

fundamental rate for Cerenkov emission of ion-acoustic plasmon by Langmuir plasmon

---

3. See also Tsytovich (1970, Eq. A.3.12). The rates for many other wave-wave mixing processes are worked out in this book, but beware: it contains a large number of typographical errors.

where

$$R_{AB\leftrightarrow C}(\mathbf{k}_A, \mathbf{k}_B, \mathbf{k}_C) = \frac{8\pi^5 \hbar e^2 (m_p/m_e)\omega_B^3}{(k_B T_e)^2 k_{ia}^2}(\hat{\mathbf{k}}_A \cdot \hat{\mathbf{k}}_C)^2.\quad (23.56b)$$

[Here we have written the ion-acoustic plasmon's wave number as k_{ia} instead of k_B, to avoid confusion with Boltzmann's constant; i.e., $k_{ia} \equiv |\mathbf{k}_B|$.]

This is the analog of the rate (23.45) for Cerenkov emission by an electron. The ion-acoustic occupation number will evolve via an evolution law analogous to Eq. (23.41) with this rate replacing W on the right-hand side, η replaced by the ion-acoustic occupation number η_B, and the electron distribution replaced by the A-mode or C-mode Langmuir occupation number; see Ex. 23.5. Moreover, there will be a similar evolution law for the Langmuir occupation number, involving the same fundamental rate (23.56); Ex. 23.6.

EXERCISES

Exercise 23.5 *Example and Challenge: Three-Wave Mixing—Ion-Acoustic Evolution*
Consider the three-wave processes shown in Fig. 23.5, with A and C Langmuir plasmons and B an ion-acoustic plasmon. The fundamental rate is given by Eqs. (23.56a) and (23.56b).

(a) By summing the rates of forward and backward reactions (Fig. 23.5a,b), show that the occupation number for the ion-acoustic plasmons satisfies the kinetic equation

$$\frac{d\eta_B}{dt} = \int W_{AB\leftrightarrow C}[(1+\eta_A+\eta_B)\eta_C - \eta_A \eta_B]\frac{d\mathcal{V}_{k_A}}{(2\pi)^3}\frac{d\mathcal{V}_{k_C}}{(2\pi)^3}. \quad (23.57)$$

[Hint: (i) The rate for $A + B \to C$ (Fig. 23.5a) will be proportional to $(\eta_C+1)\eta_A\eta_B$; why? (ii) When you sum the rates for the two diagrams, Fig. 23.5a and b, the terms involving $\eta_A\eta_B\eta_C$ should cancel.]

(b) The ion-acoustic plasmons have far lower frequencies than the Langmuir plasmons, so $\omega_B \ll \omega_A \simeq \omega_C$. Assume that they also have far lower wave numbers, $|\mathbf{k}_B| \ll |\mathbf{k}_A| \simeq |\mathbf{k}_C|$. Assume further (as will typically be the case) that the ion-acoustic plasmons, because of their tiny individual energies, have far larger occupation numbers than the Langmuir plasmons, so $\eta_B \gg \eta_A \sim \eta_C$. Using these approximations, show that the evolution law (23.57) for the ion-acoustic waves reduces to the form

$$\frac{d\eta_B(\mathbf{k})}{dt}$$
$$= \eta_B(\mathbf{k})\int R_{AB\leftrightarrow C}(\mathbf{k}'-\mathbf{k}, \mathbf{k}, \mathbf{k}')\delta[\omega_B(\mathbf{k}) - \mathbf{k}\cdot\mathbf{V}_{gL}(\mathbf{k}')]\,\mathbf{k}\cdot\nabla_{\mathbf{k}'}\eta_L(\mathbf{k}')\frac{d\mathcal{V}_{k'}}{(2\pi)^6},$$
$$(23.58)$$

where η_L is the Langmuir (waves A and C) occupation number, $\mathbf{V}_{gL}$ is the Langmuir group velocity, and $R_{C\leftrightarrow BA}$ is the fundamental rate (23.56b).

(c) Notice the strong similarities between the evolution equation (23.58) for the ion-acoustic plasmons that are Cerenkov emitted and absorbed by Langmuir plasmons, and the evolution equation (23.44) for Langmuir plasmons that are Cerenkov emitted and absorbed by fast electrons! Discuss the similarities and the physical reasons for them.

(d) *Challenge:* Carry out an explicit classical calculation of the nonlinear interaction between Langmuir waves with wave vectors $\mathbf{k}_A$ and $\mathbf{k}_C$ to produce ion-acoustic waves with wave vector $\mathbf{k}_B = \mathbf{k}_C - \mathbf{k}_A$. Base your calculation on the nonlinear Vlasov equation (23.55) and [for use in relating $\mathbf{E}$ and f_1 in the nonlinear term] the 3-dimensional analog of Eq. (23.13). Assume a spatially independent Maxwellian averaged electron velocity distribution f_0 with temperature T_e (so $\nabla f_0 = 0$). From your result compute, in the random-phase approximation, the evolution of the ion-acoustic energy density $\mathcal{E}_\mathbf{k}$ and thence the evolution of the occupation number $\eta(\mathbf{k})$. Bring that evolution equation into the functional form (23.58). By comparing quantitatively with Eq. (23.58), read off the fundamental rate $R_{C\leftrightarrow BA}$. Your result should be the rate in Eq. (23.56b).

Exercise 23.6 *Example and Challenge: Three-Wave Mixing—Langmuir Evolution*
This exercise continues the analysis of the preceding one.

(a) Derive the kinetic equation for the Langmuir occupation number. [Hint: You will have to sum over four Feynman diagrams, corresponding to the mode of interest playing the role of A and then the role of C in each of the two diagrams in Fig. 23.5.]

(b) Using the approximations outlined in part (b) of Ex. 23.5, show that the Langmuir occupation number evolves in accord with the diffusion equation

$$\frac{d\eta_L(\mathbf{k}')}{dt} = \nabla_{\mathbf{k}'} \cdot [\mathbf{D}(\mathbf{k}') \cdot \nabla_{\mathbf{k}'}\eta_L(\mathbf{k}')], \qquad (23.59a)$$

where the diffusion coefficient is given by the following integral over the ion-acoustic-wave distribution:

$$\mathbf{D}(\mathbf{k}') = \int \eta_B(\mathbf{k})\, \mathbf{k} \otimes \mathbf{k}\, R_{AB\leftrightarrow C}(\mathbf{k}'-\mathbf{k}, \mathbf{k}, \mathbf{k}')\, \delta[\omega_B(\mathbf{k}) - \mathbf{k}\cdot\mathbf{V}_{gL}(\mathbf{k}')]\frac{d\mathcal{V}_{\mathbf{k}'}}{(2\pi)^6}.$$

(23.59b)

(c) Discuss the strong similarity between the evolution law (23.59) for resonant Langmuir plasmons interacting with ion-acoustic waves, and the one for resonant electrons interacting with Langmuir waves [Eqs. (23.29c), (23.29d)]. Why are they so similar?

23.4 Quasilinear Evolution of Unstable Distribution Functions—A Bump in the Tail

In plasma physics, one often encounters a weak beam of electrons passing through a stable Maxwellian plasma, with electron density n_e, beam density n_b, and beam speed v_b much larger than the thermal width σ_e of the background plasma. When the velocity width of the beam σ_b is small compared with v_b, the distribution is known as a *bump-in-tail* distribution (see Fig. 23.6). In this section, we explore the stability and nonlinear evolution of such a distribution.

We focus on the simple case of a 1-dimensional electron distribution function $F_0(v)$ and approximate the beam by a Maxwellian distribution:

$$F_b(v) = \frac{n_b}{(2\pi)^{1/2}\sigma_b} e^{-(v-v_b)^2/(2\sigma_b^2)}, \qquad (23.60)$$

where n_b is the beam electron density. For simplicity, we treat the protons as a uniform neutralizing background.

Now, let us suppose that at time $t = 0$, the beam is established, and the Langmuir wave energy density $\mathcal{E}_k$ is very small. The waves will grow fastest when the waves' phase velocity $V_{\rm ph} = \omega_r/k$ resides where the slope of the distribution function is most positive (i.e., when $V_{\rm ph} = v_b - \sigma_b$). The associated maximum growth rate as computed from Eq. (23.12b) is

maximum growth rate for Langmuir waves

$$\omega_{i\max} = \left(\frac{\pi}{8e}\right)^{1/2}\left(\frac{v_b}{\sigma_b}\right)^2\left(\frac{n_b}{n_e}\right)\omega_p, \qquad (23.61)$$

where $e = 2.718\ldots$ is not the electron charge. Modes will grow over a range of wave phase velocities $\Delta V_{\rm ph} \sim \sigma_b$. By using the Bohm-Gross dispersion relation (21.31) rewritten in the form

$$\omega = \omega_p(1 - 3\sigma_e^2/V_{\rm ph}^2)^{-1/2}, \qquad (23.62)$$

we find that the bandwidth of the growing modes is given roughly by

$$\Delta\omega = K\omega_p \frac{\sigma_b}{v_b}, \qquad (23.63)$$

where $K = 3(\sigma_e/v_b)^2[1 - 3(\sigma_e/v_b)^2]^{-3/2}$ is a constant typically $\gtrsim 0.1$. Combining Eqs. (23.61) and (23.63), we obtain

$$\frac{\omega_{i\max}}{\Delta\omega} \sim \left(\frac{\pi}{8eK^2}\right)^{1/2}\left(\frac{v_b}{\sigma_b}\right)^3\left(\frac{n_b}{n_e}\right). \qquad (23.64)$$

Dropping constants of order unity, we conclude that the waves' growth time $\sim(\omega_{i\max})^{-1}$ is long compared with their coherence time $\sim(\Delta\omega)^{-1}$, provided that

$$\sigma_b \gtrsim \left(\frac{n_b}{n_e}\right)^{1/3} v_b. \qquad (23.65)$$

validity of quasilinear analysis

When inequality (23.65) is satisfied, the waves will take several coherence times to grow, and so we expect that no permanent phase relations will be established in

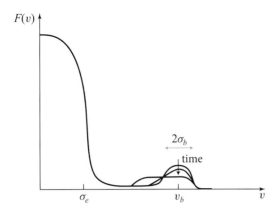

FIGURE 23.6 Evolution of the 1-dimensional electron distribution function from a bump-on-tail shape to a flat distribution function, due to the growth and scattering of electrostatic waves. The direction of evolution is indicated by the downward-pointing arrow at $v = v_b$.

the electric field and that quasilinear theory is an appropriate tool. However, when this inequality is reversed, the instability resembles more the two-stream instability of Chap. 21, and the growth is so rapid as to imprint special phase relations on the waves, so the random-phase approximation fails and quasilinear theory is invalid.

Restricting ourselves to slow growth, we use quasilinear theory to explore the evolution of the wave and particle distributions. We can associate the wave energy density $\mathcal{E}_k$ not just with a given value of k but also with a corresponding value of $V_{\text{ph}} = \omega_r/k$, and thence with the velocities $v = V_{\text{ph}}$ of electrons that resonate with the waves. Using Eq. (23.22a) for the velocity diffusion coefficient and Eq. (23.12b) for the associated wave-growth rate, we can then write the temporal evolution equations for the electron distribution function $F_0(v, t)$ and the wave energy density $\mathcal{E}_k(v, t)$ as

$$\frac{\partial F_0}{\partial t} = \frac{\pi e^2}{m_e^2 \epsilon_0} \frac{\partial}{\partial v}\left(\frac{\mathcal{E}_k}{v} \frac{\partial F_0}{\partial v}\right), \quad \frac{\partial \mathcal{E}_k}{\partial t} = \frac{\pi e^2}{m_e \epsilon_0 \omega_p} v^2 \mathcal{E}_k \frac{\partial F_0}{\partial v}. \tag{23.66}$$

evolution equations for bump-in-tail instability of electrons interacting with Langmuir waves that all propagate in the same direction

Here $v = \omega_r/k$ and, for simplicity, we have assumed a spatially homogeneous distribution of particles and waves so $d/dt \rightarrow \partial/\partial t$.

This pair of nonlinear equations must be solved numerically, but their qualitative behavior can be understood analytically without much effort; see Fig. 23.6. Waves resonant with the rising part of the electron distribution function at first grow exponentially, causing the particles to diffuse and flatten the slope of F_0 and thereby reduce the wave-growth rate. Ultimately, the slope $\partial F_0/\partial v$ diminishes to zero and the wave energy density becomes constant, with its integral, by energy conservation (Ex. 23.2), equal to the total kinetic energy lost by the beam. In this way we see that *a velocity-space irregularity in the distribution function leads to the growth of electrostatic waves, which can react back on the particles in such a way as to saturate the instability.* The

qualitative description of the evolution

net result is a beam of particles with a much-broadened width propagating through the plasma. The waves will ultimately damp through three-wave processes or other damping mechanisms, converting their energy into heat.

EXERCISES

Exercise 23.7 *Problem: Stability of Isotropic Distribution Function*
For the bump-in-tail instability, the bump must show up in the 1-dimensional distribution function F_0 (after integrating out the electron velocity components orthogonal to the wave vector $\mathbf{v}$).

Consider an arbitrary, isotropic, 3-dimensional distribution function $f_0 = f_0(|\mathbf{v}|)$ for electron velocities—one that might even have an isotropic bump at large $|\mathbf{v}|$. Show that this distribution is stable against the growth of Langmuir waves (i.e., it produces $\omega_i < 0$ for all wave vectors $\mathbf{k}$).

23.4.1 Instability of Streaming Cosmic Rays

For a simple illustration of this general type of instability (but one dealing with a different wave mode), we return to the issue of the isotropization of galactic cosmic rays, which we introduced in Sec. 19.7. We argued there that cosmic rays propagating through the interstellar medium are effectively scattered by hydromagnetic Alfvén waves. We did not explain where these Alfvén waves originated. Although often there is an independent turbulence spectrum, sometimes these waves are generated by the cosmic rays themselves.

example of bump-in-tail instability: Alfvén waves generated by Cerenkov emission from cosmic rays' motion along interstellar magnetic field

Suppose we have a beam of cosmic rays propagating through the interstellar gas at high speed. The interstellar gas is magnetized, which allows many more types of wave modes to propagate than in the unmagnetized case. It turns out that the particle distribution is unstable to the growth of Alfvén wave modes satisfying the resonance condition (23.53), modified to account for mildly relativistic protons rather than nonrelativistic electrons:

$$\omega - \mathbf{k}_\| \cdot \mathbf{v}_\| = \omega_g. \tag{23.67}$$

Here $\omega_g = \omega_{cp}\sqrt{1 - v^2/c^2}$ is the proton's gyro frequency, equal to its nonrelativistic cyclotron frequency reduced by its Lorentz factor, and we assume that the number of cyclotron (gyro) quanta fed into each emitted Alfvén plasmon is $N = 1$. For pedagogical simplicity, let the wave propagate parallel to the magnetic field so $k = k_\|$. Then, since cosmic rays travel much faster than the Alfvén waves ($v \gg a = \omega/k$), the wave's frequency ω can be neglected in the resonance condition (23.67); equivalently, we can transform to a reference frame that moves with the Alfvén wave, so ω becomes zero, altering $\mathbf{v}_\|$ hardly at all. The resonance condition (23.67) then implies that the distance the particle travels along the magnetic field when making one trip around its Larmor orbit is $v_\|(2\pi/\omega_g) = 2\pi/k$, which is the Alfvén wave's wavelength. This makes sense physically; it enables the wave to resonate with the particle's gyrational motion, absorbing energy from it.

The growth rate of these waves can be studied using a kinetic theory analogous to the one we have just developed for Langmuir waves (Kulsrud, 2005, Chap. 12; Melrose, 1984, Sec. 10.5). Dropping factors of order unity, the growth rate for waves that scatter cosmic rays of energy E is given approximately by

$$\omega_i \simeq \left(\frac{n_{cr}}{n_p}\right) \omega_{cp} \left(\frac{u_{cr}}{a} - 1\right), \quad (23.68)$$

where n_{cr} is the number density of cosmic rays with energy $\sim E$, n_p is the number density of thermal protons in the background plasma, u_{cr} is the mean speed of the cosmic ray protons through the background plasma, and a is the Alfvén speed; see Ex. 23.8. Therefore, if the particles have a mean speed greater than the Alfvén speed, the waves will grow, exponentially at first.

It is observed that the energy density of cosmic rays in our galaxy builds up until it is roughly comparable with that of the thermal plasma. As more cosmic rays are produced, they escape from the galaxy at a sufficient rate to maintain this balance. Therefore, in a steady state, the ratio of the number density of cosmic rays to the thermal proton density is roughly the inverse of their mean-energy ratio. Adopting a mean cosmic ray energy of ~ 1 GeV and an ambient temperature in the interstellar medium of $T \sim 10^4$ K, this ratio of number densities is $\sim 10^{-9}$. The ion gyro period in the interstellar medium is ~ 100 s for a typical field of strength of ~ 100 pT. Cosmic rays streaming at a few times the Alfvén speed create Alfvén waves in $\sim 10^{10}$ s, of order a few hundred years—long before they escape from the galaxy (e.g., Longair, 2011).

The waves then react back on the cosmic rays, scattering them in momentum space [Eq. (23.51a)]. Now, each time a particle is scattered by an Alfvén-wave quantum, the ratio of its energy change to the magnitude of its momentum change must be the same as that in the waves and equal to the Alfvén speed, which is far smaller than the original energy-to-momentum ratio of the particle ($\sim c$) for a mildly relativistic proton. Therefore, the effect of the Alfvén waves is to scatter the particle directions without changing their energies significantly. As the particles are already gyrating around the magnetic field, the effect of the waves is principally to change the angle between their momenta and the field (known as the *pitch angle*), so as to reduce their mean speed along the magnetic field.

Alfvén waves scatter cosmic rays until their speeds along the magnetic field are reduced to Alfvén speed, saturating the bump-in-tail instability

When this mean speed is reduced to a value of order the Alfvén speed, the growth rate diminishes, just like the growth rate of Langmuir waves is diminished after the electron distribution function is flattened. Under a wide variety of conditions, cosmic rays are believed to maintain the requisite energy density in Alfvén-wave turbulence to prevent them from streaming along the magnetic field with a mean speed much faster than the Alfvén speed (which varies between ~ 3 and ~ 30 km s^{-1}). This model of their transport differs from spatial diffusion, which we assumed in Sec. 19.7.3, but the end result is similar, and cosmic rays are confined to our galaxy for more than ~ 10 million years. These processes can be observed directly using spacecraft in the interplanetary medium.

EXERCISES

Exercise 23.8 *Challenge: Alfvén Wave Emission by Streaming Cosmic Rays*
Consider a beam of high-energy cosmic ray protons streaming along a uniform background magnetic field in a collisionless plasma. Let the cosmic rays have an isotropic distribution function in a frame that moves along the magnetic field with speed u, and assume that u is large compared with the Alfvén speed but small compared with the speeds of the individual cosmic rays. Adapt our discussion of the emission of Langmuir waves by a bump-on-tail distribution to show that the growth rate is given to order of magnitude by Eq. (23.68).

For a solution using an order-of-magnitude analysis, see Kulsrud (2005, Sec. 12.2). For a detailed analysis using quasilinear theory, and then a quasilinear study of the diffusion of cosmic rays in the Alfvén waves they have generated, see Kulsrud (2005, Secs. 12.3–12.5).

23.5 Parametric Instabilities; Laser Fusion

laser fusion

One of the approaches to the goal of attaining commercial nuclear fusion (Sec. 19.3.1) is to compress and heat pellets containing a mixture of deuterium and tritium by using powerful lasers. The goal is to produce gas densities and temperatures large enough, and for long enough, that the nuclear energy released exceeds the energy expended in producing the compression (Box 23.2). At these densities the incident laser radiation behaves like a large-amplitude plasma wave and is subject to a new type of instability that may already be familiar from dynamics, namely, a *parametric instability*.

Consider how the incident light is absorbed by the relatively tenuous ionized plasma around the pellet. The critical density at which the incident wave frequency equals the plasma frequency is $\rho \sim 5\lambda_{\mu m}^{-2}$ kg m^{-3}, where $\lambda_{\mu m}$ is the wavelength measured in microns. For an incident wave energy flux $F \sim 10^{18}$ W m^{-2}, the amplitude of the wave's electric field $E \sim [F/(\epsilon_0 c)]^{1/2} \sim 2 \times 10^{10}$ V m^{-1}. The velocity of a free electron oscillating in a wave this strong is $v \sim eE/(m_e \omega) \sim 2{,}000$ km s^{-1}, which is almost 1% of the speed of light. It is therefore not surprising that nonlinear wave processes are important.

stimulated Raman scattering of laser beam parametrically amplifies a Langmuir wave, strengthening the scattering (a parametric instability) and impeding laser fusion

One of the most important such processes is called *stimulated Raman scattering*. In this interaction the coherent electromagnetic wave with frequency ω convects a small preexisting density fluctuation—one associated with a relatively low-frequency Langmuir wave with frequency ω_{pe}—and converts the Langmuir density fluctuation into a current that varies at the beat frequency $\omega - \omega_{pe}$. This creates a new electromagnetic mode with this frequency. The vector sum of the **k** vectors of the two modes must also equal the incident **k** vector. When this condition can first be met, the new **k** is almost antiparallel to that of the incident mode, and so the radiation is backscattered.

The new mode can combine nonlinearly with the original electromagnetic wave to produce a pressure force $\propto \nabla E^2$, which amplifies the original density fluctuation. Provided the growth rate of the wave is faster than the natural damping rates (e.g.,

> **BOX 23.2. LASER FUSION**
>
> In the simplest proposed scheme for laser fusion, solid pellets of deuterium and tritium would be compressed and heated to allow the reaction
>
> $$d + t \to \alpha + n + 17.6 \text{ MeV} \qquad (1)$$
>
> to proceed. (We will meet this reaction again in Sec. 28.4.2.) An individual pellet would have mass $m \sim 3$ mg and initial diameter $r_i \sim 2$ mm. The total latent nuclear energy in a single pellet is ~ 1 GJ. So if, optimistically, the useful energy extraction were $\sim 3\%$ of this, pellets would have to burn at a combined rate of $\sim 10^5$ s^{-1} in reactors all around the world to supply, say, a fifth of the *current* (2016) global power usage ~ 15 TW.
>
> The largest program designed to accomplish this goal is at the National Ignition Facility in Lawrence Livermore National Laboratory in California. Neodymium-glass pulsed lasers are used to illuminate a small pellet from 192 directions at a wavelength of 351 nm. (The frequency is tripled from the initial infrared frequency using KDP crystals; Box 10.2.) The goal is to deliver a power of ~ 0.5 PW for a few nanoseconds. The nominal initial laser energy is ~ 3 MJ derived from a capacitor bank storing ~ 400 MJ of energy. The energy is delivered to a small metal cylinder, called a *hohlraum*, where a significant fraction of the incident energy is converted to X-rays that illuminate the fuel pellet, a technique known as *indirect drive*. The energy absorbed in the surface layers of the pellet is ~ 150 kJ. This causes the pellet to implode to roughly a tenth of its initial size and a density of $\sim 10^6$ kg m^{-3}; nuclear energy releases of 20 MJ are projected. It has recently become possible to create as much fusion energy as the energy deposited in the pellet (Hurricane et al., 2014). The shot repetition rate is currently about one per day. There is clearly a long road ahead to safe commercial reactors supplying a significant fraction of the global power usage, but progress is being made.

that of Landau damping) there can be a strong backscattering of the incident wave at a density well below the critical density of the incident radiation. (A further condition that must be satisfied is that the bandwidth of the incident wave must also be less than the growth rate. This will generally be true for a laser.) This stimulated Raman scattering is an example of a parametric instability. The incident wave frequency is called the *pump* frequency. One difference between parametric instabilities involving waves as opposed to just oscillations is that it is necessary to match spatial as well as temporal frequencies.

Reflection of the incident radiation by this mechanism reduces the ablation of the pellet and also creates a population of suprathermal electrons, which conduct heat into the interior of the pellet and inhibit compression. Various strategies, including increasing the wave frequency, have been devised to circumvent Raman backscattering (and also a related process called *Brillouin backscattering*, in which the Langmuir mode is replaced by an ion-acoustic mode).

EXERCISES

Exercise 23.9 *Example: Ablation vs. Radiation Pressure in Laser Fusion*
Demonstrate that ablation pressure is necessary to compress a pellet to densities at which nuclear reactions can progress efficiently. More specifically, do the following.

(a) Assume that a temperature $T \sim 30$ keV is necessary for nuclear reactions to proceed. Using the criteria developed in Sec. 20.2.2, evaluate the maximum density before the electrons become degenerate. Estimate the associated pressure and sound speed.

(b) Employ the Lawson criterion (19.25) to estimate the minimum size of a pellet for break-even.

(c) Using the laser performance advertised in Box 23.2, compare the radiation pressure in the laser light with the pressure needed from part (a). Do you think that radiation pressure can create laser fusion?

(d) Assume instead that the incident laser energy is carried off by an outflow. Show that the associated ablation pressure can be orders of magnitude higher than the laser light pressure, and compute an upper bound on the speed of the outflow for laser fusion to be possible.

23.6 Solitons and Collisionless Shock Waves

In Sec. 21.4.3, we introduced ion-acoustic waves that have a phase speed $V_{\rm ph} \sim (k_B T_e/m_p)^{1/2}$, determined by a combination of electron pressure and ion inertia. In Sec. 22.3.6, we argued that these waves would be strongly Landau-damped unless the electron temperature greatly exceeded the proton temperature. However, this formalism was only valid for waves of small amplitude so that the linear approximation could be used. In Ex. 21.6, we considered the profile of a nonlinear wave and found a solution for a single ion-acoustic soliton valid when the waves are weakly nonlinear. We now consider this problem in a slightly different way, one valid for a wave amplitude that is strongly nonlinear. However, we restrict our attention to waves that propagate without change of form and so will not generalize the KdV equation.

Once again we use the fluid model and introduce an ion-fluid velocity u. We presume the electrons are highly mobile and so have a local density $n_e \propto \exp[e\phi/(k_B T_e)]$, where ϕ (not Φ as above) is the electrostatic potential. The ions must satisfy equations of continuity and motion:

$$\frac{\partial n}{\partial t} + \frac{\partial}{\partial z}(nu) = 0,$$

$$\frac{\partial u}{\partial t} + u\frac{\partial u}{\partial x} = -\frac{e}{m_p}\frac{\partial \phi}{\partial z}. \quad (23.69)$$

We now seek a solution for an ion-acoustic wave moving with constant speed V through the ambient plasma. In this case, all physical quantities must be functions of a single dependent variable: $\xi = z - Vt$. Using a prime to denote differentiation with respect to ξ, Eqs. (23.69) become

two-fluid theory of a strongly nonlinear ion-acoustic soliton with speed V in a plasma with thermalized electrons: ion speed u and electrostatic potential ϕ as functions of $\zeta = z - Vt$

$$(u - V)n' = -nu',$$

$$(u - V)u' = -\frac{e}{m_p}\phi'. \quad (23.70)$$

These two equations can be integrated and combined to obtain an expression for the ion density n in terms of the electrostatic potential:

$$n = n_0[1 - 2e\phi/(m_p V^2)]^{-1/2}, \quad (23.71)$$

where n_0 is the ion density, presumed uniform, long before the wave arrives. The next step is to combine this ion density with the electron density and substitute into Poisson's equation to obtain a nonlinear ordinary differential equation for the potential:

$$\phi'' = -\frac{n_0 e}{\epsilon_0}\left\{\left(1 - \frac{2e\phi}{m_p V^2}\right)^{-1/2} - e^{e\phi/(k_B T_e)}\right\}. \quad (23.72)$$

Now, the best way to think about this problem is to formulate the equivalent dynamical problem of a particle moving in a 1-dimensional potential well $\Phi(\phi)$, with ϕ the particle's position coordinate and ξ its time coordinate. Then Eq. (23.72) becomes $\phi'' = -d\Phi/d\phi$, whence $\frac{1}{2}\phi'^2 + \Phi(\phi) = \mathcal{E}$, where $\mathcal{E}$ is the particle's conserved energy. Integrating $-d\Phi/d\phi = $ [right-hand side of Eq. (23.72)], and assuming that $\Phi \to 0$ as $\phi \to 0$ (i.e., as $\xi \to \infty$, long before the arrival of the pulse), we obtain

$$\Phi(\phi) = \frac{n_0 k_B T_e}{\epsilon_0}\left\{\left[1 - (1 - \phi/\phi_o)^{1/2}\right]M^2 - (e^{M^2\phi/(2\phi_o)} - 1)\right\}, \quad (23.73a)$$

where

$$\phi_o = m_p V^2/(2e), \quad M = [m_p V^2/(k_B T_e)]^{1/2}. \quad (23.73b)$$

We have assumed that $0 < \phi < \phi_o$; when $\phi \to \phi_o$, the proton density $n \to \infty$ [Eq. (23.71)].

The shape of this potential well $\Phi(\phi)$ is sketched in Fig. 23.7; it is determined by the parameter $M = [m_p V^2/(k_B T_e)]^{1/2}$, which is readily recognizable as the ion-acoustic Mach number (i.e., the ratio of the speed of the soliton to the ion-acoustic speed in the undisturbed medium). A solution for the soliton's potential profile $\phi(\xi)$

23.6 Solitons and Collisionless Shock Waves   1143

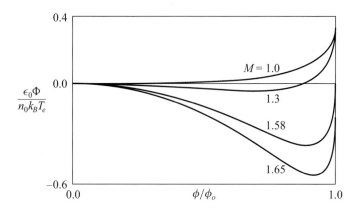

FIGURE 23.7 Potential function $\Phi(\phi)$ used for exhibiting the properties of an ion-acoustic soliton for four different values of the ion-acoustic Mach number M.

corresponds to the trajectory of the particle. Physically, a soliton (solitary wave) must be concentrated in a finite region of space at fixed time; that is, $\phi(\xi) = \phi(z - Vt)$ must go to zero as $\xi \to \pm\infty$. For the particle moving in the potential $\Phi(\phi)$ this is possible only if its total energy $\mathcal{E}$ vanishes, so $\frac{1}{2}\phi'^2 + \Phi(\phi) = 0$. The particle then starts at $\phi = 0$, with zero kinetic energy (i.e., $\phi' = 0$) and then accelerates to a maximum speed near the minimum in the potential before decelerating. If there is a turning point, the particle will come to rest, $\phi(\xi)$ will attain a maximum, and then the particle will return to the origin. This particle trajectory corresponds to a symmetrical soliton, propagating with uniform speed.

Two conditions must be satisfied for a soliton solution. First, the potential well must be attractive. This only happens when $d^2\Phi/d\phi^2(0) < 0$, which implies that $M > 1$. Second, there must be a turning point. This happens if $\Phi(\phi = \phi_o) > 0$. The maximum value of M for which these two conditions are met is a solution of the equation

$$e^{M^2/2} - 1 - M^2 = 0 \tag{23.74}$$

(i.e., $M = 1.58$). Hence, ion-acoustic-soliton solutions only exist for

Mach number for ion-acoustic soliton

$$1 < M < 1.59. \tag{23.75}$$

If $M < 1$, the particle's potential Φ acts as a barrier, preventing it from moving into the $\phi > 0$ region. If $M > 1.59$, the plasma's electrons short out the potential; that is, the electron term $(e^{M^2\phi/(2\phi_o)} - 1)$ in Eq. (23.73a) becomes so negative that it makes $\Phi(\phi_o)$ negative.

This analogy with particle dynamics is not only helpful in understanding large-amplitude solitons. It also assists us to understand a deep connection between these solitons and laminar shock fronts. The equations that we have been solving so far

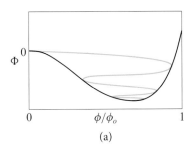 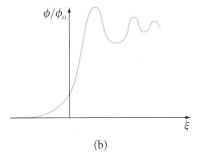

FIGURE 23.8 Ion-acoustic shock wave. (a) Solution in terms of damped particle motion $\phi(\xi)$ (blue curve) in the equivalent potential $\Phi(\phi)$ (black curve). (b) Electrostatic potential profile $\phi(\xi) = \phi(z - Vt)$ in the shock. The proton number-density variation $n(\xi)$ in the shock can be inferred from this $\phi(\xi)$ using Eq. (23.71).

contain the two key ingredients for a soliton: nonlinearity to steepen the wave profile and dispersion to spread it. However, they do not make provision for any form of dissipation, a necessary condition for a shock front, in which the entropy must increase.

In a real collisionless plasma, this dissipation can take on many forms. It may be associated with anomalous resistivity or perhaps with some viscosity associated with the ions. In many circumstances, some ions are reflected by the electrostatic potential barrier and counterstream against the incoming ions, which they eventually heat. Whatever its origin, the net effect of this dissipation will be to cause the equivalent particle to lose its total energy, so that it can never return to its starting point. Given an attractive and bounded potential well, we find that the particle has no alternative except to sink to the bottom of the well. Depending on the strength of the dissipation, the particle may undergo several oscillations before coming to rest. See Fig. 23.8a.

adding dissipation converts soliton into a laminar, collisionless shock

The structure to which this type of solution corresponds is a laminar shock front. Unlike that for a soliton, the wave profile in a shock wave is not symmetric in this case and instead describes a permanent change in the electrostatic potential ϕ (Fig. 23.8b). Repeating the arguments above, we find that a shock wave can only exist when $M > 1$, that is to say, it must be supersonic with respect to the ion-acoustic sound speed. In addition there is a maximum critical Mach number close to $M = 1.6$, above which a laminar shock becomes impossible.

What happens when the critical Mach number is exceeded? In this case there are several possibilities, which include relying on a more rapidly moving wave to form the shock front or appealing to turbulent conditions downstream from the front to enhance the dissipation rate.

This ion-acoustic shock front is the simplest example of a collisionless shock. Essentially every wave mode can be responsible for the formation of a shock. The dissipation in these shocks is still not very well understood, but we can observe them in several different environments in the heliosphere and also in the laboratory. The

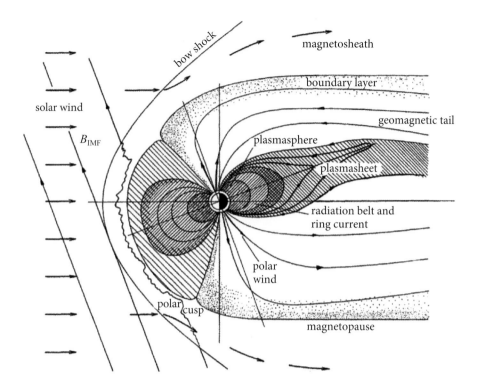

FIGURE 23.9 The form of the collisionless bow shock formed around Earth's magnetosphere. Earth's bow shock has been extensively studied using spacecraft. Alfvén, ion-acoustic, whistler, and Langmuir waves are all generated with large amplitudes in the vicinity of shock fronts by the highly nonthermal particle distributions. Adapted from Parks (2004).

collisionless bow shock at interface of Earth's magnetic field with solar wind

best studied of these shock waves are those based on magnetosonic waves, which were introduced in Sec. 19.7. The solar wind moves with a speed that is typically 5 times the Alfvén speed. It should therefore form a bow shock (one based on the fast magnetosonic mode) when it encounters a planetary magnetosphere. This bow shock forms even though the mean free path of the ions for Coulomb scattering in the solar wind is typically much larger than the thickness of the shock front. The thickness turns out to be a few ion Larmor radii. This is a dramatic illustration of the importance of collective effects in controlling the behavior of essentially collisionless plasmas (Fig. 23.9; see Sagdeev and Kennel, 1991).

EXERCISES

Exercise 23.10 *Derivation: Maximum Mach Number for an Ion-Acoustic Shock Wave*
Verify Eq. (23.73a), and show numerically that the maximum Mach number for a laminar shock front is $M = 1.58$.

Exercise 23.11 *Problem: Solar-Wind Termination Shock*
The solar wind is a quasi-spherical outflow of plasma from the Sun. At the radius of Earth's orbit, the mean proton and electron densities are $n_p \sim n_e \sim 4 \times 10^6$ m^{-3}, their temperatures are $T_p \sim T_e \sim 10^5$ K, and their common radial fluid speed is

~400 km s^{-1}. The mean magnetic field strength is ~1 nT. Eventually, the radial momentum flux in the solar wind falls to the value of the mean interstellar pressure, ~10^{-13} N m^{-2}, and a shock develops.

(a) Estimate the radius where the shock develops.

(b) The solar system moves through the interstellar medium with a speed ~30 km s^{-1}. Sketch the likely flow pattern near this radius.

(c) How do you expect the magnetic field to vary with radius in the outflowing solar wind? Estimate its value at the termination shock.

(d) Estimate the electron plasma frequency, the ion-acoustic Mach number, and the proton Larmor radius just ahead of the termination shock front, and comment on the implications of these values for the shock structure.

(e) The Voyager 1 spacecraft was launched in 1977 and is traveling radially away from the Sun with a terminal speed ~17 km s^{-1}. It was observed to cross the termination shock in 2004, and in 2012 it passed beyond the limit of the shocked solar wind into the interstellar medium. The Voyager 2 spacecraft passed through the termination shock a few years after Voyager 1. How do these observations compare with your answer to part (a)?

Exercise 23.12 *Example: Diffusive Shock Acceleration of Galactic Cosmic Rays*
Cosmic ray particles with energies between ~1 GeV and ~1 PeV are believed to be accelerated at the strong shock fronts formed by supernova explosions in the strongly scattering, local, interstellar medium. We explore a simple model of the way in which this happens.

(a) In a reference frame where the shock is at rest, consider the stationary (time-independent) flow of a medium (a plasma) with velocity $u(x)\,\mathbf{e}_x$, and consider relativistic cosmic rays, diffusing through the medium, that have reached a stationary state. Assume that the mean free paths of the cosmic rays are so short that their distribution function $f = dN/d\mathcal{V}_x d\mathcal{V}_p$ is nearly isotropic in the local rest frame of the medium, so $f = f_0(\tilde{p}, x) + f_1(\mathbf{p}, x)$, where $\tilde{p}$ is the magnitude of a cosmic ray's momentum as measured in the medium's local rest frame, $\tilde{p} \equiv |\tilde{\mathbf{p}}|$, $f_0 \equiv \langle f \rangle$ is the average over cosmic-ray propagation direction in the medium's local rest frame, and $|f_1| \ll |f_0|$. Show that in terms of the cosmic ray's momentum $p = |\mathbf{p}|$ and energy $\mathcal{E} = \sqrt{p^2 + m^2}$ as measured in the shock's rest frame, $\tilde{p} = p - u(x)\mathcal{E}\cos\theta$, where θ is the angle between $\mathbf{p}$ and the direction $\mathbf{e}_x$ of the medium's motion. (Here and throughout this exercise we set the speed of light to unity, as in Chap. 2.)

(b) By expanding the full Vlasov equation to second order in the ratio of the scattering mean free path to the scale on which $u(x)$ varies, it can be shown (e.g., Blandford and Eichler, 1987, Sec. 3.5) that, in the shock's frame,

$$u\frac{\partial f}{\partial x} - u\frac{\partial}{\partial x}\left(D\frac{\partial f}{\partial x}\right) = \frac{\tilde{p}}{3}\frac{\partial f}{\partial \tilde{p}}\frac{\partial u}{\partial x}, \tag{23.76a}$$

where $D(\tilde{p}, x) > 0$ is the spatial diffusion coefficient which arises from f_1. Explore this *convection-diffusion* equation. For example, by integrating it over momentum space in the shock's frame, show that it conserves cosmic ray particles. Also, show that, when the diffusion term is unimportant, f_0 is conserved moving with the medium's flow in the sense that $u \partial f_0/\partial x + (d\tilde{p}/dt)\partial f_0/\partial \tilde{p} = 0$, where $d\tilde{p}/dt$ is the rate of change of cosmic ray momenta due to elastic scattering in the expanding medium (for which a volume element changes as $d \ln \mathcal{V}_x/dt = \partial u/\partial x$).

(c) Argue that the flux of cosmic-ray particles, measured in the shock's frame, is given by $\int_0^\infty F(p, x) 4\pi p^2 dp$, where $F(p, x) = -D(\partial f_0/\partial x) - u(p/3)(\partial f/\partial \tilde{p})$; or, to leading order in u/v, where $v = p/\mathcal{E}$ is the cosmic ray speed,

$$F(\tilde{p}, x) = -D \frac{\partial f_0}{\partial x} - u \frac{\tilde{p}}{3} \frac{\partial f}{\partial \tilde{p}} . \tag{23.76b}$$

This F is often called the flux of particles at momentum $\tilde{p}$.

(d) Idealize the shock as a planar discontinuity at $x = 0$ in the medium's velocity $u(x)$; idealize the velocity as constant before and after the shock, $u = u_- =$ constant for $x < 0$, and $u = u_+ =$ constant for $x > 0$; and denote by $r = u_-/u_+ > 1$ the shock's compression ratio. Show that upstream from the shock, the solution to the convection-diffusion equation (23.76a) is

$$f_0(\tilde{p}, x) = f_-(\tilde{p}) + [f_s(\tilde{p}) - f_-(\tilde{p})] \exp\left[-\int_x^0 \frac{u_- dx'}{D(\tilde{p}, x')}\right] \quad \text{at} \quad x < 0. \tag{23.76c}$$

Here $f_-(\tilde{p}) \equiv f_0(\tilde{p}, x = -\infty)$, and $f_s(\tilde{p}) \equiv f_0(\tilde{p}, x = 0)$ is the value of f_0 at the shock, which must be continuous across the shock. (Why?) Show further that downstream from the shock f_0 cannot depend on x, so

$$f_0(\tilde{p}, x) = f_s(\tilde{p}) \quad \text{at} \quad x > 0 . \tag{23.76d}$$

(e) By matching the flux of particles at momentum $\tilde{p}$, $F(\tilde{p}, x)$ [Eq. (23.76b)], across the shock, show that the post-shock distribution function is

$$f_s(\tilde{p}) = q \tilde{p}^{-q} \int_0^{\tilde{p}} d\tilde{p}' \, f_-(\tilde{p}') \, \tilde{p}'^{(q-1)} , \tag{23.76e}$$

where $q = 3r/(r-1)$.

(f) The fact that this stationary solution to the convection-diffusion equation has a power-law spectrum $f_s \propto \tilde{p}^{-q}$, in accord with observations, suggests that the observed cosmic rays may indeed be accelerated in shock fronts. How, physically, do you think the acceleration occurs and how does this lead to a power-law spectrum? What is the energy source for the acceleration? It may help to explore the manner in which Eq. (23.76c) evolves an arbitrary initial distribution function $f_-(\tilde{p})$ into the power-law distribution (23.76d). [For discussion and details, see, e.g., Blandford and Eichler (1987), and Longair (2011).]

Bibliographic Note

For a concise treatment of the classical quasilinear theory of wave-particle interactions as in Sec. 23.2, see Lifshitz and Pitaevskii (1981, Sec. 49 for quasilinear theory, and Sec. 51 for fluctuations and correlations).

For more detailed and rich, pedagogical, classical treatments of quasilinear theory and some stronger nonlinear effects in plasmas, see Bellan (2006, Chaps. 14, 15), Boyd and Sanderson (2003, Chap. 10), Swanson (2003, Chaps. 7, 8), and Krall and Trivelpiece (1973, Chaps. 10, 11). For a classical treatment of quasilinear theory that is extended to include excitations of magnetized plasmas, see Stix (1992, Chaps. 16–18). For the interactions of cosmic rays with Alfvén waves, see Kulsrud (2005, Chap. 12). For wave-wave coupling parametric instabilities, see Swanson (2003, Sec. 8.4). For ion-acoustic solitons and associated shocks, see Krall and Trivelpiece (1973, Sec. 3.9.4) and Swanson (2003, Sec. 8.2).

A classic text on radiation processes in plasmas is Bekefi (1966). An extensive treatment of nonlinear plasma physics from a quantum viewpoint is contained in Melrose (2008, 2012).

For applications to astrophysical plasmas, see Melrose (1984), Parks (2004), and Kulsrud (2005), and for applications to laser-plasma interactions, see Kruer (1988).

PART VII

GENERAL RELATIVITY

We have reached the final part of this book, in which we present an introduction to the basic concepts of general relativity and its most important applications. This subject, although a little more challenging than the material that we have covered so far, is nowhere near as formidable as its reputation. Indeed, if you have mastered the techniques developed in the first five parts, the path to the Einstein field equations should be short and direct.

The general theory of relativity is the crowning achievement of classical physics, the last great fundamental theory created prior to the discovery of quantum mechanics. Its formulation by Albert Einstein in 1915 marks the culmination of the great intellectual adventure undertaken by Newton 250 years earlier. Einstein created it after many wrong turns and with little experimental guidance, almost by pure thought. Unlike the special theory, whose physical foundations and logical consequences were clearly appreciated by physicists soon after Einstein's 1905 formulation, the unique and distinctive character of the general theory only came to be widely appreciated long after its creation. Ultimately, in hindsight, rival classical theories of gravitation came to seem unnatural, inelegant, and arbitrary by comparison [see Will (1993b) for a popular account and Pais (1982) for a more scholarly treatment].

Experimental tests of Einstein's theory also were slow to come. Only since 1970 have there been striking tests of high enough precision to convince most empiricists that—in all probability and in its domain of applicability—general relativity is essentially correct. Despite these tests, it is still very poorly tested compared to, for example, quantum electrodynamics.

We begin our discussion of general relativity in Chap. 24 with a review and an elaboration of special relativity as developed in Chap. 2, focusing on those concepts that are crucial for the transition to general relativity. Our elaboration includes (i) an extension of differential geometry to curvilinear coordinate systems and general bases both in the flat spacetime of special relativity and in the curved spacetime that is the venue for general relativity; (ii) an in-depth exploration of the stress-energy tensor, which in general relativity generates the curvature of spacetime; and (iii) construction and exploration of the reference frames of accelerated observers (e.g., physicists who reside on Earth's surface).

In Chap. 25, we turn to the basic concepts of general relativity, including spacetime curvature, the Einstein field equation that governs the generation of spacetime curvature, the laws of physics in curved spacetime, and weak-gravity limits of general relativity.

In the remaining chapters, we explore applications of general relativity to stars, black holes, gravitational waves, experimental tests of the theory, and cosmology. We begin in Chap. 26 by studying the spacetime curvature around and inside highly compact stars (such as neutron stars). We then discuss the implosion of massive stars and describe the circumstances under which the implosion inevitably produces a black hole. We explore the surprising and, initially, counterintuitive properties of black holes (both nonspinning and spinning holes), and we learn about the many-fingered nature of time in general relativity. In Chap. 27, we study experimental tests of general relativity and then turn to gravitational waves (i.e., ripples in the curvature of spacetime that propagate with the speed of light). We explore the properties of these waves, and their close analogy with electromagnetic waves, and their production by binary stars and merging black holes. We also describe projects to detect them (both on Earth and in space) and the prospects and success for using them to explore observationally the "warped side of the universe" and the nature of ultrastrong spacetime curvature. Finally, in Chap. 28,[1] we draw on all the previous parts of this book, combining them with general relativity to describe the universe on the largest of scales and longest of times: cosmology. It is here, more than anywhere else in classical physics, that we are conscious of reaching a frontier where the still-promised land of quantum gravity beckons.

---

1. Chapter 28 is very different in style from the rest of the book. It presents a minimalist treatment of the now standard description of the universe at large. This is a huge subject from which we have ruthlessly excised history, observational justification, and didacticism. Our goal is limited to showing that much of what is now widely accepted about the origin and evolution of the cosmos can be explained directly and quantitatively using the ideas developed in this book.

CHAPTER TWENTY-FOUR

From Special to General Relativity

> The Theory of Relativity confers an absolute meaning on a magnitude which in classical theory has only a relative significance: the velocity of light. The velocity of light is to the Theory of Relativity as the elementary quantum of action is to the Quantum Theory: it is its absolute core.
>
> MAX PLANCK (1949)

24.1 Overview

We begin our discussion of general relativity in this chapter with a review, and elaboration of relevant material already covered in earlier chapters. In Sec. 24.2, we give a brief encapsulation of special relativity drawn largely from Chap. 2, emphasizing those aspects that underpin the transition to general relativity. Then in Sec. 24.3 we collect, review, and extend the fundamental ideas of differential geometry that have been scattered throughout the book and that we shall need as foundations for the mathematics of *spacetime curvature* (Chap. 25). Most importantly, we generalize differential geometry to encompass coordinate systems whose coordinate lines are not orthogonal and bases that are not orthonormal.

Einstein's field equation (to be studied in Chap. 25) is a relationship between the curvature of spacetime and the matter that generates it, akin to the Maxwell equations' relationship between the electromagnetic field and the electric currents and charges that generate it. The matter in Einstein's equation is described by the stress-energy tensor that we introduced in Sec. 2.13. We revisit the stress-energy tensor in Sec. 24.4 and develop a deeper understanding of its properties.

In general relativity one often wishes to describe the outcome of measurements made by observers who refuse to fall freely—for example, an observer who hovers in a spaceship just above the horizon of a black hole, or a gravitational-wave experimenter in an Earthbound laboratory. As a foundation for treating such observers, in Sec. 24.5 we examine measurements made by accelerated observers in the flat spacetime of special relativity.

24.2 Special Relativity Once Again

Our viewpoint on general relativity is unapologetically geometrical. (Other viewpoints, e.g., those of particle theorists such as Feynman and Weinberg, are quite different.) Therefore, a prerequisite for our treatment of general relativity is understanding special relativity in geometric language. In Chap. 2, we discussed the foundations of

> **BOX 24.1. READERS' GUIDE**
>
> - This chapter relies significantly on:
>   - Chap. 2 on special relativity, which now should be regarded as Track One.
>   - The discussion of connection coefficients in Sec. 11.8.
> - This chapter is a foundation for the presentation of general relativity theory and cosmology in Chaps. 25–28.

special relativity with this in mind. In this section we briefly review the most important points.

We suggest that any reader who has not studied Chap. 2 read Sec. 24.2 first, to get an overview and flavor of what will be important for our development of general relativity, and then (or in parallel with reading Sec. 24.2) read those relevant sections of Chap. 2 that the reader does not already understand.

24.2.1 Geometric, Frame-Independent Formulation

24.2.1

review of the geometric, frame-independent formulation of special relativity

In Secs. 1.1.1 and 2.2.2, we learned that *every law of physics must be expressible as a geometric, frame-independent relationship among geometric, frame-independent objects*. This is equally true in Newtonian physics, in special relativity, and in general relativity. The key difference between the three is the geometric arena: in Newtonian physics, the arena is 3-dimensional Euclidean space; in special relativity, it is 4-dimensional Minkowski spacetime; in general relativity (Chap. 25), it is 4-dimensional curved spacetime (see Fig. 1 in the Introduction to Part I and the associated discussion).

Principle of Relativity—laws as geometric relations between geometric objects

In special relativity, the demand that the laws be geometric relationships among geometric objects that live in Minkowski spacetime is the *Principle of Relativity*; see Sec. 2.2.2. Examples of the geometric objects are:

examples of geometric objects: points, curves, proper time ticked by an ideal clock, vectors, tensors, scalar product

1. A point $\mathcal{P}$ in spacetime (which represents an *event*); Sec. 2.2.1.
2. A parameterized curve in spacetime, such as the world line $\mathcal{P}(\tau)$ of a particle, for which the parameter τ is the particle's *proper time* (i.e., the time measured by an ideal clock[1] that the particle carries; Fig. 24.1); Sec. 2.4.1.

---

1. Recall that an ideal clock is one that ticks uniformly when compared, e.g., to the period of the light emitted by some standard type of atom or molecule, and that has been made impervious to accelerations. Thus two ideal clocks momentarily at rest with respect to each other tick at the same rate independent of their relative acceleration; see Secs. 2.2.1 and 2.4.1. For greater detail, see Misner, Thorne, and Wheeler (1973, pp. 23–29, 395–399).

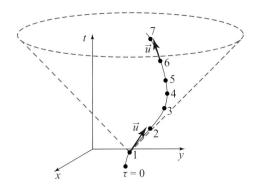

FIGURE 24.1 The world line $\mathcal{P}(\tau)$ of a particle in Minkowski spacetime and the tangent vector $\vec{u} = d\mathcal{P}/d\tau$ to this world line; $\vec{u}$ is the particle's 4-velocity. The bending of the world line is produced by some force that acts on the particle, such as the Lorentz force embodied in Eq. (24.3). Also shown is the light cone emitted from the event $\mathcal{P}(\tau = 1)$. Although the axes of an (arbitrary) inertial reference frame are shown, no reference frame is needed for the definition of the world line, its tangent vector $\vec{u}$, or the light cone. Nor is one needed for the formulation of the Lorentz force law.

3. Vectors, such as the particle's 4-velocity $\vec{u} = d\mathcal{P}/d\tau$ [the tangent vector to the curve $\mathcal{P}(\tau)$] and the particle's 4-momentum $\vec{p} = m\vec{u}$ (with m the particle's rest mass); Secs. 2.2.1 and 2.4.1.

4. Tensors, such as the electromagnetic field tensor $\mathbf{F}(\_,\_)$; Secs. 1.3 and 2.3.

Recall that a tensor is a linear real-valued function of vectors; when one puts vectors $\vec{A}$ and $\vec{B}$ into the two slots of $\mathbf{F}$, one obtains a real number (a scalar) $\mathbf{F}(\vec{A}, \vec{B})$ that is linear in $\vec{A}$ and in $\vec{B}$ so, for example: $\mathbf{F}(\vec{A}, b\vec{B} + c\vec{C}) = b\mathbf{F}(\vec{A}, \vec{B}) + c\mathbf{F}(\vec{A}, \vec{C})$. When one puts a vector $\vec{B}$ into just one of the slots of $\mathbf{F}$ and leaves the other empty, one obtains a tensor with one empty slot, $\mathbf{F}(\_, \vec{B})$, that is, a vector. The result of putting a vector into the slot of a vector is the scalar product: $\vec{D}(\vec{B}) = \vec{D} \cdot \vec{B} = \mathbf{g}(\vec{D}, \vec{B})$, where $\mathbf{g}(\_,\_)$ is the metric.

In Secs. 2.3 and 2.4.1, we tied our definitions of the inner product and the spacetime metric to the ticking of ideal clocks: If $\Delta \vec{x}$ is the vector separation of two neighboring events $\mathcal{P}(\tau)$ and $\mathcal{P}(\tau + \Delta\tau)$ along a particle's world line, then

$$\mathbf{g}(\Delta \vec{x}, \Delta \vec{x}) \equiv \Delta \vec{x} \cdot \Delta \vec{x} \equiv -(\Delta \tau)^2. \tag{24.1}$$

spacetime metric

This relation for any particle with any timelike world line, together with the linearity of $\mathbf{g}(\_,\_)$ in its two slots, is enough to determine $\mathbf{g}$ completely and to guarantee that it is symmetric: $\mathbf{g}(\vec{A}, \vec{B}) = \mathbf{g}(\vec{B}, \vec{A})$ for all $\vec{A}$ and $\vec{B}$. Since the particle's 4-velocity $\vec{u}$ is

$$\vec{u} = \frac{d\mathcal{P}}{d\tau} = \lim_{\Delta\tau \to 0} \frac{\mathcal{P}(\tau + \Delta\tau) - \mathcal{P}(\tau)}{\Delta\tau} \equiv \lim_{\Delta\tau \to 0} \frac{\Delta \vec{x}}{\Delta \tau}, \tag{24.2}$$

Eq. (24.1) implies that $\vec{u} \cdot \vec{u} = \mathbf{g}(\vec{u}, \vec{u}) = -1$ (Sec. 2.4.1).

The 4-velocity $\vec{u}$ is an example of a *timelike* vector (Sec. 2.2.3); it has a negative inner product with itself (negative "squared length"). This shows up pictorially in the

light cone; timelike, null, and spacelike vectors

fact that $\vec{u}$ lies inside the *light cone* (the cone swept out by the trajectories of photons emitted from the tail of $\vec{u}$; see Fig. 24.1). Vectors $\vec{k}$ on the light cone (the tangents to the world lines of the photons) are *null* and so have vanishing squared lengths: $\vec{k} \cdot \vec{k} = g(\vec{k}, \vec{k}) = 0$; vectors $\vec{A}$ that lie outside the light cone are *spacelike* and have positive squared lengths: $\vec{A} \cdot \vec{A} > 0$ (Sec. 2.2.3).

An example of a physical law in 4-dimensional geometric language is the Lorentz force law (Sec. 2.4.2):

Lorentz force law

$$\frac{d\vec{p}}{d\tau} = q\mathbf{F}(\_, \vec{u}). \tag{24.3}$$

Here q is the particle's charge (a scalar), and both sides of this equation are vectors, or equivalently, first-rank tensors (i.e., tensors with just one slot). As we learned in Secs. 1.5.1 and 2.5.3, it is convenient to give names to slots. When we do so, we can rewrite the Lorentz force law as

$$\frac{dp^\alpha}{d\tau} = q F^{\alpha\beta} u_\beta. \tag{24.4}$$

slot-naming index notation

Here α is the name of the slot of the vector $d\vec{p}/d\tau$, α and β are the names of the slots of $\mathbf{F}$, β is the name of the slot of $\mathbf{u}$. The double use of β with one up and one down on the right-hand side of the equation represents the insertion of $\vec{u}$ into the β slot of $\mathbf{F}$, whereby the two β slots disappear, and we wind up with a vector whose slot is named α. As we learned in Sec. 1.5, this slot-naming index notation is isomorphic to the notation for components of vectors, tensors, and physical laws in some reference frame. However, no reference frames are needed or involved when one formulates the laws of physics in geometric, frame-independent language as above.

Those readers who do not feel completely comfortable with these concepts, statements, and notation should reread the relevant portions of Chaps. 1 and 2.

EXERCISES

Exercise 24.1 *Practice: Frame-Independent Tensors*
Let $\mathbf{A}$, $\mathbf{B}$ be second-rank tensors.

(a) Show that $\mathbf{A} + \mathbf{B}$ is also a second-rank tensor.

(b) Show that $\mathbf{A} \otimes \mathbf{B}$ is a fourth-rank tensor.

(c) Show that the contraction of $\mathbf{A} \otimes \mathbf{B}$ on its first and fourth slots is a second-rank tensor. (If necessary, consult Secs. 1.5 and 2.5 for discussions of contraction.)

(d) Write the following quantities in slot-naming index notation: the tensor $\mathbf{A} \otimes \mathbf{B}$, and the simultaneous contraction of this tensor on its first and fourth slots and on its second and third slots.

24.2.2

24.2.2 Inertial Frames and Components of Vectors, Tensors, and Physical Laws

inertial reference frame

In special relativity, a key role is played by *inertial reference frames*, Sec. 2.2.1. An inertial frame is an (imaginary) latticework of rods and clocks that moves through spacetime freely (inertially, without any force acting on it). The rods are orthogonal to one another and attached to inertial-guidance gyroscopes, so they do not rotate. These rods are used to identify the spatial, Cartesian coordinates $(x^1, x^2, x^3) = (x, y, z)$ of an event $\mathcal{P}$ [which we also denote by lowercased Latin indices $x^j(\mathcal{P})$, with j running over 1, 2, 3]. The latticework's clocks are ideal and are synchronized with one another by the Einstein light-pulse process. They are used to identify the temporal coordinate $x^0 = t$ of an event $\mathcal{P}$: $x^0(\mathcal{P})$ is the time measured by that latticework clock whose world line passes through $\mathcal{P}$, at the moment of passage. The spacetime coordinates of $\mathcal{P}$ are denoted by lowercased Greek indices x^α, with α running over 0, 1, 2, 3. An inertial frame's spacetime coordinates $x^\alpha(\mathcal{P})$ are called *Lorentz coordinates* or *inertial coordinates*.

Lorentz (inertial) coordinates

In the real universe, spacetime curvature is small in regions well removed from concentrations of matter (e.g., in intergalactic space), so special relativity is highly accurate there. In such a region, frames of reference (rod-clock latticeworks) that are nonaccelerating and nonrotating with respect to cosmologically distant galaxies (and hence with respect to a local frame in which the cosmic microwave radiation looks isotropic) constitute good approximations to inertial reference frames.

Associated with an inertial frame's Lorentz coordinates are basis vectors $\vec{e}_\alpha$ that point along the frame's coordinate axes (and thus are orthogonal to one another) and have unit length (making them orthonormal); see Sec. 2.5. This orthonormality is embodied in the inner products

orthonormal basis vectors of an inertial frame

$$\boxed{\vec{e}_\alpha \cdot \vec{e}_\beta = \eta_{\alpha\beta},} \quad (24.5)$$

where by definition:

$$\boxed{\eta_{00} = -1, \quad \eta_{11} = \eta_{22} = \eta_{33} = +1, \quad \eta_{\alpha\beta} = 0 \quad \text{if } \alpha \neq \beta.} \quad (24.6)$$

Here and throughout Part VII (as in Chap. 2), we set the speed of light to unity (i.e., we use the geometrized units introduced in Sec. 1.10), so spatial lengths (e.g., along the x-axis) and time intervals (e.g., along the t-axis) are measured in the same units, seconds or meters, with $1\,\text{s} = 2.99792458 \times 10^8$ m.

geometrized units

In Sec. 2.5 (see also Sec. 1.5), we used the basis vectors of an inertial frame to build a component representation of tensor analysis. The fact that the inner products of timelike vectors with each other are negative (e.g., $\vec{e}_0 \cdot \vec{e}_0 = -1$), while those of spacelike vectors are positive (e.g., $\vec{e}_1 \cdot \vec{e}_1 = +1$), forced us to introduce two types of components: *covariant* (indices down) and *contravariant* (indices up). The covariant components of a tensor are computable by inserting the basis vectors into the tensor's slots: $u_\alpha = \vec{u}(\vec{e}_\alpha) \equiv \vec{u} \cdot \vec{e}_\alpha$; $F_{\alpha\beta} = \mathbf{F}(\vec{e}_\alpha, \vec{e}_\beta)$. For example, in our Lorentz basis the covariant components of the metric are $g_{\alpha\beta} = \mathbf{g}(\vec{e}_\alpha, \vec{e}_\beta) = \vec{e}_\alpha \cdot \vec{e}_\beta = \eta_{\alpha\beta}$. The contravariant components of a tensor were related to the covariant components via "index lowering" with the aid of the metric, $F_{\alpha\beta} = g_{\alpha\mu} g_{\beta\nu} F^{\mu\nu}$, which simply said that one reverses the sign when lowering a time index and makes no change of sign when lowering a space index. This lowering rule implied that the contravariant components of the metric in a Lorentz basis are the same numerically as the covariant

covariant and contravariant components of vectors and tensors

24.2 Special Relativity Once Again

components, $g^{\alpha\beta} = \eta_{\alpha\beta}$, and that they can be used to raise indices (i.e., to perform the trivial sign flip for temporal indices): $F^{\mu\nu} = g^{\mu\alpha} g^{\nu\beta} F_{\alpha\beta}$. As we saw in Sec. 2.5, tensors can be expressed in terms of their contravariant components as $\vec{p} = p^\alpha \vec{e}_\alpha$, and $\mathbf{F} = F^{\alpha\beta} \vec{e}_\alpha \otimes \vec{e}_\beta$, where $\otimes$ represents the tensor product [Eqs. (1.5)].

We also learned in Chap. 2 that any frame-independent geometric relation among tensors can be rewritten as a relation among those tensors' components in any chosen Lorentz frame. When one does so, the resulting component equation takes precisely the same form as the slot-naming-index-notation version of the geometric relation (Sec. 1.5.1). For example, the component version of the Lorentz force law says $dp^\alpha/d\tau = qF^{\alpha\beta} u_\beta$, which is identical to Eq. (24.4). The only difference is the interpretation of the symbols. In the component equation $F^{\alpha\beta}$ are the components of $\mathbf{F}$ and the repeated β in $F^{\alpha\beta} u_\beta$ is to be summed from 0 to 3. In the geometric relation $F^{\alpha\beta}$ means $\mathbf{F}(\_, \_)$, with the first slot named α and the second β, and the repeated β in $F^{\alpha\beta} u_\beta$ implies the insertion of $\vec{u}$ into the second slot of $\mathbf{F}$ to produce a single-slotted tensor (i.e., a vector) whose slot is named α.

component equations are same as slot-naming-index-notation equations

As we saw in Sec. 2.6, a particle's 4-velocity $\vec{u}$ (defined originally without the aid of any reference frame; Fig. 24.1) has components, in any inertial frame, given by $u^0 = \gamma$, $u^j = \gamma v^j$, where $v^j = dx^j/dt$ is the particle's ordinary velocity and $\gamma \equiv 1/\sqrt{1 - \delta_{ij} v^i v^j}$. Similarly, the particle's energy $E \equiv p^0$ is $m\gamma$, and its spatial momentum is $p^j = m\gamma v^j$ (i.e., in 3-dimensional geometric notation: $\mathbf{p} = m\gamma \mathbf{v}$). This is an example of the manner in which a choice of Lorentz frame produces a "3+1" split of the physics: a split of 4-dimensional spacetime into 3-dimensional space (with Cartesian coordinates x^j) plus 1-dimensional time $t = x^0$; a split of the particle's 4-momentum $\vec{p}$ into its 3-dimensional spatial momentum $\mathbf{p}$ and its 1-dimensional energy $\mathcal{E} = p^0$; and similarly a split of the electromagnetic field tensor $\mathbf{F}$ into the 3-dimensional electric field $\mathbf{E}$ and 3-dimensional magnetic field $\mathbf{B}$ (cf. Secs. 2.6 and 2.11).

components of 4-velocity in an inertial frame

3 + 1 split

The Principle of Relativity (all laws expressible as geometric relations between geometric objects in Minkowski spacetime), when translated into 3+1 language, says that, when the laws of physics are expressed in terms of components in a specific Lorentz frame, the form of those laws must be independent of one's choice of frame. When translated into operational terms, it says that, if two observers in two different Lorentz frames are given identical written instructions for a self-contained physics experiment, then their two experiments must yield the same results to within their experimental accuracies (Sec. 2.2.2).

Principle of Relativity restated: laws take same form in every inertial frame

The components of tensors in one Lorentz frame are related to those in another by a Lorentz transformation (Sec. 2.7), so the Principle of Relativity can be restated as saying that, when expressed in terms of Lorentz-frame components, *the laws of physics must be Lorentz-invariant* (unchanged by Lorentz transformations). This is the version of the Principle of Relativity that one meets in most elementary treatments of special relativity. However, as the above discussion shows, it is a mere shadow of the true Principle of Relativity—the shadow cast into Lorentz frames when one performs

Lorentz transformations

Principle of Relativity restated: laws are Lorentz invariant

a 3+1 split. The ultimate, fundamental version of the Principle of Relativity is the one that needs no frames at all for its expression: *all the laws of physics are expressible as geometric relations among geometric objects that reside in Minkowski spacetime.*

ultimate version of Principle of Relativity

24.2.3 Light Speed, the Interval, and Spacetime Diagrams

One set of physical laws that must be the same in all inertial frames is Maxwell's equations. Let us discuss the implications of Maxwell's equations and the Principle of Relativity for the speed of light c. (For a more detailed discussion, see Sec. 2.2.2.) According to Maxwell, c can be determined by performing nonradiative laboratory experiments; it is not necessary to measure the time it takes light to travel along some path; see Box 2.2. The Principle of Relativity requires that such experiments must give the same result for c, independent of the reference frame in which the measurement apparatus resides, so the speed of light must be independent of reference frame. It is this frame independence that enables us to introduce geometrized units with $c = 1$.

light speed is the same in all inertial frames

Another example of frame independence (Lorentz invariance) is provided by the *interval between two events* (Sec. 2.2.3). The components $g_{\alpha\beta} = \eta_{\alpha\beta}$ of the metric imply that, if $\Delta \vec{x}$ is the vector separating the two events and Δx^α are its components in some Lorentz coordinate system, then the squared length of $\Delta \vec{x}$ [also called the *interval* and denoted $(\Delta s)^2$] is given by

$$\begin{aligned}(\Delta s)^2 &\equiv \Delta \vec{x} \cdot \Delta \vec{x} = \mathbf{g}(\Delta \vec{x}, \Delta \vec{x}) = g_{\alpha\beta} \Delta x^\alpha \Delta x^\beta \\ &= -(\Delta t)^2 + (\Delta x)^2 + (\Delta y)^2 + (\Delta z)^2.\end{aligned} \quad (24.7)$$

interval between two events

Since $\Delta \vec{x}$ is a geometric, frame-independent object, so must be the interval. This implies that the equation $(\Delta s)^2 = -(\Delta t)^2 + (\Delta x)^2 + (\Delta y)^2 + (\Delta z)^2$ by which one computes the interval between the two chosen events in one Lorentz frame must give the same numerical result when used in any other frame (i.e., this expression must be Lorentz invariant). This *invariance of the interval* is the starting point for most introductions to special relativity—and, indeed, we used it as a starting point in Sec. 2.2.

invariance of the interval

Spacetime diagrams play a major role in our development of general relativity. Accordingly, it is important that the reader feel very comfortable with them. We recommend reviewing Fig. 2.7 and Ex. 2.14.

spacetime diagrams

EXERCISES

Exercise 24.2 *Example: Invariance of a Null Interval*
You have measured the intervals between a number of adjacent events in spacetime and thereby have deduced the metric $\mathbf{g}$. Your friend claims that the metric is some other frame-independent tensor $\tilde{\mathbf{g}}$ that differs from $\mathbf{g}$. Suppose that your correct metric $\mathbf{g}$ and his wrong one $\tilde{\mathbf{g}}$ agree on the forms of the light cones in spacetime (i.e., they agree as to which intervals are null, which are spacelike, and which are timelike), but they give different answers for the value of the interval in the spacelike and timelike cases: $\mathbf{g}(\Delta \vec{x}, \Delta \vec{x}) \neq \tilde{\mathbf{g}}(\Delta \vec{x}, \Delta \vec{x})$. Prove that $\tilde{\mathbf{g}}$ and $\mathbf{g}$ differ solely by

a scalar multiplicative factor, $\tilde{\mathbf{g}} = a\mathbf{g}$ for some scalar a. We say that $\tilde{\mathbf{g}}$ and $\mathbf{g}$ are *conformal to each other*. [Hint: Pick some Lorentz frame and perform computations there, then lift yourself back up to a frame-independent viewpoint.]

Exercise 24.3 *Problem: Causality*
If two events occur at the same spatial point but not simultaneously in one inertial frame, prove that the temporal order of these events is the same in all inertial frames. Prove also that in all other frames the temporal interval Δt between the two events is larger than in the first frame, and that there are no limits on the events' spatial or temporal separation in the other frames. Give *two* proofs of these results, one algebraic and the other via spacetime diagrams.

24.3 Differential Geometry in General Bases and in Curved Manifolds

The differential geometry (tensor-analysis) formalism reviewed in the last section is inadequate for general relativity in several ways.

First, in general relativity we need to use bases $\vec{e}_\alpha$ that are not orthonormal (i.e., for which $\vec{e}_\alpha \cdot \vec{e}_\beta \neq \eta_{\alpha\beta}$). For example, near a spinning black hole there is much power in using a time basis vector $\vec{e}_t$ that is tied in a simple way to the metric's time-translation symmetry and a spatial basis vector $\vec{e}_\phi$ that is tied to its rotational symmetry. This time basis vector has an inner product with itself $\vec{e}_t \cdot \vec{e}_t = g_{tt}$ that is influenced by the slowing of time near the hole (so $g_{tt} \neq -1$); and $\vec{e}_\phi$ is not orthogonal to $\vec{e}_t$ ($\vec{e}_t \cdot \vec{e}_\phi = g_{t\phi} \neq 0$), as a result of the dragging of inertial frames by the hole's spin. In this section, we generalize our formalism to treat such nonorthonormal bases.

Second, in the curved spacetime of general relativity (and in any other curved space, e.g., the 2-dimensional surface of Earth), the definition of a vector as an arrow connecting two points (Secs. 1.2 and 2.2.1) is suspect, as it is not obvious on what route the arrow should travel nor that the linear algebra of tensor analysis should be valid for such arrows. In this section, we refine the concept of a vector to deal with this problem. In the process we introduce the concept of a *tangent space* in which the linear algebra of tensors takes place—a different tangent space for tensors that live at different points in the space.

Third, once we have been forced to think of a tensor as residing in a specific tangent space at a specific point in the space, the question arises: how can one transport tensors from the tangent space at one point to the tangent space at an adjacent point? Since the notion of a gradient of a vector depends on comparing the vector at two different points and thus depends on the details of transport, we have to rework the notion of a gradient and the gradient's connection coefficients.

Fourth, when doing an integral, one must add contributions that live at different points in the space, so we must also rework the notion of integration.

We tackle each of these four issues in turn in the following four subsections.

24.3.1 Nonorthonormal Bases

Consider an n-dimensional *manifold*, that is, a space that, in the neighborhood of any point, has the same topological and smoothness properties as n-dimensional Euclidean space, though it might not have a locally Euclidean or locally Lorentz metric and perhaps has no metric at all. If the manifold has a metric (e.g., 4-dimensional spacetime, 3-dimensional Euclidean space, and the 2-dimensional surface of a sphere) it is called "Riemannian." In this chapter, all manifolds we consider will be Riemannian.

At some point $\mathcal{P}$ in our chosen n-dimensional manifold with metric, introduce a set of basis vectors $\{\vec{e}_1, \vec{e}_2, \ldots, \vec{e}_n\}$ and denote them generally as $\vec{e}_\alpha$. We seek to generalize the formalism of Sec. 24.2 in such a way that the index-manipulation rules for components of tensors are unchanged. For example, we still want it to be true that covariant components of any tensor are computable by inserting the basis vectors into the tensor's slots, $F_{\alpha\beta} = \mathbf{F}(\vec{e}_\alpha, \vec{e}_\beta)$, and that the tensor itself can be reconstructed from its contravariant components: $\mathbf{F} = F^{\mu\nu} \vec{e}_\mu \otimes \vec{e}_\nu$. We also require that the two sets of components are computable from each other via raising and lowering with the metric components: $F_{\alpha\beta} = g_{\alpha\mu} g_{\beta\nu} F^{\mu\nu}$. The only thing we do not want to preserve is the orthonormal values of the metric components: we must allow the basis to be nonorthonormal and thus $\vec{e}_\alpha \cdot \vec{e}_\beta = g_{\alpha\beta}$ to have arbitrary values (except that the metric should be nondegenerate, so no linear combination of the $\vec{e}_\alpha$s vanishes, which means that the matrix $||g_{\alpha\beta}||$ should have nonzero determinant).

We can easily achieve our goal by introducing a second set of basis vectors, denoted $\{\vec{e}^1, \vec{e}^2, \ldots, \vec{e}^n\}$, which is *dual* to our first set in the sense that

$$\boxed{\vec{e}^\mu \cdot \vec{e}_\beta \equiv \mathbf{g}(\vec{e}^\mu, \vec{e}_\beta) = \delta^\mu{}_\beta.} \quad (24.8)$$

Here $\delta^\alpha{}_\beta$ is the Kronecker delta. This duality relation actually constitutes a *definition* of the $\vec{e}^\mu$ once the $\vec{e}_\alpha$ have been chosen. To see this, regard $\vec{e}^\mu$ as a tensor of rank one. This tensor is defined as soon as its value on each and every vector has been determined. Expression (24.8) gives the value $\vec{e}^\mu(\vec{e}_\beta) = \vec{e}^\mu \cdot \vec{e}_\beta$ of $\vec{e}^\mu$ on each of the four basis vectors $\vec{e}_\beta$; and since every other vector can be expanded in terms of the $\vec{e}_\beta$s and $\vec{e}^\mu(\_)$ is a linear function, Eq. (24.8) thereby determines the value of $\vec{e}^\mu$ on every other vector.

The duality relation (24.8) says that $\vec{e}^1$ is always perpendicular to all the $\vec{e}_\alpha$s except $\vec{e}_1$, and its scalar product with $\vec{e}_1$ is unity—and similarly for the other basis vectors. This interpretation is illustrated for 3-dimensional Euclidean space in Fig. 24.2. In Minkowski spacetime, if the $\vec{e}_\alpha$ are an orthonormal Lorentz basis, then duality dictates that $\vec{e}^0 = -\vec{e}_0$, and $\vec{e}^j = +\vec{e}_j$.

The duality relation (24.8) leads immediately to the same index-manipulation formalism as we have been using, if one defines the contravariant, covariant, and mixed components of tensors in the obvious manner:

$$\boxed{F^{\mu\nu} = \mathbf{F}(\vec{e}^\mu, \vec{e}^\nu), \quad F_{\alpha\beta} = \mathbf{F}(\vec{e}_\alpha, \vec{e}_\beta), \quad F^\mu{}_\beta = \mathbf{F}(\vec{e}^\mu, \vec{e}_\beta);} \quad (24.9)$$

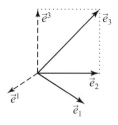

FIGURE 24.2 Nonorthonormal basis vectors $\vec{e}_j$ in Euclidean 3-space and two members $\vec{e}^1$ and $\vec{e}^3$ of the dual basis. The vectors $\vec{e}_1$ and $\vec{e}_2$ lie in the horizontal plane, so $\vec{e}^3$ is orthogonal to that plane (i.e., it points vertically upward), and its inner product with $\vec{e}_3$ is unity. Similarly, the vectors $\vec{e}_2$ and $\vec{e}_3$ span a vertical plane, so $\vec{e}^1$ is orthogonal to that plane (i.e., it points horizontally), and its inner product with $\vec{e}_1$ is unity.

see Ex. 24.4. Among the consequences of this duality are the following:

covariant and contravariant components of the metric

1. The matrix of contravariant components of the metric is inverse to that of the covariant components, $||g^{\mu\nu}|| = ||g_{\alpha\beta}||^{-1}$, so that

$$g^{\mu\beta} g_{\beta\nu} = \delta^{\mu}{}_{\nu}. \tag{24.10}$$

This relation guarantees that when one raises an index on a tensor $F_{\alpha\beta}$ with $g^{\mu\beta}$ and then lowers it back down with $g_{\beta\mu}$, one recovers one's original covariant components $F_{\alpha\beta}$ unaltered.

reconstructing a tensor from its components

2. One can reconstruct a tensor from its components by lining up the indices in a manner that accords with the rules of index manipulation:

$$\mathbf{F} = F^{\mu\nu} \vec{e}_\mu \otimes \vec{e}_\nu = F_{\alpha\beta} \vec{e}^\alpha \otimes \vec{e}^\beta = F^\mu{}_\beta \vec{e}_\mu \otimes \vec{e}^\beta. \tag{24.11}$$

component equations are same as slot-naming-index-notation equations

3. The component versions of tensorial equations are identical in mathematical symbology to the slot-naming-index-notation versions:

$$\mathbf{F}(\vec{p}, \vec{q}) = F^{\alpha\beta} p_\alpha p_\beta. \tag{24.12}$$

Associated with any coordinate system $x^\alpha(\mathcal{P})$ there is a *coordinate basis* whose basis vectors are defined by

coordinate basis

$$\vec{e}_\alpha \equiv \frac{\partial \mathcal{P}}{\partial x^\alpha}. \tag{24.13}$$

Since the derivative is taken holding the other coordinates fixed, the basis vector $\vec{e}_\alpha$ points along the α coordinate axis (the axis on which x^α changes and all the other coordinates are held fixed).

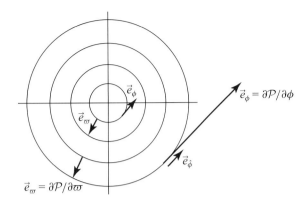

FIGURE 24.3 A circular coordinate system $\{\varpi, \phi\}$ and its coordinate basis vectors $\vec{e}_\varpi = \partial \mathcal{P}/\partial\varpi$, $\vec{e}_\phi = \partial \mathcal{P}/\partial\phi$ at several locations in the coordinate system. Also shown is the orthonormal basis vector $\vec{e}_{\hat{\phi}}$.

In an orthogonal curvilinear coordinate system [e.g., circular polar coordinates (ϖ, ϕ) in Euclidean 2-space; Fig. 24.3], this coordinate basis is quite different from the coordinate system's orthonormal basis. For example, $\vec{e}_\phi = (\partial \mathcal{P}/\partial\phi)_\varpi$ is a very long vector at large radii and a very short one at small radii; the corresponding unit-length vector is $\vec{e}_{\hat{\phi}} = (1/\varpi)\vec{e}_\phi = (1/\varpi)\partial/\partial\phi$ (i.e., the derivative with respect to physical distance along the ϕ direction). By contrast, $\vec{e}_\varpi = (\partial \mathcal{P}/\partial\varpi)_\phi$ already has unit length, so the corresponding orthonormal basis vector is simply $\vec{e}_{\hat{\varpi}} = \vec{e}_\varpi$. The metric components in the coordinate basis are readily seen to be $g_{\phi\phi} = \varpi^2$, $g_{\varpi\varpi} = 1$, and $g_{\varpi\phi} = g_{\phi\varpi} = 0$, which are in accord with the equation for the squared distance (interval) between adjacent points: $ds^2 = g_{ij}dx^i dx^j = d\varpi^2 + \varpi^2 d\phi^2$. Of course, the metric components in the orthonormal basis are $g_{\hat{i}\hat{j}} = \delta_{ij}$.

Henceforth, we use hats to identify orthonormal bases; bases whose indices do not have hats will typically (though not always) be coordinate bases.

We can construct the basis $\{\vec{e}^\mu\}$ that is dual to the coordinate basis $\{\vec{e}_\alpha\} = \{\partial \mathcal{P}/\partial x^\alpha\}$ by taking the gradients of the coordinates, viewed as scalar fields $x^\alpha(\mathcal{P})$:

$$\boxed{\vec{e}^\mu = \vec{\nabla} x^\mu.} \tag{24.14}$$

the basis dual to a coordinate basis

It is straightforward to verify the duality relation (24.8) for these two bases:

$$\vec{e}^\mu \cdot \vec{e}_\alpha = \vec{e}_\alpha \cdot \vec{\nabla} x^\mu = \nabla_{\vec{e}_\alpha} x^\mu = \nabla_{\partial \mathcal{P}/\partial x^\alpha} x^\mu = \frac{\partial x^\mu}{\partial x^\alpha} = \delta^\mu{}_\alpha. \tag{24.15}$$

In any coordinate system, the expansion of the metric in terms of the dual basis, $\mathbf{g} = g_{\alpha\beta} \vec{e}^\alpha \otimes \vec{e}^\beta = g_{\alpha\beta} \vec{\nabla} x^\alpha \otimes \vec{\nabla} x^\beta$, is intimately related to the line element $ds^2 = g_{\alpha\beta} dx^\alpha dx^\beta$. Consider an infinitesimal vectorial displacement $d\vec{x} = dx^\alpha (\partial/\partial x^\alpha)$. Insert this displacement into the metric's two slots to obtain the interval ds^2 along

orthogonal curvilinear coordinates

$d\vec{x}$. The result is $ds^2 = g_{\alpha\beta}\nabla x^\alpha \otimes \nabla x^\beta(d\vec{x}, d\vec{x}) = g_{\alpha\beta}(d\vec{x} \cdot \nabla x^\alpha)(d\vec{x} \cdot \nabla x^\beta) = g_{\alpha\beta}dx^\alpha dx^\beta$:

the line element for the invariant interval along a displacement vector

$$ds^2 = g_{\alpha\beta}dx^\alpha dx^\beta. \tag{24.16}$$

Here the second equality follows from the definition of the tensor product $\otimes$, and the third from the fact that for any scalar field ψ, $d\vec{x} \cdot \nabla\psi$ is the change $d\psi$ along $d\vec{x}$.

Any two bases $\{\vec{e}_\alpha\}$ and $\{\vec{e}_{\bar{\mu}}\}$ can be expanded in terms of each other:

transformation matrices linking two bases

$$\vec{e}_\alpha = \vec{e}_{\bar{\mu}}L^{\bar{\mu}}{}_\alpha, \quad \vec{e}_{\bar{\mu}} = \vec{e}_\alpha L^\alpha{}_{\bar{\mu}}. \tag{24.17}$$

(By convention the first index on L is always placed up, and the second is always placed down.) The quantities $||L^{\bar{\mu}}{}_\alpha||$ and $||L^\alpha{}_{\bar{\mu}}||$ are transformation matrices, and since they operate in opposite directions, they must be the inverse of each other:

$$L^{\bar{\mu}}{}_\alpha L^\alpha{}_{\bar{\nu}} = \delta^{\bar{\mu}}{}_{\bar{\nu}}, \quad L^\alpha{}_{\bar{\mu}} L^{\bar{\mu}}{}_\beta = \delta^\alpha{}_\beta. \tag{24.18}$$

These $||L^{\bar{\mu}}{}_\alpha||$ are the generalizations of Lorentz transformations to arbitrary bases [cf. Eqs. (2.34) and (2.35a)]. As in the Lorentz-transformation case, the transformation laws (24.17) for the basis vectors imply corresponding transformation laws for components of vectors and tensors—laws that entail lining up indices in the obvious manner:

transformation of tensor components between bases

$$A_{\bar{\mu}} = L^\alpha{}_{\bar{\mu}} A_\alpha, \quad T^{\bar{\mu}\bar{\nu}}{}_{\bar{\rho}} = L^{\bar{\mu}}{}_\alpha L^{\bar{\nu}}{}_\beta L^\gamma{}_{\bar{\rho}} T^{\alpha\beta}{}_\gamma,$$
and similarly in the opposite direction. $\tag{24.19}$

For coordinate bases, these $L^{\bar{\mu}}{}_\alpha$ are simply the partial derivatives of one set of coordinates with respect to the other:

transformation matrices between coordinate bases

$$L^{\bar{\mu}}{}_\alpha = \frac{\partial x^{\bar{\mu}}}{\partial x^\alpha}, \quad L^\alpha{}_{\bar{\mu}} = \frac{\partial x^\alpha}{\partial x^{\bar{\mu}}}, \tag{24.20}$$

as one can easily deduce via

$$\vec{e}_\alpha = \frac{\partial \mathcal{P}}{\partial x^\alpha} = \frac{\partial x^\mu}{\partial x^\alpha}\frac{\partial \mathcal{P}}{\partial x^\mu} = \vec{e}_\mu \frac{\partial x^\mu}{\partial x^\alpha}. \tag{24.21}$$

In many physics textbooks a tensor is *defined* as a set of components $F_{\alpha\beta}$ that obey the transformation laws

$$F_{\alpha\beta} = F_{\mu\nu}\frac{\partial x^\mu}{\partial x^\alpha}\frac{\partial x^\nu}{\partial x^\beta}. \tag{24.22}$$

This definition (valid only in a coordinate basis) is in accord with Eqs. (24.19) and (24.20), though it hides the true and very simple nature of a tensor as a linear function of frame-independent vectors.

EXERCISES

Exercise 24.4 *Derivation: Index-Manipulation Rules from Duality*

For an arbitrary basis $\{\vec{e}_\alpha\}$ and its dual basis $\{\vec{e}^\mu\}$, use (i) the duality relation (24.8), (ii) the definition (24.9) of components of a tensor, and (iii) the relation $\vec{A} \cdot \vec{B} = \mathbf{g}(\vec{A}, \vec{B})$ between the metric and the inner product to deduce the following results.

(a) The relations
$$\vec{e}^\mu = g^{\mu\alpha}\vec{e}_\alpha, \quad \vec{e}_\alpha = g_{\alpha\mu}\vec{e}^\mu. \tag{24.23}$$

(b) The fact that indices on the components of tensors can be raised and lowered using the components of the metric:
$$F^{\mu\nu} = g^{\mu\alpha} F_\alpha{}^\nu, \quad P_\alpha = g_{\alpha\beta} P^\beta. \tag{24.24}$$

(c) The fact that a tensor can be reconstructed from its components in the manner of Eq. (24.11).

Exercise 24.5 *Practice: Transformation Matrices for Circular Polar Bases*

Consider the circular polar coordinate system $\{\varpi, \phi\}$ and its coordinate bases and orthonormal bases as shown in Fig. 24.3 and discussed in the associated text. These coordinates are related to Cartesian coordinates $\{x, y\}$ by the usual relations: $x = \varpi \cos\phi$, $y = \varpi \sin\phi$.

(a) Evaluate the components ($L^x{}_\varpi$, etc.) of the transformation matrix that links the two coordinate bases $\{\vec{e}_x, \vec{e}_y\}$ and $\{\vec{e}_\varpi, \vec{e}_\phi\}$. Also evaluate the components ($L^\varpi{}_x$, etc.) of the inverse transformation matrix.

(b) Similarly, evaluate the components of the transformation matrix and its inverse linking the bases $\{\vec{e}_x, \vec{e}_y\}$ and $\{\vec{e}_{\hat{\varpi}}, \vec{e}_{\hat{\phi}}\}$.

(c) Consider the vector $\vec{A} \equiv \vec{e}_x + 2\vec{e}_y$. What are its components in the other two bases?

24.3.2 Vectors as Directional Derivatives; Tangent Space; Commutators

As discussed in the introduction to Sec. 24.3, the notion of a vector as an arrow connecting two points is problematic in a curved manifold and must be refined. As a first step in the refinement, let us consider the tangent vector $\vec{A}$ to a curve $\mathcal{P}(\zeta)$ at some point $\mathcal{P}_o \equiv \mathcal{P}(\zeta = 0)$. We have defined that tangent vector by the limiting process:

$$\boxed{\vec{A} \equiv \frac{d\mathcal{P}}{d\zeta} \equiv \lim_{\Delta\zeta \to 0} \frac{\mathcal{P}(\Delta\zeta) - \mathcal{P}(0)}{\Delta\zeta}} \tag{24.25}$$

tangent vector to a curve

[Eq. (24.2)]. In this definition the difference $\mathcal{P}(\zeta) - \mathcal{P}(0)$ means the tiny arrow reaching from $\mathcal{P}(0) \equiv \mathcal{P}_o$ to $\mathcal{P}(\Delta\zeta)$. In the limit as $\Delta\zeta$ becomes vanishingly small, these two points get arbitrarily close together. In such an arbitrarily small region of the manifold, the effects of the manifold's curvature become arbitrarily small and

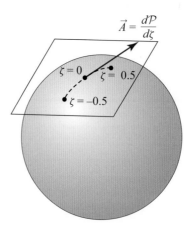

FIGURE 24.4 A curve $\mathcal{P}(\zeta)$ on the surface of a sphere and the curve's tangent vector $\vec{A} = d\mathcal{P}/d\zeta$ at $\mathcal{P}(\zeta = 0) \equiv \mathcal{P}_o$. The tangent vector lives in the tangent space at $\mathcal{P}_o$ (i.e., in the flat plane that is tangent to the sphere there, as seen in the flat Euclidean 3-space in which the sphere's surface is embedded).

negligible (just think of an arbitrarily tiny region on the surface of a sphere), so the notion of the arrow should become sensible. However, before the limit is completed, we are required to divide by $\Delta\zeta$, which makes our arbitrarily tiny arrow big again. What meaning can we give to this?

One way to think about it is to imagine embedding the curved manifold in a higher-dimensional flat space (e.g., embed the surface of a sphere in a flat 3-dimensional Euclidean space, as shown in Fig. 24.4). Then the tiny arrow $\mathcal{P}(\Delta\zeta) - \mathcal{P}(0)$ can be thought of equally well as lying on the sphere, or as lying in a surface that is tangent to the sphere and is flat, as measured in the flat embedding space. We can give meaning to $[\mathcal{P}(\Delta\zeta) - \mathcal{P}(0)]/\Delta\zeta$ if we regard this expression as a formula for lengthening an arrow-type vector in the flat tangent surface; correspondingly, we must regard the resulting tangent vector $\vec{A}$ as an arrow living in the tangent surface.

tangent space at a point

The (conceptual) flat tangent surface at the point $\mathcal{P}_o$ is called the *tangent space* to the curved manifold at that point. It has the same number of dimensions n as the manifold itself (two in the case of the surface of the sphere in Fig. 24.4). Vectors at $\mathcal{P}_o$ are arrows residing in that point's tangent space, tensors at $\mathcal{P}_o$ are linear functions of these vectors, and all the linear algebra of vectors and tensors that reside at $\mathcal{P}_o$ occurs in this tangent space. For example, the inner product of two vectors $\vec{A}$ and $\vec{B}$ at $\mathcal{P}_o$ (two arrows living in the tangent space there) is computed via the standard relation $\vec{A} \cdot \vec{B} = \boldsymbol{g}(\vec{A}, \vec{B})$ using the metric $\boldsymbol{g}$ that also resides in the tangent space. (Scalars reside in both the manifold and the tangent space.)

This pictorial way of thinking about the tangent space and vectors and tensors that reside in it is far too heuristic to satisfy most mathematicians. Therefore, mathematicians have insisted on making it much more precise at the price of greater abstraction. Mathematicians define the tangent vector to the curve $\mathcal{P}(\zeta)$ to be the derivative $d/d\zeta$

that differentiates scalar fields along the curve. This derivative operator is well defined by the rules of ordinary differentiation: if $\psi(\mathcal{P})$ is a scalar field in the manifold, then $\psi[\mathcal{P}(\zeta)]$ is a function of the real variable ζ, and its derivative $(d/d\zeta)\psi[\mathcal{P}(\zeta)]$ evaluated at $\zeta = 0$ is the ordinary derivative of elementary calculus. Since the derivative operator $d/d\zeta$ differentiates in the manifold along the direction in which the curve is moving, it is often called the *directional derivative* along $\mathcal{P}(\zeta)$. Mathematicians notice that all the directional derivatives at a point $\mathcal{P}_o$ of the manifold form a vector space (they can be multiplied by scalars and added and subtracted to get new vectors), and so the mathematicians define this vector space to be the tangent space at $\mathcal{P}_o$.

directional derivative

This mathematical procedure turns out to be isomorphic to the physicists' more heuristic way of thinking about the tangent space. In physicists' language, if one introduces a coordinate system in a region of the manifold containing $\mathcal{P}_o$ and constructs the corresponding coordinate basis $\vec{e}_\alpha = \partial \mathcal{P}/\partial x^\alpha$, then one can expand any vector in the tangent space as $\vec{A} = A^\alpha \partial \mathcal{P}/\partial x^\alpha$. One can also construct, in physicists' language, the directional derivative along $\vec{A}$; it is $\partial_{\vec{A}} \equiv A^\alpha \partial/\partial x^\alpha$. Evidently, the components A^α of the physicist's vector $\vec{A}$ (an arrow) are identical to the coefficients A^α in the coordinate-expansion of the directional derivative $\partial_{\vec{A}}$. Therefore a one-to-one correspondence exists between the directional derivatives $\partial_{\vec{A}}$ at $\mathcal{P}_o$ and the vectors $\vec{A}$ there, and a complete isomorphism holds between the tangent-space manipulations that a mathematician performs treating the directional derivatives as vectors, and those that a physicist performs treating the arrows as vectors.

"Why not abandon the fuzzy concept of a vector as an arrow, and *redefine the vector* $\vec{A}$ *to be the same as the directional derivative* $\partial_{\vec{A}}$?" mathematicians have demanded of physicists. Slowly, over the past century, physicists have come to see the merit in this approach. (i) It does, indeed, make the concept of a vector more rigorous than before. (ii) It simplifies a number of other concepts in mathematical physics (e.g., the commutator of two vector fields; see below). (iii) It facilitates communication with mathematicians. (iv) It provides a formalism that is useful for calculation. With these motivations in mind, and because one always gains conceptual and computational power by having multiple viewpoints at one's fingertips (see Feynman, 1966, p. 160), we henceforth shall regard vectors both as arrows living in a tangent space and as directional derivatives. Correspondingly, we assert the equalities:

tangent vector as directional derivative along a curve

$$\boxed{\frac{\partial \mathcal{P}}{\partial x^\alpha} = \frac{\partial}{\partial x^\alpha} \;,\quad \vec{A} = \partial_{\vec{A}},} \tag{24.26}$$

and often expand vectors in a coordinate basis using the notation

$$\boxed{\vec{A} = A^\alpha \frac{\partial}{\partial x^\alpha}.} \tag{24.27}$$

This directional-derivative viewpoint on vectors makes natural the concept of the *commutator* of two vector fields $\vec{A}$ and $\vec{B}$: $[\vec{A}, \vec{B}]$ is the vector that, when viewed

commutator of two vector fields

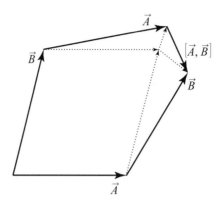

FIGURE 24.5 The commutator $[\vec{A}, \vec{B}]$ of two vector fields. The vectors are assumed to be so small that the curvature of the manifold is negligible in the region of the diagram, so all the vectors can be drawn lying in the manifold itself rather than in their respective tangent spaces. In evaluating the two terms in the commutator (24.28), a locally orthonormal coordinate basis is used, so $A^\alpha \partial B^\beta / \partial x^\alpha$ is the amount by which the vector $\vec{B}$ changes when one travels along $\vec{A}$ (i.e., it is the rightward-and-downward pointing dashed arrow in the upper right), and $B^\alpha \partial A^\beta / \partial x^\alpha$ is the amount by which $\vec{A}$ changes when one travels along $\vec{B}$ (i.e., it is the rightward-and-upward pointing dashed arrow). According to Eq. (24.28), the difference of these two dashed arrows is the commutator $[\vec{A}, \vec{B}]$. As the diagram shows, this commutator closes the quadrilateral whose legs are $\vec{A}$ and $\vec{B}$. If the commutator vanishes, then there is no gap in the quadrilateral, which means that in the region covered by this diagram, one can construct a coordinate system in which $\vec{A}$ and $\vec{B}$ are coordinate basis vectors.

as a differential operator, is given by $[\partial_{\vec{A}}, \partial_{\vec{B}}]$—where the latter quantity is the same commutator as one meets elsewhere in physics (e.g., in quantum mechanics). Using this definition, we can compute the components of the commutator in a coordinate basis:

$$[\vec{A}, \vec{B}] \equiv \left[A^\alpha \frac{\partial}{\partial x^\alpha}, B^\beta \frac{\partial}{\partial x^\beta} \right] = \left(A^\alpha \frac{\partial B^\beta}{\partial x^\alpha} - B^\alpha \frac{\partial A^\beta}{\partial x^\alpha} \right) \frac{\partial}{\partial x^\beta}. \qquad (24.28)$$

This is an operator equation where the final derivative is presumed to operate on a scalar field, just as in quantum mechanics. From this equation we can read off the components of the commutator in any coordinate basis; they are $A^\alpha B^\beta{}_{,\alpha} - B^\alpha A^\beta{}_{,\alpha}$, where the comma denotes partial differentiation. Figure 24.5 uses this equation to deduce the geometric meaning of the commutator: it is the fifth leg needed to close a quadrilateral whose other four legs are constructed from the vector fields $\vec{A}$ and $\vec{B}$. In other words, it is "the change in $\vec{B}$ relative to $\vec{A}$," and as such it is a type of derivative of $\vec{B}$ along $\vec{A}$, called the *Lie derivative*: $\mathcal{L}_{\vec{A}} \vec{B} \equiv [\vec{A}, \vec{B}]$ (cf. footnote 2 in Chap. 14).

The commutator is useful as a tool for distinguishing between coordinate bases and noncoordinate bases (also called nonholonomic bases). In a coordinate basis, the basis vectors are just the coordinate system's partial derivatives, $\vec{e}_\alpha = \partial/\partial x^\alpha$, and since partial derivatives commute, it must be that $[\vec{e}_\alpha, \vec{e}_\beta] = 0$. Conversely (as Fig. 24.5 shows), if one has a basis with vanishing commutators $[\vec{e}_\alpha, \vec{e}_\beta] = 0$, then it

coordinate bases have vanishing commutators

is possible to construct a coordinate system for which this is the coordinate basis. In a noncoordinate basis, at least one of the commutators $[\vec{e}_\alpha, \vec{e}_\beta]$ will be nonzero.

24.3.3 Differentiation of Vectors and Tensors; Connection Coefficients

In a curved manifold, the differentiation of vectors and tensors is rather subtle. To elucidate the problem, let us recall how we defined such differentiation in Minkowski spacetime or Euclidean space (Sec. 1.7). Converting to the notation used in Eq. (24.25), we began by defining the directional derivative of a tensor field $\boldsymbol{F}(\mathcal{P})$ along the tangent vector $\vec{A} = d/d\zeta$ to a curve $\mathcal{P}(\zeta)$:

$$\boxed{\nabla_{\vec{A}} \boldsymbol{F} \equiv \lim_{\Delta\zeta \to 0} \frac{\boldsymbol{F}[\mathcal{P}(\Delta\zeta)] - \boldsymbol{F}[\mathcal{P}(0)]}{\Delta\zeta}.} \qquad (24.29)$$

directional derivative of a tensor field

This definition is problematic, because $\boldsymbol{F}[\mathcal{P}(\Delta\zeta)]$ lives in a different tangent space than does $\boldsymbol{F}[\mathcal{P}(0)]$. To make the definition meaningful, we must identify some connection between the two tangent spaces, when their points $\mathcal{P}(\Delta\zeta)$ and $\mathcal{P}(0)$ are arbitrarily close together. That connection is equivalent to identifying a rule for transporting $\boldsymbol{F}$ from one tangent space to the other.

In flat space or flat spacetime, and when $\boldsymbol{F}$ is a vector $\vec{F}$, that transport rule is obvious: keep $\vec{F}$ parallel to itself and keep its length fixed during the transport. In other words, keep constant its components in an orthonormal coordinate system (Cartesian coordinates in Euclidean space, Lorentz coordinates in Minkowski spacetime). This is called the *law of parallel transport*. For a tensor $\boldsymbol{F}$, the parallel transport law is the same: keep its components fixed in an orthonormal coordinate basis.

Now, just as the curvature of Earth's surface prevents one from placing a Cartesian coordinate system on it, so nonzero curvature of any other manifold prevents one from introducing orthonormal coordinates; see Sec. 25.3. However, in an arbitrarily small region on Earth's surface, one can introduce coordinates that are arbitrarily close to Cartesian (as surveyors well know); the fractional deviations from Cartesian need be no larger than $O(L^2/R^2)$, where L is the size of the region and R is Earth's radius (see Sec. 25.3). Similarly, in curved spacetime, in an arbitrarily small region, one can introduce coordinates that are arbitrarily close to Lorentz, differing only by amounts quadratic in the size of the region—and similarly for a *local* orthonormal coordinate basis in any curved manifold.

When defining $\nabla_{\vec{A}} \boldsymbol{F}$, one is sensitive only to first-order changes of quantities, not second, so the parallel transport used in defining it in a flat manifold, based on constancy of components in an orthonormal coordinate basis, must also work in a *local* orthonormal coordinate basis of any curved manifold: In Eq. (24.29), one must transport $\boldsymbol{F}$ from $\mathcal{P}(\Delta\zeta)$ to $\mathcal{P}(0)$, holding its components fixed in a locally orthonormal coordinate basis (parallel transport), and then take the difference in the tangent space at $\mathcal{P}_o = \mathcal{P}(0)$, divide by $\Delta\zeta$, and let $\Delta\zeta \to 0$. The result is a tensor at $\mathcal{P}_o$: the directional derivative $\nabla_{\vec{A}} \boldsymbol{F}$ of $\boldsymbol{F}$.

gradient of a tensor field

Having made the directional derivative meaningful, one can proceed as in Secs. 1.7 and 2.10: define the gradient of $\mathbf{F}$ by $\nabla_{\vec{A}}\mathbf{F} = \vec{\nabla}\mathbf{F}(\_, \_, \vec{A})$ [i.e., put $\vec{A}$ in the last—differentiation—slot of $\vec{\nabla}\mathbf{F}$; Eq. (1.15b)].

As in Chap. 2, in any basis we denote the components of $\vec{\nabla}\mathbf{F}$ by $F_{\alpha\beta;\gamma}$. And as in Sec. 11.8 (elasticity theory), we can compute these components in any basis with the aid of that basis's *connection coefficients*.

In Sec. 11.8, we restricted ourselves to an orthonormal basis in Euclidean space and thus had no need to distinguish between covariant and contravariant indices; all indices were written as subscripts. Now, dealing with nonorthonormal bases in spacetime, we must distinguish covariant and contravariant indices. Accordingly, by analogy with Eq. (11.68), we define the connection coefficients $\Gamma^{\mu}{}_{\alpha\beta}$ as

connection coefficients for a basis and its dual

$$\boxed{\nabla_{\beta}\vec{e}_{\alpha} \equiv \nabla_{\vec{e}_{\beta}}\vec{e}_{\alpha} = \Gamma^{\mu}{}_{\alpha\beta}\vec{e}_{\mu}.} \tag{24.30}$$

The duality between bases $\vec{e}^{\nu} \cdot \vec{e}_{\alpha} = \delta^{\nu}{}_{\alpha}$ then implies

$$\boxed{\nabla_{\beta}\vec{e}^{\mu} \equiv \nabla_{\vec{e}_{\beta}}\vec{e}^{\mu} = -\Gamma^{\mu}{}_{\alpha\beta}\vec{e}^{\alpha}.} \tag{24.31}$$

Note the sign flip, which is required to keep $\nabla_{\beta}(\vec{e}^{\mu} \cdot \vec{e}_{\alpha}) = 0$, and note that the differentiation index always goes last on Γ. Duality also implies that Eqs. (24.30) and (24.31) can be rewritten as

$$\Gamma^{\mu}{}_{\alpha\beta} = \vec{e}^{\mu} \cdot \nabla_{\beta}\vec{e}_{\alpha} = -\vec{e}_{\alpha} \cdot \nabla_{\beta}\vec{e}^{\mu}. \tag{24.32}$$

With the aid of these connection coefficients, we can evaluate the components $A_{\alpha;\beta}$ of the gradient of a vector field in any basis. We just compute

$$A^{\mu}{}_{;\beta}\vec{e}_{\mu} = \nabla_{\beta}\vec{A} = \nabla_{\beta}(A^{\mu}\vec{e}_{\mu}) = (\nabla_{\beta}A^{\mu})\vec{e}_{\mu} + A^{\mu}\nabla_{\beta}\vec{e}_{\mu}$$
$$= A^{\mu}{}_{,\beta}\vec{e}_{\mu} + A^{\mu}\Gamma^{\alpha}{}_{\mu\beta}\vec{e}_{\alpha}$$
$$= (A^{\mu}{}_{,\beta} + A^{\alpha}\Gamma^{\mu}{}_{\alpha\beta})\vec{e}_{\mu}. \tag{24.33}$$

In going from the first line to the second, we have used the notation

$$A^{\mu}{}_{,\beta} \equiv \partial_{\vec{e}_{\beta}}A^{\mu}; \tag{24.34}$$

that is, *the comma denotes the result of letting a basis vector act as a differential operator on the component of the vector*. In going from the second line of (24.33) to the third, we have renamed some summed-over indices. By comparing the first and last expressions in Eq. (24.33), we conclude that

components of the gradient of a vector field

$$\boxed{A^{\mu}{}_{;\beta} = A^{\mu}{}_{,\beta} + A^{\alpha}\Gamma^{\mu}{}_{\alpha\beta}.} \tag{24.35}$$

The first term in this equation describes the changes in $\vec{A}$ associated with changes of its component A^{μ}; the second term *corrects for* artificial changes of A^{μ} that are induced by turning and length changes of the basis vector $\vec{e}_{\mu}$. We shall use the short-hand terminology that the second term "corrects the index μ."

By a similar computation, we conclude that in any basis the covariant components of the gradient are

$$A_{\alpha;\beta} = A_{\alpha,\beta} - \Gamma^\mu{}_{\alpha\beta} A_\mu, \tag{24.36}$$

where again $A_{\alpha,\beta} \equiv \partial_{\vec{e}_\beta} A_\alpha$. Notice that, when the index being corrected is down [α in Eq. (24.36)], the connection coefficient has a minus sign; when it is up [μ in Eq. (24.35)], the connection coefficient has a plus sign. This is in accord with the signs in Eqs. (24.30) and (24.31).

These considerations should make obvious the following equations for the components of the gradient of a second rank tensor field:

$$\begin{aligned}
F^{\alpha\beta}{}_{;\gamma} &= F^{\alpha\beta}{}_{,\gamma} + \Gamma^\alpha{}_{\mu\gamma} F^{\mu\beta} + \Gamma^\beta{}_{\mu\gamma} F^{\alpha\mu}, \\
F_{\alpha\beta;\gamma} &= F_{\alpha\beta,\gamma} - \Gamma^\mu{}_{\alpha\gamma} F_{\mu\beta} - \Gamma^\mu{}_{\beta\gamma} F_{\alpha\mu}, \\
F^\alpha{}_{\beta;\gamma} &= F^\alpha{}_{\beta,\gamma} + \Gamma^\alpha{}_{\mu\gamma} F^\mu{}_\beta - \Gamma^\mu{}_{\beta\gamma} F^\alpha{}_\mu.
\end{aligned} \tag{24.37}$$

components of the gradient of a tensor field

Notice that each index of **F** must be corrected, the correction has a sign dictated by whether the index is up or down, the differentiation index always goes last on the Γ, and all other indices can be deduced by requiring that the free indices in each term be the same and all other indices be summed.

If we have been given a basis, then how can we compute the connection coefficients? We can try to do so by drawing pictures and examining how the basis vectors change from point to point—a method that is fruitful in spherical and cylindrical coordinates in Euclidean space (Sec. 11.8). However, in other situations this method is fraught with peril, so we need a firm mathematical prescription. It turns out that the following prescription works (see Ex. 24.7 for a proof).

1. Evaluate the commutation coefficients $c_{\alpha\beta}{}^\rho$ of the basis, which are defined by the two equivalent relations:

$$[\vec{e}_\alpha, \vec{e}_\beta] \equiv c_{\alpha\beta}{}^\rho \vec{e}_\rho, \quad c_{\alpha\beta}{}^\rho \equiv \vec{e}^\rho \cdot [\vec{e}_\alpha, \vec{e}_\beta]. \tag{24.38a}$$

commutation coefficients for a basis

(Note that in a coordinate basis the commutation coefficients will vanish. Warning: Commutation coefficients also appear in the theory of Lie groups; there it is conventional to use a different ordering of indices than here: $c_{\alpha\beta}{}^\rho{}_{\text{here}} = c^\rho{}_{\alpha\beta\,\text{Lie groups}}$.)

2. Lower the last index on the commutation coefficients using the metric components in the basis:

$$c_{\alpha\beta\gamma} \equiv c_{\alpha\beta}{}^\rho g_{\rho\gamma}. \tag{24.38b}$$

3. Compute the quantities

$$\Gamma_{\alpha\beta\gamma} \equiv \frac{1}{2}(g_{\alpha\beta,\gamma} + g_{\alpha\gamma,\beta} - g_{\beta\gamma,\alpha} + c_{\alpha\beta\gamma} + c_{\alpha\gamma\beta} - c_{\beta\gamma\alpha}). \tag{24.38c}$$

formulas for computing connection coefficients

Here the commas denote differentiation with respect to the basis vectors as though the metric components were scalar fields [as in Eq. (24.34)]. Notice that the pattern of indices is the same on the gs and on the cs. It is a peculiar pattern—one of the few aspects of index gymnastics that cannot be reconstructed by merely lining up indices. In a coordinate basis the c terms will vanish, so $\Gamma_{\alpha\beta\gamma}$ will be symmetric in its last two indices. In an orthonormal basis $g_{\mu\nu}$ are constant, so the g terms will vanish, and $\Gamma_{\alpha\beta\gamma}$ will be antisymmetric in its first two indices. And in a Cartesian or Lorentz coordinate basis, which is both coordinate and orthonormal, both the c terms and the g terms will vanish, so $\Gamma_{\alpha\beta\gamma}$ will vanish.

4. Raise the first index on $\Gamma_{\alpha\beta\gamma}$ to obtain the connection coefficients

$$\boxed{\Gamma^{\mu}{}_{\beta\gamma} = g^{\mu\alpha}\Gamma_{\alpha\beta\gamma}.} \tag{24.38d}$$

In a coordinate basis, the $\Gamma^{\mu}{}_{\beta\gamma}$ are sometimes called *Christoffel symbols*, though we will use the name connection coefficients independent of the nature of the basis.

The first three steps in the above prescription for computing the connection coefficients follow from two key properties of the gradient $\vec{\nabla}$. First, the gradient of the metric tensor vanishes:

vanishing gradient of the metric tensor

$$\boxed{\vec{\nabla}\mathbf{g} = 0.} \tag{24.39}$$

Second, for any two vector fields $\vec{A}$ and $\vec{B}$, the gradient is related to the commutator by

relation of gradient to commutator

$$\boxed{\nabla_{\vec{A}}\vec{B} - \nabla_{\vec{B}}\vec{A} = [\vec{A},\vec{B}].} \tag{24.40}$$

For a derivation of these relations and then a derivation of the prescription 1–4, see Exs. 24.6 and 24.7.

The gradient operator $\vec{\nabla}$ is an example of a geometric object that is not a tensor. The connection coefficients $\Gamma^{\mu}{}_{\beta\gamma} = \vec{e}^{\mu} \cdot \left(\nabla_{\vec{e}_{\gamma}}\vec{e}_{\beta}\right)$ can be regarded as the components of $\vec{\nabla}$; because it is not a tensor, these components do not obey the tensorial transformation law (24.19) when switching from one basis to another. Their transformation law is far more complicated and is rarely used. Normally one computes them from scratch in the new basis, using the above prescription or some other, equivalent prescription (cf. Misner, Thorne, and Wheeler, 1973, Chap. 14). For most curved spacetimes that one meets in general relativity, these computations are long and tedious and therefore are normally carried out on computers using symbolic manipulation software, such as Maple, Matlab, or Mathematica, or such programs as GR-Tensor and MathTensor that run under Maple or Mathematica. Such software is easily found on the Internet using a search engine. A particularly simple Mathematica program for use with coordinate bases is presented and discussed in Appendix C of Hartle (2003) and is available on that book's website: http://web.physics.ucsb.edu/~gravitybook/.

EXERCISES

Exercise 24.6 *Derivation: Properties of the Gradient* $\vec{\nabla}$
(a) Derive Eq. (24.39). [Hint: At a point $\mathcal{P}$ where $\vec{\nabla}\mathbf{g}$ is to be evaluated, introduce a locally orthonormal coordinate basis (i.e., locally Cartesian or locally Lorentz). When computing in this basis, the effects of curvature show up only to second order in distance from $\mathcal{P}$. Show that in this basis, the components of $\vec{\nabla}\mathbf{g}$ vanish, and from this infer that $\vec{\nabla}\mathbf{g}$, viewed as a frame-independent third-rank tensor, vanishes.]
(b) Derive Eq. (24.40). [Hint: Again work in a locally orthonormal coordinate basis.]

Exercise 24.7 *Derivation and Example: Prescription for Computing Connection Coefficients*
Derive the prescription 1–4 [Eqs. (24.38)] for computing the connection coefficients in any basis. [Hints: (i) In the chosen basis, from $\vec{\nabla}\mathbf{g} = 0$ infer that $\Gamma_{\alpha\beta\gamma} + \Gamma_{\beta\alpha\gamma} = g_{\alpha\beta,\gamma}$. Notice that this determines the part of $\Gamma_{\alpha\beta\gamma}$ that is symmetric in its first two indices. Show that the number of independent components of $\Gamma_{\alpha\beta\gamma}$ thereby determined is $\frac{1}{2}n^2(n+1)$, where n is the manifold's dimension. (ii) From Eq. (24.40) infer that $\Gamma_{\gamma\beta\alpha} - \Gamma_{\gamma\alpha\beta} = c_{\alpha\beta\gamma}$, which fixes the part of Γ antisymmetric in the last two indices. Show that the number of independent components thereby determined is $\frac{1}{2}n^2(n-1)$. (iii) Infer that the number of independent components determined by (i) and (ii) together is n^3, which is the entirety of $\Gamma_{\alpha\beta\gamma}$. By somewhat complicated algebra, deduce Eq. (24.38c) for $\Gamma_{\alpha\beta\gamma}$. (The algebra is sketched in Misner, Thorne, and Wheeler, 1973, Ex. 8.15.) (iv) Then infer the final answer, Eq. (24.38d), for $\Gamma^\mu{}_{\beta\gamma}$.]

Exercise 24.8 *Practice: Commutation and Connection Coefficients for Circular Polar Bases*
Consider the circular polar coordinates $\{\varpi, \phi\}$ of Fig. 24.3 and their associated bases.
(a) Evaluate the commutation coefficients $c_{\alpha\beta}{}^\rho$ for the coordinate basis $\{\vec{e}_\varpi, \vec{e}_\phi\}$, and also for the orthonormal basis $\{\vec{e}_{\hat{\varpi}}, \vec{e}_{\hat{\phi}}\}$.
(b) Compute by hand the connection coefficients for the coordinate basis and also for the orthonormal basis, using Eqs. (24.38). [Note: The answer for the orthonormal basis was worked out pictorially in our study of elasticity theory; Fig. 11.15 and Eq. (11.70).]
(c) Repeat this computation using symbolic manipulation software on a computer.

Exercise 24.9 *Practice: Connection Coefficients for Spherical Polar Coordinates*
(a) Consider spherical polar coordinates in 3-dimensional space, and verify that the nonzero connection coefficients, assuming an orthonormal basis, are given by Eq. (11.71).

(b) Repeat the exercise in part (a) assuming a coordinate basis with

$$\mathbf{e}_r \equiv \frac{\partial}{\partial r}, \quad \mathbf{e}_\theta \equiv \frac{\partial}{\partial \theta}, \quad \mathbf{e}_\phi \equiv \frac{\partial}{\partial \phi}. \tag{24.41}$$

(c) Repeat both computations in parts (a) and (b) using symbolic manipulation software on a computer.

Exercise 24.10 *Practice: Index Gymnastics—Geometric Optics*

This exercise gives the reader practice in formal manipulations that involve the gradient operator. In the geometric-optics (eikonal) approximation of Sec. 7.3, for electromagnetic waves in Lorenz gauge, one can write the 4-vector potential in the form $\vec{A} = \tilde{\mathcal{A}} e^{i\varphi}$, where $\tilde{\mathcal{A}}$ is a slowly varying amplitude and φ is a rapidly varying phase. By the techniques of Sec. 7.3, one can deduce from the vacuum Maxwell equations that the wave vector, defined by $\vec{k} \equiv \vec{\nabla}\varphi$, is null: $\vec{k} \cdot \vec{k} = 0$.

(a) Rewrite all the equations in the above paragraph in slot-naming index notation.

(b) Using index manipulations, show that the wave vector $\vec{k}$ (which is a vector field, because the wave's phase φ is a scalar field) satisfies the geodesic equation $\nabla_{\vec{k}} \vec{k} = 0$ (cf. Sec. 24.5.2). The geodesics, to which $\vec{k}$ is the tangent vector, are the rays discussed in Sec. 7.3, along which the waves propagate.

24.3.4 Integration

Our desire to use general bases and work in curved manifolds gives rise to two new issues in the definition of integrals.

The first issue is that the volume elements used in integration involve the Levi-Civita tensor [Eqs. (2.43), (2.52), and (2.55)], so we need to know the components of the Levi-Civita tensor in a general basis. It turns out (see, e.g., Misner, Thorne, and Wheeler, 1973, Ex. 8.3) that the covariant components differ from those in an orthonormal basis by a factor $\sqrt{|g|}$ and the contravariant by $1/\sqrt{|g|}$, where

$$\boxed{g \equiv \det ||g_{\alpha\beta}||} \tag{24.42}$$

is the determinant of the matrix whose entries are the covariant components of the metric. More specifically, let us denote by $[\alpha\beta \ldots \nu]$ the value of $\epsilon_{\alpha\beta\ldots\nu}$ in an orthonormal basis of our n-dimensional space [Eq. (2.43)]:

$$[12\ldots n] = +1,$$

$$[\alpha\beta \ldots \nu] = \begin{cases} +1 & \text{if } \alpha, \beta, \ldots, \nu \text{ is an even permutation of } 1, 2, \ldots, n \\ -1 & \text{if } \alpha, \beta, \ldots, \nu \text{ is an odd permutation of } 1, 2, \ldots, n \\ 0 & \text{if } \alpha, \beta, \ldots, \nu \text{ are not all different.} \end{cases} \tag{24.43}$$

(In spacetime the indices must run from 0 to 3 rather than 1 to $n = 4$.) Then in a general right-handed basis the components of the Levi-Civita tensor are

$$\epsilon_{\alpha\beta\ldots\nu} = \sqrt{|g|}\,[\alpha\beta\ldots\nu], \quad \epsilon^{\alpha\beta\ldots\nu} = \pm\frac{1}{\sqrt{|g|}}\,[\alpha\beta\ldots\nu], \qquad (24.44)$$

components of Levi-Civita tensor in an arbitrary basis

where the $\pm$ is plus in Euclidean space and minus in spacetime. In a left-handed basis the sign is reversed.

As an example of these formulas, consider a spherical polar coordinate system (r, θ, ϕ) in 3-dimensional Euclidean space, and use the three infinitesimal vectors $dx^j(\partial/\partial x^j)$ to construct the volume element $d\Sigma$ [cf. Eq. (1.26)]:

$$dV = \epsilon\left(dr\frac{\partial}{\partial r},\, d\theta\frac{\partial}{\partial \theta},\, d\phi\frac{\partial}{\partial \phi}\right) = \epsilon_{r\theta\phi}dr d\theta d\phi = \sqrt{g}\, dr d\theta d\phi = r^2\sin\theta dr d\theta d\phi.$$
$$(24.45)$$

Here the second equality follows from linearity of ϵ and the formula for computing its components by inserting basis vectors into its slots; the third equality follows from our formula (24.44) for the components. The fourth equality entails the determinant of the metric coefficients, which in spherical coordinates are $g_{rr} = 1$, $g_{\theta\theta} = r^2$, and $g_{\phi\phi} = r^2\sin^2\theta$; all other g_{jk} vanish, so $g = r^4\sin^2\theta$. The resulting volume element $r^2\sin\theta dr d\theta d\phi$ should be familiar and obvious.

The second new integration issue we must face is that such integrals as

$$\int_{\partial\mathcal{V}} T^{\alpha\beta}d\Sigma_\beta \qquad (24.46)$$

[cf. Eqs. (2.55), (2.56)] involve constructing a vector $T^{\alpha\beta}d\Sigma_\beta$ in each infinitesimal region $d\Sigma_\beta$ of the surface of integration $\partial\mathcal{V}$ and then adding up the contributions from all the infinitesimal regions. A major difficulty arises because each contribution lives in a different tangent space. To add them together, we must first transport them all to the same tangent space at some single location in the manifold. How is that transport to be performed? The obvious answer is "by the same parallel transport technique that we used in defining the gradient." However, when defining the gradient, we only needed to perform the parallel transport over an infinitesimal distance, and now we must perform it over long distances. When the manifold is curved, long-distance parallel transport gives a result that depends on the route of the transport, and in general there is no way to identify any preferred route (see, e.g., Misner, Thorne, and Wheeler, 1973, Sec. 11.4).

As a result, *integrals such as Eq. (24.46) are ill-defined in a curved manifold. The only integrals that are well defined in a curved manifold are those such as $\int_{\partial\mathcal{V}} S^\alpha d\Sigma_\alpha$, whose infinitesimal contributions $S^\alpha d\Sigma_\alpha$ are scalars* (i.e., integrals whose value is a scalar). This fact will have profound consequences in curved spacetime for the laws of conservation of energy, momentum, and angular momentum (Secs. 25.7 and 25.9.4).

integrals in a curved manifold are well defined only if infinitesimal contributions are scalars

24.3 Differential Geometry in General Bases and in Curved Manifolds

EXERCISES

Exercise 24.11 *Practice: Integration—Gauss's Theorem*

In 3-dimensional Euclidean space Maxwell's equation $\nabla \cdot \mathbf{E} = \rho_e/\epsilon_0$ can be combined with Gauss's theorem to show that the electric flux through the surface $\partial \mathcal{V}$ of a sphere is equal to the charge in the sphere's interior $\mathcal{V}$ divided by ϵ_0:

$$\int_{\partial \mathcal{V}} \mathbf{E} \cdot d\mathbf{\Sigma} = \int_{\mathcal{V}} (\rho_e/\epsilon_0)\, dV. \tag{24.47}$$

Introduce spherical polar coordinates so the sphere's surface is at some radius $r = R$. Consider a surface element on the sphere's surface with vectorial legs $d\phi \partial/\partial\phi$ and $d\theta \partial/\partial\theta$. Evaluate the components $d\Sigma_j$ of the surface integration element $d\mathbf{\Sigma} = \boldsymbol{\epsilon}(\ldots, d\theta \partial/\partial\theta, d\phi \partial/\partial\phi)$. (Here $\boldsymbol{\epsilon}$ is the Levi-Civita tensor.) Similarly, evaluate dV in terms of vectorial legs in the sphere's interior. Then use these results for $d\Sigma_j$ and dV to convert Eq. (24.47) into an explicit form in terms of integrals over r, θ, and ϕ. The final answer should be obvious, but the above steps in deriving it are informative.

24.4 The Stress-Energy Tensor Revisited

In Sec. 2.13.1, we defined the stress-energy tensor $\mathbf{T}$ of any matter or field as a symmetric, second-rank tensor that describes the flow of 4-momentum through spacetime. More specifically, the total 4-momentum $\vec{P}$ that flows through some small 3-volume $\vec{\Sigma}$ (defined in Sec. 2.12.1), going from the negative side of $\vec{\Sigma}$ to its positive side, is

stress-energy tensor

$$\boxed{\mathbf{T}(\_, \vec{\Sigma}) = (\text{total 4-momentum } \vec{P} \text{ that flows through } \vec{\Sigma}); \quad T^{\alpha\beta} \Sigma_\beta = P^\alpha} \tag{24.48}$$

[Eq. (2.66)]. Of course, this stress-energy tensor depends on the location $\mathcal{P}$ of the 3-volume in spacetime [i.e., it is a tensor field $\mathbf{T}(\mathcal{P})$].

From this geometric, frame-independent definition of the stress-energy tensor, we were able to read off the physical meaning of its components in any inertial reference frame [Eqs. (2.67)]: T^{00} is the total energy density, including rest mass-energy; $T^{j0} = T^{0j}$ is the j-component of momentum density, or equivalently, the j-component of energy flux; and T^{jk} are the components of the stress tensor, or equivalently, of the momentum flux.

In Sec. 2.13.2, we formulated the law of conservation of 4-momentum in a local form and a global form. The local form,

local form of 4-momentum conservation

$$\boxed{\vec{\nabla} \cdot \mathbf{T} = 0,} \tag{24.49}$$

says that, in any chosen Lorentz frame, the time derivative of the energy density plus the divergence of the energy flux vanishes, $\partial T^{00}/\partial t + \partial T^{0j}/\partial x^j = 0$, and similarly for the momentum, $\partial T^{j0}/\partial t + \partial T^{jk}/\partial x^k = 0$. The global form, $\int_{\partial \mathcal{V}} T^{\alpha\beta} d\Sigma_\beta = 0$ [Eq. (2.71)], says that all the 4-momentum that enters a closed 4-volume $\mathcal{V}$ in spacetime through its boundary $\partial \mathcal{V}$ in the past must ultimately exit through $\partial \mathcal{V}$ in the future (Fig. 2.11). Unfortunately, this global form requires transporting vectorial contributions $T^{\alpha\beta} d\Sigma_\beta$ to a common location and adding them, which cannot be done in a route-independent way in curved spacetime (see the end of Sec. 24.3.4). Therefore (as we shall discuss in greater detail in Secs. 25.7 and 25.9.4), the global conservation law becomes problematic in curved spacetime.

The stress-energy tensor and local 4-momentum conservation play major roles in our development of general relativity. Almost all of our examples will entail perfect fluids.

Recall [Eq. (2.74a)] that in the local rest frame of a perfect fluid, there is no energy flux or momentum density, $T^{j0} = T^{0j} = 0$, but there is a total energy density (including rest mass) ρ and an isotropic pressure P:

$$T^{00} = \rho, \quad T^{jk} = P\delta^{jk}. \tag{24.50}$$

From this special form of $T^{\alpha\beta}$ in the fluid's local rest frame, one can derive a geometric, frame-independent expression for the fluid's stress-energy tensor $\boldsymbol{T}$ in terms of its 4-velocity $\vec{u}$, the metric tensor $\boldsymbol{g}$, and the rest-frame energy density ρ and pressure P:

$$\boxed{\boldsymbol{T} = (\rho + P)\vec{u} \otimes \vec{u} + P\boldsymbol{g}; \quad T^{\alpha\beta} = (\rho + P)u^\alpha u^\beta + Pg^{\alpha\beta}} \tag{24.51}$$

stress-energy tensor for a perfect fluid

[Eq. (2.74b)]; see Ex. 2.26. This expression for the stress-energy tensor of a perfect fluid is an example of a geometric, frame-independent description of physics.

The equations of relativistic fluid dynamics for a perfect fluid are obtained by inserting the stress-energy tensor (24.51) into the law of 4-momentum conservation $\vec{\nabla} \cdot \boldsymbol{T} = 0$, and augmenting with the law of rest-mass conservation. We explored this in brief in Ex. 2.26, and in much greater detail in Sec. 13.8. Applications that we have explored are the relativistic Bernoulli equation and ultrarelativistic jets (Sec. 13.8.2) and relativistic shocks (Ex. 17.9). In Sec. 13.8.3, we explored in detail the slightly subtle way in which a fluid's nonrelativistic energy density, energy flux, and stress tensor arise from the relativistic perfect-fluid stress-energy tensor (24.51).

These issues for a perfect fluid are so important that readers are encouraged to review them (except possibly the applications) in preparation for our foray into general relativity.

Four other examples of the stress-energy tensor are those for the electromagnetic field (Ex. 2.28), for a kinetic-theory swarm of relativistic particles (Secs. 3.4.2 and 3.5.3), for a point particle (Box 24.2), and for a relativistic fluid with viscosity and diffusive heat conduction (Ex. 24.13). However, we shall not do much with any of these during our study of general relativity, except viscosity and heat conduction in Sec. 28.5.

BOX 24.2. STRESS-ENERGY TENSOR FOR A POINT PARTICLE

For a point particle that moves through spacetime along a world line $\mathcal{P}(\zeta)$ [where ζ is the affine parameter such that the particle's 4-momentum is $\vec{p} = d/d\zeta$, Eq. (2.14)], the stress-energy tensor vanishes everywhere except on the world line itself. Correspondingly, $\mathbf{T}$ must be expressed in terms of a Dirac delta function. The relevant delta function is a scalar function of two points in spacetime, $\delta(\mathcal{Q}, \mathcal{P})$, with the property that when one integrates over the point $\mathcal{P}$, using the 4-dimensional volume element $d\Sigma$ (which in any inertial frame just reduces to $d\Sigma = dt\, dx\, dy\, dz$), one obtains

$$\int_{\mathcal{V}} f(\mathcal{P})\delta(\mathcal{Q}, \mathcal{P})\, d\Sigma = f(\mathcal{Q}). \tag{1}$$

Here $f(\mathcal{P})$ is an arbitrary scalar field, and the region $\mathcal{V}$ of 4-dimensional integration must include the point $\mathcal{Q}$. One can easily verify that in terms of Lorentz coordinates this delta function can be expressed as

$$\delta(\mathcal{Q}, \mathcal{P}) = \delta(t_Q - t_P)\delta(x_Q - x_P)\delta(y_Q - y_P)\delta(z_Q - z_P), \tag{2}$$

where the deltas on the right-hand side are ordinary 1-dimensional Dirac delta functions. [Proof: Simply insert Eq. (2) into Eq. (1), replace $d\Sigma$ by $dt_Q\, dx_Q\, dy_Q\, dz_Q$, and perform the four integrations.]

The general definition (24.48) of the stress-energy tensor $\mathbf{T}$ implies that the integral of a point particle's stress-energy tensor over any 3-surface $\mathcal{S}$ that slices through the particle's world line just once, at an event $\mathcal{P}(\zeta_o)$, must be equal to the particle's 4-momentum at the intersection point:

$$\int_{\mathcal{S}} T^{\alpha\beta} d\Sigma_\beta = p^\alpha(\zeta_o). \tag{3}$$

It is a straightforward but sophisticated exercise (Ex. 24.12) to verify that the following frame-independent expression has this property:

$$\mathbf{T}(\mathcal{Q}) = \int_{-\infty}^{+\infty} \vec{p}(\zeta) \otimes \vec{p}(\zeta)\, \delta[\mathcal{Q}, \mathcal{P}(\zeta)]\, d\zeta. \tag{4}$$

Here the integral is along the world line $\mathcal{P}(\zeta)$ of the particle, and $\mathcal{Q}$ is the point at which $\mathbf{T}$ is being evaluated. Therefore, Eq. (4) is the point-particle stress-energy tensor.

Exercise 24.12 *Derivation: Stress-Energy Tensor for a Point Particle* T2
Show that the point-particle stress-energy tensor (4) of Box 24.2 satisfies that box's Eq. (3), as claimed.

Exercise 24.13 *Example: Stress-Energy Tensor for a Viscous Fluid with Diffusive Heat Conduction*
This exercise serves two roles: It develops the relativistic stress-energy tensor for a viscous fluid with diffusive heat conduction, and in the process it allows the reader to gain practice in index gymnastics.

In our study of elasticity theory, we introduced the concept of the irreducible tensorial parts of a second-rank tensor in Euclidean space (Box 11.2). Consider a relativistic fluid flowing through spacetime with a 4-velocity $\vec{u}(\mathcal{P})$. The fluid's gradient $\vec{\nabla}\vec{u}$ ($u_{\alpha;\beta}$ in slot-naming index notation) is a second-rank tensor in spacetime. With the aid of the 4-velocity itself, we can break it down into irreducible tensorial parts as follows:

$$u_{\alpha;\beta} = -a_\alpha u_\beta + \frac{1}{3}\theta P_{\alpha\beta} + \sigma_{\alpha\beta} + \omega_{\alpha\beta}. \tag{24.52}$$

Here: (i)

$$P_{\alpha\beta} \equiv g_{\alpha\beta} + u_\alpha u_\beta \tag{24.53}$$

is a tensor that projects vectors into the 3-space orthogonal to $\vec{u}$ (it can also be regarded as that 3-space's metric; see Ex. 2.10); (ii) $\sigma_{\alpha\beta}$ is symmetric, trace-free, and orthogonal to the 4-velocity; and (iii) $\omega_{\alpha\beta}$ is antisymmetric and orthogonal to the 4-velocity.

(a) Show that the rate of change of $\vec{u}$ along itself, $\nabla_{\vec{u}}\vec{u}$ (i.e., the fluid 4-acceleration) is equal to the vector $\vec{a}$ that appears in the decomposition (24.52). Show, further, that $\vec{a} \cdot \vec{u} = 0$.

(b) Show that the divergence of the 4-velocity, $\vec{\nabla} \cdot \vec{u}$, is equal to the scalar field θ that appears in the decomposition (24.52). As we shall see in part (d), this is the fluid's rate of expansion.

(c) The quantities $\sigma_{\alpha\beta}$ and $\omega_{\alpha\beta}$ are the relativistic versions of a Newtonian fluid's shear and rotation tensors, which we introduced in Sec. 13.7.1. Derive equations for these tensors in terms of $u_{\alpha;\beta}$ and $P_{\mu\nu}$.

(d) Show that, as viewed in a Lorentz reference frame where the fluid is moving with speed small compared to the speed of light, to first order in the fluid's ordinary velocity $v^j = dx^j/dt$, the following statements are true: (i) $u^0 = 1$, $u^j = v^j$; (ii) θ is the nonrelativistic rate of expansion of the fluid, $\theta = \vec{\nabla} \cdot \vec{v} = v^j{}_{,j}$ [Eq. (13.67a)]; (iii) σ_{jk} is the fluid's nonrelativistic shear [Eq. (13.67b)]; and (iv) ω_{jk} is the fluid's nonrelativistic rotation tensor [denoted r_{ij} in Eq. (13.67c)].

(e) At some event $\mathcal{P}$ where we want to know the influence of viscosity on the fluid's stress-energy tensor, introduce the fluid's local rest frame. Explain why, in that frame, the only contributions of viscosity to the components of the stress-energy tensor are $T^{jk}_{\text{visc}} = -\zeta\theta g^{jk} - 2\mu\sigma^{jk}$, where ζ and μ are the coefficients of bulk and shear viscosity, respectively; the contributions to T^{00} and $T^{j0} = T^{0j}$ vanish. [Hint: See Eq. (13.73) and associated discussions.]

(f) From nonrelativistic fluid mechanics, infer that, in the fluid's rest frame at $\mathcal{P}$, the only contributions of diffusive heat conductivity to the stress-energy tensor are $T^{0j}_{\text{cond}} = T^{j0}_{\text{cond}} = -\kappa \partial T/\partial x^j$, where κ is the fluid's thermal conductivity and T is its temperature. [Hint: See Eq. (13.74) and associated discussion.] Actually, this expression is not fully correct. If the fluid is accelerating, there is a correction term: $\partial T/\partial x^j$ gets replaced by $\partial T/\partial x^j + a^j T$, where a^j is the acceleration. After reading Sec. 24.5 and especially Ex. 24.16, explain this correction.

(g) Using the results of parts (e) and (f), deduce the following geometric, frame-invariant form of the fluid's stress-energy tensor:

$$T_{\alpha\beta} = (\rho + P)u_\alpha u_\beta + P g_{\alpha\beta} - \zeta\theta g_{\alpha\beta} - 2\mu\sigma_{\alpha\beta} - 2\kappa u_{(\alpha} P_{\beta)}{}^\mu (T_{;\mu} + a_\mu T). \quad (24.54)$$

Here the subscript parentheses in the last term mean to symmetrize in the α and β slots.

From the divergence of this stress-energy tensor, plus the first law of thermodynamics and the law of rest-mass conservation, one can derive the full theory of relativistic fluid mechanics for a fluid with viscosity and heat flow (see, e.g., Misner, Thorne, and Wheeler, 1973, Ex. 22.7). This particular formulation of the theory, including Eq. (24.54), is due to Carl Eckart (1940). Landau and Lifshitz (1959) have given a slightly different formulation. For discussion of the differences, and of causal difficulties with both formulations and the difficulties' repair, see, for example, the reviews by Israel and Stewart (1980), Andersson and Comer (2007, Sec. 14), and López-Monsalvo (2011, Sec. 4).

24.5 The Proper Reference Frame of an Accelerated Observer

Physics experiments and astronomical measurements almost always use an apparatus that accelerates and rotates. For example, if the apparatus is in an Earthbound laboratory and is attached to the laboratory floor and walls, then it accelerates upward (relative to freely falling particles) with the negative of the "acceleration of gravity," and it rotates (relative to inertial gyroscopes) because of the rotation of Earth. It is useful, in studying such an apparatus, to regard it as attached to an accelerating, rotating reference frame. As preparation for studying such reference frames in the presence of gravity, we study them in flat spacetime. For a somewhat more sophisticated treatment, see Misner, Thorne, and Wheeler (1973, pp. 163–176, 327–332).

Consider an observer with 4-velocity $\vec{U}$, who moves along an accelerated world line through flat spacetime (Fig. 24.6) so she has a nonzero 4-acceleration:

$$\vec{a} = \nabla_{\vec{U}} \vec{U}. \quad (24.55)$$

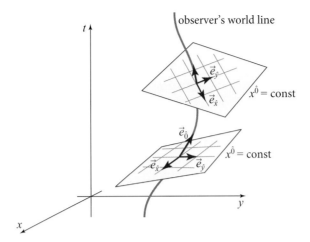

FIGURE 24.6 The proper reference frame of an accelerated observer. The spatial basis vectors $\vec{e}_{\hat{x}}$, $\vec{e}_{\hat{y}}$, and $\vec{e}_{\hat{z}}$ are orthogonal to the observer's world line and rotate, relative to local gyroscopes, as they move along the world line. The flat 3-planes spanned by these basis vectors are surfaces of constant coordinate time $x^{\hat{0}} \equiv$ (proper time as measured by the observer's clock at the event where the 3-plane intersects the observer's world line); in other words, they are the observer's slices of simultaneity and "3-space." In each of these flat 3-planes the spatial coordinates $\{\hat{x}, \hat{y}, \hat{z}\}$ are Cartesian, with $\partial/\partial\hat{x} = \vec{e}_{\hat{x}}$, $\partial/\partial\hat{y} = \vec{e}_{\hat{y}}$, and $\partial/\partial\hat{z} = \vec{e}_{\hat{z}}$.

Have that observer construct, in the vicinity of her world line, a coordinate system $\{x^{\hat{\alpha}}\}$ (called her *proper reference frame*) with these properties: (i) The spatial origin is centered on her world line at all times (i.e., her world line is given by $x^{\hat{j}} = 0$). (ii) Along her world line, the time coordinate $x^{\hat{0}}$ is the same as the proper time ticked by an ideal clock that she carries. (iii) In the immediate vicinity of her world line, the spatial coordinates $x^{\hat{j}}$ measure physical distance along the axes of a little Cartesian latticework that she carries (and that she regards as purely spatial, which means it lies in the 3-plane orthogonal to her world line). These properties dictate that, in the immediate vicinity of her world line, the metric has the form $ds^2 = \eta_{\hat{\alpha}\hat{\beta}} dx^{\hat{\alpha}} dx^{\hat{\beta}}$, where $\eta_{\hat{\alpha}\hat{\beta}}$ are the Lorentz-basis metric coefficients, Eq. (24.6); in other words, all along her world line the coordinate basis vectors are orthonormal:

proper reference frame of an accelerated observer

$$g_{\hat{\alpha}\hat{\beta}} = \frac{\partial}{\partial x^{\hat{\alpha}}} \cdot \frac{\partial}{\partial x^{\hat{\beta}}} = \eta_{\hat{\alpha}\hat{\beta}} \quad \text{at } x^{\hat{j}} = 0. \tag{24.56}$$

Moreover, properties (i) and (ii) dictate that along the observer's world line, the basis vector $\vec{e}_{\hat{0}} \equiv \partial/\partial x^{\hat{0}}$ differentiates with respect to her proper time, and thus is identically equal to her 4-velocity $\vec{U}$:

$$\vec{e}_{\hat{0}} = \frac{\partial}{\partial x^{\hat{0}}} = \vec{U}. \tag{24.57}$$

There remains freedom as to how the observer's latticework is oriented spatially. The observer can lock it to the gyroscopes of an *inertial-guidance system* that she carries (Box 24.3), in which case we say that it is "nonrotating"; or she can rotate it relative to such gyroscopes. For generality, we assume that the latticework rotates.

rotating and nonrotating proper reference frames

BOX 24.3. INERTIAL GUIDANCE SYSTEMS

Aircraft and rockets often carry inertial guidance systems, which consist of an accelerometer and a set of gyroscopes.

The accelerometer measures the system's 4-acceleration $\vec{a}$ (in relativistic language). Equivalently, it measures the system's Newtonian 3-acceleration $\mathbf{a}$ relative to inertial coordinates in which the system is momentarily at rest. As we see in Eq. (24.58), these quantities are two different ways of thinking about the same thing.

Each gyroscope is constrained to remain at rest in the aircraft or rocket by a force that is applied at its center of mass. Such a force exerts no torque around the center of mass, so the gyroscope maintains its direction (does not precess) relative to an inertial frame in which it is momentarily at rest.

As the accelerating aircraft or rocket turns, its walls rotate with some angular velocity $\vec{\Omega}$ relative to these inertial-guidance gyroscopes. This is the angular velocity discussed in the text between Eqs. (24.57) and (24.58).

From the time-evolving 4-acceleration $\vec{a}(\tau)$ and angular velocity $\vec{\Omega}(\tau)$, a computer can calculate the aircraft's (or rocket's) world line and its changing orientation.

Its angular velocity, as measured by the observer (by comparing the latticework's orientation with inertial-guidance gyroscopes), is a 3-dimensional spatial vector $\mathbf{\Omega}$ in the 3-plane orthogonal to her world line; and as viewed in 4-dimensional spacetime, it is a 4-vector $\vec{\Omega}$ whose components in the observer's reference frame are $\Omega^{\hat{j}} \neq 0$ and $\Omega^{\hat{0}} = 0$. Similarly, the latticework's acceleration, as measured by an inertial-guidance accelerometer attached to it (Box 24.3), is a 3-dimensional spatial vector $\mathbf{a}$ that can be thought of as a 4-vector with components in the observer's frame:

$$a^{\hat{0}} = 0, \quad a^{\hat{j}} = (\hat{j}\text{-component of the measured } \mathbf{a}). \tag{24.58}$$

This 4-vector is the observer's 4-acceleration, as one can verify by computing the 4-acceleration in an inertial frame in which the observer is momentarily at rest.

constructing coordinates of proper reference frame

Geometrically, the coordinates of the proper reference frame are constructed as follows. Begin with the basis vectors $\vec{e}_{\hat{\alpha}}$ along the observer's world line (Fig. 24.6)—basis vectors that satisfy Eqs. (24.56) and (24.57), and that rotate with angular velocity $\vec{\Omega}$ relative to gyroscopes. Through the observer's world line at time $x^{\hat{0}}$ construct the flat 3-plane spanned by the spatial basis vectors $\vec{e}_{\hat{j}}$. Because $\vec{e}_{\hat{j}} \cdot \vec{e}_{\hat{0}} = 0$, this 3-plane is orthogonal to the world line. All events in this 3-plane are given the same value of coordinate time $x^{\hat{0}}$ as the event where it intersects the world line; thus the 3-plane is a surface of constant coordinate time $x^{\hat{0}}$. The spatial coordinates in this flat 3-plane are ordinary, Cartesian coordinates $x^{\hat{j}}$ with $\vec{e}_{\hat{j}} = \partial/\partial x^{\hat{j}}$.

24.5.1 Relation to Inertial Coordinates; Metric in Proper Reference Frame; Transport Law for Rotating Vectors

It is instructive to examine the coordinate transformation between these proper-reference-frame coordinates $x^{\hat\alpha}$ and the coordinates x^μ of an inertial reference frame. We pick a very special inertial frame for this purpose. Choose an event on the observer's world line, near which the coordinate transformation is to be constructed; adjust the origin of the observer's proper time, so this event is $x^{\hat 0}=0$ (and of course $x^{\hat j}=0$); and choose the inertial frame to be one that, arbitrarily near this event, coincides with the observer's proper reference frame. If we were doing Newtonian physics, then the coordinate transformation from the proper reference frame to the inertial frame would have the form (accurate through terms quadratic in $x^{\hat\alpha}$):

$$x^i = x^{\hat i} + \frac{1}{2}a^{\hat i}(x^{\hat 0})^2 + \epsilon^{\hat i}{}_{\hat j\hat k}\Omega^{\hat j}x^{\hat k}x^{\hat 0}, \quad x^0 = x^{\hat 0}. \tag{24.59}$$

Here the term $\frac{1}{2}a^{\hat i}(x^{\hat 0})^2$ is the standard expression for the vectorial displacement produced after time $x^{\hat 0}$ by the acceleration $a^{\hat i}$; and the term $\epsilon^{\hat i}{}_{\hat j\hat k}\Omega^{\hat j}x^{\hat k}x^{\hat 0}$ is the standard expression for the displacement produced by the rotation rate (rotational angular velocity) $\Omega^{\hat j}$ during a short time $x^{\hat 0}$. In relativity theory there is only one departure from these familiar expressions (up through quadratic order): after time $x^{\hat 0}$ the acceleration has produced a velocity $v^{\hat j} = a^{\hat j}x^{\hat 0}$ of the proper reference frame relative to the inertial frame; correspondingly, there is a Lorentz-boost correction to the transformation of time: $x^0 = x^{\hat 0} + v^{\hat j}x^{\hat j} = x^{\hat 0}(1 + a_{\hat j}x^{\hat j})$ [cf. Eq. (2.37c)], accurate only to quadratic order. Thus, the full transformation to quadratic order is

$$x^i = x^{\hat i} + \frac{1}{2}a^{\hat i}(x^{\hat 0})^2 + \epsilon^{\hat i}{}_{\hat j\hat k}\Omega^{\hat j}x^{\hat k}x^{\hat 0},$$

$$x^0 = x^{\hat 0}(1 + a_{\hat j}x^{\hat j}). \tag{24.60a}$$

inertial coordinates related to those of the proper reference frame of an accelerated, rotating observer

From this transformation and the form of the metric, $ds^2 = -(dx^0)^2 + \delta_{ij}dx^i dx^j$ in the inertial frame, we easily can evaluate the form of the metric, accurate to linear order in **x**, in the proper reference frame:

$$\boxed{ds^2 = -(1 + 2\mathbf{a}\cdot\mathbf{x})(dx^{\hat 0})^2 + 2(\mathbf{\Omega}\times\mathbf{x})\cdot d\mathbf{x}\, dx^{\hat 0} + \delta_{\hat j\hat k}dx^{\hat j}dx^{\hat k}} \tag{24.60b}$$

metric in proper reference frame of an accelerated, rotating observer

(Ex. 24.14a). Here the notation is that of 3-dimensional vector analysis, with **x** the 3-vector whose components are $x^{\hat j}$, $d\mathbf{x}$ that with components $dx^{\hat j}$, **a** that with components $a^{\hat j}$, and **Ω** that with components $\Omega^{\hat j}$.

Because the transformation (24.60a) was constructed near an arbitrary event on the observer's world line, the metric (24.60b) is valid near any and every event on the world line (i.e., it is valid all along the world line). In fact, it is the leading order in an expansion in powers of the spatial separation $x^{\hat j}$ from the world line. For higher-order terms in this expansion see, for example, Ni and Zimmermann (1978).

Notice that precisely on the observer's world line, the metric coefficients $g_{\hat{\alpha}\hat{\beta}}$ [the coefficients of $dx^{\hat{\alpha}}dx^{\hat{\beta}}$ in Eq. (24.60b)] are $g_{\hat{\alpha}\hat{\beta}} = \eta_{\hat{\alpha}\hat{\beta}}$, in accord with Eq. (24.56). However, as one moves farther away from the observer's world line, the effects of the acceleration $a^{\hat{j}}$ and rotation $\Omega^{\hat{j}}$ cause the metric coefficients to deviate more and more strongly from $\eta_{\hat{\alpha}\hat{\beta}}$.

From the metric coefficients of Eq. (24.60b), one can compute the connection coefficients $\Gamma^{\hat{\alpha}}{}_{\hat{\beta}\hat{\gamma}}$ on the observer's world line, and from these connection coefficients, one can infer the rates of change of the basis vectors along the world line: $\nabla_{\vec{U}}\vec{e}_{\hat{\alpha}} = \nabla_{\hat{0}}\vec{e}_{\hat{\alpha}} = \Gamma^{\hat{\mu}}{}_{\hat{\alpha}\hat{0}}\vec{e}_{\hat{\mu}}$. The result is (Ex. 24.14b):

equations for transport of proper reference frame's basis vectors along observer's world line

$$\nabla_{\vec{U}}\vec{e}_{\hat{0}} \equiv \nabla_{\vec{U}}\vec{U} = \vec{a}, \tag{24.61a}$$

$$\nabla_{\vec{U}}\vec{e}_{\hat{j}} = (\vec{a} \cdot \vec{e}_{\hat{j}})\vec{U} + \epsilon(\vec{U}, \vec{\Omega}, \vec{e}_{\hat{j}}, \_). \tag{24.61b}$$

Equation (24.61b) is the general "law of transport" for constant-length vectors that are orthogonal to the observer's world line and that the observer thus sees as purely spatial. For the spin vector $\vec{S}$ of an inertial-guidance gyroscope (Box 24.3), the transport law is Eq. (24.61b) with $\vec{e}_{\hat{j}}$ replaced by $\vec{S}$ and with $\vec{\Omega} = 0$:

Fermi-Walker transport for the spin of an inertial-guidance gyroscope

$$\boxed{\nabla_{\vec{U}}\vec{S} = \vec{U}(\vec{a} \cdot \vec{S}).} \tag{24.62}$$

This is called *Fermi-Walker transport*. The term on the right-hand side of this transport law is required to keep the spin vector always orthogonal to the observer's 4-velocity: $\nabla_{\vec{U}}(\vec{S} \cdot \vec{U}) = 0$. For any other vector $\vec{A}$ that rotates relative to inertial-guidance gyroscopes, the transport law has, in addition to this "keep-it-orthogonal-to $\vec{U}$" term, a second term, which is the 4-vector form of $d\mathbf{A}/dt = \mathbf{\Omega} \times \mathbf{A}$:

transport law for a vector that is orthogonal to observer's 4-velocity and rotates relative to gyroscopes

$$\nabla_{\vec{U}}\vec{A} = \vec{U}(\vec{a} \cdot \vec{A}) + \epsilon(\vec{U}, \vec{\Omega}, \vec{A}, \_). \tag{24.63}$$

Equation (24.61b) is this general transport law with $\vec{A}$ replaced by $\vec{e}_{\hat{j}}$.

24.5.2 Geodesic Equation for a Freely Falling Particle

Consider a particle with 4-velocity $\vec{u}$ that moves freely through the neighborhood of an accelerated observer. As seen in an inertial reference frame, the particle travels through spacetime on a straight line, also called a *geodesic* of flat spacetime. Correspondingly, a geometric, frame-independent version of its *geodesic law of motion* is

geodesic law of motion for freely falling particle

$$\boxed{\nabla_{\vec{u}}\vec{u} = 0} \tag{24.64}$$

(i.e., the particle parallel transports its 4-velocity $\vec{u}$ along $\vec{u}$). It is instructive to examine the component form of this geodesic equation in the proper reference frame of the observer. Since the components of $\vec{u}$ in this frame are $u^{\alpha} = dx^{\alpha}/d\tau$, where τ is the particle's proper time (not the observer's proper time), the components $u^{\hat{\alpha}}{}_{;\hat{\mu}}u^{\hat{\mu}} = 0$ of the geodesic equation (24.64) are

$$u^{\hat{\alpha}}{}_{,\hat{\mu}}u^{\hat{\mu}} + \Gamma^{\hat{\alpha}}{}_{\hat{\mu}\hat{\nu}}u^{\hat{\mu}}u^{\hat{\nu}} = \left(\frac{\partial}{\partial x^{\hat{\mu}}}\frac{dx^{\hat{\alpha}}}{d\tau}\right)\frac{dx^{\hat{\mu}}}{d\tau} + \Gamma^{\hat{\alpha}}{}_{\hat{\mu}\hat{\nu}}u^{\hat{\mu}}u^{\hat{\nu}} = 0; \qquad (24.65)$$

or equivalently,

$$\boxed{\frac{d^2 x^{\hat{\alpha}}}{d\tau^2} + \Gamma^{\hat{\alpha}}{}_{\hat{\mu}\hat{\nu}}\frac{dx^{\hat{\mu}}}{d\tau}\frac{dx^{\hat{\nu}}}{d\tau} = 0.} \qquad (24.66)$$

Suppose, for simplicity, that the particle is moving slowly relative to the observer, so its ordinary velocity $v^{\hat{j}} = dx^{\hat{j}}/dx^{\hat{0}}$ is nearly equal to $u^{\hat{j}} = dx^{\hat{j}}/d\tau$ and is small compared to unity (the speed of light), and $u^{\hat{0}} = dx^{\hat{0}}/d\tau$ is nearly unity. Then to first order in the ordinary velocity $v^{\hat{j}}$, the spatial part of the geodesic equation (24.66) becomes

$$\frac{d^2 x^{\hat{i}}}{(dx^{\hat{0}})^2} = -\Gamma^{\hat{i}}{}_{\hat{0}\hat{0}} - (\Gamma^{\hat{i}}{}_{\hat{j}\hat{0}} + \Gamma^{\hat{i}}{}_{\hat{0}\hat{j}})v^{\hat{j}}. \qquad (24.67)$$

By computing the connection coefficients from the metric coefficients of Eq. (24.60b) (Ex. 24.14), we bring this low-velocity geodesic law of motion into the form

$$\frac{d^2 x^{\hat{i}}}{(dx^{\hat{0}})^2} = -a^{\hat{i}} - 2\epsilon^{\hat{i}}{}_{\hat{j}\hat{k}}\Omega^{\hat{j}}v^{\hat{k}}, \quad \text{that is,} \quad \frac{d^2 \mathbf{x}}{(dx^{\hat{0}})^2} = -\mathbf{a} - 2\mathbf{\Omega} \times \mathbf{v}. \qquad (24.68)$$

geodesic equation for slowly moving particle in proper reference frame of accelerated, rotating observer

This is the standard nonrelativistic form of the law of motion for a free particle as seen in a rotating, accelerating reference frame. The first term on the right-hand side is the inertial acceleration due to the failure of the frame to fall freely, and the second term is the Coriolis acceleration due to the frame's rotation. There would also be a centrifugal acceleration if we had kept terms of higher order in distance away from the observer's world line, but this acceleration has been lost due to our linearizing the metric (24.60b) in that distance.

This analysis shows how the elegant formalism of tensor analysis gives rise to familiar physics. In the next few chapters we will see it give rise to less familiar, general relativistic phenomena.

EXERCISES

Exercise 24.14 *Derivation: Proper Reference Frame*
(a) Show that the coordinate transformation (24.60a) brings the metric $ds^2 = \eta_{\alpha\beta}dx^{\alpha}dx^{\beta}$ into the form of Eq. (24.60b), accurate to linear order in separation $x^{\hat{j}}$ from the origin of coordinates.
(b) Compute the connection coefficients for the coordinate basis of Eq. (24.60b) at an arbitrary event on the observer's world line. Do so first by hand calculations, and then verify your results using symbolic-manipulation software on a computer.
(c) Using the connection coefficients from part (b), show that the rate of change of the basis vectors $\mathbf{e}_{\hat{\alpha}}$ along the observer's world line is given by Eq. (24.61).

(d) Using the connection coefficients from part (b), show that the low-velocity limit of the geodesic equation [Eq. (24.67)] is given by Eq. (24.68).

24.5.3 Uniformly Accelerated Observer

transformation between inertial coordinates and uniformly accelerated coordinates

As an important example (cf. Ex. 2.16), consider an observer whose accelerated world line, written in some inertial (Lorentz) coordinate system $\{t, x, y, z\}$, is

$$t = (1/\kappa)\sinh(\kappa\tau), \quad x = (1/\kappa)\cosh(\kappa\tau), \quad y = z = 0. \qquad (24.69)$$

Here τ is proper time along the world line, and κ is the magnitude of the observer's 4-acceleration: $\kappa = |\vec{a}|$ (which is constant along the world line; see Ex. 24.15, where the reader can derive the various claims made in this subsection and the next).

The world line (24.69) is depicted in Fig. 24.7 as a thick, solid hyperbola that asymptotes to the past light cone at early times and to the future light cone at late times. The dots along the world line mark events that have proper times $\tau = -1.2, -0.9, -0.6, -0.3, 0.0, +0.3, +0.6, +0.9, +1.2$ (in units of $1/\kappa$). At each of these dots, the 3-plane orthogonal to the world line is represented by a dashed line (with the 2 dimensions out of the plane of the paper suppressed from the diagram). This 3-plane is labeled by its coordinate time $x^{\hat{0}}$, which is equal to the proper time of the dot. The basis vector $\vec{e}_{\hat{1}}$ is chosen to point along the observer's 4-acceleration, so $\vec{a} = \kappa \vec{e}_{\hat{1}}$. The coordinate $x^{\hat{1}}$ measures proper distance along the straight line that starts out tangent to $\vec{e}_{\hat{1}}$. The other two basis vectors $\vec{e}_{\hat{2}}$ and $\vec{e}_{\hat{3}}$ point out of the plane of the figure and are parallel transported along the world line: $\nabla_{\vec{U}}\vec{e}_{\hat{2}} = \nabla_{\vec{U}}\vec{e}_{\hat{3}} = 0$. In addition, $x^{\hat{2}}$ and $x^{\hat{3}}$ are measured along straight lines, in the orthogonal 3-plane, that start out tangent to these vectors. This construction implies that the resulting proper reference frame has vanishing rotation, $\vec{\Omega} = 0$ (Ex. 24.15), and that $x^{\hat{2}} = y$ and $x^{\hat{3}} = z$, where y and z are coordinates in the $\{t, x, y, z\}$ Lorentz frame that we used to define the world line [Eqs. (24.69)].

Usually, when constructing an observer's proper reference frame, one confines attention to the immediate vicinity of her world line. However, in this special case it is instructive to extend the construction (the orthogonal 3-planes and their resulting spacetime coordinates) outward arbitrarily far. By doing so, we discover that the 3-planes all cross at location $x^{\hat{1}} = -1/\kappa$, which means the coordinate system $\{x^{\hat{\alpha}}\}$ becomes singular there. This singularity shows up in a vanishing $g_{\hat{0}\hat{0}}(x^{\hat{1}} = -1/\kappa)$ for the spacetime metric, written in that coordinate system:

singularity of uniformly accelerated coordinates

spacetime metric in uniformly accelerated coordinates

$$ds^2 = -(1+\kappa x^{\hat{1}})^2(dx^{\hat{0}})^2 + (dx^{\hat{1}})^2 + (dx^{\hat{2}})^2 + (dx^{\hat{3}})^2. \qquad (24.70)$$

[Note that for $|x^{\hat{1}}| \ll 1/\kappa$ this metric agrees with the general proper-reference-frame metric (24.60b).] From Fig. 24.7, it should be clear that this coordinate system can only cover smoothly one quadrant of Minkowski spacetime: the quadrant $x > |t|$.

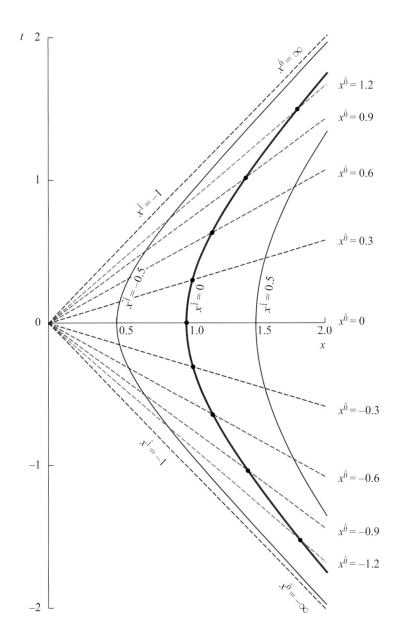

FIGURE 24.7 The proper reference frame of a uniformly accelerated observer. All lengths and times are measured in units of $1/\kappa$. We show only 2 dimensions of the reference frame—those in the 2-plane of the observer's curved world line.

24.5.4 Rindler Coordinates for Minkowski Spacetime

The spacetime metric (24.70) in our observer's proper reference frame resembles the metric in the vicinity of a black hole, as expressed in coordinates of observers who accelerate so as to avoid falling into the hole. In preparation for discussing this in

Rindler coordinates

Chap. 26, we shift the origin of our proper-reference-frame coordinates to the singular point and rename them. Specifically, we introduce so-called *Rindler coordinates*:

$$t' = x^{\hat{0}}, \quad x' = x^{\hat{1}} + 1/\kappa, \quad y' = x^{\hat{2}}, \quad z' = x^{\hat{3}}. \tag{24.71}$$

It turns out (Ex. 24.15) that these coordinates are related to the Lorentz coordinates that we began with, in Eqs. (24.69), by

$$t = x' \sinh(\kappa t'), \quad x = x' \cosh(\kappa t'), \quad y = y', \quad z = z'. \tag{24.72}$$

The metric in this Rindler coordinate system, of course, is the same as (24.70) with displacement of the origin:

spacetime metric in Rindler coordinates

$$ds^2 = -(\kappa x')^2 dt'^2 + dx'^2 + dy'^2 + dz'^2. \tag{24.73}$$

The world lines of constant $\{x', y', z'\}$ have uniform acceleration: $\vec{a} = (1/x')\vec{e}_{x'}$. Thus we can think of these coordinates as the reference frame of a family of uniformly accelerated observers, each of whom accelerates away from their *horizon* $x' = 0$ with acceleration equal to 1/(her distance x' above the horizon). (We use the name "horizon" for $x' = 0$, because it represents the edge of the region of spacetime that these observers are able to observe.) The local 3-planes orthogonal to these observers' world lines all mesh to form global 3-planes of constant t'. This is a major factor in making the metric (24.73) so simple.

horizon of Rindler coordinates

EXERCISES

Exercise 24.15 *Derivation: Uniformly Accelerated Observer and Rindler Coordinates*
In this exercise you will derive the various claims made in Secs. 24.5.3 and 24.5.4.

(a) Show that the parameter τ along the world line (24.69) is proper time and that the 4-acceleration has magnitude $|\vec{a}| = 1/\kappa$.

(b) Show that the unit vectors $\vec{e}_{\hat{j}}$ introduced in Sec. 24.5.3 all obey the Fermi-Walker transport law (24.62) and therefore, by virtue of Eq. (24.61b), the proper reference frame built from them has vanishing rotation rate: $\vec{\Omega} = 0$.

(c) Show that the coordinates $x^{\hat{2}}$ and $x^{\hat{3}}$ introduced in Sec. 24.5.3 are equal to the y and z coordinates of the inertial frame used to define the observer's world line [Eqs. (24.69)].

(d) Show that the proper-reference-frame coordinates constructed in Sec. 24.5.3 are related to the original $\{t, x, y, z\}$ coordinates by

$$t = (x^{\hat{1}} + 1/\kappa) \sinh(\kappa x^{\hat{0}}), \quad x = (x^{\hat{1}} + 1/\kappa) \cosh(\kappa x^{\hat{0}}), \quad y = x^{\hat{2}}, \quad z = x^{\hat{3}}; \tag{24.74}$$

and from this, deduce the form (24.70) of the Minkowski spacetime metric in the observer's proper reference frame.

Chapter 24. From Special to General Relativity

(e) Show that, when converted to Rindler coordinates by moving the spatial origin, the coordinate transformation (24.74) becomes (24.72), and the metric (24.70) becomes (24.73).

(f) Show that observers at rest in the Rindler coordinate system (i.e., who move along world lines of constant $\{x', y', z'\}$) have 4-acceleration $\vec{a} = (1/x')\vec{e}_{x'}$.

Exercise 24.16 *Example: Gravitational Redshift*

Inside a laboratory on Earth's surface the effects of spacetime curvature are so small that current technology cannot measure them. Therefore, experiments performed in the laboratory can be analyzed using special relativity. (This fact is embodied in Einstein's equivalence principle; end of Sec. 25.2.)

(a) Explain why the spacetime metric in the proper reference frame of the laboratory's floor has the form

$$ds^2 = (1 + 2gz)(dx^{\hat{0}})^2 + dx^2 + dy^2 + dz^2, \qquad (24.75)$$

plus terms due to the slow rotation of the laboratory walls, which we neglect in this exercise. Here g is the acceleration of gravity measured on the floor.

(b) An electromagnetic wave is emitted from the floor, where it is measured to have wavelength λ_o, and is received at the ceiling. Using the metric (24.75), show that, as measured in the proper reference frame of an observer on the ceiling, the received wave has wavelength $\lambda_r = \lambda_o(1 + gh)$, where h is the height of the ceiling above the floor (i.e., the light is *gravitationally redshifted* by $\Delta\lambda/\lambda_o = gh$). [Hint: Show that all crests of the wave must travel along world lines that have the same shape, $z = F(x^{\hat{0}} - x^{\hat{0}}_e)$, where F is some function, and $x^{\hat{0}}_e$ is the coordinate time at which the crest is emitted from the floor. You can compute the shape function F if you wish, but it is not needed to derive the gravitational redshift; only its universality is needed.]

The first high-precision experiments to test this prediction were by Robert Pound and his student Glen Rebka and postdoc Joseph Snider, in a tower at Harvard University in the 1950s and 1960s. They achieved 1% accuracy. We discuss this gravitational redshift in Sec. 27.2.1.

Exercise 24.17 *Example: Rigidly Rotating Disk*

Consider a thin disk with radius R at $z = 0$ in a Lorentz reference frame. The disk rotates rigidly with angular velocity Ω. In the early years of special relativity there was much confusion over the geometry of the disk: In the inertial frame it has physical radius (proper distance from center to edge) R and physical circumference $\mathcal{C} = 2\pi R$. But Lorentz contraction dictates that, as measured on the disk, the circumference should be $\sqrt{1 - v^2}\,\mathcal{C}$ (with $v = \Omega R$), and the physical radius, R, should be unchanged. This seemed weird. How could an obviously flat disk in flat spacetime have a curved, non-Euclidean geometry, with physical circumference divided by physical radius smaller than 2π? In this exercise you will explore this issue.

(a) Consider a family of observers who ride on the edge of the disk. Construct a circular curve, orthogonal to their world lines, that travels around the disk (at $\sqrt{x^2 + y^2} = R$). This curve can be thought of as lying in a 3-surface of constant time $x^{\hat{0}}$ of the observers' proper reference frames. Show that it spirals upward in a Lorentz-frame spacetime diagram, so it cannot close on itself after traveling around the disk. Thus the 3-planes, orthogonal to the observers' world lines at the edge of the disk, cannot mesh globally to form global 3-planes (by contrast with the case of the uniformly accelerated observers in Sec. 24.5.4 and Ex. 24.15).

(b) Next, consider a 2-dimensional family of observers who ride on the surface of the rotating disk. Show that at each radius $\sqrt{x^2 + y^2} = \text{const}$, the constant-radius curve that is orthogonal to their world lines spirals upward in spacetime with a different slope. Show this means that even locally, the 3-planes orthogonal to each of their world lines cannot mesh to form larger 3-planes—thus there does not reside in spacetime any 3-surface orthogonal to these observers' world lines. There is no 3-surface that has the claimed non-Euclidean geometry.

Bibliographic Note

For a very readable presentation of most of this chapter's material, from much the same point of view, see Hartle (2003, Chap. 20). For an equally elementary introduction from a somewhat different viewpoint, see Schutz (2009, Chaps. 1–4). A far more detailed and somewhat more sophisticated introduction, largely but not entirely from our viewpoint, will be found in Misner, Thorne, and Wheeler (1973, Chaps. 1–6). More sophisticated treatments from rather different viewpoints than ours are given in Wald (1984, Chaps. 1, 2, and Sec. 3.1), and Carroll (2004, Chaps. 1, 2). A treasure trove of exercises on this material, with solutions, is in Lightman et al. (1975, Chaps. 6–8). See also the bibliography for Chap. 2.

For a detailed and sophisticated discussion of accelerated observers and the measurements they make, see Gourgoulhon (2013).

CHAPTER TWENTY-FIVE

Fundamental Concepts of General Relativity

> The physical world is represented as a four dimensional continuum. If in this I adopt a Riemannian metric, and look for the simplest laws which such a metric can satisfy, I arrive at the relativistic gravitation theory of empty space.
>
> ALBERT EINSTEIN (1934)

25.1 History and Overview

Newton's theory of gravity is logically incompatible with the special theory of relativity. Newtonian gravity presumes the existence of a universal, frame-independent 3-dimensional space in which lives the Newtonian potential Φ and a universal, frame-independent time t with respect to which the propagation of Φ is instantaneous. By contrast, special relativity insists that the concepts of time and of 3-dimensional space are frame dependent, so that instantaneous propagation of Φ in one frame would mean noninstantaneous propagation in another.

The most straightforward way to remedy this incompatibility is to retain the assumption that gravity is described by a scalar field Φ but modify Newton's instantaneous, action-at-a-distance field equation

$$\left(\frac{\partial^2}{\partial x^2} + \frac{\partial^2}{\partial y^2} + \frac{\partial^2}{\partial z^2}\right)\Phi = 4\pi G\rho \qquad (25.1)$$

(where G is Newton's gravitation constant and ρ is the mass density) to read

$$\Box\Phi \equiv g^{\alpha\beta}\Phi_{;\alpha\beta} = -4\pi G T^\mu{}_\mu, \qquad (25.2a)$$

where $\Box \equiv \vec{\nabla} \cdot \vec{\nabla}$ is the squared gradient (i.e., d'alembertian or wave operator) in Minkowski spacetime, and $T^\mu{}_\mu$ is the trace (i.e., contraction on its two slots) of the stress-energy tensor. This modified field equation at first sight is attractive and satisfactory (but see Ex. 25.1): (i) it satisfies Einstein's Principle of Relativity in that it is expressed as a geometric, frame-independent relationship among geometric objects; and (ii) in any Lorentz frame it takes the form [with factors of $c = $ (speed of light) restored]:

$$\left(-\frac{1}{c^2}\frac{\partial^2}{\partial t^2} + \frac{\partial^2}{\partial x^2} + \frac{\partial^2}{\partial y^2} + \frac{\partial^2}{\partial z^2}\right)\Phi = \frac{4\pi G}{c^2}(T^{00} - T^{xx} - T^{yy} - T^{zz}), \qquad (25.2b)$$

which reduces to the Newtonian field equation (25.1) in the kinds of situation contemplated by Newton [energy density predominantly due to rest-mass density, $T^{00} \cong \rho c^2$;

> **BOX 25.1. READERS' GUIDE**
>
> - This chapter relies significantly on:
>   - Chap. 2 on special relativity; and
>   - Chap. 24 on the transition from special relativity to general relativity.
> - This chapter is a foundation for the applications of general relativity theory in Chaps. 26–28.

stress negligible compared to rest mass-energy density, $|T^{jk}| \ll \rho c^2$; and $1/c \times$ (time rate of change of Φ) negligible compared to spatial gradient of Φ].

Not surprisingly, most theoretical physicists in the decade following Einstein's formulation of special relativity (1905–1915) presumed that gravity would be correctly describable, in the framework of special relativity, by this type of modification of Newton's theory, or something resembling it. For a brief historical account, see Pais (1982, Chap. 13). To Einstein, by contrast, it seemed clear that the correct description of gravity should involve a generalization of special relativity rather than an incorporation into special relativity: since an observer in a local, freely falling reference frame near Earth should not feel any gravitational acceleration at all, local freely falling frames (local inertial frames) should in some sense be the domain of special relativity, and gravity should somehow be described by the relative acceleration of such frames.

key idea of general relativity: gravity described by relative acceleration of local inertial frames

Although the seeds of this idea were in Einstein's mind as early as 1907 [see the discussion of the equivalence principle in Einstein (1907)], it required 8 years for him to bring them to fruition. A first crucial step, which took half the 8 years, was for Einstein to conquer his initial aversion to Minkowski's geometric formulation of special relativity and to realize that a curvature of Minkowski's 4-dimensional spacetime is the key to understanding the relative acceleration of freely falling frames. The second crucial step was to master the mathematics of differential geometry, which describes spacetime curvature, and using that mathematics, to formulate a logically self-consistent theory of gravity. This second step took an additional 4 years and culminated in Einstein's (1915, 1916a) general theory of relativity. For a historical account of Einstein's 8-year struggle toward general relativity, see, for example, Pais (1982, Part IV). For selected quotations from Einstein's technical papers during this 8-year period, which tell the story of his struggle, see Misner, Thorne, and Wheeler (1973, Sec. 17.7). For his papers themselves with scholarly annotations, see Einstein (1989, vols. 2–4, 6).

the mathematics of general relativity: differential geometry in curved spacetime

It is remarkable that Einstein was led, not by experiment, but by philosophical and aesthetic arguments, to reject the incorporation of gravity into special relativity [Eqs. (25.2) and Ex. 25.1], and to insist instead on describing gravity by curved spacetime. Only after the full formulation of his general relativity did experiments begin to confirm that he was right and that the advocates of special-relativistic gravity were wrong, and only a half century after general relativity was formulated did the experimental evidence become extensive and strong. For detailed discussions see, for example, Will (1993a,b, 2014).

The mathematical tools, the diagrams, and the phrases by which we describe general relativity have changed somewhat in the century since Einstein formulated his theory. Indeed, we can even assert that we understand the theory more deeply than did Einstein. However, the basic ideas are unchanged, and general relativity's claim to be the most elegant and aesthetic of physical theories has been reinforced and strengthened by our growing insights.

General relativity is not merely a theory of gravity. Like special relativity before it, the general theory is a framework in which to formulate all the laws of physics, classical and quantum—but now with gravity included. However, there is one remaining, crucial, gaping hole in this framework. It is incapable of functioning—indeed, it fails completely—when conditions become so extreme that space and time themselves must be quantized. In those extreme conditions general relativity must be married in some deep, as-yet-ill-understood way, with quantum theory, to produce an all-inclusive quantum theory of gravity—a theory that, one may hope, will be a "theory of everything." To this we shall return, briefly, in Chaps. 26 and 28.

general relativity as a framework for all the laws of physics

In this chapter, we present, in modern language, the foundations of general relativity. Our presentation is proudly geometrical, as this seems to us the most powerful approach to general relativity for most situations. By contrast, some outstanding physicists, particularly Weinberg (1972, especially his preface), prefer a field-theoretic approach. Our presentation relies heavily on the geometric concepts, viewpoint, and formalism developed in Chaps. 2 and 24.

We begin in Sec. 25.2 with a discussion of three concepts that are crucial to Einstein's viewpoint on gravity: a local Lorentz frame (the closest thing there is, in the presence of gravity, to special relativity's "global" Lorentz frame), the extension of the Principle of Relativity to deal with gravitational situations, and Einstein's equivalence principle by which one can "lift" laws of physics out of the flat spacetime of special relativity and into the curved spacetime of general relativity. In Sec. 25.3, we see how gravity prevents the meshing of local Lorentz frames to form global Lorentz frames and infer from this that spacetime must be curved. In Sec. 25.4, we lift into curved spacetime the law of motion for free test particles, and in Sec. 25.5, we see how spacetime curvature pushes two freely moving test particles together or apart, and we use this phenomenon to make contact between spacetime curvature and the Newtonian "tidal gravitational field" (gradient of the Newtonian gravitational acceleration). In Sec. 25.6, we study some mathematical and geometric properties of the tensor field that embodies spacetime curvature: the Riemann tensor. In Sec. 25.7, we examine "curvature coupling delicacies" that plague the lifting of laws of physics from flat spacetime to curved spacetime. In Sec. 25.8, we meet the Einstein field equation, which

25.1 History and Overview

describes the manner in which spacetime curvature is produced by the total stress-energy tensor of all matter and nongravitational fields. In Sec. 25.9.1, we examine in some detail how Newton's laws of gravity arise as a weak-gravity, slow-motion, low-stress limit of general relativity. In Sec. 25.9.2, we develop an approximation to general relativity called "linearized theory" that is valid when gravity is weak but speeds and stresses may be high, and in Sec. 25.9.3, we use this approximation to deduce the weak relativistic gravitational field outside a stationary (unchanging) source. Finally, in Secs. 25.9.4 and 25.9.5, we examine the conservation laws for energy, momentum, and angular momentum of gravitating bodies that live in "asymptotically flat" regions of spacetime.

EXERCISES

Exercise 25.1 *Example: A Special Relativistic, Scalar-Field Theory of Gravity*
Equation (25.2a) is the field equation for a special relativistic theory of gravity with gravitational potential Φ. To complete the theory, one must describe the forces that the field Φ produces on matter.

(a) One conceivable choice for the force on a test particle of rest mass m is the following generalization of the familiar Newtonian expression:

$$\nabla_{\vec{u}} \vec{p} = -m \vec{\nabla} \Phi; \quad \text{that is,} \quad \frac{dp_\alpha}{d\tau} = -m \Phi_{,\alpha} \quad \text{in a Lorentz frame,} \quad (25.3)$$

where τ is proper time along the particle's world line, $\vec{p}$ is the particle's 4-momentum, $\vec{u}$ is its 4-velocity, and $\vec{\nabla} \Phi$ is the spacetime gradient of the gravitational potential. Show that this equation of motion reduces, in a Lorentz frame and for low particle velocities, to the standard Newtonian equation of motion. Show, however, that this equation of motion is flawed in that the gravitational field will alter the particle's rest mass—in violation of extensive experimental evidence that the rest mass of an elementary particle is unique and conserved.

(b) Show that the equation of motion (25.3), when modified to read:

$$\nabla_{\vec{u}} \vec{p} = -(\mathbf{g} + \vec{u} \otimes \vec{u}) \cdot m \vec{\nabla} \Phi;$$

$$\text{that is,} \quad \frac{dp^\alpha}{d\tau} = -(g^{\alpha\beta} + u^\alpha u^\beta) m \Phi_{,\beta} \quad \text{in a Lorentz frame,} \quad (25.4)$$

preserves the particle's rest mass. In this equation of motion $\vec{u}$ is the particle's 4-velocity, $\mathbf{g}$ is the metric, and $\mathbf{g} + \vec{u} \otimes \vec{u}$ projects $\vec{\nabla} \Phi$ into the 3-space orthogonal to the particle's world line (cf. Fig. 24.6 and Ex. 2.10).

(c) Show, by treating a zero-rest-mass particle as the limit of a particle of finite rest mass ($\vec{p} = m \vec{u}$ and $\zeta = \tau/m$ finite as τ and m go to zero), that the theory sketched in parts (a) and (b) predicts that in any Lorentz reference frame, $p^\alpha e^\Phi$ (with $\alpha = 0, 1, 2, 3$) are constant along the zero-rest-mass particle's world line. Explain why this prediction implies that there will be no gravitational deflection of light by the Sun, which conflicts severely with experiments that were done after Einstein formulated his general theory of relativity (see Sec. 27.2.3). (There was no way, experimentally, to rule out this theory in the epoch, ca. 1914, when Einstein was doing battle with his colleagues over whether gravity should be treated by adding a gravitational force to special relativity or should be treated as a geometric extension of special relativity.)

25.2 Local Lorentz Frames, the Principle of Relativity, and Einstein's Equivalence Principle

One of Einstein's greatest insights was to recognize that special relativity is valid not globally, but only locally, inside local, freely falling (inertial) reference frames. Figure 25.1 shows an example of a *local inertial frame*: the interior of a Space Shuttle in Earth orbit, where an astronaut has set up a freely falling (from his viewpoint "freely floating") latticework of rods and clocks. This latticework is constructed by all the rules appropriate to a special relativistic, inertial (Lorentz) reference frame (Secs. 2.2.1 and 24.2.2): (i) the latticework moves freely through spacetime, so no forces act on it, and its rods are attached to gyroscopes so they do not rotate; (ii) the measuring rods are orthogonal to one another, with their intervals of length uniform compared, for example, to the wavelength of light (orthonormal lattice); (iii) the clocks are densely packed in the lattice, they tick uniformly relative to ideal atomic standards (they are ideal clocks), and they are synchronized by the Einstein light-pulse process. However, there is one crucial change from special relativity: The latticework must be *small enough* that one can neglect the effects of inhomogeneities of gravity (which general relativity will associate with spacetime curvature; and which, e.g., would cause two freely floating particles, one nearer Earth than the other, to gradually move apart, even though initially they are at rest with respect to each other). The necessity for smallness is embodied in the word "local" of "local inertial frame," and we shall quantify it with ever greater precision as we move through this chapter.

local inertial (Lorentz) frame

We use the phrases *local Lorentz frame* and *local inertial frame* interchangeably to describe the above type of synchronized, orthonormal latticework. The spacetime coordinates $\{t, x, y, z\}$ that the latticework provides (in the manner of Sec. 2.2.1) we call, interchangeably, *local Lorentz coordinates* and *local inertial coordinates*.

local inertial (Lorentz) coordinates

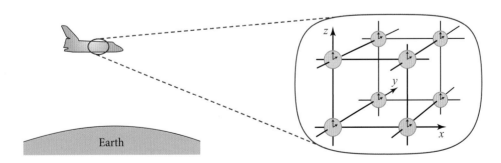

FIGURE 25.1 A local inertial frame (local Lorentz frame) inside a Space Shuttle that is orbiting Earth.

Principle of Relativity in presence of gravity

Since in the presence of gravity, inertial reference frames must be restricted to be local, the inertial-frame version of the Principle of Relativity (Sec. 2.2.2) must similarly be restricted: *all the* local, *nongravitational laws of physics are the same in every* local *inertial frame, everywhere and everywhen in the universe.* Here, by "local" laws we mean those laws, classical or quantum, that can be expressed entirely in terms of quantities confined to (measurable in) a local inertial frame. The exclusion of gravitational laws from this version of the Principle of Relativity is necessary, because gravity is to be described by a curvature of spacetime, which (by definition; see below) cannot show up in a local inertial frame. This version of the Principle of Relativity can be described in operational terms using the same language as for the special relativistic version (Secs. 2.2.2 and 24.2.2): If two different observers, in two different local Lorentz frames, in different (or the same) regions of the universe, are given identical written instructions for a physics experiment that can be performed within the confines of their local Lorentz frames, then their two experiments must yield the same results to within their experimental accuracies.

It is worth emphasizing that the Principle of Relativity is asserted to hold everywhere and everywhen in the universe: the local laws of physics must have the same form in the early universe, a fraction of a second after the big bang, as they have on Earth today, and as they have at the center of the Sun or inside a black hole.

Einstein's equivalence principle

It is reasonable to expect that *the specific forms that the local, nongravitational laws of physics take in general relativistic local Lorentz frames are the same as they take in the (global) Lorentz frames of special relativity*. This assertion is a modern version of Einstein's equivalence principle. (Einstein's original version states that local physical measurements in a uniformly accelerated reference frame cannot be distinguished from those in a uniform gravitational field. How is this related to the modern version? See Ex. 26.11.) In the next section, we use this principle to deduce some properties of the general relativistic spacetime metric. In Sec. 25.7 we use it to deduce the forms of some nongravitational laws of physics in curved spacetime, and we discover delicacies (ambiguities) in this principle of equivalence triggered by spacetime curvature.

25.3 The Spacetime Metric, and Gravity as a Curvature of Spacetime

The Einstein equivalence principle guarantees that nongravitational physics in a local Lorentz frame can be described using a spacetime metric $\mathbf{g}$, which gives for the invariant interval between neighboring events with separation vector $\vec{\xi} = \Delta x^\alpha \partial/\partial x^\alpha$, the standard special relativistic expression

$$\vec{\xi}^2 = g_{\alpha\beta}\xi^\alpha \xi^\beta = (\Delta s)^2 = -(\Delta t)^2 + (\Delta x)^2 + (\Delta y)^2 + (\Delta z)^2. \qquad (25.5)$$

Correspondingly, in a local Lorentz frame the components of the spacetime metric take on their standard special relativity values:

$$g_{\alpha\beta} = \eta_{\alpha\beta} \equiv \{-1 \text{ if } \alpha = \beta = 0, \quad +1 \text{ if } \alpha = \beta = (x, y, \text{ or } z), \quad 0 \text{ otherwise}\}.$$

$$(25.6)$$

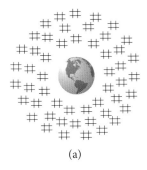
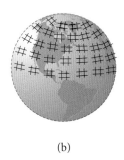

(a)   (b)

FIGURE 25.2 (a) A family of local Lorentz frames, all momentarily at rest above Earth's surface. (b) A family of local, 2-dimensional Euclidean coordinate systems on Earth's surface. The nonmeshing of Lorentz frames in (a) is analogous to the nonmeshing of Euclidean coordinates in (b) and motivates attributing gravity to a curvature of spacetime.

Turn, now, to a first look at gravity-induced constraints on the size of a local Lorentz frame. Above Earth, set up a family of local Lorentz frames scattered over the entire region from two Earth radii out to four Earth radii, with all the frames initially at rest with respect to Earth (Fig. 25.2a). From experience—or, if you prefer, from Newton's theory of gravity which after all is quite accurate near Earth—we know that, as time passes, these frames will all fall toward Earth. If (as a pedagogical aid) we drill holes through Earth to let the frames continue falling after reaching its surface, the frames will all pass through Earth's center and fly out the opposite side.

Obviously, two adjacent frames, which initially were at rest with respect to each other, acquire a relative velocity during their fall, which causes them to interpenetrate and pass through each other. Gravity is the cause of their relative velocity.

If these two adjacent frames could be meshed to form a larger Lorentz frame, then as time passes they would always remain at rest relative to each other. Thus, a meshing to form a larger Lorentz frame is impossible. The gravity-induced relative velocity prevents it. In brief: gravity prevents the meshing of local Lorentz frames to form global Lorentz frames.

nonmeshing of local Lorentz frames

This situation is closely analogous to the nonmeshing of local, 2-dimensional, Euclidean coordinate systems on the surface of Earth (Figure 25.2b): the curvature of Earth prevents a Euclidean mesh—thereby giving grief to mapmakers and surveyors. This analogy suggested to Einstein in 1912 a powerful new viewpoint on gravity. Just as the curvature of Earth prevents the meshing of local Euclidean coordinates on Earth's surface, so it must be that a curvature of spacetime prevents the meshing of local Lorentz frames in the spacetime above Earth—or anywhere else, for that matter. And since it is already known that gravity is the cause of the nonmeshing of Lorentz frames, it must be that *gravity is a manifestation of spacetime curvature.*

gravity is a manifestation of spacetime curvature

25.3 The Spacetime Metric, and Gravity as a Curvature of Spacetime

To make this idea more quantitative, consider, as a pedagogical tool, the 2-dimensional metric of Earth's surface, idealized as spherical and expressed in terms of a spherical polar coordinate system in line-element form [Eq. (2.24)]:

$$ds^2 = R^2 d\theta^2 + R^2 \sin^2\theta d\phi^2. \tag{25.7a}$$

Here R is the radius of Earth, or equivalently, the "radius of curvature" of Earth's surface. This line element, rewritten in terms of the alternative coordinates

$$x \equiv R\phi, \quad y \equiv R\left(\frac{\pi}{2} - \theta\right), \tag{25.7b}$$

has the form

$$ds^2 = \cos^2(y/R) dx^2 + dy^2 = dx^2 + dy^2 + O(y^2/R^2) dx^2, \tag{25.7c}$$

where as usual, $O(y^2/R^2)$ means "terms of order y^2/R^2 or smaller." Notice that the metric coefficients have the standard Euclidean form $g_{jk} = \delta_{jk}$ all along the equator ($y = 0$); but as one moves away from the equator, they begin to differ from Euclidean by fractional amounts of $O(y^2/R^2) = O[y^2/(\text{radius of curvature of Earth})^2]$. Thus, local Euclidean coordinates can be meshed and remain Euclidean all along the equator—or along any other great circle—but Earth's curvature forces the coordinates to cease being Euclidean when one moves off the chosen great circle, thereby causing the metric coefficients to differ from δ_{jk} by amounts $\Delta g_{jk} = O[(\text{distance from great circle})^2/(\text{radius of curvature})^2]$.

nonmeshing of local Euclidean coordinates on Earth's curved surface

Turn next to a specific example of curved spacetime: that of a $k = 0$ Robertson-Walker model for our expanding universe (to be studied in depth in Chap. 28). In spherical coordinates $\{\eta, \chi, \theta, \phi\}$, the 4-dimensional metric of this curved spacetime, described as a line element, can take the form

$$ds^2 = a^2(\eta)[-d\eta^2 + d\chi^2 + \chi^2(d\theta^2 + \sin^2\theta d\phi^2)]. \tag{25.8a}$$

Here a, the "expansion factor of the universe," is a monotonic increasing function of the "time" coordinate η (not to be confused with the flat metric $\eta_{\alpha\beta}$). This line element, rewritten near $\chi = 0$ in terms of the alternative coordinates

cosmological example of nonmeshing

$$t = \int_0^\eta a d\eta + \frac{1}{2}\chi^2 \frac{da}{d\eta}, \quad x = a\chi \sin\theta \cos\phi, \quad y = a\chi \sin\theta \sin\phi, \quad z = a\chi \cos\theta, \tag{25.8b}$$

takes the form (Ex. 25.2)

$$ds^2 = \eta_{\alpha\beta} dx^\alpha dx^\beta + O\left(\frac{x^2 + y^2 + z^2}{\mathcal{R}^2}\right) dx^\alpha dx^\beta, \tag{25.8c}$$

where $\mathcal{R}$ is a quantity that, by analogy with the radius of curvature R of Earth's surface, can be identified as a radius of curvature of spacetime:

$$\frac{1}{\mathcal{R}^2} = O\left(\frac{\dot{a}^2}{a^2}\right) + O\left(\frac{\ddot{a}}{a}\right), \quad \text{where} \quad \dot{a} \equiv \left(\frac{da}{dt}\right)_{x=y=z=0}, \quad \ddot{a} \equiv \left(\frac{d^2 a}{dt^2}\right)_{x=y=z=0}.$$

(25.8d)

From the form of the metric coefficients in Eq. (25.8d), we see that, all along the world line $x = y = z = 0$, the coordinates are precisely Lorentz, but as one moves away from that world line they cease to be Lorentz, and the metric coefficients begin to differ from $\eta_{\alpha\beta}$ by amounts $\Delta g_{\alpha\beta} = O[\text{(distance from the chosen world line)}^2 / \text{(radius of curvature of spacetime)}^2]$. This result is completely analogous to our equatorial Euclidean coordinates on Earth's surface. The curvature of Earth's surface prevented our local Euclidean coordinates from remaining Euclidean as we moved away from the equator; here the curvature of spacetime prevents our local Lorentz coordinates from remaining Lorentz as we move away from the chosen world line.

Notice that the chosen world line is the spatial origin of our local Lorentz coordinates. Thus we can think of those coordinates as provided by a tiny, spatial latticework of rods and clocks, like that of Figure 25.1. The latticework remains locally Lorentz for all time (as measured by its own clocks), but it ceases to be locally Lorentz when one moves a finite spatial distance away from the spatial origin of the latticework.

This behavior is generic. One can show [see, e.g., Misner, Thorne, and Wheeler (1973, Sec. 13.6, esp. item (5) on p. 331)] specialized to vanishing acceleration and rotation] that, if any freely falling observer, anywhere in spacetime, sets up a little latticework of rods and clocks in accord with our standard rules and keeps the latticework's spatial origin on her free-fall world line, then the coordinates provided by the latticework will be locally Lorentz, with metric coefficients

spacetime curvature forces Lorentz coordinates to be only locally Lorentz

$$g_{\alpha\beta} = \begin{cases} \eta_{\alpha\beta} + O\left(\frac{\delta_{jk} x^j x^k}{\mathcal{R}^2}\right), \\ \eta_{\alpha\beta} \quad \text{at spatial origin} \end{cases} \text{in a local Lorentz frame,} \quad (25.9a)$$

metric coefficients in a local Lorentz frame

where $\mathcal{R}$ is the radius of curvature of spacetime. Notice that, because the deviations of the metric from $\eta_{\alpha\beta}$ are of second order in the distance from the spatial origin, the first derivatives of the metric coefficients are of first order: $g_{\alpha\beta,k} = O(x^j/\mathcal{R}^2)$. This, plus the vanishing of the commutation coefficients in our coordinate basis, implies that the connection coefficients of the local Lorentz frame's coordinate basis are [Eqs. (24.38c) and (24.38d)][1]

$$\Gamma^{\alpha}{}_{\beta\gamma} = \begin{cases} O\left(\frac{\sqrt{\delta_{jk} x^j x^k}}{\mathcal{R}^2}\right), \\ 0 \quad \text{at spatial origin} \end{cases} \text{in a local Lorentz frame.} \quad (25.9b)$$

connection coefficients in a local Lorentz frame

---

1. In any manifold, coordinates for which the metric and connection have the form of Eqs. (25.9) in the vicinity of some chosen geodesic (the "spatial origin") are called *Fermi coordinates* or sometimes *Fermi normal coordinates*.

It is instructive to compare Eq. (25.9a) for the metric in the local Lorentz frame of a freely falling observer in curved spacetime with Eq. (24.60b) for the metric in the proper reference frame of an accelerated observer in flat spacetime. Whereas the spacetime curvature in Eq. (25.9a) produces corrections to $g_{\alpha\beta} = \eta_{\alpha\beta}$ of second order in distance from the world line, the acceleration and spatial rotation of the reference frame in Eq. (24.60b) produce corrections of first order. This remains true when one studies accelerated observers in curved spacetime (e.g., Sec. 26.3.2). In their proper reference frames, the metric coefficients $g_{\alpha\beta}$ contain both the first-order terms of Eq. (24.60b) due to acceleration and rotation [e.g., Eq. (26.26)], and the second-order terms of Eq. (25.9a) due to spacetime curvature.

EXERCISES

Exercise 25.2 *Derivation: Local Lorentz Frame in Robertson-Walker Universe*
By inserting the coordinate transformation (25.8b) into the Robertson-Walker metric (25.8a), derive the metric (25.8c), (25.8d) for a local Lorentz frame.

25.4 Free-Fall Motion and Geodesics of Spacetime

To make more precise the concept of spacetime curvature, we need to study quantitatively the relative acceleration of neighboring, freely falling particles. Before we can carry out such a study, however, we must understand quantitatively the motion of a single freely falling particle in curved spacetime. That is the objective of this section.

In a global Lorentz frame of flat, special relativistic spacetime, a free particle moves along a straight world line—one with the form

$$(t, x, y, z) = (t_o, x_o, y_o, z_o) + (p^0, p^x, p^y, p^z)\zeta\, ; \quad \text{that is,} \quad x^\alpha = x_o^\alpha + p^\alpha \zeta. \tag{25.10a}$$

Here the p^α are the Lorentz-frame components of the particle's 4-momentum; ζ is the affine parameter such that $\vec{p} = d/d\zeta$, so $p^\alpha = dx^\alpha/d\zeta$ [Eq. (2.10) and subsequent material]; and x_o^α are the coordinates of the particle when its affine parameter is $\zeta = 0$. The straight-line motion (25.10a) can be described equally well by the statement that the Lorentz-frame components p^α of the particle's 4-momentum are constant (i.e., are independent of ζ):

$$\frac{dp^\alpha}{d\zeta} = 0. \tag{25.10b}$$

Even nicer is the frame-independent description, which says that, as the particle moves, it parallel-transports its tangent vector $\vec{p}$ along its world line:

$$\nabla_{\vec{p}}\vec{p} = 0, \quad \text{or equivalently,} \quad p^\alpha{}_{;\beta}p^\beta = 0. \tag{25.10c}$$

For a particle with nonzero rest mass m, which has $\vec{p} = m\vec{u}$ and $\zeta = \tau/m$ (with $\vec{u} = d/d\tau$ its 4-velocity and τ its proper time), Eq. (25.10c) is equivalent to $\nabla_{\vec{u}}\vec{u} = 0$.

This is the *geodesic* form of the particle's law of motion [Eq. (24.64)]; Eq. (25.10c) is the extension of that geodesic law to a particle that may have vanishing rest mass. Recall that the word *geodesic* refers to the particle's straight world line.

This geodesic description of the motion is readily carried over into curved spacetime using the equivalence principle. Let $\mathcal{P}(\zeta)$ be the world line of a freely moving particle in curved spacetime. At a specific event $\mathcal{P}_o = \mathcal{P}(\zeta_o)$ on that world line, introduce a local Lorentz frame (so the frame's spatial origin, carried by the particle, passes through $\mathcal{P}_o$ as time progresses). Then the equivalence principle tells us that the particle's law of motion must be the same in this local Lorentz frame as it is in the global Lorentz frame of special relativity [Eq. (25.10b)]:

$$\left(\frac{dp^\alpha}{d\zeta}\right)_{\zeta=\zeta_o} = 0. \tag{25.11a}$$

More powerful than this local-Lorentz-frame description of the motion is a description that is frame independent. We can easily deduce such a description from Eq. (25.11a). Since the connection coefficients vanish at the spatial origin of the local Lorentz frame where Eq. (25.11a) is being evaluated [cf. Eq. (25.9b)], Eq. (25.11a) can be written equally well, in our local Lorentz frame, as

$$0 = \left(\frac{dp^\alpha}{d\zeta} + \Gamma^\alpha{}_{\beta\gamma} p^\beta \frac{dx^\gamma}{d\zeta}\right)_{\zeta=\zeta_o} = \left((p^\alpha{}_{,\gamma} + \Gamma^\alpha{}_{\beta\gamma} p^\beta)\frac{dx^\gamma}{d\zeta}\right)_{\zeta=\zeta_o} = (p^\alpha{}_{;\gamma} p^\gamma)_{\zeta=\zeta_o}. \tag{25.11b}$$

Thus, as the particle passes through the spatial origin of our local Lorentz coordinate system, the components of the directional derivative of its 4-momentum along itself vanish. Now, if two 4-vectors have components that are equal in one basis, their components are guaranteed [by the tensorial transformation law (24.19)] to be equal in all bases; correspondingly, the two vectors, viewed as frame-independent, geometric objects, must be equal. Thus, since Eq. (25.11b) says that the components of the 4-vector $\nabla_{\vec{p}}\vec{p}$ and the zero vector are equal in our chosen local Lorentz frame, it must be true that

$$\boxed{\nabla_{\vec{p}}\vec{p} = 0} \tag{25.11c}$$

geodesic equation of motion for a freely falling particle in curved spacetime

at the moment when the particle passes through the point $\mathcal{P}_o = \mathcal{P}(\zeta_o)$. Moreover, since $\mathcal{P}_o$ is an arbitrary point (event) along the particle's world line, it must be that Eq. (25.11c) is a geometric, frame-independent equation of motion for the particle, valid everywhere along its world line. Notice that this geometric, frame-independent equation of motion $\nabla_{\vec{p}}\vec{p} = 0$ in curved spacetime is precisely the same as that [Eq. (25.10c)] for flat spacetime. We generalize this conclusion to other laws of physics in Sec. 25.7.

Our equation of motion (25.11c) for a freely moving point particle says, in words, that the particle *parallel transports* its 4-momentum along its world line. As in flat spacetime, so also in curved spacetime, if the particle has finite rest mass, we can rewrite the equation of motion $\nabla_{\vec{p}}\vec{p} = 0$ as

geodesic equation for particle with finite rest mass

$$\boxed{\nabla_{\vec{u}}\vec{u} = 0,} \qquad (25.11d)$$

where $\vec{u} = \vec{p}/m = d/d\tau$ is the particle's 4-velocity, and $\tau = m\zeta$ is proper time along the particle's world line.

In any curved manifold, not just in spacetime, the relation $\vec{\nabla}_{\vec{u}}\vec{u} = 0$ (or $\nabla_{\vec{p}}\vec{p} = 0$) is called the *geodesic equation,* and the curve to which $\vec{u}$ is the tangent vector is called a *geodesic*. If the geodesic is spacelike, its tangent vector can be normalized such that $\vec{u} = d/ds$, with s the proper distance along the geodesic—the obvious analog of $\vec{u} = d/d\tau$ for a timelike geodesic.

On the surface of a sphere, such as Earth, the geodesics are the great circles; they are the unique curves along which local Euclidean coordinates can be meshed, keeping one of the two Euclidean coordinates constant along the curve [cf. Eq. (25.7c)]. They are also the trajectories generated by an airplane's inertial guidance system, which guides the plane along the straightest trajectory it can. Similarly, in spacetime the trajectories of freely falling particles are geodesics. They are the unique curves along which local Lorentz coordinates can be meshed, keeping the three spatial coordinates constant along the curve and letting the time vary, thereby producing a local Lorentz reference frame [Eqs. (25.9)]. They are also the spacetime trajectories along which inertial guidance systems guide a spacecraft.

The geodesic equation $\nabla_{\vec{p}}\vec{p} = 0$ for a particle in spacetime guarantees that the square of the 4-momentum will be conserved along the particle's world line; in slot-naming index notation, we have:

$$(g_{\alpha\beta}p^\alpha p^\beta)_{;\gamma} p^\gamma = 2g_{\alpha\beta} p^\alpha p^\beta{}_{;\gamma} p^\gamma = 0. \qquad (25.12)$$

Here the standard Leibniz rule for differentiating products has been used; this rule follows from the definition (24.29) of the frame-independent directional derivative of a tensor; it also can be deduced in a local Lorentz frame, where $\Gamma^\alpha{}_{\mu\nu} = 0$, so each gradient with a ";" reduces to a partial derivative with a ",". In Eq. (25.12) the term involving the gradient of the metric has been discarded, since it vanishes [Eq. (24.39)], and the two terms involving derivatives of p^α and p^β, being equal, have been combined. In index-free notation the frame-independent relation (25.12) says

$$\nabla_{\vec{p}}(\vec{p} \cdot \vec{p}) = 2\vec{p} \cdot \nabla_{\vec{p}}\vec{p} = 0. \qquad (25.13)$$

conservation of rest mass for freely falling particle

This is a pleasing result, since the square of the 4-momentum is the negative of the particle's squared rest mass, $\vec{p} \cdot \vec{p} = -m^2$, which surely should be conserved along the particle's free-fall world line! Note that, as in flat spacetime, so also in curved, for a particle of finite rest mass the free-fall trajectory (the geodesic world line) is timelike, $\vec{p} \cdot \vec{p} = -m^2 < 0$, while for a zero-rest-mass particle it is null, $\vec{p} \cdot \vec{p} = 0$. Spacetime also supports spacelike geodesics [i.e., curves with tangent vectors $\vec{p}$ that satisfy the geodesic equation (25.11c) and are spacelike, $\vec{p} \cdot \vec{p} > 0$]. Such curves can be thought of as the world lines of freely falling "tachyons" (i.e., faster-than-light particles)—though it seems unlikely that such particles exist in Nature. Note that the constancy of $\vec{p} \cdot \vec{p}$ along a geodesic implies that a geodesic can never change its character: if initially timelike, it always remains timelike; if initially null, it remains null; if initially spacelike, it remains spacelike.

The geodesic world line of a freely moving particle has three very important properties:

1. When written in a coordinate basis, the geodesic equation, $\nabla_{\vec{p}} \vec{p} = 0$, becomes the following differential equation for the particle's world line $x^\alpha(\zeta)$ in the coordinate system (Ex. 25.3):

$$\boxed{\frac{d^2 x^\alpha}{d\zeta^2} + \Gamma^\alpha{}_{\mu\nu} \frac{dx^\mu}{d\zeta} \frac{dx^\nu}{d\zeta} = 0.} \qquad (25.14)$$

coordinate representation of geodesic equation

Here $\Gamma^\alpha{}_{\mu\nu}$ are the connection coefficients of the coordinate system's coordinate basis. [Equation (24.66) was a special case of this.] Note that these are four coupled equations ($\alpha = 0, 1, 2, 3$) for the four coordinates x^α as functions of affine parameter ζ along the geodesic. If the initial position, x^α at $\zeta = 0$, and initial tangent vector (particle momentum), $p^\alpha = dx^\alpha/d\zeta$ at $\zeta = 0$, are specified, then these four equations will determine uniquely the coordinates $x^\alpha(\zeta)$ as a function of ζ along the geodesic.

2. Consider a spacetime that possesses a symmetry, which is embodied in the fact that the metric coefficients in some coordinate system are independent of one of the coordinates x^A. Associated with that symmetry there will be a so-called *Killing vector field* $\vec{\xi} = \partial/\partial x^A$ and a conserved quantity $p_A \equiv \vec{p} \cdot \partial/\partial x^A$ for free-particle motion. Exercises 25.4 and 25.5 discuss Killing vector fields, derive this conservation law, and develop a familiar example.

Killing vectors and conserved quantities for geodesic motion

3. Among all timelike curves linking two events $\mathcal{P}_0$ and $\mathcal{P}_1$ in spacetime, those whose proper time lapse (timelike length) is stationary under small variations of the curve are timelike geodesics; see Ex. 25.6. In other words, timelike geodesics are the curves that satisfy the action principle (25.19). Now, one can always send a photon from $\mathcal{P}_0$ to $\mathcal{P}_1$ by bouncing it off a set of strategically located mirrors, and that photon path is the limit of a timelike curve as the curve becomes null. Therefore, there exist timelike curves from $\mathcal{P}_0$ to $\mathcal{P}_1$ with vanishingly small length, so no timelike geodesics can be an absolute minimum of the proper time lapse, but one can be an absolute maximum.

action principle of stationary proper time lapse for geodesic motion

EXERCISES

Exercise 25.3 *Derivation and Problem: Geodesic Equation in an Arbitrary Coordinate System*

Show that in an arbitrary coordinate system $x^\alpha(\mathcal{P})$ the geodesic equation (25.11c) takes the form of Eq. (25.14).

Exercise 25.4 **Derivation and Example: Constant of Geodesic Motion in a Spacetime with Symmetry*

(a) Suppose that in some coordinate system the metric coefficients are independent of some specific coordinate x^A: $g_{\alpha\beta,A} = 0$ (e.g., in spherical polar coordinates $\{t, r, \theta, \phi\}$ in flat spacetime $g_{\alpha\beta,\phi} = 0$, so we could set $x^A = \phi$). Show that

$$p_A \equiv \vec{p} \cdot \frac{\partial}{\partial x^A} \tag{25.15}$$

is a constant of the motion for a freely moving particle [p_ϕ = (conserved z-component of angular momentum) in the above, spherically symmetric example]. [Hint: Show that the geodesic equation can be written in the form

$$\frac{dp_\alpha}{d\zeta} - \Gamma_{\mu\alpha\nu} p^\mu p^\nu = 0, \tag{25.16}$$

where $\Gamma_{\mu\alpha\nu}$ is the covariant connection coefficient of Eqs. (24.38c), (24.38d) with $c_{\alpha\beta\gamma} = 0$, because we are using a coordinate basis.] Note the analogy of the constant of the motion p_A with Hamiltonian mechanics: there, if the hamiltonian is independent of x^A, then the generalized momentum p_A is conserved; here, if the metric coefficients are independent of x^A, then the covariant component p_A of the momentum is conserved. For an elucidation of the connection between these two conservation laws, see Ex. 25.7c.

(b) As an example, consider a particle moving freely through a time-independent, Newtonian gravitational field. In Ex. 25.18, we learn that such a gravitational field can be described in the language of general relativity by the spacetime metric

$$ds^2 = -(1 + 2\Phi)dt^2 + (\delta_{jk} + h_{jk})dx^j dx^k, \tag{25.17}$$

where $\Phi(x, y, z)$ is the time-independent Newtonian potential, and h_{jk} are contributions to the metric that are independent of the time coordinate t and have magnitude of order $|\Phi|$. That the gravitational field is weak means $|\Phi| \ll 1$ (or, in conventional—SI or cgs—units, $|\Phi/c^2| \ll 1$). The coordinates being used are Lorentz, aside from tiny corrections of order $|\Phi|$, and as this exercise and Ex. 25.18 show, they coincide with the coordinates of the Newtonian theory of gravity. Suppose that the particle has a velocity $v^j \equiv dx^j/dt$ through this coordinate system that is $\lesssim |\Phi|^{\frac{1}{2}}$ and thus is small compared to the speed of light. Because the metric is independent of the time coordinate t, the component p_t of the particle's 4-momentum must be conserved along its world line. Since throughout physics, the conserved quantity associated with time-translation invariance is always the energy, we expect that p_t, when evaluated accurate to first order in $|\Phi|$, must be equal to the particle's conserved Newtonian energy, $E = m\Phi + \frac{1}{2}mv^j v^k \delta_{jk}$, aside from some multiplicative and additive constants. Show that this, indeed, is true, and evaluate the constants.

Exercise 25.5 *Example: Killing Vector Field*
A *Killing vector field*[2] is a coordinate-independent tool for exhibiting symmetries of the metric. It is any vector field $\vec{\xi}$ that satisfies

$$\xi_{\alpha;\beta} + \xi_{\beta;\alpha} = 0 \tag{25.18}$$

(i.e., any vector field whose symmetrized gradient vanishes).

(a) Let $\vec{\xi}$ be a vector field that might or might not be Killing. Show, by construction, that it is possible to introduce a coordinate system in which $\vec{\xi} = \partial/\partial x^A$ for some coordinate x^A.

(b) Show that in the coordinate system of part (a) the symmetrized gradient of $\vec{\xi}$ is $\xi_{\alpha;\beta} + \xi_{\beta;\alpha} = \partial g_{\alpha\beta}/\partial x^A$. From this infer that a vector field $\vec{\xi}$ is Killing if and only if there exists a coordinate system in which (i) $\vec{\xi} = \partial/\partial x^A$ and (ii) the metric is independent of x^A.

(c) Use Killing's equation (25.18) to show, without introducing a coordinate system, that, if $\vec{\xi}$ is a Killing vector field and $\vec{p}$ is the 4-momentum of a freely falling particle, then $\vec{\xi} \cdot \vec{p}$ is conserved along the particle's geodesic world line. This is the same conservation law as we proved in Ex. 25.4a using a coordinate-dependent calculation.

Exercise 25.6 *Problem: Timelike Geodesic as Path of Extremal Proper Time*
By introducing a specific but arbitrary coordinate system, show that among all timelike world lines that a particle could take to get from event $\mathcal{P}_0$ to event $\mathcal{P}_1$, the one or ones whose proper time lapse is stationary under small variations of path are the free-fall geodesics. In other words, an action principle for a timelike geodesic $\mathcal{P}(\lambda)$ [i.e., $x^\alpha(\lambda)$ in any coordinate system x^α] is

$$\delta \int_{\mathcal{P}_0}^{\mathcal{P}_1} d\tau = \delta \int_0^1 \left(-g_{\alpha\beta} \frac{dx^\alpha}{d\lambda} \frac{dx^\beta}{d\lambda} \right)^{\frac{1}{2}} d\lambda = 0, \tag{25.19}$$

---

2. Named after Wilhelm Killing, the mathematician who introduced it.

where λ is an arbitrary parameter, which by construction ranges from 0 at $\mathcal{P}_0$ to 1 at $\mathcal{P}_1$. [Note: Unless, after the variation, you choose the arbitrary parameter λ to be "affine" ($\lambda = a\zeta + b$, where a and b are constants and ζ is such that $\vec{p} = d/d\zeta$), your equation for $d^2x^\alpha/d\lambda^2$ will not look quite like Eq. (25.14).]

Exercise 25.7 *Problem: Super-Hamiltonian for Free Particle Motion*
(a) Show that, among all curves $\mathcal{P}(\zeta)$ that could take a particle from event $\mathcal{P}_0 = \mathcal{P}(0)$ to event $\mathcal{P}_1 = \mathcal{P}(\zeta_1)$ (for some ζ_1), those that satisfy the action principle

$$\delta \frac{1}{2} \int_0^{\zeta_1} g_{\alpha\beta} \frac{dx^\alpha}{d\zeta} \frac{dx^\beta}{d\zeta} d\zeta = 0 \qquad (25.20)$$

are geodesics, and ζ is the affine parameter along the geodesic related to the particle's 4-momentum by $\vec{p} = d/d\zeta$. [Note: In this action principle, by contrast with Eq. (25.19), the integration parameter is necessarily affine.]

(b) The lagrangian $\mathcal{L}(x^\mu, dx^\mu/d\zeta)$ associated with this action principle is $\frac{1}{2} g_{\alpha\beta} (dx^\alpha/d\zeta)(dx^\beta/d\zeta)$, where the coordinates $\{x^0, x^1, x^2, x^3\}$ appear in the metric coefficients and their derivatives appear explicitly. Using standard principles of Hamiltonian mechanics, show that the momentum canonically conjugate to x^μ is $p_\mu = g_{\mu\nu} dx^\nu/d\zeta$—which in fact is the covariant component of the particle's 4-momentum. Show, further, that the hamiltonian associated with the particle's lagrangian is

super-hamiltonian for geodesic motion

$$\mathcal{H} = \frac{1}{2} g^{\mu\nu} p_\mu p_\nu. \qquad (25.21)$$

Explain why this guarantees that Hamilton's equations $dx^\alpha/d\zeta = \partial \mathcal{H}/\partial p_\alpha$ and $dp_\alpha/d\zeta = -\partial \mathcal{H}/\partial x^\alpha$ are satisfied if and only if $x^\alpha(\zeta)$ is a geodesic with affine parameter ζ and p_α is the tangent vector to the geodesic.

(c) Show that Hamilton's equations guarantee that, if the metric coefficients are independent of some coordinate x^A, then p_A is a conserved quantity. This is the same conservation law as we derived by other methods in Exs. 25.4 and 25.5.

$\mathcal{H} = \frac{1}{2} g^{\mu\nu} p_\mu p_\nu$ is often called the geodesic's *super-hamiltonian*. It turns out that, often, the easiest way to compute geodesics numerically (e.g., for particle motion around a black hole) is to solve the super-hamiltonian's Hamilton equations (see, e.g., Levin and Perez-Giz, 2008).

25.5 Relative Acceleration, Tidal Gravity, and Spacetime Curvature

Now that we understand the motion of an individual freely falling particle in curved spacetime, we are ready to study the effects of gravity on the relative motions of such particles. Before doing so in general relativity, let us recall the Newtonian description.

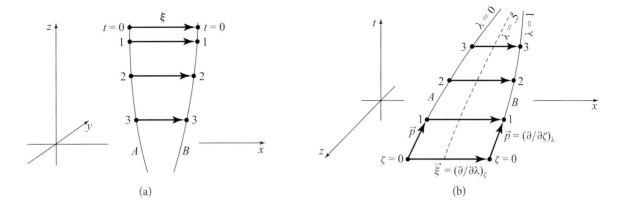

FIGURE 25.3 The effects of tidal gravity on the relative motions of two freely falling particles. (a) In Euclidean 3-space using Newton's theory of gravity. (b) In spacetime using Einstein's theory of gravity, general relativity.

25.5.1 Newtonian Description of Tidal Gravity

Consider, as shown in Fig. 25.3a, two point particles, A and B, falling freely through 3-dimensional Euclidean space under the action of an external Newtonian potential Φ (i.e., a potential generated by other masses, not by the particles themselves). At Newtonian time $t = 0$ the particles are separated by only a small distance and are moving with the same velocity: $\mathbf{v}_A = \mathbf{v}_B$. As time passes, however, the two particles, being at slightly different locations in space, experience slightly different gravitational potentials Φ and gravitational accelerations $\mathbf{g} = -\nabla\Phi$ and thence develop slightly different velocities: $\mathbf{v}_A \neq \mathbf{v}_B$. To quantify this, denote by $\boldsymbol{\xi}$ the vector separation of the two particles in Euclidean 3-space. The components of $\boldsymbol{\xi}$ on any Euclidean basis (e.g., that of Fig. 25.3a) are $\xi^j = x_B^j - x_A^j$, where x_I^j is the coordinate location of particle I. Correspondingly, the rate of change of ξ^j with respect to Newtonian time is $d\xi^j/dt = v_B^j - v_A^j$ (i.e., the relative velocity of the two particles is the difference of their velocities). The second time derivative of the relative separation (i.e., the relative acceleration of the two particles) is thus given by

$$\frac{d^2 \xi^j}{dt^2} = \frac{d^2 x_B^j}{dt^2} - \frac{d^2 x_A^j}{dt^2} = -\left(\frac{\partial \Phi}{\partial x^j}\right)_B + \left(\frac{\partial \Phi}{\partial x^j}\right)_A = -\frac{\partial^2 \Phi}{\partial x^j \partial x^k} \xi^k, \quad (25.22)$$

accurate to first order in the separation ξ^k. This equation gives the components of the relative acceleration in an arbitrary Euclidean basis. Rewritten in geometric, basis-independent language, this equation is

$$\boxed{\frac{d^2 \boldsymbol{\xi}}{dt^2} = -\boldsymbol{\mathcal{E}}(\_, \boldsymbol{\xi}); \quad \text{or} \quad \frac{d^2 \xi^j}{dt^2} = -\mathcal{E}^j{}_k \xi^k,} \quad (25.23)$$

relative acceleration of freely falling particles

Newtonian tidal gravitational field

where $\mathcal{E}$ is a symmetric, second-rank tensor, called the *Newtonian tidal gravitational field*:

$$\mathcal{E} \equiv \nabla\nabla\Phi = -\nabla\mathbf{g}; \quad \text{that is,} \quad \mathcal{E}_{jk} = \frac{\partial^2\Phi}{\partial x^j \partial x^k} \quad \text{in Euclidean coordinates.}$$

(25.24)

The name "tidal gravitational field" comes from the fact that this is the field which, generated by the Moon and the Sun, produces the tides on Earth's oceans. Note that, since this field is the gradient of the Newtonian gravitational acceleration **g**, it is a quantitative measure of the inhomogeneities of Newtonian gravity.

Equation (25.23) shows quantitatively how the tidal gravitational field produces the relative acceleration of our two particles. As a specific application, one can use it to compute, in Newtonian theory, the relative accelerations and thence relative motions of two neighboring local Lorentz frames as they fall toward and through the center of Earth (Fig. 25.2a and associated discussion).

25.5.2 Relativistic Description of Tidal Gravity

Now turn to the general relativistic description of the relative motions of two free particles. As shown in Fig. 25.3b, the particles, labeled A and B, move along geodesic world lines with affine parameters ζ and 4-momentum tangent vectors $\vec{p} = d/d\zeta$. The origins of ζ along the two world lines can be chosen however we wish, so long as events with the same ζ on the two world lines, $\mathcal{P}_A(\zeta)$ and $\mathcal{P}_B(\zeta)$, are close enough to each other that we can perform power-series expansions in their separation, $\vec{\xi}(\zeta) = \mathcal{P}_B(\zeta) - \mathcal{P}_A(\zeta)$, and keep only the leading terms. As in our Newtonian analysis, we require that the two particles initially have vanishing relative velocity, $\nabla_{\vec{p}}\vec{\xi} = 0$, and we compute the tidal-gravity-induced relative acceleration $\nabla_{\vec{p}}\nabla_{\vec{p}}\vec{\xi}$.

As a tool in our calculation, we introduce into spacetime a 2-dimensional surface that contains our two geodesics A and B, and also contains an infinity of other geodesics in between and alongside them. On that surface, we introduce two coordinates, $\zeta =$ (affine parameter along each geodesic) and $\lambda =$ (a parameter that labels the geodesics); see Fig. 25.3b. Geodesic A carries the label $\lambda = 0$; geodesic B is $\lambda = 1$; $\vec{p} = (\partial/\partial\zeta)_{\lambda=\text{const}}$ is a vector field that, evaluated on any geodesic (A, B, or other curve of constant λ), is equal to the 4-momentum of the particle that moves along that geodesic; and $\vec{\xi} \equiv (\partial/\partial\lambda)_{\zeta=\text{const}}$ is a vector field that, when evaluated on geodesic A (i.e., at $\lambda = 0$), we identify as a rigorous version of the separation vector $\mathcal{P}_B(\zeta) - \mathcal{P}_A(\zeta)$ that we wish to study. This identification requires, for good accuracy, that the geodesics be close together and be so parameterized that $\mathcal{P}_A(\zeta)$ is close to $\mathcal{P}_B(\zeta)$.

Our objective is to compute the relative acceleration of particles B and A, $\nabla_{\vec{p}}\nabla_{\vec{p}}\vec{\xi}$, evaluated at $\lambda = 0$. The quantity $\nabla_{\vec{p}}\vec{\xi}$, which we wish to differentiate a second time in that computation, is one of the terms in the following expression for the commutator of the vector fields $\vec{p}$ and $\vec{\xi}$ [Eq. (24.40)]:

$$[\vec{p}, \vec{\xi}] = \nabla_{\vec{p}} \vec{\xi} - \nabla_{\vec{\xi}} \vec{p}. \tag{25.25}$$

Because $\vec{p} = (\partial/\partial\zeta)_\lambda$ and $\vec{\xi} = (\partial/\partial\lambda)_\zeta$, these two vector fields commute, and Eq. (25.25) tells us that $\nabla_{\vec{p}} \vec{\xi} = \nabla_{\vec{\xi}} \vec{p}$. Correspondingly, the relative acceleration of our two particles can be expressed as

$$\nabla_{\vec{p}} \nabla_{\vec{p}} \vec{\xi} = \nabla_{\vec{p}} \nabla_{\vec{\xi}} \vec{p} = (\nabla_{\vec{p}} \nabla_{\vec{\xi}} - \nabla_{\vec{\xi}} \nabla_{\vec{p}}) \vec{p}. \tag{25.26}$$

Here the second equality results from adding on, for use below, a term that vanishes because $\nabla_{\vec{p}} \vec{p} = 0$ (geodesic equation).

This first part of our calculation was performed efficiently using index-free notation. The next step will be easier if we introduce indices as names for slots. Then expression (25.26) takes the form

$$(\xi^\alpha{}_{;\beta} p^\beta)_{;\gamma} p^\gamma = (p^\alpha{}_{;\gamma} \xi^\gamma)_{;\delta} p^\delta - (p^\alpha{}_{;\gamma} p^\gamma)_{;\delta} \xi^\delta, \tag{25.27}$$

which can be evaluated by using the rule for differentiating products and then renaming indices and collecting terms. The result is

$$(\xi^\alpha{}_{;\beta} p^\beta)_{;\gamma} p^\gamma = (p^\alpha{}_{;\gamma\delta} - p^\alpha{}_{;\delta\gamma})\xi^\gamma p^\delta + p^\alpha{}_{;\gamma}(\xi^\gamma{}_{;\delta} p^\delta - p^\gamma{}_{;\delta}\xi^\delta). \tag{25.28}$$

The second term in this expression vanishes, since it is just the commutator of $\vec{\xi}$ and $\vec{p}$ [Eq. (25.25)] written in slot-naming index notation, and as we noted above, $\vec{\xi}$ and $\vec{p}$ commute. The resulting equation,

$$(\xi^\alpha{}_{;\beta} p^\beta)_{;\gamma} p^\gamma = (p^\alpha{}_{;\gamma\delta} - p^\alpha{}_{;\delta\gamma})\xi^\gamma p^\delta, \tag{25.29}$$

reveals that the relative acceleration of the two particles is caused by noncommutation of the two slots of a double gradient (slots here named γ and δ). In the flat spacetime of special relativity, the two slots would commute[3] and there would be no relative acceleration. Spacetime curvature prevents them from commuting and thereby causes the relative acceleration.

Now, one can show that for any vector field $\vec{p}(\mathcal{P})$, $p^\alpha{}_{;\gamma\delta} - p^\alpha{}_{;\delta\gamma}$ is linear in p^α; see Ex. 25.8. Thus there must exist a fourth-rank tensor field $\mathbf{R}(\_,\_,\_,\_)$ such that

$$p^\alpha{}_{;\gamma\delta} - p^\alpha{}_{;\delta\gamma} = -R^\alpha{}_{\beta\gamma\delta} p^\beta \tag{25.30}$$

Riemann curvature tensor

---

3. In flat spacetime, in global Lorentz coordinates, $\Gamma^\alpha{}_{\beta\gamma} = 0$ everywhere, so $p^\alpha{}_{;\gamma\delta} = \partial^2 p^\alpha/\partial x^\gamma \partial x^\delta$. Because partial derivatives commute, expression (25.29) vanishes.

for any $\vec{p}$. The tensor **R** can be regarded as responsible for the failure of gradients to commute, so it must be some aspect of spacetime curvature. It is called the *Riemann curvature tensor*.

Inserting Eq. (25.30) into Eq. (25.29) and writing the result in both slot-naming index notation and abstract notation, we obtain

equation of geodesic deviation

$$(\xi^\alpha{}_{;\beta} p^\beta)_{;\gamma} p^\gamma = -R^\alpha{}_{\beta\gamma\delta} p^\beta \xi^\gamma p^\delta, \quad \nabla_{\vec{p}} \nabla_{\vec{p}} \vec{\xi} = -\mathbf{R}(\_, \vec{p}, \vec{\xi}, \vec{p}). \quad (25.31)$$

This is the equation of relative acceleration for freely moving test particles. It is also called the *equation of geodesic deviation*, because it describes the manner in which spacetime curvature **R** forces geodesics that are initially parallel (the world lines of freely moving particles with zero initial relative velocity) to deviate from one another (Fig. 25.3b).

EXERCISES

Exercise 25.8 *Derivation: Linearity of the Commutator of the Double Gradient*
(a) Let a and b be scalar fields with arbitrary but smooth dependence on location in curved spacetime, and let $\vec{A}$ and $\vec{B}$ be vector fields. Show that

$$(aA^\alpha + bB^\alpha)_{;\gamma\delta} - (aA^\alpha + bB^\alpha)_{;\delta\gamma} = a(A^\alpha{}_{;\gamma\delta} - A^\alpha{}_{;\delta\gamma}) + b(B^\alpha{}_{;\gamma\delta} - B^\alpha{}_{;\delta\gamma}). \quad (25.32)$$

[Hint: The double gradient of a scalar field commutes, as one can easily see in a local Lorentz frame.]

(b) Use Eq. (25.32) to show that (i) the commutator of the double gradient is independent of how the differentiated vector field varies from point to point and depends only on the value of the field at the location where the commutator is evaluated, and (ii) the commutator is linear in that value. Thereby conclude that there must exist a fourth-rank tensor field **R** such that Eq. (25.30) is true for any vector field $\vec{p}$.

25.5.3 Comparison of Newtonian and Relativistic Descriptions

It is instructive to compare this relativistic description of the relative acceleration of freely moving particles with the Newtonian description. For this purpose we consider a region of spacetime, such as our solar system, in which the Newtonian description of gravity is highly accurate; and there we study the relative acceleration of two free particles from the viewpoint of a local Lorentz frame in which the particles are both initially at rest.

In the Newtonian description, the transformation from a Newtonian universal reference frame (e.g., that of the center of mass of the solar system) to the chosen local Lorentz frame is achieved by introducing new Euclidean coordinates that are uniformly accelerated relative to the old ones, with just the right uniform acceleration to annul the gravitational acceleration at the center of the local Lorentz frame. This transformation adds a spatially homogeneous constant to the Newtonian acceleration, $\mathbf{g} = -\nabla\Phi$, but leaves unchanged the tidal field, $\mathcal{E} = \nabla\nabla\Phi$. Correspondingly, the Newtonian equation of relative acceleration in the local Lorentz frame retains its standard Newtonian form, $d^2\xi^j/dt^2 = -\mathcal{E}^j{}_k \xi^k$ [Eq. (25.23)], with the components of the tidal field computable equally well in the original universal reference frame or in the local Lorentz frame, using the standard relation $\mathcal{E}^j{}_k = \mathcal{E}_{jk} = \partial^2\Phi/\partial x^j \partial x^k$.

As an aid in making contact between the relativistic and the Newtonian descriptions, we convert from using the 4-momentum $\vec{p}$ as the relativistic tangent vector and ζ as the relativistic parameter along the particles' world lines to using the 4-velocity $\vec{u} = \vec{p}/m$ and the proper time $\tau = m\zeta$. This conversion brings the relativistic equation of relative acceleration (25.31) into the form

$$\nabla_{\vec{u}}\nabla_{\vec{u}}\vec{\xi} = -\mathbf{R}(\ldots, \vec{u}, \vec{\xi}, \vec{u}). \tag{25.33}$$

geodesic deviation for particles with finite rest mass

Because the particles are (momentarily) at rest near the origin of the local Lorentz frame, their 4-velocities are $\vec{u} \equiv d/d\tau = \partial/\partial t$, which implies that the components of their 4-velocities are $u^0 = 1$, $u^j = 0$, and their proper times τ are equal to coordinate time t, which in turn coincides with the time t of the Newtonian analysis: $\tau = t$. In the relativistic analysis, as in the Newtonian, the separation vector $\vec{\xi}$ will have only spatial components, $\xi^0 = 0$ and $\xi^j \neq 0$. (If this were not so, we could make it so by readjusting the origin of proper time for particle B, so $\vec{\xi} \cdot \vec{p} = m\vec{\xi} \cdot \vec{u} = 0$, whence $\xi^0 = 0$; Fig. 25.3b.) These facts, together with the vanishing of all the connection coefficients and derivatives of them ($\Gamma^j{}_{k0,0} = 0$) that appear in $(\xi^j{}_{;\beta} u^\beta)_{;\gamma} u^\gamma$ at the origin of the local Lorentz frame [cf. Eqs. (25.9)], imply that the local Lorentz components of the equation of relative acceleration (25.33) take the form

$$\frac{d^2\xi^j}{dt^2} = -R^j{}_{0k0}\xi^k. \tag{25.34}$$

By comparing this with the Newtonian equation of relative acceleration (25.23), we infer that, in the Newtonian limit, in the local rest frame of the two particles, we have

$$\boxed{R^j{}_{0k0} = \mathcal{E}_{jk} = \frac{\partial^2\Phi}{\partial x^j \partial x^k}.} \tag{25.35}$$

space-time-space-time components of Riemann tensor become tidal field in Newtonian limit of general relativity

Thus, the Riemann curvature tensor is the relativistic generalization of the Newtonian tidal field. This conclusion and the above equations make quantitative the statement that gravity is a manifestation of spacetime curvature.

Outside a spherical body with weak (Newtonian) gravity, such as Earth, the Newtonian potential is $\Phi = -GM/r$, where G is Newton's gravitation constant, M is the body's mass, and r is the distance from its center. If we introduce Cartesian coordinates with origin at the body's center and with the point at which the Riemann tensor is to be measured lying on the z-axis at $\{x, y, z\} = \{0, 0, r\}$, then Φ near that point is

components of Riemann tensor outside a Newtonian, gravitating body

$\Phi = -GM/(z^2 + x^2 + y^2)^{\frac{1}{2}}$, and on the z-axis the only nonzero $R^j{}_{0k0}$ components, as computed from Eq. (25.35), are

$$R^z{}_{0z0} = \frac{-2GM}{r^3}, \quad R^x{}_{0x0} = R^y{}_{0y0} = \frac{+GM}{r^3}. \tag{25.36}$$

Correspondingly, for two particles separated from each other in the radial (z) direction, the relative acceleration (25.34) is $d^2\xi^j/dt^2 = (2GM/r^3)\xi^j$ (i.e., the particles are pulled apart by the body's tidal gravitational field). Similarly, for two particles separated from each other in a transverse direction (in the x-y plane), we have $d^2\xi^j/dt^2 = -(GM/r^3)\xi^j$ (i.e., the particles are pushed together by the body's tidal gravitational field). There thus is a radial tidal stretch and a lateral tidal squeeze; the lateral squeeze has half the strength of the radial stretch but occurs in two lateral dimensions compared to the one radial dimension. This stretch and squeeze, produced by the Sun and the Moon, are responsible for the tides on Earth's oceans; Ex. 25.9.

EXERCISES

Exercise 25.9 **Example: Ocean Tides*

(a) Place a local Lorentz frame at the center of Earth, and let $\mathcal{E}_{jk}$ be the tidal field there, produced by the Newtonian gravitational fields of the Sun and the Moon. For simplicity, treat Earth as precisely spherical. Show that the gravitational acceleration (relative to Earth's center) at some location on or near Earth's surface (radius r) is

$$g_j = -\frac{GM}{r^2}n^j - \mathcal{E}^j{}_k rn^k, \tag{25.37}$$

where M is Earth's mass, and n^j is a unit vector pointing from Earth's center to the location at which g_j is evaluated.

(b) Show that this gravitational acceleration is minus the gradient of the Newtonian potential

$$\Phi = -\frac{GM}{r} + \frac{1}{2}\mathcal{E}_{jk} r^2 n^j n^k. \tag{25.38}$$

(c) Consider regions of Earth's oceans that are far from any coast and have ocean depth large compared to the heights of ocean tides. If Earth were nonrotating, then explain why the analysis of Sec. 13.3 predicts that the ocean surface in these regions would be a surface of constant Φ. Explain why this remains true to good accuracy also for the rotating Earth.

(d) Show that in these ocean regions, the Moon creates high tides pointing toward and away from itself and low tides in the transverse directions on Earth; and similarly for the Sun. Compute the difference between high and low tides produced by the Moon and by the Sun, and the difference of the total tide when the Moon and the Sun are in approximately the same direction in the sky. Your answers are reasonably accurate for deep-ocean regions far from any coast, but near a coast, the tides are typically larger and sometimes far larger, and they are shifted in phase relative to the positions of the Moon and Sun. Why?

25.6 Properties of the Riemann Curvature Tensor

We now pause in our study of the foundations of general relativity to examine a few properties of the Riemann curvature tensor **R**.

As a tool for deriving other things, we begin by evaluating the components of the Riemann tensor at the spatial origin of a local Lorentz frame (i.e., at a point where $g_{\alpha\beta} = \eta_{\alpha\beta}$ and $\Gamma^{\alpha}{}_{\beta\gamma}$ vanishes, but its derivatives do not). For any vector field $\vec{p}$, a straightforward computation reveals

$$p^{\alpha}{}_{;\gamma\delta} - p^{\alpha}{}_{;\delta\gamma} = (\Gamma^{\alpha}{}_{\beta\gamma,\delta} - \Gamma^{\alpha}{}_{\beta\delta,\gamma})p^{\beta}. \tag{25.39}$$

By comparing with Eq. (25.30), we can read off the local-Lorentz components of the Riemann tensor:

$$R^{\alpha}{}_{\beta\gamma\delta} = \Gamma^{\alpha}{}_{\beta\delta,\gamma} - \Gamma^{\alpha}{}_{\beta\gamma,\delta} \quad \text{at the spatial origin of a local Lorentz frame.} \tag{25.40}$$

From this expression we infer that, at a spatial distance $\sqrt{\delta_{ij}x^i x^j}$ from the origin of a local Lorentz frame, the connection coefficients and the metric have magnitudes

$$\Gamma^{\alpha}{}_{\beta\gamma} = O(R^{\mu}{}_{\nu\lambda\rho}\sqrt{\delta_{ij}x^i x^j}), \quad g_{\alpha\beta} - \eta_{\alpha\beta} = O(R^{\mu}{}_{\nu\lambda\rho}\,\delta_{ij}x^i x^j)$$

in a local Lorentz frame. (25.41)

influence of spacetime curvature on connection and metric in a local Lorentz frame

Comparison with Eqs. (25.9) shows that the radius of curvature of spacetime (a concept defined only semiquantitatively) is of order the inverse square root of the components of the Riemann tensor in a local Lorentz frame:

$$\boxed{\mathcal{R} = O\left(\frac{1}{|R^{\alpha}{}_{\beta\gamma\delta}|^{\frac{1}{2}}}\right) \quad \text{in a local Lorentz frame.}} \tag{25.42}$$

radius of curvature of spacetime

By comparison with Eq. (25.36), we see that at radius r outside a weakly gravitating body of mass M, the radius of curvature of spacetime is

$$\boxed{\mathcal{R} \sim \left(\frac{r^3}{GM}\right)^{\frac{1}{2}} = \left(\frac{c^2 r^3}{GM}\right)^{\frac{1}{2}},} \tag{25.43}$$

where the factor c (speed of light) in the second expression makes the formula valid in conventional units. For further discussion, see Ex. 25.10.

Using the components (25.40) of the Riemann tensor in a local Lorentz frame in terms of the connection coefficients, and using expressions (24.38) for the connection coefficients in terms of the metric components and commutation coefficients together

components of Riemann tensor in a local Lorentz frame

with the vanishing of the commutation coefficients (because a local Lorentz frame is a coordinate basis), one easily can show that

$$R_{\alpha\beta\gamma\delta} = \frac{1}{2}(g_{\alpha\delta,\beta\gamma} + g_{\beta\gamma,\alpha\delta} - g_{\alpha\gamma,\beta\delta} - g_{\beta\delta,\alpha\gamma}) \quad \text{in a local Lorentz frame.}$$

(25.44)

From these expressions, plus the commutation of partial derivatives $g_{\alpha\gamma,\beta\delta} = g_{\alpha\gamma,\delta\beta}$ and the symmetry of the metric, one readily can show that in a local Lorentz frame the components of the Riemann tensor have the following symmetries:

three symmetries of Riemann tensor

$$R_{\alpha\beta\gamma\delta} = -R_{\beta\alpha\gamma\delta}, \quad R_{\alpha\beta\gamma\delta} = -R_{\alpha\beta\delta\gamma}, \quad R_{\alpha\beta\gamma\delta} = +R_{\gamma\delta\alpha\beta}$$

(25.45a)

(antisymmetry in first pair of indices, antisymmetry in second pair of indices, and symmetry under interchange of the pairs). When one computes the value of the tensor on four vectors, $\mathbf{R}(\vec{A}, \vec{B}, \vec{C}, \vec{D})$ using component calculations in this frame, one trivially sees that these symmetries produce corresponding symmetries under interchange of the vectors inserted into the slots, and thence under interchange of the slots themselves. (This is always the case: any symmetry that the components of a tensor exhibit in any special basis will induce the same symmetry on the slots of the geometric, frame-independent tensor.) The resulting symmetries for $\mathbf{R}$ are given by Eq. (25.45a) with the "Escher mind-flip" (Sec. 1.5.1) in which the indices switch from naming components in a special frame to naming slots. The Riemann tensor is antisymmetric under interchange of its first two slots, antisymmetric under interchange of the last two, and symmetric under interchange of the two pairs.

One additional symmetry can be verified by calculation in the local Lorentz frame [i.e., from Eq. (25.44)]:[4]

a fourth symmetry of Riemann tensor

$$R_{\alpha\beta\gamma\delta} + R_{\alpha\gamma\delta\beta} + R_{\alpha\delta\beta\gamma} = 0.$$

(25.45b)

Riemann tensor has 20 independent components generically

One can show that the full set of symmetries (25.45) reduces the number of independent components of the Riemann tensor, in 4-dimensional spacetime, from $4^4 = 256$ to "just" 20.

Of these 20 independent components, 10 are contained in the *Ricci curvature tensor*, which is the contraction of the Riemann tensor on its first and third slots:

Ricci tensor and its symmetry

$$R_{\alpha\beta} \equiv R^{\mu}{}_{\alpha\mu\beta},$$

(25.46)

and which, by the symmetries (25.45) of Riemann, is itself symmetric:

---

4. Note that this cyclic symmetry is the same as occurs in the second of Maxwell's equations (2.48), $\epsilon^{\alpha\beta\gamma\delta}F_{\gamma\delta;\beta} = 0$; it is also the same as occurs in the Jacobi identity for commutators $\left[\vec{B}, [\vec{C}, \vec{D}]\right] + \left[\vec{C}, [\vec{D}, \vec{B}]\right] + \left[\vec{D}, [\vec{B}, \vec{C}]\right] = 0.$

$$R_{\alpha\beta} = R_{\beta\alpha}. \qquad (25.47)$$

The other 10 independent components of Riemann are contained in the Weyl curvature tensor:

$$C^{\mu\nu}{}_{\rho\sigma} = R^{\mu\nu}{}_{\rho\sigma} - 2g^{[\mu}{}_{[\rho} R^{\nu]}{}_{\sigma]} + \frac{1}{3} g^{[\mu}{}_{[\rho} g^{\nu]}{}_{\sigma]} R. \qquad (25.48)$$

Weyl tensor

Here the square brackets denote antisymmetrization, $A_{[\alpha\beta]} \equiv \frac{1}{2}(A_{\alpha\beta} - A_{\beta\alpha})$, and R is the contraction of the Ricci tensor on its two slots,

scalar curvature

$$R \equiv R^{\alpha}{}_{\alpha}, \qquad (25.49)$$

and is called the *curvature scalar* or *scalar curvature*. The Weyl curvature tensor $C^{\mu\nu}{}_{\rho\sigma}$ has vanishing contraction on every pair of slots and has the same symmetries as the Riemann tensor; Ex. 25.12.

One often needs to know the components of the Riemann curvature tensor in some non-local-Lorentz basis. Exercise 25.11 derives the following equation for them in an arbitrary basis:

components of Riemann tensor in an arbitrary basis

$$R^{\alpha}{}_{\beta\gamma\delta} = \Gamma^{\alpha}{}_{\beta\delta,\gamma} - \Gamma^{\alpha}{}_{\beta\gamma,\delta} + \Gamma^{\alpha}{}_{\mu\gamma}\Gamma^{\mu}{}_{\beta\delta} - \Gamma^{\alpha}{}_{\mu\delta}\Gamma^{\mu}{}_{\beta\gamma} - \Gamma^{\alpha}{}_{\beta\mu}c_{\gamma\delta}{}^{\mu}. \qquad (25.50)$$

Here $\Gamma^{\alpha}{}_{\beta\gamma}$ are the connection coefficients in the chosen basis; $\Gamma^{\alpha}{}_{\beta\gamma,\delta}$ is the result of letting the basis vector $\vec{e}_{\delta}$ act as a differential operator on $\Gamma^{\alpha}{}_{\beta\gamma}$, as though $\Gamma^{\alpha}{}_{\beta\gamma}$ were a scalar; and $c_{\gamma\delta}{}^{\mu}$ are the basis vectors' commutation coefficients. Calculations with this equation are usually long and tedious, and so are carried out using symbolic-manipulation software on a computer. See, for example, the simple Mathematica program (specialized to a coordinate basis) in Hartle (2003, Appendix C), also available on that textbook's website: http://web.physics.ucsb.edu/~gravitybook/mathematica.html.

EXERCISES

Exercise 25.10 *Example: Orders of Magnitude for the Radius of Curvature of Spacetime*
With the help of the Newtonian limit (25.35) of the Riemann curvature tensor, show that near Earth's surface the radius of curvature of spacetime has a magnitude $\mathcal{R} \sim$ (1 astronomical unit) $\equiv$ (distance from the Sun to Earth). What is the radius of curvature of spacetime near the Sun's surface? Near the surface of a white-dwarf star? Near the surface of a neutron star? Near the surface of a one-solar-mass black hole? In intergalactic space?

Exercise 25.11 *Derivation: Components of the Riemann Tensor in an Arbitrary Basis*
By evaluating expression (25.30) in an arbitrary basis (which might not even be a coordinate basis), derive Eq. (25.50) for the components of the Riemann tensor. In your derivation keep in mind that commas denote partial derivations *only* in a

coordinate basis; in an arbitrary basis they denote the result of letting a basis vector act as a differential operator [cf. Eq. (24.34)].

Exercise 25.12 *Derivation: Weyl Curvature Tensor*
Show that the Weyl curvature tensor (25.48) has vanishing contraction on all its slots and has the same symmetries as Riemann: Eqs. (25.45). From these properties, show that Weyl has just 10 independent components. Write the Riemann tensor in terms of the Weyl tensor, the Ricci tensor, and the scalar curvature.

Exercise 25.13 *Problem: Curvature of the Surface of a Sphere*
On the surface of a sphere, such as Earth, introduce spherical polar coordinates in which the metric, written as a line element, takes the form

$$ds^2 = a^2(d\theta^2 + \sin^2\theta \, d\phi^2), \tag{25.51}$$

where a is the sphere's radius.

(a) Show (first by hand and then by computer) that the connection coefficients for the coordinate basis $\{\partial/\partial\theta, \partial/\partial\phi\}$ are

$$\Gamma^\theta{}_{\phi\phi} = -\sin\theta\cos\theta, \quad \Gamma^\phi{}_{\theta\phi} = \Gamma^\phi{}_{\phi\theta} = \cot\theta, \quad \text{all others vanish.} \tag{25.52a}$$

(b) Show that the symmetries (25.45) of the Riemann tensor guarantee that its only nonzero components in the above coordinate basis are

$$R_{\theta\phi\theta\phi} = R_{\phi\theta\phi\theta} = -R_{\theta\phi\phi\theta} = -R_{\phi\theta\theta\phi}. \tag{25.52b}$$

(c) Show, first by hand and then by computer, that

$$R_{\theta\phi\theta\phi} = a^2 \sin^2\theta. \tag{25.52c}$$

(d) Show that in the basis

$$\{\vec{e}_{\hat\theta}, \vec{e}_{\hat\phi}\} = \left\{\frac{1}{a}\frac{\partial}{\partial\theta}, \frac{1}{a\sin\theta}\frac{\partial}{\partial\phi}\right\}, \tag{25.52d}$$

the components of the metric, the Riemann tensor, the Ricci tensor, the curvature scalar, and the Weyl tensor are

$$g_{\hat j \hat k} = \delta_{jk}, \quad R_{\hat\theta\hat\phi\hat\theta\hat\phi} = \frac{1}{a^2}, \quad R_{\hat j\hat k} = \frac{1}{a^2}g_{\hat j\hat k}, \quad R = \frac{2}{a^2}, \quad C_{\hat\theta\hat\phi\hat\theta\hat\phi} = 0, \tag{25.52e}$$

respectively. The first of these implies that the basis is orthonormal; the rest imply that the curvature is independent of location on the sphere, as it should be by spherical symmetry. [The θ dependence in the coordinate components of Riemann, Eq. (25.52c), like the θ dependence in the metric component $g_{\phi\phi}$, is a result of the θ dependence in the length of the coordinate basis vector $\vec{e}_\phi$: $|\vec{e}_\phi| = a\sin\theta$.]

Exercise 25.14 *Problem: Geodesic Deviation on a Sphere*

Consider two neighboring geodesics (great circles) on a sphere of radius a, one the equator and the other a geodesic slightly displaced from the equator (by $\Delta\theta = b$) and parallel to it at $\phi = 0$. Let $\vec{\xi}$ be the separation vector between the two geodesics, and note that at $\phi = 0$, $\vec{\xi} = b\partial/\partial\theta$. Let l be proper distance along the equatorial geodesic, so $d/dl = \vec{u}$ is its tangent vector.

(a) Show that $l = a\phi$ along the equatorial geodesic.

(b) Show that the equation of geodesic deviation (25.31) reduces to

$$\frac{d^2\xi^\theta}{d\phi^2} = -\xi^\theta, \quad \frac{d^2\xi^\phi}{d\phi^2} = 0. \tag{25.53}$$

(c) Solve Eq. (25.53), subject to the above initial conditions, to obtain

$$\xi^\theta = b\cos\phi, \quad \xi^\phi = 0. \tag{25.54}$$

Verify, by drawing a picture, that this is precisely what one would expect for the separation vector between two great circles.

25.7 Delicacies in the Equivalence Principle, and Some Nongravitational Laws of Physics in Curved Spacetime

Suppose that one knows a local, special relativistic, nongravitational law of physics in geometric, frame-independent form—for example, the expression for the stress-energy tensor of a perfect fluid in terms of its 4-velocity $\vec{u}$ and its rest-frame mass-energy density ρ and pressure P:

$$\mathbf{T} = (\rho + P)\vec{u} \otimes \vec{u} + P\mathbf{g} \tag{25.55}$$

[Eq. (24.51)]. Then the equivalence principle guarantees that in general relativity this law will assume the same geometric, frame-independent form. One can see that this is so by the same method as we used to derive the general relativistic equation of motion $\nabla_{\vec{p}}\,\vec{p} = 0$ for free particles [Eq. (25.11c) and associated discussion]:

1. Rewrite the special relativistic law in terms of components in a global Lorentz frame $[T^{\alpha\beta} = (\rho + P)u^\alpha u^\beta + Pg^{\alpha\beta}]$.

2. Infer from the equivalence principle that this same component form of the law will hold, unchanged, in a local Lorentz frame in general relativity.

3. Deduce that this component law is the local-Lorentz-frame version of the original geometric law $[\mathbf{T} = (\rho + P)\vec{u} \otimes \vec{u} + P\mathbf{g}]$, now lifted into general relativity.

equivalence principle implies that local nongravitational laws are the same in general relativity as in special relativity

Thus, when the local, nongravitational laws of physics are known in frame-independent form, one need not distinguish between whether they are special relativistic or general relativistic.

In this conclusion the word *local* is crucial. The equivalence principle is strictly valid only at the spatial origin of a local Lorentz frame; correspondingly, it is in danger of failure for any law of physics that cannot be formulated solely in terms of quantities that reside at the spatial origin (i.e., along a timelike geodesic). For the above example, $\mathbf{T} = (\rho + P)\vec{u} \otimes \vec{u} + P\mathbf{g}$, there is no problem; and for the local law of 4-momentum conservation, $\vec{\nabla} \cdot \mathbf{T} = 0$, there is no problem. However, for the global law of 4-momentum conservation

examples of the influence of spacetime curvature on nonlocal laws:

$$\int_{\partial \mathcal{V}} T^{\alpha\beta} d\Sigma_\beta = 0 \tag{25.56}$$

[Eq. (2.71) and Fig. 2.11], there is serious trouble. This law is severely nonlocal, since it involves integration over a finite, closed 3-surface $\partial \mathcal{V}$ in spacetime. Thus the equivalence principle fails for it. The failure shows up especially clearly when one notices (as we discussed in Sec. 24.3.4) that the quantity $T^{\alpha\beta} d\Sigma_\beta$, which the integral is trying to add up over $\partial \mathcal{V}$, has one empty slot, named α (i.e., it is a vector). This means that, to compute the integral (25.56), we must transport the contributions $T^{\alpha\beta} d\Sigma_\beta$ from the various tangent spaces in which they normally live to the tangent space of some single, agreed-on location, where they are to be added. The result of that transport depends on the route used, and in general no preferred route is available. As a result, the integral (25.56) is ill defined, and in general relativity we lose the global conservation law for 4-momentum!—except in special situations, one of which is discussed in Sec. 25.9.5.

no global conservation of 4-momentum in general relativity, generically

25.7.1

25.7.1 Curvature Coupling in the Nongravitational Laws T2

Another instructive example is the law by which a freely moving particle transports its spin angular momentum. The spin angular momentum is readily defined in the instantaneous local Lorentz rest frame of the particle's center of mass; there it is a 4-vector $\vec{S}$ with vanishing time component (so $\vec{S}$ is orthogonal to the particle's 4-velocity), with space components given by the familiar integral

$$S_i = \int_{\text{interior of body}} \epsilon_{ijk} x^j T^{k0} \, dV, \tag{25.57}$$

where the T^{k0} are components of the momentum density. In special relativity, the law of angular momentum conservation (e.g., Misner, Thorne, and Wheeler, 1973, Sec. 5.11) guarantees that the Lorentz-frame components S^α of this spin angular momentum remain constant, so long as no external torques act on the particle. This conservation law can be written in special relativistic, frame-independent notation as Eq. (24.62), specialized to a nonaccelerated particle:

$$\nabla_{\vec{u}} \vec{S} = 0; \tag{25.58}$$

that is, the spin vector $\vec{S}$ is parallel-transported along the world line of the freely falling particle (which has 4-velocity $\vec{u}$). If this were a local law of physics, it would take this same form, unchanged, in general relativity (i.e., in curved spacetime). Whether the law is local or not clearly depends on the size of the particle. If the particle is vanishingly small in its own rest frame, then the law is local, and Eq. (25.58) will be valid in general relativity. However, if the particle has finite size, the law (25.58) is in danger of failing—and, indeed, it does fail if the particle's finite size is accompanied by a finite quadrupole moment. In that case, the coupling of the quadrupole moment $\mathcal{I}_{\alpha\beta}$ to the curvature of spacetime $R^\alpha{}_{\beta\gamma\delta}$ produces a torque on the "particle," so Eq. (25.58) acquires a driving term on the right-hand side:

$$S^\alpha{}_{;\mu} u^\mu = \epsilon^{\alpha\beta\gamma\delta} \mathcal{I}_{\beta\mu} R^\mu{}_{\nu\gamma\zeta} u_\delta u^\nu u^\zeta; \qquad (25.59)$$

see Ex. 25.16. Earth is a good example: the Riemann tensor $R^\alpha{}_{\beta\gamma\delta}$ produced at Earth by the Moon and the Sun couples to Earth's centrifugal-flattening-induced quadrupole moment $\mathcal{I}_{\mu\nu}$. The resulting torque (25.59) causes Earth's spin axis to precess relative to the distant stars, with a precession period of 26,000 years—sufficiently fast to show up clearly in historical records[5] as well as in modern astronomical measurements.

tidally induced precession of Earth's spin axis

This example illustrates the fact that, if a small amount of nonlocality is present in a physical law, then, when lifted from special relativity into general relativity, the law may acquire a small *curvature-coupling* modification.

What is the minimum amount of nonlocality that can produce curvature-coupling modifications in physical laws? As a rough rule of thumb, the minimum amount is double gradients. Because the connection coefficients vanish at the origin of a local Lorentz frame, the local Lorentz components of a single gradient are the same as the components in a global Lorentz frame (e.g., $A^\alpha{}_{;\beta} = \partial A^\alpha/\partial x^\beta$). However, because spacetime curvature prevents the spatial derivatives of the connection coefficients from vanishing at the origin of a local Lorentz frame, any law that involves double gradients is in danger of acquiring curvature-coupling corrections when lifted into general relativity. As an example, it turns out that the vacuum wave equation for the electromagnetic vector 4-potential, which in Lorenz gauge (Jackson, 1999, Sec. 6.3) takes the form $A^{\alpha;\mu}{}_\mu = 0$ in flat spacetime, becomes in curved spacetime:

double gradients as a source of curvature coupling in physical laws

$$A^{\alpha;\mu}{}_\mu = R^{\alpha\mu} A_\mu, \qquad (25.60)$$

example: curvature coupling in wave equation for electromagnetic 4-potential

where $R^{\alpha\mu}$ is the Ricci curvature tensor; see Ex. 25.15. [In Eq. (25.60)—and always—all indices that follow the semicolon represent differentiation slots: $A^{\alpha;\mu}{}_\mu \equiv A^{\alpha;\mu}{}_{;\mu}$.]

The curvature-coupling ambiguities that occur when one lifts slightly nonlocal laws from special relativity into general relativity using the equivalence principle are very similar to "factor-ordering ambiguities" that occur when one lifts a hamiltonian from classical mechanics into quantum mechanics using the correspondence

curvature-coupling ambiguities in general relativity are analogous to factor-ordering ambiguities in the quantum mechanical correspondence principle

---

5. For example, Earth's north pole did not point toward the star Polaris in the era of the ancient Egyptian civilization. Hipparchus of Nicaea discovered this precession in 127 BC by comparing his own observations of the stars with those of earlier astronomers.

principle. In the case of the equivalence principle, the curvature coupling can be regarded as stemming from double gradients that commute in special relativity but do not commute in general relativity. In the case of the correspondence principle, the factor-ordering difficulties result because quantities that commute classically (e.g., position x and momentum p) do not commute quantum mechanically ($\hat{x}\hat{p} \neq \hat{p}\hat{x}$), so when the products of such quantities appear in a classical hamiltonian one does not know, a priori, their correct order in the quantum hamiltonian [does xp become $\hat{x}\hat{p}$, or $\hat{p}\hat{x}$, or $\frac{1}{2}(\hat{x}\hat{p} + \hat{p}\hat{x})$?]. (However, in each case, general relativity or quantum mechanics, the true curvature coupling or true factor ordering is unambiguous. The ambiguity is solely in the prescription for deducing it via the equivalence principle or the correspondence principle.)

EXERCISES

Exercise 25.15 *Example and Derivation: Curvature Coupling in the Electromagnetic Wave Equation* **T2**

Since Maxwell's equations, written in terms of the classically measurable electromagnetic field tensor **F** [Eqs. (2.48)] involve only single gradients, it is reasonable to expect them to be lifted into curved spacetime without curvature-coupling additions. Assume this is true. It can be shown that: (i) if one writes the electromagnetic field tensor **F** in terms of a 4-vector potential $\vec{A}$ as

$$F_{\alpha\beta} = A_{\beta;\alpha} - A_{\alpha;\beta}, \tag{25.61}$$

then half of the curved-spacetime Maxwell equations, $F_{\alpha\beta;\gamma} + F_{\beta\gamma;\alpha} + F_{\gamma\alpha;\beta} = 0$ [the second of Eqs. (2.48)] are automatically satisfied; (ii) **F** is unchanged by gauge transformations in which a gradient is added to the vector potential, $\vec{A} \to \vec{A} + \vec{\nabla}\psi$; and (iii) by such a gauge transformation one can impose the Lorenz-gauge condition $\vec{\nabla} \cdot \vec{A} = 0$ on the vector potential.

Show that, when the charge-current 4-vector vanishes, $\vec{J} = 0$, the other half of the Maxwell equations, $F^{\alpha\beta}{}_{;\beta} = 0$ [the first of Eqs. (2.48)] become, in Lorenz gauge and in curved spacetime, the wave equation with curvature coupling [Eq. (25.60)].

Exercise 25.16 *Example and Derivation: Curvature-Coupling Torque* **T2**

(a) In the Newtonian theory of gravity, consider an axisymmetric, spinning body (e.g., Earth) with spin angular momentum S_j and time-independent mass distribution $\rho(\mathbf{x})$, interacting with an externally produced tidal gravitational field $\mathcal{E}_{jk}$ (e.g., that of the Sun and the Moon). Show that the torque around the body's center of mass, exerted by the tidal field, and the resulting evolution of the body's spin are

$$\frac{dS_i}{dt} = -\epsilon_{ijk}\mathcal{I}_{jl}\mathcal{E}_{kl}. \tag{25.62}$$

Here

$$\mathcal{I}_{kl} = \int \rho \left(x_k x_l - \frac{1}{3}r^2 \delta_{kl} \right) dV \tag{25.63}$$

is the body's mass quadrupole moment, with $r = \sqrt{\delta_{ij} x_i x_j}$ the distance from the center of mass.

(b) For the centrifugally flattened Earth interacting with the tidal fields of the Moon and the Sun, estimate in order of magnitude the spin-precession period produced by this torque. [The observed precession period is 26,000 years.]

(c) Show that when rewritten in the language of general relativity, and in frame-independent, geometric language, Eq. (25.62) takes the form (25.59) discussed in the text. As part of showing this, explain the meaning of $\mathcal{I}_{\beta\mu}$ in that equation.

For a derivation and discussion of this relativistic curvature-coupling torque when the spinning body is a black hole rather than a Newtonian body, see, for example, Thorne and Hartle (1985).

25.8 The Einstein Field Equation

One crucial issue remains to be studied in this overview of the foundations of general relativity: What is the physical law that determines the curvature of spacetime? Einstein's search for that law, his *Einstein field equation*, occupied a large fraction of his efforts during the years 1913, 1914, and 1915. Several times he thought he had found it, but each time his proposed law turned out to be fatally flawed; for some flavor of his struggle, see the excerpts from his writings in Misner, Thorne, and Wheeler (1973, Sec. 17.7).

In this section, we briefly examine one segment of Einstein's route toward his field equation: the segment motivated by contact with Newtonian gravity.

The Newtonian potential Φ is a close analog of the general relativistic spacetime metric $\mathbf{g}$. From Φ we can deduce everything about Newtonian gravity, and from $\mathbf{g}$ we can deduce everything about spacetime curvature. In particular, by differentiating Φ twice we can obtain the Newtonian tidal field $\mathcal{E}$ [Eq. (25.24)], and by differentiating the components of $\mathbf{g}$ twice we can obtain the components of the relativistic generalization of $\mathcal{E}$: the Riemann curvature tensor [Eq. (25.44) in a local Lorentz frame; Eq. (25.50) in an arbitrary basis].

Newtonian gravity

In Newtonian gravity, Φ is determined by Newton's field equation

$$\nabla^2 \Phi = 4\pi G \rho, \tag{25.64}$$

which can be rewritten in terms of the tidal field, $\mathcal{E}_{jk} = \partial^2 \Phi / \partial x^j \partial x^k$, as

$$\mathcal{E}^j{}_j = 4\pi G \rho. \tag{25.65}$$

Note that this equates a piece of the tidal field—its contraction or *trace*—to the density of mass. By analogy we can expect the Einstein field equation to equate a piece of the Riemann curvature tensor (the analog of the Newtonian tidal field) to some tensor analog of the Newtonian mass density. Further guidance comes from the demand that in nearly Newtonian situations (e.g., in the solar system), the Einstein field equation

deducing the Einstein field equation from Newtonian gravity

should reduce to Newton's field equation. To exploit that guidance, we can (i) write the Newtonian tidal field for nearly Newtonian situations in terms of general relativity's Riemann tensor, $\mathcal{E}_{jk} = R_{j0k0}$ [Eq. (25.35); valid in a local Lorentz frame], (ii) then take the trace and note that by its symmetries $R^0{}_{000} = 0$ so that $\mathcal{E}^j{}_j = R^\alpha{}_{0\alpha 0} = R_{00}$, and (iii) thereby infer that the Newtonian limit of the Einstein equation should read, in a local Lorentz frame:

$$R_{00} = 4\pi G \rho. \tag{25.66}$$

Here R_{00} is the time-time component of the Ricci curvature tensor, which can be regarded as a piece of the Riemann tensor. An attractive proposal for the Einstein field equation should now be obvious. Since the equation should be geometric and frame independent, and since it must have the Newtonian limit (25.66), it presumably should say $R_{\alpha\beta} = 4\pi G \times$ (a second-rank symmetric tensor that generalizes the Newtonian mass density ρ). The obvious generalization of ρ is the stress-energy tensor $T_{\alpha\beta}$, so a candidate is

a failed field equation

$$R_{\alpha\beta} = 4\pi G T_{\alpha\beta}. \tag{25.67}$$

Einstein flirted extensively with this proposal for the field equation during 1913–1915. However, it, like several others he studied, was fatally flawed. When expressed in a coordinate system in terms of derivatives of the metric components $g_{\mu\nu}$, it becomes (because $R_{\alpha\beta}$ and $T_{\alpha\beta}$ both have 10 independent components) 10 independent differential equations for the 10 $g_{\mu\nu}$. This is too many equations. By an arbitrary change of coordinates, $x^\alpha_{\text{new}} = F^\alpha(x^0_{\text{old}}, x^1_{\text{old}}, x^2_{\text{old}}, x^3_{\text{old}})$, involving four arbitrary functions F^0, F^1, F^2, and F^3, one should be able to impose on the metric components four arbitrary conditions, analogous to gauge conditions in electromagnetism (e.g., one should be able to set $g_{00} = -1$ and $g_{0j} = 0$ everywhere). Correspondingly, the field equations should constrain only 6, not 10 of the components of the metric (the 6 g_{ij} in our example).

In November 1915, Einstein (1915), and independently Hilbert (1915) [who was familiar with Einstein's struggle as a result of private conversations and correspondence] discovered the resolution of this dilemma. Because the local law of 4-momentum conservation guarantees $T^{\alpha\beta}{}_{;\beta} = 0$ independent of the field equation, if we replace the Ricci tensor in Eq. (25.67) by a constant (to be determined) times some new curvature tensor $G^{\alpha\beta}$ that is also automatically divergence free independently of the field equation ($G^{\alpha\beta}{}_{;\beta} \equiv 0$), then the new field equation $G^{\alpha\beta} = \kappa T^{\alpha\beta}$ (with κ = constant) will not constrain all 10 components of the metric. Rather, the four equations, $(G^{\alpha\beta} - \kappa T^{\alpha\beta})_{;\beta} = 0$ with $\alpha = 0, 1, 2, 3$, will automatically be satisfied; they will not constrain the metric components in any way, and only six independent constraints on the metric components will remain in the field equation, precisely the desired number.

the need for a divergence-free curvature tensor

It turns out, in fact, that from the Ricci tensor and the scalar curvature one can construct a curvature tensor $G^{\alpha\beta}$ with the desired property:

$$\boxed{G^{\alpha\beta} \equiv R^{\alpha\beta} - \frac{1}{2}Rg^{\alpha\beta}.} \quad (25.68)$$

Einstein curvature tensor

Today we call this the *Einstein curvature tensor*. That it has vanishing divergence, independently of how one chooses the metric,

$$\boxed{\vec{\nabla} \cdot \mathbf{G} \equiv 0,} \quad (25.69)$$

Bianchi identity and contracted Bianchi identity

is called the *contracted Bianchi identity*, since it can be obtained by contracting the following *Bianchi identity* on the tensor $\epsilon_\alpha{}^{\beta\mu\nu}\epsilon_\nu{}^{\gamma\delta\epsilon}$:

$$\boxed{R^\alpha{}_{\beta\gamma\delta;\epsilon} + R^\alpha{}_{\beta\delta\epsilon;\gamma} + R^\alpha{}_{\beta\epsilon\gamma;\delta} = 0.} \quad (25.70)$$

[This Bianchi identity holds true for the Riemann curvature tensor of any and every manifold (i.e., of any and every smooth space; Ex. 25.17). For an extensive discussion of the Bianchi identities (25.70) and (25.69) and their geometric interpretation, see Misner, Thorne, and Wheeler (1973, Chap. 15).]

The Einstein field equation should then equate a multiple of $T^{\alpha\beta}$ to the Einstein tensor $G^{\alpha\beta}$:

$$G^{\alpha\beta} = \kappa T^{\alpha\beta}. \quad (25.71a)$$

The proportionality constant κ is determined from the Newtonian limit as follows. By rewriting the field equation (25.71a) in terms of the Ricci tensor

$$R^{\alpha\beta} - \frac{1}{2}g^{\alpha\beta}R = \kappa T^{\alpha\beta}, \quad (25.71b)$$

then taking the trace to obtain $R = -\kappa g_{\mu\nu}T^{\mu\nu}$ and inserting this back into (25.71a), we obtain

$$R^{\alpha\beta} = \kappa \left(T^{\alpha\beta} - \frac{1}{2}g^{\alpha\beta}g_{\mu\nu}T^{\mu\nu} \right). \quad (25.71c)$$

In nearly Newtonian situations and in a local Lorentz frame, the mass-energy density $T^{00} \cong \rho$ is far greater than the momentum density T^{j0} and is also far greater than the stress T^{jk}. Correspondingly, the time-time component of the field equation (25.71c) becomes

$$R^{00} = \kappa \left(T^{00} - \frac{1}{2}\eta^{00}\eta_{00}T^{00} \right) = \frac{1}{2}\kappa T^{00} = \frac{1}{2}\kappa\rho. \quad (25.71d)$$

By comparing with the correct Newtonian limit (25.66) and noting that in a local Lorentz frame $R_{00} = R^{00}$, we see that $\kappa = 8\pi G$, whence the Einstein field equation is

$$\boxed{G^{\alpha\beta} = 8\pi G T^{\alpha\beta}.} \quad (25.72)$$

Einstein field equation

EXERCISES

Exercise 25.17 *Derivation and Example: Bianchi Identities*

(a) Derive the Bianchi identity (25.70) in 4-dimensional spacetime. [Hint: (i) Introduce a local Lorentz frame at some arbitrary event. (ii) In that frame show, from Eq. (25.50), that the components $R_{\alpha\beta\gamma\delta}$ of Riemann have the form of Eq. (25.44) plus corrections that are quadratic in the distance from the origin. (iii) Compute the left-hand side of Eq. (25.70), with index α down, at the origin of that frame, and show that it is zero. (iv) Then argue that because the origin of the frame was an arbitrary event in spacetime, and because the left-hand side of Eq. (25.70) is an arbitrary component of a tensor, the left-hand side viewed as a frame-independent geometric object must vanish at all events in spacetime.]

(b) By contracting the Bianchi identity (25.70) on $\epsilon_\alpha{}^{\beta\mu\nu}\epsilon_\nu{}^{\gamma\delta\epsilon}$, derive the contracted Bianchi identity (25.69).

These derivations are easily generalized to an arbitrary manifold with any dimension by replacing the 4-dimensional local Lorentz frame by a locally orthonormal coordinate system (cf. the third and fourth paragraphs of Sec. 24.3.3).

25.8.1 Geometrized Units

By now the reader should be accustomed to our use of geometrized units in which the speed of light is unity. Just as setting $c = 1$ has simplified greatly the mathematical notation in Chaps. 2 and 24, so also subsequent notation is greatly simplified if we set Newton's gravitation constant G to unity. This further geometrization of our units corresponds to equating mass units to length units via the relation

geometrized units: setting Newton's gravitational constant to one

$$1 = \frac{G}{c^2} = 7.426 \times 10^{-28} \text{ m kg}^{-1}; \quad \text{so} \quad 1 \text{ kg} = 7.426 \times 10^{-28} \text{ m}. \quad (25.73)$$

Any equation can readily be converted from conventional units to geometrized units by removing all factors of c and G; it can readily be converted back by inserting whatever factors of c and G one needs to make both sides of the equation dimensionally the same. Table 25.1 lists a few important numerical quantities in both conventional units and geometrized units.

In geometrized units, the Einstein field equation (25.72) assumes the following standard form, to which we appeal extensively in the three coming chapters:

the Einstein field equation in geometrized units

$$G^{\alpha\beta} = 8\pi T^{\alpha\beta}; \quad \text{or} \quad \mathbf{G} = 8\pi \mathbf{T}. \quad (25.74)$$

25.9 Weak Gravitational Fields

All the foundations of general relativity are now in our hands. In this concluding section of the chapter, we explore their predictions for the properties of weak gravitational

TABLE 25.1: Some useful quantities in conventional and geometrized units

Quantity	Conventional units	Geometrized units
Speed of light, c	2.998×10^8 m sec^{-1}	one
Newton's gravitation constant, G	6.674×10^{-11} m^3 kg^{-1} s^{-2}	one
G/c^2	7.426×10^{-28} m kg^{-1}	one
c^5/G	3.628×10^{52} W	one
Planck's reduced constant $\hbar$	1.055×10^{-34} kg m^2 s^{-1}	$(1.616 \times 10^{-35}$ m$)^2$
Sun's mass, $M_\odot$	1.989×10^{30} kg	1.477 km
Sun's radius, $R_\odot$	6.957×10^8 m	6.957×10^8 m
Earth's mass, $M_\oplus$	5.972×10^{24} kg	4.435 mm
Earth's mean radius, $R_\oplus$	6.371×10^6 m	6.371×10^6 m

Note: 1 Mpc = 10^6 parsecs (pc), 1 pc = 3.262 light-years (lt-yr), 1 lt-yr = 0.9461×10^{16} m, 1 AU = 1.496×10^{11} m. For other useful astronomical constants, see Cox (2000).

fields, beginning with the Newtonian limit of general relativity and then moving on to other situations.

25.9.1 Newtonian Limit of General Relativity

A general relativistic gravitational field (spacetime curvature) is said to be *weak* if there exist "nearly globally Lorentz" coordinate systems in which the metric coefficients differ only slightly from unity:

$$g_{\alpha\beta} = \eta_{\alpha\beta} + h_{\alpha\beta}, \quad \text{with } |h_{\alpha\beta}| \ll 1. \tag{25.75a}$$

The Newtonian limit requires that gravity be weak in this sense throughout the system being studied. It further requires a slow-motion constraint, which has three aspects:

conditions required for general relativity to become Newton's theory of gravity (the Newtonian limit)

1. The sources of gravity must have slow enough motions that the following holds for some specific choice of the nearly globally Lorentz coordinates:

$$|h_{\alpha\beta,t}| \ll |h_{\alpha\beta,j}|; \tag{25.75b}$$

2. the sources' motions must be slow enough that in this frame the momentum density is very small compared to the energy density:

$$|T^{j0}| \ll T^{00} \equiv \rho; \tag{25.75c}$$

3. and any particles on which the action of gravity is to be studied must move with low velocities, so they must have 4-velocities satisfying

$$|u^j| \ll u^0. \tag{25.75d}$$

25.9 Weak Gravitational Fields

Finally, the Newtonian limit requires that the stresses in the gravitating bodies be small compared to their mass densities:

$$|T^{jk}| \ll T^{00} \equiv \rho. \tag{25.75e}$$

When conditions (25.75) are all satisfied, then to leading nontrivial order in the small dimensionless quantities $|h_{\alpha\beta}|$, $|h_{\alpha\beta,t}|/|h_{\alpha\beta,j}|$, $|T^{j0}|/T^{00}$, $|u^j|/u^0$, and $|T^{jk}|/T^{00}$ the laws of general relativity reduce to those of Newtonian theory.

The details of this reduction are an exercise for the reader (Ex. 25.18); here we give an outline.

details of the Newtonian limit:

The low-velocity constraint $|u^j|/u^0 \ll 1$ on the 4-velocity of a particle, together with its normalization $u^\alpha u^\beta g_{\alpha\beta} = -1$ and the near flatness of the metric (25.75a), implies that

$$u^0 \cong 1, \quad u^j \cong v^j \equiv \frac{dx^j}{dt}. \tag{25.76}$$

Since $u^0 = dt/d\tau$, the first of these relations implies that in our nearly globally Lorentz coordinate system the coordinate time is very nearly equal to the proper time of our slow-speed particle. In this way, we recover the "universal time" of Newtonian theory. The universal, Euclidean space is that of our nearly Lorentz frame, with $h_{\mu\nu}$ completely ignored because of its smallness. The universal time and universal Euclidean space become the arena in which Newtonian physics is formulated.

Equation (25.76) for the components of a particle's 4-velocity, together with $|v^j| \ll 1$ and $|h_{\mu\nu}| \ll 1$, imply that the geodesic equation for a freely moving particle to leading nontrivial order is

$$\frac{dv^j}{dt} \cong \frac{1}{2}\frac{\partial h_{00}}{\partial x^j}, \quad \text{where} \quad \frac{d}{dt} \equiv \frac{\partial}{\partial t} + \mathbf{v} \cdot \mathbf{\nabla}. \tag{25.77}$$

(Because our spatial coordinates are Cartesian, we can put the spatial index j up on one side of the equation and down on the other without creating any danger of error.)

By comparing Eq. (25.77) with Newton's equation of motion for the particle, we deduce that h_{00} must be related to the Newtonian gravitational potential by

$$h_{00} = -2\Phi, \tag{25.78}$$

so the spacetime metric in our nearly globally Lorentz coordinate system must be

spacetime metric in the Newtonian limit

$$\boxed{ds^2 = -(1+2\Phi)dt^2 + (\delta_{jk} + h_{jk})dx^j dx^k + 2h_{0j}dt\, dx^j.} \tag{25.79}$$

Because gravity is weak, only those parts of the Einstein tensor that are linear in $h_{\alpha\beta}$ are significant; quadratic and higher-order contributions can be ignored. Now, by the same mathematical steps as led us to Eq. (25.44) for the components of the Riemann tensor in a local Lorentz frame, one can show that the components of the linearized Riemann tensor in our nearly global Lorentz frame have that same form (i.e., setting $g_{\alpha\beta} = \eta_{\alpha\beta} + h_{\alpha\beta}$):

$$R_{\alpha\beta\gamma\delta} = \frac{1}{2}(h_{\alpha\delta,\beta\gamma} + h_{\beta\gamma,\alpha\delta} - h_{\alpha\gamma,\beta\delta} - h_{\beta\delta,\alpha\gamma}). \tag{25.80}$$

From this equation and the slow-motion constraint $|h_{\alpha\beta,t}| \ll |h_{\alpha\beta,j}|$, we infer that the space-time-space-time components of Riemann are

$$\boxed{R_{j0k0} = -\frac{1}{2}h_{00,jk} = \Phi_{,jk} = \mathcal{E}_{jk}.} \tag{25.81}$$

tidal gravity in the Newtonian limit

In the last step we have used Eq. (25.78). We have thereby recovered the relation between the Newtonian tidal field $\mathcal{E}_{jk} \equiv \Phi_{,jk}$ and the relativistic tidal field R_{j0k0}. That relation can now be used, via the train of arguments in the preceding section, to show that the Einstein field equation, $G^{\mu\nu} = 8\pi T^{\mu\nu}$, reduces to the Newtonian field equation, $\nabla^2 \Phi = 4\pi T^{00} \equiv 4\pi \rho$.

This analysis leaves the details of h_{0j} and h_{jk} unknown, because the Newtonian limit is insensitive to them.

EXERCISES

Exercise 25.18 *Derivation: Newtonian Limit of General Relativity*
Consider a system that can be covered by nearly globally Lorentz coordinates in which the Newtonian-limit constraints (25.75) are satisfied. For such a system, flesh out the details of the text's derivation of the Newtonian limit. More specifically, do the following.

(a) Derive Eq. (25.76) for the components of the 4-velocity of a particle.
(b) Show that the geodesic equation reduces to Eq. (25.77).
(c) Show that to linear order in the metric perturbation $h_{\alpha\beta}$, the components of the Riemann tensor take the form of Eq. (25.80).
(d) Show that in the slow-motion limit the space-time-space-time components of Riemann take the form of Eq. (25.81).

25.9.2 Linearized Theory

There are many systems in the universe that have weak gravity [Eq. (25.75a)], but for which the slow-motion approximations (25.75b)–(25.75d) and/or weak-stress approximation (25.75e) fail. Examples are high-speed particles and electromagnetic fields. For such systems, we need a generalization of Newtonian theory that drops the slow-motion and weak-stress constraints but keeps the weak-gravity constraint:

details of linearized theory (weak-gravity limit of general relativity):

$$\boxed{g_{\alpha\beta} = \eta_{\alpha\beta} + h_{\alpha\beta}, \quad \text{with } |h_{\alpha\beta}| \ll 1.} \tag{25.82}$$

spacetime metric and metric perturbation

The obvious generalization is a linearization of general relativity in $h_{\alpha\beta}$, with no other approximations being made—the so-called *linearized theory of gravity*. In this subsection we develop it.

In formulating linearized theory we can regard the metric perturbation $h_{\mu\nu}$ as a gravitational field that lives in flat spacetime, and correspondingly, we can carry out our mathematics as though we were in special relativity. In other words, linearized theory can be regarded as a field theory of gravity in flat spacetime—a variant of the type of theory that Einstein toyed with and then rejected (Sec. 25.1).

In linearized theory, the Riemann tensor takes the form of Eq. (25.80), but we have no right to simplify it further to the form of Eq. (25.81), so we must follow a different route to the linearized Einstein field equation.

Contracting the first and third indices in Eq. (25.80), we obtain an expression for the linearized Ricci tensor $R_{\mu\nu}$ in terms of $h_{\alpha\beta}$. Contracting once again, we obtain the scalar curvature R, and then from Eq. (25.68) we obtain for the Einstein tensor and the Einstein field equation:

linearized Einstein field equation

$$2G_{\mu\nu} = h_{\mu\alpha,\nu}{}^{\alpha} + h_{\nu\alpha,\mu}{}^{\alpha} - h_{\mu\nu,\alpha}{}^{\alpha} - h_{,\mu\nu} - \eta_{\mu\nu}(h_{\alpha\beta}{}^{,\alpha\beta} - h_{,\alpha}{}^{\alpha})$$
$$= 16\pi T_{\mu\nu}. \qquad (25.83)$$

Here all indices, subscript or superscript, that follow the comma are partial-derivative indices [e.g., $h_{,\beta}{}^{\beta} = (\partial^2 h/\partial x^\beta \partial x^\alpha)\eta^{\alpha\beta}$], and

$$h \equiv \eta^{\alpha\beta} h_{\alpha\beta} \qquad (25.84)$$

is the "trace" of the metric perturbation. We can simplify the field equation (25.83) by reexpressing it in terms of the quantity

trace-reversed metric perturbation

$$\boxed{\bar{h}_{\mu\nu} \equiv h_{\mu\nu} - \frac{1}{2}h\eta_{\mu\nu}.} \qquad (25.85)$$

One can easily check that this quantity has the opposite trace to that of $h_{\mu\nu}$ ($\bar{h} \equiv \bar{h}_{\alpha\beta}\eta^{\alpha\beta} = -h$), so it is called the *trace-reversed metric perturbation*. In terms of it, the field equation (25.83) becomes

$$-\bar{h}_{\mu\nu,\alpha}{}^{\alpha} - \eta_{\mu\nu}\bar{h}_{\alpha\beta}{}^{,\alpha\beta} + \bar{h}_{\mu\alpha,\nu}{}^{\alpha} + \bar{h}_{\nu\alpha,\mu}{}^{\alpha} = 16\pi T_{\mu\nu}. \qquad (25.86)$$

We can simplify this field equation further by specializing our coordinates. We introduce a new nearly globally Lorentz coordinate system that is related to the old one by

infinitesimal coordinate transformation (gauge change)

$$\boxed{x^\alpha_{\text{new}}(\mathcal{P}) = x^\alpha_{\text{old}}(\mathcal{P}) + \xi^\alpha(\mathcal{P}),} \qquad (25.87)$$

where ξ^α is a small vectorial displacement of the coordinate grid. This change of coordinates via four arbitrary functions ($\alpha = 0, 1, 2, 3$) produces a change of the functional form of the metric perturbation $h_{\alpha\beta}$ to

influence of gauge change on metric perturbation

$$\boxed{h^{\text{new}}_{\mu\nu} = h^{\text{old}}_{\mu\nu} - \xi_{\mu,\nu} - \xi_{\nu,\mu}} \qquad (25.88)$$

(Ex. 25.19), and a corresponding change of the trace-reversed metric perturbation. This is linearized theory's analog of a *gauge transformation* in electromagnetic theory. Just as an electromagnetic gauge change generated by a scalar field ψ alters the vector potential, $A_\mu^{\text{new}} = A_\mu^{\text{old}} - \psi_{,\mu}$, so the linearized-theory gauge change generated by ξ^α alters $h_{\mu\nu}$ and $\bar{h}_{\mu\nu}$; and just as the force-producing electromagnetic field tensor $F_{\mu\nu}$ is unaffected by an electromagnetic gauge change, so the tidal-force-producing linearized Riemann tensor is left unaffected by the gravitational gauge change (Ex. 25.19).

By a special choice of the four functions ξ^α (Ex. 25.19), we can impose the following four gauge conditions on $\bar{h}_{\mu\nu}$:

$$\boxed{\bar{h}_{\mu\nu,}{}^\nu = 0.} \quad (25.89)$$

gravitational Lorenz gauge condition

These are linearized theory's analog of the electromagnetic Lorenz gauge condition $A_{\mu,}{}^\mu = 0$, so they are called the *gravitational Lorenz gauge*. Just as the flat-spacetime Maxwell equations take the remarkably simple wave-equation form $A_{\mu,\alpha}{}^\alpha = 4\pi J_\mu$ in Lorenz gauge, so also the linearized Einstein equation (25.86) takes the corresponding simple wave-equation form in gravitational Lorenz gauge:

$$\boxed{-\bar{h}_{\mu\nu,\alpha}{}^\alpha = 16\pi T_{\mu\nu}.} \quad (25.90)$$

linearized Einstein field equation in Lorenz gauge

By the same method as one uses in electromagnetic theory (e.g., Jackson, 1999, Sec. 6.4), one can solve this gravitational field equation for the field $\bar{h}_{\mu\nu}$ produced by an arbitrary stress-energy-tensor source:

$$\boxed{\bar{h}_{\mu\nu}(t, \mathbf{x}) = \int \frac{4 T_{\mu\nu}(t - |\mathbf{x} - \mathbf{x}'|, \mathbf{x}')}{|\mathbf{x} - \mathbf{x}'|} dV_{x'}.} \quad (25.91)$$

Lorenz-gauge trace-reversed metric perturbation as retarded integral over stress-energy tensor

The quantity in the numerator is the stress-energy source evaluated at the "retarded time" $t' = t - |\mathbf{x} - \mathbf{x}'|$. This equation for the field, and the wave equation (25.90) that underlies it, show explicitly that dynamically changing distributions of stress-energy must generate *gravitational waves*, which propagate outward from their source at the speed of light (Einstein, 1916b; Einstein, 1918). We study these gravitational waves in Chap. 27.

EXERCISES

Exercise 25.19 *Derivation: Gauge Transformations in the Linearized Theory*
(a) Show that the "infinitesimal" coordinate transformation (25.87) produces the change (25.88) of the linearized metric perturbation and that it leaves the Riemann tensor (25.80) unchanged.
(b) Exhibit a differential equation for the ξ^α that brings the metric perturbation into gravitational Lorenz gauge [i.e., that makes $h_{\mu\nu}^{\text{new}}$ obey the Lorenz gauge condition (25.89)].

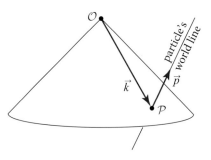

FIGURE 25.4 The past light cone of an observation event $\mathcal{O}$, the world line of a particle, and two 4-vectors: the particle's 4-momentum $\vec{p}$ at the point $\mathcal{P}$ where it passes through $\mathcal{O}$'s past light cone, and the past-directed null vector $\vec{k}$ that reaches from $\mathcal{O}$ to $\mathcal{P}$.

(c) Show that in gravitational Lorenz gauge, the Einstein field equation (25.86) reduces to Eq. (25.90).

Exercise 25.20 *Example: Gravitational Field of a Rapidly Moving Particle* T2

In this exercise we illustrate linearized theory by computing the gravitational field of a moving particle with finite rest mass and then that of a zero-rest-mass particle that moves with the speed of light.

(a) From Eq. (25.91), deduce that, for a particle with mass M at rest at the origin, the only nonvanishing component of $\bar{h}^{\mu\nu}$ is $\bar{h}^{00} = 4M/r$.

(b) Regarding $\bar{h}^{\mu\nu}$ as a field that lives in flat spacetime, show that it can be written in frame-independent, special relativistic form as

$$\bar{h}^{\mu\nu} = \frac{4 p^\mu p^\nu}{\vec{k} \cdot \vec{p}}, \qquad (25.92)$$

where $\vec{p}$ is the particle's 4-momentum, and $\vec{k}$ is the past-directed null vector that reaches from the observation event $\mathcal{O}$ to the event $\mathcal{P}$ at which the particle's world line passes through the observer's past light cone; see Fig. 25.4. Equation (25.92) is a very powerful formula. It is an explicit form of the solution (25.91) to the wave equation (25.90) not only for a particle that moves inertially (and thus could be at rest in our original reference frame) but also for an arbitrarily accelerated particle. Explain why.

(c) In the Lorentz rest frame of the observer, let the particle move along the x-axis with speed v, so its world line is $\{x = vt, y = z = 0\}$, and its 4-momentum has components $p^0 = M\gamma$ and $p^x = Mv\gamma$, with $\gamma = 1/\sqrt{1-v^2}$. Show that, expressed in terms of the observation event's coordinates (t, x, y, z),

$$\vec{k} \cdot \vec{p} = \gamma M R, \quad \text{where } R = \sqrt{(1-v^2)(y^2 + z^2) + (x - vt)^2}. \qquad (25.93)$$

Show, further, that the linearized spacetime metric is

$$ds^2 = \left(1 + \frac{2M}{\gamma R}\right)(-dt^2 + dx^2 + dy^2 + dz^2) + \frac{4\gamma M}{R}(dt - v dx)^2. \quad (25.94)$$

(d) Take the limit of a zero-rest-mass particle moving at the speed of light by sending $m \to 0$, $v \to 1$, and $M\gamma \to \mathcal{E}$ (the particle's energy). [The limit of $1/R = 1/\sqrt{(1-v^2)(y^2 + z^2) + (x - vt)^2}$ is quite tricky. It turns out to be $1/|x - t| - \delta(x - t)\ln(y^2 + z^2)$, where δ is the Dirac delta function (Aichelberg and Sexl, 1971).] Show that the resulting metric is

$$ds^2 = -dt^2 + dx^2 + dy^2 + dz^2$$
$$+ 4\mathcal{E}\left(\frac{1}{|x - t|} - \ln(y^2 + z^2)\delta(x - t)\right)(dx - dt)^2.$$

By a change of coordinates, get rid of the $1/|x - t|$ term, thereby obtaining our final form for the metric of a zero-rest-mass particle:

$$ds^2 = -dt^2 + dx^2 + dy^2 + dz^2 - 4\mathcal{E}\ln(x^2 + y^2)\,\delta(x - t)(dx - dt)^2. \quad (25.95)$$

Equation (25.95) turns out to be an exact solution of the fully nonlinear Einstein field equation for a zero-rest-mass particle; it is called the Aichelberg-Sexl *ultraboost* solution. Just as, when a charged particle is accelerated to near light speed, its electric field lines are compressed into its transverse plane, so the metric (25.95) has all its deviations from flat spacetime concentrated in the particle's transverse plane.

25.9.3 Gravitational Field outside a Stationary, Linearized Source of Gravity

Let us specialize to a time-independent source of weak gravity (so $T_{\mu\nu,t} = 0$ in our chosen nearly globally Lorentz frame) and compute its external gravitational field as a power series in 1/(distance to source). We place our origin of coordinates at the source's center of mass, so

$$\int x^j T^{00} dV = 0, \quad (25.96)$$

and in the same manner as in electromagnetic theory, we expand

$$\frac{1}{|\mathbf{x} - \mathbf{x}'|} = \frac{1}{r} + \frac{x^j x^{j'}}{r^3} + \cdots, \quad (25.97)$$

where $r \equiv |\mathbf{x}|$ is the distance of the field point from the source's center of mass. Inserting Eq. (25.97) into the general solution (25.91) of the Einstein equation and taking note of the conservation laws $T^{\alpha j}{}_{,j} = 0$, we obtain for the source's external field:

trace-reversed metric perturbation

$$\bar{h}_{00} = \frac{4M}{r} + O\left(\frac{1}{r^3}\right), \quad \bar{h}_{0j} = -\frac{2\epsilon_{jkm}J^k x^m}{r^3} + O\left(\frac{1}{r^3}\right), \quad \bar{h}_{ij} = O\left(\frac{1}{r^3}\right).$$

(25.98a)

Here M and J^k are the source's mass and angular momentum:

source's mass and angular momentum

$$M \equiv \int T^{00} dV, \quad J_k \equiv \int \epsilon_{kab} x^a T^{0b} dV$$

(25.98b)

(see Ex. 25.21). This expansion in $1/r$, as in the electromagnetic case, is a multipolar expansion. At order $1/r$ the field is spherically symmetric and the monopole moment is the source's mass M. At order $1/r^2$ there is a "magnetic-type dipole moment," the source's spin angular momentum J_k. These are the leading-order moments in two infinite sets: the "mass multipole" moments (analog of electric moments), and the "mass-current multipole" moments (analog of magnetic moments). For details on all the higher-order moments, see, for example, Thorne (1980).

The metric perturbation can be computed by reversing the trace reversal: $h_{\alpha\beta} = \bar{h}_{\alpha\beta} - \eta_{\alpha\beta}\bar{h}$. Thereby we obtain for the spacetime metric, $g_{\alpha\beta} = \eta_{\alpha\beta} + h_{\alpha\beta}$, at linear order, outside the source:

spacetime metric

$$ds^2 = -\left(1 - \frac{2M}{r}\right)dt^2 - \frac{4\epsilon_{jkm}J^k x^m}{r^3}dt\,dx^j$$
$$+ \left(1 + \frac{2M}{r}\right)\delta_{jk}dx^j dx^k + O\left(\frac{1}{r^3}\right)dx^\alpha dx^\beta.$$

(25.98c)

In spherical polar coordinates, with the polar axis along the direction of the source's angular momentum, the leading-order terms take the form

$$ds^2 = -\left(1 - \frac{2M}{r}\right)dt^2 - \frac{4J}{r}\sin^2\theta\,dt\,d\phi$$
$$+ \left(1 + \frac{2M}{r}\right)(dr^2 + r^2 d\theta^2 + r^2\sin^2\theta\,d\phi^2),$$

(25.98d)

where $J \equiv |\mathbf{J}|$ is the magnitude of the source's angular momentum.

reading off the source's mass and angular momentum from its metric

This is a very important result. It tells us that we can "read off" the mass M and angular momentum J^k from the asymptotic form of the source's metric. More specifically: (i) The mass M shows up in g_{00} in just the way we expect from the Newtonian limit—by comparing Eqs. (25.98c) and (25.79), we see that $\Phi = -M/r$, and from our experience with Newtonian gravity, we conclude that M is the mass that governs the Keplerian orbits of planets around our gravitational source. (ii) The angular momentum J^k shows up in $g_{0j} = -(2/r^3)\epsilon_{jkm}J^k x^m$. The physical manifestation of this g_{0j} is a gravitational torque on gyroscopes.

reading off mass from Keplerian orbits of planets

Consider an inertial-guidance gyroscope whose center of mass is at rest in the coordinate system of Eq. (25.98c) (i.e., at rest relative to the gravitating source). The transport law for the gyroscope's spin is $\nabla_{\vec{u}}\vec{S} = \vec{u}(\vec{a}\cdot\vec{S})$ [Eq. (24.62) boosted from special relativity to general relativity via the equivalence principle]. Here $\vec{u}$ is the gyroscope's 4-velocity (so $u^j = 0$, $u^0 = 1/\sqrt{1-2M/r} \simeq 1 + M/r \simeq 1$), and $\vec{a}$ is its 4-acceleration. The spatial components of this transport law are

$$S^j{}_{,t}u^0 \simeq S^j{}_{,t} = -\Gamma^j{}_{k0}S^k u^0 \simeq -\Gamma^j{}_{k0}S^k \simeq -\Gamma_{jk0}S^k \simeq \frac{1}{2}(g_{0k,j} - g_{0j,k})S^k. \quad (25.99)$$

Here each "$\simeq$" means "is equal, up to fractional corrections of order M/r." By inserting g_{0j} from the line element (25.98c) and performing some manipulations with Levi-Civita tensors, we can bring Eq. (25.99) into the form (cf. Ex. 26.19 and Sec. 27.2.5)

$$\boxed{\frac{d\mathbf{S}}{dt} = \boldsymbol{\Omega}_{\text{prec}} \times \mathbf{S}, \quad \text{where } \boldsymbol{\Omega}_{\text{prec}} = \frac{1}{r^3}[-\mathbf{J} + 3(\mathbf{J}\cdot\mathbf{n})\mathbf{n}].} \quad (25.100)$$

reading off angular momentum from gyroscopic precession (frame dragging)

Here $\mathbf{n} = \mathbf{e}_{\hat{r}}$ is the unit radial vector pointing away from the gravitating source.

Equation (25.100) says that the gyroscope's spin angular momentum rotates (precesses) with angular velocity $\boldsymbol{\Omega}_{\text{prec}}$ in the coordinate system (which is attached to distant inertial frames, i.e., to distant galaxies and quasars). This is sometimes called a *gravitomagnetic precession*, because the off-diagonal term g_{j0} in the metric, when thought of as a 3-vector, is $-2\mathbf{J}\times\mathbf{n}/r^2$, which has the same form as the vector potential of a magnetic dipole; and the gyroscopic precession is similar to that of a magnetized spinning body interacting with that magnetic dipole. It is also called a *frame-dragging precession*, because one can regard the source's angular momentum as dragging inertial frames into precession and regard those inertial frames as being locked to inertial-guidance gyroscopes, such as **S**. And it is sometimes called Lense-Thirring precession after the physicists who first discovered it mathematically, with Einstein's help (Pfister, 2007).

Figure 25.5 shows this frame-dragging precessional angular velocity $\boldsymbol{\Omega}_{\text{prec}}$ as a vector field attached to the source. Notice that it has precisely the same form as a dipolar magnetic field in electromagnetic theory. In Sec. 27.2.5, we discuss the magnitude of this frame dragging in the solar system and the experiments that have measured it.

For a time-independent body with *strong* internal gravity (e.g., a black hole), the distant gravitational field will have the same general form [Eqs. (25.98)] as for a weakly gravitating body, but the constants M and J^k that appear in the metric will not be expressible as the integrals (25.98b) over the body's interior. Nevertheless, they will be measurable by the same techniques as for a weakly gravitating body (Kepler's laws and frame dragging), and they can be interpreted as the body's total mass and angular momentum. We explore this in the next chapter [Eq. (26.71)], where the body will be a spinning black hole.

25.9 Weak Gravitational Fields

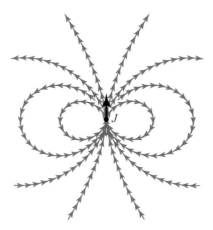

FIGURE 25.5 The precessional angular velocity $\Omega_{\rm prec}$ [Eq. (25.100)] of an inertial-guidance gyroscope at rest outside a stationary, linearized source of gravity that has angular momentum **J**. The arrows are all drawn with the same length rather than proportional to the magnitude of $\Omega_{\rm prec}$.

EXERCISES

Exercise 25.21 *Derivation and Example: External Field of a Stationary, Linearized Source* T2

Derive Eqs. (25.98a) for the trace-reversed metric perturbation outside a stationary (time-independent), linearized source of gravity. More specifically, do the following.

(a) First derive $\bar{h}_{00}$. In your derivation identify a dipolar term of the form $4D_j x^j/r^3$, and show that by placing the origin of coordinates at the center of mass, Eq. (25.96), one causes the dipole moment D_j to vanish.

(b) Next derive $\bar{h}_{0j}$. The two terms in Eq. (25.97) should give rise to two terms. The first of these is $4P_j/r$, where P_j is the source's linear momentum. Show, using the gauge condition $\bar{h}^{0\mu}{}_{,\mu} = 0$ [Eq. (25.89)], that if the momentum is nonzero, then the mass dipole term of part (a) must have a nonzero time derivative, which violates our assumption of stationarity. Therefore, the linear momentum must vanish for this source. Show that the second term gives rise to the $\bar{h}_{0j}$ of Eq. (25.98a). [Hint: You will have to add a perfect divergence, $-\frac{1}{2}(T^{0a}x^j x^m)_{,a}$ to the integrand $T^{0j}x^m$.]

(c) Finally derive $\bar{h}_{ij}$. [Hint: Show that $T^{ij} = (T^{ia}x^i)_{,a}$ and thence that the volume integral of T^{ij} vanishes; similarly for $T^{ij}x^k$.]

Exercise 25.22 *Derivation and Problem: Differential Precession and Frame-Drag Field* T2

(a) Derive the equation $\Delta\Omega_i = \mathcal{B}_{ij}\xi^j$ for the precession angular velocity of a gyroscope at the tip of $\boldsymbol{\xi}$ as measured in an inertial frame at its tail. Here $\mathcal{B}_{ij}$ is the frame-drag field introduced in Box 25.2. [For a solution, see Nichols et al. (2011, Sec. III.C).]

BOX 25.2. DECOMPOSITION OF RIEMANN: TIDAL AND FRAME-DRAG FIELDS T2

In any local Lorentz frame, and also in a Lorentz frame of the linearized theory, the electromagnetic field tensor $F_{\mu\nu}$ can be decomposed into two spatial vector fields: the electric field $E_i = F_{i0}$ and magnetic field $B_i = \frac{1}{2}\epsilon_{ipq}F^{pq}$ (Sec. 2.11). Similarly, in vacuum (for simplicity) the Riemann curvature tensor can be decomposed into two spatial tensor fields: the *tidal field* $\mathcal{E}_{ij} = R_{i0j0}$ and the *frame-drag field* $\mathcal{B}_{ij} = \frac{1}{2}\epsilon_{ipq}R^{pq}{}_{j0}$. The symmetries (25.45) of Riemann, and the fact that in vacuum it is trace-free, imply that both $\mathcal{E}_{ij}$ and $\mathcal{B}_{jk}$ are symmetric and trace-free (STF). In the 3-space of the chosen frame, they are the irreducible tensorial parts of the vacuum Riemann tensor (cf. Box 11.2).

In a local Lorentz frame for strong gravity, and also in the linearized theory for weak gravity, the Bianchi identities (25.70) take on the following Maxwell-like form [Nichols et al., 2011, Eqs. (2.4), (2.15)], in which the superscript S means to symmetrize:

$$\nabla \cdot \mathcal{E} = 0, \quad \nabla \cdot \mathcal{B} = 0, \quad \frac{\partial \mathcal{E}}{\partial t} - (\nabla \times \mathcal{B})^S = 0, \quad \frac{\partial \mathcal{B}}{\partial t} + (\nabla \times \mathcal{E})^S = 0. \tag{1}$$

This has motivated some physicists to call the tidal field $\mathcal{E}$ and the frame-drag field $\mathcal{B}$ the "electric" and "magnetic" parts of the vacuum Riemann tensor. We avoid this language because of the possibility of confusing these second-rank tensorial gravitational fields with their truly electromagnetic vector-field counterparts **E** and **B**.

The tidal and frame-drag fields get their names from the forces they produce. The equation of geodesic deviation (25.34) says that, in a local Lorentz frame or in linearized theory, the relative acceleration of two test particles, separated by the vector $\boldsymbol{\xi}$, is $\Delta a_i = -\mathcal{E}_{ij}\xi^j$. Similarly, a gyroscope at the tip of $\boldsymbol{\xi}$ precesses relative to an inertial frame at its tail with the differential frame-dragging angular velocity $\Delta\Omega_i = \mathcal{B}_{ij}\xi^j$ (Ex. 25.22). Not surprisingly, in linearized theory, $\mathcal{B}$ is the symmetrized gradient of the angular velocity Ω_{prec} of precession relative to distant inertial frames: $\mathcal{B} = \left(\nabla\Omega_{\text{prec}}\right)^S$.

Just as the electric and magnetic fields can be visualized using field lines that are the integral curves of **E** and **B**, $\mathcal{E}$ and $\mathcal{B}$ can be visualized using the integral curves of their eigenvectors. Since each field, $\mathcal{E}$ or $\mathcal{B}$, has three orthogonal eigenvectors, each has a network of three orthogonal sets of field lines. The field lines of $\mathcal{B}$ are called (Nichols et al., 2011) *frame-drag vortex lines* or simply "vortex lines"; those of $\mathcal{E}$ are called *tidal tendex lines* or simply "tendex lines"

(continued)

BOX 25.2. (continued)

(from the Latin *tendere*, meaning "to stretch" by analogy with vortex from *vertere*, meaning "to turn"). For a spinning point mass in linearized theory, $\mathcal{E}$ and $\mathcal{B}$ are given by Eqs. (25.101), and their tendex and vortex lines are as shown below (adapted from Nichols et al., 2011). The red tendex lines (left diagram) stretch; the blue squeeze. The red vortex lines (right) twist a gyroscope at a person's feet (or head) counterclockwise relative to inertial frames at her head (or feet). The blue vortex lines twist clockwise. We explore these concepts in greater detail in Box 26.3.

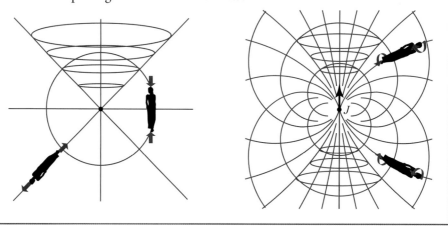

(b) Show that in linearized theory, $\mathcal{B}$ is the symmetrized gradient of the angular velocity $\mathbf{\Omega}_{\text{prec}}$ of precession of a gyroscope relative to distant inertial frames.

Exercise 25.23 *Problem: Spinning Particle in Linearized Theory: Tidal and Frame-Drag Fields* T2

Show that in linearized theory, for a spinning particle at the origin with mass M and with its spin J along the polar axis, the orthonormal-frame components of $\mathcal{E}$ and $\mathcal{B}$ are

$$\mathcal{E}_{\hat{r}\hat{r}} = -2\mathcal{E}_{\hat{\theta}\hat{\theta}} = -2\mathcal{E}_{\hat{\phi}\hat{\phi}} = -\frac{2M}{r^3}, \qquad (25.101a)$$

$$\mathcal{B}_{\hat{r}\hat{r}} = -2\mathcal{B}_{\hat{\theta}\hat{\theta}} = -2\mathcal{B}_{\hat{\phi}\hat{\phi}} = -\frac{6J\cos\theta}{r^4}, \quad \mathcal{B}_{\hat{r}\hat{\theta}} = \mathcal{B}_{\hat{\theta}\hat{r}} = -\frac{3J\sin\theta}{r^4}. \quad (25.101b)$$

What are the eigenvectors of these fields? Convince yourself that these eigenvectors' integral curves (the tidal tendex lines and frame-drag vortex lines) are as depicted at the bottom of Box 25.2.

25.9.4 Conservation Laws for Mass, Momentum, and Angular Momentum in Linearized Theory T2

Consider a static (unmoving) sphere $\mathcal{S}$ surrounding our time-independent source of gravity, with such a large radius r that the $O(1/r^3)$ corrections in $\bar{h}_{\mu\nu}$ and in the metric [Eqs. (25.98)] can be ignored. Suppose that a small amount of mass-energy E (as measured in the sphere's and source's rest frame) is injected through the sphere, into the source. Then the special relativistic law of mass-energy conservation tells us that the source's mass $M = \int T^{00} dV$ will increase by $\Delta M = E$. Similarly, if an energy flux T^{0j} flows through the sphere, the source's mass will change by

for a stationary source of gravity perturbed by an infalling or outflowing stress-energy tensor:

$$\boxed{\frac{dM}{dt} = -\int_{\mathcal{S}} T^{0j} d\Sigma_j,} \qquad (25.102)$$

mass conservation

where $d\Sigma_j$ is the sphere's outward-pointing surface-area element, and the minus sign is because $d\Sigma_j$ points outward, not inward. Since M is the mass that appears in the source's asymptotic gravitational field $\bar{h}_{\mu\nu}$ and metric $g_{\alpha\beta}$, this conservation law can be regarded as describing how the source's gravitating mass changes when energy is injected into it.

From the special relativistic law for angular momentum conservation (e.g., Misner, Thorne, and Wheeler, 1973, Box 5.6), we deduce a similar result. A flux $\epsilon_{ijk} x^j T^{km}$ of angular momentum through the sphere produces the following change in the angular momentum J_i that appears in the source's asymptotic field $\bar{h}_{\mu\nu}$ and metric:

$$\boxed{\frac{dJ_i}{dt} = -\int_{\mathcal{S}} \epsilon_{ijk} x^j T^{km} d\Sigma_m.} \qquad (25.103)$$

angular-momentum conservation

There is also a conservation law for a gravitationally measured linear momentum. That linear momentum does not show up in the asymptotic field and metric that we wrote down above [Eqs. (25.98)], because our coordinates were chosen to be attached to the source's center of mass (i.e., they are the Lorentz coordinates of the source's rest frame). However, if linear momentum P_j is injected through our sphere $\mathcal{S}$ and becomes part of the source, then the source's center of mass will start moving, and the asymptotic metric will acquire a new term:

$$\boxed{\delta g_{0j} = -4 P_j / r,} \qquad (25.104)$$

where (after the injection)

$$\boxed{P_j = P^j = \int T^{0j} dV} \qquad (25.105)$$

source's linear momentum

[see Eq. (25.91) with $\bar{h}^{0j} = -\bar{h}_{0j} = -h_{0j} = -\delta g_{0j}$; also see Ex. 25.21b]. More generally, the rate of change of the source's total linear momentum (the P_j term in the asymptotic g_{0j}) is the integral of the inward flux of momentum (inward component of the stress tensor) across the sphere:

momentum conservation

$$\boxed{\frac{dP_j}{dt} = -\int_{\mathcal{S}} T^{jk} d\Sigma_k.}$$

(25.106)

25.9.5 Conservation Laws for a Strong-Gravity Source

strong-gravity source surrounded by asymptotically flat spacetime:

For a time-independent source with strong internal gravity, not only does the asymptotic metric, far from the source, have the same form [Eqs. (25.98c), (25.98d), (25.104)] as for a weakly gravitating source, but also the conservation laws for its gravitationally measured mass, angular momentum, and linear momentum [Eqs. (25.102), (25.103), (25.106), respectively] continue to hold true. Of course, the sphere $\mathcal{S}$ must be placed far from the source, in a region where gravity is weak, so spacetime is *asymptotically flat*[6] and linearized theory is valid in the vicinity of $\mathcal{S}$. When this is done, then the special relativistic description of inflowing mass, angular momentum, and energy is valid at $\mathcal{S}$, and the linearized Einstein equation, applied in the vicinity of $\mathcal{S}$ (and not extended into the strong-gravity region), turn out to guarantee that the M, J_j, and P_j appearing in the asymptotic metric evolve in accord with the conservation laws (25.102), (25.103), and (25.106).

in asymptotically flat region, source's mass, momentum, angular momentum, and their conservation are the same as for a weak-gravity source

For strongly gravitating sources, these conservation laws owe their existence to the spacetime's asymptotic time-translation, rotation, and space-translation symmetries. In generic, strong-gravity regions of spacetime there are no such symmetries and correspondingly, no integral conservation laws for energy, angular momentum, or linear momentum.

If a strongly gravitating source is dynamical rather than static, it will emit gravitational waves (Chap. 27). The amplitudes of those waves, like the influence of the source's mass, die out as $1/r$ far from the source, so spacetime retains its asymptotic time-translation, rotation, and space-translation symmetries. These symmetries continue to enforce integral conservation laws on the gravitationally measured mass, angular momentum, and linear momentum [Eqs. (25.102), (25.103), and (25.106)], but with the new requirement that one include, in the fluxes through $\mathcal{S}$, contributions from the gravitational waves' energy, angular momentum, and linear momentum; see Chap. 27.

For a brief derivation and discussion of these asymptotic conservation laws, see Thorne (1983, Sec. 3.3.2); for far greater detail, see Misner, Thorne, and Wheeler (1973, Chaps. 18 and 19).

---

6. Our real universe, of course, is not asymptotically flat. However, nearly everywhere the distances between gravitating systems (e.g., between our solar system and the alpha centauri system) are so large that spacetime is very nearly asymptotically flat as one moves outward into the region between the systems. Correspondingly, to very high accuracy all the statements in this section remain true.

Bibliographic Note

For a superb, detailed historical account of Einstein's intellectual struggle to formulate the laws of general relativity, see Pais (1982). For Einstein's papers of that era, in the original German and in English translation, with detailed annotations and explanations by editors with strong backgrounds in both physics and the history of science, see Einstein (1989). For some key papers of that era by other major contributors besides Einstein, in English translation, see Lorentz et al. (1923).

This chapter's pedagogical approach to presenting the fundamental concepts of general relativity is strongly influenced by Misner, Thorne, and Wheeler (1973), where readers will find much greater detail. See, especially, Chap. 8 for the mathematics (differential geometry) of curved spacetime, or Chaps. 9–14 for far greater detail; Chap. 16 for the Einstein equivalence principle and how to lift laws of physics into curved spacetime; Chap. 17 for the Einstein field equations and many different ways to derive them; Chap. 18 for weak gravitational fields (the Newtonian limit and linearized theory); and Chaps. 19 and 20 for the metric in the asymptotically flat region outside a strongly gravitating source and for the source's conservation laws for mass, momentum, and angular momentum.

For an excellent, elementary introduction to the fundamental concepts of general relativity from a viewpoint that is somewhat less mathematical than this chapter or Misner, Thorne, and Wheeler (1973), see Hartle (2003). We also recommend, at a somewhat elementary level, Schutz (2009); and at a more advanced level, Carroll (2004), Straumann (2013), and Zee (2013). At a very advanced and mathematical level, we recommend Wald (1984). For a rather different approach to general relativity, one that emphasizes the connection to field theory over that to geometry, we recommend Weinberg (1972).

Our physicist's approach to differential geometry in this chapter lacks much of the rigor and beauty of a mathematician's approach, for which we recommend Spivak (1999).

CHAPTER TWENTY-SIX

Relativistic Stars and Black Holes

All light emitted from such a body would be made to return towards it by its own proper gravity.
JOHN MICHELL (1783)

26.1 Overview

Having sketched the fundamentals of Einstein's theory of gravity, general relativity, we now illustrate his theory by several concrete applications: stars and black holes in this chapter, gravitational waves in Chap. 27, and the large-scale structure and evolution of the universe in Chap. 28.

While stars and black holes are the central thread of this chapter, we study them less for their own intrinsic interest than for their roles as vehicles by which to understand general relativity. Using them, we elucidate some issues we have already met: the physical and geometric interpretations of spacetime metrics and of coordinate systems, the Newtonian limit of general relativity, the geodesic motion of freely falling particles and photons, local Lorentz frames and the tidal forces measured therein, proper reference frames, the Einstein field equation, the local law of conservation of 4-momentum, and the asymptotic structure of spacetime far from gravitating sources. Stars and black holes also serve to introduce several new physical phenomena that did not show up in our study of the foundations of general relativity: the "many-fingered" nature of time, event horizons, and spacetime singularities.

We begin this chapter, in Sec. 26.2, by studying the geometry of the curved spacetime outside any static star, as predicted by the Einstein field equation. In Sec. 26.3, we study general relativity's description of the interiors of static stars. In Sec. 26.4, we turn attention to the spherically symmetric gravitational implosion by which a nonrotating star is transformed into a black hole, and to the Schwarzschild spacetime geometry outside and inside the resulting static, spherical hole. In Sec. 26.5, we study the Kerr spacetime geometry of a spinning black hole. In Sec. 26.6, we elucidate the nature of time in the curved spacetimes of general relativity. And in Ex. 26.13, we explore the role of the vacuum Schwarzschild solution of the Einstein field equation as a wormhole.

> **BOX 26.1. READERS' GUIDE**
>
> - This chapter relies significantly on:
>   - Chap. 2 on special relativity;
>   - Chap. 24 on the transition from special relativity to general relativity; and
>   - Chap. 25 on the fundamental concepts of general relativity.
> - Portions of this chapter are a foundation for the applications of general relativity theory to gravitational waves (Chap. 27) and to cosmology (Chap. 28).

26.2 Schwarzschild's Spacetime Geometry

26.2.1 The Schwarzschild Metric, Its Connection Coefficients, and Its Curvature Tensors

On January 13, 1916, just 7 weeks after formulating the final version of his field equation, **G** = 8π **T**, Albert Einstein read to a meeting of the Prussian Academy of Sciences in Berlin a letter from the eminent German astrophysicist Karl Schwarzschild. Schwarzschild, as a member of the German army, had written from the World-War-I Russian front to tell Einstein of a mathematical discovery he had made: he had found the world's first exact solution to the Einstein field equation.

Written as a line element in a special coordinate system (coordinates $\{t, r, \theta, \phi\}$) that Schwarzschild invented for the purpose, Schwarzschild's solution takes the form (Schwarzschild, 1916a)

Schwarzschild metric

$$ds^2 = -(1 - 2M/r)dt^2 + \frac{dr^2}{(1 - 2M/r)} + r^2(d\theta^2 + \sin^2\theta d\phi^2), \quad (26.1)$$

where M is a constant of integration. The connection coefficients, Riemann tensor, and Ricci and Einstein tensors for this metric can be computed by the methods of Chaps. 24 and 25; see Ex. 26.1. The results are tabulated in Box 26.2. The key bottom line is that the Einstein tensor vanishes. Therefore, the Schwarzschild metric (26.1) is a solution of the Einstein field equation with vanishing stress-energy tensor.

roles of the Schwarzschild metric: exterior of static star and imploding star; black hole and wormhole

Many readers know already the lore of this subject. The Schwarzschild metric is reputed to represent the vacuum exterior of a nonrotating, spherical star; and also the exterior of a spherical star as it implodes to form a black hole; and also the exterior and interior of a nonrotating, spherical black hole; and also a wormhole that connects two different universes or two widely separated regions of our own universe.

How does one discover these physical interpretations of the Schwarzschild metric (26.1)? The tools for discovering them—and, more generally, the tools for interpreting

BOX 26.2. CONNECTION COEFFICIENTS AND CURVATURE TENSORS FOR SCHWARZSCHILD SOLUTION

The coordinate basis vectors for the Schwarzschild solution of Einstein's equation are

$$\vec{e}_t = \frac{\partial}{\partial t}, \quad \vec{e}_r = \frac{\partial}{\partial r}, \quad \vec{e}_\theta = \frac{\partial}{\partial \theta}, \quad \vec{e}_\phi = \frac{\partial}{\partial \phi};$$

$$\vec{e}^t = \vec{\nabla} t, \quad \vec{e}^r = \vec{\nabla} r, \quad \vec{e}^\theta = \vec{\nabla} \theta, \quad \vec{e}^\phi = \vec{\nabla} \phi. \tag{1}$$

The covariant and contravariant metric coefficients in this coordinate basis are [cf. Eq. (26.1)]

$$g_{tt} = -\left(1 - \frac{2M}{r}\right), \quad g_{rr} = \frac{1}{(1 - 2M/r)}, \quad g_{\theta\theta} = r^2, \quad g_{\phi\phi} = r^2 \sin^2\theta; \tag{2a}$$

$$g^{tt} = -\frac{1}{(1 - 2M/r)}, \quad g^{rr} = \left(1 - \frac{2M}{r}\right), \quad g^{\theta\theta} = \frac{1}{r^2}, \quad g^{\phi\phi} = \frac{1}{r^2 \sin^2\theta}. \tag{2b}$$

The nonzero connection coefficients in this coordinate basis are

$$\Gamma^t{}_{rt} = \Gamma^t{}_{tr} = \frac{M}{r^2}\frac{1}{(1-2M/r)}, \quad \Gamma^r{}_{tt} = \frac{M}{r^2}(1 - 2M/r), \quad \Gamma^r{}_{rr} = -\frac{M}{r^2}\frac{1}{(1-2M/r)},$$

$$\Gamma^r{}_{\theta\theta} = -r(1 - 2M/r), \quad \Gamma^\theta{}_{r\theta} = \Gamma^\theta{}_{\theta r} = \Gamma^\phi{}_{r\phi} = \Gamma^\phi{}_{\phi r} = \frac{1}{r}, \tag{3}$$

$$\Gamma^r{}_{\phi\phi} = -r\sin^2\theta(1 - 2M/r), \quad \Gamma^\theta{}_{\phi\phi} = -\sin\theta\cos\theta, \quad \Gamma^\phi{}_{\theta\phi} = \Gamma^\phi{}_{\phi\theta} = \cot\theta.$$

The orthonormal basis associated with the above coordinate basis is

$$\vec{e}_{\hat{0}} = \frac{1}{\sqrt{1 - 2M/r}}\frac{\partial}{\partial t}, \quad \vec{e}_{\hat{r}} = \sqrt{1 - \frac{2M}{r}}\frac{\partial}{\partial r}, \quad \vec{e}_{\hat{\theta}} = \frac{1}{r}\frac{\partial}{\partial \theta}, \quad \vec{e}_{\hat{\phi}} = \frac{1}{r\sin\theta}\frac{\partial}{\partial \phi}. \tag{4}$$

The nonzero connection coefficients in this orthonormal basis are

$$\Gamma^{\hat{r}}{}_{\hat{0}\hat{0}} = \Gamma^{\hat{0}}{}_{\hat{r}\hat{0}} = \frac{M}{r^2\sqrt{1 - 2M/r}}, \quad \Gamma^{\hat{\phi}}{}_{\hat{\theta}\hat{\phi}} = -\Gamma^{\hat{\theta}}{}_{\hat{\phi}\hat{\phi}} = \frac{\cot\theta}{r},$$

$$\Gamma^{\hat{\theta}}{}_{\hat{r}\hat{\theta}} = \Gamma^{\hat{\phi}}{}_{\hat{r}\hat{\phi}} = -\Gamma^{\hat{r}}{}_{\hat{\theta}\hat{\theta}} = -\Gamma^{\hat{r}}{}_{\hat{\phi}\hat{\phi}} = \frac{\sqrt{1 - 2M/r}}{r}. \tag{5}$$

(continued)

> **BOX 26.2. (continued)**
>
> The nonzero components of the Riemann tensor in this orthonormal basis are
>
> $$R_{\hat{r}\hat{0}\hat{r}\hat{0}} = -R_{\hat{\theta}\hat{\phi}\hat{\theta}\hat{\phi}} = -\frac{2M}{r^3}, \quad R_{\hat{\theta}\hat{0}\hat{\theta}\hat{0}} = R_{\hat{\phi}\hat{0}\hat{\phi}\hat{0}} = -R_{\hat{r}\hat{\phi}\hat{r}\hat{\phi}} = -R_{\hat{r}\hat{\theta}\hat{r}\hat{\theta}} = \frac{M}{r^3}, \quad (6)$$
>
> and those obtainable from these via the symmetries (25.45a) of Riemann. The Ricci tensor, curvature scalar, and Einstein tensor all vanish—which implies that the Schwarzschild metric is a solution of the vacuum Einstein field equation.

physically any spacetime metric that one encounters—are a central concern of this chapter.

EXERCISES

Exercise 26.1 *Practice: Connection Coefficients and the Riemann Tensor for the Schwarzschild Metric*
(a) Explain why, for the Schwarzschild metric (26.1), the metric coefficients in the coordinate basis have the values given in Eqs. (2a,b) of Box 26.2.
(b) Using tensor-analysis software on a computer,[1] derive the connection coefficients given in Eq. (3) of Box 26.2.
(c) Show that the basis vectors in Eqs. (4) of Box 26.2 are orthonormal.
(d) Using tensor-analysis software on a computer, derive the connection coefficients (5) and Riemann components (6) of Box 26.2 in the orthonormal basis.

26.2.2 The Nature of Schwarzschild's Coordinate System, and Symmetries of the Schwarzschild Spacetime

When presented with a line element such as Eq. (26.1), one of the first questions one is tempted to ask is "What is the nature of the coordinate system?" Since the metric coefficients will be different in some other coordinate system, surely one must know something about the coordinates to interpret the line element.

Remarkably, one need not go to the inventor of the coordinates to find out their nature. Instead, one can turn to the line element itself: the line element (or metric

---

1. Such as the simple Mathematica program in Hartle (2003, Appendix C), which is available on that book's website: http://web.physics.ucsb.edu/~gravitybook/mathematical.html.

coefficients) contain full information not only about the details of the spacetime geometry, but also about the nature of the coordinates. The line element (26.1) is a good example:

properties of the Schwarzschild metric and its Schwarzschild coordinates:

Look first at the 2-dimensional surfaces in spacetime that have constant values of t and r. We can regard $\{\theta, \phi\}$ as a coordinate system on each such 2-surface. The spacetime line element (26.1) tells us that the geometry of the 2-surface is given in terms of those coordinates by

$$^{(2)}ds^2 = r^2(d\theta^2 + \sin^2\theta \, d\phi^2) \qquad (26.2)$$

(where the prefix $^{(2)}$ refers to the dimensionality of the surface). This is the line element (metric) of an ordinary, everyday 2-dimensional sphere expressed in standard spherical polar coordinates. Thus we have learned that the Schwarzschild spacetime is spherically symmetric, and moreover, θ and ϕ are standard spherical polar coordinates. This is an example of extracting from a metric information about both the coordinate-independent spacetime geometry and the coordinate system being used.

spherically symmetric

Furthermore, note from Eq. (26.2) that the circumferences and surface areas of the spheres $(t, r) = \text{const}$ in Schwarzschild spacetime are given by

$$\boxed{\text{circumference} = 2\pi r, \quad \text{area} = 4\pi r^2.} \qquad (26.3)$$

radial coordinate r is circumference/2π, or square root of area/4π

This tells us one aspect of the geometric interpretation of the r coordinate: r is a radial coordinate in the sense that the circumferences and surface areas of the spheres in Schwarzschild spacetime are expressed in terms of r in the standard manner of Eq. (26.3). We must not go further, however, and assert that r is radius in the sense of being the proper distance from the center of one of the spheres to its surface. The center, and the line from center to surface, do not lie on the sphere itself, and they thus are not described by the spherical line element (26.2). Moreover, since we know that spacetime is curved, we have no right to expect that the proper distance from the center of a sphere to its surface will be given by distance = circumference/$2\pi = r$ as in flat spacetime.

26.2.3 Schwarzschild Spacetime at Radii $r \gg M$: The Asymptotically Flat Region

26.2.3

Returning to the Schwarzschild line element (26.1), let us examine several specific regions of spacetime: At "radii" r large compared to the integration constant M, the line element (26.1) takes the form

$$ds^2 = -dt^2 + dr^2 + r^2(d\theta^2 + \sin^2\theta \, d\phi^2). \qquad (26.4)$$

This is the line element of flat spacetime, $ds^2 = -dt^2 + dx^2 + dy^2 + dz^2$ written in spherical polar coordinates $\{x = r\sin\theta\cos\phi, y = r\sin\theta\sin\phi, z = r\cos\theta\}$. Thus, Schwarzschild spacetime is asymptotically flat in the region of large radii $r/M \to \infty$. This is just what one might expect physically when one gets far away from all sources of

asymptotically flat

gravity. Thus, it is reasonable to presume that the Schwarzschild spacetime geometry is that of some sort of isolated, gravitating body that is located in the region $r \sim M$.

The large-r line element (26.4) not only reveals that Schwarzschild spacetime is asymptotically flat; it also shows that in the asymptotically flat region the Schwarzschild coordinate t is the time coordinate of a Lorentz reference frame. Notice that the region of strong spacetime curvature has a boundary (say, $r \sim 100M$) that remains forever fixed relative to the asymptotically Lorentz spatial coordinates $\{x = r \sin\theta \cos\phi, \ y = r \sin\theta \sin\phi, \ z = r \cos\theta\}$. This means that the asymptotic Lorentz frame can be regarded as the body's **asymptotic rest frame**. We conclude, then, that far from the body the Schwarzschild t coordinate becomes the Lorentz time of the body's asymptotic rest frame, and the Schwarzschild $\{r, \theta, \phi\}$ coordinates become spherical polar coordinates in the body's asymptotic rest frame.

As we move inward from $r = \infty$, we gradually begin to see spacetime curvature. That curvature shows up, at $r \gg M$, in slight deviations of the Schwarzschild metric coefficients from those of a Lorentz frame: to first order in M/r the line element (26.1) becomes

$$ds^2 = -\left(1 - \frac{2M}{r}\right)dt^2 + \left(1 + \frac{2M}{r}\right)dr^2 + r^2(d\theta^2 + \sin^2\theta \, d\phi^2), \quad (26.5)$$

or, equivalently, in Cartesian spatial coordinates:

$$ds^2 = -\left(1 - \frac{2M}{\sqrt{x^2+y^2+z^2}}\right)dt^2 + dx^2 + dy^2 + dz^2$$
$$+ \frac{2M}{r}\left(\frac{x}{r}dx + \frac{y}{r}dy + \frac{z}{r}dz\right)^2. \quad (26.6)$$

It is reasonable to expect that, at these large radii where the curvature is weak, Newtonian gravity will be a good approximation to Einsteinian gravity. In Sec. 25.9.1, we studied in detail the transition from general relativity to Newtonian gravity, and found that in nearly Newtonian situations, if one uses a nearly globally Lorentz coordinate system (as we are doing), the line element should take the form [Eq. (25.79)]:

$$ds^2 = -(1 + 2\Phi)dt^2 + (\delta_{jk} + h_{jk})dx^j dx^k + 2h_{tj}dt \, dx^j, \quad (26.7)$$

where the $h_{\mu\nu}$ are metric corrections that are small compared to unity, and Φ (which shows up in the time-time part of the metric) is the Newtonian potential. Direct comparison of Eq. (26.7) with (26.6) shows that a **Newtonian description of the body's distant gravitational field** will entail a Newtonian potential given by

$$\Phi = -\frac{M}{r} \quad (26.8)$$

($\Phi = -GM/r$ in conventional units). This, of course, is the external Newtonian field of a body with mass M. Thus, the integration constant M in the Schwarzschild line element is the mass that characterizes the body's distant, nearly Newtonian gravitational field. This is an example of reading the mass of a body off the asymptotic form of the metric (last paragraph of Sec. 25.9.3).

M is the mass that characterizes the asymptotic Newtonian gravitational field

Notice that the asymptotic metric here [Eq. (26.5)] differs in its spatial part from that in Sec. 25.9.3 [Eq. (25.98d)]. This difference arises from the use of different radial coordinates here. If we define $\bar{r}$ by $r = \bar{r} + M$ at radii $r \gg M$, then to linear order in M/r, the asymptotic Schwarzschild metric (26.5) becomes

$$ds^2 = -\left(1 - \frac{2M}{\bar{r}}\right) dt^2 + \left(1 + \frac{2M}{\bar{r}}\right) [d\bar{r}^2 + \bar{r}^2(d\theta^2 + \sin^2\theta \, d\phi^2)], \quad (26.9)$$

which is the same as Eq. (25.98d) with vanishing angular momentum $J = 0$. This easy change of the spatial part of the metric reinforces the fact that one reads the asymptotic Newtonian potential and the source's mass M from the time-time components of the metric and not from the spatial part of the metric.

We can describe, in operational terms, the physical interpretation of M as the body's mass as follows. Suppose that a test particle (e.g., a small planet) moves around our central body in a circular orbit with radius $r \gg M$. A Newtonian analysis of the orbit predicts that, as measured using Newtonian time, the period of the orbit will be $P = 2\pi (r^3/M)^{\frac{1}{2}}$ (one of Kepler's laws). Moreover, since Newtonian time is very nearly equal to the time t of the nearly Lorentz coordinates used in Eq. (26.5) (cf. Sec. 25.9.1), and since that t is Lorentz time in the body's relativistic, asymptotic rest frame, the orbital period as measured by observers at rest in the asymptotic rest frame must be $P = 2\pi (r^3/M)^{\frac{1}{2}}$. Thus, M is the mass that appears in Kepler's laws for the orbits of test particles far from the central body. This quantity is sometimes called the body's "active gravitational mass," since it is the mass that characterizes the body's gravitational pull. It is also called the body's "total mass-energy," because it turns out to include all forms of mass and energy that the body possesses (rest mass; internal kinetic energy; and all forms of internal binding energy, including gravitational).

M is the mass in Kepler's laws for distant planets

We note in passing that one can use general relativity to deduce the Keplerian role of M without invoking the Newtonian limit: place a test particle in the body's equatorial plane $\theta = \pi/2$ at a radius $r \gg M$, and give it an initial velocity that lies in the equatorial plane. Then symmetry guarantees the particle remains in the equatorial plane: there is no way to prefer going toward north, $\theta < \pi/2$, or toward south, $\theta > \pi/2$. Furthermore, adjust the initial velocity so the particle remains always at a fixed radius. Then the only nonvanishing components $u^\alpha = dx^\alpha/d\tau$ of the particle's 4-velocity are $u^t = dt/d\tau$ and $u^\phi = d\phi/d\tau$. The particle's orbit is governed by the geodesic equation $\nabla_{\vec{u}} \vec{u} = 0$, where $\vec{u}$ is its 4-velocity. The radial component of this geodesic equation, computed in Schwarzschild coordinates, is [cf. Eq. (25.14) with a switch from affine parameter ζ to proper time $\tau = m\zeta$]

relativistic analysis of Keplerian orbital motion

$$\frac{d^2 r}{d\tau^2} = -\Gamma^r{}_{\mu\nu} \frac{dx^\mu}{d\tau} \frac{dx^\nu}{d\tau} = -\Gamma^r{}_{tt} \frac{dt}{d\tau} \frac{dt}{d\tau} - \Gamma^r{}_{\phi\phi} \frac{d\phi}{d\tau} \frac{d\phi}{d\tau}. \quad (26.10)$$

(Here we have used the vanishing of all $dx^\alpha/d\tau$ except the t and ϕ components and have used the vanishing of $\Gamma^r{}_{t\phi} = \Gamma^r{}_{\phi t}$ [Eq. (3) of Box 26.2].) Since the orbit is circular, with fixed r, the left-hand side of Eq. (26.10) must vanish; correspondingly, the right-hand side gives

$$\frac{d\phi}{dt} = \frac{d\phi/d\tau}{dt/d\tau} = \left(-\frac{\Gamma^r{}_{tt}}{\Gamma^r{}_{\phi\phi}}\right)^{\frac{1}{2}} = \left(\frac{M}{r^3}\right)^{\frac{1}{2}}, \quad (26.11)$$

where we have used the values of the connection coefficients from Eq. (3) of Box 26.2, specialized to the equatorial plane $\theta = \pi/2$. Equation (26.11) tells us that the amount of coordinate time t required for the particle to circle the central body once, $0 \leq \phi \leq 2\pi$, is $\Delta t = 2\pi (r^3/M)^{\frac{1}{2}}$. Since t is the Lorentz time of the body's asymptotic rest frame, observers in the asymptotic rest frame will measure for the particle an orbital period $P = \Delta t = 2\pi (r^3/M)^{\frac{1}{2}}$. This, of course, is the same result as we obtained from the Newtonian limit—but our relativistic analysis shows it to be true for circular orbits of arbitrary radius r, not just for $r \gg M$.

26.2.4
26.2.4 Schwarzschild Spacetime at $r \sim M$

Next we move inward, from the asymptotically flat region of Schwarzschild spacetime toward smaller and smaller radii. As we do so, the spacetime geometry becomes more and more strongly curved, and the Schwarzschild coordinate system becomes less and less Lorentz. As an indication of extreme deviations from Lorentz, notice that the signs of the metric coefficients,

$$\frac{\partial}{\partial t} \cdot \frac{\partial}{\partial t} = g_{tt} = -\left(1 - \frac{2M}{r}\right), \quad \frac{\partial}{\partial r} \cdot \frac{\partial}{\partial r} = g_{rr} = \frac{1}{(1 - 2M/r)} \quad (26.12)$$

outside $r = 2M$, t is a time coordinate and r a space coordinate; inside, their roles are reversed

reverse as one moves from $r > 2M$ through $r = 2M$ and into the region $r < 2M$. Correspondingly, outside $r = 2M$, world lines of changing t but constant $\{r, \theta, \phi\}$ are timelike, while inside $r = 2M$, those world lines are spacelike. Similarly outside $r = 2M$, world lines of changing r but constant $\{t, \theta, \phi\}$ are spacelike, while inside they are timelike. In this sense, outside $r = 2M$, t plays the role of a time coordinate and r the role of a space coordinate; while inside $r = 2M$, t plays the role of a space coordinate and r the role of a time coordinate. Moreover, this role reversal occurs without any change in the role of r as $1/(2\pi)$ times the circumference of circles around the center [Eq. (26.3)].

For many decades this role reversal presented severe conceptual problems, even to the best experts in general relativity. We return to it in Sec. 26.4. Henceforth we refer to the location of role reversal, $r = 2M$, as the gravitational radius of the Schwarzschild spacetime, henceforth *gravitational radius*. It is also known as the *Schwarzschild radius* and, as we shall see, is the location of an absolute *event horizon*. In Sec. 26.4, we seek a clear understanding of the "interior" region, $r < 2M$; but until then, we confine attention to the region $r > 2M$, outside the gravitational radius.

Notice that the metric coefficients in the Schwarzschild line element (26.1) are all independent of the coordinate t. This means that the geometry of spacetime itself is invariant under the translation $t \to t + $ const. At radii $r > 2M$, where t plays the role of a time coordinate, $t \to t + $ const is a time translation; correspondingly, the Schwarzschild spacetime geometry is time-translation-invariant (i.e., "static") outside the gravitational radius.

the spacetime geometry is static outside $r = 2M$

EXERCISES

Exercise 26.2 *Example: The Bertotti-Robinson Solution of the Einstein Field Equation*
Bruno Bertotti (1959) and Ivor Robinson (1959) independently solved the Einstein field equation to obtain the following metric for a universe endowed with a uniform magnetic field:

$$ds^2 = Q^2(-dt^2 + \sin^2 t \, dz^2 + d\theta^2 + \sin^2\theta \, d\phi^2). \quad (26.13)$$

Here

$$Q = \text{const}, \quad 0 \leq t \leq \pi, \quad -\infty < z < +\infty, \quad 0 \leq \theta \leq \pi, \quad 0 \leq \phi \leq 2\pi. \quad (26.14)$$

If one computes the Einstein tensor from the metric coefficients of the line element (26.13) and equates it to 8π times a stress-energy tensor, one finds a stress-energy tensor that is precisely the same as for an electromagnetic field [Eqs. (2.75) and (2.80)] lifted, unchanged, into general relativity. The electromagnetic field is one that, as measured in the local Lorentz frame of an observer with fixed $\{z, \theta, \phi\}$ (a "static" observer), has vanishing electric field and has a magnetic field directed along $\partial/\partial z$ with magnitude independent of where the observer is located in spacetime. In this sense, the spacetime metric (26.13) is that of a homogeneous magnetic universe. Discuss the geometry of this universe and the nature of the coordinates $\{t, z, \theta, \phi\}$. More specifically, do the following.

(a) Which coordinate increases in a timelike direction and which coordinates in spacelike directions?

(b) Is this universe spherically symmetric?

(c) Is this universe cylindrically symmetric?

(d) Is this universe asymptotically flat?

(e) How does the geometry of this universe change as t ranges from 0 to π? [Hint: Show that the curves $\{(z, \theta, \phi) = \text{const}, t = \tau/Q\}$ are timelike geodesics—the world lines of the static observers referred to above. Then argue from symmetry, or use the result of Ex. 25.4a.]

(f) Give as complete a characterization as you can of the coordinates $\{t, z, \theta, \phi\}$.

26.3 Static Stars

26.3.1 Birkhoff's Theorem

Birkhoff's theorem: uniqueness of Schwarzschild solution

In 1923, George Birkhoff, a professor of mathematics at Harvard, proved a remarkable theorem (Birkhoff, 1923). (For a textbook proof, see Misner, Thorne, and Wheeler, 1973, Sec. 32.2.) The Schwarzschild spacetime geometry is the unique spherically symmetric solution of the vacuum Einstein field equation $\mathbf{G} = 0$. This Birkhoff theorem can be restated in more operational terms as follows. Suppose that you find a solution of the vacuum Einstein field equation, written as a set of metric coefficients $g_{\bar{\alpha}\bar{\beta}}$ in some coordinate system $\{x^{\bar{\mu}}\}$. Suppose, further, that these $g_{\bar{\alpha}\bar{\beta}}(x^{\bar{\mu}})$ coefficients exhibit spherical symmetry but do not coincide with the Schwarzschild expressions [Eqs. (2a) of Box 26.2]. Then the Birkhoff theorem guarantees the existence of a coordinate transformation from your coordinates $x^{\bar{\mu}}$ to Schwarzschild's coordinates x^{ν} such that, when that transformation is performed, the resulting new metric components $g_{\alpha\beta}(x^{\nu})$ have precisely the Schwarzschild form [Eqs. (2a) of Box 26.2]. For an example, see Ex. 26.3. This implies that, thought of as a coordinate-independent spacetime geometry, the Schwarzschild solution is completely unique.

Now consider a static, spherically symmetric star (e.g., the Sun) residing alone in an otherwise empty universe (or, more realistically, residing in our own universe but so far from other gravitating matter that we can ignore all other sources of gravity when studying it). Since the star's interior is spherical, it is reasonable to presume that the exterior will be spherical; since the exterior is also vacuum ($\mathbf{T} = 0$), its spacetime geometry must be that of Schwarzschild. If the circumference of the star's surface is $2\pi R$ and its surface area is $4\pi R^2$, then that surface must reside at the location $r = R$ in the Schwarzschild coordinates of the exterior. In other words, the spacetime geometry will be described by the Schwarzschild line element (26.1) at radii $r > R$, but by something else inside the star, at $r < R$.

Since real atoms with finite rest masses reside on the star's surface, and since such atoms move along timelike world lines, it must be that the world lines $\{r = R, \theta = \text{const}, \phi = \text{const}, t \text{ varying}\}$ are timelike. From the Schwarzschild invariant interval (26.1) we read off the squared proper time, $d\tau^2 = -ds^2 = (1 - 2M/R)dt^2$, along those world lines. This $d\tau^2$ is positive (timelike world line) if and only if $R > 2M$. Thus, a static star with total mass-energy (active gravitational mass) M can never have a circumference smaller than $2\pi R = 4\pi M$. Restated in conventional units:

a static star must have radius R greater than $2M$

$$\frac{\text{circumference}}{2\pi} = R \equiv \begin{pmatrix} \text{radius} \\ \text{of star} \end{pmatrix} > 2M = \frac{2GM}{c^2}$$

$$= 2.953 \text{ km} \left(\frac{M}{M_\odot}\right) \equiv \begin{pmatrix} \text{gravitational} \\ \text{radius} \end{pmatrix}. \quad (26.15)$$

Here $M_\odot$ is the mass of the Sun. The Sun satisfies this constraint by a huge margin: $R = 7 \times 10^5$ km $\gg 2.953$ km. A 1-solar-mass white-dwarf star satisfies it by a smaller margin: $R \simeq 6 \times 10^3$ km. And a 1-solar-mass neutron star satisfies it by only a modest margin: $R \simeq 10$ km. For a pedagogical and detailed discussion see, for example, Shapiro and Teukolsky (1983).

EXERCISES

Exercise 26.3 *Problem: Schwarzschild Geometry in Isotropic Coordinates*

(a) It turns out that the following line element is a solution of the vacuum Einstein field equation $\mathbf{G} = 0$:

$$ds^2 = -\left(\frac{1 - M/(2\bar{r})}{1 + M/(2\bar{r})}\right)^2 dt^2 + \left(1 + \frac{M}{2\bar{r}}\right)^4 [d\bar{r}^2 + \bar{r}^2(d\theta^2 + \sin^2\theta d\phi^2)].$$

(26.16)

Since this solution is spherically symmetric, Birkhoff's theorem guarantees it must represent the standard Schwarzschild spacetime geometry in a coordinate system that differs from Schwarzschild's. Show that this is so by exhibiting a coordinate transformation that converts this line element into Eq. (26.1). [Note: The $\{t, \bar{r}, \theta, \phi\}$ coordinates are called *isotropic*, because in them the spatial part of the line element is a function of $\bar{r}$ times the 3-dimensional Euclidean line element, and Euclidean geometry picks out at each point in space no preferred spatial directions (i.e., it is isotropic).]

(b) Show that at large radii $r \gg M$, the line element (26.16) takes the form (25.98c) discussed in Chap. 25, but with vanishing spin angular momentum $\mathbf{J} = 0$.

Exercise 26.4 **Example: Gravitational Redshift of Light from a Star's Surface*
Consider a photon emitted by an atom at rest on the surface of a static star with mass M and radius R. Analyze the photon's motion in the Schwarzschild coordinate system of the star's exterior, $r \geq R > 2M$. In particular, compute the "gravitational redshift" of the photon by the following steps.

(a) Since the emitting atom is nearly an ideal clock, it gives the emitted photon nearly the same frequency ν_{em}, as measured in the emitting atom's proper reference frame (as it would give were it in an Earth laboratory or floating in free space). Thus the proper reference frame of the emitting atom is central to a discussion of the photon's properties and behavior. Show that the orthonormal basis vectors of that proper reference frame are

$$\vec{e}_{\hat{0}} = \frac{1}{\sqrt{1 - 2M/r}} \frac{\partial}{\partial t}, \quad \vec{e}_{\hat{r}} = \sqrt{1 - 2M/r} \frac{\partial}{\partial r}, \quad \vec{e}_{\hat{\theta}} = \frac{1}{r} \frac{\partial}{\partial \theta}, \quad \vec{e}_{\hat{\phi}} = \frac{1}{r \sin\theta} \frac{\partial}{\partial \phi},$$

(26.17)

with $r = R$ (the star's radius).

(b) Explain why the photon's energy as measured in the emitter's proper reference frame is $\mathcal{E} = h\nu_{em} = -p_{\hat{0}} = -\vec{p} \cdot \vec{e}_{\hat{0}}$. (Here and below h is Planck's constant, and $\vec{p}$ is the photon's 4-momentum.)

(c) Show that the quantity $\mathcal{E}_\infty \equiv -p_t = -\vec{p} \cdot \partial/\partial t$ is conserved as the photon travels outward from the emitting atom to an observer at very large radius, which we idealize as $r = \infty$. [Hint: Recall the result of Ex. 25.4a.] Show, further, that $\mathcal{E}_\infty$ is the photon's energy, as measured by the observer at $r = \infty$—which is why it is called the photon's "energy-at-infinity" and denoted $\mathcal{E}_\infty$. The photon's frequency, as measured by that observer, is given, of course, by $h\nu_\infty = \mathcal{E}_\infty$.

(d) Show that $\mathcal{E}_\infty = \mathcal{E}\sqrt{1 - 2M/R}$ and thence that $\nu_\infty = \nu_{\text{em}}\sqrt{1 - 2M/R}$, and that therefore the photon is redshifted by an amount

$$\boxed{\frac{\lambda_{\text{rec}} - \lambda_{\text{em}}}{\lambda_{\text{em}}} = \frac{1}{\sqrt{1 - 2M/R}} - 1.} \tag{26.18}$$

Here λ_{rec} is the wavelength that the photon's spectral line exhibits at the receiver, and λ_{em} is the wavelength that the emitting kind of atom would produce in an Earth laboratory. Note that for a nearly Newtonian star (i.e., one with $R \gg M$), this redshift becomes $\simeq M/R = GM/Rc^2$.

(e) Evaluate this redshift for Earth, for the Sun, and for a 1.4-solar-mass, 10-km-radius neutron star.

26.3.2 Stellar Interior

We now take a temporary detour from our study of the Schwarzschild geometry to discuss the interior of a static, spherical star. We do so less because of an interest in stars than because the detour will illustrate the process of solving the Einstein field equation and the role of the contracted Bianchi identity in the solution process.

Since the star's spacetime geometry is to be static and spherically symmetric, we can introduce as coordinates in its interior: (i) spherical polar angular coordinates θ and ϕ, (ii) a radial coordinate r such that the circumferences of the spheres are $2\pi r$, and (iii) a time coordinate $\bar{t}$ such that the metric coefficients are independent of $\bar{t}$. By their geometrical definitions, these coordinates will produce a spacetime line element of the form

$$ds^2 = g_{\bar{t}\bar{t}}d\bar{t}^2 + 2g_{\bar{t}r}d\bar{t}dr + g_{rr}dr^2 + r^2(d\theta^2 + \sin^2\theta d\phi^2), \tag{26.19}$$

with $g_{\alpha\beta}$ independent of $\bar{t}$, θ, and ϕ. Metric coefficients $g_{\bar{t}\theta}$, $g_{r\theta}$, $g_{\bar{t}\phi}$, and $g_{r\phi}$ are absent from Eq. (26.19), because they would break the spherical symmetry: they would distinguish the $+\phi$ direction from $-\phi$ or $+\theta$ from $-\theta$, since they would give nonzero values for the scalar products of $\partial/\partial\phi$ or $\partial/\partial\theta$ with $\partial/\partial t$ or $\partial/\partial r$. [Recall that the metric coefficients in a coordinate basis are $g_{\alpha\beta} = \mathbf{g}(\partial/\partial x^\alpha, \partial/\partial x^\beta) = (\partial/\partial x^\alpha) \cdot (\partial/\partial x^\beta)$.] We can get rid of the off-diagonal $g_{\bar{t}r}$ term in the line element (26.19) by specializing the time coordinate. The coordinate transformation

$$\bar{t} = t - \int \left(\frac{g_{\bar{t}r}}{g_{\bar{t}\bar{t}}}\right) dr \tag{26.20}$$

brings the line element into the form

$$ds^2 = -e^{2\Phi}dt^2 + e^{2\Lambda}dr^2 + r^2(d\theta^2 + \sin^2\theta d\phi^2).\tag{26.21}$$

coordinates and line element for interior of a static star

Here, after the transformation (26.20), we have introduced the names $e^{2\Phi}$ and $e^{2\Lambda}$ for the time-time and radial-radial metric coefficients, respectively. The signs of these coefficients (negative for g_{tt} and positive for g_{rr}) are dictated by the fact that inside the star, as on its surface, real atoms move along world lines of constant $\{r, \theta, \phi\}$ and changing t, and thus those world lines must be timelike. The name $e^{2\Phi}$ is chosen because when gravity is nearly Newtonian, the time-time metric coefficient $-e^{2\Phi}$ must reduce to $-(1 + 2\Phi)$, with Φ the Newtonian potential [Eq. (25.79)]. Thus, the Φ used in Eq. (26.21) is a generalization of the Newtonian potential to relativistic, spherical, static gravitational situations.

To solve the Einstein field equation for the star's interior, we must specify the stress-energy tensor. Stellar material is excellently approximated by a perfect fluid, and since our star is static, at any point inside the star the fluid's rest frame has constant $\{r, \theta, \phi\}$. Correspondingly, the 4-velocity of the fluid is

$$\vec{u} = e^{-\Phi}\frac{\partial}{\partial t}.\tag{26.22}$$

fluid 4-velocity inside static star

Here the factor $e^{-\Phi}$ guarantees that the 4-velocity will have $\vec{u}^2 = -1$, as it must.

Of course, this fluid is not freely falling. Rather, for a fluid element to remain always at fixed $\{r, \theta, \phi\}$ it must accelerate relative to local freely falling observers with a 4-acceleration $\vec{a} \equiv \nabla_{\vec{u}} \vec{u} \neq 0$ (i.e., $a^\alpha = u^\alpha{}_{;\mu}u^\mu \neq 0$). Symmetry tells us that this 4-acceleration cannot have any θ or ϕ components, and orthogonality of the 4-acceleration to the 4-velocity tells us that it cannot have any t component. The r component, computed from $a^r = u^r{}_{;\mu}u^\mu = \Gamma^r{}_{00}u^0 u^0$, is $a^r = e^{-2\Lambda}\Phi_{,r}$; and thus we have

$$\vec{a} = e^{-2\Lambda}\Phi_{,r}\frac{\partial}{\partial r}.\tag{26.23}$$

fluid 4-acceleration inside static star

Each fluid element can be thought of as carrying an orthonormal set of basis vectors:

$$\vec{e}_{\hat{0}} = \vec{u} = e^{-\Phi}\frac{\partial}{\partial t}, \quad \vec{e}_{\hat{r}} = e^{-\Lambda}\frac{\partial}{\partial r}, \quad \vec{e}_{\hat{\theta}} = \frac{1}{r}\frac{\partial}{\partial \theta}, \quad \vec{e}_{\hat{\phi}} = \frac{1}{r\sin\theta}\frac{\partial}{\partial \phi};\tag{26.24a}$$

$$\vec{e}^{\hat{0}} = e^{\Phi}\vec{\nabla}t, \quad \vec{e}^{\hat{r}} = e^{\Lambda}\vec{\nabla}r, \quad \vec{e}^{\hat{\theta}} = r\vec{\nabla}\theta, \quad \vec{e}^{\hat{\phi}} = r\sin\theta\vec{\nabla}\phi.\tag{26.24b}$$

orthonormal basis vectors of fluid's local rest frame (its proper reference frame)

These basis vectors play two independent roles. (i) One can regard the tangent space of each event in spacetime as being spanned by the basis (26.24), specialized to that event. From this viewpoint, Eqs. (26.24) constitute an orthonormal, noncoordinate basis that covers every tangent space of the star's spacetime. This basis is called the

26.3 Static Stars

fluid's *orthonormal, local-rest-frame basis*. (ii) One can focus attention on a specific fluid element, which moves along the world line $r = r_o$, $\theta = \theta_o$, $\phi = \phi_o$; and one can construct the proper reference frame of that fluid element in the same manner as we constructed the proper reference frame of an accelerated observer in flat spacetime in Sec. 24.5. That proper reference frame is a coordinate system $\{x^{\hat\alpha}\}$ whose basis vectors on the fluid element's world line are equal to the basis vectors (26.24):

$$\frac{\partial}{\partial x^{\hat\mu}} = \vec{e}_{\hat\mu}, \quad \vec{\nabla} x^{\hat\mu} = \vec{e}^{\hat\mu} \text{ at } x^{\hat j} = 0, \quad \text{with } \hat 1 = \hat r, \hat 2 = \hat\theta, \hat 3 = \hat\phi. \tag{26.25a}$$

More specifically, the proper-reference-frame coordinates $x^{\hat\mu}$ are given, to second-order in spatial distance from the fluid element's world line, by

coordinates of fluid element's proper reference frame

$$x^{\hat 0} = e^{\Phi_o} t, \quad x^{\hat 1} = \int_{r_o}^{r} e^{\Lambda} dr - \frac{1}{2} e^{-\Lambda_o} r_o [(\theta - \theta_o)^2 + \sin^2\theta_o (\phi - \phi_o)^2],$$

$$x^{\hat 2} = r(\theta - \theta_o) - \frac{1}{2} r_o \sin\theta_o \cos\theta_o (\phi - \phi_o)^2, \quad x^{\hat 3} = r \sin\theta (\phi - \phi_o), \tag{26.25b}$$

from which one can verify relation (26.25a) with the basis vectors given by Eqs. (26.24). [In Eqs. (26.25b) and throughout this discussion all quantities with subscripts $_o$ are evaluated on the fluid's world line.] In terms of the proper-reference-frame coordinates (26.25b), the line element (26.21) takes the following form, accurate to first order in distance from the fluid element's world line:

$$ds^2 = -[1 + 2\Phi_{,r}(r - r_o)](dx^{\hat 0})^2 + \delta_{ij} dx^{\hat i} dx^{\hat j}. \tag{26.25c}$$

Notice that the quantity $\Phi_{,r}(r - r_o)$ is equal to the scalar product of (i) the spatial separation $\hat{\mathbf{x}} \equiv (r - r_o)\partial/\partial r + (\theta - \theta_o)\partial/\partial\theta + (\phi - \phi_o)\partial/\partial\phi$ of the "field point" (r, θ, ϕ) from the fluid element's world line, with (ii) the fluid's 4-acceleration (26.23), viewed as a spatial 3-vector $\mathbf{a} = e^{-2\Lambda_o} \Phi_{,r} \partial/\partial r$. Correspondingly, the spacetime line element (26.25c) in the fluid element's proper reference frame takes the standard proper-reference-frame form (24.60b):

$$ds^2 = -(1 + 2\mathbf{a} \cdot \hat{\mathbf{x}})(dx^{\hat 0})^2 + \delta_{jk} dx^{\hat j} dx^{\hat k}, \tag{26.26}$$

accurate to first order in distance from the fluid element's world line. To second order, as discussed at the end of Sec. 25.3, there are corrections proportional to the spacetime curvature.

In the local rest frame of the fluid [i.e., when expanded on the fluid's orthonormal rest-frame basis vectors (26.24) or equally well (26.25a)], the components $T^{\hat\alpha\hat\beta} = (\rho + P) u^{\hat\alpha} u^{\hat\beta} + P g^{\hat\alpha\hat\beta}$ of the fluid's stress-energy tensor take on the standard form [Eq. (24.50)]:

stress-energy tensor in fluid's proper reference frame: fluid density and pressure

$$\boxed{T^{\hat 0 \hat 0} = \rho, \quad T^{\hat r \hat r} = T^{\hat\theta\hat\theta} = T^{\hat\phi\hat\phi} = P,} \tag{26.27}$$

corresponding to a rest-frame mass-energy density ρ and isotropic pressure P. By contrast with the simplicity of these local-rest-frame components, the contravariant components $T^{\alpha\beta} = (\rho + P)u^\alpha u^\beta + Pg^{\alpha\beta}$ in the $\{t, r, \theta, \phi\}$ coordinate basis are rather more complicated:

$$T^{tt} = e^{-2\Phi}\rho, \quad T^{rr} = e^{-2\Lambda}P, \quad T^{\theta\theta} = r^{-2}P, \quad T^{\phi\phi} = (r\sin\theta)^{-2}P. \quad (26.28)$$

This shows one advantage of using orthonormal bases: the components of vectors and tensors are generally simpler in an orthonormal basis than in a coordinate basis. A second advantage occurs when one seeks the physical interpretation of formulas. Because every orthonormal basis is the proper-reference-frame basis of some local observer (the observer with 4-velocity $\vec{u} = \vec{e}_{\hat{0}}$), components measured in such a basis have an immediate physical interpretation in terms of measurements by that observer. For example, $T^{\hat{0}\hat{0}}$ is the total density of mass-energy measured by the local observer. By contrast, components in a coordinate basis typically do not have a simple physical interpretation.

EXERCISES

Exercise 26.5 *Derivation: Proper-Reference-Frame Coordinates*
Show that in the coordinate system $\{x^{\hat{0}}, x^{\hat{1}}, x^{\hat{2}}, x^{\hat{3}}\}$ of Eqs. (26.25b), the coordinate basis vectors at $x^{\hat{j}} = 0$ are Eqs. (26.24), and, accurate through first order in distance from $x^{\hat{j}} = 0$, the spacetime line element is Eq. (26.26); that is, errors are no larger than second order.

26.3.3 Local Conservation of Energy and Momentum

Before inserting the perfect-fluid stress-energy tensor (26.27) into the Einstein field equation, we impose on it the local law of conservation of 4-momentum: $\vec{\nabla} \cdot \mathbf{T} = 0$. In doing so we require from the outset that, since the star is to be static and spherical, its density ρ and pressure P must be independent of t, θ, and ϕ (i.e., like the metric coefficients Φ and Λ, they must be functions of radius r only).

The most straightforward way to impose 4-momentum conservation is to equate to zero the quantities

$$T^{\alpha\beta}{}_{;\beta} = \frac{\partial T^{\alpha\beta}}{\partial x^\beta} + \Gamma^\beta{}_{\mu\beta}T^{\alpha\mu} + \Gamma^\alpha{}_{\mu\beta}T^{\mu\beta} = 0 \quad (26.29)$$

in our coordinate basis, making use of expressions (26.28) for the contravariant components of the stress-energy tensor, and the connection coefficients and metric components given in Box 26.2.

This straightforward calculation requires a lot of work. Much better is an analysis based on the local proper reference frame of the fluid. The temporal component of $\vec{\nabla} \cdot \mathbf{T} = 0$ in that reference frame [i.e., the projection $\vec{u} \cdot (\vec{\nabla} \cdot \mathbf{T}) = 0$ of this conservation law onto the time basis vector $\vec{e}_{\hat{0}} = e^{-\Phi}\partial/\partial t = \vec{u}$] represents energy conservation as seen by the fluid—the first law of thermodynamics:

first law of thermodynamics

$$\frac{d(\rho V)}{d\tau} = -P\frac{dV}{d\tau}. \tag{26.30}$$

Here τ is proper time as measured by the fluid element we are following, and V is the fluid element's volume. (This equation is derived in Ex. 2.26b, in a special relativistic context; but since it involves only one derivative, there is no danger of curvature coupling, so that derivation and the result can be lifted without change into general relativity, i.e., into the star's curved spacetime; cf. Ex. 26.6a.) Now, inside this static star, the fluid element sees and feels no changes. Its density ρ, pressure P, and volume V always remain constant along the fluid element's world line, and energy conservation is therefore guaranteed to be satisfied. Equation (26.30) tells us nothing new.

The spatial part of $\vec{\nabla} \cdot \mathbf{T} = 0$ in the fluid's local rest frame can be written in geometric form as $\mathbf{P} \cdot (\vec{\nabla} \cdot \mathbf{T}) = 0$. Here $\mathbf{P} \equiv \mathbf{g} + \vec{u} \otimes \vec{u}$ is the tensor that projects all vectors into the 3-surface orthogonal to $\vec{u}$, that is, into the fluid's local 3-surface of simultaneity (Exs. 2.10 and 25.1b). By inserting the perfect-fluid stress-energy tensor $\mathbf{T} = (\rho + P)\vec{u} \otimes \vec{u} + P\mathbf{g} = \rho\vec{u} \otimes \vec{u} + P\mathbf{P}$ into $\mathbf{P} \cdot (\vec{\nabla} \cdot \mathbf{T}) = 0$, reexpressing the result in slot-naming index notation, and carrying out some index gymnastics, we must obtain the same result as in special relativity (Ex. 2.26c):

force balance inside the fluid

$$(\rho + P)\vec{a} = -\mathbf{P} \cdot \vec{\nabla} P \tag{26.31}$$

(cf. Ex. 26.6b). Here $\vec{a}$ is the fluid's 4-velocity. Recall from Ex. 2.27 that for a perfect fluid, $\rho + P$ is the inertial mass per unit volume. Therefore, Eq. (26.31) says that the fluid's inertial mass per unit volume times its 4-acceleration is equal to the negative of its pressure gradient, projected orthogonally to its 4-velocity. Since both sides of Eq. (26.31) are purely spatially directed as seen in the fluid's local proper reference frame, we can rewrite this equation in 3-dimensional language as

$$\boxed{(\rho + P)\mathbf{a} = -\nabla P.} \tag{26.32}$$

A Newtonian physicist, in the proper reference frame, would identify $-\mathbf{a}$ as the local gravitational acceleration, $\mathbf{g}$, and correspondingly, would rewrite Eq. (26.31) as

Newtonian viewpoint on force balance

$$\nabla P = (\rho + P)\mathbf{g}. \tag{26.33}$$

This is the standard equation of hydrostatic equilibrium for a fluid in an Earthbound laboratory (or swimming pool or lake or ocean), except for the presence of the pressure P in the inertial mass per unit volume (Ex. 2.27). On Earth the typical pressures of fluids, even deep in the ocean, are only $P \lesssim 10^9$ dyne/cm$^2 \simeq 10^{-12}$ g cm$^{-3} \lesssim 10^{-12}\rho$. Thus, to extremely good accuracy one can ignore the contribution of pressure to the inertial mass density. However, deep inside a neutron star, P may be within a factor 2 of ρ, so the contribution of P cannot be ignored.

We can convert the law of force balance (26.31) into an ordinary differential equation for the pressure P by evaluating its components in the fluid's proper reference frame. The 4-acceleration (26.23) is purely radial; its radial component is $a^{\hat{r}} = e^{-\Lambda}\Phi_{,r} = \Phi_{,\hat{r}}$. The gradient of the pressure is also purely radial, and its radial component is $P_{;\hat{r}} = P_{,\hat{r}} = e^{-\Lambda}P_{,r}$. Therefore, the law of force balance reduces to

$$\frac{dP}{dr} = -(\rho + P)\frac{d\Phi}{dr}. \tag{26.34}$$

force balance rewritten

EXERCISES

Exercise 26.6 *Practice and Derivation: Local Conservation of Energy and Momentum for a Perfect Fluid*

(a) Use index manipulations to show that in general (not just inside a static star), for a perfect fluid with $T^{\alpha\beta} = (\rho + P)u^\alpha u^\beta + Pg^{\alpha\beta}$, the law of energy conservation $u_\alpha T^{\alpha\beta}{}_{;\beta} = 0$ reduces to the first law of thermodynamics (26.30). [Hint: You will need the relation $u^\mu{}_{;\mu} = (1/V)(dV/d\tau)$; cf. Ex. 2.24.]

(b) Similarly, show that $P_{\mu\alpha}T^{\alpha\beta}{}_{;\beta} = 0$ reduces to the force-balance law (26.31).

26.3.4 The Einstein Field Equation

Turn, now, to the Einstein field equation inside a static, spherical star with isotropic pressure. To impose it, we must first compute, in our $\{t, r, \theta, \phi\}$ coordinate system, the components of the Einstein tensor $G_{\alpha\beta}$. In general, the Einstein tensor has 10 independent components. However, the symmetries of the line element (26.21) impose identical symmetries on the Einstein tensor computed from it: The only nonzero components in the fluid's proper reference frame will be $G^{\hat{0}\hat{0}}$, $G^{\hat{r}\hat{r}}$, and $G^{\hat{\theta}\hat{\theta}} = G^{\hat{\phi}\hat{\phi}}$; and these three independent components will be functions of radius r only. Correspondingly, the Einstein equation will produce three independent differential equations for our four unknowns: the metric coefficients ("gravitational potentials") Φ and Λ [Eq. (26.21)], and the radial distribution of density ρ and pressure P.

These three independent components of the Einstein equation will actually be redundant with the law of hydrostatic equilibrium (26.34). One can see this as follows. If we had not yet imposed the law of 4-momentum conservation, then the Einstein equation $\mathbf{G} = 8\pi \mathbf{T}$, together with the Bianchi identity $\vec{\nabla} \cdot \mathbf{G} \equiv 0$ [Eq. (25.69)], would enforce $\vec{\nabla} \cdot \mathbf{T} = 0$. More explicitly, our three independent components of the Einstein equation together would imply the law of radial force balance [i.e., of hydrostatic equilibrium (26.34)]. Since we have already imposed Eq. (26.34), we need evaluate only two of the three independent components of the Einstein equation; they will give us full information.

with force balance imposed, Einstein's equation provides just two additional constraints on the stellar structure

A long and rather tedious calculation (best done on a computer), based on the metric coefficients of Eq. (26.21) and on Eqs. (24.38), (25.50), (25.46), (25.49), and (25.68), produces the following for the time-time and radial-radial components of the Einstein tensor, and thence of the Einstein field equation:

$$G^{\hat{0}\hat{0}} = \frac{1}{r^2}\frac{d}{dr}[r(1-e^{-2\Lambda})] = 8\pi T^{\hat{0}\hat{0}} = 8\pi\rho, \tag{26.35}$$

$$G^{\hat{r}\hat{r}} = -\frac{1}{r^2}(1-e^{-2\Lambda}) + \frac{2}{r}e^{-2\Lambda}\frac{d\Phi}{dr} = 8\pi T^{\hat{r}\hat{r}} = 8\pi P. \tag{26.36}$$

We can bring these components of the field equation into simpler form by defining a new metric coefficient $m(r)$ by

$$\boxed{e^{2\Lambda} \equiv \frac{1}{1-2m/r}.} \tag{26.37}$$

Note [cf. Eqs. (26.1), (26.21), and (26.37)] that outside the star, m is equal to the star's total mass-energy M. This, plus the fact that in terms of m the time-time component of the field equation (26.35) takes the form

Einstein equation for the mass inside radius r, $m(r)$

$$\boxed{\frac{dm}{dr} = 4\pi r^2 \rho,} \tag{26.38a}$$

motivates the name *mass inside radius r* for the quantity $m(r)$. In terms of m the radial-radial component (26.36) of the field equation becomes

Einstein equation for $\Phi(r)$

$$\boxed{\frac{d\Phi}{dr} = \frac{m + 4\pi r^3 P}{r(r-2m)};} \tag{26.38b}$$

combining this with Eq. (26.34), we obtain an alternative form of the equation of hydrostatic equilibrium:

TOV equation of hydrostatic equilibrium

$$\boxed{\frac{dP}{dr} = -\frac{(\rho+P)(m+4\pi r^3 P)}{r(r-2m)}.} \tag{26.38c}$$

[This form is called the Tolman-Oppenheimer-Volkoff (or TOV) equation, because it was first derived by Tolman (1939) and first used in a practical calculation by Oppenheimer and Volkoff (1939).] Equations (26.38a), (26.38b), (26.38c) plus an equation of state for the pressure of the stellar material P in terms of its density of total mass-energy ρ,

$$\boxed{P = P(\rho),} \tag{26.38d}$$

summary: the relativistic equations of stellar structure, Eqs. (26.38)

determine the four quantities Φ, m, ρ, and P as functions of radius. In other words, Eqs. (26.38) are the relativistic equations of stellar structure.

Actually, for full determination, one also needs boundary conditions. Just as the surface of a sphere is everywhere locally Euclidean (i.e., is arbitrarily close to Euclidean in arbitrarily small regions), so also spacetime must be everywhere locally Lorentz; cf. Eqs. (25.9). For spacetime to be locally Lorentz at the star's center (in particular, for circumferences of tiny circles around the center to be equal to 2π times their radii), it is necessary that m vanish at the center:

$$m = 0 \text{ at } r = 0, \quad \text{and thus} \quad \boxed{m(r) = \int_0^r 4\pi r^2 \rho \, dr} \qquad (26.39)$$

boundary conditions for relativistic equations of stellar structure

[cf. Eqs. (26.21) and (26.37)]. At the star's surface the interior spacetime geometry (26.21) must join smoothly to the exterior Schwarzschild geometry (26.1):

$$\boxed{m = M \quad \text{and} \quad e^{2\Phi} = 1 - 2M/r \text{ at } r = R.} \qquad (26.40)$$

26.3.5 Stellar Models and Their Properties

A little thought now reveals a straightforward method of producing a relativistic stellar model.

procedure for constructing a relativistic stellar model

1. Specify an equation of state for the stellar material $P = P(\rho)$, and specify a central density ρ_c or central pressure P_c for the star.

2. Integrate the coupled hydrostatic-equilibrium equation (26.38c) and "mass equation" (26.38a) outward from the center, beginning with the initial conditions $m = 0$ and $P = P_c$ at the center.

3. Terminate the integration when the pressure falls to zero; this is the surface of the star.

4. At the surface read off the value of m; it is the star's total mass-energy M, which appears in the star's external, Schwarzschild line element (26.1).

5. From this M and the radius $r \equiv R$ of the star's surface, read off the value of the gravitational potential Φ at the surface [Eq. (26.40)].

6. Integrate the Einstein field equation (26.38b) inward from the surface toward the center to determine Φ as a function of radius inside the star.

Just 6 weeks after reading to the Prussian Academy of Science the letter in which Karl Schwarzschild derived his vacuum solution (26.1) of the field equation, Albert Einstein again presented the Academy with results from Schwarzschild's fertile mind: an exact solution for the structure of the interior of a star that has constant density ρ. [And just 4 months after that, on June 29, 1916, Einstein had the sad task of announcing to the Academy that Schwarzschild had died of an illness contracted on the World-War-I Russian front.]

In our notation, Schwarzschild's solution for the interior of a star is characterized by its uniform density ρ, its total mass M, and its radius R, which is given in terms of ρ and M by

details of a relativistic star with constant density ρ

$$M = \frac{4\pi}{3} \rho R^3 \qquad (26.41)$$

26.3 Static Stars

[Eq. (26.39)]. In terms of these quantities, the mass M inside radius r, the pressure P, and the gravitational potential Φ as functions of r are (Schwarzschild, 1916b)

$$m = \frac{4\pi}{3}\rho r^3, \quad P = \rho \left[\frac{(1 - 2Mr^2/R^3)^{\frac{1}{2}} - (1 - 2M/R)^{\frac{1}{2}}}{3(1 - 2M/R)^{\frac{1}{2}} - (1 - 2Mr^2/R^3)^{\frac{1}{2}}} \right], \quad (26.42\text{a})$$

$$e^\Phi = \frac{3}{2}\left(1 - \frac{2M}{R}\right)^{\frac{1}{2}} - \frac{1}{2}\left(1 - \frac{2Mr^2}{R^3}\right)^{\frac{1}{2}}. \quad (26.42\text{b})$$

We present these details less for their specific physical content than to illustrate the solution of the Einstein field equation in a realistic, astrophysically interesting situation. For discussions of the application of this formalism to neutron stars, where relativistic deviations from Newtonian theory can be rather strong, see, for example, Shapiro and Teukolsky (1983). For the seminal work on the theory of neutron-star structure, see Oppenheimer and Volkoff (1939).

Among the remarkable consequences of the TOV equation of hydrostatic equilibrium (26.38c) for neutron-star structure are the following. (i) If the mass m inside radius r ever gets close to $r/2$, the "gravitational pull" [right-hand side of (26.38c)] becomes divergently large, forcing the pressure gradient that counterbalances it to be divergently large, and thereby driving the pressure quickly to zero as one integrates outward. This protects the static star from having M greater than $R/2$ (i.e., from having its surface inside its gravitational radius). (ii) Although the density of matter near the center of a neutron star is above that of an atomic nucleus (2.3×10^{17} kg m^{-3}), where the equation of state is ill-understood, we can be confident that there is an upper limit on the masses of neutron stars, a limit in the range $2M_\odot \lesssim M_{\max} \lesssim 3M_\odot$.[2] This mass limit cannot be avoided by postulating that a more massive neutron star develops an arbitrarily large central pressure and thereby supports itself against gravitational implosion. The reason is that an arbitrarily large central pressure is self-defeating: The "gravitational pull" that appears on the right-hand side of Eq. (26.38c) is quadratic in the pressure at very high pressures (whereas it would be independent of pressure in Newtonian theory). This purely relativistic feature guarantees that, if a star develops too high a central pressure, it will be unable to support itself against the resulting "quadratically too high" gravitational pull.

upper limit on mass of a neutron star, and how it comes about

EXERCISES

Exercise 26.7 *Problem: Mass-Radius Relation for Neutron Stars*
The equation of state of a neutron star is very hard to calculate at the supra-nuclear densities required, because the calculation is a complex, many-body problem and the particle interactions are poorly understood and poorly measured. Observations of

---

2. Measured neutron star masses range from $\sim$1.2 solar masses to more than 2.0 solar masses.

neutron stars' masses and radii can therefore provide valuable constraints on fundamental nuclear physics. As we discuss briefly in the following chapter, various candidate equations of state can already be excluded on these observational grounds.

A necessary step for comparing observation with theory is to compute the stellar structure for candidate equations of state. We can illustrate the approach using a simple functional form, which, around nuclear density ($\rho_{\text{nuc}} \simeq 2.3 \times 10^{17}$ kg m^{-3}), is a fair approximation to some of the models:

$$P = 3 \times 10^{32} \left(\frac{\rho}{\rho_{\text{nuc}}}\right)^3 \text{ N m}^{-2}. \tag{26.43}$$

For this equation of state, use the equations of stellar structure (26.38a) and (26.38c) to find the masses and radii of stars with a range of central pressures, and hence deduce a mass-radius relation, $M(R)$. You should discover that, as the central pressure is increased, the mass passes through a maximum, while the radius continues to decrease. (Stars with radii smaller than that at the maximum mass are unstable to radial perturbations.)

26.3.6 Embedding Diagrams

We conclude our discussion of static stars by using them to illustrate a useful technique for visualizing the curvature of spacetime: the embedding of the curved spacetime, or a piece of it, in a flat space of higher dimensionality.

The geometry of a curved, n-dimensional manifold is characterized by $\frac{1}{2}n(n+1)$ metric components (since those components form a symmetric $n \times n$ matrix), of which only $\frac{1}{2}n(n+1) - n = \frac{1}{2}n(n-1)$ are of coordinate-independent significance (since we are free to choose arbitrarily the n coordinates of our coordinate system and can thereby force n of the metric components to take on any desired values, e.g., zero). If this n-dimensional manifold is embedded in a flat N-dimensional manifold, that embedding will be described by expressing $N - n$ of the embedding manifold's Euclidean (or Lorentz) coordinates in terms of the other n. Thus, the embedding is characterized by $N - n$ functions of n variables. For the embedding to be possible, in general, this number of choosable functions must be at least as large as the number of significant metric coefficients $\frac{1}{2}n(n-1)$. From this argument we conclude that the dimensionality of the embedding space must be $N \geq \frac{1}{2}n(n+1)$. Actually, this argument analyzes only the local features of the embedding. If one also wants to preserve the global topology of the n-dimensional manifold, one must in general go to an embedding space of even higher dimensionality.

the problem of embedding a curved manifold inside a flat higher-dimensional manifold (the embedding space); number of dimensions needed in the embedding space

Curved spacetime has $n = 4$ dimensions and thus requires for its local embedding a flat space with at least $N = \frac{1}{2}n(n+1) = 10$ dimensions. This is a bit much for 3-dimensional beings like us to visualize. If, as a sop to our visual limitations, we reduce our ambitions and seek only to extract a 3-surface from curved spacetime and visualize it by embedding it in a flat space, we will require a flat space of $N = 6$ dimensions. This is still a bit much. In frustration, we are driven to extract from spacetime $n = 2$ dimensional surfaces and visualize them by embedding in flat spaces with $N = 3$ dimensions. This is doable—and, indeed, instructive.

embedding the equatorial plane of a static, relativistic star in a flat 3-dimensional embedding space

As a nice example, consider the equatorial "plane" through the spacetime of a static spherical star, at a specific "moment" of coordinate time t [i.e., consider the 2-surface $t = \text{const}, \theta = \pi/2$ in the spacetime of Eqs. (26.21), (26.37)]. The line element on this equatorial 2-surface is

$$^{(2)}ds^2 = \frac{dr^2}{1 - 2m/r} + r^2 d\phi^2, \quad \text{where } m = m(r) = \int_0^r 4\pi r^2 \rho \, dr \quad (26.44)$$

[cf. Eq. (26.39)]. We seek to construct in a 3-dimensional Euclidean space a 2-dimensional surface with precisely this same 2-geometry. As an aid, we introduce in the Euclidean embedding space a cylindrical coordinate system $\{r, z, \phi\}$, in terms of which the space's 3-dimensional line element is

$$^{(3)}ds^2 = dr^2 + dz^2 + r^2 d\phi^2. \quad (26.45)$$

The surface we seek to embed is axially symmetric, so we can describe its embedding by the value of z on it as a function of radius r: $z = z(r)$. Inserting this (unknown) embedding function into Eq. (26.45), we obtain for the surface's 2-geometry,

$$^{(2)}ds^2 = [1 + (dz/dr)^2] dr^2 + r^2 d\phi^2. \quad (26.46)$$

Comparing with our original expression (26.44) for the 2-geometry, we obtain a differential equation for the embedding function:

$$\frac{dz}{dr} = \left(\frac{1}{1 - 2m/r} - 1 \right)^{\frac{1}{2}}. \quad (26.47)$$

If we set $z = 0$ at the star's center, then the solution of this differential equation is

shape of embedded equatorial plane in general

$$\boxed{z = \int_0^r \frac{dr'}{[r'/(2m) - 1]^{\frac{1}{2}}}.} \quad (26.48)$$

shape outside the star where the spacetime geometry is Schwarzschild

Near the star's center $m(r)$ is given by $m = (4\pi/3) \rho_c r^3$, where ρ_c is the star's central density; and outside the star $m(r)$ is equal to the star's r-independent total mass M. Correspondingly, in these two regions Eq. (26.48) reduces to

$$z = \sqrt{(2\pi/3)\rho_c} \, r^2 \quad \text{at } r \text{ very near zero.}$$

$$z = \sqrt{8M(r - 2M)} + \text{const at } r > R, \quad \text{i.e., outside the star.} \quad (26.49)$$

Figure 26.1 shows the embedded 2-surface $z(r)$ for a star of uniform density $\rho = \text{const}$ (Ex. 26.8). For any other star the embedding diagram will be qualitatively similar, though quantitatively different.

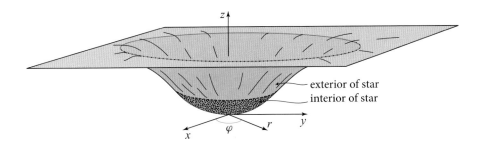

FIGURE 26.1 Embedding diagram depicting an equatorial, 2-dimensional slice $t = \text{const}, \theta = \pi/2$ through the spacetime of a spherical star with uniform density ρ and with radius R equal to 2.5 times the gravitational radius $2M$. See Ex. 26.8 for details.

The most important feature of this embedding diagram is its illustration of the fact [also clear in the original line element (26.44)] that, as one moves outward from the star's center, its circumference $2\pi r$ increases less rapidly than the proper radial distance traveled, $l = \int_0^r (1 - 2m/r)^{-\frac{1}{2}} dr$. As a specific example, the distance from the center of Earth to a perfect circle near Earth's surface is more than the circumference/2π by about 1.5 mm—a number whose smallness compared to the actual radius, 6.4×10^8 cm, is a measure of the weakness of the curvature of spacetime near Earth. As a more extreme example, the distance from the center of a massive neutron star to its surface is about 1 km greater than its circumference/2π—greater by an amount that is roughly 10% of the $\sim$10-km circumference/2π. Correspondingly, in the embedding diagram for Earth the embedded surface would be so nearly flat that its downward dip at the center would be imperceptible, whereas the diagram for a neutron star would show a downward dip about like that of Fig. 26.1.

EXERCISES

Exercise 26.8 *Example: Embedding Diagram for Star with Uniform Density*

(a) Show that the embedding surface of Eq. (26.48) is a paraboloid of revolution everywhere outside the star.

(b) Show that in the interior of a uniform-density star, the embedding surface is a segment of a sphere.

(c) Show that the match of the interior to the exterior is done in such a way that, in the embedding space, the embedded surface shows no kink (no bend) at $r = R$.

(d) Show that, in general, the circumference/(2π) for a star is less than the distance from the center to the surface by an amount of order one sixth the star's gravitational radius, $M/3$. Evaluate this amount analytically for a star of uniform density, and numerically (approximately) for Earth and for a neutron star.

26.3 Static Stars

26.4 Gravitational Implosion of a Star to Form a Black Hole

26.4.1 The Implosion Analyzed in Schwarzschild Coordinates

J. Robert Oppenheimer, on discovering with his student George Volkoff that there is a maximum mass limit for neutron stars (Oppenheimer and Volkoff, 1939), was forced to consider the possibility that, when it exhausts its nuclear fuel, a more massive star will implode to radii $R \leq 2M$. Just before the outbreak of World War II, Oppenheimer and his graduate student Hartland Snyder investigated the details of such an implosion for the idealized case of a perfectly spherical star in which all the internal pressure is suddenly extinguished (Oppenheimer and Snyder, 1939). In this section, we repeat their analysis, though from a more modern viewpoint and using somewhat different arguments.

By Birkhoff's theorem, the spacetime geometry outside an imploding, spherical star must be that of Schwarzschild. This means, in particular, that an imploding, spherical star cannot produce any gravitational waves; such waves would break the spherical symmetry. By contrast, a star that implodes nonspherically can produce a burst of gravitational waves (Chap. 27).

Since the spacetime geometry outside an imploding, spherical star is that of Schwarzschild, we can depict the motion of the star's surface by a world line in a 2-dimensional spacetime diagram with Schwarzschild coordinate time t plotted upward and Schwarzschild coordinate radius r plotted rightward (Fig. 26.2). The world line of the star's surface is an ingoing curve. The region to the left of the world line must be discarded and replaced by the spacetime of the star's interior, while the region to the right, $r > R(t)$, is correctly described by Schwarzschild.

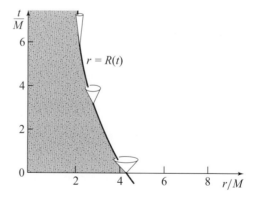

FIGURE 26.2 Spacetime diagram depicting the gravitationally induced implosion of a star in Schwarzschild coordinates. The thick solid curve is the world line of the star's surface, $r = R(t)$, in the external Schwarzschild coordinates. The stippled region to the left of that world line is not correctly described by the Schwarzschild line element (26.1); it requires for its description the spacetime metric of the star's interior. The surface's world line $r = R(t)$ is constrained to lie inside the light cones.

As for a static star, so also for an imploding one: because real atoms with finite rest masses live on the star's surface, the world line of that surface, $\{r = R(t), \theta \text{ and } \phi \text{ constant}\}$, must be timelike. Consequently, at each point along the world line it must lie within the local light cones that are depicted in Fig. 26.2.

The radial edges of the light cones are lines along which the Schwarzschild line element, the ds^2 of Eq. (26.1), vanishes with θ and ϕ held fixed:

$$0 = ds^2 = -(1 - 2M/R)dt^2 + \frac{dr^2}{1 - 2M/R}; \quad \text{that is,} \quad \frac{dt}{dr} = \pm \frac{1}{1 - 2M/R}. \quad (26.50)$$

Therefore, instead of having 45° opening angles $dt/dr = \pm 1$ as they do in a Lorentz frame of flat spacetime, the light cones "squeeze down" toward $dt/dr = \infty$ as the star's surface $r = R(t)$ approaches the gravitational radius: $R \to 2M$. This is a peculiarity due not to spacetime curvature, but rather to the nature of the Schwarzschild coordinates: If, at any chosen event of the Schwarzschild spacetime, we were to introduce a local Lorentz frame, then in that frame the light cones would have 45° opening angles.

Since the world line of the star's surface is confined to the interiors of the local light cones, the squeezing down of the light cones near $r = 2M$ prevents the star's world line $r = R(t)$ from ever, in any finite coordinate time t, reaching the gravitational radius, $r = 2M$.

This conclusion is completely general; it relies in no way on the details of what is going on inside the star or at its surface. It is just as valid for completely realistic stellar implosion (with finite pressure and shock waves) as for the idealized, Oppenheimer-Snyder case of zero-pressure implosion. In the special case of zero pressure, one can explore the details further:

Because no pressure forces act on the atoms at the star's surface, those atoms must move inward along radial geodesic world lines. Correspondingly, the world line of the star's surface in the external Schwarzschild spacetime must be a timelike geodesic of constant (θ, ϕ). In Ex. 26.9, the geodesic equation is solved to determine that world line $R(t)$, with a conclusion that agrees with the above argument: only after a lapse of infinite coordinate time t does the star's surface reach the gravitational radius $r = 2M$. A byproduct of that calculation is equally remarkable. Although the implosion to $R = 2M$ requires infinite Schwarzschild coordinate time t, it requires only a finite proper time τ as measured by an observer who rides inward on the star's surface. In fact, the proper time is

the surface of an imploding star reaches the gravitational radius $r = 2M$ in finite proper time τ, but infinite coordinate time t

$$\boxed{\tau \simeq \frac{\pi}{2}\left(\frac{R_o^3}{2M}\right)^{\frac{1}{2}} = 15\,\mu s \left(\frac{R_o}{2M}\right)^{3/2} \frac{M}{M_\odot} \quad \text{if } R_o \gg 2M,} \quad (26.51)$$

where R_o is the star's initial radius when it first begins to implode freely, $M_\odot$ denotes the mass of the Sun, and proper time τ is measured from the start of implosion. Note that this implosion time is equal to $1/(4\sqrt{2})$ times the orbital period of a test particle at the radius of the star's initial surface. For a star with mass and initial radius equal to those of the Sun, τ is about 30 minutes; for a neutron star that has been pushed over the maximum mass limit by accretion of matter from its surroundings, τ is about 50 μs. For a hypothetical supermassive star with $M = 10^9 M_\odot$ and $R_o \sim$ a few M, τ would be about a day.

26.4.2 Tidal Forces at the Gravitational Radius

What happens to the star's surface, and an imagined observer on it, when—after infinite coordinate time but a brief proper time—it reaches the gravitational radius? There are two possibilities: (i) the tidal gravitational forces there might be so strong that they destroy the star's surface and any observers on it; or (ii) the tidal forces are not that strong, and so the star and observers must continue to exist, moving into a region of spacetime (presumably $r < 2M$) that is not smoothly joined onto $r > 2M$ in the Schwarzschild coordinate system. In the latter case, the pathology is all due to poor properties of Schwarzschild's coordinates. In the former case, it is due to an intrinsic, coordinate-independent singularity of the tide-producing Riemann curvature.

To see which is the case, we must evaluate the tidal forces felt by observers on the surface of the imploding star. Those tidal forces are produced by the Riemann curvature tensor. More specifically, if an observer's feet and head have a vector separation $\boldsymbol{\xi}$ at time τ as measured by the observer's clock, then the curvature of spacetime will exert on them a relative gravitational acceleration given by the equation of geodesic deviation in the form appropriate to a local Lorentz frame:

$$\frac{d^2\xi^{\bar{j}}}{d\tau^2} = -R^{\bar{j}}{}_{\bar{0}\bar{k}\bar{0}}\xi^{\bar{k}} \tag{26.52}$$

[Eq. (25.34)]. Here the barred indices denote components in the observer's local Lorentz frame. The tidal forces will become infinite and will thereby destroy the observer and all forms of matter on the star's surface, if and only if the local Lorentz Riemann components $R_{\bar{j}\bar{0}\bar{k}\bar{0}}$ diverge as the star's surface approaches the gravitational radius. Thus, to test whether the observer and star survive, we must compute the components of the Riemann curvature tensor in the local Lorentz frame of the star's imploding surface.

The easiest way to compute those components is by a transformation from components as measured in the proper reference frames of observers who are "at rest" (fixed r, θ, ϕ) in the Schwarzschild spacetime. At each event on the world tube of the star's surface, then, we have two orthonormal frames: one (barred indices) a local Lorentz frame imploding with the star; the other (hatted indices) a proper reference frame at rest. Since the metric coefficients in these two bases have the standard flat-space form $g_{\bar{\alpha}\bar{\beta}} = \eta_{\alpha\beta}$, $g_{\hat{\alpha}\hat{\beta}} = \eta_{\alpha\beta}$, the bases must be related by a Lorentz transformation [cf. Eq. (2.35b) and associated discussion]. A little thought makes it clear that the required transformation matrix is that for a pure boost [Eq. (2.37a)]:

$$L^{\hat{0}}{}_{\bar{0}} = L^{\hat{r}}{}_{\bar{r}} = \gamma, \quad L^{\hat{0}}{}_{\bar{r}} = L^{\hat{r}}{}_{\bar{0}} = -\beta\gamma, \quad L^{\hat{\theta}}{}_{\bar{\theta}} = L^{\hat{\phi}}{}_{\bar{\phi}} = 1; \quad \gamma = \frac{1}{\sqrt{1-\beta^2}}, \tag{26.53}$$

with β the speed of implosion of the star's surface, as measured in the proper reference frame of the static observer when the surface flies by. The transformation law for the components of the Riemann tensor has, of course, the standard form for any fourth-rank tensor:

$$R_{\bar{\alpha}\bar{\beta}\bar{\gamma}\bar{\delta}} = L^{\hat{\mu}}{}_{\bar{\alpha}} L^{\hat{\nu}}{}_{\bar{\beta}} L^{\hat{\lambda}}{}_{\bar{\gamma}} L^{\hat{\sigma}}{}_{\bar{\delta}} R_{\hat{\mu}\hat{\nu}\hat{\lambda}\hat{\sigma}}. \tag{26.54}$$

The basis vectors of the proper reference frame are given by Eqs. (4) of Box 26.2, and from that box we learn that the components of Riemann in this basis are

$$R_{\hat{0}\hat{r}\hat{0}\hat{r}} = -\frac{2M}{R^3}, \quad R_{\hat{0}\hat{\theta}\hat{0}\hat{\theta}} = R_{\hat{0}\hat{\phi}\hat{0}\hat{\phi}} = +\frac{M}{R^3},$$

$$R_{\hat{\theta}\hat{\phi}\hat{\theta}\hat{\phi}} = \frac{2M}{R^3}, \quad R_{\hat{r}\hat{\theta}\hat{r}\hat{\theta}} = R_{\hat{r}\hat{\phi}\hat{r}\hat{\phi}} = -\frac{M}{R^3}. \tag{26.55}$$

These are the components measured by static observers.

By inserting these static-observer components and the Lorentz-transformation matrix (26.53) into the transformation law (26.54) we reach our goal: the following components of Riemann in the local Lorentz frame of the star's freely imploding surface:

$$R_{\bar{0}\bar{r}\bar{0}\bar{r}} = -\frac{2M}{R^3}, \quad R_{\bar{0}\bar{\theta}\bar{0}\bar{\theta}} = R_{\bar{0}\bar{\phi}\bar{0}\bar{\phi}} = +\frac{M}{R^3},$$

$$R_{\bar{\theta}\bar{\phi}\bar{\theta}\bar{\phi}} = \frac{2M}{R^3}, \quad R_{\bar{r}\bar{\theta}\bar{r}\bar{\theta}} = R_{\bar{r}\bar{\phi}\bar{r}\bar{\phi}} = -\frac{M}{R^3}. \tag{26.56}$$

These components are remarkable in two ways. First, they remain perfectly finite as the star's surface approaches the gravitational radius, $R \to 2M$; correspondingly, tidal gravity cannot destroy the star or the observers on its surface. Second, the components of Riemann are identically the same in the two orthonormal frames, hatted and barred, which move radially at finite speed β with respect to each other [expressions (26.56) are independent of β and are the same as Eqs. (26.55)]. This is a result of the very special algebraic structure that Riemann's components have for the Schwarzschild spacetime; it will not be true in typical spacetimes.

the Riemann tensor (tidal field), as measured on the imploding star's surface, remains finite as the surface approaches radius $r = 2M$

26.4.3 Stellar Implosion in Eddington-Finkelstein Coordinates

From the finiteness of the components of Riemann in the local Lorentz frame of the star's surface, we conclude that something must be wrong with Schwarzschild's $\{t, r, \theta, \phi\}$ coordinate system in the vicinity of the gravitational radius $r = 2M$: although nothing catastrophic happens to the star's surface as it approaches $2M$, those coordinates refuse to describe passage through $r = 2M$ in a reasonable, smooth, finite way. Thus to study the implosion as it passes through the gravitational radius and beyond, we need a new, improved coordinate system.

Several coordinate systems have been devised for this purpose. For a study and comparison of them see, for example, Misner, Thorne, and Wheeler (1973, Chap. 31). In this chapter we confine ourselves to one of them: a coordinate system devised for

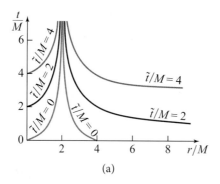
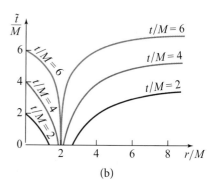

FIGURE 26.3 (a) The 3-surfaces of constant Eddington-Finkelstein time coordinate $\tilde{t}$ drawn in a Schwarzschild spacetime diagram, with the angular coordinates $\{\theta, \phi\}$ suppressed. (b) The 3-surfaces of constant Schwarzschild time coordinate t drawn in an Eddington-Finkelstein spacetime diagram, with angular coordinates suppressed.

other purposes by Arthur Eddington (1922), then long forgotten and only rediscovered independently and used for this purpose by Finkelstein (1958). Yevgeny Lifshitz, of Landau-Lifshitz fame, told one of the authors many years later what an enormous impact Finkelstein's coordinate system had on peoples' understanding of the implosion of stars. "You cannot appreciate how difficult it was for the human mind before Finkelstein to understand [the Oppenheimer-Snyder analysis of stellar implosion]," Lifshitz said. When, 19 years after Oppenheimer and Snyder, the issue of *Physical Review* containing Finkelstein's paper arrived in Moscow, suddenly everything was clear.

Finkelstein, a postdoctoral fellow at the Stevens Institute of Technology in Hoboken, New Jersey, found the following simple transformation, which moves the region $\{t = \infty, r = 2M\}$ of Schwarzschild coordinates in to a finite location. His transformation involves introducing a new time coordinate

Eddington-Finkelstein time coordinate

$$\tilde{t} = t + 2M \ln\left|[r/(2M)] - 1\right|, \tag{26.57}$$

but leaving unchanged the radial and angular coordinates. Figure 26.3 shows the surfaces of constant Eddington-Finkelstein time [3] $\tilde{t}$ in Schwarzschild coordinates, and the surfaces of constant Schwarzschild time t in Eddington-Finkelstein coordinates. Notice, as advertised, that $\{t = \infty, r = 2M\}$ is moved to a finite Eddington-Finkelstein location.

By inserting the coordinate transformation (26.57) into the Schwarzschild line element (26.1), we obtain the following line element for Schwarzschild spacetime written in Eddington-Finkelstein coordinates:

---

3. Sometimes $\tilde{t}$ is also called "ingoing Eddington-Finkelstein time," because it enables one to analyze infall through the gravitational radius.

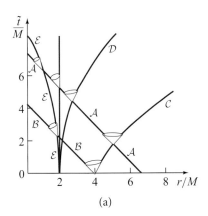
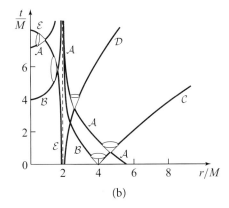

FIGURE 26.4 (a) Radial light rays and light cones for the Schwarzschild spacetime as depicted in Eddington-Finkelstein coordinates [Eq. (26.59)]. (b) The same light rays and light cones as depicted in Schwarzschild coordinates [cf. Fig. 26.2]. $\mathcal{A}$ and $\mathcal{B}$ are ingoing light rays that start far outside $r = 2M$; $\mathcal{C}$ and $\mathcal{D}$ are outgoing rays that start near $r = 2M$; $\mathcal{E}$ is an outgoing ray that is trapped inside the gravitational radius.

$$ds^2 = -\left(1 - \frac{2M}{r}\right) d\tilde{t}^2 + \frac{4M}{r} d\tilde{t}\, dr + \left(1 + \frac{2M}{r}\right) dr^2 + r^2(d\theta^2 + \sin^2\theta d\phi^2).$$

Schwarzschild metric in Eddington-Finkelstein coordinates

(26.58)

Notice that, by contrast with the line element in Schwarzschild coordinates, none of the metric coefficients diverge as r approaches $2M$. Moreover, in an Eddington-Finkelstein spacetime diagram, by contrast with Schwarzschild, the light cones do not pinch down to slivers at $r = 2M$ (compare Figs. 26.4a and 26.4b): The world lines of radial light rays are computable in Eddington-Finkelstein, as in Schwarzschild, by setting $ds^2 = 0$ (null world lines) and $d\theta = d\phi = 0$ (radial world lines) in the line element. The result, depicted in Fig. 26.4a, is

$$\frac{d\tilde{t}}{dr} = -1 \text{ for ingoing rays;} \quad \text{and} \quad \frac{d\tilde{t}}{dr} = \left(\frac{1 + 2M/r}{1 - 2M/r}\right) \text{ for outgoing rays.} \quad (26.59)$$

edges of radial light cone in Eddington-Finkelstein coordinates

Note that, in the Eddington-Finkelstein coordinate system, the ingoing light rays plunge unimpeded through $r = 2M$ and onward into $r = 0$ along 45° lines. The outgoing light rays, by contrast, are never able to escape outward through $r = 2M$: because of the inward tilt of the outer edge of the light cone, all light rays that begin inside $r = 2M$ are forced forever to remain inside, and in fact are drawn inexorably into $r = 0$, whereas light rays initially outside $r = 2M$ can escape to $r = \infty$.

Now return to the implosion of a star. The world line of the star's surface, which became asymptotically frozen at the gravitational radius when studied in Schwarzschild coordinates, plunges unimpeded through $r = 2M$ and to $r = 0$ when studied in

imploding star passes through the gravitational radius and onward to $r = 0$ in finite Eddington-Finkelstein coordinate time

26.4 Gravitational Implosion of a Star to Form a Black Hole

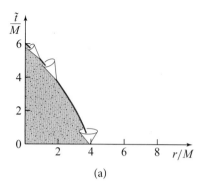
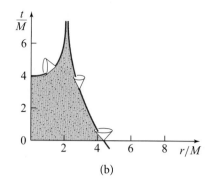

FIGURE 26.5 World line of an observer on the surface of an imploding star, as depicted (a) in an Eddington-Finkelstein spacetime diagram, and (b) in a Schwarzschild spacetime diagram; see Ex. 26.9.

Eddington-Finkelstein coordinates; see Ex. 26.9 and compare Figs. 26.5b and 26.5a. Thus to understand the star's ultimate fate, we must study the region $r = 0$.

EXERCISES

Exercise 26.9 *Example: Implosion of the Surface of a Zero-Pressure Star Analyzed in Schwarzschild and in Eddington-Finkelstein Coordinates*

Consider the surface of a zero-pressure star, which implodes along a timelike geodesic $r = R(t)$ in the Schwarzschild spacetime of its exterior. Analyze that implosion using Schwarzschild coordinates $\{t, r, \theta, \phi\}$ and the exterior metric (26.1) in those coordinates, and then repeat your analysis in Eddington-Finkelstein coordinates. More specifically, do the following.

(a) Using Schwarzschild coordinates, show that the covariant time component u_t of the 4-velocity $\vec{u}$ of a particle on the star's surface is conserved along its world line (cf. Ex. 25.4a). Evaluate this conserved quantity in terms of the star's mass M and the radius $r = R_o$ at which it begins to implode.

(b) Use the normalization of the 4-velocity to show that the star's radius R as a function of the proper time τ since implosion began (proper time as measured on its surface) satisfies the differential equation

$$\frac{dR}{d\tau} = -[\text{const} + 2M/R]^{\frac{1}{2}}, \qquad (26.60)$$

and evaluate the constant. Compare this with the equation of motion for the surface as predicted by Newtonian gravity, with proper time τ replaced by Newtonian time. (It is a coincidence that the two equations are identical.)

(c) Show from the equation of motion (26.60) that the star implodes through the gravitational radius $R = 2M$ and onward to $R = 0$ in a finite proper time given by Eq. (26.51). Show that this proper time has the magnitudes cited in Eq. (26.51) and the sentences following it.

(d) Show that the Schwarzschild coordinate time t required for the star to reach its gravitational radius, $R \to 2M$, is infinite.

(e) Show, further, that when studied in Eddington-Finkelstein coordinates, the surface's implosion to $R = 2M$ requires only finite coordinate time $\tilde{t}$; in fact, a time of the same order of magnitude as the proper time (26.51). [Hint: Derive a differential equation for $d\tilde{t}/d\tau$ along the world line of the star's surface, and use it to examine the behavior of $d\tilde{t}/d\tau$ near $R = 2M$.]

(f) Show that the world line of the star's surface as depicted in an Eddington-Finkelstein spacetime diagram has the form shown in Fig. 26.5a, and that in a Schwarzschild spacetime diagram it has the form shown in Fig. 26.5b.

26.4.4 Tidal Forces at $r = 0$—The Central Singularity

As with $r \to 2M$, there are two possibilities: either the tidal forces as measured on the star's surface remain finite as $r \to 0$, in which case something must be going wrong with the coordinate system; or else the tidal forces diverge, destroying the star. The tidal forces are computed in Ex. 26.10, with a remarkable result: they diverge. Thus, the region $r = 0$ is a *spacetime singularity*: a region where tidal gravity becomes infinitely large, destroying everything that falls into it.

$r = 0$ is a spacetime singularity; tidal gravity becomes infinite there

This conclusion, of course, is very unsatisfying. It is hard to believe that the correct laws of physics will predict such total destruction. In fact, they probably do not. As we will find in Sec. 28.7.1, in discussing the origin of the universe, when the radius of curvature of spacetime becomes as small as $L_P \equiv (G\hbar/c^3)^{\frac{1}{2}} \simeq 10^{-33}$ cm, space and time must cease to exist as classical entities; they and the spacetime geometry must then become quantized. Correspondingly, general relativity must then break down and be replaced by a quantum theory of the structure of spacetime—a quantum theory of gravity. That quantum theory will describe and govern the classically singular region at $r = 0$. Since, however, only rough hints of the structure of that quantum theory are in hand at this time, it is not known what that theory will say about the endpoint of stellar implosion.

the singularity is governed by the laws of quantum gravity

EXERCISES

Exercise 26.10 *Example: Gore at the Singularity*

(a) Show that, as the surface of an imploding star approaches $R = 0$, its world line in Schwarzschild coordinates asymptotes to the curve $\{(t, \theta, \phi) = \text{const}, r \text{ variable}\}$.

(b) Show that this curve to which it asymptotes [part (a)] is a timelike geodesic. [Hint: Use the result of Ex. 25.4a.]

(c) Show that the basis vectors of the infalling observer's local Lorentz frame near $r = 0$ are related to the Schwarzschild coordinate basis by

$$\vec{e}_{\hat{0}} = -\left(\frac{2M}{r} - 1\right)^{\frac{1}{2}} \frac{\partial}{\partial r}, \quad \vec{e}_{\hat{1}} = \left(\frac{2M}{r} - 1\right)^{-\frac{1}{2}} \frac{\partial}{\partial t},$$

$$\vec{e}_{\hat{2}} = \frac{1}{r} \frac{\partial}{\partial \theta}, \quad \vec{e}_{\hat{3}} = \frac{1}{r \sin \theta} \frac{\partial}{\partial \phi}. \quad (26.61)$$

What are the components of the Riemann tensor in that local Lorentz frame?

(d) Show that the tidal forces produced by the Riemann tensor stretch an infalling observer in the radial, $\vec{e}_{\hat{1}}$, direction and squeeze the observer in the tangential, $\vec{e}_{\hat{2}}$ and $\vec{e}_{\hat{3}}$, directions. Show that the stretching and squeezing forces become infinitely strong as the observer approaches $r = 0$.

(e) Idealize the body of an infalling observer to consist of a head of mass $\mu \simeq 20$ kg and feet of mass $\mu \simeq 20$ kg separated by a distance $h \simeq 2$ m, as measured in the observer's local Lorentz frame, and with the separation direction radial. Compute the stretching force between head and feet, as a function of proper time τ, as the observer falls into the singularity. Assume that the hole has the mass $M = 7 \times 10^9 M_\odot$, which has been measured by astronomical observations for the black hole at the center of the supergiant elliptical galaxy M87. How long before hitting the singularity (at what proper time τ) does the observer die, if he or she is a human being made of flesh, bone, and blood?

26.4.5 Schwarzschild Black Hole

Unfortunately, the singularity and its quantum mechanical structure are totally invisible to observers in the external universe: the only way the singularity can possibly be seen is by means of light rays, or other signals, that emerge from its vicinity. However, because the future light cones are all directed into the singularity (Fig. 26.5), no light-speed or sub-light-speed signals can ever emerge from it. In fact, because the outer edge of the light cone is tilted inward at every event inside the gravitational radius (Figs. 26.4 and 26.5), no signal can emerge from inside the gravitational radius to tell external observers what is going on there. In effect, the gravitational radius is an *absolute event horizon* for our universe, a horizon beyond which we cannot see—except by plunging through it, and paying the ultimate price for our momentary exploration of the hole's interior: we cannot publish the results of our observations.

As most readers are aware, the region of strong, vacuum gravity left behind by the implosion of the star is called a *black hole*. The horizon, $r = 2M$, is the surface of the hole, and the region $r < 2M$ is its interior. The spacetime geometry of the black hole, outside and at the surface of the star that creates it by implosion, is that of

gravitational radius as an absolute event horizon

black hole, its interior, surface (horizon), and exterior

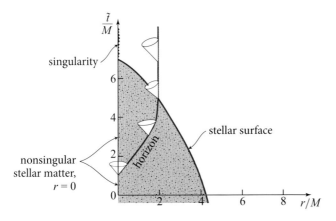

FIGURE 26.6 Spacetime diagram depicting the formation and evolution of the horizon of a black hole. The coordinates outside the surface of the imploding star are those of Eddington and Finkelstein; those inside are a smooth continuation of Eddington and Finkelstein (not explored in this book). Note that the horizon is the boundary of the region that is unable to send outgoing null geodesics to radial infinity.

Schwarzschild—though, of course, Karl Schwarzschild had no way of knowing this in the few brief months left to him after his discovery of the Schwarzschild line element.

The horizon—defined as the boundary between spacetime regions that can and cannot communicate with the external universe—actually forms initially at the star's center and then expands to encompass the star's surface at the precise moment when the surface penetrates the gravitational radius. This evolution of the horizon is depicted in an Eddington-Finkelstein-type spacetime diagram in Fig. 26.6.

evolution of the horizon as black hole is formed

Our discussion here has been confined to spherically symmetric, nonrotating black holes created by the gravitational implosion of a spherically symmetric star. Of course, real stars are not spherical, and it was widely believed—perhaps we should say, hoped—in the 1950s and 1960s that black-hole horizons and singularities would be so unstable that small nonsphericities or small rotations of the imploding star would save it from the black-hole fate. However, elegant and very general analyses carried out in the 1960s, largely by the British physicists Roger Penrose and Stephen Hawking, showed otherwise. More recent numerical simulations on supercomputers have confirmed those analyses: singularities are a generic outcome of stellar implosion, as are the black-hole horizons that clothe them.

genericity of singularities inside black holes

EXERCISES

Exercise 26.11 *Example: Rindler Approximation near the Horizon of a Schwarzschild Black Hole*

(a) Near the event $\{r = 2M, \theta = \theta_o, \phi = \phi_o, t \text{ finite}\}$, on the horizon of a black hole, introduce locally Cartesian spatial coordinates $\{x = 2M \sin\theta_o(\phi - \phi_o), y = 2M(\theta - \theta_o), z = \int_{2M}^{r} dr/\sqrt{1 - 2M/r}\}$, accurate to first order in distance from that event. Show that the metric in these coordinates has the form (accurate to leading order in distance from the chosen event):

$$ds^2 = -(g_H z)^2 dt^2 + dx^2 + dy^2 + dz^2, \quad \text{where} \quad g_H = \frac{1}{4M} \quad (26.62)$$

is the horizon's so-called *surface gravity*, to which we shall return, for a rotating black hole, in Eq. (26.90).

(b) Notice that the metric (26.62) is the same as that for flat spacetime as seen by a family of uniformly accelerated observers (i.e., as seen in the Rindler coordinates of Sec. 24.5.4). Why is this physically reasonable?

Exercise 26.12 **Example: Orbits around a Schwarzschild Black Hole*
Around a Schwarzschild black hole, spherical symmetry dictates that every geodesic orbit lies in a plane that bifurcates the $t = $ const 3-volume. We are free to orient our coordinate system, for any chosen geodesic, so its orbital plane is equatorial: $\theta = \pi/2$. Then the geodesic has three conserved quantities: the orbiting particle's rest mass μ, energy-at-infinity $\mathcal{E}_\infty$, and angular momentum L, which are given by

$$\mu^2 = -\vec{p}^2 = -g_{\alpha\beta} \frac{dx^\alpha}{d\zeta} \frac{dx^\beta}{d\zeta}, \quad \mathcal{E}_\infty = -p_t = -g_{tt} \frac{dt}{d\zeta}, \quad L = p_\phi = g_{\phi\phi} \frac{d\phi}{d\zeta}, \quad (26.63)$$

respectively. In this exercise we focus on particles with finite rest mass. Zero-rest-mass particles can be analyzed similarly; see the references at the end of this exercise.

(a) Set the rest mass μ to unity; equivalently, switch from 4-momentum to 4-velocity for the geodesic's tangent vector. Then, by algebraic manipulation of the constants of motion (26.63), derive the following orbital equations:

$$\left(\frac{dr}{d\zeta}\right)^2 + V^2(r) = \mathcal{E}_\infty^2, \quad \text{where} \quad V^2(r) = \left(1 - \frac{2M}{r}\right)\left(1 + \frac{L^2}{r^2}\right), \quad (26.64a)$$

$$\frac{d\phi}{d\zeta} = \frac{L}{r^2}, \quad \frac{dt}{d\zeta} = \frac{\mathcal{E}_\infty}{1 - 2M/r}. \quad (26.64b)$$

(b) We use a device that we have also encountered in our treatment of ion acoustic solitons in Sec. 23.6. We think of Eq. (26.64a) as an equivalent nonrelativistic energy equation with $\frac{1}{2}V^2$ being the effective potential energy and $\frac{1}{2}\mathcal{E}_\infty^2$ the effective total energy. As the energy we actually care about is $\mathcal{E}_\infty$, it is more direct to refer to $V(r)$ as our potential and investigate its properties. This $V(r)$ is plotted in Fig. 26.7a for several values of the particle's angular momentum L. Explain why: (i) Circular geodesic orbits are at extrema of $V(r)$—the large dots in the figure. (ii) Each bound orbit can be described by a horizontal line, such as the red one in the figure, with height equal to the orbit's $\mathcal{E}_\infty$; and the particle's radial motion is back and forth between the points at which the horizontal line intersects the potential.

(c) Show that the innermost stable circular orbit (often abbreviated as ISCO) is at $r = 6M$, and it occurs at a saddle point of the potential, for $L = 2\sqrt{3}\, M$. Show

Chapter 26. Relativistic Stars and Black Holes

that all inward-moving particles with $L < 2\sqrt{3}\,M$ are doomed to fall into the black hole.

(d) Show that the innermost unstable circular orbit is at $r = 3M$ and has infinite energy for finite rest mass. From this infer that there should be an unstable circular orbit for photons at $r = 3M$.

(e) The geodesic equations of motion in the form (26.64) are not very suitable for numerical integration: at each radial turning point, where $V(r) = \mathcal{E}_\infty$ and $dr/d\zeta = 0$, the accuracy of straightforward integrations goes bad, and one must switch signs of $dr/d\zeta$ by hand, unless one is sophisticated. For these reasons and others, it is preferable in numerical integrations to use the super-hamiltonian form of the geodesic equation (Ex. 25.7), or to convert Eq. (26.64a) into a second-order differential equation before integrating. Show that Hamilton's equations, for the super-hamiltonian, are

$$\frac{dr}{d\zeta} = \left(1 - \frac{2M}{r}\right) p_r, \quad \frac{dp_r}{d\zeta} = \frac{L^2}{r^3} - \frac{M}{r^2} p_r^2 - \frac{M}{(r-2M)^2} \mathcal{E}_\infty^2, \quad \frac{d\phi}{d\zeta} = \frac{L}{r^2}. \tag{26.65}$$

When integrating these equations, one must make sure that the initial value of p_r satisfies $g^{\alpha\beta} p_\alpha p_\beta = -1$ (for our unit-rest-mass particle). Show that this condition reduces to

$$p_r = \pm\sqrt{\frac{\mathcal{E}_\infty^2/(1 - 2M/r) - L^2/r^2 - 1}{1 - 2M/r}}. \tag{26.66}$$

(f) Integrate the super-hamiltonian equations (26.65) numerically for the orbit described by the red horizontal line in Fig. 26.7a, which has $L = 3.75M$ and

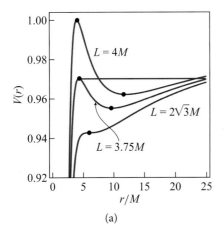
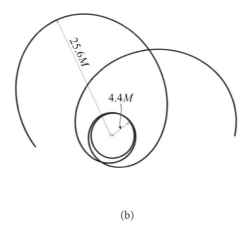

(a)   (b)

FIGURE 26.7 (a) The potential $V(r)$ for the geodesic radial motion $r(\zeta)$ of finite-rest-mass particles around a Schwarzschild black hole. Each curve is labeled by the particle's orbital angular momentum L. (b) The orbit $r(\phi)$ corresponding to the horizontal red line in (a); it has $L = 3.75M$ and $\mathcal{E}_\infty = 0.9704$. It is called a *zoom-whirl* orbit, because its particle zooms inward from a large radius, whirls around the black hole several times, then zooms back out—and then repeats.

$\mathcal{E}_\infty = 0.9704$. The result should be the zoom-whirl orbit depicted in Fig. 26.7b. (The initial conditions used in that figure were $r = 25M$, $\phi = 0$, and [from Eq. (26.66)] $p_r = 0.0339604$.)

(g) Carry out other numerical integrations to explore the variety of shapes of finite-rest-mass orbits around a Schwarzschild black hole.

Orbits around a Schwarzschild black hole are treated in most general relativity textbooks; see, for example, Misner, Thorne, and Wheeler (1973, Chap. 25) and Hartle (2003); see also Frolov and Novikov (1998), Sec. 2.8. Analytic solutions to the geodesic equation, expressed in terms of elliptic functions, are given by Darwin (1959).

Exercise 26.13 *Example: Schwarzschild Wormhole* T2
Our study of the Schwarzschild solution of Einstein's equations in this chapter has been confined to situations where, at small radii, the Schwarzschild geometry joins onto that of a star—either a static star or one that implodes to form a black hole. Suppose, by contrast, that there is no matter anywhere in the Schwarzschild spacetime. To get insight into this situation, construct an embedding diagram for the equatorial 2-surfaces $\{t = \text{const}, \theta = \pi/2\}$ of the vacuum Schwarzschild spacetime, using as the starting point the line element of such a 2-surface written in isotropic coordinates (Ex. 26.3):

$$^{(2)}ds^2 = \left(1 + \frac{M}{2\bar{r}}\right)^4 (d\bar{r}^2 + \bar{r}^2 d\phi^2). \tag{26.67}$$

Show that the region $0 < \bar{r} \ll M/2$ is an asymptotically flat space, that the region $\bar{r} \gg M/2$ is another asymptotically flat space, and that these two spaces are connected by a *wormhole* ("bridge," "tunnel") through the embedding space. This exercise reveals that the pure vacuum Schwarzschild spacetime represents a wormhole that connects two different universes—or, with a change of topology, a wormhole that connects two widely separated regions of one universe.

Exercise 26.14 *Example: Dynamical Evolution of Schwarzschild Wormhole* T2
The isotropic-coordinate line element (26.16) describing the spacetime geometry of a Schwarzschild wormhole is independent of the time coordinate t. However, because $g_{tt} = 0$ at the wormhole's throat, $\bar{r} = M/2$, the proper time $d\tau = \sqrt{-ds^2}$ measured by an observer at rest appears to vanish, which cannot be true. Evidently, the isotropic coordinates are ill behaved at the throat.

(a) Martin Kruskal (1960) and George Szekeres (1960) independently introduced a coordinate system that covers the wormhole's entire spacetime and elucidates its dynamics in a nonsingular manner. The Kruskal-Szekeres time and radial coordinates v and u are related to the Schwarzschild t and r by

$$[r/(2M) - 1] e^{r/(2M)} = u^2 - v^2, \tag{26.68}$$

$$t = 4M \tanh^{-1}(v/u) \quad \text{at } r > 2M, \quad t = 4M \tanh^{-1}(u/v) \quad \text{at } r < 2M.$$

Show that the metric of Schwarzschild spacetime written in these Kruskal-Szekeres coordinates is

$$ds^2 = (32M^3/r)e^{-r/(2M)}(-dv^2 + du^2) + r^2(d\theta^2 + \sin^2\theta d\phi^2), \quad (26.69)$$

where $r(u, v)$ is given by Eq. (26.68).

(b) Draw a spacetime diagram with v increasing upward and u increasing horizontally and rightward. Show that the radial light cones are 45° lines everywhere. Show that there are two $r = 0$ singularities, one on the past hyperbola, $v = -\sqrt{u^2 + 1}$, and the other on the future hyperbola, $v = +\sqrt{u^2 + 1}$. Show that the gravitational radius, $r = 2M$, is at $v = \pm u$. Show that our universe, outside the wormhole, is at $u \gg 1$, and there is another universe at $u \ll -1$.

(c) Draw embedding diagrams for a sequence of spacelike hypersurfaces, the first of which hits the past singularity and the last of which hits the future singularity. Thereby show that the metric (26.69) represents a wormhole that is created in the past, expands to maximum throat circumference $4\pi M$, then pinches off in the future to create a pair of singularities, one in each universe.

(d) Show that nothing can pass through the wormhole from one universe to the other; anything that tries gets crushed in the wormhole's pinch off.

For a solution, see Fuller and Wheeler (1962), or Misner, Thorne, and Wheeler (1973, Chap. 31). For discussions of what is required to hold a wormhole open so it can be traversed, see Morris and Thorne (1988); and for discussions of whether arbitrarily advanced civilizations can create wormholes and hold them open for interstellar travel, see the nontechnical discussions and technical references in Everett and Roman (2011). For a discussion of the possible use of traversable wormholes for backward time travel, see Sec. 2.9, and also Everett and Roman (2011) and references therein. For the visual appearance of wormholes and for guidance in constructing wormhole images by propagating light rays through and around them, see Thorne (2014) and James et al. (2015a).

26.5 Spinning Black Holes: The Kerr Spacetime

26.5.1 The Kerr Metric for a Spinning Black Hole

Consider a star that implodes to form a black hole, and assume for pedagogical simplicity that during the implosion no energy, momentum, or angular momentum flows through a large sphere surrounding the star. Then the asymptotic conservation laws discussed in Secs. 25.9.4 and 25.9.5 guarantee that the mass M, linear momentum P_j, and angular momentum J_j of the newborn hole, as encoded in its asymptotic metric, will be identical to those of its parent star. If (as we shall assume) our asymptotic coordinates are those of the star's rest frame ($P_j = 0$), then the hole will also be at rest in those coordinates (i.e., it will also have $P_j = 0$).

If the star was nonspinning ($J_j = 0$), then the hole will also have $J_j = 0$, and a powerful theorem due to Werner Israel (reviewed in Carter 1979) guarantees that—after it has settled down into a quiescent state—the hole's spacetime geometry will be that of Schwarzschild.

If, instead, the star was spinning ($J_j \neq 0$), then the final, quiescent hole cannot be that of Schwarzschild. Instead, according to powerful theorems due to Stephen Hawking, Brandon Carter, David Robinson, and others (also reviewed in Carter 1979), its spacetime geometry will be that described by the following exact, vacuum solution to the Einstein field equation:

the Kerr metric for a spinning black hole in Boyer-Lindquist coordinates

$$ds^2 = -\alpha^2 dt^2 + \frac{\rho^2}{\Delta} dr^2 + \rho^2 d\theta^2 + \varpi^2 (d\phi - \omega dt)^2, \qquad (26.70a)$$

where

$$\Delta = r^2 + a^2 - 2Mr, \quad \rho^2 = r^2 + a^2 \cos^2\theta, \quad \Sigma^2 = (r^2 + a^2)^2 - a^2 \Delta \sin^2\theta,$$

$$\alpha^2 = \frac{\rho^2}{\Sigma^2} \Delta, \quad \varpi^2 = \frac{\Sigma^2}{\rho^2} \sin^2\theta, \quad \omega = \frac{2aMr}{\Sigma^2}. \qquad (26.70b)$$

This is called the *Kerr solution*, because it was discovered by the New Zealand mathematician Roy Kerr (1963).

In this line element, $\{t, r, \theta, \phi\}$ are the coordinates, and there are two constants, M and a. The physical meanings of M and a can be deduced from the asymptotic form of the Kerr metric (26.70) at large radii:

$$ds^2 = -\left(1 - \frac{2M}{r}\right) dt^2 - \frac{4Ma}{r} \sin^2\theta \, d\phi \, dt$$
$$+ \left[1 + O\left(\frac{M}{r}\right)\right] [dr^2 + r^2(d\theta^2 + \sin^2\theta \, d\phi^2)]. \qquad (26.71)$$

the black hole's mass M and spin angular momentum Ma

By comparing with the standard asymptotic metric in spherical coordinates [Eq. (25.98d)], we see that M is the mass of the black hole, $Ma \equiv J_H$ is the magnitude of its spin angular momentum, and its spin points along the polar axis, $\theta = 0$. Evidently, then, the constant a is the hole's angular momentum per unit mass; it has the same dimensions as M: length (in geometrized units).

It is easy to verify that, in the limit $a \to 0$, the Kerr metric (26.70) reduces to the Schwarzschild metric (26.1), and the coordinates $\{t, r, \theta, \phi\}$ in which we have written it (called "Boyer-Lindquist coordinates") reduce to Schwarzschild's coordinates.

Just as it is convenient to read the covariant metric components $g_{\alpha\beta}$ off the line element (26.70a) via $ds^2 = g_{\alpha\beta} dx^\alpha dx^\beta$, so also it is convenient to read the contravariant metric components $g^{\alpha\beta}$ off an expression for the wave operator: $\Box \equiv \vec{\nabla} \cdot \vec{\nabla} = g^{\alpha\beta} \nabla_\alpha \nabla_\beta$. (Here $\nabla_\alpha \equiv \nabla_{\vec{e}_\alpha}$ is the directional derivative along the basis vector $\vec{e}_\alpha$.) For the Kerr metric (26.70), a straightforward inversion of the matrix

$\|g_{\alpha\beta}\|$ gives the $\|g^{\alpha\beta}\|$ embodied in the following equation:

$$\Box = -\frac{1}{\alpha^2}(\nabla_t + \omega\nabla_\phi)^2 + \frac{\Delta}{\rho^2}\nabla_r{}^2 + \frac{1}{\rho^2}\nabla_\theta{}^2 + \frac{1}{\varpi^2}\nabla_\phi{}^2. \quad (26.72)$$

the wave operator and contravariant components of the Kerr metric

26.5.2 Dragging of Inertial Frames [T2]

As we saw in Sec. 25.9.3, the angular momentum of a weakly gravitating body can be measured by its frame-dragging, precessional influence on the orientation of gyroscopes. Because the asymptotic metric (26.71) of a Kerr black hole is identical to the weak-gravity metric used to study gyroscopic precession in Sec. 25.9.3, the black hole's spin angular momentum $J_{\rm BH}$ can also be measured via frame-dragging gyroscopic precession.

This frame dragging also shows up in the geodesic trajectories of freely falling particles. For concreteness, consider a particle dropped from rest far outside the black hole. Its initial 4-velocity will be $\vec{u} = \partial/\partial t$, and correspondingly, in the distant, flat region of spacetime, the covariant components of $\vec{u}$ will be $u_t = -1, u_r = u_\theta = u_\phi = 0$.

Now, the Kerr metric coefficients $g_{\alpha\beta}$, like those of Schwarzschild, are independent of t and ϕ: the Kerr metric is symmetric under time translation (it is stationary) and under rotation about the hole's spin axis (it is axially symmetric). These symmetries impose corresponding conservation laws on the infalling particle (Ex. 25.4a): u_t and u_ϕ are conserved (i.e., they retain their initial values $u_t = -1$ and $u_\phi = 0$ as the particle falls). By raising indices—$u^\alpha = g^{\alpha\beta}u_\beta$, using the metric coefficients embodied in Eq. (26.72)—we learn the evolution of the contravariant 4-velocity components: $u^t = -g^{tt} = 1/\alpha^2$, $u^\phi = -g^{t\phi} = \omega/\alpha^2$. These in turn imply that, as the particle falls, it acquires an angular velocity around the hole's spin axis given by

$$\Omega = \frac{d\phi}{dt} = \frac{d\phi/d\tau}{dt/d\tau} = \frac{u^\phi}{u^t} = \omega. \quad (26.73)$$

frame dragging: angular velocity of a particle that falls in from $r = \infty$

(The coordinates ϕ and t are tied to the rotational and time-translation symmetries of the spacetime, so they are very special; that is why we can use them to define a physically meaningful angular velocity.)

At large radii, $\omega = 2aM/r^3 \to 0$ as $r \to \infty$. Therefore, when first dropped, the particle falls radially inward. However, as the particle nears the hole and picks up speed, it acquires a significant angular velocity around the hole's spin axis. The physical cause of this is *frame dragging*: The hole's spin drags inertial frames into rotation around the spin axis, and that inertial rotation drags the inertially falling particle into a circulatory orbital motion.

26.5.3 The Light-Cone Structure, and the Horizon [T2]

Just as for a Schwarzschild hole, so also for Kerr: the light-cone structure is a powerful tool for identifying the horizon and exploring the spacetime geometry near it.

outer and inner edges of light cone

At any event in spacetime, the tangents to the light cone are those displacements $\{dt, dr, d\theta, d\phi\}$ along which $ds^2 = 0$. The outermost and innermost edges of the cone are those for which $(dr/dt)^2$ is maximal. By setting expression (26.70a) to zero, we see that dr^2 has its maximum value, for a given dt^2, when $d\phi = \omega dt$ and $d\theta = 0$. In other words, the photons that move radially outward or inward at the fastest possible rate are those whose angular motion is that of frame dragging [Eq. (26.73)]. For these extremal photons, the radial motion (along the outer and inner edges of the light cone) is

$$\frac{dr}{dt} = \pm \frac{\alpha\sqrt{\Delta}}{\rho} = \pm \frac{\Delta}{\Sigma}. \tag{26.74}$$

Now, Σ is positive definite, but Δ is not; it decreases monotonically with decreasing radius, reaching zero at

$$\boxed{r = r_H \equiv M + \sqrt{M^2 - a^2}} \tag{26.75}$$

[Eq. (26.70b)]. (We assume that $|a| < M$, so r_H is real; we justify this assumption below.) Correspondingly, the light cone closes up to a sliver and then pinches off as $r \to r_H$; it pinches onto a null curve (actually, a null geodesic) given by

$$r = r_H, \quad \theta = \text{const}, \quad \phi = \Omega_H t + \text{const}, \tag{26.76}$$

where

$$\boxed{\Omega_H = \omega(r = r_H) = \frac{a}{2Mr_H}.} \tag{26.77}$$

This light-cone structure is depicted in Fig. 26.8a,b. The light-cone pinch-off as shown there is the same as that for Schwarzschild spacetime (Fig. 26.2) except for the light cones' frame-dragging-induced angular tilt $d\phi/dt = \omega$. In the Schwarzschild case, as $r \to 2M$, the light cones pinch onto the geodesic world lines $\{r = 2M, \theta = \text{const}, \phi = \text{const}\}$ of photons that travel along the horizon. These null world lines are called the horizon's *generators*. In the Kerr case, the light-cone pinch-off reveals that the horizon is at $r = r_H$, and the horizon generators are null geodesics that travel around and around the horizon with angular velocity Ω_H. This motivates us to regard the horizon itself as having the rotational angular velocity Ω_H.

the horizon, $r = r_H$, and its rotational angular velocity, Ω_H

When a finite-rest-mass particle falls into a spinning black hole, its world line, as it nears the horizon, is constrained always to lie inside the light cone. The light-cone pinch-off then constrains its motion asymptotically to approach the horizon generators. Therefore, as seen in Boyer-Lindquist coordinates, the particle is dragged into an orbital motion, just above the horizon, with asymptotic angular velocity $d\phi/dt = \Omega_H$, and it travels around and around the horizon "forever" (for infinite Boyer-Lindquist coordinate time t), and never (as $t \to \infty$) manages to cross through the horizon.

in Boyer-Lindquist coordinates, infalling particles asymptote to horizon and to its angular velocity

Chapter 26. Relativistic Stars and Black Holes

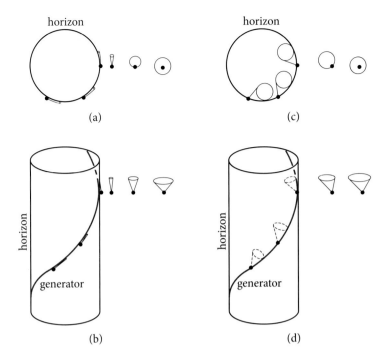

FIGURE 26.8 (a), (b): Light-cone structure of Kerr spacetime depicted in Boyer-Lindquist coordinates. Drawing (b) is a spacetime diagram; drawing (a) is the same diagram as viewed from above. (c), (d): The same light-cone structure in Kerr coordinates.

As in the Schwarzschild case, so also in Kerr: this infall to $r = r_H$ requires only finite proper time τ as measured by the particle, and the particle feels only finite tidal forces (only finite values of the components of Riemann in its proper reference frame). Therefore, as for Schwarzschild spacetime, the "barrier" to infall through $r = r_H$ must be an illusion produced by a pathology of the Boyer-Lindquist coordinates at $r = r_H$.

This coordinate pathology can be removed by a variety of different coordinate transformations. One is the following change of the time and angular coordinates:

$$\boxed{\tilde{t} = t + \int \frac{2Mr}{\Delta} dr, \quad \tilde{\phi} = \phi + \int \frac{a}{\Delta} dr.} \qquad (26.78)$$

Kerr coordinates

The new (tilded) coordinates are a variant of a coordinate system originally introduced by Kerr, so we call them "Kerr coordinates."[4] By inserting the coordinate transformation (26.78) into the line element (26.70a), we obtain the following form of the Kerr metric in Kerr coordinates:

Kerr metric in Kerr coordinates

---

4. They are often called "ingoing Kerr coordinates," because they facilitate analyzing infall through the horizon.

26.5 Spinning Black Holes: The Kerr Spacetime

$$\boxed{\begin{aligned}ds^2 = &-\alpha^2 d\tilde{t}^2 + \frac{4Mr\rho^2}{\Sigma^2}dr\,d\tilde{t} + \frac{\rho^2(\rho^2+2Mr)}{\Sigma^2}dr^2 \\ &+ \rho^2 d\theta^2 + \varpi^2\left[d\tilde{\phi} - \omega d\tilde{t} - \frac{a(\rho^2+2Mr)}{\Sigma^2}dr\right]^2\end{aligned}}$$ (26.79)

It is easy to verify that when $a \to 0$ (so Kerr spacetime becomes Schwarzschild), the Kerr coordinates (26.78) become those of Eddington and Finkelstein [Eq. (26.57)], and the Kerr line element (26.79) becomes the Eddington-Finkelstein one [Eq. (26.58)]. Similarly, when one explores the light-cone structure for a spinning black hole in the Kerr coordinates (Fig. 26.8c,d), one finds a structure like that of Eddington-Finkelstein (Fig. 26.4a): At large radii, $r \gg M$, the light cones have their usual 45° form, but as one moves inward toward the horizon, they begin to tilt inward. In addition to the inward tilt, there is a frame-dragging-induced tilt in the direction of the hole's rotation, $+\phi$. At the horizon, the outermost edge of the light cone is tangent to the horizon generators; in Kerr coordinates, as in Boyer-Lindquist, these generators rotate around the horizon with angular velocity $d\tilde{\phi}/d\tilde{t} = \Omega_H$ [cf. Eq. (26.78), which says that at fixed r, $\tilde{t} = t + $ const and $\tilde{\phi} = \phi + $ const].

in Kerr coordinates, infalling particles cross the horizon and are pulled on inward

This light-cone structure demonstrates graphically that the horizon is at the radius $r = r_H$. Outside there, the outer edge of the light cone tilts toward increasing r, and so it is possible to escape to radial infinity. Inside r_H the outer edge of the light cone tilts inward, and all forms of matter and energy are forced to move inward, toward a singularity whose structure, presumably, is governed by the laws of quantum gravity.[5]

26.5.4

26.5.4 Evolution of Black Holes—Rotational Energy and Its Extraction

When a spinning star collapses to form a black hole, its centrifugal forces will flatten it, and the dynamical growth of flattening will produce gravitational radiation (Chap. 27). The newborn hole will also be flattened and will not have the Kerr shape; but rather quickly, within a time $\Delta t \sim 100 M \sim 0.5$ ms $(M/M_\odot)$, the deformed hole will shake off its deformations as gravitational waves and settle down into the Kerr shape. This is the conclusion of extensive analyses, both analytic and numerical.

newborn black holes settle down into Kerr form

Many black holes are in binary orbits with stellar companions and pull gas off their companions and swallow it. Other black holes accrete gas from interstellar space. Any such accretion causes a hole's mass and spin to evolve in accord with the conservation laws (25.102) and (25.103). One might have thought that by accreting a large amount of angular momentum, a hole's angular momentum per unit mass a could grow larger than its mass M. If this were to happen, then $r_H = M + \sqrt{M^2 - a^2}$ would cease to

---

5. Much hoopla has been made of the fact that in the Kerr spacetime it is possible to travel inward, through a "Cauchy horizon" and then into another universe. However, the Cauchy horizon, located at $r = M - \sqrt{M^2 - a^2}$, is highly unstable against perturbations, which convert it into a singularity with infinite spacetime curvature. For details of this instability and the singularity, see, e.g., Brady, Droz, and Morsink (1998), Marolf and Ori (2013), and references therein.

be a real radius—a fact that signals the destruction of the hole's horizon: as a grows to exceed M, the inward light-cone tilt gets reduced, so that everywhere the outer edge of the cone points toward increasing r, which means that light, particles, and information are no longer trapped.

Remarkably, however, it appears that the laws of general relativity forbid a ever to grow larger than M. As accretion pushes a/M upward toward unity, the increasing drag of inertial frames causes a big increase of the hole's cross section to capture material with negative angular momentum (which will spin the hole down) and a growing resistance to capturing any further material with large positive angular momentum. Infalling particles that might try to push a/M over the limit get flung back out by huge centrifugal forces, before they can reach the horizon. A black hole, it appears, is totally resistant to having its horizon destroyed.

accretion of matter onto a black hole pushes its spin parameter a up to a maximum value that is a little less than $a = M$

In 1969, Roger Penrose discovered that a large fraction of the mass of a spinning black hole is in the form of rotational energy, stored in the whirling spacetime curvature outside the hole's horizon. Although this rotational energy cannot be localized in any precise manner, it nevertheless can be extracted. Penrose discovered this by the following thought experiment, which is called the *Penrose process*.

From far outside the hole, you throw a massive particle into the vicinity of the hole's horizon. Assuming you are at rest with respect to the hole, your 4-velocity is $\vec{U} = \partial/\partial t$. Denote by $\mathcal{E}^{\text{in}}_\infty = -\vec{p}^{\text{in}} \cdot \vec{U} = -\vec{p}^{\text{in}} \cdot (\partial/\partial t) = -p_t^{\text{in}}$ the energy of the particle (rest mass plus kinetic), as measured by you—its *energy-at-infinity* in the language of Ex. 26.4 and Sec. 4.10.2. As the particle falls, $\mathcal{E}^{\text{in}}_\infty = -p_t^{\text{in}}$ is conserved because of the Kerr metric's time-translation symmetry. Arrange that, as the particle nears the horizon, it splits into two particles; one (labeled "plunge") plunges through the horizon and the other (labeled "out") flies back out to large radii, where you catch it. Denote by $\mathcal{E}^{\text{plunge}}_\infty \equiv -p_t^{\text{plunge}}$ the conserved energy-at-infinity of the plunging particle and by $\mathcal{E}^{\text{out}}_\infty \equiv -p_t^{\text{out}}$ that of the out-flying particle. Four-momentum conservation at the event of the split dictates that $\vec{p}^{\text{in}} = \vec{p}^{\text{plunge}} + \vec{p}^{\text{out}}$, which implies this same conservation law for all components of the 4-momenta, in particular:

Penrose process for extracting spin energy from a black hole

$$\mathcal{E}^{\text{out}}_\infty = \mathcal{E}^{\text{in}}_\infty - \mathcal{E}^{\text{plunge}}_\infty. \tag{26.80}$$

Now, it is a remarkable fact that the Boyer-Lindquist time basis vector $\partial/\partial t$ has a squared length $\partial/\partial t \cdot \partial/\partial t = g_{tt} = -\alpha^2 + \varpi^2 \omega^2$ that becomes positive (so $\partial/\partial t$ becomes spacelike) at radii

$$\boxed{r < r_{\text{ergo}} \equiv M + \sqrt{M^2 - a^2 \cos^2\theta}.} \tag{26.81}$$

Notice that r_{ergo} is larger than r_H everywhere except on the hole's spin axis: $\theta = 0, \pi$. The region $r_H < r < r_{\text{ergo}}$ is called the hole's *ergosphere*. If the split into two particles occurs in the ergosphere, then it is possible to arrange the split such that the scalar product of the *timelike* vector $\vec{p}^{\text{plunge}}$ with the *spacelike* vector $\partial/\partial t$ is posi-

black hole's ergosphere

26.5 Spinning Black Holes: The Kerr Spacetime

tive, which means that the plunging particle's conserved energy-at-infinity, $\mathcal{E}_\infty^{\text{plunge}} = -\vec{p}^{\text{plunge}} \cdot (\partial/\partial t)$, is negative; whence [by Eq. (26.80)]

$$\mathcal{E}_\infty^{\text{out}} > \mathcal{E}_\infty^{\text{in}}. \tag{26.82}$$

See Ex. 26.16a.

When the outflying particle reaches your location, $r \gg M$, its conserved energy is equal to its physically measured total energy (rest-mass plus kinetic), and the fact that $\mathcal{E}_\infty^{\text{out}} > \mathcal{E}_\infty^{\text{in}}$ means that you get back more energy (rest-mass plus kinetic) than you put in. The hole's asymptotic energy-conservation law (25.102) implies that the hole's mass has decreased by precisely the amount of energy that you have extracted:

$$\Delta M = -(\mathcal{E}_\infty^{\text{out}} - \mathcal{E}_\infty^{\text{in}}) = \mathcal{E}_\infty^{\text{plunge}} < 0. \tag{26.83}$$

A closer scrutiny of this process (Ex. 26.16f) reveals that the plunging particle must have had negative angular momentum, so it has spun the hole down a bit. The energy you extracted, in fact, came from the hole's enormous store of rotational energy, which makes up part of its mass M. Your extraction of energy has reduced that rotational energy.

Stephen Hawking has used sophisticated mathematical techniques to prove that, independently of how you carry out this thought experiment, and, indeed, independently of what is done to a black hole, general relativity requires that the horizon's surface area A_H never decrease. This is called the *second law of black-hole mechanics*, and it actually turns out to be a variant of the second law of thermodynamics in disguise (Ex. 26.16g). A straightforward calculation (Ex. 26.15) reveals that the horizon surface area is given by

the second law of black-hole mechanics

$$\boxed{A_H = 4\pi(r_H^2 + a^2) = 8\pi M r_H \quad \text{for a spinning hole,}} \tag{26.84a}$$

$$\boxed{A_H = 16\pi M^2 \quad \text{for a nonspinning hole, } a = 0.} \tag{26.84b}$$

Dimitrious Christodoulou has shown (cf. Ex. 26.16) that, in the Penrose process described here, the nondecrease of A_H is the only constraint on how much energy one can extract, so by a sequence of optimally designed particle injections and splits that keep A_H unchanged, one can reduce the mass of the hole to

$$\boxed{M_{\text{irr}} = \sqrt{\frac{A_H}{16\pi}} = \sqrt{\frac{M(M + \sqrt{M^2 - a^2})}{2}},} \tag{26.85}$$

black hole's irreducible mass and rotational energy

but no smaller. This is called the hole's irreducible mass. The hole's total mass is the sum of its irreducible mass and its rotational energy M_{rot}; so the rotational energy is

$$\boxed{M_{\text{rot}} = M - M_{\text{irr}} = M\left[1 - \sqrt{\frac{1}{2}\left(1 + \sqrt{1 - a^2/M^2}\right)}\right].} \tag{26.86}$$

For the fastest possible spin, $a = M$, this gives $M_{\text{rot}} = M(1 - 1/\sqrt{2}) \simeq 0.2929M$. This is the maximum amount of energy that can be extracted, and it is enormous compared to the energy $\sim 0.005M$ that can be released by thermonuclear burning in a star of mass M.

The Penrose process of throwing in particles and splitting them in two is highly idealized, and of little or no importance in Nature. However, Nature seems to have found a very effective alternative method for extracting rotational energy from spinning black holes (Blandford and Znajek, 1977; Thorne, Price, and MacDonald, 1986; McKinney, Tchekhovskoy, and Blandford, 2012; Ex. 26.21) in which magnetic fields, threading through a black hole and held on the hole by a surrounding disk of hot plasma, extract energy electromagnetically. This process is thought to power the gigantic jets that shoot out of the nuclei of some active galaxies. It might also be the engine for some powerful gamma-ray bursts.

Blandford-Znajek process by which magnetic fields extract energy from black holes in Nature

EXERCISES

Exercise 26.15 *Derivation: Surface Area of a Spinning Black Hole* **T2**
From the Kerr metric (26.70) derive Eqs. (26.84) for the surface area of a spinning black hole's horizon—that is, the surface area of the 2-dimensional surface $\{r = r_H, t = \text{constant}\}$.

Exercise 26.16 **Example: Penrose Process, Hawking Radiation, and Thermodynamics of Black Holes* **T2**
This exercise is a foundation for the discussion of black-hole thermodynamics in Sec. 4.10.2.

(a) Consider the Penrose process, described in the text, in which a particle flying inward toward a spinning hole's horizon splits in two inside the ergosphere, and one piece plunges into the hole while the other flies back out. Show that it is always possible to arrange this process so the plunging particle has negative energy-at-infinity: $\mathcal{E}_\infty^{\text{plunge}} = -\vec{p}^{\text{plunge}} \cdot \partial/\partial t < 0$. [Hint: Perform a calculation in a local Lorentz frame in which $\partial/\partial t$ points along a spatial basis vector, $\vec{e}_{\hat{1}}$. Why is it possible to find such a local Lorentz frame?]

(b) Around a spinning black hole consider the vector field

$$\vec{\xi}_H \equiv \partial/\partial t + \Omega_H \partial/\partial \phi, \tag{26.87}$$

where Ω_H is the horizon's angular velocity. Show that at the horizon (at radius $r = r_H$) this vector field is null and is tangent to the horizon generators. Show that all other vectors in the horizon are spacelike.

(c) In the Penrose process the plunging particle changes the hole's mass by an amount ΔM and its spin angular momentum by an amount ΔJ_H. Show that

$$\Delta M - \Omega_H \Delta J_H = -\vec{p}^{\text{plunge}} \cdot \vec{\xi}_H. \tag{26.88}$$

Here $\vec{p}^{\text{plunge}}$ and $\vec{\xi}_H$ are to be evaluated at the event where the particle plunges through the horizon, so they both reside in the same tangent space. [Hint: The angular momentum carried into the horizon is the quantity $p_\phi^{\text{plunge}} = \vec{p}^{\text{plunge}} \cdot \partial/\partial\phi$. Why? This quantity is conserved along the plunging particle's world line. Why?] Note that in Sec. 4.10.2, p_ϕ is denoted $\mathbf{j} \cdot \hat{\mathbf{\Omega}}_H$—the projection of the particle's orbital angular momentum on the black hole's spin axis.

(d) Show that if $\vec{A}$ is any future-directed timelike vector and $\vec{K}$ is any null vector, both living in the tangent space at the same event in spacetime, then $\vec{A} \cdot \vec{K} < 0$. [Hint: Perform a calculation in a specially chosen local Lorentz frame.] Thereby conclude that $-\vec{p}^{\text{plunge}} \cdot \vec{\xi}_H$ is positive, whatever may be the world line and rest mass of the plunging particle.

(e) Show that for the plunging particle to decrease the hole's mass, it must also decrease the hole's angular momentum (i.e., it must spin the hole down a bit).

(f) Hawking's second law of black-hole mechanics says that, whatever may be the particle's world line and rest mass, when the particle plunges through the horizon, it causes the horizon's surface area A_H to increase. This suggests that the always positive quantity $\Delta M - \Omega_H \Delta J_H = -\vec{p}^{\text{plunge}} \cdot \vec{\xi}_H$ might be a multiple of the increase ΔA_H of the horizon area. Show that this is indeed the case:

$$\boxed{\Delta M = \Omega_H \Delta J_H + \frac{g_H}{8\pi} \Delta A_H,} \tag{26.89}$$

where g_H is given in terms of the hole's mass M and the radius r_H of its horizon by

$$\boxed{g_H = \frac{r_H - M}{2M r_H}.} \tag{26.90}$$

(You might want to do the algebra, based on Kerr-metric formulas, on a computer.) The quantity g_H (which we have met previously in the Rindler approximation; Ex. 26.11) is called the hole's "surface gravity" for a variety of reasons. One reason is that an observer who hovers just above a horizon generator, blasting his or her rocket engines to avoid falling into the hole, has a 4-acceleration with magnitude g_H/α and thus feels a "gravitational acceleration" of this magnitude; here $\alpha = g^{tt}$ is a component of the Kerr metric called the *lapse function* [Eqs. (26.70) and (26.72)]. This gravitational acceleration is arbitrarily large for an observer arbitrarily close to the horizon (where Δ and hence α are arbitrarily close to zero); when renormalized by α to make it finite, the acceleration is g_H. Equation (26.89) is called the "first law of black-hole mechanics" because of its resemblance to the first law of thermodynamics.

(g) Stephen Hawking has shown, using quantum field theory, that a black hole's horizon emits thermal (blackbody) radiation. The temperature of this "Hawking radiation," as measured by the observer who hovers just above the horizon, is proportional to the gravitational acceleration g_H/α that the observer measures, with a proportionality constant $\hbar/(2\pi k_B)$, where $\hbar$ is Planck's reduced constant and k_B is Boltzmann's constant. As this thermal radiation climbs out of the horizon's vicinity and flies off to large radii, its frequencies and temperature get redshifted by the factor α, so as measured by distant observers the temperature is

$$T_H = \frac{\hbar}{2\pi k_B} g_H. \tag{26.91}$$

This suggests a reinterpretation of the first law of black-hole mechanics (26.89) as the first law of thermodynamics for a black hole:

$$\Delta M = \Omega_H \Delta J_H + T_H \Delta S_H, \tag{26.92}$$

where S_H is the hole's entropy [cf. Eq. (4.62)]. Show that this entropy is related to the horizon's surface area by

$$S_H = k_B \frac{A_H}{4L_P^2}, \tag{26.93}$$

where $L_P = \sqrt{\hbar G/c^3} = 1.616 \times 10^{-33}$ cm is the Planck length (with G Newton's gravitation constant and c the speed of light). Because $S_H \propto A_H$, the second law of black-hole mechanics (nondecreasing A_H) is actually the second law of thermodynamics (nondecreasing S_H) in disguise.[6]

(h) For a 10-solar-mass, nonspinning black hole, what is the temperature of the Hawking radiation in Kelvins, and what is the hole's entropy in units of the Boltzmann constant?

(i) Reread the discussions of black-hole thermodynamics and entropy in the expanding universe in Secs. 4.10.2 and 4.10.3, which rely on the results of this exercise.

Exercise 26.17 *Problem: Thin Accretion Disks: Circular, Equatorial, Geodesic Orbits around a Kerr Black Hole* T2

Astronomers find and observe black holes primarily through the radiation emitted by infalling gas. Usually this gas has sufficient angular momentum to form an *accretion disk* around the black hole, lying in its equatorial plane. When the gas can cool efficiently, the disk is physically thin. The first step in describing accretion disks is to compute their equilibrium structure, which is usually approximated by assuming that the fluid elements follow circular geodesic orbits and pressure can be ignored.

---

[6]. Actually, the emission of Hawking radiation decreases the hole's entropy and surface area; but general relativity is oblivious to this, because general relativity is a classical theory, and Hawking's prediction of the thermal radiation is based on quantum theory. Thus, the Hawking radiation violates the second law of black-hole mechanics. It does not, however, violate the second law of thermodynamics, because the entropy carried into the surrounding universe by the Hawking radiation exceeds the magnitude of the decrease of the hole's entropy. The total entropy of hole plus universe increases.

The next step is to invoke some form of magnetic viscosity that catalyzes the outward transport of angular momentum through the disk and the slow inspiral of the disk's gas. The final step is to compute the spectrum of the emitted radiation. In this problem, we consider only the first step: the gas's circular geodesic orbits.

(a) Following Ex. 26.12, we can write the 4-velocity in covariant form as $u_\alpha = \{-\mathcal{E}_\infty, u_r, 0, L\}$, where $\mathcal{E}_\infty$ and L are the conserved energy and angular momentum per unit mass, respectively, and u_r describes the radial motion (which we eventually set to zero). Use $u^\alpha u_\alpha = -1$ and $u^r = dr/d\tau$ to show that, for any equatorial geodesic orbit:

$$\frac{1}{2}\left(\frac{dr}{d\tau}\right)^2 + V_{\text{eff}} = 0, \text{ where } V_{\text{eff}} = \frac{1 + u^t u_t + u^\phi u_\phi}{2g_{rr}} \quad (26.94a)$$

is an effective potential that includes $\mathcal{E}_\infty$ and L inside itself [by contrast with the effective potential we used for a Schwarzschild black hole, Eq. (26.64a), from which $\mathcal{E}_\infty$ was pulled out].

(b) Explain why a circular orbit is defined by the twin conditions, $V_{\text{eff}} = 0, \partial_r V_{\text{eff}} = 0$, and use the Boyer-Lindquist metric coefficients plus computer algebra to solve for the conserved energy and angular momentum:

$$\mathcal{E}_\infty = \frac{1 - 2M/r \pm aM^{1/2}/r^{3/2}}{(1 - 3M/r \pm 2aM^{1/2}/r^{3/2})^{1/2}}, \quad L = \frac{\pm M^{1/2}r^{1/2} - 2aM/r \pm M^{1/2}a^2/r^{3/2}}{(1 - 3M/r \pm 2aM^{1/2}/r^{3/2})^{1/2}}, \quad (26.94b)$$

where, in $\pm$, the $+$ is for a prograde orbit (same direction as hole spins) and $-$ is for retrograde.

(c) Explain mathematically and physically why the angular velocity $\Omega = u^\phi/u^t$ satisfies

$$\frac{\partial \mathcal{E}_\infty}{\partial r} = \Omega \frac{\partial L}{\partial r}, \quad (26.94c)$$

and show that it evaluates to

$$\Omega = \frac{\pm M^{1/2}}{r^{3/2} \pm aM^{1/2}}. \quad (26.94d)$$

(d) Modify the argument from Ex. 26.12 to show that the binding energy of a gas particle in its smallest, stable, circular, equatorial orbit around a maximally rotating ($a = M$) Kerr hole is $1 - 3^{-1/2} = 0.423$ per unit mass for prograde orbits and $1 - 5/3^{3/2} = 0.0377$ per unit mass for retrograde orbits. The former is a measure of the maximum power that can be released as gas accretes onto a hole through a thin disk in a cosmic object like a quasar. Note that it is two orders of magnitude larger than the energy typically recoverable from nuclear reactions.

For the visual appearance of a thin accretion disk around a black hole, distorted by gravitational lensing (bending of light rays in the Kerr metric), and for other aspects of a black hole's gravitational lensing, see, for example, James et al. (2015b) and references therein.

Exercise 26.18 *Problem: Thick Accretion Disks* T2

A quite different type of accretion disk forms when the gas is unable to cool. This can occur when its gas supply rate is either very large or very small. In the former case, the photons are trapped by the inflowing gas; in the latter, the radiative cooling timescale exceeds the inflow timescale. (In practice such inflows are likely to produce simultaneous outflows to carry off the energy released.) Either way, pressure and gravity are of comparable importance. There is an elegant description of gas flow close to the black hole as a sort of toroidal star, where the metric is associated with the hole and not the gas. In this problem we make simple assumptions to solve for the equilibrium flow.

(a) Treat the gas as a perfect fluid, and use as thermodynamic variables its pressure P and enthalpy density w. When the gas supply rate is large, the pressure is generally dominated by radiation. Show that in this case $w = \rho_o + 4P$, with ρ_o the rest-mass density of the plasma. Show, further, that $w = \rho_o + \frac{5}{2}P$ when the pressure is due primarily to nonrelativistic ions and their electrons.

(b) Assume that the motion is purely azimuthal, so the only nonzero covariant components of the gas's 4-velocity, in the Boyer-Lindquist coordinate system, are u^t and u^ϕ. (In practice there will also be slow poloidal circulation and slow inflow.) Show that the equation of hydrostatic equilibrium (26.31) can be written in the form

$$\frac{P_{,a}}{w} = u^t u_{t,a} + u^\phi u_{\phi,a}, \qquad (26.95\text{a})$$

where $a = r, \theta$, and the commas denote partial derivatives.

(c) Define for the gas the specific energy at infinity, $\mathcal{E}_\infty = -u_t$; the *specific fluid angular momentum*, $\ell = -u_\phi/u_t$ (which is $L/\mathcal{E}_\infty$ in the notation of the previous exercise); and the angular velocity, $\Omega = d\phi/dt = u^\phi/u^t$. Show that Eq. (26.95a) can be rewritten as

$$-\frac{\vec{\nabla} P}{w} = \vec{\nabla}\ln\mathcal{E}_\infty - \frac{\Omega\vec{\nabla}\ell}{1 - \Omega\ell}, \qquad (26.95\text{b})$$

where the spacetime gradient $\vec{\nabla}$ has components only in the r and θ directions, and ℓ and $\mathcal{E}_\infty$ are regarded as scalar fields.

(d) Now make the first simplifying assumption: the gas obeys a *barotropic* equation of state, $P = P(w)$ (cf. Sec. 14.2.2). Show that this assumption implies that the specific angular momentum ℓ is a function of the angular velocity.

(e) Show that the nonrelativistic limit of this result—applicable to stars like white dwarfs—is that the angular velocity is constant on cylindrical surfaces, which is von Zeipel's theorem (Ex. 13.8).

(f) Compute the shape of the surfaces on which Ω and ℓ are constant in the r-θ "plane" (surface of constant t and ϕ) for a spinning black hole with $a = 0.9\,M$.

(g) Now make the second simplifying assumption: the specific angular momentum ℓ is constant. Compute the shape of the isobars, also in the r-θ plane, and show that they exhibit a cusp along a circle in the equatorial plane, whose radius shrinks as ℓ increases.

(h) Compute the specific energy at infinity, $\mathcal{E}_\infty$, of the isobar that passes through the cusp, and show that it can vanish if ℓ is large enough. Interpret your answer physically.

Of course it is possible to deal with more realistic assumptions about the equation of state and the angular momentum distribution, but this relatively simple model brings out some salient features of the equilibrium flow.

Exercise 26.19 *Problem: Geodetic and Lense-Thirring Precession* T2

(a) Consider a pulsar in a circular, equatorial orbit around a massive Kerr black hole, with the pulsar's spin vector $\vec{S}$ lying in the hole's equatorial plane. Using the fact that the spin vector is orthogonal to the pulsar's 4-velocity (Sec. 25.7), show that its only nonzero covariant components are S_r, S_ϕ, and $S_t = -\Omega S_\phi$, where $\Omega = d\phi/dt$ is the pulsar's orbital angular velocity [Eq. (26.94d)]. Now, neglecting usually negligible quadrupole-curvature coupling forces (Sec. 25.7), the spin is parallel-transported along the pulsar's geodesic, so $\nabla_{\vec{u}}\vec{S} = 0$. Show that this implies

$$\frac{d^2 S_\alpha}{d\tau^2} = \Gamma^\delta{}_{\alpha\gamma}\Gamma^\beta{}_{\delta\epsilon}u^\gamma u^\epsilon S_\beta. \tag{26.96a}$$

(b) Next use Boyer-Lindquist coordinates and the results from Ex. (26.17) plus computer algebra to show that the spatial part of Eq. (26.96a) takes the remarkably simple form

$$\frac{d^2 S_i}{d\tau^2} = -\frac{M S_i}{r^3}, \tag{26.96b}$$

where $i = r, \phi$. Interpret this equation geometrically. In particular, comment on the absence of the hole's spin parameter a in the context of the dragging of inertial frames, and consider how two counter-orbiting stars would precess relative to each other.

(c) Explain why the rate of precession of the pulsar's spin as measured by a distant observer is given by

$$\Omega_p = \Omega - \frac{M^{1/2}}{r^{3/2} u^t}, \tag{26.96c}$$

where $u^t = dt/d\tau$ is the contravariant time component of the pulsar's 4-velocity.

(d) Use computer algebra to evaluate Ω_p, and show that a Taylor expansion for $r \gg M$ gives

$$\Omega_p = \frac{3M^{3/2}}{2r^{5/2}} - \frac{aM}{r^3} + \ldots. \qquad (26.96d)$$

The first term on the right-hand side is known as *geodetic* precession; the second as *Lense-Thirring* precession.

Exercise 26.20 *Challenge: General Orbits around a Kerr Black Hole* T2

By combining the techniques used for general orbits around a Schwarzschild black hole (Ex. 26.12) with those for equatorial orbits around a Kerr black hole (Ex. 26.17), explore the properties and shapes of general, geodesic orbits around a Kerr hole. To make progress on this exercise, note that because the Kerr spacetime is not spherically symmetric, the orbits will not, in general, lie in "planes" (2- or 3-dimensional surfaces) of any sort. Correspondingly, one must explore the orbits in 4 spacetime dimensions rather than 3. In addition to the fairly obvious three constants of motion that the Kerr hole shares with Schwarzschild—μ, $\mathcal{E}_\infty$, and L [Eqs. (26.63)]—there is a fourth called the *Carter constant* after Brandon Carter, who discovered it:

$$\mathcal{Q} = p_\theta^2 + \cos^2\theta [a^2(\mu^2 - \mathcal{E}_\infty^2) + \sin^{-2}\theta L^2]. \qquad (26.97)$$

The numerical integrations are best carried out using Hamilton's equations for the super-hamiltonian (Ex. 25.7).

Formulas for the geodesics and some of their properties are given in most general relativity textbooks, for example, Misner, Thorne, and Wheeler (1973, Sec. 33.5) and Straumann (2013, Sec. 8.4). For an extensive numerical exploration of the orbital shapes, based on the super-hamiltonian, see Levin and Perez-Giz (2008)—who present and discuss the Hamilton equations in their Appendix A.

Exercise 26.21 *Challenge: Electromagnetic Extraction of Energy from a Spinning Black Hole* T2

As discussed in the text, spinning black holes contain a considerable amount of rotational energy [Eq. (26.86)]. This exercise sketches how this energy may be extracted by an electromagnetic field: the Blandford-Znajek process.

(a) Suppose that a Kerr black hole, described using Boyer-Lindquist coordinates $\{t, r, \theta, \phi\}$, is orbited by a thick accretion disk (Ex. 26.18), whose surface near the hole consists of two funnels, axisymmetric around the hole's rotation axis $\theta = 0$ and $\theta = \pi$. Suppose, further, that the funnels' interiors contain a stationary, axisymmetric electromagnetic field described by a vector potential $A_\alpha(r, \theta)$. The surface of each funnel contains surface current and charge that keep the disk's interior free of electromagnetic field, and the disk's gas supplies pressure across the funnel's surface to balance the electromagnetic stress. Whatever plasma there may be inside the funnel has such low density that the electromagnetic contribution to the stress-energy tensor is dominant, and so we can write $F_{\alpha\beta} J^\beta = 0$ [cf. Eq. (2.81b)]. Show that the electromagnetic field tensor [Eq. (25.61)] in the Boyer-Lindquist coordinate basis can be written $F_{\alpha\beta} = A_{\beta,\alpha} - A_{\alpha,\beta}$, where the commas denote partial derivatives.

(b) Write $A_\phi = \Phi/(2\pi)$, $A_t = -V$, and by considering the electromagnetic field for $r \gg M$, interpret Φ and V as the magnetic flux contained in a circle of fixed r, θ and the electrical potential of that circle, respectively. This also remains true near the black hole, where we define the magnetic and electric fields as those measured by observers who move orthogonally to the hypersurfaces of constant Boyer-Lindquist time t, so $\vec{E}$ and $\vec{B}$ can be regarded as spatial vectors **E** and **B** that lie in those hypersurfaces of constant t. Can you prove all this?

(c) Explain why it is reasonable to expect the electric field **E** to be orthogonal to the magnetic field **B** near the hole as well as far away, and use Faraday's law to show that its toroidal component $E_{\hat\phi}$ must vanish. Thereby conclude that the equipotential surfaces $V = $ constant coincide with the magnetic surfaces $\Phi = $ constant.

(d) Define an angular velocity by $\Omega = -2\pi V_{,\theta}/\Phi_{,\theta}$, show that it, too, is constant on magnetic surfaces, and show that an observer who moves with this angular velocity $d\phi/dt = \Omega$ measures vanishing electric field. Thereby conclude that the magnetic field lines rotate rigidly with this angular velocity, and deduce an expression for the electric field **E** in terms of Φ and Ω. (You might want to compare with Ex. 19.11.)

(e) Use the inhomogeneous Maxwell equations to show that the current density J^α describes a flow of charge along the magnetic field lines. Hence calculate the current I flowing inside a magnetic surface. Express **B** in terms of I and Φ.

(f) Now sketch the variation of the electromagnetic field in the funnel both near the horizon and at a large distance from it, assuming that there is an outward flow of energy and angular momentum.

(g) Use the fact that $\partial/\partial t$ and $\partial/\partial \phi$ are Killing vectors (Ex. 25.5) to confirm that electromagnetic energy and angular momentum are conserved in the funnels.

(h) We have derived these general principles without giving an explicit solution that demonstrates energy extraction. In order to do this, we must also specify boundary conditions at the horizon of the black hole and at infinity. The former is essentially that the electromagnetic field be nonsingular when measured by an infalling observer or when expressed in Kerr coordinates, for example. The latter describes the flow of energy and angular momentum as outward. Stable solutions can be exhibited numerically and, typically, they have $\Omega \sim 0.3$ to $0.5\Omega_H$ (McKinney, Tchekhovskoy, and Blandford, 2012). These solutions demonstrate that energy and angular momentum can flow outward across the event horizon so the mass of the black hole gradually decreases. How can this be?

26.6 The Many-Fingered Nature of Time

We conclude this chapter with a discussion of a concept that John Archibald Wheeler (the person who has most clarified the conceptual underpinnings of general relativity) calls the *many-fingered nature of time*.

In the flat spacetime of special relativity there are preferred families of observers: each such family lives in a global Lorentz reference frame and uses that frame to split spacetime into space plus time. The hypersurfaces of constant time (slices of simultaneity) that result from that split are flat hypersurfaces slicing through all of spacetime (Fig. 26.9a). Of course, different preferred families live in different global Lorentz frames and thus split up spacetime into space plus time in different manners (e.g., the dotted and dashed slices of constant time in Fig. 26.9a in contrast to the solid ones). As a result, no universal concept of time exists in special relativity. But at least there are some strong restrictions on time: each inertial family of observers will agree that another family's slices of simultaneity are flat slices.

<!-- margin: in special relativity, different inertial observers slice spacetime into space plus time differently, but all slices are flat -->

In general relativity (i.e., in curved spacetime), even this restriction is gone. In a generic curved spacetime there are no flat hypersurfaces, and hence no candidates for flat slices of simultaneity. In addition, no global Lorentz frames and thus no preferred families of observers exist in a generic curved spacetime. A family of observers who are all initially at rest with respect to one another, and each of whom moves freely (inertially), will soon acquire relative motion because of tidal forces. As a result, their

<!-- margin: in general relativity, the slices of constant time have arbitrary shapes; hence, the "many-fingered nature of time" -->

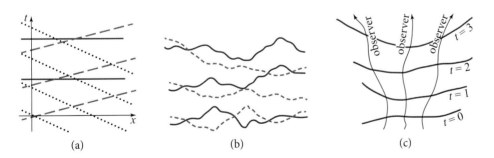

FIGURE 26.9 Spacetime diagrams showing the slices of simultaneity as defined by various families of observers. (a) Flat spacetime. The three families (those with solid slices, those with dashed, and those with dotted) are inertial, so their slices of constant time are those of global Lorentz frames. (b) Curved spacetime. The two families' slices of simultaneity illustrate the "many-fingered" nature of time. (c) Curved spacetime. The selection of an arbitrary foliation of spacelike hypersurfaces of simultaneity, and the subsequent construction of the world lines of observers who move orthogonally to those hypersurfaces (i.e., for whom light-ray synchronization will define those hypersurfaces as simultaneities).

slices of simultaneity (defined locally by Einstein light-ray synchronization, and then defined globally by patching together the little local bits of slices) may soon become rather contorted. Correspondingly, as is shown in Fig. 26.9b, different families of observers will slice spacetime up into space plus time in manners that can be quite distorted, relative to one another—with "fingers" of one family's time slices pushing forward, ahead of the other family's here, and lagging behind there, and pushing ahead in some other place.

In curved spacetime it is best to not even restrict oneself to inertial (freely falling) observers. For example, in the spacetime of a static star, or of the exterior of a Schwarzschild black hole, the family of static observers [observers whose world lines are $\{(r, \theta, \phi) = \text{const}, t \text{ varying}\}$] are particularly simple; their world lines mold themselves to the static structure of spacetime in a simple, static manner. However, these observers are not inertial; they do not fall freely. This need not prevent us from using them to split up spacetime into space plus time, however. Their proper reference frames produce a perfectly good split. When one uses that split, in the case of a black hole, one obtains a 3-dimensional-space version of the laws of black-hole physics that is a useful tool in astrophysical research; see Thorne, Price, and MacDonald (1986).

For any family of observers, accelerated or inertial, the slices of simultaneity as defined by Einstein light-ray synchronization over small distances (or equivalently by the space slices of the observer's proper reference frames) are the 3-surfaces orthogonal to the observers' world lines (cf. Fig. 26.9c). To see this most easily, pick a specific event along a specific observer's world line, and study the slice of simultaneity there from the viewpoint of a local Lorentz frame in which the observer is momentarily at rest. Light-ray synchronization guarantees that locally, the observer's slice of simultaneity will be the same as that of this local Lorentz frame. Since the frame's slice is orthogonal to its own time direction and that time direction is the same as the direction of the observer's world line, the slice is orthogonal to the observer's world line. By the discussion in Sec. 24.5, the slice is also locally the same (to first order in distance away from the world line) as a slice of constant time in the observer's proper reference frame.

If the observers rotate around one another (in curved spacetime or in flat), it is not possible to mesh their local slices of simultaneity, defined in this manner, into global slices of simultaneity (i.e., there are no global 3-dimensional hypersurfaces orthogonal to their world lines). We can protect against this eventuality by choosing the slices first: select any family of nonintersecting spacelike slices through the curved spacetime (Fig. 26.9c). Then there will be a family of timelike world lines that are everywhere orthogonal to these hypersurfaces. A family of observers who move along those orthogonal world lines and who define their 3-spaces of simultaneity by local light-ray synchronization will thereby identify the orthogonal hypersurfaces as their simultaneities. Exercise 26.22 illustrates these ideas using Schwarzschild spacetime, and Box 26.3 uses these ideas to visualize a black hole's spacetime curvature.

BOX 26.3. TENDEX AND VORTEX LINES OUTSIDE A BLACK HOLE T2

When one uses a family of spacelike slices (a *foliation*) with unit normals (4-velocities of orthogonally moving observers) $\vec{w}$ to split spacetime up into space plus time, the electromagnetic field tensor $\mathbf{F}$ splits into the electric field $E^\alpha = F^{\alpha\beta} w_\beta$ and the magnetic field $B^\beta = \frac{1}{2}\epsilon^{\alpha\beta\gamma\delta} F_{\gamma\delta} w_\alpha$. These are 3-vectors lying in the spacelike slices. In terms of components in the observers' proper reference frames, they are $E^{\hat{j}} = F^{\hat{0}\hat{j}}$ and $B^{\hat{i}} = \epsilon^{\hat{i}\hat{j}\hat{k}} F_{\hat{j}\hat{k}}$ (see Sec. 2.11 and especially Fig. 2.9). Similarly, the foliation splits the vacuum Riemann tensor into the symmetric, trace-free tidal field $\mathcal{E}_{\hat{i}\hat{j}} = R_{\hat{i}\hat{0}\hat{j}\hat{0}}$ (which produces relative accelerations $\Delta a_{\hat{i}} = -\mathcal{E}_{\hat{i}\hat{j}} \xi^{\hat{j}}$ of particles separated by $\xi^{\hat{j}}$), and the frame-drag field $\mathcal{B}_{\hat{i}\hat{j}} = \frac{1}{2}\epsilon_{\hat{i}\hat{p}\hat{q}} R^{\hat{p}\hat{q}}{}_{\hat{j}\hat{0}}$ (which produces differential frame dragging $\Delta\Omega_{\hat{i}} = \mathcal{B}_{\hat{i}\hat{j}}\xi^{\hat{j}}$). See Box 25.2.

Just as the electromagnetic field can be visualized using electric and magnetic field lines that live in the spacelike slices, so also the tidal field $\mathcal{E}_{\hat{i}\hat{j}}$ and frame-drag field $\mathcal{B}_{\hat{i}\hat{j}}$ can each be visualized using integral curves of its three eigenvector fields: the tidal field's tendex lines and the frame-drag field's vortex lines (Box 25.2). These lines lie in the space slices and are often color coded by their eigenvalues, which are called the lines' *tendicities* and *(frame-drag) vorticities*.

For a nonspinning (Schwarzschild) black hole, the frame-drag field vanishes (no spin implies no frame dragging); and the tidal field (with slicing either via Schwarzschild coordinate time t or Eddington-Finkelstein coordinate time $\tilde{t}$) has proper-reference-frame components $\mathcal{E}_{\hat{r}\hat{r}} = -2M/r^3$, $\mathcal{E}_{\hat{\theta}\hat{\theta}} = \mathcal{E}_{\hat{\phi}\hat{\phi}} = +M/r^3$ [Eq. (6) of Box 26.2]. This tidal field's eigenvectors are $\vec{e}_{\hat{r}}, \vec{e}_{\hat{\theta}}$, and $\vec{e}_{\hat{\phi}}$, and their integral curves (the tendex lines) are identical to those shown for a weakly gravitating, spherical body in the figure at the bottom left of Box 25.2.

For a black hole with large spin, $a/M = 0.95$ for example, it is convenient to choose as our space slices, hypersurfaces of constant Kerr coordinate time $\tilde{t}$ [Eq. (26.78)], since these penetrate smoothly through the horizon. The tendex and vortex lines that live in those slices have the following forms (Zhang et al., 2012).

The tendex lines are shown on the left and in the central inset; the vortex lines are shown on the right. The horizon is color coded by the sign of the tendicity (left) and vorticity (right) of the lines that emerge from it: blue for positive and red for negative.

The blue tendex lines have positive tendicity, and the tidal-acceleration equation $\Delta a_{\hat{i}} = -\mathcal{E}_{\hat{i}\hat{j}} \xi^{\hat{j}}$ says that a woman whose body is oriented along them

(continued)

BOX 26.3. (continued)

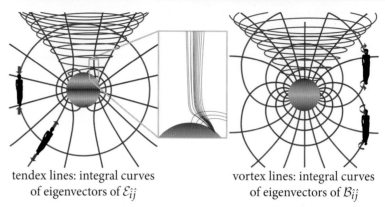

tendex lines: integral curves of eigenvectors of $\mathcal{E}_{\hat{i}\hat{j}}$

vortex lines: integral curves of eigenvectors of $\mathcal{B}_{\hat{i}\hat{j}}$

gets squeezed with relative "gravitational" acceleration between head and foot equal to the lines' tendicity times her body length. The red lines have negative tendicity, so they stretch a man with a head-to-foot relative acceleration equal to the magnitude of their tendicity times his body length. Notice that near the poles of a fast-spinning black hole, the radial tendex lines are blue, and in the equatorial region they are red. Therefore, a man falling into the polar regions of a fast-spinning hole gets squeezed radially, and one falling into the equatorial regions gets stretched radially.

For a woman with her body oriented along a vortex line, the differential frame-drag equation $\Delta \Omega_{\hat{i}} = \mathcal{B}_{\hat{i}\hat{j}} \xi^{\hat{j}}$ says that a gyroscope at her feet precesses clockwise (blue line, positive vorticity) or counterclockwise (red line, negative vorticity) relative to inertial frames at her head. And a gyroscope at her head precesses in that same direction relative to inertial frames at her feet. Thus, the vortex lines can be regarded as either counterclockwise (red) or clockwise (blue). The precessional angular velocity is equal to the line's (frame-drag) vorticity times the woman's body length.

Notice that counterclockwise (red) vortex lines emerge from the north polar region of the horizon, swing around the south pole, and descend back into the north polar region. Similarly, (blue) vortex lines emerge from the horizon's south polar region, swing around the north pole, and descend back into the south. This is similar to the vortex lines for a spinning body in linearized theory (bottom right figure in Box 25.2). The differential precession is counterclockwise along the (red) near-horizon radial lines in the north polar region, because gyroscopes near the horizon are dragged into precession more strongly by the hole's spin, the nearer one is to the horizon. This also explains the clockwise precession along the (blue) near-horizon radial lines in the south polar region.

For more details about the tendex and vortex lines of Kerr black holes, see Zhang et al. (2012).

Exercise 26.22 *Practice: Slices of Simultaneity in Schwarzschild Spacetime* T2

(a) One possible choice of slices of simultaneity for Schwarzschild spacetime is the set of 3-surfaces $\{t = \text{const}\}$, where t is the Schwarzschild time coordinate. Show that the unique family of observers for whom these are the simultaneities are the static observers, with world lines $\{(r, \theta, \phi) = \text{const}, t \text{ varying}\}$. Explain why these slices of simultaneity and families of observers exist only outside the horizon of a black hole and cannot be extended into the interior. Draw a picture of the world lines of these observers and their slices of simultaneity in an Eddington-Finkelstein spacetime diagram.

(b) A second possible choice of simultaneities is the set of 3-surfaces $\{\tilde{t} = \text{const}\}$, where $\tilde{t}$ is the Eddington-Finkelstein time coordinate. What are the world lines of the observers for whom these are the simultaneities? Draw a picture of those world lines in an Eddington-Finkelstein spacetime diagram. Note that they and their simultaneities cover the interior of the hole as well as its exterior.

Bibliographic Note

In our opinion, the best elementary textbook treatment of black holes and relativistic stars is that in Hartle (2003, Chaps. 12, 13, 15, 24); this treatment is also remarkably complete.

For the treatment of relativistic stars at an elementary level, we also recommend Schutz (2009, Chap. 10), and at a more advanced level (including stellar pulsations), Straumann (2013, Chaps. 4, 9) and Misner, Thorne, and Wheeler (1973, Chaps. 31–34).

For the study of black holes at an intermediate level, see Carroll (2004, Chaps. 5, 6), and Hobson, Efstathiou, and Lasenby (2006, Chaps. 9, 11, 13), and at more advanced levels, Wald (1984, Chap. 12), which is brief and highly mathematical, and Straumann (2013, Chap. 7) and Misner, Thorne, and Wheeler (1973, Chaps. 23, 24, 26), which are long and less mathematical.

The above are all portions of general relativity textbooks. There are a number of books and monographs devoted solely to the theory of black holes and/or relativistic stars. Among these, we particularly recommend the following. Shapiro and Teukolsky (1983) is an astrophysically oriented book at much the same level as this chapter, but with much greater detail and extensive applications; it deals with black holes, neutron stars, and white dwarf stars in astrophysical settings. Meier (2012) is a much more up-to-date and quite comprehensive treatment of the astrophysics and observations of black holes of all sizes. Frolov and Novikov (1998) is a thorough monograph on black holes, including their fundamental theory and their interactions with the rest of the universe; it includes extensive references to the original literature and readable summaries of all the important issues that had been treated by black-hole researchers as of 1997. Chandrasekhar (1983) is an idiosyncratic but elegant and complete monograph on the theory of black holes and especially small perturbations of them. Thorne, Price, and MacDonald (1986) is an equally idiosyncratic monograph that formulates the theory of black holes in 3+1 language, which facilitates physical understanding.

CHAPTER TWENTY-SEVEN

Gravitational Waves and Experimental Tests of General Relativity

> A system of moving bodies emits gravitational waves. . . . We shall assume that the speeds of all the system's bodies are small compared to the speed of light [and that the system's gravity is weak]. . . . Because of the presence of matter, the equation for the radiated waves . . . will have the form $\frac{1}{2}\Box \psi_i^k = \kappa \tau_i^k, \ldots$ where τ_i^k . . . contain, along with the [material stress-energy tensor], also terms of quadratic order from [the Einstein tensor].[1]
>
> LEV LANDAU AND EVGENY LIFSHITZ (1941)

27.1 Overview

In 1915, when Einstein formulated general relativity, human technology was inadequate for testing it definitively. Only a half century later did technology begin to catch up. In the years since then, the best experiments have improved from accuracies of a few tens of percent to a part in 10,000 or 100,000, and general relativity has passed the tests with flying colors. In Sec. 27.2, we describe some of these tests, derive general relativity's predictions for them, and discuss the experimental results.

Observations of gravitational waves are changing the character of research on general relativity. As of 2016, they have enabled observational studies of the large-amplitude, highly nonlinear vibrations of curved spacetime triggered when two black holes collide. Thereby, they have produced, for the first time, tests of general relativity when gravity is ultra strong and dynamical. They are enabling high-accuracy studies of relativistic effects in inspiraling black-hole binaries and soon should do the same for neutron stars—where they may also teach us about the equation of state of high-density nuclear matter. In the future, they will enable us to map the spacetime geometries of quiescent black holes with high precision. And they might provide a window into the physical conditions present during the first moments of the expansion of the universe (Sec. 28.7.1).

In this chapter, we develop the theory of gravitational waves in much detail and describe efforts to detect them and the sources that may be seen. More specifically, in Sec. 27.3, we develop the mathematics of gravitational waves, both classically and quantum mechanically (in the language of gravitons), and we study their propagation through flat spacetime. Then, in Sec. 27.4, we study their propagation through curved spacetime using the tools of geometric optics. In Sec. 27.5, we develop the simplest approximate method for computing the generation of gravitational waves, the "quadrupole-moment formalism." We also describe and present a few details of

---

1. By including the quadratic terms from the Einstein tensor, Landau and Lifshitz made their analysis of gravitational-wave generation valid for self-gravitating bodies such as binary stars—a remarkable achievement at this early date in the development of gravitational-wave theory.

> **BOX 27.1. READERS' GUIDE**
>
> - This chapter relies significantly on:
>   - Chap. 2 on special relativity;
>   - Chap. 24 on the transition from special relativity to general relativity;
>   - Chap. 25 on the fundamental concepts of general relativity, especially Sec. 25.9 on weak, relativistic gravitational fields;
>   - Chap. 26 on relativistic stars and black holes; and
>   - Sec. 7.3 on geometric optics.
> - In addition, Sec. 27.2.3 on Fermat's principle and gravitational lenses is closely linked to Sec. 7.3.6 on Fermat's principle, Sec. 7.6 on gravitational lenses, and Sec. 8.6 on diffraction at a caustic.
> - Portions of this chapter are a foundation for Chap. 28 on cosmology.

other, more sophisticated and accurate methods based on multipolar expansions, post-Newtonian techniques, and numerical simulations on supercomputers (numerical relativity). In Sec. 27.6, we turn to gravitational-wave detection, focusing especially on the Laser Interferometer Gravitational wave Observatory (LIGO), pulsar timing arrays, and the proposed Laser Interferometer Space Antenna (LISA).

27.2 Experimental Tests of General Relativity

In this section, we describe briefly some of the most important experimental tests of general relativity. For greater detail and other tests, see Will (1993a, 2014).

27.2.1 Equivalence Principle, Gravitational Redshift, and Global Positioning System

A key aspect of the equivalence principle is the prediction that any object, whose size is extremely small compared to the radius of curvature of spacetime and on which no nongravitational forces act, should move on a geodesic. This means, in particular, that its trajectory through spacetime should be independent of its chemical composition. This is called the *weak equivalence principle*, or the *universality of free fall*.

weak equivalence principle (i.e., universality of free fall) and tests of it

Efforts to test the universality of free fall date back to Galileo's (perhaps apocryphal) experiment of dropping objects from the leaning tower of Pisa. Over the past century, a sequence of ever-improving experiments led by Roland von Eötvös (ca. 1920), Robert Dicke (ca. 1964), Vladimir Braginsky (ca. 1972), and Eric Adelberger (ca. 2008) have led to an accuracy $\Delta a/a < 2 \times 10^{-13}$ for the difference of gravitational acceleration toward the Sun for Earthbound bodies with very different chemical compositions (Schlamminger et al., 2008). A planned atom-interferometer experiment is designed to reach $\Delta a/a \lesssim 1 \times 10^{-15}$ (Biedermann et al., 2015). A proposed space experiment called *STEP* has the prospect to increase this accuracy to the phenomenal level of $\Delta a/a \lesssim 1 \times 10^{-18}$ (Overduin et al., 2012).

General relativity predicts that bodies with significant self-gravity (even black holes) should also fall, in a nearly homogeneous external gravitational field, with the same acceleration as a body with negligible self-gravity. This prediction, sometimes called the *strong equivalence principle,* has been tested by comparing the gravitational accelerations of Earth and the Moon toward the Sun. Their fractional difference of acceleration (as determined by tracking the relative motions of the Moon and Earth using laser beams fired from Earth, reflected off mirrors that astronauts and cosmonauts have placed on the Moon, and received back at Earth) has been measured to be $\Delta a/a \lesssim 1 \times 10^{-13}$. Since Earth and the Moon have (gravitational potential energy)/(rest-mass energy) $\simeq -4 \times 10^{-10}$ and $\simeq -2 \times 10^{-11}$, respectively, this verifies that gravitational energy falls with the same acceleration as other forms of energy to within about 2.5 parts in 10,000. A recently discovered pulsar, J0337+1715, is in orbit about a white dwarf binary (Ransom et al. 2014). As the gravitational binding energy of the pulsar is roughly 10% of its mass, this will soon allow a high-accuracy test of the strong equivalence principle. For references and for discussions of a variety of other tests of the equivalence principle, see Merkowitz (2010) and Will (1993a, 2014).

tests for self-gravitating bodies

From the equivalence principle one can deduce that, for an emitter and absorber at rest in a Newtonian gravitational field Φ, light (or other electromagnetic waves) must be gravitationally redshifted by an amount $\Delta \lambda / \lambda = \Delta \Phi$, where $\Delta \Phi$ is the difference in Newtonian potential between the locations of the emitter and receiver. (See Ex. 26.4 for a general relativistic derivation when the field is that of a nonspinning, spherical central body with the emitter on the body's surface and the receiver far from the body; see Ex. 24.16 for a derivation when the emitter and receiver are on the floor and ceiling of an Earthbound laboratory.) Relativistic effects produce a correction to this shift of magnitude $\sim (\Delta \Phi)^2$ [cf. Eq. (26.18)], but for experiments performed in the solar system, the currently available precision is too poor to detect this correction, so such experiments test the equivalence principle and not the details of general relativity.

gravitational redshift, and tests of it

The highest-precision test of this gravitational redshift thus far was NASA's 1976 Gravity-Probe-A Project (led by Robert Vessot), in which several atomic clocks were flown to a height of about 10,000 km above Earth and were compared with atomic clocks on Earth via radio signals transmitted downward. After correcting for special relativistic effects due to the relative motions of the rocket's clocks and the Earth clocks, the measured gravitational redshift agreed with the prediction to within the experimental accuracy of about 2 parts in 10,000.

The Global Positioning System (GPS), by which one can routinely determine one's location on Earth to within an accuracy of about 10 m, is based on signals transmitted from a set of Earth-orbiting satellites. Each satellite's position is encoded on its transmitted signals, together with the time of transmission as measured by atomic clocks onboard the satellite. A person's GPS receiver contains a high-accuracy clock and a computer. It measures the signal arrival time and compares with the encoded transmission time to determine the distance from satellite to receiver. It uses

influence of gravitational redshift on GPS

those distances from several satellites, together with the encoded satellite positions, to determine (by triangulation) the receiver's location on Earth.

The transmission times encoded on the signals are corrected for the gravitational redshift before transmission. Without this redshift correction, the satellite clocks would quickly get out of synchronization with all the clocks on the ground, thereby eroding the GPS accuracy; see Ex. 27.1. Thus a clear understanding of general relativity was crucial to the design of GPS!

EXERCISES

Exercise 27.1 *Practice: Gravitational Redshift for GPS*
The GPS satellites are in circular orbits at a height of 20,200 km above Earth's surface, where their orbital period is 12 sidereal hours. If the ticking rates of the clocks on the satellites were not corrected for the gravitational redshift, roughly how long would it take them to accumulate a time shift, relative to clocks on Earth, large enough to degrade the GPS position accuracy by 10 m? by 1 km?

27.2.2 Perihelion Advance of Mercury

It was known at the end of the nineteenth century that the point in Mercury's orbit closest to the Sun, known as its perihelion, advances at a rate of about 575″ per century with respect to the fixed stars, of which about 532″ can be accounted for by Newtonian perturbations due to the other planets. The remaining ∼43″ per century was a mystery until Einstein showed that it can be accounted for quantitatively by the general theory of relativity.

More specifically (as is demonstrated in Ex. 27.2), if we idealize the Sun as non-rotating and spherical (so its external gravitational field is Schwarzschild), we ignore the presence of the other planets, and we note that the radius of Mercury's orbit is very large compared to the Sun's mass (in geometrized units), then Mercury's orbit will be very nearly an ellipse; and the ellipse's perihelion will advance, from one orbit to the next, by an angle

predicted perihelion or periastron advance

$$\Delta\phi = 6\pi M/p + O(M^2/p^2) \text{ radians.} \qquad (27.1)$$

Here M is the Sun's mass,[2] and p is the ellipse's *semi-latus rectum*, which is related to its semimajor axis a (half its major diameter) and its eccentricity e by $p = a(1 - e^2)$. For the parameters of Mercury's orbit ($M = M_\odot \simeq 1.4766$ km, $a = 5.79089 \times 10^7$ km, $e = 0.205628$), this advance is 0.10352″ per orbit. Since the orbital period is 0.24085 Earth years, this advance corresponds to 42.98″ per century.

Although the Sun is not precisely spherical, its tiny gravitational oblateness (as inferred from measurements of its spectrum of pulsations; Fig. 16.3) has been shown to contribute negligibly to this perihelion advance. In addition, the frame dragging

---

2. The same formula is true in a binary whose two masses are comparable, with M the sum of the masses.

due to the Sun's rotational angular momentum is also (sadly!) negligible compared to the experimental accuracy, so 42.98″ per century is the relativistic contribution to Mercury's perihelion advance. Modern observational data agree with this to within the data's accuracy of about 1 part in 1,000.

The advance of periastron has also been measured in several binary pulsars. In the double pulsar, PSR J0737-3039 (see Sec. 27.2.4), the rate is $17°$ yr^{-1}. In practice this is used to measure the masses of the binary's neutron stars, but it also validates Eq. (27.1) for the advance rate at the same level, $\sim 10^{-3}$, as the Mercury measurement, with the important difference that it is testing the influence of having comparable masses.

Gravity in the solar system is weak. Even at Mercury's orbit, the gravitational potential of the Sun is only $|\Phi| \sim 3 \times 10^{-8}$. Therefore, when one expands the spacetime metric in powers of Φ, current experiments with their fractional accuracies $\sim 10^{-5}$ or worse are able to see only the first-order terms beyond Newtonian theory (i.e., terms of *first post-Newtonian order*). To move on to second post-Newtonian order, $O(\Phi^2)$ beyond Newton, in our solar system will require major advances in technology. However, second- and higher-order effects are beginning to be measured via gravitational waves from inspiraling compact binaries, where $|\Phi|$ is $\gtrsim 0.1$.

> measurements of perihelion advance of Mercury's orbit

> measurements of periastron advance in binary pulsars

> higher-order post-Newtonian corrections to periastron advance

EXERCISES

Exercise 27.2 *Example: Perihelion Advance*
Consider a small satellite in a noncircular orbit about a spherical body with much larger mass M, for which the external gravitational field is Schwarzschild. The satellite will follow a timelike geodesic. Orient the Schwarzschild coordinates so the satellite's orbit is in the equatorial plane: $\theta = \pi/2$.

(a) Because the metric coefficients are independent of t and ϕ, the satellite's energy-at-infinity $\mathcal{E}_\infty = -p_t$ and angular momentum $L = p_\phi$ must be constants of the satellite's motion (Ex. 25.4a). Show that

$$\mathcal{E}_\infty = \left(1 - \frac{2M}{r}\right)\frac{dt}{d\tau}, \quad L = r^2 \frac{d\phi}{d\tau}. \tag{27.2a}$$

See Ex. 26.12. Here and below we take the satellite to have unit mass, so its momentum and 4-velocity are the same, and its affine parameter ζ and proper time τ are the same.

(b) Introduce the coordinate $u = r^{-1}$ and use the normalization of the 4-velocity to derive the following differential equation for the orbit:

$$\left(\frac{du}{d\phi}\right)^2 = \frac{\mathcal{E}_\infty^2}{L^2} - \left(u^2 + \frac{1}{L^2}\right)(1 - 2Mu). \tag{27.2b}$$

(c) Differentiate this equation with respect to ϕ to obtain a second-order differential equation:

$$\frac{d^2u}{d\phi^2} + u - \frac{M}{L^2} = 3Mu^2. \tag{27.2c}$$

27.2 Experimental Tests of General Relativity

By reinstating the constants G, c, and comparing with the Newtonian orbital equation, argue that the right-hand side represents a relativistic perturbation to the Newtonian equation of motion.

(d) Henceforth in this exercise, assume that $r \gg M$ (i.e., $u \ll 1/M$), and solve the orbital equation (27.2c) by perturbation theory. More specifically, at zero order (i.e., setting the right-hand side to zero), show that the Kepler ellipse (Goldstein, Poole, and Safko, 2002, Sec. 3.7),

$$u_K = \left(\frac{M}{L^2}\right)(1 + e \cos \phi), \qquad (27.2d)$$

is a solution. Here e (a constant of integration) is the ellipse's eccentricity, and L^2/M is the ellipse's *semi-latus rectum*, p. The orbit has its minimum radius at $\phi = 0$.

(e) By substituting u_K from part (d) into the right-hand side of the relativistic equation of motion (27.2c), show (to first-order in the relativistic perturbation) that in one orbit the angle ϕ at which the satellite is closest to the mass advances by $\Delta \phi \simeq 6\pi M^2/L^2$. [Hint: Try to write the differential equation in the form $d^2u/d\phi^2 + (1+\epsilon)^2 u \simeq \ldots$, where $\epsilon \ll 1$.]

(f) For the planet Mercury, using the parameter values given after Eq. (27.1), deduce that the relativistic contribution to the rate of advance of the perihelion (point of closest approach to the Sun) is 42.98″ per century.

Exercise 27.3 *Example: Gravitational Deflection of Light*
Repeat the analysis of Ex. 27.2 for a photon following a null geodesic. More specifically, do the following.

(a) Show that the photon trajectory $u(\phi)$ (with $u \equiv 1/r$) obeys the differential equation

$$\frac{d^2u}{d\phi^2} + u = 3Mu^2. \qquad (27.3)$$

(b) Obtain the zeroth-order solution by ignoring the right-hand side:

$$u = \frac{\sin \phi}{b}, \qquad (27.4)$$

where b is an integration constant. Show that this is just a straight line in the asymptotically flat region far from the body, and b is the impact parameter (projected distance of closest approach to the body).

(c) Substitute this solution (27.4) into the right-hand side of Eq. (27.3), and show that the perturbed trajectory satisfies

$$u = \frac{\sin \phi}{b} + \frac{M}{b^2}(1 - \cos \phi)^2. \qquad (27.5)$$

(d) Hence show that a ray with impact parameter $b \gg M$ will be deflected through an angle

$$\Delta\phi = \frac{4M}{b} \qquad (27.6)$$

[cf. Eq. (7.87) and associated discussion].

27.2.3 Gravitational Deflection of Light, Fermat's Principle, and Gravitational Lenses

Einstein not only explained the anomalous perihelion shift of Mercury. He also predicted (Ex. 27.3) that the null rays along which starlight propagates will be deflected, when passing through the curved spacetime near the Sun, by an angle

gravitational deflection of starlight

$$\boxed{\Delta\phi = 4M/b + O(M^2/b^2)} \qquad (27.7)$$

relative to their trajectories if spacetime were flat. Here M is the Sun's mass, and b is the ray's impact parameter (distance of closest approach to the Sun's center). For comparison, theories that incorporated a Newtonian-like gravitational field into special relativity (Sec. 25.1 and Ex. 25.1) predicted no deflection of light rays; the corpuscular theory of light combined with Newtonian gravity predicted half the general relativistic deflection, as did a 1911 principle-of-equivalence argument by Einstein that was ignorant of the curvature of space. The deflection was measured to an accuracy ~20% during the 1919 solar eclipse and agreed with general relativity rather than with the competing theories—a triumph that helped make Einstein famous. Modern experiments, based on the deflection of radio waves from distant quasars, as measured using very long baseline interferometry (interfering the waves arriving at radio telescopes with transcontinental or transworld separations; Sec. 9.3), have achieved accuracies of about 1 part in 10,000, and they agree completely with general relativity. Similar accuracies are now achievable using optical interferometers in space and may soon be achievable via optical interferometry on the ground.

measurements of gravitational light deflection

These accuracies are so great that, when astronomers make maps of the sky using either radio interferometers or optical interferometers, they must now correct for gravitational deflection of the rays not only when the rays pass near the Sun but also for rays coming in from nearly all directions. This correction is not quite as easy as Eq. (27.7) suggests, since that equation is valid only when the telescope is much farther from the Sun than the impact parameter. In the more general case, the correction is more complicated and must include aberration due to the telescope motion as well as the effects of spacetime curvature.

light deflection for large impact parameters: influences on astronomical observations

The gravitational deflection of light rays passing through or near a cluster of galaxies can produce a spectacular array of distorted images of the light source. In Sec. 7.6, we deduced the details of this gravitational lens effect using a model in which we treated spacetime as flat but endowed with a refractive index $\mathfrak{n}(\mathbf{x}) = 1 - 2\Phi(\mathbf{x})$, where $\Phi(\mathbf{x})$ is the Newtonian gravitational potential of the lensing system. This model can also be used to compute light deflection in the solar system. We now derive this model from general relativity.

gravitational lensing

The foundation for this model is the following general relativistic version of Fermat's principle [see Eq. (7.46) for the Newtonian version]: Consider any static spacetime geometry [i.e., one for which we can introduce a coordinate system in which $\partial g_{\alpha\beta}/\partial t = 0$ and $g_{jt} = 0$; so the only nonzero metric coefficients are $g_{00}(x^k)$ and $g_{ij}(x^k)$]. In such a spacetime the time coordinate t is special, since it is tied to the spacetime's temporal symmetry. An example is Schwarzschild spacetime and the Schwarzschild time coordinate t. Now, consider a light ray emitted from a spatial point $x^j = a^j$ in the static spacetime and received at a spatial point $x^j = b^j$. Assuming the spatial path along which the ray travels is $x^j(\eta)$, where η is any parameter with $x^j(0) = a^j$, $x^j(1) = b^j$, then the total coordinate time Δt required for the light's trip from a^j to b^j is

general relativistic Fermat's principle

$$\Delta t = \int_0^1 \sqrt{\gamma_{jk}\frac{dx^j}{d\eta}\frac{dx^k}{d\eta}} d\eta, \quad \text{where } \gamma_{jk} \equiv \frac{g_{jk}}{-g_{00}} \tag{27.8}$$

(computed using the fact that the ray must be null so $ds^2 = g_{00}dt^2 + g_{ij}dx^i dx^j = 0$). Fermat's principle says that *the actual spatial trajectory of the light path, in any static spacetime, is one that extremizes this coordinate time lapse.*

This principle can be proved (Ex. 27.4) by showing that the Euler-Lagrange equation for the action (27.8) is equivalent to the geodesic equation for a photon in the static spacetime with metric $g_{\mu\nu}(x^k)$.

Derivation of Index-of-Refraction Model. The index-of-refraction model used to study gravitational lenses in Sec. 7.6 is easily deduced as a special case of this Fermat principle. In a nearly Newtonian situation, the linearized-theory, Lorenz-gauge, trace-reversed metric perturbation has the form (25.91) with only the time-time component being significantly large: $\bar{h}_{00} = -4\Phi$, $\bar{h}_{0j} \simeq 0$, $\bar{h}_{jk} \simeq 0$. Correspondingly, the metric perturbation [obtained by inverting Eq. (25.85)] is $h_{00} = -2\Phi$, $h_{jk} = -2\Phi\delta_{jk}$, and the full spacetime metric $g_{\mu\nu} = \eta_{\mu\nu} + h_{\mu\nu}$ is

$$ds^2 = -(1+2\Phi)dt^2 + (1-2\Phi)\delta_{jk}dx^j dx^k. \tag{27.9}$$

This is the standard spacetime metric (25.79) in the Newtonian limit with a special choice of spatial coordinates, those of linearized-theory Lorenz gauge. The Newtonian limit includes the slow-motion constraint that time derivatives of the metric are small compared to spatial derivatives [Eq. (25.75b)], so on the timescale for light to travel through a lensing system, the Newtonian potential can be regarded as static: **Newtonian limit of Fermat's principle** $\Phi = \Phi(x^j)$. Therefore, the Newtonian-limit metric (27.9) can be regarded as static, and the coordinate time lapse along a trajectory between two spatial points, Eq. (27.8), reduces to

$$\Delta t = \int_0^L (1-2\Phi)d\ell, \tag{27.10}$$

where $d\ell = \sqrt{\delta_{jk}dx^j dx^k}$ is distance traveled treating the coordinates as though they were Cartesian in flat space and L is that total distance between the two points.

According to the relativistic Fermat principle (27.8), this Δt is extremal for light rays. However, Eq. (27.10) is also the action for the Newtonian, nongravitational version of Fermat's principle [Eq. (7.47)], with index of refraction

$$\boxed{\mathfrak{n}(x^j) = 1 - 2\Phi(x^j).} \tag{27.11}$$

refractive-index model for computing gravitational lensing when gravity is weak

Therefore, the spatial trajectories of the light rays can be computed via the Newtonian Fermat principle, with the index of refraction (27.11). ∎

Although this index-of-refraction model involves treating a special (Lorenz-gauge) coordinate system as though the spatial coordinates were Cartesian and space were flat (so $d\ell^2 = \delta_{jk} dx^j dx^k$)—which does not correspond to reality—nevertheless, this model predicts the correct gravitational lens images. The reason is that it predicts the correct rays through the Lorenz-gauge coordinates, and when the light reaches Earth, the cumulative lensing has become so great that the slight difference in the coordinates here from truly Cartesian has negligible influence on the images one sees.

EXERCISES

Exercise 27.4 *Derivation: Fermat's Principle for a Photon's Path in a Static Spacetime*
Show that the Euler-Lagrange equation for the action principle (27.8) is equivalent to the geodesic equation for a photon in the static spacetime metric $g_{00}(x^k)$, $g_{ij}(x^k)$. Specifically, do the following.

(a) The action (27.8) is the same as that for a geodesic in a 3-dimensional space with metric γ_{jk} and with t playing the role of proper distance traveled [Eq. (25.19) converted to a positive-definite, 3-dimensional metric]. Therefore, the Euler-Lagrange equation for Eq. (27.8) is the geodesic equation in that (fictitious) space [Eq. (25.14) with t the affine parameter]. Using Eq. (24.38c) for the connection coefficients, show that the geodesic equation can be written in the form

$$\gamma_{jk}\frac{d^2 x^k}{dt^2} + \frac{1}{2}(\gamma_{jk,l} + \gamma_{jl,k} - \gamma_{kl,j})\frac{dx^k}{dt}\frac{dx^l}{dt} = 0. \tag{27.12a}$$

(b) Take the geodesic equation (25.14) for the light ray in the real spacetime, with spacetime affine parameter ζ, and change parameters to coordinate time t. Thereby obtain

$$g_{jk}\frac{d^2 x^k}{dt^2} + \Gamma_{jkl}\frac{dx^k}{dt}\frac{dx^l}{dt} - \Gamma_{j00}\frac{g_{kl}}{g_{00}}\frac{dx^k}{dt}\frac{dx^l}{dt} + \frac{d^2 t/d\zeta^2}{(dt/d\zeta)^2}g_{jk}\frac{dx^k}{dt} = 0,$$

$$\frac{d^2 t/d\zeta^2}{(dt/d\zeta)^2} + 2\Gamma_{0k0}\frac{dx^k/dt}{g_{00}} = 0. \tag{27.12b}$$

(c) Insert the second of these equations into the first, and write the connection coefficients in terms of derivatives of the spacetime metric. With a little algebra, bring your result into the form Eq. (27.12a) of the Fermat-principle Euler-Lagrange equation.

27.2.4 Shapiro Time Delay

Shapiro time delay

In 1964, Irwin Shapiro proposed a new experiment to test general relativity: monitor the round-trip travel time for radio waves transmitted from Earth and bounced off Venus or some other planet, or transponded by a spacecraft. As the line-of-sight between Earth and the planet or spacecraft gradually moves nearer and then farther from the Sun, the waves' rays will pass through regions of greater and then smaller spacetime curvature, which will influence the round-trip travel time by greater and then smaller amounts. From the time evolution of the round-trip time, one can deduce the changing influence of the Sun's spacetime curvature.

One can compute the round-trip travel time with the aid of Fermat's (geometric-optics) principle. The round-trip proper time, as measured on Earth (neglecting, for simplicity, Earth's orbital motion; i.e., pretending Earth is at rest relative to the Sun while a radio-wave's rays go out and back), is $\Delta\tau_\oplus = \sqrt{1 - 2M/r_\oplus}\,\Delta t \simeq (1 - M/r_\oplus)\Delta t$, where M is the Sun's mass, $r_\oplus$ is Earth's distance from the Sun's center, Δt is the round-trip coordinate time in the static solar-system coordinates, and we have used $g_{00} = -(1 - 2M/r_\oplus)$ at Earth. Because Δt obeys Fermat's principle, it is stationary under small perturbations of the ray's spatial trajectory. This allows us to compute it using a straight-line trajectory through the spatial coordinate system. Letting b be the impact parameter (the ray's closest coordinate distance to the Sun) and ℓ be coordinate distance along the straight-line trajectory and neglecting the gravitational fields of the planets, we have $\Phi = -M/\sqrt{\ell^2 + b^2}$, so the coordinate time lapse out and back is [Eq. (27.10)]

$$\Delta t = 2 \int_{-\sqrt{r_\oplus^2 - b^2}}^{\sqrt{r_{\text{refl}}^2 - b^2}} \left(1 + \frac{2M}{\sqrt{\ell^2 + b^2}}\right) d\ell. \tag{27.13}$$

Here r_{refl} is the radius of the location at which the ray is reflected (or transponded) back to Earth. Performing the integral and multiplying by $\sqrt{g_{00}} \simeq 1 - M/r_\oplus$, we obtain for the round-trip travel time measured on Earth:

$$\Delta\tau_\oplus = 2\left(a_\oplus + a_{\text{refl}}\right)\left(1 - \frac{M}{r_\oplus}\right) + 4M \ln\left[\frac{(a_\oplus + r_\oplus)(a_{\text{refl}} + r_{\text{refl}})}{b^2}\right], \tag{27.14}$$

where $a_\oplus = \sqrt{r_\oplus^2 - b^2}$, and $a_{\text{refl}} = \sqrt{r_{\text{refl}}^2 - b^2}$.

sharply varying term in Shapiro time delay

As Earth and the reflecting planet or transponding spacecraft move along their orbits, only one term in this round-trip time varies sharply: the term

$$\boxed{\Delta\tau_\oplus = 4M \ln(1/b^2) = -8M \ln b \simeq -40\,\mu s\, \ln b.} \tag{27.15}$$

When the planet or spacecraft passes nearly behind the Sun, as seen from Earth, b plunges to a minimum (on a timescale of hours or days) and then rises back up; correspondingly, the time delay shows a sharp blip. By comparing the observed blip with the theory in a measurement with the Cassini spacecraft, this Shapiro time delay has been verified to the remarkable precision of 2×10^{-5} (Bertotti, Iess, and Tortora, 2003).

The Shapiro effect has been seen in several binary pulsar systems. The two best examples are the double pulsar PSR J0737-3039 and PSR J1614-2230, where the lines of sight pass within 1° of the orbital plane, very close to the companions. The peak Shapiro delays are 51 μs and 21 μs, respectively. Remarkably, in the second example, the neutron star has a well-measured mass of 2.0 $M_\odot$, which is large enough to rule out several candidate equations of state for cold nuclear matter (DeMorest et al., 2010) (cf. Sec. 26.3.5).

measurements of Shapiro time delay in the solar system and in binary pulsars

27.2.5 Geodetic and Lense-Thirring Precession

As we have discussed in Secs. 25.9.3 and 26.5, the mass M and the angular momentum **J** of a gravitating body place their imprint on the body's asymptotic spacetime metric:

$$ds^2 = -\left(1 - \frac{2M}{r}\right)dt^2 - \frac{4\epsilon_{jkm}J^k x^m}{r^3}dt\, dx^j + \left[1 + O\left(\frac{M}{r}\right)\right]\delta_{jk}dx^j dx^k \quad (27.16)$$

[Eq. (25.98c)]. The mass imprint can be deduced from measurements of the orbital angular velocity $\Omega = \sqrt{M/r^3}$ of an object in a circular orbit, and the angular-momentum imprint, from the precession it induces in an orbiting gyroscope [Eq. (25.100)].

As we deduced in Ex. 26.19, there are actually two precessions: geodetic precession and Lense-Thirring precession. The geodetic precession, like the orbital angular velocity, arises from the spherically symmetric part of the metric; it says that the spin of a gyroscope in a circular orbit of radius r will precess around the orbit's angular momentum vector at a rate $\frac{3}{2}(M/r)\Omega$. The Lense-Thirring precession is the average of Eq. (25.100) around the orbit, which gives $\mathbf{J}/(2r^3)$ for a polar orbit, $-\mathbf{J}/r^3$ for an equatorial orbit, and $(-J_i + \frac{3}{2}P_{ij}J_j)/r^3$ for a general circular orbit, where **J** is the central body's angular momentum, r is the orbital radius, and P_{ij} projects into the plane of the orbit.

geodetic precession of a gyroscope

Lense-Thirring precession of a gyroscope

The earth-orbiting experiment Gravity Probe B (GP-B), led by Francis Everitt, has measured these two precessions (Everitt et al., 2011). GP-B comprises four spinning spheres (gyroscopes) in one satellite on a polar orbit, and has verified the geodetic precession of 6.6″ yr^{-1} to a fractional accuracy of 0.003, and the Lense-Thirring precession of 0.040″ yr^{-1} to fractional accuracy 0.2. More recently, a combination of three satellites, LAGEOS, LAGEOS2, and LARES, has been used to measure the Lense-Thirring precession of an inclined equatorial orbit (rather than a gyroscope in orbit), where the prediction is $J/r^3 = 0.080″$ yr^{-1}. A fractional accuracy of 0.05 has been reported (Ciufolini et al., 2016).

measurements of precessions by GP-B and LAGEOS

Relativistic precession has also been observed in six binary pulsar systems. Here we must take account of the pulsar (p) and its compact companion (c). We can give a heuristic argument for the precession rate. When $m_p \ll m_c$, the dominant precession is geodetic in the gravitational field of the companion at a rate $\Omega_p = \frac{3}{2}(m_c/r)\Omega$. In the opposite limit, the companion undergoes orbital Lense-Thirring precession at a rate $\Omega_p = 2J_p/r^3$ and so the torque is $(2m_c \vec{J}_p/r) \times \vec{\Omega}$, and the reflex precession of the pulsar, about the total (orbital plus spin) angular momentum, will be at a rate $\Omega_p = (2m_c/r)\Omega$. The simplest interpolation between these two limiting cases is

precession in binary pulsars

$$\Omega_p = \Omega \frac{m_c}{r} \frac{(3m_c + 4m_p)}{2(m_c + m_p)}. \tag{27.17}$$

This turns out to be correct. If the orbit is elliptical, we (again) replace the radius r with the *semi-latus rectum p*.

The best measured example is the double pulsar PSR J0737-3039, where pulses from both pulsars were observed. One pulsar has a spin almost aligned with the orbital angular momentum, and so its precession is hard to measure. The other has an inclined spin, which has now precessed so far that the pulses are no longer detected. However, this precession changes the projected shape of the magnetosphere, which occults the pulses from the nonprecessing pulsar, allowing a measurement of the rate of precession of $4.7 \pm 0.6°$ yr^{-1}, which is consistent with the predicted value of $5.1°$ yr^{-1} (Kramer et al., 2006).

measurements of precession in double pulsars

27.2.6 Gravitational Radiation Reaction

Radio observations of binary pulsars have already provided several indirect detections of gravitational waves, via radiation reaction in the binary.

Hulse-Taylor pulsar

The first binary pulsar (PSR B1913+16) was discovered in 1974 by Russell Hulse and Joseph Taylor. One star is a pulsar that emits radio pulses with a period of $\sim$59 ms at predictable times (allowing for the slowing down of the pulsar), and their arrival times can be determined to $\sim$15 μs. Its companion is almost certainly a neutron star, but it does not pulse. The orbital period is roughly 8 hours, and the orbit's eccentricity is $\sim$0.6. The pulses are received at Earth with Shapiro time delays due to crossing the binary orbit, and with other relativistic effects.

We do not know a priori the orbital inclination or the neutron-star masses. However, we obtain one relation between these three quantities by analyzing the Newtonian orbit. A second relation comes from measuring the consequences of the combined second-order Doppler shift and gravitational redshift as the pulsar moves in and out of its companion's gravitational field. A third relation comes from measuring the relativistic precession of the orbit's periastron. From these three relations, one can solve for the stars' masses and the orbital inclination, and as a check can verify that the Shapiro time delay comes out correctly. One can then use the system's parameters to predict the rate of orbital inspiral due to gravitational radiation reaction—a phenomenon with a magnitude of $\sim |\Phi|^{2.5}$ beyond Newton (i.e., 2.5 post-Newtonian order; Sec. 27.5.3). The prediction agrees with the measurements to an accuracy of $\sim 2 \times 10^{-3}$ (Weissberg, Nice, and Taylor, 2010)—a major triumph for general relativity! The agreement is even better, $<10^{-3}$, for the double pulsar PSR J0737-3039.

measurement of gravitational radiation reaction in binary pulsars

In LIGO's observations of gravitational waves from the inspiral of binary black holes, the radiation reaction is measured with lower accuracy than this, but because the binary is so compact—with M/a increasing from ~ 0.01 to ~ 0.3—higher-order corrections to the radiation reaction are readily measured.

For reviews of other tests of general relativity using binary pulsars, see Kaspi and Kramer (2016) and papers cited therein.

27.3 Gravitational Waves Propagating through Flat Spacetime

Gravitational waves are ripples in the curvature of spacetime that are emitted by violent astrophysical events and that propagate with the speed of light. It was clear to Einstein and others, even before general relativity was fully formulated, that his theory would have to predict gravitational waves, and within months after completing the theory, Einstein (1916b, 1918) worked out those waves' basic properties.

It turns out that, after they have been emitted, gravitational waves propagate through matter with near impunity: they propagate as though in vacuum, even when other matter and fields are present. (For a proof and discussion see, e.g., Thorne, 1983, Sec. 2.4.3.) This justifies specializing our analysis to vacuum propagation.

27.3.1 Weak, Plane Waves in Linearized Theory

Once the waves are far from their source, the radii of curvature of their phase fronts are huge compared to a wavelength, as is the radius of curvature of the spacetime through which they propagate. Thus to high accuracy, we can idealize the waves as plane-fronted and as propagating through flat spacetime. The appropriate formalism for describing this is the linearized theory developed in Sec. 25.9.2.

We introduce coordinates that are as nearly Lorentz as possible, so the spacetime metric can be written as

$$g_{\alpha\beta} = \eta_{\alpha\beta} + h_{\alpha\beta}, \quad \text{with } |h_{\alpha\beta}| \ll 1 \qquad (27.18a)$$

[Eq. (25.82)], and we call $h_{\alpha\beta}$ the waves' *metric perturbation*. We perform an "infinitesimal coordinate transformation" (gauge change):

$$x^\alpha_{\text{new}}(\mathcal{P}) = x^\alpha_{\text{old}}(\mathcal{P}) + \xi^\alpha(\mathcal{P}), \quad \text{which produces } h^{\text{new}}_{\mu\nu} = h^{\text{old}}_{\mu\nu} - \xi_{\mu,\nu} - \xi_{\nu,\mu} \qquad (27.18b)$$

[Eqs. (25.87) and (25.88)], with the gauge-change generators $\xi^\alpha(\mathcal{P})$ chosen to impose the Lorenz gauge condition

$$\bar{h}_{\mu\nu}{}^{,\nu} = 0 \qquad (27.18c)$$

[Eq. (25.89)] on the new trace-reversed metric perturbation:

$$\bar{h}_{\mu\nu} \equiv h_{\mu\nu} - \frac{1}{2} h\, \eta_{\mu\nu}, \quad h \equiv \eta^{\alpha\beta} h_{\alpha\beta} \qquad (27.18d)$$

[Eqs. (25.85) and (25.84)]. In this Lorenz gauge, the vacuum Einstein field equation becomes the flat-space wave equation for $\bar{h}^{\mu\nu}$ and so also for $h^{\mu\nu}$:

wave equation

$$\boxed{\bar{h}_{\mu\nu,\alpha}{}^{\alpha} = h_{\mu\nu,\alpha}{}^{\alpha} = 0} \tag{27.18e}$$

[Eq. (25.90)]. Here all indices after a comma are partial derivatives and they are raised with the flat metric $h_{\mu\nu,\alpha}{}^{\alpha} = h_{\mu\nu,\alpha\beta}\eta^{\beta\alpha}$.

This is as far as we went in vacuum (far from the waves' source) in Chap. 25. We now go further. We simplify the mathematics by orienting the axes of our nearly Lorentz coordinates so the waves are planar and propagate in the z direction. Then the obvious solution to the wave equation (27.18e) and the consequence of the Lorenz gauge condition (27.18c) are

$$\bar{h}_{\mu\nu} = \bar{h}_{\mu\nu}(t - z), \quad \bar{h}_{\mu 0} = -\bar{h}_{\mu z}. \tag{27.19}$$

There are now six independent components of the trace-reversed metric perturbation: the six spatial $\bar{h}_{ij}$; the second of Eqs. (27.19) fixes the time-space and time-time components in terms of them.

Remarkably, these six independent components can be reduced to two by a further specialization of gauge. The original infinitesimal coordinate transformation (27.18b), which brought us into Lorenz gauge, relied on four functions $\xi_{\mu}(\mathcal{P}) = \xi_{\mu}(x^{\alpha})$ of four spacetime coordinates. A more restricted set of gauge-change generators, $\xi_{\mu}(t-z)$, that are functions solely of retarded time (and thus satisfy the wave equation) will keep us in Lorenz gauge and can be used to annul the four components $\bar{h}_{xz}, \bar{h}_{yz}, \bar{h}_{zz}$, and $\bar{h} \equiv \eta^{\mu\nu}\bar{h}_{\mu\nu}$, whence (thanks to the Lorenz conditions $\bar{h}_{\mu 0} = -\bar{h}_{\mu z}$) all the $\bar{h}_{\mu 0}$ are also annulled. See Ex. 27.5. As a result, the trace-reversed metric perturbation $\bar{h}_{\mu\nu}$ and the metric perturbation $h_{\mu\nu}$ are now equal, and their only nonzero components are $h_{xx} = -h_{yy}$ and $h_{xy} = +h_{yx}$.

further specialization to TT gauge

for locally plane waves propagating in z direction

This special new gauge has the name *transverse-traceless gauge* or *TT gauge*, because in it the metric perturbation is purely spatial, it is transverse to the waves' propagation direction (the z direction), and it is traceless. It is convenient to use the notation $h_{\mu\nu}^{\mathrm{TT}}$ for the metric perturbation in this TT gauge, and convenient to give the names h_+ and $h_\times$ to its two independent, nonzero components (which are associated with two polarization states for the waves, "+" and "×"):

the metric perturbation in TT gauge; gravitational-wave fields h_+ and $h_\times$

$$\boxed{h_{xx}^{\mathrm{TT}} = -h_{yy}^{\mathrm{TT}} = h_+(t-z), \quad h_{xy}^{\mathrm{TT}} = +h_{yx}^{\mathrm{TT}} = h_\times(t-z).} \tag{27.20}$$

The Riemann curvature tensor in this TT gauge, as in any gauge, can be expressed as

Riemann tensor

$$\boxed{R_{\alpha\beta\gamma\delta} = \frac{1}{2}h_{\{\alpha\beta,\gamma\delta\}}^{\mathrm{TT}} \equiv \frac{1}{2}(h_{\alpha\delta,\beta\gamma}^{\mathrm{TT}} + h_{\beta\gamma,\alpha\delta}^{\mathrm{TT}} - h_{\alpha\gamma,\beta\delta}^{\mathrm{TT}} - h_{\beta\delta,\alpha\gamma}^{\mathrm{TT}})} \tag{27.21}$$

[Eq. (25.80)]. Here the subscript symbol $\{\cdot\}$, analogous to $[\cdot]$ for antisymmetrization and $(\cdot)$ for symmetrization, means the combination of four terms on the right side of the $\equiv$ sign. Of particular interest for physical measurements is the relativistic tidal field $\mathcal{E}_{ij} = R_{i0j0}$, which produces a relative acceleration of freely falling particles [geodesic deviation; Eq. (25.34)]. Since the temporal components of $h_{\mu\nu}^{\rm TT}$ vanish, the only nonzero term in Eq. (27.21) for R_{i0j0} is the third one, in which the temporal components are derivatives, whence

$$\mathcal{E}_{ij} = R_{i0j0} = -\frac{1}{2}\ddot{h}_{ij}^{\rm TT}; \quad \text{or}$$

tidal fields

$$\mathcal{E}_{xx} = -\mathcal{E}_{yy} = -\frac{1}{2}\ddot{h}_+(t-z), \quad \mathcal{E}_{xy} = +\mathcal{E}_{yx} = -\frac{1}{2}\ddot{h}_\times(t-z). \tag{27.22}$$

Here the dots mean time derivatives: $\ddot{h}_+(t-x) \equiv \partial^2 h_+/\partial t^2$. A useful index-free way to write these equations is

$$\mathcal{E} = -\frac{1}{2}\ddot{h}^{\rm TT} = -\frac{1}{2}\ddot{h}_+\mathbf{e}^+ - \frac{1}{2}\ddot{h}_\times\mathbf{e}^\times, \tag{27.23a}$$

where

$$\mathbf{e}^+ = \vec{e}_x \otimes \vec{e}_x - \vec{e}_y \otimes \vec{e}_y, \quad \mathbf{e}^\times = \vec{e}_x \otimes \vec{e}_y + \vec{e}_y \otimes \vec{e}_x \tag{27.23b}$$

polarization tensors

are the *polarization tensors* associated with the $+$ and $\times$ polarizations, respectively.

It is a very important fact that the Riemann curvature tensor is gauge invariant. An infinitesimal coordinate transformation $x_{\rm new}^\alpha(\mathcal{P}) = x_{\rm old}^\alpha(\mathcal{P}) + \xi^\alpha(\mathcal{P})$ changes it by tiny fractional amounts of order ξ^α, by contrast with the metric perturbation, which is changed by amounts of order itself: $\delta h_{\mu\nu} = -2\xi_{(\mu,\nu)}$ (i.e., by fractional amounts of order unity; Ex. 25.19). This has two important consequences. (i) The gauge-invariant Riemann tensor (or its space-time-space-time part, the tidal field) is an optimal tool for discussing physical measurements (Sec. 27.3.2)—a much better tool than, for example, the gauge-dependent metric perturbation. (ii) The gauge invariance of Riemann motivates us to change our viewpoint on $h_{ij}^{\rm TT}$ in the following way.

gauge invariance of Riemann tensor

We define a dimensionless "gravitational-wave field" $h_{ij}^{\rm TT}$ to be minus twice the double time integral of the wave's tidal field:

$$h_{ij}^{\rm TT} \equiv -2 \int dt \int dt\, \mathcal{E}_{ij}. \tag{27.24}$$

alternative viewpoint: gravitational wave field defined in terms of Riemann tensor's tidal field

And we regard the computation that led to Eq. (27.22) as a demonstration that it is possible to find a gauge in which the metric perturbation is purely spatial and its spatial part is equal to this gravitational-wave field: $h_{0\mu} = 0$ and $h_{ij} = h_{ij}^{\rm TT}$.

In Box 27.2, we show that, if we have found a gauge in which the metric perturbation propagates as a plane wave at the speed of light, then we can compute the gravitational-wave field $h_{ij}^{\rm TT}$ from that gauge's $h_{\alpha\beta}$ or $\bar{h}_{\alpha\beta}$ by a simple projection process. This result is useful in the theory of gravitational-wave generation (see, e.g., Sec. 27.5.2).

BOX 27.2. PROJECTING OUT THE GRAVITATIONAL-WAVE FIELD h_{ij}^{TT}

Suppose that, for some gravitational wave, we have found a gauge (not necessarily TT) in which $h_{\mu\nu} = h_{\mu\nu}(t-z)$. Then a simple calculation with Eq. (25.80) reveals that the only nonzero components of this wave's tidal field are $\mathcal{E}_{ab} = -\frac{1}{2}\ddot{h}_{ab}$, where a and b run over x and y. But by definition, $\mathcal{E}_{ab} = -\frac{1}{2}\ddot{h}_{ab}^{TT}$. Therefore, in this gauge we can compute the gravitational-wave field by simply throwing away all parts of $h_{\mu\nu}$ except the spatial, transverse parts: $h_{xx}^{TT} = h_{xx}$, $h_{xy}^{TT} = h_{xy}$, and $h_{yy}^{TT} = h_{yy}$.

When computing the generation of gravitational waves, it is often easier to evaluate the trace-reversed metric perturbation $\bar{h}_{\alpha\beta}$ than the metric perturbation itself [e.g., Eq. (25.91)]. But $\bar{h}_{\alpha\beta}$ differs from $h_{\alpha\beta}$ by only a trace, and the gravitational-wave field h_{jk}^{TT} is traceless. Therefore, in any gauge where $\bar{h}_{\mu\nu} = \bar{h}_{\mu\nu}(t-z)$, we can compute the gravitational-wave field h_{jk}^{TT} from $\bar{h}_{\mu\nu}$ by throwing away everything except its spatial, transverse part, and by then removing its trace (i.e., by projecting out the spatial, transverse, traceless part):

$$\boxed{\begin{aligned} h_{jk}^{TT} &= \left(\bar{h}_{jk}\right)^{TT}; \quad \text{or} \\ h_+ &= h_{xx}^{TT} = \bar{h}_{xx} - \tfrac{1}{2}(\bar{h}_{xx} + \bar{h}_{yy}) = \tfrac{1}{2}(h_{xx} - h_{yy}), \\ h_\times &= h_{xy}^{TT} = \bar{h}_{xy}. \end{aligned}} \quad (1)$$

Here the symbol $\left(\bar{h}_{jk}\right)^{TT}$ means "project out the spatial, transverse, traceless part."

If we rotate the spatial axes so the waves propagate along the unit spatial vector $\mathbf{n}$ instead of along $\mathbf{e}_z$, then the "speed-of-light-propagation" forms of the metric perturbation and its trace reversal become $h_{\alpha\beta} = h_{\alpha\beta}(t - \mathbf{n}\cdot\mathbf{x})$ and $\bar{h}_{\alpha\beta} = \bar{h}_{\alpha\beta}(t - \mathbf{n}\cdot\mathbf{x})$, respectively, and the TT projection can be achieved with the aid of the transverse projection tensor:

$$\boxed{P^{jk} \equiv \delta^{jk} - n^j n^k.} \quad (2)$$

Specifically, we have

$$\boxed{h_{jk}^{TT} = (\bar{h}_{jk})^{TT} = P_j{}^l P_k{}^m \bar{h}_{lm} - \frac{1}{2} P_{jk} P^{lm} \bar{h}_{lm}.} \quad (3)$$

Here the notation is that of Cartesian coordinates with $P_j{}^k = P^{jk} = P_{jk}$.

Exercise 27.5 *Derivation: Bringing $h_{\mu\nu}$ into TT Gauge*

Consider a weak, planar gravitational wave propagating in the z direction, written in a general Lorenz gauge [Eqs. (27.19)]. Show that by appropriate choices of new gauge-change generators that have the plane-wave form $\xi_\mu(t-z)$, one can (i) keep the metric perturbation in Lorenz gauge; (ii) annul $\bar{h}_{xz}$, $\bar{h}_{yz}$, $\bar{h}_{zz}$, and $\bar{h} \equiv \eta^{\mu\nu}\bar{h}_{\mu\nu}$; and (iii) thereby make the only nonzero components of the metric perturbation be $h_{xx} = -h_{yy}$ and $h_{xy} = +h_{yx}$. [Hint: Show that a gauge change (27.18b) produces $\bar{h}^{\text{new}}_{\mu\nu} = \bar{h}^{\text{old}}_{\mu\nu} - \xi_{\mu,\nu} - \xi_{\nu,\mu} + \eta_{\mu\nu}\xi_\alpha{}^{,\alpha}$, and use this in your computations.]

27.3.2 Measuring a Gravitational Wave by Its Tidal Forces

We seek physical insight into gravitational waves by studying the following idealized problem. Consider a cloud of test particles that floats freely in space and is static and spherical before the waves pass. Study the wave-induced deformations of the cloud as viewed in the nearest thing there is to a rigid, orthonormal coordinate system: the local Lorentz frame (in the physical spacetime) of a "fiducial particle" that sits at the cloud's center. In that frame the displacement vector ζ^j between the fiducial particle and some other particle has components $\zeta^j = x^j + \delta x^j$, where x^j is the other particle's spatial coordinate before the waves pass, and δx^j is its coordinate displacement produced by the waves. By inserting this into the local-Lorentz-frame variant of the equation of geodesic deviation [Eq. (25.34)], and neglecting the tiny δx^k compared to x^k on the right-hand side, we obtain

$$\frac{d^2 \delta x^j}{dt^2} = -R_{j0k0}x^k = -\mathcal{E}_{jk}x^k = \frac{1}{2}\ddot{h}^{\text{TT}}_{jk}x^k, \quad (27.25)$$

gravitational-wave tidal acceleration

which can be integrated twice to give

$$\boxed{\delta x^j = \frac{1}{2}h^{\text{TT}}_{jk}x^k.} \quad (27.26)$$

gravitational-wave tidal displacement

Expression (27.25) is the *gravitational-wave tidal acceleration* that moves the particles back and forth relative to one another. It is completely analogous to the Newtonian tidal acceleration, $-\mathcal{E}_{jk}x^k = -(\partial^2\Phi/\partial x^j \partial x^k)x^k$, by which the Moon raises tides on Earth's oceans (Sec. 25.5.1).

Now specialize to a wave with + polarization (for which $h_\times = 0$). By inserting expression (27.20) into (27.26), we obtain

$$\delta x = \frac{1}{2}h_+ x, \quad \delta y = -\frac{1}{2}h_+ y, \quad \delta z = 0. \quad (27.27)$$

tidal displacements produced by h_+

This displacement is shown in Fig. 27.1a,b. Notice that as the gravitational-wave field h_+ oscillates at the spherical cloud's location, the cloud is left undisturbed in the z-direction (propagation direction), and in transverse planes it gets deformed into an ellipse elongated first along the x-axis (when $h_+ > 0$), then along the y-axis (when

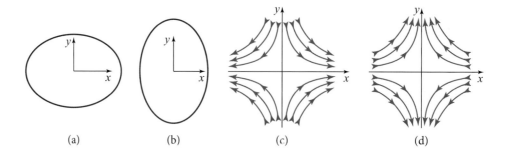

FIGURE 27.1 Physical manifestations, in a particle's local Lorentz frame, of h_+ gravitational waves. (a) Transverse deformation of an initially spherical cloud of test particles in a transverse plane at a phase of the wave when $h_+ > 0$. (b) Deformation of the cloud when $h_+ < 0$. (c) Field lines representing the acceleration field $\delta \ddot{\mathbf{x}}$ that produces the cloud's deformation, at a phase when $\ddot{h}_+ > 0$. (d) Acceleration field lines when $\ddot{h}_+ < 0$.

$h_+ < 0$). Because $\mathcal{E}_{xx} = -\mathcal{E}_{yy}$ (i.e., because $\mathcal{E}_{jk}$ is traceless), the ellipse is squashed along one axis by the same amount as it is stretched along the other (i.e., the area of the ellipse is preserved during the oscillations).

h_+ tidal acceleration described by lines of force

The effects of the h_+ polarization state can also be described in terms of the *tidal acceleration field* that it produces in the central particle's local Lorentz frame:

$$\frac{d^2}{dt^2}\delta\mathbf{x} = -\mathcal{E}_+ \cdot \mathbf{x} = \frac{1}{2}\ddot{h}_+(x\mathbf{e}_x - y\mathbf{e}_y), \tag{27.28}$$

where $\ddot{h}_+ \equiv \partial^2 h_+/\partial t^2$. Notice that this acceleration vector field $\delta\ddot{\mathbf{x}}$ is divergence free. Because it is divergence free, it can be represented by lines of force analogous to electric field lines, which point along the field and have a density of lines proportional to the magnitude of the field. When this is done, the field lines never end. Figure 27.1c,d shows this acceleration field at the phases of oscillation when $\ddot{h}_+$ is positive and when it is negative. Notice that the field is quadrupolar in shape, with a field strength (density of lines) that increases linearly with distance from the origin of the local Lorentz frame. The elliptical deformations of the spherical cloud of test particles shown in Fig. 27.1a,b are the responses of that cloud to this quadrupolar acceleration field. The polarization state that produces these accelerations and deformations is called the + state because of the orientation of the axes of the quadrupolar acceleration field (Fig. 27.1c,d).

tidal effects of $h_\times$

Next consider the × polarization state. In this state the deformations of the initially spherical cloud are described by

$$\delta x = \frac{1}{2}h_\times y, \quad \delta y = \frac{1}{2}h_\times x, \quad \delta z = 0. \tag{27.29}$$

These deformations, like those for the + state, are purely transverse; they are depicted in Fig. 27.2a,b. The acceleration field that produces these deformations is

$$\frac{d^2}{dt^2}\delta\mathbf{x} = -\mathcal{E}_\times \cdot \mathbf{x} = \frac{1}{2}\ddot{h}_\times(y\mathbf{e}_x + x\mathbf{e}_y). \tag{27.30}$$

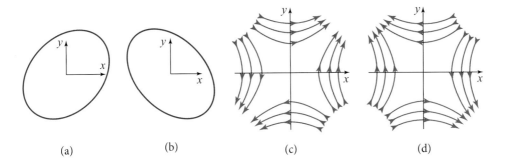

FIGURE 27.2 Physical manifestations, in a particle's local Lorentz frame, of $h_\times$ gravitational waves. (a) Deformation of an initially spherical cloud of test particles in a transverse plane at a phase of the wave when $h_\times > 0$. (b) Deformation of the sphere when $h_\times < 0$. (c) Field lines representing the acceleration field $\delta\ddot{\mathbf{x}}$ that produces the sphere's deformation, at a phase of the wave when $\ddot{h}_\times > 0$. (d) Acceleration field lines when $\ddot{h}_\times < 0$.

This acceleration field, like the one for the + polarization state, is divergence free and quadrupolar; the field lines describing it are depicted in Fig. 27.2c,d. The name "× polarization state" comes from the orientation of the axes of this quadrupolar acceleration field.

Planar gravitational waves can also be depicted in terms of the tendex and vortex lines associated with their tidal tensor field $\mathcal{E}$ and frame-drag tensor field $\mathcal{B}$; see Box 27.3.

When defining the gravitational-wave fields h_+ and $h_\times$, we have relied on a choice of (local Lorentz) reference frame (i.e., a choice of local Lorentz basis vectors $\vec{e}_\alpha$). Exercise 27.6 explores how these fields change when the basis is changed. The conclusions are simple. (i) When one rotates the transverse basis vectors $\vec{e}_x$ and $\vec{e}_y$ through an angle ψ, then h_+ and $h_\times$ rotate through 2ψ in the sense that:

behavior of gravitational wave fields under rotations and boosts

$$\boxed{(h_+ + ih_\times)_{\text{new}} = (h_+ + ih_\times)_{\text{old}} e^{2i\psi}, \quad \text{when } (\vec{e}_x + i\vec{e}_y)_{\text{new}} = (\vec{e}_x + i\vec{e}_y)_{\text{old}} e^{i\psi}.}$$

(27.31)

(ii) When one boosts from an old frame to a new one moving at some other speed but chooses the old and new spatial bases such that (a) the waves propagate in the z direction in both frames and (b) the plane spanned by $\vec{e}_x$ and $\vec{k} \equiv \vec{e}_0 + \vec{e}_z =$ (propagation direction in spacetime) is the same in both frames, then h_+ and $h_\times$ are the same in the two frames—they are scalars under such a boost! The same is true of the transverse components of the vector potential **A** for an electromagnetic wave.

EXERCISES

Exercise 27.6 *Derivation: Behavior of h_+ and $h_\times$ under Rotations and Boosts*

(a) Derive the behavior [Eq. (27.31)] of h_+ and $h_\times$ under rotations in the transverse plane. [Hint: Write the gravitational-wave field, viewed as a geometric object, as $\mathbf{h}^{\text{TT}} = \Re\left[(h_+ + ih_\times)(\mathbf{e}^+ - i\mathbf{e}^\times)\right]$, where $\mathbf{e}^+ = (\vec{e}_x \otimes \vec{e}_x - \vec{e}_y \otimes \vec{e}_y)$ and

BOX 27.3. TENDEX AND VORTEX LINES FOR A GRAVITATIONAL WAVE T2

A plane gravitational wave with + polarization, propagating in the z direction, has as its only nonzero tidal-field components $\mathcal{E}_{xx} = -\mathcal{E}_{yy} = -\frac{1}{2}\ddot{h}_+(t-z)$ [Eq. (27.22)]. This tidal field's eigenvectors are $\mathbf{e}_x$ and $\mathbf{e}_y$, so its tendex lines (Boxes 25.2 and 26.3) are straight lines pointing along these basis vectors (i.e., the solid lines in the following picture).

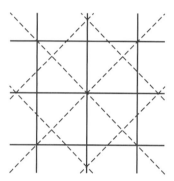

These lines' tendicities $\mathcal{E}_{xx}$ and $\mathcal{E}_{yy}$ are equal and opposite, so one set of lines stretches (red) and the other squeezes (blue). As the wave propagates, each line's tendicity oscillates as seen at fixed z, so its color oscillates between red and blue.

From the Maxwell-like Bianchi identity $\partial\mathcal{B}/\partial t = -(\nabla \times \mathcal{E})^S$ (Box 25.2)—with $\mathcal{E}$ a function of $t - \mathbf{n}\cdot\mathbf{x}$, and $\mathbf{n} = \mathbf{e}_z$ the wave's propagation direction—we infer that the wave's frame-drag field and tidal field are related by $\mathcal{B} = (\mathbf{n}\times\mathcal{E})^S$. This means that the nonzero components of $\mathcal{B}$ are $\mathcal{B}_{xy} = \mathcal{B}_{yx} = \mathcal{E}_{xx} = -\mathcal{E}_{yy} = -\frac{1}{2}\ddot{h}_+(t-z)$. Therefore, the gravitational wave's vortex lines are the dashed lines in the figure above (where the propagation direction, $\mathbf{n} = \mathbf{e}_z$, is out of the screen or paper, toward you).

Electric and magnetic field lines are generally drawn with line densities proportional to the magnitude of the field—a convention motivated by flux conservation. Not so for tendex and vortex lines, which have no corresponding conservation law. Instead, their field strengths (tendicities and vorticities) are usually indicated by color coding (see, e.g., Nichols et al., 2011).

Most discussions of gravitational waves (including the text of this chapter) focus on their tidal field $\mathcal{E}$ and its physical stretch and squeeze; they ignore the frame-drag field with its differential precession (twisting) of gyroscopes. The reason is that modern technology is able to detect and monitor the stretch and squeeze, but the precession is far too small to be detected.

$\mathbf{e}^\times = (\vec{e}_x \otimes \vec{e}_y + \vec{e}_y \otimes \vec{e}_x)$ are the polarization tensors associated with $+$ and $\times$ polarized waves, respectively [Eqs. (27.23b)]. Then show that $\mathbf{e}_+ - i\mathbf{e}_\times$ rotates through -2ψ, and use this to infer the desired result.]

(b) Show that, with the orientations of spatial basis vectors described after Eq. (27.31), h_+ and $h_\times$ are unchanged by boosts.

27.3.3 Gravitons and Their Spin and Rest Mass

Most of the abovementioned features of gravitational waves (though not expressed in this language) were clear to Einstein in 1918. Two decades later, as part of the effort to understand quantum fields, Markus Fierz and Wolfgang Pauli (1939), at the Eidgenössische Technische Hochschule (ETH) in Zurich, Switzerland, formulated a classical theory of linear fields of arbitrary spin so designed that the fields would be quantizable by canonical methods. Remarkably, their canonical theory for a field of spin two and zero rest mass is identical to general relativity with nonlinear effects removed, and the plane waves of that spin-two theory are identical to the waves described above. When quantized by canonical techniques, these waves are carried by zero-rest-mass, spin-two gravitons.

quantization of gravitational waves: gravitons with zero rest mass and spin two

One can see by simple arguments that the gravitons that carry gravitational waves must have zero rest mass and spin two. First, fundamental principles of quantum theory guarantee that any wave that propagates in vacuum with the speed of light must be carried by particles that have that same speed (i.e., particles whose 4-momenta are null, which means particles with zero rest mass). General relativity predicts that gravitational waves propagate with the speed of light. Therefore, its gravitons must have zero rest mass.

elementary explanations of the graviton rest mass and spin

Second, consider any plane-wave field (neutrino, electromagnetic, gravitational, etc.) that propagates at the speed of light in the z-direction of a (local) Lorentz frame. At any moment of time examine any physical manifestation of that field (e.g., the acceleration field it produces on test particles). Rotate that manifestation of the field around the z-axis, and ask what minimum angle of rotation is required to bring the field back to its original configuration. Call that minimum angle, θ_{ret}, the waves' *return angle*. The spin S of the particles that carry the wave will necessarily be related to that return angle by[3]

$$\boxed{S = \frac{360°}{\theta_{\text{ret}}}.} \quad (27.32)$$

This simple formula corresponds to the elegant mathematical statement that "the waves generate an irreducible representation of order $S = 360°/\theta_{\text{ret}}$ of that subgroup of the Lorentz group that leaves their propagation vector unchanged (the 'Little group'

---

3. For spin 0 this formula fails. Spin 0 corresponds to circular symmetry around the spin axis.

of the propagation vector)." For electromagnetic waves, a physical manifestation is the electric field, which is described by a vector lying in the x-y plane; if one rotates that vector about the z-axis (propagation axis), it returns to its original orientation after a return angle $\theta_{\rm ret} = 360°$. Correspondingly, the spin of the particle that carries the electromagnetic waves (the photon) is one. For neutrinos, the return angle is $\theta_{\rm ret} = 720°$; and correspondingly, the spin of a neutrino is $\frac{1}{2}$. For gravitational waves, the physical manifestations include the deformation of a sphere of test particles (Figs. 27.1a,b and 27.2a,b) and the acceleration fields (Figs. 27.1c,d and 27.2c,d). Both the deformed, ellipsoidal spheres and the quadrupolar lines of force return to their original orientations after rotation through $\theta_{\rm ret} = 180°$; correspondingly, the graviton must have spin two. This spin two also shows up in the rotation factor $e^{i2\psi}$ of Eq. (27.31).

Although Fierz and Pauli (1939) showed us how to quantize linearized general relativity, the quantization of full, nonlinear general relativity remains a difficult subject of current research.

27.4 Gravitational Waves Propagating through Curved Spacetime

Richard Isaacson (1968a,b) has developed a geometric-optics formulation of the theory of gravitational waves propagating through curved spacetime, and as a by-product he has given a rigorous mathematical description of the waves' stress-energy tensor and thence the energy and momentum carried by the waves. In this section, we sketch the main ideas and results of Isaacson's analysis.[4]

two-lengthscale expansion for gravitational waves in curved spacetime

The foundation for the analysis is a two-lengthscale expansion $\lambda/\mathcal{L}$ like we used in Sec. 7.3 when formulating geometric optics. For any physical quantity, we identify the wave contribution as the portion that varies on some short lengthscale $\bar\lambda = \lambda/(2\pi)$ (the waves' reduced wavelength), and the background as the portion that varies on a far longer lengthscale $\mathcal{L}$ (which is less than or of order the background's spacetime radius of curvature $\mathcal{R}$); see Fig. 27.3.

steady coordinates

To make this idea work, we must use "steady" coordinates (i.e., coordinates that are smooth to as great an extent as the waves permit, on lengthscales shorter than $\mathcal{L}$). In such coordinates, components of the spacetime metric $g_{\alpha\beta}$ and of the Riemann curvature tensor $R_{\alpha\beta\gamma\delta}$ split into background (B) plus gravitational waves (GW),

metric and Riemann tensor split into background and gravitational-wave parts

$$g_{\alpha\beta} = g^{\rm B}_{\alpha\beta} + h_{\alpha\beta}, \quad R_{\alpha\beta\gamma\delta} = R^{\rm B}_{\alpha\beta\gamma\delta} + R^{\rm GW}_{\alpha\beta\gamma\delta}, \quad (27.33a)$$

where the background quantities are defined as the averages (denoted $\langle \cdot \rangle$) of the full quantities over lengthscales long compared to $\bar\lambda$ and short compared to $\mathcal{L}$:

$$g^{\rm B}_{\alpha\beta} \equiv \langle g_{\alpha\beta} \rangle, \quad R^{\rm B}_{\alpha\beta\gamma\delta} \equiv \langle R_{\alpha\beta\gamma\delta} \rangle. \quad (27.33b)$$

---

4. In the 1980s and 1990s, Isaacson, as the Program Director for Gravitational Physics at the U.S. National Science Foundation (NSF), played a crucial role in the creation of the LIGO Project for detecting gravitational waves and in moving LIGO toward fruition.

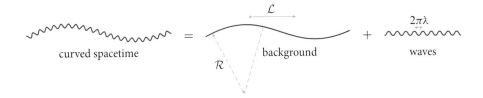

FIGURE 27.3 Heuristic embedding diagram for the decomposition of curved spacetime into a background spacetime plus gravitational waves.

To assist us in solving the Einstein equation, we treat the Einstein tensor $G_{\alpha\beta}$ a bit differently from the metric and Riemann. We begin by expanding $G_{\alpha\beta}$ as a power series in the metric perturbation $h_{\alpha\beta}$: $G_{\alpha\beta} = G^B_{\alpha\beta} + G^{(1)}_{\alpha\beta} + G^{(2)}_{\alpha\beta} + \cdots$. Here $G^B_{\alpha\beta}$ is the Einstein tensor computed from the background metric $g^B_{\alpha\beta}$, $G^{(1)}_{\alpha\beta}$ is linear in $h_{\alpha\beta}$, $G^{(2)}_{\alpha\beta}$ is quadratic, and so forth. We then split $G_{\alpha\beta}$ into its rapidly varying part, which is simply $G^{(1)}_{\alpha\beta}$ to leading order, and its smoothly varying part, which through quadratic order is $\langle G_{\alpha\beta}\rangle = G^B_{\alpha\beta} + \langle G^{(2)}_{\alpha\beta}\rangle$. The vacuum Einstein equation $G_{\alpha\beta} = 0$ will be satisfied only if the rapidly and smoothly varying parts both vanish.

In Sec. 27.4.1, by setting the fast-varying part $G^{(1)}_{\alpha\beta}$ to zero, we obtain a wave equation in the background curved spacetime for $h_{\alpha\beta}$ (the gravitational waves), which we can solve (Sec. 27.4.2) using the geometric-optics approximation that underlies this analysis. Then, in Sec. 27.4.3, by setting the slowly varying part $G^B_{\alpha\beta} + \langle G^{(2)}_{\alpha\beta}\rangle$ to zero, we obtain Isaacson's description of gravitational-wave energy and momentum.

27.4.1 Gravitational Wave Equation in Curved Spacetime

The metric perturbation $h_{\alpha\beta}$ can be regarded as a tensor field that lives in the background spacetime.[5] The rapidly varying part of the Einstein equation, $G^{(1)}_{\alpha\beta} = 0$, gives rise to a wave equation for this tensorial metric perturbation (Isaacson, 1968a; Misner, Thorne, and Wheeler, 1973, Secs. 35.13, 35.14). We can infer this wave equation most easily from a knowledge of the form it takes in any local Lorentz frame of the background (with size $\gg \lambda$ but $\ll \mathcal{L}$). In such a frame, $G^{(1)}_{\alpha\beta} = 0$ must reduce to the field equation of linearized theory [the vacuum version of Eq. (25.83)]. And if we introduce Lorenz gauge [Eq. (27.18c)], then $G^{(1)}_{\alpha\beta} = 0$ must become, in a local Lorentz frame, the vacuum wave equation (27.18e). The frame-invariant versions of these local-Lorentz-frame equations, in the background spacetime, should be obvious. The trace-reversed metric perturbation (27.18d) in frame-invariant form must become

$$\bar{h}_{\mu\nu} \equiv h_{\mu\nu} - \frac{1}{2} h\, g^B_{\mu\nu}, \quad h \equiv g_B^{\alpha\beta} h_{\alpha\beta}. \tag{27.34a}$$

The Lorenz-gauge condition (27.18c) must become

$$\bar{h}_{\mu\nu|}{}^{\nu} = 0, \tag{27.34b}$$

Lorenz-gauge condition

---

5. Actually, this characterization requires that we restrict the coordinates to be steady.

where the | denotes a gradient in the background spacetime (i.e., a covariant derivative computed using connection coefficients constructed from $g^B_{\mu\nu}$) and the index is raised with the background metric, $\bar{h}_{\mu\nu|}{}^{\nu} = \bar{h}_{\mu\nu|\alpha} g_B^{\alpha\nu}$. And the gravitational wave equation (Einstein field equation) (27.18e) must become

gravitational wave equation

$$\bar{h}_{\mu\nu|\alpha}{}^{\alpha} = 0 \tag{27.34c}$$

plus curvature coupling terms, such as $R^B_{\alpha\mu\beta\nu}\bar{h}^{\alpha\beta}$, that result from the noncommutation of the double gradients. The curvature coupling terms have magnitude $h/\mathcal{R}^2 \lesssim h/\mathcal{L}^2$ (where $\mathcal{R}$ is the radius of curvature of the background spacetime; cf. Fig. 27.3), while the terms kept in Eq. (27.34c) have the far-larger magnitude $h/\bar{\lambda}^2$, so the curvature coupling terms can be (and are) neglected.

curvature coupling is negligible

27.4.2 Geometric-Optics Propagation of Gravitational Waves

When one solves Eqs. (27.34) using the geometric-optics techniques developed in Sec. 7.3, one obtains precisely the results that one should expect, knowing the solution (27.19) for weak, planar gravitational waves in flat spacetime (linearized theory).

geometric-optics wave propagation

1. If we split $h_{\mu\nu}$ up into two polarization pieces $+$ and $\times$, each with its own rapidly varying phase φ and slowly varying amplitude $\mathsf{A}_{\mu\nu}$, then the Lorenz-gauge, trace-reversed metric perturbation for each piece takes the standard geometric-optics form [eikonal approximation; Eq. (7.20)]:

eikonal approximation

$$h_{\mu\nu} = \Re(\mathsf{A}_{\mu\nu} e^{i\varphi}). \tag{27.35a}$$

2. Because the linearized-theory waves propagate in a straight line (z direction) and travel at the speed of light, the geometric-optics waves propagate through curved spacetime on rays that are null geodesics. More specifically, the wave vector $\vec{k} = \vec{\nabla}\varphi$ is tangent to the null-ray geodesics, and φ is constant along a ray and hence is a rapidly varying function of the retarded time τ_r at which the source (in its own reference frame) emitted the ray:

rays are null geodesics

$$\varphi = \varphi(\tau_r), \quad \vec{k} = \vec{\nabla}\varphi, \quad \vec{k}\cdot\vec{k} = 0, \quad \nabla_{\vec{k}}\vec{k} = 0, \quad \nabla_{\vec{k}}\varphi = 0. \tag{27.35b}$$

3. Because the x- and y-axes that define the two polarizations in linearized theory remain fixed as the wave propagates, for each polarization we can split the amplitude $\mathsf{A}_{\mu\nu}$ up into a scalar amplitude A_+ or $\mathsf{A}_\times$, and a polarization tensor $\boldsymbol{e}^+$ or $\boldsymbol{e}^\times$ [like those of Ex. 27.6 and Eqs. (27.23b)], and the polarization tensors are parallel-transported along the rays:

polarization tensors are parallel propagated along rays

$$\mathsf{A}_{\mu\nu} = \mathsf{A}\, e_{\mu\nu}, \quad \nabla_{\vec{k}}\boldsymbol{e} = 0. \tag{27.35c}$$

graviton conservation implies amplitude scales inversely with square root of cross section of a bundle of rays

4. Because gravitons are conserved (cf. the conservation of quanta in our general treatment of geometric optics, Sec. 7.3.2), the flux of gravitons (which is proportional to the square A^2 of the scalar amplitude) times the cross sectional area $\mathcal{A}$ of a bundle of rays that are carrying the gravitons must be

constant. Therefore, the scalar wave amplitude A must die out as 1/(square root of cross sectional area $\mathcal{A}$ of a bundle of rays):

$$A \propto 1/\sqrt{\mathcal{A}}. \qquad (27.35d)$$

Now, just as the volume of a 3-dimensional fluid element, for a perfect fluid, changes at the rate $d \ln V/d\tau = \vec{\nabla} \cdot \vec{u}$, where $\vec{u}$ is the 4-velocity of the fluid [Eq. (2.65)], so (it turns out) the cross sectional area of a bundle of rays increases as $\nabla_{\vec{k}} \mathcal{A} = \vec{\nabla} \cdot \vec{k}$. Therefore, the transport law for the wave amplitude, $A \propto 1/\sqrt{\mathcal{A}}$, becomes

$$\nabla_{\vec{k}} A = -\frac{1}{2}(\vec{\nabla} \cdot \vec{k})A. \qquad (27.35e)$$

transport law for amplitude

Equations (27.35) are derived more rigorously in Isaacson (1968a) and Misner, Thorne, and Wheeler (1973, Sec. 35.14). They can be used to compute the influence of the cosmological curvature of our universe, and the gravitational fields of intervening bodies, on the propagation of gravitational waves from their sources to Earth. Once the waves have reached Earth, we can compute the measured gravitational-wave fields h_+ and $h_\times$ by projecting out the spatial, transverse-traceless parts of $h_{\mu\nu} = \Re(Ae_{\mu\nu}e^{i\varphi})$, as discussed in Box 27.2.

The geometric-optics propagation of gravitational waves, as embodied in Eqs. (27.35), is essentially identical to that of electromagnetic waves. Both waves, gravitational and electromagnetic, propagate along rays that are null geodesics. Both parallel-transport their polarizations. Both are carried by quanta that are conserved as they propagate along bundles of rays and as a result both have scalar amplitudes that vary as $A \propto 1/\sqrt{\mathcal{A}}$, where $\mathcal{A}$ is the cross sectional area of a ray bundle.

Therefore, gravitational waves must exhibit exactly the same vacuum propagation phenomena as electromagnetic waves: Doppler shifts, cosmological redshifts, gravitational redshifts, gravitational deflection of rays, and gravitational lensing!

gravitational waves exhibit same vacuum propagation phenomena as electromagnetic waves: redshifts, ray deflection, gravitational lensing

In Ex. 27.14, we illustrate this geometric optics propagation of gravitational waves by applying it to the waves from a binary system, which travel outward through our expanding universe.

Exercise 27.7 explores an application where geometric optics breaks down due to diffraction.

EXERCISES

Exercise 27.7 **Example: Gravitational Lensing of Gravitational Waves by the Sun*
Gravitational waves from a distant source travel through the Sun with impunity (negligible absorption and scattering), and their rays are gravitationally deflected. The Sun is quite centrally condensed, so most of the deflection is produced by a central region with mass $M_c \simeq 0.3 M_\odot$ and radius $R_c \simeq 10^5$ km $\simeq R_\odot/7$, and the maximum deflection angle is therefore $\Delta\phi \simeq 4M_c/R_c$ [Eq. (27.7)]. A few of the deflected rays, along which the waves propagate according to geometric optics, are shown in Fig. 27.4.

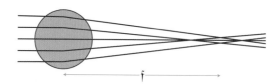

FIGURE 27.4 Some gravitational-wave rays that pass through the Sun are brought to an imperfect focus at a distance $\mathfrak{f}$, the focal length.

(a) Show that the rays are brought to an imperfect focus and thence produce caustics (Sec. 7.5) at a distance from the Sun (the focal length) $\mathfrak{f} \sim R_c^2/(4M_c) \sim 38$ AU. A calculation with a more accurate solar model gives $\mathfrak{f} \sim 20$ AU, which is near the orbit of Uranus.

(b) If the waves were to have arbitrarily small wavelength λ, then at the caustics, their wave fields h_+ and $h_\times$ would become divergently large (Sec. 7.5). Finite wavelength causes diffraction near the caustics (Sec. 8.6). Explain why the focused field thereby is smeared out over a region with transverse size $\sigma \sim [\lambdabar/(2R_c)]\mathfrak{f} \sim [\lambdabar/(8M_c)]R_c$. [Hint: See Eq. (8.9) and associated discussion.]

(c) Explain why, if $\sigma \ll R_c$ (i.e., if $\lambdabar \ll 8M_c \sim 3M_\odot$), substantial focusing occurs, and the field near the caustics is strongly amplified; but if $\sigma \gtrsim R_c$ (i.e., $\lambdabar \gtrsim 3M_\odot$), there is only slight or no focusing. Explain why it is unlikely that any discrete, strong gravitational-wave sources in the universe emit wavelengths shorter than $3M_\odot \sim 5$ km and therefore are strongly lensed by the Sun.

27.4.3 Energy and Momentum in Gravitational Waves

Now turn from the rapidly varying piece of the vacuum Einstein equation, $G^{(1)}_{\mu\nu} = 0$, to the piece that is averaged over scales long compared to $\lambdabar$ and short compared to $\mathcal{L}$:

$$G^B_{\alpha\beta} + \langle G^{(2)}_{\alpha\beta} \rangle = 0. \tag{27.36}$$

(Recall that $G^B_{\alpha\beta}$ is the Einstein tensor constructed from the slowly varying background metric, and $G^{(2)}_{\alpha\beta}$ is the piece of the full Einstein tensor that is quadratic in the rapidly varying metric perturbation $h_{\mu\nu}$ and that therefore does not average to zero.)

Notice that Eq. (27.36) can be brought into the standard form for Einstein's equation in the background spacetime,

Einstein equation for background metric in vacuum

$$\boxed{G^B_{\alpha\beta} = 8\pi T^{GW}_{\alpha\beta},} \tag{27.37}$$

by moving $\langle G^{(2)}_{\alpha\beta} \rangle$ to the right-hand side and then attributing to the waves a stress-energy tensor defined by

stress-energy tensor for gravitational waves: formal expression

$$T^{GW}_{\alpha\beta} = -\frac{1}{8\pi} \langle G^{(2)}_{\alpha\beta} \rangle. \tag{27.38}$$

Because this stress-energy tensor involves an average over a few wavelengths, its energy density, momentum density, energy flux, and momentum flux are not defined on lengthscales shorter than a wavelength. One cannot say how much energy or momentum resides in the troughs of the waves and how much in the crests. One can only say how much total energy there is in a region containing a few or more wavelengths. However, once reconciled to this amount of nonlocality, one finds that $T^{\mathrm{GW}}_{\alpha\beta}$ has all the other properties expected of any good stress-energy tensor. Most especially, in the absence of coupling of the waves to matter (the situation we are treating), it obeys the standard conservation law:

$$\boxed{T^{\mathrm{GW}\,\alpha\beta}{}_{|\beta} = 0,} \tag{27.39}$$

conservation law for gravitational-wave energy and momentum

where, as above, the symbol $|$ denotes the covariant derivative in the background spacetime (i.e., the derivative using the connection coefficients of $g^B_{\alpha\beta}$). This law is a direct consequence of the averaged field equation (27.37) and the contracted Bianchi identity for the background spacetime: $G^{\mathrm{B}\,\alpha\beta}{}_{|\beta} = 0$.

By grinding out the second-order perturbation of the Einstein tensor and inserting it into Eq. (27.38), performing several integrations by parts in the average $\langle \cdot \rangle$, and expressing the result in terms of h_+ and $h_\times$, one arrives at the following simple expression for $T^{\mathrm{GW}}_{\alpha\beta}$ in terms of the wave fields h_+ and $h_\times$:

$$\boxed{T^{\mathrm{GW}}_{\alpha\beta} = \frac{1}{16\pi} \langle h_{+,\alpha} h_{+,\beta} + h_{\times,\alpha} h_{\times,\beta} \rangle.} \tag{27.40}$$

stress-energy tensor in terms of gravitational-wave fields

[For details of the derivation, see Isaacson (1968b) or Misner, Thorne, and Wheeler (1973, Secs. 35.13, 35.15).]

Let us examine this stress-energy tensor in a local Lorentz frame of the background spacetime where the waves are locally plane and are propagating in the z direction—the kind of frame we used in Sec. 27.3.2 when exploring the properties of gravitational waves. Because in this frame, $h_+ = h_+(t-z)$ and $h_\times = h_\times(t-z)$, the only nonzero components of Eq. (27.40) are

$$\boxed{T^{\mathrm{GW}\,00} = T^{\mathrm{GW}\,0z} = T^{\mathrm{GW}\,z0} = T^{\mathrm{GW}\,zz} = \frac{1}{16\pi} \langle \dot{h}_+^2 + \dot{h}_\times^2 \rangle.} \tag{27.41}$$

This has the same form as the stress-energy tensor for a plane electromagnetic wave propagating in the z direction, and the same form as the stress-energy tensor for any collection of zero-rest-mass particles moving in the z-direction [cf. Eq. (3.32d)], as it must, since the gravitational waves are carried by zero-rest-mass gravitons just as electromagnetic waves are carried by zero-rest-mass photons.

Suppose that the waves have frequency $\sim f$ and that the amplitudes of oscillation of h_+ and $h_\times$ are $\sim h_{\mathrm{amp}}$. Then by inserting factors of G and c into Eq. (27.41) (i.e., by switching from geometrized units to conventional units) and by setting

$\langle(\partial h_+/\partial t)^2\rangle \simeq \frac{1}{2}(2\pi f h_{\text{amp}})^2$ and similarly for $h_\times$, we obtain the following approximate expression for the energy flux in the waves:

magnitude of gravitational-wave energy flux

$$T^{\text{GW 0}z} \simeq \frac{\pi}{4}\frac{c^3}{G}f^2 h_{\text{amp}}^2 \simeq \frac{0.01\,\text{W}}{\text{m}^2}\left(\frac{f}{200\,\text{Hz}}\right)^2\left(\frac{h_{\text{amp}}}{10^{-21}}\right)^2. \quad (27.42)$$

The numbers in this equation are those for the first gravitational waves ever detected: LIGO's GW150914 at its peak brightness. Those waves' observed energy flux, ~ 0.01 W m^{-2}, was several times higher than that from the full moon as seen on Earth, but the waves' source, two $\sim 30 M_\odot$ colliding black holes, is ~ 1.2 billion light years away compared to the Moon's distance of 1 light second. Of course, the moon shines steadily, while the holes' collision and this enormous flux lasted for only ~ 20 ms.

For a short gravitational wave burst such as GW150914 (only a few wave cycles, which we shall approximate as just one), the enormous energy flux (27.42) corresponds to a huge mean occupation number for the quantum states of the gravitational-wave field (i.e., a huge value for the number of spin-two, zero-rest-mass gravitons in each quantum state). To compute that occupation number, we evaluate the volume in phase space occupied by the waves and then divide by the volume occupied by each quantum state (cf. Sec. 3.2.5). At a time when the waves have reached a distance r from the source, they occupy a spherical shell of area $4\pi r^2$ and thickness $\sim c/f = 2\pi\lambdabar$, where $\lambdabar = 1/(2\pi f)$ is their reduced wavelength, so their volume in physical space is $\mathcal{V}_x \sim 8\pi^2 r^2 \lambdabar$. As seen by observers whom the waves are passing, they come from a solid angle $\Delta\Omega \sim (2\lambdabar/r)^2$ centered on the source, and they have a spread of angular frequencies $\Delta\omega \sim \omega = c/\lambdabar$. Since each graviton carries an energy $\hbar\omega = \hbar c/\lambdabar$ and a momentum $\hbar\omega/c = \hbar/\lambdabar$, the volume that they occupy in momentum space is $\mathcal{V}_p \sim (\hbar/\lambdabar)^3 \Delta\Omega$, or $\mathcal{V}_p \sim 4\hbar^3/(\lambdabar r^2)$. The gravitons' volume in phase space, then, is

$$\mathcal{V}_x \mathcal{V}_p \sim 32\pi^2 \hbar^3 \sim (2\pi\hbar)^3. \quad (27.43)$$

Since each quantum state for a zero rest-mass particle occupies a volume $(2\pi\hbar)^3$ in phase space [Eq. (3.17) with $g_s = 1$], the total number of quantum states occupied by the gravitons is of order unity! Correspondingly, for a total energy radiated like that of GW150914, $\sim M_\odot c^2 \sim 10^{47}$ J with each graviton carrying an energy $\hbar c/\lambdabar \sim 10^{-31}$ J, the mean occupation number of each occupied state is of order the total number of gravitons emitted:

mean occupation number for quantum states of gravitational waves from a strong gravitational wave burst

$$\eta \sim \frac{M_\odot c^2}{\hbar\omega} \sim 10^{78}. \quad (27.44\text{a})$$

This is the mean occupation number from the viewpoint of the emitter.

A detector on Earth has available to it only those gravitons that pass through a region with transverse size of order their wavelength $\lambdabar$—which means a fraction $(2\pi\lambdabar)^2/(4\pi r^2)$ of the emitted waves' volume. We can think of the detector as collapsing the gravitons' wave function into that volume. The number of available quantum states is still of order unity (demonstrate this!), but the number of gravitons occupying them is reduced by this factor, so from the detector's viewpoint, the mean occupation number is

$$\eta_{\text{collapsed}} \sim \frac{M_\odot c^2}{\hbar\omega} \frac{(2\pi\lambdabar)^2}{4\pi r^2} \sim \frac{T^{\text{GW0z}}}{\hbar\omega c}(2\pi\lambdabar)^3 \sim 10^{39}. \quad (27.44b)$$

mean occupation number from viewpoint of a detector on Earth

Notice that this is the number of gravitons in a cubic wavelength at the wave burst.

Whichever viewpoint one takes, the occupation number is enormous. It guarantees that the waves behave exceedingly classically; quantum-mechanical corrections to the classical theory have fractional magnitude $1/\sqrt{\eta} \sim 10^{-39}$, or $\sim 10^{-20}$.

27.5 The Generation of Gravitational Waves

When analyzing the generation of gravitational waves, it is useful to divide space around the source (in the source's rest frame) into the regions shown in Fig. 27.5.

If the source has size $L \lesssim M$, where M is its mass, then spacetime is strongly curved inside and near it, and we refer to it as a *strong-gravity source*. The region with radii (measured from its center of mass) $r \lesssim 10M$ is called the source's *strong-field region*. Examples of strong-gravity sources are vibrating or spinning neutron stars, and merging binary black holes. The region with radii $10M \lesssim r \lesssim \lambdabar =$ (the reduced wavelength of the emitted waves) is called the source's *weak-field near zone*. In this region, the source's gravity is fairly well approximated by Newtonian theory and a Newtonian gravitational potential Φ. As in electromagnetic theory, the region $\lambdabar \lesssim r \lesssim \lambda$ is called the *induction zone* or the *intermediate zone*. The *wave zone* begins at $r \sim \lambda = 2\pi\lambdabar$.

It is useful to divide the wave zone into two parts: a part near the source ($r \lesssim r_o$ for some r_o) called the *local wave zone*, in which the spacetime curvatures of external bodies and of the universe as a whole are unimportant, and the *distant wave zone* ($r \gtrsim r_o$), in which the emitted waves are significantly affected by external bodies and the external universe (i.e., by background spacetime curvature). The theory of

regions of space around a gravitational-wave source: Fig. 27.5

theory of wave generation predicts waves in source's local wave zone; geometric optics carries them onward through distant wave zone

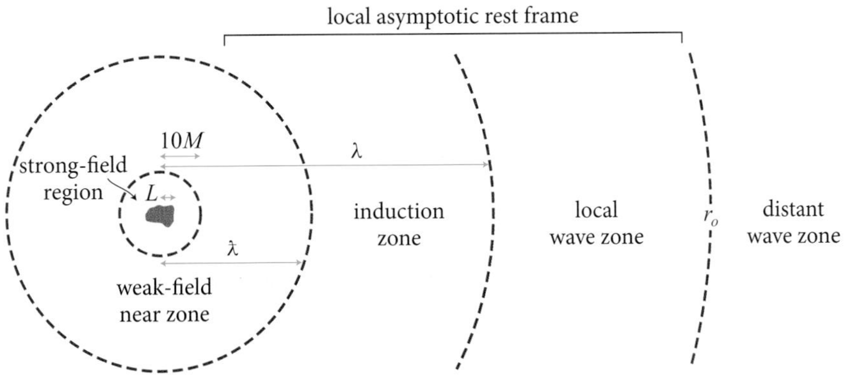

FIGURE 27.5 Regions of space around a source of gravitational waves.

gravitational-wave generation deals with computing, from the source's dynamics, the gravitational waves in the local wave zone. Propagation of the waves to Earth is dealt with by using geometric optics (or other techniques) to carry the waves from the local wave zone outward, through the distant wave zone, to Earth.

source's local asymptotic rest frame

The entire region in which gravity is weak and the spacetime curvatures of external bodies and the universe are unimportant ($10M \lesssim r \lesssim r_o$)—when viewed in nearly Lorentz coordinates in which the source is at rest—is called the source's *local asymptotic rest frame*.

27.5.1 Multipole-Moment Expansion

The electromagnetic waves emitted by a dynamical charge distribution are usually expressed as a sum over the source's multipole moments. There are two families of moments: the electric moments (moments of the electric-charge distribution) and the magnetic moments (moments of the electric-current distribution).

source's mass moments and current moments

Similarly, the gravitational waves emitted by a dynamical distribution of mass-energy and momentum can be expressed, in the local wave zone, as a sum over multipole moments. Again there are two families of moments: the *mass moments* (moments of the mass-energy distribution) and the *current moments* (moments of the mass-current distribution, i.e., the momentum distribution). The multipolar expansion of gravitational waves is developed in great detail in Blanchet (2014) and Thorne (1980). In this section, we sketch and explain its qualitative and order-of-magnitude features.

In the source's weak-gravity near zone (if it has one), the mass moments show up in the time-time part of the metric in a form familiar from Newtonian theory:

multipolar expansion of metric outside a source

$$g_{00} = -(1 + 2\Phi) = -1 \,\&\, \frac{\mathcal{I}_0}{r} \,\&\, \frac{\mathcal{I}_1}{r^2} \,\&\, \frac{\mathcal{I}_2}{r^3} \,\&\, \cdots \quad (27.45)$$

[cf. Eq. (25.79)]. Here r is radius, $\mathcal{I}_\ell$ is the mass moment of order ℓ, and "&" means "plus a term with the form" (i.e., a term whose magnitude and parameter dependence are shown but whose multiplicative numerical coefficients do not interest us, at least not for the moment). The mass monopole moment $\mathcal{I}_0$ is the source's mass, and the mass dipole moment $\mathcal{I}_1$ can be made to vanish by placing the origin of coordinates at the center of mass [Eq. (25.96) and Ex. 25.21].

In the source's weak-gravity near zone, its current moments $\mathcal{S}_\ell$ similarly show up in the space-time part of the metric:

$$g_{0j} = \frac{\mathcal{S}_1}{r^2} \,\&\, \frac{\mathcal{S}_2}{r^3} \,\&\, \cdots . \quad (27.46)$$

Just as there is no magnetic monopole moment in classical electromagnetic theory, so there is no current monopole moment in general relativity. The current dipole moment $\mathcal{S}_1$ is the source's angular momentum J_k, so the leading-order term in the expansion (27.46) has the form (25.98c), which we have used to deduce the angular momenta of gravitating bodies.

If the source has mass M, size L, and internal velocities $\sim v$, then the magnitudes of its moments are

$$\mathcal{I}_\ell \sim M L^\ell, \quad \mathcal{S}_\ell \sim M v L^\ell. \tag{27.47}$$

magnitudes of moments

These formulas guarantee that the near-zone fields g_{00} and g_{0j}, as given by Eqs. (27.45) and (27.46), are dimensionless.

As the source's moments oscillate dynamically, they produce gravitational waves. Mass-energy conservation [Eq. (25.102)] prevents the mass monopole moment $\mathcal{I}_0 = M$ from oscillating; angular-momentum conservation [Eq. (25.103)] prevents the current dipole moment $\mathcal{S}_1 =$ (angular momentum) from oscillating; and because the time derivative of the mass dipole moment $\mathcal{I}_1$ is the source's linear momentum, momentum conservation [Eq. (25.106)] prevents the mass dipole moment from oscillating. Therefore, the lowest-order moments that can oscillate and thereby contribute to the waves are the quadrupolar ones. The wave fields h_+ and $h_\times$ in the source's local wave zone must (i) be dimensionless, (ii) die out as $1/r$, and (iii) be expressed as a sum over derivatives of the multipole moments. These considerations guarantee that the waves will have the following form:

lowest-order moments that produce gravitational waves are quadrupolar

$$\boxed{h_+ \sim h_\times \sim \frac{\partial^2 \mathcal{I}_2/\partial t^2}{r} \;\&\; \frac{\partial^3 \mathcal{I}_3/\partial t^3}{r} \;\&\; \ldots \;\&\; \frac{\partial^2 \mathcal{S}_2/\partial t^2}{r} \;\&\; \frac{\partial^3 \mathcal{S}_3/\partial t^3}{r} \;\&\; \cdots}$$

multipolar expansion of gravitational waves

$$(27.48)$$

(Ex. 27.8).

The timescale on which the moments oscillate is $T \sim L/v$, so each time derivative produces a factor v/L. Correspondingly, the ℓ-pole contributions to the waves have magnitudes

$$\frac{\partial^\ell \mathcal{I}_\ell / \partial t^\ell}{r} \sim \frac{M}{r} v^\ell, \quad \frac{\partial^\ell \mathcal{S}_\ell / \partial t^\ell}{r} \sim \frac{M}{r} v^{(\ell+1)}. \tag{27.49}$$

This means that, for a *slow-motion source* (one with internal velocities v small compared to light, so the reduced wavelength $\lambdabar \sim L/v$ is large compared to the source size L), the mass quadrupole moment $\mathcal{I}_2$ will produce the strongest waves. The mass octupole (3-pole) waves and current quadrupole waves will be weaker by $\sim v \sim L/\lambdabar$; the mass 4-pole and current octupole waves will be weaker by $\sim v^2 \sim L^2/\lambdabar^2$, and so forth. This is analogous to the electromagnetic case, where the electric dipole waves are the strongest, the electric quadrupole and magnetic dipole are smaller by $\sim L/\lambdabar$, and so on.

for slow-motion source: relative magnitudes of gravitational waves' multipolar components

In the next section, we develop the theory of mass quadrupole gravitational waves. For the corresponding theory of higher-order multipoles, see, for example, Thorne (1980, Secs. IV and VIII) and Blanchet (2014). In Sec. 27.5.3, we will see that a source's mass quadruopole waves cannot, by themselves, carry net linear momentum. Net wave momentum and the corresponding recoil of the source require a beating of mass quadrupole waves against current quadrupole or mass octupole waves.

27.5 The Generation of Gravitational Waves

EXERCISES

Exercise 27.8 *Derivation: Multipolar Expansion of Gravitational Waves*
Show that conditions (i), (ii), and (iii) preceding Eq. (27.48) guarantee that the multipolar expansion of the gravitational-wave fields will have the form (27.48).

27.5.2 Quadrupole-Moment Formalism

Consider a weakly gravitating, nearly Newtonian system (which is guaranteed to be a slow-motion gravitational-wave source, since Newtonian theory requires internal velocities $v \ll 1$). An example is a binary star system. Write the system's Newtonian potential (in its near zone) in the usual way:

$$\Phi(\mathbf{x}) = -\int \frac{\rho(\mathbf{x}')}{|\mathbf{x}-\mathbf{x}'|} dV_{x'}. \tag{27.50}$$

By using Cartesian coordinates, placing the origin of coordinates at the center of mass so $\int \rho x^j dV_x = 0$, and expanding:

$$\frac{1}{|\mathbf{x}-\mathbf{x}'|} = \frac{1}{r} + \frac{x^j x^{j'}}{r^3} + \frac{x^j x^k (3x^{j'} x^{k'} - r'^2 \delta_{jk})}{2r^5} + \cdots, \tag{27.51}$$

we obtain the multipolar expansion of the Newtonian potential:

for nearly Newtonian source: multipolar expansion of Newtonian potential

$$\boxed{\Phi(\mathbf{x}) = -\frac{M}{r} - \frac{3\mathcal{I}_{jk} x^j x^k}{2r^5} + \cdots.} \tag{27.52}$$

Here

$$\boxed{M = \int \rho dV_x, \quad \mathcal{I}_{jk} = \int \rho \left(x^j x^k - \frac{1}{3} r^2 \delta_{jk} \right) dV_x} \tag{27.53}$$

are the system's mass and mass quadrupole moment. Note that the mass quadrupole moment is equal to the second moment of the mass distribution with its trace removed.

As we have discussed, dynamical oscillations of the quadrupole moment generate the source's strongest gravitational waves. Those waves must be describable, in the source's near zone and local wave zone, by an outgoing-wave solution to the Lorenz-gauge, linearized Einstein equation,

$$\bar{h}_{\mu\nu,}{}^{\nu} = 0, \quad \bar{h}_{\mu\nu,\alpha}{}^{\alpha} = 0 \tag{27.54}$$

[Eqs. (25.89) and (25.90)], that has the near-zone Newtonian limit:

$$\frac{1}{2}(\bar{h}_{00} + \bar{h}_{xx} + \bar{h}_{yy} + \bar{h}_{zz}) = h_{00} = -(\text{quadrupole part of } 2\Phi) = \frac{3\mathcal{I}_{jk} x^j x^k}{r^5} \tag{27.55}$$

[cf. Eq. (25.79)].

1330 Chapter 27. Gravitational Waves and Experimental Tests of General Relativity

The desired solution can be written in the form

$$\bar{h}_{00} = 2\left[\frac{\mathcal{I}_{jk}(t-r)}{r}\right]_{,jk}, \quad \bar{h}_{0j} = 2\left[\frac{\dot{\mathcal{I}}_{jk}(t-r)}{r}\right]_{,k}, \quad \bar{h}_{jk} = 2\frac{\ddot{\mathcal{I}}_{jk}(t-r)}{r}, \quad (27.56)$$

Newtonian potential transitions into this linearized-theory gravitational field in transition and local wave zones

where the coordinates are Cartesian, $r \equiv \sqrt{\delta_{jk}x^j x^k}$, and the dots denote time derivatives. To verify that this is the desired solution: (i) Compute its divergence $\bar{h}_{\alpha\beta}{}^{,\beta}$ and obtain zero almost trivially. (ii) Notice that each Lorentz-frame component of $\bar{h}_{\alpha\beta}$ has the form $f(t-r)/r$ aside from some derivatives that commute with the wave operator, which implies that it satisfies the wave equation. (iii) Notice that in the near zone, the slow-motion assumption inherent in the Newtonian limit makes the time derivatives negligible, so $\bar{h}_{jk} \simeq 0$ and $\bar{h}_{00}$ is twice the right-hand side of Eq. (27.55), as desired.

Because the trace-reversed metric perturbation (27.56) in the local wave zone has the speed-of-light-propagation form, aside from its very slow decay as $1/r$, we can compute the gravitational-wave field h_{jk}^{TT} from it by transverse-traceless projection [Eq. (3) of Box 27.2 with $\mathbf{n} = \mathbf{e}_r$]:

$$h_{jk}^{\mathrm{TT}} = 2\left[\frac{\ddot{\mathcal{I}}_{jk}(t-r)}{r}\right]^{\mathrm{TT}}. \quad (27.57)$$

resulting quadrupolar gravitational-wave field

This is called the *quadrupole-moment formula for gravitational-wave generation*. Our derivation shows that it is valid for any nearly Newtonian source.

Looking back more carefully at the derivation, one can see that, in fact, it relies only on the linearized Einstein equation and the Newtonian potential in the source's local asymptotic rest frame. Therefore, *this quadrupole formula is also valid for slow-motion sources that have strong internal gravity* (e.g., slowly spinning neutron stars), so long as we read the quadrupole moment $\mathcal{I}_{jk}(t-r)$ off the source's weak-field, near-zone Newtonian potential (27.52) and don't try to compute it via the Newtonian volume integral (27.53).

validity for strong-gravity slow-motion sources

When the source is nearly Newtonian, so the volume integral (27.53) can be used to compute the quadrupole moment, the computation of the waves is simplified by computing instead the second moment of the mass distribution:

$$I_{jk} = \int \rho x^j x^k dV_x, \quad (27.58)$$

which differs from the quadrupole moment solely in its trace. Then, because the TT projection is insensitive to the trace, the gravitational-wave field (27.57) can be computed as

$$h_{jk}^{\mathrm{TT}} = 2\left[\frac{\ddot{I}_{jk}(t-r)}{r}\right]^{\mathrm{TT}}. \quad (27.59)$$

for Newtonian source: gravitational-wave field in terms of second moment of mass distribution

27.5 The Generation of Gravitational Waves

27.5.3 Quadrupolar Wave Strength, Energy, Angular Momentum, and Radiation Reaction

To get an order-of-magnitude feel for the strength of the gravitational waves, notice that the second time derivative of the quadrupole moment, in order of magnitude, is the nonspherical part of the source's internal kinetic energy, $E_{\text{kin}}^{\text{ns}}$:

magnitude of slow-motion source's quadrupolar gravitational-wave field

$$\boxed{h_+ \sim h_\times \sim \frac{E_{\text{kin}}^{\text{ns}}}{r} = G\frac{E_{\text{kin}}^{\text{ns}}}{c^4 r},} \tag{27.60}$$

where the second expression is written in conventional units. Although this estimate is based on the slow-motion assumption of source size small compared to reduced wavelength, $L \ll \lambda$, it remains valid in order of magnitude when extrapolated into the realm of the strongest of all realistic astrophysical sources, which have $L \sim \lambda$. In Ex. 27.17 we use Eq. (27.60) to estimate the strongest gravitational waves that might be seen by ground-based gravitational-wave detectors.

Because the gravitational stress-energy tensor $T_{\mu\nu}^{\text{GW}}$ produces background curvature via the Einstein equation $G_{\mu\nu}^{\text{B}} = 8\pi T_{\mu\nu}^{\text{GW}}$, just like nongravitational stress-energy tensors, it must contribute to the rate of change of the source's mass M, linear momentum P_j, and angular momentum J_i [Eqs. (25.102)–(25.106)] just like other stress-energies. When one inserts the quadrupolar $T_{\mu\nu}^{\text{GW}}$ into Eqs. (25.102)–(25.106) and integrates over a sphere in the wave zone of the source's local asymptotic rest frame, one finds that (Ex. 27.11):

source's rate of change of mass, momentum, and angular momentum due to mass quadrupolar gravitational-wave emission

$$\boxed{\frac{dM}{dt} = -\frac{1}{5}\left\langle \frac{\partial^3 \mathcal{I}_{jk}}{\partial t^3} \frac{\partial^3 \mathcal{I}_{jk}}{\partial t^3} \right\rangle,} \tag{27.61}$$

$$\boxed{\frac{dJ_i}{dt} = -\frac{2}{5}\epsilon_{ijk}\left\langle \frac{\partial^2 \mathcal{I}_{jm}}{\partial t^2} \frac{\partial^3 \mathcal{I}_{km}}{\partial t^3} \right\rangle,} \tag{27.62}$$

and $dP_j/dt = 0$. It turns out (cf. Thorne, 1980, Sec. IV) that the dominant linear-momentum change (i.e., the dominant radiation-reaction "kick") arises from a beating of the mass quadrupole moment against the mass octupole moment, and mass quadrupole against current quadrupole:

$$\boxed{\frac{dP_i}{dt} = -\frac{2}{63}\left\langle \frac{\partial^3 \mathcal{I}_{jk}}{\partial t^3} \frac{\partial^4 \mathcal{I}_{jki}}{\partial t^4} \right\rangle - \frac{16}{45}\epsilon_{ijk}\left\langle \frac{\partial^3 \mathcal{I}_{jl}}{\partial t^3} \frac{\partial^3 \mathcal{S}_{kl}}{\partial t^3} \right\rangle.} \tag{27.63}$$

Here the mass octupole moment $\mathcal{I}_{jki}$ is the trace-free part of the third moment of the mass distribution, and the current quadrupole moment $\mathcal{S}_{kp}$ is the symmetric, trace-free part of the first moment of the vectorial angular momentum distribution. (See, e.g., Thorne, 1980, Secs. IV.C, V.C; Thorne, 1983, Sec. 3.)

The back reaction of the emitted waves on their source shows up not only in changes of the source's mass, momentum, and angular momentum, but also in accompanying changes of the source's internal structure. These structure changes can be deduced fully, in many cases, from dM/dt, dJ_j/dt, and dP_j/dt. A nearly Newtonian binary system is an example (Sec. 27.5.4). However, in other cases (e.g., a compact body orbiting near the horizon of a massive black hole), the only way to compute the structure changes is via a gravitational-radiation-reaction force that acts back on the system.

The simplest example of such a force is one derived by William Burke (1971) for quadrupole waves emitted by a nearly Newtonian system. Burke's quadrupolar radiation-reaction force can be incorporated into Newtonian gravitation theory by simply augmenting the system's near-zone Newtonian potential with a radiation-reaction term, computed from the fifth time derivative of the system's quadrupole moment:

$$\Phi^{\text{react}} = \frac{1}{5} \frac{\partial^5 \mathcal{I}_{jk}}{\partial t^5} x^j x^k. \qquad (27.64)$$

gravitational radiation-reaction potential

This potential satisfies the vacuum Newtonian field equation $\nabla^2 \Phi \equiv \delta^{jk} \Phi_{,jk} = 0$, because $\mathcal{I}_{jk}$ is traceless.

This augmentation of the Newtonian potential arises as a result of general relativity's outgoing-wave condition. If one were to switch to an ingoing-wave condition, Φ^{react} would change sign, and if the system's oscillating quadrupole moment were joined onto standing gravitational waves, Φ^{react} would go away. In Ex. 27.12, it is shown that the radiation-reaction force density $-\rho \nabla \Phi^{\text{react}}$ saps energy from the system at the same rate as the gravitational waves carry it away.

radiation-reaction force density in source

Burke's gravitational radiation-reaction potential Φ^{react} and force density $-\rho \nabla \Phi^{\text{react}}$ are close analogs of the radiation-reaction potential [last term in Eq. (16.79)] and acceleration [right-hand side of Eq. (16.82)] that act on an oscillating ball that emits sound waves into a surrounding fluid. Moreover, Burke's derivation of his gravitational radiation-reaction potential is conceptually the same as the derivation, in Sec. 16.5.3, of the sound-wave reaction potential.

EXERCISES

Exercise 27.9 *Problem: Gravitational Waves from Arm Waving*
Wave your arms rapidly, and thereby try to generate gravitational waves.

(a) Using classical general relativity, compute in order of magnitude the wavelength of the waves you generate and their dimensionless amplitude at a distance of one wavelength away from you.

(b) How many gravitons do you produce per second? Discuss the implications of your result.

Exercise 27.10 *Example: Quadrupolar Wave Generation in Linearized Theory*
Derive the quadrupolar wave-generation formula (27.59) for a slow-motion, weak-gravity source in linearized theory using Lorenz gauge and beginning with the retarded-integral formula:

$$\bar{h}_{\mu\nu}(t, \mathbf{x}) = \int \frac{4 T_{\mu\nu}(t - |\mathbf{x} - \mathbf{x}'|, \mathbf{x}')}{|\mathbf{x} - \mathbf{x}'|} dV_{\mathbf{x}'} \qquad (27.65)$$

[Eq. (25.91)]. Your derivation might proceed as follows.

(a) Show that for a slow-motion source, the retarded integral gives for the $1/r \equiv 1/|\mathbf{x}|$ (radiative) part of $\bar{h}_{jk}$:

$$\bar{h}_{jk}(t, \mathbf{x}) = \frac{4}{r} \int T_{jk}(t - r, \mathbf{x}') dV_{\mathbf{x}'}. \qquad (27.66)$$

(b) Show that in linearized theory using Lorenz gauge, the vacuum Einstein equation $-\bar{h}_{\mu\nu,\alpha}{}^{\alpha} = 16\pi T_{\mu\nu}$ [Eq. (25.90)] and the Lorenz gauge condition $\bar{h}_{\mu\nu,}{}^{\nu} = 0$ [Eq. (25.89)] together imply that the stress-energy tensor that generates the waves must have vanishing coordinate divergence: $T^{\mu\nu}{}_{,\nu} = 0$. This means that linearized theory is ignorant of the influence of self-gravity on the gravitating $T^{\mu\nu}$!

(c) Show that this vanishing divergence implies $[T^{00} x^j x^k]_{,00} = [T^{lm} x^j x^k]_{,ml} - 2[T^{lj} x^k + T^{lk} x^j]_{,l} + 2T^{jk}$.

(d) By combining the results of parts (a) and (c), deduce that

$$\bar{h}_{jk}(t, \mathbf{x}) = \frac{2}{r} \frac{d^2 I_{jk}(t-r)}{dt^2}, \qquad (27.67)$$

where I_{jk} is the second moment of the source's (Newtonian) mass-energy distribution $T^{00} = \rho$ [Eq. (27.58)].

(e) Noticing that the trace-reversed metric perturbation (27.67) has the "speed-of-light-propagation" form, deduce that the gravitational-wave field h_{jk}^{TT} can be computed from Eq. (27.67) by a transverse-traceless projection [Box 27.2].

Part (b) shows that this linearized-theory analysis is incapable of deducing the gravitational waves emitted by a source whose dynamics is controlled by its self-gravity (e.g., a nearly Newtonian binary star system). By contrast, the derivation of the quadrupole formula given in Sec. 27.5.2 is valid for any slow-motion source, regardless of the strength and roles of its internal gravity; see the discussion following Eq. (27.57).

Exercise 27.11 *Problem: Energy and Angular Momentum Carried by Gravitational Waves*

(a) Compute the net rate at which the quadrupolar waves (27.57) carry energy away from their source, by carrying out the surface integral (25.102) with T^{0j} being Isaacson's gravitational-wave energy flux (27.40). Your answer should be Eq. (27.61). [Hint: Perform the TT projection in Cartesian coordinates using the projection tensor, Eq. (2) of Box 27.2, and make use of the following integrals over the solid angle on the unit sphere:

$$\frac{1}{4\pi}\int n_i d\Omega = 0, \quad \frac{1}{4\pi}\int n_i n_j d\Omega = \frac{1}{3}\delta_{ij}, \quad \frac{1}{4\pi}\int n_i n_j n_k d\Omega = 0,$$

$$\frac{1}{4\pi}\int n_i n_j n_k n_l d\Omega = \frac{1}{15}(\delta_{ij}\delta_{kl} + \delta_{ik}\delta_{jl} + \delta_{il}\delta_{jk}). \tag{27.68}$$

These integrals should be obvious by symmetry, aside from the numerical factors out in front. Those factors are most easily deduced by computing the z components (i.e., by setting $i = j = k = l = z$ and using $n_z = \cos\theta$).]

(b) The computation of the waves' angular momentum can be carried out in the same way, but is somewhat delicate, because a tiny nonradial component of the energy flux, that dies out as $1/r^3$, gives rise to the $O(1/r^2)$ angular momentum flux (see Thorne, 1980, Sec. IV.D).

Exercise 27.12 *Problem: Energy Removed by Gravitational Radiation Reaction*
Burke's radiation-reaction potential (27.64) produces a force per unit volume $-\rho\nabla\Phi^{\text{react}}$ on its nearly Newtonian source. If we multiply this force per unit volume by the velocity $\mathbf{v} = d\mathbf{x}/dt$ of the source's material, we obtain thereby a rate of change of energy per unit volume. Correspondingly, the net rate of change of the system's mass-energy must be

$$\frac{dM}{dt} = -\int \rho \mathbf{v} \cdot \nabla\Phi^{\text{react}} dV_x. \tag{27.69}$$

Show that, when averaged over a few gravitational-wave periods, this formula agrees with the rate of change of mass (27.61) that we derived in Ex. 27.11 by integrating the outgoing waves' energy flux.

27.5.4 Gravitational Waves from a Binary Star System

A very important application of the quadrupole formalism is to wave emission by a nearly Newtonian binary star system. Denote the stars by indices A and B and their masses by M_A and M_B, so their total and reduced mass are (as usual)

$$\boxed{M = M_A + M_B, \quad \mu = \frac{M_A M_B}{M};} \tag{27.70a}$$

and for simplicity, let the binary's orbit be circular, with separation a between the stars' centers of mass. Then Newtonian force balance dictates that the orbital angular velocity Ω is given by Kepler's law:

dynamics of a Newtonian binary in a circular orbit

$$\boxed{\Omega = \sqrt{M/a^3},} \tag{27.70b}$$

and the orbits of the two stars are

$$x_A = \frac{M_B}{M} a \cos \Omega t, \quad y_A = \frac{M_B}{M} a \sin \Omega t,$$

$$x_B = -\frac{M_A}{M} a \cos \Omega t, \quad y_B = -\frac{M_A}{M} a \sin \Omega t.$$

(27.70c)

The second moment of the mass distribution [Eq. (27.58)] is $I_{jk} = M_A x_A^j x_A^k + M_B x_B^j x_B^k$. Inserting the stars' time-dependent positions (27.70c), we obtain as the only nonzero components:

$$I_{xx} = \mu a^2 \cos^2 \Omega t, \quad I_{yy} = \mu a^2 \sin^2 \Omega t, \quad I_{xy} = I_{yx} = \mu a^2 \cos \Omega t \sin \Omega t.$$ (27.70d)

Noting that $\cos^2 \Omega t = \frac{1}{2}(1 + \cos 2\Omega t)$, $\sin^2 \Omega t = \frac{1}{2}(1 - \cos 2\Omega t)$, and $\cos \Omega t \sin \Omega t = \frac{1}{2} \sin 2\Omega t$ and evaluating the double time derivative, we obtain:

$$\ddot{I}_{xx} = -2\mu(M\Omega)^{2/3} \cos 2\Omega t, \quad \ddot{I}_{yy} = 2\mu(M\Omega)^{2/3} \cos 2\Omega t,$$

$$\ddot{I}_{xy} = \ddot{I}_{yx} = -2\mu(M\Omega)^{2/3} \sin 2\Omega t.$$ (27.70e)

We express these components in terms of Ω rather than a, because Ω is a direct gravitational-wave observable: the waves' angular frequency is 2Ω.

To compute the gravitational-wave field (27.59), we must project out the transverse traceless part of this $\ddot{I}_{jk}$. The projection is most easily performed in an orthonormal spherical basis, since there the transverse part is just the projection onto the plane spanned by $\vec{e}_{\hat\theta}$ and $\vec{e}_{\hat\phi}$, and the transverse-traceless part has components

$$(\ddot{I}_{\hat\theta\hat\theta})^{TT} = -(\ddot{I}_{\hat\phi\hat\phi})^{TT} = \frac{1}{2}(\ddot{I}_{\hat\theta\hat\theta} - \ddot{I}_{\hat\phi\hat\phi}), \quad (\ddot{I}_{\hat\theta\hat\phi})^{TT} = \ddot{I}_{\hat\theta\hat\phi}$$ (27.70f)

[cf. Eq. (1) of Box 27.2]. Now, a little thought will save us much work: We need only compute these quantities at $\phi = 0$ (i.e., in the x-z plane), since their circular motion guarantees that their dependence on t and ϕ must be solely through the quantity $\Omega t - \phi$. At $\phi = 0$, $\vec{e}_{\hat\theta} = \vec{e}_x \cos\theta - \vec{e}_z \sin\theta$ and $\vec{e}_{\hat\phi} = \vec{e}_y$, so the only nonzero components of the transformation matrices from the Cartesian basis to the transverse part of the spherical basis are $L^x_{\hat\theta} = \cos\theta$, $L^z_{\hat\theta} = -\sin\theta$, and $L^y_{\hat\phi} = 1$. Using this transformation matrix at $\phi = 0$, we obtain: $\ddot{I}_{\hat\theta\hat\theta} = \ddot{I}_{xx} \cos^2\theta$, $\ddot{I}_{\hat\phi\hat\phi} = \ddot{I}_{yy}$, and $\ddot{I}_{\hat\theta\hat\phi} = \ddot{I}_{xy} \cos\theta$. Inserting these and expressions (27.70e) into Eq. (27.70f), and setting $\Omega t \to \Omega t - \phi$ to make the formulas valid away from $\phi = 0$, we obtain:

$$(\ddot{I}_{\hat\theta\hat\theta})^{TT} = -(\ddot{I}_{\hat\phi\hat\phi})^{TT} = -(1 + \cos^2\theta)\mu(M\Omega)^{2/3} \cos[2(\Omega t - \phi)],$$

$$(\ddot{I}_{\hat\theta\hat\phi})^{TT} = +(\ddot{I}_{\hat\phi\hat\theta})^{TT} = -2\cos\theta\,\mu(M\Omega)^{2/3} \sin[2(\Omega t - \phi)].$$ (27.70g)

The gravitational-wave field (27.59) is $2/r$ times this quantity evaluated at the retarded time $t - r$.

We make the conventional choice for the polarization tensors:

$$\mathbf{e}^+ = (\vec{e}_{\hat\theta} \otimes \vec{e}_{\hat\theta} - \vec{e}_{\hat\phi} \otimes \vec{e}_{\hat\phi}), \quad \mathbf{e}^\times = (\vec{e}_{\hat\theta} \otimes \vec{e}_{\hat\phi} + \vec{e}_{\hat\phi} \otimes \vec{e}_{\hat\theta});$$

$$\vec{e}_{\hat\theta} = \frac{1}{r}\frac{\partial}{\partial \theta}, \quad \vec{e}_{\hat\phi} = \frac{1}{r\sin\theta}\frac{\partial}{\partial \phi}. \tag{27.71a}$$

Then Eqs. (27.59) and (27.70g) tell us that the gravitational-wave field is, in slot-naming index notation:

gravitational waves from Newtonian binary

$$h^{\rm TT}_{\mu\nu} = h_+ e^+_{\mu\nu} + h_\times e^\times_{\mu\nu}, \tag{27.71b}$$

where

$$h_+ = h^{\rm TT}_{\hat\theta\hat\theta} = \frac{2}{r}[\ddot{I}_{\hat\theta\hat\theta}(t-r)]^{\rm TT} = -2(1+\cos^2\theta)\frac{\mu(M\Omega)^{2/3}}{r}\cos[2(\Omega t - \Omega r - \phi)], \tag{27.71c}$$

$$h_\times = h^{\rm TT}_{\hat\theta\hat\phi} = \frac{2}{r}[\ddot{I}_{\hat\theta\hat\phi}(t-r)]^{\rm TT} = -4\cos\theta\,\frac{\mu(M\Omega)^{2/3}}{r}\sin[2(\Omega t - \Omega r - \phi)]. \tag{27.71d}$$

We have expressed the amplitudes of these waves in terms of the dimensionless quantity $(M\Omega)^{2/3} = M/a = v^2$, where v is the relative velocity of the two stars.

Notice that, as viewed from the polar axis $\theta = 0$, h_+ and $h_\times$ are identical except for a $\pi/2$ phase delay, which means that the net stretch-squeeze ellipse (the combination of those in Figs. 27.1 and 27.2) rotates with angular velocity Ω. This is the gravitational-wave variant of circular polarization, and it arises because the binary motion as viewed from the polar axis looks circular. By contrast, as viewed by an observer in the equatorial plane $\theta = \pi/2$, $h_\times$ vanishes, so the net stretch-squeeze ellipse just oscillates along the + axes, and the waves have linear polarization. This is natural, since the orbital motion as viewed by an equatorial observer is just a linear, horizontal, back-and-forth oscillation. Notice also that the gravitational-wave frequency is twice the orbital frequency:

binary's gravitational-wave frequency is twice its orbital frequency

$$f = 2\frac{\Omega}{2\pi} = \frac{\Omega}{\pi}. \tag{27.72}$$

To compute, via Eqs. (27.61) and (27.62), the rate at which energy and angular momentum are lost from the binary, we need to know the double and triple time derivatives of its quadrupole moment $\mathcal{I}_{jk}$. The double time derivative is just $\ddot{I}_{jk}$ with its trace removed, but Eq. (27.70e) shows that $\ddot{I}_{jk}$ is already traceless so $\ddot{\mathcal{I}}_{jk} = \ddot{I}_{jk}$. Inserting Eq. (27.70e) for this quantity into Eqs. (27.61) and (27.62) and performing

27.5 The Generation of Gravitational Waves

the average over a gravitational-wave period, we find that

$$\boxed{\frac{dM}{dt} = -\frac{32}{5}\frac{\mu^2}{M^2}(M\Omega)^{10/3}, \quad \frac{dJ_z}{dt} = -\frac{1}{\Omega}\frac{dM}{dt}, \quad \frac{dJ_x}{dt} = \frac{dJ_y}{dt} = 0.}$$ (27.73)

binary's orbital inspiral due to gravitational-wave emission

This loss of energy and angular momentum causes the binary to spiral inward, decreasing the stars' separation a and increasing their orbital angular velocity Ω. By comparing Eqs. (27.73) with the standard equations for the binary's orbital energy and angular momentum [M − (sum of rest masses of stars) = $E = -\frac{1}{2}\mu M/a = -\frac{1}{2}\mu(M\Omega)^{2/3}$, and $J_z = \mu a^2 \Omega = \mu(M\Omega)^{2/3}/\Omega$], we obtain an equation for $d\Omega/dt$, which we can integrate to give

$$\boxed{\Omega = \pi f = \left(\frac{5}{256}\frac{1}{\mu M^{2/3}}\frac{1}{t_o - t}\right)^{3/8}.}$$ (27.74)

Here t_o (an integration constant) is the time remaining until the two stars merge, if the stars are thought of as point masses so their surfaces do not collide sooner. This equation can be inverted to read off the time until merger as a function of gravitational-wave frequency.

These results for a binary's waves and radiation-reaction-induced inspiral are of great importance for gravitational-wave detection (see, e.g., Cutler and Thorne, 2002; Sathyaprakash and Schutz, 2009).

EXERCISES

Exercise 27.13 *Problem: Gravitational Waves Emitted by a Linear Oscillator*
Consider a mass m attached to a spring, so it oscillates along the z-axis of a Cartesian coordinate system, moving along the world line $z = a \cos \Omega t$, $y = x = 0$. Use the quadrupole-moment formalism to compute the gravitational waves $h_+(t, r, \theta, \phi)$ and $h_\times(t, r, \theta, \phi)$ emitted by this oscillator, with the polarization tensors chosen as in Eqs. (27.71a). Pattern your analysis after the computation of waves from a binary in Sec. 27.5.4.

Exercise 27.14 **Example: Propagation of a Binary's Waves Through an Expanding Universe*
As we shall see in Sec. 28.3, the following line element is a possible model for the large-scale structure of our universe:

$$ds^2 = b^2[-d\eta^2 + d\chi^2 + \chi^2(d\theta^2 + \sin^2\theta d\phi^2)], \quad \text{where } b = b_o\eta^2,$$ (27.75)

and b_o is a constant with dimensions of length. This is an expanding universe with flat spatial slices $\eta = $ constant. Notice that the proper time measured by observers at rest in the spatial coordinate system is $t = b_o \int \eta^2 d\eta = (b_o/3)\eta^3$.

A nearly Newtonian, circular binary is at rest at $\chi = 0$ in an epoch when $\eta \simeq \eta_o$. The coordinates of the binary's local asymptotic rest frame are $\{t, r, \theta, \phi\}$, where $r = b\chi$ and the coordinates cover only a tiny region of the universe: $\chi \lesssim \chi_o \ll \eta_o$. The gravitational waves in this local asymptotic rest frame are described by Eqs. (27.71). Use geometric optics (Sec. 27.4.2) to propagate these waves out through the expanding universe. In particular, do the following.

(a) Show that the null rays along which the waves propagate are the curves of constant θ, ϕ, and $\eta - \chi$.

(b) Each ray can be characterized by the retarded time τ_r at which the source emitted it. Show that

$$\tau_r = \frac{1}{3} b_o (\eta - \chi)^3. \tag{27.76a}$$

(c) Show that in the source's local asymptotic rest frame, this retarded time is $\tau_r = t - r$, and the phase of the wave is $\varphi = 2(\Omega \tau_r + \phi)$ [cf. Eqs. (27.71c) and (27.71d)]. Because the frequency Ω varies with time due to the binary inspiral, a more accurate formula for the wave's phase is $\varphi = 2(\int \Omega \, d\tau_r + \phi)$. Using Eq. (27.74), show that

$$\varphi = 2\phi - \left(\frac{t_o - \tau_r}{5\mathcal{M}}\right)^{5/8}, \quad \Omega = \frac{d\varphi}{d\tau_r} = \left(\frac{5}{256} \frac{1}{\mathcal{M}^{5/3}} \frac{1}{t_o - \tau_r}\right)^{3/8}, \tag{27.76b}$$

where

$$\mathcal{M} \equiv \mu^{3/5} M^{2/5} \tag{27.76c}$$

(with μ the reduced mass and M the total mass) is called the binary's *chirp mass*, because, as Eqs. (27.76b) show, it controls the rate at which the binary's orbital angular frequency Ω and the gravitational-wave angular frequency 2Ω "chirp upward" as time passes. The quantity τ_r as given by Eq. (27.76a) is constant along rays when they travel out of the local wave zone and into and through the universe. Correspondingly, if we continue to write φ in terms of τ_r on those rays using Eqs. (27.76b), this φ will be conserved along the rays in the external universe and therefore will satisfy the geometric-optics equation: $\nabla_{\vec{k}} \varphi = 0$ [Eqs. (27.35b)].

(d) Show that the orthonormal basis vectors and polarization tensors

$$\vec{e}_{\hat{\theta}} = \frac{1}{b\chi} \frac{\partial}{\partial \theta}, \quad \vec{e}_{\hat{\phi}} = \frac{1}{b\chi \sin\theta} \frac{\partial}{\partial \phi},$$

$$\mathbf{e}^+ = (\vec{e}_{\hat{\theta}} \otimes \vec{e}_{\hat{\theta}} - \vec{e}_{\hat{\phi}} \otimes \vec{e}_{\hat{\phi}}), \quad \mathbf{e}^\times = (\vec{e}_{\hat{\theta}} \otimes \vec{e}_{\hat{\phi}} + \vec{e}_{\hat{\phi}} \otimes \vec{e}_{\hat{\theta}}) \tag{27.76d}$$

in the external universe: (i) are parallel-transported along rays and (ii) when carried backward on rays into the local asymptotic rest frame, become the basis vectors and tensors used in that frame's solution (27.71) for the gravitational waves. Therefore, these $e^+_{\mu\nu}$ and $e^\times_{\mu\nu}$ are the polarization tensors needed for our geometric-optics waves.

(e) Consider a bundle of rays that, at the source, extends from ϕ to $\phi + \Delta\phi$ and from θ to $\theta + \Delta\theta$. Show that this bundle's cross sectional area, as it moves outward to larger and larger χ, is $\mathcal{A} = r^2 \sin\theta \Delta\theta \Delta\phi$, where r is a function of η and χ given by

$$r = b\chi = b_o \eta^2 \chi. \tag{27.76e}$$

Show that in the source's local asymptotic rest frame, this r is the same as the distance r from the source that appears in Eqs. (27.71c) and (27.71d) for h_+ and $h_\times$.

(f) By putting together all the pieces from parts (a) through (e), show that the solution to the equations of geometric optics (27.35) for the gravitational-wave field as it travels outward through the universe is

$$\bm{h}^{\mathrm{TT}} = h_+ \bm{e}^+ + h_\times \bm{e}^\times, \tag{27.76f}$$

with $\bm{e}^+$ and $\bm{e}^\times$ given by Eqs. (27.76d), with h_+ and $h_\times$ given by

$$h_+ = -2(1+\cos^2\theta)\frac{\mathcal{M}^{5/3}\Omega^{2/3}}{r}\cos\varphi, \quad h_\times = -4\cos\theta\,\frac{\mathcal{M}^{5/3}\Omega^{2/3}}{r}\sin\varphi, \tag{27.76g}$$

and with Ω, φ, and r given by Eqs. (27.76b), (27.76a), and (27.76e). [Hint: Note that all quantities in this solution except r are constant along rays, and r varies as $1/\sqrt{\mathcal{A}}$, where $\mathcal{A}$ is the area of a bundle of rays.]

(g) The angular frequency of the waves that are emitted at retarded time τ_r is $\omega_e = 2\Omega$. When received at Earth these waves have a cosmologically redshifted frequency $\omega_r = \partial\varphi/\partial t$, where $t = (b_o/3)\eta^3$ is proper time measured at Earth, and in the derivative we must hold fixed the spatial coordinates of Earth: $\{\chi, \theta, \phi\}$. The ratio of these frequencies is $\omega_e/\omega_r = 1 + z$, where z is the so-called cosmological redshift of the waves. Show that $1 + z = (\partial\tau_r/\partial t)^{-1} = \eta^2/(\eta - \chi)^2$.

(h) Show that the information carried by the binary's waves is the following. (i) From the ratio of the amplitudes of the two polarizations, one can read off the inclination angle θ of the binary's spin axis to the line of sight to the binary. (ii) From the waves' measured angular frequency ω and its time rate of change $d\omega/dt$, one can read off $(1+z)\mathcal{M}$, the binary's redshifted chirp mass. (iii) From the amplitude of the waves, with θ and $(1+z)\mathcal{M}$ known, one can read off $(1+z)r$, a quantity known to cosmologists as the binary's *luminosity distance*. [Note: It is remarkable that gravitational waves by themselves reveal the source's luminosity distance but not its redshift, while electromagnetic observations reveal the redshift but not the luminosity distance. This complementarity illustrates the importance and power of combined gravitational-wave and electromagnetic observations.]

27.5.5 Gravitational Waves from Binaries Made of Black Holes, Neutron Stars, or Both: Numerical Relativity [T2]

Among the most interesting sources of gravitational waves are binary systems made of two black holes, a black hole and a neutron star, or two neutron stars—so-called *compact binaries*. When the two bodies are far apart, their motion and waves can be described accurately by Newtonian gravity and the quadrupole-moment formalism: the formulas in Sec. 27.5.4. As the bodies spiral inward, $(M\Omega)^{2/3} = M/a = v^2$ grows larger, h_+ and $h_\times$ grow larger, and relativistic corrections to our Newtonian, quadrupole analysis grow larger. Those relativistic corrections (including current-quadrupole waves, mass-octupole waves, etc.) can be computed using a post-Newtonian expansion of the Einstein field equations (i.e., an expansion in $M/a \sim v^2$). The accuracies of ground-based detectors such as LIGO require that, for compact binaries, the expansion be carried at least to order v^7 beyond our Newtonian, quadrupole analysis! (See Blanchet, 2014.)

compact binaries analyzed by a post-Newtonian expansion

At the end of the inspiral, the binary's bodies come crashing together. To compute the waves from this final merger with an accuracy comparable to the observations, it is necessary to solve the Einstein field equation on a computer. The techniques for doing this are called *numerical relativity* (Baumgarte and Shapiro, 2010; Shibata, 2016) and were pioneered by Bryce DeWitt, Larry Smarr, Saul Teukolsky, Frans Pretorius, and others.

merger analyzed by numerical relativity

For binary black holes with approximately equal masses, simulations using numerical relativity reveal that the total energy radiated in gravitational waves is $\Delta E \sim 0.1 M c^2$, where M is the binary's total mass. Most of this energy is emitted in the last ~ 5 to 10 cycles of waves, at wave periods $P \sim (10 \text{ to } 20) GM/c^3$ [i.e., frequencies $f = 1/P \sim 1{,}000$ Hz $(10 M_\odot/M)$]. The gravitational-wave power output in these last 5–10 cycles is $dE/dt \sim 0.1 M c^2/(100 GM/c^3) = 0.001 c^5/G$, which is roughly 10^{23} times the luminosity of the Sun, and 100 times the luminosity of all the stars in the observable universe put together! If the holes have masses $\sim 10 M_\odot$, this enormous luminosity lasts for only ~ 0.1 s, and the total energy emitted is the rest-mass energy of the Sun. If the holes have masses $\sim 10^9 M_\odot$, the enormous luminosity lasts for ~ 1 yr, and the energy emitted is the rest-mass energy of $\sim 10^8$ Suns.

gravitational-wave power output in black hole mergers: 100 universe luminosities

For the simplest case of two identical, nonspinning black holes that spiral together in a circular orbit, both waveforms (both wave shapes) have the simple time evolution shown in Fig. 27.6. As the holes spiral together, their amplitude and phase increase in accord with the Newtonian-quadrupole formulas (27.71), (27.72), and (27.74) but by the time of this figure, post-Newtonian corrections are producing noticeable differences from those formulas. When the holes merge, the gravitational-wave amplitude

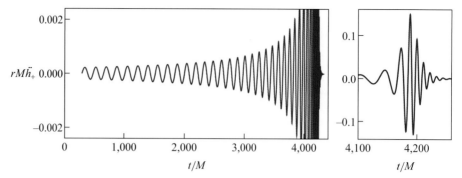

FIGURE 27.6 For a binary made of two identical, nonspinning black holes that spiral together and merge: the time evolution of the gravitational-wave tidal field $\mathcal{E}_{ij} \propto \ddot{h}_+$ [Eq. (27.22)] for the + polarization. The × polarization waveform is the same as this but with a phase shift. The right panel shows the detailed signal just prior to merger. Based on simulations performed by the Caltech/Cornell/CITA numerical relativity group (Mroué et. al. 2013). (CITA is the Canadian Institute for Theoretical Astrophysics.)

reaches a maximum amplitude. The merged hole then vibrates and the waves "ring down" with exponentially decaying amplitude.

Much more interesting are binaries made of black holes that spin. In this case, the angular momentum of each spinning hole drags inertial frames, as does the binary's orbital angular momentum. This frame-dragging causes the spins and the orbital plane to precess, and those precessions modulate the waves. Figure 27.7 depicts a generic example: a binary whose holes have a mass ratio 6:1, dimensionless spins $a_A/M_A = 0.91$, $a_B/M_B = 0.30$, and randomly chosen initial spin axes and orbital plane. Frame dragging causes the orbital motion to be rather complex, and correspondingly, the two waveforms are much richer than in the nonspinning case. The waveforms carry detailed information about the binary's masses, spins, and orbital evolution, and also about the geometrodynamics of its merger (Box 27.4).

information in gravitational waves

EXERCISES

Exercise 27.15 *Problem: Maximum Gravitational-Wave Amplitude*
Extrapolating Eqs. (27.71)–(27.73) into the strong-gravity regime, estimate the maximum gravitational-wave amplitude and emitted power for a nonspinning binary black hole with equal masses and with unequal masses. Compare with the results from numerical relativity discussed in the text.

Exercise 27.16 *Problem: Gravitational Radiation from Binary Pulsars in Elliptical Orbits*
Many precision tests of general relativity are associated with binary pulsars in elliptical orbits (Sec. 27.2.6).

(a) Verify that the radius of the relative orbit of the pulsars can be written as $r = p/(1 + e \cos \phi)$, where p is the *semi-latus rectum*, e is the eccentricity, and $d\phi/dt = (Mp)^{1/2}/r^2$ with M the total mass.

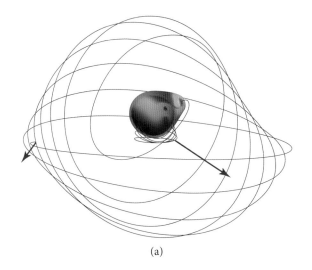

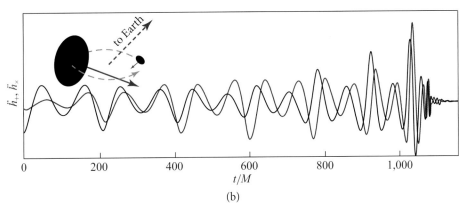

FIGURE 27.7 (a) The orbital motion of a small black hole around a larger black hole (mass ratio $M_B/M_A = 1/6$), when the spins are $a_B/M_B = 0.30$ and $a_A/M_A = 0.91$ and the initial spin axes and orbital plane are as shown in panel b. (b) The two gravitational waveforms emitted in the direction toward Earth (blue dashed line). These waveforms are from a catalog of simulations of 174 different binary-black-hole mergers, carried out by the Caltech/Cornell/CITA numerical relativity group (Mroué et al., 2013).

(b) Show that the traceless mass quadrupole moment, Eq. (27.53), in suitable coordinates, is [cf. Eq. (27.70d)]

$$\mathcal{I}_{jk} = \frac{\mu p^2}{(1+e\cos\phi)^2} \begin{pmatrix} \cos^2\phi - \frac{1}{3} & \cos\phi\sin\phi & 0 \\ \cos\phi\sin\phi & \sin^2\phi - \frac{1}{3} & 0 \\ 0 & 0 & -\frac{1}{3} \end{pmatrix}. \quad (27.77a)$$

(c) Use computer algebra to evaluate the second and third time derivatives of this tensor, and then use Eqs. (27.61) and (27.62) to calculate the orbit-averaged energy and angular momentum emitted in gravitational waves, per orbit.

27.5 The Generation of Gravitational Waves

BOX 27.4. GEOMETRODYNAMICS T2

When spinning black holes collide, they excite nonlinear vibrations of curved spacetime—a phenomenon that John Wheeler has called *geometrodynamics*. This nonlinear dynamics can be visualized using tidal tendex lines (which depict the tidal field $\mathcal{E}_{ij}$) and frame-drag vortex lines (which depict the frame-drag field $\mathcal{B}_{ij}$); see Boxes 25.2 and 26.3. Particularly helpful are the concepts of a *tendex* (a collection of tendex lines with large tendicities) and a *(frame-drag) vortex* (a collection of vortex lines with large vorticities). A spinning black hole has a counterclockwise vortex emerging from its north polar region, and a clockwise vortex emerging from its south polar region (right diagram in Box 26.3).

As an example of geometrodynamics, consider two identical black holes that collide head on, with their spins transverse to the collision direction. Numerical-relativity simulations (Owen et al., 2011) reveal that, when the holes collide and merge, each hole deposits its two vortices onto the merged horizon. The four vortices dynamically attach to each other in pairs (panel a in figure below). The pairs then interact, with surprising consequences. The blue (clockwise) vortex disconnects from the hole and forms a set of closed vortex loops that wrap around a torus (thick blue lines in panel c), and the red (counterclockwise) vortex does the same (thin red lines in panel c). This torus expands outward at the speed of light, while energy temporarily stored in near-horizon tendices (not shown) regenerates the new pair of horizon-penetrating vortices shown in panel b, with reversed vorticities (reversed directions of twist). As the torus expands outward, its motion, via the Maxwell-like Bianchi identity $\partial \mathcal{E}/\partial t = (\nabla \times \mathcal{B})^S$ (Box 25.2), generates a set of tendex lines that wrap around the torus at 45° angles to the vortex lines (dashed lines in panel c). The torus's interleaved vortex and tendex lines have become a gravitational wave, which locally looks like the plane wave discussed in Box 27.3. This process repeats, with amplitude damping, generating a sequence of expanding tori. (Figure adapted from Owen et al., 2011.)

(continued)

Your answer should be

$$\Delta E = \frac{64\pi \mu^2 M^{5/2}}{5p^{7/2}}\left(1 + \frac{73}{24}e^2 + \frac{37}{96}e^4\right), \quad \Delta J = \frac{64\pi \mu^2 M^2}{5p^2}\left(1 + \frac{7}{8}e^2\right).$$

(27.77b)

(d) Combine the results from part (c) with Kepler's laws to calculate an expression for the rate of increase of pulse period and decrease of the eccentricity (cf. Sec. 27.2.6).

BOX 27.4. (continued)

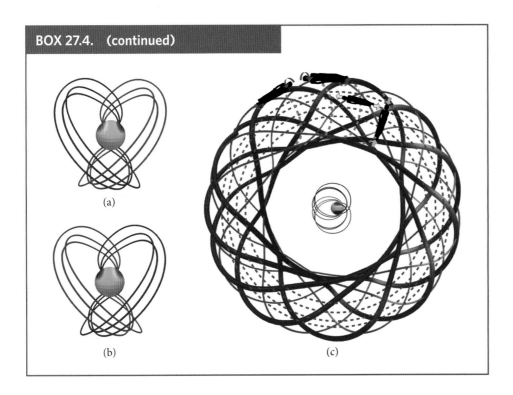

(a)

(b)

(c)

(e) Consider a parabolic encounter between two stars, and show that the energy and angular momentum radiated are, respectively,

$$\Delta E = \frac{170\pi \mu^2 M^{5/2}}{3 p^{7/2}}, \quad \Delta L = \frac{24\pi \mu^2 M^2}{p^2}. \tag{27.78}$$

27.6 The Detection of Gravitational Waves

27.6.1 Frequency Bands and Detection Techniques

Physicists and astronomers are searching for gravitational waves in four different frequency bands using four different techniques:

- In the extremely low-frequency (ELF) band, $\sim 10^{-18}$ to $\sim 10^{-15}$ Hz, gravitational waves are sought via their imprint on the polarization of the cosmic microwave background (CMB) radiation. There is only one expected ELF source of gravitational waves, but it is a very interesting one: quantum fluctuations in the gravitational field (spacetime curvature) that emerge from the big bang's quantum-gravity regime, the *Planck era*, and that are subsequently amplified to classical, detectable sizes by the universe's early inflationary expansion. We shall study this amplification and the resulting ELF gravitational waves in Sec. 28.7.1 and shall see these waves' great potential for probing the physics of inflation.

gravitational-wave frequency bands: ELF, VLF, LF, and HF; sources and detection techniques in each band

- In the very-low-frequency (VLF) band, $\sim 10^{-9}$ to $\sim 10^{-7}$ Hz, gravitational waves are sought via their influence on the propagation of radio waves emitted by pulsars (spinning neutron stars) and by the resulting fluctuations in the arrival times of the pulsars' radio-wave pulses at Earth (Sec. 27.6.6 and Ex. 27.20). The expected VLF sources are violent processes in the first fraction of a second of the universe's life (Secs. 28.4.1 and 28.7.1) and the orbital motion of extremely massive pairs of black holes in the distant universe.

- In the low-frequency (LF) band, $\sim 10^{-4}$ to ~ 0.1 Hz, gravitational waves have been sought, in the past, via their influence on the radio signals by which NASA tracks interplanetary spacecraft. In the 2020s or 2030s, this technique will likely be supplanted by some variant of the proposed LISA—three "drag-free" spacecraft in a triangular configuration with 5-km-long arms, that track one another via laser beams. LISA is likely to see waves from massive black-hole binaries (hole masses $\sim 10^5$ to $10^7 M_\odot$) out to cosmological distances; from small holes, neutron stars, and white dwarfs spiraling into massive black holes out to cosmological distances; from the orbital motion of white-dwarf binaries, neutron-star binaries, and stellar-mass black-hole binaries in our own galaxy; and possibly from violent processes in the very early universe.

- The high-frequency (HF) band, ~ 10 to $\sim 10^3$ Hz, is where Earth-based detectors operate: laser interferometer gravitational-wave detectors, such as LIGO, and resonant-mass detectors in which a gravitational wave alters the amplitude and phase of vibrations of a normal mode of a large, solid cylinder or sphere. On September 14, 2015, the advanced LIGO gravitational wave detectors made their first detection: a wave burst named GW150914 with amplitude 1.0×10^{-21}, duration ~ 150 ms, and frequency chirping upward from ~ 50 Hz (when it entered the LIGO band) to 240 Hz. By comparing the observed waveform with those from numerical relativity simulations, the LIGO-VIRGO scientists deduced that the waves came from the merger of a $29 M_\odot$ black hole with a $36 M_\odot$ black hole, 1.2 billion light years from Earth, to form a $62 M_\odot$ black hole, with a release of $3 M_\odot c^2$ of energy in gravitational waves. When this textbook went to press, additional black hole binaries were being detected. As LIGO's sensitivity improves and additional interferometers come on line, the LIGO scientists expect to see other sources: waves from spinning, slightly deformed neutron stars (e.g., pulsars) in our Milky Way galaxy, the final inspiral and collisions of binaries made from neutron stars in the more distant universe, the tearing apart of a neutron star by the spacetime curvature of a companion black hole, supernovae, the triggers of gamma-ray bursts, and possibly waves from violent processes in the very early universe.

For detailed discussions of these gravitational-wave sources in all four frequency bands, and of prospects for their detection, see, for example, Cutler and Thorne (2002)

Exercise 27.17 *Example: Strongest Gravitational Waves in HF Band*

(a) Using an order-of-magnitude analysis based on Eq. (27.60), show that the strongest gravitational waves that are likely to occur each year in LIGO's HF band have $h_+ \sim h_\times \sim 10^{-21}$—which is the actual amplitude of LIGO's first observed wave burst, GW150914. [Hint: The highest nonspherical kinetic energy achievable must be for a highly deformed object (or two colliding objects), in which the internal velocities approach the speed of light—say, for realism, $v \sim 0.3c$. To achieve these velocities, the object's size L must be of order 2 or 3 Schwarzschild radii, $L \sim 5M$, where M is the source's total mass. The emitted waves must have $f \sim 200$ Hz (the frequency at the minimum of Advanced LIGO's noise curve—which is similar to initial LIGO, Fig. 6.7, but a factor ~ 10 lower). Using these considerations, estimate the internal angular frequency of the source's motion, and thence the source's mass, and finally the source's internal kinetic energy. Such a source will be very rare, so to see a few per year, its distance must be some substantial fraction of the Hubble distance. From this, estimate $h_+ \sim h_\times$.]

(b) As a concrete example, estimate the gravitational-wave strength from the final moments of inspiral and merger of two black holes, as described by Eqs. (27.71) and (27.70b) extrapolated into the highly relativistic domain.

27.6.2 Gravitational-Wave Interferometers: Overview and Elementary Treatment

We briefly discussed Earth-based gravitational-wave interferometers such as LIGO in Sec. 9.5, focusing on optical interferometry issues. In this section we analyze the interaction of a gravitational wave with such an interferometer. This analysis will not only teach us much about gravitational waves, but it will also illustrate some central issues in the physical interpretation of general relativity theory.

idealized gravitational-wave interferometer

To get quickly to the essentials, we examine a rather idealized interferometer: a Michelson interferometer (one without the input mirrors of Fig. 9.13) that floats freely in space, so there is no need to hang its mirrors by wires; see Fig. 27.8. In Sec. 27.6.5, we briefly discuss more realistic interferometers.

If we ignore delicate details, the operation of this idealized interferometer is simple. As seen in a local Lorentz frame of the beam splitter, the gravitational wave changes the length of the x arm by $\delta x = \frac{1}{2} h_+ \ell_x$, where ℓ_x is the unperturbed length, and it changes that of the y arm by the opposite amount: $\delta y = -\frac{1}{2} h_+ \ell_y$ [Eqs. (27.29)]. The interferometer is operated with unperturbed lengths ℓ_x and ℓ_y that are nearly but not quite equal, so there is a small amount of light going toward the photodetector. The wave-induced change of arm length causes a relative phase shift of the light

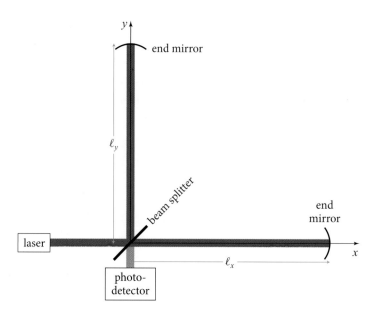

FIGURE 27.8 An idealized gravitational-wave interferometer.

returning down the two arms to the beam splitter given by $\Delta\varphi(t) = \omega_o(2\delta y - 2\delta x) = \omega_o(\ell_x + \ell_y)h_+(t)$, where ω_o is the light's angular frequency (and we have set the speed of light to unity); cf. Sec. 9.5. This oscillating phase shift modulates the intensity of the light going into the photodetector by $\Delta I_{\rm PD}(t) \propto \Delta\varphi(t)$. Setting $\ell_x \simeq \ell_y = \ell$, this modulation is

interferometer's photodetector current output

$$\Delta I_{\rm PD}(t) \propto \Delta\varphi(t) = 2\omega_o \ell h_+(t). \tag{27.79}$$

Therefore, the photodetector output tells us directly the gravitational waveform $h_+(t)$.

In the following two (Track-Two) subsections, we rederive this result much more carefully in two different coordinate systems (two different gauges). Our two analyses predict the same result (27.79) for the interferometer output, but they appear to attribute that result to two different mechanisms.

TT-gauge analysis attributes output signal to influence of waves on interferometer's light

In our first analysis (performed in TT gauge; Sec. 27.6.3), the interferometer's test mass remains always at rest in our chosen coordinate system, and the gravitational wave $h_+(t - z)$ interacts with the interferometer's light. The imprint that $h_+(t - z)$ leaves on the light causes a fluctuating light intensity $I_{\rm out}(t) \propto h_+(t)$ to emerge from the interferometer's output port and be measured by the photodetector.

proper-reference-frame analysis attributes output signal to interaction of waves with interferometer mirrors

In our second analysis (a more rigorous version of the above quick analysis, performed in the proper reference frame of the interferometer's beam splitter; Sec. 27.6.4) the gravitational waves interact hardly at all with the light. Instead, they push the end mirrors back and forth relative to the coordinate system, thereby lengthening one arm while shortening the other. These changing arm lengths cause a changing interference of the light returning to the beam splitter from the two arms, and that changing interference produces the fluctuating light intensity $I_{\rm out}(t) \propto h_+(t)$ measured by the photodetectors.

These differences of viewpoint are somewhat like the differences between the Heisenberg picture and the Schrödinger picture in quantum mechanics. The intuitive pictures associated with two viewpoints appear to be very different (Schrödinger's wave function versus Heisenberg's matrices; gravitational waves interacting with light versus gravitational waves pushing on mirrors). But when one computes the same physical observable from the two different viewpoints (probability for a quantum measurement outcome; light intensity measured by photodetector), the two viewpoints give the same answer.

analogy with Heisenberg and Schrödinger pictures in quantum mechanics

27.6.3 Interferometer Analyzed in TT Gauge T2

interferometer analyzed in TT gauge

For our first analysis, we place the interferometer at rest in the x-y plane of a TT coordinate system, with its arms along the x- and y-axes and its beam splitter at the origin, as shown in Fig. 27.8. For simplicity, we assume that the gravitational wave propagates in the z direction and has $+$ polarization, so the linearized spacetime metric has the TT-gauge form:

spacetime metric

$$ds^2 = -dt^2 + [1 + h_+(t-z)]\,dx^2 + [1 - h_+(t-z)]\,dy^2 + dz^2 \qquad (27.80)$$

[Eq. (27.20)]. For ease of notation, we omit the subscript $+$ from h_+ in the remainder of this section.

The beam splitter and end mirrors move freely and thus travel along geodesics of the metric (27.80). The splitter and mirrors are at rest in the TT coordinate system before the wave arrives, so initially, the spatial components of their 4-velocities vanish: $u_j = 0$. Because the metric coefficients $g_{\alpha\beta}$ are all independent of x and y, the geodesic equation dictates that the components u_x and u_y are conserved and thus remain zero as the wave passes, which implies (since the metric is diagonal) $u^x = dx/d\tau = 0$ and $u^y = dy/d\tau = 0$. One can also show (see Ex. 27.18) that $u^z = dz/d\tau = 0$ throughout the wave's passage. Thus, in terms of motion relative to the TT coordinate system, the gravitational wave has no influence at all on the beam splitter and mirrors; they all remain at rest (constant x, y, and z) as the waves pass.

mirrors and beam splitter do not move relative to TT coordinates

(Despite this lack of motion, the proper distances between the mirrors and the beam splitter—the interferometer's physically measured arm lengths—do change. If the unchanging coordinate lengths of the two arms are $\Delta x = \ell_x$ and $\Delta y = \ell_y$, then the metric (27.80) says that the physically measured arm lengths are

$$L_x = \left[1 + \frac{1}{2}h(t)\right]\ell_x, \qquad L_y = \left[1 - \frac{1}{2}h(t)\right]\ell_y. \qquad (27.81)$$

When h is positive, the x arm is lengthened and the y arm is shortened; when negative, L_x is shortened and L_y is lengthened.)

Next turn to the propagation of light in the interferometer. We assume, for simplicity, that the light beams have large enough transverse sizes that we can idealize them, on their optic axes, as plane electromagnetic waves. (In reality, they will be Gaussian beams, of the sort studied in Sec. 8.5.5.) The light's vector potential A^α satisfies the curved-spacetime vacuum wave equation $A^{\alpha;\mu}{}_\mu = 0$ [Eq. (25.60) with vanishing Ricci tensor]. We write the vector potential in geometric optics (eikonal-approximation) form as

$$A^\alpha = \Re(\mathsf{A}^\alpha e^{i\varphi}), \tag{27.82}$$

where A^α is a slowly varying amplitude, and φ is a rapidly varying phase [cf. Eq. (7.20)]. Because the wavefronts are (nearly) planar and the spacetime metric is nearly flat, the light's amplitude A^α will be nearly constant as it propagates down the arms, and we can ignore its variations. Not so the phase. It oscillates at the laser frequency $\omega_o/2\pi \sim 3 \times 10^{14}$ Hz [i.e., $\varphi^{\text{out}}_{x\,\text{arm}} \simeq \omega_o(x-t)$ for light propagating outward from the beam splitter along the x arm, and similarly for the returning light and the light in the y arm]. The gravitational wave places tiny deviations from this $\omega_o(x-t)$ onto the phase; we must compute those deviations.

In the spirit of geometric optics, we introduce the light's spacetime wave vector

$$\vec{k} \equiv \vec{\nabla}\varphi, \tag{27.83}$$

and we assume that $\vec{k}$ varies extremely slowly compared to the variations of φ. Then the wave equation $A^{\alpha;\mu}{}_\mu = 0$ reduces to the statement that the wave vector is null: $\vec{k} \cdot \vec{k} = \varphi_{,\alpha}\varphi_{,\beta} g^{\alpha\beta} = 0$. For light in the x arm the phase depends only on x and t; for that in the y arm it depends only on y and t. Combining this with the TT metric (27.80) and noting that the interferometer lies in the $z=0$ plane, we obtain

influence of waves on phase of light in interferometer arms

$$-\left(\frac{\partial \varphi_{x\,\text{arm}}}{\partial t}\right)^2 + [1-h(t)]\left(\frac{\partial \varphi_{x\,\text{arm}}}{\partial x}\right)^2 = 0,$$

$$-\left(\frac{\partial \varphi_{y\,\text{arm}}}{\partial t}\right)^2 + [1+h(t)]\left(\frac{\partial \varphi_{y\,\text{arm}}}{\partial y}\right)^2 = 0. \tag{27.84}$$

We idealize the laser as perfectly monochromatic, and we place it at rest in our TT coordinates, arbitrarily close to the beam splitter. Then the outgoing light frequency, as measured by the beam splitter, must be precisely ω_o and cannot vary with time. Since proper time as measured by the beam splitter is equal to coordinate time t [cf. the metric (27.80)], the frequency that the laser and beam splitter measure must be $\omega = -\partial \varphi/\partial t = -k_t$. This dictates the following boundary conditions (initial conditions) on the phase of the light that travels outward from the beam splitter:

$$\frac{\partial \varphi^{\text{out}}_{x\,\text{arm}}}{\partial t} = -\omega_o \text{ at } x=0, \quad \frac{\partial \varphi^{\text{out}}_{y\,\text{arm}}}{\partial t} = -\omega_o \text{ at } y=0. \tag{27.85}$$

It is straightforward to verify that the solutions to Eq. (27.84) (and thence to the wave equation and thence to Maxwell's equations) that satisfy the boundary conditions (27.85) are

$$\varphi_{x\,\text{arm}}^{\text{out}} = -\omega_o\left[t - x + \frac{1}{2}H(t-x) - \frac{1}{2}H(t)\right],$$

$$\varphi_{y\,\text{arm}}^{\text{out}} = -\omega_o\left[t - y - \frac{1}{2}H(t-y) + \frac{1}{2}H(t)\right], \tag{27.86}$$

where $H(t)$ is the first time integral of the gravitational waveform:

$$H(t) \equiv \int_0^t h(t')dt' \tag{27.87}$$

(cf. Ex. 27.19).

The outgoing light reflects off the mirrors, which are at rest in the TT coordinates at locations $x = \ell_x$ and $y = \ell_y$. As measured by observers at rest in these coordinates, there is no Doppler shift of the light, because the mirrors are not moving. Correspondingly, the phases of the reflected light, returning back along the two arms, have the following forms:

$$\varphi_{x\,\text{arm}}^{\text{back}} = -\omega_o\left[t + x - 2\ell_x + \frac{1}{2}H(t+x-2\ell_x) - \frac{1}{2}H(t)\right],$$

$$\varphi_{y\,\text{arm}}^{\text{back}} = -\omega_o\left[t + y - 2\ell_y - \frac{1}{2}H(t+y-2\ell_y) + \frac{1}{2}H(t)\right]. \tag{27.88}$$

The difference of the phases of the returning light, at the beam splitter ($x = y = 0$), is

$$\Delta\varphi \equiv \varphi_{x\,\text{arm}}^{\text{back}} - \varphi_{y\,\text{arm}}^{\text{back}} = -\omega_o[-2(\ell_x - \ell_y) + \frac{1}{2}H(t - 2\ell_x) + \frac{1}{2}H(t - 2\ell_y) - H(t)]$$

$$\simeq +2\omega_o[\ell_x - \ell_y + \ell h(t)] \quad \text{for Earth-based interferometers.} \tag{27.89}$$

difference between arms for output light's phase shift

In the final expression we have used the fact that for Earth-based interferometers operating in the HF band, the gravitational wavelength $\lambda_{\text{GW}} \sim c/(100\,\text{Hz}) \sim 3{,}000$ km is long compared to the interferometers' $\sim$4-km arms, and the arms have nearly the same length: $\ell_y \simeq \ell_x \equiv \ell$.

The beam splitter sends a light field $\propto e^{i\varphi_{x\,\text{arm}}^{\text{back}}} + e^{i\varphi_{y\,\text{arm}}^{\text{back}}}$ back toward the laser, and a field $\propto e^{i\varphi_{x\,\text{arm}}^{\text{back}}} - e^{i\varphi_{y\,\text{arm}}^{\text{back}}} = e^{i\varphi_{y\,\text{arm}}^{\text{back}}}(e^{i\Delta\varphi} - 1)$ toward the photodetector. The intensity of the light entering the photodetector is proportional to the squared amplitude of the field: $I_{\text{PD}} \propto |e^{i\Delta\varphi} - 1|^2$. We adjust the interferometer's arm lengths so their difference $\ell_x - \ell_y$ is small compared to the light's reduced wavelength $1/\omega_o = c/\omega_o$ but large compared to $|\ell h(t)|$. Correspondingly, $|\Delta\varphi| \ll 1$, so only a tiny fraction of the light goes toward the photodetector (it is the interferometer's "dark port"), and that darkport light intensity is

$$\boxed{I_{\text{PD}} \propto |e^{i\Delta\varphi} - 1|^2 \simeq |\Delta\varphi|^2 \simeq 4\omega_o^2(\ell_x - \ell_y)^2 + 8\omega_o^2(\ell_x - \ell_y)\ell h_+(t).} \tag{27.90}$$

photodiode output

Here we have restored the subscript $+$ onto h. The time-varying part of this intensity is proportional to the gravitational waveform $h_+(t)$ [in agreement with Eq. (27.79)]. It is this time-varying part that the photodetector reports as the interferometer output.

EXERCISES

Exercise 27.18 *Derivation and Practice: Geodesic Motion in TT Coordinates* **T2**
Consider a particle that is at rest in the TT coordinate system of the gravitational-wave metric (27.80) before the gravitational wave arrives. In the text it is shown that the particle's 4-velocity has $u^x = u^y = 0$ as the wave passes. Show that $u^z = 0$ and $u^t = 1$ as the wave passes, so the components of the particle's 4-velocity are unaffected by the passing gravitational wave, and the particle remains at rest (constant x, y, and z) in the TT coordinate system.

Exercise 27.19 *Example: Light in an Interferometric Gravitational-Wave Detector in TT Gauge* **T2**
Consider the light propagating outward from the beam splitter, along the x arm of an interferometric gravitational-wave detector, as analyzed in TT gauge, so (suppressing the subscript "x arm" and superscript "out") the electromagnetic vector potential is $A^\alpha = \Re(\mathcal{A}^\alpha e^{i\varphi(x,t)})$, with $\mathcal{A}^\alpha$ constant and with $\varphi = -\omega_o\left[t - x + \frac{1}{2}H(t-x) - \frac{1}{2}H(t)\right]$ [Eqs. (27.86) and (27.87)].

(a) Show that this φ satisfies the nullness equation (27.84), as claimed in the text—which implies that $A^\alpha = \Re(\mathcal{A}^\alpha e^{i\varphi(x,t)})$ satisfies Maxwell's equations in the geometric-optics limit.

(b) Show that this φ satisfies the initial condition (27.85), as claimed in the text.

(c) Show that, because the gradient $\vec{k} = \vec{\nabla}\varphi$ of this φ satisfies $\vec{k} \cdot \vec{k} = 0$, it also satisfies $\nabla_{\vec{k}}\vec{k} = 0$. Thus, the wave vector is the tangent vector to geometric-optics rays that are null geodesics in the gravitational-wave metric. Photons travel along these null geodesics and have 4-momenta $\vec{p} = \hbar\vec{k}$.

(d) Because the gravitational-wave metric (27.80) is independent of x, the p_x component of a photon's 4-momentum must be conserved along its geodesic world line. Compute $p_x = k_x = \partial\varphi/\partial x$, and thereby verify this conservation law.

(e) Explain why the photon's frequency, as measured by observers at rest in our TT coordinate system, is $\omega = -k_t = -\partial\varphi/\partial t$. Explain why the rate of change of this frequency, as computed moving with the photon, is $d\omega/dt \simeq (\partial/\partial t + \partial/\partial x)\omega$, and show that $d\omega/dt \simeq -\frac{1}{2}\omega_o dh/dt$.

27.6.4 Interferometer Analyzed in the Proper Reference Frame of the Beam Splitter **T2**

proper-reference-frame analysis

We now carefully reanalyze our idealized interferometer in the proper reference frame of its beam splitter, denoting that frame's coordinates by $\hat{x}^\alpha$. Because the beam splitter is freely falling (moving along a geodesic through the gravitational-wave spacetime), its proper reference frame is locally Lorentz (LL), and its metric coefficients have the form $g_{\hat\alpha\hat\beta} = \eta_{\alpha\beta} + O(\delta_{jk}\hat{x}^j\hat{x}^k/\mathcal{R}^2)$ [Eq. (25.9a)]. Here $\mathcal{R}$ is the radius of curvature of spacetime, and $1/\mathcal{R}^2$ is of order the components of the Riemann tensor, which have magnitude $\ddot{h}(\hat{t} - \hat{z})$ [Eq. (27.22) with t and z equal to $\hat{t}$ and $\hat{z}$, aside from fractional corrections of order h]. Thus we have:

$$g_{\hat{\alpha}\hat{\beta}} = \eta_{\alpha\beta} + O[\ddot{h}(\hat{t} - \hat{z})\delta_{jk}\hat{x}^j\hat{x}^k]. \qquad (27.91)$$

(Here and below we again omit the subscript $+$ on h for ease of notation.)

The following coordinate transformation takes us from the TT coordinates x^α used in Sec. 27.6.3 to the beam splitter's LL coordinates:

$$x = \left[1 - \frac{1}{2}h(\hat{t} - \hat{z})\right]\hat{x}, \quad y = \left[1 + \frac{1}{2}h(\hat{t} - \hat{z})\right]\hat{y},$$

$$t = \hat{t} - \frac{1}{4}\dot{h}(\hat{t} - \hat{z})(\hat{x}^2 - \hat{y}^2), \quad z = \hat{z} - \frac{1}{4}\dot{h}(\hat{t} - \hat{z})(\hat{x}^2 - \hat{y}^2). \qquad (27.92)$$

It is straightforward to insert this coordinate transformation into the TT-gauge metric (27.80) and thereby obtain, to linear order in h:

$$\boxed{ds^2 = -d\hat{t}^2 + d\hat{x}^2 + d\hat{y}^2 + d\hat{z}^2 + \frac{1}{2}(\hat{x}^2 - \hat{y}^2)\ddot{h}(t - z)(d\hat{t} - d\hat{z})^2.} \qquad (27.93)$$

spacetime metric in proper reference frame of beam splitter

This has the expected LL form (27.91) and, remarkably, it turns out not only to be a solution of the vacuum Einstein equation in linearized theory but also an exact solution to the full vacuum Einstein equation (cf. Misner, Thorne, and Wheeler, 1973, Ex. 35.8)!

Throughout our idealized interferometer, the magnitude of the metric perturbation in these LL coordinates is $|h_{\hat{\alpha}\hat{\beta}}| \lesssim (\ell/\lambdabar_{\rm GW})^2 h$, where $\lambdabar_{\rm GW} = \lambda_{\rm GW}/(2\pi)$ is the waves' reduced wavelength, and h is the magnitude of $h(\hat{t} - \hat{z})$. For Earth-based interferometers operating in the HF band ($\sim$10 to $\sim$1000 Hz), $\lambdabar_{\rm GW}$ is of order 50–5,000 km, and the arm lengths are $\ell \leq 4$ km, so $(L/\lambda)^2 \lesssim 10^{-6}$ to 10^{-2}. Thus, the metric coefficients $h_{\hat{\alpha}\hat{\beta}}$ are no larger than $h/100$. This has a valuable consequence for the analysis of the interferometer: up to fractional accuracy $\sim(\ell/\lambdabar_{\rm GW})^2 h \lesssim h/100$, the LL coordinates are globally Lorentz throughout the interferometer (i.e., $\hat{t}$ measures proper time, and $\hat{x}^j$ are Cartesian and measure proper distance). In the rest of this section, we restrict attention to such Earth-based interferometers but continue to treat them as though they were freely falling. (See Sec. 27.6.5 for the influence of Earth's gravity.)

Being initially at rest at the origin of these LL coordinates, the beam splitter remains always at rest, but the mirrors move. Not surprisingly, the geodesic equation for the mirrors in the metric (27.93) dictates that their coordinate positions are, up to fractional errors of order $(\ell/\lambdabar_{\rm GW})^2 h$,

27.6 The Detection of Gravitational Waves

mirror motions in proper reference frame

$$\hat{x} = L_x = \left[1 + \frac{1}{2}h(\hat{t})\right]\ell_x, \quad \hat{y} = \hat{z} = 0 \quad \text{for mirror in } x \text{ arm,}$$

$$\hat{y} = L_y = \left[1 - \frac{1}{2}h(\hat{t})\right]\ell_y, \quad \hat{x} = \hat{z} = 0 \quad \text{for mirror in } y \text{ arm.}$$

(27.94)

[Equations (27.94) can also be deduced from the gravitational-wave tidal acceleration $-\mathcal{E}_{\hat{j}\hat{k}}\hat{x}^k$, as in Eq. (27.25), and from the fact that to good accuracy $\hat{x}$ and $\hat{y}$ measure proper distance from the beam splitter.] So even though the mirrors do not move in TT coordinates, they do move in LL coordinates. The two coordinate systems predict the same time-varying physical arm lengths (the same proper distances from beam splitter to mirrors), L_x and L_y [Eqs. (27.81) and (27.94)].

As in TT coordinates, so also in LL coordinates, we can analyze the light propagation in the geometric-optics approximation, with $A^{\hat{\alpha}} = \Re(\mathcal{A}^{\hat{\alpha}} e^{i\varphi})$. Just as the wave equation for the vector potential dictates in TT coordinates that the rapidly varying phase of the outward light in the x arm has the form $\varphi_{x\,\text{arm}}^{\text{out}} = -\omega_o(t - x) + O(\omega_o \ell h_{\mu\nu})$ [Eqs. (27.86) with $x \sim \ell \ll \lambda_{\text{GW}}$, so $H(t-x) - H(t) \simeq \dot{H}(t)x = h(t)x \sim h\ell \sim h_{\mu\nu}\ell$], so similarly the wave equation in LL coordinates turns out to dictate that

$$\varphi_{x\,\text{arm}}^{\text{out}} = -\omega_o(\hat{t} - \hat{x}) + O(\omega_o \ell h_{\hat{\mu}\hat{\nu}}) = -\omega_o(\hat{t} - \hat{x}) + O\left(\omega_o \ell h \frac{\ell^2}{\lambda_{\text{GW}}^2}\right), \quad (27.95)$$

and similarly for the returning light and the light in the y arm. The term $O(\omega_o \ell h \, \ell^2/\lambda_{\text{GW}}^2)$ is the influence of the direct interaction between the gravitational wave and the light, and it is negligible in the final answer (27.96) for the measured phase shift. Aside from this term, the analysis of the interferometer proceeds in exactly the same way as in flat space (because $\hat{t}$ measures proper time and $\hat{x}$ and $\hat{y}$ proper distance): the light travels a round trip distance L_x in one arm and L_y in the other, and therefore acquires a phase difference, on arriving back at the beam splitter, given by

difference in output phase shift between arms: same as in TT gauge

$$\Delta\varphi = -\omega_o[-2(L_x - L_y)] + O\left(\omega_o \ell h \frac{\ell^2}{\lambda_{\text{GW}}^2}\right)$$

$$\simeq +2\omega_o[\ell_x - \ell_y + \ell h(\hat{t})] + O\left(\omega_o \ell h \frac{\ell^2}{\lambda_{\text{GW}}^2}\right). \quad (27.96)$$

This net phase difference for the light returning from the two arms is the same as we deduced in TT coordinates [Eq. (27.89)], up to the negligible correction $O(\omega_o \ell h \, \ell^2/\lambda_{\text{GW}}^2)$, and therefore the time-varying intensity of the light into the photodetector will be the same [Eq. (27.90)].

In our TT analysis the phase shift $2\omega_o \ell h(t)$ arose from the interaction of the light with the gravitational waves. In the LL analysis, it is due to the displacements of the mirrors in the LL coordinates (i.e., the displacements as measured in terms of proper distance) that cause the light to travel different distances in the two arms. The direct LL interaction of the waves with the light produces only the tiny correction $O(\omega_o \ell h\, \ell^2/\lambdabar_{\rm GW}^2)$ to the phase shift.

It should be evident that the LL description is much closer to elementary physics than the TT description is. This is always the case, when one's apparatus is sufficiently small that one can regard $\hat{t}$ as measuring proper time and $\hat{x}^j$ as Cartesian coordinates that measure proper distance throughout the apparatus. But for a large apparatus (e.g., planned space-based interferometers such as LISA, with arm lengths $\ell \gtrsim \lambdabar_{\rm GW}$, and the pulsar timing arrays of Sec. 27.6.6) the LL analysis becomes quite complicated, as one must pay close attention to the $O(\omega_o \ell h\, \ell^2/\lambdabar_{\rm GW}^2)$ corrections. In such a case, the TT analysis is much simpler.

27.6.5 Realistic Interferometers

For realistic, Earth-based interferometers, one must take account of the acceleration of gravity. Experimenters do this by hanging their beam splitters and test masses on wires or fibers. The simplest way to analyze such an interferometer is in the proper reference frame of the beam splitter, where the metric must now include the influence of the acceleration of gravity by adding a term $-2g\hat{z}$ to the metric coefficient $h_{\hat{0}\hat{0}}$ [cf. Eq. (24.60b)]. The resulting analysis, like that in the LL frame of our freely falling interferometer, will be identical to what one would do in an accelerated reference frame of flat spacetime, so long as one takes account of the motion of the test masses driven by the gravitational-wave tidal acceleration $-\mathcal{E}_{\hat{i}\hat{j}}\hat{x}^j$, and so long as one is willing to ignore the tiny effects of $O(\omega_o \ell h\, \ell^2/\lambdabar_{\rm GW}^2)$.

To make the realistic interferometer achieve high sensitivity, the experimenters introduce a lot of clever complications, such as the input mirrors of Fig. 9.13, which turn the arms into Fabry-Perot cavities. All these complications can be analyzed, in the beam splitter's proper reference frame, using standard flat-spacetime techniques augmented by the gravitational-wave tidal acceleration. The direct coupling of the light to the gravitational waves can be neglected, as in our idealized interferometer.

27.6.6 Pulsar Timing Arrays

A rather different approach to direct detection of gravitational waves is by the influence of the waves on the timing of an array of radio pulsars (Hobbs et al., 2010).

As we have discussed in Sec. 27.2.6, many pulsars, especially those with millisecond periods, have pulse arrival times at the radio telescope that can be predicted with high accuracy—30 to 100 ns in the best cases—after correcting for slowing down of the pulsar, propagation effects, and the motion of the telescope relative to the center of mass of the solar system. If the radio pulses travel through a gravitational wave, then they incur tiny variations in arrival time that can be used to detect the wave. This technique is most sensitive to waves with frequency $f \sim 30$ nHz (the gravitational

gravitational-wave searches via pulsar timing arrays in VLF band

VLF band; see Sec. 27.6.1), so a few periods are sampled over the duration of an observation, typically a few years. The most promising sources are supermassive binary black holes in the nuclei of distant galaxies.

Let us consider the idealized problem of radio pulses emitted at a steady rate by a stationary pulsar with position **z** relative to a stationary telescope on Earth and traveling at the speed of light. Now, suppose that there is a single, monochromatic, plane gravitational wave, $\boldsymbol{h} = \boldsymbol{h}_o \cos[2\pi f(t + \mathbf{x} \cdot \hat{\mathbf{r}})]$, where $\hat{\mathbf{r}}$ is a unit vector pointing toward the source. For the wave, adopt the metric (27.18a) and TT gauge. The pulsar and the telescope will remain at rest with respect to the coordinates (Ex. 27.18), and t measures their proper times.

The wave-induced time delay in the arrival of a pulse, called the *residual* $R(t)$, is then given by integrating along a past directed null geodesic ($ds = 0$) from the telescope to the pulsar:[6]

influence of gravitational wave on radio pulses' time delay (their "residual")

$$R(t) = -\frac{1}{2} \int_0^z dz' \hat{\mathbf{z}} \cdot \boldsymbol{h}_o \cdot \hat{\mathbf{z}} \cos\bigl(2\pi f[t - z'(1 - \hat{\mathbf{z}} \cdot \hat{\mathbf{r}})]\bigr)$$

$$= \frac{\hat{\mathbf{z}} \cdot \boldsymbol{h}_o \cdot \hat{\mathbf{z}} \{\sin(2\pi f[t - z(1 - \hat{\mathbf{z}} \cdot \hat{\mathbf{r}})]) - \sin(2\pi f t)\}}{4\pi f(1 - \hat{\mathbf{z}} \cdot \hat{\mathbf{r}})} \quad (27.97)$$

(Ex. 27.20a). Note that two terms contribute to the residual; the first is associated with the wave as it passes the pulsar, the second is local and associated with the wave passing Earth.

correlation of residuals between different pulsars

This search for gravitational waves is being prosecuted not just with a single pulsar but also with an array of tens (currently and hundreds in the future) of millisecond pulsars. Programs using different telescopes are being coordinated and combined through the International Pulsar Timing Array (IPTA) collaboration (http://www.ipta4gw.org). The local residual (27.97) is correlated between different lines of sight to different pulsars with differing polarization projections $\hat{\mathbf{z}} \cdot \boldsymbol{h}_o \cdot \hat{\mathbf{z}}$. This facilitates identifying the waves amidst different noises associated with different pulsars, and it allows a coherent addition of the data from many pulsars, which will be particularly valuable in searches for individual sources of gravitational radiation. If and when a source is located, the array can be used to fix its direction $\hat{\mathbf{r}}$ and the polarization of the waves. In addition to extracting astrophysics of the source, there should be a clear affirmation that gravitational waves have spin two and travel at the speed of light.

gravitational-wave sources in VLF band

However, unless we are very lucky, it is more likely that the first detection or the most prescriptive upper limit will relate to a stochastic background created by the superposed waves from many black hole binaries (Ex. 27.21). Although the bandwidth from an individual source will be of order the reciprocal of the time it takes the binary

---

6. Of course the actual null geodesic followed by the radio waves changes in the presence of the gravitational wave but, invoking Fermat's principle, the travel time is unchanged to $O(h)$ if we integrate along the unperturbed trajectory.

frequency to double—in the fHz to pHz range—the randomly phased signals are likely to overlap in frequency, ensuring that the gravitational radiation can be well described by a spectral energy density.

Other more speculative sources of gravitational waves in the VLF frequency band have been proposed, especially cosmic strings. There could be many surprises waiting to be discovered.

EXERCISES

Exercise 27.20 *Example: Pulsar Timing Array* [T2]
Explore how pulsar timing can be used to detect a plane gravitational wave.

(a) Derive Eq. (27.97). [Hint: One way to derive this result is from the action

$$\frac{1}{2} \int g_{\alpha\beta}(dx^\alpha/d\zeta)(dx^\beta/d\zeta)d\zeta$$

that underlies the rays' super-hamiltonian; Ex. 25.7. The numerical value of the action is zero, and since it is an extremum along each true ray, if you evaluate it along a path that is a straight line in the TT coordinate system instead of along the true ray, you will still get zero at first order in h_o.]

(b) Recognizing local and pulsar contributions to the timing residuals, explain how much information about the amplitude, direction, and polarization of a gravitational wave from a single black-hole binary can be obtained using accurate timing data from one, two, three, four, and many pulsars.

(c) Suppose, optimistically, that 30 pulsars will be monitored with timing accuracy $\sim$100 ns, and that arrival times will be measured 30 times a year. Make an estimate of the minimum measurable amplitude of the sinusoidal residual created by a single binary as a function of observing duration and wave frequency f.

(d) Using the result from part (c) and using the predicted residual averaged over the direction and the orientation of the source from Eqs. (27.71c) and (27.71d), estimate the maximum distance to which an individual source could be detected as a function of the chirp mass $\mathcal{M}$ and the frequency f.

Exercise 27.21 *Problem: Stochastic Background from Binary Black Holes* [T2]
The most likely signal to be detected using a pulsar timing array is a stochastic background formed by perhaps billions of binary black holes in the nuclei of galaxies.

(a) For simplicity, suppose that these binaries are all on circular orbits and that they lose energy at a rate given by Eqs. (27.73). Show that for each binary the gravitational wave energy radiated per unit log wave frequency is

$$\frac{dE_{GW}}{d\ln f} = -\frac{dE}{d\ln f} = -\frac{2E}{3} = \frac{\pi^{2/3}}{3}\mathcal{M}^{5/3} f^{2/3}, \qquad (27.98a)$$

where $\mathcal{M}$ is the chirp mass given by Eq. (27.76c) and E is its orbital energy given by $-\frac{1}{2}\mu M/a$.

27.6 The Detection of Gravitational Waves   1357

(b) Suppose that, on astrophysical grounds, binaries only radiate gravitational radiation efficiently if they merge in less than $t_{\text{merge}} \sim 300$ Myr. Show that this requires

$$\mathcal{M} > \left(\frac{5}{256\, t_{\text{merge}}}\right)^{3/5} (\pi f)^{-8/5}. \tag{27.98b}$$

This evaluates to ~ 1 million solar masses for $f \sim 30$ nHz. By contrast, measured masses of the large black holes in the nuclei of galaxies range from about 4 million solar masses as in our own modest spiral galaxy to perhaps 20 billion solar masses in the largest galaxies observed.

(c) Now suppose that these black holes grew mostly through mergers of holes with very different masses. Show that the total energy radiated per unit log frequency over the course of the many mergers that led to a black hole with mass M is

$$\frac{dE_{GW}}{d\ln f} \sim \frac{\pi^{2/3}}{5} M^{5/3} f^{2/3}. \tag{27.98c}$$

(d) The number density per unit mass dn/dM today of these merging black holes has been estimated to be

$$\frac{dn}{d\ln M} = \frac{\rho_{BH} M^2}{2 M_t^3} e^{-M/M_t}, \tag{27.98d}$$

where $\rho_{BH} \sim 2 \times 10^{-15}$ J m^{-3} is the contribution of black holes (i.e., of their masses) to the average energy density in the local universe and $M_t = 5 \times 10^7$ solar masses $\sim 10^{38}$ kg. Show that

$$\frac{d\rho_{GW}}{d\ln f} \sim \frac{\pi^{2/3} \Gamma(11/3)}{10} (M_t f)^{2/3} \rho_{BH} \sim 7 \times 10^{-19} \left(\frac{f}{30\,\text{nHz}}\right)^{2/3} \text{J m}^{-3}. \tag{27.99}$$

This is an overestimate, because most of the black holes were likely assembled in the past, when the universe was about three times smaller than it is today. If one thinks of the gravitational waves as gravitons that lose energy as the universe expands, then this energy density should be reduced by a factor of three. Furthermore, as the black holes are thought to grow through accretion of gas and not by mergers, this estimate should be considered an upper bound.

(e) Making the same assumptions as in the previous exercise, determine whether it will be possible to detect this background in 5 years of observation.

Bibliographic Note

For an elementary introduction to experimental tests of general relativity in the solar system, we recommend Hartle (2003, Chap. 10). For an enjoyable, popular-level book on experimental tests, see Will (1993b). For a very complete monograph on the theory underlying experimental tests, see Will (1993a), and for a more nearly up-to-date review of experimental tests, see Will (2014).

For elementary and fairly complete introductions to gravitational waves, we recommend Hartle (2003, Chaps. 16, 23) and Schutz (2009, Chap. 9). For more advanced treatments, we suggest Misner, Thorne, and Wheeler (1973, Sec. 18.2 and Chaps. 35, 36), Thorne (1983), and Straumann (2013, Secs. 5.3–5.7); but Misner, Thorne, and Wheeler (1973, Chap. 37) on gravitational-wave detection is terribly out of date and is not recommended. For fairly complete reviews of gravitational-wave sources for ground-based detectors (LIGO, etc.) and space-based detectors (LISA, etc.), see Cutler and Thorne (2002) and Sathyaprakash and Schutz (2009). For a lovely monograph on the physics of interferometric gravitational-wave detectors, see Saulson (1994).

Because gravitational-wave science is a rapidly maturing and burgeoning field, there are long, in-depth treatments that include considerable experimental detail and much detail on data analysis techniques, as well as on wave sources and the fundamental theory: Thorne, Bondarescu, and Chen (2002), Maggiore (2007), and Creighton and Anderson (2011).

CHAPTER TWENTY-EIGHT

Cosmology

The expansion thus took place in three phases, a first period of rapid expansion in which the atom universe was broken into atomic stars, a period of slowing down, followed by a third period of accelerated expansion.

GEORGES LEMAÎTRE (1933)

28.1 Overview

The extragalactic sky is isotropic and dark at night. Distant galaxies accelerate away from us. Five-sixths of the matter in the universe is in a "dark" form that we can only detect through its gravitational effects. We are immersed in a bath of blackbody radiation—the *cosmic microwave background* (CMB)—with a temperature of ~ 2.7 K that exhibits tiny fluctuations with fractional amplitude $\sim 10^{-5}$. These four profound observations lead inexorably to a description of the universe that began from a hot big bang nearly 14 billion years ago. Remarkable progress over recent years in making careful measurements of these features has led to a *standard cosmology* that involves mostly classical physics and incorporates many of the ideas and techniques we have discussed in the preceding 27 chapters.

Let us begin by describing the universe. We live on Earth, the "third rock" out from an undistinguished star located in the outskirts of a quite ordinary galaxy—the second largest in a loose federation of galaxies called the *local group* on the periphery of the *local supercluster*. Our neighbors are no more remarkable. Meanwhile, the distant galaxies and microwave background photons that we observe exhibit no preferred directions. In short, we do not appear to be special, and nowhere else that we can see appears to be special. It is therefore reasonable to assume that the universe has no "center" and that it is essentially homogeneous on the largest scales that we observe, so that all parts are equivalent at the same time. By extension, the average observed properties of the universe and its inferred history would be similar if we lived in any other galaxy.

However, it has long been known that the universe cannot also be infinite and static. If it were, then any line of sight would eventually intercept the surface of a star, and the night sky would be bright. Instead, as we observe directly, galaxies and their constituent stars are receding from us, and the speed of recession is now increasing with time, contrary to our Newtonian expectation. This expansion is observed out to distances that are so great that the recession velocities are relativistic, dimming the light from the distant stars and galaxies, and making the night sky between nearby

BOX 28.1. READERS' GUIDE

- This chapter draws on every one of the preceding chapters:
    - Chap. 1—geometry, tensors;
    - Chap. 2—stress energy tensor, Lorentz invariance;
    - Chap. 3—phase space, radiation thermodynamics, Liouville equation, stars, Boltzmann equation;
    - Chap. 4—entropy, Fermi-Dirac and Bose-Einstein distribution functions, statistical mechanics in the presence of gravity;
    - Chap. 5—chemical potential, Monte Carlo method;
    - Chap. 6—correlation functions and spectral density, Wiener-Khintchine theorem, Fokker-Planck formalism;
    - Chap. 7—paraxial optics, gravitational lenses, polarization;
    - Chap. 8—Fourier transforms, point spread function;
    - Chap. 9—coherence and random processes;
    - Chap. 10—radiation physics, parametric resonance;
    - Chap. 11—strain tensor, spherical coordinates;
    - Chap. 12—longitudinal modes, zero point fluctuations;
    - Chap. 13—relativistic fluids, stars;
    - Chap. 14—barotropic equation of state, viscosity;
    - Chap. 15—turbulence, power law spectrum;
    - Chap. 16—sound waves, nonlinear waves;
    - Chap. 17—compressible flow, 1-dimensional flow;
    - Chap. 18—heat conduction;
    - Chap. 19—nuclear reactions, magnetic stress tensor;
    - Chap. 20—Saha equation, Coulomb collisions;
    - Chap. 21—plasma oscillations, Debye screening;
    - Chap. 22—Jeans' theorem, collisionless particles, Landau damping;
    - Chap. 23—nonlinear wave dynamics, quantization of classical fields, collisionless shocks;
    - Chap. 24—differential geometry, local Lorentz frames;
    - Chap. 25—Riemann tensor, Einstein equation, geodesic deviation, geometrized units;
    - Chap. 26—horizons, pressure as source of gravitation; and
    - Chap. 27—gravitational time dilation, gravitational waves.

stars as dark as we observe. Furthermore, the amount of matter in the universe induces gravitational potential differences of order c^2 across the observed universe.

These features require a general relativistic description. However, even if we had tested it adequately in the strong field regime of black holes and neutron stars (which we have only begun to do), there is no more guarantee that general relativity describes the universe at large than that classical mechanics and electromagnetism suffice to describe atoms! Despite this, the best way to proceed is to continue to adopt general relativity and be alert to the possibility that it could fail us. What we shall discover is that it provides a sufficient and successful framework for describing the observations.

The discovery of the CMB in 1964 transformed cosmology. This radiation must have dominated the stress-energy tensor and, consequently, the geometry and dynamics of the universe in the past. More recently, it has been used to show that the spatial curvature of the universe is very small and that the CMB's tiny temperature fluctuations have simple behavior consistent with basic (mostly classical) physics. And the fluctuations can be used to argue that the universe exhibited a brief growth spurt, called *inflation,* when it was very young—a surge that established both the observed uniformity and the structure that we see around us today. No less important has been the quantitative description, mostly developed over the past 20 years, of the two invisible entities (popularly known as *dark matter* and the *cosmological constant* or, more generally, *dark energy*) that account for 95% of the modern universe. In fact, the measurements have become so good that most alternative descriptions of the universe have been ruled out and no longer need to be discussed. This is a good time to benefit from this simplification.

This chapter is longer, more ambitious, and less didactic—there are no practice and derivation exercises—than its predecessors, because we want to take advantage of the opportunity to exhibit modern classical physics in one of its most exciting and currently successful applications. The reader must expect to read slowly and carefully to "connect the dots." To keep the chapter at a manageable length, we eschew essentially all critical discussion of the observations and measurements, including errors. We also avoid the idiosyncratic terminology and conventions of astronomy. Finally, we stop short of describing the birth and development of galaxies, stars, and planets, as these phenomenological investigations would take us too far away from the direct application of general principles. Despite these limitations, it is striking how much of what is known about the universe can be calculated with passable accuracy using the ideas, principles, and techniques discussed in this book.

We begin in Sec. 28.2 by developing general relativistic cosmology, emphasizing the geometry and the kinematics before turning to the dynamics. In Sec. 28.3, we describe all the major constituents of the universe today and we follow this in Sec. 28.4 with a development of standard cosmology starting from when the temperature was roughly 10^{11} K, distinguishing seven distinct ages. In Sec. 28.5, we develop a theory to describe the growth of perturbations, which ultimately led to clustered galaxies. Most of what we know about the universe comes from observing photons, and so in

Sec. 28.6, we develop the cosmological optics that we need to draw inferences about the birth and growth of the universe. In Sec. 28.7, we conclude by discussing three foci of current research and incipient progress—attempts to understand the origin of the universe, notably inflation, theories of the creation of dark matter and baryons, and speculations about the fate of the universe involving the role of the cosmological constant.

28.2 General Relativistic Cosmology

28.2.1 Isotropy and Homogeneity

galaxies at a glance

We start by introducing *galaxies* and describing their distribution in space. A typical galaxy, like our own, comprises $\sim 10^{11}$ luminous stars (with a combined luminosity of $\sim 10^{37}$ W)[1] and gas, mostly concentrated in a sphere of radius about 3×10^{20} m $\sim$ 10 kpc located at the center of a sphere about ten times larger dominated by collisionless dark matter (Fig. 28.1). A typical galaxy mass is 10^{42} kg. There is a large range in galaxy masses from less than a millionth of the mass of our galaxy to more than a hundred times its mass.

distribution of galaxies

Roughly a trillion galaxies can be seen over the whole sky (Fig. 28.1; Beckwith et al., 2006). Their distribution appears to be quite isotropic. We can estimate their distances and study their spatial distribution. It is found that the galaxies are strongly clustered on scales of $\lesssim 3 \times 10^{23}$ m $\sim$ 10 Mpc. Especially prominent are clusters of roughly a thousand galaxies. The strength of their relative clustering diminishes with increasing linear scale to $|\delta N/N| \sim 10^{-5}$ when the scale size is comparable with the "size" of the universe (which we will define more carefully below as $\sim 10^{26}$ m $\sim$ 3 Gpc). Insofar as luminous galaxies fairly sample all material in the universe, it seems that on large enough scales, we can regard the matter distribution today as quite homogeneous.

cosmic microwave background

An even more impressive demonstration of this uniformity comes from observations of the CMB (Fig. 28.1; Penzias and Wilson, 1965). Not only does it retain a blackbody spectrum, the temperature fluctuations are also only $|\delta T/T| \lesssim 3 \times 10^{-5}$ on angular scales from radians to arcminutes. (There is a somewhat larger dipolar component, but this is attributable to a Doppler shift caused by the compounded motion of Earth, the Sun, and our galaxy.) This radiation has propagated to us from an epoch when the universe was a thousand times smaller than it is today and, as we shall demonstrate, the temperature was a thousand times greater, roughly 3,000 K. If the universe were significantly inhomogeneous at this time, then we would see far larger fluctuations in its temperature, spectrum, and polarization. We conclude that the young universe was quite homogeneous, just like the contemporary universe appears to be on the largest scales.

---

1. For calibration, the Sun has a mass of 2.0×10^{30} kg and a luminosity of 3.9×10^{26} W. The standard astronomical distance measure is 1 parsec (pc) $= 3.1 \times 10^{16}$ m.

FIGURE 28.1 Four astronomical images that illustrate recent discoveries about the universe. (a) Image of the whole sky made by the Planck satellite, exhibiting ∼10 μK fluctuations in the observed ∼3 K temperature of the CMB (Sec. 28.6.1; credit: Planck Collaboration, 2016a). These fluctuations, observed when the universe was ∼400,000 yr old, depict the seeds out of which grew the large structures we see around us today. (b) The deepest image of the sky taken by Hubble Space Telescope, roughly 2 arcmin across (Hubble eXtreme Deep Field, http://en.wikipedia.org/wiki/Hubble_eXtreme_Deep_Field; credit: NASA; ESA; G. Illingworth, and the HUDF09 Team, 2013). The light from the most distant galaxies in this image is estimated to have been emitted when the universe was only ∼0.08 of its present size and less than 500 Myr old (Sec. 28.4.5). (c) Combined ultraviolet and infrared image of the Andromeda galaxy, the nearest large galaxy to our Milky Way galaxy. (Credit: NASA/JPL-Caltech.) Orbiting hydrogen gas can be seen out to about five times the size of this image, demonstrating that the stars we see form at the bottom of a large potential well of dark matter (Sec. 28.2.1). (d) The Bullet cluster of galaxies. (Credit: X-ray: NASA/CXC/CfA/M. Markevitch et al.; lensing map: NASA/STScI; ESO WFI; Magellan/University of Arizona/D. Clowe et al.; optical: NASA/STScI; Magellan/University of Arizona/D. Clowe et al.) Two clusters, each containing roughly several hundred galaxies, are in the process of merging. The hot gas, in red, can be traced by its X-ray emission and is separated from the dark matter, in blue, which can be located by the weak gravitational lensing distortion it imposes on the images of background galaxies (Sec. 28.6.2). The separation of these concentrations of matter demonstrates that dark matter is effectively collisionless (Sec. 28.3.2).

There is only one conceivable escape from this conclusion. We could live at the origin of a spherically symmetric, radially inhomogeneous universe. The assumption that this is not the case is sometimes known as the *Copernican Principle* by analogy with the proposition that Earth is not at the center of the solar system. According to this principle, the universe would look similarly isotropic from all other vantage points at the same time. If one accepts this hypothesis, then isotropy about us and about all other points at the same time implies homogeneity. This can be demonstrated formally.

These observational data justify a procedure in modeling the average universe, which was adopted by Einstein (1917) and others with little more than philosophical justification in the early days of relativistic cosmology. Like Einstein, we assume, as a zeroth order approximation, that the universe is homogeneous and isotropic at a given time. This can be stated more carefully as follows. *There exists a family of slices of simultaneity (3-dimensional spacelike hypersurfaces; Fig. 28.2), which completely covers spacetime, with the special property that on a given slice of simultaneity (i) no two points are distinguishable from each other (homogeneity) and (ii) at a given point no one spatial direction is distinguishable from any other (isotropy).*

homogeneous and isotropic universe

So, how do we assign these slices of simultaneity? Fortunately, the universe comes with a clock, the temperature of the CMB. (For the moment, ignore the tiny fluctuations.) We then introduce a set of imaginary *fundamental observers* (FOs; Fig. 28.2) and give them the velocity that removes the dipole anisotropy in the temperature distribution.[2] (Henceforth when we talk about observations at Earth, we imagine that these are observations made by an FO coincident with Earth today. To get the actual Earth-based observations, we just add a small Doppler shift to these FO observations.) These FOs individually move on world lines that keep the CMB isotropic. We regard the 3-dimensional hypersurfaces that they inhabit when they all measure the same CMB temperature as spaces in their own right, which we approximate as homogeneous. Isotropy guarantees that the FO world lines are orthogonal to these "slices of simultaneity." Let us focus on the hypersurface that we inhabit today, formed by freezing the action when everyone measures the CMB temperature to be 2.725 K.

fundamental observers

28.2.2 Geometry

METRIC TENSOR

We want to explore the geometry of this frozen, 3-dimensional space.[3] Our first task is to deduce the form of its metric tensor. Introduce a spherical coordinate system $\{\chi, \theta, \phi\}$, centered on us. Here χ is the radius, the proper distance (measured by a rather large number of meter rules) from us to a distant point; θ, ϕ are spherical polar angles, measured from Earth's north pole and the direction of

spatial coordinates

---

2. Roughly 360 km s^{-1} for Earth.
3. For a fuller treatment see, e.g., Misner, Thorne, and Wheeler (1973), Sec. 27.6 and Box 27.2.

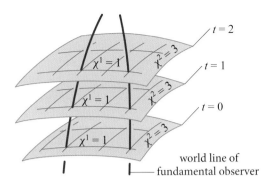

FIGURE 28.2 The slices of simultaneity, world lines of fundamental observers, and synchronous coordinate system for a homogeneous, isotropic model of the universe.

the Sun at the vernal equinox, for example.[4] We will also find it convenient to introduce an equivalent Cartesian coordinate system (Fig. 28.2): $\chi \equiv \{\chi^1, \chi^2, \chi^3\} = \{\chi \sin\theta \cos\phi, \chi \sin\theta \sin\phi, \chi \cos\theta\}$. We know that the space around us is isotropic, which implies that the angular part of the metric is $d\theta^2 + \sin^2\theta d\phi^2$ at each radius.[5] However, we cannot assume that the space is globally flat and that the area of a sphere of constant radius χ is $4\pi \chi^2$. Therefore, we generalize the metric to become

$$^{(3)}ds^2 = d\chi^2 + \Sigma^2(d\theta^2 + \sin^2\theta d\phi^2), \tag{28.1}$$

where $\Sigma(\chi)$ is a function to be determined. Now we know that $\Sigma \simeq \chi$ for small χ to satisfy the requirement that space be locally flat.

HOMOGENEOUS 2-DIMENSIONAL SPACES

Restrict attention to the 2-dimensional subspace $\phi = \text{const}$, which has metric $^{(2)}ds^2 = d\chi^2 + \Sigma^2(\chi)d\theta^2$, and consider a curve with $\theta = 0$ emanating from us at point O. This is obviously a geodesic—the shortest distance between its start at O and any point we care to choose along it. Let this geodesic pass through two galaxies labeled A, B and end at a third galaxy C (Fig. 28.3). Let the proper distances from O to A, A to B, and B to C, be χ_1, χ_2, χ_3, respectively. Now add a second geodesic, also emanating from O, and inclined to the first geodesic by a small angle θ.[6] As θ is

---

4. This is how astronomers set up their equivalent "right ascension"–"declination" coordinate system.
5. The deep, underlying reason it is not possible to cover the surface of a sphere with a metric with constant coefficients is that the vector field $(\partial/\partial\phi)_\theta$ that generates rotations about the polar axis does not commute with the vector field that generates rotations about any other axis. In principle, we could rotate our angular coordinate system from one radius to the next, but this would hide the symmetry that we are so eager to exploit!
6. This is similar in spirit to the description of gravitational lensing in Sec. 7.6.

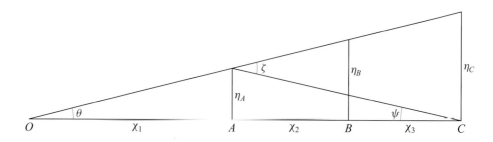

FIGURE 28.3 Geodesics in a homogeneous 2-dimensional space. We suppose that the angles θ, ψ, ζ are small and compute the transverse displacements η assuming the metric (28.1).

small, the proper separations of these geodesics (measured along short paths perpendicular to the first geodesic at A, B, and C) will be $\eta_A = \theta \Sigma(\chi_1)$, $\eta_B = \theta \Sigma(\chi_1 + \chi_2)$, $\eta_C = \theta \Sigma(\chi_1 + \chi_2 + \chi_3)$, respectively, ignoring terms $O(\theta^3)$, which will vanish in the limit $\theta \to 0$. Next, introduce a third geodesic backward from C and intersecting with the second geodesic, at a point a perpendicular distance η_A from galaxy A. Denote the (small) angle it makes with the first geodesic at C to be ψ and with the second geodesic near A to be ζ.

derivation of metric tensor

Now, by assumption, the metric must have the same form, with the same metric function $\Sigma(\chi)$ when the origin is at O, A, or C. We can use this to derive a functional relationship that must be satisfied by $\Sigma(\chi)$ by evaluating η_B in two different ways:

$$\eta_B = \theta \Sigma(\chi_1 + \chi_2) = \psi \Sigma(\chi_3) + \zeta \Sigma(\chi_2)$$
$$= \theta \left(\frac{\Sigma(\chi_3)\Sigma(\chi_1)}{\Sigma(\chi_2 + \chi_3)} + \frac{\Sigma(\chi_2)\Sigma(\chi_1 + \chi_2 + \chi_3)}{\Sigma(\chi_2 + \chi_3)} \right). \qquad (28.2)$$

(In the last equality, we have used elementary geometry to express ψ and ζ in terms of θ.) Rearranging, we obtain

$$\Sigma(\chi_1 + \chi_2)\Sigma(\chi_2 + \chi_3) = \Sigma(\chi_1)\Sigma(\chi_3) + \Sigma(\chi_1 + \chi_2 + \chi_3)\Sigma(\chi_2), \qquad (28.3)$$

for all choices of χ_1, χ_2, χ_3.[7]

This sort of nonlocal, functional relationship is probably quite unfamiliar and, in general, such equations are hard to solve. However, in this instance we can use the device of solving it in a special case and then showing that the solution satisfies the original equation. If we assume that χ_1 and χ_3 are equal and small, and we expand Eq. (28.3) to second order in χ_1 with $\chi_2 = \chi$ finite, we find that

$$\Sigma'' \Sigma - \Sigma'^2 + 1 = 0. \qquad (28.4)$$

Next, multiply the left-hand side by $2\Sigma'/\Sigma^3$, and note that it is the derivative of $(\Sigma'^2 - 1)/\Sigma^2$. Hence, we get

$$\Sigma'^2 + K\Sigma^2 - 1 = 0, \qquad (28.5)$$

---

7. The well-educated reader may notice a similarity to Ptolemy's theorem. The comparison is instructive.

where the integration constant K is known as the *Gaussian curvature* of the 2-dimensional surface, or equivalently, half the scalar curvature (Sec. 25.6). Differentiating this equation leads to the simple harmonic equation

$$\Sigma'' + K\Sigma = 0. \tag{28.6}$$

Gaussian curvature K

We need solutions that will be locally Euclidean with $\Sigma(0) = 0$, $\Sigma'(0) = 1$.

There are three cases. If $K = 0$, then $\Sigma = \chi$. If K is positive, we write it as $K = R^{-2}$, where R is the *radius of curvature* and $\Sigma(\chi) = R\sin(\chi/R)$. If $K < 0$, we write it as $K = -R^{-2}$ and $\Sigma(\chi) = R\sinh(\chi/R)$. The final step is to verify that these solutions are valid for general χ_1, χ_2, χ_3, which they are. We have therefore found three general solutions to the functional equation (28.3). Suppose that there were an additional solution. We could subject it to the same limiting procedure and recover the same nonlinear differential equation (28.4) as $\chi_1 = \chi_3 \to 0$ for all $\chi_2 = \chi$. However, we know all the locally Euclidean solutions $\Sigma(\chi)$ to this equation, and our hypothetical fourth solution would not be among them. Therefore, it cannot exist, and so we have found all possible distance relations for homogeneous 2-dimensional spaces.

We summarize these three solutions by introducing a parameter $k = \pm 1$ or 0:

$$K = k/R^2; \quad \Sigma = R\sin(\chi/R) \text{ for } k = +1, \quad \chi \text{ for } k = 0, \quad R\sinh(\chi/R) \text{ for } k = -1. \tag{28.7}$$

NON-EUCLIDEAN GEOMETRIES

The $K = 0$ solution is instantly recognizable as describing the geometry of flat, Euclidean space. The $K > 0$ solution describes the 2-dimensional surface of a sphere embedded in 3-dimensional flat space. The circumference of a circle of radius χ is $2\pi R\sin(\chi/R) < 2\pi\chi$ and vanishes at an antipodal point, where $\chi = \pi R$. The space has a natural end—we call it *closed*. The associated area of this space is $2\pi \int_0^{\pi R} d\chi\, \Sigma = 4\pi R^2$. The geodesics we have been using are simply "great circles" on the 2-sphere.[8] The theorems of spherical trigonometry, which we could have gone on to prove, are familiar to navigators, astronomers, and crystallographers.

homogeneous positive curvature space

When $K < 0$, the circumference is $2\pi R\sinh(\chi/R) > 2\pi\chi$ and the area is unbounded, so the space is *open*. This 2-dimensional space *cannot* be embedded in 3-dimensional Euclidean space. This discovery, made independently in the nineteenth century by Bolyai, Gauss, and Lobachevsky, was a source of mathematical wonder and philosophical consternation.[9]

homogeneous negative curvature space

---

8. To go from the skinny triangles we have been considering so far to large values of η, we need to construct the geodesic $\chi(\theta)$ by minimizing the length $\int d\theta (\chi'^2 + \Sigma^2)^{1/2}$ of the curve connecting the endpoints. This is a standard exercise in the calculus of variations, and carrying it out opens the door to deriving counterparts to the theorems of Euclidean geometry for non-Euclidean spaces.
9. This *hyperbolic* geometry can be embedded in 3-dimensional Minkowski space, and we could have used our understanding of special relativity to explore it by analogy with what is done with the positively curved space. Had we done so, we would have discovered that it is the same as the mass hyperboloid depicted in Fig. 3.2. However, the spirit of Riemannian geometry is not to do this but instead to explore the space through its *inner* properties.

28.2 General Relativistic Cosmology

HOMOGENEOUS 3-DIMENSIONAL SPACES

The generalization to 3 dimensions is straightforward. If space is isotropic, then the geometrical statements we made about the subspace $\phi =$ const must be true about any 2-dimensional subspace; so we arrive at the same three possibilities for Σ in the 3-dimensional metric (28.1). We replace the circumference of a circle by the area of a sphere of radius χ, which is less than (more than) $4\pi \chi^2$ for positive (negative) curvature. The total volume of the positively curved, closed 3-space is $4\pi \int_0^{\pi R} d\chi \, \Sigma^2 = 2\pi^2 R^3$.[10]

ROBERTSON-WALKER METRIC

Relativistic cosmology is transacted in spacetime, not space (Sec. 25.2). How do we generalize our understanding of these homogeneous subspaces to include a time coordinate and allow for the expansion of the universe? The first step is to assume that the 3-spaces in the past were also homogeneous and so had the same geometry as today. The only thing that can change is the radius of curvature, and so we write

scale factor

$$R(t) = a(t) R_0, \tag{28.8}$$

where R_0 is its value today when $t = t_0$ and $a = 1$. We call $a(t)$ the *scale factor*. (We handle a flat space by taking the limit $R_0 \to \infty$.) An FO then moves from one spatial hypersurface to the next carrying the same spherical coordinates $\{\chi, \theta, \phi\}$ (Fig. 28.2). We call these three coordinates *comoving coordinates*.

comoving coordinates

The basis vector $\partial_t \equiv \partial/\partial t$ is tangent to the world line of an FO, which is orthogonal to the spatial hypersurface $t =$ const. Thus, $g_{ti} \equiv \partial_t \cdot \partial_i = 0$.[11] Next consider g_{tt}. We have so far implicitly measured time using the temperature of the CMB, but as we want to be quantitative, we choose as a time coordinate the *proper time* (or *cosmic time* or simply, *time*) Δt that elapses, as measured by a clock carried by an exceedingly patient FO with $u^\alpha = (1, 0, 0, 0)$, from one hypersurface to the next. (We set $c = 1$ in all subsequent equations.) It is implicit in the assumption of homogeneity that this interval of time will be the same for all FOs, and we can add up all the intervals to make our time coordinate t the *total age of the universe* since the big bang. With this choice, $g_{tt} = -1$.[12]

time coordinate

---

10. Although homogeneity and isotropy force the cosmological model's hypersurfaces to have one of the three metrics we have described, the topologies of those hypersurfaces need not be the obvious ones adopted here. For example, a flat model could have a closed topology with finite volume rather than an open topology with infinite volume. This could come about in much the same way as a flat piece of paper can be rolled into a cylinder. The geometry is still Euclidean, but one of the two coordinates, say x, is identified with $x + C$, where C is the length of the circumference of the cylinder. In 3 dimensions we can make similar identifications for the y and z coordinates. There are no credible observations to suggest that the universe is like this, but the possibility is worth keeping in mind.
11. Such a coordinate system is termed *synchronous*.
12. Other time coordinates, such as *conformal time*, $\int dt/a$, are in common use for a variety of technical reasons, but we shall eschew these.

Our full metric is then

$$ds^2 = -dt^2 + a(t)^2[d\chi^2 + \Sigma^2(d\theta^2 + \sin^2\theta d\phi^2)]. \quad (28.9)$$

This is known as the *Robertson-Walker metric*.[13] Using this metric, we can easily verify that the acceleration $\nabla_{\vec{u}}\vec{u}$ of an FO vanishes, so FOs with fixed χ follow geodesics.

The scale factor $a(t)$ is very important. Not only does it measure the size of the universe in the past, it also measures the separation of any two FOs as a fraction of their separation today. Insofar as the universe is well described as homogeneous and isotropic, this single function tells us all we need to know about how the universe expanded and how it will continue to expand. We shall use it—actually, its logarithm—as our independent variable, because physical quantities such as densities and temperatures scale simply with it and because it is closely related to what astronomers commonly measure. Measuring $t(a)$ is the kinematic challenge to observational cosmology; explaining it is the dynamical challenge to general relativity.

meaning of $a(t)$

EINSTEIN TENSOR

Our final geometrical task is to calculate the Einstein tensor (Sec. 25.8). Although the calculation can be simplified by exploiting symmetry, it is best to use computer algebra. The only nonzero elements are

$$G^{\hat{t}\hat{t}} = -G_t{}^t = -\frac{2\Sigma''\Sigma + \Sigma'^2 - 3\Sigma^2\dot{a}^2 - 1}{\Sigma^2 a^2},$$

$$G^{\hat{\chi}\hat{\chi}} = G_\chi{}^\chi = \frac{\Sigma'^2 - \Sigma^2(2\ddot{a}a + \dot{a}^2) - 1}{\Sigma^2 a^2},$$

$$G^{\hat{\theta}\hat{\theta}} = G_\theta{}^\theta = G^{\hat{\phi}\hat{\phi}} = G_\phi{}^\phi = \frac{\Sigma'' - \Sigma(2\ddot{a}a + \dot{a}^2)}{\Sigma a^2}. \quad (28.10)$$

components of Einstein tensor in orthonormal basis

Here each dot or prime denotes a derivative with respect to t or χ; the hatted components are in the proper reference frame of an FO [with orthonormal basis vectors $\vec{e}_{\hat{t}} = \partial_t, \vec{e}_{\hat{\chi}} = a^{-1}\partial_\chi, \vec{e}_{\hat{\theta}} = (a\Sigma)^{-1}\partial_\theta, \vec{e}_{\hat{\phi}} = (a\Sigma\sin\theta)^{-1}\partial_\phi$; cf. the metric (28.9)]; the unhatted components are those of the coordinate basis. The mixed-coordinate components (one index up, the other down) are the easiest to evaluate mathematically, but the orthonormal components are the best for physical interpretation, since they are what an FO measures. The two sets of components are equal up to a sign, because the metric is diagonal.

---

13. Historically, the three possible choices for the geometry of a homogeneous, isotropic cosmological model were discovered by the Russian meteorologist Alexander Friedmann (1922). These solutions were found independently by a Belgian priest, Georges Lemaître (1927), who included a cosmological constant and discussed the growth of perturbations. The first proof that these three choices are the only possibilities was due to Robertson (1935, 1936a,b) and Walker (1935).

inclusion of geometry

If we now substitute our geometrical conditions (28.4) and (28.5) with $K = kR_0^{-2}$ [Eqs. (28.7) and (28.8)], we obtain

$$G^{\hat{t}\hat{t}} = 3\frac{\dot{a}^2 + kR_0^{-2}}{a^2}, \quad G^{\hat{\chi}\hat{\chi}} = G^{\hat{\theta}\hat{\theta}} = G^{\hat{\phi}\hat{\phi}} = -\frac{2\ddot{a}a + \dot{a}^2 + kR_0^{-2}}{a^2}. \tag{28.11}$$

Note that the spatial part of the Einstein tensor is proportional to the 3-metric $g^{\hat{i}\hat{j}} = \delta^{ij}$. In other words, it is isotropic, as we expect.

EXERCISES

Exercise 28.1 *Example: Alternative Derivation of the Spatial Metric*

Not surprisingly, there are several other approaches to deriving the possible forms of $\Sigma(\chi)$.[14] Another derivation exploits the symmetries of the Riemann tensor.

(a) The 3-dimensional Riemann curvature tensor of the hypersurface must be homogeneous and isotropic. Explain why it should therefore only involve the metric tensor and not any other tensor, for example, not the Levi-Civita tensor.

(b) Show that the only combination of these quantities that exhibits the full symmetry properties of the Riemann tensor is

$$^{(3)}R_{ijkl} = K(g_{ik}g_{jl} - g_{il}g_{jk}), \tag{28.12}$$

where K is a constant.

(c) Write a computer algebra routine to evaluate the Riemann tensor from the metric tensor (28.1) directly. By equating it to Eq. (28.12), show that two differential equations must be satisfied and these are identical to Eqs. (28.5) and (28.6).

(d) Compute the Ricci tensor, and compare it with the metric tensor. Comment.

Exercise 28.2 *Problem: Area of a Spherical Triangle*

Consider the triangle formed by the three geodesics in Fig. 28.3. In a flat space, the exterior angle ζ must equal $\theta + \psi$. However, if the space is homogeneous and positively curved, then the angle deficit $\Delta \equiv \theta + \psi - \zeta$ will be positive.

(a) By considering the geometry of the 2-dimensional surface of a sphere embedded in 3-dimensional Euclidean space, show that the area of the triangle is Δ/K. [Hint: We know that the area of a *lune* lying between two lines of longitude separated by an angle ϕ is $2\phi/K$.]

(b) Make a conjecture (or, better still, devise a demonstration) as to the formula for the area of a triangle in a negatively curved homogeneous space.

These results are special cases of the famous *Gauss-Bonnet* theorem,[15] which allows for the possibility that the topology of the space might not be simple.

---

14. For example, there is an elegant, group-theoretic approach (e.g., Ryan and Shepley, 1975).
15. See, e.g., Peacock (1999) and Carroll (2004).

28.2.3 Kinematics

RAYS AND WORLD LINES

We are interested in events which lie at the intersection of our past light cone and the world lines of other FOs. Imagine a single wavelength of light emitted in the χ direction at time t_e. Let there be two FOs at either end of the wavelength, and let their comoving radii be χ_e and $\chi_e + d\chi_e$. Their physical separation at the time of emission is λ_e, the emitted wavelength. Their separation in terms of the comoving coordinate is therefore $d\chi = \lambda_e/a(t_e)$. Let the light be observed by us today. Let the front of the wavelength be observed at time t_0 and the end of the wavelength at $t_0 + P_0$, where P_0 is the observed period. Using the metric (28.9) with $ds = d\theta = d\phi = 0$, we see that $\chi_e = \int_{t_e}^{t_0} dt/a$ and $\chi_e + \lambda_e/a(t_e) = \int_{t_e}^{t_0+P_0} dt/a$, from which we deduce that $\lambda_e/a(t_e) = P_e/a(t_e) = P_0$ or in terms of the frequencies, $\nu_0 = a(t_e)\nu_e$.[16] Put another way, as a photon crosses the universe, its frequency, as measured by the FOs, satisfies $\nu \propto a^{-1}$. So, if we observe a spectral line that was emitted by the stars in a distant galaxy with frequency ν_e and we observe it today with frequency ν_0, then we can deduce immediately that the scale factor a at the time of emission was ν_0/ν_e. The first stars and galaxies to form, and therefore the most distant ones that we can see, emitted their light when $a \sim 0.1$, so that a Lyman α spectral line of hydrogen emitted with a wavelength of $\lambda_e = 122$ nm is observed in the infrared today with $\lambda_0 \sim 1.2 \, \mu$m.

cosmological redshift

THERMODYNAMICS

Construct a small, imaginary sphere around us with radius χ and let the sphere's surface be carried by a population of FOs, so it comoves and expands with them. These observers each see isotropic blackbody radiation. Every time a photon leaves the sphere, it will be replaced, on average, by an entering photon. The FOs could therefore construct a spherical mirror, perfectly reflecting on both sides and expanding with the universe. Nothing would change. The radiation inside this sphere will undergo slow adiabatic expansion. If it starts off as thermal blackbody radiation, it will remain so. The blackbody temperature T_γ, photon number density n_γ, pressure P_γ, and energy density ρ_γ will therefore vary as (cf. Secs. 3.2.4 and 3.5.5)

$$T_\gamma \propto a^{-1}; \quad n_\gamma \propto a^{-3}; \quad \rho_\gamma, P_\gamma \propto a^{-4} \propto n_\gamma^{4/3}, \qquad (28.13)$$

radiation in an expanding universe

and the specific heat ratio is 4/3. If the radiation is isotropic but not blackbody, its distribution function $\eta_\gamma(p, a)$ will not change along a trajectory in phase space (Sec. 3.6), and so $\eta_\gamma(p', a') = \eta_\gamma(p = p'a'/a, a)$. Here $p = |\mathbf{p}|$ is the magnitude of the photon's momentum, or equally well its energy $\mathcal{E} = p$.

By parallel arguments, the individual momenta p of massive particles vary as $p \propto a^{-1}$—their de Broglie wavelengths expand with the universe just like photon

---

16. This argument neatly avoids a generally unhelpful separation into the Doppler shift and gravitational redshift. The quantity *redshift*, $z = (\nu_e - \nu_0)/\nu_0 = 1/a - 1$, is in common use by astronomers. However, we shall continue to work with the scale factor $a(t)$, as we are focusing on the kinematics of the expansion and not the observations through which this has been inferred.

gas in an expanding universe

wavelengths—so long as their behavior is adiabatic. When nonrelativistic, their kinetic energies and temperatures vary according to $T \propto p^2 \propto a^{-2}$, and their pressure P scales with their number density n as $P \propto nT \propto a^{-5} \propto n^{5/3}$, recovering the familiar specific heat ratio for a monatomic gas.

EXPANSION, DECELERATION, AND JERK

kinematics of the expansion

The kinematic behavior of the homogeneous universe is fully described by the single function $a(t)$. Not surprisingly, its derivatives turn out to be very useful. Adopting standard (if archaic) conventions, we define the *expansion rate* $H(t)$, the *deceleration function* $q(t)$, and the *jerk function* $j(t)$ by

expansion rate H, deceleration q and jerk j

$$H = \frac{\dot{a}}{a} \equiv \left(\frac{dt}{d\ln a}\right)^{-1}, \quad q = -\frac{\ddot{a}a}{\dot{a}^2}, \quad j = \frac{\dddot{a}a^2}{\dot{a}^3}, \tag{28.14}$$

respectively. Note that $H^{-1} = dt/d\ln a$ converts a derivative with respect to time (denoted with a dot) to a derivative with respect to $\ln a$ (denoted with a prime), which is our preferred independent variable. The expansion rate H is a reciprocal time; q and j are dimensionless. A galaxy at small radius χ emitted its photons when the scale factor was $a \simeq 1 - \chi \dot{a}$, and so the shift in the wavelengths of spectral lines observed on Earth will satisfy to first order $\delta\lambda/\lambda \equiv v = \dot{a}\chi \equiv H_0\chi$,[17] where v is the inferred recession speed, and the *Hubble constant*[18] $H_0 \equiv H(t_0)$ is the contemporary value of H and is one of the key parameters of observational cosmology. It measures the age and size of the universe. We now know that the universe is accelerating $q_0 \equiv q(t_0) < 0$ and $j_0 \equiv j(t_0) > 0$ (Riess et al., 1998; Perlmutter et al., 1999).

Hubble constant H_0

DISTANCE MEASUREMENT

There are two common measures of distance in cosmology. The first is based on observing a small source of known physical size, η, perpendicular to the line of sight at radius χ. The source's angular size measured at Earth will be $\theta = \eta/(a\Sigma)$ (Fig. 28.3, where O at $\chi = 0$ is Earth's location). This motivates us to define the angular diameter distance $d_A \equiv a\Sigma$.

angular diameter distance d_A

The second measure is based on an isotropic source of known luminosity L. If the universe were flat and static, the source's measured flux would be $F = L/(4\pi\chi^2)$. Relativistic cosmology introduces three modifications. First, the area of a sphere centered on the source with comoving radius χ, at the time of observation (today), is $4\pi \Sigma(\chi)^2$. Second, the source emits photons, and their individual energies $h\nu$ as observed at O will be reduced by a factor a from their energies when emitted. Third,

---

17. The quantity χ must be large enough for the recession speed to exceed the random motion with respect to a local FO.
18. The Hubble "law," that the recession velocity is proportional to the distance, was published in 1929 by Edwin Hubble, following theory by Lemaître and observations by himself, Slipher, and others. It provided the first compelling demonstration that the universe was expanding, although Hubble's inferred constant of proportionality, $H_0 \sim 500$ km s^{-1} Mpc^{-1}, was over seven times larger than the contemporary measurement.

the time it takes to emit a fixed number of photons will be shorter by another factor a than the time it takes these same photons to pass the observer on Earth. The flux will therefore be modified to $F = a^2 L/(4\pi \Sigma^2)$. This motivates us to define a *luminosity distance* $d_L = \Sigma/a = a^{-2} d_A$.

luminosity distance d_L

Making cosmological distance measurements is challenging. There are many astronomical candidates for sources of known size or luminosity for use in measuring d_A or d_L. However, they should all be greeted with suspicion, as it is not just the universe that evolves with time; its contents can likewise change. The most natural distance measures use variable and exploding stars. As we discuss below, statistical measures in the CMB and distribution of galaxies as well as gravitational lenses are also important. Many methods are anchored by local trigonometric surveys. Inevitably, there are systematic errors (*inaccuracy*), which are now more limiting than random error (*imprecision*). Remarkably, the spread in measurements of H_0 has been reduced to roughly 10%. For specificity, we shall use cosmological parameters from Planck Collaboration (2016b), starting with $H_0 = 67.7$ km s^{-1} Mpc$^{-1} \equiv (4.56 \times 10^{17}$ s$)^{-1} \equiv (14.4$ Gyr$)^{-1} \equiv (1.37 \times 10^{26}$ m$)^{-1} \equiv (4.43$ Gpc$)^{-1}$.

measurement of H_0

HORIZON

The final kinematic quantity we introduce is the *horizon*[19] radius $\chi_H(t)$ of the event on our past world line at time t. This is the comoving radius of a sphere, centered on a point on our past world line, that contains the region of the universe that can have sent signals to the event since the big bang. We also find it useful to introduce the *acoustic horizon radius* $\chi_A = C/\dot{a}$. This is the comoving distance a sound wave can travel at the local sound speed C during one expansion time scale $a/\dot{a} = H^{-1}$:

horizon radius χ_H

acoustic horizon radius χ_A

$$\chi_H(t) = \int_0^t \frac{dt'}{a(t')} = \int_0^{a(t)} \frac{da'}{a'^2 H(a')}; \quad \chi_A = \frac{C}{\dot{a}}. \quad (28.15)$$

When the fluid is ultrarelativistic, C is simply $3^{-1/2}$ and $\chi_A \equiv \chi_R = 3^{-1/2} H^{-1}$.

EXERCISES

Exercise 28.3 *Example: Mildly Relativistic Particles*

Suppose that the universe contained a significant component in the form of isotropic but noninteracting particles with momentum p and rest mass m. Suppose that they were created with a distribution function $f(p, a) \propto p^{-q}$, with $4 < q < 5$ extending from $p \ll m$ to $p \gg m$ (Sec. 3.5.3).

(a) Show that the pressure and the internal energy density (not including rest mass) of these particles are both well defined.

(b) Show that the adiabatic index of this component is $q/3$ and interpret your answer.

(c) Explain how P and ρ for this component should vary with the scale factor a.

---

19. Or more properly the *particle horizon*, to distinguish it from an *event horizon* (Sec. 26.4.5).

Exercise 28.4 *Problem: Observations of the Luminosity Distance*

Astronomers find it convenient to use the redshift $z = 1/a - 1$ to measure the size of the universe when the light they observed was emitted.

(a) Perform a Taylor expansion in z to show that the luminosity distance of a source is given to quadratic order by
$$d_L = H_0^{-1}[z + 1/2(1 - q_0)z^2 + \cdots].$$

(b) The cubic term in this expansion involves the curvature K_0 and the jerk j_0 today. Calculate it.

28.2.4 Dynamics

FRIEDMANN EQUATIONS

Ultimately, we must understand how the universe's individual constituents—matter, photons, neutrinos—behave and interact. However, for the initial purposes of providing an idealized description of the average contents of the universe that can match our idealized description of its geometry and kinematics, we approximate everything as a homogeneous perfect fluid at rest in the FOs' proper reference frame, with pressure $T^{\hat{x}\hat{x}} = T^{\hat{\theta}\hat{\theta}} = T^{\hat{\phi}\hat{\phi}} = P$ (Sec. 2.13). This is related to the Einstein tensor through the Einstein field equation (Sec. 25.8). We start with the time-time part and employ geometrized units (Sec. 25.8.1): $G^{\hat{t}\hat{t}} = 8\pi T^{\hat{t}\hat{t}}$, which becomes, with the aid of Eq. (28.11):

equation of motion

$$\dot{a}^2 = \frac{8\pi}{3}\rho a^2 - \frac{k}{R_0^2}. \tag{28.16}$$

This equation of motion is the foundation of relativistic cosmology. It relates the universe's expansion rate $\dot{a}$ to its mean mass-energy density ρ and curvature k/R_0^2.

It is illuminating to write Eq. (28.16) in the form

$$\frac{1}{2}(\dot{a}\chi)^2 - \frac{4\pi\rho(a\chi)^3}{3(a\chi)} = -\frac{k\chi^2}{2R_0^2} = \text{const}, \tag{28.17}$$

for fixed χ. Again imagine that we are at the center of a small, imaginary, expanding sphere of radius $a\chi$ carried by FOs. We can regard the first term as the kinetic energy of a unit mass resting on the surface of the sphere and the second as its Newtonian gravitational potential energy. So this looks just like an equation of Newtonian energy conservation. Could we not have written it down without recourse to general relativity? The answer is "No." Equation (28.16) addresses three crucial issues on which Newtonian cosmology must remain silent. First, it neatly handles the influence of the exterior matter. Second, it relates the total energy of the expansion, the right-hand side of Eq. (28.17), to the spatial curvature. Third, it has taken account of the pressure, which as we saw with neutron stars [(Eqs. 26.38)], can contribute an active gravitational mass. The absence of pressure in Eq. (28.16) is correct but does not

inadequacy of Newtonian treatment

have a simple Newtonian explanation. These three features demonstrate the power of general relativistic cosmology and the inadequacy of a purely Newtonian approach.

Next, consider the spatial part of the Einstein equation $G^{\hat{i}\hat{j}} = 8\pi T^{\hat{i}\hat{j}} = 8\pi P \delta^{\hat{i}\hat{j}}$. With the aid of Eqs. (28.11) and (28.16), this gives a second-order differential equation:

$$\ddot{a} = -\frac{4\pi}{3}(\rho + 3P)a. \qquad (28.18)$$

cosmic acceleration

The divergence of the stress-energy tensor must vanish. We have seen many times in this book [Ex. 2.26, Eqs. (13.86), Secs. 13.8.1, 26.3.3] that for any perfect fluid, in the fluid's rest frame (which for cosmology is the same as the FO rest frame), the time component $T^{\hat{t}\hat{\alpha}}{}_{;\hat{\alpha}} = 0$ is energy conservation, or equivalently, the first law of thermodynamics. For a fluid element with unit comoving volume, the physical volume is a^3, so the first law states

$$d(\rho a^3) = -P d(a^3). \qquad (28.19)$$

first law of thermodynamics

If the fluid comprises two or more independent components that do not exchange heat and evolve separately, then Eq. (28.19) must apply to each component.

We have also seen many times [Ex. 2.26, Eqs. (13.86), Sec. 26.3.3] that in the fluid's rest frame, the spatial component $T^{\hat{i}\hat{\alpha}}{}_{;\hat{\alpha}} = 0$ is force balance; it equates the fluid's pressure gradient ∇P to its inertial mass per unit volume $(\rho + P)$ times its 4-acceleration. However, homogeneity and isotropy guarantee that in our cosmological situation, the fluid's pressure gradient and 4-acceleration vanish, so $T^{\hat{i}\hat{\alpha}}{}_{;\hat{\alpha}} = 0$ is satisfied identically and automatically. It teaches us nothing new.

Equations (28.16), (28.18), and (28.19) are called the *Friedmann equations* in honor of Alexander Friedmann. They are not independent; the contracted Bianchi identity guarantees that from any two of them, one can deduce the third (Sec. 25.8).

CRITICAL DENSITY

Not only does the Hubble constant define a scale of length and of time, it also defines a critical density, ρ_{cr}—the value of ρ for which the universe's spatial curvature $K = k/R_0^2$ vanishes today [Eqs. (28.16) and (28.14)]:

$$\rho_{cr} = \frac{3H_0^2}{8\pi} = 8.6 \times 10^{-27} \text{ kg m}^{-3} \equiv 7.7 \times 10^{-10} \text{ J m}^{-3}. \qquad (28.20)$$

It is conventional to express the energy density of constituent i today (e.g., radiation or baryonic matter) as a fraction by introducing

density fraction

$$\Omega_i = \frac{\rho_i}{\rho_{cr}}; \quad \Omega = \frac{\rho}{\rho_{cr}} = \sum_i \Omega_i = 1 - \Omega_k, \qquad (28.21)$$

where Ω refers to the total energy density, and

$$\Omega_k = -\frac{k}{H_0^2 R_0^2} \qquad (28.22)$$

takes account of curvature.

expansion rate, deceleration function, and jerk function

The kinematic quantities, H, q, j [Eq. (28.14)] are then given by

$$H = H_0 \left(\frac{\rho}{\rho_{cr}} + \frac{\Omega_k}{a^2} \right)^{1/2} \; ; \quad q = \frac{1}{2} \frac{\rho + 3P}{\rho + \frac{\Omega_k \rho_{cr}}{a^2}} ; \quad j = \frac{\rho - \frac{3}{2}\frac{dP}{d \ln a}}{\rho + \frac{\Omega_k \rho_{cr}}{a^2}}, \quad (28.23)$$

respectively. Note that if the curvature is negligible, $q = \frac{1}{2}$ when $P = 0$ and $j = 1$ when P is constant.

A FLAT UNIVERSE

At this point we introduce a key simplification. Essentially geometrical arguments, based largely on observations of CMB fluctuations, have shown that the radius of curvature today is $|R_0| \gtrsim 14/H_0$. Equivalently, $\Omega_k \lesssim 0.005$. This is so small, and was even smaller in the past (when we replace R_0 by R) that henceforth we shall set $\Omega_k = 0$ and $\Sigma = \chi$. If significant spatial curvature is measured one day, then small corrections will be required to what follows, while the principles are unaffected. However, none of this absolves us from the obligation to explain *why* the universe is so flat, an issue to which we shall return in Sec. 28.7.1.

SUMMARY

We have described an idealized universe that is homogeneous and isotropic everywhere. We have shown its spatial geometry can take one of three different forms and have invoked observation to restrict attention to the spatially flat case. We have also introduced some useful cosmological measures, most notably the scale factor $a(t)$, to characterize the kinematics of the universe. We have married the universe's geometry to its kinematics to calculate the two independent components of the Einstein tensor, which we then combined with the volume-averaged stress-energy tensor of the contents of the universe to derive the Friedmann equations. We now turn to cataloging the contributions to the stress-energy tensor today.

EXERCISES

Exercise 28.5 *Example: Einstein–de Sitter Universe*
Suppose, as was once thought to be the case, that the universe today is flat and dominated by cold (pressure-free) matter.

(a) Show that $a \propto t^{2/3}$ and evaluate the age of the universe assuming the Hubble constant given in Sec. 28.2.3: $H_0 = 67.7$ km s^{-1} Mpc^{-1}.

(b) Evaluate an expression for the angular diameter distance as a function of a and find its maximum value.

(c) Calculate the comoving volume within the universe back to very early times.

28.3 The Universe Today

28.3.1 Baryons

Among all the constituents in our universe, baryonic matter is the easiest form to identify, because it is capable of radiating when hot and absorbing when cool. The baryons from the very early universe are mostly in the form of hydrogen, with mass fraction 0.75, and helium, with mass fraction $0.25 \equiv Y_{He}$. Some of these baryons found their way into stars, which produced more helium and created about 1% by mass, on average, of heavier elements. Stars have masses ranging from a tenth to more than 100 times that of the Sun. The most massive stars shine for roughly a million years, while the least massive ones will last more than a trillion years, much longer than the age of the universe.

stars at a glance

Very roughly 10^{-4} of the total mass of a typical galaxy is found in a massive, spinning black hole residing in its nucleus. When this black hole is supplied with gas, roughly 10% of the rest mass energy of this gas is converted into radiation. The galaxy is then said to have an *active galactic nucleus*, and when this outshines the entire galaxy, it is called a *quasar*.

active galactic nuclei and quasars

It is common practice to use the luminosity of a galaxy as a measure of its stellar mass. However, this must be done with care, because the answer depends on the relative proportions of low- and high-mass stars. The latter are far more luminous per unit mass. It also depends on the time that has elapsed since the stars were formed. If the stars are old, the luminous high-mass ones will have consumed all their nuclear fuel and evolved to form neutron stars and black holes. Absorption is also an issue. Despite all these difficulties, astronomers estimate that the average fraction of stellar mass today is $\Omega_* \sim 0.005$.

The baryons that are not contained in stars exist as gas. We can assay this gas fraction by measuring the X-ray emission from clusters of galaxies and find that $\Omega_{gas} \sim 0.05$. More accurate measurements of the total baryon fraction Ω_b are made possible by measuring the tiny fraction of deuterium and other light elements that are created in trace amounts in the early universe (Sec. 28.4.2) and through interpreting the spectrum of CMB fluctuations (Sec. 28.6.1). The best estimate today is $\Omega_b = 0.049$, implying that the mean energy and number densities of baryons in the universe are $\rho_{b0} = 3.8 \times 10^{-11}$ J m^{-3} = 4.2×10^{-28} kg m^{-3} and $n_{b0} = 0.25$ m^{-3}. The equivalent proton, electron, and helium densities (Sec. 28.4.2) are $n_{p0} = 0.19$ m^{-3}, $n_{e0} = 0.22$ m^{-3}, and $n_{\alpha 0} = 0.016$ m^{-3} (Table 28.1). Of course, these densities are substantially higher in galaxies.

measuring Ω_b

EXERCISES

Exercise 28.6 *Example: Stars and Massive Black Holes in Galaxies*
Assume that a fraction ~ 0.2 of the baryons in the universe is associated with galaxies, split roughly equally between stars and gas. Also assume that a fraction $\sim 10^{-3}$ of the baryons in each galaxy is associated with a massive black hole and that most of the radiation from stars and black holes was radiated when $a \sim 0.3$.

TABLE 28.1: The universe today

Geometric	$\Omega_k = 0$		
Kinematic	Hubble constant	Deceleration parameter	Jerk parameter
	$H_0 = 67.7$ km s^{-1} Mpc^{-1}	$q_0 = -0.54$	$j_0 = 1$
Derived	Hubble time	Hubble distance	Critical density
	$t_H = H_0^{-1} = 4.6 \times 10^{17}$ s	$d_H = ct_H = 1.4 \times 10^{26}$ m	$\rho_{cr} = 7.7 \times 10^{-10}$ J m^{-3}
Constituent	Energy fraction	Energy density (J m^{-3})	Number density (m^{-3})
Baryons	$\Omega_b = 0.049$	$\rho_{b0} = 3.8 \times 10^{-11}$	$n_{b0}, n_{e0} = 0.25, 0.22$
Dark matter	$\Omega_D = 0.26$	$\rho_{D0} = 2.0 \times 10^{-10}$	$n_{D0} = 0.0013 m_{D,12}^{-1}$
Cosmological	$\Omega_\Lambda = 0.69$	$\rho_\Lambda = 5.3 \times 10^{-10}$	
Photons	$\Omega_\gamma = 5.4 \times 10^{-5}$	$\rho_{\gamma 0} = 4.2 \times 10^{-14}$	$n_{\gamma 0} = 4.1 \times 10^{8}$
Neutrinos	$\Omega_\nu = 0.0014 m_{\nu,-1}$	$\rho_{\nu 0} = 1.1 \times 10^{-12} m_{\nu,-1}$	$n_{\nu 0} = 3.4 \times 10^{8}$

Note: The notation is explained in the text.

(a) The current energy density in light from all the galaxies is $\sim 4 \times 10^{-16}$ J m^{-3}. Estimate what fraction of stellar hydrogen has been converted by nuclear reactions into helium. (You may assume that heat is created by these reactions at a rate of 6.4 MeV per nucleon.)

(b) The current energy density of light radiated by observed accreting black holes—dominated by quasars—is $\sim 6 \times 10^{-17}$ J m^{-3}. Estimate the actual efficiency with which the rest mass energy of the accreting gas is released.

(c) Estimate the total entropy per baryon associated with the horizons of massive black holes today and compare with the entropy per baryon associated with the microwave background and the intergalactic medium (see Sec. 4.10.2).

28.3.2 Dark Matter

After studying the Coma cluster of galaxies, Fritz Zwicky (1933)[20] argued that most of the gravitational mass in clusters of galaxies is neither in the form of stars nor of gas but in a *dark* form that has only been detectable through its gravity (Fig. 28.1c). In effect, the gas and the galaxies move in giant gravitational potential wells formed by this dark matter. Careful measurements of the motions of gas and stars have led to an estimate of the dark matter density about five times the baryon density. Again, CMB observations improve the accuracy, and today we find that the dark matter fraction is $\Omega_D = 0.26$ so that $\Omega_b + \Omega_D = 0.31$ is the total matter fraction. It is widely suspected that dark

---

20. Oort (1932) realized that there is invisible matter in the solar neighborhood. Important subsequent evidence came from optical and radio observations of nearby spiral galaxies by Rubin, Roberts, and others; e.g., Rubin and Ford (1970); and Roberts and Whitehurst (1975).

matter mostly comprises new elementary particles. Furthermore, the development of gravitational clustering on small linear scales points to this matter being collisionless and having negligible pressure when the perturbations grow.[21] The inferred energy and number density of putative dark matter particles is $\rho_{D0} = 2.0 \times 10^{-10}$ J m^{-3}, $n_{D0} = 0.0013 m_{D,12}^{-1}$ m^{-3}, where $m_{D,12}$ is the mass of the hypothesized particle in units of 1 TeV. Of course, the local dark matter density in our galaxy is much larger than this.

cold dark matter

Exercise 28.7 *Challenge: Galaxies*

Make a simple (numerical) model of a spherical galaxy in which the dark matter particles moving in the (Newtonian) gravitational field they create behave like collisionless plasma particles moving in an electromagnetic field. Ignore the baryons.

(a) Adopt the fluid approximation, treat the pressure P as isotropic and equal to $K\rho^{4/3}$, and use the equation of hydrostatic equilibrium (Sec. 13.3) to solve for the mass and radius. Comment on the answer you get, and make a suitable approximation to define an effective mass and radius.

(b) Consider a line passing through the galaxy with impact parameter relative to the center given by b. Solve for the dark matter particles' rms velocity along this line as a function of b.

(c) It is observed that the central densities of galaxies scale as the inverse square of the rms velocity for $b = 0$. How does the mass scale with the velocity?

(d) Solve for the distribution function of the dark matter particles assuming that it is just a function of the energy (Sec. 22.2).

EXERCISES

28.3.3 Photons

The next contributor to the energy density of the universe is the CMB. With its measured temperature of $T_{\gamma 0} = 2.725$ K, the energy density is $\rho_{\gamma 0} = 4.2 \times 10^{-14}$ J m^{-3} (Sec. 3.5). Equivalently, $\Omega_\gamma = 5.4 \times 10^{-5}$. This is dynamically insignificant today but was very important in the past. We can also compute the number density of photons to be $n_{\gamma 0} = 4.1 \times 10^8$ m$^{-3} = 1.6 \times 10^9 n_{b0}$. An equivalent measure, which we need below, is the photon entropy per baryon (Secs. 4.8 and 4.10):

photon number density

$$\sigma_{\gamma 0} = \frac{\rho_{\gamma 0} + P_{\gamma 0}}{n_{b0} T_{\gamma 0}} = \frac{32\pi^5}{45 n_{b0}} \left(\frac{k_B T_{\gamma 0}}{h}\right)^3 k_B = 5.9 \times 10^9 k_B. \quad (28.24)$$

entropy of the CMB

The photon entropy per baryon—the entropy in a box that contains, on average, one baryon and expands with the universe—is therefore conserved to high accuracy.

---

[21]. It is often called *cold dark matter*. However, "cold" is inappropriate at early times, when the particles are probably relativistic, and at late times, when they virialize in dark matter potential wells with thermal speeds $\sim$100–1,000 km s^{-1}.

28.3.4 Neutrinos

There is also a (currently) undetectable background of neutrinos that was in thermal equilibrium with the photons at early epochs and then became thermodynamically isolated with an approximate Fermi-Dirac distribution. As we show in the next section, the current total neutrino density is $n_{\nu 0} = 3.4 \times 10^8$ m^{-3}. There are three *flavors* of neutrinos (ν_e, ν_μ, ν_τ) plus their antiparticles, and they are known to have small masses. The contemporary total neutrino energy density and mass fraction are then $\rho_{\nu 0} = 1.8 \times 10^{-12} m_{\nu, -1}$ J m^{-3}, $\Omega_\nu = 0.0023 m_{\nu, -1}$, respectively, where $m_{\nu, -1} = (m_{\nu_e} + m_{\nu_\mu} + m_{\nu_\tau})/100$ meV. This total m_ν is bounded below through neutrino oscillation measurements (e.g., Cahn and Goldhaber, 2009) at 60 meV and above by cosmological observations at 230 meV. For illustration purposes, we shall assume that there is a single, dominant neutrino of mass 100 meV (i.e., 0.1 eV).

28.3.5 Cosmological Constant

In 1998, it was discovered that the universe is accelerating. This was first inferred from observations of exploding stars called *supernovae*,[22] which turn out to be surprisingly good distance indicators. As we know the scale factor at the time of the explosion, we can measure $\chi(a)$ and infer contemporary values for the deceleration and jerk functions: $q_0 = -0.54$ and j_0 consistent with 1. Using Eq. (28.23), we infer that the pressure is also negative, $P = -0.69\rho$, where ρ includes the matter density.

As we have mentioned, Einstein anticipated this possibility in 1917 when he noted that the field equations would remain "covariant"—in our language, would continue to obey the Principle of Relativity (be expressible in geometric language)—if an additional term proportional to the metric tensor $\boldsymbol{g}$ were added to the field equation (Sec. 25.8):[23]

$$\boldsymbol{G} + \Lambda \boldsymbol{g} = 8\pi \boldsymbol{T}. \tag{28.25}$$

The constant of proportionality Λ is known as the *cosmological constant*. As the Einstein tensor involves second derivatives of the metric tensor, a cosmological constant term that is significant on a cosmological scale should be undetectable on the scale of the solar system or a compact object.

Although we might interpret Eq. (28.25) as a modification to the field equation, we can instead incorporate Λ into the stress-energy tensor $\boldsymbol{T}$ [move $\Lambda \boldsymbol{g}$ to the right-hand side of Eq. (28.25) and absorb it into $8\pi \boldsymbol{T}$], resulting in a *cosmological energy density* ρ_Λ and *cosmological pressure* P_Λ satisfying

$$\rho_\Lambda = -P_\Lambda = \frac{\Lambda}{8\pi}. \tag{28.26}$$

---

22. Perlmutter et al. (1999), Riess et al. (1998).
23. Einstein then went on to propose a static universe, made possible by the cosmological constant Λ. But it was proved wrong when observations revealed the Hubble expansion of the universe, and it was also unstable; so he renounced Λ (Einstein, 1931).

In the absence of another major component of the universe, we deduce that $\Omega_\Lambda = 1 - \Omega_b - \Omega_D - \Omega_\gamma - \Omega_\nu = 0.69$ and $\rho_\Lambda = -P_\Lambda = 5.3 \times 10^{-10}$ J m^{-3}.

Although the large negative pressure may seem strange, it is precedented in classical electromagnetism. Consider a uniform magnetic field B permeating a cylinder plus piston and aligned with their axis. The energy density is $\rho_{\text{mag}} = B^2/(2\mu_0)$, and the total stress acting on the piston—the combination of the isotropic magnetic pressure and the magnetic tension—is $P_{\text{mag}} = -B^2/(2\mu_0) = -\rho_{\text{mag}}$. When we withdraw the piston, ρ_{mag} and P_{mag} will not change, in agreement with the first law of thermodynamics, just like the cosmological constant. What is different about the cosmological constant is that the stress is isotropic.

Λ as density and negative pressure

The cosmological constant contribution to the stress-energy tensor is, by assumption, constant on a hypersurface of simultaneity—it is *ubiquitous*. Applying the first law of thermodynamics [Eq. (28.19)], we see that it was the same in the past and will be the same in the future—it is (past and future) *eternal*. In addition, as the contribution to the stress-energy tensor is directly proportional to the metric tensor, it takes the same form in all frames—it is *invariant*. Combining these features, we see that from a relativistic perspective, it is *universal*.

Λ is universal

28.3.6 Standard Cosmology

This cosmological energy density ρ_Λ and pressure P_Λ are the most conservative explanation for the universe's acceleration. They motivate us to define a *standard cosmology*, in which there are no other major components of the universe besides ρ_Λ, P_Λ, baryonic matter, dark matter, photons, and neutrinos, and in which the spatial geometry is flat.

EXERCISES

Exercise 28.8 *Example: The Pressure of the Rest of the Universe*
Calculate three contributions to the pressure of the contemporary universe.

(a) Baryons. Assume that most of the baryons in the universe outside of stars make up a uniform, hot intergalactic medium with temperature 10^6 K.

(b) Radiation.

(c) Neutrinos. Assume that almost all the neutrino pressure is associated with one flavor with an associated mass of 100 meV.

28.4 Seven Ages of the Universe

Given this description of the present contents of the universe, we are now in a position to describe its history. As our main purpose is to use cosmology to illustrate and apply many features of classical physics, we do not retrace the tortuous, inferential path that led to the standard description of the universe. Instead, we proceed more or less deductively from the earliest time and describe many of the essentially classical processes that must have occurred along the way. To help us do this, we use the device of dividing the history into seven ages, each dominated by different physical processes.

28.4.1 Particle Age

CONSTITUENTS

As explained at the start of this chapter, the universe began expanding from a very hot, dense state; we commence our story at a time when we still have confidence in essentially classical principles. To be specific, this is when $a \sim 10^{-11}$, $T \sim 10$ MeV $\sim 10^{11}$ K, and $t \sim 10$ ms. At earlier times, thermodynamic equilibrium required the presence of a large density of nucleon-antinucleon pairs and pions at near-nuclear density. A description in terms of quarks and gluons is needed (Cahn and Goldhaber, 2009). However, for $a \gtrsim 10^{-11}$, only a tiny but permanent residue of neutrons n and protons p was present, and it was dynamically and thermodynamically irrelevant. We use the comoving density of these baryons as a reference:

baryon density
$$n_b \equiv n_p + n_n = n_{b0}a^{-3} = 2.5 \times 10^{32} a_{-11}^{-3} \text{ m}^{-3}; \quad a \gtrsim 10^{-11}, \quad (28.27)$$

where $a_{-11} \equiv 10^{11}a$.

There were also muons and τs along with their antiparticles, but like the pions, they disappeared from thermodynamic equilibrium and quickly decayed. However, when $a \sim 10^{-11}$, the much lighter electrons and positrons were present with combined density n_e comparable to that of the photons n_γ. The photons and pairs interacted electromagnetically and remained in thermodynamic equilibrium with each other, as well as with the baryons, at a common temperature $T_\gamma(a)$. Neutrinos, with combined density n_ν, were also present at early times. Each of the three neutrino flavors had an approximate Fermi-Dirac distribution with zero chemical potential (Secs. 4.4.3, 5.5.3); and as they were effectively massless, like the photons, the neutrinos' temperature T_ν decreased $\propto a^{-1}$ until comparatively recently, when their finite masses became significant. However, unlike the photons, they did so in thermodynamic isolation (they were *decoupled* from other particles) as the rate at which they self-interacted or exchanged energy with the other constituents quickly became less than the rate of expansion, as we demonstrate below.

pair density

neutrino decoupling

ENTROPY AND TEMPERATURE

Prior to neutrino decoupling, the photons and neutrinos were in equilibrium at the same temperature. However, when $T_\gamma \lesssim m_e/k_B \sim 10^{10}$ K, the pairs became nonrelativistic; they annihilated, and their number (and entropy) density, relative to that of the photons, quickly declined to a very small value—ultimately just that required to equalize the proton charge density—while the photon entropy [Eq. (28.24)] was augmented. Now, the photon and pair entropies per (conserved) baryon, σ_γ and σ_e, are given by Eq. (28.24) with $T_{\gamma 0}$ replaced by T_γ and n_{b0} by n_b, and by

pair annihilation

$$\sigma_e = \frac{\rho_e + P_e}{n_b T_\gamma} = \frac{16\pi a^3}{n_{b0} h^3 T_\gamma} \int_0^\infty \frac{dp \; p^2(E + p^2/(3E))}{e^{E/(k_B T_\gamma)} + 1} = \frac{7}{4}\sigma_\gamma f_{\sigma e}(y), \quad (28.28)$$

where $E = (p^2 + m_e^2)^{1/2}$, $y = m_e/(k_B T_\gamma)$, and

$$f_{\sigma e}(y) = \frac{90 y^4}{7\pi^4} \int_0^\infty \frac{dx \; x^2(1 + 4x^2/3)}{(1+x^2)^{1/2}(e^{y(1+x^2)^{1/2}} + 1)} \quad (28.29)$$

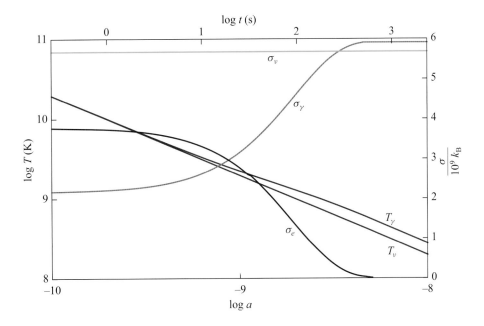

FIGURE 28.4 Temperature T and entropy σ variation during the particle age. At early times t (or equivalently, small scale factors a), the photons, electron-positron pairs, and neutrinos (designated by subscripts γ, e, and ν, respectively) were in equilibrium with a common temperature. However, when the temperature approached m_e/k_B, the pairs began to annihilate, and their contribution to the total entropy was taken up by the photons. (The neutrinos had decoupled by this time, and their entropy per baryon remained constant.) The photon temperature T_γ increased relative to the neutrino temperature T_ν by a factor of 1.4.

varies between 1 when $y \to 0$ ($T_\gamma \to \infty$) and 0 when $y \to \infty$ ($T_\gamma \to 0$) (Secs. 4.8, 4.10.3). Entropy conservation dictates that $\sigma_e + \sigma_\gamma = \sigma_{\gamma 0}$. We can therefore solve for the scale factor $a = (y/y_0)(1 + 7f_{\sigma e}(y)/4)^{-1/3}$, where $y_0 = m_e/(k_B T_{\gamma 0}) = 2.2 \times 10^9$, and hence for the photon temperature as a function of a. At early times and high temperatures, $\sigma_\nu = (4/11)\sigma_{\gamma 0}$, and so $T_\nu/T_\gamma = (4/11)^{1/3} = 0.71$ at later times.[24] The neutrino and photon densities satisfy $n_\nu = \frac{9}{4}(T_\nu/T_\gamma)^3 n_\gamma \propto n_b$, and so n_ν/n_γ decreased from $\frac{9}{4}$ to $\frac{9}{11}$, whence $n_\nu = 3.4 \times 10^{41} a_{-11}^{-3}$ m^{-3} (Fig. 28.4).

neutrino density

ENERGY DENSITY

We can now sum ρ_γ, ρ_e, ρ_ν, ρ_D, ρ_b, and ρ_Λ to give the total energy density for $T_\gamma \lesssim 10^{11}$ K (Fig. 28.5; Secs. 3.5.4, 3.5.5)

$$\rho = a_B T_\gamma^4 + \frac{7}{4} a_B T_\gamma^4 f_{\rho e}(y) + \frac{21}{8} a_B T_\nu^4 + \rho_{D0} a^{-3} + \rho_{b0} a^{-3} + \rho_\Lambda, \qquad (28.30)$$

variation of energy density over cosmic time

where

$$f_{\rho e} = \frac{120 y^4}{7\pi^4} \int_0^\infty \frac{dx\, x^2 (1+x^2)^{1/2}}{(e^{y(1+x^2)^{1/2}} + 1)}. \qquad (28.31)$$

24. As neutrinos have mass yet do not interact, they do not maintain a thermal distribution function when they eventually become nonrelativistic, and so $T_{\nu 0}$ is meaningless.

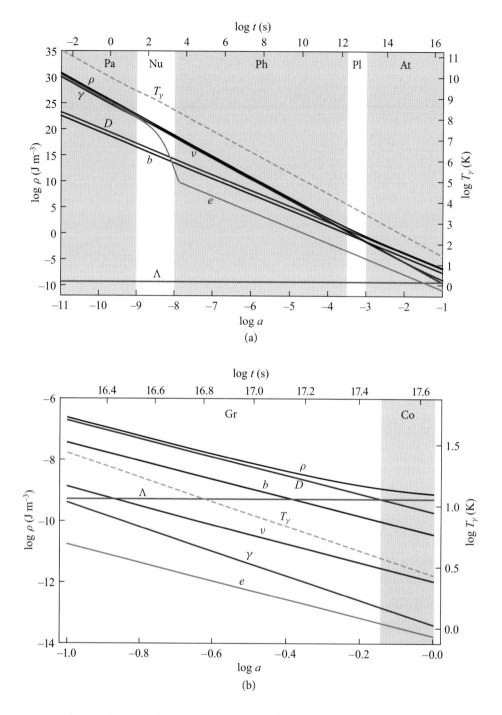

FIGURE 28.5 Energy density in the expanding universe. The energy density ρ (in J m^{-3}) for photons (γ), neutrinos (ν), dark matter (D), baryons (b), and electrons (e) as a function of the scale factor a while the universe evolved. (a) Evolution through the particle (Pa), nuclear (Nu), photon (Ph), plasma (Pl), and atomic (At) ages ending at the *epoch of reionization*, when $a \sim 0.1$. (b) Evolution through the gravitational (Gr) and cosmological (Co) ages. The total energy density is depicted by a thick black line. Also displayed is the cosmic time (proper time) t and the photon temperature T_γ.

Only the first three terms in Eq. (28.30) are significant at early times, but we give the complete expression here.[25] We also make a correction for the (unknown) neutrino rest mass density as discussed in Sec. 28.3.4; it has small, observable effects at late times.

AGE

Given the density, we can use the equation of motion to compute the age of the universe (cosmic time) t as a function of the scale factor a:

$$t(a) = \left(\frac{3}{8\pi}\right)^{1/2} \int_0^a \frac{da'}{a'} \rho(a')^{-1/2} \qquad (28.32)$$

time as a function of scale factor

[Eq. (28.16)]. The age today ($a = 1$) is $t_0 = 4.4 \times 10^{17}$ s $= 13.8$ Gyr.

HORIZON

At early times when matter is ultrarelativistic with $T_\gamma \gtrsim 10^{11}$ K, t is proportional to a^2, ignoring a weak dependence on the number of types of particle contributing to the expansion. In this case, using Eq. (28.15), we find for the horizon radius $\chi_H = 2t/a \propto a$. This tells us that the horizon sphere gets smaller and smaller in terms of comoving baryon mass $M_{Hb} = 4\pi \chi_H^3 \rho_{b0}/3 \propto (t/a)^3$ as we go back in time. When $a \sim 10^{-11}$, $\chi_H \sim 3 \times 10^{16}$ m and $M_H \sim 10^{25}$ kg (roughly an Earth mass). As we have emphasized, the universe that we see around us today grew deterministically out of a hot big bang and is believed to be homogeneous as far as we can see. Yet the regions that were able to establish this smoothness through some form of mixing were comparatively tiny at any time when this could have happened (Fig. 28.6). A plausible resolution of this longstanding paradox, *inflation*, will be discussed in Sec. 28.7.1.

horizon problem

28.4.2 Nuclear Age

28.4.2

NEUTRON-PROTON RATIO

During the early particle age, neutrons and protons were able to establish thermodynamic equilibrium primarily through the weak reactions $n + \nu \leftrightarrow p + e^-$ and $n + e^+ \leftrightarrow p + \bar{\nu}$. When 10^{12} K $\gtrsim T_\gamma \gtrsim 10^{11}$ K, $n_n \sim n_p \sim 0.5 n_b$. As the universe cooled, a Boltzmann distribution was established: $n_n/n_p = \exp[-\gamma_{np} m_e/(k_B T)]$, where we write the mass difference $m_n - m_p$ as $\gamma_{np} m_e = 2.53 m_e = 1.29$ MeV. However, when $T_\gamma \lesssim 3 \times 10^{10}$ K, the reaction rate became slower than the expansion rate, and n_n/n_p declined slowly to ~ 0.14. If these were the only possible interactions, then the neutrons would have eventually undergone β-decay, $n \to p + e^- + \bar{\nu}$, when the universe had expanded further and the age was of order the mean life of a neutron, ~ 15 min. ($\gamma_{np} m_e$ is the maximum electron energy in this decay process.) However, before this happened, strong interactions intervened, and helium was formed.

neutron freeze out

---

25. Although the free electrons are dynamically insignificant at late times, they are very important for the evolution of the microwave background, and their density evolution is discussed below.

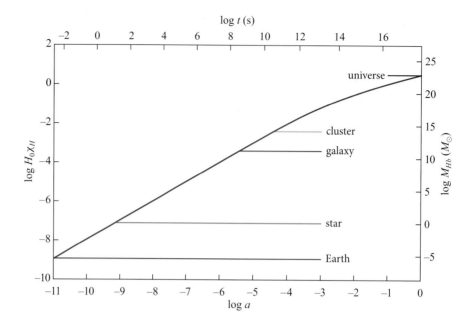

FIGURE 28.6 Comoving horizon radius χ_H as a function of the scale factor a. Also displayed is the associated baryon mass M_{Hb} measured in solar masses (see footnote 1 in this chapter). Note that the mass of baryons in causal contact when $a \sim 10^{-11}$ was equal to only an Earth mass. The mass associated with a cluster of galaxies entered our horizon when $t \sim 300$ yr. The horizon mass when the microwave background was last scattered, $a \sim 10^{-3}$, is equivalent to $\sim 10^4$ galaxy clusters but is only $\sim 10^{-5}$ of the total mass that is observed today, which appears to be impressively homogeneous.

Let us begin analyzing this process by ignoring the expansion of the universe and considering the conversion of n to p through the single reaction $n + \nu \to p + e^-$ (cf. Sec. 5.5.3). Let us define the *rate constant* $\lambda_{n\nu}$ by the expression

forward reaction

$$\dot{n}_p = -\dot{n}_n = \lambda_{n\nu} n_n, \tag{28.33}$$

where a dot denotes differentiation with respect to t. To compute $\lambda_{n\nu}$, we make some well-justified simplifications. First, we ignore the motion of the nucleons and the mass of the neutrinos, which is justified when $T \sim 10^9$ K. Second, we suppose that all distribution functions are isotropic, which is ensured by frequent scattering. Third, we suppose that the electrons and positrons have a common Fermi-Dirac distribution function $f_e = (e^{\gamma m_e/(k_B T_\gamma)} + 1)^{-1}$ with zero chemical potential and temperature T_γ, which is reasonable, as there were many more electrons than nucleons at this time and they were strongly coupled to the photons through Compton scattering. Likewise, the neutrinos and antineutrinos maintain a similar Fermi-Dirac distribution f_ν with an independent temperature T_ν.

On general grounds [cf. Eqs. (3.28), (23.50), and (23.56)], we expect to be able to write the rate coefficient in the form:

$$\lambda_{n\nu} = \int \left(\frac{d^3 p_\nu}{h^3} f_\nu\right) \left(\frac{h^6 W}{m_e^5} \delta(E_\nu - \gamma m_e + \gamma_{np} m_e)\right) \left(2\frac{d^3 p_e}{h^3}(1 - f_e)\right). \tag{28.34}$$

The first term in this expression describes the number density of neutrinos. The second is the *rate per state*, defining W and including a delta function to ensure energy conservation, while the third describes the electron density of states, taking account of both spins and including a blocking factor when the states are occupied. The quantity W can be calculated using the theory of electro-weak interactions and can be measured experimentally. It has the value $W = 2.0 \times 10^{-6}$ s^{-1} (e.g., Weinberg, 2008). As everything is isotropic, we can write $d^3p = 4\pi(E^2 - m^2)^{1/2} E dE$ and integrate over the delta function to obtain

$$\lambda_{n\nu} = 32\pi^2 W \int_{\gamma_{np}}^{\infty} d\gamma \frac{\gamma(\gamma^2 - 1)^{1/2}(\gamma - \gamma_{np})^2}{(e^{(\gamma - \gamma_{np})m_e/(k_B T_\nu)} + 1)(e^{-\gamma m_e/(k_B T_\gamma)} + 1)}. \quad (28.35)$$

integration over phase space

The integration is over the range of γ permitted by energy conservation.

Note that if we set $T_\nu = T_\gamma = T$, the neutrons and protons will attain Boltzmann equilibrium (Sec. 4.4), and so the rate constant λ_{pe} for the inverse reaction will be $e^{-\gamma_{np} m_e/(kT)} \lambda_{n\nu}$.[26] This, in turn, implies that W, in the integral corresponding to Eq. (28.34), is the same for the forward and backward reactions. Furthermore, the theory of electro-weak interactions informs us that W is the same for the $n + e^+ \leftrightarrow p + \bar{\nu}$ and $n \to p + e^- + \bar{\nu}$ reactions. The five relevant rate constants are exhibited in Fig. 28.7(a).

inverse reaction

Including the expansion of the universe is a straightforward matter. We replace the number density of neutrons and protons with the number contained in a fixed volume expanding with the universe. If we set this volume as n_b^{-1}, recalling that baryons are conserved, then we can allow for the expansion by the device of replacing n_p with the proton fraction $X_p \equiv n_p/n_b$, and similarly treating n_n in Eq. (28.33).

including expansion

DEUTERIUM AND HELIUM FORMATION

Neutrons and protons can combine to form deuterons (^{2}H $\equiv d$) through the reactions $p + n \leftrightarrow d + \gamma$ (e.g., Cyburt et al., 2016, and Ex. 4.10). If we just consider the forward reaction and, temporarily, ignore the expansion, d would have formed at the rate $\dot{n}_d = n_p n_n \langle \sigma v \rangle_{np}$. Here the product of the cross section σ and the relative speed v of the nucleons, averaged over the velocity distribution at a given temperature and over the evolution of the universe when these reactions are most significant, is $\langle \sigma v \rangle_{np} = 5 \times 10^{-26}$ m^3 s^{-1}.

deuterium formation

We can now include the inverse reaction by observing that, in equilibrium, we can relate the density n_i of each species i to the chemical potential μ_i through

including inverse reaction

$$n_i = g_i \int \left(\frac{d^3p}{h^3}\right) e^{(\mu_i - E)/(k_B T_\gamma)} = g_i \left(\frac{2\pi m_i k_B T_\gamma}{h^2}\right)^{3/2} e^{\mu_i/(k_B T_\gamma)} \quad (28.36)$$

---

26. As we could have inferred quantum mechanically.

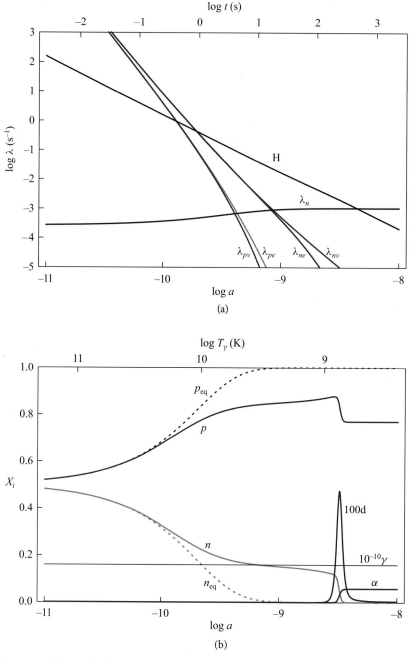

FIGURE 28.7 (a) Rate coefficients for the five reactions that determine the neutron-proton ratio n/p; the subscripts on the rates λ correspond to the left-hand side of the reaction. Note that when time $t \gtrsim 300$ ms, the equilibration rate is slower than the Hubble expansion rate H, and the neutron fraction is stabilized. When $t \gtrsim 600$ s, the few remaining neutrons are able to undergo β-decay before being incorporated in αs. (b) Number fractions (number of particles per baryon) for photons γ, protons p, neutrons n, deuterons d, and alpha particles α. When $T_\gamma \lesssim 2 \times 10^{10}$ K, the nucleons depart from thermodynamic equilibrium. The deuterons are produced when $T_\gamma \sim 10^9$ K before being quickly incorporated into αs. These two panels demonstrate that the current deuterium and helium fractions depend sensitively on the expansion rate and the nuclear physics details; the agreement with observation is a nontrivial validation of standard cosmology.

Chapter 28. Cosmology

(Sec. 5.2), where g_i is the degeneracy of species i, and we treat the nucleons as nonrelativistic particles with $E = p^2/(2m_i) + \ldots$. Under equilibrium conditions, the relativistic chemical potentials satisfy $\tilde{\mu}_p + \tilde{\mu}_n = \tilde{\mu}_d + \tilde{\mu}_\gamma = \tilde{\mu}_d$, and so we can eliminate the chemical potentials to obtain the Saha-like (Ex. 5.10) relation:

$$n_p n_n = S(T_\gamma) n_d, \tag{28.37a}$$

where

$$S(T_\gamma) = 2^{1/2} 3^{-1} \left(\frac{2\pi m_p k_B T_\gamma}{h^2} \right)^{3/2} e^{-\gamma_d m_e/(k_B T_\gamma)}, \tag{28.37b}$$

we have used $g_p = g_n = 2$ and $g_d = 3$ and $\gamma_d m_e = m_n + m_p - m_d = 4.34\, m_e$ is the deuteron binding energy. The combined forward-reverse reaction rate must then be described by $dn_d/dt = \langle \sigma v \rangle_{np} [n_p n_n - S(T_\gamma) n_d]$, as this is the only way that we can maintain equilibrium at all temperatures. At early times, $S \langle \sigma v \rangle_{np} \gg H$ and equilibrium must have been maintained.

The final step is to explain how the tritons (nuclei of $^3\text{H} \equiv t$), helions (nuclei of $^3\text{He} \equiv h$), and alpha particles (nuclei of $^4\text{He} \equiv \alpha$) were formed. There are two important pathways: $d + d \rightarrow t + p, t + d \rightarrow \alpha + n$; and $d + d \rightarrow h + n, h + d \rightarrow \alpha + p$ (cf. Sec. 19.3.1). The effective, combined forward reaction rate is calculated using $\langle \sigma v \rangle_{dd} = 7 \times 10^{-24}$ m^3 s^{-1}. The reverse reaction rates can be calculated using a simple generalization of the argument leading to Eq. (28.37) for reactions involving four or more nucleons. We can now allow for the expansion as before, introducing the d number fraction $X_d = n_d/n_b$, to obtain three coupled rate equations for the proton, neutron, and deuteron number fractions $X_p = n_p/n_b$, $X_n = n_n/n_b$, and $X_d = n_d/n_b$:

deuterium reactions

$$H X'_p = (\lambda_{nv} + \lambda_{ne} + \lambda_n) X_n - (\lambda_{pe} + \lambda_{pv}) X_p - \langle \sigma v \rangle_{np} (n_b X_p X_n - S(T_\gamma) X_d),$$

$$H X'_n = (\lambda_{pe} + \lambda_{pv}) X_p - (\lambda_{nv} + \lambda_{ne} + \lambda_n) X_n - \langle \sigma v \rangle_{np} (n_b X_p X_n - S(T_\gamma) X_d),$$

$$H X'_d = \langle \sigma v \rangle_{np} (n_b X_p X_n - S(T_\gamma) X_d) - 2 \langle \sigma v \rangle_{dd} n_b X_d^2, \tag{28.38}$$

rate equations

where a prime denotes differentiation with respect to $\ln a$. It is easiest to solve these equations by assuming the neutrons and protons are in equilibrium until $a \sim 10^{-11}$ and the deuterons are in equilibrium until $a \sim 10^{-9}$.

The numerical solution to these equations (Fig. 28.7) shows that the neutrons and protons remained in thermodynamic equilibrium until the temperature fell to $\sim 2 \times 10^{10}$ K, at which time the neutron fraction declined slowly, reaching a value ~ 0.15 by the time the temperature had fallen to $\sim 10^9$ K. At this temperature, the deuteron equilibrium fraction climbed quickly, and the deuteron density became large enough for the $d + d$ reactions to produce α at $T_\gamma = 8 \times 10^8$ K, and then declined. Most neutrons were incorporated into helium, though a small fraction was left to decay freely. The final helium fraction is[27] $X_\alpha = n_\alpha/n_b = \frac{1}{4}(1 - X_p) = 0.058$ (Fig. 28.7), slightly smaller than the (observed) value $X_\alpha = 0.062$ obtained by more

numerical solution

nucleosynthesis

final helium and deuterium fractions

---

27. Astronomers reserve the symbol X for hydrogen and use the symbol $Y = 4 X_\alpha$ for the helium mass fraction and Z for the mass fraction of all other elements which they call "metals"!

detailed calculations that include about ten more reactions, temperature dependence of the reaction rates, and other refinements. The late-time deuterium number fraction computes to be $X_d = 2.3 \times 10^{-5}$, consistent with the detailed calculations and with observations. Yields of h, ^{6}Li, and ^{7}Li are also computed and can be reconciled with observations if one takes account of large astrophysical uncertainty. The t decays.

It is the absence of stable nuclei with $A = 5, 8$ that prevented the build-up of elements beyond helium in the early universe. Instead, the synthesis of these elements had to await the formation and evolution of stars.

28.4.3 Photon Age

Primordial nucleosynthesis was complete by the time the scale factor had increased to $a \sim 10^{-8}$ and the temperature had fallen to $T_\gamma \sim 3 \times 10^8$ K. This marks the beginning of the photon age, during which the energy density of the universe was still dominated by photons (and neutrinos) and $\rho_\gamma \propto a^{-4}$, $a \propto t^{1/2}$. Now is a good time to examine the implicit assumption that we have been making that, with the conspicuous exception of the neutrinos, all major constituents of the universe are maintained in thermal equilibrium during the nuclear and photon ages.

PLASMA EQUILIBRATION

Let us first consider protons and electrons. The plasma frequency $\omega_P = [ne^2/(m\epsilon_0)]^{1/2} \propto a^{-3/2}$ (Sec. 20.3) is very high in comparison with the expansion rate. In fact, $\omega_p H \sim 10^{17} a_{-8}^{1/2}$, where $a_{-8} = a/10^{-8}$. Likewise, the ratio of the Debye length $\lambda_D = [\epsilon_0 k_B T/(ne^2)]^{1/2}$ to the horizon radius is $\lambda_D \chi_H \sim 10^{-18} a_{-8}^{-1}$. Therefore, the use of fluid mechanics in place of plasma physics is amply justified for the whole expansion of the universe.

validity of the fluid approximation

Now let us question our implicit assumption of thermodynamic equilibrium. Using Eqs. (20.23), (28.27), and (28.13), we find that the ratios of the e-e, p-p, and p-e equilibration timescales ($\propto T^{3/2} n^{-1} \propto a^{3/2}$) to the expansion timescale ($\propto a^2$) are given by $H\{t_{ee}, t_{pp}, t_{ep}\} \sim \{1, 40, 1800\} \times 10^{-10} a_{-8}^{-1/2}$ throughout the photon age, ensuring that baryons remained in thermal equilibrium.

equilibration timescales

COMPTON SCATTERING

Next consider the interaction between the plasma and the photons, which is mediated by Compton scattering. If there had been no such interaction, then the (nonrelativistic) plasma would have maintained its Maxwellian distribution with a temperature that would have fallen as $T \propto a^{-2}$ (Sec. 28.2.3). However, each Compton scattering will lead to an energy transfer from the photons to the electrons through a combination of the Doppler effect and Compton recoil. This energy change will then be quickly shared with the protons, so that the plasma is only allowed to become a tiny bit cooler than the radiation. To be quantitative, the mean fractional energy exchange per scattering is $|\Delta \ln E| \sim k_B T/m_e$. Therefore, it will take $\sim m_e/(k_B T)$ scatterings to equilibrate the plasma with the far more numerous photons. The scattering time is $t_{e\gamma} \sim (n_\gamma \sigma_T)^{-1}$, where $\sigma_T = 8\pi r_e^2/3 = 6.65 \times 10^{-29}$ m^{-2} is the *Thomson cross* section, and $r_e = 2.8$ fm is the classical electron radius (Sec. 3.7.1). The ratio of the timescale for the plasma to equilibrate with the radiation, to the expansion timescale, is $Ht_{e\gamma} = Hm_e/(\rho_\gamma \sigma_T) \sim 2 \times 10^{-16} a_{-8}^2$. The electrons therefore exchange energy with the radiation field even faster than they share it among themselves and with protons. This justifies our assumption of a common matter and radiation temperature in the early universe. Insofar as the universe is homogeneous, the far more numerous photons will automatically maintain their initial Planck distribution without energy redistribution by electrons. Their response to small, inhomogeneous perturbations is considered below.

electron-photon coupling

28.4.4 Plasma Age

Eventually, the rest-mass energy density of matter (mostly dark matter) exceeded that of radiation and neutrinos, and the plasma age began. This happened when $a = a_{eq} = 0.00030$, $t = t_{eq} = 52$ kyr, and $T_\gamma = T_{eq} = 9{,}100$ K. Thereafter, according to Eq. (28.30), $\rho \propto a^{-3}$ and $a \propto t^{2/3}$, approximately.[28] The Friedmann equation (28.16) can be integrated in this plasma age to give, more precisely,

onset of matter dominance

$$\frac{t}{t_{eq}} = \left(1 + 2^{-1/2}\right)\left[2 - \left(2 - \frac{a}{a_{eq}}\right)\left(1 + \frac{a}{a_{eq}}\right)^{1/2}\right], \qquad (28.39)$$

which remains valid until the cosmological age.

Next the helium ions became singly ionized through capturing one electron and then became neutral by taking on a second electron, leaving a proton-electron plasma. After a further interval, the hydrogen recombined.[29] Unlike the helium recombination, the details of this process are highly significant.

helium recombination

The total (atomic plus ionized) hydrogen density is $n_{H+p} = 1.9 \times 10^8 a_{-3}^{-3}$ m^{-3}, where $a_{-3} = a/10^{-3}$. In equilibrium, the atomic fraction would satisfy the Saha equation (5.68). However, just as happened with nucleosynthesis, the universe expanded too fast for the reactions to keep up. The basic problem is that when an electron and a proton recombine, they emit one or more photons that have a short mean free path and are mostly reabsorbed by neighboring atoms, leading to no net change in ionization. This is especially true of the Lyman α photons emitted with frequency $\nu_\alpha = 2.47 \times 10^{15}$ Hz when a hydrogen atom transitions from its first excited state, designated by quantum number $n = 2$, to its ground state with $n = 1$. A good approximation is to treat the $n = 2$ level as the effective ground state, changing the effective ionization potential from $I_1 = 13.6$ eV to $I_2 = 13.6/4 = 3.4$ eV and modifying the degeneracy in the Saha equation, and then to allow for the slow permanent population of the true, $n = 1$, ground state.

hydrogen recombination

---

28. The thermal energy density of the plasma was only 6×10^{-10} times the radiation energy density at this time.
29. The use of the term *recombination* is conventional but misleading, because it is the first occurrence of this process.

If we denote by X_1 and X_2 the fraction of the hydrogen in the $n = 1$ and $n = 2$ states, respectively, and ignore higher energy levels,[30] then the *ionization fraction* is $x = n_e/n_{H+p} = 1 - X_1 - X_2$. The rate equations analogous to Eqs. (28.38) are

rate equations

$$HX_2' = \alpha_2 \left(n_{H+p} x^2 - \frac{1}{4} X_2 \left(\frac{2\pi m k_B T_\gamma}{h^2} \right)^{3/2} e^{-I_2/(k_B T_\gamma)} \right) - \lambda_{21} X_2,$$

$$HX_1' = \lambda_{21} X_2. \tag{28.40}$$

The use of fractional densities once again takes account of the expansion of the universe. The quantity $\alpha_2 = 2.8 \times 10^{-19} T_4^{-1/2}$ m³ s⁻¹ is the *recombination coefficient* into the $n = 2$ level, which is computed by summing over all pathways, excluding transitions to $n = 1$. The $\frac{1}{4}$ takes into account the 2-fold degeneracy of the $n = 1$ level

allowing for inverse reactions

and 8-fold degeneracy of the $n = 2$ level, and the "Saha" factor ensures equilibrium in the absence of transitions to $n = 1$, just like the factor S in Eq. (28.37). The rate constant λ_{21} describes the permanent transitions to the ground state, corrected for reverse transitions.

We compute λ_{21} as follows. The sublevels of the $n = 2$ level are well mixed by collisions, so one-quarter of the excited atoms will be in the 2s sublevel, while three-

two-photon deexcitation

quarters will be in the 2p sublevel. The 2s atoms can permanently deexcite by emitting two photons, neither of which is reabsorbed. This process has a *forbidden* spontaneous rate $A_{2s} = 8.2$ s⁻¹, and the inverse process can be ignored. The 2p atoms create Lyman α photons with a *permitted* rate $A_{2p} = 4.7 \times 10^8$ s⁻¹. The spectral line will have a small, combined natural and Doppler width. The cross section for absorbing a Lyman α photon in the low-frequency wing of the line decreases with decreasing frequency and eventually becomes small enough that the expansion of the universe allows the photon to avoid absorption altogether and therefore leads to the forma-

escape probability

tion of a hydrogen atom in its ground state. Let us define by P_{esc} the probability that one of these photons avoids absorption in this manner. Next let us describe the line profile by $P_\nu(\nu)$, the cumulative probability that an emitted Lyman α photon has frequency less than ν. (See Ex. 28.9 for this and some other details of the analysis.)

Kirchhoff's law of radiation (e.g., Sec. 10.2.1) ensures that the net absorption cross section has the same frequency dependence as the emissivity. Using the Einstein coefficients, we can show that the net absorption cross section associated with the same atoms as those that are emitting the Lyman α photons is $\sigma_\alpha = [A_{2p}/(8\pi \nu_\alpha^2)] dP_\nu/d\nu$.[31] The probability that a Lyman α photon with any frequency will escape due to expansion is then given by

---

30. The fractional occupancy of the $n = 2$ level never exceeds $\sim 10^{-13}$, and so this is a good approximation.
31. Another way of expressing this is $\int d\nu \sigma_\alpha = \pi r_e f_\alpha$, where $f_\alpha = 0.42$ is the *oscillator strength* (Cohen-Tannoudji, Diu, and Laloë, 1977).

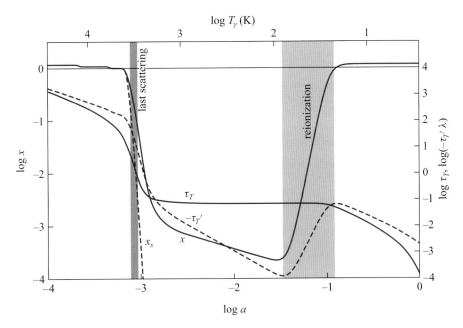

FIGURE 28.8 Ionization fraction of the universe, x, plotted as a function of the scale factor a. The blue dashed curve shows the result from assuming Saha equilibrium x_S. The solid blue curve shows the actual ionization fraction, x, deduced from Eqs. (28.40) after including in λ_{21} the atomic processes that delay recombination of the electrons. The solid red curve shows the Thomson optical depth τ_T from now back to scale factor a, while the dashed red curve is its derivative with respect to $-\ln a$: $-\tau_T' = n_e \sigma_T H^{-1}$. The shaded region to the left delineates the short interval when τ_T fell from ~ 3 to ~ 1. The shaded region to the right delineates the poorly understood epoch of reionization (Sec. 28.4.5), when newly formed stars and black holes are thought to have created sufficient ultraviolet photons to change most hydrogen in the universe back to a plasma. The ionization fraction adopted for $0.05 < a < 1$ is $x = 1.16[1 + 1.5(a/0.1)^{-8.7}]^{-1}$, which allows for the presence of helium.

$$P_{\text{esc}} = \int_0^1 dP_\nu e^{-n_1 \int dt\sigma_\alpha} = \int_0^1 dP_\nu e^{-\frac{n_1}{H\nu_\alpha} \int d\nu \sigma_\alpha} = \int_0^1 dP_\nu e^{-\frac{n_1 A_{2p} P_\nu}{8\pi \nu_\alpha^3 H}} \simeq \frac{8\pi \nu_\alpha^3 H}{n_1 A_{2p}}, \tag{28.41}$$

where n_1 is the number density of H atoms in the $n = 1$ state, which is effectively constant in the short time it takes the universe to Doppler shift the frequency of the photon by enough to escape absorption. Therefore $\lambda_{21} = \frac{1}{4} A_{2s} + \frac{3}{4} A_{2p} P_{\text{esc}}$, independent of A_{2p}.

To solve Eqs. (28.40), note that $X_2 \ll X_1$, and so X_2' can be set to zero. The resulting ionization x (solid blue curve in Fig. 28.8) follows the full Saha evolution as long as $T \gtrsim 4{,}000$ K (Fig. 28.8). Thereafter the ionization fraction is significantly larger. The universe is half-ionized when $T_\gamma \sim 3{,}500$ as opposed to $\sim 3{,}700$ K, according to the Saha equation.

We will need the Thomson optical depth τ_I on our past light cone:

$$\tau_T(\ln a) = \int_{\ln a}^0 (d \ln a) n_e \sigma_T / H. \tag{28.42}$$

28.4 Seven Ages of the Universe

last scattering surface

According to this equation, the average last scattering surface, where $\tau_T = 2/3$, occurred when $a = a_{ls} = 0.00093$, $T_\gamma = 2{,}920$ K, $t = 1.2 \times 10^{13}$ s $= 380$ kyr. These values are in good agreement with more careful calculations.

EXERCISES

Exercise 28.9 *Problem: Spectral Line Formation*

In our discussion of recombination, we related the emission of Lyman α photons to their absorption. This involves some important ideas in the theories of radiation and thermodynamics.

(a) Consider a population of two-state atoms. Let the number of atoms in the lower state be N_1 and in the upper state N_2. The probability per unit time of an upper-state atom changing to a lower state and releasing a photon of energy $h\nu$ equal to the energy difference of the states is denoted by A. We expect that the rate of upward, $1 \to 2$ transitions is proportional to the occupation number η_γ of the photons with frequency ν (Sec. 3.2.5). Call this rate $K_u \eta_\gamma$. By requiring that the atoms should be able to remain in Boltzmann equilibrium with the Planckian radiation field of the same temperature, show that there must also be downward, *stimulated emission* at a rate per state-2 atom of $K_d \eta_\gamma$, and show that $K_u = K_d = A$.

(b) The absorption cross section σ for an atom at rest can be written in the *Lorentz* or *Breit-Wigner* form as:

$$\sigma = \frac{\pi A^2}{(\nu - \nu_0)^2 + A^2}.$$

Either make a classical model of an atom as an electron oscillator with natural frequency ν_0, or use time-dependent perturbation theory in quantum mechanics to justify the form of this formula.

(c) Identify the frequency probability function P_ν introduced in Sec. 28.4.4, and plot the natural line profile for an emission line.

(d) Atoms also have thermal motions, which Doppler shift the photon frequencies. Modify the line profile by numerically convolving the natural profile with a 1-dimensional Gaussian velocity distribution and replot it, drawing attention to its behavior when A is much more than the thermal Doppler shift.

(e) We have restricted our attention to a two-state system. It is usually the case that we are dealing with energy levels containing several distinct states, and the formalism we have described has to be modified to include the *degeneracies* g_i of these levels. Make the necessary corrections and recover the formulas used in the text.

(f) A second complication is polarization (Sec. 7.7). Discuss how to include this.

28.4.5 Atomic Age

The next age is the atomic age.[32] If atomic hydrogen were completely decoupled from the radiation field, then it would cool with temperature $T_H \propto a^{-2}$ and eventually become cryogenic! However, this is not what happened, as the electrons that remained were still able to keep in thermal contact with the radiation and with the protons and atomic hydrogen. The plasma was maintained at roughly the electron temperature. Eventually, when $a \sim 0.03$, the temperature had fallen to $T \sim 100$ K, and molecular hydrogen appeared. However, this was also about the time when the very first self-luminous stars and black holes formed and emitted ultraviolet radiation, which caused the molecular hydrogen to dissociate and the atoms to ionize. This is known as the *epoch of reionization* and must have continued until $a \sim 0.14$, because neutral hydrogen is actually detected at this time through Lyman α absorption of quasar light. Characterizing the epoch of reionization is a major goal of modern research and involves many considerations that lie beyond the scope of this book. The evolution adopted in Fig. 28.8 and Sec. 28.5 is consistent with current observations but is not yet well constrained.

Exercise 28.10 *Problem: Reionization of the Universe*

(a) Estimate the minimum fraction of the rest mass energy of the hydrogen that must have undergone nuclear reactions inside stars to have ionized the remaining gas when $a \sim 0.1$.

(b) Suppose that these stars radiated 30 times this minimum energy at optical frequencies. Estimate the energy density and frequency of this stellar radiation background today.

You may find Exercises 4.11 and 28.6 helpful.

28.4.6 Gravitational Age

SCALE FACTOR

As we discuss further in Sec. 28.5, after recombination small inhomogeneities in the early universe grew under the influence of gravity to form galaxies and larger-scale structures. The influence of stars was supplemented by that of accreting massive black holes that formed and grew in the nuclei of galaxies. After reionization this radiative onslaught kept most baryons in the universe in a multiphase, high-temperature state.

However, this is also the time when the influence of the cosmological constant started to become significant. If we ignore photons, neutrinos, and spatial curvature, then Eq. (28.16) describing the expansion of the universe in the gravitational age has

---

32. Sometimes called the *dark age*.

a simple analytical solution derived and discussed by Bondi (1952a):

$$t = \frac{2}{3H_0(1-\Omega_M)^{1/2}} \sinh^{-1}\left[(\Omega_M^{-1}-1)^{1/2} a^{3/2}\right], \qquad (28.43)$$

where Ω_M is the current matter density in units of the critical density. We see that the influence of the cosmological constant is negligible at early enough times, $t \ll t_0$, and the energy density of the matter dominates the expansion. It also dominates any curvature that there might be today. Such a universe is usually called an *Einstein–de Sitter universe* and has $a = (3H_0 t/2)^{2/3}$. Similarly, at late times, $t \gg t_0$, we can ignore the matter and find that $a \propto \exp(Ht)$, where $H = (1-\Omega_M)^{1/2} H_0$. This is called a *de Sitter universe*. The kinematic properties of our standard universe relative to an Einstein–de Sitter universe and an open ($k = -1$) one are best expressed using the deceleration parameter q [Eqs. (28.14) and (28.23)], which is exhibited in Fig. 28.9. Note that the jerk parameter is $j = 1$ throughout the gravitational and cosmological ages.

Einstein–de Sitter universe

DISTANCE AND VOLUME

Excepting the CMB, most cosmological measurements are made when $a \gtrsim 0.1$. Therefore, now is a good time to calculate distance and volume. The comoving distance to a source whose light was emitted when the expansion parameter was a, $\Sigma(a)$ (equal to the radius $\chi(a)$ in our flat universe), can be computed from the Friedmann equations and is exhibited in Fig. 28.10. We can also compute the angular diameter distance $d_A = a\chi(a)$, which is seen to reach a maximum value of $0.41/H_0 = 5.6 \times 10^{25}$ m at $a = 0.39$. More distant sources of fixed physical length will actually appear progressively larger. By contrast, the luminosity distance $d_L = a^{-1}\chi(a)$ increases rapidly with distance, and individual sources become unobservably faint. The total distance to the early universe is $\chi_H(t_0) = 3.2/H_0 = 14$ Gpc $= 4.4 \times 10^{26}$ m and the associated comoving volume is $V_H(t_0) = 4\pi \chi_H(t_0)^3/3 = 140 H_0^{-3} = 1.2 \times 10^4$ Gpc$^3 = 3.6 \times 10^{80}$ m^3. The universe is a big place!

angular diameter and luminosity distances

volume of observable universe

EXERCISES

Exercise 28.11 *Problem: Type 1a Supernovae and the Accelerating Universe*
Rather surprisingly, it turns out that a certain type of supernova explosion (called "Type 1a" and associated with detonating white-dwarf stars) has a peak luminosity L that can be determined by studying the way its brightness subsequently declines. Astronomers can measure the peak fluxes F for a population of supernovae at a range of distances.

Calculate the flux measured at Earth for a given L as a function of the scale factor $0.3 < a < 1$ at the time of emission for the following.

(a) An Einstein–de Sitter universe.

(b) A nonaccelerating universe with $a \propto t$.

(c) Our standard model universe.

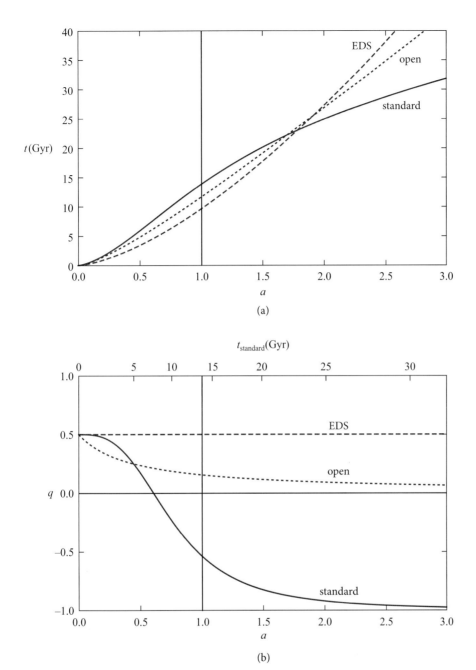

FIGURE 28.9 Expansion of the universe during the gravitational and cosmological ages. (a) The variation of age t with scale factor a is shown for standard cosmology (solid curve). Note that the expansion of the universe initially decelerated under the pull of gravity, but then began to accelerate at age ~6 Gyr under the influence of the cosmological constant. If this continues for the next ~15 Gyr, then the universe will embark on an exponential growth. (b) This exponential growth is brought out in a plot of the associated deceleration parameter q versus a. Also shown in both panels are the solutions for a (dashed) Einstein–de Sitter (EDS) model, which is flat and matter dominated, with an identical Hubble constant as standard cosmology, and a negatively curved (dotted) open model with the same contemporary density parameter as used in standard cosmology. Neither of these models exhibits acceleration; the former decelerates forever, the latter eventually expands with constant speed.

28.4 Seven Ages of the Universe

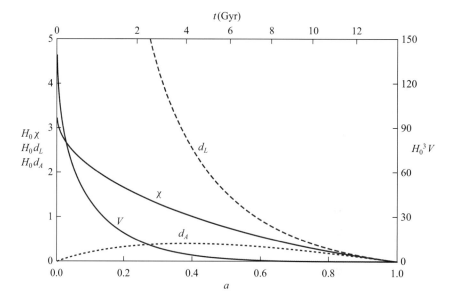

FIGURE 28.10 Comoving distance χ (in units H_0^{-1}) to a source whose light was emitted when the scale factor was $a < 1$. Also shown are the luminosity distance d_L, the angular diameter distance d_A, and the comoving volume V associated with a sphere of radius χ (red).

Assume the same Hubble constant today, H_0. How accurately must F be measured to confirm the prediction of the standard model with an error of ~ 0.1 in a single measurement at $a \sim 0.7$? Astronomers do not, in practice, measure the total flux but the flux in a specific spectral band, but this adjustment can be made if the spectrum is known.

28.4.7 Cosmological Age

The final age, which began about 5 Gyr ago, when $a \sim a_\Lambda \sim 0.7$, is called the cosmological age because the energy density is thereafter dominated by the cosmological constant. We are entering a phase of exponential, de Sitter expansion, presaging a future dominated by dilution and decay—an agoraphobic's worst nightmare! Operationally, the acceleration slows the development of large-scale structure in the distribution of galaxies, which provides one way to measure the value of the cosmological constant. Of course, as a pure cosmological constant is still a weakly constrained fit to the observations, the future could be more subtle, as we discuss in Sec. 28.7.3.

eschatology!

EXERCISES

Exercise 28.12 *Problem: Future Evolution of the Universe*
Assume that the universe will continue to expand according to Eq. (28.43).

(a) Calculate the behavior of the angular diameter distance and the associated volume as a function of the scale factor for the next 20 billion years.

(b) Interpret your answer physically.

(c) Explain qualitatively what will happen if the universe accelerates even faster than this.

We return to this topic in Sec. 28.7.3.

28.5 Galaxy Formation T2

The universe we have described so far is homogeneous and isotropic and completely ignores the large density fluctuations on small scales, observed today as clustered galaxies. Our task now is to set up a formalism to describe the growth under gravity of the perturbations that produce this structure as the universe ages. Much of this problem can be handled using Newtonian physics, but as the most interesting questions are intrinsically relativistic and as we have already developed the necessary formalism, we shall dive right into a fully relativistic analysis (Peebles and Yu, 1970; Sunyaev and Zel'dovich, 1970).

28.5.1 Linear Perturbations T2

METRIC

We generalize the Robertson-Walker metric [Eq. (28.9)] to include linear perturbations in a manner inspired by our discussions of weak fields (Sec. 25.9) and the Schwarzschild spacetime (Sec. 26.2):

$$ds^2 = -(1+2\Phi)dt^2 + a^2(1-2\Psi)\delta_{ij}d\chi^i d\chi^i. \tag{28.44}$$

relativistic perturbation theory

Here a is the same function of t as in the unperturbed model, and FOs (by definition) continue to move with fixed comoving coordinate χ. The changes to the spacetime geometry are all contained in the *curvature perturbation* Ψ and in the *potential perturbation* Φ, which agrees with its Newtonian counterpart when it is small in magnitude (relative to unity) and inhomogeneity scale[33] (relative to the horizon).[34]

curvature and potential perturbations

KINEMATICS

An FO no longer follows a timelike geodesic. We use $\vec{u} \cdot \vec{u} = -1$ to evaluate the components of its 4-velocity and 4-acceleration $a_\alpha = u^\beta u_{\alpha;\beta}$ to first order in the perturbation:

$$u^t = 1 - \Phi, \quad u^j = 0; \quad a^t = 0, \quad a_i = \partial_i \Phi. \tag{28.45}$$

---

33. Note that Φ does not include the Newtonian potential difference $2\pi\rho r^2/3$ that we might be tempted to associate with two points separated by r in the background medium if we had not appreciated the way that general relativity neatly resolves this ambiguity.
34. The coordinate choice is sometimes called a *gauge* [cf. Eqs. (25.87), (25.88), and (27.18)] by analogy with classical electromagnetism and particle physics and is often motivated by considerations of symmetry. Physical observables should not (and do not) depend on the coordinate/gauge choice. Our choice is known as the *Newtonian gauge* and is useful for perturbations that can be expressed as scalar quantities. (We will encounter tensor perturbations in Sec. 28.7.1.) This gauge choice is appropriate because, when Φ and Ψ are small, it becomes the weak-gravitational-field limit of general relativity (Sec. 25.9), which we need to interpret the actual cosmological observations we make today and which can be used to discuss the initial conditions, as we shall see in Sec. 28.7.1.)

These expressions exhibit gravitational time dilation—$dt/d\tau = 1 - \Phi$ (Sec. 27.2.1)—and Einstein's equivalence principle (Sec. 25.4). A particle that moves with small 3-velocity **v**, as measured in an FO's local Lorentz frame, has a 4-velocity $u^t = 1 - \Phi$, $u^i = v_i/a$ and a 4-acceleration $a_t = 0$, $a_i = d(av_i)/dt + \partial_i \Phi$. In the limit $\Phi = 0$, $\mathbf{v} \propto a^{-1}$ if the particle is freely moving (Sec. 28.2.3).

acceleration with respect to FO

FOURIER MODES

We are interested in the linear evolution of perturbations, and it is convenient to work with the spatial Fourier transform of the perturbed quantities (Sec. 8.3). For the potential, which is our primary concern, we write

$$\Phi(t, \chi) = \int \frac{d^3k}{(2\pi)^3} e^{i\mathbf{k}\cdot\chi} \tilde{\Phi}(t, \mathbf{k}), \tag{28.46}$$

potential oscillations

where the $\sim$ denotes spatial Fourier transform, the comoving wave vector **k** does not change with time, and we treat t and a as interchangeable coordinates. As Φ is real, $\tilde{\Phi}(t, -\mathbf{k}) = \tilde{\Phi}^*(t, \mathbf{k})$. Implicit in this Fourier expansion is a box, which we presume is much larger than the current horizon but which will not feature in our development. The actual modes can be considered as traveling waves moving in antiparallel directions or as standing modes in quadrature, which is a better way to think about their nonlinear development. In what follows, we shall consider the temporal development of linear perturbations, which can be thought of as either individual wave modes or as continuous Fourier transforms.

PERTURBED EINSTEIN EQUATION

Now focus on a single Fourier oscillation and use computer algebra to evaluate the nonzero, linear perturbations to the Einstein field equations in a local orthonormal basis:

$$\tilde{G}^{\hat{t}\hat{t}} = -2[3H(\dot{\tilde{\Psi}} + H\tilde{\Phi}) + (k/a)^2 \tilde{\Psi}] = 8\pi \tilde{T}^{\hat{t}\hat{t}} = 8\pi \tilde{\rho}, \tag{28.47a}$$

$$\tilde{G}^{\hat{t}\hat{\|}} = -\frac{2ik}{a}(\dot{\tilde{\Psi}} + H\tilde{\Phi}) = 8\pi \tilde{T}^{\hat{t}\hat{\|}} = 8\pi(\rho + P)\tilde{v}, \tag{28.47b}$$

perturbed field equations

$$\tilde{G}^{\hat{\|}\hat{\|}} = 2[\ddot{\tilde{\Psi}} + H(\dot{\tilde{\Phi}} + 3\dot{\tilde{\Psi}}) + (1 - 2q)H^2\tilde{\Phi}] = 8\pi \tilde{T}^{\hat{\|}\hat{\|}}, \tag{28.47c}$$

$$\tilde{G}^{\hat{\perp}\hat{\perp}} = \tilde{G}^{\hat{\|}\hat{\|}} + \frac{k^2(\tilde{\Psi} - \tilde{\Phi})}{a^2} = 8\pi \tilde{T}^{\hat{\perp}\hat{\perp}}, \tag{28.47d}$$

where $\|$ and $\perp$ are components parallel and perpendicular to **k**. Equation (28.47b) defines the mean velocity perturbation, which is purely parallel, $\tilde{v} = \tilde{v}^{\|}$. Note that if the cosmological fluid is perfect, there is no shear stress ($\tilde{T}^{\hat{\|}\hat{\|}} = \tilde{T}^{\hat{\perp}\hat{\perp}} = \tilde{P}$) and, consequently, $\tilde{\Psi} = \tilde{\Phi}$. This is a major simplification, echoing our treatment of Schwarzschild spacetime (Sec. 26.2). However, when neutrinos or photons have decoupled from matter and free stream through primordial (dark-matter) density perturbations, their stresses become sufficiently anisotropic to produce a measurable distinction between $\tilde{\Phi}$ and $\tilde{\Psi}$.

Because these equations only involve the component of mean velocity parallel to **k**, they describe longitudinal waves, generalizations of the sound waves discussed in Sec. 16.5. This mean velocity is irrotational and can therefore be written as a carefully chosen function of time, multiplied by the gradient of a velocity potential $\tilde{\psi}$ (Sec. 13.5.4):[35]

$$\tilde{\mathbf{v}} = i\boldsymbol{\beta}\tilde{\psi}. \qquad (28.48a)$$

velocity potential

Here we introduce a *scaled wave vector*

$$\boldsymbol{\beta} = \frac{\mathbf{k}}{3^{1/2}\dot{a}}, \qquad (28.48b)$$

scaled wave vector

which can be regarded as a function of either t or a. (Note that $q \equiv (\partial \ln \beta / \partial \ln a)_\mathbf{k}$, which is a useful relation.)

We also define the *total relative density perturbation*:

$$\tilde{\delta} \equiv \tilde{\rho}/\rho = (\rho_b \tilde{\delta}_b + \rho_D \tilde{\delta}_D + \rho_\gamma \tilde{\delta}_\gamma + \rho_\nu \tilde{\delta}_\nu)/\rho, \quad \text{with} \quad \tilde{\delta}_{b,D,\gamma,\nu} \equiv \tilde{\rho}_{b,D,\gamma,\nu}/\rho_{b,D,\gamma,\nu}, \qquad (28.49)$$

relative density perturbation

where the subscripts b, D, γ, and ν continue to refer to baryons, dark matter, photons, and neutrinos. Equation (28.47a) then becomes

$$\tilde{\Psi}' + \beta^2 \tilde{\Psi} + \tilde{\Phi} = -\frac{1}{2}\tilde{\delta}, \qquad (28.50)$$

potential evolution equation

where the prime denotes a derivative with respect to $\ln a$.

EARLY EVOLUTION

Before neutrino decoupling, when $a \lesssim 10^{-11}$, the stress-energy tensor **T** is dominated by a well-coupled, relativistic fluid consisting of radiation, neutrinos, and elementary particles with $P = \rho/3$ and $a \propto t^{1/2}$, so that $q = 1$, $\tilde{\Psi} = \tilde{\Phi}$, and $\tilde{T}^{\hat{i}\hat{i}} = \tilde{\rho}/3$. This remains a pretty good approximation until dark matter dominates the density during the plasma era when $a \sim a_{\text{eq}}$, and it can be used to bring out some important features of the general evolution. We can identify β as the ratio of the (relativistic) acoustic horizon χ_R [Eq. (28.15) and subsequent line] to the size of the perturbation, measured by $1/k$, both in comoving coordinates. Further simplification results from changing the independent variable to $\beta \propto a$ and combining Eqs. (28.47b) and (28.47d) to obtain a single second-order, homogeneous differential equation:[36]

$$\frac{d^2\tilde{\Phi}}{d\beta^2} + \frac{4}{\beta}\frac{d\tilde{\Phi}}{d\beta} + \tilde{\Phi} = 0. \qquad (28.51)$$

evolution under radiation dominance

---

35. Actually there are modes with vorticity (embodied in the perpendicular part of the velocity $\tilde{\mathbf{v}}^\perp$), and they evolve according to the relativistic generalization of the equations discussed in Sec. 14.2.1. In principle, they could have been created by some sort of primordial turbulence. However, in practice they decay quickly as the universe expands and so will be ignored.
36. This equation, like many other equations describing the evolution of perturbations, has the form of a damped simple harmonic oscillator equation. The first derivative term is then often called a "friction" term. However, this is only a mathematical analogy. Physically, it represents the loss of energy in work done on the expanding medium, not a true dissipation.

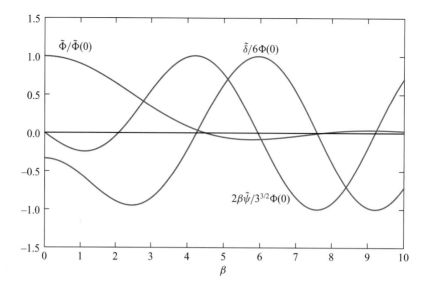

FIGURE 28.11 Early growth of a single spatial Fourier component of the perturbations. The amplitude of the potential $\tilde{\Phi}$, relative density $\tilde{\delta}$, and velocity potential $\tilde{\psi}$ perturbations are shown as functions of $\beta = k\chi_R$ for a single spatial Fourier component. The perturbations are frozen until they "enter the horizon" when $\beta \sim 1$. The perturbations then convert oscillations in which the amplitude of $\tilde{\delta}$ is constant while $\tilde{\Phi}$ is in antiphase and $\tilde{\psi}$ is in quadrature.

This has a unique solution, nonsingular as $\beta, a \to 0$ and valid for all scales:

$$\tilde{\Phi}(t, \mathbf{k}) = \frac{3\tilde{\Phi}(0, \mathbf{k})}{\beta^2} \left(\frac{\sin \beta - \beta \cos \beta}{\beta} \right) = \tilde{\Phi}(0, \mathbf{k})(1 - \beta^2/10 + \ldots). \quad (28.52)$$

The mode does not evolve significantly until it is contained by the acoustic horizon.[37]

Using Eq. (28.50), the relative density and velocity potential perturbations are then given by

$$\tilde{\delta} = -6\tilde{\Phi}(0) \left(\frac{2(\beta^2 - 1)\sin \beta - \beta(\beta^2 - 2)\cos \beta}{\beta^3} \right) = -2\tilde{\Phi}(0)(1 + 7\beta^2/10 + \ldots),$$

$$\tilde{\psi} = \frac{-3^{3/2}\tilde{\Phi}(0)}{2} \left(\frac{(\beta^2 - 2)\sin \beta + 2\beta \cos \beta}{\beta^3} \right) = \frac{-3^{1/2}\tilde{\Phi}(0)}{2}(1 - 3\beta^2/10 + \ldots)$$

$$(28.53)$$

evolution after entering horizon

(see Fig. 28.11). When $\beta \gg 1$, the wavelength is smaller than the acoustic horizon and the amplitude of the velocity perturbation is $3^{1/2}\delta/4$, in agreement with the expectation for a sound wave in a stationary, relativistic fluid. The angular frequency of the

---

37. What this really means is that we have chosen a coordinate system in which the expectation is clearly expressed that a physical perturbation not change significantly until a signal can cross it. Even in the absence of a genuine physical perturbation, we could have created the illusion of one simply by changing to a "wrinkled" set of coordinates.

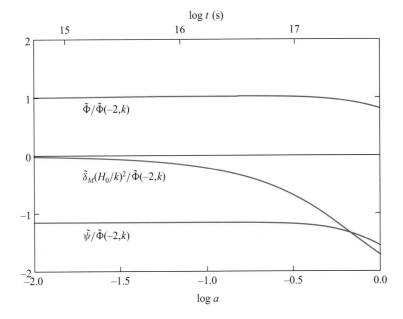

FIGURE 28.12 Evolution of the potential $\tilde{\Phi}$, the matter density perturbation $\tilde{\delta}_M$, and the velocity potential $\tilde{\psi}$ for $-2.0 \leq \log a \leq 0.0$. Note that $\tilde{\Phi}$ and $\tilde{\psi}$ only change slowly during the cosmological era when the cosmological constant is significant. By contrast, the density perturbation $\tilde{\delta}_M$ is $\propto k^2$ and grows rapidly to create the structure we observe today.

wave is $\omega = d\beta/dt = k/(3^{1/2}a)$, just what is expected for a sound wave (Sec. 16.5) in a relativistic fluid. The amplitude of the potential Φ decays as $\sim 3\tilde{\Phi}_0/\beta^2$, in accord with Poisson's equation.

The constancy of the wave amplitude $\tilde{\delta}$ turns out to be a nice illustration of adiabatic invariance (cf. Ex. 7.4). The locally measured wave energy in a single wavelength is $\propto \tilde{\delta}^2 \rho a^3$. This should scale with the wave frequency, implying that $\tilde{\delta}$ is constant (since $\omega \propto a^{-1}$).

LATE EVOLUTION

We can also describe the evolution during the gravitational and cosmological ages, when photons, neutrinos, and pressure can be ignored.[38] The space-space part of the Einstein tensor [Eq. (28.47d)] gives

$$\tilde{\Phi}'' + (3-q)\tilde{\Phi}' + (1-2q)\tilde{\Phi} = 0, \quad (28.54)$$

where $q = \frac{1}{2} - 3\rho_\Lambda/2\rho$ is initially $\sim \frac{1}{2}$. This says that the Fourier transform of the potential $\tilde{\Phi}$ is almost constant until the cosmological constant becomes important; then it decreases to ~ 0.80 times its starting value (Fig. 28.12). The density and velocity

slow potential evolution

---

38. Photons and neutrinos contribute small corrections right after recombination, and there are transients associated with the sudden decrease of the coupling of baryons to photons. These are preserved in the full solution below but can be ignored in this approximate treatment.

growth of density perturbations

potential perturbations for $\beta \gg 1$ [cf. Eqs. (28.53)] are

$$\tilde{\delta}_M \equiv \frac{\tilde{\rho}_M}{\rho_M} = \frac{\tilde{\rho}_D + \tilde{\rho}_b}{\rho_D + \rho_b} \sim -\frac{3\beta^2 \tilde{\Phi}}{1+q}; \quad \tilde{\psi}_M \sim -\frac{3^{1/2}}{1+q}(\tilde{\Phi}' + \tilde{\Phi}) \qquad (28.55)$$

for short wavelengths $\beta \gg 1$, where we have introduced the relative matter perturbation $\tilde{\delta}_M$, which grows $\propto \beta^2 \propto a$ in accord with Poisson's equation until ρ_Λ takes over and the growth rate is reduced. The velocity potential $\tilde{\psi}_M$ is that of the matter (baryons and dark matter). Note also that, although this potential is always small, the density fluctuation $\tilde{\delta}_M$ becomes nonlinear for large β. The resulting corrections must be computed numerically.

SUMMARY

We have described the early evolution of linear perturbations that were frozen until they entered the horizon and became sound waves with constant $\tilde{\delta}$ and the universe could no longer be approximated as a single relativistic fluid. We have also outlined the growth of matter perturbations when galaxies are visible—specifically when $0.1 \lesssim a < 1$. To connect these two limits, we must examine the behavior of the separate perturbations to dark matter, neutrinos, photons, and baryons.

28.5.2 Individual Constituents

DARK MATTER

dark matter density perturbation and velocity potential

Because the nongravitational interactions of dark matter, neutrinos, baryons, and photons are negligible during the epoch of galaxy formation, we handle the evolution of the different constituent perturbations by equating the 4-divergence of their individual stress-energy tensors to zero in the given spacetime.[39] Dark matter has no pressure, so the nonzero, mixed, orthonormal stress-energy tensor components are $\tilde{T}_D^{\hat{i}\hat{i}} = \rho_D \tilde{\delta}_D$, $\tilde{T}_D^{\hat{i}\hat{\parallel}} = \rho_D \tilde{v}_D = i\beta \rho_D \tilde{\psi}_D$. Setting its divergence to zero leads to two independent equations:

$$\tilde{\delta}_D' - 3\tilde{\Psi}' - 3^{1/2}\beta^2 \tilde{\psi}_D = 0; \quad \tilde{\psi}_D' + (1+q)\tilde{\psi}_D + 3^{1/2}\tilde{\Phi} = 0, \qquad (28.56)$$

where we have used the conservation law $\rho_D' + 3\rho_D = 0$ [cf. Eq. (28.19)].

Importantly for what follows, we can derive the first of Eqs. (28.56) from the flux of dark matter particles, $(\rho_D/m_D)[1 + \tilde{\delta}_D, \tilde{\mathbf{v}}_D]$ in orthonormal coordinates, setting its divergence to zero.

What is the initial dark matter density perturbation? It could have been quite independent of the perturbation to the photons, neutrinos, pairs, etc. However, the simplest assumption to make is that it just depended on local physics and that equilibrium was established on a timescale short compared with the expansion time.[40]

---

39. The derivations in this section are only sketched. Confirming them will take some work.
40. It is not necessary that the initial conditions be established simultaneously—for example, during inflation—only that they be fixed before the neutrinos start to decouple (when $a \sim 10^{-11}$) and before the pairs annihilate ($a \sim 10^{-9}$).

Equivalently, the number of photons per dark matter particle $\propto \rho_\gamma^{3/4}/\rho_D$ was a fixed number and so, using Eqs. (28.53), $\tilde{\delta}_D(0) = \frac{3}{4}\tilde{\delta}_\gamma(0) = -\frac{3}{2}\tilde{\Phi}(0)$. Using a similar argument we can deduce that $\tilde{\delta}_b(0) = \tilde{\delta}_D(0)$, $\tilde{\delta}_\nu(0) = \tilde{\delta}_\gamma(0)$. This type of perturbation is called *adiabatic* and is found to describe the observations very accurately, vindicating our trust in basic principles. The initial velocity potential perturbation, common to all constituents, is $\tilde{\psi}_D(0) = -(3^{1/2}/2)\tilde{\Phi}$, from Eqs. (28.53).

adiabatic perturbations

NEUTRINOS

Neutrinos add pressure but are effectively massless and travel at the speed of light until recent epochs, and are collisionless after decoupling. Let us throw caution to the winds and follow our treatment of a warm plasma (Sec. 22.3.5) to develop a fluid model. We set

$$\tilde{T}_\nu^{\hat{t}\hat{t}} = \rho_\nu \tilde{\delta}_\nu, \quad \tilde{T}_\nu^{\hat{t}\hat{\parallel}} = 4\rho_\nu \tilde{v}_\nu/3, \quad \tilde{T}_\nu^{\hat{\parallel}\hat{\parallel}} = \tilde{T}_\nu^{\hat{\perp}\hat{\perp}} = \rho_\nu \tilde{\delta}_\nu/3. \qquad (28.57)$$

fluid approximation

When we set the divergence of the neutrino stress-energy tensor to zero, we obtain two equations for the perturbations. These equations are the same as those that we would have gotten if we had treated neutrinos as collisional and would lead to oscillations after the mode entered the horizon. However, the neutrinos can free stream through the mode to damp δ_ν and ψ_ν. If we ignore the potential $\tilde{\Phi}$ and imagine starting with a simple sine wave, we can use Jeans' theorem to solve approximately for the time evolution of the distribution function. We find that the wave will decay in a time $\sim 2a/k$. A neutrino wave initialized in this fashion and with no other perturbations would then decay according to $\delta'_\nu \sim -\beta\delta_\nu, \delta\psi'_\nu \sim -\beta\delta\psi_\nu$. We therefore add these terms to the perturbation equations to account for free-streaming. The final results are:

$$\tilde{\delta}'_\nu - 4\tilde{\Psi}' - (4/3^{1/2})\beta^2\tilde{\psi}_\nu = -\beta\tilde{\delta}_\nu; \quad \tilde{\psi}'_\nu + q\tilde{\psi}_\nu + 3^{1/2}\tilde{\Phi} + (3^{1/2}/4)\tilde{\delta}_\nu = -\beta\tilde{\psi}_\nu.$$

$$(28.58)$$

BARYONS AND PHOTONS

Prior to recombination at the end of the plasma age, the photons and baryons were tightly coupled by Thomson scattering and behaved as a single fluid with the photons dominating the density. The baryons were therefore prevented from falling into dark-matter gravitational potential wells. However, around the time of recombination, the baryon density grew larger than the photon density, and the photon mean free paths lengthened, causing the photon-baryon fluctuations to damp through heat conduction and viscosity (Sec. 18.2). This effect is known as Silk damping (Silk, 1968). If there were no cold dark matter, structure would have been erased on small scales and we would have had to find some other explanation for galaxy formation. Instead, as baryons released themselves from photons, they fell into potential wells formed by the dark matter. It is this complex evolution that we must now try to address.

baryon-photon coupling

Silk damping

The baryons are relatively easy. They can be treated as a cold fluid, just like the dark matter (because their pressure is never significant in the linear regime), with a single fluid velocity $\tilde{\mathbf{v}}_b = i\boldsymbol{\beta}\tilde{\psi}_b$ parallel to $\mathbf{k}$ (because the ions and electrons must have net zero charge density on all scales larger than the Debye length; Sec. 20.3). Their conservation law can be obtained by setting the divergence of the flux of baryons to zero just like we did for dark matter, Eq. (28.56):

baryon density evolution

$$\tilde{\delta}_b' - 3\tilde{\Psi}' - 3^{1/2}\beta^2\tilde{\psi}_b = 0. \tag{28.59}$$

Now turn to the photons. Just as with the neutrinos, the photons contribute $\rho_\gamma \tilde{\delta}_\gamma$ to the energy density and $\frac{1}{3}\rho_\gamma\tilde{\delta}_\gamma$ to the pressure. Initially, they shared the baryon velocity and so also contributed a term $\frac{4}{3}\rho_\gamma\tilde{v}_b$ to the momentum density/energy flux. Under the diffusive approximation, their heat flux in the baryon rest frame is $-ik\rho_\gamma\tilde{\delta}_\gamma/(3n_e\sigma_T a)$, where we do not have to worry about frequency shifts of the photons. We now define a photon velocity potential, $\tilde{\psi}_\gamma$, in the frame of the FOs, by equating the photon heat flux to $(4i/3)\beta\rho_\gamma(\tilde{\psi}_\gamma - \tilde{\psi}_b)$ [cf. Eq. (28.48a)]. However, this relation breaks down when the photon mean free path approaches the wavelength of the perturbation. The heat flux will then be limited by $\sim \rho_\gamma\tilde{\delta}_\gamma/3$, and we simply modify the photon velocity potential by adding a flux-limiter:

heat flux

$$\tilde{\psi}_\gamma = \tilde{\psi}_b + 3^{1/2}\tilde{\delta}_\gamma/4(\tau_T' - 3^{1/2}\beta), \tag{28.60}$$

substituting the Thomson optical depth from Eq. (28.42). The combined baryon-photon energy flux in the FO frame is then $\tilde{T}_{b\gamma}^{\hat{i}\hat{\parallel}} = i\beta(\rho_b\tilde{\psi}_b + 4\rho_\gamma\tilde{\psi}_\gamma/3)$.

We can now take the divergence of the stress-energy tensor to obtain

$$\tilde{\delta}_\gamma' - 4\tilde{\psi}' - 4\beta^2\tilde{\psi}_\gamma/3^{1/2} = 0,$$

$$\rho_b[\tilde{\psi}_b' + (1+q)\tilde{\psi}_b + 3^{1/2}\tilde{\Phi}] + (4/3)\rho_\gamma[\tilde{\psi}_\gamma' + q\tilde{\psi}_\gamma + 3^{1/2}\tilde{\Phi} + 3^{1/2}\tilde{\delta}_\gamma/4] = 0. \tag{28.61}$$

need for kinetic treatment

As with the neutrinos, accurate calculation mandates a kinetic treatment (Ex. 28.12; Sec. 28.6.1), but this simplified treatment captures most of the kinetic results.

SUMMARY

We have now derived a complete set of linear equations describing the evolution of a single mode with wave vector **k**. These equations can be used over the observable range, $10^{-4} \lesssim k/H_0 \lesssim 0.3$, and for the whole range of evolution for $10^{-11} < a < 1$, although different terms are significant during different epochs, as we have described. We next turn to the solution of these equations.

EXERCISES

Exercise 28.13 *Challenge: Kinetic Treatment of Neutrino Perturbations*
The fluid treatment of the neutrino component would only be adequate if the neutrinos were self-collisional, which they are not.[41] The phenomenon of Landau damping (Sec. 22.3) alerts us to the need for a kinetic approach. We develop this in stages.

---

41. If we were to introduce shear stress, then viscous damping should also be included (cf. the discussion of stars in Sec. 3.7.1).

Following the discussion in Sec. 3.2.5, we introduce the neutrino distribution function $\eta_\nu(t, x^i, p_j)$, where x^i is the (contravariant) comoving (spatial) coordinate, and p_j is the (covariant) conjugate 3-momentum (Sec. 3.6, Box 3.2, and Ex. 4.1). The function η_ν satisfies the collisionless Boltzmann equation [Eq. (3.65)]:

$$\frac{\partial \eta_\nu}{\partial t} + \frac{dx^i}{dt}\frac{\partial \eta_\nu}{\partial x^i} + \frac{dp_j}{dt}\frac{\partial \eta_\nu}{\partial p_j} = 0. \tag{28.62}$$

We work with this equation to linear order.

(a) Show that the neutrino equation of motion in phase space can be written as

$$\frac{dx^i}{dt} = \frac{p_i}{a(t)(p_k p_k)^{1/2}}(1 + \Phi + \Psi); \quad \frac{dp_j}{dt} = -\frac{(p_k p_k)^{1/2}}{a(t)}\frac{\partial(\Phi + \Psi)}{\partial x^j}, \tag{28.63}$$

and explain why it is necessary to express the right-hand sides in terms of t, x^i, and p_i.

(b) Interpret the momentum equation in terms of the expansion of the universe (Sec. 28.2.3) and gravitational lensing (Sec. 7.6.1).

(c) Introduce locally orthonormal coordinates in the rest frame of the FOs, and define $p^{\hat{\alpha}} = \{\mathcal{E}, p^{\hat{1}}, p^{\hat{2}}, p^{\hat{3}}\}$. Carefully interpret the density, velocity, and pressure in this frame, remembering that it is only necessary to work to linear order in Φ.

(d) Multiply the Boltzmann equation successively by 1, $\mathcal{E}$, and $p^{\hat{i}}$, and integrate over the momentum space volume element $dp^{\hat{1}} dp^{\hat{2}} dp^{\hat{3}}$ to show that

$$\tilde{\delta}'_{n_\nu} - 3\tilde{\Psi}' + 3^{1/2} i\boldsymbol{\beta} \cdot \tilde{\mathbf{S}}_\nu = 0; \quad \tilde{\delta}'_\nu - 4\tilde{\Psi}' + \frac{4i\beta}{3^{1/2}}\tilde{v}_\nu = 0;$$

$$i\tilde{v}'_\nu - 3^{1/2}\beta\left(\frac{\tilde{\delta}_\nu}{4} + \tilde{\Phi} + \tilde{\Psi}\right), \tag{28.64}$$

where $\tilde{\delta}_{n_\nu}$ is the fractional fluctuation in neutrino number density, and $\tilde{\mathbf{S}}_\nu$ is the associated number flux. In deriving these equations, it is necessary to impose the same closure relation (cf. Sec. 22.2.2) as was used to derive the fluid equations (28.57).[42] These kinetic equations have the same form as the fluid equations, although the coefficients are different and would change again if we changed the closure relation. This demonstrates that fluid equations can only be approximate, even when derived using the Boltzmann equation.

(e) One standard way to handle the neutrino perturbations accurately is to expand the distribution function in spherical harmonics. Outline how you would carry this out in practice, and how you would then use the more accurate neutrino distribution to improve the evolution equation for the dark-matter, baryon, and photon components.

---

42. Thus the trace of the stress-energy tensor vanishes in all Lorentz frames.

Exercise 28.14 *Problem: Neutrino Mass*

Neutrinos have mass, which becomes measurable at late times through its influence on the growth of structure.

(a) Explain how the expansion of the universe is changed if there is a single dominant neutrino species of mass 100 meV.

(b) Modify the equations for neutrino phase-space trajectories described in the preceding problem to allow for neutrino rest mass, and outline how this will affect the growth of perturbations.

(c) Describe how you could, in principle, measure the individual neutrino masses using cosmological observations. (In practice, this would be extremely difficult.)

28.5.3 Solution of the Perturbation Equations

GROWTH VECTOR

If we ignore shear stress and equate Ψ to Φ [as discussed after Eq. (28.47d)], we have eight first-order, coupled, linear differential equations [Eqs. (28.50), (28.56), (28.58), (28.59), and (28.61)] plus an algebraic equation (28.60), describing the evolution of small perturbations in the presence of a single wave mode. A simple reorganization gives

$$\tilde{\Upsilon}' = \mathcal{M}\tilde{\Upsilon}, \tag{28.65}$$

where $\tilde{\Upsilon} = \{\tilde{\Phi}, \tilde{\delta}_D, \tilde{\psi}_D, \tilde{\delta}_\nu, \tilde{\psi}_\nu, \tilde{\delta}_b, \tilde{\Psi}_b, \tilde{\delta}_\gamma, \tilde{\psi}_\gamma\}$ is dimensionless, and the elements of the matrix $\mathcal{M}$ are all real functions of a that we have already calculated and that describe the unperturbed universe. The solution of Eq. (28.65) can be written as $\tilde{\Upsilon} = \Gamma(a, \mathbf{k})\tilde{\Phi}(0, \mathbf{k})$, where Γ is the *growth vector* for adiabatic perturbations, initialized by

$$\Gamma(0, k) = \left\{1, -\frac{3}{2}, -\frac{3^{1/2}}{2}, -2, -\frac{3^{1/2}}{2}, -\frac{3}{2}, -\frac{3^{1/2}}{2}, -2, -\frac{3^{1/2}}{2}\right\}. \tag{28.66}$$

[Note that we here switch from the scaled wave vector $\boldsymbol{\beta}$, Eq. (28.48b), to its unscaled, comoving form $\mathbf{k}$, as that is what we will need in using the growth vector.]

In the above equations and analysis, we have neither allowed for uncertainty in the governing parameters nor the addition of speculative physical processes (Ex. 28.26). We have also ignored nonlinear corrections (cf. Sec. 23.2, Ex. 28.15). Nonetheless, in their domain of applicability, these equations allow us to exhibit much that is observed and measured (Fig. 28.13).

POTENTIAL POWER SPECTRUM

What determined the initial amplitudes $\tilde{\Phi}(0, \mathbf{k})$? It turns out that a simple early conjecture[43] accounts very well for a large number of independent cosmological mea-

---

43. First proposed by Harrison (1970), who considered the mode amplitudes when they entered the horizon, not when they were initialized much earlier, as we shall do. The idea was put on a more general footing by Zel'dovich (1972); see also Peebles and Yu (1970).

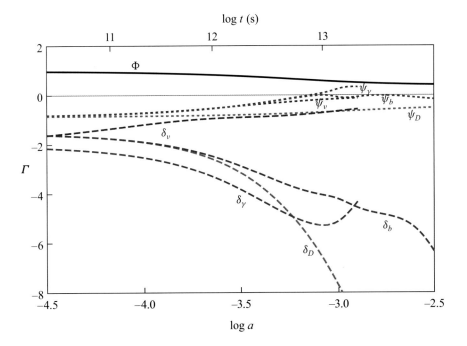

FIGURE 28.13 Variation of the growth vector Γ for one choice of wave number $k_{eq} = 78H_0$, which enters the horizon, $\beta = 1$, when the universe becomes matter dominated ($a = a_{eq} = 0.00030$; $\log a_{eq} = -3.52$). The solid black line is the potential perturbation; the dashed lines are the density perturbations, δ; the dotted lines are the velocity potential perturbations. The potential decreases slowly as the universe expands through the radiation, plasma, and atomic ages. The dark matter density perturbation starts to grow $\propto a$ after it enters the horizon. By contrast, the neutrino and photon perturbations, being hot, do not fall into the dark matter potential wells. The baryon perturbations are initially coupled to the photon perturbations, as can be seen by their common velocity potentials, but after recombination, they are released to fall into the dark matter potential wells. The model for the neutrino and photon perturbations is decreasingly realistic after recombination but irrelevant for our purpose here, and so is not shown.

surements. The conjecture is that the initial metric perturbations were *scale invariant and isotropic*. To explain what this means, adopt the formalism developed in Sec. 6.4, and imagine an ensemble of universes[44] defining a dimensionless *power spectrum* of potential fluctuations $P_\Phi(a, k)$. Specifically, we define [cf. Eq. (6.31)]

$$\langle \tilde{\Phi}(a, \mathbf{k}) \tilde{\Phi}^*(a, \mathbf{k}') \rangle = (2\pi)^3 P_\Phi(a, k) \delta(\mathbf{k} - \mathbf{k}') = (2\pi)^3 \Gamma_1(a, k)^2 P_\Phi(0, k) \delta(\mathbf{k} - \mathbf{k}'). \tag{28.67}$$

Henceforth, we use $\langle \cdot \rangle$ to denote an ensemble average; all other averages that can be computed by integrating known functions over position, angle, frequency, etc. will

ensemble average

---

44. Of course, we only have one universe to observe, but when we study many small regions, the average properties are well defined. By contrast, when we examine large regions, the *cosmic variance* is also large, and no matter how precisely we make our measurements, there is a limit to how much we can learn about the statistical properties of the ensemble.

be denoted by an overbar. Scale invariance is the assertion that $P_\Phi(0, k) \propto k^{-3}$.[45] Equivalently, the primordial contribution to $\langle \Phi^2 \rangle$ from each octave of k is a constant:

$$\langle \Phi^2 \rangle_k \equiv \left(\frac{k}{2\pi}\right)^3 P_\Phi(k) \equiv \mathcal{Q} = \text{const.} \tag{28.68}$$

initial cosmic noise

Observations of the CMB and galaxies imply that the dimensionless quantity $\mathcal{Q}$, the *initial cosmic noise*, is 1.8×10^{-10}.[46]

In general, a power spectrum does not capture all possible statistical properties of noise. However, in this case a stronger statement was conjectured—that the initial amplitudes of individual Fourier modes had a Gaussian distribution with zero mean, constant variance, and random phases (implying no covariance and no need for any other independent statistical measures beyond $\mathcal{Q}$; cf. Secs 6.3.2 and 6.3.3).

Gaussian fluctuation spectrum

We can now employ the Wiener-Khintchine theorem (Sec. 6.4) to obtain a symmetric *correlation matrix*:

correlation matrix

$$\mathcal{C}_{ij}(\mathbf{s}, a) \equiv \langle \Upsilon_i(a, \boldsymbol{\chi} + \mathbf{s}) \Upsilon_j(a, \boldsymbol{\chi}) \rangle = \mathcal{Q} \int \frac{d^3k}{4\pi k^3} e^{i\mathbf{k}\cdot\mathbf{s}} \Gamma_i(a, \mathbf{k}) \Gamma_j(a, \mathbf{k}) \tag{28.69}$$

(cf. Ex. 9.8). Many entries in this matrix have been verified observationally, thereby validating the remarkably simple physical model that we have outlined. The birth of the universe was accompanied by a hum, not a fanfare!

SUMMARY

We have set up a general formalism for describing, approximately and linearly, the growth of perturbations under general relativity in the expanding universe all the way from neutrino decoupling to the present day. The principal output is the evolution of the potential functions Ψ, Φ and the accompanying density perturbations through recombination and after reionization. This provides a basis for a more careful treatment of the photons, to which we turn in Sec. 28.6, and the observed clustering of galaxies, which we now address.

28.5.4 Galaxies

SURVEYS

observations of galaxies

Much of what we have learned about cosmology has come from systematic surveys of distant galaxies over large areas of sky. As we have emphasized (cf. Sec. 28.2.1), galaxies are not well standardized; they are more like people than elementary particles! However, it is possible to average over this diversity to study their clustering. To date,

---

45. This spectrum may be thought of as the (3-dimensional) spatial generalization of flicker or "$1/f$" noise that is commonly measured in time series, such as music (Sec. 6.6.1). The common property is that there are no characteristic spectral features, and the power diverges logarithmically at both small and large scales. The measured spectrum has slightly more power at small k, as we discuss in Sec. 28.7.1.

46. For reasons that make perfect sense to astronomers, the conventional normalization is expressed as the rms relative density fluctuation in a sphere of radius 11 Mpc, assuming linear evolution of the perturbations. This quantity, known as σ_8, has the value ~ 0.82.

roughly a billion galaxies (out of the roughly one trillion that are observable) have had their positions, shapes, and fluxes measured in a few spectral bands. Over a million of these have had their spectra taken, so their distances can be determined using the Hubble law.

GALAXY POWER SPECTRUM

The power spectrum of density perturbations in the gravitational and cosmological ages, $P_M(a, k)$, defined by $\langle \tilde{\delta}_M(a, \mathbf{k}) \tilde{\delta}_M^*(a, \mathbf{k}') \rangle = (2\pi)^3 P_M(a, k) \delta(\mathbf{k} - \mathbf{k}')$, can be simply related to the power spectrum for the gravitational perturbations $P_\Phi(a, k)$ [the evolved Eq. (28.68)] using Eqs. (28.55). If we also assume that, despite the different evolution of baryons and dark matter, the space density of galaxies is directly proportional to their combined density,[47] then we can use galaxy counts to measure the total matter power spectrum. The computed power spectrum is exhibited in Fig. 28.14, and its high-k region is explored in Ex. 6.7. Note that after recombination, the small-scale (large k) structure was relatively more important than the larger-scale structure in comparison with today. In other words, smaller *groups* of protogalaxies merged to create larger groups as the universe expanded. This principle is also apparent before reionization, when smaller, pregalactic dark matter *halos* merged to form larger halos, which eventually made observable galaxies when the gas was able to cool and form luminous stars.

merging of galaxies and groups

To describe merging requires that we handle the perturbations nonlinearly. Mild nonlinearity can be handled by a variety of analytical techniques, but the best approach is to assume that the dark matter is collisionless and to perform N-body numerical simulations, which can now (2016) follow over a trillion test particles. These calculations can then be supplemented with prescriptions for handling the nongravitational behavior of the baryons on the smallest length scales as they cool to form stars and massive black holes. This happened most vigorously when $a \sim 0.3$. The current incidence of small structure appears to be significantly less than expected. This may be a result of nongravitational effects, or it could signify a high-k cutoff in the initial potential power spectrum [Eq. (28.68)].

nonlinearity of evolution

BARYON ACOUSTIC OSCILLATIONS

The perturbations at recombination are basically sound waves whose amplitude depends on the phase of the oscillation measured since the time when the wave mode entered the horizon. As a result, at recombination (380 kyr), the amplitude oscillates with the (comoving) wavelength. These oscillations are observed directly in angular fluctuations of the CMB (see Fig. 28.15), where they are known as *acoustic peaks*. Baryons were released from the grip of photons during recombination and fell into dark matter potential wells. There should thus be preferred scales imprinted on the

---

47. This turns out to be a better approximation than might be imagined, but it fails on small scales, where cooling and stellar activity become more important than gravity in determining what we see. These effects are addressed by attempting to compute *bias* factors.

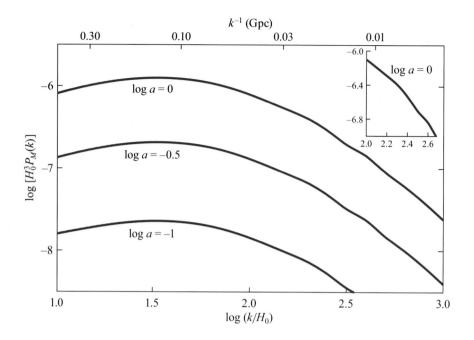

FIGURE 28.14 Matter fluctuation power spectrum $P_M(k)$ for $\log a = -1$, -0.5, and 0, corresponding to recombination, reionization, the epoch of maximal galaxy and star formation, and today, respectively. The spectrum demonstrates clearly that structure grows first on small scales, and it provides a fair description of the results of large-scale galaxy surveys. Note the wiggles in the power spectrum for $2 < \log(k/H_0) < 2.7$, shown more clearly in the inset for $\log a = 0$. These are baryon acoustic oscillations—echoes of the acoustic oscillations observed using CMB measurements at recombination.

distribution of galaxies we see around us today. These *baryon acoustic oscillations* are very important, because they allow astronomers to follow the expansion of a comoving ruler over time—in other words, to measure $a(t)$. As we shall see, the calibration length at recombination is very well determined by observations of the radiation (Sec. 28.6.1). Baryon acoustic oscillations can be seen in the angular correlation functions measured in large galaxy surveys and also in studies of the radial velocities of these galaxies.

EXERCISES

Exercise 28.15 *Example: Nonlinearity*
Explore nonlinear effects in the growth of perturbations in the gravitational age—when radiation and the cosmological constant can be ignored—by considering the evolution of a sphere in which the matter density is uniform and exceeds the external density by a small quantity.

(a) Use the Friedmann equations (28.16), (28.18), and (28.19) to show that the sphere behaves like a universe with density greater than the critical density and stops expanding when its density exceeds the external value by a factor of $9\pi^2/16$.

(b) Assume that the perturbation remains strictly spherical, and determine by what additional scale factor the external universe will have expanded when the perturbation collapses to a point.

(c) Argue that realistic perturbations behave quite differently, that non-spherical perturbations grow during the collapse, and that the infall kinetic energy effectively randomizes during the collapse. Show that the collapse stops when the radius of the sphere is roughly half its maximum value and that this occurs when the average density exceeds that in the still-expanding external universe by a factor of ~ 150.

28.6 Cosmological Optics

28.6.1 Cosmic Microwave Background

OVERVIEW

So far, we have emphasized the dynamical effects that govern the evolution of small perturbations in the expanding universe and have shown how these lead to a statistical description of the potential, density, and velocity perturbations. We now consider the effect of these perturbations on extragalactic observations where the radiation propagates passively through them. We start with the CMB. As we have explained in Sec. 28.4.4, most of the action happens at recombination—over an interval of time short compared with the age of the universe—when free electrons are rapidly captured and retained by protons and the rate of Thomson scattering plummets. We need to describe photons as they transition from belonging to a perfect fluid to uninterrupted, free propagation along null geodesics. We then discuss the statistical properties of the relative temperature and polarization fluctuations.

radiative transfer

MONTE CARLO RADIATIVE TRANSFER

The standard way to compute the radiative transfer is to generalize the moments of the Boltzmann equation (28.64) to include baryon motion and the potentials $\tilde{\Phi}$ and $\tilde{\Psi}$. Hundreds of spherical harmonics are necessary to achieve the requisite accuracy. As we have already discussed many calculations of this general character in the preceding chapters, we shall elucidate the underlying physical processes by using a Monte Carlo description (cf. Sec. 5.8.4). Monte Carlo methods are often used for problems that are too complex for a Boltzmann approach.

To do this, we first ignore the perturbations and follow backward in time a photon observed by us, today (at time t_0), with initial direction $\mathbf{n}$ and polarization (electric field unit vector) $\hat{\mathbf{E}}$. (The backward transition probabilities are just the same as for the actual, forward path.) We then assign the Thomson optical depth to the first scattering (going backward) according to $\tau_T = -\ln R_1$, where R_1 is a random number distributed uniformly in [0, 1]. Using the discussion in Sec. 28.4.4, we associate this optical depth with the location of the scatterer in spacetime, which is

Thomson optical depth

T2

scattering probability

specified by t_1 and $\chi_1 = \mathbf{n} \int_{t_1}^{t_0} dt/a$. The differential cross section for electron scattering into direction $\mathbf{n}_1$ is $d\sigma/d\Omega = r_e^2(1 - \mu_s^2)$, where r_e is the classical electron radius, and $\mu_s = \hat{\mathbf{E}} \cdot \mathbf{n}_1$ (e.g., Jackson, 1999).[48] The cumulative probability distribution for μ_s is $(3\mu_s/4 - \mu_s^3/4 + 1/2)$, and so we equate this to another uniformly distributed random number R_2 and solve for μ_s. The scattered photon's azimuth is likewise assigned as $2\pi R_3$, with R_3 a third random number uniform on [0, 1], and

polarization

the new polarization vector $\hat{\mathbf{E}}_1$ is along the direction of $(\hat{\mathbf{E}} \times \mathbf{n}_1) \times \mathbf{n}_1$. We iterate and trace the photon path backward until the scatterings were so frequent that the evolution of the radiation can be treated as adiabatic. This happened at time t_{ad} and location χ_{ad}.

Having determined the path, we consider a photon traveling forward along it. According to Eq. (28.63) with $\Phi = \Psi = 0$, the covariant momentum p_i is constant between scatterings, and so the photon's frequency is $\nu \propto a^{-1}$. This frequency is unchanged by scattering, as the electrons are assumed to be at rest with respect to the FOs. When we repeat this exercise many times, we find no net polarization to statistical accuracy, as we must.

We now switch on a single perturbation with wave vector $\mathbf{k}$. [It is helpful to express the perturbation's Ψ and Φ as standing waves rather than running waves; cf. the passage following Eq. (28.46).] And we make the approximation that $\Psi = \Phi$ [i.e., we neglect gravitational effects of anisotropies in the free-streaming photon and neutrino stresses; see passage following Eq. (28.47d).] We need to calculate the linear relative frequency shift $\delta_\nu \equiv \delta\nu/\nu$ induced by the perturbation for a photon propagating forward in time along the above path, superposing four effects. The first

starting frequency shift

is that the perturbation mode changes the *initial* frequency through the radiation density perturbation, the Doppler shift associated with the baryon velocity, and the gravitational frequency shift (Sec. 25.9). Specifically, $\delta_\nu = \frac{1}{4}\tilde{\delta}_\gamma(t_{\text{ad}}) - \mathbf{n}_{\text{ad}} \cdot \tilde{\mathbf{v}}_b(t_{\text{ad}}) + \tilde{\Phi}(t_{\text{ad}})$. The second effect is more subtle. The proper time (age of the universe)

starting time shift

at the start of the path differs from that of an FO in a homogeneous universe by $\delta\tilde{t} = \int_0^{t_{\text{ad}}} dt\,\tilde{\Phi}(t)$ (cf. Eq. 28.44). Since the universe is expanding according to local laws, the local scale factor (as measured, for example, by the local temperature) is modified by $\delta\tilde{a}/a = H\delta\tilde{t}$; and correspondingly, the frequency of the photon moving along the above Monte Carlo path is modified by $\delta_\nu = -H(t_{\text{ad}}) \int_0^{t_{\text{ad}}} dt\,\tilde{\Phi}(t)$.[49]

The third effect is related to the second one. The scattering rate is $n_e\sigma(1 - \mathbf{n}_i \cdot \tilde{\mathbf{v}}_b)$, and its perturbation leads to a change in the propagation time and consequently to

---

48. The Compton recoil is ignorable, and so the Thomson cross section suffices.
49. This is known as the Sachs-Wolfe effect (Sachs and Wolfe, 1967). The largest influence on the photon frequency comes from long-wavelength (low-k) gravitational perturbations, for which $\tilde{\Phi}$—which arises from dark matter—is nearly time independent during $0 < t < t_{\text{ad}}$. Combining this with the value $H(t_{\text{ad}}) = 2/(3t_{\text{ad}})$ in the plasma age, we obtain for the second effect $\delta_\nu = -\frac{2}{3}\tilde{\Phi}(t_{\text{ad}})$. And adding this to our first effect's gravitational frequency shift $\delta_\nu = +\tilde{\Phi}(t_{\text{ad}})$, we obtain a combined direct gravitational frequency shift $\delta_\nu = +\frac{1}{3}\tilde{\Phi}(t_{\text{ad}})$.

the time and frequency at the start of the path. The baryon density perturbations $\tilde{\delta}_b$ induce relative electron density perturbations $\tilde{\delta}_e$. They can be estimated using

$$\tilde{\delta}_e = \left(\frac{\partial \ln(n_p x)}{\partial \ln n_b}\right)\tilde{\delta}_b + \frac{1}{4}\left(\frac{\partial \ln(n_p x)}{\partial \ln T_\gamma}\right)\tilde{\delta}_\gamma, \qquad (28.70) \qquad \text{ionization shift}$$

where x is the ionization fraction (Sec. 28.4.4), and the partial derivatives are computed at the recombination surface using the formalism of Sec. 28.4.4. (A more careful treatment includes many more atomic processes.) When the scattering rate increased and the total duration of the path decreased, the universe became colder, which contributes a negative frequency shift $\delta_\nu = -\int_{t_{\text{ad}}}^{t_0} dt\, H\tau_T'[\tilde{\delta}_e + \mathbf{n}(t)\cdot\tilde{\mathbf{v}}_b]/\tau_T'(t_{\text{ad}})$. The fourth and final effect is the Doppler shift applied at each scattering: $\delta_\nu = \sum_i (\mathbf{n}_i - \mathbf{n}_{i-1})\cdot\tilde{\mathbf{v}}(t_i)$. $\qquad$ Doppler shift

Now, the point of the Boltzmann equation is that the photon distribution function $\eta_\gamma = \{\exp[h\nu/(k_B T_\gamma)] - 1\}^{-1}$ is conserved along a trajectory in phase space (Sec. 3.6). Furthermore, it is not changed by scattering in the electron rest frame or by Lorentz transformation into and out of this frame. Therefore, the relative temperature fluctuation, which is what is actually measured, satisfies $\delta_{T_\gamma} \equiv \delta_T = \langle\delta_\nu\rangle$, averaging over the sum of the four contributions. $\qquad$ temperature fluctuation

We can also consider the effect on the polarization. The natural basis for the electric vector is $\mathbf{e}_a = \mathbf{k}\times\mathbf{n}/|\mathbf{k}\times\mathbf{n}|$ and $\mathbf{e}_b = \mathbf{n}\times\mathbf{e}_a$, and we expect any measured polarization to be perpendicular or parallel to the projection of $\mathbf{k}$ on the sky. (See Ex. 28.16.) However, $\mathbf{v}_b$ is along $\mathbf{k}$, and transforming into and out of the electron rest frame does not rotate the polarization vector in the $\{\mathbf{e}_a, \mathbf{e}_b\}$ basis. And the influence of gravitational deflections on the polarization is also negligible. $\qquad$ polarization

SPHERICAL HARMONIC EXPANSION

The Monte-Carlo calculation that we have just outlined allows us to compute the expected temperature fluctuation and write it in the form

$$\tilde{\delta}_T(\mathbf{n}, \mathbf{k}) = \mathcal{T}_k(\mathbf{n}\cdot\hat{\mathbf{k}})\tilde{\Phi}(0, \mathbf{k}) \qquad (28.71)$$

where $\mathcal{T}_k$ is the *initial potential-temperature transfer function*. As $\mathcal{T}_k$ is defined on a sphere, it is natural to expand it in Legendre polynomials, the functional equivalent of a Fourier series:

$$\mathcal{T}_k(\mathbf{n}\cdot\hat{\mathbf{k}}) = \sum_{l=0}^{\infty} \mathcal{T}_{kl} P_l(\mathbf{n}\cdot\hat{\mathbf{k}}), \quad \text{where} \quad \mathcal{T}_{kl} = \frac{(2l+1)}{2}\int_{-1}^{1} d(\mathbf{n}\cdot\hat{\mathbf{k}})\mathcal{T}_k P_l(\mathbf{n}\cdot\hat{\mathbf{k}}).$$

$$(28.72)$$

We are interested in the cross correlation of the temperature fluctuations $\langle\delta_T(\mathbf{n})\delta_T(\mathbf{n}')\rangle$, where the average is over all directions $\mathbf{n}, \mathbf{n}'$ separated by a fixed angle and we take the ensemble average over perturbations using Eqs. (28.67), (28.68). This must depend only on that angle and can therefore be expanded as another sum over Legendre polynomials:[50]

expansion in Legendre polynomials

$$\langle \delta_T(\mathbf{n})\delta_T(\mathbf{n}')\rangle = \int \frac{d^3k}{(2\pi)^3} P_\Phi(0,k) \sum_{l,\,l'} \mathcal{T}_{kl}\mathcal{T}^*_{kl'} \langle P_l(\hat{\mathbf{k}}\cdot\mathbf{n}) P_{l'}(\hat{\mathbf{k}}\cdot\mathbf{n}')\rangle$$

$$= \int \frac{d^3k}{(2\pi)^3} P_\Phi(0,k) \sum_{l=0}^{\infty} |\mathcal{T}_{kl}|^2 \frac{P_l(\mathbf{n}\cdot\mathbf{n}')}{2l+1}. \qquad (28.73)$$

This is the functional equivalent of the Wiener–Khintchine theorem (Sec. 6.4.4). It is conventional to express $\langle \delta_T(\mathbf{n})\delta_T(\mathbf{n}')\rangle$ in terms of the total *multipole coefficient* C_l:

$$\langle \delta_T(\mathbf{n})\delta_T(\mathbf{n}')\rangle = \sum_{l=0}^{\infty} \frac{2l+1}{4\pi\,T_{\gamma o}^2} C_l\, P_l(\mathbf{n}\cdot\mathbf{n}') \;; \qquad (28.74a)$$

$$C_l = \frac{4\pi T_{\gamma 0}^2}{(2l+1)^2} \int \frac{d^3k}{(2\pi)^3} P_\Phi(0,k) |\mathcal{T}_{lk}|^2$$

$$= \frac{16\pi^2\, \mathcal{Q}\, T_{\gamma 0}^2}{(2l+1)^2} \int_0^\infty d\ln k\, |\mathcal{T}_{lk}|^2 , \qquad (28.74b)$$

where we have used the scale-invariant form (28.68) of the initial perturbation spectrum $P_\Phi(0,k)$.

REIONIZATION SCATTERING

influence of intervening electrons

The rapid increase in electron density and Thomson scattering following reionization at the end of the atomic age (see Fig. 28.8) can actually be detected in the observations of the CMB and is described by essentially the same equations that we used for recombination. The Thomson optical depth, backward in time from us through reionization (as defined in Sec. 28.4.4) averages to $\tau_T \sim 0.066$.

INTEGRATED SACHS-WOLFE EFFECT

influence of intervening structure

Another late-time effect is better described in configuration space. Consider a photon crossing a large negative gravitational potential well, associated with an excess of matter during the cosmological age. The cosmological constant causes the potential perturbation to decrease [cf. Eq. (28.54)] while the photon crosses it and so the photon loses less energy climbing out of it than it gained by falling into it, so there is a net positive temperature fluctuation. This is most apparent at long wavelengths and is most easily detected by cross-correlating the matter distribution with the temperature fluctuations.

GRAVITATIONAL LENSING

As we describe in more detail in Sec. 28.6.2, the gravitational deflection of rays crossing the universe leads to the distortion of the images of background sources. Of

---

50. To verify this identity, consider the special case $\mathbf{n} = \mathbf{n}'$.

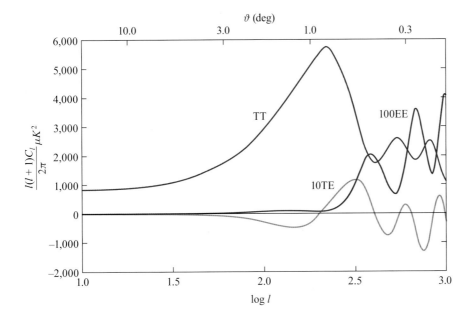

FIGURE 28.15 Theoretical spectra for anisotropy of the microwave background fluctuations as measured by the coefficients C_l, where l is the spherical harmonic quantum number as defined in Eqs. (28.74). The curve labeled TT shows the C_ls for the temperature fluctuations [Eqs. (28.74); multiplied by $l(l + 1)/(2\pi)$]; that labeled EE shows the C_ls for the E-mode polarization fluctuations; and that labeled TE shows the C_ls for the temperature-polarization cross correlation. These curves are adapted from theoretical calculations by Planck Collaboration (2016b) for parameter values that best fit the observations. The low-l portion of the TT curve can be reproduced with the formalism presented in this section. The fluctuations' angular scale on the top axis is $\vartheta = 180°/l$. Note the prominence of the first "acoustic peak" at $l \sim 200$, which corresponds to waves that have reached maximum amplitude at recombination. The large-angle fluctuations with $l \lesssim 70$ are basically gravitational redshifts associated with perturbations that have not yet entered the horizon at the time of recombination. Modes with $l \gtrsim 300$ are dominated by density changes. Velocity effects contribute heavily to intermediate l harmonics. Ten of the predicted acoustic peaks have been measured in the TT spectrum out to $l \sim 3,000$. The observations agree extremely well with these predictions after correcting for some additional effects listed in the text.

course this makes no difference to the appearance of a uniform background radiation. However, it will change the fluctuation spectrum. The formalism used is an adaption of that developed below. This has turned out to be a powerful probe of intervening structure and, consequently, an important consistency check on the standard model.

TEMPERATURE FLUCTUATION SPECTRUM

The theoretical temperature fluctuation spectrum computed using more detailed calculations than ours (Planck Collaboration, 2016b) is shown in Fig. 28.15. It fits the observational data extremely well with only six adjustable parameters. Our Monte Carlo results (not shown) roughly recover this spectrum for low l. The detailed fit of the observations to the theory is responsible for many of the features of the standard model described earlier in this chapter.

the spectrum of temperature fluctuations

POLARIZATION

A very important recent development has been the calculation and measurement of polarization in the CMB. The calculation outlined above predicts that a single wave perturbation will produce roughly 10% polarization along the projected direction of **k** on the sky and also predicts that the polarization observed along neighboring directions will be correlated (Ex. 28.22). When we sum over all modes, expand in spherical harmonics, and average over the sky, there is a net polarization signal of a few percent and a cross correlation with the temperature spectrum (Fig. 28.15). Measurements of these effects are used to refine the standard model. A simple generalization of Eq. (28.74b) gives the multipole coefficients for the polarization as well as the cross correlation with the temperature fluctuations (Fig. 28.15).

Because this polarization arises from photon scattering in the presence of density fluctuations, it turns out to have a pattern described by tensor spherical harmonics that are double gradients (on the sky) of scalar spherical harmonics; these are sometimes called "electric-type" spherical harmonics, and polarization patterns constructed from them are called *E-modes*. Primordial gravitational waves, interacting with the plasma during recombination, can catalyze a second type of polarization pattern called *B-modes*, whose "magnetic-type" spherical harmonics are constructed by operating on a scalar spherical harmonic with one gradient ∇ and one angular momentum operator $\mathbf{L} = \mathbf{e}_{\hat{r}} \times \nabla$. For some details of E-modes, B-modes, and other aspects of the predicted polarization, see Ex. 28.22.

RADIATION STRESS-ENERGY TENSOR

The least satisfactory aspect of our treatment of the growth of perturbations is the approximation of the radiation as a perfect fluid. Our treatment of radiative transfer allows one to refine this approximation by including in Eq. (28.61) an estimate of the anisotropic part of the photon stress-energy tensor.

EXERCISES

Exercise 28.16 *Example: Stokes' Parameters*

There are many ways to represent the polarization of electromagnetic radiation (Sec. 7.7). A convenient one that is used in the description of CMB fluctuations was introduced by Stokes.

(a) Consider a monochromatic wave propagating along $\mathbf{e}_z$ with electric vector $\mathbf{E} = \{E_x, E_y\}e^{i\omega t}$, where the components are complex numbers. Explain why this wave is completely polarized, and introduce the *Stokes' parameters* (following Jackson, 1999) $I = E_x E_x^* + E_y E_y^*$, $Q = E_x E_x^* - E_y E_y^*$, $U = 2\Re(E_x E_y^*)$, and $V = -2\Im(E_x E_y^*)$. Sketch the behavior of the electric vector as the complex ratio $r = E_y/E_x$ is varied, and hence associate Q, U, and V with different states of polarization.

(b) Derive the transformation laws for the Stokes' parameters if we rotate the $\mathbf{e}_x$, $\mathbf{e}_y$ directions about $\mathbf{e}_z$ through an angle ψ.

(c) Show that $Q^2 + U^2 + V^2 = I^2$ and that the polarization of the wave may be represented as a point on a sphere, which you should identify.

(d) Now suppose that the wave is polychromatic and partially polarized, and replace the definitions of I, Q, U, and V with the time averages $\langle E_x E_x^* \rangle$, and so forth. Show that $Q^2 + U^2 + V^2 < I^2$, and give expressions for the degree of polarization and the associated position angle in terms of I, Q, U, and V.

Exercise 28.17 *Problem: Cosmic Variance* T2

The precision with which the low-l spherical harmonic power spectrum can be determined observationally is limited because of the low number of independent measurements that can be averaged over. Give an approximate expression for the *cosmic variance* that should be associated with the CMB fluctuation spectrum.

Exercise 28.18 *Problem: Acoustic Peaks* T2

We have explained how the peaks in the CMB temperature fluctuation spectrum arise because the sound waves all began at the same time and are all effectively observed at the same time, while they entered the horizon at different times. Suppose that the universe had been radiation dominated up to recombination, so that Eq. (28.51) is valid and oscillatory waves of constant amplitude were created with different values of k. Calculate the total relative density perturbation $\tilde{\delta}(k)$ at recombination. Describe the main changes in standard cosmology that are introduced.

Exercise 28.19 *Example: Cosmic Dawn* T2

The cosmic dawn that preceded the epoch of reionization can be probed by low-frequency CMB observations using a special radio hyperfine line emitted and absorbed by hydrogen atoms. This line is associated with a flip in direction of the magnetic dipole associated with the central proton relative to the magnetic field created by the orbiting electron. The line's frequency and strength can be calculated using quantum mechanics. For our purposes, all that we need to know is that the frequency associated with the transition is $\nu_H = 1.42$ GHz, the degeneracy of the ground/upper state is 1/3, and the rate of spontaneous transition is $A = 3 \times 10^{-15}$ s^{-1} (Ex. 28.9).

(a) Explain why these hyperfine transitions should produce a change in the measured CMB spectrum over a range of frequencies $\sim$20–150 MHz, where the lower limit is due to the practicality of making the measurement.

(b) Consider hydrogen atoms with total number density n_1 in the ground state and $n_2 = 3n_1 \exp(-h\nu_H/k_B T_S)$ in the excited state, where T_S defines the *spin temperature*, and we measure frequencies and rates locally. Show that the net creation rate of photons per unit volume and frequency can be written as $3n_1 A\delta(\nu - \nu_H)(1 - T_\gamma/T_S)$. [Hint: $T_S, T_\gamma \gg h\nu_H/k_B$.]

(c) Hence, show that the CMB temperature fluctuation at frequency $\nu_0 = \nu_H/a$, produced when the expansion factor was a, is

$$\delta_T = \left(\frac{3}{32\pi}\right)\left(\frac{n_H}{\nu_H^3}\right)\left(\frac{A}{H(a)}\right)[1 - T_{\gamma 0}/T_s(a)],$$

where n_H is the atomic hydrogen density, $H(a)$ is the expansion rate, and $T_{\gamma 0}$ is the CMB temperature today while T_γ is the CMB temperature at the point of emission.

(d) It is predicted that $T_S \sim 10$ K when $a \sim 0.05$. Estimate the associated temperature perturbation, δ_T.

The spin temperature T_s will follow the nonrelativistic gas and fall faster than the radiation temperature T_γ (Sec. 28.2.3), creating absorption above $\sim$10 MHz. The first stars will heat the gas and create Lyman alpha photons (Sec. 28.4.4), which end up populating the upper hyperfine state (cf. Sec. 10.2.1), increasing the spin temperature and reducing the absorption above $\sim$50 MHz. Black holes are expected to create highly penetrating X-rays, which may make $T_s > T_\gamma$ and lead to emission above 100 MHz. Eventually the gas will be fully ionized, so that the spectrum above $\sim$200 MHz should be unaffected.

At the lowest frequencies observable today, the radiation from our galaxy is $\sim$300 times brighter than the CMB and has to be carefully removed, along with the influence of the ionosphere (cf. Sec. 21.5.4). Most attention is now (2016) focused on measuring a signal associated with the growing density perturbations from the time just before reionization. If the measurements are successful, we will have another powerful probe of the growth of matter perturbations (Sec. 28.5.3).

28.6.2 Weak Gravitational Lensing

NULL GEODESIC CONGRUENCE

We introduced strong gravitational lensing in Sec. 7.6. Such lensing is important for rare lines of sight where the galaxy-induced gravitational deflections of light rays are strong enough to image background sources more than once. There is a complementary effect called *weak gravitational lensing*, which is a consequence of the growth of perturbations in the universe and is present for all images (e.g., Schneider, Ehlers, and Falco, 1992). Basically, the tidal actions of gravitational perturbations distort galaxy images, inducing a correlated ellipticity that we can measure if we assume that the galaxies' intrinsic shapes are randomly oriented on the sky. To quantify this effect, we need to consider the propagation of neighboring rays through the inhomogeneous universe, under the geometrical optics approximation.

What we actually do is a little more subtle and much more powerful. We consider one *fiducial* ray and a *congruence* of rays that encircle it—a generalization of the paraxial optics developed in Sec. 7.4. We imagine this congruence as propagating backward in time from us, now (in a scholastically correct manner!), toward a distant galaxy. We label rays that belong to the congruence by the vectorial angle $\pmb{\psi}$ they make with the fiducial ray here and now—what an astronomer observes. The fiducial ray will follow a crooked path, but we are concerned with the proper transverse separations of neighboring rays $\pmb{\xi}(\chi;\pmb{\psi})$. This is a job for the equation of geodesic deviation [Eq. (25.31)]. (For a more detailed analysis along the lines of the following, see Blandford et al., 1991.)

ray congruence

As we have discussed in Sec. 25.4, we parameterize distance along a null geodesic using an affine parameter ζ, which must satisfy $dt/d\zeta \propto p^0$ [cf. Eq. (2.14)]. Now, p^0 is the energy of a photon measured by an FO; in the homogeneous universe, this will vary $\propto a^{-1}$. A convenient choice for ζ is therefore

affine parameter

$$\zeta(a) = \int_{t(a)}^{t_0} dt'\, a(t') = \int_a^1 \frac{da'}{H(a')} = \int_0^{\chi(a)} d\chi'\, a(\chi')^2. \qquad (28.75)$$

Note that we use the scale factor appropriate to the unperturbed universe in defining ζ, because the overall expansion of the universe is dictated by the behavior of the stress-energy tensor $\pmb{T}$ on the largest scales where it is, by assumption, homogeneous. The associated tangent vector to use in the equation of geodesic deviation is $dx^\alpha/d\zeta = \{a^{-1}, 0, 0, a^{-2}\}$.

The Riemann tensor for the perturbed metric (28.44), like the Einstein tensor (28.47), is easily computed to linear order in the perturbations and then inserted into the equation of geodesic deviation (25.31). In contrast to our treatment of the CMB, for weak lensing we explicitly assume that the relevant perturbations are of short wavelength and are effectively static when crossed by the photons we see today, allowing us to use local Lorentz coordinates parallel-propagated along the ray with $\hat{e}_3$ aligned along the ray. Assuming that $\Phi = \Psi$ as dictated by the Einstein equations (28.47d) for a perfect fluid, and just retaining lowest order terms, the equation of geodesic deviation becomes

$$a^2 \frac{d^2\xi^i}{d\zeta^2} - \dot{H}\xi^i = -\left(\Phi_{,3}{}^3 \xi^i + 2\Phi_{,j}{}^i \xi^j\right) = -\left(4\pi\delta\rho_M \xi^i + 2\bar{\Phi}_{,j}{}^i \xi^j\right),$$

equation of geodesic deviation

for $i, j = 1, 2,$ (28.76)

where we have used Poisson's equation $\Phi_{,k}{}^k + \Phi_{,3}{}^3 = 4\pi\delta\rho_M$ and have introduced the trace-free *tidal tensor* $\bar{\Phi}_{,j}{}^i \equiv \Phi_{,j}{}^i - \frac{1}{2}\Phi_{,k}{}^k \delta_j{}^i$; and where spatial indices are raised and lowered with the flat metric, and all indices following a comma represent partial derivatives.

CONVERGENCE AND SHEAR

It is instructive and helpful to express this equation in terms of comoving coordinates. Substituting $\boldsymbol{\xi} \to a\boldsymbol{\eta}$ and $d\zeta \to a^2 d\chi$, we obtain[51]

$$\frac{d^2\eta^i}{d\chi^2} = -\left(4\pi\delta\rho_M \delta_j{}^i + 2\bar{\Phi}_{,j}{}^i\right) a^2 \eta^j. \tag{28.77}$$

The right-hand side vanishes in the absence of perturbations, which is what we expect, as the 3-space associated with the homogeneous universe is flat (Sec. 28.2.2). Equation (28.77), made linear by inserting the unperturbed η^i on the right-hand side, admits a Green function solution:

$$\boldsymbol{\eta} = \frac{\boldsymbol{\xi}}{a} = \chi[(1-\kappa)\mathbf{I} - \boldsymbol{\gamma}] \cdot \boldsymbol{\psi};$$

$$\{\kappa, \gamma_j{}^i\} = \int_0^\chi d\chi' \left(\chi' - \frac{\chi'^2}{\chi}\right) a(\chi')^2 \left\{4\pi\delta\rho_M(\chi'), 2\bar{\Phi}_{,j}{}^i(\chi')\right\}. \tag{28.78}$$

Here $\boldsymbol{\psi} = (d\boldsymbol{\xi}/d\zeta)_0 = (d\boldsymbol{\xi}/d\chi)_0$ is the observed angular displacement from the fiducial ray, and $\mathbf{I}$ is the 2-dimensional unit tensor (metric). Also, we introduce the **convergence** κ; it is the analog of the expansion that appears in the theory of elastostatics (Sec. 11.2), and it is produced by matter density perturbations inside the congruence. The quantity $\boldsymbol{\gamma}$ is the trace-free *cosmic shear tensor* produced by matter distributed anisotropically outside the congruence.

If we replace χ with d_A/a, Eq. (28.78) appears to be a simple linear generalization of the formalism introduced in our discussion of strong gravitational lensing (Sec. 7.6). However, the equation of geodesic deviation is necessary to justify the neglect of the cosmological constant, to handle the potential derivatives along the ray, and to include large density perturbations.

Now we jump into **k**-space, write $\bar{\Phi}$ in terms of the Fourier transform $\tilde{\Phi}$, and express the two components of the **distortion** as:

$$\{\kappa, \gamma_j{}^i\} = -\int_0^\chi d\chi' \left(\chi' - \frac{\chi'^2}{\chi}\right) \int \frac{d^3k}{(2\pi)^3} \left\{k^2, k_j k^i - \frac{1}{2} k_l k^l \delta_j{}^i\right\} \tilde{\Phi} e^{i\mathbf{k}\cdot\boldsymbol{\chi}'}. \tag{28.79}$$

The first component is a scalar; the second is a tensor.[52]

CORRELATION FUNCTIONS

The ensemble average distortion vanishes, because the wave phases are random. When we consider the convergence and shear along two congruences labeled 1, 2,

---

[51]. The cosmological constant does not contribute to the focusing, but we should include radiation and neutrinos when considering weak lensing of the CMB.

[52]. Formally, κ should be a tensor with an antisymmetric part describing image rotation. In the case of elastostatics, this term is dropped, because it induces no stress. Weak lensing rotation vanishes in the linear, scalar approximation, but there is a tiny nonlinear contribution. In principle, an antisymmetric part of κ could be created by a hypothetical cosmic torsion field, which has been sought (unsuccessfully!) using polarization observations.

and separated by an angle ϑ, the contribution from the products of different waves likewise vanishes. Even the contribution from a single wave is small except when it is directed almost perpendicular to the line of sight and we are integrating along its crest or trough. We formalize these expectations by using the potential power spectrum [Eq. (28.67)] and keeping faith with the magic of delta functions. Let us assume that $\vartheta \ll 1$, so the sky can be treated as flat, and introduce local basis vectors $\mathbf{e}_\parallel$ parallel to the separation of the congruences and $\mathbf{e}_\perp$ perpendicular to the separation. We then define two independent components of shear by $\gamma_+ \equiv \gamma_\parallel{}^\parallel - \gamma_\perp{}^\perp$ and $\gamma_\times \equiv 2\gamma_\perp{}^\parallel$. The convergence correlation function is $C_{\kappa\kappa} = \langle \kappa_1 \kappa_2 \rangle$, where the average is over all pairs of rays on the sky separated by an angle ϑ. We can likewise define correlation functions involving the shear. Then the values of the nonzero correlation functions are

$$\begin{pmatrix} C_{\kappa\kappa} \\ C_{\kappa+} \\ C_{++} \\ C_{\times\times} \end{pmatrix} = \int_0^\chi d\chi' \int_0^\chi d\chi'' \left(\chi' - \frac{\chi'^2}{\chi} \right) \left(\chi'' - \frac{\chi''^2}{\chi} \right) \times$$

$$\int \frac{d^3k}{(2\pi)^3} e^{i(k_\parallel \vartheta \chi' + k_3(\chi' - \chi''))} P_\Phi(a,k) \begin{pmatrix} k^4 \\ k^2(k_\parallel^2 - k_\perp^2) \\ (k_\parallel^2 - k_\perp^2)^2 \\ 4k_\parallel^2 k_\perp^2 \end{pmatrix}. \quad (28.80)$$

(The vanishing of the two remaining averages provides a check on the accuracy of the measurements.) As $k\chi \gg 1$, we approximate the integral over χ'' by an infinite integral over $\chi'' - \chi'$ to produce a delta function, which can then be integrated over to obtain

$$\begin{pmatrix} C_{\kappa\kappa} \\ C_{\kappa+} \\ C_{++} \\ C_{\times\times} \end{pmatrix} = \int_0^\chi d\chi' \left(\chi' - \frac{\chi'^2}{\chi} \right)^2 \int \frac{dk}{2\pi} k^5 P_\Phi(a,k) \int_0^{2\pi} \frac{d\phi_k}{2\pi} e^{ik\vartheta \chi' \cos\phi_k} \begin{pmatrix} 1 \\ \cos 2\phi_k \\ \cos^2 2\phi_k \\ \sin^2 2\phi_k \end{pmatrix}$$

$$= (2\pi)^2 \mathcal{Q} \int_0^\chi d\chi' \left(\chi' - \frac{\chi'^2}{\chi} \right)^2 \int dk\, k^2 \Gamma_1^2(a,k) \begin{pmatrix} J_0(k\chi\vartheta) \\ -J_2(k\chi\vartheta) \\ \tfrac{1}{2}[J_0(k\chi\vartheta) + J_4(k\chi\vartheta)] \\ \tfrac{1}{2}[J_0(k\chi\vartheta) - J_4(k\chi\vartheta)] \end{pmatrix},$$

$$(28.81)$$

where we have substituted the initial cosmic noise [Eq. (28.68)] for P_Φ, $\Gamma_1(a,k)$ is the first component of the growth vector Eq. (28.66), and J_0, J_2, J_4 are Bessel functions. The remaining integrals must be performed numerically; see Fig. 28.16. Note that

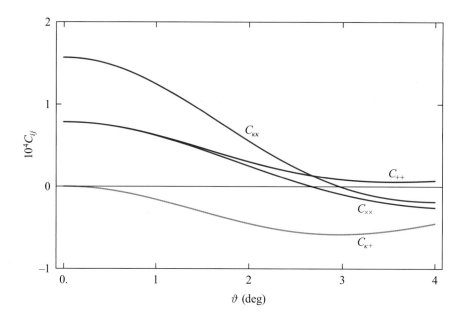

FIGURE 28.16 Two-point correlation functions for weak gravitational lensing for sources located at $\log a = -0.5$, calculated from Eq. (28.81) and adopting the growth factor $\Gamma_1(a, k)$ calculated numerically from Eq. (28.65). The cross correlation of magnification fluctuations is $C_{\kappa\kappa}$. The correlation function for parallel stretching of galaxy images is C_{++}. The correlation function for stretching along a direction inclined at 45° to the separation vector is $C_{\times\times}$. The magnification-shear cross correlation is $C_{\kappa+}$.

shear correlation

$C_{++} + C_{\times\times}$, the total shear cross correlation, equals $C_{\kappa\kappa}$, as can be seen directly from Eq. (28.80) if we recall that $k_3 \sim 0$. Note also that $C_{\times\times}(\vartheta)$ can change sign. This effect has been observed. The calculation is performed here in the linear approximation, which breaks down for small scales, and so numerical simulations must be used in practice.

tomography

These four correlation functions refer to galaxies at a fixed distance. In practice, a given survey averages over a range of distances and, by varying this range, can make a *tomographic* examination of the growth of the potential power spectrum P_Φ.

STATISTICAL MEASUREMENT

How do we actually use observations of galaxies to measure convergence and shear? Weak lensing does not change the intensity of radiation. Therefore a source's flux F

flux perturbation

is determined by the solid angle it subtends. Using Eq. (28.78) to linear order, we see that $\Delta F/F = 2\kappa$. Now, we do not know the intrinsic galaxy luminosities. However, their distribution should be the same in different directions at the same distance if the universe is homogeneous, and the predicted magnifications have been measured. Unfortunately, absorption in our galaxy and other problems limit the utility of this approach. Conversely, $C_{\kappa\kappa}(0)$ is a measure of the variance of a flux measurement and a limit on how precisely a measurement need be made in a single case when attempting to use "standard candles" (sources of known luminosity) to measure the universe.

Statistical shear measurements are more useful. We expect the orientations of galaxies to be uncorrelated,[53] and so, if we associate the fiducial ray with the center of an observed galaxy, the tensor $\langle \boldsymbol{\xi} \otimes \boldsymbol{\xi} \rangle$ should be proportional to the unit tensor. Inverting the first part of Eq. (28.78) then provides a linear estimator of the shear:

intrinsic galaxy alignment

shear estimator

$$\{\gamma_+, \gamma_\times\} = \frac{\langle \{\psi_\parallel^2 - \psi_\perp^2, 2\psi_\parallel \psi_\perp\} \rangle}{\langle \psi_\parallel^2 + \psi_\perp^2 \rangle}, \qquad (28.82)$$

where the averages are over the photons associated with observed images of individual galaxies and then over galaxies adjacent on the sky but at differing distances. This estimator is quite general, and different types of source[54] and weighting can be employed in evaluating it observationally.

There are many observational challenges. For example, the dominance of Fourier modes with **k** almost perpendicular to the line of sight means that occasional structures elongated along the line of sight can hinder the measurement of an unbiased estimate of the ensemble-averaged correlations. Despite this, cosmic shear measurements are consistent with the predicted perturbation spectrum (28.68) and its evolution and show great promise for future surveys that will image 20 billion galaxies.

observing cosmic shear

EXERCISES

Exercise 28.20 *Example: Weak Lensing in an Empty Universe*
Consider an extended congruence propagating through an otherwise homogeneous universe, from which all matter has been removed. Show that the affine distance functions as an effective angular diameter distance in this congruence. Now reinstate the matter as compact galaxies and modify Eq. (28.79) for the convergence and shear. Bookkeep the average density of matter as purely positive density perturbations. Explain qualitatively how the convergence and shear are changed when the ray passes through one or more galaxies, when it misses all of them, and on average.

Exercise 28.21 *Problem: Mean Deviation*
Consider a single light ray propagating across the universe from a source at $\log a = -0.5$ to us. The cumulative effect of all the deflections caused by large-scale inhomogeneities makes the observed direction of this ray deviate by a small angle from the direction it would have had in the absence of the inhomogeneities. Estimate this deviation angle. Is it large enough to be observable?

---

53. In fact, neighboring galaxies are systematically aligned, but provided we average over large enough volumes, this is ignorable. In addition, systematic biases associated with telescope and atmospheric distortion must be removed carefully.
54. For example, the CMB fluctuations and the galaxy correlation function at the smallest angular scales can be used.

Exercise 28.22 *Challenge: CMB Polarization*

Polarization observations of the CMB provide an extremely important probe of fluctuations in the early universe.

(a) By invoking the electromagnetic features of Thomson scattering by free electrons, give a heuristic demonstration of why a net linear polarization signal is expected.

(b) Using the Monte Carlo formalism sketched in Sec. 28.6.1, calculate the polarization expected from a single fluctuational mode $\tilde{\Phi} e^{i\mathbf{\kappa}\cdot\mathbf{x}}$ of given amplitude.

(c) The description of polarization has many similarities with the formalism we have outlined to describe cosmic shear. Linear polarization is unchanged by rotation through π just like a shear deformation. In addition, the polarization pattern that should be seen from a single inhomogeneity mode should have an electric vector that alternates between parallel and perpendicular to the projection of the mode's wave vector $\mathbf{k}$ on the sky, just like the elongation of the images of background galaxies in weak gravitational lensing. We do not expect to produce a signal in either case along a direction at 45° to the projection of $\mathbf{k}$. These predicted polarization/shear patterns are commonly called "E-modes." However, as we discuss in Sec. 28.7.1, primordial gravitational wave modes may also be present. Explain qualitatively how a single gravitational wave mode can produce a "B-mode" polarization pattern with electric vectors inclined at ±45° to the direction of the projection of $\mathbf{k}$ on the sky.

(d) When one sums over inhomogeneity modes, and over gravitational-wave modes, the resulting polarization E-modes and B-modes have distinctive patterns that differ from each other. In what ways do they differ? Read, explain, and elaborate the discussion of E-modes and B-modes near the end of Sec. 28.6.1.

(e) Outline how our perturbed metric, Eq. (28.44), would have to be modified to accommodate the presence of primordial gravitational waves.

28.6.3 Sunyaev-Zel'dovich Effect

When we discussed CMB radiative transfer in Sec. 28.6.1, we assumed that the plasma was cold. This was appropriate when the temperature was $T_e \sim 3{,}000$ K. However, the gas that settles in rich clusters of galaxies (see Sec. 28.2.1 and Fig. 28.1) has a temperature of $\sim 10^8$ K, and thermal effects are very important. To understand these important effects in general, consider a homogeneous and isotropic radiation field with distribution function $\eta_\gamma(\nu)$. Every time a photon is scattered by an electron, its energy changes through small increments by Doppler shifting and Compton recoil. This problem is ideally suited for a Fokker-Planck treatment (Sec. 6.9.1). If we ignore emission and absorption, photons are conserved, and the Fokker-Planck equation must have the form

$$\frac{\partial \eta_\gamma}{\partial t} + \frac{1}{\nu^2}\frac{\partial}{\partial \nu}\left(\nu^2 \mathcal{F}_\gamma\right) = 0, \tag{28.83}$$

where the flux $\mathcal{F}_\gamma$ in frequency space depends on η_γ and its frequency derivative. When the electrons are very cold, each scattering produces a Compton recoil with average frequency change $\langle \Delta \nu \rangle = -h\nu^2/m_e$, and so $\mathcal{F}_\gamma = -(n_e \sigma_T h\nu^2/m_e)\eta_\gamma$. Now, our experience with the Fokker-Planck equation suggests that we add a term proportional to $\partial \eta_\gamma / \partial \nu$ to account for the heating of the photons by hot electrons with temperature T_e. However, $\mathcal{F}_\gamma$ must vanish when η_γ is Planckian with temperature T_e. A quick inspection shows that to deal with this, we must also add a term quadratic in η_γ:

$$\mathcal{F}_\gamma = -\frac{n_e \sigma_T h \nu^2}{m_e}\left(\eta_\gamma + \eta_\gamma^2 + \frac{k_B T_e}{h}\frac{\partial \eta_\gamma}{\partial \nu}\right). \tag{28.84}$$

Kompaneets equation

This $\mathcal{F}_\gamma$ vanishes when η_γ has the Bose-Einstein form $\{\exp[(h\nu - \mu)/(k_B T_e)] - 1\}^{-1}$, with μ the chemical potential. This is entirely reasonable, as the total photon number density is fixed in pure electron scattering and the equilibrium photon distribution function under these conditions has the Bose-Einstein form, not the Planck form (Sec. 4.4). The kinetic equation (28.83), adopting the flux $\mathcal{F}_\gamma$ of Eq. (28.84), is known as the *Kompaneets* equation (Kompaneets, 1957). The third term in parentheses in Eq. (28.84) describes the diffusion of photons in frequency space due to the Doppler shift $\Delta \nu \sim \nu(k_B T_e/m_e)^{1/2}$, leading to the familiar second-order frequency increase. The diffusion coefficient in frequency space, $D_{\nu\nu} = n_e \sigma_T \nu^2 k_B T_e / m_e$, can be calculated explicitly by averaging over all electron velocities, assuming a Maxwell-Boltzmann distribution function.

The quadratic term in parentheses in Eq. (28.84) describes the *induced Compton effect*, in which the scattering rate between two photon states (from unprimed to primed) is $\propto \eta_\gamma(1 + \eta_\gamma')$. To lowest order, the product term is canceled by inverse scattering. However, when we make allowance for the electron recoil, there is a finite effect, which is important at low frequency. This term is also derivable from classical electrodynamics without recourse to quantum mechanics.[55]

induced Compton scattering

Now return to what happens in a galaxy cluster. The radiation is still effectively isotropic, and time should be interpreted as path length through the cluster. The derivative term in $\mathcal{F}_\nu$ dominates at high T_e, and so the observed CMB relative temperature change measured at a fixed frequency is given by

frequency shift

$$\delta_{T\,SZ} = \left(\frac{\partial \eta_\gamma}{\partial \ln T_\gamma}\right)_\nu^{-1}\left(\frac{k_B T_e \tau_T}{m_e \nu^2}\right)\left[\frac{\partial}{\partial \nu}\left(\nu^4 \frac{\partial \eta_\gamma}{\partial \nu}\right)\right]_{T_\gamma}$$

$$= \left(\frac{k_B T_e \tau_T}{m_e}\right)\left[\left(\frac{h\nu}{k_B T_\gamma}\right)\coth\left(\frac{h\nu}{2k_B T_\gamma}\right) - 4\right]. \tag{28.85}$$

---

55. The first realization that *induced* Compton scattering could be important was by Kapitsa and Dirac (1933).

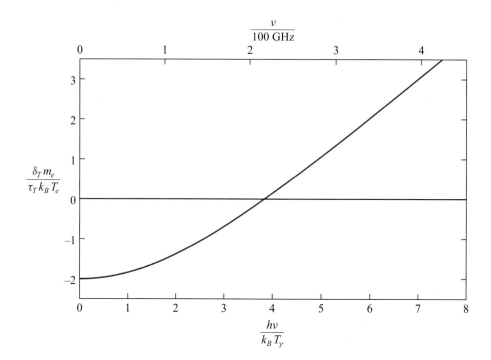

FIGURE 28.17 Sunyaev-Zel'dovich effect: Distortion of the CMB blackbody spectrum by passage of radiation through a cluster of galaxies, where the hot electrons Doppler boost (on average) the frequencies of individual photons. At frequencies well below the peak in the spectrum, the intensity decreases, while at high frequencies, it increases.

This is known as the *Sunyaev-Zel'dovich* (1970) *effect* (Fig. 28.17). In a typical cluster, the optical depth is $\tau_T \sim 0.1$ and $\delta_{T\,SZ} \sim 10^{-3}$. The photon temperature decreases by a fractional amount $\delta_T = -2k_B T_e \tau_T / m_e$ in the Rayleigh-Jeans part of the spectrum and increases in the Wien part. The crossover occurs at $\nu = 3.83 k_B T_\gamma / h = 217$ GHz, as observed.

EXERCISES

Exercise 28.23 *Example and Challenge: The Music of the Sphere*
We have hitherto focused on the statistical properties of the cosmological perturbations as probed by a variety of observations. However, we on Earth occupy a unique location in a specific realization of wave modes that we have argued are drawn from a specific set of waves with particular amplitudes and phases, despite these supposedly being drawn from a statistical distribution (in much the same way that different pieces of music are distinct despite having similar power spectra). It should not be too long before we can produce a 3-dimensional "map" of our universe out to recombination based on a suite of observations, presuming that our standard cosmology and theory for the evolution of perturbations is correct.

(a) Calculate the comoving radii of recombination, reionization, and the most distant galaxies and quasars.

(b) Suppose that we have noise-free CMB fluctuation maps in temperature and polarization up to spherical harmonic quantum number $l = 100$. How many numbers would be contributed by these measurements?

(c) How many numbers would we need to measure to describe the potential and the associated density perturbations out to the radius of recombination with comparable resolution?

(d) Would these modes still be in the linear regime today?

(e) *Challenge.* Explore some of the practical challenges of carrying out this program, paying attention to the investigations we have described in Sec. 28.6 and assuming them to be carried out over the whole sky.

28.7 Three Mysteries

The cosmology that we have described depends on the application of basic physical laws, mostly classical—including (among others) general relativity with a cosmological constant—in locales where the laws are not independently tested; and it also depends on the "simplest" assumptions, including flatness and an initial, scale-free spectrum of adiabatic fluctuations. In this cosmology's most elementary version, on which we have focused, what we observe follows from just four dimensionless parameters, which we can choose to be Ω_b, Ω_D, Ω_γ, and $\mathcal{Q}$, plus a single length/timescale which we can choose to be t_0; and we learn these parameters by fitting observations. In a more comprehensive version of this cosmology, one must also fit additional astrophysical parameters associated with reionization and galaxy formation.

We conclude this chapter with a brief summary of contemporary views on the more fundamental processes that presumably determine this cosmology's assumptions and parameters.

28.7.1 Inflation and the Origin of the Universe

KINEMATICS

Early in our discussion of cosmology, we introduced the horizon and emphasized that it was smaller and smaller when the universe was younger and younger, so less and less of the universe was in causal contact at early times (Fig. 28.6). However, the universe we observe today is nearly homogeneous and isotropic. In particular, large-scale, spatially coherent fluctuations in the CMB are seen at recombination, when the horizon was less than 1% of their size. We have argued that the universe began in thermodynamic equilibrium at a very high temperature and that key properties (such as the net number of baryons and dark matter particles per photon and the amplitude of the potential perturbations) were determined by specific, though unknown, local physical laws. If this is so, how can there be spacelike slices of simultaneity that are homogeneous on the largest scale? It is also surprising that the universe is as flat as

horizon problem

it appears to be.[56] As was realized surprisingly early in the development of modern cosmology, these mysteries may have a common explanation called *inflation*.[57]

Under the inflationary scenario, it is proposed that the material of the universe we see today was initially in causal contact and was well mixed and therefore homogeneous. There followed an epoch of runaway expansion—inflation—when parts of the universe on different world lines from our own exited our horizon and lost contact with us. After this phase ended, these parts independently followed the evolution we described in Sec. 28.4 and, primed with the same features as us, their world lines then reentered our horizon, the last to leave being the first to return. "Hello, goodbye, hello!"

We have not demonstrated that the universe had such an origin, but all recent observations are consistent with the simplest version of inflation, and nothing that we have learned is in conflict with it. At the very least, the theory of inflation illustrates some fascinating features of general relativity, bringing out clearly the challenge that will have to be met by any rational description of the very early universe.[58]

We do not understand the physics that underlies inflation, but let us make a guess, inspired by observation. We have explained that the inferred initial amplitude of a potential perturbation is almost independent of its comoving length scale (scale invariance). We presume that each perturbation was laid down by local physics just prior to that scale exiting the horizon and argue that the physical conditions were therefore constant in time. The one homogeneous cosmology that has this character is a de Sitter expansion with $\rho = \text{const}$ and $a \propto e^{H_i t}$ for $t_i < t < t_h$, with a constant *inflation Hubble constant* H_i and where the subscript h denotes the end of inflation.

It is commonly supposed that inflation began during the epoch of "grand unification" of the electroweak and strong forces, when $t \sim t_i \sim H_i^{-1} \sim 10^{-36}$ s. If we also assume that the universe was homogeneous on the scale of the horizon $\sim t_i$, and inflation went on for long enough to encompass our horizon, then $a_i \sim t_i/\chi_{0H} \sim 3 \times 10^{-55}$. However, we also know that $a_h \sim a_i \exp(H_i t_h)$; and if we denote the start of the particle era by $a_p \sim 10^{-11}$ when $t_p \sim 3$ ms, then we also have that $a_h \sim a_p (t_h/t_p)^{1/2}$. These two relations and the above numbers imply that $t_h \sim 64 t_i$, $a_h \sim 2 \times 10^{-27}$. In this simple example, 64 e-foldings of inflationary expansion suffice to explain the homogeneity of our universe today.

Furthermore, any significant curvature that may have been present when the universe started to inflate would have made a fractional contribution to the Friedmann equation (28.16) $\sim (a_i/a_h)^2 \sim 10^{-56}$ at t_h. This fractional contribution would then

---

56. There are additional quantum field theory puzzles, especially the apparent scarcity of topological defects like monopoles, strings, and domain walls that are addressed by this theory.
57. Pioneers of these ideas included Kazanas (1980), Starobinsky (1980), Guth (1981), Sato (1981), Albrecht and Steinhardt (1982), and Linde (1982).
58. We exclude "Just So" stories—for example, flood geology—that assert that the world began at a specific time in the relatively recent past with just the right initial conditions necessary to evolve to the world of today.

grow by a factor $\sim(a_{eq}/a_h)^2 \sim 10^{46}$ by the end of the radiation era and by a further factor $\sim a_\Lambda/a_{eq} \sim 3{,}000$ by the start of the cosmological era. In this example it is only $\sim 10^{-56} \times 10^{46} \times 3{,}000 \sim 3 \times 10^{-7}$ in the early cosmological age (today). However, with other plausible assumptions, it may just be detectable. Therefore, the observed flatness, which otherwise requires very careful fine tuning of the initial conditions, also has a natural explanation.

CLASSICAL ELECTROMAGNETIC FIELD THEORY

The constant energy density that we have argued is needed to drive inflation is commonly associated with a classical scalar field, sometimes called the *inflaton*. To describe its properties necessitates a short digression into classical field theory (cf. Ex. 7.4).

inflaton field

Lagrangian methods were devised to solve problems in celestial mechanics and turned out to be useful for a larger class of classical problems (Goldstein, Poole, and Safko, 2002). To summarize the approach (with which the reader is presumed to be quite familiar), the coordinates **x** of all particles are replaced by a sufficient number of generalized coordinates $\mathbf{q}(\mathbf{x}, t)$—for example, three Euler angles for a spinning top—to describe the system. A scalar *lagrangian* $L(\mathbf{q}, \dot{\mathbf{q}}, t)$ is introduced as the difference of the kinetic and potential energies, and the system evolves so as to make the *action* $\int dt\, L$ stationary.[59] This implies the Euler-Lagrange equations $\partial_\mathbf{q} L = d(\partial_{\dot{\mathbf{q}}} L)/dt$, where $\partial_\mathbf{q}$ is shorthand notation for $\partial/\partial q^i$. A *hamiltonian*, $\partial_{\dot{\mathbf{q}}} L \cdot \dot{\mathbf{q}} - L$, is introduced, which equals the conserved energy if the system does not interact with its environment and evolve explicitly with time.

Lagrangian dynamics

This Lagrangian approach was generalized to describe the classical electromagnetic field[60] (where it is not very much used in practice; Jackson, 1999).[61] For this, three changes need to be made to the particle lagrangian. First, as we are dealing with a relativistic theory, we work in spacetime coordinates. Second, we use the lagrangian density $\mathcal{L}$, so that the action is $\int dt\, dV\, \mathcal{L}$; $\mathcal{L}$ must be a Lorentz-invariant scalar (cf. Sec. 2.12 and Ex. 7.4). Third, we treat the electromagnetic field itself as a generalized coordinate. However, instead of being an N-dimensional vector, we take the limit $N \to \infty$ and treat it as a continuous variable.

The natural choice of field coordinate is the 4-vector potential A^α constrained by the Lorenz gauge condition $\partial_\alpha A^\alpha = 0$. (We use indices here to avoid notational confusion, we use ∂_α for partial derivatives, $\partial_\alpha A^\alpha \equiv \partial A^\alpha/\partial x^\alpha$, and for simplicity we assume spacetime is flat.) For the free electromagnetic field, the only choice for the

---

59. In a phrase that captures the philosophical, political, theological, and literary context in which this revolutionary approach to physics was created, we live "in the best of all possible worlds."
60. Faraday first conceptualized a field description of electromagnetism and gravity in the 1820s, contrasting it with the Newtonian "action at a distance" and gradually developed this idea (Faraday, 1846). Although Maxwell and others sought a variational description of electromagnetism, it was the astronomer Karl Schwarzschild (1903) who first got it right, 2 years before the advent of special relativity.
61. It is, of course, indispensible for understanding quantum mechanics, quantum electrodynamics, and quantum field theory and is extremely useful for probing the fundamental character of general relativity.

T2

lagrangian density for electromagnetic field

lagrangian density (except for an additive or multiplicative constant) consistent with these requirements is $\mathcal{L}_{EM} = \frac{1}{4\pi}\partial^{[\alpha}A^{\beta]}\partial_{[\beta}A_{\alpha]}$, in Gaussian units. Comparing with the particle lagrangian, we recognize this as kinetic energy–like with no potential energy–like contribution. Varying the action leads to the Euler-Lagrange equations:

$$\frac{\delta\mathcal{L}_{EM}}{\delta A_\alpha} \equiv -\partial_\beta\left(\frac{\partial\mathcal{L}_{EM}}{\partial(\partial_\beta A_\alpha)}\right) = 0 \Rightarrow F^{\alpha\beta}{}_{,\beta} = 0 \quad \text{where } F_{\alpha\beta} = 2\partial_{[\beta}A_{\alpha]}. \quad (28.86)$$

These are the free-field Maxwell equations.[62]

The electromagnetic stress-energy tensor is a natural generalization of the hamiltonian for particle dynamics:[63]

$$T_\beta{}^\alpha = \mathcal{L}_{EM}\delta_\beta{}^\alpha - \left(\frac{\partial\mathcal{L}_{EM}}{\partial(\partial_\alpha A_\gamma)}\right)\partial_\beta A_\gamma = \frac{1}{4\pi}\left(\partial^{[\gamma}A^{\delta]}\partial_{[\delta}A_{\gamma]}\delta_\beta^\alpha - 4\partial^{[\alpha}A^{\gamma]}\partial_{[\gamma}A_{\beta]}\right).$$

(28.87)

This agrees with the standard form, Eq. (2.75).

SCALAR FIELD THEORY

The observed isotropy of the universe suggests that the fundamental inflaton field we seek—henceforth designated as φ—is a real scalar and not a vector field (as in electromagnetism) or a tensor field (as in general relativity) or a complex or spinorial quantum field (like the Higgs field). In this section we set $G = c = 1$, so that φ is dimensionless and we deal with the real universe where spacetime is curved. The simplest form for the lagrangian density, by analogy with $\mathcal{L}_{EM}$, is $\mathcal{L} = -\frac{1}{2}\vec{\nabla}\varphi \cdot \vec{\nabla}\varphi -$

lagrangian density for scalar field

$V(\varphi)$ (where we no longer need to use indices, as the field is a scalar). The first term is the only invariant choice we have for the kinetic energy–like part (except that the $-\frac{1}{2}$ is a convention); the second term is the simplest potential energy–like part, which is absent for classical electromagnetism but necessary here. Continuing the analogy, the stress-energy tensor is given by

stress-energy tensor

$$\boldsymbol{T} = \mathcal{L}\boldsymbol{g} - \frac{\partial\mathcal{L}}{\partial(\vec{\nabla}\varphi)} \otimes \vec{\nabla}\varphi = \mathcal{L}\boldsymbol{g} + \vec{\nabla}\varphi \otimes \vec{\nabla}\varphi. \quad (28.88)$$

potential minimum

Now consider a harmonic potential $V = \frac{1}{2}\omega_h^2(\varphi - \varphi_h)^2$, in which φ_h is the (vacuum) value of the field, where the potential vanishes. The (Euler-Lagrange) field equation for our lagrangian density is $\nabla_\alpha\nabla^\alpha\varphi = dV/d\varphi$. If we seek a wave solution in a local Lorentz frame, then we recover the dispersion relation: $\omega^2 - k^2 = \omega_h^2$, where ω_h is Lorentz invariant. This describes a longitudinal scalar wave propagating in vacuo but has a similar dispersion relation to a transverse vector wave propagating in an unmagnetized plasma [cf. Eq. (21.24)]. Note that we can Lorentz transform into a frame

---

62. Including a current source requires the addition of an interaction lagrangian, which is straightforward but need not concern us here.
63. The formal justification of this heuristic argument hinges on the celebrated theorem (Noether, 1918) that relates conserved quantities to symmetry.

in which $\omega = \omega_h$ and $k = 0$, and the wave becomes a pure oscillation with no spatial gradients. In this frame, the field is directly analogous to a classical particle moving in a stationary potential well. In a general frame, the stress tensor for an individual wave is anisotropic and oscillatory.

Next, consider a potential maximum $V = V_i - \frac{1}{2}\omega_i^2\varphi^2$, and transform into the frame where there are no spatial gradients. The equation of motion for the field is $\ddot{\varphi} - \omega_i^2\varphi = 0$. It is simplest to imagine the field as starting with $\varphi(0) = 0$ and a small positive velocity $\dot{\varphi}(0)$, so that $\dot{\varphi}(t) = \dot{\varphi}(0)\cosh\omega_i t$, which soon increases exponentially with time.

potential maximum

SLOW-ROLL INFLATION

We have now introduced all the ideas we need to design a potential that allows the universe to *slow roll* and inflate for ~ 64 *e*-foldings before transitioning classically to a decelerating expansion. Qualitatively, we require a potential maximum with small enough curvature for sufficient inflation to take place, joined to a potential minimum into which the field can settle and allow fundamental particles to take over the dynamics.[64]

A convenient and illustrative choice for the potential is $V = V_i[1 - (\varphi/\varphi_h)^2]^2$ (Fig. 28.18a). The two parameters V_i and φ_h measure the height and the width of the potential and ought to be derivable from basic physics. Let us work in the frame in which $\varphi = \varphi(t)$ with no spatial gradients, which will define and evolve into the sequence of homogeneous spatial hypersurfaces that we have been using. Using Eq. (28.88) and comparing with the perfect fluid stress-energy tensor, we can identify the energy density ρ and the pressure P:

model potential

$$\rho_\varphi = V + \frac{1}{2}\dot{\varphi}^2, \quad P_\varphi = -V + \frac{1}{2}\dot{\varphi}^2. \tag{28.89}$$

density and pressure for field

If we ignore all other possible contributions and insert these expressions into the first law of thermodynamics [Eq. (28.19)], we obtain the cosmological scalar field equation:

evolution of the field

$$\ddot{\varphi} + 3H\dot{\varphi} + dV/d\varphi = 0, \tag{28.90}$$

where $H^2 = 8\pi\rho_\varphi/3$. We set $V_i = 3H_i^2/8\pi = 1.6 \times 10^{98}$ J m^{-3} and we choose $\varphi_h = 1.5$ and the value of the scalar field $\dot{\varphi}(0)$ so as to prolong inflation for 64 *e*-foldings and to match the expansion during the particle age (Fig. 28.18).

Now, this model thus far is seriously incomplete, because its evolution asymptotes to an empty, static universe with $H = 0$ and $a = $ const. We have completely ignored the relativistic, fundamental particles that drive the post-inflationary expansion. These cannot be primordial, because their contribution would have inflated away; instead, they must have been generated at the end of inflation, specifically, as the field oscillates in the potential well. From a quantum mechanical perspective, this is quite reasonable,

particle production

---

[64]. In the original theory, this transition—called the *graceful exit*—was attributed, unsuccessfully, to quantum mechanical tunneling.

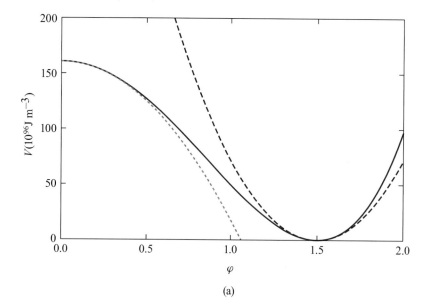
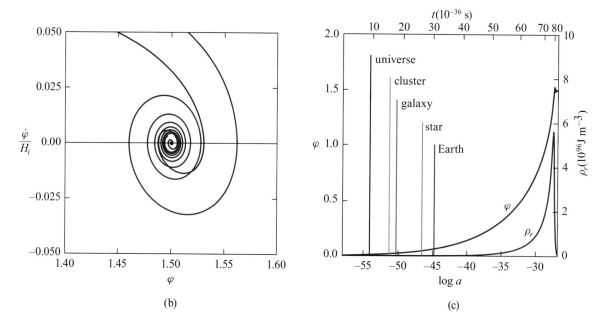

FIGURE 28.18 (a) The three types of potential considered in the text. The dashed line is a (shifted) potential minimum, the dotted line is a potential maximum, and the solid line is the simple model inflaton potential that joins these two solutions. (b) Variation of φ as it oscillates about the potential minimum exhibited in the φ–$\dot{\varphi}$ plane. The red curve is the free field variation; the blue curve includes the ad hoc particle production. (c) Variation of the field φ, the scale factor a, and the relativistic particle energy density ρ_r as a function of a. The rapid decline in ρ_r at late time is due to expansion. The times at which perturbations, on the comoving scale of various cosmic structures, leave the horizon are indicated by vertical lines.

though the detailed mechanisms are not understood.[65] The production of particles is irreversible, and this is where the entropy of today's universe, mostly carried by CMB photons, would have originated.

So (inspired by the treatment of collisionless shocks in Sec. 23.6), let us make an ad hoc modification to Eq. (28.90) by changing the "friction" term from $3H\dot{\varphi}$ to $(3H + k_r H_i)\dot{\varphi}$ with $k_r > 0$, to include some genuine dissipation. The equation for the production of relativistic particle energy density ρ_r is then

$$\dot{\rho}_r + 4H\rho_r = k_r H_i \dot{\varphi}^2, \qquad (28.91)$$

particle production ansatz

where the source term on the right is the negative of the dissipative rate of loss of inflaton energy, $\dot{\rho}_{\varphi\,\mathrm{diss}}$. The results from adding this particle production to our evolution equation are shown in Fig. 28.18b. We see that a choice of $k_r = 0.2$ leads to a scale factor evolution similar to what we suggest may have happened.

One of the more serious issues glossed over in this abbreviated account of inflation is the value of the potential at its minimum. Our model potential has $V(\varphi_h) = 0$. What this really means is that after inflation is over, the finite scalar field contributed nothing to the stress-energy tensor and generated no spacetime curvature. This must be essentially forever, because all other contributions to ρ diminish with time, at least until the cosmological age. Quantum mechanically, the issue is even murkier. We return to this point when we discuss the fate of the universe.

potential must asymptote to 0

PERTURBATIONS

We have not yet explained why the primordial potential perturbations have their inferred amplitude. It is widely argued that these perturbations are quantum mechanical in origin, and to describe the associated theory goes beyond the scope of this book. See, for example, Mukhanov (2005) for details. However, we can give a heuristic semiclassical argument that captures the essence of the calculations.

In the early stages of inflation, when the perturbations that we measure leave our horizon, the inflaton field φ is very small, V and H_i are almost constant, and the end of inflation is far in the future. There is a limit to how far a photon can travel in comoving coordinate χ, since $dt = a\,d\chi \propto e^{H_i t} d\chi$. The de Sitter spacetime therefore has an effective *event horizon*. It should therefore come as no surprise to discover that there is an associated temperature similar to the Hawking temperature for a black hole, $T_i = \hbar H_i/(2\pi k_B) \sim 10^{24}$ K as H_i is the effective acceleration (Secs. 4.10.2 and 4.10.3). This temperature is too low for excited states to be important, so we are only interested in zero-point fluctuations in φ (Sec. 12.5). We can estimate the magnitude $\delta\varphi$ of the fluctuations by equating the energy associated with it, $\sim(H_i\delta\varphi)^2(1/H_i)^3/G$, to the zero-point energy in a mode $\sim\hbar H_i$, temporarily abandoning geometrized units and reinstating G for clarity. We deduce that $\delta\varphi \sim (G\hbar)^{1/2} H_i$.

de Sitter horizon

relation to Hawking radiation

---

65. The production of electron-positron pairs by a rapidly varying electromagnetic field is an analogous process. The most developed explanations involve parametric resonance, as discussed in Secs. 23.5 and 10.7.2.

Planck units

The quantity $(G\hbar)^{1/2}$ is known as the *Planck length*, denoted $L_P \sim 10^{-35}$ m, and is equivalent to the Planck time $t_P \sim 10^{-43}$ s. It marks the scale when gravitational, relativistic, and quantum mechanical effects should be comparably important and it makes no sense to invoke classical physics alone. The physics must be described by some future complete theory of quantum gravity, perhaps derived from *string theory*. Associated with the Planck length/time is a *Planck mass* $m_P \sim (\hbar/G)^{1/2} \sim 10^{-8}$ kg $\sim 10^{28}$ eV, and a *Planck energy density* $\sim (G^2\hbar)^{-1} \sim 10^{114}$ J m^{-3}. The associated Planck temperature is $\sim 10^{32}$ K. In our model for inflation, $t_{\text{start}} \sim 10^7 t_P$, and this justifies a semi-classical treatment.

The way these field fluctuations $\delta\phi$ are transformed into density and potential fluctuations is a bit more subtle. Spatial gradients can no longer be ignored when there are perturbations. Positive fluctuations develop faster than negative fluctuations, just as we saw in the gravitational age. However, we can ignore $\ddot{\varphi}$ so the rate of change of φ is $\dot{\varphi} \sim -(dV/d\varphi)/(3H_i)$ from Eq. (28.90), and the time interval to change the field is $\delta t \sim \delta\varphi/\dot{\varphi}$. The absolute values of the relative density and potential fluctuations on a given scale are then estimated by $\delta \sim \Phi \sim H_i \delta t$ (Peacock, 1999, pp. 338–340). Now allowing φ to be dynamical and to evolve means that the production of the potential fluctuations from quantum fluctuations is not completely scale free, and the detailed shape of the potential $V(\varphi)$ should be reflected in the observed fluctuation spectrum.

We mentioned at the start of our discussion of perturbations (Sec. 28.5.1) that we were confining our attention to scalar perturbations. If the inflationary mechanism for creating them is basically correct, then they should also be accompanied by tensor perturbations, which take the form of gravitational wave modes whose wavelengths expand with the universe in just the same way. Repeating the zero-point energy calculation using an estimate of the gravitational wave energy of $\sim (H_i \delta h)^2 (1/H_i)^3/G$, where δh is the amplitude of the wave (cf. Sec. 27.4.3), we find that $\delta h \sim H_i t_P$. A full discussion of the growth of the tensor perturbations requires the addition of extra terms to the perturbed Robertson-Walker metric, and care must be taken to deal only with physical quantities that are independent of the choice of coordinate system/gauge.

TESTING INFLATION

We have addressed several cosmological riddles by invoking an inflaton scalar field accompanied by a specific potential for which we have no experimental evidence. Although this may seem quite unscientific, there are many precedents for astronomical observations anticipating laboratory measurement, including the deduction of an inverse square law of gravitation, the discovery of helium, and the prediction of the excited state of carbon that allowed stellar nucleosynthesis to proceed. Although there is a limit to what we can hope to learn, the recent determination that the spectrum of potential perturbations was not quite scale-free—specifically, that $P^\Phi(0, k) \propto k^{-3.04}$, so the longer-wavelength modes have slightly larger amplitudes—is encouraging. As can be appreciated from the preceding discussion, this is a fairly general feature of simple, slow-roll potentials, because the vacuum fluctuations diminish as the field "rolls" down the potential.

As outlined in Ex. 28.22, gravitational waves are detectable through the polarization patterns that they produce in the CMB. Specifically, the B-modes, which have specific handedness locally, have been sought observationally and have been found with levels that are attributable to gravitational lensing and interstellar dust. More sensitive measurements are planned to seek the primordial signal.

INITIAL CONDITIONS

Even if we accept uncritically—on theoretical or observational grounds—that inflation happened roughly as just described,[66] the mechanism has not explained the birth of the universe. It has only translated the discussion of this question to earlier times. The simplest explanation of the origin of the universe is that there is only one universe that began as quantum mechanical decay of a Planck-scale entity.[67] Discussion of "outside" or "before" is relegated to metaphysics. More appealing ideas are that the initial state was chaotic, that it involved extra spatial dimensions, or that the evolution of the inflaton was not semi-classical.

However, the most discussed and intriguing possibility is that our universe is just one member of an ensemble of universes—the *multiverse*—that may be both past and future eternal. Members of this ensemble—also called the *landscape*—correspond to a large number[68] of possible vacuum states that are continuously created and destroyed—most on Planckian timescales. The initial conditions (Kachru et al., 2003; Susskind, 2005) mandated by the existence of our universe (e.g., that it has existed for 14 billion years) are highly specialized and will arise extremely rarely. These unusual circumstances may also extend beyond the starting requirements to the actual laws of physics that come into play as the universe expands. They may need to be fine-tuned as well to allow dark matter, baryons, nuclei, atoms, molecules, galaxies, stars, planets, and life to have developed.

multiverse
landscape

The apparent improbability of this outcome has been turned on its head by conjecturing that our very existence actually selects, as opposed to just allows us to infer, the initial conditions and perhaps also the laws of physics. This is known as the *anthropic principle* (Dicke, 1961). Things were as they were because we are as we are! Many physicists are repelled by such a principle, arguing that it is basically

anthropic principle

---

66. A minority of cosmologists argue that the inflation theory creates more problems than it solves.
67. If we try to do away with inflation altogether and assume that there were only relativistic particles present, then the scale factor at the Planck time is $a \sim 10^{-32}$, and we would have to explain how a region $\sim 1\,\mu$m (or $\sim 10^{29}$ Planck lengths) across could be homogenized and synchronized!
68. Over 10^{500} in some formulations! This is sufficient for adopting the precepts of statistical mechanics. The problem is that we do not have a Liouville equation for assigning a priori probabilities to individual states.

teleological—a belief in the existence of final causes—and an abandonment of the quest to understand the foundations of science. Others see it as an important extension of the scientific method to accommodate the discussion of hypotheses that cannot be tested experimentally[69] and as the only practical way to advance our understanding of physics at energies well beyond current experimental reach. It will be interesting to see which of these viewpoints holds sway as our measurement of cosmological parameters continues to improve. What is undeniable, though, is that the contemplation of a universe in which, for example, P_Φ or Ω_b have very different values or a physics under which planets obey an inverse cube law of gravitation, nuclear interactions fail to produce carbon or even helium, or the fine structure constant is 0.1[70] greatly sharpens our appreciation of the laws that we otherwise take for granted.

EXERCISES

Exercise 28.24 *Example: Blind Cosmologist*
A blind (but hearing) cosmologist observed the radiation-dominated universe. He detected faint tones and noted that their frequencies declined as $t^{-1/2}$ and believed (correctly) that the sound speed is constant. As he was blind, he knew nothing of photons but did understand classical scalar field theory. What did he conclude about $V(\varphi)$?

28.7.2 Dark Matter and the Growth of Structure

The material content of the modern universe is dominated by dark matter and baryons. The existence of each of these constituents is puzzling. Let us consider them in turn.

DARK PARTICLES

It is widely conjectured that dark matter comprises one or more new elementary particles. For specificity, let us suppose that there is a single particle that is neutral (like a neutron), fermionic with spin $\frac{1}{2}$ (like an electron), its own antiparticle (like a photon), weakly interacting (like a neutrino), and stable (like a proton appears to be).[71] Introduce $Z_D \equiv n_D/n_\nu$, the ratio of its density n_D to the neutrino density n_ν, and let its mass be $m_D \equiv m_{D,12}$ TeV. When $T \gg m_D/k_B \sim 10^{16} m_{D,12}$ K, the dark particles were presumably created and destroyed freely in thermal equilibrium with other particles and with zero chemical potential, so that $Z_D = \frac{1}{3}$, as there are three

---

69. It can be argued that many of the most important modern applications of classical physics (e.g., geophysics, climate change, and astrophysics) are of this character, although when we understand the governing principles, we substitute simulation for experiment and Bayesian likelihood for frequentist statistics.
70. In some extreme formulations, even the laws of arithmetic are allowed to vary!
71. Such dark matter particles are often called Weakly Interacting Massive Particles or WIMPs!

neutrino species. Using the current dark matter and neutrino densities from Table 28.1, we find that today $Z_D = 4 \times 10^{-12} m_{D,12}^{-1}$ (Table 28.1).

So, most of the dark matter particles must have vanished. The simplest explanation is that during binary collisions they annihilated to form less massive, relativistic particles when they became mildly relativistic, like electrons and positrons annihilated, consistent with thermodynamic equilibrium (cf. Sec 28.4.1). Eventually the annihilation rate fell below the Hubble expansion rate, and a small relict density was left over, out of equilibrium, constituting the dark matter we see today (Lee and Weinberg, 1977). Using the principles we established in our discussion of nucleosynthesis (Sec. 28.4.2), we can write down immediately a kinetic equation for the dark particles that balances annihilation with the reverse reactions:

relict density

$$\dot{Z}_D = n_\nu \langle \sigma_{DD} v_D \rangle \left(\frac{n_{D\text{th}}^2}{n_\nu^2} - Z_D^2 \right), \quad (28.92a)$$

$$\text{where} \quad n_{D\text{th}}(T, m_D) = 2 \int_0^\infty \frac{4\pi p^2 dp}{h^3} \frac{1}{e^{(m_D^2 + p^2)^{1/2}/(k_B T)} + 1} \quad (28.92b)$$

[cf. Eqs. (28.38)], and where $\langle \sigma_{DD} v_D \rangle$ is the annihilation rate, $n_{D\text{th}}(m_D, T)$ is the density the dark particles would have in thermal equilibrium, and the neutrino density is $n_\nu(T) = 3 n_{D\text{th}}(0, T)$. (The particle speeds v_D are nonrelativistic when they stop annihilating and typically, $\sigma_{DD} \propto v_D^{-1}$.) We assume that the dark particles are kept at the same temperature as the rest of the universe and that the expansion obeyed the same $t \propto a^2$ law as at the start of the photon age. Eq. (28.92a) is now easily solved numerically for different masses to derive the relation $\sigma_{DD}(m_D)$ needed to produce the contemporary value of Z_D. More simply, though less accurately, we note that $\langle \sigma_{DD} v_D \rangle \sim H/n_D \propto m_D a_D$ when $a \sim a_D$ and annihilation ceased. Guided by pair annihilation, we estimate that the associated temperature was $T_D \sim 0.1 m_D / k_B \propto a_D^{-1}$. In order for this to be the origin of dark matter, we require that $\langle \sigma_{DD} v_D \rangle \sim 3 \times 10^{-32}$ m^3 s^{-1}, which is also the result from the more careful calculation and is pretty insensitive to the more detailed assumptions and entirely consistent with the expectations from particle physics.

"WIMP miracle"

We do not understand the properties of the dark particle, and the choices we have just made look like they have been delivered by a committee! (Indeed, every single one of them can be negated and still lead to a viable explanation.) However, these choices do describe the most widely supported explanation for dark matter, namely, that it is the lightest supersymmetric particle.[72] *Supersymmetry* is a promising extension of the standard model of particle physics that postulates the existence of fermionic partners to the bosons of the standard model and vice versa. A major experimental program

supersymmetry

---

72. Alternatives that have been seriously considered and sought include axions (Peccei and Quinn, 1977) and sterile neutrinos (Pontecorvo, 1968).

dark matter searches

under way at the Large Hadron Collider seeks evidence for it. Many *direct* searches are also being conducted in deep mines (to filter out the cosmic ray background) for very rare collisions between dark particles and atomic nuclei. Finally, the small rate of annihilations still going on today might lead to γ-rays and positrons that can be seen, indirectly, by astronomers. All three searches—below, on, and above ground—are under way or are being undertaken in 2016. What is interesting and encouraging is that the sensitivity attainable, in each case, is roughly compatible with the value of $\langle \sigma_{DD} v \rangle$ inferred on the basis of cosmology observations and theoretical calculations.

To date, despite exquisite experiments, no convincing evidence for dark particles has been found. Instead, significant constraints on their properties are being measured by all three techniques, and improvements in sensitivity should be forthcoming over the next several years. Only Nature knows whether we are now on the threshold of identifying most of the matter in the universe and exploring a second standard model of particle physics or if we must look elsewhere for an explanation, but the hunt is on.

BARYOGENESIS

The puzzle over baryons is why there are any of them at all! Protons have antiparticles with which they can annihilate with very large cross sections, so symmetry would seem to suggest that none should have been left over after the temperature fell far below $\sim m_p / k_B \sim 10^{13}$ K. The only way residual baryons could have been created is if some imbalance were created between baryons and antibaryons.[73] The protons (and neutrons, which decay into protons) derived from quarks and primordial asymmetry can be preserved through the quark phase of the universe by a conserved quantum number—the *baryon number*. Essentially, what we seek is some process occurring during the very early universe and involving a significant constituent that creates such an asymmetry. Any serious discussion of this topic involves highly speculative high-energy-physics considerations that we cannot go into here. However, there are essentially classical considerations that should underlie any proposed mechanism, and these we now discuss.

baryon number

Let us suppose that a particle—one with a distinct antiparticle—undergoes a decay that creates a net baryon number. Such a possibility is permitted—and indeed, suggested—by the existence of finite neutrino masses, though it has never been observed. The problem is that on quite general grounds, an antiparticle will create the opposite baryon number and so we have to explain why there is a difference between the amounts of matter and antimatter. Now, the laws of classical physics are tacitly assumed to be the same when the spatial coordinates are inverted: $\mathbf{x} \to -\mathbf{x}$. In the language of particle physics, we say that they are invariant under a *parity* (P) transformation. Likewise, the laws of classical electromagnetism are unchanged if we change

---

73. Equally effective would be a process that created and maintained an asymmetry between electrons and positrons, so that a baryon asymmetry would be needed to preserve charge neutrality. However, baryogenesis is thought to be more likely than *leptogenesis*.

the signs of all the charges. This is invariance under a *charge* (C) transformation. Finally, the microscopic classical laws (e.g., those describing collisions of particles or the evolution of an electromagnetic field) are invariant under time reversal—a *time* (T) transformation. This last observation does not imply that all physical phenomena are reversible. As discussed in Sec. 4.7, the act of averaging (or "coarse graining") introduces the arrow of time, embodied in the second law of thermodynamics.

This need not have been the case and is not true in particle physics. Specifically, it was shown in the 1950s that β decay—a weak interaction—produces left-handed (not right-handed) neutrinos, which violates P symmetry. This was a surprise to many physicists. Likewise, if we were to change the signs of all charges in this decay, we would expect to produce a left-handed antineutrino, and this is not seen either, so C symmetry is violated as well. However, combined C and P transformations lead to the preservation of CP symmetry in β decay. This is important, because a fundamental theorem states that CPT symmetry must be respected, and so a violation of CP would imply a violation of T.

It therefore came as a second, even greater, shock when in the 1960s, experimental measurements of neutral K mesons showed that their decay into two channels that were CP equivalent occurred at different rates. CP symmetry is violated, and Nature can distinguish matter from antimatter and forward from backward in time. For the universe to have actually made this distinction also requires a departure from thermodynamic equilibrium in the past. If this did not happen, the particles and antiparticles would have had identical distribution functions with zero chemical potentials, and no net baryons would have been made. Now, thermodynamic equilibrium would not have been maintained had the expansion been too fast for the particle reactions (cf. Secs. 28.4.1 and 28.4.2). An alternative possibility is that the particles did not scatter, so that their momenta evolved according to $p \propto a^{-1}$ (cf. Sec. 28.2.3). This automatically leads to a nonthermal distribution function[74] and allows baryon asymmetry to proceed. We are probably a long way from understanding the detailed particle physics of baryogenesis, but determining that its general requirements could plausibly have been satisfied is a good first step.

Exercise 28.25 *Challenge: Including More Details*

We have made many simplifying assumptions in this chapter to demonstrate the strong connection to the principles and techniques developed in the preceding 27 chapters. It is possible to improve on our standard cosmological model by being more careful without introducing anything fundamentally new. (The research frontier, of course, is advancing fast and contains much we have not attempted to explain.) Consider how to implement the following corrections to standard cosmology that

---

74. That all three violations—baryon number, C/CP symmetry, and thermal equilibrium—were necessary was first recognized by Sakharov (1965).

are mandated by observation, and then repeat the calculations with these changes, comparing with the research literature.

(a) Changing the slope of the initial cosmic noise spectrum (Sec. 28.5.3).

(b) Including more spherical harmonics in the description of the radiation field and using the Boltzmann equation instead of a Monte Carlo simulation (Sec. 28.6.1).

(c) Including tensor perturbations in the Robertson-Walker metric (Secs. 28.5.1, 28.5.3, and 28.7.1).

(d) Accounting for neutrino mass (Sec. 28.3.4).

Exercise 28.26 *Challenge: Testing Standard Cosmology*
There are many elaborations of standard cosmology either involving new features following from known physics or involving new physics. While no convincing evidence exists for any of them as of this writing, they are all being actively sought. Explain how to generalize standard cosmology to accommodate these possibilities and to test for them, repeating calculations, where possible.

(a) Space curvature (Sec. 28.2.2).

(b) Dark energy. Adopt an empirical equation of state with the parameter w being constant and slightly greater than -1 (Sec. 28.7.3).

(c) Additional neutrino flavors, including *sterile* neutrinos (Secs. 28.4.1 and 28.4.2).

(d) Nonadiabatic and non-Gaussian initial fluctuations (Sec. 28.5.3). [Hint: Consulting an advanced textbook is recommended.]

28.7.3 The Cosmological Constant and the Fate of the Universe

We have explained (Sec. 28.3.5) that the observed acceleration of the universe requires an effective negative pressure in the stress-energy tensor, and how such a possibility was anticipated by Einstein (1917). Furthermore, we have explained that Einstein's cosmological constant is consistent with the data. This is important, because it is a simple prescription whose consequences can be computed without too many extraneous assumptions and is therefore readily falsifiable. However, there are also theoretical reasons for questioning this interpretation; they, too, lie at the interface between quantum and classical physics.

The earliest view of the cosmological constant (implicitly, that of Einstein) was that its presence in the field equations is an expression of the true law of gravitation necessary on a large scale, in an analogous fashion to Newton proposing the inverse square law for planets and expanding our understanding of gravitational force beyond what is needed on the surface of Earth. The ultimate connection between the general relativity of, say, stellar-mass black holes and that of the universe at large has yet to be divined, but it should exist independent of the messy details of our cosmic environment. Many interesting ideas have been proposed, in particular those involving extra spatial dimensions and oscillating universes. There have even been proposals to dispense with dark matter particles altogether and to interpret the observed motions of stars and galaxies in terms of modified Newtonian gravity or entanglement.

Today it is more common to view the field equations (Sec. 25.8) as providing a complete framework for describing gravitation and the cosmological constant as being just one of several contributions to the stress-energy tensor. This is the approach we followed in Sec. 28.3.5. Its most natural identification is with the quantum mechanical vacuum. Having $\boldsymbol{T}_\Lambda \propto \boldsymbol{g}$ ensures that the vacuum looks the same in all local Lorentz frames, consistent with the principle of relativity. However, attempts to develop this relationship quantitatively have mostly foundered on the tiny size of Λ relative to the Planck scale of quantum gravity[75]—$G^2 \hbar \rho_\Lambda \sim 10^{-122}$. Either this represents an unprecedented degree of fine tuning, or the cosmological constant has even less to do with quantum mechanics than, say, oceanography, and it is instead some ungrasped expression of the "fabric" of spacetime on a supraclassical scale.[76] If so, it throws down a challenge to the *reductionist* view of physics, under which physics at essentially all scales (especially classical physics) is viewed as being derivative of physics at the smallest lengthscales and highest energies. The properties of materials depend on the behavior of atoms and molecules, which depend on electrons and nuclei, which depend on quantum electrodynamics and the interactions of quarks, and so on. This does not preclude the existence, interest, or importance of *emergent* phenomena (e.g., ferromagnetism, shock fronts, or astrophysical black holes) that require appealing to the properties of matter in bulk, but it is a statement of faith that ultimately, the governing principles of these phenomena are reducible to the physics of the smallest scales, even if this is not useful in describing what we observe. In this sense Λ might be like ferromagnetism.

However, the square peg of quantum field theory need not be forced into the round hole of Riemannian geometry. The very existence of something like a cosmological constant should also cause us to inquire whether some physics is derivative of the largest scales and lowest energies instead of the smallest scales and the highest energies. Consider, allegorically, a bug in a still pond, living its low-Reynolds-number life. A fish swims by, and the bug finds itself in the fish's turbulent wake. It observes that its food is moving more rapidly with an average speed that increases as the cube root of its distance [Eq. (15.23)] and, if it is very sensitive, it might find the water

---

75. A more intriguing possibility is that the natural, quantum mechanical mass scale for Λ is $\sim (\hbar^3 \rho_\Lambda / c^3)^{1/4} \sim 10$ meV, similar to the neutrino mass scale. However, there is no known good reason for this.
76. It is interesting that the cosmological constant was taken very seriously before the 1980s (e.g., Lemaître, 1934; Tolman, 1934; Bondi, 1952a; Zel'dovich, 1968) because it allowed the universe to be older than its contents and because it provided a scale distinct from those associated with atoms, nuclei, and elementary particles (e.g., Eddington, 1933). However, the challenge of reconciling it with quantum field theory then led to its near abandonment. The subsequent discovery of cosmic acceleration therefore came as a surprise to many physicists.

heating up a little. While it is true that the Navier-Stokes equations can be derived by suitable averaging of a kinetic theory, if the bug desires to reconcile its observations with their causes, it is to the outer scale of a turbulent spectrum that it should turn, not the properties of water molecules. The speculations involving the anthropic principle and the multiverse contain some of this spirit. Perhaps Λ is "situational"—our first glimpse of physics on the large scale—just like blackbody radiation and the photo-electric effect opened the door to the physics of small scales and quantum mechanics more than a century ago. Of course, this physics, dependent on conditions at and beyond our current horizon, might reflect what happened immediately prior to our observable universe leaving the horizon during inflation; alternatively, it might reflect the accidental properties of our contemporary cosmological environment.

A more pragmatic approach is to consider generalizations of the cosmological constant. These include *dark energy* (or *quintessence*; Perlmutter et al., 1999). In particular, the behavior of the stress-energy tensor has been parameterized by introducing an equation of state $P_\Lambda = w\rho_\Lambda$, where w is negative, although there is no compelling theoretical reason for doing this. Several observational studies have concluded that $|w + 1| \lesssim 0.08$, and more accuracy is in the works. Another common approach is to invoke classical field theory (just as we did for inflation) and to use the same formalism to develop a description of contemporary inflation. It is tempting to suppose that there have been many instances of inflation over the history of the universe, and that Nature has managed to find graceful exits from every one of these expansions in the past and, in the fullness of time, it will do so again. However, there is no requirement that this will ever happen. Another way to connect to inflation is to suppose that the bottom of the inflaton potential well is associated with some sort of quantum mechanical zero-point energy or a classical offset, and that this is what the universe is experiencing now.

A particularly interesting outcome would be if future observations demonstrated that $-P_\Lambda > \rho_\Lambda > 0$; for example, if $w < -1$.[77] This condition corresponds to negative cosmological enthalpy and negative kinetic energy for a scalar field (cf. Sec. 28.7.1). Not only would it exclude a simple scalar field; it would also, if taken literally, predict an unusual fate for the universe. If $w = $ const, then the first law of thermodynamics implies that eventually $\rho \propto a^{-3(w+1)}$. The energy density increases as the universe expands if $w < -1$; the universe reaches infinite size in finite time, while the horizon shrinks and closer and closer neighbors disappear, a behavior dubbed "the big rip." Despite its eschatological fascination, many cosmologists reject this outcome on grounds of physics inconsistency and regard $w \geq -1$ as a prediction.

As with the experimental searches for dark matter, the prospects for learning more about the expansion and fate of the universe from astronomical studies over the next decade are bright, and the observations will presumably continue to corral speculative

---

77. This is sometimes known as a violation of the *weak energy condition*.

interpretation. At present, progress in cosmology seems to be following that in the standard model of particle physics, where a relatively simple physical model suffices to account for essentially all data pertaining to questions addressed by the model, while leaving other questions to be confronted by new physics and future experiments. Three familiar constituents, baryons, neutrinos, and photons—supplemented by dark matter, zero curvature, and a cosmological constant and imprinted with an almost scale-free spectrum of adiabatic, Gaussian potential perturbations—suffice to account for most of what we measure. It will be fascinating to see whether this apparent simplicity is maintained through the next phase of cosmological exploration, when the true nature of inflation, dark matter, and the cosmological constant will likely be a focus of attention.

Bibliographic Note

The literature on cosmology is enormous, and it is not hard to find excellent texts and research papers to elaborate on the many topics we have touched on in this chapter. Arguably, the most useful advanced text is Weinberg (2008). Other books that are helpful on the physics but are rather out of date observationally are Padmanabhan (1993); Peebles (1993); Kolb and Turner (1994); Peacock (1999); Dodelson (2003); and Hobson, Efstathiou, and Lasenby (2006). Texts that emphasize the early universe include Liddle and Lyth (2000) and Mukhanov (2005). An especially lucid and up-to-date textbook accessible to undergraduates that emphasizes observational cosmology is Schneider (2015). Excellent discussions of cosmology from a more elementary standpoint include Ryden (2002) and Hartle (2003). Among the most important contemporary observations are the CMB results from the Planck (Planck Collaboration, 2016a, 2016b) and Wilkinson Microwave Anisotropy Probe (WMAP) (Komatsu et al., 2011) satellites, one of several careful analyses of weak-lensing observations (Heymans et al., 2012), and the preliminary results on galaxy clustering from the Baryon Oscillation Spectroscopic Survey (BOSS) galaxy survey (Anderson et al., 2014).

REFERENCES

Abbott, B. P., R. Abbott, T. D. Abbott, M. R. Abernathy, et al. (2016a). Observation of gravitational waves from a binary black hole merger. *Physical Review Letters* **116**, 061102.

——— (2016b). GW150914: The advanced LIGO detectors in the era of first discoveries. *Physical Review Letters* **116**, 131103.

Abbott, B. P., R. Abbott, R. Adhikari, P. Ajith, et al. (2009a). Observation of a kilogram-scale oscillator near its quantum ground state. *New Journal of Physics* **11**, 1–13.

——— (2009b). LIGO: the Laser Interferometer Gravitational-wave Observatory. *Reports on Progress in Physics* **72**, 076901.

Abernathy, F. (1968). National Committee for Fluid Mechanics Films movie: Fundamentals of boundary layers.

Ablowitz, M. J. (2011). *Nonlinear Dispersive Waves: Asymptotic Analysis and Solitons*. Cambridge: Cambridge University Press.

Acheson, D. J. (1990). *Elementary Fluid Dynamics*. Oxford: Clarendon Press.

Adair, R. K. (1990). *The Physics of Baseball*. New York: Harper and Row.

Ade, P. A. R., R. W. Aikin, D. Barkats, S. J. Benton, et al. (2014). BICEP2 I: Detection of B-mode polarization at degree angular scales. *Physical Review Letters* **112**, 241101.

Aichelberg, P. C., and R. U. Sexl (1971). On the gravitational field of a massless particle. *General Relativity and Gravitation* **2**, 303–312.

Albrecht, A., and P. Steinhardt (1982). Cosmology for grand unified theories with radiatively induced symmetry breaking. *Physical Review Letters* **48**, 1220–1223.

Alfvén, H. (1970). Plasma physics, space research and the origin of the solar system. Nobel lecture. Available at http://www.nobelprize.org/nobel_prizes/physics/laureates/1970/alfven-lecture.pdf. Chapter 19 epigraph reprinted with permission of The Nobel Foundation.

Alligood, K. T., T. D. Sauer, and J. A. Yorke (1996). *Chaos, an Introduction to Dynamical Systems*. Berlin: Springer-Verlag.

Allis, W. P., S. J. Buchsbaum, and A. Bers (1963). *Waves in Anisotropic Plasmas*. Cambridge, Mass.: MIT Press.

Almheiri, A., D. Marolf, J. Polchinski, and J. Sully (2013). Black holes: Complementarity or firewalls? *Journal of High Energy Physics* **2**, 62–78.

Anderson, J. D. (2003). *Modern Compressible Flow: With Historical Perspective*. New York: McGraw-Hill.

Anderson, L., E. Aubourg, S. Bailey, F. Beutler, et al. (2014). The clustering of galaxies in the SDSS-III baryon oscillation spectroscopic survey: Baryon acoustic oscillations in the data releases 10 and 11 galaxy samples. *Monthly Notices of the Royal Astronomical Society* **441**, 24–62.

Anderson, M. H., J. R. Ensher, M. R. Matthews, C. E. Wieman, and E. A. Cornell (1995). Observation of Bose-Einstein condensation in a dilute atomic vapor. *Science* **269**, 198–201.

Andersson, N., and G. L. Comer (2007). Relativistic fluid dynamics: Physics for many different scales. *Living Reviews in Relativity* **10**, 1.

Archimedes (ca 250 BC). Exclamation (perhaps apocryphal) upon discovering a variant of Archimedes' Law, which is presented in his book *On Floating Bodies*.

Arfken, G. B., H. J. Weber, and F. E. Harris (2013). *Mathematical Methods for Physicists*. Amsterdam: Elsevier.

Armenti, A. J. (1992). *The Physics of Sports*. New York: American Institute of Physics.

Arnol'd, V. I. (1992). *Catastrophe Theory*. Cham, Switzerland: Springer.

Ashby, N., and J. Dreitlein (1975). Gravitational wave reception by a sphere. *Physical Review D* **12**, 336–349.

Bachman, H. E. (1994). *Vibration Problems in Structures*. Basel: Birkhauser.

Baker, G. L., and J. P. Gollub (1990). *Chaotic Dynamics, An Introduction*. Cambridge: Cambridge University Press.

Basov, N. G., and A. M. Prokhorov (1954). First Russian ammonia maser. *Journal of Experimental and Theoretical Physics* **27**, 431–438.

——— (1955). Possible methods for obtaining active molecules for a molecular oscillator. *Journal of Experimental and Theoretical Physics* **28**, 249–250.

Batchelor, G. K. (2000). *An Introduction to Fluid Dynamics*. Cambridge: Cambridge University Press.

Bateman, G. (1978). *MHD Instabilities*. Cambridge, Mass.: MIT Press.

Båth, M. (1966). Earthquake energy and magnitude. In L. H. Ahrens, F. Press, S. K. Runcorn, and H. C. Urey (eds.), *Physics and Chemistry of the Earth*, pp. 115–165. Oxford: Pergamon.

Baumgarte, T. W., and S. Shapiro (2010). *Numerical Relativity: Solving Einstein's Equations on the Computer*. Cambridge: Cambridge University Press.

Beckwith, S. V., M. Stiavelli, A. M. Koekemoer, J. A. R. Caldwell, et al. (2006). The Hubble Ultra Deep Field. *Astronomical Journal* **132**, 1729–1755.

Bejan, A. (2013). *Convection Heat Transfer*. New York: Wiley.

Bekefi, G. (1966). *Radiation Processes in Plasmas*. New York: Wiley.

Bekenstein, J. (1952). Black holes and the second law. *Lettre al Nuovo Cimento* **4**, 737–740.

Bellan, P. M. (2000). *Spheromaks*. London: Imperial College Press.

——— (2006). *Fundamentals of Plasma Physics*. Cambridge: Cambridge University Press.

Bennett, C. A. (2008). *Principles of Physical Optics*. New York: Wiley.

Bennett, C. L., M. Halpern, G. Hinshaw, and N. E. Jarosik (2003). First-year Wilkinson Microwave Anisotropy Probe (WMAP) Observations: Preliminary Maps and Basic Results. *Astrophysical Journal Supplement* **148**, 1–28.

Bernstein, I. B., J. M. Greene, and M. D. Kruskal (1957). Exact nonlinear plasma oscillations. *Physical Review* **108**, 546–550.

Bernstein, I. B., E. A. Frieman, M. D. Kruskal, and R. M. Kulsrud (1958). An energy principle for hydromagnetic stability problems. *Proceedings of the Royal Society A* **244**, 17–40.

Berry, M. (1990). Anticipations of the geometric phase. *Physics Today* **43**, 34–40.

Berry, M. V., and C. Upstill (1980). Catastrophe optics: Morphologies of caustics and their diffraction patterns. *Progress in Optics* **18**, 257–346.

Bertotti, B. (1959). Uniform electromagnetic field in the theory of general relativity. *Physical Review* **116**, 1331–1333.

Bertotti, B., L. Iess, and P. Tortora (2003). A test of general relativity using radio links with the Cassini spacecraft. *Nature* **425**, 374–376.

Biedermann, G. W., X. Wu, L. Deslauriers, S. Roy, C. Mahadeswaraswamy, and M. A. Kasevich (2015). Testing gravity with cold-atom interferometers. *Physical Review A* **91**, 033629.

Binney, J. J., and S. Tremaine (2003). *Galactic Dynamics*. Princeton, N.J.: Princeton University Press.

Birkhoff, G. (1923). *Relativity and Modern Physics*. Cambridge, Mass.: Harvard University Press.

Birn, J., and E. Priest (2007). *Reconnection of Magnetic Fields*. Cambridge: Cambridge University Press.

Bittencourt, J. A. (2004). *Fundamentals of Plasma Physics*. Berlin: Springer-Verlag.

Black, E. D. (2001). An introduction to Pound–Drever–Hall laser frequency stabilization. *American Journal of Physics* **69**, 79–87.

Blanchet, L. (2014). Gravitational radiation from post-Newtonian sources and inspiraling compact binaries. *Living Reviews in Relativity* **17**, 2.

Blandford, R. D., and D. Eichler (1987). Particle acceleration at astrophysical shocks: A theory of cosmic ray origin. *Physics Reports* **154**, 1–75.

Blandford, R. D., and R. Narayan (1992). Cosmological applications of gravitational lensing. *Annual Reviews of Astronomy and Astrophysics* **30**, 311–358.

Blandford R. D., and M. J. Rees (1974). A "twin-exhaust" for double radio sources. *Monthly Notices of the Royal Astronomical Society* **169**, 395–415.

Blandford, R. D., and R. L. Znajek (1977). The electromagnetic extraction of energy from Kerr black holes. *Monthly Notices of the Royal Astronomical Society* **179**, 433–456.

Blandford, R. D., A. B. Saust, T. G. Brainerd, and J. Villumsen (1991). The distortion of distant galaxy images by large-scale structure. *Monthly Notices of the Royal Astronomical Society* **251**, 600–627.

Boas, M. L. (2006). *Mathematical Methods in the Physical Sciences*. New York: Wiley.

Bogolyubov, N. N. (1962). Problems of a dynamical theory in statistical physics. In J. de Boer and G. E. Uhlenbeck (eds.), *Studies in Statistical Mechanics*, p. 1. Amsterdam: North Holland.

Bondi, H. (1952a). *Cosmology*. Cambridge: Cambridge University Press.

—— (1952b). On spherically symmetric accretion. *Monthly Notices of the Royal Astronomical Society* **112**, 195–204.

Boresi, A. P., and K. P. Chong (1999). *Elasticity in Engineering Mechanics*. New York: Wiley.

Born, M., and H. S. Green (1949). *A General Kinetic Theory of Liquids*. Cambridge: Cambridge University Press.

Born, M., and E. Wolf (1999). *Principles of Optics: Electromagnetic Theory of Propagation, Interference and Diffraction of Light*. Cambridge: Cambridge University Press.

Boyd, R. W. (2008). *Nonlinear Optics*. New York: Academic Press.

Boyd, T. J. M., and J. J. Sanderson (2003). *The Physics of Plasmas*. Cambridge: Cambridge University Press.

Brady, P. R., S. Droz, and S. M. Morsink (1998). The late-time singularity inside non-spherical black holes. *Physical Review D* **58**, 084034–084048.

Braginsky, V. B., and F. Y. Khalili (1992). *Quantum Measurement*. Cambridge: Cambridge University Press.

Braginsky, V. B., M. L. Gorodetsky, and S. P. Vyatchanin (1999). Thermodynamical fluctuations and photo-thermal shot noise in gravitational wave antennae. *Physics Letters A* **264**, 1–10.

Brenner, M. P., S. Hilgenfeldt, and D. Lohse (2002). Single bubble sonoluminescence. *Reviews of Modern Physics* **74**, 425–484.

Brooker, G. (2003). *Modern Classical Optics*. Oxford: Oxford University Press.

Brown, L., and G. Gabrielse (1986). Geonium theory: Physics of a single electron or ion in a Penning trap. *Reviews of Modern Physics* **58**, 233–311.

Brown, R. (1828). XXVII. A brief account of microscopical observations made in the months of June, July and August 1827, on the particles contained in the pollen of plants; and on the general existence of active molecules in organic and inorganic bodies. *Philosophical Magazine* **4**, 161–173.

Bryson, A. (1964). National Committee for Fluid Mechanics Films movie: Waves in fluids.

Burke, W. L. (1970). Runaway solutions: Remarks on the asymptotic theory of radiation damping. *Physical Review A* **2**, 1501–1505.

——— (1971). Gravitational radiation damping of slowly moving systems calculated using matched asymptotic expansions. *Journal of Mathematical Physics* **12**, 402–418.

Cahn, R., and G. Goldhaber (2009). *The Experimental Foundations of Particle Physics*. Cambridge: Cambridge University Press.

Callen, H. B., and T. A. Welton (1951). Irreversibility and generalized noise. *Physical Review* **83**, 34–40.

Carroll, S. M. (2004). *Spacetime and Geometry. An Introduction to General Relativity*. New York: Addison-Wesley.

Carter, B. (1979). The general theory of the mechanical, electromagnetic and thermodynamic properties of black holes. In S. W. Hawking and W. Israel (eds.), *General Relativity, an Einstein Centenary Survey*. Cambridge: Cambridge University Press.

Cathey, W. T. (1974). *Optical Information Processing and Holography*. New York: Wiley.

Caves, C. M. (1980). Quantum-mechanical radiation-pressure fluctuations in an interferometer. *Physical Review Letters* **45**, 75–79.

——— (1981). Quantum mechanical noise in an interferometer. *Physical Review D* **23**, 1693–1708.

Chan, J., T. M. Alegre, A. H. Safavi-Naeini, J. T. Hill, et al. (2011). Laser cooling of a nanomechanical oscillator into its quantum ground state. *Nature* **478**, 89–92.

Chandler, D. (1987). *Introduction to Modern Statistical Mechanics*. Oxford: Oxford University Press.

Chandrasekhar, S. (1939). *Stellar Structure*. Chicago: University of Chicago Press.

——— (1961). *Hydrodynamics and Hydromagnetic Stability*. Oxford: Oxford University Press.

——— (1962). *Ellipsoidal Figures of Equilibrium*. New Haven, Conn.: Yale University Press.

——— (1983). *The Mathematical Theory of Black Holes*. Oxford: Oxford University Press.

Chelton, D., and M. Schlax (1996). Global observations of oceanic Rossby waves. *Science* **272**, 234–238.

Chen, F. F. (1974). *Introduction to Plasma Physics*. New York: Plenum Press.

——— (2016). *Introduction to Plasma Physics and Controlled Fusion*, third edition. Heidelberg: Springer.

Chew, G. F., M. L. Goldberger, and F. E. Low (1956). The Boltzmann equation and the one-fluid hydromagnetic equations in the absence of particle collisions. *Proceedings of the Royal Society A* **236**, 112–118.

Chou, C. W., D. B. Hume, T. Rosenband, and D. J. Wineland (2010). Optical clocks and relativity. *Science* **329**, 1630–1633.

Chu, S., C. Cohen-Tannoudji, and W. D. Phillips (1998). Nobel lectures. *Reviews of Modern Physics* **70**, 685–742.

Ciufolini, I., et al. (2016). A test of general relativity using the LARES and LAGEOS satellites and a GRACE Earth gravity model. Measurement of Earth's dragging of inertial frames. *The European Physical Journal C* **76**, 120.

Clash, J. (2012). An interview with the late Sally Ride. Available at http://www.askmen.com/entertainment/right-stuff/sally-ride-interview.html. Chapter 17 epigraph reprinted with permission of Jim Clash.

Clayton, D. D. (1968). *Principles of Stellar Evolution and Nucleosynthesis*. Chicago: University of Chicago Press.

Clemmow, P. C., and J. P. Dougherty (1969). *Electrodynamics of Particles and Plasmas*. New York: Addison-Wesley.

Cohen-Tannoudji, C., B. Diu, and F. Laloë (1977). *Quantum Mechanics*. New York: Wiley.

Cole, J. (1974). *Perturbation Methods in Applied Mathematics*. New York: Blaisdell Publishing.

Coles, D. (1965). National Committee for Fluid Mechanics Films movie: Channel flow of a compressible fluid.

Constantinescu, A., and A. Korsunsky (2007). *Elasticity with Mathematica*. Cambridge: Cambridge University Press.

Copson, E. T. (1935). *An Introduction to the Theory of Functions of a Complex Variable*. Oxford: Oxford University Press.

Cornell, E. (1996). Very cold indeed: The nanokelvin physics of Bose-Einstein condensation. *Journal of Research of NIST* **101**, 419–434.

Cox, Arthur N., ed. (2000). *Allen's Astrophysical Quantities*. Cham, Switzerland: Springer.

Creighton, J. D. E., and W. G. Anderson (2011). *Listening to the Universe*. New York: Wiley.

Crookes, W. (1879). The Bakerian lecture: On the illumination of lines of molecular pressure, and the trajectory of molecules. *Philosophical Transactions of the Royal Society* **170**, 135–164.

Cundiff, S. T. (2002). Phase stabilization of ultrashort optical pulses. *Journal of Physics D* **35**, 43–59.

Cundiff, S. T., and J. Ye (2003). Colloquium: Femtosecond optical frequency combs. *Reviews of Modern Physics* **75**, 325–342.

Cushman-Roisin, B., and J.-M. Beckers (2011). *Introduction to Geophysical Fluid Dynamics*. New York: Academic Press.

Cutler, C., and K. S. Thorne (2002). An overview of gravitational wave sources. In N. Bishop and S. D. Maharaj (eds.), *Proc. GR16 Conference on General Relativity and Gravitation*, pp. 72–111. Singapore: World Scientific.

Cyburt, R. H., B. D. Fields, K. A. Olive, and T.-H. Yeh (2016). Big bang nucleosynthesis: Present status. *Reviews of Modern Physics* **88**, 015004.

Dalfovo, F., S. Giorgini, L. P. Pitaevskii, and S. Stringari (1999). Theory of Bose-Einstein condensation in trapped gases. *Reviews of Modern Physics* **71**, 463–512.

Darwin, C. (1959). The gravity field of a particle. *Proceedings of the Royal Society A* **249**, 180–194.

Dauxois, T., and M. Peyrard (2010). *Physics of Solitons*. Cambridge: Cambridge University Press.

Davidson, P. A. (2001). *An Introduction to Magnetohydrodynamics*. Cambridge: Cambridge University Press.

——— (2005). *Turbulence: An Introduction for Scientists and Engineers*. Oxford: Oxford University Press.

Davidson, R. C. (1972). *Methods in Nonlinear Plasma Theory*. New York: Academic Press.

Davies, P. C. W. (1977). The thermodynamic theory of black holes. *Proceedings of the Royal Society A* **353**, 499–521.

Davison, L. (2010). *Fundamentals of Shock Wave Propagation in Solids*. Berlin: Springer-Verlag.

DeMorest, P. B., T. Pennucci, S. M. Ransom, M. S. E. Roberts, et al. (2010). A two solar mass neutron star measured using Shapiro delay. *Nature* **467**, 1081–1083.

Dicke, R. H. (1961). Dirac's cosmology and Mach's principle. *Nature* **192**, 440–441.

Dodelson, S. (2003). *Modern Cosmology*. New York: Academic Press.

Doob, J. L. (1942). The Brownian movement and stochastic equations. *Annals of Mathematics* **43**, 351–369.

Dorf, R. C., and R. H. Bishop (2012). *Modern Control Systems*. Upper Saddle River, N.J.: Pearson.

Drazin, P. G., and R. S. Johnson (1989). *Solitons: An Introduction*. Cambridge: Cambridge University Press.

Drazin, P. G., and W. H. Reid (2004). *Hydrodynamic Stability*. Cambridge: Cambridge University Press.

Duderstadt, J. J., and L. J. Hamilton (1976). *Nuclear Reactor Analysis*. New York: Wiley.

Dyson, F. (1986). Quoted in T. A. Heppenheimer, After the Sun dies. *Omni* **8**, no. 11, 38. Chapter 23 epigraph reprinted with permission of Freeman Dyson.

Eckart, C. (1940). The thermodynamics of irreversible processes, III: Relativistic theory of the simple fluid. *Physical Review* **58**, 919–924.

Eddington, A. S. (1919). The total eclipse of 1919 May 29 and the influence of gravitation on light. *Observatory* **42**, 119–122.

——— (1922). *The Mathematical Theory of Relativity*. Cambridge: Cambridge University Press.

——— (1927). March 1927 Gifford lecture at the University of Edinburgh. As published in Arthur S. Eddington: *The Nature of the Physical World* Gifford Lectures of 1927: An Annotated Edition by H. G. Callaway. Cambridge: Cambridge Scholars Publishing (2014). Chapter 16 epigraph reprinted with permission of the publisher.

——— (1933). *The Expanding Universe*. Cambridge: Cambridge University Press.

Einstein, A. (1907). Über das Relativitätsprinzip und die ausdemselben gesogenen Folgerungen. *Jahrbuch der Radioaktivität und Elektronik* **4**, 411–462.

——— (1915). Die Feldgleichungen der Gravitation. *Sitzungsberichte der Preussischen Akademie* **1915**, 844–847.

——— (1916a). Die Grundlage der allgemeinen Relativitätstheorie. *Annalen der Physik* **49**, 769–822.

——— (1916b). Näherungsweise Integration der Feldgleichungen der Gravitation. *Sitzungsberichte der Preussischen Akademie der Wissenschaften* **1916**, 688–696.

——— (1917). Kosmologische Betrachtungen zur allgemeinen Relativitätstheorie. *Sitzungsberichte der Preussischen Akademie der Wissenschaften* **1917**, 142–152.

——— (1918). Über Gravitationswellen. *Sitzungsberichte der Preussischen Akademie der Wissenschaften* **1918**, 154–167.

——— (1925). Quantum theory of ideal gases. *Sitzungsberichte der Preussischen Akademie der Wissenschaften* **3**, 18–25.

——— (1931). Zum kosmologischen Problem der allgemeinen Relativitätstheorie. *Sitzungsberichte der Preussischen Akademie der Wissenschaften* **1931**, 235–237.

——— (1934) On the method of theoretical physics. *Philosphy of Science* **1**, 163–169. Published version of Einstein's Herbert Spencer Lecture, delivered at Oxford, June 10, 1933. Chapter 25 epigraph reprinted with permission of the University of Chicago Press.

——— (1989). *The Collected Papers of Albert Einstein*. Princeton, N.J.: Princeton University Press.

Eisenstein, D. J., I. Zehavi, D. W. Hogg, R. Scoccimarro, et al. (2005). Detection of the baryon acoustic peak in the large-scale correlation function of SDSS luminous red galaxies. *Astrophysical Journal* **633**, 560–574.

Ensher, J. R., D. S. Jin, M. R. Matthews, C. E. Wieman, and E. A. Cornell (1996). Bose-Einstein condensation in a dilute gas: Measurement of energy and ground-state occupation. *Physical Review Letters* **77**, 4984–4987.

Eringen, A. C., and E. S. Suhubi (1975). *Elastodynamics, Vol. II: Linear Theory*. New York: Academic Press.

Everett, A., and T. Roman (2011). *Time Travel and Warp Drives: A Scientific Guide to Shortcuts through Time and Space*. Chicago: University of Chicago Press.

Everitt, C. W. F., D. B. DeBra, B. W. Parkinson, J. P. Turneaure, et al. (2011). Gravity Probe B: Final results of a space experiment to test general relativity. *Physical Review Letters* **106**, 221101.

Faber, T. E. (1995). *Fluid Dynamics for Physicists*. Cambridge: Cambridge University Press.

Faraday, M. (1846). Thoughts on ray vibrations. *Philosophical Magazine* **140**, 147–161.

Farquhar, I. E. (1964). *Ergodic Theory in Statistical Mechanics*. London: Interscience.

Feigenbaum, M. (1978). Universal behavior in nonlinear systems. *Journal of Statistical Physics* **19**, 25–52.

Fenstermacher, P. R., H. L. Swinney, and J. P. Gollub (1979). Dynamical instabilities and the transition to chaotic Taylor vortex flow. *Journal of Fluid Mechanics* **94**, 103–128.

Feynman, R. P. (1966). *The Character of Physical Law*. Cambridge, Mass.: MIT Press.

——— (1972). *Statistical Mechanics*. New York: Benjamin.

Feynman, R. P., R. B. Leighton, and M. Sands (1964). *The Feynman Lectures on Physics*. Reading, Mass.: Addison-Wesley. Chapter 14 epigraph reprinted with permission of Caltech.

Fierz, M., and W. Pauli (1939). On relativistic wave equations for particles of arbitrary spin in an electromagnetic field. *Proceedings of the Royal Society A* **173**, 211–232.

Finkelstein, D. (1958). Past-future asymmetry of the gravitational field of a point particle. *Physical Review* **110**, 965–967.

Flanders, H. (1989). *Differential Forms with Applications to the Physical Sciences*, corrected edition. Mineola, N.Y.: Courier Dover Publications.

Fletcher, C. A. J. (1991). *Computational Techniques for Fluid Dynamics, Vol I: Fundamental and General Techniques*. Berlin: Springer-Verlag.

Forbes, T., and E. Priest (2007). *Magnetic Reconnection*. Cambridge: Cambridge University Press.

Fortere, Y., J. M. Skothelm, J. Dumals, and L. Mahadevan (2005). How the Venus flytrap snaps. *Nature* **433**, 421–425.

Francon, M., and I. Willmans (1966). *Optical Interferometry*. New York: Academic Press.

Franklin, G. F., J. D. Powell, and A. Emami-Naeini (2005). *Feedback Control of Dynamic Systems*. Upper Saddle River, N.J.: Pearson.

Fraunhofer, J. von (1814–1815). Determination of the refractive and color-dispersing power of different types of glass, in relation to the improvement of achromatic telescopes. *Denkschriften der Königlichen Academie der Wissenschaften zu München* **5**, 193–226.

Frautschi, S. (1982). Entropy in an expanding universe. *Science* **217**, 593–599.

Friedman, J., and A. Higuchi (2006). Topological censorship and chronology protection. *Annalen der Physik* **15**, 109–128.

Friedmann, A. A. (1922). Über die Krümmung des Raumes. *Zeitschrift für Physik* **10**, 377–386.

Frolov, V. P., and I. D. Novikov (1990). Physical effects in wormholes and time machines. *Physical Review D* **42**, 1057–1065.

——— (1998). *Black Hole Physics: Basic Concepts and New Developments*. Dordrecht: Kluwer.

Frolov, V. P., and D. N. Page (1993). Proof of the generalized second law for quasistationary semiclassical black holes. *Physical Review Letters* **71**, 3902–3905.

Frolov, V. P., and A. Zelnikov (2011). *Introduction to Black Hole Physics*. Oxford: Oxford University Press.

Fuller, R. W., and J. A. Wheeler (1962). Causality and multiply connected spacetime. *Physical Review* **128**, 919–929.

Fultz, D. (1969). National Committee for Fluid Mechanics Films movie: Rotating flows.

Galleani, L. (2012). The statistics of the atomic clock noise. In L. Cohen et al. (eds.), *Classical, Semi-classical and Quantum Noise*, pp. 63–77. Cham, Switzerland: Springer Science + Business Media.

Genzel, R., F. Eisenhauer, and S. Gillessen (2010). The galactic center massive black hole and nuclear star cluster. *Reviews of Modern Physics* **82**, 3121–3195.

Ghatak, A. (2010). *Optics*. New Delhi: McGraw-Hill.

Ghez, A. M., S. Salim, N. N. Weinberg, J. R. Lu, et al. (2008). Measuring distance and properties of the Milky Way central supermassive black hole with stellar orbits. *Astrophysical Journal* **689**, 1044–1062.

Gibbs, J. W. (1881). Letter accepting the Rumford Medal. Quoted in A. L. Mackay, *Dictionary of Scientific Quotations*. London: IOP Publishing.

——— (1902). *Elementary Principles in Statistical Mechanics*. New York: Charles Scribner's Sons.

Gill, A. E. (1982). *Atmosphere-Ocean Dynamics*. New York: Academic Press.

Gladwell, G. M. L. (1980). *Contact Problems in the Classical Theory of Elasticity*. Alphen aan den Rijn: Sijthoff and Noordhoff.

Goedbloed, J. P., R. Keppens, and S. Poedts (2010). *Advanced Magnetohydrodynamics*. Cambridge: Cambridge University Press.

Goedbloed, J. P., and S. Poedts (2004). *Principles of Magnetohydrodynamics, with Applications to Laboratory and Astrophysical Plasmas*. Cambridge: Cambridge University Press.

Goldstein, H., C. Poole, and J. Safko (2002). *Classical Mechanics*. New York: Addison-Wesley.

Gollub, J. P., and S. V. Benson (1980). Many routes to turbulent convection. *Journal of Fluid Mechanics* **100**, 449–470.

Goodman, J. W. (1985). *Statistical Optics*. New York: Wiley.

—— (2005). *Introduction to Fourier Optics*. Englewood, Colo.: Roberts and Company.

Goodman, J. J., R. W. Romani, R. D. Blandford, and R. Narayan (1987). The effect of caustics on scintillating radio sources. *Monthly Notices of the Royal Astronomical Society* **229**, 73–102.

Goodstein, D. L. (2002). *States of Matter*. Mineola, N.Y.: Courier Dover Publications.

Goodstein, D. L., and J. R. Goodstein (1996). *Feynman's Lost Lecture: The Motion of Planets around the Sun*. New York: W. W. Norton.

Gordon, J. P., H. J. Zeiger, and C. H. Townes (1954). Molecular microwave oscillator and new hyperfine structure in the microwave spectrum of NH_3. *Physical Review* **95**, 282–284.

—— (1955). The maser—new type of microwave amplifier, frequency standard, and spectrometer. *Physical Review* **99**, 1264–1274.

Gorman, M., and H. L. Swinney (1982). Spatial and temporal characteristics of modulated waves in the circular Couette system. *Journal of Fluid Mechanics* **117**, 123–142.

Gourgoulhon, E. (2013). *Special Relativity in General Frames: From Particles to Astrophysics*. Berlin: Springer-Verlag.

Grad, H. (1958). *Principles of the Kinetic Theory of Gases*. Cham, Switzerland: Springer.

Greenspan, H. P. (1973). *The Theory of Rotating Fluids*. Cambridge: Cambridge University Press.

Griffiths, D. J. (1999). *Introduction to Electrodynamics*. Upper Saddle River, N.J.: Prentice-Hall.

—— (2004). *Introduction to Quantum Mechanics*. Upper Saddle River, N.J.: Prentice-Hall.

Grossman, S. (2000). The onset of shear flow turbulence. *Reviews of Modern Physics* **72**, 603–618.

Guth, A. H. (1981). Inflationary universe: A possible solution to the horizon and flatness problems. *Physical Review D* **23**, 347–356.

Gutzwiller, M. C. (1990). *Chaos in Classical and Quantum Mechanics*. New York: Springer Verlag.

Hafele, J. C., and R. E. Keating (1972a). Around-the-world atomic clocks: Predicted relativistic time gains. *Science* **177**, 166–168.

—— (1972b). Around-the-world atomic clocks: Observed relativistic time gains. *Science* **177**, 168–170.

Hariharan, P. (2007). *Basics of Interferometry*. New York: Academic Press.

Harrison, E. R. (1970). Fluctuations at the threshold of classical cosmology. *Physical Review D* **1**, 2726–2730.

Hartle, J. B. (2003). *Gravity: An Introduction to Einstein's General Relativity*. San Francisco: Addison-Wesley.

Hassani, S. (2013). *Mathematical Physics: A Modern Introduction to Its Foundations*. Cham, Switzerland: Springer.

Hawking, S. W. (1975). Particle creation by black holes. *Communications in Mathematical Physics* **43**, 199–220.

—— (1976). Black holes and thermodynamics. *Physical Review D* **13**, 191–197.

Hawking, S. W., and R. Penrose (2010). *The Nature of Space and Time*. Princeton, N.J.: Princeton University Press.

Heaviside, O. (1912). *Electromagnetic Theory*, Volume III, p. 1. London: "The Electrician" Printing and Publishing.

Hecht, E. (2017). *Optics*. New York: Addison-Wesley.

Heisenberg, W. (1969). Significance of Sommerfeld's work today. In Bopp F., and H. Kleinpoppen (eds.), *Physics of the One and Two Electron Atoms*, p. 1. Amsterdam: North Holland. Chapter 18 epigraph reprinted with permission of the publisher.

Hénon, M. (1982). Vlasov equation? *Astronomy and Astrophysics* **114**, 211–212.

Heymans, C., L. van Waerbeke, L. Miller, T. Erben, et al. (2012). CFHTLenS: The Canada-France-Hawaii telescope lensing survey. *Monthly Notices of the Royal Astronomical Society* **427**, 146–166.

Hilbert, D. (1915). Die Grundlagen der Physik. *Königliche Gesellschaft der Wissenschaften zu Göttingen. Mathematische-physikalische Klasse. Nachrichten* **1917**, 53–76.

Hobbs, G., A. Archibald, Z. Arzoumanian, D. Backer, et al. (2010). The International Pulsar Timing Array project: Using pulsars as a gravitational wave detector. *Classical and Quantum Gravity* **27**, 8, 084013.

Hobson, M. P., G. P. Efstathiou, and A. N. Lasenby (2006). *General Relativity: An Introduction for Physicists*. Cambridge: Cambridge University Press.

Hooke, R. (1678). Answer to the anagram "ceiiinosssttuv," which he had previously published, to establish his priority on the linear law of elasticity. De Potentia, or of spring explaining the power of springing bodies, Hooke's Sixth Cutler Lecture, R. T. Gunther facsimile reprint. In *Early Science in Oxford*. Vol. 8. London: Dawsons of Pall Mall (1968).

Hubble, E. P. (1929). A relation between distance and radial velocity among extragalactic nebulae. *Proceedings of the National Academy of Sciences* **15**, 169–173.

Hull, J. C. (2014). *Options, Futures and Other Derivatives*, ninth edition. Upper Saddle River, N.J.: Pearson.

Hurricane, O. A., D. A. Callahan, D. T. Casey, P. M. Cellers, et al. (2014). Fuel gain exceeding unity in an inertially confined fusion explosion. *Nature* **506**, 343–348.

Iizuka, K. (1987). *Engineering Optics*. Berlin: Springer-Verlag.

Illingworth, G. D., and the HUDF09 team (2013). The HST extreme deep field (XDF): Combining all ACS and WFC3/IR data on the HUDF region into the deepest field ever. *Astrophysical Journal Supplement Series* **209**, 6.

Iorio, L., M. L. Ruggiero, and C. Corda (2013). Novel considerations about the error budget of the LAGEOS-based tests of frame-dragging with GRACE geopotential models. *Acta Astronautica* **91**, 141–148.

Isaacson, R. A. (1968a). Gravitational radiation in the limit of high frequency. I. The linear approximation and geometrical optics. *Physical Review* **166**, 1263–1271.

——— (1968b). Gravitational radiation in the limit of high frequency. II. Nonlinear terms and the effective stress tensor. *Physical Review* **166**, 1272–1280.

Israel, W., and J. M. Stewart (1980). Progress in relativistic thermodynamics and electrodynamics of continuous media. In A. Held, (ed.), *General Relativity and Gravitation. Vol. 2. One Hundred Years after the Birth of Albert Einstein*, p. 491. New York: Plenum Press.

Jackson, J. D. (1999). *Classical Electrodynamics*. New York: Wiley.

James, O., E. von Tunzelmann, P. Franklin, and K. S. Thorne (2015a). Gravitational lensing by spinning black holes in astrophysics, and in the movie *Interstellar*. *Classical and Quantum Gravity* **32**, 065001.

——— (2015b). Visualizing *Interstellar*'s Wormhole. *American Journal of Physics* **83**, 486–499.

Jeans, J. H. (1929). *Astronomy and Cosmology*, second edition. Cambridge: Cambridge University Press.

Jeffrey, A. and T. Taniuti, eds. (1966). *Magnetohydrodynamic Stability and Thermonuclear Confinement: A Collection of Reprints*. New York: Academic Press.

Jenkins, F. A., and H. E. White (1976). *Fundamentals of Optics*. New York: McGraw-Hill.

Johnson, J. B. (1928). Thermal agitation of electricity in conductors. *Physical Review* **32**, 97–109.

Johnson, K. L. (1985). *Contact Mechanics*. Cambridge: Cambridge University Press.

Johnson, L. R. (1974). Green's function for Lamb's problem. *Geophysical Journal of the Royal Astronomical Society* **37**, 99–131.

Kachru, S., R. Kallosh, A. Linde, and S. Trivedi (2003). De Sitter vacua in string theory. *Physical Review D* **68**, 046005.

Kapitsa, P. L., and P. A. M. Dirac (1933). The reflection of electrons from standing light waves. *Proceedings of the Cambridge Philosophical Society* **29**, 297–300.

Kapner, D. J., T. S. Cook, E. G. Adelberger, J. H. Gundlach, et al. (2008). Tests of the gravitational inverse-square law below the dark-energy length scale. *Physical Review Letters* **98**, 021101.

Kardar, M. (2007). *Statistical Physics of Particles*. Cambridge: Cambridge University Press.

Kaspi, V., and M. Kramer (2016). Radio pulsars: The neutron star population and fundamental physics. In R. D. Blandford, D. Gross, and A. Sevrin (eds.), *Proceedings of the 26th Solvay Conference on Physics, Astrophysics and Cosmology*, pp. 21–62. Singapore: World Scientific.

Kausel, E. (2006). *Fundamental Solutions in Elastodynamics*. Cambridge: Cambridge University Press.

Kay, B. S., M. J. Radzikowski, and R. M. Wald (1997). Quantum field theory on spacetimes with a compactly generated Cauchy horizon. *Communications in Mathematical Physics* **183**, 533–556.

Kazanas, D. (1980). Dynamics of the universe and spontaneous symmetry breaking. *Astrophysical Journal* **241**, L59–L63.

Keilhacker, M., and the JET Team (1998). Fusion physics progress on JET. *Fusion Engineering and Design* **46**, 273–290.

Kerr, R. P. (1963). Gravitational field of a spinning mass as an example of algebraically special metrics. *Physical Review Letters* **11**, 237–238.

Kim, S.-W., and K. S. Thorne (1991). Do vacuum fluctuations prevent the creation of closed timelike curves? *Physical Review D* **43**, 3929–3949.

Kirkwood, J. G. (1946). Statistical mechanical theory of transport processes. I. General theory. *Journal of Chemical Physics* **14**, 180–201.

Kittel, C. (2004). *Elementary Statistical Physics*. Mineola, N.Y.: Courier Dover Publications.

Kittel, C., and H. Kroemer (1980). *Thermal Physics*. London: Macmillan.

Klein, M. V., and T. E. Furtak (1986). *Optics*. New York: Wiley.

Kleppner, D., and R. K. Kolenkow (2013). *An Introduction to Mechanics*. Cambridge: Cambridge University Press.

Kolb, E. W., and M. S. Turner (1994). *The Early Universe*. New York: Addison-Wesley.

Kolsky, H. (1963). *Stress Waves in Solids*. Mineola, N.Y.: Courier Dover Publications.

Komatsu, E., K. M. Smith, J. Dunkley, C. L. Bennett, et al. (2011). Seven-year Wilkinson Microwave Anisotropy Probe (WMAP) observations: Cosmological interpretation. *Astrophysical Journal Supplement* **192**, 18–35.

Kompaneets, A. (1957). The establishment of thermal equilibrium between quanta and electrons. *Journal of Experimental and Theoretical Physics* **4**, 730–737.

Krall, N. A., and A. W. Trivelpiece (1973). *Principles of Plasma Physics*. New York: McGraw-Hill.

Kramer, M., I. H. Stairs, R. N. Manchester, M. A. McLaughlin, et al. (2006). Tests of general relativity from timing the double pulsar. *Science* **314**, 97–102.

Kravtsov, Y. A. (2005). *Geometrical Optics in Engineering Physics*. Oxford: Alpha Science International.

Kruer, W. L. (1988). *The Physics of Laser-Plasma Interactions*. New York: Addison-Wesley.

Kruskal, M. D. (1960). The maximal extension of the Schwarzschild metric. *Physical Review* **119**, 1743–1745.

Kulsrud, R. M. (2005). *Plasma Physics for Astrophysics*. Princeton, N.J.: Princeton University Press.

Kundu, P. K., I. M. Cohen, and D. R. Dowling (2012). *Fluid Mechanics*. New York: Academic Press.

La Porta, A., R. Slusher, and B. Yurke (1989). Back-action evading measurements of an optical field using parametric down conversion. *Physical Review Letters* **62**, 28–31.

Lagerstrom, P. (1988). *Matched Asymptotic Expansions: Ideas and Techniques*. Berlin: Springer-Verlag.

Lamb, H. (1882). On the vibrations of an elastic sphere. *Proceedings of the London Mathematical Society* **13**, 189–212.

Landau, L. D. (1944). On the problem of turbulence. *Doklady Akademii Nauk SSSR* **44**, 311–314.

——— (1946). On the vibrations of the electronic plasma. *Journal of Physics USSR* **10**, 25–37.

Landau, L. D., and E. M. Lifshitz (1941). Teoriya Polya. Moscow: Gosudarstvennoye Izdatel'stvo Tekhniko-Teoreticheskoi Literaturi. First Russian edition of Landau and Lifshitz (1951).

——— (1951). *The Classical Theory of Fields*, first English edition. Cambridge, Mass.: Addison-Wesley.

——— (1959). *Fluid Mechanics*. Oxford: Pergamon.

——— (1975). *The Classical Theory of Fields*, fourth English edition. Oxford: Butterworth-Heinemann.

——— (1976). *Mechanics*. Oxford: Butterworth-Heinemann.

——— (1986). *Elasticity*. Oxford: Pergamon.

Landau, L. D., L. P. Pitaevskii, and E. M. Lifshitz (1979). *Electrodynamics of Continuous Media*. Oxford: Butterworth-Heinemann.

Landauer, R. (1961). Irreversibility and heat generation in the computing process. *IBM Journal of Research and Development* **5**, 183–191.

——— (1991). Information is physical. *Physics Today* **44**, 23–29.

Langmuir, I. (1928). Oscillations in Ionized Gases. *Proceedings of the National Academy of Sciences* **14**, 627–637. Chapter 22 epigraph reprinted with permission of the publisher.

Lautrup, B. (2005). *Physics of Continuous Matter*. Bristol and Philadelphia: Institute of Physics Publishing.

Lax, M. J., W. Cai, M. Xu, and H. E. Stanley (2006). *Random Processes in Physics and Finance*. Oxford: Oxford University Press.

Lee, B. W., and S. Weinberg (1977). Cosmological lower bound on heavy-neutrino masses. *Physical Review Letters* **39**, 165–167.

Lemaître, G. (1927). Un univers homogène de masse constante et de rayon croissant rendant compte de la vitesse radiale des nébuleuses extra-galactiques. *Annales de la Société Scientifique Bruxelles A* **47**, 49–59.

——— (1933). La formation des nebuleuses dans l'univers en expansion. *Comptes Rendus* **196**, 903–904. Translated in *Cosmology and Controversy: The Historical Development of Two Theories of the Universe* by Helge Kragh. Copyright © 1996 by Princeton University Press. Chapter 28 epigraph reprinted with permission of Princeton University Press.

——— (1934). Evolution of the expanding universe. *Proceedings of the National Academy of Sciences* **20**, 12–17.

Levin, J., and G. Perez-Giz (2008). A periodic table for black hole orbits. *Physical Review D* **77**, 103005–103023.

Levin, Y. (1998). Internal thermal noise in the LIGO test masses: A direct approach. *Physical Review D* **57**, 659–663.

Lewin, L. (1981). *Polylogarithms and Associated Functions*. New York: North Holland.

Libbrecht, K. G., and M. F. Woodard (1991). Advances in helioseismology. *Science* **253**, 152–157.

Libchaber, A., C. Laroche, and S. Fauve (1982). Period doubling cascade in mercury, a quantitative measurement. *Journal de Physique—Lettres* **43**, L211–L216.

Liddle, A., and D. Lyth (2000). *Cosmological Inflation and Large Scale Structure*. Cambridge: Cambridge University Press.

Liepmann, H., and A. Roshko (2002). *Compressible Gas Dynamics*. Mineola, N.Y.: Courier Dover Publications.

Lifshitz, E. M., and L. P. Pitaevskii (1980). *Statistical Physics, Part 1*. Oxford: Pergamon.

——— (1981). *Physical Kinetics*. Oxford: Pergamon.

Lighthill, M. J. (1952). On sound generated aerodynamically. I. General theory. *Proceedings of the Royal Society A* **211**, 564–587.

——— (1954). On sound generated aerodynamically. II. Turbulence as a source of sound. *Proceedings of the Royal Society A* **222**, 1–32.

——— (1986). *An Informal Introduction to Theoretical Fluid Mechanics*. Oxford: Oxford University Press.

——— (2001). *Waves in Fluids*. Cambridge: Cambridge University Press.

Lightman, A. P., W. H. Press, R. H. Price, and S. A. Teukolsky (1975). *Problem Book in Relativity and Gravitation*. Princeton, N.J.: Princeton University Press.

LIGO Scientific Collaboration (2015). Advanced LIGO. *Classical and Quantum Gravity* **32**, 074001.

Linde, A. (1982). A new inflationary universe scenario: A possible solution to the horizon, flatness, homogeneity, isotropy and primordial monopole problems. *Physics Letters B* **108**, 389–393.

Liu, Y. T., and K. S. Thorne (2000). Thermoelastic noise and thermal noise in finite-sized gravitational-wave test masses. *Physical Review D* **62**, 122002–122011.

Longair, M. S. (2011). *High Energy Astrophysics*. Cambridge: Cambridge University Press.

Longhurst, R. S. (1973). *Geometrical and Physical Optics*. London: Longmans.

López-Monsalvo, C. S. (2011). *Covariant Thermodynamics and Relativity*, PhD thesis, University of Southampton. Available at https://arxiv.org/pdf/1107.1005.pdf.

Lorentz, H. A. (1904). Electromagnetic phenomena in a system moving with any velocity smaller than that of light. *Proceedings of the Royal Netherlands Academy of Arts and Sciences (KNAW)* **6**, 809–831.

Lorentz, H. A., A. Einstein, H. Minkowski, and H. Weyl (1923). *The Principle of Relativity: A Collection of Original Memoirs on the Special and General Theory of Relativity*. Mineola, N.Y.: Courier Dover Publications.

Lorenz, E. N. (1963). Deterministic nonperiodic flow. *Journal of Atmospheric Sciences* **20**, 130–141.

Love, A. E. H. (1927). *A Treatise on the Mathematical Theory of Elasticity*. Mineola, N.Y.: Courier Dover Publications.

Lynden-Bell, D. (1967). Statistical mechanics of violent relaxation in stellar systems. *Monthly Notices of the Royal Astronomical Society* **136**, 101–121.

Macintosh, B., J. R. Graham, P. Ingraham, Q. Konopacky, et al. (2014). First light of the Gemini planet imager. *Proceedings of the National Academy of Sciences* **111**, 12661–12666.

Mack, J. E. (1947). *Semi-Popular Motion Picture Record of the Trinity Explosion*. University of Michigan Library, Ann Arbor.

Maggiore, M. (2007). *Gravitational Waves. Volume 1: Theory and Experiment*. Oxford: Oxford University Press.

Maiman, T. H. (1960). Stimulated optical radiation in ruby. *Nature* **187**, 493–494.

Majda, A. J., and A. L. Bertozzi (2002). *Vorticity and Incompressible Flow*. Cambridge: Cambridge University Press.

Marion, J. B., and S. T. Thornton (1995). *Classical Dynamics of Particles and Systems*. Philadelphia: Saunders College Publishing.

Maris, H. J., and L. P. Kadanoff (1978). Teaching the renormalization group. *American Journal of Physics* **46**, 653–657.

Marko, J. F., and S. Cocco (2003). The micro mechanics of DNA. *Physics World* **16**, 37–41.

Marolf, D., and A. Ori (2013). Outgoing gravitational shock-wave at the inner horizon: The late-time limit of black hole interiors. *Physical Review D* **86**, 124026.

Maroto, J. A., V. Perez-Munuzuri, and M. S. Romero-Cano (2007). Introductory analysis of Benard-Marangoni convection. *European Journal of Physics* **28**, 311–320.

Marsden, J. E., and T. J. Hughes (1986). *Mathematical Foundations of Elasticity*. Upper Saddle River, N.J.: Prentice-Hall.

Martin, R. F. (1986). Chaotic particle dynamics near a two-dimensional neutral point with application to the geomagnetic tail. *Journal of Geophysical Research* **91**, 11985–11992.

Mather, J. C., E. S. Cheng, D. A. Cottingham, R. E. Eplee Jr., et al. (1994). Measurement of the cosmic microwave background spectrum by the COBE FIRAS instrument. *Astrophysical Journal* **420**, 439–444.

Mathews, J., and R. L. Walker (1970). *Mathematical Methods of Physics*. New York: Benjamin.

Maxwell, J. C. (1873). Letter to William Grylls Adams (3 Dec 1873). In P. M. Harman (ed.). (1995). *The Scientific Letters and Papers of James Clerk Maxwell, Vol 2, 1862–1873*, pp. 949–950. Cambridge: Cambridge University Press.

McClelland, D. E., N. Mavalvala, Y. Chen, and R. Schnabel (2011). Advanced interferometry, quantum optics and optomechanics in gravitational wave detectors. *Lasers and Photonics Reviews* **5**, 677–696.

McEliece, R. J. (2002). *The Theory of Information and Coding*. Cambridge: Cambridge University Press.

McKinney, J. C., A. Tchekhovskoy, and R. D. Blandford (2012). General relativistic magnetohydrodynamical simulations of magnetically choked accretion flows around black holes. *Monthly Notices of the Royal Astronomical Society* **423**, 3083–3117.

Meier, D. L. (2012). *Black Hole Astrophysics: The Engine Paradigm*. Cham, Switzerland: Springer.

Melrose, D. B. (1980). *Plasma Astrophysics*. New York: Gordon and Breach.

—— (1984). *Instabilities in Space and Laboratory Plasmas*. Cambridge: Cambridge University Press.

—— (2008). *Quantum Plasmadynamics, Vol 1: Unmagnetized Plasmas*. Cham, Switzerland: Springer.

—— (2012). *Quantum Plasmadynamics, Vol 2: Magnetized Plasmas*. Cham, Switzerland: Springer.

Merkowitz, S. M. (2010). Tests of gravity using lunar laser ranging. *Living Reviews in Relativity* **13**, 7.

Messiah, A. (1962). *Quantum Mechanics, Volume II*. New York: North Holland.

Metropolis, N., A. Rosenbluth, M. Rosenbluth, A. Teller, and E. Teller (1953). Combinatorial minimization. *Journal of Chemical Physics* **21**, 1087–1092.

Michell, J. (1783). On the means of discovering the distance, magnitude, etc., of the fixed stars, in consequence of the diminution of their light, in case such a diminution should be found to take place in any of them, and such other data should be procured from observations, as would be further necessary for that purpose. *Philosophical Transactions of the Royal Society of London* **74**, 35–57; presented to the Royal Society on November 27, 1783.

Michelson, A. A., and F. G. Pease (1921). Measurement of the diameter of α Orionis with the interferometer. *Astrophysical Journal* **53**, 249–259.

Mikhailovskii, A. B. (1998). *Instabilities in a Confined Plasma*. Bristol and Philadelphia: Institute of Physics Publishing.

Miles, J. (1993). Surface-wave generation revisited. *Journal of Fluid Mechanics* **256**, 427–441.

Millikan, R. A. (1938). Biographical Memoir of Albert Abraham Michelson, 1852–1931. *Biographical Memoirs of the National Academy of Sciences of the United States of America* **19**, 121–146. Chapter 4 epigraph reprinted with permission of the publisher.

Minkowski, H. (1908). Space and time. Address delivered at the 80th Assembly of German Natural Scientists and Physicians, at Cologne, Germany, September 21, 1908. First German publication: *Jahresbericht der Deutschen Mathematiker-Vereinigung* **1909**, 75–88. English translation in Lorentz et al. (1923).

Misner, C. W., K. S. Thorne, and J. A. Wheeler (1973). *Gravitation*. San Francisco: Freeman.

Morris, M., and K. S. Thorne (1988). Wormholes in spacetime and their use for interstellar travel: A tool for teaching general relativity. *American Journal of Physics* **56**, 395–416.

Morris, M. S., K. S. Thorne, and U. Yurtsever (1988). Wormholes, time machines, and the weak energy condition. *Physical Review Letters* **61**, 1446–1449.

Mroué, A. H., M. A. Scheel, B. Szilagyi, H. P. Pfeiffer, et al. (2013). Catalog of 174 black hole simulations for gravitational wave astronomy. *Physical Review Letters* **111**, 241104.

Mukhanov, V. (2005). *Physical Foundations of Modern Cosmology*. Cambridge: Cambridge University Press.

Munson, B. R., D. F. Young, and T. H. Okiishi (2006). *Fundamentals of Fluid Mechanics*. New York: Wiley.

NIST (2005). *Final Report on the Collapse of the World Trade Center Towers*. National Institute of Standards and Technology Report Number NIST NCSTAR 1. Washington, D.C.: U.S. Government Printing Office.

——— (2008). *Final Report on the Collapse of the World Trade Center Building 7*. National Institute of Standards and Technology Report Number NIST NCSTAR 1A. Washington, D.C.: U.S. Government Printing Office.

Nelson, P. (2008). *Biological Physics*. San Francisco: Freeman.

Newton, I. (1687). *Philosophiae Naturalis Principia Mathematica*. London: Royal Society. English translation by I. B. Cohen and A. Whitman. Berkeley: University of California Press (1999).

Ni, W.-T., and M. Zimmermann (1978). Inertial and gravitational effects in the proper reference frame of an accelerated, rotating observer. *Physical Review D* **17**, 1473–1476.

Nichols, D., R. Owen, F. Zhang, A. Zimmerman, et al. (2011). Visualizing spacetime curvature via frame-drag vortexes and tidal tendexes: General theory and weak-gravity applications. *Physical Review D* **84**, 124014.

Noether, E. (1918). Invariante Variationenprobleme. *Nachrichten von der Gesellschaft der Wissenschaften zu Göttingen* **1918**, 235–257.

Northrop, T. (1963). *Adiabatic Motion of Charged Particles*. New York: Interscience.

Nye, J. (1999). *Natural Focusing and Fine Structure of Light*. Bristol and Philadelphia: Institute of Physics Publishing.

Nyquist, H. (1928). Thermal agitation of electric charge in conductors. *Physical Review* **32**, 110–113.

Oelker, T. I., T. Isogai, J. Miller, M. Tse, et al. (2016). Audio-band frequency-dependent squeezing for gravitational-wave detectors. *Physical Review Letters* **116**, 041102.

Ogorodnikov, K. F. (1965). *Dynamics of Stellar Systems*. Oxford: Pergamon.

Onsager, L. (1944). Crystal statistics. I. A two-dimensional model with an order-disorder transition. *Physical Review* **65**, 117–149.

Oort, J. H. (1932). The force exerted by the stellar system in the direction perpendicular to the galactic plane and some related problems. *Bulletin of the Astronomical Institute of the Netherlands* **238**, 249–287.

Oppenheimer, J. R., and H. Snyder (1939). On continued gravitational contraction. *Physical Review* **56**, 455–459.

Oppenheimer, J. R., and G. Volkoff (1939). On massive neutron cores. *Physical Review* **55**, 374–381.

Ott, E. (1982). Strange attractors and chaotic motions of dynamical systems. *Reviews of Modern Physics* **53**, 655–671.

——— (1993). *Chaos in Dynamical Systems*. Cambridge: Cambridge University Press.

Overduin, J., F. Everitt, P. Worden, and J. Mester (2012). STEP and fundamental physics. *Classical and Quantum Gravity* **29**, 184012.

Owen, R., J. Brink, Y. Chen, J. D. Kaplan, et al. (2011). Frame-dragging vortexes and tidal tendexes attached to colliding black holes: Visualizing the curvature of spacetime. *Physical Review Letters* **106**, 151101.

Padmanabhan, T. (1993). *Structure Formation in the Universe*. Cambridge: Cambridge University Press.

Page, D. N., F. Weinhold, R. L. Moore, F. Weinhold, and R. E. Barker (1977). Thermodynamic paradoxes. *Physics Today* **30**, 11.

Pais, A. (1982). *Subtle Is the Lord. . . . The Science and Life of Albert Einstein*. Oxford: Oxford University Press.

Panton, R. L. (2005). *Incompressible Flow*. New York: Wiley.

Parker, E. N. (1979). *Cosmical Magnetic Fields*. Oxford: Clarendon Press.

Parker, L., and D. Toms (2009). *Quantum Field Theory in Curved Spacetime: Quantized Fields and Gravity*. Cambridge: Cambridge University Press.

Parks, G. K. (2004). *Physics of Space Plasmas: An Introduction*. Boulder: Westview Press.

Pathria, R. K., and P. D. Beale (2011). *Statistical Mechanics*, third edition. Amsterdam: Elsevier.

Paul, W., and J. Baschnagel (2010). *Stochastic Processes: From Physics to Finance*. Cham, Switzerland: Springer.

Peacock, J. A. (1999). *Cosmological Physics*. Cambridge: Cambridge University Press.

Peccei, R. D., and H. R. Quinn (1977). CP conservation in the presence of pseudoparticles. *Physical Review Letters* **38**, 1440–1443.

Pedlosky, J. (1987). *Geophysical Fluid Dynamics*. Berlin: Springer-Verlag.

Pedrotti, F. L., L. S. Pedrotti, and L. M. Pedrotti (2007). *Introduction to Optics*. Upper Saddle River, N.J.: Pearson.

Peebles, P. J. E. (1993). *Principles of Physical Cosmology*. Princeton, N.J.: Princeton University Press.

Peebles, P. J. E., and J. T. Yu (1970). Primeval adiabatic perturbation in an expanding universe. *Astrophysical Journal* **162**, 815–836.

Pellew, A., and R. V. Southwell (1940). On maintained convective motion in a fluid heated from below. *Proceedings of the Royal Society A* **176**, 312–343.

Penrose, O. (1960). Electrostatic instabilities of a uniform non-Maxwellian plasma. *Physics of Fluids* **3**, 258–265.

Penrose, R. (1999). *The Emperor's New Mind: Concerning Computers, Minds, and the Laws of Physics*. Oxford: Oxford University Press.

——— (2016). *Fashion, Faith and Fantasy in the New Physics of the Universe*. Princeton, N.J.: Princeton University Press.

Penzias, A. A., and R. W. Wilson (1965). A measurement of excess antenna temperature at 4080 Mc/s. *Astrophysical Journal* **142**, 419–421.

Perlmutter, S., M. Turner, and M. White (1999). Constraining dark energy with Type Ia supernovae and large-scale structure. *Physical Review Letters* **83**, 670–673.

Perlmutter, S., G. Aldering, G. Goldhaber, R. A. Knop, et al. (1999). Measurements of Ω and Λ from 42 high-redshift supernovae. *Astrophysical Journal* **517**, 565–586.

Petters, A. O., H. Levine, and J. Wambsganss (2001). *Singularity Theory and Gravitational Lensing*. Cham, Switzerland: Springer.

Pfister, H. (2007). On the history of the so-called Lense-Thirring effect. *General Relativity and Gravitation* **39**, 1735–1748.

Phillips, O. M. (1957). On the generation of waves by turbulent wind. *Journal of Fluid Mechanics* **2**, 417–445.

Pierce, J. R. (2012). *An Introduction to Information Theory: Symbols, Signals and Noise*. Mineola, N.Y.: Courier Dover Publications.

Pines, D., and J. R. Schrieffer (1962). Approach to equilibrium of electrons, plasmons and phonons in quantum and classical plasmas. *Physical Review* **125**, 804–812.

Planck, M. (1949). *Scientific Autobiography and Other Papers*. New York: Philosophical Library. Chapter 24 epigraph reprinted with permission of the publisher.

Planck Collaboration (2016a). Planck 2015 results. I. Overview of products and scientific results. *Astronomy and Astrophysics* **594**, A1.

——— (2016b). Planck 2015 results. XIII. Cosmological parameters. *Astronomy and Astrophysics* **594**, A13.

Pontecorvo, B. (1968). Neutrino experiments and the problem of conservation of leptonic charge. *Soviet Physics JETP* **26**, 984–988.

Pop, I., and D. B. Ingham (2001). *Convective Heat Transfer: Mathematical Computational Modelling of Viscous Fluids and Porous Media.* Amsterdam: Elsevier.

Pope, S. B. (2000). *Turbulent Flows.* Cambridge: Cambridge University Press.

Poruchikov, V. B., V. A. Khokhryakov, and G. P. Groshev (1993). *Methods of the Classical Theory of Elastodynamics.* Berlin: Springer-Verlag.

Poston, T., and I. Stewart (2012). *Catastrophe Theory and Its Applications.* Mineola, N.Y.: Courier Dover Publications.

Potter, M. C., D. C. Wiggert, and B. H. Ramadan (2012). *Mechanics of Fluids.* Stamford, Conn.: Cengage Learning.

Press, W. H. (1978). Flicker noises in astronomy and elsewhere. *Comments on Astrophysics* **7**, 103–119.

Press, W. H., S. A. Teukolsky, W. T. Vetterling, and B. P. Flannery (2007). *Numerical Recipes: The Art of Scientific Computing.* Cambridge: Cambridge University Press.

Purcell, E. M. (1983). The back of the envelope. *American Journal of Physics* **51**, 205.

Raisbeck, G. (1963). *Information Theory.* Cambridge, Mass.: MIT Press.

Ransom, S. M., I. H. Stairs, A. M. Archibald, J.W.T. Hessels, et al. (2014). A millisecond pulsar in a stellar triple system. *Nature* **505**, 520–524.

Rashed, R. 1990. A pioneer in anaclastics: Ibn Sahl on burning mirrors and lenses. *Isis* **1**, 464–491.

Reichl, L. E. (2009). *A Modern Course in Statistical Physics.* London: Arnold.

Reif, F. (2008). *Fundamentals of Statistical and Thermal Physics.* Long Grove, Ill.: Waveland Press.

Rezzolla, L., and O. Zanotti (2013). *Relativistic Hydrodynamics.* Oxford: Oxford University Press.

Richardson, L. (1922). *Weather Prediction by Numerical Process.* Cambridge: Cambridge at the University Press.

Richter, C. F. (1980). Interview with Henry Spall. *Earthquake Information Bulletin*, January–February. Chapter 12 epigraph reprinted with permission of the publisher.

Ride, S. (2012). Interview with Jim Clash. Available at http://www.askmen.com/entertainment/right-stuff/sally-ride-interview.html. Reprinted by permission of Jim Clash.

Riess, A. G., A. Filippenko, P. Challis, A. Clochiatti, et al. (1998). Observational evidence from supernovae for an accelerating universe and a cosmological constant. *Astronomical Journal* **116**, 1009–1038.

Roberts, M. S., and R. Whitehurst (1975). The rotation curve and geometry of M31 at large galactocentric distances. *Astrophysical Journal* **201**, 327–346.

Robertson, H. P. (1935). Kinematics and world structure I. *Astrophysical Journal* **82**, 248–301.

——— (1936a). Kinematics and world structure II. *Astrophysical Journal* **83**, 187–201.

——— (1936b). Kinematics and world structure III. *Astrophysical Journal* **83**, 257–271.

Robinson, I. (1959). A solution of the Maxwell-Einstein equations. *Bulletin of the Polish Academy of Sciences* **7**, 351–352.

Roddier, F. (1981). The effects of atmospheric turbulence in optical astronomy. *Progress in Optics* **19**, 281–376.

Rohrlich, F. (1965). *Classical Charged Particles*. New York: Addison-Wesley.

Rosenbluth, M. N., M. MacDonald, and D. L. Judd (1957). Fokker-Planck equation for an inverse square force. *Physical Review* **107**, 1–6.

Rouse, H. (1963a). University of Iowa movie: Introduction to the study of fluid motion. Available at
http://www.iihr.uiowa.edu/research/publications-and-media/films-by-hunter-rouse/.

—— (1963b). University of Iowa movie: Fundamental principles of flow. Available at http://www.iihr.uiowa.edu/research/publications-and-media/films-by-hunter-rouse/.

—— (1963c). University of Iowa movie: Fluid motion in a gravitational field. Available at http://www.iihr.uiowa.edu/research/publications-and-media/films-by-hunter-rouse/.

—— (1963d). University of Iowa movie: Characteristics of laminar and turbulent flow. Available at
http://www.iihr.uiowa.edu/research/publications-and-media/films-by-hunter-rouse/.

—— (1963e). University of Iowa movie: Form, drag, lift, and propulsion. Available at http://www.iihr.uiowa.edu/research/publications-and-media/films-by-hunter-rouse/.

—— (1963f). University of Iowa movie: Effects of fluid compressibility. Available at http://www.iihr.uiowa.edu/research/publications-and-media/films-by-hunter-rouse/.

Rubin, V. C., and W. K. Ford, Jr. (1970). Rotation of the Andromeda nebula from a spectroscopic survey of emission regions. *Astrophysical Journal* **159**, 379–403.

Ruelle, D. (1989). *Chaotic Evolution and Strange Attractors*. Cambridge: Cambridge University Press.

Ryan, M., and L. Shepley (1975). *Homogeneous, Relativistic Cosmology*. Princeton, N.J.: Princeton University Press.

Ryden, B. S. (2002). *Introduction to Cosmology*. New York: Addison-Wesley.

Sachs, R. K., and A. M. Wolfe (1967). Perturbations of a cosmological model and angular variations of the microwave background. *Astrophysical Journal* **147**, 73–90.

Sagdeev, R. Z., and C. F. Kennel (1991). Collisionless shock waves. *Scientific American* **264**, April issue, 106–113.

Sagdeev, R. Z., D. A. Usikov, and G. M. Zaslovsky (1988). *Non-linear Physics from the Pendulum to Turbulence and Chaos*. Newark, N.J.: Harwood Academic Publishers.

Sakharov, A. D. (1965). The initial stage of an expanding universe and the appearance of a nonuniform distribution of matter. *Journal of Experimental and Theoretical Physics* **49**, 345–358.

Sahl, I. (984). *On Burning Mirrors and Lenses*. Discussed in Rashed (1990).

Saleh, B. E., and M. C. Teich (2007). *Fundamentals of Photonics*. New York: Wiley.

Sathyaprakash, B. S., and B. F. Schutz (2009). Physics, astrophysics and cosmology with gravitational waves. *Living Reviews in Relativity* **12**, 3.

Sato, K. (1981). Cosmological baryon number domain structure and the first order phase transition of the vacuum. *Physics Letters B* **33**, 66–70.

Saulson, P. (1994). *Fundamentals of Interferometric Gravitational Wave Detectors*. Singapore: World Scientific.

Saunders, P. T. (1980). *An Introduction to Catastrophe Theory*. Cambridge: Cambridge University Press.

Schlamminger, S., K.-Y. Choi, T. A. Wagner, J. H. Gundlach, and E. G. Adelberger (2008). Test of the equivalence principle using a rotating torsion balance. *Physical Review Letters* **100**, 041101.

Schmidt, G. (1979). *Physics of High Temperature Plasmas*. New York: Academic Press.

Schneider, P. (2015). *Extragalactic Astronomy and Cosmology*. Heidelberg: Springer.

Schneider, P., J. Ehlers, and E. Falco (1992). *Gravitational Lensing*. Berlin: Springer-Verlag.

Schneier, B. (1997). *Applied Cryptography: Protocols, Algorithms and Source Code in C*. New York: Wiley.

Schrödinger, E. (1944). *What Is Life?* Cambridge: Cambridge University Press.

Schutz, B. (2009). *A First Course in General Relativity*. Cambridge: Cambridge University Press.

Schwarzschild, K. (1903). Zur Elektrodynamik. 1. Zwei Formen des Princips der Action in der Elektrontheorie. *Nachrichten von der Gesellschaft der Wissenschaften zu Göttingen* **1903**, 126–131.

——— (1916a). Über das Gravitationsfeld eines Massenpunktes nach der Einsteinschen Theorie. *Sitzungsberichte der Preussischen Akademie der Wissenschaften* **1916**, 189–196.

——— (1916b). Über das Gravitationsfeld einer Kugel aus Inkompressibler Flüssigkeit nach der Einsteinschen Theorie. *Sitzungsberichte der Preussischen Akademie der Wissenschaften* **1916**, 424–434.

Scott-Russell, J. (1844). Report on waves. *British Association for the Advancement of Science* **14**, 311–390, Plates XLVII–LVII.

Sedov, L. I. (1946). Propagation of strong blast waves. *Prikhladnaya Matematika i Mekhanika* **10**, 241–250.

——— (1957). Russian Language Fourth Edtion of Sedov (1993): *Metody podobiya i razmernosti v mekhanike*. Moskva: Gostekhizdat.

——— (1993). *Similarity and Dimensional Methods in Mechanics*. Boca Raton, Fla.: CRC Press.

Sethna, J. P. (2006). *Statistical Mechanics: Entropy, Order Parameters, and Complexity*. Oxford: Oxford University Press.

Shannon, C. E. (1948). A mathematical theory of communication. *Bell System Technical Journal* **27**, 379–423.

Shapiro, A. (1961a). National Committee for Fluid Mechanics Films. Available at web.mit.edu/hml/ncfmf.html.

——— (1961b). National Committee for Fluid Mechanics Films movie: Vorticity.

Shapiro, S. L., and S. A. Teukolsky (1983). *Black Holes, White Dwarfs and Neutron Stars: The Physics of Compact Objects*. New York: Wiley.

Sharma, K. (2006). *Optics: Principles and Applications*. New York: Academic Press.

Shearer, P. M. (2009). *Introduction to Seismology*. Cambridge: Cambridge University Press.

Shercliff, J. A. (1965). National Committee for Fluid Mechanics Films movie: Magnetohydrodynamics.

Shibata, M. (2016). *Numerical Relativity*. Singapore: World Scientific.

Shkarofsky, I. P., T. W. Johnston, and M. P. Bachynski (1966). *The Particle Kinetics of Plasmas*. New York: Addison-Wesley.

Silk, J. (1968). Cosmic black-body radiation and galaxy formation. *Astrophysical Journal* **151**, 459–471.

Slaughter, W. S. (2002). *The Linearized Theory of Elasticity*. Boston: Birkhäuser.

Southwell, R. V. (1941). *An Introduction to the Theory of Elasticity for Engineers and Physicists.* Oxford: Clarendon Press.

Spitzer, Jr., L. (1962). *Physics of Fully Ionized Gases.* New York: Interscience. Chapter 20 epigraph reprinted with permission of the publisher.

Spitzer, Jr., L., and R. Harm (1953). Transport phenomena in a completely ionized gas. *Physical Review* **89**, 977–981.

Spivak, M. (1999). *A Comprehensive Introduction to Differential Geometry,* Volumes 1–5. Houston: Publish or Perish.

Stacey, F. D. (1977). *Physics of the Earth.* New York: Wiley.

Stanyukovich, K. P. (1960). *Unsteady Motion of Continuous Media.* Oxford: Pergamon.

Starobinsky, A. (1980). A new type of isotropic cosmological model without singularity. *Physics Letters B* **91**, 99–102.

Stein, S., and M. Wysession (2003). *An Introduction to Seismology, Earthquakes and Earth Structure.* Oxford: Blackwell.

Stewart, R. W. (1968). National Committee for Fluid Mechanics Films movie: Turbulence.

Stix, T. H. (1992). *Waves in Plasmas.* New York: American Institute of Physics.

Straumann, N. (2013). *General Relativity.* Cham, Switzerland: Springer.

Strogatz, S. H. (2008). *Nonlinear Dynamics and Chaos: With Applications to Physics, Biology, Chemistry and Engineering.* Boulder: Westview Press.

Sturrock, P. A. (1994). *Plasma Physics: An Introduction to the Theory of Astrophysical, Geophysical and Laboratory Plasmas.* Cambridge: Cambridge University Press.

Sunyaev, R. A., and Ya. B. Zel'dovich (1970). Small-scale fluctuations of relic radiation. *Astrophysics and Space Science* **7**, 3–19.

Susskind, L. (2005). *The Cosmic Landscape: String Theory and the Illusion of Intelligent Design.* New York: Little, Brown.

Swanson, D. G. (2003). *Plasma Waves.* Bristol and Philadelphia: Institute of Physics Publishing.

Szekeres, G. (1960). On the singularities of a Riemann manifold. *Publicationes Mathematicae Debrecen* **7**, 285–301.

Tanimoto, T., and J. Um (1999). Cause of continuous oscillations of the earth. *Journal of Geophysical Research* **104**, 28723–28739.

Taylor, E. (1968). National Committee for Fluid Mechanics Films movie: Secondary flow.

Taylor, E. F., and J. A. Wheeler (1966). *Spacetime Physics,* first edition. San Francisco: Freeman.

—— (1992). *Spacetime Physics,* second edition. San Francisco: Freeman.

Taylor, G. (1950). The formation of a blast wave by a very intense explosion. II. The atomic explosion of 1945. *Proceedings of the Royal Society A* **201**, 175–186.

—— (1964). National Committee for Fluid Mechanics Films movie: Low Reynolds number flows.

Tennekes, H., and J. L. Lumley (1972). *A First Course on Turbulence.* Cambridge, Mass.: MIT Press.

ter Haar, D. (1955). Foundations of statistical mechanics. *Reviews of Modern Physics* **27**, 289–338.

Thom, R. (1994). *Structural Stability and Morphogenesis.* Boulder: Westview Press.

Thompson, P. A. (1984). *Compressible Fluid Dynamics.* Boulder: Maple Press.

Thorne, K. S. (1973). Relativistic shocks: The Taub adiabat. *Astrophysical Journal* **179**, 897–907.

——— (1980). Multipole expansions of gravitational radiation. *Reviews of Modern Physics* **52**, 299–340.

——— (1981). Relativistic radiative transfer—moment formalisms. *Monthly Notices of the Royal Astronomical Society* **194**, 439–473.

——— (1983). The theory of gravitational radiation: An introductory review. In N. Dereulle and T. Piran (eds.), *Gravitational Radiation*, pp. 1–57. New York: North Holland.

——— (1994). *Black Holes and Time Warps: Einstein's Outrageous Legacy*. New York: W. W. Norton.

——— (2014). *The Science of Interstellar*. New York: W. W. Norton.

Thorne, K. S., M. Bondarescu, and Y. Chen (2002). Gravitational waves: A web-based course. Available at http://elmer.caltech.edu/ph237/.

Thorne, K. S., and J. Hartle (1985). Laws of motion and precession for black holes and other bodies. *Physical Review D* **31**, 1815–1837.

Thorne, K. S., R. H. Price, and D. A. MacDonald (1986). *Black holes: the Membrane Paradigm*. New Haven, Conn.: Yale University Press.

Timoshenko, S., and J. N. Goodier (1970). *Theory of Elasticity*. New York: McGraw-Hill.

Todhunter, I., and K. Pearson (1886). *A History of the Theory of Elasticity and of the Strength of Materials, from Galilei to the Present Time*. Cambridge: Cambridge University Press.

Tolman, R. C. (1934). *Relativity, Thermodynamics and Cosmology*. Oxford: Oxford University Press.

——— (1938). *The Principles of Statistical Mechanics*. Mineola, N.Y.: Courier Dover Publications.

——— (1939). Static solutions of Einstein's field equations for spheres of fluid. *Physical Review* **55**, 364–373.

Toro, E. F. (2010). *Riemann Solvers and Numerical Methods for Fluid Dynamics: A Practical Introduction*. Berlin: Springer-Verlag.

Townes, C. H. (2002). *How the Laser Happened: Adventures of a Scientist*. Oxford: Oxford University Press. Chapter 10 epigraph reprinted with permission of the publisher.

Townsend, A. A. (1949). The fully developed turbulent wake of a circular cylinder. *Australian Journal of Scientific Research* **2**, 451–468.

Tranah, D., and P. T. Landsberg (1980). Thermodynamics of non-extensive entropies II. *Collective Phenomena* **3**, 81–88.

Tritton, D. J. (1987). *Physical Fluid Dynamics*. Oxford: Oxford University Press.

Tsytovich, V. (1970). *Nonlinear Effects in Plasma*. New York: Plenum.

Turco, R. P., O. B. Toon, T. B. Ackerman, J. B. Pollack, and C. Sagan (1986). Nuclear winter: Global consequences of multiple nuclear explosions. *Science* **222**, 1283–1292.

Turcotte, D. L., and G. Schubert (1982). *Geodynamics*. New York: Wiley.

Turner, J. S. (1973). *Buoyancy Effects in Fluids*. Cambridge: Cambridge University Press.

Ugural, A. C., and S. K. Fenster (2012). *Advanced Mechanics of Materials and Applied Elasticity*. Upper Saddle River, N.J.: Prentice-Hall.

Unruh, W. G. (1976). Notes on black hole evaporation. *Physical Review D* **14**, 870–892.

Vallis, G. K. (2006). *Atmospheric and Oceanic Fluid Dynamics*. Cambridge: Cambridge University Press.

Van Dyke, M. (1982). *An Album of Fluid Flow*. Stanford, Calif.: Parabolic Press.

Van Kampen, N. G. (2007). *Stochastic Processes in Physics and Chemistry*. New York: North Holland.

Verhulst, P. F. (1838). Notice sur la loi que la population poursuit dans son accroissement. *Correspondance Mathématique et Physique* **10**, 113–121.

Vogel, S. (1994). *Life In Moving Fluids: The Physical Biology Of Flow,* 2nd Edition, Revised and Expanded by Steven Vogel, Illustrated by Susan Tanner Beety and the Author.

Wagner, T., S. Schlamminger, J. Gundlach, and E. Adelberger (2012). Torsion-balance tests of the weak equivalence principle. *Classical and Quantum Gravity* **29**, 1–15.

Wainstein, L. A., and V. D. Zubakov (1962). *Extraction of Signals from Noise*. London: Prentice-Hall.

Wald, R. M. (1984). *General Relativity*. Chicago: University of Chicago Press.

——— (1994). *Quantum Field Theory in Curved Spacetime and Black Hole Thermodynamics*. Chicago: University of Chicago Press.

——— (2001). The thermodynamics of black holes. *Living Reviews in Relativity* **4**, 6.

Walker, A. G. (1935). On Milne's theory of world structure. *Proceedings of the London Mathematical Society* **42**, 90–127.

Weber, J. (1953). Amplification of microwave radiation by substances not in thermal equilibrium. *IRE Transactions of the Professional Group on Electron Devices* **3**, 1–4.

Weinberg, S. (1972). *Gravitation and Cosmology. Principles and Applications of the General Theory of Relativity*. New York: Wiley.

——— (2008). *Cosmology*. Oxford: Oxford University Press.

Weiss, R. (1972). Electromagnetically coupled broadband gravitational antenna. *Quarterly Progress Report of the Research Laboratory of Electronics, M.I.T.,* **105**, 54–76.

Weissberg, J. M., D. J. Nice, and J. H. Taylor (2010). Timing measurements of the relativistic binary pulsar PSR B1913+16. *Astrophysical Journal* **722**, 1030–1034.

Weld, D. M., J. Xia, B. Cabrera, and A. Kapitulnik (2008). A new apparatus for detecting micron-scale deviations from Newtonian gravity. *Physical Review D* **77**, 062006.

Welford, W. T. (1988). *Optics*. Oxford: Oxford University Press.

Wheeler, J. A. (2000). *Geons, Black Holes, and Quantum Foam: A Life in Physics*. New York: W. W. Norton.

White, F. M. (2006). *Viscous Fluid Flow*. New York: McGraw-Hill.

——— (2008). *Fluid Mechanics*. New York: McGraw-Hill.

Whitham, G. B. (1974). *Linear and Non-linear Waves*. New York: Wiley.

Wiener, N. (1949). *The Extrapolation, Interpolation, and Smoothing of Stationary Time Series with Engineering Applications*. New York: Wiley.

Will, C. M. (1993a). *Theory and Experiment in Gravitational Physics*. Cambridge: Cambridge University Press.

——— (1993b). *Was Einstein Right?* New York: Basic Books.

——— (2014). The confrontation between general relativity and experiment. *Living Reviews in Relativity* **17**, 4.

Wolgemuth, C. W., T. R. Powers, and R. E. Goldstein (2000). Twirling and whirling: Viscous dynamics of rotating elastic filaments. *Physical Review Letters* **84**, 1623–1626.

Xing, X., P. M. Goldbart, and L. Radzihovsky (2007). Thermal fluctuations and rubber elasticity. *Physical Review Letters* **98**, 075502.

Yariv, A. (1978). Phase conjugate optics and real time holography. *IEEE Journal of Quantum Electronics* **14**, 650–660.

——— (1989). *Quantum Electronics*. New York: Wiley.

Yariv, A., and P. Yeh (2007). *Photonics: Optical Electronics in Modern Communications*. Oxford: Oxford University Press.

Yeganeh-Haeri, A., D. J. Weidner, and J. B. Parise (1992). Elasticity of α-Cristobalite: A silicon dioxide with a negative Poisson's ratio. *Science* **257**, 650.

Young, T. (1802). On the theory of light and colours (read in 1801). *Philosophical Transactions* **92**, 34.

Yvon, J. (1935). *La Théorie des Fluides et l'Équation d'État*. Paris: Hermann.

Zangwill, A. (2013). *Modern Electrodynamics*. Cambridge: Cambridge University Press.

Zee, A. (2013). *Einstein Gravity in a Nutshell*. Princeton, N.J.: Princeton University Press.

Zel'dovich, B. Ya., V. I. Popovichev, V. V. Ragul'skii, and F. S. Faizullov (1972). Connection between the wavefronts of the reflected and exciting light in stimulated Mandel'shtem-Brillouin scattering. *Journal of Experimental and Theoretical Physics Letters* **15**, 160–164.

Zel'dovich, Ya. B. (1968). The cosmological constant and the theory of elementary particles. *Soviet Physics Uspekhi* **11**, 381–393.

——— (1972). A hypothesis, unifying the structure and the entropy of the universe. *Monthly Notices of the Royal Astronomical Society* **160**, 1p–3p.

Zel'dovich, Ya. B., and Yu. P. Raizer (2002). *Physics of Shock Waves and High Temperature Hydrodynamic Phenomena*. Mineola, N.Y.: Courier Dover Publications.

Zhang, F., A. Zimmerman, D. Nichols, Y. Chen, et al. (2012). Visualizing spacetime curvature via frame-drag vortexes and tidal tendexes II. Stationary black holes. *Physical Review D* **86**, 084049.

Zipf, G. K. (1935). *The Psycho-Biology of Language*. Boston: Houghton-Mifflin.

Zurek, W. H., and K. S. Thorne (1985). Statistical mechanical origin of the entropy of a rotating, charged black hole. *Physical Review Letters* **54**, 2171–2175.

Zwicky, F. (1933). Die Rotverschiebung von extragalaktischen Nebeln. *Helvetica Physics Acta* **6**, 110–127.

NAME INDEX

Page numbers for entries in boxes are followed by "b," those for epigraphs at the beginning of a chapter by "e," those for figures by "f," and those for notes by "n."

Abbé, Ernst, 439
Adelberger, Eric, 1300
Albrecht, Andreas, 1432n
Alfvén, Hannes, 943e
Anderson, Wilhelm, 127
Appleton, Edward Victor, 1058n
Arago, François, 436n
Archimedes of Syracuse, 675e

Bacon, Roger, 347
Basov, Nicolay Gennadiyevich, 517
Bekenstein, Jacob, 206
Bernstein, Ira B., 981, 1101
Berry, Michael, 406
Bertotti, Bruno, 1249
Birkhoff, George, 1250
Bogolyubov, Nikolai Nikolayevich, 1103
Bohr, Niels, 1039n
Boltzmann, Ludwig, 181n, 182n
Bondi, Hermann, 890, 1398, 1445n
Born, Max, 1103
Braginsky, Vladimir Borisovich, 1300
Brown, Robert, 283e, 313
Burke, William, 869, 1333

Callen, Herbert, 331
Carlini, Francesco, 358n
Carter, Brandon, 1278, 1291
Cauchy, Augustin-Louis, 565, 588n
Christodoulou, Demetrios, 1284
Chu, Steven, 340
Ciufolini, Ignazio, 1309

Clough, Ray W., 565
Cohen-Tannoudji, Claude, 340
Cornell, Eric, 193

Davies, Paul, 204
De Laval, Gustaf, 887n
DeWitt, Bryce, 1341
Dicke, Robert, 1300
Dirac, Paul, 438n, 1429n
Doob, Joseph, 295n
Drever, Ronald W. P., 498, 503
Dyson, Freeman, 1111e

Eddington, Arthur, 396, 835e, 1268
Einstein, Albert, 41, 42, 51, 53, 140n, 193, 396, 398n, 1151,
    1191e, 1192, 1193, 1194, 1195, 1197, 1221, 1222, 1228,
    1233, 1239, 1242, 1259, 1299, 1302, 1305, 1311, 1319,
    1366, 1382, 1382n, 1444
Emden, Robert, 688
Eötvös, Roland von, 1300
Euclid, 9n
Euler, Leonhard, 371n, 565, 600, 602, 603, 697n
Everitt, Francis, 1309

Faraday, Michael, 5n, 1433n
Feigenbaum, Mitchell, 828
Feynman, Richard P., 14n, 35, 438n, 729e
Fierz, Marcus, 1319, 1320
Finkelstein, David, 1268
Fraunhofer, Joseph von, 411e
Fresnel, Augustin-Jean, 436n
Friedmann, Alexander, 1371n, 1377

1473

Gabor, Dennis, 521n
Galileo Galilei, 347, 565, 1300
Genzel, Reinhard, 471b
Germain, Marie-Sophie, 565
Ghez, Andrea, 471b
Gibbs, J. Willard, 5n, 155e, 160, 219e, 219
Goldreich, Peter, 702
Goldwasser, Samuel M., 554f
Gordon, James P., 517
Green, George, 358n
Green, Herbert S. 1103
Greene, John M., 1101
Grimaldi, Francesco Maria, 347
Guth, Alan, 1432n

Hafele, Josef, 70
Hall, John, 498
Hamilton, William Rowan, 5n, 347, 1062
Hanbury Brown, Robert, 509, 511
Hänsch, Theodor, 498
Harrison, Edward R., 1410
Hawking, Stephen, 204, 206, 1273, 1278, 1284, 1286, 1287n
Hazard, Cyril, 433
Heaviside, Oliver, 5n, 995, 1033e, 1058n
Hecht, Eugene, 434f
Heisenberg, Werner, 788n, 917e
Hertz, Heinrich, 347, 502
Hipparchus of Nicaea, 1219n
Hooke, Robert, 565, 567e
Hubble, Edwin, 1374n
Hulse, Russell, 502, 1310
Huygens, Christiaan, 347, 411

Isaacson, Richard, 1320, 1320n, 1321
Ising, Ernst, 272n

Jeans, James, 1071n
Jeffreys, Harold, 358n
Johnson, John, 327

Kapitsa, Pyotr, 1429n
Kazanas, Demosthenes, 1432n
Keating, Richard, 70
Kennelly, Arthur, 995, 1058e
Kerr, Roy, 1278
Ketterle, Wolfgang, 193
Killing, Wilhelm, 1205n
Kirkwood, John, 1103
Kruskal, Martin, 1101, 1276

Lagrange, Joseph-Louis, 5n, 14, 565
Landau, Lev Davidovich, 1080, 1299e

Lane, Jonathan Homer, 688
Langmuir, Irving, 1044, 1069e
Laplace, Pierre-Simon, 155e
Leibniz, Gottfried Wilhelm, 371n
Lele, Sanjiva K., 807f
Lemaître, Georges, 1371n, 1374n, 1445n
Lense, Josef, 1233
Levin, Yuri, 334
Libchaber, Albert, 830
Lifshitz, Evgeny Mikhailovich, 1268, 1299e
Linde, Andrei, 1432n
Liouville, Joseph, 358n
Lorentz, Hendrik, 41n, 88
Lorenz, Edward, 834
Love, Augustus Edward Hugh, 565
Lynden-Bell, Donald, 113n

Maiman, Theodore H. 517
Marriotte, Edme, 565
Martin, H. C., 565
Maupertuis, Pierre Louis, 371n
Maxwell, James Clerk, 5n, 95e, 155e, 347, 1433n
Metropolis, Nicholas, 280
Michell, John, 1241e
Michelson, Albert A., 455, 464, 470b, 474, 479, 483
Millikan, Robert, 155e, 749
Minkowski, Hermann, 1n, 37e, 88, 1192
Morley, Edward, 474, 483

Navier, Claude-Louis, 565, 588n
Nelson, Jerry, 609
Newton, Isaac, 5e, 5n, 14, 41, 347, 398n, 690, 712, 1151, 1444
Noether, Emmy, 1434n
Nyquist, Harry, 327, 1091

Onsager, Lars, 273, 275, 278
Oort, Jan, 1380n
Oppenheimer, J. Robert, 1258, 1260, 1264, 1268
Oseen, Carl Wilhelm, 754

Page, Don, 206
Pauli, Wolfgang, 160, 1319, 1320
Pease, Francis, 464, 470b
Penrose, Oliver, 1091
Penrose, Roger, 210n, 1273, 1283
Penzias, Arno, 1363, 1364
Phillips, William D., 340
Planck, Max, 1153e
Poisson, Siméon, 436n
Pound, Robert, 498, 1189
Pretorius, Frans, 1341
Prokhorov, Alexander Mikhailovich, 517

Rayleigh, Lord (John William Strutt), 899n
Richardson, Bernard, 529f
Richardson, Lewis, 787e
Richter, Charles, 629e
Ride, Sally, 875e
Rittenhouse, David, 422n
Roberts, Morton S., 1380n
Robertson, Howard P., 1371n
Robinson, David, 1278
Robinson, Ivor, 1249
Rouse, Hunter, 727, 731b
Rubin, Vera, 1380n
Ryle, Martin, 480n

Sahl, Ibn, 351e
Saint-Venant, Barré de, 590bn
Sakharov, Andrei Dmitrievich, 532n, 959n, 1443n
Sato, Katsuhiko, 1432n
Schmidt, Maarten, 433
Schrödinger, Erwin, 210n
Schwarzschild, Karl, 935, 1242, 1259, 1273, 1433n
Scott-Russell, John, 850, 855
Sedov, Leonid I., 911, 912
Seurat, Georges, 427, 427f
Shannon, Claude, 212
Shapiro, Ascher, 731b
Shapiro, Irwin, 1308
Slipher, Vesto, 1374n
Smarr, Larry, 1341
Snell (Willebrord Snellius), 347
Snyder, Hartland, 1264, 1268
Spitzer, Lyman, 997e
Starobinsky, Alexei Aexandrovich, 1432n
Steinhardt, Paul, 1432n
Stevenson, David, 816n, 843n
Stewart, Potter, 788n
Stokes, George G., 486, 754, 899n
Stoner, Edmund, 127

Sunyaev, Rashid, 1430
Szekeres, George, 1276

Taam, Igor Yevgenyevich, 959n
Taylor, Geoffrey Ingram, 458, 746, 911, 912
Taylor, Joseph, 502, 1310
Teukolsky, Saul, 1341
Thirring, Hans, 1233
Tolman, Richard Chace, 1258, 1445n
Topp, L. J., 565
Townes, Charles H. 513e, 517
Turner, M. J., 565
Twiss, Richard Q., 509, 511

Unruh, William, 204

Vessot, Robert, 1301
Vlasov, Anatoly Alexandrovich, 1071n
Volkoff, George, 1258, 1264

Walker, Arthur Geoffrey, 1371n
Weber, Joseph, 517
Weinberg, Steven, 1153, 1193
Weiss, Rainer, 503
Welton, Theodore A., 331
Wheeler, John Archibald, 88, 153, 1293, 1344b
Whitehorse, R. N., 1380n
Wieman, Carl, 193
Wiener, Norbert, 319, 438n
Wilson, Robert, 1363, 1364

Yariv, Amnon, 532
Young, Thomas, 347, 455e
Yvon, Jacques, 1103

Zeiger, Herbert, 517
Zel'dovich, Boris Yakovlevich, 532, 532n
Zel'dovich, Yakov Borisovich, 532n, 916, 1410n, 1430, 1445n
Zwicky, Fritz, 1380

SUBJECT INDEX

Second and third level entries are not ordered alphabetically. Instead, the most important or general entries come first, followed by less important or less general ones, with specific applications last.

Page numbers for entries in boxes are followed by "b," those for epigraphs at the beginning of a chapter by "e," those for figures by "f," for notes by "n," and for tables by "t."

$\mathbf{E} \times \mathbf{B}$ drift, 1025, 1026f, 1039
Θ-pinch for plasma confinement, 961f, 962–963, 971
    stability of, 978–979
Θ-pinch, toroidal, for plasma confinement, 978
    flute instability of, 978–979, 978f

3+1 split
    of spacetime into space plus time, 60, 1158
    of electromagnetic field tensor, 72–74
    of stress-energy tensor, 82–84, 120
    of 4-momentum conservation, 60, 85–88
4-acceleration related to acceleration that is felt, 69, 1182
4-force
    as a geometric object, 51
    orthogonal to 4–velocity, 52
4-momentum
    as a geometric object, 50
    components in Lorentz frame: energy and momentum, 58–59
    and affine parameter, 51
    related to 4–velocity, 50
    related to quantum wave vector, 50
    related to stress-energy tensor, 82–84
4-momentum conservation (energy-momentum conservation)
    3+1 split: energy and momentum conservation, 60, 85–88
    expressed in terms of stress-energy tensor, 84–85
    global, for asymptotically flat system, 1237–1238
    global version fails in generic curved spacetime, 1177, 1218
    for particles, 51, 52f, 60
    for perfect fluid, 86–87
    for electromagnetic field and charged matter, 88
4-momentum density, 82
4-vector. *See* vector in spacetime
4-velocity
    as a geometric object, 49
    3+1 split: components in Lorentz frame, 58

Abbé condition, 373n
aberration of photon propagation direction, 107, 1305
aberrations of optical instruments, 395–396
    of Hubble space telescope 426–427
absorption of radiation, 115–116, 131, 260, 341–342, 515–516, 937, 1017, 1054, 1057, 1126–1130, 1394–1397
accelerated observer
    proper reference frame of, 1180–1186, 1181f, 1200, 1254, 1274
    uniformly, 1186–1189, 1187f
acceleration of universe, 1382, 1398–1401, 1444, 1445n
accretion disk around spinning black hole, 784, 969
    thin, 1287–1289
    thick, 1289–1290
accretion of gas onto neutron star or black hole, 205, 890–891, 1266, 1282–1283

acoustic horizon radius, χ_A, 1375
  ultrarelativistic, χ_R, 1375, 1403–1404
acoustic peaks, in CMB anisotropy spectrum, 1413, 1419f, 1421
action principles
  Hamilton's, in analytical mechanics, 15
  for geodesic equation, 1203, 1205–1206, 1357
  for rays in geometric optics, 371–373
  for elastic stress, 584
  for eigenfrequencies of normal modes, 980–981
active galactic nuclei, 1379
adaptive optics, 470b–471b, 472
adiabatic index
  definition, 243, 724b
  polytropic, 878
  for ideal gas: ratio of specific heats, 244, 678, 681b, 724b
    influence of molecules' internal degrees of freedom, 879–880
  for air, as function of temperature, 880f
  in plasma: anisotropic, 1020–1024
adiabatic invariants
  accuracy of, 1030
  failure of, 1030
  for charged particle in magnetic field, 1028–1030
  wave action of a classical wave, 365–366
advective (convective) time derivative, 32, 692, 724b, 892
affine parameter, 50–51, 99–100, 133, 136b, 1178b, 1200, 1203, 1206, 1208, 1247, 1303, 1307, 1423
Aichelberg-Sexl ultraboost metric of a light-speed particle, 1231
airplane wing or airfoil, lift on, 743f, 743–744, 824
Airy diffraction pattern for circular aperture, 426–428, 437, 442
Airy disk
  for diffraction pattern of a circular aperture, 426–427, 437, 471
  for low-pass filter to clean laser beam, 441
  in phase-contrast microscopy, 442–443, 442f
  in Strehl ratio for a telescope's performance, 472
Airy function, for diffraction near a caustic, 451–454, 452f
Alcator C-Mod, 963
Alfvén waves, 354, 990–991, 1053f
  two-fluid analysis of, 1055–1058
  dispersion relation of, 354, 990
  phase velocity of, 354, 990
  group velocity of, 355f, 356
  polarization of, 405
  relativistic corrections for, 1055–1056
  as plasma-laden, plucked magnetic field lines, 990, 1056–1057
  in magnetosphere, 370f

  in solar wind, 970–971
  generated near shock fronts, 1146f
  interaction with cosmic rays. See cosmic rays
Allan variance of clocks, 310f, 320–21
allometry, 587, 609
ALMA (Atacama Large Millimeter Array), 480, 482
Andromeda galaxy, 305, 1365f
angular momentum
  and moment of inertia tensor, 6
  of fundamental particles (spin), 22
  of a Kerr black hole, 204–205, 226n, 1278, 1282–1283, 1285–1286, 1342. See also frame dragging by spinning bodies
  of a relativistic, spinning body, 1218, 1220, 1232–1234, 1237–1238, 1328
  in accretion disks, 1287–1292
  carried by gravitational waves, 1332–1333, 1335, 1338, 1345
  in statistical mechanics, 169, 172–173, 179
  in elastodynamics, 644, 661–662
  in fluid mechanics, 702, 729, 732, 733, 784, 826, 849
angular momentum conservation, Newtonian, 14–15
  in circulating water, 729, 732–733
angular momentum conservation, relativistic
  global, for asymptotically flat system, 1237–1238
  for geodesic orbits around a black hole, 1274, 1303
angular-diameter distance, d_A, 1374, 1378, 1398–1400, 1427
anthropic principle, 1439–1440, 1446
aperture synthesis, in radio astronomy, 480
Appleton-Hartree dispersion relation, 1059–1060
Archimedes law, 675, 684–685, 692
artery, blood flow in, 716–719
astigmatism, 390, 395
astronomical seeing, 425, 464–472, 481, 511
asymptotic rest frame, 1237, 1246–1248
  local, 1328, 1331, 1332, 1339–1340
asymptotically flat system in general relativity, 1194, 1238, 1238n
  imprint of mass and angular momentum on exterior metric, 1232–1233, 1238
  conservation laws for mass and angular momentum, 1237–1238, 1332, 1338
atmosphere of Earth. See also tornados; winds
  structure of, 683–684, 684f
  chemical reactions in, 256–258
  and greenhouse effect, 138, 748–749
  storms in, 768, 769b
    excitation of ocean waves by, 783
  billow clouds in, 783, 783f
  sedimentation in, 748–755

turbulence of
    and astronomical seeing, 425, 464–472, 481, 511
    excitation of earth's normal modes by, 816–817
atomic bomb, 153, 912–914, 1009

B-modes, of CMB polarization, 1420, 1428, 1439
Babinet's principle, 428–429, 523
bacterium, swimming, 747b–748b, 756–757
baffles, to control scattered light, 448–451
balls, physics of flight, 817, 823–825
bandwidth of a filter, 315–318
barotropic fluid, 681b, 724b
baryogenesis, 1442–1443
baryons in universe
    origin of: baryogenesis, 1442–1443
    evolution of, 1407–1408, 1410–1414
    observations today, 1379
basis vectors in Euclidean space
    orthogonal transformation of, 20–21
    Cartesian, 16–17
    spherical and Cartesian, orthonormal, 614
basis vectors in spacetime
    dual sets of, 1161
    coordinate, 1162–1163, 1167
    orthonormal (Lorentz), 54, 1157
        Lorentz transformation of, 63–65
    nonorthonormal, 1160–1163
        transformation between, 1164
baths for statistical-mechanical systems
    concept of, 160
    general, 172
    tables summarizing, 160t, 221t, 251t
BBGKY hierarchy of kinetic equations, 803, 1103–1106, 1109
BD, 530, 531
beam, bent. *See* bent beam
bending modulus (flexural rigidity), 594
bending torque, 593f, 594, 596, 600, 602, 603, 608, 611, 612
bent beam, elastostatics of, 592–596
    elastostatic force-balance equation for, 595
    solutions of force-balance equation
        for clamped cantilever pulled by gravity, 596–597
        for Foucault pendulum, 597–598
        for elastica, 600–601, 601f
bent plate, elastostatics of, 609–613
    elastostatic force-balance equation: shape equation, 610
Bernoulli function, 698–700, 702, 721, 722, 724b
Bernoulli's theorem
    most general version: for any ideal fluid, 698
    for steady flow of an ideal fluid, 698, 700
    for irrotational flow of an ideal, isentropic fluid, 701
    relativistic, 722

Berry's phase (geometric phase), 406–409
BGK waves in a plasma, 1100–1101
Bianchi identities, 1223–1224
    in Maxwell-like form, 1235b, 1318b
bifurcation of equilibria
    formal mathematical foundations for, 384
    onset of dynamical instability at bifurcation
        in general, in absence of dissipation, 648
        for beam under compression, 647–648
    for convection, Rayleigh-Bénard, 931
    for beam under compression, 603–605
    for rapidly spinning star, 607
    for rotating Couette flow, 826–827
    for Venus fly trap, 607
    for whirling shaft, 607
big rip, 1446
biharmonic equation, 589, 610, 754, 756
billow clouds, 783, 783f
binary black holes, 1341–1342, 1342f, 1343f, 1344b–1345b
binary pulsars
    Hulse-Taylor: B1913+16, 502, 1310
    J0337+715, 1301
    J0737+3039, 1303, 1309, 1310
    J1614–2230, 1309
    observation of gravitational radiation reaction in, 1310–1311
    tests of general relativity in, 1301, 1303, 1311
binary star system. *See also* binary pulsars
    gravitational waves from, 1335–1341
    tidally locked, shapes of stars, 691
bird flight
    V-shaped configuration, 744
    wingtip vortices, 734f, 739, 744, 744f
birefringent crystals, 541b–542b, 546, 547. *See also* nonlinear crystals
    phase-matching via birefringence, 546–548
    three-wave mixing in. *See* three-wave mixing in nonlinear crystals
Birkhoff's theorem, 1250–1251, 1264
black holes. *See also* horizon, black-hole event; Kerr metric; Schwarzschild metric
    nonspinning, Schwarzschild, 1272–1276. *See also* Schwarzschild metric
        geodesic orbits around, 1274–1276, 1275f
    spinning, Kerr, 1277–1293. *See also* Kerr metric
    laws of black-hole mechanics and thermodynamics, 205–209, 1284–1287
    statistical mechanics of, 204–206
    entropy of, 205–209, 1287
    irreducible mass of, 1284
    inside a box: thermal equilibrium, 206–209

Subject Index   **1479**

black holes *(continued)*
    rotational energy and its extraction, 1282–1287, 1291–1293
    evolution of, 1282–1287
    quantum thermal atmosphere of, 204–205
    Hawking radiation from, 204–205, 1286–1287
    accretion of gas onto, 205, 784, 890–891, 969, 1282–1283, 1287–1290
    binary, 1341–1342, 1342f, 1343f, 1344b–1345b
    collisions of and their gravitational waves, 1341–1342, 1342f, 1343f, 1344b–1345b
    in the universe, 1379–1380, 1397
blackbody (Planck) distribution and specific intensity, 113, 128, 132
Blandford-Znajek process, 1285, 1291–1293
Blasius profile for laminar boundary layer, 758–764
blast wave. *See* explosion and blast wave
blood flow in arteries, 717–719
boat waves, 846–848
boat, stability of, 685–686
Boltzmann distribution (mean occupation number), 113, 177
    entropy of, 187
Boltzmann equation, collisionless, 134–135, 167, 169. *See also* Vlasov equation
    derivation from Hamiltonian, 136b–137b
    implies conservation of particles and 4-momentum, 135
Boltzmann transport equation, 135, 139
    for photons scattered by thermalized electrons, 144–148
    accuracy of solutions, 140–141
    order-of-magnitude solution, 143–144
    solution via Fokker-Planck equation, 343
    solution via Monte Carlo methods, 1415–1418, 1428
    solution via two-lengthscale expansion, 145–148
boost, Lorentz, 64–65
Bose-Einstein condensate, 193–201
    condensation process, 193, 196, 197f, 198–200
    critical temperature, 196
    specific heat change, 200
    in cubical box, 201
Bose-Einstein ensemble
    probabilistic distribution function for, 176
    mean occupation number of, 112–113, 176–177
    entropy of, 187
bosons, 110
boundary layers
    laminar, 757–766
        Blasius profile, 758–764
        sublayer of turbulent boundary layer, 818
        Ekman, for rotating flow, 772–777
        vorticity creation in, 758
        diffusion of vorticity in, 741–742, 741f, 758
        instability of, 822–823
        in a pipe, 766
        near curved surface, 764–765
        separation from boundary, 764, 793
    turbulent, 817–825
        profile of, 818–820, 818f
        separation from boundary, 821f, 820–821
    thermal, 923
    influence on flight of sports balls, 823–825
Boussinesq approximation, 923–925
bow shock around Earth, 876f, 957, 1090, 1146–1147, 1146f
bremsstrahlung, 142, 260, 1009, 1017
brightness temperature, 482
Brillouin scattering, 1142
Brownian motion, 296, 309, 313–315. *See also* random walk
    spectral density and correlation function for, 313–314
    relaxation time for, 328
    fluctuation-dissipation theorem applied to, 327–329
Brunt-Väisälä frequency for internal waves in stratified fluid, 941
bubbles
    in water, collapse of, 703
    in water, rising, 937
    soap, 846
buckling of compressed beam or card, 602
    onset of buckling at bifurcation of equilibria, 603–605
    onset of elastostatic instability at bifurcation of equilibria, 648
    elementary theory of, 602–605, 608–609
    free energy for, 604
    applications
        collapse of World Trade Center buildings, 605–607
        mountain building, 609
        thermal expansion of pipes, 609
bulk modulus, for elasticity, 581
    values of, 586t, 651t
    atomic origin of, 649f
    relation to equation of state, 650
bump-in-tail instability, 1136–1137, 1138–1139
bunching of bosons, 511, 511n, 1117

canonical ensemble, 160t, 169–172, 221t
    distribution function, 171, 173
canonical transformation, 162, 164, 166
cantilever, 566, 592–593, 596–597
capillary waves, 844–848
Cartesian coordinates, 16, 26, 28
    local, on curved surface, 1198
catastrophe theory
    caustics as examples of catastrophes, 384

state variables, 385
control parameters, 386
five elementary catastrophes
   fold catastrophe, 386–388, 393f
   cusp catastrophe, 389, 392, 391f, 393f
   swallowtail catastrophe, 389–390, 391f, 393f
   hyperbolic umbillic catastrophe, 389–390, 391f, 393f
   elliptic umbillic catastrophe, 391f, 392, 393f
applications
   to caustics of light propagation, 384–394
   to elliptic gravitational lens, 403
   to van der Waals equation, 394–395
   to buckling of a beam, 606–607
caustics, 351, 384–394. See also catastrophe theory
   diffraction near, 451–454
   examples
      sunlight on bottom of swimming pool, 384, 384n, 385f
      sunlight reflected onto bottom of a cup, 385f
cavitation, 702–703
CD, 530, 531
central limit theorem, 292–294
   examples and applications of, 261, 294–295, 322, 465, 510
centrifugal acceleration, 689, 767
Cerenkov emission of plasmons by fast electrons in a plasma, 1127–1129, 1131, 1133, 1138
CGL equations of state, 1024
chaos in dynamical systems, 832–834
   Lyapunov exponent, 833
   Lyapunov time, 832
   strange attractors, 833–834
   examples of, 832, 1030
   quantum chaos, 832
chaos, onset of in dynamical systems, 825–833. See also turbulence, onset of
   in idealized equations and mathematical maps
      logistic equation and Feigenbaum sequence, 828–831
      Lorenz equations, 834
      universality of routes to chaos, 830
Chapman-Kolmogorov equation. See Smoluchowski equation
characteristics of a dynamical fluid flow, 852, 892–893, 893f, 894–896
charge density
   as integral over plasma distribution function, 1072
   as time component of charge-current 4-vector, 74
charge-current 4-vector
   geometric definition, 78
   components: charge and current density, 78
   local (differential) conservation law for, 79
   global (integral) conservation law for, 79, 79f
      evaluation in a Lorentz frame, 81

relation to nonrelativistic conservation of charge, 81
charged-particle motion in electromagnetic field, 1024–1032
chemical free energy (Gibbs potential), 246–249. See also under fundamental thermodynamic potentials; fundamental thermodynamic potentials out of statistical equilibrium
chemical potential, excluding rest mass, μ, 112, 173
chemical potential, including rest mass, $\tilde{\mu}$, 112, 172–173
chemical reactions, including nuclear and particle, 256
   direction controlled by Gibbs potential (chemical-potential sum), 256–258
   partial statistical equilibrium for, 256
   examples
      water formation from hydrogen and oxygen, 256–257
      electron-positron pair formation, 258–259, 1001
      emission and absorption of photons, 115–116
      ionization of hydrogen: Saha equation, 259–260, 998–1000
      controlled thermonuclear fusion, 959–960, 1141b
      nucleosynthesis in nuclear age of early universe, 192–193, 1387–1392
      recombination in early universe, 1393–1396
      annihilation of dark-matter particles, 1440–1442
Christoffel symbols, 1172
chromatic resolving power, 424, 496
chronology protection, 69
circular polar coordinates, 1163, 1163f, 1165, 1173. See also cylindrical coordinates
circulation, 729, 733, 734, 739–740
   and lift on airplane wing, 743
   as flux of vorticity, 739
   evolution equations and Kelvin's theorem, 740
Clausius-Clapeyron equation, 254–256
climate change, 748–749, 755, 958, 1440n. See also greenhouse effect
clocks
   ideal, 39, 39n, 49, 1154n
   frequency fluctuations of, 310f, 310n, 320–321
closure phase, in multiple-element interferometry, 481
closure relation, in plasma kinetic theory, 1074, 1105, 1409
clouds, billow, 783, 783f
CMA diagram for waves in cold, magnetized plasma, 1062–1065, 1064f
CMB. See cosmic microwave background
Coanda effect, 809f, 809–810, 820, 821f
coarse graining, 183–185, 184f, 206, 210–211, 1443
COBE (Cosmic Background Explorer), 476
coherence length
   longitudinal or temporal, 472–473
   spatial or lateral, 462–463
   volume of coherence, 477

coherence of radiation
    qualitative description, 437–438
    perfect coherence, 459
    incoherent superposition of radiation, 460
    spatial coherence, 456–464
    temporal (longitudinal) coherence, 472–474, 458n
    degree of coherence, 461n
        lateral, 460–461
        spatial, 462
        longitudinal (temporal), 472–474, 458n
        3-dimensional, 477
        applications of, 463–465, 466b–471b, 474
        fringe visibility, 461–463, 475
coherent state, quantum mechanical, for light, 518
collective effects in plasmas, 907, 943, 1003–1006, 1016, 1020, 1070, 1146
collisionless shocks, 907, 1145–1147
communication theory, 211–217
commutation coefficients, 1171, 1215
commutator
    of two vector fields, 735n, 1167–1169, 1172, 1209, 1214n
comoving coordinates, in cosmology, 1370
component manipulation rules
    in Euclidean space, 16–19
    in spacetime with orthormal basis, 54–57
    in spacetime with arbitrary basis, 1161–1165
components of vectors and tensors. See under vector in Euclidean space; vector in spacetime; tensor in Euclidean space; tensor in spacetime
compressible fluid flow
    equations for, 877–879
    1-dimensional, time-dependent, 891–897
        Riemann invariants for, 891–895
        nonlinear sound wave, steepening to form shock, 894, 894f
        in shock tube. See shock tube, fluid flow in
    transonic, quasi-1-dimensional, steady flow, 884f, 880–891
        equations in a stream tube, 880–882
        properties of, 882–883
        relativistic, 890
Compton scattering, 1388, 1392–1393, 1428–1430
conductivity, electrical, κ_e, 139
    in plasmas with Coulomb collisions, 1015, 1018
    in magnetized plasma, tensorial, 1022–1023, 1036
conductivity, thermal, κ, 139
    energy flux for, 714
    for photons scattered by thermalized electrons, 148,
        derivation from Boltzmann transport equation, 144–148
conformally related metrics, 1159–1160
congruence of light rays, 1423–1424

connection coefficients
    for an arbitrary basis, 1171–1173
    for orthonormal bases in Euclidean space, 615
        pictorial evaluation of, 616f
        used to compute components of gradient, 617
        for cylindrical orthonormal basis, 615
        for spherical orthonormal basis, 616
conservation laws. See also specific conserved quantities
    differential and integral, in Euclidean 3-space, 28
    differential and integral, in spacetime, 79
    related to symmetries, 1203–1205
contact discontinuity, 953
continental drift, 932b
contraction of tensors
    formal definition, 12–13, 48
    in slot-naming index notation, 19
    component representation, 17, 56
controlled fusion. See fusion, controlled thermonuclear
convection
    onset of convection and of convective turbulence, 830–831, 931
    Boussinesq approximation for, 924–925
    between two horizontal plates at different temperatures: Rayleigh-Bénard convection, 925–933
        Boussinesq-approximation analysis, 925–928, 930
        critical Rayleigh number for onset, 930, 930f, 933
        pattern of convection cells, 930–931, 931f
        toy model, 929
    in a room, 931
    in Earth's mantle, 932
    in a star, 933–937
    in the solar convection zone, 936
convergence of light rays, 1424
convolution theorem, 421–422
coordinate independence. See geometric principle; principle of relativity
coordinates. See specific names of coordinates
Copernican principle, 1366
Coriolis acceleration, 735, 767–768, 1185
    as restoring force for Rossby waves, 858
Cornus spiral, for Fresnel integrals and Fraunhofer diffraction, 431f
correlation functions
    for 1-dimensional random process, 297
        correlation (relaxation) time of, 297
        value at zero delay is variance, 297
    for 2-dimensional random process, 306–308
        cross correlation, 307
    for 3-dimensional random process
        cosmological density fluctuations (galaxy distribution), 304–306, 306f

for many-particle system, 1104–1106
　　　　　two-point and three-point, 1104–1106
　　applications of
　　　　Brownian motion, 314
　　　　cosmological density fluctuations, 303–306, 1414
　　　　distortion of galaxy images due to weak lensing, 1424–1427
　　　　angular anisotropy of cosmic microwave background, 1417–1420
correlation (relaxation) time, 297, 298f
Cosmic Background Explorer (COBE), 476
cosmic dawn, 1421–1422
cosmic microwave background (CMB)
　　evolution of in universe
　　　　before recombination, 1384–1387, 1407–1408
　　　　during and since recombination, 1415–1422
　　　　redshifting as universe expands, 1373
　　observed properties today, 1381, 1419f
　　　　Doppler shift of due to Earth's motion, 116–117
　　　　isotropy of, 1364
　　　　map of, by Planck, 1365f
　　frequency spectrum of, today
　　　　measured by COBE, 476
　　　　Sunyaev-Zeldovich effect on, 1428–1430
　　anisotropies of, today
　　　　predicted spectrum, 1419f
　　　　acoustic peaks, 1413, 1419f, 1421
　　polarization of, today, 1416, 1417, 1420, 1428
　　　　E-mode 1419f, 1420, 1428
　　　　B-mode, 1420, 1428, 1439
cosmic rays
　　spectrum of, 988
　　ultra-high-energy, 1024
　　anisotropy of arrival directions, 992–993
　　acceleration of in strong shock fronts, 1147–1148
　　interaction with Alfvén waves, 992–993, 1138–1139
　　　　Cerenkov emission of Alfvén waves by, 1138–1139
　　　　observational evidence for scattering of, 989
　　　　scattering of, by Alfvén waves, 992–993, 1138–1139
cosmic shear tensor, 1424, 1427
cosmic strings, 1357, 1432n
cosmic variance, 1411n, 1421
cosmological constant
　　observational evidence for, 1382–1383
　　history of ideas about, 1382n, 1444–1445, 1445n
　　as energy density and negative pressure, 1282–1383, 1445
　　as a property of the vacuum, 1445
　　as a "situational" phenomenon, 1446
　　as an emergent phenomenon, 1445
cosmology, standard, 1383
Coulomb correction to pressure in plasma, 1108

Coulomb logarithm, 1008–1009
Coulomb scattering
　　Rutherford scattering analysis, 1006–1007
　　Fokker-Planck analysis of, 1013–1015
　　deflection times and frequencies, 1008
　　energy equilibration times, 1010–1012, 1002t
Cowling's theorem, for dynamos, 984
critical density for universe, 1377
critical point of transonic fluid flow, 883, 886, 891
Crocco's theorem, 702, 742
Cross correlation, 306–308
cross product, 25–26
Cross spectral density, 307–308
cruise-control system for automobiles, 1097–1098
crystals, nonlinear. See nonlinear crystals
curl, 25–26
current density
　　as spatial part of charge-current 4-vector, 74
　　as integral over plasma distribution function, 1072
current moments, gravitational, 1328–1332
curvature coupling in physical laws, 1219–1221
curvature drift, 1026, 1027f
curve, 9, 49, 1154–1155
cutoff, in wave propagation, 1049–1050, 1050f
cyclic symmetry, 1214n
cyclotron frequencies, 1019, 1002t
　　relativistic, 1024
Cygnus X-1, 111
cylindrical coordinates
　　related to Cartesian coordinates, 614
　　orthonormal basis and connection coefficients for, 614–615
　　expansion and shear tensor in, 617, 618
　　coordinate basis for, 1163, 1163f

d'Alembertian (wave operator), 71, 1191, 1434
d'Alembert's paradox, for potential flow around a cylinder, 765
dam, water flow after breaking, 857–858, 897
dark energy, 1363, 1444, 1446. See also cosmological constant
dark matter
　　observational evidence for, 1380–1381
　　physical nature of, 1440–1442
　　searches for dark-matter particles, 1442
　　evolution of, in early universe, 1406–1407, 1411f
de Broglie waves, 44b
De Laval nozzle, 887
de Sitter universe or expansion, 1398, 1400, 1432, 1437
Debye length, 1002t, 1004
Debye number, 1002t, 1004
Debye shielding, 1003–1004

decay time for magnetic field, in MHD, 949, 950t
deceleration function $q(t)$ for the universe, 1374, 1378
    value today, 1382
decibel, 865
decoherence, quantum, and entropy increase, 185, 186b–187b, 190–191
deflection of starlight, gravitational, 1304–1307. *See also* gravitational lensing
degeneracy, of gas, 122–124, 122f, 1000, 1002
    relativistic, 122f, 125, 127
degree of coherence. *See under* coherence of radiation
density fractions, Ω_k, for cosmology, 1377–1378
density of states (modes)
    for free particles, 108–110
    in statistical mechanics, 162–163
density operator (matrix), in quantum statistical mechanics, 165b–166b
derivatives of scalars, vectors, and tensors
    directional derivatives, 22–23, 70, 1167, 1169
    gradients, 23, 70–71, 617, 1170–1171, 1173
    Lie derivative, 735n
deuterium formation in early universe, 1389–1392
dielectric tensor, 1036
    in nonlinear crystal, 537
    in cold, magnetized plasma, 1051–1052
differential forms, 78
    one-forms used for 3-volumes and integration, 77n
    and Stokes' theorem, 78
diffraction grating, 422–424, 524–529
diffraction: scalar, Helmholtz-equation formalism for, 413–436. *See also under* Fraunhofer diffraction; Fresnel diffraction
    propagator through an aperture, 416–417
    Fresnel and Fraunhofer regions defined, 417–419, 418f
    Fraunhofer diffraction, 420–429
    Fresnel diffraction, 429–436
    failure near edges of apertures, 415
    application to weak sound waves in a homogenous medium, 413
    application to electromagnetic waves in vacuum or a homogeneous dielectric medium, 413
    failure due to polarization effects, 413, 413n, 416n
diffusion. *See also* Boltzmann transport equation; diffusion coefficient; diffusion equation
    approximation: criteria for validity, 140
    conditional probability for, 291–292
    of neutrons in a nuclear reactor, 151–153
diffusion coefficient
    defined, 139
    for particle diffusion through thermalized scatterers, 150–151
    for temperature, in thermally conducting fluid, 142, 920
    for vorticity, in viscous fluid, 741
    for electrons interacting with electrostatic waves in plasma, 1118–1119, 1123
    for magnetic field, in MHD, 948
diffusion equation, 140
    solution in infinite, homogenous medium, 141, 291
    and random walk, 140, 140n
    Fokker-Planck equation as, 339
    for temperature in homogenous medium, 142, 920
    for vorticity, in viscous fluid, 741
    for magnetic field, in MHD, 948
    in nonlinear plasma physics, 1118–1119, 1135, 1148
dimensional analysis for functional form of a fluid flow, 790–791
dimensional reduction in elasticity theory, 590b
    for bent beam, 592–595
    for bent plate, 609–613
Dirac equation, 44b
    energy eigenstates (modes) of, 175n
directional derivative, 22–23, 70, 1167, 1169
dispersion relation, 353. *See also under* geometric optics; *specific types of wave*
    as Hamiltonian for rays, 361, 367
displacement vector, in elasticity, 570
    gradient of, decomposed into expansion, shear, and rotation, 570–571
    Navier-Cauchy equation for, 587. *See also* Navier-Cauchy equation for elastostatic equilibrium
dissipation, 724b. *See also* fluctuation-dissipation theorem
distortion of images, 1424
distribution function. *See also under specific ensembles*
    as a geometrical object, 162
    Newtonian number density in phase space, 99
    relativistic number density in phase space, 104
    statistical mechanical, number density of systems in phase space, 163
    statistical mechanical, probabilistic, ρ, 161
        in statistical equilibrium, general, 173
        normalization, 163
    mean occupation number, 108–110. *See also* occupation number, mean
    N-particle, 1102–1103
    isotropic, 120–121
    integrals over momentum space, 117–121
    evolution of. *See* Boltzmann equation, collisionless; Boltzmann transport equation; Vlasov equation
    for photons, 106–108
        in terms of specific intensity, 107
    for particles with range of rest masses, 104–105

for particles in a plasma, 105–106, 1071
    N-particle, 1102–1103
divergence, 24, 71
DNA molecule, elastostatics of, 599–600
domain walls, 1432n
Doob's theorem, 295–296
    proof of, 298–299
Doppler shift, 62
    of temperature of CMB, 116–117
double diffusion, 937–940
drag force and drag coefficient, 792
    at low Reynolds number (Stokes flow), 753–754
    influence of turbulence on, 792f, 794, 820–821
    on a flat plate, 763–764
    on a cylinder, 792–794, 792f
    on an airplane wing, 820–821
    on sports balls, 825
    on fish of various shapes, 797–798
drift velocities
    for charged particles in electromagnetic field, 1025–1027
    for electron and ion fluids, 1038–1039
drift waves, 1067–1068
DVD, 530, 531
dynamos, 984–988

E-modes, of CMB polarization, 1419f, 1420, 1428
Earth. *See also* atmosphere of Earth; elastodynamic waves in Earth
    internal structure of, 651t
    pressure at center of, 649
    Moho discontinuity, 650
    mantle viscosity, 755–756
    mantle convection, 932
    continental drift, 932
    normal modes excited by atmospheric turbulence, 816
eddies
    in flow past a cylinder, 791f, 793–794
    in turbulence, 798–800, 802, 804–807, 811–814
eikonal approximation. *See* geometric optics
Einstein curvature tensor, 1223
    contracted Bianchi identity for, 1223
    components in specific metrics
        static, spherical metric, 1258
        linearized metric, 1228
        Robertson-Walker metric for universe, 1371–1372
        perturbations of Robertson-Walker metric, 1401–1402
Einstein field equation, 1223, 1224
    derivation of, 1221–1223
    Newtonian limit of, 1223, 1226–1227
    linearized, 1229
    cosmological perturbations of, 1402

solutions of, for specific systems. *See under* spacetime metrics for specific systems
Einstein summation convention, 16, 55
Einstein–de Sitter universe, 1378, 1398, 1399f
Ekman boundary layer, 772–777
Ekman number, 768
Ekman pumping, 773
elastic energy density, 583–584
    elastic physical free energy density, 584
elastic limit, 580, 581f
elastic moduli, 580–582
    physical origin of, and magnitudes, 585–586, 586t
    for anisotropic solid, 580
    for isotropic solid
        shear and bulk, 581–582
        Young's, 591
        numerical values, 586t
    in Earth, 651t
elastodynamic waves in a homogeneous, isotropic medium, 630–642
    influence of gravity at ultralow frequencies, 639
    wave equation, 635–636
    energy density, energy flux, and Lagrangian for, 641–642
    decomposition into longitudinal and transverse, 636, 636f, 640
    Heaviside Green's functions for, 658–660
    longitudinal waves, 637–638
        displacement is gradient of scalar, 637, 639–640
        sound speed, dispersion relation, group and phase velocities, 637–638
    transverse waves, 638–639
        displacement is curl of a vector, 637, 639–640
        sound speed, dispersion relation, group and phase velocities, 638–639
    Rayleigh waves at surface, 654–657
elastodynamic waves in Earth
    body waves, 650–654
        P-modes and S-modes, 650
        wave speeds at different depths, 651t
        geometric optics ray equation, 652
        junction conditions and mixing of, at discontinuities, 651–654, 651f, 653f
        rays inside Earth, 653f
    edge waves, 654
        Rayleigh wave at Earth's surface, 654–657
        Love waves at Earth's surface, 658
        internal waves, 941
elastodynamic waves in rods, strings, and beams, 642–648
    waves on a string under tension, 644–645
    flexural waves in a stretched or compressed beam, 645–646
    torsion waves in a circular rod, 643–644

elastodynamics. *See also* elastodynamic waves in Earth; elastodynamic waves in a homogeneous, isotropic medium; elastodynamic waves in rods, strings, and beams
   force density, 587
      in cylindrical coordinates, 624
   when gravity can be ignored, 631
   momentum conservation, 631
   wave equation for displacement vector, 635–636
   quantization of, 667–670
elastostatic force balance. *See* Navier-Cauchy equation for elastostatic equilibrium
electric charge. *See* charge density
electromagnetic field. *See also* electromagnetic waves; Maxwell's equations
   electromagnetic field tensor, 52, 53, 72
   electric and magnetic fields, 72
      as 4–vectors living in observer's slice of simultaneity, 72–73, 73f
   4–vector potential, 74–75
      scalar and 3–vector potentials, 75
   electric displacement vector, 536, 1036
   stress tensor, 33
   stress-energy tensor. *See under* stress-energy tensor
electromagnetic waves
   vacuum wave equation for vector potential, 75, 1219–1220
   in curved spacetime: curvature coupling, 1219–1220
   in nonlinear dielectric medium, 536–564. *See also* wave-wave mixing; three-wave mixing in nonlinear crystals; four-wave mixing in isotropic, nonlinear media
      wave equation, 537
   in anisotropic, linear dielectric medium, 551, 1035–1037
      wave equation and dispersion relation, 551, 1037
   in cold plasma, 1035–1068
electron microscope, 444–445
electron motion in electromagnetic field, 1024–1032
electro-optic effects, 539
electrostatic waves. *See also* Langmuir waves; ion-acoustic waves; Landau damping
   dispersion relation for, 1083–1084
      for weakly damped or growing modes, 1085–1086
   kinetic-theory analysis of, 1077–1079
   stability analysis of, 1090–1092, 1095–1098
   quasilinear theory of, 1113–1135
   nonlinear: BGK waves, 1100–1101
embedding diagram, 1261–1263, 1276–1277, 1321f
emission
   spontaneous, 115
      in lasers, 515
      of plasmons, in a plasma, 1126–1128
   stimulated, 115
      in lasers, 496, 515–516, 516f
      of plasmons, in a plasma, 1134
energy conservation, Newtonian, 359, 695
energy conservation, relativistic
   differential, 85, 1176
   integral (global) in flat spacetime, 84, 86
   global, in curved, asymptotically flat spacetime, 1237–1238
   global, in generic curved spacetime: fails!, 1177, 1218
energy density, Newtonian, U
   deduced from lagrangian, 365
   as integral over distribution function, 121
   for prototypical wave equation, 365
energy density, relativistic
   as component of stress-energy tensor, 83
   as integral over distribution function, 120, 126
energy flux, Newtonian, $\mathbf{F}$,
   deduced from lagrangian, 365
   in diffusion approximation, 147–148
   for prototypical wave equation, 365
energy flux, relativistic
   as integral over distribution function, 120
   as component of stress-energy tensor, 83
energy potential. *See under* fundamental thermodynamic potentials
energy principle for perturbations of magnetostatic equilibria, 980–982
energy, relativistic, 34, 59
   as inner product of 4-momentum and observer's 4-velocity, 60–61
   for zero-rest-mass particle, 60, 106
   kinetic, 34, 59
   Newtonian limit, 34, 112
engine, adiabatic, 241
engine, isothermal, 241
ensemble average
   in statistical mechanics, 163–164
   in theory of random processes, 287
ensemble of systems, 160. *See also specific ensembles*
   in statistical equilibrium, 160–161, 172–177
      general, 172–173
      tables summarizing, 160t, 221t
   out of statistical equilibrium, 248–270
      table summarizing, 251t
enthalpy ensemble, 221t, 245
enthalpy, 174
entrainment of one fluid by another
   laminar, 796–797
   turbulent, 806, 809–810
entropy, 181
   additivity of, derived, 185

estimates of, 185
maximized in statistical equilibrium, 183
per particle, 187, 191–192
increase of. See thermodynamics, second law of
of specific entities
  general ensemble, 181
  microcanonical ensemble, 182
  thermalized mode, 187
  thermalized radiation, 188
  classical, nonrelativistic, perfect gas, 188–190
  mixing of two different gases, 190
  black hole, 206
  black hole and radiation inside a box, 206–209
  the universe, 209–210
  information, 211–217
Eötvös experiment, 1300
equations of state
  computed from kinetic theory, 121
  polytropic, 681b, 687, 726b, 878
    thermodynamic quantities in terms of sound speed, 878–879
  for ideal or perfect gas, 228, 675n
  for fluids, 680b–681b, 725b
  for nonrelativistic hydrogen gas, 122–125
  for van der Waals gas, 234
  for thermalized radiation, 128–129
equipartition theorem, 177–178
equivalence principle
  weak, 1300–1301
  Einstein's, 1196, 1217
    delicacies of, 1218–1221
    used to lift laws of physics into curved spacetime, 1217–1218
ergodic hypothesis,
  in statistical mechanics, 180–181
  in theory of random processes, 288–289
ergodic theory, 181n
ergosphere of black hole, 1283–1284
eschatology of universe, 1400–1401
etalon, 483–486, 489
  reflection and transmission coefficients, 484, 485, 486–488
  power reflectivity and transmissivity, 486
Euler equation of fluid dynamics, 697, 725b
Euler's equation (relation) in thermodynamics, 226–227, 231, 240, 247, 256–257
Eulerian changes, 725b
Eulerian perturbations, 971–972
evanescent wave, 654
event, 40
expansion, in elasticity theory, Θ, 571, 572b, 574, 577
  temperature change during, 585

expansion rate of fluid, θ, 693, 725b
expansion rate of universe, $H(t)$, 1374
explosions and blast waves
  in atmosphere or interstellar space, 908–914
  into stellar wind, 915–916
  underwater, 914–915
extensive variables, 169, 172, 221
  complete set of, for a closed system, 222
extraordinary waves
  in nonlinear crystals, 546–548, 551
  in a cold, magnetized plasma, 1057–1058, 1060, 1063, 1064f

Fabry-Perot interferometer (cavity), 490, 491–502
  with spherical mirrors: modes of, 491b–492b
  finesse, 493
  free spectral range, 493
  Gouy phase, 493
  reflection and transmission coefficients, 490
  power transmissivity and reflectivity, 494f
  resonance FWHM, 493
  Bose-Einstein behavior on resonance, 495
  applications of
    laser stabilization, 497–498, 501
    lasers, 496
    mode cleaner for laser beam, 496–497
    reshaping light beam, 497
    spectrometer, 496
    optical frequency comb, 498–501
factor ordering in correspondence principle, 1219–1220
Faraday rotation, 1053–1054, 1060–1061
feedback-control system, 1093b
  stability analysis of, 1093b–1095b, 1098
Feigenbaum sequence and number, 828–831
Fermat's principle, 371
  for dispersionless waves, 372–373
  and Feynman sum over paths, 372
  for general relativistic light rays, 1306–1307. See also gravitational lensing
Fermi momentum, 124
Fermi-Dirac distribution
  probabilistic distribution function for, 176
  mean occupation number of, 112–113, 176
    near-degenerate, 124f, 125
  entropy of, 187
Fermi-Walker transport, 1184
fermion, 110
Ferraro's law of isorotation for magnetosphere, 970
filtering of images. See image processing

filtering of random processes (noise), 311–313
    types of filters
        differentiation and integration, 311
        averaging, 317–318
        band-pass, 315–317
        high-pass, 441
        low-pass, 441
        notch, 441
        finite-Fourier-transform, 317–318
        Wiener's optimal, 318–320
finesse, of a Fabry-Perot interferometer, 493
finite-element methods, 565, 590b, 606f
fish
    streamlining and drag coefficients, 797–798
    swimming, 744, 747b–748b
flexural rigidity (bending modulus), 594
flexural waves on a beam or rod, 353, 355f, 356, 645–646
flows, fluid. *See* fluid flows
fluctuation-dissipation theorem
    Langevin equation, 324–325
    physics underlying, 323–325
    elementary version of, 325–326
        derivation of, 326–327
    generalized version of, 331–334
        derivation of, 334–335
    applications of
        Johnson noise in a resistor, 327
        thermal noise in an oscillator, 329–330. *See also* noise, thermal
        laser-beam measurement of mirror position, 331–334
fluctuations away from statistical equilibrium. *See under* statistical equilibrium
fluid, 677. *See also* fluid dynamics, fundamental equations; fluid dynamics, relativistic; fluid flows; fluid-flow instabilities
    thermodynamics for, 679b–681b
    perfect (ideal), 675, 675n
    Newtonian, 712, 726b
    non-Newtonian, 712f
fluid dynamics, fundamental equations. *See also* fluid dynamics, relativistic
    terminology, 724b–726b
    mass density and flux, 708t
    mass conservation, 32–33, 692–693
    for ideal fluid in external gravitational field
        momentum density and flux, 708t
        Euler equation (momentum conservation), 696–697
        energy density and flux, 704, 708t
        energy conservation, 707
        entropy conservation, 697, 707
    for self-gravitating ideal fluid, 705b–707b, 709
    for viscous, heat-conducting fluid in external gravitational field, 710–719, 919–920
        momentum and energy densities and stress tensor, 715t
        viscous stress tensor, 712
        Navier-Stokes equation (momentum conservation), 712–713. *See also* Navier-Stokes equation
        total energy flux, 715t
        viscous and thermal-conductive energy flux, 714
        entropy evolution (dissipative heating), 715–716
    for viscous, heat-conducting, incompressible flow with negligible dissipation, 919–920
        Boussinesq approximation, 924–925. *See also* convection
    in rotating reference frame, 767–768, 770
fluid dynamics, relativistic, 719–724
    fundamental equations, 719–720
    nonrelativistic limit, 723–724
    relativistic Bernoulli equation and theorem, 721–722
    application to steady, relativistic jet, 721–722
    application to relativistic shock wave, 902–903
fluid flows. *See also* fluid-flow instabilities; fluid dynamics, fundamental equations; fluid dynamics, relativistic
    between two plates, steady, 718
    through a pipe
        laminar, 716–717, 766
        onset of turbulence, 787
    around a body at low Reynolds number: Stokes flow, 749–754, 749f
    around a cylinder: high-Reynolds-number, potential flow, 765, 789–794
    types of
        barotropic, inviscid, 736–738, 740
        viscous, 710–716
        high-Reynolds-number, 757–766
        low-Reynolds-number, 746–757. *See also* low-Reynolds-number flow
        irrotational (potential), 701
        irrotational, incompressible, 837
        incompressible, 709–710
        compressible, 875–916. *See also* compressible fluid flow
        laminar, 716–717. *See also under* boundary layers; wakes; jets
        turbulent, 787–834. *See also under* boundary layers; wakes; jets
        nearly rigidly rotating, 766–768
        geostrophic, 770–777
        self-similar, 759. *See* self-similar flows
fluid-flow instabilities. *See also* fluid flows
    convective. *See* convection
    density inversion: Rayleigh-Taylor instability, 783–784

shear flows
    Kelvin-Helmholtz instability, 778–782
    influence of gravity and density stratification, 782–783, 784–786
    laminar boundary layer, 822–823
    rotating Couette flow, 784, 785f, 825–828
flute instability for toroidal Θ-pinch, 978–979, 978f
Fokker-Planck equation
    in one dimension, 335–337
        as a conservation law for probability, 339
        derivation of, 337–338
        for a Gaussian, Markov process, 338–339
        for detailed-balance processes, 339–340
        time-independent, 338
    in multiple dimensions, 343
    for Coulomb collisions, 1013–1016, 1032
    for Doppler cooling of atoms by laser beams, 340–343, 341f
    for electrons interacting with plasmons, 1123, 1130–1131
    solutions of
        for Brownian motion of a dust particle, 340
        for Doppler cooling of atoms by laser beams, 340–343, 341f
        for photon propagation through intergalactic gas (Sunyaev-Zel'dovich effect), 1428–1430
        for thermal noise in an oscillator, 344
force density, as divergence of stress tensor, 578
force-free magnetic field, 964
Foucault pendulum, 407, 597–598
Fourier transform, conventions for
    in theory of random processes, 299
    in theory of diffraction, 420
    in plasma physics, 1115
Fourier-transform spectroscopy, 474–476
four-wave mixing in isotropic, nonlinear media, 540, 558–564
    specific nonlinear materials used, 559t
    fully degenerate, evolution equations for, 561
    resonance conditions, 560, 562
    phase conjugation via, 559–562, 560f
    squeezing via, 562
    optical Kerr effect in an optical fiber, 562–564
fracture, criterion for safety against, 621
frame dragging by spinning bodies, 1233–1236, 1279–1282, 1295b–1296b, 1342
frame-drag field, 1235b–1236b
frame-drag vortex lines, 1235b–1236b
    around a linearized, spinning particle, 1236b
    around a Kerr black hole, 1295b–1296b
    around colliding black holes, 1344b–1345b
    in a gravitational wave, 1318b, 1345b

Fraunhofer diffraction, 420–429
    diffracted field as Fourier transform of aperture's transmission function, 420
    convolution theorem applied to, 422, 423f
    Babinet's principle for, 428–429
    specific diffraction patterns
        slit, 421, 434
        diffraction grating, 422, 423f, 424
        circular aperture: Airy pattern, 425–427
free energy, 241n
    chemical (Gibbs) free energy. See under fundamental thermodynamic potentials
    physical (Helmholtz) free energy, 241–246
        physical meaning of, 241, 241f
        for elastic medium, 584, 603–604
free spectral range, of a Fabry-Perot interferometer, 493
free-fall motion and geodesics, 1200–1203
frequency and time standards, 310f, 310n
frequency doubling in nonlinear optics, 545–546, 553–555
Fresnel diffraction, 429–436. See also paraxial Fourier optics
    Fresnel integrals for, 430–431
        Cornu spiral, 431f
    specific diffraction patterns
        unobscured plane wave, 432
        straight edge, 432–434
        aperture with arbitrary shape, 430, 433–434, 434f
        rectangular aperture, 430–431
        circular aperture: Fresnel zones and zone plates, 434–436
        near a caustic, 451–454, 452f
Fresnel integrals, 430–431
Fresnel length, 418
Fresnel zones and zone plates, 435–436
Fried parameter, 468b
Friedmann equations for expansion of the universe, 1376–1377
fringe visibility, 461–463
fringes, interference. See interference fringes
fundamental observers (FOs), in cosmology, 1366–1367
fundamental thermodynamic potentials. See also under thermodynamics
    energy potential
        for energy representation of thermodynamics, 222–223
        for nonrelativistic, classical, perfect gas, 227
    Gibbs potential, 246–247
        physical interpretation as chemical free energy, 247–248
        computed by a statistical sum, 246, 248
    grand potential, 229–230
        computed by a statistical sum, 230, 232–238
        for relativistic, perfect gas, 238–239
        for van der Waals gas, 234

fundamental thermodynamic potentials *(continued)*
    physical-free-energy potential, 239–240
        computed by a statistical sum, 239, 242–243
        for ideal gas with internal degrees of freedom, 243
        for elastic medium, 584, 603–604
    enthalpy potential, 244–245
fundamental thermodynamic potentials out of statistical equilibrium
    Gibbs potential, 248–250, 251t
        minimum principle for, 249, 251t
    other potentials and their extremum principles, 250, 251t
    used to analyze fluctuations away from statistical equilibrium, 260–270
fusion, controlled thermonuclear, 959–964, 999f, 1001, 1002t, 1140–1142
    motivation for, 958–959
    Lawson criterion for, 960
    d, t fusion reaction, 959
    magnetic confinement for, 960–964.
    laser fusion, 1140–1142

g modes of sun, 837, 849–850
gain margin, for stability of control system, 1095b, 1098
galaxies
    structures of, 201–202
    observed properties of, 1364, 1365f, 1412–1413
    distortion of images by gravitational lensing, 1424–1427
    spatial distribution of, 1364
        power spectrum for, 1412–1415, 1414f
        correlation function for, 306f
    statistical mechanics of, 202–204
    formation of in early universe, 210–211, 1401–1406
    dark matter in, 201–204, 1076n, 1364, 1365f, 1381
    mergers of, 1413
galaxy clusters
    dark matter in, 1380–1381
    hot gas in, and Sunyaev-Zel'dovich effect, 1428–1430
    merging, image of, 1365f
gas, 678, 725b
    perfect gas, nonrelativistic, 121, 188–189, 726b
    perfect, relativistic, 127, 238–239
    ideal, 242–244, 725b
    hydrogen, 122–123, 122f, 127, 999f
    degenerate, 127–128
gas discharge, in laboratory, 999f, 1001, 1002t
gauge transformations and choices
    in linearized theory of gravity, 1228–1229, 1312
    in cosmological perturbations, 1401n
Gauss's theorem
    in Euclidean 3-space, 27
    in spacetime, 78

Gaussian beams, 445–448
    in Fabry-Perot cavity with spherical mirrors, 491b–492b
    in interferometric gravitational wave detectors (LIGO), 447–448
    manipulation of, 447, 448
Gaussian random process, 292–294. *See also* Markov, Gaussian random process
general relativity, 1191–1224
    some history of, 1191–1193
    linearized approximation to, 1227–1231
    Newtonian limit of, 1225–1227
    experimental tests of, 1299–1311
geodesic deviation, equation of, 1210
    for light rays, 1423
    on surface of a sphere, 1217
geodesic equation
    geometric form, 1201–1202
    in coordinate system, 1203
    conserved rest mass, 1202
    super-hamiltonian for, 1206, 1357
    action principles for
        stationary proper time, 1203, 1205–1206
        super-Hamiltonian, 1357
    conserved quantities associated with symmetries, 1203–1205
geodetic precession, 1290–1291, 1309–1310
geometric object, 1, 5, 41
geometric optics, 357–375, 1174. *See also* Fermat's principle; paraxial ray optics
    as two-lengthscale expansion for a wave, 357, 359, 360
    limitations (failure) of, 369–371
    for a completely general wave, 366–368
    for prototypical wave equation, 358–366
    eikonal approximation, 359
    connection to quantum theory, 362–365
    rays and Hamilton's equations for them, 361
    dispersion relation as Hamiltonian, 361, 367
    amplitude and its propagation, 359, 361, 364–365, 368
    phase and its propagation, 359, 362, 367
    angular frequency and wave vector, 359
    energy density, U, 359
    energy flux, $\mathbf{F}$, 359
    quanta, 363
        conservation of, 364, 365, 368
        Hamiltonian, energy, momentum, number density, and flux of, 363
    polarization vector, for electromagnetic waves, 405–406
        propagation of: parallel transport along rays, 406–409
        geometric phase, 406–409
    eikonal equation (Hamilton-Jacobi equation), 362

for dispersionless waves in time-independent medium
  index of refraction, 372
  ray equation, 373
  Fermat's principle: rays have extremal time, 372
  Snell's law, 373, 374f
  examples
    light propagating through lens, 370f,
    light rays in an optical fiber, 374–375
    flexural waves in a tapering rod, 368–369
    spherical sound waves, 368, 369
    Alfvén waves in Earth's magnetosphere, 370f
    gravitational waves, 1320–1324, 1338–1341
geometric phase, 406–409
geometric principle, 1, 6–7, 10
  examples, 28, 29
geometrized units, 33–34, 35, 1157, 1224
  numerical values of quantities in, 1225t
geometrodynamics, 1344b–1345b
geostrophic flow, 770–777
Gibbs ensemble, 160t, 173–174, 221t. *See also under* fundamental thermodynamic potentials; thermodynamics
  distribution function, 174
Gibbs potential. *See under* fundamental thermodynamic potentials; fundamental thermodynamic potentials out of statistical equilibrium
global positioning system, 1301–1302
global warming, 748–749, 755, 958, 1440n
globular star cluster, energy equilibration time for, 1012–1013. *See also* stellar dynamics
Gouy phase
  for freely propagating Gaussian beam, 446
  for mode of a Fabry-Perot interferometer, 493
gradient drift, 1027, 1027f
gradient operator, 23, 70–71, 617, 1170–1171, 1173
Gran Telescopio Canarias, 609
grand canonical ensemble, 160t, 174, 221t, 229–239. *See also under* fundamental thermodynamic potentials; thermodynamics
  distribution function, 174
grand partition function. *See also* fundamental thermodynamic potentials, grand potential
  as log of grand potential, 229–230
grand potential. *See under* fundamental thermodynamic potentials
gravitation theories
  general relativity, 1191–1224
  Newtonian theory. *See* gravity, Newtonian
  relativistic scalar theory, 53, 1194–1195
gravitational drift of charged particle in magnetic field, 1026
gravitational fields of relativistic systems. *See* spacetime metrics for specific systems
gravitational lensing, 396–404, 1305–1307, 1422–1427. *See also* deflection of starlight, gravitational
  refractive index models for, 396–397
    derivation of, 1305–1307
  Fermat's principle for, 396–397, 1306–1307
  microlensing by a point mass, 398–401
    Einstein ring, 399, 400f
    time delay in, 401
  lensing by galaxies, 401–404, 404f
  lensing of gravitational waves, 1323–1324
  weak lensing, 1422–1427
gravitational waves, 1321f. *See also* gravitons
  speed of, same as light, 45b
  stress-energy tensor of, 1324–1326
  energy and momentum carried by, 1324–1326
  dispersion relation for, 354
  generation of, 1327–1345
    multipole-moment expansion, 1328–1329
    quadrupole-moment formalism, 1330–1335
    radiation reaction in source, 1333, 1338
    numerical relativity simulations, 1341–1342
    energy, momentum, and angular momentum emitted, 1332, 1334–1335
    mean occupation number of modes, 1326–1327
  propagation through flat spacetime, 1229, 1311–1320
    h_+ and $h_\times$, 1315–1316
    behavior under rotations and boosts, 1317, 1319
    TT gauge, 1312–1315
    projecting out TT-gauge field, 1314b
    Riemann tensor and tidal fields, 1312–1313
    deformations, stretches and squeezes, 1315–1317
    tidal tendex and frame-drag vortex lines for, 1318b
  propagation through curved spacetime (geometric optics), 1320–1327, 1338–1341
    same propagation phenomena as electromagnetic waves, 1323
    gravitational lensing of, 1323–1324
  penetrating power, 1311
  frequency bands for: ELF, VLF, LF, and HF, 1345–1347
  sources of
    human arm waving, 1333
    linear oscillator, 1338
    supernovae, 111
    binary star systems, 1335–1342
    binary pulsars in elliptical orbits, 1342–1345
    binary black holes, 1341–1342, 1342f, 1343f, 1344b–1345b

gravitational waves (continued)
  sources of (continued)
    stochastic background from binary black holes, 1356–1358,
    cosmic strings, 1357
   detection of, 1345–1357
    gravitational wave interferometers, 1347–1355. See also LIGO; laser interferometer gravitational wave detector
    pulsar timing arrays, 1355–1357
gravitons
  speed of, same as light, 45b, 1319
  spin and rest mass, 1319–1320
gravity, Newtonian
  gravitational potential, Φ, 682
  field equation for Φ, 682
  gravitational acceleration **g**, 682
  gravitational stress tensor, 705b
  gravitational energy density, 706b
  gravitational energy flux, 706b
  total gravitational energy, 709
gravity probe A, 1301
gravity probe B, 1309
gravity waves on water, 353, 355f, 356, 837–843
  arbitrary depth, 837–840
  shallow water, 840–843
    dam breaking: water flow after, 857
    nonlinear, 840–841, 843, 850–858, 897
    solitary waves (solitons) and KdV equation, 850–858
  deep water, 353, 355f, 356, 840
    viscous damping of, 842
  capillary (with surface tension), 844–848
Green's functions
  for wave diffraction, 417
    in paraxial optics, 438
  for elasticity theory, 590b
  for elastostatic displacement, 626–627
  for elastodynamic waves, 658–661, 660f
  in Fokker-Planck theory, 343
greenhouse effect, 135, 137–138, 748, 958. See also climate change
group velocity, 355
guide star, for adaptive optics, 470–471
guiding-center approximation for charged-particle motion, 1025–1030, 1055
gyre, 773, 775–776, 805
gyro frequencies. See cyclotron frequencies
gyroscope, propagation of spin
  in absence of tidal gravity
    parallel transport if freely falling, 1218–1219
    Fermi-Walker transport if accelerated, 1184
  precession due to tidal gravity (curvature coupling), 1219–1221
gyroscopes
  inertial-guidance, 1182
  used to construct reference frames, 39, 1156, 1180–1182, 1195
  precession of due to frame-dragging by spinning body, 1232–1236, 1279, 1296b, 1309, 1318
  laser, 501, 502f, 520
  on Martian rover, 409

Hamilton-Jacobi equation, 362, 375
Hamilton's equations
  for particle motion, 136b
  for particle motion in curved spacetime, 1206, 1275, 1291
  for rays in geometric optics, 361–363, 367
  for plasmons, 1124
  in statistical mechanics, 158
Hamilton's principal function, 362, 375
hamiltonian, constructed from lagrangian, 1433
hamiltonian for specific systems
  harmonic oscillator, 159
  L-C circuit, 332
  crystal
    fundamental mode, 159
    all modes, weakly coupled, 159
  damped system, 159n
  star moving in galaxy, 159
  particle motion in curved spacetime. See also geodesic equation
    super-hamiltonian, 1206, 1357
Hanbury Brown and Twiss intensity interferometer, 509–511
harmonic generation by nonlinear medium, 537, 545–546, 553–55
harmonic oscillator
  hamiltonian for, 159
  complex amplitude for, 344
  thermal noise in, 344
Harriet delay line, 381
Hartmann flow, 965–969
Hartmann number, 968
Hawking radiation
  from black holes, 204–205, 1286–1287
  from cosmological horizon, 1437
heat conduction, diffusive. See also conductivity, thermal; diffusion; diffusion equation; random walk
  in a stationary, homogeneous medium, 141–142
  in a star, 142–148
  in the sun, 937
  in a flowing fluid, 920
    fluid flow equations with heat conduction. See under fluid dynamics, fundamental equations

Heaviside Green's functions, 658–660, 660b
helicity
    hydrodynamic, 985
    magnetic, 965
helioseismology, 848–850
helium formation in early universe, 192–193, 1387–1392
Helmholtz equation, 413
Helmholtz free energy. See under free energy
Helmholtz-Kirchhoff integral for diffraction, 414, 415f
high-Reynolds-number flow, 757–766
    boundary layers in. See boundary layers
Hilbert space, 18b
hologram, 522–531. See also holography
holography, 521–531
    recording hologram, 522–525, 530
    reconstructing 3-dimensional image from hologram, 525–527, 530
    secondary (phase conjugated) wave and image, 525b, 527, 535
    types of
        simple (standard) holography, 521–528
        reflection holography, 528, 530
        white-light holography, 528
        full-color holography, 528–529
        phase holography, 528
        volume holography, 528
    applications of
        holographic interferometry, 529, 529f
        holographic lenses, 529, 530–531
homogeneity of the universe, 1364–1366
homogeneous spaces
    2-dimensional, 1367–1370
    3-dimensional, 1370, 1372
Hooke's law, 568f, 591
    realm of validity and breakdown of, 580, 581f
horizon problem in cosmology, 1387, 1388f, 1431–1432
horizon radius of universe, χ_H, 1375
horizon, black-hole event
    nonrotating (Schwarzschild), 1272
        formation of, in imploding star, 1273, 1273f
        surface gravity of, 1274
    rotating (Kerr), 1279–1280
        generators of, 1280, 1281f, 1282
        angular velocity of, 1280
        surface gravity of, 1286
        surface area of, 1284, 1285
horizon, cosmological, 1375
    horizon radius, χ_H, 1375
    horizon problem, 1387, 1388f, 1431–1432
    and theory of inflation, 1437–1438
    acoustic horizon and radius, χ_A, 1375

Hubble constant, H_0, 1374
    measurements of, 1375
Hubble law for expansion of universe, 1374
Hubble Space Telescope
    images from, 400, 404, 1365
    spherical aberration in, and its repair, 426–427
Huygen's model for wave propagation, 411, 417
hydraulic jump, 903–904, 904f
hydrogen gas. See gas, hydrogen
hydromagnetic flows, 965–971
hydrostatic equilibrium
    in uniform gravitational field
        equation of, 681
        theorems about, 682–683
    of nonrotating stars and planets, 686–689
    of rotating stars and planets, 689–691
        barotropic: von Zeipel's theorem, 702
        centrifugal flattening, 690, 691
    of spherical, relativistic star, 1258
hydrostatics, 681–691

ideal fluid. See perfect fluid
ideal gas. See gas, ideal
image processing
    via paraxial Fourier optics, 436–437, 441–445
        low-pass filter: cleaning laser beam, 441
        high-pass filter, accentuating features, 441
        notch filter: removing pixellation, 441
        convolution of two images or functions, 443–444
        phase-contrast microscopy, 442–443
        transmission electron microscope, 444–445
    speckle, 470b, 472
impedance
    acoustic, 654
    complex, for fluctuation-dissipation theorem, 332
incompressible approximation for fluid dynamics, 709–710, 725b
index gymnastics. See component manipulation rules
index of refraction, 372
    numerical values, 541b–542b, 547f, 559t
    for axisymmetric optical systems, 377
    for optical elements, 483, 486–489, 497n
    for optical fiber, 374, 447, 534
    for anisotropic crystals, 546
    for plasma waves, 1052, 1057, 1058f
    for Earth's atmosphere, 466b–469b, 814–815
    for seismic waves in Earth, 652
    for model of gravitational lensing, 396, 1307
induction zone, 1327f
inertial (Lorentz) coordinates, 41, 54, 1157
inertial-guidance system, 1182b

inertial mass density (tensorial)
    definition, 87
    for perfect fluid, 87
inertial reference frame. See Lorentz reference frame
inflation, cosmological, 1431–1440
    motivation for, 1431–1432
    theory of, 1434–1438
    particle production at end of, 1435, 1437
    tests of, 1438–1439
inflaton field, 1433
    potential for, 1435, 1436b
    energy density and pressure of, 1435
    evolution of, 1435, 1436b, 1437
    dissipation of, produces particles, 1437
information
    definition of, 212
    properties of, 216
    statistical mechanics of, 211–218
    per symbol in a message, 214, 215
    gain defined by entropy decrease, 211–212
inner product
    in Euclidean space, 10–12, 17
    in spacetime, 48, 56
    in quantum theory, 18b
instabilities in fluid flows. See fluid-flow instabilities
integrals in Euclidean space
    over 2-surface, 27
    over 3-volume, 27
    Gauss's theorem, 27, 1176
integrals in spacetime, 75–78, 1174–1176
    over 3-surface, 77, 80–81, 1175
    over 4-volume, 75, 1175
    Gauss's theorem, 27, 78
    not well defined in curved spacetime unless infinitesimal contributions are scalars, 1175
intensive variables, 172, 221–222
interference by division of the amplitude, 473
interference by division of the wavefront, 458
interference fringes
    for two-beam interference, 457f, 458, 458n
    for perfectly coherent waves, 459
    for waves from an extended source, 460
    fringe visibility, 460–463, 475
    in Fresnel diffraction, 419f,
    near a caustic, 452f, 453
interferogram, 475, 476
interferometer
    Fabry-Perot, 490–495. See also Fabry-Perot interferometer
    gravitational wave. See LIGO; laser interferometer gravitational wave detector
    Michelson, 474, 475f

    Michelson stellar, 464, 465f
    Sagnac, 501–502
    radio-telescope, 479–483
    very long baseline (VLBI), 482
    intensity, 509–511
    stellar intensity, 511
interferometric gravitational wave detector. See laser interferometer gravitational wave detector; LIGO
interferometry, multiple-beam, 483–486
intergalactic medium, 999f, 1002, 1002t
intermittency in turbulence, 798–799, 807
internal waves in a stratified fluid, 941
international pulsar timing array (IPTA), 1356
interstellar medium, 891, 914–916, 950t, 989, 992, 999f, 1001, 1002t, 1012, 1060–1061, 1138–1139, 1146–1147
interval
    defined, 45, 1159
    invariance of, 45–48, 1159–1160
    spacelike, timelike, and null (lightlike), 45
inviscid, 725b
ion-acoustic waves. See also electrostatic waves; plasmons
    in two-fluid formalism, 1046–1050
    in kinetic theory, 1088–1090
    Landau damping of, 1088
    nonlinear interaction with Langmuir waves, 1132–1135
    solitons, 1142–1146
ionosphere, 999f, 1001, 1002t, 1059
    radio waves in, 1058–1062
irreducible mass of black hole, 1284–1287
irreducible tensorial parts of second-rank tensor, 572b–574b, 577, 711
irrotational flow (vorticity-free), 701, 725b, 837
isentropic, 725b
Ising model for ferromagnetic phase transition, 272–282
    1-dimensional Ising model, 278–279
    2-dimensional Ising model, 272–273
        solved by Monte Carlo methods, 279–282
        solved by renormalization group methods, 273–278
isobar, 725b
isothermal engine, 241
isotropy of the universe, 1364–1366
ITER (International Thermonuclear Experimental Reactor), 963

James Webb Space Telescope, 427
Jeans' theorem, 169, 1074–1077, 1100, 1407
jerk function $j(t)$ for universe, 1374, 1378
    value today, 1382
JET (Joint European Torus), 963
jets
    laminar, 796–797

turbulent, 809f, 810
Johnson noise in a resistor, 327
Joukowski's (Kutta-Joukowski's) theorem, 743
Joule-Kelvin cooling, 708, 708f
junction conditions
    elastodynamic, 588–589, 651, 654
    hydraulic jump, 903–904
    MHD, 953–956
    shock front. *See* Rankine-Hugoniot relations for a shock wave
Jupiter, 455, 687, 689, 702, 801, 1100
JVLA (Jansky Very Large Array), 480

Kármán vortex street, 791f, 794
KDP nonlinear crystal, 541b, 546–548, 1141
Keck telescopes, 609–611
Kelvin-Helmholtz instability in shear flow, 778–782
    influence of gravity on, 782–783
    onset of turbulence in, 801b
Kelvin's theorem for circulation, 740, 746, 824
Kepler's laws, 14, 691, 784, 1232–1233, 1247, 1304, 1335, 1344
kernel of a filter, in theory of random processes (noise), 311–313
Kerr metric. *See also* black holes; horizon, black-hole event
    in Boyer-Lindquist coordinates, 1277–1279
    in (ingoing) Kerr coordinates, 1281–1282, 1281n
    geodesic orbits in, 1291
    dragging of inertial frames in, 1279, 1290–1291
    precession of gyroscope in orbit around, 1290–1291
    tidal tendex lines and frame-drag vortex lines in, 1295b–1296b
    light-cone structure of, 1279–1282
    event horizon of, 1280
    Cauchy horizon of and its instability, 1282n
Killing vector field, 1203–1205
kink instability, of magnetostatic equilibria, 977–978
Knudsen number, 755n
Kolmogorov spectrum for turbulence, 467, 810–815
    phenomena missed by, 814
    derivation of, 810–812
    for transported quantities, 467, 814–816
    in Earth's atmosphere, 466b–471b
Kompaneets equation, 1429
Korteweg–de Vries equation and soliton solutions, 850–856, 858, 1048–1049
KTP nonlinear crystal, 542, 554–555
Kutta-Joukowski's theorem, 743

Lagrange multiplier, 183
Lagrangian changes, 725b

lagrangian methods for dynamics, 1433
    lagrangian density
        energy density and flux in terms of, 365, 642, 1434
        for prototypical wave equation, 365
        for scalar field, 1434
        for electrodynamics, 1433–1434
        for elastodynamic waves, 642
Lagrangian perturbations, 971–972
Lamé coefficients, 582
laminar flow, 716–717. *See also under* boundary layers; jets; wakes
Landau contour, 1082f, 1083
Landau damping
    physics of: particle surfing, 1046, 1069–1070, 1098–1099
    for ion-acoustic waves, 1088–1090
    for Langmuir waves, 1086–1088
    in quantum language, 1127–1129
Landauer's theorem in communication theory, 217–218
Langmuir waves. *See also* electrostatic waves; plasmons
    in two-fluid formalism, 1044–1047
    in kinetic theory, 1086–1088, 1090
    in quasilinear theory, 1115–1123
        summary of, in one dimension, 1120
        evolution of electron distribution, 1118–1120
        evolution of wave spectral density, 1118
        in three dimensions, 1122–1123
    Landau damping of, 1086–1088, 1098–1099
    particle trapping in, 1098–1100
    nonlinear interaction with ion-acoustic waves, 1132–1135
Laplace transform, 1081, 1084
    used to evolve initial data, 1081
laplacian, 24, 402, 665
Larmor radius, 1019, 1002t
laser
    principles of, 515–519
    light in quantum coherent state, 518
    pump mechanisms, 519
    types of
        continuous wave, 517, 519
        pulsed, 517, 519
        Nd:YAG, 553
        mode-locked, 520–521. *See also* optical frequency comb
        free electron, 521
        nuclear powered X-ray, 521
laser frequency stabilization
    locking to atomic transition, 519
    locking to mode of an optical cavity: PDH locking, 497–498, 501, 519
laser gyroscope, 501, 502f, 520

laser interferometer gravitational wave detector. See also LIGO
  spectral density of noise, 302
    in initial LIGO detectors, 302f
  sensitivity: weakest detectable signal, 505
  order-of-magnitude analysis of, 503–505
  general relativistic analyses of
    in proper reference frame of beam splitter, 1347–1349, 1352–1355
    in TT gauge, 1347–1352
    for more realistic interferometer, 1355
  Gaussian beams in, 447–448
  power recycling in, 505
  signal recycling in, 506
  phase shift in arm, 506–507
  photon shot noise in, 507–509
  scattered light in, 448–451
  experimental challenges, 505
laser pointer, 554–555
latent heat, 252, 254, 255, 270
Lawson criterion for controlled fusion, 960
least action, principle of, 371, 371n
left modes, for plasma electromagnetic waves parallel to magnetic field, 1052–1056
lens, thin: light propagation through
  geometric optics description of, 378–379, 379f
  paraxial Fourier optics description of (Abbé's theory), 439–441
  and optical Fourier transforms, 439–441
Lense-Thirring precession, 1233, 1290–1291, 1309–1310
Lenz's law, 945
Levi-Civita tensor in Euclidean space, 24–26
  product of two, 25
Levi-Civita tensor in spacetime, 71, 1174–1175
Lie derivative, 735n
light cones, 1155–1156, 1155f, 1159, 1186–1187, 1230, 1230f
  near Schwarzschild black hole, 1264–1265, 1269, 1272
  near Kerr black hole, 1279–1283
LIGO (Laser Interferometer Gravitational-Wave Observatory). See also laser interferometer gravitational wave detector
  discovery of gravitational waves, 506, 1326, 1346
  initial LIGO detectors (interferometers), 447–448, 503
    noise in, 302f, 323, 334, 448–451, 507–509, 626
    order-of-magnitude analysis of, 503–506
    schematic design of, 503f
  advanced LIGO detectors (interferometers), 448, 503, 506, 1346–1347
  signal recycling in, 506
  Gaussian beams in, 447–448
  laser frequency stabilization in, 519
  signal processing for, 320, 329–330, 1341
  squeezed states of light in, 557
LINAC Coherent Light Source (LCLS), 521
line element, 57, 1163–1164
linearized theory (approximation to general relativity), 1227–1231
Liouville equation, in statistical mechanics, 167
  quantum analog of, 165b–166b
Liouville's theorem
  in kinetic theory, 132–134, 133f
  in statistical mechanics, 166, 168f
liquid, 678, 726b
  bulk modulus for, 678
liquid crystals and LCDs, 539, 712
Lithium formation in early universe, 1392
local Lorentz reference frame and coordinates, 1195–1196, 1195f
  connection coefficients in, 1199–1200
    influence of spacetime curvature on, 1213
  metric components in, 1196–1200
    influence of spacetime curvature on, 1213
  Riemann tensor components in, 1214
  nonmeshing of neighboring frames in curved spacetime, 1197–1199, 1197f
logistic equation, 828–831
Lorentz contraction
  of length, 66–67
  of volume, 99
  of rest-mass density, 81, 723
Lorentz coordinates, 41, 54, 1157
Lorentz factor, 58
Lorentz force
  in terms of electromagnetic field tensor, 53, 71, 1156
  in terms of electric and magnetic fields, 6, 14, 72
  geometric derivation of, 52–53
Lorentz group, 64
Lorentz reference frame, 39, 39f, 1156–1157. See also local Lorentz reference frame and coordinates
  slice of simultaneity (3-space) in, 58, 59f
Lorentz transformation, 63–65, 1158
  boost, 64, 65f
  rotation, 65
Lorenz equations for chaotic dynamics, 834
Lorenz gauge
  electromagnetic, 75, 760, 1219–1220
  gravitational, 1229–1230
low-pass filter, optical, 441
low-Reynolds-number flow, 746–757
  nearly reversible, 746
  pressure gradient balances viscous stress, 746

regimes of: small-scale flow, or very viscous large-scale flow, 746
luminosity distance, d_L, 1375–1376
Lyapunov time and exponent, 832–833
Lyman alpha spectral line, 1373, 1393–1396

Mach number, 882
magnetic bottle, 1028, 1029f
magnetic confinement of plasma, for controlled fusion, 960–964
magnetic field diffusion in MHD, 948–950, 950t, 956
magnetic field interaction with vorticity, 957–958, 958f
magnetic field-line reconnection, 950, 986–988, 987f
magnetic force density on fluid, in MHD, 951–952
magnetic lenses for charged particles, 381–383
magnetic materials, 270–282
    paramagnetism and Curie's law, 271–272
    ferromagnetism, 272–282
        phase transition into, 272–273. See also Ising model for ferromagnetic phase transition
magnetic Reynolds number, 950, 950t
magnetization
    in magnetic materials, 270
    in magnetized plasma, 1039
magnetohydrodynamics (MHD), 943–993
    conditions for validity, 1020
    fundamental equations of
        electric field, charge, and current density in terms of magnetic field, 947–948
        magnetic-field evolution equation, 948
        freezing-in of magnetic field, 948
        magnetic field diffusion, 948
        fluid equation of motion, 951
        magnetic force density on fluid, 951–952
        boundary conditions and junction conditions, 953–956
        momentum conservation, 951
        energy conservation, 952
        entropy evolution, 953
        ohm's law, 947
    generalizations of, 1020–1022
magnetoionic theory, for radio waves in ionosphere, 1058–1062
magnetosonic waves, 988–992
    Alfvén waves (intermediate magnetosonic mode), 354–355, 990–991, 1053f. See also Alfvén waves
    fast magnetosonic mode, 990–992
    slow magnetosonic mode, 990–992
magnetosphere, 999f, 1001, 1002t
    rotating, 969–970
    in binary pulsars, 1310
    Alfvén waves in, 370
    interaction of solar wind with Earth's, 950, 957, 987–988, 1090, 1146–1147, 1146f
magnetostatic equilibria, 958–965, 971, 1030–1031
    equations of, 960, 962
    perturbations and stability of, 971–984
        dynamical equation, 973
        boundary conditions, 974
        energy principle, 980–982
        virial theorems, 982–984
magnification of images
    by thin lens, 379
    near a caustic, 390
    in gravitational microlensing, 399, 400f, 401
Maple, 129, 132, 431, 619, 647, 691, 858, 1172
Markov random process, 289–291
Markov, Gaussian random process
    probabilities for (Doob's theorem), 295–296, 298–299
    spectral density for, 303, 304f
    correlation function for, 297, 304f
    and fluctuation-dissipation theorem, 325
    Fokker-Planck equation for, 336–338, 343
mass conservation, 32–33, 80, 692–693
mass density
    rest-mass density, 81
    as integral over distribution function, 121
mass hyperboloid, 100–101, 100f
mass moments, gravitational, 1328–1332
mass-energy density, relativistic
    as component of stress-energy tensor, 83, 85
    as integral over distribution function, 126
matched asymptotic expansions, 874
    in Stokes flow, 753–754
    in theory of radiation reaction, 869–871, 872f
Mathematica, 129n, 132, 431, 619, 647, 691, 858, 1172
Matlab, 129n, 132, 431, 619, 647, 691, 858, 1172
Maxwell relations, thermodynamic, 227–228, 232, 240, 247
    as equality of mixed partial derivatives of fundamental thermodynamic potential, 227–228
Maxwell velocity distribution for nonrelativistic thermalized particles, 113–114, 114f
Maxwell-Jüttner velocity distribution for relativistic thermalized particles, 114–115, 114f
Maxwell's equations
    in terms of electromagnetic field tensor, 73–74
    in terms of electric and magnetic fields, 74, 946
    in linear, polarizable (dielectric) medium, 1036
    in nonlinear, polarizable (dielectric) medium, 536–537
mean free path, 140, 143–145, 146b, 149
mean molecular weight, 680b, 726b
Mercury, perihelion advance of, 1302–1304
method of moments. See moments, method of

metric perturbation and trace-reversed metric perturbation, 1227–1228, 1311
metric tensor
　in Euclidean space
　　geometric definition, 11–12
　　components in orthonormal basis, 17
　in spacetime, 48, 1155
　　geometric definition, 48, 1155
　　components in orthonormal basis, 55, 1157
metrics for specific systems. See spacetime metrics for specific systems
Metropolis rule in Monte Carlo computations, 280
MHD electromagnetic brake, 966, 967f
MHD electromagnetic pump, 966–968, 967f
MHD flow meter, 966, 967f
MHD power generator, 966, 967f
Michelson interferometer, 474, 475f, 476
　application to Fourier-transform spectroscopy, 475–476
Michelson stellar interferometer, 464, 465f
Michelson-Morley experiment, 474, 483
microcanonical ensemble, 160t, 178–180, 221–228, 221t
　correlations of subensembles in, 179
　distribution function for, 179
　and energy representation of thermodynamics, 221–228
microcantilever, 597
microscope
　simple, 380f
　　rays traveling through, 380, 380f
　phase-contrast, 442–443, 442f
　transmission electron, 444–445
Minkowski spacetime, 1–2
mirror machine for confining plasma, 963, 1030–1031
mirror point for particle motion in magnetic field, 1027f, 1028–1029, 1029f
mixing length for convection in a star, 935–936
modes (single-particle quantum states), 174–176
　for Bose-Einstein condensate, 194–195
Moho discontinuity, 648, 650
moments, method of: applications
　solving Boltzmann transport equation, 147n
　dimensional reduction in elasticity theory, 594–595, 612–613
　constructing fluid models from kinetic theory, 1074
momentum, relativistic, 34, 59
　relation to 4-momentum and observer, 59, 61
　of a zero-rest-mass particle, 60, 106
　Newtonian limit, 34
momentum conservation, Newtonian
　differential, 32, 694–695
　integral, 32

momentum conservation, relativistic
　for particles, 60
　differential, 85, 1176–1177
　global, for asymptotically flat system, 1237–1238
　global, fails in generic curved spacetime, 1177, 1218
momentum density
　as component of stress-energy tensor, 83
　as integral over distribution function, 118
momentum space
　Newtonian, 98, 98f
　relativistic, 100–101, 100f
monopoles, 1432n
Monte Carlo methods
　origin of name, 279n
　for 2-dimensional Ising model of ferromagnetism, 279–282
　for radiative transfer, 1415–1419, 1428
Morse theory, 384
multiplicity factor for states in phase space, $\mathcal{M}$, 163
multiplicity for particle's spin states, g_s, 109
multipole moments
　in sound generation, 865–867
　gravitational, 1232, 1328–1334
　of CMB anisotropy, 1418, 1419f
　method of moments. See moments, method of

National Ignition Facility, 519, 664, 1141
Navier-Cauchy equation for elastostatic equilibrium, 588
　in cylindrical coordinates, 624
　displacement is biharmonic, 589
　boundary conditions for, 588–589
　methods for solving, 590b
　　simple methods, 619–622
　　separation of variables, 624–626
　　Green's function, 626–627
Navier-Stokes equation
　general form, 712
　for incompressible flow, 713, 726b
　in rotating reference frame, 767
Nd:YAG crystal and laser, 447, 553–554, 561
Nd:YVO$_4$ crystal, 554
near zone, 1327f
nephroid, 384n
neutral surface, in elasticity theory, 592–593
neutrinos
　chirality of, 109n
　spin of, deduced from return angle, 1319–1320
　spin-state multiplicity, 109
　from supernovae, 914
　in universe today, 1380t, 1382
　in universe, evolution of, 1384, 1385f

temperature and number density compared to photons, 1385, 1385n
decoupling in early universe, 1384, 1385f, 1406n
thermodynamically isolated after decoupling, 192, 209, 1384
influence of rest mass, 1385n, 1410
free streaming through dark matter potentials, 1407–1409
neutron stars. See also binary pulsars; pulsars; stars, spherical, in general relativity
birth in supernovae, 111, 914
equation of state, 125, 1257
structures of, 579, 734, 1258–1260
upper limit on mass of, 1260
magnetospheres of, 969–970
r-modes of oscillation, 860
accretion of gas onto, 784, 890–891
neutrons in early universe, 1384, 1387–1392, 1390f
noise. See also fluctuation-dissipation theorem; LIGO; random process; spectral density
as a random process, 308–313
types of spectra (spectral densities)
flicker noise, 308–310, 323
random-walk noise, 308–310
white noise, 308–310
information missing from spectral density, 310–311
filtering of, 311–313
Johnson noise in a resistor, 327
shot noise, 321–323, 507–509
thermal noise, 302f, 329–330, 334, 343–345, 448, 505, 598, 622–623, 626. See also fluctuation-dissipation theorem
nonlinear crystals
dielectric tensor for, 537
dielectric susceptibilities of, 536–540
for isotropic crystal, 538, 538n, 539–540
specific crystals and their properties, 541b–542b
wave-wave mixing in. See three-wave mixing in nonlinear crystals; wave-wave mixing
nonlinear media. See nonlinear crystals; wave-wave mixing; three-wave mixing; four-wave mixing in isotropic, nonlinear media
normal modes
Sturm-Liouville, 974–975
of elastic bodies, 661–662, 664–668
quantization of, 668–669
of elastic, homogeneous sphere, 661–662, 664–667
radial, 661, 664–665
ellipsoidal, 662, 666–667
torsional, 661–662, 665–666
of Earth. See under Earth

of sun, 848–850
of magnetostatic equilibria, 975–976, 980–981
nuclear reactions. See chemical reactions, including nuclear and particle
nuclear reactor
neutron diffusion in, 151–153
cooling of, 922–923
xenon poisoning in, 153
nucleosynthesis, in nuclear age of early universe, 192–193, 1387–1392
number density
as time component of number-flux 4-vector, 79–80
as integral over distribution function, 117, 119, 121, 126
number flux
as spatial part of number-flux 4-vector, 79–80
as integral over distribution function, 117
number-flux 4-vector
geometric definition, 79–80
as integral over distribution function, 118, 119
components: number density and flux, 79–80
conservation laws for, 79–80
Nyquist diagram and method for analyzing stability, 1091–1098, 1092f

observer in spacetime, 41
occupation number, mean
defined, 110
ranges, for fermions, bosons, distinguishable particles, and classical waves, 110, 111
for plasmon modes in a plasma, 1123–1124
for cosmic X-rays, 111
for astrophysical gravitational waves, 111, 1326–1327
ocean currents
surface currents driven by winds, 775–776
deep currents driven by gyre pressure, 768, 775–776
ocean tides, 1212–1213
ocean waves
generation by atmospheric pressure fluctuations in storms, 783
damping by turbulent viscosity, 842
breaking near shore, 903–904
Ohm's law, 139. See also conductivity, electrical
in magnetohydrodynamics, 946–947
tensorial in magnetized plasmas, 1036
ohmic dissipation, 945, 949–950, 953, 957, 966, 987–988
Olber's paradox, 138–139
Onsager relation, 1018
optical cavity
paraxial (ray) optics of, 380–381, 381f
modes of, 491b–492b
optical depth, 1395

Subject Index   **1499**

optical fiber
　　light rays in, 374
　　Gaussian beams in, 447, 448
　　image distortions in, 534, 534f
　　geometric phase in, 406–409
　　four-wave mixing in, 562–564
　　solitons in, 564
optical frequency comb, 310n, 498–501, 512, 520–521
optical Kerr effect, 562–564
optimal filtering, 318–320
ordinary waves
　　in nonlinear crystals, 546–548, 551
　　in a cold, magnetized plasma, 1057–1058, 1060, 1063, 1064f
orthogonal transformation, 20–21

p modes of sun, 849–850
pairs, electron-positron
　　thermal equilibrium of, 258–259, 1001
　　temperature-density boundary for, 259f, 999f
　　annihilation of, in early universe, 1384, 1385f
parallel transport
　　for polarization vector in geometric optics, 406–409
　　for 4-vectors in curved or flat spacetime, 1169
parametric amplification, 555–558, 1140–1142
parametric instability, 1140–1141
paraxial Fourier optics, 436–451
　　point spread functions for, 438–439
　　image processing using. See image processing
paraxial ray optics, 375–384. See also catastrophe theory, caustics
　　ray equation, 376
　　transfer matrices, 377–378
　　　　for optical elements: straight section, thin lens, spherical mirror, 378
　　conjugate planes, 378
　　applications
　　　　thin lens, 378, 379f
　　　　microscope, 380, 380f
　　　　refracting telescope, 379f, 379–380
　　　　optical cavity, 380–381, 381f
　　　　magnetic lenses, 381–383
Parseval's theorem, 300, 303, 478, 811, 1116–1117
particle conservation law
　　Newtonian, 28
　　relativistic, 80
　　in plasma, 1071
particle density. See number density
particle kinetics
　　in Euclidean space
　　　　geometric form, 13–15
　　　　in index notation, 19–20

　　in flat spacetime
　　　　geometric form, 49–52, 1154–1156, 1178b
　　　　in index notation, 57–62
　　in Newtonian phase space, 97–99
partition function, in statistical mechanics. See also fundamental thermodynamic potentials, physical-free-energy potential
　　as log of physical free energy, 239
Pascal, unit of stress, 578
path integrals in quantum mechanics, 371–372, 438n
path of particle (Newtonian analog of world line), 9–10
paths, for visualizing fluid flows, 699b
Péclet number, 921
Penning trap, 1031–1032
Penrose process for black holes, 1283–1285
Penrose stability criterion for electrostatic waves, 1097
perfect fluid (ideal fluid), 30, 675, 675n
　　Euler equation for, 33, 697
　　stress tensor for, 30–31, 32
perfect gas. See gas, perfect
perfect MHD, 950
perihelion and periastron advances due to general relativity, 1302–1304
perturbations in expanding universe
　　origin of, 1437
　　initial spectrum of, 1410–1412
　　evolution of, 1401–1422
phase conjugation of an optical wave, 531–535
　　and time reversal, 535
　　in holographic secondary image, 527
　　used to remove wave-front distortions, 532–535, 533f, 534f
　　produced by phase-conjugating mirror, 532
　　　　implemented via four-wave mixing in a nonlinear medium, 559–562
phase margin, for stability of control system, 1098
phase matching, in nonlinear optics, 543, 548–549
phase mixing in statistical mechanics, 184, 184f, 210–211
phase of a wave. See under geometric optics
phase space
　　Newtonian, 98–99
　　relativistic, 101–105
　　in statistical mechanics, 161–163
phase transitions, 251
　　governed by Gibbs potential, 251–254
　　first-order, 252
　　　　Clausius-Clapeyron equation for, 254–255
　　second-order, 253
　　　　specific heat discontinuity in, 200, 254
　　triple point, 254–255, 255f
　　specific examples
　　　　water-ice, 251–252, 254–255

water vapor–water, 255
van der Waals gas, 266–268
crystal structure change, 253–254
Bose-Einstein condensation, 196–197, 197f, 254
ferromagnetism, 272–282. See also Ising model for ferromagnetic phase transition
phase velocity, 352
phonons
modes for, 175, 175n
for modes of an elastic solid, mathematical theory of, 667–670
momentum and energy of, 363
specific heat of in an isotropic solid, 131–132
photography, 522, 522f
photon, gravitational field of in linearized theory, 1231
physical laws
frameworks and arenas for, 1–3
geometric formulation of. See geometric principle; principle of relativity
piezoelectric fields, 586
pipe
stressed, elastostatics of, 619–621
fluid flow in, 716–717, 766, 787
pitch angle, 1028
Pitot tube, 700, 701f
Planck energy, 580, 1438
Planck length, 579, 580, 1287, 1438, 1439
Planck satellite, 1365f
Planck time, 209, 1438, 1439
Planck units, 1438
planets. See also Earth; Jupiter
Mercury, perihelion advance of, 1302–1304
plasma electromagnetic waves
validity of fluid approximation for, 1020, 1392
in unmagnetized plasma, 1042–1044, 1050
in magnetized plasma, parallel to magnetic field: left and right modes, 1052–1066
plasma frequency, 1005, 1002t, 1041
plasma oscillations
elementary analysis of, 1005
in two-fluid formalism, 1041–1042
in moving reference frame, 1065
two-stream instability for, 1065–1067
plasma waves
in an unmagnetized, cold plasma, 1040–1044. See also plasma electromagnetic waves; plasma oscillations
in a magnetized, cold plasma, 1050–1065. See also Alfvén waves; ordinary waves; extraordinary waves; magnetosonic waves; plasma oscillations; whistler wave in plasma
in an unmagnetized, warm plasma, 1044–1050. See also ion-acoustic waves; Langmuir waves

plasmas. See also magnetohydrodynamics; plasma waves; plasmons
summary in density-temperature plane, 999
summary of parameter values for, 1002t
electron correlations (antibunching) in, 1106
electron-positron pair production in, 1001
ionization of, 999–1000
degeneracy of, 1000, 1002
examples of, 999f, 1001–1002, 1002t
relativistic, 1000
plasmons. See also ion-acoustic waves; Langmuir waves; quasilinear theory in plasma physics
mean occupation number for, 1124
master equation for evolution of, 1126
fundamental emission rates, 1127, 1133–1134
interaction with electrons, 1124–1131
nonlinear interaction with each other, 1132–1135
plate, bent. See bent plate, elastostatics of
Pockels cell, 497n, 539
point spread functions for paraxial Fourier optics, 438–439
pointillist paintings, 427
Poiseuille flow (confined laminar, viscous flow)
between two plates, 718–719
with MHD magnetic field, 965–969
down a pipe, 717, 922
Poiseuille's law, for laminar fluid flow in a pipe, 717
Poisson distribution, 264, 505
Poisson's equation, 686, 705, 1003, 1078
Poisson's ratio, 591–592, 586t
polarization of charge distribution
in a linear dielectric medium, 1036
in a nonlinear dielectric medium, 536
dielectric tensor, 537
energy density, 538
nonlinear susceptibilities, 536–540. See also susceptibilities, dielectric; nonlinear crystals
in a plasma, 1035–1036
polarization of electromagnetic waves
for CMB radiation, 1415–1416, 1417, 1419f, 1420–1421, 1428, 1439
Stokes parameters for, 1420–1421
polarization vector in geometric optics, 405–409
polarization of gravitational waves, 1312–1313, 1316–1317
polytrope, 687–689
population inversion, 516
creation of by pumping, 517 517f
and lasing, 513, 518–519
post-Newtonian approximation to general relativity, 1303, 1310, 1341
potential flow (irrotational flow), 701
Prandtl number, 920, 921t

Subject Index **1501**

pressure, 30
    as component of stress tensor, 30–31
    as component of stress-energy tensor, 85
    as integral over distribution function, 121, 126
pressure ratio β in MHD, 959
pressure self-adjustment in fluid dynamics, 742
primordial nucleosynthesis, 192–193, 1387–1392
principle of relativity, 42, 1154, 1158–1159
    in presence of gravity, 1196
probability distributions, 286–288
    conditional, 287
projection tensors
    into Lorentz frame's 3-space, 61
    for TT-gauge gravitational waves, 1314b
proper reference frame of accelerated, rotating observer, 1180–1186, 1181f
    metric in, 1183
    geodesic equation in, 1185
    for observer at rest inside a spherical, relativistic star, 1253–1254
proper time, 49, 1154
proportionality limit, in elasticity, 580, 581f
PSR B1913+16 binary pulsar, 502, 1310. *See also* binary pulsars
pulsar. *See also* binary pulsars; neutron stars
    radio waves from, 1060–1061
    timing arrays for gravitational wave detection, 1355–1357

quadratic degree of freedom, 177
quantum state
    single-particle (mode), 174–175
        for Bose-Einstein condensate, 194–195
    many-particle, 175
        distribution function for, 175
quantum statistical mechanics, 165b–166b
quasars, 193, 309, 396, 403, 404f, 433, 482, 969, 995, 1233, 1288, 1305, 1379, 1380, 1397, 1430
quasilinear theory in plasma physics. *See also* ion-acoustic waves; Langmuir waves
    in classical language, 1113–1123
        in three dimensions, 1122–1123
    in quantum language, 1123–1135
quintessence, 1446

radiation, equation of state for thermalized, 128–129, 132
radiation reaction, gravitational: predictions and observations
    predictions of, 1333, 1335
    measurements of, in binary pulsars, 1310
    measurements of, in binary black holes, by LIGO, 1311

radiation reaction, theory of
    slow-motion approximation, 871
    matched asymptotic expansion, 871
    radiation-reaction potential, 871, 1333, 1335
    damping and energy conservation, 872, 873, 1335
    runaway solutions, their origin and invalidity, 872–873
    examples
        electromagnetic waves from accelerated, charged particle, 873
        sound waves from oscillating ball, 869–874
        gravitational waves from any slow-motion, gravitating system, 1333, 1335
radiative processes. *See also under* chemical reactions
    in statistical equilibrium, 115–116
    bremsstrahlung, 142, 260, 1009, 1017
    Thomson scattering, 142–144, 937, 1407–1408, 1415, 1416n, 1418, 1428
    Compton scattering, 1388, 1392–1393, 1428–1430
    Raman scattering, 1140
    Rayleigh scattering, 471
radiative transfer, Boltzmann transport analysis of
    by two-lengthscale-expansion, 145–148
    by Monte Carlo methods, 1415–1418, 1428
radio waves: AM, FM, and SW, 1060
RadioAstron, 482
radius of curvature of spacetime, 1213
Raman scattering, stimulated, 1140
random process, 1-dimensional, 285
    stationary, 287–288
    ergodic, 288–289
    Gaussian, 292–294
    Markov, 289–290
    Gaussian, Markov. *See* Markov, Gaussian random process
random process, 3-dimensional
    complex, 478–479
    cosmological density fluctuations, 304–306
random process, 2-dimensional, 306–308
random variable, 285
random walk, 139, 140, 140n, 141, 286f, 291–292, 294–295, 309, 310, 314–315, 320, 321, 465, 1007. *See also* diffusion
random-number generator, 279n, 294, 294n
random-phase approximation, 1116–1117, 1137
rank of tensor, 11
Rankine-Hugoniot relations for a shock wave, 900
    derivation from conservation laws, 898–900
    physical implications of, 900–902
    for polytropic equation of state, 905
    for strong polytropic shock, 905
    relativistic, 902–903
rarefaction wave, 895–896

Rayleigh criterion for instability of rotating flows, 784
Rayleigh number, 928
Rayleigh principle, 980. *See also* action principles
Rayleigh scattering, 471
Rayleigh waves at surface of a homogeneous solid, 654–657, 659, 661, 839–840, 941
Rayleigh-Bénard convection. *See under* convection
Rayleigh-Jeans spectrum, 482, 1430
Rayleigh-Taylor instability, 783–784
reciprocity relations for reflection and transmission coefficients, 485, 486–488
recombination in early universe, 1392–1396
redshift, cosmological, 1373
redshift, gravitational
    in proper reference frame of accelerated observer, 1189
    from surface of spherical star, to infinity, 1251–1252
    influence on GPS, 1301–1302
    experimental tests of, 70, 1301
reference beam for holography, 525, 525f
reflection and transmission coefficients
    reciprocity relations for, 485, 486–488
    for an interface between dielectric media, 489
    for a locally planar optical device, 486–488
    for an etalon, 484, 485, 486–488
    for a Fabry-Perot interferometer, 490
    modulus squared: power reflectivity and transmissivity, 486
refractive-index surface, 1062–1063, 1062f
reionizaton of universe, 193, 1386f, 1395f, 1397, 1418, 1431
relaxation (correlation) time, 297, 298f
renormalization group
    idea of, 273
    applied to 2-dimensional Ising model for ferromagnetism, 273–278
    applied to the onset of chaos in the logistic equation, 831
resistance
    in electrical circuit, 324
    in an oscillator, 326
    in Stokes fluid flow, 328, 753
    as real part of complex impedance, 332
resonance conditions in wave-wave mixing, 542, 543–544, 1132
rest frame
    momentary, 49
    local, 85, 86, 366, 677, 719–720
    asymptotic, 1237, 1246–1248
    local asymptotic, 1327f, 1328, 1331, 1332, 1339–1340
rest mass, 34, 58–59
    global and local conservation laws for, 80, 82
rest-mass density, relativistic, 81

rest-mass-flux 4-vector
    geometric definition of, 80
    components: rest-mass density and flux, 81
Reynolds number, 716, 726b
    as ratio of inertial to viscous acceleration, 746
    magnetic, 950, 987
Reynolds stress for turbulence, 802
    and turbulent viscosity, 804
Ricci (curvature) tensor, 1214–1215
Richardson criterion for instability of shear flows, 785–786
Richardson number, 785–786
Riemann curvature tensor
    definition, 1209
    symmetries of, 1214
    components of
        in an arbitrary basis, 1215–1216
        in local Lorentz frame, 1214
    Bianchi identity for, 1223
    decomposition into tidal and frame-drag fields, in vacuum, 1235b–1236b
    components in specific spacetimes or spaces
        surface of a sphere, 1216
        general linearized metric, 1227
        Schwarzschild metric, 1244b, 1267
    Newtonian limit of, 1227
        magnitude of, 1213
        outside Newtonian, gravitating body, 1212–1213
Riemann invariants, 852, 891–897, 901–902
right modes, for plasma electromagnetic waves parallel to magnetic field, 1052–1056
rigidly rotating disk, relativistic, 1189–1190
Rindler approximation, 1273–1274
Rindler coordinates
    in flat spacetime, 1187–1189
    near black-hole horizon, 1273–1274
Robertson-Walker metric for a homogeneous, isotropic universe, 1371
    coordinates for, 1370
    derivation of, 1366–1372
    Einstein tensor for, 1371–1372
    perturbations of, and their evolution, 1401–1422
rocket engines, fluid flow through, 887–890
rod. *See* bent beam, elastostatics of
Rossby number, 768
Rossby waves in rotating fluid, 858–861
rotating disk, relativistic, 1189–1190
rotating reference frame, fluid dynamics in, 766–777
rotation, rate of, in fluid mechanics, 711, 726b
    as vorticity in disguise, 711
rotation group, 21, 572b–574b
rotation matrix, 21, 65

rotation tensor and vector, in elasticity theory, 571, 573b–574b, 575–576, 577
rupture point, in elasticity, 580, 581f
Rutherford scattering, 1006–1007

Sackur-Tetrode equation for entropy of a perfect gas, 189
Sagnac interferometer, 501–502, 502f
Saha equation for ionization equilibrium, 259–260, 999–1000
Saint-Venant's principle, for elastostatic equilibrium, 590b
salt fingers due to double diffusion of salt and heat, 937–940
sausage instability, of magnetostatic equilibria, 977
scale factor, in cosmology, 1370
    as a function of time, 1387, 1388f, 1390f, 1399f, 1400f
scaling relations in fluid flows
    between similar flows, 791–792
    for drag force on an object, 765
    for Kolmogorov turbulence spectrum, 789, 810–814
scaling relations near a catastrophe (caustic), 392
scattering of light. See also under radiative processes
    by large, opaque particle, 429
    in LIGO, 448–451
Schrödinger equation
    energy eigenstates (modes) of, 175n, 194–195, 848–849
    and Coulomb wave functions, 1009
    propagation speed of waves, 44b
    geometric optics for, 375, 409
    nonlinear variant of, and solitons, 856–857
Schwarzschild criterion for onset of convection in a star, 935, 935f
Schwarzschild metric, 1242. See also black holes; horizon, black-hole event; stars; wormhole
    uniqueness of: Birchoff's theorem, 1250
    in Schwarzschild coordinates, 1242
        bases, connection coefficients, and Riemann tensor, 1243b–1244b
        Schwarzschild coordinate system and symmetries, 1244–1249
    in isotropic coordinate system, 1251
    in ingoing Eddington-Finkelstein coordinates, 1269
    gravitational (horizon) radius of, 1250
    Rindler approximation near horizon, 1273–1274
    geodesic orbits in, 1247–1248, 1274–1276
    Newtonian limit of, 1246
    roles of
        exterior metric of static star, 1250–1252
        exterior metric of imploding star, 1264–1266, 1269
        metric of nonspinning black hole, 1272–1276
        metric of wormhole, 1276–1277
second quantization, 175
secondary fluid flows, 775–776
sedimentation, 749, 754–755

Sedov-Taylor blast wave, 909–912
seeing, atmospheric, 425, 465, 466b–471b, 471–472
seismic waves. See elastodynamic waves in Earth
self-gravity, in fluid dynamics, 705b–706b, 709
self-similar flows
    boundary layer near flat plate: Blasius profile, 758–763
    Sedov-Taylor blast wave, 909–912
    underwater blast wave, 914–915
    flow in shock tube, 916
    stellar wind, 915–916
    water flow when dam breaks, 857–858
separation of variables for Navier-Cauchy equation, 590b, 624–625
Shapiro time delay, 1009, 1308–1309
shear, rate of, 711, 726b
shear modulus, for elasticity, 581, 586t, 651t
shear tensor, in elasticity theory, 571, 572b–574b, 574–577
    stretch and squeeze along principal axes, 574–575, 575f
shock fronts. See shock waves in various media
shock tube, fluid flow in
    analyzed using similarity methods, 916
    analyzed using Riemann invariants, 895–896
    shock front in, 906
shock waves (shock fronts) in a fluid, adiabatic
    terminology for, 898, 900f
    inevitability of, 897
    Rankine-Hugoniot relations for, 900. See also Rankine-Hugoniot relations for a shock wave
    shock adiabat, 900–901, 901f
    properties of, 900–901
    internal structure of, 898, 906–907
        role of viscosity in, 898
    patterns of
        around a supersonic aircraft, 876f
        bow shock in solar wind around Earth, 876f, 957, 1146–1147, 1146f
        Mach cone, 907–908, 907f
        Sedov-Taylor blast wave, 909–912
    sonic boom, 908, 908f
    acceleration of cosmic rays in, 1145–1148
shock waves in an elastic medium, 663–664
shock waves in a plasma, collisionless, 907, 1145–1147, 1146f
shot noise, 321–323, 504–505, 506–509, 557
signal-to-noise ratio
    for band-pass filter, 317
    for Wiener's optimal filter, 319–320
similarity methods in fluid mechanics. See self-similar flows
simultaneity in relativity
    breakdown of, 66
    slices of, 58, 73, 73f, 1181f, 1293–1294, 1293f, 1297
single-particle quantum states (modes), 174–175, 194–195

singularity, spacetime
    at center of Schwarzschild black hole, 1271–1272, 1273f
    generic, inside all black holes, 1273, 1282n
    for Schwarzschild wormhole, 1277
SLAC National Accelerator Laboratory, 521
slot-naming index notation, 19–20, 23, 25, 56–57, 70, 1156
smoke rings, 744f
Smoluchowski equation, 290
    applications of, 291–292, 337
soap film, shapes of, 846
solar dynamo, 985–986
solar furnace, 138–139
solar wind, 875, 876f, 970–971, 988, 999f, 1001, 1002t
    two-stream instability in, 1066–1077
    collisionless shocks in, 907, 1146
    bow shock at interface with Earth, 876f, 957, 1090, 1146–1147, 1146f
    termination shock with interstellar medium, 1146–1147
solid-body normal modes. See normal modes, of elastic bodies
solitons
    balance of nonlinearity against dispersion in, 853–854
    equations exhibiting, 856–857
    Korteweg–de Vries equation and solutions, 852–856, 858
    venues for, 856–857
        in optical fiber, 564
        in ion-acoustic waves in plasma, 1142–1146
        in nonlinear gravity waves on water, 852–856, 858
sonic boom, 889b, 907–908
sound speed in elastic solid, C_L, 586t, 638
sound waves in a fluid
    wave equation, 862
    sound speed, 862
    analysis of, 862–865
    dispersion relation, 353
    phase velocity and group velocity, 355
    energy density, 864
    energy flux, 865
    in inhomogeneous fluid: example of prototypical wave equation, 863
    generation of, 865–869
        radiation reaction on source, 869–874
    attenuation of, 868
    nonlinearity of, and shock formation, 894
    propagating in a horizontal wind with shear, 366
    quanta: phonons, 363. See also phonons
space shuttle, 889b–890b
    sonic boom from, 889b, 907–908
space telescope. See Hubble Space Telescope; James Webb Space Telescope

spacetime diagram, 40–41
    for Lorentz boost, 65–67, 65f
spacetime metrics for specific systems
    for a spherical star, 1250, 1253, 1258–1260. See also stars, spherical in general relativity
    for a moving particle, linearized, 1230–1231
    for a photon: Aichelberg-Sexl ultraboost metric, 1231
    for exterior of any weak-gravity stationary system, 1231–1234, 1236
        conservation of mass and angular momentum: influence on, 1237–1238
        reading off source's mass and angular momentum from exterior metric, 1232–1233
    for exterior of any asymptotically flat, strong-gravity, stationary system, 1238
    for gravitational waves in flat spacetime, 1311–1314
    Schwarzschild metric for a spherical star, black hole, or wormhole, 1242. See also Schwarzschild metric
    Robertson-Walker metric for a homogeneous, isotropic universe, 1371, 1366–1372. See also Robertson-Walker metric for a homogeneous, isotropic universe
    Bertotti-Robinson metric, for a homogeneous magnetic universe, 1249
specific heats, C_p, c_p, C_V, and c_V, 244, 678. See also adiabatic index
    for ideal gas with internal degrees of freedom, 879–880
    for nonrelativistic, degenerate electrons, 130–131
    for phonons in an isotropic solid, 131–132
specific intensity (spectral intensity) of radiation, 107, 107f
speckle image processing, 470b, 472
speckles, in light images, 466bf, 469b–470b
spectral density
    for a 1-dimensional random process, 299–300
        as mean of square of Fourier transform, 303, 304
        double-sided vs single-sided, 300
        for sum of two random processes, 308
        integral of is variance, 300
        physical meaning of, 301–302
        rms oscillations in terms of, 301
    for a 2-dimensional random process, 307–308
        cross spectral density, 307
    for a 3-dimensional random process,
        cosmological density fluctuations (galaxy distribution), 304–306, 306f
    Wiener-Khintchine theorem for. See Wiener-Khintchine theorem
    influence of filtering on, 312. See also filtering of random process
    types of spectral densities. See under noise

spectral density (continued)
 applications of
  Brownian motion, 314
  light, 301
  LIGO gravitational wave detector, 302
  noise, 308–311, 321–323, 334. See also noise
  cosmological density fluctuations, 304–306
spectral energy density $\mathcal{E}_k$, in quasilinear theory of plasma waves, 1117
spectral energy flux (or spectrum), F_ω, 473, 473n
spectral intensity, I_ν or I_ω. See specific intensity
spectrometer, Fabry-Perot cavity as, 496
spectroscopy, Fourier-transform using Michelson interferometer, 474–476
spectrum (spectral energy flux), 473, 473n
spectrum of light related to spectral density, 301
speed of light
 constancy of, 34, 42, 1159
 measuring without light, 43b
 in geometrized units, 34
 contrasted with speeds of other waves, 44b
spherical coordinates
 related to Cartesian coordinates, 614
 orthonormal bases and connection coefficients, 614, 616
  expansion and shear tensor in, 617, 618b
spherical triangle, 1372
spheromak, 964
sports, physics of, 823–825
spy satellite, 436
squeezed vacuum and squeezed states of light, 556–558, 558f
stagnation pressure, 700, 792
standard cosmology, 1383
Star Wars (Strategic Defense Initiative), 521
stars. See also neutron stars
 formation of first stars in early universe, 1397
 observed properties of, 1379
 diffusive heat conduction in, 142–148
 spherical, in general relativity, 1250–1263
  equations of stellar structure, 1258–1259
  interior metric, 1253, 1258–1259
  exterior spacetime metric: Schwarzschild, 1250
  embedding diagram for, 1262–1263, 1263f
  star with constant density, full structure, 1260
 implosion to form black holes, 1264–1272
  in Schwarzschild coordinates, 1264–1267, 1270–1271
  in ingoing Eddington-Finkelstein coordinates, 1267–1271
statistical equilibrium, 168–178
 ensembles in, 168–177. See also specific ensembles
  general, 172–173
  tables summarizing, 160t, 221t

fluctuations away from, 260–270
 for ensemble of closed systems, 260–261
 for particle distribution in a closed box, 262–263
 for particle number in an open box, 263–264
 for temperature and volume of an ideal gas, 264–265
 for van der Waals gas: volume fluctuations, 266–268
 for volume of a thermally isolated gas (constant-pressure balloon), 265–266
statistical equilibrium for fundamental particles
 for identical bosons, Bose-Einstein distribution, 112–113
 for identical fermions, Fermi-Dirac distribution, 112
 for identical classical particles, Boltzmann distribution, 113
statistical independence, 170
steady fluid flow, 726b
stellar dynamics. See also under galaxies
 statistical mechanics of galaxies and star clusters, 201–204
 Jeans' theorem in, 1076–1077
 evolution of cluster due to ejection of stars, 203–204
 equilibration time for stars in cluster, 1012–1013
 violent relaxation of star clusters, 113n
stochastic differential equations, 325
Stokes flow, 749–754, 749f, 766
 drag force in: Stokes' law, 753
Stokes parameters for polarization of radiation, 1420–1421
Stokes' paradox for fluid flow past a cylinder, 754
Stokes' theorem for integrals, 27
storms, fluid dynamics of, 768, 769b, 842
strain tensor, in elasticity theory, 576
strange attractors, 833–834
Strategic Defense Initiative, 521
stratosphere, 683, 684f, 731, 748, 755, 784–786
streaks, for visualizing fluid flows, 699b
stream function for 2-dimensional incompressible flow
 in Cartesian coordinates, 759
 in any orthogonal coordinate system, 760b–761b, 766
stream tube, in fluid dynamics, 699–700, 700f, 721–722, 721f
streamlines, for visualizing fluid flows, 698, 699f
stress polishing mirrors, 609–611, 611f, 613
stress tensor
 geometric definition of, 29, 577
 components, meaning of, 30
 symmetry of, 30
 as integral over distribution function, 118
 as spatial part of relativistic stress-energy tensor, 83
 for specific entities
  electromagnetic field, 33
  perfect fluid, 30–31, 32, 696
  strained elastic solid, 581, 584
  strained and heated elastic solid, 584
 magnitudes of, 578–580

stress-energy tensor
    geometric definition of, 82, 1176
    constructed from Lagrangian, 1434
    components of, 82–83, 120, 1176
    symmetry of, 83–84
    as integral over distribution function, 118, 120
    and 4-momentum conservation, 84–85, 1176–1177
    for electromagnetic field
        in terms of electromagnetic field tensor, 86
        in terms of electric and magnetic fields, 88
        in terms of vector potential, 1434
    for perfect fluid, 85, 720, 1177
        nonrelativistic limit, 723–724
    for point particle, 1178b, 1179
    for viscous, heat-conducting fluid, 1179–1180
structure function, for fluctuations, 467b, 815
Sturm-Liouville equation and theory, 974–975, 980–981
subensemble, 170
sun. See also solar dynamo; solar wind
    core of, 933f, 937, 999f, 1001, 1002t
    convection zone, 933f, 936
    disturbances on surface, 1035, 1065
    normal modes of, 848–850. See also normal modes
Sunyaev-Zel'dovich effect, 1428–1430
superfluid, rotating, 733–734
supernovae
    neutron stars produced in, 111, 914
    as gravitational-wave sources, 111
    Sedov-Taylor blast wave from, 914–915
    observations of reveal acceleration of the universe, 1398, 1400
supersymmetry, 1441
surface tension, 844b–845b
    force balance at interface between two fluids, 846
surfing of electrons and protons on electrostatic waves, 1046, 1069–1070, 1098–1099. See also Landau damping
susceptibilities, dielectric
    linear, 536–537, 1036
    nonlinear, 536–540
        isotropic, 538, 538n, 539–540
    magnitudes of, 539
swimming mechanisms, 744, 747b–748b, 756–757
symmetries and conservation laws, 1203–1205
system, in statistical mechanics
    defined, 157
    closed, 158–159
    semiclosed, 157–158

tangent space, 9, 1160, 1165–1169, 1166f, 1175, 1218, 1253
tangent vector, 9, 49, 1155, 1155f, 1165–1166
    as directional derivative, 1167

Taylor rolls, in rotating Couette flow, 826, 826f
Taylor-Proudman theorem for geostrophic flow, 771
tea cup: circulating flow and Ekman boundary layer, 776–777
telescopes, optical. See also adaptive optics; astronomical seeing; Gran Telescopio Canarias; Hubble Space Telescope; James Webb Space Telescope; Keck telescopes
    simple refracting, and light rays, 379f, 379–380
    angular resolution of, 425–427
    aberrations in, 395
telescopes, radio, 479–483
    angular resolution of, 479, 480, 482
temperature
    definition, 168, 168n, 171
    measured by idealized thermometer, 223–224
temperature diffusion equation, 142, 920
tensor in Euclidean space
    definition and rank, 11
    algebra of without coordinates or bases, 11–13
    expanded in basis, 16
    component representation, 17–19
tensor in quantum theory, 18b
tensor in spacetime. See also component manipulation rules; specific tensors
    definition and rank, 48
    bases for, 55
    components of, 54–57
        contravariant, covariant, and mixed components, 55, 1157–1158, 1161–1162
        raising and lowering indices, 55, 1164
    algebra of
        without coordinates or bases, 48, 61–62
        component representation in orthonormal basis, 54–57, 1157–1158
        component representation in arbitrary basis, 1162–1165
tensor product, 12, 48
thermal diffusivity, 920, 921t
thermal equilibrium. See statistical equilibrium
thermal noise. See under noise
thermal plume, 933
thermodynamics. See also equations of state; Euler's equation; fundamental thermodynamic potentials; Maxwell relations, thermodynamic
    representations of, summarized, 221t, 228
        Legendre transformation between representations, 230–232, 240, 244, 247
    energy representation, and microcanonical ensemble, 221–229
    enthalpy representation, 244–246

thermodynamics *(continued)*
 grand-potential representation and grand canonical ensemble, 229–239
 physical-free-energy representation, and canonical ensemble, 239–244
 Gibbs representation and Gibbs ensemble, 246–260
 first law of, 225
  in all representations, 221t
  as mnemonic for deducing other relations, 227–228
  for fluid, in fluid dynamics, 679b–680b
  for black hole, 205
 second law of, 182
  underlying physics of: coarse graining and discarding correlations, 183–185, 184f, 186b–187b
  underlying quantum physics of: discarding correlations (quantum decoherence), 185, 186b–187b, 190–191
  in theory of information: when information is erased, 217–218
 of black holes, 204–209, 1286–1287
thermoelastic noise in mirrors, 623, 626
thermoelasticity, 584–585
thermoelectric transport coefficients, 1017–1018
Thomson scattering of photons by electrons, 142–144, 937, 1407–1408, 1415, 1416n, 1418, 1428
three-point correlation function, 1105
three-wave mixing in nonlinear crystals, 540–558
 polarization for, 540, 542
 evolution equations
  for birefringent crystal, 546–552
  for medium that is linearly dispersion-free and isotropic, 544–546
 phase matching for, 543, 548–549
 applications of
  frequency doubling, 545–546, 553–555
  optical parametric amplification, 555–557
  degenerate optical parametric amplification, 556–558
  squeezing, 556–558, 558f
three-wave mixing in plasmas
 driving term for, in Vlasov equation, 1114
 quasilinear theory of, 1132–1135
tidal gravitational field
 Newtonian, 1207–1208
 relativistic, 1211–1212, 1235b–1236b
tidal gravity
 Newtonian description, 1207–1208
 relativistic description, 1208–1210
 comparison of Newtonian and relativistic descriptions, 1210–1212, 1227
tidal tendex lines, 1235b–1236b
 around a linearized, spinning particle, 1236b
 around Kerr black hole, 1295b–1296b
 around colliding black holes, 1344b–1345b
 in a gravitational wave, 1318b, 1345b
time. *See also* clocks, ideal; simultaneity in relativity, slices of
 coordinate, of inertial frame, 39–40
 proper, 49
 imaginary, 54
 in cosmology, 1370
 in general relativity: many-fingered nature of, 1293–1294, 1297
time and frequency standards, 310f, 310n
time derivative
 advective (convective), 32, 692, 724b, 892
 fluid, 736
 with respect to proper time, 49, 52
time dilation, 66
 observations of, 70
time travel, 67–70
tokamak, 963
 MHD stability of, 979
Tollmien-Schlichting waves, 823
tomography, seismic, 663
topological defects, 1432n
tornado, 738, 739f
 pressure differential in, 702, 738
torsion pendulum, elastostatics of, 621–622
TOV equation of hydrostatic equilibrium, 1258, 1260
trace-reversed metric perturbation, 1228, 1311
transformation matrices, between bases, 1164
 orthogonal, 20–21
 Lorentz, 63–65, 65f, 1158
transport coefficients, 139. *See also* conductivity, electrical; conductivity, thermal; diffusion coefficient; viscosity
 in plasmas, 1015–1018
 thermoelectric, 1017–1018
triple point for phase transitions, 254–255, 255f
trumpet, sound generation by, 868
tsunamis, 841b, 843, 922
TT gauge, 1312–1315
turbulence, 787–834
 weak and strong, 800
 characteristics of
  3-dimensional, 794
  disorder, 798
  irregularly distributed vorticity, 799
  wide range of interacting scales, 798–799
  eddies, 798–800, 802, 804–807, 811–814
  efficient mixing and transport, 799
  large dissipation, 799
  intermittency, 798–799, 807, 814, 831
 onset of. *See* turbulence, onset of

vorticity in, 799–800
    drives energy from large scales to small, 799f
semiquantitative analysis of, 800–817
    Kolmogorov spectrum, 467, 810–813, 813f, 815. *See also* Kolmogorov spectrum for turbulence
    weak turbulence formalism, 800–810. *See also* weak turbulence formalism
generation of sound by, 869
turbulence, 2-dimensional analog of, 801b
    inverse cascade of energy, 799n
    transition to (3-dimensional) turbulence, 800
turbulence, onset of. *See also* chaos, onset of in dynamical systems
    critical Reynolds number for, 787, 794, 822, 826
    in convection, 830, 831
    in flow past a cylinder, 789–794, 800
    in rotating Couette flow, 825–828
    routes to turbulence
        one frequency, two frequencies, turbulence, 825–828
        one frequency, two frequencies, phase locking, turbulence, 831
        one frequency, two frequencies, three frequencies, turbulence, 831
        period doubling sequence, 830–831
        intermittency, 831
twins paradox, 67–70
two-fluid formalism, for plasma physics, 1037–1068
    fundamental equations, 1037–1038
    deduced from kinetic theory, 1073–1074
two-lengthscale expansion, 146b
    bookkeeping parameter for, 360b
    and statistical independence in statistical mechanics, 170n
    for solving Boltzmann transport equation, 145
    for geometric optics, 357–358, 359–360
    for quasi-linear theory in plasma physics, 1113–1114
    for gravitational waves in curved spacetime, 1320–1321, 1321f
two-point correlation function, 305, 1104–1107, 1424–1426
    for Coulomb corrections to pressure in a plasma, 1107–1108
    for electron antibunching in a plasma, 1104–1107
    for galaxy clustering, 305, 306f
    for weak gravitational lensing, 1424–1426
two-stream instability
    in two-fluid formalism, 1065–1068
    in kinetic theory, 1079–1080, 1137

ultrasound, 663–664
universe, evolution of
    expansion, kinematics of, 1373–1376
    evolution of radiation and gas properties during, 1373–1375
    expansion, dynamics of, 1376–1378
        Friedman equations, 1376–1377
    graphical summaries of
        entire life: distances as functions of scale factor, 1400f
        entire life: energy densities of constituents, 1386f
        particle age: temperatures and entropies of particle constituents, 1385f
        nuclear age: reaction rates; nuclear and particle abundances, 1390f
        plasma and atomic ages: ionization fraction and optical depths, 1395f
        gravitational and cosmological ages: scale factor and deceleration function, 1399f
    perturbations, evolution of, 1404f, 1405f, 1411f, 1414f
    formation of structure
        origin of primordial perturbations, 1437–1440
        perturbations, initial spectrum, 1410–1412
        evolution of perturbations, 1401–1422
        statistical mechanics of, 210–211
    seven ages
        before the particle age, 1431–1440
        particle age, before nucleosynthesis, 1384–1387
        nuclear age, primordial nucleosynthesis, 192–193, 1387–1392
        photon age, from nucleosynthesis to matter dominance, 1392–1393
        plasma age, from matter dominance through recombination, 1393–1396
        atomic age, from recombination through reionization, 193, 1394
        gravitational age, from reionization to dark-energy influence, 1394–1400
        cosmological age, the era of dark-energy influence, 1400–1401
        galaxy formation, 210–211, 1401–1415
universe, observed properties of
    isotropy and homogeneity, 1364–1366
    spatial flatness, 1378
    parameter values today, 1380t
    age of, 1387
    volume of, 1398
    constituents of
        baryons, 1379. *See also* baryons in universe
        neutrinos, 1382
        photons: cosmic microwave background, 1381. *See also* cosmic microwave background
        dark matter, 1380–1381. *See also* dark matter
        dark energy or cosmological constant, 1382–1383, 1444–1447

universe, observed properties of *(continued)*
   constituents of *(continued)*
      galaxies. *See* galaxies
      black holes, 1379–1380, 1397
      acceleration of, 1382, 1398, 1400, 1444
      spectral line formation, 1396
universe, statistical mechanics of, 209–211

van Allen belts, 1028, 1029f
van Cittern-Zernike theorem
   for lateral coherence, 461, 463
   for temporal coherence, 473
   3-dimensional, 477–478
   relation to Wiener-Khintchine theorem, 478–479
van der Waals gas
   equation of state for, 234
   grand potential for, 234
      derivation of, 232–238
   phase transition for, 266–268
   volume fluctuations in, 266–268
   catastrophe theory applied to, 394–395
variance, 287
vector
   as arrow, 8, 40, 1166
   as derivative of a point, 9, 49, 1165
   as differential operator, 1167–1169
vector in Euclidean space (3-vector): components, 16
vector in quantum theory, 18b
vector in spacetime (4-vector)
   contravariant and covariant components of, 55
   raising and lowering indices of, 55
   timelike, null, and spacelike, 47, 1155–1156
velocity
   Newtonian, in Euclidean space, 9
   ordinary, in relativity, 58, 59f, 61–62. *See also* 4-velocity
velocity potential for irrotational flow, 701
   in cosmological perturbations, 1403
violent relaxation of star distributions, 210
violin string, sound generation by, 868
virial theorems
   for any system obeying momentum conservation, 982–984
   for self-gravitating systems, 984
   MHD application, 982, 984
virtual image vs real image, 527
viscosity, bulk, coefficient of, 712, 724b
viscosity, molecular origin of, 713–714
viscosity, shear, coefficient of, 139, 712, 726b
   dynamic, η, 713, 724b
   kinematic, ν, 713, 725b
   values of, for various fluids, 713t, 921t
   for monatomic gas, 149–150, 714

VLA (Jansky Very Large Array), 480
Vlasov equation, in plasma kinetic theory, 1071–1072
   solution via Jeans' theorem, 1075
VLBI (very long baseline interferometry), 482
volcanic explosions, 748, 755
volume in Euclidean space
   2-volume (area), 26
   vectorial surface area in 3-space, 27
   3-volume, 27
   n-volume, 24
   differential volume elements, 28
volume in phase space
   Newtonian, 98
   relativistic, 102–104
      Lorentz invariance of, 103–104, 105f
volume in spacetime, 75–77
   4-volume, 75
   vectorial 3-volume, 76–77, 77f
      positive and negative sides and senses, 76
   differential volume elements, 77
volume of coherence, 477
von Zeipel's theorem, 702
vortex. *See also* vortex lines; vorticity
   diffusive expansion of, 742
   above a water drain, 729, 732, 777
   vortex sheet, 782, 801b
   vortex ring, 744
   starting vortex, 824, 825b
   vortex street, Kármán, 791f, 794
   wingtip vortex, 734f, 739, 744, 744f
   tornado, 702, 732, 738, 739f
   vortex generated by spatula, 745–746
   vortex generators, on airplane wing, 821–822, 821f
vortex cores, in superfluid, 733–734
vortex generator on airplane wing, 821–822, 821f
vortex lines, 734, 734f. *See also* frame-drag vortex lines
   diffuse due to viscosity, 741–742
   frozen into fluid, for barotropic, inviscid flows, 736–738, 737f
vortex rings, 744, 744f
vorticity, 697, 731–732, 732f
   relation to angular velocity of a fluid element, 697–698
   measured by a vane with orthogonal fins, 732
   sources of, 744–746
   evolution equations for, 735–738, 741
      diffusion of vorticity, 741
      frozen into an inviscid, barotropic flow, 736–737
   interaction with magnetic field, 957–958
   delta-function: constant-angular-momentum flow, 732–733
Voyager spacecraft, 1147

wakes
    2-dimensional, behind cylinder
        laminar, 794–795
        turbulent, 805–810
    3-dimensional, behind sphere
        laminar, 796
        turbulent, 810
water waves. *See* gravity waves on water; sound waves in a fluid
wave equations
    prototypical, 358
        lagrangian, energy density flux, and adiabatic invariant for, 365–366
    algebratized, 1037
    for electromagnetic waves. *See* electromagnetic waves
    for elastodynamic waves, 635. *See also* elastodynamic waves *in various media*
    for gravitational waves, 1312, 1322. *See also* gravitational waves
    for sound waves, 862, 863
wave packet, 354–356
    Gaussian, 356–357
    spreading of (dispersion), 356–357
wave zone, 1327f
wave-normal surface, 1062–1063, 1062f, 1064f
wave-wave mixing. *See also* three-wave mixing *in various media*; four-wave mixing in isotropic, nonlinear media
    in nonlinear dielectric media, 540–564
    in plasmas, 1132–1135
waves, monochromatic in homogeneous medium. *See also* sound waves in a fluid; gravity waves on water; flexural waves on a beam or rod; gravitational waves; Alfvén waves; Rossby waves in rotating fluid
    dispersion relation, 353
    group velocity, 355
    phase velocity, 352
    plane, 352
weak turbulence formalism, 800–810
    Reynolds stress and turbulent viscosity, 802, 803–804
    turbulent diffusion coefficient, 805
    turbulent thermal conductivity, 805
    correlation functions in, 802–803
    spatial evolution of turbulent energy, 804, 808f, 808–809
Weyl (curvature) tensor, 1215, 1216
whistler wave in plasma, 1053f, 1054–1055, 1062f, 1146f
Wiener-Khintchine theorem
    for 1-dimensional random process, 303
    for 2-dimensional random process, 307–308
    for complex 3-dimensional random process (van Cittert-Zernike theorem), 478–479
Wiener's optimal filter, 318–320
WIMPs, 1440–1441
winds
    around low-pressure region, 770
    drive ocean's surface currents, 772–776, 805
    in stratosphere, 784–785
    lee waves in, 821n
    propagation of sound waves in, 366
wingtip vortices, 734f, 739, 744, 744f
WKB approximation, as example of eikonal approximation, 358
Womersley number, 719
world line, 49, 59f, 1155f
World Trade Center buildings, collapse of, 605–607
world tube, 49n, 68f, 69
wormhole, 68–69, 68f
    as time machine, 69
    Schwarzschild, 1276–1277

yield point, in elasticity, 580, 581f
    origin of yield: dislocations, 586, 587f
    yield strains for various materials, 586t
Young's modulus, 582, 589–592
    values of, for specific materials, 586t
Young's slits, 456–458

zero point energy, 175n, 669, 1002, 1437–1438, 1446
zone plate, 435–436
Z-pinch for plasma confinement, 960–962, 961f
    stability of, 975–978